If you put your motor home through what we put ours through, you'd be on one lousy vacation.

No motor home can survive everything. But you'll be glad to know every motor home built by Winnebago® Industries is subjected to tests that are often as hard on a motor home as the worst of Mother Nature. Frankly, our testing procedures are so uncompromising they even challenge the intestinal fortitude of our engineering department. We start with a drench test from 160 spray heads that simulates rain. Then there's our familiar drop test and computerized analysis. Our bone-jarring shake machine. Our fabric wear index. And our laser-guided chassis alignment check. Just to name a few. Quite simply, for a motor home to survive the short, inspection-laden trip down our assembly lines is as much a testament to its overall durability as the thousands of miles you put on it once you own one. But just to be safe, we take motor homes where the going is even tougher, like our regular trip out to Death Valley for our air conditioning test. Or our hard-hitting question and answer sessions with customers like you. And since we build 80% of everything that goes into the motor homes we sell, we can apply what we learn right away to make our motor homes even better. We're proud builders of Winnebago, Itasca,® Luxor,™ Vectra® and Rialta® motor homes. We make one that's right for you. For one of our fine Winnebago or Itasca dealers across North America, call 1-800-643-4892, ext. G018WC. (Our extensive dealer network and unsurpassed warranty package mean we're behind you no matter where you are.) Then give us a test of your own. Say, a short test drive. Rest assured, no matter which model you're interested in, you can feel confident it will stand up to any vacation you can imagine. As well as one or two you can't.

Interstate America

Letter from the Publisher

Dear Fellow Traveler,

We are pleased to present the 1998 Exit Authority — a complete guide to Every Business and Service at Every Exit on Every U.S. Interstate!*

You can use Exit Authority to plan all of your interstate travel stops well in advance, and:

- Plan to stop at your favorite restaurants, shopping outlets, and hotels along the way.
- Find special RV services, like RV Dump Stations, RV Camps, Truck Stops or Rest Areas with Overnight Parking.
- Easily identify stops with easy RV and bus access and parking by looking for businesses highlighted in red!
- Find fuel stops by brand or type (diesel) or with repair, car wash or convenience store facilities.
- Choose the safest, busiest exits in unfamiliar areas.
- Find medical services, including pharmacies, hospitals, and minor emergency centers.
- and much more!

Just as we did for our previous editions, we traveled every interstate and stopped at every exit to record business names and services, additional facilities, and even the size of the parking lot.

In the 1998 Exit Authority, you'll find we've added thousands of new businesses, updated name changes and new services, and listed new exits and facilities. This year you'll find more listings for major beltways and connectors, in addition to the complete listing of all U.S. interstates. Also new for 1998 are our Travel Saver Coupons, which can save you money at your favorite hotels, outlet malls, and other travel stops.

Thank you for purchasing the 1998 Exit Authority, and we wish you safe and pleasant travels in 1998.

Happy Motoring!

Dennis

Dennis Robbins
Publisher

*Excludes Alaska

Interstate America • 5695 Oakbrook Pkwy. • Suite G • Norcross, GA 30093
Questions or comments? Call 1-800-683-3948

Printed in U.S.A. Entire Contents ©1997–1998 by Interstate America • All Rights Reserved ISBN # 1-880477-14-9

INTERSTATE EXIT AUTHORITY

PRESIDENT: Ron Peterson

V.P./PUBLISHER: Dennis Robbins

MARKETING DIRECTOR: Bonnie Elmore

ASSOCIATE EDITOR: Suellen Stroup

PRODUCT FULFILLMENT: Mark Patton

PREMIUM/DATA SALES: Les Helms

ADVERTISING SALES:

Jeff Mitchell - Audrey Rawls - Cory Davis

Paige Goff - Mike Richards - Gina Sinnett

Hesta Baker

How to Use the Exit Authority... in three easy steps...

1 Turn to the Interstate.

(Interstates are listed in numerical order)

2 Find the State.

(Organized North to South or East to West)

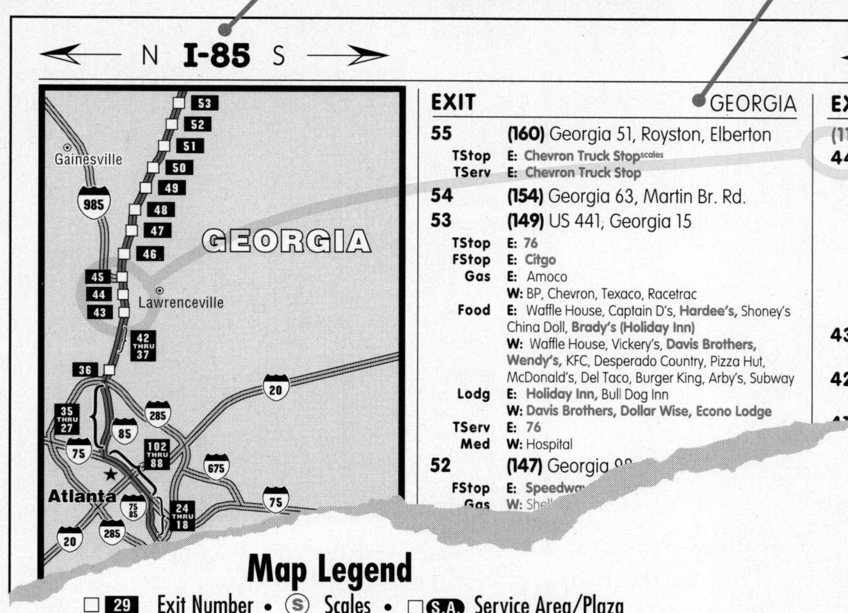

← N **I-85** S →

EXIT	GEORGIA
55	**(160)** Georgia 51, Royston, Elberton
TStop	E: Chevron Truck Stop scales
TServ	E: Chevron Truck Stop
54	**(154)** Georgia 63, Martin Br. Rd.
53	**(149)** US 441, Georgia 15
TStop	E: 76
FStop	E: Citgo
Gas	E: Amoco W: BP, Chevron, Texaco, Racetrac
Food	E: Waffle House, Captain D's, **Hardee's,** Shoney's China Doll, **Brady's (Holiday Inn)** W: Waffle House, Vickery's, **Davis Brothers, Wendy's,** KFC, Desperado Country, Pizza Hut, McDonald's, Del Taco, Burger King, Arby's, Subway
Lodg	E: **Holiday Inn,** Bull Dog Inn W: **Davis Brothers, Dollar Wise, Econo Lodge**
TServ	E: 76
Med	W: Hospital
52	**(147)** Georgia 9...
FStop	E: Speedwa...
Gas	W: Shell...

← N **I-85** S →

EXIT	GEORGIA
(112)	Rest Area (northbnd)-RR, phone
44	**(111)** Georgia 317
FStop	E: Phillips 66 W: Shell
Gas	E: BP, Amoco W: Chevron, Exxon
Food	E: Waffle House, Wendy's, **Justin's (Holiday Inn),** W: **Waffle House,** McDonald's, Porkland BBQ, Denny's (Days Inn)
Lodg	E: **Holiday Inn, Comfort Inn** W: **Days Inn,** Falcon Inn (Best Western)
43	**(109)** Old Peachtree Rd.
Gas	E: Texaco
42	**(107)** Georgia 120...
FStop	E: ...

Map Legend

29 Exit Number • (S) Scales • S.A. Service Area/Plaza

3 Follow Directional Arrows at top of Page to Search For Exits in the Desired Direction.

Legend For Abbreviations & Symbols:

TStop	24-Hour diesel fuel location w/ full service restaurant and/or store
FStop	Diesel fuel location with large vehicle clearance
Food	Food Outlets (fast food, restaurants, cafeterias, etc.)
Lodg	Hotel/Motel Accommodations
Gas	Automobile fueling locations
AServ	Facilities with Automotive Service
TServ	Commercial truck/diesel engine service facilites
TWash	Commercial vehicle wash facilites
RVCamp	Camping sites, RV service or supply facilities
Parks	National, state or local forest, parks, preserves & lakes
ATM	Banks or other services services with cash machine
Med	Hospital, emergency treatment or other medical facility
Other	Useful services, sites or attractions near interstate
(MM)	Numbers in ( ) indicate mile markers when not same as exit #.
N-S-E-W	Indicates side of the highway that services are located
I-O	Three-digit perimeter hwys – (I) inside or (O) outside perimeter
AAA	— AAA Official Listing Requirements
♦	— Meets AAA Listing Requirements

Rest Area & Welcome Center Facilities: (P) Overnight parking allowed at rest area

RR – Rest rooms	**RV Dump** – Sanitary Waste Dump
Vending – Snack & Drink Machine	**RV Water** – RV Water Hookup
Phones – Public Pay Phones	**HF** – Handicapped Facilities
Grills – Outdoor Cooking Grills	**Grills** – Outdoor Cooking Grills
Picnic – Picnic Tables	**Pet Walk** – Designated Area for Pets

Superscripts: (*) Convenience store; (cw) Automatic car wash; (D) Diesel Fuel Available; (LP) Liquid propane gas

- Every Business at every exit for every Interstate in the U.S. (excluding Alaska) within one quarter mile.
- Areas with adequate RV & Bus Parking are highlighted in Red (Parking may be nearby, but provides easy access)
- Services grouped by category (see legend) and direction from exit (N-S-E-W)
- RV Dump Stations highlighted in Red
- Additional notations for Convenience stores, Diesel Fuel, Propane, and Automatic Car Washes (see legend at left)

Upfit conversion package shown is from an independent sup

NOW, IMPROVE T
RAISING T

INTRODUCING THE 1998 CONVERSION VAN FROM CHEVY. It's designed to be the best van we've ever built. The ride is smooth and comfortable, the next best thing to fresh blacktop. Its foundation is a rugged, full-length chassis with welded ladder-frame construction that adds tons of structural rigidity. Independent short- and long-arm front suspension allows front wheels to act individually over bumps. The extended model has

WALMART INTERSTATE LOCATIONS

Interstate	Exit
Alabama	
I-10	15AB
I-20	133
I-20	185
I-59	108
I-59	218
I-65	3AB
I-65	130
I-65	255
I-65	271
I-85	79
Arkansas	
I-30	73
I-30	118
I-40	13
I-40	58
I-40	84
I-40	125
I-40	216
I-40	241A
I-40	276
I-55	67
Arizona	
I-19	4
I-19	69
I-40	51
I-40	253
California	
I-10	75
I-15	8
I-15	73
I-15	112
I-40	1
I-5	111B
I-5	603
I-5	647A
I-5	775
I-8	24A
I-99	21
I-99	120
I-99	160
I-99	227
I-99	243
I-99	271
I-99	302
Colorado	
I-25	15
I-25	150A
I-70	167
I-70	203
I-70	264
Connecticut	
I-395	80
I-395	97
I-95	81
Florida	
I-10	12
I-10	14
I-10	18
I-10	30
I-10	48
I-275	23AB
I-295	2
I-295	13
I-75	46
I-75	82
I-95	35AB
I-95	88
I-95	91C
Georgia	
I-20	3
I-20	10
I-20	36
I-75	5
I-75	18
I-75	33
I-75	62
I-75	118
I-85	9
I-85	13
I-85	40
Iowa	
I-35	92
Idaho	
I-15	93
I-84	208
I-86	61
Illinois	
I-255	13
I-280	2
I-55	52
I-55	197
I-57	54AB
I-57	71
I-57	95
I-57	116
I-57	160
I-57	190AB
I-57	315
I-64	14
I-70	61
I-72	141AB
I-74	95A
I-74	181
I-80	19
I-80	56
I-80	75
I-80	90
I-80	112AB
I-80	130AB
I-88	41
NWTL	63
TSTL	74
Indiana	
I-465	27
I-64	105
I-65	4
I-65	29
I-65	172
I-69	3
I-69	112AB
I-69	129
I-70	104
I-74	116
I-74	134AB
I-94	34AB
Kansas	
I-135	60
I-135	89
I-35	71
I-35	128
I-35	183AB
I-70	53
I-70	159
I-70	298
Kentucky	
I-24	4
I-64	53AB
I-64	137
I-71	22
I-75	38
Louisiana	
I-10	64
I-10	82
I-10	109
I-10	151
I-10	157B
I-10	163
I-10	266
I-12	7
I-12	63B
I-20	10
I-20	114
I-20	138
I-49	18
I-55	31
Massachusetts	
I-195	1
I-195	18
I-495	18
I-495	38
I-95	9
Maryland	
I-68	40
I-70	54
I-81	5
Maine	
I-95	33
I-95	39
I-95	49
I-95	62
METNPK	4
Michigan	
I-196	20
I-69	13
I-69	61
I-69	141
I-75	282
I-75	392
I-94	29
I-94	98AB
I-94	181AB
I-96	30AB
Minnesota	
I-35	42B
I-35	56
I-35	131
I-35E	97AB
I-35E	115
I-90	119
I-94	54
I-94	103
Missouri	
I-255	2
I-270	29
I-35	16
I-35	54
I-44	77
I-44	80AB
I-44	100
I-44	129
I-44	208
I-44	226
I-44	261
I-55	96
I-55	129
I-55	191
I-57	10
I-70	15AB
I-70	124
I-70	193
I-70	208
I-70	227
Mississippi	
I-10	34AB
I-20	42AB
I-55	18
I-55	40
I-55	61
I-55	206
I-55	243AB
I-55	291
I-59	4
I-59	154AB
Montana	
I-15	192AB
I-90	306
North Carolina	
I-26	18AB
I-40	103
I-40	151
I-40	214
I-85	21
I-85	45AB
I-85	75
I-85	91
I-85	141
I-85	164
I-85	204
I-85	213
I-95	173
North Dakota	
I-29	64
I-94	61
I-94	258
Nebraska	
I-80	177
New Hampshire	
I-93	20
New Jersey	
I-295	47AB
New Mexico	
I-25	3
I-40	160
Nevada	
I-80	301
New York	
I-81	45
I-87	6
I-88	15
NYTH	59
Ohio	
I-270	15
I-270	32
I-275	33
I-275	63
I-70	36
I-70	91AB
I-70	218
I-71	8
I-71	234
I-75	22
I-75	74AB
I-75	92
I-77	1
I-77	81
I-77	120
OHTNPK	8
Oklahoma	
I-35	109
I-35	116
I-40	82
I-40	123
I-40	136
I-40	185
I-40	311
I-44	80
I-44	108
I-44	196
Oregon	
I-5	21
I-5	55
I-5	174
I-84	62
I-84	376AB
Pennsylvania	
I-70	7AB
I-76	3
I-79	25
I-80	19
I-90	6
Rhode Island	
I-295	4
South Carolina	
I-26	21AB
I-26	103
I-26	111AB
I-26	199AB
I-85	92
South Dakota	
I-29	77
I-29	132
I-90	12
I-90	59
Tennessee	
I-24	4
I-24	114
I-24	152
I-40	12
I-40	80AB
I-40	82AB
I-40	287
I-40	379
I-40	435
Texas	
I-10	11
I-10	28B
I-10	696
I-10	720
I-10	747
I-10	780
I-20	42
Texas (continued)	
I-20	343
I-20	408
I-27	49
I-30	7A
I-30	68
I-30	93AB
I-30	124
I-30	201
I-30	220A
I-30	223AB
I-35	3B
I-35	186
I-35	205
I-35	241
I-35	250
I-35	251
I-35	261
I-35	339
I-35	368A
I-35E	414
I-35E	415
I-35E	452
I-35E	463
I-40	72B
I-410	13B
I-410	21A
I-45	25
I-45	59
I-45	88
I-45	116
I-45	251
I-635	2
I-635	23
I-820	10AB
I-820	20B
I-820	27
Utah	
I-15	272
I-15	297
I-15	334
I-15	342
I-80	145
Virginia	
I-64	55
I-64	255AB
I-64	263AB
I-64	290AB
I-66	47
I-81	7
I-85	12
I-95	53
I-95	143AB
Washington	
I-5	79
Wisconsin	
I-43	43
I-43	96
I-43	149
I-90	4
I-90	25
I-90	171C
I-94	2
I-94	41
I-94	70
I-94	116
I-94	143
I-94	287
West Virginia	
I-64	15
I-64	169
I-77	138
I-79	99
I-79	119
I-79	132
I-81	12
I-81	13
Wyoming	
I-80	5
I-80	102
I-90	25
I-90	126

FLYING J is RV Ready

...a Highway Haven for Travelers

Designed to serve interstate traffic, Flying J's easy access Travel Plazas and Fuel Stops offer every amenity a traveler could imagine:

- Free RV dump stations
- Special RV fueling islands
- RV Real Value membership club
- Highest quality fuel
- Restaurants & fast food operations
- Well-stocked convenience stores
- Open 24-hours
- ATM's
- Daily U.S. Postal pickup
- UPS/FedEX boxes
- Laundry facilities
- Barber/hair salons

And more . . .

J Care Service Centers provide:
(at select locations)

- RV wash
- Oil change and lube
- Tire repair
- Roadside services

Flying J · PO Box 678
Brigham City, UT 84302
801-734-6400

R.V. Real Value CLUB
0123 4567 8900
NAME HERE
ROADLINK
PREPAID CALLING CARD PRIVILEGES
FLYING J

Join the Real Value Club and receive:

- 1¢ per Gallon Fuel Discount* (gas or diesel)
- 5¢ per Gallon Propane Discount*
- Discounts on Selected Products & Services
- Prepaid Calling Card Privileges (5 minutes free included)
- Lifetime Membership to Audio Adventures Book on Tape Rental Program

Fill out the attached application and mail. You'll then receive your personal membership card and benefits information. It's that simple!

*Discounts will be granted unless prohibited by law.

FLYING J

is RV

Ready

Flying J Real Value Membership Application

(Please print clearly)

FOR OFFICE USE ONLY	Card No. ______________
Branch No. ______________	Emp. No. ______________

Membership Information

Customer Name: ____

Customer Address: ____

City: ______________________ State: ______________ Zip: ______________

Home Telephone: (   ) ______________ Work Telephone: (   ) ______________

Are you a member of. . . ? (Check applicable box(es)

- ❑ Good Sam Club
- ❑ FMCA (Family Motor Coach Association)
- ❑ ICC (International Coachmen Caravan)
- ❑ Other: ______________

Have you ever applied for a Real Value membership before?: ❑ Yes ❑ No

Type of RV (Check applicable box(es)

- ❑ Trailers
  - ❑ Less than 25 ft.
  - ❑ Over 25 ft.
- ❑ Folding Trailers
- ❑ 5th Wheel
  - ❑ Less than 30 ft.
  - ❑ Over 30 ft.
- ❑ Motorhomes
  - ❑ Less than 30 ft. Class A
  - ❑ Over 30 ft.
  - ❑ Mini Motor Homes
  - ❑ Micro Mini
- ❑ Bus Conversions
- ❑ Campers
  - ❑ Truck Campers
  - ❑ Camping Vans

EA

BUSINESS REPLY MAIL
FIRST-CLASS MAIL PERMIT NO. 60 BRIGHAM CITY UT

POSTAGE WILL BE PAID BY ADDRESSEE

NO POSTAGE
NECESSARY
IF MAILED
IN THE
UNITED STATES

FLYING J
PO BOX 678
BRIGHAM CITY UT 84302-9975

You're in Flying J Country

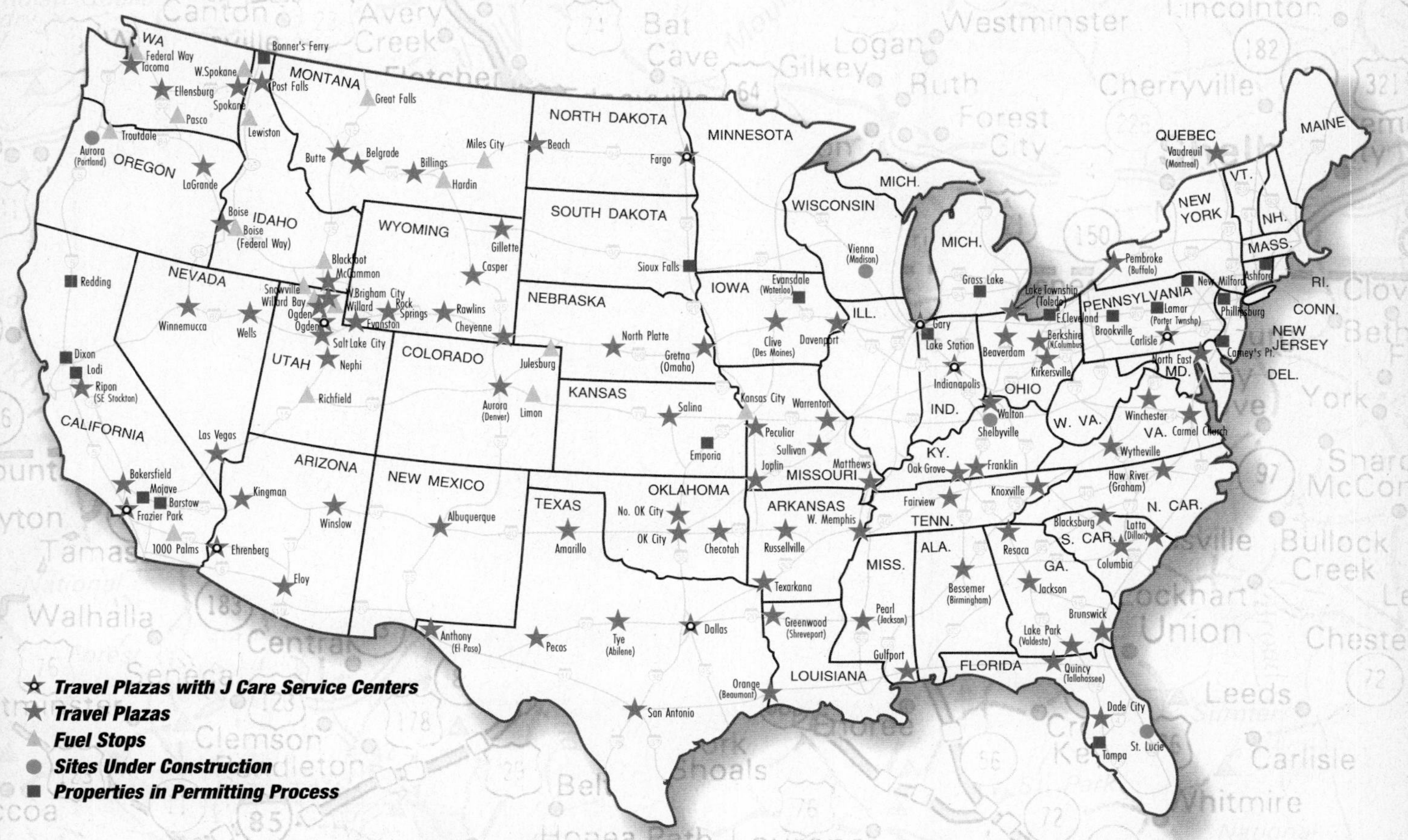

★ U.S. TRAVEL PLAZAS		
AL	Bessemer (Birmingham)	I-20 & I-59, Exit 104
AR	Russellville	I-40 Exit 84
AR	Texarkana	I-30 Exit 7
AR	West Memphis	I-40 Exit 280 & I-55 Exit 4
AZ	Ehrenberg	I-10 Exit 1
AZ	Eloy	I-10 Exit 208A
AZ	Kingman	I-40 Exit 53
AZ	Winslow	I-40 Exit 255
CA	Bakersfield	Hwy 99 Exit Merced Ave.
CA	Frazier Park	I-5 Frazier Park Exit
CA	Ripon (S.E. Stockton)	Hwy 99 Exit Jacktone Rd
CO	Aurora (Denver)	I-70 Airport Blvd. So. Exit 285
FL	Dade City	I-75 Exit 59
FL	Quincy (Tallahassee)	I-10 Exit 27
GA	Brunswick	I-95 Exit 6
GA	Jackson	I-75 Exit 66
GA	Lake Park (Valdosta)	I-75 Exit 1
GA	Resaca	I-75 Exit 133
IA	Clive (Des Moines)	I-80 & I-35 Exit 125
IA	Davenport	I-80 Exit 292
ID	Boise	I-84 Exit 50
ID	McCammon	I-15 & Hwy 30 Exit 47
ID	Post Falls	I-90 Exit 2
IN	Gary	I-94 & I-80 Exit 9A
IN	Indianapolis	I-465 Exit 4
KS	Salina	I-70 Exit 253
KY	Franklin	I-65 @ 31 W. Exit 2
KY	Oak Grove **(Opens 1/98)**	I-24 Exit 86
KY	Walton	I-75 Exit 171
LA	Greenwood (Shreveport)	I-20 Exit 3
MD	*North East (Elkton)	I-95 Exit 100
MO	Joplin	I-44 U.S. 71 Exit 11A
MO	Matthews (Sikeston)	I-55 Exit 58
MO	Peculiar	U.S. 71 Exit Hwy J
MO	Sullivan	I-44 & Hwy 185 Exit 226
MO	Warrenton	I-70 Exit 188
MS	Gulfport	I-10 Exit 31
MS	Pearl (Jackson)	I-20/I-55 Exit 47
MT	*Belgrade (BAIR'S)	I-90 Exit 298
MT	Billings	I-90 Exit 455
MT	Butte	I-90 Exit 122
NC	Haw River (Graham)	I-85 & I-40 Exit 150
ND	Beach	I-94 & Hwy 16 Exit 1
ND	Fargo	I-29 & I-94 Exit 62
NE	Gretna (Omaha)	I-80 Exit 432
NE	North Platte **(Opens 11/97)**	I-80 Exit 179
NM	Albuquerque	I-40 & 98th Street Exit 153
NV	*Las Vegas	I-15 Cheyenne Exit
NV	Wells	I-80 & Hwy 93 Exit 352
NV	Winnemucca	I-80 Exit 176
NY	Pembroke (Buffalo) **(Opens 12/97)**	I-90 Exit 48A
OH	Beaverdam	I-75 Exit 135
OH	Berkshire (N. Columbus)	I-71 Sunbury Exit 131
OH	Kirkersville	I-70 State Rd. 158 Exit 122
OH	Lake Township (Toledo)	I-280 Exit 1B
OK	Checotah	I-40 Exit 264B & Jct. Hwy 69 & Hwy 266
OK	Oklahoma City	I-40 Exit 140, Morgan Rd.
OK	Oklahoma City (N.) **(Opens 2/98)**	I-35 & N.E. 122nd St.
OR	*LaGrande	I-84 Exit 265
PA	Carlisle	I-81 Exit 17/17A
SC	Blacksburg	I-85 Exit 102
SC	Columbia	I-20 Exit 70
SC	Latta (Dillon)	I-95 Exit 181
TN	Fairview	I-40 & Hwy 96 Exit 182
TN	Knoxville	I-40 & I-75 Exit 369
TX	Amarillo	I-40 Exit 76
TX	Anthony (El Paso)	I-10 Exit 0
TX	Dallas	I-20 & Exit 472
TX	Orange (Beaumont)	I-10 Exit 873
TX	Pecos	I-20 & Exit 42
TX	San Antonio	I-10 Exit 583
TX	Tye (Abilene)	I-20 FM 707 Exit 277
UT	Nephi	I-15 Exit 222
UT	Ogden	I-15 Exit 346
UT	Salt Lake City	I-15 Exit 21st So. Wstbnd
UT	Willard Bay	I-15 Exit 360
VA	Carmel Church	I-95 Exit 104
VA	Winchester	I-81 Exit 323
VA	Wytheville	I-77 & I-81 Exit 77
WA	*Ellensburg	I-90 Canyon Rd. Exit
WA	*Spokane	I-90 Exit 286
WA	Tacoma	I-5 Exit 136
WY	Casper	I-25 Exit 185
WY	Cheyenne	I-25 Exit 7
WY	*Evanston	I-80 Exit 3
WY	*Gillette	I-90 & Hwy 59
WY	Rawlins	I-80 Exit 209
WY	Rock Springs	I-80 Exit 104
CANADAN TRAVEL PLAZAS		
QC	Vaudreuil (Montreal)	NW quadrant of intersection of Hwy 540 & Cité des Jeunes Blvd.
▲ U.S. FUEL STOPS		
CA	Thousand Palms	I-10 Ramon Exit
CO	Julesburg	I-76 Hwy 385
CO	Limon	I-70 Exit 361 Main Intrchng
ID	Blackfoot	I-15 Exit 93
ID	Boise	I-84 Exit 54 (Federal Way)
ID	Lewiston	Jct Hwy 12 & 95
MO	Kansas City	I-435 Exit 57 Front St.
MT	Great Falls	I-15 Exit 280
MT	*Hardin (Bair's)	I-90 Exit 495
MT	Miles City	I-94 & Baker Exit
OR	Troutdale	I-84 Exit 17
UT	Ogden	I-15 Exit 347 (12th St.)
UT	Richfield	I-70 Exit 40
UT	Snowville	I-84 Exit 7
UT	West Brigham	I-15 Exit 364
UT	Willard	Hwy 89
WA	*Federal Way	I-5 Exit 142B
WA	*Pasco	U.S. Hwy 395
WA	*West Spokane	I-90 Exit 276

*Franchised Facilities

Tourism

State Tourism Offices ~ Nationwide

Alabama Bureau of Tourism and Travel
Alabama - Unforgettable
401 Adams Avenue, Suite 126 P.O. Box 4927
Montgomery, AL 36103-4309
www.state.al.us
1-800-ALABAMA

Alaska Division of Tourism
If you've got what it takes, you can make it to the top. Alaska Ô All Things Wild and Wonderful
P.O. Box 11081
Juneau, AK 99811-0801
www.state.ak.us/tourism
1-800-76-ALASKA;1-800-667-8489 (ALASKA STO)

Arizona Office of Tourism
Arizona ~ Grand Canyon State
2702 N. 3rd Street, Suite 4015
Phoenix, AZ 85004
www.arizonaguide.com
1-800-842-8257

Arkansas Department of Parks and Tourism
Arkansas...The Natural State
One Capitol Mall
Little Rock, AR 72201
www.state.ar.us/html/ark_parks.html
1-800-NATURAL

California Division of Tourism California.
Everything Under The Sun.
801 K St., Suite 1600
Sacramento, CA 95814
gocalif.ca.gov
1-800-TO-CALIF

Colorado Travel & Tourism Authority
P.O. Box 3524
Englewood, CO 80155
www.colorado.com
1-800-COLORADO

Connecticut Department of Economic Development, Tourism Division
Connecticut. We're full of surprises.
865 Brook St.
Rocky Hill, CT 06067
www.state.ct.us/tourism.htm
1-800-CT-BOUND

Delaware Tourism Office
Delaware, Small Wonder; The First State of America
99 Kings Highway, Box 1401
Dover, DE 19903
www.state.de.us/tourism/intro.htm
1-800-441-8846

Washington, DC Convention and Visitors Association
Capital Region USA
1212 New York Ave., NW
Washington, DC 20005
www.washington.org
1-800-422-8644

Florida Tourism Industry Marketing Corporation
Return to your senses in Florida. One Florida, Many Faces!
661 East Jefferson Street, Suite 300
Tallahassee, FL 32301
www.flausa.com
1-904-487-1462

Georgia Department of Industry, Trade & Tourism
Georgia On My Mind
285 Peachtree Center Ave. Marquis Tower Two, 10th Floor
Atlanta, GA 30303
www.Georgia-On-My-Mind.org/code/welcome.html
1-800-VISIT-GA

Hawaii State Tourism Office
Aloha and Diversity; Islands of Aloha
PO Box 2359
Honolulu, HI 96804
www.gohawaii.com
1-800-GOHAWAII

Idaho Division of Tourism Development
Department of Commerce
Idaho is what America used to be!
700 West State St.
Boise, ID. 83720
www.visitid.org
1-800-635-7820

Illinois Bureau of Tourism
Illinois. A Million Miles From Monday!
State of Illinois Center
100 W. Randolph, Suite 3-400
Chicago, IL 60601
www.enjoyillinois.com
1-800-2-CONNECT

Indiana Tourism Division
Department of Commerce
You Could Use a Little Indiana
One North Capitol, Suite 700
Indianapolis, IN 46204-2288
www.ai.org/tourism
1-800-289-6646

Iowa Division of Tourism Department of Economic Development
Iowa. You make me Smile.
200 East Grand Ave.
Des Moines, IA 50309
www.state.ia.us/tourism/
1-800-345-IOWA (For ordering vacation kit only) (U.S. only);1-800-528-5265 (Special Events Calender)

Continues . . .

INTRODUCING THE MOTORCOA INTO A NEW

THE WANDERLODGE LXi,
CH THAT TRANSPORTS YOU
DIMENSION OF LUXURY.

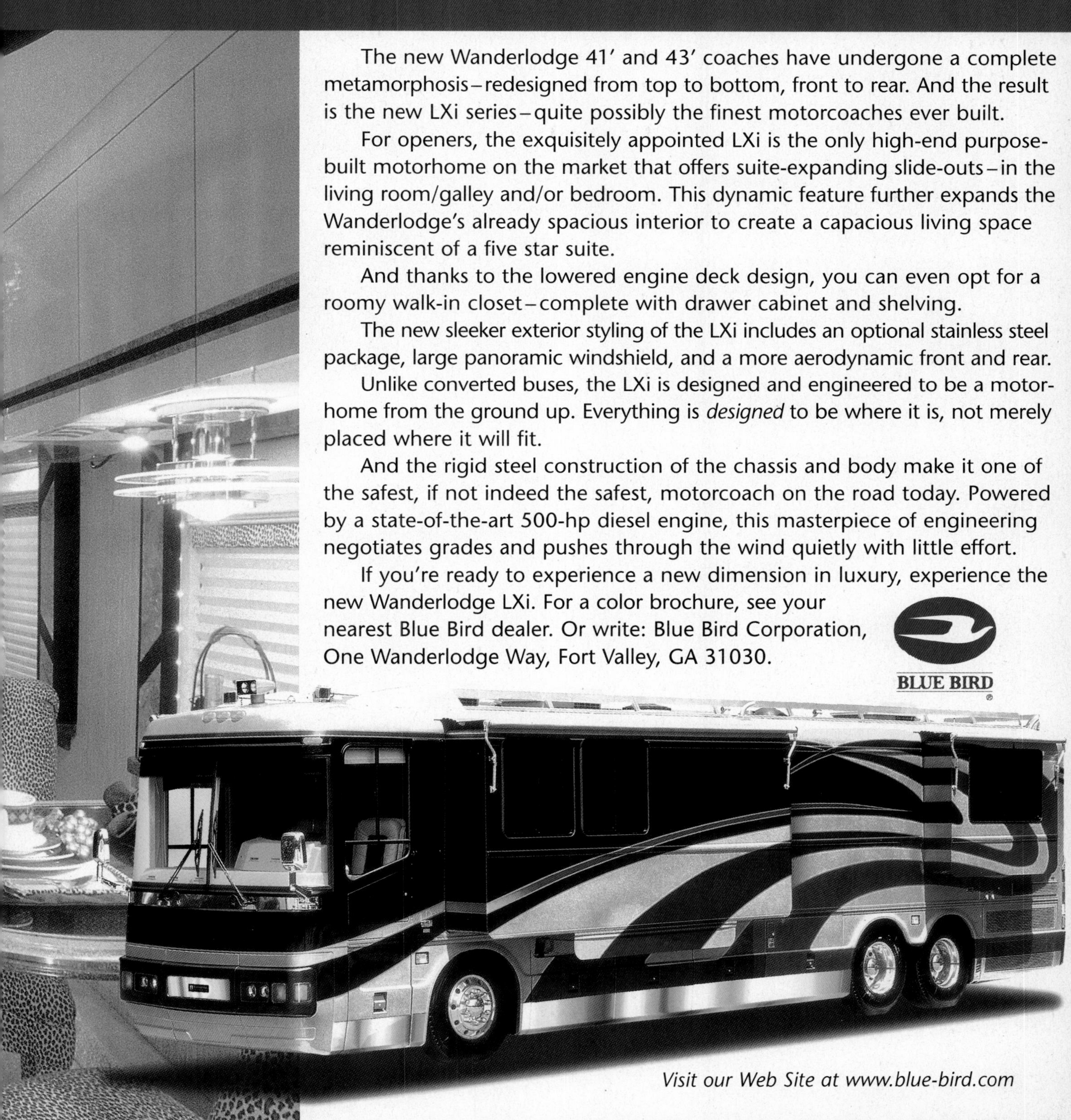

The new Wanderlodge 41′ and 43′ coaches have undergone a complete metamorphosis–redesigned from top to bottom, front to rear. And the result is the new LXi series–quite possibly the finest motorcoaches ever built.

For openers, the exquisitely appointed LXi is the only high-end purpose-built motorhome on the market that offers suite-expanding slide-outs–in the living room/galley and/or bedroom. This dynamic feature further expands the Wanderlodge's already spacious interior to create a capacious living space reminiscent of a five star suite.

And thanks to the lowered engine deck design, you can even opt for a roomy walk-in closet–complete with drawer cabinet and shelving.

The new sleeker exterior styling of the LXi includes an optional stainless steel package, large panoramic windshield, and a more aerodynamic front and rear.

Unlike converted buses, the LXi is designed and engineered to be a motorhome from the ground up. Everything is *designed* to be where it is, not merely placed where it will fit.

And the rigid steel construction of the chassis and body make it one of the safest, if not indeed the safest, motorcoach on the road today. Powered by a state-of-the-art 500-hp diesel engine, this masterpiece of engineering negotiates grades and pushes through the wind quietly with little effort.

If you're ready to experience a new dimension in luxury, experience the new Wanderlodge LXi. For a color brochure, see your nearest Blue Bird dealer. Or write: Blue Bird Corporation, One Wanderlodge Way, Fort Valley, GA 31030.

Visit our Web Site at www.blue-bird.com

Tourism

State Tourism Offices ~ Nationwide

Kansas Travel & Tourism Division
Kansas. Simply Wonderful.
700 S.W. Harrison Street, Suite 1300
Topeka, KS, 66603-
3957kicin.cecase.ukans.edu/kdoch/html/tour
1.html
1-800-2KANSAS

Kentucky Department of Travel Development
Kentucky - Always in Season.
2200 Capital Plaza Tower 500 Mero St.
Frankfort, KY, 40601
www.state.ky.us/tour/tour.htm
1-800-225-TRIP

Louisiana Office of Tourism Department of Culture, Recreation & Tourism
PO Box 94291
Baton Rouge, LA, 70804-9291
www.crt.state.la.us/crt/tourism.htm
1-800-334-8626

Maine Office of Tourism Department of Economic & Community Development
Maine, The Way Life Should Be
189 State St., Station 59
Augusta, ME 04333
www.visitmaine.com
1-800-533-9595

Maryland Office of Tourism Development Department of Economic & Employment Development
So many things to do. So close together.
217 East Redwood, 9th Floor
Baltimore, MD 21202
www.mdisfun.org
1-800-543-1036 (For vacation kit only)

Massachusetts Office of Travel & Tourism
We'd Love to Show You Around
100 Cambridge St., 13th Floor
Boston, MA, 02202
www.mass-vacation.com
1-800-447-MASS (For vacation kit only)

Michigan Jobs Commission/Travel Michigan
Michigan: Great Lakes Ö Great Times
PO Box 30226
Lansing, MI 48909
www.michigan.org
1-888-78-GREAT

Minnesota Office of Tourism
Explore Minnesota
100 Metro Square 121 Seventh Place East
St. Paul, MN, 55101-2112
www.tccn.com/mn.tourism/mnhome.html
1-800-657-3700

Mississippi Tourism Development Division
Department of Economic Development
Mississippi, *The South's Warmest Welcome*
PO Box 849 Jackson, MS, 39205
www.decd.state.ms.us/tourism.htm
1-800-WARMEST

Missouri Division of Tourism
Wake Up to Missouri
PO Box 1055
Jefferson City, MO 65102
www.ecodev.state.mo.us/tourism/
1-800-877-1234

Montana Department of Commerce Tourism Development & Promotion
Big Sky Country
PO Box 200533
Helena, MT 59620-0533
travel.mt.gov
1-800-VISIT-MT

Nebraska Divsion of Travel & Tourism
Department of Economic Development
Send "A Postcard from Nebraska"
PO Box 94666 301 Centennial Mall South
Lincoln, NE 68509
www.ded.state.ne.us/tourism.html
1-800-228-4307

Nevada Commission on Tourism
Discover Both Sides of Nevada
Capitol Complex
Carson City, NV 89710
www.travelnevada.com/
1-800-NEVADA-8

New Hampshire Office of Travel & Tourism Development
It's Right In New Hampshire
172 Pembroke Rd. PO Box 1856
Concord, NH 03302-1856
www.visitnh.gov
1-800-FUN-IN-NH

New Jersey Division of Travel & Tourism
Department of Commerce & Economic Development New Jersey -
What A Difference A State Makes
20 West State St., CN826
Trenton, NJ 08625-0826
www.state.nj.us/travel
1-800-JERSEY-7

New Mexico Department of Tourism
New Mexico, USA
491 Old Santa Fe Trail
Santa Fe, NM, 87503
www.newmexico.org

Continues . . .

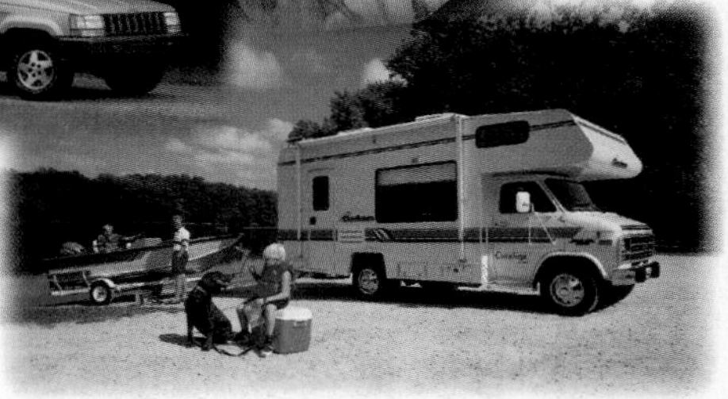

Tourism

State Tourism Offices ~ Nationwide

New York State Division of Tourism Department of Economic Development
I LOVE NY
One Commerce Plaza
Albany, NY 12245
www.iloveny.state.ny.us
1-800-CALL-NYS

North Carolina Travel & Tourism Division
Department of Economic and Community Development
301 N. Wilmington
Raleigh, NC 27601-2825
www.visitnc.com
1-800-VISIT-NC

North Dakota Tourism Department
Discover the Spirit
Liberty Memorial Bldg.
604 East Boulevard
Bismarck, ND 58505
www.glness.com/tourism/
1-800-HELLO-ND

Ohio Division of Travel & Tourism Department of Development
Ohio...The Heart Of It All
PO Box 1001
Columbus, OH 43216-1001
www.ohiotourism.com
1-800-BUCKEYE (Continental U.S. & Canada)

Oklahoma Tourism & Recreation Department
Travel and Tourism Division
Oklahoma - Native America
15 N. Robinson, Suite 100
Oklahoma City, OK 73102-5403
www.otrd.state.ok.us
1-800-652-6552 (Information Requests Only)

Oregon Tourism Commission
Oregon. Things Look Different Here.
775 Summer St., NE
Salem, OR 97310
www.traveloregon.com
1-800-547-7842

Pennsylvania Office of Travel, Tourism & Film Promotion
Pennsylvania Memories Last a Lifetime
453 Forum Building
Harrisburg, PA 17120
www.state.pa.us/visit
1-800-VISIT-PA (For ordering visitors guide only)

Rhode Island Economic Development Corporation
America's First Resort
One West Exchange Street
Providence, RI 02903
www.visitrhodeisland.com
1-800-556-2484

S. Carolina Dept. of Parks, Recreation & Tourism
Smiling Faces, Beautiful Places
1205 Pendleton St., Suite 106
Edgar A. Brown Bldg.
Columbia, SC 29201
www.sccsi.com/sc/

South Dakota Department of Tourism
Great Faces. Great Places
Capitol Lake Plaza 711 E. Wells Ave.
Pierre, SD 57501-3369
www.state.sd.us/tourism/
1-800-SDAKOTA

Tennessee Department of Tourist Development
Tennessee Sounds Good to Me
PO Box 23170
Nashville, TN 37202-3170
www.state.tn.us/tourdev/
1-800-836-6200

Texas Department of Commerce, Tourism Division
Texas. *It's Like A Whole Other Country.*
Texas. "To do un Poco y Mas."
PO Box 12728
Austin, TX 78711-2728
www.traveltex.com
1-800-88-88-TEX

Utah Travel Council
Utah!
Council Hall
Salt Lake City, UT 84114
www.utah.com
1-800-200-1160

Vermont Department of Tourism & Marketing
It Will Change The Way You Look At Things!
134 State St.
Montpelier, VT 05602
www.travel-vermont.com
1-800-VERMONT

Virginia Tourism Corporation
Virginia Is For Lovers
901 E. Byrd St. 19th Floor
Richmond, VA 23219
www.virginia.org
1-800-VISIT-VA

Washington State Tourism Development Division
Department of Trade & Economic Development
Washington State: The Place You've Been Trying To Get To!
PO Box 42500
Olympia, WA, 98504-2500
www.tourism.wa.gov
1-800-544-1800

West Virginia Division of Tourism & Parks
2101 Washington St., East
Charleston, WV 25305
www.state.wv.us/tourism
1-800-225-5982

Wisconsin Division of Tourism Department of Development
Wisconsin. You're Among Friends
123 West Washington Ave. PO Box 7970
Madison, WI 53707
tourism.state.wi.us
1-800-432-TRIP (Out-of-State);
1-800-372-2737(In-State)

Wyoming Division of Tourism
Wyoming. Like No Place on Earth.
Frank Norris Jr. Travel Center I-25 & College Dr.
Cheyenne, WY, 82002-0660
www.state.wy.us/state/tourism/tourism.html
1-800-225-5996

Continues

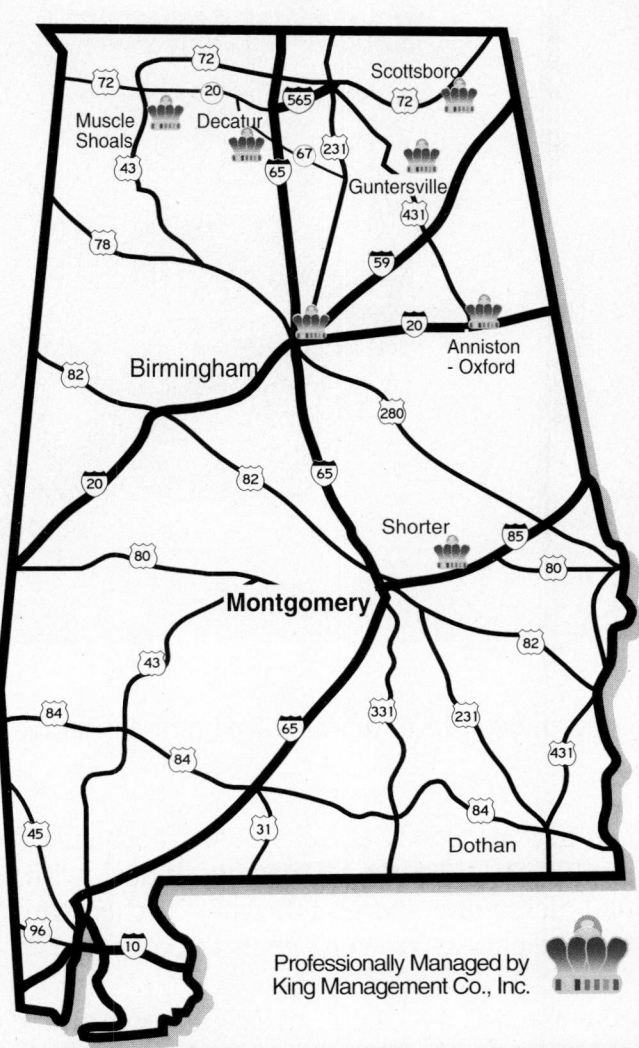

You have to see it to believe it!

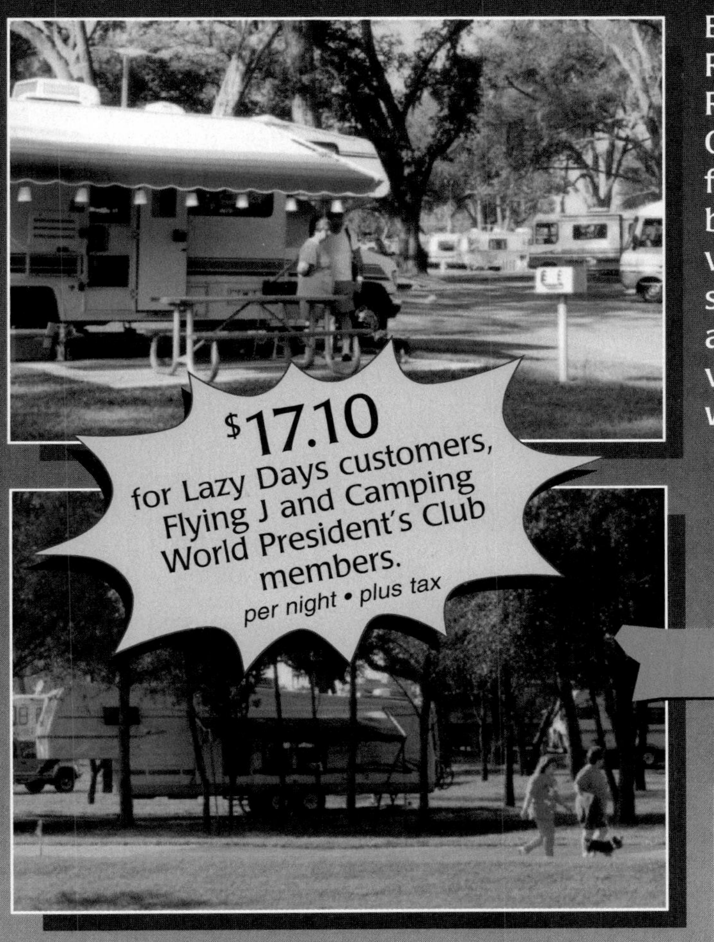

Everyone's talking about the new Lazy Days RV Resort designed especially for you, the RV enthusiast. Nestled in a canopy of Live Oak trees draped with southern moss and filled with southern hospitality surrounded by over 800 RVs on display. For your convenience you'll also find a completely stocked 40,000 sq. ft. Camping World and an on-site Cracker Barrel Restaurant. One visit to Lazy Days and you'll understand why everyone's saying...

"You have to see it to believe it!"

Special Rally Rates Available

Call for information or registration:

1-800-626-7800 • Ext. 4777

- 16,000 sq. ft. rec. building
- Complete laundry facilities
- Heated swimming pool
- Indoor showers
- Full banquet facilities
- Full service, pull-thru 30 amp parking
- Acres of flowered walking and jogging paths
- Rally parking available
- Cracker Barrel Restaurant
- RV rentals available
- Central Florida location is only minutes away from Disney World, Epcot, Sea World, Universal Studios, MGM Studios, Busch Gardens and spring training baseball.
- Coming soon... Flying J RV Travel Plaza including a hotel, convenience store, buffet style restaurant and an RV quick lube.

Cracker Barrel Old Country Store

Your Home Away From Home!

Woodall's Campground Rating 

6190 Lazy Days Boulevard • Seffner (Tampa), FL • Intersection of I-4 and County Road 579 • Exit 8
Just 1 mile east of I-4 and I-75 interchange • Phone: 1-800-626-7800; Ext. 4777
Website: www.lazydays.com

National Parks

State Park Offices ~ Nationwide

Acadia National Park
P.O. Box 177
Bar Harbor, ME 04609-0177
Headquarter Phone: 207-288-9561
Fax Number: 207-288-5507

Arches National Park
P.O. Box 907
Moab, UT 84532
Headquarter Phone: 801-259-8161
Fax Number: 801-259-3411

Badlands National Park
P.O. Box 6
Interior, SD 57750
Headquarter Phone: 605-433-5361
Fax Number: 605-433-5404

Big Bend National Park
P.O Box 129
Big Bend National Park, TX 79834
Headquarter Phone: 915-477-2251
Fax Number: 915-477-2357

Biscayne National Park
P.O. Box 1369
Homestead, FL 33030
Headquarter Phone: 305-230-1144
Fax Number: 305-230-1190

Bryce Canyon National Park
P.O. Box 170001
Bryce Canyon, UT 84717-0001
Headquarter Phone: 801-834-5322
Fax Number: 801-834-4102

Canyonlands National Park
2282 S. West Resource Blvd.
Moab, UT 84532-3298
Headquarter Phone: 801-259-3911
Fax Number: 801-259-8628

Capitol Reef National Park
HC 70 Box 15
Torrey, UT 84775-9602
Headquarter Phone: 801-425-3791
Fax Number: 801-425-3026

Carlsbad Caverns National Park
3225 National Parks Highway
Carlsbad, NM 88220
Headquarter Phone: 505-785-2232
Fax Number: 505-785-2133

Channel Islands National Park
1901 Spinnaker Drive
Ventura, CA 93001
Headquarter Phone: 805-658-5700
Fax Number: 805-658-5799

Crater Lake National Park
P.O. Box 7
Crater Lake, OR 97604
Headquarter Phone: 541-594-2211
Fax Number: 541-584-2299

Death Valley
National Park
Death Valley, CA 92328
Headquarter Phone: 619-786-2331
Fax Number: 619-786-3283

Denali National Park & PRES
National Park Service
Denali Park, AK 99755
Headquarter Phone: 907-683-2294
Fax Number: 907-683-9612

Dry Tortugas National Park
40001 State Road 9336
Homestead, FL 33034-6733
Headquarter Phone: 305-242-7700
Fax Number: 305-242-7711

Everglades National Park
40001 State Road 9336
Homestead, FL 33030
Headquarter Phone: 305-242-7700
Fax Number: 305-242-7728

Gates of the Arctic
National Park & PRES
201 First Avenue
P.O. Box 74680
Fairbanks, AK 99707
Headquarter Phone: 907-456-0281
Fax Number: 907-456-0452

Glacier Bay National Park & PRES
P.O. Box 140
Gustavus, AK 99826-0140
Headquarter Phone: 907-697-2230
Fax Number: 907-697-2654

Glacier National Park
P.O. Box 128
West Glacier, MT 59936-0128
Headquarter Phone: 406-888-7800
Fax Number: 406-888-7808

Grand Canyon National Park
P.O. Box 129
Grand Canyon, AZ 86023
Headquarter Phone: 520-638-7888
Fax Number: 520-638-7797

Grand Teton National Park
P.O. Box 170
Moose, WY 83012-0170
Headquarter Phone: 307-739-3300
Fax Number: 307-739-3438

Great Basin National Park
c/o Great Basin National Park
Baker, NV 89311
Headquarter Phone: 702-234-7331
Fax Number: 702-234-7269

Great Smoky Mountain National Park
107 Park Headquarters Road
Gatlinburg, TN 37738
Headquarter Phone: 423-436-1200
Fax Number: 423-436-1220

Guadalupe Mountains National Park
HC 60, Box 400
Salt Flat, TX 79847-9400
Headquarter Phone: 915-828-3251
Fax Number: 915-828-3269

Haleakala National Park
P.O. Box 369
Makawao, HI 96768
Headquarter Phone: 808-572-9306
Fax Number: 808-572-1304

Hawaii Volcanoes National Park
P.O. Box 52
Volcanoes, HI 96718
Headquarter Phone: 808-967-7311
Fax Number: 808-967-8186

Hot Springs National Park
P.O. Box 1860
Hot Springs, AR 71902
Headquarter Phone: 501-624-3383
Fax Number: 501-624-1536

Continues

"With Camping World, the honeymoon never ends."

It's not every day a commitment like this comes along. We're talking about a 100% Satisfaction

Guarantee on every item and service Camping World offers. No Hassle Refunds & Exchanges. And a history of quality that goes all the way back to 1966. At Camping World, they make sure every customer is completely happy. Not just for a day, but for a lifetime. We recommend you take the plunge, too. Browse through their latest catalog or take a walk down the aisles of one of Camping World's 27 stores nation-wide. A very promising relationship awaits.

Your RV Traveling Companion
Since 1966

Call toll-free 1-800-845-7875 24 hours a day for a FREE catalog and mention code EY.
Visit our Internet site: www.campingworld.com

FREE! SeaLand Toilet Tissue (4-pak) with any purchase of $25 or more.

SeaLand Toilet Tissue #6667 (reg. $2.48). This offer is not retroactive and cannot be used in combination with any other offer.

Present coupon at time of purchase. If ordering by phone, mention discount code below. Limit one per customer.

Offer expires 11/30/98.

Cashier: Ring as discount 8823.

National Parks

State Park Offices ~ Nationwide

Isle Royale National Park
800 E. Lakeshore Drive
Houghton, MI 49931-1895
Headquarter Phone: 906-482-0986
Fax Number: 906-482-8753

Joshua Tree National Park
74485 National Park Drive
Twentynine Palms, CA 92277
Headquarter Phone: 619-367-6376
Fax Number: 619-367-6392

Katmai National Park & PRES
202 Center Ave. Suite 201
Kodiak, AK 99615
Headquarter Phone: 907-486-6730
Fax Number: 907-486-3331

Kenai Fjords National Park
P.O. Box 1727
Seward, AK 99664
Headquarter Phone: 907-224-3175
Fax Number: 907-224-2144

Kings Canyon National Park
Sequoia and Kings Canyon Natl Pks
Three Rivers, CA 93271
Headquarter Phone: 209-565-3341
Fax Number: 209-565-3730

Kobuk Valley National Park
P.O. Box 1029
Kotzebue, AK 99752
Headquarter Phone: 907-442-3890
Fax Number: 907-442-8316

Lake Clark National Park & PRES
4230 University Drive, Suite 311
Anchorage, AK 99508
Headquarter Phone: 907-271-3751
Fax Number: 907-271-3707

Lassen Volcanic National Park
P.O. Box 100
Mineral, CA 96063
Headquarter Phone: 916-595-4444
Fax Number: 916-595-3262

Mammoth Cave National Park
Mammoth Cave National Park
Mammoth Cave, KY 42259
Headquarter Phone: 502-758-2251
Fax Number: 502-758-2349

Mesa Verde National Park
P.O. Box 8
Mesa Verde National Park, CO 81330-0008
Headquarter Phone: 970-529-4465
Fax Number: 970-529-4498
Mount Rainier National Park
Tahoma Woods, Star Route
Ashford, WA 98304-9751
Headquarter Phone: 360-569-2211
Fax Number: 360-569-2170

North Cascades National Park
2105 State Route 20
Sedro Woolley, WA 98284-9314
Headquarter Phone: 360-856-5700
Fax Number: 360-856-1934

Olympic National Park
600 East Park Avenue
Port Angeles, WA 98362-6798
Headquarter Phone: 360-452-4501
Fax Number: 360-452-0335

Petrified Forest National Park
Box 2217
Petrified Forest, AZ 86028
Headquarter Phone: 520-524-6228
Fax Number: 520-524-3567

Redwood National Park
1111 Second Street
Crescent City, CA 95531
Headquarter Phone: 707-464-6101
Fax Number: 707-464-1812

Rocky Mountain National Park
c/o Rocky Mountain National Park
Estes Park, CO 80517-8397
Headquarter Phone: 970-586-1399
Fax Number: 970-586-1310

Saguaro National Park
3693 South Old Spanish Trail
Tucson, AZ 85730-5601
Headquarter Phone: 520-733-5153
Fax Number: 520-733-6681

Sequoia & Kings Canyon National Park
c/o Sequoia & Kings Canyon
National Park
Three Rivers, CA 93271
Headquarter Phone: 209-565-3341
Fax Number: 209-565-3730

Shenandoah National Park
Route 4Box 348
Luray, VA 22835
Headquarter Phone: 540-999-3500
Fax Number: 540-999-3601
Theodore Roosevelt National Park
315 Second Avenue
P.O. Box 7
Medora, ND 58645-0007
Headquarter Phone: 701-623-4466
Fax Number: 701-623-4840

Voyageurs National Park
3131 Highway 53
International Falls, MN 56649-8904
Headquarter Phone: 218-283-9821
Fax Number: 218-285-7407

Wind Cave National Park
RR #1, Box 190 - WCNational Park
Hot Springs, SD 57747-9430
Headquarter Phone: 605-745-4600
Fax Number: 605-745-4207

Wrangell-St Elias National Park & PRES
P.O. Box 439
Mile 105.5 Old Richardson Hwy.
Copper Center, AK 99573
Headquarter Phone: 907-822-5234
Fax Number: 907-822-7216

Yellowstone National Park
P.O. Box 168
Yellowstone, WY 82190
Headquarter Phone: 307-344-7381
Fax Number: 307-344-2005

Yosemite National Park
P.O. Box 577
Yosemite, CA 95389
Headquarter Phone: 209-372-0201
Fax Number: 209-372-0220

Zion National Park
Springdale, UT 84767-1099
Headquarter Phone: 801-772-3256
Fax Number: 801-772-3426

Cracker Barrel
Old Country Store

Cracker Barrel Old Country Store

Convenient Locations in 33 States

ALABAMA
ATHENS*
CULLMAN
GARDENDALE
GREENVILLE*
MADISON*
MOBILE*
MONTGOMERY
MOODY*
OPELIKA*
OXFORD
PELHAM
PRATTVILLE*
TUSCALOOSA*

ARIZONA
FLAGSTAFF*
GOODYEAR*
MARANA*
PEORIA*
YUMA

ARKANSAS
ALMA*
BRYANT*
CONWAY*
RUSSELLVILLE*

COLORADO
COLORADO SPRINGS*
LOVELAND*
NORTHGLENN*
PUEBLO*

CONNECTICUT
MILFORD

FLORIDA
BOYNTON BEACH*
BRADENTON*
BROOKSVILLE*
CRESTVIEW*
DAYTONA BEACH*
FT. MEYERS*
FT. PIERCE*
GAINESVILLE*
JACKSONVILLE*
KISSIMMEE*
LAKE CITY*
LAKELAND*
MELBOURNE*
NAPLES*
OCALA*
ORANGE PARK*
ORLANDO*
PALM COAST*
PENSACOLA*
PORT CHARLOTTE*
SANFORD*
SEFFNER*
ST. AUGUSTINE
STUART*
TALLAHASSEE
TITUSVILLE*
VENICE
VERO BEACH*
WESLEY CHAPEL*
WEST PALM BEACH*

GEORGIA
AUGUSTA *
BRUNSWICK*
CARTERSVILLE*
COMMERCE*
CONYERS*
DALTON*
DOUGLASVILLE*
KENNESAW
LAKE PARK*
MACON (2)
MARIETTA
MORROW*
NORCROSS
PERRY*
SAVANNAH*
SUWANEE*
TIFTON*
UNION CITY
VALDOSTA

ILLINOIS
BLOOMINGTON*
BRADLEY
CASEYVILLE*
DECATUR*
EFFINGHAM*
ELGIN*
GURNEE*
JOLIET*
MARION*
MATTESON*
MORTON*
MT VERNON*
NAPERVILLE*
ROCKFORD*
ROMEOVILLE
SPRINGFIELD*
TINLEY PARK*
TROY*
URBANA*

INDIANA
ANDERSON
EDINBURGH*
ELKHART*
EVANSVILLE*
FISHERS*
FT. WAYNE
INDIANAPOLIS (3)
LAFAYETTE
MERRILLVILLE*
PLAINFIELD*
RICHMOND*
SEYMOUR*
TERRE HAUTE*

IOWA
CLIVE*
COUNCIL BLUFFS
DAVENPORT*

KANSAS
KANSAS CITY
OLATHE*
TOPEKA*

KENTUCKY
BOWLING GREEN*
CORBIN*
ELIZABETHTOWN
FLORENCE*
FRANKLIN*
GEORGETOWN*
JEFFERSONTOWN*
LAGRANGE
LEXINGTON (2)*
MT. STERLING*
PADUCAH
RICHMOND
SHEPHERDSVILLE*

LOUISIANA
BATON ROUGE*
GONZALES*
HAMMOND*
LAFAYETTE*
SHREVEPORT*
SLIDELL*
SULPHER*
WEST MONROE

MARYLAND
BELCAMP*
FREDERICK*

MICHIGAN
BATTLE CREEK*
BELLEVILLE*
BRIGHTON
FLINT*
GRANDVILLE*
GRAND RAPIDS*
JACKSON*
KALAMAZOO*
LANSING*
MONROE*
PORT HURON*
SAGINAW*
STEVENSVILLE*

MINNESOTA
BROOKLYN CENTER*
LAKEVILLE*
WOODBURY*

MISSISSIPPI
BATESVILLE
HATTIESBURG
HORN LAKE*
JACKSON*
MERIDIAN*
MOSS POINT
PEARL
VICKSBURG

MISSOURI
BRANSON*
CAPE GIRARDEAU*
COLUMBIA*
FENTON*
INDEPENDENCE*
JOPLIN*
LIBERTY*
SPRINGFIELD*
ST. CHARLES
ST. JOSEPH
ST. LOUIS*

MONTANA
BILLINGS*

NEW JERSEY
PENNSVILLE*

NEW MEXICO
ALBUQUERQUE*
LAS CRUCES*

NEW YORK
BLASDELL*
CICERO
E. GREENBUSH
LANCASTER
WATERTOWN

NORTH CAROLINA
ASHEVILLE*
BURLINGTON*
CHARLOTTE
CLEMMONS*
CONCORD*
DURHAM*
FAYETTEVILLE
GASTONIA
GREENSBORO*
HENDERSON*
HENDERSONVILLE
HICKORY*
JONESVILLE*
LEXINGTON*
LUMBERTON*
MOORESVILLE*
ROANOKE RAPIDS*
SMITHFIELD*
STATESVILLE*
WILSON

OHIO
AKRON*
AUSTINTOWN*
CAMBRIDGE*
CINCINNATI*
COLUMBUS*
DAYTON*
FINDLAY*
FOREST PARK
GROVE CITY*
LIMA*
MANSFIELD*
MEDINA*
MIDDLETON*
N. CANTON*
PERRYSBURG*
PINKERINGTON
SPRINGFIELD

OKLAHOMA
EDMOND*
MIDWEST CITY
NORMAN*
OKLAHOMA CITY*
TULSA

PENNSYLVANIA
CHAMBERSBURG*
ERIE*
FOGELSVILLE
HAMBERG
NEW STANTON
YORK*

SOUTH CAROLINA
ANDERSON*
COLUMBIA
FLORENCE
GAFFNEY*
GREENVILLE
HILTON HEAD
MURRELLS INLET*
N. CHARLESTON*
N. MYRTLE BEACH*
ROCK HILL*
SANTEE*
SPARTANBURG

TENNESSEE
CHATTANOOGA (3)*
CLARKSVILLE
CLEVELAND
COOKEVILLE
CROSSVILLE*
DICKSON*
FARRAGUT*
FRANKLIN*
HARRIMAN
JACKSON
JOHNSON CITY
KINGSPORT*
KNOXVILLE (3)*
LAKE CITY*
LEBANON*
MANCHESTER*
MEMPHIS
MT. JULIET*
MURFREESBORO (2)
NASHVILLE (6)
NEWPORT
PIGEON FORGE*
SWEETWATER*

TEXAS
AMARILLO*
ARLINGTON (2)*
BAYTOWN*
BENBROOK
BURLESON*
CONROE
CORPUS CHRISTI*
DENTON*
DESOTO*
EL PASO*
FT. WORTH*
LEAGUE CITY*
LEWISVILLE*
MESQUITE*
ROUND ROCK*
SAN ANTONIO (2)*
WACO*
W. HOUSTON*

UTAH
LAYTON
SPRINGVILLE*
WEST VALLEY

VIRGINIA
ABINGDON*
ASHLAND*
CHESTER*
CHRISTIANSBURG*
DUMFRIES*
FREDERICKSBURG*
HARRISONBURG*
MANASSAS*
MECHANICSVILLE*
NEWPORT NEWS*
STAUNTON*
TROUTVILLE
WILLIAMSBURG*
WINCHESTER*

WEST VIRGINIA
BARBOURSVILLE*
BECKLEY*
MARTINSBURG
MINERAL WELLS*
PRINCETON

WISCONSIN
GERMANTOWN*
JANESVILLE*
KENOSHA*
MADISON*

*Designated RV and bus parking

We hope you'll visit your nearest Cracker Barrel for a detailed map of our store locations.

Favorite Outlet Stops

Nationwide

Location	Number of Stores
Orlando, Florida	**233**
Belz Factory Outlet World	122
International Designer Outlets	45
Kissimmee Manufacturer's Mall	25
Lake Buena Vista Factory	24
Quality Outlet Centers	17
I-35 Austin to San Antonio, Texas	**177**
San Marcos Factory Shops	104
New Braunfels Mill Store Plaza	39
Tanger Factory Outlet Center	34
I-45 Dallas to Galveston, Texas	**80**
*Factory Stores of America at La Marque	35
*Factory Stores of America at Corsicana	4
Conroe Outlet Center	26
*Factory Stores of America at Hempstead	5
*Factory Stores of America at Sulphur Springs	11
Mineral Wells, Texas	**4**
*Factory Stores of America	4
Lancaster, Pennsylvania	**171**
Rockvale Square Outlets	115
Tanger Factory Outlet Center at Millstream	50
Quality Outlets	6
Georgetown, Kentucky	**32**
*Factory Stores of America	32

Location	Number of Stores
Birch Run, Michigan	**169**
Outlets at Birch Run	169
Humbolt, Tennessee	**3**
*Dan's Factory Outlet	3
Williamsburg, Virginia	**158**
Berkeley Commons Outlet Center	75
Williamsburg Outlet Mall	49
Williamsburg Pottery	25
Patriot Plaza	9
Reading, Pennsylvania	**157**
Reading Outlet Center	72
Reading Station Outlet Center	19
VF Outlet Village/Desingner Place	66
Pigeon Forge, Tennessee	**155**
Belz Factory Outlet World	68
Pigeon Forge Outlet Mall	40
Five Oaks Factory Stores	23
Tanger Factory Outlet Center	24
Salt Lake City, Utah	**37**
*Factory Stores of America in Draper	36
Patagonia (Single Outlet)	1
Central Valley, New York	**146**
Woodbury Common Factory Outlets	146
Jeffersonville, Ohio	**142**
Ohio Factory Shops	78
Jeffersonville Outlet Center	64
Blountsville, Tennessee	**20**
*Factory Stores of America	20
Crossville, Tennessee	**23**
*Factory Stores of America	23
Kenosha, Wisconsin	**141**
Lakeside Marketplace	71
Kenosha Outlet Centre	70

** See Travel Saver Coupons Section for Special coupon offer*

Continues . . .

Favorite Outlet Stops

Nationwide

Location	Number of Stores
Lake Park, Georgia	**32**
*Factory Stores of America	20
Lake Park Factory Stores	12
Nashville, Tennessee	**103**
100 Oaks Mall	33
*Factory Stores of America	70
Branson, Missouri	**138**
Factory Merchants Branson	75
Factory Stores of America	11
Tanger Factory Outlet Center	52
Rehoboth Beach, DE	**129**
Ocean Outlets Bayside/Seaside	92
Rehonoth Outlet Center	37
Gilroy, California	**124**
Outlets at Gilroy	124
Foley, Alabama	**122**
Riviera Centre Factory Stores	122
Kittery, Maine	**120**
Centers on Coastal Rte. 1 in Kittery	120
Barstow, California	**112**
*Factory Merchants Barstow	89
Tanger Factory Outlet Center	23
Michigan City, Indiana	**113**
Lighthouse Place	112
Jaymar Ruby Factory Outlet	1
Las Vegas, Nevada	**112**
Belz Factory Outlet World	66
*Factory Stores of America	45
Boaz, Alabama	**110**
Boaz Outlet Center	51
Fashion Outlets of Boaz	29
Tanger Factory Outlet Center	18
*Factory Stores of America	12
Livingston, Texas	**4**
Factory Stores of America	4
Secaucus, New Jersey	**110**
Harmon Cove Outlet Center	41
Outlets at the Cove	10
Other Centers & Freestanding Stores	59
Osage Beach, Missouri	**104**
Factory Outlet Village	104
Castle Rock, Colorado	**101**
Castle Rock Factory Shops	101
Flemington, New Jersey	**91**
Liberty Village Factory Outlet	56
Circle Outlet Center	13
Other Centers & Freestanding Stores	22
Max Meadows, Virginia	**33**
*Factory Merchants Fort Chiswell	33
St. Augustine, Florida	**95**
*St. Augustine Outlet Center	95
Jackson, New Jersey	**49**
*Six Flags Factory Outlet	49
Vacaville, California	**112**
*Factory Stores of America	112
Raleigh, North Carolina	**65**
*Factory Stores of America in Smithfield	40
Tower Shopping Complex	25
Phoenix, Arizona	**38**
*Factory Stores of America in Mesa	36
Converse Factory Outlet	1
Allen-Edmonds Company Shoe Store	1

** See Travel Saver Coupons Section for Special coupon offer*

DAYS INN®

Follow the Sun

No matter where life's road takes you, chances are there's a Days Inn nearby. In fact, with 1,800 locations you're sure to find a clean, comfortable room at a great value. Days Inn offers special travel programs like September Days Club® for seniors and the Inn-Credible Card Plus® club for business travelers. And try one of our Days Business Place℠ hotels featuring Days Work Zone℠ rooms which include a large work desk, in-room coffee, data ports and even a snack pack for every night you stay. So Follow the Sun to Days Inn. For reservations, call your travel agent or **1-800-DAYS INN.**

Every amenity. *Absolutely* free. *Everywhere* we are.

That's Wingate Inn.

Finally, a hotel chain that offers all the essential amenities necessary to wash away the stress of life on the road. All Wingate Inn® hotels are newly constructed and offer all-inclusive amenities, like a fitness room with whirlpool, in-room cordless phones and an expanded continental breakfast—at rates lower than you might expect.

At every Wingate Inn hotel you'll find:

Complimentary:

- expanded continental breakfast
- local calls and long-distance access
- fitness room with whirlpool
- 24-hour business center with fax, copier and printer

Plus Standard In-Room Amenities Like:

- two-line speakerphone with data port, conference call and voice mail capabilities
- coffee maker, iron, ironing board and safe
- 900 megahertz cordless phone
- comfortable, well-lighted work space

And:

- boardroom and meeting rooms
- automated check-in and check-out
- 100% Satisfaction Guarantee

So get off the road and check in to a Wingate Inn hotel. You'll see why we're built for the road-weary traveler.

BUILT FOR BUSINESS®

For Reservations, Contact Your Travel Professional Or Call

1.800.228.1000

www.wingateinns.com

Travelodge® Makes Life On The Road Easier To Bear.

When you're on the road and need a comfortable place to rest your heels (and wheels) for the night, follow Sleepy Bear® into the nearest Travelodge® location. We start with a great room at an affordable rate, then add special extras* like: **Free fresh-brewed in-room coffee**, a **free weekday lobby newspaper**, **no long distance access charges**, and **free cable TV channels** including **movies, news and sports**. Plus, with our Travelodge Miles℠ Program** you can earn free night stays,† airline frequent flyer miles, or your choice of dozens of other guest rewards like watches, cameras, luggage and more. With all that, it's easy to see why so many travelers like to turn in with us. Call for reservations today.

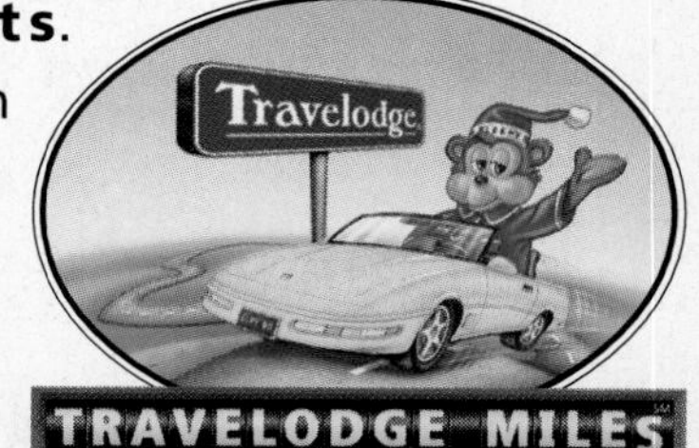

TRAVELODGE MILES℠
The Guest Rewards Program

For reservations at our more than 500 locations throughout North America call:

1-800-578-7878

www.travelodge.com

*Special amenities available only at Travelodge® locations; not available at Thriftlodge® locations. Weekday newspaper available while supplies last. **To earn Travelodge Miles℠ you must pay Regular Rate, Travelodge Miles Guest Rewards Rate or higher. One Travelodge Mile will be awarded for each lodging dollar spent, not for dollars spent on incidentals or taxes. The Travelodge Miles program is subject to change or discontinuation without notice. Rewards vary by country of residence; see Official Brochure for details. †Free Night Certificate valid at any participating Travelodge or Thriftlodge location throughout North America. Sleepy Bear® is a registered trade-mark in Canada of Travelodge Hotels, Inc. Sleepy Bear is a registered service mark in the United States of TM Acquisition Corp., an affiliate of Travelodge Hotels, Inc.

CALL TOLL FREE 1-800-843-5644
1-800-TheKnights
Defend your savings.
Capture the value.
When it comes to defending your dollars with legendary value, the Knights is your "Exit Authority!"
Coast-to-coast, we keep you rolling with clean, comfortable accommodations at true budget rates. With over 200 convenient locations, Knights makes it easy for you to get a good night's rest and be on the road again – with more money to spend on the things great travels are made of.
Take the exit to one of the best values in budget lodging – just look for the shield that defends your dollars all along the way and back again!
Knights Inn Royalty Club
Save more with a "car load" of member benefits every time you travel... and even when you don't. Join the Royalty® Club today!
Knights Inn
Spend less here.
Spend more there.
www.knightsinn.com
Knights Inn®

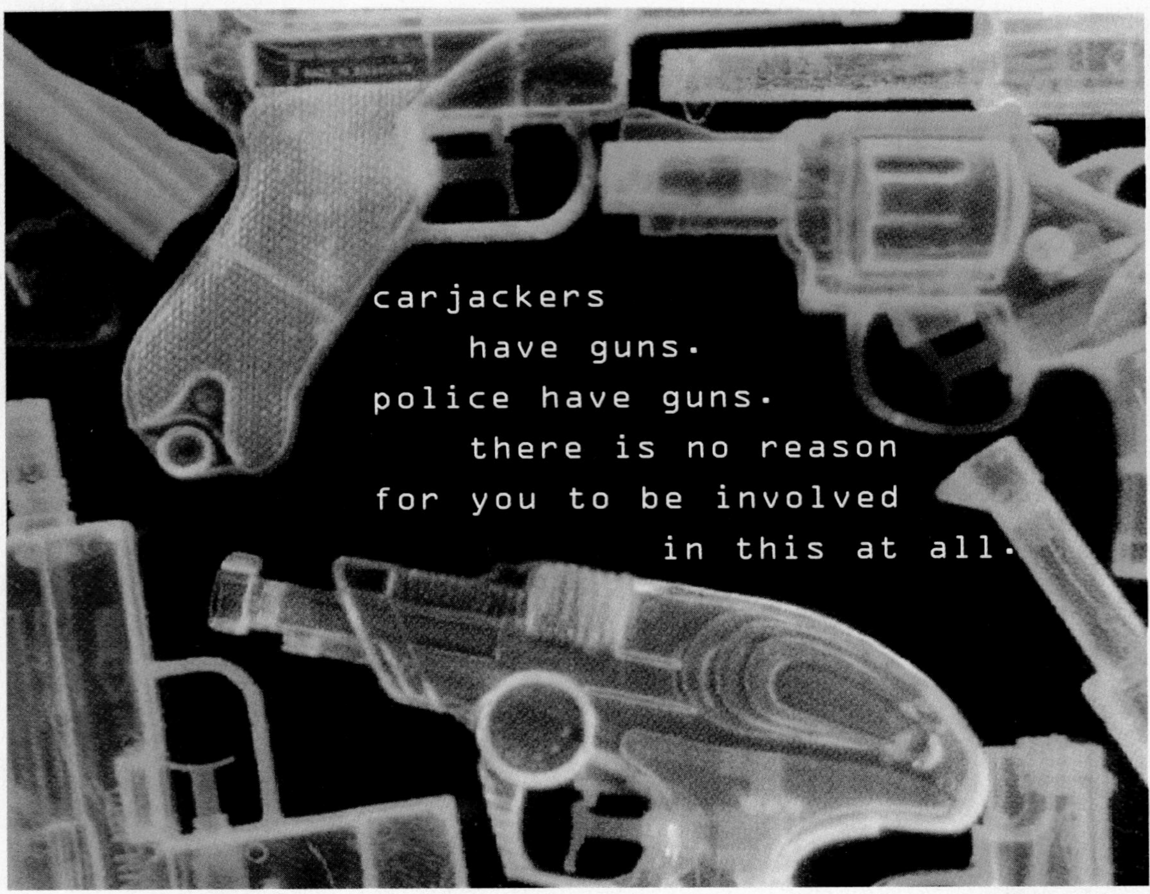

carjackers
have guns.
police have guns.
there is no reason
for you to be involved
in this at all.

> When you have The CERES System protecting you and your car, you just press a button at the first sign of trouble. Immediately, a CERES 24-hour operator is alerted, satellites pinpoint your car's location and the police are on their way.

And as thorough as our carjacking feature is, it is but one of many. Because CERES is designed to give you complete protection and peace of mind wherever you travel with your car.

For example, in a medical emergency, press a button and the operator sends medical help to you. If your car breaks down, CERES will find you and send roadside assistance. If you're lost, it will help you find your way.

You'll also have a panic button for your Remote Key. Press it and your car's horn will sound, the light will flash and one of our operators will be alerted.

CERES will even protect your car when you're not in it. Because your car won't start unless the correct personal ID number is entered. And if your car does get stolen, the satellites will locate it and an operator will alert the police.

USES SATELLITE TECHNOLOGY.

CERES has been designed with your total security in mind. It's less than $1000 installed, plus a low monthly monitoring fee, and it's worth it. Because it guards you against everything from carjackers to dishonest valets and makes criminals who are thinking about hurting you think about something else. Like finding a good lawyer. For more information or for installation, call 1-800-934-0546.

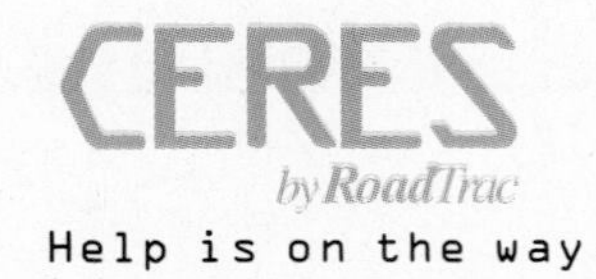

CERES
by RoadTrac

Help is on the way.

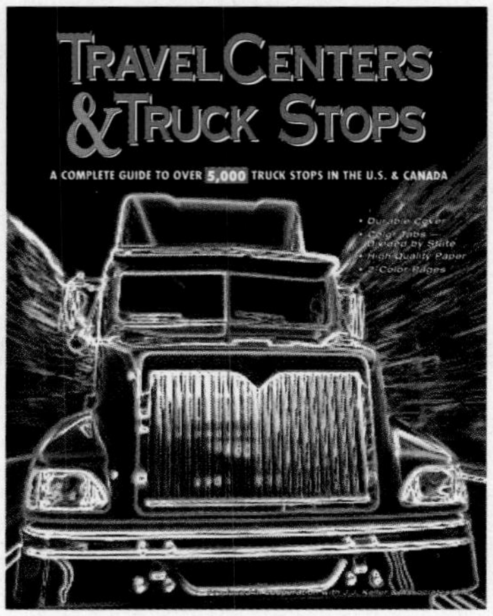

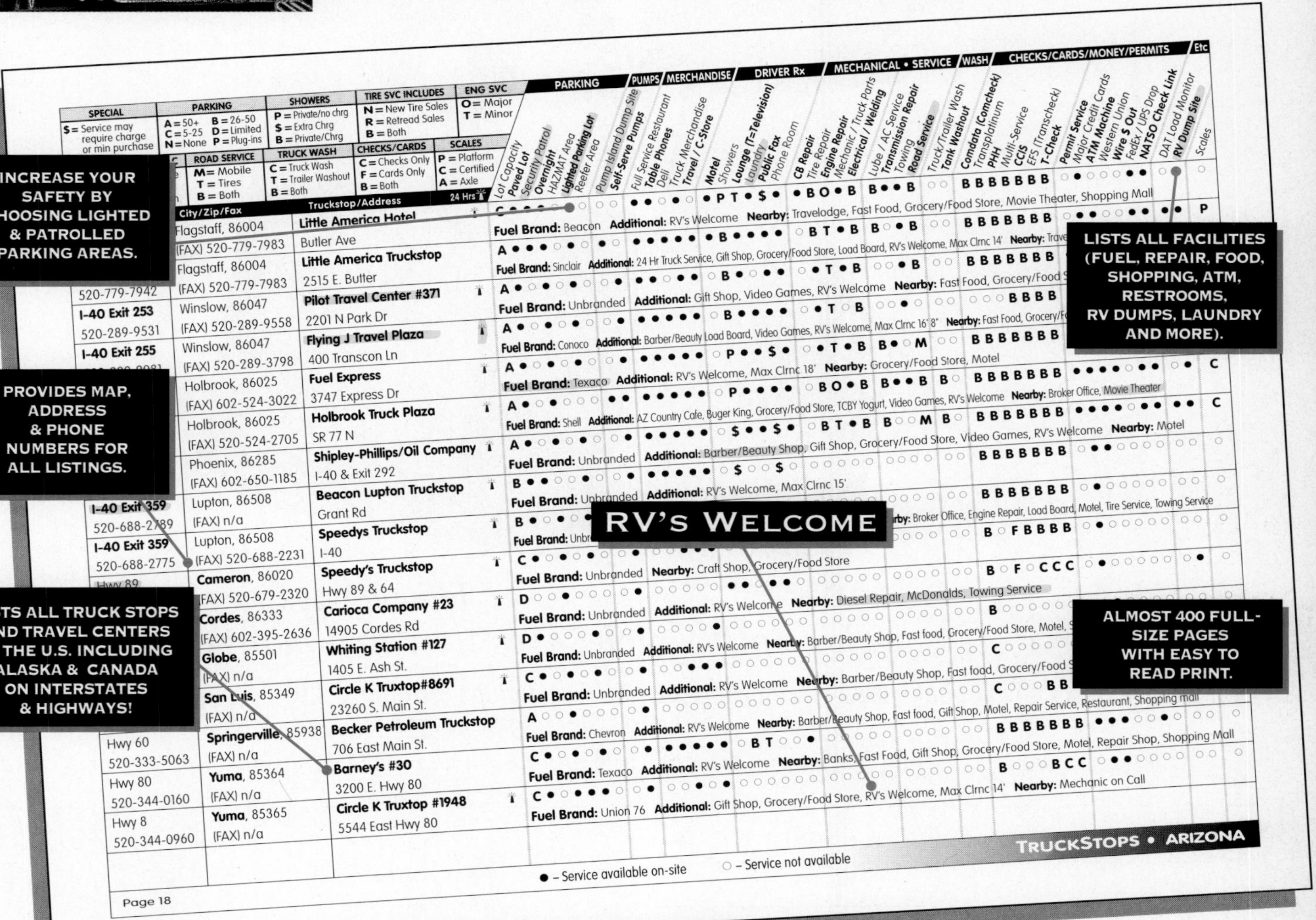

The Perfect Gift for *your* Special Traveler!

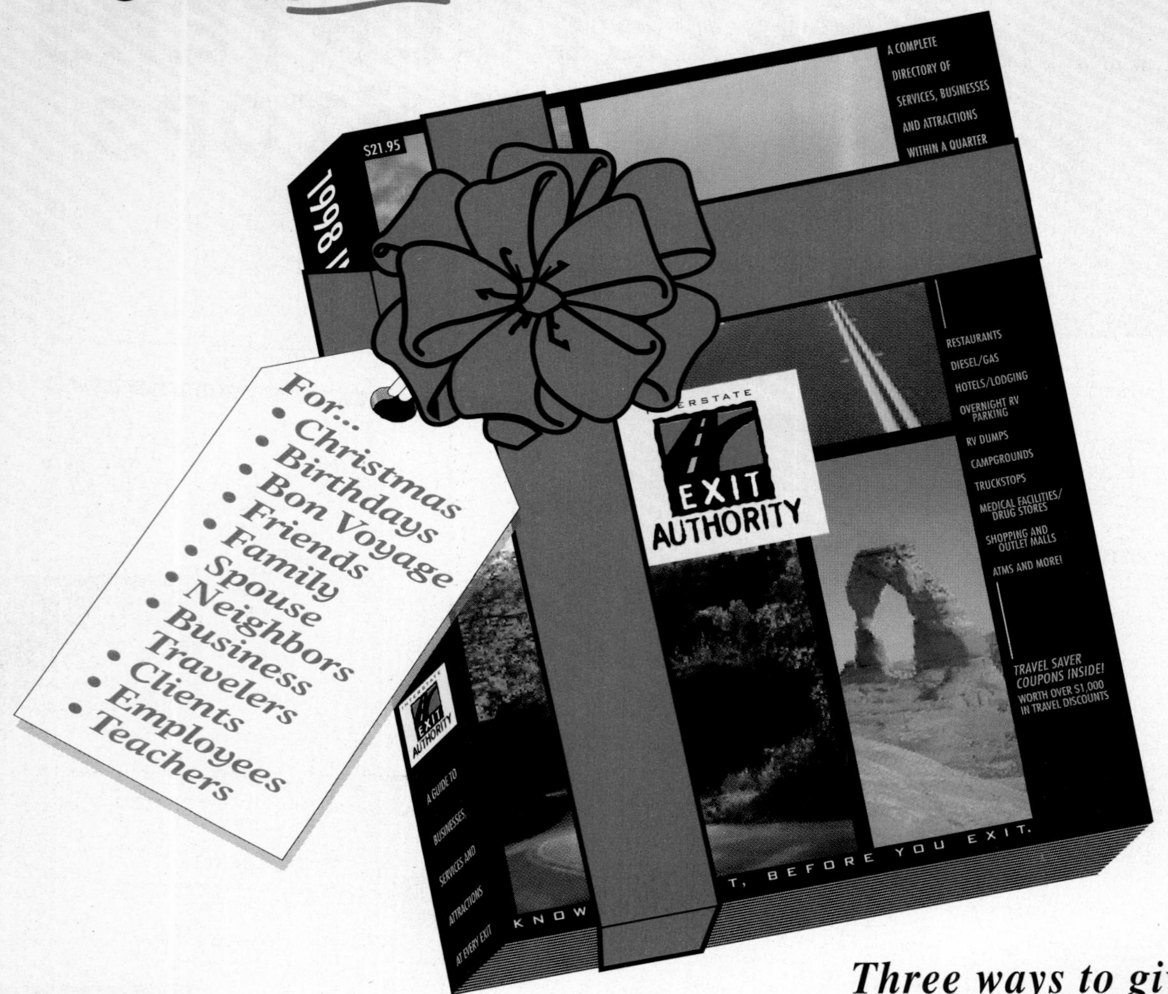

Three ways to give someone a unique gift which saves time and money all year long!

1. Pick up a copy at bookstores, AAA locations or RV Dealerships for only $21.95 or let us deliver one for you!
2. Call 1-800-494-5566 ext. 59 with your credit card number.
3. Clip or photocopy this order form and mail with payment to Interstate America.

Bill To:

Name ______________________

Address ______________________

City ____________ State ______ Zip ________

Phone ______________________

Shipping: ☐ Regular ☐ Priority

☐ Enclosed is my check for $21.95 ($23.95 for priority shipping) payable to Interstate America.

☐ Bill my credit card ☐ MC ☐ VISA

Card # ______________ Exp. date: ______

Signature ______________________

Deliver to: ☐ Same Address ☐ Address Below

Name ______________________

Address ______________________

City ____________ State ______ Zip ________

Phone ______________________

Gift Message: ______________________

Mail to: Interstate America • Dept. 59 • 5695G Oakbrook Parkway • Norcross, GA 30093

Villager® Value

There's more in it for you

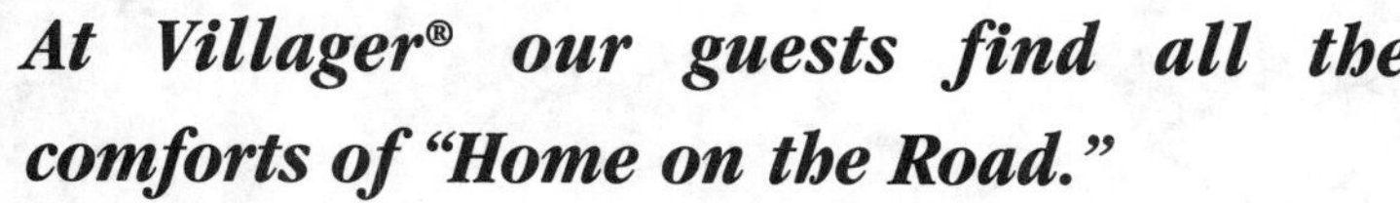

At Villager® our guests find all the comforts of "Home on the Road."

- By offering great rooms that feature mini-kitchenettes, Cable TV, a premium movie channel and with free hot coffee waiting in the lobby, there really is more in it for you.

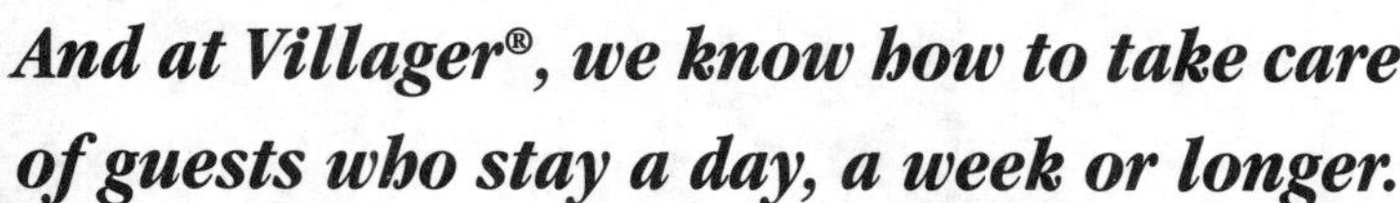

And at Villager®, we know how to take care of guests who stay a day, a week or longer.

- With extended stay rates you will really see the value of "Extended Stay Living℠."
- Or take advantage of our affordable daily rates for stays of six days or less. Either way, you will find the Villager Value.

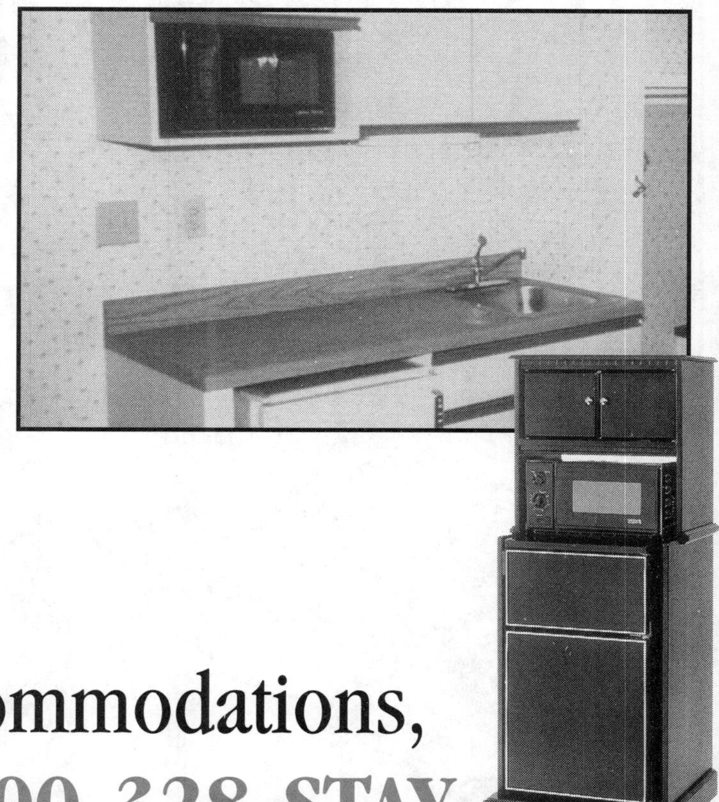

For accommodations, call 1-800-328-STAY or your travel professional.

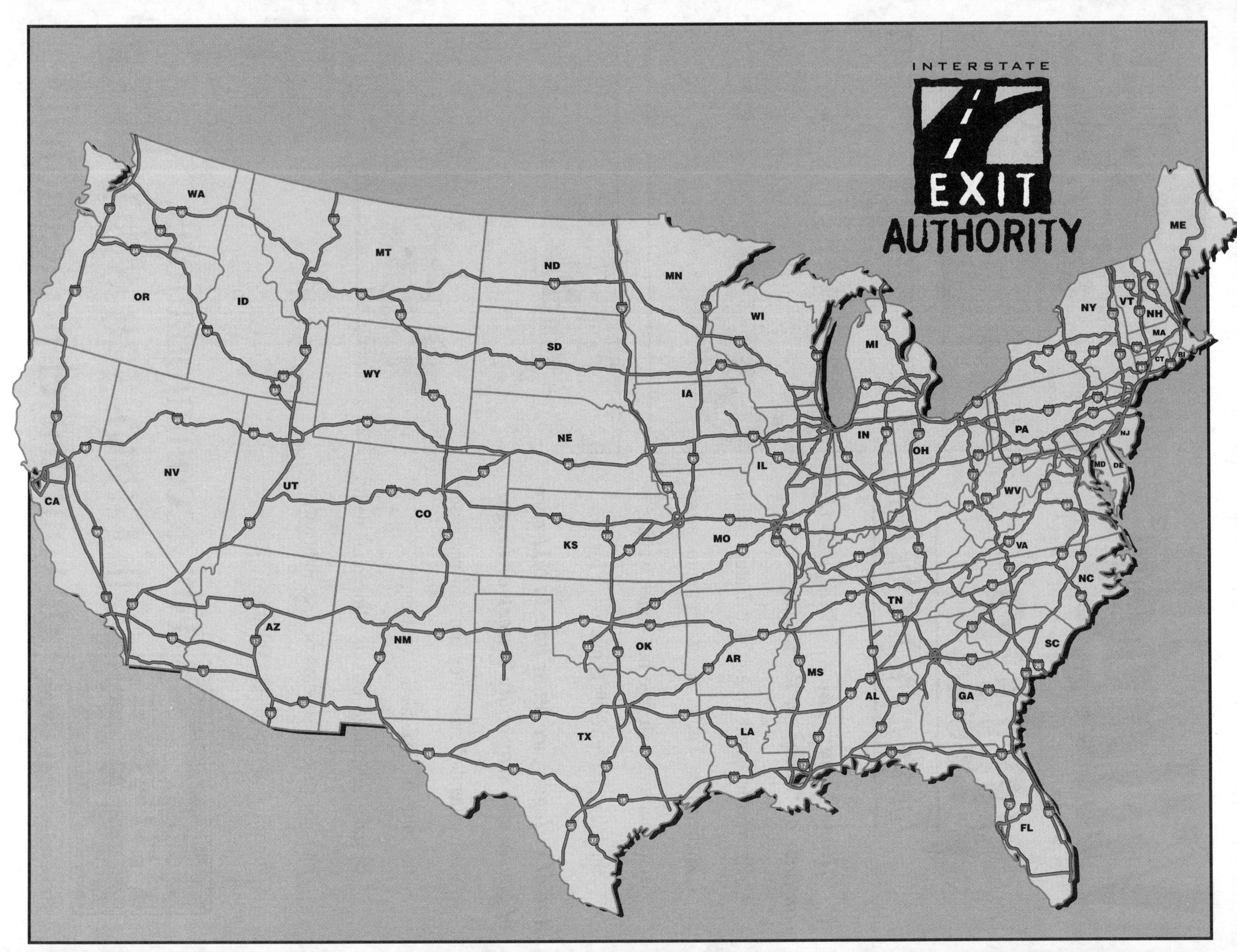

INTERSTATE
EXIT
AUTHORITY
WA
OR
ID
MT
ND
MN
SD
WY
NV
CA
UT
CO
NE
IA
KS
MO
IL
IN
OH
MI
WI
NY
VT
NH
ME
MA
CT
RI
PA
NJ
MD
DE
WV
VA
NC
SC
GA
FL
AL
MS
TN
AR
LA
OK
TX
NM
AZ

U.S. Time Zones & Telephone Area Codes

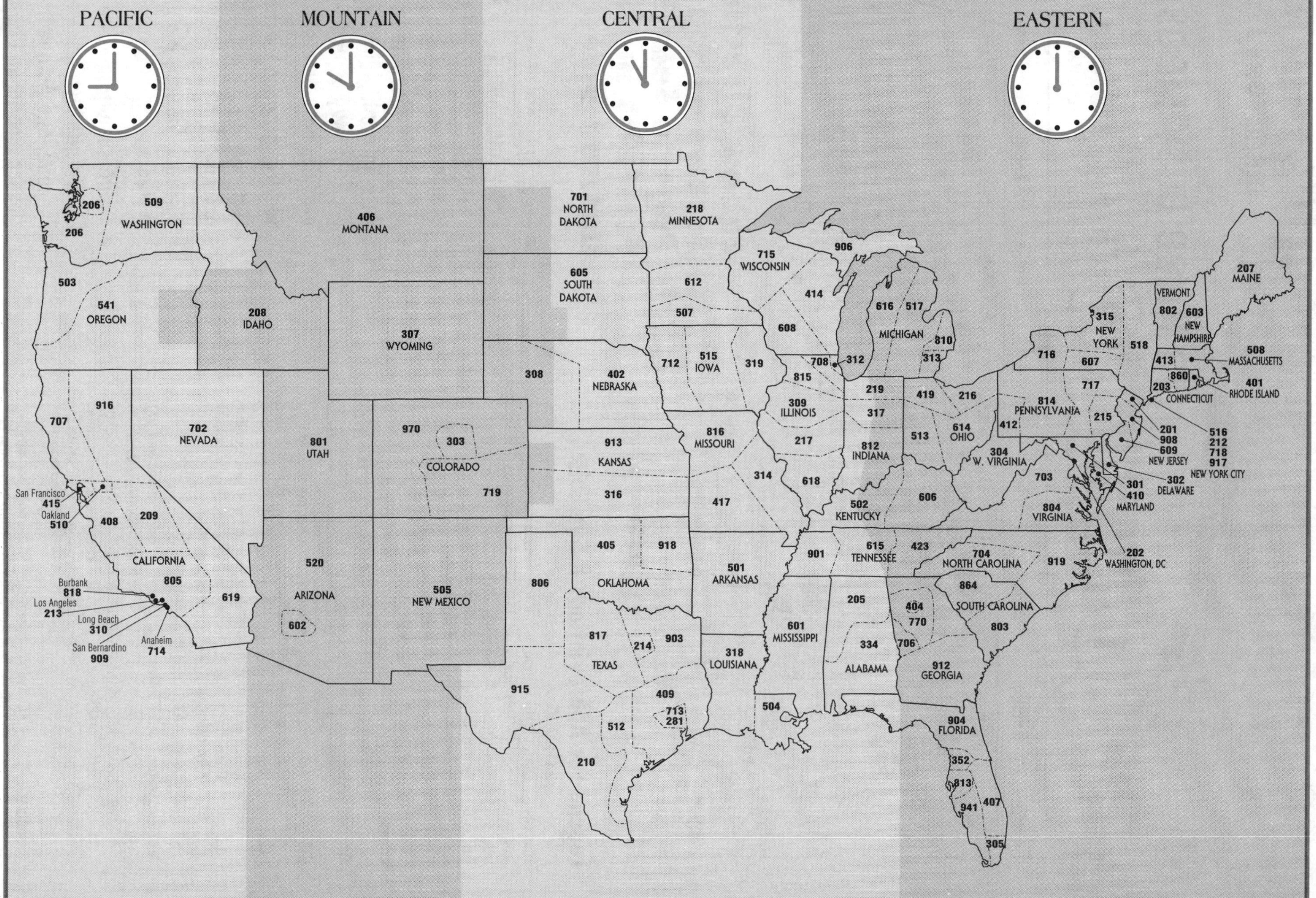

PACIFIC
MOUNTAIN
CENTRAL
EASTERN

206
206
509
WASHINGTON
503
541
OREGON
707
916
408
209
CALIFORNIA
805
619
San Francisco
415
Oakland
510
Burbank
818
Los Angeles
213
Long Beach
310
San Bernardino
909
Anaheim
714
702
NEVADA
208
IDAHO
406
MONTANA
307
WYOMING
801
UTAH
970
303
COLORADO
719
520
ARIZONA
602
505
NEW MEXICO
701
NORTH DAKOTA
605
SOUTH DAKOTA
308
402
NEBRASKA
913
KANSAS
316
405
918
OKLAHOMA
806
817
214
903
TEXAS
915
409
713
281
512
210
218
MINNESOTA
612
507
712
515
IOWA
319
816
MISSOURI
314
417
501
ARKANSAS
318
LOUISIANA
504
715
WISCONSIN
414
608
906
815
708
312
309
ILLINOIS
217
618
616
517
MICHIGAN
810
313
219
317
812
INDIANA
419
216
614
OHIO
513
502
KENTUCKY
606
901
615
TENNESSEE
423
601
MISSISSIPPI
205
334
ALABAMA
404
770
706
912
GEORGIA
904
FLORIDA
352
813
941
407
305
864
SOUTH CAROLINA
803
704
NORTH CAROLINA
919
804
VIRGINIA
703
304
W. VIRGINIA
412
814
PENNSYLVANIA
717
215
716
607
315
NEW YORK
518
VERMONT
802
603
NEW HAMPSHIRE
207
MAINE
413
508
MASSACHUSETTS
860
203
CONNECTICUT
401
RHODE ISLAND
516
212
718
917
NEW YORK CITY
201
908
609
NEW JERSEY
302
DELAWARE
301
410
MARYLAND
202
WASHINGTON, DC

How to Use the Exit Authority...

SEE BACK PAGE OF DIRECTORY FOR INDEX

in three easy steps...

1 Turn to the Interstate.

(Interstates are listed in numerical order)

2 Find the State.

(Organized North to South or East to West)

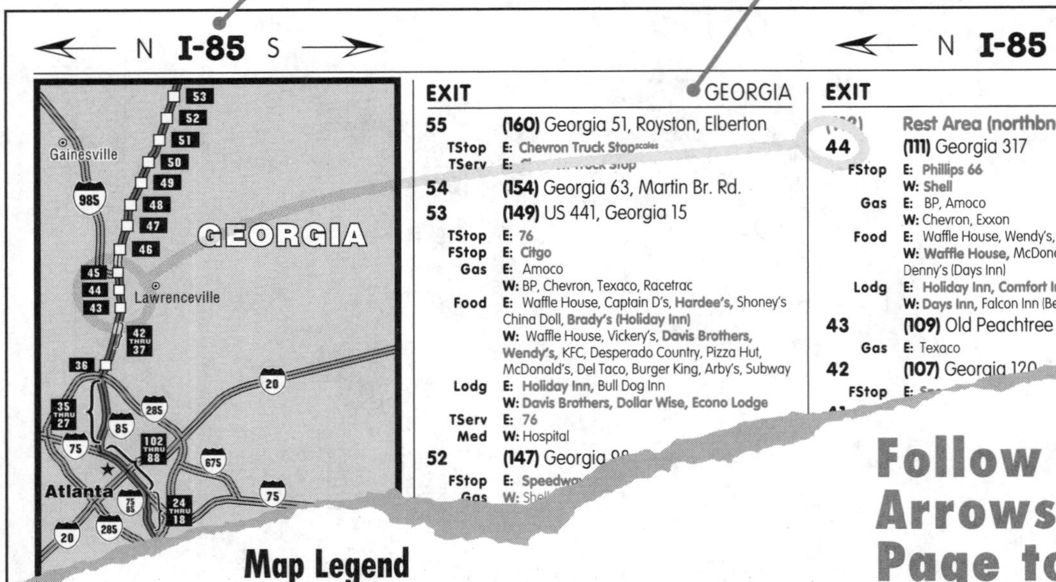

3 Follow Directional Arrows at top of Page to Search For Exits in the Desired Direction.

Legend For Abbreviations & Symbols:

TStop	24-Hour diesel fuel location w/full service restaurant and/or store
FStop	Diesel fuel location with large vehicle clearance
Food	Food Outlets (fast food, restaurants, cafeterias, etc.)
Lodg	Hotel/Motel Accommodations
Gas	Automobile fueling locations
AServ	Facilities with Automotive Service
TServ	Commercial truck/diesel engine service facilites
TWash	Commercial vehicle wash facilites
RVCamp	Camping sites, RV service or supply facilities
Parks	National, state or local forest, parks, preserves & lakes
ATM	Banks or other services services with cash machine (please use caution - many banking institutions cannot accommodate large RV's. For RV access check other listings at the exit)
Med	Hospital, emergency treatment or other medical facility
Other	Useful services, sites or attractions near interstate
(MM)	Numbers in ( ) indicate mile markers when not same as exit #.
N-S-E-W	Indicates side of the highway that services are located
I-O	Three-digit perimeter hwys – (I) inside or (O) outside perimeter
AAA	AAA Official Listing Requirements
◆	Meets AAA Listing Requirements

Rest Area & Welcome Center Facilities: P Overnight parking allowed at rest area

RR – Rest rooms	RV Dump – Sanitary Waste Dump
Vending – Snack & Drink Machine	RV Water – RV Water Hookup
Phones – Public Pay Phones	HF – Handicapped Facilities
Grills – Outdoor Cooking Grills	Grills – Outdoor Cooking Grills
Picnic – Picnic Tables	Pet Walk – Designated Area for Pets

Superscripts: (*) Convenience store; (cw) Automatic car wash; (D) Diesel Fuel Available; (LP) Liquid propane gas

- Every Business at every exit for every Interstate in the U.S. (excluding Alaska) within one quarter mile.
- Areas with adequate RV & Bus Parking are highlighted in Red (Parking may be nearby, but provides easy access)
- Services grouped by category (see legend) and direction from exit (N-S-E-W)
- RV Dump Stations highlighted in Red
- Additional notations for Convenience Stores, Diesel Fuel, Propane, and Automatic Car Washes (see legend at left)

EXIT FLORIDA

Begin I-4

↓FLORIDA

1 FL 585, 22nd Street, 21st Street
- **Gas** **N:** BP*, Fina*(D)
  - **S:** Amoco*
- **Food** **S:** Burger King, Hardee's
- **Other** **S:** Ybor City Historic District Museum

2 **(1)** FL 569, 40th Street
- **FStop** **N:** Texaco*
- **Gas** **N:** Citgo*(D)
- **Food** **N:** Budget Inn
- **Lodg** **N:** Budget Inn
- **AServ** **N:** Penske Auto Service
- **ATM** **N:** Texaoc

3 **(2)** U.S. 41, 50th Street, Columbus Dr.
- **FStop** **S:** Speedway*(LP), United 500*
- **Gas** **N:** Texaco*(D)
  - **S:** Exxon(D)(CW)
- **Food** **N:** Subway (Texaco)
  - **S:** Burger King (Playground), Church's Fried Chicken, Days Inn, Eggroll King Chinese, Pizza Hut, Taco Bell, Wendy's
- **Lodg** **N:** Econolodge, ◆ Holiday Inn, AAA La Quinta Inn
  - **S:** Days Inn Days Inn

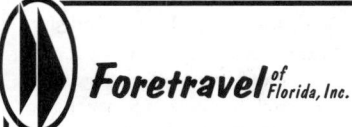

EXIT FLORIDA

- **ATM** **N:** Texaco
- **Other** **S:** Eckerd Drugs

4 **(5)** FL 574, Martin Luther King Jr. Blvd
- **Food** **N:** McDonalds
  - **S:** Wendy's
- **Lodg** **S:** Economy Inns
- **TServ** **S:** Great Dane Trailers, Kenworth

5 **(6)** Orient Road (Eastbound, Westbound Reaccess)

6AB **(7)** U.S. 92, to U.S. 301, Riverview

6C **(8)** U.S. 92 West, Hillsborough Ave
- **TStop** **N:** Cigar City Auto Truck Stop(SCALES)
- **Gas** **N:** Circle K*, Citgo*(CW)
  - **S:** Racetrac*
- **Food** **N:** Cigar City Family Restaurant (Cigar City TS), Sky Box Sports Cafe
  - **S:** Denny's
- **Lodg** **N:** Cigar City Truck Stop, Eastlake Inn, Four Points Motel
  - **S:** Budgetel Inn, Red Roof Inn
- **TServ** **N:** Cigar City Truck Service TS, Freightliner Dealer
- **TWash** **N:** Cigar City Truck Stop
- **RVCamp** **N:** Beaudry's RV & Travel Trailer Service
  - **S:** Holiday RV (see our ad this page), Foretravel, Inc. (see our ad this page)
- **ATM** **N:** Circle K
- **Other** **S:** State Fair Grounds

7 **(9)** Junction I-75, Ocala, Naples

8 **(10)** CR 579, Mango, Thonotosassa
- **FStop** **S:** Texaco*
- **Gas** **N:** Amoco*
- **Food** **N:** Cracker Barrel Cracker Barrel

EXIT FLORIDA

- **S:** Denny's, Hardee's, Wendy's
- **Lodg** **S:** Master's Inn
- **RVCamp** **N:** Camping World (see our ad this page), Lazy Days RV Park, Lazy Days RV Supercenter Sales & Service
- **ATM** **S:** Texaco FS

9 **(14)** McIntosh Road
- **Gas** **N:** Amoco
  - **S:** Racetrac*, Texaco*
- **Food** **S:** McDonalds
- **Other** **S:** Bates RV Exchange

10 **(17)** Branch Forbes Road
- **Gas** **N:** Shell* (see our ad this page), Spur*(D)

(19) Weigh Station (Both Directions)

11 **(19)** Thonotosassa Road
- **Gas** **N:** Amoco*
- **Food** **S:** Buddy Freddy's Southern Family Dining, McDonalds (Playground), Sonny's BBQ

12 **(21)** Alexander Street (Westbound, Reaccess Eastbound)
- **Med** **S:** ✚ Hospital

13 **(21)** FL. 39, Zephyrhills, Plant City
- **FStop** **N:** Texaco*
- **Gas** **S:** Amoco*(D), Shell* (24 Hrs) (see our ad this page)
- **Food** **S:** 1776 Restaurant

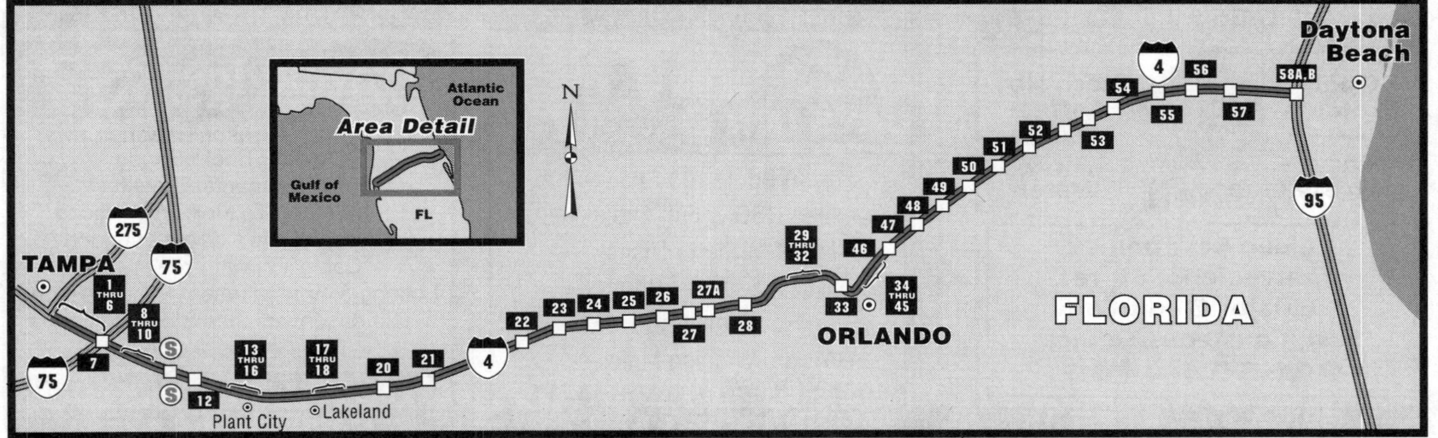

Bold red print shows RV & Bus parking available or nearby

EXIT FLORIDA

Lodg S: Days Inn ◆Days Inn (see our ad this page), ◆ Ramada Inn (see our ad this page)

14 **(23)** Plant City, Park Road
- Gas S: Shell* (see our ad this page), Texaco*(D)
- Food S: Arby's, Burger King, Denny's, Subway (Shell)

15 **(26)** County Line Road
- Gas S: Shell* (see our ad this page)
- ATM S: Shell
- Other S: Sun N Fun Aviation Center

16 **(28)** U.S. 92, Lakeland (Eastbound, Reaccess Westbound)

17 **(31)** FL. 539, Kathleen, Lakeland
- Other S: Munn Park Historic District

18 **(32)** U.S. 98, Lakeland, Dade City
- FStop S: Citgo*
- Gas N: Chevron*(D), Texaco*
  S: Amoco(CW), Coastal*(D), Racetrac*
- Food N: Bennigan's, Blimpie's Subs (Texaco), Chuck E. Cheese's, Don Pablo's, Hooters, IHOP, Lone Star Steakhouse, McDonalds (Play Land), Olive Garden, Shoney's, Steak & Shake
  S: Bob Evans Restaurant, Burger King (Play Land), Denny's, Grenson's (Citgo), Roadhouse Grill, Waffle House, Wendy's
- Lodg N: Royalty Inn, AAA Wellesley Inn & Suites
  S: ◆ Best Western, Motel 6, Ramada, Travelodge
- AServ N: Auto Express
  S: Amoco
- RVCamp N: Tiki Village Resort
- ATM N: Chevron, Texaco

EXIT FLORIDA

- S: Coastal
- Other N: Lakeland Square Mall, Wal-Mart
  S: Winn Dixie Supermarket

19 **(34)** CR. 582, FL. 33, to Lakeland.
- FStop S: Amoco*
- Gas N: Exxon*(CW)
- Food N: Cracker Barrel Cracker Barrel
  S: Icecream Churn (Amoco FS), Waffle House
- Lodg N: ◆ Budgetel Inn, Quality Inn
- Med S: ✚ Hospital
- ATM S: First Federal Florida

(34) Rest Area (Eastbound)

(35) Rest Area (Westbound)

20 **(38)** FL. 33

EXIT FLORIDA

21 **(44)** FL. 559, Auburndale, Polk City
- TStop S: Dixie Boy Fuel Center*(SCALES)
- Gas S: Amoco*
- Food S: Dixie Boy Restaurant (Dixie Boy TS)
- AServ S: Amoco
- TServ S: Dixie Boy Tire Service (Dixie Boy TS)
- ATM S: Dixie Boy TS

(46) Rest Area - Picnic, RR, HF, Phones (Eastbound)

(47) Rest Area - Picnic, RR, HF, Phones (Westbound)

22 **(48)** CR. 557, Lake Alfred, Winter Haven
- Gas S: BP*(D)

23 **(55)** U.S. 27, Haines City, Clermont
- TStop S: Commercial 76 A/T Stop*(SCALES) (On U.S.27N)
- FStop N: Speedway*
- Gas N: Chevron* (24 Hrs)
  S: Amoco*(D), Exxon*, Shell*, Texaco(D)
- Food N: Burger King, McDonalds, Shoney's, Waffle House, Wendy's
  S: Bob Evans Restaurant, Hardee's, Perkins Family Restaurant (24 Hrs)
- Lodg N: AAA Comfort Inn, ◆ Florida Southgate Inn
  S: Days Inn ◆ Days Inn (see our ad this page), Holiday Inn
- AServ S: Texaco

24 **(58)** CR. 532, Kissimmee (Eastbound)

25AB **(64)** U.S. 192, Kissimmee, Disney World (Access To Theme Parks)
- Food S: Atlantic Bay Seafood, Charly's Steakhouse, Kobe Japanese Steakhouse, Pizza Lover Italian, Shoney's, Summer House Breakfast Buffet
- Lodg N: Econolodge Main (see our ad opposite page), Ramada Inn Resort (see our ad opposite page)
  S: Hampton Inn Maingate (see our ad opposite page), Holiday Inn (see our ad opposite page), Homewood Suites (see our ad opposite page), Howard Johnson, Hyatt Hotel, Larson's Lodge, Radisson, Travelodge (see our ad this page)

Bold red print shows RV & Bus parking available or nearby

EXIT		FLORIDA
RVCamp		S: **Camping World (see our ad this page)**
Med		S: **Medi Clinic**
Other		N: **Disney-MGM Studios Theme Park, Magic Kingdom, Walt Disney World**
26AB		**(68)** FL. 536, Epcot Village, Disney Village
27		**(70)** FL. 535, Lake Buena Vista, Kissimmee
Gas		N: Chevron*(D) (24 Hrs), Citgo*, Texaco*(CW) S: Citgo*, Shell*(D)(CW)
Food		N: Baskin Robbins, Burger King, Chevy's Mexican, China Town Buffet, Dunkin' Donuts, IHOP, McDonalds, Miami Subs, Perkins Family Restaurant, Pizza Hut, Red Lobster, Shoney's, T.G.I. Friday's, Taco Bell, Uno Pizzaria S: Landry's Seafood, Lone Star Steakhouse, Wendy's
Lodg		N: AAA Days Inn (see our ad this page), Walt Disney World Resort S: AAA Holiday Inn
Med		N: **Centra Care (Walk-In Clinic)**
Other		N: **Gooding's Supermarket (24 Hrs, Pharmacy), Walt Disney World Resort (Access to Theme Parks)** S: **WalGreens (Pharmacy)**
(70)		**Rest Area - RR, Phones, Vending, Picnic**
27A		**(71)** Sea World (Eastbound, Reaccess Westbound)
Med		N: **Sandlake Hospital**
Other		S: **Sea World of Florida Marine Park**

EXIT		FLORIDA
28		**(72)** FL. 528 East Toll Rd., Cape Canaveral, Titusville
29		**(528)** FL 482, Sand Lake Road (2701.281)
Gas		N: Chevron* (24 Hrs), Citgo* S: Exxon*, Shell*(CW) (24 Hrs), Texaco*(D)
Food		N: Lakeside Restaurant & Cafeteria (Days Inn), Wendy's S: Burger King, Charlie's Steakhouse, Checkers Burgers, Chuck E. Cheese's Pizza, Denny's, Dunkin' Donuts, Fish Bones Restaurant, Golden Corral, IHOP, Italianni's, Kobe Japanese Steakhouse, McDonalds (Play Land), Morrisons Cafeteria, Olive Garden, Perkins Family Restaurant, Popeye's Chicken, The Crab House
Lodg		N: AAA Days Inn S: ◆ Embassy Suites, ◆ Fairfield Inn, AAA Hampton Inn, AAA Inns of America, AAA La Quinta Inn, ◆ Marriot, ◆ Quality Inn, AAA Radisson, AAA Ramada, ◆ Residence Inn
ATM		N: Citgo, NationsBank S: Shell, Texaco
RVCamp		S: **Holiday RV (see our ad this page)**
Other		N: **K-Mart, Publix Supermarket (Pharmacy), Universal Studios Florida** S: **Ripley's Believe-it-or-Not, WalGreens (Pharmacy)**
30AB		**(75)** FL 435, Kirkman Road
Gas		N: Chevron*(D) (24 Hrs), Mobil* S: Shell*(CW)
Food		N: Hooter's, Kobe Japanese Steakhouse,

EXIT FLORIDA

McDonalds (Playground), Ponderosa, Shoney's, Waffle House, Wendy's
S: Bagel King, Kenny Rogers Roasters, Perkins Family Restaurant, Pizza Hut, Steak & Ale, Subway, Wendy's, Western Steer Family Steakhouse, Wing Ding Chinese

Lodg **N:** Hampton Inn, Holiday Inn (see our ad this page), Radisson
S: Days Inn (Exit 30A -- see our ad this page), Days Inn (Exit 30B -- see our ad this page), Gateway Inn, Hampton Inn, Holiday Inn Express (see our ad this page), Universal Inn

ATM **N:** Amsouth Bank

Other **N:** Universal Studios Florida, WalGreens (Pharmacy)
S: Wet N Wild Water Park

31 **(76)** Junction Florida Turnpike, to Miami (Toll Road)

32 **(80)** 33rd Street, John Young Pkwy

Gas **S:** Citgo*, Racetrac*, Texaco*(D)(CW)

Food **N:** McDonalds, Sub Station
S: IHOP

Lodg **N:** Catalina Inn
S: Days Inn

AServ **S:** Texaco

33 **(80)** U.S. 441, U.S. 17, U.S. 92

Gas **N:** Amoco*
S: Hess*, Racetrac*

Food **N:** China Palace
S: Checkers, Denny's, Gary's Restaurant, Ryan's Steakhouse, Tam Tam Restaurant, Waffle House, Wendy's

Lodg **S:** Regency Inn

AServ **N:** Cruz Muffler Shop, F.L.M. Automotive Inc., General Tire, Larry's Discount Auto Parts

34 **(81)** Michigan Street (Northbound)

Gas **N:** Citgo*(D)(CW)

35AB **(81)** West Kaley Avenue

Gas **S:** Mobil(D), Texaco*(D)

AServ **S:** Dop Auto Repair, Mobil

Med **S:** ✚ Hospital

Other **S:** Coin Laundry

36 **(82)** Junction East - West Expressway Fl. 408 (Toll Road)

37 **(82)** Gore Street

Gas **N:** Onmark*

38 **(83)** Anderson Street

39 **(84)** South Street

40 **(84)** FL. 526, Robinson Street (Eastbound, Reaccess Westbound)

AServ **S:** Goodyear Tire & Auto

ATM **S:** NationsBank

41 **(84)** U.S. 17, U.S. 92, FL 50, Colonial Dr., Naval Training Center

Food **N:** Angel's Diner, Steak & Ale
S: Mama B's Giant Subs

Lodg **N:** ◆ Holiday Inn, Howard Vernon Motel
S: Knights Inn (see our ad this page)

42 **(85)** FL. 50, Ivanhoe Blvd

Lodg **S:** Radisson Hotel

43 **(85)** Princeton Street, Orlando

Gas **S:** Shell, Spur*

Food **S:** Chester Fried Chicken (Spur)

AServ **S:** Shell

Med **S:** ✚ Hospital

44 **(86)** Par Ave. (Eastbound, Westbound Reaccess)

Gas **S:** Texaco*

AServ **N:** Link's Auto Repair & Towing

ATM **S:** Texaco

45 **(87)** Fairbanks Ave.

Gas **S:** Amoco* (24 Hrs)

Food **S:** Sam's Subs

Other **S:** Medicine Shop Pharmacy

46 **(88)** FL. 423, Lee Road

Gas **N:** Amoco(CW) (24 Hrs), Citgo*(D), Shell
S: Chevron* (24 Hrs), Mobil*(CW)

Food **N:** Amoco, Arby's, Burger King, China Buffet, Del Frisco's Steak & Lobster, McDonalds, Shell's Seafood, Straub's Steaks & Seafood, Waffle House

Lodg **N:** AAA Comfort Inn, ◆ Holiday Inn, In Town Suites, Motel 6

AServ **N:** AAMCO Transmission, Firestone Tire & Auto, Shell

47AB **(89)** FL 414, Maitland

Gas **N:** Citgo*

Food **N:** Pizza Hut

Lodg **N:** Sheraton

ATM **N:** SunTrust

Bold red print shows RV & Bus parking available or nearby

EXIT FLORIDA

48 **(92)** FL. 436, Altamonte Springs, Apopka
- **Gas** **N:** Citgo*, Exxon*(D), Mobil(CW), Shell, Texaco*(D)(CW)
  S: Amoco*(CW), Chevron* (24 Hrs), Texaco*
- **Food** **N:** Angel's Diner, Bangkok Thai & Oriental, Bennigan's, Crickets, Long John Silvers, McDonalds, Olive Garden, Perkins Family Restaurant, Pizza Hut, Rio Bravo, Steak & Ale, Waffle House
  S: Denny's, Fuddrucker's, Steak & Shake
- **Lodg** **N:** Altamonte Springs Best Western, Days Inn, AAA Hampton Inn, AAA Holiday Inn, Howard Johnson (see our ad opposite page), AAA La Quinta Inn, Residential Inn, Travelodge
  S: ◆ Embassy Suites
- **AServ** **N:** Firestone Tire & Auto, Shell
  S: Tune Up Clinic, Xpress Lube
- **Other** **N:** West Monte Animal Clinic

49 **(93)** FL. 434, Longwood, Winter Springs
- **Gas** **N:** Chevron* (24 Hrs), Exxon*(D)(CW), Mobil*(D)
  S: Mobil*(D)(CW), Shell*
- **Food** **N:** Burger King, Denny's, Kyoto Japanese Steakhouse, Miami Subs, Pizza Hut, Wild Jack's Steak & BBQ
  S: Baskin Robbins, Black Eyed Pea, Boston Market
- **Lodg** **N:** ◆ Ramada Inn
- **ATM** **N:** First Union, Sun Trust
- **Parks** **N:** Wekiwa Springs State Park

(94) **Rest Area - Picnic, RR, HF, Phones, RV Dump (Westbound)**

(96) **Rest Area - RR, Phones, Vending, Picnic (Eastbound)**

50 **(98)** Lake Mary, Heathrow
- **Gas** **N:** Exxon*(LP)
  S: Amoco*(CW) (24 Hrs), Citgo* (24 Hrs)
- **Food** **N:** Liguino's Pasta & Steaks
  S: Arby's, Baskin Robbins, Bob Evans Restaurant, Chick-fil-A, Chili's, Dunkin Donuts, KFC, Krystal, Longhorn Steaks, Osaka Japanese Steakhouse, Wendy's
- **Lodg** **N:** Marriot Courtyard
- **AServ** **N:** Exxon
- **ATM** **S:** NationsBank
- **Other** **N:** Eckerd Drugs
  S: Albertson's Grocery (Pharmacy), Book Rack Bookstore, General Cinema, K-Mart (Pharmacy), Lake Emma Animal Hospital, Lake Mary Car Wash, Old Time Pottery

51 **(102)** FL 46, to Florida Toll Rd. 417, Sanford, Mount Dora
- **FStop** **N:** Amoco* (24 Hrs)
- **Gas** **S:** Chevron, Speedway*(LP), Texaco*
- **Food** **N:** Pizza Hut
  S: Burger King, Denny's, Don Pablo's, McDonalds, Red Lobster, Steak & Shake, Waffle House

EXIT FLORIDA

- **Lodg** **S:** AAA Days Inn (see our ad this page), ◆ Super 8 Motel
- **AServ** **S:** Chevron, Texaco Lube Express
- **Other** **N:** Central Florida Zoological Park
  S: Seminole Town Center Mall

52 **(104)** U.S. 17, U.S. 92, Sanford (Difficult Reaccess)
- **Gas** **S:** Citgo*
- **Food** **S:** Subway (Citgo)
- **AServ** **S:** Fred Tire Service
- **Med** **S:** ✚ Hospital
- **ATM** **S:** Citgo
- **Other** **S:** Central Florida Zoological Park

53 **(108)** De Bary, Deltona
- **Gas** **S:** Lil' Champ*
- **Food** **N:** McDonalds
  S: Burger King, Shoney's
- **Lodg** **N:** Best Western
  S: ◆ Hampton Inn
- **Parks** **S:** Gemini Springs Park

53CA **(109)** Deltona

53CB **(110)** Orange City, Deltona (Services Accessible To 53CA)
- **Gas** **S:** Chevron*
- **Other** **S:** Albertson's Grocery

54 **(114)** FL. 472, Orange City, DeLand
- **RVCamp** **S:** KOA Campground, Village Park RV Park
- **Parks** **S:** Blue Springs State Park

55 **(116)** Lake Helen, DeLand

56 **(118)** FL. 44, DeLand, New Smyrna Beach
- **Gas** **N:** Texaco*(D)
  S: Shell*
- **Lodg** **N:** AAA Quality Inn
- **Med** **N:** ✚ Hospital

(127) **Parking Area (Eastbound, No Facilaties)**

57 **(129)** U.S. 92 East

58AB **(132)** Junction I-95, Jacksonville, Miami, Daytona

↑FLORIDA

Begin I-4

EXIT WASHINGTON

Begin I-5

↓WASHINGTON

276 WA 548S., Blaine City Center, Peace Arch Park
- **Gas** **E:** 76* (24 Hrs), Arco*, Blain*, Border Gas, Exxon*(D), MP*, USA* (24 Hrs)
  W: Chevron*, Exxon*, Shell*
- **Food** **E:** Denny's, Paso Del Norte
  W: Costa Azul Mexican, Subway, Vista Pizza, Ribs, & Steaks, Wheel House Grill
- **Lodg** **E:** AAA Northwoods Motel
  W: Book Inn, International Motel
- **AServ** **W:** Chevron
- **ATM** **E:** MP
  W: Exxon, Exxon, Key Bank, Seafirst Bank, U.S. Bank
- **Parks** **E:** Peace Arch State Park
- **Other** **E:** Duty Free Store
  W: Paul's Antique Store, Peace Arch Veterinary Clinic, Tourist Information

275 WA 543 North, Truck Customs (Northbound, Reaccess Southbound Only)
- **FStop** **E:** Exxon*
- **Gas** **E:** BP*, Texaco*(D)(CW)
- **Food** **E:** Burger King, Dairy Queen, McDonalds
- **AServ** **E:** NAPA Auto Parts
- **ATM** **E:** Exxon
- **Other** **E:** International Supermarket, Payless Drugs

274 Peace Portal Drive, Semiahmoo (Southbound Reaccess Only)
- **Gas** **W:** Chevron*
- **Parks** **W:** Birch Bay Semiahmoo Resort
- **Other** **W:** Visitor Information

270 Lynden, Birch Bay
- **Gas** **W:** Arco*, Texaco*(D) (24 Hrs)
- **Food** **W:** TCBY Yogurt (Peace Arch Factory Outlets), Taco Bell
- **RVCamp** **W:** Camping (5 Miles)
- **ATM** **W:** Peace Arch Factory Outlets
- **Other** **W:** Peace Arch Factoru Outlet

(269) **Rest Area P (Southbound)**

(268) **Rest Area - RR, Picnic, Vending P (Northbound)**

266 WA 548, Custer, Grandview Road
- **Gas** **W:** Arco*
- **Parks** **W:** Birch Bay State Park

263 Portal Way
- **FStop** **E:** Pacific Pride Commercial Fueling
- **Gas** **E:** Texaco*(D) (24 Hrs)
- **RVCamp** **E:** Cedar's RV Resort Camping

262 Ferndale, Axton Road, Main St.
- **FStop** **E:** BP*(LP), Chevron*(SCALES) (24 Hrs)
- **Gas** **W:** Exxon*
- **Food** **E:** McDonalds
  W: Bob's Burger's, Dairy Queen, Great Wall Restaurant
- **Lodg** **E:** Super 8 Motel
  W: Scottish Motel
- **AServ** **E:** Michelin Tire Service
  W: Martin's Radiator & Exhaust, NAPA Auto Parts
- **Other** **E:** Northwest Propane(LP), Whatcon Veterinary Hospital
  W: Glacier View Animal Hospital

EXIT	WASHINGTON
260	Lummi Island, Slater Road
Gas	**E:** Arco*
Food	**E:** Lummi Cafe
TServ	**E:** International Dealer (Frontage Rd.)
RVCamp	**E:** El Monty RV Service, Unity RV Service
	W: Eagle RV Park
Other	**E:** Antique Mall
258	Bakerview Road, Bellingham International Airport
FStop	**W:** Exxon*(D) (RV Dump)
Gas	**W:** BP*
Lodg	**W:** AAA Hampton Inn, Shamrock Motel
257	Northwest Ave
AServ	**E:** GM Auto Dealership
256AB	WA. 539 North, Meridian St.
Food	**E:** Burger King, Denny's, Godfather's Pizza, McDonalds, Mi Mexico Mexican Restaurant, Mitzel's American Kitchen, Red Robin, Shari's Restaurant, Sizzler, Taco Time, Taste of India, Thai House
Lodg	**E:** Best Western
	W: Comfort Inn, DAYS INN Days Inn, Holiday Inn Express, Quality Inn, Rodeway Inn, Travelers Inn
ATM	**E:** Key Bank
Other	**E:** Bellis Fair Mall, Eyesrite Optical, Payless Drugs, Vitamin's
	W: State Patrol Post
255	WA 542 East, Sunset Drive, Mt. Baker
Gas	**E:** Chevron*(D), Exxon*, Sunset Self Serve
Food	**E:** Daruma Thai Food, Jack-In-The-Box, Marker Restaurant, McDonalds, Round Table Pizza, Scotty B Bagel, Slow Pitch Pub & Eatery, Starbuck's Coffee, Taco Bell
AServ	**E:** K-Mart
Med	**W:** Hospital
ATM	**E:** 7-11 Convenience Store, The Fair
Other	**E:** 7-11 Convenience Store*, Dental Clinic, GNC, K-Mart, Maytag Laundry, Payless Drugs, Sunset Car Wash, The Fair Supermarket (24 Hrs), U.S. Post Office
254	State St., Iowa St.
Gas	**E:** BP*, Texaco*(LP)
	W: 76, Chevron*
Food	**W:** Dairy Queen, McDonalds
AServ	**E:** Cooper Tires, Dr. John's Auto Clinic, Honda, Mitsubishi Dealer, Nissan Dealer, Oldmobile Dealer, Volvo Dealer
	W: 76, Bill Bailey Tire's, Cascade Performance Center, Ford Dealership, Les Schwab Tires, Schuck's Auto Supply & Service
RVCamp	**E:** RV Center, Vacationland RV Sales & Service
Other	**W:** Coin Car Wash
253	Lakeway Drive
Gas	**E:** Self Service Gas
Food	**E:** Pizza Hut, Port of Subs, Ricky's Restaurant
Lodg	**E:** AAA Best Western, Val-U-Inn Motel
AServ	**E:** Discount Tire Co
ATM	**E:** U.S. Bank
Other	**E:** Coin Laundry, Fred Meyer Grocery, Visitor Information
252	Samish Way, West Washington University (Difficult Reaccess)
Gas	**W:** 76(LP), Arco*, Chevron*, Texaco*(D)
Food	**W:** Arby's, Black Agnes, Burger King, Cristos Pizza, Denny's, Diega's Authentic Mexican, Friday's, Godfather's Pizza, IHOP, McDonalds, New Peking Restaurant, Rib'n Reef, Starbuck's Coffee, Subway
Lodg	**E:** Evergreen Motel
	W: Aloha Motel, AAA Bay City Motor Inn, Cascade Inn, Coachman Inn Motel, Macs Motel, AAA Motel 6, AAA Ramada, ◆ Travelodge, Villa Inn
AServ	**W:** 76, Jeep Eagle Dealer
ATM	**W:** Chevron, Seafirst Bank
Other	**W:** Haggen Grocery Store, Payless Drugs
250	WA 11 South, Chuckanut Drive, Old Fairhaven Parkway
Gas	**W:** Chevron
AServ	**W:** Chevron
Other	**W:** Albertson's Grocery (Pharmacy), Old Fairhaven Historic District
246	North Lake Samish
Gas	**E:** Texaco*(D)
RVCamp	**W:** Camping
Parks	**W:** Lake Padden & Recreation
242	Nulle Road, South Lake Samish
240	Alger
RVCamp	**E:** Sudden Valley RV Park
(238)	**Rest Area - RR, Phones, Picnic P (Both Directions)**
236	Bow Hill Road
Food	**E:** Harah's Skagit Casino & Restaurant
RVCamp	**E:** Thousand Trails RV Park
(235)	**Weigh Station (Southbound)**
232	Cook Road, Sedro - Woolley
FStop	**E:** Texaco*(CW)(LP)
Gas	**E:** Shell*
Food	**E:** Blimpie's Subs (Shell), Burger King, Dairy Queen, Iron Skillet
RVCamp	**E:** KOA Campground
Med	**E:** Hospital (3 Miles)
231	WA 11 North, Chuckanut Drive
Parks	**W:** Larabee State Park
230	WA 20, Burlington, Anacortes
FStop	**E:** Pacific Pride
Gas	**E:** Exxon*, Texaco*(D)
	W: Arco*, Holiday*
Food	**E:** Burger King, China Wok, Hooligan's, Pizza Factory, Red Robin, Saucy's Pizza, Skagit's Bagels, Subway, The Galley Restaurant
	W: McDonalds
Lodg	**E:** AAA Cocusa Motel, AAA Sterling Motor Inn
	W: Mark II Motel
AServ	**E:** Les Schwab Tires, Oil Well Fast Lube, Robinson's Auto Rebuild, Skagit Transmission, Texaco
RVCamp	**E:** RV Service & RV Dealer
ATM	**E:** Seafirst Bank
	W: Holiday
Other	**E:** Burlington Coin Laundry, Cascade Mall, Fred Meyer Grocery (Pharmacy), Kwik-N-Clean Car Wash, Office Max, Target Department Store
229	George Hopper Rd
Gas	**E:** Arco* (24 Hrs), USA Mini-Mart*(D)
Food	**E:** Little Caesars Pizza (K-Mart), Shari's Restaurant, The Sports Keg, Wendy's
AServ	**E:** Auto Beauty, Penzoil
	W: Chrysler Auto Dealer, Dodge Dealer, Honda, Nissan
RVCamp	**E:** Camping
ATM	**E:** Pavilion
Other	**E:** Coin Laundry, K-Mart, Liz Claiborne Outlet Mall, Lock Smith, Pavilion Grocery* (24 Hrs), Pets-R-Us, Skagit Animal Clinic
	W: Tourist Information
227	WA 538, East College Way
Gas	**E:** 76(LP), Exxon*
	W: Shell*(D), Texaco*(CW)
Food	**E:** Cascade Pizza Inn, Jack-In-The-Box, KFC, McDonalds, On The Road Espresso, Port Of Subs, Round Table Pizza, Shakey's Pizza, Skipper's Seafood & Chowder House, Starbuck's Coffee, Subway, Taco Bell, Taco Time
	W: Arby's, Best Western, Burger King, Cranberry Tree Restaurant, Dairy Queen, Drumman's Restaurant, Fortune Mandarin Restaurant, Mitzel's American Kitchen, Royal Fork Buffet, Sub Shop (Shell)
Lodg	**E:** DAYS INN ◆ Days Inn, West Winds Motel
	W: AAA Best Western, AAA Comfort Inn, Tulip Inn
AServ	**E:** 76, Goodyear Tire & Auto, Schuck's Auto Supply, Short Stop Tune Up & Brake Center
	W: Quaker State Lube, Walt's Radiator, Muffler & Brakes
RVCamp	**W:** Valley RV Service(LP)
ATM	**E:** Interwest Bank, People's Bank
	W: Texaco
Other	**E:** Al's Auto Supply, Albertson's Grocery, Coin Laundry, Payless Drugs, Safe Way Grocery
	W: Eagle Hardware, Sears (Frontage Rd.), Visitor Information
226	WA 536 West, Kincaid St
Food	**W:** Coffee Corner
AServ	**W:** LAE Auto Part Store
RVCamp	**W:** Camping
Med	**E:** Hospital
ATM	**W:** Seafirst Bank
Other	**W:** Red Apple Market, Vaux Drugs
225	Anderson Rd
FStop	**W:** Truck City*
Gas	**E:** Amerigas(LP) (Frontage Rd.), BP*
Food	**W:** Crane's Restaurant (Truck City FS)
AServ	**W:** Iversen Auto Body
TServ	**W:** International Dealer, The CB Shop
224	South Mount Vernon (Northbound, Reaccess Southbound Only)
221	WA 534 East, Lake McMurray
Gas	**E:** Texaco*(D)(LP)
RVCamp	**W:** Blakes RV Park
ATM	**E:** Texaco
218	Starbird Road
Lodg	**W:** Hillside Motel (W. Frontage Rd.)
215	300th St N.W.
Gas	**W:** MiniMart*
AServ	**W:** MiniMart
212	WA 532 West, Stanwood, Camano Island
Gas	**W:** Shell*(D)(LP), Texaco*(D)(LP) (24 Hrs)
RVCamp	**W:** Camping
ATM	**W:** Shell
Parks	**W:** Camano Island State Park (19 Miles)
210	236th St N.E.
208	WA 530, Silvana, Arlington
FStop	**W:** Exxon*
Gas	**E:** Arco* (24 Hrs), BP*(LP), Chevron*(LP), Texaco*
Food	**E:** Denny's, O'Brien Turkey House

Bold red print shows RV & Bus parking available or nearby

EXIT	WASHINGTON
Lodg	E: AAA Arlington Motor Inn
Med	E: Hospital
ATM	W: Exxon FS
(207)	**Rest Area - Phones, Coffee, Info, RV Dump, RR, Picnic P (Both Directions)**
206	**WA 531, Lakewood, Smokey Point**
Gas	E: Arco*, Chevron*(CW)(LP), Citgo*, Texaco*
	W: Chevron* (24 Hrs)
Food	E: Baskin Robbins, Buzz Inn Steakhouse, Jack-In-The-Box, KFC, McDonalds, Olympia Pizza House, Pizza & Teriyaki, Taco Time, Wendy's
	W: Nick's Place Restaurant
Lodg	E: ◆ Smokey Point Motor Inn
AServ	E: Al's Auto Supply, Chevron, GM Auto Dealership, Q-Lube
RVCamp	W: Camping
ATM	E: Key Bank, Safe Way Grocery, Seafirst Bank, U.S. Bank
Parks	W: State Park (7 Miles)
Other	E: Pavillion Grocery Store (Pharmacy, 24 Hrs), Payless Drugs, Safe Way Grocery (24 Hrs), Sears (Frontage Rd.), Western Union (Payless Drugs)
202	**116th St N.E.**
TStop	W: Chevron*(SCALES)
Gas	E: Texaco* (24 Hrs)
TServ	W: Kenworth
Other	W: State Patrol Post, Visitor Information
200	**88th St. NE., Quill Teda Way**
199	**WA 528 East, Marysville, Tulalip**
Gas	E: Arco*, BP*, Chevron*, Texaco*(D)
	W: 76(LP)
Food	E: 4th St. Market & Deli, Burger King, Conto's Pizza & Pasta, Dairy Queen, Don's Restaurant, Jack-In-The-Box, Las Margaritas Mexican Restaurant, Village Restaurant
	W: Arby's, McDonalds, Restaurant (Best Western), Taco Time, Wendy's
Lodg	E: AAA Motor Inn
	W: AAA Best Western, AAA Holiday Inn Express
AServ	E: BP, Fast Lube Oil
	W: 76, GM Auto Dealership
RVCamp	E: RV Dump
	W: Roy Robinson RV Service
ATM	E: Seafirst Bank
Other	E: Albertson's Grocery, Coin Car Wash, JC Penny, Lamont's, Morning Star Books, Payless Drugs
198	**WA 529 South, North Broadway, Port of Everett (Southbound, Reaccess Northbound Only)**
195	**Port of Everett, Marine View Drive (Northbound, Southbound Reaccess Only)**
194	**U.S. 2 East, Snohomish, Wenatchee**
Gas	W: Texaco*(D)
AServ	W: Les Schwab Tires
TServ	W: Les Schwab Tires
193	**Pacific Ave, City Center, WA 529**
Gas	W: Texaco*
Food	W: Burger King, Casbah, Denny's, KFC, McDonalds, Starbuck's Coffee, Wendy's
Lodg	W: AAA Best Western, AAA Howard Johnson, ◆ Travelodge
AServ	W: Les Schwab Tires, Meineke Discount Mufflers

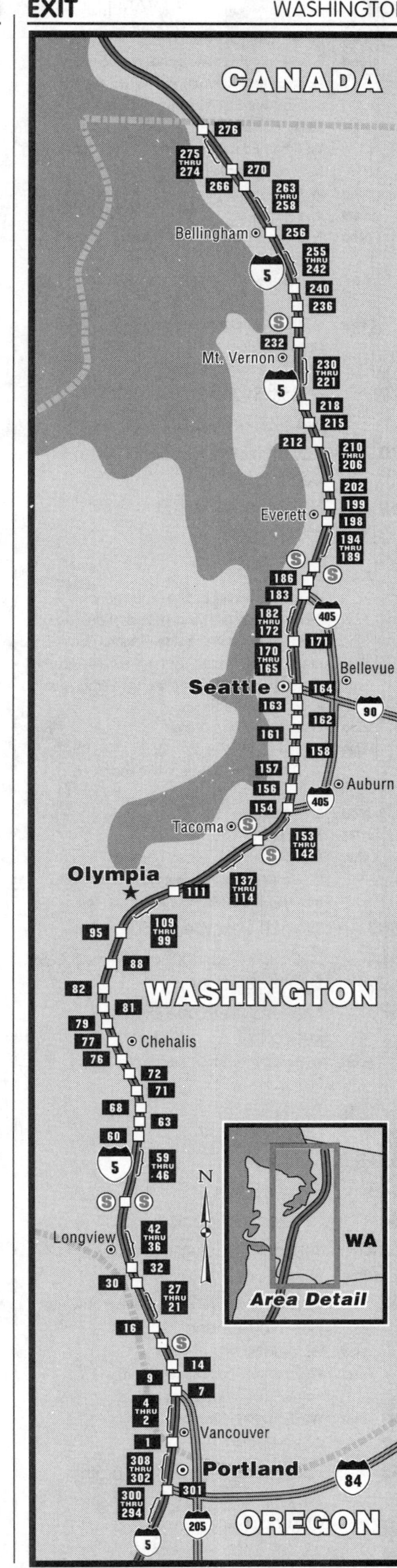

EXIT	WASHINGTON
Med	W: Hospital
ATM	W: Texaco
Other	W: QFC Grocery (24 Hrs, Pharmacy)
192	**Broadway, Navel Station, Port of Everett (Left Lane Exit Only)**
Gas	W: Arco*, BP*(CW), Chevron* (24 Hrs), Texaco
Food	W: Buzz In Steak House, China Doll Restaurant, Petosa's Restaurant, Taco Bell
AServ	W: Pacific Power Battery(LP), Quaker State Lube
TServ	W: Morgan -Johnson
RVCamp	W: Morgan-Johnson RV Service
Med	W: Hospital
189	**WA 527 N., To WA 99, Broadway, Everett Mall Way**
Gas	E: Arco*, Texaco*
	W: Shell
Food	E: Alfy's Pizza, Buzz Inn, McDonalds, Orchid Thai Cuisine
Lodg	E: Ramada Inn
	W: Motel 6
AServ	E: Texaco*
Med	E: Silver Lake Medical Center
Other	E: Coin Car Wash, Dental Clinic
(188)	**Weigh Station, Rest Area - RR, HF, Phones, Picnic, RV Dump P**
186	**128th St S.W., WA 96**
Gas	E: BP*(LP), Texaco*(D)
	W: 7-11 Convenience Store*, Arco*, Chevron* (24 Hrs), Texaco*(CW)
Food	W: Burger King, Dairy Queen, Denny's, Great American Bagel, Great River Restaurant Chinese, KFC, McDonalds, Mitzel's American Kitchen, Pizza Hut, Seattle Fish & Chips, Subway, Taco Bell, Taco Time
Lodg	E: Express Inn, AAA Holiday Inn
	W: Motel 6, Quality Inn
AServ	W: Auto Part Store, Goodyear Tire & Auto, Q Lube, Tony's Auto Care & Repair
RVCamp	W: Lake Side RV Park
ATM	W: U.S. Bank
Other	E: 24 Hr. Market/Deli, Northcreek Dental Care, Visitor Information
	W: Albertson's Grocery (Pharmacy), Coin Laundry
183	**164th St S.W.**
Gas	E: 76, Arco*, Chevron*, Texaco*(D)(CW)
Food	E: Jack-In-The-Box, Subway, Taco Time
Lodg	W: ◆ Residence Inn
AServ	E: 76, Premium Tune & Lube, Q-Lube
RVCamp	W: Twin Cedars RV Park
ATM	E: Wells Fargo
182	**WA 525 North, to WA 99, I-405 South, Belleview, Renton**
181B	**44th Ave, Lynnwood, WA 524 (Northbound, Difficult Northbound Reaccess)**
Gas	E: Arco*
	W: Arco*, BP*(D)(CW), Chevron*(LP), Texaco*(D), USA*
Food	E: Pizza Place
	W: Alfy's Pizza, Applebee's, Arby's, Black Angus Steakhouse, Burger King, China Kitchen, Chuck E. Cheese's Pizza, Denny's, Donuts, Eastside Mario's Italian, El Torito Mexican Restaurant, IHOP, Jack-In-The-Box, KFC, Kostalee's Family Pizza, Macheezmo Mouse, McDonalds, Red Lobster, Roaster's, Starbuck's Coffee, Taco Bell,

Bold red print shows RV & Bus parking available or nearby

EXIT WASHINGTON

The Bagel Baking Company, The Great China Restaurant, The Olive Garden, The Yankee Diner, Tony Roma's, Tropicana Restaurant, Wendy's, Wong's Authenic Cuisine
Lodg **W:** Holiday Inn Express
AServ **W:** Discount Tire Company, Goodyear Tire & Auto, K C Automotive, Oldmobile Dealer, Precision Tune & Lube, Super Shops, Texaco
Med **W:** + **Chec Medical Center**
ATM **W:** First Interstate Bank
Other **W:** **Alderwood Veterinary Clinic, Eye's Rite Optical, Fred Meyer Grocery, K-Mart, Rocky Grocery & Deli, Sportmart**

181A To Lynnwood, To WA. 524
Gas **E:** Arco* (24 Hrs)
W: BP(D)(CW), Chevron*(LP), Fred Meyer, Texaco
Food **E:** Darn Good Pizza
W: Bagel Baking Co., Bagel Co., Barbara's Family Dining, Black Angus, Burger King, Cafe Taka, Chi Chi's Mexican, China Kitchen, Dave's Burger's, Denny's, El Torito, Evergreen Donuts, Great China Restaurant, Happy Teriyaki, IHOP, Indian Cuisine, Jack-in-the-Box, Japanese Restaurant, Just Between Friends (Best Western), KFC, Kafe Athens, Kenny Rogers Roasters, Mario's Italian, McDonalds, Old Country Buffet, Olive Garden, Red Lobster, Starbuck's Coffee, Subway, Tony Roma's Ribs, Wendy's, Yankee Diner
Lodg **E:** Embassy Suites
W: Best Western, Holiday Inn Express
AServ **W:** Bucky's Muffler & Brakes, Centers Collision Center, Chevron, Discount Tire Co., Goodyear Tire & Auto, Oldmobile Dealer, Precision Tune & Lube, Super Shops, Texaco
ATM **W:** Key Bank, U.S. Bank, Wells Fargo
Other **E:** **K-Mart**
W: **Eyes Rite Optical, Fred Meyer Grocery (Pharmacy, Optical), Kinko's Copies, Sport Mart**

179 220th St. Southwest
Gas **W:** Shell*, Texaco*(D)(CW)
Food **W:** Azteca Mexican, Bento Teriyaki, China Passage Restaurant, Port of Subs, Subway, Thai Terrace
Med **W:** + **Hospital,** + **Mt Lake Medical Immediate Care Center**
Other **W:** **Animal Care Vet. Hosp., Dental Clinic, National Forest Headquarters, Super Deli Mart, Treasure Thoughts Bookstore**

178 236th ST. W., Mt. Lake Terrace

177 WA 104, NE 205th St, Edmonds, Lake Forest Park
Gas **E:** Chevron* (24 Hrs), Chevron*(CW)(LP), Texaco*(D)
Food **E:** Buzz Inn Steak House, Kings III Chinese Restaurant, McDonalds, Subway, Taco Bell (Texaco), The CookHouse, Toshi's Teriyaki, Winchell's Donuts, Zoopa Italian
Lodg **E:** (DAYS INN) Days Inn, K&D Motor Inn, (AAA) Travelodge
AServ **E:** Chevron, Q-Lube, Shuck's, The Auto Clinic
ATM **E:** Chevron, City Bank
Other **E:** **Terrace Pharmacy, Thrift Way**

176 NE 175th St, Shoreline

175 WA 523, NE 145th St, 5th Ave

174 NE 130th St, Roosevelt Way (Northbound)

173 Northgate Way, 1st Ave NE

EXIT WASHINGTON

Gas **W:** BP*, Chevron*, Texaco*(D)
Food **E:** Baskin Robbins, Fresh Choice, Mocha Tree, North China Restaurant, Round Table Pizza, Sbarro Italian Eatery, Starbuck's Coffee, Sub Shop, Subway, Toshi's Teriyaki
W: Arby's, Barnaby's, Denny's, Family Donut, McDonalds, Teriyaki Plus
Lodg **W:** (AAA) Ramada Inn
AServ **W:** BP, Texaco*
Med **E:** + **North Gate Medical Hospital,** + **Seattle Chinese Med Center**
ATM **E:** Seafirst Bank, U.S. Bank
W: BP
Other **E:** **Mall, Pace Dental, Payless Drugs, Western Optical**
W: **7-11 Convenience Store**

172 N. 85th St, Aurora Ave. North

171 WA 522, Bothell, Lake City Way

170 Ravenna Blvd, NE 65th St (Northbound)

169 NE 45th St, NE 50th St
Gas **E:** 76, Texaco*(CW)
W: Texaco*(D) (24 Hrs)
Food **E:** Subway
W: Boston Market, Dick's Hamburgers, Excaliburs (University Plaza Hotel), Fudge Bros. Pizza, Juli's Restaurant, Kabul Afghan Cuisine, Moon Temple Restaurant, Nikolas Cafe, Sea-Thia Restaurant, Starbuck's Coffee, Taco Time, Tatto Pizza, Winchell's Donuts
Lodg **W:** University Plaza Motel
AServ **E:** 76
W: Quaker State Q Lube, Walt's Radiator & Brakes
Med **E:** + **Hospital**
ATM **E:** Key Bank
Other **E:** **Burke Museum, Kinko's Copies**
W: **Bartell Drugs, Heathway Natural Foods, Natural Food Store, Payless Drugs, Zoo**

168B WA 520, Bellevue, Kirkland

168A Lakeview Blvd, Roanoke St (Reaccess Southbound Only)

167 Mercer St. (Left Lane Exit, No Reaccess)
Food **W:** Benjamin's Fish & Chips, Burger King, Denny's, Home Deli
AServ **W:** Lincoln Towing

166 Olive Way

165AB Seneca St.

164B 4th Ave South

164 Junction I-90, Spokane

163A Spokane St, Columbian Way
Other **W:** **King Dome**

162 Corson Ave, Michigan St (Left Hand Exit - Northbound)
Gas **W:** Texaco*(D)(CW)
Food **W:** Arby's, CP Thai, Daimonji, Herfy's Hamburgers, Original Deli
Lodg **W:** (AAA) Georgetown Inn

161 Albro Place, Swift Ave

158 Pacific Hwy South, East Marginal Way

157 WA 900 East, ML King Way

EXIT WASHINGTON

(Reaccess Northbound Only)

156 WA 599 Tukwila, West Marginal
FStop **E:** **Pacific Pride Commercial Fueling**
Gas **W:** 76, BP*, Texaco*(CW)(LP) (ATM)
Food **W:** Denny's, Jack-In-The-Box
Lodg **W:** Silver Cloud Motel
AServ **W:** 76, BP*
TServ **W:** **Husky International**
ATM **W:** Texaco

154B WA 518 West

154A I-405

153 Southcenter Parkway & Mall (Northbound. Difficult Reaccess)
Gas **E:** Chevron(D)
Food **E:** Houlihan's, Mayflower of China, McDonalds, Sizzler Steak House, Tony Roma's Ribs, Wendy's
Lodg **E:** (AAA) Doubletree Inn (Restaurant), Doubletree Suites
AServ **E:** Chevron, Firestone Tire & Auto
Other **E:** **JC Penny, Ranier Court Mall, Target Department Store**

152 Sea - Tac Airport, South 188th St.
Gas **W:** BP*
Lodg **W:** Motel 6
RVCamp **E:** **KOA Campground**

151 South 200th St, Military Road
Gas **W:** BP*, Chevron*, Citgo*
Food **W:** Bullpin, Gikan 5 Teriyaki
Lodg **E:** **Motel 6**
W: Comfort Inn, Hampton Inn, Holiday Inn Express, Howard Johnson, Super 8 Motel
TServ **W:** **Kenworth Northwest**

149 WA 516, Kent, DesMoines
Gas **W:** Texaco*(D)(CW)
Food **W:** Big Apple Bagels, Blockhouse Restaurant, Burger King (Playground), Dunkin Donuts, KFC, McDonalds (Playground), Meal Time Drive In, Pizza Hut, Skipper's, Starbuck's Coffee, Subway, Sunrise Diner, Taco Bell, Wendy's
Lodg **E:** Century Motel
W: King's Arms Motor Inn, New Best Inn
AServ **E:** Lloyd's Automotive Repair, Timlick's Auto Repair
W: General Transmission, Midas Muffler & Brakes, Quaker State Lube, Schuck's Auto Supply
Other **E:** **Valley I-5 RV Center**
W: **Book World, EZ-Clean, Kinko's**

147 South 272nd St
Gas **E:** BP*
Other **W:** **Saltwater State Park**

143 Federal Way, South 320th St
Gas **W:** Arco, BP*(CW)
Food **W:** A-1 Teriyaki House, Applebee's, Azteca Mexican Restaurant, Billy McHales Restaurant, Black Angus Steakhouse, Burger King, Cafe Arizona, Denny's, Duncan Donut, Emerald City Smoothie, Godfather's Pizza, Ivan's Seafood Bar, Marlene's, McDonalds, Outback Steakhouse, Pho Hoang, Red Lobster, Red Robin, TCBY Yogurt, Taco Time, Tony Roma's, Wendy's
Lodg **W:** (AAA) Best Western
Med **W:** + **Hospital**
ATM **W:** Key Bank, Tacoma Mutual
Other **W:** **Christian Book House, Dash Point State Park, Kinko's, Pier 1 Imports, Seattle Tacoma Mall, Subway, Ynot Magizines and**

Bold red print shows RV & Bus parking available or nearby

EXIT WASHINGTON

Paperbacks

142AB WA 18 East, Federal Way, Auburn
- **TStop** W: Flying J Travel Plaza*(LP)(SCALES)
- **Gas** W: Chevron*, Texaco*(D)(CW)
- **Food** W: Burger King, Dairy Queen, Denny's, Dillano's Coffee, McDonalds, Olive Garden, Pacific Highway Diner, Popeye's Chicken, Sharis Restaurant, Subway
- **Lodg** W: Evergreen Motel, Holiday Inn Express, Super 8 Motel
- **AServ** W: Eagle Tire and Automotive, Federal Way Radiator and Muffler, Les Schwab Tires, Pyramid Tire, Safe Lite Auto Glass
- **Med** W: Hospital
- **Other** W: The Trucker's Store

(141) Weigh Station, Rest Area - Picnic, Phones, RR, RV Dump P (Both Directions)

137 WA 99, Junction, Fife, Milton
- **Gas** E: Arco*, BP*(D), Chevron*(D), Texaco*(CW)
  W: BP*(CW)
- **Food** E: Dairy Queen, Johnny's
  W: Arby's, Baskin Robbins, Burger King, Denny's, German Deli, KFC, King's Palace Restaurant, McDonalds, Mitzel's, Poodle Dog, Skipper's, Taco Bell, Wendy's
- **Lodg** E: Motel 6
  W: AAA Best Western, AAA Comfort Inn, Kings Motor Inn, Paradise Motel, AAA Royal Coachman
- **RVCamp** W: Camping World (see our ad this page)
- **AServ** E: Nissan, Volvo Dealer
  W: Fife National Auto Parts
- **ATM** W: Seafirst Bank, US Bank
- **Other** W: Coin Laundry, Fife Valley Pharmacy, United Drugs

136AB Port of Tacoma, 20th St
- **TStop** W: Flying J Travel Plaza*(LP)(SCALES)
- **FStop** E: CFN Fuel Stop
  W: CFN
- **Gas** W: Chevron*, Texaco*(CW)
- **Food** E: Joe's Deli
  W: Fife City Bar & Grill (Days Inn), Jack-In-The-Box, La Casa Real, Thads Restaurant(SCALES) (Flying J Travel Plaza)
- **Lodg** W: Days Inn Days Inn, Econolodge, Glacier Motel, Holiday Inn Express, Hometel Inn, Sunshine Motel, Travlers Inn
- **AServ** W: Goodyear Tire & Auto, Walt's Radiator & Muffler
- **TServ** E: Western Peterbilt Inc.
  W: Cummins Diesel

135 WA 167 Junction, Puyallup, Portland Ave.

134 Portland Ave (Northbound)
- **FStop** W: BP*
- **Gas** W: Arco*
- **Food** W: Dairy Queen, Pegasus Restaurant, Western Cafe
- **Lodg** W: La Quinta Inn (Restaurant), Valley Motel
- **AServ** W: Phil's Auto Care

133 I-705 North, South WA 7, City Center

132 WA 16 West, South 38th St, Cig Harbor Bremerton
- **Gas** E: Texaco*
  W: Arco*, BP(CW)

EXIT WASHINGTON

- **Food** E: Kyoto Stop Teriyaki
  W: Arby's, Blimpie's Subs, Burger King, Great Wall of China, KFC, La Donut, McDonalds, Outback Steakhouse, Shanghai Restaurant, Starbuck's Coffee, Taco Bell, Tokyo Teriyaki
- **AServ** E: Auto Lube, Sione's Auto Repair, Texaco
  W: Firestone Tire & Auto, Precision Tune & Lube, Walt's
- **Med** E: Hospital
- **Other** E: 38th St. Laundry Mat, Borders Books and Music, Petco, Safeway Drug

130 South 56th St, Tocoma Mall Blvd.
- **Gas** W: Texaco*(CW)
- **Food** W: Chuck E. Cheese's Pizza, Mexican Azteca, Subway, Tony Roma's

129 South 72nd St, South 84th St
- **Gas** E: Arco*, Exxon*
- **Food** E: Applebee's, Burger King, Dairy Queen, Elmer's, IHOP, Jack-In-The-Box, Mitzel's American Kitchen, Mongolian Grill, Olive Garden, Red Lobster, Round Table Pizza, Shari's Restaurant, Starbuck's Coffee, Taco Bell, Teriyaki, Zoopa
  W: Yankee Diner
- **Lodg** E: Motel 6, Shilo Inn Motel
  W: Days Inn ◆ Days Inn
- **AServ** E: Precision Tune & Lube
- **Other** E: Drug Emporium

128 South 84th St
- **Gas** E: BP, Chevron*, Texaco*(D)(CW)
  W: Arco*
- **Food** E: Alligater Pub and Grill, Copperfield's Restaurant, Crocodile Pub & Grill, Denny's, Raintree Restaurant, Subway
- **Lodg** E: Best Western, Copperfield's, Holiday Inn Express, Howard Johnson, AAA King Oscar Motel, Rothem Inn, AAA Travelodge
- **AServ** E: BP
  W: Discount Tire Company

127 East WA 512, Puyallup, Mt Rainier
- **Gas** W: Arco*, BP*, Chevron*, Texaco*(D)(CW)
- **Food** W: Billy McHales Restaurant, Burger King (Playground), Dairy Queen, Ivan's Seafood, Jim Moore's Restaurant, Korean Restaurant, Lessie's Southern Quisine, Mazatlan Mexican Restaurant, McDonalds (Playground), Sizzler Steak House, Subway, Taco Time, Wendy's
- **Lodg** W: Budget Inn, Vagabond Motel, Western Inn
- **AServ** W: Northwest Transmissions

125 Lakewood, McChord A.F.B.
- **Gas** W: 76(D), Texaco(D)
- **Food** W: Al's BBQ and Burgers, Doughnuts, Happy Day's Diner, Ho Ban Restaurant, Hogie's Sub Express, KFC, Mory's Family Dining
- **Lodg** W: Colonial Motel, Fort Lewis Motel, Home Motel, Lakewood Lodge, Rose Garden Motel
- **AServ** W: 76, AAMCO Transmission, Aces High, Lakewood Car Wash, Lakewood Transmission, Rich's Auto Service, Walt's
- **Med** W: Hospital

EXIT WASHINGTON

- **Other** E: McChord Air Force Base Air Museum
  W: Bridgeport Way Animal Hosp.

124 Gravelly Lake Drive
- **Gas** W: Arco (24 Hrs), BP
- **AServ** W: BP

123 Thorne Lane
- **AServ** E: A-1 Transmission

122 Madigan Hospital, Camp Murray
- **Gas** W: Chevron
- **Food** W: Baskin Robbins, Ching Ha, Gertie's Grill, House Of Teriyaki, KFC, McDonalds, Subway, Taco Bell
- **AServ** W: An Honest Mechanic, Chevron, Coin Car Wash
- **ATM** W: 7-11 Convenience Store
- **Other** E: Fort Lewis Military Reservation
  W: 7-11 Convenience Store

120 Fort Lewis, North Fort Lewis
- **Other** W: Fort Lewis Military Museum (3 Miles)

119 Dupont, Steilacoom

(117) Weigh Station (Northbound)

116 Mounts Road

114 Old Nisqually
- **Gas** E: Chevron(LP), Texaco*
- **Food** E: Nisqually Pub, Tiny's Burger House
- **AServ** E: Chevron
- **RVCamp** E: Nisqually

111 WA 510, Yelm, Marvin Road
- **FStop** E: BP* (ATM)
  W: Pacific Pride Commercial Fueling
- **Gas** E: Chevron*, Texaco*(D)(CW)
- **Food** E: Blimpie's Subs, Burger King, Dairy Queen (Playground), Godfather's Pizza, Hawk's Prairie, McDonalds (Playground), Pizza Hut (Texaco), Taco Bell
  W: Country Junction Family Restaurant
- **Lodg** E: King Oscar Motel
- **AServ** E: Les Schwab Tires, Q Lube
- **ATM** E: BP
- **Other** E: Payless Drugs, Safe Way Grocery

109 Martin Way, Sleater - Kinney Road
- **Gas** W: BP*, Texaco(D)(CW)
- **Food** E: Mitzel's, Taco Bell
  W: Denny's, Expresso and Deli, Red Lobster, Sharis Restaurant
- **Lodg** W: AAA Comfort Inn, Days Inn AAA Days Inn, AAA Holiday Inn Express, Super 8 Motel
- **AServ** E: Discount Tire Company
- **ATM** E: US Bank
  W: Texaco
- **Other** E: Top Food Supermarket

108 Sleater - Kinney Road, South College St. (Difficult Northbound Reaccess)
- **Gas** E: Texaco*
  W: 76, Arco*, Shell*
- **Food** E: Arby's, Baskin Robbins, Double Joy Teriyaki, Godfather's Pizza, Kenny Rogers Roasters, McDonalds (Playground), Red Corral, Starbucks, Wendy's, Winchell's Donuts
  W: Burger King, Casa Mia, Dirty Dave's Pizza and Spaghetti Parlor, El Sarape Mexican, Jack-In-The-Box, Mandarin House, Subway
- **AServ** E: Firestone Tire & Auto, Q Lube, Sears
  W: 76, Gary's Tire Factory, K-Mart
- **Med** W: Hospital
- **ATM** E: Sea First Bank, Wells Fargo

EXIT		WASHINGTON
	Other	E: Fred Meyer Grocery, Lacey Civic Plaza, Mervyn's, Payless Drugs, Sears W: K-Mart
107	Pacific Ave	
	Gas	E: Chevron*, Texaco*(D)
	Food	E: Izzy's Pizza, Sharis Restaurant, Sizzler Steak House, Taco Time, Yukio's Teriyaki
	AServ	W: Ford Dealership, US Auto Glass
	RVCamp	W: Coumbs RV Service
	Med	E: Hospital
	ATM	E: Texaco
	Other	E: Albertson's Grocery, Animal Care Vet. Clinic, Kinko's
105	State Capitol, City Center, Port of Olympia (Difficult Reaccess)	
	Gas	W: Chevron*
	Food	W: Original Steak House
	Lodg	W: AAA Carriage Inn Motel
	AServ	W: Chevron*, Saturn Dealership
104	U.S. 101 North, Aberdeen, Port Angeles	
103	2nd Ave, Deschutes Way (Reaccess Northbound Only)	
	Food	E: Buffalo Bill's Burgers, Falls Terrace Restaurant, Red Barn Tavern
102	Trosper Road, Black Lake	
	Gas	E: Citgo*, Texaco*(D) W: BP*, Chevron*
	Food	E: Arby's, Brewery City Pizza, Burger King, Cattin's Family Restaurant, Domino's Pizza, Happy Teriyaki V, Jack-In-The-Box, Jim's Diner, KFC, McDonalds (Playground), Pizza Hut, Subway, Taco Bell W: Blimpie's Subs, Iron Skillet, Nickelby's Restaurant, Seasons Teriyaki Too
	Lodg	E: AAA Best Western, Books, Motel 6 W: AAA Tyee Hotel
	AServ	E: Al's, Brad's Auto Parts, Goodyear Tire & Auto, Poages Auto Repair, Tumwater Auto Parts W: Cut Rate Auto Parts
	ATM	E: Citgo, First Community Bank, Key Bank, Texaco W: BP
	Other	E: Southgate Drug W: Albertson's Grocery, Megafoods Supermarket
101	Airdustrial Way	
99	93rd Ave, Scott Lake	
	TStop	W: Restover TStop*(LP)(SCALES) (ATM)
	Food	W: Hannah's Pantry (Restover TStop)
	RVCamp	E: American Heritage Campground, Olympia Campground
	ATM	W: Restover
	Other	E: Washington State Patrol
95	WA 121, Littlerock Maytown	
	Food	W: Farm Boy Restaurant
	RVCamp	E: Deep Lake Resort
(93)	Rest Area - RR, Phones, Info, Picnic, Vending P (Southbound)	
(91)	Rest Area - Picnic, RR, Phones, Vending, Info, RV Dump P (Northbound)	
88AB	Jct. U.S. 12, Aberdeen, Tenino	
	FStop	W: Pacific Pride Commercial Fueling, Texaco*
	Gas	W: Arco*, Shell(D)(LP)
	Food	W: Burgermaster, Dairy Queen, Golden Chinese (Texaco), Little Red Barn

EXIT		WASHINGTON
	AServ	W: Fred's Discount Tire, Shell
	ATM	W: Key Bank, Texaco
82	Harrison Ave.	
	Gas	E: Shell*(D) W: BP*, Chevron*, Texaco*
	Food	E: Arco*, Bambino, Burger King, Burgerville USA, Casa Ramos Mexican, China Dragon, Dairy Queen, Godfather's Pizza, Pizza Hut, Sharis Restaurant, TCBY Yogurt, Wendy's, Yum Yum Teryaki W: Andree's, Arby's, Country Cousin Family Restaurant, Denny's, Jack-In-The-Box, PaPa Murphy's Pizza, Subway, Taco Bell
	Lodg	E: DAYS INN AAA Days Inn, AAA Ferryman's Inn, Riverside Motel, AAA Travel Lodge W: AAA Motel 6, Park Motel
	AServ	E: Quaker State Lube W: Al's Auto Supply, Les Schwab Tires
	ATM	W: Centenial Bank
	Other	E: Payless Drugs W: Factory Outlet Center, Safe Way Grocery
81	WA 507 North, Mellen St Bucoda	
	Gas	E: Shell*(LP), Texaco*
	Food	E: Buchanan's, King Solomon Restaurant
	Lodg	E: Lakeshore Motel, Peppertree West Motel
	AServ	E: Shell
	Med	W: Hospital
	ATM	E: Texaco
79	Chamber Way, City Center	
	Gas	E: Texaco*(D) W: BP (Burger King, Taco Bell)
	Food	E: Plaza Jalisco Mexican W: Burger King (BP), McDonalds (Wal-Mart), Taco Bell (BP)
	AServ	E: Ford Dealership, Goodyear Tire & Auto, State Ave Auto & Muffler, The Gear Box W: Wal-Mart
	TServ	E: Fleetguard Cummins Dealership, Uhlmann Motors
	RVCamp	E: Uhlman's RV Center(LP)
	Other	E: Visitor Information W: K-Mart, Wal-Mart (McDonald's, Tire & Lube)
77	WA 6 West, Pe Ell, Raymond	
	Gas	E: Arco*
	AServ	E: Chehalis Muffler
	Other	W: State Patrol Post
76	13th St	
	Gas	E: Arco*, Chevron*
	Food	E: Denny's, Jack-In-The-Box, Kit Carson Restaurant
	Lodg	E: Nendels Inn, Relax Inn
72	Rush Road, Napa Vine	
	TStop	E: Shell (Taco Bell)
	FStop	W: Pacific Pride Commercial Fueling, Texaco* (ATM)
	Gas	W: Chevron*(D) (Hot Stuff Pizza, Cinnamon Street, Smash Hit Subs)
	Food	E: McDonalds, Rib Eye Restaurant, Subway, Taco Bell (Shell) W: Cinnamon Street (Chevron), Hot Stuff Pizza (Chevron), Smash Hit Subs
	ATM	W: Texaco
71	WA 508 East, Napavine, Onalaska	
	TStop	E: 76*(SCALES)
	FStop	E: CFN Card Lock
	TServ	E: 76 Auto/Truck Plaza, KC Truck Parts

EXIT		WASHINGTON
	ATM	E: 76
68	U.S. 12 East, Mt St Helens, Morton, Yakima	
	FStop	W: Texaco*(LP)
	Gas	E: Arco*, BP*(LP)
	Food	E: Spiffy's W: Mustard Seed Restaurant
	RVCamp	E: KOA RV Park(LP)
	ATM	E: BP
	Parks	E: Louis & Clark State Park, Mount Ranier National Park
	Other	E: Visitor Information
63	WA 505, Winlock	
	Gas	W: Shell*(D)(LP) (ATM)
	Lodg	W: Sunrise Motel
	ATM	W: Shell
60	Toledo	
59	WA 506 West, Vader, Ryderwood	
	FStop	E: BP*
	Gas	W: Texaco*(D)(LP)
	Food	E: Mrs. Beesley's W: Country House Restaurant, Grandma's In'n'Out
	Lodg	E: Cowlitz Motel
	RVCamp	E: Cowlitz RV Park W: River Oaks
57	Jackson Highway	
	TStop	W: Texaco(LP)(SCALES)
	Food	W: Restaurant (Texaco TStop)
	TServ	W: Texaco Truck Stop, Western Star Trucks
	RVCamp	W: Camping
(55)	Rest Area - Picnic, Phones, RR P (Both Directions)	
52	Barnes Dr, Toutle Park Road	
	RVCamp	E: Fox Park RV Park
49	WA 411S, WA 504E, Toutle, Castle Rock	
	Gas	E: Arco*
	Food	E: 49er Restaurant, Papa Pete's, Rose Tree Restaurant
	Lodg	E: Motel 7 West, Mount St Helens Motel, Timberland Inn and Suites
	RVCamp	E: Mount St Helens RV Park (2 Miles), Silver Lake Camping Area (6 miles)
	ATM	E: Texaco
	Other	E: Mount St Helens Visitor Info Ctr
48	Bus Loop 5, Castle Rock	
46	Headquarters Road, Pleasant Hill Road	
	RVCamp	E: Cedars RV Park
(44)	Weigh Station (Southbound)	
42	Ostrander Road, Pleasanthill Road	
40	North Kelso Ave, South to WA 4, Longview, Long Beach	
39	WA 4W, Kelso, Longview	
	Gas	E: Arco*, Shell*
	Food	E: Denny's, Hilander Restaurant, Jitters, McDonalds, Sharis Restaurant W: Azteca, Burger King, Dairy Queen, Izzy's Pizza, Red Lobster, Taco Bell
	Lodg	E: Double Tree Hotel, AAA Motel 6, Super 8 Motel W: AAA Comfort Inn
	AServ	W: Quaker State Lube, Sears
	Other	E: Payless Drugs, Safe Way Grocery, Tam-O-

Bold red print shows RV & Bus parking available or nearby

EXIT	WASHINGTON
	Shanter Park, Tourist Information
	W: Sears, Target Department Store, Three Rivers Mall, Top Foods Supermarket
36AB	WA 432W, to WA 4, Longview, Long Beach Peninsula (Difficult Reaccess)
Med	**E:** Hospital
Other	**E:** State Patrol Post
32	Kalama River Road
RVCamp	**E:** Camp Kalama RV Park
30	Kalama (Difficult Northbound Reaccess)
Gas	**E:** BP*
Food	**E:** Bob Paul's Cafe, Burger Bar, Columbia Inn Restaurant, Kalama Cafe, Poker Pete's Pizza Parlor, The Key Tavern
Lodg	**E:** Columbia Inn
AServ	**E:** Big A Auto Repair, Phil Poage's Transmission
RVCamp	**W:** RV Park
Other	**E:** Post Office
27	Todd Road, Port of Kalama
FStop	**E:** Texaco*(LP)
AServ	**E:** Louis River Tire, Rebel Tire Repair. (Truck & Car)
TServ	**E:** Rebel Tire Repair Truck & Car
RVCamp	**W:** Campground
22	Dike Access Road
RVCamp	**W:** Columbia River Front RV Park (2 Miles)
21	WA 503 East, Mt Saint Helens, Woodland, Cougar
Gas	**E:** Arco*, Chevron*, Shell*(D)(CW)(LP), Texaco*
	W: Quick Stop*, Shell*(D)(LP)
Food	**E:** Brock's Oak Tree Restaurant, Burgerville USA, Casa Maria's Mexican, Dairy Queen, Figaro's Pizza, Raitano's Italian Eaterty, Rosie's Restaurant, Subway
	W: Eager Beaver Drive In
Lodg	**E:** Louis River Inn Motel, (AAA) Woodlander Inn
	W: Hansen's Motel, Lakeside Motel
AServ	**W:** Louis River Motor Co., Napa Auto Parts, Shell
RVCamp	**E:** Woodland Shores RV Park
ATM	**W:** Columbia Bank, Sea First Bank, U.S. Bank
Other	**E:** Clean Wash Coin Laundry, Coin Car Wash, Hi-School Pharmacy
	W: Bookstore, Horseshoe Lake Park
16	N.W. 319th St, La Center
FStop	**E:** Texaco* (ATM)
ATM	**E:** Texaco Fuel Stop
(16)	Weigh Station (Northbound)
14	WA 501 West, NW 269th St, Ridgefield
FStop	**W:** Chevron*(D)
Gas	**E:** Arco*, BP*
Food	**E:** Country Junction Restaurant
RVCamp	**E:** Big Fir RV Camp (4 Miles)
(13)	Rest Area - RR, Phones, Vending, RV Dump, Picnic P (Southbound)
(11)	Rest Area - RR, Phones, Vending, RV Dump, Picnic P (Northbound)
9	WA 502E, Northeast 179th St, Battle Ground
FStop	**W:** Chevron*(D)
Food	**E:** Jollie's Restaurant
	W: Northwest BBQ
RVCamp	**E:** U-Neek RV Sales & Service(LP)
ATM	**W:** First Independent Bank

EXIT	WASHINGTON
7	Jct I-205S, to WA 14, I-84, Salem, Northeast 134th St
Gas	**E:** 76, Citgo* (ATM), Trail Mart*
	W: Expressway Market*(D)
Food	**E:** Burger King, Burgerville USA, JB's Roadhouse, McDonalds, Round Table Pizza, Taco Bell
Lodg	**E:** Shilo Motel
AServ	**E:** 76
ATM	**E:** Citgo
Other	**E:** Hi-School Pharmacy
5	Northeast 99th Street
Gas	**E:** Arco*, Citron*
	W: Arco*, Chevron*
Food	**E:** Bernabe's, Burgerville USA, Domino's Pizza
	W: Bortolami's Pizzaeria, Clancy's Seafood Mexican, Vancouver Pizza Co.
ATM	**W:** Northwest National Bank
Other	**E:** Cub Foods
	W: Albertson's Grocery, Hi-School Pharmacy
4	N.E. 78th St
Gas	**E:** 76*, Exxon*(CW), Texaco*(D)(CW)
	W: Arco*, Shell(D), Texaco*(D)
Food	**E:** 78th St. Barista, Akura, Burger King, Dragon King, Edelweiss Inn, Izzy's Pizza, KFC, Little Caesars Pizza, Perry's Restaurant, The New Hong Kong Lounge, Totem Pole
	W: City Grill, Denny's, Kenny Rogers Roasters, Round Table Pizza, Sunrise Bagels, Wendy's, YansingChinese
Lodg	**E:** Value Motel
	W: (AAA) Best Western
AServ	**E:** Battery X-Change, Dusty's Auto Parts, Firestone Tire & Auto, Hazel Dell Muffler
	W: Shell
ATM	**E:** First Independent Bank, Riverview Savings Bank, Seafirst Bank
	W: Wells Fargo
Other	**E:** Evergreen Animal Hospital, Fred Meyer Grocery, Petco
	W: Bookstore, Payless Drugs, Safe Way Grocery
3	Hazel Dell, Northeast Hwy 99, Main Street
Gas	**E:** Chevron*, Citgo*, Texaco(D)(CW)
Food	**E:** A & W Drive-In, Baskin Robbins, Bob's Supper Club, McDonalds, Peachtree Restaurant & Pie House, Pizza Hut, Ragmuffin's Deli, Skipper's Seafood, Smokey's Hot Oven Pizza, Steakburger, Subway, Taco Bell, Taco Time
	W: Hideaway Tavern
Lodg	**E:** Kay's Motel, Quality Inn
AServ	**E:** Chevron, Goodyear Tire & Auto, Les Schwab Tires, Midas Muffler & Brakes, My Daddy's Muffler, Oil Can Henry's, Precision Tune & Lube, Schuck's Auto Parts, Tire Factory
	W: Tech-tune Transmission
ATM	**E:** U.S. Bank
2	WA 500E, 39th St, Orchards
1D	WA 505, East - 4th Plain Blvd, Port of Vancouver
Food	**W:** Dairy Queen
AServ	**W:** Clark's European Car Service, Larson Tire Co, Main Street Auto Care
Other	**W:** Hi-School Pharmacy, Save Drugs
1C	Mill Plain Blvd, City Center (Watch For One Ways)
Gas	**W:** Chevron*

EXIT	WASHINGTON/OREGON
Food	**W:** Burgerville USA, Denny's
AServ	**W:** General Tire Company, Pinkerton Auto Mechanics, Steven's Auto Service
Other	**W:** Kinko's
(0)	Rest Area - RR, HF, Phones P
1AB	WA 14 East, Camas, Yakima

↑WASHINGTON

↓OREGON

EXIT	WASHINGTON/OREGON
308	Jantzen Beach Center
Gas	**E:** Chevron(LP)
	W: 76
Food	**E:** Bayou Side Coffee Shop, Burger King, Taco Bell, Waddles
	W: Chang's Mongolian, Denny's, McDonalds, Newport Bay Restaurant, Pietro's Engine House Pizza, Stanford's
Lodg	**E:** (AAA) Double Tree Hotel, (AAA) Oxford Suites
	W: (AAA) Double Tree
AServ	**E:** Chevron, Jantzen Automatic Car Wash
	W: 76, Montgomery Ward
ATM	**E:** Wells Fargo
Other	**E:** Payless Drugs, Safe Way Grocery
	W: K-Mart, Montgomery Ward, Visitor Information
307	OR 99E, Martin Luther King Jr Blvd, Marine Drive
306B	Interstate Ave, Delta Park
Gas	**E:** 76*(LP)
Food	**E:** Burger King, Burito House, Elmer's, Mar's Medows, Shari's
Lodg	**E:** Best Western, (AAA) Delta Inn
AServ	**E:** Baxter Auto Parts, Eric's Oilery
306A	Columbia Blvd (Northbound, Reaccess Northbound Difficult)
305AB	U.S. 30, Lombard St (Northbound, Reaccess Southbound Only)
Gas	**W:** 76*, Interstate*(D), Texaco*(D)
Food	**W:** Cam Ranh Bay Restaurant, KFC, Lung Fung Chinese, Taco Time, Wendy's, Winchell's Donuts
ATM	**W:** 7-11 Convenience Store
Other	**W:** 7-11 Convenience Store* (ATM), Fred Meyer Grocery
304	Portland Blvd
Gas	**W:** Arco*
Food	**W:** Nite Hawk Restaurant, Swan Garden
Lodg	**W:** Viking Motel
303	Killingsworth St, Swan Island
Food	**W:** Alibi Restaurant, Shamrock Restaurant, Subway, Taco Time
Lodg	**W:** Crown Motel, Knickerbocker Motel, Marco Polo Motel, Mel's Motor Inn, Monticello Motel, Westerner Motel
AServ	**W:** Interstate Automotive
ATM	**W:** U.S. Bank
Other	**W:** Plaid Pantery
302C	Greeley Ave, Swan Island (Difficult Reaccess)
302B	I-405, U.S. 30, Beaverton, St Helens
302A	Coliseum, Broadway - Weidler Street (Watch For One Ways)
Gas	**E:** BP*(CW)(LP), Texaco*(D)
Food	**E:** Burger King, Coffee People, Denny's, Golden

Bold red print shows RV & Bus parking available or nearby

EXIT OREGON

Palace Restaurant, KFC, McDonalds, Taco Bell, Wendy's

Lodg E: AAA Ramada Plaza Hotel, Roadway Inn

AServ E: Car Washman, Les Schwab Tires, Oil Can Henry's

ATM E: 7-11 Convenience Store

Other E: **7-11 Convenience Store**

300B Jct I-84, U.S. 30E, Portland Airport, The Dalles

299B I-405, U.S. 26, City Center, Beaverton

299A U.S. 26 East, OR 43, Ross Island Bridge, Macadam Ave

298 Corbett Ave (Northbound, No Reaccess)

297 Terwilliger Blvd

Food W: Burger King, KFC, La Costa Mexican, Norm's Garden Chinese

AServ W: Kaddy Car Wash

Med W: **+ VA Hospital**

ATM W: Bank of America, U.S. Bank

Other E: **Tryon Creek State Park**
W: **Fred Meyer Grocery**

296B Multnomah Blvd (Southbound, Reaccess Northbound Only)

296A Barbur Blvd (Southbound, Reaccess Northbound Only)

Gas W: Chevron*, Texaco

Food W: Golden Touch Family Restaurant, Manana Mexican Restaurant, Original Pancake House, Pizza Hut, Subway, Szechuan Restaurant, The Onion Deli

Lodg W: Capital Hill Motel, King's Row Motel, Portland Rose Motel

AServ W: Chevron, Jiffy Lube, Q-lube

ATM W: 7-11 Convenience Store

Other W: **7-11 Convenience Store* (ATM), Capitol Hill Veterinary Hospital, Safe Way Grocery**

295 Capitol highway, Taylors Ferry Road (Difficult Reaccess)

FStop W: **Pacific Pride Commercial Fueling**

Gas W: Texaco(LP)

Food W: Boston Market, Dunkin Donuts, IHOP, McDonalds, Round Table Pizza, Taco Time, The Old Barn Restaurant, Wendy's

Lodg W: Hospitality Inn

AServ W: Kaddy's Automatic Car Wash, Metro Tire and Auto Repair, Oil Can Henry's, Texaco, The Master Wrench

294 Barbur Blvd

Gas W: BP*, Texaco(CW)

Food E: Angelo & Rose's
W: Arby's, Banning's Restaurant, Burger King, Buster's Texas BBQ, Carrows Restaurant, Dimsum, Hi Hat Northern Cantonese & American, KFC, Mazatland Mexican, New Port Bay Restaurant, Pizza Caboose, Sante Fe Burrito, Skipper's Seafood, Subway, Taco Bell, Tang's Garden Restaurant, Tarra Thai

Lodg W: Days Inn Days Inn, AAA Wayside Motor Inn

AServ W: BP, Baxter Auto Parts, Les Schwab Tires, Tigard Transmission Center

ATM W: U.S. Bank

Other E: **Barbur Blvd. Vet. Hospital**
W: **Fred Meyer**

293 Haines St

EXIT OREGON

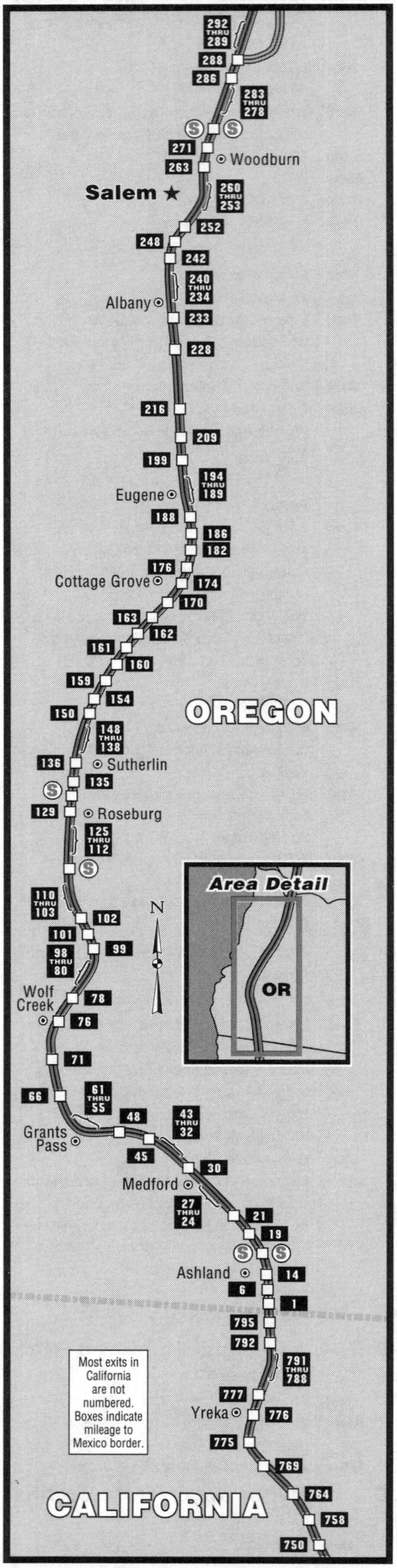

Most exits in California are not numbered. Boxes indicate mileage to Mexico border.

EXIT OREGON

292 OR 217, Tigard, Beaverton

Gas E: Texaco*(D)

Food E: Applebee's, Chevy's Mexican, Chili's, Hunan Pearle, Kobo's Expresso, Olive Garden, Stafford's Restaurant, Taco Bell

Lodg E: Crown Plaza Hotel (see our ad this page), Marriott, AAA Phoenix Inn

ATM E: US Bank

Other E: **Deseret Book**

291 Carman Drive, King City

Gas W: Chevron(LP), Shell*(LP) (ATM)

Food W: Burgerville USA

Lodg W: AAA Best Western, Courtyard Marriott

AServ W: Chevron

ATM W: Shell

290 Durham, Lake Oswego

Gas E: 76, BP(D)(LP), Chevron*(LP)
W: Arco, Texaco*(D)(LP)

Food E: Arby's, Burger King, China Cafe, Dalton's Steakhouse, Denny's, Fuddrucker's, Miller's Homestead Restaurant, Skippers Seafood, Taco Bell
W: Koon Lok Chinese Restaurant, Pin'n Pancakes, Village Inn

Lodg E: Motel 6
W: AAA Best Western, Quality Inn

AServ E: 76, Lake Oswego Transmission
W: Arco, Texaco

Other E: **Cat Care Veterinary Clinic, Safe Way Grocery**

289 Tualatin, Sherwood

Gas E: Arco*(CW), Shell(LP), Texaco*(CW)

Food E: El Sol De Mexican, McDonalds
W: Jiggle's Tavern, Pogys Subs, Taco Bell, Wendy's

Lodg E: The Sweetbrier Inn

AServ E: Instant Oil Change, Shell(LP)
W: K-Mart

RVCamp E: **Trailer Park of Portland (.5 Miles)**

Med E: **+ Legacy Meridian Hospital**

ATM E: 7-11 Convenience Store
W: US Bank, Wells Fargo

Other E: **7-11 Convenience Store* (ATM)**
W: **Fred Meyer Grocery, K-Mart, Safe Way Grocery**

288 to I-84, Jct I-205, Dalles, Seattle

286 Stafford, North Wilsonville

TStop E: **Burns Bros Travel Stop(SCALES)**

Gas E: BP*

Food E: Mrs. B's Homestyle

Lodg E: Burns Bros, **Burns West Motels**, AAA Motel Orleans, Super 8 Motel
W: AAA Holiday Inn (see our ad this page)

Bold red print shows RV & Bus parking available or nearby

EXIT OREGON

AServ E: BP, Burns Bros
TServ E: **Burns Bros**

283 Wilsonville
FStop E: **Pacific Pride Commercial Fueling**
Gas E: 76*
W: Chevron*(CW), Shell*
Food E: Arby's, Boston's Pub & Grill, Brew Ha Ha Coffee, Club House Deli, Denny's, Izzy's Pizza, Josh's Cafe, La Isla Bonita, McDonalds, Portlandia Pizza, Royal Panda Restaurant, Scotty's, Sharis, Subway, Taco Bell, Wendy's
W: Baskin Robbins, Burger King, Chili's, Doughnuts and Yogurt, Izzy's Pizza, Kathy's Expresso, New Century Chinese
Lodg E: (AAA) Comfort Inn, **Snooz Inn**
W: (AAA) Phoenix Inn
RVCamp W: **Camping World (see our ad this page)**
AServ E: 76
W: Chevron, Shell
ATM E: Bank of America
W: 7-11 Convenience Store, US Bank, Wells Fargo
Other E: **Payless Drugs, Thriftway Supermarket,** Town Center Car Wash, **Town Center Veterinary Clinic**
W: **7-11 Convenience Store (ATM), Lowries I.G.A.**

282A Canby, Hubbard

282B Charbonneau District, Canby

(282) **Rest Area - RR, Phones, Picnic, Info, RV Dump (P) (Both Directions)**

278 Donald, Aurora
TStop W: **76 Auto/Truck Plaza(LP)(SCALES)**
FStop W: **Texaco***
Gas E: **Tesoro Alaska*(D)(LP) (RV Dump)**
Food W: **Restaurant (76 TStop)**
TServ W: **76 Auto/Truck Plaza**
RVCamp E: **Isberg RV Park**
ATM W: Texaco

(274) **Weigh Station (Both Directions)**

271 OR 214, Woodburn, Newberg
Gas E: 76(LP), Arco*, Chevron*, Shell(CW)
W: Texaco*(D)(LP)
Food E: Burger King, Dairy Queen, Denny's, KFC, McDonalds, Oregon Berry Restaurant, Patterson's, Sandwich Express, Taco Bell, Wendy's
Lodg E: **Fairway Inn Motel,** (AAA) Holiday Inn Express, Super 8 Motel
W: (AAA) Comfort Inn
AServ E: 76, Shell
W: GM Auto Dealership, Hillyer's Ford Dealership
RVCamp W: **Woodburn I-5 RV Park**
ATM E: First Security Bank
Other E: **Fairway Drugs, Visitor Information**

263 Brooks, Gervais
TStop W: **Pilot(LP)(SCALES) (Taco Bell, Subway)**
Food W: **Subway (Pilot), Taco Bell (Pilot)**
Parks W: **Williamette Mission State Park (4 Miles)**

260AB OR 99E, Keizer, North Chemawa Road, Salem Parkway

258 OR 99E, Pacific Highway East
Gas E: BP*
W: Pacific Pride Commercial Fueling*
Food E: Figaro's, McDonalds, The Original Pancake House
Lodg E: (AAA) Best Western, ◆ Sleep Inn
AServ E: Superior Tire Service
Med W: **+ Hospital**
Other E: **Thriftway Market Place**
W: **Circle K Food Store, State Patrol Post**

256 Market St, Silverton
FStop W: **Pacific Pride Commercial Fueling**
Gas W: Arco*, Texaco
Food E: Best Teriyaki, Blue Willow Chinese Restaurant, Carl's Jr Hamburgers, Chalet Restaurant, Chinese Restaurant, Denny's, Don Bedro, Dunkin Donuts, Elmer's Pancake & Steakhouse, Los Baez Mexican, Lucky Fortune Restaurant, Nacho's Mexican Restaurante, Olive Garden, Santa Fe Burrito Company, Skipper's Seafood, Subway, Taco Bell
W: Baskin Robbins, Canton Garden Chinese, McDonalds, Newport Bay Restaurant, O'Callahan's Restaurant (Quality Inn), Pietro's Engine House Pizza, Richard's Restaurant, Rock-In- Rogers, Village Inn
Lodg E: (AAA) Best Western, Tiki Lodge Motel, (AAA) Travelodge
W: **Holiday Lodge,** Motel 6, Quality Inn Salem Inn, ◆ Shilo Inn, **Super 8 Motel**
AServ E: Les Schwab Tires, Magic Touch Machine Car Wash, McEwen's Car Wash, Midas Muffler & Brakes
W: Auto Glass Express, Market Street Mazda & Isuzu, Texaco
TServ W: **Bratan International**
ATM E: First Security Bank
W: Wells Fargo
Other E: **Albertson's Grocery, Fred Meyer Grocery, Suds City Depot**

253 OR 22 North, Santiam Hwy, Stayton, Detroit Lake
Food W: Denny's
Lodg W: (AAA) Best Western
AServ W: Robertson Chrysler/Plymouth
Med W: **+ Hospital**
Other W: **Costco, Visitor Information**

252 Kuebler Blvd

249 Salem (Northbound. Reaccess Southbound Only.)
Med W: **+ Hospital**
Other W: **Salem Historic Museum**

248 Sunnyside, Turner
FStop W: **Pacific Pride Commercial Fueling**
AServ W: Ole's Towing

244 Jefferson

243 Ankeny Hill

242 Talbot Road

(241) **Rest Area - RR, Picnic, Phones, Vending, RV Dump (P) (Both Directions)**

240 Hoefer Road

239 Dever - Conner

238 Jefferson, Scio

237 Viewcrest (Southbound, Reaccess Southbound Only)

235 Viewcrest (Northbound, Reaccess Northbound Only)

234AB Albany, Knox, Butte
Gas W: Arco*, Chevron*
Food W: **Burger King,** McDonalds, Taco Bell, **Tom Tom Restaurant,** Yaquina Bay Restaurant
Lodg W: Budget Inn, Comfort Inn
RVCamp W: **Camper Shoppe RV Center**
Other W: **K-Mart, Mervyn's**

233 U.S. 20, Santiam Hwy, Lebanon, Albany
TStop E: **Beacon(D) (ATM), Jack's Truck Stop*(SCALES) (ATM)**
FStop E: **Chevron*(LP)**
Gas W: Leathers Oil Co, Plaza Gas, Texaco(LP)
Food E: Burgundy's Restaurant, **T & R Restaurant**
W: Abby's Legendary Pizza, Appletree Family Restaurant, Baskin Robbins, Burger King, BurgerVille USA, Cameron's Restaurant, Carl's Jr Hamburgers, Chan Kam Kee Chinese, Cork's Donuts, Elmer's Pancake & Steakhouse, Skipper's Seafood, Taco Time, Tequila's
Lodg E: (AAA) Holiday Inn Express, Motel Orleans
W: Valu-Inn
AServ E: Doug's, GM Auto Dealership
W: Leathers Oil, Les Schwab Tires, Mark Thomas Plymouth, Chrysler, Dodge, Jeep Eagle & Subaru, Texaco
TServ E: **Battam International, Brattain International and Caterpillar Dealership**
RVCamp E: **Blue Ox RV Park**
ATM E: Beacon, T & R Restaurant
W: U.S. Bank
Other E: **Albany Municipal Airport**
W: **Albertson's Grocery, Fred Meyer Grocery, Payless Drugs**

228 OR 34, Lebanon, Corvallis
FStop E: **Texaco*(D) (Commercial Fueling)**
Gas E: Hers
W: 76*(LP), Arco*, BP*(D), Cash and Credit*(LP)
Food E: **Pine Cone Cafe (Texaco FS)**
AServ E: Hers

216 OR 228, Brownsville, Halsey
TStop E: **BP* (Blimpie)**
FStop W: **Texaco* (Subway & Taco Bell)**
Food E: Blimpie's Subs (Pioneer Villa), **Restaurant* (Pioneer Villa)**
W: **Subway (Texaco), Taco Bell (Texaco)**
Lodg E: Pioneer Villa (ATM)
AServ W: Larry's Auto Parts and Repair
ATM E: Pioneer Villa

209 Harrisburg, Junction City
Food W: **The Hungry Farmer**
TServ W: **Diamond Hill Truck & RV Repair**
RVCamp W: **Diamond Hill**

(206) **Rest Area - RR, Phones, Picnic, Information (P) (Both Directions)**

199 Coburg
TStop W: **TravelCenters of America*(SCALES) (RV Dump, Motel)**
Gas W: BP*(LP), Shell*, Texaco(LP)
Food E: **LB (Country Squire Inn)**
W: **Country Pride (T/C of America),** The Hillside

EXIT OREGON

Grill
Lodg E: Country Squire Inn
AServ W: Road Runner Tire, Texaco
TServ E: Road Runner Electronics
W: Basin Tire, Cummins Diesel, Kelly Trailer Repair
RVCamp W: Eugene Kamping World

195AB Junction City, Florence
Gas E: Arco*, Chevron*
Food E: Denny's, Gateway Chinese Buffet, IHOP, KFC, McDonalds, Sharis Restaurant, Shilo Inn, Taco Bell, Taco Time
Lodg E: Best Western, Double Tree Hotel, Gateway Inn, Marriott, Motel 6, Motel Orleans, Pacific 9, Roadway Inn, Shilo Inn
TServ E: Stalick International Trucks
ATM E: Western Bank
Other E: Gateway Mall

194AB OR 126, I-105, Springfield, Eugene

192 OR 99, Eugene, Univ of Oregon (Northbound, Reaccess Southbound Only)
Gas W: BP*, Space Age
Food W: Barrons (Travel Lodge), Black Angus, Burger King, Deb's Family Restaurant, Domino's Pizza, House of Chen Restaurant, Kim's Restaurant, Lai Lai Mandarin & Chinese, Lyon's Restaurant, McCallums, Tracktown Pizza, Wendy's
Lodg W: Best Western, Franklin Inn, Quality Inn, Travel Lodge (Barrons)
AServ W: GM Auto Dealership, German Auto Service
Med W: Hospital
Other W: Hirxons Pharmacy

191 Glenwood, Springfield
Gas W: BP[LP], Texaco*[D][LP]
Food W: Denny's
Lodg W: Motel 6
TServ E: Pape Caterpillar Dealership

189 30th Ave, South Eugene
Gas E: Texaco*[D]
W: Exxon[LP]
Food E: Taco Bell (Texaco), Taco Bell (Texaco), The Road House Pub and Grill
W: The Old Smokehouse (Exxon)
RVCamp E: Eugene Mobile Home Park (1.75 Miles), Interstate 5 RV Service

188B OR 99 South, Goshen
TStop W: Goshen Truck Stop
FStop W: Pacific Pride
Food W: Summers Cafe
TServ W: Road-Runner Tire
Other W: R&D Propane[LP]

188A OR 58, Willamette Hwy, Oakridge
FStop W: Goshen Truck Stop & Cafe
Food W: Goshen Truck Stop & Cafe
TServ W: Goshen Truck Stop Tire Service

186 Goshen (Northbound, Reaccess Southbound Only)

182 Creswell
Gas W: Arco*, BP*[LP], Chevron, Texaco*[D][LP]
Food W: Creswell Cafe and Deli, Dairy Queen, Los Cabos, Mr. Macho's Pizza, TJ's Family Restaurant, The Pizza Station (BP)
Lodg W: AAA Motel Orleans
AServ W: Chevron, Knecht's Discount Auto Parts, Ruiz Repairs Auto Trucks & RVs
RVCamp W: KOA Campground
ATM W: Siuslaw Bank
Other W: Century Grocery Store, Coin Laundry, Creswell Pharmacy, Creswell Vet. Clinic

(178) Rest Area - RR, Phones, Picnic, Tourist Info P (Both Directions)

176 Saginaw

174 Cottage Grove, Dorena Lake
FStop W: Chevron*[D] (ATM)
Gas E: Chevron[D][LP], Shell[LP]
Food E: Subway, Taco Bell
W: Burger King, Carl's Jr Hamburgers, KFC, McDonalds, Vintage Inn Restaurant
Lodg E: AAA Best Western
W: AAA Comfort Inn, AAA Holiday Inn Express
AServ E: Chevron, GM Auto Dealership, Lowther Chrysler Plymouth Dodge Jeep Eagle, Shell, Wal-Mart
RVCamp E: Village Green RV Park
ATM W: Chevron
Other E: Sud-n-Shine, Wal-Mart

170 Bus OR 99, Cottage Grove (Difficult Northbound Reaccess)
Med W: Hospital

163 Curtin
FStop E: Shell*[D][LP]
Gas W: 76*
Food E: Curtin Kitchen
W: The Coach House
Lodg E: Stardust Motel
RVCamp W: Pass Creek Park
Other E: US Post Office

162 Drain, Elkton

161 Anlauf, Lorane (Northbound, No Reaccess)

160 Salt Springs Road

159 Elk Creek, Cox Road

154 Scotts Valley, Elkhead

150 OR 99, to OR 38 Yoncalla, Drain
RVCamp W: Trees of Oregon RV Park

148 Rice Hill
TStop E: Texaco Truck Stop*[SCALES]
FStop E: Commercial Fueling Station
Gas E: Chevron*
Food E: Meggy's Restaurant, Quickies Drive In, Ranch Restaurant
W: K & R Drive-in
Lodg E: Ranch Motel
AServ E: Carl's Towing, Jim's Towing & Garage
TServ E: Northwest Diesel Service

146 Rice Valley

(143) Rest Area - RR, Picnic, Phones P (Both Directions)

142 Metz Hill

138 Oakland (Northbound, Reaccess Southbound Only)

136 OR 138, Sutherlin, Elkton
Gas E: 76, Chevron, Road Runner*[D][LP], Texaco[D] (Subway)
W: BP[LP]
Food E: Abby's Legendary Pizza, Bosman's Burger Barn, Dory's Restaurante Italiano, Hong Kong Chinese & American, Lettuce Inn Hamburgers, McDonalds, Papa Murphy's, Subway (Texaco), Tequilas
W: Dairy Queen, Taco Bell, West Winds
Lodg E: Penny Wise Motel, Town & Country Motel
W: Best Budget Inn
AServ E: 76, Big A Auto Parts, Chevron, Coin Car Wash, Kress-Vacken, Les Schwab Tires
W: BP
ATM E: Douglas National Bank

135 Sutherlin, Wilbur

(130) Weigh Station (Southbound)

129 Winchester, Wilbur
TServ E: Southern Oregon Diesel

127 N. Roseburg, Stewart Pkwy

125 Roseburg, Garden Valley Blvd
Gas E: BP*, Fireball, Texaco*[D]
W: 76, Chevron* (24 Hrs), Shell
Food E: Casey's Restaurant, KFC, McDonalds, Papa Murphy's Pizza, Purple Parrot Deli, Sandpiper (Windmill Inn), Shumart Pizza and Subs, Taco Bell
W: Arby's, Burger King, Carl's Jr Hamburgers, Fox Den, IHOP, Izzy's Pizza, La Hacienda Mexican Restaurant, Round Table Pizza, Sizzler Steak House, Tom Tom Restaurant, Waldron's, Wendy's
Lodg E: ◆ Comfort Inn, Hotel Orleans, AAA Windmill Inn
W: AAA Best Western
AServ E: Champion Car Wash, Fireball, Pennzoil Lube, Texaco
W: Schuck's Auto Supply, Shell
Med W: Hospital
ATM E: Douglas National Bank
W: Bank of America, South Umpqua Bank
Other E: Albertson's Grocery, Coin Opp Laundry, Payless Drugs
W: Fred Meyer Grocery, Garden Valley Shopping Center

124 OR 138, City Center, Diamond Lake
Gas E: BP*[D][LP]
W: Astro[D] (24 Hrs), Chevron, Texaco
Food E: Denny's
W: Gay 90's Ice Cream, KFC, Subway, Taco Time
Lodg E: Dunes Motel, AAA Holiday Inn Express, Holiday Motel, AAA Travelodge
AServ E: BP
W: Speedy Lube, Texaco
Med W: Douglas Community Hospital
Other E: Visitor Information
W: Books and Expresso, Grocery Outlet

123 Fairgrounds, Umpqua Park
RVCamp E: Fairgrounds RV Park
Other E: County Museum

121 McLain Ave

120 OR 99N, Roseburg (Difficult Northbound Reaccess)
FStop W: Bassett-Hyland Fuel Center
Lodg E: AAA Shady Oaks Motel
TServ W: Transit Support Services

119 OR 99, OR 42W, Winston Coos Bay
FStop W: Texaco*[D][CW]
RVCamp W: RV Camp

113 Round Prairie, Clarks Branch Road
Lodg W: Quick Stop Motel & Market
TServ W: Doug's Diesel Inc
RVCamp W: On the River RV Park (2 Miles)

Bold red print shows RV & Bus parking available or nearby

EXIT OREGON

112 OR 99, OR 42, Dillard, Coos Bay
- RVCamp **E:** RV Park
- Other **E:** Rest Area - RR, Picnic, Phone

(112) Rest Area - RR, Phones, Picnic P (Southbound)

(111) Weigh Station (Northbound)

110 Boomer Hill Road

108 Myrtle Creek
- Food **E:** Cross Creek Restaurant, Dairy Queen, Fat Elk Deli Cafe
- Lodg **E:** Rose Motel, South Umpqua Inn
- AServ **E:** Car Quest Auto Center, Chevron
- RVCamp **E:** Myrtle Creek RV Park
- ATM **E:** U.S Bank
- Other **E:** Mike's Books and Expresso

106 Weaver Road

103 Tri - City, Myrtle Creek
- FStop **W:** Chevron*(D)
- Food **W:** McDonalds

102 Gazley Road

101 Riddle

99 North Canyonville, Stanton Park
- TStop **W:** Fat Harvey's Travel Center*(LP)(SCALES) (RV Dump, ATM)
- Food **E:** Burger King
  W: Restaurant (Fat Harvey's TStop)
- Lodg **E:** Riverside Lodge Motel, Valley View Motel
- ATM **W:** Fat Harvey's
- Other **E:** Crow Creek Indian Gaming Center, Stanton Park

98 OR 99, Canyonville, Days Creek
- Gas **E:** BP*(D)(LP), Texaco*(D)
- Food **E:** BJ's Place Restaurant, Bella Donna, Bob's Country Junction, Feed Lot Family Restaurant, Kathy's Kitchen Cafe
- Lodg **E:** Leisure Inn
- AServ **E:** Dick's Pacific Hwy Garage & Towing, Jake's Auto Center, Napa Auto Parts
  W: Bill's Tire Towing & Automotive(LP)
- ATM **E:** South Umpqua Bank
- Other **E:** Rexall Drugs (.5 Miles), Visitor Information
  W: Dave's Book Barn, Pioneer Indian Museum

95 Canyon Creek

88 Azalea, Galesville Reservoir

86 Quines Creek Road
- Gas **E:** Texaco*(D)(LP)
- Food **E:** Heaven On Earth Restaurant
- Other **E:** Galesville General Store & Tourist Information

83 Barton Road (Northbound, Reaccess Southbound Only)
- RVCamp **E:** Meadow Wood RV Park (1 Mile)

(82) Rest Area - RR, Phones, Picnic P (Both Directions)

80 Glendale
- Gas **W:** T & T
- Food **W:** Lynn's Drive Inn(LP)

78 Speaker Road, Glendale (Southbound, Reaccess Northbound Only)
- RVCamp **W:** Creekside*(LP)

76 Wolf Creek (Difficult Northbound Reaccess)
- Gas **W:** Texaco*(LP)
- Food **W:** The Barrr, Wolf Creek Tavern
- Lodg **E:** Irish Sheperds Inn
- Other **W:** U.S. Post Office, Wolf Creek General Store

71 Sunny Valley, Leland
- Gas **E:** BP*(D)(LP)
- Food **W:** Covered Bridge Crossing
- Lodg **E:** Sunny Valley Motel
- RVCamp **W:** KOA Camp
- Other **E:** Covered Bridge Country Store

66 Hugo
- RVCamp **E:** Joe Creek Waterfalls RV Park*(LP)

(63) Rest Area - RR, Phones, Picnic, RV Dump P (Both Directions)

61 Merlin
- Gas **W:** Texaco(D)
- AServ **W:** MR Auto Parts
- ATM **W:** Colonial Bank
- Other **W:** Ray's (Grocery)

58 OR 99, OR 199, N. Grants Pass
- Gas **W:** 76, BP(D), Chevron, Gas 4 Less(LP), Shell, Texaco*(D), Towne Pump(CW)
- Food **W:** Baskin Robbins, Bee Gee's Restaurant, Burger King, Carl's Jr Hamburgers, Dairy Queen, Della's Restaurant, Denny's, Maggie's Pizza, Manning's, Matsukaze, McDonalds, Pizza Hut, Pongsri's Thai-Chinese, Royal Vue, Skipper's Seafood, Subway, Taco Bell, The Jelly Doughnut, The Lantern Grill, Wendy's
- Lodg **W:** Golden Inn, Hawk's Inn Motel, (AAA) Motel 6, Motel Orleans, Royal Vue, Shilo Inn Motel, Super 8 Motel, (AAA) Sweet Breeze Inn
- AServ **W:** 76, Caveman Towing, Chevron, Gas 4 Less, Jim Sigel Chevrolet Nissan/Honda, Les Schwab Tires, Texaco(D)(LP)
- RVCamp **W:** Rogue Valley Overniters
- ATM **W:** Valley Of The Road Bank, Wells Fargo, Western Bank
- Other **W:** Coin Laundry, Visitor Information

55 U.S. 199, Grants Pass, Crescent City
- Gas **W:** Arco*
- Food **W:** Abby's Pizza Inn, Elmer's Pancake & Steakhouse, Hamilton House Restaurant (Fred Meyer), J.J. North's Grand Buffet, McDonalds, Sharis, Taco Bell
- Lodg **W:** (AAA) Best Western, (AAA) Holiday Inn Express
- ATM **W:** Fred Meyer
- Other **W:** Fred Meyer Grocery (ATM), Wal-Mart

48 City of Rogue River
- Gas **E:** Arco*, Exxon, Texaco*(D)(LP)
- Food **E:** Abby's Legendary Pizza, Cattleman's Saloon & Rib House, China House, Suzu-Ya Japanese, The Wright Place
  W: Branding Iron Steakhouse & Mexican, Karen's Kitchen
- Lodg **W:** (AAA) Best Western
- AServ **E:** Exxon
- RVCamp **W:** Circle W RV Park (1 Mile), River Park RV Resort (6.25 miles), Whispering Pines RV Park
- ATM **E:** Valley of the Rogue Bank
- Other **E:** Dove Book Center, Rogue River Vet. Hospital

45B Valley of the Rogue Park & Rest Area - Picnic, HF, RR, Camping

45A OR 99, Savage Rapids Dam and Rogue River Route
- Food **E:** Rogue Riviera Supper Club
- Lodg **E:** Reeves Flycaster Motel
- RVCamp **E:** Cypress Grove RV Park(LP)
- Other **E:** Valley of the Rogue Park

43 OR 99, Rogue River Rt, Rock Point
- Food **E:** Rock Point Bistro
- Lodg **E:** Rock Point Motel & RV Park, Rogue River Guest House

40 OR 99, OR 234, Gold Hill, Scenic Byway
- Gas **W:** BP*
- Food **W:** Gato Gordo Mexican
- RVCamp **E:** KOA Campground, Lazy Acres
  W: Dardanelle's RV Park (.25 Miles)

35 OR 99, Blackwell Road

33 Central Point
- TStop **E:** Pilot(SCALES) (Taco Bell, Subway)
- Gas **E:** Chevron*(LP)
  W: BP*(D)(LP), Texaco*(D)(LP) (ATM)
- Food **E:** Burger King, Subway (Pilot), Taco Bell (Pilot)
  W: Bee Gee's Restaurant, Berg's Old Fashoned Bakery, McDonalds, Pappy's Pizza Inn
- ATM **W:** Texaco

30 OR 62, Medford, Crater Lake
- TStop **W:** Witham*(SCALES) (TServ, Restaurant)
- Gas **E:** Chevron*, Gas 4 Less(LP)
  W: Chevron*
- Food **E:** Arby's, Coyote Grill (Reston Hotel), Dairy Queen, Denny's, Elmer's Pancake & Steakhouse, IHOP, Pizza Hut, Taco Delite
  W: Restaurant (Witham TStop)
- Lodg **E:** (AAA) Best Western, Howard Johnson, (AAA) Reston Hotel, (AAA) Windmill Inn
  W: (AAA) Comfort Inn, Motel 6, (AAA) Rogue Regency Inn
- AServ **E:** Mastercraft Tires
- TServ **W:** Freightliner Dealer(SCALES), Witham TStop
- Other **E:** Barnes & Noble, Food 4 Less Supermarket, Kinko's

27 Medford, Barnett Road
- Gas **E:** Exxon*(D), Shell*
  W: Astro(D), BP(CW), Chevron, Exxon*(D), Texaco*
- Food **E:** Apple Annie's Family Restaurant, Kopper Kitchen
  W: Abby's Pizza, Apple Annie's, Burger King, Home Town Buffet, Jack-In-The-Box, KFC, Kenny Rogers Roasters, Kim's, McDonalds, McGrath's Fish House, Pizza Hut, Roosters Homestyle Cooking, Senor Sam's Mexican Grill, Sharis Restaurant, Strayhorn, Subway, Wendy's
- Lodg **E:** (DAYS INN) Days Inn, (AAA) Horizon Motor Inn, Motel 6
  W: Capri Motel, Comfort Inn, (AAA) Medford Inn, Royal Crest Motel
- AServ **W:** Big O Tires, Chevron, Hand Car Wash, Quality Tire
- Med **E:** + Rogue Valley Medical Hospital
- ATM **W:** Wells Fargo, Western Bank
- Other **W:** Grocery Outlet, Visitor Information

24 Phoenix
- TStop **E:** Petro(LP)(SCALES)
- Gas **E:** Texaco*(LP)
  W: Exxon*
- Food **E:** Iron Skillet (Petro)
  W: Courtyard Cafe, McDonalds
- Lodg **E:** (AAA) Pear Tree Motel
- TServ **E:** DSU Peterbilt
- RVCamp **W:** Holiday RV Park
- Other **W:** Factory Outlet Center

Bold red print shows RV & Bus parking available or nearby

EXIT	OREGON

(22) **Rest Area - RR, Phones, Picnic P (Southbound)**

21 Talent
- **TStop** W: **Talent Truck Stop**(D)
- **Gas** W: Arco*, Lewis River Shell* (see our ad this page), Woodland Shell* (see our ad this page)
- **Food** W: Cafe Electra Expresso, Figaro's, **Talent Truck Stop Restaurant**
- **AServ** W: Wal-Mart
- **TWash** W: **B & H Truck Wash**
- **RVCamp** W: **Oregon RV Roundup**
- **Other** W: **Wal-Mart**

19 Valley View Road, Ashland
- **FStop** W: **Pacific Pride Commercial Fueling**
- **Gas** W: Exxon(LP), Texaco*
- **Food** W: Burger King, Regency Inn
- **Lodg** W: (AAA) Best Western, (AAA) Regency Inn
- **AServ** W: Exxon, Texaco
- **RVCamp** W: **Regency Inn**

(18) **Weigh Station (Both Directions)**

14 OR 66, Ashland, Klamath Falls
- **Gas** E: BP(LP), Chevron*(D), Texaco*(D)(LP)
  W: 76, Arco*, Exxon
- **Food** W: All American Ice Cream and Yogurt, Chubby's, Ezteca, Hot Shot Expresso, Jade Dragon, McDonalds, Oak Tree Restaurant, Pizza Hut, Round Table Pizza, Taco Bell
- **Lodg** E: (AAA) Ashland Hills Inn, Flagship Inn, Vista Motel
  W: (AAA) Knight's Inn, Super 8 Motel
- **AServ** E: BP
  W: 76, Exxon, Les Schwab Tires

EXIT	OREGON/CALIFORNIA

- **ATM** W: Bank Of America
- **Other** W: **Albertson's Grocery, Oregon Welcome Center, Payless Drugs**

11 OR 99, Siskiyou Blvd, Ashland (Northbound. Reaccess Southbound Only)

9 Runaway Truck Ramp (Northbound)

7 Runaway Truck Ramp (Northbound)

6 Mt Ashland
- **Food** E: Callahan's Dinner House & Country Store

1 Siskiyou Summit (Northbound, Reaccess Southbound Only)

↑OREGON

↓CALIFORNIA

795 Hilt
- **Gas** W: Texaco*
- **Food** W: Hamburger Cafe (Texaco)

792 Bailey Hill Road

(791) Agricultural Inspection station (Southbound)

790 Hornbrook Hwy, Ditch Creek Road

788 Henley, Hornbrook
- **FStop** E: **Chevron***(D)(LP)
- **Other** E: **Iron Gate Recreation Area (8 Miles)**

786 CA 96 Klamath River Hwy, Rest Area Inside Exit (Steep Off Ramp)

782 Vista Point (Southbound)

777 Yreka, Montague
- **FStop** W: **USA Gasoline***(D)
- **Food** W: Claim Jumper's Family Restaurant, Ma & Pa's Restaurant, Sterlings Take Out BBQ
- **Lodg** W: Gold Pan Motel, Super 8 Motel
- **RVCamp** W: **Tandy's RV and Auto Service**
- **ATM** W: Ray's
- **Other** E: **Airport**
  W: **Amerigas Propane**(LP), **Coin Laundry, Ray's Food Place Grocery (ATM)**

776 Central Yreka
- **FStop** E: **Pacific Pride Commercial Fueling**
- **Gas** W: BP, Chevron, Texaco*(D)
- **Food** W: A & W Drive-In (Texaco), China Dragon, Grandma's House, Hunan Villa, Log Cabin Club, Miner Street Backery, Ming's, Sue's Coffee and Cones, Taco Time (Texaco), The Daily Grind Coffee, The Deli
- **Lodg** W: Best Western, Heritage Inn, Roadway Inn, Yreka Motel
- **AServ** W: All Pro Auto Parts, BP, Chevron, Clayton Tire Center, Weldon's Tire
- **ATM** W: Tri Counties Bank, US Bank
- **Other** W: **US Post Office**

775 CA 3, Fort Jones, Yreka
- **Gas** W: Chevron*
- **Food** W: Burger King, **Jerry's Restaurant**, McDonalds, Papa Murphy's Pizza, Pizza Hut, Subway, The Old Boston Chef Restaurant, Yogurt Etc.
- **Lodg** W: Amerihost Inn, Motel 6, Motel Orleans
- **AServ** E: Les Schwab Tires (access from West frontage road)
  W: Jim Wilson Ford Mercury & Lincoln, NAPA

EXIT	CALIFORNIA

Auto Parts
- **RVCamp** E: **Waiiaka Trailer Haven & RV Park (access from West frontage road)**
- **Med** W: ✚ **Hospital**
- **Other** E: **Siskiyou Golden Fair**
  W: **JC Penny, Raley's Supermarket, Wal-Mart (Pharmacy)**

769 Easy St., Shamrock Rd.
- **Gas** W: Beacon*(D)(LP)
- **RVCamp** W: **Campground**

764 Montague, Grenada
- **FStop** E: **Shell***
- **AServ** E: Billy Tanner's Repair & Towing

758 Louie Road

752 Weed Airport Road (Rest Area Inside Exit)
- **Food** W: **Porky Bob's BBQ**
- **Other** E: **Weed Airport**

750A Stewart Springs Road, Edgewood, Gazelle
- **RVCamp** E: **Campground**
- **Other** E: **Lake Shastina Recreation Area**

746A US 97, Klamath Falls, Central Weed, North Weed Blvd.
- **Gas** E: BP*(LP), Chevron*, Shell*(D), Texaco*
- **Food** E: **Hi-Lo Restaurant**, Y Restaurant
- **Lodg** E: **Hi-Lo Motel**, Motel 6, Town House Motel, Y Motel
- **AServ** E: Bill's Garage & Towing, Chevron(CW)(LP), Napa Auto Parts
- **RVCamp** E: **RV Park (Hi-Lo Motel)**
- **Other** E: **E-Z Wash Coin Laundry**

745 South Weed Blvd
- **FStop** E: **CFN**
- **Gas** E: BP*, Chevron*(D)(CW)
- **Food** E: **Burger King, McDonalds, Silva's Family Restaurant, Taco Bell**
- **Lodg** E: **Comfort Inn**, Holiday Inn Select, **Sis-Q Inn Motel**
- **Other** E: **Kellogg Ranch**

743 Summit Drive, Truck Village Drive
- **FStop** E: **Commercial Fueling Network**
- **Other** W: **Grand Rental Station**(LP)

741 Abrams Lake Road
- **RVCamp** W: **Camping**

740 Mount Shasta City (Southbound, Reaccess Northbound Only)

739 Central Mount Shasta
- **Gas** E: BP*(D), Chevron*, Shell(LP), Texaco*(CW)
- **Food** E: Black Bear Diner (Best Western), Burger King, Little Caesars Pizza (Texaco), Round Table Pizza, Say Cheese Pizza (Best Western), Shasta's Family Restaurant, Subway (Best Western), Taco Shop
- **Lodg** E: Best Western
- **AServ** E: Kragen
- **RVCamp** W: **Lake Siskiyou Campground, Mount Shasta RV Resort**
- **Med** E: ✚ **Hospital**
- **ATM** E: Ameican Savings Bank
- **Parks** W: **Lake Siskiyou**
- **Other** E: **Coin Laundry, Golden Boug Book Store, Payless Drugs, Ray's Food Place Grocery, Thrifty Drugs, Tourist Information, Wings**

Bold red print shows RV & Bus parking available or nearby

Book Store

737 Mount Shasta City (Northbound. Difficult Reaccess)
Lodg E: Finlandia Motel

736 CA 89, McCloud, Mount Shasta City
Food E: Wayside Inn
Lodg E: Swiss Holiday Lodge
RVCamp E: **McCloud Dance Country RV Park (10 Miles)**

(734) **Weigh Station, Inspection Center (Southbound)**

733 Mott Road, Dunsmuir Ave

731 Dunsmuir Ave, Siskiyou Ave
Food E: Nicole's (Best Choice Inn)
Lodg E: Best Choice Inn
W: Cedar Lodge, Garden Motel

730 Central, Dunsmuir
Gas E: Shell*
W: BP[LP], Texaco*[D]
Food W: Don's Frosties (Tacos & Hamburgers), Hitching Post Cafe
Lodg E: Dunsmuir Inn, Travelodge
W: Kay Springs Motel
AServ E: Napa Auto Parts, Shasta Ford & Mercury Dealer
W: BP
Other E: **US Post Office**
W: **Botanical Gardens, Dunsmuir City Park**

728 Dunsmuir
Gas E: 76*[D] (ATM)
Lodg E: Oak Tree Inn
ATM E: 76

727B Cragview Drive, Railroad Park Road
Food W: Diner House
Lodg W: **Caboose Motel**
RVCamp E: **Rustic Park*[LP]**
Parks W: **Railroad Park**

727 Crag View Drive (Northbound)
RVCamp E: **Cedar Pines RV Park**

726 Soda Creek Road
Other W: **Pacific Crest Trail**

724 Castella
Gas W: Chevron*[D]
Food W: Castle Crag Tavern & Cafe (Chevron)
RVCamp E: **Cragview Valley Camp RV Park**
Parks W: **Castle Crags State Park**
Other W: **U.S. Post Office**

723 Vista Point (Northbound, Observation Area)

722 Sweetbrier Ave

721 Conant Road

720 Flume Creek Road

718 Sims Road
RVCamp W: **Campground**

713 Gibson Road

711 Pollard Flat
FStop E: **Exxon*[LP] (ATM)**
Food E: **Pollard Flat USA Restaurant**
ATM E: Exxon Fuel Stop

710 La Moine, Slate Creek Road

706 Vollmers, Dog Creek Road, Delta Road

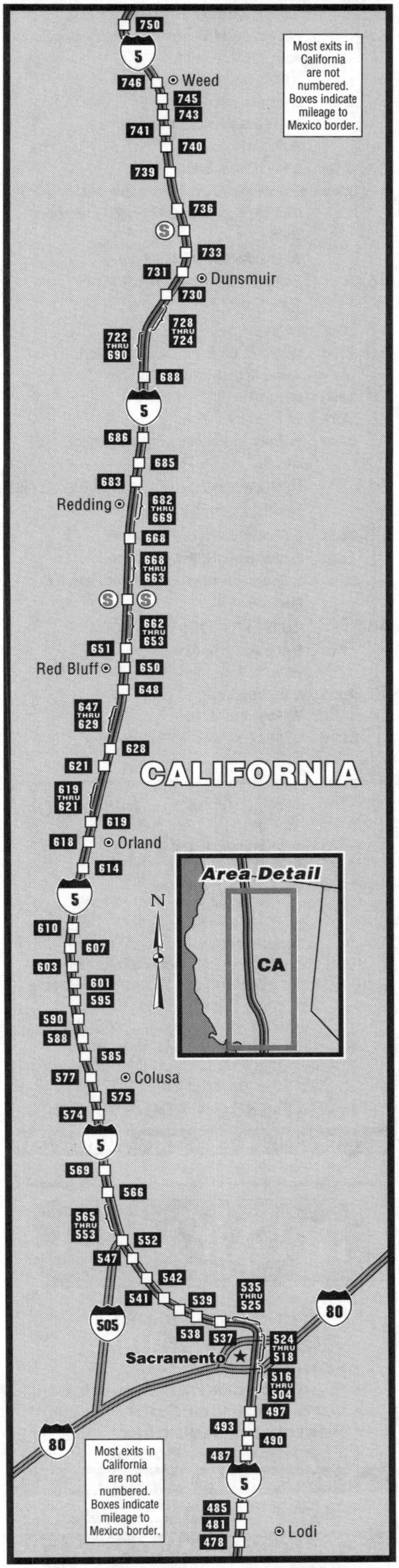

Most exits in California are not numbered. Boxes indicate mileage to Mexico border.

CALIFORNIA

Area Detail

CA

Most exits in California are not numbered. Boxes indicate mileage to Mexico border.

(704) **Rest Area - RR, Phones, Picnic (Southbound)**

703 Lakehead, Riverview Drive
Food E: Klub Klondike
Lodg E: Yukon Motel

702 Lakeshore Drive, Antlers Road
Gas E: Chevron*[D]
W: 76*
Food E: Top Hat Cafe
W: Bass Hole[LP], Kanyon Kettle, Pizza Station
Lodg E: Neu Lodge Motel
W: Shasta Lake Motel
AServ E: Shasta Lake Auto Repair
Other E: **U.S. Post Office**
W: **Lake Villa Coin Laundry (1.2 Miles)**

698 Salt Creek Road, Gilman Road
RVCamp W: **Cascade Cove Resort RV Park (2.6 Miles), Salt Creek RV Park & Campground (.9 Miles), Treail In RVPark (1 Mile)**

694 O'Brien, Shasta Caverns Road
RVCamp W: **Campground**

693 Packers Bay Road

(693) **Rest Area - RR, Phones, Picnic (Northbound)**

692 Turntable Bay Road

690 Bridge Bay Road (No Trucks)
Food W: Bridge Bay Restaurant
Lodg W: Bridge Bay Motel
Other W: **Bridge Bay Resort & Stores**

688 Fawndale Road, Wonderland Blvd
Lodg E: Fawndale Lodge[LP]
RVCamp E: **Fawndale Oaks RV Park**
W: **Wonderland RV Park[LP]**

686 Mountain Gate, Wonderland Blvd
Gas E: Exxon*
Food E: Mountain Gate Family Take-Out
RVCamp E: **Lake Shasta's Bear Mtn RV Resort, Mountain Gate RV Park**
W: **Campground**
Other E: **Visitor Information**

685 Shasta Dam Blvd, Shasta Dam
Gas W: Chevron*[D] (ATM), Texaco[D] (AServ)
Food W: McDonalds, Taco Den, Take'N Bake Pizza, The Stage Stop Cafe
Lodg W: Shasta Dam Motel
AServ W: Coin Car Wash, Texaco
ATM W: Chevron

683 Pine Grove Ave
Gas W: Exxon*[D][LP]

682 Oasis Road
Gas W: Arco*
TServ E: **Anderson's Truck Repair, Peterbilt Dealer**
RVCamp E: **California RV Supply & Repair**

681 Market St, Redding (Southbound, Northbound Reaccess Only)
FStop W: **Exxon* (RV Dump)**
AServ W: Tire Cobbler
TServ W: **CAT, Redding Freightliner**

680B Twin View Blvd
FStop W: **Pacific Pride Commercial Fueling**
Gas E: BP*
Lodg E: Motel 6
W: Holiday Inn Express (see our ad this page)

680A CA 299 East, Burney, Alturas, Lake

EXIT CALIFORNIA

Blvd.
Gas W: Texaco*
Food W: Bartel's Giant Burger
Lodg E: Bel-Air Motel (see our ad this page)
AServ W: J&K Muffler
RVCamp W: KOA Campground, Redding RV Park

678B CA 299 West, CA 44, Central Redding, Eureka

678A CA 299W, Hilltop Drive, Central Drive, Weaverville, Eureka (Difficult Northbound Reaccess)
Gas E: BP[D][LP] (AServ), Chevron*
W: 76*
Food E: Applebee's, Chevy's Mexican, Italian Cottage Restaurant, Marie Callender's Restaurant, Pizza Hut
W: Country Kitchen, Far East Cafe
Lodg E: Comfort Inn, Double Tree Hotel, Motel 6, Oxford Suites
AServ E: BP, Chevron
W: Les Schwab Tires, Sears Auto Center
ATM W: Bank Of America
Other E: Albertson's Grocery, Coin Laundry, Payless Drugs, Petco
W: Sears, Thrifty Drug Store

677 Redding, Cypress Ave
Gas E: 76, Chevron*, Exxon*
W: 76, Beacon*, Shell*, USA Gasoline[D]
Food E: Black Bear Restaurant, Burger King, CR Gibbs (Best Western), Carl's Jr Hamburgers, Cattlemens, Denny's, Dreyer's Tacos and Shakes, Hoang's Chinese, IHOP, KFC, McDonalds, Taco Bell, Wendy's
W: California Cattle Company Steaks, Denny's, Lyon's Restaurant
Lodg E: Best Western, Hotel Orleans/ Motel Orleans, La Quinta Inn
W: Vagabond Inn
AServ E: 76, Grand Auto Supply
W: Big Tires, Midas Muffler & Brakes, Redding Toyota Volvo, Shasta Nissan
ATM E: North Valley Bank, Wells Fargo Bank
Other E: Asher Animal Hospital, Longs Drugs

675 Bechelli Lane, Churn Creek Road, S. Bonnyview Road
Gas E: Arco*, BP*, Shell*[D][LP]
Lodg E: Super 8 Motel
Other E: Traveler's Hill RV Service[LP]
W: Amerigas[LP]

672 Knighton Road, Redding Airport
TStop E: 76 Auto/Truck Plaza[LP][SCALES]
Food E: Night N' Day (76 TStop)
TServ E: 76
RVCamp W: Sacramento River RV Park (2 Miles)

669 Riverside Ave

668 Anderson (Southbound, Difficult Southbound Reaccess)
Gas W: Beacon*, Exxon*[D] (24 Hrs)
Food E: Burger King, McDonalds, The Big Taco, Walkabout Creek
W: Bartell's Giant Burger, Golden Room Steakhouse
AServ W: Texaco
ATM W: Bank of America

667B Central Anderson, Lassen Park, Bells Ferry Road, North St (Northbound)
Gas E: 76*[D], Beacon*
Food E: Bill's Take 'N Bake Pizza, Burger King, Denny's, Donut Time, Furrow's, McDonalds, Papa Murphy's, Perko's Family Restaurant, Round Table Pizza, Silver Star Chinese American, Subway, Taco Bell, Twinkle's
W: Frank's Old Fashoned Creamery, Good Times Pizza, KFC
Lodg E: Best Western, Valley Inn
AServ E: Franklin Auto Parts, Les Schwab Tires, Napa Auto Parts, Quality Lube & Oil[CW]
W: Rick's Place
ATM E: North Valley Bank
Other E: Gateway Animal Clinic, Launder Land Coin Laundry, Safe Way Grocery (& Pharmacy), Thrifty
W: Holiday Supermarket

667A Deschutes Road, Anderson, Junction 273
Gas W: Shell[D][CW]
Food W: Arby's, California Expresso Cafe, Long John Silvers, Rocky Mountain Chocolate Factory.
Lodg W: AmeriHost
ATM W: Bank of America
Other W: Book Warehouse, Shasta Factory Stores Outlet

664 Cottonwood (Southbound, Reaccess Northbound Only)
Food E: Shasta Livestock Restaurant
Lodg E: Alamo Motel, Travelers Motel
Other E: Shasta Western Shop (Shasta Livestock Restaurant)

663 Balls Ferry, Gas Point Road
Gas E: Chevron*[D], Shell*
W: 76*[D][LP], Arco*[LP] (RV Dump)
Food E: Lilly's Donuts
W: Eagles Nest Pizza
AServ E: HHH Auto Parts, Jim's Auto Repair

ATM W: Sentry Supermarket
Other E: Cottonwood Drugs, Visitor Information
W: Sentry Supermarket (ATM), Wash & Dry Coin Laundry

662 Bowman Road, Cottonwood
FStop E: Texaco*

(661) Weigh Station (Both Directions)

659 Snively Road

657 Auction Yard Road, Hooker Creek Road

(654) Rest Area - RR, Phones, Picnic (Both Directions)

653 Jellys Ferry Road
RVCamp E: Bend RV Park
Other E: Wildlife Viewing

651 Wilcox Golf Road

650 Red Bluff, Bus. 5 (Southbound, Reaccess Northbound Only)

648 CA 36, CA 99S, Red Bluff, Chico, Lassen Park
Gas E: 76*, Payless Drugs, Shell[LP] (AServ)
W: BP*, Gas 4 Less*, USA*
Food E: Burger King, Golden Corral, KFC, McDonalds, Perko's
W: Carl's Jr Hamburgers, Denny's, Egg Roll King, Round Table Pizza, Shari's Restaurant, Subway
Lodg E: Best Western, Motel 6, Super 8 Motel
W: Goodnite Inn, Kings Lodge, Red Bluff Inn
AServ E: Shell
W: Kragen Auto Supply
RVCamp W: Idlewheels RV Park (GoodSam Park)
Other W: Antelope Veterinary Hospital, Coin Laundry, Food 4 Less

647B Diamond Ave. (Southbound, Northbound Reaccess Only)

647A South Main St, Red Bluff
Gas E: Exxon*
W: Chevron*[CW]
Food E: La Corona, The Outpost
W: Arby's, Italian Cottage, Jack-In-The-Box, Papa Murphy's, Pizza Hut, Yogurt Alley
Lodg E: Motel Orleans
W: Triangle Motel
AServ W: Wal-Mart
Parks E: Ide Adobe Historical Park
Other W: Raley's Supermarket, Wal-Mart (Tire & Lube Express)

642 Flores Ave, Proberta, Gerber (No Trucks Over 25 Tons)

636 Tehama, Los Molinos, Gyle Road

633 Richfield, Finnel Road

(632) Rest Area - Picnic, RR, Phones (Both Directions)

631 Corning
Gas E: 76*, Chevron*, Jiffy*, Shell[D][LP] (AServ)
Food E: Broasted Chicken Deli, Burger King, Corning Restaurant, Francisco's Fine Mexican, Marco's Pizza, Rancho Grande Mexican American, Roadhouse Family Restaurant, Round Table Pizza, Taco Bell
W: Bartels Giant Burger
Lodg E: 7 Inn, Corning Olive Inn Motel, Economy Inn, Olive City Inn, Shastaway Motel
AServ E: Cardon's Tires Plus, Corning Ford Dealership, Lyndon Johnson Dealership, Napa Auto Parts, Shell

EXIT CALIFORNIA

RVCamp E: Heritage RV Park(LP)
Other E: Clark's Drug Store, Coin Laundry, Safe Way Grocery
629 South Ave
TStop E: Burns Bros Travel Stop*(LP)(SCALES), Petro*(SCALES)
Food E: Bonnie's Restaurant and Steakhouse, Iron Skillet* (Petro TStop), McDonalds, Ms. B's Homestyle*(LP)(SCALES) (Burns Bros. TStop), Subway (Burns Bros), TCBY (Burns Bros)
Lodg E: Days Inn, Olive Tree Motel, Shilo Inn
AServ E: Linnet's Tire Shop
TServ E: Burns Bros TStop, Corning Truck and Radiator Service, Petro
TWash E: Blue Beacon Truck Wash, Corning Truck Wash
RVCamp E: Heritage
Other E: Ace Hardware
628 Road 99W, Liberal Ave
621 Road 7
619 CA 32, Chico
Gas E: Exxon*
W: 76*(D), Sportsman's GasMart*(LP)
Food E: Berry Patch Restaurant, Burger King
Lodg E: AmberLight Inn Motel, Orlanda Inn Motel
AServ E: Mike's Mufflers
RVCamp W: Green Acres, Old Orchard RV Park
618 Orland
FStop E: Beacon*, CFN
Food E: Pizza Factory
Lodg E: Orland Inn
Other E: Coin Laundry, Long's Drugs
614 Road 27
610 Artois
(608) Rest Area - Picnic, RR, Phones, RV Dump, RV Water (Both Directions)
607 Bayliss - Blue Gum Road
603 CA 162, Willows, Glenn Elk Creek, Oroville
Gas E: Arco*, Chevron*, Shell (AServ)
Food E: Burger King, Denny's, Eagle's Garden Chinese, Java Jim's Ice Cream, Jerry's Restaurant, KFC, McDonalds, Round Table Pizza, Taco Bell, Willow Brook Restaurant (Best Western)
W: Nancy's 24 Hour Cafe
Lodg E: Best Western, Days Inn, Super 8 Motel
AServ E: Express Lube, Shell
Med E: + Hospital
Other E: Hand Coin Car Wash, Highway Patrol
W: Wal-Mart (& Pharmacy), Willows Airport (& Pharmacy)
601 Willows, Rd. 57
595 Norman Road, Princeton
Parks E: Sacramento National Wildlife Refuge
590 Delevan Road
588 Maxwell (Southbound)
585 Stonyford, Maxwell
Food W: Chateau Basque Restaurant, Maxwell Inn
AServ W: Napa Auto Parts
Parks E: Delevan National Wildlife Refuge
(583) Rest Area - Picnic, RR, Phones (Both Directions)
577 CA 20, Colusa, Clear Lake
Med E: + Hospital
575 CA Bus 20, Williams
FStop W: CFN Comercial Fueling Network
Gas W: Arco*, BP*, Chevron*(D), Shell*(D)
Food E: Carl's Jr Hamburgers, Taco Bell

EXIT CALIFORNIA

W: A & W Drive-In, Caliente Mexican Drive-in, Dairy Queen, Denny's, Granzella's Deli (Granzella's Inn), McDonalds, Top Hat Cafe, Wendy's
Lodg W: Granzella's Inn, Motel 6, Stage Stop Motel, Travelers Motel, Woodcrest Inn
Other W: Post Office
574 Husted Road
FStop E: BC Petroleum (Automated Fueling)
569 Hahn Road, Grimes
566 Arbuckle
Gas E: Shell*(D)
565 Arbuckle, College City (Difficult Northbound Reaccess)
FStop E: Pacific Pride
Gas E: 76*, Chevron*(D)
W: Beacon* (ATM)
AServ E: DeMarchi's
ATM W: Beacon
Other E: U.S. Post Office
559 County Line Road
(557) Rest Area - RR, Picnic, Phones (Both Directions)
556 Dunnigan
Gas E: Chevron*(CW), Shell*(D)(LP)
W: BP*
Food E: Bill & Kathy's
Lodg E: Best Western, Value Lodge
553 Road 8
TStop E: BP*(D)(LP)(SCALES)
FStop W: Beacon*(LP)(SCALES)
Food E: Aladero Coffee Shop (BP Truck Stop), Judy's Country Cafe
Lodg E: Budget Motel
RVCamp E: Happy Time RV Park
552 Jct I-505, Winters, San Francisco (Southbound)
547 Zamora
Food E: Zamora Mini Mart & Deli
542 Yolo
541 CA 16 West, Road E7
539 West St
Food W: Denny's
538 CA 113 North, Yuba City
Gas W: Shell
Food W: Denny's
Lodg E: Valley Oaks Inn
W: Best Western
537 CA 113 South, Davis, Woodland, Main St (Difficult Reaccess)
Gas W: Exxon*, Union 76*
Food W: Burger King, Denny's, McDonalds, Primo's Restaurant, Taco Bell
Lodg W: Comfort Inn, Motel 6, Phoenix Inn
AServ E: Lasher Auto Center
W: Art's Automotive, Bee Line Service
TServ W: J & J Trucker Repair
Other W: Canned Foods Grocery Outlet, K-Mart
535 Road 102
Gas E: Arco*
AServ E: Art's Automotive, Blakely Auto Repair, Rohwer Bros. Repair
Other E: Hays Antique Truck Museum
531 Road 22 West, Sacramento, Elkhorn
(530) Rest Area - RR, Phones, Picnic (Southbound)

EXIT CALIFORNIA

528 Airport, Garden Hwy, Sac Metro
525 CA 99, to CA 70, Yuba City, Marysville (Difficult Reaccess)
524 Del Paso Road
522 Jct I-80
521 West El Camino Ave (Northbound. Southbound Reaccess Only)
Gas E: Exxon*
Food E: A-Mart Deli, Gourmet Wok
W: Jack-In-The-Box, The Greek Connection
Lodg E: Residence Inn
ATM W: Golden One Credit Union
520B Garden Hwy
Lodg W: Courtyard by Marriott
520A Richards Blvd
Gas E: Chevron* (24 Hrs)
W: Arco*, Shell*(CW)
Food E: Buttercup Pantry, Hungry Hunter, Lyon's Restaurant (24 Hrs), McDonalds, Monterey Seafood
W: Las Altos, Perko's Cafe
Lodg E: Days Inn, Fountain Suites, Governor's Inn, Super 8 Motel
W: Best Western, Capital Inn, Crossroads Inn, La Quinta Inn, Motel 6
519 J St, Downtown Sacramento
Food E: China Moon Cafe, Denny's, Hong King Lum Restaurant, Lu Shan Chinese Buffet
Lodg E: Vagabond Inn
Other E: Downtown Plaza
W: Old Sacramento South Historic Park
518B Q St
518A U.S. 50, Business I-80, San Francisco, CA 99 Jct.
516 Sutterville Rd
514B Fruitridge Rd, Seamas Ave
514A 43rd Ave (Southbound)
512 Florin Rd
Gas E: Chevron*, Shell*(CW) (24 Hrs)
Food E: Alicia's Mexican, Round Table Pizza, Yogurt Shop
W: Burger King, Round Table Pizza, Sheri's Restaurant, Sonoma Valley Bagel
AServ E: Kragen Auto Parts, Shell
ATM W: Bank of America, Cal Fed
Other E: Bel Air Grocery, Long's Drugs
W: Payless Drugs
511 Pocket Rd, Meadowview Rd
Gas E: Shell*(CW)(LP) (24 Hrs, RV Dump Station)
Food E: McDonalds
507 Laguna Blvd
506 Elk Grove Blvd
504 Hood Franklin Road
497 Twin Cities Road, Walnut Grove, Galt
493 Thornton, Walton Grove
FStop E: BP*
AServ E: C & N Repair & Road Service
490 Peltier Road
487 Turner Road
485 CA 12, Lodi, Fairfield
TStop E: Texaco Truck & Auto Plaza
Gas E: Chevron*(D)(LP), Exxon(LP), Texaco(D)
Food E: Baskin Robbins (Exxon), McDonalds, Rocky's (24 Hrs), Subway (Chevron), Taco Bell, Wendy's

Bold red print shows RV & Bus parking available or nearby

EXIT	CALIFORNIA
TServ	**E:** Texaco Truck Plaza
RVCamp	**W:** Tower Park Marina Camping (5 Miles)
ATM	**E:** Texaco TStop
481	8 Mile Road
RVCamp	**E:** KOA Campground (5 Miles)
478	Hammer Lane
Gas	**E:** Arco* (24 Hrs), BP*(CW)
	W: Exxon* (24 Hrs)
Food	**E:** Adalbella Mexican, Bangkok, KFC, Little Caesars Pizza, Western Garden Chinese Deli
	W: Arby's, Burger King, Chava's Taco House, Jack-In-The-Box, Round Table Pizza, Sacramento Baking Company, Subway, Taco Bell
Lodg	**W:** Inn Cal
AServ	**E:** Auto Parts Express
Other	**E:** S-Mart Foods (& Pharmacy)
477	Benjamin Holt Drive
Gas	**E:** Arco*, Chevron*(CW) (24 Hrs)
	W: Shell*
Food	**E:** Pizza & Deli
	W: Lyon's, McDonalds, Round Table Pizza, Subway, Wong's Deli
Lodg	**E:** Motel 6
Other	**W:** 7-11 Convenience Store, Village Veternarian Hospital
475	March Lane
Gas	**E:** Citgo*
	W: 76*(D) (24 Hrs)
Food	**E:** Black Angus Steakhouse, Boston Market, Cactus Bay, Carl's Jr Hamburgers, Denny's, El Torito Mexican, Honey Treat Yogurt, Jack-In-The-Box, Marie Callender's Restaurant, Red Lobster, Safari Coffee & Tea, Taco Bell, Tony Roma's, Toot Sweets Bakery Cafe, Wendy's
	W: Carrows Restaurant, In-N-Out Hamburgers
Lodg	**E:** Traveler's Inn
	W: La Quinta Inn, Super 8 Motel
Other	**E:** Longs Drugs
474	Alpine Ave, Country Club Blvd
Gas	**E:** Shell*(LP)
	W: Citgo*, Exxon*(D)(LP)
Food	**W:** Baskin Robbins (Exxon), Subway (Exxon)
AServ	**E:** Shell
ATM	**W:** Citgo
Other	**W:** Safe Way Grocery (& Pharmacy)
473	Monte Diablo Ave
AServ	**W:** AAMCO Transmission
472	Oak St, Fremont St (Southbound)
Gas	**W:** Arco*
Lodg	**W:** Fremont Inn
471C	Persian Ave. (Northbound)
471B	Downtown Stockton, Fresno Ave
471A	CA 4, Charter Way
TStop	**W:** Vanco Truck/Auto Stop*(LP)(SCALES)
Gas	**E:** Chevron*, Shell*(CW), Texaco(D)(CW)
Food	**E:** Burger King, Denny's, McDonalds
	W: Carrows Restaurant, Taco Bell (24 Hrs)
Lodg	**E:** Best Western, Motel 6
	W: Motel 6
AServ	**E:** Transmission Service Center
TServ	**W:** International Trucks, Peterbilt Dealer
470	8th St
Gas	**W:** Texaco*(LP)
Lodg	**W:** Econolodge
469	Downing Ave

EXIT	CALIFORNIA
468	French Camp, Stockton Airport
FStop	**E:** Exxon (Pacific Pride Commercial Fueling)
Food	**E:** Togo's (Exxon)
467	Mathews Road
Med	**W:** ✚ Hospital
Other	**W:** County Jail
466	El Dorado St. (Northbound)
Gas	**E:** Exxon*(D)
Food	**E:** Chinese Deli
AServ	**E:** Dunlap Tire, Exxon
465	Sharpe Depot, Roth Road
TServ	**E:** Freightliner Dealer, Kenworth Trucks
463	Lathrop Road
Gas	**E:** Chevron*, Exxon*
Food	**E:** Bella's Bakery, Express Grill (Chevron), Papa's Pizza, Yan Yan Deli
Lodg	**E:** DAYS INN Days Inn
462	Louise Ave
Gas	**E:** Arco*, BP
Food	**E:** Carl's Jr Hamburgers, Denny's, Jack-in-the-Box, McDonalds (Playground), Taco Bell
Lodg	**E:** Holiday Inn Express
AServ	**E:** BP
460	CA 120E, Modesto, Fresno
459	Manthey Road (Southbound)
458	Jct I-580, Jct I-205, San Francisco (Southbound)
457A	Defense Depot, Tracy (Southbound)
455B	Mossdale Rd. (Northbound)
Gas	**E:** Texaco*
456	Kasson Road
FStop	**W:** Commercial Fueling Network (24 Hrs)
452	CA 33, Vernalis
449	CA132, San Francisco, Modesto
448	Jct I-5 & 80
(447)	**Rest Area - RR, Picnic, RV Water, RV Dump (Both Directions)**
437	Westley
TStop	**E:** Westley Truck Stop(SCALES)
Gas	**E:** 76*, BP* (24 Hrs), Chevron*
	W: Shell*
Food	**E:** Bobby Ray's Restaurant, McDonalds, Pepper Tree Restaurant
	W: Ingram Creek Coffee Shop (24 Hrs)
Lodg	**E:** Budget Inn, DAYS INN Days Inn
TServ	**E:** Westley TS
433	Patterson
Med	**E:** ✚ Hospital
430	Vista Point (Northbound)
427	Crows Landing
422	Newman
419	Vista Point (Southbound)
417	CA140, Gustine, Merced

EXIT	CALIFORNIA
Gas	**E:** Shell*(D) (24 Hrs), Texaco(D)
AServ	**E:** Auto & Tire Service Center, Texaco
(409)	**Weigh Station (Both Directions)**
407	CA 33, Santa Nella Blvd
TStop	**E:** Mid Cal 76 Auto/Truck Stop(SCALES)
FStop	**W:** Rotten Robby's Auto TruckPlaza*(SCALES)
Gas	**E:** Texaco*
	W: Beacon*, Shell*
Food	**E:** 76 Auto/Truck Plaza, Burger King (Playground), Carl's Jr Hamburgers, Pea Soup Anderson
	W: Denny's (24 Hrs), McDonalds (Playground), Orient, Taco Bell
Lodg	**E:** Best Western, Holiday Inn Express, Pea Soup Anderson
	W: Motel 6, Ramada Inn
TServ	**E:** 76 Auto/Truck Plaza
TWash	**E:** 76 Auto/Truck Plaza
RVCamp	**W:** Santa Nella RV Park
Other	**W:** Sheriff's Department
402	Jct. CA 152, CA 33
RVCamp	**W:** Camping World (see our ad this page)
391	CA165, Mercy Springs Rd
Gas	**W:** Chevron*
Med	**E:** ✚ Hospital
390	Dos Amigos Vista Pointe, Rest Area (Northbound)
(387)	**Rest Area - RR, Phones, Picnic, RV Water (Both Directions)**
384	Nees Ave, Firebaugh
FStop	**W:** Texaco*
379	Shields Ave, Mendota
371	Russell Ave
368	Panoche Rd
Gas	**W:** Chevron* (24 Hrs), Shell*, Texaco*(D) (24 Hrs)
Food	**W:** Apricot Tree Restaurant, Foster's Freeze Hamburgers
Lodg	**W:** Shilo Inn
AServ	**W:** Chevron
364	Manning Ave, San Joaquin
357	Kamm Ave
349	CA 33N, Derrick Ave
337	CA 33 S, Fresno, Coalinga, CA 145 N.
Med	**W:** ✚ Hospital
Other	**W:** Highway Patrol
333	CA198, Lemoore, Hanford
Gas	**E:** Shell*, Texaco*(LP)
	W: 76* (24 Hrs), Chevron*(D) (24 Hrs), Mobil(D)(LP)
Food	**E:** Harris Ranch Restaurant (Harris Ranch Inn)
	W: Burger King, Carl's Jr Hamburgers (Playground), Denny's (24 Hrs), Garden Room (Big Country Inn), McDonalds, Oriental Express, Red Robin Restaurant, Taco Bell (24 Hrs)
Lodg	**E:** Harris Ranch Inn
	W: Big Country Inn, Motel 6 (#1), Motel 6 (#2)
AServ	**E:** Texaco
	W: Chevron
Med	**W:** ✚ Hospital
ATM	**E:** Shell
	W: Mobil
324	Jayne Ave
Gas	**W:** Arco*
RVCamp	**W:** Sommerville's Almond Tree RV Camp(LP)
Med	**W:** ✚ Hospital
Other	**W:** Highway Patrol

Bold red print shows RV & Bus parking available or nearby

EXIT CALIFORNIA

(322) **Rest Area - RR, Phones, Picnic (Both Directions)**

319 CA.269, Lassen Ave, Avenal

308 Kettleman City, Fresno, Jct. Ca. 41

- **FStop** E: **Beacon*(SCALES)**
- **Gas** E: 76, Arco*, Chevron*(D), Exxon*(D), Shell*
- **Food** E: **Burger King, Carl's Jr Hamburgers (Playground, 24 Hrs)**, In-N-Out Hamburgers, **Jack-In-The-Box, Major's Bros. Farms Restaurant (24 Hrs), McDonalds (Playground)**, Subway (Exxon), TCBY Yogurt, **Taco Bell**
- **Lodg** E: **Best Western, Olive Tree Inn**
- **AServ** E: 76, Chevron, Kettleman City Tire & Repair Service (24 Hr repairs)
- **TServ** E: **Beacon, Kettleman City Tire Service**
- **RVCamp** E: **Kettleman RV Park**

304 Utica Ave (Airport)

287 Twisselman Rd

282 CA 46, Lost Hills, Wasco

- **TStop** W: **Burns Brothers Travel Stop(LP)**
- **Gas** E: Texaco*(D)(LP)
  W: 76*, BP* (24 Hrs), Beacon*(D), Chevron*
- **Food** W: **Carl's Jr Hamburgers, Denny's (24 Hrs), Jack-In-The-Box, Mrs. B's (Burns Bro.)**
- **Lodg** W: **Economy Motels Of America, Motel 6**
- **AServ** W: 76
- **RVCamp** W: **Koa RV Camp(LP)**

267 Lerdo Hwy, Shafter

265B Buttonwillow, McKittrick (Southbound)

265 7th Rd. (Northbound)

- **RVCamp** E: **Camping World (see our ad this page)**

(264) **Rest Area - Picnic, RR, Phones, Vending (Both Directions)**

263 Buttonwillow, McKittrick, Jct. Ca. 58, Bakersville

- **TStop** E: **76(SCALES) (Restaurant, TServ), Bruce's*(SCALES) (24 Hrs)**
- **FStop** W: **Exxon*(D)**

EXIT CALIFORNIA

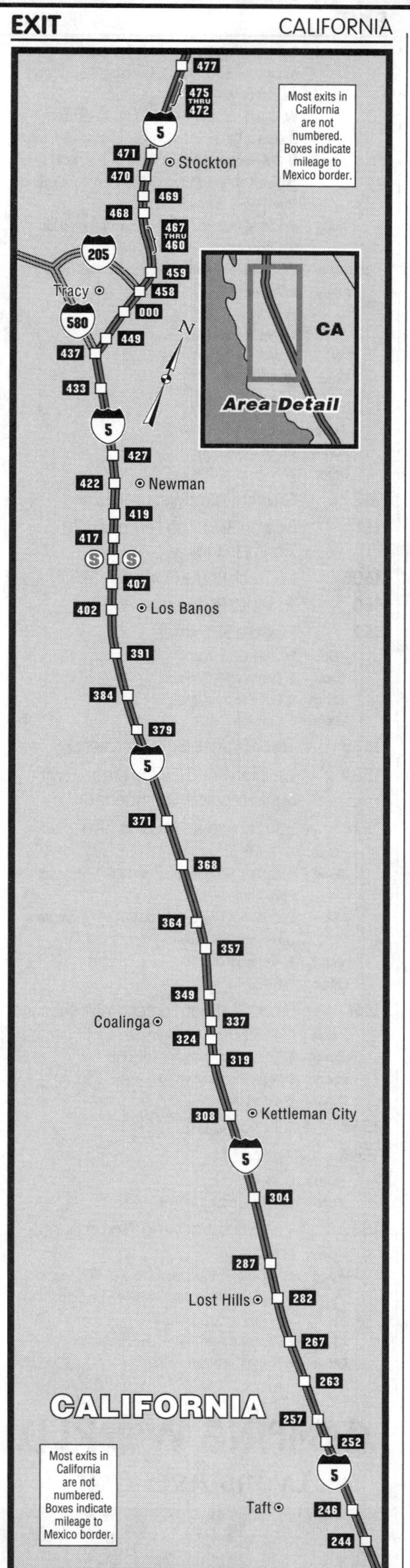

EXIT CALIFORNIA

- **Gas** E: Arco* (AServ), Chevron* (AServ, 24 Hrs), Mobil*(D), Shell*, Texaco*
- **Food** E: **Burger King, Carl's Jr Hamburgers, Cotton Country, Country Kettle Restaurant, Denny's (24 Hrs), McDonalds (Playground)**, Sweetwater Saloon
- **Lodg** E: Good Nite Inn, **Motel 6, Sixpence Inn, Super 8 Motel**
- **AServ** E: Chevron
- **TServ** E: **76**
- **TWash** E: **Bruce's**
- **ATM** E: Mobil

252 Stockdale Hwy

- **FStop** E: **Exxon*(LP) (Restaurant)**
- **Gas** E: British Petroleum* (24 Hrs), Shell
- **Food** E: **Jack-In-The-Box, Palm Tree Restaurant, Perko's Family Dining (Exxon FS)**
- **Lodg** E: **Best Western, Econolodge**
- **Parks** W: **Tule Elk State Reserve**
- **Other** E: **Highway Patrol**

246 CA 43, Shaftes, Wasco

244 CA 119, Pumpkin Center

238 CA 223, Bear Mountain Blvd, Arvin

234 Old River Rd

228 Copus Rd

225 CA166, Maricopa

- **Lodg** W: Santa Marin Inn (west on CA 166, see our ad this page)

217B Junction CA.99 Bakersfield

217 Laval Rd - Southbound, Lamont Lake Isabella (Northbound)

- **TStop** E: **TravelCenters of America(SCALES)**
- **Food** E: Burger King, Country Pride Restaurant, Subway, Taco Bell
- **TServ** E: **Blue Beacon Truck Wash**

216B Lamont Lake Isabella

(216) **Weigh Station (Southbound)**

215 Grapevine

- **Gas** E: Shell (AServ)
  W: 76*, Arco(LP), Texaco*(CW)(LP)
- **Food** E: Denny's, Ranch House Coffee Shop
  W: Farmer's Table Restaurant, Taco Bell
- **AServ** E: Shell
  W: Texaco
- **ATM** W: Arco

(214) Runaway Truck Ramp (Northbound)

210 Fort Tejon (Water & Phone Available)

- **Parks** W: **Fort Tejon State Historical Park**

209 Lebec

(208) **Rest Area - RR, Phones, Picnic, Vending, RV Dump (Both Directions)**

205 Frazier Park

- **TStop** W: **Flying J Travel Plaza*(LP)(SCALES) (Motel & Restaurant)**
- **Gas** W: **Exxon***, Texaco(D)
- **Food** W: **Jack-In-The-Box**, Mark's Coffee Shop Restaurant, **Thads Restaurant (Flying J)**, Wahoo's Family Restaurant
- **Lodg** W: **Best Rest Inn (Flying J Truck Stop)**
- **AServ** W: High Tech Transmissions, Texaco, Tony's Auto Repair
- **TServ** W: **J Care Service Center**

EXIT CALIFORNIA

(204) Truck Brake Inspection Area (Southbound)

202 Gorman
- **Gas** E: Chevron*(D), Texaco(D)
  W: Mobil*
- **Food** E: Brian's Diner (Sizzler Motor Lodge), Carl's Jr Hamburgers, Sizzler Steak House (Sizzler Motor Lodge)
  W: McDonalds
- **Lodg** E: Sizzler Motor Lodge
- **AServ** E: Texaco
  W: Mobil

198 Quail Lake Rd

199 CA 138, Lancaster, Palmdale

194 Smoky Bear Road
- **RVCamp** W: Pyramid Lake RV Campground

190 Vista Del Lago Road

(184) Truck Brake Inspection Area (Southbound)

182 Templin Hwy (Steep Climb)

177 Lake Hughes Rd., Castaic (Steep Climb)
- **TStop** E: Giant Truckstop*(SCALES)
- **Gas** E: Citgo*
  W: 76, Mobil*
- **Food** E: Burger King, Casa Lupe, Domino's Pizza, Foster Freeze, McDonalds (24 Hrs), Subway, Zorbas
  W: Del Taco, Jack-In-The-Box
- **AServ** W: S Star Auto Clinic
- **ATM** E: McDonalds
  W: 76
- **Other** E: Payless Drugs, Ralph's Grocery

176 Parker Rd, Castaic (Norhtbound)
- **FStop** E: Village Fuel Stop
- **Gas** E: Arco*, Shell*(CW)
- **Food** E: Cafe Mike Mexican American (24 Hrs), Carl's Jr Hamburgers, Teddi's Donuts, Tommie's Hamburgers
- **Lodg** E: Castaic Inn, Comfort Inn, Econolodge
- **AServ** E: Canyon Tire
- **RVCamp** E: Castaic Lake RV Park
- **Other** E: U.S. Post Office

173 Hasley Canyon Rd
- **Parks** W: Wayside Honor Rancho Valverde Park

172 CA 126, Ventura

171 Rye Canyon Rd (Southbound)
- **Gas** W: Texaco*

(170) Weigh Station (Northbound)

170 CA126, Saugus, Magic Mtn Pkwy
- **Gas** W: Chevron* (24 Hrs)
- **Food** E: Dillions Steakhouse
  W: El Torito Mexican, Hamburger Hamlet, Marie Callender's Restaurant, Red Lobster, Wendy's
- **Lodg** E: Ranch House Inn
  W: Hilton/Garden Inn
- **ATM** W: Chevron
- **Other** W: Six Flags California

169 Valencia Blvd (No Southbound Reaccess)

168 McBean Pkwy
- **Med** E: ✚ Hospital

167 Lyons Ave, Pico Canyon Rd
- **Gas** E: 76, Arco*, Texaco*(D)(CW)
  W: Arco*, Mobil*, Shell* (24 Hrs)

EXIT CALIFORNIA

- **Food** E: Burger King (Playground), Genghis Khan Mongolian BBQ, Pizza Esperienza, Tiny Naylors Restaurant (24 Hrs)
  W: Blimpie's Subs (Mobil), Carl's Jr Hamburgers, Del Taco, Denny's, El Pollo Loco Mexican, IHOP, In-N-Out Hamburgers, Jack-In-The-Box, Kelly's Coffee & Fudge, McDonalds (Play Area), Outback Steakhouse, Subway, Taco Bell
- **Lodg** W: Comfort Suites, Fairfield Inn, Hampton Inn, Residence Inn
- **RVCamp** W: Camping World (see our ad this page)
- **AServ** E: 76, Arco
  W: Jiffy Lube, Shell
- **ATM** W: McDonalds, Mobil
- **Parks** E: William S. Hart Park
- **Other** W: Petsmart

165 Calgrove Blvd
- **Food** E: Carrows Restaurant
- **Parks** W: Ed Davis Park
- **Other** W: U.S. Post Office

162 CA14N, Palmdale, Lancaster

161 Balboa Blvd, San Fernando Rd (Southbound)

160B I-5 Truck Route North, CA 14

160 Jct I-210, Pasadena

159 Roxford St, Sylmar
- **Gas** E: Chevron*(D), Mobil*
- **Food** E: Denny's, McDonalds
- **Lodg** E: Good Nite Inn, Motel 6
- **AServ** E: Chevron

158B Truck Bypass (Both Directions)

158A Junction I-405 S., San Diego Frwy, Santa Monica (Southbound)

157 San Fernadino Mission Blvd
- **Gas** E: Mobil*
- **Food** E: Carl's Jr Hamburgers, In-N-Out Hamburgers, Pollo Gordo
- **AServ** E: Chief Auto Parts, Jiffy Lube, Tune Up Masters, Uniroyal Tire & Auto
- **Med** E: ✚ Hospital
- **Other** E: K-Mart

156C Brand Blvd. San Fernando (Northbound)
- **Gas** E: Arco (24 Hrs), Chevron*, Shell
- **Food** E: Carnitas Michoacan, Taco Bell
- **AServ** E: Arco, Mission Hills Ford Truck
- **Other** E: Coin Laundry

156B Junction CA118

156A Paxton St
- **Gas** E: Shell*
- **Food** E: Royal Donuts, Teriyaki House

154C Van Nuys, Pacoima (Northbound)
- **Gas** E: 76
- **Food** E: Birrieria, El Mero Mero, Jack-In-The-Box, KFC, McDonalds, Pizza Hut, Popeye's Chicken
  W: Carnitas 3, Domino's
- **AServ** E: 76, Chief Auto Parts, Tune Up Masters
- **Other** E: Fiesta Car Wash

EXIT CALIFORNIA

154B Terra Bella St. (Northbound, No Reaccess)

154 Osborne St
- **Gas** E: 76, Arco*, Shell*
  W: 76
- **Food** E: La Pizza Logo, New York Pizza, Papas & Tacos, Peter Piper's Pizza
  W: Yummy Donuts (24 Hrs)
- **AServ** E: 76, A&J Tires, Joe's Auto Repair
  W: 76
- **Other** E: Target Department Store
  W: Auto Spa Car Wash, Coin Laundry

153C Branford St. (Northbound)

153B CA170 S, Hollywood Frwy (Southbound)

153A Sheldon St
- **Med** W: ✚ Hospital

152 Lankershim Blvd
- **FStop** E: Texaco(D)(LP)(SCALES)
- **Food** E: The Donut Baker
- **AServ** E: Tune Up Masters

151 Penrose St

150 Sunvalley Blvd.
- **Gas** E: 76*, Mobil*
  W: Shell*
- **Food** E: Acapulco Mexican, China King, Donut Star, El Pollo Loco Mexican, Georgio's Pizza, Thai-Chinese
  W: Dimions, El Mexicano Mexican, Los Hermones, McDonalds, Tortellini's
- **Lodg** E: Scottish Inns
- **AServ** E: Mobil
  W: Round's Auto Parts
- **Other** E: 7-11 Convenience Store, Coin Laundry

149B Glen Oaks Blvd (Northbound)
- **Gas** E: Arco(LP)
- **Food** E: Cheo's Taco (Arco)
- **Lodg** E: Willow's Motel
- **AServ** E: Arco

149 Hollywood Way
- **Gas** W: Texaco*(D)
- **AServ** W: J.A. Auto Center

148 Buena Vista St
- **Gas** W: Exxon
- **Food** W: Jack-In-The-Box
- **Lodg** W: Buena Vista Motel, Ramada

147 Scott Rd
- **Gas** E: Sevan
- **Food** E: Carl's Jr Hamburgers, Eiffel Bakery, Mr Big Burger, Sun Moon Chinese
- **Lodg** E: Bell Vista Motel, Scott Motel
- **AServ** E: European Car Service, Sevan
- **Other** E: In-Out Hand Wash

146B Burbank Blvd.
- **Gas** E: 76
- **Food** E: Baskin Robbins, Chevy's, El Mexicano, Harry's Family Restaurant, IHOP, Kenny Rogers Roasters, McDonalds, Mipace Italian, Popeye's, Taco Bell
  W: Cody's Landing, El Burrito Loco, Granada Mexican, Kim Hua Chinese, McDonalds, Subway
- **AServ** E: 76
  W: Discount Tire Center
- **RVCamp** W: Fitzpatrick Trailer Service
- **ATM** E: Home Savings Of America, Wells Fargo
- **Other** E: Barnes & Noble, K-Mart, Vons Grocery
  W: Burbank Animal Hospital

146A Olive Ave, Burbank Ave.
- **Food** E: Black Angus, Bobby McGee's, Bombay Bicycle

Bold red print shows RV & Bus parking available or nearby

EXIT	CALIFORNIA
	Club, Crocadile Cafe, Wild Thyme
	W: Frank's Coffee Shop, Submarine King, Subway
Lodg	**E:** Holiday Inn
AServ	**E:** Southern Cal Tire Co., Valley Tire
	W: Community Chevrolet, Jay's Auto Clinic, Jiffy Lube, Pep Boys Auto Center
RVCamp	**W: Metro RV Inc.**
Other	**W: 7-11 Convenience Store (24 Hrs)**
145	Alameda Ave
Gas	**W:** Arco, Chevron*, Shell*(CW), United Gas(D)
Food	**W:** Willie's Grill
Lodg	**W:** Burbank Inn & Suites
AServ	**W:** Arco, United Gas
144B	Western Ave
144A	Junction CA134, Ventura, Pasadena
142	Colorado St (Northbound)
141B	Griffith Park Dr (Southbound)
141A	Los Feliz Blvd
Gas	**E:** Arco*
Food	**E:** Del Taco, La Fogata, Sizzler, Sun Hai Chinese, Swedish Table Restaurant, Tamo'shantern
Lodg	**E:** Los Feliz Motel, Travelodge (in Hollywood, see our ad this page)
AServ	**E:** 76
Other	**E: Animal Clinic,** Darla's Dog Wash, Los Feliz Car Wash
140B	Glendale Blvd
Gas	**E:** 76* (24 Hrs), Shell*
AServ	**E:** Shell
140	Fletcher Dr
139	Junction CA2, Glendale Freeway
138B	Stadium Way
137	Junction CA110, Pasadena Frwy
136	North Broadway, Pasadena Ave
135F	Main St. (Northbound)
AServ	**E:** Tune Up Masters
135D	Mission Rd. (Southbound)
135B	Junction I-10, San Bernardino Freeway
135A	Cesar Chavez
Gas	**W:** Shell
Food	**E:** El Pollo Loco, Shakeys Pizza
AServ	**E:** Chief Auto Parts
	W: Shell
Med	**W: ✚ White Memorial Hospital**
134	4th St
Gas	**E:** Shell*
	W: 76*(D)

EXIT CALIFORNIA

CALIFORNIA

Area Detail

Most exits in California are not numbered. Boxes indicate mileage to Mexico border.

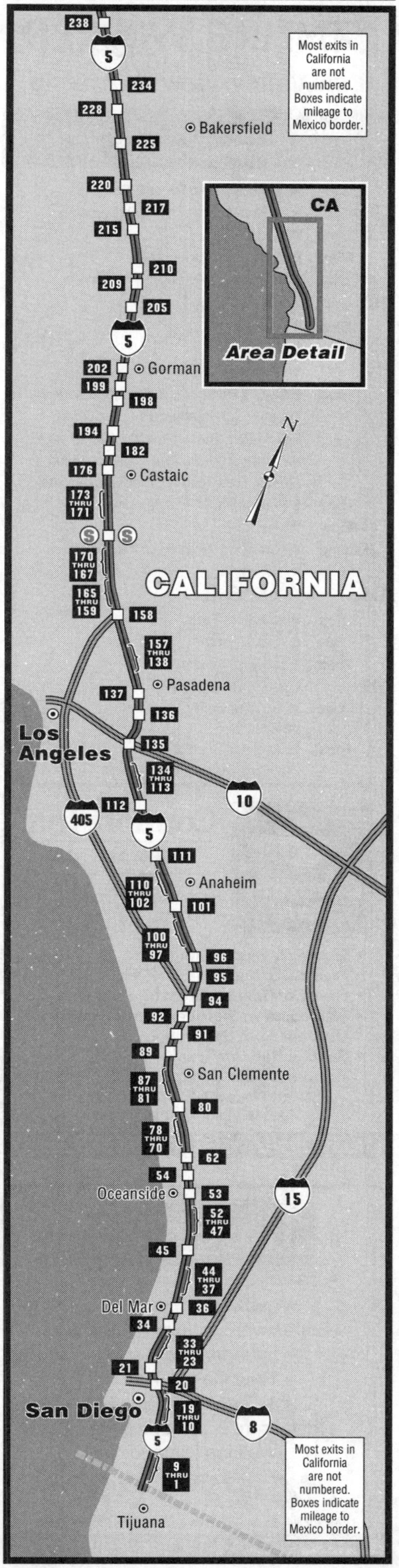

EXIT	CALIFORNIA
AServ	**E:** Shell, Trejo's Radiator
	W: Hygrade Transmissions
133A	Jct I-10E, CA 60, Santa Monica Frwy
133C	Seventh St., Junction I-10W, CA 60, Santa Monica Frwy.
133B	Soto St (Difficult Reaccess)
Gas	**E:** Mobil*
	W: 76*(D), Cardlock, Shell*
Food	**E:** China Town Express, El Ranchito Taco Shop, Jim's Burgers, Pioneer Chicken
	W: El Pollo Loco, Jack-In-The-Box, McDonalds, Seafood Bay, Subway (76), Tacos El Unico, Yoshinoya Rice Bowl
AServ	**E:** Tijuana Tire Center
ATM	**W:** Bank Of America, Coast Federal Bank
Other	**E:** Coin Car Wash, **La Familia Pharmacy**
132	Grande Vista Ave
131B	Calzona
131	Indian St.
FStop	**W: Mobil***(D)(CW)
Gas	**E:** Arco*, Texaco*(D)
Food	**W:** Jim's Hamburgers
AServ	**E:** Arco, Texaco
130B	Eastern Ave. (Northbound)
130	Jct I-710, South To Long Beach (Left Hand Exit For Northbound)
129B	Triggs St
129A	Atlantic Blvd
FStop	**W: Arco**
Food	**W:** Sergio's Tacos, Steven's Steakhouse
Lodg	**E:** Wyndham Garden Hotel
	W: Bamby's Motel, Bob's Motel, Hines Motel, Rodeo's Hotel
AServ	**W:** Paul's Automatic Transmission
Other	**E: Citadel Mall**
128	Washington Blvd
Gas	**E:** Chevron(LP)
Food	**E:** JC's Bar & Grill (Commerce Hotel)
Lodg	**E:** Commerce Hotel
AServ	**E:** Chevron, Firestone Tire & Auto
TServ	**W: California Transport Refrigeration**
ATM	**E:** Banko Popular
	W: Sanwa Bank
Other	**E: Commerce Casino**
127	Garfield Ave, Bandini Blvd
126B	Slauson Ave, to Montebello
Gas	**E:** Shell
	W: Arco*
Food	**E: Ozzie's Diner**

EXIT CALIFORNIA

W: Denny's
Lodg E: Four Corners Motel, Howard Johnson, **Super 8 Motel**
W: **Ramada (see our ad this page)**, Travelodge
AServ E: Shell
126A Paramount Blvd, Downey
Food E: Florence Pizza, Granny's Donuts, Teriyaki Inn No. 8
125 CA 19, Lakewood Blvd, Rosemead Blvd
Gas E: Mobil, Shell*, Thrifty
Food E: Chateau Brian Restaurant, Del Taco, El Pedregal Mexican, Foster's Freeze, Pepe's, Sam's Burgers, Sports Bar & Restaurant
W: Chris & Pitt's BBQ Steaks, Little Caesars Pizza, McDonalds
Lodg E: Econolodge
AServ E: Mobil, R & A Auto Service, Windshields Of America (Coin Car Wash & Auto Service)
W: Pacific Lincoln Mercury
Med E: ✚ **Hospital**
Other E: Best Car Wash Full Service
W: **Ralphs Grocery**
123 Jct I-605
122 Imperial Hwy, Pioneer Blvd
Gas E: 76, Chevron*
W: 76(CW), Shell*(CW)
Food E: Frantone's Italian, IHOP, Jack-In-The-Box, McDonalds, Mexican, Subway, Wendy's
W: Denny's, Panda King, Pizza Hut, Rally's Hamburgers, Sizzler Steak House, Taco Bell, Weinerschnitzel
Lodg E: **Comfort Inn (see our ad this page)**
W: Anchor's Inn, Comfort Inn, Towne House, Westland Motel
AServ E: Firestone Tire & Auto, **Wards Automotive**
ATM E: Bank of America
Other E: **Montgomery Ward, Thrifty Drug, U.S. Post Office**
122B Florence Ave, Downey (Difficult Northbound Reaccess)
Gas E: Mobil
Food E: Billie's, Birds Nest Chinese, Cherri's Donuts, Dragon Express, Express Pizza
AServ E: Chief Auto Parts, Mobil
W: Discount Auto Service, Massey Geo-Chevrolet-Cadillac
Other W: **Sams Club**
121 Norwalk Blvd, San Antonio Drive
Gas E: 76, Chevron*, Shell
W: 76
Food E: IHOP, Jack-In-The-Box, McDonalds, Wendy's
Lodg E: Marriott
AServ E: Firestone Tire & Auto, Shell
W: 76, AC-DC Auto Repair, Jimmy's Auto & Smog Repair Service, Keystone Ford Dealership
Other W: **Coin Laundry**
120 Rosecrans Ave, La Mirada
Gas E: Mobil(D), Shell* (24 Hrs)
W: Arco*
Food E: Jim's Burger, Las Aguilas Bakery, Taco Joe's
W: El Pollo Loco, Salt & Pepper Family Restaurant
AServ E: Mobil
W: Norwalk Lube, Sante Fe Nissan, Tune Up Masters
119 Carmenita Rd
Gas E: 76, Vic Oil Company
W: Arco*, Mobil
Food E: Jack-In-The-Box
W: Paul's Place Hamburger's
Lodg E: **Motel 6**

EXIT CALIFORNIA

W: Dynasty Suites, **Super 8 Motel**
AServ E: 76, Figueroa Tire Co., Vic Oil Company
W: Mobil
TServ E: **Carmenita Truck Center.**
RVCamp E: **Cal-Western RV Supplies & Service**
118 Valley View Rd.
FStop W: **Texaco**
Gas E: Arco*
W: Chevron*
Food E: Carl's Jr Hamburgers, Clarman's North Woods Inn, In-N-Out Hamburgers, Le Diplomate (Holiday Inn), The Elephant Bar Restaurant
W: Denny's, El Pollo Loco, Fast Food China, Mama Bear's, Taco Tio, Winchell's-N-More
Lodg E: Holiday Inn, Residence Inn
AServ W: Texaco
RVCamp W: **Camping World (in La Mirada, see our ad this page)**
116 Artesia Blvd
FStop W: **Cardlock Fuels**
Gas E: Arco, Texaco(D)
Food E: Subway, Taco Time
115 CA 39, Beach Blvd
Gas E: Arco, Chevron*
W: Chevron
Food E: Szechuan Oriental
W: Arby's, Denny's, Fuddrucker's, KFC, Karuta

EXIT CALIFORNIA

Japanese Cuisine, Korean BBQ, Pizza Hut, Soot Bul Restaurant (Korean BBQ)
Lodg E: Copper Barrel Motel, Coral Hotel, Travelodge (see our ad this page)
W: Franklin Motel, Holiday Inn
AServ E: Arco, Buena Park Tire, Ken Grody Ford, Toyota
W: Chevron
ATM E: Bank of America
Other W: **Buena Animal Clinic, Coin Laundry**, Full Car Wash
114C Manchester Ave. (Northbound)
AServ E: Lincoln Mercury
114B Jct CA 91E, Riverside
114A Magnolia Ave, Orangethorpe Ave
Gas E: Mobil*(D)
W: 76*, Arco
Food E: Burger Town, Cancun Mexican, Dieter's VW, Pizza Shack, Taco Bell (24 Hrs)
W: Del Taco, El Polo Loco, Keno's Family Restaurant, Peking Bowl, Star Indian Restaurant, Wienerschnitzel
Lodg E: Fullerton Inn Motel
AServ E: Firestone Tire & Auto
W: Arco
Other E: **Self Car Wash.**
W: **Lens Crafters**
113 Brookhurst St, La Palma Ave
Gas E: Chevron*
W: Arco*
Food E: Four Seasons Burgers, Jahan
W: La Estralla
Lodg W: Kona Motel
AServ E: Guerrero
W: Brookhurst Motor Company
Other W: **VCA Animal Hospital**
112B Crescent Ave (Northbound)
112A Euclid St (Southbound)
Gas W: 76, Arco, Mobil, Texaco(CW)
Food E: Chris & Pitts BBQ, Marie Callender's Restaurant, McDonalds, Taco Bell, Yoshinoya Beef Bowl Restaurant
W: Denny's
AServ W: 76, Arco, GM Auto Dealership, Jin's Transmission, Mobil
ATM E: Home Savings of America
Other E: **Anaheim Plaza, Mervyn's, Petco**
W: **Coin Laundry, Sav-On Drugs**
111B Loara St (Northbound)
AServ E: **Wal-Mart**
Other E: **Anaheim Plaza, Petco, Wal-Mart**
111A Lincoln Ave
Gas E: Texaco*(D)
Food E: El Nopal #2
AServ W: Action Mufflers, Cal-Best Transmission, California Japanese Auto Service, Discount Tire & Auto Repair, Neil's Ford
Other E: Lincoln West Coin Car Wash
W: **Anaheim Animal Hospital**
110C Ball Rd. (Southbound)
110B South St., West St. (Northbound, No Trucks)
110A Harbor Blvd
Gas E: Shell*
Food E: Carrows Restaurant, Shakey's Pizza
W: Aculpoco Mexican, Coco's, Denny's, IHOP, McDonalds, Millie's Restaurant, Tony Roma's
Lodg E: Anaheim Motel, Anaheim Tropic Motel, Courtesy Lodge, DAYS INN Days Inn, Frontier Motel, Holiday Inn, Ramada, Tops Motel
W: Best Western, Carousel Inn & Suites, DAYS INN Days Inn, Desert Inn & Suites, Econolodge, Fairfield Inn, Howard Johnson, Park Inn International, Penny Sleeper Inn, Ramada, Saga Inn, Tropicana

Bold red print shows RV & Bus parking available or nearby

EXIT CALIFORNIA

AServ E: Harbor Auto Care
RVCamp E: **Anaheim Harbor, Travelers Lodge**
Med E: **+ Hospital**
Other W: **Disneyland**

109 Katella Ave, Anaheim Blvd
Gas E: 76
W: Exxon
Food E: Courtyard Cafe, Fritz, Mr. Stox Fine Dining, Tweed's
W: Chinese Food, Cowboy Boogie, Del Taco, Donuts, Flaky Jake's Burgers, Ming Delight Restaurant, Persepolis, Thai Thai Restaurant
Lodg E: Comfort Inn, Ramada Inn
W: Alender Motel, Arenan, Crystal Suites, Little Boy Blue Motel, Peacock Suites, Rip Van Winkle Motel, Riviera Motel, Samoa Motel, Super 8 Motel
AServ E: 76
W: Cowboy Automotive, Exxon
TServ E: **LA Freightliner**
Other W: **7-11 Convenience Store (24 Hrs), Coin Laundry, Disneyland**

107 CA. 57 N. to Pamona

106B Chapman Ave (Difficult Reaccess)
Food E: Burger King, Denny's
W: City Club
Lodg E: **Motel 6**
W: Countryside Inn, Doubletree Hotel

106A CA 22, Garden Grove Fwy, to Orange

105B Broadway, Main St
Gas W: 76 (Full Service Car Wash)
Food E: California Pizza Kitchen, Polly's Tasty Food & Pies, Spoons
W: El Polo Loco, Topaz Cafe
Lodg E: Red Roof Inn
W: Golden West Hotel, Travel Inn
Other E: **Main Place Santa Anna (Mall)**

105A 17th St
Gas E: Shell*(D)(CW)
W: Chevron
Food E: Country Harvest Buffet, IHOP, Lampost Pizza, McDonalds
W: Las Palmas, Norms (24 Hrs), Ruby Tuesday, Yum Yum Donuts
Lodg E: Howard Johnson
W: Aqua Motel
AServ E: GM Auto Dealership
W: Chevron
ATM W: Great Western Bank

104 Santa Ana Blvd, Grand Ave

103 4th St , 1st St.
Lodg W: Villager Lodge (see our ad this page)

102B Jct CA 55, Newport Beach

102A Newport Ave (Southbound)

101 Red Hill Ave
Gas E: Arco*, Arco(CW), Mobil(D), Shell
W: 76, Arco*, Chevron*, Circle K*
Food E: D&N Donuts, Denny's, Mario's Pizza, Seoul BBQ Buffet
W: Taco Bell
Lodg E: Key Inn
AServ E: Arco, Mobil, Shell
W: 76, Arco
Other E: **Coin Laundry, Drug Emporium (24 Hrs)**
W: **Stater Grocery**

100B Tustin Ranch Road
AServ E: Buick Dealer, Cadillac-Oldsmobile-GMC, Costco, Dodge, Geo-Ford, Infiniti-Acura-Chevrolet, Lincoln Mercury, Mazda, Pontiac, Toyota & Lexus Dealer, Tustin Auto Center, Tustin Ford-GM
Other E: **K-Mart**

EXIT CALIFORNIA

100A Jamboree Road
Food E: Black Angus Steakhouse, Busy B Oriental Food, Carl's Jr Hamburgers, Dairy Queen, El Pollo Loco, In-N-Out Hamburgers, Juice Club, Koo-Koo-Roo, Northwood Pizza, Red Robin, Sub Station II, Taco Bell
Other E: **Tustin Marketplace**

99 Culver Drive
Gas E: Shell
W: Chevron
Food W: Ace Donuts, Bread Smith, Burger King, Carrows, Caspian Restaurant, China Oasis, Denny's, Diho Bakery, Domino's Pizza, East Coast Bagel, India Cook House, Japanese Restaurant, Korean Restaurant, Lampost Pizza, Los Primos, Milano's Italian, Mitsuba, Seattle's Best Coffee, Simply Pasta, Sizzler, Tai Nam, Wendy's
AServ E: Shell
ATM W: Bank of America, First Bank & Trust, Wells Fargo
Other W: **Sav-On Drugs**

97 Jeffrey Road
Food W: A&J Restaurant, Champagne Cuisine, China Express, Irvine Pizza, Jenny's Donuts & Croissants, O'Shine Cafe, Ra, Taiko, Wienerschnitzel
AServ W: 76
ATM W: Great Western Bank
Other W: **Arbor Animal Hospital, Camino Pet Hospital, Coin Laundry, Irvine Eye Care, Lucky Food Center, TS Emporium, The Arbor (Shopping Center), Thifty Drugs**

96 Sand Canyon Ave
Gas W: 76(CW)
Food W: Fast Break Food, Knowlwood, Shark's, Ti

EXIT CALIFORNIA

Juana's
Lodg W: La Quinta Inn
RVCamp W: **El Toro Gen. Store(LP), Traveland USA**
ATM W: 76
Other W: **U.S. Post Office**

95 Jct 133S, Laguna Beach (Southbound)

94 Alton Pkwy
Gas E: Texaco
Food E: Carl's Jr. Hamburger, Subway (Texaco), Taco Bell (Texaco)
W: Bertolini's Authentic Trattoria, Champps Americana, Sloppy Joe's Bar
Lodg E: Homestead Village
Med W: **+ Hospital**
Other E: **Irvine Spectrum Pavilion (Shopping Mall)**
W: **Irvine Spectrum**

93 Jct. I-405 Long Beach

92B Blake Pkwy., I-5 Truck By - pass

92 Lake Forest Drive
Gas E: Chevron*, Texaco*(D)(LP)
W: Shell*(CW)
Food E: Black Angus, Burger King, Captain Creme, Country Rock, Del Taco, Hungry Hunter, IHOP, Mandarin Taste, McDonalds, Nory's Restaurant, Pho Bo Vang, Subway, Taco Bell, Teriyaki House
W: Carl's Jr Hamburgers, Coco's Restaurant, Gyro Hobbies, N.Y. C. Cafe, Snooty Fox Cafe
Lodg E: Best Western, Irvine Suites Hotel, Travelodge (see our ad this page)
W: Comfort Inn
AServ E: Chevron, Econo Lube & Tune, Mercedes, Suzuki
ATM E: Farmers & Merchants Bank
Other W: **Kinko's Copies**

91 El Toro Road
Gas E: Arco*, Chevron, Mobile*, Shell, Texaco(D), USA Gas*(D)
W: 76, Chevron(D) (24 Hrs), Shell(CW)
Food E: Arby's, Baker's Square, Baskin Robbins, Carmels Burger's & Burrito, Fuddrucker's, I Love Bagels, Jack-In-The-Box, McDonalds, Mega Burger's, Mr. Wok, Numero Uno, Red Lobster, Scarantino's Italian, Spoons
W: Carrows, Don Jose, Koo-Koo-Doo
Lodg W: Laguna Hills Lodge
AServ E: Chevron, Firestone Tire & Auto, Shell, Texaco
W: 76, Just Tires, Shell, Tire Station
ATM E: Bank of America, Wells Fargo
W: American Savings Bank, Coast Federal Bank, Great Western Bank
Other E: **Petco, Sav-On Drugs (24 Hrs), Thrifty Drugs**
W: **JC Penny, Macy's**

90 Alicia Pkwy
Gas E: 76(D)
W: 76(LP) (24 Hrs), Chevron (24 Hrs)
Food E: Del Taco, Denny's, Mission Donut, Piccadilly Pizza, Subway
W: Carl's Jr Hamburgers, Golden Baked Hams, Natraj Cuisine of India, New York Deli Restaurant, Pasta Palace, Royal Donuts, Thai Rice Chinese, Time Square Pizza, Togo's Eatery, Wendy's
AServ E: Kragen, Parnelli Jones Firestone
W: 76, Buick Pontiac Mazda
Other E: **Mervyn's, Target Department Store**
W: Laguna Hills Car Wash (Chevron)

89 La Paz Road
Gas E: Arco*, Beacon, Mobil
W: 76, Chevron
Food E: Diedrich Coffee, Donuts & Deli, Taco Bell
W: China Palms, Claimjumper, Del Taco,

Bold red print shows RV & Bus parking available or nearby

EXIT CALIFORNIA

Elephant Bar, Haddleburg Pastry, McDonalds, Outback Steakhouse, Roma D Italian, Shakeys Pizza, Spasso's Family Italian, Wienerschnitzel

Lodg **W:** Holiday Inn

AServ **E:** Beacon, Mobil

W: 76, Chevron, Winston Tire Company

ATM **E:** Downey Savings Bank

Other **E: Animal & Bird Clinic, Lucky Food Center, Sav-On Drugs**

W: Lapaz Animal Clinic

87 Pacific Park Drive, Oso Pkwy

Gas **E:** 76(D) (24 Hrs), Chevron (24 Hrs)

Lodg **E:** Fairfield Inn

AServ **E:** 76

86 Crown Valley Pkwy

Gas **E:** 76, Arco (24 Hrs), Chevron

W: Chevron*(D)(CW)

Food **E:** Coco's Restaurant, TJ's Mexican

AServ **E:** 76, Arco, Chevron

W: Econo Lube & Tune, Tucker Tire

ATM **E:** Bank of America, Coast Federal Bank

Other **E: Parkway Animal Hospital**

85 Avery Pkwy

Gas **E:** Texaco*

W: Shell*(D)

Food **E:** Albertaco's Mexican, Booster's, Carrows Restaurant, Del Taco, Geckos, Jack-In-The-Box, Manderin Dynasty, Subway

W: A's Burgers, Buffy's Family Restaurant

Lodg **W:** Laguna Inn & Suites

AServ **E:** Land Rover, Lexus, Mercedes Benz, Preway Auto Supply

W: Allen Oldsmobile-Cadillac-GMC Trucks

ATM **E:** Wells Fargo

Other **E: Kinko's Copies**

84 CA 73 N. to Long Beach (Toll Rd, Northbound)

83 Junipero Serra Road

Gas **W:** Shell, Ultramar(D)

AServ **W:** Shell, Ultramar

82 CA 74, Ortega Hwy, to San Juan Capistrano

Gas **E:** 76, Chevron, Shell

W: Arco*, Chevron

Food **E:** Cafe Capo (Best Western), Denny's, Parino's Trattoria

W: Carl's Jr Hamburgers, Cedar Creek Inn, Jack-In-The-Box, Marie Callender's Restaurant, McDonalds, Sizzler Steak House, Walnut Grove Restaurant

Lodg **E:** Best Western

W: Mission Inn

AServ **E:** 76, Chevron, Shell

W: Arco, Chevron

ATM **W:** Farmer's & Merchant Bank

Other **W: Capistrano Historic Mission, Ralph's (Grocery Store)**

81 Camino Capistrano, San Juan Creek Rd., Valle Rd.

Gas **W:** Chevron

Food **W:** Baskin Robbins, Donut Hut, Eng's, Harry's Family Restaurant, India Kitchen Cuisine, Pizza Hut, Pranzare Italian

AServ **E:** Capistrano Volkswagon Peugeot

W: Chevron, Goodyear Tire & Auto

ATM **W:** El Dorado Bank, Glendale Federal Savings, Wells Fargo Bank

EXIT CALIFORNIA

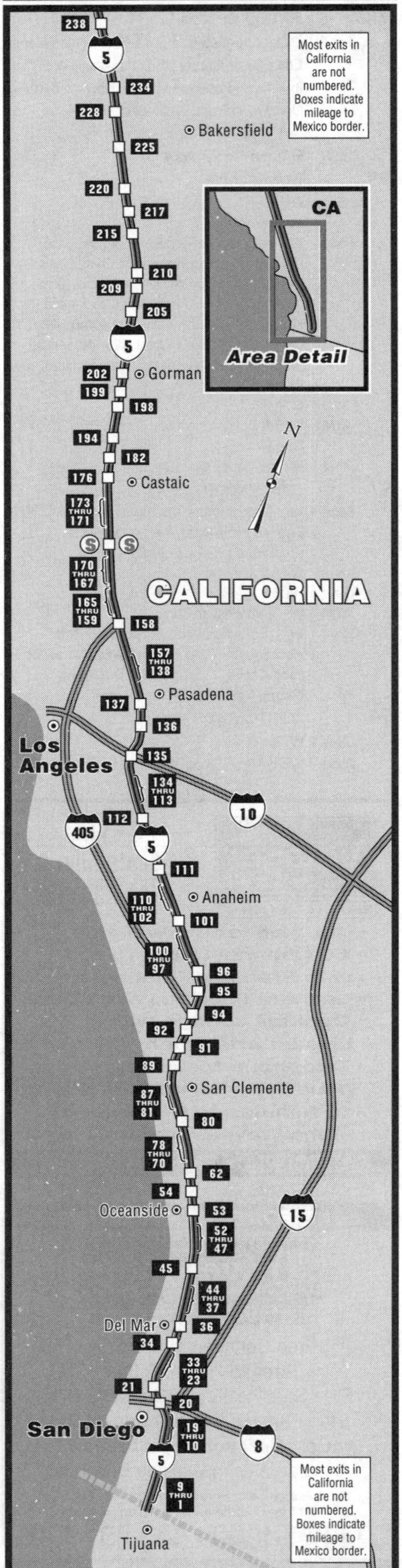

Most exits in California are not numbered. Boxes indicate mileage to Mexico border.

EXIT CALIFORNIA

Other **W: CA Highway Patrol, Petco, Thrifty Drug Store**

80 CA 1 Beach Cities & Camino Las Ramblos

78 Casino de Estrella

Gas **E:** 76

W: Thrifty Gas*

Food **E:** Bakers Square Restaurant & Pies, Carl's Jr Hamburgers, China Well Restaurant, Golden Chicken, Gourmet Bagel, Juice Time, Ralph's, Round Table Pizza, Starbuck's Coffee, Subway, Trader Joe's

W: Japanese Restaurant, Kultured Kitchen, Thai Paradise

AServ **E:** 76

ATM **E:** Bank of America, Home Savings Of America, Wells Fargo Bank

W: Bank of America, California Federal Bank, Great Western Bank

Other **E: Lucky Food Center, Ocean View Plaza, Sav-On Drugs**

W: Jim's Pharmacy

77 Avenue Pico (Northbound)

Gas **E:** Mobil*(CW)

W: Chevron*, Exxon*, Texaco*(D)

Food **E:** Carrows Restaurant, McDonalds

W: Burger Stop, Caterina's Yogurt, Denny's (24 Hrs), Donuts, Pick-Up-Stix, Stuffed Pizza, Subway

Lodg **W:** Countryside Inn

AServ **W:** Midas Muffler

ATM **W:** Del Taco, El Dorado

Other **W: Petco, Sav-On Drugs, U.S. Post Office**

76 Ave Palizada (Southbound)

Gas **W:** Ultramar

AServ **W:** Ultramar

ATM **W:** Wells Fargo

75 Ave Prasedio, San Clemente

Other **E: Civic Center, San Clemente Police Headquarters**

74 El Camino Real

Gas **E:** Chevron (24 Hrs)

W: 76(D), Mobil

Food **E:** Burrito Basket, Mariachi, The Shack Restaurant

W: FatBurger, Love Burger, Taco Bell, Taste of China, Tommy's Family Restaurant

Lodg **E:** Budget Lodge, Tradewinds Motel

AServ **E:** Chevron, San Clemente Foreign Car Repair

W: Kragen Auto, Mobil, Top Tune

Other **E: Coin Laundry, San Clemente Veterinary Clinic & Boarding**

73 Ave Magdalena

Gas **E:** 76 (24 Hrs), Shell

Food **E:** China Beach Cantina, Coco's Bakery & Restaurant, Jack-In-The-Box, Pedro's Tacos, Sugar Shack Cafe, The Beefcutter Prime Rib

Lodg **E:** C-Vu Motel, El Rancho, La Vista Inn, Quality Suites, San Clemente Motor Lodge, Travelodge

W: San Clemente Inn

AServ **E:** 76, Dick Watson Auto Service, Winston Tire Company

Other **E:** Buggy Bath Car Wash

72 Christianitos Rd.

Food **E: Carl's Jr Hamburgers**

Lodg **E:** Carinelo Motel, Comfort Suites

Parks **W: San Clemente State Park**

71 Basilone Rd, San Onofre

Bold red print shows RV & Bus parking available or nearby

EXIT CALIFORNIA

70 Scenic Vista

(67) **Weigh Station (Both Directions)**

(63) Vista Point (Southbound, Ocean Lookout)

62 Las Pulgas Road

(57) **Rest Area - RR, Picnic, Vending, Phones (Both Directions)**

54 Oceanside Harbor Drive, Camp Pendleton East Side
- **Gas** W: Mobil(LP)
- **Food** W: Chart House, Del Taco, Denny's, Jolly Roger
- **Lodg** W: Sandman Motel, Travelodge
- **AServ** W: Mobil
- **Other** E: **Camp Pendleton Marine Base**

53B Bonsall, Coast Hwy. CA 76E, Oceanside, Hill St.
- **Gas** W: Arco*
- **Food** W: Angelo's Burgers, **Carrows Restaurant**, Hamburger Heaven, Wienerschnitzel
- **Lodg** W: Miramar Motel, Motel 9, **The Bridge Motor Inn**
- **Other** W: **Visitor Information**

53A Mission Ave, Downtown Oceanside
- **Gas** E: Arco*
- **Food** E: Burger King, El Charrito, Grandee Family Restaurant, Jack-in-the-Box, Mission Donut House, Rally's Hamburgers, Taco Shop (24 Hrs)
  W: Burger Joint, El Pollo Loco, Long John Silvers
- **Lodg** E: Grandee Inn, Welton Inn
- **AServ** E: Bussey's Auto Service, Econo Lube & Tune
- **Other** W: **Thrifty Drugs**

52 Oceanside Blvd
- **Food** E: IHOP, Marsha's Deli, McDonalds, New Hong Kong, Pizza Hut, Pizza Place, Super Bronco
- **Lodg** W: Best Western
- **AServ** W: Texaco
- **Other** E: **California Highway Patrol Post, Coin Laundry, Longs Drugs, Ralph's, Sav-on-Drugs, Vons Grocery**

51C Oceanside, Vista Way
- **Food** W: Hungry Hunter

51B Jct CA 78, to Escondido
- **RVCamp** E: **Camping World (see our ad this page)**
  W: **Foretravel, Inc (see our ad this page)**

51A Las Flores Drive

50 Carlsbad Village Drive, Downtown Carlsbad
- **Gas** E: Shell
  W: 76(CW), Arco*, Chevron(D), Texaco*(D) (24 Hrs)
- **Food** E: Fresco's, Lotus Thai, Pizzaria
  W: Alberto's Mexican, Carl's Jr Hamburgers, Jack-in-the-Box, KFC, Kahala Cafe, Mikko, Taco Bell
- **Lodg** W: Motel 6
- **AServ** E: Shell
  W: Chevron

EXIT CALIFORNIA

- **ATM** W: Union Bank of California, Wells Fargo
- **Other** W: **Albertson's Grocery**

49 Tamarack Ave
- **Gas** E: Chevron (24 Hrs), Texaco(D) (24 Hrs)
  W: 76(LP) (24 Hrs)
- **Food** E: Pizza Shuttle & Subs, Village Kitchen & Pie Shop
  W: Gerico's Family Restaurant
- **AServ** E: Chevron, Texaco
  W: 76
- **Other** E: **Coin Laundry, Thrifty Drug Store, Vons Grocery**

48 Cannon Road
- **AServ** E: Acura, Buick, Chrysler Auto Dealer, Ford Dealership, Hoehn Porche-Audi, Lexus Dealer, Lincoln Mercury, Mercedes Benz, Pontiac-GMC Trucks-Mazda, Rorick Buick, Toyota Dealer, Weseloh Chevrolet, Worthington Dodge

47 Carlsbad Blvd, Palomar Airport Road
- **Gas** E: Chevron*(D) (24 Hrs), Citgo*, Mobil*(CW)
  W: Texaco*(D) (24 Hrs)
- **Food** E: Creek Village Restaurant, Denny's, Fogg's (Best Western), Haedrich's Tip Top Meats, Pea Soup Anderson, Subway, Taco Bell
  W: California Woodfired Pizza, Claimjumper Restaurant, In-N-Out Hamburgers, Marie Callender's Restaurant, McDonalds
- **Lodg** E: Best Western, Motel 6
- **AServ** E: Jiffy Lube
  W: Texaco
- **ATM** E: Citgo
- **Other** E: **Costco,** Hand Car Wash

45 Poinsettia Ln.
- **Food** W: El Pollo Loco, It's Coffee Time, Jack-In-The-Box, Rain Tree Restaurant (Inns of America), Subway
- **Lodg** W: Inns of America, Motel 6, Ramada Inn, Travel Inn
- **Other** W: **All Cats Hospital, Payless Drugs, Ralph's Grocery (24 Hrs), Vision Center**

44 La Costa Ave
- **Gas** W: Chevron (24 Hrs)
- **AServ** W: Chevron

43 Leucadia Blvd
- **Gas** W: Shell, Texaco(D)(LP)
- **Lodg** E: Holiday Inn Express
- **AServ** W: Shell, Texaco

41 Encinitas Blvd
- **Gas** E: Chevron*, Mobil, Texaco*(LP)

EXIT CALIFORNIA

W: Shell(CW)(LP)
- **Food** E: Coco's Restaurant, Del Taco, Honey Baked Ham, Red Oak Steakhouse
  W: Denny's (see our ad this page), Italian (Radisson), Wendy's
- **Lodg** W: Budget Motels, Radisson
- **AServ** E: Mobil, Mossy Nissan, Texaco
- **Other** E: **Kinko's Copies, Sav-On Drugs**
  W: **Med-Rx Prescription Drugs, Petco**

40B Santa Fe Drive, to Encinitas
- **Gas** E: Shell
  W: 76(D)
- **Food** E: Carl's Jr Hamburgers, El Nopalito, Encinitas Donuts, Papa Tony's Pizza, Shide Away Cafe
  W: Burger King (76), California Yogurt Company, Great Wall
- **Med** W: **✚ Scripps Memorial Hospital**
- **ATM** W: Downey Savings Bank
- **Other** E: **Coin Laundry**
  W: **Animal Clinic, Lauderland Coin Laundry, Thrifty Drug Store, Vons Grocery (24 Hrs)**

40A Birmingham Drive
- **Gas** E: Chevron* (24 Hrs), Texaco*(CW)(LP)
  W: Thrifty*(D)
- **Food** E: Glenn's Pancake House
- **AServ** E: Texaco

38 Manchester Ave
- **Gas** E: 76*
- **AServ** E: 76

37 Lomas Santa Fe Drive, to Solana Beach
- **Gas** W: Mobil*
- **Food** E: Baskin Robbins, Boulangerie, Donna Lee's, Einstein Bros. Bagels, Road Side Grill, Samurai Restaurant
  W: Cafe Europa, Carl's Jr Hamburgers, Golden Bowl Chinese, Sante Fe Yogurt
- **AServ** W: Texaco
- **ATM** E: Mission Federal Credit Union, Wells Fargo Bank
  W: Great Western Bank
- **Other** E: **Vons Grocery**
  W: **SAV On Drugs, Vons Grocery (& Pharmacy)**

36 Via de la Valle, Del Mar
- **Gas** E: Mobil*
  W: Arco*, Texaco(D)(CW)
- **Food** E: Chevy's, Koo-Koo-Doo, McDonalds, Milton's Deli, Pancho Chino, Pasta Pronto, Pick-up-Stix, Tony Roma's
  W: Denny's, The Fish Market
- **Lodg** W: Hilton

EXIT		CALIFORNIA
	ASserv	W: Texaco
	ATM	E: Grossmont Bank, Wells Fargo Bank
	Other	E: **Alberson's Supermarket, Blaine's Drugs - N- Such, Pet Cove**
		W: **Del Mar Fairgrounds & Racetrack**
34		Del Mar Heights Road
	Gas	E: Citgo*(D), Texaco*
	Food	W: Bellisario's Pizza, Golden Bagel Cafe, Jack-In-The-Box, Le Bambou, Mexican Grill, Mucho Gusto, Rotisserie Chicken
	AServ	E: Texaco
	ATM	E: Citgo, Wells Fargo
		W: Bank of America
	Other	W: **Coin Laundry, Longs Drugs, Vons Grocery**
33		Sorrento Valley Road (Northbound, Difficult Reaccess)
32		Caramel Valley, Sorrento Valley Road (Difficult Reaccess)
	Gas	E: Arco* (24 Hrs), Shell
	Food	E: A & W Drive-In (Arco), Taco Bell
	Lodg	E: Doubletree Hotel
	AServ	E: Shell
30		Jct I-805
29		Genesee Ave
	Med	E: ✚ **Scripps Memorial Hospital**
28		La Jolla Village Drive
	Gas	W: Mobil*(CW)
	Food	W: El Torito Mexican
	Lodg	E: Hyatt
		W: Radisson
27B		Noble Rd. (No Northbound Reaccess)
	Food	W: BJ's Pizza & Grill, California Pizza Kitchen, Garden State Bagels, Islands, Jamha Juice, Mrs. Gooch's Cafe, Pasta Bravo, Pick-Up Stix, Rubio's Baja Grill, Samson's Deli & Bakery, T.G.I. Friday's
	ATM	W: Bank Of America, Glendale Federal Bank
	Other	W: **Kinko's Copies (24 Hrs), La Jolla Village Square Mall, Ralph's Big Grocery Store, SAV On Drugs, Vision Center**
27		Gilman Drive, La Jolla Colony Drive
26		CA 52E, San Clemente Canyon
25		Ardath Rd.
24		Balboa Ave. Ca. 274 (Reaccess Difficult)
	Gas	W: 76, Citgo*, Mobil
	Food	W: Dragon House, In-N-Out Hamburgers, Sheldon's Cafe, Wienerschnitzel, Yum-Yum Donuts
	Lodg	W: Comfort Inn
	AServ	W: 76, B-Line Alignment, Cal's Radiator, Discount Tire Co., Dualton Mufflers & Brakes, Econo Lube & Tune, Jiffy Lube, Mobil, Pacific Beach Transmission, Quiki Oil Change, Winston Tire Company
	Med	W: ✚ **Hospital**
	ATM	W: Citgo, Home Savings of America
	Other	W: **Clairmont Animal Hospital**
23		CA 274, Garnet Ave, Beaches, Grand Ave. (No Northbound Reaccess)

EXIT		CALIFORNIA
	Gas	W: Shell
	Food	W: Kolbeh Restaurant
	Lodg	W: Sleepy Time Motel, Super 8 Motel, Trade Winds Motel, Western Shores Motel
	AServ	W: Guy Hill Cadillac, Massy Ford, Mossy Ford, Pacific Nissan, Shell
	Med	W: ✚ **Mission Bay Hospital**
21		Clairemont Drive, Mission Bay Drive
	Gas	E: 76
	Food	E: JR's, Jack-In-The-Box, Petricca's Pizza
	Lodg	E: DAYS INN Days Inn
	AServ	E: 76, Grease Monkey, Shell
	ATM	E: Wells Fargo
	Other	E: **Mission Bell Pharmacy**
20B		Sea World Drive, Tecolote Road
	Gas	E: Shell (AServ)
	Lodg	E: Motel
	AServ	E: Shell, World Wide Auto Parts
	Other	W: **Highway Patrol, Sea World**
20A		Jct I-8, CA 209, El Centro, Rosecrans St
	Lodg	S: Super 8 Motel (San Diego, see our ad this page)
19		Old Town Ave. (Marine Base)
	Gas	E: Arco*, Shell
	Food	E: El Agave, Old Town Pizza
	Lodg	E: Ramada, Travel Lodge, Vacation Inn, Western Inn
	Other	E: **Old Town Ave. State Park**
		W: **Marine Base**
17		Washington St (Northbound)
	Gas	E: Chevron*
	Med	E: ✚ **Hospital**
16B		Front St, to downtown San Diego, Civic Center (Southbound)
	Gas	W: Shell* (AServ)
	Lodg	W: Holiday Inn, Radisson, Super 8 Motel
	AServ	W: Speedo Lube
	Med	W: ✚ **Hospital**
16A		CA163N, Escondido
16		Pacific Hwy. (Northbound)
15C		Sassafras St., Indian Street (Northbound)
15B		Hawthorne St, San Diego Airport (Difficult Reaccess)
15		Sixth Avenue Downtown (Northbound)
	Lodg	E: New Palace Hotel

EXIT		CALIFORNIA
14D		CA 94E, Pershing Dr, B St
14C		Imperial Ave (Northbound)
14		Jay Street (Northbound)
14B		Crosby St
	Food	W: El Sarape, Imperial Express, Panchita's, Taco Shop
	Lodg	W: Econolodge
	AServ	W: Ace Radiator
	Other	W: **Coin Laundry**
14A		Jct CA 75, Cornado (Toll Crossing)
13		National Ave, 28th St
	Gas	W: Texaco*
	Food	W: Burger King, El Pollo Loco, Long John Silvers, McDonalds, Roberto's
	AServ	E: Chief Auto Parts
		W: Texaco
12		Jct CA15, Riverside
11B		Main St, National City Blvd (Difficult Southbound Reaccess)
	Gas	E: Mobil(LP), Shell
		W: Chevron*(D)
	Food	E: Keith's
	Lodg	E: Budget Inn, Ramada Limited
	AServ	E: Hopsing, Mobil, Shell
		W: Sergio's Auto Electric, Welch Tire
11A		8th St, National City
	Gas	E: Thrifty(D)
	Lodg	E: Holiday Inn, Radisson
	ATM	E: Union Bank
10C		Plaza Blvd., Downtown (No Reaccess)
10B		Civic Center Drive (No Northbound Reaccess)
10A		24th St
	Food	E: Denny's, In-N-Out Hamburgers
9		Jct CA 54E
8		E St
	Gas	E: 76*, Arco*, Mobil
	Food	E: Aunt Emma's, Black Angus Steakhouse, Denny's, Little China Club, McDonalds, New Day (Royal Vista Inn), Pizza Junction, Royal China Palace, Taco Bell, Wendy's, Yum Yum Donuts
		W: Anthony's Fish Gratto
	Lodg	E: **Best Western**, DAYS INN Days Inn, Motel 6, Royal Vista Inn, Traveler Motel Suites
		W: Good Nite Inn
	AServ	E: Chief Auto Service, Hydro Spray Car Wash, Mobil
	Other	E: **IGA Grocery**
7		H St
	Gas	E: Arco*, Chevron(CW), World*
	Food	E: BaJa Lobster, Casa Salsa, El Pollo Loco, Jack-In-The-Box, La Tostada
	Lodg	E: Early California Motel
	AServ	E: Chevron
6C		J St, to Chula Vista Harbor

Bold red print shows RV & Bus parking available or nearby

EXIT		CALIFORNIA
6B		L St
6A		Palomar St
	Gas	E: Arco*
	Lodg	E: Palomar Inn
5		Main St
4B		CA 75, Imperial Beach, Palm Ave
	Gas	E: Arco (AServ)
		W: Arco*, Shell
	Food	E: Armando's, Fresh Doughnuts
		W: Alberto Jr., Carrows Restaurant, Evergreen Garden, Oriental Express
	Lodg	W: Imperial Palm Motel, **Silverado**
	AServ	E: Arco, Gators Auto Parts, Midas Muffler & Brakes, Mike's Tires
		W: Shell, Smitty's Tires, Summerset Transmission, Taylors
	ATM	W: Wells Fargo
	Other	W: **Sav-on Drugs, Vons Grocery**
4A		Coronado Ave
	Gas	E: Chevron, Shell(LP)
		W: Arco*, Texaco*(D) (AServ)
	Food	E: Denny's, Dos Panchos
		W: Mike's Giant New York Pizza
	Lodg	E: E-Z 8 Motel, San Diego Inn Motel, Travlers Motel
		W: South Bay Lodge
	AServ	E: Chevron, Leo's Garage, Shell
		W: Texaco
	Other	W: **Nestor Pharmacy**
2B		Jct CA 905, Tocayo Ave
2A		Dairy Mart Road
	Gas	E: Arco*
	Food	E: Burger King, Carl's Jr Hamburgers, Coco's, McDonalds
	Lodg	E: Americana Inn & Suites, Motel 66, Valli-Hi Motor Hotel
	RVCamp	E: **LA Pacific**
1C		Via de San Ysidro (Difficult Southbound Reaccess)
	Gas	E: 76, Chevron*, Mobil*, Shell* (ATM)
		W: Mini Mart Gas*
	Food	E: Amigos Tacos
		W: KFC
	Lodg	W: Economy Motels of America, Motel 6
	AServ	E: 76
		W: Chevron*
	ATM	E: Bank of America, Shell
1B		Jct I-805
1A		Camino de la Plaza
	Food	E: Burger King, El Pollo Authentic Mexican Chicken, Jack-In-The-Box, McDonalds, Mercado Sonora, Subway
		W: Taco Bell
	Lodg	E: Gateway Inn, Holiday Lodge Motel, Travelodge
	AServ	W: Border Tire
	ATM	E: Wells Fargo Bank
	Other	W: **Five Star Border Parking, San Diego Factory Outlet Center**

↑CALIFORNIA

Begin I-5

EXIT		CALIFORNIA
		Begin I-8
		↓CALIFORNIA
0		Junction I-5 South
	Lodg	S: **Super 8 Motel (see our ad this page)**
3		West Mission Bay Drive Sports Arena Blvd.
2		Junction I-5
4B		Morena Blvd
4		Taylor St, Hotel Circle
	Food	N: D.W. Ranch, Hunter Steakhouse
	Lodg	N: **Best Western, Comfort Inn, Motel 6,** Premier Inn
5A		Hotel Circle
	Gas	S: Chevron
	Food	N: **Kelly's Steak House (Town & Country Hotel),** Restaurant (Handlery Hotel)
		S: **Albi's Beef Inn (Travelodge),** King's Grille, **Pam Pam Cafe (Econolodge),** Ricky's (Hotel Circle Inn), Tickled Trout (Ramada Plaza), Valley Kitchen
	Lodg	N: **Handlery Hotel, Town & Country Hotel**
		S: **Best Western, Days Inn, Econolodge, Holiday Inn Select,** Hotel Circle Inn (Ricky's), Howard Johnson, **Kings Inn (Kings Grille), Quality Resort,** Ramada Plaza Hotel (Tickled Trout), Regency Plaza Hotel, **TraveLodge,** Vagabond Inn
	Other	S: **Emergency Animal Hospital**
5B		Junction 163, Escondido, Downtown
5C		Mission Center Road
	Gas	S: Arco*
	Food	N: Hooters, Mandarin Cuisine, Rusty Pelican
		S: Benihana, Denny's, Hayama Restaurant, Love's (The Great Rib Restaurant), **Padre's Pub (Hilton),** Taco Mucho, Wendy's
	Lodg	S: **Budget Motel, Comfort Suites, Hilton, Radisson**
	AServ	N: Montgomery Ward
		S: Marvin Brown Cadillac, Buick, GMC Truck, & Hummer, Trevellyan Oldsmobile, Trevellyan Subaru
	ATM	N: Great Western Bank, Home Savings of America
	Other	N: **Mission Valley Center Mall, Montgomery Ward**
6A		Texas St, Stadium Way
	Gas	N: Chevron(LP) (AServ)
	Food	N: Bennigan's, Hogi Yogi
		S: Bully's

EXIT		CALIFORNIA
	AServ	N: Chevron
	ATM	N: Bank of America
6B		Junction I-805
8A		CA 15, 40th St Junction
8B		Mission Gorge Road, Fairmont Ave
	Gas	N: Mobile(D)(CW), Shell
	Food	N: Bagel King, Boll Weevil, Chili's, Dublin Pub, Osaka, Szechuan, Taco Bell
	Lodg	N: Budget Motel
	AServ	N: Quiki Oil Change
	Med	N: **+ Hospital**
9		Waring Road
	Food	N: Nicolosi's, Patches (Good Nite Inn)
	Lodg	N: Good Nite Inn, Madrid Suites, Motel 7
10		College Ave (San Diego State University)
11		70th St, Lake Murray Blvd
	Gas	N: Shell*(CW)
		S: Texaco(D)(LP) (AServ)
	Food	N: Peppers, Subway, The Chickenest
		S: D.Z. Akins, Denny's, Marie Callender's Restaurant
	AServ	N: Prestigous Auto Service
		S: Texaco
	Med	N: **+ Hospital**
13A		Fletcher Parkway
	Food	N: Baker's Square, Boston Market, McDonalds
	Lodg	N: **Comfort Inn, E-Z 8 Motel**
		S: **Motel 6**
	AServ	N: Costco Tire Service
	RVCamp	S: **Camperland, RV Peddler**
	ATM	N: Western Financial
	Other	N: **Coin Laundry**
13B		El Cajon, Spring St.
	Food	S: Country Kitchen and Grill, La Salsa, Togo's Eatery
	Lodg	S: **Hitching Post Motel, Travelodge**
14A		Jackson Drive, Grossmont Blvd.
	Gas	N: Chevron(CW)
	Food	N: Arby's, Fuddrucker's, KFC, Olive Garden, Red Lobster, Trophy's Sports Grill
		S: Chilli Bandito, Honey Baked Ham, Jack-In-The-Box
	AServ	N: Shell
		S: Firestone Tire & Auto
	ATM	N: Home Savings of America
	Other	N: **Grossmont Center Mall, Longs Drugstore**
		S: **Coin Laundry, Pet Emergency Clinic**
14B		La Mesa Blvd., Grossmont Center Dr., El Cajon Blvd. (No Reacces)
	Gas	S: Circle K*
	Food	N: Carlos Muphy's Pizzas, Claim Jumper, Pizza Nova, Rubio's Baja Grill, Schlotzkys Deli, Submarina
		S: Marieta's
	AServ	S: Discount Tire Co., Drew Volkswagon, Hyundai
	Med	N: **+ Grossmont Hospital**
	ATM	N: Bank of America, Great Western Bank, Grossmont Bank, Vallede Oro Bank
	Other	N: **Grossmont Center Mall**
15B		Junction CA125 S & CA125 N.
15A		Severin Dr., FuerteDr.
	Food	S: Briganteen Seafood
16		El Cajon Blvd (Eastbound Access Only, No Reaccess)

Bold red print shows RV & Bus parking available or nearby

EXIT	CALIFORNIA
AServ	**S:** Cunningham BMW
17	Main St
Gas	**N:** Arco*
	S: 76 (AServ), Chevron (AServ)
Food	**N:** Denny's, Sombrero
Lodg	**N:** **Thriftlodge**
AServ	**N:** Banny Hom
	S: 76, Chevron, Import Auto Repair, Village Auto Repair
18A	Johnson Ave (Eastbound Only, No Eastbound Reaccess)
Food	**N:** Burger King, El Cotixan 3, Japenese Restaurant, Subway
AServ	**N:** Batteries Plus, Big O Tires, Dorman's, Sear's Auto Center
	S: AAMCO Transmission, Green's Brake and Alignment, Greg's Automotive, Saturn of El Cajon, Summit Transmissions
TServ	**S:** **Cummins Diesel**
ATM	**N:** Home Savings of America
Other	**S:** **El Cajon Valley Animal Hospital**
18C	Magnolia Ave, CA 67
Gas	**S:** Texaco (AServ)
Food	**S:** Perry's Cafe, The Best Chinese, Wienerschnitzel
Lodg	**S:** Midtown Motel, **Motel 6, TraveLodge**
AServ	**S:** Tire Stop
Other	**S:** **Nudo's Pharmacy**
19	Mollison Ave
Gas	**N:** Chevron (AServ), Shell*(CW), Thrifty*
	S: Arco*
Food	**N:** Arby's, Denny's (see our ad this page), Wendy's, Winchell's Donuts
	S: Taco Bell
Lodg	**N:** **Best Western, Plaza International Inn**
	S: Super 8 Motel
AServ	**N:** Chevron
Other	**N:** **Valley Coin Laundry**
20	Second St, CA 54
Gas	**N:** Arco*, Mobil(LP) (AServ)
	S: 76* (AServ), Shell
Food	**N:** Julian's BBQ, Marechiaro's Pizza, Taco Shop
	S: Arby's, Baskin Robbins, Boll Weevil, Carl's Jr Hamburgers, Chinese Cuisine, Cotixan Taquerias, El Compadre, Finest Donut, Golden Coral, IHOP, Jack-In-The-Box, KFC, Kips, McDonalds, Pizza Hut, Subway, Taco Bell, Tyler's, Wings-N-Things
AServ	**N:** Atlantic Auto Repair, Auto Doc, Foreign Auto Supply, Geni Car Wash, Midas Muffler & Brakes, Run Rite Auto Service, Sun Auto Repair, Winston Tire Co
	S: 76, El Cajon, Firestone Tire & Auto, Instant Oil Change, The Clutch Exchange, World Wide Auto Parts
ATM	**N:** Bank Of America
	S: Union Bank of California
Other	**S:** **Petco, Thrifty Pharmacy**
20B	East Main St. (Westbound, No Westbound Reaccess)
Gas	**S:** Arco* (AServ)
Food	**N:** Bailey's Pancake Restaurant, Pernicano's
Lodg	**N:** **Budget Inn, Embasadora Motel, Fabulous 7 Motel, Ha'Penney**
	S: Best Western, St Francis Motel
AServ	**N:** Decker's Auto Service, Ford Dealership
	S: Arco, De La Fuente (Cadillac, Pontiac Dealership), Touchless Car Wash
RVCamp	**N:** **Rick's RV Center, Vacationer's Travel**
	S: **El Capitan**
21	Greenfield Drive
Gas	**N:** Exxon (AServ), Shell(LP) (AServ)
	S: Mobil*(LP)
Food	**N:** Janet's Cafe, Los Panchos Taco Shop, Lost Dutchman, McDonalds, Yogurt Gallery, Yum Yum Donuts
AServ	**N:** Chief Auto Parts, Shell(LP)
Med	**N:** **+ Hospital**
Other	**N:** **State Patrol Post**
24A	Los Coches Road, Lakeside
Gas	**N:** Mobil(LP) (AServ)
	S: Shell* (Blimpie)
Food	**S:** Blimpie's Subs (Shell), McDonalds (Wal-Mart), McDonalds, Subway, Taco Bell
AServ	**N:** Mobil
	S: Tire & Lube Express (Wal-Mart)
Other	**S:** **Vons Grocery, Wal-Mart (Tire & Lube Express)**
24B	Lake Jennings Park Road
Gas	**S:** Citgo*
Food	**S:** Jilberto's Taco Shop, Marechiaro's (Italian)
26	Map Stop - Phones (Westbound)
28	Harbison Canyon, Dunbar Lane
31	Alpine, Tavern Road
FStop	**N:** **Texaco*(CW)(LP) (AServ)**
Gas	**N:** Alpine Express*(D)
	S: Circle K*, Shell (AServ)
Food	**S:** Alpine Frontier Deli, Carl's Jr Hamburgers, Glacier's Frozen Yogurt, Little Caesars Pizza, Mananas, Steph's Donut Shop
Lodg	**S:** **Countryside Inn**
AServ	**S:** Alpine Moutain Garage, Shell
ATM	**S:** Great Western Bank
34	Alpine Blvd, Willows Road
Other	**S:** **Ranger Station**
37	East Willows Road
40	CA 79, Julian, Japatul Road
46	Pine Valley
47	Sunrise Highway
(51)	**Rest Area - RR, Phones, Picnic (Both Directions)**
51	Buckman Springs Road
Other	**S:** **Rest Area (Picnic, RR, Phones, RV Dump)**
53	Cameron Station, Kitchen Creek Road
61	Crestwood Road, Live Oak Springs
Other	**S:** **Tourist Information**
66	CA 94, Campo
Lodg	**S:** Buena Vista
70	Jacumba
Gas	**S:** Shell*(LP), **Texaco*(LP)**
(77)	Truck Brake Inspection Area (Eastbound)
78	In - Ko - Pah Park Road
81	Mountain Springs Road
88	CA 98, Calexico
90	Ocotillo Imperial Highway
FStop	**N:** **OTU Fuel Mart***
	S: **Desert FStop***

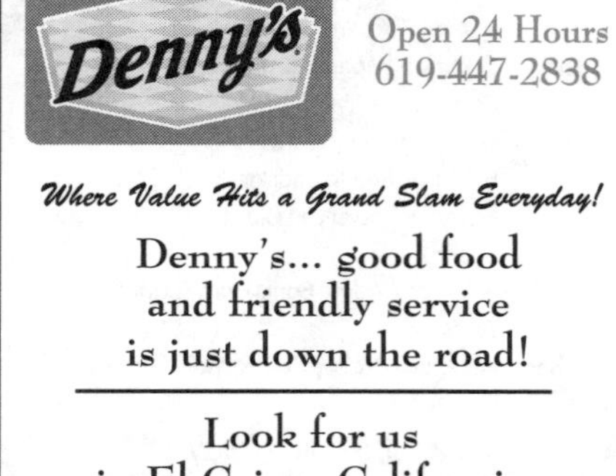

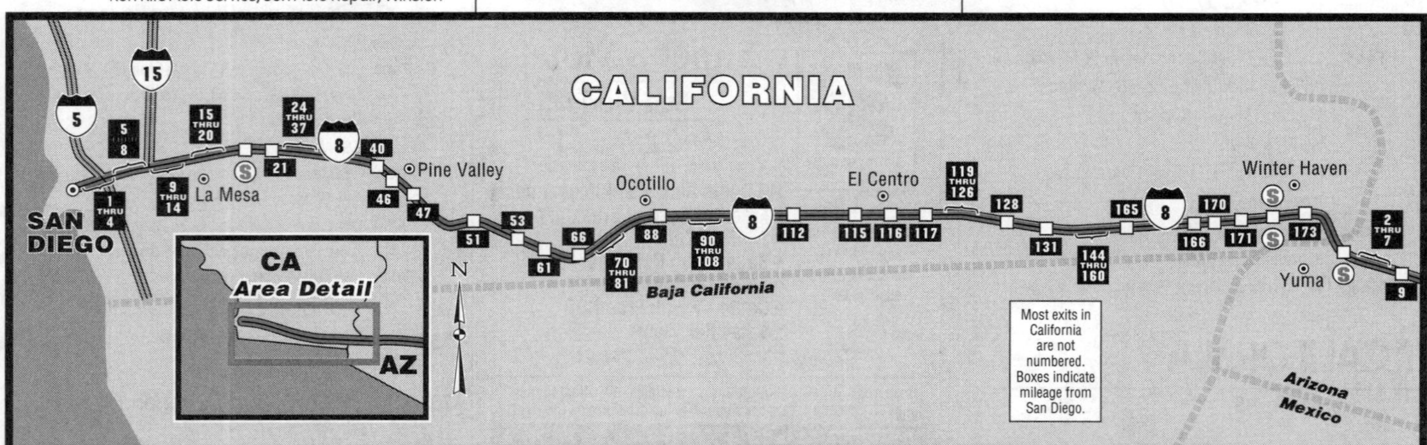

Bold red print shows RV & Bus parking available or nearby

EXIT	CALIFORNIA
Food	N: Lazy Lizard Saloon
	S: **Desert Kitchen**
Lodg	N: **Ocotillo Motel**
RVCamp	N: **Ocotillo RV Park (Laundry Facility)**
Other	N: **US Post Office**
101	Dunaway Road
108	Drew Road, Seeley
RVCamp	N: **Sunbeam Lake**
	S: **Rio Bend**
(109)	**Rest Area - RR, Picnic, RV Dump, Phones (Both Directions)**
112	Forrester Road
115	Imperial Ave, El Centro
Gas	N: Chevron(CW) (AServ), Citgo*(D) (ATM)
Food	N: Del Taco, Denny's (Ramada Inn), KFC, **Vacation Inn**
Lodg	N: Ramada Inn, Vacation Inn
AServ	N: Chevron
RVCamp	N: **Vacation Inn (Restaurant)**
ATM	N: Citgo
116	CA 86, 4th St, El Centro
TStop	S: **Mobil*(SCALES) (Millie's Kitchen, ATM)**
Gas	N: Arco*, Chevron* (AServ), Citgo*(D) (ATM), Texaco*(D)
	S: Mobil*
Food	N: Carl's Jr Hamburgers, Jack-In-The-Box, La Fonda, McDonalds, Rally's Hamburgers
	S: **Millie's Kitchen (Mobile)**, Taco Bell
Lodg	N: **Motel 6**
	S: **Best Western, E-Z 8**
AServ	N: Chief Auto Parts, Guerrereo's Shop
	S: Gene's Auto Service, Thomas Motor Company
RVCamp	S: **Desert Trails**
ATM	N: Citgo
	S: Mobil
117	Dogwood Road
119	CA 111, Indico Calexico
TStop	N: **Texaco*(LP)(SCALES) (Express Lube)**
TServ	N: **Express Lube (Texaco TStop)**
RVCamp	N: **Country Boys**
120	Bowker Road
126	Holtville, Orchard Road
128	Bonds Corner Road
131	CA 115, Holtsville
144	CA 98 Calexico
147	Brock Research Center Road
151	Gordons Well

EXIT	CALIFORNIA/ARIZONA
(152)	**Rest Area - Parking Only (Both Directions)**
153	Grays Well Road
160	Ogilby Road, Blythe
164	Sidewinder Road
Gas	S: Shell*(LP) (RV Dump)
Other	N: **California Highway Patrol Post**
165	California Agriculture Inspection & Weigh Station (Both Directions, Weigh Station Westbound Only)
166	Algodones Road, Andrade
170	Winterhaven Drive
RVCamp	S: **Rivers Edge (RV Resort)**
171	Winterhaven 4th Ave.
Gas	S: Circle K*
Food	S: Jack-In-The-Box, Taco Mi Rancho, Yuma Landing
Lodg	S: Best Western, Regal Lodge, Yuma Inn Motel

↑CALIFORNIA

↓ARIZONA

EXIT	ARIZONA
1	Giss Pkwy., Yuma
Food	S: California Bakery
Lodg	S: Lee Hotel
Other	S: **US Post Office**
(1)	**Weigh Station (Both Directions)**
2	US 95, 16th St, San Luis
Gas	N: 76*
	S: Texaco*(D)

INTERSTATE EXIT AUTHORITY

EXIT	ARIZONA
Food	N: **Cracker Barrel**, Denny's
	S: Burger King (ATM), Jonny's Sports Bar and Grill, McDonalds, Shoney's
Lodg	N: Best Western, **Days Inn, La Fuente Inn, Motel 6**, Shilo Inn Hotel
	S: Comfort Inn, Motel 6, Super 8 Motel
AServ	S: Apple's Garage
ATM	S: Burger King, Marine Air Federal Credit Union
Other	S: **Gila Veterinarian Clinic**
3	AZ 280, Ave 3E, Yuma International Airport
FStop	N: **Texaco*(SCALES)**
Food	N: **Dairy Queen**
TServ	S: **Diesel Componets and Fuel Injection**
TWash	N: **Texaco**
7	Araby Road
TServ	S: **Yuma Diesel Service**
RVCamp	S: **RV Merchant Sales and Service**
9	Business Loop 8
12	Fortuna Road
TStop	N: **Barney's Truck Stop(SCALES) (Hot Stuff Pizza, Smash Hit Subs)**
Gas	N: Chevron* (ATM)
	S: Texaco*(LP) (Hardee's)
Food	N: **Copper Miner's (Barney's TStop), Hot Stuff Pizza (Barney's TStop)**, Jack-in-the-Box, Pizza Hut, **Smash Hit Subs (Barney's TStop)**
	S: Candy Land, Don Quijotoe, Hardee's (Texaco)
Lodg	N: **Caravan Oasis(LP)**
TServ	N: **Goodyear Tire & Auto (Barney's TStop)**
RVCamp	S: **Al's RV**
ATM	N: Chevron
	S: Bank One
Other	S: **Coin Laundry, Eckerd Drugs**
14	Foothills Blvd
Food	S: Domino's Pizza, Foothills Restaurant (Closed Every Summer)
AServ	S: Foothill Auto Parts
RVCamp	S: **S & H RV Parts and Service**
ATM	S: National Bank Of Arizona
Other	N: **Sundance RV Park**
(17)	Inspection Station (Eastbound)
21	Dome Valley
(22)	**Parking Area**
30	Ave 29 East, Wellton
37	Ave 36 East, Roll
42	Ave 40E, Tacna
Gas	N: Chevron*
Food	N: Alamo Saloon, Basque Etchea Restaurant,

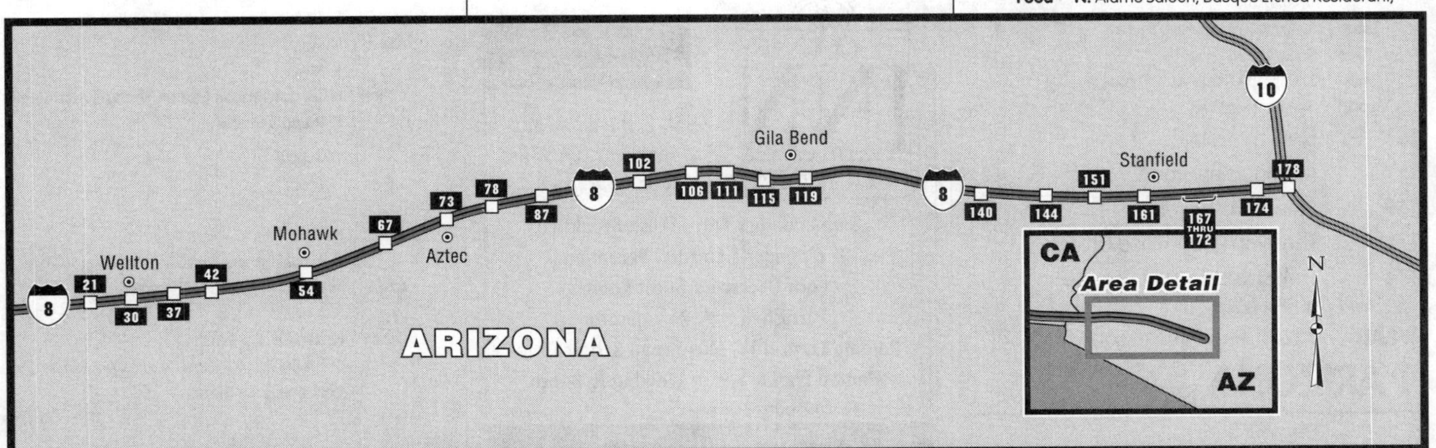

Bold red print shows RV & Bus parking available or nearby

EXIT		ARIZONA
		Patio Cafe
	Lodg	N: Chapparal Motel
	AServ	N: 2 Bay Hand Car Wash, Tipton Auto Parts
	TServ	N: Pilo Verde (Scale Co.)
	Other	N: True Value Hardware(LP), U.S. Post Office
54		Ave. 52 E, Mount Mohawk Valley
(56)		Rest Area - Phones, RR, Vending, Picnic, HF P (Both Directions)
67		Dateland
	Gas	S: Exxon*(D)(LP) (AServ)
	Food	S: Dateland Cafe* (RV Dump, ATM)
	AServ	S: Exxon(LP)
73		Aztec
78		Spot Road
(84)		Rest Area - RR, Vending, Picnic P (Eastbound)
(85)		Rest Area - Vending, RR, Picnic P (Westbound)
87		Sentinel, Hyder, Agua Caliente
	Gas	N: No Name*
102		Painted Rock Road
106		Paloma Road
111		Citrus Valley Road
115		Business 8, AZ 85, Gila Bend, Phoenix
	Gas	N: Circle K*, Exxon*, Gas For Less*(D)(LP), Texaco* (ATM)
	Food	N: A & W Drive-In, Burger King, El Taco Kid, McDonalds
	Lodg	N: Desert Gem
	AServ	N: Bill Henry's Auto Supply, Goodyear Tire & Auto
	Med	N: + Gila Bend Primary Care
	ATM	N: Texaco
	Other	N: Wheel Inn (RV Camp)
119		Business 8 West, Gila Bend
	TStop	N: Texaco(SCALES) (RV Dump)
	Food	N: Exit West Cafe
	Lodg	N: Motel 8
(123)		Rest Area - Picnic P (Both Directions)
140		Freeman Road
144		Vekol Road
(148)		Rest Area - Picnic P (Both Directions)
151		AZ 84 East, Maricopa Road, Stanfield
	TStop	S: Truck 19 Stop
	Gas	S: Gas N Go*
	Food	S: Pullman Restaurant (Truck 19)
	Lodg	S: Pullman Motel
161		Stanfield Road
167		Montgomery Road
169		Bianco Road
172		Thornton Road, Casa Grande
174		Trekell Road, Casa Grande
	Med	N: + Hospital
178AB		Jct I-10, Phoenix, Tucson

↑ ARIZONA

Begin I-8

EXIT		CALIFORNIA
		Begin I-10

↓ CALIFORNIA

EXIT		CALIFORNIA
1		Lincoln Blvd, CA1S
	Gas	S: 76, Chevron*, Shell(CW)
	Food	N: Denny's, Norm's
		S: Big Bowl Express Chinese, Donut King, Jack-In-The-Box, Subway, Taco Bell
	AServ	N: Lincoln Auto Center, Tune Up Masters
		S: Firestone Tire & Auto
	Lodg	N: Country Inn (see our ad this page)
1A		4th St, 5th St
	Gas	N: 76
		S: Shell
	Food	N: Denny's, Norm's Burgers
		S: Subway, Tommy's Burgers
	AServ	S: Lincoln Auto Repair, Shell
	Other	N: Car Wash
		S: JP's Market
1B		20th ST. (Difficult Eastbound Reaccess)
	Gas	S: 76
	Food	N: Big Joe's
		S: Burger King, Campo's Mexican
	AServ	N: Bug Alley
1C		Cloverfield Blvd., 26th St (Westbound)
	Gas	N: Shell
	AServ	N: Barry Slawter, Shell
	Med	N: + Hospital
2		Centinela Ave
2A		Bundy Dr. (Westbound)
	Other	W: 9-11 Food Mart, Coin Laundry
3		Junction I-405
4		Overland Ave, National Blvd., Palms & Rancho Park
	Gas	S: Arco*, Mobil*(D)
	ATM	N: Citi Bank
5		National Blvd
6		Roberson Blvd, Culver City
	Food	S: Adam's Restaurant, Del Taco, Golden China Restaurant, Grandma's Lucia's Pizza, Noah's Bagels, Starbucks Coffee, Tom's #5 Chili Burgers
	Lodg	S: Golden Gate Lodge
	AServ	S: Fox Hills Buick Dealership, Mike Miller Toyota Dealership
	Med	S: + Hospital
	Other	S: Save-On Drugs

EXIT		CALIFORNIA
7		La Cienega Blvd, Culver City
	Gas	N: Chevron, Mobil*
		S: Arco
	Food	N: Pantry's Pizza, Yum Yum Donuts
		S: Subway, Venice Grill
	AServ	N: Firestone Tire & Auto
		S: Davis Brothers Tires, Delta Battery, La Cienaga Auto Center
	Med	N: + Hospital
	Other	S: Coin Laundry
7B		Fairfax
8		La Brea Ave, Englewood
	Gas	N: Gas 4 You, Shell, Shell
		S: Chevron*
	AServ	N: G&M Transmission
		S: Chief Auto Parts
	Other	N: Coin Laundry
9		Crenshaw Blvd, Englewood
	Gas	S: Chevron*, Thrifty
	Food	S: El Pollo Loco, Leo's Barbeque, McDonalds (Outdoor Playground), Tasty BBQ
	AServ	S: Mobil
	ATM	S: Bank of America
	Other	S: Ace Hardware
10		Arlington Ave
	Gas	E: Mobil*
		N: Chevron
	AServ	E: Jeff's Auto & Lube
		N: Chevron
11A		Western Ave
	Gas	N: Chevron
		S: Chevron
	Food	N: Chabelita Ice Cream, Chabelita Tacos, Daily Donuts, Kay Kay Chinese, McDonalds (Outdoor Playground), The Seafood Outlet, Two For One Pizza, Winchell's Donuts
	AServ	N: Best Transmission, Chevron, Chief Auto Parts, Modern Auto Center, Modern Auto Specialist
		S: Auto Haus, Chevron, King's Transmission, New Gold Automotive
	Med	S: + Hospital
	Other	N: Coin Laundry, Food For Less, Sav On Drugs
		S: Coin Laundry, Triangle Market
11B		Normandie Ave
	Gas	S: Chevron(CW), Shell*, Texaco*
	AServ	S: Chevron(CW), Texaco
11C		Vermont Ave
	Gas	N: Arco
		S: Mobil, Shell*
	Food	N: Burger King
		S: El Unico Tacos, Jack-In-The-Box
	AServ	N: Arco
		S: Mobil
	Other	N: LA Convention Center, Thrifty Supermarket
		S: Ralph's Grocery
12		Hoover St
	Gas	N: Mobil
		S: Arco
	Food	N: King Donuts, Lucy's Pastrami
		S: Pete's Burgers, Vera Cruz Mexican
	AServ	N: Mobil, Pep Boys Auto Center
		S: Arco
	Other	N: Mid City Car Wash
		S: Full Service Car Wash, Jerry's Market, Mana Christian Book Store

Bold red print shows RV & Bus parking available or nearby

EXIT CALIFORNIA

13B Maple Ave.
Food N: Mr Steve's Donuts
S: El Pollo Loco
AServ S: Maple Auto Center, Napa Auto Parts, Super Trak
Other S: **99 Cents Only Grocery (Pharmacy), Launderland Coin Laundry, Thrifty Grocery Store**

13 Jct. I-110 S. San Pedro, CA 110N Pasadena
Lodg: N: Oxnard Hilton (see our ad this page), Casa Sirena (see our ad this page),

13A Central Ave.
Gas N: Shell
S: Shell(CW)
Food S: King Taco, L.A. Hamburgers
AServ N: Tune Up Masters

15 Santa Fe Ave.
Gas S: Arco, Chevron, Shell
Food S: Big Bear Convenient Store and Fast Food
AServ S: Sunland Tires

16 Alameda
TStop S: **Alameda Petroleum Truck Stop***(D)(LP)(SCALES)
Gas N: 76(LP), Fuel Stop
S: Mobil(CW)
Food N: TV Cafe
S: Georgia's Hamburgers
AServ S: Berlin Tire Center
TServ S: **J&J Truck and Bus Repair**
TWash S: **Alameda Truck Wash**
Med S: **+ Alameda Medical Group**
Other N: **LA Wholesale Produce Market**
S: **Ryder Truck Rental**

16B Santa Fe
Gas S: Shell
Food S: Mini Market Fast Food(LP), Tacos D.F.
AServ S: Shell, Sunland Tire
Other S: **Kelco's Oil & Propane**

16A 4th St.
Gas S: Arco, Shell
Food S: Olympic Donuts, Tom's Burgers

17 Jct I-5, CA 60E to Pamona , Santa Anna

17A Boyle Ave. (Difficult Eastbound Reaccess)
Food S: Chinese Fast Food, Domino's Pizza, Homestyle Donuts
Other S: **Coin Laundry**

18 Soto St, State St.

EXIT CALIFORNIA

Gas S: Mobil, Shell
Food S: Orozcos Tacos
AServ S: Shell
Other S: **Clean King Laundry**

19 City Terrace Dr

19A Eastern Ave.
Gas S: Chevron, Mobil
Food S: McDonalds

20 Jct I-710, Long Beach, Valley Blvd

21 Fremont Ave
Food S: Kings Donuts

21A Atlantic Blvd, Monterey Park
Gas N: 76, Mobil
Food N: Del Taco, Pizza Hut, Popeye's
S: Ocean Star Restaurant

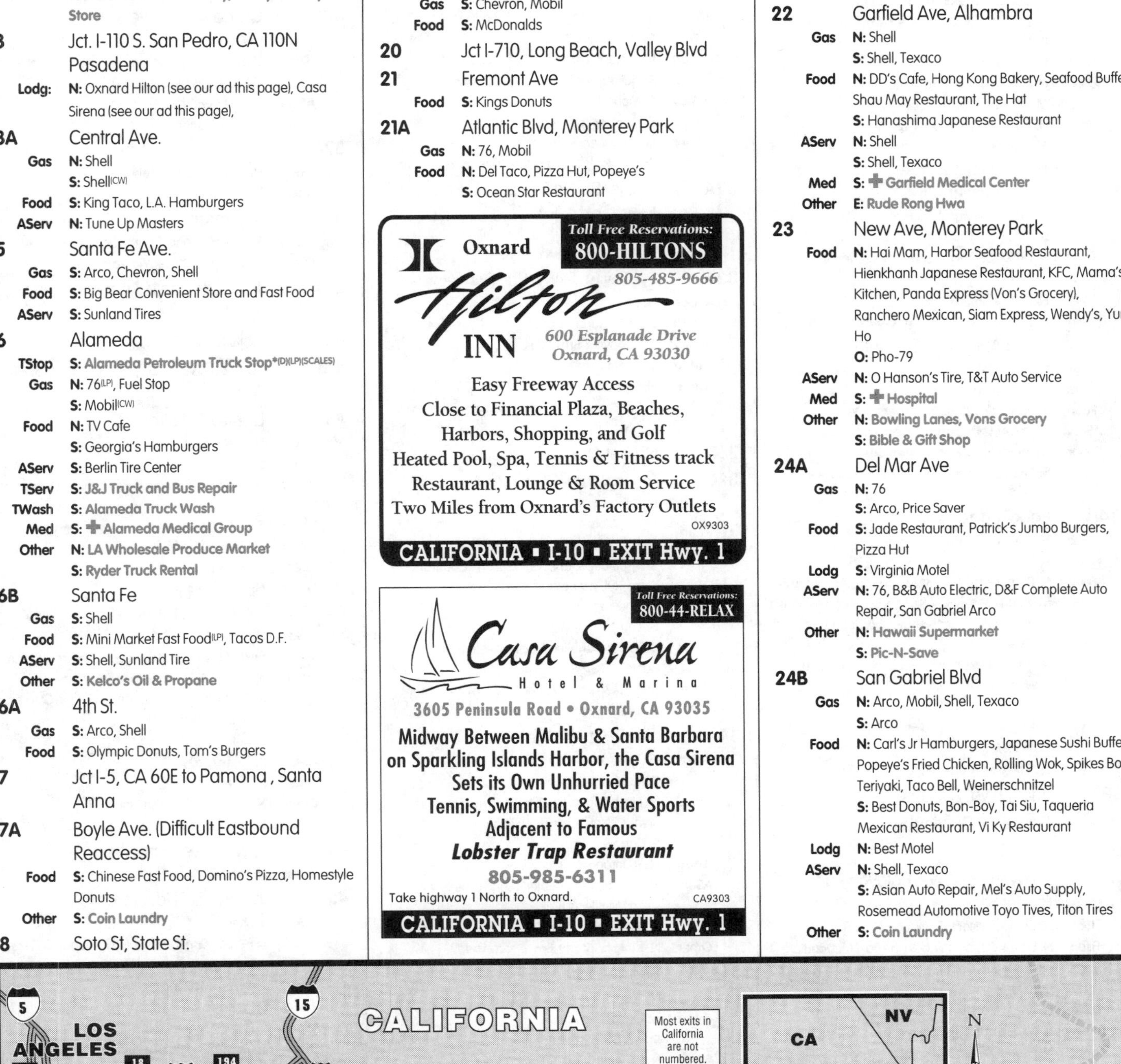

EXIT CALIFORNIA

Lodg S: Monterey Park Inn
AServ S: Econolube & Tune, Firestone Tire & Auto
Med N: **+ Hospital**
ATM S: Bank of America, First Continental Bank, San Wa Bank
Other S: **Hughes Family Market**, Monterey Car Wash, **Shun Fat**

22 Garfield Ave, Alhambra
Gas N: Shell
S: Shell, Texaco
Food N: DD's Cafe, Hong Kong Bakery, Seafood Buffet, Shau May Restaurant, The Hat
S: Hanashima Japanese Restaurant
AServ N: Shell
S: Shell, Texaco
Med S: **+ Garfield Medical Center**
Other E: **Rude Rong Hwa**

23 New Ave, Monterey Park
Food N: Hai Mam, Harbor Seafood Restaurant, Hienkhanh Japanese Restaurant, KFC, Mama's Kitchen, Panda Express (Von's Grocery), Ranchero Mexican, Siam Express, Wendy's, Yung Ho
O: Pho-79
AServ N: O Hanson's Tire, T&T Auto Service
Med S: **+ Hospital**
Other N: **Bowling Lanes, Vons Grocery**
S: **Bible & Gift Shop**

24A Del Mar Ave
Gas N: 76
S: Arco, Price Saver
Food S: Jade Restaurant, Patrick's Jumbo Burgers, Pizza Hut
Lodg S: Virginia Motel
AServ N: 76, B&B Auto Electric, D&F Complete Auto Repair, San Gabriel Arco
Other N: **Hawaii Supermarket**
S: **Pic-N-Save**

24B San Gabriel Blvd
Gas N: Arco, Mobil, Shell, Texaco
S: Arco
Food N: Carl's Jr Hamburgers, Japanese Sushi Buffet, Popeye's Fried Chicken, Rolling Wok, Spikes Bowl Teriyaki, Taco Bell, Weinerschnitzel
S: Best Donuts, Bon-Boy, Tai Siu, Taqueria Mexican Restaurant, Vi Ky Restaurant
Lodg N: Best Motel
AServ N: Shell, Texaco
S: Asian Auto Repair, Mel's Auto Supply, Rosemead Automotive Toyo Tives, Titon Tires
Other S: **Coin Laundry**

CALIFORNIA
Most exits in California are not numbered. Boxes indicate mileage from San Diego.
Los Angeles · Santa Monica · Long Beach · San Bernardino · Riverside · Palm Springs · Desert Center · Blythe · Quartzsite · Arizona
Area Detail: CA · NV · AZ

Bold red print shows RV & Bus parking available or nearby

EXIT		CALIFORNIA
25A	Walnut Grove Ave	
	Food	S: Cafe Benh Minh
	Other	S: Coin Laundry, Tony's Mini Market
25B	CA19, Temple City, Rosemead Blvd, Pasadena	
	Gas	N: Mobil
	Food	N: China Way, Denny's, Love's The Great Rib Restaurant, McDonalds, Subway, Winsel's Donuts
	Lodg	N: Vagabond Inn
	AServ	N: Econo Lube and Tune, Goodyear Tire & Auto, Montgomery Ward, Super Trak
	Other	N: Country Village Car Wash, Kids R Us, Michael's Craft Shop, Radio Shack, Rosemead Square Shopping Center
26	Baldwin Avenue	
	Gas	N: Arco, USA Gasoline
		S: Arco
	Food	N: Donuts, El Taco Riendo, Thai Food
		S: Denny's
	Lodg	N: Eunice Plaza Motel
	Other	N: Baldwin Car Wash, Coin Laundry
27	Santa Anita Ave, El Monte	
	Gas	N: Texaco*(D)
		S: 76, Ultra Mar*
	Food	S: Little Caesars Pizza, Universal Donuts
	AServ	N: GM Auto Dealership, Geo Dealership
	Other	N: Max's Hardware, Metro Bus Terminal
		S: Casa Camacho Grocery, Emergency Pet Clinic
28A	Peck Road S	
	Gas	N: Chevron, Mobil(CW)
		S: 76, Shell
	Food	N: Burger King, Denny's, Gardunos American & Mexican Food, KFC, Tommy's Burgers, Yoshinoya
		S: McDonalds, Taco Ready, USA Donuts
	Lodg	N: Motel 6
	AServ	N: Auto Clinic, Chevron, Goodyear Tire & Auto
		S: Winston Tires
	ATM	N: Wells Fargo Bank
	Other	N: EZ Car Wash, Radio Shack, Star Car Wash, Thrifty Drugs
		S: ETD Food Mart
28B	Valley Blvd, El Monte	
29	Garvey St, Duirfield St	
30	Jct I-605	
31A	Baldwin Park Ave, Baldwin Park, Frazier St	
	Gas	N: Chevron, Mobil, Shell, USA Gasoline
	Food	N: Baskin Robbins, Jack in the Box, McDonalds, Rosa Italian, Taco Bell, Taco Ready, Yum Yum Donuts
	Lodg	N: Angel Motel
	AServ	N: Mobil, Transmission Specialist
	Other	N: Food For Less, Office Max, Target Department Store
		S: UPS Drop Box
31B	Francisquito Ave, La Puente	
	Gas	N: V&G Gasoline
	Food	N: Hong Kong Restaurant, In & Out Burger, Taco Ready
		S: La Fogata Mexican, Paqueria Los Amigos, Wienerschnitzel
	Lodg	S: Travelodge
	AServ	N: V&G Gasoline
		S: Chief Auto Parts
	Other	N: Shopping Mall
		S: Francisquito Laundry Mat
32	Puente Ave	
	Gas	S: 76, Mobil, Texaco
	Food	N: China Palace, Denny's, Guadalajara Grill, Happy Wok, McDonalds (Outdoor Playground), Milano Pizza, Taco Las Palmas
		S: Carl's Jr Hamburgers, Papa Joe's Pizza, The Green Burrito
	Lodg	N: Motel 6, Queen Lodge Motel, Radisson
		S: Holiday Lodge
	AServ	S: Mobil, Texaco
	Med	S: Hospital
	Other	N: Car Wash, Staples
		S: Tomson's Grocery, U-Haul Center(LP)
32A	Frazier St. (No Eastbound Reaccess)	
	Gas	N: Arco, Shell
	Lodg	N: Crazy 8
	Other	N: 7-11 Convenience Store
33A	W. Covina Pkwy, Pacific Ave	
	Gas	N: 76
	Food	S: La Passada
	AServ	S: Goodyear Tire & Auto
	Med	S: Hospital
	Other	N: Coin Laundry
		S: Dry Cleaners, JC Penny, Sears, Shopping Center
33B	Sunset Ave, W.Covina	
33C	Vincent Ave, Glendora Ave.	
	Gas	N: Chevron
		S: 76*, Mobil
	Food	N: Hungry Al's
		S: Applebee's, Chevys Mexican Restaurant, Pizza Hut, Wienerschnitzel
	AServ	N: Chevron
		S: Comp Auto, Mobil
	ATM	S: Bank of America, Pacific Western National Bank
	Other	S: Barnes & Noble, Big & Tall Mens Clothing, Coin Car Wash
34	CA39, Azusa Ave	
	Gas	N: 76, Arco*
		S: Mobil
	Food	N: Black Angus, La Zona Rosa, McDonalds, Steak Corral, Taqueria La Fogata
		S: Carrows
	Lodg	N: El Dorado
	AServ	N: Chrysler Auto Dealer, Econo Lube & Tune, Midas Muffler & Brakes
		S: Mobil, Nissan Dealership, Toyota Dealership
	Other	N: Kinko's, Stater Bro. Grocery
36A	Barranca St.	
	Gas	N: 76
	Food	N: Chili's, The Safari
		S: In & Out Burgers
	AServ	N: Full Auto Service
	Other	N: Eastland Mall, Office Depot, Ross, Toys "R" Us
36	Citrus St	
	Gas	N: Chevron, Shell
		S: 76
	Food	N: Burger King (Indoor Playground), IHOP, T.G.I. Friday's
		S: Casa Jimenez, Jackie's Cafe
	AServ	N: Chevron, Citrus Brake & Automotive, Lincoln Mercury Dealership, Shell, Trojan Tire, Ultimate Auto Repair
		S: 76
	Med	N: Hospital
	ATM	N: Coast Federal Bank, San Wa Bank
	Other	N: Lolman's, Lucky Food Center, Marshall's, Mervyn's, Old Navy, Ross, Target Department Store
		S: Trader Joe's
36B	Grand Ave., Olive St.	
	Gas	N: Chevron, Shell
	Food	N: Emporer Mongolian BBQ, House of Louie Chinese Restaurant, Magic Recipe
	AServ	N: Chevron, Shell
	Other	N: Mail Station
37	Holt Ave, Barracan St	
	Food	N: Blakes, Sizzling Kabob
	Lodg	N: Embassy Suites
39	Via Verde	
	Food	S: Arties Pizzaria, Donuts Galore
	Other	S: U.S. Post Office, Via Verde Plaza
39A	JCT I-210, Santa Anna, San Dinas, CA 71	
41A	Kellogg Dr	
41B	Jct I-210W, CA57, Pasadena, Santa Anna	
43	Dudley St, Fairplex Dr	
	Gas	N: 76(LP), Texaco(LP)
		S: Mobil*
	Food	N: Denny's, Pamona Valley Mining, Quinta Real Restaurant
		S: McDonalds
	Lodg	N: Lemon Tree Mt
	AServ	S: Mobil
	Other	N: Bonnilli Park
44	Garey Ave, Pamona	
	Gas	N: Thrifty(D)
		S: 76, Advanced Auto Repair, Chevron, Exxon
	Food	N: Donuts & Burgers, Mandarin Express
		S: Bravo Burgers, Donahoo's Golden Chicken
	Lodg	N: Sheriton Suites
		S: Auto Lodge
	AServ	S: 76, Texaco
	Med	N: Hospital
	Other	N: Ganesha County Park
		S: Coin Car Wash, Stater Bros. Grocery
45	Towne Ave, Pamona	
	Food	N: The Jelly Donut
46	Indian Hill Blvd, Claremont	
	Gas	S: Shell
	Food	S: Burger King (Outdoor Playground), Carl's Jr Hamburgers, In-N-Out Hamburgers, Juanita's Mexican, McDonalds
	Lodg	S: Ramada Inn
	AServ	S: Parnelli Jones Tires
	Other	E: Greyhound Bus Terminal
		S: Claremont Auto Wash, Radio Shack, Vons Grocery
48A	Monte Vista Ave	
	Gas	N: Texaco
	Food	N: Black Angus Steakhouse, Olive Garden, Toni Roma's Ribs
	Med	S: Hospital
48B	Montclare, Chino	
	Food	N: El Pollo Loco, McDonalds, Theo's Cafe Greek & American
		S: All You Can Eat Pizza, Jack-In-The-Box, Jonny's

Bold red print shows RV & Bus parking available or nearby

Donuts, Long John Silvers, Wienerschnitzel
AServ **N:** America's Tire Company, Firestone Tire & Auto, Hi-Lo Auto Supply
Med **S:** ✚ **Hospital**
ATM **N:** American Savings Bank
Other **N:** **Kids R Us, Office Depot**

49 Mountain Ave, Mount Baldy
Gas **N:** Arco, Mobil, Texaco*
S: 76
Food **N:** China Gate Restaurant, Denny's, El Torito, Happy Wok, Honey Baked Ham, Mi Taco, San Biagio Pizza, Starbucks Coffee, Subway
S: Carl's Jr Hamburgers, Celia's El Loco
Lodg **N:** Comfort Inn
ATM **N:** Wells Fargo Bank
Other **N:** **BBQ Galore BBQ Unit Sales, Big & Tall Mens Clothing, Long's Drugstore, Mervyn's, Miller's Outpost Western Wear, Staples Office Superstore, Trader Joe's Grocery Store**
S: **Food For Less Grocery**

50 CA83, Euclid Ave, to Upland, to Ontario (Am Track Metro Link)
Gas **S:** Arco, Chevron, Mobil
Food **S:** IHOP
AServ **S:** Arco, Mobil
Med **S:** ✚ **Hospital**
Other **S:** **Ace Hardware, Overdale Park (5 Blocks)**

52 4th St
Gas **N:** Arco*, Chevron, Shell
S: 76, Arco*, Fast Fuel, Texaco*
Food **N:** Burger King (Indoor Playground), Carl's Jr Hamburgers, Domino's Pizza, Fine Mexican Food, Jack-In-The-Box, Star Donut
S: Denny's, Gordos Mexican Restaurant, KFC
Lodg **N:** Motel 6, Quality Inn, Red Carpet Motel
S: DAYS INN Days Inn, West Coast Inn
Other **N:** **K-Mart**
S: **Jax Market**

53 Vineyard Ave, Ontario Airport
Gas **S:** Arco, Mobil, Texaco
Food **S:** Bombay Bicycle Club, In-N-Out Hamburgers, Marie Callender's Restaurant, Michael J's Restaurant, Rosa's Fine Italian Restaurant, Spires
Lodg **S:** Best Western, Country Suites, Countryside Suites, Double Tree Hotel, Express Inn, Good Nite Inn, Marriott, Super 8 Motel (see our ad this page)
AServ **S:** Mobil, Texaco

54 Archibald Ave
Food **N:** Burger Town USA, Joey's Pizza, Teriyaki Champ

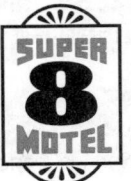

EXIT CALIFORNIA

S: Marie Callender's Restaurant
AServ **S:** GM Auto Dealership

55 Haven Ave
Gas **N:** Mobil*(CW)
Food **N:** Black Angus, El Torito, Plaza Cafe, Spoons Ribs Burgers & Tacos, Tony Roma's Ribs
S: Chevy's Restaurant, T.G.I. Friday's
Lodg **N:** Extended Stay America, Hilton
S: Fairfield Inn, Panda Inn
Other **N:** **Ontario Mills Mall, Plaza Continental Factory Stores**

56 Milliken Ave
TStop **S:** **76 Auto/Truck Plaza*(LP)(SCALES), TravelCenters of America*(SCALES)**
Food **N:** Coco's, Wolfgang Puck Cafe
S: **Burger King (TravelCenters of America)**, Farmer's Boy Hamburgers, Sbarro Pizza
AServ **S:** Bridgestone Tire & Auto, Parts & Service
TServ **S:** **76 Truck Lube**
Other **N:** **Burlington Coat Factory, Foozles Bookstore, JC Penny, Marshall's, Mi Casa, Ontario Mills Shopping Center**

57 Jct I-15 , Barstow, S to San Diego

58 Etiwanda

60 Cherry Ave
TStop **S:** **Three Sisters Truck Stop*(D)(SCALES)**
Gas **N:** Mobil*
S: Circle K*
Food **N:** Jack-in-the-Box
S: Farmer Boy's Hamburgers
Lodg **N:** **Circle Inn Motel**
AServ **S:** John's Car Care Service, Nelson's
TServ **N:** **Big Rig Truck Sales**
S: **Rush Truck Center, Western States Truck Repair**

62 Citrus Ave
Gas **N:** Ultramar
S: Arco
Food **N:** Burger Delight, Superstar Donuts
AServ **N:** Repo Ranch Car Sales
TServ **N:** **California Car & Truck Wash**
Other **N:** **Brothers Market**

63A Sierra Ave (Many Services Not Listed)
Gas **N:** Arco*, Mobil(CW), Texaco
Food **N:** Applebee's, Arby's, Asaderos Mexican Restaurant, Bakers, Burger King, China Cook Chinese Restaurant, China Panda, Coco's, Country Harvest Buffet Restaurant, Dairy Queen, Denny's, McDonalds, Millie's, Sub Stop, Subway, Taco Bell, Wienerschnitzel, Yoshinoya
Lodg **N:** Comfort Inn, Econolodge
AServ **N:** Pep Boys Auto Center, Texaco
Other **N:** **Food For Less, K-Mart, Kids R Us, Pick & Save Drugs, Toys "R" Us, Vons Grocery**

63B Sierra Ave, Fontana

65 Cedar Ave, Bloomington
Gas **N:** Chevron, Mobil*
S: 7-11 Convenience Store

EXIT CALIFORNIA

Food **N:** El Mezonizquite Mariscos, El Rancho Restaurant, Porky's Pizza, Star Donut
S: El Burrito
AServ **N:** Mobil

67 Riverside Ave, Rialto
TStop **N:** **I-10 Truckstop*(D)**
Gas **N:** Arco*, Chevron
S: Amerigas(LP), Mobil, Poma (Commercial Fueling Network)
Food **N:** Hometown Buffet, **I-10 Restaurant (I-10 TS), Taco-Jo's**
Lodg **N:** Rialto Motel
AServ **N:** Chevron
TServ **N:** **P&S Truck Supply**
TWash **N:** **I-10 TS**
RVCamp **S:** **Sierra RV Repair**

68 Pepper Ave
Gas **N:** Shell*(LP)
Food **N:** Bakers
AServ **N:** LJ Snow Ford, Shell
Med **N:** ✚ **Hospital**
Other **N:** **Movie Land Fronteer Town**

69 Rancho Ave
Gas **N:** Arco*
Food **N:** Antoniou's Pizza, Auction Chinese Food, Del Taco, Diane's Coffee Shop, El Rancho Chinese Mexican Food, Leno's Tacos, Star Donuts, Tom's Hamburgers, Wienerschnitzel

70A 9th St, Downtown
Gas **N:** Mobil(CW)
S: Arco, Valley Colton Fuel(LP)
Food **N:** Burger King, Carrows (Open 24 Hours), Denny's, Jeremiah's Steak House
S: Pepito's
Lodg **N:** Thrift Lodge
S: Rio Inn
AServ **N:** Big A Auto Parts, Colton Radiator, Murray's BF Goodwrench
S: A-1 Auto Glass & Tires, Colton Muffler & Brake
RVCamp **N:** **Alpac RV Service Center**
ATM **N:** Citizens Business Bank
Other **N:** **Stater Bros. Grocery**

70B Sperry Road, Mount Vernon Ave.
TStop **N:** **Colton(SCALES)**
FStop **N:** **Valley Colton*(SCALES)**
Gas **N:** Arco*
Food **N:** Pepito's Tostados and Fries, Pepper Steak
Lodg **N:** **Colony Inn, Howard Johnson**, Rio Inn
AServ **N:** Colton Muffler and Brake, Pro Auto repair
TServ **N:** **Colton Truck Terminal**

71 Jct I-215, San Bernardino, Riverside

72 Waterman Ave, Tippecanoe
Gas **N:** 76*(LP), Shell*
Food **N:** Bobby McGee's, Coco's, Hilton, Lacinta, Stewart Anderson's Black Angus, T.G.I. Friday's, Yamazoto of Japan
S: Carl's Jr Hamburgers, McDonalds, Popeye's Chicken
Lodg **N:** Hilton, Super 8 Motel, Travelodge
S: Motel 6
RVCamp **S:** **Camping World (see our ad this page)**
Other **S:** **Oriental Food Mart**

73 University, Lomalinda, Tippecanoe Ave, Anderson St.
Gas **N:** Arco, Shell
Food **N:** Romberto's Taco Shop
S: BK Subs and Sandwiches, **KFC**, Kool Kactus

EXIT CALIFORNIA

Cafe, Napoli Italian
ASserv N: Arco, Lomalinda Auto Repair

74 Mountain View Ave, Bryn Mawr
Gas N: Mobil*(D), Shell
S: Arco*
Food N: Applebee's, GI Yogi, Jose's Mexican Food, Mr. You Chinese Food
S: Farmer Boy's, Lupe's
RVCamp S: **Camping**
Other N: Car Wash, **Wal-Mart**

75 California St
Gas S: Arco*, Shell*(LP)
Food S: Applebee's, Hosea's Mexican Food, Jack-In-The-Box
Other N: **Pharaoh's Water Park**
S: **Food 4 Less, Wal-Mart**

76A Tennesse St, Alabama St (Services More Than A Mile Away)
Gas S: Texaco*(D)(LP)
Food N: Tom's World Famous Hamburgers
S: Baker's, Burger King, Carl's Jr Hamburgers, El Pollo Loco, Foster's Donuts, Long John Silvers, Marie Callender's Restaurant, McDonalds (Outdoor Playground), Mr. J's Donuts, Taco Bell
Lodg N: Motel 7 West, Redlands Motor Lodge, Super 8 Motel
S: Best Western, Hanson's Motel, Starlite Motel, Sunrise Motel
AServ S: Boyd's Parts & Auto Service, Econolube & Tune, Nissan, Purrfect
Med S: **+ Hospital**
Other S: **Cask-N-Cleaver, Longs Drug Store, Mervyn's, Pick & Save, Tri City Mall**

76B Jct CA30, Highland, Running Springs

78 CA38, 6th St, Downtown, Orange St
Gas N: Chevron
Food N: Antonio's Pizza, Donut n Burger
S: China Town Kitchen, Pizza Stop, Subway, Subway
AServ N: Reid's Auto Service, Salerno's Tire, Super Track, Warehouse Auto Parts
Lodg S: Milner Hotel (see our ad this page)
Other N: **Stater Brothers Grocery**
S: **Albertson's Grocery**

79 University St, Cypress St
Med S: **+ Hospital**

80 4th St, Redlands Blvd, Ford Street
Gas S: 76*
Food S: Dillion's Steak & Seafood, Griswolds Seafood
AServ S: 76

82 Yucaipa Blvd, Oak Glen
Gas N: Arco*, Chevron
Food N: Bakers Hamburgers, La Mexicana
AServ N: Chevron, Deshler's Specialties, European Auto Repair, Gateway Tires and service
Parks N: **Yucaipa Regional Park**

84A Live Oak Canyon Road
Food N: Cedar Hills Steaks, Seafood & Italian

(85) **Wildwood Safety Rest Area - Picnic, RR, Phones, Pet Walk**

84B Calimesa, County Line Rd.
Gas N: Fastrip*, Texaco(LP)
Food N: CNI Coffee, Del Taco, Plaza Cafe
AServ N: Calimesa Tire Center, Dinosaur Tire, House of Quality Auto Parts, Texaco, Troyce's Automotive Shop, Truck and Tire Repair
Other N: **Coast to Coast Hardware, Coin Laundry, Sunrise Foods Health Store**

87 Calimesa Blvd
Gas N: Arco*, Shell*(CW)(LP)

EXIT CALIFORNIA

Food N: Johnnies Roasted Chicken, McDonalds, New York Pizzeria, Sea Kettle Fish and Chips, Taco Bell
Lodg N: Calimesa Inn Motel
ATM N: American Savings Bank, Redlands Federal Bank
Other N: **Starter Brothers Grocery**

88 Singleton Road (Westbound)

89 Cherry Valley Blvd, Desert Lawn Dr.

(91) **Rest Area - RR, Picnic (Westbound)**

92 San Timoteo Canyon Road, 14th Street

93 CA60, Riverside, Beaumont

94A CA79 Beaumont Ave, Hamet
Gas N: 76, Arco*
Food N: Baker's Old Fashioned Ice Cream and Shakes, Denny's, Donald's Burgers, Tacos, El Rancho, Frijoles Mexican, McDonalds (Outdoor Playground), Taco Shop
S: Denny's
Lodg N: Best Western, Budget Host Inn
AServ N: D&S Auto Service & Towing, Dick's Repair
RVCamp S: **Camping**
ATM N: CA State Bank, North County Bank Plus

94B Pennsylvania Ave

95 Highland Springs Ave, Banning
Gas N: Arco*, Chevron*, Ultramar(LP)
S: Mobil*(CW)
Food N: Burger King (Indoor Playground), Denny's, Espresso Donuts, Farmhouse Restaurant, Pizza-N-Pasta Lovers, Ramsey Burger, Subway
S: Carl's Jr Hamburgers
AServ N: Highland Springs Express Lube & Car Wash
S: Mobil
ATM S: Wells Fargo Bank
Other N: **CA Highway Patrol, Food For Less, Radio Shack, Stater Brothers Grocery**
S: **Albertson's Grocery**

97 Sunset Ave
Gas N: Chevron
Food N: Billy T's Restaurant, Chinese Table Restaurant, Gus Jr's Famous Burgers
AServ N: Auto Zone Auto Parts, Chief Auto Parts, Mountain Air Auto Care
RVCamp N: **Ray's RV Sales & Repairs**
ATM N: Redmonds Federal Bank
Other N: **Sav-U-Foods, Thrifty Drugs**

98 22nd St, Banning
Gas N: 7-11 Convenience Store*, Arco*, Mobil*
Food N: Carl's Jr Hamburgers, Del Taco, Gramma's Kitchen, KFC, Raliderto Mexican, Ramsey Burger, Sizzler Steak House, Taco Bell, Wendy's
Lodg N: Sands Motel, Sunset Motel, Super 8 Motel, Travelodge

EXIT CALIFORNIA

AServ N: Auto Repair Service, Chrysler Auto Dealer, Ford Dealership
ATM N: North County Bank

99 CA243, 8th St, Idlewild
Gas N: Ultra Mar
Food N: Banning Burgers, Banning Donut, Chinese Cafe, Paradise Pizza
Lodg N: Hacienda Inn, Peach Tree Motel
AServ N: Dick's Tire Mart
RVCamp S: **RV Repair**
ATM N: American Savings Bank
Other N: **Coin Laundry**

100 Hargrave St, Idlewild
FStop N: **Shell*(CW), Texaco(LP)**
Gas N: A Z Mini Mart & Gas*, Shell(CW)
Food N: Consuelo's Mexican, San Gorgonia
Lodg N: **Desert Star Motel,** San Gorgonio Inn, Stage Coach Motor Inn
AServ N: Parts Plus Auto Store
Other S: **Airport**

(15) **Weigh Station**

101 Ramsey St (No Westbound Reaccess)

(102) **Weigh Station (Both Directions)**

103 Sports Road

104 Apache Trail
Food N: Hadley's Kitchen

106 Fields Rd.
Parks N: **Morongo Indian Reservation & Casino**
Other N: **Factory Outlet Center**

105 Cabazon, Apache Trail
TStop N: **Truck stop***
Parks S: **Cabazon Park and Communtiy Center**
Other N: **Travel Center**

111 Verbenia Ave
TStop N: **Texaco***
Food N: Cabazon Country Store & Grill
Lodg N: **Wheel Inn**
S: Cabazon Inn
AServ N: Drive Train LTD Tires
Other S: **Cabazon Amusement Area**

112 Palm Springs, Hwy 111 (Service In The Town)

(112) **Rest Area - Picnic, HF, RR, RV Park (Westbound)**

113 White Water

115 CA62, 29 Palms, Yucca Valley, Joshua Tree (Difficult Reaccess)
Parks N: **Joshua Tree National Park**

119 Indian Ave, N Palm Springs
TStop S: **Pilot*(D)(SCALES)**
Gas N: 76*, Arco*(CW), Shell*(LP)
Food N: Johnson's Donuts, Special Chinese Food
S: **Dairy Queen (Pilot)**
AServ N: NAPA Auto Parts
RVCamp N: **Camping**
Med S: **+ Hospital**

122 Palm Dr, Desert Hot Springs, Gene Autry Trail
Lodg S: **Travelodge (see our ad oppsite page), Super 8 (see our ad opposite page)**

125 Date Palm Dr, Catherdral City, Rancho Mirage
Gas S: Mobil*(D)(CW)

129 Thousand Palms, Rancho Mirage, Palm Desert

Bold red print shows RV & Bus parking available or nearby

EXIT	CALIFORNIA
TStop	N: Flying J Travel Plaza[D][SCALES]
Gas	N: Chevron, Mobil, Ultramar
Food	N: Burger King (Playground Inside), Carl's Jr Hamburgers, Del Taco, Denny's, Flying J Travel Plaza, In-N-Out Hamburgers, McDonalds (Playground)
Lodg	N: Travel Inn[LP]
AServ	N: Park House Tire
TServ	N: Little Sisters Truck Wash
Med	S: ✚ Hospital
Other	N: Animal Emergency Clinic, Animal Samaritans
130	Palm Desert, Monterey Ave
Food	S: Bubba Bears Pizza, IHOP, Taco Bell
AServ	S: Price Club Auto Service
131	Cook St
136	Washington St, Country Club Dr, Palm Desert, Bermuda Dunes
Gas	N: Arco*
	S: Unical 76
Food	N: Del Taco
	S: Carl's Jr Hamburgers, Carmen's Mexican, Goody's Cafe, Lilly's Chinese, Pizza & Pasta, Pizza Hut, Subway, Swiss Donut, Teddy's Donuts, Tuttis Pasta
Lodg	N: Motel 6
AServ	S: Certified Auto Service, Goodyear Tire & Auto, Unical 76
138	Jefferson St
Other	S: Airport, Highway Patrol Post
140	Monroe St, Central Indo
Food	S: Alicia's, Carro's Cafe, Jerry's
Lodg	S: Comfort Inn
AServ	S: Anaya's Transmission and Auto Repair
TServ	S: Smitty's Truck Service
RVCamp	N: Camping
Med	S: ✚ Hospital
141	Indio, Jackson St
RVCamp	N: Camping
142	CA111, Palm Desert, Indio Auto Center Dr, CA 86
Lodg	N: Holiday Inn Express
Other	N: Fantasy Springs Casino
143	CA86, CA111S, Dillon Road, Coachella, Indio, Elcentro
Gas	N: Arco
Food	N: Country Kitchen
Lodg	N: Pinta Inn
	S: Super 8 (see our ad this page)
AServ	N: Padilla's Tire Service
RVCamp	N: Camping
(143)	Rest Area - RR, HF, Phones, Picnic, RV Dump (Eastbound)
(144)	Rest Area - RR, HF, Phones, Picnic, RV Dump (Westbound)
166	Frontage Road
168	Mecca, Twenty - Nine Palms
Parks	N: Joshua Tree National Park
170	Chiriaco, Summit
Food	N: Chiriaco Summit Coffee Shop
Lodg	N: Chiriaco Motel
AServ	N: Chavez Truck & Tire Repair
Other	N: General Patton Memorial Museum, U.S. Post Office
175	Hayfield Road
180	Red Cloud Road
187	Eagle Mountain Road

EXIT	CALIFORNIA/ARIZONA
190	CA177, Rice Rd., Desert Center , Parker, Needles
Gas	N: Stanco*
Food	N: Desert Center Cafe
Parks	N: Lake Tamarisk
Other	N: Desert Center Market, US Post Office
199	Corn Springs Road
215	Ford Dry Lake Road
(216)	Rest Area - RR, Phones, HF
220	Wileys Well Road
(224)	Weigh Station (Westbound)
230	Airport, Mesa Dr
TStop	N: 76 Auto/Truck Plaza*[SCALES]
FStop	S: Mesa Verde Truck Stop*[D]
TServ	N: Tire Senter, Lube, RV Service
234	CA78, Novis Blvd, Brawley, Neighbors Blvd.
Gas	N: Texaco*
237	Paulo Baird College, Lovekin Blvd
Gas	N: Mobil*[D], Shell*[CW]
	S: 76, Arco, Chevron*, Mobil
Food	N: Carl's Jr Hamburgers, Del Taco, La Casitados Mexican Food
	S: Blythe Phoenix Pizza, Del Taco, KFC, McDonalds (Playground Outside), Steak House, Subway, Town's Square Cafe
Lodg	N: E-Z Motel, Travelodge
	S: Holiday Inn Express (see our ad this page), Travel Lodge
AServ	N: Mobil, Shell
Med	S: ✚ Hospital
238	Bus10, Sutton St, Blythe, 7th Street
Gas	N: Chevron*
Food	N: Blythe Inn, Dean's Paradise Chinese, Foster's Freeze Hamburgers, Pizza Place
Lodg	N: Astro Motel, Blue Line Motel and Trailer Park, Blythe Inn, Budget Inn, Comfort Inn, Sea Shell Motel
AServ	N: Blythe Ford, Del Rio Chrysler
Other	N: Coin Laundry
239	U.S. 95, Intake Blvd, Needles
Gas	N: Cal West[D][LP], Gas Mart, Shell*, Texaco*
Food	N: Chinese Food, Judy's, Ruperto's Cafe, Steaks n' Cakes
Lodg	N: Best Western, Desert Winds Motel, El Rancho Verde Motel, Flying Inn
AServ	N: Bays II Auto Repair, Mr Engine, Taylor Automotive
TServ	N: Ramsey Truck & Trailer Repair
RVCamp	N: Burton's Mobil Home & RV Park
Parks	N: Mayflower Park
	S: MacIntire Park
241	Inspection Station (Westbound)
243	Riviera Dr.
RVCamp	S: Riveria RV Resort and Marina

↑CALIFORNIA

↓ARIZONA

1	Ehrenburg, Parker
TStop	S: Flying J Travel Plaza*[D][CW][LP][SCALES]
Gas	S: Flying J Travel Plaza
Food	N: Silly Al's Pizza & Burgers
	S: Thads Restaurant (Flying J), Wendy's
Lodg	N: River Lagoon Resort
	S: (AAA) Best Western (Flying J)

EXIT	ARIZONA
TServ	S: Flying J Travel Plaza, Two Way CB Shop
RVCamp	N: Villa Verde RV Park
Other	N: Landromat
(2)	Inspection Station (Eastbound)
(3)	**Weigh Station Scales (Westbound)**
(4)	**Rest Area - RR, HF, Phones, Picnic P (Eastbound)**
(5)	**Rest Area - RR, HF, Phones, Picnic P (Westbound)**
5	Tom Wells Road
TStop	N: Beacon(SCALES)
FStop	N: Beacon*
Food	N: Beacon
TWash	N: Beacon

EXIT	ARIZONA
11	Dome Rock Road
17	U.S. 95, AZ95, Bus10, Quartzsite
Gas	N: Chevron, Mini Mart(LP), Mobil Mart
Food	N: Best Chinese, Burger King, Ted's Bullpen Restaurant
AServ	N: Big A Auto Service, Ed's Complete Auto Repair, **Express Center Tire & Truck Repair**, Richard's Custom Tire
RVCamp	N: **Desert Trails, RV Dump(LP)**
	S: **Buzzard Gardens RV Park(LP), RV Lifestyles Parts and Service**
Other	N: **Library, Police Station, Q Landromat Coin, Showers(LP), Road Runner Market**
18	Unnamed
19	U.S. 95, AZ95, Bus10, Yuma, Parker,

EXIT	ARIZONA
	Quartzsite
FStop	N: **Chevron(SCALES), Texaco**
RVCamp	S: **Clouds Trailer Park, Desert Edge RV Park**
Other	N: **Quartzsite General Store**
26	Gold Nugget Road
31	U.S. 60E, Wickenburg, Prescott (Difficult Reaccess)
45	Vicksburg Road, AZ72W, Vicksburg, Parker
TStop	S: **Tomahawk TStop***
TServ	S: **Jobski's Towing**
Other	S: **Tomahawk Trading Post**
(52)	**Rest Area P (Westbound)**
(53)	**Rest Area - RR P (Eastbound)**
53	Havotter Road
69	Ave 75E
81	Salome Road, Harquahala Valley Rd.
(85)	**Rest Area - RR, Picnic, RV Parking, Truck Parking P (Westbound)**
(86)	**Rest Area - RR, Picnic, RV Parking, Truck Parking P (Eastbound)**
94	411st Ave, Tonopah
FStop	S: **Texaco*(LP)**
Food	S: **Joe & Alice's Restaurant (ATM), Subway (Texaco)**
TServ	S: **Tonopah Joe's & Alice's Tires, Towing**
RVCamp	S: **Saddle Mountain RV Park**
Other	S: **U.S. Post Office**
98	Wintersburg Road
TStop	S: **AmBest Truck Stop***
TServ	S: **Rip Griffin Travel Center**
99	Oglesby Rd.
RVCamp	N: **Camping**
Other	N: **Tourist Information**
103	339th Ave
TStop	S: **Rip Griffin Travel Center(SCALES) (RV Dump)**
Food	S: Pizza Hut, **Restaurant (Rip Griffin)**, Subway
109	Sun Valley Parkway, Palo Verde Road
112	AZ85, Yuma, San Diego
114	Miller Road, Buckeye
TStop	S: **Love's Truck Stop*(D)(SCALES)**
Food	S: **Baskin Robbins (Love's), Dairy Queen (Love's), Taco Bell (Love's)**
Other	S: **Tourist Information (Love's)**

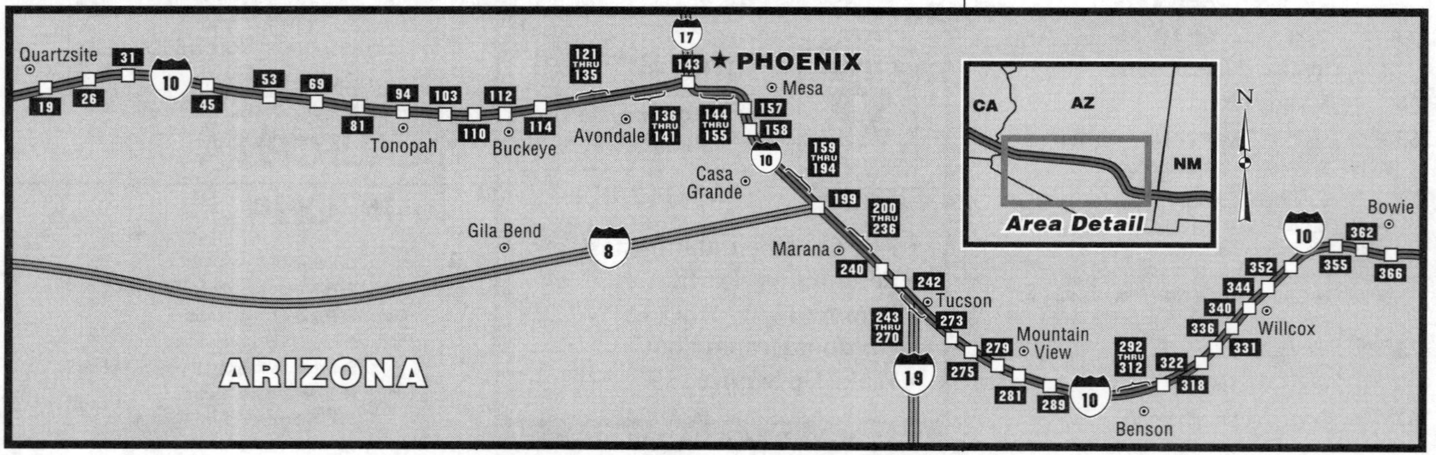

Bold red print shows RV & Bus parking available or nearby

EXIT ARIZONA

121 Jackrabbit Trail
- **Gas** S: Circle K*
- **Food** N: Waddel's Longhorn Corral

124 Cotton Lane, AZ Loop 303
- **RVCamp** S: **KOA Campground**

126 Pebble Creek Pkwy, Estrella Pkwy
- **Parks** S: **Estrella Park**

128 Litchfield Road, Goodyear, Luke Air Force Base
- **Gas** S: Chevron, Mobil
- **Food** N: Burger King, Cracker Barrel, JB's, Taco Bell, Wendy's
  S: Arby's, Ramira's Mexican Food, Schlotzkys Deli
- **Lodg** N: AAA Best Western, Holiday Inn Express
- **AServ** S: Checker Auto Parts, GM Auto Dealer
- **Med** S: **+ Hospital**
- **ATM** S: Bank One Express
- **Parks** N: **Litchfield Parks**
- **Other** N: **Wig Wam Outlet Store**
  S: **Albertson's Grocery, Goodyear Airport, Landromat, Radio Shack**

129 Dysart Road, Avondale
- **Food** S: Jerry's, Mazzone's Plaza, McDonalds (Playground), Waffle House
- **Lodg** S: Comfort Inn, **Super 8 Motel**
- **AServ** S: S&S Tires
- **Other** N: **Veterinary Clinic**

131 115th Ave, Cashion

132 107th Ave

133 99th Ave , 107th Ave., to AZ 101 Loop

134 91st Ave, Tolleson
- **Other** N: **K-Mart**

135 83rd Ave
- **Gas** N: Circle K*
- **Food** N: Arby's, Burger King (Playground inside), Jack-In-The-Box, Waffle House
- **Lodg** N: AAA **Econolodge**
- **AServ** N: Car Wash
- **Other** N: **Super K-Mart**

136 75th Ave
- **Gas** N: Chevron, Circle K*
  S: Arco*
- **Food** N: Denny's, Taco Bell
- **AServ** N: Pioneer Ford
- **Other** N: **Home Depot, Wal-Mart (Photo)**

137 67th Ave
- **Gas** N: Circle K*, Texaco*(CW)

138 59th Ave
- **TStop** S: **Circle K***(D)
- **Gas** N: Circle K*
  S: Circle K
- **Food** N: Los Armandos Mexican (24 Hrs)
  S: Waffle House, What A Burger
- **AServ** N: Auto Zone Auto Parts
- **TServ** S: **Danny's**(CW), **Inland Kenworth**
- **ATM** S: Bank One

139 51st Ave
- **Gas** N: 7-11 Convenience Store
- **Food** N: Domino's Pizza, La Cazuela, Sonic, Waffle House
- **Lodg** N: **Motel 6**, Red Roof Inn
  S: Fairfield Inn, Hampton Inn, **Travelers Inn**
- **AServ** N: Discount Tire Company, Firestone Tire & Auto, Q-Lube
- **ATM** N: Wells Fargo Bank
- **Other** N: **Medical Care Center, Southwest Supermarkets**

140 43rd Ave
- **Gas** N: 7-11 Convenience Store, Exxon(D)
- **Food** N: Dunkin Donuts, Hong Kong Express, KFC, Little Caesars Pizza, Mixteca, Subway, Wendy's
- **AServ** N: Big O Tires, Midas Muffler & Brakes
- **ATM** N: Bank of America
- **Other** N: **Coin Laundry/ Dry Cleaning, Dental Clinic, Omego Foods, Radio Shack, Smitty's Supermarket, Walgreens**

141 35th Ave
- **Gas** S: Circle K*, Texaco*
- **Food** N: Jack-In-The-Box, Rada's Mexican
- **AServ** N: Loyd Smith's Garage, Paul Johnson Garage
  S: Kar-Life Battery Company, Latino's Car Wash, S&G Auto Repair
- **Other** N: **Southwest Supermarket**

143C U.S. 60, 19th Ave

143AB Jct I-17, Flagstaff

144 7th Ave
- **Gas** N: Circle K*
  S: Circle K*
- **Food** N: Buffalo Brown's Wings and Things

- **Other** S: Car Wash

145 Seventh St.
- **Gas** N: Circle K* (ATM)
  S: Circle K*, Texaco*
- **Food** N: Chico's Tacos, Coco's, Einstein Brothers Bagels, McDonalds, Starbuck's Coffee, Subway, What A Burger
  S: Taco Juarez Mexican
- **Other** N: **ABCO Foods**

146 16th St (No Eastbound Reaccess)
- **Gas** N: Circle K*
  S: Circle K*
- **Food** N: Brookshire's, Gourmet House of Hong Kong, KFC, Maricio's Mexican, Seville's
  S: Church's Fried Chicken, Taqueria Three Amigo's Restaurant
- **AServ** S: K-Mart
- **Other** S: **K-Mart**

147 AZ51N, Loop 202, Squaw Peak Pkwy

148 Jefferson St (East), Washington St (West) (Each Street Is One Way)
- **Gas** N: Chevron, Tiemco*(D)
  S: Circle K
- **Food** N: Carl's Jr Hamburgers, McDonalds, Rally's Hamburgers
- **AServ** S: **Bill's Truck & Auto Repair**
- **Med** S: **+ Hospital**

149 Buckeye Road, Sky Harbor Int Airport (Difficult Reaccess, No Eastbound Reaccess)
- **Food** N: Road House
- **Lodg** N: **Roadway Inn**
- **ATM** S: Bank of America

150A Jct I-17N, Flagstaff

150B 24th St (No Westbound Reaccess)

151 University Dr

152 40th St

153A AZ143N, Sky Harbor Airport

153B Broadway Road, 52nd St (Difficult Reaccess)

154 U.S. 60, Scottsdale, Mesa
- **RVCamp** S: **Camping World (see our ad this page)**

155 Baseline Road
- **Other** N: **Mesa Factory Stores (see our ad this page)**

157 Elliot Road, Guadalupe

158 Warner Road

159 Ray Road

160 Chandler Blvd, Chandler

162 Maricapa Road

164 Queen Creek Road

167 Riggs Road, Sun Lakes

175 AZ587N, Chandler, Gilbert, Casa Blanca Road, Sacaton
- **RVCamp** S: **Camping**

(182) **Rest Area - RR, Picnic P**

(183) **Rest Area - RR, Picnic P (Westbound)**

185 AZ187, AZ387, Sacaton, Florence

190 McCartney Road

194 AZ287, Bus10, Casa Grande,

Bold red print shows RV & Bus parking available or nearby

EXIT	ARIZONA
	Coolidge
Other	**S:** Casa Grande Factory (see our ad this page)
198	AZ 84, Casa Grande
199	Jct I-8W, AZ84, San Diego, Casa Grande
200	Sunland Gin Road, Arizona City
TStop	**N:** Petro*(SCALES), Pilot Travel Center(SCALES) (Subway)
Gas	**S:** Shell, Texaco
Food	**N:** Burger King, Eva's Mexican, Iron Skillet (Petro), Subway(SCALES)
	S: Golden (Motel 6)
Lodg	**N:** Days Inn, Sunland Inn
	S: AAA Motel 6
AServ	**N:** Southwest Towing Truck and Trailor Repair
	S: Texaco
TServ	**N:** Petro Lube, Red Baron
ATM	**N:** Petro TStop
203	Toltec Road
FStop	**N:** Circle K
Gas	**N:** Exxon
	S: Exxon*
Food	**N:** Carl's Jr Hamburgers, McDonalds, Mexican Restaurant, Waffle House
	S: Country Pride (TravelCenters of America), Taco Bell (TravelCenters of America)
Lodg	**N:** Super 8 Motel, Tol Tec Inn
AServ	**S:** Exxon
TServ	**N:** Blue Beacon*, Cactus Country Tires
	S: TravelCenters of America(SCALES)
TWash	**N:** Blue Beacon Truck Wash*
	S: Little Sisters
ATM	**S:** TravelCenters of America
208	Eloy, St John Blvd, Sunshine Blvd.
TStop	**S:** Flying J Travel Plaza*(D)(LP)(SCALES) (RV Dump)
Food	**S:** Cookery (Flying J TStop)
211B	AZ84W, AZ87N, Coolidge, Florence
Parks	**N:** Casa Grande Ruins National Monument
212	Picacho
Gas	**N:** Exxon(LP) (AServ)
Food	**N:** Picacho Restaurant
Lodg	**N:** Motel 9, ◆ Picacho Motel
AServ	**N:** Exxon
RVCamp	**S:** KOA Campground*(LP)
Other	**N:** U.S. Post Office
(217)	Rest Area - Picnic, RR P (Eastbound)
219	Picacho Peak Road
Gas	**N:** Citgo*
Food	**N:** Dairy Queen (Citgo)
	S: Arizona Nut House, Long's Family Restaurant
AServ	**S:** No Name
Parks	**S:** Picacho Peak State Park
226	Red Rock
232	Pinal Air Park Road
236	Marana Road
Gas	**S:** Chevron*(D)(LP) (Restaurant), Circle K*(LP)
Food	**S:** Don's, Green Garden, Restaurant (Chevron)
RVCamp	**S:** Valley Of the Sun
ATM	**S:** Bank One
240	Tangerine Road
RVCamp	**N:** Camping
243	Cortaro, Avra Valley Road, Rillito
246	Cortaro Road

EXIT	ARIZONA
Food	**S:** Burger King, McDonalds
248	Ina Road
Gas	**S:** Exxon*(LP), Texaco*(D) (ATM)
Food	**N:** Long John Silvers, Perkins Family Restaurant
	S: Denny's (Holiday Inn Express)
Lodg	**N:** Motel 6
	S: ◆ Comfort Inn, ◆ Holiday Inn Express, AAA Red Roof Inn
ATM	**S:** Texaco
249	Orange Grove
253	Sunset Road, El Camino Del Cerro
TStop	**N:** National Truck Stop(SCALES) (TServ)
FStop	**N:** Pacific Pride
Food	**N:** Mr. Catfish Restaurant
AServ	**N:** Coin Car Wash, Star Automotive and Machine, T-N-S Automobile Service
254	Prince Road
Gas	**N:** Circle K*, Diamond Shamrock*(D) (ATM)
	S: Texaco(LP)
AServ	**S:** Texaco
TServ	**N:** Cummins Diesel
ATM	**N:** Diamond Shamrock
Other	**N:** Deaert Small Animal Hospital, Silent Wheels (RV Park)
255	AZ77N, Bus10, Miracle Mile
Gas	**N:** Circle K*
Food	**N:** EL Toro, Overpass Cafe
Lodg	**N:** Vista del Sol
AServ	**N:** S&J Garage
RVCamp	**N:** Sandy's RV Center
256	Grant Road
Gas	**S:** Exxon
Food	**N:** Sonic Drive In
	S: Sonora Sam's, Waffle House
Lodg	**S:** AAA Hampton Inn, AAA Roadway Inn
AServ	**N:** Desert Dog Automotive, Precision Engineering
	S: Exxon
257	Speedway Blvd.
Gas	**S:** Arco*
AServ	**N:** Gin's Oil Express
257A	St. Mary's Rd.
Gas	**N:** 76*(LP)
	S: Texaco(D) (AServ)
Food	**S:** Burger King, Denny's, Furr's Family Dining
Lodg	**S:** La Quinta Inn
AServ	**S:** Texaco
Other	**S:** St. Mary's Animal Clinic
258	Convention Center, Congress Street, Broadway

EXIT	ARIZONA
Gas	**N:** Circle K* (ATM)
Food	**N:** Carlos Murphy's Mexican
	S: Carl's Jr Hamburgers
Lodg	**N:** ◆ Holiday Inn
	S: Days Inn
ATM	**N:** Circle K
Other	**N:** Visitor Information
259	22nd St, Starr Pass Blvd.
Food	**S:** Ho Jo, Kettle (Holiday Inn Express), Waffle House
Lodg	**S:** Comfort Inn, E-Z 8 Motel, AAA Holiday Inn Express, Howard Johnson (Restaurant), Motel 6, Motel 6, Super 8 Motel, Travel Inn
AServ	**N:** 786 Auto Repair
Other	**N:** Coin Car Wash
260	I-19S, Nogales
261	4th Ave, 6th Ave, Bus 19
Gas	**N:** Chevron*
Food	**N:** Carrows Restaurant, Flanagan's San Diego
	S: Arby's, Bagels and Doughnuts, Burger King, El Indio, Jack-In-The-Box, McDonalds
Lodg	**N:** Budget Inn, AAA Econolodge, El Comino Motel
AServ	**N:** Discount Tire Co., Super Wheels and Tires Plus
	S: Pete's Auto Repair
Med	**S:** ✚ VA Medical Center
Other	**S:** Coin Laundry, Southgate Shoping Center (No Trucks)
262	Park Ave
Gas	**S:** Arco*, Chevron*, Texaco*(D)
Food	**S:** JB'S (Ho Jo), LCL Restaurant, Waffle House
Lodg	**S:** AAA Howard Johnson*(see our ad this page), Motel 6, AAA Roadway Inn
AServ	**S:** Chevron, Texaco
TServ	**S:** Peterbilt Dealer
263	Ajo Way, Kino Blvd (Difficult Reaccess)
Med	**N:** ✚ Hospital
264	Irvington Road, Palo Verde Road
Food	**N:** Carl's Jr Hamburgers, Denny's (Red Roof Inn), Old Town Restaurant and Saloon
	S: Arby's, McDonalds
Lodg	**N:** ◆ Days Inn, Fairfield Inn, Holiday Inn, Red Roof Inn
	S: Motel 6, Ramada Inn
RVCamp	**S:** Beaudry RV Center, La Mesa
Med	**N:** ✚ Hospital
Other	**S:** Tucson Outlet Center (see our ad this page)
265	Alvernon Way, Davis Monthan Air

Bold red print shows RV & Bus parking available or nearby

EXIT	ARIZONA
	Base
267	Bus Loop10, Tucson Int Airport, Valencia Road
268	**(6)** Craycroft Road
TStop	**N:** Triple T Truck Stop*(SCALES) (RV Dump)
FStop	**N:** Mr. T's(LP) (RV Dump)
Gas	**N:** Circle K*
Food	**N:** Hi-Way Chef (Triple T Truck Stop)
TServ	**N:** Fruehauf, Triple T Truck Stop
	S: Daffy's Diesel Truck repair
TWash	**N:** Triple T Truck Stop
RVCamp	**N:** Crazy Horse
ATM	**N:** Mr. T's, Triple T Truck Stop
269	Wilmot Road
TStop	**N:** Texaco
Food	**N:** Jason's (Texaco Truck Stop)
Lodg	**N:** Saguaro Inn
AServ	**S:** Wilmot Auto World
RVCamp	**N:** Camping
Other	**N:** Pima Air Museum
270	Kolb Road
RVCamp	**S:** Voyager
273	Rita Road
Parks	**S:** Pima County Fair Grounds
275	Houghton Road
Parks	**N:** Saguaro National Monument
279	Vail Road, Wentworth Road
Parks	**N:** Colossal Cave
281	AZ83S, Sonoita, Patagonia
289	Marsh Station Road
292	Empirita Road
297	J - Six, Ranch Road, Mescal Road
FStop	**N:** Quick Pic*(LP)
Food	**N:** Mr. Easy's Grill
299	Skyline Road
302	AZ90, Fort Huachuca, Sierra Vista
Gas	**S:** Shell* (Subway)
Food	**S:** McDonalds, Subway
Parks	**S:** Fort Huachuca National Historic Sight
304	Ocotillo St, Benson Road
Gas	**S:** Chevron*
Food	**N:** Denny's
	S: Bru's Hitchin' Post, Burger King, Plaza Restaurant
Lodg	**N:** Days Inn Days Inn, ◆ Super 8 Motel
	S: AAA Best Western
AServ	**S:** Dutchman Diesel Service

EXIT	ARIZONA
RVCamp	**N:** KOA Campground
	S: Butterfield R.V. Resort (see our ad this page)
Med	**S:** ✚ Hospital
Parks	**N:** Benson I-10, Red Barn RV Park(LP)
Other	**S:** Safe Way Grocery, Tourist Information
306	AZ80, Bus10, Pomerene Road, Benson Road
Parks	**S:** Douglas National Monument
312	Sibyl Road
Gas	**S:** Stuckey's*
318	Dragoon Road
(320)	**Rest Area - Vending, RR, Picnic P (Eastbound)**
(321)	**Rest Area - RR, Phones, Picnic P (Westbound)**
322S	Johnson Road
Gas	**S:** Citgo* (Dairy Queen)
Food	**S:** Dairy Queen (Citgo)
Other	**S:** The Thing? Museum
331	U.S. 191S, Douglas, Sunsites
Parks	**S:** Cochise Strong Hold
336	Bus10, Willcox
FStop	**S:** Texaco(LP)
AServ	**S:** Texaco
340	AZ186, Rex Allen Dr, Fort Grant
TStop	**N:** Rip Griffin Travel Center*(SCALES) (RV Dump, Restaurant, Taco Bell, Subway)
	S: Chevron*
FStop	**S:** Mobil*
Gas	**S:** Circle K*
Food	**N:** Restaurant (Rip Griffin's TStop), Subway (Rip Griffen's), Taco Bell (Rip Griffin's TStop)
	S: Burger King, KFC, McDonalds, Pizza Hut, R & R Pizza Express, Solarium (Best Western)
Lodg	**N:** ◆ Super 8 Motel
	S: AAA Best Western (Solarium), Days Inn AAA Days Inn, Motel 6
AServ	**S:** Dick's Tire and Auto
TServ	**S:** Mark's Tire Stop
TWash	**N:** 340 Truck Wash
RVCamp	**N:** Magic Circle
	S: Grande Vista
Med	**S:** ✚ Northern Cochise Hospital
Parks	**N:** Chiri Cahua National Monument
Other	**N:** Tourist Information
	S: IGA, Safe Way Grocery (Pharmacy)
344	Bus10 loop, Willcox (Difficult Reaccess)

EXIT	AIZONA/NEW MEXICO
TServ	**S:** Willcox Diesel Service
352	U.S. 191N, Safford
355	U.S. 191N, Safford
362	Bus10, Bowie
Gas	**N:** Gas Station
RVCamp	**N:** RV Park(D)(LP)
366	Bus10 Loop , Bowie
Gas	**N:** Texaco*
Food	**N:** Baskin Robbins (Texaco), Subway (Texaco)
RVCamp	**S:** Alaskan
Parks	**N:** Fort Bowie Historic Sight
378	Bus Loop 10, San Simon
TStop	**N:** 4-K*(SCALES) (RV Dump)
Gas	**N:** Texaco (AServ)
Food	**N:** Restaurant (4-K Truck Stop)
AServ	**N:** Texaco
TServ	**N:** Car RV & Truck Repair
382	Bus10, Portal Road, San Simon
(383)	**Weigh Station (Eastbound)**
(384)	**Weigh Station (Westbound)**
(388)	**Rest Area - RR, Picnic, Grills P (Eastbound)**
(389)	**Rest Area - RR, Vending, Picnic, Grills P (Westbound)**
390	Cavot Road

↑ARIZONA
↓NEW MEXICO

EXIT	NEW MEXICO
3	Steins
5	NM80S, Road Forks, Douglas, Agua Prieta
TStop	**S:** U.S.A. Truck Stop*(SCALES)
Gas	**S:** Fina*
Food	**S:** Road Kill Saloon (USA Truck Stop), Shady Grove (USA Truck Stop)
Lodg	**S:** AAA Desert West (Restaurant)
11	NM338S, Animas
15	Gary
20	West Motel Dr
TStop	**N:** Love's Truck Stop*(SCALES) (Taco Bell & ATM)
Food	**N:** Taco Bell (Love's Truck Stop)
ATM	**N:** Love's Truck Stop
(20)	**Info Center, Rest Area - RR, Phones, RV Dump P (Eastbound)**

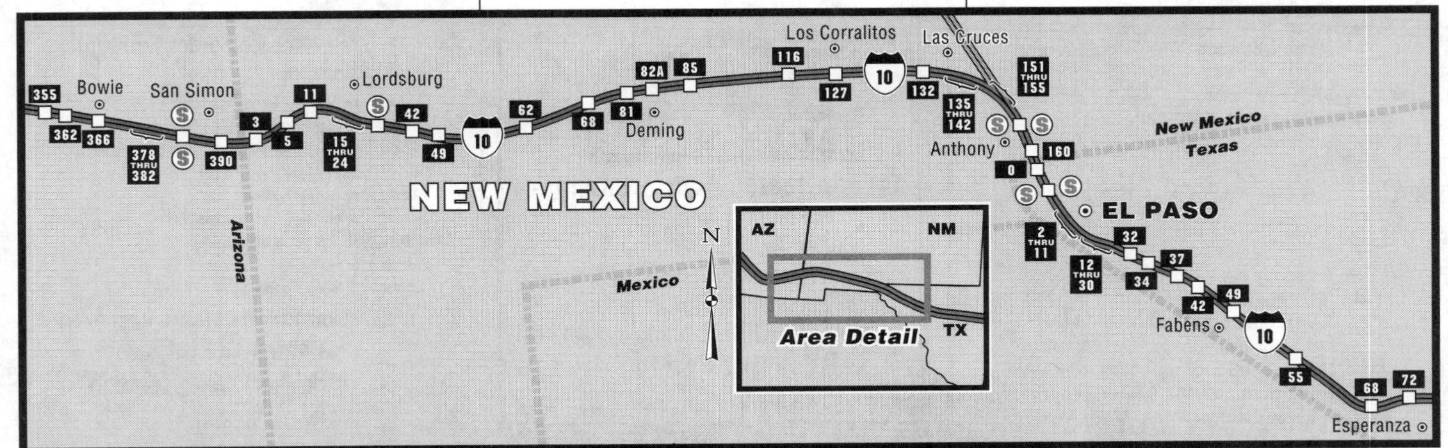

Bold red print shows RV & Bus parking available or nearby

EXIT — NEW MEXICO

22 U.S. 70, NM494, Main St, Silver City
- **Gas** N: Save
  S: Chevron, Diamond Shamrock*, Exxon* (AServ), Snappy Mart, Texaco*(D)
- **Food** N: Dairy Queen, Hot Spot Cafe, McDonalds, Subway
  S: KFC, Kranberry's, Pizza King
- **Lodg** S: AAA Best Western, Motel 10, AAA Super 8 Motel
- **AServ** N: Hand Car Wash, Raise Tire
  S: Exxon
- **TServ** N: Oscar's Truck Service
- **RVCamp** S: KOA Campground
- **Parks** N: Gilla Cliff Dwellings

(24) Weigh Station (Both Directions)

24 Bus Loop 10, E Motel Dr
- **Gas** N: Chevron (AServ), Texaco*
- **Food** N: AJ's Cafe, Best Western
- **Lodg** N: AAA Best Western (Restaurant)
- **AServ** N: Auto Diagnostic Center, Chevron, Turn A Tire
- **RVCamp** N: Camping

29 Unnamed

34 NM113, Playas

42 Separ
- **FStop** S: Separ Gas
- **Gas** S: Chevron
- **Food** S: Windmill Diner
- **TServ** S: Tire Shop

49 NM146, Hachita, Playas, Antelope Wells

51 Continental Divide

(53) Rest Area - RR, Picnic, RV Dump P (Eastbound)

55 Quincy

(61) Rest Area - RR, Picnic P (Westbound)

62 Cage
- **FStop** S: Citgo*
- **Food** S: Dairy Queen
- **RVCamp** S: Butterfield Station

68 NM418
- **TStop** S: Savoy Truck Stop* (Restaurant)
- **Gas** S: Stuckey's*
- **Food** S: Restaurant (Savoy TStop)

81 Deming
- **TStop** S: Deming Truck Terminal (Restaurant)
- **FStop** S: Fina
- **Gas** S: Chevron, Diamond Shamrock*(D), Texaco
- **Food** S: Arby's, Burger King, Burger Time, El Camino Real, Restaurant, Sonic
- **Lodg** S: AAA Best Western, Budget Motel, Super 8 Motel, AAA Wagon Wheel Motel, Western Motel
- **AServ** S: Chevron, Texaco
- **TServ** S: El Pavo Real, Truck Wash
- **RVCamp** S: 81 Palms
- **ATM** S: Deming Truck Stop

82A U.S. 180, NM26, NM11, Silver City, Hatch
- **Gas** N: Fina*, Texaco*(D)(LP)
  S: Exxon (AServ), Fina, Phillips 66*, Sav-o-mat
- **Food** N: Lotaburger
  S: Betty's Ice Cream, Burger King, Cactus Cafe, China Restaurant, Domino's Pizza, K-Bob's Steakhouse, KFC, La Fonda, Long John Silvers, Mucho Gusto, Pizza Hut, Si Senor, Subway
- **Lodg** S: Butterfield Stage Motel, Mirador Motel

EXIT — NEW MEXICO

- **AServ** N: Texaco
  S: Checker Auto Parts, Exxon, Harry's Auto Electric, On Sale Tire Stores Inc., Penzoil PDQ Lube, Rex Null's
- **Other** N: Servi Gas(LP)
  S: Furr's, Rockhound State Park

85 NM11, Deming, Motel Dr, Bus Loop 10
- **Gas** S: Chevron* (AServ), Texaco*(D)
- **Lodg** S: ◆ Holiday Inn, Motel 6
- **RVCamp** S: Dream Catcher

102 Akela
- **Gas** N: Conoco*
- **Parks** S: Rock Hound State Park

(111) Rest Area P (Westbound)

116 NM549
- **Gas** S: Gas Station*
- **Food** S: Old Fashioned Hamburgers

(121) Inspection Point (Westbound)

127 Corralitas Road
- **Gas** N: Chevron*

132 Las Cruces International Airport
- **TStop** S: Loves Truck Stop(SCALES) (Taco Bell)
- **Food** S: Taco Bell (Loves Truck Stop)

(135) Rest Area - RR, HF, Phones, Picnic, RV Dump P

135 U.S. 70E, Las Cruces, Alamogordo
- **RVCamp** N: Camping

139 NM292, Motel Blvd, Amador Ave
- **TStop** N: Pilot Travel Center*(SCALES), TravelCenters of America*(SCALES)
- **Gas** N: Shell*
  S: Chevron (AServ)
- **Food** N: Country Pride (TravelCenters of America), Subway (Pilot Travel Center)
  S: Dick's Restaurant (Coachlight Motel)
- **Lodg** S: Coachlight Motel
- **AServ** S: Chevron
- **TServ** N: Las Cruces Truck Services

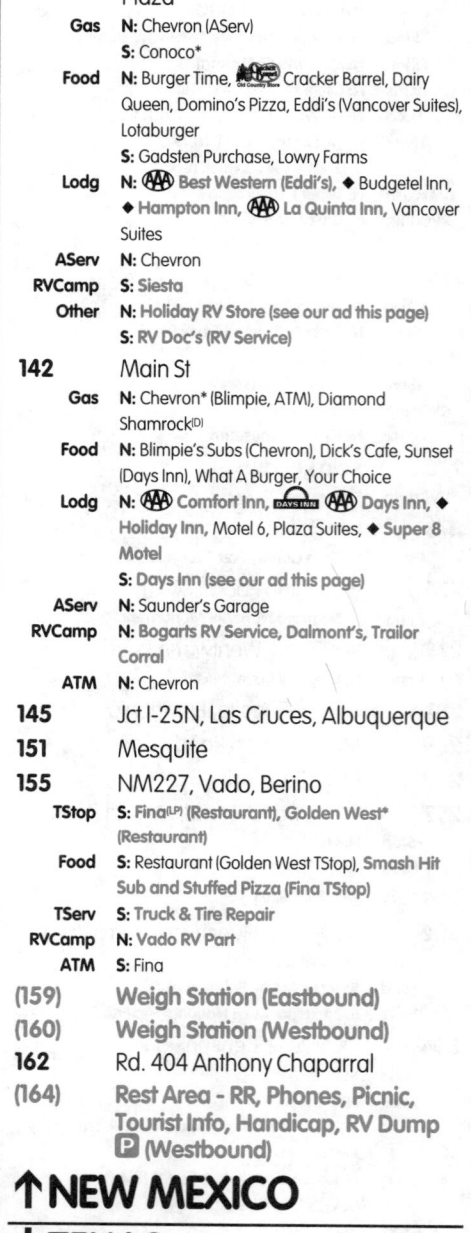

EXIT — NEW MEXICO/TEXAS

- **TWash** N: TravelCenters of America(SCALES)
- **ATM** N: Pilot Travel Center, TravelCenters of America

140 NM28, Mesilla, Historic Mesilla Plaza
- **Gas** N: Chevron (AServ)
  S: Conoco*
- **Food** N: Burger Time, Cracker Barrel, Dairy Queen, Domino's Pizza, Eddi's (Vancover Suites), Lotaburger
  S: Gadsten Purchase, Lowry Farms
- **Lodg** N: AAA Best Western (Eddi's), ◆ Budgetel Inn, ◆ Hampton Inn, AAA La Quinta Inn, Vancover Suites
- **AServ** N: Chevron
- **RVCamp** S: Siesta
- **Other** N: Holiday RV Store (see our ad this page)
  S: RV Doc's (RV Service)

142 Main St
- **Gas** N: Chevron* (Blimpie, ATM), Diamond Shamrock(D)
- **Food** N: Blimpie's Subs (Chevron), Dick's Cafe, Sunset (Days Inn), What A Burger, Your Choice
- **Lodg** N: AAA Comfort Inn, Days Inn, AAA Days Inn, ◆ Holiday Inn, Motel 6, Plaza Suites, ◆ Super 8 Motel
  S: Days Inn (see our ad this page)
- **AServ** N: Saunder's Garage
- **RVCamp** N: Bogarts RV Service, Dalmont's, Trailor Corral
- **ATM** N: Chevron

145 Jct I-25N, Las Cruces, Albuquerque

151 Mesquite

155 NM227, Vado, Berino
- **TStop** S: Fina(LP) (Restaurant), Golden West* (Restaurant)
- **Food** S: Restaurant (Golden West TStop), Smash Hit Sub and Stuffed Pizza (Fina TStop)
- **TServ** S: Truck & Tire Repair
- **RVCamp** N: Vado RV Part
- **ATM** S: Fina

(159) Weigh Station (Eastbound)

(160) Weigh Station (Westbound)

162 Rd. 404 Anthony Chaparral

(164) Rest Area - RR, Phones, Picnic, Tourist Info, Handicap, RV Dump P (Westbound)

↑NEW MEXICO

↓TEXAS

0 New Mexico border, Fm1905, Anthony
- **TStop** N: Flying J Travel Plaza(LP)(SCALES)
- **Gas** S: Chevron*(D), Exxon(D) (Burgar King)
- **Food** N: Thads Restaurant (Flying J)
  S: Burger King (Exxon)
- **Lodg** N: Super 8 Motel
  S: Holiday Inn Express
- **TServ** N: S & W Refrigeration
- **TWash** S: No Name
- **Other** S: Big 8 Foods

(0) Information Bureau, Rest Area - RR, Phones P (Eastbound)

(1) Truck Check/Weigh Station

2 Westway, Vinton

Bold red print shows RV & Bus parking available or nearby

EXIT TEXAS

TStop N: Petro 2*(SCALES) (RV Dump)
Gas S: Circle N*
Food N: Blimpie's Subs (Petro 2), Rinoncito (Mexican)
AServ S: Venton Auto Repair
TServ N: Petrolube

6 Loop375, Trans Mountain Road, Canutillo Road
Parks N: Franklin Mountain State Park

8 Artcraft Rd.

9 Redd Road

11 TX20, Mesa St
Gas N: Diamond Shamrock*
S: Diamond Shamrock*
Food N: Basken Robbin's, Carrows, Chili's, CiCi's Pizza, Cracker Barrel, Denny's, K-Bob's Steak House, Long John Silvers, Mei Li's Chinese Food, Outback Steakhouse, Wendy's
S: Burger King, KFC, Lubby's Cafeteria, McDonalds, Peter Piper Pizza, Pizza Hut, Restaurant (Village Inn), Taco Cabana
Lodg N: ◆ Budgetel Inn, ◆ Comfort Inn, AAA La Quinta Inn, ◆ Red Roof Inn
S: Motel 6, AAA Travelers Inn
AServ S: Martin TIre Co., Master Care Auto Service, Pronto Lube
ATM N: Texas Commerce Bank
Other N: Promenade Mall
S: Big 8 Foods, Sam's Club

12 Resler Dr (No Westbound Reacces)

13 Sunland Park Dr, U.S. 85
Gas S: Diamond Shamrock* (AServ), Shell*(CW) (AServ, Taco Bell)
Food N: Carino's, Chuck E. Cheese's Pizza, Olive Garden, Red Lobster, T.G.I. Friday's
S: La Malinche, McDonalds, Taco Bell*
Lodg S: AAA Holiday Inn (Restaurant)(see our ad this page), AAA Best Western, Comfort Inn, Sleep Inn
AServ N: Auto Express Montogomery Ward
ATM N: Nations Bank, Sunwest Bank
Other N: Barnes & Noble, Office Depot, Pier 1 Imports, Sunland Park Mall, Sunland Plaza
S: Sunland Park Race Track

16 Executive Center Blvd
Food N: The Cafe (Ramada Inn)
Lodg N: Ramada Inn (The Cafe)

18A Schuster Ave. UT El Paso

18B Porfirio Diaz St
Gas N: Circle K* (ATM)
ATM N: Circle K*

19A Mesa St, TX20 (Westbound)
Gas S: Chevron (AServ)

Lodg S: International Hotel
AServ S: Chevron

19B Dowtown, to Convention Center (Eastbound, Difficult Reaccess)
Lodg N: Travel Lodg
Med N: ✚ Hospital
ATM N: Sunwest Bank

20 Cotton St (Westbound)
Gas N: Texaco

21 Piedras St
Food N: Burger King, McDonalds

22A Loop478, Copia St, US24 , Patriot HWY
Gas N: Diamond Shamrock*, Fina*
Food N: KFC
AServ N: Copia Auto Service
S: R.D. Tires
ATM N: Fina

23A Raynolds St, U.S. 64, Patrick Freeway, Alamogordo Juarez, Fort Bliss
Food S: Arby's
Lodg S: Motel 6
AServ S: Central Auto Supply
Med S: ✚ Hospital

23B Carlsbad, U.S. 62, U.S. 180 , Paisano Dr
Food N: Alexandro's Fine Mexican Food
Lodg N: Budget Inn, Knights Inn (see our ad this page)
AServ N: Auto West
Other S: El Paso Zoo

24 Geronimo Dr, Trowbridge Dr
Gas S: Exxon*, Phillips 66*
Food N: Seafood Galley, Yellow Rose Cafe (Quality Inn)
S: Bombay Bicycle Club, Denny's
Lodg N: Quality Inn (Yellow Rose Cafe)
S: Embassy Suites, La Quinta Inn, Sumner Suites
AServ N: Bassett Tire and Service Store, Chevron
S: Art's Muffler and Brake Shop
ATM N: Norwest Bank
Other N: Bassett Center Shopping Mall, Office Depot

25 El Paso Airport
TStop S: El Paso Truck Terminal(SCALES)
Gas N: Shell*(D) (see our ad this page)
Food N: Jaxon's, Landry's Seafood House, Teddy's Flame Room Grill (Holiday Inn)
Lodg N: ◆ Hampton Inn and Suites, ◆ Holiday Inn
AServ N: Mercedes Volvo VW Dealer
TServ S: El Paso Tire Center, Great Basin GMC Truck Dealership

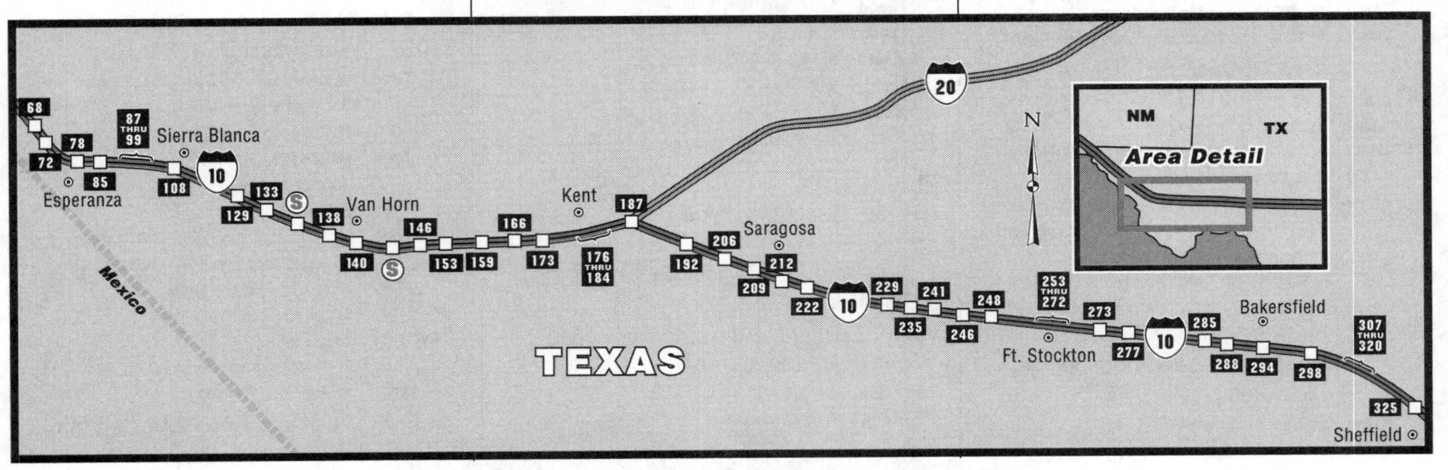

EXIT		TEXAS
26	Hawkins Blvd.	
	Gas	N: Chevron (AServ), Diamond Shamrock*(D), Shell S: Diamond Shamrock*(D)
	Food	N: Arby's, Burger King, IHOP, Lubby's Cafeteria, Olive Garden S: Billy Gin Restaurant (Best Western), China King's, McDonalds
	Lodg	N: ◆ Howard Johnson S: AAA Best Western
	AServ	N: Montgomery Ward, Vintage Full Service Car Wash S: Spitzer Electrical Service
	Other	N: Cielo Bista Mall, Mission Plaza Mall (Shell Gas)
27	Hunter Drive, Viscount Blvd (Eastbound)	
	Gas	N: Diamond Shamrock*(D)(CW), Fina* (24 Hr) S: Exxon(D) (Subway), Shell (AServ)
	Food	N: Carrows, Doc's BBQ, Garden Buffet, Red Lobster, Taco Bell S: Subway (Exxon), What A Burger
	Lodg	N: Days Inn
	AServ	S: Shell
	Other	S: Food City, Ranchland Laundry, Ranchland Village (Mall)
28A	Giles Road, Andrew Dr, Fm2316, Fort Bliss, Hunter Drive	
	Gas	N: Chevron* S: Chevron(CW), Fina*, Phillips 66*(CW)
	Food	N: Chico's Tacos, Jack-In-The-Box S: Gabriel's Cafe
	AServ	N: K-Mart, Tune Up S: Diesel Truck & Car Wash, Saturn Dealership
	ATM	N: Texas Commerce Bank S: Fina*
	Other	N: K-Mart (Pharmacy), Office Max, WalGreens S: Mcrae Animal Hospital
28B	Yarbough, Sumac Dr	
	Gas	N: Chevron (AServ) S: Diamond Shamrock*
	Food	N: Baskin Robbins, Bennigan's, Bobby Q's BBQ and Mexican Ribs, Dunkin Donuts, Grandy's, Long John Silvers, McDonalds, Pistol Pete's Pizza (Wal-Mart), Sonic Drive-In, Wendy's, What A Burger, Wyatt's Cafeteria S: Applebee's, La Malinche Cafe, Pizza Hut, Quizno's Subs
	Lodg	N: Travelers Inn S: ◆ Budgetel Inn, Comfort Inn
	AServ	N: Chevron, Midas Break and Muffler, Pep Boys Auto Center
	Med	N: ✚ Columbia Medical Center
	Other	N: Town East Plaza, Wal-Mart (Pistol Pete's Pizza), Yarborough Plaza
30	Lee Trevino Dr, Lomand Dr	
	Gas	S: Diamond Shamrock*
	Food	N: Mi Pueblito, Oriental Hut, Smokie's Pit Stop BBQ, Subway
	AServ	N: Firestone Tire & Auto, Martin Tire Co., National Tire and Battery, Northside Tire, Shamaley Ford & Isuzu, Travino Transmissions S: Story Nissan, Vista Chevrolet
	ATM	N: Sunwest Bank
	Other	N: Kinko's, Trevino Mall
32	Fm659, Zaragosa Road, George Dieter Road	
	FStop	S: Diamond Shamrock*
	Gas	N: Chevron*(D) (ATM) S: Chevron*, Conoco
	Food	N: What A Burger
	AServ	S: International Dealership
	ATM	N: Cheveron
	Other	S: Supro Energy(LP), Ysleta Vet. Clinic
34	Loop375, Americas Ave	
	FStop	N: Texaco* (Taco Bell)
	Gas	N: Chevron*(D)
	Food	N: Taco Bell (Texaco)
	RVCamp	N: Mission
37	Fm1281, Horizon Blvd	
	TStop	N: 76 Auto/Truck Plaza*, Loves*(SCALES) (Pizza Hut, Subway, Taco Bell) S: Petro(SCALES) (Iron Skillet, TServ)
	Gas	S: Mobile (Blimpie)
	Food	N: 76 Auto/Truck Plaza (Love's TStop), Pizza Hut (Love's TStop), Subway (Love's TStop), Taco Bell (Love's TStop) S: Blimpie's Subs (Mobile), McDonalds
	Lodg	N: AAA Americana S: Deluxe Inn
	TServ	N: 76 Auto/Truck Plaza, Cummins Diesel, El Paso Thermo King S: Channel 17 Electronic Repair, Chrome TNT Accessories, Expert Trailor Repair, P&I Fleet Service Truck Repair, West Texas Great Dane Trailors
	TWash	S: Blue Beacon Truck Wash(SCALES), Horizon
	RVCamp	S: Sampson's RV Park
42	Fm1110, Clint, San Elizario	
	Gas	S: Exxon*(D)
	Food	S: Boll Weevil Restaurant
	Lodg	S: Motel
	RVCamp	S: Cotton Valley
49	Fm793, Fabens	
	TStop	S: Texas Red's Truck Stop* (Restaurant)
	Gas	S: Texaco(D)
	Food	N: Cattleman's Steakhouse S: Hilltop Cafe, Texas Red's Express
	Lodg	S: Hilltop Motel, Motel
(50)	Rest Area - RR P (Eastbound)	
(51)	Rest Area - RR, Picnic P (Westbound)	
55	Tornillo	
68	Acala Road	
72	Spur148, Fort Hancock	
	FStop	S: Shell*
	Gas	S: Texaco*
	Food	S: Argie's
	Lodg	S: ◆ Motel
	AServ	S: McNary Garage, Shell
	RVCamp	S: RV Park
	Other	S: Sheiff's Office
78	TX20W, McNary	
81	Fm2217	
85	Esperanza Road	
87	Fm34	
	TStop	S: Desert Outpost Truck Stop* (Restaurant)
	TServ	S: Hot Spot CB Repair
95	Frontage Road (No Westbound Reaccess)	
	Gas	S: No Name*
	Lodg	S: Motel (no name)
(98)	Rest Area - Picnic P (Eastbound)	
(99)	Rest Area P (Westbound)	
99	Alsca Road	
(102)	Inspection Station (Eastbound)	
106	Bus Loop I-10	
	Gas	N: No Name
	Food	N: Chuckwagon Restaurant
107	Fm 1111, Sierra Blanca Ave	
	FStop	N: Exxon*
	Gas	N: Texaco S: Chevron*
	Food	N: Best Cafe, Chonas Parlor, Dude's Restaurant, Michael's Restaurant S: Hole In The Wall Cafe
	Lodg	N: El Camino Motel, Sierra Motel
	AServ	N: Ellison Motor Co., Texaco
	TServ	N: Wayne's Truck Shop
	RVCamp	N: Siera Blanca
	Other	N: Blanca Coin Laundry, Hueco Tanks State Park, RR Depot- Hudspeth County Museum, US Post Office
108	Bus10, Sierra Blanca (Caution Low Clearance 13'6")	
129	Allamore, Hot Springs	
133	Frontage Road	
136	Scenic Overlook - Picnic (Westbound)	
(137)	Weigh Station (Eastbound)	
138	Golf Course Dr.	
	FStop	S: Chevron (Restaurant)
	Gas	N: Phillip 66* S: Exxon*
	Food	N: Dairy Queen, Pizza Hut, Subway S: Don's (Chevron FStop), McDonalds
	Lodg	N: AAA Best Western, AAA Howard Johnson, AAA Motel 6 S: AAA Holiday Inn Express, AAA Super 8 Motel
	Other	N: Special Events Tourist Info.
140A	U.S. 90, TX54, Van Horn Drive, Marfa, Del Rio	
	Gas	N: Texaco* (Burgar King) S: Conoco*(D), HH Gas*
	Food	N: Burger King (Texaco), Leslie's BBQ S: Rosa's Cafe
	Lodg	S: AAA Freeway Inn Motel
	AServ	N: Car Quest Auto Center, Napa Auto Parts, Ramirez RV and Auto Shop, Texaco
	Med	N: ✚ Hospital
	Parks	N: Carlsbad Caverns, Guadelupe Mountains
140B	Bus10, Ross Drive, Van Horn	
	TStop	N: Love's Country Store (Taco Bell) S: Conoco* (Restaurant)
	Gas	N: Chevron*, Exxon*(D)(LP)
	Food	N: Iron Rail, Okey D's Family Restaurant, Sands, Taco Bell S: Restaurant (Conoco TStop)
	Lodg	N: Bells Motel (Okey D's), Days Inn ◆ Days Inn (Iron and Rail), Sands Motel (Sands Restaurant)
	TServ	N: Ana Repair, Tire Shop S: Conoco Truck Stop
	RVCamp	N: El Campo S: Camping (Mountain View RV Park)
	Other	N: Van Horn Pharmacy
(145)	Rest Area - RR, Picnic P (Both	

Bold red print shows RV & Bus parking available or nearby

EXIT	TEXAS
	Directions)
(146)	**Weigh Station (Westbound)**
146	Wild Horse Road
153	Michigan Flat
159	Plateau
TStop	N: **Citgo* (Restaurant)**
166	Boracho Station
173	Hurds Draw Road
176	TX118, Fm2424, Kent, Fort Davis
Gas	N: Chevron*(D)
Parks	S: **Davis Mountains State Park**
Other	N: **Post Office**
	S: **McDonald Observatory**
181	Cherry Creek Road
FStop	S: **Chevron***
184	Springhills
(185)	**Rest Area - Picnic, Grills, HF P (Eastbound)**
(186)	**Rest Area - Picnic, Grills, HF P (Westbound)**
187	Junction I-20, Pecos
188	Giffin Road
192	TX 3078, Toyahvale
206	TX 2903, U.S. 10, Toyah, Balmorhea
209	Bus 10 West, TX 17 South , Bal Morhea
212	TX 17, Pecos
FStop	S: **Fina***
Food	S: **Circle Bar, Restaurant (Fina)**
Parks	N: **Picnic Area**
214	Fm 2448 (Difficult Reaccess)
222	Hoefs Road
229	Hovey Road
(233)	**Rest Area - Picnic, Grills, RR, Phones P (Both Directions)**
235	Mendel Road
241	Kennedy Road
246	Firestone Road
248	U.S. 67 South, Alpine, FM1776, Big Bend
253	FM 2037, Belding
256	Fort Stockton
FStop	N: **Shell***
	S: **Texaco***
Food	S: Alpine Lodge, Brazen Bean
Lodg	S: AAA **Best Western**, ◆ **Comfort Inn**
257	U.S. 285, Pecos, Sanderson
259	TX 18, FM 1053, Monahan's
Gas	N: Shell*(D) (see our ad this page)
261	U.S. 10, U.S. 385, Marathon, Fort Stockton
Gas	N: Exxon*
RVCamp	S: **Big Ben Park**
Med	S: **+ Hospital**
Other	S: **Fort Stockton Historical District**
264	Warnock Road
272	University Road
273	U.S. 67, U.S. 385, McCarmey
(273)	**Rest Area P (Westbound)**
277	FM 2023
(279)	**Rest Area - Picnic, Grills, RR P (Eastbound)**
285	McKenzie Road
288	Ligon Road
294	TX 11, Bakersfield
Gas	N: Exxon*
	S: Chevron*(D)
298	FM 2886
307	U.S. 190, Iraan, FM305, McCarney
(308)	**Rest Area - Picnic, Phones, Grills, HF P**

EXIT	TEXAS
314	Frontage Road
320	Frontage Road
325	TX 290, 349, Iraan, Sheffield
Parks	S: **Wild Life Viewing Area**
328	River Road
337	Live Oak Road
343	TX 290, Sheffield
(346)	**Parking Area (Eastbound)**
350	FM 2398, Howard Draw
361	FM 2083, Pandale Road
363	RD 466, Business Ozona
365	TX 163, Sterling City, Comstock
Gas	N: Chevron*(D)(CW), Texaco(D)
	S: Chevron*, Phillips 66*
Food	N: Dairy Queen, The Cafe Next Door
Lodg	N: **Comfort Inn**
RVCamp	N: **Economy Inn**
ATM	N: Motor Bank
368	Loop 466, Ozona
372	Taylor Box Road
TStop	N: **76 Auto/Truck Plaza*(SCALES) (RV Dump)**
Food	N: **Circle Bar (76 Truck Stop)**
Lodg	N: **Circle Bar Hotel**
RVCamp	N: **Camping (Showers & Grills)**
ATM	N: 76 Auto/Truck Plaza
381	RC 1312
388	Fm 1312 (Westbound, Difficult Reaccess)
392	Road 1989, Caverns of Sonora Road
(394)	**Rest Area - RV Dump P (Both Directions)**
399	TX 467, Sonora
400	U.S. 277, San Angelo, Del Rio
Gas	N: Texaco*
	S: Chevron*(D), Conoco*(D)(CW)(LP), Exxon*, Fina*, Phillips 66, Shell(D) (AServ)
Food	N: Sutton County Steak House
	S: Big Tree Restaurant, Country Fried Chicken & Fish, Dairy Queen, La Mexicana, **Pizza Hut**, Sonic Drive In
Lodg	N: **Devils River Inn**
	S: AAA **Twin Oaks Motel, Zola's Motel**
AServ	N: Atkinson Chevrolet, Buick, I-10 Auto Ranch GM
	S: Phillips 66, Shell
ATM	S: Sutton County Bank

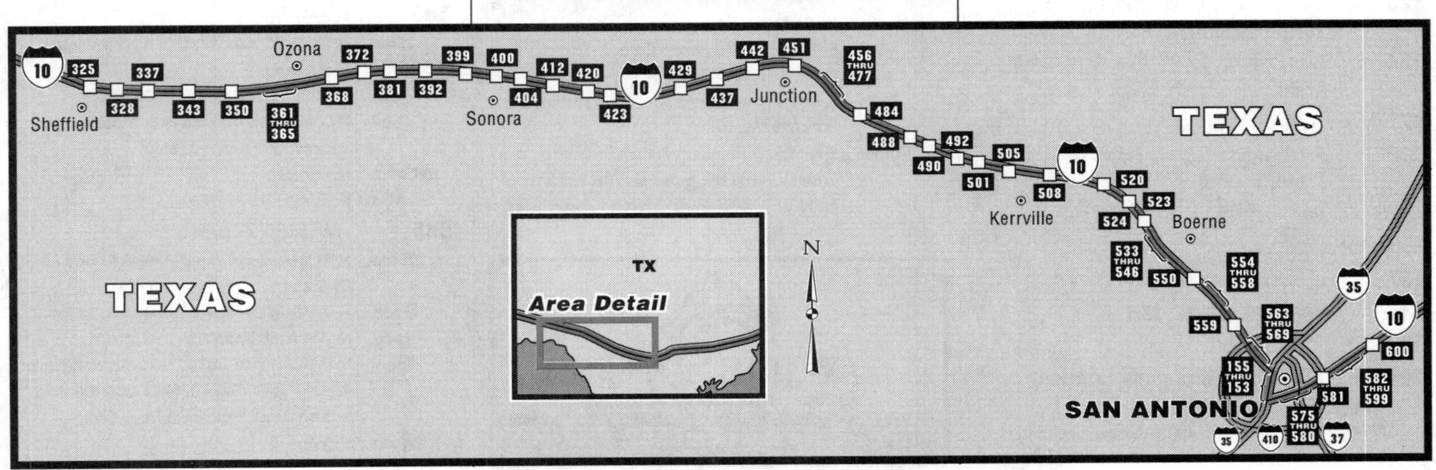

Bold red print shows RV & Bus parking available or nearby

EXIT TEXAS

404 Road 864, Fort McKavatt , Sonora
412 RD. 3130, Allison Road
420 Baker Road, RM 3130
423 Parking Area (Both Directions)
429 Harrell Road , FR3130
437 TX 291, Roosevelt (Difficult Reaccess)
442 TX 291, Fort McKavatt
- **Parks** **S:** **Fort McKavatt State Historic Park**

445 Road 1674
451 Cleo Road
456 U.S. 83 North, Menard, U.S. 377, Mason
- **FStop** **N:** **Texaco (AServ)**
- **Gas** **N:** Chevron*
  S: Exxon*, Phillips 66*
- **Food** **N:** Tastee Freeze
  S: Come-n-Git-It, **Dairy Queen**
- **Lodg** **N:** **Comfort Inn**
  S: **Slumber Inn**
- **AServ** **N:** Texaco
- **RVCamp** **S:** **Camping**
- **Med** **S:** **✚ Hospital**

457 Martinez St, FM 2169 Junction
460 Bus Loop
461 Picnic (Eastbound)
462 U.S. 83, Uvalde
465 Road 2169, Segovia
472 Old Segovia Road, FM 479, 2169
477 U.S. 290, Fredericksburg
484 Midway Road
488 TX 27, Mountain Home, Ingram
490 TX 41, Mountain Home, Rocks Springs
492 RD 479
- **Gas** **S:** Loan Oak*[D]
- **RVCamp** **N:** **Camping**
  S: **Camping**

(497) **Rest Area - Picnic, Grills** P
501 Road 1338
505 Road 783, Harper, Kerrville, Ingram
508 TX 16, Kerrville
(513) **Rest Area - Picnic, Grills, RR, RV Dump** P
520 Road 1341, Cypress Creek Road
523 U.S. 87 North, Comfort, San Angelo, Fredericksburg
524 Bus U.S. 87, TX 27, Comfort, Center Point
527 FM 1631 Waring
(531) **Rest Area - Picnic, Grills** P **(Westbound)**
(529) **Rest Area - Picnic, Grills** P
533 Welfare
537 U.S. 87, Boerne
538 Ranger Creek Road
539 John's Road
540 TX 46, New Braunfels, Bandera
- **Gas** **N:** Texaco*[CW]
- **Food** **N:** Church's Fried Chicken, Dairy Queen, Pizza Hut Carry Out
- **Lodg** **N:** **Best Western**
- **Other** **N:** **HEB Food And Drugs**

542 Bus 87, Boerne (No Westbound Reaccess)
543 Boerne Stage Road, Cascade Caverns Road
546 Fair Oaks Pkwy, Tarpon Drive
550 Ralph Fair Road, Boerne Stage Road
551 Boerne Stage Road
554 Camp Bullis Road
555 La Cantera Pkwy, Fiesta
556A 1604 Anderson Loop
556B Frontage Road
557 TX 53, Univ. of Texas San Antonio
- **Lodg** **N:** ◆ **Best Western, Bradford Inn, Howard Johnson, Super 8 Motel**
- **AServ** **N:** Brake Check, Smith Chevrolet
- **ATM** **N:** Frost Bank

558 De Zavala
559 U.S. 87, Fredericksburg Road
560A Huebner Rd. (Eastbound)
561 Huebner Road
562 Wurzbach Road
- **Lodg** **N:** Amerisuites (see our ad this page)

563 Callaghan Road
564 Junction I-410
565A Crossroads Blvd, Balcones Heights
565C Vance Jackson Rd.
565B West Ave
566A Fresno Drive
566B Hildebrand Ave, Fulton Ave
567 Loop 354, Fredericksburg Road, Woodlawn Ave
568 Culebra Ave, Bandera, Spur 421
569C Santa Rosa St

Note: I-10 runs concurrent below with I-35. Numbering follows I-35.

156 Junction N. U.S. 87 El Paso & W. 10
155B Durango Boulevard (Difficult Reaccess)
- **Food** **E:** Bill Miller BBQ
- **Lodg** **E:** Courtyard by Marriott, Fairfield Inn, Holiday Inn, Residence Inn
  W: Radisson Hotel
- **Med** **E:** **✚ Santa Rosa Health Care**
- **ATM** **E:** Stopn Go, Wells Fargo
- **Parks** **E:** **San Antonio Missions National Historical Park (2.8 Miles)**
- **Other** **E:** **K-Mart (Pharmacy), RxCare Pharmacy, San Antonio Police HQ, Stopn Go, The Market Square**

EXIT TEXAS

  W: **Fire Station**

154B South Laredo St, Cevallos St
- **Gas** **E:** Texaco*
- **Food** **E:** Church's Fried Chicken, Eddie's Taco House, McDonalds (Playground), Pizza Hut Delivery & Carry out, Wendy's
  W: Piedras Negras
- **Lodg** **E:** DAYS INN Days Inn
- **Other** **W:** **Gleason Veterinary Hospital**

154A San Marcos St, Nogalitos St (Difficult Reaccess)
- **Gas** **E:** J & L Food & Gas*
- **Food** **E:** Maria's Cafe, Tommy's Old San Antonio Cafe
  W: Henry's Tacos To Go, Midtown Grill, Simon's Bakery
- **AServ** **E:** General Brake & Alignment, Hernandez Tire & Muffler
- **Other** **W:** **Collins Branch Garden Library, Fire Station, H-E-B Grocery**

153 Junction I-35, I-37, U.S. 281, Austin, Corpus Christi

Note: I-10 runs concurrent above with I-35. Numbering follows I-35.

573 Probadt St
575 Pine St
576 New Brownfels Road, Gevers St
577 U.S. 87 South, Roland Ave, Victoria
578 Pecan Valley Drive, MLK Drive
- **Gas** **N:** Phillips 66*
  S: Shell*
- **AServ** **N:** General Repairs

579 Houston St, Commerce St.
580 W.W. White Road, Loop 13
581 I-410, Connally Loop
582 Ackerman Road, Kirby
- **TStop** **N:** **Pilot Travel Center*[SCALES] (24 Hrs)**
  S: **Petro*[SCALES] (24 Hrs, Mobil)**
- **Food** **N:** **Arby's (Pilot Travel Center TStop, 24 Hrs), Miss B's Cafe**
  S: KFC
- **Lodg** **S:** **Relay Station Motel**
- **TServ** **N:** **Buddy Storbeck's Diesel Service, Fruehouf (Pilot TS), Janke Engines, Santex International Trucks (Pilot TS), Stewart & Stevenson San Antonio Branch, Toyo Tires, W W Tank & Trailor Parts, Western Star Trucks**
- **TWash** **S:** **Blue Beacon Truck Wash[SCALES] (Petro TS)**

583 Foster Road
- **TStop** **N:** **Flying J Travel Plaza*[LP][SCALES] (RV Dump, FedEx Drop)**
- **Gas** **N:** Diamond Shamrock*[CW] (24 Hrs)
- **Food** **N:** **Jack-in-the-Box (24 Hrs), Thads Restaurant[LP] (Flying J TStop)**
- **RVCamp** **N:** **Camping**
- **ATM** **N:** **Flying J Travel Plaza**

585 FM 1516, Converse
- **TStop** **N:** **Mobil Travel Center*[LP][SCALES] (Laundromat, CB Shop)**
- **Food** **N:** **Winfield's (Mobil Travel Center, 24 Hrs)**
- **Lodg** **N:** **Winfield's (Mobil TS)**
- **TServ** **N:** **Performance Trailor, Texas Truck & Trailer Repair (Mobil TStop), Utility Trailor Sales**
  S: **Rush GMC Truck, Rush Truck Center**
- **TWash** **N:** **Mobil TS**

Bold red print shows RV & Bus parking available or nearby

EXIT TEXAS

ATM N: Mobil TS

587 Loop 1604, Anderson Loop, Randolph AFB, Universal City

589 Graytown Road, Pfeil Road

(590) **Rest Area - RR, Phones, Picnic, Grills, RV Dump, 24 Hr Park P (Both Directions)**

591 Schertz, Fm1518

TStop **N: Exxon Travel Center*(LP)(SCALES) (24 Hrs)**

Food **N: Farmers Restaurant (Exxon TStop)**

Lodg **N: Motel (Exxon TStop)**

TServ **N: Exxon TS**

TWash **N: Exxon TS**

ATM **N: Exxon TS**

593 FM 2538, Trainer Hale Road

TStop **N: Citgo Truck/Auto Plaza***
S: Diamond Shamrock*

Food **N: Dairy Queen (Citgo TStop, Playground)**
S: Restaurant (Diamond Shamrock TStop)

TServ **S: Alamo City Truck & Equip. Service (Diamond Shamrock TS)**

ATM **N: Citgo TS**
S: Diamond Shamrock TS

Other **S: 4 Watt CB Shop**

595 Zuehl Road

597 Santa Clara Road

TServ **N: Brown Truck & Equip. Sales**

599 FM 465, Marion

600 Schwab Road

601 FM 775, New Berlin, La Vernia

FStop **N: Mobil* (24 Hrs)**

Food **N: Arby's (Mobil, 24 Hrs)**

603 East U.S. 90, Alt. U.S. 90, Seguin (Eastbound, No Eastbound Reaccess)

RVCamp **S: RV Park (1.5 Miles)**

604 FM 725

AServ N: Quality Service Automotive

Other **N: Allens Boats & Motors**

605 FM 464

RVCamp **N: Camping**

Other **S: Public Boat Ramp**

607 TX 46, FM 78, New Braunfels, Lake McQueeney

Gas S: Chevron*(D), Exxon*(D)

Food S: Bart's Old Town BBQ, Kettle, **McDonalds**

Lodg **S: (AAA) Best Western, Super 8 Motel**

AServ S: Exxon

609 Bus. TX 123, Austin St

FStop **S: Mobil***

RVCamp **S: Shady Lane Mobil Home & RV Park**

Other **S: Sharp Propane Inc.(LP)**

610 TX 123, San Marcos, Stockdale

TStop **N: Chevron Travel Center* (24 Hrs)**

Gas N: Exxon*(D)

Food **N: Chester Fried Chicken (Chevron TS), Hot Stuff Pizza (Chevron TS), K&G Steakhouse, Long Horn BBQ, Subway (Chevron TS)**

Lodg **N: Econolodge, Holiday Inn**

AServ N: Carter's Tire Service

TServ **N: Carter's Tire Center**

ATM **N: Chevron TS, Exxon**

612 U.S. 90, Seguin

RVCamp **N: Dusty Oaks RV Park (3 Miles, Complete Facility)**

(615) **Weigh Station (Westbound)**

EXIT TEXAS

(616) **Weigh Station (Eastbound)**

617 FM 2438

620 FM 1104, Kingsbury

(621) **Rest Area - RR, Phones, Vending, Picnic, Grills, RV Dump P (Both Directions)**

625 Darst Field Road

TServ **S: Trucks of Texas**

628 TX 80, Nixon, San Marcos

FStop **N: Diamond Shamrock*(D)**

Gas S: Exxon

Med **N: + Hospital**

632 U.S. 90, U.S. 183, Gonzales, Cuero, Luling, Lockhart

TStop **N: Love's* (24 Hrs)**

Food **N: Ice Cream Shoppe (Love's TS)**

ATM **N: Love's TS**

Parks **S: Palmetto State Park (5 Miles)**

637 FM 794, Harwood

642 TX 304, Bastrop, Gonzales

RVCamp **N: Noah's Land RV Park (5.5 Miles)**

Parks **N: Noah's Land Wild Life Park**

649 TX 97, Waelder, Gonzales

653 U.S. 90, Waelder

Gas N: Texaco*(D)

(657) Picnic - Table (Both Directions)

661 TX 95, FM 609, Flatonia, Smithville, Moulton, Shiner

TStop **N: Conoco*, Mobil* (24 Hrs, Showers, 24 Hr Roadside Assistance)**

Gas **S: Exxon* (No Truck Fuel), Shell***

Food **N: Joel's BBQ (Conoco TS)**, Mi Ranchito,

EXIT TEXAS

Stockman's Restaurant (Mobil TS)
S: Dairy Queen, **Grumpy's Restaurant (Grumpy's Motor Inn), Homestyle Fried Chicken (Exxon)**

Lodg **S: Grumpy's Motor Inn, Sav-Inn Antlers Inn**

AServ N: Flatonia Auto & Truck Repair

TServ **N: Flatonia Auto & Truck Repair, Mobil TS**

ATM N: Conoco TS
S: Flatonia State Bank

Other **S: Maytag Washateria (Coin Laundry)**

668 FM 2238, Engle

674 U.S. 77, Schulenburg, La Grange, Hallettsville

FStop **S: Phillips 66*(D)(LP)**

Gas N: Chevron*(D)(CW) (24 Hrs), Exxon*(D)
S: Diamond Shamrock*, Exxon*, Texaco

Food N: Bar-B Q Smoke House, McDonalds (Playground), **Oakridge Restaurant**
S: Burger King, Dairy Queen, Diamond S Family Dining, **Frank's Restaurant**

Lodg **N: Oakridge Motor Inn**

AServ S: Texaco

RVCamp **S: Schulenberg RV Park (see our ad this page)**

ATM N: Chevron
S: Instant Cash

677 U.S. 90

Food **N: Nannie's Biscuit and Bakery**

Other **N: Fostoria Glass Outlet**

682 FM 155, Weimar

Gas N: Exxon(D), Texaco(D)

Food N: Dairy Queen, Fishbeck's BBQ (Texaco)

AServ N: Exxon, Weimar Ford

Med **N: + Hospital**

689 U.S. 90, Hattermann Lane

(692) **Rest Area - Picnic, Grills, RR, Phones, Vending P (Both Directions)**

693 FM 2434, Glidden

695 West TX 71, La Grange, Austin

696 S. TX 71, Bus. TX 71, Columbus, EL Campo

FStop **S: Diamond Shamrock***

Gas N: Chevron*(D)(CW) (24 Hrs), Texaco*
S: Phillips 66*

Food N: Burger King, Denny's, Guadalajara Restaurant, Pizza Hut, Schobels' Restaurant, What A Burger (24 Hrs, Playground)
S: Chester Fried Chicken (Exxon), Church's Fried Chicken (Phillips 66), **McDonalds (Playground)**, Sonic Drive In, Subway

Lodg **N: ◆ Columbus Inn Motel**
S: ◆ Country Hearth Inn

AServ N: Columbus Ford Mercury
S: Exxon

RVCamp **S: Columbus RV Park & Campground**

Med **N: + Hospital**

ATM S: Exxon

Other **N: H-E-B Pantry Foods, Voelkel's Pharmacy, Wal-Mart**
S: Columbus Mini Golf

698 Alleyton Road, US 90, Columbus

Gas N: Augie's*
S: Shell*(D) (see our ad this page)

Food N: Augie's (Augie's Gas), **Jerry Mikeska's BBQ**
S: Baskin Robbins (Shell), Cattlemen's Restaurant, Grandy's (Shell), Taco Bell (Shell)

TServ **S: Bob's Tire & Truck Service**

EXIT TEXAS

ATM N: Shell

699 (29) FM 102 Eagle Lake
RVCamp N: **Happy Oaks RV Park (1 Mile)**
Parks S: **Wild Life Viewing Area**

704 FM 949

709 FM 2761, Bernardo Road
Food N: **Texas Travel Stop and Grill**

713 Beckendorff Road

716 Pyka Road
TStop N: **Sealy***
Food N: **Restaurant (Sealy TS)**
TServ N: **Sealy TS**

718 U.S. 90 (Eastbound, No Eastbound Reaccess, Reaccess Via US 90 E.)

720 TX 36, Sealy, Rosenberg, Bellville, Eagle Lake
FStop S: **Mobil***
Gas N: Exxon(D), Texaco*(D)
S: Chevron*(D)(CW), Shell*(D)(CW)
Food N: Dairy Queen, Hartz Chicken, McDonalds (Playground), Sonic Drive In, **Tony's Restaurant**
S: Buffet House (Chinese), Hinze's BBQ, KFC, **Omar's**, Pizza Hut, Subway, Taco Bell, What A Burger (24 Hrs)
Lodg S: AAA Best Western, **Holiday Inn Express (see our ad this page), Rodeway Inn**
AServ N: Cliff Jones GM, Exxon, Hi/Lo Auto Supply
S: Dowell Plymouth, Dodge, Eagle
RVCamp N: **Airstream RV Sales & Service**(LP)
ATM N: Austin County State Bank
Other S: **Bill's Supermarket, Wal-Mart**

721 U.S. 90 (Westbound, No Westbound Reaccess, Reaccess Via US 90)
AServ N: West 10 Ford Mercury
Other N: **Outlet Center**

723 FM1458, San Felipe, Frydek
TStop N: **Knox***(SCALES) **(Coastal, 24 Hrs)**
Food N: **Subway (Knox TS)**
AServ S: **Riverside Tire Center**
TServ N: **Brown Truck Tire Center**
S: **Riverside Tire Center**
ATM N: **Knox TS**
Parks N: **Stephen F. Austin State Park (3 Miles)**

(726) **Weigh Station (Westbound)**

725 Mlcak Rd. (Westbound)

730 Peach Ridge Road

(731) Picnic - Tables, No Services (Left Exit, Both Directions)

EXIT TEXAS

731 FM 1489, Koomey Road, Simonton
FStop N: **Exxon***
Gas N: Shell*(CW)
S: Diamond Shamrock
Food N: **El Vaquero (Travelodge)**, Villa Fuentes Mexican Restaurant
S: Berg-N-Moore's Country Fried Chicken
Lodg N: **Brookshire Hotel, Travelodge**
AServ S: Diamond Shamrock
RVCamp N: **KOA Kampground (RV Dump)**
Other S: **Brookshire Animal Clinic**

732 FM 359, Brookshire
TStop N: **76 Auto/Truck Plaza***(SCALES) **(Phillips 66, Showers, Laundry), U.S. Truxtop* (Citgo)**
FStop S: **Conoco***(SCALES)
Gas N: Texaco*
S: Chevron*(CW), Coastal*
Food N: Charlie's Hamburgers, Joel's BBQ, **Seventy-Six Restaurant (76 TS)**
S: Burger King (Coastal), Jack-in-the-Box (24 Hrs)
Lodg N: DAYS INN **Days Inn**
TServ N: **76 Auto/Truck Plaza**(SCALES)
S: **Brookshire Truck and Trailor Parts, Kunkel Motors**
TWash S: **Conoco**
ATM N: **76 Auto/Truck Plaza, U.S. Truxtop**
S: Chevron

737 Pederson Road

740 FM 1463, Pin Oak Road
Gas S: Chevron*
Food S: Calli Chocolates & Coffee House, Truffle Hound Gourmet

EXIT TEXAS

Med S: **Columbia Katy Medical Center**

742 West US 90, Katy; Katy - Fort Bend County Road
FStop S: **Mobil***
Gas N: Citgo
ATM N: Citgo
Other N: **Police Station**

743 Grand Pkwy, TX 99
AServ S: Westside Chevrolet

745 Mason Road
Gas S: Diamond Shamrock*, Shell*(CW) (24 Hrs)
Food S: Black-Eyed Pea, Chick-fil-A (Playground), Chili's, China Inn Cafe, Hartz Chicken, Johnny Carino's Italian Kitchen, Landry's Seafood, Luby's Cafeteria, Marble Slab Creamery Ice Cream, Marco's, McDonalds (Playground), Papa John's Pizza, Pizza Hut, Southern Maid Donuts, Taco Cabana
Lodg S: Holiday Inn Express, Ramada Limited, Super 8 Motel
AServ S: Discount Tire Company, Jiffy Lube
ATM S: Gerland's Food Store, Texas Commerce Bank
Other S: **Briar Toys, Clubhouse Books, Eckerd Drugs, Fed Ex Drop Box, Gerland's Food Store (24 Hrs), Katy Vision Center, Petco, Rejoice Christian Bookstore, Stop-N-Fast Food Store**

747 Fry Road
Gas N: Phillips 66*, Shell*(CW)
S: Chevron*(CW), Diamond Shamrock*(D), Shell*(CW)(LP) (24 Hrs)
Food N: Blimpie's Subs, Burger King (Playground), Dairy Queen, Daylight Donuts, Hartz Chicken Buffet, McDonalds (Playground), Pizza Hut, Sonic Drive In, Taco Bell, Texas Borders, Victor's Casa Garcia
S: Bo's Best BBQ (Shell), Boston Market, El Chico Mexican, Fazoli's Italian Food, IHOP, Jack-in-the-Box, Omar's, Orient Express, Outback Steakhouse, Shipley Do-nuts, The Original Pasta Company, Wendy's
AServ N: Econo Lube N Tune, Wal-Mart
ATM N: **Kroger Supermarket**
S: Albertson's Grocery, Community Bank, Diamond Shamrock, NationsBank, Shell
Other N: **Garden Ridge Pottery, Kroger Supermarket (24 Hrs, Pharmacy)**, Kwik Kar Wash, **Mail Boxes Etc, Wal-Mart (Vision Center, 1-Hr Photo, Pharmacy), WalGreens (24 Hrs, 1-Hr Photo)**
S: **Albertson's Grocery (Pharmacy), Fox Photo 1-Hr Lab, Gerland's Food Fair (24 Hrs, Pharmacy), Postmark Mail & Copies, Target**

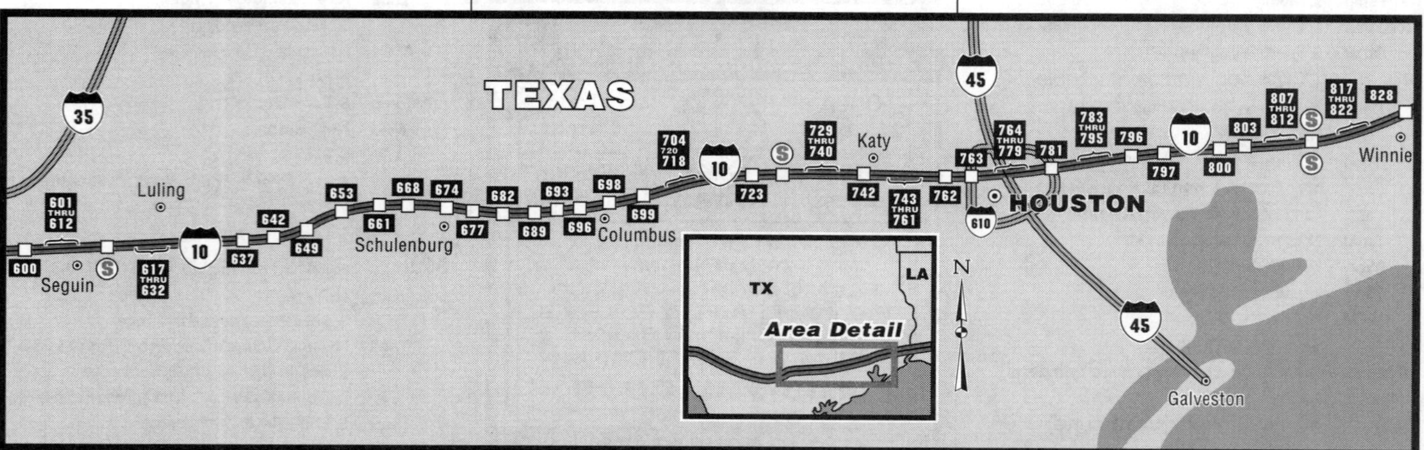

Bold red print shows RV & Bus parking available or nearby

EXIT TEXAS

Department Store

748 Barker - Cypress Road
- **Gas** N: Shell*
- **Food** S: Blue Corn Grill, Cracker Barrel

750 Park Ten Blvd. (Westbound)
- **Food** N: Freddy's Breakfast & Lunch
- **Other** N: US Post Office

751 TX 6, Addicks
- **Gas** S: Chevron* (24 Hrs), Conoco*(D) (CNG Natural Gas), Texaco*(D)
- **Food** N: Cattleguard Restaurant, The Park Cafe (Holiday Inn Select), Waffle House
  S: Burger Tex Burgers & Steaks, Charlie's Hamburger Joint, Denny's, Durango's, El Yucatan, Jack-in-the-Box (24 Hrs), Kim House Hunan Cuisine, Lupe Tortilla (Playground), Paddington Station, Pasta Lo Monte's, Red Barn Creamery
- **Lodg** N: ◆ Drury Inn, ◆ Holiday Inn Select, Homestead Village, ◆ Red Roof Inn
  S: Fairfield Inn by Marriott, La Quinta Inn & Suites, Motel 6
- **AServ** S: Texaco
- **ATM** S: Wells Fargo
- **Other** N: Sam's Club

753A Eldridge Parkway, Addicks - Fairbanks Rd.
- **Gas** N: Conoco*(D)(CW)
  S: Diamond Shamrock*(CW)
- **Food** N: Shipley Do-nuts (Conoco)
- **Lodg** N: AAA Marriott
- **AServ** N: Conoco
  S: Kwik Kar Lube & Tune

753B Dairy - Ashford Road
- **Gas** S: Exxon*(D)(CW)
- **Food** S: Becks Prime, Cliff's & Otto's Burgers & BBQ, Ryan's Family Steakhouse, Shoney's (Shoney's Inn), Subway, What A Burger (24 Hrs)
- **Lodg** S: ◆ Hilton Inn, Shoney's Inn & Suites
- **AServ** S: Don McGill Porsche, Audi, Ernie Guzman Pontiac GMC Trucks, McGinnis Cadillac, McGinnis Mitsubishi
- **Other** S: Golf Warehouse

754 Kirkwood
- **Gas** S: Chevron(D), Shell*(CW)
- **Food** S: Carrabba's Italian Grill, Dirty's Restaurant Bar, IHOP, Taco Cabana, The Original Pasta Co.
- **Lodg** S: ◆ Hampton Inn
- **AServ** N: Crown Dodge, Discount Tire Co., Don McGill Toyota, Nils Sefeldt Volvo, Professional Car Care, Saturn of Houston, West Point Buick, West Point KIA, West Point Lincoln Mercury, West Side Lexus
  S: Chevron, Mac Haik Chevrolet GEO, Mac Haik Subaru
- **Other** S: Kirkwood Pharmacy, Mountasia Fantasy Golf

755 Wilcrest Dr., E. & W. Beltway 8, Frontage Rd
- **FStop** S: Star Parts* (24 Hrs)
- **Gas** N: Gas*
  S: Exxon* (24 Hrs), Sunny's*, Texaco*
- **Food** N: Sam's Deli Diner
  S: China Plaza (La Quinta Inn), Jo Jos (24 Hrs), McDonalds (24 Hrs, Playground, Exxon), Shipley Do-nuts (Exxon), Steak & Ale
- **Lodg** S: AAA La Quinta Inn

EXIT TEXAS

- **AServ** N: Discount Collision, Jeff Haas Mazda, NTB
  S: Star Parts, Texaco
- **TServ** S: Star Parts
- **ATM** S: Exxon, Star Parts, Sunny's
- **Other** N: Washateria
  S: Joe's Golf House, Sunny's Food Store, Town & Country Mall

756A Beltway 8, N. & S. Frontage Roads (Westbound)

756B N. & S. Sam Houston Toll Way

757 Gessner Road
- **Gas** N: Chevron(CW) (Mr. Car Wash), Exxon*
  S: Shell*, Texaco(D)(CW)
- **Food** N: Asiana Garden, Ci Ci's Pizza, Golden King Chinese Buffet, Goodson's Cafe, Luther's BBQ, Schlotzkys Deli, ThunderCloud Subs, What A Burger (24 Hrs)
  S: Champs Country Breakfast, Fuddruckers, Jason's Deli, Olive Garden, Pappa Deaux Seafood, Pappasitos Cantina, Romano's Macaroni Grill, Taste of Texas Restaurant
- **AServ** N: CSI Collision, Exxon, Pep Boys Auto Center
  S: Firestone Tire & Auto, Mac Haik Ford, Montgomery Ward, Sears
- **Med** S: ✚ Memorial Hospital Memorial City
- **ATM** S: Wells Fargo
- **Other** N: Boater's World Discount Marine, Eckerd Drugs, Memorial Market, U-Haul Center(LP)
  S: Discovery Zone, Memorial City Shopping Mall, Office Depot, Sony Theatres, WalGreens

758A Bunker Hill Road, Memorial City Way
- **Gas** S: Diamond Shamrock*(D), Texaco(D)
- **Food** S: Charlie's Hamburger Joint, Cobe Japanese, Hunan Inn, Hunan Village Mongolian B-B-Q, Prince's Hamburgers, Shipley Do-nuts, Subway
- **Lodg** S: DAYS INN ◆ Days Inn, Quality Inn
- **AServ** N: Spring Branch Honda
  S: Goodyear Tire & Auto, Texaco
- **ATM** S: Compass Bank
- **Other** S: American Pro-Line Golf, Bunker Hill Bowl, Eyemasters, Target Department Store, Toys "R" Us

758B Blalock Road, Campbell Road, Echo Lane
- **Gas** S: Chevron*(CW)
- **Food** N: Sonic Drive In
  S: Baskin Robbins, Frankie and Johnnies, McDonalds (Playground, 24 Hrs, Chevron), Ninfas, Three E's, Villages Cafe (Kroger)
- **AServ** N: Adams Automotive, Lube Stop
  S: Robert's Auto Works
- **Med** S: ✚ Columbia West Houston Medical Center
- **ATM** S: Stopn Go
- **Other** N: Fiesta Supermarket, Key Rexall Drugs, The Photo Lab
  S: AAA Auto Club, Echo Lane Animal Clinic, Factory Eyeglass Outlet, Kroger Supermarket (24 Hrs, Pharmacy, 1/2 Hr Photo), Postmark Mail, Stopn Go, United Optical, WalGreens

759 Campbell Rd. (Westbound)
- **Gas** N: Phillips 66*
- **Food** N: Bel Ami Croissant, Ciro's Cibi Italiani, General Joe's Chopstix, The Great Charcoal Chicken, Yummies & Co.
- **Med** N: ✚ Hospital
- **Parks** N: City Hall (Playground)

EXIT TEXAS

- **Other** N: Mail It!

760 Bingle Road, Voss Road
- **Gas** S: Exxon(D), Texaco(D)
- **Food** S: Good Co. Texas Bar-B-Q, Las Alamedas, Pappy's, Saltgrass Steak House, Southwell's Hamburger Grill, The Mason Jar, Ugo's Italian Grill
- **AServ** S: Exxon, Texaco
- **Med** N: ✚ Hospital
- **ATM** S: NationsBank
- **Other** S: Academy (Outdoors), Animal Emergency Clinic, Scuba, United Pharmacy

761A Chimney Rock Road, Wirt Road
- **Gas** S: Chevron*(D)(CW), Exxon*(CW), Shell*(CW)
- **Food** S: 59 Diner, Dixie's Red Hot Roadhouse, McDonalds, TCBY (Exxon)
- **Lodg** S: ◆ La Quinta Inn
- **ATM** S: Exxon, Stopn Go, Sunbelt National Bank
- **Other** S: Stopn Go*

761B Antoine Drive (Eastbound)
- **Gas** S: Stopn Go*
- **Food** S: Denny's, Fuzzy's Pizza & Cafe, Mandola's Italian Restaurant, Papa John's Pizza, What A Burger (24 Hrs)
- **AServ** S: NTB
- **ATM** S: Memorial Market
- **Other** S: Hospital for Animals, Mail Boxes Etc, Memorial Market, Rei Sports Supplies, West Marine

762 Silber Road, Post Oak Road, Antoine Dr.
- **Gas** N: Chevron, Stopn Go*
  S: Shell*(CW)
- **Food** N: Hunan Chef All Day Buffet, Steak Kountry Buffet, Wheel Burger
  S: Aubrey's Ribs, Jack-in-the-Box, Shipley Do-nuts
- **Lodg** S: ◆ Holiday Inn, Ramada Plaza Hotel
- **AServ** N: Chevron, Courtesy Geo Chevrolet, Katy Freeway Tire (Firestone)
  S: Courtesy Chevrolet
- **Med** N: ✚ Family Practice & Walk-In Clinic
- **ATM** N: Stopn Go
- **Other** S: Country Club Car Wash

763 I-610 North & South

764 Westcott St, Washington Ave, Katy Road
- **Gas** S: Chevron* (24 Hrs)
- **Food** N: Denny's
  S: McDonalds, Sam's Place
- **Lodg** N: Roadway Inn
  S: Star Motel
- **AServ** S: Cooksey's West End Radiator
- **Parks** S: Memorial Park

765A TC Jester Blvd
- **Gas** S: Exxon*, Texaco*(D)
- **Food** N: Perfecto Espresso & Tea
- **AServ** N: Comet Tire, Discount Auto Parts Exchange, TC Automotive Service
  S: Reliable Battery Company, Texaco
- **Other** N: City of Houston Fire Station, Texas Scuba, Washateria (Coin-Op Laundry)
  S: Metro Propane(LP)

765B N. Shepherd Dr., Patterson St., N. Durham Dr. (Caution - One Way Streets)

Bold red print shows RV & Bus parking available or nearby

EXIT		TEXAS
	Gas	**N:** Stewart's Walco Service*
		S: Diamond Shamrock*
	Food	**N:** Wendy's
		S: Jax Grill, Pizzitola's Bar-B-Cue
	Lodg	**N:** (AAA) Howard Johnson (see our ad this page)
	AServ	**N:** ABC Auto Parts, Castl Transmissions, Mexico Auto, Midas Muffler & Brakes, Stewart's Walco Service
	ATM	**N:** NationsBank
		S: Diamond Shamrock
	Other	**N:** **Pet Vet Animal Hospital**
766		Studemont St., Heights Blvd. , Yale St. (Eastbound Exit Is Ex. 767A)
	Gas	**N:** Chevron* (24 Hrs)
		S: Exxon*(D)
	AServ	**N:** Kelley's Complete Car Care
		S: Exxon
767A		Yale St, Height Blvd, Studemont St.
	Gas	**S:** Exxon*(D)
	AServ	**S:** Exxon
767B		Taylor St
768AB		768A - I-45 N. Dallas, 768B - I-45 S. Galveston (Westbound Exit 768B Is A Left Exit Eastbound Exit 768A Is Left Exit)
769A		Downtown, Smith St
	Food	**S:** All Street Deli, Longhorn Cafe, Palace Cafe
	ATM	**S:** NationsBank, Texas Commerce Motor Bank
	Other	**S:** **Alley Theater, The Jesse Jones Hall for the Performing Arts, The Lyric Center, The Theater District**
769C		McKee St, Hardy St, Nance St (Difficult Eastbound Reaccess)
	Parks	**S:** **James Bute Park**
769B		San Jacinto St., Main St. (No Westbound Reaccess)
770A		US 59S Victoria, Dowtown (Westbound Is A Left Exit, Limited Access Hwy)
770C		US 59 N Cleavand
770B		Jensen Dr, Gregg St, Meadow St. (Difficult Reaccess)
	TServ	**N:** **Daniel Radiator**
771A		Waco St
	Food	**N:** Frenchy's Chicken
	Lodg	**N:** Sunrise Motel, Waco Motel
	Other	**N:** **Washateria**
		S: **American Food Store & Washateria**
771B		Lockwood Drive
	FStop	**S:** **Texaco***
	Gas	**N:** Chevron*(D)(CW) (24 Hrs), Mobil*
		S: King Truck Stop*(D)
	Food	**N:** McDonalds (Playground)
		S: Kozy Kitchen BBQ
	Lodg	**S:** Palace Inn
	AServ	**N:** Auto Zone Auto Parts, Western Auto
	ATM	**N:** Texas Commerce Bank
	Other	**N:** **Eckerd Drugs, Fiesta Grocery, Walgreens (Pharmacy, 1-Hr Photo)**
772		Kress St, Lathrop St (Difficult Reaccess)
	Gas	**N:** Citgo*, Conoco, Exxon*, Franco's Drive-In Grocery*
		S: Best Buy Food Mart
	Food	**N:** Dixie Maid Hamburger, Domino's Pizza, Popeyes Chicken, Restaurant El Pariente, Samburger
		S: Burger King (Playground), El Chinero, Ostioneria 7 Mares #5, Porras Prontito
	AServ	**N:** Anderson Motors, Conoco, Cowles Bros. Tire Store, Denver Harbor Auto Clinic, Parts USA
	TServ	**N:** **D H Tire**
	ATM	**N:** Citgo
	Other	**N:** Car Wash, **Fire Dept., Matamoros Meat Market, Public Library, Rice Supermarket,** Wash & Vac Truck & Car Wash (Light Trucks), **You Do Washateria (Coin Op Laundry)**
		S: **Kress Food Store, Lathrop Washateria, Russo's Food Market**
773A		U.S. 90 Alt, N Wayside Drive (No Westbound Reaccess)
	FStop	**S:** **Coastal*, Mobil***
	Gas	**N:** Diamond Shamrock*
		S: Shell*(CW)
	Food	**N:** Jack-in-the-Box, What A Burger (24 Hrs)
		S: Church's Fried Chicken, Cozumel Refresqueria, Mucho Mexico !!
	AServ	**N:** Wayside Auto & Truck Parts
		S: Buddy Doyle's Diagnostic Car Clinic, G & G Body Shop, J & T Parts & Repair, Quality Tire Service
	TServ	**N:** **Houston Truck Parts, Ogburn's Brake, Wayside Truck Parts (& Auto Parts)**
		S: **Buddy Doyle's, Mobil**
	ATM	**N:** Diamond Shamrock
		S: Mucho Mexico
773B		McCarty Drive
	Gas	**N:** Exxon*(D), Shell*(CW)
	Food	**S:** Don Chile Mexican, Subway
	AServ	**N:** Harbor Dale Radiator, McCarty Auto Parts
	TServ	**N:** **Depco Diesel Engine, Harbor Dale Radiator, Houston Freightliner**
		S: **A-1 Public Scales (24 Hrs), A-1 Truck Parts & Equip., Amigo Truck Parts(SCALES), Cat Scales, Nick's Diesel Service, Truck Wash 24 Hr (Oil Changes, etc.), Trucks Of Texas**
	TWash	**S:** **Truck Wash 24 Hr**
	Other	**N:** **East Freeway Animal Clinic**
774		Gellhorn Drive
775AB		Junction I-610
	Lodg	**S:** (DAYS INN) (AAA) Days Inn, (see our ad this page)
776A		Mercury Drive, Jacinto City, Galena Park
	Gas	**N:** Citgo*
		S: Diamond Shamrock* (24 Hrs), Mr. Mercury*, Texaco*(D)(CW)
	Food	**N:** Burger King, Jalisco, McDonalds (Playground), Pizza Inn, Taqueria Arandas
		S: Dairy Queen, Donald Donuts, Luby's Cafeteria, Pizzini's, Steak & Ale, Taqueria Mi Jalisco, Tony's Seafood Galley
	Lodg	**N:** ◆ Fairfield Inn, Hampton Inn
	AServ	**N:** Citgo
	TServ	**N:** **Lakeview Volvo**
	ATM	**S:** Capital Bank, Diamond Shamrock, Prime Bank
	Other	**S:** **Fire Dept., Mercury Dr. Pharmacy**
776B		John Ralston Road, Holland Ave, Jacinto City, Galena Park
	Gas	**N:** Chevron*, Mobil*(D)
		S: Eddie's Food Store*
	Food	**N:** Denny's, East China Restaurant, Golden River Chinese Buffet, Pappasitos, **The Raintree**
	Lodg	**N:** (DAYS INN) ◆ **Days Inn, La Quinta Inn**
	AServ	**N:** Chevron, Discount Tire Company, Western Auto
		S: Discount Auto Service Center Inc., Parker Automotive Inc.
	Other	**N:** **Army Surplus Command Post, Fiesta Grocery, Kroger Supermarket (24 Hrs), Sunny's Food Store, Target Department Store**
		S: **Jacinto City Animal Hospital**
778A		FM 526, Pasadena, Federal Road (Westbound Says Normandy Street)
	FStop	**N:** **Conoco***
	Gas	**N:** Diamond Shamrock*(D)
		S: Chevron, Conoco*, Diamond Shamrock* (24

Bold red print shows RV & Bus parking available or nearby

EXIT TEXAS

Hrs), Exxon, Shell*

Food **N:** Casa Ole, Champ's, El Imperial, Express Subs, Golden Corral, KFC, Little Caesars Pizza, Long John Silvers, Luby's Cafeteria, Panaderia Bakery, Pizza Hut, Popeyes Chicken, Taco Bell, Wendy's
S: Bennigan's, Bradley's Steakhouse, Catfish Kitchen, Church's Fried Chicken, Jack-in-the-Box, **James Coney Island, McDonalds (Playground)**, Pappas Seafood House, Peking Bo, Taqueria Los Reyes

AServ **N:** Andrell Automotive
S: Chevron, Exxon, Hi/Lo Auto Supply, Midas Muffler Shops, Romeo's Auto Repair

TServ **S:** **Romeo's Truck Repair**

TWash **S:** **Truck Wash**

ATM **N:** Conoco, Lone Star Bank, NationsBank
S: Diamond Shamrock

Parks **S:** **J.P. White Park (Playground)**

Other **N:** **Eckerd Drugs, H-E-B Pantry Foods, Office Depot**, Sam Car Wash
S: **Auto Wash at the Truck Wash, Eastway Cinema 4**

778B Normandy St.

FStop **S:** **Eastop*, Texaco*[D]**

Food **S:** Taqueria Guanajuato

Lodg **S:** Bayou Motel, Normandy Inn

AServ **S:** Carquest Auto Parts, Fain's Auto Service & Parts, Jimmy's Paint & Body Shop, Market St. Used Auto Parts, Mauricio's, Spurlock's Garage, Texaco

TServ **S:** **Spurlock's Garage**

Other **S:** **Animana Veterinary Center, Garcia & Associates Eye Clinic**

779A Westmont St. (Difficult Westbound Reaccess)

Gas **N:** Chevron*

Lodg **N:** Interstate Motor Lodge

Med **N:** **✚ Columbia East Houston Medical Center**

779 Uvalde Road, Freeport St, Market St. (Westbound Lists Exit As Exit 779B)

Gas **N:** Exxon[D], Gas-N-Go II*, Mobil*

Food **N:** Baskin Robbins, Frey's, HoBo's, IHOP, Jack-in-the-Box, KFC, Poncho's Mexican Buffet, Shipley Do-nuts (24 Hrs), Subway, Szechuan Wok, Uvalde Malt & Burger

AServ **N:** Exxon, Meineke Discount Mufflers, Uvalde Mufflers

ATM **N:** Wood Forest National Bank

Other **N:** **ACE Hardware, Dollar Cinema, Fire Dept., Sellers Bros., Texas State Optical, Walgreens (Pharmacy)**

780 Market St Road, Uvalde Rd, Freeport St

Gas **N:** Mobil*, Texaco*

Food **N:** Nuevo Leon Meat Market
S: Black-Eyed Pea, Joe King Steakhouse, Marco's Mexican, **McDonalds (Wal-Mart)**, What A Burger (24 Hrs)

AServ **N:** Crazy Don's Used Tires (24 Hrs), Freeport Tire Service
S: NTB, Sam's Club, Wal-Mart

ATM **N:** Texaco
S: **Wal-Mart**

Other **N:** **Bi-Rite Supermarket**
S: **Sam's Club, The Home Depot[LP], U-Haul Center[LP], Wal-Mart (24 Hrs, Optical, Pharmacy, 1-Hr Photo)**

781A Beltway 8, Frontage Rd (Westbound Exit Says Sam Houston Pkwy, Limited Access Hwy)

FStop **S:** **Mobil***

AServ **S:** Baker Tire Co., Hanvy Auto Electric

TServ **S:** **Baker Tire Co.**

781B Beltway 8, Sam Houston Pkwy (Westbound Exit Says Market St.)

Food **N:** Brewsky's (Holiday Inn)

Lodg **N:** Holiday Inn

Med **N:** **✚ Columbia East Houston Medical Center**

Other **N:** **North Channel Area Chamber of Commerce**

782 Dell Dale Ave.

Gas **N:** Harper's*[D] (Washateria)

Food **N:** Helen's Malts "N" Hamburgers, Mom's Kountry Cafe

Lodg **N:** Days Inn, Dell Dale Motel
S: Shady Glen Motel

RVCamp **N:** **Bob's LP Gas[LP], Channelview Supply Company (parts,supplies,service)**
S: **Channelview Supply Co.[LP]**

Other **N:** **Bob's LP Gas[LP] (Kerosene), Bud's LP Gas[LP], Dell Dale Superstore, Let Carter's Cleaners[CW]**

783 Sheldon Road, Channelview

FStop **N:** **Coastal*[D]**

Gas **N:** Diamond Shamrock*, Shell*[CW] (24 Hrs)
S: Chevron* (24 Hrs), Shell

Food **N:** Burger King, Jack-in-the-Box (24 Hrs), KFC, **Nick's (Best Western)**, Pizza Hut, Pizza Inn, Popeyes Chicken, Subway, Taco Bell, What A Burger (24 Hrs)
S: Captain D's Seafood, El Tejano, McDonalds (Playground), Wendy's

Lodg **N:** ◆ **Best Western**, I-10 Motel, Super 8 Motel, Travelodge Suites, Days Inn (see our ad this page)

AServ **N:** Auto Zone Auto Parts, Channelview Auto Supply, Discount Tire Co., Western Auto
S: Lyall Bros. Collision, Shell, Spiller's Automotive

TServ **N:** **DFW Truck Sales**

ATM **N:** Diamond Shamrock, Highlands State Bank
S: Chevron

Other **N:** **Eckerd Drugs, Gerland's Food Fair (24 Hrs), Living Word Book & Gift Ctr., US Post Office**

784 Cedar Lane, Bayou Drive

Gas **N:** Citgo*

Food **S:** I-10 Mexican Restaurant

Lodg **N:** Magnolia Motel, **Ramada Limited**

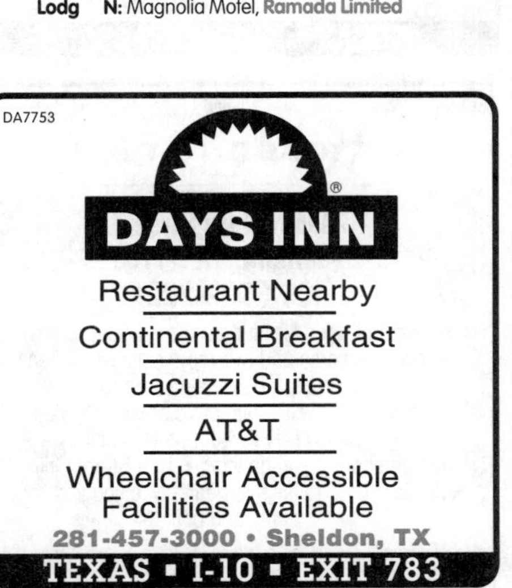

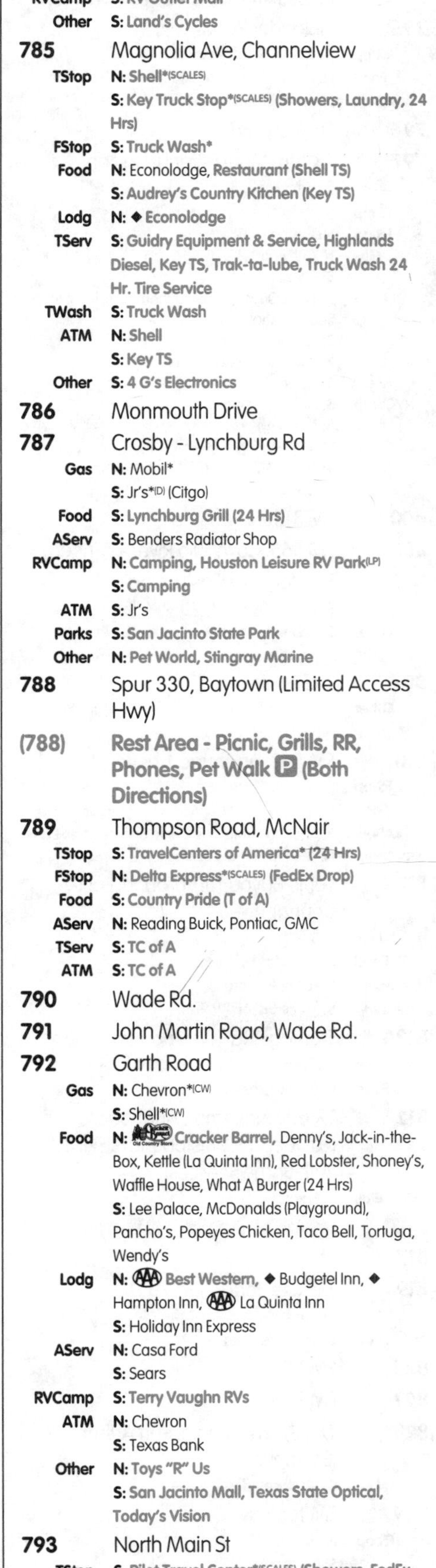

TServ **S:** **Channelview, Griffith Truck & Equip, Tommy's Truck Sales & Equip**

RVCamp **S:** **RV Outlet Mall**

Other **S:** **Land's Cycles**

785 Magnolia Ave, Channelview

TStop **N:** **Shell*[SCALES]**
S: **Key Truck Stop*[SCALES] (Showers, Laundry, 24 Hrs)**

FStop **S:** **Truck Wash***

Food **N:** Econolodge, **Restaurant (Shell TS)**
S: **Audrey's Country Kitchen (Key TS)**

Lodg **N:** ◆ **Econolodge**

TServ **S:** **Guidry Equipment & Service, Highlands Diesel, Key TS, Trak-ta-lube, Truck Wash 24 Hr. Tire Service**

TWash **S:** **Truck Wash**

ATM **N:** **Shell**
S: **Key TS**

Other **S:** **4 G's Electronics**

786 Monmouth Drive

787 Crosby - Lynchburg Rd

Gas **N:** Mobil*
S: Jr's*[D] (Citgo)

Food **S:** **Lynchburg Grill (24 Hrs)**

AServ **S:** Benders Radiator Shop

RVCamp **N:** **Camping, Houston Leisure RV Park[LP]**
S: **Camping**

ATM **S:** Jr's

Parks **S:** **San Jacinto State Park**

Other **N:** **Pet World, Stingray Marine**

788 Spur 330, Baytown (Limited Access Hwy)

(788) Rest Area - Picnic, Grills, RR, Phones, Pet Walk P (Both Directions)

789 Thompson Road, McNair

TStop **S:** **TravelCenters of America* (24 Hrs)**

FStop **N:** **Delta Express*[SCALES] (FedEx Drop)**

Food **S:** **Country Pride (T of A)**

AServ **N:** Reading Buick, Pontiac, GMC

TServ **S:** **TC of A**

ATM **S:** **TC of A**

790 Wade Rd.

791 John Martin Road, Wade Rd.

792 Garth Road

Gas **N:** Chevron*[CW]
S: Shell*[CW]

Food **N:** **Cracker Barrel**, Denny's, Jack-in-the-Box, Kettle (La Quinta Inn), Red Lobster, Shoney's, Waffle House, What A Burger (24 Hrs)
S: Lee Palace, McDonalds (Playground), Pancho's, Popeyes Chicken, Taco Bell, Tortuga, Wendy's

Lodg **N:** AAA **Best Western**, ◆ Budgetel Inn, ◆ Hampton Inn, AAA La Quinta Inn
S: Holiday Inn Express

AServ **N:** Casa Ford
S: Sears

RVCamp **S:** **Terry Vaughn RVs**

ATM **N:** Chevron
S: Texas Bank

Other **N:** **Toys "R" Us**
S: **San Jacinto Mall, Texas State Optical, Today's Vision**

793 North Main St

TStop **S:** **Pilot Travel Center*[SCALES] (Showers, FedEx**

Bold red print shows RV & Bus parking available or nearby

EXIT	TEXAS
	Drop)
Food	S: KFC Express (Pilot TS), Subway (Pilot TS)
AServ	N: Baytown Nissan, Ron Craft Chrysler
795	Sjolander Road
TStop	N: Pappa Truck Stop* (Showers)
Food	N: Convenient Cafeteria (Pappa Truck Stop)
AServ	N: Auto Supply (Pappa Truck Stop)
796	Frontage Rd
797	TX 146, Mont Belvieu, Dayton, Baytown
TStop	S: Mobil*
FStop	N: Conoco*, Texaco*
Gas	N: Chevron*
	S: Exxon
Food	N: Dairy Queen, IHOP, Iguana Joe's, Pizza Inn, Subway, Waffle House
	S: Jack-in-the-Box, Popeyes (Mobil TS)
Lodg	N: Motel 6, Scanadian Inn
AServ	S: Exxon
RVCamp	S: KOA Kampground
Med	S: Hospital
Other	N: Stop-N-Good Food Market
	S: Able LP Gas Inc.(LP), Abshier-Meuth Animal Hospital
800	FM 3180
803	FM 565, Cove, Old River - Winfree
806	Frontage Rd (Eastbound)
Parks	S: Wallisville Heritage Park
Other	S: US Post Office, Wallisville Lake Project Ranger Station
807	Wallisville
Other	S: Wallisville Heritage Park, Wallisville Ranger Station
810	FM 563, Anahuac, Liberty
FStop	S: Chevron* (24 Hrs)
Food	S: Blimpie's Subs (Chevron)
AServ	S: Anahuac Auto, Bumper to Bumper Auto Parts
RVCamp	N: Mobile Home Parking RV (8 Miles)
811	Turtle Bayou Turnaround (Eastbound)
Gas	S: Shell*(D)
Food	S: Grandma's Diner
AServ	S: Ed Santos Ford
RVCamp	S: Turtle Bayou RV Park
812A	Frontage Rd. (Westbound)
Gas	S: Shell*
Food	N: Bayou Crossing
812	TX 61, Hankamer, Anahuac (Eastbound, Westbound Is Exit 813 With Same Services)
Gas	S: Exxon*(D)
(815)	Weigh Station (Both Directions)
817	FM 1724
819	Jenkins Road
Gas	S: Texaco*
Food	S: Stuckey's (Texaco)
822	FM 1410
827	FM 1406
828	TX 73, 124, Winnie, Port Arthur (Eastbound)
Med	S: Hospital
829	FM 1663
FStop	N: Exxon*
	S: Phillips 66*
Gas	N: Mobil*(D), Texaco*(D)
Food	N: Burger King (Mobil), McDonalds, Taco Bell
	S: AL T's Cajun, Pizza Inn, Subway (Phillips 66), Waffle House
Lodg	S: AL T's Inn, Best Western
AServ	S: Vick's Road Service (24 Hr)
RVCamp	N: RV Park
	S: AL T's RV Park
ATM	N: Texaco
	S: Phillips 66
833	Hamshire Road
FStop	N: Chevron*
Food	N: Bergeron's (Chevron)
(837)	Picnic - Shelters, Grills (Both Directions)
838	FM 365
Gas	N: Shell*
RVCamp	N: Campground & RV Park
843	Smith Road
845	FM 364, Major Dr
846	Brooks Road
848	Walden Road
TStop	S: Petro*(SCALES) (Mobil, Petro)
Gas	N: Texaco*(D)
	S: Chevron* (24 Hrs)
Food	N: Pappadeaux Seafood Kitchen
	S: Iron Skillet (Petro TS), Jack-in-the-Box, Subway (Chevron)
Lodg	N: ◆ Holiday Inn
	S: Ramada Limited
TServ	N: Performance Truck
	S: Petro-lube (Petro TS)
TWash	S: Blue Beacon Truck Wash (Petro TS)
Parks	S: Tyrrell Park

EXIT	TEXAS
Other	N: US Post Office
849	U.S. 69 South, Washington Blvd, Port Arthur
850	Washington Blvd.
Gas	N: Exxon*(D), Mobil*
Food	N: Don's Seafood & Steakhouse, Outback Steakhouse
Lodg	N: Beaumont Hilton (see our ad this page)
TServ	N: Smart's Truck & Trailor Equip
851	U.S. 90, College St
Gas	N: Birdsong's(D)(LP), Chevron(CW), Racetrac*
	S: Exxon*(CW), Mobil, Shell*(D)
Food	N: Carrabba's, Fujiyama, Golden Corral, Waffle House
	S: Baskin Robbins, Burger King, Catfish Kitchen, China Hut, Dairy Queen (Playground), Happy Buddha, IHOP, Jason's Deli, Pig Stands Restaurant (24 Hrs), Pizza Hut, Poncho's, Shoney's, Taste of China
Lodg	N: AAA Best Western, Roadrunner Motor Inn
	S: Econolodge
AServ	N: Binswanger Glass, Birdsongs, Hi/Lo Auto Parts, Magic Car & Lube Service
	S: Firestone Tire & Auto, Mobil, Super Shops
RVCamp	N: New World RV Sales & Service
ATM	N: Racetrac
	S: Market Basket, Texas Commerce Bank
Other	N: Dawson Marine, M & D Supply (Hardware)
	S: Cinema I & II, Market Basket, Office Depot, Texas State Optical, U-Haul Center(LP), Walgreens (24-Hr Pharmacy, 1-Hr Photo)
852A	Laurel Ave (Difficult Reaccess)
Gas	N: Chevron*(CW) (24 Hrs), Texaco*(D)(CW), Willie Ray's BBQ Co.
Food	N: Chili's, Suchun
AServ	N: Joel's Paint & Body Shop
ATM	N: Texaco
852B	Calder Ave, Harrison Ave, Gladys Ave
Gas	N: Conoco*
	S: Chevron, E-Z Mart*
Food	N: Bennigan's, Casa Ole, La Suprema, Olive Garden, Steak & Ale
	S: Church's Fried Chicken, Great China Restaurant, McDonalds, The Sports Page
Lodg	S: DAYS INN Days Inn, AAA La Quinta Inn
AServ	S: American Transmission, Chevron, ChevyLand
Med	S: St. Elizabeth Hospital
ATM	N: Community Bank, PrimeBank, Texas Commerce Bank
Other	S: Art Museum of S.E. Texas, StatCare Pharmacy, Texas Marine
853A	U.S. 69 North, Lufkin (Limited Access Hwy)
853B	11th St
Gas	N: Fina*
	S: Conoco*, E-Z Mart*
Food	N: 11th Street Grill (Holiday Inn), Denny's, Hoffbrau Steaks, Ninfa's, Red Lobster, Restaurant (Ramada Inn), Waffle House
	S: Chula Vista, Luby's Cafeteria, Restaurant (Castle Motel)
Lodg	N: AAA Best Western, ◆ Holiday Inn (see our ad this page), Ramada Inn, Scottish Inns, AAA Super 8 Motel, Travel Inn
	S: Castle Motel (Laundromat), AAA Quality Inn
Other	N: Market Basket
853C	7th St (Eastbound)
Gas	S: Shop N Go*
854	Spur 380, Babe Zaharias Museum,

Bold red print shows RV & Bus parking available or nearby

EXIT	TEXAS
	Fairgrounds (Mlk Pkwy)
Gas	N: Chevron*
Food	N: Burger King (Playground)
AServ	S: Fasulo Paint & Body, Neal Jacobs Automotive
RVCamp	S: **Morgan RVs Service & Sales**(D)
Other	N: **Babe Didrikson Zaharias Museum (Playground)**
855B	Pine St., Magnolia Ave. (Difficult Reaccess)
AServ	N: Automatic Transmission Service
855A	US 90, Dowtown, Civic Center, Port of Beaumont
856	Old Highway
857A	Rose City West
857B	The Workman Turnaround
AServ	N: Moore Auto Care
858A	Rose City East (Eastbound)
FStop	S: **E-Z Mart***
AServ	S: Freeway Auto Parts
Other	S: **Progas**(LP)
858B	The Asher Turnaround
TStop	S: **Spindletop***(SCALES) **(Chevron)**
Food	N: The Riverfront Restaurant
	S: **24-Hr Family Restaurant (Spindletop TS)**
TServ	S: **Spindletop TS**
RVCamp	N: **Boomtown USA RV & Fishing Park**
859	Bonner Turnaround
860A	Dewitt Rd.
AServ	N: ACE Auto Repair
860B	West Vidor
Lodg	N: Greenway Motel
AServ	N: Hubcaps, Morris Moore
861A	FM 105, Vidor
Gas	N: Diamond Shamrock*, Fina*(D), Shell*(D)
	S: Conoco*, Exxon*(D)
Food	N: Casa Ole, Dairy Queen (Playground), Domino's Pizza, Jody Leigh's Restaurant & Ice Cream, Little Caesars Pizza, McDonalds (Playground), Steve's Country Kitchen, The Donut Place, Waffle House
	S: Blimpie's Subs (Exxon), Burger King (Playground), Great Wall Chinese, KFC, Pizza Hut, Sonic Drive In, Taco Bell, What A Burger (24 Hrs)
AServ	N: Ray's 24-Hr Towing
	S: B & B Paint & Body, Gunstream's Tire & Alignment, Herrera's Collision & Towing
ATM	N: Orange Savings Bank
	S: Vidor State Bank

EXIT	TEXAS
Other	N: **Eckerd Drugs, Price-Lo Grocery**
	S: **Maytag Laundry, T and T Pharmacy**
861B	Lamar St.
FStop	N: **Conoco***
861C	Denver St.
861D	TX 12 Deweyville (Left Exit)
862A	Railroad Ave
862B	Old Hwy (Westbound)
RVCamp	N: **Hodge RV**
862C	Timberlane Dr. (Eastbound)
864	FM 1132, 1135
Lodg	N: Cacey Motel
865	Doty Rd. (Westbound)
RVCamp	N: **Claiborne West Park**
867	Frontage Road (Eastbound)
(867)	**Rest Area - RR, Phones, Vending, Picnic, Grills P (Both Directions)**
869	FM 1442, Bridge City
RVCamp	S: **Bridge City RV, Inc.**(LP)
870	FM 1136
Gas	N: Stuckey's*
Food	N: Stuckey's
873	TX 62, TX 73, Bridge City, Port Arthur
TStop	N: **Flying J Travel Plaza***(LP)(SCALES) **(RV Dump)**
FStop	S: **Delta Express***(SCALES) **(FedEx Drop)**
Gas	N: Chevron*(D)
	S: Diamond Shamrock*(D), Texaco*(D)
Food	N: Hot Stuff Pizza (Chevron), Subway (Chevron), **Thads Restaurant (Flying J TStop)**
	S: Jack-in-the-Box, **McDonalds,** Waffle House
TServ	N: **M & M Tire (24 Hr Rd Service), Vic's Road Service (24 Hr)**
TWash	N: **M & M**
RVCamp	N: **Oak Leaf RV Park Full Service Campground**
ATM	S: **Delta Express,** Texaco
874A	U.S. 90 Bus., Orange (Eastbound)
Food	S: JB's BBQ
RVCamp	N: **Oak Leaf RV Park**
Med	S: **+ Hospital**
874B	Womack Rd. (Westbound)
875	FM 3247, M. L. King Jr Drive
Food	N: Luby's Cafeteria
Med	S: **+ Hospital**
876	Adams Bayou, Frontage Rd
Gas	N: Racetrac*
Food	N: Cajun Cookery, Gary's Coffee Shop (24 Hrs),

EXIT	TEXAS/LOUISIANA
	Polo's (Ramada Inn), Waffle House
Lodg	N: AAA Best Western, Days Inn, ◆ **Holiday Inn Express,** Kings Inn, Motel 6, AAA Ramada Inn
AServ	N: Austin Lee GM
TServ	N: **Bennett's Authorized Diesel Pump**
877	TX 87, 16th St , Port Arthur
Gas	N: Chevron*(CW), Diamond Shamrock*
	S: Chevron*, Shell*(D)
Food	N: Little Caesars Pizza, Pizza Hut, Subway, Wyatt's Cafeteria
	S: Cody's, Dairy Queen (Playground), Taco Bell
AServ	N: Chevron, Jerry's Specialty
	S: Hi-Lo Auto Supply (Parts)
RVCamp	S: **Costello RV Sales**
ATM	N: Market Basket, Orange Bank
	S: Shell
Other	N: **California Rib House,** Car Wash, **Eckerd Drugs, Market Basket Grocery, North Orange Veterinary Clinic**
	S: **H-E-B Pantry Foods**
878	U.S. 90 Bus, Orange (Livestock Inspection Station)
FStop	N: **Mobil***
AServ	S: Dave's Auto Clinic
RVCamp	N: **Cypress Lake RV Resort**
Other	N: **Air Boat Rides**
	S: **Pottery World**
(879)	TX Travel Info Center - RR, Phones, Pet Rest (Westbound)
880	Sabine River Turnaround (Eastbound)
Other	S: **Public Boat Ramp**

↑TEXAS

↓LOUISIANA

EXIT	LOUISIANA
1	Sabine River Turnaround (Westbound)
(1)	**LA Tourist Info center, Rest Area - RR, Picnic, Phones, RV Dump (Eastbound)**
(2)	**Weigh Station (Both Directions)**
4	U.S. 90, LA 109, Toomey, Starks
TStop	N: **Cajun Auto/Truck Plaza* (Fuel-Shell), The Lucky Longhorn***(SCALES) **(RV Dump, 24 Hrs, Fuel-Diamond Shamrock)**
	S: **Delta Truck Plaza* (Showers)**

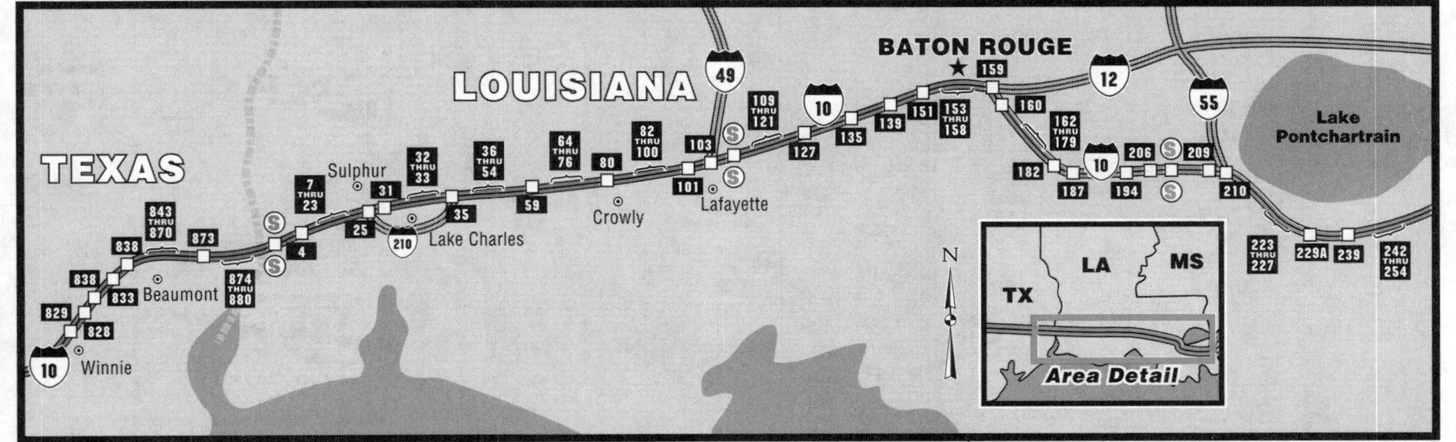

Bold red print shows RV & Bus parking available or nearby

EXIT LOUISIANA

- **FStop** N: Exxon*
  - S: Exxon*
- **Food** N: Bayou Gold, Nevada Magic (No Minors), The Hard Country Cafe & Lounge (The Lucky Longhorn)
  - S: Restaurant (Best Western, Delta Truck Plaza), Texas Pelican
- **Lodg** N: Lucky Longhorn Motel
  - S: Best Western (Delta Truck Plaza)
- **TServ** N: The Lucky Longhorn
- **TWash** N: The Luck Longhorn
- **RVCamp** S: Texas Pelican RV Park (Washateria, Bath)
- **Parks** N: Niblet Bluff Park

7 LA 3063, Vinton
- **FStop** N: Exxon*(CW)
- **Food** N: Champ's Fried Chicken, Dairy Queen, The Lucky Delta
- **AServ** N: BJ's Auto Supply, Delta Tire & Lube, Exxon

8 LA 108, Vinton
- **FStop** N: E-Z Mart*
- **Gas** N: Chevron*(D)
- **Food** N: Glenn's Mart Restaurant
- **RVCamp** N: KOA Kampground

(15) Rest Area - RR, Picnic, Shelters (Westbound)

20 LA 27, Sulphur, Cameron
- **FStop** S: Speedway*
- **Gas** N: Canal*, Circle K*, Exxon*, Shell* (24 Hrs)
  - S: Texaco*(D)(CW)
- **Food** N: Bonanza Steak House, Burger King (Playground), Cajun Charlie's Seafood, Checkers Burgers, Hong Kong Chinese, Mr. Gatti's, Schillileagh's, Shoney's, Subway, Taco Bell, Wendy's
  - S: Joey's Italian Grill (24 Hrs), Pitt Grill, Pizza Hut, Sunshine Frozen Yogurt, Waffle House
- **Lodg** N: Chateau Motor Inn, Holiday Inn (see our ad this page)
  - S: (AAA) La Quinta Inn, (AAA) Microtel Inn
- **AServ** N: Insta Lube
- **RVCamp** N: Southern Mobile Home & RV Supply
- **ATM** N: First National Bank
- **Other** N: La Fleur Roller Rink
  - S: Propane(LP), Sabine Migratory Water Fowl Refuge (22 Miles)

21 LA 3077, DeQuincy
- **FStop** S: USA Super Stop*
- **Gas** S: Chevron*
- **Food** N: The Boiling Point
  - S: Cajun Deli USA (USA Super Stop)
- **RVCamp** S: Hidden Ponds RV
- **ATM** S: USA Super Stop

23 LA 108, Industries, Sulfur (Difficult Reacces)

EXIT LOUISIANA

- **TStop** S: Texaco*
- **Gas** S: Exxon*(CW)
- **Food** S: Cracker Barrel, Little Caesars Pizza (K-Mart), McDonalds (Playground), Waffle House, Winners Choice (Texaco TS, 24 Hrs)
- **Lodg** S: Comfort Suites
- **TServ** S: Southern Tire Mart (Texaco TS)
- **ATM** S: Hibernia
- **Other** S: K-Mart (Pharmacy)

25 I-210 East, Lake Charles Loop

26 Southern Road, Columbia, US 90W
- **FStop** N: Chateau Charles
- **Food** N: Chateau Charles
- **Lodg** N: Chateau Charles Hotel & Suites

27 LA 378, Westlake (Difficult Reaccess)
- **FStop** S: Conoco*
- **Gas** N: Circle K*, Fina*(D)
- **Food** N: Dairy Queen
- **Parks** N: Sam Houston Jones State Park (6 Miles)

29 LA 385, Business District, Tourist Bureau
- **Lodg** S: (AAA) Holiday Inn
- **Other** S: Tourist Information

30A LA 385, N. Lakeshore Dr.
- **Gas** N: Chevron*
- **Food** N: Cafe Du Lac (Travel Inn), Steamboat Bill's
- **Lodg** N: Lakeview Motel, Travel Inn
- **ATM** N: Hibernia

30B Ryan St., Business District, Lake Charles Civic Center

31A US 90 Bus, Enterprise Blvd (Westbound Exit Is Closed For Construction)

EXIT LOUISIANA

- **Gas** S: Texaco*(CW)
- **Food** S: Popeyes Chicken
- **Other** S: Pryce's Pharmacy

31B U.S. 90 East To LA 14, Shattuck St.
- **TStop** N: Kings Ransom*(SCALES) (Fuel-Texaco)
- **FStop** S: Shell* (24 Hrs)
- **Food** N: Restaurant (Kings Ransom TS)
- **TServ** N: Kings Ransom TS

32 Opelousas St
- **Gas** N: Conoco*, Exxon*
- **Lodg** S: Motel 6
- **Other** N: Nash's Superette, Willis Laundry Center

33 U.S. 171 North, DeRidder To LA 14
- **Gas** N: Chevron*(CW), Racetrac* (24 Hrs), Texaco*(D)(CW)
- **Food** N: Burger King (Playground), Taco Bell
- **Lodg** N: (AAA) Best Western, Comfort Inn, (DAYS INN) Days Inn
- **AServ** N: R & D Automotive
- **ATM** N: First National Bank, Market Basket
- **Other** N: Eckerd Drugs, Market Basket Grocery, Walgreens

34 I-210 West, Lake Charles Loop

35 **(34)** Junction I-210, Lake Charles Loop

36 LA 397, Creole, Cameron
- **TStop** S: Conoco Travel Plaza*(SCALES) (FedEx Drop)
- **Food** S: Restaurant (Conoco TS)
- **TServ** S: Conoco
- **RVCamp** S: James Mobile Campground (1 Mile, Full Hookups, Laundry, Bath House)
- **Other** S: Double Diamond Casino (Conoco TStop)

43 LA 383, Iowa
- **Gas** S: Citgo*, Shell*(LP), Texaco*(D)(CW)
- **Food** S: E T Express, Emery's Family Restaurant, Fausto's Fried Chicken, Gulf Fresh Seafood
- **AServ** S: Shell
- **ATM** S: Hibernia (Fact. Stores of Amer.)
- **Other** S: Factory Outlet Center

44 U.S. 165, Alexandria
- **RVCamp** N: Mobile City Campground & RV Park
  - S: A OK Campground

48 LA 101, Lacassine
- **Gas** S: Diamond Shamrock*

54 LA 99, Welsh
- **Gas** S: Chevron*, Citgo*, Exxon*, Texaco*
- **Food** S: Cajun Tales, Dairy Queen, Fannie's Hot Tamales
- **AServ** S: Texaco
- **Other** S: Welsh Municipal Airport

59 LA 395, Roanoke

Bold red print shows RV & Bus parking available or nearby

EXIT	LOUISIANA
64	LA 26, Elton, Jennings, Lake Arthur
FStop	S: Shell*
Gas	S: Fina*, Texaco*(D)(CW)(LP)
Food	N: Golden Dragon
	S: Blimpie's Subs (Fina), Burger King (Playground), McDonalds (Playground), Shoney's, Sugar Mill (Holiday Inn), Taco Bell, Waffle House
Lodg	S: ◆ Holiday Inn, Thrifty Inn
AServ	S: Texaco, Wal-Mart
Med	S: ✚ Jennings Hospital
ATM	S: Wal-Mart
Parks	N: Louisiana Oil & Gas Park
Other	N: Visitor Information
	S: Strand Theatre, Tupper Museum, Wal-Mart Supercenter (1-Hr Photo)
65	LA 97, Evangeline, Jennings (S.W. La State School)
TStop	S: Reed's I-10 Auto/Truck Plaza* (Fuel-Shell)
Lodg	S: DAYS INN ◆ Days Inn
TServ	S: Reed's I-10 Auto/Truck Plaza
(67)	Rest Area - RR, Phones, Picnic, RV Dump, Pet Rest (Both Directions)
72	Egan
Gas	N: Cajun Cafe(D)
Food	N: Cajun Cafe
	S: Cajun Connection Restaurant
RVCamp	N: Cajun Haven RV Park
	S: Trails End Campground (1.2 Miles)
76	LA 91, Iota, Estherwood
TStop	S: Manuel's (24 Hrs, Fuel-Conoco)
TServ	S: Eddie's Tire Service (Manuel's TS), Manuel's TS (24 Hrs)
80	LA 13, Eunice, Crowley
TStop	N: Citgo Truck/Auto Plaza*(SCALES)
Gas	N: Texaco*(D)(CW)
	S: Chevron*(D), Circle K Express* (24 Hrs), Exxon*(CW) (24 Hrs)
Food	N: Brodie's Place (Crowley Inn), Rice Palace (Citgo TStop), Waffle House
	S: Burger King (Playground), Chef Roy's, KFC, Mr. Gatti's, P.J.'s Grill, Sonic Drive In, Taco Bell
Lodg	N: Best Western, Crowley Inn
AServ	N: Thibodeaux Auto Repair
	S: Auto Zone Auto Parts, Chevron
TServ	N: Dubus Engine Co. Inc.
ATM	N: Citgo TS
	S: Chevron, First Bank
Other	N: Crowley Veterinary Hospital
	S: Cinema IV, DelChamps Supermarket
82	LA 1111, East Crowley
Food	S: McDonalds (Wal-Mart)
AServ	S: Wal-Mart
Med	S: ✚ Hospital
ATM	S: Wal-Mart
Other	S: Wal-Mart Supercenter (24 Hrs, Pharmacy)
87	LA 35, Church Point, Rayne
FStop	S: Mobil*
Gas	N: Chevron* (24 Hrs), Exxon(LP)
	S: Diamond Shamrock*(CW), RK's*, Shell*, Texaco*(CW)
Food	N: McDonalds
	S: Dairy Queen, Gabe's Cajun Food, Popeyes Chicken, Roland's Cajun Grill (Mobil, 24 Hrs), Subway (Shell)
AServ	N: Exxon
	S: NAPA Auto Parts
ATM	S: Shell
Other	S: City Police Dept., Eckerd Drugs, The Rayne Chamber of Commerce & Agriculture, Winn Dixie Supermarket
92	LA 95, Mire, Duson
97	LA 93, Cankton, Scott

EXIT	LOUISIANA
Gas	S: Texaco*(D)
Food	S: Church's Fried Chicken (Texaco), Miss Helen's 100% Cajun
AServ	S: Gerald's Towing & Recovery
TServ	S: Gerald's Towing & Recovery
RVCamp	S: KOA Kampground*(LP) (Showers, Laundry, Playground)
Other	S: Cajun Propane(LP)
100	Ambassador Caffery Pkwy (Cajun Dome)
Gas	N: Exxon*(D)
	S: Chevron*(CW) (24 Hrs), Racetrac* (24 Hrs), Texaco*(D)(CW)
Food	S: Burger King (Playground), Taco Bell, Waffle House, Wendy's
AServ	N: Sam's Club
	S: Specialty Automotive & Muffler
TServ	S: LaFayette Tire & Service, Treadco, Inc.
Med	S: ✚ Hospital
Other	N: Sam's Club
101	LA 182, Carencro, Lafayette
TStop	N: 76 Auto/Truck Plaza*(SCALES) (Fuel-Mobil)
FStop	N: Speedway*
Gas	N: Chevron*, Texaco*(D)(CW)
	S: Diamond Shamrock*(CW), Shell*(CW) (24 Hrs), Texaco*(D)(CW)
Food	N: McDonalds (Chevron), Robichaux's (76 TS), Waffle House

EXIT	LOUISIANA
	S: Cracker Barrel
Lodg	N: ◆ Red Roof Inn
	S: DAYS INN Days Inn, Quality Inn, St. Francis Motel
TServ	N: 76 Auto/Truck Plaza
ATM	N: Speedway
Other	N: Honda of Lafayette (Motorcycles)
	S: Diamond Touch Detailing
103AB	Junction I-49, U.S. 167, Opelousas, LaFayette
Gas	S: Shell*, Texaco*(D)
Food	S: Cajun Chicken, Kettle, McDonalds, Mr. Gatti's, Pizza Hut, Popeye's Chicken, Shoney's, Waffle House
Lodg	S: Best Western (see our ad this page), Courtyard by Marriott (see our ad this page), Fairfield Inn, Holiday Inn, La Quinta Inn, Rodeway Inn, Shoney's Inn, Super 8 Motel
AServ	S: A-Abal Auto Service, Firestone Tire & Auto
Other	S: K&B Drug, Northgate Mall
(104)	Rest Area - Picnic, RR, HF, Phones, RV Dump (Eastbound)
(106)	Rest Area - Picnic, RR, HF, Phones, RV Dump (Westbound)
(108)	Weigh Station (Both Directions)
109	LA. 328, Breaux Bridge
Gas	N: Shell*
	S: Delta Express*, Mobil*(D), Texaco*
Food	N: Blimpie's Subs (Shell), Crawfish Kitchen Restaurant
	S: Baskin Robbins (Mobil), Burger King (Playground), Burger Tyme
RVCamp	S: Pioneer Campground
ATM	S: Delta Express
115	LA 347, Cecilia, Henderson
FStop	N: Exxon*
	S: Mapco Express*
Gas	N: Texaco*
	S: Exxon*
Food	N: Exxon, Landry's Seafood
Other	N: Coin Laundry (At Exxon)
(121)	Rest Area - Picnic, RR, HF, Phones, RV Dump (South Side Of Exit 121)
121	Butte, La Rose
RVCamp	S: Camping
Other	N: Tourist Information (Picnic Area, Facilities)
127	LA 975, Whiskey Bay
135	LA 3000, Ramah, Maringouin
Gas	N: PJ's Bait Shop, Texaco*
Food	N: Icecream Churn
(137)	Rest Area - RV Dump, Picnic, RR, HF, Phones (Both Directions)
139	LA 77, Rosedale, Grosse Tete, Maringuin
TStop	S: Tiger Truckstop*(D)
FStop	N: Texaco*
Food	N: Restaurant (Texaco FS)
	S: Tiger Cafe (Tiger TS)
AServ	N: Brock Hoeft Chevrolet GEO
TServ	S: Tiger TS
151	LA. 415, to U.S. 90, Lobdell, Alexandria
FStop	N: Delta Express*(SCALES)
	S: Shell* (24 Hrs)
Gas	N: Chevron*, Exxon*(D), Racetrac*, Texaco*
	S: Texaco*
Food	N: Burger King, KFC, McDonalds, Popeye's Chicken, Shoneys, Taco Bell, Waffle House
Lodg	N: DAYS INN Days Inn, Holiday Inn Express (see our ad this page), Shoney's Inn
	S: Motel 6, Ramada, Super 8 Motel
ATM	N: Racetrac
Other	N: Wal-Mart (Pharmacy)

Bold red print shows RV & Bus parking available or nearby

EXIT LOUISIANA

S: LA. Tourist Information

153 LA 1, Port Allen, Plaquemine
- **Gas** N: Chevron*, Exxon*, Texaco*(D)(CW)
- **AServ** N: Russo's Auto Parts

155A LA 30, Nicholson Dr., Highland Dr. (Eastbound)

155B I-110N, Business Dist., Metro Airport

155C Louise St.

156A Washington St (Eastbound)

156B Dairymple Drive, LSU
- **Parks** S: City Park Lake
- **Other** S: LSU

157A Perkins Road (Eastbound)
- **Gas** S: Phillips 66*, Texaco*
- **Food** S: O' Brian Crab House

157B Acadian Thruway
- **Gas** N: Citgo*, Shell*(CW)
  S: Exxon*, Mobil*, Texaco*
- **Food** N: Cafe Louisiane, Denny's, Ribs
  S: Outback Steakhouse
- **Lodg** N: (AAA) Comfort Inn, (AAA) La Quinta Inn
- **AServ** S: Mobil
- **ATM** S: Hancock Bank
- **Other** S: Books-A-Million, Calibans Books, Eckerd Drugs, Wal-Mart

158 College Drive
- **Gas** N: Raceway*
  S: Chevron* (24 Hrs), Texaco*(CW)
- **Food** N: Capital City Brewery, Grady's, Macaroni Grill, Ruby Tuesday, Steamroom Grill, Waffle House
  S: Applebee's, Burger King (Playground), Chili's, Great Wall Restaurant, Koto Oriental, McDonalds, Shoney's
- **Lodg** N: Corporate Inn, (AAA) Hilton, Residence Inn
  S: ◆ Hampton Inn
- **AServ** N: Midas Muffler & Brakes
- **ATM** N: Bank One
  S: City National Bank, Regions Bank
- **Other** N: Esplanade Shopping Mall
  S: Albertson's Grocery (Pharmacy), K&B Drugs

159 Junction I-12, Hammond

160 LA 3064, Essen Lane
- **Gas** S: Chevron*, Exxon*(D)(CW), Racetrac*
- **Food** S: Louisiana Pizza Kitchen, McDonalds, Wendy's
- **Med** S: ✚ Our Lady of the Lake Regional Medical Center

162 Blue Bonnet Road
- **Gas** S: Exxon*, Racetrac* (24 Hrs)
- **Food** N: Albasha Greek, Cadillac Cafe
  S: Glynwood, Ralph & Kacoos Seafood
- **Lodg** N: (AAA) Quality Suites
  S: (AAA) AmeriSuites (see our ad this page)
- **Other** N: Bluebonnet Veterinary Hospital

163 Siegen Lane
- **Gas** N: Chevron*(CW) (24 Hrs), Racetrac* (24 Hrs), Shell*(CW)
- **Food** N: Burger King, Chili's, Fantastic Sams, Joe's Crab Shack, Kabob's Greek, Shoney's, Subway, Waffle House, Wendy's
  S: McDonalds (Wal-Mart)
- **Lodg** N: ◆ Budgetel Inn, Courtyard by Marriott, Holiday Inn, Motel 6
- **AServ** N: Western Auto
- **Other** N: K-Mart (Pharmacy)
  S: Wal-Mart (Pharmacy)

EXIT LOUISIANA

166 LA 42, 427, Highland Road, Perkins Road
- **Gas** S: Texaco*(CW)
- **ATM** S: Hancock Bank
- **Other** N: Water Amusement Park
  S: Louisiana State Police Troop A

173 LA 73, Geismar, Prairieville
- **FStop** N: Texaco*
- **Gas** S: Exxon*(CW), Mobil*
- **Food** S: Subway (In Mobil)
- **RVCamp** S: Twin Lakes RV Park

177 **(178)** LA 30, Gonzales, St. Gabriel
- **TStop** N: USA Auto Truck Plaza*(SCALES) (RV Dump)
- **Gas** N: Shell*, Texaco*
  S: Citgo*, Texaco*(CW)
- **Food** N: Church's Fried Chicken, McDonalds (Play Place), Shoney's, Taco Bell, USA Auto Truck Plaza, Waffle House
  S: Blimpie's Subs (Citgo), Popeye's Chicken, Wendy's
- **Lodg** N: Budget Inn, Cajun Inn, Holiday Inn
  S: Comfort Inn
- **TServ** N: USA Auto Truck Plaza
- **Med** N: ✚ Hospital
- **ATM** S: Citgo
- **Other** S: Tanger Factory Outlet

179 LA 44, Gonzales , Burnside
- **TStop** N: Red Man of Louisiana*
- **FStop** N: Mobil*
- **Food** N: Red Man of Louisiana
- **TServ** N: Red Man of Louisiana

(180) **Rest Area - RV Dump, Picnic, RR, HF, Phones (Both Directions)**

182 LA 22, Sorrento, Donaldsonville
- **TStop** S: Square Deal*(SCALES)
- **FStop** S: PS Save*
- **Food** S: McDonalds (In Square Deal)
- **Other** S: Tourist Information

187 U.S. 61, Gramercy, Sorrento (No Westbound Reaccess)

EXIT LOUISIANA

194 LA 641 South, Gramercy, Lutcher

206 LA 3188, La Place
- **Gas** S: Shell*
- **Food** S: McDonalds (Play Place)
- **AServ** S: Goodyear Tire & Auto

(207) **Weigh Station - Both Directions**

209 U.S. 51, LaPlace, Hammond
- **FStop** S: Speedway*
- **Gas** S: Chevron* (24 Hrs), Citgo*, Shell*(CW)
- **Food** S: McDonalds (Play Place), Shoney's, Waffle House, Wendy's
- **Lodg** S: ◆ Holiday Inn
- **ATM** S: Speedway

210 Junction I-55 North, Hammond (Westbound)

220 Jct. 310S Boutte, Houma

221 Loyola Dr
- **Gas** N: Circle K*, Citgo*, Exxon*(D), Shell*(CW), Texaco*(D)(CW), Timesaver*
  S: Chevron*, Conoco*, Speedway*
- **Food** N: Blimpie's Subs, Church's Fried Chicken, Kenner's Seafood, McDonalds, Popeye's Chicken, Rally's Hamburgers, Subway, Taco Bell, Tastee Donuts
  S: Toddle House, Wendy's
- **AServ** N: Shell
- **ATM** N: Circle K
- **Other** N: Sam's Club, Super Foods Deli and Bakery Grocery Store
  S: Tourist Information

223A LA 49, Williams Blvd, New Orleans International Airport

224 Power Blvd. (No Westbound Reaccess)

225 Veteran's Blvd
- **Lodg** S: Holiday Inn (see our ad this page)

226 Clearview Pkwy, Huey Long Bridge

228 Causeway Blvd, Downtown Metairie, Mandeville

229 Bonnabel Blvd

230 Junction I-610, Downtown New Orleans

231B Florida Blvd, West End Blvd

231A Metairie Road, City Park Ave

232 U.S. 61, Airline Highway, Tulane Ave, Carrollton Ave, Downtown New Orleans (Difficult Reaccess)

234A Manhattan Rd.
- **Lodg** S: Travelodge Hotel (see our ad opposite page)

235C U.S. 90 West, Claiborne Ave, Downtown New Orleans

235B U.S. 90 Business West, Canal St, Superdome

235A Orleans Ave, Vieux Carre
- **Other** S: Casino, Police Station

236C Saint Bernard Ave, Downtown New Orleans (No Reaccess)

237 Elysian Fields Ave (Difficult Reaccess)
- **TStop** N: Texaco
- **TServ** N: Texaco

Bold red print shows RV & Bus parking available or nearby

EXIT LOUISIANA

TWash N: Texaco

238B Junction I-610 West, Baton Rouge, New Orleans Int'l Airport

238A Franklin Ave (No Reaccess)

239 Louisa St, Almonaster Blvd. (Difficult Reaccess)

- **FStop** S: **Fleetman Automated Fueling Station**
- **Gas** N: Chevron*, Exxon*(CW)
  S: Fina*(D), Fuelman
- **Food** N: Burger King, Daiquire's, McDonalds, Pizza Hut, Popeye's Chicken, Rallys Hamburgers
- **Lodg** N: Budget Inn, Scotish Inns
- **AServ** N: Firestone Tire & Auto, Goodyear Tire & Auto, Precision Tune & Lube, Spee-Dee Oil Change
- **ATM** N: First NBC
- **Other** N: Coin Car Wash, **K-Mart, WalGreens**

240B U.S. 90, Chef Highway

241 Morrison Road (Reaccess Eastbound Only)

242 Crowder Blvd

244 Read Blvd

- **Lodg** S: **Days Inn (see our ad this page)**

245 Bullard Ave

246AB Junction I-510 South, LA 47, Chalmette (No Reaccess)

248 Michoud Blvd (No Reaccess)

254 U.S. 11, Irish Bayou, North Shore Dr.

261 Oak Harbor Blvd, Eden Isles

263 **(264)** LA 433, Slidell

EXIT LOUISIANA/MISSISSIPPI

- **FStop** S: **Texaco***
- **Gas** N: Citgo*, Exxon*, Shell*(CW)
  S: Speedway*
- **Food** N: Pitt Grill (24 Hrs), Waffle House
  S: **Blimpie's Subs (Texaco FS)**, McDonalds (Play Place), Wendy's
- **Lodg** N: AAA Comfort Inn
- **AServ** N: Exxon, Levi's Chevrolet
- **Other** N: **Slidell Factory Outlets**

(265) **Weigh Station - Both Directions**

266 U.S. 190, Slidel

- **TStop** N: **76 Auto/Truck Plaza*(SCALES) (RV Dump)**
- **Gas** N: Exxon*(D), Shell*(CW), Texaco*(CW), USA*
  S: Chevron*(D) (24 Hrs), Racetrac*
- **Food** N: Arby's, Baskin Robbins, Burger King (Play Land), China Wok, Happy Dragon Chinese, KFC, Lone Star Steakhouse, McDonalds (Play Place), **Pete's Family Restaurant (76 Auto / Truck Stop)**, Pizza Hut, Shoney's, Taco Bell, Wendy's
  S: Applebee's, **Cracker Barrel**, Luther's Smokehouse & Grill, Outback Steakhouse, Waffle House
- **Lodg** N: Budget Host Inn, AAA Days Inn, Motel 6
  S: AAA Econolodge, King's Guest Lodge, AAA La Quinta Inn, ◆ Ramada, Value Travel Inn
- **AServ** S: Ernie's, Meineke Discount Mufflers
- **TServ** N: **76 Auto/Truck Plaza**
- **Other** N: **Schwegmann (Pharmacy)**
  S: **Wal-Mart (Pharmacy)**

267AB Junction I-59 North, Jct 267B, Jct I-12W, Baton Rouge, Hammond

(269) **LA Tourist Info Center, Rest Area - RV Dump (Full Facilities)**

↑LOUISIANA

↓MISSISSIPPI

(1) **Weigh Station (Both Directions)**

(2) MS Welcome Center - RV Dump, Phones, Picnics, RR, HF

2 MS. 607, NASA John C. Stennis Space Center

- **Other** N: **Naval Oceanographic Center, Stennis Space Center**

(8) **Parking Area (Eastbound, No Services)**

13 MS 43, MS 603, Bay Saint Louis,

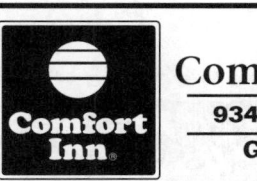

EXIT MISSISSIPPI

Picayune

- **Gas** S: Chevron*, Conoco*, Exxon*
- **Food** S: Subway (Exxon), Waffles (Conoco)
- **RVCamp** S: **Bay Marina RV Park, Casino Magic RV Park, KOA Campground**
- **Med** S: **✚ Hospital**
- **ATM** S: Conoco, Exxon
- **Parks** N: **McLeod State Park**
  S: **Buccaneer State Park**

16 Diamondhead

- **Gas** N: BP*, Chevron*(CW)
  S: Texaco*
- **Food** N: Blimpie's Subs (Chevron), Burger King (Playground), Dairy Queen, Domino's Pizza (BP), Robert's, Waffle House
  S: Pawpaw's (Texaco)
- **Lodg** N: Diamond Head Resort
- **ATM** N: Chevron, The People's Bank, Whitney

20 DeLisle

- **Gas** N: Shell*, Spur*
- **Food** N: **GiGi's**

24 **(25)** Menge Ave

- **TStop** N: **Amoco*(SCALES)**
- **Gas** S: Texaco*
- **Food** S: Stuckey's (Texaco)
- **TServ** N: **Amoco TS**
- **TWash** N: **Amoco TS**

28 Long Beach, Pass Christian

- **Gas** S: Chevron*, Citgo*
- **TServ** S: **Coastal Energy Tires**
- **RVCamp** S: **Magic River Camping**

31 Canal Road, Gulfport

- **TStop** S: **Flying J Travel Plaza*(D)(SCALES) (RV Dump, Travel Store)**
- **FStop** S: **BP***
- **Food** S: **Thads Restaurant (Flying J)**
- **Lodg** S: ◆ Crystal Inn
- **TServ** N: **61 Tire Service**
  S: **Flying J Travel Plaza**
- **TWash** N: **Russel's Truck Wash**

34AB U.S. 49, Gulfport, Hattiesburg

- **FStop** N: **Speedway***
  S: **Shell*(D)**
- **Gas** N: Amoco*(CW), Texaco*(D)
  S: Chevron*(CW), Fast Lane*, Texaco*
- **Food** N: Hardee's, Waffle House
  S: Applebee's, Arby's, Burger King, KFC, Krispy Kreme Donuts, McDonalds (Wal-Mart), Morrison's Cafeteria, Randy's, **Shiners Real Pit BBQ**, Shoney's, **Waffle House**, Wendy's
- **Lodg** S: Comfort Inn (see our ad this page), Hampton Inn, Holiday Inn Express, Motel 6, **Seaway Inn Best Western**, Shoney's
- **AServ** S: 10 Min. Oil Change, Pat Peck Nisson, **Wal-Mart**
- **Med** S: **✚ Hospital**
- **ATM** N: Texaco
- **Other** S: **Wal-Mart Supercenter (Pharmacy)**

38 Lorraine - Cowan Road

- **FStop** N: **Chevron*(CW), Speedway*(LP)**
- **Gas** N: Exxon*, Texaco*(D)
- **Food** N: Domino's Pizza, Subway (Exxon)
- **RVCamp** S: **Gaywood RV Camp**

41 MS. 67, Woolmarket

- **Gas** N: **BP*(LP)**
- **RVCamp** S: **Mazlalea Travel Park, Parker's Landing RV Park**

EXIT — MISSISSIPPI

44 Cedar Lake Road, Coast Coliseum
- **FStop** S: Shell*(LP) (RV Dump)
- **Gas** S: Exxon*(D)
- **AServ** S: Cedar Lake Auto Parts
- **Med** S: ✚ Hospital
- **ATM** S: Shell
- **Other** S: Cedar Lake Animal Hospital

46AB I-110, MS 15, Biloxie, Keesler Air Force Base
- **Gas** N: Texaco*(D)

50 MS. 609, Ocean Springs
- **FStop** S: Speedway* (24 Hrs)
- **Gas** N: Chevron*, Texaco*(CW)
  S: Amoco* (24 Hrs)
- **Food** N: Pizza Inn (Texaco), Taco Bell (Texaco)
  S: Blimpie's Subs (Speedway FS), Denny's, McDonalds (Amoco), Waffle House
- **Lodg** N: Super 8 Motel
  S: Hampton Inn, (AAA) Holiday Inn Express, ◆ Sleep Inn (see our ad this page)
- **ATM** N: Texaco
- **Other** S: Blue Ridge Veterinary Clinic

57 MS. 57, Fontainbleau, Vancleave
- **FStop** S: Texaco*
- **Gas** S: BP*
- **RVCamp** N: KOA Campground
- **Other** S: Gulf Island National Seashore

61 Gautier, Vancleave
- **RVCamp** N: Wonderland Park
  S: Bluff Creek Camping
- **Other** N: Sand Hill Crane Wildlife Refuge Area

(64) Rest Area - Phones, Picnics, RR, HF, RV Dump (P) (Both Directions)

68 MS 613, Moss Point, Pascagoula
- **FStop** N: Citgo*(D), Conoco*
- **Gas** N: Texaco*(D)
  S: Chevron*
- **Food** N: KFC (Texaco)
- **Lodg** N: Super 8 Motel

69 MS. 63, East Moss Point, East Pacagoula

EXIT — MISSISSIPPI/ALABAMA

- **TStop** S: Chevron*(SCALES)
- **FStop** S: Amoco*
- **Gas** N: Texaco*(CW)
  S: Exxon*
- **Food** N: Blimpie's Subs (Texaco), Pizza Inn (Texaco)
  S: Burger King (Playground), Cracker Barrel, Hardee's, J.J.'s (24 Hrs, Chevron TS), McDonalds (Play Place), Pizza Hut, Quincy's Family Steakhouse, Subway (Exxon), Waffle House
- **Lodg** N: Ashbury Suites
  S: (AAA) Best Western, ◆ Comfort Inn, Days Inn, ◆ Hampton Inn, ◆ Holiday Inn, ◆ Shular Inn, Holiday Inn (see our ad this page)
- **TServ** S: Chevron
- **TWash** S: Chevron TS
- **ATM** S: Amoco FS, Exxon

(74) Weigh Station (Eastbound)

(75) Welcome Center - RV Dump, RR, Phones, Picnic (Westbound)

75 Franklin Creek Road

(77) Weigh Station (Westbound)

↑ MISSISSIPPI

↓ ALABAMA

(1) Welcome Center - Picnic, RR, HF, RV Dump, Phones (Eastbound)

4 AL. 188, Grand Bay

EXIT — ALABAMA

- **TStop** N: 76 Auto/Truck Plaza*(SCALES) (Travel Store, 24 Hrs)
- **Gas** N: Citgo*, Texaco*
  S: Chevron*(CW) (24 Hrs)
- **Food** N: 76 Auto/Truck Plaza, Dairy Queen (Texaco), Stuckey's (Texaco)
  S: Hardee's
- **TServ** N: 76 Auto/Truck Plaza

13 (14) Theodore, Dawes
- **TStop** N: Pilot Travel Center*(D)(SCALES)
- **Gas** N: Amoco* (24 Hrs), Conoco*, Shell*(CW)
  S: Chevron*
- **Food** N: Krispy Kreme (Pilot TS), McDonalds, Subway (Shell), Waffle House, Wendy's (Pilot TS)
- **RVCamp** S: I-10 Campground
- **ATM** N: Amoco
- **Other** N: Mobile Greyhound Park

15AB U.S. 90, Theodore, Historic Mobile Pkwy., Tillman's Corner
- **Gas** N: Chevron* (24 Hrs), Racetrac* (24 Hrs), Shell*(CW)
  S: Chevron*(CW), Texaco*(D)
- **Food** N: Burger King, Checkers Burgers, Coffee Kettle, Godfather's Pizza, McDonalds (Play Place), Papa John's Pizza, Pizza Inn, Popeye's Chicken, Quincy's Family Steakhouse, Shoney's, Subway, Taco Bell, Waffle House
  S: Waffle House
- **Lodg** N: Best Western, Comfort Inn, Days Inn, Hampton Inn, Holiday Inn, Motel 6, Shoney's Inn, Super 8 Motel, Holiday Inn (see our ad this page)
- **ATM** N: Regions Bank, SouthTrust Bank
- **Other** N: Coin Laundry, Del Champ Supermarket, K & B Drugs, Wal-Mart (W/ Pharmacy, 24 Hrs), Winn Dixie Supermarket (W/ Pharmacy)

17AB AL 193, Dauphin Island, Tillman's Corner (Difficult Reaccess)
- **Gas** N: Citgo*
- **Food** N: Blimpie's Subs (Citgo)
- **Med** S: ✚ Hospital

20 Junction I-65 N, Montgomery

22AB Alabama 163, Dauphin Island

Bold red print shows RV & Bus parking available or nearby

EXIT ALABAMA

Pkwy
- **Gas** S: Bubba's*, Exxon*
- **Food** S: Checkers, Gone Fishin Catfish Restaurant, Subway (Exxon), Waffle House
- **Lodg** N: Motel 6
- **AServ** N: I-10 Auto Parts; S: Cooper Tires
- **ATM** S: Regions Bank
- **Other** S: **Revco Drugs**

23 Michigan Ave.
- **Gas** N: Exxon*
- **Lodg** N: Rodeway Inn

24 Broad St., Duval St.
- **Gas** N: Chevron*(CW) (24 Hrs), Citgo*(D)

25 Canal St. (Eastbound)

25B Virginia St
- **Gas** N: Texaco*(D)
- **AServ** S: Tennessee Tom's Tires

25A Texas St. (Difficult Reaccess)

26 Water St, Downtown Mobile
- **Lodg** N: Holiday Inn Express

27 U.S. 90, U.S. 98, Government St.
- **Food** S: Captain's Table
- **Lodg** S: Battleship Inn Best Western
- **TServ** N: **Dixie Nationwide Truck Service**

30 U.S. 31, U.S. 90, U.S. 98, Battle-ship Pkwy
- **FStop** S: **Citgo***
- **Gas** S: Texaco*(D)
- **Food** S: Kimberly's (Ramada), Pier 4 Restaurant, Poor Man's
- **Lodg** S: Ramada
- **Parks** S: **Battleship USS Alabama Park**

35 U.S. 98, U.S. 90, Daphne - Fairhope Spanish Fort
- **Gas** N: Amoco*; S: Chevron*(D) (24 Hrs), Exxon*(D), Shell*, Texaco*
- **Food** S: Burger King (Playground), Checkers Burgers, Guido's, Hardee's, IHOP, McDonalds (Playground), Nautilus, Quincy's Family Steakhouse, Shoney's, South China, Taco Bell, Waffle House, Wendy's
- **Lodg** S: Eastern Shore Motel, ◆ Hampton Inn
- **AServ** N: Bubba's Service Center, Discount Auto Parts
- **ATM** N: Amoco

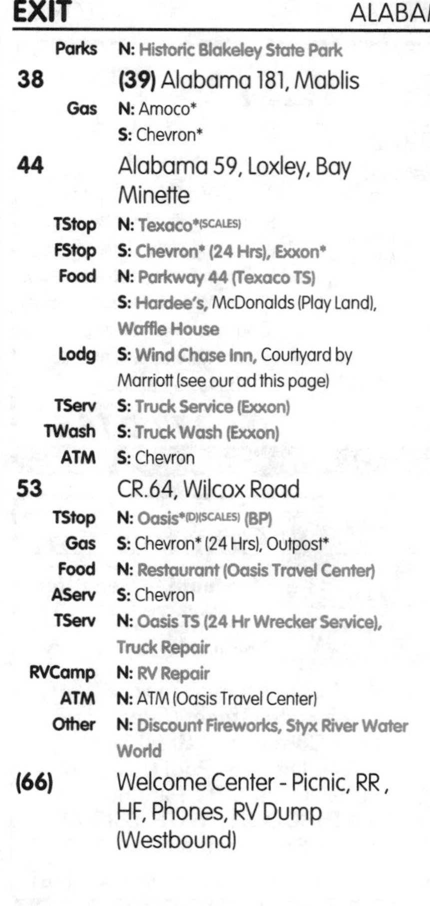

EXIT ALABAMA

- **Parks** N: **Historic Blakeley State Park**

38 **(39)** Alabama 181, Malbis
- **Gas** N: Amoco*; S: Chevron*

44 Alabama 59, Loxley, Bay Minette
- **TStop** N: **Texaco***(SCALES)
- **FStop** S: **Chevron* (24 Hrs), Exxon***
- **Food** N: **Parkway 44 (Texaco TS)**; S: **Hardee's,** McDonalds (Play Land), **Waffle House**
- **Lodg** S: **Wind Chase Inn,** Courtyard by Marriott (see our ad this page)
- **TServ** S: **Truck Service (Exxon)**
- **TWash** S: **Truck Wash (Exxon)**
- **ATM** S: Chevron

53 CR.64, Wilcox Road
- **TStop** N: **Oasis***(D)(SCALES) **(BP)**
- **Gas** S: Chevron* (24 Hrs), Outpost*
- **Food** N: **Restaurant (Oasis Travel Center)**
- **AServ** S: Chevron
- **TServ** N: **Oasis TS (24 Hr Wrecker Service), Truck Repair**
- **RVCamp** N: **RV Repair**
- **ATM** N: ATM (Oasis Travel Center)
- **Other** N: **Discount Fireworks, Styx River Water World**

(66) Welcome Center - Picnic, RR , HF, Phones, RV Dump (Westbound)

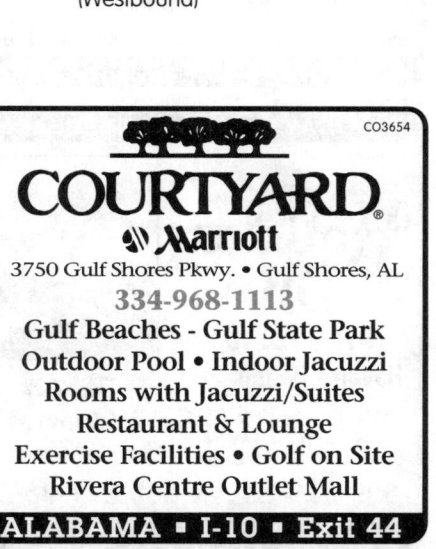

EXIT ALABAMA/FLORIDA

↑ALABAMA
↓FLORIDA

(5) Welcome Center - Phones, RR, HF, Picnic (Eastbound)

(4) **Weigh Station (Both Directions)**

1 **(6)** U.S. 90 Alternate
- **FStop** N: **Speedway***(LP)(SCALES)
- **Food** N: **Subway (Speedway)**
- **RVCamp** N: **Tall Oaks Campground**
- **ATM** N: Speedway

2 **(7)** FL 297, Pine Forest Road, Pensacola Naval Air Station
- **Gas** S: BP*, Citgo*, Texaco*(D)
- **Food** S: Burger King, Hardee's, Little Caesars Pizza, McDonalds (Play Land), Waffle House
- **Lodg** N: ◆ Rodeway Inn
- **RVCamp** N: **Tall Oaks Campground**
- **Other** S: **Food World Supermarket, Naval Aviation Museum**

3AB **(10)** U.S. 29, Pensacola, Cantonment
- **FStop** N: **Mapco Express***(SCALES)
- **Gas** N: Exxon*, Parade*; S: Racetrac*, Shell*
- **Food** N: **Hardee's,** Waffle House; S: Burger King, **Denny's,** McDonalds (Playground), Waffle House, Wendy's
- **Lodg** S: Comfort Inn, Days Inn **Days Inn,** Econolodge, Executive Inn, Howard Johnson
- **ATM** S: Racetrac

4 **(12)** Junction I-110, Pensacola

5 **(13)** FL 291, Pensacola, University of West Florida
- **Gas** N: Amoco(CW) (24 Hrs), Chevron*(CW), Texaco*(D)
- **Food** N: Arby's, Barnhills Buffet, Burger King, Captain D's Seafood, Denny's, McDonalds (Playground), Peking Garden Chinese, Shoney's, Taco Bell, Waffle

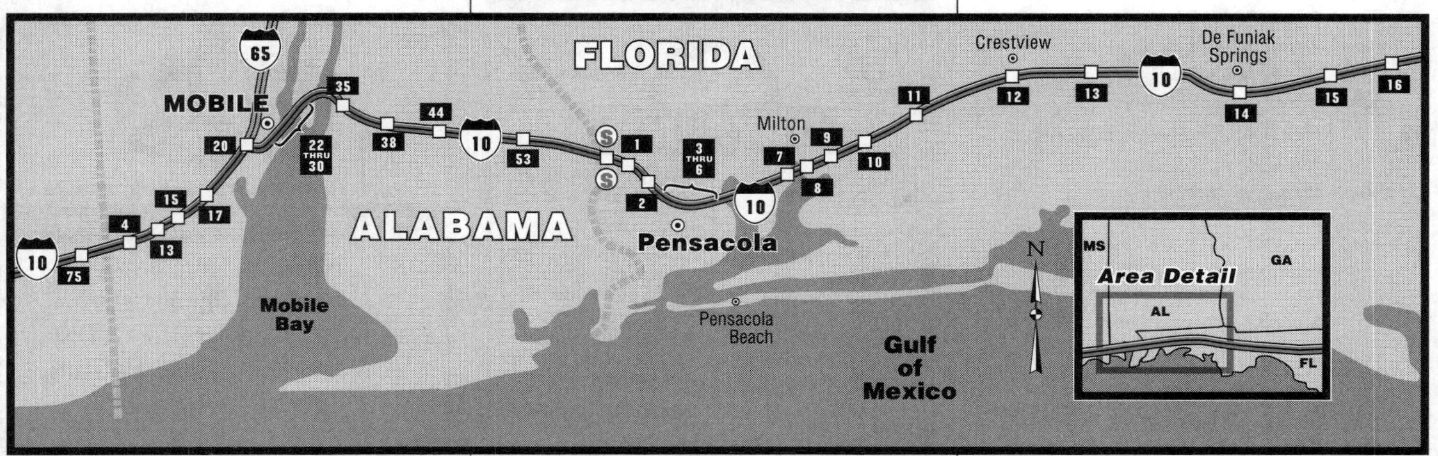

EXIT FLORIDA

House
S: Chuck E. Cheese's Pizza, Cuco's Restaurant, Pizza Hut, Popeye's Chicken, Steak & Ale, Waffle House, Wendy's
Lodg N: Knight's Inn, (AAA) La Quinta Inn, Motel 6, (AAA) Shoney's Inn
S: ◆ Fairfield Inns, ◆ Holiday Inn, ◆ Red Roof Inn
AServ N: Amoco, Auto Pit Stop, Auto Zone Auto Parts, Super Lube
S: Big 10 Tires, Master Care Auto Service, Mr. Transmission, Sears Auto Center, Super Lube, Tires Inc.
Med N: ✚ **Hospital**
ATM S: Barnett Bank, SunTrust
Other N: **Big B Drugs, Food World Supermarket**
S: **Eckerd Drugs, United Artist Theaters, University Mall**

6 **(16)** U.S. 90, Pensacola
Gas N: BP*
Food S: Ramada
Lodg S: (AAA) Ramada (see our ad this page)

(20) Picnic

7 **(22)** Avalon Blvd, Fl. 281
Gas N: Shell*
Food N: McDonalds
RVCamp S: **By The Bay RV Park**
Med N: ✚ **Hospital**

8 **(26)** CR.191, Milton, Bagdad
FStop S: **T&B Food Store***
Gas S: Citgo*
Food S: Dairy Queen (Citgo), Stuckey's (Citgo)
RVCamp S: **Sunny Acres RV Resort**
ATM S: Citgo

9 **(28)** CR. 89, Milton
RVCamp S: **Evans Cedar Lake RV Campground**
Med N: ✚ **Hospital**

(30) **Rest Area - Picnic, RR, Phones, HF (Both Directions)**

10 **(31)** CR87, Milton, Navarre
TStop N: **Rolling Thunder Truck Stop***
Gas S: Amoco*, Chevron* (24 Hrs), Express*
Food N: **Restaurant (Rolling Thunder TS)**
S: Red Carpet Inn Restaurant
Lodg S: Red Carpet Inn, **Comfort Inn (see our ad this page)**, Holiday Inn (see our ad this page)
TServ N: **Ray Rich CB Shop, Wooten's Road Service**
RVCamp N: **Gulf Pines Resort**

11 **(45)** CR189, Holt
FStop S: **Shell***(LP)
RVCamp N: **Log Lake Road RV Park**
Parks N: **BlackWater River State Park**

12 **(56)** FL 85, Crestview, Eglin AFB, Niceville, Ft. Walton Beaches
Gas N: Racetrac*, Shell*(D)
S: Amoco*(LP) (24 Hrs), Citgo*(D), Exxon*
Food N: Cajun's Fried Chicken, Great China, Mc Lain's Family Restaurant, McDonalds, Taco Bell
S: Arby's, Baskin Robbins (Exxon), (Cracker Barrel logo) **Cracker Barrel, Hardee's,** Nim's Garden Chinese, Porch BBQ, Shoney's, Subway, Waffle House, **Wendy's**
Lodg N: ◆ **Econolodge**

EXIT FLORIDA

S: (Days Inn) (AAA) Days Inn (see our ad this page), Hampton Inn (see our ad this page), ◆ Holiday Inn, (AAA) Super 8 Motel
AServ N: Discount Auto Parts
S: Auto Value Auto Parts, Hub City Ford Mercury
RVCamp S: **Holiday Lake Travel Park**
Med N: ✚ **Hospital**
Other N: **Wal-Mart (24 Hrs, Pharmacy)**
S: **Air Force Armament Museum**

(58) **Rest Area - Picnic, RR, HF, Phones (Eastbound)**

(60) **Rest Area - Picnic, RR, HF, Phones (Westbound)**

13 **(70)** FL 285, Niceville, Eglin Air Force Base, Ft Walton Bches
TStop S: **Lucky 13 Truck/Auto Plaza***
FStop N: **The Cola Cafe**
Gas N: RaceWay
Food N: Cola Cafe
S: **Restaurant (Lucky 13)**
Lodg S: (Days Inn) (AAA) Days Inn (see our ad this page)
TServ S: **Lucky 13 TS**

14 **(85)** U.S. 331, De Funiak Springs, Freeport
FStop S: **Exxon***
Gas N: Amoco*, Chevron*(CW)
S: Shell*, Texaco*
Food N: Burger King, McLain's Family Steakhouse, Waffle House

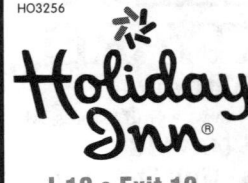

Bold red print shows RV & Bus parking available or nearby

EXIT	FLORIDA
	S: Hardee's, McDonalds (Playground), Taco Bell (Exxon FS)
Lodg	N: ◆ Comfort Inn, Days Inn, AAA Days Inn, Hampton Inn (see our ad opposite page)
	S: AAA Best Western
15	**(96)** FL 81, Ponce de Leon
FStop	S: Exxon*, Texaco*
Gas	S: Amoco*
Food	S: Subway (Texaco)
Lodg	S: Ponce de Leon Motor Lodge, Sleep Inn (see our ad this page)
RVCamp	N: Ponce de Leon Campground, Vortex Springs Resort
Parks	N: Ponce de Leon Springs State Recreation Area (No Camping)
(96)	**Rest Area - Picnic, RR, HF, Phones (Southside Of Exit 15)**
16	**(104)** CR. 279, Caryville
17	**(111)** FL 79, Bonifay, Panama City Beach
TStop	N: Citgo Truck/ Auto Plaza*
FStop	N: Exxon*
Gas	N: Chevron*
Food	N: Blitch's Family Restaurant, Hardee's, McDonalds (Playground), Pizza Hut, Simbo's Steaks & Seafood (Citgo TS), Subway, Waffle House
Lodg	N: Econolodge, Tivoli Inn Best Western
AServ	N: Hightower Auto Parts
TServ	N: Citgo TS

EXIT	FLORIDA
TWash	N: Citgo TS
RVCamp	N: Hidden Lake Campground
	S: Cypress Springs Campground
Med	N: ✚ Hospital
18	**(119)** FL 77, Chipley, Panama City
Gas	N: Exxon*, Texaco
	S: Shell*
Food	N: Coffee House (Days Inn), KFC, TCBY Treats (Exxon), Taco Bell (Texaco), Wendy's
	S: Icecream Churn (Shell)
Lodg	N: Days Inn (see our ad this page), Holiday Inn Express, AAA Super 8 Motel
AServ	N: Wal-Mart
Med	N: ✚ Hospital
Other	N: Wal-Mart (Pharmacy)
	S: Falling Water State Recreation Area
19	**(130)** U.S. 231, Panama City, Cottondale
FStop	N: Amoco*
Gas	N: BP*, Chevron*
Food	N: Hardee's, Subway (BP)
TServ	N: Plea's Truck Service
(133)	**Rest Area - Picnic, RR, HF, Phones (Both Directions)**
20	**(136)** FL 276, Marianna
FStop	N: Chevron*
Med	N: ✚ Hospital
21	**(142)** FL 71, Marianna, Blountstown
TStop	N: Pilot Travel Center*(D)(SCALES)

EXIT	FLORIDA
	S: 76 Auto/Truck Plaza*(SCALES) (BP)
FStop	N: Amoco* (24 Hrs)
	S: Speedway*
Gas	N: Shell*, Texaco*(D)
Food	N: Arby's (Pilot Truckstop), Baskin Robbins, Shoney's, Waffle House
	S: Courtesy House Restaurant (76 Auto/Truck Stop), McDonalds
Lodg	N: AAA Comfort Inn, ◆ Hampton Inn, Holiday Inn
	S: AAA Best Western
TServ	S: 76 Auto/Truck Plaza
RVCamp	N: Arrowhead Campsite
	S: Dove Nest RV Park
ATM	N: Texaco
Parks	N: Florida Caverns State Park
22	**(152)** FL 69, Grand Ridge, Blountstown
FStop	N: Chevron*
Gas	N: Exxon*, Texaco*
Food	N: Icecream Churn (Texaco), The Golden Lariat
Lodg	N: Durden's Family Inn
(155)	**Weigh Station - Both Directions**
23	**(158)** CR. 286, Sneads
Other	N: Three Rivers State Recreation Area
(161)	**Rest Area - Phones, RR, HF, Picnic (Both Directions)**
24	**(165)** CR.270 - A, Lake Seminole, Chattahoochee
Gas	S: Texaco*
Other	N: Lake Seminole
25	**(174)** FL.12, Quincy, Greensboro, Gretna
FStop	N: Amoco*, Texaco*
Food	N: Burger King (Amoco)
RVCamp	N: Beaver Lake Campground
26	**(181)** FL 267, Quincy
Gas	N: Texaco*
	S: BP*(D), Shell*
Food	N: Icecream Churn (Texaco)
Lodg	S: Holiday Inn Express
Med	N: ✚ Hospital
Other	S: Lake Talquin Recreation Area

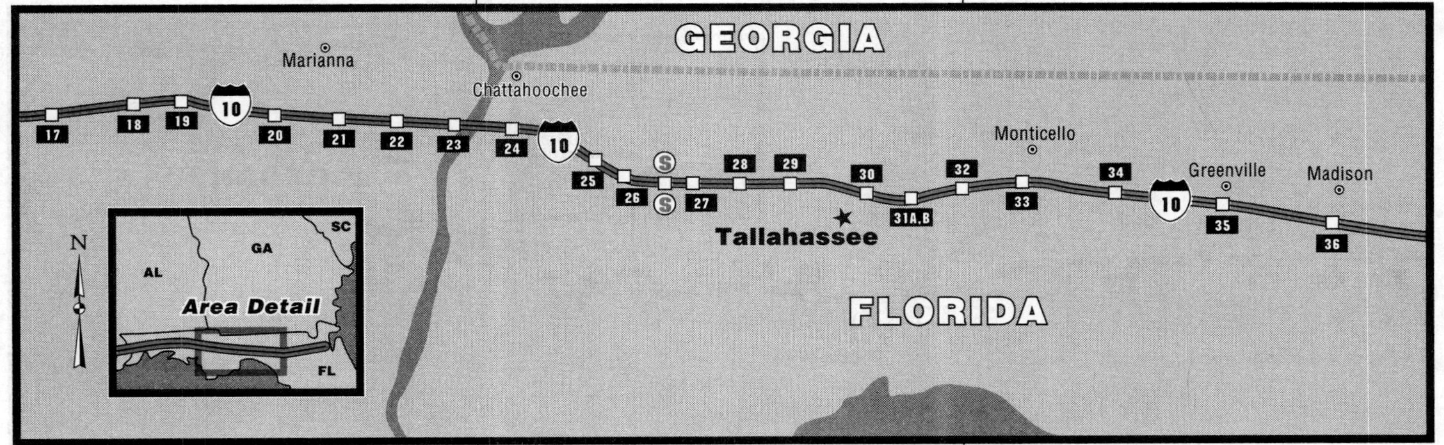

EXIT	FLORIDA
27	**(192)** U.S. 90, Quincy, Midway
TStop	N: **Flying J Travel Plaza*(LP)(SCALES) (Travel Store, RV Dump)**
Gas	N: BP*(LP), Gas N Go
Food	N: **Country Market (Flying J TS), Magic Dragon (Flying J TS), Pepperoni Super Slice (Flying J TS)**
(194)	**Rest Area - Picnic, RR, HF, Phones (Both Directions)**
28	**(195)** FL 263, Capitol Circle, Regional Airport
Gas	S: Chevron*(CW), Dixie*(D), Shell*(D) (see our ad this page)
Food	S: Waffle House
Lodg	S: ◆ Sleep Inn, Days Inn (see our ad this page)
Other	S: **Tallahassee Museum of History & Natural Science**
29	**(199)** U.S. 27, Havana, Tallahassee
Gas	S: Texaco*, Shell (see our ad this page)
Food	N: Burger King, Waffle House
	S: Blimpie's Subs, Cracker Barrel, Julie's Place, Red Lobster, Shoney's, Sonny's BBQ
Lodg	N: AAA Best Inns, ◆ Comfort Inn, ◆ Hampton Inn, ◆ Holiday Inn(see our ad this page), Thrifty Inn
	S: ◆ Cabot Lodge, Days Inn, AAA Howard Johnson, AAA La Quinta Inn, Motel 6, AAA Ramada, AAA Shoney's Inn
30	**(202)** U.S. 319, FL. 61, Thomasville, Tallahassee
Gas	N: Amoco* (24 Hrs), BP, Dixie*(D), Shell*(D) (see our ad this page), Texaco*
Food	N: Applebee's, Bagel Peddler, Baskin Robbins, Fuddrucker's, KFC, King House Buffet Chinese, Manhattan Bagel, McDonalds (Play Land), Philly Connection, Pizza Hut, Popeye's Chicken, Quincy's Family Steakhouse, Subway, TCBY Yogurt, Taco Bell, Veranda's Restaurant, Village Pizza, Wendy's
	S: Boston Market, Outback Steakhouse, Steak & Shake, T.G.I. Friday's
Lodg	N: Motel 6
	S: ◆ Cabot Lodge, Residence Inn
AServ	N: Auto Super Service, Super Lube
Med	N: **+ Hospital**
ATM	N: Ameri South Bank, Barnett Bank, Capital City Bank, FMB Bank, First Union, NationsBank, SunTrust, Tallahassee State Bank
Parks	N: **Alfred B. Maclay State Gardens**
Other	N: **Albertson's Grocery, Books-A-Million, Eckerd Drugs, My Favorite Books, Northeast Animal Hospital, Publix Supermarket,** Super Suds Car Wash, **Wal-Mart, Winn Dixie Supermarket (Pharmacy)**

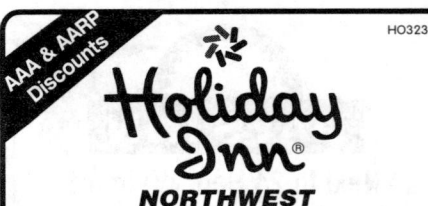

Bold red print shows RV & Bus parking available or nearby

EXIT		FLORIDA
31AB		(208) U.S. 90, Tallahassee, Monticello
	FStop	S: Citgo*(LP)
	Food	S: Cross Creek Restaurant
	Lodg	S: Seminole Inn Best Western, Ramada Limited (see our ad this page)
	RVCamp	S: Tallahassee RV Camping
	ATM	S: Citgo FS
	Other	S: Cross Creek Driving Range & Golf Course
32		(216) FL. 59
	TStop	S: Big Bend Auto/Truck Plaza*(SCALES)
	FStop	S: Amoco*, Citgo*
	Food	S: Big Bend Restaurant (Big Bend TS), Joyner's (Amoco), Subway (Amoco FS)
	Lodg	S: The Big Bend Auto/Truck Plaza Motel
	TServ	S: Big Bend Auto/Truck Plaza
	TWash	S: I-10 Truck Wash (Big Bend Truck Wash)
33		(225) Monticello, Thomasville
	FStop	S: Texaco*, United 500*
	Gas	S: Chevron*, Exxon*, Shell*
	Food	S: Arby's (Texaco Fs), Huddle House, Taco Bell (Texaco FS), Wendy's
	Lodg	S: Days Inn, ◆ Super 8 Motel (see our ad this page)
	RVCamp	N: Campers World Campground
		S: KOA Campground

EXIT		FLORIDA
34		(233) CR. 257
	Gas	N: Chevron*, Fina*
(235)		Rest Area - Picnic, Phones, RR, HF (Both Directions)
35		(242) U.S. 221, Greenville, Perry
	Gas	N: Texaco*
		S: BP*
	Food	N: Dairy Queen (Texaco)
36		(252) FL. 14, Madison, Perry
	Gas	N: Texaco*
	Food	N: Arby's (Texaco), Icecream Churn (Texaco)
	Med	N: ✚ Hospital
37		(258) FL 53, Madison
	TStop	N: Texaco*(SCALES), Jimmies Firestone (OPENING SOON -- see our ad this page)
	FStop	N: Amoco*, Citgo*(D)
	Food	N: Burger King (Amoco FS), Dairy Queen (Texaco TS), Jimmy's Restaurant, Latrelle's Family Restaurant (see our ad this page), Subway (Texaco), Taco Bell (Texaco TS)
	Lodg	N: ◆ Days Inn (see our ad this page), Holiday Inn Express, ◆ Super 8 Motel (see our ad this page)
		S: Deerwood Inn
	TServ	N: Texaco TS
	RVCamp	S: Frontier Town (see our ad this page), Madison Campground
	Med	N: ✚ Hospital
	Other	S: Frontier Territory
38		(262) CR. 255, Lee
	FStop	N: Exxon*
		S: Penn Oil*(D)
	Gas	S: Chevron*
	Food	S: Icecream Churn (Chevron)
(265)		Rest Area - Picnic, RR, HF, Phones (Both Directions)
(271)		Weigh Station, Agricultural Inspection (Both Directions)
39		(275) U.S. 90, Live Oak, Lee

EXIT FLORIDA

Gas S: United 500*

AServ S: McCook Auto Center

40 **(384)** U.S. 129, Live Oak, Jasper

FStop N: **Spur*(D)**

Gas S: BP*, Chevron*, Shell*, Texaco*

Food S: Huddle House, McDonalds, Taco Bell, Wendy's

Lodg S: AAA **Econolodge,** Suwannee River Inn

AServ S: Chevron

RVCamp S: **Spirit of the Suwanne Campground**

Med S: **Hospital**

ATM S: Shell

(284) **Parking Area (Westbound)**

41 **(293)** CR.137, Wellborn

(294) **Rest Area - Picnic, RR, HF, Phones (Eastbound)**

(295) **Rest Area - Pincic, RR, HF, Phones (Westbound)**

42AB **(297)** Junction I-75, Valdosta, Tampa

43 **(302)** U.S. 41, Lake City, White Springs

Gas N: Amoco*

S: Texaco*

Parks N: **Stephen Foster Folk Center**

44 **(304)** U.S. 441, Lake City, Fargo

FStop S: **Joy***

Gas N: BP*, Chevron*(D)

AServ N: BP

RVCamp S: **KOA Campground**

Med S: **Hospital**

ATM S: Texaco FS

(318) **Rest Area - Picnic, RR, HF, Phones (Both Directions)**

45 **(324)** U.S. 90, Sanderson, Olustee

Other S: **Olustee Battlefield, Osceola National Forest**

46 **(327)** CR. 229, Sanderson, Raiford

47 **(333)** FL.125, Glen, St. Mary

FStop N: **Exxon*(D)**

ATM N: Exxon

48 **(336)** FL 121, Macclenny, Lake Butler

FStop N: **BP*(D)**

S: **Exxon***

Gas N: Exxon*

Food N: Hardee's, KFC, McDonalds (Play Land), Pizza Hut, Subway, Taco Bell, Waffle House

EXIT FLORIDA

S: Burger King (Playground), China Garden Restaurant

Lodg N: **Days Inn**

S: AAA **Econolodge,** Expressway Motel

AServ N: Discount Auto Parts, Jiffy Lube, Xpress Lube

Med N: **Hospital**

ATM N: BP FS, SouthTrust, Winn Dixie Supermarket

Other N: **Food Lion Supermarket, Okefenokee National Wildlife Refuge, Wal-Mart, Winn Dixie Supermarket (24 Hrs, Pharmacy)**

49 **(337)** FL 228, Macclenny, Maxville

Med N: **Hospital**

Other N: **Phantom Fireworks**

50 **(344)** U.S. 301, Baldwin, Starke

TStop S: **Jacksonville Unocal 76*(SCALES) (Amoco)**

FStop S: **Speedway*(SCALES)**

Gas S: Exxon*, Shell*, Texaco*

Food S: Burger King, **McDonalds, Town & Country (Unocal 76 TS),** Waffle House

Lodg S: Best Western

(351) **Rest Area - Picnic, RR, HF, Phones (Westbound)**

(352) **Rest Area - Picnic, RR, HF, Phones (Eastbound)**

EXIT FLORIDA

51 **(352)** Whitehouse, Cecil Field

Gas N: Chevron*, Lil' Champ*(CW)

S: Shell*(D), Texaco*(LP)

RVCamp N: **Rivers RV Camp**

ATM N: Chevron

52 **(357)** Marietta

Gas N: Gate*, Texaco*

S: Amoco*

Other S: Coin Car Wash

53 **(357)** Junction I-295, Savannah, St Augustine

54 **(358)** FL 103, Lane Ave

FStop N: **Fleet Card**

Gas S: Chevron*(CW) (24 Hrs), Shell*, Texaco*(D)(CW)

Food N: Andy's Sandwich Shop, Tad's Restaurant (Quality Inn)

S: Cross Creek BBQ, Denny's, Hardee's, KFC, Lee's Dragon Chinese, Linda's Seafood, Piccadilly Cafeteria, Shoney's

Lodg N: **Days Inn (see our ad this page),** Quality Inn

S: Executive Inn, Paradise Inn, **Super 8 Motel**

AServ S: Pep Boys Auto Center

ATM S: SunTrust Bank, Texaco

Other S: **Jax Lane's West Bowling Alley**

55 **(359)** FL 111, Cassat Ave, Edgewood Ave

Gas N: Amoco*(D), Hess(D), Texaco*

S: Racetrac*

Food N: Pat's Subs

S: Andy's Sandwich Shop, Krispy Kreme Donuts

AServ N: Auto Zone Auto Parts, Fred's Service Center, Mr. Transmission

S: Xpress Lube

Other N: **Murray Hill Animal Hospital, Winn Dixie Supermarket**

S: **Winn Dixie Supermarket (Pharmacy)**

56 **(360)** Lenox Ave, Edgewood Ave (Westbound, Difficult Reaccess)

57 **(361)** FL129, McDuff Ave.

Gas S: Amoco*(D) (24 Hrs), Chevron*

Food S: Popeye's Chicken

AServ S: Amoco, McDuff Auto Center

Other N: **Trinity Baptist College**

58 **(362)** U.S. 17, Roosevelt Blvd

59 **(363)** Stockton St.

Gas S: Amoco*, Gate*(D)

Med S: **Hospital**

60 **(363)** Junction I-95, Savannah, St Augustine

↑FLORIDA

Begin I-10

EXIT	LOUISIANA

Begin I-12

↓ LOUISIANA

1A Junction I-10

1B LA 3064, Essen Ln.
- **Med** S: ✚ **Hospital**
- **ATM** N: Union Planters Bank

2AB (3) U.S. 61S., Airline Hwy
- **Gas** N: Exxon*(D), Shell* (24 Hrs)
  S: Circle K*, Speedway*, Texaco*(CW)
- **Food** N: Applebee's, McDonalds, New York Deli, Shoney's, Taste of China
  S: Bonanno's Seafood, McDonalds, Waffle House
- **Lodg** N: Hampton Inn, Holiday Inn, Motel 6, Shoney's
  S: Days Inn, Host Inns
- **AServ** N: Pep Boys Auto Center
  S: Duplessis Cadillac
- **ATM** S: Circle K, Regions Bank

4 Sherwood Forest Blvd
- **Gas** N: Exxon*(CW), Jet 24*, Texaco*(D)(CW)
  S: Chevron*(CW), Shell*(CW)
- **Food** N: Bamboo House, Burger King (Playground), Chuck E. Cheese's Pizza, Coriu Greek, Lebanese, Italian, Denny's, Eggroll King Chinese, KFC, McDonalds (Play Place), Popeye's Chicken, Subway, Waffle House
  S: Pasta Garden, Pizza Hut, Podnuh's BBQ, Taco Bell, The Ground Pati
- **Lodg** N: ◆ Red Roof Inn, Super 8 Motel
  S: Quality Inn
- **AServ** N: Car Quest Auto Center, Firestone Tire & Auto, Goodyear Tire & Auto
  S: Precision Tune
- **Other** N: **Eckerd Drugs, K & B Drugs**

6 Millerville Rd
- **Gas** S: Citgo*
- **Food** S: Chris' Specialty Meats
- **AServ** S: Penske Auto Center
- **Other** N: **K-Mart (24 Hrs, Pharmacy)**

7 O'Neal Ln.
- **Gas** N: Mobil*(CW), Shell*(CW) (24 Hrs)
  S: Chevron*(CW) (24 Hrs), Exxon*(D), Texaco*(D)(CW)
- **Food** S: Blimpie's Subs, **Little Caesars Pizza (Wal-Mart)**, Popeye's Chicken, Subway, Taco Bell, Waffle House
- **AServ** S: Exxon

EXIT	LOUISIANA

- **RVCamp** N: **Knight's RV Park**
- **Med** S: ✚ **Hospital**
- **ATM** N: Union Planter's Bank
  S: Regions Bank
- **Other** S: **Wal-Mart Supercenter (Pharmacy)**

10 LA 3002, Denham Springs
- **FStop** S: **Speedway*(SCALES)**
- **Gas** N: Chevron* (24 Hrs), Circle K*, Exxon*, Racetrac* (24 Hrs), Texaco*(CW)
  S: Texaco*
- **Food** N: Burger King (Playground), Chinese Inn, Crawford's Family Restaurant (Catfish), Golden Corral, **McDonalds (Play Place)**, Popeye's Chicken, Subway, Waffle House, Wendy's
  S: Church's Fried Chicken (Speedway FS), Las Vegas Deli, Piccadilly Cafeteria, Shoney's
- **Lodg** N: ◆ **Denham Springs Best Western**
- **AServ** N: Firestone Tire & Auto, Michelin
  S: All Star Dodge
- **RVCamp** S: **KOA Campground**
- **ATM** N: Exxon, Hancock Bank
- **Other** N: **Del Champs Supermarket, K & B Drugs, Louisiana Fireworks**

15 LA 447, Walker, Port Vincent
- **FStop** S: **Chevron*, Fina***
- **Gas** N: Shell*(CW)
- **Food** N: Burger King (Playground), McDonalds (Play Place)
- **ATM** S: **Chevron**

19 Satsuma

22 LA. 63, Livingston, Frost

EXIT	LOUISIANA

- **Gas** N: Conoco*, Texaco(D)
- **AServ** N: Sibley's Car Care, Texaco
- **RVCamp** S: **RV Park**

(27) **Rest Area - RR, Phones, RV Dump, Picnic (Both Directions)**

29 LA 441, Holden
- **Gas** N: Coastal*(D)

32 LA 43, Albany, Springfield
- **Gas** N: Chevron*
  S: Citgo*
- **Food** S: Levi's Cafeteria

35 Baptist Pumpkin Center
- **Gas** N: Exxon*
  S: Chevron*
- **AServ** S: C&S Automotive & Transmission Repair
- **RVCamp** N: **Pumpkin Park Campground**

(36) **Weigh Station (Both Directions)**

38AB Junction I-55, New Orleans, Jackson

40 U.S. Hwy 51 Bus., Hammond, Ponchatoula
- **TStop** S: **Petro*(SCALES), Pilot Travel Center***
- **Gas** N: Chevron*, Racetrac*, Shell*, Texaco*(CW)
- **Food** N: Burger King, Denny's, McDonalds (Playground), Morrisons Cafeteria, Pizza Hut, Ryan's Steakhouse, Taco Bell, Wendy's
  S: **Arby's (Pilot), Iron Skillet (Petro), Waffle House**
- **Lodg** N: ◆ Comfort Inn
  S: **Colonial Inn**
- **TServ** S: **Petro**
- **TWash** S: **Blue Beacon Truck Wash (Petro)**
- **ATM** N: Racetrac
- **Other** N: **Hammond Square Mall, K & B Drugs**

42 LA. 3158, Airport
- **FStop** N: **Chevron***
- **Lodg** N: **Friendly Inn, The Lucky Dollar (Chevron)**
- **TWash** N: **Hammond Truck Wash**

47 LA. 445, Robert
- **RVCamp** N: **Hidden Oaks Campground, Jellystone Park Campground, Sunset Campground**
- **Parks** N: **Zemmurray Gardens**

57 Goodbee, Madisonville, LA1077
- **Parks** S: **Fairview Riverside State Park**

59 LA21, Covington, Madisonville
- **Gas** N: Shell*(CW) (see our ad this page), Spur*
- **Food** N: Burger King (Spur), McDonalds

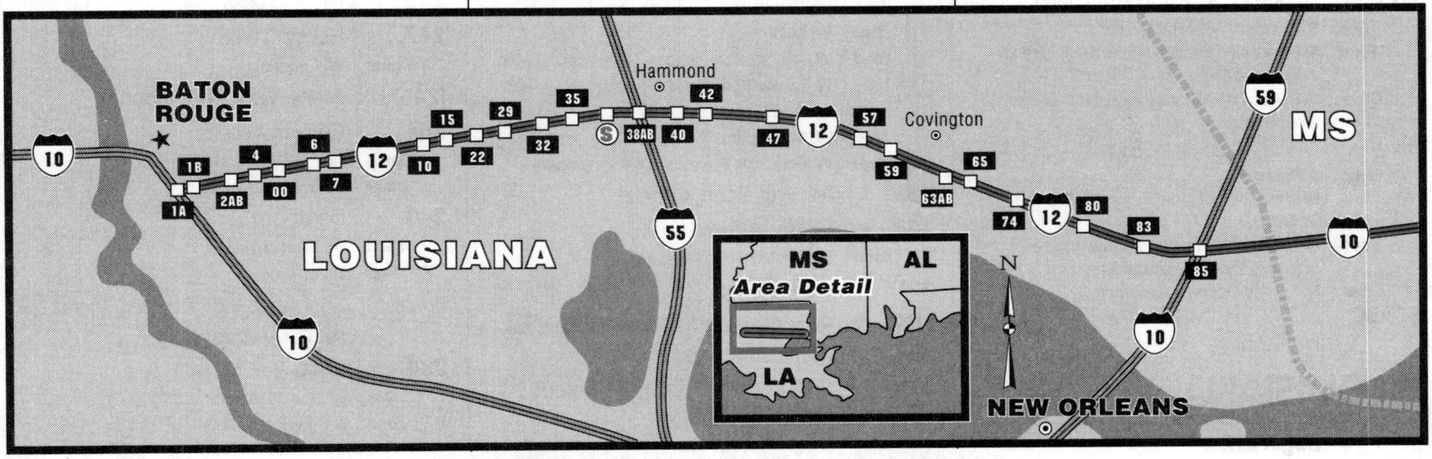

Bold red print shows RV & Bus parking available or nearby

EXIT LOUISIANA

Med N: ✚ Hospital

(60) **Rest Area - Picnic, RR, HF, Phones, RV Dump (Both Directions)**

63A New Orleans via Causeway Toll Bridge

63B U.S. 190, Covington, Mandeville.

Gas N: Exxon*(CW), Racetrac* (24 Hrs), Shell*, Texaco*(D)(CW)

Food N: Applebee's, Burger King (Playground), Pasta Kitchen, Shoney's, Subway

Lodg N: (AAA) Best Western, Courtyard by Marriott, ◆ **Holiday Inn (see our ad this page)**, (AAA) Northpark Inn

Other N: **Albertson's Grocery, Books-A-Million, Wal-Mart**

65 LA59, Abita Springs, Mandeville

Gas N: Chevron*(CW), Shell*(CW) (see our ad this page)

Food N: Subway (Chevron)

Parks S: **Fountain Bleau State Park**

Other N: **Tourist Information**

74 LA. 434, St. Tammany, Lacombe

80 Airport Drive, North Shore Blvd

Gas S: Chevron*(CW) (24 Hrs), Exxon*

Food S: Burger King (Playground), McDonalds (Play Place), Taco Bell

Other S: **Del Champs Supermarket, Northshore Square Mall**

83 U.S. 11, Pearl River, Slidell

FStop S: **Speedway***

Gas N: Chevron*(CW), Exxon*(D)
S: Shell*(CW)

Food N: McDonalds (Play Place), Waffle House
S: **Starvin Marvin (Speedway)**

Other N: **County Market Supermarket**

85ABC I-10 W. to New Orleans, I-59 N. to Hattiesburg

↑LOUISIANA

Begin I-12

EXIT MONTANA

Begin I-15

↓MONTANA

397 Sweetgrass

Gas W: Conoco*, Sinclair*, Speedy Mart*

Food W: Glocca Morra Motel, Paper Dollar Restaurant

Lodg W: Glocca Morra Motel

Other W: **Rest Area - RR, Picnic, Phone**

394 Ranch Access

389 CR 552, Sunburst

FStop W: **Suta's Supply***

Food W: Cafe, Chuck Wagon

AServ W: Suta's Supply

Other W: **Coin Laundry, Martin's Market, U.S. Post Office**

385 Swayze Road

379 CR 215, CR 343, Kevin, Oilmont

Food W: 4 Corners Cafe

373 Potter Road

369 Bronken Road

(367) **Weigh Station - RV Dump (Southbound)**

364 Bus Loop 15, Shelby

RVCamp E: **Camping**

363 U.S. 2, Shelby, Cut Bank, Bus I-15

TStop E: **Exxon**

Gas E: Gasamat*

Food E: **Exxon**, Fine Mexican Food, Pizza Hut, Ron's Cafe, **The Dixie Inn**
W: **McDonalds**

Lodg E: (AAA) Comfort Inn, Crossroads Inn, Glacier Motel

TServ E: **Exxon**

RVCamp E: **Camping, Glacier Park**
W: **Glacier RV Park**

Med E: ✚ **Hospital**

Parks W: **Glacier National Park**

Other W: **Pamida Discount Center (General merchandise & pharmacy)**

358 Marias Valley Road, Golf Course Road

RVCamp E: **Camping**

352 Bullhead Road

348 MT 44, Valier Road

Parks W: **Lake Francis Recreation Area**

345 CR 366, Ledger Road, Tiber Dam

339 Bus Loop I-15, Conrad

FStop W: **Sinex***(LP)

Gas W: Exxon*(D)

Food W: Arby's (Exxon), Pizza Hut, The Main Drive Inn

Lodg W: **Lowest Rate Motel**, Northgate Motel, (AAA) **Super 8 Motel**

Med W: ✚ **Hospital**

Other W: Car Wash, **Econo-Wash Coin Laundry**

335 Midway Road, Conrad

328 CR 365, Brady

Food W: The Red Onion Restaurant

321 Collins Road

(319) **Rest Area - RR, Phones, Picnic (P) (Both Directions)**

313 CR 221, CR 379, Dutton, Choteau

FStop W: **Conoco***

Food W: **Cafe Dutton**

305 Bozeman

302 CR 431, Power

297 Gordon

290 U.S. 89N, MT 200W, Missoula, Choteau

RVCamp W: **Camping**

286 Manchester

282 U.S. 87N, N.W. Bypass (Southbound, No Southbound Reaccess)

TStop E: **Teton Truckstop**(SCALES)

Lodg E: **Evergreen Motel**

TServ E: **Metco Kenworth**

280 U.S. 87 N, Bus Loop 15, Central Ave West

TStop E: **Flying J Travel Plaza**

Gas E: Mini Mart*, Sinclair*

Food E: Arizona Grill & Seafood, Dairy Queen, Ford's Drive-in, Hilltop Cafe, **Jack's Club Restaurant**

Lodg E: Alberta Motel, (AAA) Central Motel, **Starlit Motel**

AServ E: Kelly Tires, Major Motors, Wayland Tire Service

TServ E: **Cummins Diesel, Flying J Travel Plaza**(LP), **Truck Rate**

ATM E: First Liberty Bank

278 Bus Loop 15, U.S. 89S, MT 200E, 10th Ave South (Loads Over 12' Wide Not Permitted)

Gas E: Exxon*, Sinclair*(LP)

Food E: China Town Restaurant, Dairy Queen, New York Pizzeria

Lodg E: Airway Motel, Best Western, Budget Inn, Motel 9, Rendevous Inn, **Wright Nites Inn**

AServ E: Sinclair

RVCamp E: **Dick's RV Park, KOA Campgrounds**

Med E: ✚ **Hospital**

Other E: **Fox Farm Foods, Visitor Information**

277 International Airport

TStop E: **Conoco**

Food E: **Conoco**

Other W: **Airport**

(275) **Weigh Station (Northbound)**

270 CR 330, Ulm

Food W: Village Inn

Other E: **Pishklin State Park, U.S. Post Office**

256 MT 68, Cascade

254 MT 68, Cascade

250 Local Access

247 Hardy Creek

RVCamp W: **Camping**

(246) Scenic View (Southbound)

244 Canyon Access

RVCamp W: **Camping (2 Miles)**

Other W: **Boating (2 Miles)**

240 Dearborn

Food W: Dearborn Inn

RVCamp E: **Camping**

(240) **Rest Area - RR, Phones, Picnic (P) (Both Directions)**

234 Craig

Gas E: O'Connell's Store*

Food E: Trout Shop Cafe

Bold red print shows RV & Bus parking available or nearby

EXIT		MONTANA
	Lodg	E: Missouri River Trout Shop & Lodge
	AServ	E: Rod's Motor Service
	RVCamp	E: **Camping**
228		U.S. 287N, Augusta, Choteau
226		CR 434, Wolf Creek
	Gas	E: Exxon*[LP]
	Food	E: Oasis Family Restaurant
		W: The Frenchman Cafe & Saloon
	Lodg	E: Frenchy's Motel
	RVCamp	E: **Camping**
	Other	E: **Holter Lake**
		W: **U.S. Post Office**
(222)		**Parking Area - RR, Phones, Picnic (Both Directions)**
219		Spring Creek (Northbound, No Northbound Reaccess)
	RVCamp	E: **Camping**
216		Sieben
209		Gates of the Mountains
	Other	E: **Recreation Area**
(202)		**Weigh Station (Southbound)**
200		CR 279, CR 453, Lincoln Road
	Gas	W: Conoco
	Food	W: Grub Stake Restaurant, Keno's Steakhouse
	RVCamp	W: **Lincoln Park Camp Ground**
	Parks	E: **Houser Lake**
	Other	W: **Ski Area**
193		Bus Loop I-15, Cedar Street
	Gas	W: Conoco*[D], Exxon*, Jolly's, Sinclair*
	Food	W: Mother Lode Casino & Restaurant, Mother Lode Restaurant, Pizza Hut, Subway, Uncle Ron's Breakfast & Steakhouse
	AServ	W: Exxon, Pennzoil Lube, Power Chevrolet
	TServ	E: **Whalen Tire**
	RVCamp	E: **Northern Energy Propane Sales**[LP]
		W: **Branding Iron, Camping**
	Other	E: **Helena Regional Airport, National Forrest Ranger Station, Visitor Information**
		W: **Animal Center, Champion Auto Stores, K-Mart, U.S. Post Office**
192AB		U.S. 12, U.S. 287, Townsend, Capitol Area
	FStop	E: **Conoco***[LP]
	Gas	E: Conoco*[D]
		W: Exxon[D], Mini Mart*, Sinclair*
	Food	E: Burger King
		W: Bullseye Casino & Restaurant, Burger King, Country Kitchen, Dairy Queen, JB's Restaurant, Jimmy B's Casino & Restaurant, KFC, Little Caesars Pizza, McDonalds, Overland Express, Pasta Pantry, Rack's, TCBY Yogurt, Taco Bell, Taco Treat Mexican, Village Inn Pizza, Wendy's
	Lodg	W: Best Western, Comfort Inn, Days Inn, Holiday Inn Express, Jorgenson's Holiday Motel, Motel 6, Shilo Inn Motel, Super 8 Motel
	AServ	E: Gary Auto Plaza, Northwest Battery & Electric
		W: Goodyear Tire & Auto, Muffler Shop, Toyota Dealer
	RVCamp	E: **RV Center**
	Med	W: **+ Helena Medical Plaza**
	ATM	W: First Security Bank
	Other	E: Car & Truck Wash, **State Patrol Post, Tourist Information, Truck Permits, Wal-Mart (Pharmacy)**
		W: **Albertson's Grocery, Bergum Drug's, Capital City Mall, Coin Laundry, Pet Supply Store, Safe Way Grocery, State Museum, Terry's Food Store**
187		CR 518, Montana City
	Gas	W: Conoco*
	Food	E: Hugo's Pizza
		W: Jackson Creek Saloon, The Exchange
	AServ	E: Certified Transmission
	Other	E: **Strawberry Mountain**
		W: Coin Car Wash
182		Clancy
	Food	W: Legal Tender Restaurant
	Other	E: **Camping**[LP]
(178)		**Rest Area - RR, Phones, Picnic P (Both Directions)**
176		Jefferson City
	Other	W: **U.S. Post Office**
164		MT 69, Boulder
	FStop	E: **Exxon***
	Gas	E: R&D (Food mart)
	Food	E: **Dairy Queen (Exxon FS)**, Mountain Good Restaurant, **Pizza Parlor (Exxon FS), Stage Stop Steak House (Exxon FS)**
	Lodg	E: O E Motel, Sunshine Health Mine
	AServ	E: El Dorado Tire's
	RVCamp	E: **RC RV Camping**
	ATM	E: Exxon
	Other	E: **Almon Drugs, Boulder Cash Grocery Store,** Coin Car Wash, **Coin Laundry, IGA Store, Post Office**
160		High Ore Road
156		Basin
	RVCamp	E: **Camping**
151		Bernice
	Other	W: **Camping**
138		Elk Park
134		Woodville
(131)		Scenic Overlook (Southbound)
129		Jct I-90, Billings, Butte
127AB		Harrison Ave
	Gas	E: Conoco* (#1 of 1), Conoco* (#2 of 2), Exxon*[D][CW]
		W: Cenex*[D][CW], Conoco*, Sinclair* (RV Dump)
	Food	E: 4 B's Restaurant (24Hr), Arby's, Burger King, Godfather's Pizza, Joey's Seafood, KFC, McDonalds, Perkins Family Restaurant, Pizza Hut, Plaza Royale Restaurant, Ray's Place, Red Rooster Supper Club, Restaurant (Motel), Silver Bow Pizza, Subway, TCBY Yogurt, Taco Bell, The Ponderosa Cafe, Uno's, Wendy's
		W: Arctic Circle Hamburgers, Dairy Queen, Denny's, Derby Steakhouse, Domino's Pizza, El Taco Mexican Food, Hardee's, Hot Stuff Pizza (Cenex), John's, Little Caesars Pizza, Restaurant (War Bonnet Inn), Smash Hit Subs (Cenex), Taco John's, Top Deck
	Lodg	E: Best Western, Comfort Inn, Days Inn, Mile High, Motel, Super 8 Motel
		W: Holiday Inn Express, War Bonnet Inn
	AServ	E: American Car Care Centers, Checker Auto Parts, Ford Dealership, GM Auto Dealership, Glenn's, Honda Dealer, Pennzoil Lube, **Wal-Mart**
		W: Bob's Fast Lube, Champion Auto Parts Store, Nissan Dealer, Uniroyal Tire & Auto
	ATM	E: American Federal Savings Bank, Conoco (#1 of 1), Conoco (#2 of 2), Exxon, First Citizens Bank, First National Bank, Norwest Bank

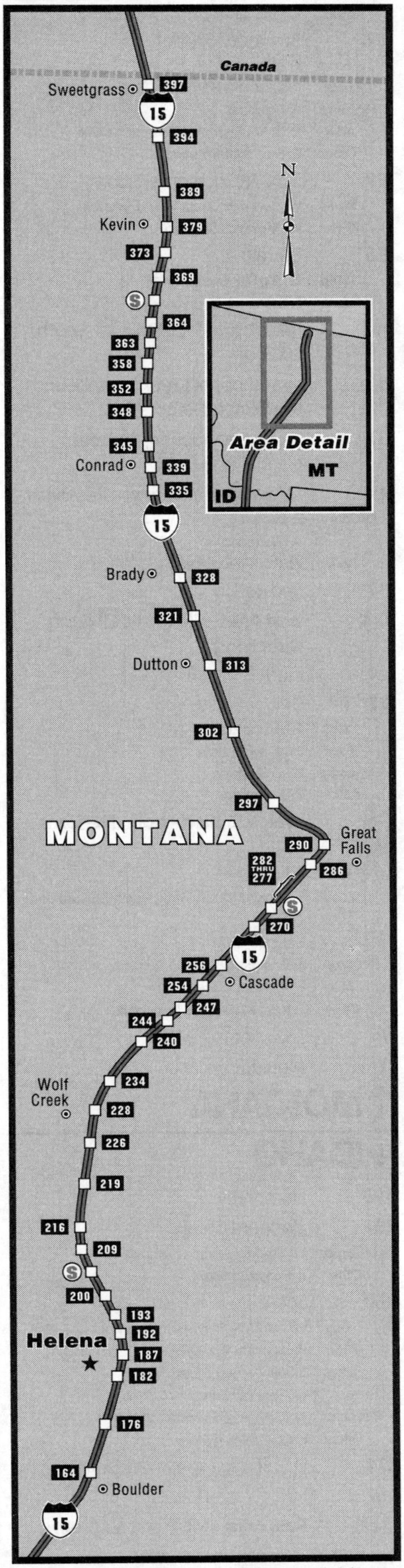

Bold red print shows RV & Bus parking available or nearby

EXIT MONTANA

W: Cenex, Conoco
Other E: **Butte Plaza Mall, Buttrey Food Grocery (Pharmacy), K-Mart, Optical, Wal-Mart (Pharmacy)**
W: **Downey Drug, Natural Healing, Safe Way Grocery (24 Hrs, W/Pharmacy)**

126 Montana Street
Gas E: Conoco*, Exxon*(D)
W: Cenex*, Sinclair
Food E: Muzz & Stan's Food
W: Jokers Wild Casino & Restaurant (Chicken/BBQ Ribs), Winter Garden Lanes
Lodg W: Eddy's Motel
AServ W: Sinclair
RVCamp W: **KOA Campground**
Med W: ✚ **Hospital**
ATM E: Conoco
Other W: **Dental Clinic, Safe Way Grocery, Tourist Information**

124 Jct I-115, Butte City Center

122 Rocker, Weigh Station (Both Directions)
TStop E: **Conoco*(LP)**
W: **Flying J Travel Plaza(LP)**
Food E: **4 B's Restaurant (Conoco), Arby's (Conoco TS)**
W: **Flying J Travel Plaza**
Lodg W: (AAA) **Rocker Inn**
AServ E: **Rocker Repair (24 Hr)**
TServ E: **Rocker Repair (24 Hr)**
ATM E: **Conoco**
W: **Cash Machine (Flying J TS)**

121 Jct I-90W, Missoula

119 Hub Access, Silver Bow

116 Buxton

111 Feely

(109) **Rest Area - RR, Phones, Picnic P (Both Directions)**

102 MT 43, Divide, Wisdom
Other W: **Big Hole Battlefield Monument**

99 Moose Creek Road

93 Melrose
FStop W: **Melrose**
Food W: Melrose Bar & Cafe
Lodg W: Melrose Motel, Sportsman Motel
RVCamp W: **Camping**
Other W: **Campcreek Trailer Court Laundry, U.S. Post Office**

85 Glen

74 Apex, Birch Creek

63 MT 41, Dillon, Twin Bridges
FStop E: **Cenex(LP)**
Gas E: Conoco(CW), **Exxon(D)**, Finex (RV Dump)
Food E: KFC, Lion's Den Supper Club, **McDonalds, Pizza Hut**, Subway
Lodg E: (AAA) Best Western, (AAA) **Comfort Inn**, Sacajewea Motel, (AAA) **Sundowner Motel**, ◆ **Super 8 Motel**
AServ E: Auto Dealers - All Makes, Johnson's Motors, Lisac's Tires, Mark's Fast Lube, Montano Motor Parts, Uniroyal Tire & Auto, Wayland Tire
RVCamp E: **Campground**
Med E: ✚ **Hospital**
ATM E: Exxon

EXIT MONTANA/IDAHO

Other E: **Amerigas(LP), Tourist Information**

62 Bus Loop 15, Dillon
Food E: Taco John's
Lodg E: Cross Winds Motel
RVCamp E: **Camping**
Med E: ✚ **Southwestern Montana Clinic**
Other E: **Tourist Information**

59 MT 278, Jackson, Wisdom
Parks W: **Bannack Ghost Town State Park**
Other W: **Maverick Ski Area**

56 Barretts
TStop W: **Big Sky Truck Stop*(LP)**
RVCamp E: **Camping**

(55) **Rest Area - RR, Picnic P (Southbound)**

52 Grasshopper Creek (Northbound, Northbound Reaccss)

51 Dalys (Southbound, Reaccess Northbound Only)

44 MTD 324, Clark Canyon Reservoir
RVCamp E: **Camping**
W: **Camping**
Parks W: **Recreation Area**

37 Red Rock

(34) **Rest Area - RR, Picnic P (Both Directions)**

29 Kidd

23 Dell
Gas E: Dell Mercantile*
Food E: Yesterday Cafe
Lodg E: Red Rock Inn
AServ E: Dell Garage

(16) **Weigh Station - both directions**

15 Lima
Gas E: Exxon*(D)(LP)
Food E: Cook Your Own Steak, **Home Cooked Grub Cafe**
AServ E: Big Sky Country
RVCamp E: **Camping**
Med E: ✚ **Ambulance Service**
Other E: **R & J Market, US Post Office**

9 Snowline

0 Monida

↑MONTANA

↓IDAHO

190 Humphrey

184 Stoddard Creek Area
RVCamp W: **Stoddard Creek Campground**
Other E: **Historical Site**

180 Spencer
Gas W: Opal Mtn Showroom, **Spencer Bar & Grill**
Food W: **Spencer Bar & Grill**
Lodg E: Spencer Bed & Breakfast
W: **Spencer Camping Cabins**
RVCamp W: **Camping, Spencer RV Park**
Other E: **Opal Mining Store**

172 U.S. Sheep Experiment Station

167 I-22, CR A2, Dubois

(167) **Rest Area - RR, Picnic P (Both Directions)**

EXIT IDAHO

150 Hamer

143 ID 28, ID 33, Salmon, Rexburg

(143) **Weigh Station - both directions**

135 ID 48, Roberts
FStop E: **Amoco***
Food E: **Restaurant (Amoco)**
Other E: **Thriftway Grocery, True Value Hardware(LP)**

128 Osgood Area
FStop E: **Sinclair***

119 U.S. 20 East, Rigby, West Yellowstone
FStop E: **Texaco*(D)(LP)**
Gas W: **Texaco*(D)(LP) (RV Dump)**
Food E: Denny's, Frontier Pies (Quality Inn), Jaker's Steak, Ribs, Fish, Shilo Inn, The Sandpiper Restaurant
Lodg E: ◆ Ameritel Inn, (AAA) Best Western, Holiday Inn, Quality Inn, ◆ Shilo Inn, **Super 8 Motel**
AServ E: Automotive Electric
W: Texaco
TServ E: **Rumble's Diesel**
RVCamp E: **KOA Campground**
Other W: **Idaho State Police**

118 Bus. Loop 15, U.S. 20, Broadway St. (Closed For Construction)
FStop E: **Phillips 66**
Gas E: Chevron
W: **Amoco***, Exxon*, Flying J Travel Plaza*(CW), Phillips 66*(CW), Sinclair*(CW)
Food E: Arctic Circle Hamburgers, Holiday Inn, JB's Restaurant, Smitty's Pancake & Steak House
W: Burger King, Dairy Queen, Hong Kong Restaurant, McDonalds, O'Brady's Family Restaurant, Pizza Hut, Subway
Lodg E: Ameritel Inn, (AAA) Best Western, (AAA) Holiday Inn
W: Comfort Inn, Motel 6, **Motel West**
AServ E: Broadway Ford, Buick Dealer, Chevron, Phillips 66
W: **Amoco**, Pennzoil Lube
RVCamp W: **Amoco (RV Service)**
ATM E: Key Bank
W: West One Bank
Other E: **The Parts Place, Tourist Information**
W: **Albertson's Grocery, Coin Laundry, Payless Drugs**

113 Bus. Loop 15, U.S. 26, Jackson, Idaho Falls
TStop E: **Yellowstone Truck Stop(SCALES)**
Lodg E: **Yellowstone Motel**
TServ E: **Lake City International, Lindsay Truck & Towing, Peterbilt Dealer**
TWash E: **Yellowstone TS**
RVCamp E: **Sunnyside RV Camping (4.5 Miles)**
Med E: ✚ **Hospital**
Parks E: **Grand Teton National Park**

108 Shelley

(101) **Rest Area - RR, Phones, Info, Picnic, Vending P (Both Directions)**

98 Rose - Firth

93 U.S. 26, ID 39, Blackfoot, Arco
Gas E: Chevron*(CW) (24 Hrs), Flying J Travel Plaza*(D)(LP), Maverick*

Bold red print shows RV & Bus parking available or nearby

EXIT	IDAHO
Food	E: Arctic Circle Hamburgers, Betty's Cafe, Hogi Yogi, Homestead Family Restaurant, KFC, Little Caesars Pizza, **McDonalds (Playground)**, Papa Murphy's Pizza, Pizza Hut, Subway, Taco Bell, Taco Time, Wendy's
Lodg	E: AAA Best Western, **Riverside Inn**
AServ	E: Ford Dealership, Les Schwab Tires, Maverick, Pennzoil Lube
Med	E: **+ Hospital**
ATM	E: Key Bank
Other	E: **Albertson's Grocery, General Denistry, Grocery Outlet, Kesler's Supermarket, Kings Department Store, Payless Drugs, Potato Expo, Sears,** Super Car Wash, **Wal-Mart (Pharmacy)**
	W: **Craters of the Moon National Monument**
89	Bus Loop I-15, U.S. 91, Blackfoot
80	Fort Hall
FStop	W: **Sinclair*(LP)**
Food	W: **Oregon Trail Restaurant**
ATM	W: **Sinclair**
Other	W: **Shoshone-Bannock Tribal Museum, Trading Post Grocery**
72	Jct I-86 West, Twin Falls
71	Pocatella Creek Road
Gas	E: Chevron(LP), Circle K*, Phillips 66*
	W: Exxon*, Sinclair
Food	E: Frontier Pies (Best Western), Holiday Inn, Jack-in-the-Box, Perkins, Subway, The Porterhouse Grill (Quality Inn), The Sandpiper
	W: Chang's Garden, Dairy Queen, Pier 49 San Francisco Pizza, Sizzler Steakhouse
Lodg	E: ◆ AmeriTel Inn, AAA Best Western, ◆ Comfort Inn, AAA Holiday Inn, AAA Quality Inn, AAA Super 8 Motel
AServ	E: Chevron
	W: Sinclair
Other	W: **Alameda Pet Hospital, Alameda Vision Center, Idaho Museum of Natural History, Payless Drugs**
69	Clark St (No Trucks)
Med	E: **+ Pocatello Regional Medical Center**
	W: **+ Bannock Regional Medical Center**
67	Bus Loop 15, U.S. 30, U.S. 91, Pocatello (One Way Streets)
FStop	W: **Cowboy Oil Company(LP), Sinclair**
Gas	E: Exxon*
Lodg	W: Rainbow Motel
RVCamp	W: **Cowboy RV Park**
Parks	E: **Constitution Park**
Other	W: **Fort Hall Replica**
63	Portneuf Area, Mink Creek Recreation Area
RVCamp	E: **Camping**
(60)	**Weigh Station, Rest Area - RR, Picnic, Phones, Vending P (Both Directions)**
58	Inkom, Pebble Creek Ski Area, Bus I-15 (Southbound)
Gas	W: Saloman's(D)(LP), Sinclair*
AServ	W: Saloman's(D)(LP)
ATM	W: Sinclair
Other	W: **Post Office**
57	Bus Loop I-15, Inkom (Northbound)
Gas	W: Salomon's(D)(LP)

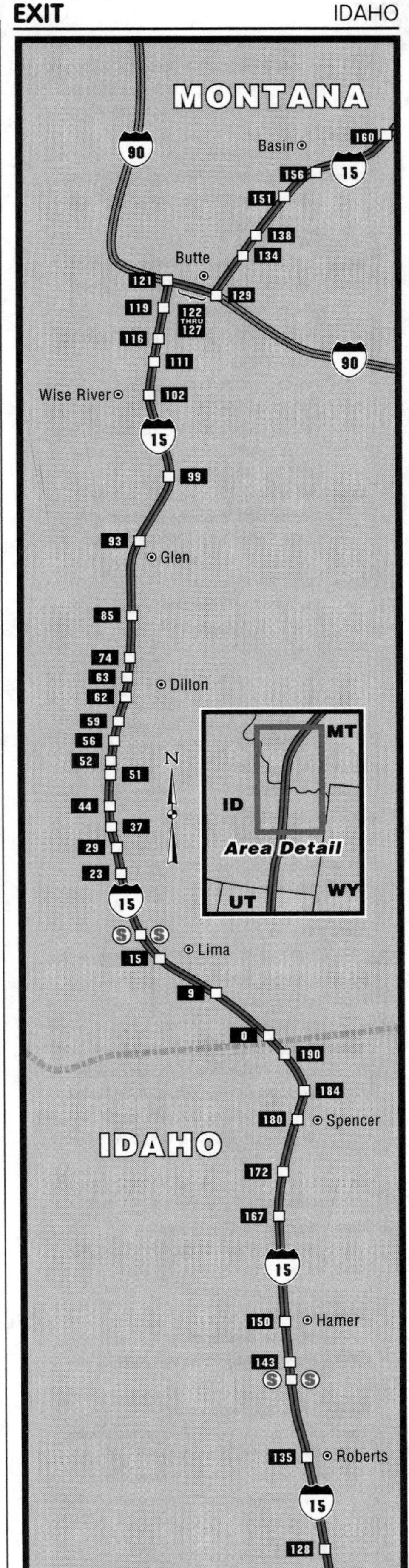

EXIT	IDAHO/UTAH
Food	W: El Rancho
AServ	W: Salomon's
Other	W: **Post Office**
47	U.S. 30, Lava Hot Springs, Soda Springs, to Jackson, WY
TStop	E: **Flying J Travel Plaza(LP)(SCALES)**
Food	E: **Thads Restaurant (Flying J TStop)**
44	Bus Loop I-15, Caccammon. Jensen Rd.
40	Arimo
36	U.S. 91, Downey, Preston
31	Downey, Preston
TStop	E: **Flags West(LP) (RV Dump)**
Food	E: **Flags West**
Lodg	E: **Flags West Motel**
TServ	E: **Flags West TS**
RVCamp	E: **Flags West(LP)**
ATM	E: **Flags West TStop**
Other	E: **Coin Laundry**
(25)	**Rest Area - RR, Phones, Picnic, Vending P (Southbound)**
22	Devil Creek Reservoir
RVCamp	E: **Camping**
17	ID 36, Weston, Preston
Food	W: Gary's Deep Creek Inn
13	ID 38, Malad
Gas	W: Phillips 66*(D)
Lodg	W: **Chantilly Bed and Breakfast,** Village Inn Motel
AServ	W: E-Z Coin & Automatic Car Wash
TServ	W: **3 R's Country Tire**
Med	W: **+ Hospital**
ATM	W: US Bank
(7)	**Info Center, Rest Area - RR, Phones, Picnic P (Northbound)**
3	Woodruff, Samaria

↑IDAHO

↓UTAH

EXIT	UTAH
402	Portage
394	UT 13 South, Plymouth
TStop	E: **Tri Valley Truck Stop***
Food	E: **Subway (Tri Valley TStop)**
392	I-84 West, Tremonton, Boise (Northbound)
387	UT 30 East, Riverside, Logan, Fielding
RVCamp	E: **Camping**
383	I-15 Loop, Tremonton, Garland
Med	E: **+ Hospital**
382	Jct I-84W, Boise (Southbound)
379	UT 13, I-15 Loop, Bus I-84, Tremonton, Garland
375	UT 240, to UT 13, Honeyville, Bear River
RVCamp	E: **Crystal Hot Springs Camping**
(370)	**Info Center, Rest Area - RR, Phones, Picnic, Vending P (Southbound)**
368	UT 13, 900 North St.
Parks	W: **Bear River Bird Refuge**

Bold red print shows RV & Bus parking available or nearby

EXIT UTAH

366 Forest St.
- **Parks** **W:** Bear River Bird Refuge

364 I-15 Loop, U.S. 89, 1100 South St.
- **Med** **E:** ✚ Hospital

(363) **Rest Area - Picnic, Vending, RR, Phones P (Northbound)**

361 Port of Entry & Weigh Station (Both Directions)

360 UT 315, Willard, Perry
- **FStop** **E:** Flying J Travel Plaza*(D)(LP) (Country Market)
- **Food** **E:** Country Market (Flying J)
- **RVCamp** **W:** Camping

354 UT 126, U.S. 89, Willard

352 UT 134, Farr West, Plain City, North Ogden, Pleasant View
- **Gas** **E:** Maverick* (Playground)
- **Food** **E:** McDonalds, Subway
- **ATM** **E:** Maverick

349 Defense Depot, Harrisville
- **Gas** **W:** Texaco*
- **TServ** **W:** JM Bodily & Sons TService

347 UT 39, 12th Street
- **TStop** **W:** Pilot Travel Center*(SCALES)
- **FStop** **E:** Flying J Travel Plaza*
- **Gas** **E:** Phillips 66*(CW)
- **Food** **E:** Jeremiah's Restaurant
  W: Country Kitchen (Pilot), Dairy Queen (Pilot), Subway (Pilot), Taco Bell (Pilot)
- **Lodg** **E:** (AAA) Best Western
  W: ◆ Sleep Inn
- **AServ** **E:** Steve's Car Care
- **TServ** **W:** General Diesel
- **Parks** **E:** Ogden Canyon Recreation Area

346 UT 104, 21st Street, Wilson Lane
- **TStop** **E:** Flying J Travel Plaza(SCALES) (RV Dump)
- **FStop** **E:** Chevron*(CW)
  W: Texaco*(D) (RV Dump)
- **Food** **E:** Arby's (Chevron), Big Z Restaurant, Cactus Reds (Comfort Suites), The Cookery Buffet
  W: Freeway Cafe
- **Lodg** **E:** Best Rest Inn, (AAA) Big Z Motel, ◆ Comfort Suites
  W: ◆ Super 8 Motel
- **TServ** **E:** Flying J Travel Plaza, Lake City International, Ogden Diesel
- **RVCamp** **W:** Century Home & RV Park
- **ATM** **E:** Flying J Travel Plaza
- **Other** **E:** Flying J Coin Laundry

345 UT 53, 24th Street, City Center, Fort Buena Ventura (Northbound, No Reaccess)
- **FStop** **E:** Sinclair*
- **Gas** **E:** Texaco*

344B UT. 79 West

344 UT 79 East

343 I-84 East, Cheyenne (Southbound)

342 UT 26, I-84 East, Riverdale (Northbound, No Northbound Reaccess)
- **Gas** **E:** Conoco(D)
  W: Phillips 66*(D)(CW)(LP)
- **Food** **E:** Applebee's, Boston Market, Chili's, McDonalds (Wal-Mart)
  W: ABC Chinese Cuisine, Arby's Roast Town, Burger Bar, Chinese Gourmet, Dairy Queen, Gade Terrace, JB's Restaurant, KFC, Lee's Fish & Rice, Lucky China, Star Burger, Subway, Taco Time, Tafoya Brother's Pizza, Warrens
- **Lodg** **E:** Motel 6
  W: Circle R Motel
- **AServ** **W:** Big O Tires, Buffalo Bros Tire Outfitters, Discount Tire Company, Midas Muffler & Brakes, Q-Lube
- **ATM** **W:** Key Bank, Phillips 66
- **Other** **E:** Target Department Store, Wal-Mart
  W: Animal Care Vet, Standard Optical, Super Serve Coin Car Wash

341 UT 97, Roy (Difficult Northbound Reaccess)
- **Gas** **W:** Citgo* (ATM), Phillips 66, Sinclair
- **Food** **W:** Arctic Circle, B.C. Chicken, Blimpie's Subs, Burger King, Chico & Wongs, Denny's, Hap-e-Trails, PaPa Murphy's Pizza, Pizza Hut, Taco Bell, Taco Time, Wendy's
- **AServ** **W:** Auto Zone Auto Parts, David Early's, Master Lube, Phillips 66, Precision Tune & Lube, Sinclair, Tubbs Goodyear Service Center
- **ATM** **W:** Bank of Utah, Citgo, First Security Bank
- **Other** **E:** Air Force Museum
  W: Albertson's Grocery, Payless Drugs

338 UT 103, Clearfield, Sunset, Hill AFB, Clinton
- **Gas** **W:** Chevron*(CW), Conoco*(CW)
- **Food** **W:** Arby's, Carl's Jr Hamburgers, KFC, McDonalds, Subway, Taco Bell
- **Lodg** **W:** (AAA) Cottage Inn, ◆ Super 8 Motel
- **AServ** **W:** Big O Tires
- **ATM** **W:** Conoco

336 UT 193, Freeport Center, Clearfield, Hill AFB, Truck Gate
- **Gas** **E:** Circle K*, Texaco*(D)(CW)(LP)
  W: Circle K*, Sinclair*
- **Food** **W:** Domino's Pizza
- **AServ** **E:** Texaco

335 UT 108, Freeport Center, Syracuse
- **FStop** **E:** Phillips 66*(D)(CW)(LP)
- **Gas** **E:** Chevron*(CW)
  W: Citgo*
- **Food** **E:** Applebee's, (Old Country Store) Cracker Barrel, Golden Corral, JB's Restaurant, Marie Callender's Restaurant, Outback Steakhouse, Pier 49 Sour Dough Pizza, Quiz No's, Red Robbin
  W: Arby's, Burger King, Central Park, Crown Burgers, Subway, The Taco Maker
- **Lodg** **E:** ◆ Fairfield Inn, ◆ Holiday Inn Express, (AAA) La Quinta Inn
- **AServ** **E:** Grandma's Tires, Tunex
  W: Checker Auto Works, Ford Dealership
- **Med** **E:** ✚ IHC Healthcare
  W: ✚ Davis Hospital
- **ATM** **E:** Phillips 66
  W: Citgo, Zion's Bank
- **Other** **W:** Albertson's Grocery, K-Mart

334 UT 232, Layton
- **TStop** **W:** Flying J Travel Plaza*(D)(CW)
- **Food** **E:** Denny's, Garcia's Mexican, McDonalds, Olive Garden, Red Lobster, Sizzler Steak House, Training Table Burgers, Wendy's
  W: Beech Boys Deli, Blimpie's Subs, Burger King, Fuddrucker's, Godfather's Pizza, KFC, Kenny Rogers Roasters, Lone Star Steakhouse, McDonalds (Wal-Mart), Seafood and Chowder House, Shoney's, Taco Bell
- **Lodg** **E:** (AAA) Travelers Inn
- **AServ** **W:** Layton Dodge Dealer, National Tire and Battery, Safelite Auto Glass, Simmons Auto Repair, Young Chevrolet/Pontiac
- **ATM** **E:** Bank One, Key Bank
  W: Barnes Banking Co.
- **Other** **E:** Layton Hills Shopping Mall
  W: Kinko's, Petsmart, Sam's Club, Wal-Mart

332 North I-15 Loop, UT 126, Layton (Northbound)
- **Gas** **W:** Chevron*, Phillips 66
- **Food** **E:** Little Orient Restaurant, Preformance Truck Accessories
  W: Doug's & Emmy's Family Restaurant, Sill's Cafe
- **AServ** **W:** Allco Auto Parts, Elmer's Auto Clinic, Phillips 66
- **Other** **E:** Coin Car Wash

331 UT 273, Kaysville
- **Gas** **E:** Chevron*(D) (McDonalds, Playground), Citgo*, Phillips 66*(CW), Sinclair
- **Food** **E:** Jakes Over the Top, Subway, The Taco Maker
- **AServ** **E:** Dick's Tire and Auto, Main St. Lube & Oil, Xpress Lube
- **RVCamp** **W:** Blaine Jensen and Son RVs
- **Other** **E:** Albertson's Grocery, Post Office

329 Truck Parking Area (Both Directions)

327 UT 225, Lagoon Drive, Farmington

326 U.S. 89 North, Ogden (Northbound)
- **RVCamp** **E:** Lagoons Pioneer Village Campgrounds

325 UT 227, Lagoon Drive, Farmington (Northbound)
- **RVCamp** **E:** Lagoon's Pioneer Village Campground

322 Centerville
- **FStop** **E:** Phillips 66*(D)(CW)
- **Gas** **E:** Chevron*
- **Food** **E:** Arby's, Arctic Circle Hamburger, Burger King, Dairy Queen, Domino's Pizza, McDonalds, Subway, Taco Bell
- **AServ** **E:** Big O Tires, Ray's Muffler Service, Super Wash Coin & Automatic Car Wash
- **ATM** **E:** Phillips 66
- **Other** **E:** Albertson's Grocery, Target Department Store

321 South U.S. 89, UT 131, 400 North, Bountiful
- **Gas** **E:** Chevron*(CW), Phillips 66, Texaco*
- **Food** **E:** Cafe' Alicia, The Fifth
- **AServ** **E:** Action Alternator and Starters, Phillips 66, Q Lube, Walton's Brake & Tire Co.
- **Other** **E:** Bob's Spot Free Coin Car Wash

320AB UT 68, 500 South, Bountiful
- **FStop** **W:** Phillips 66*
- **Gas** **E:** Amoco*(CW), Texaco*(D)
- **Food** **E:** Blimpie's Subs, Burger King, Chuck-A-Rama Buffet, Galaxy Diner, KFC, McDonalds,

EXIT UTAH

Pine Garden, Renny's Charbroiled, Sizzler Steak House, Subway, Taco Bell, Winger's American Diner

AServ E: AAMCO Transmission, Auto Zone Auto Parts, Checker Auto Parts, David Early Tires, Express Automotive, Firestone Tire & Auto, Lodder's Auto Repair, Meineke Discount Mufflers, Midas Muffler & Brakes, Ray's Muffler Service, Triangle Tire Service

W: Dave Roberts Auto Service

Med E: **Benchmark Regional Hospital**

ATM E: Barnes Bank, US Bank

Other E: **Barnes & Noble, Shopko Grocery,** Super Wash Coin Car Wash, **TJ Maxx**

W: **U.S. Post Office**

318 UT 93, North Salt Lake, Woods Cross

FStop E: **Chevron***

Gas E: Sinclair*, Texaco*

Food E: Arby's, Atlantis Burgers, Cutler's Grill, McDonalds, Pizza Hut, Subway, Village Inn

W: Denny's, Lorena's Restaurant

Lodg W: Hamilton Inn, Motel 6

AServ E: Ken Garff Mazda, Menlove Dodge and Toyota, Mercury Lincoln, Nation Wide Auto Glass, Tunex

ATM E: Key Bank, Wells Fargo, Zions First National Bank

Other E: **K-Mart, Smith's**

317 Cudahy Lane, North Salt Lake (Southbound, No Southbound Reaccess)

316 I-215, Belt Route, Salt Lake International Airport (Southbound)

315 U.S. 89 South, Beck

314 2300 North, Warm Springs Road

313 900 West (Southbound, No Southbound Reaccess)

312AB 600 North, 6th North (Closed)

Gas E: Conoco*

Lodg E: Gateway Motel

RVCamp W: **Camping**

Med W: **Hospital**

311 Junction I-80W, Airport, Reno

310 6th South St, City Center

Gas E: Circle K, Phillips 66*, Sinclair*

Food E: Cafe Olympus, Denny's, Hilton, McDonalds, Sophi Garcia's Mexican

Lodg E: Best Western, Crystal Inn, Little America Hotel, Motel 6, Olympus Hotel, Quality Inn

AServ E: Miller Toyota

ATM E: Cash Machine (Sinclair)

309 13th St, 21st St (Difficult Reacess)

TStop W: **Flying J Travel Plaza**(LP)(SCALES)

FStop E: **Cash Saver***

Food E: Atlantis Burgers

W: Wendy's

AServ E: **Truckland Truck Service**

W: Dick's Check 'n Lube, Seiner Chevrolet, Seiner Mitsubishi, Val's

TServ E: **Cummins Diesel**

W: **Flying J Travel Plaza, Great Dane Trailer Service, Semi Service, Utility Trailer Service, West Valley Tire**

TWash W: **Blue Beacon Truck Wash**

ATM E: Key Bank

Other E: **U-Haul Center**(LP)

EXIT UTAH

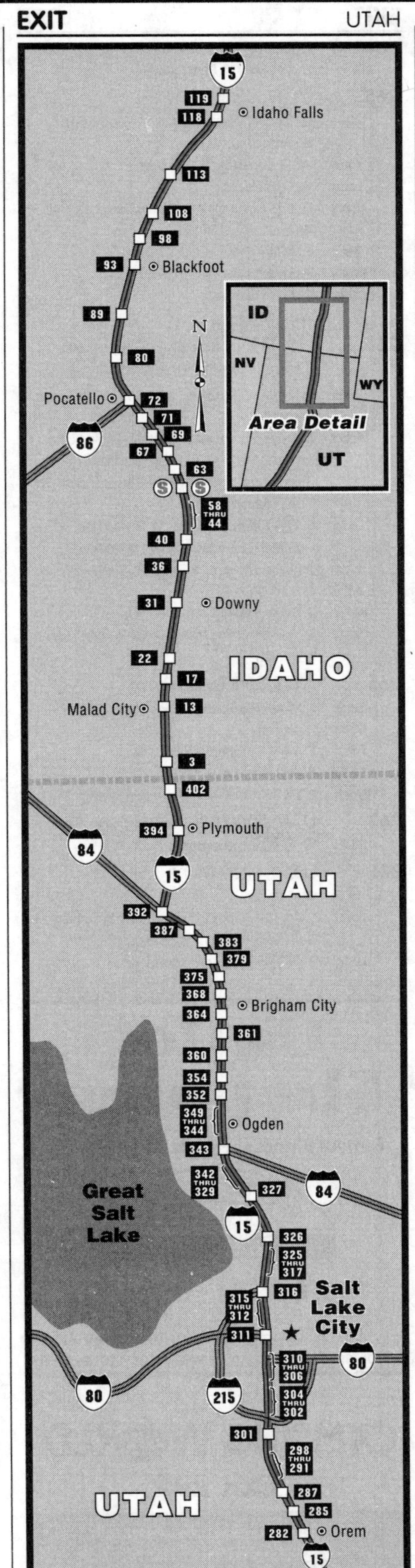

EXIT UTAH

308 Hwy 201

TServ E: **Rick Warner**

307 Junction I-80E, to U.S. 40, Denver, Cheyenne

306 UT171, 33rd S St, South Salt Lake

Gas E: AAMCO Transmission, Chevron*(CW), Citgo*, Sinclair

Food E: Central Park, McDonalds, Subway, Taco Bell, Wendy's

Lodg E: ◆ Days Inn

AServ E: Jiffy Lube, Keith's Automotive

ATM E: Brighton Bank, First Interstate Bank

304 UT 266, 45th S St, Murray, Kearns

Gas E: Chevron, Phillips 66*(CW)

W: Shell, Texaco*

Food E: Imperial Garden, McDonalds, Super Grinders Sandwiches

W: Denny's, Rocky's Deli

Lodg W: ◆ Hampton Inn, AAA Quality Inn

AServ E: Checker's Auto Parts, Midas Muffler

Other E: **Zim's Craft Supply**

W: **Eagle Hardware & Garden**

303 UT173, 53rd S St, Murray, Kearns

Gas W: Chevron(CW), Conoco*(CW)

Food W: Papa Murphy's, Reston Hotel, Schlotzkys Deli, Tio's Fine Mexican

Lodg W: AAA Reston Hotel

TServ E: **Sam's Truck Accessories**

Med E: **Hospital**

ATM W: Handy Bank, Smith's Grocery

Other W: **Smith's Grocery**

302 Junction I-215

301 UT48, Midvale, 72nd S St

Gas E: Hart's, Phillips 66*, Standard(CW), Texaco*(LP)

Food E: Cafe Silvestre Mexican, Denny's, KFC, McDonalds, Midvale Mining Company

Lodg E: AAA Best Western, AAA Days Inn, AAA Discovery Inn, AAA La Quinta Inn, Motel 6

298 UT209, 90th S St, Sandy, West Jordan

Gas E: Sinclair*

W: Maverick*, Texaco*(LP)

Food E: Arby's, Sandy's Station, The Scone Cutter

Lodg E: Comfort Inn

AServ E: David Early Tires, Ford Dealership

W: AAMCO Transmission, Advantage Auto Service, National Tire & Battery

Med E: **Hospital**

ATM E: Gardian State Bank

Other E: **Classic Skating & Water Slide, Eagle Hardware & Garden**

297 106th St, Sandy, South Jordan

Gas E: Conoco*, Phillips 66*(CW)

Food E: Carver's Prime Rib, Jade Garden Chinese, Shoney's, T.G.I. Friday's, **Village Inn**

W: Denny's

Lodg E: Best Western, Courtyard by Marriott, Hampton Inn

W: ◆ Sleep Inn, ◆ Super 8 Motel

AServ E: Conoco, Utah Auto Mall Chevrolet Olds

ATM E: Bank One, Zions First National Bank

Other E: **Locksmith, Shopping Mall**

295 UK.S. 89N, Sandy

Lodg E: Holiday Inn Express

Other E: **Factory Outlet Center**

294 UT 71, Draper, Riverton

Gas E: Holdiay Gas*, Texaco*(CW)

Bold red print shows RV & Bus parking available or nearby

EXIT UTAH

- Food E: Guadalahonky's Mexican, Neal's Broiler
- AServ E: Economy Auto Repair, Richin's Auto Service
- TWash E: **Draper Wash**
- RVCamp E: **Camping, Holiday Camping, Quality RV Service, Camping World (see our ad this page)**
- Other E: **Factory Outlet Center (see our ad this page)**

291 State Prison, Bluffdale, Draper
- AServ E: Beus Isuzu

287 UT 92, Highland, Alpine
- Gas W: Amoco(CW)
- Food W: **Burger King (Amoco)**
- Parks E: **Timpanogos Cave**

285 UT 73, 11th West
- AServ E: Gear's Tramission Repair
- TServ E: **Gear's**

282 UT73, Lehi
- Gas W: Chevron(CW), Walker's
- Food E: Carnivors Sandwiches, One Man Band Diner
  W: KFC Express (Chevron), Mayberries, McDonalds, Pap Murphy's Pizza, Subway, The Arctic Circle Burgers, Walker's
- Lodg W: ◆ Best Western, Super 8 Motel
- AServ E: Rex's Deisel
  W: Chevron
- ATM W: Zions Bank
- Other E: **Kurt's Propane**

281 Main St
- FStop E: **Hart***
- AServ E: Dodge, Plymoth Dealer

279 5th St, Main St., American Fork (Service More Then 1 Mile Away)
- Food E: Taco Bell
- Lodg E: Quality Inn
- AServ W: Ford Dealership
- RVCamp E: **Stuarts RV Parts & Service**
- Med E: **Hospital**
- Other E: **Wal-Mart**

(278) Rest Area - Parking P (Both Directions)

276 Lindon, Pleasant Grove Road
- TStop W: **Conoco(SCALES)**
- FStop W: **Texaco(LP)**
- Food W: **Frontier Cafe (24 Hrs)**
- TServ W: **Certified Transmission & Drivetrain, RJ's Truck Service Tire Towing, Auto, Utah Truck Systems(D)**

275 UT 52, U.S. 189, 8th St North
- Gas E: Phillips 66*(CW)
- Food E: Red Hot Chili Peppers Mexican
- AServ E: Advanced Automotive Specialist, Brimhall & Brimhall Auto Service
- Parks E: **Provo Canyon, Sundance Recreation Park**
- Other E: **Highway Patrol Post**

274 Center St
- Gas E: 7-11 Convenience Store*, Conoco(CW), Stesans's Fuel Stop*
- Food E: The Pizza Pipe Line
- AServ W: Alpine Auto
- Med E: **Hospital**
- Other E: **Trafalga Fun Center- Water Sports, Miniature Golf**
  W: Coin Car Wash

272 Orum, University Parkway
- Gas E: Amoco(CW), Shell* (24 Hrs), Sinclair*
- Food E: McDonalds, Subway

EXIT UTAH

- AServ E: Saturn Dealership
- Other E: **Wal-Mart (& Pharmacy)**

268 UT114, Center St
- Gas E: Conoco*, Sinclair*, Walker's Food & Fuel*(CW)
  W: Amoco, Conoco*
- Food W: The Great Steak Sandwich
- Lodg W: Econolodge
- AServ E: Jim's Automotive, SDS Auto Center, Sinclair
  W: Tyacke Auto Service
- Med E: **Hospital**
- Parks W: **Utah Lake**
- Other E: **Ream Grocery**
  W: **Pharmacy**

266 U.S. 189N, University Ave, Provo (Difficult Westbound Reaccess)
- Gas E: Chevron*(CW), Circle K*, Phillips 66*(CW), Texaco(CW)
- Food E: Arby's, Burger King, Chuck-a-Rama Buffet, Jovera's, McDonalds, Ruby River Steak House (Holiday Inn), Shoney's, Sizzler Steak House, Taco Time, Wendy's
- Lodg E: AAA Colony Inn Suites, ◆ Fairfield Inn, ◆ Hampton Inn, ◆ Holiday Inn, ◆ Howard Johnson, **Motel 6**, ◆ Sleep Inn, Super 8 Motel
- ATM E: Bank One
- Parks E: **Seven Peaks Recreation Area**
- Other E: **K-Mart, Office Max, Sam's Club**, Self Serve Car Wash

265 UT75, Springville, Provo
- TStop E: **Mountain Springs Truck Stop(LP)(SCALES)**
- Gas E: Amoco
- Food E: Mountain Springs Restaurant
- Lodg E: Best Western
- TWash E: **T & T Truck Wash (Mountain Springs TS)**

263 UT77, Springville, Mapleton
- Food W: Cracker Barrel

261 U.S. 6, Price, Manti
- Gas W: Chevron
- Food E: Arby's, Chappy's, Cinnamon Street Bakery, Hot Stuff Pizza, McDonalds, Smash Hit Subs
- Lodg W: AAA Holiday Inn Express

EXIT UTAH

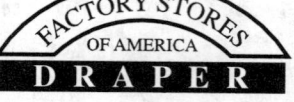

- ATM W: Chappy's
- Other W: **K-Mart**

260 U.S. 6E, Price, Spanish Fork
- Gas E: Amoco*(D)(CW), Chevron, Conoco, Phillips 66, Texaco*(D)(CW)
- Food E: Artic Circle, Bob's Express Pizza, Hickory Kist, JB's Restaurant, KFC, Little Caesars Pizza, Taco Bell
- AServ E: Checker Auto Works, GM Auto Dealership, Ken's Service, Texaco
- ATM E: Amoco, Bank One
- Other E: **Dr Jenson Veterinary Clinic, Macy's food, Bakery Drugs, Seagull Books & Tapes, Shopko Grocery**

256 UT164, Spanish Fork
- Other E: **County Fairgrounds**

254 UT115, Payson, Benjamin
- TStop E: **Phillips 66(CW)(LP)(SCALES) (RV Dump)**
- Gas E: Texaco*
- Food E: **McDonalds**, Subway, The Cobblestone Restaurant
- Lodg E: ◆ Comfort Inn
- TServ E: **Pason Diesel(CW)(SCALES)**
- Med E: **Hospital**
- Other E: **Pason Market**

252 Payson, Salem
- Gas E: **Chevron- RV Dump*(LP)**

248 U.S. 6W, Santaquin
- Gas W: Conoco*, Jr's Express Amoco, Texaco*(D)(CW)
- Food W: Hot Stuff Pizza (Amoco), Lenny's Dollar and More
- AServ W: Rocky Mountain Tire

245 Santaquin

236 UT 54, Mona

228 Nephi
- Med W: **Hospital**

225 UT132, Nephi, Manti, Moroni, Ephriam (Many Services On Into The Town)
- Gas E: AAMCO Transmission(CW)
  W: **Phillips 66***
- Food E: Taco Time
  W: **Wendy's (Phillips 66)**
- RVCamp E: **KOA Campground**
- Med W: **Hospital**
- ATM W: Phillips 66

222 I-15 Loop, UT 28, Salina, Nephi
- TStop E: **Circle C Truck Stop(SCALES) (RV Dump)**
  W: **Flying J Travel Plaza*(D)(LP)(SCALES) (RV Dump)**
- FStop E: **Texaco***
- Gas E: **Chevron***
- Food E: Burger King, Denny's, Hogi Yogi, J. C. Mickelson Restaurant, **Subway**
  W: Pepperoni's, **Thads Restaurant**
- Lodg E: Motel 6, **Roberto's Cove Motor Inn**, ◆ **Super 8 Motel**
  W: AAA Best Western
- TServ E: **Doyles Diesel**
- Parks E: **Mt Nebo Scenic Loop**

207 UT 78E, Mills, Lavan

202 Yuba Lake
- RVCamp E: **Camping**

188 U.S. 50E, Scipio, to I-70, Salina
- FStop E: **Chevron*(D)**
- Gas E: Amoco

Bold red print shows RV & Bus parking available or nearby

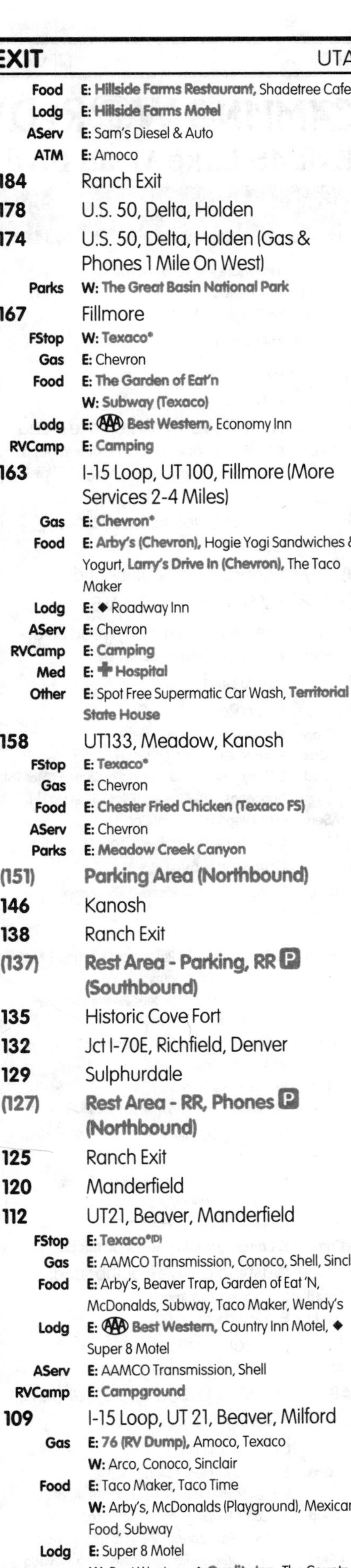

EXIT	UTAH
Food	E: **Hillside Farms Restaurant,** Shadetree Cafe
Lodg	E: **Hillside Farms Motel**
AServ	E: Sam's Diesel & Auto
ATM	E: Amoco
184	Ranch Exit
178	U.S. 50, Delta, Holden
174	U.S. 50, Delta, Holden (Gas & Phones 1 Mile On West)
Parks	W: **The Great Basin National Park**
167	Fillmore
FStop	W: **Texaco***
Gas	E: Chevron
Food	E: **The Garden of Eat'n**
	W: **Subway (Texaco)**
Lodg	E: (AAA) **Best Western,** Economy Inn
RVCamp	E: **Camping**
163	I-15 Loop, UT 100, Fillmore (More Services 2-4 Miles)
Gas	E: **Chevron***
Food	E: **Arby's (Chevron),** Hogie Yogi Sandwiches & Yogurt, **Larry's Drive In (Chevron),** The Taco Maker
Lodg	E: ◆ Roadway Inn
AServ	E: Chevron
RVCamp	E: **Camping**
Med	E: ✚ **Hospital**
Other	E: Spot Free Supermatic Car Wash, **Territorial State House**
158	UT133, Meadow, Kanosh
FStop	E: **Texaco***
Gas	E: Chevron
Food	E: **Chester Fried Chicken (Texaco FS)**
AServ	E: Chevron
Parks	E: **Meadow Creek Canyon**
(151)	**Parking Area (Northbound)**
146	Kanosh
138	Ranch Exit
(137)	**Rest Area - Parking, RR** P **(Southbound)**
135	Historic Cove Fort
132	Jct I-70E, Richfield, Denver
129	Sulphurdale
(127)	**Rest Area - RR, Phones** P **(Northbound)**
125	Ranch Exit
120	Manderfield
112	UT21, Beaver, Manderfield
FStop	E: **Texaco***(D)
Gas	E: AAMCO Transmission, Conoco, Shell, Sinclair
Food	E: Arby's, Beaver Trap, Garden of Eat 'N, McDonalds, Subway, Taco Maker, Wendy's
Lodg	E: (AAA) **Best Western,** Country Inn Motel, ◆ Super 8 Motel
AServ	E: AAMCO Transmission, Shell
RVCamp	E: **Campground**
109	I-15 Loop, UT 21, Beaver, Milford
Gas	E: **76 (RV Dump),** Amoco, Texaco
	W: Arco, Conoco, Sinclair
Food	E: Taco Maker, Taco Time
	W: Arby's, McDonalds (Playground), Mexican Food, Subway
Lodg	E: Super 8 Motel
	W: Best Western, ◆ **Quality Inn,** The Country Inn

EXIT UTAH

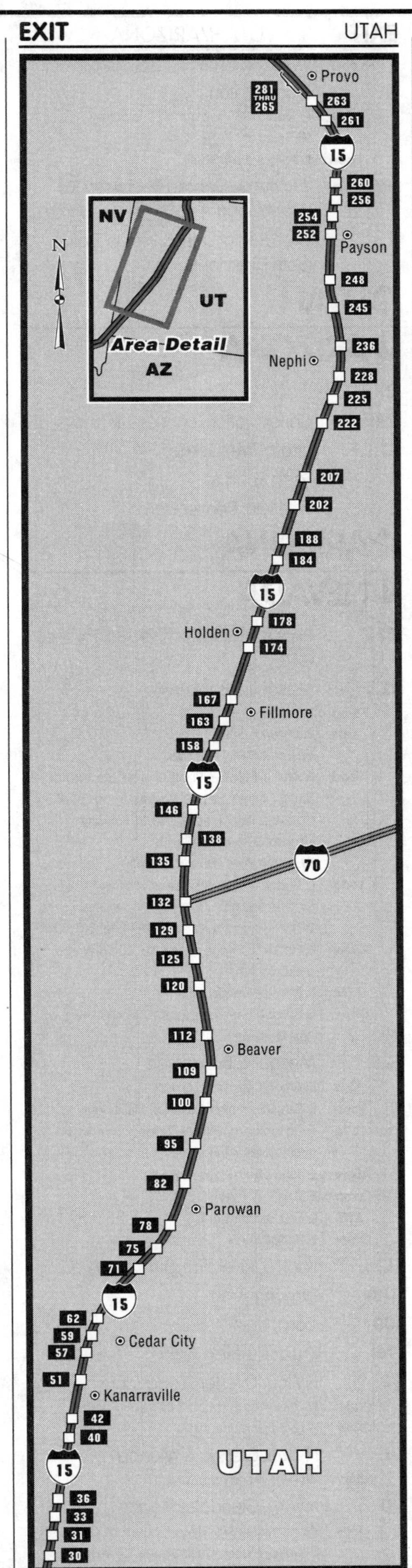

EXIT	UTAH
AServ	E: Goodyear Tire & Auto, NAPA Auto Parts, Texaco
	W: Arco, Chevron(D), Conoco
RVCamp	E: **76, Campground**
	W: **RV, and Auto and Diesel Repair & Towing (559-5376)**
Med	E: ✚ **Hospital**
Parks	E: **Great Basin National Park, Minersville State Park**
100	Ranch Exit
95	UT20, U.S. 89, Panguitch, Circleville
(88)	**Rest Area - RR, Picnic, Phones** P **(Both Directions)**
82	UT 271, Paragonah
78	Business 15, UT143, Parowan, Paragonah
75	I-15 Loop, UT 143, Parowan
Other	E: **Bryan Head Ski Resort, Cedar Brakes, Fair Grounds, Petroglyths**
71	Summit
TStop	W: **Sunshine**
Food	W: **Sunshine Restaurant**
TServ	W: **Sunshine**
62	UT 130, Minersville, Milford
Gas	E: Phillips 66, Trimart
	W: Texaco*
Food	E: Cinnamon Street Bakery (Phillips 66), Hot Stuff Pizza (Phillips 66)
	W: Steaks n Stuff
Lodg	W: (AAA) Travel Lodge
AServ	W: Texaco
Other	E: **Highway Patrol**
59	UT56, Cedar City
Gas	W: Texaco*(D)(LP)
Food	W: Bristlecone Restaurant, Subway
Lodg	W: (AAA) Holiday Inn, Motel 6, ◆ Super 8 Motel
AServ	W: Texaco
Parks	E: **Iron Mission State Park**
57	I-15 Loop, to UT 14, Cedar City
Gas	E: Chevron, Shell*, Texaco*
Food	E: Taco Time
Lodg	E: Village Inn Motel
AServ	E: Allied Discount Tires & Brakes, Chevron, Condie's Auto Service, Desert Pines, Superior Service Center, Tires & Tools
RVCamp	E: **KOA Campground**
Med	E: ✚ **Hospital**
Parks	E: **Bryce Canyon, Cedar Brakes , Navajo Lakes, Ducks Creek**
Other	E: **K-Mart, Albertson's Grocery, Wal-Mart**
51	Hamilton Fort
(45)	**Rest Area - Picnic, RR, HF, Phones** P **(Both Directions)**
42	New Harmony, Kanarraville
40	Kolob Canyons
Parks	E: **Zion National Park**
Other	E: **Visitor Information Center 5 Miles**
36	Ranch Exit
33	Ranch Exit
31	Pintura
30	Browse
27	UT17, Toquerville

EXIT UTAH

23 Leeds, Silver Reef (Difficult Reaccess)
- RVCamp E: **Camping**
- Other E: **Silver Reef Museum**

16 UT 9, Hurricane, Grand Canyon, Lake Powell
- RVCamp E: **Harrisburg RV Park**
- Parks E: **Grand Canyon, Lake Powell**
- Other E: **Quail Creek Amusement and Recreation**

10 Washington City, Middleton Dr
- Gas E: Phillips 66*
- Food E: Burger King, Juice Company, Pizza Pizza, Red Cliffs Inn
- Lodg E: ◆ Red Cliffs Inn
- RVCamp E: **Redlands RV Camp**
- ATM E: First Security Bank
- Other E: **Albertson's Grocery**

8 Business 15, UT18, St George Blvd, Santa Clara
- Gas E: Shell, Texaco*
  W: Amoco, Chevron*, Maverick Country Store, Phillips 66, Shell, Sinclair
- Food E: Carl's Jr Hamburgers, Chili's, Chuck-A-Rama Buffet, Red Lobster, Shoney's, Subway (Texaco), Village Inn
  W: Burger King, Denny's, Hogie Yogie, Hong Kong Dynasty, KFC, Larsen's Frost Stop, Luigi's Italian Restaurant, McDonalds (Playground), Taco Bell, Taco Time
- Lodg E: AAA Hampton Inn, AAA Ramada, ◆ Shoney's Inn
  W: Cinnamon Hills, Howard Johnson, Southside Inn, The Chalet Motel, The Sands Motel
- AServ E: Sunrise Tire
  W: ABC Auto Parts, Big O Tires, Checker Auto Parts, Shell, Sinclair
- RVCamp W: **St George RV Service**
- Other E: **Zion Factory Stores**

6 Jct I-15, Bus., UT18, Bluff St
- Gas W: Texaco*(D)
- Food W: Burger King, Dairy Queen, **Denny's**, Flame Jumper Steak House, Halley's, JB's Restaurant (Comfort Suites), Jimmy John's Gourmet Subs, Pizza Hut, Poncho & Lefty's, The Palms Restaurant
- Lodg W: ◆ Bluff Motel, **Budget 8 Motel**, ◆ **Budget Inn**, Claridge Inn, AAA Comfort Suites, AAA Holiday Inn
- AServ W: Big O Tires, Dixie Lube Auto Center
- RVCamp W: **Vacation World**
- ATM W: US Bank

EXIT UTAH/ARIZONA/NEVADA

- Other W: **Albertson's Grocery**, St George Car Wash

4 Bloomington
- Gas E: **Shell*(D)(CW) (see our ad this page)**
  W: Texaco*
- Food E: **Burger King (Shell)**

(2) **Welcome Center, Rest Area** P **(Northbound)**

1 Utah Port of Entry & Weigh Station (Both Directions)

↑UTAH

↓ARIZONA

27 Black Rock Road

(24) Check Station for Buses & Trucks

18 Virgin River Gorge

9 Farm Road

8 Littlefield, Beaver Dam

↑ARIZONA

↓NEVADA

(123) **Rest Area - RR, HF, Picnic, Phones** P

122 Mesquite, Bunkerville
- TStop W: **76*(D)**
- Gas E: Chevron, Shell, Texaco
  W: Virgin River Food Mart
- Food E: Arby's, Bulls Eye Pizza, Chalet Cafe, Cielito Lindo, Denny's, Jesse's Restaurant, KFC, Panda Gardens, Polar Freeze Drive In, Subway (Chevron), Taco Time
  W: **Virgin River Hotel & Casino**
- Lodg E: Budget Inn, Desert Palms, Hardys, State Line Motel & Casino
  W: Holiday Inn, ◆ **Virgin River Hotel & Casino**
- AServ E: Harley's Garage, Mesquite Lube & Wash, Mesquite Tire
- ATM E: Bank of America
- Other E: Bulldog Car Wash, **Smiths Food & Drug, US Post Office**

120 Mesquite, Bunkerville
- Gas E: Arco*, Chevron*, Texaco*
- Food E: Baskin Robbins, Carollo's, McDonalds
- Lodg E: Casablanca Hotel & Casino, **Oasis Resort & Casino**, Valley Inn Motel
- AServ E: John's Auto Repair Service
- RVCamp E: **Oasis RV Park**
- ATM E: First Security Bank
- Other E: **Peggy Sue's**

112 NV 170, Riverside, Bunkerville

(110) **Parking Area**

100 Carp, Elgin

(96) **Truck Parking Area**

93 NV169, Overton, Logandale
- Parks E: **Lake Mead National Recreation Area**
- Other E: **Lost City Museum**

91 NV168, Glendale, Moapa
- AServ W: Arrowhead Auto Service

90 NV 161, Glendale, Moapa
- Gas W: Arrowhead Food Mart, Chevron*, Gas & Goodies, Moapa Oasis General Store*

EXIT NEVADA

(Hardware, Grocerys, Gasoline)
- Food W: Chevron
- Lodg W: Glendale Motel

88 Hidden Valley

84 Byron

80 Ute

75 NV 169E, Valley of Fire, Lake Mead
- Other E: **Casino, Moapa Band of Paiutes, Moapa Tribal Enterprises (C-Store), Valley of Fire State Park**

64 U.S. 93N, Pioche, Ely, Alamo, Caliente

(61) Check Station (Southbound)

58 NV 604, Apex, Nellis AFB

54 Speedway Blvd, Hollywood Blvd
- Other E: **Las Vegas Motor Speedway**

50 Lamb Blvd

48 Craig Road
- TStop E: **Pilot Travel Center*(SCALES)**
- Gas E: Arco*, Chevron*, Citgo*
- Food E: **Burger King (Pilot TS), Dairy Queen (Pilot TS)**, Finnegans, **KFC Express (Pilot TS)**, Pizza Hut
- AServ E: Las Vegas Home for Car
- TServ E: **First Gear Transmission Service**

46 Cheyenne Ave
- TStop W: **Flying J Travel Plaza(LP) (RV Dump)**
- Gas E: Arco
  W: Citgo*(D)
- Food E: Cheyenne Hotel Restaurant & Lounge, Mario's Market, Sunrise Donuts
  W: **McDonalds**, Muleteer, **Restaurant (Flying J Travel Plaza)**
- Lodg E: Cheyenne Hotel Restaurant & Lounge
  W: **Comfort Inn**
- AServ W: Joe Rossi Tires
- TServ W: **Flying J Travel Plaza, Joe Rossi Tires**

45 Lake Mead Blvd
- Gas E: 76, Citgo*
- Food E: Arby's, Burger King, JSS Fish & Chips, McDonalds, Wendy's
- AServ E: Pep Boys Auto Center
- RVCamp E: **Camping World (see our ad this page)**

44 Washington Ave (Southbound)
- Food E: Daniel's Restaurant
- Lodg E: Best Western

43 D Street
- AServ E: Complete Automotive Center

42AB Jct I-515S, U.S. 93, U.S. 95, Phoenix, Mercury

41 Charleston Blvd
- Gas E: 7-11 Convenience Store, Arco
  W: Texaco*(D) (#1 of 2), Texaco*(D)(LP) (#2 of 2)
- Food W: Carl's Jr Hamburgers, Del Taco, McDonalds (Outdoor Playground), Real Donuts, Wendy's
- AServ E: Bob's Garage, Econo Lube & Tune, Nevada Transmission

Bold red print shows RV & Bus parking available or nearby

EXIT — NEVADA

W: Texaco (#2)
Med W: + **University Medical Center**
Other W: **Genesis Books**

40 Sahara Ave
Gas E: **76**, Texaco
W: Texaco*(D)(LP), Texaco*(D)(LP) (2 of 2)
Food E: El Azteca, Golden Steer Steakhouse
W: Boston Market, Carrows Restaurant, Landry's Seafood, Macaroni Grill, Palace Station Casino
Lodg E: Sahara Hotel & Casino, Travelodge (see our ad this page)
W: Palace Station Casino
AServ E: Nevada Tire City, Texaco
W: Texaco (#1 of 2)
TServ E: **Nevada Tire City**
ATM W: Bank West of Nevada
Other E: **Circus Circus Casino, General Store Mini Market, Grand Slam Canyon Adventure**

39 Spring Mtn Road

38AB Dunes Flamingo Road
Gas E: Arco
W: Arco, Texaco*
Food E: **Bally's Hotel & Casino, Barbary Coast Hotel & Casino,** Battista's Italian Restaurant, **Caesars Palace Hotel & Casino**
W: Del Taco, Outback Steakhouse
Lodg E: **Bally's Hotel & Casino, Barbary Coast Hotel & Casino, Caesars Palace Hotel & Casino,** Comfort Inn, **Flamingo Hilton Hotel & Casino,** Super 8 Motel (see our ad this page), Crowne Plaza (see our ad this page)
W: **Gold Coast Hotel & Casino, Rio Hotel & Casino**
AServ W: Terrible Herbst Lube & Car Wash
ATM W: Sun State Bank
Other E: **Bourban St. Casino, Ellis Island Casino**

37 Tropicana Ave
TStop W: **King 8 Truck Plaza**(LP)(SCALES)
Gas W: 76, Arco*, Shell, Standard*(D)
Food E: Carrows Restaurant, **Excalibur Hotel & Casino,** Jeremiah's Steakhouse, **MGM Grand Hotel & Casino, Tropicana Hotel & Casino**
W: Burger King, IHOP, In-N-Out Hamburgers, Jack-In-The-Box, KFC, McDonalds, Taco Bell, Taco Cabana, Wendy's
Lodg E: **Excalibur Hotel & Casino, MGM Grand Hotel & Casino,** Motel 6, New York New York, Roadway Inn, San Remo Hotel, **Tropicana Hotel & Casino**
W: Best Western, Budget Suites, Heritage Inn, **King 8 Hotel & Casino**
Other W: **King 8 Casino**

36 Russell Road
Gas E: Texaco*
W: Terrible's
Lodg E: Cardinal Inn, Casa Malaga, Diamond Inn, Fuzz Motel, Klondike Hotel, Pollyanna Motel, Super 8 Motel, Warren Motel
TServ W: **E & R Truck Service**

34 Jct I-215E, McCarran Airport

33 NV160, Blue Diamond, Pahrump
TStop W: **76 Auto/Truck Plaza**(LP)(SCALES) **(RV Dump)**
Food W: **Restaurant (76)**
Lodg W: **Firebird Motel (76 TS)**
AServ W: 76 Auto/Truck Plaza
RVCamp E: **Oasis RV Camp**
Other E: **Belz Factory Outlet, Las Vegas Factory Outlet**

EXIT — NEVADA

Stores (see our ad this page)

27 NV146, Henderson, Lake Mead

25 Sloan

(13) Truck Check Station (Northbound)

12 NV161, Jean, Goodsprings
FStop E: **Texaco***

EXIT — NEVADA/CALIFORNIA

Gas W: Shell*
Food E: **Burger King, Gold Strike Hotel & Casino**
W: **Nevada Landings Hotel & Casino**
Lodg E: **Gold Strike Hotel & Casino**
W: **Nevada Landings Hotel & Casino**
Other E: **Nevada Welcome Center; RR, Phones, Tourist Information, US Post Office**

(1) **Rest Area - RR, HF, Phones** (P)

1 State Line, Primm
TStop W: **Whiskey Pete's Truck Stop**(SCALES)
FStop W: **Texaco**(SCALES)
Gas E: 76
Food E: **McDonalds, Primadonna Resort & Casino**
W: **Whiskey Pete's Casino**
Lodg E: **Buffalo Bill's Resort & Casino, Primadonna Resort & Casino**
W: **Whiskey Pete's Casino**
RVCamp E: **Primadonna RV Village**

↑NEVADA

↓CALIFORNIA

289 Yates Well Road

284 Nipton Road

279 Bailey Road

(278) Truck Brake Inspection Area

270 Cima Road
Gas E: Mobil*

(269) **Rest Area (Both Directions)**

263 Halloran Summit Road
FStop E: **Paso Alto Fuel Stop**
AServ E: Mr. D's Tire Service
TServ E: **Paso Alto**

257 Halloran Springs Road
FStop E: **Lo-Gas**
TServ E: **Halloran Springs Towing**

246 Baker
Gas W: 76, Mobil, Texaco, XCel
Food W: Arby's, Bun Boy, Burger King
Lodg W: Wells Fargo Motel
TServ W: **The Baker Garage**

244 CA127, Death Valley, Kelbaker Road
Gas W: Arco*, Mobil*(D), Texaco, XCel
Food W: Arby's, **Bun Boy,** Burger King, Del Taco, Denny's, The Mad Greek
Lodg W: Royal Hawaiian Motel, Wells Fargo Motel
AServ W: Baker Garage, NAPA Auto Parts
TServ W: **A-1 Towing, The General Store**
Other W: **Bakers Hardware Complex**

238 Zzyx Road

232 Rasor Road
FStop E: **Gas Fuel Stop***
AServ E: Gas FS
TServ E: **Gas FS**

228 Basin Road

219 Afton Rd, Dunn

(218) **Rest Area (Both Directions)**

211 Field Road

204 Harvard Road
RVCamp E: **Twin Lakes RV Park (5 Miles)**
Other E: **Harvard Station Farmers Market**

196 Minneola Rd

EXIT CALIFORNIA

TStop **W:** Rustlers*
Gas **W:** Mobil
Food **W:** Minneola Mini Mart

(195) Agricultural Inspection Station (Southbound)

194 Yermo Road

192 Calico Road, Yermo Road
Food **E:** International Chinese & American Cafe

189 Ghost Town Road
TStop **E:** Vegas Truck Stop*[D]
FStop **W:** Shell*
Gas **E:** Circle C
W: Texaco
Food **E:** Peggy Sue's Diner
W: Jenny Rose Cafe
Lodg **E:** Calico Motel
RVCamp **W:** KOA Campground

187 Fort Irwin Road
FStop **W:** 76*
RVCamp **W:** Camping

184 CA 58W, Bakersfield

182 East Main St
Gas **E:** Mobil*, Shell, Terrible Herbst
W: 76*, Arco, Chevron, Shell*
Food **E:** Barstow Station, Big Nick's Restaurant, Donut Star Donuts & Chinese Food, McDonalds, Taco Bell, Tom's Burgers
W: Burger King (Indoor Playground), Cactus Club Grill, Carl's Jr Hamburgers, China Gormet Restaurant, IHOP, Jack-In-The-Box, Sizzler Steak House, Taco Bell, The Green Burrito
Lodg **E:** Gateway Motel
W: Astro Budget Motel, Best Motel, Comfort Inn, Econolodge, Executive Inn, Holiday Inn, Quality Inn
AServ **E:** Mobil, Shell
W: Shell
Other **W:** Vons Grocery

181B Jct I-40E, Needles

181 CA 247, Barstow Road, to Down-town Barstow
Gas **E:** Ultra Mart
Food **E:** Pizza Hut
W: Little Caesars Pizza
AServ **W:** Chief Auto Parts
Other **W:** Food 4 Less, K-Mart

179 W Main St
TStop **W:** Burns Bros Travel Stop*, Heartland Truck Stop[SCALES]
Gas **W:** Arco*, Thrifty*
Food **W:** Bun Boy Restaurant, Country Meats, Robertio's Mexican Food, Straw Hat Pizza
Lodg **W:** Desert Lodge Motel, Holiday Inn Express
AServ **W:** Barstow Automotive, Barstow Tire #2 (Towing)
TWash **W:** Burns Bros Travel Stop, Heartland Truck Stop, Vernon Truck Wash
RVCamp **W:** Desert Moon RV Camp
ATM **W:** Cash Machine (Burns Bros TS)

176 Lenwood Road
TStop **W:** Rip Griffin Travel Center
FStop **W:** Arco* (RV Dump), Pilot Travel Center[SCALES]
Gas **E:** 76*, Chevron*, Exxon, Shell*
Food **E:** Arby's, Baskin Robbins, Carl's Jr Hamburgers, Del Taco, Denny's, Hanna Grill, Harvey House, In-N-Out Hamburgers, KFC, Little Caesars Pizza,

EXIT CALIFORNIA

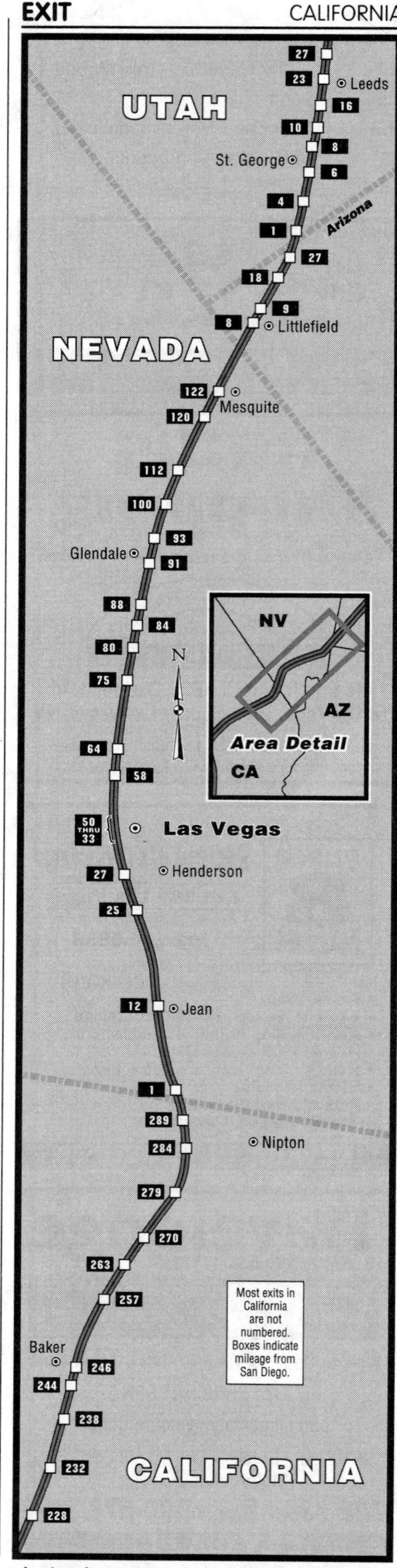

EXIT CALIFORNIA

Panda Express, Pollo Loco, Quigley's Restaurant, Subway, Wendy's (Exxon)
W: McDonalds
Lodg **W:** Good Nite Inn
TServ **W:** Rip Griffin Travel Center[SCALES]
TWash **W:** Little Sisters Truck Wash
Other **E:** Tanger Factory Outlet
W: Factory Outlet Center

173 Outlet Road

168 Hodge Rd

163 Wild Wash Road

159 Boulder Road

152 Bell Mountain, Stoddard Wells Road

151C Stoddard Wells Road
Gas **W:** 76[D], Mobil*[D], US Gas
Food **E:** Peggy Sue's Restaurant
W: Denny's
Lodg **W:** Motel 6, Queens Motel, Travelodge
AServ **E:** Complete Auto Repair
RVCamp **E:** Camping

151B E Street, Oroville

151A CA 18, D Street, Oro Grande

150 Mojave Drive
Gas **E:** 76*[D]
W: Chevron, Kwik Stop*
Food **W:** Mollies Country Kitchen, Sunset Inns
Lodg **E:** Budget Inn
W: Economy Inns of America, Sunset Inns
AServ **E:** 76
W: Central Transmission, Chevron, Quality Automotive Repair

149 Roy Rogers Drive
Gas **E:** Chevron
Food **E:** Dairy Queen, Hometown Buffet, Jack-In-The-Box
AServ **E:** Goodyear Tire & Auto
ATM **E:** Desert Community Bank
Other **E:** Cosco, Food 4 Less

148 Palmdale Rd, CA 18
Gas **E:** Texaco*[D]
W: Mobile[CW], Shell*[CW], Thrifty*
Food **E:** A&C Bar-B-Que, Anton's Cafe, Best Western, Billy T's, Burger King, Casa Delicias Mexican, Don's Coffee Shop, Los Roberto's, StarWok Express
W: Coco's, Del Taco, Donuts, Hole in One Donuts, Long John Silvers, Oriental Gardens Chinese, Taco Bell, The Grumpy Englishman, The Oak Room (Holiday Inn), Tom's #21 Burgers
Lodg **E:** Best Western
W: Budget Inn, EZ8 Motel, Holiday Inn
AServ **E:** Cadallic, Oldsmobile, GMC Dealer, Chief Auto Parts, Chrysler Auto Dealer, Pep Boys Auto Center
W: Ford Dealership, Lovelands Automotive Servuce
TServ **W:** Custom Truck Warehouse
Other **E:** Smile Dental Lab, Thrifty
W: Target

146 Lucerne Valley, Bear Valley Rd.
Gas **E:** Arco*, Citgo*, Mobil*[CW], Shell*[CW]
Food **E:** Bagel Bakery & Deli, Burger King, Carl's Jr Hamburgers, Del Taco, John's Pizza, KFC, Long John Silvers, McDonalds, Panda Express, The Baganos Italian, Wienerschnitzel

EXIT CALIFORNIA

W: El Tio Pepe Mexican, Olive Garden, Red Lobster, Tony Roma's Ribs
Lodg E: Days Inn, Econolodge, Super 8 Motel
AServ E: American TransWorld, Bear Valley Tire, Firestone Tire & Auto
RVCamp E: **Range RV Sales & Service**
Other W: **The Mall**

145 Hesperia, Main St
Gas E: Phillips 66, Texaco Star Mart*
W: Arco*
Food E: Burger King, Dairy Queen (Phillips 66), E-Z Burger (Phillips 66), In-N-Out Hamburgers
W: Bob's Big Boy
RVCamp W: **Camping***

139 Jct U.S. 395, Bishop, Adelanto
TStop W: **Outpost**(LP)(SCALES)
Food W: **Outpost Cafe**
TServ W: **Firestone Tire & Auto**
TWash W: **Outpost**

136 Oak Hill Road
Gas E: Arco*, Texaco
Food E: Sub Works (Arco), Summit Inn Cafe
AServ E: Texaco
RVCamp E: **Oak Hill RV Village**

131 CA 138, Palmdale, Silverwood Lake
Gas W: 76*(LP), Texaco*
Food W: Del Taco (76)
Lodg W: Econo-Inn
AServ W: 76, Texaco
Parks W: **Silverwood Lake (10 Miles)**

(129) **Weigh Station (Both Directions)**

127 Cleghorn Dr

123 Jct I-215S, San Bernardino

119 Sierra Ave
Gas W: **Shell***(LP)

115 CA 30, Highland Ave

113 Baseline

112 CA 66, Foothill Blvd
Food E: In-N-Out Hamburgers, Stuffed Bagel, Subway, The Cinnamon Cafe, Wienerschnitzel, Wok In
AServ E: Cosco, Wal-Mart
Other E: **Cosco, Food 4 Less, Mail Boxes Etc, Wal-Mart**

110 4th St
Gas W: Mobil Mart
Food W: Burger King
Other W: **Foozles Bookstore, Mall, Pick & Save**

109 Jct I-10, Los Angeles, San Bernardino

108 Jurupa Ave
Gas W: Arco* (ATM)
Food W: Carl Jr.'s Hamburger
AServ E: The Ontario Auto Center - All Makes
W: Citrus Ford

106 CA 60, Riverside, Los Angeles

103 Limonite Ave

100 6th St, Norco
Gas E: Chevron*(CW)
W: Arco*, Old Town Arco
Food E: 6th Street Deli, Country Kitchen, Jack-In-The-Box, McDonalds, Pats Kitchen
W: Baskin Robbins (Arco)
Other E: **Arco Animal Hospital**

98 Second St, Norco
Gas W: Shell(CW)
Food W: Burger King, Dominio's, Donuts, Norco Subs, Pizzeria, Sizzler Steakhouse, Weinerschnitzel
Lodg W: Howard Johnson
AServ W: Norco Auto Mall - All Makes
RVCamp W: **RV Parts & Supplies**
Med W: **+ Hospital**
Other W: Coin Car Wash

96 Jct CA 91, Riverside, Beach Cities
Lodg E: **Holiday Inn (north on Riverside Frwy to Market St Exit -- see our ad this page)**

95 Magnolia Ave
Gas W: Shell*(CW)
Food W: Balia's Pizza, Baskin Robbins, Burger King, Coffee Town, Donut Star, Little Caesars Pizza, Lotus Garden Chinese, McDonalds, Pizza Palace, Sizzler Steak House, Subway, Zendejas Mexican
AServ W: Kragen Auto Parts
TServ W: **American Diesel Performance**
ATM W: Redlands Federal Bank, **Union Bank of CA (Ralphs)**
Other W: **Payless Drugs, Post Office Plus, Ralph's Grocery (ATM), Save on Drugs, Stater Grocery Store**

93 Ontario Ave
Gas W: Arco*
AServ W: Pete's Road Service

92 El Cerrito Road
Gas E: 76, Circle K*
Food E: Don Pepe's, El Rodeo Cafe, Pheo's Pizza, **Pizza & Deli**
AServ E: Hubbstead Transmissions, Ontario's Auto Repair

91 Cajalco Road

90 Weirick Road

86 Temescal Canyon Road, Glen Ivy
Gas W: Arco*
Food W: Carl's Jr Hamburgers, **Tom's Farm Hamburgers**

85 Indian Truck Trail

81 Lake St, Alberhill

78 Nichols Road
Other W: **Outlet Stores**

77 CA 74, Central Ave
Gas E: Arco*, **Chevron***, Mobil(CW), Phillips 66
Food E: Douglas #23 Burgers, Off Ramp Cafe

75 Rt 15 Bus, Main St
Gas W: 76(LP), Circle K, Elsinore (ATM), Texaco(LP)
Food W: Family Basket Restaurant
Lodg W: Elsinore Lodge
AServ W: 76, Approved Mechanical Auto Repair, Elsinore Auto Service
Other W: **Food Smart Groceries, Van's Coin Laundry**

73 Railroad Canyon Road
Gas E: Union 76 (ATM)
W: Arco*, Chevron, Mobil*(D)

Food E: Denny's, Donut Depot (WalMart), In-N-Out Hamburgers, KFC (WalMart), **McDonalds (WalMart)**
W: Burger King, Carl's Jr Hamburgers, Del Taco, Gill's Chinese, Los Roberto's, Manny's Restaurant, McDonalds, My Buddies Pizza, Pizza Hut, Sizzler Steak House, Subway, Taco Bell, The Green Burrito, Vista Donuts
Lodg W: Lake View Inn
AServ E: Wal-Mart
W: A&M Auto Repair, Auto Aide Parts & Advice, Firestone Tire & Auto
ATM W: Bank of America
Other E: **Vons Grocery, Wal-Mart**
W: **Dental Clinic, Save on Drugs**

71 Bundy Canyon Road, Sedco Hills
Gas W: Arco*(CW) (ATM)

69 Baxter Road

68 Clinton Keith Road
Med E: **+ Inland Valley Hospital**

65 California Oaks Rd., Kalmia St
Gas E: Mobil(D), Shell(CW), Texaco
W: **Arco* (ATM)**
Food E: Burger King, Carl's Jr Hamburgers, Dairy Queen, KFC, Numero Pizzeria, Taco Bell (Texaco)
AServ E: Express Auto Service, Mobil
Other E: **Mall**

64 Murrieta Hot Springs Road
Food W: McDonalds
Med E: **+ Hospital**
Other W: **Shopping Mall**

62 Jct to I-215

61 CA 79, Winchester Rd, Hamet
Gas W: Arco* (ATM), Chevron, Mobil
Food E: Baskin Robbins, Carl's Jr Hamburgers, Godfather's Pizza, Taco Bell
W: Dairy Queen, Del Taco, El Pollo Loco, Filippi's Italian, Jack-In-The-Box, Little Chung King, Margarita Grill, Pasquale Italian, Richie's All American Diner, Stadium Pizza, Tony Roma's Ribs, Verf's
Lodg W: **Comfort Inn**
AServ E: Kragen Auto Works
Med E: **+ Walk in Medical Clinic**
ATM E: Home Savings of America, Union Bank of CA
W: Well Fargo Bank
Other E: **Food 4 Less**
W: **Lady of the Lake Books**, Rancho Car Wash

59 Rancho California Road, Temecula
Gas E: Mobil(CW)
W: 76, Chevron, Mobil*
Food E: Black Angus Steakhouse, Chesapeake Bagel, Claimjumper Restaurant, Dragon's, El Rancho Taco Shop, Great Grains Bread, Hunan Garden, Kelly's Coffee & Fudge, Little Caesars Pizza, Raised Donuts, Round Table Pizza, Rubo's Bahia Grill
W: Coco's, Denny's, Dominio's Pizza, KFC, McDonalds, Mexico Chiquito, Nicks Super Burgers, Olde Town Donuts & Pastry, Steak Ranch Restaurant, Taco Bell
Lodg E: **Embassy Suites**
W: Best Western, Motel 6
AServ W: 76, All Import Auto Parts, Chevron
ATM E: Great Western Bank
Other E: **Coin Laundry, Discovery Zone, Little Professor Bookstore, Sav-On Drugs, Target**

Bold red print shows RV & Bus parking available or nearby

Department Store
W: Tru Value Hardware(LP), US Post Office

58 CA 79S, Temecula, Indio, Warner Springs
- **Gas** E: Mobil*
  W: Chic-n Station* (Texaco)
- **Food** E: Carl's Jr Hamburgers, Donut Depot
  W: Alberto's Mexican, European Deli, Sunrise Market
- **Lodg** W: Ramada
- **AServ** W: Complete Auto Care, Express Tire

(53) Inspection & Weigh Station (Northbound)

52 Rainbow Valley Blvd

(52) Weigh Station (Southbound)

51 Mission Road, Fallbrook
- **Med** W: ✚ Hospital

46 CA 76, Pala, Oceanside
- **Gas** W: Mobil*
- **Food** W: Nessy Burgers, Southwest Cafe
- **Lodg** W: La Estancia Inn, Pala Mesa Resort
- **Other** W: Pala Mesa Market

43 Old Hwy 395

41 Gopher Canyon Road, Old Castle Road
- **RVCamp** E: Campground

37 Deer Springs Road, Mountain Meadow Road
- **Gas** W: Arco* (ATM)

34 Centre City Pkwy, Escondido

33 El Norte Pkwy, Escondido
- **Gas** E: Mobil(CW), Shell*(CW)
  W: Circle K*
- **Food** W: Big Apple Bagels, Donut Star, Graciano's Italian, Rosa #1 Pizza, Royal Dragon Chinese, Ruperto's Mexican, Safari Coffee, Wendy's
- **Lodg** E: Best Western
- **AServ** E: Express Tire Goodyear, Precision Tune & Lube
- **RVCamp** E: Camping
- **Other** E: Mr. Sudz Car Wash (Mobil)
  W: Mail Boxes Etc, Vons Grocery (& Pharmacy)

32 CA 78, Oceanside, Ramona
- **RVCamp** W: Camping World (see our ad this page)

31 Valley Pkwy, Downtown, Escondido
- **Food** E: Chili's, McDonalds, The Olive Garden, Yoshinoya Beef Bowl
  W: Applebees, Burger King, Coco's Bakery & Restaurant, Del Taco, PGI Steaks, Ribs, Seafood, Taco Bell
- **Lodg** W: Comfort Inn, Holiday Inn
- **AServ** E: Econo Lube & Tune
- **Med** E: ✚ Hospital
  W: ✚ Escondido Family Medical
- **Other** E: Barnes & Noble
  W: Shopping Center, Target Department Store (C-Store)

30 9th Ave, Auto Park Way, Escondido
- **Food** W: Applebee's, Peking Panda Chinese, Subway, Taco Bell
- **Other** W: Target Department Store

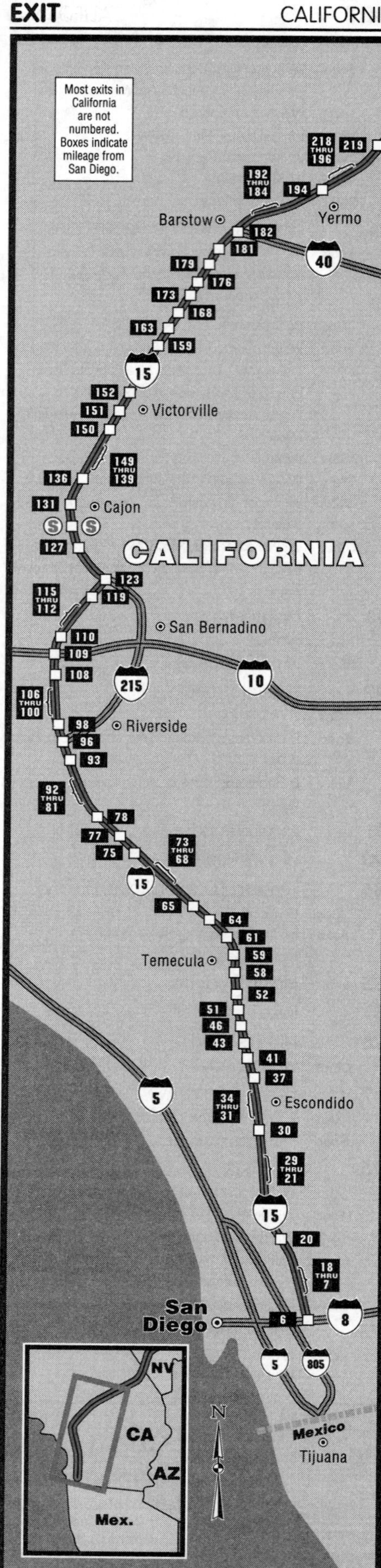

29 Bus Route 15, Center Center Pkwy

27 Via Rancho Pkwy
- **Gas** E: Chevron*, Shell*(CW)
  W: Texaco*
- **Food** W: McDonalds, Tony Spunky Steer
- **Other** E: North County Fair Shopping Mall

25 W Bernardo Drive, Pomerado Road

24 Rancho Bernardo Road, Lake Poway
- **Gas** E: Arco*, Mobil
  W: 76, Shell*(CW), Texaco*
- **Food** E: Submarines, Taco Shop, The Incredible Egg
  W: The Elephant Bar
- **Lodg** W: Econolodge, Holiday Inn, Radisson Suite Hotel
- **AServ** W: Shell
- **ATM** E: Wells Fargo Bank
- **Other** E: Mail Boxes Etc, Mercado Shopping Center, Thrifty Jr. Drugs

23 Bernardo Center Drive
- **Food** E: Burger King, Chinese Fortune Cookie, El Torito, Passage to India, Sesame Donuts, Spices Thia, Taco Bell
- **AServ** E: Firestone Tire & Auto, Goodyear Tire & Auto
- **ATM** E: American Savings Bank, Rancho Bernado Community Bank

22 Camino del Norte
- **Med** E: ✚ Hospital

21 Carmel Mountain Road
- **Gas** E: Chevron*(CW)(LP)
  W: Chevron*(CW)(LP)
- **Food** E: Athens Market Cafe, Baskin Robbins, Boston Market, Cafe Luna, California Pizza Kitchen, Carl's Jr Hamburgers, Chevy's Mexican, Heidi's Frozen Yogurt, Little Tokyo, Planet Wraps, Pogo's Eatery, Schlotzkys Deli, Sesame Donuts, Sombero Mexican, T.G.I. Friday's, Taco Bell, The Olive Garden, Wendy's
  W: Dragon Wok
- **Lodg** E: Residence Inn

EXIT CALIFORNIA

W: Doubletree Motel
ATM E: USA Federal Credit Union
Other E: **Borders Bookstore**

20 Jct CA 56W, Ted Williams Pkwy

18 Poway Rd, Rancho Penasquitos Blvd
Gas W: Mobil*(CW), Shell*(CW), Unocal 76
Food W: Burger King, IHOP, Little Caesars Pizza, McDonalds, Subway, Taco Bell
Lodg W: La Quinta Inn, Ramada Limited
AServ W: Unocal 76
Other W: **7-11 Convenience Store (C-Store)**

17 Mercy Road, Scripps Poway Pkwy

16 Mira Mesa Blvd
Gas W: Shell(CW)
Food E: C's Ice Cream, Chinese Restaurant, Chuck E. Cheese's Pizza, Denny's, Filippis Pizza Grotto, Sesame Donuts, Taco Bell
W: Applebees, Burger King, Dairy Queen, Golden Steak Seafood, Hungry Howie's Pizza, In-N-Out Hamburgers, Jack-In-The-Box, Starbuck's Coffee, Subway
Lodg E: Quality Suites

EXIT CALIFORNIA

ATM W: Bank of America
Other W: **Mail Boxes Etc, Pick & Save, Ralph's Grocery**

15 Carroll Canyon Road
Food E: Canyon Grill, Mieki's Sushi

14 Palmerado Road, Miramar Road
Gas W: Chevron, Mobil*(CW), Shell*(CW), Texaco*(D), Thrifty
Food W: Days Inn, Holiday Inn, Keith's Restaurant, Maxwell
Lodg W: Best Western, Days Inn, Holiday Inn
AServ W: NAPA Auto Parts
ATM W: Bank of America

13 Miramar Way, Miramar NAS

12 Jct CA 52

11 Clairmont Mesa Blvd

10 Clairmont Mesa Blvd
Food W: Boll Weevil Hamburgers, Carl's Jr Hamburgers, Chop Chop Pastaria, Hsu's Sezchwan Cuisine, Mexican Buffet, Subway, Sushi Bros., Taco Bell, Taco Shop, Wendy's

EXIT CALIFORNIA

ATM W: First Interstate Bank

9 Balboa Ave, CA 274, Tierrasanta Blvd
Gas E: Chevron
Food E: Ambio Italian, Boston Market, Coffee Bean, Round Table Pizza, Subway
ATM E: Home Savings of America

8 Aero Drive
Gas W: Arco*, Shell(CW)
Food E: Popeyes
W: Del Monto Italian, Del Taco, El Berto's, Jack-In-The-Box, McDonalds, Sizzler Steak House, Star Bucks Coffee, Submarina, TCBY Treats
Lodg W: Holiday Inn
Other W: **Wal-Mart**

7 Friars Rd, San Diego Stadium
Med E: **+ Hospital**

6 Junction I-8

↑CALIFORNIA

Begin I-15

I-16 E →

EXIT GEORGIA

Begin I-16

↓GEORGIA

1 **(0)** To I-75 S., Valdosta (Westbound)

2 **(1)** U.S. 23, U.S. 129, GA 49, Spring St
Gas N: Fina(D)
S: Amoco*, Conoco*(D)(LP), Exxon*
Food N: El Soubrero, Krispy Kreme Donuts, McDonalds, Nu-Way Hamburgers, Papa John's Pizza, Popeye's Chicken, Subway
S: Burger King, Checkers Burgers, Hardee's, KFC, Krystal, Pizza Hut, Satterfield's BBQ, Waffle House, Wendy's
Lodg S: Macon Inn
AServ N: Attaway Tire, Fina
S: Five Star Suzuki & Hyundai
Med S: **+ Georgia Medical Center**
ATM N: **NationsBank (Kroger)**
Other N: **Baconfield Pharmacy, Kroger Supermarket**
S: **Greyhound Bus Station**

3 **(1)** GA 22, to U.S. 129, to GA 49, 2nd Street, Macon (Westbound)
Lodg S: Crowne Plaza Hotel
Med N: **+ Coliseum Hospital**
S: **+ Georgia Medical Center, + Middle Georgia Hospital**

EXIT GEORGIA

4 **(2)** U.S. 80, GA 87, Martin Luther King Blvd., Coliseum Drive
Gas S: Fina*(D)(LP)
Food S: Green Jacket Steak & Seafood
Med N: **+ Coliseum Hospital**
Other N: **Ocmulgee National Monument**
S: **Georgia Music Hall of Fame, Tourist Information**

5 U.S. 23, U.S. 129, Golden Isles Highway, Ocmulgee East Blvd
FStop S: **Exxon*(LP)**
Gas S: Texaco*(D)
Food S: **Sam's Country Kitchen (Exxon)**, Subway (Texaco)
Other N: **Ocmulgee National Monument**

6 **(11)** Sgoda Road, Huber
FStop N: **Fina***

7 **(18)** Bullard Road, Jeffersonville

8 **(24)** GA 96, Jeffersonville, Tarversville
FStop N: **Citgo*(LP)**
S: **BP*(D)(LP)**
Food N: **Chester Fried Chicken (Citgo)**
S: Huddle House
Lodg S: **Days Inn**
AServ N: Crossroads Auto Service
ATM N: **Citgo**

EXIT GEORGIA

9 **(27)** GA 358, Danville

10 **(32)** GA 112, Montrose, Allentown
TStop S: **Amoco* (24 Hrs)**
FStop S: **Chevron***
Food S: Me-Ma's
TServ S: **Amoco**

11 **(39)** GA 26, Cockran, Montrose

12 **(42)** GA 338, Dudley, Dexter

(44) **Rest Area - RR, Phones, Picnic, Vending, RV Dump (Eastbound)**

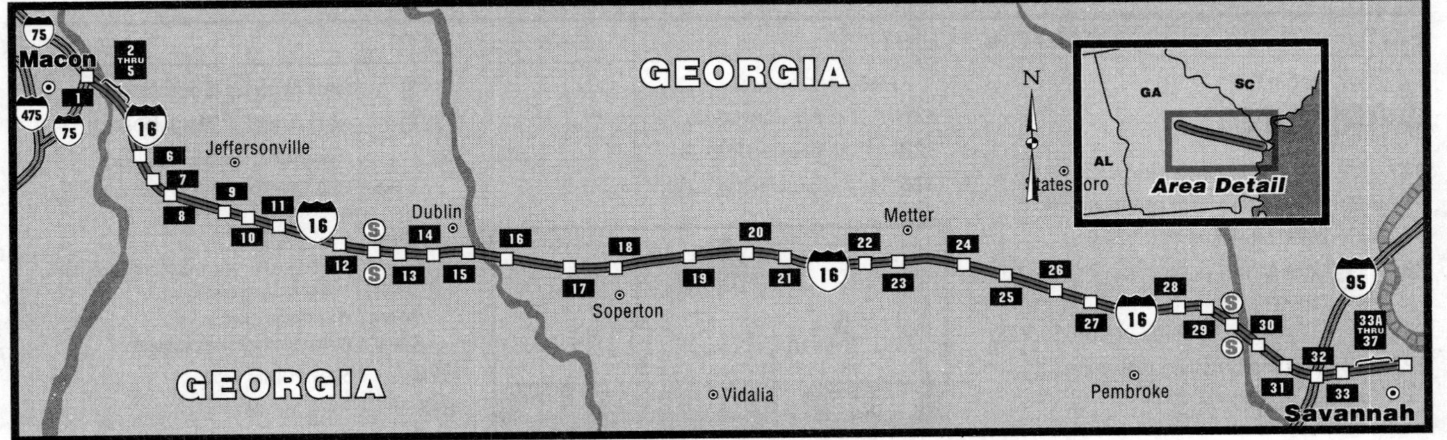

Bold red print shows RV & Bus parking available or nearby

EXIT — GEORGIA

(46) **Rest Area - RR, Phones, Picnic, Vending, RV Dump (Westbound)**

13 **(49)** GA 257, Dexter
- TStop — S: **Barney's Truck Stop* (Fina Gas)**
- Gas — N: Chevron*
- Food — S: **Barney's (Barney's Tstop)**
- TServ — S: **Barney's Truck Service, LJL Truck Center**
- TWash — S: **Barney's**

14 **(51)** U.S. 319, U.S. 441, McRae, Dublin
- FStop — N: **Amoco***, **Exxon*(D)(SCALES)**, **Speedway*(LP)**
- Gas — N: BP*, Texaco*(CW)
- Food — N: Arby's (Texaco), Burger King, KFC (24 Hrs Amoco), McDonalds (Playground), **Pizza Hut Express (Holiday Inn)**, Shoney's, Starvin Marvin, Subway (BP), TCBY Yogurt, Taco Bell, Waffle House, Wendy's (Texaco)
- Lodg — N: AAA **Comfort Inn (see our ad this page)**, DAYS INN Days Inn, Hampton Inn, AAA **Holiday Inn**, Jameson Inn, Shoney's
- Med — N: + **Hospital**
- ATM — N: Exxon, Texaco
- Parks — S: **Little Ocmulgee State Park**
- Other — N: **Georgia State Patrol Post**

15 **(54)** GA 19, Dublin
- Med — N: + **Hospital**

16 **(59)** GA 199, East Dublin

17 **(67)** GA 29
- FStop — S: **Texaco*(D)**
- Gas — S: Chevron*
- Food — S: **Huddle House**, Icecream Churn
- Other — S: **Tourist Information**

18 **(71)** GA 15, GA 78, Soperton, Adrian
- Gas — N: Chevron*
- Food — N: Front Porch BBQ

19 **(78)** U.S. 221, GA 56, Soperton

20 **(84)** GA 297, Vidalia
- TServ — N: **I-16 Truck Sales & Equipment**

21 **(89)** U.S. 1, Swainsboro, Lyons
- TStop — N: **Ed's Truckstop-Phillips 66***
- Gas — N: BP*
- Food — N: Barbecue Bill's, Ed's Truckstop Restaurant, Seafood Sally's
- TServ — N: **I-16 Service, Truck Tire Service (Ed's Tstop)**

22 **(98)** GA 57, Swainsboro, Stillmore
- FStop — S: **Amoco*(D)**
- Gas — S: Chevron*(D)
- ATM — S: Chevron

23 **(104)** GA 23, GA 121, Metter, Reidsville
- FStop — N: **BP*(D)(LP) (24 Hrs), Phillips 66**
- S: **Texaco***
- Gas — N: Exxon*, Shell*(D)(LP)
- Food — N: Dairy Queen, Hardee's (Playground), Huddle House, Market Place BBQ, **McDonalds (Playground)**, Subway, Taco Bell, Village Pizza, Waffle House, **Western Steer Family Steakhouse**
- Lodg — N: AAA **Comfort Inn**, DAYS INN Days Inn, **Holiday Inn**, Metter Inn
- AServ — N: Jimmy Franklin Chevrolet, Oldsmobile, Phillips 66, Williams Tire Center
- Med — N: + **Hospital**
- ATM — N: BP, Metter Banking Co.
- Parks — N: **George L. Smith State Park**
- Other — N: **Rite Aide Pharmacy**

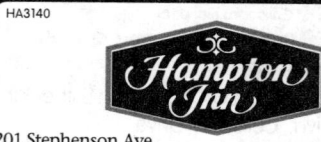

24 **(111)** Pulaski - Excelsior Road
- TStop — S: **Citgo***
- Food — S: **Grady's Grill (Citgo)**
- TServ — S: **Citgo**
- RVCamp — S: **Beaver Run**

25 **(117)** U.S. 25, U.S. 301, Statesboro, Claxton
- FStop — N: **Chevron*(SCALES) (24 Hrs)**; S: **El Cheapo*(LP) (Shell)**
- Food — S: **Huddle House (El Cheapo)**
- Lodg — S: Red Carpet Inn

26 **(126)** GA 67, Statesboro
- FStop — S: **Chevron***
- Food — S: Huddle House
- RVCamp — S: **Oasis Campground**

27 **(132)** Ash Branch Church Road

28 **(137)** GA 119, Springfield, Pembrooke, Ft. Stewart
- Other — N: **Historic Guyton**

29 **(143)** U.S. 280, to U.S. 80
- FStop — S: **BP***
- Gas — S: Chevron*(D)
- Food — S: Country Cafe (Chevron)
- ATM — S: **BP**

(144) **Weigh Station (Both Directions)**

30 **(148)** Old River Road, to U.S. 80

31 **(152)** GA 17, Bloomingdale Rd.

32A **(157)** I-95 (S To Jacksonville, Brunswick)

32B **(157)** North, Florence

33 **(160)** GA 307, Dean Forest Rd
- FStop — N: **Speedway***
- Gas — N: Chevron*(D)
- Food — N: Ronnie's, Waffle House
- Other — S: **Georgia State Patrol**

33A **(162)** Chatham Parkway

34AB **(164)** Jct.I-516, Lynes Pkwy, U.S. 80, U.S. 17, GA 21 (To Limited Access)
- Lodg — S: Hampton Inn (see our ad this page)

35 **(165)** GA 204, 37th Street (Eastbound)

36 **(166)** U.S. 17, GA 25, Gwinnett Street, Louisville Rd (Eastbound)

37AB **(166)** Martin Luther King Blvd, Montgomery St (Eastbound)
- Gas — N: Enmark, Speedway*
- Food — S: Burger King, Popeye's Chicken, Wendy's

↑GEORGIA

Begin I-16

I-17 S →

EXIT — ARIZONA

Begin I-17

↓ARIZONA

341 McConnel Drive, Flagstaff
- Lodg — W: Amerisuites (see our ad this page)

340AB Jct I-17, I-40

339 Lake Mary Rd, Mormon Rd
- Gas — E: Pick Quic*

337A U.S. 89A S, Flagstaff Airport, Sedona

333 Kachina Blvd, Mountainaire Rd
- Gas — W: 76*(LP)
- Food — W: Subway (76)

331 Kelly Canyon Rd

328 Newman Park Rd

326 Willard Springs Rd

(325) **Rest Area P (Southbound)**

(324) **Rest Area P (Northbound)**

322 Pinewood Rd, Munds Park
- Gas — E: 76, Chevron(LP) (ATM); W: Exxon*(LP)
- Food — E: Pines Rest; W: Pinewood Pizza, Woody's Pizza & Subs
- Lodg — E: AAA Motel In The Pines
- AServ — W: Old Town Garage
- RVCamp — W: **Munds Park Campground**

320 Schnebly Hill Rd

317 Fox Ranch Rd

EXIT ARIZONA

315 Rocky Park Rd

313 Scenic Views (Available To Trucks)

306 Stoneman Lake Rd

298 AZ179, Sedona, Oak Creek Canyon

(297) **Rest Area P (Southbound)**

(296) **Rest Area - RR, Picnic, HF P (Northbound)**

293 Cornville, McGuireville, Lake Montezuma
- **Gas** **E:** Bever Creek Auto Service
  W: 76
- **Food** **W:** Beaver Hollow Cafe
- **AServ** **E:** American Automotive, Beaver Creek Auto Service

289 Camp Verde, Middle Verde Rd, Camp Rd

287 AZ260, to U.S. 89A, Clarkedale, Jerome
- **Gas** **E:** Mobil, Texaco[D] (ATM)
  W: Chevron*
- **Food** **E:** Burger King, Dairy Queen, Denny's, McDonalds, Subway, Taco Bell
- **Lodg** **E:** Microtel Inn, AAA **Super 8 Motel (see our ad this page)**
- **TServ** **E: Valley Pro**
- **RVCamp** **E: Crazy K RV Park**

285 Camp Verde

278 AZ169, Cherry Rd, Prescott

EXIT ARIZONA

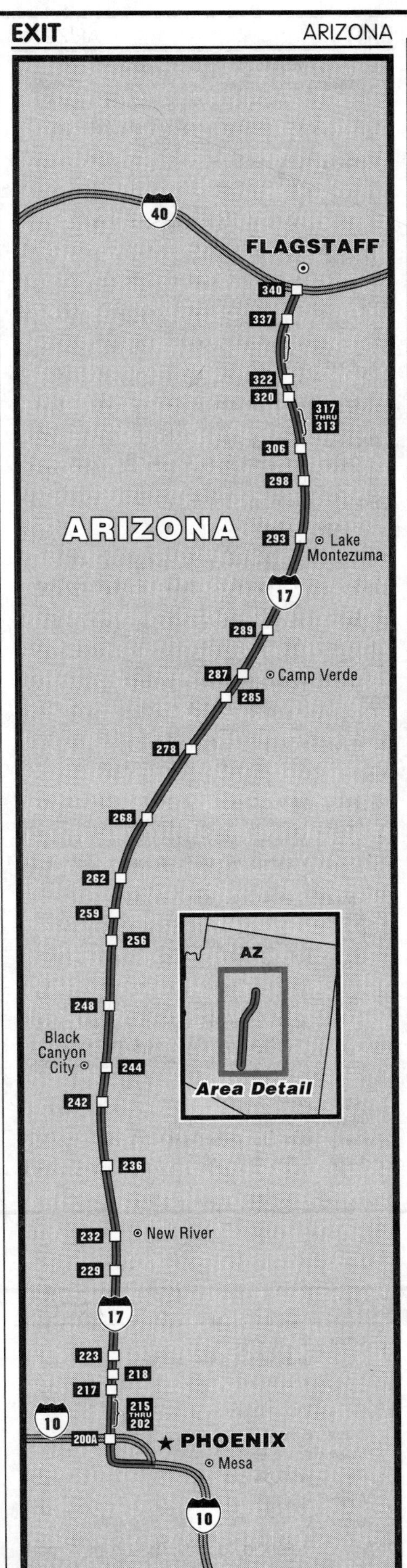

EXIT ARIZONA

268 Dugas Rd, Orme Rd

262A AZ69N, Cortes Junction Rd, Prescott
- **Gas** **E:** Exxon, **Texaco*[LP] (ATM, Subway)**
- **Food** **E:** McDonalds
- **Lodg** **E:** Lights On Motel

262B AZ69N, Prescott

259 Bloody Basin Rd, Crown King

256 Badger Springs Rd

(253) **Rest Area - Picnic, Phones, RR, HF P (Southbound)**

(252) **Rest Area - RV Dump, RR, Phones, HF, Vending, Picnic P (Northbound)**

248 Bumble Bee

244 Bus17, Black Canyon City, School Valley Rd
- **Food** **E:** Squaw Peak Steak House

242 Bus17, Rock Springs, BlackCanyon City
- **Gas** **E:** Texaco (1 Mile)
  W: Chevron (Towing)
- **Food** **W:** Rock Springs Cafe
- **RVCamp** **E: KOA Campground**

236 Table Mesa Rd

232 New River
- **Food** **E:** Road Runner Cafe

229 Desert Hills Rd
- **Other** **W: Factory Outlet Center (Food Court)**

225 Pioneer Rd
- **RVCamp** **E: Pioneer RV Park**

223 AZ78, Carefree Hwy, Wickenburg

218 Happy Valley Rd

217 Pinnacle Peak Rd

215A Deer Valley Rd E, Airport
- **Gas** **E:** Circle K 76, Exxon, Texaco Star Mart*[D]
  W: Circle K 76[LP], Texaco[D]
- **Food** **E:** Burger King, Kasha Chicken, McDonalds, Sonic, Taco Bell, Wendy's
  W: Denny's, Mia Homestyle Pizza
- **AServ** **E:** Car Go, Lyon's Auto Service, RS Auto Parts
  W: NAPA Auto Parts, Texaco
- **Other** **W: Grimes Coin Car Wash (RV Accessible)**

215B Deer Valley Rd W, Beardsley

214 Yorkshire Dr, Union Hills Dr
- **Gas** **E:** Circle K, **Diamond Shamrock**
- **Food** **E:** Andy's Family Restaurant, Burrito Bandito, KFC, Little Caesars Pizza, Orient Express, Sardello's Pizza & Wings, What A Burger
- **Lodg** **W:** Homestead Village, Wyndham Garden Hotel
- **ATM** **E:** Bank One
- **Other** **E: Fry's Food & Drug, Mail Boxes Etc, Safe Way Grocery, WalGreens**

212AB Bell Rd W, Sun City
- **Gas** **E:** Chevron, Exxon, Mobil, Texaco
  W: Mobil[CW]
- **Food** **E:** Blimpie's Subs, Coco's, IHOP, Long John Silvers, McDonalds, Waffle House, Wendy's
  W: Chill Out Yogurt, Denny's, Hometown Buffet, Kyoto Bowl, Santisa Brothers Pizza, Sizzler Steak House, The Good Egg
- **Lodg** **E:** Best Western, Fairfield Inn, Motel 6

EXIT	ARIZONA
	W: Red Roof Inn
ASErv	**E:** Big O Tires, Exxon, Ford Dealership, GM Auto Dealership, Meineke Discount Mufflers
ATM	**E:** Wells Fargo Bank
	W: Thunderbird Bank
Other	**E:** **Smith's Food & Drugs**
	W: **Albertson's Grocery,** Danny's Family Car Wash, **Mr. Books**
211	Greenway Rd
Gas	**W:** Texaco Star Mart
Food	**W:** Cousins Subs, R.K. Cafe, Rancho & Pinto, Schlotzkys Deli, Top Shelf Mexican, Wendy's
Lodg	**E:** Embassey Suites, La Quinta Inn
Med	**E:** **+ Deer Valley Medical Center**
Other	**W:** **Abco Grocery Store, K-Mart**
210	Thunderbird Rd
Gas	**E:** Arco, Circle K*, Exxon, Texaco
	W: Exxon
Food	**E:** Burger King, Dairy Queen, Jack-In-The-Box, Kenny Rogers Roasters, Magic Bowl Chinese, McDonalds, Streets of NY Pizza & Subs, Subway, Taco Bell, Wendy's
ASErv	**E:** Penzoil Quick Lube
	W: Exxon[LP]
ATM	**E:** Bank Of America
Other	**E:** **Home Depot[LP], Osco Drugs, Safe Way Grocery, WalGreens**
209	Cactus Rd
Gas	**W:** Chevron
Food	**W:** Cousins Subs, Denny's, Olivia's Mexican, Shangra La Chinese
Lodg	**W:** Ramada Plaza
ASErv	**W:** Chevron, Tire Outlet
208	Peoria Ave
Food	**E:** Fajitas, Lone Star Steakhouse, The Rusty Pelican
	W: Burger King, Coco's, Island Hamburgers, Poncho's Mexican Buffet, Sizzler Steak House, Wendy's
Lodg	**E:** Comfort Suites, Holiday Inn
ASErv	**W:** Master Care Auto Service, Q-Lube
ATM	**W:** Bank One
Other	**W:** **Barnes & Noble**
207	Dunlap Ave
Gas	**E:** 76
	W: Exxon
Food	**E:** Blimpie's Subs, Fuddrucker's
	W: Denny's, Schlotzkys Deli
Lodg	**E:** Sheraton, Travelodge
ASErv	**E:** AAMCO Transmission
	W: Midas Muffler
Med	**E:** **+ Dunlap Medical**
Other	**W:** **Toys "R" Us**
206	Northern Ave
Gas	**E:** Exxon

EXIT	ARIZONA
	W: Arco
Food	**E:** Burger King, Carl's Jr Hamburgers, Denny's, Los Compadres, Maria Calenders, Prickly Pair
	W: Dairy Queen, First Cafeteria, Village Inn Restaurant, Winchell's Donuts
Lodg	**E:** Hampton Inn
	W: Travelers Inn
ASErv	**E:** Exxon
	W: Hamilton Auto Parts, Smart Mart
ATM	**E:** Wells Fargo Bank
Other	**E:** **Desert Dove Books**
	W: Car Wash, **K-Mart**
205	Glendale Ave
Gas	**E:** 7-11 Convenience Store* (Citgo), Mobil
	W: Circle K*, **Exxon[LP]**
Food	**E:** Circle K
	W: Alberto's, Jack-In-The-Box
ASErv	**E:** Sun Transmission
	W: Checker's Auto Parts, Herb's
RVCamp	**W:** **RV Camping**
Other	**E:** **Ace Hardware, Animal Clinic,** Coin Car Wash, **Community Pharmacy**
204	Bethany Home Rd
Gas	**E:** Arco*
	W: Chevron, Exxon
Food	**E:** Brad's Fish & Chips, McDonalds
	W: Burger King, Lazy Lou's Fish & Shrimp, Pier De'Orleans Steaks, The Bagel Hut
ASErv	**W:** Chevron, Jake's Auto Service, Q-Lube, Tune Up While You Wait
Med	**E:** **+ Pheonix Baptist Hospital**
Other	**W:** **Southwest Grocery Store**
203	Camelback Rd
Gas	**W:** Mobil, Texaco*[D]
Food	**E:** Burger King, Pizza Hut
	W: Dairy Queen, David Kwan's Chinese, McDonalds, Taco Bell
Lodg	**W:** Comfort Inn
ASErv	**E:** Discount Tire Company, NAPA Auto Parts, Ric's Auto Care, Volvo Dealer, Wing's Auto Service
	W: Auto Zone Auto Parts, Danny's Car Wash, GM Auto Dealership
Med	**E:** **+ Phoenix Baptist Hospital**
Other	**E:** **WalGreens**
202	Indian School Rd
Gas	**E:** Arco*
	W: Exxon, Texaco
Food	**E:** Alberto's Mexican, Albertson's Grocery, Filberto's Mexican, Jimmy's Family Restaurant, Pizza Hut, Pizza Mia, Subway, Yin Chinese
	W: Hunter Phoenix Steak Out, J.B.'s Restaurant, Wendy's
Lodg	**W:** Motel 6, Super 8 Motel
ASErv	**E:** Sage Auto Service
RVCamp	**E:** **Arizona Trailer Sales**
Other	**E:** **Ace Hardware**

EXIT	ARIZONA
	W: **Wide World of Maps**
201	U.S. 60, Thomas Rd
Food	**E:** Arby's, Circle K, Denny's, Jack-In-The-Box, Taco Bell
	W: Burger King, Hilberto's Mexican
Lodg	**E:** Days Inn, La Quinta Inn
ASErv	**W:** NAPA Auto Parts, Pioneer Ford
200A	Jct. I-10, Tucson, Los Angeles, Central Phoenix
199B	Adams St, Van Buren St
Gas	**E:** Circle K*
	W: Circle K*
Food	**E:** Jack-In-The-Box
	W: Asia Express, Burger Shop, Pete's Fish & Chips
Lodg	**E:** Coconut Grove Motel, K Motel
ASErv	**W:** Bill's Radiator, Penny Pincher Auto Parts, Pep Boys Auto Center
TServ	**W:** **Truck Stuff**
ATM	**E:** Bank One
Other	**W:** **Southwest Grocery Store**
199A	Grant
Gas	**E:** Circle K*
ASErv	**E:** Joe's Auto Service
TServ	**W:** **Ron's Used Pick Up Parts**
198	Buckeye Rd
Gas	**W:** Circle K*
Food	**E:** Horseshoe Restaurant, Raliberto's Mexican
ASErv	**E:** Arizona Auto Parts
TServ	**E:** **American Truck Sales & Salvage, Carrillo's Truck Parts, Ray & Bob's Truck Sales**
	W: **Joplin Trailer Sales & Service, Lugo's Tire Service**
Other	**E:** **Southwest Supermarket**
197	19th Ave, State Capital
TStop	**E:** **AFCO***
Food	**E:** Jack-In-The-Box, What A Burger
TServ	**E:** **Mac's Trucks**
	W: **Arizona Truck & Trailer, Williams Detroit Diesel**
196	Central Ave
Food	**E:** Burger King
Other	**E:** Car Wash
195B	7th Ave, Central Ave
Gas	**E:** Circle K, Exxon[LP], Trailside General Store*
Food	**E:** McDonalds, **Taco Bell**
Lodg	**E:** **EZ 8 Motel**
ASErv	**W:** Big A Auto Parts
TServ	**E:** **Arizona Truck & Equipment Repair (Frontage road)**
RVCamp	**E:** **Desert West Coach**

↑ARIZONA

Begin I-17

I-19 S →

EXIT	ARIZONA
	Begin I-19
	↓ARIZONA
99	AZ 86, Ajo Way
Gas	**E:** Circle K*, Union 76*
	W: Chevron
Food	**E:** Eegee's, Original Hamburger Stand, Pied Piper Pizza, Pizza Hut, Taco Bell
ASErv	**E:** El Campo Tires
ATM	**E:** Norwest Bank, Wells Fargo

EXIT	ARIZONA
Other	**E:** **Coin Laundry**
	W: **Desert Museum Old Tuson, Santa Cruse River Park**
98	Irvington Rd.
Gas	**E:** Arco*
Food	**E:** Little Mexico (Mexican)
	W: McDonalds
ASErv	**E:** Super Wash
Other	**E:** **Smith's (Grocery With Pharmacy)**
95A	Valencia Rd. East, Tucson Int. Airport

EXIT	ARIZONA
95B	Valencia Rd. West
92	San Xavier Rd.
Other	**W:** **San Xavier Mission**
87	Papago
80	Pima Mine Rd.
75	Helmet Peak Rd. Sahuarita
69	Bus 19 N. Duval Mine Rd. Sahuarita
Gas	**W:** 76[CW], Citgo*
Food	**E:** Bashas', Bumpers, Denny's, Pizza Hut, Subway
	W: Arby's, Coach's All American Bar and Grill,

Bold red print shows RV & Bus parking available or nearby

EXIT ARIZONA

Dairy Queen
Lodg W: ◆ Holiday Inn Express
AServ W: Jim Click Ford Mercury Lincoln
ATM E: World Savings
Other E: **Wal-Mart**
W: **Greenvalley RV Resort, Titan Missile Landmark**

65 Esperanza Blvd. Green Valley
Gas E: Texaco (AServ)
W: Exxon*(LP)
Food W: Arizona Family Restaurant, Los Amigos Mexican
AServ E: Texaco
W: Exxon
ATM W: Arizona Bank, Bank Of America, Bank One
Other W: **Walgreens**

63 Continental Rd.
Gas W: Exxon* (Hot Stuff Pizza)
Food W: Carlos Grill, China Ying, Hot Stuff Pizza (Exxon), KFC, Kelly's Ice Cream, McDonalds, Safe Way Grocery
AServ W: Cobre Tire, Goodyear Tire & Auto (Exxon), Merle's Service
ATM W: Arizona Bank, Bank One, Wells Fargo Bank
Other E: **Madera Canyon, US Post Office**
W: **Arizona Tourist Information, Osco Drugs**

56 Canoa Rd.

(54) **Rest Area - Phones, RR, Picnic P (Both Directions)**

48 Arivaca Rd. Amado
Other E: **Mountain View RV Park**

42 Aqua Linda Rd.

40 Chavez Siding. Tubac

34 Tubac.
Other E: **Tubac Presidio State Park**

29 Tumacacori Carmen
Other E: **Tumacacori National Historical Park**

25 Palo Parado Rd.

22 Peck Canyon Rd.

17 Rio Rico Rd. Yavapai Dr.
Gas W: Chevron (AServ)
Food W: IGA (ATM), The Bandits Rendezdous
AServ W: Chevron
ATM W: Bank Of America, IGA
Other W: **Rio Rico Laundry Mat, Rio Rico Plaza (No Big Trucks)**

12 AR. 289 Ruby Rd
Other E: **Nogales Ranger Station**
W: **Pina Blanca Lake Recreational Area**

8 NM82 Bus. Loop 19 (No Southbound Reaccess)
RVCamp E: **Mi Casa**

4 AZ189 Marivosa Rd.
Gas E: 76(D), Chevron*
W: Shell*(D) (ATM)
Food E: Arby's, **Barrow's (Super 8)**, Dairy Queen, Fomosa, KFC, McDonalds
W: Carl's Jr Hamburgers
Lodg E: ◆ **Super 8 Motel**
AServ W: Ed Moss Chrysler Dealer
ATM W: Shell
Other E: **K-Mart, Wal-Mart, Walgreens**

1 Western Ave. Nogales
Med W: ✚ **Carondelet Holy Cross Hospital**
Other E: **Coin Laundry**

↑ARIZONA

Begin I-19

EXIT ARIZONA

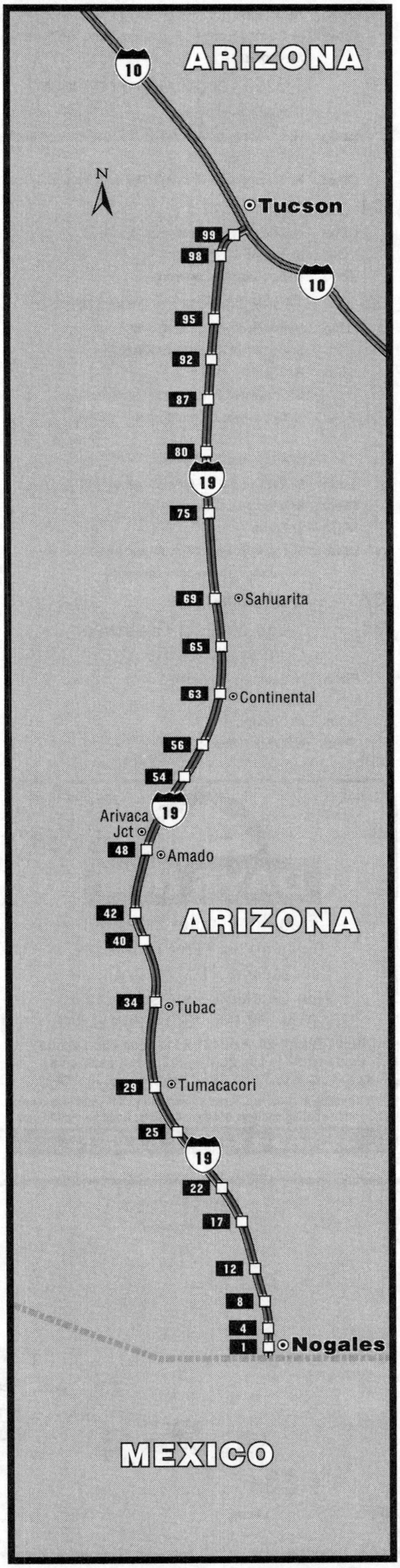

EXIT TEXAS

Begin I-20

↓TEXAS

3 Stocks Road

7 Johnson Road

13 McAlpine Road

22 Fm2903, Toyah
TStop N: **Toyah Auto/Truck Stop**
Gas S: Texaco*(D)
Food N: Roses (24 Hr), **Toyah Truck Stop Restaurant**
TServ N: **Toyah Mechanical Truck Service**
S: **Pete's Garage and Tire Service**
Other N: **Post Office**

(24) Picnic (Westbound)

(25) Picnic (Eastbound)

29 Shaw Road

33 Fm869

37 Bus20E, Pecos

39 TX17, Fort Davis, Balmorhea
AServ S: Eagle Tire Service
TServ N: **C&L Deisel**
Med N: ✚ **Hospital**

40 Country Club Dr, Pecos
Gas S: Chevron* (Subway)
Food S: **Alpine Lodge (Best Western)**, Subway (Chevron)
Lodg S: AAA **Best Western (Restaurant)**
RVCamp S: **Camping**
Other N: **Dept. Of Public Safety**

42 U.S. 285, Carlsbad, Fort Stockton
TStop N: **Flying J Travel Plaza(SCALES) (RV Dump, Thad's,ATM)**
Gas N: Exxon, Texaco*
Food N: McDonalds, **Thads Restaurant**
Lodg N: **Motel 6**, ◆ **Quality Inn**
TServ N: **I-20 Truck Service (24 Hr)**
ATM N: **Flying J Travel Plaza**
Other N: **Wal-Mart**

44 Pecos, Collie Rd

49 Fm516, Barstow

52 Barstow, Bus20W

58 Frontage Road

66 Fm1927, TX115, Pyote, Kermit
FStop N: **Citgo* (Restaurant)**
Food N: **Restaurant (Citgo)**

(68) **Rest Area - RR, Picnic P (Westbound)**

(69) **Rest Area - RR, Picnic P (Eastbound)**

70 Spur65

73 Fm1219, Wickett
TStop S: **National Truck Stop*(D) (Restaurant)**
FStop N: **Fina***
Food S: Restaurant (Natonal Truck Stop)
AServ S: Jim's Auto & Truck Repair
TServ N: **Tire Shop**

76 Bus20E, Monahans

79 Loop464

80 TX18, Kermit, Fort Stockton
FStop N: **Chevron(D)**
S: **Fina* (Taco Bell, Country Kitchen), Phillips**

EXIT TEXAS

66*
- **Gas** N: Chevron*[D], Exxon (Auto Service)
  S: Diamond Shamrock*, Texaco
- **Food** N: Dairy Queen, McDonalds
  S: Country Kitchen (Fina), Restaurant (Best Western), Taco Bell (Fina)
- **Lodg** S: (AAA) Best Western (Restaurant), Texan Inn
- **AServ** N: Ted's Car Wash
- **Other** N: Lowe's

83 Bus20W, Monahans

86 TX41, Monahans Sandhill State Park

93 Fm1053, Fort Stockton

101 Fm1601, Penwell
- **TStop** S: Texas Interstate Truck Stop (Country Kitchen Cafe)
- **Food** S: Country Kitchen Cafe (TStop)
- **TServ** S: Penwell Enterprises

(103) Parking Area (Both Directions)

104 Fm866, Goldsmith, Meteor Crater Road

112 Fm1936, Odessa
- **FStop** N: Citgo*

113 TX302, Loop338, Kermit, Meteor Crater Road

115 Fm1882, County Road W
- **Gas** N: Fina*
- **Other** N: Texas Department Of Public Safety

116 U.S. 385, Andrews, Crane
- **FStop** S: Phillips 66*
- **Gas** N: Chevron*, Fina*
  S: Texaco*
- **Food** N: Dairy Queen, Garden Buffet Restaurant (Best Western)
- **Lodg** N: (AAA) Best Western (Garden Buffet Restaurant), ◆ Econolodge, (AAA) Villa West Inn
  S: Motel 6
- **Med** N: ✚ Hospital

118 Fm3503, Grandview Ave
- **Gas** N: Shell*[D]
- **TServ** N: Cummins Diesel, Goodyear Truck Tires and Alignment, West Texas Peterbilt-Odessa
- **RVCamp** N: Miller's RV Specialists

121 Loop338, Odessa

126 Fm1788, Midland Int Airport
- **TStop** N: Warfield Truck Terminal*[CW][SCALES] (Restaurant)
- **Gas** N: Chevron*, Fina*[D]

EXIT TEXAS

- **Food** N: Warfield Truck Termanal
- **Other** N: Confederate Air Force Museum, Vietnam Memorial

131 TX1250, Loop 250 (for northbound trucks), TX158
- **Lodg** N: Economy Inns of America, Days Inn (see our ad this page)
- **Other** N: Texas Department Of Public Safety

134 Midkiff Road
- **FStop** N: Chevron* (Subway), Texaco*
- **Gas** S: Patriot Oil*[D]
- **Food** N: Subway (In Chevron)

136 TX349, Midland, Rankin, Lamesa
- **FStop** N: Pacific Pride, Phillips 66
  S: Exxon* (ATM, Burger King)
- **Gas** N: Shell*
  S: Chevron[D], Fina*
- **Food** N: Cowboy's 2 Cafe (Super 8 Motel), McDonalds, Minie's Pizza, Sonic Drive In
  S: Burger King (Exxon)
- **Lodg** N: (AAA) Super 8 Motel (Cowboy's 2)
- **TServ** S: Eddins-Walcher[D][LP]
- **ATM** S: Exxon
- **Other** N: Fiesta Foods, I G A Supermarket, Pharmacy, Petroleum Museum

137 Old Lamesa Rd.

138 TX158, Midland, Greenwood, Garden City, Fm715
- **FStop** N: Eddins-Walcher[LP] (24 Hr)
  S: Fina*
- **Gas** N: Chevron
- **Food** N: What A Burger

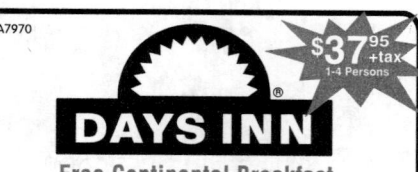

EXIT TEXAS

- **AServ** N: Big 3 Tire And Automotive Center

(141) Rest Area - RV Dump (P) (Eastbound)

(142) Rest Area - RV Dump (P) (Westbound)

143 Frontage Road (Eastbound)

144 Bus20, Midland, TX250 Loop

151 Fm829

154 Loop 214E, BS20E

156 TX137, Stanton, Lamesa
- **TStop** S: Diamond Shamrock*[D] (Country Kitchen, ATM, Subway)
- **FStop** S: Fina*[LP]
- **Gas** N: Chevron*
  S: Exxon*[D]
- **Food** N: Guys
  S: Dairy Queen, Subway (Diamond Shamrock)
- **ATM** S: Diamond Shamrock

158 Loop154, Stanton, BUS20W

165 Fm818

167 Picnic (Eastbound)

(168) Parking, Picnic (Westbound)

169 Fm2599

171 Moore Field Road

172 Cauble Road

174 Bus20E, Big Spring
- **Gas** S: Exxon*
- **Food** S: Sonic

176 Andrews, Link TX176

177 U.S. 87, San Angelo, Lamesa
- **TStop** N: Rip Griffin Travel Center*[SCALES] (Country Fare, Subway)
- **FStop** S: Fina*
- **Gas** N: Texaco* (Burger King)
  S: Chevron*[D]
- **Food** N: Burger King (Texaco), Country Fare (Rip Griffen), Subway, Sunrise Grill (Econolodge)
  S: Casa Blanca, Dairy Queen, McDonalds
- **Lodg** N: (AAA) Best Western, Econolodge (Sunrise Grill), Motel 6
- **AServ** S: Cross Roads Tires

178 TX350, Snyder
- **Gas** N: Texaco*

179 Bus20, Big Spring
- **Gas** S: Fina* (ATM), Phillips 66[D], TNT Hickory House

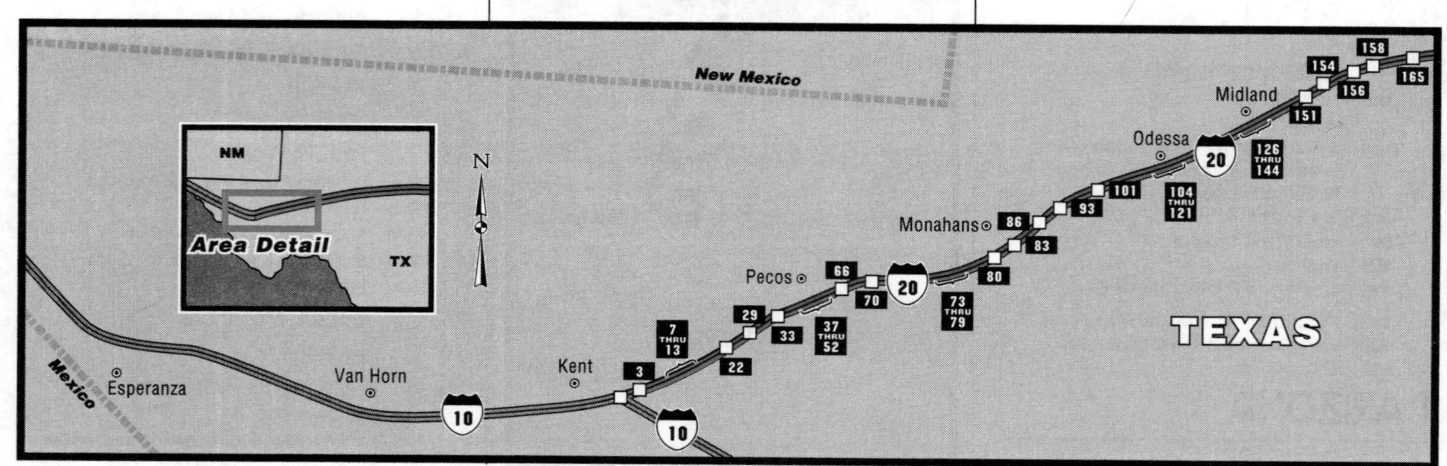

Bold red print shows RV & Bus parking available or nearby

EXIT TEXAS

BBQ, Texico*

Food **S:** College Park Cafe, Dairy Queen, Dell's Cafe, Denny's

Lodg **S:** ◆ Comfort Inn, Days Inn (see our ad this page), Super 7 Motel

ATM **S:** Fina

Other **S:** College Automatic Laundry, Eckerd Drugs, IGR Store

181A Fm700, Airport

AServ **N:** Big Springs Auto Electric

RVCamp **N:** Hillside RV Park

181B Refinery Road

Parks **N:** Big Springs State Park

182 Midway Road

184 Moss Lake Road, Sand Springs

Gas **N:** Phillips 66*(D)

S: Lakeway Grocery*

RVCamp **N:** Whip-In Campground

186 Salem Road, Sand Spring

188 Fm820, Coahoma Road

Gas **N:** Chevron*, Fina(D)

Food **N:** Dairy Queen, Kelly's Cafe

Lodg **N:** Motel

AServ **N:** Joe's Tire Service

Other **N:** US Post Office

189 McGregor Road

(190) Rest Area P (Eastbound)

190 Snyder Field Road

(191) Rest Area - Picnic, Phones, RR P (Eastound)

192 Fm821

194A East Howard Field Road

195 Frontage Road

199 Iatan Rd.

200 Conaway Road

(204) Rest Area - Picnic, RR, Phones P (Westbound)

206 Fm670, Westbrook

FStop **N:** Citgo*

207 Bus20, Westbrook

Gas **S:** Conoco*

209 Dorn Road

210 Fm2836

Parks **S:** Lake Colorado City State Park

212 Fm1229

213 Bus20E, Enderly Road, Colorodo City

215 Fm3525

216 TX208N, Snyder, Colorado City

Gas **N:** Chevron*, Exxon*(D) (ATM, McDonald's), Texaco*

S: Henderson's*, Phillips 66* (Burger King)

Food **N:** Dairy Queen, McDonalds (Exxon), Subway

S: Burger King (Phillips 66), Pizza Hut, Sonic Drive In, Villa Restaurant (Villa Inn)

Lodg **N:** Days Inn

S: Villa Inn Motel (Restaurant)

AServ **N:** Cooper Tires

Med **S:** ✚ Hospital

ATM **N:** Exxon

S: First Bank of West Texas

Other **S:** Coin Car Wash, The Medicine Place

EXIT TEXAS

217 TX208S, San Angelo

TStop **S:** Homeward Bound

Food **N:** Platter Restaurant

S: Collum's Restaurant, Feed Store BBQ

RVCamp **S:** Camping

219A Country Club Road

219B Bus20, Colorado City

220 Fm1899

221 Lasky Road

223 Lucas Road

224 Bus20, Lorraine

225 Fm644S

226A Fm644N

226B Bus20, Loraine , Wimberly Road

227 Narrell Road

(228) Picnic (Eastbound)

(229) Picinc Area (Westbound)

230 Fm1230

235 Bus20 , Roscoe

236 Fm608, Roscoe

Gas **N:** Conoco*, Texaco*(D)

Food **S:** Dairy Queen

TServ **S:** Audio Plus CB Radio Repairs and Sales

237 Cemetery Road

238A U.S. 84W, Roscoe, Lubbock, Snyders

238B Blackland Road

239 May Road

240 TX170, City Airport170 loop

241 Bus20, Sweetwater

242 Hopkins Road

TStop **S:** Sweetwater Travel Center

Gas **S:** 76, Conoco

Food **S:** 76 Restaurant, Pizza Hut (76)

TServ **S:** Jim's Truck and Alignment Service*, Truck Lube Center(SCALES)

TWash **S:** Texas West TruckWash

Other **S:** Dandy Western Wear

243 Robert Lee St, Hillsdale Road

Other **N:** Bentty RV Service

244 TX70S, TX70Bus, Sweetwater, San Angelo

FStop **N:** Fina*

Gas **N:** Diamond Shamrock (AServ), Fina*, Texaco*(D)

S: Chevron*, Fina*, Texaco*(D)

Food **N:** Dairy Queen, Kettle, McDonalds

EXIT TEXAS

S: Best Western, Buck's Steak and BBQ, Golden Fried Chicken, Jack's Family Steak House, Mona's Authentic Mexican Restaurant, Ranch House Motel, Schlotzkys Deli, Taco Bell

Lodg **N:** ◆ Holiday Inn, Motel 6

S: Best Western (Restaurant), Comfort Inn, AAA Ranch House Motel (Restaurant)

AServ **N:** Bledsoe Ford, Diamond Shamrock

S: GM Auto Dealership, Laser Wash

RVCamp **N:** Sweetwater RV Park

S: Chabarral, Sweetwater RV Park

Med **N:** ✚ Rolling Plains Memorial Hospital

ATM **S:** First National Bank

Other **N:** Rolling Plains Pharmacy, Texas Department Of Public Safety

S: IGA Store, K-Mart

245 Arizona Ave.

246 Alabama Ave

247 Bus20, TX70N, Sweetwater, Roby

249 Fm1856

251 Eskota Road

255 Adrian Road

256 Stink Creek Road

(256) Rest Area - RR, Phones P (Westbound)

(257) Rest Area - RR, Phones P (Eastbound)

258 White Flat Road

259 Sylvester Road

261 Bus20

FStop **N:** Texaco*

262 Fm1085

FStop **S:** Fina*

AServ **N:** Lindsey Garage

263 Bus20, Trent

264 Noodle Dome Road

266 Derstine Road

267 Bus20, Merkel

269 Fm126

FStop **S:** Fina* (Open24 Hours)

Gas **S:** Texaco

Food **N:** Pug and Leo's, Subway

S: Dairy Queen, Merkel Restaurant, Mesquite Bean Barbeque

Lodg **N:** Save Inn

S: Merkle Motel (Restaurant)

AServ **N:** Fred Hughes Chevrolet, Fred Hughes GMC Truck

S: Badger Lube, Texaco

Other **S:** Lemens LP Gas

270 Bus20, Fm1235, Merkel

TStop **N:** Wiley's Texaco

Gas **N:** Diamond Shamrock*

272 Wimberly Road

274 Wells Lane

277 Fm707, Bus20

TStop **N:** Flying J Travel Plaza* (Propane, Restaurant)

Gas **S:** Fina*, Texaco*(D)

Food **N:** Flying J Travel Plaza

278 Bus20 , Tye

TStop **S:** Wes-T-Go*(SCALES) (Restaurant)

FStop **N:** Conoco* (ATM)

EXIT TEXAS

Food S: Wes-T-Go Restaurant
TServ N: 278 Alternators-Starter
S: Hughes GMC Volvo
TWash S: 278 Truck Wash, Wes-T-Go
ATM N: Conoco
Other N: Camping

280 Fulwiler Road

282 Shirley Road
Lodg S: Motel 6
RVCamp S: Camping, KOA Campground
Other S: Dyess Airforce Base

283AB U.S. 83, 277S, Ballinger, San Angelo
Lodg S: Court by Marriott (see our ad this page)

285 Old Anson Road, Impact
Gas N: Texaco*
S: Texaco*
Food N: Charolette Belle (Travel Inn)
Lodg N: Travel Inn (Charolette Belle)
Other N: Coin Truck or Car Wash (Open 24 Hours)

286AB Abilene
Gas S: Fina*, Fisca*
Food S: Dairy Queen
Lodg S: Best Western
AServ S: Fina

286C Fm600
FStop N: Diamond Shamrock*, Fina*
Gas S: Chevron*(D)
Food N: Denny's
S: Bobbie's Home Cooking
Lodg N: AAA La Quinta Inn

288 TX351, Albany
FStop N: Diamond Shamrock*, Fina* (Subway)
Gas N: Chevron
Food N: Dairy Queen, Kettle, Subway (Fina)
Lodg N: Comfort Inn, Days Inn (see our ad this page), AAA Holiday Inn Express (see our ad this page), Roadway Inn
S: AAA Super 8 Motel

290 TX36, Loop322, Cross Plains, Reg Airport
RVCamp N: Abeline Zoo
Other N: Airport

292A Bus20, Abilene (Left Side Exit Only)

292B Elmdale Road
RVCamp S: Abeline Campground

294 Buck Creek Road

(296) Rest Area - Phones P (Eastbound)

(297) Rest Area - Picnic, Phones, RR P (Westbound)

297 Fm603

299 Fm1707

300 Fm604N
Gas S: Fina*
RVCamp S: White's Shady Oak

301 Fm604S, Spur 189S, Clyde, Cherry Lane
FStop S: Texaco(D)
Gas N: Chevron* (Open 24 Hours)
S: Fina*, Texaco*
Food N: Pizza House
S: Dairy Queen
TWash S: 301 Truck Wash
Other S: Falk Pharmacy

EXIT TEXAS

303 Union Hill Road
Food S: Ann's Country Kitchen

306 Bus20, Fm2047
AServ N: Hanner Chevrolet, Pontiac, GMC, Geo, Hanner Jeep and Eagle

307 U.S. 283, Albany, Coleman
FStop S: Conoco*
Gas S: Fina*
Food N: Dairy Queen

308 Bus20, Baird

310 Finley Road

313 Fm2228

316 Brushy Creek Road

319 Fm880S, Putnam, Cross Plains
Gas N: Conoco*(D)
Food N: Spur and Sportsman
Other N: US Post Office

320 Fm880N, Fm2945, Moran

322 Cooper Creek Road

324 Scranton Road

EXIT TEXAS

(326) Rest Area - No Services P (Eastbound)

(327) Picnic (Westbound)

(328) Parking Picnic (Westbound)

330 TX206, Cross Plains, Coleman
FStop N: Chevron
Food N: White Elephant
Lodg N: Roadway Inn
TServ N: Deans Radio Shop

332 U.S. 183, Cisco, Brownwood, Breckenridge
Gas N: Phillips 66*
Food N: Binger's Ranch House Restaurant, Dairy Queen, Sisco Steak House, Sonic Drive In
Lodg N: Oak Motel
AServ N: T&G's Auto Repair
S: Hanner Dealership
RVCamp N: Everett RV Park
Other N: Sisco Car Wash

337 Spur490

340 TX6, Gorman, Breckenridge
FStop N: Texaco*
S: Texaco*
Gas N: Fina
Food N: Ramona's Mexican
AServ N: Mangum Automotive and Electric
Med N: ✚ Hospital

343 Fm570, Eastland
Gas N: Fina* (Subway), Texaco*(D)
Food N: Dairy Queen, Happy Jordan Steak House, Home Cooked BBQ, Ken's Chicken-N-Fish, McDonalds, Sonic Drive In, Subway (Fina), Taco Bell, The Asia Restaurant
S: Pulido's Mexican
Lodg N: ◆ Super 8 Motel
S: ◆ Econolodge (Restaurant), ◆ Ramada (Restaurant)
AServ S: Davis Chrysler Plymouth Jeep Eagle Dodge
Med N: ✚ Hospital
Other N: Wal-Mart

347 Fm3363, Olden

349 Fm2461, Lake Leon, Ranger College
FStop N: Love's Country Store*
Gas S: Fina*(D)
Food N: Dairy Queen, Dan's BBQ and Steakhouse, Pizza Hut Express (Love's FS), Subway (Love's FS)
S: Last Chance Chicken and Fish
Lodg N: Days Inn

351 Desdemona Blvd. (Eastbound)

352 Blundell Street

354 Loop 254W, Ranger
AServ N: Freddy's Grage and Wrecker Service

361 TX16, Strawn, De Leon

(362) Rest Area - Parking, Picnic P (Westbound)

(363) Rest Area - Parking, Picnic P (Eastbound)

363 Tudor Road
Other N: Picnic Area

367 TX108, Thurber, Mingus
Gas N: Thurber Station*(D)
Food N: Smokestack Restaurant

Bold red print shows RV & Bus parking available or nearby

EXIT TEXAS

S: New York Hill
AServ **N:** Thurber Station

370 TX 108, Gordon, Stephenville
TStop **N:** Conoco(SCALES) (Restaurant, ATM)
FStop **S:** Citgo
Gas **N:** Texaco*
Food **N:** Conoco
Lodg **S:** Longhorn Inn
TServ **N:** Clay's Installation Center, Conoco, I-20 Diesel Service Truck Repair
TWash **N:** Conoco
ATM **N:** Conoco

373 TX 193, Gordon
FStop **N:** Chevron*
AServ **S:** Turners Garage

376 Panama Road, Blue Flat Road

380 Fm4, Palo Pinto, Lipan - Santo
Food **S:** Roadrunner Cafe
RVCamp **S:** Windmill Acres RV Park

386 U.S. 281, Stephenville, Mineral Wells
FStop **N:** Chevron*
Gas **N:** Diamond Shamrock*
S: Texaco* (ATM)
Food **N:** Billy Jean's
ATM **S:** Texaco
Other **N:** Brazos Rattlesnake Ranch and Petting Zoo, Mineral Wells Factory (see our ad this page)

(389) Rest Area - RR, Phones P (Eastbound)

(390) Rest Area - RR, Phones P (Westbound)

391 Gilbert Pit Road

394 Fm113, Millsap
FStop **N:** Fina*

397 Fm1189, Brock
Gas **N:** Diamond Shamrock*(D) (Open 24 Hours)

402 Weatherford, Spur312E (Eastbound)

403 Dennis Road (Westbound)

406 Old Dennis Road
TStop **N:** Conoco(SCALES) (RV Dump)
Food **N:** Wayside
Lodg **N:** Wayside Motel
RVCamp **S:** Safari Camping

407 FM1884, Tin Top Road (Westbound)

408 TX 171, FM 51, FM 1884, Tin Top Rd.
Gas **N:** Mobil* (ATM)
S: Exxon*(D), Texaco*(D) (Burger King)

EXIT TEXAS

Food **N:** Golden Corral, Grandy's, McDonalds, Taco Bell
S: Burger King (With Texaco)
Lodg **S:** Hampton Inn, Holiday Inn Express, Super 8 Motel
AServ **N:** Kwik Lube & Tune
ATM **N:** Mobile
Other **N:** San's Propane Inc., Wal-Mart

409 Clear Lake Road
TStop **N:** Petro(SCALES)
Gas **N:** Chevron*
Food **N:** Armondos Mexican Food, Domino's Pizza, Iron Skillet (Petro), Jimmy's Pancake-N-Waffle House (Best Western)
Lodg **N:** Bed and Bath, ◆ Best Western
TServ **N:** Blue Beacon Truck Wash, Weatherford
Med **N:** ✚ Hospital

410 Bankhead Hwy
TStop **S:** Love's Truck Stop*(D) (Subway and Baskin Robbins)
Food **S:** Baskin Robbins (Love's), Subway (Love's)

414 U.S. 180, Weatherford, Mineral Wells, Hudson Oaks
Gas **N:** Phillips 66* (ATM), Racetrac*, Texaco*
S: Chevron*
Food **N:** Cowboy's BBQue and Rib Co., Jack's Restaurant, Sonic
AServ **N:** Auto Dock, Hook's Lincoln Mercury, Jerry's Chevrolet, Jerry's Nissan, Oldmobile Dealer, Southwest Ford, Southwest Jeep Eagle
ATM **N:** Phillips 66

415 Mikus Road Annetta Road
FStop **S:** Citgo

EXIT TEXAS

Gas **S:** Texaco* (ATM)
Food **S:** Drivers Diner
AServ **S:** 415 RV Service
ATM **S:** Texaco

(417) Weigh station (Westbound)

418 Willow Park, Ranch House Road
FStop **N:** Citgo*
Gas **S:** Texaco*
Food **N:** Pizza Hut, Subway, That BBQue Place
S: McDonalds
Lodg **S:** AAA Ramada Limited
RVCamp **S:** Count Down* (Propane)
ATM **N:** Texas Bank
Other **N:** Brookshire's, Trinity Medows Raceway
S: Aledo Veterinarian Clinic, Red River M-Trailer Service and Repair, Winn Dixie Supermarket

(419) Rest Area P (Westbound)

420 Fm1187, Aledo Farm Road, Parking Area

421 Jct I-30 (Eastbound)

425 Markum Ranch Road (Eastbound)

426 Chapin School Road

428 Jct. I-820N

429A U.S. 377, Granbury
Gas **S:** Chevron*(CW), Exxon*(CW), Racetrac*
Food **S:** Dairy Queen, Domino's Pizza, McDonalds, Waffle House, What A Burger
ATM **S:** Alertsons
Other **S:** Albertson's Grocery (ATM), Animal Hospital

429B Winscott Road
Gas **S:** Diamond Shamrock*, Texaco*
Food **N:** Cracker Barrel
ATM **S:** First National Bank of Texas
Other **S:** Food Lion

431 Bryant Irving Rd.
Gas **N:** Mobil*(CW)
S: Chevron*(CW) (McDonald's Food Express), Texaco*
Food **S:** Black Eyed Pea, Colonial Cafe, Lone Star Oyster Bar, McDonalds (Chevron), Outback Steakhouse, Razzoo's Cajun Cafe, Sharky's Texas Sports Bar, Subway
Lodg **S:** La Quinta Inn
AServ **S:** Charlie Hillard Ford Dealership, City Garage, Cityview Shop, Goodyear Tire & Auto, Hillard Buick, Kwic Kar Lube And Tune, Lexus Hillard Of Fort Worth, Saturn Of Fort Worth

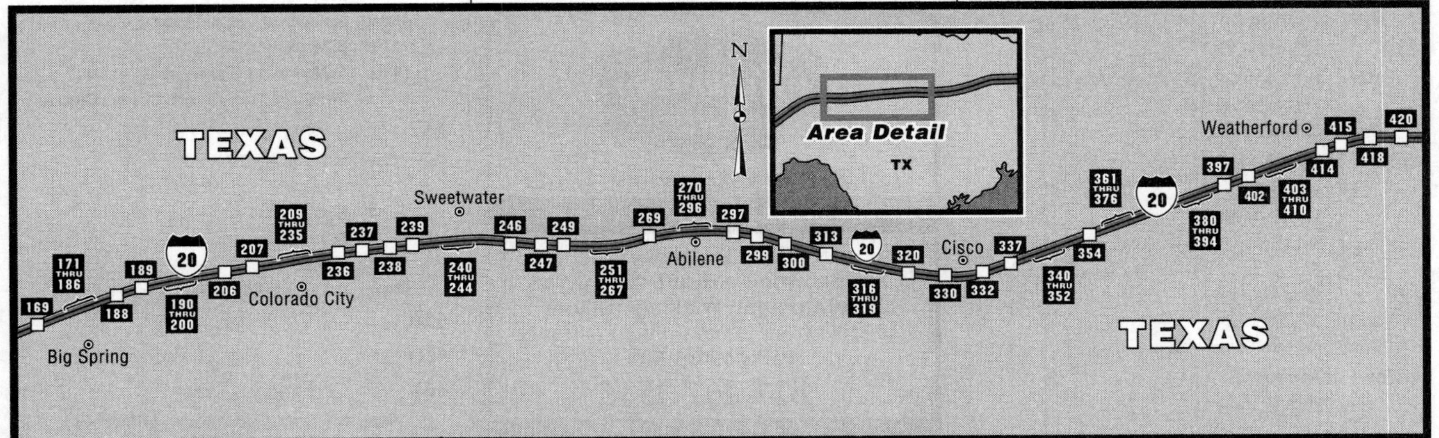

EXIT TEXAS

ATM N: Citizens National Bank
S: Bank One
Other N: **Country Day Pharmacy**
S: **Tom Thumb Grocery Store**

432 TX 183, Southwest Blvd.

433 Hulen St
Gas N: Texaco*[D]
Food N: Bagel Chain, Grady's, Honey Baked Ham, Olive Garden, Souper Salad, Subway, Tia's Texan Mexican Food
S: Bennigan's, Chili's, Colter's Barbeque, JO JO's, Jack-In-The-Box, McDonalds, Red Lobster
AServ S: Montgomery Ward
ATM N: Overton Bank and Trust
S: Bank One, Bank of America
Other N: **Albertson's Grocery, Office Depot, Petsmart**
S: **Hulen Mall, Pearl Vision Center**

434A Granbury Drive, South Drive
Gas S: Citgo*
Food S: Charbrolier Steakhouse, Dairy Queen, Ming Wok, New Great Wall Chinese Food, Poncho's Mexican Buffet
AServ S: Econo Lube N Tune, Hi/Lo Auto Supply
ATM S: Fort Worth Federal Credit Union
Other S: **Eckerd Drugs, US Post Office, Wedgewood Animal Hospital**

434B Trail Lake Drive
Gas S: Citgo*, Exxon*, Texaco
AServ S: Citgo, Texaco
ATM S: Southwest Bank

435 McCart Ave, West Creek Dr
Gas N: Conoco*[D], Fina*[LP] (ATM)
S: Fina*, Mobil*
Food S: Busy B's Bakery
AServ N: Bolen's Automotive
S: Discount Tire Company
ATM N: Fina
Other N: Quick Wash, **South Hills Animal Hospital,** Speedway Coin Car Wash

436A Fm731, Crowley Road, James Ave
Gas N: Conoco*[D]
S: Chevron*[CW]
Food S: Dairy Queen, Taco Bell
AServ S: Transmission Masters

436B Hemphill St
Gas N: Texaco*[D][CW] (ATM)
ATM N: Texaco

437 Jct. I-35W, Fort Worth, Waco

438 Oak Grove Road
Gas S: Conoco*, Texaco*
Other S: Car Wash

439 Campus Dr
AServ N: Meador Chrysler Plymoth, Meador Oldsmobile, Nichol's Ford
Other S: **Sam's Club**

440A Wichita St
Gas N: Chevron*
S: Fina*, Total*
Food N: McDonalds, Taco Bell, Wendy's
S: Braum's, Jack-In-The-Box, Luby's Cafeteria, What A Burger
Lodg S: Comfort Inn
ATM N: Texas Commerce Bank

440B Forest Hill Dr
Gas S: Citgo*, Texaco*

EXIT TEXAS

Food S: Captain D's Seafood, Ci Ci's Pizza, Dairy Queen, Golden Brown Chicken, Subway
AServ S: Chief's Auto Parts
Other S: **Eckerd Drugs, Kroger Supermarket (Pharmacy)**

441 Anglin Drive Hartman Lane
Gas N: Texaco*[CW]
S: Conoco*[LP]
AServ S: Automobile and Diesel Doctor, Clark's Garage Foreign and Domestic

442A 287Bus, Mansfield Hwy, Kennedale

442B Jct. I-820, 287N, Downtown Fort Worth

443 Bowman Springs Road (Westbound)

444 U.S. 287, Waxahachie

445 Green Oaks Blvd, Little Road , Kennedale
Gas N: Chevron*[CW], Mobile*, Texaco*[CW]
S: Chevron, Citgo*
Food N: Braum's, Burger Street, Colter's Barbeque, Dunkin Donuts, Grandy's, KFC, Khaki's Burgars and Pizza, Pizza Hut, Sacred Grounds Coffee House, Taco Bell, Tai-Pan Chinese Buffet, Wendy's
S: Hibachi 93 Japanese Steakhouse, McDonalds, Pancho's Mexican Food, Pasta Oggi & Pizza, Schlotzkys Deli, Southern Maid Donuts, Taco Bueno
AServ S: Chief Auto Parts, Jiffy Lube
ATM N: Nations Bank
S: Security Bank Of Arlington
Other N: **Eckerd Drugs, Minyard Supermarket**
S: **Gateway Auto Supply, Speed Queen Coin Laundry, Winn Dixie Supermarket**

447 Park Springs Road, Kelly - Elliott Road
Gas S: Exxon, Fina*[CW]
Food S: Blimpie's Subs (Fina)

448 Bowen Road
Gas N: Chevron*[CW], Conoco*[LP] (ATM)
S: Texaco*[D][CW]
Food N: Bobby Valentine Sports Gallery, **Cracker Barrel**
ATM N: Conoco[LP]

449 Fm157, Cooper St, Arlington
Gas N: Mobil*[CW], Shell*[CW], Texaco*[CW]
S: Chevron*[CW] (Open 24 Hours), Conoco*
Food N: Bennigans, Chili's, Eastside Mario's, Grandy's, IHOP, McDonalds (Indoor Playground), Outback Steakhouse, Owen's Restaurant, Pizza Hut, Red Lobster, Schlotzkys Deli, Starbucks Coffee
S: Applebee's, Arby's, Burger King, Burger Street, Chick-fil-A, Denny's, Little Caesars Pizza (K-Mart), Long John Silvers, Macaroni Grill, Olive Garden, Subway, T G I Friday's, Taco Bueno
Lodg N: AAA Best Western, ◆ Holiday Inn Express, Homestead Vllage
AServ N: Discount Tire Co, Sears
ATM N: Comerica Bank
S: Texas Commerce Bank
Other N: **Parks Mall, Target Department Store**
S: **Circut City, K-Mart, Office Max, Wal-Mart**

EXIT TEXAS

450 Matlock Road
FStop N: **Fina*[D]**
Gas N: Citgo*
S: Citgo*, Diamond Shamrock*[CW], Mobile*[CW], Shell[CW] (Blimpie, TCBY), Texaco*[CW]
Food N: Iron Skillet, Mercado Juarez Cafe, Tony Roman, **Wendy's**
S: Blimpie's Subs (Shell), Domino's Pizza, Joe's Pizza & Pasta, TCBY (Shell)
Lodg N: ◆ Hampton Inn (see our ad this page), La Quinta Inn
AServ S: Gateway Auto Supply, Texas Oil Xchange
Med N: **+ Hospital**
ATM N: Compass Bank, Overton Bank and Trust
Other S: **Pet Pause Grooming**

451 Collins St New York Ave.
Gas N: Exxon*[CW] (ATM), Mobil*[CW], Racetrac*
S: Diamond Shamrock, Texaco*
Food S: McDonalds
AServ N: Hiley Mazda, Saturn Dealership
ATM N: Exxon
Other S: **Davis Nissan, Post Office**

453A TX 360, Dallas, Fort Worth, Airport

453B S H 360S Watson Road

454 Great Southwest Pkwy
FStop N: **Conoco***
Gas N: Chevron*[CW]
S: Diamond Shamrock*, Texaco*[D][CW]
Food N: McDonalds, Taco Bell, Waffle House, Wendy's
S: Arby's (Stop & Go), Burger King
Lodg N: Days Inn
S: Comfort Inn
ATM S: Independent National Bank
Other N: **Lynn Creek Park**

456 Carrier Pkwy
Gas S: Mobil*[CW] (ATM), Texaco[CW]
Food S: Boston Market, Brass Bean Coffee Shop, Chili's, Denny's, Little Caesars Pizza, McDonalds, Subway
AServ S: Alex's Car Wash, City Garage
ATM S: Mobil
Other S: **Albertson's Grocery, Animal Clinic, Methodist Family Health Ctr., WalGreens**

457 Grand Prairie, Cedar Hill
Gas N: Texaco*[D] (ATM), Total* (ATM)
S: Racetrac* (Open 24 Hours)
Food N: Waffle House
S: Jack-In-The-Box
ATM N: Texaco, Total
Other N: **Cedar Hill State Park & Joe Pool Lake**

458 Mt. Creek Parkway

460 U.S. 408 (Difficult Reaccess)

461 Cedar Ridge Drive
Gas S: Diamond Schamroc*[CW] (ATM), LP Gas, Racetrac*

Bold red print shows RV & Bus parking available or nearby

EXIT	TEXAS
ATM	S: Diamond Shamroc
462B	North Main St
Gas	S: Exxon, Fina, Texaco*(D), Total*(D)
Food	S: Arby's, Brenda's Place, Captain D's Seafood, Chow-Line Buffet, Church's Fried Chicken, Judy's Cafe, K.C. Doughnuts, Los Lupes, O Doughnuts, Taco Bell, What A Burger
Lodg	S: Motel 6
AServ	S: Car Quest Auto Center, Exxon, Firestone Tire & Auto, Petrik's, Powell's Battery Exchange
Other	S: **Byron's Auto Supply, Eckerd Drugs, Jiffy Wash Coin Laundry, Knick Knacks, Kroger Supermarket, Kwik Wash Coin Laundry,** Kwik Kar Wash
463	Cockrell Hill Road , Camp Wisdom Road
Gas	N: Chevron*(CW)
	S: Diamond Shamrock*
Food	N: Bennigan's, Catfish King, Denny's, Miss Behavin, Papa John's Pizza, Taco Bueno, Taco Cabana Mexican Patio Cafe
	S: Burger King, Donut Palace, Olive Garden, Owen's Restaurant, Red Lobster, Subway
Lodg	N: (AAA) Hampton Inn, ◆ Holiday Inn, Lexington Hotel Suites, Motel 6, Royal Inn
AServ	N: Midas Muffler & Brakes, Pep Boys Auto Center, Red Bird Jeep and Eagle, Red Bird Nissan, RedBird Ford
	S: Red Bird Pontiac, Red Bird Toyota
Other	N: **Texas Drug Warehouse Pharmacy**
	S: **K-Mart, Medifirst Walk-in Clinic, Target Department Store**
464A	JCT I-76, Dallas
464B	U.S. 67 Cleburne
465	Hampton Road, Wheatland Road
Gas	N: Texaco*(D)(LP)
	S: Chevron*(CW), Racetrac* (Multi Store, ATM)
Food	S: Arby's, Cheddar's Casual Cafe, Jack-in-the-Box, McDonalds (Chevron), Sonic Drive In, Spring Creek BBQ, Wendy's
AServ	S: Davis Dealership,GMC Trucks,& Buick, Hyundai Davis Dealership, Lincoln Mercury Dealer, Sam's Club, Saturn of Duncanville Dealership, Village Mazda
Med	S: **✚ Hospital**
ATM	S: Racetrac
Other	S: **Home Depot, Sam's Club**
466	South Polk St
Gas	N: Citgo*, Texaco*(D)
Food	N: Dairy Queen, W.B.S.Catfish and Hamburgers, Western Barbeque
Other	N: Coin Car Wash, **Sun Brite Coin Laundry**
467A	Jct. I-35E, Downtown Dallas
467B	I-35E
468	Houston School Road
470	TX 342, Lancaster Road
TStop	N: **Pilot Travel Center*(D)(CW)(SCALES) (RV Dump, Kerosene)**
Gas	N: Chevron*(CW)
	S: Racetrac*(SCALES), Texico*
Food	N: **Dairy Queen (Pilot), Wendy's (Pilot)**
	S: **McDonalds,** Williams Chicken
Lodg	S: (DAYS INN) ◆ **Days Inn**
TServ	S: **Dallas Truck Wash, Goodyear Tire & Auto**

EXIT	TEXAS
472	Bonnie View Road
TStop	N: **Flying J Travel Plaza(LP)(SCALES)**
Gas	N: Conoco*(D) (Flying J Travel Plaza)
Food	N: **Layover Restaurant, The Cookery (Flying J)**
Lodg	N: Ramada Limied
TServ	N: **Chrome Plus, Flying J Travel Plaza, United Truck Wash (24 Hrs)**
TWash	N: **Flying J Travel Plaza**
473AB	473A - N I-45 Dallas, 473B - S I-45 Houston
474	TX 310 N
476	Dowdy Ferry Road (No Westbound Reaccess)
477	St. Augustine Road
479AB	US 175, Dallas, Kaufman
480	I-635 North
481	Seagoville Road
Gas	N: Beasley's Grocery*, Total*
Food	N: Smith Donuts, Taco Bell (Total)
	S: **Lindy's Family Restaurant**
ATM	N: Total
482	Belt Line Road
RVCamp	S: **Countryside RV & Mobile Home Park (RV Dump)**
483	Lawson Road., Lasater Road
487	FM 740, Forney
490	FM 741
491	FM 2932, Helms Tr. Road
493	FM 1641
498	FM 148
499A	To US 80W Dallas (Limited Access Hwy)
499B	Rose Hill Road
501	TX 34, Terrell, Kaufman
FStop	S: **Total* (24 Hrs)**
Gas	N: Chevron* (24 Hrs), Exxon*
	S: Citgo*, Overland Express*
Food	N: Burger King, **Charlotte Plumers Seafare,** Puerto Escondido
	S: Arby's (Overland Express), **McDonalds (Playground), Wendy's**
Lodg	N: (AAA) Best Western, ◆ **Classic Inn,** (DAYS INN) Days Inn, Holiday Inn Express
	S: ◆ Comfort Inn
AServ	N: Exxon
Med	N: **✚ Hospital**

INTERSTATE EXIT AUTHORITY

EXIT	TEXAS
ATM	S: Citgo, **Total**
Other	S: **Tanger Factory Outlet**
503	Wilson Road
TStop	S: **Rip Griffin Travel Center*(SCALES) (Fuel-Texaco, Barber, RV Dump, Showers, Laundry)**
Food	S: **Pizza Hut Express (Rip Griffin TS), Restaurant (Rip Griffin TS), Subway (Rip Griffin TS)**
AServ	S: **Rip Griffin Travel Center**
TServ	S: **Rip Griffin Travel Center**
RVCamp	S: **Rip Griffin Travel Center**
ATM	S: **Rip Griffin Travel Center**
506	FM 429, FM 2728, College Mound Road
509	Hiram Road
FStop	S: **McDonald's Truck Center (Fuel - Fina)**
Food	S: **McDonald's Bar-B-Q**
(510)	**Rest Area - RR, Picnic, Grills, Vending (P) (Both Directions)**
(511)	**Weigh Station (Both Directions)**
512	FM 2965, Hiram - Wills Point Road
516	FM 47, Lake Tawakoni
Gas	S: Diamond Shamrock*(D), Robertson's*(D)
Food	S: **Cafe (24 Hrs, Interstate Motel),** Hamburgers & Pizza (Diamond Shamrock)
Lodg	S: **Interstate Motel**
519	Turner - Hayden Road
521	Myrtle Cemetery Road., Myrtle Springs Road
Food	S: **Canton Jubilee Superb Buffet**
AServ	S: Tommy's Garage and Wrecker Service
RVCamp	S: **Marshall's RV Centers Inc.**
523	TX 64, Wills Point
RVCamp	N: **Action RV Park (1 Mile), Canton Campground & RV Park(LP) (1 Mile)**
526	FM 859, Edgewood
Food	S: Crazy Cow BBQ & Steakhouse
Parks	N: **Edgewood Heritage Park**
	S: **First Monday Park**
527	TX 19, Emory, Canton
FStop	S: **Fina*, Overland Express***
Gas	N: Citgo* (24 Hrs), Texaco*(D)
	S: Chevron*(D)(CW) (24 Hrs), Shell*
Food	N: Burger King (Playground), Jewel's Restaurant & Steakhouse, Ranchero Restaurant, **What A Burger**
	S: **Arby's (Overland Express FStop),** Dairy Palace (24 Hrs), Dairy Queen, Jerry's Pizza, McDonalds (Playground), Taco Bell, Two Senorita's Mexican
Lodg	N: All Suites Inn, **Super 8 Motel**
	S: ◆ **Best Western,** (DAYS INN) **Days Inn**
AServ	S: Chevron
RVCamp	S: **RV Park (Best Western, RV Dump), RV Service**
ATM	N: Citgo
	S: Chevron, Shell
Other	S: Jerry's Car Wash
528	FM 17
530	FM 1255
533	Colfax Oakland Road
TStop	N: **Citgo**
TServ	N: **Van Zandt Tire & Service Center (Citgo TS)**
ATM	N: **Citgo TS**
536	Tank Farm Road

EXIT TEXAS

537 FM 773, FM 16

(537) **Rest Area - RR, Phones, Picnic, Grills, RV Dump, Vending P (Both Directions)**

540 FM 314, Van
- **Gas** N: Exxon*(D)
- **Food** N: Dairy Queen, **Rail Head Cafe**
- **Lodg** N: Van Inn
- **AServ** N: The Van Service Center (Exxon)

544 Willowbranch Road
- **TStop** N: **Conoco***
- **TServ** N: **Conoco TS (24 Hrs)**
- **RVCamp** N: **Willowbranch RV Park (1 Mile, RV Dump)**

(546) **Weigh Station (Both Directions)**

548 TX 110, Grand Saline
- **FStop** S: **Chevron***
- **Gas** N: Chevron
- **AServ** N: Chevron
- **Other** N: **YWAM Go Center**

552 FM 849
- **Gas** N: Chevron* (24 Hrs)
- **ATM** N: Chevron
- **Other** N: **Elliott's Pharmacy, Hide-A-Way Small Animal Clinic, The Travel Store**

554 Harvey Road
- **RVCamp** S: **554 Campground**

556 U.S. 69, Lindale, Mineola
- **FStop** N: **Total* (FedEx Drop)**
- **Gas** N: Racetrac*
  S: Chevron*, Texaco*(D)
- **Food** N: **Burger King,** Juanita's Family Mexican, **McDonalds (Playground),** Paco's Mexican, Subway, Taco Bell
  S: Dairy Queen (Texaco), **Terry's Deli & Bar-B-Q (Coachlight Inn)**
- **Lodg** N: **Comfort Inn,** Executive Inn, **Days Inn (see our ad this page)**
  S: **Coachlight Inn**

557 Jim Hogg Road
- **FStop** N: **Texaco***
- **Food** N: Tio's Restaurant

560 Lavender Road

562 TX 14, Tyler State Park, Hawkins
- **Food** N: Bodacious Bar-B-Q
- **RVCamp** S: **Blue Jay Camp (4 Miles)**
- **Parks** N: **Tyler State Park (2 Miles)**

565 FM 2015, Driskill, Lake Road

EXIT TEXAS

567 TX 155, East Texas Center, Gilmer, Winona, Big Sandy
- **Lodg** S: Days Inn
- **Med** S: **Univ. Texas Health Center**

571A U.S. 271, Tyler
- **Gas** S: Chevron*(D)

571B FM 757, Starrville, Omen Road

(573) Picnic (Both Directions, No Services)

575 Barber Road

579 Joy - Wright Mountain Road
- **TStop** N: **Diamond Shamrock***
- **Gas** S: Fina*
- **Food** N: Hickory Ridge Cafe, **The Truck Stop Cafe (Diamond Shamrock TS)**
- **TServ** N: **Diamond Shamrock TS**

EXIT TEXAS

- **TWash** N: **Red Devil Truck Wash (24 Hrs)**
- **ATM** S: Fina

582 FM 3053, Liberty City, Overton
- **Gas** N: Exxon*(D), Shell*(D), Texaco*
- **Food** N: Dairy Queen, Pizza Boy, Subway (Exxon)
- **Lodg** S: Thrifty Inn
- **AServ** N: Hog-Eye Auto Parts
- **Other** N: Aquajet Car Wash, **Security Mail Service, Soapbox Laundry,** Spot Free Rinse Car Wash

583 TX 135, Gladewater, Overton
- **Gas** N: E-Z Mart*
  S: Phillips 66*
- **Food** N: Liberty City Cafe

587 TX 42, Kilgore, White Oak
- **FStop** S: **Chevron***
- **Gas** N: Diamond Shamrock*(D)
- **Food** N: Bodacious Bar-B-Q, Maxie Brothers Grill
  S: **Piccadilly Pizza & Subs (Chevron)**

589 U.S. 259, TX 31, Kilgore, Henderson, Longview (Westbound Is Left Exit)
- **RVCamp** S: **Foretravel, Inc. (see our ad this page)**

591 FM 2087, FM 2011
- **TServ** S: **Master Aligners of 18 Wheelers**
- **RVCamp** S: **Pine Meadows Campground (RV Dump)**

595 TX 322, FM 1845, Loop 281, Estes Parkway, Longview
- **Gas** N: Chevron*(D), Exxon(D)(LP), Exxon*(D), Texaco*(D)(LP)
  S: Chevron*(D), Fina*(D)
- **Food** N: Dairy Queen, Denny's (Longview Inn), Drillers Grill (Econolodge), KFC, Lupe's Mexican, McDonalds (Playground), Pizza Hut, Ribs, Subway (Chevron), Waffle House
  S: The Hot Biscuit (La Quinta Inn)
- **Lodg** N: Days Inn AAA Days Inn, AAA Econolodge, Holiday Inn, ◆ Longview Inn, Stratford House Inns
  S: Hampton Inn, AAA La Quinta Inn, Motel 6
- **AServ** N: Exxon, Mike London

596 U.S. 259 North, TX 149, Eastman Road
- **FStop** N: **Texaco***
- **Gas** N: Exxon*(D)
  S: Total*
- **Food** N: Burger King, Grandy's* (Exxon), **Taco Bell (Texaco),** What A Burger (24 Hrs)
  S: Arby's (Total)
- **Lodg** N: **Economy Inn,** Ramada Limited
- **TServ** N: **Twin States Truck Inc.**
- **ATM** N: Exxon

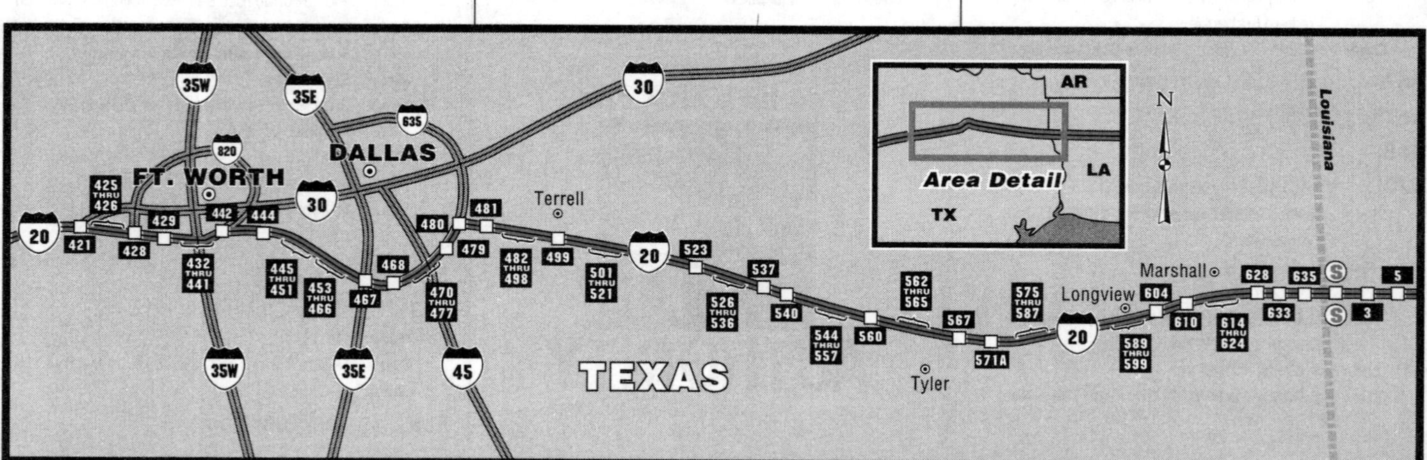

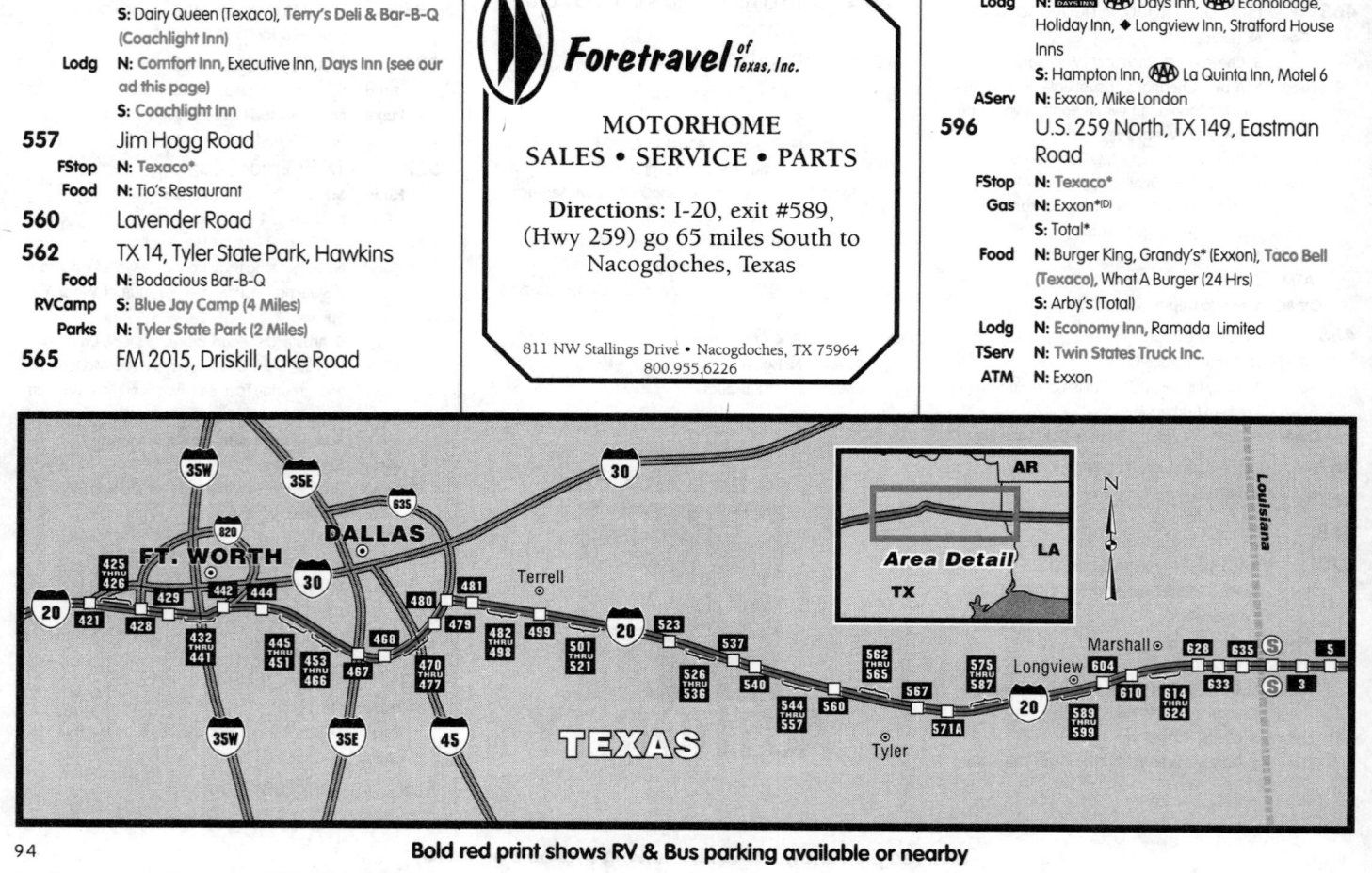

Bold red print shows RV & Bus parking available or nearby

EXIT	TEXAS
Parks	S: Martin Lake State Park
Other	N: Bullwinkles CB Sales & Repair, Lake O the Pines
599	Loop 281, FM 968
TStop	S: National Truck Stop (SCALES) (Fina Fuel)
Gas	S: Exxon*
Food	S: Bar-B-Q Deli (Exxon), Restaurant (National Truck Stop)
TServ	N: Fluid Power Service, Longview Truck & Trailor, Texas Kenworth Co., Truck Parts World
	S: National Truck Stop
TWash	S: Studebaker's Truck Wash
RVCamp	N: Gum Springs RV Travel Trailor Park (Showers, Laundromat, 1 Mile)
	S: RV Park, Wanda's Kountry Korner RV Park
ATM	S: National Truck Stop
Other	S: Laundromat (Exxon)
604	FM 450, Hallsville
Gas	N: Chevron*
RVCamp	N: 450 RV Park Hitchin' Post (Laundry, Showers), Lake O the Pines
(608)	Rest Area - RR, Phones, Picnic, Grills, Vending P (Closed For Repairs)
610	FM 3251
614	TX 43, Marshall, Henderson
Parks	S: Martin Creek Lake State Park (18 Miles)
617	U.S. 59, Marshall, Carthage
TStop	S: Exit 617 Pony Express* (SCALES) (Texaco, Laundry, FedEx Drop)
FStop	S: Diamond Shamrock*
Gas	N: Exxon* (24 Hrs), Texaco*
	S: Chevron* (D), Total*
Food	N: Elmwood Cafeteria (Ramada Inn), Golden Corral, McDonalds (Playground), The Hot Biscuit, Waffle House
	S: Hungri Maverick (24 Hrs)
Lodg	N: Best Western, Holiday Inn Express, Ramada Inn
	S: Comfort Inn, ◆ Days Inn, Motel 6, Super 8 Motel
AServ	N: Exxon
TServ	S: Exit 617 Pony Express (Texaco)
RVCamp	N: Camping, Hilltop RVPark
ATM	S: Exit 617 Pony Express (Texaco)
620	FM 31, Elysian Fields
624	FM 2199, Scottsville
628	U.S. 80
633	FM 134, Caddo Lake (Longhorn Army Ammunition Plant)
Gas	N: Texaco* (24 Hrs)
Food	N: Robertson's Hams
RVCamp	S: Camping (.7 Miles)
635	Spur 156, FM 9, Waskom
Gas	N: Chevron* (D), Texaco* (D)
Food	N: Dairy Queen, Jim's Bar-B-Q & Catfish
AServ	N: Buddy's Auto Service
ATM	N: Chevron, First State Bank
Other	N: Tiller Veterinary Clinic
(636)	TX Welcome Center - RR, Info, Phones, Snacks, Souvenirs, RV Dump (RR-8:30AM To 5:00PM Central Time)

EXIT	TEXAS/LOUISIANA

↑ TEXAS

↓ LOUISIANA

EXIT	TEXAS/LOUISIANA
(1)	Weigh Station (Both Directions)
(2)	LA Welcome Center - RR, Info, Phones, Picnic, RV Dump (Eastbound)
3	South U.S. 79, LA 169, Mooringsport, Carthage Texas
TStop	S: Flying J Travel Plaza (SCALES) (RV Dump)
Food	S: Thads Restaurant (Flying-J Truck Plaza)
ATM	S: Flying J Travel Plaza
Other	N: Boothill Speedway
5	North U.S. 79, U.S. 80, Greenwood
TStop	N: Kelly's Truck Terminal (LP)(SCALES) (CB Shop, FedEx Drop), The Derrick* (Overnight Parking)
Food	N: The Derrick Restaurant
Lodg	N: Country Inn (The Derrick TS), Mid Continent Inn
TServ	N: Kelly Truck Terminal
TWash	N: Kelly's Truck Terminal
RVCamp	N: Kelly's RV Park
ATM	N: Kelly's Truck Terminal
8	U.S. 80, LA 526 East, Industrial Loop
TStop	S: Petro* (SCALES) (24 Hrs, Auto Fuel-Mobil, FedEx Drop)
FStop	S: Speedway*
Gas	S: Chevron* (D)
Food	S: Iron Skillet Restaurant (Petro TS), Jack-In-The-Box
Lodg	N: ◆ Red Roof Inn
TServ	N: Detroit Diesel United Engines, Shreveport Truck Center
	S: Petro (Bridgestone)
TWash	S: Blue Beacon Truck Wash
RVCamp	N: Randy's Travel Town
	S: Campers RV Center Campgrounds (2 Miles), KOA Kampground
10	Pines Road
Gas	N: Fina* (D) (24 Hrs)
	S: Exxon (D), Shell* (D)(CW)
Food	N: Dairy Queen, Little Caesars Pizza, Pizza Hut, Subway, Western Sizzlin'
	S: Baskin Robbins, Blimpie's Subs (Exxon), Burger King (Playground), Church's Fried Chicken, Cracker Barrel, Grandy's, KFC, Mamie's Rib Shack, McDonalds (Playground), Nicky's Mexican Restaurant, TCBY (Exxon), Taco Bell, Wendy's
Lodg	S: ◆ Fairfield Inn by Marriott, La Quinta Inn & Suites
AServ	S: Exxon, Minit Oil Change, Xpress Lube
RVCamp	S: KOA Kampground (1.7 Miles)
ATM	N: Hibernia National Bank
	S: Deposit Guaranty, Exxon
Other	N: Brookshire's Grocery Store
	S: Eckerd Drugs, Eyes & Eyewear, K&B Drugs, Kroger Supermarket, Public Library, US Postal Service, Wal-Mart (Pharmacy, 24 Hrs)
11	I-220 East, LA 3132 East, Inner Loop Expwy
13	Monkhouse Dr., Airport
Gas	S: Chevron* (CW) (24 Hrs), Exxon* (D), Texaco (D)
Food	N: Denny's, Kettle Restaurant, Leon 's Bar-B-Que
	S: Restaurant (Ramada Inn), Restaurant (24 Hrs, Pelican Inn Airport Station), Subway (Exxon), Waffle House
Lodg	N: Best Value Inn & Suites, ◆ Days Inn, Econolodge, ◆ Holiday Inn Express
	S: Best Western, Pelican Inn Airport Station, Ramada Inn, ◆ Super 8 Motel
AServ	S: Texaco
ATM	N: Regions Bank
Other	S: Shreveport Airport

EXIT	LOUISIANA
14	Jewella Ave
Gas	N: Mobil (D), Texaco* (D)(CW)
Food	N: Burger King (Playground), Church's Fried Chicken, John's Seafood, McDonalds, Popeye's Chicken, Subway, What A Burger
Lodg	N: Jo-dan Motel
AServ	N: Auto Zone Auto Parts, Brake-O Brake Shop, Pennzoil Lube, Western Auto
ATM	N: Bank One, Hibernia, K & B Drugs
Other	N: County Market Total Discount Foods, Eckerd Drugs, Fred's Discount Drug Store, K&B Drugs (24 Hrs, Pharmacy), Louisiana State Fairgrounds, Medic Pharmacy, Super 1 Foods
	S: Libbey Factory Outlet Store (Glassware)
16A	U.S. 171, Hearne Ave., State Fairgrounds
Gas	N: Texaco*
	S: Exxon*, Fina*, Texaco* (D), Triple JJJ*
Food	N: Hot Donuts
	S: KFC
AServ	N: Computer Tune-A-Car, Service Tire Inc.
	S: Exxon
Med	N: ✚ LSU Medical Center, ✚ WK Medical Center
Other	N: Caddo Animal Clinic, Fairgrounds Field (Shreveport Captains - Minor League Baseball), Fire Dept., Louisiana State Exhibition Museum, Louisiana State Fairgrounds, Medic Pharmacy, Spar Planetarium
	S: Superior Hand Wash & Detail
16B	U.S. 79, U.S. 80, Greenwood Road
Gas	S: Mobil*
Food	S: Bobbie's Cafe Shreve, El Chico's Mexican, Restaurant (24 Hrs, Sundowner West)
Lodg	S: Sundowner West
AServ	S: Ed's Muffler Shop, H & M Generator & Alternator Service, Randy's Automotive
Med	N: ✚ WK Medical Center
ATM	N: Hibernia Bank
Other	S: Fed Ex Drop Box, US Post Office
17A	Linwood Ave, Lakeshore Dr.
Gas	S: Mobil*
AServ	N: Jordan Auto Parts
	S: Dixie Imported Auto Parts, Used Tires, Windshields Of America (Mobil)
Other	S: Arkla Natural Gas Vehicle Refueling Station (.46 Miles), Davis Tool & Tarp, Scuba Training School
17B	I-49 South, Alexandria
18A	Common St., Line Ave. (Eastbound, Difficult Westbound Reaccess)
Food	S: Monjuni's, The Village Grill
Lodg	S: Mid-City Motor Hotel
AServ	S: Joe Rachal Transmissions
Med	S: ✚ Schumpert Medical Center
ATM	S: Hibernia
18C	Fairfield Ave (Westbound, Difficult Westbound Reaccess)
Gas	N: Circle K*

EXIT LOUISIANA

Food N: Fertitta's Deli
AServ N: Under Car Service Specialists, Uniroyal Tire & Auto, Walker's Auto Spring
Med N: ✚ **Schumpert Medical Center**
ATM N: Circle K

18D Common St, Louisiana Ave (Westbound)
AServ N: Caddo Radiator

19A North U.S. 71, North LA 1, Spring St, South LA 1, Market St. (One Way Streets)
Lodg N: AAA Best Western, Holiday Inn Downtown
Other N: **Car Museum, Visitor Information**

19B Traffic St (One Way Streets)
Food N: Jacob's Well Coffee House, L'Italiano Restaurant
AServ N: A & R Automotive Service, Parker's Automotive, Red River GM Service Center
Other N: **Barksdale Drugstore, East Bank Theater & Gallery, Municipal Building Historical Site**

20C US 71 South, Barksdale Blvd (Eastbound)
Gas S: Mobil*
Food S: Southern Maid Donuts Cafe
AServ S: Martin's Auto Electric

20B LA 3, Benton Rd (Eastbound, Reaccess Eastbound Via Old Minden)
Gas S: Exxon*(CW)
Food S: Blimpie's Subs, Burger King, Johnny's Pizza House, McDonalds, Podnuh's Bar-B-Q, Poncho's Mexican Buffet, Posados Cafe, Ralph & Kacoo's Seafood, Shoney's (Shoney's Inn), Subway, TCBY, What A Burger (24 Hrs)
Lodg S: Shoney's Inn
AServ S: Doug's Pit Stop Import Service, Mazda, Volks, Audi, Porsche, Western Auto
ATM S: Bank One, Exxon, Hibernia
Other S: **Book Rack, Inspiration Christian Bookstore, US Post Office**

20A Isle of Capri Blvd, Hamilton Rd
Gas N: Circle K*(LP), Speedway*, Texaco*
S: Chevron
Food N: Cobb's Bar-B-Q
S: Imperial Garden Chinese (Ramada Inn)
Lodg N: Comfort Inn
S: Ramada Inn
AServ N: AC Delco, Davis Car Clinic, Don's Auto Repair, John's Paint & Body, Jr.'s & Tommy's Alternators & Starters, Parker's Automotive, Regency Tire
S: Chevron
ATM N: Circle K
Other N: **Martin Animal Hospital, Peter's Grocery Market**

21 LA 72, To U.S. 71South, Old Minden Rd
Gas N: Circle K Express (24 Hrs), Exxon*(CW)
S: Racetrac*
Food N: Adam's Restaurant (Holiday Inn), El Chico Mexican, Kettle Restaurant (24 Hrs, La Quinta Inn), Shoney's (Shoney's Inn), TCBY
S: **Dragon House (Chinese)**
Lodg N: ◆ Hampton Inn, Holiday Inn, La Quinta Inn, ◆ Residence Inn by Marriott, ◆ Shoney's Inn
S: DAYS INN **Days Inn**, Motel 6, **Days Inn** **(see our ad this page)**

EXIT LOUISIANA

AServ N: Mazda, Volks, Audi, Porsche
ATM N: Exxon
Other S: **Visitor Information**

22 Airline Dr., Barksdale AFB (Bossier Parish Community Coll.)
Gas N: Chevron*, Exxon(D), Mobil*(CW)
S: Circle K, Exxon*(D)(CW), Texaco*(D)(CW)
Food N: Casino Cafe (Best Western), Grandy's, Jack-In-The-Box, Luby's Cafeteria, McDonalds (Chevron), Pizza Hut, Red Lobster, Taco Bell, **Waffle House**
S: Darrell's Grill & Restaurant (Red Carpet Inn), David Beard's Catfish & Seafood, Outback Steakhouse
Lodg N: ◆ Best Western, Isle Of Capri Hotel, ◆ Rodeway Inn
S: Red Carpet Inn
AServ N: Exxon, **K-Mart**, NAPA Auto Parts
Med N: ✚ **Hospital**
ATM S: Bank One, Bossier Federal Credit Union
Parks N: **Cyprus Black Bayou Recreation Area**
Other N: **K-Mart (Pharmacy), Pierre Bossier Mall**
S: **Barksdale Airforce Base, Eight Airforce Museum, Tourist Information**

23 Industrial Dr
FStop S: **Chevron***(D)
Gas N: Phillips 66(D), Texaco*(D)
S: Road Mart*
Food N: Burger King (Playground), Country Kitchen, Greenway's Harvest Buffet, McDonalds (Playground), Popeye's Chicken (Texaco), Taco Bell
S: Restaurant (Road Mart)
Lodg N: Horseshoe Le Boss'ier Hotel

EXIT LOUISIANA

S: Quality Inn
Other S: **State Police Troop G**

26 I-220 West Bypass, Louisiana Downs Racetrack

33 LA 157, Haughton, Fillmore, Doyline
Gas N: Phillips 66*
AServ S: Haughton Auto Repair
RVCamp N: **Hilltop Campground (1.9 Miles)**
Parks N: **Bodcau Recreational Area (10.75 Miles)**
S: **Lake Bistineau State Park**

(36) **Rest Area - RR, Phones, Picnic, Grills, RV Dump, Pet Rest (Both Directions)**

38 Goodwill Road, Ammunition Plant
TStop S: Fina*
Food S: **Rainbow Diner (Fina Truck Stop)**
TServ S: **ACE Truck & Trailer Service Center**

44 LA 7, Cotton Valley, Springhill
Gas N: Bud's, Hamburger Happiness, Roscoe's Grocery*
Food N: **Earl's Bayou Inn, Hamburger Happiness, Southern Maid Donuts**
Other N: **Rick's Marine**

47 LA 7, Minden, Sibley
Gas N: Chevron*(D), Mobil*, Texaco*
Food N: Golden Biscuit (24 Hrs)
Lodg N: Best Western, AAA Exacta Inn, AAA Southern Inn
Parks S: **Lake Bistineau State Park (17 Miles)**
Other N: Clark's Car Wash

49 LA 531, Dubberly, Minden
TStop N: **Truckers Paradise* (24 Hrs, Washeteria, Laundry)**
Food N: **Oasis (Truckers Paradise)**
TServ N: **Petchak's Truck & Tire Repair, Trucker's Pardise TS**

52 LA 532, Dubberly, To U.S. 80
FStop S: **Chevron***
Gas S: Texaco*

55 U.S. 80, Ada, Taylor

(58) **Rest Area - RR, Phones, Picnic, RV Dump, Pet Rest (Both Directions)**

61 LA 154, Gibsland, Athens
Parks N: **Lake Claiborne State Park (17 Miles)**

67 LA 9, Arcadia, Homer
Gas S: Texaco
Food S: J&J Express
AServ S: Major Muffler (Texaco)
Parks N: **Lake Claiborne State Park (16 Miles)**

69 LA 151, Arcadia, Dubach
FStop S: **Fina***
Gas S: Exxon*, Texaco*
Food S: Country Folks Kitchen, **Restaurant (Nob Hill Inn)**, Snuffy's Pizza, Sonic Drive In, Subway
Lodg S: DAYS INN Days Inn, **Nob Hill Inn**
AServ S: Arcadia Tire & Service, Big A Auto Parts
ATM S: Texaco
Other S: **Brookshire's Grocery Store**, Car Wash, **Factory Outlet Center, Freds Drug Store (Pharmacy)**

76 LA 507, Simsboro

78 LA 563, Industry

81 Grambling, LA 149, Grambling State University

Bold red print shows RV & Bus parking available or nearby

EXIT LOUISIANA

Gas S: Exxon*, Texaco*

84 LA 544, Ruston, LA Tech Univ.
- **Gas** N: Mobil*
  S: Chevron (24 Hrs), Citgo[CW] (24 Hrs), Exxon*, Texaco*[D]
- **Food** S: Dairy Queen, Johnny's Pizza, Pizza Inn, Subway, TCBY
- **Lodg** S: AAA Super 8 Motel
- **AServ** S: Chevron
- **ATM** S: Citgo
- **Other** N: **Acorn Creek Antiques & Old Books**
  S: **Louisiana Tech University**

85 U.S. 167, Ruston, Dubach
- **Gas** N: Chevron*[CW] (24 Hrs), Citgo*[CW], Shell*[CW]
  S: Phillips 66[D], Texaco*[D]
- **Food** N: Baskin Robbins, Burger King (Playground), Captain D's Seafood, Huddle House, Maxwell's (Holiday Inn), McDonalds (Playground), Peking, Pizza Inn, Shoney's, Subway (Citgo), Wendy's
  S: Pizza Hut
- **Lodg** N: DAYS INN AAA **Days Inn**, Hampton Inn, ◆ Holiday Inn
  S: AAA Best Western
- **AServ** N: ACME Glass & Mirror
  S: Snappy Lube, Taylor's Tire
- **Med** S: ✚ **Hospital**
- **ATM** N: Bank One, CentralBank, Community Trust Bank, **Super 1 Foods**
- **Other** N: **Super 1 Foods (Pharmacy)**
  S: **Lincoln Parish Museum & Historical Society**

86 LA 33, Ruston, Farmerville
- **Gas** N: Mini Mart*, Shell*[CW]
  S: Texaco* (24 Hrs)
- **Food** N: Cowboy's BAR-B-Q, Log Cabin Smokehouse
- **Lodg** N: AAA Comfort Inn, Ramada Limited
- **AServ** N: Lyndsey's Auto Care Inc., Ruston Chrysler Plymouth Dodge Jeep Eagle, Steve Graves Pont, Cadillac, Chev, Geo, Xpress Lube
- **RVCamp** N: **Tri-Lake Marine & RV**
- **ATM** N: Shell
- **Other** N: **All Creatures Animal Hospital**, Do-It Center Car Wash, **Pack & Mail**
  S: **The Soap Opera Coin Laundry**

93 LA 145, Choudrant, Sibley
- **FStop** N: **Texaco***
- **Gas** S: Chevron*[CW]
- **TServ** N: **Texaco**
- **Other** N: O'Nealgas[LP]

(95) **Rest Area - RR, Phones, Picnic, Pet Rest, RV Dump (Eastbound)**

EXIT LOUISIANA

(97) **Rest Area - RR, Phones, Picnic, Pet Rest, RV Dump (Westbound)**

101 Calhoun, Downsville, LA 151
- **TStop** S: **Shell***
- **Gas** N: Texaco[D]
- **Food** S: **Royal Flush (Shell TStop)**
- **TServ** S: **Ogden's Diesel & Tire Repair**
- **Other** S: **Coastal Truck Driver Training College**

103 U.S. 80, Calhoun
- **TStop** N: **USA Truck Stop Chevron***[SCALES]
- **Gas** N: Buckshot's CB Shop, Shell*
- **Food** N: **Restaurant (Chevron TStop)**
- **Lodg** N: Avant West Motel
- **TServ** N: **Ogden's I-20 Tire Repair**
  S: **24 Hr. Truck Repair, Coleman's Truck City**

EXIT LOUISIANA

- **TWash** N: **Calhoun Super Truck Wash**
- **ATM** N: Chevron Truck Stop
- **Other** N: **103 Calhoun CB Shop**

107 Camp Road, Ouachita Parish 25

108 LA 546, Cheniere, To U.S. 80
- **Gas** N: Texaco*
- **RVCamp** N: **Carters Camping Center Sales & Service**

112 Well Road
- **Gas** N: Chevron*, Shell*[CW], Texaco*[D][CW] (24 Hrs)
- **Food** N: Chester Fried Chicken (Texaco), Dairy Queen, **McDonalds**
- **AServ** N: Big A Auto Parts Claiborne, Ray's Garage & Tire Service, Xpress Lube
- **RVCamp** S: **I-20 Trailor Camp**
- **ATM** N: CentralBank
- **Other** N: **Claiborne IGA, Claiborne Veterinary Clinic & Pet Grooming, Cooper Veterinary Hospital, Home & Ranch Hardware**

114 LA 617, Thomas Road
- **FStop** S: **Exxon, Texaco***
- **Gas** N: Racetrac* (24 Hrs), Texaco*
  S: Citgo*[CW]
- **Food** N: Bonanza Steak House, Chick-fil-A, **Chile Verde Mexican Bar & Grill (Super 8 Motel)**, KFC, **McDonalds (Wal-Mart Supercenter)**, McDonalds (Playground), Pizza Q, Popeye's Chicken, Shoney's, Subway, TCBY, Taco Bell, Waffle House
  S: Baskin Robbins (Citgo), Chili's, **Cracker Barrel**, Donut Shop, Kettle, Lone Star Steakhouse, Outback Steakhouse, Peking Chinese, Sonic Drive In, Waffle House, Western Sizzlin'
- **Lodg** N: **Shoney's Inn**, ◆ **Super 8 Motel**
  S: Best Western, Holiday Inn Express, ◆ Red Roof Inn
- **AServ** N: **Wal-Mart Supercenter**, Xpress Lube
  S: **K-Mart**, Roger Pinton Tire & Auto Center
- **Med** N: ✚ **Glenwood Regional Medical Center**
- **ATM** N: Hibernia Bank, **Wal-Mart Supercenter**
  S: **First Republic Bank (Exxon), K-Mart**
- **Other** N: **Ouachita Public Library, Wal-Mart Supercenter (24 Hrs)**
  S: **K-Mart (24 Hrs, Pharmacy)**

115 LA 34, Stella St, Mill St
- **Lodg** E: Holiday Inn (see our ad this page)

116B Bus. US 165, LA 15, to Civic Center, Jackson Street (Eastbound)
- **Food** S: Popeye's Chicken
- **Med** S: ✚ **St. Francis Medical Center**
- **ATM** S: Bank One
- **Other** S: **Monroe Medical Clinic (Pharmacy)**

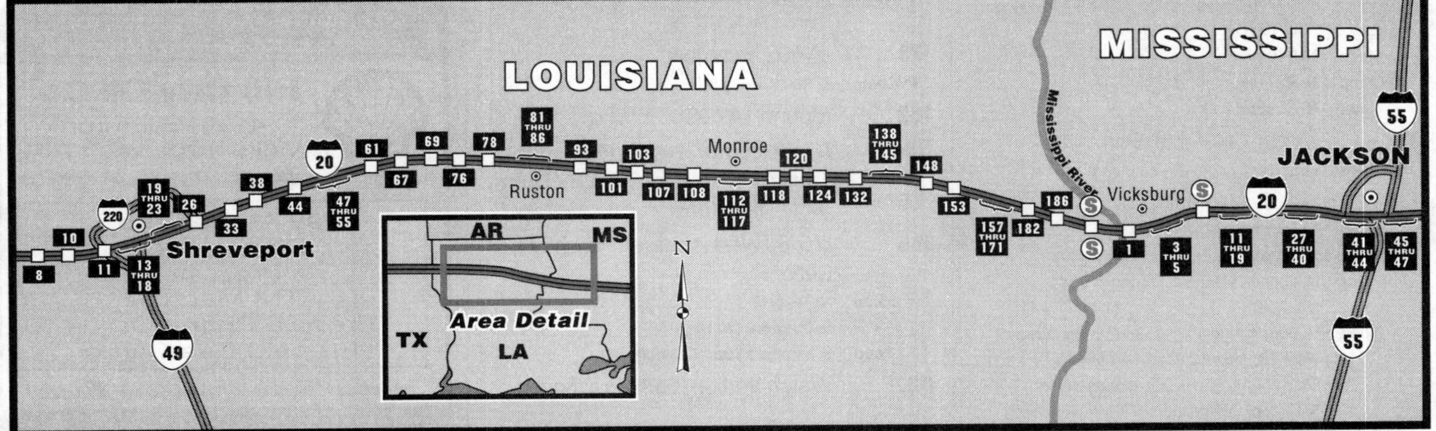

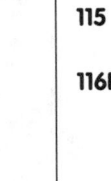

EXIT LOUISIANA

116A Fifth St
- **Gas** N: Circle K*
- **AServ** N: Hank Wallace High Performance Exhaust, Welker Auto Inc. Diagnostics
- **Other** N: US Post Office

117A Hall St (Eastbound)
- **Food** S: Popeyes Chicken
- **Med** S: ✚ St. Francis Medical Center
- **Other** S: Monroe Medical Clinic (Pharmacy)

117D South 2nd St, Layton Ave (Westbound)
- **Food** N: Popeye's Chicken, Restaurant (Guesthouse Inn)
- **Lodg** N: Guesthouse Inn
- **Other** N: Louisiana Purchase Museum & Zoo

117C Bus. US 165, LA 15, Civic Center (Westbound)
- **Food** N: Restaurant (Guesthouse Inn)
- **Lodg** N: Guesthouse Inn
- **Med** N: ✚ St. Francis Medical Center
- **Other** N: Monroe Civic Center, Police Dept

117B LA 594, Texas Ave
- **Food** S: McCoy's BAR-B-QUE
- **Other** S: Corner Grocery

118AB U.S. 165, Bastrop, Columbia (118B - Northeast Louisiana Univ.)
- **Gas** S: Exxon*, Shell*(CW)
- **Food** N: Bobbisox (Holiday Inn), Kettle (La Quinta)
  S: Haskell's Donuts, McDonalds, Podnuh's BBQ
- **Lodg** N: Holiday Inn, La Quinta Inn, Stratford House Inns
  S: Hampton Inn, Motel 6, Ramada Limited
- **AServ** N: Big A Auto Parts, Cooper Jeep Eagle Buick GMC, Interstate Dodge Dodge Trucks
- **TServ** N: Consolidated Truck Parts Inc., Monroe Mack Sales Louisiana Kenworth, Scott Truck & Tractor
- **Other** N: United Parcel Service
  S: Tourist Information

120 Garrett Rd, Monroe Airport
- **FStop** N: Citgo Travel Plaza*
- **Gas** N: Shell*
  S: Exxon*(D)
- **Lodg** S: Best Western, DAYS INN AAA Days Inn
- **AServ** N: Sears Auto Center
- **RVCamp** S: Monroe's Shiloh Camp Resort (1.5 Miles), Pecanland RV Park
- **ATM** N: Citgo Travel Plaza
- **Other** N: Pecanland Mall
  S: Sam's Club

124 LA 594, R Sage Wildlife Area, Millhaven
- **FStop** N: E-Z Mart*
- **RVCamp** N: RV Park
- **Other** N: LA State Police

132 LA 133, Columbia, Start
- **Gas** N: Exxon*

138 LA 137, Archibald, Rayville
- **TStop** N: Pilot Travel Center*(SCALES)
- **FStop** S: Citgo*
- **Gas** N: Cannon's Handee Korner*(LP)
  S: Exxon*
- **Food** N: Burger King (Playground), Dairy Queen (Pilot TS), Johnny's Pizza, McDonalds (Playground), Sports Grill, Wendy's (Pilot TS)
  S: Cottonland Restaurant (Cottonland Inn), Popeye's Chicken, Waffle House
- **Lodg** N: DAYS INN Days Inn
  S: AAA Cottonland Inn
- **AServ** N: Classic Autoplex Chev, Pont, Olds, Buick, Geo, Express Lube
  S: R.C. Tires Inc.
- **TServ** N: Lee's Truck Service (24 Hrs)
- **RVCamp** S: Cottonland Campground (Cottonland Inn)
- **Med** N: ✚ Richardson Medical Center
- **ATM** N: Republic Bank, Richland State Bank
  S: Exxon
- **Other** N: Car Wash, Kaye's Food Market, Wal-Mart

141 LA 583, Bee Bayou Road
- **FStop** N: Texaco* (C C's CB Shop)
- **TServ** N: Atkins Tire Repair, Bee Bayou Truck Service (Texaco, 24 Hrs)

145 LA 183, Richland Parish 202, Holly Ridge

148 LA 609, Dunn

(150) **Rest Area - RR, Picnic, Pet Rest, Phones, RV Dump (Both Directions)**

153 LA 17, Winnsboro, Delhi
- **TStop** N: Jubilee Truck Stop*(CW) (24 Hrs, Showers, Laundry, Mechanic On Call 24 Hrs)
- **Gas** N: Chevron (Jubilee Truck Stop), Exxon(D)(LP), Texaco
  S: Texaco*
- **Food** N: Dairy Queen, Pizza Hut Dine-In, Carry-Out, Subway, Taco Bell (Jubilee Truck Stop)
  S: Restaurant (Best Western)
- **Lodg** S: AAA Best Western
- **AServ** N: Exxon, Texaco
- **TServ** N: Interstate Engine (24 Hr Service)
- **TWash** N: Jubilee Truck Stop
- **Other** N: Brookshire's Grocery, Freds Drug Store, Poverty Point State Comm. (18 Miles), Singley's Pharmacy

157 LA 577, Waverly
- **TStop** N: Tiger Truck Stop* (24 Hrs)
- **Food** N: Great Cajun Cookin'

171 U.S. 65, Vidalia, Tallulah, Newellton
- **TStop** S: LA I-20 East Travel Center*(SCALES) (Mobil Fuel, 76 A/T S), Phillips 66(SCALES)
- **FStop** N: Shell*
- **Gas** N: Citgo*
  S: Texaco*
- **Food** N: McDonalds (Playground), Wendy's
  S: Brushy Bayou (Phillips 66 TS), Louisianne Restaurant (LA I-20 East Travel Center)
- **Lodg** N: DAYS INN ◆ Days Inn, Super 8 Motel
- **TServ** S: Bridgestone Tire & Auto (LA I-20 East Travel Center), Phillips 66 TS

173 LA 602, Richmond
- **RVCamp** S: Roundaway RV Park

182 Mound, LA 602

(184) **LA Welcome Center - RR, Phones, Tourist Info, Picnic, Grills, RV Dump (Both Directions)**

186 U.S. 80 West, Madison Parish 193, Delta
- **FStop** N: Texaco*
  S: Chevron* (24 Hrs)
- **Food** S: Blimpie's Subs (Chevron)

(187) **Weigh Station (Both Directions)**

EXIT LOUISIANA/MISSISSIPPI

↑LOUISIANA

↓MISSISSIPPI

1A Washington St, Warrenton Rd, MS Welcome Center
- **Gas** N: Exxon*(D), Shell*(D)(LP)(see our ad this page)
- **Food** N: Delta Point River Restaurant, Goldie's Trail BAR-B-QUE
  S: Waffle House
- **Lodg** S: Ramada Limited, Ridgeland Inn
- **RVCamp** N: Isle of Capri RV Park
- **ATM** N: Exxon
- **Parks** N: Navy Circle & Grant's Canal (2 Miles)
- **Other** N: Kar Kleen Car Wash, Mississippi Welcome Center - RR, Phones, Dog Trail, Rainbow Casino (see our ad this page)
  S: Gray Oaks Tour Home

1B U.S. 61 South, Natchez (Westbound Left Exit, Limited Access Highway)

1C Halls Ferry Rd.
- **Gas** N: Chevron* (24 Hrs), Exxon*(D)(LP)
  S: Fast Lane*(D)(CW), Shell*(D)(CW)
- **Food** N: Burger King
  S: Baskin Robbins (Playground), Captain D's Seafood, China Doll Restaurant, Goldies Express BAR-B-QUE, Hardee's, Little Caesars Pizza, Popeye's Chicken, Restaurant (Kroger), Shipley's Do-Nuts, Shoney's, TCBY, Taco Bell, Taco Casa, Wendy's
- **Lodg** N: Guesthouse Inn, Hampton Inn
  S: DAYS INN Days Inn, Econolodge, ◆ Fairfield Inn by Marriott (see our ad this page)
- **AServ** S: Halls Ferry Auto Supply, Pennzoil Lube, Saxton's Tire Barn, Xpress Lube
- **Med** N: ✚ Columbia Vicksburg Medical Center, ✚

Bold red print shows RV & Bus parking available or nearby

EXIT MISSISSIPPI

Hospital
- **ATM** S: Deposit Guaranty, Fast Lane, Merchants Bank, Trustmark National Bank
- **Other** N: **Big Wheelie Roller Skates, Durst Discount Drugs**
  S: **FabraCare, K-Mart (Pharmacy)**, Kar-Kleen Car Wash, **Kroger Supermarket (Pharmacy, 24 Hrs), SAVE-A-CENTER Food Market, The Book Rack, US Post Office**

3 Indiana Ave
- **Gas** N: Texaco*(D)(CW)
  S: Chevron*(D)(CW), Shell*
- **Food** N: Blimpie's Subs (Texaco), Krystal, McDonalds (Playground), Pizza Hut Delivery & Carry Out, Sun Garden Restaurant, Taco Casa, Waffle House
  S: Heavenly Ham, KFC, Ponderosa (Steak House)
- **Lodg** N: Best Western
  S: (AAA) Quality Inn
- **AServ** N: Atwood Chevrolet, Olds, Geo, Delta Ford, George Carr Buick Pontiac Cadillac GMC Trucks Subaru
- **ATM** N: Texaco, Trustmark Bank
  S: Shell
- **Other** N: **Battlefield Discount Drugs, Del Champs Superstore, Eyeline Optical, K & B Drugs**
  S: **Sears**

4A East Clay St. (Westbound, Same Services Of Exit 4B)

4B West Clay Street, Downtown Vicksburg
- **Gas** N: Chevron* (24 Hrs)
  S: Texaco*
- **Food** N: Garnett's (Holiday Inn), Restaurant (Park Inn Intern.), Waffle House
  S: **Cracker Barrel**, Pizza Inn, The Dock Seafood
- **Lodg** N: Hampton Inn, Holiday Inn, (AAA) Park Inn

EXIT MISSISSIPPI

International, ◆ Super 8 Motel
  S: Comfort Inn, **Scottish Inns**
- **AServ** S: Rivertown Toyota Jeep Eagle Lincoln Mercury
- **RVCamp** N: **Battlefield Kampground (Laundry, Store, Pool, Hookups)**
- **Parks** N: **Vicksburg National Military Park**
- **Other** N: **Vicksburg Convention & Visitors Bureau**
  S: **Sun-Up Laundry, Vicksburg Factory Outlets**

5AB U.S. 61 North, MS 27 South, Rolling Fork, Utica, Yazoo City (5A Eastbound Is Left Exit)
- **FStop** N: **Shell***
- **Gas** N: Conoco*(D), Fina*
  S: Texaco*(CW)
- **Food** S: Beechwood Restaurant (Beechwood Inn), Blimpie's Subs(D) (Texaco), Maxwell's Restaurant & Lounge, **Rowdy's Restaurant**, Taco Bell (Texaco)
- **Lodg** S: Beechwood Inn, Hillcrest Inn
- **AServ** N: Conoco
- **ATM** N: Shell, Trustmark National Bank
  S: Texaco
- **Other** N: **ACME Hardware, Vicksburg Factory (see our ad this page)**

(10) **Weigh Station (Westbound)**

11 Bovina
- **FStop** N: **Texaco*(D)(LP) (RV Dump)**
- **Food** S: Restaurant
- **AServ** S: Import Automotive Specialist
- **RVCamp** N: **Clear Creek Campground RV Park (Laundromat)**
- **Other** N: **Volunteer Fire Dept**

15 Flowers
- **Other** N: **Truck Parking Area**
  S: **Truck Parking Area**

EXIT MISSISSIPPI

19 MS 22, Flora, Edwards
- **FStop** S: **Conoco***
- **Gas** S: Texaco*
- **Food** S: Dairy Queen (Texaco), Stuckey's (Texaco)
- **Lodg** S: Relax Inn
- **RVCamp** N: **Askew's Landing Campground (2.5 Miles)**
- **Other** S: **Cactus Plantation (1.5 Miles), Petrified Forest, US Post Office**

27 Bolton
- **Gas** N: Chevron*

34 Natchez Trace Parkway (No Commercial Vehicles)

35 Clinton - Raymond Road
- **TStop** S: **Conoco**
- **Gas** N: Chevron*(LP), Phillips 66*(D)(CW)
- **Food** S: **Restaurant (Conoco TS)**
- **AServ** N: Chevron
- **TServ** S: **Conoco**
- **Other** N: **Mississippi College**

36 Springridge Road (Mississippi College)
- **Gas** N: BP*, Exxon*, Shell*(LP)
  S: Texaco*(D)(CW)
- **Food** N: Burger King (BP), Captain D's Seafood, China Garden Restaurant, Dragon Palace, J B's BBQ, McDonalds (Playground), Papa John's Pizza, Restaurant (Clinton Inn), Shipley Do-nuts, Smoothie King, Subway, USA Bumpers Drive-In, Waffle House, Wendy's
  S: Blimpie's Subs (Texaco), Popeye's Chicken, Shoney's
- **Lodg** N: Clinton Inn, (DAYS INN) Days Inn (see our ad this page)
  S: ◆ Comfort Inn, Holiday Inn Express
- **AServ** N: Pennzoil Lube, **Wal-Mart**
- **RVCamp** S: **Spring Ridge**
- **ATM** N: BP, Bank of Miss., Kroger Supermarket, Merchant & Planters Bank
- **Other** N: **Kroger Supermarket, Mississippi College**, Quality Car Wash, **Wal-Mart (Pharmacy, 24 Hrs)**
  S: **Animal Hospital of Clinton, Book Rack, United Artist Theaters**

40AB MS 18 West, Raymond, Robinson Road
- **Gas** N: Phillips 66, Spur*(D)
- **Food** N: Mazzio's Pizza
  S: Waffle House
- **Lodg** S: Comfort Inn
- **Med** S: ✚ **Hospital**
- **ATM** S: Trustmark Bank
- **Other** N: **The Home Depot(LP)**

41 I-220, U.S. 49, N. Jackson, Yazoo City (Limited Access Hwy)

42AB Ellis Ave (Jackson State Univ., Jackson Zoo)
- **Gas** N: BP*(D)(CW), Shell*(D)(CW)
  S: Amoco*(D), Conoco*(D)(CW), Exxon*(D), Pump & Save Gas
- **Food** N: Burger King, Captain D's Seafood, Country Fisherman Buffet, Denny's, Hunan Garden, Kai Hwang Chinese, Little Caesars Pizza, McDonalds (Playground), Ponderosa (Steak House), Popeye's Chicken, Rally's Hamburgers, Sonny's Real Pit Bar-B-Q, Wendy's, Western Sizzlin'
  S: Dairy Queen, Pizza Hut
- **Lodg** N: Best Western, (DAYS INN) Days Inn, Econolodge, Holiday Motel, Ramada Inn, Red Carpet Inn, Scottish Inns, Sleep Inn, Stonewall Jackson Motor Lodge, Holiday Inn (see our ad this page)
- **AServ** N: Auto Zone Auto Parts, Automuff, Carquest Auto Parts, Coleman Taylor Transmission, Howard Wilson Chrysler Plymouth, Master Care Auto Service, Midas Mufflers & Brakes, Mr.

EXIT MISSISSIPPI

Transmission, Peach Auto Painting & Collision, Tire Zone, Wilson Dodge
S: Amoco

Med **S:** ✚ **Hospital**

Other **N:** **MediSave Drug Center, U-Haul Center**(LP)
S: **Eyeline Optical Discount Eyewear, Joe's Super Discount Drugs, Sac & Save Food & Drug (Pharmacy), Wal-Mart**

43AB Terry Road

Gas **N:** A & M Food Mart*, Citgo*, Shell*, Texaco(D)

Food **N:** George's Cafeteria, Kim's Seafood Restaurant, Krystal, Tai Hong Restaurant

Lodg **N:** Tarrymore Motel
S: **La Quinta Inn**

AServ **N:** Capitol Body Shop, Delta Muffler & Exhaust, Odell's Carburetor, S & S Auto Parts, Southern Auto Supply, Texaco
S: Dixie Glass

RVCamp **N:** **S & S Apache Camping Center Sales & Service**

ATM **N:** Deposit Guaranty

Other **N:** **Mooreco Drugs, Powell Animal Clinic**

44 Junction I-55 South, McComb, New Orleans (Westbound Is A Left Exit)

45A Gallatin St.

Gas **N:** Chevron*, Dixie Gas*, S & S Mini Service, Texaco*

Lodg **N:** Crossroads Inn

AServ **N:** Anglin Tire Co, Cowboys Used Tires, Gentry's Body Shop, Smith Automatic Transmission, Xpress Lube
S: Regency Toyota, Nissan, Mitsubishi

Other **N:** **Camping, Duncan's Discount Marine, Jackson Animal Clinic**
S: **Animal Health Products, Magnolia Animal Hospital**

45B US 51N, State St

AServ **N:** AAMCO Transmission, Fowler Buick, Fowler GMC Truck, Freeman Auto Service, Paul Moak Pontiac, Volvo, Honda

46 Junction I-55 North, Grenada, Memphis

47AB U.S. 49 South, Hattiesburg, U.S. 80 Flowood

TStop **N:** **Flying J Travel Plaza***(LP)(SCALES) **(RV Dump, FedEx Drop)**

Food **N:** **Country Market (Flying J TS)**, Western Sizzlin'

Lodg **N:** Days Inn (see our ad this page)

ATM **N:** **Flying J Travel Plaza**

48 MS 468, Pearl

FStop **N:** **Speedway***(LP)
S: **Chevron* (24 Hrs)**

Gas **N:** BP*(CW), Conoco*(LP), Shell*(CW)
S: Texaco*(D)

Food **N:** Arby's, Burger King, **Cracker Barrel,** Domino's Pizza, El Charro Mexican, Frisco Deli, McDonalds (Playground), Popeyes Chicken, Schlotzkys Deli, Shoney's, Waffle House
S: Pearl Cafe

Lodg **N:** AAA Best Western, ◆ Comfort Inn, **Econolodge**, Holiday Inn Express

AServ **N:** E-Z Lube(CW), High Tech Transmission

ATM **N:** Conoco, Deposit Guaranty National Bank

Other **N:** **Discount Drugs, Fire Station**
S: **Pearson Road Laundromat**

52 MS 475, Jackson Int'l Airport, Whitfield (Central Miss. Correctional Facility)

Gas **N:** Exxon*(D)(LP)

Food **N:** Waffle House
S: Taylor's Fish House

TServ **N:** **Peterbilt Dealer**

54 Crossgates Blvd, Greenfield Road

Gas **N:** Amoco*(LP), BP*(CW), Exxon*

Food **N:** Domino's Pizza, KFC, Kismet's, Papa John's Pizza, Pizza Hut, Tropical Sno, Waffle House, Wendy's

Lodg **N:** Ridgeland Inn

AServ **N:** Crossgates Tire & Auto, Goodyear Tire & Auto, Gray-Daniels Ford, Jet-Lube, Van-Trow Oldsmobile

Med **N:** ✚ **Rankin Medical Center**

ATM **N:** **Kroger Supermarket**

Other **N:** **Kroger Supermarket (Pharmacy)**

56 U.S. 80, Downtown Brandon

Gas **S:** Exxon*, Texaco*(D)(CW)(LP)

Food **N:** McDonalds (Playground), Taco Bell
S: Blimpie's Subs (Texaco), Brandon Seafood, Dairy Queen (Playground), Sonic Drive In

Lodg **S:** Days Inn

AServ **N:** Auto Zone Auto Parts, Brandon Radiator Supply, Ozene Cumberland Body Shop
S: Brandon Discount Tire, Delta Muffler & Brake, Texaco Express Lube

ATM **S:** Texaco

Other **N:** **Critters Pet Grooming & Supplies**

59 U.S. 80, East Brandon

68 MS 43, Pelahatchie, Puckett

FStop **N:** **Chevron* (24 Hrs), Conoco***(CW)(LP) **(RV Dump, 24 Hrs)**

S: BP*(LP)

Gas **S:** Ol' Yeller*

Food **N:** **Stuckey's Express (Chevron), Subway (Chevron)**

TWash **N:** **Conoco**

RVCamp **N:** **Camping**

ATM **N:** **Chevron**

(75) **Rest Area - RR, Phones, Vending, RV Dump P (Westbound)**

Other **N:** **RV Dumping Station**

77 MS 13, Morton, Puckett

TStop **N:** **Phillips 66 (CB Shop, 24 Hrs)**

Gas **N:** Phillips Farm & Garden*(D)

Food **N:** **Restaurant (24 Hrs, Phillips 66 TS)**, Southern BBQ & Stuff (Phillips Farm & Garden)

TServ **N:** **77 Truck Repair Garage (24 Hr Towing)**

RVCamp **S:** **Cooper Lake Camping (5 Miles)**

Parks **N:** **Roosevelt State Park**

Other **N:** **National Forest Trail**

80 MS 481, Morton, Raleigh

88 MS 35, Forest, Raleigh

FStop **N:** **Amoco***
S: **Chevron* (24 Hrs)**

Gas **N:** BP*(D)

Food **N:** Restaurant (Best Western), Wendy's
S: **Santa Fe Steak House, Stuckey's Express (Chevron)**

Lodg **N:** AAA Best Western, ◆ Comfort Inn, **AAA Days Inn (see our ad this page)**

Med **N:** ✚ **Hospital**

ATM **S:** **Chevron**

Other **N:** **Marathon Lake, Shell* (No Gas Pumps)**
S: **Bienville National Forest Info Center**

(91) **Rest Area - RR, Phones, RV Dump, Picnic, Grills, Vending, Pet P (Eastbound)**

96 Lake

100 U.S. 80, Lake, Lawrence

TStop **N:** **BP* (Showers, 24 Hrs)**

ATM **N:** BP

109 MS 15, Newton, Philadelphia, Decatur, Union (East Central Community College)

FStop **N:** **Texaco***
S: **Citgo* (No Pump Handles), Conoco***(CW)(LP)

Gas **S:** Chevron*(D) (24 Hrs)

Food **N:** **BoRo Family Restaurant**, Wendy's (Texaco)
S: **Hardee's**, McDonalds (Playground)

Lodg **N:** Thrifty Inn
S: AAA Days Inn

ATM **N:** **Texaco**

115 MS 503, Hickory, Decatur

121 Chunky

Food **N:** Chunky Junction Seafood (Thurs.-Sun.)

129 U.S. 80 West, Lost Gap, Meehan Junction

Gas **S:** Spaceway*

(130) Unnumbered Exit; I-59 South, Laurel

Note: I-20 runs concurrent below with I-59. Numbering follows I-59.

150 South U.S. 11, North MS 19, Philad. Meridian Airport

TStop **W:** **Phillips 66***(LP) **(Restaurant)**

FStop **E:** **Amoco*, Stuckey's Express***

Bold red print shows RV & Bus parking available or nearby

EXIT		MISSISSIPPI
	Food	E: Subway (Stuckey's Express)
	TServ	W: International, Phillips 66, Tire Centers Inc.
151		49th Avenue, Valley Road, Meridian
152		29th Avenue
	FStop	W: Amoco*
	Gas	W: Chevron*(D)
	Food	E: Restaurant (Royal Inn)
	Lodg	E: Royal Inn Motel
		W: Ramada Limited
	ATM	W: Chevron
153		MS 145 South, 22nd Avenue, Quitman, Downtown Meridian
	Gas	E: Chevron*(D) (24 Hrs), Conoco*(D)(CW)(LP) (RV Dump), Exxon*(D)(LP), Shell
		W: Amoco* (24 Hrs), BP*
	Food	E: Depot (Best Western), Waffle House
		S: Morrison's Cafeteria
		W: Burger King, Captain D's Seafood, El Chico Mexican, Hardee's, KFC, Quincy's Family Steakhouse, Shoney's, Wendy's
	Lodg	E: Astro Motel, Best Western, Budget 8 Motel, ◆ Budgetel Inn, Econolodge, Holiday Inn Express, Motel 6, Sleep Inn
	AServ	E: Nelson Hall Hyundai Chevrolet, Shell
		W: Firestone Tire & Auto, Goodyear Tire & Auto
	ATM	E: Conoco
		W: BP, Citizens National Bank, Trustmark National Bank, Union Planters Bank
	Other	W: Cinema 5, Foodmax Grocery, Fred's, Mall, Medi Save Pharmacy, Optical
154AB		MS 19 South, MS 39 North, De Kalb
	FStop	W: Texaco
	Gas	E: Chevron*(CW), Conoco*(D)(CW)(LP)
		W: Amoco*(D), BP*(D)(CW), Food Shop 7*, Shell*(CW) (24 Hrs)
	Food	E: McDonalds, Popeyes Chicken, Red Lobster, Ryan's Steakhouse, Taco Bell
		W: Applebee's, Backyard Burgers, Cracker Barrel, Denny's (Holiday Inn), Greenbriar (Howard Johnson), Nelva Restaurant, Waffle House, Western Sizzlin'
	Lodg	E: Comfort Inn, Pine Haven Motel
		W: Days Inn, Economy Inn, Hampton Inn, Holiday Inn, Howard Johnson, Relax Inn, Rodeway Inn, Super 8 Motel, Super Inn, Western Motel
	AServ	E: Express Lube
		W: Bill Ethridge Lincoln Mercury Isuzu, Branning Auto Supply, Johnson Dodge, Meridian Honda, Sellers Olds Cadillac, Texaco
	TServ	E: Truckers Supply Co. (Parts)

EXIT		MISSISSIPPI/ALABAMA
		W: Stribling Equipment Empire Truck
	RVCamp	W: Ethridge RV Center
	ATM	E: Chevron, Conoco
		W: Howard Johnson, Shell
	Other	E: K-Mart (Pharmacy), Wal-Mart (Pharmacy)
		W: Old South Antique Mall, U-Haul Center(LP)
157AB		U.S. 45, Macon, Quitman (Limited Access Hwy)
160		Russell
	TStop	E: Amoco*(LP)(SCALES) (CB Shop)
		W: 76 Auto/Truck Plaza*(LP)(SCALES) (24 Hrs, FedEx Drop)
	Food	E: Restaurant (Amoco TS)
		W: Restaurant (76 Auto/Truck Stop)
	TServ	E: Amoco TS
		W: Truck Lube Center (76 TS)
	RVCamp	W: Nanabe Creek Campground (1 Mile)
	ATM	E: Amoco TS
		W: 76 Auto/Truck Plaza
(164)		Rest Area - RR, Phones, RV Dump P (Westbound)
165		Toomsuba
	Gas	N: Shell*, Texaco*(D)
	Food	N: Travla Restaurant (Texaco)
	RVCamp	S: KOA Campground
	Other	N: Coin Car Wash
169		Kewanee
	TStop	S: Kewanee One Stop* (Restaurant)
	FStop	N: Dixie
		S: Red Apple*
	Food	S: Restaurant (Kewanee One Stop)
	Other	S: Speedy's CB Shop
(170)		Weigh Station - both directions

↑MISSISSIPPI

↓ALABAMA

EXIT		ALABAMA
(1)		Welcome Area - RR, Phones, RV Dump (Eastbound)
1		U.S. 80 East, Cuba
	TStop	N: Phillips 66
	Gas	S: Chevron*, Dixie*
8		AL 17, York
	TStop	S: BP*(SCALES)
	Food	S: Restaurant (BP TS)
	Lodg	S: Days Inn
	TServ	S: Dun-Rite Truck Wash (BP)
17		AL 28, Livingston, Boyd
	TStop	S: Noble Truck Stop*(SCALES)
	FStop	S: Texaco*
	Gas	S: Chevron*
	Food	S: Burger King, Janet's Restaurant (Country Cookin'), Rose's Country Kitchen, Royal Waffle King, Subway (Chevron)
	TServ	S: Dickey's 24hr Shop Lube & Repair (Noble)
23		Gainesville, Epes
32		Boligee
	FStop	N: BP* (Restaurant)
	Gas	S: Chevron*
	Food	N: Restaurant (BP)
		S: Subway (Chevron)
(38)		Rest Area - RR, Phones, RV Dump (Eastbound)
(39)		Rest Area - RR, Phones (Westbound)
40		AL 14, Eutaw, Aliceville
	Med	E: ✚ Hospital
	Other	W: Visitor Information
45		Union
	FStop	S: Amoco*
	Food	N: Cotton Patch Restaurant
		S: Hardee's, Southfork Restaurant, Wishbone Restaurant
	Lodg	S: Western Inn
	AServ	S: Auto Value (Parts)
52		U.S. 11, U.S. 43, Knoxville
	FStop	N: SpeedStop*
62		Fosters
71A		AL 695, Moundville (Difficult Reaccess)
	Gas	E: Exxon(D)
	Food	E: Arby's, Lone Star Steakhouse, Outback Steakhouse, Pizza Hut, Wendy's
	Lodg	E: Courtyard by Marriott, Fairfield Inn
	Other	E: Kmart
71B		Junction I-359, AL 69 North, Tuscaloosa (Difficult Reaccess)
73		U.S. 82, McFarland Blvd
	FStop	N: BP*
	Gas	N: Chevron*(D)(CW), Parade*, Phillips 66*, Racetrac*, Shell*(CW)
		S: Texaco*(CW)
	Food	N: Burger King, Captain D's Seafood, Denny's, Ezell's Catfish Cabin, Fortune Garden, Krystal, Long John Silvers, O'Charley's, Shoney's, Waffle House

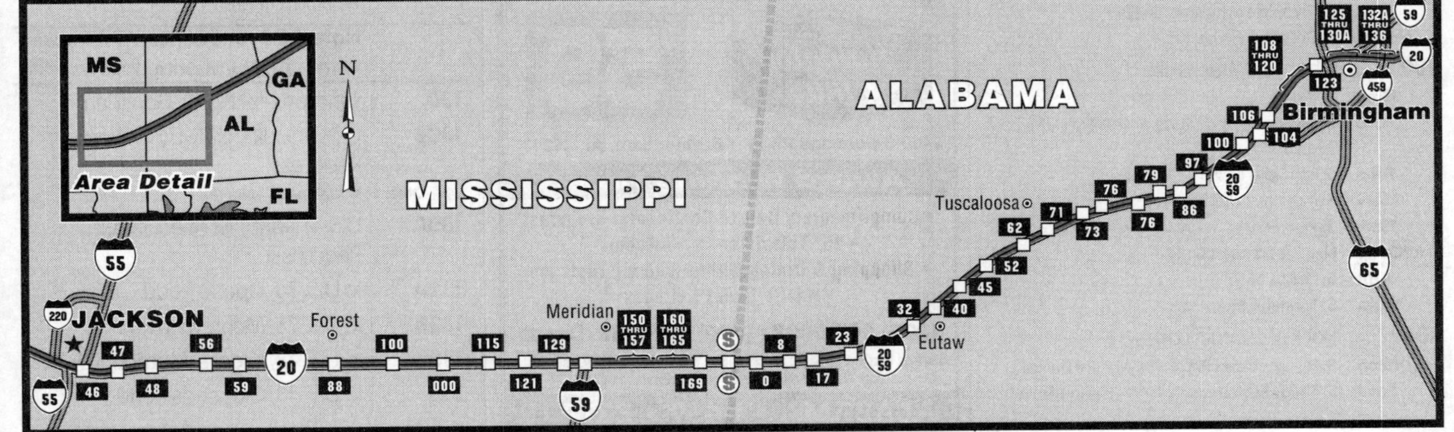

EXIT ALABAMA

S: Chili's, Guthrie's Chicken, Hardee's, Huddle House, KFC, McDonalds, Piccadilly Cafeteria, Pizza Hut, Quincy's Family Steakhouse, Sonic Drive In, Subway, Taco Bell, Taco Casa, Trey Yuen, Wendy's
Lodg N: AAA Best Western, Holiday Inn, Master's Inn
S: Comfort Inn, Days Inn, La Quinta Inn, LaQuinta Inn, Motel 6, Ramada Inn, Super 8 Motel
AServ N: Barkley's Pontiac, GMC, O.K. Tire Service, Postle's Auto Service, Tuscaloosa Auto Parts
S: Carport Auto Service, Townsend Ford, Tuscaloosa Isuzu
ATM S: AM South
Parks S: Lake Lurene State Park
Other S: DelChamps Supermarket, Harkco Super Drug, McFarland Mall

76 U.S. 11, East Tuscaloosa, Cottondale
TStop S: 76 Auto/Truck Plaza(SCALES)
Gas N: Citgo*, Shell*
S: Texaco*(D)(LP)
Food N: Cracker Barrel
S: Baggett's Restaurant (76 Auto/TS)
Lodg N: Knight's Inn
S: Sleep Inn
AServ S: Texaco
TServ S: 76 Auto/Truck Plaza(SCALES)
TWash S: 76 Auto/Truck Plaza
RVCamp N: Sunset II Travel Park

77 Cottondale
TStop N: TravelCenters of America*(SCALES) (RV Dump)
Gas N: BP*(D)(SCALES) (TravelCenters of America)
Food N: Country Pride (TravelCenters of America), McDonalds, Subway (TravelCenters of America), Taco Bell (Truck Stop)
Lodg N: Hampton Inn
AServ S: Troy's Honda Parts
TServ N: TravelCenters of America
TWash N: TravelCenters of America

79 U.S. 11, University Blvd, Coaling

(85) Rest Area - RR, Phones, RV Dump

86 Brookwood, Vance
TStop N: Shell* (Restaurant)
Food N: Restaurant (Shell TS)
TServ N: Shell
RVCamp S: Lakeside RV Park
Other S: Crawford Groceries

89 Mercedes Drive

97 U.S. 11 South, AL 5 South, West Blocton, Centreville
FStop S: Amoco*, Texaco*(LP)
Gas S: Complete Mini Mart*, Shell*(LP)
Food S: Dot's Farmhouse, KFC

100 Abernant, Bucksville
TStop S: Petro
Gas S: Chevron (Petro TStop), Phillips 66* (ATM), Texaco*(LP)
Food S: Iron Skillet (Petro)
AServ S: McKinney Wrecker 24 Hr.
TServ S: Petro(SCALES)
RVCamp N: KOA Campground
ATM S: Phillips 66*(LP)
Parks S: Tannehill State Park

104 Rock Mountain Lakes
FStop S: Flying J Travel Plaza*(LP)(SCALES) (RV Dump)
Food S: Thads Restaurant*(LP)(SCALES) (Flying J Travel

EXIT ALABAMA

Plaza)

106 Junction I-459 North, Gadsden, Atlanta

108 U.S. 11, AL 5, Academy Drive
Gas S: BP*(CW)
Food S: Little Caesars Pizza
AServ S: Tire & Lube Express (Wal-Mart)
Other S: Baskin Robbins (Wal-Mart), Drugs for Less, Wal-Mart (Tire Lube Express & Pharmacy), Winn Dixie Supermarket

112 18th St, 19th St, Bessemer
Gas N: Amoco*
S: Chevron*(CW)
Food N: Jack's Hamburgers
AServ N: O.K. Tire & Battery Co.
S: City Auto Parts
Other N: Mick's True Value Hardware(LP)
S: B&C Rental(LP), FMS Pharmacy

113 18th Avenue, Brighton
FStop S: Speedway*
Food S: McDonalds

115 Jaybird Road, Midfield, Pleasant Grove

118 Valley Road, Fairfield
Lodg S: Villager Lodge
AServ S: Alton Jones Auto Service, Fairfield Transmission, NTB Warehouse, Scogin Bros Auto Repair, Valley Road Auto Parts
Other S: Home Depot(LP), Miles College Historic District

119A Lloyd Nolan Pkwy
Gas N: Amoco*(D), Chevron*
S: Exxon, Texaco*
Food N: Burger King, McDonalds, Mrs Winner's Chicken, Subway, Taco Bell, Wingo's Buffalo Wings
S: Omelet Shoppe
AServ N: CloverLeaf Auto Center
S: B&D Wrecker, Big Moe Spring & Alignment, Exxon, Mr. T's, Northside Auto Service, Sharp Auto Supply
TServ S: B&D Wrecker
Med S: Hospital
Other N: Food Fair Grocery Store, S&J Wash House (Coin Laundry)
S: Dolly Madison Thrift Store, Jim Clay Optician's

119B Avenue I (Westbound, Difficult Reaccess)

EXIT ALABAMA

120 20th St, Ensley Avenue, Alabama State Fair Complex (Eastbound)
Gas N: Crown*(D) (ATM)
S: BP*, Jiffy Mart #4*
Food N: KFC
AServ N: Wayne Gargus GMC (Truck & Auto Service)
S: Jim Burke Chevrolet, Limbaugh Toyota
ATM N: Crown*(D)
Other S: Washing Well Coin Laundry

121 Bush Blvd, Ensley (Southbound)
Gas N: Citgo*, Exxon*(LP), Pure
Food N: Fat Burger
AServ N: Pure

123 U.S. 78, Arkadelphia Road
Gas N: Amoco*, Chevron*, Shell*(CW)
Food N: Charley's Cafe, Denny's, Popeye's Chicken
Lodg N: AAA La Quinta Inn
AServ N: AAMCO Transmission
Med W: Hospital
Other W: Birmingham S. College

124A Junction I-65 South, Montgomery

124B Junction I-65 North, Nashville

125A 17th St, Downtown Birmingham
Gas E: Gem*
Food E: Burger King
Other E: AL School of Fine Arts, Fire Dept.

125B 22nd St, Downtown Birmingham

126A U.S. 31 South, U.S. 280 East, Carraway Blvd
Gas W: Citgo
Food W: Church's Fried Chicken, Ed's Diner, Rally's Hamburgers
AServ W: Citgo

126B 31st St, Sloss Furnaces
Gas W: Circle K*, Texaco*
Food W: McDonalds, Sol's Hotdogs
TServ E: Kurt's Truck Parts, Liberty Truck Sales & Service
Other W: Sani-Clean Laundromat

128 AL 79, Tallapoosa St

129 Airport Blvd (Eastbound, Reaccess Both Directions)
Gas E: Exxon*, Shell*(CW), Texaco*(D)
Food E: Blimpie's Subs (Texaco), Hardee's (24 Hrs), Huddle House, Sammy's Sandwich Shop
Lodg E: ◆ Days Inn, Holiday Inn
W: Ramada Inn
ATM E: Shell
Other E: Woodlawn Mart
W: Airport Car Wash

Note: I-20 runs concurrent above with I-59. Numbering follows I-59.

130 Junction I-59 North, Gadsden

130B U.S. 11 South, 1st Ave (No West Reaccess)
Other N: Colonial Center (RV Parts & Service)

130A U.S. 11 North, 1st Ave (Difficult Reaccess)

132A to U.S. 78, Oporto Road

132B to U.S. 78, Montevallo Road

133 to U.S. 78, Kilgore, Memorial Dr

135 to U.S. 78, Old Leeds Road

EXIT ALABAMA

136 Junction I-459, Montgomery, Tuscaloosa, Gadsden
- **Lodg** S: Hampton Inn (take I-459 to Exit 19 -- see our ad this page)

140 U.S. 78, Leeds
- **Gas** S: Chevron*(CW) (24 Hrs)

144AB U.S. 411, Leeds, Moody
- **FStop** S: **Speedway*(LP)**
- **Gas** N: BP*(CW), Racetrac* (24 Hrs), Shell*
  S: Exxon*(CW)(LP), RaceWay*
- **Food** N: Arby's, Bamboo House Chinese, **Cracker Barrel**, Pizza Hut, Waffle House, Wendy's
  S: Captain D's Seafood, Dairy Queen, Guadalajara Jalisco Mexican, Hardee's, KFC, Little Caesars Pizza, McDonalds (Play Place), Quincy's Family Steakhouse, Taco Bell, Waffle House
- **Lodg** N: Comfort Inn
  S: **Days Inn**
- **AServ** N: Xpress Lube
  S: Advance Auto Parts, Auto Zone Auto Parts
- **RVCamp** N: **Holiday Travel**
- **ATM** N: Regions Bank
  S: **Starvin Marvin (Speedway)**
- **Other** N: **Open Book Bookstore, Winn Dixie Supermarket (Pharmacy)**
  S: **Eckerd Drugs, Showtime Cinema, Wal-Mart (Pharmacy)**

147 Brompton
- **Other** S: **Alabama Outdoors (RV Parts, Sales, & Service)**

152 Cook Springs

153 U.S. 78 West, Chula Vista

156 Eden, Odenville
- **FStop** S: **Exxon***
- **Gas** S: Chevron*(LP) (24 Hrs)

158 U.S. 231, Ashville, Pell City
- **Gas** S: BP*(CW)
- **Food** S: **Burger King, Hardee's,** Waffle House
- **Lodg** S: AAA **Ramada Limited**
- **AServ** S: Pell City Ford Lincoln Mercury
- **RVCamp** S: **Lakeside Landing RV Park & Campground**
- **Med** S: ✚ **Hospital**

162 U.S. 78, Riverside, Pell City
- **Gas** S: Phillips 66*
- **Food** S: Hungry Bear (Best Western), **Pancake House,** Suzy Lou's Open Pit BBQ
- **Lodg** S: AAA Best Western
- **RVCamp** N: **Safe Harbor RV Park**

165 Embry, Cross Roads
- **TStop** S: **I-20 Texaco Truckstop*(SCALES)**
- **FStop** S: **Speedmart***
- **Food** S: **Broiler Room (I-20 TS), Huddle House**
- **Lodg** S: McCaig's Motel
- **TServ** S: **Bobby Orr Tire Shop & Garage**

168 AL. 77, Talladega, Lincoln
- **FStop** N: **Citgo*, Conoco*(LP)**
- **Gas** S: Bill's Gas(LP), Chevron*(D), Texaco*
- **Food** N: **Jack's (Playground)**
  S: Baskin Robbins, Burger King (Playground Inside, Texaco), **Gateway Restaurant (24 Hr), McDonalds (Play Place),** Rippin Ribs (Chevron)
- **Lodg** S: **McCaig Motel**
- **RVCamp** N: **Dogwood Campground**
- **Other** N: **Coin Laundry (Citgo)**

EXIT ALABAMA

173 Eastaboga
- **Gas** S: Shell*, Texaco*
- **Food** S: Dairy Queen* (Texaco), Stuckey's (Texaco)
- **Other** S: **International Motor Sports Hall of Fame, Talladega Superspeedway**

179 Munford, Coldwater
- **Other** N: **Anniston Army Depot**

185 AL 21, Oxford, Anniston
- **FStop** S: **Mapco Express*(SCALES), Texaco***
- **Gas** N: BP*(D)(CW), Chevron(D)(CW) (24 Hrs), Shell*, Texaco*(D)
  S: Exxon*(LP), Racetrac* (24 Hrs), Shell*
- **Food** N: Applebee's, Arby's, Burger King (Playground), Captain D's Seafood, Diamond Dave's Cafe, Domino's Pizza, McDonalds (Play Place), Morrison's Cafeteria, O'Charley's, Pizza Boy, **Quincy's Family Steakhouse,** Red Lobster, Shoney's, Taco Bell, Waffle House, Western Sizzlin'
  S: **Baskin Robbins (Wal-Mart), Baskin Robbins (Texaco),** Chick-fil-A, **Food Outlet of Oxford, McDonalds (Wal-Mart), Subway (Texaco),** Waffle House
- **Lodg** N: Days Inn, AAA Holiday Inn, AAA Howard Johnson, Red Carpet Inn
  S: AAA Comfort Inn, ◆ Econolodge, Hampton Inn, **Motel 6,** Travelodge
- **AServ** N: Firestone Tire & Auto, Texaco
  S: Cobb Automotive
- **TServ** S: **Cobb Automotive Truck Center**
- **Med** N: ✚ **Hospital**
- **ATM** N: AmSouth, Colonial Bank, Compass Bank, Regions Bank
  S: **Colonial Bank (Wal-Mart)**
- **Other** N: **Anniston Museum of Natural History, Book Land, Books-A-Million, Crazy Joe's Fireworks, Gregerson's Supermarket, Martin's Pharmacy, Quintard Mall**
  S: **Jolly Joe's Fireworks, Oxford Discount Drugs, U.S. Post Office, Wal-Mart Supercenter (Vision Center, Pharmacy)**

188 **(189)** to U.S. 78 , Oxford, Anniston
- **FStop** N: **Skinner's Super Stop*(LP)**
- **Gas** N: Texaco*
- **Food** N: **Cracker Barrel,** Lone Star Steakhouse, Wendy's
- **Lodg** N: Jameson Inn
- **AServ** N: **Golden Springs Tire & Service Center**

191 to U.S. 78, U.S. 431, Talladaga Scenic Hwy
- **Parks** S: **Cheaha State Park**

EXIT ALABAMA/GEORGIA

199 AL 9, Heflin, Hollis
- **FStop** N: **Texaco*(D)(LP)**
  S: **Citgo***
- **Gas** S: Chevron*(CW) (24 Hrs)
- **Food** N: **Hardee's, Icecream Churn (Texaco FS), Pop's Char Burgers, Subway* (Texaco), Taco Bell (Texaco FS)**
  S: **Huddle House, Rippin Ribs (Citgo)**
- **Lodg** N: ◆ **Howard Johnson (see our ad this page)**
- **ATM** N: **Texaco FS**

205 AL 46, Ranburne
- **FStop** N: **BP*, Texaco***
- **TServ** N: **BP Truck Tire Service**
- **ATM** N: BP

(208) **Weigh Station (Westbound)**

210 **(211)** Abernathy
- **Other** S: **Stateline Fireworks**

(214) AL Welcome Center - Picnic, RR, HF, Phones, RV Dump (Westbound)
- **Other** N: **Phones, Picnic Tables, RV Site, Rest Rooms**

↑ALABAMA

↓GEORGIA

(1) **GA Welcome Center - Picnic, RR, HF, Phones, RV Dump (Eastbound)**

1 **(5)** GA 100, Tallapoosa, Bowdon
- **TStop** S: **Noble Truck Plaza*(SCALES) (Citgo Gas)**
- **FStop** N: **Phillips 66* (24 Hrs)**
  S: **Big O Tires(D)**
- **Gas** S: BP*
- **Food** N: **Robinsons (Phillips 66 FS)**
  S: Big "O" BBQ, Dairy Queen, Huddle House, **Janet's Country Restaurant (Noble),** Waffle House
- **Lodg** S: Comfort Inn, **Noble Auto/Truck Plaza**
- **TServ** S: **Noble TS**
- **TWash** S: **Big O FS, Noble TS**
- **ATM** N: Phillips 66

2 **(9)** Waco Rd

3 **(12)** U.S. 27, Bremen, Carrollton
- **FStop** N: **Texaco*(LP)**
  S: **BP*(LP)**
- **Gas** S: Amoco*(D)
- **Food** N: Arby's, **McDonalds (Play Place),** Wendy's
  S: Waffle House
- **Lodg** N: ◆ **Days Inn,** Hampton Inn
- **TServ** S: **BP*(LP)**
- **Med** N: ✚ **Hospital**
- **ATM** N: Texaco FS
  S: Amoco
- **Parks** S: **John Tanner State Park**

(15) **Weigh Station (Westbound)**

4 **(19)** GA 113, Temple, Carrollton
- **Food** N: **Hardee's**

5 **(24)** GA.61, GA. 101, Villa Rica, Carrollton
- **FStop** N: **Shell***
- **Gas** N: Amoco*(D)(CW), Chevron*(LP), Racetrac* (24 Hrs)
- **Food** N: Arby's, Big Chick (Amoco), Domino's Pizza, El Torito, Gondolier Pizza, **Hardee's,** KFC, Krystal, McDonalds (Play Place), New China Buffet, Pizza Hut, Subway, Taco Bell
- **Lodg** N: ◆ Comfort Inn, AAA Super 8 Motel
- **Med** N: ✚ **Hospital**

EXIT GEORGIA

ATM N: Carrollton Federal Bank, West Ga. National Bank
Other N: **Ingles Supermarket, Revco Drugs, Winn Dixie Supermarket (24 Hrs, Pharmacy)**

6 **(26)** Liberty Rd, Villa Rica
TStop S: **Leathers Truck Stop*(SCALES)**
Food S: **Restaurant (Leathers Truck Stop)**
TServ S: **Leathers TS**
TWash S: **Leathers TS**

7 **(31)** Post Rd
FStop S: **Texaco***
ATM S: **Texaco FS**
Other N: **U.S. Post Office***

8 **(34)** GA 5, Douglasville
Gas N: Fina*(LP)
S: Chevron*(D) (24 Hrs)
Food N: Huddle House
S: Baskin Robbins, Chili's, Chinese Pagoda, Dunkin' Donuts, El Rodeo Mexican, Folks, KFC, Long John Silvers, McDonalds (Play Place), Miano's Pasta, Monterrey Restaurant, Philly Connection, Pizza Inn, Red Lobster, Ruby Tuesday, Shoney's, Subway, Szechuan Village, Topps Bar & Grill, Waffle House, Wendy's
Lodg N: ◆ Quality Inn, Sleep Inn
AServ N: Willet Honda, Isuzu
S: Chevron, Xpress Lube
ATM S: NationsBank, SunTrust
Other S: **Big B Drugs, Eckerd Drugs, K-Mart, Kroger Supermarket (w/pharmacy), Pearl Vision Center (Optometry), Revco Drug, Wolf Camera**

9 **(36)** Chapel Hill Rd.
Gas S: Amoco*(CW) (24 Hrs), Citgo*, Conoco*(LP), QuikTrip*
Food S: McDonalds (Amoco), Outback Steakhouse
Med N: ✚ **Douglas Hospital**
Other S: **Eckerd Drugs**

10 **(38)** GA 92, Fairburn Rd, Douglasville
Gas N: BP*(CW), Chevron (24 Hrs), Racetrac*, Texaco*(D)(CW)
S: Amoco*, Shell*(D)
Food N: Burger King (Playground), Captain D's Seafood, Chick-fil-A, Cracker Barrel, House Of Ming, KFC, Krystal, Little Caesars Pizza, Martin's, Mazzio's Pizza, Monterrey Mexican, Pagoda Express, Pizza Hut, Shoney's, Taco Bell
Lodg N: **Best Western,** Days Inn, Holiday Inn Express, Ramada
AServ N: Advance Auto Parts, Chevron, Douglas County Dodge, Midas Muffler & Brakes
S: AAMCO Transmission, ABS Automotive, Amoco
Med N: ✚ **Hospital**
ATM N: Regions Bank, Suntrust, Texaco NationsBank
S: Shell
Other N: **Wal-Mart (Pharmacy)**

11 **(42)** Lee Rd., Lithia Springs
FStop N: **Citgo*(LP)**
S: **Speedway*(LP)**
Food N: **Hardee's, Icecream Churn (Citgo FS)**
S: Waffle House
AServ N: Lee Road Transmission
ATM S: **Speedway**

(42) **Weigh Station (Eastbound)**

12 **(43)** GA 6, Thorton Rd, Austell
Gas N: Amoco*(CW) (24 Hrs), Phillips 66*(CW), Racetrac*, Wallace*

EXIT GEORGIA

Food N: Blimpie's Subs, Burger King (Playground), Chick-fil-A, Dunkin Donuts, **Hardee's,** International House of Pancakes, McDonalds (Play Place), Shoney's, Subway, Taco Bell, Waffle House
Lodg N: **Hampton Inn,** Shoney's Inn & Suites
S: Fairfield Inn
AServ N: Thornton Chevrolet, GMC, Mazda
S: Westside Mitsubishi, Toyota
RVCamp N: **Atlanta Camp RV Park**
Med N: ✚ **Parkway Medical Center**
ATM N: Douglas County Bank, Douglas Federal Bank, Racetrac, Suntrust
Parks S: **Sweetwater Creek State Park**
Other N: **Revco Drugs**

(43) **Weigh Station**

13BA Six Flags Dr., Riverside Pkwy.
Gas N: Amoco*(LP), Conoco*, QT*
S: Coastal*
Food N: Church's Fried Chicken (Amoco), Denny's, Waffle House
Lodg N: La Quinta Inn
S: Holiday Inn
RVCamp N: **Arrowhead RV Park**
ATM N: Amoco
Other S: **Sam's Club, Six Flags Over Georgia**

13C **(47)** Riverside Pkwy, Six Flags Pkwy (West Access By Access Rd.)
FStop N: **Amoco***
Gas N: Mark Inn
Lodg N: Mark Inn
S: Days Inn Days Inn
TServ N: **Sunbelt Power Cat. Truck Engine Parts & Service**
RVCamp N: **Arrowhead Campground**
Other S: **Sam's Club, Six Flags Over Georgia**

14 **(49)** GA 70, Fulton Industrial Blvd, Fulton County Airport
FStop N: **Happy Stores***
S: **Mapco Express*, UFO**
Gas N: Racetrac*
S: Amoco*, Chevron(CW) (24 Hrs), Racetrac*, Texaco*
Food N: Baskin Robbins, Captain D's Seafood, Checkers Burgers, Dairy Queen, Dunkin' Donuts, EJ's, Hardee's, Krystal, Mrs. Winner's, Subway, Waffle House, Wendy's
S: Arby's, Blimpie's Subs, Burger King (Playground), Golden Palace, McDonalds, Roma Sub Express, Waffle House, Wency's Cafe

EXIT GEORGIA

Lodg N: Fulton Inn, AAA Masters Inn, Ramada Inn (see our ad this page), Suburban Lodge
S: ◆ Comfort Inn, Executive Inn, Mark Inn, ◆ Red Roof Inn, Super 8 Motel
AServ S: Chevron
TServ S: **Great American Inc. Truck Parts**
ATM N: NationsBank (RaceTrac)
S: Amoco, Mapco Express, Racetrac
Other N: **Fulton County Airport**
S: **U-Haul Center(LP)**

15A **(49)** Junction I-285 South, Macon, Montgomery

15B **(49)** Junction I-285 North, Chattanooga, Greenville

16 **(50)** GA 280, Hightower Rd
Gas S: Exxon*(D)

17 **(51)** GA 139, Martin Luther King Jr. Dr
Gas N: Citgo*, Texaco(D)
S: The Right Stuff*
Food S: Unity Restaurant
AServ N: Citgo, Texaco
S: Holt Auto Parts
Other N: **Coin Laundry**

18 **(52)** Langhorn Street, Cascade Road (Westbound, Reaccess Eastbound)

19 **(52)** Ashby St., West End (Difficult Reaccess)
Gas S: BP*(CW), Exxon*
Food S: Chinese Kitchen, Church's Fried Chicken, Dipper Dan Icecream, Italian Sub & Salad, Krispy Kreme Donuts, Momo's Pizza & Pasta, New York Subs, Pizza Hut, Popeye's Chicken, Taco Bell, West Inn Cafeteria
ATM S: Capitol City Bank, First Union, NationsBank, SouthTrust, Wachovia Bank
Other S: **A & P Supermarket, Hardy's Market, West End Mall, Western Union**

20 **(53)** Lee St. , Ft. McPherson, Atlanta University Center (Westbound)
Other S: **Fed Ex Drop Box**

21 **(54)** U.S. 19, U.S. 29, McDaniel St., Whitehall St. (Eastbound, Reaccess Westbound)

22 **(55)** Windsor St., Spring St., Stadium, Georgia World Congress Center
Food N: Anns Fine Foods

23 **(56)** Junction I-75, I-85

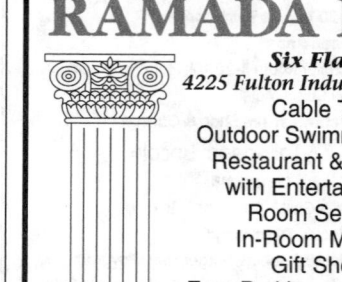

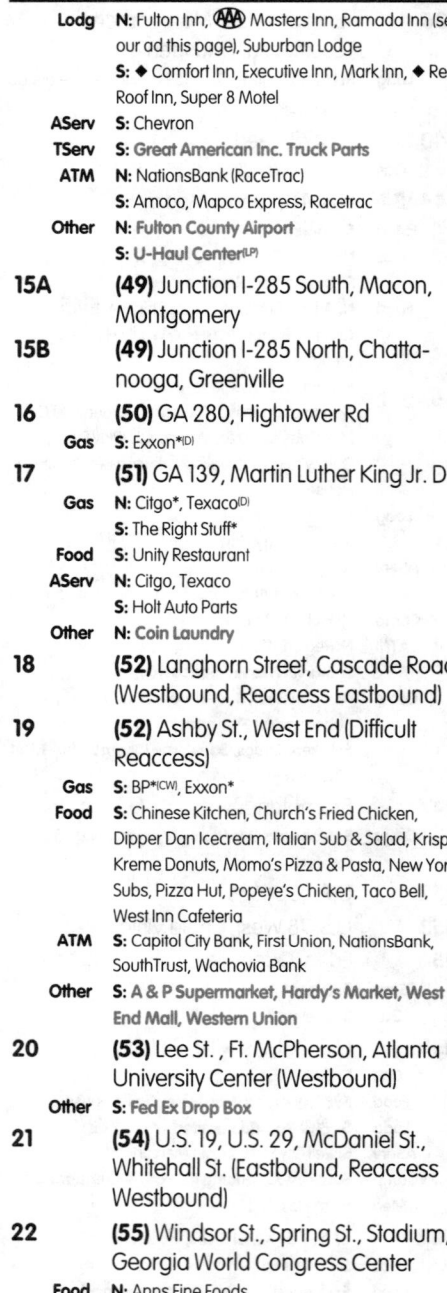

Bold red print shows RV & Bus parking available or nearby

EXIT GEORGIA

24 **(57)** Capitol Ave., Downtown Atlanta (Westbound, Reaccess Eastbound)

25 **(58)** Hill St., Stadium (Westbound)
- **Gas** **N:** Amoco*, Fina, Texaco*
- **Food** **N:** Mrs Winner's Chicken
- **AServ** **N:** Fina

26 **(59)** Boulevard Rd, Zoo Atlanta, Cyclorama
- **Gas** **N:** Chevron*
- **Food** **N:** Blimpie's Subs, Tangie's Hot Wings
- **Other** **N: Martin Luther King National Historic Site**
  S: Cyclorama (see our ad opposite page), Zoo Atlanta (see our ad this page)

27 Memorial Dr., Glenwood Ave.

28AB **(60)** U.S. 23, Moreland Ave.
- **Gas** **N:** Exxon*
  S: Fina*(D), Shell*
- **Food** **S:** Checkers Burgers, KFC, Krystal, Long John Silvers, McDonalds, Mrs Winner's Chicken, Taco Bell, Wendy's
- **Lodg** **N:** Atlanta Motel
- **AServ** **N:** ATZ Tire Service, Gormans Auto Parts, Lo's Garage, Nice People Auto Service, TNT Auto Repair, Tune-Up Clinic, X-Press Lube
  S: Downey's Auto Parts, Moreland Auto Center (BMW other foreign cars)
- **ATM** **S:** NationsBank
- **Other** **S: Coin Laundry**

29 **(61)** Maynard Terrace, Memorial Dr (Westbound Reaccess Only)
- **Gas** **N:** Amoco*(CW), Crown* (Open 24 Hours)
- **Food** **N:** Checkers Burgers, Wyatts Country BBQ
- **Other** **N: Coin Laundry**

30 **(62)** Glenwood Ave., GA 260
- **Gas** **N:** Amoco*
- **Food** **N:** KFC

31 **(63)** Flat Shoals Rd (Reaccess Westbound Only)

32A **(64)** Gresham Rd (Difficult Reaccess)

EXIT GEORGIA

- **Gas** **S:** Amoco*(LP), Citgo*
- **Food** **S:** Church's Fried Chicken, G&G BBQ
- **AServ** **S:** Smith's Auto Repair
- **Med** **S: + Southside Health Care Inc.**
- **Other** **S: Buy Rite Pharmacy, Coin Laundry,** Super Soaper Coin Car Wash

32B **(64)** Flat Shoals Rd, Brannen Rd, Gresham Rd (Westbound)
- **Gas** **S:** Shell*
- **AServ** **S:** Brake & Wheels

33 **(65)** GA 155, Candler Rd, Decatur
- **Gas** **N:** Amoco*, BP*, Hess*
  S: Amoco*(CW), Chevron* (Open 24 Hours), Circle K*, Conoco*(CW), Shell*(D)
- **Food** **N:** Dundee's Cafe, Dynasty Chinese, Long John

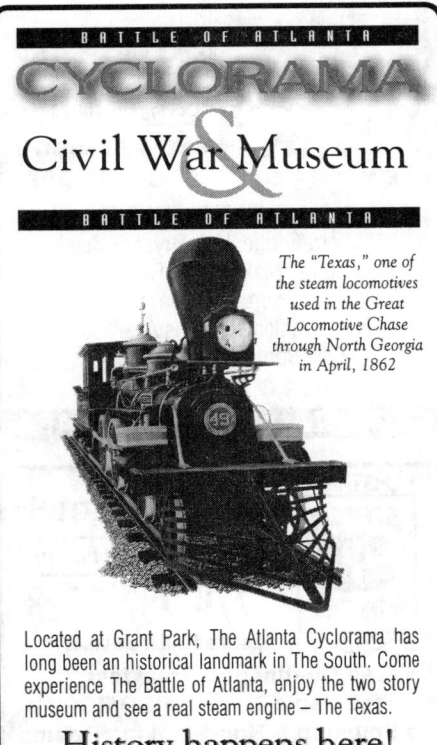

EXIT GEORGIA

Silvers, Pizza Hut, Red Lobster, Wendy's
  S: Arby's, Checkers Burgers, China Cafeteria, Church's Fried Chicken, Dairy Queen, Dunkin Donuts, Home Box, J.R. Crickets, KFC, McDonalds, Taco Bell, W.K. Wings, Waffle King
- **Lodg** **N:** Discover Inn, (AAA) Econolodge, Howard Johnson
  S: Candler Inn, Gulf American Inns, Sun Set Inn
- **AServ** **N:** Tune-Up Clinic
  S: Advance Auto Parts, Firestone Tire & Auto, Mitchell Tire Company, Pep Boys Auto Center, Q-Lube, Rich's Auto Center
- **Med** **S: + Emory Clinic, + Grady Medical Center**
- **ATM** **S:** Circle K, Conoco, First Southern Bank, First Union Bank, SunTrust Bank, Wachovia Bank
- **Other** **N:** Car Wash, **U-Haul Center**
  S: Big B Drugs, Coin Laundry, Kroger Supermarket, Rainbow Coin Laundry, Revco, South Dekalb Mall, U.S. Post Office, Winn Dixie Supermarket

34 **(66)** Columbia Drive (Eastbound, Reaccess Westbound)
- **Gas** **N:** Phillip 66*(LP), Texaco*
  S: Fina*
- **AServ** **N:** Columbia Emissions Inspection
- **ATM** **N:** Texaco
- **Other** **N: New Deal Convient Store**
  S: Coin Laundry

35AB **(67)** Junction I-285, Macon, Greenville

36 **(68)** Wesley Chapel Rd, Snapfinger Rd
- **Gas** **N:** Exxon*(D), Racetrac*
  S: Chevron* (Open 24 Hours), Crown*(CW)(LP), Speedway*(D)(LP)
- **Food** **N:** Baskin Robbins, Blimpie's Subs, Captain D's Seafood, Checkers Burgers, Chick-fil-A, China Cafeteria, Church's Fried Chicken, Happy Dragon Chinese, Hardee's, KFC, Long John Silvers, Martino's Pizza, Popeye's, Shoney's, Subway, Taco Bell, Waffle House, Wendy's
  S: Burger King, Dairy Queen, Dragon Chinese Restaurant, McDonalds, Mr. Philly's, Supreme Fish Delight
- **Lodg** **N:** Motel 6
  S: (DAYS INN) Days Inn, (AAA) Holiday Inn Express
- **AServ** **N:** Q Lube
  S: Jiffy Lube, Precision Tune
- **ATM** **N:** Citizen Trust Bank, NationsBank, Racetrac, SouthTrust Bank, Wachovia Bank

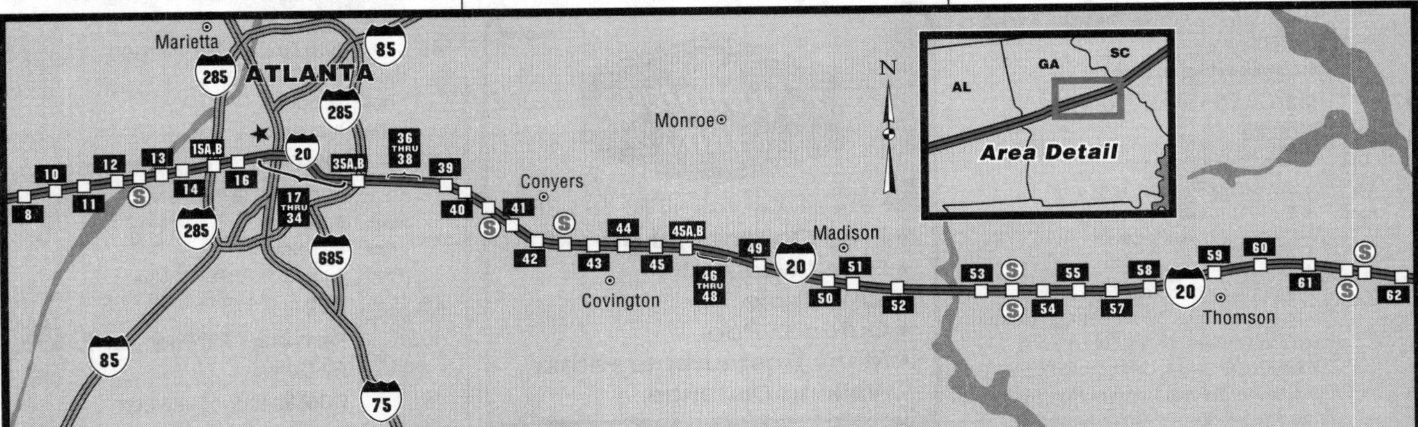

EXIT GEORGIA

S: Crown

Other N: **Big B Drugs, Coin Laundry, Ingles Supermarket, K-Mart (Pharmacy), Kroger Supermarket, Revco, Wal-Mart**
S: **Brandon Animal Clinic, Mr. Quick Copy**

37 Panola Rd
Gas N: QuikTrip*(D) (Open 24 Hours), Texaco*(D)(CW)
S: Exxon, Stop N Shop*
Food N: Brandon's Restaurant, Burger King, Checkers Burgers, McDonalds, Waffle House
S: Domino's Pizza
Lodg N: (AAA) La Quinta Inn, Super 8 Motel (see our ad this page)
S: Sleep Inn
AServ S: Exxon
ATM N: First Southern Bank, Suntrust

38 Evans Mill Rd, Lithonia
Gas N: Amoco*(CW), Chevron*(CW), Texaco*(CW)
S: Speedway*(D)(LP)
Food N: Captain D's Seafood, KFC, Mama's Kitchen, McDonalds, Pizza Hut, Taco Bell, Wendy's
S: Dairy Queen, Golden Palace 2, Jazz Cafe, Krystal, Snuffy's, Waffle House
Other N: **Coin Laundry**, Eastgate Car Wash
S: **Ace Hardware**, Coin Car Wash, **Piggly Wiggly Supermarket, Revco, Western Union**

39 **(73)** GA 124, U.S. 278, Turner Hill Rd
FStop N: **US Discount***
Gas N: Amoco*

40 **(76)** Sigman Rd
FStop N: **Circle K***(LP)
Food N: Waffle House
AServ N: J.R.'s Auto & Tire
Med N: **+ Hospital**
ATM N: Circle K

(76) **Weigh Station, Rest Area - Phones (Eastbound)**

41 **(78)** West Ave., Conyers
Gas N: Shell*(CW), Speedway*(D)(LP), Texaco*(D)(CW)
S: Exxon*
Food N: American Deli, Dairy Queen, Domino's Pizza, Donna Marie's Pizza, Golden Palace Chinese, Mrs Winner's Chicken, Subway
S: Longhorn Steaks, McDonalds, Quik-Chick, Waffle House
Lodg N: Holiday Inn, Rich Field
S: (AAA) Comfort Inn
AServ N: Midas Muffler, Statham Tire Co.
S: Conyers Nissan, Conyers Toyota, Courtesy Ford
ATM N: FNB Bank, NationsBank
Other N: **Coin Laundry, Conyers Pharmacy, Family Foods Grocery**
S: **Conyers Animal Hospital**

42 **(82)** GA 20, GA 138, Conyers, Monroe
Gas N: Amoco*, BP*(D)(CW), Speedway*(D)
S: Chevron*(CW), Racetrac*, Texaco*(D)(LP)
Food N: Brusters Ice Cream, Carrabba's, Crabbydill's, **Cracker Barrel**, IHOP, O'Charley's, Outback Steakhouse
S: Applebee's, Arby's, Bagel Barn, Blimpie's Subs, Boston Market, Burger King, Captain D's Seafood, Checkers Burgers, Chianti Iltalian Restaurant, Chick-fil-A, Chili's, Crown Palace, DJ's Kitchen, Daruma Japanese, Folks, Frontera Tex-Mex Grill, Hardees, Hickory Hams Cafe, Hooters, Huddle House, I Love Sushi, KFC, Krystal, Long John Silvers, Mandarin Gardens, McDonalds, Morrison's, Papa John's Pizza, Philly Connection, Pizza Hut, Popeye's Chicken, Provino's Iltalian Restaurant, Ruby Tuesday, Ryan's Steakhouse, Schlotzkys Deli, Seven Gables Restaurant, Shades, Shoney's, Silver Dragon Chinese Restaurant, Sonny's BBQ, Subway, TCBY Yogurt, Taco Bell, The Italian Oven Restaurant, Waffle House, Waffle Shop, Wendy's, Yenshing Garden Chinese
Lodg N: Conyers Motor Inn, (Days Inn) ◆ Days Inn, Hampton Inn (see our ad this page), (AAA) Ramada, The Jameson Inn
S: Intown Suites, Suburban Lodge
AServ S: Big Ten Tires, Conyers Honda, Good Year, Jiffy Lube, Master Care Auto Service, Pep Boys Auto Center, Precision Tune, Tune-Up Clinic
ATM S: First Union Bank, Main St. Bank, Nations Bank, Racetrac, Regions Bank, South Bank, **Suntrust Bank (Publix)**, Texaco
Other N: Buddy's Bubble Car Wash, **PetsMart, Wal-Mart Supercenter**
S: **A & P Supermarket, Big B Drugs, Cin Star Theaters, Cinema 8, Coin Laundry, Drug Emporium, Eckerd Drugs, K-Mart (Pharmacy), Kroger Supermarket (Pharmacy), Mail Boxes Etc, Patient's Pharamcy, Publix Supermarket (Pharmacy), Quality Food Depot, Scotty's Pet Center, Target Department Store**, Wash Lube, **Wolf Camera Shop**

(82) **Weigh Station, Rest Area - Phones (Westbound)**

43 **(84)** GA 162, Salem Rd
Gas S: Amoco*(CW), BP*, Chevron* (Open 24 Hours), Citgo*, Racetrac*
Food S: Burger King, Hardee's, Highland House Restaurant, La Bravo's, Subway, Waffle House
AServ S: BJ McDaniel Auto Service
ATM S: Chevron, Suntrust
Other S: **Eckerd Drugs, Little Henry's Food Store, Rockdale Animal Hospital, The Medicine Shop, Winn Dixie Supermarket (Pharmacy)**

44 **(88)** Almon Rd, Porterdale
Gas N: Chevron*(LP)
RVCamp S: **Riverside Estate**

45 **(90)** U.S. 278 East, Covington, Oxford
Gas S: Citgo*, Exxon*(CW), Prue(D), Racetrac*, Shell*(D), Texaco*
Food S: Arby's, Burger King, Captain D's Seafood, Checkers Burgers, Chen's Chinese, Donut King, El Arrollo Mexican, KFC, Shoney's, Sports Bar & Grill, Taco Bell, Waffle House
Lodg S: The Crest
AServ S: Advance Auto Parts, Ginn Chevrolet Oldsmobile, Henry's Fast Lube
ATM S: Newton Federal, Racetrac
Other S: **Eckerd Drugs, Ingles Supermarket, K-Mart (& pharmacy), The Soap Opera Laundromat, Winn Dixie Supermarket**

45AB **(92)** Alcovy Rd
FStop N: **Circle K***(LP)
Gas N: Chevron*(LP)
Food N: Crowe Nest & Bar, Pippin's BBQ, Subway (Chevron), Waffle House
Lodg N: **Best Western**, Econolodge, **Holiday Inn (see our ad this page)**
Med S: **+ Hospital**
ATM N: **Circle K**

46 **(93)** GA 142, Hazelbrand
FStop S: **Exxon***(LP)
Gas S: Texaco*(LP)
ATM S: **Exxon**

47 **(98)** GA 11, Monroe, Monticello, Social Circle, Mansfield
FStop S: **Citgo***(LP)
Gas S: Chevron*(LP)
Food S: Log Cabin Restaurant (BBQ)

48 **(101)** U.S. 278

(103) **Rest Area - RR, Phones, HF, Picnic, RV Dump**

49 **(106)** Rutledge, Newborn
Gas N: Chevron*(D)

Bold red print shows RV & Bus parking available or nearby

EXIT GEORGIA

Parks N: Hard Labor Creek State Park

(108) Rest Area - RR, Phones, RV Dump (Westbound)

50 **(113)** GA 83, Madison, Monticello

Gas N: BP*

Med N: ✚ Hospital

Other N: Georgia State Patrol Post

51 **(115)** U.S. 441, GA 129, Rock Eagle, Madison, Eatonton

TStop S: Madison 20 Auto Truck Plaza BP*(LP)(SCALES)

FStop N: Shell Fuel Center*(CW) (Open 24 Hours)
S: Fuel Mart

Gas N: Amoco* (Open 24 Hours), Chevron*(LP), Racetrac* (Open 24 Hours)
S: Texaco*(LP)

Food N: Arby's, Burger King, Hardee's, KFC, McDonalds, Pizza Hut, Subway, Waffle House, Wendy's
S: Country Kettle Restaurant (BP), Taco Bell (BP), Waffle House

Lodg N: Comfort Inn, Days Inn (see our ad this page), Hampton Inn
S: Holiday Inn Express (see our ad this page), AAA Ramada Inn, Super 8 Motel (see our ad this page)

AServ N: Phil Cook Chevrolet

TServ S: Truck Service, Firestone Tires, Truck Lube Center (BP)

ATM N: Racetrac

52 **(121)** Buckhead

Gas S: Phillip 66*(D)(LP)

53 **(130)** Greensboro, Eatonton

FStop S: Chevron*(LP) (Open 24 Hours)

Gas N: Amoco*(D)(LP) (Open 24 Hours), BP*

Food N: Pizza Hut, Subway, Waffle House

Lodg N: Jameson Inn, AAA Microtel Inn

ATM N: BP
S: Chevron

(130) Weigh Station - both directions

54 **(137)** GA 77, GA 15, Siloam, Union Point, Sparta

Gas S: Amoco*(D)(LP), Exxpon*(LP)

Med S: ✚ Hospital

55 **(148)** GA 22, Crawfordville, Sparta

FStop N: Amoco*
S: Chevron*(LP)

TServ S: Chevron

ATM S: Chevron

Parks N: A.H. Stevens State Park

Other S: Coin Laundry (Chevron)

56 **(154)** U.S. 278, Warrenton, Washington

57 **(159)** Norwood

58 **(165)** GA 80, Camak

59 **(172)** U.S. 78, GA 17, Thomson, Washington

FStop S: Citgo*(D), Fuel City*(SCALES) (24 Hrs)

Gas N: BP, Chevron*(D)
S: Amoco(D)(CW), Racetrac*(LP), Texaco*

Food S: Blimpie's Subs, Burger King, Dairy Queen (Amoco), Long John Silvers, McDonalds, Pizza Hut, Plantation House (Best Western), Shoney's, Taco Bell, Waffle House, Wendy's, Western Sizzlin'

Lodg S: AAA Econolodge, AAA Ramada, AAA White Columns Inn (Best Western)

AServ N: BP, Thomson Chrysler Plymouth, Dodge, Jeep Eagle

Med N: ✚ Hospital

ATM S: Fuel City

60 **(175)** GA 150

Parks N: Mistletoe State Park

(182) Rest Area - RR, Phones, RV Dump

61 **(183)** U.S. 221, Appling, Harlem

(187) Weigh Station (Eastbound)

(188) Weigh Station (Westbound)

62 **(190)** GA 388, Grovetown

FStop N: Fuel Stop*(D) (24 Hrs)

Gas S: BP*(LP)

Other N: Fort Gordon

63 **(194)** GA 383, Belair Rd, Evans

FStop N: Texaco*
S: Smile Gas*(LP), Speedway*(LP)(SCALES)

Gas N: Circle K*(LP), Exxon*(LP)
S: Amoco*(D) (24 Hrs)

Food N: Hardee's, Taco Bell, Waffle House
S: Blimpie's Subs, Cracker Barrel, Dairy Queen (Amoco), Huddle House, TCBY Treats (Smile Gas), Waffle House

Lodg N: National Inn 9
S: Econolodge, National Inn 9

ATM N: Southtrust, Texaco
S: Smile Gas

Other N: Coin Car Wash

64AB **(198)** Junction I-520, GA 232, Bobby Jones Expwy.

Gas N: Racetrac* (24 Hrs), Smile Gas* (24 Hrs)

Food N: Applebee's, Blimpie's Subs (Smile Gas), Burger King, Checkers Burgers, China Pearl, Golden Corral Steakhouse, Krispy Kreme, Krystal, Neighborhood Restaurant, Ruby Tuesday, Ryan's Steakhouse, Shoney's, Taco Bell, Tasty Freez Food, Waffle House

Lodg N: Howard Johnson
S: Holiday Inn (see our ad this page)

AServ N: B. C. Tire & Auto Service, Q Lube

Med N: ✚ Martinez Urgent Care

ATM N: Bi-Lo Supermarket, NationsBank, Racetrac, SunTrust

Other N: Bi-Lo Supermarket, Eckerd Drugs, Eye Glass World Express, Home Quarters Warehouse, Sams Club Warehouse, Villa Plaza Shopping Center, Wal-Mart (24 Hrs, Pharmacy)

65 **(200)** GA 28, Washington Rd, Augusta

Gas N: BP*(CW), Racetrac* (24 Hrs)
S: Amoco*(CW) (24 Hrs), Crown*(LP), Hess(D), Shell*(LP), Smile Gas*, Texaco*(D)(LP)

Food N: Applebee's, Burger King, Captain D's Seafood, Chick-fil-A, Dairy Queen, Damon's Ribs, Empress Chinese Supper Buffet, Huddle House, McDonalds, Mikoto Japanese, Picadilly Cafeteria, Pizza Hut, Rhinehart's Oysters & Seafood, Sho Gun Japanese Steakhouse, Shoney's, Veracruz Mexican Restaurant, Waffle House
S: Arby's, Blimpie's Subs, Bojangle's, Church's

EXIT GEORGIA

Fried Chicken, Fat Tuesday Restaurant, Fazoli's Italian Food, Hooter's, Krispy Kreme Donuts, Kyota Japanese Steak & Seafood, Lone Star Steakhouse, Long John Silvers, McDonalds, Michael's Fine Food, Olive Garden, Outback Steakhouse, Peking Chinese, Red Lobster, Rio-Bravo, Shanri-La Chinese, Smiley's Basketball, Subway, T.G.I. Friday's, TBONEZ Steakhouse, Taco Bell, Thai Jong, The China Restaurant, The Plum Crazy, Vallarta Mexican Restaurant, Waffle House, Wendy's

Lodg **N:** ◆ Courtyard by Marriott, Days Inn, Hampton Inn, Holiday Inn, Homewood Suites Hotel, **La Quinta**, Master's Inn, Radisson Suites, Scottish Inns, Shoney's Inn, Sunset Inn, Travelodge
S: Econolodge, ◆ Fairfield Inn, Knight's Inn, Motel 6, West Bank Inn

AServ **N:** Raders Mazda, Xpress Lube
S: Midas Mufflers &Brakes, National Hills Tire & Service, Your Home Town Firestone Tire & Auto

ATM **N:** Racetrac, SouthTrust
S: First Union Bank, Georgia Bank & Trust Co., NationsBank, Smile Gas, SouthTrust, Suntrust Bank, Texaco

Other **N:** **Redwing Roller Way**
S: Car Wash, **Drug Emporium, Kroger Supermarket (& pharmacy)**, National Pride Coin Car Wash, **Publix Supermarket (Pharmacy), Steinmart Mall, Thrifty Rexall Drugs, Wolf Camera**

66 **(200)** GA 104, River Watch Pkwy, Augusta

Lodg **N:** Ameri Suites (see our ad this page)

AServ **S:** Steve's Auto Service

EXIT GEORGIA/SOUTH CAROLINA

(201) GA Welcome Center - RR, Phones, RV Dump (Westbound)

↑GEORGIA

↓SOUTH CAROLINA

(1) SC Welcome Center - RR, Phones (Eastbound)

1 SC 230, N. Augusta

FStop **S: Smile Gas*(LP)**

Food **S: Blimpie's Subs (Smile Gas)**, Tastee Freez, Waffle House

AServ **S:** Kinsey's Auto

Other **S: Wacky Wayne's Fireworks**

5 U.S. 25, SC 121, Edgefield, Johnston

FStop **N: Fuel City*(LP)(SCALES)**
S: Speedway*(LP)

Gas **N:** BP(D), Shell*(D)
S: Exxon*(LP)

Food **N:** Bojangle's (Shell), Burger King, **Hardee's, Huddle House**, Smiley's Deli, Sonic American
S: Taco Bell (Exxon), **Waffle House**

AServ **N:** BP

RVCamp **N: Mobile Home Estates (RV hookup & overnite camping)**

ATM **N: Fuel City**, Shell, **Winn Dixie Supermarket**

Other **N: Fed Ex Drop Box (Fuel City), Winn Dixie Supermarket (Pharmacy)**
S: GA - SC Souvenir

11 Road 144, Graniteville

18 SC 19, Aiken, Johnston

FStop **S: Fuel Stop*, Shell***

Food **S: Blimpie's Subs (Shell), Subway (Fuel Stop)**, Waffle House

Lodg **S: Deluxe Inn**, Ramada Inn

Med **S: + Hospital**

ATM **S:** Fuel Stop

(20) **Parking Area (Both Directions)**

22 U.S. 1, Aiken, Ridge Spring

Gas **S:** Amoco*(LP), BP*(LP), Racetrac*

Food **S:** Hardee's, **McDonalds, Waffle House**

Lodg **S:** ◆ **Days Inn**, ◆ **Holiday Inn**

ATM **S:** Racetrac

29 Road 49

33 SC 39, Wagener, Monetta

FStop **N: Shell***

(35) **Weigh Station (Eastbound)**

EXIT SOUTH CAROLINA

39 U.S. 178, Batesburg, Pelion

TStop **S: Citgo Hill View Truck Stop***

FStop **N: Exxon***

Food **S:** The Fatback Connection

44 Road 34, Gilbert

TStop **N: 44 Truck Stop***

Food **N: 44 Restaurant**

TServ **N: Truck Service (Road service)**

ATM **N: 44 TStop**

(48) **Parking Area (Both Directions)**

51 Road 204

TStop **N: Texaco*(LP)**

FStop **S: Amoco* (24 Hrs)**

Food **N: Subway (Texaco)**

AServ **S:** International Garage (Truck tire service)

ATM **N: Texaco**
S: Amoco

(53) **Weigh Station (Westbound)**

55 SC 6, Lexington, Swansea

FStop **S: Exxon*(D)(LP)**

Gas **N:** Texaco*

Food **N: Hardee's**
S: Golden Town Chinese, **McDonalds**, Subway, Waffle House

ATM **N:** BB&T Bank
S: Exxon, Wachovia Bank

Other **N: Car Quest Auto Center**
S: Revco Drugs, Winn Dixie Supermarket (Pharmacy)

58 U.S. 1, Columbia Airport, West Columbia, Univ of South Carolina

FStop **N: Texaco***

Food **N:** Subway, Waffle House
S: Meat'n Place Ribs & BBQ

ATM **N:** Texaco

Other **S: Four Oaks Farm**

61 U.S. 378, West Columbia

FStop **S: Amoco***

Gas **S:** Phillips 66*(LP)

Food **S:** Waffle House

63 Bush River Road

Gas **N:** BP*(CW)
S: Exxon*(CW)

Food **N:** Cracker Barrel, Steak-Out, Subway, Wings & Ale
S: El Chico Mexican, Fuddrucker's, Key West Grill, The Villa (Italian), Waffle House

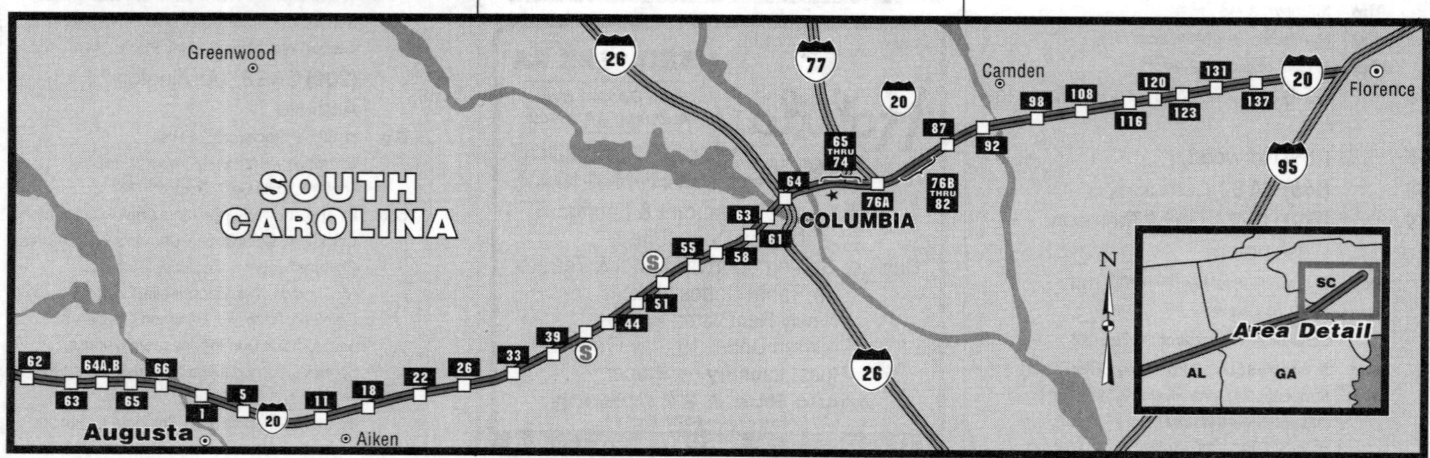

Bold red print shows RV & Bus parking available or nearby

EXIT SOUTH CAROLINA

Lodg N: ◆ Travelodge
S: Best Western, AAA Knight's Inn (see our ad this page), AAA Sheraton, Sleep Inn
Other N: Market Point Mall
S: Kinko's Copies, Major League Lanes (Bowling)

64A Junction I-26, U.S. 76 East, Columbia

64B Junction I-26, U.S. 76 West, Spartanburg

65 U.S. 176, Broad River Road
Gas N: BP*(CW)
S: Racetrac*
Food N: Applebee's, Bojangles, Chopstix Chinese, Donut Hole, Great Wall Chinese, Rush's Fabulous Food, Subway, Waffle House
S: Annabelle's Restaurant & Pub, Arby's, Burger King, Captain Tom's Seafood, Chicago Smokehouse, Chungking Chinese Restaurant, Church's Fried Chicken, Cracker Jack's Food & Drink, Decision's Bar & Grill, Hooter's, KFC, Lizard's Thicket Country Cookin, McDonalds, Morrison's Cafeteria, Pizza Hut, Taco Bell, Taco Cid, The Red Checker Sandwich Shop
Lodg N: Quality Inn
S: American Inn, ◆ La Quinta Inn, Royal Inn
AServ N: AAMCO Transmission, Yokohama Tires
S: Goodyear Tire & Auto, Pep Boys Auto Center, Precision Tune & Lube, Q Lube
ATM N: First Citizens Trust
S: MBSC, Racetrac, Wachovia Bank
Other N: National Pride Car Wash, Snappy Carwash, U.S. Post Office
S: A & P Supermarket, Dutch Square Mall, Eckerd Drugs

68 SC 215, Monticello Road, Jenkinsville
FStop N: Exxon*(D)(LP)
Gas N: Amoco*(D), Texaco*(D)(LP)
S: Citgo*(D)
Food N: Culler's Restaurant, Virginia's Grill
S: Sunrise
AServ N: Andy's Express Tire & Lube

70 U.S. 321, Fairfield Road, Winnsboro
TStop S: Flying J Travel Plaza*(D)(LP)(SCALES) (RV Dump)
Gas S: Exxon*
Food S: Hardee's, Thads Restaurant
Lodg S: Super 8 Motel
ATM S: Exxon, Flying J Travel Plaza

71 U.S. 21, North Main St, Blythewood
TStop N: Columbia Travel Center Amoco*(SCALES)
FStop N: United*(D) (24 Hrs)
Gas N: BP*(D)(CW)
S: Shell* (see our ad this page)
Food N: Carolina Country (Firestone), Pizza Hut (Columbia Travel Center), Restaurant (United), Subway (Columbia Travel Center)
S: Bert's Grill and Diner
Lodg N: DAYS INN Days Inn
TServ N: Firestone Tire & Auto*
TWash N: National Truck Wash
ATM N: Columbia Travel Center, United
S: First Citizens Bank
Other N: Fed Ex Drop Box (Columbia)
S: A Plus Locksmith

72 SC 555, Farrow Road

73A SC 277, Columbia (Eastbound, Reaccess On Both Sides)

73B Junction I-77, SC 277, Charlotte

74 U.S. 1, Two Notch Road, Fort Jackson
Gas N: BP(CW)
S: Citgo*(CW)(LP) (24 Hrs), Hess(D)

EXIT SOUTH CAROLINA

Food N: Chili's, Fazoli's Italian Food, International House of Pancakes, Lizard's Thicket Country Cookin, Mc Kenna's, Outback Steakhouse, Waffle House
S: Applebee's, Baskin Robbins, Bojangle's, Captain D's Seafood, Godfather's Pizza, Hardee's, Maurice's BBQ, McDonalds, Quincy's Family Steakhouse, Roadhouse Grill, Shoney's, Sub Station II, Sub's Miami Grill, Taco Bell, Wendy's
Lodg N: AAA AmeriSuites (see our ad this page), ◆ Budgetel Inn, AAA Comfort Inn, Hampton Inn, ◆ Holiday Inn, AAA Quality Inn, ◆ Red Roof Inn
S: DAYS INN ◆ Days Inn
AServ N: BP
S: Midas Muffler, NAPA Auto Parts, Penske Auto Services, Precision Tune, Q Lube

EXIT SOUTH CAROLINA

ATM S: Citgo, First Citizen's Bank, NBSC, NationsBank
Other N: Sesqui Centennial State Park
S: Columbia Mall, K-Mart (Pharmacy), Kinko's Copies, Parts America, Pearl Vision Center, Snappy's Car Wash, Winn Dixie Supermarket

76A Alpine Rd., Fort Jackson, Charlotte
Gas N: BP*(CW), Exxon*(CW)
Food N: Waffle House

76B Alpine Road

80 Clemson Road
Gas N: BP*(CW), Exxon*(D)(CW)
Food N: McDonalds, Waffle House
ATM N: Exxon
Other N: CVS Pharmacy

82 Road 53, Pontiac

87 Road 47, Elgin
FStop N: BP*(D)
Gas N: Texaco*
AServ N: Regal Auto Care

92 U.S. 601, Lugoff, Camden
TStop N: Pilot Travel Center*(SCALES)
Gas N: Texaco*
Food N: Dairy Queen, Hardee's, Subway, Waffle House
Lodg N: DAYS INN Days Inn
ATM N: Pilot Travel Center

(93) Rest Area - RR, Phones, Vending, Picnic (Both Directions)

98 U.S. 521, Camden, Sumpter
Gas N: Exxon*
Med N: ✚ Hospital
Parks N: Revolutionary War Park

101 Road 91, Road 189

108 SC 34, Road 31, Manville
FStop S: Shell*(D)
Gas S: Citgo*(D)(CW)
AServ S: Frankie's Garage & Wrecker Service

116 U.S. 15, Bishopville, Hartsville, Sumter
TStop N: Shell*(D) (24 Hrs)
Food N: KFC, McDonalds
Lodg N: AAA Econolodge

120 SC 341, Lynchburg, Elliot
FStop S: Citgo*(D)(LP)
Food S: A Taste of Country (Citgo)
Lodg S: Bishopville Motel (24 Hrs)

123 Road 22, Lamar, Lee State Park
FStop N: Amoco*
Parks N: Lee State Park

(129) Parking Area (Both Directions)

131 U.S. 401, SC 403, Timmonsville, Darlington
FStop N: Phillips 66*
Gas N: Exxon*(D) (24 Hrs)
Food N: Restaurant (Phillips 66)
TServ N: Handee's Truck Shop
ATM N: Exxon

137 SC 340, Darlington
Gas S: BP*
AServ N: B&B Auto Shop

141AB Junction I-95, Fayetteville, Savannah

↑SOUTH CAROLINA

Begin I-20

EXIT	ILLINOIS/KENTUCKY

Begin I-24

↓ILLINOIS

7 Goreville, Tunnel Hill
- FStop S: **Citgo***(D)
- RVCamp S: **Camping**

14 U.S. 45, Vienna, Harrisburg

16 IL 146, Vienna, Golconda
- Gas S: Amoco*, Citgo*, Shell*
- Food S: Dairy Queen
- Lodg N: **Budget Inn**
  S: Ramada
- Other S: Car Wash, **South Illinois Propane Gas**

27 New Columbia, Big Bay

37 U.S. 45, Metropolis, Brookport
- TStop N: **Veach's Service***(D)
- FStop S: **BP***(D)(SCALES)
- Food N: **Veach's Service**
  S: Pizza Hut, Ponderosa Steakhouse, **Waffle Hut (BP, Open 24 Hours)**
- Lodg S: AAA Best Inns of America, AAA Comfort Inn, Metropolis Inn
- TServ S: **BP**

(37) IL Welcome Center - RR, Vending, Phones

↑ILLINOIS

↓KENTUCKY

3 Kentucky 305, Paducah
- TStop N: **Citgo**(D) **(Truck Fuel Only)**
  S: **Pilot Travel Center***
- FStop S: **Pilot Travel Center***(CW)(SCALES)
- Gas N: Ashland*
  S: BP*
- Food N: El Maguey Mexican, Leroy & Lita's Restaurant, Slim's BBQ
  S: Salubre' Pizza (BP), **Subway (Pilot FS)**, Waffle Hut (Open 24 Hours)
- Lodg N: AAA **Comfort Inn**, Ramada Limited, ◆ Super 8 Motel
  S: ◆ Budgetel Inn

4 U.S. 60, Bus. Loop 24, Wickliffe, Paducah
- Gas N: Shell*
  S: Ashland*, BP*, Petro*
- Food N: Applebee's, Burger King, Denny's (Peartree Inn), Double Happiness Chinese, O'Charley's, Outback Steakhouse
  S: Chong's Chinese, Cracker Barrel, El Chico Mexican, Fazoli's Italian Food, Godfather's Pizza, Hardee's, Olive Garden, Pizza Hut, Red Lobster, Ryan's Steakhouse, Shoney's, Taco Bell (Outdoor Playground)
- Lodg N: **Courtyard by Marriott (see our ad this page)**, Days Inn ◆ **Days Inn**, ◆ Drury Inn, **Holiday Inn Express, Westowne Inn**
  S: AAA Best Inns of America, AAA Comfort Suites (see our ad this page), ◆ Hampton Inn, Peartree Inn
- AServ S: Goodyear Tire & Auto
- ATM S: Ashland, Citizens Bank and Trust, Peoples First
- Other S: **Kentucky Oaks Mall, Sam's Club, Wal-Mart Supercenter (w/Pharmacy, Auto Service)**

7 U.S. 62, U.S. 45, Mayfield, Bardwell
- Gas N: Ashland*, Shell* (Open 24 Hours)
  S: BP* (US 62), Old Fashioned Deli (BP, US 62), Scot Market*, Shell*(CW) (US 62)
- Food N: Burger King (Indoor Playground), **McDonalds (Outdoor Playground)**, Subway (Ashland), Taco Bell
  S: KFC, Little Caesars Pizza
- Lodg S: AAA Denton Motel, Quality Inn (see our ad this page)
- Med N: **+ Hospital**
- ATM N: Paducah Bank
  S: BP (US 62), Citizens Bank and Trust, Republic Bank, Shell
- Other N: Car Wash

(7) KY Welcome Center - RR, Phones, Vending, Tourist Info (Truck Parking)
- Gas N: **Ashland***(D), Shell*(D)
- Food N: Burger King (Ashland), Subway (Ashland), Taco Bell (Ashland)
  S: Arby's, McDonalds, Sonic
- Lodg S: Sunset Inn
- AServ S: K-Mart
- Med N: **+ Hospital**
- Other S: Coin Car Wash, **K-Mart, White Haven Welcome Center**

11 KY 1954, Husband Road, Paducah
- TServ N: **Hartman's Truck & Equip. Service**

16 U.S. 68, Business 24, Paducah
- TStop S: **BP/AmBest***(D)(SCALES)
- Gas S: Citgo*
- Food S: **Southern Pride (BP/AmBest)**
- AServ S: Citgo
- ATM S: **BP/AmBest**

25AB Purchase Pkwy (toll), Fulton, Calvert City

27 U.S. 62, Calvert City, KY Dam
- TStop S: **I-24 76 Auto/Truck Stop***(SCALES)
- Gas N: Ashland, **BP***
- Food N: KFC, McDonalds (Playground), Willow Pond Rest. (Foxfire)
  S: 76 Auto/Truck Plaza
- Lodg N: Foxfire Motor Inn
- TServ N: **Freightliner Dealer**
  S: **Bridgestone Tire & Auto**

31 KY 453, Grand Rivers, Smithland

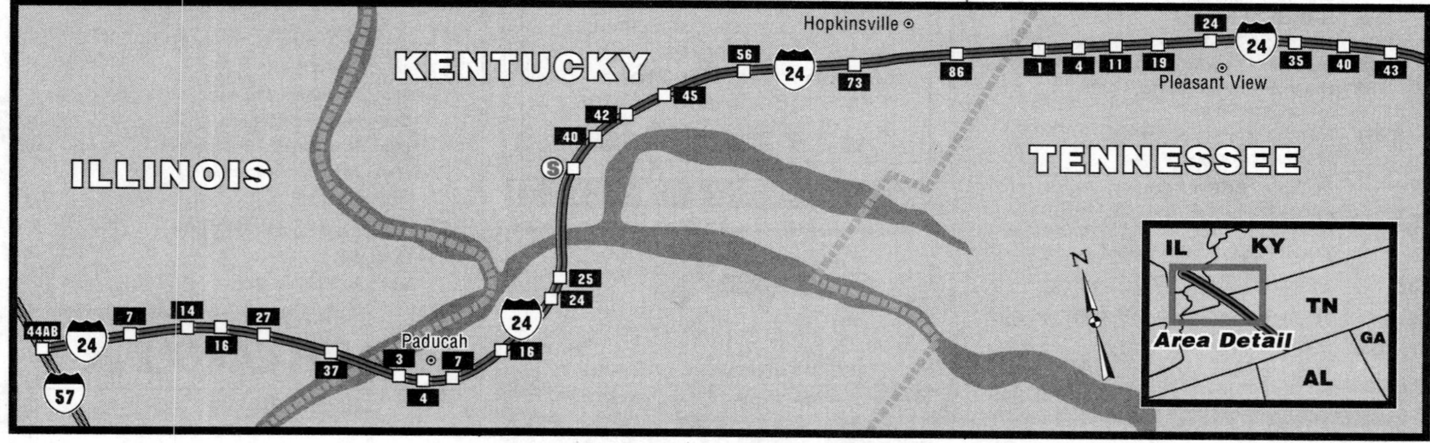

Bold red print shows RV & Bus parking available or nearby

EXIT KENTUCKY

Gas S: BP*
Food S: Mrs. Scarletts Restaurant
Lodg S: AAA Best Western
(34) **Rest Area (Westbound)**
(36) **Weigh Station (Both Directions)**
40 U.S. 62, U.S. 641 Kuttawa, Eddyville
FStop S: **Sunshine Travel Center*(D)**
Gas S: BP*, Shell*
Food S: Burger King (Shell), Taco Bell (Shell)
Lodg N: Relax Inn
S: DAYS INN ◆ **Days Inn, Hampton Inn**
42 West KY Pkwy. East, Princeton, Elizabethtown
45 KY 293, Princeton
Gas S: Chevron*(D)
Food S: Smash Hit Subs (Chevron)
RVCamp N: **R.V. Park**
56 KY 139, Cadiz, Princeton
FStop S: **Chevron***
65 U.S. 68, KY 80, Cadiz, Hopkinsville (Kentucky State Park Resort And Recreational Area)
FStop S: **Coastal*(D) (Souvenir)**
Gas S: BP*, Phillips 66*(D), Shell*
Food S: Broad Bent Cafe (Coastal)
Lodg S: AAA Country Inn, Holiday Inn Express, **Knight's Inn (see our ad this page)**, Lakeway Motel
Med S: **+ Hospital**
ATM S: Shell
Other S: **Tourist Information**
73 KY 117, Newstead, Gracey
86 U.S. Alt. 41, Hopkinsville, Ft.

EXIT KENTUCKY/TENNESSEE

Campbell, Pennyrile
TStop S: **Pilot Travel Center(SCALES)**
FStop N: **Chevron*(SCALES)**
S: **Pilot Travel Center*(D)**
Gas S: Amoco*
Food N: **Taco Bell (Chevron)**
S: **Burger King, Subway (Pilot), Waffle House**
Lodg S: ◆ **Budgetel Inn**, Comfort Inn (see our ad this page)
TServ S: **Pilot Travel Center**
ATM N: Chevron
89 KY 115, Oak Grove, Pembroke (Jefferson Davis Monument)
(93) **KY Welcome Center - RR, Phones, Vending, Truck Parking, RV Parking, HF (Westbound)**

↑KENTUCKY

↓TENNESSEE

(1) **TN Welcome Center - RR, Phones, Truck Parking (Eastbound)**
1 TN 48, Clarksville, Trenton
Gas N: Shell*
S: Exxon*
Food S: Ice Cream Corner (Exxon)
Lodg S: **Radisson (see our ad this page)**
ATM S: Acme Credit, Exxon
Other N: **Clarksville Campground**
4 U.S. 79, Guthrie, Clarksville (Fort Campbell Army Post, Austin Peay State University)
FStop N: **BP***
S: **Speedway***
Gas N: Exxon
S: Amoco*(CW), Texaco*
Food S: Applebee's, Arby's, Baskin Robbins, Hunan Garden Chinese Buffet, Krystal, Little Caesars Pizza, Loco Lupe's, Logan's Roadhouse Steaks & Ribs, Long John Silvers, McDonalds, O'Charley's, Olive Garden, Outback, Ponderosa, Rafferty's, Red Lobster, Rio Bravo, Ryan's Steak House, Speedway, Taco Bell, Waffle House, Wendy's
Lodg S: AAA Best Western, AAA Comfort Inn, DAYS INN AAA Days Inn (see our ad this page), AAA Econolodge (see our ad this page), ◆ Fairfield Inn, AAA Hampton Inn, ◆ Holiday Inn, ◆ **Ramada Limited**, Royal Inn, AAA Shoney's Inn, AAA Super 8 Motel (see our ad this page), ◆ Travelodge, Comfort Inn (see our ad this page)
AServ N: Texaco
S: Q-Lube, Western Auto
TServ S: **Goodyear Tire & Auto**
RVCamp N: **Clarksville RV Sales and Service Center**
ATM N: Exxon
S: First Union Bank, Heritage Bank, NationsBank
Other S: Coin Car Wash, **Governor's Square Mall, Hampton Plaza, K-Mart (Pharmacy), Sam's Club, Wal-Mart (Open 24 Hours, Food Court)**

EXIT TENNESSEE

Bold red print shows RV & Bus parking available or nearby

EXIT TENNESSEE

8 TN 237, Rossview Rd.

11 TN 76, Adams, Clarksville
- **FStop** N: Texaco*
  S: Amoco* (Open 24 Hours)
- **Gas** S: Citgo*, Phillips 66* (Days Inn)
- **Food** S: Homeplace Restaurant (Days Inn), Waffle House
- **Lodg** S: Comfort Inn (see our ad this page), ◆ Days Inn, Holiday Inn Express (see our ad this page)
- **ATM** S: Citgo

19 TN 256, Maxey Road, Adams
- **FStop** N: Phillips 66
- **Gas** N: Dixieland
  S: Shell (see our ad this page)
- **Food** N: Phillips 66
- **AServ** N: Truck & Trailer Repair (Phillips 66)
- **TServ** N: Phillips 66
- **TWash** N: Phillips 66

24 TN 49, Springfield, Ashland City
- **FStop** N: Delta Express*
- **Gas** N: Amoco*, Texaco*
  S: Shell* (see our ad this page)
- **ATM** N: Delta Express

31 TN 249, New Hope Road (Nashville Zoo)
- **TStop** S: BP*
- **FStop** S: BP* (Bubba's Truck Stop)
- **Gas** N: Shell* (see our ad this page)
- **Food** S: Buddy's Chop House Steak, Chicken and Fish
- **AServ** S: BP
- **TServ** S: BP

35 U.S. 431, Joelton, Springfield
- **Gas** S: Amoco*, BP*
- **Food** S: Country Junction, McDonalds, Subway (BP)
- **Lodg** S: Days Inn
- **RVCamp** S: OK Campground
- **Med** N: + Hospital
- **Other** S: Coin Car Wash

40 TN 45, Old Hickory Blvd
- **Gas** N: Citgo Quick Mart, Phillips 66*
- **Food** N: Family Restaurant (Super 8 Motel), Subway (Citgo)
- **Lodg** N: Super 8 Motel
- **ATM** N: Citgo

43 TN 155, Brick Church Pike

44AB Junction I-65, Nashville

Note: I-24 runs concurrent below with I-65. Numbering follows I-65.

EXIT TENNESSEE

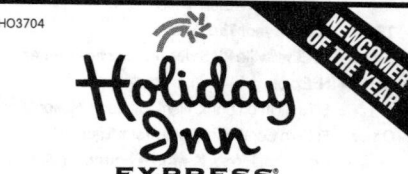

87AB U.S. 431, Trinity Lane
- **TStop** E: Pilot Travel Center[SCALES]
- **Gas** E: Circle K*[LP]
  W: Amoco*[D][CW] (Kerosene), BP*[D][CW], Exxon*[D], Texaco*[D]
- **Food** E: Arby's (Pilot), Candle Light Restaurant, Chuch's Chicken, Krystal (24 hrs), Sonic, TJ Restaurant, White Castle Restaurant
  W: Burger King, Captain D's Seafood, Chugers Restaurant, Club Paradise, Denny's (24 hrs), Gabe's Lounge, Jack's BAR-B-QUE, Lockers Sports Grill, McDonalds, Ponderosa, Shoney's, Taco Bell, The Broken Spoke Cafe (Ramada Inn), Track 1 Cafe (Rain Tree Inn), Waffle House
- **Lodg** E: Cumberland Inn (Pilot TS), Key Motel, Red Carpet Inn, Savoy Motel, Scottish Inns, Trinity Inn
  W: Comfort Inn, Days Inn, Econolodge, Hallmark Inn, Hampton Inn, Holiday Inn Express, Knights Inn, Liberty Inn, Motel 6, Oxford Inn, Rain Tree Inn, Ramada
- **AServ** E: Bobby's Tire Service, Gary's, The Tire Store
  W: Exxon
- **RVCamp** E: RV and Camper Corral Truck Accessories
- **ATM** W: BP
- **Parks** E: Trinity Park (Mobil Home Park)
- **Other** E: Coin Car Wash, National Car Wash, Sweeney's Food Town Gocery Store, Trinity Gas Co. Inc[LP], US Post Office

86 Junction I-265 Memphis

85B Jefferson St
- **Gas** W: Spur
- **Lodg** W: Best Western, ◆ Days Inn

85A U.S. 31 E, Ellington Pkwy, Spring St
- **FStop** W: Cone*
- **Gas** W: Spur*[D]
- **Lodg** W: Best Western (Metro Inn), Days Inn
- **AServ** W: Jefferson Street Car Care, NAPA Auto Parts, Nashville Tansmission Parts, Todd's Auto Parts
- **Other** W: Burnette's Truck Wash, U-Haul Center[LP]

85 James Robertson Parkway, State Capitol (Difficult Reaccess)
- **TStop** W: TravelCenters of America[SCALES] (RV Dump)
- **Gas** E: Mapco Express, Shell*[CW]
  W: Cone*
- **Food** W: Country Pride (Truckstop), Gersthaus Restaurant, Mrs. C's, Shoney's, Subway (Truckstop)
- **Lodg** W: Best Host Motel, Econolodge, Ramada Limited
- **AServ** E: Amoco, Coin Car Wash
  W: Vogely & Todd Collision Repair Experts
- **TServ** W: TravelCenters of America
- **ATM** E: Mapco Express
- **Other** E: Main Street Auto Wash
  W: Country Hearth Bread, Bakery Thrift Store, Performing Arts Center, State Capitol, Tourist Information

84 Shelby Ave
- **Gas** W: Exxon*
- **Food** W: Gersthaus
- **Lodg** W: Econolodge, Ramada Limited
- **AServ** W: Napa Auto Parts, Standard Motor Parts
- **Other** E: Lynn Drugs, Martin's Grocery
  W: Prop Shop & More

83B I-40 West, South I-65, Memphis, Birmingham

Bold red print shows RV & Bus parking available or nearby

EXIT	TENNESSEE
83A	I-24 East, I-40 East, South I-65, Nashville, Chattanooga

Note: I-24 runs concurrent above with I-65. Numbering follows I-65.

Note: I-24 runs concurrent below with I-40. Numbering follows I-40.

211A Junction I-65 South, Junction I-40 West, Memphis, Birmingham

212 Hermitage Ave (Westbound, Difficult Reaccess, Reaccess Via I-40)

- **Gas** **S:** **Citgo*(D)**
- **Food** **N:** El Captain Bar and Grill
  S: Hermitage Cafe Hamburgers & Chili, Lady J's Gingerbread House
- **AServ** **N:** Martin's Wrecking Service
- **TServ** **N:** **Goodyear Tire & Auto, Neely Coble Sunbelt Truck Center, TIP Trailer Rental, Leasing, Sales Service, Tennessee Truck Sales**
  S: **Rawlings Truck Rental and Leasing**
- **Med** **N:** **+ Hospital**
  S: **+ Hospital**
- **Other** **N:** **Hermitage Ave Deli & Market, J C Napier Community Center Swimming Pool**

213A Junction I-24 East, Junction I-40

213B Junction I-24 West, Junction I-40 West

Note: I-24 runs concurrent above with I-40. Numbering follows I-40.

52AB 52A - I-24W, I-40W, Nashville 52B - I-40E, Knoxville

52 U.S. 41 South, Murfreesboro Road

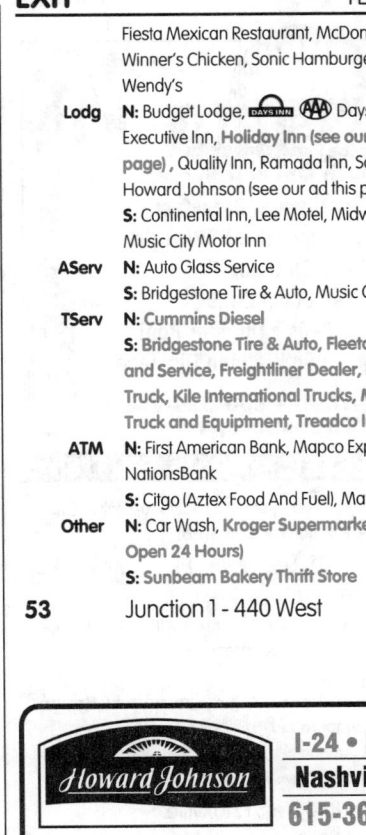

- **FStop** **S:** **Aztex Food and Fuel Center***
- **Gas** **N:** BP*(CW), Exxon*, Mapco Express*, Texaco*
  S: Amoco*, Citgo* (Aztex Food and Fuel), Mapco Express*, Texaco*(CW)
- **Food** **N:** Baskin Robbins, Burger King, Chili's, Dad's Place (Ramada Inn), Denny's, Domino's Pizza, Dunkin Doughnuts, Emperor of China Restaurant, Fifth Quarter Steak House, Hunan Chinese, KFC, Lin's Garden Chinese Restaurant (Quality Inn), Los Reyes, O' Charley's Restaurant, PJ's Sports Bar, Piccadilly Cafeteria, Pizza Hut, Ramada, Santa Fe Cantina and Cattlemans Club, Shoney's, Super Buffet, Taco Bell, The Down Under Bar & Grill (Days Inn), The Peddler Steakhouse and Lounge, Waffle House
  S: Aregis Restaurant, Burger King, J&R Family Restuaurant, Krystal (Aztex Food and Fuel), La Fiesta Mexican Restaurant, McDonalds, Mrs Winner's Chicken, Sonic Hamburgers, Teasers, Wendy's
- **Lodg** **N:** Budget Lodge, DAYS INN (AAA) Days Inn, Executive Inn, **Holiday Inn (see our ad this page)**, Quality Inn, Ramada Inn, Scottish Inns, Howard Johnson (see our ad this page)
  S: Continental Inn, Lee Motel, Midway Motel, Music City Motor Inn
- **AServ** **N:** Auto Glass Service
  S: Bridgestone Tire & Auto, Music City Dodge Inc.
- **TServ** **N:** **Cummins Diesel**
  S: **Bridgestone Tire & Auto, Fleetco Truck Parts and Service, Freightliner Dealer, International Truck, Kile International Trucks, Music City Truck and Equiptment, Treadco Inc.**
- **ATM** **N:** First American Bank, Mapco Express, NationsBank
  S: Citgo (Aztex Food And Fuel), Mapco Express
- **Other** **N:** Car Wash, **Kroger Supermarket (Pharmacy, Open 24 Hours)**
  S: **Sunbeam Bakery Thrift Store**

53 Junction 1 - 440 West

54AB TN 155, Briley Parkway (Ltd. Access Road North, Difficult Truck Reacess Both Directions)

- **Gas** **S:** Shell*
- **AServ** **S:** Shell
- **Parks** **S:** **Grassmere Wildlife Park**

56 Route 255, Harding Place

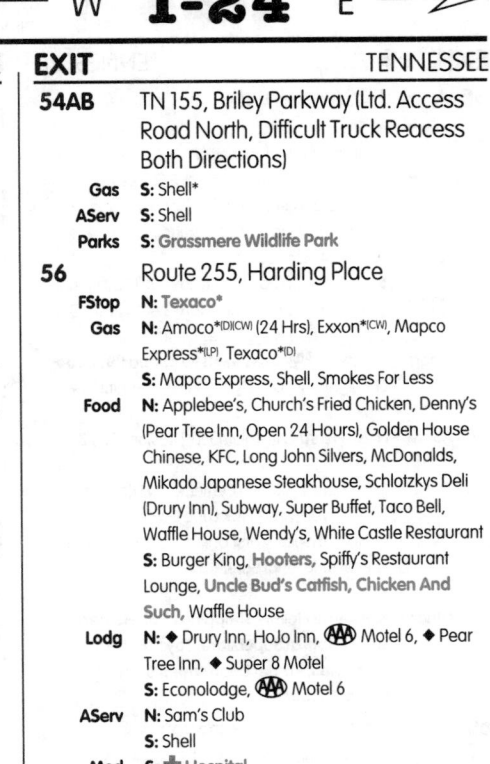

- **FStop** **N:** **Texaco***
- **Gas** **N:** Amoco*(D)(CW) (24 Hrs), Exxon*(CW), Mapco Express*(LP), Texaco*(D)
  S: Mapco Express, Shell, Smokes For Less
- **Food** **N:** Applebee's, Church's Fried Chicken, Denny's (Pear Tree Inn, Open 24 Hours), Golden House Chinese, KFC, Long John Silvers, McDonalds, Mikado Japanese Steakhouse, Schlotzkys Deli (Drury Inn), Subway, Super Buffet, Taco Bell, Waffle House, Wendy's, White Castle Restaurant
  S: Burger King, **Hooters**, Spiffy's Restaurant Lounge, **Uncle Bud's Catfish, Chicken And Such**, Waffle House
- **Lodg** **N:** ◆ Drury Inn, HoJo Inn, (AAA) Motel 6, ◆ Pear Tree Inn, ◆ Super 8 Motel
  S: Econolodge, (AAA) Motel 6
- **AServ** **N:** Sam's Club
  S: Shell
- **Med** **S:** **+ Hospital**
- **ATM** **N:** Amoco, Mapco Express, Sam's Club, Texaco
- **Other** **N:** **Sam's Club, YMCA**
  S: Car Wash

57AB Antioch, Haywood Lane

- **Gas** **N:** Amoco(D) (Kerosene), Kwik Sak*(LP)
  S: Exxon*, Phillips 66*(CW)
- **Food** **N:** BBQ, Fat Mo's, Georgia's Coast Pizza, Steak, & Burgers, Gold-N-Nut, Golden Wok, Hardee's, Lambino's Pizzeria, Pizza Hut Delivery, Rio Grande Mexican & Salvadorian Restaurant, Waffle House, Whitt's BBQ
  S: Chanello's Pizza, TCBY (Exxon), Taco Bell (Exxon)
- **AServ** **N:** Amoco, Car Quest Auto Center, Mike Jones Antioch Auto Center
- **TServ** **N:** **Peterbilt Dealer**
- **Med** **N:** **+ Antioch Medical Center (Walk-Ins Welcome)**
- **Other** **N:** **Food Lion, JML Market*, Jumbo Washette Coin Laundry**, Pop's Antioch Car Wash, **Soap Opera Laundry, The Book Trader**, U-Wash Car Wash, **WalGreens**
  S: **Apache Trail Animal Hospital, Brunswick Bowling Alley, Quick Stop Food Mart*(LP)**

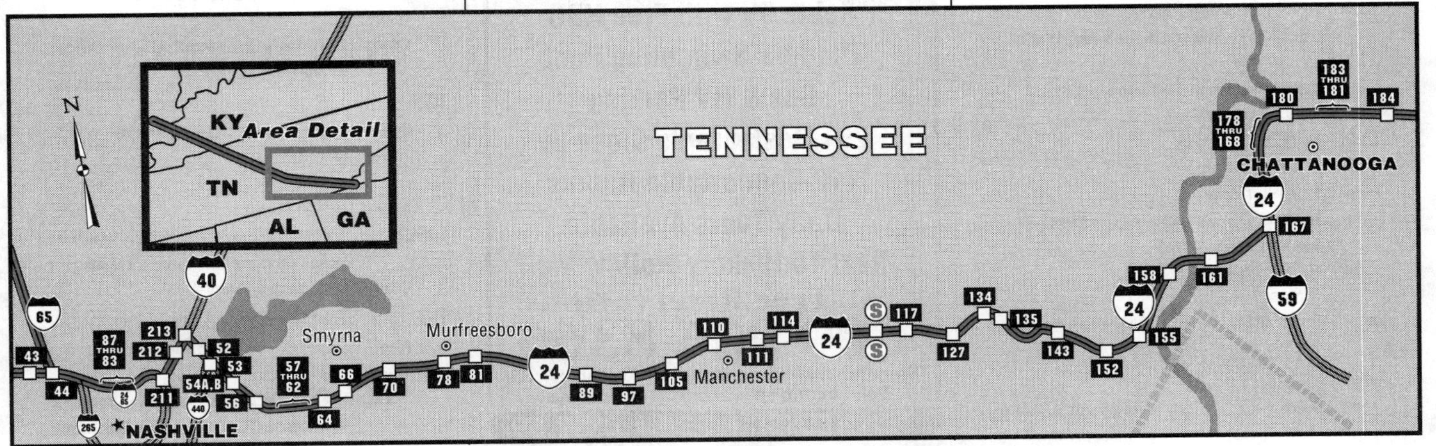

Bold red print shows RV & Bus parking available or nearby

EXIT		TENNESSEE
59		TN 254, Bell Road
	Gas	N: Shell*
		S: Amoco*[LP] (Kerosene), BP*[D][CW], Texaco*[D][CW][LP]
	Food	N: Applebee's, Arby's, Chuck E Cheese Pizza, Cracker Barrel, KFC, Logan's Roadhouse Steaks, McDonalds (Playground), O'Charley's, Pizza Hut, Red Lobster
		S: Casa Fiesta Mexican Restaurant (Quality Inn), Evergreen Chinese, LongHorn Steakhouse, Olive Garden, Shoney's, Waffle House
	Lodg	N: AAA Days Inn (see our ad this page)
		S: Knights Inn, Quality Inn, AAA The Quarters Motor Inn
	AServ	N: Bill Heard Chevrolet/Geo, Firestone Tire & Auto
		S: Bill Heard Collision Center, Gary Force Acura, Texaco, Valvoline Quick Lube Center
	ATM	N: First American, First Union, SunTrust, Union Planter's National Bank
		S: Texaco
	Other	N: **Hickory Hollow Shopping Mall, Horner-Rauch Optical Superstore, Toys "R" Us**
		S: **Pet Med, Target Department Store, The Home Depot**[LP]
60		Hickory Hollow Parkway
	Food	N: Arby's Roast Town, Chuck E Cheese Pizza, Logan's Roadhouse, Outback Steakhouse, Panda Garden Chinese, T.G.I. Friday's, Wendy's
	Lodg	N: Hampton Inn, ◆ Holiday Inn
	AServ	N: Bill Heard Chevrolet, Hickory Hollow Mazda, Master Care Auto Service, Mr. Transmission
	ATM	N: AMSouth Bank, First American
	Other	N: **Discovery Zone (Children's Play Place), Hickory Hollow Cinemas, Media Play Bookstore, Etc., Super Petz Pet Owner's Warehouse, US Post Office, Valu Vision Optometrist**
62		TN 171, Old Hickory Boulevard
	TStop	N: **Music City Auto/Truck Stop***[SCALES] **(Auto Fuel-BP, CB Shop)**
	FStop	N: **Speedway***[LP]
	Gas	N: Amoco*[D][CW], Texaco*
	Food	N: Blimpie's Subs* (Amoco), **Country Fiddler (Music City TS)**, Waffle House
	Lodg	N: AAA Best Western (Music City Inn)
	TServ	N: **76 (Music City TS)**
	ATM	N: **Music City TS**, Texaco
	Other	N: **Starwood Amphitheater**
64		Waldron Road, La Vergne
	FStop	N: **Speedway***[SCALES] **(RV Dump)**
	Gas	N: Exxon*, Kwik Sak*[LP]
		S: Mapco Express* (Kerosene)
	Food	N: Arby's, **Hardee's, Krystal**, McDonalds (Playground), **Subway (Speedway), Waffle House**
		S: **Daisy's Kitchen (Driftwood Inn)**, Pizza Neatza, Rice Bowl II
	Lodg	N: AAA Comfort Inn
		S: **Driftwood Inn**
	TServ	N: **International**
	RVCamp	N: **Music City Campground (1.9 Miles), RV Service (1 Mile, 7 days)**
	ATM	N: Exxon, **Speedway**
		S: Mapco Express
	Other	S: **La Vergne Fire Dept.**, New Wave Auto Wash
66AB		TN 266, Sam Ridley Pkwy, Smyrna
	Gas	N: Texaco*
	Food	N: Baskin Robbins (Texaco), Subs & Salads (Texaco), The Catfish House Seafood-Chicken-Steaks
	RVCamp	N: **R.V. Service (1.25 Miles)**
70		TN 102, Lee Victory Pky, Almaville Rd, Smyrna
	Gas	S: Amoco*, Citgo*[LP], Exxon[D] (Kerosene), Mapco Express*[D][LP] (Kerosene)
	AServ	S: Exxon
	RVCamp	N: **Nashville I-24 Campground, New Smyrna Campground**
	ATM	S: Amoco
	Other	N: **Sam Davis Home**
74		TN 840 East, Knoxville
78AB		TN 96, Franklin, Murfreesboro (Tn Technology Center At Murfreesboro)
	Gas	N: Phillips 66*[D], Shell, Texaco*
		S: Amoco*[CW][LP], BP*[CW] (Kerosene), Chevron* (24 Hrs), Exxon*[D]
	Food	N: Baskin Robbins, Burger King (Texaco), **Cracker Barrel**, KFC, McDonalds, Murfree's, Outback Steakhouse, Subway, Waffle House, Wendy's
		S: Corkey's Ribs & BBQ, Hardee's, Waffle House
	Lodg	N: Best Western, Comfort Inn, Country Inn & Suites, Garden Plaza Hotel, Hampton Inn, Holiday Inn, Microtel Inn, Motel 6, Wyngate Inn
	AServ	N: Parkway Auto & Tire Service, Shell, **Wal-Mart**
		S: **Sam's Club**
	Med	N: **+ Hospital**
	ATM	N: SunTrust Bank
		S: Cavalry Bank
	Other	N: **Fed Ex Drop Box, Golf USA Fun Park, Oaklands State Historical Site, Office Max, Wal-Mart Supercenter (24 Hrs), William's Animal Hospital, YMCA**
		S: Auto Pride Car Wash, **Nashville Pet Products Center, Outlets LTD. Mall, Sam's Club**
81AB		U.S. 231, Murfreesboro, Shelbyville (81B-Middle TN State Univ.)
	FStop	N: **Uncle Sandy's BP**
		S: **Phillips 66***[CW][LP][SCALES]
	Gas	N: Amoco[D], Racetrac*
		S: Citgo*[LP], Mapco Express*[D][LP] (Kerosene), Texaco*[LP]
	Food	N: Burger King (Playground), Cracker Barrel, Hunan Chinese, Shoney's, The Parthenon Steakhouse, Waffle House, Wendy's
		S: McDonalds (Playground), Q's Country Cookin' (Howard Johnson), Subway, Whitt's BBQ
	Lodg	N: **Days Inn (see our ad this page), Scottish Inns**, Shoney's Inn, Travelodge
		S: Howard Johnson, Quality Inn, Safari Inn
	AServ	S: Neill Sandler Toyota
	Med	N: **+ Hospital**
	ATM	S: Citgo, **Phillips 66**, Texaco
	Other	N: Car Wash, **Direct Factory Outlet Fireworks Supermarket, Keepsake Antique Mall Center, Oakland State Historic Site**
		S: **Laundry (Coin Laundry)**
89		Buchanan Road
	FStop	S: **Citgo Danny's Food & Fuel Plaza* (Kerosene)**
	Gas	N: Texaco*
	Food	S: Crazy J's Fireworks & Hot Grill, **Huddle House (Citgo)**
	Other	N: **Crazy J's Fireworks & Hot Grill**
97		TN 64, Beechgrove Road, Shelbyville
	Gas	N: Beech Grove General Store*
	AServ	N: The Import Clinic
	Other	N: **Parker's Quick Mart*, US Post Office**
		S: **Wartrace Bell Buckle**
105		U.S. 41, Manchester
	TStop	N: **Busy Corner Truck Stop* (BP)**
	Gas	N: Texaco*[D]
		S: Amoco*[LP]
	AServ	N: Best Tire Co.
	RVCamp	S: **KOA Campground (8 Miles), Whispering Oaks Campground (Approx. 1 Mile)**
	Other	S: **Normandy Dam**
110		TN 53, Manchester, Woodbury
	FStop	S: **Texaco*[D] (FedEx Drop)**
	Gas	N: BP*, Exxon*, Phillips 66*
	Food	N: **Cracker Barrel, Davy Crockett's Roadhouse**, TCBY (Exxon), The Oak Family

Bold red print shows RV & Bus parking available or nearby

EXIT	TENNESSEE
	Restaurant
	S: Los 3 Amigos, Waffle House
Lodg	N: Ambassador Inn & Luxury Suites, ◆ Days Inn, Hampton Inn
AServ	S: Hulletts Service Center
TServ	S: Hulletts Service Center
Med	N: ✚ Hospital
ATM	N: Exxon, Phillips 66
Parks	S: Old Stone Fort State Park
111	TN 55, Manchester, McMinnville
Gas	N: Amoco*(D)(LP) (24 Hrs, Kerosene), Chevron*(D) (Kerosene), Citgo*(LP)
	S: BP*(D) (Kerosene)
Food	N: Cafe (Amoco)
	S: J & G Pizza & Steakhouse, Porky's Pit BBQ
AServ	S: Gateway Tire & Service Center
Med	N: ✚ Hospital
	S: ✚ Hospital
ATM	N: Amoco, Citgo
Parks	N: Rock Island State Park (37 Miles)
	S: Old Stone Fort State Park (3 Miles), Tim's Ford State Park (23 Miles)
Other	N: All Creature's Veterinary Clinic, Antique Mall, Coffee Veterinary Hospital, The Corn Crib
114	U.S. 41, Manchester
TStop	N: Jiffy Truck/Auto Plaza*(SCALES) (Showers, Kerosene, FedEx Drop)
Gas	N: Phillips 66*(D), Texaco*(D)
	S: BP*, Exxon*(LP), RaceWay* (24 Hrs)
Food	N: Longfellow's Family Restaurant (Ramada Inn), Trucker's Inn Cafe (Jiffy T/A Plaza)
	S: Arby's, Burger King (Playground), Captain D's Seafood, KFC, McDonalds (Playground), Shoney's, TCBY (Exxon), Taco Bell, Waffle House, Wendy's
Lodg	N: AAA Ramada Inn, Scottish Inns, AAA Super 8 Motel, Trucker's Inn (Jiffy T/A Plaza)
	S: Budget Motel, ◆ Holiday Inn, Red Carpet Inn
AServ	N: Roberts Nissan/Toyota
	S: Auto Zone Auto Parts, Bobby Vann Chevrolet/Geo
RVCamp	N: KOA Kampground(LP)
ATM	S: Bi-Lo Supermarket
Other	N: Arrowhead Aerospace Museum, Train Toy WWII Museum
	S: Bi-Lo Supermarket, Wal-Mart
(116)	Weigh Station (Both Directions)
117	U.S. Air Force, Arnold Center, Tullahoma, UT Space Institute
(120)	Parking Area - No Facilities (Both Directions)
127	US 64, TN 50, Pelham, Winchester
Gas	N: Amoco*(D), Phillips 66*(D), Texaco*
	S: Exxon*(D)
Food	N: Stuckey's (Texaco)
Parks	S: Tim's Ford State Park
Other	N: Wonder Cave (5 Miles)
(134)	Rest Area - RR, Vending, Phones (Both Directions, Under Construction)
134	U.S. 41A, Monteagle, Sewanee
Gas	N: Exxon, The Depot*(LP)
	S: Chevron*(D), Citgo*(LP), Texaco*
Food	N: Papa Ron's Pizza
	S: Hardee's (Playground), Jim Oliver's Smokehouse Restaurant & Trading Post, Tony's Deli (Chevron), Tubby's Diner, Waffle House
Lodg	S: Budget Host Inn, Jim Oliver's Smokehouse Lodge
AServ	N: Exxon
	S: Monteagle Firestone
TServ	S: Monteagle Firestone
ATM	S: Citgo
Parks	N: South Cumberland State Park
Other	S: Monteagle Wedding Chapel, Piggly Wiggly Grocery, Sewanee University of the South
135	U.S. 41 North, Monteagle, Tracy City
TStop	N: 76 Monteagle Truck Plaza(SCALES)
FStop	N: Citgo*(D), Phillips 66*(LP)
Food	N: Deli (Phillips 66), Pop's Happyland Restaurant & Truck Stop, Unocal (Monteagle Truck Plaza)
	S: Spanky's
Lodg	S: Days Inn
AServ	N: Tate's Garage
TServ	N: Monteagle Truck Plaza
(136)	Truck Inspection Station (Eastbound)
(137)	Left Runaway Truck Exit
(139)	Left Runaway Truck Ramp
143	Martin Springs Road
FStop	N: Chevron*
152	U.S. 41, U.S. 64, U.S. 72, Kimball, S. Pittsburg
Gas	N: Amoco*, Phillips 66* (TN/AL Fireworks), Racetrac*, Shell
Food	N: Arby's, China Inn, KFC, Krystal, Long John Silvers, McDonalds (Playground), Pizza Hut, Shoney's, Subway, Taco Bell, Waffle House
	S: Bubba's Down Home Pizza, Oscar's BBQ
Lodg	N: Budget Host Inn Kimball Motor Lodge, AAA Days Inn
AServ	N: Jentry Geo/GM/Chevrolet, Shell
Med	S: ✚ Hospital
ATM	N: Citizen's State Bank, Pioneer Bank (In front of Wal-Mart)
Other	N: Bi-Lo Supermarket (Pharmacy), Chattanooga State Technical Community College, Eyear Optical, Laundromat, New Tennessee Alabama Fireworks, Russel Cave National Monument, Space Age Factory Outlet (Fireworks, Amoco), Wal-Mart
	S: Bean Raulston Graveyard (7 Miles)
155	TN 28, Jasper, Dunlap, Whitwell
Gas	N: Amoco*(D)
Food	N: Hardee's, Western Sizzlin'

EXIT	TENNESSEE/GA
Lodg	N: Acuff Country Inn
Other	N: Fireworks Supermarket (24 Hrs)
158	TN 27, Nickajack Dam, Powells Crossroads
FStop	N: Phillips 66* (Accessible Via Both Sides)
	S: Phillips 66* (Accessible Via Both Sides)
Gas	N: Texaco* (Kerosene)
RVCamp	S: Camping (3 Miles)
Other	N: Tennessee/Alabama Fireworks (Accessible Via Both Sides)
	S: Tennessee/Alabama Fireworks (Accessible Via Both Sides)
(160)	Rest Area - RR, HF, Phones, Picnic, RV Parking, No Trucks (Both Directions)
161	TN156, Haletown, New Hope
Gas	S: Chevron* (Big Daddy's Fireworks)
AServ	S: Glenn's Wrecker Service (24 Hrs)
RVCamp	S: Camp on the Lake
Other	S: Big Daddy's Fireworks, TVA Maple View Recreation Area, Wholesale Fireworks
167	Junction I-59 South, Birmingham

↑TENNESSEE

↓GEORGIA

EXIT	GEORGIA
169	Georgia 299 to U.S. 11
FStop	N: $av-A-Ton*
	S: Amoco*(LP)(SCALES) (24 Hrs, Kerosene), Cone Auto/Truck Plaza*

↑GEORGIA

↓TENNESSEE

EXIT	TENNESSEE
(172)	TN Welcome Center - RR, Vending, Phones, Dog Walk
174	Junction U.S. 11, U.S. 41, U.S. 64, Lookout Valley & Mtn.
FStop	S: Amoco*(LP)
Gas	S: BP*, Exxon*
Food	N: Waffle House
	S: C Bar-B-Q, Cracker Barrel, Taco Bell, Waffle House
Lodg	N: AAA Days Inn, ◆ Holiday Inn
	S: AAA Best Western, Econolodge, Fricks Motel, AAA Hampton Inn, Ramada Limited (see our ad this page), ◆ Super 8 Motel
RVCamp	N: Barren Mountain (1.9 Miles)
	S: Lookout Valley RV Park (.5 Miles), Raccoon Mtn, Crystal Caverns, Campground (1.3 Miles)
ATM	S: AMSouth Bank
Other	N: Cravens House
	S: Battle Above the Clouds (Civil War Battle), Tiftonia True Value Hardware
175	Browns Ferry Road, Lookout Mtn (Difficult Reaccess)
FStop	S: Conoco*(LP) (Kerosene, 24 Hrs)
Gas	N: Citgo*(LP), Exxon* (Kerosene)
	S: Strobel Oil (Kerosene), Texaco*(D)(LP) (FedEx)
Food	N: Mike's Pizza & Ice Cream, Mountain Manor (Knights Inn)
	S: Hardee's, Subway
Lodg	N: Knights Inn (see our ad this page)
	S: Comfort Inn
AServ	S: Strobel Oil
ATM	S: SunTrust Bank

EXIT TENNESSEE

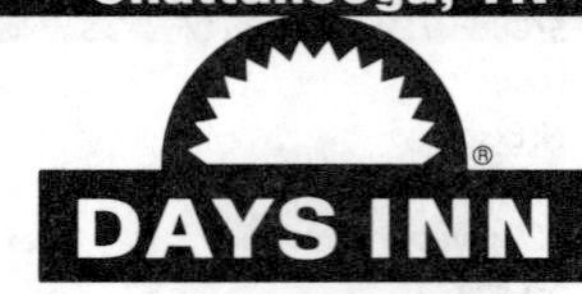

Other N: **Food Lion Supermarket, Lookout Valley Post Office, Revco Drugs**
S: **Bi-Lo Supermarket (Pharmacy)**

178 U.S. 27 N, US 11, US 64, U.S. 41, Lookout Mtn, Downtown Chattanooga, Tennessee Aquarium
- **FStop** N: **Amoco**
- **Gas** N: Amoco(D) (Kerosene), BP* (Kerosene)
- **Food** N: Chatt's Restaurant (Guesthouse Inn), Restaurant (Knights Inn)
  S: KFC
- **Lodg** N: Days Inn, Guesthouse Inn, Knights Inn (see our ad this page)
  S: ◆ Comfort Suites, ◆ Hampton Inn, Microtel Inn
- **AServ** N: Amoco, Ford Truck Center, Grant Auto Glass Co., Midas Muffler & Brakes, Mountain View Ford, Mountain View Nissan
  S: Auto Care
- **Other** N: **Mr. C's Meats & Things**
  S: **Bits of History Antique Mall, Historical Chattanooga Choo-Choo**

180A TN 8 North, Rossville Blvd, Central Ave (Chattanooga, Difficult Westbound Reaccess)
- **Gas** N: Ted's
- **Food** N: Chattanooga Fisheries Seafood & Deli, Master Blasters Barbeque
- **AServ** N: Brakes & More, Computer Car Care, Discount Tire, Foreign Car Parts, Ken Smith Auto Parts, TNT Auto Parts, Ted's, Tire Centers Inc., Universal Tire & Auto
- **Other** N: **Coin Laundry**, Hy's Car Wash, **Korner Market**

180B U.S. 27 South, Rossville Blvd (Difficult Westbound Reaccess)
- **FStop** S: **Speedway* (Deli)**
- **Gas** S: Phillips 66*(LP), Racetrac*
- **Food** S: Juke Box Players Cafe, Long John Silvers
- **AServ** S: Baker's Transmission & Muffler, Roy's Tire Service, Scenic City Auto Repair (NAPA), Tennessee Automotive, Willie's Auto Parts

EXIT TENNESSEE

- **Med** S: **+ Medical Care Center (Walk-Ins Welcome)**
- **Other** S: **Automotive Part Exchange, Eyear One Hour Optical**

181 Fourth Ave
- **Gas** N: Amoco*, Exxon(D) (24 Hrs)
  S: Citgo*
- **Food** N: Burger King, Central Park Burgers, Hardee's, Krystal, **Mick's 24 Hour Diner**, The Jones Family BBQ, **Waffle House**
  S: Ricky's Restaurant
- **Lodg** N: (AAA) Quality Inn
- **AServ** N: Lancaster Auto Parts, Mr. Transmission
- **TServ** N: **Doug Yate's Wrecker Service (24 Hrs, Diesel Repair), Mack Trucks Cummins, Cat, & Mercedes**
- **Med** N: **+ Hospital**
- **ATM** N: First Tennessee Bank
- **Other** N: **Animal World Boarding & Supplies, Tennessee Temple University**
  S: **Don's Fish Market, Stewart's Coin Laundry**

181A U.S. 41 South, East Ridge (Eastbound)
- **Gas** N: Amoco*
- **Food** S: Ridge Cut Cafe (King's Lodge), Westside Grille & Bar
- **Lodg** S: Gateway Motel, King's Lodge

183A **(184)** Belvoir Ave, Germantown Road (Westbound)

184 Moore Road, Chattanooga (Westbound 184 Is Closed, Difficult Reaccess If Eastbound)
- **AServ** S: **K-Mart**
- **Med** S: **+ Hospital**
- **Other** S: **K-Mart (Pharmacy), Winn Dixie Supermarket (Pharmacy)**

185AB Junction I-75

↑TENNESSEE

Begin I-24

I-25 S →

EXIT WYOMING

Begin I-25

↓WYOMING

(300) Jct I-90, Gillette, Billings

299 U.S. 16, Buffalo
- **TStop** W: **Big Horn Travel Plaza**
- **Gas** E: Conoco*(D)(LP), Texaco*
- **Food** W: Col. Bozeman's Steakhouse, Crossroads Inn, Dash Inn Restaurant, Hardee's, McDonalds, Pizza Hut, Subway, Taco John's, The Breadboard Sub Shop
- **Lodg** E: ◆ Comfort Inn, Stage Brush Inn
  W: Crossroads Inn, Econolodge, (AAA) Super 8 Motel, (AAA) Wyoming Motel
- **AServ** E: Conoco
- **RVCamp** E: **KOA Campground**
  W: **Camping**
- **Other** W: Squeeky Klean Car Wash, **Tourist Information**

298 Business 25, Business U.S. 87, Buffalo
- **Other** W: **Visitor Information**

291 Trabing Road

280 Middlefork Road

EXIT WYOMING

265 Reno Road

254 Kaycee
- **FStop** E: **Texaco*(CW)**
  W: **Sinclair***
- **Food** E: Country Inn Dining, The Feed Rack Cafe, The Sundance Grill
- **Lodg** E: Cassidy Inn Motel, Ciesta Motel
- **AServ** E: Texaco
- **Other** E: **Grocery Store, Kaycee Meat Market, Kaycee Police Department, US Post Office**
  W: **Rest Area-Picnic, RR, Grills, RVDump, RVDrinking Water (Phones)**

(254) **Rest Area - RR, Phones, Picnic, RV Dump**

249 TTT Road

246 Powder River Road

235 Tisdale, Mountain Road

227 WY387, Midwest, Edgerton

223 Kaycee

(219) **Parking Area (Both Directions)**

216 Ranch Road

210 Horse Ranch Creek Road

EXIT WYOMING

197 Ormsby Road

191 Wardwell Road, Bar Nunn
- **Gas** W: Texaco*
- **Other** W: **Antelope Campground, KOA Campground**

189 U.S. 20, U.S. 26, Shoshoni (Port Of Entry)
- **Other** W: **Wyoming Port of Entry**

188B WY220, Poplar St
- **Gas** E: Rider(D)
  W: Exxon*
- **Food** E: El Jarro, Hilton Inn, JB's Restaurant
  W: Casper's Family Dining, Dairy Queen
- **Lodg** E: Hampton Inn, ◆ Hilton Inn, Holiday Inn, ◆ Kelly Inn, Motel 6
- **RVCamp** W: **Camping**
- **Other** E: **Casper Planetarium**
  W: **Fort Casper**

188A Bus Loop I-25, U.S. 87 Bus, Center St, Downtown Casper
- **FStop** E: **Texaco*(CW)**
- **Gas** E: Conoco(D), Mini Mart, Texaco
  W: Cenex*
- **Food** E: Benham's Great Food, Taco John's
  W: Denny's (Parkway Plaza Motel), Subway
- **Lodg** E: Holiday Inn, National 9 Inn/Showboat Motel

Bold red print shows RV & Bus parking available or nearby

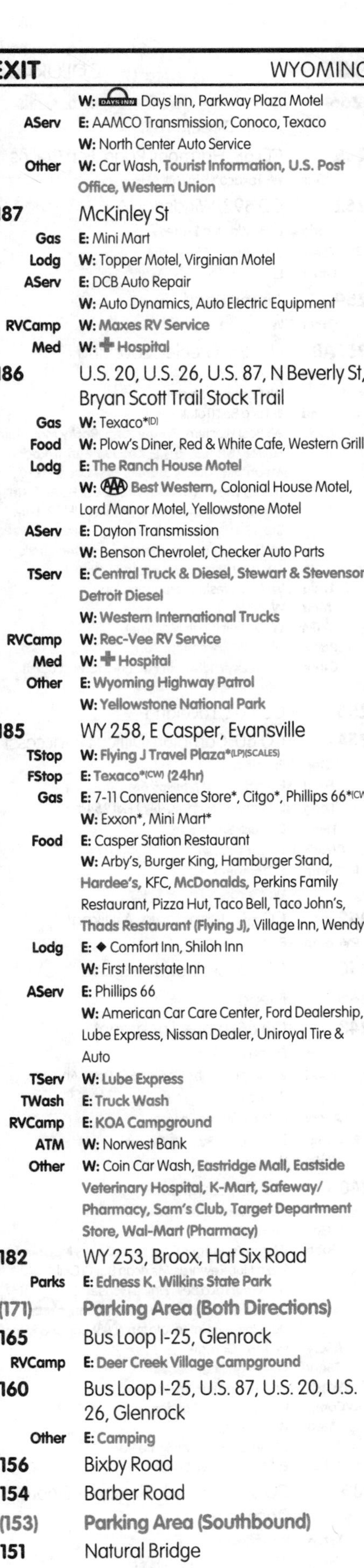

EXIT WYOMING

	W: Days Inn, Parkway Plaza Motel
AServ	**E:** AAMCO Transmission, Conoco, Texaco
	W: North Center Auto Service
Other	**W:** Car Wash, **Tourist Information, U.S. Post Office, Western Union**
187	McKinley St
Gas	**E:** Mini Mart
Lodg	**W:** Topper Motel, Virginian Motel
AServ	**E:** DCB Auto Repair
	W: Auto Dynamics, Auto Electric Equipment
RVCamp	**W: Maxes RV Service**
Med	**W: ✚ Hospital**
186	U.S. 20, U.S. 26, U.S. 87, N Beverly St, Bryan Scott Trail Stock Trail
Gas	**W:** Texaco*(D)
Food	**W:** Plow's Diner, Red & White Cafe, Western Grill
Lodg	**E: The Ranch House Motel**
	W: AAA **Best Western,** Colonial House Motel, Lord Manor Motel, Yellowstone Motel
AServ	**E:** Dayton Transmission
	W: Benson Chevrolet, Checker Auto Parts
TServ	**E: Central Truck & Diesel, Stewart & Stevenson Detroit Diesel**
	W: Western International Trucks
RVCamp	**W: Rec-Vee RV Service**
Med	**W: ✚ Hospital**
Other	**E: Wyoming Highway Patrol**
	W: Yellowstone National Park
185	WY 258, E Casper, Evansville
TStop	**W: Flying J Travel Plaza*(LP)(SCALES)**
FStop	**E: Texaco*(CW) (24hr)**
Gas	**E:** 7-11 Convenience Store*, Citgo*, Phillips 66*(CW)
	W: Exxon*, Mini Mart*
Food	**E:** Casper Station Restaurant
	W: Arby's, Burger King, Hamburger Stand, **Hardee's,** KFC, **McDonalds,** Perkins Family Restaurant, Pizza Hut, Taco Bell, Taco John's, **Thads Restaurant (Flying J),** Village Inn, Wendy's
Lodg	**E:** ◆ Comfort Inn, Shiloh Inn
	W: First Interstate Inn
AServ	**E:** Phillips 66
	W: American Car Care Center, Ford Dealership, Lube Express, Nissan Dealer, Uniroyal Tire & Auto
TServ	**W: Lube Express**
TWash	**E: Truck Wash**
RVCamp	**E: KOA Campground**
ATM	**W:** Norwest Bank
Other	**W:** Coin Car Wash, **Eastridge Mall, Eastside Veterinary Hospital, K-Mart, Safeway/ Pharmacy, Sam's Club, Target Department Store, Wal-Mart (Pharmacy)**
182	WY 253, Broox, Hat Six Road
Parks	**E: Edness K. Wilkins State Park**
(171)	**Parking Area (Both Directions)**
165	Bus Loop I-25, Glenrock
RVCamp	**E: Deer Creek Village Campground**
160	Bus Loop I-25, U.S. 87, U.S. 20, U.S. 26, Glenrock
Other	**E: Camping**
156	Bixby Road
154	Barber Road
(153)	**Parking Area (Southbound)**
151	Natural Bridge
150	Inez Road

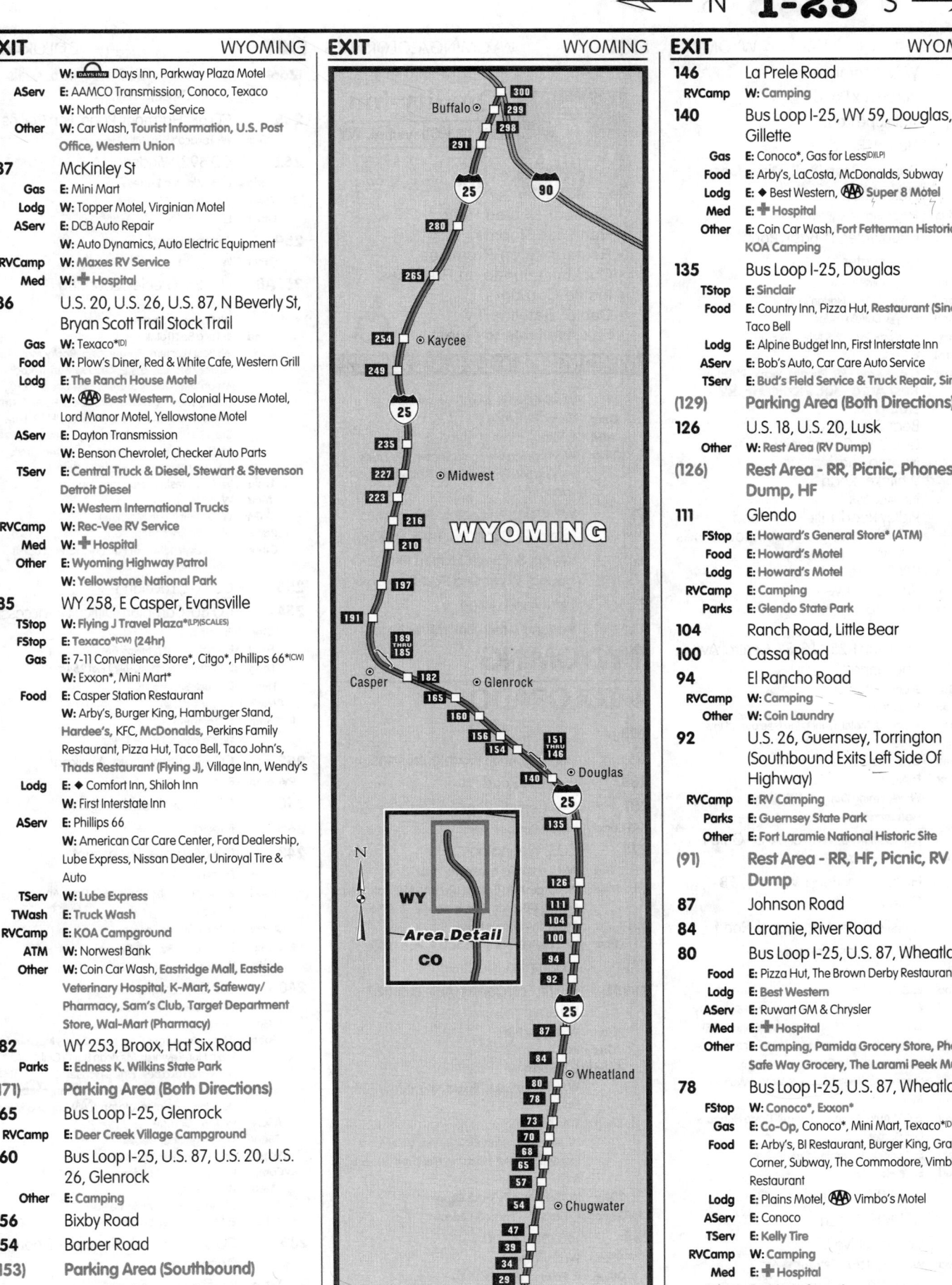

EXIT WYOMING

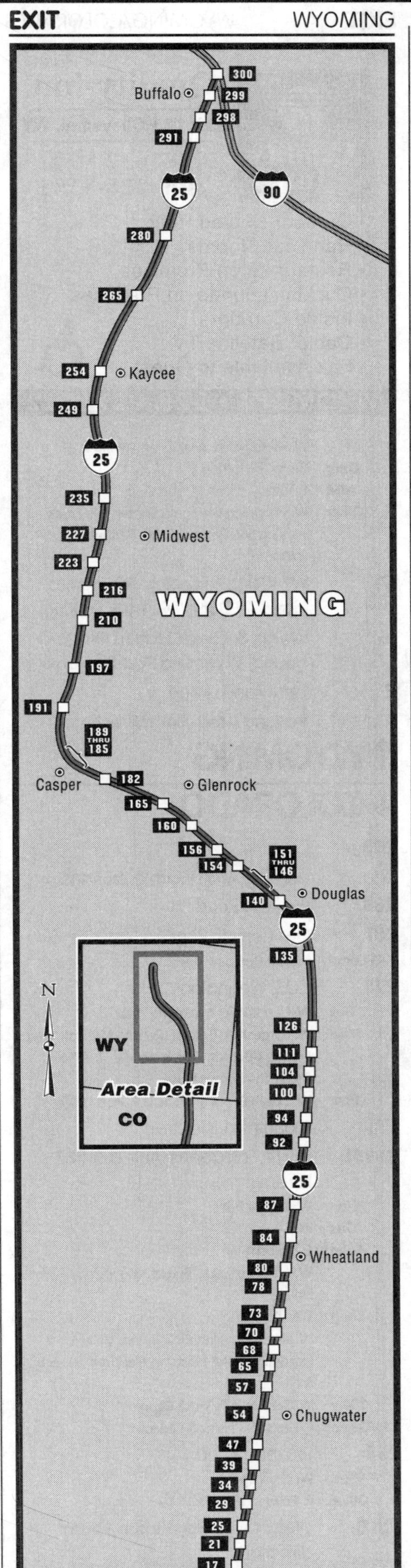

EXIT WYOMING

146	La Prele Road
RVCamp	**W: Camping**
140	Bus Loop I-25, WY 59, Douglas, Gillette
Gas	**E:** Conoco*, Gas for Less(D)(LP)
Food	**E:** Arby's, LaCosta, McDonalds, Subway
Lodg	**E:** ◆ Best Western, AAA **Super 8 Motel**
Med	**E: ✚ Hospital**
Other	**E:** Coin Car Wash, **Fort Fetterman Historic Site, KOA Camping**
135	Bus Loop I-25, Douglas
TStop	**E: Sinclair**
Food	**E:** Country Inn, Pizza Hut, **Restaurant (Sinclair),** Taco Bell
Lodg	**E:** Alpine Budget Inn, First Interstate Inn
AServ	**E:** Bob's Auto, Car Care Auto Service
TServ	**E: Bud's Field Service & Truck Repair, Sinclair**
(129)	**Parking Area (Both Directions)**
126	U.S. 18, U.S. 20, Lusk
Other	**W: Rest Area (RV Dump)**
(126)	**Rest Area - RR, Picnic, Phones, RV Dump, HF**
111	Glendo
FStop	**E: Howard's General Store* (ATM)**
Food	**E: Howard's Motel**
Lodg	**E: Howard's Motel**
RVCamp	**E: Camping**
Parks	**E: Glendo State Park**
104	Ranch Road, Little Bear
100	Cassa Road
94	El Rancho Road
RVCamp	**W: Camping**
Other	**W: Coin Laundry**
92	U.S. 26, Guernsey, Torrington (Southbound Exits Left Side Of Highway)
RVCamp	**E: RV Camping**
Parks	**E: Guernsey State Park**
Other	**E: Fort Laramie National Historic Site**
(91)	**Rest Area - RR, HF, Picnic, RV Dump**
87	Johnson Road
84	Laramie, River Road
80	Bus Loop I-25, U.S. 87, Wheatland
Food	**E:** Pizza Hut, The Brown Derby Restaurant, Timber
Lodg	**E: Best Western**
AServ	**E:** Ruwart GM & Chrysler
Med	**E: ✚ Hospital**
Other	**E: Camping, Pamida Grocery Store, Pharmacy, Safe Way Grocery, The Larami Peek Museum**
78	Bus Loop I-25, U.S. 87, Wheatland
FStop	**W: Conoco*, Exxon***
Gas	**E: Co-Op,** Conoco*, Mini Mart, Texaco*(D)
Food	**E:** Arby's, Bl Restaurant, Burger King, Granny's Corner, Subway, The Commodore, Vimbo's Restaurant
Lodg	**E:** Plains Motel, AAA Vimbo's Motel
AServ	**E:** Conoco
TServ	**E: Kelly Tire**
RVCamp	**W: Camping**
Med	**E: ✚ Hospital**
Other	**E: Visitor Information**
	W: Coin Laundry (Conoco FS)

EXIT	WYOMING
73	WY 34, Laramie
70	Bordeaux Road
68	Antelope Road
66	Hunton Road
(65)	**Parking Area (Both Directions)**
65	Slater Road
Other	**E:** Parking Area (Both Sides)
57	Chugwater, TY Basin Road
54	Chugwater
FStop	**E:** Sinclair*(LP)
Food	**E:** Buffalo Grill, Horton's Corner
Lodg	**E:** AAA Buffalo Lodge
RVCamp	**W:** Camping (12.5 Miles May-Oct)
ATM	**E:** Sinclair
Other	**E:** Rest Area - RR, Phones, Picnic, RVDump
(54)	**Rest Area - RR, Picnic, HF, RV Dump**
47	Bear Creek Road
39	Little Bear Community
34	Nimmo Road
29	Whitaker Road
25	No Access
21	Ridley Road, Little Bear Road
17	U.S. 85, Torrington (Southbound Exits Left Side Of Highway)
16	WY 211, Horse Creek Road
13	Vandehei Ave
Gas	**E:** Mini Mart*
	W: Total*
12	Bus Loop I-25, U.S. 87, Central Ave, Cheyenne
Gas	**E:** Conoco*
Food	**E:** Central Cafe
Lodg	**E:** Central Motel, Quality Inn (see our ad this page)
Med	**E:** ✚ Hospital
Other	**E:** Airport
	W: Wyoming Game & Fish Commission Headquarters
11	Randall Ave, Warren AFB - Gate 1
Other	**W:** Warren AFB
10B	Happy Jack Road, Warren AFB - Gate 2
10D	Missile Drive, Happy Jack Road
9	U.S. 30, West Lincolnway
TStop	**W:** Little America*
FStop	**E:** Conoco(SCALES)
Food	**E:** Denny's, G & C Repair (Conoco FS), Luxury Diner (Conoco FS), Restaurant (Conoco FS)
Lodg	**E:** Best Western, DAYS INN Days Inn, Econolodge, Firebird Motel, La Quinta Inn, Motel 6, Stagecoach Motel, Super 8 Motel, Wyoming Motel
AServ	**E:** GM Auto Dealership, Tyrrell Doyle Honda & GM Dealer
TServ	**E:** Conoco, Wyoming Cat
Med	**E:** ✚ Hospital
8B	Jct I-80W, Laramie
8D	Jct I-80E, Omaha
7	College Drive
TStop	**W:** Flying J Travel Plaza(LP)(SCALES)
FStop	**E:** Total*(LP)
Gas	**W:** Conoco* (Flying J)
Food	**E:** Subway (Total)

EXIT	WYOMING/COLORADO
	W: McDonalds, Restaurant (Flying J)
Lodg	**W:** ◆ Comfort Inn
ATM	**E:** Total
Other	**W:** Wyoming Welcome Center - No Truck Parking (RR, Phones, Grills, Information, RVDump)
(7)	**WY Welcome Center - RR, Phones, Picnic, RV Dump (No Truck Parking)**
(6)	**Weigh & Check Station (North-bound, Wyoming Port Of Entry)**
2	Terry Ranch Road
(0)	**Parking Area (Both Directions)**

↑WYOMING

↓COLORADO

EXIT	COLORADO
293	Carr
(296)	**Parking Area (Both Directions)**
288	Buckeye Road
281	Owl Canyon Road
RVCamp	**E:** KOA Campground
278	CO 1, Wellington
Gas	**W:** Conoco*(CW), Minimart*, Total
Food	**W:** Burger King (Total), El Rancho Mexican, Pizza Palace, T-Bar-Inn Cafe & Lounge, Taco Bell (Total), The Coffee House
Other	**W:** Susy's Grocery Store, U.S. Post Office
271	Mountain Vista Drive
269AB	CO 14, Fort Collins, Ault, U.S. 287 Laramie
FStop	**W:** Phillips 66*
Gas	**W:** Conoco*
Food	**E:** Mulberry Inn
	W: Al Fresco Italian, Burger King, Denny's, Waffle House
Lodg	**E:** Mulberry Inn
	W: DAYS INN Days Inn, Holiday Inn, Motel 6, National 9 Motel, Plaza Inn, Sleep Inn, Super 8 Motel
TServ	**W:** Schwinvell's Truck Repair
RVCamp	**E:** Lee's RV Center & Service
268	Prospect Road
Med	**W:** ✚ Hospital
Other	**E:** Powder River RV Dealer
(267)	**Weigh & Check Station - both directions**
(266)	**Rest Area - RR, Picnic, Info, Grills (Both Directions)**
265	CO 68, Harmony Road, Fort Collins
Gas	**W:** Texaco*(LP) (ATM)
262	CO 392, Windsor
Lodg	**E:** Holiday Inn Express
TServ	**W:** WhiteGMC Dealer
Other	**E:** Airport
259	Airport Road
Other	**W:** Airport
257AB	U.S. 34, Greeley, Loveland
Gas	**E:** Total*
	W: Conoco*, Sinclair*
Food	**E:** Taco Bell (Total)
	W: Best Western, Burger King (Rocky Mountain Factory Stores), Old Country Store Cracker Barrel (Rocky Mountain Factory Stores), East Coast Pizza (Rocky Mountain Factory Stores), Food Court (Rocky Mountain Factory Stores), IHOP, Lone Star (Rocky Mountain Factory Stores), McDonalds (Rocky Mountain Factory Stores), Subway (Rocky Mountain Factory Stores)
Lodg	**W:** Best Western, Hampton Inn
Med	**W:** ✚ Hospital
ATM	**W:** Sinclair (Cash Machine)
Parks	**W:** Boyd Lake State Park
Other	**W:** Rocky Mountain Factory Stores, Tourist Information
255	CO 402, Loveland
254	CO 60, Campion (Difficult Reaccess)
TStop	**E:** Phillips 66
Food	**E:** Restaurant (Phillips 66)
Lodg	**E:** Budget Host Inn, AAA Exit 254 Inn
TServ	**E:** Truck Service
TWash	**E:** Truck Wash
RVCamp	**E:** Camping
	W: Murdock Trailer Sales
252	CO 60, Johnstown, Milliken
RVCamp	**E:** Phillips 66(LP)
250	CO 56, Berthoud
245	Mead
243	CO 66, Lyons, Longmont
Gas	**E:** Phillips 66*(D)
Food	**E:** Blimpie's Subs (Phillips 66), Peter Angelo's (Texaco)
AServ	**E:** Michelin Tire Service
RVCamp	**E:** K & C RV Sales & Service
Parks	**W:** Rocky Mountain National Park
240	CO 119, Longmont
FStop	**W:** Conoco*(SCALES), Texaco*(SCALES)
Gas	**W:** Total*(D)
Food	**W:** Arby's, Burger King, Little Oasis, McDonalds, Pizza Hut, Rewards Mexican Bar & Grill, Subway, Taco Bell, Waffle House
Lodg	**W:** Budget Host Inn, ◆ Comfort Inn, DAYS INN Days Inn, First Interstate Inn, AAA Super 8 Motel
AServ	**W:** M&S Garage
TServ	**W:** Freightliner Dealer
TWash	**W:** Del Camino Truck Wash (Showers)
RVCamp	**W:** RV Sales & Service
Med	**W:** ✚ Hospital
Parks	**W:** Barbour Ponds State Park
Other	**E:** Stevinson RV Center
235	CO 52, Frederick, Firestone, Eldora, Dacono
TStop	**W:** Phillips 66
Food	**W:** Restaurant (Phillips 66)
232	Erie

Bold red print shows RV & Bus parking available or nearby

EXIT		COLORADO
229		CO 7, Lafayette, Brighton
	RVCamp	**E: Camping**
223		CO 128, 120th Ave
	Gas	E: Diamond Shamrock*(CW), Sinclair, Texaco(CW)(LP)
		W: Total*(CW)(LP)
	Food	E: Applebee's, Damon's, Daybreak, Fazoli's Italian Food, Holiday Inn, Lone Star, Outback Steakhouse
		W: **Cracker Barrel**, Jade City Chinese Cafe, La Casa Loma Cafe, Perkins Family Restaurant, Radisson, The Village Inn, Wendy's
	Lodg	E: Days Inn, AAA Days Inn, ◆ Hampton Inn, AAA Holiday Inn, AAA Radisson
		W: ◆ La Quinta Inn, Super 8 Motel
	AServ	E: Brakes Plus, Discount Tire Company
	Other	**E: Albertson's Grocery, Checker Auto Parts**
221		104th Ave, Northglenn
	Gas	E: Conoco*(CW), Phillips 66*(CW), Self Service Gas(CW), Total*
		W: 7-11 Convenience Store*, Conoco*(CW), Texaco*(CW)
	Food	E: Burger King, Denny's, HoHo Chinese, IHOP, Subway
		W: Applebee's, Black Eyed Pea, Coco's, McDonalds, Taco Bell
	Lodg	W: ◆ Ramada Limited
	AServ	E: Midas Muffler & Brakes
		W: Omeara Ford Isuzu
	ATM	E: Total
	Other	**E: Dr. Scott's Optical, King's Super(LP) (Pharmacy, ATM), Target Department Store**
		W: Albertson's Grocery, Northglenn Mall, Payless Drugs

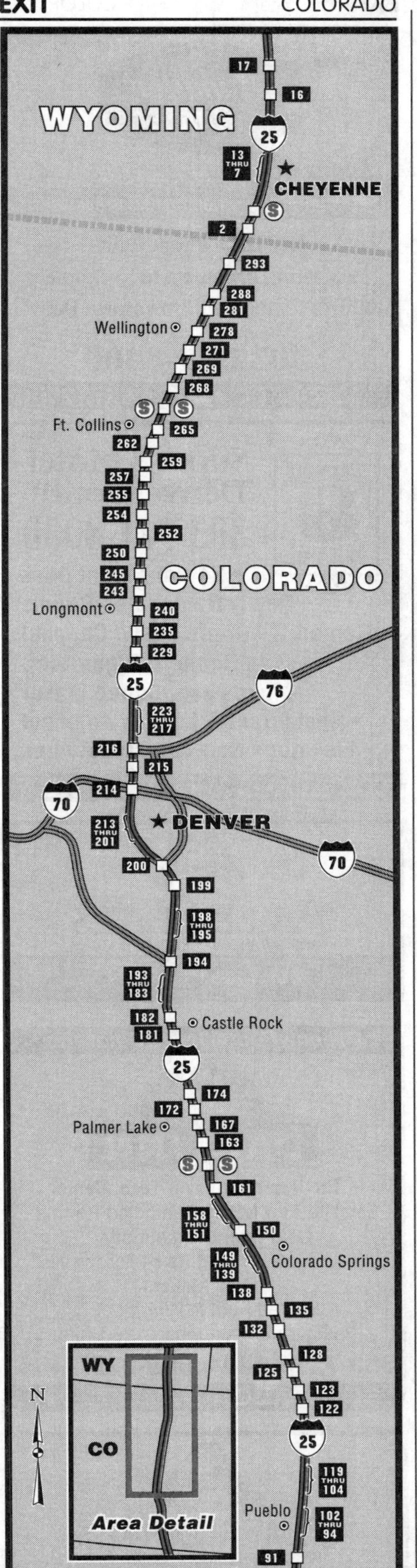

EXIT		COLORADO
220		Thornton Pkwy
	Med	**E: Hospital**
219		Federal Heights, 84th Ave
	Gas	E: Conoco*(CW)
		W: Amoco*(CW), Coastal*, Total*(CW)
	Food	E: Arby's, Bonanza Steak House, Dan & Elanore's Mexican, Goodtime's, Waffle House
		W: Burger King, Dairy Queen, Dunkin Donuts, Pizza Hut, Toyoko Bowl, Village Inn
	Lodg	W: Motel 6
	AServ	E: Firestone Tire & Auto
		W: Amoco, Auto Zone Auto Parts, Car Quest Auto Center, Discount Tire Company, Econo Lube & Tune, Meineke Discount Mufflers, Nationwide Transmissions
	Other	E: Car Wash (North Valley Mall), **Montgomery Ward, WalGreens (North Valley Mall)**
		W: Inside Optical, Sears, The Vet Animal Hospital
217		U.S. 36, Westminster, Boulder
	Food	E: Subway
	AServ	E: Chesrown's Chevrolet
216		Jct I-76, Grand Junction, Fort Morgan
215		58th Ave
	Lodg	W: Super 8 (see our ad this page)
214AB		Jct I-70, 48th Ave, Grand Junction, Limon (Difficult Reaccess)
	Food	W: Holiday Inn, Travelodge, Village Inn
	Lodg	W: Holiday Inn, Travelodge, Village Inn
213		Park Ave, W 38th Ave, Coors Field
	Gas	E: Amoco*(CW)
		W: Citgo*, Texaco
	Food	E: Denny's, McDonalds
		W: Carbone's Sausage House, Chubby's, Regency Hotel, Subway
	Lodg	E: ◆ La Quinta Inn
		W: Regency Hotel
	AServ	W: Goodyear Tire & Auto
212C		20th St
212A		Speer Blvd, Downtown Denver
	Gas	W: Conoco, Texaco*
	Lodg	W: Budgethost, Residence Inn, Super 8 Motel (see our ad this page)
	AServ	W: Conoco
211		23rd Ave
	Food	E: Maxfield's, Shakespear's
		W: Baby Does
210C		CO33, Auraria Pkwy

EXIT COLORADO

210B 17th Ave
- **Food** W: Denny's, Fan's Fine Food, KFC, Pizza Hut
- **Lodg** W: ◆ Ramada
- **AServ** W: F&M Auto Parts
- **Other** W: **Denver Sports Complex**

210A U.S. 40, Colfax Ave, Downtown Denver (Difficult Northbound Reaccess, No Southbound Reaccess)

209C Zuni Street
- **Lodg** E: Motel 7
- **AServ** W: Bridgestone Tire & Auto

209AB U.S. 6, 6th Ave
- **Lodg** W: Days Inn (see our ad this page)

208 CO 26W, Alameda Ave
- **FStop** W: **Conoco***(D)(CW)
- **Gas** E: Amoco*(CW)(LP), Total*(D)
- **Food** E: Burger King, Denny's
  W: Valverde Yacht Club
- **Lodg** W: Motel 5
- **AServ** W: High Country Transmissions

207B U.S. 85S, Santa Fe Drive

207A Broadway
- **Gas** W: Texaco*
- **Food** E: Friends Cafe, Griff's Hamburgers
  W: Sunny Chine Cafe
- **AServ** E: Econolube, Zimmer Auto Parts
  W: Denver Discount Tire Service, General Tires, Jim Paris Tire City
- **ATM** W: Colorado National Bank
- **Other** E: **True Value Hardware**

206B Emerson St, Washington St
- **Food** E: **Wild Oats Market & Cafe**
- **Med** E: **✚ Hospital**
- **Other** E: **Wild Oats Market & Cafe**

206A Downing St (No Reaccess Northbound)

205AB University Blvd
- **Gas** W: Conoco*(LP)
- **Food** W: Black Jack Pizza, Bruger's Bagels, Colorado Espresso, Dunkin Donuts, Isle Of Singapore, Juice Stop, Keefan Mediterranean Food, Mustards Last Stand, Red Dragon Chinese, Starbuck's Coffee, Subway, The Border Mexican Restaurant, University Park Cafe
- **Med** W: **✚ University Park Medical Center**
- **Other** W: **WalGreens**

204 CO 2, Colorado Blvd
- **Gas** E: Conoco*(CW)
  W: Total*(D)
- **Food** E: Arby's, Black Eyed Pea, Grisanti's Italian, Le Peep Restaurant, Little Caesars Pizza, Subway, Village Inn
  W: Baskin Robbins, Denny's, Genroku, KFC, Perkins Family Restaurant, Taco Stop, Wang's Chinese Kitchen
- **Lodg** E: ◆ Fairfield Inn, Windam Garden Hotel
  W: ◆ La Quinta Inn
- **ATM** E: Vectra Bank
- **Other** W: **Albertson's Grocery, Colorado Center, Radio Shack, US Post Office**

203 Evans Ave
- **Food** W: Chinese Kitchen, Wang's Reastuarnt
- **Lodg** W: (AAA) Cameron Motel
- **AServ** E: House of Mufflers, Pro Star Inc
  W: Car Quest Auto Center, Discount Muffler & Brake, Express Lube, Freeway Ford, Windshields Of America

202 Yale Ave
- **Gas** E: Diamond Shamrock*, Sinclair

EXIT COLORADO

- W: Total*(D)
- **Food** E: Bagel Stop, Fazoli's Italian Food, Pizza Hut, Taco Bell
  W: Sam Wilsons Restaurant
- **Other** E: **Safe Way Grocery**
  W: **King Sooper's Grocery**

201 U.S. 285, CO 30, Hampden Ave, Englewood, Aurora
- **Gas** E: Amoco(CW), Conoco*(CW)(LP), Texaco
  W: Conoco*, Texaco
- **Food** E: Piccolo's, Quality Inn
  W: Burger King, Starbucks Coffee
- **Lodg** E: ◆ Marriott, (AAA) Quality Inn
- **AServ** E: Just Brakes
- **Other** E: **King Soopers Grocery**
  W: **Safe Way Grocery**

200 Jct I-225 to I-70, Limon (Southbound Exits Left Side Of Highway)
- **Lodg** W: Amerisuites (see our ad this page)

199 CO 88, Belleview Ave, Littleton
- **Gas** E: Phillips 66(LP)
- **Food** E: Bagels & Beyond, Kinko's Copies, Sandwich Plus, The Harvest Restaurant, Wendy's
  W: European Cafe, McDonalds, Taco Bell
- **Lodg** W: Days Inn (see our ad this page), Holiday Inn Express, Home Gate Inn, Super 8 Motel (see our ad this page)
- **AServ** E: Phillips 66
  W: Grease Monkey, Midas Muffler & Brakes
- **Other** W: **Mountain View Miniature Golf**

198 Orchard Road
- **Gas** W: Texaco*(D)
- **Food** W: Bayou Bob's, Great Bagel (Texaco), Taco Bell
- **Lodg** W: ◆ Hilton, Lucky Inn

197 Centennial Airport, CO 88, Arapahoe Road
- **Gas** E: Amoco
  W: Phillips 66*(CW)
- **Food** E: Bennigan's, Denny's, El Parral Mexican, Gunther Toody's, Ho Ho Chinese Food, Taco Cabana, Wendy's
  W: Arby's, Baja Fresh Mexican Grill, Black Eyed Pea, Boston Market, Brookes Steakhouse, Bruegger's Bagel Bakery, Chevy's Mexican, Dairy Queen, Grisanti's Italian, McDonalds, Red Robin Hamburgers, Ruby Tuesday, Subway, Taco Bell
- **Lodg** E: ◆ Courtyard by Marriott, (AAA) Radisson, ◆ Woodfield Suites
- **AServ** W: Firestone Tire & Auto
- **ATM** W: Big O Tires, First Bank
- **Other** E: **Target Department Store**
  W: **Albertson's Grocery (C-Store), Barnes & Noble**

196 Dry Creek Road

195 County Line Road
- **Food** E: The 19th Hole, The Columnbine, The Garden Terrace, The Swan (Inverness Hotel)
  W: Alexander's, Burger King, Walnut Brewery
- **Lodg** E: (AAA) Inverness Hotel
- **ATM** W: Community First National Bank
- **Other** W: **Borders Bookstore, Centennial Promenade, Hallmark Cards, Michael's Crafts, Office Max, Park Meadows Shopping Center, Toys "R" Us**

194 CO 470, Tollway 470, Grand Junction
- **Lodg** W: Amerisuites (see our ad this page), Super 8

193 Lincoln Ave, Parker

191 Unnamed

190 Surrey Ridge

188 Castle Pines Pkwy
- **Gas** W: Total*(LP)
- **Food** W: Pizza (Total)

Bold red print shows RV & Bus parking available or nearby

EXIT		COLORADO
	Other	W: US Mail Drop Box (Total)
187		Happy Canyon Road
184		Founders Pkwy, Meadows Pkwy
	Other	W: Factory Outlet Center
183		U.S. 85, Sedalia, Littleton
182		WIlcox St., Wolfensberger Rd., CO86, Franktown
	Gas	E: Amoco, Phillips 66*, Standard, Western* W: Diamond Shamrock*, Texaco*(D)(CW)
	Food	E: Bagel Stop, Little Caesars Pizza, Mexacalie Mexican Food W: Burger King, KFC, McDonalds (Outdoor Playground), Sharis Restaurant, Taco Bell, Village Inn, Wendy's
	Lodg	E: Castle Pines Motel W: ◆ Comfort Inn, Holiday Inn Express, Super 8 Motel
	AServ	E: Amoco W: NAPA Auto Parts, Plymouth Jeep Dodge Dealership
	Other	W: Malibu Car Wash
181		Wilcox St, Plum Creek Pkwy
	Gas	E: Bradley, Citgo*, Total*(CW), Western
	Food	E: Dairy Queen, Nicolo's Pizza, Pizza Hut, Subway
	Lodg	E: Castle Rock Motel
	AServ	E: Big O Tires, Big A Auto Parts
	Other	E: Cindy's Hallmark Cards, Hooked on Books, Radio Shack, Safe Way Grocery (Pharmacy), U.S. Post Office
174		Tomah Road
	RVCamp	W: KOA Campground
173		Larkspur, Palmer Lake (Southbound, No Southbound Reaccess)
172		Gulch Road
(170)		Rest Area - RR, Phones, Picnic, Vending (Both Directions)
167		Greenland Road
163		County Line Road
(161)		Weigh Station - both directions
161		CO 105, Woodmoor Dr., Monument, Palmer Lake
	Gas	E: Amoco W: 7-11 Convenience Store*, Conoco, Texaco(D)(LP)
	Food	W: Boston Market, Burger King, Daylight Donuts, Lots A Bagels, McDonalds (Indoor Playground), Monument Pizza, Pizza Hut, Subway, Taco Bell, Village Inn
	Lodg	E: Falcon Inn
	AServ	E: Amoco W: Red Mountain Oil Change Center, Texaco
	ATM	W: Norwest Bank
	Other	E: Coin Car Wash W: Safe Way Grocery (& Pharmacy)
158		Baptist Road
	FStop	W: Total*(D)(SCALES)
156B		North Entrance to Air Force Academy
	Other	E: Air Force Academy Visitor Center, Mining Museum
156A		Gleneagle Drive
(152)		Parking Area (Southbound)
151		Briargate Pkwy, Black Forest
	Other	E: Focus on the Family Visitor Center
150B		South Entrance to Air Force Academy
150A		CO 83, Academy Blvd
	Gas	E: Texaco*
	Food	E: Arby's, Baskin Robbins, Burger King, Captain D's Seafood, Chevy's Mexican, Cracker Barrel, Denny's, Eat at Joe's Crab Shack, Houlihan's, IHOP, KFC, McDonalds, Pizza Hut, Red Robin, Subway, Taco Bell, Village Inn, Wendy's
	Lodg	E: Comfort Inn, ◆ Drury Inn, (AAA) Radisson, (AAA) Red Roof Inn, Sleep Inn, Super 8 Motel
	AServ	E: Big O Tires, Firestone Tire & Auto, Jiffy Lube, Midas Muffler & Brakes, Sear's, Western Auto
	ATM	E: Colorado National Bank, Security Service Bank
	Other	E: Chapel Hills Mall, JC Penny, K-Mart, Kinko's, Sear's, Wal-Mart (& Pharmacy)
149		Woodmen Road
	Food	E: Carradda's, Hardee's W: Old Chicago Pasta & Pizza, Outback Steakhouse, T.G.I. Friday's
	Lodg	W: (AAA) Embassy Suites, ◆ Fairfield Inn, ◆ Hampton Inn (see our ad this page)
148A		Business 25, Nevada Ave, University of Colorado (Southbound Exits Left)
	Gas	E: Texaco
	Food	E: Squatting Chicken
	Lodg	E: B-N-B Motel, Howard Johnson
	AServ	E: Front Range Radiator, Mountain View Motors, Sheltons Transmissions
	RVCamp	E: Morris RV Sales & Service, Peak View Camp
148B		Corporate Center Drive
	Food	W: New South Wales Restaurant
	Lodg	W: Comfort Inn
147		Rockrimmon Blvd
	Gas	W: Texaco*(CW)
	Food	W: Marriott
	Lodg	W: Marriott
	AServ	W: Texaco
	ATM	W: UMB Bank
	Other	W: Prorodeo Hall of Champions
146		Garden of the Gods Road
	FStop	W: Citgo*
	Gas	E: Amoco, Texaco*(CW) W: Conoco*(CW), Phillips 66(CW)
	Food	E: Antonio's Pizza & Seafood, Denny's, Hardee's, McDonalds (Outdoor Playground), Taco John's W: Alpine Chalet, Dunkin Donuts, Freshens Yogurt, Garfield's, Holiday Inn, Quiznos Classic Subs, Sien Sien Chinese Food, Subway, Taco Bell, The Hungry Farmer, Village Inn Restaurant, Wendy's
	Lodg	E: ◆ La Quinta Inn W: (Days Inn) ◆ Days Inn (see our ad this page), (AAA) Holiday Inn, (AAA) Quality Inn, Ramada, (AAA) Super 8 Motel
	Med	E: ✚ EC Emergency Care Family Medical Center
145		CO 38, Filmore St
	Gas	E: 7-11 Convenience Store*, Citgo, Diamond Shamrock*, Texaco, Total* W: Conoco*, Texaco*(D)(CW)
	Food	E: Burger King, Dairy Queen, Lucky Dragon Chinese, Pizza Hut W: Waffle House
	Lodg	E: ◆ Ramada Inn W: (AAA) Best Western, (AAA) Motel 6, Super 8 Motel
	AServ	E: Action Car Care, Auto Tech Plaza, Firestone Tire & Auto, Jack's Alignment Service, Texaco W: Hertz Penske Super Lube
	RVCamp	E: Adventures In RV's (1 Mile, Sales & Rental)
	Med	E: ✚ Hospital
	Other	E: Car Wash, Coin Car Wash, Coin Laundry
144		Fontanero St (Difficult Reaccess)
143		Uintah St
	Gas	E: 7-11 Convenience Store*
	Other	E: Fine Arts Center, U.S. Olympic Center
142		Bijou St, Central Business District (East Access Very Congested)
	Gas	W: Total
	Food	W: Best Western, Denny's
	Lodg	W: (AAA) Best Western, (AAA) Le Baron Hotel
	AServ	E: Firestone Tire & Auto
	ATM	E: Peoples National Bank, Western National Bank
141		U.S. 24W, Cimarron St, Manitou Springs, Pikes Peak
	Gas	E: Texaco W: Conoco*, Phillips 66, Texaco
	Food	E: Chuck's Stop Diner W: Billy's Pizza, Burger King, Captain D's Seafood, McDonalds (Indoor Playground), Popeye's Chicken, Shelden's Luncheonette, Sonic, Subway, Taco John's, Waffle House, Western Sizzlin'
	Lodg	W: Holiday Inn Express, J.P.McGills Hotel Casino (see our page this page)
	AServ	W: Auto Tech Plaza, Auto Zone Auto Parts, Grease Monkey
	Other	W: Wal-Mart

EXIT COLORADO

140B U.S. 85S, Tejon St, Nevada Ave
Food E: Grindelwald, Luigi's Italian Restaurant, Nemeth El Tejon, Old Hiedleburg, Thrift House
W: Jamie's English Connection
AServ E: Jordan's Auto Service, Peerless Tire
W: Allied Transmissions, Colorado Springs Landrover, Ellegard Lincoln Mercury, Faricy Jeep Eagle, Heuberger Subaru VW Saab, Liberty Toyota, Long Ford Isuzu Suzuki, Long KIA, Long Mercedes Audi, Long Mitsubishi, Long Nissan, Noland Cadillac, Penkhus Volvo, Perkins Chrysler Plymouth, Rielly Buick GMC, Saturn Dealership, Sports Car City 2 Foriegn Car Repair

140A Business 25, Nevada Ave
Gas W: Amoco, Citgo*, Texaco*, Total*
Food W: Asian Garden Restaurant, Burger King, China Kitchen, Dunkin Donuts, Fazoli's Italian Food, Japanese Steakhouse, KFC, La Casita, Little Caesars Pizza, McDonalds, Red Top Restaurant, Rudy's Little Hideaway, Schlotzkys Deli, Subway, Taco Bell, Taco Bell, Wendy's
Lodg E: Big Horn Lodge, Chateau Motel, AAA Economy Inn, Four U Motel
W: American Inn, AAA Chief Motel, Circle S Motel, AAA Econolodge, Shamrock Hotel, Western Inn
AServ E: JW Brewer, Tire King
W: Checkers Auto Works, Gray's Tire & Service, Meineke Discount Mufflers, Midas Muffler & Brakes, Texaco
Other W: **Ace Hardware, Coin Laundry, King Sooper's Grocery, One Hour Dry Cleaners, Sear's, Southgate Mall, WalGreens**

139 U.S. 24, Limon

138 CO 29, Airport Circle Dr
Gas E: Conoco*, Texaco*(D)
W: 7-11 Convenience Store*
Food E: Days Inn (see our ad this page), Las Palmeras, Sheration
W: Arby's, Burger King, Carrabbas Italian Restaurant, Chesapeake Bagel, Denny's, Fazoli's Italian Food, Hardee's, Outback Steakhouse
Lodg E: Days Inn, ◆ Sheraton, Springs Motor Inn
W: Budget Inn, Double Tree Motel, Hampton Inn
Other W: **Cheyenne Mountain Center**

135 CO 83, Academy Blvd, Fort Carson, Security

132 CO 16, Widefield, Security
RVCamp E: **KOA Campground**

128 Fountain
TStop W: **Texaco**(SCALES)
Gas E: Citgo*
Food W: **Restaurant (Texaco)**
Lodg E: Southwinds
W: **First Interstate Inn**
AServ E: A&B Auto Body Shop, First Stop Auto Repair
TServ W: **Texaco**
TWash W: **Texaco**
ATM E: Plus Center
Other E: **Coin Laundry**

125 Ray Nixon Road

123 Unnamed

122 Pikes Peak Meadows
Other W: **Pikes Peak International Raceway**

119 Midway

116 County Line Road

(115) **Rest Area - RR, Phones, Picnic, Vending (Northbound)**

114 Young Hollow

EXIT COLORADO

(112) **Rest Area - RR, Phones, Picnic, Vending, RV Dump (Southbound)**

110 Pinon
TStop W: **Sinclair**(SCALES)
Food W: **Restaurant/Pinon Truckstop**
Lodg W: **Pinon Inn (Sinclair TS)**
TServ W: **Sinclair TS**

108 Bragdon
Other W: **KOA Campground**(LP)

106 Porter Draw

104 Eden
TServ W: **Wagner Caterpillar (1 Mile)**
RVCamp E: **Empiregas**(LP)

102 Eagleridge Blvd
Gas E: Total
Food W: Country Kitchen, Cracker Barrel, DJ's Steakhouse, IHOP
Lodg W: Comfort Inn (see our ad this page), Days Inn, AAA **Hampton Inn**, Motel 6, National 9 Motel, Wingate (see our ad this page)
TServ W: **Kenworth Dealer**
Med W: ✚ **Emergicare**
Other E: **Sam's Club**

EXIT COLORADO

101 U.S. 50, Canon City, Royal Gorge
Gas W: Diamond Shamrock, Total
Food E: Captain D's Seafood, Denny's
W: Applebee's, Arby's, Bamboo Garden, Black Eyed Pea, Boston Market, Burger King, Chinese Express, Dairy Queen, Fazoli's Italian Food, Gaetano's Supper Club, Hardee's, McDonalds, Southwest Grill, Subway, Wendy's, Western Sizzlin'
Lodg E: ◆ Sleep Inn
W: ◆ **Holiday Inn**, Motel 6, Pueblo Motor Inn, Super 8 Motel
AServ W: Batterys Plus, Discount Tire, Goodyear Tire & Auto, Western Auto
ATM W: Minnequa Bank, Total
Other E: **Target Department Store**
W: Car Wash, **Coin Laundry, Galleria West, K-Mart (& Pharmacy), Tourist Information**

100B 29th St
Gas W: Amoco, Sinclair*
Food E: Country Buffet, Mandarin, Panda Buffet, Peter Piper Pizza
W: Submarine Sandwiches, Taco Casa
Lodg W: Bel Mar Motel
AServ W: Amoco, Checker Auto Parts, Grease Monkey, Hamid's Complete Auto Service
Other E: **King Sooper's Grocery, Pueblo Mall**
W: Full Service Car Wash, **Safe Way Grocery**

100A U.S. 50, La Junta

99B 13th St, Santa Fe Ave
Food W: Cornell Cafe, Siagon Garden
Lodg W: Travelers Motel
AServ W: Diodosio Isuzu Pontiac, Freedom Ford Subaru, Spradley Chevrolet
Med W: ✚ **Hospital**

99A CO 96, 6th St
Food W: Best Western, Wendy's
Lodg W: Best Western
AServ W: Auto Tech Auto Parts, Jiffy Lube, Vidmar Honda Mazda

98B CO 96, 1st St
Gas W: Total*
Food W: Hardee's, Quizno's Subs
AServ W: Pueblo Auto Parts
ATM W: Colorado National Bank, Minnequa Bank

98A U.S. 50, La Junta
Gas W: Texaco*(D)
Food W: Sonic Drive In

97B Abriendo Ave
Gas W: Texaco*
Food W: Passkey, Pizza King & Subs, Subway, Taco Bell

97A Central Ave
Gas W: Total*
Food W: Taco Bell
AServ W: Larry's Transmission
ATM W: Total

96 Indiana Ave (Difficult Reaccess)
Gas W: Texaco*
Food W: Kozy Korner Cafe & Bar
AServ W: Cobra Auto Repair

Bold red print shows RV & Bus parking available or nearby

EXIT	COLORADO
Med	W: Hospital
95	Illinois Ave (Southbound, No Reaccess)
94	CO 45, Pueblo Blvd
Gas	W: Standard, Total*
Food	W: Old Coral Cafe, Taco Bell (Total)
AServ	E: J-S Auto Parts
RVCamp	W: Forts RV & Mobil Home Park
Other	W: Dog Track, Go Cart Track
91	Stem Beach
Gas	W: Country Side Mini Mart*
Lodg	W: Country Bunk Inn
Other	W: Antique Shop (Country Side Mini Mart)
88	Burnt Mill Road
87	Verde Road
83	Unnamed
(82)	Parking Area - Picnic (Both Directions)
77	Cedarwood
74	CO165, Colorado City, Rye, San Isabel
Gas	E: Total(D)
Food	E: Cafe (Total), Taco Bell
	W: Greenhorn Inn
Lodg	W: Greenhorn Inn
Other	W: Tourist Information
71	Graneros Road
67	Apache
64	Lascar Road
60	Huerfano
59	Butte Road
56	Redrock Road
55	Airport Road
52	Bus Loop I-25S, U.S. 160W, CO 69W, Walsenburg, Gardner
FStop	W: Phillips 66*
Gas	W: Diamond Shamrock*
Food	W: Best Western, Pizza Hut
Lodg	W: AAA Best Western, Budget Host Inn
AServ	W: Phillips 66
RVCamp	W: Camping
Parks	W: Lathrup State Park
Other	W: The Great Sand Dunes National Monument (45 Miles)
50	CO 10, El Junta
Lodg	W: Days Inn
Med	W: Hospital
Other	W: Tourist Information
49	Bus Loop I-25, U.S. 160, Walsenburg, Alamosa
42	Rouse Road
41	Rugby Road
(37)	Rest Area - RR, Picnic, Grills, Vending, RV Dump (Both Directions)
34	Aguilar
FStop	E: Amoco
Food	E: Amoco Fuel Stop
ATM	E: Amoco
30	Aguilar Road
27	CR 440, Ludlow
Other	W: Point of Historical Interest
23	Hoehne Road
18	El Moro Road
15	U.S. 160E, Kit Carson Trail, La Junta
Gas	W: Texaco*
Food	E: Burger King
	W: Frontier Cafe
Lodg	E: ◆ Super 8 Motel

EXIT COLORADO

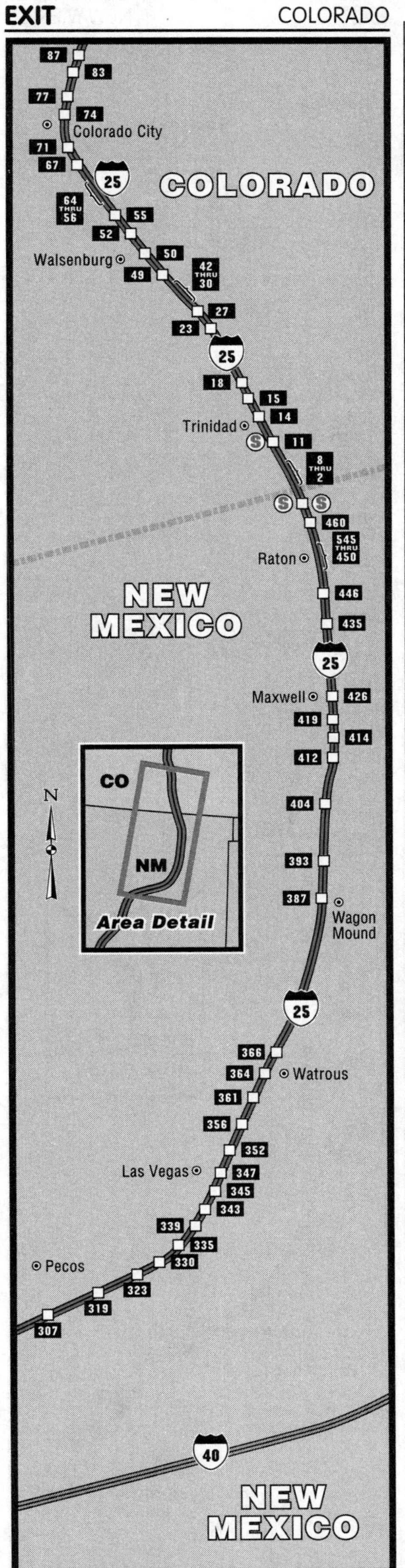

EXIT	COLORADO/NEW MEXICO
	W: Columnbine Motel, Frontier Motel
14B	Bus Loop I-25, Trinidad
Gas	W: Phillips 66, Texaco
AServ	W: Kadad General Tire
Other	W: Spot Free Car Wash
14A	CO 12W, Cuchara, La Vetta
Gas	E: Conoco, Texaco*
	W: Conoco, Diamond Shamrock*
Food	E: McDonalds, Pizza Hut, Subway, Taco Bell
	W: Dairy Queen, El Capitan Italian & Mexican
Lodg	E: Days Inn, Santa Fe Trail Inn
AServ	E: Car Quest Auto Center
Med	E: Hospital
Other	E: Safe Way Grocery
13B	U.S. 160, U.S. 350, Main St
Gas	E: Conoco*(CW)
Food	E: Days Inn, Golden Nugget Restaurant
Lodg	E: AAA Days Inn
AServ	E: Trinidad Tire
Med	E: Hospital
Other	E: Safe Way Grocery
13A	Trinidad
Gas	E: Phillips 66
Food	E: Chef Liu's Chinese Food, Restaurant/Best Western
Lodg	E: AAA Best Western
RVCamp	E: Campground
Other	E: Car Wash
11	Weigh & Check Station, Starkville, Santa Fe Trail (Both Directions)
FStop	E: Texaco*
Food	E: Bob & Earl's Family Restaurant
	W: Holiday Inn
Lodg	E: Budget Host Inn, Budget Inn
	W: ◆ Holiday Inn
AServ	W: Circle Chevrolet Buick
RVCamp	E: Budget Inn
8	Spring Creek
6	Gallinas
2	Wootton

↑COLORADO

↓NEW MEXICO

(461)	Weigh Station - both directions
(460)	Rest Area - RR, Picnic P
460	Unnamed
454	Bus Loop I-25, U.S. 87, Raton
452	NM 72, Raton
Gas	W: Conoco*
Lodg	W: Mesa Vista Motel
Med	W: Hospital
Parks	W: Sugarite State Park
451	U.S. 64, U.S. 87, Raton, Clayton
FStop	E: Texaco*
Gas	E: Chevron, Diamond Shamrock, Total*
	W: Conoco*, Texaco*
Food	E: Hooter Brown's Restaurant (Total), Subway (Diamond Shamrock)
	W: All Seasons Family Restaurant, Arby's, Dairy Queen, KFC, McDonalds, Sands Restaurant
Lodg	W: ◆ Comfort Inn, El Kapp Motel, AAA Harmony Manor Motel, Holiday Classic Motel, Motel 6, Sands Motel, ◆ Super 8 Motel, Texan Motel, Travel Motel
AServ	E: Chevron, Raton Auto Truck Service
	W: NAPA Auto Parts
Parks	W: Capulin Volcano

Bold red print shows RV & Bus parking available or nearby

EXIT NEW MEXICO

450 Business 25, Raton
- **Gas** W: Diamond Shamrock, Phillips 66
- **Food** W: Oasis Restaurant, Sonic Drive In, Thia Country Kitchen
- **Lodg** W: Maverick Motel, Oasis Motel, Robin Hood Motel, Westerner Motel
- **Med** W: ✚ Hospital
- **Other** W: **Mesa Vista Veterinary Hospital**

446 U.S. 64, Taos, Cimarron
435 Tinaja
(434) **Rest Area - RV Dump, RR, HF, Picnic P (Both Directions)**
426 NM505, Maxwell
- **Other** W: **Maxwell National Wildlife Refuge (4 Miles)**

419 NM58, Cimarron
- **RVCamp** W: **RV Park**

414 Springer
- **Gas** E: Conoco*, Diamond Shamrock, Phillips 66[D], Texaco*
- **Food** E: El Taco Mexican, Minnie's Dairy Delite, Smokeys Cafe, Stockman's Cafe
- **Lodg** E: Cozy Motel, Oasis Motel, The Brown Hotel
- **AServ** E: Road King Tires
- **Other** E: **Tourist Information**

412 Springer, U.S. 56, NM21, NM468
404 NM569, Colmor
393 Levy
387 NM120, Wagon Mound, Roy, Ocate
- **Gas** E: Chevron*, Phillips 66*
- **Food** E: Levi's Cafe
- **Lodg** E: Mat Travel Center
- **AServ** E: Phillips 66

(374) **Rest Area - RV Dump, RR, HF, Picnic P (Both Directions)**
366 NM 161, NM97, Watrous, Valmora
- **Other** W: **Fort Union National Monument**

364 NM97, NM161, Watrous, Valmora
361 Unnamed
(360) **Rest Area - no services P (Both Directions, No Facilaties)**
356 Onava
352 Airport
347 NM518N, Las Vegas , Taos
- **Gas** W: Fina, Texaco*
- **Food** W: Pino's Restaurant
- **Lodg** W: (AAA) Comfort Inn, (DAYS INN) **(AAA) Days Inn (see our ad this page, Inn of Las Vegas, ◆ Super 8 Motel**
- **AServ** W: Pennzoil Lube[LP]

345 NM65, NM104, University Ave
- **Gas** W: Allsups*, Bell, Fina*, Save Mat, Texaco*
- **Food** W: 85 Coffee Shop, Burger King, Dairy Queen, Golden Dragon, Hillcrest Restaurant, KFC, Mexican Kitchen, Pizza Hut, State Cafe, Taco Bell
- **Lodg** W: (AAA) El Camino Motel, El Fidel, Knight Rest, Palamina Motel, Plaza Hotel, Santa Fe Trail Inn, Scottish Inns
- **AServ** W: GM Auto Dealership, NAPA Auto Parts
- **Med** W: ✚ **Hospital**
- **ATM** W: Allsups
- **Other** W: **Coin Laundry, Historic Olds Towns Plaza**

343 Business 25, Las Vegas, NM518
- **Gas** W: Conoco, Ferrel Gas[LP], Texaco
- **Food** W: Marion's El Camino Restaurant, Subway, Taco Bell, Teresa's Restaurant
- **Lodg** W: (AAA) Plaza Hotel, Thunderbird Motel
- **AServ** E: Garcia Tire
- **Other** W: **Suds Laundromat**

339 U.S. 84, Romeroville, Santa Rosa
- **RVCamp** E: **KOA Campground**

335 Tecolote
330 Bernal

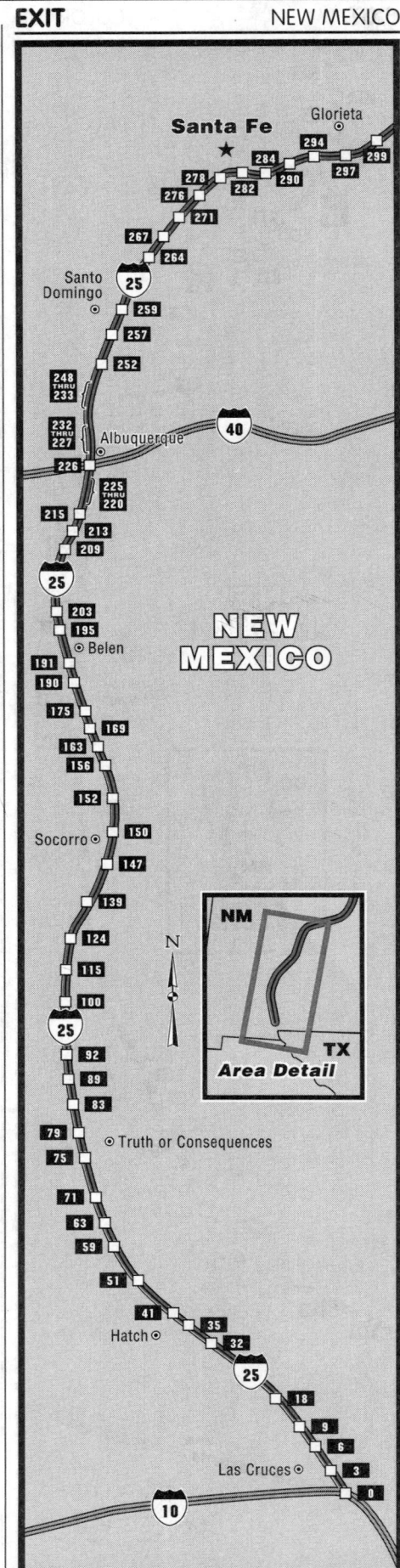

EXIT NEW MEXICO

EXIT NEW MEXICO

(325) **Rest Area P (Both Directions)**
323 NM3, Villanueva
319 San Juan, San Jose
307 NM63, Rowe, Pecos
- **Other** E: **Pecos National Monument**

299 NM50, Glorieta, Pecos
297 Valencia
294 Canoncito at Apache Canyon
290 U.S. 285, Clines Corners, Eldorado, Lamy
284 Old Pecos Trail, NM466
- **Med** W: ✚ **Hospital**

282 U.S. 84, U.S. 285, St. Francis Dr, Sante Fe Plaza, Los Alimos
278 Business 25, NM14, Cerrillos Road, Madrid
276AB NM599, NM14, Madrid
271 NM587, La Cienega
(269) **Rest Area - RV Dump, RR, HF, Phones, Picnic P (Northbound)**
264 NM16, Cochiti, Pueblo
259 NM22, Santo Domingo, Pueblo
257 Budaghers
- **Gas** W: Outlet Mall
- **Food** W: Red River Grill

252 San Felipe, Pueblo
248 Algodones
242 NM44, NM164, Rio Rancho, Bernalillo, Farmington, Placitas
240 NM473, Bernalillo
- **RVCamp** W: **KOA Campground**

234 NM556, Tramway Road
233 Alameda Blvd
232 Paseo del Norte
231 San Antonio Blvd, Ellison, San Antonio, Osuna
- **Lodg** N: Amberly Suite Hotel (see our ad opposite page), Howard Johnson (see our ad opposite page), Hampton Inn (see our ad opposite page)

230 San Mateo Blvd
229 Jefferson St
228 Montgomery Blvd, Montano
227B Comanche Blvd, Griegos
- **Gas** W: Phillips 66*
- **ATM** W: Phillips 66
- **Other** W: Comanche Car Wash

227A Candelaria Road, Menaul Blvd
- **TStop** E: **76 Auto/Truck Plaza**
- **FStop** W: **Chevron***[CW]
- **Gas** E: Shell
- **Food** E: 76, Liberty Cafe (Motel 76), Milly's Restaurant
- **Lodg** E: **Motel 1**, Motel 76
  W: ◆ Travelodge

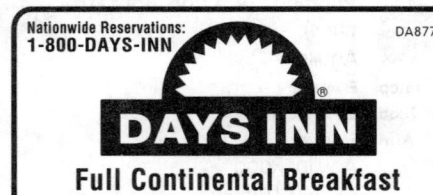

Bold red print shows RV & Bus parking available or nearby

EXIT NEW MEXICO

ATM E: Western Bank
226 Junction I-40
225 Lomas Blvd
Food E: JB's Restaurant (Plaza Inn)
Lodg E: Plaza Inn
AServ E: Bob Turner's Ford Country
224B Dr. Martin Luther King Jr. Ave.
224A Lead - Coal Central Ave. (One Way Streets)
223 Stadium Blvd
Gas W: Chevron*(D)(CW), Plateau*
Lodg E: **Motel 6**
222B Gibson Blvd
222A Gibson Blvd, Airport
Gas E: Chevron, Circle K, Phillips 66
Food E: Applebee's, Goddy's, Waffle House
Lodg E: Budgetel Inn, Courtyard by Marriott, Fairfield Inn, La Quinta Inn, Radisson Inn, Ramada Limited, Sleep Inn
Med E: **Hospital**
220 Rio Bravo Blvd
215 NM47, Broadway
Gas E: Conoco*(D)
RVCamp E: **RV Park**
213 NM314, Isleta Blvd.
Gas W: Chevron*(D)
Food W: Subway (Chevron)
209 NM45, Isleta, Pueblo
203 NM6, Los Lunas
Gas E: Chevron*(CW), Diamond Shamrock*
Food E: Arby's, McDonalds
Lodg E: Days Inn
195 Bus25, Belen
191 NM548, Sosima, Padilla Blvd
Gas E: Fina(D)
Lodg E: Budgetel
W: Best Western
190 Bus25, S Belen
Gas E: AKIN(D)(LP), Conoco, Gasman*
Food E: Golden Coral, KFC, McDonalds
Lodg E: Freeway Inn, Mountainview Motel, Super 8 Motel
RVCamp E: **KOA Campground, La Marada**
175 U.S. 60E, Bernardo, Mountainaire
Gas E: Fuel*
169 La Joya State Game Refuge
(167) **Rest Area - RV Dump, RR, HF, Phones, Picnic P (Southbound)**
(166) **Rest Area P (Northbound)**
163 San Acacia
156 Lemitar
FStop W: **M Gas*(D)**
152 Escondida
150 U.S. 60W, Bus25, Socorro, Magdelena
Gas W: Chevron*(CW), Circle K*, Exxon*
Food W: Blakes Hamburgers, Burger King, Denny's, El Camino, K-Bob's Steakhouse, KFC, McDonalds, Pizza Hut, Sonic Drive In, Subway, Taco Bell
Lodg W: AAA Econolodge, Holiday Inn Express, Motel Vagabond, San Miguel Motel, ◆ **Super 8 Motel**
AServ W: Bob's Auto Machine
147 Bus25, Bus.60W, Socorro, Magdalena
Gas W: Conoco(LP), Exxon, Fina*(LP), Phillips 66, Shell, Texaco*(D)
Food W: Arby's, Denny's, KFC, McDonalds, Taco Bell
Lodg W: Economy Inn Motel, Holiday Inn Express, Motel 6, Super 8 Motel
RVCamp W: **KOA Campground(LP)**
Other W: **NAPA Auto Parts**

EXIT NEW MEXICO

139 U.S. 380E, San Antonio, Carrizozo
RVCamp E: **Night Owl RV Park**
124 San Marcial
115 NM107
TStop E: **Santa Fe Diner & Truck Stop (RV Dump)**
Gas E: Santa Fe
Food E: **Santa Fe**
AServ E: Sante Fe(LP)
RVCamp E: **Sante Fe RV Park**
(114) **Rest Area - RR, RV Dump P (Southbound)**
(113) **Rest Area - RR, HF, Phones, Picnic P (Northbound)**
100 Red Rock
92 Mitchell Point
89 NM181, NM52, Cuchillo, Monticello

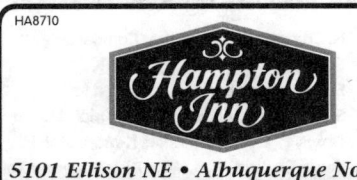

EXIT NEW MEXICO

83 NM181 NM 52 Cuchillo, Montecello (No Northbound Reaccess)
Food E: Butte Restaurant
Other E: **Marina Suites Motel**
79 Bus25, Truth or Consequences
Gas E: Chevron(D)
Food E: **K-Bobs, McDonalds**
Lodg E: AAA Best Western
75 Bus25, Truth or Consequences (Difficult Reaccess Northbound)
Gas E: Chevron, Conoco(D), Diamond Shamrock, Fast Stop
Food E: La Pinata, Los Artos, McDonalds
Lodg E: Ace Lodge, Red Haven Motel, Rio Grande Motel, Travelers Lodge
AServ E: Borman Geo
RVCamp E: **RV Park**
Med E: **Hospital**
Other E: **County Fair Grounds, The Elks Club**
71 Las Palomas
(69) **Rest Area - no facilities P (Southbound)**
(68) **Rest Area - no facilities P (Northbound)**
63 NM152, Hillsboro
Gas E: Texaco(LP)
RVCamp E: **KOA Campground**
59 Caballo and Percha State Parks
51 Garfield
41 NM26, Hatch, Deming
Gas W: Conoco*
Food W: Dairy Queen, Exxon
RVCamp W: **Happy Trails RV Park**
35 Rincon
32 Upham
(26) Inspection Station (Northbound)
(24) **Rest Area - RR, HF, Phones, Picnic P (Southbound)**
(23) **Rest Area - RR, Vending, HF, Picnic P (Northbound)**
18 Radium Springs
RVCamp W: **Camping**
9 Dona Ana
Gas W: Chevron*, Diamond Shamrock*
RVCamp W: **Camping**
6A U.S. 70, U.S. 82E, Alamogordo, Las Cruces
Gas E: Texaco*
Food E: IHOP, Silver Streak Hamburgers
Lodg E: ◆ Super 8 Motel
Med E: **Hospital**
Other E: **K-Mart**
6B U.S. 70, U.S. 82W, Las Cruces
3 Lohman Ave
Gas W: Conoco*
Food E: Applebee's, Burger King, Chili's, Jack-In-The-Box, KFC, Red Lobster, Village Inn Restaurant
W: McDonalds, Taco Bell, Wienerschnitzel
Lodg E: AAA Hilton
ATM W: Nations Bank
Other E: **Office Max**
W: **Wal-Mart**
1 University Ave
Food W: **Dairy Queen,** Furr's Family Dining
Lodg W: Days Inn, Holiday Inn, La Quinta Inn
Med E: **Memorial Medical Hospital**
0 Jct. I-10, El Paso, Tucson

↑NEW MEXICO

Begin I-25

Bold red print shows RV & Bus parking available or nearby

EXIT	NORTH CAROLINA

Begin I-26

↓NORTH CAROLINA

1 Junction I-40, Canton, Knoxville, NC 191 to Blue Ridge Pwky.

2 NC 191, to Blue Ridge Parkway
- **Food** S: Cookies N Creme, Harbor Inn, Little Caesars Pizza, Long John Silvers, Ryan's, Subway, Taco Bell, Waffle House
- **Lodg** N: AAA Hampton Inn, Wingate Inn
  S: ◆ Comfort Suites
- **AServ** N: Jim Barkley Toyota, Saturn of Ashville
- **RVCamp** N: **Bear Creek Campground**
  S: **North Mills Recreation Area, Powhaton Camping**
- **Other** N: **Biltmore Estates, Super Petz (Pet owners warehouse)**
  S: **Biltmore Square Mall, Ingles Supermarket, K-Mart, Ridgefield Mall, U.S. Post Office**

6 NC 280, Skyland
- **Gas** N: Amoco*(D)(LP)
- **Food** N: Eat Rite Cafe, Hardee's, Shoney's, Shoney's Inn, Waffle House
  S: Cider Mill Restaurant & Pub
- **AServ** N: Airport Connection Auto Service
- **Other** N: **W.N.C. Vetninarian Hospital**

9 NC 280, Brevard, Arden, Asheville Regional Airport
- **Gas** N: BP*(D)(CW), Texaco*
  S: Amoco*(D)(CW)
- **Food** N: Arby's, McDonalds, Waffle House
- **Lodg** N: Budget Motel, Days Inn **Days Inn (see our ad this page)**, AAA **Econolodge**, AAA Holiday Inn
  S: ◆ Fairfield Inn
- **AServ** N: Appletree Honda, L.B. Smith Volvo
- **ATM** N: Texaco
- **Other** S: **Ashville Regional Airport**

(10) **Rest Area - RR, Phones, Vending, Picnic (Both Directions)**

13 U.S. 25, Fletcher, Mountain Home
- **TStop** S: **Skyway***(SCALES)
- **Gas** S: Exxon*, Phillips 66
- **Food** N: Hardee's
  S: Burger King, **Huddle House, Skyway Restaurant**
- **AServ** S: Phillips 66
- **TServ** S: **Skyway TStop, Will's CD Shop**
- **Med** S: ✚ **Park Ridge Hospital**
- **ATM** S: Skyway

(15) **Weigh Station (Both Directions)**

18AB U.S. 64, Hendersonville, Bat Cave
- **FStop** N: **Shell***(D)
- **Gas** N: Chevron*, Texaco*(D)(LP)
  S: Exxon*(D)
- **Food** N: A Day in the Country (Bakery & Coffee shop), Fireside Restaurant, Waffle House
  S: Applebee's, Arby's, Baskin Robbins, Bojangles, Bunyan's Road House, Burger King, Checkers Burgers, Denny's, Hardee's, KFC, Lon Sen Chinese, Long John Silvers, McDonalds, Savannah Beach Grill, Schlotzkys Deli, Subway, Taco Bell
- **Lodg** N: **Best Western**, Hampton Inn, Quality Inn, Ramada Limited
  S: Comfort Inn, Days Inn Days Inn
- **AServ** N: Chevron, Hendersonville Tire & Oil, I-26 Auto Supply
  S: Carolina Tire Co., Penske Auto Service, Qlube

EXIT	NORTH/SOUTH CAROLINA

- **RVCamp** N: **Campground, Mountain Camper Sales & Servoce**
- **ATM** S: First Citizen's Bank
- **Other** S: **Allan's Pharmacy, Blue Ridge Mall, Eckerd Drugs, Ingles Supermarket, K-Mart, Mail Boxes Etc, Wal-Mart, Winn Dixie Supermarket**

22 Upward Road, Hendersonville
- **FStop** S: **Shell***(D)
- **Gas** N: Texaco*(D)(LP)
- **Food** S: Cracker Barrel **Cracker Barrel**, McDonalds
- **Lodg** S: Holiday Inn Express
- **RVCamp** N: **Lakewood RV Resort, Rite-Way Services (RV Parts & Repairs), Twin Ponds (Open year round)**
- **ATM** S: Exxon
- **Other** N: **Country Village Crafts**

23 U.S. 25, to U.S. 176 (Closed)

28 Saluda
- **FStop** S: **Amoco***
- **Gas** S: Texaco*(LP)
- **Lodg** N: Heaven's View Motel
- **AServ** S: **Hipps Garage**
- **TServ** S: **Hipps Garage**
- **Other** S: **The Apple Mill (Cidar Mill Museam)**

36 NC 108, Hwy. 74
- **Gas** N: Texaco*
  S: Amoco*(D)(LP)
- **Food** N: Blimpie's Subs, Chester Fried Chicken, Hardee's, McDonalds, Papa's Pizza, Subway, Waffle House
  S: KFC, Sow Sun Chinese
- **Lodg** S: Days Inn Days Inn
- **ATM** S: Bi-Lo Supermarket
- **Other** N: **Food Lion Supermarket, Revco Drugs**
  S: **Bi-Lo Supermarket**

(36) NC Welcome Center - RR, Vending, Phones, Picnic

↑NORTH CAROLINA

↓SOUTH CAROLINA

1 SC 14, Landrum
- **Gas** S: BP(D)
- **Food** S: Denny's
- **AServ** S: BP
- **Other** S: **Ingles Supermarket**

(3) SC Welcome Center - RR, Vending, Phones, Tourist Info (Eastbound)

5 SC 11, Campobello, Chesnee

EXIT	SOUTH CAROLINA

- **FStop** N: **Citgo***
- **Gas** S: Phillips 66* (24 Hrs), Spur(D) (Fireworks Superstore)
- **Food** N: **Aunt M's (Citgo FStop)**
- **ATM** N: Citgo

(9) **Rest Area (Both Directions)**

10 SC 292, Inman
- **FStop** N: **Conoco***(LP)(SCALES)
- **Food** N: Hope's Home Cooked Meals, **Krispy Kreme (Conoco), Subway (Conoco FStop)**
- **ATM** N: Conoco

15 U.S. 176, Inman
- **FStop** N: **Fuel City***(SCALES)
  S: **Exxon***
- **Gas** N: Fast Stop*
- **Food** N: Waffle House
  S: **Blimpie's Subs (Exxon FStop)**
- **AServ** N: NIX Tires, Alignment, & Brakes
- **ATM** N: Fast Stop, Fuel City FStop

16 John Dodd Road, Wellford
- **FStop** N: **Citgo***
- **Food** N: **Aunt M's (Citgo FStop)**
- **ATM** N: Citgo
- **Other** N: **Red Star Fireworks**

17 New Cut Road, Sigsbee
- **Gas** S: Citgo*, Speedway*(LP)
- **Food** S: Burger King, **Hardees**, McDonalds, Waffle House
- **Lodg** S: AAA Comfort Inn, Days Inn ◆ **Days Inn**, Econolodge, ◆ Howard Johnson
- **RVCamp** S: **Spantanburg Cunninghan Camping**
- **Other** S: **Foothills Factory Stores**

19AB Junction I-85, Greenville, Charlotte

21B U.S. 29, Greer

21A U.S. 29, Spartanburg
- **Gas** N: Crown* (24 Hrs), Exxon*
  S: Citgo*(D), Texaco(D)
- **Food** N: Aloha Oriental Dining, Carfields, Chick-fil-A, Ci Ci's Pizza, Hardee's, Long John Silvers, Pizza Hut, Quincy's Family Steakhouse, Sub Station II, Subway (Exxon), Wendy's
  S: Applebee's, Aunt M's, McDonalds, Piccadilly, Steakout, Taco Bell, Vola's Restaurant
- **AServ** N: Firestone Tire & Auto, Goodyear Tire & Auto, Jiffy Lube, Sears
  S: Dayton Tires, Spartan Automotive, Western Auto
- **ATM** N: BB&T Bank, NationsBank, Wachovia Bank
- **Other** N: **Eckerd Drugs, K-Mart (w/pharmacy 24 Hrs), Petsmart Vetinarian**
  S: **Companion Animal Clinic, Ingles Supermarket, Noah's Ark Kennels, Sam's Club, Wal-Mart**

22 SC 296, Reidville Road, Spartanburg
- **Gas** S: BP*(CW), Hess*(D)
- **Food** N: Arby's, Capri Italian, Carolina BBQ, Hong Kong Express, Little Caesars Pizza, McDonalds, Outback Steakhouse, Pizzazz Pizza, Waffle House
  S: Burger King, Denny's, Domino's Pizza, Hardee's, Hunank Chinese, TCBY Yogurt, Waffle House
- **Lodg** S: ◆ Sleep Inn
- **AServ** N: B.F. Goodrich
  S: Dave Edward's Toyota, Reidville Road Auto Service, Rick's Auto Service, Xpress Lube(CW)
- **ATM** N: American Federal Bank, NVSC, NationsBank, Palmetto Bank, Spartanbug National Bank
  S: Bi-Lo Supermarket, First Union Bank
- **Other** N: **Eckerd Drugs, Mail Boxes Etc, Putt-Putt & other games, Wynnsong Cinema's**
  S: **Bi-Lo Supermarket (w/pharmacy), Mail**

Bold red print shows RV & Bus parking available or nearby

EXIT SOUTH CAROLINA

Center (FedEx, Western Union, UPS), Red Star Fireworks, Revco

28 U.S. 221, Spartanburg, Moore
- **Gas** N: Conoco*(D)(LP)
- **Food** N: Subway (Conoco), Walnut Grove Seafood (Open evening hrs. Thurs. - Sun.)
- **RVCamp** N: Pine Ridge Campground
- **Med** N: ✚ Hospital
- **ATM** N: Conoco
- **Other** N: Walnut Grove Plantation

35 Road 50, Woodruff
- **FStop** S: Citgo*
- **Food** S: Family Restaurant (Citgo, 24 Hrs)
- **Med** S: ✚ Hospital

38 SC 146, Woodruff (Westbound)
- **Food** N: Big Country Restaurant

41 SC 92, Enoree
- **Gas** S: Phillips 66*(LP)

(43) Rest Area - Picnic (Both Directions)

44 SC 49, Cross Anchor, Union

51 Junction I-385, Laurens, Greenville

52 SC 56, Cross Anchor, Clinton
- **FStop** N: Speedway*(SCALES)
- **Gas** S: Phillips 66*(D)
- **Food** N: McDonalds, Waffle House
  S: Hardee's, Waffle House, Wendy's
- **Lodg** N: AAA Comfort Inn
  S: DAYS INN AAA Days Inn, AAA Holiday Inn, Traveler's Inn
- **TServ** N: Carolina Transportation Inc. (24 Hr.Truck Wrecker & Truck Trailer Service, Truck wash)

54 SC 72, Clinton
- **FStop** N: Citgo*
  S: Texaco*(LP)

60 SC 66, Whitmire, Joanna
- **FStop** S: BP*(D)
- **Food** S: 60 Truck Stop Restaurant
- **AServ** S: BP
- **RVCamp** S: Joanna Koa Campground(LP)

(63) Rest Area - RR, Phones, Vending, Picnic (Both Directions)

66 Road 32, Jalapa

72 SC 121, Whitmire, Union
- **Gas** N: Exxon*
- **Med** S: ✚ Hospital

74 SC 34, Newberry, Winnsboro
- **FStop** N: Shell*
  S: Texaco*
- **Gas** N: BP*(D)
- **Food** N: Bill & Fran's, Mickey's

EXIT SOUTH CAROLINA

- S: Waffle House
- **Lodg** N: AAA Best Western
  S: DAYS INN Days Inn
- **Med** S: ✚ Hospital
- **ATM** N: Shell

76 SC 219, Newberry

(81) Weigh Station (Eastbound)

82 SC 773, Pomaria, Prosperity
- **FStop** N: Amoco Travel Plaza*(SCALES)
- **Food** N: Subway
- **AServ** N: Craig Hipps Auto Service
- **TServ** N: Craig Hipps Truck Service
- **RVCamp** N: Flea Market Campground
- **Other** N: Jockey Lot Flea & Farmers Market

(84) Parking Area (Eastbound)

85 SC 202, Pomaria, Little Mountain

EXIT SOUTH CAROLINA

(88) Parking Area (Westbound)

91 Road 48, Chapin
- **FStop** S: Texaco*(D)(LP)
- **Gas** S: Amoco*
- **Food** S: Capital City Subs & Salad, McDonalds
- **ATM** S: Texaco
- **Other** S: Dreher Island State Park

(94) Weigh Station (Westbound)

97 U.S. 176, Peak
- **FStop** S: Amoco* (24 Hrs)
- **Food** N: Kathryn's Whales Tail
- **RVCamp** S: Smokewood Campground

101 U.S. 76, U.S. 176, Ballentine, White Rock
- **Gas** S: Amoco*, BP(CW)
- **Food** S: Baskin Robbins, Burger King
- **AServ** S: BP
- **ATM** S: BB&T Bank (24 Hr. banking)

102 SC 60, Irmo (Westbound, Eastbound Reaccess)
- **Food** S: Cookies by Design, Papa John's Pizza, TCBY Yogurt
- **Other** S: Piggly Wiggly Supermarket, Revco

103 Harbison Blvd
- **Gas** S: Amoco*(CW), Exxon*, Hess*(D), Corner Pantry (see our ad this page)
- **Food** N: Applebee's
  S: Bagle Works, Bailey's Sports Grill, Blimpie's Subs, Chick-fil-A, Chili's, Fazoli's Italian Food, McDonalds, Olive Garden, Outback Steakhouse, Palmetto Best Chinese, Ruby Tuesday, Rush's Hamburgers, Shoney's, Sonic Drive In, Subway, Taco Bell
- **Lodg** N: ◆ Hampton Inn
- **AServ** S: Speedlee lube & tune up
- **Med** S: ✚ Baptist Medical Center
- **ATM** S: BB&T Bank, First Union Bank, South Carolina Federal Credit Union, SouthTrust, Wachovia
- **Other** S: Bi-Lo Supermarket, Carmike Cinemas, Columbiana Centre Mall, Phar Mor Drugs, U.S. Post Office, Wal-Mart Supercenter (Pharmacy, 24 Hrs)

104 Piney Grove Road
- **TStop** N: Texaco*
- **FStop** N: Speedway*(LP)
- **Gas** S: Exxon*(CW), Shell* (see our ad this page)
- **Food** N: Lizard's Thicket Country Cookin (TS), Quincy's Family Steakhouse, Waffle House
- **Lodg** N: AAA Comfort Inn, Econolodge
- **AServ** S: Saturn of Columbia
- **ATM** N: Speedway
  S: Exxon, Shell

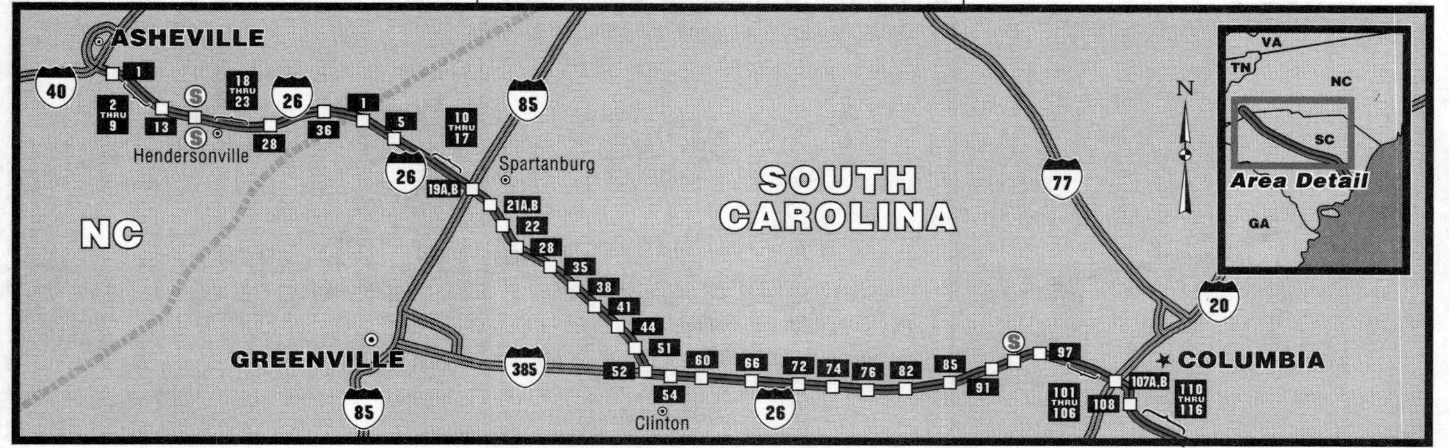

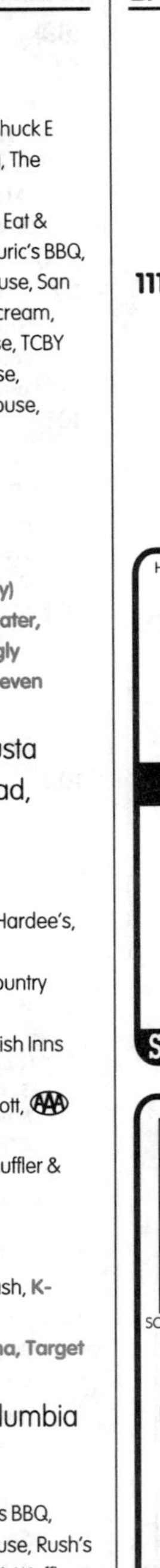

EXIT SOUTH CAROLINA

106AB St Andrews Rd.
- **Gas** N: Exxon*
  S: BP*(D)(CW), Circle K*(LP), Hess*, Shell*
- **Food** N: Blimpie's Subs (Exxon), Chick-fil-A, Chuck E Cheese Pizza, IHOP, Papa John's Pizza, The Cobblestone
  S: Bojangles, Burger King, Cussin' Bill's Eat & Drink, Domino's, Lowcountry Cafe, Mauric's BBQ, McDonalds, Pizza Hut, Ryan's Steakhouse, San Jose Mexican, Sandy's Hot Dogs & Icecream, Sub Station II, Sushi & Inakaya Jepenese, TCBY Yogurt, Taste of China Hut, Waffle House, Wendy's, Western Steer Family Steakhouse, Yamato Japanese
- **Lodg** N: Motel 6
  S: Holiday Inn, Red Roof Inn
- **AServ** S: Tire America
- **ATM** S: Nations Bank
- **Other** N: **Kroger Supermarket (w/pharmacy)**
  S: **Coin Laundry, Comedy House Theater, Eckerd Drugs, Food Lion, Piggly Wiggly Supermarket, Rite Aide Pharmacy, Seven Oaks Animal Hospital**

107AB Junction I-20, Florence, Augusta

108 Junction I-126, Bush River Road, Columbia
- **Gas** N: Citgo*(D), Texaco*(D)(CW)
  S: RaceWay*(LP), Speedway*(D)(LP)
- **Food** N: Captain D's Seafood, China Town, Hardee's, Schlotzkys Deli, Shoney's
  S: El Chico Mexican, Lizard's Thicket Country Cookin, The Villa Italian
- **Lodg** N: Budgetel Inn, DAYS INN Days Inn, Scottish Inns (see our ad this page)
  S: Best Western, ◆ Courtyard by Marriott, AAA Howard Johnson, Junction Inn
- **AServ** N: Master Care Auto Service, Midas Muffler & Brakes, Penske
- **ATM** N: First Union Bank
  S: Speedway
- **Other** N: **Dutch Square Mall,** Franks Car Wash, **K-Mart (Pharmacy)**
  S: **Corley Lock & Key, General Cinema, Target Department Store**

110 U.S. 378, Lexington, West Columbia
- **Gas** N: Phillips 66*(D)(LP), Texaco*(D)
  S: BP*(CW)
- **Food** N: Burger King, McDonalds, Millinder's BBQ, Pizza House, Quincy's Family Steakhouse, Rush's Hamburgers, The Plantation (Ramada), Waffle House
  S: Bojangles, Hardee's, Pizza Hut
- **Lodg** N: ◆ Hampton Inn (see our ad this page), AAA Ramada
- **AServ** N: Phillips 66
- **Med** S: ✚ **Lexington Medical Center**
- **ATM** N: BB& T
  S: Wachovia Bank
- **Other** N: **Sunset Drugs**

111AB U.S. 1, Lexington, West Columbia
- **Gas** N: BP*(D)(CW)(LP), Racetrac* (24 Hrs)
- **Food** N: Domino's Pizza, **Hardee's,** Magnolia's, Sonic Drive In, Subway, Terry's Diner, Waffle House
- **Lodg** N: Delta Motel, Holiday Inn, Super 8 Motel
- **AServ** N: Jiffy Lube, Nasco Auto Service
- **TServ** N: **Perkins Detroit Diesel & Alison Transmissions**

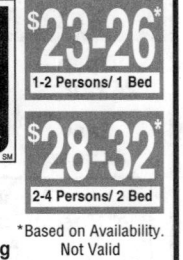

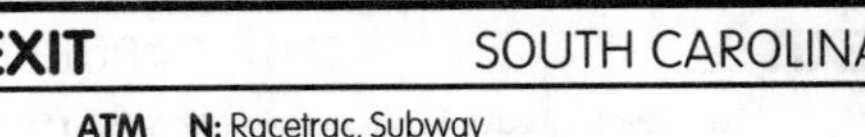

EXIT SOUTH CAROLINA

- **ATM** N: Racetrac, Subway
- **Other** N: **Bi-Lo Supermarket, Kroger Supermarket (w/pharmacy), Wal-Mart (Pharmacy)**

113 SC 302, Columbia Airport, Cayce
- **Gas** N: Amoco*
  S: Racetrac* (24 Hrs), Smile Gas* (24 Hrs)
- **Food** S: Burger King, Kettle, Shoney's, Subway, Waffle House
- **Lodg** N: AAA Best Western, Knight's Inn, **Masters Inn**
  S: AAA Comfort Inn, DAYS INN **Days Inn**
- **AServ** N: Peake's Auto & Tire
  S: NAPA Auto Parts
- **TServ** N: **Kenworth**
- **ATM** S: Racetrac, Smile Gas

115 U.S. 321, U.S. 21, U.S. 176, Cayce, Columbia
- **TStop** S: **Pilot Travel Center*(SCALES)**
- **FStop** S: **United***
- **Gas** N: Racetrac* (24 Hrs)
- **Food** N: **Waffle House**
  S: **Dairy Queen (TS),** Fried Chicken Deli, Great China Restaurant, Hardee's, **McDonalds,** Shooters Grill & Pub, Steak & Eggs House, Subway, **Wendy's (TS)**
- **AServ** N: Scott's Automotive
  S: Firestone Tire & Auto
- **ATM** S: United
- **Other** S: **Piggly Wiggly Supermarket**

116 I-77 to Charlotte, Fort Jackson

119 U.S. 21, U.S. 176, Dixiana
- **TStop** S: **Citgo*(LP)**
- **Gas** N: BP*(D)

(122) **Rest Area - RR, Phones, Vending, Picnic (Both Directions)**

125 Road 31, Gaston
- **TServ** N: **Trailer Services, Wolfe's Truck & Trailer Repair**

129 U.S. 21, Orangeburg
- **Gas** N: Shell*

136 SC 6, North Swansea St.
- **FStop** N: **Exxon***
- **Gas** N: BP*
- **Food** N: **Briar Batch (Exxon)**
- **AServ** N: Expressway Tire & Auto
- **ATM** N: Exxon

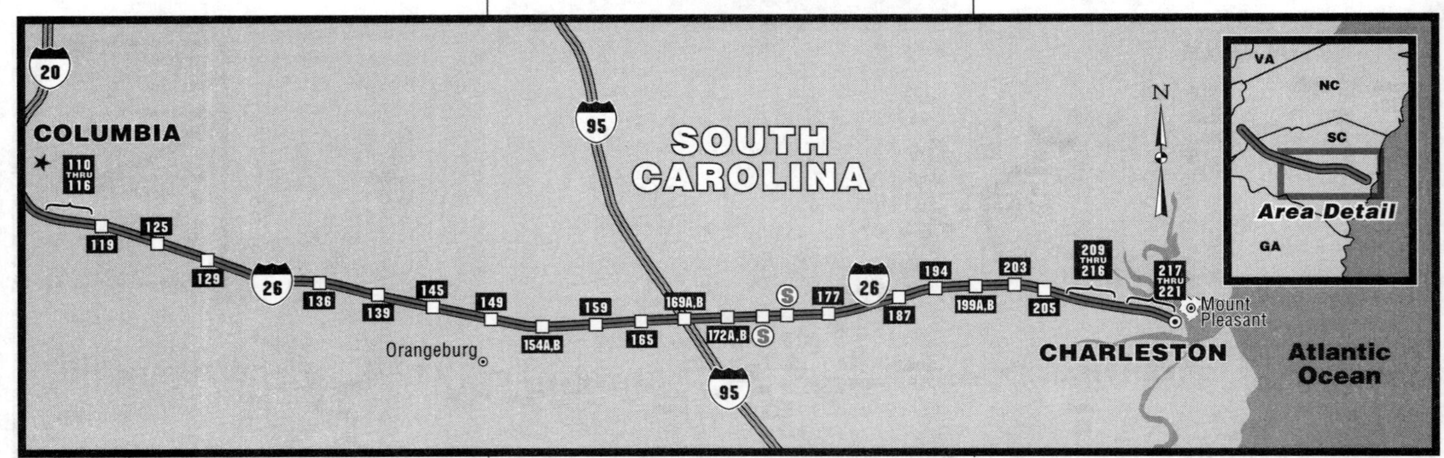

Bold red print shows RV & Bus parking available or nearby

EXIT SOUTH CAROLINA

139 Road 22, St Matthews
- **Gas** S: Citgo*
- **TServ** N: **Truck Tire & Service**
- **RVCamp** S: **Campground**
- **ATM** S: Citgo

145AB U.S. 601, St Matthews, Orangeburg
- **FStop** S: **Speedway***(LP)
- **Gas** S: Amoco*, Exxon*
- **Food** S: **Hardee's**, KFC, **McDonalds**, Waffle House
- **Lodg** S: Carolina Lodge, Comfort Inn, Days Inn, Hampton Inn (see our ad this page), Howard Johnson, **Southern Lodge**
- **ATM** S: Exxon

149 SC 33, Orangeburg, Cameron

(150) **Rest Area - RR, Phones, Picnic, Vending (Eastbound)**

(151) **Rest Area - RR, Phones, Picnic, Vending (Westbound)**

154AB U.S. 301, Orangeburg, Santee
- **FStop** S: **BP***, **Texaco***
- **Food** N: **Days Inn**
- **Lodg** N: **Days Inn, Villager Lodge (see our ad this page)**
- **AServ** S: S&R Garage
- **ATM** S: Texaco

159 Road 36, Bowman
- **TStop** N: **Speedway***(SCALES)
- **Food** N: **Starvin Marvin (Speedway)**
- **TServ** N: **Speedway**
  - S: **Abe's Truck Service (24 Hr. Service)**

165 SC 210, Bowman
- **TStop** S: **Texaco***(SCALES)
- **FStop** N: **Exxon***
  - S: **BP***
- **Food** S: Blimpie's Subs, Carolina Cafe
- **TServ** S: **Texaco**
- **ATM** S: BP, Texaco

169AB Junction I-95, Savannah, Florence

172AB U.S. 15, Holly Hill, St George
- **TStop** S: **Amoco***(SCALES) **(24 Hrs)**
- **Food** N: I-26 Coffee Shop
  - S: Dixie Boy
- **TServ** S: **J&L Tire & Diesel Service**
- **ATM** S: Amoco

(173) **Weigh Station (Eastbound)**

(174) **Weigh Station (Westbound)**

177 SC 453, Holly Hill, Harleyville
- **FStop** S: **BP***(LP)
- **ATM** S: BP
- **Other** S: **Francis Beidler Forest**

187 SC 27, Ridgeville

194 Road 16, Jedburg, Pinopolis

199AB Alternate U.S. 17, Mocks Corner, Lincolnville
- **TStop** N: **Citgo***(CW)(LP)(SCALES)
- **FStop** N: **Speedway***(LP)
- **Gas** N: Hess*(D)
  - S: Amoco*(D)(LP), BP*(D)(CW)
- **Food** N: Pizza Hut, **Ruthy's Diner (Citgo)**
  - S: Blimpie's Subs, Bojangles, Burger King, China Eight, Fazoli's Italian Food, Hardee's, Huddle House, Perkins Family Restaurant, **Quincy's Family Steakhouse**, Shoney's, TCBY Yogurt (Amoco), Waffle House
- **Lodg** S: **Comfort Inn**, Econolodge, **Economy Inn**, Hampton Inn, Holiday Inn Express
- **AServ** N: Advance Auto Parts, McElveen Pontiac, Buick, GMC Trucks
  - S: Park's Auto Parts, Parts America
- **TServ** N: **FWS Truck Service & Repair (Citgo)**
- **TWash** N: **Citgo TStop**
- **ATM** N: Citgo, Speedway
  - S: Amoco, First Citizens Bank, First Federal of Charleston
- **Other** N: **Bi-Lo Supermarket, Eckerd Drugs, Food Lion Grocery, Revco Drugs, Sangaree Animal Hospital**
  - S: **North Main Market Mall, Super Petz, Wal-Mart (Pharmacy), Winn Dixie Supermarket (Pharmacy)**

(202) **Rest Area - RR, Phones, Picnic, Vending (Westbound)**

203 College Park Road, Ladson
- **FStop** N: **Speedway***(LP)
- **Gas** N: BP*(CW)
- **Food** N: **McDonalds**, Waffle House
- **Lodg** N: ◆ **Days Inn (see our ad this page)**
- **Other** S: **College Park Veterinarian**

(204) **Rest Area - RR, Phones, Vending, Picnic (Eastbound)**

205 U.S. 78, to U.S. 52, Goose Creek
- **Gas** N: Shell*(LP)
- **Food** N: Subway, Waffle House
- **Med** N: ✚ **Columbia Trident Med Center**
- **ATM** N: First Citizens Bank

208 U.S. 52, Goose Creek, Moncks Corner

209 Ashley Phosphate Road
- **Gas** N: Exxon*(CW), Shell*(LP)
  - S: Citgo*(LP), Gate*(LP), Racetrac*
- **Food** N: Blimpie's Subs (Exxon), Fazoli's Italian Food, Hardee's, Krispy Kreme Donuts, McDonalds, Noisy Oyster, Pizza Hut, Quincy's Family Steakhouse, Ryan's Steakhouse, Taco Bell, Waffle House, Wendy's
  - S: Bojangle's, **Cracker Barrel**, Domino's Pizza, McDonalds, Old Fashioned Ice Cream, Pizza Hut, Shoney's, Waffle House, Wong's Palace
- **Lodg** N: Best Western, ◆ Red Roof Inn, ◆ Residence Inn, ◆ Super 8 Motel
  - S: Fairfield Inn, Howard Johnson, La Quinta Inn, Motel 6
- **AServ** N: Jiffy Lube, Master Care Auto Service
- **ATM** N: First Citizen's Bank, South Trust Bank, Wachovia Bank
  - S: Racetrac
- **Other** N: Cactus Car Wash, **Northwoods Mall**

211AB Aviation Ave, Air Force Base, Remount Rd.
- **FStop** S: **Speedway***(LP)
- **Gas** N: Amoco*, Exxon*(CW), Hess(D), Shell*
- **Food** N: Arby's, Burger King, Captain D's Seafood, China Garden, China Town,

Bold red print shows RV & Bus parking available or nearby

EXIT	SOUTH CAROLINA
	Church's Fried Chicken, Grandy's, Huddle House, KFC, La Hacienda, Mongolian BBQ, Old Country Buffet, Oriental Cuisine, Pizza Hut, Pizza Inn, Schlotzkys Deli, Sea Fare, Shoney's, Subway, Taco Bell, Wendy's
	S: Veranda's Restaurant (Holiday Inn), Waffle House
Lodg	**N:** **Masters Inn**, Radisson
	S: Budget Inn, Holiday Inn, **Knight's Inn**
AServ	**N:** Advance Auto Parts, Amoco, Goodyear Tire & Auto, Grease Monkey, Pep Boys Auto Center
ATM	**N:** BB& T, SC Federal Credit Union
Other	**N:** **Sam's Club**
	S: **United Artist Theaters**
212A	Hanahan (Westbound, Reaccess Eastbound, Difficult Reaccess)
Gas	**N:** Hess[D]
Food	**N:** Alex's Restaurant, Marie's Diner, McDonalds, Tom Portaro's Italian
AServ	**N:** Advance Auto Parts, Altman Dodge, Auto Zone Auto Parts, Carolina Transmission, Goodyear Tire & Auto, Grease Monkey, Jones Ford, Muffler Shop
ATM	**N:** Wachovia
Other	**N:** Alex Car Wash, **Coin Laundry**
212BC	Junction I-526, Savannah, Mt. Pleasant
213AB	Montague Ave
Gas	**S:** Ashley's*[D] (24 Hrs), Citgo*
Food	**N:** Piccadilly Cafeteria, Red Lobster
	S: Waffle House

EXIT	SOUTH CAROLINA
Lodg	**N:** Hilton
	S: Comfort Inn, Days Inn, Extended StayAmerica, Hampton Inn (see our ad this page), Orchard Inn, Quality Suites, Ramada
AServ	**S:** Ashley's
ATM	**N:** First Federal of Charleston, NBSC
Other	**N:** **Charles Towne Square**
215	SC 642, Dorchester Road
Gas	**N:** Hess[D]
	S: Texaco*[D][CW]
Food	**N:** Hardee's
Lodg	**N:** Howard Johnson
	S: ◆ Econolodge, Super 8 (see our ad this page)
Other	**N:** **Coin Laundry, Greyhound Bus Station**
216AB	SC 7, Cosgrove Ave, US 17S.
Med	**N:** ✚ **Roper Medical Center**
217	North Meeting St, Charleston (Eastbound)
218	Spruill Ave, Naval Base
219	Rutledge Ave , The Citadel, Morrison Dr. (Eastbound, Reaccess Westbound)
220	Romney Street (Westbound, No Reaccess)
221AB	U.S. 17, King St, Meeting St
Med	**S:** ✚ **Hospital**

↑SOUTH CAROLINA

Begin I-26

I-27 S →

EXIT	TEXAS
	Begin I-27

↓TEXAS

EXIT	TEXAS
124	Junction I-40 , Amarillo
123	34th Ave, Tyler St
Gas	**E:** Conoco*
Food	**W:** Burritos El Gordo, D.J.'s
122B	Fm1541, 34th Ave, Tyler St
122A	Fm 1541, Washington St
Gas	**E:** Texaco* (ATM)
	W: Texaco
Food	**E:** Sonic
	W: Taco Bell
Lodg	**E:** Amarillo Motel
AServ	**W:** Texaco
ATM	**E:** Boatmen's First, Texaco
Other	**W:** **Maxor**
121B	Hawthorne Dr, Austin St (Southbound)
Gas	**W:** Texaco* (ATM)
AServ	**W:** Scotties Transmission
ATM	**W:** Texaco
Other	**W:** Coin Car Wash
121A	Georgia St
Gas	**E:** Phillips 66[CW]
Food	**E:** Jeff's Grand Burger
AServ	**E:** GMC Truck, Honda Dealership, Nissan Dealership

EXIT	TEXAS
Other	**W:** **Texas Department Of Public Safety**
120B	45th Ave
FStop	**E:** **Fina***[D]
Gas	**W:** Diamond Shamrock*
Food	**W:** Abuelo's Mexican, Hardee's, Lubby's Cafeteria, The Doughnut Stop
Lodg	**E:** American Motel
	W: Travler Motel
AServ	**W:** Farnsworth Transmission
TServ	**W:** **Roberts Ford Truck, Roberts Truck Center**
ATM	**W:** Diamond Shamrock
Other	**W:** Coin Car Wash, **Colonial Vet. Clinic**
120A	Republic Ave
119B	Western St, 58th Ave
Gas	**E:** Phillip 66* (ATM)
	W: Texaco*[D]
Food	**W:** Blimpie's Subs, Ming Palace, Pizza Hut
ATM	**E:** Phillip 66
Other	**E:** **Homeland**
	W: **Dickie Stout RV Service**
119A	Hillside Rd W
117	Bell St
Gas	**W:** Fina*, Texaco*
Food	**W:** Long John Silvers, Mr. Burger, Sonic
AServ	**W:** Super Shops Automotive Performance Center
ATM	**W:** Texaco
Other	**E:** Crystal Car Wash
116	Loop 335, Hollywood Rd
FStop	**W:** **Love's Country Store***[D][CW]

EXIT	TEXAS
Food	**W:** McDonalds, Waffle House
Lodg	**W:** Days Inn
115	Sundown Lane
113	McCormick Rd
Gas	**W:** Texaco
AServ	**E:** Ron Clark Ford
112	Fm 2219
FStop	**W:** **Citgo**[D]
111	Rockwell Rd
Gas	**E:** Fina*
AServ	**E:** Midway Chevorlet
110	U.S. 87S, U.S. 60W, Canyon, Hereford
109	Buffalo Stadium Rd
108	Fm3331, Hunsley Rd
106	Canyon, W Texas A & M University
103	Fm1541N, Cemetary Rd
99	Hungate Rd
(98)	**Parking Area P (Both Directions)**
96	Dowlen Rd
94	Fm285, Wayside
92	Haley Rd
90	Fm1075, Happy
88	U.S. 87N, Fm1881
83	Fm2698
82	Fm214
77	U.S. 87, Tulia
75	N W 6th St

EXIT TEXAS

74 TX86, Tulia
- **TStop** **W:** Rip Griffin Travel Center*(D) (Subway, Grandy's)
- **Food** **W:** Grandy's, Subway
- **Lodg** **W:** Best Western

(70) Parking Area (No Facilaties)
68 Fm928
63 Fm145, Kress
61 U.S. 87, County Rd, Kress
56 Fm788
54 Fm3184
53 Bus27, Plainview
51 Quincy St
50 TX194
- **Med** **E:** Hospital

49 U.S. 70
- **Gas** **E:** Diamond Shamrock*, Texaco* (Subway)
  W: Chevron(CW), Shell*
- **Food** **E:** Cotton Patch, Far East Restaurant, Furr's, Kettle, Leal's Mexican, Long John Silvers, Mrs. K's Bakery, Subway
  W: Arby's, Blimpie's Subs (Chevron), Burger King, Casa Ole, Chicken Express, McDonalds, Mr Gatti's Pizza, Sabrie's Seafood, Subway, Taco Villa
- **Lodg** **E:** ◆ Best Western, Days Inn
  W: AAA Holiday Inn Express (see our ad this page)
- **AServ** **E:** Ford Dealership
- **ATM** **E:** American State Bank
- **Other** **E:** Revco
  W: Eye Wear Outlet, K-Mart, Planet Express Car Wash, Wal-Mart

48 Fm3488
45 Bus27, Plainview
43 Fm2337
41 County Rd
38 Main St, Hale
37 Fm1914, Cleveland St, Hale Center
- **Food** **E:** Dairy Queen
- **Med** **W:** Hospital
- **Other** **W:** Post Office

36 Fm1424N
32 Fm37W
31 Fm37E
(29) Rest Area - Picnic, Grills, RR, Phones P
(28) Rest Area - Picnic, Grills, RR, Phones P
27 County Rd
24 Fm54
22 Loop369, Abernathy
- **Gas** **E:** Phillips 66(D)
  W: Phillips 66*
- **AServ** **E:** Phillips 66

21 Fm597, Main St, Abernathy

EXIT TEXAS

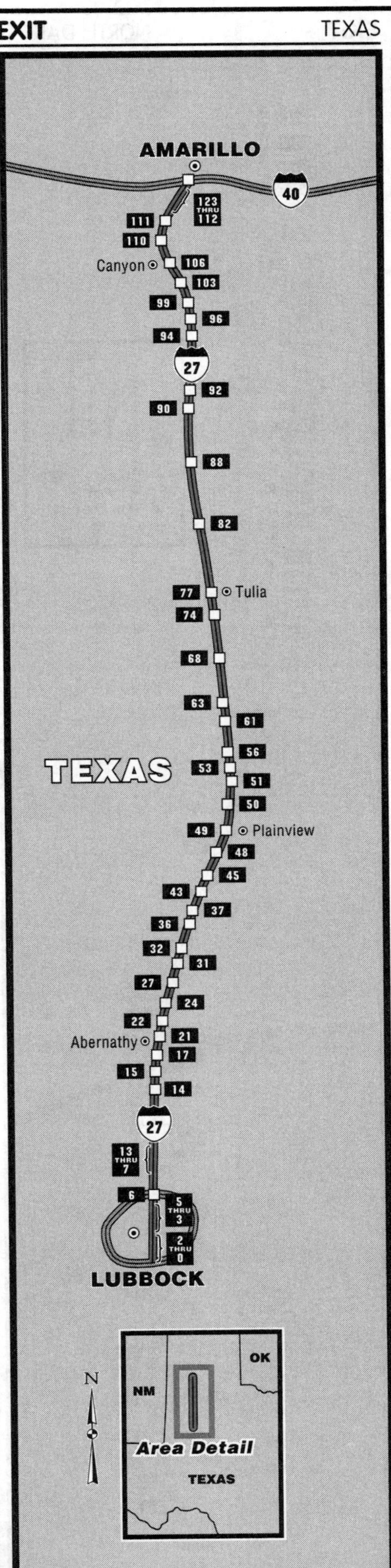

EXIT TEXAS

- **FStop** **W:** Co-op(D)
- **Gas** **W:** Conoco*
- **Food** **W:** Dairy Queen
- **AServ** **W:** Co-op, Gordon Automotive
- **ATM** **W:** First State Bank
- **Other** **W:** Pinson Pharmacy, Post Office

17 County Rd 53
15 Loop461, New Deal
14 Fm1729
13 Loop461, New Deal
11 Fm1294, Shallowater
10 Keuka St
9 Lubbock Int Airport
8 Fm2641, Regis St, Airport Passenger Terminal
7 Yucca Lane
6B Loop289
- **Lodg** **W:** Texas Motor Inn
- **RVCamp** **E:** Pharr Rv

6A Spur326, Ave Q
- **Lodg** **W:** El Tejas Motel

5 Ave H, Municipal Dr
4 Bus87, U.S. 82, U.S. 87, 4th St, Crosbyton
3 U.S. 62, TX114, Floydada, Levelland, 19th St
- **Gas** **E:** Diamond Shamrock
- **Food** **W:** Stubb's BBQ, The Depot Restaurant
- **AServ** **W:** Lubbock Carburetor and Electric

2 34th St, Ave H
1C 50th St
- **Gas** **E:** Buddy's*, Conoco*, Pronto Mart*
- **Food** **W:** Bryan's Steaks, Country Plate Diner, Dairy Queen, KFC, Long John Silvers, Subway, Wienerschnitzel
- **AServ** **E:** Beebers Radiator
- **Other** **W:** Post Office

1B U.S. 84, Loop289E, Post (Souhtbound)
- **Lodg** **W:** AAA Super 8 Motel, Best Western (see our ad this page)

1A 50th St, Loop289W
1 Bus87, 82nd St
- **Gas** **W:** Chevron*, Conoco*(D)
- **Food** **W:** Charley's BBQ

0 98th St
- **FStop** **W:** Diamond Shamrock*(D)
- **Food** **W:** Texas Tony's
- **AServ** **W:** Ben's Muffler

↑TEXAS

Begin I-27

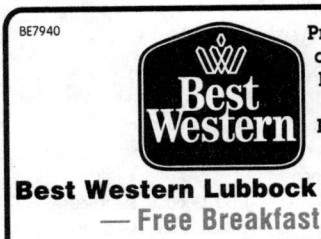

Bold red print shows RV & Bus parking available or nearby

Begin I-29

↓NORTH DAKOTA

215 ND 59, County 55, Neche, Pembina
- **FStop** E: **Gastrak*(SCALES) (RV Dump), The Trad'N' Post***
- **Food** E: **Restaurant (Trade'N'Post)**, Tastee Freeze, The Depot Cafe
- **Other** E: **Amex Duty Free Shop, The Pembina Museum, Tourist Information**

212 Unnamed

208 CR1, Bathgate

203 U.S. 81, ND 5, Hamilton, Cavalier
- **TStop** W: **Joliette Express Truck Stop* (24 Hr)**
- **Food** W: **Restaurant* (Joliette)**
- **Parks** W: **Icelandic State Park (25 Mi)**
- **Other** W: **Weigh Station**

(203) Weigh Station (Both Directions)

200 Unnamed

196 CR 3

193 Unnamed

191 CR 11, St. Thomas

187 ND 66, Drayton
- **TStop** E: **Cub Int'l**
- **FStop** E: **Amoco, Cenex*, Fur***
- **Food** E: Eat, **Hot Stuff Pizza (Cenex), Smash Hit Subs (Cenex)**
- **Lodg** E: **Motel 66**
- **RVCamp** E: **Catfish Haven Campground**
- **Other** E: Car & Truck Wash, **Coin Laundry**

184 Drayton
- **RVCamp** E: **Camping**

180 Unnamed

(179) Rest Area - RR, HF, Phones, Picnic, RV Dump (Both Directions, Left Sided Exit, Closed Due To Construction)

176 ND 17, Grafton
- **RVCamp** W: **Campground**
- **Med** W: ✚ **Unity Medical Center**

172 Unnamed

168 Warsaw, Minto

164 Unnamed

161 ND 54, CR 19, Oslo, Ardoch

157 Unnamed

152 U.S. 81, Manvel, Gilby
- **FStop** W: **Towne Mart***

145 N. Washington St

141 U.S. 2, Gateway Drive
- **TStop** W: **Phillips 66(LP)(SCALES), Stamart*(LP)(SCALES) (RV Dump, Showers, Coin Laundry)**
- **Gas** E: Amoco*, Phillips 66*, Sinclair*, Surburan Propane(LP)
- **Food** E: Burger King, G.F. Goodribs Steakhouse, **Highway Host Restaurant**, Holiday Inn, McDonalds, Ramada Inn, Village Inn
  W: Perkins Family Restaurant, **Restaurant (Stamart), Restaurnat (Phillips 66)**
- **Lodg** E: ◆ Best Western, ◆ Econolodge, ◆ **Holiday Inn**, ◆ Ramada, (AAA) Roadking Inn, (AAA) Select Inn, Super 8 Motel
  W: Prairie Inn

- **AServ** E: Amoco, Big A, Big Wheel Auto Service, Smitty's Transmission
  W: GM Auto Dealership
- **TServ** E: **Superior Body Works**
  W: **Cummins Diesel, Phillips 66, Scott's Inc, Stamart**
- **RVCamp** W: **Budget RV Service, Key RV**
- **ATM** E: First Nat'l Bank, Sinclair
  W: **Stamart**
- **Parks** W: **Turtle River State Park (19 Miles)**
- **Other** E: Cleanerama Operated Car Wash, **Tourist Information**
  W: **Coin Laundry (Stamart)**

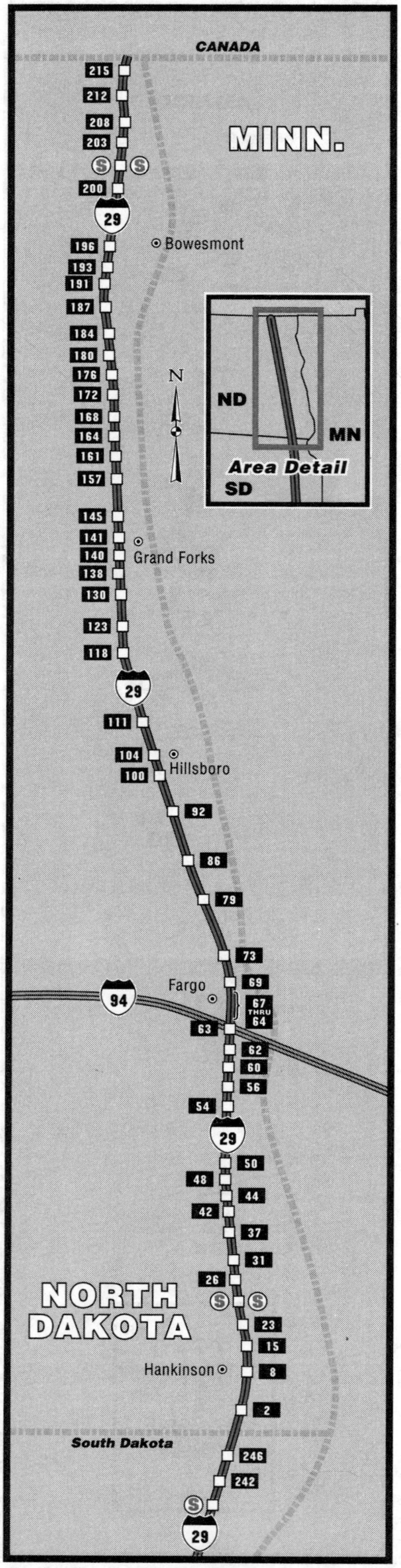

140 DeMers Ave, City Center
- **Gas** E: Amoco*(CW)(LP)
- **AServ** E: Amoco
- **Med** E: ✚ **Hospital**
- **ATM** E: Amoco

138 U.S. 81 Bus, 32nd Ave South
- **TStop** W: **Conoco*(LP) (RV Dump)**
- **Gas** E: Holiday*(CW), Super Pumper*(CW)
- **Food** E: Holiday Inn, Subway (Holiday Gas)
  W: **I-29 Cafe (Conoco), Subway (Conoco)**
- **Lodg** E: C'mon Inn, ◆ Comfort Inn, Country Inn Suite, ◆ Fairfield Inn, Road King Inn
- **AServ** E: National Muffler
- **RVCamp** W: **Campground**
- **ATM** E: Holiday Gas
- **Other** E: **Columbia Mall, Menard's**
  W: **Coin Laundry (Conoco), Kindess Animal Hospital**

130 CR81, ND 15, Thompson

123 Reynolds

(120) Weigh Station (Both Directions)

118 Buxton

111 ND 200 West, Mayville, Cummings

104 Hillsboro
- **FStop** E: **Cenex*(LP)**
- **Gas** E: Amoco*
- **Food** E: **Burger King (Cenex)**, Country Hearth Family Restaurant, Food Court (Amoco), Hot Stuff Pizza (Amoco)
- **AServ** E: Hillsboro Auto
- **RVCamp** E: **Hillsboro Campground**
- **Med** E: ✚ **Hospital**
- **ATM** E: Amoco, Goose River Bank
- **Other** E: **Coin Laundry**

100 ND 200 E&W, Mayville, Cummings

(99) Rest Area - RR, HF, PHones, Picnic, RV Dump (Both Directions)

92 Grandin
- **FStop** E: **CoOp(LP)**
- **Gas** W: Stop-N-Shop*
- **Food** E: **Rendezvous Grill (CoOp)**
  W: Stop-N-Shop Restaurant
- **Other** W: **Halstead Park**

86 Gardner
- **FStop** W: **Sinclair***
- **Food** W: **Cafe (Sinclair)**
- **AServ** W: Sinclair

79 Argusville

(74) Rest Area - RR, HF, Phones, Picnic, RV Dump, Pet Walk (Both Directions)

Bold red print shows RV & Bus parking available or nearby

EXIT NORTH DAKOTA

73 Harwood
- **FStop** E: Cenex*(LP)
- **Food** E: Cafe*(LP) (Cenex)

69 County Route 20

67 Bus U.S. 81, 19th Ave North
- **Med** W: ✚ VA Hospital

66 12th Ave North
- **Gas** E: Stop 'N Go
- **TServ** W: Interstate Detroit Diesel, Northern Truck & Trailer, OK Tire Shop, Power Brake Midwest Inc.
- **ATM** E: Stop 'N Go
- **Other** E: Whale of A Wash (Car Wash)

65 U.S. 10, Downtown Fargo, West Fargo
- **FStop** E: Sta-Mart*
  W: Cenex*
- **Gas** E: Amoco*, Suburban City(LP)
  W: Mobil, Stop 'N Go*
- **Food** E: Highway Host Restaurant, Valley Kitchen
  W: Embers, Hardee's, Subway (Cenex)
- **Lodg** W: ◆ Best Western
- **AServ** E: A Transmission City, Amoco, Big Wheel, Dave's I-29 Service Ctr
  W: Car Quest Auto Center, Ford Dealership, Lincoln, Mercury, Mobil
- **TServ** W: American Truck, Cummins Diesel, Ford Dealership, Geller's, Hall Truck Service, Nelson International, RDO Mack, Volvo Dealer, Wallwork Truck Center
- **RVCamp** E: Bernie's Camper Corral (RV Service)
- **ATM** E: Amoco, Sta-Mart
- **Other** E: NAPA Auto Parts, Valley Veterinary Hospital

64 13th Ave South
- **FStop** E: Sinclair*(D), Sta-Mart*
- **Gas** E: Amoco, Holiday*
  W: Amoco*(CW)
- **Food** E: Best Western, Burger King, Dairy Queen, Deflies, Giovanni's Pizza, Happy Chef, Mr. Steak, Nine Dragons (Chinese), Perkins Family Restaurant, Ponderosa
  W: Arby's, Blimpie's Subs, Chi Chi's Mexican, Chili's, Country Harvest, Denny's, Fuddrucker's, Godfather's Pizza, Grandma's, Lone Star Restaurant, McDonalds, Paradiso Mexican, Pizza Hut, Randy's Restaurant, Red Lobster, Royal Fork Buffet, T.G.I. Friday's, The Olive Garden, Valentino's
- **Lodg** E: (AAA) Americinn Motel, (AAA) Best Western, ◆ Comfort Inn, Country Hospitality Suites, ◆ Econolodge, ◆ Hampton Inn, Motel 6, ◆ Super 8 Motel
  W: ◆ Comfort Inn, (DAYS INN) ◆ Days Inn, ◆ Fairfield Inn, ◆ Holiday Inn, ◆ Kelly Inn, (AAA) Ramada Plaza Suites, (AAA) Select Inn
- **AServ** E: Goodyear Tire & Auto, Jiffy Lube
  W: Amoco, GM Auto Dealership, Midas, NAPA Auto Parts, Sears Auto Center
- **ATM** E: Amoco, Sinclair
  W: Amoco, Community First, First Bank, Gate City Federal Savings Bank
- **Other** E: Cash Wise Foods Grocery, Dental Clinic, Menard's, Pearle Vision, Sterling Optical, White Drug Pharmacy
  W: Car Wash, General Nutrition, Hornbacher's Grocery Store, Kohl's Dept Store, Mall, Office Max, Target Department Store, Toys "R" Us, WalGreens

EXIT NORTH/SOUTH DAKOTA

63AB Jct I-94, Fargo, Bismarck

62 32nd Ave South
- **TStop** W: Flying J Travel Plaza(SCALES)
- **FStop** E: Fleet Farm*(CW)
- **Gas** E: Amoco*(CW)
- **Food** E: Country Kitchen
  W: Windbreak (Flying J)
- **Lodg** W: (AAA) Flying J Inn
- **AServ** E: ABRA, Fleet Farm
  W: Goodyear Tire & Auto, Thermo-King Radiator
- **TServ** E: Butler CAT
  W: Flying J Travel Plaza, Goodyear Tire & Auto, Johnsen Trailer, Peterbilt Dealer, Timpke Trailer Co, Truck Licensing & Service, Utility Trailer Shop
- **TWash** W: Flying J Travel Plaza
- **ATM** E: Amoco
  W: Flying J Travel Plaza
- **Other** E: Rent-All Hardware(LP)

60 CR 6, Frontier
- **Other** E: Fireworks

56 Wild Rice, Horace

54 CR 16, Oxbow, Davenport

50 CR 18, Hickson

48 ND 46, Kindred

44 Christine

42 CR 2, Walcott

(40) Rest Area - RR, HF, Phones, Picnic, RV Dump (Both Directions)

37 CR 4, Colfax, Abercrombie
- **Other** E: Fort Abercrombie Historic Site (7 miles)

31 CR 8, Galchutt

26 Dwight

(24) Weigh Station (Both Directions, Left Lane Exit)

23 ND 13, Mooreton, Wahpeton
- **Med** E: ✚ Hospital
- **Other** W: Bagg Bonanza Farm Interpretive Ctr (2 Miles)

15 CR 16, Great Bend, Mantador

8 ND 11, Hankinson, Fairmount
- **FStop** E: Moblie*
- **Other** W: Camping (3 Miles)

(3) Rest Area - RR, HF, Phones, Picnic, RV Dump (Northbound)

2 CR 22

1 Ct 1E
- **Other** E: The Dakota Magic Casino

↑NORTH DAKOTA

↓SOUTH DAKOTA

(251) Rest Area - RR, HF, Phones, Picnic, RV Dump, Tourist Info (Southbound)

246 SD 127, New Effington, Rosholt
- **Parks** W: Sica Hollow State Park (24 Miles)

242 SD

(235) Weigh Station (Southbound)

232 SD 10, Sisseton, Browns Valley

EXIT SOUTH DAKOTA

- **FStop** E: Phillips 66* (Casino)
- **RVCamp** W: Camping
- **Med** W: ✚ Hospital
- **Parks** W: Fort Sisseton (25 miles), Roy Lake State Park (25 miles)

224 Peever
- **FStop** E: Cenex*
- **Parks** W: Pickerel Lake (16 Miles)
- **Other** W: Sioux Tribal Headquarters

(213) Rest Area - RR, HF, Phones, Picnic, RV Dump
- **Parks** E: Hartford Beach State Park (17 Miles)
- **Other** E: Highway Patrol, Rest Area - RR, Picnic, Phone

213 SD 15, Wilmot, Rest Area Eastbound
- **Parks** E: Hartford Beach State Park
- **Other** E: Highway Patrol, Rest Area-Full Facilities

207 U.S. 12, Summit, Aberdeen
- **FStop** E: Conoco*
- **Food** E: Hot Stuff Pizza (Conoco)
- **Other** W: Blue Dog State Fish Hatchery, Waubay National Wildlife (19 Miles)

201 Twin Brooks

193 SD 20, South Shore, Stockholm
- **RVCamp** E: Camping (6 Miles)

185 Waverly

180 Bramble Park Zoo, Municipal Airport
- **Other** W: Bramble Park Zoo (5 Miles), Municipal Airport (6 Miles)

177 U.S. 212, Watertown, Kranzburg
- **TStop** E: Sinclair*(SCALES)
- **FStop** W: Conoco*
- **Food** E: Sinclair
  W: Country Kitchen (Conoco), Hot Stuff Pizza (Conoco), McDonalds
- **Lodg** E: (AAA) Stone's Inn
  W: (AAA) Comfort Inn, (DAYS INN) ◆ Days Inn, Drake Motor Inn
- **TServ** E: Sinclair
- **Med** W: ✚ Hospital
- **ATM** E: Sinclair
- **Other** E: Lew's Fireworks
  W: Sandy Shore Recreation Area (10 Miles), Tourist Information (10 Miles)

164 SD 22, Castlewood, Clear Lake
- **Gas** E: Cenex (9 Miles)
- **Med** E: ✚ Hospital (10 Miles)

(160) Rest Area - RR, HF, Phones, Picnic, RV Dump (Both Directions)

157 Brandt

150 SD 15, 28, Toronto, Estelline
- **Parks** W: Lake Poinsett Rec Area (24 miles)

140 SD 30, White, Bruce
- **Parks** W: Oakwood Lakes State Park (12 Miles)

133 US 14 Bypass, Volga, Arlington
- **Other** W: Laura Ingles Wilder Homestead

132 I-29 Bus, U.S. 14, Brookings, Huron
- **FStop** E: Conoco*
- **Gas** W: Amoco*(D)(CW) (24 Hr), Citgo*(CW)
- **Food** E: Burger King (Conoco)
  W: Burger King, Country Kitchen, Dairy Queen, Hardee's, KFC (Best Western), Little Caesars Pizza (K-Mart), McDonalds, Perkins Family Restaurant, Steak & Buffet

Bold red print shows RV & Bus parking available or nearby

EXIT SOUTH DAKOTA

Lodg E: ◆ Fairfield Inn, ◆ Super 8 Motel
W: Best Western, AAA Comfort Inn, Holiday Inn
AServ W: Amoco, Car Quest Auto Center, GM Auto Dealership, Husky Auto Repair
RVCamp E: A&D Campers
Med W: ✚ Hospital
ATM E: Conoco
W: First National Bank
Other W: Econo Mart Grocery Store, K-Mart (Pharmacy), South Dakato State Museum, South Dakota State University, Visitor Information, Wal-Mart (Pharmacy)

127 SD 324, Elkton, Sinai

(121) **Rest Area - RR, Phones, Picnic, Vending, RV Dump (Both Directions)**

121 Nunda, Ward, Rest Area - Full Facilities, Pet Walk (Both Directions)

114 SD 32, Flandreau

109 SD 34, Madison, Colman
Gas W: Sinclair*, Texaco*(CW)
Parks W: Lake Herman State Park (24 Miles)
Other W: Dakota State University, South Dakota State Museum

104 Trent

(103) **Parking Area (Southbound)**

(102) **Parking Area (Northbound)**

98 SD115, Dell Rapids, Chester

94 Baltic, Colton

86 Renner, Crooks
TServ E: Harmon Trucking, Northern Truck Equipment Inc.

84AB Jct. I-90

83 SD 38, 60th Street
TServ E: Holcomb Freightliner Inc.
Other W: Fireworks

81 SD 38, Airport, Russell St (Bridge 15ft 4in)
Food E: Best Western, Oaks (Howard Johnson), Rollin' Pin (Kelly Inn)
Lodg E: ◆ Best Western, AAA Brimark Inn, Excel Inn, Howard Johnson, ◆ Kelly Inn, Motel 6
AServ E: Sioux Brake
TServ E: Sioux Brake
RVCamp E: Schaap's Travel land
Other E: Highway Patrol

79 I-29 Bus, SD 42, 12th St
FStop W: Conoco*
Gas E: Amoco* (24 Hr)
W: Citgo*
Food E: Chris' Restaurant, Dairy Queen, Taco Bell, Taco Villa, Wendy's
W: Casino West (Casino Inside), Golden Crust (Conoco), Hardee's, Victory Lane Homecooked Food
Lodg E: Ramada Limited
W: Pinecrest (Rents By Month), Sunset Motel, West Wick Motel (RV Hook-Up)
AServ E: Cartiva, GM Auto Dealership, K-Mart
W: Mr Tire Service, Performance Pros, Select Motors, Sioux Falls Tires, Wheel City Motors, Wheelco Brake
RVCamp W: Tower Campground
ATM E: Amoco

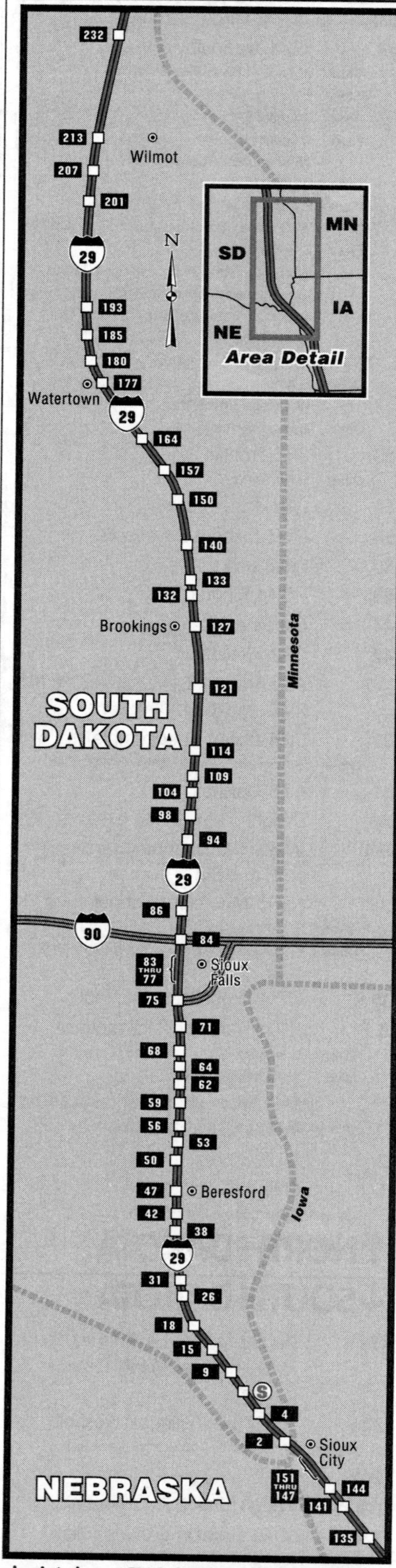

EXIT SOUTH DAKOTA

W: Conoco
Parks E: Sioux River Waterslide Park
Other E: Great Plains Zoo & Museum (1 Mile), K-Mart (1 Mile), USS South Dakota Battleship Memorial (1 Mile)
W: Stop N Cart Food Mart

78 26th St
Lodg E: Hampton Inn

77 41st St
Gas E: Amoco*(CW), Conoco(CW), Super America*
Food E: Applebee's (Target Shop Ctr), Arby's, Burger King, Carlos O'Kelly's Cafe (Mexican), Chi Chi's Mexican, Dots & Eddy's, Embers, Fryin' Pan, Fuddrucker's, Hooters, KFC, McDonalds, Olive Garden, Perkin's Family Restaurant, Pizza Hut, Pizza Inn, Red Lobster, Spaghetti Jack's, Subway, Szechuan Chinese, T.G.I. Friday's, Taco Bell, Timber Lodge Steakhouse, Wendy's
W: Denny's, Perkins Family Restaurant
Lodg E: ◆ Best Western, ◆ Comfort Suites, ◆ Fairfield Inn, AAA Radisson, ◆ Super 8 Motel
W: ◆ AmericInn, Bugetel Inn, DAYS INN ◆ Days Inn, AAA Select Inn
AServ E: Ben Hur Ford, Ford Dealership, GM Auto Dealership, Jiffy Lube, Schultz Automotive, Tires Plus, Western Auto
ATM E: Bank First, First Premier Bank, Valley Bank
Parks E: Falls Park
Other E: Fast Car Wash, Food Surgical Ctr, Harold's Car Wash, Kinko's, Kohl's Dept Store, Manard's, Randal's Foods (Pharmacy), Sam's Club, Sunshine Supermarket, Target Department Store, The Empire Mall, Toys "R" Us, Wal-Mart (Pharmacy)

75 Jct. I-229

73 Tea
FStop E: Larry's Texaco Truck Stop*
Food E: Larry's Texaco, Pat's Steakhouse
AServ E: Midwest Tire & Muffler, Quaker State
TServ E: Larry's Texaco
RVCamp W: Red Barn Campground
Other E: S.D. Fireworks
W: Fireworks Outlet, Rich Bros Fireworks, Sioux Falls Fireworks

71 Tea, Harrisburg
AServ E: Repairables
W: Kars & Kustom Car Service
TServ E: Lund Truck Parts
RVCamp W: Camping

68 Parker

64 SD 44, Worthing, Lennox
AServ W: GM Auto Dealership

62 U.S. 18E, Canton (Low Clearance Under Bride 14ft.7in)
FStop E: Phillips 66*
Lodg E: Charlie's Motel
AServ W: Star Auto Repairables

59 U.S. 18, Davis, Hurley

56 Fairview
Parks E: Newton Hills State Park (12 Miles)

53 Viborg

50 Centerville, Hudson

47 SD 46, Irene, Beresford
TStop W: Cenex*(LP)
FStop E: Sinclair* (Casino)

Bold red print shows RV & Bus parking available or nearby

EXIT SOUTH DAKOTA/IOWA

Gas E: Ampride*(D), Casey's Gen Store*, Cenex*
Food E: Emily's Cafe (Speciality Chicken Fried Steak), Hot Stuff Pizza (Cenex)
W: Cafe (Cenex)
Lodg E: Crossroads Motel
AServ E: Ampride, GM Auto Dealership
TServ E: Joe's Service Ctr
W: Cenex
TWash W: Cenex*(LP)
RVCamp E: Windmill RV Campground (Laundry Facilities)
ATM E: Cenex, Sinclair
W: Cenex*
Other E: Car Wash, Jubilee Foods Supermarket, Laundry Facilities (Windmill RV Camp)
W: Travel Store (Cenex)

42 Alchester, Wakonda

(41) **Parking Area (Southbound)**

38 Volin
Parks E: Union County State Park (3 Miles)

31 SD 48, Spink, Akron

(26) **Rest Area - RR, Phones, Picnic, RV Dump (Both Directions)**

26 SD 50, Vermillion, Yankton, Rest Area Eastbound
FStop W: Conoco
Food W: Coffee Cup Restaurant (Conoco)
Lodg W: ◆ Comfort Inn
Other E: Rest Area, Full Facilities
W: Shrine Music Museum (7 Miles)

18 Elk Point, Burbank
Gas E: Amoco*(CW)(LP) (Casino), Phillips 66*
Food E: Dairy Queen, Hot Stuff Pizza (Phillips 66)
Lodg E: Hometown Inn
ATM E: Amoco

15 I-29 Bus, Elk Point

9 SD105, Jefferson
FStop E: Amoco*(D)(LP)
Food E: Countryside (Amoco)
Other E: Sav-N-Sam Fireworks, South Dakota Fireworks

4 McCook
RVCamp W: KOA Campground
Other E: Discount Fireworks

(3) **Weigh Station (Northbound)**

2 SD 105N, North Sioux City
FStop E: 76*
Gas E: Amoco* (Casino)
W: Casey's Gen Store*
Food E: Cafe (76), McDonalds
W: Picadilly's Pizza (Casey's Gen Store)
Lodg W: Apple Inn, Comfort Inn, ◆ Super 8 Motel
RVCamp W: Cott's Campground
ATM E: Amoco
Other E: Fireworks, Ike's Casino, The Gateway Factory Outlet, The King Sea Casino

1 Dakota Dunes
Food W: The First Edition 2 (Beef & Seafood)
Lodg W: (AAA) Country Inn
ATM W: First Financial Bank

↑SOUTH DAKOTA

↓IOWA

EXIT IOWA

151 IA 12, Riverside Boulevard

(149) **Rest Area - RR, Phones, Picnic, Vending, Info (Both Directions)**

149 Hamilton Boulevard
Gas E: Conoco*(CW) (24 Hr)
Food E: Horizon Restaurant
Lodg E: Holiday Inn (see our ad this page)
AServ E: Jiffy Lube
Parks W: Chris Larsen Jr Park
Other W: Iowa Tourist Information (Full Facilities), Riverboat Museum

148 Wesley Way, S. Sioux City, U.S. 77S

147B U.S. 20, IA 12S, Bus District (Difficult Reaccess, Southbound Reaccess Only)
Gas E: Amoco*(CW), Holiday*(LP)
Food E: Arby's, Burger King, Hardee's, Pepperdines Restaurant (Riverboat Inn), Perkins Family Restaurant
Lodg E: Best Western, Hilton, ◆ Riverboat Inn
AServ E: GM Auto Dealership, Midas, Speedy Lube
Med E: ✚ Hospital
Other E: Staple's (Office Supply Store), WalGreens

147A Floyd Boulevard, Stockyards

144B U.S. 20W, US 75S, South Sioux City (Divided Hwy)

144A U.S. 20, Fort Dodge (Divided Hwy)

143 U.S. 75, Singing Hills Blvd
TStop E: Texaco*(SCALES)
W: Amoco*(SCALES)
FStop E: Cennex*
Food E: McDonalds
W: Restaurant (Amoco), Wendy's
Lodg E: Budgetel Motel, DAYS INN ◆ Days Inn, Haven Inn Motel
W: Amoco
TServ E: Texaco
W: Amoco, Bridgestone Tire & Auto, International, Kenworth White GMC Volvo, Peterbilt Dealer
TWash E: Texaco
Parks E: Lewis & Clark State Park
Other E: CB Shop (Texaco), Sam's Club

141 CR D 38, Sioux Gateway Airport
Gas E: Phillips 66*(CW), Texaco*
W: Amoco*(LP)
Food E: Cheers, Godfather's Pizza, Subway
Lodg E: Rath Inn
W: Motel 6

EXIT IOWA

ATM E: Pioneer Bank
Other E: Bluff Stop Grocery Store
W: Air Museum

(139) **Rest Area - RR, Phones, Picnic, Vending, RV Dump, Weather Radio, Newspaper, Pet Area (Both Directions)**

135 Port Neal Landing

134 Salix
Gas E: Citgo*
AServ E: Citgo
RVCamp W: Camping

(132) **Weigh Station - both directions**

127 IA 141, Sloan
Gas E: Phillips 66*
Lodg E: Homestead Inn, ◆ Rodeway Inn
Other W: Winnebago Indian Reservation

120 Whiting
RVCamp W: Camping

112 IA 175, Oanwa, Decatur
TStop E: Conoco* (24 Hr)
FStop E: Phillips 66*
Food E: Jan's Restaurant (Conoco), McDonalds, Michael's Restaurant, Oehler Brothers Food, Subway (Phillips 66)
Lodg E: Super 8 Motel
AServ E: Buick Dealer, Don's Exhaust
TServ E: Conoco
RVCamp E: Interchange RV Camp
Med E: ✚ Hospital
Parks W: Lewis & Clark State Park
Other E: Pamida Grocery
W: Indian Reservation, Keel Boat Exhibit

(109) **Rest Area - RR, Phones, Picnic, RV Dump, Weather Radio (Both Directions)**

105 CR E 60, Blencoe

95 IA 301, Little Sioux
Gas E: Citgo
Food E: Stuckey's (Citgo)
RVCamp W: Woodland Campground

(92) **Rest Area - Picnic, RR, HF, Phones, RV Dump (Parking Only)**

89 IA 127, Mondamin

82 IA 300, CR F50, Modale
FStop W: Cenex*

(79) **Rest Area - RR, Phones, Picnic, RV Dump, Grills, Weather Radio, Newspaper (Both Directions)**

75 U.S. 30, Missouri Valley, Blair NE
FStop E: Texaco* (24 Hr)
Gas E: Phillips 66*(CW)
W: Conoco*
Food E: McDonalds, Subway
W: Burger King, Copper Kettle, Oehler Brother
Lodg W: DAYS INN Days Inn, Rath Inn, Super 8 Motel
AServ E: Bob Anderson Ford Mercury
W: A & G Auto And Truck Repair, GM Auto Dealership
TServ W: A&G Auto And Truck Repair
Med E: ✚ Hospital (2 Miles)
ATM E: Texaco (24 Hr)
Other E: Iowa Welcome Center (5 Miles)

EXIT IOWA

W: DeSoto Refugee, Steamboard Exhibit
(74) **Rest Area - Picnic, RR, HF, Phones, RV Dump (Southbound)**
(73) **Weigh Station (Northbound)**
72 IA 362, Loveland, Weigh Station Both Directions
FStop E: Conoco*
71 Jct. I-680 East, Des Moines
66 Honey Creek
FStop W: Phillips 66*(LP)
Food W: Iowa Feed & Grain Company
RVCamp W: Camping
61AB Jct. I-680, IA 988, Omaha Crescent Council Bluffs
56 Council Bluffs (Southbound)
55 North 25th St
Gas E: Andrew Lounge*, Pump N Munch*, Sinclair*
Food E: Dairy Fair
Lodg E: Ramada Inn
Other E: Coin Car Wash
54B North 35th St
54A G. Ave., Council Bluffs
53B Jct. I-480, US 6, Omaha (Left Lane Exit)
53A 9th Ave, Harvey's Blvd
Gas E: Conoco*(CW), Phillips 66*, Total*
Food E: Country Kitchen
Lodg E: Days Inn ◆ Days Inn
W: Harvey's Hotel & Casino
52 Nebraska Ave, Dog Track Casino
Gas E: Conoco*(D)(CW)
Food E: Franny's Subs (Conoco)
Lodg W: AmeriStar Hotel, ◆ Holiday Inn
ATM E: Conoco
Other E: Casino, Dog Track
W: AmeriStar Casino
51 Jct. I-80W. Omaha

Note: I-29 runs concurrent below with I-80. Numbering follows I-80.

1B South 24th St, Council Bluffs
3 IA 192, Council Bluffs, Lake Manawa
5 Madison Avenue, Council Bluffs
Gas N: Amoco*
S: Phillips 66* (ATM), Texaco*(CW)
Food N: Burger King, Great Wall, Pizza Hut, Subway, The Garden Cafe
S: Dairy Queen, Shoney's
Lodg N: ◆ Heartland Inn
S: AAA Western Inn
ATM N: First Bank
Other N: Coin Laundry, Council Bluffs Medical Mall, Hy-Vee Food Store (Pharmacy), Mall of the Bluffs, Target Department Store
S: Scrub & Dub Auto Wash

Note: I-29 runs concurrent above with I-80. Numbering follows I-80.

48 Jct. I-80 East, Des Moines
47 U.S. 275, IA 92, Lake Manawa
42 IA 370, Bellevue
(37) **Rest Area - RR, Phones, Picnic, RV Dump (Both Directions)**
35 U.S. 34, Glenwood, Red Oak
FStop W: Amoco*
Lodg W: AAA Bluff View Motel
AServ W: Amoco
TServ W: Amoco

EXIT IOWA

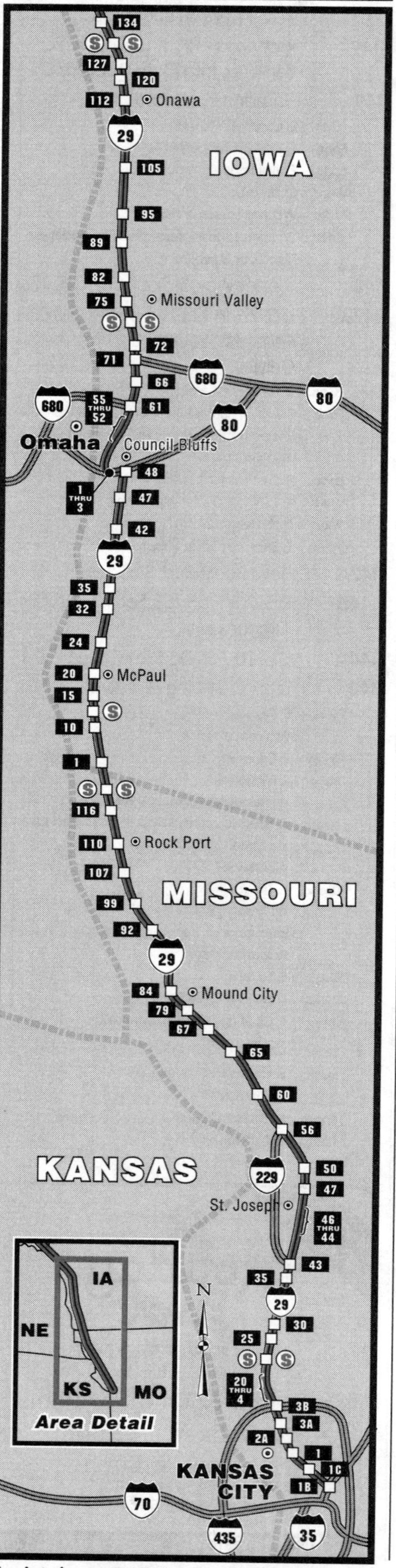

EXIT IOWA/MISSOURI

32 U.S. 34, Plattsmouth, Pacific Jct
24 Cnty Road L31, Bartlett, Tabor
20 IA 145, McPaul, Thurman
AServ E: J&K Tires
15 CR J26, Percival
(11) **Weigh Station (Northbound)**
10 IA 2, Sidney, Nebraska City
3 IA192, Lake Manawa
1 IA 383, Hamburg
Gas W: Fox Lake Citgo*(D)

↑IOWA

↓MISSOURI

(121) **Weigh Station (Both Directions)**
116 CR A & B, Watson
AServ E: Don's Auto & Truck Salvage
110 U.S. 136, Rock Port, Phelps City
TStop W: Amoco*(SCALES), Phillips 66*(SCALES)
Gas E: Conoco*, Texaco*(D)
Food E: Hardee's
W: McDonalds, Trails End Restaurant
Lodg E: Rockport Inn
W: Oak Grove Inn
TServ W: Amoco
(110) **Rest Area - RR, Phones, Picnic, Vending P (Southbound)**
107 MO 111, Rock Port, Langdon
Food W: Elk Plaza Cafe
Lodg W: Elk Inn
99 CR W: Corning
92 U.S. 59, Craig, Fairfax
84 MO 118, Mound City
FStop W: King*
Gas E: Conoco*, Phillips 66*(D)(LP), Total*(D)
Food E: Hardees, Karnes Cafe
Lodg E: Audreys Motel
ATM E: UMB Bank
(81) **Rest Area - RR, Phones, Picnic, Vending P (Both Directions)**
79 U.S. 159, Rulo
TStop E: Phillips 66*(SCALES) (RV Dump)
Lodg E: Plaza Motel
TServ E: Phillips 66
75 U.S. 59, Oregon
67 U.S. 59, Oregon
65 U.S. 59, Fillmore, CR RA, Savannah
60 CR CC, K, Amazonia
56AB Jct I-229, U.S. 71, 59, Maryville, St. Joseph
53 Bus. 51, US 59, Bus. I-29 Savannah, St.Joseph
RVCamp E: KOA Campground
50 U.S. 169, St. Joseph, King City
47 MO 6, Fredrick Boulevard, Clarksdale
Gas E: Conoco* (ATM)
W: Amoco*(CW), Sinclair, Texaco*(D), Total*
Food E: Country Kitchen, Golden Grill
W: Applebee's, Boston Market, Burger King, Denny's, Dunkin Donuts, Fazoli's Italian Food, Godfather's Pizza, Hardee's, McDonalds, Perkins, Pizza Hut, Red Lobster, Sizzler Steak House, Sonic Drive In, Taco Bell, The Ground Round, Winsteads, Wyatt Cafeteria
Lodg E: Days Inn AAA Days Inn, ◆ Drury Inn

Bold red print shows RV & Bus parking available or nearby

EXIT		MISSOURI
		W: Budget Inn, Motel 6, Pony Express Motel, AAA Ramada, Super 8 Motel
	ASers	W: Firestone Tire & Auto, Goodyear Tire & Auto, Midas Muffler & Brakes, Texaco
	ATM	W: Bank Midwest, Merchantile Bank, North American Savings Bank
	Other	W: **Osco Drugs**
46AB		U.S. 36, St. Joseph, Cameron (Difficult Reaccess)
44		I-29 Bus, U.S. 169, St. Joseph, Gower
	TStop	W: **Texaco*(SCALES)**
	Gas	E: King*
	Food	E: St.Joe Prime Meat, Subway, The Grill
		W: **McDonalds, Texaco**
	Lodg	E: AAA Best Western
		W: **Texaco**
	TServ	E: **Dave's Truck Service**
		W: **Texaco**
	Other	E: Car & Truck Wash, **Suburband Propane**
43		Jct. I-229
35		CR DD, Faucett
	TStop	W: **Farris Truckstop(LP)(SCALES)**
	Food	W: **Subway (Farris TStop)**
	Lodg	W: **Farris Truckstop**
	TServ	W: **Farris Truckstop**
	TWash	W: **Farris Truckstop**
30		MO Z, H, Dearborn, New Market
	FStop	E: **Conoco*(D) (ATM)**
(27)		**Rest Area - RR, Vending, Phones, Picnic P (Both Directions)**
25		MO U & E, Camden Point
	Gas	E: KCL Express, Phillips 66*
(24)		**Weigh Station - both directions**
20		MO 92, 371, Atchison, Weston, Leavenworth
	Parks	W: **Weston Bend State Park (5 Miles)**
19		MO HH, Platte City (Difficult Northbound Reaccess)
	Gas	W: Phillips 66
	Food	W: Dairy Queen, Red Dragon 2
	Lodg	W: AAA Best Western, Comfort Inn
	AServ	W: Mike's Auto Parts
	Other	W: **IGA Food Store**
18		MO 92, Platte City, Leavenworth, Weston
	TStop	W: **QuikTrip*(SCALES)**
	Food	W: Dairy Queen, **McDonalds,** Subway, Taco Bell
	ATM	W: Platte City Bank
17		Jct. I-435, Topeca
15		Mexico City Ave
	Lodg	W: Marriott

EXIT		MISSOURI
	Other	W: **Kansas City Airport**
14		Jct. I-435E.
13		KCI Airport, Jct. I-435
	Lodg	E: Club House Inn, ◆ Fairfield Inn
12		Northwest 112th St
	Gas	E: Amoco*(CW)
	Lodg	E: AAA Hilton, AAA Holiday Inn Express, Hampton Inn (see our ad this page)
		W: ◆ Econolodge
10		Tiffany Springs Parkway
	Gas	E: Texaco*(D)(CW)
	Food	E: Jade Garden
	Lodg	E: ◆ Embassy Suites, Homewood Suites Hotel
		W: ◆ Courtyard by Marriott, Drury Inn, ◆ Residence Inn
9AB		MO 152, Liberty, Topeca (Difficult Reaccess)
8		MO 9, Rt T, Northwest Barry Road
	Gas	W: Phillips 66*(CW), QuikTrip*
	Food	E: 54th Street Grill, Applebee's, Bagel & Bagel, Boston Market, Carlito's Mexican, Chili's, China Wok, Lamar's Donuts, Lone Star Steakhouse, Subway, Taco Bell, Waids, Wendy's, Winsteads
		W: Daylight Donuts, Hardee's, Long John Silvers, McDonalds, Minsky's Pizza, Outback Steakhouse, Rainbow Restaurant, Taco John's
	Lodg	W: **Motel 6,** ◆ Super 8 Motel
	Med	E: **+ St. Luke's Northland Hospital**
	ATM	E: Merchantile Bank
		W: UMB Bank
	Other	E: **Eckerd Drugs**
6		Northwest 72nd St, Platte Woods
	Gas	E: Sinclair*
		W: QuikTrip*
	Food	W: KFC*, Papa John's Pizza, Pizza Hut
	AServ	E: Sinclair
	Other	E: **Animal Hospital**
		W: **K-Mart**
5		MO 45 North, Northwest 64th St
	Gas	W: Texaco*(CW)
	Food	W: IHOP, Little Caesars Pizza, McDonalds, Paradise Grill
	ATM	W: Midland Bank
	Other	W: **Osco Drugs**
4		Northwest 56th St
3B		Jct. I-635, U.S. 169 South , Kansas
3C		CR.A, Riverside (Southbound, Difficult Reaccess)
3A		Cty. Road AA, Waukomis Drive (Northbound Only)
2A		U.S. 169, Smithville (Difficult Reaccess)
1		U.S. 69, Vivion Road (Southbound)
	Food	E: Lamar's Donuts
	AServ	E: Courtsey Chevrolet, Cadillac, Jack Miller Chrysler Plymouth, Northtown Mercury Lincoln, Northtown Mitibushi
	Other	E: **Price Chopper Supermarket, Venture Discount Store**
1D		MO 283S, Oak Trfwy
1C		Old Trfwy (Southbound, Difficult Reaccess)
	Food	E: Deli Depot, Lemar's Donuts
	Other	E: **Price Chopper Supermarket, Venture**
1B		Jct. I-35 North, Des Moines

↑ MISSOURI

Begin I-29

EXIT		TEXAS

Begin I-30

↓ TEXAS

EXIT		TEXAS
1B		Linkcrest Dr.
	FStop	S: **Mobil*(D)**
	Gas	S: Chevron*, Fina*(D), Phillips 66*
	Food	S: Joseifinas Mex Cafe, Wet Willies
2		Spur 580
3		RM2871, Chapell, Creek Blvd
5A		Alemeda St.
5BC		Junction I-820
6		Las Vegas Trail
	Gas	N: Chevron*(CW), Highway Oil
		S: Citgo*, Diamond Shamrock*, Exxon*(D)(CW), Texaco*(D)(CW)
	Food	N: Laudmark Cafe, McDonalds (Chevron), Waffle House
		S: Denny's, Vegas Grill
	Lodg	N: DAYS INN Days Inn
		S: Comfort Inn, Motel 6, ◆ Royal Western Suites
	AServ	S: Cheif Auto Parts
	Other	N: **Rigmar Animal Hospital (24 Hrs)**
		S: **Coin Laundry**
7A		Cherry Lane
	Gas	N: Conoco*(D), Texaco*(CW)
	Food	N: Chuck E. Cheese's, Ci Ci's Pizza, IHOP, Luby's, Papa John's Pizza, Ryan's Steakhouse, Subway, Taco Bell
		S: Partons Pizza
	Lodg	N: AAA La Quinta Inn, Super 8 Motel
		S: ◆ Hampton Inn (see our ad on this page)
	AServ	N: Texas Motors Ford Dealership
	Other	N: **K-Mart, Wal-Mart**
		S: **Target Department Store**
7B		TX183, Spur 341, NAS Fort Worth, JRB, Green Oaks Rd.
8B		Ridgmar Blvd, Ridglea Ave
	Gas	N: Texaco*
8C		Green Oaks Rd
	Food	N: Applebee's, T.G.I. Friday's
		S: Chinese Cuisine Buffet, Tommy's Hamburgers
	AServ	N: National Tire and Battery, Sear's Auto Care
	Other	N: **Albertson's Grocery, Joshua's Book Store, Nieman Marcus, Ridgmar Mall**
8D		TX183S
	Lodg	S: Best Western
9A		Bryant - Irvin Rd
	Gas	S: Texaco*

EXIT TEXAS

9B Horne St
- **Gas** S: Citgo* (ATM), Fina*(D)
- **Food** S: Dunkin Donut, Edmondson's Fried Chicken, Tejano Cafe
- **ATM** S: Citgo

10 Hulen St

11 Montgomery St
- **Gas** S: Texaco*(CW)
- **Food** N: Cafe Jazz
  S: What A Burger
- **Other** N: **Coin Laundry**

12A University Dr

12B Rosedale St
- **Med** S: **Hospital**

12C Forest Park Blvd

13 Summit Ave

14 TX199, Henderson St

14A Junction I-35W, Denton, Waco

15 U.S. 287, TX 180

16C Beach St
- **Lodg** S: (AAA) Comfort Inn, (AAA) Holiday Inn
- **ATM** S: Bank One

18 Oakland Blvd
- **Gas** N: Phillips 66*(D), Texaco*(D)(CW)
- **Food** N: Burger King, Waffle House
- **Lodg** N: Motel 6

19 Brentwood Stair Rd
- **Gas** N: Chevron*(CW), Texaco*
  S: Citgo*, Texaco
- **Food** N: Chuy's, Italy Pasta and Pizza, Steak & Ale
  S: Fast Freddy's, Lino's Mexican
- **AServ** S: Texaco, Transmission Depot
- **Other** S: Coin Car Wash

21AB Junction I-820

21C Bridgewood Dr
- **Gas** S: Citgo*(D)
- **Food** N: Luby's Cafeteria, Subway
  S: What A Burger
- **Other** N: **Woodhaven Pet Clinic**

23 Cooks Lane

24 Eastchase Pkwy
- **Gas** S: Chevron*(CW), Texaco*
- **Food** S: Baskin Robbins, Dunkin Donuts, IHOP, McDonalds (Playground), Schlotzkys Deli, Subway, Taco Bell, Wendy's, What A Burger
- **ATM** S: First National Bank of Texas
- **Other** S: **Crown Books, Petsmart, Target Department Store**

26 Fielder Rd

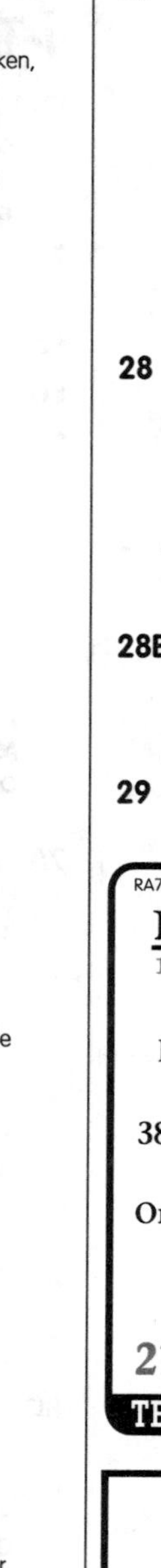

EXIT TEXAS

27 Lamar Blvd, Cooper St
- **Gas** N: Mobil(CW)
  S: Citgo* (ATM), Texaco(CW)
- **Food** N: Jack-in-the-Box, Subway, Tao Yen Chinese
  S: Burger King (Indoor Playground)
- **AServ** N: Kwik Lube
- **ATM** N: Bank One
  S: Citgo
- **Other** N: Coin Car Wash, **Eckerd Drugs, Kroger Supermarket**
  S: **Arlington North Animal Clinic, Coin Laundry**

28 FM157, Collins St
- **Gas** N: Mobil*, Shell*
- **Food** N: Waffle House
  S: Royal Panda
- **Lodg** N: Ramada
- **AServ** N: Shell
- **Other** N: **Six Flags Over Texas, Wet N' Wild Water Park**

28B Pennant Dr, Arlington Stadium, Nolan Ryan Expressway
- **Food** S: Cozymels, Eat At Joe's, Pappadeaux's
- **Other** S: **Arlington Stadium**

29 Ball Park Way
- **Gas** S: Fina*(D)(CW)

EXIT TEXAS

- **Food** N: Macroni Grill
  S: On the Border Cafe
- **Lodg** S: Marriott, Roadway Inn
- **ATM** N: Bank Of America

30 TX360, Six Flags Dr
- **Gas** S: Texaco*(D)(CW), Total
- **Food** N: McDonalds
  S: Bennigan's, Cheddar's, Grand Slam Pizza, Luby's, McDonalds, Ninfa's, Owen's Restaurant, Steak & Ale, Subway
- **Lodg** S: ◆ Budgetel Inn, (AAA) La Quinta Inn, (AAA) Sleep Inn, Amerisuites (see our ad this page)
- **Other** S: **Six Flags Mall, Six Flags Over Texas**

32 NW19th St
- **Gas** N: Exxon
  S: Fina(D), Mobil(CW)
- **Food** S: Denny's, Long John Silvers, McDonalds, Pizza Hut, Subway, Taco Bell, What A Burger
- **Lodg** S: (AAA) La Quinta Inn
- **ATM** S: Central Bank and Trust

34 Belt Line Rd
- **Gas** N: Chevron*(CW)
  S: Diamond Shamrock*(CW), Fina*(CW) (Blimpie), Racetrac* (ATM), Texaco (AServ)
- **Food** S: Blimpie's Subs (Fina), McDonalds, Schlotzkys Deli
- **Lodg** N: Ramada Inn
- **AServ** S: Texaco
- **ATM** S: Racetrac
- **Other** N: **Palace Of Wax, Ripley's Believe It or Not**
  S: **Animal Clinic**

36 MacArthur Blvd.

38 Loop12

41 Hampton Rd South
- **Food** S: Luby's Cafeteria

42 Hampton Rd North

44A Beckley Ave

45 Junction I-35

45A Griffin/Lamar
- **Lodg** N: Ramada Plaza Dallas (see our ad this page)

46A Downtown Central Expressway

46B Junction I-45, Houston Sherman, U.S. 75

47A 2nd Ave
- **Parks** S: **Fair Park**

47B !st Ave. Fair Park

48A TX78, East Grand Ave, Barry Ave, Munger Blvd

48B Winslow St
- **Gas** N: Citgo*, Phillips 66*(CW), Texaco*

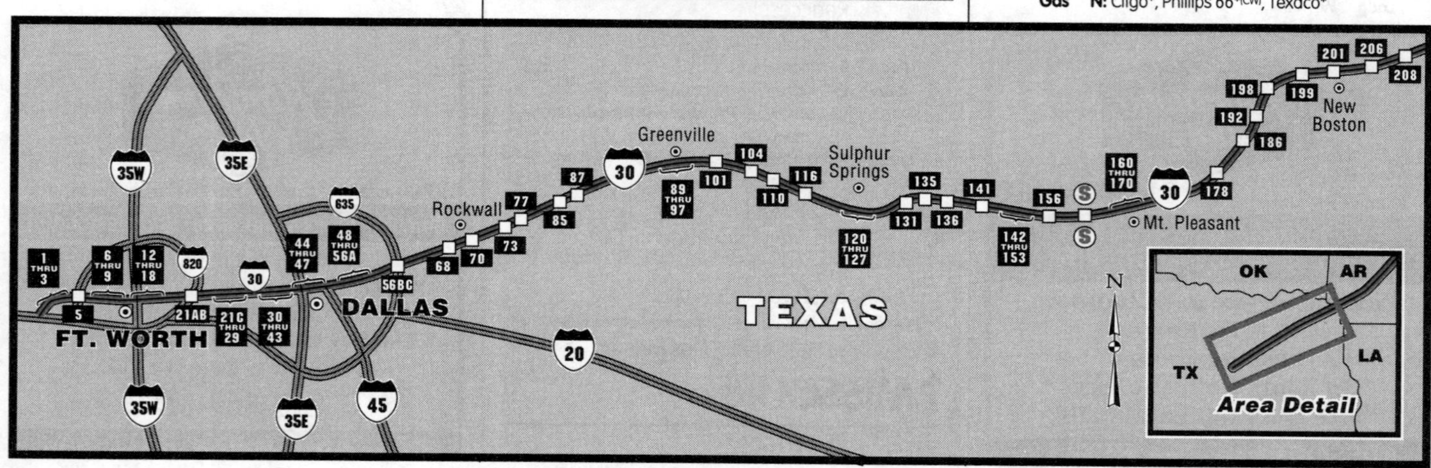

Bold red print shows RV & Bus parking available or nearby

EXIT	TEXAS
	S: Phillips 66*, Shell*
Food	N: Church's Fried Chicken
49A	Dolphin Rd
Gas	N: Texaco*
Lodg	N: Eastern Hills Motel, Welcome Inn
AServ	N: Ken's Muffler's and Shocks
49B	Lawnview Ave, Samuell Blvd
Gas	N: Diamond Shamrock, Phillips 66, Texaco*
Food	N: Keller's Hamburgers
Lodg	N: Howard Johnson
	S: Motel
AServ	N: Phillips 66
	S: Brako
51	Highland Rd, Jim Miller Rd
Gas	N: Exxon(D)
	S: Racetrac, Texaco(CW)
Food	N: Denny's, McDonalds, Owen's Restaurant
	S: Burger King, Captain D's Seafood, Casa Cavasas, Grandy's, KFC, Pizza Hut, Pizza Inn, Subway, Taco Bell, Wendy's
Lodg	N: La Quinta Inn
	S: ◆ **Red Roof Inn**
AServ	N: Exxon
	S: Chief Auto Parts, Jim Miller
Other	S: **Eckerd Drugs**
52A	Saint Francis
52B	Loop 12
53	U.S. 80, Terrell
54	Big Town Blvd
TStop	S: **TravelCenters of America*(D)(SCALES) (Country Pride)**
Gas	N: Citgo*, Shell*, Texaco*
Food	N: Her Donut and Bakery Shop
	S: **TravelCenters of America**
Lodg	S: **Villa Inn**
TServ	S: **Cummins Diesel**
TWash	S: **Big Town Truck Wash**
RVCamp	S: **Holiday RV Center**
55	Motley Dr
Gas	S: Chevron(CW)
Med	S: ✚ **Hospital**
56A	Gus Thomasson Rd, Galloway Ave
Gas	N: Diamond Shamrock*, Exxon*
	S: Citgo* (ATM)
Food	N: KFC, Taco Bell
	S: Baskin Robbins, China 1st, Dickey's BBQ, Domino's Pizza, Donna's Kitchen, Grandy's, Jack-In-The-Box, Luby's Cafeteria, Subs Miami Grill, Subway, The Feedbag, Wendy's
Lodg	S: Delux Inn
AServ	N: Chief Auto Parts, Classic Auto Care
	S: Firestone Tire & Auto, Jiffy Lube, Midas Muffler & Brakes, Tune-Up Masters
RVCamp	S: **Mesquite Camper Sales & Service**
ATM	S: Citgo
Other	N: Coin Car Wash, **Dave's Books, WalGreens**
	S: **Kroger Supermarket, Spiegler Eye Care**
56BC	Junction I-635
57	Northwest Dr
Gas	S: Fina*(D)
Food	S: Jack-In-The-Box
58	Garland, Beltline Rd
Gas	N: Mobil* (ATM), Total*
	S: Exxon(D), Shell*(CW)
Food	N: China City, Denny's, Donuts, KFC, Little Caesars Pizza, McDonalds, Subway, Taco Bell
	S: Baker's Ribs, Buck's Pizza, Burger King, Dairy Queen, Grandy's, Waffle House, Wok One
Lodg	S: Days Inn, Motel 6
AServ	N: Chief Auto Parts
	S: Exxon
ATM	N: Mobil
	S: Bank Of America, Bank of Texas
Other	S: **Kroger Supermarket, Petmobile Pet Hospital**
59	Rose Hill Rd
60	Bobtown Rd
61	Zion Rd
62	Chaha Rd
Gas	S: Diamond Shamrock*(LP), Texaco*
Lodg	N: Best Western
64	Dalrock Rd
Gas	N: Conoco, Diamond Shamrock(D)(CW), Mobil, Texaco, Total
	S: Exxon(D), Shell
Food	N: China City Restaurant, McDonalds
	S: Burger King, Dairy Queen, Grandy's, WOK One, Waffle House
AServ	N: Chief Auto Parts, Diamond Shamrock
ATM	N: Mobil
Other	N: **K-Mart**
67A	Village Dr, Horizon
Gas	N: Chevron*(CW), E-Z Mart
	S: Total*
Food	N: Arby's, Burger King, Cajun Catfish, El Trebino's, Grandy's, Hamby's Sports Grill, McDonalds, Waffle House, Wendy's
	S: Boston Market, Chili's, Culpeppers, Donut Palace, El Chico Mexican, Jack-In-The-Box, Pizza Getti, Rockwall Country Cafe, Subway, The Oar Restaurant
AServ	N: Goodyear Tire & Auto
ATM	N: Bank of America
	S: American National Bank, Lakeside National Bank
Other	N: Coin Car Wash
	S: **Albertson's Grocery, Eckerd Drugs, Kroger Supermarket, Ridge Road Animal Hospital, Roma's Pre-read Books, Target Department Store**
68	TX205, Rockwall, Terrall
TStop	S: **76 Auto/Truck Plaza*(LP)(SCALES)**
Gas	N: E-Z Mart*, Racetrac*, Shell, Texaco
	S: Total*(D) (Taco Bell)
Food	N: Braums Restaurant, Dairy Queen, Donna's Kitchen, Granma's Fried Chicken, KFC, Luigis, Pizza Hut, Wan Fu, What A Burger/24 Hrs
	S: **76**, Taco Bell (Total)
Lodg	N: Super 8 Motel
AServ	N: Napa Pro Lube Center, Shell, Texaco
TServ	N: **R-Tex Auto and Truck Parts**
	S: **76**
TWash	N: **Truck and Car Coin Car Wash**
ATM	S: Total
Other	N: **Rockwall Veterinary Hospital, The Eyes of Texas**
70	FM549
73	FM551, Fate
77A	FM548, Royse City
Gas	N: Texaco*(D)(CW) (Pizza Inn)
	S: Conoco*
Food	N: **Kountry Charm Cafe**, Pizza Inn (Texaco)
Lodg	N: **Sun Royse Inn**
AServ	N: Crumpton Auto Repair, Royse City Tire Co.
77B	FM35
TStop	N: **Knox*(D)(SCALES) (ATM)**
FStop	N: **Coastal*(D)(SCALES), Exxon***
Food	N: DK's Drive In, **Soul Man's BBQ**, Subway (Knox)
AServ	N: Johnson's Auto & Wrecker Repair
ATM	N: Knox
79	FM1565, FM2642
83	FM1565N
85	FM36, Caddo Mills
AServ	S: Cams Auto Service
87	FM1903, Caddo Mills
FStop	S: **Fina*(D)**
Food	S: **Fina**
TServ	S: **T & B Truck Repair**
89	FM1570
Lodg	S: Meadowview Motel
93AB	TX34, Terrell, Greenville
Gas	N: Jim Dandy
	S: Chevron*(CW)
Food	N: Applebee's, Ci Ci's Pizza, IHOP, Jack-In-The-Box, Ryan's Steak House, Schlotzkys Deli, Subway, TCBY Yogurt, Ta Mally's, Taco Bell, Taco Bueno, Wendy's, What A Burger
	S: Denny's, Luby's, Red Lobster, The Spare Rib
AServ	S: Wal-Mart
Med	N: ✚ **Hospital 3 miles**
ATM	N: Aliance Bank
Other	N: Coin Car Wash
	S: **Wal-Mart**
94A	U.S. 69, U.S. 380, Denison, McKinney
Gas	N: Chevron*(D)(CW), Exxon
	S: Fina*
Food	N: Arby's, McDonalds
	S: IHOP
Lodg	N: Comfort Inn, Economy Inn, Holiday Inn, **Motel 6**
ATM	N: Lone Oak State Bank
94B	Bus. US 69, Emory
FStop	N: **Total* (ATM)**
Gas	N: Fisca
	S: Chevron*(CW) (ATM), Exxon
Food	N: El Sombrero, Kettle, Taco Bell (Total)
	S: **Arby's**, Catfish King, McDonalds
Lodg	N: ◆ Best Western, Days Inn, Royal Inn
	S: ◆ Comfort Inn, Economy Inn, ◆ Holiday Inn, Motel 6, Super 8 Motel
AServ	S: Exxon
ATM	N: Total
	S: Chevron
Other	S: Coin Car Wash
95	Frontage Rd
Gas	S: Fina
Lodg	N: Dream Lodge Motel
	S: Sunrise Motel
96	U.S. 67, Loop 302
97	Lamar St
Gas	N: Exxon
	S: Fina*
Lodg	S: Sunrise Motel
AServ	N: Exxon
TServ	N: **Truck Tire Repair**
RVCamp	S: **Sunrise**

EXIT		TEXAS
	Other	S: Animal Hospital
101		Cooper Lake, Commerce, Paris
	FStop	N: Phillips 66*
	Food	N: Phillips 66
	Lodg	N: Cooper Motel
	Other	N: Pippin's Propane Gas(LP)
104		Lake Tawakoni, Campbell, Lone Oak
110		FM275, Cumby
	Gas	N: Citgo*(D), Phillips 66*(D)
		S: Texaco*
	AServ	N: Auto Repair Co.
		S: Morgan's Auto Repair
116		U.S. 67, FM2653, Brashear Rd
120		U.S. 67
122		TX19, Emory
	FStop	S: Phillips 66(D) (Restaurant)
	Gas	N: Chevron*
	Food	S: Phillips 66
	RVCamp	S: I-30 RV Sales and Service
123		FM2297, League St
	Gas	N: Chevron*, Phillips 66(CW), Texaco*
124		TX11, TX154, Sulphur Springs, Quitman
	Gas	N: Diamond Shamrock, Exxon, Phillips 66*, Total*
		S: Exxon*, Shell*(D) (ATM) (see our ad this page), Texaco*
	Food	N: Catfish King, KFC, North Cut Donuts, Pizza Hut, Sonic Drive In, Stromboli's Pizza, Subway, Ta Molly's, Western Sizzlin'
		S: Baskin Robbins, Bodacious BBQ, Braum's, Burger King, Furr's Cafeteria, Grandy's, K-Bobs Steakhouse, McDonalds, McDonalds (Wal-Mart), Pizza Inn, Taco Bell
	Lodg	N: Royal Inn
		S: Holiday King Motel
	AServ	N: Chief Auto Parts, Diamond Shamrock, Exxon, I-30 Radiator Shop, Phillips 66
		S: 10 Minute Oil Change, Discount Wheel and Tire Co., Tire Town, Wal-Mart
	ATM	N: Aliance Bank
		S: Shell
	Other	N: Factory Stores of America (see our ad this page), Suburban Propane(LP)
		S: Eckerd Drugs, Lum's Car Wash, Wal-Mart, Winn Dixie Supermarket
125		Frontage Rd
126		FM1870, College St
	Gas	S: Exxon*
	Lodg	S: Country Folks Inn
	AServ	S: Firestone Tire & Auto
127		Loop 301
	FStop	S: Chevron*(D)
	Gas	S: Texaco*
	Food	N: Kettle
		S: Burton's Family Restaurant
	Lodg	N: Budget Inn, ◆ Holiday Inn
		S: ◆ Best Western
	AServ	S: Chevron
	TServ	S: Chevron
	Med	N: ✚ Hospital
131		FM69
	AServ	S: Smith's Garage
135		U.S. 67
136		FM269, Weaver Rd
	AServ	N: Weaver Automotive

EXIT		TEXAS
141		FM900, Saltillo Rd
	FStop	N: Exxon*(D)
142		County Line Rd
(143)		Rest Area - Picnic, Phones, RR, Vending, Grills P (Both Directions)
146		Mt. Vernon, TX37, Clarksville, Winnsboro
	FStop	S: Texaco*(D) (Restaurant)
	Gas	N: Exxon*
		S: Chevron*
	Food	S: Barnstormers Restaurant, Burger King (Playground), Dairy Queen, Restaurant
	Lodg	S: AAA Motel, AAA Super 8 Motel
	AServ	N: Crowston's Auto Service
		S: Perry's Tires, Quick Lube
	Med	N: ✚ Hospital
	ATM	S: Franklin National Bank
	Other	S: Coin Car Wash
147		Lupe 423
	TStop	N: Love's Truck Stop* (Subway, Taco Bell)
	Food	N: Subway (Love's), Taco Bell (Love's)
	Lodg	N: Mount Vernon Motel
	AServ	S: Big A Auto Parts
	Other	S: Shelton's Propane
150		Ripley Rd
153		Spur 185, Winfield, Millers Cove
	TStop	N: Winfield Travel Center*(D) (Restaurant)
	Gas	S: Miller's Cove*
	Food	N: Citgo
	TServ	N: Cummins Diesel, Winfield
	TWash	N: Winfield
	Other	N: Coin Laundry

EXIT		TEXAS
156		Frontage Rd
(157)		Weigh Station - both directions
160		U.S. 67, U.S. 271, TX49, Mt Pleasant, Pittsburg
	FStop	S: Texaco*(D) (24 hrs)
	Gas	S: Exxon*(D)
	Food	N: Two Senoritas
		S: El Chico Mexican, Peach Tree Cafe (Holiday Inn), The Hot Biscuit, Western Sizzlin'
	Lodg	S: AAA Comfort Inn, DAYS INN AAA Days Inn, Holiday Inn
	ATM	S: Texaco
162		U.S. 271, FM1402, FM2152, Mt. Pleasant
	FStop	S: Total(D) (Subway)
	Gas	N: Chevron*(CW) (ATM)
		S: Shell*, Texaco*(CW)
	Food	N: Alps Restaurant, Blaylocks BBQ
		S: Burger King, McDonalds, Subway (Total)
	Lodg	S: AAA Best Western
	RVCamp	N: KOA Campground
	Med	S: ✚ Hospital
	ATM	N: Chevron
165		FM1001
	FStop	S: Chevron*(D)
170		FM1993
	Gas	N: Mobil*(D)
178		U.S. 259, De Kalb, Daingerfield
186		FM561
(190)		Rest Area P (Both Directions)
192		FM990
	FStop	N: Fina*
	Food	N: Pitt Grill
	AServ	N: Burkett Tire
198		TX98
199		U.S. 82, New Boston
	Gas	N: Phillips 66*
201		TX8
	FStop	N: Texaco*(D), Total*(D)
		S: Overland Express*(D) (Arby's, ATM)
	Gas	S: E-Z Mart*, Exxon
	Food	N: Pitt Grill
		S: Arby's (Overland), Burger King, Dairy Queen, McDonalds, Pizza Hut, Taco Delite
	Lodg	N: Tex Inn
		S: AAA Best Western, Bostonian Motor Inn
	Med	S: ✚ Hospital
	ATM	S: Overland Express
206		Spur 86, Red River, Army Depot
208		FM560, Hooks
	TStop	N: Hooks Truckstop*(SCALES), Love's Country Store*(SCALES) (Restaurant)
	FStop	S: Total*(D) (Taco Bell, Restaurant)
	Gas	S: Pic 'n' Pac*
	Food	N: Brownings Main Street Cafe, Love's*(D)(SCALES), Restaurant (Hooks)
		S: Dairy Queen (Playground), Taco Bell (Total)
	Lodg	N: Motel (Hooks)
	AServ	S: Hook's Tire Service, KRP Tire Repair
	TServ	N: Hooks Truckstop
	Other	S: Coin Car Wash, Coin Laundry
212		Spur74, Lonestar Army Ammunition Plant
	Gas	S: Texaco*

Bold red print shows RV & Bus parking available or nearby

EXIT TEXAS

213 FM2253, Leary

218 FM989, Nash
- **Gas** **S:** Exxon*(D) (Burger King, TCBY, Playground)
- **Food** **S:** Burger King (Exxon), TCBY (Exxon)
- **AServ** **S:** Volvo Dealer
- **Other** **S:** **Post Office**

220A U.S. 59S
- **Gas** **S:** Exxon* (Wendy's)
- **Food** **S:** Wendy's (Exxon)
- **Other** **S:** **Wal-Mart**

220B FM559, Richmond Rd
- **Gas** **N:** E-Z Mart*, Exxon*(CW), Texaco
  S: Chevron, Total*(D)
- **Food** **N:** Bob's Smokehouse BBQ, Burger King, Dairy Queen, Domino's Pizza, Pancho's Mexican, Pizza Hut, Popeye's (Texaco), Taco Bell
  S: Arby's, Baskin Robbins, Chili's, Grandy's, McDonalds, Subway
- **AServ** **S:** Chevron, Jiffy Lube, Sear's
- **Med** **N:** + **Hospital**
- **ATM** **S:** First National Bank, Hibernia, Teacher's Credit Union
- **Other** **E:** Coin Car Wash, **Pleasant Grove Animal Hospital**
  N: **Super 1**
  S: **Albertson's Grocery, Books-A-Million, Central Mall, Fant Optical, Target Department Store**

222 TX93, FM1397, Summerhill Rd
- **Gas** **N:** Shell*(CW), Total*
  S: Texaco*(CW)
- **Food** **N:** Applebee's*, Branson's, CJ's, McDonalds, Old Tyme Burger Shop (Shell), Subway (Total), Waffle House
  S: Bryce's Cafeteria
- **Lodg** **N:** Motel 6
  S: Ramada
- **TServ** **N:** **Texarkana**
- **RVCamp** **N:** **Andrews Recreation Center**
- **Med** **N:** + **Hospital**
- **ATM** **N:** State First National Bank

(223) **Welcome Center, Rest Area - RR, Picnic P (Westbound)**

223AB U.S. 59, U.S. 71, State Line Ave, Ashdown
- **TStop** **S:** **Texarkana Auto/Truck Stop**
- **Gas** **N:** E-Z Mart, Shell*, Texaco
  S: E-Z Mart*, Exxon*(CW), Mobil, Racetrac*, Total*(D)
- **Food** **N:** Red Lobster (Holiday Inn), Shoney's, Waffle House
  S: Arby's, Backyard Burgers, Bennigan's, Burger King, Cattleman's Steakhouse, Chi Chi's Pizza, Country Boy Buffet, Doc Alexander's, El Chico Mexican, Harriett's Restaurant, KFC, Little Caesars Pizza, Long John Silvers, Mandarian House, McDonalds, Pizza Hut, Pizza Inn, TCBY, Taco Bell, Taco Tico, The Kettle, Tombo's BBQ, Western Sizzlin', What A Burger
- **Lodg** **N:** Best Western, Budgetel Inn, Four Points By Sheraton, Hampton Inn, **Holiday Inn (see our ad on this page)**, Holiday Inn Express, Shoney's Inn, Super 8 Motel
  S: Best Western, **Comfort Inn**, Days Inn, Econolodge, Knights Inn, La Quinta Inn, Twin City Inn
- **AServ** **N:** Cooper Tire, Texaco
- **TServ** **N:** **Jo Esco Truck Tire Co., State Line Tire**
  S: **Kenworth**
- **TWash** **S:** **Texarcana**
- **RVCamp** **N:** **Northgate RV**
- **Med** **N:** + **Hospital (3 Miles)**
- **ATM** **S:** Hibernia
- **Other** **S:** **Wal-Mart**

↑TEXAS

↓ARKANSAS

EXIT TEXAS/ARKANSAS

1 Jefferson Ave
- **Lodg** **S:** Motel 6
- **AServ** **S:** Po Auto Glass

(1) **Rest Area - RR, Phones, Picnic, Vending, Info P (Eastbound)**

EXIT ARKANSAS

2 AR245, Airport
- **FStop** **N:** **Overland Express*, Total*(D)**
  S: **BP*(D)**
- **Food** **N:** **A & W Drive-In (Overland)**, Arby's (Total)
  S: Brian's Restaurant

7 AR108, Mandeville
- **TStop** **N:** **Flying J Travel Plaza(D)(LP)(SCALES) (RV Dump)**
- **Food** **N:** **Restaurant (Flying J)**
- **TServ** **N:** **Tire and AC Shop**
- **TWash** **N:** **TWA Truck Wash**

12 U.S. 67, Fulton

18 Fulton
- **FStop** **N:** **BP(D)**
- **Food** **N:** Exit 18 Quick Stop
- **AServ** **N:** BP
- **TServ** **N:** **BP**

(25) **Weigh Station (Both Directions)**

30 AR4, Hope, Nashville
- **Gas** **N:** Phillips 66
  S: Shell*(CW), Texaco*(D)
- **Food** **N:** Little B's, **Western Sizzlin'**
  S: Burger King, Catfish King, McDonalds, Pitt Grill (Days Inn), Pizza Hut, Taco Bell
- **Lodg** **N:** AAA **Best Western**, Holiday Inn Express, ◆ **Super 8 Motel**
  S: Days In (see our ad this page)
- **AServ** **S:** Ford Dealership, Plymouth, Eagle Dealership, Smith Tire Co.
- **Med** **S:** + **Hospital**
- **Other** **S:** **Wal-Mart**

31 AR29, Hope
- **FStop** **N:** **Shell*(D), Texaco*(D)**

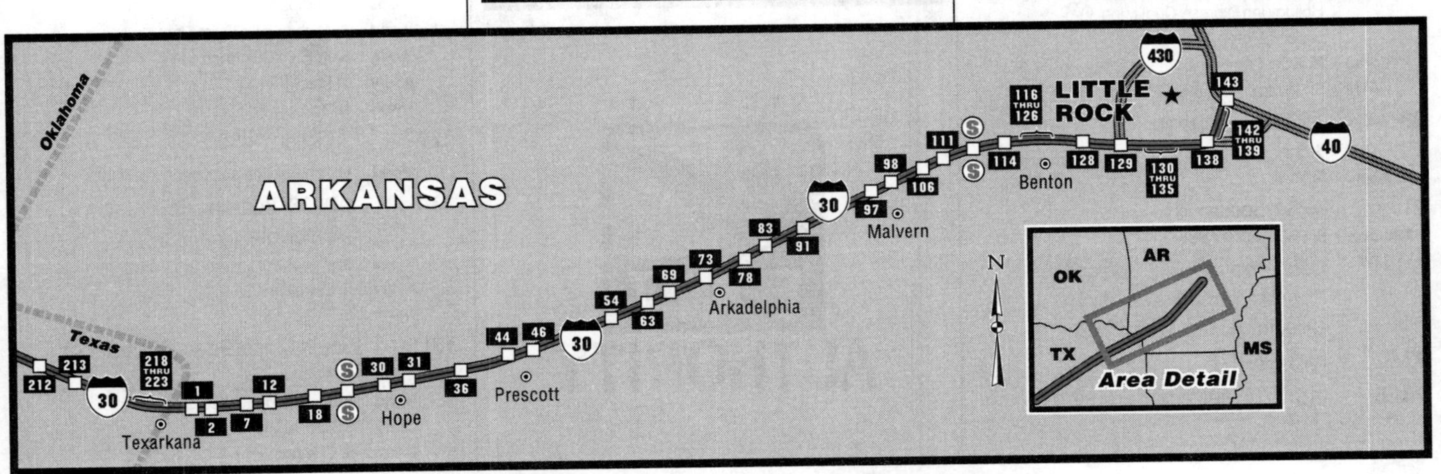

Bold red print shows RV & Bus parking available or nearby

EXIT ARKANSAS

Gas S: Exxon*, Texaco*, Total*
Food N: Pitt Grill, Quality Inn
S: B & H Cafe, KFC
Lodg N: AAA Quality Inn
S: King's Inn
AServ S: Exxon
Other N: Coin Car Wash

36 AR299, Emmet

44 US 371, AR24, Prescott
TStop S: Exit 44 Truck Stop*(SCALES) (ATM)
Food S: Norman's 44, Split Rail Cafe
Lodg S: Econolodge, Split Rail Inn
TWash S: Exit 44 Truck Wash
ATM S: Exit 44

46 AR19, Prescott
FStop N: Exxon*(D)(SCALES) (24 Hrs), Phillips 66*(D)
Gas S: Texaco*(D)
AServ S: Texaco
TServ N: Phillips 66 Tire Service
Parks N: Crater of Diamonds State Park

54 AR51, Okolona, Gurdon

(56) Rest Area - RR, Picnic, Vending P (Both Directions)

63 AR53, Gurdon
TStop N: Southfork Truck Stop(D) (ATM)
Gas S: Shell*(D)
Food N: Restaurant (Southfork)
AServ S: I-30 Automotive
TServ N: Southfork
ATM N: Southfork

69 AR26, Gum Springs

73 AR8, AR26, AR51, Arkadelphia
Gas N: Citgo*(D), Shell*
S: Exxon*, Texaco*, Total*
Food N: Stuckey's (Shell), Western Sizzlin'
S: Andy's Restaurant, Burger King, Kregs Family Restaurant, Mazzio's Pizza, Subway, TCBY, Taco Tico
AServ S: Auto Zone Auto Parts
Med S: + Baptist Medical Center
Other N: Coin Car Wash

78 AR7, Caddo Valley, Hot Springs
TStop N: Fina*(D)(LP)(SCALES)
FStop N: Shell(D), Total*(D)
Gas S: Exxon(CW)
Food N: Fina, Pizza Inn (Shell)
S: McDonalds, Pig Pit BBQ, Shoney's, Subway, Taco Bell, Waffle House, Wendy's
Lodg S: AAA Best Western, DAYS INN ◆ Days Inn, Holiday Inn Express, Quality Inn, AAA Super 8 Motel
RVCamp N: KOA Campground
ATM S: Citizen's First Bank, Elkhorn Bank

83 AR283, Friendship
Gas S: Shell(D)
AServ S: Shell

91 AR84, Social Hill
RVCamp S: Social Hill RV Park

(93) Rest Area - Picnic, Grills, RR, Phones, Vending P (Both Directions, Left Hand Exit)

97 AR84, AR171

98AB U.S. 270, Malvern, Hot Springs

EXIT ARKANSAS

FStop S: Fina*, Shell
Gas S: Total*
Food S: Blimpie's Subs (Shell), Waffle House
Lodg N: Super 8 Motel
S: Economy Inn
ATM S: Shell

106 Old Military Rd
TStop N: JR's Truck Stop*(SCALES)
Food N: JR's
Lodg N: 106 Motel
TServ N: JR's
RVCamp N: J.B.

111 U.S. 70, Hot Springs

(112) Truck Inspection Station

114 U.S. 67, Benton Services Center

116 Benton, Sevier Rd
Gas N: Shell*, Sinclair*, Texaco*
S: BP*, Fina, Sinclair, Texaco
Food N: Chile & Co., Hunan Place, Tastee Freeze
S: BBQ, C.K. Family Restaurant, El Cena Casa
Lodg N: AAA Troutt Motel
S: Capri Motel
AServ S: Benton Transmissions, Btitt's Tire, Sinclair, Texaco

EXIT ARKANSAS

ATM N: Shell
Other S: Coin Car Wash

117 AR5, AR35
Gas N: Conoco*(D), Shell*
Food N: Denny's (Ramada), Pizza Hut, Waffle House
S: McDonalds
Lodg N: AAA Econolodge, AAA Ramada
Med S: + Hospital
Other N: Coin Car Wash

118 Congo Rd
Gas S: Exxon, Shell*
Food S: Burger King, Dairy Queen, Sergio's Pizza, Shoney's, Tia Wanda Mexican
Lodg N: Scottish Inns
S: DAYS INN ◆ Days Inn
AServ N: Donnie's Foreign Car Service, Precision Brake, Williams Tire Service
S: Exxon, Whitfield Tire
Other S: Wal-Mart

121 Alcoa Rd
Other S: Branch Hollow RV Park

123 AR183, Bryant, Bauxite, Reynolds Rd.
FStop S: Total*(D)
Gas N: Phillips 66(CW) (Sinnamon Street)
S: Exxon(D), Phillips 66
Food N: Cracker Barrel
S: Little Caesars Pizza, McDonalds, Ole South, Pizza Hut, Sergio's Pizza, Sonic, Subway, Taco Bell, Wendy's
Lodg N: Holiday Inn Express
S: Super 8 Motel
AServ S: Exxon, Reade's Automotive
ATM S: Union Bank of Bryant
Other S: Animal Clinic, Coin Car Wash, Foster's, Post Office, R & R Full Service Car Wash

126 AR111, County Line Rd, Alexander
FStop N: Shell*(D)(CW)
Gas N: Citgo
S: Delta Express*
Food N: KFC, Taco Bell
AServ N: Eddie's Auto Parts
Other N: Woody's RVs
S: Moix RV

128 Mabelvale West, Otter Creek Rd
Gas S: Phillips 66*
Food S: Michael's (La Quinta Inn)
Lodg S: AAA La Quinta Inn
AServ S: Purcell Tire Co.
Med N: + Hospital

129 Junction I-430
Food N: Amerisuites (see our ad this page)

130 AR338, Baseline Rd, Mabelvale
Gas S: Phillips 66*
AServ S: Uniroil Tire Co.
ATM S: Phillips 66

131 South Chicot Road (Eastbound)
Food S: Waffle House
Lodg S: ◆ Ramada Limited, Super 7 Inn
AServ S: Terry Auto Care
Other S: Camper Caps

Bold red print shows RV & Bus parking available or nearby

EXIT ARKANSAS

132 U.S. 70B, University Ave

133 Geyer Springs Rd

Gas N: Exxon*, Fuel Mart*, Jack Pot
S: Phillips 66*, Total*(D)

Food N: Church's Fried Chicken, Cuisine of China, Simm's BBQ, Subway, Tony's Chinese Restaurant
S: Arby's, Backyard Burgers, Burger King, El Chico Mexican, KFC, Mazzio's Pizza, McDonalds, Pizza for Less, TCBY, Taco Bell, The Dixie Cafe, Wendy's, Western Sizzlin'

Lodg S: Comfort Inn, ◆ Hampton Inn (see our ad opposite page)

AServ N: Razorback Transmission
S: Goodyear Tire & Auto, Jiffy Lube

ATM S: Boatmen's Bank, First Commercial

Other N: **Post Office**
S: **Kroger Supermarket, Mega Pantry**

134 Scott Hamilton Dr

Gas S: Exxon*(D)

Food S: Waffle House

Lodg S: Motel 6, ◆ Red Roof Inn

135 65th St

Gas N: Exxon*, Mapco*, Texaco*(D)(CW)

Food S: Denny's

Lodg N: DAYS INN AAA Days Inn
S: AAA La Quinta Inn

AServ S: Transmission Express

TServ N: **Rawlings Truck Rental & Leasing**

ATM N: Nations Bank

Other N: **Hertz Truck Rental**

138A Junction I-440

138B U.S. 65, U.S. 167, Pine Bluff

139 AR365, Roosevelt Rd

Gas N: Exxon, Unnamed
S: Texaco*

Food S: Catfish Young's

AServ N: Exxon
S: Napa Auto & Truck Parts

ATM S: Boatman's Bank

Other N: Coin Car Wash
S: **Kroger Supermarket**

139B Junction I-630

140 9th St, 6th St

Gas N: Exxon*, Shell*(CW), Texaco*
S: No Name, Phillips 66*

Food N: Pizza Hut, Waffle House

Lodg N: Best Western
S: Masters Inn

ATM S: First Commercial Bank

Other S: **Rexall Drugs**

141A AR10, Cantrell Rd, Markham St

141B U.S. 70, Broadway

Gas N: Exxon*(CW), Shell*
S: Total*

Food N: Burger King
S: Arby's, KFC, McDonalds, Popeye's Chicken, Wendy's

AServ N: Three Star Muffler Shop
S: Ed's Brake Service

142 15th St

Gas S: Phillips 66*

143AB Junction I-40EW

↑ARKANSAS

Begin I-30

EXIT MINNESOTA

Begin I-35

↓MINNESOTA

258 21st Ave E

Gas E: Spur*

Food E: Burger King, McDonalds

Lodg E: AAA Best Western, Chalet Motel

256AB Mesaba Ave, Superior St, Lake Ave

Lodg W: Holiday Inn, The Radisson

Other W: **Hospital**

255B I-535

255A U.S. 53, 21st Ave West (Left Lane Exit)

254 27th Ave West

Gas W: Amoco*(CW), Spur*(D)

Food W: Duluth Grill, Subway

Lodg W: Motel 6

AServ W: Amoco, Ford Dealership

253B 40th Ave

253A U.S. 2, Wisconsin

252 Central Ave

Gas W: Conoco*(D), Holiday*(LP)

Food W: Domino's Pizza, Jade Fountain Restaurant (Chinese & American), KFC, McDonalds, Sammy's Pizza & Restaurant, Subway, Taco Bell (Conoco), The Gopher Restaurant (Italian & American)

AServ W: Bumper To Bumper Auto Parts

Med W: ✚ **West Duluth Med-Dental**

ATM W: Conoco, First Bank, Holiday, Western National Bank

Other W: **Gentle Dentist, K-Mart (Pharmacy), Laundromat, Super 1 Foods Grocery, Veterinary Clinic, WalGreens**

251B MN 23 South, Grand Ave

251A Cody St (Southbound Reaccess Only)

Other E: **Lake Superior Zoo**

250 U.S. 2 (Southbound, Divided Hwy)

249 Boundary Ave, Skyline Parkway

Gas E: Holiday*
W: Phillips 66*

Food E: McDonalds
W: Country Kitchen

Lodg E: Country Inn & Suites, Sundown Motel (Access Rd)
W: Spirit Mountain Lodge

RVCamp E: **Campground**

ATM E: Holiday
W: Phillips 66

Other E: **Spirit Mountain Recreation Area**
W: **Rest Area, Full Facilities (Senic Overlook Of Lake Superior)**

246 CR13, Midway Road, Nopeming

245 CR61

Food E: Buffalo House Restaurant

RVCamp E: **Campground**

242 CR1, Esko, Thomson

Gas E: Amoco*

ATM E: Amoco

239 MN 45, Scanlon, Cloquet

Gas W: Conoco*(CW)

Food W: Blimpie's Subs (Conoco), **Golden Gate Motel**, The Pantry Restaurant

Lodg W: **Golden Gate Motel**

RVCamp E: **KOA Campground**

EXIT MINNESOTA

ATM W: Conoco

237 MN 33, Cloquet (Divided Hwy)

(236) **Weigh Station - Both Directions**

235 MN 210, Cromwell, Carlton

TStop E: **76***

FStop E: **Amoco***

Food E: **Junction Oasis (76), Stage Stop Restaurant (24Hr)**
W: Black Bear Casino

Lodg E: Americinn Motel, **Royal Pines Motel**
W: Black Bear Casino

TServ E: **Truck Center**

Parks E: **Jay Cook State Park (5 Miles)**

227 CR4, Mahtowa

RVCamp E: **Campground**

(225) **Rest Area - RR, HF, Phones, Picnics (Northbound)**

220 CR6, Barnum

Gas W: Amoco* (24Hr)

Food W: Restaurant (Amoco)

RVCamp E: **Bear Lake County Campground**

Other W: **Munger Trail**

216 MN 27, Moose Lake (Southbound, Reaccess Northbound Only)

214 MN 73, Moose Lake

FStop W: **Conoco***

Food W: **Subway (Conoco)**

RVCamp E: **Camping**
W: **Red Fox Campground**

Med W: ✚ **Hospital (2 Miles)**

ATM W: Conoco

Parks E: **Moose Lake State Park**

Other W: **Sioux, Moose Munger Trails**

209 CR46, Sturgeon Lake

Gas E: Phillips 66*(LP)

Food E: Cafe (Phillips 66)

Lodg W: Sturgeon Lake Motel

RVCamp E: **Campground**
W: **Campground**

ATM E: Phillips 66

(208) **Rest Area - Picnic, RR, HF, Phones (Southbound)**

205 CR43, Willow River, Bruno

FStop W: **Citgo***(D)

Food W: **Cafe (Citgo)**

AServ W: Citgo

RVCamp W: **Campground**

(198) **Rest Area - Picnic, RR, HF, Phones (Northbound)**

195 MN 18, MN 23E, Finlayson, Askov

Gas E: **76***

Food E: **Cafe (76)**

Lodg E: ◆ **Super 8 Motel**

RVCamp E: **Camping**
W: **Camping**

Parks E: **Banning State Park**

Other E: **Tourist Information**

191 MN 23E, Sandstone

FStop E: **Conoco***(CW)(LP)

Food E: **Hot Stuff Pizza (Conoco)**

Lodg E: Sandstone 61 Motel

Med E: ✚ **Hospital**

183 MN 48, Hinckley

Gas E: Conoco*, Holiday*(D)(LP) (24Hr)

Bold red print shows RV & Bus parking available or nearby

EXIT MINNESOTA

W: Amoco*(CW), Little Store*

Food E: Dairy Queen, Grand Hinckley Inn, **Hardee's**, Subway, Taco Bell, Tobies Bakery, **Tobies Restaurant**

W: Cassidy's Restaurant, White Castle Restaurant (Little Store)

Lodg E: Days Inn, **Econolodge**, ◆ Grand Hinckley Inn (Casino), Holiday Inn Express

W: Best Western

AServ W: Hinckley Automotive

RVCamp E: **Camping**

ATM E: Conoco, Holiday

W: Little Store

Parks E: **St. Croix State Park**

Other E: **Casino (Grand Hinckley Inn), Coin Laundry**

W: **Hinckley Fire Museum**

180 MN 23, CR61, Mora

175 CR 14, Beroun

Note: I-35 runs concurrent below with I-35E. Numbering follows I-35E.

171 **(172)** CR 11

FStop E: **Super America***(D)(LP)

Food E: **McDonalds**

RVCamp W: **Camping**

ATM E: Super America

Note: I-35 runs concurrent above with I-35E. Numbering follows I-35E.

169 MN 324, CR 7, Pine City

Gas E: Amoco*(D)(LP), Holiday*

Food E: Hardee's, KFC, Pizza Hut, **Pizza Pub**, Red Shed Restaurant, Subway

AServ E: Mercury

ATM E: Amoco, Holiday

Other E: **Jubilee Grocery, Pamida Grocery (Pharmacy), Public Beach, Tourist Information, Wal-Mart**

165 MN 70, Rock Creek, Grantsburg

FStop E: **Tim's Crossroads Truck Stop***

W: **Total***

Food E: **Tim's Crossroads**

W: **Rock Creek Cafe (Total)**

Lodg E: Chalet Motel

TServ E: **Tim's Crossroads**

RVCamp E: **Campground**

Other W: **Federated Propane**(LP)

159 MN 361, County 1, Rush City

FStop E: **Tank & Tackle***

Gas E: 76*

RVCamp W: **Campground (2 Miles)**

Med E: ✚ **Hospital**

ATM E: Tank & Tackle

Other E: **Coin Laundry, Super Fair Foods**

(154) **Rest Area - Picnic, RR, HF, Phones, Pet Walk (Northbound)**

152 CR10, Harris

147 MN 95, Cambridge, North Branch

Gas E: Casey's General Store*, Conoco*(D)(CW)(LP), Holiday*(D)(LP), Phillips 66*(D)(LP)

Food E: Homemade Pizza To Go (Casey's), McDonalds, Subway, The Oak Inn Restaurant

Lodg E: Crossroads Motel

AServ E: Auto Care, Car Quest Radiator Repair, GM Auto Dealership (Frontage Rd)

EXIT MINNESOTA

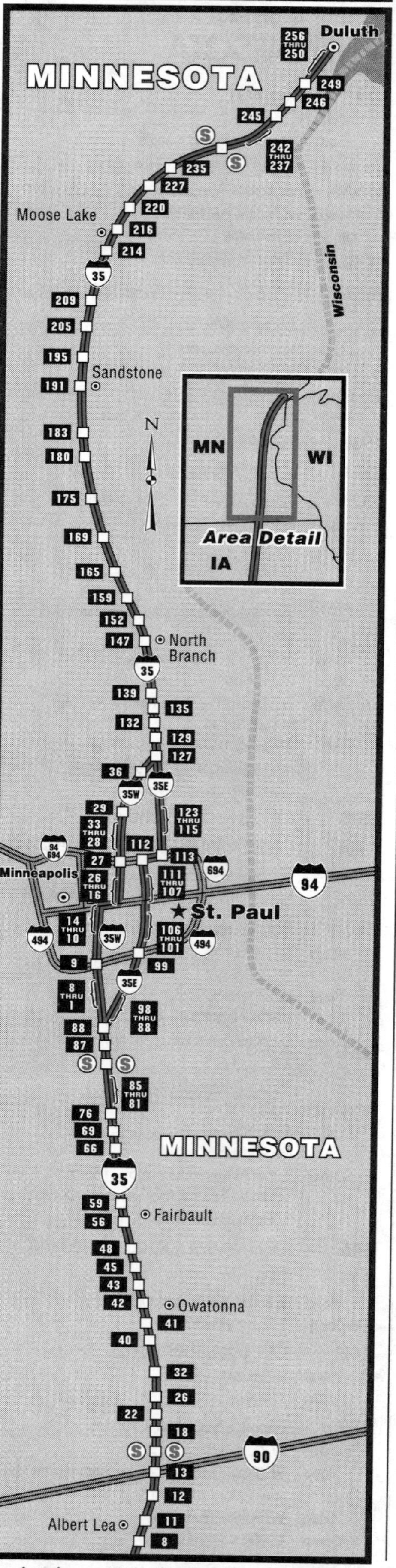

EXIT MINNESOTA

W: Ford Dealership

TServ E: **Freightliner Dealer (Frontage Rd)**

RVCamp W: **Campground (6 Miles)**

Med E: ✚ **Wild River Med Clinic**

ATM E: Conoco, Holiday

Parks E: **Wild River Park**

Other E: Happy Wheels Car Wash, **Super Value Grocery Store**

W: **Tanger Factory Outlet**(LP)

139 County 19, Stacy

Gas W: Conoco*

Food W: Hot Stuff Pizza (Conoco)

Other W: **IGA Grocery Store**

135 U.S. 61, CR22, Wyoming

FStop E: **76***(LP)

W: **Fina***(CW)(LP)

Food E: Dairy Queen, Subway

W: **Village Inn Restaurant**

AServ E: Auto Glass Plus, Car Quest Auto Center

RVCamp E: **Campground (2 Miles)**

W: **Campground (10 Miles)**

ATM W: **Fina**

Other E: **Alumacraft Boat Service**

132 U.S. 8, Taylors Falls, Difficult Reaccess (Divided Hwy, Northbound, Reaccess Southbound Only)

131 CR2, Forest Lake

Gas E: Amoco, Holiday*(D)(LP), Super America*(CW)(LP)

Food E: Arby's, Burger King, Cheung Sing Chinese, Hardee's, Hot Stuff Pizza (Holiday), KFC, McDonalds, McDonalds (Wal-Mart), Perkins Family Restaurant, Sparro's Pizza & Subs, Subway

W: Little Caesars Pizza (K-Mart), Taco Bell

Lodg E: Americinn Hotel

AServ E: Amoco, Big Wheel Auto Parts, Champion Auto Stores, Crown Auto, Jems Service Center, Tire Plus, Wal-Mart

W: Chrysler Auto Dealer, Ford Dealership, GM Auto Dealership, Novack GM Dealer, Quaker State, Valvoline Quick Lube Center

Med E: ✚ **Hospital**

ATM E: Amoco, Holiday, Super America, TCF Bank Of Minnesota

Other E: 12st Corner Car Wash, Broadway Car Wash, **Rainbow Foods Grocery, Wal-Mart (Pharmacy & Optical)**

W: **K-Mart (Pharmacy)**

129 MN 97, County 23

FStop W: **Conoco***

Food W: **Hot Stuff Pizza (Conoco)**, Smash Hit Subs, Trout Air Restaurant

TServ E: **A+ Radiator Service, Columbus Truck Service**

RVCamp E: **Campground**

W: **Campground**

128 Weigh Station (Both Directions)

127 I-35E Merges With I-35W, East St Paul, West Minneapolis (Southbound)

Note: I-35 runs concurrent below with I-35E. Numbering follows I-35E.

123 CR14

Gas W: Texaco*(LP)

Food W: Pizza & Subs

Bold red print shows RV & Bus parking available or nearby

EXIT	MINNESOTA
RVCamp	E: Otter Lake RV Center
ATM	W: Texaco
Other	W: Super Car Wash
120	CRJ (Reaccess Southbound Only)
117	CR 96
Gas	E: Total*
	W: Amoco*(CW)(LP), PDQ*
Food	E: Burger King
	W: Applebee's, Arby's, Boston Market, Hardee's, Little Caesars Pizza, McDonalds, Sparro's, Subway, Tia Pan
AServ	E: Goodyear Tire & Auto
	W: Valvoline Quick Lube Center
Med	W: + Northeast Medical
ATM	W: Cub Foods, PDQ
Other	W: Cub Foods (24 Hrs, Pharmacy), Northeast Dental, WalGreens (Pharmacy)
115	CRE
Gas	E: Amoco*(CW), Super America*, Total*
Food	E: Perkins Family Restaurant
	W: McDonalds (Wal-Mart)
AServ	W: Wal-Mart
ATM	E: Amoco, Super America
Other	W: Optical (Wal-Mart), Target Department Store (Pharmacy), Wal-Mart (Pharmacy)
113	I-694W, U.S. 10 (Left Lane Exit)
112	Little Canada Road
Gas	E: 76*(LP)
	W: Sinclair*
Food	W: Rocco's Pizza
AServ	E: 76
Other	W: Dental Clinic, Tom Thumb C-Store
111AB	MN 36, Stillwater, Minneapolis (Divided Hwy)
110B	Roselawn Ave
110A	Wheelock Parkway, Larpenteur Ave
Gas	E: Amoco*, Total*(LP)
	W: Sinclair*
Food	E: Roadside Pizza, Subway
	W: Chanpp's
AServ	E: Amoco
	W: Sinclair
ATM	E: Amoco, Total
109	Maryland Ave
Gas	E: 76*, Super America*
Food	W: Wendy's
AServ	E: 76, Pete's Auto Repair, Tires Plus Performance Center
	W: K-Mart
ATM	E: Super America
Other	E: NHIA Grocery Store (Asian & Mexican Foods)
	W: K-Mart
108	Pennsylvania Ave
107A	I-94E Junction
106C	11th St, State Capitol (Northbound, Reaccess Southbound Only)
106B	Kellogg Blvd (Southbound Reaccess Only)
Lodg	E: Days Inn
Other	E: Civic Center
106A	Grand Ave (Southbound Reaccess Only)
Gas	E: Mobil*(D)

EXIT	MINNESOTA
AServ	E: Mobil
Med	E: + Hospital
105	St. Clair Ave (Southbound, Reaccess Northbound Only)
104C	Victoria St, Jefferson Ave (Southbound, Reaccess Northbound Only)
104A	Randolph Ave
Gas	W: Total*
103B	MN 5, West 7th St
Gas	E: Holiday*(D)
	W: Super America*
Food	E: Burger King
AServ	E: Champion Crown Auto Service, Crown Auto
	W: Midas
ATM	E: Holiday
	W: Super America
Other	W: US Post Office
103A	Shepard Road (Northbound, Southbound Reaccess Only)
102	MN 13, Sibley Highway
Gas	W: Holiday*(LP)
Food	E: China Delight, Moose Country Restaurant
ATM	W: Holiday
Parks	E: City Park
Other	E: Just Paws Pet Store, The Parkview Cat Clinic
101AB	MN 110
Gas	W: General Tire Company
AServ	W: General Tire Company
99AB	I-494
98	CR26, Lone Oak Road
Gas	W: Fina*(CW)(LP)
Food	W: Costello Coffee, Cracker Barrel
Lodg	W: Hampton Inn, Residence Inn
ATM	W: Fina
Other	W: Dental Clinic
97AB	CR 31, CR 28, Pilot Knob Road, Yankee Doodle Road
Gas	E: Citgo(CW), Holiday*, Phillips 66*(CW), Super America*(D)
	W: Amoco*(CW), Conoco*, Super America*(LP)
Food	E: Applebee's, Arby's, Big Apple Bagels, Chili's, Hardee's, He He Chinese Food, KFC, McDonalds, Perkins Family Restaurant, Pizza Hut, Schlotzkys Deli, Taco Bell, Wendy's
	W: Al Baker's, Arby's, Burger King, Dragon Palace Chinese Food, Italian Pie Shoppe, John Hardy BBQ, The Best Steakhouse
Lodg	E: Fairfield Inn
	W: Best Western
AServ	E: Abra, All Imports & Domestic Auto Service, Auto Maul, Car X Muffler & Brakes, Firestone Tire & Auto, Goodyear Tire & Auto, Jiffy Lube(CW), Kennedy Transmission, Phillips 66, Precision Tune, Tires Plus
	W: Amoco, NAPA Auto Parts
Med	E: + Eagan Medical Center
ATM	E: First Bank, Holiday
	W: Firststar Bank, Norwest Bank, Signal Bank, Super America
Other	E: Companion Animal Hospital, Kinko's Copies, Kohl's Department Store, Rainbow Foods Grocery, Yankee Eye Clinic
94	CR30, Diffley Road
Gas	W: Conoco*(CW)(LP)
93	CR32, Cliff Road

EXIT	MINNESOTA
Gas	W: Total*(CW)
Food	W: Anna Chung (Chinese), Baker's Square, Baskin Robbins, Boston Market, Broadway Pizza, Burger King, Dairy Queen, Davanni's Pizza & Hoagie's, Green Mill (Holiday Inn Express), KFC, McDonalds, Pizza Hut, Taco Bell
Lodg	W: Holiday Inn Express
AServ	W: Big Wheel Auto Store, Meineke Discount Mufflers, Tires Plus, Valvoline Quick Lube Center
Med	E: + Park Nicollett Medical Clinic
ATM	W: Dakota Bank, First American Bank, Firstar Bank
Other	W: Cub Foods (24Hr), Dental Clinic, Dental Clinic, GNC Nutrition, Now Med Ctr, Pet Clinic, Post Office, Target Department Store (Pharmacy), WalGreens
92	MN 77, Cedar Ave (Divided Hwy)
Other	E: Zoo
90	CR11
Gas	E: PDQ*, QuikTrip*(D)(LP)
ATM	E: PDQ
Other	E: Valley View Pet Hospital
88B	CR42
Gas	W: Amoco*(CW), PDQ* (ATM)
Food	W: Arby's, Burger King, China Seas, Fuddruckers, Marie Callender's Restaurant, McDonalds, Old Country Buffet, Sbarro Pizza, The Ground Round
Lodg	W: Country Inn, Fairfield Inn, Holiday Inn
AServ	W: Midas Muffler & Brake
Med	W: + Ridge Point Medical
ATM	W: The Richfield Bank
Other	W: Animal Hospital
	Note: I-35 runs concurrent above with I-35E. Numbering follows I-35E.
	Note: I-35 runs concurrent below with I-35W. Numbering follows I-35W.
36	CR23
FStop	E: Fina*(LP)
ATM	E: Fina
33	CR17, Lexington Ave (Closed Due To Construction)
32	CR52, 95th Ave NE
Other	W: National Sports Center
31B	Lake Drive, CRI, 85th Ave (Northbound)
	Note: I-35 runs concurrent above with I-35W. Numbering follows I-35W.
31A	CR J, MN118 (Divided Hwy, Northbound)
	Note: I-35 runs concurrent below with I-35W. Numbering follows I-35W.
30	MN 118 (Difficult Reaccess)
29	CRI
28C	U.S. 10, CRH (Southbound, Reaccess Southbound Only)
Gas	W: Amoco*(CW)
Food	W: KFC, McDonalds, Perkins Family Restaurant, Taco Bell

Bold red print shows RV & Bus parking available or nearby

EXIT MINNESOTA

Lodg W: Mounds View Inn
Other W: Car Wash

28B U.S. 10, St. Paul (Southbound, Reaccess Northbound Only)

28A MN 96
Parks W: **Long Lake Regional Park**

27AB Junction I-694

26 CRE2
Gas W: Tank-n-Tummy*(D)

25AB MN 88, CRD (Southbound, Reaccess Northbound Only)
Gas E: Spur
W: Total*(D)
Food W: Kin Jing Chinese Restaurant, Main Event Restaurant, McDonalds, Perkins Family Restaurant, Picadilly Pizza (Total), Subway
ATM W: Total
Other E: **Ken's Market**

24 CRC
Food E: Burger King
Lodg E: Ramada
W: Comfort Inn
AServ W: GM Auto Dealership
TServ E: **Cummins Diesel**
W: **Mack Truck Dealer**

23B MN 36, Cleveland Ave (Left Lane Exit, Divided Hwy)

23A MN 280 (Divided Hwy)

22 Industrial Blvd, St. Anthony Blvd

21A Broadway St, County 88, Stinson Blvd
Food W: Country Kitchen, McDonalds, Taco Bell
Other W: **Target Department Store**

21B Johnson St (Northbound, Reaccess Southbound Only)

19 E. Hennepin Ave

18 4th St SE, University Ave

17C Washington Ave

17B I-94W, 11th Ave

17A MN 55, Hiawatha Ave. (Southbound, Reaccess Northbound Only)
Lodg E: Holiday Inn

16A I-94 (Northbound)

14 35th St, 36th St
Gas W: Super America*

13 46th St
Gas W: 76*, Amoco*(CW)
Food W: Bruegger's Bagels, Fresh Wok Chinese, KFC, Papa John's Pizza, Subway
AServ E: Mobil*, Mobil
W: 76, Hawkins Service Inc

12B Diamond Lake Road
Gas W: Holiday*(LP)
Food W: Chinese Restaurant, Steep & Brew, The Best Steakhouse
AServ W: Big Wheel Auto
ATM W: Holiday
Other W: **ACE Hdwe(LP), Towns Edge Cleaners/Coin Laundry**

12A 60th St (Reaccess Northbound Only)
Food W: Hong Kong Restaurant (Chinese), Perkins Family Restaurant

EXIT MINNESOTA

AServ W: Meineke Discount Mufflers
ATM W: Cub Foods
Other W: **Cub Foods (Pharmacy)**, Nicollet Car Wash

11B MN 62E (Left Lane Exit)

11A Lyndale Ave (Reaccess Northbound Only)
Gas E: Food-N-Fuel*
Food E: Mister Donut, Pizza Hut
AServ E: Champion, Nelvin's Auto Body
ATM E: Food-N-Fuel
Other E: **Dental Clinic, Northstar Paint & Body Supply, Wood Ave Veterinary**

10B MN 62, I-35W (Difficult Reaccess)

10A CR53, 66th St

9C 76th St (Southbound, Reaccess Northbound Only)

9AB Jct I-494

8 82nd St
AServ W: GM Auto Dealership (Frontage Rd)

7B 90th St

7A 94th St
Gas E: Amoco*(CW)
Lodg W: Holiday Inn
AServ E: Amoco
ATM E: Amoco
Other E: **Bloomington Rental Ctr(LP)**

6 98th St
Gas E: Indian Joe's*
W: Super America*(D)
Food E: Baker's Square, Chinese Restaurant
W: Denny's
AServ E: Champion Auto, Ford Dealership
W: A-1 Body
ATM E: Firstar Bank, Norwest Bank
Other E: **Bloomington Drugs**

5 106th St

4B Black Dog Rd
AServ W: Burnsville Volkswagon

4A Cliff Road
AServ E: Dodge of Burnsville
W: Volkswagen Dealer
Other E: **Mystic Lake Casino, Zoo**

3AB MN 13, Shakopee
Other W: **Best Buy**

2 Burnsville Parkway
Gas E: 76*(LP), Fina*(D)
W: Standard*(D)
Food E: Benchwarmer Bob's, Denny's, Hardee's, Little Caesars Pizza
W: Chinese Gourmet, Embers, Timberlodge Steakhouse
Lodg W: Days Inn, Prime Rate Motel, Red Roof Inn, Super 8 Motel
AServ E: 76
W: Standard
ATM E: Fina, Firstar Bank
Other W: **Quik Mart**

1 County 42, Crystal Lake Road (Southbound, Reaccess Northbound Only)
Gas E: Amoco*(CW), PDQ*
W: Sinclair*(D)
Food E: Arby's, Burger King, China Seas, Ciatti's Italian, Dakota County (Holiday Inn), Fuddrucker's, Marie

EXIT MINNESOTA

Callender's Restaurant, McDonalds, Old Country Buffet, Sparro Pizza, Taco Bell, The Ground Round
W: Applebee's, Asia Grille, Baker's Square, Champps, Chi Chi's Mexican, Chili's, Chung King Garden, Macaroni Grill, Olive Garden, Red Lobster
Lodg E: Country Inn, Holiday Inn
AServ W: Car X Muffler & Brakes, Ford Dealership, Goodyear Tire & Auto, Grossman Chevrolet, Jiffy Lube, Sinclair
Med E: **+ Hospital**
ATM E: PDQ, Richfield Bank & Trust
W: Norwest Bank
Other E: **Benson Optical, Dental Clinic, Petsmart**
W: **Kinko's Copies, Kohl's Department Store, Office Max, Target Department Store (Pharmacy), Toys "R" Us**

Note: I-35 runs concurrent above with I-35W. Numbering follows I-35W.

88 Junction I-35EW

87 Crystal Lake Road

(86) **Weigh Station - both directions**

85 MN50
Gas E: Amoco*(CW), Sinclair*(LP), Super America*(D)(CW), Tom Thumb*(D)
W: Holiday*(D)(CW)(LP)
Food E: Burger King, Dairy Queen, Pizza Hut, Taco Bell
W: **Cracker Barrel**
Lodg E: ◆ Comfort Inn
W: ◆ Americinn
AServ E: Goodyear Tire & Auto, Rapid Oil Change
ATM E: Marquette Lakeville
Other E: Mr. Sparkle Truck & Car Wash

84 185th St W

81 CR70, Lakeville, Farmington
FStop E: **MegaStop*(CW)**
Food E: **JL Pizza, McDonalds, Subway,** Tacoville
Lodg E: **Motel 6,** Super 8 Motel
TWash E: **Mega Stop**
ATM E: MegaStop

76 CR2, Elko, New Market (Exit Closed Due To Construction)
FStop E: **Phillips 66*(LP)**
TServ E: **Phillips 66**

(76) **Rest Area - RR, Phones, Picnic, Vending (Southbound)**

69 MN19, Northfield, New Prague
FStop W: **Conoco(CW)(LP)(SCALES)**
Food E: Bridgeman's Soda Fountion
W: **Big Steer/Conoco**
Med E: **+ Hospital (7 miles)**

66 CR1, Dundas, Montgomery

59 MN21, Fairbault, Le Center
FStop E: **Texaco***
Food E: Country Kitchen, Restaurant (Lavender Inn), Restaurant/Texaco, Truckers Inn Restaurant (Lavender Inn)
Lodg E: ◆ Americinn Motel, Best Western, ◆ Lavender Inn, Super 8 Motel

56 MN60, Fairbault, Waterville
FStop W: **Conoco* (ATM)**
Gas E: Amoco*
Food E: Bernie's Grill (Amoco), China Cafe

Bold red print shows RV & Bus parking available or nearby

EXIT	MINNESOTA
	W: Dairy Queen, Happy Chef
Lodg	W: AAA Select Inn
ATM	E: State Bank of Fairbault
Other	E: 4 More Grocery Store, Fairbault West Mall, Wal-Mart
48	CR12, CR23, Medford
Food	W: McDonalds
Other	W: Outlet Center
45	CR9, Clinton Falls
43	26th St, Airport Road
42B	U.S. 14 CR 45, Waseca
FStop	W: Amoco*(CW)(LP)(SCALES)
Gas	E: Sinclair*(D)
Food	E: Kernel Restaurant, McDonalds
	W: Happy Chef, Perkins Family Restaurant, Restaurant (Ramada Inn)
Lodg	E: Budget Host Inn
	W: Best Western, AAA Ramada Inn, Super 8 Motel
AServ	E: Champion Auto Stores
Parks	E: Rice Lake State Park
Other	E: Wal-Mart
41	Bridge St
FStop	W: Budget Oil Company Commerical Fueling*
Gas	E: Citgo*(D)(CW)(LP)
Food	E: Applebee's, Arby's, Burger King, KFC, Subway, Taco Bell
Lodg	E: ◆ Americinn, ◆ Country Inn & Suites
ATM	E: Citgo
Other	W: Target Department Store
40	U.S. 218, U.S. 14, Owatonna Rochester
RVCamp	E: Camping
Med	E: ✚ Hospital
(35)	Rest Area - RR, Phones, Picnic, Vending (Both Directions)
32	CR 4, Hope
Other	E: Camping
26	MN30, Blooming Prairie, New Richland
FStop	W: Texaco
Gas	E: Amoco*(LP)
Food	E: Restaurant/Amoco
	W: Restaurant/Texaco
TWash	W: Texaco
22	CR35, Geneva, Hartland
18	MN251, Hollandale, Clarks Grove
FStop	W: Phillips 66*(LP)
(17)	Weigh Station - both directions
14	35th St, 36th St
Gas	W: Super America*
13AB	Junction I-90, Sioux Falls, La Crosse, 46th St
Gas	E: Mobil*
	W: 76*, Amoco*(CW)
Food	W: Bruegger's Bagels, KFC, Papa John's, Subway
AServ	E: Mobil
	W: 76, Hawkins Serv Inc
12B	Diamond Lake Rd
Gas	W: Holiday*(LP)
Food	W: Chinese Restaurant, Steep & Brew, The Best Steakhouse
AServ	W: Big Wheel
ATM	W: Holiday

EXIT	MINNESOTA/IOWA
Other	W: ACE Hdwe, Towns Edge Cleaner/Coin Laundry
12A	60th St (Reaccess Northbound Only)
Food	W: Hong Kong Restaurant, Perkin's Family Restaurant
AServ	W: Meineke Discount Mufflers
ATM	W: Cub Foods
Other	W: Cub Foods (Pharmacy), Nicolet Carwash
12	U.S. 65, I-35 Business, Albert Lea
11A	International Airport
11	CR46
RVCamp	W: KOA Campground
Med	W: ✚ Hospital
Parks	E: Myre-Big Island State Park
8	U.S. 65 Bus. 35, Albert Lea
10B	Minnesota 62 W
5	CR13, Twin Lakes Glenville
RVCamp	W: Camping
2	CR5
(1)	Welcome Center, Rest Area - RR, Phones, Picnic, Vending, Info (Both Directions)

↑ MINNESOTA

↓ IOWA

EXIT	IOWA
214	CR105, Lake Mills, Northwood
(213)	IA Welcome Center, Rest Area - RR, Phones, Picnic, Info (Southbound)
(212)	Weigh Station - both directions
208	CR A38, Joice, Kensett
203	IA9, Manly, Forest City
FStop	W: Amoco*(D)
RVCamp	W: Winnebago (see our ad this page)
197	CR B20
Parks	E: Lime Creek Nature Center (8 miles)
(196)	Rest Area - RR, Phones, Picnic, Vending, RV Dump (Both Directions)
194	U.S. 18, Clear Lake, Mason City
FStop	E: Coastal*(D) (RV Dump)
	W: Conoco*(D)(SCALES)
Gas	W: Kum & Go*
Food	W: Bennigan's, Burger King, Country Kitchen, KFC, McDonalds, Papa Johns Pizza (Conoco), Perkins Family Restaurant, Pizza Hut
Lodg	W: Best Western, Budget Inn Motel
Med	E: ✚ Hospital
193	CR B35, Clear Lake, Mason City
FStop	E: Amoco*(D)(LP)
Gas	W: Phillips 66*
Food	E: Happy Chef
Lodg	E: Super 8 Motel
AServ	E: Amoco
188	CR B43, Burchinal
182	CR B60, Swalendale, Rockwell
180	CR B65, Thornton
RVCamp	W: Campground
176	CR C13, Sheffield Belmond

EXIT — IOWA

170 CR C25, Alexander

165 IA3, Hampton, Clarion
- **FStop** E: **Texaco*(D)**
- **Food** E: Texaco
- **ATM** E: Texaco

159 CR C47, Dows

151 CR R75, Woolstock

147 CR D20

144 Blairsburg
- **TStop** E: **Phillips 66 (RV Dump)**
  W: **AM Best*(SCALES) (Amoco)**
- **FStop** E: **Phillips 66***
- **Food** E: **Boondocks Motel***, Dairy Queen
  W: **Trump Restaurant (Amoco)**
- **Lodg** E: (AAA) Best Western
- **TServ** E: **Phillips 66**
- **ATM** E: Phillips 66

142 U.S. 20E, Waterloo

139 CR D41, Kamrar

133 IA175, Jewell Eldora

128 CR D65, Randall, Stanhope
- **Parks** W: **Little Wall State Park (5 Miles)**

124 Story City
- **FStop** W: **Phillips 66*(D)**, **Texaco***
- **Food** W: Dairy Queen, Dairy Queen, Godfather's Pizza, Happy Chef, **McDonalds**, Pizza Hut, Subway, Valhalla Restaurant
- **Lodg** W: ◆ **Super 8 Motel (see our ad this page)**, (AAA) Viking Motor Inn
- **RVCamp** W: **Gookin RV Center**
- **ATM** W: Story County Bank & Trust

123 IA221, CR E18, Roland, McCallsburg

(120) **Rest Area - Phones, RR, Picnic, Vending, RV Dump (Northbound)**

116 CR E29

113 13th St, U.S. DA Veterinary Lab, Ames
- **Gas** W: Amoco*, Kum and Go(D)
- **Food** W: Burger King
- **Lodg** W: (AAA) Best Western

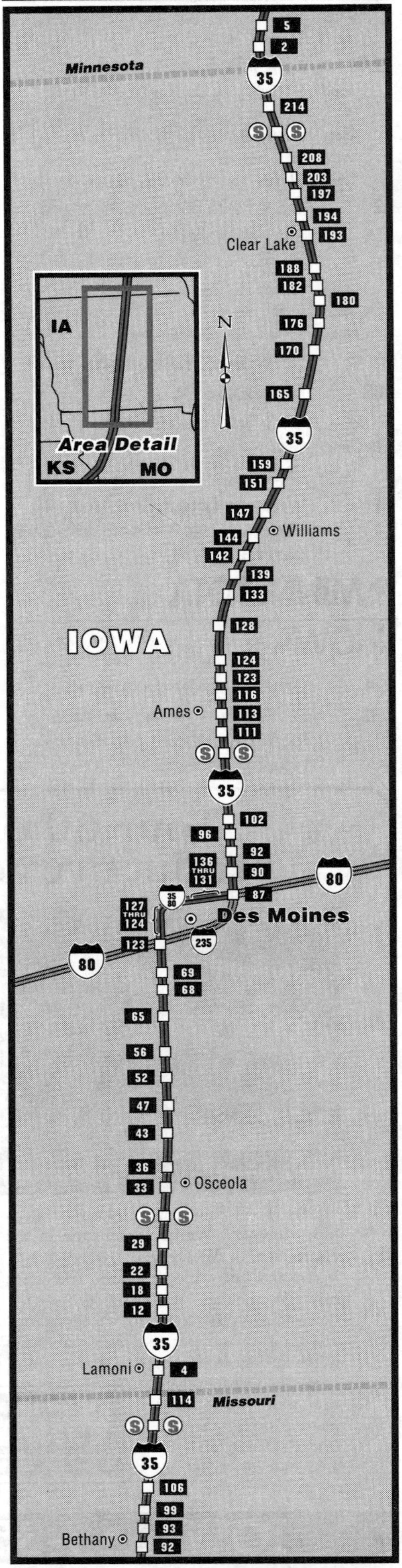

111AB U.S. 30, Ames, Nevada
- **FStop** W: **Texaco***
- **Food** W: Happy Chef
- **Lodg** W: Hampton Inn (see our ad this page), Heartland Inn, **Super 8 Motel (see our ad this page)**
- **TServ** W: **Texaco**
- **RVCamp** E: **Campground**

(106) **Weigh Station - both directions**

102 IA210, Slater, Maxwell

96 Elkhart

(94) **Rest Area - RR, Phones, Picnic, Vending (Both Directions)**

92 1st St
- **Gas** W: Amoco*(D), Kum and Go*, QuikTrip* (Open 24 Hours)
- **Food** W: Applebee's, Arby's, Burger King, Fazoli's Italian Food, Happy Chef, KFC, Stuffy's Restaurant
- **Lodg** W: (AAA) Best Western, (DAYS INN) Days Inn, ◆ Heartland Inn, Super 8 Motel
- **AServ** W: Amoco
- **ATM** W: Brenton Bank, Kum and Go
- **Other** W: Super Wash, **Wal-Mart Supercenter**

Bold red print shows RV & Bus parking available or nearby

EXIT	IOWA
90	IA 160, Bondurant, Ankeny Industrial Area
Gas	W: Casey's*
Food	W: **Burger King (Playground), McDonalds (Indoor Playground)**
Lodg	E: Country Inn, Holiday Inn Express (see our ad opposite page)
87B	Junction I-80W
	Note: I-35 runs concurrent below with I-80. Numbering follows I-80.
136	U.S. 69, East 14th St, Ankeny
FStop	S: **QuikTrip*(SCALES)**
Gas	N: Amoco*(D)(CW), Phillips 66*(CW), Sinclair*(D) S: QuikTrip
Food	N: Bonanza Steak House, Country Kitchen, Okoboji S: **Burger King**
Lodg	N: (AAA) Best Western, Motel 6 S: 14th Street Inn
AServ	N: Sinclair
TServ	S: **Housby Mack Trucks**
Other	N: **K-Mart**
135	IA 415, 2nd Ave, Polk City
Gas	S: Coastal*, QuikTrip*
TServ	N: **Interstate Detroit Diesel** S: **Freightliner Dealer**
131	IA 28, Merle Hay Road, Urbandale
Gas	N: Coastal*(D), QuikTrip* S: Amoco*(CW), QuikTrip*, Sinclair*
Food	N: North Inn Diner (The Inn) S: Burger King, Denny's, Embers, Hostetler's BBQ, McDonalds, Perkins Family Restaurant, Wendy's
Lodg	N: (AAA) Best Inns of America, (AAA) The Inn S: ◆ Comfort Inn, (DAYS INN) (AAA) Days Inn, Holiday Inn, (AAA) Howard Johnson, Roadway Inn, ◆ Super 8 Motel
Other	S: **Touchless of Merle Hay(CW)**
129	NW 86th St., Camp Dodge
127	IA 141, Grimes, Perry
FStop	N: **Phillips 66*(D)**
Food	N: Subway
126	Douglas Ave, Urbandale
TStop	N: **Pilot Travel Center(SCALES)**
Gas	S: Phillips 66*(CW)
Food	S: Dragon House
Lodg	S: (DAYS INN) **Days Inn, Econolodge**
TServ	N: **Pilot Travel Center(D)(SCALES) (ATM)**
TWash	N: **Pilot Travel Center**
ATM	S: Magna Bank
125	U.S. 6, Hickman Road, Adel
Food	S: Abigails (Best Western), Iowa Machine Shed (Comfort Suites)
Lodg	S: Best Western, (AAA) Comfort Suites
AServ	S: Goodyear Tire & Auto
124	University Ave, Clive
Gas	N: Texaco*(CW)
Food	S: Chili's
Lodg	S: ◆ Courtyard by Marriott, ◆ Fairfield Inn, ◆ Holiday Inn, ◆ Residence Inn
AServ	N: Texaco*(CW)
123	Junction I-80, I-235, Downtown Des Moines
	Note: I-35 runs concurrent above with I-80. Numbering follows I-80.

EXIT	IOWA
69AB	West Des Moines, Grand Ave
68	IA 5, Des Moines Airport
Parks	E: **Blank Park Zoo (2 Miles)**
Other	E: **Des Moines Airport (7 Miles)**
65	CR G14, Cumming, Norwalk
56	IA 92, Indianola, Winterset
FStop	W: **Kum & Go*(D) (Texaco)**
Food	W: Piccadilli Pizza (Texaco)
52	CR G50, St. Charles, St. Marys
Other	W: **Madison County Museum (14 Miles)**
(51)	**Rest Area - Parking area (Southbound)**
47	CR G64, Truro
43	IA 207, New Virginia
Gas	E: Conoco*(D)(LP) W: Texaco*
36	IA 152
33	U.S. 34, Osceola, Creston
TStop	E: **AM Pride*(D)(SCALES)**
FStop	E: **AM Pride*(D)(SCALES) (Open 24 Hours)**
Gas	E: Texaco*
Food	E: AM Pride, Byers Restaurant (Texaco), McDonalds, Pizza Hut, Subway
Lodg	E: (AAA) Best Western, Holiday Inn Express, **Super 8 Motel (see our ad this page)**
AServ	E: Goodyear Tire & Auto
TServ	E: **AM Pride**
TWash	E: **AM Pride**
ATM	E: Am Pride, American State Bank
(33)	**Rest Area - RR, Phones, Picnic, Vending, RV Dump (Both Directions)**
(31)	**Weigh Station - both directions**
29	CR H45
22	IA 258, Van Wert
18	CR J20, Grand River, Garden Grove
12	IA 2, Leon, Mount Ayr
FStop	E: **Texaco*(LP)**
Food	E: **Country Corner Restaurant (Texaco)**
(7)	**Rest Area - Phones, RR, Picnic, Vending, HF, RV Dump**
4	U.S. 69, Lamoni, Davis City
Gas	W: Amoco*(D)(LP), Kum & Go*
Food	W: Quilt Country Restaurant
Lodg	W: Chief Lamoni Motel, Super 8 Motel
Other	W: **Iowa Welcome Center**

EXIT	IOWA/MISSOURI
	↑IOWA
	↓MISSOURI
114	U.S. 69 Lamoni
FStop	W: **Conoco*(D) (RV Dump)**
TServ	W: Conoco
(110)	**Weigh Station - both directions**
106	CR N, Blythedale, Eagleville
TStop	E: **Eagleville Truckstop*(D)(SCALES)**
FStop	W: **Texaco*(D)**
Gas	E: Conoco*
Lodg	E: Eagles Landing Motel
TServ	E: **Eagleville Truckstop**
99	CR A, Ridgeway
RVCamp	W: **Eagle Ridge**
93	U.S. 69, Spur Bethany
RVCamp	W: **Camping (RV Dump)**
92	U.S. 136, Bethany, Princeton
FStop	E: **Conoco*(D)**
Gas	E: **Chick's Mini Mart(D)** W: **Amoco*(D)**, QuikTrip*(D)
Food	E: **McDonalds** W: Country Kitchen, Dairy Queen, Hardee's, Subway, **Toot Toot Restaurant**, Wendy's
Lodg	E: (AAA) **Best Western** W: Super 8 Motel
Other	E: **24 Hour Laundromat (Chick's Mini Mart)** W: **Wal-Mart Supercenter**
88	MO 13, Bethany, Gallatin
84	CR H, CR AA, Gilman City
(81)	**Rest Area - RR, Phones, Picnic (P) (Both Directions)**
80	CR B, CR N, Coffey
78	CR C, Pattonsburg
72	CR DD
68	U.S. 69, Pattonsburg
64	MO 6, Gallatin, Maysville
61	U.S. 69, Winston, Jamesport
FStop	E: **Texaco*(LP)**
Food	E: Restaurant (Texaco)
54	U.S. 36, Busi Loop 35, Hamilton, Cameron
TStop	E: **Total*(SCALES)**
FStop	E: **Amoco***
Gas	W: Total*
Food	E: Country Kitchen, **McDonalds, Subway (Amoco), Total**, Wendy's W: Breadeaux Pisa, Burger King, Chinese Chef, Dairy Queen, KFC
Lodg	E: (AAA) Best Western, Comfort Inn, **Crossroads Inn** W: (DAYS INN) Days Inn, ◆ Holiday Inn Express, Super 8 Motel
TServ	E: **Total**
TWash	E: **Total**
ATM	W: Farmer State Bank
Other	W: **Wal-Mart Supercenter**
52	Bus Loop 35, CR BB, Cameron (Northbound, Southbound Reaccess)
48	U.S. 69, Cameron, Wallace State Park

Bold red print shows RV & Bus parking available or nearby

EXIT		MISSOURI
	Gas	W: Total*
	Parks	E: State Park (2 Miles)
40		MO 116, Polo, Lathrop
	FStop	E: Phillips 66*(D)
	Gas	E: Total*
	Food	E: Country Cafe (Phillips 66)
	ATM	E: Phillips 66
(35)		Rest Area - RR, Phones, Picnic P (Southbound)
(34)		Rest Area - RR, Phones, Picnic P (Northbound)
33		CR PP, Lawson, Holt
	FStop	W: Conoco*(D), Phillips 66*(D)
	Lodg	W: Holt Steel Inn
26		MO 92, Kearney, Excelsior Springs
	TStop	W: Kearney Truck Plaza*(D)(SCALES)
	Gas	E: Phillips 66*, Texaco Star Mart*(D)
	Food	E: Sonic Drive In, Taco Bell (Texaco Star Mart)
		W: Donuts Plus, GB's Grill, Hardee's, Hunan Garden Chinese, Kearney Truck Plaza, Legends Sports Grill, Pizza'N More, Son Light, Subway
	Lodg	E: ◆ Super 8 Motel
		W: Econolodge
	TServ	W: Kearney Truck Plaza
	ATM	E: Big V Country Mart
		W: Pony Express Bank
	Other	E: Big V Country Mart*
(23)		Weigh Station - both directions
20		U.S. 69, MO 33, Excelsior Springs
	Gas	E: Blue Light*
	Med	E: ✚ Liberty Hospital
17		MO 291, CR A, Liberty, KCI Airport
	Gas	W: Phillips 66*(CW)
	ATM	W: Phillips 66
	Lodg	E: Best Western (see our ad this page)
16		MO 152, Liberty
	Gas	E: Conoco*(D), Texaco*
		W: Phillips 66*(CW)
	Food	E: Boston Market, Carlitos Mexican Restaurant, El Cerro Grande Mexican Restaurant, Godfathers Pizza, Greaser's Burgers, Hunan Garden Chinese, Lamar's Doughnuts, Perkins Family Restaurant, Pizza Hut, Ponderosa Steak House, Wendy's
		W: Applebee's, Country Kitchen, Cracker Barrel, Golden Corral, McDonalds
	Lodg	E: AAA Best Western
		W: ◆ Fairfield Inn
	AServ	E: Conoco, Firestone Tire & Auto, Parts America Auto Parts
	RVCamp	E: Camping
	ATM	E: Commerce Bank, Nations Bank, Platte Valley Bank, UMB Bank
	Other	E: Hy Vee Grocery and Pharmacy, K-Mart
		W: Wal-Mart
13		U.S. 69, Pleasant Valley
	Gas	E: Sinclair*, Texaco*(CW), Total*(D)
		W: Conoco*
	Food	E: Hot Stuff Pizza (Sinclair)
	AServ	E: Frank's Tow & Service, Outback Body Shop
	RVCamp	E: I-35 Parts & Accessories
12A		Jct. I-435S, St. Louis
11		U.S. 69, Vivian Road
	Gas	W: QuikTrip*, Total*
	Food	W: Big Burger, Church's Fried Chicken, Sonic Drive In, Strouds Restaurant, Subway
	ATM	W: Mark Twain Bank
	Other	W: Bob's Automatic Car Wash, Foxwood Drugs

EXIT		MISSOURI
		Pharmacy, Timber Ridge Park
10		North Brighton Ave
	Gas	W: Quick Trip*, Total*
	Food	W: Big Burger, Church's Fried Chicken, Sonic Drive-In, Subway
	AServ	W: O' Reilly Auto Parts
	Other	W: Foxwood Drugs
9		MO 269, Chouteau Trafficway
	Gas	E: Phillips 66*(CW), Sinclair*
	AServ	E: Sinclair
8C		MO 1, Antioch Road
	Gas	E: Citgo*, Sinclare*(LP)
		W: Phillips 66*(CW)
	Food	E: Country Girl, Domino's Pizza
		W: Waffle House
	Lodg	E: ◆ Best Western, Inn Towne
8B		Jct. I-29 North US Highway 71 North
8A		Parvin Road
	Gas	E: Texaco*(CW)(LP), Total*(D)
	AServ	E: O' Reilly Auto Parts
	Other	E: Thrifty Coin Laundry
6AB		Armour Road, North Kansas City MO Hwy.210
	Gas	E: Total*(D)
		W: Phillips 66*(CW), QuikTrip*, Texaco*(CW)
	Food	E: Arby's, Captain D's Seafood, McDonalds, Shoney's
		W: Church's Fried Chicken (White Castle Hamburgers), Long John Silvers, Pizza Hut, Taco Bell, Wendy's, White Castle Restaurant
	Lodg	E: Budgetel Inn
		W: American Inn
	Med	W: ✚ Omni Family Medicine Health South
	ATM	E: NationsBank
		W: First Federal Bank
5B		16th Ave. (Northbound)
5A		Levee Rd., Bedford St. (Difficult Reaccess)
4B		Front St
4A		U.S. 24, Independence Ave
	Gas	E: Citgo*, Phillips 66
		W: Phillips 66*
	Food	E: Chubbies
		W: McDonalds
	Lodg	E: Capri Motel
		W: Admiral Motel, Royale Inn
	AServ	E: Phillips 66, Thoroughbred Ford
3		Junction I-70
2F		Oak St, Grand - Walnut
2D		Maine - Delaware, Wyandotte St

EXIT		MISSOURI/KANSAS
2C		Broadway, US 169N
	Food	E: Broadway Deli
	Lodg	E: HoJo
2W		12th St
2G		JCT 29
2H		US 24, MO 9
2U		Junction I-70E, I-670, St.Louis
1D		20th St
1C		Pennway (Northbound Only, Difficult Reaccess)
1B		27th St, Broadway
1A		SW Trafficway

↑ MISSOURI

↓ KANSAS

EXIT		KANSAS
235		Cambridge Circle
234		U.S. 169, 7th St Trafficway, Rainbow Blvd
	Food	E: Dairy Queen, Katz', Shoney's, Sonic Drive-In
233		SW Blvd, Mission Road
233B		37th Ave
232B		18th St Expressway, Roe Ave, US69N
232A		Lamar Ave
	Gas	E: QuikTrip*(LP), Total*(LP)
231AB		Junction I-635, U.S. 69, Metcalf Ave
230		Antioch Road (Reaccess Northbound Only)
229		Johnson Dr
	Gas	W: Conoco*(D)
	ATM	W: Conoco
228B		U.S. 56, U.S. 69, Shawnee, Mission Pkwy
	Gas	E: Texaco*(CW)
	Food	E: Checkers Burgers, IHOP, Shoneys, Taco Bell, Winstead's
		W: Baskin Robbins, Furr's Family Buffet, Hardee's, Long John Silvers, Perkin's Family Restaurant, Red Lobster, Texas Tom's Tacos
	Lodg	E: ◆ Comfort Inn (see our ad this page), ◆ Drury Inn
	AServ	W: AAMCO Transmission, Car X Muffler & Brakes, Firestone Tire & Auto, Goodyear Tire & Auto
	Other	E: K-Mart (Auto Service)

Bold red print shows RV & Bus parking available or nearby

EXIT KANSAS

228A 67th St
- **Gas** W: Phillips 66*(CW)
- **Food** E: Burger King (Playground)
- **Lodg** E: ◆ Fairfield Inn

227 75th St
- **Gas** E: Circle K*(LP)
  W: Citgo*, Total*(CW)
- **Food** E: McDonalds, Mr. Goodcents Subs, Perkins Family Restaurant
  W: Sonic Drive In, Subway, Wendy's
- **Med** E: ✚ **Shawnee Mission Medical Center**
- **Other** E: **Georgetown Pharmacy, Gerry Optical, Wal-Mart**

225B U.S. 69, Overland Pkwy

225A 87th St Pkwy
- **Gas** E: Amoco*(CW)
  W: Kicks 66*(CW), Total*(LP)
- **Food** E: Phoenix Chinese, Raddison Inn, Shoney's, Tippins, Wendy's
  W: KFC, Longbranch Steakhouse, Zarda BBQ
- **Lodg** E: Raddison Hotel
- **AServ** E: Amoco
- **ATM** E: UMV Bank
  W: Premier Bank
- **Other** E: **JM Bauersfeld's Grocery**

224 95th St
- **Gas** E: Phillips 66*
  W: Conoco*, JB's One Stop* (Open 24 Hours)
- **Food** E: Applebee's, Bagel and Bagel, Denny's (La Quinta Inn), McDonalds, Ming Palace Restaurant, Steak & Ale, Winsteade's
- **Lodg** E: Days Inn, ◆ Holiday Inn, La Quinta Inn, ◆ Super 8 Motel
- **ATM** E: NationsBank
- **Other** E: **Sam's Club**

222AB Junction I-435

220 119th St
- **Food** E: **Cracker Barrel,** Kansas Machine Shed, McDonalds (Playground), Subway, Tres Hombres Mexican Resetaurant
- **Lodg** E: ◆ Comfort Suites, ◆ Fairfield Inn, Hampton Inn
- **Other** E: **Target Department Store**

218 Santa Fe
- **Gas** E: Amoco*(CW), Circle K*

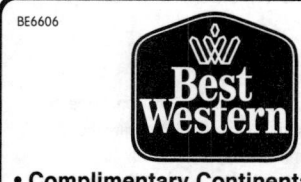

EXIT KANSAS

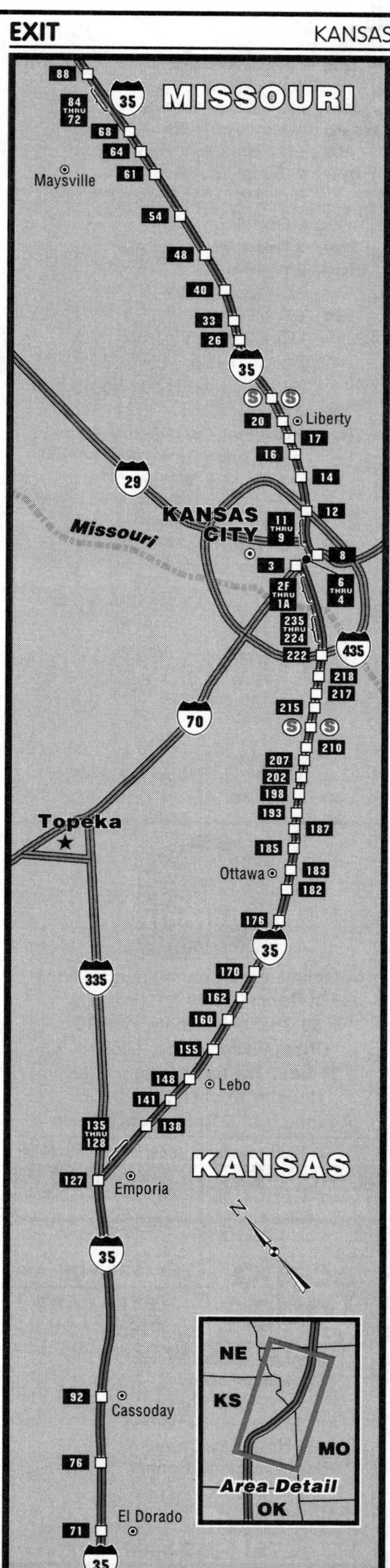

EXIT KANSAS

- W: Amoco*(CW), Phillips 66*(CW), QuikTrip*(CW) (Open 24 Hours)
- **Food** E: Burger King, Golden Bowl Chinese, Perkins Family Restaurant
  W: Denny's, Domino's Pizza, Ponderosa, Taco Bell, Wendy's
- **Lodg** W: AAA **Best Western (see our ad this page)**
- **AServ** E: Instant Oil Change, Midas Muffler & Brakes
  W: Parts of America
- **ATM** E: NationsBank
- **Other** E: **K-Mart (Auto Service), Osco Drugs, Shopping Mall**

217 Old Hwy 56 (Difficult Reaccess)
- **Food** W: JC's Bar and Grill

215 U.S. S169, KS7, Paola
- **TStop** W: **Texaco*(SCALES) (Open 24 Hours)**
- **Gas** E: Phillips 66*(CW) (Open 24 Hours)
- **Food** E: Red Lobster
  W: **Blimpie's Subs (Texaco),** Burger King, **McDonalds (Playground),** Wendy's
- **Lodg** W: Microtel Inn
- **Med** W: ✚ **Olaphe Medical Center**

(213) Motor Vehicle Inspection Station

210 U.S. 56, Gardner
- **Gas** W: Kicks 66*(CW)
- **Food** W: McDonalds, Waffle House
- **Lodg** W: Super 8 Motel

207 Gardner Road
- **Gas** W: Conoco*(D)
- **RVCamp** E: **Olathe RV Center**

202 Edgerton

198 KS33, Wellsville
- **Gas** W: Total*

193 Le Loup Road, Baldwin
- **Food** W: McKenzie's Country Cafe

187 U.S. Bus.50, KS68, Ottawa, Paola
- **FStop** W: **Phillips 66*(D)**
- **Gas** W: Fill R Up*

185 15th St, Ottawa
- **Med** W: ✚ **Hospital**
- **Lodg** W: Days Inn (see our ad this page)

183AB U.S. 59, Ottawa, Garnett
- **FStop** W: **Phillips 66***
- **Gas** W: Amoco*(CW), Conoco* (Open 24 Hours),

EXIT KANSAS

Woods Mini Mart* (Open 24 Hours)
- **Food** **W:** Country Kitchen, Long John Silvers, **McDonalds (Playground),** Sirloin Stockade, Taco Bell, Wendy's
- **Lodg** **W:** Best Western (see our ad this page), Holiday Inn Express, Econo Lodge (see our ad this page)
- **AServ** **W:** Big A Auto Parts
- **Other** **W: Country Mart Grocery, Wal-Mart (Pharmacy, Tire and Lube Service)**

182 U.S. 50, Ottawa, County Road (Difficult Reaccess)

176 Homewood

(175) **Rest Area - RR, Phones, Picnic, Vending, RV Dump P (Both Directions)**

170 KS 273, Williamsburg, Pomona
- **Gas** **W:** Davey and Gails Convenience Store

162 KS 31S, Waverly

160 KS 31N, Melvern

155 U.S. 75, Burlington, Lyndon
- **TStop** **E: Beto Junction***(SCALES)
- **FStop** **E: Total***
- **Food** **E:** Country Pride Restaurant
- **TServ** **E: Lebo Garage**

148 KS 131, Lebo
- **FStop** **E: Coastal***
- **Food** **E:** Universal Inn
- **Lodg** **E:** Universal Inn

141 KS 130, Neosho Rapids, Hartford

138 County Road

135 Thorndale
- **RVCamp** **W: Camping**

133 U.S. 50, 6th Ave
- **Food** **E:** McDonalds, Pizza Hut
- **Lodg** **E:** AAA Budget Host Inn (.75 Miles)

131 Burlingame Road
- **Gas** **E:** Amoco*, Coastal*
- **Food** **E: Burger King (Playground, Bus Parking),** Hardee's (Playground)
- **AServ** **E:** Amoco
- **ATM** **E:** Admire Bank, Coastal, NationsBank

130 KS 99, Merchant St
- **Gas** **E:** Phillips 66(D)
- **Lodg** **E:** Ramada Inn
- **AServ** **E:** Phillips 66

128 Industrial Blvd
- **Gas** **E:** Conoco*, Total* **W:** Texaco*(CW)
- **Food** **E:** Amigos Mexican, Burger King **W:** McDonalds, Taco Bell
- **Lodg** **E:** AAA Comfort Inn, Motel 6, ◆ Ramada Inn **W:** ◆ Fairfield Inn
- **AServ** **E:** Goodyear Tire & Auto, Western Auto **W:** Texaco
- **ATM** **E:** NationsBank **W:** Texaco
- **Other** **E: Food 4 Less Grocery (Open 24 Hours)** **W: Wal-Mart (Pharmacy, Vision Center, Tire and Lube Service)**

127 Jct I-335, U.S. 50W, KS 57, Newton, Imporia Ottawa
- **TStop** **W: Martin's Truck Service**(SCALES)
- **FStop** **E: Phillips 66***(D)(LP)
- **Gas** **E:** Amoco, Conoco*(D)
- **Food** **E:** Arby's, Hardee's (Playground), Mazzio's Pizza, Western Sizzlin' **W:** Ranch House Cafe

EXIT KANSAS

- **Lodg** **E:** Best Western, DAYS INN Days Inn, Super 8 Motel **W:** Ranch House
- **AServ** **E:** Amoco, Napa Auto Parts
- **RVCamp** **W: KOA Campground**
- **ATM** **E:** First National Bank, Phillips 66
- **Other** **E:** Car Wash, **Laundromat (Phillips 66)**

(97) Mattfield Green Service Area (Left Lane Exit)
- **FStop** **B: Coastal***(D)
- **Food** **B: Hardee's**

92 KS 177, Cassoday
- **Gas** **E:** Phillips 66

76 U.S. 77, El Dorado North, Nick Badwey Plaza

71 KS 254, KS 196, El Dorado, Augusta
- **Gas** **E:** Phillips 66
- **Food** **E:** Burger King, Chin's Chinese Restaurant, Golden Corral, KFC, McDonalds, Moe's BBQ, Pizza Hut, Sub & Stuff Sandwich Shop
- **Lodg** **E:** Best Western, Heritage Inn, **Super 8 Motel**
- **AServ** **E:** All Pro Auto Parts, Jarvis Auto Supply
- **ATM** **E:** Commerce Bank, NationsBank
- **Other** **E: Coin Laundry, Wal-Mart (Pharmacy, Auto Service)**

(65) Towanda Service Area (Both Directions)
- **FStop** **B: Coastal***(D)
- **Food** **B: Hardee's**
- **Other** **B: Tourist Information**

57 Andover

53 KS 96W, Wichata (Difficult Reaccess)

50 U.S. 54, U.S. 400, Kellogg Ave
- **Gas** **W:** Coastal*

EXIT KANSAS/OKLAHOMA

- **Food** **E:** Shoney's **W:** Denny's, Grandy's, KFC, Long John Silvers, McDonalds, Pizza Hut, Red Lion Sports Grill, Spangles, Steak & Ale, Taco Bell, Wendy's
- **Lodg** **E:** Clubhouse Inn, Comfort Inn, Fairfield Inn, Hampton Inn, Hampton Inn, Marriott, Residence Inn, Scottsman Inn, Super 8 Motel **W:** DAYS INN Days Inn, Harvey Hotel, La Quinta Inn, Mark 8 Lodge, Roadway Inn
- **AServ** **W:** National Tire and Battery, Pep Boys Auto Center, Splash and Dash Buggy Bath
- **ATM** **E:** AACU Credit Union **W:** Coastal
- **Other** **W: Barnes & Noble, K-Mart (& Pharmacy), Osco Drugs, Town East Square Shopping Mall**

45 KS 15, Wichita K - 15

42 Jct I-135, I-235, U.S. 81, South Wichita (Toll Plaza)
- **Gas** **W:** Coastal
- **Lodg** **W:** Comfort Inn, Holiday Inn Express

39 U.S. 81, Haysville, Derby

33 U.S. 81, KS 53, Mulvane

(26) Belle Plain Service Area (Both Directions)
- **FStop** **B: Coastal***(D)
- **Food** **B: Hardee's**

19 U.S. 160, Wellington
- **RVCamp** **W: KOA Campground**

(19) Toll Plaza

4 U.S. 166, U.S. 81, Arkansas City, South Haven
- **FStop** **E: Total***(D)
- **Food** **E: Piccadilly Circus Pizza & Subs (Total)**
- **Lodg** **E:** Economy Inn
- **AServ** **E:** Strickland Road Service

(1) **KS Welcome Center, Weigh Station, Rest Area - Full Facilities, Truck Parking P (Northbound)**

↑KANSAS

↓OKLAHOMA

231 U.S. 177, Braman
- **FStop** **E: Conoco***(D)(SCALES)
- **Food** **E: Kanza Cafe**
- **Lodg** **E: Motel**

230 Braman Road
- **FStop** **W: Texaco***(D)

(226) **Park Area, Rest Area - RR, Phones, Picnic, Vending, RV Dump P (Both Directions)**

222 **(219)** OK11, Blackwell, Medford
- **FStop** **E: Conoco***(D)
- **Gas** **E:** Phillips 66*, Texaco*(D) **W:** Texaco*
- **Food** **E: Braum's Ice Cream, Days Inn,** Kettle Restaurant (Comfort Inn), **McDonalds,** Plainsman Restaurant, Subway
- **Lodg** **E:** AAA Comfort Inn, DAYS INN **Days Inn,** Super 8 Motel
- **AServ** **E:** Texaco
- **Med** **E:** ✚ **Hospital**
- **Other** **E: Top Of Oklahoma Museum, Whites Factory Outlet Center**

218 Hubbard Road

(217) **Weigh Station - both directions**

214 **(212)** U.S. 60, Lamont, Ponca City.

EXIT	OKLAHOMA
FStop	W: Phillips 66*(D)
Food	W: Conetoga Restaurant, Grill (Phillips 66)
Lodg	W: Motel
AServ	W: 24 Hr Tire
TServ	W: 24 Hr Tire
RVCamp	W: Woodland Campground
211	Fountain Road
TStop	E: Love's Country Store*(D)
Food	E: Deli (Love's), Pizza Hut Express (Love's FS)
TServ	E: Oklahoma Truck Supply, Wilkins Truck Supply (Truck Chrome & Acc)
ATM	E: Love's Country Store*(D)
(210)	Parking Area - both directions
203	OK15, Billings, Marland
TStop	E: Conoco*(D)
Food	E: Dairy Queen (Conoco)
ATM	E: Conoco*(D)
Other	E: Coin Operated Laundry*(D) (Conoco), Fax*(D) (Conoco)
(196)	Parking Area (Northbound)
(195)	Parking Area (Southbound)
194AB	U.S. 64, U.S. 412W, Enid, Cimarron Turnpike (Difficult Reaccess)
186	U.S. 64E, Perry, Fir St
TStop	W: Conoco*(SCALES)
FStop	E: Diamond Shamrock*(D)
Food	E: Braum's Ice Cream, McDonalds, Subway* (Diamond Shamrock), Taco Mayo
	W: Restaurant (Conoco FS)
Lodg	W: Days Inn (AAA) Days Inn
Med	E: + Hospital
Other	E: Cherokee Strip Museum
185	U.S. 77, Perry, Covington
TStop	W: Texaco*(LP)
Gas	E: Phillips 66*
Food	E: Cherokee Strip (At Best Western)
	W: Texaco
Lodg	E: (AAA) Best Western
	W: Sooners Corner Motel, Texaco Truck Stop
AServ	E: Perry Tire Serv (Goodyear)
	W: Sooners Corner Auto
TServ	E: Perry Tire Serv (Goodyear)
	W: Texaco Truck Stop
RVCamp	W: Sooner's RV Park
ATM	W: Texaco
Other	W: Coin Operated Laundry (Texaco), Horse Stalls (Texaco)
180	Orlando Road
174	OK 51, Stillwater, Hennessey. (Oklahoma State University)
FStop	E: Phillips 66(D), Texaco*(D)
Other	E: Oklahoma State University
(173)	Parking Area (Southbound)
(171)	Parking Area (Northbound)
170	Mulhall Road
FStop	E: Sinclair(D)
AServ	E: Sinclair FS*
157	OK 33, Guthrie, Cushing
FStop	E: Bowman's(D)(LP)
	W: Texaco*(D)
Gas	W: Phillip 66*(D)
Food	W: Burger King, Chester Fried Chicken (Texaco FS), Dairy Queen, J & L Restaurant (Texaco FS)
Lodg	W: (AAA) Best Western, Interstate Motel
AServ	E: Bowman's
	W: Ron's Tire Service (Phillips 66)
RVCamp	W: Cedar Park Recreation Camping
Med	W: + Hospital (4 Miles)
Other	W: Oklahoma State Museum
153	U.S. 77, Guthrie (Northbound,

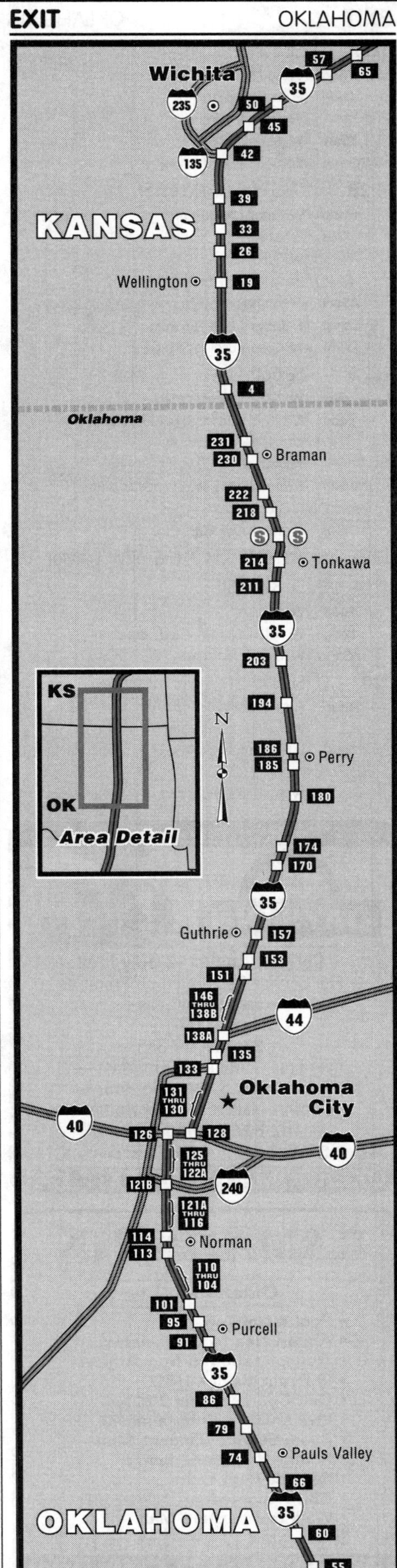

EXIT	OKLAHOMA
	Reaccess Southbound)
AServ	W: Chrysler Auto Dealer, Ford Dealership
151	Seward Road
FStop	E: Citgo*(D)
Food	E: Stuckey's (Citgo)
RVCamp	E: Camping
(149)	Rest Area - RR, Phones, Picnic P (Both Directions)
(148)	Weigh Station Both Directions
146	Waterloo Road
Gas	E: Conoco*(D)
Food	E: County Line Cafe
AServ	E: Stouts Body Shop
	W: Tire Store
TServ	W: Tire Store
Other	E: County Line Cafe C-Store, Royal Coach RV Service
143	Covell Road
142	Danforth Rd (Difficult Reaccess, Use W Sooner Dr For Reaccess)
141	US77S, Second St, OK66E, Edmond, Tulsa
Gas	W: Texaco*
Parks	E: Central State Park, Edmonds Park
Other	E: Arcadia Lake
140	SE 15th St
Parks	E: Arcadia Spring Creek Park
139	SE 33rd St
138D	Memorial Road
138C	Sooner Road
138B	Kilpatrick Turnpike
138A	Jct I-44E, Tulsa
137	NE 122nd St
FStop	E: Texaco
	W: Love's Country Store*(SCALES) (Taco Bell and Subway inside), Texaco*(D) (Hardees)
Gas	E: Total*(D)
Food	E: Hardee's (Texaco FS), Kettle (Texaco FS)
	W: McDonalds, Waffle House
Lodg	E: (AAA) Motel 6
	W: Comfort Inn, Days Inn (AAA) Days Inn (see our ad this page), (AAA) Motel 6, ◆ Quality Inn, Red Carpet Inn
AServ	W: Interstate Auto Repair
RVCamp	W: Camping
136	Hefner Road
Gas	N: Conoco*
Food	N: Papa's Lil Italy Restaurant (Phillips 66)
Lodg	N: Ramada Inn, The Grand Hotel

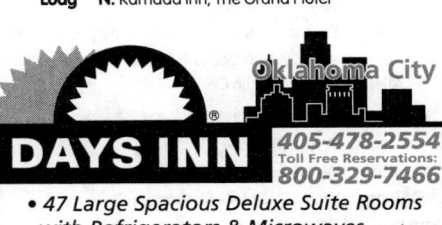

Bold red print shows RV & Bus parking available or nearby

EXIT OKLAHOMA

Other N: Frontier City (Amusement Park)

135 Britton Road, NE 93rd St

134 Wilshire Blvd, NE 78th St

TStop W: Texaco*(SCALES)

TServ W: Texaco Truck Stop

TWash W: Blue Beacon(SCALES)

ATM W: Texaco

Other W: Blue Beacon Truck Wash

132B NE63rd St.

Gas E: Conoco*(D)

Food E: Braum's, Derby's

Lodg E: Super 8 Motel

RVCamp W: ACME RV Service

Other E: Animal Clinic

133 Jct I-44W, Lawton, Amarillo

132A NE 50th St

131A NE 36th St

130 U.S. 62E, NE 23rd St

Gas E: Conoco*, Total*

Food E: Country Club BBQ

Other W: State Capitol Building

129 NE 10th St

128 Jct I-40E, Fort Smith (Southbound Exits Left Side Of Highway)

127 Reno Ave, Martin Luther King Ave, Eastern Ave

TStop W: Petro Travel Center*(D)(SCALES)

Gas W: Conoco*, Texaco

Food W: Gary Dale's BBQ, Waffle House

Lodg W: (AAA) Best Western, Central Place Hotel

TWash W: Blue Beacon (Petro Travel Center)

RVCamp W: Lewis RV Sales & Service

126 Jct I-40, Jct I-235N, Edmond, Amarillo

125D SE 15th St

Gas E: Conoco*, Save a Stop, Total*

Food E: Choice Restaurant, Skyliner Restaurant, Subway (Save a Stop)

Lodg E: (AAA) Howard Johnson

TServ W: Oklahoma Truck Supply

125A SE 25th St

Gas W: Phillips 66* (ATM)

Food E: Black Gold Restaurant

124B SE 29th St

Food E: Days Inn, Denny's, McDonalds, Taco Bell, Waffle House

W: Mama Lou's

Lodg E: (Days Inn) Days Inn (see our ad this page), Plaza Inn, Royal Inn, Super 8 Motel

AServ W: Gary's Auto Center

124A Grand Blvd

Food W: Drover's Inn

Lodg W: Drover's Inn, Hampshire Inn

RVCamp W: McClain's RV Superstore

123B SE 44th St

FStop E: Conoco*

Gas E: Stax Gas*

Food E: Domino's Pizza, Lucy's Restaurant, Sonic Drive In

W: Dairy Queen, Little Mermaid Seafood, Pizza 44, Sunset Grill, Taco Mayo

Lodg E: Econolodge

ATM E: Cash Machine (Stax Gas)

Other W: Sav-A-Lot Grocery

EXIT OKLAHOMA

123A SE 51st St

Food W: Southgate Inn

Lodg E: Best Value Inn

W: Southgate Inn

TServ W: Cherokee Tire

RVCamp E: Roadrunner RV Park

122B SE 59th St, Hillcrest St

FStop W: Total* (RV Dump)

Gas E: Texaco*

W: Stax Gas*

Food W: Taco Bell (Total FS)

AServ E: Championship Transmission Auto Repair

RVCamp E: Brisko's RV & Fun Park

ATM W: Cash Machine (Stax Gas)

122A SE 66th St

Gas W: 7-11 Convenience Store

Food E: Burger King, Lubby's Cafeteria, McDonalds, Ramada Inn, Taco Bell

W: Arby's

Lodg E: Ramada Inn (see our ad this page)

AServ E: Cooper Tires

Other E: Crossroads Mall

121B Jct I-240, OK 3, U.S. 62W, Lawton

121A SE 82nd St

Food W: Denny's

Lodg W: ◆ La Quinta Inn, Red Carpet Inn

AServ W: Classic Auto Parts

120 SE 89th St

FStop E: Total*(SCALES)

W: Love's Country Store*

Food E: Hot Stuff Pizza (Total FS), Smash Hit Subs (Total FS)

W: Hardee's, Mom's Bar-B-Que, Subway

EXIT OKLAHOMA

(Love's Country Store), Taco Bell (Love's Country Store)

119B N 27th St

FStop E: Conoco

Gas E: Texaco*

W: Phillips 66*

Food E: Best Western

W: Pickle's Restaurant

Lodg E: (AAA) Best Western

AServ E: Conoco

W: Goodyear Tire & Auto

119A Shields Blvd (Northbound)

118 N12th St, Main St, N 5th St

Gas E: Stax Gas*(CW), Total*

W: Texaco*(CW)

Food E: Chinese Wok, Golden Apple Restaurant, Harry Bear's Restaurant, Sonic Drive In

W: Arby's, Braum's, Fajitas Mexican, Grandy's, KFC, Long John Silvers, McDonalds, PaPa John's, Pizza Hut, Wendy's, Western Sizzlin'

Lodg E: (AAA) Super 8 Motel

W: (Days Inn) (AAA) Days Inn

Other W: Auto Zone Auto Parts

117 OK37, S 4th St

Gas E: Texaco

W: Total*(CW)

Food E: China Doll Chinese

W: Jerry's Foods, Mr Burger, The Donut Delight

AServ E: Moore Kar Kare, Texaco

W: Total

Med W: ✚ Moore Medical Center

Other E: US Post Office

W: Moore Veterinary Hospital, Southgate Drug

116 S 19th St

Gas E: Conoco*

Food E: Braum's Hamburgers, Dairy Queen, HB Burgermaker, Hardee's, McDonalds

AServ E: Moore Auto

Other E: Wal-Mart Supercenter

114 Indian Hill Road, 179th St

RVCamp E: Kerr Country

113 U.S. 77S, Franklin Road (Southbound Exits Left Side Of Highway)

110 Robinson St

Gas E: Petro - Self-Service, Sinclair

W: 7-11 Convenience Store*

Food E: Hardees, Sonic Drive In

W: Arby's, Braums, (Cracker Barrel) Cracker Barrel, Hunan Chinese, Jama's Italian, Pizza Hut, Pizza Shuttle, Ryan's Steakhouse, Subway, Waffle House

Lodg W: ◆ Holiday Inn

AServ E: Fastland Oil Change, Lincoln Mercury Dealer

W: Auto Valet

Other E: Albertson's Grocery (& Pharmacy)

109 Main St

Gas E: Phillips 66*(CW), Texaco*(D)(CW)

Food E: Denny's, Furr's Cafeteria, Golden Corral, Kettle Restaurant, Little Caesars Pizza, Waffle House

W: Burger King, Chili's, Don Pablo, McDonalds, Olive Garden, Red Lobster

Lodg E: Econolodge, (AAA) Guest Inn, Super 8 Motel, Travel Lodge

W: ◆ Fairfield, Hampton Inn

AServ E: Cooper Buick Mazda, Hibdon Tire, Pemberton

Bold red print shows RV & Bus parking available or nearby

EXIT — OKLAHOMA

Olds Cadillac Nissan, Spencer Chevrolet
W: Wal-Mart
ATM W: Bank of Oklahoma
Other E: Buy for Less Grocery, Target Department Store
W: Borders Bookstore, Shopping Mall, Wal-Mart

108B OK 9E, OK 74A, Lindsey St, Tecumseh, Univ of Oklahoma
Gas E: Conoco*(CW)(LP), Mr. Short Shop (ATM), Phillips 66, Sinclair
Food E: Arby's, Arthur's, Bo Bo's Chinese, Braum's, Del Rancho's Hamburger, KFC, Schlotzkys Deli, Taco Bell, Taco Bueno
Lodg E: (AAA) Ramada Inn (see our ad this page)
AServ E: Fast Lube, Midas Muffler Shop, O'Rilley Auto Parts, Ron's Automotive
Other E: Coin Car Wash

108A OK 9E, Tucumseh
Gas E: Phillips 66*(D)
Lodg E: ◆ Residence Inn, ThunderBird Lodge

106 OK 9W, Chickasha
FStop E: Mid-Continent Truck Sales
W: Conoco, Love's Country Store*
Food W: Burger King (Conoco), Taco Bell (Love's Country Stores)

104 OK 74S, Goldsby, Washington
FStop W: Crawford's Grocery* (Total)
Gas W: Texaco*(D)
Food W: Hydes Smokehouse BBQ
TServ W: Andrew's Service Center - Trucks, Cars and Foreign (Behind Total)
RVCamp E: Floyd's RV Center

101 Ladd Road

98 Johnson Road
FStop E: Texaco*
RVCamp E: Alamo RV Park

95 U.S. Hwy 77, Purcell, Lexington (Southbound Exits Left Side Of Highway)
Med E: ✚ Hospital

91 OK 74, Purcell, Maysville
FStop W: Texaco*(D)
Gas E: Phillips 66*(D), Total*(D)
W: Texaco*
Food E: McDonalds
W: Chickasha Hills Restaurant
Lodg E: (AAA) Econolodge, Horse Country Inn, Ruby's Inn

EXIT — OKLAHOMA

AServ W: Texaco
TWash E: J&M Truck Wash

86 OK 59, Wayne, Payne

79 OK145E, Paoli

74 Kimberlin Road

72 OK 19, Pauls Valley, Maysville
FStop W: Love's Country Store*, Phillips 66
Gas W: Phillips 66
Food E: Ballard's Drive In, Braum's, Hardee's, KFC, McDonalds, Punkin's BBQ, Subway, Taco Mayo
W: Mindy's Family Steakhouse
Lodg E: (AAA) Amish Inn Motel, (Days Inn) (AAA) Days Inn, Garden Inn Motel, Park Inn International
TServ E: Total FS
TWash W: Rusty's Truck Wash
RVCamp E: Pecan RV Park
Other E: Bi-Lo Supermarket

70 Airport Road
Med W: ✚ Hospital

66 OK 29, Wynnewood, Elmore City
FStop E: Kent's Truck Stop - Citgo*(D)
Gas W: Texaco*
Lodg E: Kent's Motel
TServ E: JJ's Tire Service, Southern Leap Supply Truck & Trailer Parts

64 OK 17A East, Wynnewood
FStop W: Fina*
Food W: Washita Crossing Restaurant(D)(LP) (Fina FS)

60 Ruppe Road

(57) Rest Area - RR, Phones, RV Dump, Picnic, HF (P) (Northbound)

(56) Rest Area - RR, Phones, RV Dump (P) (Southbound)

55 OK 7, Davis, Duncan, Sulphur
FStop E: Chickasaw Trading Post*(D) (Total), Phillips 66* (A&W Root Beer)
RVCamp W: RV Park
Parks E: Chickasaw National Recreation Area
Other E: Arbuckle Lake

(53) Weigh Station - both directions

51 U.S. 77, Turner Falls Area
Gas W: Honey Creek Station (Sinclair)
Food W: The Trout Place
Lodg E: Arbuckle Motel, Mountain View Inn
W: Canyon Breeze Motel (& RV Hookups)
RVCamp E: Martins Rv Park
Other E: Arbuckle Wilderness (1 Mile)
W: Cedarville Botanical Gardens (1 Mile)

(49) Scenic View (Both Directions)

47 U.S. 77, Turner Falls Area

(46) Scenic View (Both Directions)

43 Dougherty, Springer

42 OK53W, Comanche
FStop W: Diamond Shamrock*

40 U.S. 77, OK 53, Autrey, Ardmore Airport
FStop E: Total*(D)(LP) (RV Dump)

33 OK 142, Ardmore
Gas E: Fuelman Automated Fueling(D), Phillips 66*, Total Rapliss
Food E: Ponders, Ryan's Family Steakhouse
Lodg E: (Days Inn) ◆ Days Inn, Guest Inn, Regency Inn,

EXIT — OKLAHOMA/TEXAS

Super 8 Motel
TServ E: Alternative Truck Parts Specialists
Other E: Oklahoma Highway Patrol

32 12th St
FStop W: Love's Country Store*(D)(LP)(SCALES)
Food W: Baskin Robbins (Love's)
Med E: ✚ Memorial Hospital (1 Mile)

31B U.S. 70W, OK 199, Waurika, Lone Grove, Ardmore
Gas W: Conoco*, Total*
Food W: Fin & Feather
AServ W: Billingsly Ford Mercury Nissan, McCullough Olds Cadillac

31A OK199E, Ardmore
Gas E: Conoco*, Sinclair, Texaco*
Food E: Black Eyed Pea, Boston Market, Burger King, Cattle Rustler's, Denny's, El Chico Mexican, KFC, Kettle, Pizza Hut, Shoney's, Sirlion Stockade
Lodg E: (AAA) Best Western, ◆ Comfort Inn, Hampton, ◆ Holiday Inn, Motel 6, Norwest
Other E: Car Wash

29 U.S. 70E, Madill, Ardmore

24 OK 77S
FStop E: Total*(D)
Food E: Buffalo Burgers
Parks E: Lake Murray State Park (3 Miles)
Other E: KOA Campground, Wilderness Way

21 Oswalt Road
RVCamp W: KOA Campground(LP)

15 OK 32, Marietta, Ryan
TStop E: Total*
Gas E: Sinclair
W: Texaco*
Food E: Denim's Restaurant, Hardee's, Robertson's Bisquits & Sandwiches, Sonic Drive In
W: Hickory House BBQ
AServ E: Love County Ford, Sinclair
TServ E: Advantage
RVCamp W: Grey Rock
ATM E: Bank of Love Co.
Parks E: Lake Texoma State Park
Other E: Harper's Western Store, Winn Dixie Supermarket

5 OK153, Thackerville
RVCamp W: Indian Nation RV Park

(4) Rest Area - RR, Phones, Picnic, HF (P) (Southbound)

(3) OK Info Center, Rest Area - RR, Phones, RV Dump (P) (Northbound)

1 U.S. 77N

↑OKLAHOMA

↓TEXAS

504 Frontage Road

(503) Rest Area - no facilities (P) (Northbound)

(502) TX Info Center, Rest Area - RR, Phones (P) (Southbound)

501 TX 1202
Gas W: Conoco*
Food W: Hilltop Cafe

500 Fm 372, Gainesville

EXIT TEXAS

TStop W: Texaco* (SCALES) (ATM)
FStop E: Diamond Shamrock* (Frontage road)
Food E: Restaurant (Diamond Shamrock Fuel Stop)
W: Harper's Restaurant
TServ W: Texaco Truck Stop
TWash E: Truck Wash
Other W: Outlet Mall

499 Frontage Road
Gas E: Exxon*, Little Red's
W: Exxon* (D)
Food E: Denny's
W: Peking Chinese Buffet
Lodg E: Budget Host Inn, Comfort Inn
W: Texas Motel

498B U.S. 82W, Wichita Fallls
Gas E: Citgo* (D), Conoco, Exxon*, Texaco*
W: Lone Star
Food E: Denny's, Waffle Inn
W: Peking Chinese
Lodg E: 12 Oaks Inn, Bed & Bath, Budget Host Inn, Comfort Inn
W: AAA Best Western, DAYS INN Days Inn, Texas Motel
AServ E: CBJ Tire & Alignment Firestone
W: Chrysler Auto Dealer
Med E: ✚ Hospital
Other E: Coin Laundry

498A U.S. 82E, Sherman (See Services On 498B)

497 Frontage Road
Gas E: Conoco*

496B TX 51, California St, Decatur
Gas E: Conoco* (D), Diamond Shamrock*, Mobil*
W: Texaco
Food E: Braum's, Burger King, Holiday Inn, McDonalds, Taco Bell, Taco Mayo, Wendy's
Lodg E: Holiday Inn
AServ E: Chrysler Auto Dealer, Oldmobile Dealer
W: Texaco
Other E: Scivally's Grocery (& Pharmacy)
W: Frank Buck Zoo

496A Weaver St, Gainsville (Northbound)

495 Frontage Road (Southbound)

494 TX1306

(492) Rest Area - RR, Phones, Picnic, Grills, Vending P (Southbound)

491 Spring Creek Road

(490) Rest Area - RR, Phones, Picnic, Grills, Vending P (Northbound)

488 Hockley Creek Road

487 TX 922, Valley View, Hood (Difficult Reaccess)
FStop W: Fina*
Food W: Dairy Queen, Suzy Q's
Other W: U.S. Post Office

486 Frontage Road
AServ W: Frank's Tire & Auto

485 Frontage Road (Southbound, Reaccess Southbound Only)

483 TX 3002, Lone Oak Road

482 Chisam Road

EXIT TEXAS

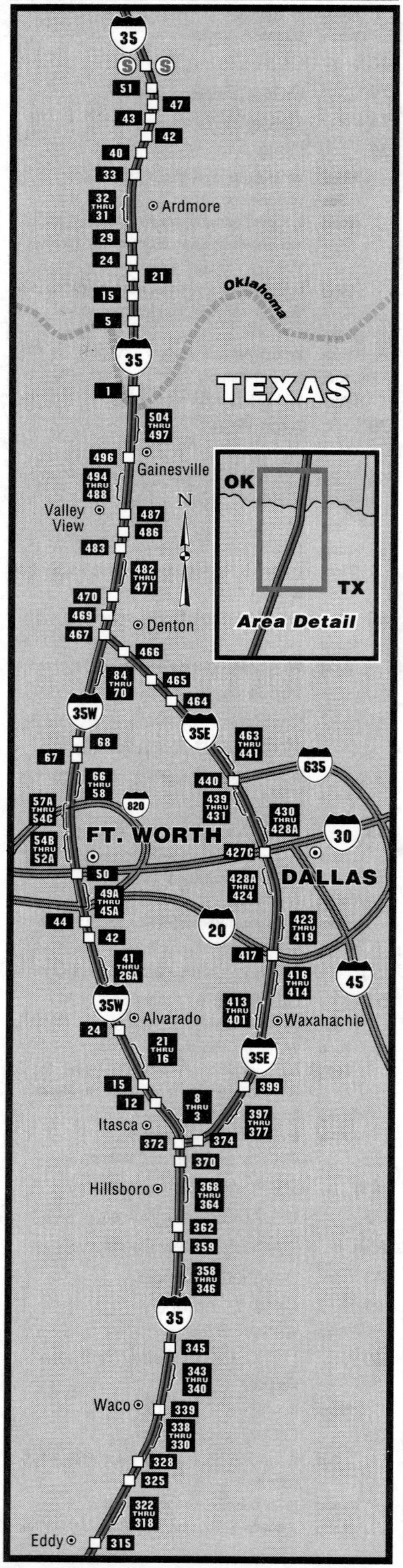

481 View Road

480 Lois Road

479 Belz Road
Lodg E: Sanger Inn
AServ E: Beard's Motors
RVCamp E: North Village Campground

478 Fm 455, Pilot Point, Bolivar
Gas W: Chevron* (D)
Food W: Subway
TServ E: Cowboy Truck Repair
ATM W: GNB Bank
Parks E: Ray Roberts State Park
Other E: Caty's Wash n Go Coin Laundry, Coin Car Wash
W: Bruss IGA Grocery Store, Post Office

477 Bus 35, Keeton Road
Gas E: Phillips 66*

475B Rector Road

475A TX 156, Krum (Southbound, Reaccess Southbound Only)

473 TX 3163, Milam Road
FStop E: Citgo, Love's Country Store*

472 Ganzer Road

471 U.S. 77, Fm1173, Loop 288, Denton, Krum
TStop W: Citgo Truckstop
FStop E: 76
W: Fina*
Gas W: First and Last* (Citgo)
Food E: 76
TServ E: 76
W: ARS Truck Repair (SCALES), Citgo Truckstop* (D)(SCALES)
TWash W: Citgo Truckstop
Med W: ✚ Hospital
Other E: Denton Factory Stores

470 Loop 288 (Northbound)
Food E: Good Eats
RVCamp W: Camping World (see our ad this page)
Other E: Denton Factory Store

469 U.S. 380, Decatur, McKinney
TStop W: Conoco* (D)
Gas E: Chevron* (ATM)
W: Diamond Shamrock*, Shell* (CW), Texaco*
Food E: Cracker Barrel, McDonalds
W: Dairy Queen, Denny's, Waffle House
Lodg W: Econolodge, Excel Inn, Motel 6
Med W: ✚ Hospital

468 Fm 1515, Airport Road, West Oak St (Southbound, Reaccess Northbound Only)

Note: I-35 runs concurrent below with I-35E. Numbering follows I-35E.

Bold red print shows RV & Bus parking available or nearby

EXIT TEXAS

467 Jct I-35W South, Fort Worth

466B Ave D
- **Gas** **E:** Citgo*, Exxon*, Phillips 66
- **Food** **E:** Burger King, IHOP, McDonalds, New York Subway, Poncho's Mexican Buffet, Winston's Ice Cream
  W: Outback Steakhouse
- **Lodg** **E:** Royal Hotel Suite
  W: Days Inn

466A McCormick St (One Way Streets)
- **Gas** **E:** Citgo*, Phillips 66*, Texaco*(D)(CW)
  W: Fina*
- **Food** **E:** IHOP, Pancho's Mexican Buffet
- **Lodg** **E:** Royal Inn & Suites

465B U.S. 377, Fort Worth Drive
- **Gas** **E:** Diamond Shamrock*
  W: Citgo, Conoco*(D), Phillips 66, Total*(D)
- **Food** **E:** Kettle Restaurant, Taco Bueno, What A Burger
  W: Frosty, MiRanchito, Mr. Pockets, Smoke House BBQ, Sonic Drive In
- **Lodg** **E:** La Quinta Inn
  W: Days Inn
- **AServ** **E:** Payne Tire Co.
- **ATM** **E:** Texas Bank

465A Fm 2181, Teasley Lane
- **Gas** **E:** Citgo* (ATM)
  W: Exxon*(LP), Fina*(D), Shell*, Texaco*
- **Food** **E:** Applebee's, Braum's Restaurant, Golden China, KFC, Little Caesars Pizza, Pizza Hut
  W: Olive 'O Branch Pizza, Red Peppers Chinese Buffet
- **Lodg** **W:** **Ramada Inn (see our ad this page)**, Super 8 Motel
- **AServ** **W:** Shell
- **Other** **E:** **Chief Auto Parts**
  W: Car Wash

464 Pennsylvania Drive, US77 Denton
- **Gas** **E:** 7-11 Convenience Store (Citgo)
- **Food** **E:** JJ Pizza
- **Lodg** **E:** Holiday Inn
- **AServ** **E:** NAPA Auto Parts, Pep Boys Auto Center
- **ATM** **E:** First State Bank

463 Loop 288
- **Gas** **W:** Mobil*(D)(CW)
- **Food** **E:** Arby's, Burger King, Cafe China, Ci Ci's Pizza, Colters BBQ, Grandys, Jason's Deli, Long John Silvers, Old Country Buffet, Taco Bell, Wendy's
  W: Black Eyed Pea, Chili's, Jack-In-The-Box, Lubby's Cafeteria, Red Lobster, Schlotzkys Deli, Tia's Texmex
- **AServ** **E:** Discount Tire, Goodyear Tire & Auto, Midas Muffler & Brakes
- **ATM** **E:** Bank One
- **Other** **E:** **Golden Triangle Mall**, Heritage Car Wash, **Kroger Supermarket, Target Department Store**

462 State School Road, Mayhill Road
- **Gas** **E:** Fina*
  W: Exxon*
- **Food** **W:** **Brier Inn**

461 Shady Shores Road, Post Oak Dr.
- **RVCamp** **E:** **McClain's RV SuperStore**

460 Corinth
- **Gas** **E:** Chevron*
- **Food** **W:** Cattle Company Steaks & Seafood
- **AServ** **E:** Chevron

EXIT TEXAS

- **RVCamp** **W:** **KOA Campground**

459 Frontage Road
- **RVCamp** **W:** **KOA Campground**

458 Fm 2181, Swisher Road, Lake Dallas, Hickory Creek
- **Gas** **E:** Overland*
- **Food** **E:** Arby's Express (In Overland)

457 Lake Dallas (Northbound)
- **Gas** **E:** Chevron*, Fast Brake*, Phillips 66*
- **Food** **E:** Ozzie's Donuts, Sonic Drive in, Subway

455 McGee Lane

454B Highland Village, Rest Stop, Park, Playground (Southbound)

454A Road 407, Justin
- **Gas** **E:** Citgo*, Diamond Shamrock
- **Food** **E:** Al's Chuck Wagon BBQ
- **RVCamp** **E:** **Maze's RV**

453 Valley Ridge Blvd
- **RVCamp** **E:** **Buddy Gregg's Motor Homes**

452 Fm 1171, Flower Mound
- **Gas** **E:** Mobil*(CW)
  W: Exxon*(D), Shell*
- **Food** **E:** Taco Bueno
  W: Baskin Robbins, Blimpie's Subs, Chick-fil-A, Church's Fried Chicken, Ci Ci's Pizza, Denny's, Golden Corral, Grandy's, Hardee's, McDonalds, Pizza Inn, Taco Bell, Taco John's, Wendy's, What A Burger
- **Lodg** **E:** Holiday Inn Express
- **AServ** **E:** Lewisville Quick Stop
  W: Atlas Transmission, Chief Auto Parts, KClinic
- **Med** **E:** **+ Hospital**
  W: **+ Prime Care Medical Clinic**
- **ATM** **E:** Bank One, First Federal Bank
- **Other** **W:** **Sam's Club, Wal-Mart**

451 Fox Ave
- **Gas** **E:** Texaco*(D)(CW)
  W: Chevron*(CW), Diamond Shamrock*
- **Food** **E:** Braum's Ice Cream
  W: Black Eyed Pea, **Cracker Barrel**, El Chico Mexican
- **Lodg** **W:** Hampton Inn
- **AServ** **W:** Atlas Transmission, Midas Muffler & Brakes

450 TX 121, Grapevine, McKinney
- **Gas** **E:** 7-11 Convenience Store, Citgo*, Racetrac*
  W: Chevron*(CW), Conoco*, Fina*, Texaco*(D)(CW)
- **Food** **E:** Bob Evans, Marshalls BBQ

EXIT TEXAS

  W: Burger King, Chili's, Church's Fried Chicken, Furr's Family Dining, Long John Silvers, Ming Garden Chinese, Roma Pasta House, Taco Bell, Tia's Tex-Mex, Waffle House
- **Lodg** **E:** Motel 6, Pines Motel, Ramada Inn
  W: J & J Motel, Spanish Trails Inn
- **AServ** **E:** Oliver's
  W: Firestone Tire & Auto, Kwik Kar Lube & Tune
- **Other** **W:** **Kroger Supermarket (& Pharmacy)**

449 Corporate Drive
- **Gas** **W:** Phillips 66*
- **Food** **E:** On the Border Cafe, Zane's
  W: Good Eats Grill, IHOP, Kettle, Outback Steakhouse, The Italian Oven
- **Lodg** **E:** Motel 6
  W: La Quinta Inn
- **AServ** **W:** Discount Tire Co.
- **Other** **W:** **Target Department Store**

448A Fm 3040, Round Grove Road, Hebron Pkwy
- **Gas** **W:** Exxon*(CW)
- **Food** **E:** Bennigan's, Chuck E. Cheese's Pizza, Olive Garden, Saltgrass Steakhouse, Souper Salad
  W: Applebee's, Carino's, Chick-fil-A, Cotton Patch Cafe, Einstein Bagels, IHOP, Jason's Deli, McDonalds, Schlotzkys Deli, Spring Creek BBQ, TCBY, Taco Cabana, Wendy's
- **Lodg** **E:** Homewood Suites Hotel
- **AServ** **E:** Auto Express
  W: Discount Tire Company
- **Med** **E:** **+ Metro Medical Associates**
- **ATM** **W:** Compass Bank
- **Other** **W:** **Border's Books, Target Department Store**

448B TX553 (Southbound)
- **Food** **W:** Don Pablos Mexican, Red Lobster, T.G.I. Friday's
- **Lodg** **W:** Comfort Inn, Country Inn Suites

446 Frankford Rd., Trinity Mills Rd.
- **RVCamp** **E:** **North Dallas RV Sales & Service**

445 Trinity Mills Road

444 Whitlock Lane, Sandy Lake Road
- **Gas** **E:** Texaco*(CW)
  W: Chevron*(CW)
- **Food** **E:** Taco Bell
  W: McDonalds
- **Lodg** **W:** Deluxe Inn
- **AServ** **E:** Stanley's Garage
- **TServ** **E:** **Stanley's Garage**
- **RVCamp** **W:** **Camping**

Note: I-35 runs concurrent above with I-35E. Numbering follows I-35E.

443B Belt Line Road, Crosby Road
- **Gas** **E:** Fina*, Racetrac*, Shell*(CW)
  W: Texaco
- **Food** **E:** Blimpie's Subs, Mom's Kitchen, The Bakery Sandwich Shop
  W: The Pocket Place Sandwich Shop
- **AServ** **E:** Old Town Automotive
  W: B&B Radiator & Muffler, Big A Auto Parts, Texaco
- **Other** **E:** **Animal Medical Center**

Note: I-35 runs concurrent below with I-35E. Numbering follows I-35E.

EXIT TEXAS

443A Crosby Road

442 Valwood Parkway
- **Gas** E: Highway Oil*, Superfuels, Texaco*(D)
  W: Fina*
- **Food** E: Dairy Queen, Denny's, El Chico Mexican, Elena's Tortillas, Grandy's, Jack-In-The-Box, Red Line Burgers, Waffle House
- **Lodg** E: Carrolton Inn, Comfort Inn, Guest Inn
- **AServ** E: Hill Tire Co. (Firestone), Texaco
  W: American Transmisions, Top Lube

441 Valley View Lane
- **Gas** E: Mobil*(CW)
  W: Exxon*
- **Food** E: The China Restaurant
  W: JoJo's Bakery and Restaurant
- **Lodg** W: Best Western, Econolodge (see our ad this page), La Quinta Inn
- **AServ** E: Advanced Auto Repair, Bob Hackler Transmission
- **ATM** W: Bank One

440BC Junction I-635 East & West
- **Lodg** E: Amerisuites (see our ad this page)

439 Royal Lane
- **Gas** E: Fina*, Texaco*
  W: Chevron*(CW)
- **Food** E: McDonalds, New Palace Chinese (At Fina), Royal Wok Chinese (At Fina), Wendy's
  W: Royal BBQ, The Family Ocean Seafood
- **AServ** E: Ken's Muffler Shop, Quaker State Lube, Tune Up Masters

438 Walnut Hill Lane
- **Gas** E: Mobil*(CW) (ATM), Shell*, Texaco*(CW)
- **Food** E: Bennigan's, Burger King, Denny's, Old San Francisco Steak House, Paul's Porterhouse Steaks and Seafood, Red Lobster, Steak & Ale, Taco Bell, Tony Roma's Ribs, Trail Dust Steak House
- **Lodg** E: Drury Inn, Hampton Inn
- **AServ** E: Midas Muffler & Brakes, Shell
- **RVCamp** W: **Morgan RVs**
- **Other** W: **Amusement Park**

436 Loop 12, Spur 348
- **Lodg** E: Days Inn (see our ad this page)

434B Regal Row
- **Gas** E: Chevron*(CW)
- **Food** E: Denny's, What A Burger
- **Lodg** E: La Quinta Inn
  W: Fairfield Inn Marriott

434A Empire Central
- **Gas** E: Shell*(CW)
  W: Exxon*
- **Food** E: McDonalds, Wendy's
- **AServ** W: Exxon

433B Mocking Bird Lane, Dallas, Love Field Airport
- **Gas** E: Mobil*
- **Food** E: Jack-In-The-Box
  W: Church's Fried Chicken
- **Lodg** E: Clarion Motel, Holiday Inn (see our ad this page)

432B Road 356, Commonwealth Dr.
- **Gas** E: Texaco*(CW)
- **Food** W: Villas Dal-Mex
- **Lodg** E: Harvey Hotel, Residence Inn Marriot
  W: Delux Inn

EXIT TEXAS

432A Inwod Road
- **Gas** E: Exxon*(CW)
  W: Fina*, Texaco*
- **Food** W: Minfa's Mexican Food
- **AServ** W: Texaco
- **Med** E: ✚ **South Western Medical Center**
- **Other** W: Alladin Car Wash

EXIT TEXAS

431 Motor St
- **Gas** E: Mobil*(CW)
  W: Shell*(CW)
- **Food** E: JoJo's
  W: Stemons Wok
- **Lodg** W: Embassy Suites
- **Med** E: ✚ **Hospital**

430C Wycliff Ave.
- **Lodg** E: Renaissance Hotel

430B Market Center Blvd
- **Food** W: Denny's, Georgeo's (Best Western)
- **Lodg** W: Best Western, Courtyard by Marriott, Fairfield Inn, Holiday Inn, Quality Hotel, Ramada, Sheraton Suites, Wyndham Garden Hotel

430A Oak Lawn Ave
- **Gas** W: Texaco(D)
- **Food** W: Medieval Times (Food & Entertainment)
- **AServ** W: Texaco

429D Tollway North (Northbound, No Reaccess)

429C Hi Line Drive (Northbound, Difficult Reaccess)

429B To I-45, US 75 Sherman

429A Continental Ave, Commerce St
- **Gas** W: Exxon*(D), Shell*(CW)
- **Food** E: Planet Hollywood
  W: Burger King, Clover Restauarant, McDonalds, Popeye's Chicken
- **Lodg** E: Amerisuites (see our ad this page)

428 I-30W, Ft.Worth, I-30E Texarkana, Commerce Street, Reunion

428B Junction I-45

427A Colorado Blvd

427C Junction I-30

427B Industrial Blvd

426C Jefferson Ave

426B East 8th St, TX 180W
- **Gas** E: Texaco*(D)
  W: Texaco*
- **Food** W: James Grill
- **Lodg** W: La Santa
- **AServ** W: USA Tire & Incorp.

426A Ewing Ave
- **Food** E: Roy's Place
- **Lodg** E: Courtsey Inn
  W: Circle Inn
- **AServ** E: State Radiator & Air Conditioning Service
  W: Good Taylor Honda, Pontiac, Trucks
- **Parks** E: **Dallas Zoo**

425C Marsalis Ave, Ewing Ave
- **Gas** E: Chevron* (24 Hrs)
  W: Diamond Shamrock*, Mobil*
- **Lodg** E: Dallas Inn Motel
- **AServ** W: Mobil
- **Parks** E: **Dallas Zoo**

425B Beckley Ave, 12th St

Bold red print shows RV & Bus parking available or nearby

EXIT TEXAS

425A Zang Blvd., Beckley Ave
- **Gas** **W:** Chevron (24 Hrs), Exxon*
- **ATM** **W:** Nations Bank

424 Illinois Ave
- **Gas** **E:** Chevron(CW) (24 Hrs)
  W: Exxon*
- **Food** **E:** Dairy Queen, Theo's Smokehouse BBQ, Williams Fried Chicken
  W: Church's Fried Chicken, IHOP, Jack-In-The-Box, Manna Donut, Red Line Burgers, Taco Bell
- **Lodg** **W:** Oak Tree Inn
- **AServ** **E:** Bud and Ben Mufflers
  W: Exxon, Factory Brake Centers, Just Brake's, Midas Muffler & Brakes
- **ATM** **W:** Bank One
- **Other** **E:** Coin Car Wash

423B Saner Ave
- **AServ** **W:** Express Tire
- **Other** **W:** **Optical Clinic**

423 U.S. 67, Cleburne

422BA Beckley Ave, Overton Rd., Kiest
- **Gas** **W:** Shell*
- **Food** **W:** McDonalds
- **Lodg** **E:** Interstate Motel
  W: Dallas Inn
- **ATM** **W:** Main Bank

421AB Junction Loop 12, Ann Arbor Ave.
- **Gas** **E:** Racetrac*, Texaco*(D)
- **Food** **W:** Little Bob's BBQ
- **Other** **W:** Coin Car Wash

421 Ann Arbor Ave

421A Ann Arbor Ave, RD 12 Loop

420 Laureland Road
- **Gas** **E:** Fina*
  W: Fina* (ATM), Mobil*
- **Lodg** **E:** Crawford Motel, Master's Suite Motel
  W: Embassy Motel, Holiday Motel

419 Camp Wisdom Road
- **Gas** **E:** Exxon*
  W: Chevron*(CW), Shell*(CW)
- **Food** **W:** McDonalds
- **Lodg** **E:** Oak Cliff Inn
  W: Sun Crest Inn

418 Junction I-20, Fort Worth, Shrieveport

417 Wheatland Road, Danieldale Road

416 Wintergreen Road
- **Gas** **W:** Citgo*
- **Food** **W:** **Cracker Barrel**
- **Lodg** **E:** Royal Inn
  W: Holiday Inn, Red Roof Inn

415 Pleasant Run Road
- **Gas** **E:** Chevron*(CW), Racetrac*, Texaco*(D)
  W: Exxon*, Mobil*(CW)
- **Food** **E:** Benavides Restaurant, Blimpie's Subs (With Texaco), Chow Line (With Texaco), Chubby's, Grandy's, Subway, Waffle House
  W: Burger King (Playground), El Chico Mexican, Golden Corral, KFC, Long John Silvers, Lubby's Cafeteria, McDonalds (Playground), Schlotzkys Deli, Taco Bueno, Wendy's
- **Lodg** **E:** Comfort Inn, Great Western Inn, Royal Inn, Spanish Trails Motel
  W: Best Western
- **AServ** **E:** Midas Muffler & Brakes, Napa Auto Parts, Wal-Mart
  W: Firestone Tire & Auto
- **ATM** **E:** Racetrac
- **Other** **E:** **Wal-Mart (Tire Service, Optical)**

EXIT TEXAS

- **W:** **Doctor's First Medical Care, K-Mart, Kroger Supermarket, The Crossing Shopping Mall**

414 Road 1382, De Soto, Belt Line Road
- **Gas** **W:** Conoco*, Texaco*(D)
- **Food** **E:** What A Burger (24 Hrs)
  W: Joe's Pizza
- **AServ** **E:** Wal-Mart
  W: Texaco
- **Other** **E:** **Wal-Mart**

413 Parkerville Road
- **Gas** **E:** Fina
  W: Total*
- **Food** **W:** Taco Bell (With Total)
- **AServ** **W:** SWS Trucks & Sales Service
- **TServ** **W:** **SWS Truck Service**

412 Bear Creek Road , Glen Heights
- **FStop** **W:** **Texaco***
- **Food** **W:** Dairy Queen (Texaco)

411 FM 664, Ovilla Road
- **Gas** **E:** Exxon*(D)(CW)
  W: Exxon* (ATM), Phillips 66*, Texaco*(D)
- **Food** **E:** Church's Fried Chicken (Exxon), McDonalds (Playground), TCBY Yogurt (Exxon), Taco Bell (Exxon)
  W: Smash Hit Subs (Exxon)
- **ATM** **E:** Main Bank
- **Other** **E:** **Eckerd Drugs**

410 Red Oak Road
- **FStop** **E:** **Coastal***
- **Food** **E:** **Baskin Robbins (Coastal)**, Denny's, **Subway (Coastal)**
- **Lodg** **E:** Days Inn
- **RVCamp** **W:** **Hilltop Travel Trailors**

408 TX 342, Red Oak

406 Sterrett Road
- **TServ** **E:** **Frontera Truck Parts**

405 Road 387
- **FStop** **E:** **Prime Travel Stop***(D)

404 Lofland Road

403 U.S. 287, Corsicana, Fort Worth
- **AServ** **E:** GM Auto Dealership
  W: Buick Pontiac GMC Trucks, Thornbill Ford, Mercury

401B U.S. 287 Business, Road 664, Waxahachie
- **Food** **E:** Restaurant
- **Lodg** **E:** Comfort Inn
- **Med** **E:** **✚ Hospital**

401A Brookside Road
- **Lodg** **E:** Ramada Limited, TraveLodge

399B Road 1446

399A Road 66, 876, CR 1446
- **Gas** **E:** Diamond Shamrock*
  W: Fina*, Texaco*(D)
- **Lodg** **E:** Texas Inn Motel
- **AServ** **E:** Diamond Shamrock, Total Automotive

397 U.S. 77, Waxahachie

(393) **Rest Area - RR, Phones, Vending, Picnic, Grills P (Both Directions)**

391 Road 329, Forreston Road

386 TX 34, Italy, Ennis
- **FStop** **E:** **Texaco***
- **Food** **E:** **Blimpie's Subs (With Texaco)**, Dairy Queen

EXIT TEXAS

- **AServ** **E:** Chris'v&v Auto Service

384 County Road

381 Road 566, Milford Road

377 Road 934

374 Road 2959, Carl's Corner
- **FStop** **W:** **Fina***

Note: I-35 runs concurrent above with I-35E. Numbering follows I-35E.

Note: I-35 runs concurrent below with I-35W. Numbering follows I-35W.

85A Jct I-35, Dallas/Ft. Worth

84 TX 1515, Bonnie Brae St

82 TX 2449, Ponder

79 Crawford Road

76 TX 407, Argyle, Justin
- **Gas** **W:** Citgo*(D)

(75) **Rest Area - Picnic P (Both Directions)**

74 Fm1171, Lewisville, Flower Mound

70 TX114, Dallas, Bridgeport
- **Other** **W:** **Texas Motor Speedway**

68 Eagle Pkwy

67 Alliance Blvd
- **Gas** **W:** Mobil
- **Food** **W:** Metro Bagel, Wendy's
- **Other** **W:** **Fort Worth Alliance Airport**

66 Haslet, Westport Pkwy
- **Gas** **W:** Mobil*(D)
- **Food** **W:** Subway, Wendy's

65 TX170E

64 Golden Triangle Blvd, Heller Hicks Road

63 Park Glen Blvd.

60 Jct U.S. 287N, U.S. 81N, Decatur

59 Jct I-35W North, Denton, US 287

58 Western Center Blvd
- **Gas** **E:** Citgo*
- **Food** **E:** Domino's Pizza

57A Jct I-820E

56B Jct I-820W, Melody Hills Dr

56A Meacham Blvd
- **Gas** **W:** Mobil*(CW)
- **Food** **W:** **Cracker Barrel**, Holiday Inn, McDonalds
- **Lodg** **E:** Best Western, Comfort Inn, La Quinta Inn
  W: Hampton Inn, Holiday Inn
- **ATM** **E:** Mercantile Bank of Ft. Worth

54C 33rd St, Long Ave
- **FStop** **W:** **Circle K***(D)
- **Gas** **W:** Circle K*
- **Lodg** **W:** Motel 6
- **AServ** **E:** Goodyear Tire & Auto
  W: Dick Smith Auto Parts
- **TServ** **E:** **Cummins Diesel**
  W: **Fort Worth Gear and Axle, ThermoKing, WF Truck and Trailor Sales**
- **Med** **E:** **✚ Puelma Medical Clinic**
- **Other** **E:** **Discount Auto Parts Exchange**
  W: **Haygood Truck & Trailer Park**

54B TX183W, Papurt St

EXIT TEXAS

TStop W: Circle K Truck Stop*(D)
Lodg W: Motel 6
TServ E: Bridgestone Tire & Auto
W: Red Oval Truck Sales & Service

54A TX183E, NE 28th St, Papurt St.
FStop W: Mid Continent Truck Plaza
Gas E: 7-11 Convenience Store*, Phillips 66*, Texaco(D)
W: Diamond Shamrock* (ATM), Quik Stop* (ATM)
Food E: Dairy Queen
W: Big Boss's, Pink Poodle Coffee Shop
Lodg W: Motel Classic Inn
TServ E: Bridgestone Tire & Auto, Treadco Inc
W: Mid Continent Truck & Auto
ATM W: The National Bank of Texas

53 Northside Dr, Yucca Ave
Gas E: Texaco*(D)
Other W: Camping

52D Carver St

52C Pharr St
TServ W: Freightliner Dealer (24 Hrs), Southwest International, CAT,Cummings

52B Spur347W, Belknap St

52A U.S. 377N, U.S. 121N, Belknap St

50 Jct I-30, U.S. 287S, U.S. 377S

49A Allen St., Rosedale St (Difficult Reaccess)
Food W: Drake's Cafeteria, Little John's BBQ
Med W: Hospital

49B Rosedale St
Food E: Frank's BBQ, Little John's BBQ
AServ E: Low Cost Auto Tire Center
W: George's Automotive

48B Morning Side Dr

48A Berry St
Gas E: Chevron*, Citgo*
W: Racetrac*
Food E: McDonalds
AServ E: Economy Tires

47 Ripy St
Lodg W: Astro Inn
AServ E: AAMCO Transmission
W: A-1 Automotive

46B Seminary Drive
Gas E: Racetrac*
W: Diamond Shamrock*, Texaco*(D)(CW)
Food E: Grandy's, Jack-In-The-Box, Long John Silvers, Sonic Drive In, Taco Bell, What A Burger (24 Hrs)
W: Arby's, Denny's, IHOP, Thrifty Bake Store, Wendy's
Lodg E: Days Inn, Delux Inn, Motel 6
AServ W: Firestone Tire & Auto, Pep Boys Auto Center, Sears Auto Center
ATM E: Racetrac
W: Nations Bank
Other W: Optical Clinic, Town Center Mall

46A Felix St
Gas E: Mobil*
Food E: Pulido's Mexican
W: Burger King, McDonalds
Lodg E: Southoaks Motel
AServ W: Dodge Dealer
Other W: Kroger Supermarket, Post Office

45B Junction I-20 East, Abilene, Dallas

45A Junction I-20 West

EXIT TEXAS

44 Altamesa Blvd
Gas W: Citgo*
Food W: The Rig Steak House, Waffle House
Lodg E: Holiday Inn
W: Ramada Limited
ATM E: Fidelity Bank & Trust

43 Sycamore School Rd
Gas W: Exxon*(CW)
Food W: Beefer's Breakfast & Burgers (24 Hrs), Joe's Pizza, Subway, Taco Bell, What A Burger (24 Hrs)

42 Everman Parkway
Gas E: Exxon*(CW)
W: Chevron*(CW) (24 Hrs)
ATM E: Bank One

41 Risinger Road
RVCamp W: Morgan's RV

40 Garden Acres Drive
FStop E: Love's Country Store*
RVCamp W: C&S RV World

39 TX 1187, McAllister Road, Rendon - Crowley Road
Gas W: Citgo*(D)(CW) (24 Hrs), Diamond Shamrock*(D)
Food W: Taco Bell
Med E: Huguley Hospital

38 Alsbury Blvd
Gas E: Chevron*(D)(CW)
W: Texaco*(D)(CW)
Food E: Cracker Barrel, JB's BBQ, McDonalds (Playground), Old Country Steak House
W: Arby's, Burger King, Denny's, Grandy's, Pancho's
AServ W: Kwik Kar Lube and Tune
ATM W: First National Bank
Other W: K-Mart

37 TX 174, Cleburne, Wilshire Blvd

36 Spur 50, Renfroe St, FM 3391
Gas E: Citgo*(D), Mobil*(CW)
W: Diamond Shamrock*, Fina*(D), Texaco*
Food E: Waffle House
ATM E: Burleson State Bank

(34) Rest Area - RR, Phones, Picnic, Grills P (Southbound)

34 Briar Oaks Rd
RVCamp W: Camping

32 Bethesda Road
Gas E: Citgo*
Lodg E: Five Star Inn
RVCamp W: Country Junction RV Park

(31) Rest Area - RR, Phones, Picnic, Grills P (Northbound)

30 TX 917, Joshua, Mansfield
Gas W: Citgo*(D), Texaco*
RVCamp E: Capri Campers & RV
Other W: Discount Auto Salvage Exchange

27 CR707/604
RVCamp E: Anacira Pace, Arrow

26A U.S. 67, Cleburne, Dallas
Gas E: Phillips 66*(D)
Food E: Pop's Honey Fried Chicken

26B Bus 35 W, Alvarado

24 U.S. 35 West, TX 1706, 3136, Alvarado, Maple Ave
Gas E: Chevron*

EXIT TEXAS

Food E: Alvarado House

21 Greensfield, Barnesville Road

17 TX 2258
RVCamp E: Dotsco Parts & Service

16 TX 81 S, Grandview
RVCamp W: Camping

15 TX 916, Grandview, Maypearl
Gas W: Diamond Shamrock*(D)(CW)
Food W: Rick's Texas BBQ

12 TX 67

(8) Rest Area - RR, Phones, Picnic, Grills P (Southbound)

8 TX 66, Itasca
FStop E: Fina*(D)
Gas W: Citgo*(D) (ATM)
Food W: Dairy Queen

(7) Rest Area - RR, Phones, Picnic, Grills P (Northbound)

7 TX 934
Other E: Picnic Area (Northbound)

3 TX 2959, Hillsboro Airport

Note: I-35 runs concurrent above with I-35W. Numbering follows I-35W.

371 Junction I-35E Dallas, I-35W Ft. Worth (Northbound I-35W Is A Left Exit)

370 North U.S. 77 , Spur 579

368B FM 286 (Southbound)
Gas W: Chevron*, Diamond Shamrock*(D)(CW)
Food W: Dairy Queen, El Conquistador, Pizza Hut
Lodg W: Best Western, Comfort Inn
ATM W: Chevron

368A TX 22, TX 171, FM 286, Whitney, Corsicana
FStop E: Love's*(SCALES) (Citgo)
W: Fina*(D)
Gas W: Citgo*, Exxon*, Mobil*(CW)
Food E: Arby's, Burger King, Golden Corral, Grandy's, McDonalds (Playground), Roze's Cafe (Ramada Inn), Taco Bell (Love's)
W: KFC, Restaurant (Thunderbird Motel), Schlotzkys Deli, What A Burger
Lodg E: Holiday Inn Express, ◆ Ramada Inn
W: Thunderbird Motel
AServ W: Dobbs & Co. Pontiac, Exxon, Westside Motors Chevrolet
Med W: Hospital
ATM W: Fina, Norwest Bank
Other E: Hillsboro Outlet Center
W: Wal-Mart (Pharmacy)

367 FM 3267, Old Bynum Rd

364B TX 81, Hillsboro (Northbound Is Left Exit)

364A FM 310
FStop W: Knox Coastal*(LP)(SCALES)
Gas E: Texaco*
ATM W: Knox Coastal

362 Chatt Road

359 FM 1304
FStop W: Mobil*
Food W: Arby's (Mobil, 24 Hrs)

Bold red print shows RV & Bus parking available or nearby

EXIT TEXAS

TServ W: Mobil (24 Hrs)
ATM W: Mobil

358 FM 1242 East, Abbott
Gas E: Exxon*
Food E: Turkey Shop Cafeteria
TWash W: Abbott Truck Wash

356 Abest Road

355 County Line Road
RVCamp E: KOA Kampground
Other E: RV & Car Wash

354 Marable St
RVCamp E: KOA Kampground
Med E: Hospital
Other E: West Gas Service(LP)

353 FM 2114, West
Gas E: Circle K*, Citgo*, Fina*, Shell*
W: Citgo*(D), Exxon
Food E: Dairy Queen, Jaime's Restaurant, Little Czech Bakery, Old Czech Smoke House, Pizza House
AServ E: Sykora Ford Dealership
W: Alvin's Body Shop & Supply, Exxon, Gerrel Bolton Geo Chevrolet, Goodyear Tire & Auto (Citgo)
Med E: Hospital
ATM E: Shell

351 FM 1858
Food E: Cafe Magaritas

349 Wiggins Road

347 FM 3149, Tours Road
AServ W: Auto & Truck Repair Emergency Service
TServ W: Auto & Truck Repair Emergency Service

346 Ross Road
TStop E: Exxon* (CB Shop, 24 Hrs)
W: Fina (Showers)
Food E: Chicken Express (Exxon TS), Dogs N Suds Express (Exxon TS)
W: Restaurant (Fina TS, 24 Hrs)
Lodg W: Motel (Fina TS)
AServ W: EZ Pickens Auto Ranch Inc.
TServ W: Fina TS, M and M Tire Sales, Pickens Truck Sales & Parts
TWash W: Fina Truck Wash
ATM E: Exxon TS
W: Fina TS

(345) Picnic - Shelters (Northbound)

345 Old Dallas Road
TServ E: Texas Truck Body Works
W: Tate's Truck Repair
Other E: Heart of Texas Speedway

343 FM 308, Elm Mott
TStop E: Texaco*(SCALES)
Gas E: Fina*(D)
W: Chevron*(D)
Food E: Charlie's BBQ House, Country Cafe (Texaco TS), Dairy Queen (Playground), Eddie Ray's Smokehouse
W: Heitmiller Family Steakhouse
AServ E: Rogers Automotive
TServ E: Texaco TS
TWash E: Texaco TS
ATM E: Texaco TS
Other W: Elm Mott Grocery, US Post Office

342B Bus. South U.S. 77

342A FM 2417, Northcrest
FStop W: Diamond Shamrock*

EXIT TEXAS

Gas E: Citgo*
Food E: Wimpy's Hamburgers
W: Dairy Queen
Lodg W: Every Day Inn
AServ W: Chuck's Automotive Repair, Chuck's Muffler & Exhaust
Other E: TX State Technical College - Waco, Texas Department Of Public Safety
W: Wash-N-Dry Laundromat (Diamond Shamrock)

341 Craven Ave, Lacy, Lakeview
Gas E: Chevron*
W: Shell*
AServ W: Chief Auto Parts
ATM W: Shell

340 Meyers Lane (Northbound)

339 FM 3051 To S. TX 6, Loop 340, Lake Waco, Lakeshore Dr
FStop E: Texaco*(D)
Gas E: Diamond Shamrock*(D)
Food E: Casa Ole, El Conquistador, Luby's Cafeteria, Pizza Hut, Shipley Do-nuts, Sonic, Wendy's, What A Burger (24 Hrs)
W: Burger King, Cracker Barrel, KFC, McDonalds (Playground)
AServ E: Kwik Kar Oil & Lube, Wal-Mart
W: Carquest Auto Parts
ATM E: American Bank, Texaco
W: Winn Dixie Supermarket
Other E: Genie Car Wash, Pharmacy Plus, US Post Office, Wal-Mart Supercenter (Vision Center, 1-Hr Photo, Pharmacy, 24 Hrs)
W: Brown's Hardware, Cinema, Winn Dixie Supermarket

338B Behrens Circle, Bell Mead
FStop W: Texaco*(D)(LP)
Gas E: Gas Station*
Food W: Blue Bonnet Cafe (24 Hrs), Neighbor's Restaurant
Lodg W: Days Inn, Delta Inns, Knights Inn, Motel 6, Royal Inn
AServ E: Superior Car Care
Other E: Mrs. Baird's Thrift Store
W: Eckerd Drugs, Food Basket IGA

338A U.S. 84, To TX 31, Waco Drive, Bellmead
Gas E: Fina*, H-E-B
W: Texaco*
AServ E: Auto Zone Auto Parts, Pickens Auto Parts, Sam's Club
ATM E: Fina
Other E: H-E-B Grocery Store (Pharmacy, 1-Hr Photo), Lynn's La Vega Pharmacy, Sam's Club

337B Business North U.S. 77, Elm Ave
Gas W: Texaco*
AServ W: Bill's Discount Tire Service, Busy Bees Auto Care Center

337A Business South U.S. 77

336 Forrest St

335C MLK Jr. Blvd (Lake Brazos Pkwy) (No Southbound Reaccess)
Gas E: Texaco*(D)(LP)
Food E: Mickey's (Howard Johnson), Paquito's Cantina & Grill
W: Dock's River Front
Lodg E: ◆ Holiday Inn (see our ad this page), Howard Johnson
W: Econolodge, Victorian Inns
Other E: Brazos Golf Range
W: Cameron Park Zoo, The Waco Suspension Bridge

EXIT TEXAS

335B FM 434, Fort Fischer, University Parks Drive
Food E: Baskin Robbins, Eggroll House, Grind 'N Stone Coffee & Gifts, Shoney's, Thai Orchid
W: Arby's, Jack-in-the-Box, Tanglewood Farms
Lodg W: ◆ Lexington Inn
RVCamp E: Fort Fischer Park
ATM W: Bank of America
Other E: Convenient Food Mart, Gov. Bill & Vera Daniels Village, Texas Ranger Hall of Fame, The Ferrell Center (Sports Arena for U. of Baylor), The Texas Sports Hall of Fame, The University of Baylor Info Center, Tourist Information
W: Fire Dept., Waco Convention Center

335A 4th - 5th Sts.
Gas E: Exxon*(D), Texaco*(D)
W: Diamond Shamrock*(D), Shell*(CW)
Food E: Cafe China Super Buffet, IHOP, Subway (Exxon), TCBY (Exxon), Taco Bueno
W: Gold-N-Crisp Chicken, Long John Silvers, McDonalds (Playground), Sonic, Taco Bell, Taco Cabana (24 Hrs), The Pineapple Grill (Clarion), Wendy's, What A Burger (24 Hrs)
Lodg E: AAA Best Western
W: Clarion
AServ E: Exxon
ATM E: Exxon, Texaco
W: Shell
Other E: The University of Baylor
W: The Dr. Pepper Museum

334B 8th St (Under Construction)
Food E: Asian Restaurant, Denny's, Pizza Hut
Lodg E: La Quinta Inn
ATM E: Bank of America
Other E: Rother's Bookstore, The University of Baylor

334A 18th - 19th Streets (Under Construction)
Gas E: Chevron*, Shell*, Texaco*(D)
Food E: Burger King (Playground), La Jaivita, Schlotzkys Deli
Lodg E: Budget Inn, Comfort Inn, Super 8 Motel
Other E: Harley Davidson Motorcycles

334 US 77 South, 17th - 18th Sts.
Gas W: Chevron*, Phillips 66*(D), Texaco
Food W: Dairy Queen, George's Steaks, Seafood, BBQ,

EXIT TEXAS

Mexico Lindo

AServ **W:** Makowski Automotive, Texaco

Med **W:** ✚ Hillcrest Baptist Medical Center

Other **W:** Startex Propane(LP)

333A Loop 396, La Salle Ave, Valley Mills Dr (Southbound Only Due To Construction)

Gas **W:** Racetrac* (24 Hrs)

Food **W:** Gold-N-Crisp Chicken, Mexican Chicken In A Tortilla Factory, Papa John's Pizza

Lodg **W:** Mardi Gras Motel

AServ **W:** Cen-Tex Brake & Spring, Jenkins Radiator Shop, Marsteller Mercury Lincoln Ford Dodge Jeep Eagle, Precision Tune

TServ **W:** Driveline Shop, Duncan Truck Sales (Freightliner, 24 Hrs)

331 New Road (Southbound Only Due To Construction)

Lodg **E:** Best 4 Less Motel, New Road Inn, Relax Inn

Med **W:** ✚ VA Hospital

Other **E:** Central Texas International Inc.

W: Cottonwood Country Club Golf Course, Heart of Texas Fairgrounds

330 TX 6, Loop 340, Meridian, Robinson (Limited Access Hwy)

328 FM 2063, FM 2113, Hewitt

Gas **W:** Diamond Shamrock*(D), Texaco*(CW)

ATM **W:** Diamond Shamrock

325 FM 3148, Robinson Road

TStop **W:** Exxon* (24 Hrs, Showers)

Food **W:** Pizza Inn (Exxon TS), RBD's (Exxon TS, 24 Hrs), Subway (Exxon TS), TCBY (Exxon TS)

TServ **W:** US Tire Co. (24 Hr. Road Service)

ATM **W:** Exxon TS

323 FM 2837 (Southbound)

Food **W:** Chicken Express, Coyote Ranch Cafe

Other **W:** US Post Office

322 FM 2837, Lorena

Gas **E:** Diamond Shamrock*

W: Phillips 66*(D)

Food **E:** Miller's Steakhouse

319 Woodlawn Road

(318) Picnic - Shelters, Grills (Southbound)

318B Bruceville, Picnic, Camping - Shelters, Grills (Services Are At Southbound Exit, Picnic At Northbound)

Gas **W:** Fina*(D)

RVCamp **W:** KOA Kampground (27 Miles)

Other **W:** Coin Laundry

318A Bruceville, Frontage Rd.

Gas **E:** Chevron

TServ **W:** Joe KinCannon Truck Tire Retreading

315 TX 7, FM 107, Moody Marlin

FStop **W:** Texaco*

Gas **W:** Fina, Shell*

Food **W:** I Can't Believe It's Yogurt (Texaco), Lil' Orky's Cafe, Pizza Hut Express (Texaco), Red Line Burgers (Texaco)

AServ **W:** Eddy Engine Repair, Fina

ATM **W:** Shell, Texaco

Parks **W:** Mother Neff State Park

Other **W:** Pelzel Foods

314 Old Blevins Road

EXIT TEXAS

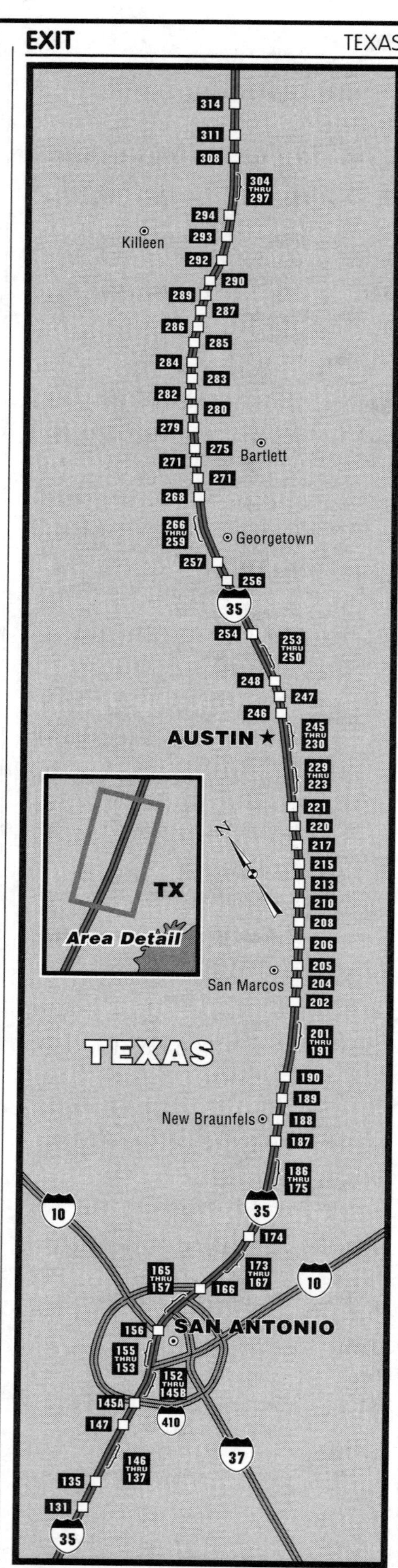

EXIT TEXAS

311 Big Elm Road

308 FM 935, Troy

Gas **E:** Texaco

W: Diamond Shamrock*, Exxon*

Food **E:** Phelan's Restaurant

W: Hot Stuff Pizzeria (Exxon)

AServ **E:** Texaco

Other **E:** Coin Operated Laundry, Payne Grocery Market

306 FM 1237, Pendleton

FStop **E:** Love's* (Citgo Fuel, RV Dump)

Food **E:** A & W Drive-In (Love's TS), Subway (Love's TS)

ATM **E:** Love's

305 Berger Road

TStop **W:** Willie's 305 Truck Stop(SCALES) (Fina)

FStop **E:** Phillips 66*

TServ **E:** Gray International

W: Goodyear Tire & Auto, Temple Diesel Service (24 Hrs)

TWash **W:** Willie's Truck Wash (Willie's 305 TS)

RVCamp **W:** Camping

304 Loop 363

Gas **W:** Diamond Shamrock*(D)

TServ **W:** Temple Freightliner

303B Mayborn Civic & Convention Center (Northbound)

Other **E:** Cultural Activities Center, Frank W. Mayborn Civic & Convention Center

303A Spur 290, FM 1143, Industrial Blvd, North 3rd St (Same As Exit 303 Southbound)

Gas **E:** Citgo*, Diamond Shamrock*

Food **E:** Why Poncho's?

Lodg **E:** Texas Sands Motel

W: Continental Motel

ATM **E:** Cash Card, Citgo

302 Nugent Ave

FStop **W:** Mobil

Gas **E:** Exxon*(D), Texaco*(D)

W: Chevron*

Food **W:** Denny's, Drake's (Ramada Inn)

Lodg **E:** Econolodge, (AAA) Holiday Inn Express

W: (DAYS INN) Days Inn, (AAA) Motel 6, ◆ Ramada Inn, Stratford House Inns

RVCamp **W:** Bird Creek Mobile Home Parking (Overnight)

ATM **E:** Texaco

W: Chevron

Other **W:** Harley-Davidson Motorcycles

301 TX 53, FM 2305, Central Ave, Adams Ave

Gas **E:** Diamond Shamrock*(D), Texaco

Food **E:** Arby's, Grandy's, KFC, Long John Silvers, McDonalds (Playground), Mr. Gatti's Pizza, Naranjo's Restaurante, Pizza Hut, Schlotzkys Deli, Shoney's, Subway, Taco Bell, Wendy's, What A Burger (24 Hrs)

W: Czech Heritage Bakery

Lodg **E:** (AAA) La Quinta Inn

AServ **E:** Griffin Body Shop, Johnson Bros. Ford, Texaco

W: Olson Auto Clinic, Temple Alternator & Starter Service, Temple Mercury Jeep Lincoln Dealer

Med **E:** ✚ Hospital

ATM **E:** Temple Santa Fe Credit Union

W: Cash Card

Bold red print shows RV & Bus parking available or nearby

EXIT — TEXAS

Parks — W: **Sammon's Park & Golf Course**
Other — E: **Eckerd Drugs, Gils Auto Shine**(CW)**, H-E-B (Pharmacy, 1-Hr Photo)**
W: **Albertson's Grocery (Pharmacy, 24 Hrs), Temple Veterinary Hospital of Western Hills**

300 — Ave H, 49th - 57th Streets
Gas — E: Shell*, Texaco, Texaco*
W: Fina*(D), Texaco*
Food — E: Clem Mikeska's BBQ, DeVinchi's Steak & Hoagie, Jack 'N' Jill Donuts, Las Casas, RDJ's Burger In A Flash, Rylander's Hamburgers, Sorge's Italian
Lodg — E: Oasis Motel
W: **Temple Inn**
AServ — E: Ave M Auto Trans & Towing, Big 3 Auto Supply, Bostick Tire & Auto Repair, Cooper Tires, Frost Automotive, Harmon Glass, Temple Collision Center, Texaco, Young Dodge Dealer
W: Carquest Auto Parts, Temple Auto Glass, Temple Grinding Co. Auto Parts
TServ — E: **Husky Trailor Parts**
ATM — E: Texaco
Parks — W: **Sammons Park & Golf Course**
Other — E: **Railroad Pioneer Museum,** Temple Wonder Wash, Triple Clean Car Wash(CW)

299 — East U.S. 190, Loop 363, TX 36, Gatesville, Cameron
Gas — E: Diamond Shamrock*, Phillips 66*, Shell*, Texaco*(D)(CW)
W: Exxon*, Texaco*
Food — E: Doyle Phillips Steak House, El Conquistador, Jack-in-the-Box, Luby's Cafeteria
W: Burger King (Exxon)
Lodg — E: **Budget Inn,** Regency Inn
W: A Rebel Inn
AServ — E: Ringler Chevrolet Toyota Geo, Saturn of Temple
W: Friend Tire, J & S Custom Paint & Body
ATM — E: Bank One, Bank of America, Phillips 66
Other — E: **K-Mart (Pharmacy), Putter Golf (Mini-Golf)**
W: **Marine Outlet**

298 — Frontage Road (Northbound)
RVCamp — E: **Ancira Motor Homes**
Other — E: **Scuba Plus**

297 — FM 817, Midway Drive
FStop — E: **Fina**
Gas — E: Citgo*(D)
W: Diamond Shamrock*(D)
Food — E: Midway Junction Temple Texas
Lodg — E: AAA Super 8 Motel
AServ — E: Chief E-Z Liner II, Shelton Volvo, Nissan Dealership
W: Tranum VW GMC Truck Dealer
Other — E: **Snippers Pet Grooming**
W: **Fed Ex Drop Box,** Laser Wash(CW) (24 Hrs)

294B — FM 93, 6th Ave, University of Mary Hardin Baylor
Gas — W: Shell*(D)(CW) (Laundromat)
Food — E: McDonalds (Playground)
W: Subway
Lodg — W: ◆ Best Western
AServ — W: M & B Auto Repair
Parks — W: **Heritage Park**
Other — E: **Mike's Complete Auto Detail**(D)**, Totally Scuba Training Center**
W: **Bell County Veterinary Hospital, University of Mary Hardin Baylor**

294A — Spur 253, Central Ave, Downtown
Gas — W: Diamond Shamrock*, Exxon*, Texaco*(D)
Food — W: Bobby's Burgers, J. Pleasant's Eatery, Pizza Plus, Restaurant (Ramada Limited), Schoepf's Old Time BBQ, Sonic Drive In, Stir Chicken Broster, What A Burger
Lodg — W: Ramada Limited
AServ — W: Auto Zone Auto Parts, Big 3 Auto Supply, Shine Bros. & Associate Goodyear, Western Auto
ATM — W: Compass Bank
Other — W: **Bell County Museum, The Medicine Shoppe (Pharmacy)**

293B — TX 317, FM 436, Main St (Southbound)
Gas — W: Citgo*
Food — W: Ernie's Fried Chicken
Other — W: Sea Horse Car Wash, **South Main Washateria**

293A — West U.S. 190, TX 317, FM 436, Killeen, Fort Hood
TStop — E: **Diamond Shamrock* (24 Hrs, FedEx Drop)**
Gas — E: Exxon*(D), Fina*(D), Smith Bros. Miller Hts. Grocery*
Food — E: Pat's Cafe
Lodg — E: **Budget Host Belton Inn**
TServ — E: **Diamond Shamrock TS**
TWash — E: **Diamond Shamrock TS**

292 — Loop 121
Gas — E: Mobil*(D)
Food — E: **Bakery Street (Mobil), Blimpie's Subs (Mobil), I Can't Believe It's Yogurt (Mobil),** Oxbow Steakhouse & BBQ, **Tia Rita's Tacos (Mobil)**
Lodg — E: Bloom's Motel
AServ — E: Fisher-Vincent Ford
RVCamp — E: **KOA Kampground, Sunbelt RV Center Service & Sales**
ATM — E: Mobil
Other — E: **Brown Boat & Motor Repair, Expo Center**

290 — Shanklin Road
TServ — W: **Capital Truck & Equipment Co.**
Other — E: **Mr. Marine Boat Sales**

289B — Frontage Road (Northbound)

289A — Tahuaya Road
AServ — E: Hi-Way Auto Parts

287 — Amity Road

286 — FM 1670, Stillhouse Hollow Lake
Other — W: **Salado Veterinary Hospital**

285 — FM 2268, Salado
Gas — W: Chevron*
Food — E: Samm's Family Restaurant
W: Cowboy's BBQ, Robertson's Choppin Block
AServ — E: Pronto Auto Service
ATM — E: Brookshire Bros., Compass Bank
W: Chevron
Other — E: **Brookshire Bros. Grocery, Fletcher's Books & Antiques, Salado Antique Mall, Village Pharmacy**

284 — Stage Coach Road
Gas — E: Exxon*
W: Mobil*(D)
Food — E: Browning's Courtyard Cafe, Burger King (Exxon), Cathy's Boardwalk Cafe, Old Town Salado & Patio Cafe, Pietro's Italian, Pink Rose Tea Room, Salado Square Cafe, Strawberry Patch, The Barton House, The Salado Mansion
W: Dairy Queen
Lodg — E: Osage Guest Cottage, Stage Coach Inn, The Inn at Salado, The Reue House
W: AAA Salado Inn
ATM — E: Exxon, First State Bank
W: Mobil
Other — E: **Central Texas Area Museum, Public Library, Wells Gallery**

283 — FM 2268, Salado, Holland

282 — FM 2115, FM 2843
TStop — E: **Texaco***(LP)
Food — E: **Restaurant (Texaco TS)**
TServ — E: **Texaco TS**
RVCamp — E: **Camping (2 Miles)**
ATM — E: **Texaco TS**
Other — E: **Sky Dive**

282A — Rest Area - RR, Vending, Picnic, Grills, Phones (Southbound)

(281) — **Rest Area - RR, Phones, Picnic, Vending, Grills (P) (Northbound)**

280 — Frontage Road

279 — Prairie Dell
RVCamp — W: **Emerald Lake RV Park (Approx. 1 Mile)**

275 — FM 487, Jarrell, Bartlett, Schwertner
TStop — E: **Citgo***(LP)(SCALES) **(CB Shop)**
FStop — E: **Exxon**
Gas — W: Texaco*
Food — E: **Doc's Cafe (Citgo TS), Joe's Country BBQ**
AServ — E: Exxon
W: Central Texas Chevy Parts
RVCamp — W: **Camping**
ATM — E: **Citgo TS**
Other — W: **Jarrell Country Market, Post Office**

271 — Theon Road
FStop — W: **Texaco***
AServ — E: RPM Tires
TServ — E: **RPM Tires**
ATM — W: **Texaco**

268 — FM 972, Walburg

266 — TX 195, Florence, Killeen
FStop — E: **Mobil* (FedEx Drop)**
Food — E: **Arby's (Mobil)**
AServ — W: Georgetown Interstate Transmission
ATM — E: **Mobil**

264 — Bus Loop I-35, Lakeway Drive, Austin Ave
AServ — E: Family Car Care Center
RVCamp — E: **Camping**

262 — Road 2338, Lake Georgetown, Andice, FM 971, Granger (Northbound Exit Number Is 261A)
Gas — E: Diamond Shamrock*, Gas Station*(D)
W: Phillips 66, Texaco*
Food — E: McDonalds (Playground), Pizza Hut, Quizno's Classic Sub, Schlotzkys Deli, Short Stop, Sonic
W: Chuck Wagon, Dairy Queen, Little Caesars Pizza, Popeyes Chicken, Taco Bueno, What A Burger (24 Hrs)
Lodg — W: DAYS INN **Days Inn,** La Quinta Inn
AServ — E: Fox Auto Parts, The Georgetown Tire Center
W: Phillips 66
ATM — E: Albertson's Grocery, Austin Area Teachers Federal Credit Union
W: Texaco, Union State Bank
Other — E: **Albertson's Grocery (Pharmacy), Eckerd Drugs, Fast Forward Mail, Tim's Book Shop**
W: **Georgetown Maritime (1 Mile)**

EXIT		TEXAS
261	TX 29, Taylor, Burnet	
FStop	**W:** **Chevron**(D)	
Gas	**E:** Texaco*(D)(CW)	
Food	**E:** Taco Bell (Playground)	
AServ	**E:** **Wal-Mart**	
	W: Chevron	
ATM	**E:** Texaco	
Other	**E:** **H-E-B Grocery Store (Pharmacy), Wal-Mart, (Pharmacy), Georgetown CVB (see our ad this page)**	
260	RM 2243, Leander	
Gas	**W:** Chevron*, Circle K*, Diamond Shamrock*	
Food	**W:** Lupita's Mexican, Page House Bed & Breakfast, Sverre's Donuts	
Lodg	**W:** ◆ **Comfort Inn**, Page House Bed & Breakfast	
AServ	**E:** A-1 Automotive	
Med	**E:** ✚ **Georgetown Medical Center**	
ATM	**W:** Chevron, Circle K	
Other	**E:** **Gus's Drugs, US Post Office**	
259	I-35 Bus Loop, Georgetown (Difficult Northbound Reaccess)	
RVCamp	**W:** **RV Outlet Mall**	
Other	**E:** **Candle Factory**	
	W: **Inner Space Cavern**	
257	Westinghouse Road	
(256)	**Rest Area - RR, Picnic, Grills, Vending, Phones P (Southbound)**	
256	RM 1431, Chandler Road	
Other	**W:** **Lake Travis (16 Miles)**	
(255)	**Rest Area - RR, Phones, Picnic, Grills, Vending, Pet Area P (Northbound)**	
254	I-35 Bus Loop, FM 3406	
Gas	**E:** Chevron	
	W: Phillips 66*(D)	
Food	**E:** Giovanni's Italian Restaurant, Hayloft Steakhouse, Lone Star Cafe, Pizzeria & Deli	
	W: **Cracker Barrel**, Denny's	
AServ	**E:** Chevron, Classic Pontiac	
ATM	**W:** Texas Heritage Bank	
253A	Frontage Rd (Southbound)	
Gas	**W:** Exxon*(D), Shell*(CW) (24 Hrs)	
Food	**W:** Hunan Lion, K-Bobs, Poke Jo's BBQ, Popeyes, Taco Bell, Thundercloud Subs	
Lodg	**W:** The Inn at Brushy Creek	
ATM	**W:** Shell, Texas Commerce Bank	
Other	**W:** **Detail Pro, Eckerd Drugs, TX State Optical, US Post Office**	
253	US 79, Taylor	
FStop	**E:** **Shell**(D)	
Gas	**E:** Diamond Shamrock*(D)(CW)	
	W: Texaco*(D)(CW)	
Food	**E:** Arby's, Baskin Robbins, Castaways Steakhouse, Dairy Queen, Damon's, KFC, Long John Silvers, Short Stop, Sirloin Stockade	
	W: K-Bob's Steakhouse, La Margarita	
Lodg	**E:** Best Western	
	W: La Quinta Inn, Ramada Limited, Sleep Inn	
AServ	**E:** Austin Datsun Repair, Auto Zone Auto Parts, Auto, Bus, & Truck Air Conditioning, Hester Auto Radiator	
	W: Bishop Automotive Machine, Longhorn Foreign & Domestic	
ATM	**E:** Bank One, Compass Bank, Heartland Bank, Shell	
Other	**E:** **H-E-B (Pharmacy)**	

EXIT		TEXAS
252B	RM 620	
Gas	**E:** Texaco*(D)(CW)	
	W: New York Deli*(D)	
Food	**E:** Golden Fried Chicken, Ole Taco	
	W: Bruegger's Bagels, Grandy's, Little Caesars Pizza, McDonalds (Playground), New York Deli, Peking Palace, Wendy's	
AServ	**E:** Kwik Kar Oil & Lube	
	W: K&S Automotive	
Med	**W:** ✚ **Hospital**	
ATM	**E:** First Texas Bank, Norwest Bank	
	W: Randall's	
Other	**E:** **Round Rock Travel & Tours, Visitor Information**	
	W: **Mail Center USA, Pro Photo Inc., Putt-Putt Golf & Games, Randall's Food & Pharmacy (24 Hrs, 1 Hr Photo)**	
252A	McNeil Road	
Gas	**W:** Citgo*, Phillips 66*(CW)	
Food	**E:** Bob's BBQ etc., CiCi's Pizza, Delaware Sub Shop, Golden Palace, What A Burger (24 Hrs)	
	W: Mr. Gatti's Pizza	
ATM	**E:** Austin Area Teachers Federal Credit Union	
Other	**E:** **Antique Mall of Texas, Mrs. Baird's Thrift Store, Pet Hospital**	
	W: **Animal Hospital**	
251	Bus Loop I-35, Round Rock	
Gas	**E:** Citgo*, Diamond Shamrock*(CW)	
	W: Citgo*, Exxon*	
Food	**W:** Burger King (Playground), China Wall Restaurant, Jack-in-the-Box, Jambalaya, Little Tony's, Luby's Cafeteria, TCBY, Taco Cabana (24 Hrs), The Bagel Tree	

EXIT		TEXAS
Lodg	**W:** Rodeway Inn & Suites	
AServ	**E:** AAMCO Transmission, Auto Tech, Brake Check, Genie Lube Express, Meineke Discount Mufflers	
	W: Brake Specialists, E & T Automotive, Precision Tune	
ATM	**W:** Albertson's Grocery, Citgo, First State Bank	
Other	**E:** Car Wash	
	W: **Albertson's Grocery (Pharmacy, 24 Hrs), Fed Ex Drop Box, Hastings, Optical Outlets, Walgreens (Pharmacy, 1 Hr Photo)**	
250	FM 1325	
Gas	**E:** Mobil*(D)	
Food	**E:** Applebee's, Boston Market, Chick-fil-A (Playground), Chili's, Dickey's BBQ, El Chico, McDonalds (Playground), Short Stop, Subway, Taco Bell	
AServ	**E:** Lamb's Tire & Auto, Pro Auto Body (Towing), **Wal-Mart**	
	W: Xpress Lube	
RVCamp	**E:** **Texas Camper Corral, Travel Town Texas**	
ATM	**E:** NationsBank, **Wal-Mart**	
Other	**E:** **Emeral Green's Golf Complex, Target Department Store, The Home Depot**(LP), **Wal-Mart Supercenter (Pharmacy, 24 Hrs)**	
	W: Doc Holiday's Car Wash, **Garden Ridge Shopping Mall**	
248	Grand Ave Parkway	
FStop	**E:** **Circle K***	
ATM	**E:** Circle K	
Other	**E:** **Certified Propane Inc. & Truck Wash**(LP), **Village at 3 Points Movies 12 Cinema**	
247	FM 1825, Pflugerville	
Gas	**E:** H-E-B (H-E-B Grocery), Racetrac*	
	W: Exxon*	
AServ	**W:** Purcell Tire Co.	
TServ	**W:** **Diesel Tech Parts & Service, Purcell Auto/Truck Service Center**	
ATM	**W:** Texas Heritage Bank	
Other	**E:** **H-E-B Grocery Store (1-Hr Photo, Pharmacy, 24 Hrs)**	
246	Dessau Road, Howard Lane	
Gas	**E:** Shell*(D), Texaco*	
	W: Diamond Shamrock*(D)(CW)	
ATM	**E:** Texaco	
245	FM 734, Parmer Lane, Yager Lane (Southbound Sign Info Includes Yager Lane)	
Gas	**W:** Citgo*	
AServ	**W:** Van's Auto Parts	
Med	**W:** ✚ **Hospital**	
244	Yager Lane (Northbound)	
243	Braker Lane	
Gas	**E:** Diamond Shamrock*	
	W: Texaco*	
Food	**E:** C & B BBQ, E and T BBQ, Gunther's Restaurant, Jack-in-the-Box, Subway, Tropical Heat Cafe & Bar, What A Burger (24 Hrs)	
Lodg	**E:** Walnut Forest Motel	
	W: Austin Motor Inn	
AServ	**E:** Austin Braker Automotive	
	W: AusTex, Lamar Body Shop, Tire Mart	
ATM	**E:** Diamond Shamrock, Randalls	
	W: Norwest Bank	
Other	**E:** **Animal Clinic, Randalls (Pharmacy)**	
241	Rutherford Lane, Rundberg Lane	
Gas	**E:** Diamond Shamrock*, Exxon(CW), Phillips 66*	

Bold red print shows RV & Bus parking available or nearby

EXIT TEXAS

W: Chevron* (24 Hrs), Conoco*, Texaco*(CW)

Food E: Dairy Queen, Golden China Restaurant, Jack-in-the-Box, Mr. Gatti's Pizza, Old San Francisco Steakhouse, Taquerias Arandas
W: Mr. India Palace

Lodg E: ◆ Ramada Inn
W: Howard Johnson, Motel 6, Park Inn International, Texas Village Motor Inn

AServ E: Exxon, Henna Chevrolet, **Wal-Mart**
W: A-One Discount Tire, I.H.-35 Automotive, J.C.'s Motor World, Nortek Auto Repair, North Austin P.A.C., Xpress Lube

ATM E: Diamond Shamrock, Exxon
W: Chevron, Conoco

Other E: Koin Wash, **Wal-Mart (24 Hrs, Pharmacy, 1-Hr Photo)**
W: Affordable Car Wash, **Golf**

240B N. U.S. 183, Research Blvd. (Limited Access Hwy)

240A S. U.S. 183, Anderson Ln., Lampasas, Lockhart

Gas E: Fisco*
W: Conoco*(D), Diamond Shamrock*, Exxon*(CW)

Food E: Chili's
W: American Grill Bandstand, Bennigan's, Burger Tex, Cafe Serranos, Denny's, Koffee 'N Ice Kream, Ninfa's, Outback Steakhouse, Rocky's Grill (Red Roof Inn), Souper Salads, Taj Palace

Lodg E: DAYS INN ◆ Days Inn, ◆ **Hampton Inn (see our ad this page)**
W: Best Western, Chariot Inn, Comfort Inn, Four Points Motel by Sheraton, AAA Holiday Inn Express (see our ad opposite page), La Quinta Inn, Motel 6, Red Roof Inn, Sumner Suites, Travelers Inn, Travelodge Suites

AServ E: Dyer Automotive
W: Henna Motor Co., Longhorn Glass

TServ W: **Austin Drive Train**

ATM W: Diamond Shamrock, Frost Bank

Other W: **General Cinema, Lucky 7 Food Mart**

240 US 183

Gas E: Fisco*

Food E: Chili's
W: Bennigan's, Jim's, Red Lobster

Lodg E: DAYS INN Days Inn, Hampton Inn
W: Best Western, Four Points Motel, Holiday Inn Express

ATM W: Frost Bank of N. Austin

239 St. Johns Ave (Westside Services Are The Same As On Exit 240A)

Food E: Fuddruckers, Owens Restaurant, Pappadeaux, Pappasito's, Savannah's (Holiday Inn), Shoney's

Lodg E: Doubletree Hotel, ◆ Drury Inn, Hawthorn Inn, AAA Holiday Inn, Homestead Village

AServ E: Heart of Texas Dodge

Other E: **Malibu Grand Prix, Post Office, The Home Depot(LP)**

238B East U.S. 290, West RM 2222, Houston

Food E: Cafe Allegri (Econo Lodge), El Torito, Palmeras, Texas Land & Cattle Steakhouse
W: Bombay Bicycle Club, Carrows Restaurant, China Cafe, IHOP, Lonestar Cafe, Ma Ferguson's, TJ Cinnamons Bakery

Lodg E: Econolodge (see our ad this page), ◆ Embassy Suites, AAA Red Lion Hotel, Fairfield Inn (see our

EXIT TEXAS

EXIT TEXAS

ad this page)
W: Drury Inn, Highland Village Motel, Hilton, La Quinta Inn, ◆ Quality Inn, Super 8 Motel

AServ E: American Collision Center, Roger Beasley Volvo

Other E: **News**
W: **Highland Mall, Lincoln Theatre, Office Depot Tech Center, OfficeMax, Presidio Theaters**

238A Reinli St, Clayton Ln (Southbound Reads 51st St)

Gas E: Chevron* (24 Hrs)
W: Texaco*

Food E: Amaya's Taco Village, CiCi's Pizza, Grandy's, McDonalds (Playground)
W: Baby Acapulco, Bad Griesbach, El Paraiso, The Captain's

Lodg W: Austin Rio Motel, Bad Griesbach, Courtyard by Marriott, Fairfield Inn by Marriott, Motel 6, Ramada Limited, Rodeway Inn

AServ E: Brake Check, Firestone Tire & Auto, Western Auto
W: International Auto Repair, Interstate Automotive, Trans-Care Transmission Parts

ATM E: Bank One
W: Bank One

Other E: **OfficeMax, Texas State Optical, Toys "R" Us, WalGreens (Pharmacy, 1-Hr Photo)**

237B 51st St, Cameron Road (Upper Level)

Gas E: Frisco Petroleum, Shell*

ATM E: Shell

237A Airport Blvd, 51st St, Cameron Rd, 38 1/2 St (Upper Level - No Services, Lower Level - Has Services)

Gas W: 1st Ave*, Chevron, Diamond Shamrock*(D)

AServ W: A-Z Auto, Airport Auto Supply, Airport Tire, Chevron, Sears Auto, The Oil Works

ATM W: Diamond Shamrock

Other E: **Academy, U-Haul Center(LP)**

236B 38 1/2 St

Gas E: Chevron*, Gas*
W: Texaco*(D)(CW)

Food E: Little Caesars Pizza, Pato's, Short Stop Hamburger
W: Jades Chinese

AServ E: Hi/Lo Auto Supply, Van's Auto Parts
W: Dura-Tune

Med W: ✚ **Hospital**

ATM E: Fiesta
W: NationsBank

Other E: **Austin Outdoor Gear & Guidance, Fiesta Grocery (24 Hrs), Monarch Food Mart**
W: **Hancock Golf Course**

236A 26th - 32nd Streets

Food E: Stars Cafe
W: Enchiladas Y Mas

Lodg E: DAYS INN AAA Days Inn (see our ad this page)
W: Rodeway Inn

Med W: ✚ **St. David's Medical Center**

Other W: **Austin Veterinary Hospital**

235C University of Texas, 15th & MLK Blvd, State Capitol

235B Manor Road, 26th St

235A 15th Street - Martin Luther King Blvd, State Capitol

Med W: ✚ **Brackenridge Hospital**, ✚ **Children's**

Hospital of Austin (Upper Level Access)
Other W: Lyndon Baines Johnson Library & Museum, The University of Texas

234C 6th - 12th Streets, State Capitol (Westside Services Are Via Upper Level)
Gas E: Chevron, Exxon*, Texaco*
W: Exxon, Texaco*(D)
Food E: Angie's, Ben's Long Branch BBQ, El Mollino
W: Capitol Cafe (La Quinta Inn), Serranos
Lodg E: Super 8 Motel
W: La Quinta Inn, Marriott
AServ E: Armadillo Auto Repair, Chevron
W: Exxon
Med W: Hospital
ATM E: Chevron
W: Austin Municipal Federal Credit Union, Texaco
Other E: Eckerd Drugs
W: Historic Sites and Museums

234B Cesar Chavez St, 2nd - 4th Sts., 8th - 3rd Sts. (8th-3rd Sts. Are Southbound Only)
Gas W: Mobil*
Food W: High Lite Cafe, O'Shucks Tomales, The Boiling-Pot Restaurant, The Pit
Lodg W: Omni Hotel
AServ E: East First Auto Supplies, Eastside Lawnmower Rentals
W: Bethke, Carter's Auto Repair
ATM W: Mobil
Other E: Public Library
W: Fire Dept., O. Henry House & Museum, Police HQ

234A Cesar Chavez St, Holly St (Southbound)
Food W: IHOP, Iron Works
AServ W: Toyo Tires
Other W: Austin Convention Center, Visitor Information

233A 1st - 4th Streets

233BC Unnamed
Gas E: Shell*
ATM E: Shell

233 Town Lake, Riverside Drive
Gas E: Diamond Shamrock*
W: Chevron*(CW) (24 Hrs)
Food E: Wok 'N Gold
W: The Pecan Tree (Holiday Inn)
Lodg E: HomeGate Studios & Suites
W: AAA Holiday Inn
Other W: Public Boat Ramp

232B Woodland Ave
Gas E: Custom Auto Detail
AServ E: Custom Auto Detail

232A Oltorf St., Live Oak
Gas E: Citgo*, Fisco*, Texaco*(CW)
W: Chevron*(CW), Exxon, Texaco
Food E: Carrows Restaurant, Kettle, Luby's Cafeteria, Mr. Gatti's Pizza, Sonic
W: Denny's, Marco Polo Restaurant (Quality Inn)
Lodg E: Austin Motel, AAA Exel Inn, ◆ La Quinta Inn, Motel 6, Park West Inn, Ramada Limited, Super 8 Motel
W: AAA Quality Inn
AServ E: Tuneup Masters
W: Chevron, Exxon, Texaco
Other W: Dallas Cowboy's Training Camp, Quality Vision Eyeware, Whip In

231 Woodward St., St. Edwards Univ.
Food E: Country Kitchen (Holiday Inn)
Lodg E: ◆ Holiday Inn

230A St. Elmo Rd (Southbound)
Food W: Furr's Family Dining
AServ W: Beasley Mazda Kia, Cen-Tex Nissan, Easy Wheels, Hendrix GMC Trucks, McMorris Ford, NTB, Red McCombs Toyota, South Point Jeep, Eagle, Lincoln, Mercury, South Point Pontiac, Cadillac, Southstar Dodge

230B US 290 West, TX 71, Ben White Blvd. (Southbound)
Food W: Bill Miller BBQ, Burger King, IHOP, Pizza Hut, Taco Cabana
Lodg W: Days Inn, Hawthorn Suites, ◆ La Quinta Inn
AServ W: Howdy Honda
Med W: Hospital

230 U.S. 290 West, TX 71, Ben White Blvd, St Elmo Rd (Northbound)
Gas E: Texaco*(D)(CW)
Food E: Celebration Station, Saigon Kitchen (Best Western), Subway (Best Western)
Lodg E: Best Western, Courtyard by Marriott, Fairfield Inn (see our ad page 165), Hampton Inn, Residence Inn by Marriott, The Omni Hotel
Other E: Celebration Station

229 Stassney Lane
Gas E: Citgo*(D), Diamond Shamrock*, Exxon*(CW)
W: Texaco*(D)
Food E: Applebee's, McDonalds, Subway, Taco Bell
W: Burger King, KFC, Long John Silvers, Mr. Gatti's Piza, Taco Cabana, Wendy's, What A Burger
AServ E: Chrysler Auto Dealer
W: Western Auto
RVCamp E: KOA Campground
Other E: H.E.B. Grocery Store, K-Mart, Target Department Store

228 William Cannon Drive
Gas E: Citgo*, Diamond Shamrock*, Exxon*(CW), Shell*(D)
W: Texaco*(D)
Food E: Applebee's, I Can't Believe It's Yogurt (Shell), McDonalds (Playground), Subway, Taco Bell, TiaRita's Tacos (Shell)
W: Burger King (Playground), Delaware Sub Shop, KFC, Long John Silvers, Mr. Gatti's Pizza, Taco Cabana, Thai Kitchen, Wendy's, What A Burger (24 Hrs)
Lodg E: Austin Airport Inn, Courtyard by Marriott (see our ad this page)
AServ E: Chief Auto Parts, Gillman Mitsubishi, Prestige Chrysler
W: Capitol Chevrolet Geo, Hackney Automotive & Truck Service, Western Auto
TServ W: Hackney Automotive & Truck Service
RVCamp E: KOA Kampground(LP)
ATM E: Diamond Shamrock, Shell
W: Texaco
Other E: H-E-B (24 Hrs, Pharmacy, 1-Hr Photo), K-Mart (Pharmacy), Mail & Box, Target Department Store, Texas State Optical
W: Academy Sports & Outdoors, Century Veterinary Hospital, Eckerd Drugs, Medi-Save Optical Outlet

227 Slaughter Lane, TX Loop 275, South Congress Ave.
Gas W: Diamond Shamrock*(D)
TServ E: Holt Cat
ATM W: Diamond Shamrock

226 FM 1626, Slaughter Creek Overpass
Gas E: Texaco*(D)
RVCamp W: Marshall's Traveland Sales

225 Onion Creek Parkway
Gas E: Diamond Shamrock*

223 FM 1327

221 Loop 4, Buda
TStop E: Dorsett 221(SCALES) (Phillips 66)
Gas W: Chevron* (24 Hrs), Texaco*(D)(CW)
Food E: Dorsett 221 Cafe (Phillips 66 TS)
W: Taco El Paso
Lodg E: Interstate Inn (Phillips 66 TS)
AServ E: 221 Fleet Service (Phillips 66 TS)
W: A-Line Auto Parts
TServ E: 221 Fleet Service (Phillips 66 TS), Carlstead Truck & Bus, Ford Truck City
ATM W: Austin National Bank

220 FM 2001, Niederwald
Food E: Hinojosa Express
RVCamp E: First RV Service
W: Crestview RV Sales & Service, & Camp

217 Loop 4, Buda
FStop E: Conoco*
Gas W: Diamond Shamrock*(D)
Food E: Burger King (Conoco)
TServ E: K D Truck Parts
RVCamp W: Interstate RV
ATM E: Conoco

215 Bunton Overpass

213 FM 150, Kyle
Gas E: Diamond Shamrock*(D)
W: Conoco*(D)
Food E: Dairy Queen
W: Blanco River Pizza Co., Chicken Willies, Panaderia Mexicana Kyle Bakery, Railroad BBQ, Scoops!
AServ W: 4 Way Auto Repair Shop (24 Hrs), Carquest Auto Center (Adolph's Auto Parts), J & R Tire Service (24 Hrs)
ATM E: Dairy Queen, Diamond Shamrock

EXIT TEXAS

W: Balcones Bank, Conoco

Other E: **Mr. T Washateria (Diamond Shamrock)**

W: **Bon-Ton Grocery, Center Grocery**

211 Weigh Station (Northbound)

(211) **Rest Area - RR, Vending, Picnic, Grills, HF P (Both Directions)**

210 Yarrington Road

208 Blanco River

206 Loop 82, Aquarena Springs Dr

Gas E: Conoco*

W: Diamond Shamrock*(D), Payless Drugs*(D), Phillips 66*(CW), Shell*(CW), Texaco*

Food W: Casa Ole, Kettle Restaurant, Popeye's Chicken, Shoney's, Sonic

Lodg W: AAA Comfort Inn, Executive House Hotel (see our ad this page), ◆ Howard Johnson, La Quinta Inn, Motel 6, Ramada Limited & Suites, Stratford House, ◆ Super 8 Motel, University Inn

RVCamp E: **United Campground RV Park**(LP)

ATM W: Diamond Shamrock

Other E: **Texas Natural Aquarium Aquarena Springs (.5 Miles)**

W: **Super Klean Laundromat, Tourist Information, Washouse Coin-Op Laundry**

205 TX 80 To TX 21 Luling, TX 12 To TX 142 Wimberley, Lockhart.

Gas E: Conoco*(D), Diamond Shamrock*(D), Exxon*, Racetrac* (24 Hrs), Shell*(D)

W: Diamond Shamrock*

Food E: Arby's, Baysea's Fish Market III & Seafood Restaurant, Dairy Queen, Fushak's BBQ, J J's Flame-Broiled Hamburgers, Pantera's Pizza, Schlotzkys Deli, Subway, Zapata's Cafe

W: Burger King (Playground), Church's Fried Chicken, Donut Palace, Gill's Fried Chicken, KFC, Long John Silvers, **McDonalds (Playground),** Pizza Hut, Sirloin Stockade, Taco Cabana, Wendy's

Lodg W: AAA Best Western, DAYS INN AAA Days Inn, Microtel Inn, Rodeway Inn, Southwest Motor Lodg

AServ E: Grease Monkey

W: Starr Lube, The Brake Shop

ATM E: Conoco, Exxon, Frost Bank, Shell

Other E: **Cinema 5, Eckerd Drugs, Hastings Bookstore, S.W. Vision Optical, Tickle Blagg Animal Hospital**

W: **H-E-B Grocery,** Laser Wash

204 West Loop 82, East TX 123, San Marcos, Seguin (Texas State University)

FStop E: **Texaco***

W: **Texaco***(LP)

Gas E: Conoco*, Phillips 66*(D)(CW)

W: Conoco*(D) (Kerosene), Diamond Shamrock*, Quick Align

Food E: Burger King, Chili's, Golden Corral, Luby's Cafeteria, McDonalds, Red Lobster, What A Burger (24 Hrs)

W: Chili Dog, Dairy Queen, Guadalupe Street Smokehouse, Hacienda Gonzalez, Hong Kong Restaurant, Panda King, Sonic, Taqueria La Fonda

Lodg E: Holiday Inn Express

W: AAA Econolodge

AServ E: Chuck Nash Jeep Eagle Dodge, Pennzoil Lube, Red Simon Ford Mercury, Rudy's Automotive

EXIT TEXAS

W: Auto Zone Auto Parts, Carquest Auto Parts, Cartek Auto Body, Conoco, Ellison's Auto Repair & Windshield, Fritz's Muffler Shop, NAPA Auto Parts, Saucedo's Wrecker Service (24 Hrs)

TServ W: **Texaco**

Med E: **✚ Hospital**

Other E: **La Palma Grocery, San Marcos Skate Center**

W: **Budget Opticals,** Car Wash

202 FM 3407, Wonder World Dr

Gas W: Diamond Shamrock*(D)

Food W: Pizza Hut Express (Diamond Shamrock)

Med E: **✚ Central TX Medical Center**

Other E: **Lowe's Home Improvement Wharehouse**(LP)

W: **Wonder World**

201 McCarty Lane

200 Center Point Road

Gas W: Diamond Shamrock*(D)

Food E: Food Court (San Marcos Factory Shops), Lone Star Cafe, Subway, Taco Bell, Wendy's

W: Centerpoint Station

Lodg W: AmeriHost Inn

ATM E: Food Court (San Marcos Factory Shops)

W: Diamond Shamrock

Other E: **San Marcos Factory Shops, Tanger Factory Outlet**

W: **Antique Outlet Center**

199A Posey Road

196 FM 1102, York Creek Rd.

Lodg W: Acapulco Motel & Restaurant

AServ W: Arsenio's Automotive 24 Hr

195 Watson Lane, Old Bastrop Rd.

193 Kohlenberg Rd., Conrads Rd.

TStop W: **Rip Griffin Travel Center***(LP)(SCALES) **(Shell, Playground)**

Food W: **A & W Drive-In (Rip Griffin TS), Subway (Rip Griffin TS)**

TServ W: **Rip Griffin Travel Center**

191 TX 306, TX 483, Canyon Lake

FStop W: **Exxon***

Gas E: Fina*(D)

W: Mobil*(CW)

Food W: I Can't Believe It's Yogurt (Mobil), Mesquite Smoked BBQ (Mobil), Pizza Hut (Mobil)

RVCamp W: **Maricopa Ranch Resort**

ATM W: Mobil

190C Post Rd (Southbound)

Other E: **Tourist Information**

190B S. Bus. Loop I-35, New Braunfels (Southbound)

EXIT TEXAS

Gas E: Conoco*(D)

AServ E: Conoco

RVCamp E: **Wayside RV Park**

190A Frontage Rd (Southbound)

Lodg E: Best Western, ◆ Comfort Suites

190 North U.S. 35, South Bus Loop

189 TX 46, Loop 337, Seguin, Boerne

Gas E: Conoco*(D), Texaco*(D)(CW) (24 Hrs)

W: Chevron*(D) (24 Hrs), Diamond Shamrock*

Food E: Luby's Cafeteria, Oma's Haus

W: Breustedt Haus, Kettle, Longhorn Grill, McDonalds (Playground), Molly Joe's, New Braunfels Smokehouse, TCBY, Taco Bell, Wendy's

Lodg E: Oak Wood Inn, ◆ Super 8 Motel

W: Dwight's Motel, Edelweiss Inn, Fountain Motel, ◆ Holiday Inn, Old Town Inn, AAA **Rodeway Inn**

ATM E: Texaco

Other E: **Comal Animal Clinic, K-Mart (Pharmacy)**

188 Frontage Road

Food W: Applebee's, IHOP, Ryan's, Taco Cabana (24 Hrs)

Lodg W: DAYS INN **Days Inn,** Hampton Inn, Lucky Star Motel

AServ W: Best Deal Tires 24-Hr Road Service

RVCamp W: **New Braunfel's RV Park**

ATM W: New Braunfel Factory Stores

Other W: **Book Warehouse, New Braunfel's Factory Stores**

187 FM 725, Seguin Ave, Lake McQueeny

Gas E: Chevron* (24 Hrs), Diamond Shamrock*

W: Exxon, Phillips 66* (Kerosene)

Food E: Arby's, Burger King (Playground), Cancun Cafe, China Kitchen Restaurant & Bar, CiCi's Pizza, Donut Palace, Long John Silvers, What A Burger (24 Hrs)

W: Adobe Cafe, Dairy Queen, Jack-in-the-Box, Keno's BBQ & Smokehouse, Peking Restaurant, Rally's Hamburgers, Restaurant (Budget Host)

Lodg W: Budget Host Inn, Budget Inn

AServ E: AAMCO Transmission, Precision Tune & Lube, Q Lube

W: B & C Service Center (NAPA), Delux Glass & Mirror, Exxon

RVCamp W: **Mobile Home & RV Parts**

Med W: **✚ Hospital (2 Miles)**

Other E: E-Z Wash #2, **Handy Andy Supermarket, Polly's Pet Shop, Texas State Optical**

186 Walnut Ave

Gas E: Diamond Shamrock*(D)(CW), Exxon*

W: Conoco*, H-E-B, Texaco*(D)(CW)

Food E: **McDonalds (Wal-Mart), McDonalds (Playground),** Papa Dante's, Popeye's Chicken, Subway (Exxon)

W: KFC, Mr. Gatti's, Papa John's Pizza, Schlotzkys Deli, Shanghai Inn, Shipley Do-nuts

Lodg E: **Ramada Limited**

AServ E: Maxwell Chevrolet, **Wal-Mart Supercenter**

W: Auto Zone Auto Parts, Brinkkoeter's, Morris Glass Co., Rick's Muffler & Hitch Center, Xpress Lube

RVCamp E: **RV Service (Wal-Mart)**

ATM E: Exxon, **Wal-Mart Supercenter**

W: Texaco, Texstar Bank

Other E: **Wal-Mart Supercenter (Pharmacy, 1-Hr Photo, 24 Hrs, Vision Center)**

W: **Country Clean Coin-Op Laundry, H-E-B**

EXIT TEXAS

Grocery Store (Pharmacy), Mail It Plus, Schlitterbahn Water Park, Target Department Store, True Value Hardware, Vivroux Toy & Sporting Goods, Walnut 6 Cinema

185 S Bus. Loop 35, FM 1044, New Braunfels
- Food W: Butcher Boy Meat Market
- AServ W: Z's 24-Hr Tow Service
- Other W: Car Wash, **Coin Laundry**

184 Loop 337, FM 482, Ruekle Road
- Gas E: Texaco*[D]
- Food E: Blimpie's Subs (Texaco)
- RVCamp E: **Hill Country RV Resort**
- ATM E: Texaco

183 Solms Road
- Gas W: Exxon*

182 Engel Road
- Food W: Mesquite Pit Junction
- RVCamp E: **Stahmann RV Sales**
- Other W: **Snake Farm**

180 Schwab Road

(179) **Rest Area - RR, Phones, Vending, Picnic, Grills, RV Dump, Pet Rest P (Both Directions)**

178 FM 1103, Cibolo, Hubertus Road
- Gas E: Conoco*[D]
- RVCamp E: **Happy Camper RV Sales, Rancho Vista Park**

177 FM 482, FM 2252
- RVCamp W: **Stone Creek RV Park**

175 FM 3009, Natural Bridge Caverns Road (Southbound Ramp Closed For Construction)
- FStop W: **Diamond Shamrock***
- Gas E: Diamond Shamrock*[D][CW], H-E-B
  W: Texaco*[D][CW]
- Food E: Bill Miller BBQ, La Pasadita, McDonalds (Playground)
  W: Abel's Diner, Arby's, Denny's, **Jack-in-the-Box, Subway (Diamond Shamrock)**, Wendy's
- Lodg W: ◆ Ramada Limited
- ATM E: Diamond Shamrock, State Bank & Trust
  W: **Diamond Shamrock**, Texaco
- Other E: **Animal Hospital, H-E-B Grocery (Pharmacy)**
  W: **Natural Bridge Caverns (8 Miles), Natural Bridge Wildlife Ranch (8 Miles), Texas Pecan Candy**

174B Schertz Parkway
- RVCamp W: **Beryl's RV Sales Center, Crestview RV Sale**

174A FM 1518, Selma, Schertz
- AServ E: Gillman Honda, Saturn of San Antonio N.E.
- Other W: **Ronnie's Marine, Tex-All Boat Co.**

173 Olympia Parkway, Old Austin Road

172 Loop 1604, Anderson Loop

171 TX 218, Pat Booker Road, Universal City, Randolph AFB (Northbound)
- AServ E: Red McCombs Nissan

170B Toepperwein Road
- Food E: Kettle Restaurant, Sasha India Cuisine, Subway, What A Burger (24 Hrs)
- Lodg E: La Quinta Inn
- AServ E: Gunn Chevrolet
- Med E: ✚ **Northeast Methodist Hospital**
- Other E: **Live Oak Pharmacy, My Buddy, Village Oaks Pharmacy**

EXIT TEXAS

170A Judson Road
- Gas E: Texaco*[D][CW]
  W: Exxon*[CW]
- Lodg W: Holiday Inn Express
- AServ E: Gunn Auto Park, Universal Toyota
  W: **Sam's Club**, Universal Mazda Subaru
- ATM E: Texaco
- Other W: **Sam's Club**

169 O'Connor Road
- FStop E: **Conoco* (24 Hrs)**
  W: **Texaco*[SCALES] (FedEx Drop)**
- Gas E: Gas*
  W: Diamond Shamrock*[D][CW], Racetrac* (24 Hrs)
- Food W: **Chester Fried Chicken (Texaco)**, Jim's Restaurant, **Little Caesars Pizza (K-Mart)**
- Lodg W: Econolodge
- AServ W: **K-Mart**, Kwik Kar Lube & Tune, Martinez Tire & Muffler Shop
- RVCamp W: **Camping**
- ATM E: Conoco
  W: Diamond Shamrock
- Other W: **K-Mart (24 Hrs, Pharmacy)**

168 Weidner Road
- Gas E: Bank Card Self Service, Diamond Shamrock*[D], Exxon*
  W: Chevron*
- Lodg E: DAYS INN AAA Days Inn
  W: ◆ Quality Inn, Super 8 Motel
- AServ E: Gunn Dodge
  W: Carburetor Shop, Kyongs Auto Center
- TServ W: **Grande Trucks**
- RVCamp W: **Interstate RV**
- ATM E: Diamond Shamrock
- Other E: **Sailboat Shop**
  W: **Harley Davidson**

167A Randolph Blvd (Southbound)
- Food W: The Jalapeno Pancake
- Lodg W: AAA Best Western, DAYS INN AAA Days Inn, **Howard Johnson**, AAA Motel 6, Ruby Inn (Playground)
- AServ W: Yang's Auto Repair

167B Starlight Terrace (Southbound)

167 Starlight Terrace (Northbound)
- Gas E: Stopn Go*
- AServ E: Automotive & Performance, Performance Automotive & Transmission, Uncle Sam's Auto Repair
- ATM E: Stopn Go

166B Randolph Blvd, Windcrest (Northbound)
- Gas E: Diamond Shamrock*[CW]
- Food E: Mongolia Restaurant, Saigon Gardens
- RVCamp E: **Traveltown Texas**
- ATM E: Diamond Shamrock
- Other E: **Christ the King Bookstore, Joe Harrison Motor Sports, Skateland**

166 I-410 west, South Loop 368

165 FM 1976, Walzem Road, Windcrest

EXIT TEXAS

- Gas E: Texaco*[D][CW]
  W: Mobil
- Food E: Applebee's, Burger King, China Cafe, Chuck E. Cheese's, Church's Fried Chicken, Jailhouse Cafe, Jim's, Marie Callender's Restaurant, Mr. Goodcents Subs & Pastas, Olive Garden, PoFolks, Red Line Hamburgers, Red Lobster, Shoney's, **Taco Bell (Wal-Mart)**, Taco Cabana (24 Hrs), Wendy's
  W: Sonic
- Lodg E: Drury Inn, ◆ Hampton Inn
- AServ E: AutoExpress (Montgomery Ward), Firestone Tire & Auto, **Wal-Mart**
  W: Mobil, NTB
- ATM E: Frost Bank, Security Service Federal Credit Union
  W: Randolph-Brooks Federal Credit Union
- Other E: **OfficeMax, PETsMART, Pearle Vision Express, The Home Depot[LP], Toys "R" Us, Wal-Mart (1 Hr Photo , Vision Center, Pharmacy), Windsor Optical, Windsor Park Mall**
  W: **Baseball City, La Placida Food Mart, NorthEast Animal Hospital**

164 Eisenhauer Road
- Gas E: Exxon*[D][CW]

164B **(163)** Rittiman Road
- Gas E: Exxon*[CW], Racetrac*, Texaco*[D][CW]
  W: Chevron*[D], Diamond Shamrock*[D], Otto Food Mart*
- Food E: Burger King, Church's Fried Chicken, **Cracker Barrel**, Denny's, Kettle Restaurant (24 Hrs), McDonalds (Playground), Red Line Burgers, Sam Won Garden, Taco Cabana (24 Hrs), What A Burger (24 Hrs)
  W: Bill Miller BBQ, Cristan's Tacos #5, Edelweiss Restaurant, Wendy's
- Lodg E: Comfort Suites (Playground), La Quinta Inn, Motel 6, Motel 6, Ramada Limited, Scotsman Inn, Stratford House Inn
- AServ E: MAACO Auto Body, Tire Shop
- TServ E: **Onan Cummins**
- ATM E: Exxon
  W: Diamond Shamrock
- Other E: **Rittiman Animal Hospital**

163C Holbrook Rd, Binz - Englemann Rd (Southbound, Unnumbered Exit)
- Lodg W: Holiday Inn Hotel & Suites, Quality Inn & Suites, Villager Lodge
- Med W: ✚ **Brook Army Medical Center (Ft. Sam Houston)**

163AA South I-410 (Southbound, Left Exit, Unnumbered Exit)

163BB Petroleum Drive (Unnumbered Exit)

162 I-410 South, Loop 13, W.W. White, FM 78, Kirby (Northbound)
- Lodg W: **Comfort Inn Airport (see our ad this page)**

161 Binz - Englemann Road
- AServ E: Jim Buffaloe Automotive, Red Line Truck Accessories (Pick-Ups)

160 Splash Town Drive, Coliseum Rd (Access To Services Varies On Direction Traveled)
- Food W: Casey's BBQ, Los Pinos
- Lodg E: Delux Inn
  W: DAYS INN ◆ Days Inn, ◆ Super 8 Motel, Travelodge

Bold red print shows RV & Bus parking available or nearby

EXIT TEXAS

TWash E: Truck Wash
Other E: Splash Town Water Park

159B Walter St., Fort Sam Houston, Colliseum Rd. (Access To Services Varies On Direction Traveled)
FStop E: Diamond Shamrock (24 Hrs)
Food E: C & C Tacos, J & E Drive-Inn, McDonalds (Playground)
W: Drop Zone Cafe, Flores Drive-In
Lodg W: Howard Johnson
AServ E: Aarco Transmission
RVCamp E: KOA Kampground
ATM E: Diamond Shamrock
Other E: Alamo Auto Detail
W: Fort Sam Houston

159A New Braunfels Ave
Gas E: Exxon*(D), Texaco*(D)(CW)
W: Chevron*, Diamond Shamrock*
Food W: Bill Miller BBQ, Johnny's Mexican Restaurant, Johnny's Seafood, TNK Oriental Restaurant
AServ W: Atlas Body Shop
ATM W: Diamond Shamrock
Other W: Historical Fort Sam Houston, San Antonio Botanical Center (1 Mile)

158C Loop 368, North Alamo St, Broadway (Southbound)
Food W: Olgita's Molino

158B South I-37, South US 281, Corpus Christi (Southbound)

158 I-37, U.S. 281 S, U.S. 281 N, Corpus Christi, Johnson City (Northbound Upper Level)

157C St. Mary's St, Loop 368, Broadway (Northbound)
Food E: El Nogal
Lodg E: Super 8 Motel
AServ E: Drive Shaft & Auto Air, Gene's Brake & Alignment, Malin's Auto Repair
Other E: Fire Station, Hickey Animal Clinic Emergency Hospital, San Antonio Museum of Art

157B McCullough Ave, Brooklyn Ave (Upper Level)
Gas E: Diamond Shamrock*
Food E: Audry's Mexican
AServ E: H H Roper Auto Parts
Med E: ✚ Baptist Medical Center
W: ✚ Metropolitan Methodist Hospital
ATM E: Diamond Shamrock
W: NationsBank
Other E: San Antonio Museum of Art
W: Your Eyes Optical

157A San Pedro Ave., Main Ave., Lexington Ave.
Gas W: Diamond Shamrock*(CW)
Food W: Cristan's Tacos, Jack-in-the-Box, Luby's Cafeteria, McDonalds (Playground), Pete's Tako House, Pizza Hut, Restaurant (Rodeway Inn), Taco Bell, Wendy's
Lodg W: Rodeway Inn
AServ W: Fred Luderus Tire Service, Quality Paint & Body, Tuneup Masters
Med E: ✚ Baptist Medical Center
ATM E: Compass Bank
W: Diamond Shamrock
Other E: Madison Square Medical Building Pharmacy

EXIT TEXAS

W: Polo's Photo 1-Hr Photo, Walgreens (Pharmacy)

156 West I-10, McDermott Freeway, North US 87 El Paso (Upper Level)

155C West Houston St, Commerce St, Market Square (Southbound, Upper Level)
Gas W: Pik Nik*
Food W: Golden Star Cafe, Jailhouse Cafe, McDonalds, Pico de Gallo
Lodg W: Motel 6, Radisson Hotel
Med W: ✚ University Health Center Downtown
ATM W: San Antonio City Employees Federal Credit Union

155B Durango Boulevard, Downtown, Frio St. (Access To Services Is Limited To The Direction Traveled)
Food E: Bill Miller BBQ
Lodg E: Courtyard by Marriott, Fairfield Inn (Marriott), Holiday Inn, Residence Inn (Marriott)
W: Radisson Hotel
Med E: ✚ Santa Rosa Health Care
ATM E: Stopn Go, Wells Fargo
Parks E: San Antonio Missions National Historical Park (2.8 Miles)
Other E: K-Mart (Pharmacy), RxCare Pharmacy, San Antonio Police HQ, Stopn Go, The Market Square
W: Fire Station

155A Spur 536, South Alamo St. (Upper Level Southbound)
Gas E: Datafleet(D) (Datafleet Credit Card Only, No Attendant)
Food E: Wen Wah's Chinese
Lodg E: ◆ Comfort Inn, Ramada Limited
W: River Inn Motel
AServ E: Speed & Sport
Other E: US Post Office

154B South Laredo St, Cevallos St (Upper Level)
Gas E: Texaco*
Food E: Church's Fried Chicken, Eddie's Taco House, McDonalds (Playground), Pizza Hut Delivery & Carry out, Wendy's
W: Piedras Negras
Lodg E: Days Inn Days Inn
Other W: Gleason Veterinary Hospital

154A San Marcos St, Nogalitos St, Loop 353 (Upper Level Southbound)
Gas E: J & L Food & Gas*
Food E: Maria's Cafe, Tommy's Old San Antonio Cafe
W: Henry's Tacos To Go, Midtown Grill, Simon's Bakery
AServ E: General Brake & Alignment, Hernandez Tire & Muffler
Other W: Collins Branch Garden Library, Fire Station, H-E-B Grocery

153 West U.S. 90, Del Rio (Upper Level Southbound)

152B Malone Ave, Theo Ave
Gas W: Diamond Shamrock*, Texaco*(D)
Food E: Taco Cabana (24 Hrs)
AServ E: Liberty Transmission & Gear Co.
W: Mack's Transmission Service
ATM W: Diamond Shamrock, Texaco

EXIT TEXAS

152A Division Ave
Gas E: Chevron*(CW)
W: Exxon*(CW), Phillips 66*
Food E: Bill Miller BBQ, What A Burger (24 Hrs)
W: Tortilleria Los Hermanos, Victoria Tortilla & Tamale Factory
Lodg E: Holiday Inn Express
AServ W: Backus Radiator Works, Coxco Transmission, Guerrero's Tire Shop
Other E: Car Wash, Pan Am Pharmacy
W: Pablo's Grocery

151 Southcross Blvd
Gas E: Citgo*, Exxon, Shop "N" Save*
W: Texaco*(D)(CW)
Food E: Centeno Market
W: Taco Jalisco
AServ E: Exxon
RVCamp E: Camping
ATM E: Citgo
Other E: Centeno Market (Flea Market)

150B Loop 13, Military Dr, Kelly AFB, Lackland AFB
Gas E: Chevron*(CW) (24 Hrs), Texaco*
W: Exxon*(CW), Mobil*
Food E: Casa Dos Pedros, Denny's (24 Hrs), Jang's Chinese, Pizza Hut, Pizza Rio Buffet, Sonic, Subway, Taco Cabana (24 Hrs)
W: Hungry Farmer Steakhouse, Hungry Italian Steakhouse, Jack-in-the-Box, Long John Silvers, Luby's Cafeteria, McDonalds (Playground), Popeyes Chicken, Red Line Hamburgers, Shoney's, Southfork Restaurant, Uncle Barney's Old Fashioned Hamburgers
Lodg E: La Quinta Inn
AServ E: Art's Tires & Mufflers, Auto Zone Auto Parts, Brake Check, Discount Tire, Eagle Auto Glass, Macias Tire Shop, Thad Ziegler
Med E: ✚ Family Pracice Minor Emergency
ATM E: Chevron
W: Bank One
Other E: Alamo City Optical, Century Plaza Theatres, EyeMasters, Ram's Texas Car Wash, South San Pharmacy, U-Haul Center(LP)
W: Eckerd Drugs, Office Depot, Southpark Mall, Toys "R" Us

150A Zarzamora St
Gas E: Diamond Shamrock*
Food E: Berta's Mexican Cafe
AServ W: Firestone Tire & Auto, Sears Auto Center
ATM E: Diamond Shamrock
Other E: Buck Shop Hunter's HQ(LP), Rios Meat Market #4
W: Skate Time, South Park Mall

149 Hutchins Blvd. (Southbound)
Med W: ✚ Hospital

148B Palo Alto Road (Northbound Is A Left Exit)
FStop W: Phillips 66*(D)
Gas W: Saven' Go*
Other W: Kwik Wash Laundry

148A Spur 422, TX 16 South, Poteet

147 Somerset Road

146 Cassin road

145B North Loop 353 (Northbound Is A Left Exit, Difficult Northbound Reaccess)
AServ W: A-1 Auto Parts, A-1 Imports, A-Alpha Import

EXIT TEXAS

Co., ABA Parts, Alamo City Imports, Alamo Imports, All Foreign Auto Parts, Apache Auto Parts, Auto World, Benzes & BMW Recyclers, Crystal Ball Auto Parts, E & D Complete Auto Parts, J & M Auto Parts, Laredo Auto & Truck Parts, Million Auto Parts, NICA Motors & Parts

TServ **W: A-1 Truck Salvage, A-Alpha Import Co., ABA Parts, International Truck Parts, Inc., Interstate Truck Sales, Laredo Auto & Truck Parts**

145A I-410, TX 16

144 Fischer Road

TStop **E: Diamond Shamrock*(SCALES) (Fax Machine)**

Food **E: Subway (Diamond Shamrock TS)**

Lodg **E: D&D Motel**

RVCamp **E: Hidden Valley RV Park & Camp (Store, Laundry, Full Hookups, 0.5 Miles)**

142 Medina River Turnaround

TServ **E: Gene's Truck Parts**

RVCamp **E: Uresti's Camper Sales & Truck Accessories**

141 Benton City Road, Von Ormy

TStop **W: Texaco* (24 Hrs)**

Food **W: Mario's Cafe (Texaco TS)**

ATM **W: Texaco TS**

Other **E: AJM Food Store, Blossoms & Bows, US Post Office**

140 Loop 1604, Anderson Loop, Somerset, Sea World, Fiesta

FStop **E: Exxon* (24 Hrs)**

Food **E: Burger King (Exxon)**

ATM **E: Exxon**

139 Kinney Road

AServ **E:** A & J's Auto Parts

RVCamp **E: Camping**

137 Shepherd Road

Gas **W:** JC's Food Mart*(D)

135 Luckey Road

131 FM 3175, FM 2790, Benton City Rd, Lytle

Gas **W:** Conoco*(D)

Food **W:** Mr. Pizza

Lodg **W:** Best Western

AServ **E:** Tires(LP)

TServ **W: Tires(LP)**

ATM **E:** Lytle State Bank

W: Conoco

(129) **Rest Area - RR, Phones, Picnic, Grills, Vending, RV Dump P (Both Directions)**

127 FM 471, Natalia

124 FM 463, Bigfoot Road

AServ **E:** Chaparral Ford

EXIT TEXAS

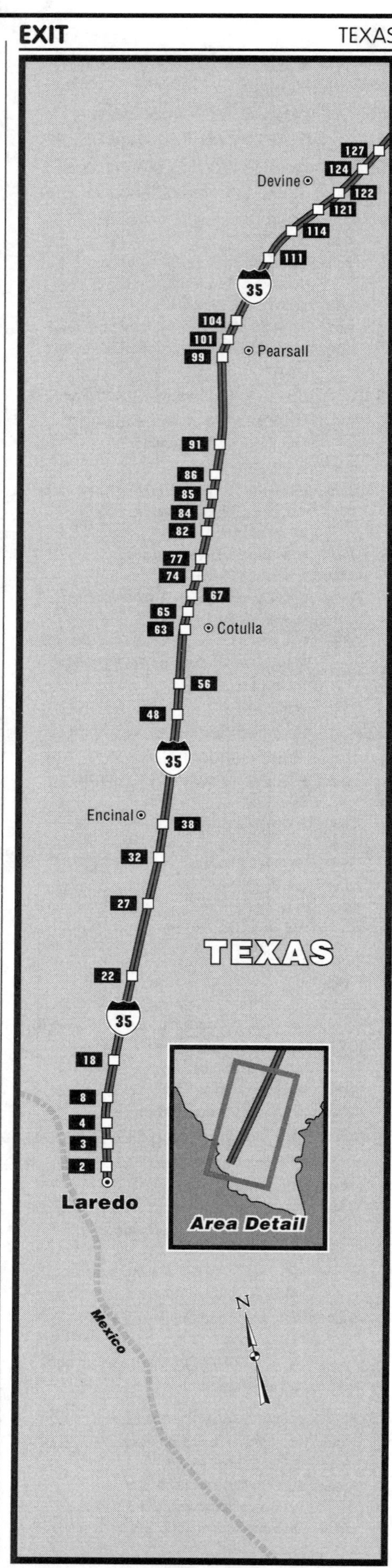

EXIT TEXAS

122 TX 173, Hondo, Jourdanton

FStop **E: Calame Store*(D), Exxon*(D)**

Gas **W:** Chevron*(D)

Food **E:** Bob's BBQ

W: Subway (Chevron), **Triple-C Steak House (Country Corner Motel)**

Lodg **W: Country Corner Motel**

AServ **E:** Brown Chevrolet Geo

RVCamp **E: Nine Oaks Camper Park**

ATM **E: Calame Store**

W: Triple-C Steak House

121 North TX 132, Devine

(118) **Weigh Station (Both Directions)**

114 FM 462, BigFoot, Yancey, Moore

Gas **E:** Diamond Shamrock*(D)

W: The Moore Store*

Food **E:** Orditas Tacos

W: The Moore Store

AServ **E:** Auto Service

W: The Pit

Other **W: Frio County Sheriff, US Post Office**

111 U.S. 57, Eagle Pass

104 South Bus Loop I-35

101 FM 140, Charlotte, Uvalde

FStop **W: Conoco*(D) (24 Hrs)**

Gas **E:** Chevron*

Food **E:** Cowpokes BBQ, Marty's (Chevron)

W: Blimpie's Subs (Conoco), Chester Fried Chicken (Conoco), Restaurant (Porter House Inn)

Lodg **E:** Rio Frio Motel, **Royal Inn**

W: Porter House Inn

AServ **E:** Brooks GMC Trucks

W: Vega Tire & Road Service

TServ **W: Jacks 24-Hr Tire Service & Sales, Vega Tire & Road Service**

ATM **W: Conoco**

99 Bus Loop I-35, FM 1581, Pearsall, Divot

(93) Picnic - Shelters, Grills, No Services (Both Directions)

91 North Spur 581, FM 1583, Derby

86 South I-35 Bus. Loop

85 FM 117, Batesville

FStop **W: Exxon*(D)**

Food **E: Pacho Garcia**

W: Dairy Queen, **La Pasadita (Safari Motel)**

Lodg **E: Pacho Garcia Motel**

W: Safari Motel

84 TX 85, Charlotte, Carrizo Springs

FStop **W: Diamond Shamrock*, Texaco* (Cleo's Travel Center)**

Food **W:** Campbell House Restaurant, **Chester**

Bold red print shows RV & Bus parking available or nearby

EXIT TEXAS

Fried Chicken (Diamond Shamrock)

Lodg W: Executive Inn

AServ E: Dilley Parts House, Tindall Chevrolet Pontiac Geo

Med E: + Hospital

ATM W: Diamond Shamrock, Texaco

Other E: Super S Foods

82 I-35N Bus Loop, County Line Road, Dilley

76 FM 469, Millett

74 Gardendale

67 FM 468, Big Wells

FStop E: Diamond Shamrock* (24 Hrs, Cleo's Travel Center)

Gas E: Conoco*, Exxon*(D)

Food E: Church's Fried Chicken (Diamond Shamrock), Dairy Queen, Hot Stuff Pizza (Exxon), Log Cabin Restaurant, The Country Store Restaurant (Conoco), Wendy's (Exxon)

Lodg E: Cotulla Executive Inn, Rodeway Inn

TServ E: Valentine's (24 Hr)

TWash E: Catulla Truckwash

RVCamp W: RV Park

ATM E: Diamond Shamrock, Exxon

Other E: Cotulla-La Salle County Chamber of Commerce

65 North Bus. Loop I-35, Cotulla

63 Elm Creek Interchange

(59) Picnic - Shelters, No Facilities (Both Directions)

56 FM 133, Artesia Wells

Gas E: Adam's Grocery*

48 Caiman Creek Interchange

38 TX 44, Bus Loop I-35, Encinal

32 San Roman Interchange

27 Callaghan Interchange

22 Webb Interchange

18 U.S. 83N, Carrizo Springs, Uvalde

(14) Inspection Station - US Border Patrol Station, All Traffic (Northbound, Southbound Is Parking Area)

13 Uniroyal Interchange

TStop E: Pilot*(SCALES)

Food E: Country Cooker (Pilot TS), Subway (Pilot)

TWash E: Blue Beacon (Pilot TS)

ATM E: Pilot TS

8 FM 3464; To FM 1472, TX 20, Bob Bullock Loop To Mines Rd

7 FM 1472, Frontage Rd (Southbound)

4 FM 1472, Del Mar Blvd, Santa

EXIT TEXAS

Maria Ave.

FStop W: Citgo*(D)

Gas E: Chevron, Coastal* (24 Hrs), Exxon

W: Texaco*(D) (24 Hrs)

Food E: Applebee's, CiCi's Pizza, Dolce Vita, El Metate, Las Asadas (24 Hrs), McDonalds (Playground), Tokyo Garden

W: Danny's Restaurant, Ernie's Smokehouse, Mi Tierra Bakery

Lodg E: (AAA) Hampton Inn

W: Executive House Hotel, (AAA) Motel 6

AServ E: Chevron, Exxon

RVCamp E: Casa Norte Trailor Park, RV Camp, Town North RV Park (1 Mile)

ATM E: Albertson's Grocery

W: International Bank of Commerce

Other E: Albertson's Grocery (Pharmacy), Book Mark Books, Border Patrol Sector HQ, J & A Pharmacy, Mail Boxes Etc, North Creek United Artists Theatres, Target Greatland

W: AAA Auto Club

3B Mann Road

Gas E: Exxon(CW)

Food E: El Taco Tote, Long John Silvers, Luby's Cafeteria, Sirloin Stockade, Tony Romas Ribs

W: Chili's, Danny's Restaurant, Golden Corral, Kettle, Pancake House, Subway, Taco Palenque Jr. (Playground)

Lodg W: Family Gardens Inn, Motel 6, Motel 6

AServ E: Montgomery Ward, Powell-Watson Oldsmobile, Sames Honda Subaru, Sames Mazda

W: Wal-Mart

Med E: + Columbia Doctor's Hospital

ATM E: IBC, NBC Bank of Laredo

W: Loredo National Bank, South Texas National Bank

Other E: Mall Del Norte, Maverick Market

W: OfficeMax, Toys "R" Us, Wal-Mart (24 Hrs, 1-Hr Photo, Pharmacy)

3A San Bernardo Ave, Calton Road

Gas E: Texaco*

W: Phillips 66(D)

Food E: Emperor Garden, Long John Silvers, Luby's Cafeteria, Peter Piper Pizza, Pizza Chef Gourmet, Sirloin Stockade, Subway

W: Acapulco Steak & Shrimp, Baskin Robbins, Burger King (Playground), Dunkin Donuts, El Pollo Loco (Playground), McDonalds (Playground), Ming Dynasty,

EXIT TEXAS

Pizza Hut, Popeyes Chicken, Taco Bell (Playground), Taco Palenque (Playground, 24 Hrs), The Big Red Line (Playground), Wendy's

Lodg W: Best Western, Days Inn, Gateway Inn, La Hacienda Motor Hotel, ◆ Red Roof Inn, Siesta Motel

AServ E: NAPA Auto Parts, Pep Boys Auto Center

W: Goodyear Tire & Auto, Sam's Club

ATM E: Norwest Banks

W: International Bank of Commerce

Other E: H.E.B. Mercado Grande, K-Mart (Pharmacy, 1-Hr Photo), Mail Boxes Etc, TX State Optical, Tran Vision Center, Yo Books & Games

W: Car Wash, Fire Dept., Sam's Club

2 U.S. 59, Freer, Corpus Christi, Houston, Laredo Intnl. Airpt

Gas E: Conoco*, Diamond Shamrock*, Texaco*

W: Coastal*, Exxon, Phillips 66*(D)

Food E: Gallegos, Jack-in-the-Box (24 Hrs), Mariscos El Pescador

W: Church's Fried Chicken, Denny's, Pan American Courts Cafe, Quick Bite, The Holiday Bakery, The Shrimp Royal, The Unicorn Restaurant, Vallarta Restaurante

Lodg W: Econolodge, El Courtez Motel, Haynes Motel, La Fonda Motel, La Quinta Inn, Mayan Inn

AServ E: Frontera Auto Parts, Laredo Auto Air & Radiator Shop, Lopez Transmission, Rodriguez Paint & Body Shop

W: Auto Zone Auto Parts, Bumper to Bumper, Exxon, Phillips 66, Ramirez Tire Center, Rodriguez Tire Shop

ATM E: Conoco

Parks E: Lake Casa Blanca State Park

Other E: Chito's, Tiburon Car Wash & Detail

W: Coin Laundry

1B Park St., Sanchez St.

Gas W: Conoco*(D)

Food W: Pizza Hut, Popeye's Chicken

AServ W: Tire Center of Laredo

Med E: + Hospital

1A Scott St., Washington St. (Southbound, Closed For Construction)

Gas W: Conoco*(D)

Food W: Favarato's, La Casita (24 Hrs), La Mexicana, La Paisana II, La Siberia, Oasis BBQ, Pizza Hut, Popeyes

Lodg W: Courtyard (Marriott), Holiday Inn

AServ W: Auto Electrical Shop, City Radiator, Conoco, Guzman Auto Parts, Perez Body Shop, Ranger Automotive, Roy's Auto Center, Tire Center of Laredo Inc., Villarreal Auto Service

RVCamp E: Camping World (see our ad this page)

Med E: + Hospital

ATM W: Conoco

Other W: Los Gueros, Rainbow Bakery Store

↑TEXAS

Begin I-35

EXIT TEXAS

Begin I-37

↓TEXAS

142AB 142A - N. I-35, Austin, 142B - S. I-35, Laredo (Northbound)

141C McCullough Ave, Nolan Street (Southbound)
- **Gas** W: Chevron*(CW)
- **Food** W: Oasis Cafe
- **Lodg** W: The Painted Lady Inn on Broadway
- **AServ** W: ACE Break Service, Auto Air & Axle, Cavender Cadillac, Downtown Auto Parts, Downtown Auto Repair, Federated Auto Parts, Grayson St. Garage, Nix Alignment

141B Houston Street, The Alamo (Southbound)
- **Gas** W: Coastal*
- **Food** W: Burger King, Haagen-Dazs, Morton's, Pizza Hut, Wendy's
- **Lodg** W: Crockett Hotels, DAYS INN Days Inn, Downtowner Motel, Hampton Inn, Holiday Inn, Ramada Inn, Residence Inn by Marriott, The Emily Morgan Hotel, The Hilton, The Menger Hotel
- **AServ** W: Bob's Auto Service
- **Other** W: **AMC Theaters, Alamo Visitor Center, Fire Station, IMAX Theater, King's X Toy Soldiers, Antiques, Ripley's Believe It or Not, Rivercenter, The Alamo**

141A Commerce Street, Downtown (Southbound)
- **Food** W: Denny's, Landry's Seafood House, The Tower of the Americas
- **Lodg** W: La Quinta Inn, Marriott, Marriott
- **Other** W: **AMC Rivercenter Theaters, Rivercenter, Rivercenter Drugstore, The Hemisfair Plaza, The Tower of the Americas**

141 Commerce St, Downtown, The Alamo (Northbound)
- **Food** E: Aldaco's, The Coffee Gallery
- **Lodg** E: Red Roof Inn
- **AServ** E: Theo's Brake & Tire
- **Other** E: **The Alamodome**

140B Durango Blvd, Alamodome, La Vallita (Northbound)
- **Food** E: Bill Miller BBQ, Ray's Mexican
- **Other** E: **Alamodome, Piknik Foods**

140A Florida Street, Carolina Street
- **FStop** W: **Datafleet (Datafleet Credit Cards Only)**
- **Gas** E: Fina*(D)
- **Food** W: Eagle's Nest Cafe

139A I-10, US 90, US 87, El Paso, Houston, Victoria del Rio

139 Fair Ave (Unnumbered)
- **Gas** E: Chevron; W: Exxon, Texaco*
- **Food** E: Dairy Queen, KFC, Peter Piper Pizza, Red Line Burgers, Taco Bell
- **AServ** E: Brake Check, Chevron; W: Exxon
- **ATM** W: Texaco
- **Other** E: **The Home Depot(LP)**

138A New Braunfels Ave, Southcross Blvd
- **Gas** W: Exxon*(CW)
- **Food** E: Hong Kong Buffet Restaurant, Luby's Cafeteria, McDonalds (Playground), Pizza Hut Carry-Out

EXIT TEXAS

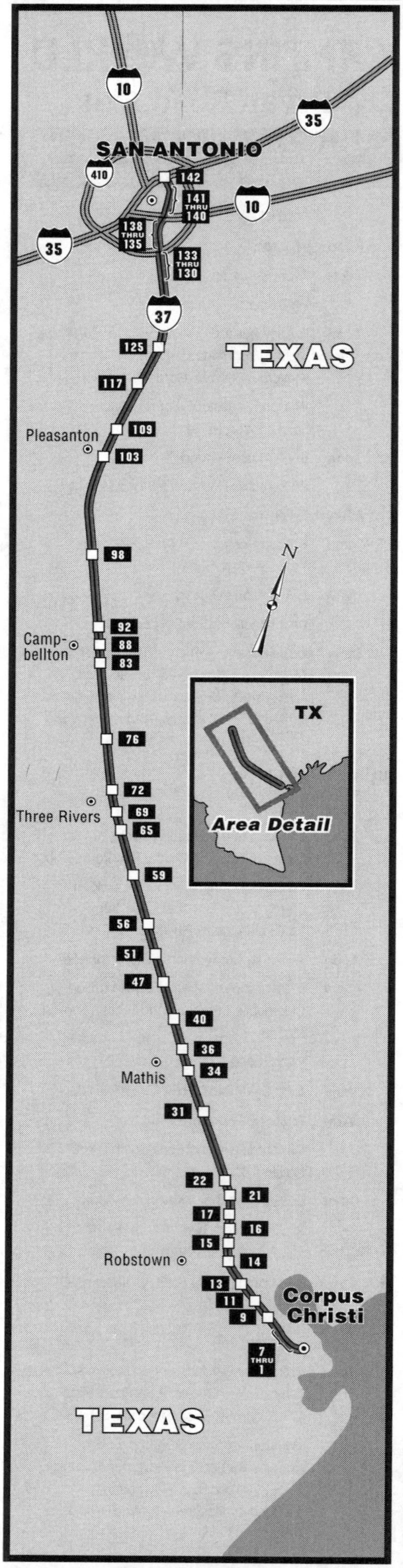

EXIT TEXAS

Delivery, Shoney's (Westside Access), Taco Cabana (24 Hrs), Wendy's; W: Burger King (Playground), Sonic
- **AServ** E: Auto Express (Montgomery Ward), Full Service Auto Parts Wharehouse
- **ATM** E: Bank of America
- **Other** E: **Handy Andy Grocery, McCreless Mall**

137 Hot Wells Blvd (Unnumbered)
- **Gas** E: Exxon
- **Food** E: Taco Hut; W: IHOP
- **AServ** E: Exxon; W: Hot Wheels Automotive Repair & Sales
- **ATM** E: NationsBank
- **Other** E: **Fire Station**

136 **(135)** Pecan Valley Drive (Unnumbered)
- **Gas** E: Coastal*(D), Exxon
- **Food** E: Baysea's Seafood, Beijing Express, Church's Fried Chicken, KFC, Neptunes Seafood House, Pizza Hut
- **AServ** E: Exxon, Joe's Tires & Wheels, One Stop Car Care
- **ATM** E: Handy Andy
- **Other** E: Delco Coin Vehicle Wash, **Eckerd Drugs, For Your Convenience Store, Handy Andy Grocery, Mail Pack, Mission Cleaning Center (Laundromat), Ponderosa Bowl**

135 Loop 13, Military Dr, Brooks AFB
- **Gas** E: Diamond Shamrock*; W: Diamond Shamrock* (24 Hrs)
- **Food** E: New China Restaurant; W: Burger King (Playground)
- **AServ** E: Goodspeed's Collision
- **ATM** E: Diamond Shamrock; W: Diamond Shamrock
- **Other** E: Car Wash; W: **H-E-B Grocery Store (Pharmacy, 24 Hrs, 1-Hr Photo), K-Mart (Pharmacy)**

133 I-410, South U.S. 281, Connally Loop

132 South US 181, Floresville (Southbound, No Southbound Reaccess)
- **Gas** E: Fina*, Stopn Go*, Texaco*

132A Spur 122 (Northbound)
- **Gas** E: Diamond Shamrock*(D)
- **ATM** E: Diamond Shamrock

130 Southton Road, Donop Road, Braunig Lake
- **TStop** E: **Diamond Shamrock Travel Center*(SCALES) (RV Dump)**
- **Food** E: **Braunig Lake Cafe (Diamond Shamrock TS)**
- **Other** E: **Braunig Lake (2 Miles), I-37 Flea Market**

127 San Antonio River Turnaround, Lake Braunig (Northbound)

125 Loop1604, Anderson Loop, Elmendorf
- **FStop** E: **Conoco***
- **Gas** W: Fina
- **Food** E: **Burger King (Conoco)**, Mi Reina

122 Priest Road, Mathis Road
- **Other** E: **Jack's Corner Store**

120 Hardy Road

117 FM 536

113 FM 3006, Pleasanton

(112) Picnic - Tables, Shelters, Grills, No

Bold red print shows RV & Bus parking available or nearby

EXIT	TEXAS
	Other Services (Both Directions)
109	TX 97, Pleasanton, Floresville
FStop	E: JP's Superstop* (Chevron)
Food	E: Shorty's Place #3
RVCamp	E: Camping
Other	W: Tourist Information
106	Coughran Road
104	Spur 199, Leal Road (Difficult Southbound Reaccess)
TStop	E: Kuntry Korner (Diamond Shamrock, 24 Hrs)
FStop	E: Bubba's Fuel Stop* (Citgo)
Food	E: Dairy Queen, Kuntry Korner
Lodg	E: Kuntry Inn
AServ	E: Bubba's FStop (24 Hrs)
TServ	E: Bubba's FStop (Truck Tires, 24 Hrs)
ATM	E: Kuntry Korner TStop
Other	E: RV Service (Bubba's FStop)
103	North US 281, Pleasanton
98	FM 541, McCoy, Poth
92	Alt. U.S. 281, Campbellton, Whitsett (No Southbound Reaccess)
88	FM 1099 To FM 791
83	FM 99, Whittsett, Karnes City, Peggy
FStop	E: Texaco*
	W: Exxon* (Deli)
Gas	W: Chevron*(D)
Food	E: The Sandwich Shop (Texaco)
AServ	E: Cypret's Garage
ATM	W: Chevron
(82)	Rest Area - RR, Phones, Picnic, Grills P (Southbound)
(78)	Rest Area - RR, Phones, Picnic, Grills P (Northbound)
76	FM 2049
(75)	Weigh Station (Southbound)
(74)	Weigh Station (Northbound)
72	U.S. 281 South, Three Rivers, Alice
69	TX 72, Three Rivers, Kenedy
FStop	W: Wolffs Travel Stop* (Daimond Shamrock, 24 Hrs)
Food	W: Andy's BBQ, Cafe (24 Hrs, Wolffs TS)
TServ	W: Wolffs Tire Center
ATM	W: Wolffs Travel Stop
Parks	W: Choke Canyon State Park
65	FM 1358, Oakville
Gas	E: Texaco*
	W: Chevron*
Food	E: Van's BBQ (Texaco)
Other	W: US Post Office (Chevron)
59	FM 799
56	U.S. 59, George West, Beeville
FStop	E: Texaco*
	W: Chevron*, Exxon*
Food	W: Burger King (Exxon), Manster's Roadhouse (Chevron)
ATM	E: Texaco
	W: Chevron
51	Hailey Ranch Road
47	FM 3024, FM 534 (No Reaccess)
RVCamp	W: KOA Kampground (4 Miles)
(43)	Parking Only - No Services P (Southbound)

EXIT	TEXAS
(41)	Parking Area - No Services (Northbound)
40	FM 888
RVCamp	W: KOA Kampground
36	TX 359, Skidmore, Alice
Gas	W: Exxon*(D)
Food	W: El Taco Loco, Pizza Hut, Ranch Motel Restaurant
Lodg	W: Mathis Motor Inn, Ranch Motel
AServ	W: Compos New & Used Tires, Exxon
RVCamp	W: Camping, KOA Kampground (9 Miles), Mathis RV Park
Parks	W: Lake Corpus Christi State Recreational Area
34	TX 359 West
Food	W: CW Sandwich Mart
AServ	E: C&C Auto Service
Parks	W: Lake Corpus Christi State Recreational Area
Other	W: Wash & Vac (Car Wash)
31	TX 188, Sinton, Rockport
22	FM 796, Edroy, Odem
20B	Cooper Road
(19)	Rest Area - RR, Phones, Vending, Picnic, Grills P (Both Directions)
17	U.S. 77 North, Victoria
16	Nueces River Park
Parks	W: Nueces River Park
Other	E: Public Boat Ramp
	W: City of Corpus Christi Tourist Info Center
15	Sharpsburg Road, Red Bird Lane
14	U.S. 77, Brownsville, Rio Grande Valley, Kingsville
TStop	W: Sun Mart #114*(SCALES) (in Robstown on U.S.77)
13B	Sharpsburg Rd. (Northbound)
13A	Callicoatte Rd, Leopard St
Food	E: Mother's Pizza
AServ	E: Don's Automotive Service
Other	W: Hilltop Bowl
11	Hart Road, TX 24, Violet Road, McKinsey Road, Carbon Plant Road
Gas	E: Conoco*
	W: Exxon(D)(LP), H-E-B
Food	W: McDonalds (Playground), What A Burger (24 Hrs)
Lodg	W: Best Western
AServ	W: Exxon
Other	W: H-E-B Grocery (w/Pharmacy, One-Hr Photo)

EXIT	TEXAS
10	Carbon Plant Rd
9	FM 2292, Up River Rd, Rand Morgan Rd
Gas	W: Texaco*
7	Tuloso Road, Suntide Road
RVCamp	W: Camping
6	Southern Minerals Road
5	Corn Products Road
FStop	W: Petro Fleet (Pay w/credit card only.)
Food	W: Kettle Restaurant, Parkside Inn, Restaurant (Travelodge)
Lodg	W: Parkside Inn, ◆ Red Roof Inn, AAA Travelodge
TServ	E: French-Ellison Truck Center*(D)
ATM	W: Travelodge
4B	Lantana Street, McBride Lane (Southbound)
Lodg	W: Motel 6
4A	TX 358, NAS - CCAD Padre Island (Limited Access Hwy, Southbound)
4	TX 44, C.C. International Airport, TX 358, Padre Island (Northbound)
3B	McBride Lane, Lantana Street (Northbound)
Lodg	E: Hampton Inn
3A	Navigation Blvd
FStop	E: Texaco*
	W: Coastal Datafleet (24 Hr, Credit Card Only, No Attendant), Petro Fleet (Automated Fueling, 24 Hrs, Credit Cards Only)
Gas	W: Exxon*(D)(CW)
Food	W: Bill Miller BBQ, Denny's, Restaurant (Days Inn)
Lodg	E: Val-U Inn Motel
	W: DAYS INN AAA Days Inn, AAA La Quinta Inn
AServ	W: Creager Tire, Richard's Auto, Winston Wharehouse Auto & Truck Parts
TServ	W: Creager Tire, Winston Wharehouse
ATM	W: Exxon
2	Up River Road
RVCamp	W: Camping
1E	Lawrence Drive, Neuces Bay Blvd
Gas	E: Citgo*, Texaco*
	W: Citgo*
1D	Port Ave , Brown Lee Blvd, Port of Corpus Christi (Southbound)
Gas	W: Coastal*(D) (Kerosene)
Food	W: Vick's Hamburgers
AServ	W: Carl Kuehn's Central Auto Body, Gulf Radiator, Victoria Tire Repair
ATM	W: Frost Bank
1C	TX 286, Crosstown Expressway (Northbound Is A Left Exit)
1B	Brownlee Blvd, Port Ave (Northbound)
Gas	W: Chevron*
Food	W: Taqueria Banda's
AServ	W: Eddie Villarreal & Son, Shaffer's Muffler
1A	Buffalo Street, City Hall (Southbound)
Gas	W: Texaco* (Parking Garage)
AServ	W: Texaco (Parking Garage)
ATM	W: Mercantile Bank
Other	W: Public Library, US Post Office
1	**(0)** U.S. 181, TX 35, Portland (Southbound, Unnumbered Exit)
Other	W: CC Beach, Texas State Aquarium, USS Lexington

↑TEXAS

Begin I-37

INTERSTATE EXIT AUTHORITY

Bold red print shows RV & Bus parking available or nearby

I-39 S →

EXIT ILLINOIS

Begin I-39

↓ILLINOIS

2 (4) Business U.S. 51, Bloomington Normal

1 Jct. i-55, I-39

5 (7) Hudson

8 IL 251, Kappa, Lake Bloomington Rd
- RVCamp W: Camping

14 U.S. 24, Peoria, El Paso
- FStop E: Shell*(D) (ATM)
- Gas E: 76*(D)(LP) (ATM)
- Food E: Dairy Queen, Hardee's, McDonalds, Subway (Shell)
  W: Monical's Pizza
- Lodg E: Days Inn
  W: Super 8 Motel
- AServ E: NAPA Auto Parts

22 IL 116, Peoria, Pontiac

27 Minonk

35 IL 17, Lacon, Wenona
- FStop E: Amoco* (ATM)
- Food E: Burger King, Buster's Family Restaurant
- Lodg E: Super 8 Motel
- ATM E: Amoco

41 IL 18, Henry, Streador

48 Tonica
- FStop E: Village Inn Cafe
- Gas E: Amoco*
- Food E: Village Cafe
- TServ E: I-39 Truck Repair, J & J Truck Repair

51 IL 71, Oglesby, Hennepin

52 IL 251, La Salle - Peru

54 Oglesby
- Gas E: Amoco*, Shell* (ATM)
- Food E: Baskin Robbins, Dunkin Donuts, Hardee's, McDonalds, Subway
- Lodg E: AAA Days Inn, AAA Holiday Inn Express

57 U.S. 6, Ottawa, La Salle - Peru
- Gas W: Casey's General Store*
- RVCamp W: Camping

59AB I-80 Junction to Chicago & Des Moines

66 U.S. 52, Troy Grove
- RVCamp E: KOA Campground

72 U.S. 34, Mendota, Earlville
- FStop E: Amoco*(SCALES), Shell*(D) (ATM)
- Food E: Buster's Family Restaurant, McDonalds
- Lodg E: Super 8 Motel

82 Paw Paw
- RVCamp E: Smith's Stonehouse Park (10 Miles)

(86) Rest Area - Picnic, RR, HF, Phones, RV Dump P

87 U.S. 30, Sterling, Rock Falls, Aurora
- Parks W: Shabbona Lake State Park

93 Steward

(108) Tollaway Oasis
- Gas B: Mobil
- Food B: McDonalds

EXIT ILLINOIS

EXIT ILLINOIS/WISCONSIN

(107) IL38, IL23, Annie Gidden Rd., Dekalb
- Lodg N: Super 8 Motel

97AB I-88 Tollway, Moline, Rock Island

99 IL 38, Rochelle, DeKalb
- TStop W: Petro*(CW)(SCALES)
- Gas W: Amoco* (Petro), Mobil (Petro)
- Food W: The Iron Skillet (Petro), Wendy's (Petro)
- Lodg W: Amerihost Inn, Super 8 Motel
- Med W: Hospital

104 IL 64, Oregon Sycamore

111 IL 72, Byron, Genoa

115 (121) Baxter Rd

122A Harrison Ave
- FStop W: Marathon
- Gas W: Citgo*(CW)
- Food W: Kegel's Diner, T.G.I. Friday's
- Other W: Shopping Mall

122B U.S. 20E, Belvidere

↑ILLINOIS

↓WISCONSIN

92 US51, Portage
- Gas E: Kwik Trip* (ATM)
- Food E: Culver's Frozen Custard, Hardees, KFC, McDonalds, Taco Bell
- Lodg E: Best Western, Ridge Motor Inn, Super 8 Motel
- ATM E: First Star
- Other E: K-Mart, Wal-Mart

89AB WI16, WI127, Portage, WI Dells
- FStop N: 76*(D)

87 WI 33, Portage

85 Cascade Mt. Rd.

100 WI23, CR P, Endeavor
- Gas E: US

104 CR D, Packwauklee

106 WI82, WI 23 Oxford, Montello
- Food E: Rissen's Family Restaurant
- Lodg E: Crossroads Motel

113 CR E, CR J, Westfield
- FStop W: Mobil*
- Gas W: Amoco* (ATM), Union 76*(CW)
- Food W: Bob's & Kay Family Inn, Brakebush Chicken, Hardees, McDonalds, Pioneer Restaurant, Subway
- Lodg E: Sandman Motel
  W: Pioneer Motor Inn
- ATM W: Montello State Bank

(117) Rest Area - Picnic, RR, Vending, RV Dump

124 WI 21, Coloma, Necedah
- Gas E: Mobil*(D)

(126) Truck Weigh Station (Left Lane)

131 CR V, Hancock
- Gas E: Phillips 66*

136 WI 73, Plainfield, WI Rapids
- TStop E: Plainfield Truck Stop*
- FStop E: Amoco*(D)
- Lodg E: R&R Motel
- AServ E: BF Goodrich
- TServ E: Amoco

Bold red print shows RV & Bus parking available or nearby

EXIT		WISCONSIN
138		CR D, Almond
143		CR W Bancroft, Wisconsin Rapids
	Gas	E: Citgo*(D)
151		WI 54, Wisconsin Rapids, Waupaca
	Gas	E: Phillips 66*(D)
	Food	E: Four Star Family Restaurant, Shooters
	Lodg	E: Days Inn, Elizabeth Inn
153		CR B Clover Amherst
	Gas	W: Amoco*, Mobil*(D)(CW)
	Food	W: Blake's Family Restaurant, Burger King (Amoco)
	Lodg	W: Americinn
	ATM	W: Wood County National Bank
156		CR HH Whiting
158		US 20 Stevens Point, Waupaca
	Gas	E: Mills Fleet Farms(CW), Mobil*(D)(CW) (ATM)
		W: Amoco*(CW)
	Food	E: Applebees, Colver's Frozen Dessert, McDonalds, Shoney's, Taco Bell, Wendy's
		W: Cafe Royal, Hilltop Grill
	Lodg	E: Fairfield Inn
		W: Best Western, Budgetel Inn
	AServ	E: Mills Fleet Farms
	Other	E: Target Department Store, Wal-Mart
159		WI 66 Stevens Point, Rosholt
	Gas	W: Kwik Trip* (ATM)
	TServ	W: Mock Truck Repair Service
161		Bus 51, Stevens Point
	Gas	W: Amoco*, Kwik Trip* (ATM), Super America*(D) (ATM)
	Food	W: Burger King, China Garden, Country Kitchen, Dominio's Pizza, Fender's, Hardees, Hunan Chinese, KFC, McDonalds, Michelle Restaurant, Olympic Family Restaurant, Perkin's Family Restaurant, Pizza Hut, Pointer's Family Restaurant, Ponderosa Steakhouse, Rocky Rococo Italian, Subway, Taco Bell, Topper's Pizza
	Lodg	W: Comfort Suites, AAA Holiday Inn, Point Motel, Super 8 Motel, The Road Star Inn
	Other	W: Coin Laundry, Kmart (Pharmacy)
164		Truck Weigh Station (Both Directions)
165		CR X
171		CR DB Knowlpon DuBay
175		WI 34, Knowlton, Wisconsin Rapids
(178)		Rest Area - No Services (Northbound)
179		WI 53, Mosinee, Elderon
	Gas	W: Amoco* (ATM), Shell*(D)(CW)
	Food	W: Hardees, McDonalds, The Stage's
	Lodg	W: AAA Amerihost Inn
	ATM	E: River Valley State Bank
	Other	E: Middle Wisconsin Airpot
181		Maple Ridge Road
(183)		Rest Area - No Services (Southbound)
185		Bus 151, Rothschild Wausau
	Gas	E: Mobil*(CW)
	Food	E: Covert's Custards, Denny's, Tony Roma's
	Lodg	E: Comfort Inn, Stoney Creek Inn
	ATM	E: M&I Bank
	Other	E: Cedar Creek Factory Outlet Stores, Gander Mountain Outdoor Sportsmen Center
187		WI 29, GreenBay, US51

EXIT		WISCONSIN
188		CR N
	TStop	E: Amoco*(SCALES)
	Gas	E: Spur*
	Food	E: Java City Bagels, McDonalds, Rosati's Pizza, Wendy's
	Lodg	E: Days Inn
	AServ	E: Exhaust Pros
	TServ	E: Cummins Diesel, Ryder Truck Rental
	RVCamp	E: Kings Campers & Service
	ATM	E: Marathon Savings Bank
		W: M&I State Bank
	Parks	W: Rib Mountain State Park
190		CR NN
	Gas	W: Citgo*(D)(CW) (ATM), Mobil*
	Food	E: Emma Krumbees
		W: Shakey's Pizza, Subway
	Lodg	E: Wausau Inn
		W: Best Western, Hoffman's House
	ATM	E: First Star Bank
	Other	E: Dental Clinic
		W: Wisconsin Station Patrol
191		Sherman Street (Difficult Northbound Reaccess)
	TServ	W: Mike's Trucks Accessories, Northwest Trucks International Dealer
192		WI59, WI 52, Wassau, Abbotsford (Difficult Reaccess, Low Bridges)
	Gas	E: 29 Super, Mobil*(D)
	Food	E: American Country Cafe, Applebees, Big Apple Bagels, Captain's Steak, Counsin's Subs, George's Restaurant, Little Caesars Pizza, McDonalds, Pizza Chef, Subway
		W: Burger King, Fender's Frozen Custard, Hardees, Schlotzkys Deli, The 25 10 Family Restaurant
	Lodg	E: Budgetel Inn, Exel Inn, Hampton Inn, Marlene Motel, Ramada Inn, Super 8 Motel
	AServ	E: Horak's
		W: Penzoil 10 Minute Lube
	ATM	E: Security Bank
	Other	E: Midwest Dental
		W: White Water Car Wash
193		Bridge Street (Low Bridge)
194		CR U, US51, CR K, Wassau (Low Bridges)
	Gas	E: Kwik Trip* (ATM)
	Food	E: McDonalds, Philly's Subs, Taco Bell
	Other	E: Car Wash
197		CR WW, Brokaw (Low Bridge)
	Gas	W: Citgo*
205		Bus 51, Merrill
	TStop	E: Union 76 Truck Stop*(D)
208		WI 64, WI 17, Merrill, Antigo (Low Bridges)
	Gas	W: Mobil*(D)
	Food	W: 3's Company, Burger King, Diamond Dave's Taco, McDonalds, Pine Ridge Restaurant
	Lodg	W: Best Western, Super 8 Motel
	AServ	W: Pine Ridge Quick Lube
	RVCamp	W: Camping
	Med	W: + Hospital
	ATM	W: River Valley State Bank
	Other	W: BP Car Wash, Wal-Mart

↑WISCONSIN

Begin I-39

EXIT		CALIFORNIA

Begin I-40

↓CALIFORNIA

EXIT		CALIFORNIA
1		I-15, E. Main St., Montaro Road
	Gas	N: Mobil* (ATM), Shell*, Terrible Herbst*
		S: Texaco*(D)
	Food	N: Big Mick's, Donut Star, Frosties Donuts, McDonalds, Mega Tom's Burgers, Star Wok Express, Straw Hat Pizza
	Lodg	N: Best Western, Gateway Motel
	AServ	N: Mobil
		S: Texaco(LP), Wal-Mart
	Other	N: Barstow Mall
		S: Wal-Mart (Pharmacy)
3		Marine Corps Logistics Base
	Lodg	N: Roadway Inn
	TServ	S: Woodard Deisel Service
7		Daggett
	Gas	N: Dagget Fuel
	AServ	N: Interstate Fleet Service(LP) (Towing)
	RVCamp	N: Desert Springs RV Park*(LP) (2 Miles)
12		Barstow - Daggett Airport
19		National Trails Hwy
	FStop	N: NewBerry Truck Stop
	Gas	S: Kelly's Market*
	AServ	S: Oasis Auto Parts
	RVCamp	S: Twin Lake RV Park
23		New Berry Springs
	FStop	N: Wesco*
	AServ	S: Silver Valley Auto Repair (2 Miles)
	RVCamp	S: Twin Lakes RV Park (3 Miles)
(27)		Rest Area - Phones, RR, HF, Phones (Eastbound)
(28)		Rest Area - Picnic, RR, HF, Phones (Westbound)
32		Hector Road
50		Ludlow, Amboy, 29 Palms
	TStop	S: Ludlow Truck Stop
	Gas	N: Texaco*
		S: Chevron*(D) (ATM)
	Food	S: Ludlow Coffee Shop
	Lodg	S: Ludlow Motel
78		Amboy, Kelso, Kelbaker Road
	RVCamp	N: Camping
100		Essex, Essex Road
	Parks	N: Providence Mtn. State Rec. Area
(104)		Rest Area - Picnic, Phones, RR, HF (Eastbound)
(105)		Rest Area - Picnic, RR, Phones, HF (Westbound)
107		Goffs Road, Essex
	FStop	N: Naja's Fuel Stop*(D)
115		Mountain Spring Road, Amboy, 29 Palms
120		Water Road
133		U.S. 95, Searchlight, Las Vegas
141		Bus40, W Broadway, River Road
	FStop	S: Arco, Chevron, Texaco(D)
	Gas	S: Mobil*
	Food	S: California Restaurant, Carl's Jr Hamburgers, Hui's Chinese American Cuisine, Taco Bell, Wagon Wheel Restaurant

Bold red print shows RV & Bus parking available or nearby

EXIT CALIFORNIA/ARIZONA

- **Lodg** S: Best Chalet, **Best Western**, Lablum Motel, Royal Inn, Star Dust Motel
- **AServ** S: Broadway Tire, GMC, Oldsmobile, Chevrolet, Pontiac Dealers, Mobil, Texaco, West Point Auto
- **TServ** N: **Southwestern Truck RV**(LP)
- **RVCamp** N: **KOA Campground**

142 Needles, J Street
- **Gas** N: 76, Shell*(CW)
- **Food** N: Jack-In-The-Box, McDonalds; S: Denny's
- **Lodg** N: **Travelers Inn**; S: Motel 6
- **AServ** N: 76
- **Med** N: **Hospital**
- **Other** N: **City of Needles Aquatics Center**

143 U.S. 95, E Broadway
- **Gas** N: Arco*, Chevron, Shell*, Texaco*(D)
- **Food** N: Burger King, Vito's Pizza
- **Lodg** S: **Super 8 Motel**
- **AServ** N: Texaco(D); S: TransTech Transmission
- **TServ** S: **Trans Tech Transmission**
- **Other** N: **Bas Ha's Supermarket, Thrifty Food Store**

147 Five Mile Road

148 Inspection Stop (All Vehicles Eastbound)

(149) Inspection Stop (All Vehicles Westbound)

153 Park Moabi Road

155 Topock

↑CALIFORNIA

↓ARIZONA

1 Topock, Lochlin , Bull Head City

2 Needle Mountain Road

3 Weigh Station (Both Directions)

9 AZ95S, Lake Havasu City, Parker
- **Parks** S: **Lake Havasu State Park**

13 Franconia Road

20 Gem Acres Road

(23) **Rest Area - RR, Phones, HF, Picnic, Vending P (Eastbound)**

(24) **Rest Area - Picnic, Phones, RR, HF, Vending P (Westbound)**

25 Alamo Road
- **FStop** N: **Micro-Mart***

26 Proving Ground Road
- **Lodg** S: Whiten Brothers Motel

EXIT ARIZONA

EXIT ARIZONA

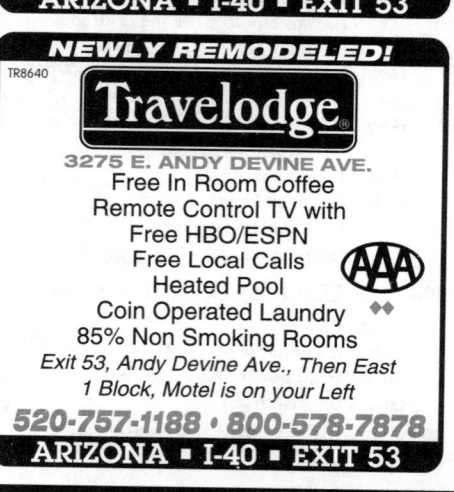

28 Old Trails Road

37 Griffith Road

44 AZ66E, Oatman Road, McConnico Rd.
- **TStop** S: **Crazy Fred's***

48 U.S. 93 to AZ66 & AZ68, Beale St, Las Vegas
- **TStop** N: **76 Auto/Truck Plaza***(SCALES)
- **FStop** N: **Pilot Travel Center*, Shell***(D)(LP)
- **Gas** N: **Exxon***(D) **(RV Dump)**; S: Chevron*(D)
- **Food** N: House of Chan (KFC Express); S: Blimpie's Subs (In Chevron), Calico's Restaurant, Carl's Restaurant
- **Lodg** S: **Arizona Inn, AAA Motel 6**
- **AServ** S: Chevron
- **Other** S: **AZ Tourist Information, Mohave Museum of Art and History**

51 Stockton Hill Road
- **Gas** N: Chevron*; S: Circle K*, Texaco*(D)
- **Food** N: **Hing's Chinese**, KFC, Subway, Taco Bell; S: Dairy Queen, Golden Corral, Little Caesars Pizza
- **AServ** N: Checker's Auto Parts, Wal-Mart
- **Med** N: **Kingman Regional Medical Center**
- **ATM** N: Arizonia Bank
- **Other** N: **Albertson's Grocery, Smith's Grocery, Wal-Mart**; S: **Hastings Bookstore, Safe Way Grocery**

53 AZ66, Bus40, Andy Devine Ave, Kingman
- **TStop** N: **Flying J Travel Plaza**(LP) **(RV Dump)**; S: **Texaco**
- **Gas** N: Terrible Herbst*, **Shell (see our ad this page)**; S: Diamond Shamrock, Exxon, Sinclair*, Texaco*
- **Food** N: **Arby's, Burger King**, Denny's, Jack-In-The-Box, **McDonalds**, Pizza Hut, Taco Bell; S: JB's Restaurant, Low's Restaurant
- **Lodg** N: DAYS INN AAA Days Inn, **First Value Inn, Silver Queen Motel**, AAA Super 8 Motel, Travelodge (see our ad this page); S: AAA Best Western, Comfort Inn, AAA High Desert Inn, AAA Holiday Inn, Lido Motel, Mohave Inn, Route 66 Motel, Wayfair Lodge
- **AServ** N: Goodyear Tire & Auto, Tire World; S: B&J Service Center Uniroyal, Exxon, Quality Tires
- **RVCamp** S: **Quality Stars, Circle S (see our ad this page)**
- **Other** N: **Bas Ha's Supermarket, K-Mart**; S: **Uptown Pharmacy**

59 D W Ranch Road

66 Blake Ranch Road
- **TStop** N: **Petro***(SCALES)

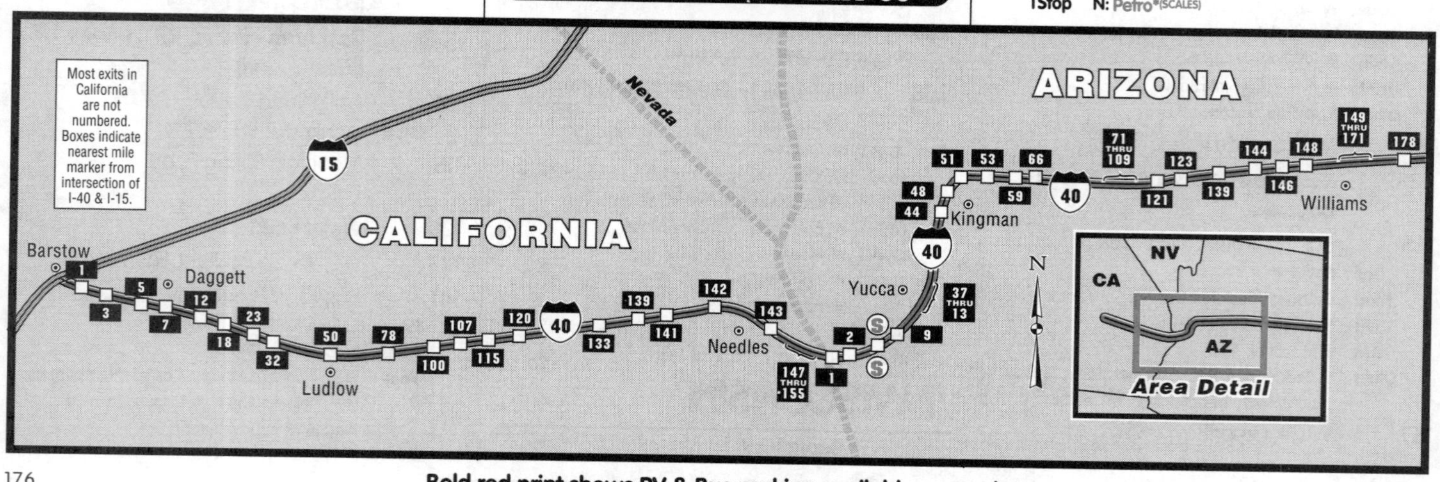

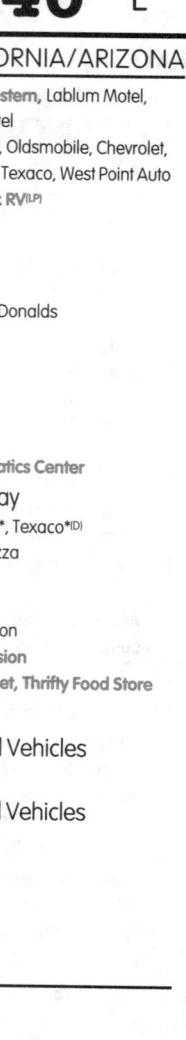

Bold red print shows RV & Bus parking available or nearby

EXIT ARIZONA

S: Beacon
Food N: Iron Skillet, Stage Stop (Petro)
S: Beacon Cafe
TServ N: Bob's Truck Refrigeration, Capital Diesel Repair, Petro
TWash N: Blue Beacon (In Petro)
RVCamp N: RV Park
71 Jct. U.S. 93S, Wickenburg, Phoenix
79 Silver Springs Road
87 Willows Ranch Road
91 Fort Rock Road
96 Cross Mountain Road
103 Jolly Road
109 Anvil Rock Road
121 AZ66, Bus40, Seligman
123 AZ66, Bus40, Seligman, Peach Springs
FStop N: Shell
S: Mobil*
Food S: Subway (In Mobil)
RVCamp N: KOA Campground (1 Mile)
Parks N: Grand Canyon Caverns Supai
139 Crookton Road
144 Bus40, Ash Fork
FStop S: Chevron, Exxon*(LP)
Gas S: Chevron*(D)
Lodg N: AAA Ash Fork Inn, Stage Coach Motel
AServ S: Chevron
TServ N: Diesel Repair
RVCamp N: KOA Campground (Behind Ashfork Inn)
S: Hillside RV Park (RV Dump)
Other S: Coin Laundry (In Exxon)
146 AZ89, Bus40, Ash Fork, Prescott
TStop N: Ted's Truck Center*(LP)
Gas N: Bell*, Mobil
Food N: Ted's Bull Pen, Ternow's Ranchouse Cafe
Lodg N: Ashfork Inn, Highline Motel
AServ N: Gary's Automotive, Mobil
Other N: Zettlers Market
148 County Line Road
149 Monte Carlo Road
TStop N: Monte Carlo Truck Stop
TServ N: Monte Carlo Truck Stop
151 Welch Road
157 Devil Dog Road
161 Bus40, Williams, Grand Canyon, Country Club Road
Gas S: Shell*
Food S: Denny's
Lodg S: AAA Best Western, DAYS INN ◆ Days Inn
RVCamp N: Camping
Parks N: Cateract Lake
163 Williams, Grand Canyon
Gas N: Chevron*
S: Mobil*, Texaco*(D)
Food N: Subway
S: Doc Holiday (Holiday Inn), Jack in the Box, McDonalds
Lodg N: AAA Fairfield Inn
S: AAA Holiday Inn, AAA Howard Johnson, Roadway Inn
Parks S: Grand Canyon Railway Depot
Other N: Tourist Information
165 AZ64, Bus40, Red Lake, Williams, Grand Canyon
Gas S: Malone's* (1 Mile)
Lodg S: ◆ Super 8 Motel (1 Mile)

EXIT ARIZONA

AServ S: Malone's
167 Garland Prairie Road, Circle Pines Road
RVCamp N: KOA Campground
171 Pittman Valley Road, Deer Farm Road
Food S: Quality Inn*
Lodg S: AAA Quality Inn
178 Parks Road
Gas N: Parks in the Pine 76 Gas* (1/2 Miles)
RVCamp N: Ponderosa RV Park
(181) Rest Area - Phones, RR, HF, Picnic P (Eastbound)
(182) Rest Area - RR, Phones, HF, Picnic P (Westbound)
185 Bellemont, Transwestern Road
TStop N: Texaco*
Food N: Country Host Restaurant, Subway (in Texaco)
TServ N: Wertz Goodyear Tires & Interstate Battery
190 A - 1 Mountain Road
191 U.S. 89N, Bus40, Flagstaff, Grand Canyon
RVCamp N: Woody's Camp (2 Miles)
192 Flagstaff Ranch Road
195B AZ89N, Flagstaff, Grand Canyon
Gas N: Circle K*, Exxon, Mobil, Texaco*
Food N: Arby's, Burger King, Buster's Restaurant, Carl's Jr Hamburgers, Chili's, Choco's, Denny's, ElChilito, KFC, McDonalds, Perkins Family Restaurant, Pizza Hut, Red Lobster, Roma Pizza, Ruby Lew's Home Cooking, Sizzler Steak House, Strombolli's, Wendy's
Lodg N: AAA Econolodge, Fairfield Inn, Quality Inn, Amerisuites (see our ad this page)
Other N: Basha's Grocery, Hasting's Books, Target Department Store, Wal-Mart
195A Jct. I-17S, U.S. 89A, Sedona, Phoenix
198 Butler Ave, Flagstaff
TStop S: Little America* (RV Dump)
Gas N: Exxon*, Giant Convenient Store*, Shell
S: Mobil*(LP)

EXIT ARIZONA

Food N: Blackbart's Steakhouse, Country Host Restaurant, Denny's, McDonalds, Taco Bell
Lodg N: Flag Staff Inn, AAA Holiday Inn, Motel 6, AAA Ramada Limited
AServ N: NAPA Auto Parts
S: Mobil Auto & RV Service
TServ N: Little America
RVCamp S: Blackbart's RV Park
Other N: Sam's Club
201 U.S. 89N, Bus40, Page, Grand Canyon, US180W, Flag Staff
Gas S: Mobil*
Lodg N: Super 8 (see our ad this page)
Med N: ✚ Hospital
204 To U.S. 89N, Walnut Canyon National Momument
Parks S: Walnut Canyon National Monument
207 Cosnino Road
211 Winona
FStop N: Texaco*
219 Twin Arrows
TStop S: Twin Arrows Trading Post*
Food S: Twin Arrows
225 Buffalo Range Road
230 Two Guns
FStop S: Shell*(LP)
RVCamp S: Two Guns RV Camp
233 Meteor Crater Road
Gas S: Mobil*
RVCamp S: Meteor Crater RV Park
(235) Rest Area - Phones, RR, HF, Picnic P (Eastbound)
(236) Rest Area - RR, HF, Picnic, Phones P (Westbound)
239 Meteor City Road, Red Gap Ranch Road
245 AZ99, Leupp
252 AZ87S, Bus40, Winslow, Payson
FStop S: Texaco*(D)
Gas S: Mobil*, Shell*(D)
Food S: Burger King, Chinese, Joe's Cafe, Mexican American
Lodg S: Best Inn, AAA Best Western, DAYS INN AAA Days Inn (AAA), Delta Motel, Mayfair Motel, AAA Super 8 Motel
AServ S: A&M Auto Repair
RVCamp S: Glen's RV Shop
253 North Park Dr, Winslow
TStop N: Pilot Travel Center*(LP)(SCALES) (RV Dump)
FStop N: Chevron*(CW) (Quick Lube & O'Hara's Tires)
Gas S: 76*
Food N: Arby's, Captain Toni's Pizza, Denny's, Pizza Hut, Senoir D's Mexican
S: KFC, McDonalds, Subway, Taco Bell
Lodg S: AAA Best Western, AAA Comfort Inn, ◆ Econolodge
AServ S: NAPA Auto Parts
TWash N: United Truck Wash
Med N: ✚ Hospital
Other N: Tourist Information, Wal-Mart
S: Basha's Grocery, Coin Laundry & Car Wash, Safe Way Grocery (Pharmacy)
255 Bus40, Winslow, 87S, Payson
TStop S: Flying J Travel Plaza(LP)(SCALES) (RV Dump)
Gas N: Shell*
Food N: Freddie's Burger Shack

EXIT ARIZONA

- **RVCamp** N: Freddie's RV Park
- **257** AZ87N, Second Mesa
  - **Parks** N: Homolovi Ruins State Park
- **264** Hibbard Road
- **269** Jackrabbit Road
  - **Gas** S: 76*(D)
- **274** Bus40, Joseph City
- **277** Bus40, Joseph City
- **280** Hunt Road, Geronimo Road
- **283** Perkins Valley Road, Golf Course Road
  - **FStop** S: Texaco Truck Stop*(LP) (24hr mechanic)
  - **Food** S: Country Host (Texaco)
- **285** AZ77, Bus40, Holbrook
  - **Food** S: Butterfield Stage Co. Steak House, Cholla, R&R Pizza Express, The Plainsman Family Restaurant
  - **Lodg** S: AAA Adobe Inn, Golden Inn Motel, Sandman Motel, StarInn, Sunn' Sand Motel, Wig Wam Motel
  - **Other** S: Rest Area (picnic, truck & car parking), Safe Way Grocery
- **286** AZ77S, Holbrook, Bus.40, US180E, Show Low
  - **Gas** N: Mobil*
    S: Chevron(D), Exxon*(D), Unical 76*
  - **Food** N: Burger King, KFC, McDonalds, Pizza Hut, Road Runner Cafe, Taco Bell
    S: Dairy Queen, Mr. Maestes
  - **Lodg** N: ◆ Holiday Inn Express (see our ad this page), Super 8 Motel
    S: Budget Inn, ElRancho, Meon Kopi Motel, Western Holiday
  - **AServ** S: Chevron, Crabtree Auto Center, Tate's Auto Center Chrysler Dodge Ford
  - **RVCamp** N: OK RV Park
  - **Other** N: Coin Laundry, Osco Drugs
- **289** Bus40, Holbrook
  - **Gas** N: Chevron*(D), Shell*
  - **Food** N: Alley's Subs & More, Denny's, Maza Restaurant
  - **Lodg** N: AAA Best Western, ◆ Comfort Inn, DAYS INN AAA Days Inn, ◆ Econolodge, AAA Rainbow Inn, AAA Ramada Limited, Sahara Inn, ◆ Travelodge
  - **AServ** N: Alley's Tire and Lube
  - **RVCamp** N: KOA Campgrounds
- **292** AZ77N, Keams Canyon
  - **TStop** N: Holbrook Truck Plaza(SCALES) (RV Dump)

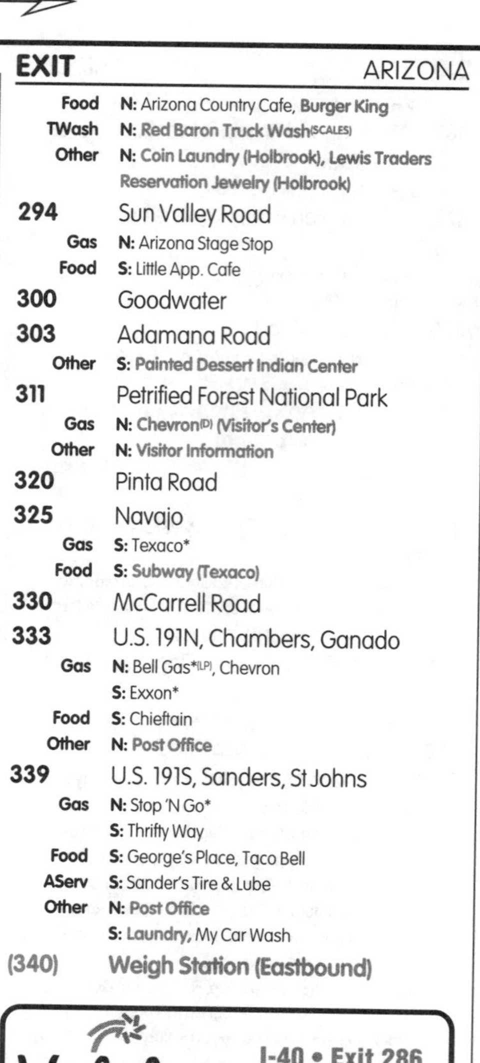

EXIT ARIZONA

  - **Food** N: Arizona Country Cafe, Burger King
  - **TWash** N: Red Baron Truck Wash(SCALES)
  - **Other** N: Coin Laundry (Holbrook), Lewis Traders Reservation Jewelry (Holbrook)
- **294** Sun Valley Road
  - **Gas** N: Arizona Stage Stop
  - **Food** S: Little App. Cafe
- **300** Goodwater
- **303** Adamana Road
  - **Other** S: Painted Dessert Indian Center
- **311** Petrified Forest National Park
  - **Gas** N: Chevron(D) (Visitor's Center)
  - **Other** N: Visitor Information
- **320** Pinta Road
- **325** Navajo
  - **Gas** S: Texaco*
  - **Food** S: Subway (Texaco)
- **330** McCarrell Road
- **333** U.S. 191N, Chambers, Ganado
  - **Gas** N: Bell Gas*(LP), Chevron
    S: Exxon*
  - **Food** S: Chieftain
  - **Other** N: Post Office
- **339** U.S. 191S, Sanders, St Johns
  - **Gas** N: Stop 'N Go*
    S: Thrifty Way
  - **Food** S: George's Place, Taco Bell
  - **AServ** S: Sander's Tire & Lube
  - **Other** N: Post Office
    S: Laundry, My Car Wash
- **(340)** Weigh Station (Eastbound)

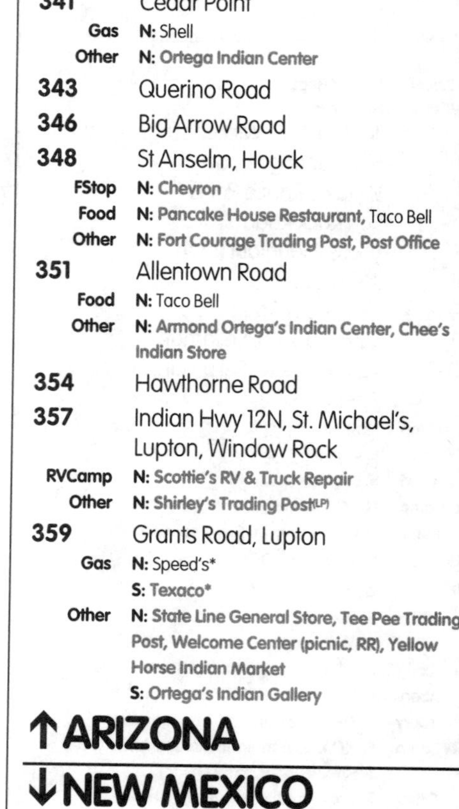

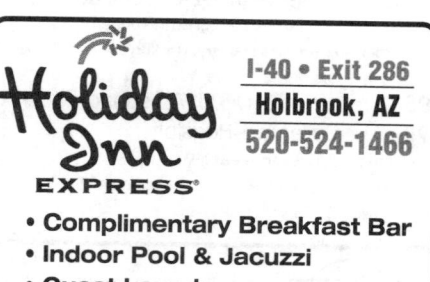

EXIT ARIZONA/NEW MEXICO

- **(341)** AZ Inspection Station (Westbound)
- **341** Cedar Point
  - **Gas** N: Shell
  - **Other** N: Ortega Indian Center
- **343** Querino Road
- **346** Big Arrow Road
- **348** St Anselm, Houck
  - **FStop** N: Chevron
  - **Food** N: Pancake House Restaurant, Taco Bell
  - **Other** N: Fort Courage Trading Post, Post Office
- **351** Allentown Road
  - **Food** N: Taco Bell
  - **Other** N: Armond Ortega's Indian Center, Chee's Indian Store
- **354** Hawthorne Road
- **357** Indian Hwy 12N, St. Michael's, Lupton, Window Rock
  - **RVCamp** N: Scottie's RV & Truck Repair
  - **Other** N: Shirley's Trading Post(LP)
- **359** Grants Road, Lupton
  - **Gas** N: Speed's*
    S: Texaco*
  - **Other** N: State Line General Store, Tee Pee Trading Post, Welcome Center (picnic, RR), Yellow Horse Indian Market
    S: Ortega's Indian Gallery

↑ARIZONA

↓NEW MEXICO

- **(2)** Rest Area P (Eastbound)
- **(3)** Rest Area - RV Dump P (Westbound)
- **8** Manuelito
- **(9)** Weigh Station (Westbound)
- **(11)** Weigh Station (Eastbound)
- **16** Bus. 40, W Gallup
  - **TStop** N: 76 Auto/Truck Plaza*(SCALES) (RV Dump), Loves*, TravelCenters of America(SCALES)
  - **Gas** S: Chevron, Gasman*, Phillips 66*, Texaco*
  - **Food** S: Carriage Inn, Foster's Market, New Mexico Steakhouse, Ranch Kitchen, Taco Bell, Westend Donut & Deli
  - **Lodg** N: Howard Johnson (TA)
    S: AAA Best Western, AAA Budget Inn, Carriage Inn, AAA Comfort Inn, DAYS INN AAA Days Inn,

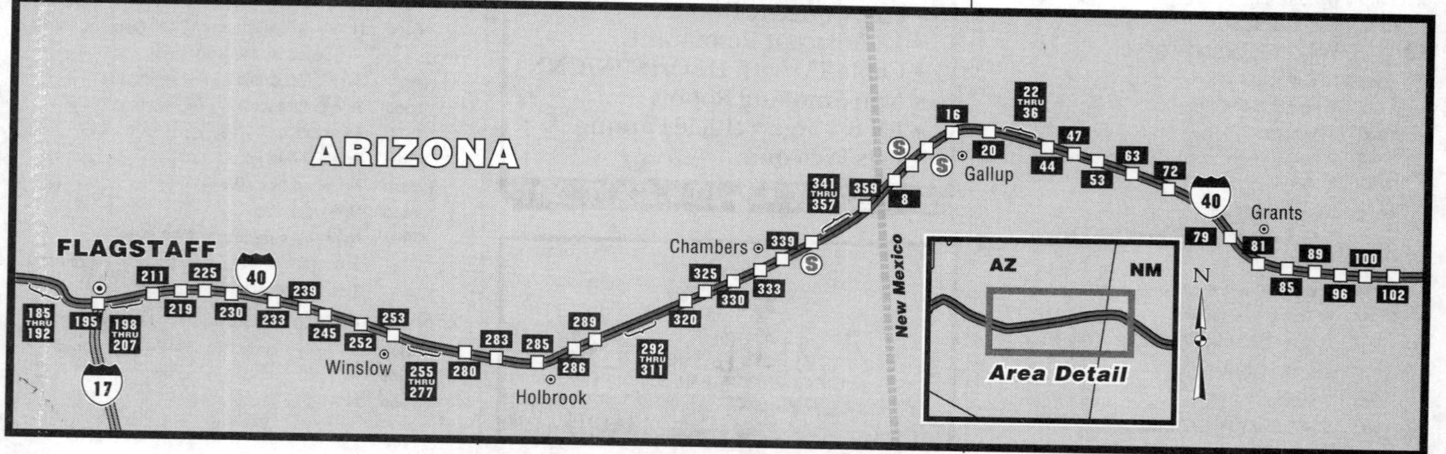

Bold red print shows RV & Bus parking available or nearby

EXIT NEW MEXICO

AAA Econolodge, AAA Holiday Inn, AAA Microtel Inn, Motel 6, Travellers Inn, ◆ Travelodge

TServ N: Augie Truck Service, Bridgestone Tire & Auto

TWash N: American Truck & Car Wash, Blue Beacon

RVCamp S: KOA Campground

Other S: Airport

20 Munoz Blvd., U.S. 666N, NM602S, Shiprock, Zuni

Gas N: Giant*(D), Malco*, Shell, Texaco

Food N: Arby's, Burger King, Captain D's Seafood, Church's Fried Chicken, Dairy Queen, Furr's Cafeteria, KFC, McDonalds, Pizza Hut, Sizzler Steak House, Sonic, Taco Bell, Wendy's

AServ N: Auto Zone Auto Parts, Car Quest Auto Center, Midas Muffler & Brakes, NAPA Auto Parts, Pep Boys Auto Center

RVCamp S: Chapperal RV Park

Med S: ✚ Hospital

Other N: Car Wash, Coin Laundry, Safe Way Grocery, Smith Food & Drug Center, Wal-Mart

22 Miyamura Dr, Montoya Blvd

Gas S: Chevron*, Gasmat, Shell, Texaco, Thrift Way

Food S: Avalon Chinese, Baskin Robbins, Beijing, Church's Fried Chicken, KoKo's, Kristy's Coffee Shop, La Barraca, Panz Alegra Restaurant, Pedro's Mexican, Pizza Hut, Taco Bell

Lodg S: AAA Blue Spruce, El Coronado Motel, El Rancho Hotel & Motel, ElCapatine, Lariat Lodge, Redwood Lodge, Zia Motel

AServ S: Big O Tires, Plaza Lube Center, Smith's Auto Service

Other S: Mission Auto Wash, The Market Scene Supermarket

26 East Gallup, Bus. 40

FStop N: Chevron*

Gas S: Gasman*, Shell*, Texaco*

Food N: Denny's

Lodg N: AAA Sleep Inn
S: AAA Red Rock Inn, AAA Road Runner Motel

AServ S: Bill's Radiator, Shell

Med S: ✚ Gallup Hospital

Parks N: Red Rock State Park

Other N: Gallup Indian Plaza, State Patrol Post
S: Gilbert Ortega's Indian Jewlery Store, Indian America Shops

31 Fort Wingate Army Depot, Church Rock

33 NM400, McGaffey, Red Rock State Park

36 Iyanbito

39 Refinery

TStop N: Giant (RV Dump)

FStop N: Giant

Food N: Giant

TWash N: Giant

(39) Rest Area - RV Dump P (Both Directions)

44 Coolidge

47 Continental Divide

Gas N: Chevron*

Other N: Indians Arts & Crafts Store
S: Post Office

53 Crownpoint, Chaco Canyon

EXIT NEW MEXICO

Farmington, Thoreau

Food N: Tri Pizza

AServ N: Herman's Garage

RVCamp N: St. Bonaventure Mission RV Park

63 NM412, NM612, Prewitt

RVCamp N: Grants West Campground

Parks S: Bluewater State Park

72 Bluewater Village

Gas N: Citgo Travel Center

79 NM122, NM605, Milan, San Mateo

TStop N: Loves*

FStop S: Petro

Food N: Dairy Queen
S: Iron Skillet Restaurant (Petro), Petro

Lodg N: Crossroads Motel

AServ N: Mr. Alternator

TServ S: Baker's Diesel Repair, Petro(SCALES)

Parks S: Chaco Culture State Park

Other N: Milan Super Mart, State Patrol Post

81 Bus.40, NM53S, Grants San Rafael

Food N: Burger King, McDonalds

RVCamp S: Blue Spruce

Parks S: Zuni Sands Canyon

85 NM122, NM547, Bus40, Grants, Mt Taylor

Gas N: Chevron*, Pump & Save, Shell, Texaco*(D)

Food N: 4 B's Restaurant, House of Pancakes, New Mexica Steakhouse, Subway

Lodg N: AAA Best Western, DAYS INN AAA Days Inn, Econolodge, AAA Holiday Inn Express, Motel 6, Super 8 Motel, ◆ Travel Lodge

AServ N: Pump & Save, Shell

RVCamp S: Lavaland

89 NM117, Quemado

Gas N: Diamond Shamrock, Stuckey's

Food N: Stuckey's (Diamond Shamrock)

Other S: El Malpais National Conservation Area

(93) Rest Area - RV Dump P (Eastbound)

96 NM124, McCartys

100 San Fidel

102 Acomita, Sky City

Med S: ✚ Hospital

Other N: Sky City Casino
S: Rest Area - RR, Picnic, Phone

(103) Rest Area P (Westbound)

104 Cubero, Budville

108 Casa Blanca, Paraje

FStop S: Conoco*(D)(LP)

Other S: Casa Blanca Commerical Center

(113) Rest Area - No facilities P (Both Directions)

114 NM124, Laguna

117 Mesita

126 NM6, Los Lunas

131 Canoncito

140 Rio Puerco

FStop N: Chevron*

Gas N: Citgo

Food N: Dairy Queen, Stuckey's

AServ N: Chevron

149 Paseo, Del Volcan, Central Ave,

EXIT NEW MEXICO

Albuquerque

FStop S: Chevron*, Oasis Truck Stop

TServ N: Freightliner Dealer

TWash S: Truck & RV Wash

RVCamp N: Enchanted Trails(LP)
S: American RV Park

Other N: Double Eagle Airport

(151) Rest Area P (Eastbound, No Services)

(152) Rest Area - no services P (Westbound)

153 98th St

TStop S: Flying J Travel Plaza*(LP)(SCALES) (RV Dump)

Food S: Country Buffet (Flting J), Magic Dragon Chinese (Flting J)

154 Unser Blvd, NM345

Gas S: Diamond Shamrock*

155 Rio Rancho, Coors Road

Gas N: Chevron*(CW), Diamond Shamrock* (ATM), Giant*(D)
S: Chevron*(CW)

Food N: Applebees, Cuco's Kitchen, Epip's Mexican, McDonalds
S: Denny's, Furr's Family Dining, Gino's Family Restaurant & Steakhouse, McDonalds, Pizza Hut, Taco Bell

Lodg S: AAA Comfort Inn, DAYS INN ◆ Days Inn, AAA Holiday Inn Express, LaQuinta, AAA Motel 6, Super 8 Motel, Village Inn

AServ N: Jiffy Lube

RVCamp N: Albuquerque West Campground

ATM N: Nations Bank (Shamrock)

Other N: Albuquerque Tourist Information

157A NM194, Rio Grande Blvd

Gas N: Diamond Shamrock*
S: Texaco*(CW)

Food S: Albuquerque Grill (Best Western), Blake's Hamburgers, Burger King, La Hacinda

Lodg S: AAA Best Western, Sherton Inn

157B 12th St

Gas N: Diamond Shamrock
S: Standard, Texaco*

Food N: Burger King
S: Blake's WhataBurger, Maria Teresa

Lodg N: Best Western
S: Sheraton

AServ S: AAPete's Tire Service, RioGrande Automotive

ATM S: Nations Bank

Other S: Old Town Tourist Center

158 8th St, 6th St

FStop N: Love's Country Store*

Food N: Dominio's Pizza

AServ N: Ken's Auto Service

TServ N: Big West Trucks

Other N: WalGreens

159A 2nd St, 3rd St, 4th St

Gas N: Chevron*(CW), Conoco*, Diamond Shamrock*, Economy*

Food N: Furr's Family Dining, What A Burger
S: Village Inn & Pancake House

Lodg S: Interstate Inn, Traveller's Inn

AServ N: Car Quest Auto Center, General Tire, Perfection Plus
S: Firestone Tire & Auto, Ray's Automotive & Truck Service

Parks S: Corondo Park

EXIT NEW MEXICO

159B Jct. I-25N, Santa Fe

160 Carlisle Blvd

- **Gas** N: Texaco
  - S: Circle K*, Conoco, Texaco(LP)
- **Food** N: Blake's Hamburgers, China Dragon, Country Harvest Buffet, Pizza Hut, Rudy's, Sonic, What A Burger
  - S: **Burger King**, Donuts, Subway (Texaco)
- **Lodg** N: (AAA) Budget Inn, (AAA) Radisson
  - S: **Rodeway Inn (see our ad this page)**
- **AServ** N: Sam Boren Tire, United Glass Auto Repair
  - S: Conoco, K-Mart, Texaco
- **Other** N: **Smith's Food & Drug, Wal-Mart (Pharmacy)**
  - S: **Coin Laundry, K-Mart, WalGreens**

161 San Mateo Blvd

- **Gas** N: Giants*(D), Phillips 66(D)(LP), Texaco*
  - S: Diamond Shamrock, Plateau*
- **Food** N: Boston Market, Burger King, K-Bob's Seafood and Steaks, KFC, StarBucks Coffee, Subway, Taco Bell
- **Lodg** N: (AAA) La Quinta Inn
- **Other** N: **Coin Laundry**

162B Louisiana Blvd

- **Food** N: Allie's American Grill, Bennigan's, Garduno's, Japanese Kitchen, LePep, Macaroni Grill, Steak & Ale, T.G.I. Friday's
  - S: Burger King
- **Lodg** N: Best Western, Marriot, Amerisuites (see our ad this page)
  - S: Barcelona Suites
- **Other** N: **Border's Bookstore, Lens Crafters, Shopping Mall**
  - S: **National Atomic Museum**

164AB Wyoming Blvd

- **Gas** N: Circle K*, Phillips 66*(CW)
- **Food** N: MaBarkares Deli, Pizza Hut
- **AServ** N: NAPA Auto Parts
  - S: Rich Ford, Rich Mazda, KIA, Zangara Dodge
- **RVCamp** N: **Travelland RV Service**

164C Lomas Blvd, Wyoming Blvd.

165 Eubank Blvd

- **Gas** N: Chevron, Diamond Shamrock*, Phillips 66, Texaco*
  - S: U-Pump-It*
- **Food** N: Applebees, JB's Restaurant, Owl Cafe, Sonic Drive In
  - S: Boston Market, Burger King, Taco Bell
- **Lodg** N: (DAYS INN) ◆ Days Inn, ◆ Econolodge, (AAA) Holiday Inn Express, ◆ Howard Johnson, ◆ **Ramada Inn (see our ad this page)**

EXIT NEW MEXICO

- **AServ** S: Dan's Foreign Car Service
- **Other** N: **Target Department Store**
  - S: **Sam's Club**

166 Juan Tabo Blvd

- **Gas** N: Chevron*, Phillips 66(CW)
- **Food** N: Atlantic Chinese, Burger King, Carrows, Dominio's Pizza, Lin's, Long John Silvers, Lubby's Cafeteria, Mack's Steak In The Rough, McDonalds, Paul Montrey's Mexican, Village Inn, Wendy's
  - S: Chen's, Dragon City Chinese, Little Caesars Pizza, Wienerschnitzel
- **Lodg** N: (AAA) Park Inn, Super 8 Motel
- **AServ** N: BrakeMasters, Lube & go, Midas Muffler, United Trasmission
- **RVCamp** S: **Action RV Repair, All Systems RV Repair, KOA Campground, Myers RV Center**
- **ATM** N: First Security Bank, Nations Bank
  - S: Norwest Bank
- **Other** N: **Hastings Books, Osco Drugs**
  - S: Citation Car Wash, **Furr's Grocery Store (ATM), WalGreens**

167 Central Ave, Tramway Blvd

- **FStop** S: **Texaco***
- **Gas** S: Chevron*, Fina*
- **Food** S: Burger King, Einstein Brothers Bagels, IHOP, KFC, McDonalds, Sirloin Stockade, Starbuck's Coffee, TCBY Yogurt, Taco Bell, The Four Hills Cafe, Waffle House
- **Lodg** S: ◆ Best Western, Canyon Motel, (AAA) **Comfort Inn**, (DAYS INN) (AAA) **Days Inn (see our ad this page)**, (AAA) Econolodge, **Motel 6**, Roadway Inn, (AAA) Travelodge
- **Other** S: **Mail Boxes Etc, Smith's Food and Drugs**

170 Carnuel

175 NM337, NM14, Cedar Crest, Tijeras

- **RVCamp** N: **Camping**
- **ATM** S: Norwest Bank
- **Parks** N: **Sandia Crest Recreation Area**
- **Other** N: **Sandia Ski Area**
  - S: **National Forest Information, U.S. Post Office**

176 NM14, San Antonio, Tijeras

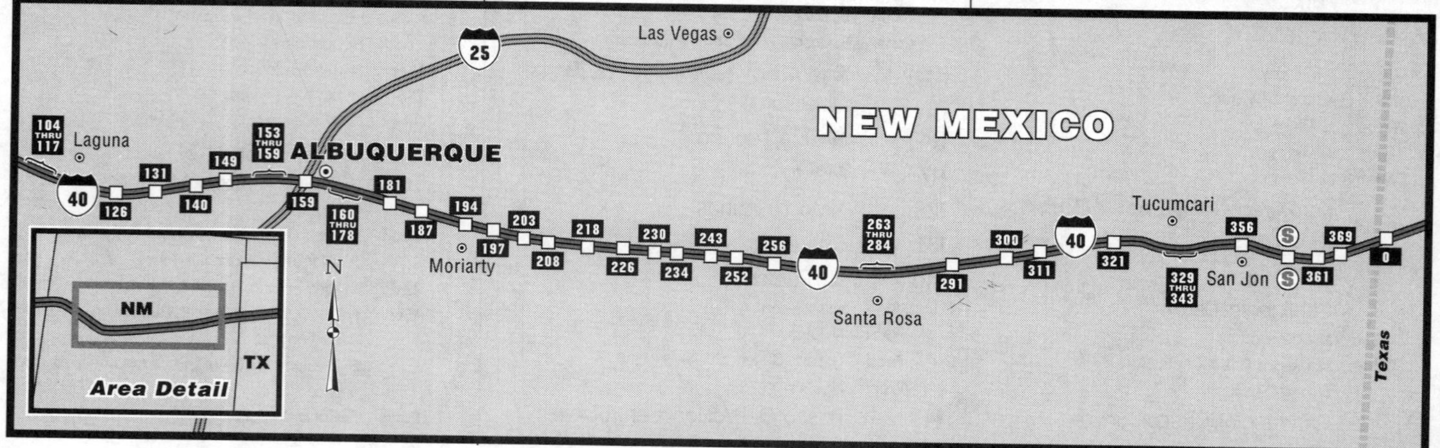

Bold red print shows RV & Bus parking available or nearby

EXIT	NEW MEXICO
178	Zuzax
Gas	S: Chevron
RVCamp	S: Camping (Summer only)
181	NM217, Sedillo
187	NM3344, Edgewood
FStop	S: Phillips 66
Gas	N: Citgo, Conoco
	S: Fina*, Stuckey's*
Food	N: Dairy Queen
	S: Homerun Pizza, Homestead, Stuckey's
AServ	N: Conoco
RVCamp	S: Red Arrow
ATM	N: Nations Bank
	S: Norwest Bank (Stuckey's)
194	Bus 40, Morieto, Rt 66
TStop	S: Rip Griffin Travel Center*(SCALES)
Gas	S: Chevron*
Food	S: McDonalds, Pizza Hut (Rip Griffin), Subway, Taco Bell, Touch Down Pizza
Lodg	S: Days Inn (AAA) Days Inn (AAA), GMC, Chevrolet Dealer, (AAA) Howard Johnson, (AAA) Super 8 Motel
RVCamp	S: Texaco Gas
ATM	S: Sunset Bank
Other	S: Fed Ex Drop Box, IGA Grocery
196	NM41, Howard Cavasos
Gas	S: Circle K, Phillips 66
Food	S: El Comedor
Lodg	S: Larriet Motel, Lazy J Motel, Ponderosa Motel, Sands Motel, Siesta Motel
AServ	S: Jr.'s Tire & Auto Parts, Phillips 66
ATM	S: Norwest Bank
Other	S: Post Office
197	NM41, Bus40, Moriarty, Escancia
FStop	S: Ted's Truck Center(LP)
Gas	S: Vikings Brothers
Food	S: Blake's What a Burger, Bull Pen, Chubbie's Restaurant, Mama Rosa's
Lodg	S: Sunset
AServ	S: Circle O Tire Service, East Route 66 RV & Auto Repair, NAPA Auto Parts
TServ	S: Brippons Complete Truck Service, West Truck Service
RVCamp	S: Camping
203	Frontage Road
Food	S: El Vaquero, Long Horn Ranch
Lodg	S: El Vaquero Motel, Long Horn Ranch Hotel
RVCamp	N: Zia RV Camp
(206)	Rest Area (P) (Eastbound)
(207)	Rest Area - RV Dump, HF, RR (P) (Westbound)
208	Wagon Wheel
Gas	N: Texaco
Food	N: Wagon Wheel Bar and Grill
Lodg	N: Wagon Wheel Motel
AServ	N: Texaco
218	U.S. 285, Vaughn, Santa Fe, Clines Corners, Encino
FStop	N: Chevron*, Shell*
Food	N: Clines Corner Restaurant
226	Local road
230	NM3, Encino, Villanueva
234	Flying C Ranch
FStop	N: Bowlin's Flying C Ranch*
Food	N: Dairy Queen

EXIT	NEW MEXICO
239	Unnamed
243	Milagro
FStop	N: Chevron*
Food	N: Milagro Cafe
(250)	Rest Area (P) (Eastbound)
(251)	Rest Area - RV Dump, RR, Picnic (P) (Westbound)
252	Local road
256	U.S. 84N, Las Vegas, NM219S, Pastura
263	San Ignacio
267	Colonias
FStop	N: Texaco
Food	N: Stuckey's
273	U.S. 54, Bus40, Santa Rosa, Vaughn
Gas	S: Chevron, Conoco*
Food	S: Mateo's
Lodg	S: Oasis Motel
AServ	S: Chevron, Mac's Auto and RV Service, Ortega's RV & Auto Repair
RVCamp	S: Sundown RV Paark
275	U.S. 54, Bus40, Santa Rosa
Gas	N: Phillips 66, Texaco*
	S: Diamond Shamrock, Exxon*, Fina
Food	N: Burger King (Texaco), McDonalds, Santa Fe Grill
	S: Joseph's Restaurant, Pizza Hut
Lodg	N: (AAA) Best Western, Days Inn (AAA) Days Inn, Ramada, Shawford Motel
	S: American Inn, Comfort Motel, La Loma Motel, Sun 'N Sand Motel, Sunset Motel, ◆ Super 8 Motel
AServ	S: Fina, Goodyear Tire & Auto
RVCamp	N: Donnie's RV Park
Med	S: ✚ Hospital
Other	N: NM State Police
	S: Coin Laundry
277	U.S. 84, Fort Sumner, Santa Rosa
TStop	S: TravelCenters of America(SCALES)
FStop	N: Phillips 66*, Texaco*
Food	N: Dairy Queen, Denny's, Silver Moon
	S: Country Pride, Subway
Lodg	N: (AAA) Best Western, ◆ Comfort Inn, (AAA) Holiday Inn, (AAA) Motel 6, Silver Moon Motel
AServ	N: Texaco
TServ	S: Big Rig Truck Service
TWash	S: Joe's
RVCamp	N: KOA Campground
284	Local road
291	Cuervo
Gas	N: Exxon*
AServ	N: 24 Hour Auto Repair
300	NM129, Newkirk
FStop	N: Phillips 66
Gas	S: Vista Mini Mart*(D)
Parks	N: Conchas Lake State Park
Other	N: U.S. Post Office
(301)	Rest Area - RV Dump (P) (Eastbound)
(302)	Rest Area (P) (Westbound)
311	Montoya
321	Palamos
FStop	S: Texaco

EXIT	NEW MEXICO/TEXAS
Food	S: Dairy Queen, Stuckey's
329	U.S. 54E, Bus40, Tucumcari
TStop	N: Shell*
Food	N: Shell Cafe
331	Camino del Corando
Food	N: Cactus Pete's Diner
Other	N: Conway Oil Co.(LP)
332	U.S. 54, NM209, Tucumcari, Quay
Gas	N: Exxon*, Phillips 66, Texaco
Food	N: Blake's Lot A Burger, KFC, McDonalds, Subway
Lodg	N: (AAA) Best Western, Days Inn Days Inn
Med	N: ✚ Hospital
Other	N: NM State Police
333	U.S. 54, Mountain Road
TStop	N: Loves
Food	N: Taco Bell
AServ	N: Bill's Radiator Shop
RVCamp	N: KIVA Park
ATM	N: Loves
335	U.S. 54, Bus40, Tucumcari
Gas	N: Chevron*, Conoco(D)(LP), Shell, Texaco*
Food	N: Western Sizzlin'
Lodg	N: Econolodge, (AAA) Holiday Inn, (AAA) Howard Johnson, Motel 6, ◆ Super 8 Motel
AServ	N: Shell
RVCamp	S: KOA Campground
339	NM278
343	Local road
356	NM469, San Jon
TStop	N: Citgo*(D) (ATM)
Gas	S: Texaco*
Food	N: Burger King (Citgo), KFC (Citgo)
	S: Burger King (Texaco), The Rustler (Texaco)
Lodg	S: San Jaun Hotel
AServ	S: Terry's Service Center
TServ	N: Old Route 66 Truck and Auto Service
Other	S: U.S. Post Office
(358)	Weigh Station (Both Directions)
361	Bard
369	NM93, NM392, Endee
(373)	NM Welcome Center - RV Dump, HF, Phones (Westbound)

↑NEW MEXICO

↓TEXAS

EXIT	TEXAS
0	Bus40, Glen Rio, Old Route 66
(12)	Rest Area - RR, HF, Phones, Picnic (P)
18	Gruhlkey Road, FM2815
FStop	S: Stuckey's*
Food	S: Stuckey's
22	TX214, Bus40, Adrian
Gas	S: Phillips 66*(D)
Food	N: Fabulous 40 Cafe, Route 66 Midway Cafe
Other	N: Loveless Propane(LP), U.S. Post Office
28	Landergin
Food	N: Granny's Kitchen
(31)	Rest Area - RR, HF, Phones, Picnic (P) (Eastbound)
(32)	Rest Area - RR, HF, Phones, Picnic (P) (Westbound)

EXIT	TEXAS
35	Bus40, Vega, Old Route 66
Lodg	**S:** Best Western
36	U.S. 385, Dalhart, Hereford
FStop	**N:** **Conoco***
Gas	**N:** Phillips 66*
	S: Texaco* (ATM)
Food	**N:** Dairy Queen
37	Bus40W, U.S. 66, Vega, U.S. 385
FStop	**N:** **Phillips 66***
Gas	**N:** Diamond Shamrock
Food	**N:** Hamburger House, Hickory Cafe
Lodg	**N:** Bonanza Motel, Vega Motel
AServ	**N:** Texas Auto Supply
	S: NAPA Auto Parts
Other	**N:** Coin Car Wash, **Walnut RV Park**
42	Everett Road
49	Fm809, Wildorado
FStop	**S:** **Diamond Shamrock*(LP)**
Food	**S:** Jesse's Cafe
Lodg	**S:** **Texan Motel**
(54)	**Rest Area - RR, HF, Phones, Picnic P (Eastbound)**
54	Adkisson Road
RVCamp	**S:** **Adkisson**
(58)	**Parking Area**
57	Fm2381, Bushland
RVCamp	**S:** **Camping**
60	Arnot Road
FStop	**S:** **Loves**
Food	**S:** **Loves**
Other	**S:** **Texas Trading Company**

EXIT	TEXAS
62A	Hope Road
RVCamp	**S:** **Camping**
62B	Bus40E, Amarillo Blvd, Old Route 66 (No Eastbound Reaccess)
64	Loop 335, Soncy Rd, Helium Rd
Food	**S:** Applebee's, Doodles, Legends Steakhouse, McDonalds, New York Bagel, Ruby Tuesday
AServ	**N:** Mr. Automotive
Other	**S:** **Barn's & Noble Books, Circut City, Office Max, Target Department Store**
65	Coulter Dr
Gas	**N:** Phillips 66*(CW)
	S: Texaco*
Food	**N:** Arby's, Donut Shop, Golden Corral, Lubby's Cafeteria, My Thai Chinese, Waffle House
	S: Donut Stop, Hud's Hamburger, Taco Bell, Wendy's, What A Burger
Lodg	**N:** AAA Best Western, Days Inn, AAA La Quinta Inn
	S: ◆ Comfort Inn, Interstate Motel
AServ	**N:** Firestone Tire & Auto, West Gate Chevrolet, GEO, KIA
RVCamp	**S:** **Interstate Motel**
Med	**N:** **+ Midstate Medical**
ATM	**N:** Amarillo National Bank
Other	**S:** **West Gate Mall**
66	Bell St, Wolflin Ave
Gas	**S:** Chevron, Diamond Shamrock*, Taylor*(CW), Texaco*(CW)
Food	**S:** Dale's Hamburgers, King and I Thai Food, Subway
Lodg	**N:** ◆ Fairfield Inn, Marriott, Motel 6

EXIT	TEXAS
AServ	**S:** Big A Auto Parts, Chevron
Med	**S:** **+ Family Medical Clinic**
ATM	**S:** BankOne
Other	**S:** **Albertson's Grocery**
67	Western St
Gas	**N:** Citgo, Phillips 66*, Texaco*
	S: Diamond Shamrock*
Food	**N:** Black Eyed Pea, Blimpie's Subs, Braum's, Burger King, Chili's, McDonalds, Sonic, Taco Bell
	S: Bagel Place, Catfish Shack, First Cafeteria, IHOP, Olive Garden, Taco Cabana, Vince's Pizza, Wienerschnitzel
AServ	**S:** Firestone Tire & Auto, Montgomery Ward
ATM	**N:** Texaco
	S: Plus ATM
Other	**N:** **Amarillo Bowl**
68A	Julian Blvd, Paramount Blvd
Food	**S:** Arby's, Bless Your Heart, Buns Over Texas, Cajun Magic, Calico County, Chinese Golden Chinese, Denny's, El Chico, Long John Silvers, Malcolm's Ice Cream, Orient Express, Peking Chinese, Poncho's, Ruby Tequila's, Steak & Ale, The Kettle, Wagon Wheel Steakhouse
Lodg	**S:** Comfort Inn, Motel 6, AAA Travelodge (see our ad this page)
AServ	**S:** GMC, Cadallic Dealer, Mr. Muffler, Penzoil Pit Stop
68B	Georgia St
FStop	**N:** **Shell*(CW), Texaco***
Food	**N:** Dyer's BBQ, Gardski's, Schlotzkys Deli, Subway
	S: Burger King, Casual Gourmet, Church's Fried Chicken, Dyer's BBQ, Furr's Family Dining, Taco Villa, What A Burger

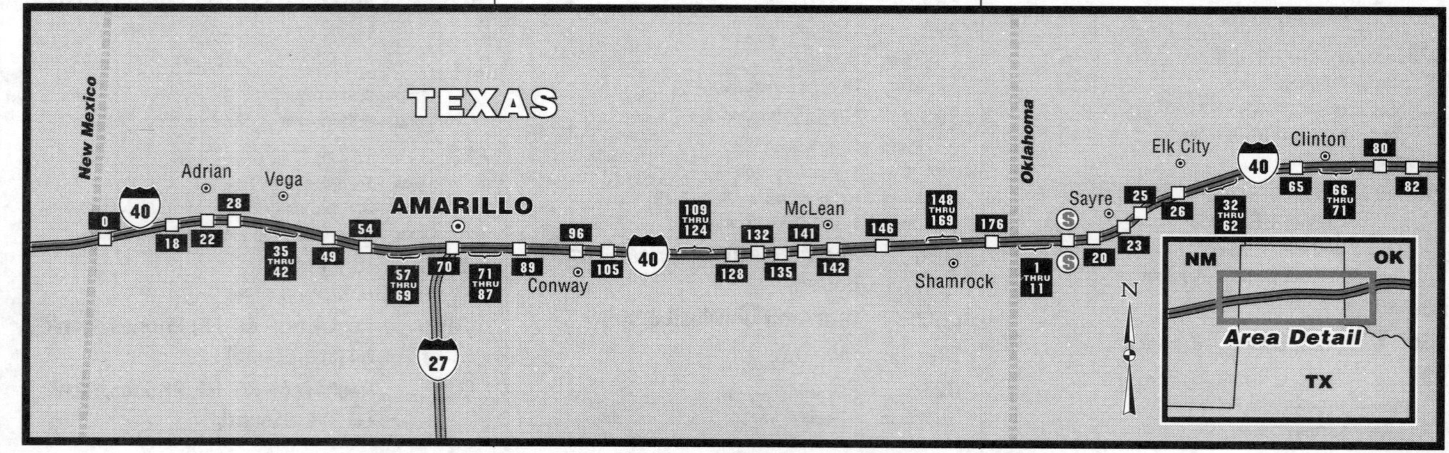

Bold red print shows RV & Bus parking available or nearby

EXIT — TEXAS

AServ S: Uniroyal Tire & Auto
ATM N: Amirello National Bank
S: Plains National Bank
Other S: **K-Mart, Office Depot, Wilson Square**

69A Crockett St

69B Washington St, Amarillo College
Gas N: Diamond Shamrock* (ATM)
S: Phillips 66*(CW), Texaco*
Food N: Arnold's Burger, Dairy Queen
S: Blimpie's Subs (In Texaco), The Donut Stop (In Texaco)
AServ S: Jiffy Discount Auto Parts
Other N: **Albertson's Grocery**
S: **Eckerd Drugs**

70 Jct I-27, U.S. 60E, Pampa, U.S. 287, Dumas, Downtown Amarila

71 Ross - Osage St, Arthur St
Gas N: Diamond Shamrock*(CW), Shell*, Texaco*
Food N: Burger King, Grandy's, Long John Silvers, McDonalds, O Kelly's Grill, Popeye's, Schlotzkys Deli, Shoney's, Taco Garcia, Thai Restaurant, Wienerschnitzel
S: Arby's, Denny's, La Fiesta, Taco Bell, Wendy's
Lodg N: **Coach Light Inn,** Days Inn, ◆ Holiday Inn (see our ad on opposite page)
S: ◆ Hampton Inn (see our ad opposite page), (AAA) La Quinta Inn, Traveler's Inn
AServ N: Texaco Express Lube
S: Chandler Ford
Other N: Coin Car Wash
S: **Post Office, Sam's Club**

72A TX395, Nelson St, Tee Anchor Blvd, Quarter Horse Dr.
Gas S: Chevron, Diamond Shamrock*, Shell
Food N: Carolyn's Place, **Cracker Barrel,** KFC
Lodg N: Budget Host, ◆ Ramada Inn, (AAA) Sleep Inn, Super 8 Motel, **Travelodge**
S: **Camelot Inn**
AServ S: Chevron
Other N: **American Quarter Horse Museum**

72B Grand St, Bolton St
Gas N: Texaco*
S: Phillips 66*(CW), Yukon*
Food N: Chicho's Cafe, Henks BBQ, The Donut Stop
S: McDonalds, Pizza Hut, Sonic, Subway, Taco Cabana, Taco Villa
Lodg S: Motel 6
AServ N: Grand Battery Electeric, Grand Street Auto Service, Napa Auto Parts
S: **Wal-Mart,** Western Auto
ATM N: Boatmens Bank
Other S: Coin Car Wash, **Wal-Mart**

73 Eastern St, Bolton St
FStop S: **Pacific Pride**
Gas N: Diamond Shamrock*(D) (ATM)
S: Chevron*
Lodg S: (AAA) Best Western
TServ S: **Cummins Diesel, Treadco**
RVCamp N: **Village East**
Other N: **Pioneer Trucker Store**

74 Whitaker Road
TStop S: **76 Auto/Truck Plaza*(D)(SCALES), Love's Country Store**
Food N: **Big Texan Steak Ranch**
S: **76 Auto/Truck Plaza**
Lodg N: **Big Texan Steak Ranch Motel**
S: **Coachlight Inn**

EXIT — TEXAS

TServ S: **Amarillo Truck Center, West Texas Peterbilt**
TWash S: **Blue Beacon**

75 Loop335, Lakeside Dr, Airport
TStop S: **Petro*(SCALES)**
Gas N: Phillips 66*, Shell*
Food N: Country Barn, Waffle House
S: **Iron Skillet (Petro)**
Lodg N: (AAA) Econolodge, ◆ **Radisson Inn, Super 8 Motel**
TWash S: **Blue Beacon**
RVCamp N: **Camping**

76 Spur468, Int'l Airport
TStop N: **Flying J Travel Plaza(D)(SCALES) (RV Dump)**
Food N: **Cookery (Flying J)**
Other N: **Texas Travel Information**

77 Fm1228, Pullman Road
TStop N: **Pilot Travel Center(SCALES) (Arby's)**
S: **Texaco Truck & Auto*(SCALES)**
Food N: **Arby's (Pilot)**
S: **Pepper Mill Restaurant (Texaco)**
AServ N: Texas Truck Parts
TServ S: **Fleet Refrigeration, Stewart & Stevenson, Utility Trailor**
Other S: **Custom RV**

78 U.S. 287, Fort Worth (Eastbound)

80 Spur228

81 Fm1912
TStop S: **TravelCenters of America(SCALES)**
FStop N: **Texaco*(D)(LP)**
Food S: **Country Fresh Restaurant (Truckstop), Subway (Truckstop)**
TServ S: **West Texas Caterpillar**

85 U.S. 66, Bus 40, Amarillo Blvd

87 Fm2373

(87) **Rest Area P (Eastbound)**

(88) **Rest Area - Picnic, Grills, RR P (Westbound)**

89 Fm2161, Old Route 66

96 TX207, Conway, Panhandle
FStop N: **Love's Country Store*(D)**
Gas S: Shell
Food N: **J & R's**
S: **S&S Restaurant**
Lodg N: LA Motel
S: **S&S Motel**
AServ S: Shell

98 South TX 207, Claude

105 Fm2880

(107) **Parking Area P (Westbound)**

109 Fm294

110 Bus40E, Groom

112 Fm295, Groom

113 Fm2300, Groom
Gas S: Texaco*(D)
Food S: **Dairy Queen**
Lodg S: (AAA) Chalet Inn

114 Bus.40, Groom
TServ N: **Tower Truck and Auto Service(D)**

121 TX70N, Pampa

124 TX70S, Clarendon
Gas N: Conoco(D)
RVCamp N: **Groom McLean Camping**

EXIT — TEXAS/OKLAHOMA

128 Fm2477, Lake McClellan
Parks N: **Wild Life Viewing Area**

(130) Picnic (Eastbound)

(131) Picnic - Phones (Westbound)

132 Johnson Ranch Road

135 Fm291, Alanreed, Old Route 66
Gas S: Conoco(D)
Lodg S: Texas Motel
AServ S: Bill's Auto Repair
Other S: **U.S. Post Office**

141 Bus40, McClean (No Reaccess)

142 TX273, Fm3143, McLean
FStop N: **Texaco***
Food N: North 40 Steakhouse and Grill
AServ N: Kirk's Auto Service and Supply

143 Bus 40, McLean
Other N: **Veterinary Clinic**

146 County Line Road
RVCamp S: **Camping**

148 Fm1443, Kellerville
RVCamp S: **Camping**

(150) **Rest Area - Picnic, Grills, RV Dump, RR P (Both Directions)**

152 Fm453, Pakan Road
Gas S: Gas Station*

157 Fm1547, Fm2474, Fm3075, Lela
FStop S: **Texaco**
AServ N: JD's Service Center
TServ N: **JD's Service Center**
RVCamp S: **Camping**

161 Bus40, Shamrock

163 U.S. 83, Bus40, Shamrock, Wheeler, Wellington
FStop S: **Phillips 66*(D)(CW) (ATM), Texaco*(D)(LP)**
Gas N: Chevron, Conoco, **Texaco*(D)**
Food N: Irish Inn, Mitchell's Family Restaurant, Pizza Hut
S: **Dairy Queen, McDonalds (In Phillips 66), Subway (In Phillips 66)**
Lodg N: (AAA) **Best Western (Irish Inn)**
S: Western Motel
AServ N: Chevron, Conoco, Texaco
TServ S: **JD Truck and Tire Co.**
Med S: ✚ **Hospital**

164 Bus 40 West Shamrock

167 Fm2168, Daberry Road
FStop N: **Texaco*(D)**
Food N: **Butch's Restaurant**

169 Fm1802, Carbon Black Road

(173) **Rest Area - Picnic P (Eastbound)**

(176) Livestock Inspection Area (Westbound)

176 Spur36E, Texola (No Reaccess)

↑TEXAS

↓OKLAHOMA

1 Texola
FStop S: **Diamond Shamrock*(D)**
Food S: **The Windmill Restaurant**
Other S: **US Post Office**

Bold red print shows RV & Bus parking available or nearby

EXIT		OKLAHOMA
5		Bus40E, Honeyfarm Road, Hollis
7		OK30, Erick, Sweetwater
	FStop	N: **Texaco*(D)**
		S: **Love's Country Store**
	Food	N: **Cal's**
		S: A & W Drive-In (Love's), **Cowboys**, Taco Bell (Love's)
	Lodg	N: (AAA) **Comfort Inn**
		S: (Days Inn) (AAA) **Days Inn (see our ad this page)**
(8)		**Welcome Center, Rest Area - RV Dump, RR, Picnic (P) (Eastbound)**
(9)		**Rest Area - RV Dump, Picnic, RR, Phones (P) (Westbound)**
11		Bus40, Erick
(13)		Check Station (Both Directions)
14		Hext Road
20		U.S. 283, Bus40E, S 4th St, Mangum, Sayre
	Parks	N: **Washita Battlefield National Landmark**
23		OK152, Main St, Cordell
	FStop	N: **Phillips 66*(D)(LP)**
		S: **Texaco***
	Food	S: **Junction Restaurant***
	AServ	N: Phillips 66
		S: Texaco
	TServ	S: **Texaco**
25		Bus40W, N 4th St
	FStop	N: **Diamond Shamrock*(D)**
	RVCamp	N: **Camping**
	Med	N: ✚ **Hospital**
26		Cemetery Road
	TStop	S: **Simon's Oasis Travel Center(LP)**
	FStop	N: **Citgo*(D)**
	Gas	S: Texaco
	Food	N: **Bud's American Cafe (Citgo)**
		S: **Hardee's**
32		OK34S, Bus40E, Mangum
34		Merritt Road, Elk City
	TServ	N: **DARR Power Systems**
38		OK6, S Main St, Altus, Elk City
	Gas	N: Citgo*, Conoco*(D)
		S: Phillips 66*(D), Total*(D)
	Food	N: Arby's, **Lennie's (Ramada)**, Long John Silvers, McDonalds, Taco Mayo, Western Sizzlin'
	Lodg	N: (AAA) **Ramada Inn**
		S: (AAA) **Econolodge**, (AAA) Holiday Inn, (AAA) **Quality Inn**

EXIT		OKLAHOMA
	AServ	S: Trop Artic
	RVCamp	N: **Elk Creek**
	ATM	N: Conoco
	Parks	S: **Quartz Mountain State Park**
	Other	S: Coin Car and Truck Wash
40		E 7th St
41		OK34N, Elk City, Woodward
	FStop	N: **Love's Country Store***
	Gas	N: Conoco*(D)
	Food	N: **Home Cooking Cafe**, Kettle
	Lodg	N: **Budget Host Inn**, (Days Inn) (AAA) **Days Inn**, (AAA) **Knight's Inn**, Motel 6, Super 8 Motel, (AAA) Travelodge
	RVCamp	N: **Camping**
	Med	N: ✚ **Hospital**
47		Canute
	Gas	S: Citgo*
	Food	S: **Sister's Cafe**
	Lodg	S: Economy Inn
	Other	S: **Post Office**
50		Clinton Lake Road
	Gas	N: KOA Camp(D)(LP)
	RVCamp	N: **KOA Campground*(D)**
53		OK 44, Foss, Altus
	FStop	S: **Texaco*(D)**
	TServ	S: **Pendleton's Tire Service**
	Parks	N: **Foss Resevoire**
57		Stafford Road
61		Haggard Road
62		Parkersburg Road
	Gas	N: Conoco(D)(LP)

EXIT		OKLAHOMA
65		Bus40, Gary BlvdClinton (No Eastbound Reaccess)
	Food	N: Hardees, Long John Silvers, McDonalds, Pancake Inn, Taco Mayo
	Lodg	N: (AAA) Best Western, Comfort Inn, (AAA) Travel Lodge, Treasure Inn
	AServ	N: Jay's Tires
	Other	N: **K-Mart**
65A		10th St, Neptune Dr
	FStop	S: **Phillips 66*(D)**
	Food	N: Braums Burgers and Ice Cream, KFC, Lupita's Mexican, Pizza Hut
	Lodg	N: (AAA) Holiday Inn, Relax Inn, (AAA) Super 8 Motel
		S: Budget Inn
	AServ	N: Pontiac Buick GMC
		S: L & M Garage, Phillips 66
	RVCamp	S: **Camping**
66		U.S. 183, S 4th St, Cordell
	AServ	S: Ballard Ford
69		Bus 40, West Clinton
71		Custer City Road
	FStop	N: **Love's Country Store*(D)**
	Food	N: **Taco Bell (Love's), Trading Post Restaurant**
80		OK54, Thomas (No Eastbound Reaccess)
	Gas	S: Conoco*(D)
	Food	N: Little Mexico
	Lodg	N: Econolodge
		S: Texaco(LP)
	AServ	S: Texaco
82		Bus40W, E Main Street
	Gas	N: Citgo*(CW), Conoco*(D), Phillips 66*, Texaco*(CW)
	Food	N: Baskin Robbins, Brickhouse BBQ, City Diner, Hardee's, Jerry's Restaurant, KFC, Mazzio's Pizza, McDonalds, Pizza Hut, Starvin' Marvin Pizza, Subway, T-Bone Steakhouse, The Dutchman
	Lodg	N: (AAA) Best Western, (Days Inn) Days Inn, Scottish Inns
	AServ	N: AA Auto, Auto Zone Auto Parts, B & D Automotive, Conoco, Duaine's Auto Clinic, O' Reilly's Auto Parts, Southwest Tire, Texaco
	Med	N: ✚ **Hospital**
	ATM	N: First National Bank, Midfirst Bank
	Other	N: Impreial Coin Car Wash, **Puckett 's Foods, Wal-Mart**
84		Airport Road
	Gas	N: Total

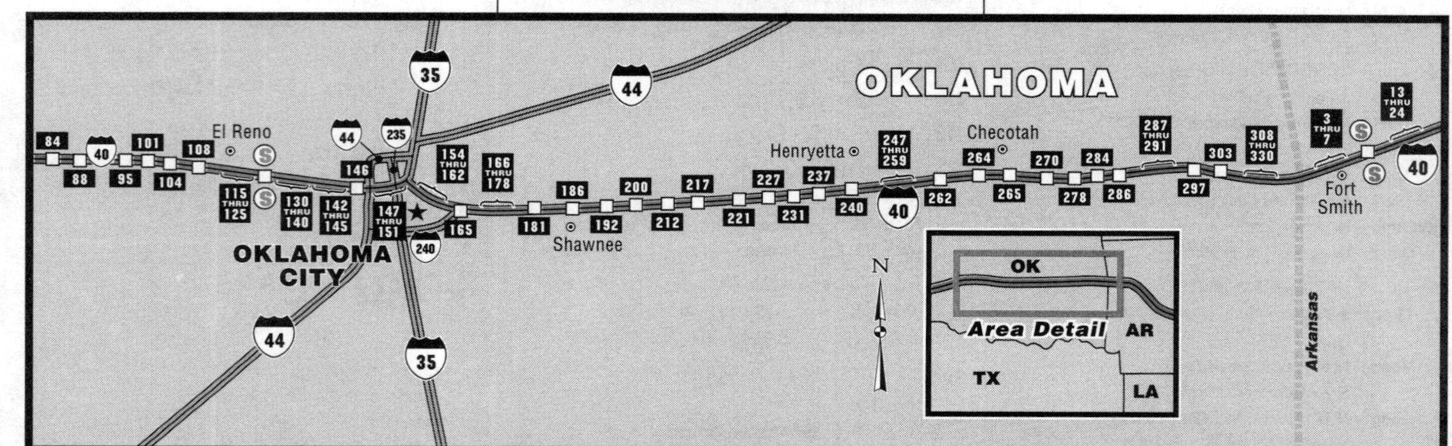

Bold red print shows RV & Bus parking available or nearby

EXIT OKLAHOMA

Lodg N: Travel Inn
AServ N: Buick & GM Dealer, Cummins Chrysler Plymouth Dodge Dealer, Jameson Chevorlet Oldsmobile, Tunes & Tint

88 OK58, Hydro, Carnegie
FStop N: **Texaco (westbound only)**

93 Picnic (Westbound)

95 Bethel Road

101 U.S. 281, OK8, Anadarko, Watonga, Hinton, Geary
TStop S: **Biscuit Hill Truck Stop*, Phillips 66*(D)**
Food S: **Biscuit Hill Restaurant**
Lodg S: **Motel (With Phillips 66)**
Parks N: **Roman Nose State Park**
Other N: **Public Picnic Area**

104 Methodist Road

108 Spur U.S. 281N, Geary, Watonga
FStop N: **Cherokee Texaco*(D)(CW) (RV Dump)**
S: **Love's Country Store***
Food N: **Cherokee Restaurant**
S: **Indian Trading Post Restaurant**
Lodg N: **Cherokee Motor Inn**
TServ S: **Exit 108 Tire Shop**
RVCamp N: **KOA Campground (Cherokee Texaco)**
S: **Good Life RV Camp**
Other N: **I-40 Lighthouse Bookstore**

(111) **Rest Area P (Eastbound)**

115 U.S. 270, Calumet

119 Bus40E, El Reno

123 Country Club Road, El Reno
FStop S: **Phillips 66*(D)(LP)**
Gas N: Citgo*
S: Texaco
Food N: Baskin Robbins, Braum's Hamburgers, Hardee's, KFC, Little Caesars Pizza, Long John Silvers, McDonalds, Pizza Hut, Subway, Taco Bell
S: Denny's, Sirloin Stockade
Lodg S: AAA **Best Western**, DAYS INN AAA Days Inn (see our ad this page), Red Carpet Inn
AServ N: **Wal-Mart**
S: Texaco
Med N: **+ Hospital**
ATM N: American Heritage Bank, The Bank of Union
Other N: Coin Car Wash, **Wal-Mart**

125 U.S. 81, Chickasha, El Reno
Gas N: Love's Country Store*
Food N: Brass Apple, Porky's Counrty Kitchen
Lodg N: Big 8 Motel, Ramada Limited, AAA **Super 8 Motel**, Western Sands Motel
AServ N: Diffee Ford, Lincoln, Mercury, Frontier Chrysler, Plymouth, Dodge

129 Weigh Station (Both Directions)

130 Banner Road, Union City
Gas N: Texaco*

132 Cimarron Road

136 OK92, Yukon, Mustang
Gas N: Texaco*(CW)
Food N: Braum's, Ci Ci's Pizza, Hardee's, Harry's Grill and Bar, KFC, Long John Silvers, Wendy's
AServ N: Auto Zone Auto Parts
ATM N: Bank of Oklahoma
Other N: **Wal-Mart**

138 OK4, Yukon, Mustang Road
Gas S: Stax*

EXIT OKLAHOMA

Food N: Denny's, Miller's Crossing
S: Sonic Drive In
Lodg N: AAA **Comfort Inn**
Other S: **Food Lion**

140 Morgan Road (Eastbound)
TStop N: **76*(CW)(SCALES) (Salad Bar Buffet)**
S: **Flying J Travel Plaza*(D)(LP)(SCALES) (RV Dump)**
FStop S: **Love's Country Store***
Food N: **Salad Bar Buffet (76)**
S: **Grandy's, Subway (Love's), Thads Restaurant (Flying J)**
TServ N: **LL Car and Truck Tires**
TWash N: **USA Truck Wash**
ATM S: Love's Country Store

142 Council Road
TStop S: **Oklahoma Transport Refidgeration, TravelCenters of America*(SCALES) (RV Dump)**
Gas N: Sinclair*, Texaco*
Food N: Braum's, McDonalds, Stan's BBQ, Taco Bell, Waffle House, Wendy's
S: **Burger King, Country Pride**
Lodg N: Best Budget Inn
S: **Econolodge**
AServ N: Goodyear Tire & Auto
RVCamp N: **Motley RV Repair**
Med N: **+ Hospital**

143 Rockwell Ave
Gas S: Interstate Gas*(D)(LP)
Lodg N: Rockwell Inn
S: **Sands Motel**
RVCamp N: **McClain's RV**
S: **Rockwell RV Park, Sands RV Park**

144 MacArthur Blvd
FStop S: **Sinclair(D)**
Gas N: Texaco*(CW)
S: Texaco*
Food S: Ruby's
Lodg S: Econolodge, Quality Inn, Super 10
TServ N: **Cummins Diesel, Fruehauf Dealership, Oklahoma City Freightliner, United Engines**

EXIT OKLAHOMA

Detroit Diesel Dealership
S: **Kenworth Dealership, Peterbilt Dealer**

145 Meridian Ave
Gas N: Conoco(CW) (ATM), Texaco*(D)
S: Phillips 66*(CW), Total*(D)
Food N: Boomerang Grill, Crocket's, Denny's, Fannies Cafe on Reno, Gator Trappers, Jimmy's Egg, McDonalds, On The Border, Outback Steakhouse, Shorty Small's Ribs, Steak & Ale, Texanna Red's, West End Iguana
S: Bennigan's, Burger King, Calhoun's Bar and Grill, **Cracker Barrel**, Doc's Subs, Edsel's, Kettle, Logan's Road House, Pepper Tree (Clarion), Shoney's, Singapore Chinese, Wendy's
Lodg N: AAA Best Western, Extended Stay America, Howard Johnson, ◆ Marriott, Radisson, Shoney's Inn, ◆ Super 8 Motel, AAA Travelers Inn
S: Clarion Hotel, Comfort Inn, ◆ Courtyard by Marriot, DAYS INN Days Inn, ◆ Holiday Inn Express, La Quinta Inn, Lexington Hotel Suites, AAA Motel 6, Park Inn Intl., Sleep Inn, Amerisuites (see our ad this page)
ATM N: Texaco
S: Banc First, First Oklahoma

146 Portland Ave (No Eastbound Reaccess)

147A Jct I-44W, OK3E, Lawton, Dallas, Wichita (Exit To The Left)

147B Jct I-44E, OK3W

147C May Ave, Fair Park (Westbound)

148A Agnew Ave, Villa Ave
FStop N: **Fuel Man**
Gas S: Total(CW)
Food S: **Braum's**, Taco Bell, The Rib Stand
AServ N: Delta Transmissions, Sooner State Ford
S: Freddie's Discount Tires, Steering Column Specialist
ATM S: Total

148B Pennsylvania Ave (Eastbound)
Gas N: Diamond Shamrock(D), Total(D)(LP)
AServ N: Diamond Shamrock, Total
TServ S: **HD Copland International**

148C Virginia Ave. (Westbound)
TServ S: **H. D. Copeland Intl. Trucks**

149A Western Ave, Reno Ave (Difficult Eastbound Reaccess)
Gas N: Total*(D)(CW)
S: Phillips 66*
Food N: McDonalds, Taco Bell
S: Burger King
AServ S: Tune Up

149B Classen Blvd. (Westbound)

150A Walker Ave

150B Harvey Ave (Eastbound, Difficult Reaccess)
AServ N: Fred Jones Ford

150C Downtown, Robinson Ave (Westbound, Difficult Reaccess)
TServ N: **Fred Jones Ford Truck**

151A Jct I-235, Lincoln Blvd (Eastbound)

151B Jct I-35S, Dallas

151C Jct I-235N , State Capital, Edmond

Note: I-40 runs concurrent below

Bold red print shows RV & Bus parking available or nearby

EXIT — OKLAHOMA

with I-235. Numbering follows I-235.

127 Eastern Ave
- TStop: **N: Petro*(SCALES) (Iron Skillet), Pilot*(SCALES) (Wendy's)**
- Gas: N: Texaco*
- Food: N: **Iron Skillet Restaurant**, Waffle House, **Wendy's (Pilot)**
- Lodg: N: **Best Western**, Travel Lodge
- TWash: N: **Blue Beacon**
- RVCamp: N: **Lewis RV Center**

128 I-35N, Tulsa, U.S. 62E, to I-44

Note: I-40 runs concurrent above with I-235. Numbering follows I-235.

154 Scott St, Reno Ave.
- Gas: N: Conoco*
  S: 7-11 Convenience Store*, Phillips 66*(CW)

155A Sunnylane Road, Del City
- FStop: N: **Total*(D)(LP)**
- Gas: S: Texaco*, Total*
- Food: S: Braum's, Pizza Hut
- AServ: N: Jim's Auto Center
  S: Stoval Auto Repair
- Med: S: **+ Lassater Clinic**
- Other: N: Murphy's Car Wash

155B SE15th St, Del City, Midwest City
- Gas: N: Texaco*(D)(CW) (ATM)
  S: Express Stop*
- Food: S: Ashley's Country Kitchen
- AServ: N: Express Automotive
- ATM: N: Megamarket
- Other: N: **Dental Clinic, Mega Market Supermarket**

156A Sooner Road
- Gas: N: Conoco*(D)
- Food: N: Bristol Station, Cracker Barrel, Kettle
- Lodg: N: Clarion Inn, Comfort Inn, **Hampton Inn**, ◆ La Quinta Inn, Sixtence Inn

156B Hudibird Dr
- Lodg: S: Motel 6
- AServ: S: Hudiburg Chevrolet, CEO, Buick

157A SE 29th St, Mid West City

157B Air Depot Blvd
- FStop: N: **Star Mart Texaco***
- Food: N: **A & W Drive-In (Texaco)**, Mr. Spriggs BBQ, Pizza Inn
- AServ: N: Reilly Auto Parts
- Med: N: **+ Hospital, + Hospital**
- ATM: N: Nations Bank (Texaco)

159A Gate 7

157C Industrial Entrance
- Lodg: N: Stratford House Inn

159B Douglas Blvd
- Gas: N: Conoco*(D), Phillips 66*(CW)
  S: Conoco*
- Food: N: Acropolis Greek Restaurant, Casey's Cajun Catfish, Denny's, LaGreek, McDonalds (playground), Pizza Hut, Sonic, Subs, Subway, Taco Bell
- AServ: N: Conoco
- RVCamp: N: **Lee's RV Service, Universal RV**
- Other: N: **Highlander Coin Laundry**

162 Anderson Road
- Other: N: **Lundy's(LP)**

EXIT — OKLAHOMA

165 Jct I-240

166 Choctaw Road
- TStop: N: **Love's Country Store**
  S: **Texaco Auto/Truck Plaza(SCALES)**
- FStop: N: **Love's***
- Food: N: **Taco Bell (Loves)**
  S: **ProAm Restaurant**
- TServ: S: **Texaco**
- RVCamp: N: **KOA Campground**

169 Peebly Rd.

172 Harrah , Newalla Road

176 OK102N, McLoud Road
- FStop: S: **Carl and Velma's General Store***
- Food: S: **Curtis Watson Restaurant**, Love's Country Store

178 OK102S, Bethel Acres, Dale
- FStop: S: **Kerr McGee***

181 U.S. 177, U.S. 270, OK3W, Stillwater, Tecumseh, Shawnee
- Gas: S: Conoco*(D)
- Food: S: Mia Casa Mexican

185 OK3E, Shawnee, Kickapoo St, Airport
- Gas: S: Phillips 66*(CW) (ATM)
- Food: N: Lubby's Cafeteria, Taco Bueno
  S: McDonalds, Shoney's
- AServ: N: **Wal-Mart**
- Med: N: **+ Hospital**
- Other: N: **Shawnee Mall, Wal-Mart**

186 OK18, Shawnee, Meeker
- Gas: N: Citgo*, Texaco*
- Food: N: Denny's
- Lodg: N: **American Inn**, Days Inn, Holiday Inn, **Motel 6**, Super 8 Motel
  S: ◆ Travel Lodge
- ATM: S: American Bank & Trust

192 OK9A, Earlsboro
- Gas: S: Texaco*(D)
- Food: S: Biscuit Hill Restaurant
- Lodg: S: **Roadway Inn**

(197) **Rest Area - RR, Phones, Picnic, RV Dump P (East Directions)**

(198) **Rest Area - RV Dump, RR, Phones, Picnic, HF P (Westbound)**

200 U.S. 377, OK99, Prague, Seminole
- TStop: N: **Phillips 66(SCALES) (RV Dump)**
- FStop: S: **Citgo*(D), Love's Country Store***
- Gas: N: Phillips 66
- Food: N: I-40 Pit Stop (Phillips 66)
  S: Robertson's Hams, Round Up Cafe
- RVCamp: S: **Round Up Campground**
- Other: S: **Post Office**

212 OK56, Wewoka, Cromwell
- Gas: N: Conoco*
  S: Texaco
- Food: S: Stuckey's
- AServ: N: Danny's Auto Center
- RVCamp: S: **Camping**

217 OK48, Bristow, Holdenville
- FStop: S: **Total(D)**
- Gas: N: Citgo*(D)(LP), Total*
- Food: S: **Derrick Cafe (Total)**
- AServ: N: Joe's Muffler & Tire Shop
  S: Total
- RVCamp: N: **Camping**

EXIT — OKLAHOMA

221 OK27, Okemah, Wetumka
- TStop: N: **Total*(D)**
- Gas: N: Conoco*(D) (ATM)
  S: Ampride*(D), Love's*(D), Texaco*
- Food: N: Aspen Fine Food, Mazzio's Pizza, Sonic
  S: John's BBQ, Subway (Love's), Taco Bell
- Lodg: N: **OK Motor Lodge**
- AServ: N: Big A Auto Parts
- TServ: S: **24 Hour Truck Repair, Ampride**
- RVCamp: S: **Blecker's RV Park**
- Med: S: **+ Hospital**
- Other: N: Coin Car Wash

227 Clearview Road

231 U.S. 75S, Weleetka
- FStop: N: **Ampride***
- Food: N: **Cow Poke's Cafe**
- AServ: N: Ampride

237 U.S. 62, To U.S. 75, Henryetta, I-40 Bus. Loop
- TStop: N: **Cow Creek Ranch House* (ATM)**
- Gas: N: Ampride*
- Food: S: **Hungry Traveler**
- Lodg: N: Sleepy Traveler, **Trail Motel**
  S: **Super 8 Motel**
- Med: N: **+ Hospital**
- Other: N: **Camping**

240B U.S. 62E, U.S. 75N, Henryetta, Okmulgee
- FStop: N: **Citgo**
- Gas: N: Conoco, Phillips 66(LP)
- Food: N: Mazzio's Pizzas, Simple Simon Pizza
- Lodg: N: Gate Way Inn, The Baron Motel, The Relax Inn
- AServ: N: Ford Dealership
- TServ: N: **J&L Diesel Service (New & Used Truck Service)**
- Med: N: **+ Hospital**
- Other: N: **Wal-Mart**

247 Tiger Mt Road

(251) **Rest Area - no facilities P (Eastbound)**

(252) **Rest Area - no facilities P (Westbound)**

255 Pierce Road
- RVCamp: N: **KOA Campground**

259 OK150, Fountainhead Road
- FStop: S: **AmPride(LP)**
- Gas: S: Texaco*
- Lodg: S: **Best Western**
- TServ: S: **J&B's Service Center**
- Parks: S: **Fountain Head State Park**

262 To U.S. 266, Lotawatah Road
- Gas: N: Citgo*, Phillips 66*
- Food: N: **Stuckey's**

264A U.S. 69S, Eufaula, Muskogee

264B U.S. 69N, Muskagee

265 Bus69, Checotah
- FStop: N: **Texaco**
  S: **Citgo*, Phillips 66*(LP)**
- Gas: N: Total*
- Food: N: Pizza Hut, Sonic
- Lodg: S: **Budget Host Inn**
- AServ: N: Texaco
- RVCamp: N: **Camping**

270 Texanna Road, Porum Landing

Bold red print shows RV & Bus parking available or nearby

EXIT OKLAHOMA

FStop S: Ten*

278 U.S. 266, OK2, Warner, Muskogee

FStop N: Texaco*
Food N: Big Country Restaurant
Lodg N: Western Sands Motel

(282) **Scenic Turnout (Truck Parking)**

284 Ross Road

286 Muskagee Tnpk, Muskagee, Tulsa

287 OK100N, Webber's Falls, Gore

FStop N: Love's Country Store* (ATM)
Food N: Charlie's Chicken, Subway, Taco Bell
Lodg N: Super 8 Motel
TServ N: Service & Tire

291 OK10N, Gore, Carlile Road

Parks N: Green Leaf Recreation Area, Ten Killer Reacreation Area
Other N: Cherokee Nation Information Center

297 OK82N, Vian, Tahlequah

Gas N: Phillips 66*(CW)
Food N: Simple Simon Pizza
AServ N: Mike's Brake & Alignment
Parks S: Sequoia Wild Life Area
Other N: Coin Car Wash

303 Dwight Mission Road, Blue Ribbon Downs

308 U.S. 59, Bus40, Sallisaw, Poteau

FStop S: Texaco*
Gas N: Phillips 66*, Total*
S: Texaco
Food N: Braum's, Dana's Restaurant, JR's Dinner Bell, Mazzio's Pizza, McDonalds, Western Sizzlin'
Lodg N: AAA Best Western, Magnolia Inn, McKnight's Motel, AAA Super 8 Motel
RVCamp S: Camping
ATM S: Texaco
Parks N: Brushy Lake State Park
Other N: Tourist Information

311 U.S. 64, Sallisaw, Stilwell

FStop N: Fina, Phillips 66, Texaco
Gas N: Texaco(D)
Food N: Hardees, KFC, Pizza Hut, Simple Simon's Pizza, Sonic, Taco Mayo
Lodg N: Econolodge
AServ N: Auto Zone Auto Parts, Bill's New and Used Tires, Wal-Mart
Parks N: Brushy Lake State Park, Sequoyah's Cabin
Other N: Wal-Mart

(314) **OK Welcome Center - Picnic, Phones, HF, RV Dump (Westbound)**

(315) **Rest Area - RR, Phones, HF, Picnic P (Eastbound)**

EXIT OKLAHOMA/ARKANSAS

321 OK64N, Muldrow

FStop N: Texaco*
S: Curt's*
Food N: Bit"O"Country
S: Carolyn's Restaurant, Flip's Cafe
Lodg S: Gold Crown Motel
TWash S: Circle H

325 U.S. 64, Roland, Fort Smith

TStop N: Total*(D)(SCALES) (ATM)
FStop S: Conoco*(CW)
Gas S: Citgo, Fina*
Food S: Hardees, Mazzio's Pizza, McDonalds, Sonic
Lodg N: Days Inn
S: Interstate Inn
AServ S: Fina*
ATM S: First National Bank
Parks S: Fort Smith National Historic Site
Other S: Marvin's IGA

330 OK64S, Dora, Fort Smith (No Eastbound Reaccess)

↑OKLAHOMA

↓ARKANSAS

(2) **Welcome Center, Rest Area P (Eastbound)**

3 Lee Creek Road

5 AR59, Van Buren, Siloam Springs (AR59 Closed To Trucks)

TStop S: Unical 76
FStop S: Shell*(D)(SCALES), Texaco (Towing)
Gas N: Citgo, Exxon*, Total
S: Citgo*, Shell, Texaco*
Food N: Pizza Country
S: 76, Big Jake's Cattle Co. Steakhouse, Braum's, Hardee's, Subway, Taco Rio, Waffle House, Wendy's
Lodg N: ◆ Holiday Inn Express
S: AAA Motel 6, ◆ Super 8 Motel
AServ N: Citgo*
S: Ford Dealership, Pennzoil Lube
TServ N: Arkansas Kenworth
RVCamp S: Outdoor Living Center
ATM S: Citizen's Bank
Other S: Kwikwi Car Wash, Pharmacy Express

EXIT ARKANSAS

7 U.S. 540S, U.S. 71S, Van Buren, Fort Smith

(9) **Weigh Station - both directions**

12 AR540N, Fayetteville, Lake Ft. Smith State Park

13 U.S. 71, Alma, Fayetteville

TStop S: Gold Truck Stop - Sinclair*(D)
Gas N: BP*, Exxon*, Phillips 66*(D), Shell*(D)
S: Citgo*(D), Total*
Food N: Burger King, Cracker Barrel, Golden Dragon Chinese, KFC, Taco Bell
S: Braum's, Market Place Cafe, McDonalds, Piccadilly Circus Pizza & Subs, Sonic, Subway
Lodg N: Alma Inn Motel, Meadors Inn, Holiday Inn Express (see our ad this page)
S: Days Inn
AServ N: Exxon
RVCamp N: KOA Campground
Med S: ✚ Quick Care Immediate Medical Care
ATM N: Bank of the Ozarks
S: Citizen's Bank
Other S: Alma Car Wash, O'Relly Auto Parts

20 Dyer, Mulberry

FStop S: Citgo
Gas N: Conoco*(D)
AServ S: Citgo*(D)(LP)
TServ S: Citgo

24 AR215, Mulberry

Parks S: Vine Prairie Park

35 AR23, Ozark, Huntsville

Med S: ✚ Hospital
Parks S: Aux Arc Park

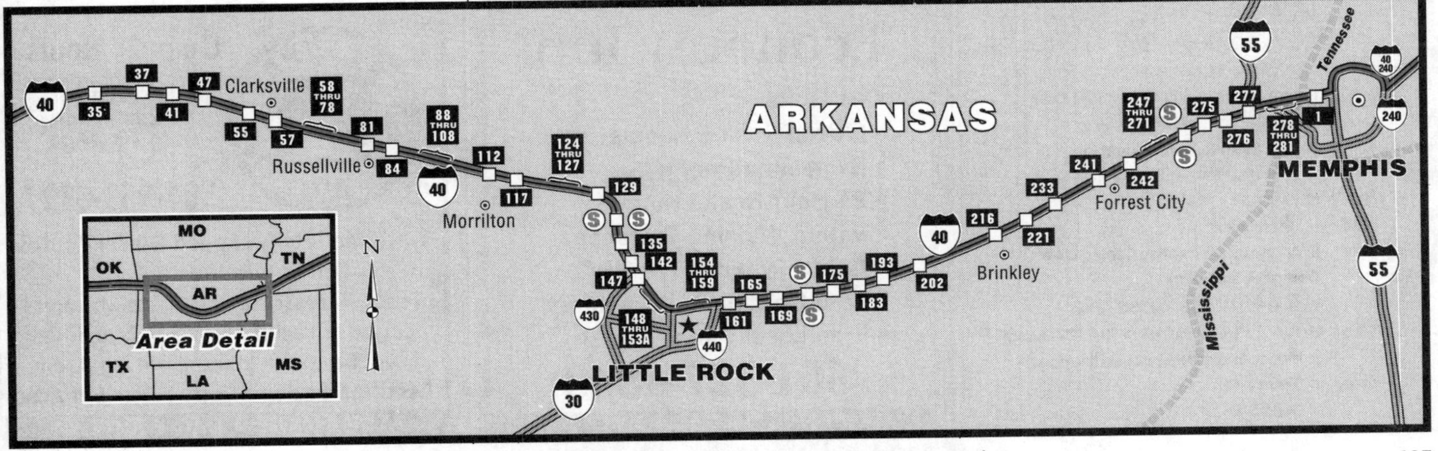

Bold red print shows RV & Bus parking available or nearby

EXIT ARKANSAS

(36) Rest Area - RR, HF, Grills, Phones (P) (Eastbound)
37 AR219, Ozark
FStop S: Love's Country Store*
Food S: A & W Drive-In (In Love's), Subway (In Love's)
Med S: ✚ Hospital
(37) Rest Area - RR (P) (Westbound)
41 AR186, Altus
RVCamp S: Camping
47 AR164, Coal Hill, Hartman
55 U.S. 64, Clarksville, AR109, Scranton
FStop S: Exxon*(D)
Gas N: Citgo*(D), Phillips 66*
Food N: BBQ Steaks, Hardees, Lotus Chinese, Peppy's Grill
S: Western Sizzlin'
Lodg N: Days Inn, (AAA) Hampton Inn
AServ N: Phillips 66
Other S: AR State Police
57 AR109, Clarksville, Scranton
58 AR103, AK21, Clarksville
TStop S: Texaco
FStop S: Texaco
Gas N: Phillips 66*, Texaco*
S: Texaco*
Food N: Emerald Dragon Chinese, KFC, Mazzio's Pizza, McDonalds, Old South Pancake House, Pizza Hut, Pizza Pro, Sonic, Taco Bell, Wendy's, Woddard's Family Restaurant
S: South Park Restaurant (Texaco)
Lodg N: (AAA) Best Western, Comfort Inn, Economy Inn, Super 8 Motel
AServ N: J&S Tire Center, Wal-Mart
S: Whitson Mortan Ford
ATM N: Bank of the Ozarks, Farmer's Bank (Piggy Wiggley)
Other N: Coin Laundry, Piggly Wiggly Supermarket, Wal-Mart
64 U.S. 64, Lemar
RVCamp S: Dad's Dream RV Park
67 AR315, Knoxville
(68) Rest Area - RR, Phones, Picnic, Grills, Vending (P) (Eastbound)
(69) Rest Area - RR (P) (Westbound)
74 AR333, London, Russellville
78 U.S. 64, Russellville (No Eastbound Reaccess)
Med S: ✚ Hospital
Parks S: Dardanelle State Park
81 AR7, Russellville
Gas N: Conoco(D)(LP)
S: Exxon*, Phillips 66*, Texaco*(CW)
Food N: Waffle House
S: Burger King, Chili's, Cracker Barrel, New China, Pancake House, Pizza Inn, Scoops Ice Cream, Shoney's, Subway, Waffle House
Lodg N: Motel 6, Ramada Limited, Sunrise Inn
S: (AAA) Best Western, Budget Inn, Hampton Inn, ◆ Holiday Inn, ◆ Super 8 Motel
AServ N: Conoco
Med N: ✚ Hospital
Other S: Arkansas Information Center, Lake Darnadell State Park
84 U.S. 64, AR331, Russellville
TStop N: Flying J Travel Plaza*(LP)(SCALES) (RV Dump)
S: Mapco Express (see our ad this page)
FStop N: Texaco*(D)
S: Delta Express*

EXIT ARKANSAS

Gas N: Texaco*
S: Exxon*, Phillips 66*
Food N: Hob Nob Restaurant, Thads Restaurant
S: Ci Ci's Pizza, Hardees, Hunan Chinese, McDonalds, Subway
Lodg S: (AAA) Comfort Inn
AServ S: Wal-Mart
TServ N: Freightliner Dealer
S: International
TWash N: Southern Shine Truck Wash
ATM S: Simmon First
Other S: Coin Car Wash, Wal-Mart
88 Pottsville
TServ S: Gala Creek Truck Service and Truck Wash
94 AR105, Atkins
Gas N: BP*(D)(LP), Citgo*(D)
S: Phillips 66*(D)
Food N: Chicken & More, Hardee's, Sonic Drive In, Tastee Burger, The Pizza Meister
AServ S: Ron's Auto Parts
ATM S: Bank of Atkins
Other S: Durb's Auto Parts, General Jack's Car Wash, Medicine Shoppe, Sav-On Foods
101 Blackwell
107 AR95, Morrilton
TStop S: Love's Country Store*(D)
Gas N: Citgo*(D)(CW)(LP)
S: Phillips 66*(D)(LP), Shell*
Food N: Chester Fried Chicken
S: Ice Cream Churn, Mom and Pops Waffles, Morralton Drive-In, Yesterdays
Lodg N: (AAA) Best Western
S: Days Inn
RVCamp N: KOA Campground
108 AR9, Morrilton
Gas S: Phillips 66*, Shell*(LP)
Food S: Bonanza Steak House, KFC, McDonalds, Pizza Hut, Pizza Pro, Subway, Sweden's Better Burgers, Waffle House
Lodg S: Super 8 Motel
Med S: ✚ Conway Co. Hospital
ATM S: Morrilton Security Bank
Other N: Kroger Supermarket, Wal-Mart
(110) Rest Area - HF, RR, Phones, Picnic (P) (Both Directions)
(111) Rest Area - RR (P) (Westbound)
112 AR92N, Plumerville
Gas N: Total
117 Menifee
124 AR25N, Conway (No Eastbound Reaccess)

EXIT ARKANSAS

Gas S: Mocking Bird*, Texaco*
Med S: **Hospital**

125 U.S. 65N, U.S. 65B S, Conway, Greenbrier, Harrison
Gas N: Conoco*(D), Shell*, Texaco*(D)(CW)
S: Citgo*(CW), Exxon*
Food N: China Buffet, Cracker Barrel, El Chico Mexican, Hardee's, McDonalds
S: Baskin Robbins, Fazoli's Italian Food, Waffle House, Wendy's
Lodg N: AAA Comfort Inn(see our ad opposite page)
S: Holiday Inn, Motel 6
AServ S: Kelly Tire
TServ S: **Kelly's Tires**
Other N: **Town Center Shopping**

127 U.S. 64, Conway, Vilonia, Beebe
Gas N: Citgo(CW)
S: Racetrac, Texaco*(CW), Total*
Food N: Bowen's Restaurant, Peppercorns Restaurant, Waffle House
S: Bamboo Garden, Burger King, Dillion's Steakhouse & Grill, Hardees, KFC, Long John Silvers, McDonalds, Shipley Donuts, Shoney's, Subway (Texaco), Taco Bell, Wendy's, Western Sizzlin'
Lodg N: AAA Best Western, DAYS INN AAA Days Inn (see our ad opposite page), ◆ **Economy Inn (see our ad opposite page)**, Hampton Inn, AAA **Ramada (see our ad opposite page)**
S: **Knight's Inn**
AServ N: Brown Pontiac, GMC Trucks, Car Pro Auto Store, Smith Ford
S: Auto Zone Auto Parts, Muffler Shop
RVCamp N: **Moix RV**
ATM S: First National Bank
Other S: **Kroger Supermarket**

129 U.S. 65B, AR286, Conway, Central Arkansas Univ
FStop S: **Delta Express***
Gas S: BP, Exxon*, Texaco
AServ S: BP, Midas Muffler & Brakes
ATM S: Boatman's Bank

(133) Truck Inspection Station (Both Directions)

135 AR89, AR365, Mayflower
Gas N: Phillip 66 - Bates Field & Stream*(LP)
S: Fina*(LP), Sinclair*(CW)
Food S: Cracker Box, Farmers Market Restaurant, Glory Be's, Mayflower Diner
TServ S: **R's Tires All Sizes**
Other N: **Camp Robinson**

142 AR365, Morgan, Maumelle
FStop N: **Shell, Total***
S: **Texaco*(CW)(SCALES)**
Food S: BJS Restaurant, **I-40 Restaurant**, KFC, McDonalds, Subway, Waffle House (In Total)
Lodg N: DAYS INN Days Inn
S: Comfort Suites (see our ad this page)
RVCamp N: **Camping**
Other S: Coin Car Wash (Texaco)

147 Jct I-430S

148 Crystal Hill Road
FStop S: **Citgo*(D)**
Gas N: Texaco*
S: Total*
RVCamp S: **Camping**

150 AR176, Burns Park, Camp Robinson
RVCamp N: **Campground**

EXIT ARKANSAS

S: **Camping**
Other N: **North Little Rock Animal Shelter**
S: **Tourist Information**

152 AR365, Levy (No Eastbound Reaccess)
Med S: **Baptist Memorial Medical Center**

153A AR107, JFK Blvd, Main St
Gas N: Exxon, Mapco Express, Shell*(CW)
S: Exxon*(CW)
Food N: Schlotzkys Deli
S: Bonanza Steak House, Hardees, Shipley Donuts, Waffle House
Lodg N: Masters Inn, Ramada Inn
S: Holiday Inn, Hampton Inn (see our ad this page)
ATM N: Nations Bank

153B Jct. 30 US 65S Little Rock

EXIT ARKANSAS

154 Lakewood (No Eastbound Reaccess)

155 U.S. 67, U.S. 167N, Jacksonville

157 AR161, Prothro Road
TStop N: **Mid-State Truck Plaza(SCALES)**
FStop S: **Citgo***
Gas N: Conoco*, Exxon
S: Texaco*(D), Total*(D)
Food S: Burger King, Fred's Place, **McDonalds**, Taco Bell, Waffle House
Lodg S: **Masters Inn**, ◆ **Super 8 Motel**
AServ N: Exxon
TWash N: **Mid-State**

159 Jct I-440W, LR River Port, Texakana

161 AR391, Galloway
TStop S: **76 Auto/Truck Plaza*(D)(LP)(SCALES) (Showers), Pilot Travel Center(SCALES) (Cars Only), TruckoMat(D)(SCALES)**
FStop N: **Love's Country Store*(D)**
Gas S: Shell
Food N: **A & W Drive-In (Love's), Taco Bell (Love's)**
S: **Dairy Queen (Pilot), Subway (Pilot)**
TServ S: **76**
TWash S: **76, Blue Beacon Truck Wash**

165 Kerr Road

169 AR15, Remington Road, Cabot

(170) Truck Inspection Station - both directions

175 AR31, Lonoke
FStop N: **Citgo*(D)**
S: **Total*(D)**
Gas S: Shell*(D)
Food N: **McDonalds (Outdoor Playground)**
S: KFC, Taco Bell
Lodg N: DAYS INN Days Inn
S: **Economy Inn**
ATM N: Citgo

183 AR13, Carlisle
FStop S: **Conoco*(D), Phillips 66*(D)**
Gas S: Exxon, Texaco*
Food S: **KFC (Phillips 66)**, Nick's BBQ & Catfish, Pizza & More, **Pizza Masters Express (Conoco)**
Lodg S: AAA **Best Western**
ATM S: Conoco

193 AR11, Hazen, Des Arc, Stuttgart, De Vaees, Clarendon
TStop S: **Shell*(D) (Showers, Overnight Parking)**
Gas N: Exxon*
Food N: Exxon Restaurant
Lodg S: ◆ Super 8 Motel

(198) **Rest Area - RR, Phones P (Both Directions)**

(199) **Rest Area - RR, Vending P (Westbound)**

202 AK33, Biscoe

216 U.S. 49, AR17, Brinkley, Cotton Point, Helena, Jonesboro
FStop N: **Exxon*(D), Texaco***
Gas N: Citgo*
S: BP* (ATM), Exxon*, Mapco Express, Shell
Food N: **Four Seasons Restaurant (Super 8 Motel)**, Smoothe Hill Subs, Western Sizzlin'
S: Gene's Pit BBQ, **McDonalds**, Simons Ole South Pancake House (Open 24 Hours), Subway, Taco Bell, Waffle House
Lodg N: DAYS INN Days Inn, Econolodge, AAA **Super 8 Motel (see our ad this page)**
S: ◆ Best Western, AAA Heritage Inn
RVCamp S: **GoodSam RV Park**
AServ S: NAPA Auto Parts
Other S: **Kroger Supermarket, Radio Shack, Wal-**

EXIT | ARKANSAS

Mart (Open 24 Hours)

221 AR78, Wheatley, Marianna
- TStop N: Holmes Truckstop*(D), Phillips 66*(D)
- FStop S: Delta Express*(D), Shell Super Stop*(D)
- Gas N: Exxon*
- Food S: KFC (Shell), Subway (Delta Express)
- TServ N: Holmes Auto Service and Truck Service, I-40 Truck and Tire RepairShop, Tire Center (Phillips 66)
- TWash N: Holmes

233 AR261, Palestine
- FStop S: Citgo*(D), Fina*
- Gas N: Loves*
- Food N: Subway
  S: Citgo Restaurant
- AServ S: 24 Hour Tire Shop
- RVCamp S: Camping

(235) Rest Area - RR, Phones P (Eastbound)

(236) Rest Area - RR, Phones P (Westbound)

241A AR1S, Forrest City, Wynne
- Gas S: Citgo*(CW), Exxon*, Phillips 66*, Texaco*
- Food S: Bonanza Steak House, Hardees, McDonalds, McDonalds (Wal-Mart), Pizza Hut, Simons Old South Pancake House, Subway, Taco Bell, Waffle House
- Lodg S: AAA Best Western, AAA Colony Inn
- ATM S: First National Bank
- Other S: Edward's Food Giant, Fred's Discount Pharmacy, Kroger Supermarket, Pay Less Shoe Store, Wal-Mart Supercenter

241B AR1N, Wynne
- FStop N: Total*(D)
- Gas N: Shell*
- Food N: Ho Ho Restaurant, Krystal (Shell), Wendy's
- Lodg N: DAYS INN Days Inn (see our ad this page), AAA Comfort Inn, ◆ Luxury Inn, Super 8 Motel

242 AR284, Crowley's Ridge Road
- Med N: + Hospital
- Parks N: Village Creek State Park

(243) Rest Area - RR, Phones, HF, Phones, Picnic, Vending, Tourist Info P

247 AR38E, Widener, Hughes
- TServ N: Truck Outlet Goodrich Tires

256 AR75, Parkin
- FStop N: Delta Express*(D)
- Parks N: Parkin Archeological State Park

260 AR149, Earle
- TStop N: TravelCenters of America(SCALES)
- FStop N: BP, Phillips 66*
- Gas S: Texaco*(CW)
- Food N: Country Pride, Subway
- Lodg N: Best Western
- TServ N: TA(SCALES)
- TWash N: BP
- RVCamp N: KOA Campground*(SCALES)
- Other S: Coin Car Wash

265 Marianna Elaina Rd.

271 AR147
- FStop S: P J's*
- Gas S: Exxon*(D), Texaco*(D)
- Food S: PJ's

(273) Weigh Station - both directions

(274) Rest Area - RR, Phones, HF, Picnic, Tourist Info P (Westbound)

275 AR118, Airport Road, West Memphis Airport

276 AR76, Rich Road, Missouri Street (Difficult Reaccess)
- Food S: McDonalds (In Walmart)
- AServ S: Wal-Mart

EXIT | ARKANSAS

- Other S: Wal-Mart

277 Junction I-55, N. Blytheville, St Louis (Difficult Reaccess)

278 7th St, West Memphis, AR191
- TStop S: Memphis Gateway Auto/Truck Plaza*(SCALES)
- FStop S: Love's* (24 hrs)
- Gas S: Citgo*
- Food N: Blimpie's Subs (Citgo)
  S: Catfish Island, I Can't Believe It's Yogurt (Memphis Gateway TS), McDonalds (Wal-Mart)
- Lodg S: Ramada Inn
- TServ S: Memphis Gateway Auto/Truck Plaza
- RVCamp S: Tom Sawyer's RV Park
- ATM S: Love's
- Other S: Wal-Mart (Open 24 Hours)

279A Ingram Blvd (Difficult Reaccess)
- Gas N: BP*
  S: BP, Citgo*(CW), Shell*
- Food N: Delta Point (Days Inn), Grandy's (Days Inn), Little Italy Pizza (Days Inn)
  S: Cinnamon Street Bakery (Citgo), Earl's Restaurant, Eddi Pepper's Mexican (Citgo), Holiday Inn Restaurant (Holiday Inn), Hot Stuff Pizza (Citgo), Smash Hit Subs (Citgo), Waffle House, Waffle House
- Lodg N: Classic Inn, AAA Comfort Inn, DAYS INN Days Inn
  S: Classic Inn, DAYS INN Days Inn, ◆ Econolodge, ◆ Hampton Inn, ◆ Holiday Inn, ◆ Holiday Inn, AAA Howard Johnson, Motel 6, Motel 6, Relax Inn, Relax Inn
- AServ N: BP
- Other N: Southwind Greyhound Park
  S: Coin Laundry

279B Junction I-55, South Memphis, Jackson, MS

280 Dr Martin Luther King Jr Dr (S. Side Services Are Same As Those On I-55, Ex. 4 East Side)
- TStop S: Flying J Travel Plaza*(LP)(SCALES) (RV Dump),

EXIT | ARKANSAS/TENNESSEE

Petro*(SCALES) (Showers, CB Shop, MS Carriers Recruiting Office)
- FStop N: Fina* (Showers), Mapco Express
  S: Pilot Travel Center*(D)(SCALES)
- Food S: Dairy Queen (Pilot TS), Iron Skillet Restaurant (Petro TS), KFC, McDonalds, Subway (Pilot TS), Taco Bell
- Lodg S: Best Western, Deluxe Inn, Express Inn, ◆ Super 8 Motel
- AServ N: Goodyear Tire & Auto
- TServ N: Goodyear Tire & Auto, Riggs Cat Truck Engine Service
  S: Bridgestone Tire & Auto (Petro TS), Jim's & Son Truck Service & Parts, Landstar Service Center
- TWash S: Blue Beacon Truck Wash (Petro TS), Mr. Clean Truck Wash, XVIII Wheeler's Truck Wash (24 Hrs)
- Other N: Best Tarps
  S: CB Shop

281 AR 131, Mound City Road (Westbound)

↑ARKANSAS

↓TENNESSEE

1 Riverside Dr, Front St, Welcome Center
- Food N: High Point Pinch Bar & Grill, TJ Mulligan's, The North End Restaurant
  S: China Restaurant, George & David's Greek Island Restaurant, Gridiron Restaurant, Wall Street Deli
- Lodg S: AAA Comfort Inn, Crowne Plaza Hotel, ◆ Sleep Inn, Days Inn (see our ad this page)
- Med S: + Hospital
- ATM S: First Commercial Bank
- Parks S: Confederate Park, Jefferson Davis Park
- Other N: Main Street Trolley, The Pyramid
  S: Easy-way Food Store, Jack's Food Store #1, Memphis Cook Convention Center, Memphis Police Station, Mud Island, Welcome Center

1A 2nd Street, 3rd Street
- Med N: + Hospitals

1B US 51, Danny Thomas Blvd
- Med N: + St. Joseph Hospital, + St. Jude Children's Research Hospital
- Other S: Sam's Cash & Carry Market

1C US 51 South, Danny Thomas Blvd (Services Same As Previous Exit B)

1D US 51 N, Danny Thomas Blvd (Services Same As Previous Exit B)

1E I-240 S, Jackson MS, Madison Ave

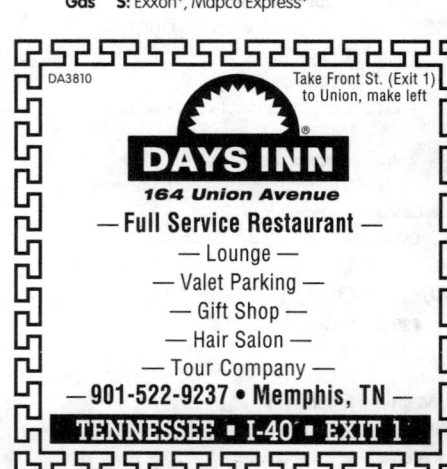

1FG TN 14, Jackson Ave
- Gas S: Exxon*, Mapco Express*

EXIT TENNESSEE

Food **N:** Jango's Bar-B-Q, Pager's Bar-B-Q, Playerettes Club
S: Randy's Ranch Burger, Southern Experience
Lodg **N:** Rainbow Inn
AServ **N:** Cowboy Used Tires
S: 1217, Donn's Transmission, Pigues Tire Co
Other **N:** **Piggly Wiggly Supermarket**
S: Car Wash, **Carol's Clean Coin-Op Laundry, Safari #1 Grocery Store**

2 Chelsea Ave, Smith Ave (Difficult Reaccess)
AServ **N:** Memphis Trucks Parts and Equipment Inc
S: Mr. Complete Used & Recycled Auto Parts
Other **N: K Corner Grocery**

2A U.S. 51, Millington

3 Watkins St
Gas **N:** Amoco*, Citgo*, Coastal, Oil City USA*(D), Total*
Food **N:** Oriental Market Restaurant
AServ **N:** The Shop, Wardlaw's Garage
Other **N:** Car Wash, **Coin-Op Laundry, U-Haul Center**(LP)

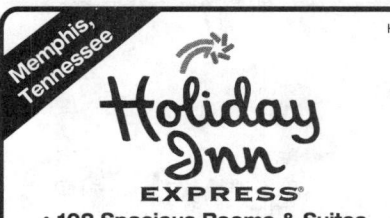

5 Hollywood St
Gas **N:** Mapco Express*
Food **N:** Hardee's, Moma's Bar-B-Q
ATM **N:** First Bank Tennessee
Other **N: Colonial Bakery Store, Stop N Shop Grocery, The Big One Memphis Flea Market, WalGreens**

6 Warford St, New Allen Road

8 TN 14, Jackson Ave, Austin Peay Hwy
Gas **S:** Texaco*(CW)
Food **S:** Atienda Mexicana Azteca, Knight's Cafe
AServ **S:** Transmission Service, Uneeda Auto Glass
Med **N:** ✚ **Raleigh Methodist Hospital**
Other **S: Peoples' Pharmacy, Rub A Duds Laundromat**

10 TN 204, Covington Pike
Gas **N:** Pump & Save
Food **N:** Hardee's
AServ **N:** Auto Parts for Imports, Covington Pike Dodge, Covington Pike Honda, Covington Pike Hyundai, Covington Pike Toyota, Dayton Tires, Gossett Jeep/Eagle, Gossett Mitsubishi, Gossett VW, Gwatney Olds Chev, Homer Skelton Mazda, Kia, Jim Keras Buick & Nissan, Pat Patterson Volvo, Sunrise Pontiac/GMC
Other **N: Mega Market Supermarket, Sam's Club, The Golf Academy Aqua Golf Center**

12A Summer Ave, U.S. 64, 70, 79

EXIT TENNESSEE

Gas **N:** Mapco Express*(D), Texaco*(CW)
S: Amoco, Exxon*, Texaco
Food **N:** Luby's Cafeteria, Pappy & Jimmie's Restaurant, Waffle House
S: Arby's, GoodTime Charlies, Great China Restaurant, Jack's Bar-B-Q Rib Shack, Lipsey Seafood Center, McDonalds, Nam King, Pancho's, Pizza Hut, Russ's Bar & Grill, Wendy's, Western Sizzlin', Young's Deli
Lodg **N: Holiday Inn (see our ad this page)**
AServ **N:** Lewis Ford, Mr. D's Service
S: Amoco, Bill King's Brake-O, Dayton Tires, Firestone Tire & Auto, Goodyear Tire & Auto, K-Mart, Texaco
ATM **S:** National Bank of Commerce
Other **N: Malco Summer Drive-In, Office Max, U-Haul Center**(LP)
S: Bookcase Paper Back Books, Cinema Showcase 12, Cloverleaf Animal Clinic, Cloverleaf Pharmacy, Hot Spot Food Store, K-Mart, Piggly Wiggly, Skateland

12B Sam Cooper Blvd (Difficult Reaccess)

12C Junction I-240, Nashville (Eastbound Must Exit To Stay On I-40)
Lodg **N: Comfort Inn (see our ad this page)**, Days Inn (Take I-240 to I-55, then south to Exit 5B -- see our ad this page), Hampton Inn

12 Sycamore View Road, Bartlett
Gas **N:** Texaco*(CW)(LP), Total*(D)
S: Amoco*(LP), BP*(CW), Exxon*, Mapco Express*
Food **N:** Cracker Barrel, McDonalds, Perkins Family Restaurant, The Garden Room (Holiday Inn), Waffle House
S: All American Eagle's Nest Sports Grill, Blimpie's Subs (Mapco), Burger King, Celebration Station Food & Fun, Delhi Palace, Denny's, Fortune Inn Chinese, Gridley's BBQ, Old Country Buffet, Sakura Japanese, Subway, Tops Bar-B-Q, Wendy's
Lodg **N:** ◆ Budgetel Inn, ◆ Drury Inn, Hampton Inn (see our ad this page), AAA Holiday Inn, AAA Howard Johnson, ◆ Red Roof Inn
S: AAA Best Western, AAA Comfort Inn, DAYS INN ◆ Days Inn, ◆ Fairfield Inn (see our ad this page), AAA La Quinta Inn, AAA Memphis Inn, Motel 6, Super 8 Motel
AServ **S:** Express Lube, Wal-Mart
ATM **S:** Mapco
Other **N: Wimbleton Sportsplex**
S: Wal-Mart (w/ pharmacy)

14 Whitten Road
Gas **N:** BP*(CW)
S: Citgo* (24 Hrs)
Food **S:** Stuckey's (Citgo)
Lodg **S:** Villager Lodge
Other **N: Wild Water & Wheels (Amusement Park)**

15AB Appling Rd.

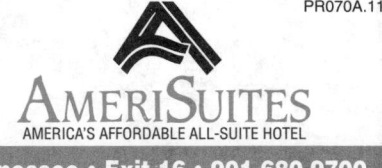

16 TN 177, Germantown
Food **N:** Alexander's Restaurant, Chik-fil-A, Chili's, Gam-ma's Gourmet Foods, Logan's Roadhouse, Luby's Cafeteria, On the Border Cafe, Ruby Tuesday, The Fine Grind
Lodg **N:** AAA AmeriSuites (see our ad this page)
ATM **N:** First Tennessee Bank
Other **N: Baptist Bookstore, Barnes & Noble, Home Depot**(LP)**, Toys "R" Us, Wolfchase Galleria Shopping Mall**

EXIT TENNESSEE

18 U.S. 64, Somerville, Bolivar
- **Gas** **N:** Exxon*(CW)
  - **S:** Amoco*(LP), Circle K*(LP)
- **Food** **N:** Bresler's Ice Cream (Exxon), Burger King (Exxon)
  - **S:** Backyard Burgers, Pizza Hut
- **Lodg** **N:** AAA Holiday Inn Express (see our ad page 191)
- **ATM** **N:** Exxon
- **Other** **S:** **Fire Station, Kroger Supermarket, Pet Health Center (Veterinary Clinic)**

20 Canada Road, Lakeland
- **FStop** **N:** **BP*(D)**
- **Gas** **N:** Discount Gas(D)(LP)
  - **S:** Exxon*
- **Food** **S:** Subway (Exxon), TCBY (Exxon), The Cotton Cabin
- **Lodg** **N:** Days Inn ◆ Days Inn, **Relax Inn**, AAA **Super 8 Motel**
- **ATM** **N:** The People's Bank
- **Other** **S:** **Belz Factory Outlet**

25 TN 205, Airline Road, Arlington, Collierville
- **Gas** **N:** Discount Gas*(LP), Exxon*
- **Food** **N:** Subway (Exxon)

35 TN 59, Covington, Somerville
- **FStop** **S:** **BP***
- **Food** **S:** **Longtown Cafe (BP)**
- **Other** **S:** **The CB Shop**

42 TN 222, Stanton, Somerville
- **Gas** **S:** Exxon*, Phillips 66(LP)
- **Lodg** **N:** Countryside Inn
- **AServ** **S:** Phillips 66

47 TN 179, Stanton - Dancyville Road
- **TStop** **S:** **PTP Stop***
- **TServ** **S:** **PTP Stop***

(49) **Weigh Station (Eastbound)**

(50) **Weigh Station (Westbound)**

52 TN 76, TN 179, Koko Road, Whiteville
- **Gas** **S:** KoKo Community Market*, Phillip's Quick Stop*

56 TN 76, Brownsville, Somerville
- **FStop** **S:** **Exxon***
- **Gas** **N:** Amoco*, Citgo* (24 hrs)
  - **S:** BP*
- **Food** **N:** Dairy Queen, Fiesta Garden, **KFC, McDonalds**
  - **S:** **Huddle House (Exxon)**
- **Lodg** **N:** Best Western, AAA Comfort Inn, Days Inn AAA **Days Inn**, ◆ **Holiday Inn Express**

EXIT TENNESSEE

- **Other** **N:** **Humboldt Fish Hatchery & Davy Crockett Lake**
  - **S:** **Headquarters for the Hatchie National Wildlife Refuge**

60 TN 19, Mercer Road
- **AServ** **N:** Convenient Store
- **Other** **N:** **Convenient Store**

66 U.S. 70
- **FStop** **S:** **Fuel Mart*(SCALES)**
- **Gas** **S:** Citgo*
- **Food** **S:** **Blimpie's Subs (Fuel Mart)**
- **Lodg** **S:** Relax Inn

68 TN 138, Providence Road
- **TStop** **S:** **76 Auto/Truck Plaza(SCALES) (BP Fuel)**
- **Gas** **S:** Citgo*
- **Food** **S:** **Union Restaurant (76 TS)**

EXIT TENNESSE

- **Lodg** **N:** Econolodge
- **TServ** **S:** **76**
- **TWash** **S:** **Al's Truck Wash**
- **RVCamp** **S:** **Joy-O RV Park**
- **Other** **S:** **Cotton Patch Grocery(LP)**

(73) **Rest Area - RR, Phones, Vending, Pet Walk, Picnic (Eastbound)**

(74) **Rest Area (Westbound)**

74 Lower Brownsville Road

76 TN 223 South, McKellar - Sipes Regional Airport

79 U.S. 412, Jackson
- **FStop** **S:** **Citgo*(CW)**
- **Gas** **S:** BP, Exxon*, PRM
- **Food** **S:** CG's Restaurant
- **Lodg** **S:** Days Inn ◆ Days Inn
- **AServ** **S:** BP
- **RVCamp** **S:** **Jackson Mobile Village & RV Park**
- **Other** **S:** **Madison Wholesale Co.**

80AB 80A - US 45 S. Bypass, Jackson 80B N Bypass, Milan, Humboldt
- **Gas** **S:** Amoco*(CW), BP*(CW), Citgo*(CW), Citgo*, Phillips 66*(D)(LP)
- **Food** **N:** Chik-fil-A, KFC, Lone Star Steakhouse, McDonalds
  - **S:** Applebee's, Arby's, Barn Hill's Country Cafe, Baudo's Italian American, Burger King, Burgers Up, Celebrations Restaurant, Cookies In Bloom, Dunkin' Donuts, El Chico Mexican Restaurant, Great Wall Chinese, Heavenly Ham, Hugh's Pit Bar-B-Q, Logan's Roadhouse, Madison's, McDonalds, Mrs Winner's Chicken, O'Charley's, Old Town Spaghetti Store, Pizza Hut Delivery, Shirley's Bakery, Sonic Drive-In, Subway, Taco Bell, Village Inn Pizza, Waffle House
- **Lodg** **S:** Best Western, Budget Inn, Casey Jones Statio Inn, Comfort Inn, Days Inn Days Inn, Econolodge, Fairfield Inn, Garden Plaza Hotel, Hampton Inn, Holiday Inn
- **AServ** **S:** Classic Car Care, K-Mart, King Tire Co, Midas Muffler & Brakes
- **ATM** **S:** Central State Bank, First American, First South Bank, First Tennessee Bank, The Bank of Jackso Volunteer Bank
- **Other** **N:** **Sam's Club, Wal-Mart, Dan's Factory Outle** **(see our ad on this page)**
  - **S:** **Casey Jones Museum, Casey Jones Village & Railroad Museum, Foto Express, Jackson's Pharmacy, K-Mart, Kroger Supermarket**

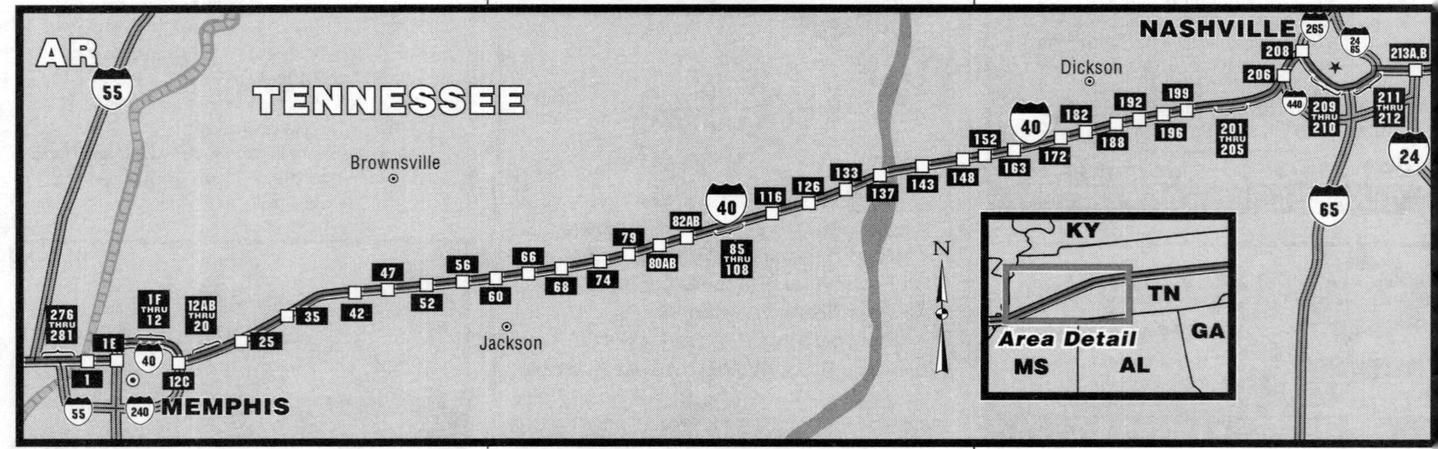

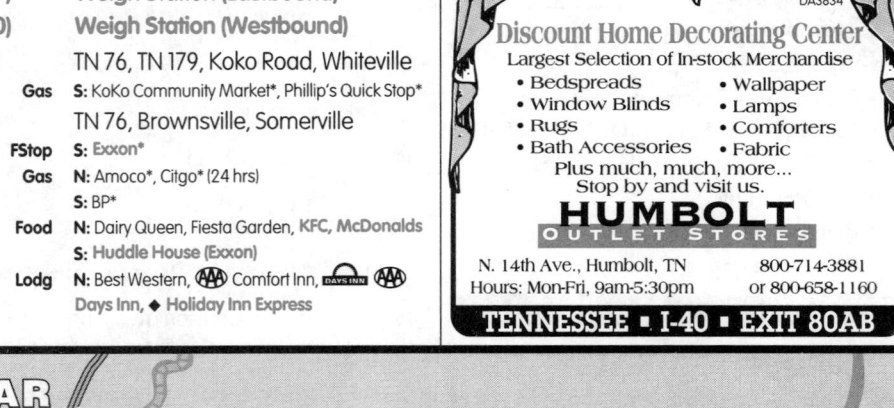

Bold red print shows RV & Bus parking available or nearby

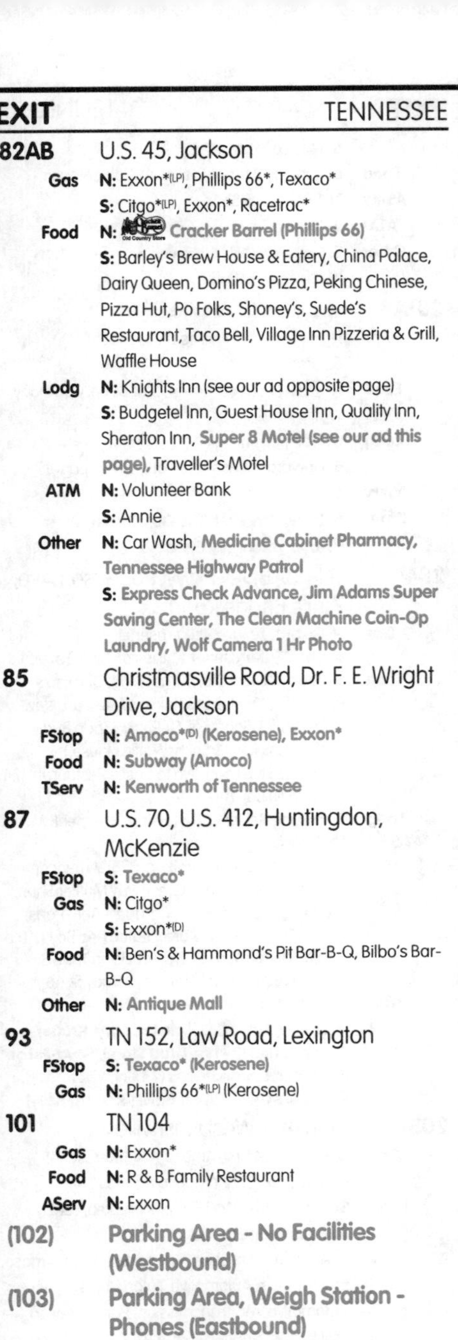

EXIT	TENNESSEE
82AB	U.S. 45, Jackson
Gas	N: Exxon*(LP), Phillips 66*, Texaco*
	S: Citgo*(LP), Exxon*, Racetrac*
Food	N: Cracker Barrel (Phillips 66)
	S: Barley's Brew House & Eatery, China Palace, Dairy Queen, Domino's Pizza, Peking Chinese, Pizza Hut, Po Folks, Shoney's, Suede's Restaurant, Taco Bell, Village Inn Pizzeria & Grill, Waffle House
Lodg	N: Knights Inn (see our ad opposite page)
	S: Budgetel Inn, Guest House Inn, Quality Inn, Sheraton Inn, Super 8 Motel (see our ad this page), Traveller's Motel
ATM	N: Volunteer Bank
	S: Annie
Other	N: Car Wash, Medicine Cabinet Pharmacy, Tennessee Highway Patrol
	S: Express Check Advance, Jim Adams Super Saving Center, The Clean Machine Coin-Op Laundry, Wolf Camera 1 Hr Photo
85	Christmasville Road, Dr. F. E. Wright Drive, Jackson
FStop	N: Amoco*(D) (Kerosene), Exxon*
Food	N: Subway (Amoco)
TServ	N: Kenworth of Tennessee
87	U.S. 70, U.S. 412, Huntingdon, McKenzie
FStop	S: Texaco*
Gas	N: Citgo*
	S: Exxon*(D)
Food	N: Ben's & Hammond's Pit Bar-B-Q, Bilbo's Bar-B-Q
Other	N: Antique Mall
93	TN 152, Law Road, Lexington
FStop	S: Texaco* (Kerosene)
Gas	N: Phillips 66*(LP) (Kerosene)
101	TN 104
Gas	N: Exxon*
Food	N: R & B Family Restaurant
AServ	N: Exxon
(102)	Parking Area - No Facilities (Westbound)
(103)	Parking Area, Weigh Station - Phones (Eastbound)
108	TN 22, Huntington, Lexington, Parkers Crossroads
FStop	N: Citgo*(LP) (Kerosene), Phillips 66*
Gas	N: Amoco*, BP* (Kerosene)
	S: Duckies* (Fireworks), Exxon*, Texaco*
Food	N: Bailey's Restaurant (24 Hrs)
	S: Cotton Patch Resturaunt (Tourist Info)
Lodg	N: Knights Inn, Hampton Inn (see our ad this page)
	S: AAA Best Western
AServ	N: Tire Shop
RVCamp	N: Beach Lake Campground
Med	S: + Hospital
ATM	N: BP, First Bank, Phillips 66
Parks	S: Pickwick Landing (65 Miles), Shiloh Park (51 Miles)
Other	N: Civil War Tour, US Post Office
116	TN 114
RVCamp	S: Bucksnort Campground
Parks	S: Natchez Trace State Park
Other	N: Maple Lake (3.5 Miles), Spring Valley Golf (7 Miles)

EXIT	TENNESSEE
126	TN 69, U.S. 641, Camden, Paris, Parsons
TStop	N: North 40 Auto/Truckstop*(SCALES), Sugar Tree
	S: Shell*
FStop	S: BP*(LP) (Kerosene)
Gas	N: Amoco*(LP), Exxon*(D), Texaco*(D)
	S: Citgo*
Food	N: Subway (Exxon), TCBY (Exxon)
	S: Apple Annie's (Shell TS), Dairy Queen (Citgo)
Lodg	S: Apple Annie's Motel (Shell TS)
TServ	N: Sugar Tree, Wilson's Repair
ATM	N: North 40 Auto/Truckstop
Parks	N: Kentucky Lake, Lake Barkley, Land Between the Lakes, Nathan Bedford Forest, Paris Landing State Parks
Other	N: Patsy Cline Memorial Plane Crash
	S: Dottie's Marine (BP)
(130)	Rest Area - RR, Phones, Vending, Picnic (Eastbound)
(131)	Rest Area - RR, Phones, Vending, Picnic
133	TN 191, Birdsong Road
RVCamp	N: Birdsong Resort & Marina
137	Cuba Landing
FStop	S: Citgo*
Food	S: Citgo Restaurant
RVCamp	N: Camping
Other	N: Cuba Landing Marina (Full Service Marina Store & Deli)
143	TN 13, Linden, Waverly, Lobelville, Buffalo
FStop	N: Pilot Travel Center*(SCALES)
Gas	N: BP(D), Phillips 66*(D), Shell*(D)
	S: Exxon*, Texaco*
Food	N: Buffalo River Grill, J J's Home Cookin', Log Cabin Restaurant (Best Western), Loretta Lynn's Kitchen, McDonalds, Subway (Pilot), The Sweet Shop
	S: South Side Restaurant
Lodg	N: ◆ Best Western, DAYS INN AAA Days Inn (see our ad this page), AAA Super 8 Motel
	S: Best Budget Motel
AServ	N: Ashton's Auto Service, BP
	S: Barnett's Towing
TServ	N: Ashton's Truck Service
	S: Barnett's Towing
RVCamp	N: KOA Kampground(LP)
ATM	N: Phillips 66
148	TN 229, Barren Hollow Road, Turney Center
Gas	S: Texaco*
152	TN 230, Bucksnort
FStop	S: Texaco*
Gas	N: BP* (Kerosene)
	S: Phillips 66*(LP)
Food	N: Rudy's Restaurant
Lodg	N: Country Inn USA
AServ	N: W & L (24 Hr Wrecker)
RVCamp	S: Camping Guide Service
Other	S: Bucksnort Trout Ranch (1 Mile)
163	TN 48, Centerville, Dickson
Gas	N: Citgo* (Kerosene), Phillips 66*(D)(LP)
	S: Shell* (Kerosene)(see our ad this page)
AServ	N: Citgo
RVCamp	S: Tanbark Campground (.3 Miles)
Other	S: CB Shop

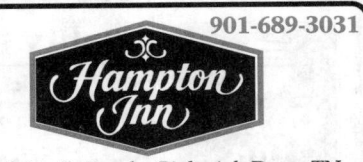

Bold red print shows RV & Bus parking available or nearby

EXIT TENNESSEE

(170) **Rest Area - RR, Phones, Picnic, Vending, Pet Walk (Both Directions)**

172 TN 46, Centerville, Dickson

- **TStop** N: **Mapco Express(SCALES)**
- **Gas** N: Amoco*(D)(LP) (24 Hrs), BP*(LP), Shell* (see our ad this page), Texaco* (Playground)
  S: Chevron* (24 hrs), Exxon*(LP) (Kerosene)
- **Food** N: Arby's, Baskin Robbins (Burger King, Playground), Burger King (Texaco, Playground), **Cracker Barrel**, Key West Cafe, McDonalds (Playground), **Steak 'N Shake (Mapco Express TS)**, Subway (BP), Waffle House, Wang's China
- **Lodg** N: ◆ Comfort Inn, **Econolodge**, Knights Inn, ◆ Quality Inn (see our ad this page)
  S: **Days Inn (see our ad this page)**, Holiday Inn
- **RVCamp** N: **KOA Kampground**
- **Med** N: **Hospital**
- **ATM** N: Mapco Express TS
- **Parks** N: **Montgomery Bell State Park**
- **Other** N: **Seven Flags Race Park - Put-Put, Go-Karts**

182 TN 96, Dickson, Fairview, Franklin

- **TStop** N: **Citgo - New Hwy 96 Truck Stop*(LP)**
- **FStop** S: **BP*(D)**
- **Gas** N: BP*
- **Food** N: **Country Cooking (Citgo TS)**
- **Lodg** N: **Dickson Motel**
- **TServ** N: **GTF Tire & Maintenance Service, Robert's Truck & Tire Repair (Citgo TS)**
- **Parks** N: **Montgomery Bell State Park (16 Miles)**

188 TN 249, Kingston Springs Road, Ashland City

- **TStop** S: **Petro-2*(LP)(SCALES)**
- **Gas** N: BP* (Kerosene), Mapco Express*(LP) (Kerosene), Shell*(CW) (see our ad this page)
- **Food** N: Arby's (Shell), Harpeth Restaurant, Miss Sadie's Diner
  S: **Quick! Skillet (Petro-2)**
- **Lodg** N: Best Western, **Econolodge**, Scottish Inns
- **TServ** S: **Petro Lube**
- **ATM** N: Cheatham State Bank, Mapco Express
  S: **Petro-2**
- **Other** N: **Public Library, US Post Office, Vegas Coin Laundry**
  S: **Kingston Springs Animal Hospital**

192 McCrory Lane, Pegram, Natchez Trace Parkway

- **Parks** N: **Newsoms Mill State Park (2 Miles)**

EXIT TENNESSEE

196 U.S. 70 South, Bellevue, Newsom Station

- **Gas** N: Mapco Express*(LP)
  S: BP*(CW)(LP), Shell, Texaco*(D)(CW)
- **Food** N: Shoney's
  S: McGillicudy's Restaurant, Sir Pizza, Subway, Taco Bell, Waffle House, Wendy's
- **AServ** S: Firestone Tire & Auto, Shell
- **Med** S: **Baptist Bellevue Medical Center**
- **ATM** N: Mapco Express
  S: First Union, Texaco
- **Other** N: **Regal Cinemas**
  S: **Bellevue Center Mall, Rent All(LP), The Home Depot**

199 TN 251, Old Hickory Blvd

- **Gas** N: Texaco*(D)(CW)(LP)

EXIT TENNESSEE

- S: Mapco Express*(LP)
- **Food** S: Subway (Mapco Express), Waffle House
- **AServ** N: Pug's Body Shop
- **ATM** S: Mapco Express
- **Other** S: **Putt-Putt Golf & Games**, Rojo Red's Car Wash, **Sam's Club**

201AB U.S. 70E, Charlotte Pike

- **Gas** N: Exxon*(CW)(LP), Phillips 66*, Texaco*
  S: Citgo
- **Food** N: **Cracker Barrel**, Star Bagel Company, Waffle House
- **Lodg** N: Hallmark Inn
  S: Howard Johnson (see our ad this page)
- **AServ** S: Citgo
- **Other** N: **Drew's Market, The Golf Center, West Meade Veterinary Clinic**

204 TN 155, Briley Pkwy, Robertson Ave, White Bridge Road

- **Gas** N: Exxon*, Texaco*(D) (Kerosene)
  S: Citgo*(D) (Kerosene), Phillips 66*(D)(LP), Texaco
- **Food** S: Bobby's Dairy Dip, Burger King, Domino's Pizza, KFC, Krystal, Las Palmas Mexican, New China Buffet, Shoney's, Sir Pizza, Uncle Bud's Catfish, Chicken, & Such, Waffle House, Wendell's Fine Food, White Castle Restaurant (24 hrs), Whitts Bar-B-Q
- **Lodg** S: ◆ Budgetel Inn, Daystop Inn, ◆ Super 8 Motel
- **AServ** N: Texaco
  S: 76 Covington Transmission (76), American Tire Co., Auto Value Parts Store, Batey Muffler Shop, Bradley's Import Service, Carquest Auto Parts, Michel Tire, NTB, NTW Tire Center, Pep Boys Auto Center, Performance Ford, Precision Tune & Lube, Q Lube, Texaco, Whaley's Body Shop
- **ATM** S: First Union, WalGreens
- **Other** S: **Bradley Drug Co., Coin Laundry, Kroger Supermarket (24 hrs, Drug Store), Save-A-Lot Food Stores, WalGreens (24 hrs, Pharmacy), Wendell's Deli(LP), West Bowl (Bowling Alley)**

205 46th Ave, West Nashville

- **Gas** S: Amoco (24 hrs), Amoco*(D)(CW) (Kerosene), Mapco Express*(LP), Mapco Express
- **Food** S: McDonalds, Mrs Winner's Chicken, Lee's Famous Recipe Chicken
- **AServ** S: Allpro Transmission, Engine Specialist, Amoco, Bradley Bros. Automotive, Laton Dress Automotive, Mr. Transmission, Parts America, Quick Lube, Stephens Service Center, Wilkie's
- **ATM** S: First American, McDonalds, SunTrust
- **Other** S: **Dark Horse Theatre, Public Library, US Post Office**

206 Junction I-440 East, Knoxville

Bold red print shows RV & Bus parking available or nearby

EXIT	TENNESSEE

207 Jefferson St. (Reaccess Westbound Only)
- **Gas** N: Circle K*
- **Food** N: Lee's Famous Recipe Chicken, Mrs Winner's, Marie's Country Kitchen, San Juan's Seafood King, Wendy's

208 Junction I-265, Louisville

209 Charlotte Ave, Church St
- **Gas** N: Exxon*(CW)
  S: Amoco
- **Food** N: McDonalds
- **Lodg** N: Clubhouse Inn & Conference Center, ◆ Union Station Hotel
  S: Days Inn
- **AServ** N: Exxon, Hippodrome Nissan Trucks, Hippodrome Olds, Nissan, Jack Morris Auto Glass, Master Care Auto Service
  S: Beaman Pontiac GMC Trucks, Import Auto Maintenance, Jim Reed Chevrolet, McDowell General Tire, Rally Mitsubishi
- **Other** N: **Baptist Bookstore**
  S: **The Parthenon**

209A US 70S, US 431 Broadway (Westbound)
- **Gas** N: Exxon*(D)(CW)
  S: Texaco*(CW)
- **Food** N: McDonalds
- **Lodg** S: AAA Days Inn
- **AServ** N: Exxon, Hippodrome Nissan Trucks, Hippodrome Olds Nissan, Jack Morris Auto Glass, Master Care Auto Service
  S: Beaman Pontiac, GMC Trucks, Suzuki, Import Auto Maintenance, Jim Reed Chevrolet & Truck Center, McDowell General Tire, Rally Mitsubishi, Toyota
- **Other** S: Car Wash, Sure Brite Care Wash, **The Parthenon**

209B U.S. 70 South, U.S. 431, Demonbreun St
- **Gas** N: Exxon*
- **Food** N: McDonalds, The Piewagon Restaurant
  S: Country Star Cafe
- **AServ** N: Car Wash, Hippodrome Nissan Trucks, Hippodrome Oldsmobile Nissan, Kelly Tire, Master Care Auto Service, Nashville Auto Service, Truck Tire Service
  S: Auto Care Center, Toyota
- **Other** N: **Tourist Information**
  S: 1 Hr Moto Photo, **Cars of the Stars Museum, Country Music Hall Of Fame, Hank Williams Jr Museum, Legends and Superstar Hall of Fame, The Country Music Wax Museum, Wild West Indian Shop**

210A Junction I-65 South, Birmingham (Eastbound)

210B Junction I-65 North (Eastbound)

210C U.S. 31 A South, U.S. 41 A, 4th Ave, 2nd Ave (Difficult Reaccess)

- **Med** N: Hospital
- **Other** S: **Cumberland Science Museum**

211B Junction I-24 West, Junction I-65 North, Clarksville, Louisville

212A Fesslers Lane (Eastbound)
- **FStop** S: **Citgo*(LP), Texaco*(CW)**
- **Food** S: **Burger King**, Elvy's Deli Etc., **Krystal (Citgo)**, McDonalds, Wendy's
- **AServ** N: Goodyear Tire & Auto, TCI Tire Centers
- **TServ** N: **Goodyear Tire & Auto, Neeley Coble Co, Tennessee Truck Sales Inc.**
  S: **PM Truck Service, Treadco Inc. Truck Tire Sales & Service**
- **RVCamp** N: **Neeley Coble Company**
- **ATM** S: Citgo, Texaco
- **Other** S: **CB Radio Sales & Service**

211A Junction I-65 South, Junction I-40 West, Memphis, Birmingham

212 Hermitage Ave (Westbound, Difficult Reaccess)
- **Gas** S: **Citgo*(D)**
- **Food** N: El Captain Bar and Grill
  S: Hermitage Cafe Hamburgers & Chili, Lady J's Gingerbread House
- **Lodg** N: Drake Inn (see our ad this page)
- **AServ** N: Martin's Wrecking Service
- **TServ** N: **Goodyear Tire & Auto, Neely Coble Sunbelt Truck Center, TIP Trailer Rental, Leasing, Sales Service, Tennessee Truck Sales**
  S: **Rawlings Truck Rental and Leasing**
- **Med** N: **Hospital**
  S: **Hospital**
- **Other** N: **Hermitage Ave Deli & Market, J C Napier Community Center Swimming Pool**

213A Junction I-24 East, West I-440, Memphis, Chattanooga

213B Junction I-24 West, Junction I-40 West, Nashville

213 U.S. 41, Spence Lane, Murfreesboro Rd (Difficult Reaccess)
- **Gas** N: Texaco*(D)(CW) (Kerosene)
- **Food** S: Dad's Place (Ramada Inn), The Down Under Bar & Grill (Days Inn), Waffle House
- **Lodg** S: Days Inn, Ramada Inn
- **TServ** N: **Kenworth of Tennessee**
  S: **Cummins Diesel**

215 TN 155, Briley Pkwy, Opryland
- **Gas** S: Phillips 66*, Shell*
- **Food** S: Denny's (24 Hrs, Ramada Inn), McRedmond's Southern Dining (Howard Johnson)
- **Lodg** S: AAA Howard Johnson, Ramada Inn, Villager Lodge, Days Inn (see our ad this page)
- **RVCamp** N: **Camping World (see our ad this page)**
- **Other** N: **Nashville Factory (see our ad this page)**

216 TN 255, Donelson Pike, International Airport
- **Gas** N: Amoco*(D)(CW), BP*(CW), Racetrac*, Texaco*
- **Food** N: KFC, McDonalds, Shoney's, Subway (Texaco), Waffle House, Wendy's
- **Lodg** N: Budgetel Inn, Hampton Inn & Suites (see our ad this page), Holiday Inn Express, Red Roof Inn, Super 8 Motel, Wyndham Garden Hotel
- **AServ** N: Amoco, **K-Mart**
- **ATM** N: Racetrac
- **Other** N: **K-Mart, Piggly Wiggly Supermarket**

Bold red print shows RV & Bus parking available or nearby

EXIT TENNESSEE

S: Airport

219 Stewart's Ferry Pike, J. Percy Priest Dam

- **Gas** N: Mapco Express*[LP] (Kerosene)
  S: Mapco Express*, Texaco*[D][CW]
- **Food** S: **Cracker Barrel,** Dragon Palace, New York Experience, Subway (Mapco Express), Uncle Bud's Catfish, Chicken, & Such, Waffle House
- **Lodg** S: (AAA) Best Western, ◆ Days Inn, Family Inns of America, Howard Johnson, ◆ Sleep Inn
- **ATM** S: Texaco
- **Other** S: **J. Percy Priest Lake & Dam, McCormick Animal Clinic,** Touchless Automatics Car Wash

221 TN 45, Old Hickory Blvd, The Hermitage

- **Gas** N: Mapco Express*[LP] (Kerosene), Racetrac*, Shell*
  S: Chevron[D][LP], Kwik Sak[LP] (Racing Gasoline with Lead), Texaco*[D][LP]
- **Food** N: Dairy Queen (Playground), Hardee's, K.O.'s Pizza & Sandwiches, Waffle House
  S: Express Pizza
- **Lodg** N: Comfort Inn, (AAA) Holiday Inn Express, ◆ Ramada Limited
- **AServ** S: Chevron
- **Med** N: **Columbia Hospital**
- **Other** N: Car Wash

226 TN 171, Mount Juliet Road

- **Gas** N: BP*[LP], Citgo*, Mapco Express*[D][LP] (Kerosene), Texaco*[D][LP]
  S: Mapco Express*[D][LP] (Kerosene)
- **Food** N: Arby's, Austin's Tex Mex (Citgo), Captain D's Seafood, McDonalds (BP)
  S: **Cracker Barrel,** Waffle House
- **AServ** N: David Harris Automotive Specialist Corvette, Cobra
- **ATM** N: Mapco Express, McDonalds
  S: Mapco
- **Other** N: **Squeeky Clean Coin Laundry**

(227) **Weigh Station, Rest Area (Eastbound)**

(228) **Parking Area, Weigh Station (Westbound)**

232 TN 109, Gallatin

- **TStop** N: **76 Auto/Truck Plaza*[SCALES] (Auto Fuel-Citgo)**
- **Gas** N: Amoco* (Kerosene)
- **Food** N: Granny's (Amoco), **Restaurant (76 Auto/Truck Stop)**
- **TServ** N: 76
- **ATM** N: **70 Auto/Truck Stop,** Delta Express

235 TN 840 West, Chattanooga (Limited Access Highway)

238 U.S. 231, Lebanon, Murfreesboro (Cumberland University)

- **TStop** S: **Pam's Truckstop, Mapco Express (see our ad this page)**
- **FStop** S: **Cone***
- **Gas** N: Amoco*, BP*, Exxon*, Pilot Travel Center*[D][LP] (Kerosene), Shell*[CW] (see our ad this page)
- **Food** N: Arby's (Amoco), **Cracker Barrel, Gondola, Hardee's, McDonalds (Playground),** Mrs Winner's Chicken, Pizza Hut, Ponderosa, Shoney's, Subway (Hampton Inn), Sunset Family Restaurant, Taco Bell, Uncle Bud's Catfish, Chicken, & Such, Waffle House, Wendy's
  S: O'Charley's Restaurant, **Restaurant (Pam's TS)**
- **Lodg** N: (AAA) Best Western, (AAA) Comfort Inn, ◆ Hampton Inn, ◆ Holiday Inn Express, Scottish Inns, (AAA) Shoney's Inn
  S: ◆ **Days Inn (see our ad this page),** (AAA) **Knights Inn,** ◆ **Super 8 Motel**
- **TServ** S: **G & R Diesel Repair (Pam's TS)**
- **RVCamp** S: **Shady Acres (2 Miles), Timberline Campground (.7 Miles)**
- **ATM** N: Suntrust Bank (Hampton Inn)
- **Parks** S: **Cedars of Lebanon State Park (6 Miles)**
- **Other** N: Car Wash, **Cumberland University, Holiday Washateria, Sweet Tooth Candy Store, Wal-Mart Supercenter (24 Hrs)**

239 U.S. 70, Lebanon, Watertown

- **TStop** S: **Uncle Pete's Super Truck Stop (Phillips 66, Showers, 24 Hrs)**
- **Gas** N: Amoco* (Kerosene), Raceway*
- **Food** S: **Four Winds Truckstop**
- **AServ** S: Kings Wrecker Service (24 Hr Service)
- **TServ** S: **Uncle Pete's TS**
- **TWash** S: **Uncle Pete's TS**
- **Med** N: **Hospital**
- **ATM** N: Raceway
- **Other** N: **Sport Outfitters**

245 Linwood Road

- **FStop** N: **BP*[LP]**

(252) **Parking Area, Weigh Station (Both Directions)**

254 TN 141, Alexandria

258 TN 53, Carthage, Gordonsville, Hartsville

- **Gas** N: Shell (see our ad this page)
  S: BP[D] (Kerosene), Citgo*[D], Keystop[D]
- **Food** N: McDonalds (Playground), Timberloft
  S: Restaurant (Keystop)
- **Lodg** N: **Comfort Inn (see our ad this page)**
- **AServ** S: BP
- **Other** N: **Cordell Hull Dam**

(267) **Rest Area - Picnic, RR, HF, Phones, RV Dump**

268 TN 96, Buffalo Valley Road, Center Hill Dam

- **Food** N: Buffalo Bill's Restaurant
- **Lodg** N: Buffalo Bill's Motel
- **Parks** S: **Edgar Evins State Park (4 Miles)**
- **Other** N: **Buffalo Bill's Market & General Store**

Bold red print shows RV & Bus parking available or nearby

EXIT	TENNESSEE
	S: Center Hill Dam, US Post Office
273	**TN 56, Smithville, McMinnville**
Gas	S: BP[D], Phillips 66*
Food	S: Rose Garden Restaurant
AServ	S: BP
Parks	S: Edgar Evins State Park
Other	S: Appalachian Center for Crafts (6 Miles), Center Hill Dam (8 Miles), US Post Office
276	**Old Baxter Road**
Gas	N: T-Tommy's*[D]
	S: Texaco*
AServ	N: T Tommy's Auto & 24-Hour Wrecker Service
280	**TN 56 North, Baxter, Gainesboro**
TStop	N: Mapco Express (see our ad this page)
Gas	N: Shell*[D] (see our ad this page)

EXIT	TENNESSEE
RVCamp	N: Camp Discovery (24 Miles)
Parks	N: Burgess Falls State Park
286	**TN 135, Burgess Falls Road**
FStop	N: Exxon*[CW], Racetrac* (24 Hrs), Shell*[CW] (see our ad this page)
	S: Texaco*[D][LP]
Gas	N: BP*
	S: Amoco*[LP] (Kerosene)
Food	N: China One Express & Gifts, Hardee's, Waffle House
	S: Rice's Restaurant (Star Motor Inn)
Lodg	S: Star Motor Inn
AServ	N: Auto Glass Service, Cookeville Nissan, Cumberland Chrysler Center, Cumberland Toyota, Cumberland Truck Center, Mike Williams Tire & Auto
TServ	N: Walker Diesel

EXIT	TENNESSEE
Med	N: Hospital
ATM	N: First American
Parks	S: Burgess Falls State Park
Other	N: Family World Bookstore
	S: Touchless Automatic Car & RV Wash
287	**TN 136, Cookeville, Sparta**
TStop	S: Exxon* (24 Hrs, Showers, Kerosene)
FStop	S: Pilot Travel Center*
Gas	N: Amoco*[CW], Chevron[D], Shell*[CW] (see our ad this page), Texaco*[D][LP] (Tailor Entrance)
Food	N: Baskin Robbins, Burger King, Captain D's Seafood, China Star Chinese Restaurant, Cracker Barrel, Dairy Queen, El Tapatio, Finish Line Subs & Such (Amoco), Long John Silvers, Louie's (Holiday Inn), McDonalds, Nick's (Executive Inn), Pizza Hut, Ponderosa, Schlotzkys Deli, Shoney's, Subway (24 Hrs), TCBY (Amoco), Uncle Budd's Catfish Chicken & Such, Waffle House, Wendy's
	S: B & J BBQ, Gondola Pizza, KFC, Waffle House
Lodg	N: AAA Best Western, Comfort Suites, AAA Days Inn (see our ad this page), Executive Inn, AAA Holiday Inn (see our ad this page), Ramada Limited, AAA Super 8 Motel
	S: ◆ Econolodge
AServ	N: Cookeville Transmission, Pennzoil Lube
ATM	S: Exxon TS
Parks	S: Fall Creek Falls (43 Miles), Rock Island State Park (29 Miles)
Other	N: Horner-Rausch Optical Co. Superstore, Mall, TN Highway Patrol
	S: Camp Clements

Bold red print shows RV & Bus parking available or nearby

EXIT TENNESSEE

288 TN 111, Livingston, Sparta
- **TStop** S: Mid Tenn. Auto/Truck Plaza*(SCALES) (24 Hrs, RV Dump)
- **Gas** S: Shell*(D)(Kerosene) (see our ad this page)
- **Food** S: General Lee (Mid Tenn. TS), Huddle House (24 Hrs), Subway (Mid Tenn. TS)
- **Lodg** S: Quality Inn
- **TServ** S: I-40 Tires (Mid Tenn TS)
- **Parks** N: Standing Stone State Park (31 Miles)
  S: Fall Creek Falls, Rock Island
- **Other** N: Tennessee Bible College

290 U.S. 70 North, Cookeville
- **Gas** N: Amoco*(D)(LP)
- **Food** S: Grady's Resturaunt (Howard Johnson)
- **Lodg** S: AAA Howard Johnson
- **Other** N: Hidden Hollow (Picnic, Hiking, Fishing, Swimming, Petting Z00, 2.5 Mi)

300 U.S. 70 North, TN 84, Monterey, Livingston
- **FStop** N: Citgo*(D) (Kerosene)
- **Gas** N: BP*(D), Phillips 66(D)
- **Food** N: Dairy Queen, Hardee's
- **AServ** N: Monterey Muffler Shop, NAPA Auto Parts, Phillips 66
  S: J.T. McCormick & Sons Complete Car & Truck Repair
- **TServ** S: J.T. McCormick & Sons Complete Car & Truck Repair
- **Other** N: ACE Hardware(LP), Monterey CB Foods

301 U.S. 70 North, TN 84, Monterey, Livingston
- **Gas** N: Amoco*(D) (Kerosene)
- **Food** N: Mountain Top Grill (Amoco), Subway
- **AServ** N: LTL Discount Tire Auto & Truck Service (24 Hr. Emergency Road Service)
- **TServ** N: LTL Discount Tire Auto & Truck Tire Service (24 Hr. Emergency Road Service)
- **Other** N: Monterey Food Center

(307) Parking Area/Weigh Station - No Services

311 Plateau Road
- **Gas** N: Chevron*(D), Plateau Grocery*
  S: BP*, Exxon*
- **RVCamp** S: Plateau Recreation Vehicle Service

317 U.S. 127, Crossville, Jamestown, Pikeville
- **Gas** N: BP(D), Citgo*(D), Exxon*(D)(CW), Shell*(CW) (see our ad this page)
  S: Amoco*(D) (Kerosene), Phillips 66* (CB Shop, Kerosene), Texaco*
- **Food** N: Blimpie's Subs (Shell), Huddle House (24 Hrs), Restaurant (Ramada Inn)
  S: Cracker Barrel, Shoney's, Waffle House
- **Lodg** N: Best Western, Hampton Inn, ◆ Ramada Inn
  S: Guesthouse Inn, Heritage Inn, Scottish Inns
- **AServ** N: BP
- **ATM** N: Exxon
- **Parks** N: Pickett State Park (46 Miles)
  S: Cumberland Mountain State Park (7 Miles)
- **Other** N: Alvin C. York Historical Site (42 Miles), Big South Fork National Recreation Area (42 Miles)
  S: Cumberland County Playhouse

320 TN 298, Genesis Road, Crossville
- **TStop** N: BP*
- **Gas** N: Shell*
- **Food** N: Catfish Cove, Dairy Queen (BP TS), Halcyon Days, Pizza Hut (BP TS)
- **AServ** N: Lacks Auto Repair
- **TServ** N: Universal Tire
- **Other** N: Antique Village Mall & Shops, Book Outlet, CrossvilleFactory Outlet Center see our ad this page), US Golf & Tennis Center Inc.

322 TN 101, Peavine Road, Crossville
- **Gas** N: BP*(LP) (Kerosene), Marathon*
- **Food** N: Bean Pot (BP), Hardee's, McDonalds (Playground)
- **RVCamp** N: Beanpot Trav-L-Park (1.5 Miles), Roam & Roost RV Campground (4.5 Miles)
  S: KOA Kampground (7.1 Miles)
- **Parks** S: Cumberland Mtn State Park
- **Other** N: Fairfield Glade Resort (6 Miles)

(324) Rest Area (Eastbound)

(327) Rest Area - Picnic, RR, HF, Phones, RV Dump (Westbound)

329 U.S. 70, Crab Orchard
- **FStop** N: BP*
- **Gas** N: Exxon
- **AServ** N: Exxon

(336) Parking Area, Weigh Station (Eastbound)

338 TN 299 South, Westel Road
- **TStop** S: East-West Truckstop*(LP) (Texaco, Kerosene, 24 Hr Mechanic)
- **Gas** N: BP(D)
- **Food** S: Restaurant (East-West Truckstop, 24 Hrs)
- **AServ** N: BP
- **TServ** S: East-West Truckstop
- **TWash** S: East-West Truckstop
- **RVCamp** S: Campground

340 TN 299 North, Airport Road

347 U.S. 27, Harriman, Rockwood (TN Technology Center At Harriman 5.6 Miles)
- **Gas** N: Phillips 66*(D) (24 Hrs)
  S: BP*, Exxon*(D), Shell*
- **Food** N: Cancun Mexican Restaurant, Cracker Barrel, Hardee's, KFC, Long John Silvers, McDonalds (Playground), Pizza Hut, Taco Bell, Wendy's
  S: Dairy Queen (Shell, 24 Hrs), Shoney's, TCBY (Exxon)
- **Lodg** N: ◆ Best Western
  S: Holiday Inn Express, Scottish Inns
- **Med** N: ✚ Harriman City Hospital, ✚ Roane Medical Center (24 Hr Emergency Room, 2 Miles)
- **ATM** S: Shell
- **Parks** N: Frozen Head State Park
- **Other** N: Big South Fork National Recreation Area, Rugby

350 TN 29, Midtown (No Westbound Reaccess Yet)
- **AServ** S: C & S Tune-Up, Midtown Tire

352 TN 58 South, Kingston (Watts Bar Lake)
- **Gas** S: Exxon* (24 Hrs, Kerosene), RaceWay*, Shell*(CW)(LP) (Kerosene)
- **Food** N: D. J.'s Restaurant
  S: Hardee's, Kimlien's Chinese Restaurant, McDonalds (Playground), Pizza Hut, Subway
- **Lodg** N: HoJo Inn
  S: ◆ Comfort Inn
- **AServ** N: The Auto Place (NAPA Auto Parts)
- **ATM** S: NationsBank, Shell
- **Other** S: Car Wash, Foodliner Grocery, Harold's Deli & Bakery

355 Lawnville Road
- **FStop** N: Texaco*(LP) (Kerosene)
- **Other** N: Golf Driving Range by Henley Golf Co.

356AB TN 58N, Oak Ridge, Gallaher Road
- **Gas** N: BP*(D)
- **Food** N: Huddle House
- **Lodg** N: Days Inn, Family Inns of America
- **RVCamp** N: 4 Seasons Campground(LP)
- **ATM** N: BP
- **Other** S: Harris Marine Boat Sales & Services

360 Buttermilk Road

Bold red print shows RV & Bus parking available or nearby

EXIT		TENNESSEE
RVCamp		N: Soaring Eagle Campground(LP) (Full Hookups, Laundry, Hot Showers)
(363)	Parking Area - No Services (Both Directions)	
364	U.S. 321, TN 95, Lenoir City, Oak Ridge	
Gas		N: Shell*
Food		N: Restaurant (Shell)
Lodg		N: Days Inn
RVCamp		N: Cross-Eyed Crickett Campground (2 Miles)
368	Junction I-75 South, Chattanooga	
369	Watt Road	
TStop		N: Flying J Travel Plaza*(LP)(SCALES), Knoxville Travel Center, 76*(SCALES) (TFuel-Unocal 76, AFuel-BP)
		S: Petro Travel Plaza*(SCALES)
Food		N: The Cookery (Flying J Travel Plaza)
		S: Burger King (Knoxville Travel Center, 76), Iron Skillet Restaurant (24 Hrs, Petro TS), Perkins (Knoxville Travel Center TS), Pizza Hut (Knoxville Travel Center, 76)
TServ		S: Knoxville Travel Center, 76, Petro TS
TWash		S: Blue Beacon Truck Wash (Petro TS)
Other		N: Fireworks Supermarket
(372)	Weigh Station - both directions	
373	Campbell Station Road, Farragut	
Gas		N: Amoco*, Texaco*(D)
		S: BP*(CW), Pilot Travel Center*(D), Speedway*
Food		S: Applecake Tearoom, Applewood, Cracker Barrel, Hardee's
Lodg		N: (AAA) Comfort Suites, (AAA) Super 8 Motel
		S: Budgetel Inn, (AAA) Holiday Inn Express
RVCamp		N: Buddy Gregg Motor Homes
Other		S: Harvest Moon Fresh Fruits & Vegetables
374	TN 131, Lovell Rd	
TStop		N: TravelCenters of America*(SCALES)
		S: Pilot Travel Center*(SCALES)
Gas		N: Amoco*(D) (24 Hrs), BP*(CW), Citgo, Marathon*(D), Texaco*(CW)
		S: Speedway*
Food		N: Country Pride (TravelCenters of America), McDonalds (Playground), Taco Bell (TravelCenters of America), Waffle House
		S: Arby's, Krystal (24 Hrs), Shoney's, Wendy's (Pilot Travel Center)
Lodg		N: (AAA) Best Western, ◆ Knights Inn, Travelodge (TravelCenters of America)
		S: ◆ Days Inn, (AAA) Motel 6
AServ		N: Citgo
TServ		N: TravelCenters of America
Other		N: All Kreatures Pet Supplies, Etc
376B	I-140E, Maryville	
376A	TN 162 North, Oak Ridge, Pelissippi State Tech Col., American Museum of Sci. & Energy (Reaccess Eastbound Only)	
Lodg		N: Days Inn
Med		N: + Hospital
378AB	Cedar Bluff Road	
Gas		N: Texaco*
Food		N: Cracker Barrel, McDonalds, Restaurant (Ramada Inn), Stefano's Chicago Style Pizza
		S: Applebee's, Bob Evans Restaurant, Corky's Ribs & BBQ, Denny's, Grady's, Outback Steakhouse
Lodg		N: Hampton Inn, Holiday Inn Select (see our ad on this page), Ramada Inn, Sleep Inn
		S: Clubhouse Inn, La Quinta Inn, Microtel Inn, Red Roof Inn, Signature Inn
AServ		N: Texaco
		S: Harry Lane KIA Chrysler Plymouth
Med		N: + Hospital
ATM		N: First American, First Tennessee Bank, SunTrust, SunTrust Bank
		S: Union Planter's National Bank
Other		N: All Ears Audio Books, Bel Air Grill, El Mercado, Food City Grocery(LP), Nevada Bob's Golf & Tennis
379	Gallaher View Rd, Walker Springs Road	
FStop		N: Citgo*(D)(LP)
Gas		N: Exxon*
		S: BP*(D)(CW)(LP), Pilot*
Food		N: Subway (Exxon)
		S: Buddy's bar-b-q, Burger King, Can Ton Restaurant, Huck Finn's Catfish Chicken & Steaks, Joe's Crab Shack, Kyoto Japanese, Logan's Roadhouse, Mrs Winner's Chicken, Old Country Buffet, Roger's Place, Shoney's, Time-Out Deli, Wendy's
Lodg		S: Nation's Best Hotel
AServ		N: Sam's Club, Wal-Mart
		S: Auto Zone Auto Parts, Firestone Tire & Auto, Goodyear Tire & Auto, Rice Mitsubishi, Rice Oldsmobile Inc GMC Trucks, Rodgers Cadillac
Med		S: + Park Med Ambulatory Care Walk-In Medical Center
ATM		N: Citgo
		S: First Tennessee Bank, Pilot, Union Planters Bank
Other		N: Sam's Club, Wal-Mart (24 Hrs)
		S: Books-A-Million, Joshua's Christian Books, Music, & Video, Revco, Winn Dixie Supermarket (24 Hrs, Pharmacy)
380	U.S. 11, U.S. 70, West Hills	
Gas		S: BP(CW), Conoco*(LP), Pilot*(LP)
Food		S: Big Orange Sports Bar & Deli, Cancun Mexican Restaurant & Cantina, Chili's, Copper Cellar Seafood & Steaks, Cozymel's, KFC, Korean Restaurant, Pizza Palace Theater & Entertainment Co., Romano's Macaroni Grill, Steak-Out Char-Broiled Delivery, Subway, Taco Bell
Lodg		S: (AAA) Comfort Hotel, Howard Johnson
AServ		S: BP, K-Mart, NTB
ATM		S: Conoco, First American Bank, First Tennessee Bank, SunTrust
Other		S: Fire Dept, Food Lion, Jeffrey Photo Lab (1 Hr Photo), K-Mart, Kingston Pike Pet Hospital, Mail Store & More, State Hwy Patrol, West Town Mall
383	Papermill Dr	
FStop		S: Pilot*(LP) (Kerosene)
Gas		N: Spur*
		S: BP*(CW), Citgo*
Food		S: Waffle House
Lodg		N: ◆ Holiday Inn
		S: ◆ Super 8 Motel
ATM		S: Pilot
Other		N: Fire Dept
385	North Junction I-75, East Junction I-640, Lexington	
386A	University Ave, Middlebrook Pike (Eastbound, Services Same As 386B)	
386B	U.S. 129, Alcoa Hwy, Airport, Smoky Mountains	
Lodg		S: Expo Inn
TServ		S: Post & Co. The Truck Body People
Other		S: Airport
387	TN 62, 17th St - 21st St, Western Ave (Convention Ctr, Museum Of Art, Knoxville College)	
Gas		N: Western Ave Tire & Service Center
AServ		N: Western Ave Tire & Service Center
Med		S: + Fort Sanders Regional Medical Center
Parks		N: World's Fair Park
Other		N: Convention Ctr, Museum of Art, Revco
		S: 17th Street Market & Deli, Coin Laundry (24 hrs), Fire Dept., University of TN, Water Works Car Wash
387A	Junction I-275 North, Lexington	
388A	to US 441 South, James White Pkwy, Downtown, Univ of TN (Limited Access Hwy)	
389A	U.S. 441 North, Broadway (Limited Access Hwy)	
389B	Fifth Avenue (Westbound)	
390	Cherry St	
FStop		N: Texaco*(CW)
Gas		N: Citgo*, Pilot*
		S: Buy Quick Market*, Exxon*(CW)(LP)
Food		N: Country Table Restaurant, Hardee's, Taco Bell (Citgo)
AServ		N: David's Tire Service, Goodyear Tire & Auto
		S: Advance Auto Parts, All Right Service Center, Ron's Import Auto Parts
ATM		N: Citgo, Pilot
Other		N: Doan's Market & Deli
		S: Knoxville Zoo, Mailboxes and More
392AB	U.S. 11 W, Rutledge Pike	
Gas		N: Spurr*(D)
		S: BP*(CW)
Food		N: The Cookhouse
		S: Hardee's, Shoney's, The Lunch House Restaurant
Lodg		S: Family Inns of America
AServ		N: Hancock Tire
		S: Atkins & Son Transmission, Auto Value the Parts Place, Quicko Mufflers, Shocks
TServ		N: Cummins Diesel, Disney Tire Company, East Tennessee Tire, Everything Chrome, Freightliner Dealer, Kenworth, Landmark International Trucks, Wheels & Brakes Inc.

EXIT TENNESSEE

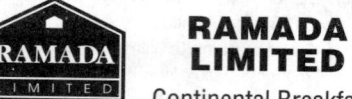

Parks S: Chilhowee Park & Zoo
Other S: Car Wash, Knoxville Zoo, TVA and I Fairgrounds

393 West I-640 to North I-75, Lexington

394 U.S. 11 East, U.S. 25 West, U.S. 70, Asheville Highway

Gas N: BP*(CW)
S: Phillips 66*, Texaco*
Food S: Waffle House
Lodg S: Days Inn AAA Days Inn
AServ N: K-Mart
ATM S: Texaco
Other N: Food Lion Supermarket, K-Mart

398 Strawberry Plains Pike (Johnson Bible College)

FStop N: Texaco* (Kerosene)
S: Pilot*(SCALES) (Showers, Kerosene), Speedway*(SCALES)
Gas N: Amoco*(D)(CW) (24 Hr Car Wash), BP*(CW), Shell*
S: Citgo*
Food N: Baskin Robbins (Amoco), McDonalds (Playground), TCBY (BP), Waffle House, Wendy's
S: Cracker Barrel, Krystal (24 Hrs), Perkins Family Restaurant
Lodg N: AAA Comfort Inn, AAA Country Inn & Suites, AAA Hampton Inn, AAA Holiday Inn Express, AAA Ramada Limited (see our ad this page), AAA Super 8 Motel
S: ◆ Fairfield Inn
AServ N: Doug's Garage
ATM N: Amoco
S: Citgo
Other N: Johnson Bible College
S: Tenn. Dept. of Trans.; Dept. of Safety

402 Midway Road

407 TN 66, Pigeon Forge, Gatlinburg, Sevierville (Dollywood)

Gas N: Texaco*
S: Amoco* (Fireworks Supercenter), BP*, Texaco*(CW)
Food S: Dairy Queen (BP), Ole Southern Pancakes, Subway, Wendy's
Lodg N: Econolodge, Super 8 (see our ad this page)
S: Holiday Inn Express (see our ad this page), AAA Best Western, Super 8 (see our ad this page), Holiday Inn Express (see our ad this page), Ramada Inn (see our ad this page), AAA Comfort Inn, Days Inn ◆ Days Inn, AAA Quality Inn
RVCamp N: KOA Kampground
S: Foretravel Sales (see our ad this page), Services, Parts, Smoky Mountain Campground

EXIT TENNESSEE

EXIT TENNESSEE

(.4 Miles)
Parks S: Great Smoky Mountains National Park
Other S: Dollywood Theme Park, Monster Fireworks, Phantom Fireworks

412 Deep Springs Road, Douglas Dam

TStop N: Sunshine Travel Center*(LP)(SCALES) (Kerosene)
S: TR Auto/Truck Plaza*(SCALES) (CB Shop)
Food N: Apple Valley Cafe
S: Rick's (TR Truck Stop, 24 hrs)
TServ S: TR Truck Service Center (TR Truck Stop)

415 U.S. 25 West, U.S. 70, Dandridge

FStop S: Texaco* (Kerosene)
Food S: Wild Bill's Texas BBQ (Thurs. thru Sun.)

417 TN 92, Dandridge, Jefferson City (Cherokee Dam, Carson-Newman College)

TStop N: 417 Travel Center(LP)(SCALES) (Exxon)
Gas N: Marathon*
S: Amoco*(D)(LP) (Kerosene), Shell*(D) (24 Hrs)
Food N: Anna Lee's Restaurant, Baskin Robbins (417 Travel Center TS), Hardee's, McDonalds (Playground), Perkins (24 Hrs), Subway (417 Travel Center), Taco Bell (417 Travel Center TS)
S: Wendy's (Shell)
Lodg N: AAA Tennessee Mountain Inn
TServ N: 417 Travel Center (24 Hrs)
ATM S: Shell
Other N: Goose Creek Mini Golf, Jefferson County Chamber of Commerce

(419) Rest Area - Picnic, RR, HF, Phones, RV Dump (Eastbound)

421 I-81 North, Bristol

424 TN 113, Dandridge, White Pine

Bold red print shows RV & Bus parking available or nearby

EXIT	TENNESSEE
FStop	N: Marathon*
Gas	N: Amoco* (Kerosene)
Food	N: Bubba's Rib Shack
RVCamp	S: Douglas Lake Campground (1 Mile), Fancher's Campground
Other	N: L & N Produce
(425)	**Rest Area - RR, Picnic, Vending, Pet Walk, Phones (Westbound)**
432AB	432A - US 411 S, Sevierville 432B - US 25 W, US 70 E, Newport
FStop	N: Exxon*(D) (Kerosene)
Gas	S: Amoco*, Citgo* (Kerosene), Shell*(CW) (Kerosene) (see our ad this page), Texaco*(D)(CW)
Food	N: 25/70 Truck Stop, Lois's Country Kitchen Restaurant
Lodg	N: Relax Inn S: Family Inns of America
AServ	N: GM Auto Dealer, Liberty Ford Mercury, Newport Tire Center
RVCamp	N: TMC Campground
ATM	N: Exxon S: Amoco
Other	N: Woodzo Drive-In (Movie Theater)
435	U.S. 321, TN 32, Newport, Gatlinburg
Gas	N: Exxon(D), Marathon* (Kerosene), Texaco*(CW) S: Amoco*, BP*, Bi-Lo Supermarket*
Food	N: Arby's, Baskin Robbins(LP) (Exxon), Burger King (Playground), Hardee's, KFC, La Carreta, McDonalds (Playground), Pizza Hut, Pizza Inn (Exxon), Shoney's, Taco Bell S: Cracker Barrel, Shiners Bar-B-Q (Best Western), Waffle House, Wendy's
Lodg	N: Bryant Town Motel

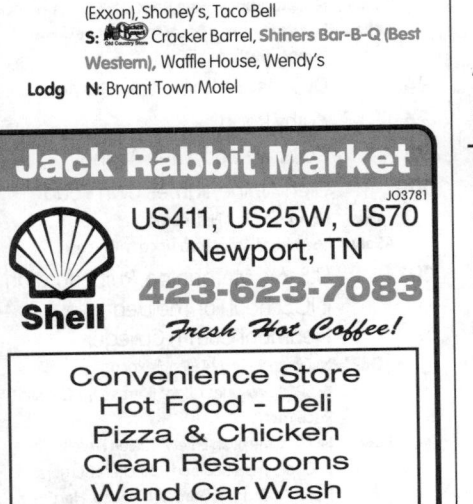

EXIT	TENNESSEE/NORTH CAROLINA
	S: (AAA) Best Western, Family Inns of America, ◆ Holiday Inn
AServ	N: Cosby Highway Service Center S: Calibur Express Lube, Wal-Mart Supercenter
ATM	N: First Union Bank, National Bank of Newport
Other	N: Cinema 4 S: Revco Drugs, Save-A-Lot Food Stores, The Clean Machine Car Wash, Wal-Mart Supercenter
440	U.S. 321 South, TN 73, Wilton Springs Road, Gatlinburg
FStop	N: Coastal* (24 Hrs)
Gas	S: Amoco*
Food	N: Mountain View Restaurant (24 Hrs)
Lodg	N: Mountain View Motel
TServ	N: Frank's Truck Repair
(441)	**Weigh Station - both directions**
443	Foothills Pkwy, Great Smoky Mtns National Park, Gatlinburg
(446)	**TN Welcome Center - Picnic, RR, HF, Phones, RV Dump (No Trucks)**
447	Hartford Road
Gas	S: Exxon* (Kerosene)
AServ	S: Exxon
Other	S: Big Pigeon Rafting, Pigeon River Outdoors, Rapid Descent River Company (White Water Rafting), Rip Roaring (Rafting), Smoky Mtn River Company (White Water Rafting), Smoky Outdoor, The Hartford Outpost*, The White Water Co, U.S. Post Office, USA Raft, Wahoo's (Rafting), Wild Water Rafting, Pigeon River Rafting Center
451	Waterville Road

↑TENNESSEE

↓NORTH CAROLINA

EXIT	NORTH CAROLINA
7	Harmon Den
(11)	**Rest Area - RR, Phones**
15	Fines Creek
20	U.S. 276, Maggie Valley
Gas	S: Amoco*(D)(LP) (Kerosene), Exxon*(D) (Kerosene)
AServ	S: Amoco
RVCamp	S: Creekwood Farm RV Park (1 Mile)
ATM	S: Amoco
Parks	S: Great Smoky Mountain National Park
24	NC 209, Lake Junaluska Hot Springs
TStop	N: Pilot Travel Center*(SCALES) S: 76 Auto/Truck Plaza*(SCALES)
Food	N: Country Cooker (Pilot TS)
Lodg	N: Pilot Travel Center Motel
AServ	S: Boyd's Truck & Auto Repair
TServ	N: Pilot Travel Center S: Boyd's Truck & Auto Repair
ATM	S: 76 Auto/Truck Plaza
27	U.S. 19, U.S. 23, U.S. 74, Clyde, Waynesville (Cherokee, Bryson City, Franklin, Difficult Reaccess)
Med	S: Hospital
Parks	S: Great Smoky Mountains National Park
Other	S: Ghost Town, Maggie Valley, Cherokee
31	NC 215, Canton
FStop	S: Amoco* (Kerosene)
Gas	S: BP(D), Texaco*(D)
Food	N: Burger King, Western Steer Family Steakhouse S: Arby's (Texaco)
Lodg	N: ◆ Econolodge S: ◆ Comfort Inn
AServ	N: An-ton Chevrolet Pontiac Oldsmobile Buick S: BP, Ken Wilson Ford
ATM	S: Amoco, Texaco
Other	S: Auto Pride Car Wash
33	Newfound Rd., Leicester
37	East Canton
TStop	N: Travelport*(SCALES)
Gas	N: Amoco* S: Exxon*(D)(LP) (Kerosene)
Food	N: Buck Horn Restaurant (Travelport TS)
Lodg	S: Days Inn (see our ad this page)
AServ	N: Brad Ragan Inc. Truck Tire Center

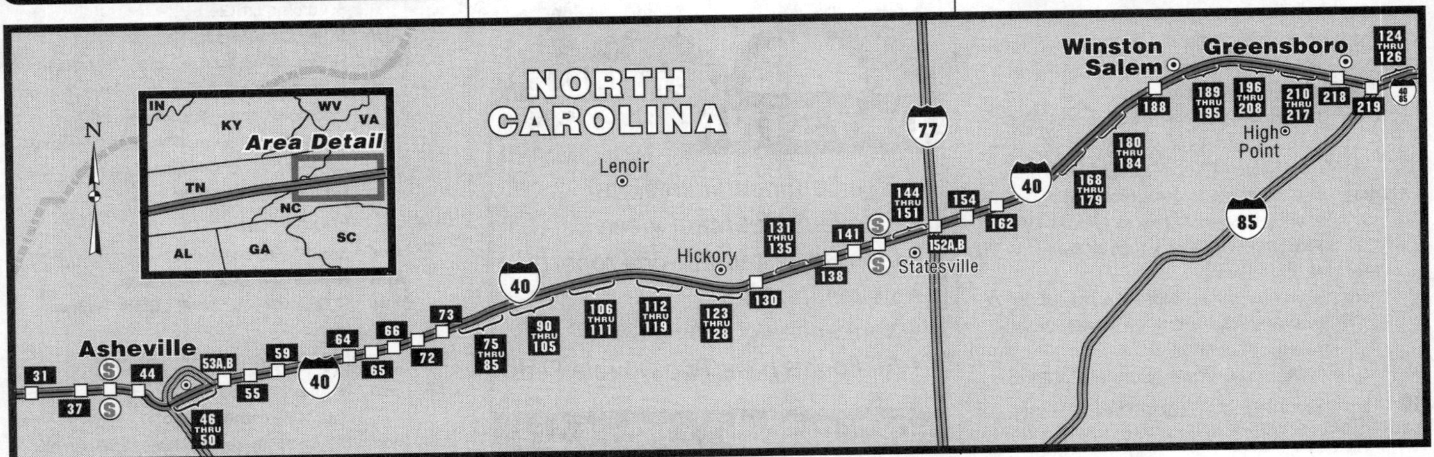

Bold red print shows RV & Bus parking available or nearby

EXIT NORTH CAROLINA

TServ **N: Brad Ragan Inc. Truck Tire Center, Travelport**
RVCamp **S: Big Cove Campground (2.7 Miles), KOA Kampground (.7 Miles)**

(41) **Weigh Station - both directions**

44 U.S. 19, U.S. 23, US 74, West Asheville, Enka - Candler
Gas **N:** Amoco*(CW), Chevron(D), Exxon*(LP) (Kerosene), Shell*(D)
Food **N: Burger King (Playground), Cracker Barrel,** El Chapala, Hardee's (24 Hrs), Sub Station II (Shell), Waffle House, Wendy's
S: McDonalds (Playground), Ramada Plaza Hotel, Restaurant (Ramada Plaza Hotel), Shoney's
Lodg **N:** ◆ Comfort Inn, ◆ Hotel West, ◆ Red Roof Inn
S: Budget Motel
AServ **N:** Auto Repair, Chevron
ATM **N:** Amoco, Exxon, Shell
Other **N: Big D's Golf Range**

46B Junction I-240, Asheville (Eastbound)

46A Junction I-26 East, East US 74, Hendersonville, Spartanburg

47 NC 191, West Asheville, Farmer's Market
Gas **S:** Phillips 66* (Kerosene)
Food **S:** Moose Cafe
AServ **S:** Carolina Collision, The Tire Station
RVCamp **N: Bear Creek RV Park & Campground**
Other **S: US Post Office, WNC Farmer's Market**

50AB U.S. 25, Asheville, Biltmore Estate
Gas **N:** Texaco*(D)
Food **N:** Arby's, Biltmore Dairy Bar & Grill, Bruegger's Fresh Bagel Bakery, Chatt's Restaurant (Howard Johnson), The Criterion Grill, The Lobster Tail
S: Apollo Flame Pizza & Subs, Huddle House (24 Hrs)
Lodg **N:** Holiday Inn Express, Howard Johnson, **Quality Inn, Sleep Inn**
ATM **N:** NationsBank
Other **N: Doctor's Vision Center**
S: Office of the Sheriff Buncombe County, The Book Rack

53AB 53A - US 74A E, Blue Ridge Pky; 53B - I-240, US 74AW, Asheville (Wnc Nature Center, Unc Asheville)

55 to U.S. 70, East Asheville (Billy Graham Training Center - The Cove, Folk Art Center)
Gas **N:** Amoco*, Citgo*(CW), Conoco*
Food **N:** Arby's (Playground), Bojangles, Hardee's, Poseidon Steak & Seafood, Shoney's, Subway (Citgo), Waffle House
Lodg **N:** (AAA) **Best Inns of America,** (AAA) Days Inn (see our ad this page), (AAA) Econolodge, ◆ Holiday Inn, (AAA) Motel 6, (AAA) Super 8 Motel
RVCamp **N: Azalea Country Campground (Full Hookups), Boulder Creek RV Park, RC RV Propane Filling Station, Taps RV Park**
Med **N: ✚ VA Hospital**
Other **N: Flowers Bakery Thrift Store, Folk Art Center, Fun Wheels Grand Prix, Go Grocery Outlet,** Sure Shine Car Wash
S: Billy Graham Training Center, The Cove

59 Swannanoa (Warren Wilson College)

EXIT NORTH CAROLINA

FStop **N: Exxon*(D) (Kerosene)**
Gas **N:** Citgo*, Texaco
Food **N:** Athen's Pizza, Burger King (Playground), Perry's Famous Southwestern BBQ, Subway (Citgo), TCBY (Citgo)
AServ **N:** Texaco
RVCamp **N: Miles Motors RV Center(LP)**
Med **N: ✚ St Joseph's Urgent Care (7 Days a Week)**
ATM **N:** Nations Bank
Other **N: Black Mountain Center, Coin Laundry, Ingles Supermarket, Kerr Drug, US Post Office, Warren Wilson College**

64 NC 9, Black Mountain, Montreat
Gas **N:** Texaco*
S: Phillips 66*
Food **N:** Pizza Hut, Subway (Texaco)
S: Campfire Steaks Buffet, Denny's, Huddle House, KFC, McDonalds (Playground), Mountain BBQ, Taco Bell, Wendy's
Lodg **S:** (AAA) Comfort Inn
AServ **N:** Black Mountain Chevrolet Geo
ATM **N:** First Union Bank
Other **N: Black Mountain Antique Mall, Visitor Information**
S: Chimney Rock (20 Miles), Eckerd Drugs, Ingles Supermarket, Lake Lure (21 Miles), Shadowbrook Golf & Games

65 U.S. 70, Black Mountain (Eastbound Reaccess Only)
Gas **N:** Chevron*(D)(CW) (Kerosene), Citgo*
Food **N:** Cong Sing Chinese, Franks Roman Pizza, H.T. Papas P.D., Hardee's, No. 1 China, Olympic Flame Restaurant, Pizza Express
Lodg **N:** Super 8 Motel
Other **N: Coin Laundry, Food Lion, Whatever Rinse(LP)**

66 Ridgecrest

72 U.S. 70, Old Fort (Eastbound)

73 Old Fort
Gas **N:** BP* (Kerosene)
S: Exxon, K-Max*, SuperTest*
Food **N:** Hardee's
S: McDonalds (Playground), Mustard's Last Stand
AServ **N:** Big Rig Tire & Brake, Old Fort Auto Parts (NAPA)
S: Exxon, Parts Plus
RVCamp **S: Camping**
Other **N: ACE Hardware, Mountain Gateway Museum**

75 Parker Padgett Road
FStop **S: Citgo***
Food **S: Dairy Queen (Citgo)**

EXIT NORTH CAROLINA

81 Sugar Hill Road
TStop **N: Chevron*(LP) (24 hrs)**
Gas **N:** Amoco* (Kerosene)
AServ **N: F F Auto Repair & Truck Service,** Stamey Chrysler Dodge Jeep
TServ **N: F F Auto Repair & Truck Service**
Med **N: ✚ McDowell Hospital**

(82) **Rest Area - Picnic, RR, HF, Phones, RV Dump (Both Directions)**

83 Ashworth Road

85 U.S. 221, Marion, Rutherfordton
FStop **S: Shell*(D) (Kerosene)**
Gas **S:** BP*(D)
Food **S:** Cafe in the Park (Park Inn), Carolina Chocolatiers Ice Cream, **Western Steer Family Steakhouse**
Lodg **S:** Park Inn, Scenic Inn
Parks **N: Mt. Mitchell State Park (approx 20 Miles)**

86 NC 226, Marion, Shelby (Mcdowell Tech Comm College)
Lodg **N:** Carolina Motel
RVCamp **N: Hidden Valley Campground & Recreation Park (2 Miles)**
Other **S: Highway Patrol, McDowell County HQ of the NC Forest Service**

90 Nebo - Lake James
FStop **N: Amoco*(LP) (Kerosene), Exxon***
Food **N: Restaurant**
Parks **N: Lake James State Park**
Other **N: Blackbeard's Boats Inc., National Forest Ranger Station**

94 Dysartsville Road, Lake James

96 Kathy Road

98 Causby Road, Glen Alpine

100 Glen Alpine, Jamestown Road
Gas **N:** Amoco* (24 Hrs, Kerosene)
AServ **N:** Martin Family Ford Lincoln Mercury

103 U.S. 64, Morganton, Rutherfordton (NC School For The Deaf, West Piedmont Comm College)
Gas **N:** Exxon*(CW)(LP) (24 Hrs), Texaco*(D)
S: BP* (Kerosene), Citgo* (Kerosene), Coastal*(D), Racetrac*
Food **N:** Max' Mexican Eatery, Tastee Freeze
S: Captain D's Seafood, Checkers Burgers, Denny's, Dragon Garden Chinese, **Hardee's,** KFC, Long John Silvers, Subway, Taco Bell
AServ **N:** Xpress Lube
S: Carolina Tire Co.
ATM **S:** First Citizens Bank
Other **N: Sen. Sam J. Ervin Jr. Library & Museum, Sunbeam Bread Bakery Thrift Store, Western Piedmont Community College**
S: Food Lion Supermarket, Ingles Supermarket, Living Water Christian Bookstore, Lowe's Hardware(LP), Wal-Mart

104 Enola Road
Gas **S:** BP*(LP) (Kerosene)
ATM **N:** State Employees Credit Union
Other **N: Burke County Sheriff's Office, Highway Patrol**

105 NC 18, Morganton, Shelby
Gas **N:** Amoco*, Exxon(D)
S: QM* (Kerosene), Texaco*(LP)
Food **N:** Arby's (Playground), Hardee's, Mr. Omelet,

EXIT NORTH CAROLINA

Peking Express Chinese, Pizza Inn, Quincy's Family Steakhouse, Shoney's, Wendy's
S: Harbor Inn Seafood Restaurant, Sagebrush Steakhouse

Lodg N: Red Carpet Inn
S: Days Inn, ◆ Holiday Inn, ◆ Sleep Inn

AServ N: Exxon, Fastway Oil Change, Rooster Bush Pontiac, Cadillac, GMC Trucks

Med N: ✚ Broughton Hospital, ✚ Grace Hospital

ATM N: Shoney's
S: Texaco

Parks S: South Mountain State Park

106 Bethel Road

FStop S: Exxon*(LP) (24 Hrs)

Food S: Timberwood (Rainbow Motel)

Lodg S: Rainbow Motel

AServ N: I-40 Auto Parts

107 NC 114, Drexel

111 Valdese

112 Mineral Springs Mtn Road, Valdese

113 Rutherford College, Connelly Springs

FStop N: Phillips 66*

AServ N: Paramount Ford, Paramount Pontiac

Med N: ✚ Hospital

Other N: Speedway

116 Icard

FStop S: Conoco*(LP) (Kerosene)

Food S: Burger King (Playground), Knotty Pine No. 2

Lodg S: Icard Inn

ATM S: Conoco

Other S: Icarda Wash

118 Old NC 10

Gas N: Exxon*(LP) (Kerosene), Texaco*(D)(LP)

ATM N: Texaco

Other N: Western Piedmont Community College East Burke Center

119 Hildebran, Henry River

Gas N: Texaco*

Food N: Hardee's, Subway (Texaco)

123 U.S. 70, U.S. 321, NC 127, Hickory, Lenoir (Appalachian State University, Limited Access Hwy)

125 Hickory (Lenoir Rhyne College)

Gas N: Exxon*(LP)
S: Servco*(CW)(LP), Shell*

Food N: Bojangles, Dragon Inn, Golden Corral, Rock-ola Cafe, Subway (Exxon), Tripps, Western Steer Family Steakhouse
S: Cracker Barrel, Fuddruckers, Hardee's, J & S Cafeteria, Krispy Kreme Donuts, Outback Steakhouse, Ragazzi's, Red Lobster, Sagebrush Steakhouse, Steak & Ale, The Stockyard & Co. Grille, Waffle House

Lodg N: ◆ Red Roof Inn
S: ◆ Comfort Suites, ◆ Fairfield Inn, Hampton Inn, ◆ Holiday Inn Select, ◆ Sleep Inn

AServ N: Hendricks Motors BMW, Lube Works, Saturn of Hickory
S: Armstrong Ford

ATM S: First Citizen's Bank

Other N: Arts & Science Centers of Catawba Valley, Waterworks Car Wash
S: Food Lion Supermarket (24 Hrs), Hickory Furniture Mart, Media Play, The Home Depot

126 Unnamed

128 Fairgrove Church Road (Catawba Valley Community College)

FStop S: Citgo*(D)(LP) (Kerosene)

Gas N: Amoco*(LP), Texaco*(LP)
S: Phillips 66*(D)

Food N: McDonalds, Waffle House
S: Arby's (Playground), Bennett's Smokehouse & Saloon, Burger King, Harbor Inn Seafood, Mr. Omelet, Shoney's, Thai Orchid, Wendy's

Lodg S: ◆ Best Western, Days Inn

TServ N: WhiteGMC Hickory

Med N: ✚ Hospital

Other S: Highway Patrol

130 Old U.S. 70

Food N: Domino's Pizza, No.1 Kitchen, Subway

Other N: City of Conover Fire Dept, Countryside Pet Hospital, Firefighters Museum, K-Mart (Pharmacy), Lowe's Foods(LP) (24 Hrs), US Post Office

131 NC 16, Conover, Taylorsville

Gas N: Amoco*(LP), Texaco*(LP)

RVCamp N: Lake Hickory RV Resort (9.6 Miles)

Other N: Jet-Kleen Car Wash

133 to U.S. 70 to U.S. 321, Rock Barn Road, Newton

TStop S: Wilco Travel Plaza*(LP)(SCALES) (Showers)

Gas N: Exxon*(CW)(LP) (Kerosene)

Food S: Subway (Wilco TS), Taco Bell (Wilco TS)

135 Claremont

Gas S: Texaco*(LP) (24 Hrs)

Food S: Flapjacks Country Cookin

Lodg S: Super 8 Motel

Other S: The Dive Shop (Scuba Diving)

(136) **Rest Area - Picnic, RR, HF, Phones, RV Dump (Both Directions)**

138 Oxford School Road, Catawba, West NC 10

FStop N: Exxon*(LP)
S: Texaco*(D) (Kerosene)

ATM S: Texaco

Other S: Murray's Mill State Historic Site

141 Sharon School Road

Gas N: Shiloh Mini Mall*(LP)

Other S: Buffalo Shoals Golf

(143) **Weigh Station - both directions**

144 Old Mountain Road

FStop N: Chevron*(LP) (Kerosene)
S: Texaco*(D)(LP) (Kerosene)

Gas N: BP(D)
S: Grand Prix*(D) (Kerosene)

Food N: Troy's 50's

AServ N: BP

ATM S: Texaco

146 Stamey Farm Road (Statesville Municipal Airport)

TStop S: Homer's Truck Stop*(SCALES) (CB Shop)

TServ S: Homer's

ATM S: Homer's Truck Stop

148 US 64, NC 90, West Statesville, Taylorsville

Gas N: Citgo*(D), Exxon* (Kerosene)
S: Traveller's

Food N: Arby's, Burger King (Playground), Country Cafe, McDonalds, Prime Sirloin, Subway, Village Inn Pizza

Lodg N: Economy Inn

AServ S: Traveller's

ATM N: Citgo

Other N: Ingles Supermarket, Revco Drugs, Southern States Farm & More(LP)

150 NC 115, Statesville Downtown, In Wilkesboro (Mitchell Community College)

FStop N: BP* (Kerosene)

Gas N: Citgo*(D), Texaco*
S: Amoco*

Food N: Little Caesars Pizza

RVCamp N: Jerry Lathan's RV World(LP)

Other N: Animal Hospital of Statesville Northside, Arts & Science Center, Revco Drugs, Riverfront

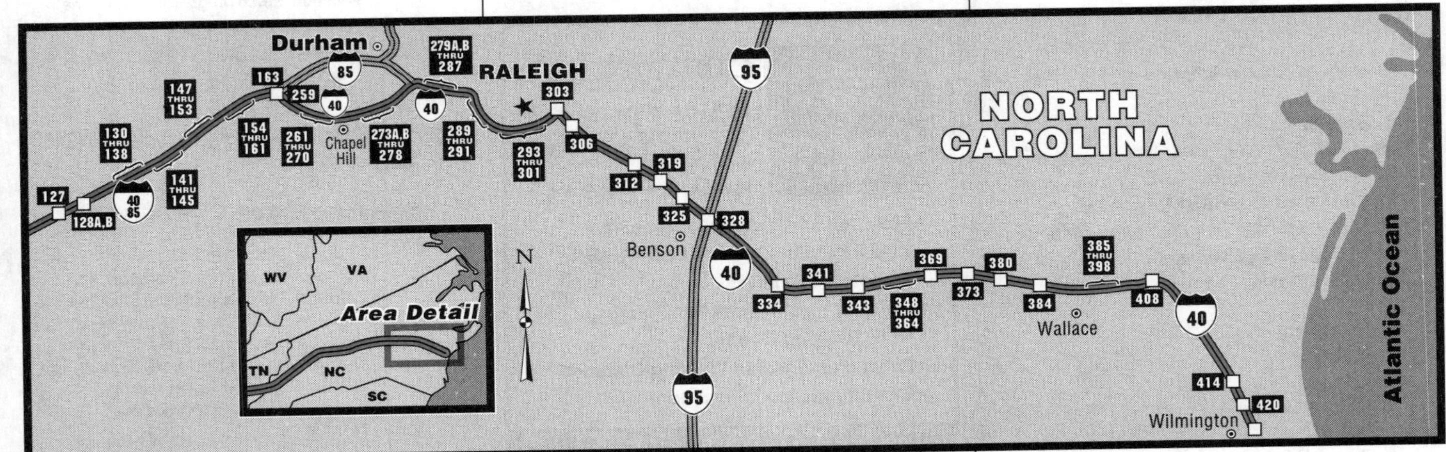

EXIT	NORTH CAROLINA

Antique Mall

151 U.S. 21 East, Statesville, Harmony

Gas N: Citgo*
S: BP*[D][CW], Exxon*[LP]

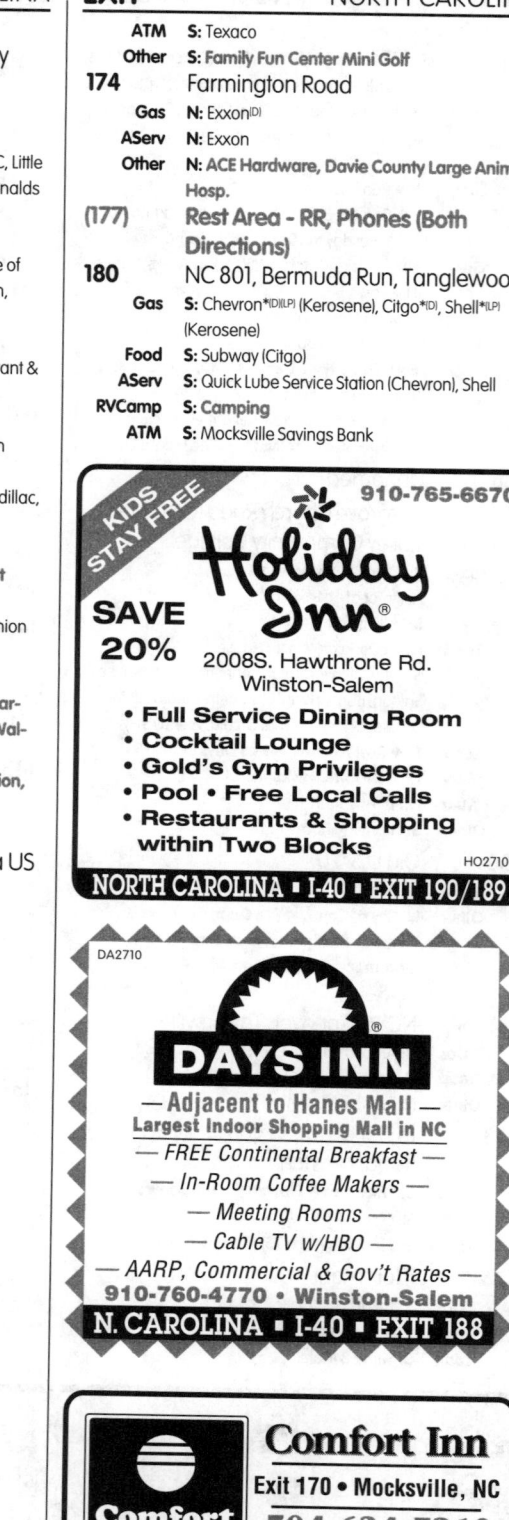

Food N: Applebee's, Bojangle's, **Cracker Barrel,** Golden Corral, K & W Cafeteria, KFC, Little China Restaurant, Long John Silvers, McDonalds (Playground), Pizza Inn, Quincy's Family Steakhouse, Red Lobster, Sagebrush Steakhouse, Sakura Seafood & Steakhouse of Japan, Shoney's, Taco Bell, The Italian Oven, Wendy's
S: Carolina Cafe & Donut, **Hardee's (Playground),** Huddle House, Lotus Pier Restaurant & Lounge, Szechuan Chinese, Waffle House

Lodg N: **Days Inn**
S: Econolodge, Hampton Inn, ◆ Holiday Inn Express, Masters Inn Economy

AServ N: Bell & Howard Chevrolet Oldsmobile Cadillac, The Lube Doctor
S: Cannon Motors Inc. Volkswagon

Med S: **✚ Iredell Memorial Hospital, ✚ Urgent Care Family Practice (7 days a week)**

ATM N: **Bi-Lo Supermarket,** Cash Points, First Union Bank
S: BB&T Bank

Other N: **Arts & Science Museum, Bi-Lo Supermarket, Lowes Foods (24 Hrs), Revco Drugs, Wal-Mart[LP] (24 Hrs, One Hr Photo)**
S: **Gateway Cinema, Highway Patrol Station, Statesville Coin Laundry**

152AB Junction I-77, Elkin, Charlotte

153 U.S. 64 (Eastbound, Reaccess Via US 64)

Gas S: Rickie's*[LP]

Food S: Ice Cream Churn

Lodg S: Hallmark Inn

TServ S: **Carolina Truck & Tractor**

ATM S: Rickie's

Other S: **Pla Mor Bowling Lanes**

154 Old Mocksville Rd

Gas S: Citgo*[D][LP]

Food S: Jay Bee's

AServ S: Bill Martin Tire

Med N: **✚ Columbia Davis Medical Center**

162 U.S. 64

Gas S: Texaco*

RVCamp N: **Lake Myers RV Resort**
S: **Midway Campground Resort**

168 U.S. 64, Mocksville

Gas N: Mobil*[LP] (Kerosene)
S: Amoco*[LP]

RVCamp N: **Lake Myers RV Resort**

170 U.S. 601, Mocksville, Yadkinville

TStop N: **76 Auto/Truck Plaza*[SCALES] (FedEx Drop Box)**

FStop S: **BP***

Gas N: Citgo[LP], Shell (76 Auto TS)
S: Amoco*, Texaco*[LP]

Food N: **Good Kitchen Rest. (76 Auto TS)**
S: Burger King, Pizza Hut, Wendy's, **Western Steer Family Steakhouse**

Lodg S: ◆ **Comfort Inn (see our ad this page)**

AServ N: Citgo
S: BP

TServ N: 76

TWash N: 76

Med S: **✚ Hospital**

EXIT	NORTH CAROLINA

ATM S: Texaco

Other S: **Family Fun Center Mini Golf**

174 Farmington Road

Gas N: Exxon[D]

AServ N: Exxon

Other N: **ACE Hardware, Davie County Large Animal Hosp.**

(177) **Rest Area - RR, Phones (Both Directions)**

180 NC 801, Bermuda Run, Tanglewood

Gas S: Chevron*[D][LP] (Kerosene), Citgo*[D], Shell*[LP] (Kerosene)

Food S: Subway (Citgo)

AServ S: Quick Lube Service Station (Chevron), Shell

RVCamp S: **Camping**

ATM S: Mocksville Savings Bank

EXIT	NORTH CAROLINA

Other S: **ACE Hardware, Eckerd Drugs, Food Lion Grocery, Revco Drugs, Village Way Veterinary, Webb Heating & Air[LP]**

182 Tanglewood (Reaccess Eastbound)

184 Lewisville, Clemmons

Gas N: Shell* (Kerosene), Texaco*
S: Amoco*[CW], BP*[CW][LP], Citgo[D][LP], Etna*[D][LP], Exxon*

Food N: Captain's Gallery, KFC, Quincy's Family Steakhouse
S: Arby's, Baskin Robbins, Burger King, Cherries Cafe, **Cracker Barrel,** Dockside Seafood Rest, Dominoes Pizza, Dunkin' Donuts (Etna), Frauenhofer's Ice Cream, Grecian House Rest, Illiano's Pizza, Krispy Kreme (Lowes), Little Caesars Pizza (K-Mart), Little Richard's Lexington BBQ, Lotus Pond Chinese, Mandarin Chinese, McDonalds, Mi Pueblo Mexican, Mountain Fried Chicken, Pizza Hut, Sagebrush Steakhouse, Sonic, Subway, Taco Bell, Waffle House

Lodg S: ◆ Ramada

AServ N: Shell
S: Advance Auto Parts, BF Goodrich, Car Quest Auto Center, Citgo, George's Village Garage, Q-Lube

ATM N: Nations Bank
S: BB&T, Central Carolina Bank, First Citizens Bank, First Union Bank

Other N: **Animal Hosp of Clemmons,** Car Wash
S: **Clemmons Coin Laundromat, Clemmons Vet Clinic, Food Lion Grocery Store (24 hrs), K-Mart, Lowes Foods (24 hrs), Mail Boxes Etc, Revco Drugs,** Streetcar's Full Service Car Wash (Etna)

188 Junction I-40 East Business, U.S. 421, Winston - Salem, Yadkinville, Wilkesboro (Difficult Reaccess)

Lodg S: Days Inn (see our ad this page)

189 U.S. 158, Stratford Road to Hanes Mall Blvd

Gas S: BP*

Food N: Bojangles, Chili's, Olive Garden, Taco Bell, The Honey Baked Ham Co., Western Steer Family Steakhouse
S: Applebee's, Burger King, New York Bagels & Deli, Romanos Macaroni Grill, Subway, The Oyster Bay

Lodg S: Extended Stay America, ◆ Hampton Inn, Holiday Inn (see our ad this page)

AServ N: Tire America

Med S: **✚ Forsyth Memorial Hospital**

ATM S: NationsBank

Other N: **Cinema, Hanes Mall, Putt Putt Golf & Games**
S: **Discovery Zone, Food Lion Grocery, PetSmart, Revco Drugs, Sam's Club, The Home Depot**

190 Hanes Mall Blvd (Westbound, Difficult Reaccess Via Exit 189)

Food N: Outback Steakhouse

192 Peters Creek Pkwy, Downtown, NC 150

Gas N: Wilco*[D]
S: BP*[CW]

Food N: Arby's, Bojangles, Boston Market, Burger King, Char's, Checkers Burgers, China Buffet, China City, Dragon Inn, IHOP, La Carreta, Little Caesars Pizza, Mayberry, Monterrey, Mr. Barbeque, **Old Country Buffet,** Omega House Family Restaurant, Quincy's Family Steakhouse, Red Lobster, Sam Pan Chinese, Shoney's, Shuckers Pelican Point, Subway, TJ's Deli, Taco Bell, The Red Pepper Tavern, Wendy's
S: Libbey Hill Seafood, McDonalds

Lodg N: Innkeeper

Bold red print shows RV & Bus parking available or nearby

EXIT NORTH CAROLINA

AServ N: Baity's Discount Tire Sales, Bob King Mazda, Brown's Quality Cars, Car Quest Auto Center, Dunite, Liberty Lincoln Mercury, NAPA Auto Parts, Parkway Ford, Shell[LP] (Kerosene)
S: Advance Auto Parts, Flow Buick BMW, Forsyth Honda, Saturn of Winston-Salem
ATM N: BB&T Bank, NationsBank
Other N: **1 Hr Photo, AMF Parkway Lanes, American Check Cashiers, Cinema 6, Eckerd Drugs, Lowes Foods (24 hrs)**, Magnum Auto Wash, **Office Depot, Phar Mor Drugs**
S: **Abri Vet Hosp, Harris Teeter Supermarket, K-Mart**

193C Silas Creek Pkwy, S. Main St (Westbound)
193AB U.S. 52, NC 8, Mount Airy, Lexington
195 U.S. 331 NC 109, Thomasville, Clemonsville Road
Gas N: Easton Grocery*
196 U.S. 311 South, Highpoint (Difficult Reaccess)
201 Union Crossroad
Food N: China Cafe, Jerry's Pizza
ATM N: Cash Points
Other N: **Food Lion Supermarket, Revco Drugs**
203 NC 66, Kernersville, Highpoint
Gas N: Amoco*[CW], Citgo*
S: Shell*[LP] (Kerosene)
Food N: McDonalds (Citgo)
Lodg N: Sleep Inn
AServ N: Merchant's Tire & Auto
ATM S: Shell
205 U.S. 421, Colfax (Eastbound)
206 North US 421, Business Loop 40, Kernersville, Downtown W - S (Westbound)
208 Sandy Ridge Road, Farmers Market
FStop S: **Citgo*[D] (Kerosene, FedEx)**
Gas S: Texaco*
ATM S: Citgo
210 NC 68, Highpoint, Piedmont Triad International Airport
FStop N: **Phillips 66*[D], U.S. Fleet[D]**
Gas S: Exxon*[D], Mobil*
Food N: Arby's, **Hardee's**, Mr. Omelet, Wendy's (Days Inn)
S: Bojangles, McDonalds, Shoney's, Subway
Lodg N: AAA Best Inns of America, DAYS INN Days Inn
S: Comfort Suites, ◆ Hampton Inn, Motel 6, ◆ Ramada, ◆ Red Roof Inn
AServ N: Piedmont Automotive Center
S: Exxon, Foxworth's Auto Repair
TServ N: **Piedmont Automotive Center**
S: **Piedmont Peterbuilt Inc**
ATM N: Phillips 66
S: Central Carolina Bank, Mobil
212 Chimney Rock Road
TServ S: **Carolina Tractor, Covington Diesel**
213 Guilford College, Jamestown
Gas N: BP[D]
S: Exxon*, Pace Olive Mtn Halal Market*
Food N: Damon's (Radisson)
Lodg N: Radisson Inn
AServ N: BP
Other S: **Buffalo Lanes**
214 Wendover Avenue
Gas N: Exxon*[D], Texaco*
Food N: Blimpie's Subs, Burger King, Hardee's, K & W Cafeteria, Queen's Gardens Chinese, TCBY, Waffle House
S: Applebee's, Bojangles, Chick-fil-A, **Cracker Barrel**, Fuddruckers, Golden Corral, IHOP, McDonalds, Red Lobster, Shoney's, Steak 'N Shake, Subway, T.G.I. Friday's, Taco Bell, Tripps, Wendy's
Lodg N: ◆ **Innkeeper**, AAA Microtel
S: AAA AmeriSuites (see our ad this page), ◆ Courtyard by Marriott, Greensboro StudioPlus, AAA Shoney's Inn
AServ N: Bob Dunn Isuzu
S: Econo Lube N Tune, Foreign Cars Italia, Sam's Club
ATM S: Wachovia
Other S: **Sam's Club, The Home Depot, Wal-Mart**
216 Greensboro, Colliseum Area, NC 6. (Eastbound, Reaccess Westbound)
217AB Coliseum Area, High Point Road
Gas N: Exxon*
S: Crown*[CW][LP], Texaco[CW]
Food N: Bennigan's, Biscuitville, Blue Marlin Seafood, Burger King, Chili's, Grady's American Grill, Hooter's, Lone Star Steakhouse, Olive Garden, Po Folks
S: Brewbakers Bagel Cafe, Cherry On Top, Darryl's, Houlihan's Rest & Bar, KFC, McDonalds, Quincy's Steak House, Waffle House (24 Hrs), Wendys
Lodg N: Howard Johnson, Park Lane Hotel, Red Roof Inn, Super 8 Motel, Travelodge
S: Best Western, Comfort Inn, DAYS INN Days Inn, Econolodge, Fairfield Inn, Hampton Inn, Holiday Inn, Marriot Residence Inn
AServ S: Southern Airbrake & Equipment Co, Tire America
TServ S: **Four Seasons Animal Hosp**
ATM N: Blue Marlin Seafood
S: First Union, NationsBank
Other N: **Better Vision Optometrist, Office Depot, Pet Supplies "Plus"**
S: **Four Seasons Town Center, Kid's Clubhouse, Kroger Supermarket (24 hrs, Pharmacy)**
218 218A - US 220 to I-85 Asheboro 218B - Freeman Mill Rd
219 Junction I-85 South, US 29 South, US 70 West, Charlotte

Note: I-40 runs concurrent below with I-85. Numbering follows I-85.

124 To Jct. I-40 West, Randleman Road
Gas E: BP*[CW], Exxon, Texaco*[LP]
W: Amoco* (ATM), Shell*[LP]
Food E: Cafe 212, Cookout Hamburgers, Mayflower Seafood, Quincy's Family Steakhouse, **Waffle House**, Wendy's
W: Arby's, Ben's Diner, Burger King, Captain D's Seafood, China Town Express, Dairy Queen, Jed's BBQ, KFC, McDonalds, Pizza Hut, Sub Station 2, Taco Bell
Lodg W: Budget Motel
AServ E: Exxon
W: Grease Monkey, Meineke Discount Mufflers, Precision Tune & Lube
ATM E: Machine (Texaco)
W: Wachovia Bank
Other E: **The Optical Place**
W: **Dry Clean America, Eckerd Drugs**, New Glow Car Wash

EXIT NORTH CAROLINA

125 South Elm - Eugene Street, Downtown Greensboro
Gas E: Amoco*[D][CW], Texaco*[D]
W: Crown*
Food W: BT Sandwich Express, VIP Express Chinese
Lodg E: **Cricket Inn**, DAYS INN AAA Days Inn, Howard Johnson, Super 8 Motel
W: Homestead Lodge, Ramada
AServ W: Auto Zone Auto Parts, Superior Auto Parts Store
ATM W: Cash Point
Other W: **Coint Laundry, Food Lion Supermarket, Kerr Pharmacy, Visitor Information**
126 U.S. 421 South, Sanford, MLK Jr. Drive
Gas E: Exxon*
Food E: Burger King, Domino's, McDonalds, Pizza Pronto, Szechuan Kitchen, Wendy's
AServ E: Advantage Auto Stores, Goodyear Tire & Auto, Hall Tire Company
ATM E: BB & T
Other E: **A-1 Convenience Store, Bi-Lo Supermarket, Buchman's Discount Drugs, Food Lion Supermarket, Post Office, Revco Drugs**
127 U.S. 29 North, U.S. 70 East, U.S. 220, U.S. 421, Reidsville, Danville (Northbound, No Reaccess)
128AB NC 6, East Lee St, US 29 to US 220 North, North US 421
130 McConnell Road
Gas E: Texaco*
AServ E: Texaco* (Towing)
132 McLeansville, Mtn Hope Church Road
Gas W: Handi Pik*, Shell*[CW][LP], Texaco*[LP] (Towing)
AServ W: Texaco*
135 Rock Creek Dairy Road
Gas W: Citgo*, Exxon*
Food W: McDonalds
AServ W: Exxon
138 NC 61, Gibsonville
TStop W: **TravelCenters of America[SCALES]**
Gas W: **BP (TravelCenters of America)**
Food W: **Burger King (TravelCenters of America), Country Pride Restaurant (TravelCenters of America)**
Lodg W: **Day Stop (TravelCenters of America)**
TServ W: **Truck Service (TravelCenters of America)**
TWash W: **Piedmont Truck Wash**
ATM W: **Machine (TravelCenters of America)**
(140) **Rest Area - RR, Phones, Picnic, HF**
141 Elon College
Gas E: Amoco*, BP*
W: Phillips 66*
Food E: IHOP
W: Applebee's, Arby's, Bojangles, Burger King, Chick-fil-A, **Cracker Barrel**, Golden Corral, The Summit, The Village Grill
Lodg E: Hampton Inn
W: **Best Western**, Super 8 Motel
AServ W: Jiffy Lube, Phillips 66
Other W: **K-Mart, Wal-Mart**
143 NC 62, Burlington, Downtown Alamance
145 NC 49, Liberty, Burlington
147 NC 87, Graham, Pittsboro
148 NC 54, Chapel Hill, Carrboro
150 Haw River
152 Trollingwood Road
153 NC 119, Mebane
154 Mebane - Oaks Road
157 Buckhorn Road

EXIT	NORTH CAROLINA
(159)	**Weigh Station - both directions**
160	Efland
161	To U.S. 70, NC 86
	Note: I-40 runs concurrent above with I-85. Numbering follows I-85.
259	Junction I-85 , North Durham
261	Hillsborough
Other	N: **Hillsborough Visitor Information**
263	New Hope Church Road
266	NC 86
270	U.S. 15, U.S. 501, Chapel Hill, Durham
Food	N: Bob Evans, Outback Steakhouse
Lodg	N: ◆ Comfort Inn
	S: ◆ Red Roof Inn
Med	N: **Hospital**
	S: **Hospital**
Other	N: **Duke University**
273AB	NC 54, Durham, Chapel Hill
Gas	S: BP*(CW), Texaco*(D)(CW)(LP)
Food	S: Hardee's, New China, Pizza A Mante
ATM	S: NationsBank
Other	S: **The Pet Store**
274	NC 751, Jordan Lake Road
276	Fayetteville Road
Gas	N: Exxon*, Kick's 66*
Food	N: China Cafe, Extrusion Pasta, Little Caesars Pizza, Subs Ect, TCBY, Waffle House, Wendy's
ATM	N: CCB Bank
Other	N: **Eckerd Drugs, Eye Care Center, Harris Teeter Supermarket**
278	NC 55 to NC 54, Apex
Gas	S: Crown*, Exxon*(CW)
Food	N: China One Restaurant, Doubletree Guests Suites, Lil' Dino Subs, **Waffle House**
	S: Arby's, Bojangles, Briggs, Burger King, Country Junction, El Dorado Mexican, Ginger Inn Chinese, Golden Corral, Hunan Express, It's Weiner, Jamacian Restaurant, Jersey Mike's Subs, KFC, McDonalds, Miami Grill, Philly Subs, Pizza Hut, Schlotzkys Deli, Shor Sezchuan, Subway, Taco Bell, Wendy's
Lodg	N: Doubletree Guests Suites, (AAA) Fairfield Inn, Innkeeper Motel, Red Roof Inn
	S: Hawthorne Suites Hotel, Homestead Village, ◆ Residence Inn
AServ	N: RTP Auto Service
	S: A & A Tire & Automotive, Best Transmission, Ingold Tire, Precision Tune & Lube
ATM	S: CCB Bank, Cashpoint, South Bank
Other	S: **Coin Laundry, Eckerd Drugs, Food Lion Supermarket, Hope Oriental Market, Parkwood Animal Hospital, Revco Drugs, Winn Dixie Supermarket**
279AB	North Carolina 147, Durham
Med	N: **Hospital**
280	Davis Dr
Lodg	S: (AAA) Radisson
ATM	S: NationsBank
281	Miami Blvd
Gas	S: BP*, Shell*(CW)(LP)
Food	N: Krispy Kreme (Shell), Marriot
	S: Ping Pong Cafe, The Deli, Tico Pizza
Lodg	N: ◆ Marriot, ◆ Wyndham Garden Hotel
AServ	S: Park Auto Service
Med	S: **Park Medical Center**
282	Page Road
Food	S: Holiday Inn, Sheraton
Lodg	S: (AAA) Holiday Inn, Sheraton
283B	Jct I-540, to U.S. 70
284	Airport Blvd
Gas	S: Amoco*(D), Exxon*(D)
Food	S: Jersey's Mike's Subs, Munchies Grill, Triangle Factory Food Court, Waffle House, Wendy's
Lodg	S: ◆ Budgetel Inn, ◆ Courtyard by Marriott, (DAYS INN) ◆ Days Inn, Hampton Inn, Innkeeper, ◆ La Quinta Inn, (AAA) Microtel

EXIT	NORTH CAROLINA
Other	S: **Triangle Factory Shops**
285	Morrisville, Aviation Pkwy
Lodg	N: Hilton Garden Inn
RVCamp	S: **Lake Crabtree Country Park**
287	Cary, Harrison Avenue
Gas	S: Texaco*
Food	S: Burger King, Manhattan Gourmet Deli, McDonalds, NY Pizza, Newton's West, Shanghai Garden Chinese, Wendy's
Lodg	S: Embassey Suites
AServ	S: Goodyear Tire & Auto
Med	S: **Med Stop Walk-In (Harrison Square)**
ATM	S: Cash Points, Wachovia
Parks	N: **William B Umstead State Park**
Other	S: **Harrison Square Shopping Center, Sam's Club**
289	to U.S. 1 North, Raleigh, Wade Avenue
Lodg	S: Ramada Inn
Med	N: **Hospital**
Other	N: **NC Highway Patrol**
290	NC 54, Cary
Gas	E: Circle K*, Circle K(LP), Crown*, Servco*(CW)
Food	E: Anedeo Italian, BoJangles, Dairy Queen, McDonalds, Mike's Subs, Pizza Hut, Subway, Taco Bell, Ten Ten Chinese, Village Inn Pizza Parlor, Wendy's
AServ	E: Clark West Auto Parts, Jiffy Lube, Meineke Discount Mufflers
	W: Peniske Oil
ATM	E: First Union
	W: Wachovia
Other	E: **Coin Laundry, Econo Food Mart**
	W: **Kmart**
291	Cary
293	U.S. 1, U.S. 64 West, Sanford, Asheboro
Other	S: **South Hills Mall**
295	Gorman St
297	Lake Wheeler Road
Gas	N: Exxon*
	S: Citgo*, Coin Car Wash (Citgo)
Food	N: Burger King, Farmers Market Restaurant
Med	N: **Hospital**
298	U.S. 401, U.S. 70, NC 50, Fayetteville
Gas	S: Citgo*, Crown*, Servco*(CW)(LP)
Food	S: Domino's Pizza, KFC, Shoney's
Lodg	S: Claremont Inn, Inn Keeper, Days Inn (see our ad this page)
AServ	S: Car Quest Auto Center

EXIT	NORTH CAROLINA
Other	S: **Sam's Club**
299	Person St, Hammond Road
300AB	Rock Quarry Road
Gas	S: Texaco*(LP)
Food	S: Burger King, Hardee's, PTA Pizza, Subway
ATM	S: M&S Bank
Other	N: **MLK Gardens**
	S: **Coin Laundry, Eckerd Drugs, Winn Dixie Supermarket**
301	East U.S. 64, Jct I-440N (Left Exit Eastbound)
303	Jones Sausage Road
FStop	S: **Texaco***
306AB	U.S. 70, Smithfield, Garner
Gas	S: Phillips 66*(LP)
312	NC 42, Clayton, Fuquay - Varina
TStop	N: **Wilco Travel Plaza***
FStop	S: **Amoco*(CW), Citgo*(LP), Texaco*(LP)**
Gas	S: BP*
Food	N: **Smithfield's BBQ & Chicken, Wendy's**
	S: Bojangles (BP), Burger King (BP), **Huddle House (BP)**, Krispy Kreme Donuts, **McDonalds**, Shane's Pizza (BP), Subway (BP)
AServ	S: Mike's Transmission, Village Auto Repair
ATM	S: Cash Point, First Citizens Bank, Four Oaks Bank & Trust, Wachovia
Other	S: **Animal Hospital**, Car Wash, **Coin Laundry, Eye Deals, Food Lion Grocery, Revco Drugs**
319	NC 210, Smithfield, Angier
(324)	**Rest Area - RR, Phones, Picnic**
325	NC 242, to U.S. 301, Benson
328AB	Junction I-95, Benson, Smithfield
334	NC 96, Meadow
341	NC 50, NC 55, to U.S. 13, Newton Grove
Gas	S: BP*(CW)
Food	S: **McDonalds (BP)**
ATM	S: BP
343	U.S. 701, Clinton
Other	N: **Bentonville Civil War Battle Ground**
348	Suttontown Road
355	NC 403, Faison, Goldsboro, Clinton
364	NC 24 to NC 50, Warsaw, Clinton, Turkey
FStop	S: **Amoco*, BP*, Texaco*(LP)**
Gas	S: Citgo*(CW), Phillips 66*
Food	S: **Bojangles, McDonalds**, Smithville Chicken & B-Que, Subway, Waffle House, **Wendy's**
(364)	**Rest Area - RR, Phones, Vending, Picnic**
369	U.S. 117, Warsaw, Magnolia
373	NC 903, Magnolia, Kenansville
Med	N: **Hospital**
Other	N: **Liberty Hall/Cowan Museum**
380	Rose Hill, Greenevers
Gas	S: Citgo*(CW)
384	NC 11, Wallace, Teachey, Kenansville
385	NC 41, Wallace, Beulaville, Chinquapin
RVCamp	N: **Campground**
390	U.S. 117, Wallace, Burgaw
398	NC 53, Burgaw, Jacksonville
Med	S: **Hospital**
408	NC 210, Hampstead, Rocky Point, Topsail Island Beaches
FStop	S: **Exxon***
Food	S: **Paul's Place Restaurant***
RVCamp	S: **Campground**
Med	S: **Medical Care Center**
Other	S: **Farmer's Market**
414	Castle Hayne, Brunswick County Beaches.
420	NC 132 North, Gordon Road
Lodg	S: Howard Johnson Inn (see our ad this page)
RVCamp	S: **Camelot Campground**

↑NORTH CAROLINA

Begin I-40

Bold red print shows RV & Bus parking available or nearby

EXIT		WISCONSIN
	Begin I-43	

↓WISCONSIN

EXIT		
192AB		Jct 41 U.S. 141
189		Atkinson Dr
187		Webster Ave, East Shore Dr
	Gas	W: Shell*(CW) (ATM)
	Food	W: **McDonalds**, Wendy's
185		U.S. 54, 57, Sturgeon Bay, Algoma (Difficult Reaccess)
183		WI V, Mason St
	Lodg	E: Amerihost
	Med	W: **+ Hospital**
181		WI JJ, Eaton road, Manitowoc Rd
	Gas	E: Citgo*(D) (ATM)
	Food	E: Blimpie's Subs, GiGi's, Hardee's, Rosati's Pizza
	ATM	E: First National Bank
180		WI 172, U.S. 41, A Straubel Airport (Difficult Reaccess)
178		U.S. 141, WI 29, Cnty Road MM, Bellevue, Kewaunee
	TStop	W: **Amoco(SCALES)**
	Food	E: Redwood Inn Food & Cocktails
		W: Restaurant
	TServ	W: **Tielens Repair Service**
171		WI 96, Greenleaf, Denmark
	FStop	E: **Union 76**
	Food	E: **McDonalds**
(169)		**Rest Area - RR, Phones, Picnic, Vending, RV Dump (Both Directions)**
164		WI 147, CR Z, Maribel, Two Rivers
	TStop	W: **Citgo***
	Food	W: **Cedar Ridge Restaurant**
160		CR K, Kellnersville
157		CR V, Mishicot Francis Creek Hillcrest Road
	FStop	E: **BP**
	Gas	E: Amoco*(D), Union 76
	Food	E: Sherry's Country Cafe
154		WI 310, U.S. 10, Two Rivers, Appleton
152		WI 42, U.S. 10, CR JJ Manitowoc, Sturgeon Bay
149		U.S. 51, WI 42, Manitowoc Fond du Lac
	Food	E: Applebees, Perkins Family Restaurant, Wendy's
	Lodg	E: ◆ Comfort Inn, ◆ Holiday Inn, Westmoor Hotel
	AServ	E: AllCar Automotive Center
	Other	E: **Wal-Mart**
144		CR C, St Nazianz, Newton
(142)		**Weigh Station (Southbound)**
137		Cnty Road XX, Kiel, Cleveland
	Gas	E: Citgo*(D) (ATM)
128		WI 42, Sheboygan, Howards Grove
	FStop	W: **Citgo(D)**
	Food	E: **Hardee's**
		W: Cousin's Sub's
	Lodg	E: ◆ **Comfort Inn**
	Other	W: **Joch's Oasis (Food Mart)**
126		WI 23, Sheboygan, Kohler
	Gas	W: Citgo(D)
	Food	E: Annie's, Applebees, Cousin's Subs, Culver's Ice Cream, Hardees, Imperial Palace, McDonalds, New China Restaurant, Nino's Steak Roundup, Ponderosa Steak House, Rococo Pizza, Taco Bell
		W: Dairy Queen, McDonalds
	Lodg	E: ◆ Budgetel Inn, ◆ Super 8 Motel
	AServ	E: Firestone Tire & Auto
	ATM	E: Community Bank
	Other	E: **Memorial Mall, Shopko Grocery (Pharmacy), Wal-Mart**

EXIT — WISCONSIN

EXIT		WISCONSIN
123		WI 28, Sheboygan, Sheboygan Falls
	FStop	E: **Citgo**
	Gas	E: Mobil
	Food	E: McDonalds, Shoney's, Wendy's
	Lodg	E: ◆ Americinn, ◆ Holiday Inn
120		Cnty Road V, OK, Waldo Sheboygan
	FStop	E: **76***
	Food	E: **Hill Farm Restaurant**
	Lodg	E: The Parkway Motel
116		Cnty Road AA, Foster Road, Oostburg
113		WI32, Cnty Road LL, Cedar Grove
	TStop	W: **76 Auto/Truck Plaza***
	TServ	W: **76**
107		Cnty Road D, Belgium, Lake Church
	TStop	W: **Howdea Truck Stop(SCALES) (Showers)**
	Gas	W: Amoco*(CW), Mobil*
	Food	W: Dairy Queen, **Hobos Korner Kitchen (Howdea Truck Stop)**
	Lodg	W: Americinn Motel
100		WI 32, 84, Fredonia, Port Washington
	Gas	E: Citgo*(D)
	Food	E: Arby's, Burger King, McDonalds, Subway
		W: Niesletts Country Inn
	AServ	E: Goodyear Tire & Auto
	Other	E: **Sentry Foods Grocery Store**
97		Jct WI-57N, to Plymouth
96		WI 33, Saukvill, Port Washington
	Gas	E: Mobil
		W: Amoco*
	Food	W: **Bublitz's Family Restaurant**, Hardee's, Subway
	Lodg	W: **Super 8 Motel**
	Other	E: Squeaky Clean Car Wash, **Wal-Mart**
		W: **Coin Laundry**
93		WI 32, 57, Port Washington, Grafton
92		WI 60, Grafton
	Food	E: McCormick 2 Dining
89		Cnty Road C, Cedarburg
	Gas	W: Mobil*
	Med	W: **+ Hospital**
	ATM	W: Grafton State Bank (Mobil)
85		WI 167, Mequon Road
	Gas	W: Amoco*, Citgo*(D), Mobil*(CW)
	Food	W: Dairy Queen, Damon's, Eistein Brothers Bagels, Freddy's, Leonardio's, McDonalds
	AServ	W: Express Lube (Citgo)
	ATM	W: First Star, Mutual Savings Bank
	Other	W: **Kohl's Food**
83		Port Washington Rd
82B		WI 100, Brown Deer Road
82A		WI 32 East
	Gas	E: PDQ*
	Food	E: Baskin Robbins, Cousins Submarine, McDonalds, Pizza Hut
	ATM	E: Bank One, First Star, Great Midwest Bank, MI Bank
	Other	E: **Kohl's Food Emporium, Osco Drugs**
80		Good Hope Road
	Lodg	E: ◆ Marriott
	Med	E: **+ Hospital**
78		Silver Spring Dr
	Gas	E: Mobil
		W: Amoco
	Food	E: Annie's American Cafe, Boston Market, Burger King, Cousin's, Denny's, Hardee's, Little Caesars Pizza, McDonalds, Schlotzkys Deli, Subway
		W: Taco Bell, The Ground Round
	Lodg	E: Budgetel Inn, Exel Inn, Woodfield Suites
	AServ	E: Car X Muffler & Brakes, Mobil, Quaker State Lube
		W: Parts Plus
	ATM	E: Security Bank

Bold red print shows RV & Bus parking available or nearby

← N I-43

EXIT WISCONSIN

77AB Hampton Ave.
- Food E: Sally's Coffee Shop
- Lodg E: Hilton Inn, North Shore Inn

76AB WI 57, 190, Green Bay Ave, Capitol Dr

75 Atkinson Ave, Keef Ave
- Gas E: Mobil* W: Amoco
- AServ E: Pennzoil Lube

74 Locust St
- Gas E: Amoco
- Food E: Burger King
- AServ E: Uniroyal Tire
- Other E: Milwaukee Police Station

73C North Ave
- Food W: Wendy's

73A WI 145, 4th St, Broadway (No Reaccess To Northbound)

72C Civic Center, Wells St, Kilbourn Ave
- Med W: + Hospital

72B Jct I-94 West Madison

72A Jct I-794 East

Note: I-43 runs concurrent below with I-94. Numbering follows I-94.

311 WI 59, National Ave, 6th St

312A Lapham Blvd, Mitchell St

313 Becher St, Lincoln Ave

314A Holt Ave
- Gas N: Citgo*(D)
- Other N: Pik & Save

314B Howard Ave
- Gas N: Clark*, Mobil*, Union 76(CW)
- Food N: Copper Kitchen
- AServ N: Mobil
- ATM N: Mutual Savings Bank
- Other N: WalGreens

316 Junction I-43 & I-894, Chicago, Beloit

Note: I-43 runs concurrent above with I-94. Numbering follows I-94.

Note: I-43 runs concurrent below with I-894. Numbering follows I-894.

9AB U.S. 41, 27th St
- Gas E: Amoco, Citgo*, Mobil*
- Food E: Arby's, Burger King, Chancery, Cousins Submarines, Pizza Hut, Spring Garden Restaurant W: Hardee's, McDonalds, Mexican Restaurant
- Lodg E: Suburban Motel W: Hospitality Inn
- AServ E: Amoco W: Midas Muffler & Brakes, Quaker State Lube
- ATM E: Guaranty Bank W: First Star
- Other E: K-Mart, Target Department Store

8A WI 36, Loomis Road
- Gas E: Amoco(CW), Citgo*(D)(CW) W: Citgo
- Food E: Los Mariachoi's W: George Webb Restaurant
- AServ W: Citgo, Penzoil
- Other E: WalGreens*

7 60th St
- Gas E: Super America*(LP) (ATM)
- Food E: Mineo's Italian
- Other E: WalGreens

5B 76th St to 84th St

5A WI 24, Forest Holm Ave (Difficult Reaccess)
- Gas E: Amoco*, Citgo*(D)
- AServ E: Citgo, QLube Quaker State
- Other E: Cub Food Grocerystore

4 Junction I-43 & I-894

2 National Avenue

1 WI59
- Gas W: SuperAmerica*(D) (ATM)
- Food E: Dicken's Grille W: Burger King, McDonalds, Steakhouse 100, Yick-Sing
- AServ E: Wisconsin Auto Parts W: Meineke Discount Mufflers
- Other W: WalGreens

Note: I-43 runs concurrent above with I-894. Numbering follows I-894.

60 U.S. 45, WI 100, 108th St

57 Moorland Road

54 Cnty Road Y, Racine Ave
- FStop E: Citgo*(D) (ATM)
- Food E: Subway

50 WI 164, Waukesha, Big Bend
- Food W: McDonalds

43 WI 83, Waterford, Mukwonago
- FStop E: Mobil*
- Gas W: Citgo*
- Food W: Dairy Queen, Taco Bell
- Med W: + Childcare Clinic
- Other E: Camping W: Wal-Mart

38 WI 20, East Troy Waterford
- TStop W: Citgo
- Food W: Burger King, Subway
- TServ W: Citgo
- Other W: Drive-N-Shine (Car Wash)

36 WI 120, East Troy, Lake Geneva
- Gas W: Side View Plaza(D)

33 Bowers Road

(33) Rest Area - RR, Phones, Picnic, RV Dump (Both Directions)

29 WI 11, Elkhorn, Burlington

27AB U.S. 12, Lake Geneva, Madison
- Med E: + Hospital

25 WI 67, Elkhorn, Williams Bay
- FStop E: Mobil*(CW)
- Lodg E: ◆ AmericInn W: Lakeland Motel

21 WI 50, Delavan
- Gas W: Mobil*(LP)
- Food W: Burger King, Cousin's Subs, KFC, McDonalds, Pizza Hut, Shoney's, Subway, Taco Bell, Wendy's
- Lodg W: Super 8 Motel
- Other W: K-Mart, WalGreens

17 Highway X, Delevan, Darien

15 U.S. 14, Darien Whitewater, Janesville
- Food E: West Wind Diner

6 WI 140, Clinton, Avalon

2 Highway X, Hart Road

1AB Jct I-90, Madison, Chicago

↑ WISCONSIN

Begin I-43

I-44 E →

EXIT TEXAS

Begin I-44

↓ TEXAS

1A U.S. 277S, Abeline (Westbound)
- Food N: Arby's
- Lodg N: Econolodge, Four Points by Sheraton S: Holiday Inn
- AServ S: Action Battery Center
- Other N: City Police, Hospital

1C Texas Trav Info Center

1B Scottland (Eastbound)
- Gas S: Citgo*
- Lodg N: Four Points Motel by Sheraton S: Scotland Park Motel
- Other N: Camping

1D Bus287
- Gas N: Conoco*, Texaco
- Food S: Pioneer
- Lodg N: River Oaks Inn
- TServ S: International Harvester

2 Maurine St
- Gas N: Conoco*, Texaco*(D) S: Chevron*(D)
- Food N: China Star, Dairy Queen, Denny's, El Chico Mexican, Long John Silvers, Pioneer Restaurant
- Lodg N: ◆ Hampton Inn, (AAA) La Quinta Inn, River Oaks Inn, Super 8 Motel, (AAA) Travelers Inn S: ◆ Comfort Inn, (DAYS INN) ◆ Days Inn (Restaurant inside), Motel 6
- AServ S: Phillips 66

3B Spur325, Sheppard AFB

3A US287 Vernon Amarillo

3C Municipal Airport (Eastbound)
- Gas N: Diamond Shamrock*(D) (ATM inside)
- Other S: Food Lion (open 24 hr.)

4 City Loop

5 Access Rd. (Eastbound)

5A Missile Rd
- Food S: Tee Pee Restaurant

6 Bacon Switch Rd

7 East Rd

(9) Rest Area - RR, Phones (P) (Westbound)

(10) Rest Area - RR, Phones (P) (Eastbound)

11 Fm3429, Daniels Rd

12 TX240 Burke Burnett
- Gas N: Diamond Shamrock*, Texaco*(D)(CW) S: Conoco*, Phillips 66*
- Food N: Braum's, Golden Fried Chicken, Hardee's, Mazzio's Pizza, McDonalds, What A Burger
- AServ N: Bill's Auto Parts

Bold red print shows RV & Bus parking available or nearby

EXIT	TEXAS/OKLAHOMA
3	TX 383, Glendale St.
Food	N: Circle BBQ
ATM	N: First National Bank of Texas
4	TX267, E Third St., Burkburnett, Glendale St
Other	N: **KOA Campground**

↑ TEXAS

↓ OKLAHOMA

EXIT	OKLAHOMA
	OK36, Grandfield
5	U.S. 277, U.S. 281, US.70 Waurika
(20)	Walters Service Plaza - Picnic, Phones, RR HF
20	OK5, Walters
Gas	S: **Texaco**
Food	S: **McDonalds**
30	OK36W, Geronimo, Faxon, Frederick
33	US Bus281, 11th St
Other	N: **Hospital**
36AB	OK7, Lee Blvd, Duncan
TStop	N: **Leo & Ken's Truck Stop Restaurant**
FStop	N: **Fina*(D), Sun Country***
Gas	N: Citgo*, Texaco*
Food	N: Big Chef Coffee Shop, KFC, Salas Mexican, Sonic Drive In
Lodg	N: **Hospitality Inn**, Melart Motel
Other	N: **Beavers Animal Hospital, Coin Laundry**
37	Gore Blvd
Food	N: Mike's Sports Grill, Tweetie Sweeties
Lodg	S: ◆ Howard Johnson
39A	US Hwy.62 Cache Altus
Gas	N: Conoco*, Phillips 66
Lodg	N: Ecmomy Inn, Super 9 Motel, Travel Inn
AServ	N: Midas Muffler
39B	Bus281, Lawton
40A	Rogers Lane
40C	Gate Two
41	Fort Sill Key Gate
45	OK49, Medicine Park, Carnegie
FStop	N: **Love's***
Food	N: A & W Drive-In, Burger King, Subway
46	US 62, US277, US281, Elgin - Apachee Anadarko (Eastbound, Difficult Reaccess)

EXIT	OKLAHOMA
53	U.S. 277, Elgin Fletcher
Gas	N: **Phillips 66*(D)**
	S: Conoco*(CW), Fina*(D)
(60)	**Rest Area - no facilities P (Eastbound)**
62	Fletcher, Elgin, Sterling
(63)	**Rest Area - no facilities P (Westbound)**
(79)	Toll Booth
80	U.S. 81, Chickasha, Duncan
Gas	N: Conoco*, Love's Country Store*(CW), Texaco*(D)
Food	N: Arby's, Braum's, Cleghorn Deli, ElRancho, Hardee's, KFC, Long John Silvers, Mazzio's Pizza, McDonalds, Munchie's, Peking Dragon, Pizza Hut, Taco Bell, Taco Mayo
	S: Burger King, Cutting Horse Restaurant (Days Inn), Jake's Rib
Lodg	N: **Best Western**, Ranch House Motel, U Lodge Motel
	S: DAYS INN AAA Days Inn, Deluxe Inn, Royal American Inn, Super 8 Motel
AServ	N: Conoco, O'Riley's Auto Parts
	S: Wal-Mart
ATM	N: Chickasha Bank & Trust, Superior Federal Bank
Other	N: **Eckerd Drugs**
	S: **Wal-Mart**
83	U.S. 62, Chickasha (Toll)
Gas	N: Total*
Food	N: Subway
Lodg	N: King's Inn
AServ	N: Ray's Radiator Shop
(85)	Service Area
(96)	**Rest Area - no facilities P (Westbound)**
(98)	Toll Plaza
(100)	**Rest Area - no facilities P (Eastbound)**
107	U.S. 62W, U.S. 277S, Newcastle, Blanchard
Gas	S: Conoco*(D)
Food	S: Burger King
107A	Indian Hills Rd
108	OK37W, Tuttle, Minco
Gas	N: Phillips 66*(CW)
Food	N: Braum's, Hardee's, Mazzio's Pizza, Subway

EXIT	OKLAHOMA
AServ	N: Tri-City Quick Lube
ATM	N: Sooner State Bank
Other	N: **Wal-Mart**
108A	Frontage Rd (Westbound)
109	SW 149th St
Food	S: JR's Pub and Grill
110	OK37E, Moore, SW 134th St
Gas	S: Diamond Shamrock*
111	SW 119th St
AServ	N: Lindsay's Auto Services
112	SW 104th St
FStop	N: **Texaco*(D)**
Food	N: **Hi-Way Grill (Texaco)**
113	SW 89th St
FStop	S: **Love's Country Store*(CW)**
Gas	S: Total*(D)
114	SW 74th St
Gas	S: 7-11 Convenience Store*, Sinclair (ATM), Stax* (ATM), Texaco*, Total*(D)(CW) (ATM)
Food	S: Braum's, Burger King, Captain D's Seafood, Crocketts Smokehouse, Perry's Restaurant, Taco Bell
Lodg	S: Cambridge Inn
AServ	S: Sinclair
115A	Jct I-240E, US 62E, OK 3E, Fort Smith
116A	SW59th St
Gas	S: Conoco*, Stax's* (ATM)
Food	S: Pizza Inn, Pizza King, Sonic Drive in, Taco Mayo, Winchells Donuts
ATM	S: Bank One
116B	Airport Rd (Difficult Reaccess)
117B	SW44th St
118	OK152W, SW29th St
Gas	N: Texaco*
Food	S: Burger King, Captain D's Seafood, Grandy's, KFC, McDonalds, Pizza House, Sonic, Taco Bell, Taco Bueno
Lodg	N: Bel-Aire
AServ	N: Keith's Transmission Repair
	S: Auto Zone Auto Parts
Other	S: **WalGreens**
119	SW 15th St
120	Jct with I-40 Amarillo Fort Smith
121	10th St, Fair Park
Gas	N: Stax* (ATM)
122	23rd St

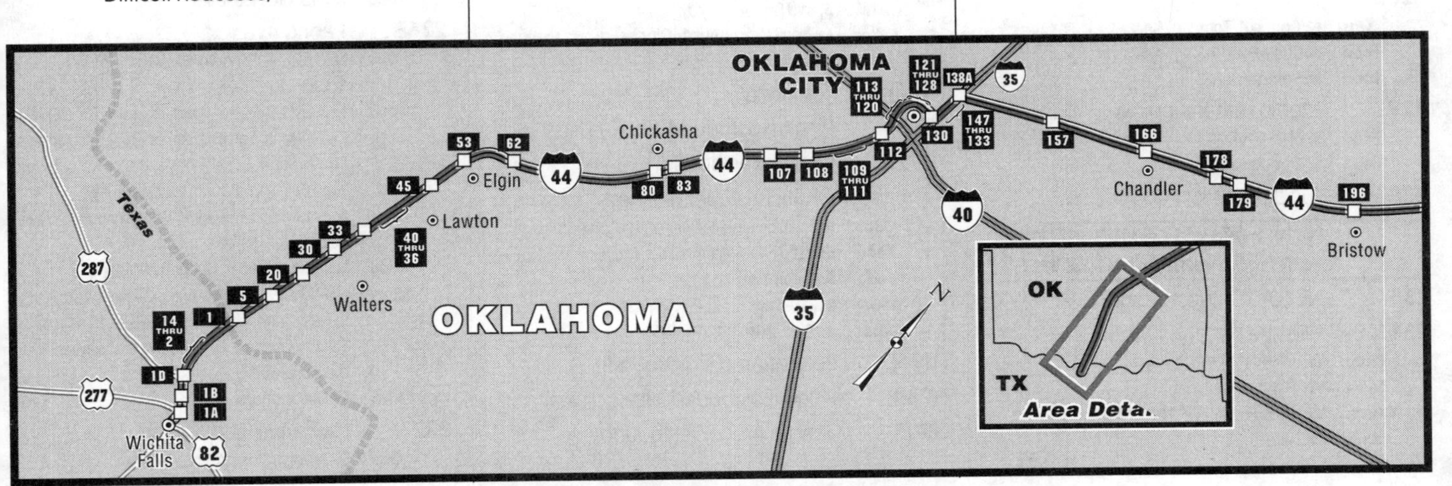

Bold red print shows RV & Bus parking available or nearby

EXIT OKLAHOMA

Gas N: 7-11 Convenience Store*, Conoco*(CW)
S: Conoco*(CW)
Food N: Church's Fried Chicken, Egg Roll King, Long John Silvers, Taco Mayo
S: Arby's, Sonic
AServ N: Hibdon Tire Center, Midwest Auto Supply
S: Big O Tires

123A NW 36th St
Gas N: Stax*(D) (ATM), Total* (ATM)

123B OK66W, Warr Acres Bethany

124 North May Ave
Gas N: Texaco*(D) (Bank IV ATM)
S: Conoco*, Texaco*
Food N: Dairy Queen, Rain Tree (Days Inn), Subway
S: Dunkin Donuts, Mr. Spriggs BBQ, Pizzeria, Wendy's, What A Burger
Lodg N: Days Inn, ◆ Super 8 Motel
AServ N: O'Reilly Auto Parts
ATM S: Bank One

125A Penn Ave to OK 3A, NW Expressway
Gas N: Conoco*(CW)
S: Texaco
Food N: Burger King (Conoco)
S: Braum's, Coit's Cafe
Lodg S: Habana Inn (Restaurant)
AServ S: Joe Esco Tire Co.

125B Classen Blvd.
Gas N: Texaco(CW)
Food N: Beverly Pancake Corner
S: Big Beef Bar-B-Q, McDonalds, Patio Cafe, Restaurant
Lodg S: Guest House Motel, Richmond Suites
AServ N: Montegomery Ward
S: Tune Up & Oil Change
Other N: **Fifty Penn Place Shopping Center, Penn Square Mall**

126 Western Ave
Gas S: Conoco*
Food N: Flip's Restaurant, Split-T Charcoal Hamburgers
Lodg S: Guest House Motel

127 Junction I-235, U.S. 77 to Edmond, Broadway Extension

128A Lincoln Blvd, State Capitol
Gas S: Sinclair, Total
Food S: Pizza Hut
Lodg S: Holiday Inn Express, Oxford Inn
AServ S: Total

128B Kelly Ave.
Gas N: Conoco*, Total*(D)(CW)
Food N: Sonic Drive In
Lodg N: Ramada Inn

129 Martin Luther King Ave.
Food S: **McDonalds**
Lodg S: Ramada Inn

130 Junction I-35S, I-40S Dallas

Note: I-44 runs concurrent below with I-35. Numbering follows I-35.

133 NE 63rd St (Westbound)

134 Wilshire Blvd, NE 78th St
TStop W: **Texaco***(SCALES)
TServ W: **Texaco Truck Stop**
TWash W: **Blue Beacon**(SCALES)
ATM W: Texaco

EXIT OKLAHOMA

135 Britton Rd

136 Heffner Rd.
Gas N: Conoco*
Food N: Papa's Lil Italy Restaurant
Lodg N: Ramada Inn, The Grand Hotel
Other N: **Frontier City (Amusement Park)**

137 122nd St.
FStop E: **Texaco**
W: **Love's Country Store***(SCALES) **(Taco Bell and Subway inside), Texaco***(D) **(Hardees)**
Gas E: Total*(D)
Food E: **Hardee's (Texaco FS),** Kettle (Texaco FS)
W: **McDonalds,** Waffle House
Lodg E: (AAA) Motel 6
W: Comfort Inn, (AAA) Days Inn, (AAA) Motel 6, ◆ Quality Inn, Red Carpet Inn
AServ W: Interstate Auto Repair
RVCamp W: **Camping**

138A Jct I-44E, Tulsa Turnpike

138B Kilpatrick Turnpike

Note: I-44 runs concurrent above with I-35. Numbering follows I-35.

(153) **Rest Area - Phones, Picnic (P) (Both Directions)**

(156) Service Area (Southbound)

(157) Service Area (Eastbound)

157 OK66, Wellston (Eastbound)

166 OK18, Chandler, Cushing
Gas S: Phillips 66
Food S: Tastee Freeze
Lodg S: Econolodge (Food Pumps Cafe)
AServ S: Chandler Tire Center

(167) Service Area (Eastbound)

(171) **Rest Area - Picnic, Phones (P) (Eastbound)**

(178) Service Area (Both Directions)

179 OK99, Stroud, Drumright
Gas N: Citgo*
S: Conoco(D), Phillips 66*(D), Stax*, Texaco*, Total*(D)
Food N: Best Western, Wendy's
S: Mazzio's Pizza, **McDonalds,** Sonic Drive In, Speciality House, Taco Mayo, Wright's Restaurant
Lodg N: **Best Western**
S: Sooner Motel
AServ S: Conoco
ATM S: Mid First Bank, Stax
Other N: **Book Warehouse, Tanger Factory Outlet**
S: Route 66 Car Wash

(183) Toll Plaza

(189) Picnic (Eastbound, No Facilities)

(192) Picnic (Westbound)

196 OK48, Bristow, Lake Keystone
Gas S: 7-11 Convenience Store*, Phillips 66*(D)
Food S: Pizza Hut, Steer Inn, Taco Mayo
Lodg S: Carolyn Inn
AServ S: Wal-Mart
Other S: **Wal-Mart (Pharmacy)**

(196) Service Area (Eastbound)

(204) Picnic (Eastbound, No Facilities)

(205) Service Area (Westbound)

EXIT OKLAHOMA

(207) Service Area (Westbound)

211 OK33, Kellyville, Sapulpa
Gas S: Phillips 66*(D)

215 OK97, Sapulpa, Sand Springs
Gas S: 7-11 Convenience Store*, Phillips 66*(D)
Food S: Marge's Restaurant
Other S: **Hospital**

222A 49th W Ave
FStop S: **QuikTrip**(SCALES), **Texaco***(CW)(SCALES)
Food N: Hardee's, Mama Lou's Restaurant, Monterey's Texaco Mexican, The Green Burrito
S: Arby's, McDonalds, Shoney's, **Subway (Texaco),** Waffle House (Travelers Inn), Wendy's
Lodg N: Motel 6
S: Days Inn, Ecomony Inn, Travelers Inn
AServ S: Service Tire & Battery
TServ N: **Mac Trucks of Tulsa**
S: **Freightliner Dealer, Kenworth**

222B 55th Place (Eastbound, Difficult Reaccess)
Food S: **Restaurant (Days Inn)**
Lodg S: (AAA) **Days Inn**

222C 56th Street (Westbound)
Food N: Mama Lou's, Montery's
Lodg N: Crystal Motel, Gateway Motor Inn, Super 9 Motel, Tulsa Inn, Western Capri Motel
TServ N: **Oklahoma Truck Supply**

223A Junction I-244, Downtown Tulsa

223C 33rd W Ave
Gas S: Phillips 66*
Food N: Braum's, Little Caesars Pizza
ATM S: Phillips 66

224 U.S. 75, Okmulgee, Bartlesville
Gas N: Citgo*, QuikTrip*, Total*
Food N: Arnold's, KFC, Linda-Mar Drive-In, Mazzio's Pizza, Subway
S: The Last Great American Diner
Lodg S: Rio Motel
AServ N: Firestone Tire & Auto, QuikTrip
RVCamp S: **Warrior Campground**
Other N: **Coin Laundry, Warehouse Supermarket,** Warrior Car Wash

225 Elwood Ave (Eastbound Exit 224, Westbound Exit At 225)
Lodg S: Old Capital Motel
AServ N: GM Auto Dealership

226A Riverside Dr
Lodg S: Stratford House Inn

226B Peoria Ave
Gas N: Citgo(CW), Conoco*(D)(CW), QuikTrip*, Total*
S: QuikTrip*, Texaco*(CW)(LP)
Food N: Arby's, Burger's Street, Church's Fried Chicken, Ci Ci's Pizza, Egg Roll Express, KFC, McDonalds, Panda Super Buffet, Philly's, Pizza Hut, Subway, Taco Bell, Taco Bueno, Waffle House
S: Braum's, Burger King, Charlie's Bakery, Kelly's Restaurant, Po Folks
Lodg N: Parkside Motel, Super 8 Motel
AServ N: Midas Muffler & Brakes, O'Riley's Auto Parts
S: Auto Zone Auto Parts, Gary's Automotive Service, Leon's Tulsa Wheel & Brake, Meineke Discount Mufflers
ATM N: Nations Bank

227 Lewis Ave, Oral Roberts Univ
Gas S: Conoco, Stax* (ATM)

Bold red print shows RV & Bus parking available or nearby

EXIT OKLAHOMA

Food **S:** Arizona Mexican, Big Al's Subs, El Chico Mexican, Goldie's Grill, Steak Stuffers USA
AServ **S:** Conoco, Hibdon Tires
RVCamp **S:** **Interstate RV Park**(LP)
Other **S:** **Coin Laundry**

228 Harvard Ave
Gas **N:** Texaco*(D)
S: Phillips 66*(D)
Food **N:** Shoney's, Subway
S: Baskin Robbins, Blimpie's Subs, Chili's, Diamond Jacks, Grill 51, Jimmy's Egg, Little Caesars Pizza, Lone Star, Long John Silvers, Mario's Pizza, McDonalds, Oska Steakhouse of Japan, Papa John's, Perry's Restaurant, Rick's Cafe
Lodg **N:** AAA Best Western, Towers Hotel & Suites
S: AAA Holiday Inn Express
AServ **N:** Jiffy Lube
S: K-Mart, Tire Plus, Tulsa Automotive Repair
Med **N:** **S. Harvard Medical Clinic**
ATM **N:** Nations Bank
Other **S:** **K-Mart**

229 Yale Ave St. Francis Hospital
Gas **S:** Phillips 66, QuikTrip*
Food **S:** Applebee's, Arby's, Braum's, Burger King, Denny's, Don Pablo Mexican, Joseph's Steak House & Seafood, Outback Steakhouse, Steak & Ale, Subway, Taco Bell
Lodg **S:** ◆ Budgetel Inn, AAA Comfort Inn, AAA Howard Johnson, Ramada
AServ **N:** Texaco
S: Fast Lube
Med **S:** **St. Francis Children Hospital**
Other **S:** Spot-Not Car Wash

230 41st St, Sheraton Rd
Gas **N:** Texaco*(CW)(LP) (ATM)
S: Citgo*
Food **N:** Bamboo Garden Chinese, Hardee's, Ricardo's Mexican, What A Burger
S: Christopher Place French Bakery, Jumbo's Hamburgers, Mayberry's Diner
Lodg **N:** Skyline East
S: ◆ Quality Inn
AServ **N:** Goodyear Tire & Auto

231AB U.S. 64, OK51, Muskogee, Broken Arrow

232 Memorial Drive, E31 St.
Gas **N:** Phillips 66
S: Texaco*
Food **N:** Country Inn, Perry's, What A Burger

EXIT OKLAHOMA

S: A & W Drive-In (Texaco), Rex Express (Texaco), Village Inn (Texaco)
Lodg **N:** Days Inn, George Town Inn, Holiday Inn
S: Embassy Suites, Hampton Inn, **Courtyard by Marriott (see our ad this page)**
AServ **N:** Phillips 66(LP)
S: Chrysler Auto Dealer, GM Dealer
Other **N:** **Burlington Coat Factory Outlet**
S: Car Wash (Behind Texaco)

233 East 21 Street (Westbound, No Reaccess)
Food **S:** El Chico
RVCamp **S:** **Dean's RV Super Store**
Other **S:** **Kmart**

234 U.S. 169, Broken Arrow, Owasso

234B Garnett Rd.
Gas **S:** QuikTrip*
Food **N:** Hardee's, Mazzio's Pizza
S: Braum's
Lodg **N:** Garnett Inn, Motel 6, Statford House Inn
AServ **N:** Jiffy Lube
Other **N:** **May's Drugstore, Red Bud Supermarket**

235 East 11th Street (Westbound Reaccess Is At Exit 234B)
Food **N:** Fajita Rita's, Hardee's/Green Burrito, Mazzio's Pizza, Sonic Drive In
S: Denny's, Taco Bueno
Lodg **N:** Garnett Inn, Motel 6, Stratford House Inn, Super 8 Motel
S: AAA Econolodge, National Inn
AServ **N:** Jiffy Lube, O'Riley Auto Parts

236A Admiral Place , 129th E. Ave.

236B 129th E Ave

238 161st E Ave
TStop **N:** **Texaco**(SCALES)
FStop **S:** **QuikTrip**(SCALES)
Gas **S:** QuikTrip*(D)(SCALES)
Food **N:** **Country Kettle/Texaco**
S: Arby's

EXIT OKLAHOMA

AServ **S:** Brad's Auto Parts Center
TServ **N:** **Onan Cummings Diesel, Texaco**

240A OK167, 193rd E Ave
TStop **N:** **Sunmart #46***(SCALES)
FStop **N:** **Diamond Shamrock, Phillips 66***
Gas **S:** Citgo*, Phillips 66*(CW), Texaco*(D)(CW)
Food **N:** **McDonalds,** Pizza Hut, **Sunny Farms* (Diamond Shamrock),** Taco Mayo, Waffle House
S: Coney Island Hot Dogs, Firehouse Bar-B-Q, Lotaburger, Mazzio's Pizza, Sonic Drive In, Subway, Sunrise Donut
Lodg **N:** **Travelers Inn**
AServ **S:** Cooper Tires
RVCamp **N:** **KOA Campground**
Other **S:** **Homeland Grocer Store**

240B U.S. 412, Choteao

241 I-44 E Turnpike to Joplin, OK66, Catossa

255 OK20, Claremore, Pryor
Gas **W:** Phillips 66*(D), Texaco
Food **W:** Hugo's Family
Lodg **W:** Motel Claremore, Walker Motel
AServ **W:** Texaco

(256) **Parking, Picnic (Both Directions)**

(269) Picnic (Eastbound)

269 OK28, Adair, Chelsea

(271) Picnic - No facilities (Westbound)

283 U.S. 69, Big Cabin
Food **N:** **Restaurant/Texaco**
Lodg **N:** **Bates Motel/Texaco**
TServ **N:** **Texaco**

(286) Toll Plaza

(288) Service Area

289 U.S. 66, U.S. 69, Vinita
Gas **N:** Total*(D)
S: Sinclair(D)
Food **N:** Braum's, McDonalds, Pizza Hut, Subway
Lodg **N:** Lewis Motel, Vinita Inn
Other **N:** **Wal-Mart**

(299) Picnic - No facilities (Eastbound)

302 U.S. 59, U.S. 66, Afton, Fairland Grove

(310) Picnic (Westbound)

(312) Picnic (Eastbound)

313 OK10, Miami
Gas **N:** Citgo*(D), Love's Country Store*(D), Texaco*,

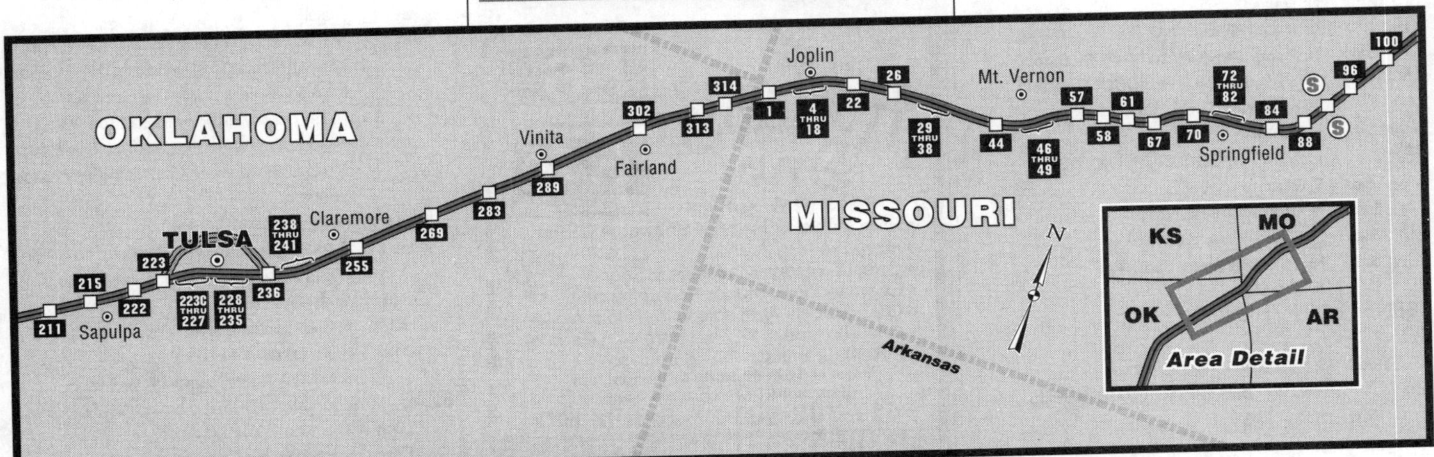

Bold red print shows RV & Bus parking available or nearby

EXIT OKLAHOMA/MISSOURI

Total*(D)

Food N: Thunderbird

Lodg N: Best Western, ◆ **Super 8 Motel**, Thunderbird Motel, Townsmen

RVCamp N: **Miami RV Park**

Other N: Coin Car Wash

(314) Service Area (Westbound)

(315) **OK Information Center, Rest Area - RR P (Westbound)**

↑ OKLAHOMA

↓ MISSOURI

1 U.S. Hwy. 400, U.S. 166, Baxter Springs, Kansas

(1) **Rest Area - RR, Phones, Picnic, Vending, Info P (Both Directions)**

(3) **Weigh Station - both directions**

4 MO43, Seneca, Joplin

TStop S: **Petro(SCALES), Pilot Travel Center(SCALES)**

Food S: **Blimpie's Subs (Pilot), Dairy Queen (Pilot), McDonalds (Pilot), Wendy's (Pilot)**

Lodg S: AAA Sleep Inn

TServ N: **Peterbilt Dealer**

S: **4 State Trucks**

RVCamp S: **KOA Campground**

6 MO43, MO86, Business 44, Joplin, Racine

Gas N: Citgo*

S: Texaco*

AServ S: Texaco

8AB U.S. 71, Neosho, Joplin

FStop S: **Phillips 66*(D)(LP)**

Gas N: Amoco*(CW), Conoco*, Phillips 66*, Texaco*

Food N: Applebee's, Bob Evans Restaurant, Braum's, Country Kitchen, Crazy Horse Steakhouse, Denny's, Iron Horse, Luby's, McDonalds, Olive Garden, Red Lobster, Ruby's, Steak 'N Shake, Subway, Waffle House, Western Sizzlin'

S: **Cracker Barrel**

Lodg N: Best Inns of America, Best Western (see our ad this page), Comfort Inn, Days Inn, Drury Inn, Fairfield Inn, Hampton Inn, Holiday Inn (see our ad this page), Howard Johnson, Motel 6, Ramada, **Riviera Roadsite Motel, Super 8 Motel**

S: Microtel Inn & Suites

AServ N: NAPA Auto Parts

TServ N: **Freightliner Dealer (24hr)**

RVCamp S: **Wheelen**

ATM N: Merchantile Bank

Other N: **Sam's Club**

11A US71S, Fayetteville

TStop N: **Flying J Travel Plaza*(SCALES) (RV Dump)**

15 MO66, Joplin (Westbound, Eastbound Reaccess Only)

RVCamp N: **RV Park**

18A U.S. 71 Alt, Carthage, Diamond

Gas S: Citgo*

Food N: Ozark Land Restaurant

AServ N: Colar RV Sales & Services

RVCamp N: **4 State RV Sales, Coachlight**

S: **RV Park(LP)**

18B U.S. 71, to Joplin

22 Cty Rd 10

FStop S: **Xpress*(D)**

26 MO 37, Bus 44, Sarcoxie, Reeds

Gas S: Standard*

EXIT MISSOURI

Other S: **Indian Gift Shop**

29 MO U, Sarcoxie, La Russell

FStop S: **Texaco***

Lodg S: **Sarcoxie Motel**

RVCamp N: **W.A.C. RV Park**

Other N: **Ozark Village**

33 MO 97, Pierce City

FStop S: **Sinclair***

Food S: **Restaurant (Sinclair)**

38 MO 97, Stotts City

Gas N: Massie's*

AServ N: Massie's

44 MO H, Bus 44, Mt. Vernon, Monett

AServ N: Rinker's Radiator Service

Other N: **Dental Clinic**

46 MO 39 Mount Vernon Aurora

TStop N: **Truckstops of America(SCALES)**

Gas N: Phillips 66*, Sinclair*(D) (ATM)

S: Total*

Food N: Dalmas Family Restaurant, Hardee's, KFC, **McDonalds**, Simple Simons Pizza, Sonic Drive in, **Tin Lizzy Restaurant (Sto's)**

Lodg N: AAA Best Western, Budget Host Inn, Mid-West Motel

S: Comfort Inn

AServ N: O'Riley Auto Parts

49 MO 174, Cty Rd CC, Chesapeake

(52) **Rest Area - RR, Phones, HF, Picnic P (Both Directions)**

57 MO O, MO 96, Halltown (Westbound, Eastbound Reaccess Only)

58 Cty Rd Z, Cty Rd O, Halltown (Difficult Eastbound Reaccess)

FStop S: **Conoco***

EXIT MISSOURI

Lodg S: **Scandinavian Motel**

AServ N: Owen's Auto Service

S: Larry's Auto Repair

61 Cty Rd K, Cty Rd PP

TStop N: **Total*(D)(SCALES)**

Gas N: Amoco*

Food N: **Hood's Restaurant (Total)**

Lodg N: AAA Motel

TServ N: **Total**

67 Cty Rd T, Cty Rd N, Bois D'Arc, Republic

FStop S: **Citgo*(D)**

Gas S: Conoco*

Food S: Ozark Travel Restaurant

70 Cty Rd B, Cty Rd MM

Gas S: Total*(D)

Other N: **Fireworks Supermarket, Western Wear Boot Outlet**

72 MO 266, Bus Loop 44, Chestnut Express

FStop S: **Seven Gables***

Food S: Seven Gables

Lodg S: Best Budget Inn

AServ N: Bridgestone Tire & Auto

TServ S: **Wilson Truck Repair**

RVCamp N: **Travellers RV Park**

Other S: **KOA Campground**

75 U.S. 160, West Bypass, Willard

Gas S: Conoco*

77 MO 13, Kansas Expressway, Bolivar

Gas N: Total*

S: Citgo* (ATM), Gas +, Phillips 66*, QuikTrip*

Food S: **McDonalds (Walmart)**, PaPa John's Pizza, Pizza Inn, Rallys Hamburgers, **Subway (Walmart)**, Tiny's Smokehouse, Waffle House

Lodg N: Interstate Inn

S: AAA Econolodge

AServ S: Express Lube (Walmart)

ATM S: Empire Bank

Other S: **Coin Laundry, Wal-Mart (Pharmacy)**

80AB Cty Rd H, Bus Loop 44, Glenstone Ave, Pleasant Hope

Gas N: Conoco*

S: Amoco*, Coastal, Conoco, Phillips 66*(D), Texaco, Total*(D)

Food N: Waffle House

S: Bob Evans Restaurant, Burger King, China Express, Country Kitchen, **Cracker Barrel**, Denny's, Fazoli's Italian Food, Hardee's, KFC, Long John Silvers, Maple Restaurant, McDonalds, Pizza Hut, Shoney's, Western Sizzlin'

Lodg N: Microtel Inn, Motel 6, Quality Inn, Super 8 Motel

S: Bass County Inn, Best Inns of America, Best Western, Budget Host Inn, Days Inn, Drug Inn & Suites, Economy Inn, Fairfield Inn, Flagship Motel, Holiday Inn, Howard Johnson, Maple Inn, Motel 6, Ramada Inn, Red Roof Inn, Satellite Motel, Scottish Inns, Sherton Motel, Skyline Motel, Village Inn

AServ N: Interstate Auto Service

S: Auto Zone Auto Parts, Goodyear Tire & Auto, O'Riley Auto Parts, Precision Tune, Texaco

Med S: ✚ **Doctors Hospital**

ATM S: Great Southern Bank, UMB

Other N: **RV Caravan Mobil Villa**

S: **K-Mart, North Town Mall, Wal-Mart**

82AB U.S. 65, Branson, Sedelia

Gas S: Amoco*(D) (ATM), Conoco*(D)

Food S: **Waffle House**

Bold red print shows RV & Bus parking available or nearby

XIT | MISSOURI

Lodg S: American Inn
TServ S: **Kenworth, Peterbilt Dealer, Southwest Missouri Truck Center**
Other S: **Highway Patrol Headquaters**
4 MO 744
8 MO 125, Fair Grove, Strafford
TStop N: **Conoco/Union 76*(SCALES) (ATM)**
FStop N: **Phillips 66*(SCALES)**
Gas S: Citgo*
Food N: **McDonalds**
S: Mekong
Lodg S: (AAA) **Super 8 Motel**
AServ N: Conoco
TWash N: **18 Wheeler Truck Wash**
89) **Weigh Station - both directions**
6 Cty Rd B, Northview
RVCamp N: **Oak Rest RV Campground**
00 Cty Rd W, MO 38, Marshfield
Gas N: Amoco*
S: Conoco*, Phillips 66*, Texaco*(D), Total*
Food N: **Tiny's Smokehouse**
S: Arcadia Restaurant, Country Kitchen, Dairy Queen, KFC, McDonalds, Pizza Hut, Sonic Drive In, Subway, Taco Bell
Lodg N: **Fair Oaks Motel,** Plaza Motel
S: ◆ Holiday Inn Express
AServ S: Break Time, Conoco, Goodyear Tire & Auto, O'Riley Auto Parts
ATM S: Citizen's Bank, Merchantile Bank
Other N: **Empire Gas(LP)**
S: **Wal-Mart**
07 Cty Rd
110) **Rest Area - RR, Phones, Picnic P (Both Directions)**
13 Cty Rd Y, Cty Rd J, Conway
FStop N: **Shell***
Gas N: Phillips 66*
Food N: McShane's Restaurant
18 Cty Rd A, Cty Rd C, Phillipsburg
Gas N: Conoco*
S: Texaco*
23 Cty Rd
Gas S: Cathyville Store*(D)
RVCamp S: **KOA Campground**
27 Bus Loop 44, Lebanon
TStop N: **B&D Truck Port*(SCALES)**
Gas N: Citgo*(D), Phillips 66*
Food N: Ranch House, **Waffle House**
Lodg N: Bestway Inn, ◆ Econolodge, **Hampton Inn,** ◆ Quality Inn, Scottish Inns, Super 8 Motel
AServ N: Citgo, Kelly Tires
TWash N: **Texaco Truck & Auto (ATM)**
29 MO 5, MO 32, MO 64, Lebanon, Hartville
Gas N: Conoco*(D), Gas+
S: Amoco*, Conoco*(D)
Food N: Baskin Robbins, Chi Time Chinese, Chinese Chef Buffet, Corner Stone Subs & Pizza, Country Kitchen, Dairy Queen, KFC, Long John Silvers, McDonalds, Shoney's, Sonic Drive-In, Subway, Taco Bell, Wendy's, Western Sizzlin'
S: Captain D's Seafood, Hardee's, Pizza Hut, Stonegate Station Restaurant
Lodg S: (AAA) Brentwood Motel
AServ N: Auto Zone Auto Parts, Brake Time, NAPA Auto Parts, O'Riley Auto Parts
S: **Wal-Mart**
TServ S: **Goodyear Tire & Auto**
ATM S: Central Bank, Commerce Bank
Other N: **K-Mart**

EXIT | MISSOURI

S: **All Star Gas(LP), Wal-Mart**
130 Bus 44, Cty Rd MM
Gas N: Gas+*, Phillips 66*, Total*
Food N: **Bell Restaurant,** What's Cookin
Lodg N: Best Budget Inn, (AAA) Best Western, Holiday Motel, Munger Moss Motel
135 Cty Rd F, Sleeper
140 Cty Rd T, Cty Rd N, Stoutland
FStop S: **Phillips 66***
Food S: **Midway Restaurant**
145 MO 133, Cty Rd A & B, Richland
FStop N: **Conoco***
RVCamp S: **Gasconade RV Park (1 1/2 Miles), RV Parking**
150 MO 7, Cty Rd P, Richland, Laquey
Gas N: Amoco*
153 MO 17, Buckhorn
FStop N: **Shell***

EXIT | MISSOURI

Gas N: Phillips 66*
S: Texaco*(D)
Food N: **Witmor Farms Restaurant**
Lodg N: **Scottish Inns**
S: S&G Motel
RVCamp S: **Campground**
156 Cty Rd H, Bus Loop 44, Waynesville
Gas N: Citgo*
Food N: McDonalds
RVCamp N: **Covered Wagon RV Park (1 3/4 Miles)**
Med N: ✚ **Roubidox Clinic**
159 MO 17, Bus Loop 44, Saint Robert
Gas N: Conoco
S: MFA Oil*(CW), Phillips 66*, Texaco*(D)
Food N: Aussie Jack's Steak and Sea, Dairy Queen, Sonic
Lodg N: Star Motel
S: Alpine Haus, Deville Motel, Motel Ozark
AServ N: American Car Care Center, Firestone Tire & Auto, O'Riley Auto Parts
Med S: ✚ **Mercy Medical Group**
ATM N: First State Bank
161 MO Y, Bus 44, Fort Leonard Wood
FStop S: **Total*(D)**
Gas N: Conoco*
S: MFA Oil
Food S: Arby's, Captain D's Seafood, Great Wall Grand Chinese, KFC, Lilly's, McDonalds, Pizza Hut, Taco Bell, Waffle House, Wendy's
Lodg S: (AAA) Econolodge, Motel DeVille, Ramada Inn, Ranch Motel
Med S: ✚ **Mercy Medical Group**
163 MO 28, Dixon
TStop N: **28 Truck Port* (Phillips 66)**
Gas S: Amoco, Conoco*(D)
Food N: JB's Market, **Restaurant (28 Truck Port)**
S: Country Cafe
Lodg S: (AAA) **Best Western,** (DAYS INN) (AAA) Days Inn (see our ad this page), **Super 8 Motel**
AServ S: Amoco
TServ N: **28 Truck Port(LP)**
RVCamp S: **Camping, Covered Wagon RV Park**
169 Cty Rd J
172 Cty Rd D, Jerome
Other N: **Camping**
176 Sugar Tree Rd
Lodg N: Vernelle's Motel
RVCamp S: **Camping**
(178) **Rest Area - RR, Phones P (Both Directions)**
179 Cty Rd C & T, Doolittle, Newburg
FStop S: **Citgo*(D)**
Food S: **Cookin' from Scratch (With Citgo)**
ATM S: Citizen's Bank
184 U.S. 63, Bus Loop 44, Rolla
Gas S: Conoco(D)(CW), Delano's(D), Gas+, Mobil*, Phillips 66*, Shell*
Food S: Arby's, Burger King, KFC, Lucky House Chinese, McDonalds, Pizza Hut, Shoney's, Sirloin Stockade, Subway, Taco Hut, Wendy's, Zeno's Steakhouse
Lodg S: Best Way Motel, (AAA) Best Western, (DAYS INN) (AAA) **Days Inn (see our ad this page),** (AAA) Econolodge, (AAA) Howard Johnson, ◆ Ramada Inn, Super 8 Motel, **Wayfarer Inn,** (AAA) Western Inn, (AAA) Zeno's Motel
AServ S: Conoco, Mobil
Med S: ✚ **Hospital**
185 Cty Rd E, Rolla
Food S: Dairy Queen, Hardee's, Taco Bell

EXIT		MISSOURI
	Med	S: Hospital
186		U.S. 63, to MO 72, Rolla , Jefferson City
	Gas	N: Sinclair*
		S: Amoco*, Conoco*, Mobil, Ozark, Shell*
	Food	N: Steak 'N Shake
		S: Denny's, Dunkin Donuts, Lee's Famous Recipe Chicken, Pizza Inn, Waffle House
	Lodg	N: ◆ Drury Inn, Sooter Inn
		S: American Motor Inn, Budget Delux Motel
	AServ	N: Plaza Tire Service
	Other	**S: Coin Laundry**
189		Cty Rd V
	FStop	**S: Conoco*(D)**
	RVCamp	**N: RV Park**
195		MO 68, MO 8, Saint James
	Gas	N: Conoco*, Mobil*(D)(CW)
		S: Delano*, Phillips 66*(CW)
	Food	N: **McDonalds**, Northside Restaurant, Pizza Hut, Ruby's Ice Cream, Subway, Taco Bell
	Lodg	N: (AAA) **Comfort Inn, Economy Inn**, Ozarks Inn
		S: Finn's Motel
	AServ	N: Conoco*(D)
		S: Phillips 66(D)
	TServ	**N: Rays Tire and Service Center**
	ATM	N: Citgo
	Other	S: Car Wash
203		Cnty Rd F & ZZ
	RVCamp	**N: Eagle Wood RV Park (1/4 Mile)**
208		(6) MO 19, Cuba, Owensville
	TStop	**N: Texaco Truck & Auto*(SCALES)**
	Gas	N: Citgo*
		S: Mobil*
	Food	**N: Restaurant (Texaco Truck Stop)**
		S: AJ's Family Restaurant, Hardee's, McDonalds, Subway, **Taco Bell**
	Lodg	N: (AAA) Best Western, **Super 8 Motel**
	AServ	S: NAPA Auto Parts
	TServ	**N: Express Lube (Texaco Truck Stop)**
	Other	**S: Wal-Mart**
210		Cty Rd UU
214		Cty Rd H, Leasburg
	Gas	N: Mobil*(D)
	Parks	**S: Onondaga Cave State Park (7 Miles)**
218		Cty Rd C & J & N, Bourbon
	Gas	N: Citgo*(D)
		S: Mobil*(LP) (ATM)
	Food	S: Henhouse Restaurant
	Lodg	N: (AAA) Budget Inn
	AServ	N: Auto Zone Auto Parts
		S: Mobil (ATM)
	RVCamp	**S: Boubon RV Center (6 Miles)**
225		MO 185, Cty Rd D, Sullivan
	TStop	**S: Bobber Travel Center(SCALES)**
	Gas	N: Mobil*(CW), Shell*
	Food	N: Domino's Pizza
		S: Captain D's Seafood, Jack-In-The-Box, Lucky House Chinese, McDonalds, Pizza Hut, Sassafras BBQ, Shoney's, Subway
	Lodg	N: ◆ Best Western, (AAA) Family Motor Inn, ◆ Ramada, Sunrise Motel, **Super 8 Motel**
		S: Sullivan Motel
	Other	S: **Bud's Outdoor Gear Warehouse,** Cooperative Car Wash
226		MO 185, Oak Grove, Sullivan
	TStop	**N: Conoco*(D)(LP) (Country Market Buffett)**
	Gas	S: Citgo*, Gas Mart
	Food	S: Beijing, Burger King, Golden Corral, Hardees, KFC, **McDonalds**, Pizza Hut, Shoneys, Steak 'N Shake, Subway, Taco Bell
	AServ	S: Wal-Mart
	TServ	**S: Goodyear Tire & Auto**
	ATM	S: Bank of Sullivan
	Other	**S: Wal-Mart**
230		Cty Rd JJ & W, Stanton
	Gas	N: Citgo*(D)
		S: Phillips 66*
	Food	N: Steakhouse
	Lodg	N: Stanton Motel
		S: Delta Motel
	RVCamp	**S: KOA Campground (3 1/2 Miles)**
	Other	**S: Jessy James Wax Museum**
(235)		**Rest Area - RR, Phones P (Both Directions)**
(238)		**Weigh Station - both directions**
239		MO 30, Cty Rd AB & WW, St. Clair
	TStop	**N: Coibion Truck Center, Robbin's Truck Center**
	Gas	S: Shell*(D)
	Food	S: Jim's Country Diner
	Lodg	N: Arch Motel
	AServ	S: Federal Auto Parts
240		MO 47, St. Clair, Union
	Gas	N: Phillips 66*(D)
		S: Mobil*
	Food	N: Burger King
		S: Hardee's, **McDonalds**
	Lodg	S: Super 8 Motel
	AServ	S: BF Goodrich, Mobil
242		Cnty Rd AH
247		U.S. 50, Cty Rd AT & O, Union (Difficult Eastbound Reaccess)
	Parks	**N: Camping**
		S: Robertsville State Park (5 Miles)
251		MO 100, Washington
	FStop	**N: Fuel Mart*(D)(SCALES), Mr. Fuel*(D)(SCALES)**
	Gas	N: Citgo*, Phillips 66*
	Food	N: Ike's
	AServ	N: I-44 Automotive Service Center
	TServ	**N: McCoy Truck Wash**
253		MO 100, Gray Summit
	Gas	S: Phillips 66*(LP)
	Lodg	S: (AAA) **Best Western**
	Other	**S: Coin Laundry**
257		Bus Loop 44, Pacific
	TStop	**N: Texaco Truck & Auto*(D) (Buffalo's)**
	Gas	S: Auto Mart*(CW), Citgo*, Conoco*, Mobil*(CW)
	Food	S: Hardee's, McDonalds, Pizza Hut, Taco Bell
	Lodg	S: (AAA) Holiday Inn
	AServ	S: NAPA Auto Parts
	Med	**S: Pacific Health Center**
	ATM	S: Missions Bank, Security Pacific Bank
	Other	S: Pacific Auto Wash
261		Allenton
	FStop	**S: Phillips 66*(D)(LP)**
	Gas	S: Shell*(CW)
	Food	N: Lion Choice Roast Beef, Mr. Goodsense, Steak & Shake
	Lodg	N: (AAA) Oak Grove Inn, (AAA) Ramada, Red Carpet Inn
	AServ	N: Auto Zone Auto Parts, Wal-Mart
	Parks	**N: Six Flags Over Mid America**
	Other	**N: Wal-Mart**
264		MO 109, Cty Rd W, Eureka
	Gas	N: Citgo*, Phillips 66, Texaco*
	Food	N: Burger King, Phil's BBQ
	Lodg	N: DAYS INN Days Inn
265		Williams Rd
266		Louis Rd
269		Antire Rd, Beaumont
272		MO 141, Fenton, Valley Park
	FStop	**S: Citgo***
	Gas	S: Shell*(CW)
	Food	S: Burger King, Golden City Chinese, Imo's Pizza, McDonalds, Steak & Shake, Subway, Taco Bell
	AServ	S: Pennzoil Lube
274		Bowles Ave

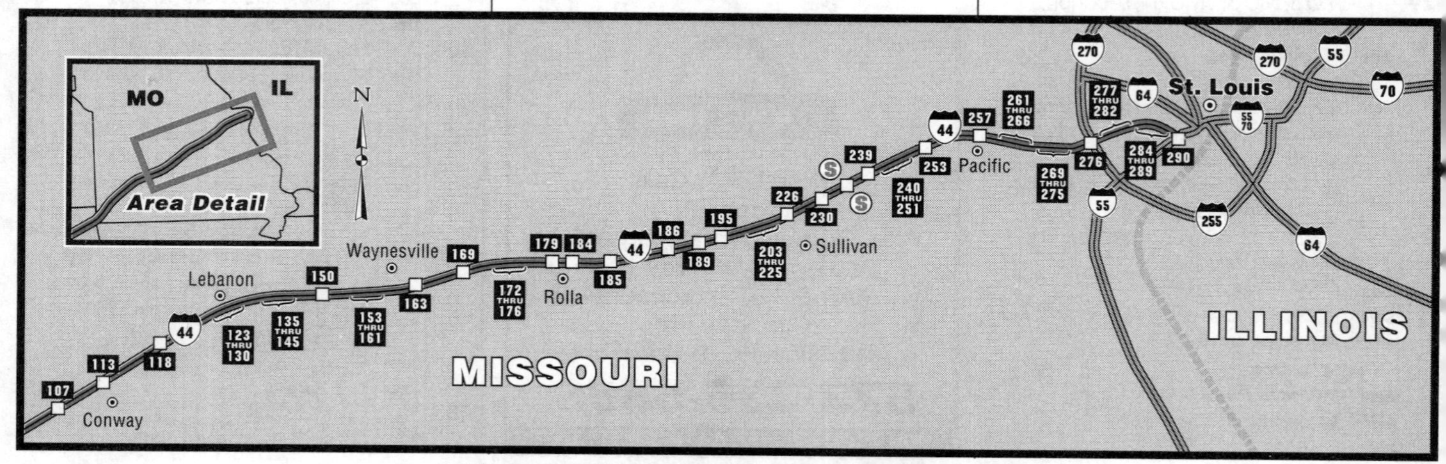

Bold red print shows RV & Bus parking available or nearby

EXIT MISSOURI

Gas S: Citgo*, QuikTrip*[D]
Food S: Cracker Barrel, Denny's, White Castle Restaurant
Lodg S: Dury Inn, Holiday Inn Express, AAA Motel 6, ◆ **Pear Tree Inn**
ATM S: Commerce Bank

275 Yarnell Rd

276 Junction I-270, Memphis, Chicago

277A MO 366, Watson Rd

277B U.S. 61, U.S. 67, U.S. 50, Lindbergh Blvd
Gas N: Shell*
S: Shell*[CW], Vickers Gas & Food Mart*
Food N: Hardees, Richard Ribs, Sunny China Buffet, The Steak & Race Chinese
S: Bob Evans Restaurant, Burger King, Denny's, Helen Fitzgerald's, House of Hunan, Steak & Shake, Viking Restaurant
Lodg N: AAA Best Western, Howard Johnson
S: Comfort Inn, Days Inn, ◆ Holiday Inn, Ramada Inn
AServ N: Jiffy Lube
S: Empire Tire Center, Kelly Auto Tire Center, Midas Muffler
Med N: **Hospital**
Other N: **Venture Discount Store**

278 Big Bend Rd
Gas S: Conoco[D], QuikTrip*, Sinclair*
AServ S: Conoco
Other S: **Animal Hospital**, Hydrojet Coin Car Wash

280 Elm Ave
Gas N: Amoco*
AServ N: Amoco

282 Murdoch Ave, Laclade Station Rd (Eastbound Difficult Reaccess)
Gas N: Amoco, Clark's*, Mobil*
Food N: Dairy Queen, Hunan Vu's, Pizza Hut
AServ N: Amoco, Jiffy Lube, Mobil

283 Shrewsberry Ave. (Westbound)

284A Jamieson Ave

286 Hampton Ave
Gas N: Amoco*, Clark, Convenience Store*, Mobil*
S: Shell*[CW], Sinclair*
Food N: Chinese Express, Denny's, Hardees, Imo's Pizza, Jack-In-The-Box, McDonalds, Steak 'N Shake, Subway, Taco Bell
S: Hardee's, Hunan Cafe, Single Wok Chinese
Lodg S: AAA Holiday Inn, ◆ Red Roof Inn
AServ N: Amoco
Other N: **St. Louis Zoo**
S: Car Wash

287 Kings Highway
AServ N: Amoco*[CW]

288 Grand Blvd., Downtown St. Louis (Difficult Westbound Reaccess)
Med N: **Incarnate Word Hospital**

289 Jefferson Ave
Gas N: Citgo*
Food N: Subway
S: McDonalds, Taco Bell

290A Junction I-55, Memphis

290B Junction I-55, 18th St, I-70 West, I-64E, US 40

↑MISSOURI

Begin I-44

EXIT TEXAS

Begin I-45

↓TEXAS

284A I-30

283 US 175 East, Kaufman, MLK Jr. Blvd, Fair Park (Southbound)
Gas W: Diamond Shamrock*, Shell
Food W: Hardeman's Bar-B-Q, KFC, Taco Bell
Lodg W: Motel Winnway
AServ W: Chief Auto Parts, House of Parts, P D's Tire Service, Shell
ATM W: Diamond Shamrock
Other W: **Friedman's Pharmacy**

283A Lamar Street
Gas W: Gas*[D]
Food W: Gilbert's Fast Food & Grocery
Other E: **Hasty Grocery**

283B Fair Park, Pennsylvania Ave, MLK Jr Blvd (Northbound)

281 Overton Road (Southbound, Reaccess Both Directions)
AServ E: Taylor Auto Supply (IAPA)

280 Linfield Street, Illinois Avenue
Gas W: Texaco*[D]
Lodg E: Linfield Motel, Star Motel
AServ E: International Paint & Body
TServ E: **Body Shop, DIS 24 Hr Tire & Break Service**
ATM W: Texaco

279 Loop 12 (Limited Access Highway, Northbound Is 279AB)

277 Simpson Stuart Road
Gas W: Chevron

276AB 276A - I-20 West Fort Worth 276B - I-20 East Shreveport

274 Dowdy Ferry Road, Hutchins
FStop W: **Afco (Automated Fueling)**
Gas E: Exxon, Texaco*[D][LP]
Food E: **Tooters Diner (Gold Inn)**
W: Dairy Queen, **The Smith's Restaurant**
Lodg E: **Gold Inn**
AServ E: Exxon
ATM W: NationsBank
Other W: **Eagle Bus Parts**

273 Wintergreen Road
TServ W: **Frank Prasifka & Sons Tire Service**

272 Fulgham Road Truck Check Station
Food E: Texas Rose Restaurant & Club
Other W: **Weigh Station**

271 Pleasant Run Road

270 Belt Line Road, Wilmer
FStop E: **Texaco*, Total***
Gas W: Exxon*[CW]
Food W: Dairy Queen, Delightful Donuts, Monopoly Pizza, Subway (Exxon)
AServ E: **Texaco**
ATM E: **Total**
Other W: Car Wash, **Laundromat, Mac's Supermarket, Police Station, US Post Office**

269 Mars Road
TServ E: **Industrial Truck Parts (Texas Star Truck Sales), Texas Star Truck Sales**

268 Bus. I-45, Malloy Bridge Rd, Ferris

267 Frontage Road

EXIT TEXAS

266 FM 660, 5th Street
Gas E: Fina*[D]
W: Diamond Shamrock*
Food W: Dairy Queen
ATM E: Commercial State Bank, Fina
W: Diamond Shamrock
Other E: **Get 'N Go Food Store**

264 Wester Road
TServ E: **T & H Bus Repair**
W: **G & S Truck & Equipment**

263A Loop 561, Trumbull

262 Risinger Rd.

260 Bus. I-45, Palmer (Northbound Is Called Hampel Rd)
Other E: **Tri-State Semi-Tractor Driving School**

259 FM 813 to FM 878
TServ W: **I-45 Truck & Trailer Repair Inc. (Goodyear)**
Other W: **US Post Office**

258 Parker Hill Road
TStop W: **Knox*[SCALES] (Fuel-Coastal, FedEx Drop), Mobil* (FedEx Drop)**
Food W: **BLT's Breakfast, Lunch & Treats (Mobil TS), Chicken Express (Mobil TS), Dogs 'N Suds Express (Mobil TS), Subway (Knox TS)**
Lodg W: **Palmer House Motel**
ATM W: **Knox TS, Mobil TS**

255 FM 879, Garrett
FStop W: **Chevron***
Gas E: Diamond Shamrock (Northbound Access Only)
ATM W: **Chevron**

253 Bus. Loop I-45, Ennis
Gas W: Texaco*[D]
AServ W: JR Auto Ranch
Med W: **Hospital**
Other W: **Veterinary Clinic**

251 TX 34, Kaufman Italy
Gas E: Fina*, Texaco*[D]
W: Chevron*[D], Exxon*[CW]
Food E: Donuts, McDonalds (Playground)
W: Arby's, Braum's Hamburgers & Ice Cream, Dairy Queen, Golden Corral, Jack-in-the-Box, KFC, Mr. Gatti's Pizza (Chevron), Taco Bell, Waffle House, What A Burger
Lodg E: Holiday Inn Express
W: **Ennis Inn**, AAA Quality Inn
AServ W: Allen Samuel's Autoplex, Terry Gregory Ford, Mercury, **Wal-Mart**
RVCamp W: **Jeff's RV Campground (Showers)**
ATM W: Chevron, Exxon
Other E: **Golf Store**
W: **Chamber of Commerce, Wal-Mart (Pharmacy)**

249 Unnamed
TStop W: **Ennis Truck Stop* (CB Shop)**
FStop W: **Total***
Food W: **Ennis Truck Stop Restaurant (Ennis Truck Stop)**
TServ W: **Doherty's Bros. Truck (24 Wrecker Service & Truck Repair)**
TWash W: **Blue Beacon Truck Wash (Total)**
ATM W: **Ennis Truck Stop, Total**

247 US 287N, Waxahachie, Ft Worth (Limited Access Highway)

246 FM 1183, Alma
Gas W: Chevron*[D], Phillips 66*

EXIT — TEXAS

ASserv — W: Tom's Bait & Auto Parts

244 — FM 1182

243 — Frontage Road

242 — Rice

AServ — W: Gary's Auto Parts, OLJ Tire & Brake

Other — W: Car Wash, **Mini Mart, Police Dept, US Post Office**

239 — FM 1126

Gas — W: Phillips 66*

238 — FM 1603

TStop — E: **Fina* (24 Hrs)**

Food — E: **Blimpie's Subs (Fina TS)**

AServ — E: H & H Auto Parts

TServ — E: **C&E Truck Tire & Road Service (24 Hrs)**

RVCamp — E: **RV Park (Fina TS)**

ATM — E: Fina Truck Stop

237 — Frontage Road

235B — Bus. I-45, Corsicana (Southbound, No Exit 235 B Northbound, Only Exit 235)

235A — Frontage Road (Southbound, Known As Only Exit 235 Northbound)

232 — Roane Road

231 — TX 31, Waco, Athens (Navarro College)

Gas — E: Mobil, Texaco*
W: Chevron*, Exxon*(D)

Food — E: **Sandy's Restaurant (Colonial Inn)**
W: **Bill's Fried Chicken,** Charlie's Bar-B-Q, **Cisco's Mexican,** Dairy Queen, McDonalds (Playground)

Lodg — E: **Colonial Inn**

AServ — E: Carl White Chevrolet, Mobil
W: Berry Chrys, Plym, Dodge, Jeep, Eagle, Brinson Ford

Med — W: **+ Hospital**

Other — E: **Navarro Pecan Company with Factory Outlet Store**

229 — US 287 South, Palestine (Richland Chambers Reservoir)

Gas — E: Shell*(D)

Food — W: **El Patio (Ramada Inn),** Fish Camp Catfish & Seafood Dinners, **Hallmark Restaurant (Days Inn),** Waffle House

Lodg — W: **DAYS INN Days Inn, Ramada Inn, Royal Inn, Travelers Inn**

Other — E: **Factory Outlet Center (see our ad this page), Texas Department Of Public Safety**

228B — Frontage Road (Northbound Reads North Bus. I-45, Corsicana)

FStop — E: **Phillips 66***

228A — 15th Street (Northbound)

FStop — E: **Phillips 66***

225 — FM 739, Angus, Mustang

FStop — W: **Citgo*(D)**

Gas — E: Bennies, Chevron*, Exxon*(D)
W: Shell*

Food — E: Bennies Old Fashioned Hamburgers (Bennies Gas Station)

RVCamp — W: **Camper Depot (1 Mile), Richland Chambers Reservoir**

221 — Unnamed

220 — Frontage Road

EXIT — TEXAS

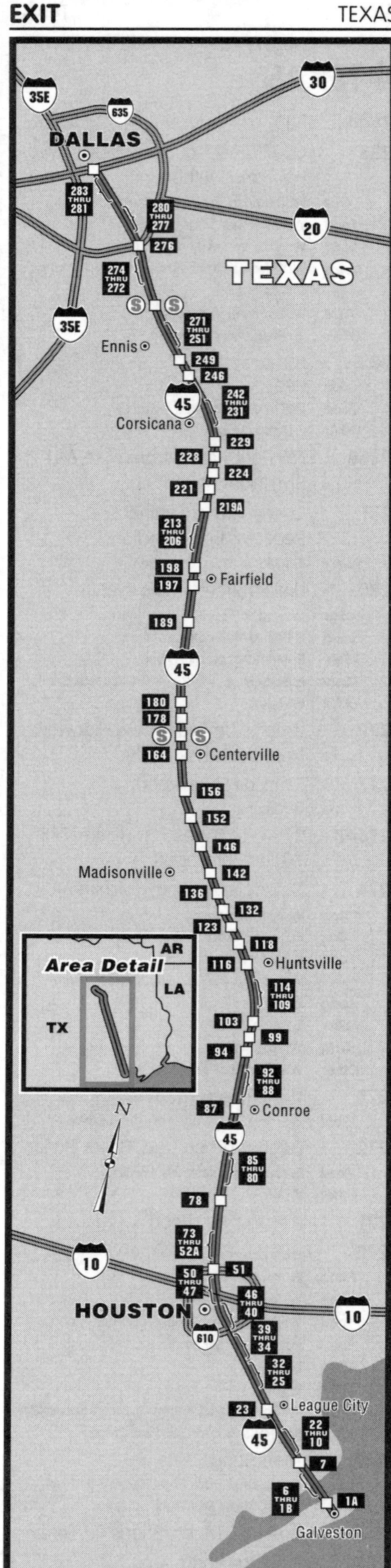

EXIT — TEXA

219B — Frontage Rd

219A — TX 14, Richland, Mexia (Southbound

Gas — W: Shell*

Food — W: B & J's Cafe

ATM — W: Shell

218 — FM 1394, Richland (Northbound)

(217) — **Rest Area - RR, Phones, Vending, Picnic, Grills, RV Dump P (Both Directions)**

213 — TX 75, FM 246, Worthman, Streetman

Gas — W: **Exxon* (No Diesel Fuel for Trucks),** Texaco*

211 — FM 80, Kirvin

206 — FM 833

198 — FM 27, Wortham, Fairfield

Gas — E: Shell*
W: Exxon*

Food — E: Real Pit Bar-B-Q (Shell)
W: First Stop Food & Grill (Exxon)

Lodg — W: **Budget Inn**

RVCamp — W: **Budget Inn RV Park**

Med — E: **+ Hospital**

197 — US 84, Fairfield, Teague

FStop — W: **Shell*, Texaco*(D)**

Gas — E: Exxon*, Fina*, Texaco*(D)

Food — E: Dairy Queen, Jack-in-the-Box, **McDonalds (Playground),** Ponte's Diner, Sam's Chicken & Bar-B-Q, **Sam's Restaurant,** Something Differen Restaurant, Subway, Texas Burger
W: **I-45 Coffee Shop, Pizza Hut, Sammy's Restaurant**

Lodg — W: Regency Inn, **Sam's Motel**

AServ — E: Bossier Chev, Chev. Trucks, Chrys, Plym, Dodge, NAPA Auto Parts (Bossier), Rutherford 24 Hr. Mechanic, Tire, Service, Towing
W: Fairfield Ford, Mercury

TServ — W: **Sandra's Tire Service (Shell)**

Other — E: Roadster Car Wash

189 — TX 179

TStop — E: **Conoco* (Showers), Fina* (24 Hrs)**

Gas — E: Chevron*

Food — E: **Dew Cafe (24 Hrs, Fina TS), Restaurant (Conoco)**
W: **Tatum's Bar-B-Q**

TServ — E: **Conoco**

ATM — E: **Conoco**

(187) — Picnic - Shelters, No Services (Both Directions)

180 — TX 164, Groesbeck

178 — US 79, Buffalo

EXIT	TEXAS
FStop	W: Diamond Shamrock* (FedEx Drop), Mobil* (24 Hrs)
Gas	E: Conoco, Texaco*
	W: Chevron*
Food	E: Pizza Inn, Weathervane Restaurant
	W: Arby's (Mobil), Dairy Queen, Pitt Grill, Pop's BBQ, Rainbow Restaurant, Texas Burger
Lodg	W: Best Western, Economy Inn
AServ	E: Buffalo Body Shop & Car Wash
	W: Triangle Tire
ATM	W: Diamond Shamrock
Other	E: Brookshire Bros. Grocery
(166)	Weigh Station (Southbound, Under Construction)
164	TX 7, Centerville
FStop	W: Woody's Diesel Express (Showers, No Auto Fuel)
Gas	E: Diamond Shamrock*, Texaco(D)
	W: Exxon*, Shell*
Food	E: Country Cousin's Barbecue, Texas Burger, Woody's Smoke House (Diamond Shamrock)
	W: Dairy Queen, Jack-in-the-Box, Mama Mike's, Piccadilly Circus Pizza (Shell), Woody's Smoke House & Bakery (Shell)
Lodg	E: Days Inn
AServ	E: Texaco
ATM	W: Exxon
(159)	Picnic - Shelters, No Services
156	FM 977, Leona
FStop	W: Diamond Shamrock*
Gas	W: Exxon*(D)
Food	E: Crystal's Skillet
	W: D-K Diner
(155)	Picnic - Shelters, No Services (Northbound)
152	TX OSR, Normangee
FStop	W: Chevron*(D)
Food	W: Chuckwagon, Yellow Rose Cattle Co.
Other	W: Old West Town (Yellow Rose Cattle Co.), Yellow Rose Arena
146	TX 75, Madisonville
142	US 190, TX 21, Bryan, Crockett, Madisonville
FStop	E: Chevron*(D), Exxon*
	W: Diamond Shamrock*(D) (24 Hrs)
Gas	E: Texaco*(D)(CW)
	W: Shell*
Food	E: Church's Fried Chicken, Corral Cafe, Eddie Pepper's Great Mexican Food (Exxon), Hot Stuff Pizza (Exxon)
	W: J W Cattleman's, McDonalds, Pizza Hut, Sonic, Subway (Shell), Texas Burger
Lodg	E: Best Western
	W: Budget Motel
AServ	E: Chevron
	W: Carquest, Drake's Service Center, Madisonville Ford, Mercury, & Ford Trucks
TServ	E: Chevron
Med	W: + Hospital
ATM	E: Chevron, Exxon, Texaco
136	Spur 67
132	FM 2989
(126)	Rest Area - RR, Picnic, Grills, RV Dump, Pet Rest, Vending P (Southbound)
(124)	Rest Area P (Northbound)
123	FM 1696
(121)	Weigh Station - Phones (Northbound)
118	TX 75, Huntsville (Sam Houston State Univ. & Museum)

EXIT	TEXAS
TStop	W: Citgo Truck/Auto Plaza*(SCALES) (FedEx Drop)
Food	E: Hitchin' Post Restaurant (Texaco TS)
	W: Blimpie's Subs (Citgo Truck/Auto Plaza), Burger King (Citgo T/A Plaza), Country Kitchen (Citgo T/A Plaza)
AServ	E: Texaco TS (24 Hrs)
TServ	E: Texaco TS (24 Hrs)
	W: Citgo Truck/Auto Plaza
TWash	E: Truck Wash (Texaco TS)
Other	E: Huntsville Municipal Airport
116	TX 30, East US 190, Huntsville (Lake Livingston)
Gas	E: Diamond Shamrock*(D)
	W: Chevron*(CW), Exxon*(D), Shell*, Texaco*(CW), Texaco*(D)
Food	E: Church's Fried Chicken, El Chico Mexican, Golden Corral, Imperial Garden, Kettle, McDonalds (Playground), Mr. Gatti's Pizza, Popeye's Chicken, TCBY, Tejas Cafe & Bar, Texas Burger, The Junction Steaks & Seafood, What A Burger (24 Hrs)
	W: Bullwinkle's Restaurant, Burger King, Chili's, Denny's, Jose's Restaurant, KFC, Shipley Do-nuts, Subway
Lodg	E: Comfort Inn, ◆ Econolodge, AAA La Quinta Inn, Motel 6
AServ	E: Huntsville Chrysler, Dodge, Jeep, & Dodge Trucks
Med	W: + Hospital
ATM	E: Citizen's Bank
	W: Exxon, First National Bank West, Guaranty Federal Bank, NationsBank, Texaco
Other	W: Hastings Books, Music, Video, Kroger Supermarket (Pharmacy, 24 Hrs), Mail Boxes Etc, Maurel Car Wash, Sam Houston Optical, Today's Vision, WalGreens, West Hill Mall
114	FM 1374, Huntsville
Gas	E: Corner Pantry*(D), Fina*(D)
	W: Chevron, Diamond Shamrock*
Food	E: Dairy Queen (Playground)
	W: Casa Tomas Mexican Restaurant, Country Inn III Steakhouse
Lodg	E: ◆ Rodeway Inn
	W: Sam Houston Inn
AServ	E: Hillcrest Ford, Lincoln, Mercury Dealership
	W: Chevron
Med	W: + Huntsville Memorial Hospital
ATM	E: Corner Pantry, Fina
Other	W: Slaughter Drugs
112	TX 75
Gas	E: Citgo*

EXIT	TEXAS
Other	E: Sam Houston Statue Visitor Center
113	TX 19, Huntsville, Crockett (Northbound)
109	Park 40, Huntsville State Park
RVCamp	W: Huntsville State Park (RV Dump)
Parks	W: Huntsville State Park
Other	E: Sam Houston Statue with Visitor Center
(105)	Picnic - Grills, Phones, No Other Services
103	TX 150, FM 1374, FM 1375, New Waverly (Southbound)
FStop	W: Citgo*
Gas	W: Texaco*
AServ	W: Texaco
Other	W: Sam Houston National Forest District Office
102	TX 150, FM 1374, FM 1375, New Waverly (Northbound)
98	(99) Danville Rd., Shepard Hill Rd.
Food	W: Squeeky's Place BBQ
RVCamp	E: The Fish Pond RV Park (RV Dump)
94	FM 1097, Willis
Gas	W: Chevron*, Texaco*(D)
Food	E: Jack-in-the-Box, Sonic
	W: McDonalds
AServ	E: Isaacks Auto Diagnostics & Car Exhaust
92	FM 830, Seven Coves Rd, Panorama Village
RVCamp	W: Camping (RV Dump)
91	League Line Road
Gas	W: Citgo*
Food	E: Wendy's
	W: Cracker Barrel
Other	E: Conroe Texas Outlet Center
90	FM 3083, Teas Nursery Road
Parks	E: Montgomery County Park (3 Miles)
88	Wilson Road, Loop 336 To Navasota, Cleveland
Gas	E: Diamond Shamrock*(D), Exxon*(CW) (24 Hrs), Shell*(CW)
	W: Chevron*(CW) (24 Hrs)
Food	E: Arby's, Burger King (Playground), China Delight, Domino's Pizza, Donut Shoppe, Grandy's, Hofbrau Steaks, Hooks Casual Seafood Restaurant, Little Caesars Pizza, Margarita's Mexican Restaurant, McDonalds (Playground), Papa John's Pizza, Rainbow Cafe, Shipley Do-nuts, Sonic, Subway, What A Burger (24 Hrs)
	W: Casa Ole Mexican, Honey-B-Ham, Hunan Village Restaurant, I Can't Believe It's Yogurt, KFC, Leaf N' Ladle Garden Cafe, McDonalds (Wal-Mart Supercenter), Ryan's Steakhouse, The Black-Eyed Pea
AServ	E: Discount Brake & Muffler, Discount Tire Co., Goodyear Tire & Auto, Pennzoil Lube, Professional Car Care Express Tire Service
	W: Sam's Club, Wal-Mart
ATM	E: Bank of America, Diamon Shamrock, Guaranty Federal Bank, Klein Bank
	W: First Bank of Conroe, Klein Bank, Randall's Food & Drugs, Wal-Mart Supercenter
Other	E: Animal Hospital Of Conroe Inc., Cinema 6, Hastings Books, Music, & Video, Kroger Supermarket (Pharmacy, 24 Hrs), Montgomery College, Timber Lane Washateria, WalGreens (1-Hr Photo)
	W: Academy Sports & Outdoor Supplies Store, Eye Land, For Heaven's Sake Christian Bookstore, Pet's Paw Animal Hospital, Randall's Food & Drugs (1 Hr Photo ,

Bold red print shows RV & Bus parking available or nearby

EXIT TEXAS

Pharmacy), Sam's Club, Wal-Mart Supercenter (Pharmacy, Vision Center, 1-Hr Photo)

87 TX 105, FM 2854, Conroe

Gas E: Chevron*(D) (24 Hrs), Diamond Shamrock*, Exxon*
W: Coastal*(D), Exxon*(CW), Texaco

Food E: Burger King (Playground), Church's Fried Chicken, Ci Ci's Pizza, Donut Wheel, ElT'son's Chicken Tender & Tasty, Jack-in-the-Box (24 Hrs), Kettle, La Carreta Taqueria, Luther's Hamburgers Bar-B-Q, McDonalds (Playground), Panda Chinese Buffet, Popeye's Chicken, Schlotzkys Deli, Taco Bell, Taqueria Vallarta, Village Restaurant (24 Hrs), Wyatt's Cafeteria
W: Luby's Cafeteria

AServ E: Bill Blankenship, Ford Dealership, Fred's Auto Parts, Sunset Auto Repair, Tom Kelley Tire Co., Tune Up Masters
W: Xpress Lube

ATM E: Bank One
W: NationsBank, Texas Commerce Bank

Other E: Car Service Center, Eckerd Drugs (24 Hrs, 1-Hr Photo), Kroger Supermarket (Pharmacy), Lifechek Drug, Mail Depot, Texas State Optical, US Post Office
W: Maurel Car Wash

85 Gladstell Street

Gas W: Diamond Shamrock*(D), Texaco*(D)

Food W: Malt-N-Burger

Lodg W: Days Inn, Motel 6

AServ E: AAA Tire Service (24 Hr Service)
W: DeMontrond Chrys, Plym, Jeep, Eagle, Dodge, Cadillac, Olds, Gullo Toyota

Med W: ✚ Hospital

ATM E: Texas Commerce Bank
W: A & H Foodstore

Other E: Barnacle Bill's Marine
W: A & H Foodstore, Texas Department Of Public Safety

84 North TX 75, Loop 336, Frazier St.

Gas E: Exxon*(D), Texaco(D)
W: Shell*(CW)

Food E: San Miguel Taqueria Mexican
W: IHOP, Pizza Hut Carry Out & Delivery, Shoney's, Taco Cabana (24 Hrs)

Lodg E: AAA Holiday Inn, Ramada Limited

AServ E: Exxon, Texaco

Med W: ✚ Columbia Conroe Regional Medical Center

ATM W: Albertson's Grocery

Other E: U-Haul Center(LP)
W: Albertson's Grocery (24 Hrs, Pharmacy), K-Mart (Pharmacy), Mail Boxes Etc

83 Crighton Road, Camp Strake

Other W: Animal Emergency Clinic

81 FM 1488, Magnolia, Hempstead

Gas W: Diamond Shamrock*(D)

RVCamp W: Camper Land Sales & Service

ATM W: Diamond Shamrock

Other W: Hempstead Outlet (see our ad this page)

79 Needham Rd, TX 242

Gas E: Texaco*(D)

Food E: McDonalds (Texaco)

ATM E: Texaco

Other W: Montgomery College

78 Tamina Rd, Research Forest Dr, Shenandoah (Services Same On Exit 77 Northbound)

FStop E: Conoco*

Gas W: Diamond Shamrock*(D), Mobil*

Food E: Cafe China Buffet, Luther's Bar-B-Q, Papa John's Pizza, Poncho's Mexican Buffet, Red Lobster, Pasta Kitchen
W: Denny's, Win's Chinese Restaurant

Lodg W: ◆ Hampton Inn, AAA La Quinta Inn, Roadrunner Motor Inn

AServ E: Knee's Automotive Volkswagon & American

Med W: ✚ Hospital

ATM E: Compass Bank
W: Diamond Shamrock

Other E: A-Med Medical Supply, Cinema 4, Office Depot, Woodland's Bowl
W: Mail Boxes Etc, Shenandoah Animal Clinic

76 Woodlands Pkwy, Robinson Rd., Oak Ridge North, Chateau Woods

Gas W: Texaco*(D)(CW)

Food E: Long John Silvers
W: Chili's, Guadalajara Mexican Grill & Bar, Jack-in-the-Box, Luby's Cafeteria, The Black-Eyed Pea

Lodg W: Courtyard by Marriott, Drury Inn

AServ E: Bill Blankenship (Firestone)

ATM W: Compass Bank, NationsBank

Other E: Drive Pro Golf Shop
W: Barnes & Noble, Target Department Store, The Woodlands Mall, Toys "R" Us, US Post Office

73 Rayford Road, Sawdust Road

Gas E: Conoco*, Shell*
W: Shell*(CW) (24 Hrs), Texaco*(D)(CW)

Food E: Baskin Robbins, Hartz Chicken Buffet, Hydens Restaurant & Oyster Bar, Jack-in-the-Box, Manuel's Mexican, McDonalds (Playground), Popeye's Chicken, Sonic, Tam's Restuarant, Thomas Bar-B-Q
W: 19th Hole Grill & Bar, Bagel Express, Bei Jing Chinese Restaurant, Ci Ci's Pizza, Giuseppe's Restaurant, Grandy's, Greek Tony's, Ninfa's, Oscar's Creamery, Papa's Icehouse, Fishfry, Rookies Sports Bar & Grill, Sam's Cafe & Restaurant, Shipley's Do-Nuts, Smoothie King, Subway, Sundale Donuts, Supreme Soup & Salad, What A Burger, Woodlands House Chinese, Z K's Cafe

Lodg W: Red Roof Inns

AServ E: AAMCO Transmission, Auto Zone Auto Parts, Hi-Lo Auto Supply, Kelly Tire, Meineke Discount Mufflers, Quick Lube & Sticker
W: Discount Tire Co., Express Lube, Fast Lube

Med E: ✚ Primary Medical Clinic (Minor Emergency/7 Days A Week)

ATM E: Conoco, Woodforrest National Bank
W: Bank of America, Kroger Supermarket, Texas Commerce Bank

Other E: Academy (Outdoor Store), Boats Unlimited, Golf Balls Unlimited, News/Craft Books, Photo Hut, Quick Clean Car Wash, Swing Masters Go[lf] Shop
W: Golf USA, H-E-B Pantry Foods, Just Cats Veterinary Hospital, Kroger Supermarket (24 Hrs, Pharmacy), Nesbit's Cleaners & Laundry, Rainbow Car Wash, Today's Vision, Walgreens 24-Hr Pharmacy, Woodlands Pet Clinic

73A (72) Frontage Road, Hardy Toll Rd South (Southbound)

TStop W: Mobil* (1-Hr Photo)

Food W: Restaurant (Mobil TS)

TServ W: King of the Road Heavy Repair & Tires

ATM W: Mobil TS

72 Unnamed (Southbound)

AServ W: Bruce Automotive

70B Spring - Stuebner Road

Other E: Island Mist Spas(LP)

70A Spring - Cypress Road, FM 2920, Tomball

Gas E: Exxon*(CW)
W: Chevron*(D), Racetrac*, Shell*(CW) (24 Hrs), Texaco*(D)

Food E: Charlie's Neighborhood Grill, El Palenque Mexican, Hartz Chicken Buffet, McDonalds (Playground), Pizza Hut, Subway, Sun Li Chinese Restaurant, Wendy's
W: Burger King, Chinese Wok, Taco Bell, What A Burger (24 Hrs)

AServ E: C & L Tire Co. Inc., Hi-Lo Auto Supply
W: Spring Body Shop

RVCamp E: Campground (RV Dump)

ATM E: Wells Fargo
W: Compass Bank

Other E: Birds of Paradise Pet Store, Kroger Supermarket (24 Hrs, Pharmacy), Northland Optical, Splash Town USA (Water Theme Park), Spring Center Animal Clinic, US Post Office, WalGreens (1-Hr. Photo)
W: Spring Animal Hospital, U-Haul Center(LP)

69 Louetta Rd., Holzwarth Rd

Gas W: Chevron*

Lodg W: Motel 6, Travelodge

AServ W: Planet Ford

Other W: The Home Depot(LP)

66 FM 1960, Addicks, Humble

Gas E: Chevron*, Racetrac*, Texaco*(CW)
W: Exxon*(CW)

Food E: Chinese Restaurant East Bow, Pour House Restaurant
W: Bennigan's, Champs (24 Hrs), Grandy's, Hooters, James Coney Island, Lasagna House III Marco's Mexican, McDonalds (Playground), Ninfa's Mexican, Outback Steakhouse, Pizza Hu[t] Red Lobster, Steak & Ale, Subway, The Original Pasta Co.

Lodg W: HomeStead Village Efficiencies, AAA La Quinta Inn

AServ E: AutoNation USA, Detail Lube Center
W: All Tune & Lube, NTB

XIT TEXAS

ATM E: First Educator's Credit Union, Texaco
W: Kroger Supermarket, Wells Fargo
Other E: Golf World Driving Range, Mr. Car Wash (Chevron)
W: Book Nook, Golfsmith Golf Center, Kroger Supermarket (24 Hrs, Pharmacy), Mail Boxes Etc, OfficeMax, Pet City Discount Center, Venture, WalGreens (24 Hr.)

4 Richey Road
TStop W: 45 Travel Center* (Citgo, Truckers Lounge)
Food W: Restaurant (45 Travel Center, 24 Hrs)
Lodg E: AAA Holiday Inn, ◆ Lexington Hotel Suites
W: Bunkhouse (45 Travel Center)
AServ E: Carmax, Discount Tire Co. of Texas, Sam's Club
W: Gunther's Auto Repair
TWash W: 45 Travel Center
ATM W: 45 Travel Center
Other E: Sam's Club

3 Airtex Drive, Rankin Rd
Food E: Atchafalaya River Cafe (Cajun & Creole Seafood), Pappasito's Cantina
AServ E: Tom Peacock Cadillac & Nissan
Other E: Garden Ridge Pottery, Louis DelHomme Marine, Marine Sports

2 Kuykendahl, Rankin Road
Lodg E: Super 8 (see our ad this page)

1 Greens Road
Gas E: Mobil*
Food E: Blimpie's Subs, Brown Sugar's Bar-B-Q, Chef Lin's Hunan, Fajitas A Sizzlin' Celebration, Fuddruckers, IHOP, Imperial Dragon Chinese Restaurant, Jack-in-the-Box, McDonalds, Monterey's Tex Mex Cafe, Pizza Inn, Souper Salad, Wendy's, Zero's Sandwich Shop
W: Burger King (Playground), Chinese Palace Restaurant, Ginger Chinese Cafe, Kiraku Sushi Bar, Luby's Cafeteria, Marco's Mexican, Rocky's Subs, Subway, Supreme Soup & Salad, TCBY, The Black-Eyed Pea
Lodg E: DAYS INN ◆ Days Inn, Wyndham Garden Hotel
W: Budgetel Inn
AServ E: Auto Express Montgomery Ward
RVCamp E: Morgan RV's (1 Mile)
ATM E: NationsBank, NationsBank
W: Bank United
Other E: Greenspoint Mall, Today's Vision Optometrist
W: Kroger Supermarket (24 Hrs), Office Depot, Petco Supplies & Fish, Target Department Store

0B Beltway 8, West Sam Houston Tollway

0A Beltway 8, Frtg Rd, Intercontinental Airport

0D Beltway 8, FM 525, Airport
Food W: Marco's Mexican Restaurant, Supreme Soup & Salad, TCBY
Other E: Greenspoint Mall
W: Office Depot, Petco Supplies & Fish

9 West Road
Gas E: E-Z Serve*, Shell(CW)
W: Exxon*(CW), Shell*(CW)
Food E: McDonalds (Playground)
W: Blimpie's Subs (Exxon), Little Caesars Pizza (K-Mart), McDonalds (Wal-Mart), TCBY (Exxon),

EXIT TEXAS

Taco Bell, What A Burger (24 Hrs)
AServ E: Bob Lunsford's Honda, Carquest Auto Parts, Midas, Shell
W: K-Mart, Pep Boys Auto Center, Wal-Mart
Other E: Pier 45 Marine
W: Academy Outdoors (Outdoor Store), K-Mart (Pharmacy), Wal-Mart (24 Hrs)

57B TX 249, West Mount Houston Rd
Gas E: Exxon*(D), Mobil*
W: Shell*(CW)
Food E: Church's Fried Chicken (Mobil), Taquito Joe
W: Pizza Inn, Sonic
Lodg W: Green Chase Inn
AServ E: Discount Tire Co.
W: Landmark Chevy Trucks, Northwood Lincoln Mercury Dealer
Other E: Kroger Supermarket
W: Eckerd Drugs (24 Hrs), Stop n Get*, Washateria

57A Gulf Bank Road
FStop E: Mobil*
Gas E: Conoco*
W: Diamond Shamrock*
Food E: Ricardo's Mexican
Lodg W: DAYS INN AAA Days Inn
AServ E: Hidden Valley Wrecker, In & Out Auto & Marine Repair, Joey Co Engines, New & Used Reproduction Parts
W: Archer Chrysler Plymouth Dealer, Archer Jeep Eagle, Landmark Geo Chevrolet
RVCamp E: Smith RV Rentals
ATM E: Conoco, Texas Commerce Bank
W: Diamond Shamrock
Other E: Cahill Veterinary Clinic, In & Out Auto & Marine Repair
W: Barber Boats

56B Spur 261, Shepherd Dr, Little York Rd (Southbound)
Gas W: Mobil*, Shell*(CW)
Food W: Arby's (Mobil), Denny's, Luby's Cafeteria
Lodg W: Gulf Wind Motel, Houston Motor Inn
AServ W: All Tune & Lube, Archer Mazda, Brandt Auto Truck & Equipment Sales, Duron Tires, Fred Haas Toyota, Lifetime Warranty Transmission, Lone Star Ford, Q-Lube, Saturn of Houston North Frwy, Western Auto
Other W: Walgreens

56A Canino Rd (Northbound)
RVCamp E: Best Buy RV Sales & Service, Holiday World RV Center

EXIT TEXAS

55B Little York Rd (Northbound)
FStop E: Coastal*(CW)
Gas E: Chevron*, Exxon, Fina*
Food E: Church's Fried Chicken, Ocean Fish Market & Restaurant, Sam's Country Bar-B-Q, What A Burger (24 Hrs)
AServ E: Exxon, Rick's Auto, Wholesale Wheel & Tire
TWash E: Coastal
ATM E: Chevron, Coastal
Other E: Price Buster Foods, Travis Boats

55A Parker Rd., Yale St.
Gas E: Mobil*, Northtown Service(D)(LP), Shell*
W: Shell
Food E: Schlotzkys Deli
W: Hartz Chicken, Long John Silvers
Lodg E: Olympic Motel
AServ E: Bap Geon Imported Car & Truck Parts, Northtown Service
W: Dave Cory Trucks, Fact-O-Bake Body & Paint Shop, Shell
Med E: ✚ Columbia North Houston Medical Center
ATM W: Merchant's Bank
Other E: ACE Pharmacy, Army Surplus, Parkway Optical, Parkway Pharmacy
W: Parker Food Mart

54 Tidwell Road
Gas E: Best Food Market*, Exxon*(CW)
W: Chevron* (24 Hrs), Circle A*, Shell*(CW)
Food E: Aunt Bea's Restaurant (24 Hrs), China Border (Chinese Buffet), Connie's Seafood, Lee's Inn Chinese & Vietnamese, Pancho's Mexican Buffet, Pizza Inn, Subway, Taco Cabana, Thomas Bar-B-Q
Lodg W: Guest Motel, Scottish Inns, South Wind Motel
AServ E: Tex-Star Motors
W: AAMCO Transmission, North Freeway Body Shop
TServ W: Rush Truck Center
RVCamp W: Ron Hoover Co.'s RV Center
ATM E: J.M. Grocery Store & Deli, Municipal Employees Credit Union
W: Chevron
Other E: Bakery Outlet, Eckerd Drugs, J. M. Grocery Store & Deli, Washaway Laundry (Coin Laundry)
W: Red Stone Food Mart, U-Haul Center(LP)

53 Airline Drive
Gas E: Chevron*(D), Conoco
W: Citgo*
Food W: What A Burger (24 Hrs)
Lodg W: Villa Provencial Motor Inn
AServ E: Conoco, Discount Tire Center, Montgomery Ward, Stop n Go, Tuneup Masters
ATM E: Chevron
Other E: Fiesta Grocery (Pharmacy), Northline Mall
W: Car Care Car Wash (Hand Wash), First Stop Store, Glen Burnie Washateria

52B Crosstimbers Road
Gas E: Shell*(CW), Texaco*(D)
W: Chevron*, Exxon*, Mobil*, N.N. Food Store*
Food E: Baskin Robbins, China Inn Restaurant, CiCi's Pizza, Denny's, Hungry Farmer Bar-B-Q, Jack-in-the-Box, James Coney Island, KFC, Long John's Donuts, McDonalds (Playground), Piccadilly Cafeteria, Pizza Hut, Sonic, Taco Bell, Texas Seafood Chinese Restaurant
W: Monterey's, Restaurant (Howard Johnson),

EXIT TEXAS

Wendy's
Lodg **W:** Economy Lodge, Howard Johnson, Luxury Inn, The Silver Glo Motel
AServ **E:** 3-A Auto Service, AJ's Grease Rack, **Montgomery Ward,** Tire Station, Western Auto
W: Crosstimbers Muffler & Brake Service, Llantas Tires, M & G Used Auto Sales, Meineke Discount Mufflers, National Engine & Supply, Nick's International Auto Care
ATM **E:** Navigation Bank Northline Branch, Shell, Texas Commerce Bank
Other **E: Cinema I-IV, Foot Locker Outlet, Northline Mall, Price Buster Foods (Grocery), Texas State Optical**

52A Frontage Road (Southbound)
Lodg **W:** Guest Inn, Super 8 Motel

51 Junction I-610

50 Cavalcade Street, Patton Street (Northbound Is Exit 50AB)
TStop **E: Pilot***(SCALES)
Gas **E:** Chevron* (24 Hrs), Diamond Shamrock*, Exxon, Texaco*(CW)
Food **E:** Casa Garay Carnes Asadas, Dairy Queen, Laredo Taqueria, Sol Y Mar Restaurant, **Wendy's (Pilot TS)**
Lodg **W:** DAYS INN Days Inn
AServ **E:** Reparacion de Radiadores
W: National Auto Parts, Unidos Auto Service
TServ **E: Embry Isuzu Truck Center, Pilot TS**
TWash **E: Pilot TS**
ATM **E: Pilot TS**
Other **E: 39 Cent Bakery Outlet Store, Grocery World (Pharmacy),** O'Banions Car Wash, **Quick Food Store**
W: Car Wash, **Snap Groceries, Texas Groceries, Washateria (Coin-Op Laundry)**

49B North Main Street, Houston Avenue
Gas **W:** Exxon*(D)(CW), Texaco*
Food **E:** Casa Grande Mexican Restaurant & Club
W: Domino's Pizza, Food Court Express (Exxon), Hunan Bo Restaurant, KFC, McDonalds (Playground), Shipley Do-Nuts, Subway, What A Burger (24 Hrs)
AServ **W:** Hi/Lo Auto Supply
Other **W: La Fiestita II Food Market, Simon's United Drugs**

49A Quitman Street (Southbound, No Southbound Reaccess)
Gas **W:** Stop N Go*
AServ **W:** Illusion Mechanic Shop

48AB 48A - I-10 East, Beaumont, 48B - I-10 West, San Antonio (48A-Southbound Is Left Exit, 48B-Northbound Is Left Exit)

47D Dallas Ave, Pierce Ave (Southbound)

47C McKinney Street (Southbound, Left Exit, Difficult Reaccess)
Food **W:** Downtown Soup & Salad, Foley's Deli Express, James Coney Island, Luther's Bar-B-Q, Mandarin Hunan Cuisine Restaurant, Massa's Restaurant, McDonalds, Souper Sandwich, Zero's Sandwich Shop
Lodg **W:** Doubletree Hotel, Hyatt Regency
Other **W: Fox Photo 1-Hr Lab, Houston Public Library, Music Hall, Prescription House, The Heritage Society, The Jesse Jones Hall for the Performing Arts, Theatre District, United Parcel Service**

EXIT TEXAS

47B Houston Ave, Memorial Dr, Theatre District (Northbound)
AServ **E:** Firestone Tire & Auto, Knapp Chevrolet

47A Allen Parkway (Left Exit Both Directions, Difficult Reaccess)

46AB US 59, TX 288, Lake Jackson, Freeport, Cleveland (Northbound - Left Exit, Southbound - Left Exit, Separated Ex)
Other **E: Livingston Outlet Mall (see our ad this page)**

45A Scott Street (Northbound Is Exit 45)
Food **E:** Blimpie's Subs (Texaco), Creole Fried Chicken
AServ **E:** University Firestone
ATM **E:** Texaco
Other **E: Joe's Washateria, Lavenderia**

44C Cullen Blvd (Southbound)
Food **W:** Burger King, McDonalds (24 Hrs), Wendy's
Other **W: University of Houston**

44B Calhoun Street (Southbound)
Other **W: University of Houston**

44A Elgin - Lockwood Cullen Blvd. (Northbound)
Gas **E:** Diamond Shamrock*(D)
ATM **E:** Diamond Shamrock

43B Tellepsen St, Schlumberger St (Northbound)

43A Phones Road
Gas **W:** 6-Eleven*
Food **E:** Fiesta The Original Loma Linda, Luby's Cafeteria, Tel-45 Kitchen
W: Panaderia Taqueria
AServ **W:** Auto-Air Wholesale Parts, Sales, Service, Galvan's Tire Service, Hernan's Transmissions, Juan Tire Service, Lidstone Garage, M & N Mechanic Shop

42 US 90 Alt., S. Wayside Drive (Southbound)
Gas **W:** Exxon*
Food **W:** Jack-in-the-Box, **Little Caesars Pizza (K-Mart),** McDonalds (Playground), **Monterey's Tex-Mex Cafe**
Lodg **W:** Josephine Motel, Sunset Inn
AServ **W:** Galvan's Tire Service, Juarez Tire Service
Other **W: K-Mart (Pharmacy)**

41B Broad Street, Griggs Road, Alt US 90, S. Wayside Dr (Southbound)
Gas **E:** Diamond Shamrock*, Shell*(CW)
W: Conoco*, Jack's Mini Mart* (24 Hrs), Mobil*
Food **E: Swifty's Bar-B-Que (Red Carpet Inn),** Taqueria Arandas
W: Little Caesars Pizza (K-Mart), Mary Lee

EXIT TEXA

Donuts (24 Hrs), Morelia Taqueria y Refresquer
Lodg **E:** Gulf Freeway Inn, Houtex Inn, Red Carpet Inn
AServ **E:** Fiesta Paint, Body, & Auto Glass
W: Casey's Paint & Body Shop, Don Rucker Tire Co.
Med **E: ✚ Clinical Santa Ana (Minor Emergencies Clinic)**
Other **W: K-Mart (Pharmacy), Sellers Bros. Grocery Store**

41A Woodridge Dr (Southbound)
Gas **W:** Shell*(CW)
Food **W:** Church's Fried Chicken, Dot Coffee Shop (24 Hrs), Grandy's, King Hoagie N' Rice, McDonald (Playground), Pappas Bar-B-Q, Pizza Hut, Schlotzkys Deli, Taqueria Mexico, Wendy's
AServ **W:** Bayway Lincoln, Mercury, Subaru, Body Sho Cabell Chevrolet, Mazda, Geo
ATM **W:** Savings of America, Shell, Texas Commerce Bank
Other **W: 4 Less Food Store, Gulfgate Animal Hospital, Gulfgate Mall, Snow White Laundromat, Stop N Go**

40 Jct.I-610

40A Frontage Rd (Northbound)

40B E.I-610 To TX 225, Pasad., S. TX 35, W.I-610 Pearland, Alvin (San Jacinto Battleground State Park, Battleship Texas)

40C I-610 West (Northbound, Left Exit)

39 Park Place Blvd, Broadway Blvd
Gas **E:** Gas Station*, Texaco*
W: Phillips 66*, Shell*, Texaco*
Food **E:** Dante Italian Cuisine, Del Sol Bakery, Ice Cream, Deli, & Gifts, Kim Long Restaurant, La Ojarasca Bakery, Nuevo Mexico, Taquerias Del Sol, Tony Mandola's Blue Oyster Bar
W: Blimpie's Subs (Texaco), Kelley's Country Cookin', Mac's Chicken & Rice, May Moon Cafe Noemi's Tacos, Sun Sai Gai Restaurant
AServ **E:** Crafts Car Center Auto Parts, Gulf Gate Auto Sales, Jimenez Auto Repair, Texaco, Trans Auto Body Complete Auto Repair, Tygco Cars N' Vett
W: Earl Scheib Paint & Body, Kelly's Auto Repair Tire, Texaco
Other **E: Food Valley Grocery, P & P Washateria, Public Library**
W: Food Picante, Park Place Pharmacy, US Post Office, Washateria

38B Bellfort Avenue, Howard Dr. (Southbound, East Services Are At Exit 38)
Gas **W:** Citgo*(D) (24 hrs)
Food **W:** Don Tako, Luby's Cafeteria, Taj Mahal Restaurant
AServ **W:** Pep Boys Auto Center
Other **W: A&M Grocery,** Car Wash, **Food Spot, Hobb Airport, St. Mary's Washateria**

38 TX 3, Monroe Road, Howard Dr.
Gas **E:** Chevron*(D)(CW), Diamond Shamrock*, Exxon*(Shell*, Stopn Go*
W: Chevron*, Get nGo*, Texaco*(D)
Food **E:** Dairy Queen, Halbrook's Bar & Grill, Jack-in-the-Box, Keiko Seafood, Lam's Chinese Restaurant, Ninfa's Mexican, Old Galveston Seafood Restaurant, Snowflake Donuts, The Captain's Half Shell Oyster Bar, Tijuana's Tex-Mex Grill, Wendy's
W: Luby's Cafeteria, Luther's Bar-B-Q, Omega

Bold red print shows RV & Bus parking available or nearby

EXIT TEXAS

Restaurant (24 Hrs), Oriental Gourmet, Subway
Lodg W: ◆ Quality Inn & Suites, Smile Inn
AServ E: Carquest Parts Center, Chevron
W: Firestone Tire & Auto
RVCamp E: **Best Buy RV**
ATM E: Bank One, Diamond Shamrock
W: Chevron
Other E: **Glenbrook Animal Clinic, Hobby Skate, Wikler Food Market, Winkler Food Mart & Deli**
W: **Best Marine, Hobby Airport**

36 College Avenue, Airport Blvd
Gas E: Shell*(CW) (24 Hrs)
W: Exxon*(CW), Texaco*(D)(CW)
Food E: Aranda's Panaderia Bakery, Dairy Queen, Shipley Do-nuts, Taqueria Arandas, Thirst Parlor, Waffle House, Yummy Chow
W: Damon's (Holiday Inn), Denny's, Kettle, Lucky Dragon Restaurant, McGrath's, Taco Cabana
Lodg E: Airport Inn, DAYS INN ◆ Days Inn
W: ◆ Comfort Inn, ◆ Courtyard by Marriott, ◆ Drury Inn, ◆ Holiday Inn, AAA La Quinta Inn, ◆ Red Roof Inn, AAA Sumner Suites, Super 8 Motel, Travel Inn
AServ E: Castl Transmission & Auto Care, Kari's Garage, Trans City Transmission
W: Fleet Transmission Exchange & Repair, Gulf Freeway Auto Service, Lone Star Automotive, Millennium Autohouse
RVCamp E: **Eastex Camper Sales**
ATM W: Comerica Bank
Other E: **Boots Follmar Marine, Davis Food City, Union Drive-In Grocery**
W: **Boat Stuff, Hobby Airport, R & M The Car Specialist (Car Detail), Stopn Go*, Supertrac Games**

35 Edgebrook Dr, Clearwood Dr
Gas E: Chevron* (24 Hrs), Exxon, Racetrac*
W: Shell*
Food E: Burger King, Burger Mart, Chine One Restuarant, Grandy's, Jack-in-the-Box (24 Hrs), Popeyes, Subway, Taco Bell
W: James Coney Island, Mary Lee Donuts, McDonalds (Playground), Pizza Hut, What A Burger (24 Hrs)
AServ E: David McDavid KIA, Exxon, Gulf Freeway Pontiac GMC Trucks, Master Care Auto Service
RVCamp E: **Terry Vaughn RV's, Thompson RV**
W: **Lone Star RV**(LP)
ATM E: Bank One, Bank of America
W: Shell
Other E: **Animal Emergency Clinic S.E., Eckerd Drugs, Fiesta, Office Depot, See-N-Focus, WalGreens**

34 South Shaver Rd, Almeda - Genoa Rd
Gas E: Conoco
W: Chevron*(CW) (24 Hrs), Exxon*, Mobil*
Food W: Burger King (Playground), General Joe's Chopstix, Pancho's Mexican Buffet, Sciortino's Deli, Wendy's
AServ E: Carlie Thomas Chrys, Plymouth, Jeep, Eagle, Charlie Thomas Acura, Charlie Thomas Hyundai, Isuzu, Conoco, Houston-Pasadena Foreign Car Service, Jay Marks Toyota, McDavid Honda, Saturn of Houston
W: David McDavid Olds, Nissan, Discount Tire Co., NTB, Western Auto
RVCamp W: **Holiday World**
ATM W: Wells Fargo

EXIT TEXAS

Other E: **Almeda Super Rink, Roy's Gas Grills**(LP)
W: **Almeda Mall, Baptist Bookstore,** Coin Car Wash, **OfficeMax, R H Tropical Fish & Pets, Toys "R" Us, Venture**

33 Fuqua Street, Beltway 8 Frtg Rd.
Gas E: Stopn Go*
W: Texaco*(D)(CW)
Food E: Blimpie's Subs (Diamond Shamrock), Chili's, Fuddruckers, King Bo, Luby's Cafeteria, Mexico Lindo, Olive Garden, Schlotzkys Deli, T.G.I. Friday's
W: Black-Eyed Pea, Boston Market, Brown Sugar Bar-B-Q, Casa Ole, CiCi's Pizza, Golden Corral, Joe's Crab Shack (Playground), Lillie's Cantina, Little Caesars Pizza, Long Horn Cafe, Loong Wah, Seafood Plus, Sing Lee Restaurant, Steak & Ale, Subway, TCBY, Taco Bell, Tejas Cafe, What A Burger
AServ E: Bayway Lincoln Mercury
W: AutoNation USA, **Sam's Club,** The Tire Station
ATM E: Diamond Shamrock, Stopn Go, Texas Commerce Bank
W: Bank of America
Other E: **Medicine Man Pharmacy, Sony Theaters, United Optical**
W: **Almeda Mall, Eckerd Drugs, Golf Superstore, Longhorn Golf, Randalls Food & Drugs (24 Hrs), Sam's Club, Sears Hardware**(LP)**, Texas State Optical, The Home Depot**(LP)

32 Sam Houston Tollway
31 FM 2553, Scarsdale Blvd, Beltway 8 Frontage Rd.
Gas W: Exxon*, Shell*(CW) (24 Hrs, CNG Natural Gas)
Food W: Dairy Queen (Playground), Danny's Donuts and Kolaches, McDonalds (Playground), Perry's Grill & Steakhouse, Pho Cong Ly
ATM W: Exxon, My Quang Market
Other E: **Stop & Sock Driving Range**
W: **My Quang Market, Scarsdale Pharmacy, Stop & Gone Groceries**

30 FM 1959, Ellington Field, Dixie Farm Road
Gas E: Conoco*
AServ E: Clear Lake Dodge, Dodge Trucks
RVCamp W: **Morgan Building & Spas**
Med W: ✚ **Hospital**
ATM W: First Community Bank
Other E: **Laundromat**

29 FM 2351, Friendswood, Clear Lake City Blvd (Space Center Houston)
Parks E: **Clear Lake Recreational Center**
Other E: **Space Center Houston**

27 El Dorado Blvd
Gas E: Texaco*(CW)
Food E: Dairy Queen (Playground)
ATM E: Texaco
Other E: Car Wash, **Fire Station, Space Center Houston**
W: **Texas Ice Stadium (Public Ice Skating)**

26 Bay Area Blvd
Gas W: Shell*
Food E: Kettle, Pappas Seafood House, Red Lobster, The Original Taco Cabana
W: Bennigan's, Burger King, Chuck E. Cheese's, Denny's, Marie Callender's Restaurant, McDonalds, Steak & Ale
Lodg E: AAA Best Western

EXIT TEXAS

AServ E: Discount Tire Co., Pro Tech Collision Repair Ctr
W: Sears
Med E: ✚ **Columbia Clear Lake Regional Medical Center**
ATM W: NationsBank, Savings of America
Other E: **Barnes & Noble, Boaters World, Discovery Zone, Sony Theatres, Space Center Houston, Venture**
W: **Baybrook Mall, Bookstop, Eyemasters, OfficeMax, PETsMART, Pet Vet Animal Hosp., Target Department Store, Toys "R" Us**

25 NASA Rd. 1, FM 528, NASA, Alvin
Gas E: Conoco*(CW), Stopn Go*
Food E: Cesar's Cantina, Chili's, CiCi's Pizza, Crazy Cajun Bayou Steakhouse, Enzo's, King Food, Logan Farms Honey-Glazed Hams, Marco's Mexican Restaurant, Mason Jar, Saltgrass Steakhouse, Simply Yogurt, Waffle House
W: Dairy Queen, Hans Mongolian BBQ, Hooters, Luther's Bar-B-Q, Subway, The Boat
Lodg E: ◆ Comfort Inn, **Motel 6**
AServ E: Western Auto
W: Stickerstop, **Wal-Mart**
ATM E: Bank of America, Stopn Go
W: **Wal-Mart**
Parks W: **Challenger 7 Memorial Park**
Other E: **Dollar Cinema, General Cinema, Joshua's Christian Bookstore, Office Depot, Scuba, Space Center Houston, Space Center Souvenirs, The Home Depot**
W: **Wal-Mart (24 Hrs, Pharmacy)**

23 FM 518, Kemah, League City
Gas E: Racetrac*, Texaco(D) (24 Hrs)
W: Exxon*, Mobil*
Food E: Ashley's Donuts, Burger King, Jack-in-the-Box (24 Hrs), KFC, Little Caesars Pizza, New Hunan Buffet, Pancho's Mexican, Subway, Taqueria Arandas, The Pardise Restaurant
W: Bonny's Donuts, **Cracker Barrel,** Hartz Chicken, McDonalds (Playground), Taco Bell, Waffle House, Wendy's
Lodg W: ◆ Super 8 Motel
AServ E: Auto Air, Q-Lube, Texaco
W: Auto Chek
ATM W: Merchant's Bank
Other E: **Academy (Outdoor Store), Box Get, Eckerd Drugs, Kroger Supermarket (Pharmacy, 24 Hrs), M & H One-Hr Photo, Mass Mailing Systems, Pets To Luv, WalGreens**
W: **Discount Mini Mart, New Concept Veterinary Clinic**

22 Calder Drive, Brittany Bay Blvd
RVCamp W: **Camping**
Other W: **Lazer Marine Inc.**

20 FM 646, Santa Fe
Other W: **Public Golf Course**

19 FM 517, Dickinson, San Leon Hughes Rd
Gas E: Diamond Shamrock*, Shell (24 Hrs)
W: Exxon*(CW), Texaco*(D)(CW)
Food E: Jack-in-the-Box, Monterey's Tex-Mex Cafe, Pizza Inn, Szechuan Garden Chinese
W: KFC, Kettle (24 Hrs), McDonalds (Playground), Pizza Hut, Subway, Taco Bell, Wendy's, What A Burger (24 Hrs)
Lodg W: El Rancho Motel
AServ E: Gay Pontiac, GMC Trucks, Shell
ATM E: Guaranty Federal Bank

EXIT — TEXAS

W: Exxon
Other E: Eckerd Drugs, Food King Grocery Store, Speed Queen Laundry (Coin-Op)
W: Animal Care Clinic of Dickinson, Kroger Supermarket (24 Hrs), Mail Boxes Etc, Oasis Pets, Walgreens (1-Hr Photo)

17 Holland Road

16 FM 1764 East, Texas City, College of the Mainland (Southbound)
Med W: ✚ Hospital

15 FM 1764 West, FM 2004, Hitchcock Mall of the Mainland Pkwy
Gas E: Shell*
W: Mobil*, Texaco*(D)
Food E: Jack-in-the-Box
W: Waffle House, What A Burger (24 Hrs)
Lodg E: ◆ Fairfield Inn, ◆ Hampton Inn
RVCamp W: Camping
Med E: ✚ Hospital
ATM W: Mobil, Texaco, Texas First Bank
Other E: Mall of the Mainland

13 Johnny Palmer Rd, Delany Rd
Lodg W: (AAA) Holiday Inn, Pelican Inn
RVCamp W: Camping (RV Dump)
ATM W: Factory Outlet Center
Other W: Factory Outlet Center (see our ad this page)

12 FM 1765, LaMarque
Gas W: Chevron*(CW) (24 Hrs), Texas Discount Gasoline*
Food W: Jack-in-the-Box, Kelley's Country Cookin', Sonic
AServ W: All Pro Automotive, Bunch's Engine Machine Service, Robby's Full Service Auto Repair
RVCamp W: Camping, Mainland Camper Sales

Other W: LaMarque Police Dept.

11 Vauthier Road

10 FM 519, Main Street
FStop W: Texaco*
Gas E: Stopn Go*
Food E: McDonalds (Playground)
AServ E: MCH Truck & Auto Repair
ATM E: Stopn Go
Parks W: Highland Bayou Park, Mahan Park
Other E: LaMarque Police Dept.

9 Frontage Rd (Southbound)

8 Frontage Rd (Northbound)

7 TX3, TX 146, TX 6, Texas City, Bayou Vista, Hitchcock (Northbound Exit Is Ex. 7AB, 7B Is A Left Exit)
Gas W: Diamond Shamrock*, Texaco*

7C Frontage Rd (Northbound)

6 Frontage Road (Southbound)

5 Frontage Road

4 Village of Tiki Island
Gas W: Conoco*(D) (24 Hrs)
Food W: The Grill (Teakwood Marina)
ATM W: Conoco
Other E: Salty's Bait, Tackle
W: Public Boat Ramp, Teakwood Marina

1C FM 188, TX 275, Harborside Drive, Teichman Rd. (Port Of Gavelston)
FStop E: Chevron*(D), Texaco*
Gas E: Conoco*(D)
AServ E: Bob Pagan Toyota, Ford, Rick Perry's Nissan, Plym, Dodge Trucks
TServ E: NCI Diesel Service
ATM E: Texaco
Other W: Payco Marina

1B 71st Street
Lodg E: Motel 6
AServ E: Galveston Auto Center
Other W: Galveston Marine Center

1A Spur 342, 61st Street, West Beach
Gas W: Exxon*(D)
Food W: The Diner On 61st Street
Lodg W: (DAYS INN) Days Inn
AServ W: Hicks Automotive
RVCamp W: Camping
ATM W: Exxon
Parks W: Washington Park Recreation Area
Other W: Galveston County 61st Street Public Boat Ramp)

↑ TEXAS

Begin I-45

I-49 S →

EXIT — LOUISIANA

Begin I-49

↓ LOUISIANA

206 I-20 Junction, Dallas, Monroe

205 Kings Hwy
Gas W: Diamond Shamrock*(CW)
Food W: Long John Silvers, Subway, Taco Bell
AServ E: Sears
Med W: ✚ Louisiana State University Medical Center, ✚ Shriners Hospital for Children
ATM E: Bank One
Other E: Mall St. Vincent, Tourist Information
W: Biomedical Research Institute

203 Hollywood Ave, Pierremont Rd.
Gas W: Fina*
Food W: Lee's Country Kitchen
Other E: Boyce's Pet Land

202 LA 511E, 70th Street
Gas W: Chevron*, Circle K*
Food E: Mr. Swiss Hot Dog Shop
W: Sonic
AServ E: Grant's Automotive
W: J & J Auto Repairs, Jones Garage, Kaddo Alternator & Starter Shop, Radiator
ATM W: Circle K
Other E: Get 'N Go, Medic Pharmacy
W: Fire Station

201 LA 3132 To Dallas To Texarkana (Limited Access Hwy)

199 LA 526, Bert Kouns - LSU Shreveport
FStop E: Chevron*(D)
Gas E: Citgo*
Food E: Burger King (Playground), KFC, Taco Bell (Chevron), Wendy's
ATM E: Mansfield Bank
Other E: The Home Depot(LP)

191 DeSoto Parish16, Stonewall

186 LA 175, Kingston, Frierson

177 LA 509, Carmel
TStop E: Country Auto Truck Stop* (Showers)
Food E: Restaurant (Country Auto Truck Stop)

172 U.S. 84, Mansfield, Grand Bayou

169 Asseff Road

162 US 371, LA 177, Pleasant Hill, Coushatta

155 LA 174, Ajax, Lake End
FStop W: Spauldings*

148 LA 485, Allen, Powhatan

142 Natchitoches Parish 547, Posey Road

138 LA 6, Many, Natchitoches
TStop W: Lott's O' Luck Truck Stop* (Fuel-Chevron, FedEx Drop)
Gas E: Exxon*, Shell*
W: Texaco*(D)
Food E: Shoney's
W: Burger King (Playground), Lott's O' Luck (24 Hrs, Lott's O' Luck TS), McDonalds
Lodg E: Best Western
W: Comfort Inn
RVCamp W: Nakatosh Campground
Med E: ✚ Hospital
Other W: Kisatchie National Forest District Office

132 LA 478, Cnty. 620

127 LA 120, Flora, Cypress

119 LA 119, Derry Gorum, Cloutierville

113 LA 490, Chopin

107 Lena

103 LA 8 West, Flatwoods
FStop E: Texaco* (24 Hrs)
Other E: Cotile Lake (8.6 Miles)
W: Youth Rodeo

99 LA 8 East, LA 1200, Boyce

98 LA 1, Boyce

94 Rapides Station Rd
Gas E: Shari Sue's*
Other E: Louisiana Boating Center

90 LA 498, Alexandria Intern'l Airport
FStop W: Chevron* (24 Hrs)
Food W: Burger King (Playground), McDonalds (Playground)
Lodg W: Guesthouse Inn, La Quinta Inn
RVCamp W: Cabana RV Park

86 US 71, US 165, MacArthur Dr

85B Monoe St., Medical Center Drive
Med E: ✚ Rapides Regional Medical Center

85A 10th St., Downtown Alexandria, To

Bold red print shows RV & Bus parking available or nearby

XIT LOUISIANA

LA 1, Elliot St
AServ **E:** M & T Automotive
Other **E:** Alexandria Fire Dept., Alexandria Library, Alexandria Police Dept.

4 North US 167, LA 28, LA 1, Pineville Expwy, Casson St.

3 Broadway Avenue
Gas **E:** Fina*(D), Lesser*
AServ **E:** Auto Service Shop (Unnamed)

0 U.S. 71, U.S. 167, MacArthur Drive (Lsu At Alexandria)
Lodg **W:** Best Western (see our ad this page), Rodeway Inn (see our ad this page)

3 LA 3265, Woodworth, Rapides Parish 22
FStop **W:** Exxon* (24 Hrs)
Food **W:** Blimpie's Subs (Exxon)
ATM **W:** Exxon

6 LA 112, Forest Hill, Lecompte
Food **W:** Nell's Cajun Kitchen (Thurs.-Sun.)

U.S. 167, Turkey Creek, Meeker (Loyd Hall Plantation)

6 LA 181, Cheneyville

3 LA 115, Bunkie
Med **E:** ✚ Hospital (6 Miles)

6 LA 106, St. Landry

0 LA 29, Ville Platte
Gas **E:** Texaco

4) Rest Area - RR, Phones, Picnic, Boat Ramp, Pet Rest, Grills (Both Directions)

7 LA 10, Lebeau

5 LA 103, Washington, Port Barre

3 U.S. 167, LA 744, Ville Platte
TStop **E:** Gold Rush Truck Stop* (Fuel-Texaco)
Med **W:** ✚ Hospital

AB U.S. 190, Baton Rouge, Opelousas
Food **W:** Mikey's Donut King
Med **W:** ✚ Opelousas General Hospital
Other **W:** Jim Bowie Museum, Ray's Grocery, Tourist Information

LA 31, Cresswell Lane
Gas **W:** Shell(CW) (24 Hrs)
Food **E:** The Back Porch Cafe
W: Burger King (Playground), Cresswell Lane Restaurant, McDonalds (Playground), Mr. Gatti's Pizza, Pizza Hut, Ryan's Steakhouse, Subway, The Big Scoop, Wendy's
Lodg **W:** ◆ Best Western
AServ **E:** Sterling Plymouth
W: Diesi Pontiac, Firestone Tire & Auto, Shell, Top Dog Exhaust Center, Wal-Mart

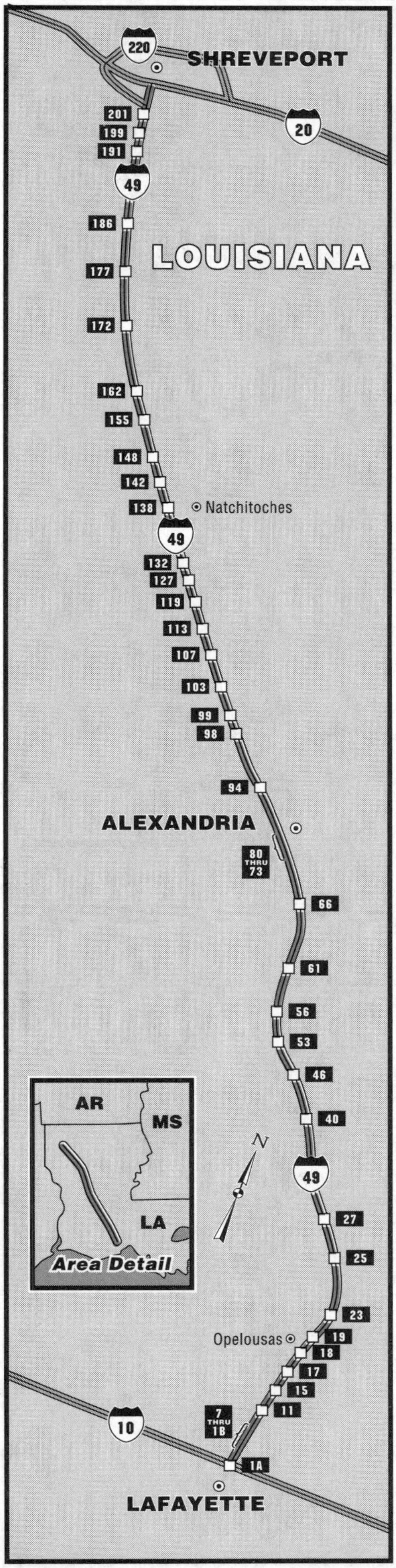

EXIT LOUISIANA

ATM **W:** American Bank
Other **W:** Cinema-IV, DelChamps Supermarket, Eckerd Drugs, K & B Drugs, Vo-Tech Office Supply, Wal-Mart (Pharmacy, 1-Hr Photo), Whipps Family Pharmacy

17 Judson Walsh Drive
Gas **E:** Texaco*(D)(CW)(LP)
Food **E:** Hil-Win Crawfish, Restaurant (Texaco)
ATM **E:** Texaco

15 LA 3233, Harry Guilbeau Road
Lodg **W:** Quality Inn
Med **W:** ✚ Columbia Doctors Hospital of Opelousas

11 LA 93, Sunset, Grand Coteau
TStop **E:** Citgo Truck/Auto Plaza*
Gas **E:** Chevron*
Food **E:** Beau Chene (Citgo TS, 24 Hrs)
RVCamp **W:** Courvelle RV
Other **E:** Baronne Veterinary Clinic, St. Cyr Clinic & Pharmacy

7 LA 182
Food **E:** Prudhomme's Cajun Cafe
Lodg **W:** Acadian Village Inn
AServ **W:** Victory

4 LA 726, Carencro
Gas **W:** Chevron(CW)
Food **W:** Burger King (Playground)
AServ **W:** PPG Collision Repair Center
TServ **W:** Acadiana Mack & LA Kenworth
ATM **W:** First National Bank
Other **W:** Carencro Eye Clinic, US Post Office

2 LA 98, Gloria Switch Road
Gas **E:** Chevron*, Citgo*, Shell*
W: Texaco*
Food **E:** Blimpie's Subs (Citgo)
W: Church's Fried Chicken (Texaco), Picante
AServ **E:** All American & Foreign Used Parts & Cars, Ben's Auto Repair, Joe's Transmission Service
RVCamp **E:** Floyd's RV Park, Jackie Edgar RV Center Inc., Trade Winds Camping Center
ATM **E:** Citgo
W: Texaco
Other **E:** LA Museum of Military History, Stemmans Pet Supplies, The Book Rack

1B Point Des, Mouton RD
TStop **W:** Golden Palace Truck Stop*(D), International Truck
Gas **E:** Exxon*(D), Texaco*(D)
Food **E:** Subway (In Texaco)
Lodg **E:** Plantation Motor Inn
AServ **W:** M&J Valve Services, Ronnie's Car Care
RVCamp **E:** Gauthier Mobile Homes (Parts & Service)
Other **E:** LA State Police

1A I-10 West, Lake Charles
TStop **W:** Golden Palace Truck Stop* (24 Hrs)
Gas **E:** Exxon*(D), Texaco*(D)
Food **E:** Subway (Texaco), The Tropics (Plantation Motor Inn)
W: Golden Palace (Golden Palace TS)
Lodg **E:** Plantation Motor Inn
AServ **W:** Ronnie's Car Care, Stelly's Auto Repair
TServ **W:** International Trucks, Stelly's Auto & Truck Repair
RVCamp **E:** Gauthier Homes & RV Center
ATM **W:** Golden Palace TS
Other **E:** Louisiana State Police Region II HQ

(0) I-10 East, Baton Rouge (Unnumbered Exit)

↑LOUISIANA

EXIT — ILLINOIS

Begin I-55

↓ILLINOIS

293D Martin Luther King Dr
- Gas E: Amoco
- Med E: ✚ Hospital

293B Junction I-90/94 East, Indiana (Southbound)

293A Cermak Road, Chinatown

292 Junction I-90/94 (Westbound)

290 Damen Ave, Ashland Ave.

289 California Ave (Northbound)
- FStop E: **Speedway* (RV Dump)**

288 Kedzie Ave (Southbound)

287 Pulaski Road

286 Cicero Ave., IL 50
- Gas E: Amoco

285 Central Ave
- Gas E: Gas Center*(D), Union 76*
- Food E: Donald's Hot Dogs
- TServ E: **International Truck**

283 IL 43, Harlem Ave
- Gas E: Shell*
- Food E: Arby's, Baskin Robbins, Beefy's Burgers, Burger King, Dunkin Donuts, Fox's Pizza, Marisco's Mexican, Subway, Zigs Italian Sausage & Burgers
- Other E: **WalGreens**

282AB IL 171, 1st Ave. (Difficult Reaccess)

279AB U.S 12, U.S 20, U.S 45, LaGrange Road (Difficult Reaccess)

277AB Junction 294 , Wisconsin, Indiana (Toll Road)

276C Joliet Road

276AB County Line Road
- Food E: China King Restaurant, Mack's & Erma's Restaurant
- Lodg E: Best Western
- ATM E: Harris Bank
- Other E: **Ceebee's Grocery**

274 IL 83, Kingery Ave
- Gas E: Shell*
  W: Mobil*(CW)
- Food E: Dunkin Donuts, Joel Falco's Pizza
  W: Denny's
- Lodg W: Budgetel Inn, ◆ Fairfield, AAA Holiday Inn
- AServ E: Buridge Car Care
- ATM E: State Bank of Countryside

273AB Cass Ave
- Gas W: Shell*(CW)
- Food W: Ripples Dining

271AB Lemont Road
- Gas W: Shell*

269 Junction I-355 North , S Joliet Road (Toll Road)

267 IL 53, Bolingbrook
- TStop E: **76 Auto/Truck Plaza**(SCALES)
- Gas E: Amoco*
  W: Shell*, Speedway*
- Food E: Bob Evans Restaurant, McDonalds, Restaurant (76)
  W: Baskin Robbins, Denny's, Family Square

EXIT — ILLINOIS

EXIT — ILLINOIS

Restaurant, Golden Chopsticks, Hardee's, Popeye's Chicken, Salsa & Tortillas, Subway
- Lodg E: ◆ Comfort Inn, Ramada Limited, ◆ Super 8 Motel
- AServ W: Merlin Muffler & Brake
- Med W: ✚ **Ogdon Medical Clinic**
- Other W: **WalGreens**
- RVCamp W: **Camping World (see our ad this page)**

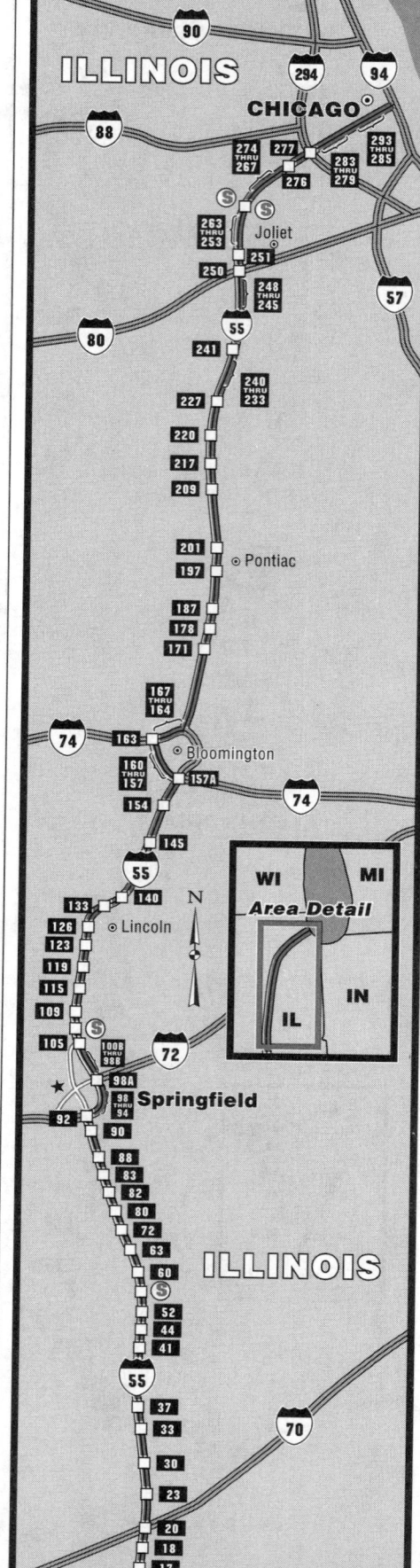

(265) **Weigh Station (Both Directions)**

263 Weber Road
- Gas W: Shell*(CW) (ATM)
- Food W: Cracker Barrel, Wendy's
- Lodg W: ◆ Country Inn & Suites, Howard Johnson

261 IL 126 West, Plainfield (Southbound, Northbound Reaccess)

257 U.S 30, Aurora, Joliet
- TStop W: **Phillips 66**(SCALES)
- Gas E: Shell*
  W: Amoco*
- Food E: Applebee's, Burger King, Diamans Family Restaurant, Lone Star Steakhouse, McDonalds, Old Country Buffet, Pizza Hut, Red Lobster, Steak & Shake, Taco Bell, Wendy's
  W: **Restaurant (Phillips 66), Subway (Phillips 66)**
- Lodg E: ◆ **Comfort Inn,** ◆ Fairfield Inn, Hampton Inn, ◆ **Ramada,** ◆ Super 8 Motel
- AServ E: Discount Tire Co.
- Other E: **Joliet Mall, Target**

253 U.S 52, Joliet, Shorewood
- FStop W: **Gas Center Phillips 66***(D)
- Gas E: Amoco*, Shell*
- Food E: Dakota Steakhouse, **McDonalds,** Restaurant (Fireside Resort), Wendy's
  W: Burrito King
- Lodg E: DAYS INN **Days Inn,** Elks Motel, Fireside Resort
- AServ E: Amoco (ATM)
- RVCamp E: **Rick's RV**
- Other E: **Dental Clinic**

251 IL 59, Shorewood, Plainfield (Northbound, Reaccess Southbound)

250AB Junction I-80, Des Moines, Toledo (Difficult Reaccess)

248 U.S 6, Moriss, Joliet
- Gas E: Speedway*(LP)
  W: Amoco*
- Food W: McDonalds (Amoco), Pit Stop Fast Food
- Lodg E: **Manor House,** Manor Motel
- ATM E: Tri-County Bank

247 Bluff Road, Channahon

245 Arsenal Road

241 Wilmington

240 Lorenzo Road
- TStop W: **Citgo* (ATM)**
- FStop E: **Phillips 66**
- Food W: **River Restaurant (Citgo)**
- Lodg W: Motel 55

Bold red print shows RV & Bus parking available or nearby

EXIT	ILLINOIS
238	IL 129 South, Braidwood
236	IL 113, Coal City, Kankakee
Food	E: The Good Table
233	Reed Road
Gas	E: Marathon*
Lodg	E: Sands Motel
227	IL 53, Gardner
Gas	W: Amoco*
220	IL 47, Morris, Dwight
FStop	E: Amoco* (ATM), Clark*
Gas	E: Fuel 24*(D)
Food	E: Burger King (Amoco), Dwight's Restaurant, Hardee's, McDonalds, Pete's Pancake House, Subway
Lodg	E: Super 8 Motel, Traffic Inn
217	IL 17, Streator, Kankakee
TStop	E: Shell* (ATM)
Gas	E: Marathon
Food	E: Harvest Table Restaurant
AServ	E: NAPA Auto Parts
TServ	E: Bridgestone Tire & Auto (Shell), Shell
209	Odell
Gas	E: Amoco
AServ	E: Amoco
201	IL 23, Pontiac, Streator
RVCamp	E: 4H Camping (April - Nov)
197	IL 116, Pontiac, Flanagan
FStop	E: Amoco*(D)
Food	E: Arby's, Burger King, Busters Family Restaurant, Taco Bell, Wendy's
Lodg	E: AAA Comfort Inn, AAA Holiday Inn Express, Super 8 Motel
AServ	E: Auto Zone Auto Parts
Other	E: Wal-Mart
(194)	Rest Area - RR, Phones P (Southbound)
(193)	Rest Area - RR, Phones, Picnics, HF P (Northbound)
187	U.S 24, Chenoa, El Paso
Gas	E: Amoco*(D), Apollo Mart*, Casey's General Store, Shell* (ATM)
Food	E: Chenoa Family Restaurant, Dairy Queen (Shell), McDonalds (Apollo)
Other	E: Coin Car Wash
178	Lexington
171	Towanda
167	I-55 Bus Veteran's Pkwy, Airport
165	Bus U.S 51, Bloomington, Normal
Gas	E: Amoco*(CW), Shell*
Food	E: Casey's Restaurant, Denny's
Lodg	E: Best Western, Holiday Inn, Motel 6, Super 8 Motel
TServ	W: International Truck
164	I-39, U.S 51, Rockford
163	Junction I-74 West, Champaign, Peoria
160AB	U.S 150, Il 9, Pekin, Market St
TStop	E: 76 Auto/Truck Plaza*, Pilot Travel Center(SCALES)
FStop	E: Speedway*(D)
Gas	E: Amoco*, Freedom*(D)(LP), Phillips 66, Shell*
Food	E: Arby's, Burger King, Cracker Barrel, KFC, McDonalds (Indoor Playground), Restaurant (76), Subway, Taco Bell, The Summit, Wendy's (Pilot Travel Center)
	W: Country Kitchen, Steak 'N Shake
Lodg	E: Best Inns of America, Comfort Inn, Days Inn, Howard Johnson
AServ	E: Shell
TWash	E: Pilot Travel Center
ATM	E: Bank One, Phillips 66
157B	Bus I-55, U.S 51, Veteran's Pkwy, Airport
Gas	E: Shell*
Food	E: McDonalds, Restaurant (Parkway Inn)
Lodg	E: Parkway Inn
157A	Jct I-74 E, U.S 51, Indianapolis, Decatur
154	Shirley
(149)	Rest Area - RR, Phones, Vending, Picnic, Playground P (Large Truck Parking)
145	US 36 McLean, Heyworth
TStop	W: AM Best*(SCALES) (Phillips 66, Mobil)
Gas	W: Citgo*(LP)
Food	W: AM Best, McDonalds, Restaurant (Citgo)
TServ	W: AM Best
ATM	W: Citgo
140	Atlanta, Lawndale
FStop	W: Phillips 66(D)
Gas	W: Phillips 66*(D)
Food	W: Country Aire Restaurant (Phillips 66)
Lodg	W: ◆ Park Inn
AServ	W: NAPA Auto Parts
RVCamp	E: Campground
133	Lincoln, I-55 Bus, Landale
RVCamp	E: Camping
Med	E: ✚ Hospital
127	Junction I-155 , Peoria, Hartsburg
126	IL 10, IL 121, Lincoln, Mason City
FStop	E: Phillips 66*(D)(SCALES)
Gas	E: Ayeroco*, Clark*, Phillips 66
Food	E: A & W Drive-In, Bob's Roast Beef, Bonanza Steakhouse, Burger King, Cafe Woodlawn, Dairy Queen, Daphne's Restaurant, Hardee's, Long John Silvers, Maverick Steakhouse, Taco Bell (Phillips 66 TruckStop), The Tropics, Wendy's
Lodg	E: ◆ Comfort Inn, Crossroads Motel, Holiday Inn Express, Lincoln Motel, Red Wood Motel, Super 8 Motel, The Lincoln Country Inn
AServ	E: Auto Zone Auto Parts, Pennzoil Lube, Phillips 66
ATM	E: Illini Bank, State Bank of Lincoln
Other	E: Car Wash, Eagle Country Supermarket, Kroger Supermarket, Wal-Mart Supercenter (Auto Service)
123	Lincoln, Bus I-55
Med	E: ✚ Hospital
119	Broadwell
RVCamp	E: Camping
115	Elkhart
Gas	W: Shell*
AServ	W: Shell
109	Williamsville, New Salem St
AServ	E: Shell
(107)	Weigh Station (Southbound)
105	I-55 Bus., Sherman
Gas	W: Amoco, Shell*
Food	W: Antonio's Pizza, Cancun Mexican, Sam's Too Italian, Subway
AServ	W: Amoco
Other	W: Car Wash
(104)	Rest Area - RR, Picnic, HF, Phones P (Southbound)
(102)	Rest Area - RR, Picnic, HF, Phones P (Northbound)
100B	Sangamon Ave., Clinton, Capitol Airport, Springfield
Gas	W: Shell*(D)(CW)(LP), Speedway*(D)(LP)
Food	E: Citgo
	W: Arby's, Burger King, Hardee's, Parkway Cafe
Lodg	W: Ramada Inn Limited
TServ	E: Kenworth(SCALES)
Other	W: Ferrellgas Liquid Propane
100A	IL 54, Clinton, Junction Sangamon Avenue
98B	IL 97 West, Clear Lake Avenue
98A	Junction I-72 East, U.S 36 East, Decatur, Champaign, Urbana
96A	IL 29 South, Taylorville
96B	IL 29 North, S Grand Avenue, Springfield
94	Stevenson Road, Eastlake Dr, Springfield
92AB	U.S 36 West, I-55 Business, 6th St., Jacksonville, Junction
Gas	W: Shell
Food	W: Heritage House Smorgasbord, McDonalds
Lodg	W: Holiday Inn, Park Inn, Super 8 Motel
AServ	W: Shell
Med	W: ✚ Hospital
90	Toronto Road, Springfield
Gas	E: Amoco*, Shell*
Food	E: Antonio's Pizza, Cracker Barrel, Hen House, Hen House Restaurant, Muchachos Mexican, Subway
	W: Johnny's Pizza
Lodg	E: ◆ Budgetel Inn, Motel 6, ◆ Ramada
Other	E: Car Wash
88	Eastlake Dr., Chatham
RVCamp	E: KOA Campground
	W: Holiday RV Park
Other	E: Lincoln Memorial Garden & Nature Center, State Patrol Post, State Police Training Academy
83	Glenarm
82	IL 104, Pawnee, Auburn
FStop	W: Mobil*(D)
Food	W: Trucker's Homestead Restaurant
TServ	W: Goodyear Truck Service
80	Divernon
Gas	W: Citgo*, Phillips 66*
Food	W: Restaurant (Citgo)
72	Farmersville, Girard
FStop	W: Mobil*(D)
Gas	W: Shell*
Food	W: Art's Restaurant, Subway (Mobil)
Lodg	W: Art's Motel
Other	W: Truck & Car Wash (Mobile)
(65)	Rest Area - RR, Phones P

Bold red print shows RV & Bus parking available or nearby

EXIT ILLINOIS

63 IL 48, IL 127, Raymond, Waggoner, Morrisonville, Taylorville

60 IL 108, Carlinville
- FStop **W: Shell***
- Gas W: Shell*
- Food W: Moonlight Cafe
- Lodg W: ◆ **Holiday Inn,** Moonlight Hotel (Moonlight Cafe)
- RVCamp **E: Kamper Kampground**

(56) **Weigh Station (Northbound)**

52 IL 16, Gillespie, Litchfield, Hillsborough, Mattoon
- Gas E: Amoco*(CW), Casey's General Store, Coastal*, Shell*
- Food E: Ariston, Dairy Queen, Hardee's, Jubelt's Bakery, **Long John Silvers (Bus Parking Available),** McDonalds, Pizza Hut, **Ponderosa,** Subway, Taco Bell, Wendy's
- Lodg E: 66 Motel Court, **Best Western, Super 8 Motel**
- AServ E: McKay Auto Parts, Neal Tire & Battery, Snappy Lube
- TServ **E: Goodyear Tire & Auto**
- ATM E: Bank & Trust Company
- Other **E: Kroger Supermarket (ATM), Wal-Mart**
  W: State Patrol Post

44 IL 138, Benld, Mount Olive, White City
- Food **E: CrossRoads Restaurant,** Scheepps
- Lodg **E: Budget 10 Motel**

41 Staunton
- FStop **W: Phillips 66***
- Food **W: Dairy Queen, Diamond's Cafe (Phillips 66)**
- Lodg W: ◆ Super 8 Motel

37 Livingston, New Douglas
- FStop **W: Amoco***
- Lodg W: Country Inn
- TServ **W: Truck Service (Amoco)**
- RVCamp **E: Shady Oak Campground**
- ATM W: First National Bank
- Other W: Coin Car Wash (Amoco)

33 IL 4, Staunton, Lebanon
- Gas W: Schlechte's
- Food W: Shooter's Grill
- AServ W: Schlechte's

30 IL 140, Alton, Greenville
- Gas W: 76*(D), Shell*
- Food E: Inn Keeper
- Lodg E: Inn Keeper
- ATM W: Hamel State Bank

23 IL 143, Edwardsville, Marine
- FStop **E: Citgo***
- Gas E: Mobil*

20AB Junction I-70, I-270, Kansas City, Indianapolis

18 IL 162, Troy
- TStop **E: Shell(SCALES)**
- Gas W: Mobil*(LP)
- Food E: Burger King, Dairy Queen, Hardee's, Jack-In-The-Box, **McDonalds,** Pizza Hut
  W: China Garden, **Cracker Barrel,** Randy's Restaurant, Taco Bell
- Lodg E: Relax Inn
  W: Scottish Inns, ◆ Super 8 Motel
- TServ **E: Amoco Auto/Truck Stop*(SCALES), Freightliner Dealer, Speedco Truck Lube**
- TWash **E: 18 Wheeler's Truck Wash**
- Med **E: ✚ Hospital**
- ATM E: Magna Bank
- Other **E: Super Valu Grocery**

17 U.S 40 East, Saint Jacob (Northbound)

15AB IL 159, Collinsville, Maryville
- Gas W: Conoco*
- Lodg **W: Best Western**
- AServ W: Conoco

(14) **Weigh Station (Southbound)**

11 IL 157, Colinsville, Edwardsville
- Gas E: Amoco*, Mobil*, Shell*(CW)
  W: Moto Mart*(D)
- Food E: China Palace, Denny's (Pear Tree Inn), Hardee's, Long John Silvers, McDonalds, Pizza Hut, The Pub Lounge & Grill, Waffle House, Wendy's
  W: Arby's, Bob Evans Restaurant, Boston Market, Burger King, Cancun Mexican, Dairy Queen, Ponderosa, Porter's (Holiday Inn), Shoney's, Steak & Shake, White Castle Restaurant
- Lodg E: ◆ Best Western, (AAA) **Days Inn (see our ad this page),** Howard Johnson, Motel 6, (AAA) Pear Tree Inn, Travelodge (see our ad this page)
  W: Comfort Inn (see our ad this page, ◆ Drury Inn, Fairfield Inn, Hampton Inn (see our ad this page), Holiday Inn, (AAA) Ramada Limited (see our ad this page), Super 8 Motel
- AServ E: Amoco, Midas Muffler & Brakes
- ATM W: Magna Carta
- Other E: Car Wash
  W: 4 Seasons Auto Wash, **Convention Center, State Patrol Post**

10 Junction I-255, I-270, Memphis

9 Black Lane (Northbound)

6 IL 111, Great River Road, Wood River, Washington, Park
- Gas E: Clark*(D)
- Food E: Rainbo Motel
- Lodg E: Rainbo Court Motel
- AServ E: B&E Auto Repair
- TServ **E: Custom Auto & Truck Accessories**
- RVCamp **E: Safari RV Park**
- Other **E: Cahokia Mounds Historic Sight, FC Grocery & Pharmacy, Foodland Supermarket (Pharmacy), Venture**

4AB IL 203, Granite City, Colinsville Road
- TStop **W: America's Best Truck Stop*(SCALES), Texaco*(SCALES)**
- Food W: Big Duga's Restaurant(SCALES)
- TServ **W: Texaco**
- Other **W: Gateway International Raceway, Volvo Dealer**

3 Exchange Ave (Southbound, Reaccess Northbound Only)

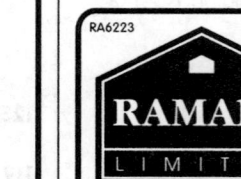

2 Junction I-64, IL 3, St Clair Avenue,

EXIT ILLINOIS/MISSOURI

Louisville
A ML King Bridge, Downtown St Louis (Southbound, Reaccess Northbound Only)
B 3rd Street (Southbound, Reaccess Northbound Only)
IL 3, Sauget

↑ILLINOIS

↓MISSOURI

09AB Junction I-70 West, I-64
08 7th St (Difficult Reaccess)
- **Other** E: Farmer's Market

07B West I-44, Lafayette St
07A Gravois Ave, 12th St (207A Is The Same As 290C On I-44)
- **Gas** E: Citgo*(D)
- **Food** E: A-1 Wok Restaurant, Jack-In-The-Box

90C 12th St, Gravois Ave
- **Gas** E: Citgo*(D)
- **Food** E: A-1 Wok Restaurant, Jack-In-The-Box (24 hrs)

06C Arsenal St
- **Gas** W: Shell*
- **Food** W: Gus' Pretzel Shop
- **ATM** W: Shell
- **Other** W: Gross Variety Inc., Metro. Police Station (Shell)

06B Broadway 3200 So. (Northbound)
- **Gas** E: Citgo*(LP) (Kerosene)
- **AServ** E: Broadway Auto Radiator, Broner Generator Service

06A Potomac St (Northbound, No Northbound Reaccess)
05 Gasconade St (Southbound, Reaccess Northbound Only)
- **Gas** W: 7-11 Convenience Store*
- **Food** W: Don's Meat Market, Gateway Sandwich Shop, Hardee's, Ho's Cuisine, Wendy's
- **Med** W: ✚ Alexian Bros. Hospital
- **Other** W: Coin Laundry, Coin Laundry (24 hrs), Pharmacy, Schnucks Budget Bakery, WalGreens (24 hrs)

04 Broadway
- **Gas** E: Mobil*
  W: Clark*(LP), Sinclair*
- **Food** E: Fat Richie's
  W: McDonalds
- **AServ** E: Mobil

03 Bates St
- **Gas** W: Clark*, Standard Amoco
- **Food** W: Paul's Pizza, Rosario's Pizza & Deli
- **AServ** W: Amoco (Standard)
- **ATM** W: 7-11 Convenience Store
- **Other** W: 7-11 Convenience Store

02C Loughborough Avenue
- **Other** W: Schnucks 24 hr Supercenter

202B Germania Avenue (Reaccess Northbound Only)
202A Carondolet Blvd (Northbound, Reaccess Southbound Only)
201B Weber Road (Southbound, Reaccess Northbound Only)
201A Bayless Avenue
- **Gas** E: Amoco* (24 hrs), Crystal Clean Car Wash & Auto Care(CW)
  W: Amoco (24 hrs), Citgo* (711 Convenience Store), Mobil*(D), Shell (24 hrs)
- **Food** E: McDonalds

EXIT MISSOURI

- W: Baskin Robbins, China Wok, Dairy Queen, Fu Wah Chinese, Jack-In-The-Box, Subway, Taco Bell, The Donut House
- **AServ** W: A to Z Auto Parts, Amoco, BF Goodrich, Shell
- **ATM** W: NationsBank, Southside National Bank
- **Other** E: Bayless Hardware(LP)
  W: Eberhart Pharmacy, Garner's Market, Maytag Laundry (Coin Laundry), Sav-A-Lot Grocery, Weber Road Branch - St. Louis County Library

200 Union Road (Southbound, Reaccess Northbound Only)
199 Reavis, Barracks Road
- **Gas** E: Shell*, Sinclair(LP)
- **Food** E: China Express Restaurant*, Papa's Steakhouse
  W: Imo's Pizza
- **AServ** E: Reavis Auto Repair, Shell
- **ATM** E: Mercantile Bank
- **Other** E: Plaza Pharmacy, V & E Pet Grooming & Supplies

197 U.S 61, U.S 50, U.S 67, Lindbergh Blvd
- **Gas** E: Amoco*, Phillips 66*(CW)
- **Food** E: Honey-Baked Ham, Hot Shots Sports Bar & Grill, KFC, R.T. Furr's Restaurant, Ruby Tuesday, Tucker's Place
  W: Bob Evans Restaurant, Casa Gallardo, Denny's, Ponderosa
- **Lodg** E: Holiday Inn
  W: (AAA) Motel 6, ◆ Oak Grove Inn
- **AServ** E: Dobbs Tire & Auto, South Town Dodge Dealership
  W: Saturn of South County
- **ATM** E: Roosevelt Bank
- **Other** E: South Country Center, Walden Books
  W: Aldi Grocery

196AB 196A - I-255E, Chicago, US 61, 67, 190B - I-270W, Kansas
- **Lodg** W: Days Inn (see our ad this page)

195 Butler Hill Road
- **Gas** E: Amoco*, Citgo*, Phillips 66*(CW) (Kerosene)
  W: Sinclair
- **Food** E: Ditto's Pizza, Lucky's Chinese Restaurant
  W: Burger King, Frailey's Southtown Grill, Hardee's, Taco Bell, Waffle House
- **Lodg** E: ◆ Holiday Inn
- **AServ** W: 55 Tires & Service, Sinclair
- **ATM** E: Phillips 66
  W: Magna Bank, Schnucks Grocery Store
- **Other** W: Schnucks Grocery Store(LP) (24 hrs, 1 hr photo)

EXIT MISSOURI

193 Meramec Bottom Road
- **Gas** E: QuikTrip*
- **Food** E: Cracker Barrel
- **Lodg** E: 55-South (Best Western)
- **AServ** E: Extra Mile Service
- **ATM** E: South Commercial Bank

191 MO 141, Arnold Fenton
- **Gas** E: Amoco*, Citgo*, QuikTrip*, Shell* (24 hrs)
  W: Citgo (Kerosene), Phillips 66*(D)
- **Food** E: Applebee's, Baskin Robbins, Burger King, Captain D's Seafood, Denny's, Fazoli's Italian Food, Hardee's, Lee's Famous Recipe Chicken, McDonalds, Pizza Hut Delivery, Rally's Hamburgers, Ruma's Deli, Schnucks Restaurant (Schnucks Supercenter), Shoney's, Steak 'N Shake, Taco Bell, The Pasta House Company, Wendy's
- **Lodg** E: ◆ Drury Inn
- **AServ** E: Instant Oil Change, Jefferson County Auto Parts, Jerry's Auto Repair, Jiffy Lube, Wal-Mart
  W: Best Auto, Brakes
- **ATM** E: Citgo, Commerce Bank, Grandpa's, Schnucks Supercenter
  W: 7-11 Convenience Store, UMB Bank
- **Other** E: CPI Photo Finish (1-Hr. Photo), Grandpa's(LP), K-Mart, National Grocery (24 hrs, w/ pharmacy), Pam's Books, Schnucks Supercenter (w/pharmacy, 24 Hrs), Shop'n Save, Wal-Mart
  W: 7-11 Convenience Store (24 hrs), Warson Coin Laundry

190 Richardson Road
- **Gas** E: Citgo*(D)(CW)(LP), Shell*(D)(CW)
  W: Citgo* (711 Convenience Store), Shell*(D)(CW) (Kerosene)
- **Food** E: Sonic
  W: Blimpie's Subs (Shell), Don's Donuts, McDonalds, Pizza Plus
- **AServ** E: Reuther Ford & Mazda
- **ATM** E: Heartland Bank, Lemay Bank
  W: Citgo, Shell
- **Other** E: Car Wash, Sav-A-Lot Grocery

186 Imperial, Kimmswick
- **Gas** E: Shell*(D)(CW)
- **Food** E: Main Street Bar & Grill
- **AServ** E: Lambert's Custom Muffler
- **ATM** E: Shell, Southern Commercial Bank
- **Other** E: Herrel's Bestway, Imperial Produce
  W: Mastadon Historic Site

185 Barnhart, Antonia, Rt. M
- **Gas** W: Citgo* (711), Citgo*
- **Food** E: Tressel Cafe
  W: Ronni's Pizza & Grill
- **AServ** W: Wylde Automotive
- **RVCamp** W: KOA Campground (2.5 Miles)
- **ATM** W: Citgo
- **Other** W: US Post Office

(185) Weigh Station (Both Directions)
180 Rt Z, Hillsboro, Pevely
- **TStop** W: McStop*(LP)(SCALES) (Phillips 66)
- **FStop** W: Mr Fuel*
- **Gas** E: Coastal*, Sinclair(LP)
- **Food** E: Bobby Tom's BBQ, Burger King, Domino's Pizza, Subway, The Kitchen (24 hrs)
  W: McDonalds
- **Lodg** W: Gateway Inn
- **AServ** E: Pevely Plaza Auto Parts, Sinclair
- **TServ** W: I-55 Towing & Repair
- **RVCamp** W: KOA Campground (2.5 Miles)
- **ATM** E: NationsBank

Bold red print shows RV & Bus parking available or nearby

EXIT MISSOURI

W: McStop
Other E: Oakie's Coin Laundry, Queen's IGA Store
178 Business Loop I-55, Herculaneum
TStop E: QT*(SCALES)
Food E: Wendy's (Truck Stop)
AServ W: Sapaugh Chev, Olds, Cad, Buick, Pont, GMC Trucks
Other W: Animal Clinic
175 Festus, Crystal City, Route A
FStop E: Phillips 66*(CW)
W: Coastal*(D)(LP) (24 hrs)
Gas W: Citgo*(LP) (711)
Food E: Bob Evans Restaurant, Captain D's Seafood, Dohack's Family Restaurant, Fazoli's Italian Food, Imo's Pizza, McDonalds, Steak 'N Shake, Taco Bell, White Castle Restaurant
W: Domino's Pizza (Citgo), Hardee's, Hot Stuff Pizza
Lodg E: ◆ Drury Inn, (AAA) Holiday Inn
W: ◆ Budgetel Inn
AServ E: Parts America, Twin City Toyota
ATM E: Commerce Bank, NationsBank
W: Citgo, Coastal
Other E: Eye Clinic Optical, Schnucks 24 Hr Supercenter(LP)
W: Rainbo Car Wash
174B US 67 South, Park Hills
AServ W: Pippin Towing, Auto & Truck Repair (24 hrs)
Med E: ✚ Hospital
174A U.S 67 North, U.S 61 Crystal City, Festus
Med E: ✚ Hospital
170 U.S 61
FStop W: Citgo*(LP) (Kerosene)
Food W: Tim and Terry's 61 Steak House
162 Rt DD, Rt OO
Food W: The Corner Grill
(160) Rest Area - RR, Phones, Picnic (P) (Both Directions)
157 Rt Y, Bloomsdale
FStop W: Texaco*(LP)
Gas E: Texaco*(CW)(LP)
AServ E: Bloomsdale Tire
Other E: Bi-Rite Food Stores, US Post Office
154 Rt O, St Genevieve, Rocky Ridge
150 MO 32, Rt B, Rt A, St Genevieve, Farmington
Gas E: Amoco* (24 hrs)
Food E: Dairy Queen (Playground)
Med E: ✚ Hospital
Parks W: Hawn State Park (11 Miles)
Other E: Felix Valle Historic Site
143 Rt M, Rt N, Ozora
TStop W: J & N Truck Stop*(LP)(SCALES) (Showers, Laundromat)
Lodg W: (AAA) Family Budget Inn (J & N Truck Stop)
AServ W: J & N Truck Stop
TServ W: J & N Truck Stop
TWash W: J & N Truck Stop (24 hrs)
ATM W: J & N Truck Stop
141 Rt Z, St Mary
135 Rt M, Brewer
Gas E: Empire Gas(LP)
129 MO 51, Perryville
Gas E: Amoco (24 Hrs), Conoco*(D) (24 Hrs)
W: Shell*(LP)
Food E: Burger King, KFC, McDonalds (Playground), Taco Bell
W: Dairy Queen (Playground), Jer's Restaurant
Lodg W: Colonial Inn Best Western

EXIT MISSOURI

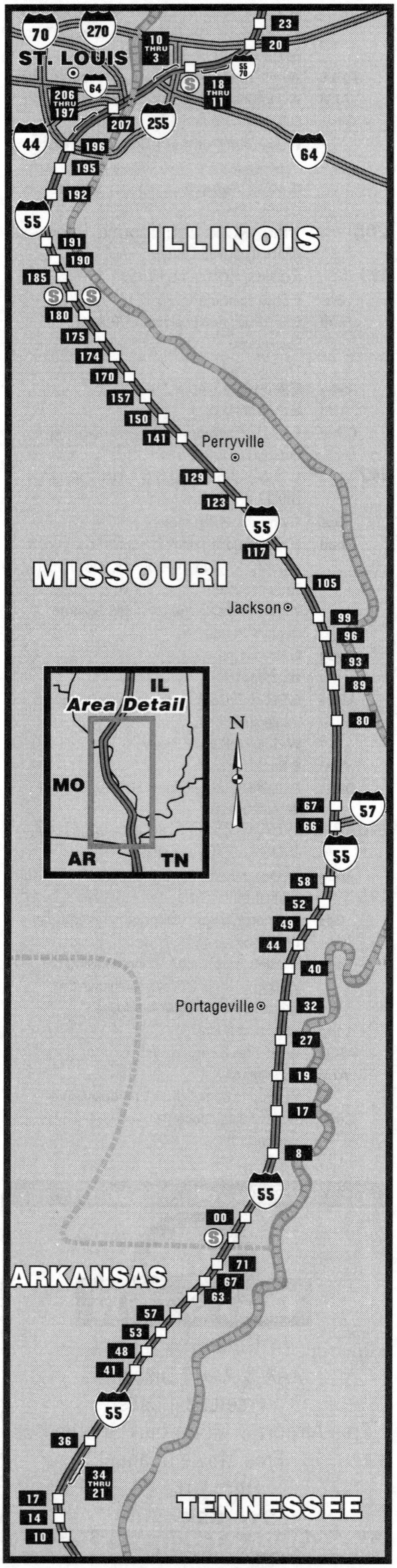

EXIT MISSOURI

AServ E: Amoco, McDowell Ford Dealership
W: Kellar GM Dealership
RVCamp W: KOA Campground (1 Mile)
ATM E: Conoco, Union Planters Bank
W: Shell
Other E: Wal-Mart
W: Heartland Outdoors Bass Pro Shop
123 Rt B, Biehle
FStop W: Phillips 66*(LP)
Food W: Country Kettle
117 Rt KK, Appleton
Gas E: Sewing's
Food E: Sewing's
AServ E: Sewing's
(110) Rest Area - RR, Phones, Picnic, Vending (P) (Both Directions)
105 U.S 61, Business Loop I-55, Fruitland, Jackson
FStop E: Phillips 66*
W: Citgo*
Gas W: Spee D's(LP)
Food W: Bavarian Haus, Bert's B-B-Q, Dairy Queen
Lodg W: ◆ Drury Inn & Suites
99 US 61, I-55 Bus., MO 34, Cape Girardeau, Jackson
Other E: Missouri Veterans Home
96 Rt K, Cape Girardeau, Gordonville
Gas E: Citgo*
W: Shell* (24 Hrs)
Food E: Blimpie's Subs, Burger King, Cedar Street Restaurant & Bar (Drury Lodge), Cracker Barrel, El Chico, Pizza Inn, Red Lobster, Ruby Tuesday, Ryan's, Steak 'N Shake, Taco Bell
W: Hardee's (Playground), Heavenly Ham, McDonalds (Playground), Outback Steakhouse
Lodg E: ◆ Drury Lodge, ◆ Holiday Inn, ◆ Pear Tree Inn by Drury, ◆ Victorian Inn
W: ◆ Drury Suites, ◆ Hampton Inn
AServ W: Sam's Club, Wal-Mart Supercenter
Med E: ✚ St Francis Medical Center
ATM E: Capaha Bank
W: Shell
Other E: Barnes & Noble, Valu Vision, West Park Mall
W: Sam's Club, Target Department Store, Wal-Mart Supercenter (McDonald's, 24 hrs, pharmacy)
93 MO74, Dutchtown, U.S 61, Cape Girardeau (Northbound)
91 Rt AB, Cape Girardeau, Airport
TStop E: Phillips 66*(SCALES) (Radio Shop CB Repair)
TServ E: Carrier Transicold (Transport Refrigeration Services), Marmon Trucks, Midwest Diesel, Raben, Sam's Ready & Repair Service (Phillips 66 TS)
TWash E: Phillips 66
RVCamp W: Cape Town RV Sales
89 US 61, MO M, MO K, Scott City, Chaffee
Gas E: Citgo*, Conoco*, Texaco*
Food E: Burger King, Dairy Queen
AServ E: Auto Parts, S & S Automotive
ATM E: Security Bank & Trust
Other E: IGA Foodliner Grocery Store, Medicap Pharmacy, Riverside Regional Library
80 MO 77, Benton, Diehlstadt
FStop W: Sinclair*
67 U.S 62, Sikeston, Bertrand
FStop E: Phillips 66*, Shell*(LP)
Gas W: Amoco, Citgo*(LP)
Food E: JD's Steakhouse & Saloon

Bold red print shows RV & Bus parking available or nearby

XIT MISSOURI

W: Hardee's, Lambert's, Queen House Chinese
Lodg E: AAA Coach House Inn & Suites (Best Western), Red Carpet Inn
W: ◆ Drury Inn, Hatfield Inn, ◆ Holiday Inn Express, Ramada, Super 8 Motel (see our ad this page)
AServ W: Amoco, Berry's Exhaust Shop, Jarvis Motor Co. Geo, Chev, Olds, Cad
TServ E: Gadberry, Griffin Truck Center Volvo
RVCamp E: Hinton Park Camping, Town & Country RV Park (1.5 Miles)
ATM W: Sikeston Factory Outlet Stores
Other E: Fed Ex Drop Box
W: MFA Propane(LP), Miner Fire Station, Piggly Wiggly Supermarket, Police Dept., Semo Kart Raceway, Sikeston Factory Outlet Stores, Sikeston Trade Fair

6B I-57S, U.S 60W, Poplar Bluff
TServ W: Duckett Truck Center (Freightliner Services)
Lodg W: Super 8 (see our this page)

6A Junction I-57 North, East U.S 60, Chicago

8 MO 80, Matthews, East Prairie, Toll Ferry to Kentucky
TStop E: Eagle Landing Travel Center*(SCALES) (Citgo)
W: Flying J Travel Plaza*(LP)(SCALES) (RV Dump)
Food E: Restaurant (Citgo TS), Taco Bell (Citgo TS)
TServ E: M. Cranford(SCALES), Truck Lube Center
TWash E: Cranfords Truck and Tank Wash

2 Rt P, Kewanee
FStop E: Amoco*(LP)
W: Shell*

9 U.S 61, U.S 62, Business Loop I-55, New Madrid

EXIT MISSOURI/ARKANSAS

FStop E: Citgo* (24 hrs)

44 U.S 61, U.S 62, New Madrid, Howardville

(42) Rest Area - RR, Phones, Vending, Picnic P (Both Directions)

40 Rt EE, Marston, St Jude Road
TStop E: Pilot Travel Center(SCALES) (24 hrs)
FStop W: Amoco(CW), Phillips 66*
Food E: Arby's (Pilot TS)
W: Jerry's Cafe (Amoco)
Lodg E: Super 8 Motel
W: Cottonboll Inn
AServ E: NAPA Auto Parts (Pilot TS)
W: Jerry's Repair (Amoco), Larry's Auto Sales & Services
TServ E: Pilot Travel Center
W: Jerry's Repair (Amoco)
ATM E: Pilot Travel Center
Other W: The Fishing Shed Crappie Specialist

32 MO 162, Portageville
FStop W: Amoco* (24 hrs)
Food W: Gary's Bar-B-Q, Restaurant (Amoco)
AServ W: Hay's Tires, Parts Towing Service

27 MO K, MO A, Rt BB, Wardell
RVCamp E: KOA Kampground (3 Miles)

19 U.S 412, MO 84, Hayti, Caruthersville
FStop E: Conoco*, Phillips 66*
W: Total*(SCALES)
Gas E: Exxon*
W: Amoco*, Conoco*, R & P*
Food E: Blimpie's Subs (Conoco), Hardee's, McDonalds, Pizza Hut
W: Bessie's Smorgasbord, Boudreaux's Cafe, Chubby's Bar-B-Q, Dairy Queen, Subway (Amoco)
Lodg E: Comfort Inn, ◆ Holiday Inn Express
W: Pear Tree Inn by Drury
RVCamp W: KOA Kampground (6 Miles)
Med W: ✚ Hospital
ATM E: Conoco

17A Junction I-155, U.S 412, Dyersburg, TN

14 Rt J, Rt H, Rt U, Caruthersville, Braggadocio

(10) Weigh Station - both directions

8 MO 164, U.S 61, Steele
Gas E: Duckie's Phillips 66*
ATM E: Duckie's Phillips 66

4 Rt E, Cooter, Holland

(3) Rest Area - RR, Phones, Vending, Picnic P (Both Directions)

1 U.S 61, Rt O, Holland
FStop E: Raceway*(LP)
Gas W: Coastal*

↑MISSOURI

↓ARKANSAS

(71) Weigh Station (Southbound)

71 AR 150
Gas W: Citgo*

(68) Welcome Center, Rest Area - RR, Phones, Vending, Picnic P (Southbound)

67 AR 18, Armorel/Hickman Rd., Blytheville
Gas W: Coastal*(CW), Exxon*(D), Texaco*
Food E: Harry's Steaks, Seafoods

EXIT ARKANSAS

W: Bonanza, Grecian Steak House, Hardee's, KFC, Mazzio's Pizza, McDonalds (Playground), Olympia, Pancho's (Comfort Inn), Perkin's Family Restaurant, Pizza Inn, Shoney's, Sonic Drive In, Subway, Taco Bell, Wendy's
Lodg E: DAYS INN Days Inn
W: AAA Comfort Inn, ◆ Drury Inn, Econolodge, ◆ Holiday Inn
AServ W: Turner Muffler Shop, Wal-Mart
Med W: ✚ Hospital
ATM W: Coastal, Farmer's Bank & Trust, First National Bank
Other W: Big Lake WMA (16 Miles), Crossroads Videos, Music, & Books, Malco Cinema, Wal-Mart (Pharmacy)

63 U.S 61, Blytheville
TStop W: Citgo Travel Center(SCALES) (RV Dump, CB Repair)
FStop E: Total*
Gas W: Dodge's Store* (Kerosene), Texaco
Food W: Subway (Texaco)
Lodg W: Best Western, ◆ Delta K Motel
AServ W: Texaco
TServ W: Citgo Travel Center
RVCamp W: Knights of the Road RV Park
Med W: ✚ Hospital

57 AR 148, Burdette

53 AR 158, Luxora, Victoria

48 AR 140, Osceola
FStop E: Texaco*
Gas E: Total*(D)
Food E: Baskin Robbins (Texaco), Chester Fried Chicken (Texaco), Pizza Inn (Texaco), Restaurant (Best Western)
Lodg E: AAA Best Western, Holiday Inn Express (see our ad this page)
Med E: ✚ Hospital
ATM E: Texaco

(45) Rest Area - RR, Picnic P (Northbound)

44 AR 181, Wilson, Keiser

41 AR 14, Marie, Lepanto

36 AR 181, Bassett, Evadale

(35) Rest Area - RR, Phones, Picnic P (Southbound)

34 AR 118, Joiner, Tyronza

23AB 23A - AR 77 S, Turrell; 23B - US 63 N, Marked Tree, Jonesboro

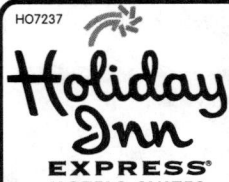

21 AR 42

EXIT ARKANSAS

FStop W: Fuelmart*[LP]
Food W: Subway (Fuelmart)
17 AR 50, Jericho
14 Crittendon Co. AR 4
FStop E: Citgo*[SCALES]
Gas W: Citgo*
Food E: Flashmart Travel Stop Restaurant (Citgo)
W: Stuckey's (Citgo)
RVCamp W: Best Holiday Trav-L-Park Camping[LP]
ATM E: Citgo
10 US 64W, Marion, Sunset, Wynne
FStop E: Delta Express*
W: Total*
Gas E: Avery's*, BP[D], Citgo*[CW][LP], Exxon*[D]
W: Citgo*[LP], Shell*[CW]
Food E: Big John's Restaurant, Ford's Family Restaurant, Hardee's, Hot Stuff Pizza (Citgo), Smash Subs, Sonic Drive In
W: Chester Fried Chicken (Shell), Restaurant (Best Western), Taco Bell (Total)
Lodg E: Hallmarc Inn
W: ◆ Best Western, Scottish Inns
AServ E: BP, Cliff's Auto & Tractor (Auto Parts), Exxon
ATM E: Citgo, Citizen's Bank, Fidelity National Bank, Union Planter's Bank
Parks W: Parkin Archaeological State Park (23 Miles)
Other E: Big Star, Marion Discount Pharmacy, US Post Office
(9) Weigh Station - both directions
8 West I-40, Little Rock
7 AR 77, Missouri St.
Gas W: Exxon*[CW], Mapco*[LP], Texaco, Total*[LP]
Food W: Baskin Robbins, Bonanza, Burger King, Domino's Pizza, Howard's Donuts, Krystal, McDonalds, Mrs Winner's Chicken, Pizza Inn, Shoney's, Subway
Lodg W: Ramada Inn
AServ W: Don Gage's Auto, Goodyear Tire & Auto, Mr. Fast Lube, Texaco
ATM W: Fidelity National Bank, Union Planter's Bank
Other W: Holiday Plaza Lanes Bowling, Kroger Supermarket, WalGreens, Wonder Hostess Bakery Thrift Shop

Note: I-55 runs concurrent below with I-40. Numbering follows I-40.

278 AR 191, 7th St
TStop S: Memphis Gateway Auto/Truck Plaza*[SCALES]
FStop S: Love's* (24 hrs)
Gas S: Citgo*
Food N: Blimpie's Subs (Citgo)
S: Catfish Island, I Can't Believe It's Yogurt (Memphis Gateway TS), McDonalds (Wal-Mart)
Lodg S: Ramada Inn
TServ S: Memphis Gateway Auto/Truck Plaza
RVCamp S: Tom Sawyer's RV Park
ATM S: Love's
Other S: Wal-Mart (Open 24 Hours)
279A Ingram Blvd
Gas N: BP*
S: BP, Citgo*[CW], Shell*
Food N: Delta Point (Days Inn), Grandy's (Days Inn), Little Italy Pizza (Days Inn)
S: Cinnamon Street Bakery (Citgo), Earl's Restaurant, Eddi Pepper's Mexican (Citgo), Holiday Inn Restaurant (Holiday Inn), Hot Stuff Pizza (Citgo), Smash Hit Subs (Citgo), Waffle House, Waffle House

EXIT ARKANSAS/TENNESSEE

Lodg N: Classic Inn, (AAA) Comfort Inn, (DAYS INN) Days Inn
S: Classic Inn, (DAYS INN) Days Inn, ◆ Econolodge, ◆ Hampton Inn, ◆ Holiday Inn, ◆ Holiday Inn, (AAA) Howard Johnson, Motel 6, Motel 6, Relax Inn, Relax Inn
AServ N: BP
Other N: Southwind Greyhound Park
S: Coin Laundry
279B I-40E, Memphis, Nashville, I-55S, Memphis, Jackson Miss
280 Jct I-40 East
TStop S: Flying J Travel Plaza*[LP][SCALES] (RV Dump), Petro*[SCALES] (Showers, CB Shop, MS Carriers Recruiting Office)
FStop N: Fina* (Showers), Mapco*[LP]
S: Pilot Travel Center*[D][SCALES]
Food S: Dairy Queen (Pilot TS), Iron Skillet Restaurant (Petro TS), KFC, McDonalds, Subway (Pilot TS), Taco Bell
Lodg S: Best Western, Deluxe Inn, Express Inn, ◆ Super 8 Motel
AServ N: Goodyear Tire & Auto
TServ N: Goodyear Tire & Auto, Riggs Cat Truck Engine Service
S: Bridgestone Tire & Auto (Petro TS), Jim's & Son Truck Service & Parts, Landstar Service Center
TWash S: Blue Beacon Truck Wash (Petro TS), Mr. Clean Truck Wash, XVIII Wheeler's Truck Wash (24 Hrs)
Other N: Best Tarps
S: CB Shop

Note: I-55 runs concurrent above with I-40. Numbering follows I-40.

4 Dr Martin Luther King Jr Drive, Southland Drive
TStop E: Flying J Travel Plaza*[LP][SCALES] (RV Dump), Petro*[SCALES] (Showers)
FStop E: Pilot Travel Center*[SCALES]
Gas W: Citgo*
Food E: Dairy Queen (Pilot), KFC, McDonalds (Playground), Subway (Pilot), Taco Bell
W: Pancho's Mexican
Lodg E: Best Western, Deluxe Inn, Express Inn, Super 8 Motel
AServ W: Bridgestone Tire & Auto, Howard Ford Truck & Car Center
TServ E: Petro
TWash E: Mr. Clean Truck Wash, Petro, XVIII Wheeler's Truck Wash (24 Hrs)
Other E: CB Shop
3B US 70, Broadway Blvd (Northbound)
3A AR 131, Mound City Rd (Northbound)
1 Bridgeport Road (Northbound)

↑ARKANSAS

↓TENNESSEE

12C Delaware St, Metal Museum Dr.
Food W: River Bluff Inn (Best Western)
Lodg W: River Bluff Inn (Best Western)
RVCamp W: Mississippi River RV Park*
12B Riverside Drive
Gas E: Texaco*
12A US 61, 64, 70, 79 E Crump Blvd

EXIT TENNESSEE

TServ E: G & W Diesel Services Inc.
Other E: Fast Check Food Store & Deli
11 McLemore Ave, Presidents Island
10 South Parkway, Memphis
Food E: Morrison's Restaurant
Parks W: Dr Martin Luther King Jr Riverside Park
9 Mallory Ave., Memphis
TServ E: Great Dane Trailors Sales & Service
8 Horn Lake Rd (Difficult Reaccess Southbound)
7 US 61S, Vicksburg
Gas E: Exxon*
Food E: Interstate Bar-B-Q, Lot-a-burger
AServ E: Anderson[CW]
6AB 6A - E I-240, Nashville, 6B - N I-240

Bold red print shows RV & Bus parking available or nearby

EXIT TENNESSEE/MISSISSIPPI

5AB U.S 51, Elvis Presley Blvd. South, Brooks Road, Graceland

FStop **W:** **Mapco Express*(LP)**
Gas **E:** Mapco Express*(LP)
W: Texaco*(CW)
Food **E:** **Dad's Place (Ramada Inn), Grid Iron Restaurant (Days Inn, 24 Hrs),** Restaurant (Howard Johnson)
W: Captain D's Seafood, Crumpy's Hot Wings, Exlines Best Pizza In Town, Hernando's Hideaway, **Kettle Restaurant (24 Hrs),** Luby's Cafeteria, Peking Inn, Tastee Bar-B-Q
Lodg **E:** Comfort Inn, **Days Inn (see our ad opposite page),** Howard Johnson, **Ramada Inn (see our ad opposite page)**
W: "Q" Inn, Graceland Inn, Motel 6, **Days Inn (see our ad opposite page)**
AServ **E:** Peach Auto Painting & Collision
W: Advance Muffler, Bob's Honda Dealership, Coleman Taylor Transmission, Graceland Dodge, NAPA Auto Parts, Pyramid Pontiac GMC Dealer, Whitehaven Car Care Center
TServ **E:** **Barton Parts & Service (Freightliner), Clarke Detroit Diesel**
W: **Star Truck & Trailor Sales**
RVCamp **W:** **D&N Camper Sales, Davis Motor Home Mart**
Med **W:** **Hospital**
ATM **E:** Mapco Express
W: **Mapco Express**
Other **E:** **Fire Station**
W: **Graceland**

(3) TN Welcome Center - RR, Vending, Phones (Northbound)

2AB U.S. 5, TN 175, Shelby Dr, White Haven, Capleville

Gas **E:** Amoco* (24 Hrs), Circle K*
W: BP*(CW), Circle K* (Gas STP), The Working Man's Friend
Food **E:** Pollard's Bar-B-Que
W: Adams Family Restaurant, Burger King (Playground), CK's Coffee Shop (24 Hrs), Central Park, F & D Sizzlin' Steakout, Hot Wings Express, Mrs Winner's Chicken, Pizza Inn, Ruby's Family Restaurant
Lodg **E:** Knights Inn, **Super 8 Motel**
AServ **W:** Goodyear Tire & Auto, Shaw's Auto Glass Co., Xpert Tune
ATM **W:** First American
Other **E:** **Dean's Coin-Op Laundry**
W: Car Wash, **Hubbard's Hardware, Leith's Coin Laundry, Piggly Wiggly Supermarket (Western Union), Super D Drugs**

↑ TENNESSEE

↓ MISSISSIPPI

291 Southaven, State Line Rd

EXIT MISSISSIPPI

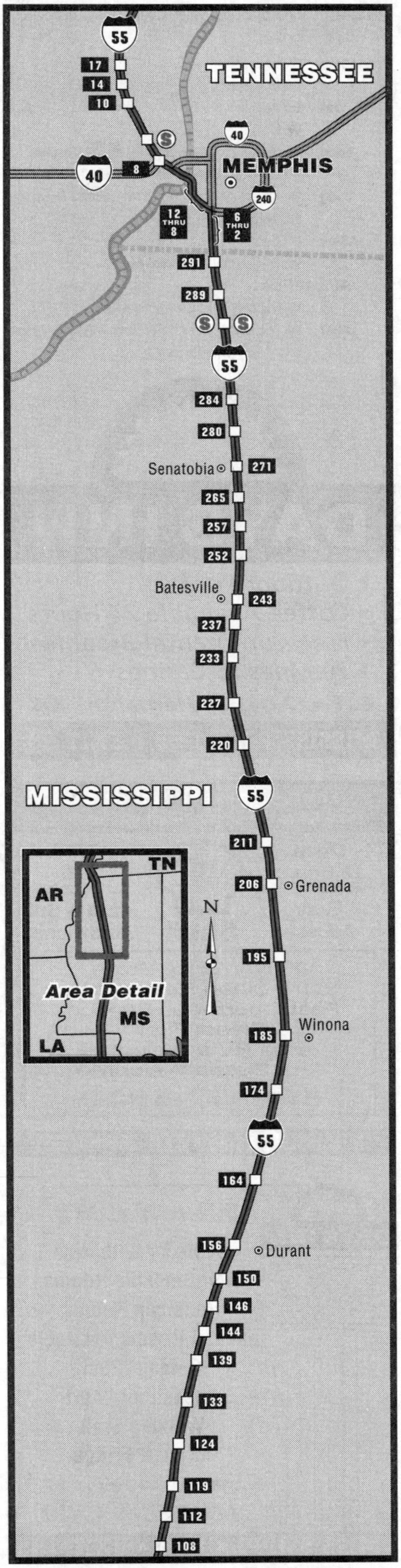

EXIT MISSISSIPPI

Gas **E:** BP*(CW), Pump 'n Save
W: Amoco*, Exxon*(D), Texaco(D)
Food **E:** Baskin Robbins, Burger King (Playground), CK's Coffee Shop, Chinese Chef, Exlines, Golden Corral, Harlow's Donuts, Little Caesars Pizza, McDonalds (Playground), Pizza Hut, Poncho's Mexican, **Restaurant (Kroger),** Shoney's, Subway, Tops Bar-B-Q, Waffle House
W: El Porton Mexican, KFC, Mrs Winner's Chicken, Rally's Hamburgers, Wendy's
Lodg **E:** AAA **Best Western,** ◆ Comfort Inn, ◆ Holiday Inn Express, Shoney's Inn
AServ **E:** American Tire Repair 'LLC, Firestone Tire & Auto, K-Mart, Mathis Tire Center
W: Amoco, Glen's Garage (24 Hr Towing), Speedy Auto Glass, Texaco, Valvoline Quick Lube Center, Xpert tune
ATM **E:** BP, K-Mart, **Kroger Supermarket,** Trustmark National Bank
W: Bank of Mississippi, The People's Bank, Union Planter's Bank
Other **E:** **Bobby's Pet A Groom (Pet Supplies & IAMS Products), Bookstore (Christian), K-Mart (24 Hrs, Pharmacy, One Hr Photo), Kroger Supermarket (w/pharmacy, 24 Hrs), Optometrist, Wal-Mart (w/ pharmacy, 24 Hrs), WalGreens**
W: **Freds Drug Store, K & B Drugs, Seessel's Grocery, Southaven Police Dept**

289 MS 302, Horn Lake, Olive Branch, S. Southaven, Goodman Rd

FStop **W:** **Phillips 66*(CW) (Kerosene)**
Gas **E:** BP*(CW), Shell*(D)(CW)
W: Shell*
Food **E:** Burger King (Playground), Krystal (24 Hrs)
W: Applebee's, Arby's, **Cracker Barrel,** Hardee's, Popeye's Chicken, Ryan's Steakhouse, Taco Bell, The Great Wall Restaurant, Waffle House, Wendy's
Lodg **E:** ◆ Hampton Inn
W: Ramada Limited, AAA Sleep Inn
AServ **E:** **E & W Auto & Truck Service Inc.,** Ferguson's Auto Service Inc. (NAPA Auto Care Service Center), Valvoline Quick Lube Center
W: Abra Auto Body, Bridgestone Tire & Auto, Gateway Tire & Service Center, Xpress Lube
TServ **E:** **E & W Auto & Truck Service**
Med **E:** **Baptist Memorial Hospital De Soto**
ATM **E:** Shell, Trustmark National Bank
W: **Kroger Supermarket**
Other **E:** **Mississippi Treasures**
W: **Kroger Supermarket (Pharmacy), The Home Depot(LP)**

287 Church Rd

(285) **Weigh Station - both directions**

284 Nesbit Road

Gas **W:** BP*, Citgo*
Food **W:** Happy Daze Dairy Bar
AServ **E:** Jimmy Gray Oldsmobile/Chevrolet/Geo
Other **W:** Car Wash, **US Post Office**

Bold red print shows RV & Bus parking available or nearby

EXIT MISSISSIPPI

280 Hernando, Arkabutla Lake
- **Gas** E: BP
  W: Chevron*(LP), Shell*(D)(CW), Total*(LP) (24 Hrs)
- **Food** E: The Sports Cafe (Hernando Inn)
  W: Church's Fried Chicken, Colemans Bar-B-Q, McDonalds, Pizza Hut, Pizza Inn, Sonic, Subway (Chevron), Taco Bell (Total)
- **Lodg** E: ◆ Days Inn (see our ad this page), Hernando Inn
- **AServ** E: BP
  W: Bryant Tire & Service Center, Delta Muffler, NAPA Auto Parts
- **RVCamp** W: Memphis South Campground & RV (1.7 Miles)
- **ATM** W: Deposit Guarantee National Bank, Piggly Wiggly Supermarket, Trustmark National Bank
- **Other** W: Arkabutla Lake, Don's Car Wash, Piggly Wiggly Supermarket

(279) MS Welcome Center - RR, Picnic, Phones, Dog Trail, RV Dump (Southbound)

(276) Rest Area - RR, Picnic, Phones, RV Dump P (Northbound)

271 MS 306, Coldwater, Independence
- **Gas** W: Amoco*
- **RVCamp** W: Arkabutla Lake (14 Miles, Camping)

265 MS 4, Senatobia, Holly Springs (Northwest Ms Community College)
- **FStop** W: Fuel Mart*
- **Gas** W: Comet(CW), Exxon*, Texaco*(CW)(LP)
- **Food** W: Colemans Bar-B-Q, Domino's Pizza, KFC, McDonalds, Pizza Hut, Senatobia Inn, Subway, Taco Bell, Wendy's, Western Sizzlin'
- **Lodg** W: AAA Comfort Inn, Senatobia Inn
- **AServ** W: Clear Vision Auto Glass, Texaco, Tommy Heafner Pontiac Buick GMC Trucks
- **TServ** W: BP(SCALES)
- **Med** W: + Senatobia Community Hospital
- **ATM** W: Senatobia Bank
- **Other** W: City Drugs, Coin Operated Laundry, Community Discount Pharmacy, Freds Drug Store, Piggly Wiggly Supermarket, Senatobia Food Market

257 MS 310, Como (North Sardis Lake)
- **Gas** W: Bob Payne's Como Marine*
- **Other** E: North Sardis Lake

252 MS 315, Sardis
- **FStop** E: Chevron*(CW)
- **Gas** E: Amoco*(D), Texaco*(D)(LP)
  W: BP(D), JW's Suds & Duds(CW) (Coin Laundry), Phillips 66*(LP)
- **Food** E: Pop's Piggy Place Bar-B-Q
  W: Happy Days, Sonic
- **Lodg** E: Lake Inn, AAA Super 8 Motel
  W: Best Western
- **AServ** E: Amoco, NAPA Auto Parts, Simmerman Auto Service
  W: BP
- **Med** E: + Hospital
- **ATM** E: Texaco
- **Other** W: Freds Discount Drug Store, Mette$ave Discount Drugs, Piggly Wiggly Supermarket, Save-A-Lot Food Stores

246 MS 35, N Batesville
- **FStop** W: Texaco*
- **Other** E: Sardis Dam (8 Miles)

EXIT MISSISSIPPI

243AB MS 6, Batesville, Oxford (South Sardis Lake)
- **FStop** E: Texaco*(D)(LP)
  W: Chevron*
- **Gas** E: Citgo
  W: Exxon*(D)
- **Food** W: Burger King (Playground), Cracker Barrel, Dairy Queen, Sonic Drive In, Taco Bell
- **Lodg** W: AmeriHost Inn, Comfort Inn, Days Inn, Hampton Inn
- **AServ** E: Citgo
  W: Tri-County Glass, Wal-Mart
- **ATM** E: Texaco
  W: Factory Outlet Center, Food World
- **Other** W: Factory Outlet Center, Food World (24 Hrs), Wal-Mart (w/pharmacy)

EXIT MISSISSIPPI

(239) Rest Area - RR, Picnic, RV Dump, Pet Walk P (Both Directions)

237 Pope, Courtland
- **FStop** E: Amoco*(LP)

233 Enid
- **Gas** W: Benson's Grocery Store*
- **RVCamp** E: Enid Lake (1 Miles)

227 MS 32, Oakland, Water Valley
- **FStop** W: Chevron*(LP)
- **Food** W: Country Catfish
- **AServ** W: 32/55 Service, Ashmore's Wrecker Service
- **RVCamp** E: George Payne Cossar State Park
- **Parks** E: George Payne Cossar State Park

220 MS 330, Coffeeville, Tillatoba Rd
- **TStop** E: Conoco*
- **Food** E: All American Restaurant (Conoco TS)
- **TServ** E: Conoco
- **ATM** E: Conoco

211 North MS 7, Scenic Route 333, Coffeeville
- **FStop** W: Texaco*(LP)
- **Food** W: Chester Fried Chicken (Texaco)
- **AServ** W: Interstate Sales & Service
- **RVCamp** E: Oxbow RV Park

208 N. Grenada - Papermill Rd.

206 South MS 7, MS 8, Grenada, Greenwood
- **Gas** E: Exxon*, Racetrac*, Shell*(D) (see our ad this page), Texaco*
  W: Texaco (24 Hrs, Hilltop Inns)
- **Food** E: Best Western, Boo-Ray's Cafe (Holiday Inn), Cobb's Seafood & Steak, Domino's Pizza, Hot Stuff Pizza (Shell), Jake & Rip's Bar-B-Q, Steaks, & Catfish, McDonalds (Wal-Mart Supercenter), Ragtime Bar & Grill, Shoney's, TCBY, Wendy's, Western Sizzlin'
- **Lodg** E: Best Western, AAA Comfort Inn, AAA Days Inn, ◆ Hampton Inn, AAA Holiday Inn, Ramada Limited
  W: Hilltop Inns
- **AServ** E: Kirk Auto Co. Merc/Lincoln/Ford/Toyota, Wal-Mart
- **RVCamp** E: Lake Grenada
- **Med** E: + Hospital
- **ATM** E: Exxon, Shell
- **Other** E: US Post Office, Wal-Mart Supercenter (24 Hrs, Pharmacy)
  W: Delta State University

(203) Parking Area - Phones, No Other Facilities (Southbound)

(202) Parking Area - Phones No Other Services (Northbound)

195 MS 404, Duck Hill

185 US 82, Winona, Greenwood (MS State Univ., MS Valley State Univ., MS Univ For Women)
- **FStop** E: Texaco*
- **Gas** E: Exxon*, Shell*, Simply Irresistable*
- **Food** E: McDonalds, Sharon's Family Restaurant (Relax Inn), TCBY (Exxon)
- **Lodg** E: ◆ Days Inn (see our ad this page), Relax Inn

174 MS 430, MS 35, Vaiden, Carrollton
- **TStop** E: 35-55 Truck Plaza(SCALES) (Chevron)

Bold red print shows RV & Bus parking available or nearby

EXIT	MISSISSIPPI
FStop	**E:** Shell*(D)
Gas	**E:** BP*, Comet
	W: Texaco*
Food	**E:** 35-55 Restaurant (35-55 Truck Plaza)
	W: Chester Fried Express (Texaco), Stuckey's (Texaco)
Lodg	**E:** 35-55 Motel (35-55 Truck Plaza)
AServ	**E:** NAPA Auto Parts
RVCamp	**E:** Vaiden Campground
Other	**E:** CB Shop, T & H Car Wash (BP), T & H Laundromat (BP)
(173)	Rest Area - RR, Phones, RV Dump, Dog Trail, Picnic P (Southbound)
164	West
FStop	**W:** West Truck Stop*
TServ	**W:** West Truck Stop
Other	**W:** Bobs 99 Grocery
(163)	Rest Area - RR, Phones, RV Dump P (Northbound)
156	MS 12, Durant, Lexington
FStop	**E:** Texaco*(LP)
Lodg	**E:** Super 8 Motel
Med	**E:** + Hospital
150	Holmes County State Park
Parks	**E:** Holmes County State Park (1.5 Miles)
Other	**E:** Camping
146	MS 14, Goodman, Ebenezer (Holmes Community College)
Other	**W:** Little Red School House, Racetrac
144	MS 17, Pickens, Lexington
TStop	**E:** Texaco(LP)(SCALES) (24 Hrs)
FStop	**W:** BP Fuel Center*(LP)
Food	**W:** Lakeside Restaurant, MGM Restaurant (BP Fuel Center)
Lodg	**W:** Motel (BP Fuel Center)
TServ	**E:** Texaco TS
ATM	**E:** Texaco TS
	W: BP Fuel Center
139	MS 432, Pickens, Yazoo City, Benton
133	Vaughan
Other	**E:** Casey Jones Museum
124	MS 16, North Canton, Yazoo City
(120)	Parking Area - No Services (Southbound)
119	MS 22, Canton, Flora
TStop	**E:** Chevron* (24 Hrs, RV Dump)
Gas	**E:** Amoco*(D), BP*, Exxon*(D), Shell*, Texaco*(D)
Food	**E:** McDonalds (Playground), Nancy's (Chevron TS), Pizza Hut, Popeye's Chicken, Two Rivers Restaurant, Wendy's
Lodg	**E:** [Days Inn logo] ◆ Days Inn, ◆ Econolodge (see our ad this page), Holiday Inn Express
AServ	**E:** Briggs Ford
Med	**E:** + Hospital
ATM	**E:** Chevron TS, Texaco
Other	**E:** Ross Barnett Reservoir Upper Lake
(117)	Parking Area - No Services (Northbound)
112	Gluckstadt
FStop	**E:** Amoco*(LP)
Gas	**E:** Mac's Gas*(D)
RVCamp	**W:** Camper Corral (Sales & Service)
ATM	**E:** Amoco
108	MS 463, Madison

EXIT	MISSISSIPPI
Gas	**E:** Shell*, Texaco*(CW)
Food	**E:** Blimpie's Subs (Texaco)
ATM	**E:** Shell, Texaco
Other	**W:** PGA Tour Golf Classic
105B	Ridgeland, Old Agency Road (Holmes Comm. College Ridgeland Campus)
FStop	**E:** Speedway*(LP)
Gas	**E:** Chevron*(D)
Lodg	**E:** (AAA) Comfort Inn (see our ad this page)
Other	**E:** Holmes Comm. College Ridgeland Campus
105A	Natchez Trace Pkwy
104	Junction I-220, West Jackson
103	County Line Road, Ridgeland (Tougaloo College, Ross Barnett Reservoir)
Gas	**E:** Exxon*(D)
Food	**E:** Applebee's, Chick-fil-A (Playground), Hardee's (24 Hrs), Pickles & Ice Cream, Ralph & Kacoo's Restaurant, Santa Fe Grill (Plaza Hotel), Shoney's, Stockyard Steaks, The Honey Baked Ham Co.
	W: Kim Long Chinese Cuisine, Olive Garden, Smoothie Q, Subway
Lodg	**E:** Cabot Lodge Bed & Breakfast, ◆ Courtyard by Marriott, ◆ Plaza Hotel, ◆ Red Roof Inn, Shoney's Inn, Studio PLUS
	W: Comfort Suites, Motel 6
AServ	**E:** Furr Transmission, Midas Mufflers & Brakes, North Park Acura Mazda, The Pit Stop Auto Service & Repair, Watson Ford
Other	**E:** AAA Auto Club, Barnes & Noble, Edwin Watts Golf Shop, Office Max, Ross Barnett Reservoir, Sam's Club
	W: Drugs for Less (1 Hr Photo), Eyemart Express (1 Hr Service), Office Depot, PETsMart, Target Department Store, The Home Depot(LP)
102B	Beasley Road, Adkins Blvd.
Gas	**E:** Amoco(CW), Chevron*(D), Chevron*
Food	**E:** [Cracker Barrel logo] Cracker Barrel, Kenny Rogers Roasters, Lone Star Restaurant, Outback Steakhouse, Subway, The Wok Shoppe

EXIT	MISSISSIPPI
	W: McDonalds (Playground)
Lodg	**W:** Econolodge
AServ	**E:** Hallmark Toyota Kia BMW, North Jackson Nissan, Tom Wimberley Jeep Eagle Lincoln Mercury, Watson Ford
	W: K-Mart
Med	**E:** + Methodist Medical Center (No Emergency Room)
ATM	**E:** Chevron
Other	**E:** Fire Dept.
	W: K-Mart (w/pharmacy, 24 Hrs)
102A	Briarwood Dr., Jackson
Gas	**E:** Amoco*
Food	**W:** Chili's Grill & Bar, Chuck E. Cheese's, Perkins Family Restaurant, Steak & Ale
Lodg	**E:** La Quinta Inn
	W: (AAA) Best Suites of America, Hampton Inn
AServ	**E:** Van-Trow Oldsmobile Volkswagon
	W: Blackwell Chevrolet, Blackwell Import Motors Merc Porsche
100	Northside Dr, Meadowbrook Rd
Gas	**E:** Chevron*, Sprint Mart*
	W: Amoco*, Exxon*(D)(CW), Shell*
Food	**E:** Golden Dragon Restaurant, Logan Farms Honey Hams, McDonalds (Playground), Olde Tyme Restaurant, Pizza Inn, Steak-Out Charbroiled Delivery, Subway, We Love Yogurt
	W: Bennigan's, Domino's Pizza, Marcel's Steak & Seafood, Perry's Soul Food (Northside Inn), Sam's Westside Restaurant & Bar, Waffle House
Lodg	**W:** Northside Inn
AServ	**E:** Chevron
	W: Car Quest Auto Center, Safelite Auto Glass
ATM	**E:** Deposit Guaranty National Bank, Trustmark Bank
	W: Amoco, Shell
Other	**E:** ACE Hardware & Kitchen Shop, Beemon Drugs, Jasper Ewing & Sons One Hr Photo, Optical Outlet, Pack & Mail, Pro Golf Discount, Sunflower Grocery Store, Super D Drug Store
	W: Eyecare Plus, Highlander Coin-Op Laundry, Minit Mart Groceries, Northside Library, The Dog Wash Pet Supplies
99	Meadowbrook Road, Northside Dr East (Northbound)
98BC	MS 25 North, Carthage, Lakeland Dr
Food	**E:** Crawdad Hole
Med	**W:** + St Dominic-Jackson Health Services (Hospital, Med Offices, Mental Health, MS Heart Center)
Parks	**E:** LeFleur's Bluff State Park (Camping)
Other	**E:** Agriculture & Forestry Museum, Farmer's Market Museum, LeFleur's Bluff State Park Golf Course/Driving Range, Mississippi Music Hall of Fame, Mississippi Sports Hall Of Fame, Smith-Wills Stadium for Jackson Generals (Minor League Baseball), Tourist Information
98A	Woodrow Wilson Dr
Med	**W:** + Blair E. Batson Hospital for Children, + Univ of Mississippi Medical Center, + VA Medical Center
Other	**W:** Miss. Highway Patrol
96C	Fortification St (Belhaven College)
Food	**W:** The Bel-Haven Bar & Grill
Lodg	**E:** ◆ Residence Inn by Marriott
Other	**W:** Community Children's Theater, New Stage Theater

EXIT MISSISSIPPI

96B High St., State Capitol
- **Gas** W: Shell*, Texaco*(D)(CW)
- **Food** W: Blimpie's Subs (Texaco), Burger King, Chimneyville Station, Dairy Queen, Dennery's Restaurant, Dunkin' Donuts, Popeye's Chicken, Shoney's, Taco Bell, The Emporium (Ramada Inn), Waffle House, Wendy's
- **Lodg** W: Days Inn, ◆ Hampton Inn, ◆ Holiday Inn Express, Ramada Inn, ◆ Red Roof Inn
- **AServ** E: Herrin-Gear Autoplex Saturn, Chev, Infin, Lexus
- **ATM** W: Shell, Texaco
- **Other** W: **Animal Emergency Clinic, Miss. State Fairgrounds**

96A Pearl St. (Difficult Reaccess)
- **Other** W: **Miss. Museum of Art, Natural Science Museum, Old Capitol Museum**

94 Jct I-20E, US49S, Meridian, Hattiesburg

Note: I-55 runs concurrent below with I-20. Numbering follows I-20.

45A Gallatin St.
- **Gas** N: Chevron*, Dixie Gas*, S & S Mini Service, Texaco*
- **Lodg** N: Crossroads Inn
- **AServ** N: Anglin Tire Co, Cowboys Used Tires, Gentry's Body Shop, Smith Automatic Transmission, Xpress Lube
  S: Regency Toyota, Nissan, Mitsubishi
- **Other** N: **Camping, Duncan's Discount Marine, Jackson Animal Clinic**
  S: **Animal Health Products, Magnolia Animal Hospital**

45B US 51N, State St
- **AServ** N: AAMCO Transmission, Fowler Buick, Fowler GMC Truck, Freeman Auto Service, Paul Moak Pontiac, Volvo, Honda

Note: I-55 runs concurrent above with I-20. Numbering follows I-20.

92C West I-20, U.S 49, Vicksburg, Yazoo City (Left Exit)

92B U.S 51 North, State St, Gallatin St

92A McDowell Road
- **Gas** W: Dixie*, Fleet Morris Petroleum*, Shell*(CW) (24 Hrs), Texaco*(D)
- **Food** W: China Wok, Golden Glazed Donuts, Thai House Restaurant, Waffle House
- **Lodg** W: (AAA) Super 8 Motel
- **AServ** W: Shell
- **TServ** E: **Jackson's Truck & Trailor Repair**
- **Other** W: **United States Post Office**

90B Daniel Lake Blvd, Cooper Rd. (Southbound)
- **Gas** W: Shell*(LP)
- **Other** W: **Harley-Davidson Dealer**

90A Savanna St
- **Gas** W: Spur*
- **Food** W: Bo Don's Seafood Restaurant
- **Lodg** E: **Rodeway Inn**
- **AServ** E: Auto Crafters, Hardy Bros. Inc. Paint & Body Shop
  W: Spur, Stegall Auto Body
- **TServ** E: **Gray's Truck Service**
- **RVCamp** W: **Camping**

88 Elton Road

EXIT MISSISSIPPI

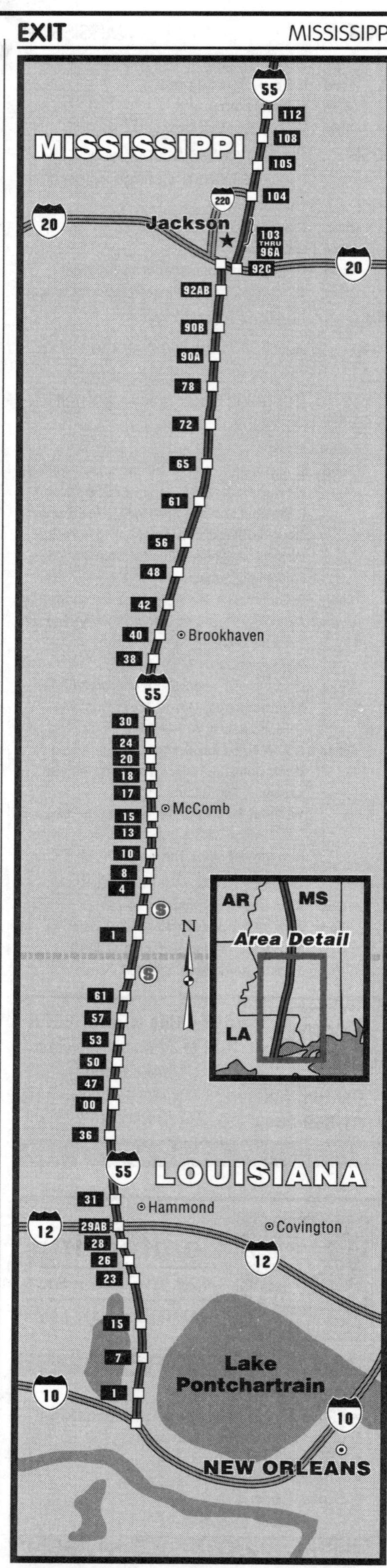

EXIT MISSISSIPPI

- **Gas** W: Chevron*(CW)(LP), Exxon*(D)
- **Food** W: Chester Fried Chicken (Exxon), Subway (Exxon)
- **Other** W: Car Wash

85 Byram
- **Gas** E: Chevron*(LP) (24 Hrs)
  W: Bill's*, Pump & Save, Texaco*(D)(CW)(LP)
- **Food** E: Krystal (Chevron)
  W: Fernando's, Golden Glazed Donuts, Hong Kong Restaurant, McDonalds (Playground), Pizza Hut, Popeyes Chicken, Subway, Taco Bell, Wendy's
- **Lodg** W: **Days Inn**
- **AServ** W: Davis Tire & Auto, Jerry Gould Auto Repair, Pennzoil Lube, Steve's Auto Parts
- **ATM** W: Bank of Mississippi, Texaco
- **Other** W: **Jitney Jungle, Super D Drug Store**

81 Wynndale Road
- **AServ** E: **Fulgham's Auto & Truck Repair**
- **TServ** E: **Fulgham's Auto & Truck Repair**

78 Terry
- **FStop** E: **Texaco***
- **Gas** E: Conoco*(D)
  W: Mac's*
- **Food** E: **Church's Fried Chicken (Texaco)**
- **AServ** W: Tannehill's Auto Repair
- **ATM** W: Mac's Gas
- **Other** E: Car Wash
  W: **Terry Animal Clinic**

72 US 51, MS 27, North Crystal Springs, Vicksburg
- **FStop** E: **Phillips 66***
- **Food** E: McDonalds (Playground)
- **AServ** E: Crystal Springs Ford Dealership

68 South Crystal Springs
- **Other** E: **Copiah Animal Hospital**

65 Gallman
- **Gas** E: Citgo*
- **Food** E: Dairy Queen (Citgo), Stuckey's (Citgo)

61 MS 28, Hazlehurst, Fayette
- **Gas** E: Exxon*(CW), Phillips 66*(CW), Pump 'n Save, The Store*
- **Food** E: J & B BBQ, KFC, McDonalds (Playground), Pizza Hut, **Restaurant (Ramada Inn), Stark's Restaurant,** Subway (Exxon), T & T Fresh Catfish, Seafood, Wendy's
- **Lodg** E: ◆ **Ramada Inn,** Western Inn Express
- **AServ** E: Quality Oil & Lube, The Little Engun
- **Med** E: ✚ **Hospital**
- **Other** E: **Wal-Mart (Pharmacy)**

59 South Hazlehurst

56 Martinsville

(54) **Rest Area - RR, Phones, RV Dump, Picnic, Grills (P) (Both Directions)**

51 Wesson (Copiah-lincoln Community College)
- **TStop** W: **Country Junction Truckstop*(LP) (Fuel-Chevron)**
- **TServ** W: **Country Junction TS**
- **RVCamp** W: **Country Junction Truck Stop (Full Hookups)**

48 Mt. Zion Road

42 North Brookhaven
- **TStop** E: **Phillips 66* (FedEx Drop)**
- **FStop** E: **Shell(LP) (see our ad opposite page)**
- **Food** E: **Ernie's Diner (Phillips 66 TS)**
- **AServ** E: Shell

Bold red print shows RV & Bus parking available or nearby

EXIT MISSISSIPPI

TServ E: Phillips 66 TS
RVCamp E: Circle N Campground (Hookups)

40 To MS 550, Downtown Brookhaven
Gas E: BP*(D), Chevron*(CW) (24 Hrs), Exxon*(D), Texaco*(D)
Food E: Bowie BBQ, Dairy Queen, Little Caesars Pizza, McDonalds (Playground), Pizza Hut, Popeyes Chicken, Restaurant (Ramada Inn), Shoney's, Subway, Taco Bell
Lodg E: ◆ Best Western, DAYS INN ◆ Days Inn, Hampton Inn, Ramada Inn
AServ E: Allbritton Sullivan Ford, Wal-Mart
ATM E: Wal-Mart
Other E: DelChamps Supermarket, Wal-Mart Supercenter (Vision Center, Pharmacy)
W: Animal Medical Center

38 U.S 84, South Brookhaven, Natchez, Monticello, Meadville
FStop W: Chevron*(LP) (24 Hrs), Hickory Hill Grocery
Food W: Busy B's (Hickory Hill Grocery)
AServ W: Chevron

30 Bogue Chitto, Norfield
FStop E: Shell* (see our ad this page)
Med W: ✚ Southwest Mississippi Regional Medical Center

(26) Parking Area - No Services (Northbound)

24 Lake Dixie Springs, Johnstons Station

(23) Parking Area - No Services (Southbound)

20AB U.S 98 West, Summit, Natchez, Meadville (SW Miss Community College)
FStop E: Texaco*(D), Shell (see our ad this page)
W: Phillips 66*(LP)
Gas E: BP*(D)
W: Exxon*(D)
Food W: Chester Fried Chicken (Exxon), Summit Smokehouse & Casey's Kitchen
ATM W: Phillips 66

18 MS 570, Smithdale Road, North McComb
FStop W: Chevron*(CW)
Food E: McDonalds
AServ E: Wal-Mart
W: Trademark Ford Dealership
Med E: ✚ Hospital

EXIT MISSISSIPPI

ATM W: Chevron
Other E: Edgewood Mall, Wal-Mart (24 Hrs, Pharmacy)

17 Delaware Ave, Downtown McComb
Gas E: BP*(D), Chevron(D), Citgo*, Shell (see our ad this page)
Food E: Dairy Queen, Huddle House, Little Joe's Family Restaurant, Restaurant (Holiday Inn), Wendy's
Lodg E: ◆ Comfort Inn, ◆ Holiday Inn, Super 8 Motel
W: AAA Best Western (see our ad this page)
AServ E: Chevron, Quality Oil & Lube
W: Delaware Motors Chrysler Plymouth Toyota
ATM E: Deposit Guaranty
Other E: Chamber of Commerce Visitor Info, Eckerd Drugs, Kroger Supermarket

15AB U.S 98 East, MS 24 West, Tylertown,

EXIT MISSISSIPPI/LOUISIANA

Liberty
FStop W: BP*

13 Fernwood Road
TStop W: Fernwood Truckstop*(LP)(SCALES) (Conoco)
Lodg W: Rodeway Inn (Fernwood TS)
TServ E: Interstate Supply
W: Fernwood TS
Parks W: Percy Quin State Park

10 MS 48, Magnolia
RVCamp E: Camping

8 MS 568, Gillsburg, Magnolia

4 Chatawa

(3) MS Welcome Center (Northbound)

(2) Weigh Station (Both Directions)

1 MS 584, Osyka, Gillsburg
RVCamp W: Camping

↑MISSISSIPPI

↓LOUISIANA

(65) LA Tourist Info, Rest Area - RR, Picnic, Phones, RV Dump (Southbound)

(64) Weigh Station (Northbound)

61 LA 38, Kentwood, Liverpool
FStop E: Chevron*
W: Texaco*
Gas E: Texaco(D)
Food E: Chicken Little, Kentwood Donut Shop, Kentwood Plaza, Pizza Shack, Sonic Drive In
AServ E: Kentwood Ford Mercury Dealership, Texaco
W: Gill Motor Co. Dodge Chrysler
RVCamp W: Great Discovery Camp (2.7 Miles)
ATM E: Service24
Other E: Bill's, Brown Morris Pharmacy, Car Wash, Sunflower Grocery

(59) Weigh Station (Southbound)

57 LA 440, Tangipahoa
RVCamp E: Camp Moore
Other E: Camp Moore Museum

(54) Rest Area - RR, Picnic (Northbound)

53 LA 10, Fluker, Greensburg

50 LA 1048, Arcola, Roseland
FStop E: Texaco*

47 LA 16, Amite, Montpelier

EXIT MISSISSIPPI

TStop **W:** Citgo Auto/Truck Plaza*(SCALES)
Gas **E:** Citgo*, Conoco*
W: Shell*, Texaco*
Food **E:** KFC, Master Chef, McDonalds (Playground), Mike's Catfish Inn, Original Hi-Ho BBQ, Popeyes Chicken, Subway
W: Stuckey's Express (Citgo TS)
Lodg **W:** Colonial Inn Motel (Citgo TS)
RVCamp **E:** Camping
Med **E:** ✚ Hospital
ATM **E:** Winn Dixie Supermarket
Other **E:** Amite Pets, Coin Laundry, Public Library, Speedy Car Wash, TG & Y, Thrift Town Pharmacy, Winn Dixie Supermarket (Pharmacy, 24 Hrs)
W: Car Wash

41 LA 40, Independence
Gas **E:** Conoco*
Food **E:** The Kingfish
RVCamp **W:** Indian Creek Campground (1.7 Miles)
Other **E:** Super Suds Laundromat

36 LA 442, Tickfaw
FStop **E:** Chevron* (Kerosene)
Food **E:** Restaurant (Chevron)
Other **E:** The Global Wildlife Center (15.3 Miles)

(33) **Rest Area - RR, Phones, Picnic, Grills, Pet Rest (Both Directions)**

32 LA 3234, Wardline Rd., Southeastern

EXIT MISSISSIPPI

LA Univ.
Gas **E:** Citgo*
Food **E:** Blimpie's Subs (Citgo), Burger King (Playground)

31 U.S 190, Albany, Hammond
TStop **E:** Fleet Truck Stop*
FStop **E:** Speedway*
Gas **E:** Chevron*(D), Exxon*
Food **E:** Cypress Palace Restaurant (Best Western Inn, Fleet TS), Fleet Truckstop Restaurant (24 Hrs), Shorty's Real Pit BBQ (Fleet TS), Subway (Exxon), TCBY (Exxon), Waffle House
Lodg **E:** Best Western (Fleet TS), Econo Motel, Super 8 Motel
W: Motel
AServ **E:** Chevron, Community Motors Chrysler Dodge, Expert CV Joint, James Tire Serv. (Fleet TS), Tire Centers Inc.
TServ **E:** James Tire Service (24 Hr Service, Fleet TS)
ATM **E:** Exxon, Speedway
Other **E:** Econo Laundry Mat, Ferrellgas(LP), Laundry (Fleet TS)

29AB Junction I-12, Baton Rouge, Slidell

28 U.S. 51 North, Hammond
Lodg **E:** Ramada Inn
TServ **E:** Big Wheel Diesel Repair
RVCamp **E:** KOA Campground
Med **E:** ✚ Hospital

EXIT MISSISSIPP

Other **E:** Sub Station Sheriff Department, Tourist Information, Watkins Car Wash

26 LA 22, Ponchatoula, Springfield
Gas **E:** Chevron*, Conoco*, Shell*
Food **E:** All Sport Pizza, Burger King, Catfish Charlie's, KFC, McDonalds (Play Land), Popeye's Chicken, Sonic Drive In, Wendy's
AServ **E:** Gateway Ford, USA Auto Service
Med **E:** ✚ Hospital
ATM **E:** Hancock Bank
Other **E:** Eckerd Drugs, Tony Turbo Car Wash, Winn Dixie Supermarket

23 U.S 51 Bus, Ponchatoula

22 Frontage Road (Southbound, Difficu Reaccess)

15 Manchac

7 Ruddock

1 U.S. 51, Junction I-10, La Place, Bator Rouge, New Orleans
FStop **E:** Speedway*
Gas **E:** Chevron*, Citgo*, Shell*(CW)
Food **E:** Shoney's, Waffle House, Wendy's
ATM **E:** Hibernia Bank, Speedway

↑LOUISIANA

Begin I-55

I-57 S →

EXIT ILLINOIS

Begin I-57

↓ILLINOIS

359 Junction I-94

357 IL 1, Halsted St
Gas **E:** Amoco*(CW), Mobil*
W: 76*, Shell*
Food **W:** McDonalds

355 111th St
Gas **W:** Amoco*

354 119th St

353 127th St, Burr Oak Ave
Gas **E:** Amoco*, Mobil*, Shell*
W: Amoco*, Gas City*(D)

EXIT ILLINOIS

Food **E:** Burger King, Dillinger's Gyros Etc., McDonalds, Subway
Lodg **E:** Super 8 Motel (see our ad this page)
AServ **E:** Firestone Tire & Auto, Goodyear Tire & Auto, Mobil
W: Meineke Discount Mufflers
Other **E:** Auto Parts Inc.

350 IL 83, Sibley Blvd, 147th St
Gas **E:** Marathon*, Mobil
W: BP*
Food **E:** Hugo's Gyro's Hamburgers & Hot Dogs
AServ **W:** Fair Muffler

348 U.S. 6, 159th St
Gas **E:** King*(D)
W: Citgo*(CW)
Food **E:** Burger King, Kickoff Grill, McDonalds, Subway, Taco Bell, USA #1 Family Restaurant, White Castle Restaurant
Lodg **E:** Hi-Way Motel
AServ **E:** Arrow Transmission, Firestone Tire & Auto, Rowe's Auto Repair, Value Plus Mufflers
W: Quik Lube (Citgo)
Other **E:** Markham Animal Clinic, WalGreens

346 167th St
Gas **E:** Amoco*(CW), Minuteman*(LP)

345AB Junction I-80, Indiana, Iowa (Difficult Reaccess)

342AB Vollmer Road
Gas **E:** Shell*

340AB U.S. 30, Lincoln Highway

EXIT ILLINOI

Gas **E:** Mobil*
Food **E:** Applebees, Baker's Square, Chuck E. Cheese' Pizza, Cracker Barrel, Fuddrucker's, J.N. Michael's, Old Country Buffet, Olive Garden, Pizza Hut
Lodg **E:** Budgetel Inn (see our ad opposite page), Hampton Inn, Holiday Inn (see our ad this page)
AServ **E:** Goodyear Tire & Auto
Other **E:** Lincoln Shopping Mall, Target Department Store, Venture, Wal-Mart

339 Sauk Trail
Gas **E:** Amoco*(CW), Shell*
Food **E:** Bozo's Hot Dogs, McDonalds, White Hen Pantry

335 Monee, Manhattan

EXIT ILLINOIS

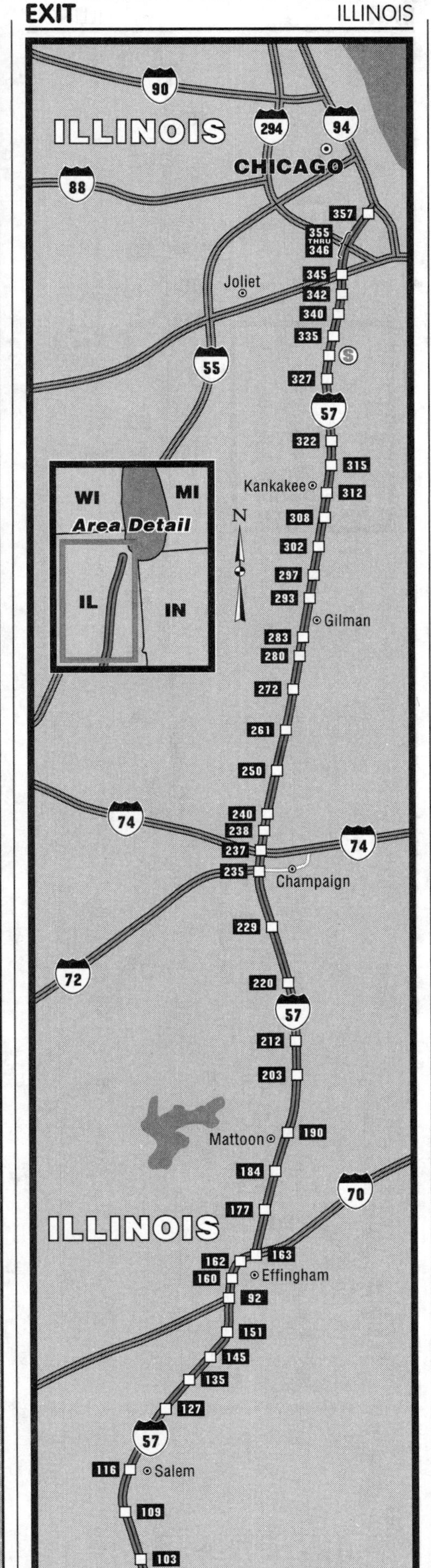

EXIT ILLINOIS

TStop	**E:** 76 Auto/Truck Plaza(SCALES)
FStop	**E:** Speedway*(SCALES) (ATM)
Gas	**E:** Amoco*
Food	**E:** Burger King (Amoco), Hardees, Leo's Gyros & Ribs, McDonalds (Amoco), Pizza Hut (Amoco)
Lodg	**E:** Country Host Inn
AServ	**E:** Amoco
TServ	**E:** 76 Auto/Truck Plaza
TWash	**E:** Blue Beacon Truck Wash

(332) Rest Area - RR, Phones, RV Dump, Picnic P (Both Directions)

(330) Weigh Station - both directions

327 Peotone, Wilmington

Gas	**E:** Shell*
Food	**E:** Dairy Queen (Shell), McDonalds, Taco Bell

322 Manteno

Gas	**E:** Amoco*, Phillips 66*
Food	**E:** Donut Land (Phillips 66), Hardee's, McDonalds (Amoco), Subway
Lodg	**E:** AAA Comfort Inn
ATM	**E:** Machine (Phillips 66)

315 IL 50, Bradley, Kankakee, Bourbonnais

Gas	**E:** Shell*
	W: Amoco*(D)
Food	**E:** Burger King, Cracker Barrel, Lone Star, Old Country Buffet, Pizza Hut, Red Lobster, White Castle Restaurant
	W: Applebee's, Bakers Square Restaurant, Boston Market, Denny's, Hardee's, Mancino's Pizza, McDonalds, Mongolian Buffet, Sirloin Stockade, Steak & Shake, Subway, Taco Bell, Wendy's
Lodg	**E:** ◆ Fairfield Inn, Hampton Inn, AAA Lees Inn
	W: Motel 6, ◆ Ramada Inn, ◆ Super 8 Motel
AServ	**E:** Midas Muffler & Brakes
	W: K-Mart
Other	**E:** Northfield Square Mall, Target Department Store
	W: K-Mart (w/Pharmacy), Wal-Mart (w/ Pharmacy)

312 IL 17, Kankakee, Momence

Gas	**W:** Clark*, Shell*
Food	**W:** McDonalds, River Oaks Restaurant, Subway, Wendy's
Lodg	**W:** DAYS INN ◆ Days Inn
AServ	**W:** Just Lube
TServ	**W:** International Truck
Other	**W:** WalGreens

308 U.S. 45, U.S. 52, Kankakee

Gas	**W:** Phillips 66*(D)
Food	**E:** Redwood Inn
Lodg	**W:** Fairview Motel
Other	**E:** Camping

302 Chebanse

Gas	**W:** 76*
TServ	**W:** Ken's Truck Repair

297 Clifton

Gas	**W:** Phillips 66* (ATM)
Food	**W:** Taco John's Express

293 IL 116, Pontiac, Ashkum

TServ	**W:** Bridgestone Tire & Auto

283 U.S. 24, Gilman, Chatsworth

TStop	**E:** Amoco(SCALES), Apollo Travel Center Citgo*(D) (ATM)

Bold red print shows RV & Bus parking available or nearby

EXIT		ILLINOIS
	Food	E: Dairy Queen, Gillman Restaurant, K & H Truck Plaza Resturaunt(D)(SCALES), McDonalds
	Lodg	E: Budget Host Inn, Days Inn ◆ Days Inn, AAA Super 8 Motel
	TServ	E: Falls Mastercraft Tires
	Other	E: Coin Car Wash
280		Onarga, Roberts, IL 54
	Gas	E: Phillips 66*(CW)
272		Buckley, Roberts
	Gas	E: Brad's Gas Station*
(269)		**Rest Area - RR, Phones, Vending, RV Dump P (Southbound)**
(268)		**Rest Area - RR, Phones, Vending, RV Dump P (Northbound)**
261		IL 9, Paxton, Gibson City
	Gas	W: Amoco*
	Food	E: Hardees, Monical's Pizza, Pizza Hut, Subway
		W: Country Garden Restaurant
	Lodg	W: Paxton Inn
	Other	E: Coin Car Wash
250		U.S. 136, Rantoul, Fisher
	Gas	E: Shell* (ATM)
	Food	E: Long John Silvers, McDonalds, Red Wheel Pancake & Steak House, Taco Bell
	Lodg	E: Best Western, Days Inn Days Inn, Super 8 Motel
	TServ	E: Grand Tool Equipment
	ATM	E: Central Illinois
240		Market St
	TStop	E: Amoco
	TWash	E: Amoco
238		Olympian Dr., Champaign
	Gas	W: Shell*
	Food	W: Dairy Queen
237AB		Junction I-74 Peoria, Indianapolis (Difficult Reaccess)
235AB		Junction I-72 (Difficult Reaccess)
229		Savoy, Monticello
	FStop	E: Speedway*(D)
(222)		**Rest Area - RR, Phones P (Southbound)**
(221)		**Rest Area - RR, Phones, Vending, Picnic, Playground P (North-bound)**
220		U.S. 45, Pesotum, Tolono
	Gas	W: Citgo*, Gasland*
	Food	W: Lakeside Resturaunt
	Other	E: State Patrol Post
212		U.S. 36, Tuscola, Newman
	TStop	E: Fuel Mart*(D)
		W: Dixie Truckers Home*(D)(SCALES), Pilot Travel Center*(D)(SCALES)
	Food	W: Burger King, Dixie Resturaunt, Hardee's, McDonalds (Outdoor Playground), Subway
	Lodg	W: Ameri Host Inn, Cooper Motel, Super 8 Motel
	TServ	W: Dixie Truckers Home
	RVCamp	E: Campground
	ATM	W: Dixie Truckers Home, Tuscola National Bank
	Other	W: Coin Car Wash, Factory Outlet Center
203		IL 133, Arcola, Paris
	FStop	E: Gasland*(D)
	Gas	W: Marathon*(LP), Phillips 66*, Shell*, Union 76
	Food	W: Dairy Queen, Hardee's (Outdoor Playground), Hen House
	Lodg	W: ◆ Amish Country Inn, Budget Host Inn, ◆ Comfort Inn
	AServ	E: Jerry's Radiator Shop
	TServ	E: Truck Repair
	RVCamp	W: Campground
	Other	W: Car Wash
190AB		IL 16, Mattoon, Charleston
	Food	W: Alamo Steak House, Cody's, Country Kitchen, Little Caesars Pizza (K-Mart), McDonalds (Indoor Playground), Pondorosa, Steak & Shake,

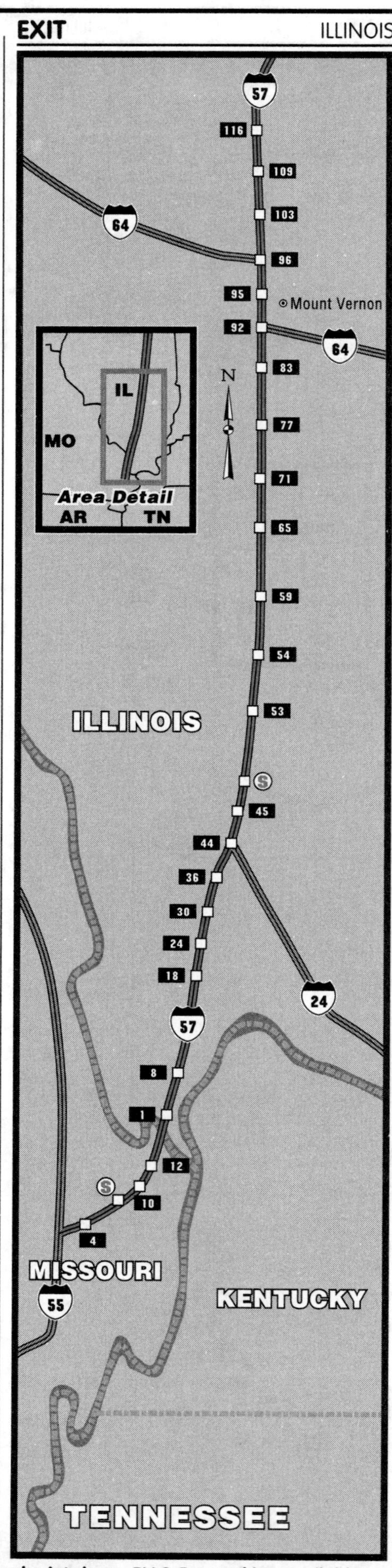

EXIT		ILLINOIS
		Taco Bell, Wendy's
	Lodg	W: Fairfield Inn, Hampton Inn, Ramada Inn, Super 8 Motel
	AServ	W: Sear's
	Med	E: ✚ Hospital
	ATM	W: Central National Bank
	Parks	E: State Park
	Other	W: JC Penny, K-Mart (w/Pharmacy, Auto Service), Mattoon Dental, Osco Pharmacy, Sear's (Express Eye Care), Wal-Mart Supercenter (w/Pharmacy, Wal-Mart Optical Tire Center, Express Lube), WalGreens
184		U.S. 45, IL 121, Toledo, Mattoon
	Gas	W: Marathon*(D), Shell*
	Food	W: McDonalds (Outdoor Playground), Old Country Restaurant (HoJo)
	Lodg	W: Budget Inn (Outdoor Playground), Howard Johnson
177		U.S. 45, Neoga
	FStop	E: Shell*(D)
	Gas	E: Citgo*
	Food	E: Shell
	RVCamp	W: Over Night Campground
(167)		**Rest Area - RR, Phones, RV Dump, Picnic, Playground P (South-bound)**
(166)		**Rest Area - RR, Phones, RV Dump, Picnic, Playground P (North-bound)**
163		Junction I-70 East, Indianapolis
162		U.S. 45, Sigel, Effingham
	Gas	E: Moto Mart*
		W: Citgo*, Shell*(D)
	Food	W: Trailways Restaurant
	Lodg	W: Budget Host Inn
	AServ	W: Shell
	RVCamp	W: Campground
	ATM	W: Citgo
160		IL 32, IL 33
	TStop	W: Bobbers Truck Plaza*(SCALES) (RV Dump), Trucks America Truck Plaza*(SCALES)
	Gas	E: Amoco*, Shell
		W: Phillips 66, Shell*(D)
	Food	E: Dixie Cream Donut Shop, Little Caesars Pizza, Papa John's Pizza, Pizza Hut
		W: Arby's, Blimpie's Subs, Bobber Restaurant, Bonanza Steak House, Burger King (Outdoor Playground), Cracker Barrel, Denny's, El Rancherito, K Square Food Court, KFC, Long John Silvers, McDonalds (Inside Wal-Mart), McDonalds, Ramada Inn, Ryan's Steakhouse, Steak & Shake, Stix B B Q, Subway (Shell), T.G.I. Friday's, Taco Bell, Trucks America, Wendy's
	Lodg	E: AAA Amerihost Inn, ◆ Hampton Inn
		W: Best Inns of America, Budgetel Inn, Econolodge, ◆ Ramada
	AServ	E: Auto Zone Auto Parts, Shell
		W: Ken Diepholz Ford Licoln Mercury, Phillips 66
	TServ	W: Bobbers Truck Plaza, Speedco Truck Service, Trucks America
	TWash	W: Trucks America
	Med	E: ✚ Hospital
	ATM	E: Amoco, Crossroads Bank, First Mid- Illinois
		W: Bobbers Truck Plaza, Illinois Community Bank
	Other	E: Aldi Grocery, Effingham Veterinary Clinic, Ever Clean Car Wash, K-Mart, Kroger Supermarket (Open 24 Hours), Rollin Hill Laundromat, Super X Pharmacy
		W: Factory Outlet Center
159		Effingham
	TStop	E: 76 Auto/Truck Plaza(LP)(SCALES)
		W: Petro(SCALES) (Open 24 Hours), Truck-O-Mat(SCALES)

Bold red print shows RV & Bus parking available or nearby

XIT ILLINOIS

FStop E: **Speedway*(SCALES) (RV Dump)**
Gas E: Amoco*(CW), Clark, Phillips 66*(D)(CW), Shell* (Open 24 Hours), **Speedway*(D)(SCALES)**
Food E: Domino's Pizza, G Wilkes Bar and Grill, Golden China Restaurant, Hardee's, Little Caesars Pizza, Niemerg's Family Dining, Spaghetti Shop, Subway, The China Buffet
W: Petro
Lodg E: Abe Lincoln Hotel, Comfort Suites, ◆ **Days Inn**, Holiday Inn, **Howard Johnson**, Paradise Inn, ◆ **Quality Inn**
W: (AAA) **Best Western**
AServ E: Amoco, Effingham Tire Center, Firestone Tire & Auto, Pennzoil Lube, Shell
TServ E: **76 Auto/Truck Plaza, Effingham/International Truck Sales, Firestone Tire & Auto**
W: **Petro**
TWash E: **76 Auto/Truck Plaza**
W: **Truck-O-Mat**
Other E: Car Wash, **Sav-A-Lot Grocery**
57 I-70 West, St Louis, Indianapolis
51 Watson, Mason (Difficult Reaccess Southbound)
45 Edgewood
Gas E: Citgo*(LP)
W: Shell*
Other E: **American Legion (Picnic, Playground)**
35 IL 185, Farina, Vandalia
FStop E: **76 Auto/Truck Plaza**
Food E: **Kountry Kitchen (76)**
27 Patoka, Kinmundy
Gas E: Marathon
AServ E: Kinmundy Diesel & Auto Repair
TServ E: **Kinmundy Diesel & Auto Repair**
16 U.S. 50, Salem, Sandoval (Steven Forbes State Park)
Gas E: Amoco*, Clark*(D), Shell*
W: Phillips 66*
Food E: Austin's Fried Chicken, Burger King, Golden Corral, Hunan Gardens Chinese Restaurant, KFC, Kim's Chinese Restaurant, Long John Silvers, Pizza Hut, Pizza Man, Subway, Taco Bell, Wendy's
W: Denny's (see our ad this page)
Lodg E: Budget Inn
W: ◆ **Holiday Inn**, ◆ **Super 8 Motel**
AServ E: Auto Zone Auto Parts, Shell, Ten Minute Oil Change
W: Salem Tire Center
TServ W: **Salem Tire Center**
Med E: **+ Hospital**
ATM E: NationsBank
Other E: **Wal-Mart (Pharmacy)**, Westgate Car Wash
W: **Amerigas Liquid Propane**
114) **Rest Area - RR, Phones, Picnic, Playground, Vending (P) (Both Directions)**
109 IL 161, Centralia
Gas W: Phillips 66*
Food W: Phillips 66
103 Dix
Gas E: **Marathon(D)**
Food E: Austin's Restaurant
Lodg E: **Scottish Inns**
96 I-64 (Difficult Reaccess)
95 IL 15, Mt. Vernon, Ashley
TStop W: **76 Auto/Truck Plaza(SCALES) (Open 24 Hours)**
FStop W: **Huck's*(SCALES)**
Gas E: Amoco*, Hucks Food Store*, Phillips 66*(CW), Shell, Speedway*
W: Mount Vernon Center (76 Auto/Truck Stop), Shell*
Food E: Bonanza Steak House, Burger King, China

EXIT ILLINOIS

Buffet, Denny's, El Rancherito, Fazoli's Italian Food, KFC, Little Caesars Pizza, Long John Silvers, McDonalds, Pizza Hut, **Pizza Pro (Kroger)**, Shoney's, Steak & Shake, Subway, Wendy's, Western Sizzlin'
W: 76 Auto/Truck Plaza, Applebee's, Arby's, Burger King, **Cracker Barrel**, Hucks Country Cooking, McDonalds (Indoor Playground)
Lodg E: Best Inns of America, **Best Western**, ◆ **Drury Inn**, Motel 6, Ramada Inn, ◆ Super 8 Motel, Thrifty Inn
W: (AAA) Comfort Inn, Holiday Inn
AServ E: Auto Zone Auto Parts, Midas Muffler and Brakes, Shell
TServ W: **76 Auto/Truck Plaza**
Med E: **+ Hospital**
ATM E: Bank of Illinois, Citizen's Bank, Merchantile Bank, NationsBank
W: 76 Auto/Truck Plaza
Other E: **K-Mart (Pharmacy), Kroger Supermarket**
W: **Wal-Mart Supercenter (Pharmacy, Tire Center, Open 24 Hours)**
92 I-64 East, Louisville (Difficult Reaccess)
83 Ina
FStop E: **BP*(LP)**
(79) **Rest Area - RR, Phones (P) (Southbound)**
77 IL 154, Sesser
RVCamp E: **Holiday Travel Park**
Other W: **Wind Lake State Park Recreation Area and Resort**
(74) **Rest Area - RR, Phones, Picnic, Playground, Pet Walk (P) (Northbound)**
71 IL 14, Benton, Christopher
Gas E: 76*, Citgo*, Coastal*, Han-dee Mart(LP)
W: Shell*(D)
Food E: Hardee's, Pizza Hut, The Gardens Restaurant (Days Inn), Wendy's
W: Bonanza Steak House, **McDonalds (Outdoor Playground)**, Subway, Taco John's
Lodg E: ◆ Days Inn, Gray Plaza Motel, ◆ Super 8 Motel
AServ S: Southern Illinois Tire
Med E: **+ Hospital**
ATM E: NationsBank
Other E: Car Wash, **Laundromat**
W: **Big John's, Revco, Wal-Mart (w/Pharmacy)**
65 IL 149, West Frankfort, Zeigler
Gas E: Amoco*, Citgo*(D), Clark On The Go*, Shell* (Ice Cream Churn, 24 Hrs)
Food E: Dixie Creme Doughnuts, Hardee's, Hungry's Pancake House, KFC, Long John Silvers, Mike's

EXIT ILLINOIS

Drive-In, Pizza Inn
W: McDonalds (Indoor Playground)
Lodg E: Gray Plaza
W: HoJo Inn
AServ E: Big A Auto Parts, Speed Lube
ATM E: Banterra Bank
Other E: Car Wash
W: **K-Mart, Kroger Supermarket, Revco Drugs**
59 Johnston City, Herrin
Gas E: Shell* (Open 24 Hours, No Trucks)
Food E: Hardee's
AServ E: KIP's Tires
RVCamp E: **Campground**
Med W: **+ Hospital**
54AB IL 13, Carbondale, Murphysboro, Marion
TStop W: **Marion Truck Plaza*(D)(SCALES) (BP)**
Gas E: Circuit Independent, Citgo*(D)(CW), Phillips 66*(CW)
W: Amoco*
Food E: Arby's, Bantera Bank, Fazoli's Italian Food, Hardee's, KFC, La Fiesta Mexican Restaurant, Little Caesars Pizza, Long John Silvers, Monster Spud Baked Potatoes, Nong Chen Chinese, Papa John's Pizza, Pizza Hut, Shoney's, Subway, Tequilas Mexican Restaurant, Wendy's
W: Bob Evans, Burger King, Grand China Restaurant, Holiday Inn, Marion Truck Plaza, **McDonalds (Indoor Playground)**, Ryan's Steakhouse, Sonic Drive In, Steak 'N Shake, Taco Bell
Lodg E: Gray Plaza Motel, Red Lion Inn, Shoney's Inn
W: **Best Inns of America**, Drury Inn, Holiday Inn, **Motel 6, Super 8 Motel**
AServ E: Instant Oil Change
TServ W: **Vernell's Intestate Service (Marion Truck Plaza, Open 24 Hours)**
Med E: **+ Columbia Eye Surgery Center of Illinois, + Twenty-first Century Dental**
ATM E: Bank of Marion, Central Bank, Charter Bank, NationsBank
Other E: **Kroger Supermarket, Marion Center, Marion Plaza Mall, Optometrist, Pet Wellness Center, Town & Country Plaza**
W: **Sam's Club, Wal-Mart**
53 Main St, Marion
Gas E: 76*(D), Coastal*, Shell*
W: Motomart*
Food E: Dairy Queen
W: **20's Hideout Steakhouse, Cracker Barrel**
Lodg E: Motel Marion
W: (AAA) Comfort Inn, (AAA) **Comfort Suites**
AServ E: 76
Med E: **+ Hospital, + VA Hospital**
(46A) Wiegh Station (Southbound)
(46) **Weigh Station - both directions (Northbound)**
45 IL 148, Herrin
TServ W: **Doc's Diesel Repair**
RVCamp E: **Camping**
44 Junction I-24 East, Nashville
40 Goreville Road
FStop W: **Scenic Ridge Plaza (Citgo)**
Food W: **Scenic Ridge Plaza**
RVCamp E: **Camping, Hilltop Campgrounds**
Parks E: **Fern Cliff State Park**
Other E: **Scenic Overlook**
36 Lick Creek Road
Other E: **Lick Creek General Store**
(32) **Rest Area - RR, Phones (P) (Southbound)**
(31) IL Welcome Center - RR, Phones, Vending, Playground (Both Direc-

← N I-57

EXIT		ILLINOIS
		tions)
30		IL 146, Anna, Vienna
	FStop	W: Shell*(LP)
	Food	W: Sam's Sizzlin's Smokehouse (Shell)
	AServ	W: Martin's Auto Service
	Other	E: Vienna Ranger Station (14 Miles)
		W: Joneboro Ranger Station (9 Miles)
24		Dongola Road
	Gas	W: Shell*(LP)
18		Ullin Road
	Gas	W: BP
	Food	W: Cheeko's
	Lodg	W: (AAA) Cheekwood Inn (Best Western)
	AServ	W: BP
	Other	E: Olmsted Locks & Dam
		W: State Police Headquarters
8		Mounds Road
	FStop	E: K & K Auto & Truck Stop*
	AServ	E: K&K
	TServ	E: K&K
	Other	E: Olmsted Locks & Dam
1		IL 3 to U.S. 51, Cairo

EXIT		ILLINOIS/MISSOURI
	TStop	E: Cairo Truck Stop (BP)
	Gas	E: Amoco*(D) (24 hrs)
	Lodg	E: (Days Inn) Days Inn
	TServ	E: Cairo Truck Stop (BP)
	RVCamp	E: Camping
		W: Camping
	Med	E: + Medical Care Center
	ATM	E: Cairo Truck Stop (BP)

↑ ILLINOIS

↓ MISSOURI

(18)		Weigh Station - both directions
12		U.S. 62, US 60, MO 77, Charleston, Wyatt
	TStop	E: Sunshine Travel Center*(SCALES)
	FStop	E: Phillips 66*
		W: The Flagstop
	Food	E: Apple Valley Cafe (Sunshines Travel Center)
		W: Charleston Restaurant (Charleston Inn Motel), KFC
	Lodg	E: Economy Motel
		W: Charleston Inn Motel (The Flagstop TS)
10		MO 105, Business Loop I-57,

EXIT		MISSOURI
		Charleston, East Prairie
	TStop	E: Pilot*(SCALES) (24 hrs)
	FStop	E: MFA Oil(LP)
	Gas	E: Boom-land* (Fireworks)
		W: Casey's General Store*
	Food	E: Subway (Pilot), Wally's Chew-Chew (Boomland)
		W: China Buffet, Dairy Queen, McDonalds, Pizza Hut
	Lodg	W: (AAA) Comfort Inn
	AServ	W: Automobile Mechanic Shop
	TServ	E: Pilot
	TWash	E: Pilot
	RVCamp	E: Camping, Sam's Camping
	ATM	E: Boom-land, Pilot
		W: Union Planter's Bank
	Other	W: Town & Country Supermarket(LP), Wal-Mart
4		Route B, Bertrand
1AB		1A - I-55 South, Memphis, 1B - I-55 North, St Louis
0		Junction I-55, I-57

↑ MISSOURI

Begin I-57

I-59 S →

EXIT		GEORGIA
		Begin I-59

↓ GEORGIA

4		(19) West I-24 Nashville
3		(18) Slygo Rd, New England
	Gas	W: Citgo*(D)(CW)
	RVCamp	W: KOA Kampground (1.7 Miles)
	ATM	W: Citgo
2		(12) GA 136, Trenton (Covenant College)
	Gas	E: Chevron(CW), Citgo*(LP) (Kerosene), Exxon*(D)
		W: Amoco*(LP) (Kerosene), Citgo*(D)(LP)
	Food	E: Burger King (Playground), Deli Dipper, Hardee's, McDonalds (Playground), Pizza Hut, Subway
		W: Huddle House, TCBY (Amoco), Taco Bell
	AServ	E: Chevron, NAPA Auto Parts, Smith's Auto Parts
	RVCamp	E: Lookout Lake Camping (7 Miles)
	ATM	E: Citgo, Citizens Bank & Trust Inc., Exxon
		W: Citgo
	Parks	E: Cloudland Canyon State Park
	Other	E: Bi-Lo Supermarket, Covenant College, Dade

EXIT		GEORGIA/ALABAMA
		County Chamber of Commerce Welcome Center, Eye Care Optical, Ingles Supermarket, Ponder Pharmacy, Revco, Tourist Information, Trenton Christian Books & More
(11)		Parking Area, Scenic View, Scales - No Services (Both Directions)
1		(4) Rising Fawn
	FStop	E: Citgo*(LP) (Kerosene)
		W: Mapco Express (see our ad this page)
	Gas	W: Amoco*(D)(LP) (Kerosene)
	Food	W: Rising Fawn Cafe
	ATM	E: Citgo Fuel Stop
	Other	W: Fox Mountain Trout Farm (1.5 Miles)

↑ GEORGIA

↓ ALABAMA

(241)		AL Welcome Center - Picnic, RR, HF, Phones, RV Dump (Southbound)
239		To U.S 11, Sulphur Springs Rd
	Lodg	E: Freeway Motel
	RVCamp	E: Sequoyah Caverns & Campground (3.5 Miles)
	Other	E: Sequoyah Caverns & Campground
231		AL 40, AL 117, Hammondville, Vly Head, Stevenson, Bridgeport
	Gas	W: Texaco
	AServ	W: Texaco
	RVCamp	E: Sequoyah Caverns Campground (5.3 Miles)
	Other	E: Sequoyah Caverns
		W: I-59 Flea Market
222		U.S 11, Fort Payne
	Gas	E: Shell
		W: Exxon*(D)(LP), Texaco*(D) (Kerosene)
	Food	W: Royal Waffle King (24 Hrs)
	AServ	W: Airport Tire & Auto Service, Harper Valley Auto Parts
218		AL 35, Fort Payne, Rainsville,

EXIT		ALABAMA
		Scottsboro
	TStop	W: Pure Truck Plaza
	Gas	E: Conoco*(D)(LP) (Kerosene)
		W: Exxon*(CW), Phillips 66* (Harco), Shell*
	Food	E: Captain D's Seafood, Central Park, Pizza Hut, Shoney's, Taco Bell
		W: Burger King (Playground), Hardee's, Huddle House, Little Caesars Pizza (K-Mart), McDonalds (Wal-Mart), Old Times Cafe & Family Restaurant, Quincy's Family Steakhouse, The Valley Cafe, Waffle House (Quality Inn)
	Lodg	W: (Days Inn) (AAA) Days Inn, (AAA) Quality Inn (FedEx Drop) (see our ad this page)
	AServ	W: Wal-Mart
	TServ	W: Pure Truck Plaza
	Med	W: + Dekalb Baptist Medical Center
	ATM	W: Exxon
	Parks	E: De Soto State Park
	Other	E: De Soto Falls, Little River Canyons
		W: Abbott Marine Sales & Service, Food World, Harco Pharmacy Express, K-Mart (w/ pharmacy), Wal-Mart Supercenter (24 Hrs, Pharmacy, One Hr Photo)

Bold red print shows RV & Bus parking available or nearby

EXIT	ALABAMA
205	AL 68, Collinsville, Crossville
FStop	W: Pure (24 Hrs), Texaco*(LP)
Gas	W: Chevron*, Shell* (24 Hrs)
Food	W: Big Valley Restaurant (HoJo Inn, 24 Hrs), Jack's (Playground), The Ice Cream Churn (Texaco)
Lodg	W: ◆ HoJo Inn
AServ	W: Pure (24 Hrs)
TServ	W: Pure (24 Hrs)
188	AL 211, To U.S 11, Noccalula Falls, Gadsden, Reece City
Gas	E: Chevron*
	W: Amoco*(D)(LP), I-59 Echo*(LP) (Kerosene)
Food	E: Noccalula Station Restaurant
RVCamp	E: Noccalula Campgrounds
183	U.S 278, U.S 431, Gadsden, Attalla (Robert Trent Jones Golf Trail)
FStop	E: Amoco* (Showers), BP(D)
Gas	E: Shell*, Texaco*(D)(LP) (Kerosene)
	W: Amoco*, Chevron*(LP) (24 Hrs), Exxon*(CW)(LP), Texaco*(LP)
Food	E: Hardee's, Magic Burger, The Ice Cream Churn (Texaco), Waffle House, Wendy's
	W: KFC, McDonalds (Playground), Pizza Hut, Quincy's Family Steakhouse, Shoney's, Subway, Taco Bell
Lodg	E: Columbia Inn, (AAA) Holiday Inn Express (see our ad this page), Rodeway Inn
	W: Econolodge
AServ	E: Attalla Auto Parts, BP
	W: Cook's Auto Body Shop, Exxon
ATM	E: The Exchange Bank of AL
Other	E: AL State Troopers Gadsden Post

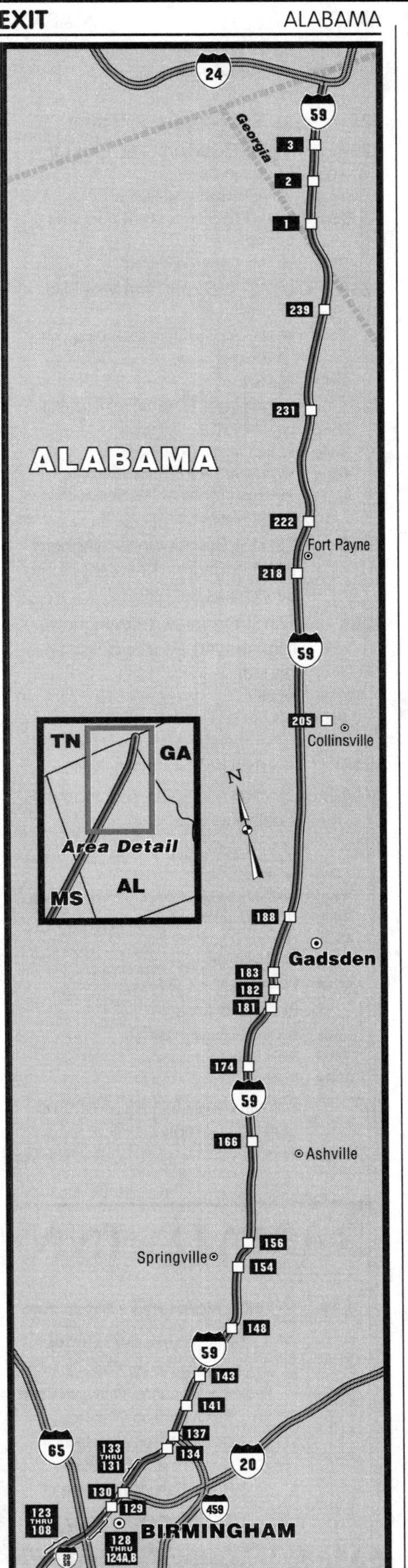

EXIT	ALABAMA
	W: Sunbeam Bakery Thrift Store
182	Junction I-759, Gadsden
Lodg	E: Super 8 (see our ad this page)
181	AL 77, Attalla, Rainbow City
FStop	W: BP*
Food	W: Hardee's, Subway
Lodg	E: (DAYS INN) Days Inn
Other	E: Flowers Bakery Thrift Store
174	Steele
TStop	W: Steele City Truck Stop* (CB Shop, Showers, Spur Fuel)
FStop	W: Chevron*(LP) (Kerosene)
Food	W: Restaurant (Steele City Truck Stop)
TServ	E: Truck Tire Service
	W: Steele City Truck Stop (24 Truck & Tire Service)
TWash	W: Steele City Truck Wash
ATM	W: Steele City Truck Stop
Other	E: AL International Dragway
(167)	**Rest Area - RR, Picnic, Phones, RV Dump, Pet Walk, Vending (Southbound)**
166	U.S. 231, Oneonta, Ashville
FStop	E: Exxon*
	W: Texaco*
Gas	E: Chevron*(D)(LP)
Food	W: Taco Bell (Texaco)
Other	W: Horse Pins Forty
(165)	**Rest Area - Picnic, RR, HF, Phones, RV Dump (Northbound)**
156	AL 23, Saint Clair Springs, Springville
154	AL 174, Springville, Odenville
Gas	W: Chevron*(CW)(LP) (24 Hrs), Texaco*(CW)
Food	W: Jack's Hamburgers (Playground), McDonalds
ATM	W: Texaco
148	To U.S. 11, Argo
Gas	E: BP*(LP) (Kerosene), Citgo*(D)
Food	E: Argo Cafe
AServ	E: Code 3 Automotive
Other	E: Argo Animal Clinic, Argo Optical, Argo True Value Hardware
143	Deerfoot Pkwy, Mt Olive Church Rd
141	Trussville, Pinson
FStop	E: Amoco(D)(CW)(LP)
Gas	E: BP*(D)(CW), Shell*(D)(CW) (Kerosene)
	W: Chevron*(CW) (24 Hrs), Shell*
Food	E: Big Dragon Chinese Restaurant, McDonalds (Playground), Pizza Hut, Subway (Shell), Taco Bell, Waffle House, Wendy's
	W: Burger King, Dairy Queen, Little Caesars Pizza, Paul's Hot Dogs, Ruby Tuesday
AServ	W: Express Lube
ATM	W: Compass Bank, SouthTrust Bank, Western Supermarket
Other	W: Big "B" Drugs, K-Mart (w/pharmacy), Tee Time Golf, Total Pet Care Hospital, Western Supermarket
137	Junction I-459 South, Tuscaloosa, Montgomery
134	To AL 75, Roebuck Pkwy
Gas	W: BP*(D), Chevron*(CW)(LP) (24 Hrs), Citgo
Food	W: Baskin Robbins, Chick-fil-A, Chuck E. Cheese's Pizza, Denny's, Guadalajara Mexican Restaurant, Johnny's Ray's, Krystal (24 Hrs),

EXIT	ALABAMA
	McDonalds (Playground), Milo's Hamburgers, Mrs Winner's Chicken, Pioneer Cafeteria, Waffle House, Wok Cuisine
Lodg	**W:** Parkway Inn
AServ	**W:** Citgo, Goodyear Tire & Auto, **K-Mart**, Marty's Transmission, NTB Complete Car Care, Peach Auto Painting & Collision, Roebuck Honda, Roebuck Parts & Auto Service
Med	**W:** **Hospital**
ATM	**W:** AM South Bank, Colonial Bank, Compass Bank, SouthTrust Bank
Other	**W:** **Alabama Medical (Medical Supply), Cara's Coin Laundry, Carol & Paul's One Hr Photo, EyeCare, Jefferson State College, Joshua's Christian Bookstore, K-Mart (Pharmacy), Polly's Pet Parlor, Putt America Carpet Golf, Revco (24 Hrs, One Hr Photo), The Mail Center**
133	4th Ave S (Northbound)
Gas	**E:** Phillips 66(CW)
Food	**E:** Arby's, Catfish Cabin Seafood Family Restaurant, Papa John's Pizza, Pasquale's Pizza
AServ	**E:** Phillips 66, Roebuck Chrysler Plymouth Jeep Eagle, Roebuck Mazda Kia
Parks	**E:** **Don A. Hawkins Park**
Other	**E:** **Bruno's Supermarket**, Car Wash, **Drugs for Less Pharmacy (1 Hr Photo), US Post Office**
132	U.S 11, 1st Ave N
Gas	**W:** Exxon
AServ	**W:** East Lake Auto Parts, Exxon
RVCamp	**E:** **Dandy RV Parts & Sales (1/3 Mile)**
131	Oporto - Madrid (Northbound, Difficult Reaccess Northbound)
Gas	**E:** Exxon*(CW)
Food	**E:** Andrew's Bar-B-Q, Burger King, Church's Fried Chicken, Joe & Deb's, Rally's Hamburgers, Subway, Taco Bell
AServ	**E:** East Lake Tire Center
RVCamp	**E:** **Dandy RV Parts & Service**
Other	**E:** **Southern Museum of Flight, U-Move U-Store(LP), Wood's V & S Drugs**
130	Junction I-20 East, Atlanta
Gas	**S:** Exxon*, Shell(CW), Texaco*
Food	**N:** Restaurant (Ramada Inn)
	S: Blimpie's Subs, Huddle House
Lodg	**N:** Ramada Inn (Restaurant)
	S: Days Inn, Holiday Inn
AServ	**S:** Airport Car Wash (Self Wash)
130A	US 11 South, 1st Ave N (Northbound, Exit 130 Ramp)
Food	**E:** Captain D's Seafood, McDonalds, Mrs. Winner's Famous Recipe Chicken & Biscuits
Lodg	**E:** Motel American, Relax Inn, Sky Inn
Other	**E:** **Merita Bakery Thrift Store**
130B	US 11 North, 1st Ave N (Northbound, Exit 130 Ramp)
Gas	**E:** Amoco, Chevron* (24 Hrs)
AServ	**E:** Amoco, Auto Zone Auto Parts, Eastern Alternator & Starter Service, Western Auto, Westwood Auto Parts
RVCamp	**E:** **Colonial RV**
Other	**E:** **Fire Station, Food Fair**
129	Airport Blvd
Gas	**E:** Exxon*, Shell*(CW), Texaco*(D)
Food	**E:** Blimpie's Subs (Texaco), Hardee's (24 Hrs), Huddle House, Sammy's Sandwich Shop
Lodg	**E:** ◆ Days Inn, Holiday Inn
	W: Ramada Inn (see our ad this page)

EXIT	ALABAMA
ATM	**E:** Shell
Other	**E:** **Woodlawn Mart**
	W: Airport Car Wash
128	AL 79, Tallapoosa St, Tarrant
126B	31st St, Sloss Furnaces, Civic Ctr
Gas	**W:** Circle K*, Texaco*
Food	**W:** McDonalds, Sol's Hotdogs
TServ	**E:** **Kurt's Truck Parts, Liberty Truck Sales & Service**
Other	**W:** **Sani-Clean Laundromat**
126A	U.S. 31, U.S 280, Carraway Blvd
Gas	**W:** Citgo
Food	**W:** Church's Fried Chicken, Ed's Diner, Rally's Hamburgers
AServ	**W:** Citgo
125	22nd Street, Downtown (Left Exit)
Food	**W:** Sophia's Deli
Lodg	**W:** Best Western, Sheraton
Other	**W:** **AL Sports Hall of Fame, Birmingham Jefferson Civic Center, Boutwell Auditorium, Tourist Information**
125B	22nd St, Downtown Birmingham (Northbound, No Reaccess, Same Services As Exit 125)
125A	17th St, Downtown Birmingham (Northbound, Reaccess Southbound)
Gas	**E:** Gem*
Food	**E:** Burger King
Other	**E:** **AL School of Fine Arts, Fire Dept.**
124B	Junction I-65 North, Nashville
124A	Junction I-65 South, Montgomery (Southbound Left Exit)
123	U.S 78, Arkadelphia Rd
Gas	**N:** Amoco*, Chevron*, Shell*(CW)
Food	**N:** Charley's Cafe, Denny's, Popeye's Chicken
Lodg	**N:** AAA La Quinta Inn
AServ	**N:** AAMCO Transmission
Med	**W:** **Hospital**
Other	**W:** **Birmingham S. College**
121	Bush Blvd, Ensley
Gas	**N:** Citgo*, Exxon*(LP), Pure
Food	**N:** Fat Burger
AServ	**N:** Pure
120	20th St, Ensley Avenue, Alabama State Fair Complex (Difficult Reaccess)

EXIT	ALABAMA
Gas	**N:** Crown*(D) (ATM)
	S: BP*, Jiffy Mart #4*
Food	**N:** KFC
AServ	**N:** Wayne Gargus GMC (Truck & Auto Service)
	S: Jim Burke Chevrolet, Limbaugh Toyota
ATM	**N:** Crown*(D)
Other	**S:** **Washing Well Coin Laundry**
119B	Avenue 1 (Westbound, Difficult Reaccess)
119A	Lloyd Nolan Pkwy
Gas	**N:** Amoco*(D), Chevron*
	S: Exxon, Texaco*
Food	**N:** Burger King, McDonalds, **Mrs Winner's Chicken**, Subway, Taco Bell, Wingo's Buffalo Wings
	S: Omelet Shoppe
AServ	**N:** CloverLeaf Auto Center
	S: B&D Wrecker, Big Moe Spring & Alignment, Exxon, Mr. T's, Northside Auto Service, Sharp Auto Supply
TServ	**S:** **B&D Wrecker**
Med	**S:** **Hospital**
Other	**N:** **Food Fair Grocery Store**, S&J Wash House (Coin Laundry)
	S: **Dolly Madison Thrift Store, Jim Clay Optician's**
118	Valley Rd, Fairfield
Lodg	**S:** Villager Lodge
AServ	**S:** Alton Jones Auto Service, Fairfield Transmission, NTB Warehouse, Scogin Bros Auto Repair, Valley Road Auto Parts
Other	**S:** **Home Depot(LP), Miles College Historic District**
115	Jaybird Rd, Midfield, Pleasant Rd
113	18th Avenue, Brighton
FStop	**S:** **Speedway***
Food	**S:** **McDonalds**
112	18th St, 19th St, Bessemer
Gas	**N:** Amoco*
	S: Chevron*(CW)
Food	**N:** Jack's Hamburgers
AServ	**N:** O.K. Tire & Battery Co.
	S: City Auto Parts
Other	**N:** **Mick's True Value Hardware(LP)**
	S: **B&C Rental(LP), FMS Pharmacy**
108	U.S 11, AL 5, Academy Drive
Gas	**S:** BP*(CW)
Food	**S:** Little Caesars Pizza
AServ	**S:** Tire & Lube Express (Wal-Mart)
Other	**S:** **Baskin Robbins (Wal-Mart), Drugs for Less, Wal-Mart (Tire Lube Express & Pharmacy), Winn Dixie Supermarket**
106	Junction I-459 North, Gadsden, Atlanta
104	Rock Mountain Lakes
FStop	**S:** **Flying J Travel Plaza*(LP)(SCALES) (RV Dump)**
Food	**S:** **Thads Restaurant*(LP)(SCALES) (Flying J Travel Plaza)**
100	Abernant, Bucksville
TStop	**S:** **Petro**
Gas	**S:** **Chevron (Petro TStop)**, Phillips 66* (ATM), Texaco*(LP)

Bold red print shows RV & Bus parking available or nearby

EXIT ALABAMA

Food S: Iron Skillet (Petro)
AServ S: McKinney Wrecker 24 Hr.
TServ S: Petro(SCALES)
RVCamp N: KOA Campground
ATM S: Phillips 66*(LP)
Parks S: Tannehill State Park

97 U.S 11 South, AL 5 South, West Blocton, Centerville
FStop S: Amoco*, Texaco*(LP)
Gas S: Complete Mini Mart*, Shell*(LP)
Food S: Dot's Farmhouse, KFC

86 Brookwood, Vance
TStop N: Shell* (Restaurant)
Food N: Restaurant (Shell TS)
TServ N: Shell
RVCamp S: Lakeside RV Park
Other S: Crawford Groceries

(85) Rest Area - RR, Phones, HF, RV Dump

79 U.S 11, University Blvd, Coaling

77 Cottondale
TStop N: Truck Stops Of America*(SCALES) (RV Dump)
Gas N: BP*(D)(SCALES) (Truck Stops Of America)
Food N: Country Pride (Truck Stops Of America), McDonalds, Subway (Truck Stops Of America), Taco Bell (Truck Stop)
Lodg N: Hampton Inn
AServ S: Troy's Honda Parts
TServ N: Truck Stops of America
TWash N: Truck Stops Of America

76 U.S 11, East Tuscaloosa, Cottondale
TStop S: 76 Auto/Truck Plaza(SCALES)
Gas N: Citgo*, Shell*
S: Texaco*(D)(LP)
Food N: Cracker Barrel
S: Baggett's Restaurant (76 Auto/TS)
Lodg N: Knight's Inn
S: Sleep Inn
AServ S: Texaco
TServ S: 76 Auto/Truck Plaza(SCALES)
TWash S: 76 Auto/Truck Plaza
RVCamp N: Sunset II Travel Park

73 U.S 82, McFarland Blvd
FStop N: BP*
Gas N: Chevron*(D)(CW), Parade*, Phillips 66*, Racetrac*, Shell*(CW)
S: Texaco*(CW)
Food N: Burger King, Captain D's Seafood, Denny's, Ezell's Catfish Cabin, Fortune Garden, Krystal, Long John Silvers, O'Charley's, Shoney's, Waffle House
S: Chili's, Guthrie's Chicken, Hardee's, Huddle House, KFC, McDonalds, Piccadilly Cafeteria, Pizza Hut, Quincy's Family Steakhouse, Sonic Drive In, Subway, Taco Bell, Taco Casa, Trey Yuen, Wendy's
Lodg N: AAA Best Western, Holiday Inn, Master's Inn
S: Comfort Inn, Days Inn, La Quinta Inn, LaQuinta Inn, Motel 6, Ramada Inn, Super 8 Motel

EXIT ALABAMA

AServ N: Barkley's Pontiac, GMC, O.K. Tire Service, Postle's Auto Service, Tuscaloosa Auto Parts
S: Carport Auto Service, Townsend Ford, Tuscaloosa Isuzu
ATM S: AM South
Parks S: Lake Lurene State Park
Other S: DelChamps Supermarket, Harkco Super Drug, McFarland Mall

71B Junction I-359, AL 69 North, Tuscaloosa (Difficult Reaccess)

71A AL 695, Moundville (Difficult Reaccess)
Gas E: Exxon(D)
Food E: Arby's, Lone Star Steakhouse, Outback Steakhouse, Pizza Hut, Wendy's
Lodg E: Courtyard by Marriott, Fairfield Inn
Other E: Kmart

62 Fosters

52 U.S 11, U.S 43, Knoxville
FStop N: SpeedStop*

45 Union
FStop S: Amoco*
Food N: Cotton Patch Restaurant
S: Hardee's, Southfork Restaurant, Wishbone Restaurant
Lodg S: Western Inn
AServ S: Auto Value (Parts)

40 AL 14, Eutaw, Aliceville
Med E: Hospital
Other W: Visitor Information

(39) Rest Area - RR, Phones, Picnic, RV Dump, HF (Westbound)

(38) Rest Area - RR, Phones, HF, Picnic, RV Dump (Eastbound)

32 Boligee
FStop N: BP* (Restaurant)
Gas S: Chevron*
Food N: Restaurant (BP)
S: Subway (Chevron)

23 Gainesville, Epes

17 AL 28, Livingston, Boyd
TStop S: Noble Truck Stop*(SCALES)
FStop S: Texaco*
Gas S: Chevron*
Food S: Burger King, Janet's Restaurant (Country Cookin'), Rose's Country Kitchen, Royal Waffle King, Subway (Chevron)
TServ S: Dickey's 24hr Shop Lube & Repair (Noble)

8 AL 17, York
TStop S: BP*(SCALES)
Food S: Restaurant (BP TS)
Lodg S: Days Inn
TServ S: Dun-Rite Truck Wash (BP)

1 U.S 80 East, Cuba
TStop N: Phillips 66
Gas S: Chevron*, Dixie*

(1) Rest Area - RR, Phones, RV Dump, Picnic, HF (Eastbound)

EXIT ALABAMA/MISSISSIPPI

↑ALABAMA
↓MISSISSIPPI

(170) Weigh Station - both directions

169 Kewanee
TStop S: Kewanee One Stop* (Restaurant)
FStop N: Dixie
S: Red Apple*
Food S: Restaurant (Kewanee One Stop)
Other S: Speedy's CB Shop

165 Toomsuba
Gas N: Shell* (see our ad on this page), Texaco*(D)
Food N: Travla Restaurant (Texaco)
RVCamp S: KOA Campground
Other N: Coin Car Wash

(164) Rest Area - RR, Phones P (Westbound)

160 Russell
TStop E: Amoco*(LP)(SCALES) (CB Shop)
W: 76 Auto/Truck Plaza*(LP)(SCALES) (24 Hrs, FedEx Drop)
Food E: Restaurant (Amoco TS)
W: Restaurant (76 Auto/Truck Stop)
TServ E: Amoco TS
W: Truck Lube Center (76 TS)
RVCamp W: Nanabe Creek Campground (1 Mile)
ATM E: Amoco TS
W: 76 Auto/Truck Plaza

157AB U.S 45, Macon, Quitman

154AB Hwy 19 south, Hwy 39 North, Naval Air Station, De Kalb
FStop W: Texaco
Gas E: Chevron*(CW), Conoco*(D)(CW)(LP)
W: Amoco*(D), BP*(D)(CW), Food Shop 7*, Shell*(CW) (24 Hrs) (see our ad on this page)
Food E: McDonalds, Popeyes Chicken, Red Lobster, Ryan's Steakhouse, Taco Bell
W: Applebee's, Backyard Burgers, Cracker Barrel, Denny's (Holiday Inn), Greenbriar (Howard Johnson), Nelva Restaurant, Waffle House, Western Sizzlin'
Lodg E: Comfort Inn, Pine Haven Motel
W: Days Inn, Economy Inn, Hampton Inn, Holiday Inn, Howard Johnson, Relax Inn, Rodeway Inn, Super 8 Motel, Super Inn, Western Motel
AServ E: Express Lube

Bold red print shows RV & Bus parking available or nearby

EXIT		MISSISSIPPI
		W: Bill Ethridge Lincoln Mercury Isuzu, Branning Auto Supply, Johnson Dodge, Meridian Honda, Sellers Olds Cadillac, Texaco
	TServ	E: Truckers Supply Co. (Parts)
		W: Stribling Equipment Empire Truck
	RVCamp	W: Ethridge RV Center
	ATM	E: Chevron, Conoco
		W: Howard Johnson, Shell
	Other	E: K-Mart (Pharmacy), Wal-Mart (Pharmacy)
		W: Old South Antique Mall, U-Haul Center(LP)
153		U.S 45, 22nd Avenue, Quitman
	Gas	E: Chevron*(D) (24 Hrs), Conoco*(D)(CW)(LP) (RV Dump), Exxon*(D)(LP), Shell
		W: Amoco* (24 Hrs), BP*
	Food	E: Depot (Best Western), Waffle House
		S: Morrison's Cafeteria
		W: Burger King, Captain D's Seafood, El Chico Mexican, Hardee's, KFC, Quincy's Family Steakhouse, Shoney's, Wendy's
	Lodg	E: Astro Motel, Best Western, Budget 8 Motel, ◆ Budgetel Inn, Econolodge, Holiday Inn Express, Motel 6, Sleep Inn
	AServ	E: Nelson Hall Hyundai Chevrolet, Shell
		W: Firestone Tire & Auto, Goodyear Tire & Auto
	ATM	E: Conoco
		W: BP, Citizens National Bank, Trustmark National Bank, Union Planters Bank
	Other	W: Cinema 5, Foodmax Grocery, Fred's, Mall, Medi Save Pharmacy, Optical
152		29th Avenue, Meridian (Difficult Reaccess Eastbound)
	FStop	W: Amoco*
	Gas	W: Chevron*(D)
	Food	E: Restaurant (Royal Inn)
	Lodg	E: Royal Inn Motel
		W: Ramada Limited
	ATM	W: Chevron
151		49th Avenue, Valley Rd, Meridian
150		U.S 11, MS 19, Philadelphia
	TStop	W: Phillips 66*(LP) (Restaurant)
	FStop	E: Amoco*, Stuckey's Express*
	Food	E: Subway (Stuckey's Express)
	TServ	W: International, Phillips 66, Tire Centers Inc.
148		I-20 West, Jackson (Unnumbered Exit, Northbound Is Left Exit)
142		Savoy, Dunn's Falls
137		North Enterprise
134		MS 513, South Enterprise, Stonewall
	FStop	E: BP*
126		MS 18, Pachuta, Rose Hill, Quitman
	FStop	E: Amoco*, BP*
118		Vossburg, Paulding, Stafford Springs, Waukaway Springs
	Other	E: S & L Country Store
113		MS 528, Heidelberg, Bay Springs
	TStop	E: J.R.'s Truckstop* (RV Dump, Fuel-BP)
	FStop	E: Stuckey's Express* (Off-Road Diesel)
	Gas	E: Exxon*(D), Shell*
		W: Texaco(D)(LP)
	Food	E: Hot Stuff Pizza (Stuckey's Express), J.R.'s Restaurant (J.R.'s Truck Stop), Subway (Exxon)
	AServ	W: Texaco
	ATM	E: Stuckey's Express
	Other	E: Hwy.-528-Car Wash
(109)		Parking Area - No Services

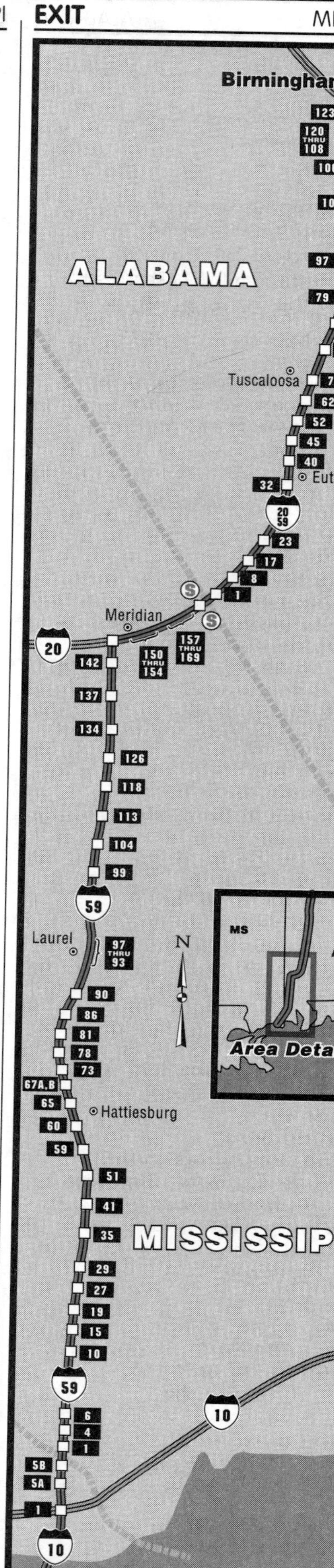

EXIT		MISSISSIPPI
		(Southbound)
(106)		Parking Area - No Services (Northbound)
104		Sandersville, Sharon
99		U.S. 11, Laurel
	FStop	E: T & B Curb*
	Food	E: Restaurant (Days Inn)
	Lodg	E: Days Inn, Magnolia Motel
	TServ	E: T & B Curb
97		U.S. 84 East, Waynesboro, Chantilly St.
	TStop	E: Doc's Food 'N Fuel Center* (Fuel-Exxon)
	FStop	E: Chevron*
	Gas	W: BP*(D), Texaco*(CW)
	Food	W: KFC, Out Back BBQ & Grill, Vic's Biscuits & Burgers
	Lodg	E: Motel (Exxon TS)
	Other	E: CB Store (Exxon TS)
96B		MS 15 South, Cook Ave
96A		4th Ave, Masonite Rd
	Gas	E: Blues*
95C		Beacon St, Downtown Laurel
	Gas	E: BP*(D) (Kerosene), Chevron*(D), Texaco*(CW)
		W: Pump & Save
	Food	E: Le's New Orleans Style Po-Boys (Chevron)
		W: Burger King, Church's Fried Chicken, McDonalds (Playground), Old Mexico Restaurant, Popeyes Chicken
	Lodg	W: Town House Motel
	Other	W: Diket's Drugs, Lauren Rogers Museum of Art, Sawmill Square Mall, Synergy Gas(LP)
95AB		U.S 84 West, MS 15 North, 16th Ave (Southeastern College)
	Gas	W: Amoco*, Exxon*(D)(CW), Shell(D)
	Food	W: Shoney's, Taco Bell, Waffle House, Wendy's
	Lodg	W: Econolodge, Quality Inn
	AServ	W: KIA of Laurel, Shell
	Med	W: + South Central Regional Medical Center
	ATM	W: Bank of Mississippi, Trustmark National Bank
	Other	E: Blossman Inc. Propane(LP)
93		South Laurel, Fairgrounds, Industrial Park
	TStop	W: American Food Truck Stop (Fuel-Citgo)
	FStop	W: Shell*(CW) (see our ad this page)
	Gas	W: Exxon*(D)
	Food	W: Hardee's, Restaurant (American Foods Truck Stop), Subway (Exxon)
	AServ	E: Paul's Discount Glass, Tires, Muffler, Rebel Alternator Service

Bold red print shows RV & Bus parking available or nearby

EXIT MISSISSIPPI

- **ATM** W: Trustmark National Bank
- **Other** E: **Mississippi National Guard**
  - W: **Smith's Sunbeam Bread Store**

90 U.S 11, Ellisville Blvd

- **FStop** W: **Dixie Gas* (24 Hrs)**
- **Gas** E: Joe's Service Center[LP] (Motor Fuel Grade Propane)
- **Food** E: Charlie's Catfish House
- **AServ** E: Joe's Service Center, Poole Wheel Service, Roger's Auto Parts
  - W: White's Auto Electric
- **TServ** W: **Industrial Services Diesel Repair**
- **Other** E: **Shaffer's Optical Express**

88 MS 29, MS 588, Ellisville

- **FStop** E: **Chevron*[D]**
- **Gas** W: Shell (see our ad opposite page), Woody's Total Discount Fuel & Tires*
- **Food** E: Country Girl Kitchen, KFC, McDonalds (Playground), Subway
  - W: Po' Boy!
- **AServ** E: Cooksey's Used Tire, Ellisville Auto Parts (NAPA)
  - W: Hanna's Auto Repair, Shell, Woody's
- **ATM** E: Union Planters Bank
- **Other** E: **Food Tiger**

85 MS 590, Ellisville, State School

- **FStop** W: **Triple T Junction* (Spur)**
- **AServ** W: Triple T Automotive
- **Other** E: **Jones County Jr. College**

80 Moselle

78 Sanford Rd

76 Hattiesburg - Laurel Regional Airport

73 Monroe Rd

69 Eatonville Rd, Glendale

67AB U.S. 49, Jackson, Hattiesburg

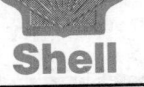

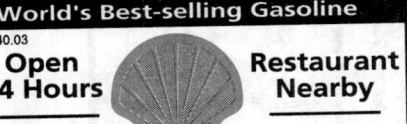

- **FStop** W: **Amoco*, Smith Quick Shop*, Speedway***
- **Gas** E: Exxon*, Shell* (see our ad this page)
  - W: RaceWay*, Shell* (see our ad this page), Stuckey's Express*
- **Food** E: Arby's, Burger King (Playground), **Cracker Barrel**, KFC, **Krystal (24 Hrs)**, McDonalds (Playground), Pizza Hut, The Hickory House Buffet, Waffle House
  - W: **Northgate Diner (Best Western)**, Stuckey's Deli (Stuckey's), Waffle House, Ward's Hamburgers
- **Lodg** E: **Budget Inn**, Cabot Lodge, Comfort Inn, Days Inn, Holiday Inn, Howard Johnson (see our ad this page), **Motel 6**, Quality Inn, Scottish Inns, Super 8 Motel
  - W: **Best Western**
- **ATM** E: Shell
  - W: Deposit Guaranty, Stuckey's

65AB U.S. 98 West, Hardy St, Columbia

- **Gas** E: Chevron*[D] (24 Hrs), Shell*[CW] (see our ad this page), Texaco*
  - W: Amoco*[CW], Pump & Save, Shell*[CW] (see our ad this page),
- **Food** E: Baskin Robbins, Crescent City Grill, Hub City Coffee Co., Mr Gatti's, Mr. Jim's Pizza, Pizza Hut, Purple Parrot Cafe, South China Restaurant, Ward's
  - W: Chili's, It's Yogurt, La Fiesta Brava, Little Caesars Pizza, Lone Star Steakhouse, Mandarin House, O'Charley's, Outback Steakhouse, Red Lobster, Shipley Do-nuts, Spuds Deli, Waffle House, Wendy's
- **Lodg** E: Fairfield Inn by Marriott
  - W: Comfort Suites, Hampton Inn
- **AServ** E: Carquest Auto Parts, NAPA Auto Parts, University Quick Lube, University Tire
  - W: **K-Mart**
- **Med** E: **+ Immediate Care**
- **ATM** E: Community Bank, Deposit Guaranty, Great Southern National Bank, Union Planters Bank
  - W: Amoco, Bank of Mississippi, Citizens National Bank
- **Other** E: **20/20 Eyecare, Angie's Pet Palace, Eckerd Drugs, Lighthouse Christian Bookstore, Stewart's Camera & 1-Hr Photo, Sunflower Food Store, U.S. Post Office**
  - W: **Books-A-Million, DelChamps Supermarket, Golf USA, K & B Drugs, K-Mart (24 Hrs, Pharmacy), Sac & Save Foods, Toys "R" Us**

60 U.S. 11, South Hattiesburg (William Carey College)

- **FStop** W: **Amoco*[D], Chevron* (24 Hrs)**
- **AServ** W: B & B Road Service (24 Hrs)
- **TServ** W: **B & B Road Service (24 Hrs), K & K Truck Trailor Parts, Inc.**
- **Other** W: **Exit 60 CB Shop**

59 U.S. 98 East, Lucedale, Mobile (Limited Access Hwy)

51 MS 589, Purvis

41 MS 13, Lumberton

- **FStop** W: **Conoco***
- **Gas** W: BP[LP]
- **Food** W: Alma's I-59 Restaurant
- **AServ** W: BP
- **Other** W: **Bass Pecan Co.**

35 Hillsdale Rd

- **FStop** E: **Pure***
- **Food** E: **Deli Express (Pure)**
- **Lodg** E: Georgetowne Inn
- **Other** E: **Hillsdale Golf Course (Georgetowne Inn)**

29 MS 26, Poplarville, Wiggins

27 MS 53, Poplarville, Necaise

19 Millard

15 McNeill

- **TStop** W: **McNeill Truck Stop* (24 Hrs)**
- **Food** W: **Restaurant (McNeill TS)**
- **Lodg** W: **Motel (McNeill Truck Stop)**

(13) **Rest Area - No Services** P **(Southbound)**

Bold red print shows RV & Bus parking available or nearby

← N I-59

EXIT — MISSISSIPPI

10 Carriere
- TStop E: Hilda's I-59 Truckstop* (Fuel-Phillips 66)
- Food E: Hilda's Cafe (Hilda's TS)
- TServ W: McNeill Service Center
- ATM E: Hilda's I-59 TS

(8) Rest Area - No facilities (P) (Northbound)

6 MS 43 North, North Picayune
- Gas W: Chevron*
- Lodg W: Majestic Inn
- ATM W: Bank Plus, First National Bank
- Other W: Berry Veterinary Clinic, Cinema IV, Winn Dixie Supermarket (Pharmacy)

4 MS 43 South, Picayune, Kiln
- Gas W: Shell*(D), Spur*, Texaco*(D)
- Food E: McDonalds (Wal-Mart)
  W: Burger King (Playground), Hardee's, Shoney's, Waffle House
- Lodg W: Comfort Inn, Days Inn, Heritage Inn
- AServ E: Bill Garrett Chev Dealer, Greg's Tire & Service Center, Wal-Mart
  W: Auto Zone Auto Parts, Dub Herring Chrysler
- RVCamp W: Paw-Paw's Camper City
- ATM E: Wal-Mart
  W: Bank Plus, First National Bank of Picayune
- Other E: The Crosby Arboretum, Wal-Mart Supercenter (Pharmacy)
  W: Eye Center, Pet Depot

(2) MS Welcome Center, Rest Area - RR, Picnic, Phones, RV Dump, Grills (P) (Northbound)

(1) Weigh Station (Both Directions)

1 U.S. 11, MS, 607, Nicholson, NASA John C. Stennis Space Ctr.
- FStop W: Chevron*(D)
- AServ W: Croney's Auto Repair
- Other E: John C. Stennis Space Ctr. (6 Miles), Naval Oceanographic Center

↑MISSISSIPPI

EXIT — LOUISIANA

↓LOUISIANA

11 Pearl River Turnaround

5B Honey Island Swamp
- Other E: Pearl River Wildlife Mngmt Area

5A LA 41 Spur, Pearl River (Southbound Only Due To Construction)
- Gas W: Citgo*
- Food W: Corner Cafe
- Other W: Wally's Gators

3 U.S. 11 South, Pearl River Town, S. LA 1090

(1) LA Tourist Info, Rest Area - RR, Picnic

1BC I-10, New Orleans, Bay, St. Louis (Southbound)

1A I-12 West, Hammond (Southbound)

↑LOUISIANA

Begin I-59

I-64 E →

EXIT — ILLINOIS

Begin I-64

↓MISSOURI

40C West I-44, South I-55

↑MISSOURI

↓ILLINOIS

1 IL 3 South, Sauget, West I-70, Great River Rd IL, Cahokia

2A Downtown St. Louis via MLK Bridge

2B 3rd Street
- Med S: Hospital
- ATM S: First Illinois Bank

3 North I-55, East I-70, Chicago, Indianapolis (Difficult Reaccess)

4 Baugh Avenue, 15th Street (Difficult Reaccess)

5 25th Street

6 Kingshighway, IL 111
- Gas N: Amoco* (24 hrs), Shell
- Food N: Popeye's Chicken
  S: Stoplight Restaurant
- AServ N: Shell

  S: Norton's Automotive
- Parks S: Frank Holten State Park (w/ Golf Course)
- Other N: Fatman's Car Wash & Barber Shop

7 I-255, US 50, Chicago, Memphis

9 IL 157, Caseyville, Centreville
- FStop S: Shell(D)(LP)
- Gas N: Phillips 66*
  S: Amoco* (24 hrs)
- Food N: Hardee's
  S: Cracker Barrel, Domino's Pizza, McDonalds, Taco Bell
- Lodg S: AAA Best Inns of America, AAA Days Inn (see our ad this page)
- AServ N: Vehicle Doctor

12 IL 159, Belleville, Collinsville, Fairview Heights
- Gas S: Amoco*(CW), Mobil*
- Food N: 20's Hideout Steak and Bar, Applebee's, Blimpie's Subs, Carlos O'Kelly's Mexican Cafe, Damon's, Houlihan's Restaurant & Bar, Lotawata Creek Southern Grill, Olive Garden, Red Lobster,

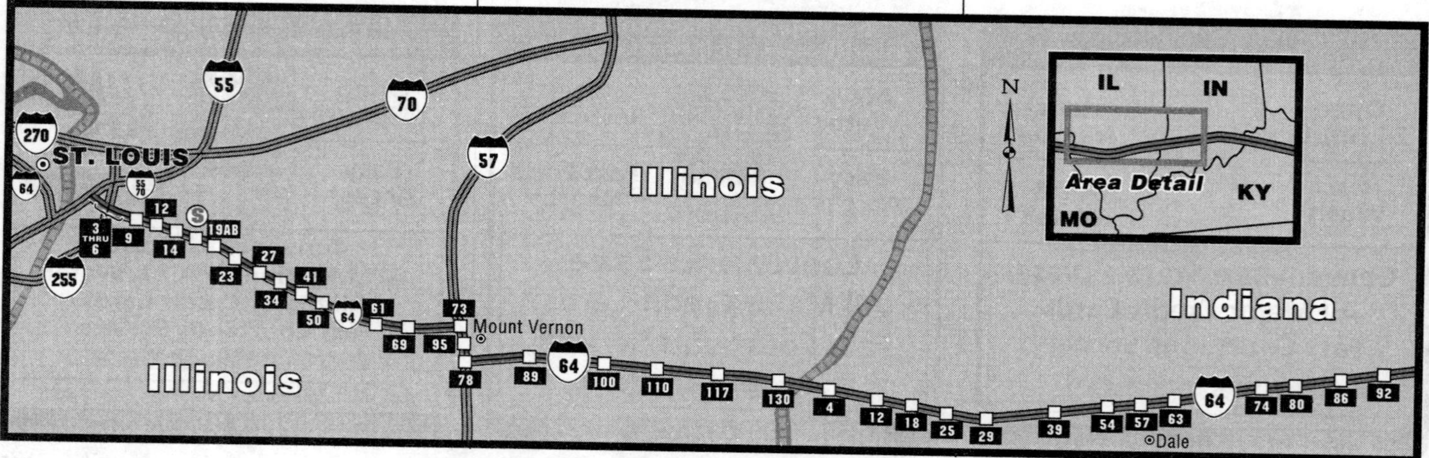

Bold red print shows RV & Bus parking available or nearby

EXIT	ILLINOIS
	Shoney's (Drury Inn), T.G.I. Friday's (Ramada)
	S: Boston Market, Casa Gallardo, Pasta House, Ruby Tuesday, The Honey Baked Ham Co.
Lodg	N: Camelot Inn Best Western, ◆ Drury Inn, ◆ Fairfield Inn, (AAA) Ramada, ◆ Super 8 Motel
AServ	N: Saturn of Metro East
	S: Firestone Tire & Auto, In & Out Auto Center, Jiffy Lube, National Tire & Battery, Sears
ATM	S: Central Bank, First Financial Bank
Other	N: **Antique Mall, St. Clair 10 Cine'**
	S: **Office Max, Pet Care Super Store, St Clair Square Mall, Venture**
14	O'Fallon
Gas	N: Shell*(CW)
Food	N: Rookie's Sports Bar & Grill, Steak 'N Shake
	S: Dairy Queen, Hardee's, Jack-In-The-Box, Lion's Choice Roast Beef, McDonalds, Taco Bell, Western Sizzlin'
Lodg	N: Holiday Inn Express
	S: Econolodge
AServ	N: Jack Schmitt Nissan Olds, Schantz Ford
	S: **Sam's Club, Wal-Mart**
ATM	N: Union Bank of IL
	S: MidAmerica Bank
Other	S: **O'Fallon 15 Cine, PETsMart, Sam's Club, The Home Depot, Wal-Mart**
(17)	**Weigh Station (Eastbound)**
19AB	US 50, IL 158, O'Fallon, Scott AFB
FStop	S: **Citgo**
Gas	N: Motomart*(D) (24 hrs)
Food	N: Hero's Pizza & Subs (Motomart), Schiappa's Pizza
	S: **Ivory Chopsticks (Citgo)**
Lodg	N: Comfort Inn
Med	S: **+ Hospital**
ATM	N: Motomart
23	IL 4, Mascoutah, Lebanon
(25)	**Rest Area - RR, Phones, Vending, Picnic, Pet Walk, Playground (P) (Both Sides)**
27	IL 161, New Baden
Gas	N: Shell*(D) (24 hrs)
ATM	N: Shell
34	Albers, Damiansville
41	IL 177, Okawville
TStop	S: **Amoco***
Food	S: Dairy Queen, **Hen House (Amoco TS)**
Lodg	S: **Super 8 Motel**
AServ	S: Exit 41 Service Center
TServ	S: **Exit 41 Service Center**
Other	S: **Heritage House Museum**
50	IL 127, Carlyle, Nashville
TStop	S: **Conoco (24 hrs)**
Food	S: **Little Nashville (Conoco TS), McDonalds**
Lodg	S: **U.S. Inn (Conoco TS)**
TServ	N: **Norrenberns Truck Service**
TWash	S: **Conoco**
ATM	S: **Conoco**
Other	N: **Carlyle Lake**
	S: **Washington County Conservation Area**
61	US 51, Centralia, Ashley
69	Woodlawn
73	I-57 North, Chicago (Eastbound, Exit 73 Is The Same As 96 On I-65/57 West/North)

EXIT	ILLINOIS
	Note: I-64 runs concurrent below with I-57. Numbering follows I-57.
95	IL 15, Mt Vernon, Ashley
TStop	W: **76 Auto/Truck Plaza(SCALES) (Open 24 Hours)**
FStop	W: **Huck's*(SCALES)**
Gas	E: Amoco*, Hucks Food Store*, Phillips 66*(CW), Shell, Speedway*
	W: Mount Vernon Center (76 Auto/Truck Stop), Shell*
Food	E: Bonanza Steak House, Burger King, China Buffet, Denny's, El Rancherito, Fazoli's Italian Food, KFC, Little Caesars Pizza, Long John Silvers, McDonalds, Pizza Hut, Pizza Pro (Kroger), Shoney's, Steak & Shake, Subway, Wendy's, Western Sizzlin'
	W: 76 Auto/Truck Plaza, Applebee's, Arby's, Burger King, **Cracker Barrel,** Hucks Country Cooking, McDonalds (Indoor Playground)
Lodg	E: Best Inns of America, **Best Western,** ◆ **Drury Inn,** Motel 6, Ramada Inn, ◆ Super 8 Motel, Thrifty Inn
	W: (AAA) Comfort Inn, Holiday Inn
AServ	E: Auto Zone Auto Parts, Midas Muffler and Brakes, Shell
TServ	W: **76 Auto/Truck Plaza**
Med	E: **+ Hospital**
ATM	E: Bank of Illinois, Citizen's Bank, Merchantile Bank, NationsBank
	W: 76 Auto/Truck Plaza
Other	E: **K-Mart (Pharmacy), Kroger Supermarket**
	W: **Wal-Mart Supercenter (Pharmacy, Tire Center, Open 24 Hours)**
	Note: I-64 runs concurrent above with I-57. Numbering follows I-57.
78	I-57 South, Memphis (Westbound)
80	IL 37, Mt Vernon
RVCamp	S: **Sherwood Camping**
(82)	**Rest Area (P) (Eastbound)**
(86)	**Rest Area - RR, Vending, Picnic, Playground, Phones (P) (Westbound)**
89	Belle Rive, Bluford
100	IL 242, Wayne City, McLeansboro
FStop	N: **Marathon***
110	US 45, Norris City, Fairfield
RVCamp	S: **Barnhill Campground**
117	Burnt Prairie

EXIT	ILLINOIS/INDIANA
FStop	S: **BP***
Food	S: **Restaurant (BP)**
130	IL 1, Grayville, Carmi, Mt Carmel
FStop	N: **Shell* (24 hrs)**
	S: **Phillips 66***
Lodg	N: (AAA) **Windsor Oaks Inn (Best Western)**
RVCamp	N: **Beall Woods State Park (10 Miles)**
Parks	N: **Beall Woods State Park (10 Miles)**
(131)	**Rest Area - RR, Phones, Info, Picnic (P) (Westbound)**

↑ILLINOIS

↓INDIANA

EXIT	INDIANA
4	Griffin
Gas	N: Griffin Depot*(D) (Kerosene), Phillips 66
Food	N: Depot Diner (Griffin Depot)
AServ	N: Phillips 66, Posey County Co-op
Other	N: **Posey County Co-op(LP)**
(8)	**Rest Area - RR, Phones (Eastbound)**
12	IN 165, Poseyville
Food	S: **T's Restaurant**
AServ	S: Broerman Geo Chevrolet
Parks	S: **Harmonie State Park**
Other	S: **New Harmony Historic Area**
18	IN 65, Evansville, Cynthiana
FStop	S: **Moto Mart***
25AB	US 41, Terre Haute, Evansville
TStop	N: **Pilot*(SCALES) (24 hrs)**
	S: **Pennzoil Busler*(SCALES) (24 hrs)**
Food	N: **Wendy's (Pilot TS)**
	S: Arby's, **Hardee's (Pennzoil Busler), McDonalds**
Lodg	S: **Comfort Inn,** Holiday Inn Express, **Pennzoil Busler, Pennzoil Busler,** Super 8 Motel, Days Inn (see our ad this page)
TServ	S: **Pennzoil Busler**
TWash	S: **Pennzoil Busler**
ATM	N: **Pennzoil Busler, Pilot TS**
Other	S: **IN State Police**
29A	I-164, IN 57 South, Evansville, Henderson KY
TStop	N: **Amoco Fuel Plaza*(SCALES) (in Washington on IN 57)**
Food	S: **Cracker Barrel (south on I-164 to Exit 7)**
29B	IN 57 North, Petersburg
39	IN 61, Boneville, Lynnville
FStop	N: **76(CW)**
Gas	N: Rocket Wholesale, Shell*
AServ	N: Lynville Auto Service Center (76), Remington Tire
TServ	N: **Lynnville Auto Service Center, Rita's Repairs**
TWash	N: **76**
RVCamp	N: **Camping**
ATM	N: Lynnville National Bank, Shell
Other	N: **Barnie's Market, Fire Dept, Main Street Pizza, US Post Office**
54	IN 161, Tennyson, Holland
RVCamp	S: **Camping**
57	US - 231, Dale, Jasper, Huntingburg
TStop	N: **AmBest*(SCALES)**
Gas	S: Shell (24 hrs)
Food	S: **Denny's**
Lodg	N: **Scottish Inns (Am Best)**

Bold red print shows RV & Bus parking available or nearby

EXIT — INDIANA

TServ N: AmBest
ATM N: AmBest
Parks S: Lincoln State Park
Other S: Amphitheater, Lincoln Boyhood National Memorial, Mega Maze Water Wars, Mini Golf

(59) Rest Area - RR, Phones, Weather Info (Both Directions)

63 IN 162, Santa Claus, Ferdinand
RVCamp S: Camping
Parks N: Ferdinand State Forrest
Other N: Tourist Information

72 IN 145, Bristow, Birdseye
Gas S: 76*(D)

79 IN 37 South, Tell City
RVCamp S: Camping
Parks S: Hoosier National Forrest Recreational Facilities
Other S: Goffinet's General Store

(80) Rest Area - No Services Available, Parking Only (Eastbound)

(81) Rest Area - No Services Available, Parking Only (Westbound)

86 IN 37N, Sulphur, English
Parks N: Patoka Lake

92 IN 66, Leavenworth, Marengo
TStop S: 76 Auto/Truck Plaza*(LP)(SCALES), Citgo*, Country Style Plaza*(SCALES) (24 hrs)
Food S: Country Style Restaurant, Days Inn (24 hrs), Taco Bell (76 TS)
Lodg S: Cavern Inn (76 TS), DAYS INN Days Inn (76 TS)
TServ S: 76 Auto/Truck Plaza
ATM S: 76 Auto/Truck Plaza

105 IN 135, Palmyra, Corydon
Gas N: Citgo*(D), Shell*
S: Fivestar*
Food N: Frisch's Big Boy
S: Arby's, Burger King, Country Folk's Buffet, Lee's Famous Recipe Chicken, Long John Silvers, McDonalds, Papa John's Pizza, Ponderosa, Subway, Taco Bell, Waffle & Steak, Wendy's
Lodg N: Old Capital Inn (Best Western)
S: ◆ Budgetel Inn
AServ S: Wal-Mart
Other S: Wal-Mart Supercenter

113 Lanesville
Gas N: Ashland*(LP) (Kerosene)

(115) IN Welcome Center, Rest Area - RR, Phones, Picnic (Westbound)

118 IN 62, IN 64W, Georgetown
FStop N: Marathon*(LP)
Gas N: Shell*
S: Marathon

EXIT — INDIANA/KENTUCKY

Food N: Korner Kitchen, McDonalds, Pizza King
AServ S: Marathon
ATM N: Harrison County Bank
Other N: Mr. Hardware, Thriftway Supermarket

119 US 150 West, Greenville, Paoli

121 Junction I-265 East

123 IN 62 East, New Albany
Gas N: Amoco*
S: BP*(LP)
Food S: Waffle & Steak
Lodg S: ◆ Holiday Inn Express
AServ N: Certified Auto Service, Consolidated Tire & Automotive Center, Guarantee Auto Supply, Mike Smith Firestone
Med N: ✚ Hospital
ATM N: Bank One
Other N: Public Library

↑ INDIANA

↓ KENTUCKY

1 Junction East I-264, Shively

3 US 150 East, 22nd St.
Gas S: Chevron*, Dairy Mart*, Shell (24 hrs)
Food S: Dairy Queen, McDonalds
AServ S: Shell

4 9th Street, Roy Wilkins Avenue

5A Junction I-65, Nashville, Indianapolis

5B 3rd Street, Downtown Louisville (Westbound, Difficult Reaccess)
Food S: City Wok, Marketplace Bar & Grill, Subway
Other S: Louisville Science Center, Louisville Slugger

EXIT — KENTUCKY

Museum & Visitor Center

6 Junction I-71, Cincinnati

7 U.S. 42, U.S. 60, Mellwood Ave
Gas N: Speedway*(D)(LP) (Kerosene)
Food N: J.K.'s Corner Grill, Picasso, Sugardoe Cafe
S: Hall's Cafeteria
AServ N: Bob Collett Auto Wreckers, Collett Bros. Auto Parts
Other N: Thomas Edison Museum

8 Grinstead Drive
Gas S: $ave, BP*, Chevron
Food S: August Moon, K.T.'s Restaurant & Bar
AServ S: BP, Chevron
Other S: Pets Galore

10 Cannons Ln. Bowman Field Airport

12 Junction I-264, Watterson Expwy

15 KY 1747, Jeffersontown, Middletown, Hurstbourne Lane
Gas N: Amoco*(CW), BP*(CW), Shell*(CW)(LP)
S: BP*(D)(CW), Thornton's Food Mart*(D)
Food N: Arby's, Benihana, Bob Evans Restaurant, Burger King, Joe's O.K. Bayou, McDonalds, Olive Garden, Sichuan Gardens Chinese, Skyline Chili, Steak-out, Subway, T.G.I. Friday's, Tumbleweed Mexican Food
S: Applebee's, Au Bon Pain (DoubleTree Club Hotel), Balihai Chinese Restaurant, Blimpie's Subs, Dairy Queen, Damon's, Hardee's, O'Charley's, Shalimar Indian Restaurant, Shoney's, Wendy's, Yen Ching Chinese
Lodg N: Blair Wood, ◆ Courtyard by Marriott, ◆ Fairfield Inn, AAA Holiday Inn, Red Roof Inn, Amerisuites (see our ad this page)
S: DAYS INN Days Inn, DoubleTree Club Hotel (see our ad this page), ◆ Hampton Inn (see our ad this page), ◆ Marriott, Red Carpet Inn,

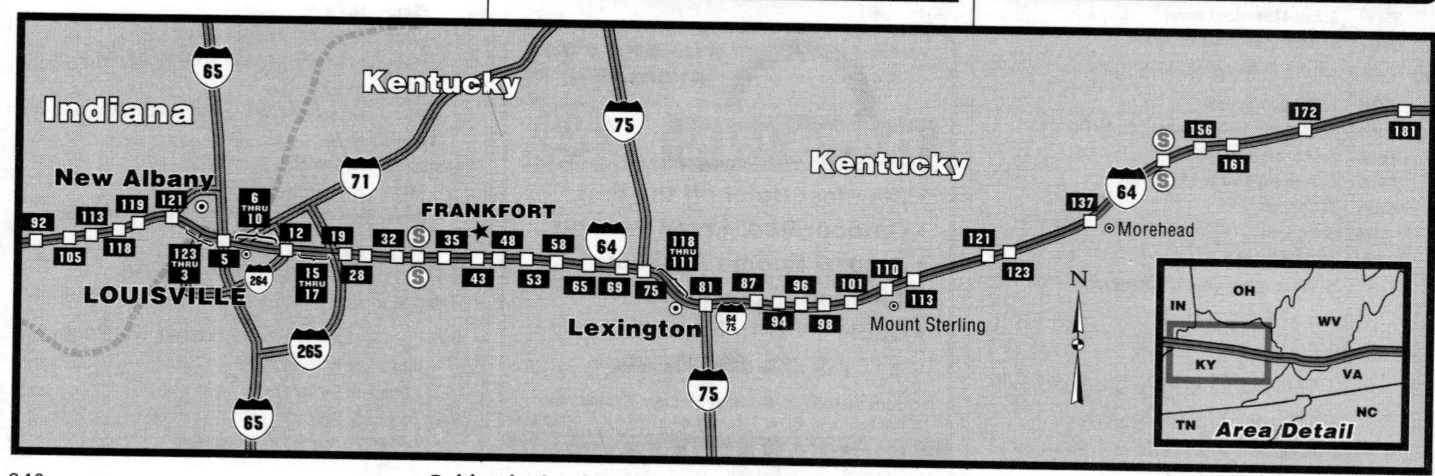

Bold red print shows RV & Bus parking available or nearby

EXIT	KENTUCKY
	StudioPLUS, Wilson Inn & Suites
AServ	N: BP, Chevron
	S: Swobe Auto Center
ATM	N: Bank One, National City Bank, PNC Bank
Other	N: **60 Min. Photo, AAA Auto Club - Kentucky, Fragile Pack (Shipping), Kroger Supermarket, Revco Drugs**
	S: **EYE Mart, Wal-Mart**
17	Blankenbaker Road, Industrial Park
Gas	S: BP*, Chevron*(CW) (24 hrs), Thornton's Food Mart*(D)
Food	S: Arby's, Burger King, **Cracker Barrel**, HomeTown Buffet, McDonalds, Subway (Thornton's), Waffle House
Lodg	S: Candlewood, ◆ Comfort Suites, AAA Microtel Inn, Regency (Best Western), ◆ Sleep Inn
ATM	S: BP, Bank of Louisville
Other	S: **Sam's Club**
19AB	19A - S.I-265, KY 841, G.Snyder Fwy 19B - N.I-265, KY 841, G.Snyder e
(24)	**Rest Area - RR, Vending, Phones (Eastbound)**
28	Veechdale Road, Simpsonville
TStop	N: **Pilot Travel Center***(SCALES)
Gas	S: BP*
Food	N: **Subway (Pilot, 24 hrs)**
(29)	**Rest Area - RR, Vending, Phones (Westbound)**
32AB	KY 55, Taylorsville, Shelbyville
Gas	N: Shell*
Food	N: Arby's, Parkside Deli & Grocery (Shell)
Lodg	N: Best Western, **Days Inn**
AServ	N: Brinkhaus Buick, Pontiac, GMC Trucks (light)
Med	N: **+ Hospital**
Parks	S: **Taylorsville Lake State Park**
Other	N: **Shelby Lanes**
35	KY 53, Shelbyville
Gas	N: BP*(D)(LP), Chevron(D)
AServ	N: Chevron
(38)	**Weigh Station (Both Directions)**
43	KY 395, Waddy, Peytonia
TStop	S: **76 Auto/Truck Plaza***(LP)(SCALES)
Food	S: **Hot Stuff Pizza (76), Restaurant (76 Auto/ TruckStop)**
TServ	S: 76
TWash	S: 76
Other	S: **Laundromat (76)**
48	KY 151, US 127 S, Lawrenceburg, Grafenburg
Gas	S: Chevron*
53AB	U.S. 127, Frankfort, Lawrenceburg
Gas	N: Chevron*, Shell*(D)
Food	N: Applebee's, Baskin Robbins, Columbia Steak House, Frisch's Big Boy, New China, Sassy's Subs & Stuff (Chevron)
AServ	N: Wal-Mart
Med	N: **+ Hospital**
ATM	N: Farmer's Bank, State National Bank
Other	N: **Advantage Optical, K-Mart (w/ Pharmacy), Kentucky State Police, Kentucky Vietnam Memorial, Kroger Supermarket (W/ Pharmacy), Mail Boxes Etc, Rite Aide Pharmacy (1-Hr Photo), Wal-Mart Supercenter, Winn Dixie Supermarket**
58	U.S. 60, Frankfort, Versailles
Gas	N: Chevron*(D)(LP), Marathon*(D), Shell*

EXIT	KENTUCKY
Lodg	N: AAA Best Western
AServ	N: Trans-A-Matic Transmissions
(60)	**Rest Area - RR, Vending, Phones, Dog Walk, Picnic (Both Directions)**
65	To KY 341, U.S. 62, Midway, Versailles
FStop	S: **76***(LP)
69	U.S. 62E, Georgetown
75	Junction I-75, Cincinnati
	Note: I-64 runs concurrent below with I-75. Numbering follows I-75.
115	KY 922, Airport, Lexington, to Bluegrass Parkway
FStop	E: **Exxon***(D)(CW)
Gas	E: Shell*(CW)(LP)
	W: Chevron*
Food	E: **Cracker Barrel**, McDonalds, Subway (Shell), Waffle House, Wyndham Garden Hotel
	W: Denny's, The Mansion Restaurant, The Post Restaurant (Holiday Inn)
Lodg	E: Knights Inn, La Quinta Inn, Wyndham Garden Hotel
	W: Holiday Inn, Marriott
AServ	W: Chevron
Other	E: **Kentucky Horse Park**
	W: **Griffin Gate Golf Club (Marriot)**
113	U.S. 27, U.S. 68, Paris, Lexington
Gas	E: BP*(D)(LP), Marathon*(LP), SuperAmerica*(LP) (Kerosene)
	W: Chevron*(D)(LP), Shell*
Food	E: Sigee's Restaurant (Harley Hotel), Waffle House
	W: Fazoli's Italian Food, Howard Johnson, Subway (Chevron)
Lodg	E: Harley Hotel
	W: **Days Inn**, Howard Johnson
AServ	E: BP
Med	W: **+ Hospital**
ATM	E: BP, SuperAmerica, Whitaker Bank
	W: ATM (Shell)
Other	E: **Joyland Bowl, Tourist Information**
	W: **Rupp Arena, Tourist Information, Transylvania University, University of Kentucky**
	Note: I-64 runs concurrent above with I-75. Numbering follows I-75.
81	Junction I-75 South, Richmond, Knoxville
87	KY 859, Lexington Army Depot, Blue

EXIT	KENTUCKY
	Grass Station
94	KY 1958, Van Meter Rd, Winchester
TStop	N: **Shell***(D)(SCALES) **(Showers, Fax Machine, 24 hrs)**
FStop	S: **Speedway***(D)(LP) **(24 hrs)**
Gas	N: Chevron (24 hrs)
	S: BP*(D), Marathon*, Somerset (Kerosene)
Food	N: **Restaurant (Shell)**
	S: Applebee's, Arby's, Captain D's Seafood, Domino's Pizza, Fazoli's Italian Food, Gators, Hardee's, KFC, Kentuckee Diner Inc., Lee's Famous Recipe Chicken, Little Caesars Pizza, Long John Silvers, Mac's Restaurant, Magarita's Traditional Mexican, McDonalds, Pizza Hut, Rally's Hamburgers, Sal's Pizza, Pasta, Subs, Shoney's, Sir Pizza, Subway, Tacos Too, Time Out Steakhouse, Waffle House, Wendy's
Lodg	N: ◆ Days Inn
	S: AAA Best Western, Red Carpet Inn
AServ	S: Auto Zone Auto Parts, Bypass Auto Hardware, Hatfields Plymouth Jeep Dodge Trucks (Light Trucks), Paynter Tire & Service Center, Pro Lube Plus(CW), S & S Tire, T & D Auto, Ted's Collision Center
TWash	N: **Truck Wash**
Med	S: **+ Hospital**
ATM	S: PNC Bank, People's Commercial Bank
Parks	S: **Fort Boonesbough State Park**
Other	S: Blue Line Car Wash, **Cridder Outfitters (Pet Supplies), K-Mart**(LP)**, Maytag Laundry, Pharmacy (K-Mart), Rite Aide Pharmacy, Sav-A-Lot Grocery**
96AB	96A - KY 627S, Winchester, 96B - KY 627N, Paris
TStop	N: **96 Truck Stop***
Gas	S: Ashland
Food	S: Holiday Inn Restaurant (Holiday Inn)
Lodg	S: **Hampton Inn,** Holiday Inn
AServ	S: Ashland, Casey Geo Chevrolet Olds
Other	N: **Fire Station**
	S: **Tourist Information**
98	Campton, Mountain Pkwy, Prestonsburg, Junction (Eastbound, Difficult Reaccess, Reaccess Westbound Only)
(98)	**Rest Area - RR, Vending, Phones (Eastbound)**
101	U.S. 60
(108)	**Rest Area - RR, Vending, Phones, Dog Walk (Westbound)**
110	U.S. 460, KY 11, Mt. Sterling, Paris
Gas	N: Chevron (24 hrs)
	S: Ashland*, BP*(D), Dairy Mart*, Exxon*, Marathon*(D) (Kerosene), Rich, Shell*(CW), Somerset(D), SuperAmerica* (Kerosene)
Food	S: Arby's, Burger King, Golden Corral, Hardee's, Jerry's Restaurant, KFC, Lee's Famous Recipe Chicken, Little Caesars Pizza, Long John Silvers, McDonalds, Peking Chinese, Pizza Hut, **Rio's Steakhouse & Lounge (Days Inn)**, Shoney's, Subway (Exxon), Taco Bell, Taco Tico, Wendy's
Lodg	N: Fairfield Inn
	S: Budget Inn, ◆ **Days Inn (see our ad this page)**, Scottish Inns
AServ	N: Chevron
	S: Advance Auto Parts, Big A Auto Parts, Car Quest Auto Center, Cole Ford Parts, Pennzoil Lube
Med	S: **+ Hospital**
ATM	S: ATM (Shell), Community Trust Bank, Mt. Sterling National Bank, Traditional Bank

EXIT KENTUCKY

Other S: Check Holders, Crystal Clean Car Wash, Food Lion, Pharmacy (Winn Dixie), Rite Aide Pharmacy, Six Bay Car Wash, Southern States(LP), The Medicine Shoppe Pharmacy, Veterinary Clinic, Winn Dixie Supermarket

113 U.S. 60, Mt. Sterling
FStop N: Citgo Truck Auto Plaza* (Kerosene)
Food N: Super Express Stop (Citgo)

121 KY 36, Frenchburg, Owingsville
FStop N: Citgo*
Gas N: BP*(D)
Food N: Dairy Queen, Kountry Kettle, McDonalds, Toms' Pizza Shop
ATM N: Owingsville Banking Co.
Other N: Maytag Laundry (Coin Operated), No Touch Automatic Car Wash, Owingsville Drug, Town & Country Car Wash

123 U.S. 60, Owingsville, Salt Lick
Gas N: 76(D)
AServ N: 76
RVCamp S: Cave Run Lake (12 Miles)

133 KY 801, Farmers, Sharkey
FStop S: BP*(LP)

137 KY 32, Morehead, Flemingsburg
FStop S: Exxon*
Gas N: Shell*(D)
S: Chevron*(D), Citgo
Food N: Boomerang's Bar & Grill, China Garden, Cutter's Roadhouse, Tacos Too
S: Baskin Robbins (Exxon), Hardee's, Jimbos Big Boy, KFC, Lee's Famous Recipe Chicken, Ponderosa, Shoney's, Subway
Lodg S: Days Inn ◆ Days Inn, ◆ Holiday Inn, Mountain Lodge, Plaza Inn (Best Western), Shoney's Inn & Suites, ◆ Super 8 Motel
AServ N: Trademore Tire Service
S: Major Brands Tire & Auto Service
Med S: ✚ Hospital
ATM S: Trans Financial Bank
Other N: Check-Advance, Kroger Supermarket, Pharmacy (Wal-Mart), Tuttle's Coin Laundry
S: All American Car Wash, Cave Run Lake (11 Miles), Food Lion, KY State Police, Visitor Information, Wal-Mart

(141) Rest Area - RR, Vending, Phones (Both Sides)

(148) Weigh Station (Both Directions)

156 KY 2 To KY 59, Olive Hill, Vanceburg
FStop S: Sunoco
Gas S: Citgo
Food S: Restaurant (Sunoco)
AServ S: Citgo
TServ S: Sunoco (Tire Repair)
Other S: CB Repair (Sunoco)

161 U.S. 60, Olive Hill
FStop S: Exxon*(SCALES) (Kerosene)
Gas N: Citgo
Parks N: Carter Caves State Resort Park (5 Miles)
Other N: U.S. Post Office
S: Car Wash, Truck Wash

172 KY 1, KY 7, Grayson
TStop N: Super Quick*(CW)(LP) (24 hrs), Truck Stop*(LP) (Showers)
S: Pilot Travel Center*(SCALES)
Gas S: Ashland*, BP*, Certified, Chevron(D)(LP), Citgo, Exxon, Rich, SuperAmerica*
Food N: Cafe Restaurant (TS), Hot Shoppe Deli (Super Quick), KFC, Long John Silvers, Shoney's, Western Steer Family Steakhouse
S: A & W Drive-In (SuperAmerica), Arby's, Baskin Robbins (SuperAmerica), Burger King, Dairy Queen, Little Caesars Pizza, McDonalds, Pizza Hut, Pizza Wizard, Sam's Hotdogs (BP), Subway, Taco Bell, Wendy's (Pilot TS, 24 hrs)
Lodg N: Country Squire Inn, AAA Econolodge, Holiday Inn Express
AServ N: Adkins-Slone Chrysler Jeep Eagle Dodge, Adkins-Slone Ford, Clay Tire Sales
S: Advance Auto Parts, Big A Auto Parts, Carquest Auto Parts, Chevron, Pennzoil Lube
TServ N: Interstate Truck Supply Inc
RVCamp S: Camp Webb (10 Miles)
Parks N: Greenbow Lake State Resort Park
S: Grayson Lake State Park
Other N: K-Mart, Pharmacy (K-Mart)
S: Bowling Center, Grayson Fire Dept, Kentucky Christian College, Kwik Photo 30 Min Processing, Messer's Army Navy Store, Ralphs Supermarket, True Value Farmer's Hardware(LP)

(173) KY Welcome Center - RR, Vending, Phones (Westbound)

181 U.S. 60
FStop N: Citgo*(LP)
Gas S: Rich*
AServ N: Tires "R" Us

185 KY 180, Cannonsburg, Ashland
Gas N: Chevron*
AServ N: Chevron
Lodg N: Days Inn Days Inn (see our ad this page), Knights Inn (see our ad this page)
Other N: KY State Patrol Post

191 U.S. 23, Ashland, Louisa,

EXIT KENTUCKY/WEST VIRGINIA

Catlettsburg
Med N: ✚ Hospital

↑KENTUCKY

↓WEST VIRGINIA

1 U.S. 52 South, Kenova, Ceredo
Gas N: Sunoco
AServ N: Queen's Automotive, Sunoco
Other S: Glass Factory Tours, Tri State Airport

6 U.S. 52 North, West Huntington, Chesapeake (OH) (Exit To Ltd Access)

8 WV 152 South, WV 527 North, 5th St East
Food S: Dairy Queen, Laredo Texas Steakhouse
Other N: Huntington Museum of Art

(10) WV Welcome Center - RR, Vending, Phones (Eastbound)

11 WV 10, Hal Greer Blvd, Downtown
RVCamp S: Beach Fork State Camp (8.5 miles South)
Med N: ✚ Hospital

15 U.S. 60, 29th St East
FStop N: Go Mart*
Gas S: Exxon*
Food N: Jolly Pirate Donuts (24 hrs), Nanna's Country Restaurant (Go Mart), Omelette Shoppe
S: Fazoli's Italian Food, Golden Corral, Golden Corral, McDonalds, Orient Express Chinese, Shoney's, Taco Bell (24 hrs)
Lodg N: Colonial Inn Motel, Traveler's Motel
S: Days Inn ◆ Days Inn, ◆ Red Roof Inn
AServ S: Davis Dodge, Guardian Auto Glass, Mitsubishi Motors, Moses Buick Pontiac, Moses Isuzu, Moses Nissan and VW, Subaru, Wal-Mart
Med N: ✚ Hospital
ATM S: Bank One
Other S: One Hr Photo, Revco Drugs, Wal-Mart (24 hrs)

20 U.S. 60, Barboursville, Mall Road
Gas N: Exxon*(D)
S: BP*, Exxon*
Food N: Applebee's, Bob Evans Restaurant, Burger King, Logan's Roadhouse, McDonalds, Olive Garden, Wendy's
S: Billy Bob's Pizza Wonderland, Cozumel Mexican Restaurant, Cracker Barrel, Gino's Pizza and Spaghetti, Johnny's Pizza, Lonestar, Outback Steakhouse, Sonic Drive In, Subway, TCBY
Lodg S: Ramada Ltd
AServ N: Firestone Tire & Auto, Sears Auto Center, Tire America
S: B & S Fast Lube, Safelite Auto Glass Replacement Repairs, Superior Oldsmobile Cadillac
RVCamp S: Foxfire Resort Camping (5.6 Miles)
ATM S: BP
Other N: Drug Emporium, Huntington Mall
S: Maytag Coin Laundry, Moore's Car Wash, Sturgeon's Optical

28 U.S. 60, Milton
Gas S: Ashland* (Kerosene), Chevron*(D), Exxon*, Go Mart*, Rich*
Food S: Bubba's Deli & Roadkill, Dairy Queen, Gino's Pizza & Spaghetti House, Granny K's Restaurant, McDonalds, Pizza Hut, Subway, Tudor's Biscuit

Bold red print shows RV & Bus parking available or nearby

EXIT		WEST VIRGINIA
		World, WV Fried Chicken (Go Mart)
	AServ	S: Advance Auto Parts, Exxon, Hubie's Auto Repair, Milton Motor Supply, Mohr's Tire Farm, Omni Muffler Center, Quality Exhaust & Tire
	RVCamp	S: Foxfire Resort (3.4 Miles)
	ATM	S: Bank One, Huntington Banks WV
	Other	S: County Lanes, Fire Dept, Foodland Grocery, Fruth Pharmacy, I Love Hardware, Milton Car Wash, Nicholas Cleaners & Coin Laundry, Save-A-Lot Food Stores
34		Hurricane
	Gas	S: Chevron(D), Exxon, Go Mart*, Sunoco(D)(CW) (Kerosene)
	Food	S: China Town, Dwight's Family Restaurant, Gino's Pizza & Spaghetti, McDonalds, Pizza Hut, Subway, Tudor's Biscuit World
	Lodg	S: AAA Ramada
	AServ	N: Ford Dealership, R.H. Peters Trucks GM Chevrolet (Light Trucks), Saturn of Charleston-Huntington
		S: Autoworks, Chevron, Exxon, Hurricane Motors(CW), Sunoco
	ATM	S: OneValley Bank, Putnam County Bank
	Parks	S: Valley Park
	Other	S: Fruth Pharmacy, Putnam Animal Hospital, Rite Aide Pharmacy
(35)		Rest Area - RR, Vending, Phones, RV Dump (Both Directions)
(38)		Weigh Station (Both Directions)
39		WV 34, Winfield, Scott Depot
	TStop	S: 76 Auto/Truck Plaza*(SCALES) (Showers)
	FStop	N: Go Mart*
	Gas	N: BP*
		S: Chevron*(D), Exxon*(D), Go Mart*
	Food	N: A Taste of Italy, Applebee's, Arby's (BP), Baskin Robbins (BP), Bleachers Sport Bar & Grill, Bob Evans Restaurant, Griff's Restaurant & Tavern, Hardee's
		S: Blimpie's Subs (Go Mart), Burger King, Captain D's Seafood, China Chef, Fazoli's Italian Food, Gino's Pizza & Spaghetti, KFC, McDonalds, Paul Bunyan Restaurant (76 Auto/Truck Stop), Schlotzkys Deli, Shoney's, Subway, TCBY Yogurt, Taco Bell, Tudor's Biscuit World, Wendy's
	Lodg	N: DAYS INN AAA Days Inn, Red Roof Inn
	AServ	N: Advance Auto Parts
		S: Cooper's Car Care Center
	TServ	S: 76 Auto/Truck Plaza
	Med	S: ✚ Health Plus, ✚ Hospital
	ATM	N: Big Bear Food, Huntington Federal Savings

EXIT		WEST VIRGINIA
		Bank
		S: OneValley Bank
	Other	N: Big Bear Food & Grocery (w/pharmacy), Putnam County Library, Putnam Village Cinemas, Teays Optometry Group, Teays Valley Laundry Center (Coin Laundry), Tee to Green (Golf Shop), US Post Office
		S: K-Mart (w/pharmacy), Kroger Supermarket (24 hrs), Pharmacy (K-Mart), Pharmacy (Kroger), Rite Aide Pharmacy, Village Pets
44		U.S. 35, St Albans
45		WV 25, Nitro
	Gas	S: BP*, Pennzoil Lube*
	Food	S: Checkers Burgers, Gino's Pizza, McDonalds, Subway, Tudor's Biscuit World
	Lodg	S: AAA Best Western
	AServ	N: Turnpike Chevrolet Geo
		S: Marty's Tire Store, Turnpike Chevrolet
	Other	S: Town & Country Supply(LP)
47		WV 622, Cross Lanes, Goff Mtn Road
	Gas	N: Chevron*(CW) (24 hrs), Exxon*, Go Mart*, SuperAmerica*
	Food	N: Bob Evans Restaurant, Captain D's Seafood, Dunkin Donuts, Gino's Pizza & Spaghetti House, Graziano's Pizza, Papa John's Pizza, Pizza Hut, Rice Bowl Chinese, Taco Bell, Tudor's Biscuit World, Wendy's
		S: KFC, Shoney's
	Lodg	N: Motel 6
		S: Comfort Inn
	AServ	N: Jiffy Lube, NAPA Auto Parts
	ATM	N: City National Bank, Machine (SuperAmerica), OneValley Bank
	Other	N: Maytag Equipped Laundry (Coin Laundry), Rite Aide Pharmacy
		S: United States Armed Forces Reserve Center
50		WV 25, Institute (Difficult Reaccess)
	Gas	S: Go Mart*
	Food	S: Andy's Grill
	Parks	S: Shawnee Regional Park
	Other	S: US Post Office
53		WV 25/25, Roxalana Rd, Dunbar
	Gas	S: Go Mart
	Food	S: Captain D's Seafood, China Inn Restaurant, Graziano's Pizza (Go Mart), Hoots Bar & Grill, Krispy Kreme (Go Mart), McDonalds, Shoney's, Subway (Go Mart), Wendy's
	Lodg	S: Super 8 Motel, ◆ Travelodge
	AServ	S: Amarica Auto Parts Pros, Earl's Auto Parts, Gail's Auto Service, NAPA Auto Parts, Tire America, Tune Up Plus

EXIT		WEST VIRGINIA
	ATM	S: Kroger Supermarket, United National Bank
	Other	S: Dunbar Animal Hospital, Economy Laundromat, Fast Chek Grocery, Fire Station, Kroger Supermarket, One Hour Photography, Police Dept, Public Library, Revco Drugs, Rite Aide Pharmacy, Wanda's Car Wash
54		U.S. 60, MacCorkle Ave, South Charleston
	Gas	S: Ashland*
	Food	N: TCBY
		S: Bob Evans Restaurant, Husson's Pizza, KFC, McDonalds, Taco Bell, Wendy's
	Lodg	S: ◆ Red Roof Inn
	AServ	S: AAMCO Transmission, Charleston Tire
	Med	S: ✚ Thomas Memorial Hosp
	Other	S: Sav-A-Lot Grocery, South Charleston Community Center, West Va State Police HQ
55		Kanawha Turnpike (Westbound)
56		Montrose Dr, South Charleston
	Gas	N: 7-11 Convenience Store*, Chevron*(CW), Exxon*(D), SuperAmerica* (24 hrs)
	Food	N: Baskin Robbins (Chevron), Blimpie's Subs (Chevron), Dairy Queen, Hardee's ' (24 hrs), Shoney's
	Lodg	N: Ramada
	AServ	N: Allen's Tune-Up, Express Lube, Harrah's South Charleston Garage, Quality Exhaust
	Other	N: Avalon Dog & Cat Hospital
58A		U.S. 119 South, Oakwood Road
58B		U.S. 119 North, Virginia Street Civic Center (Eastbound)
	Gas	N: BP
	Food	S: Captain D's Seafood, Long John Silvers, Mack's Restaurant & Lounge, Shoney's, Wendy's
	Lodg	S: Elk River Town Center Inn, Hampton Inn, ◆ Holiday Inn
	AServ	N: BP, Jiffy Lube, Sportcar Clinic
		S: Barracks Automotive
	ATM	N: City National Bank, Huntington Banks WV
	Other	N: Mountaineer Pride Car Wash, Swan Cleaners (Coin Laundry), US Post Office, Valley West Veterinary Hosp
		S: Phillip's Animal Center
58C		U.S. 60, Washington Street Civic Center
	Gas	N: Chevron(CW), Exxon, Go Mart*, Rich
	Food	N: Country Junction Rest, Dutchess Bakeries, The Grill, Tudor's Biscuit World
		S: Allie's American Grill, Cagney's Seafood

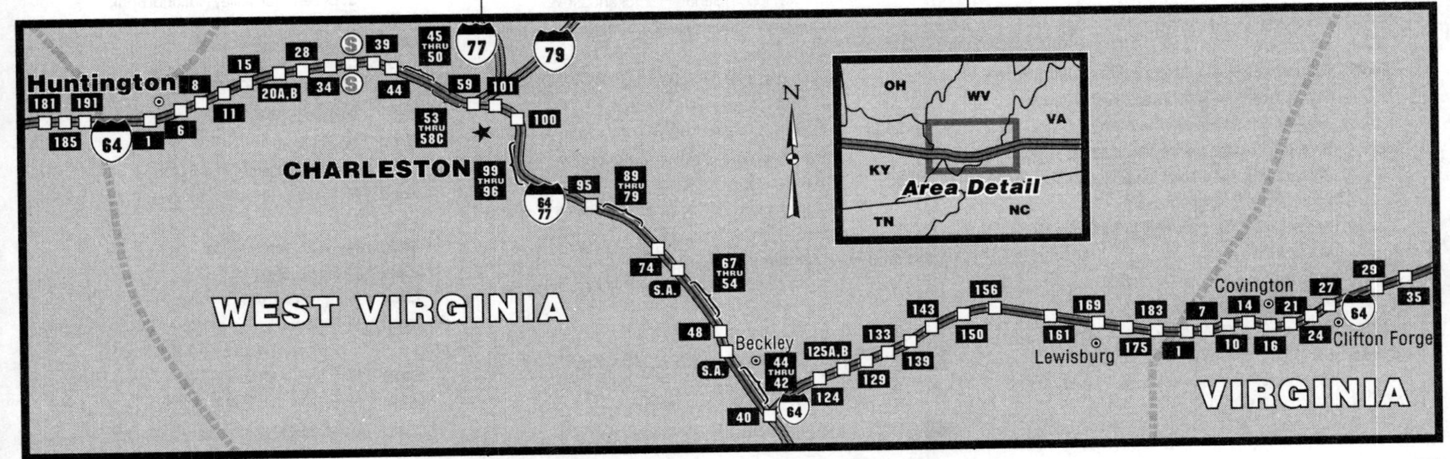

Bold red print shows RV & Bus parking available or nearby

EXIT	WEST VIRGINIA
	Steaks, Fifth Quarter Steak House
Lodg	**S:** Holiday Inn, ◆ Marriot
AServ	**N:** Exxon
	S: Sears
Med	**S:** ✚ Columbia St Francis Hospital
ATM	**N:** OneValley
	S: OneValley Bank
Other	**S:** Charleston Civic Center, Charleston Town Center Mall, Phillips Animal Hospital
59	Junction I-77, Junction I-79, Huntington, Parkersburg
	Note: I-64 runs concurrent below with I-77. Numbering follows I-77.
101	Junction I-64 West, Huntington
100	Broad St, Capitol St (Difficult Reaccess Northbound)
Gas	**W:** Chevron(D)
Food	**W:** Chesapeake Bagel Bakery, Graziano's Pizza, Lee Street Deli & Lounge, Murfy's Coffee Shop, Pavilion Cafe (Heart O' Town), Ponderosa, Tudor's Biscuit World
Lodg	**W:** Heart O' Town
AServ	**W:** A & S Automotive, Big A Auto Parts, Chevron, Firestone Tire & Auto, Goodyear Tire & Auto, NAPA Auto Parts, Ziebart TidyCar
Med	**W:** ✚ Charleston Area Medical Center, ✚ Columbia St Francis Hosp
ATM	**W:** Huntington Banks
Other	**W:** Capitol Market, Fire Dept, Kanawha County Public Library, Mail Boxes Etc, Park Place Cinema 7, Revco Drugs, Rite Aide Pharmacy, US Post Office
99	WV 114, Greenbrier St, State Capital
Gas	**W:** Chevron* (711 Store), Exxon*(D)
Food	**W:** Applegarth Cafe, Capital Lounge Rest, Domino's Pizza, New China Restaurant, Rally's Hamburgers, Subway, Wendy's
ATM	**W:** Bank One
Other	**W:** US Post Office, West Virginia State Capitol Building
98	WV 61, 35th Street Bridge (Southbound)
97	West US 60, Kanawha Blvd, Midland Trail
96	U.S. 60 East, Belle, Midland Trail (Difficult Reaccess)
Food	**E:** Gino's Italian, Tudor's Biscuit House
Lodg	**W:** Budget Host Inn
95	WV 61, MacCorkle Ave
FStop	**E:** Go Mart*
Gas	**W:** Exxon*(D)
Food	**E:** Bob Evans Restaurant, Graziano's Italian, Lone Star Steakhouse, McDonalds, Wendy's, West Virginia Fried Chicken (Go Mart)
	W: Burger King, Cancun Mexican Rest, Captain D's Seafood, Ponderosa Steak House, Taco Bell, Tony's Place Gold Dome BBQ
Lodg	**E:** DAYS INN ◆ Days Inn, AAA Knights Inn, Motel 6, ◆ Red Roof Inn
AServ	**E:** Advance Auto Parts, K-Mart
	W: W.H. Service Center
ATM	**W:** Bank One, One Valley Bank
Other	**E:** K-Mart, Pharmacy (K-Mart)
	W: Kanawha Mall, Kroger Supermarket (24 hrs), Pharmacy (Kroger)

EXIT	WEST VIRGINIA
89	WV 61, WV 94, Marmet, Chesapeake
FStop	**E:** Exxon*(D) (Kerosene)
Gas	**E:** BP(D), Go Mart*
Food	**E:** Gino's Pizza, Hardees (24 hrs), Subway, Tudor's Biscuit World
AServ	**E:** BP, Hudson's Auto Repair, NAPA Auto Parts
Other	**E:** Kroger Supermarket (24 hrs), Pharmacy (Kroger), Rite Aide Pharmacy
85	Chelyan, US 60, WV 61, Cedar Grove (Pay Toll To Reaccess)
AServ	**E:** Paul White Chevrolet Geo
(83)	Toll Plaza
79	Sharon, Cabin Creek Road
74	WV 83, Paint Creek Road
(72)	Travel Plaza (Northbound)
FStop	**N:** Exxon(D) (RV Dump)
Food	**N:** Roy Rogers (Travel Plaza), TCBY Yogurt (Travel Plaza)
(70)	Rest Area - RR (Southbound)
66	WV 15, Mahan
60	WV 612, Mossy, Oak Hill
Gas	**E:** Exxon*(D)
Food	**E:** Miss Ann's Fancy Food
Lodg	**E:** Motel (Exxon)
AServ	**E:** Exxon
RVCamp	**E:** Camping (Full Hookup)
(55)	Toll Plaza (Both Directions)
54	Pax, Mount Hope, 23/2
Gas	**E:** BP*
	W: Exxon*(D)
Food	**W:** Long Branch (Kerosene)
Other	**E:** Plum Orchard Lake Public Fishing Area
48	U.S. 19, Summersville, North Beckley (Pay Toll To Exit)
(45)	Tamarack the Best of West Virginia Travel Plaza (Southbound)
FStop	**W:** Exxon (Tamarack TP, RV Dump)
Gas	**W:** Exxon (Tamarack TP)
Food	**W:** Mrs Fields Cookies (Tamarack TP), Sbarro Italian (Tamarack TP), TCBY, Taco Bell (Tamarack TP)
Other	**W:** Tamarack Huge Craft Center (Tamarack TP)
44	WV 3, Harper Road, Beckley
FStop	**E:** Go Mart*
Gas	**E:** BP*, Chevron*(D), Exxon*
Food	**E:** Applebees, Baskin Robbins (BP), Beckley Pancake House, Bennetts Smokehouse & Saloon, Dairy Queen, Omelet Shoppe, Pizza Hut, Subway (BP)
	W: Bob Evans Restaurant, Pasquale Mira (Days Inn), Texas Steakhouse, Wendy's
Lodg	**E:** AAA Best Western, AAA Comfort Inn, Fairfield Inn, ◆ Holiday Inn, AAA Howard Johnson, Super 8 Motel
	W: Country Inn and Suites, DAYS INN Days Inn, ◆ Hampton Inn, ◆ Shoney's Inn
RVCamp	**W:** Lake Stephens
Med	**E:** ✚ Doctors Immedia Care (7 Days/Week, No Appt), ✚ Hospital
Other	**E:** Beckley Vision Center
42	WV 16, WV 97, Mabscott, Robert C. Byrd Drive (Difficult Reaccess)
40	Junction I-64 East, Louisburg

EXIT	WEST VIRGINIA
	Note: I-64 runs concurrent above with I-77. Numbering follows I-77.
121	Junction I-77 South, Bluefield (Southbound)
124	U.S. 19, East Beckley (Exit To Ltd Access)
125	WV 307, WV 9/9, Beaver, Airport Road
FStop	**N:** Shell* (24 hrs)
Food	**N:** Bojangles (Shell), TCBY (Shell)
Lodg	**N:** Sleep Inn
ATM	**N:** Shell
129	WV 9, Shady Spring, Grandview Road
Parks	**N:** New River Gorge National River Grandview
	S: Little Beaver State Park
133	WV 27, Bragg, Pluto Road
(136)	Run - away Truck Ramp (Eastbound)
(137)	Run - away Truck Ramp (Westbound)
139	WV 20, Sandstone, Hinton
Gas	**S:** Ashland*(D)(LP)
Parks	**S:** Bluestone Pipestem Resort Park
Other	**S:** US Post Office (Ashland), Visitor Information
143	WV 20, Green Sulphur Springs, Meadowbridge
Gas	**N:** Bore's Sporting Goods*
Other	**N:** U.S. Post Office
150	WV 29/4, Dawson
156	U.S. 60, Midland Trail, Sam Black Church, Rupert
Gas	**N:** Exxon, Shell*(D)(LP) (Kerosene)
AServ	**N:** Exxon
161	WV 12, Alta, Alderson
Gas	**S:** Exxon*(D) (Kerosene)
Food	**S:** Grandpaw's Rest
RVCamp	**S:** Campground (14.5 Miles)
169	U.S. 219, Lewisburg, Ronceverte
Gas	**S:** Exxon*(D) (Kerosene), Shell*
Food	**S:** Hardee's, **McDonalds (Wal-Mart)**, Shoney's, Subway, TCBY (Shell), Western Sizzlin'
Lodg	**N:** DAYS INN AAA Days Inn
	S: Greenbriar Inn, AAA Super 8 Motel
AServ	**S:** Wal-Mart
Med	**S:** ✚ Hospital
ATM	**N:** Bank of White Sulphur Springs
	S: Greenbriar Valley National Bank
Other	**S:** Historic District, Pharmacy (Wal-Mart), Wal-Mart (24 hrs)
175	U.S. 60, Caldwell, White Sulphur Springs, WV 92
FStop	**N:** Shell*
Gas	**N:** Chevron*(D) (24 hrs), Exxon
Food	**N:** Granny's House Restaurant, McDonalds, Wendy's
Lodg	**N:** The Sleeper Motel
AServ	**N:** Exxon, Shell
TServ	**N:** Dixon's Truck Service
RVCamp	**S:** Greenbriar Mountain Air Camp Ground, Greenbriar State Forest (2 Miles)
Parks	**N:** Greenbriar State Forest Cavern Tours
Other	**N:** Organ Caverns (2 Miles)
(179)	WV Welcome Center - RR, Phones,

Bold red print shows RV & Bus parking available or nearby

EXIT		WEST VIRGINIA/VIRGINIA
		RV Dump (Westbound)
181		U.S. 60, WV 92, White Sulphur Springs (Westbound, Reaccess Eastbound)
	Gas	N: Pennzoil Lube(D)
	Food	N: April's Pizzeria Rest, Blake's Rest & Lounge, **Hardee's**, Pizza Hut Carryout
	Lodg	N: (AAA) Budget Inn, The Old White Motel
	AServ	N: Pennzoil Lube
	RVCamp	S: **Twilight Overnight Camping (.9 Miles)**
	Other	N: **Food Lion Supermarket, Hardware Auto Service NAPA Auto Parts, Rite Aide Pharmacy, True Value Hardware**
183		VA 311, Crows (Eastbound, Reaccess Westbound)

↑ WEST VIRGINIA

↓ VIRGINIA

EXIT		VIRGINIA
1		Jerry's Run Trail
	Other	N: **Allegheny Trail**
(2)		**VA Welcome Center - RR, Phones, 2 Hr Parking, Pet Rest (Eastbound)**
7		CR 661 (Reaccess Southbound On Midland Trail East (3 Miles))
10		U.S. 60 East, VA 159 to South VA 311, Callaghan
	Gas	S: Midway
14		VA 154, Covington, Hot Springs
	Gas	N: Coastal*, Exxon*
	Food	N: Arbys (Exxon), Baskin Robbins (Exxon), Great Wall, KFC, Little Caesars Pizza, Subway, Wendy's
	AServ	N: Advance Auto Parts
	Other	N: **Food Lion Supermarket, Horizon Pharmacy, Kroger Supermarket (Pharmacy), Revco Drugs,** Stable Wash Too(CW)
16		U.S. 60 West, U.S. 220 North, Covington, Hot Springs
	Gas	N: Exxon, Texaco*
	Food	N: Burger King, Jigg's Drive-In, Marion's Cafe, Mountain View Best Western
		S: Choice Picks Food Court, Comfort Center, **Long John Silvers, McDonalds, Restaurant (K-Mart),** The Painted Elephant
	Lodg	N: Kings Court, Pinehurst Motel
		S: Comfort Center, ◆ Comfort Inn
	AServ	N: Buick Pontiac, Exxon
		S: K-Mart
	ATM	S: First Virginia Bank
	Other	N: **Fore Mountain Trail Parking, James River Ranger Station**
		S: **K-Mart (w/pharmacy), Mallow Lanes**
21		VA 696 , Low Moor
	Med	S: ✚ **Allegheny Regional Hospital**
24		U.S. 60 East, U.S. 220 South, Clifton Forge
	Other	S: **Allegheny Highlands Arts & Crafts Center**
27		U.S. 60 West Bus., U.S. 220 South, CR 629, Clifton Forge
	Gas	S: Citgo*(D)(LP) (Kerosene), Texaco*
	AServ	N: NAPA Auto Parts
	TServ	N: **Forrest Park Truck & Equip Inc**
	Parks	N: **Douthat State Park**
	Other	S: **Cliftondale Country Club Golf**
29		VA 269 East, VA 42 North
	Gas	S: Exxon*(D)(LP)
	Food	S: Triangle Restaurant (Exxon)
	AServ	S: Triangle Auto Tires
35		VA 269, CR 850, Longdale Furnace
	Food	S: Old Stacks
43		CR 780, Goshen
	Other	S: **National Forest Scenic Byway**
50		U.S. 60, CR 623, Kerrs Creek, Lexington
55		U.S. 11 to VA 39, Lexington
	Gas	N: Exxon(D) (Kerosene Varsol)
		S: Texaco*(D)
	Food	N: **Burger King**
		S: Golden Corral, Redwood Family Restaurant, Shoney's
	Lodg	N: Colony House Motel, (AAA) Inn at Hunt Ridge (Best Western), Super 8 Motel
		S: (AAA) Comfort Inn, Econolodge, (AAA) **Holiday Inn**
	AServ	S: Oil Exchange & Lube
	RVCamp	N: **Long's Campground**
	Med	S: ✚ **Hospital (2.5 Miles)**
	ATM	S: Bank Of Rockbridge
	Other	N: **Blue Ridge Animal Clinic, Wal-Mart**
		S: **Dabney S. Lancaster Rockbridge Center, George C. Marshall Museum, Kroger Supermarket, Lexington Historical Shop, Revco Drugs, Stonewall Jackson House, Tourist Information**
56		Junction I-81 South, Roanoke (Eastbound)
75		7th St to Norfolk
		Note: I-64 runs concurrent below with I-81. Numbering follows I-81.
191		Junction I-64 West, Lexington, Charleston
195		U.S. 11, Lexington
	TStop	W: **Shell*(D)**
	Gas	W: Citgo
	Food	E: Maple Hall
		W: Aunt Sara's
	Lodg	E: ◆ Maple Hall
		W: (AAA) Howard Johnson, (AAA) **Ramada**, Red Oaks Inn
(199)		**Rest Area - RR, Phones (Southbound)**
200		VA 710, Fairfield
205		VA 606, Raphines, Steeles Tavern
213AB		U.S. 11, Greenville
217		VA 654, Mint Spring, Stuarts Draft
220		VA 262, U.S. 11, Staunton
221		Junction I-64 East
		Note: I-64 runs concurrent above with I-81. Numbering follows I-81.
87		Junction I-81 North, Staunton, Lexington (North To Winchester, South Roanoke, Harrisonburg)
	Other	N: **Museum Of American Frontier Culture, Robert E. Lee Burial Grounds, Staunton Historic Dist, Stonewall Jackson Burial Grounds, Woodrow Wilson Birthplace**
91		VA 608, Fisherville, Stuarts Draft
	Gas	S: Exxon* (24 Hr)
	Food	S: McDonalds
	AServ	N: Eddie's Tire
		S: Eddie's Tire
	Med	N: ✚ **Augusta Medical Ctr**
	Other	N: **Woodrow Wilson Rehab Ctr**
94		U.S. 340, Waynesboro, Stuarts Draft
	Gas	N: Citgo*, Exxon(D)
		S: Pennzoil Lube*
	Food	N: KFC, Shoney's, Wendy's, Western Sizzlin'
	Lodg	N: Best Western, (Days Inn) **Days Inn**, (AAA) Holiday Inn Express, ◆ **Super 8 Motel**
	AServ	N: Exxon
		S: Ladd Auto Repair
	Other	S: **Moss Museum, Waynesboro Outlet Village**
96		VA 624, Waynesboro, Lyndhurst

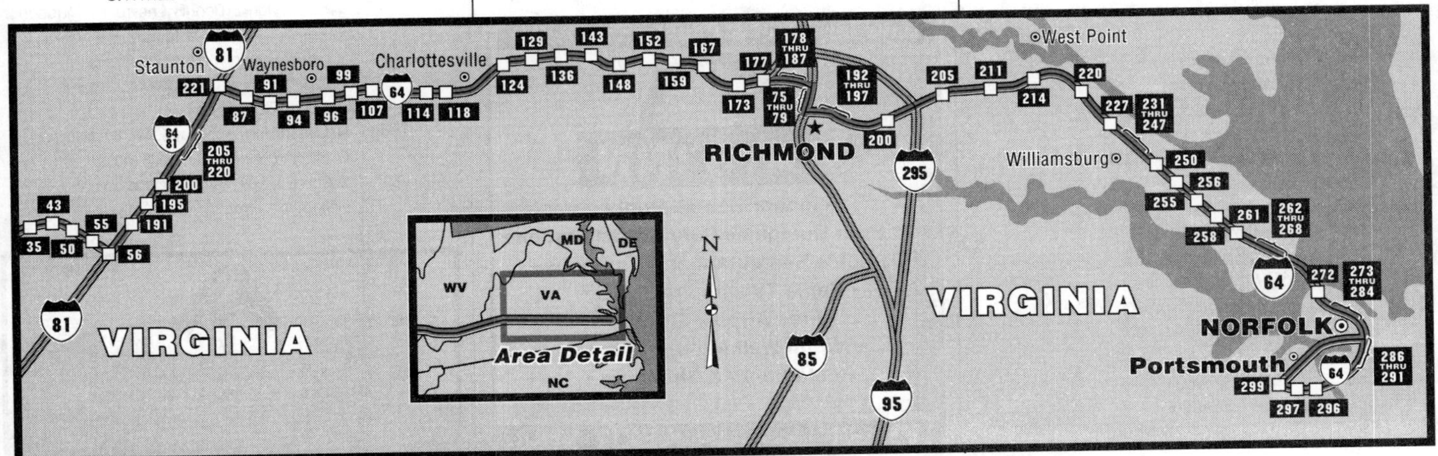

Bold red print shows RV & Bus parking available or nearby

EXIT VIRGINIA

(Trucks Use Lower Gear)
TServ S: Overnight Truck Facility
RVCamp N: The North Forty RV Campground
Other N: Sherando Lake, George Washington Nat'l Forrest

99 U.S. 250, Afton, Waynesboro
Gas S: Chevron*
Food S: Howard Johnson
Lodg N: Colony House Motel, Redwood Lodge
S: Howard Johnson, The Inn at Afton
Other N: Appalachian Trail Crossing
S: Blue Ridge Parkway, Shenandoah National Park, Tourist Information, Wintergreen Ski

(105) Rest Area - RR, Phones, Picnic, Vending (Eastbound)

107 U.S. 350, Crozet
Gas S: BP
Other S: Village Mkt

(113) Rest Area - RR, Phones, Picnic, Vending (Westbound)

114 VA 637, Ivy

118AB U.S. 29, Charlottesville, Lynchburg (Culpepper)
Lodg N: Days Inn Conference (see our ad this page), Holiday Inn
Other N: University Info Ctr

120 VA 631, 5th St, Charlottesville
Gas N: Exxon*, Texaco*(LP)
Food N: Domino's Pizza, Hardee's, Henry's Restaurant, Jade Garden (Chinese)
Lodg N: (AAA) Holiday Inn (see our ad this page)
AServ N: Texaco
ATM N: F&M Bank, Nationsbank (Texaco)
Other N: Coin Car Wash, Pet Motel & Salon

121 VA 20, Charlottesville, Scottsville
Gas N: Amoco*, Chevron*
Food N: Blimpie's Subs (Amoco), Holiday Food & Deli, Moore's Creek Restaurant
AServ N: Chevron, Shull's Wrecker Service
RVCamp S: KOA Campground (9.6 Miles)
Med S: + Hospital
Other S: Ashlawn (James Monroe's Home), Jefferson Vinyard Tour, Monticello (Thomas Jefferson's Home), Monticello Visitor Center

124 U.S. 250, Charlottesville, Shadwell, Jefferson Hwy
Food S: Ramada Inn
Lodg S: Ramada Inn
Med N: + Hospital

129 VA 616, Keswick, Boyd Tavern

136 U.S. 15, Gordonsville, Palmyra (Zions Cross Roads & Fork Union)
FStop S: Amoco*, Citgo
Gas S: Exxon*(D), RaceWay*(LP), Texaco*
Food S: Burger King (Exxon), Crescent Inn Restaurant, McDonalds (Amoco), Zion Cross Roads Market & Deli
Lodg S: Zion Cross Roads Motel
Other S: Crossroads Market

143 VA 208, Louisa, Ferncliff
Gas S: Citgo*(D), Exxon(D)(LP)
RVCamp N: Small Country Camping (7 Miles)
ATM S: Citgo
Other N: Ferncliff Market
S: Ferncliff Market (Citgo)

149 VA 605, Shannon Hill

152 VA 629, Hadensville

159 U.S. 522, Gum Spring, Mineral
FStop N: Exxon*(D)(CW) (Picnic Area)
Gas S: Citgo*
Food N: Ernhardt Hot Dogs (Exxon)
S: Junction Restaurant
AServ N: Gum Springs Auto Service

EXIT VIRGINIA

S: Excel
TServ N: Gum Springs Truck Service
ATM S: Citgo
Other S: Perish Grocery

167 VA 617, Oilville, Goochland
FStop S: Amoco*(LP) (Picnic)
Food S: Bullet's (Amoco)
ATM S: Amoco

(168) Rest Area - RR, Phones, Picnic, Vending (Westbound)

(169) Rest Area - RR, Phones, Picnic, Vending (Eastbound, No Night Parking Allowed)

173 VA 623, Rockville, Manakin
FStop S: Texaco*

177 Junction I-295, Washington, Norfolk, Williamsburg

178AB US 250, Richmond, Short Pump
FStop N: Texaco(CW) (Kerosene)
Gas S: Amoco*
Food N: Boychick's Deli, Chesapeake Bagel, Crab House, Dairy Queen, DeFazio Of Innsbrook Steaks, Pasta), Hickory Ham, Hunan Express, Leonardo's Pizza, Manhattan Bagel, Mulligan's, Parkside Cafe, Sharkie's, Starbuck's Coffee, Thai Garden, Zack's Frozen Yogurt
S: Casa Grande, Grill & Cafe, Hot Stuff Pizza (Amoco), McDonalds (Walmart), Taco Bell, Wendy's
Lodg N: AmeriSuites (see our ad this page), Hampton Inn, Homestead Villas
AServ N: Auto Import Serv All In One Detail, Carmax, Firestone Tire & Auto (Master Care), Import Auto House, Jiffy Lube, Universal Ford
S: Car Quest Auto Center, Richard's Auto Repair, Short Pump Tire & Exhaust

EXIT VIRGINIA

Med S: + Patient Walk In Clinic
ATM N: Fidelity Federal, First Union Bank, Jefferson National Bank, Signet Bank
S: Commerce Bank, Crestar Bank
Other N: Auto Port Shopping Ctr, Dog's & Cat's World, Fed Ex Drop Box, Great Land Hess Pharmacy, Innsbrook Pavalion, Innsbrook Shopping Ctr, Mail Room
S: Target Department Store, Wal-Mart

180AB Gaskins Road
FStop N: East Coast*(D)(LP) (Kerosene)
Gas N: Amoco*, Exxon*, Texaco*
S: Citgo(D)(LP), Texaco*
Food N: Applebee's, Bistro's Pantry, Blimpie's Subs (East Coast), China Express, Golden Corral, Jimmy's American Deli, McDonalds, Ruby Tuesday
AServ N: Costco, Goodyear Tire & Auto, Haynes Jeep & Eagle
S: Citgo
ATM N: Central Fidelity Bank, First Virginia Bank, Nationsbank
Other N: Circuit City Corral Plaza, Commonwealth Eyeglasses, Deep Run Animal Clinic, LeGourmet Bakery, Lexington Commons Shopping Ctr, Pet Center, Phar Mor Drugs, UKrops (Pharmacy & Supermarket)
S: 7-11 Convenience Store, Breeze Convience Store, Coin Operated Laundry

181AB Parham Road
Med N: + Hospital
S: + Patient First Walk-In
ATM N: First Virginia Bank, Jefferson Nat'l Bank

183 U.S. 250, Broad St, Glenside Dr
Gas N: Amoco, Citgo*(CW), Crown*(CW)(LP), Exxon*(D), Merritt, Shell(CW)
S: Amoco*, Chevron*, Shell, Texaco*(D)(CW)
Food N: Aunt Sara's Pancake House, Awful Arthur's Seafood, Bennigan's, Blue Marlin Seafood, Bob Evans Restaurant, Bojangle's, Brothers Italian Restaurant & Pizza, Bullets' Burgers & More, Burger King, Casa Grande Mexican Restaurant, China Buffet, China Inn, Friendly's, Fuddrucker's, Hooters, Indian Foods & Flavor, Italian Kitchen West, Kabuto Japanese Steakhouse, Little Caesars Pizza, Lone Star Steakhouse, McDonalds, Morrisons Cafeteria, Mozzarella's Cafe, Old Country Buffett, Olive Garden, Outback Steakhouse, Peking Garden, Piccadilly Cafeteria, Pizza Hut, Red Lobster, Shoney's, Steak & Ale, Subway, T.G.I. Friday's, Taco Bell, Thai Dynasty (Thai/American), The Wood Grill (Steaks & Seafood), Vietnam Harbor Restaurant, Waffle House, Wendy's
S: Arby's, Bill's BBQ, Chinese Restaurant, Denny's, Dunkin Donuts (Chevron), El Maddador Restaurant, Full Kee Restaurant, Holiday Inn, House of Hunan, McDonalds, Me-Kong Restaurant, Mexico Restaurant, Taco Bell, Toppings Pizza & Subs
Lodg N: Comfort Inn, Embassy Suites, Fairfield Inn, Quality Inn, Shoney's, Super 8 Motel
S: Courtyard by Marriott, (Days Inn logo) Days Inn, Holiday Inn, Hyatt

Bold red print shows RV & Bus parking available or nearby

EXIT	VIRGINIA
AServ	N: Amoco, Capitol Ford Lincoln, Mercury, Exxon, George Toyoto, Haywood Clark Pontaic Buick, Hutch's Body Shop, Jiffy Lube, K-Mart, Lawrence Dodge, Merchant's Tire & Auto, Midas, Mitsubishi/Honda Dealer, Royal Oldsmobile, Saturn, Trak Auto, Volvo Dealer
	S: Advance Auto Parts, Alan Tire Goodyear, All Tune & Lube, Amoco, Baugh Auto Body, Dominion Chevrolet, Import Auto Haus, Muffler Shop, Shell
ATM	N: Central Fidelity Bank, First Union Bank, First Virginia Bank, Hanover Bank, Jefferson Nat'l Bank, Nationsbank, Signet Bank
	S: Crestar Bank, First Virginia Bank
Other	N: **7-11 Convenience Store, A&A Convience, Allied Animal Hospital, CVS Pharmacy, Coin Operated Laundry, Fountain Square Shopping, K-Mart, Krops Pharmacy, Merchant's Walk Shopping Ctr, Old Town Shopping Ctr, Petsmart, Richmond Lock & Safe,** Western Wash Car Wash, **Winn Dixie Supermarket**
	S: **7-11 Convenience Store, Ambassador Dog & Cat Hospital, Coin Laundry (Horsepin Convience), Crestview Food Store, Horsepin Convience, Rolling Hills Super Center, UniMart Convience Store, WalGreens, Westwood Shopping Ctr**
185	U.S. 33, Staples Mill Road, Dickens Road (Difficult Reaccess, Dickens Rd)
Gas	S: Shell
AServ	S: Shell
Med	S: **Hospital**
ATM	S: Central Fidelity Bank
Other	S: **7-11 Convenience Store**
186	Junction I-195 South
187	Junction I-95 North, Washington (Left Exit Eastbound)

Note: I-64 runs concurrent below with I-95. Numbering follows I-95.

EXIT	VIRGINIA
79	Junction I-64 West, I-195 South
78	Boulevard (Difficult Reaccess)
Gas	E: Citgo
	W: Amoco*(D), Lucky*(LP)
Food	E: Holiday Inn, Zippy's BBQ
	W: Bill's Virginia BBQ, Taylor's Family Restaurant (Days Inn)
Lodg	E: Diamond Lodge and Suites, Gadnes Restaurant, Holiday Inn
	W: Days Inn, AAA Days Inn
TServ	W: **Dolan International**
ATM	W: Jefferson National Bank (Lucky)
Other	W: **Tourist Information, US Marine Museum**
76B	U.S. 1, U.S. 301, Belvidere
Med	W: **Belvidere Medical Center**
Other	W: **Maggie Walker Historical Site, VA War Memorial**
76A	Chamberlayne Ave (Northbound)
Gas	E: Citgo
Food	E: Burger King, Captain D's Seafood, Dunkin Donuts, Hawks Bar-B-Que And Seafood, McDonalds
Lodg	E: Belmont Motel
AServ	E: Emrick Chevrolet, Napa Auto Parts, Texaco*(D) (Kerosene)
Other	E: **7-11 Convenience Store, Easters Convience Store**

Note: I-64 runs concurrent above with I-95. Numbering follows I-95.

EXIT	VIRGINIA
190	Junction I-95 South
192	U.S. 360, Mechanicsville
Gas	N: Amoco(CW) (24 Hr), Lucky Convience, Texaco(D) (Wrecker Service)
	S: Chevron(D)(CW)
Food	N: Church's Fried Chicken, KFC, McDonalds, Murrey's Steak & Seafood, Ocean Seafood Market
	S: Stuarts Seafood Take-Out
AServ	N: Texaco
	S: Tuffy, Weaver Transmission
Other	N: **Community Pride Food Store, Lucky Coin Operated Laundry, Lucky Convience Store, Mathematics and Science Center, Repair and Towing, Strawberry Hill, Western Union**
	S: **Cheek & Shockley RV, Coin Operated Laundry (Chevron)**
193AB	VA 33, Nine Mile Road
Gas	N: Amoco*(D), Exxon*
AServ	N: Amoco
TServ	N: **Capitol Freightliner Sales and Services**
Med	S: **Hospital**
195	Laburnum Ave, VA 33, US60, VA5
Gas	N: Mobil, Shell
Food	S: Applebee's, China King, Subway, Taco Bell
AServ	N: Mobil, Shell
ATM	S: Central Fidelity Bank, Jefferson National Bank
Other	N: **James River Plantation**
	S: **Eye Care Center, Fish Tail Pet Store, Puppy Love Pet Care Center, Rite Aide Pharmacy**
197AB	VA 156, Airport Dr, Highland Springs,

EXIT	VIRGINIA
	Sandston
FStop	S: **East Coast*(LP)**
Gas	S: BP*
Food	S: Best Western, Bullets, Burger King, Ma & Pa's Country Diner, Mexico Restaurant, Pizza Hut, Waffle House
Lodg	S: Best Western, Days Inn, Days Inn, Econolodge, Hampton Inn, Holiday Inn, Legacy Inn, Motel 6, Super 8 Motel
AServ	S: **Central Virginia Auto Sales,** Town & Country Auto Sales and Truck Center
TServ	N: **Penske Truck Center**
ATM	S: Central Fidelity
Other	S: **7-11 Convenience Store* (24 Hr), VA Aviation Museum**
200	Junction I-295, U.S. 60, Washington D.C., Rocky Mount NC
FStop	N: **Exxon* (24 Hr)**
Gas	N: Amoco*(LP)
	S: Texaco*(LP)
Food	N: Fast Mart Cafe, Nava's Pizza
	S: McDonalds (Texaco)
AServ	N: Amoco
ATM	N: Citizens and Farmers Bank
Other	N: **Food Lion Grocery, New Kent Crossing Supermarket, Post Office (Amoco), Revco Drugs**
	S: **Lipscomes Hardware, Winn Dixie Supermarket (Pharmacy)**
(203)	**Weigh Station (Both Directions)**
205	VA 33E, VA 249W, to U.S. 60, Bottoms Bridge, Quinton
FStop	N: **Exxon*(D) (24 Hr)**
Gas	N: Amoco*(D)(LP) (Wrecker Service)
	S: Texaco*(LP)
Food	N: Nava's Pizza
	S: McDonalds (Texaco)
AServ	S: Texaco(LP)
ATM	N: Citizens & Farmers Bank
Other	N: **Food Lion Grocery Store, New Kent Crossing Supermarket, Revco Drugs**
	S: **Lipscomes Hardware Store, U.S. Post Office, Winn Dixie Supermarket (Deli & Pharmacy)**
211	VA 106, Talleysville, Prince George
(213)	**Rest Area - RR, Vending, Phones, Picnic**
214	VA 155, New Kent, Providence Forge
RVCamp	S: **Ed Allen's Campground (9.9 Miles)**
220	VA 33 East, West Point
Gas	N: Exxon*(D) (Kerosene)
227	VA 30, to VA 60, Toano, Williamsburg
FStop	S: **Texaco***
Gas	S: Shell(LP)
Food	S: **McDonalds (Texaco FS), Stuckey's(LP) (Shell)**
RVCamp	S: **Williamsburg Campsites (6.5 Miles)**
231AB	VA 607, Croaker, Norge
Gas	N: Citgo*
Parks	N: **York River State Park**
234AB	199 &VA 646, Lightfoot
Lodg	N: **Days Inn (see our ad this page)**
RVCamp	N: **Camp Skimono, KOA Campground**
	S: **Fair Oaks Campground, Kin Kaid Kampground**

Bold red print shows RV & Bus parking available or nearby

EXIT		VIRGINIA
	Other	N: Old Dominion Oprey, Williamsburg Pottery (see our ad this page)
238		VA 143, Camp Peary, Colonial Williamsburg, to U.S. 60
	Lodg	N: Williamsburg Hotel (see our ad this page)
		S: Holiday Inn (see our ad opposite page)
	RVCamp	S: Anvil Campground
	Med	N: Hospital
	Parks	S: Water Mill Park
	Other	N: Cheatham Annex
242AB		VA 199, Colonial Pkwy, Williamsburg
	Food	N: Days Inn
	Lodg	N: Days Inn (see our ad this page)
		S: Red Carpet Inn (see our ad this page)
	Other	N: Water Country USA Water Park, Yorktown
		S: Busch Gardens
243		VA 143 West, Williamsburg (Difficult Reaccess, Left Exit Westbound)
247AB		VA 238, Yorktown, Lee Hall (Westbound)
	Gas	N: BP*(D), Citgo*
	AServ	N: BP
	ATM	N: BP
250AB		VA 105, Fort Eustis Blvd, Ft Eustis, Yorktown
	RVCamp	N: Campground
	Parks	N: Newport News Park
	Other	N: US Army & Transportation Museum
255AB		VA 143, Jefferson Ave, Newport News, Williamsburg
	FStop	N: Mobil*(D)

Bold red print shows RV & Bus parking available or nearby

EXIT VIRGINIA

Gas S: Citgo*(D), Exxon*(D)

Food S: Applebee's, Burger King, Cheddar's Casual Cafe, Chick-Fil-A, Cracker Barrel, Don Pablo's Mexican Kitchen, KFC, Krispy Kreme Donuts, McDonalds, Nappo Suschi Bar Restaurant (Japanese), Outback Steakhouse, Samuri Steaks & Seafood, Subway, Waffle House, Wendy's

Lodg S: Comfort Inn, Hampton Inn, Days Inn (see our ad this page)

AServ N: AAMCO Transmission, Hall Acura, Suttle Cadillac, Oldsmobile, GMC Trucks
S: Exxon, Tires of America

ATM S: Crestar Bank, NationsBank (FedEx/UPS)

Other N: Newport New Int'l Airport, Sam's Club, Wal-Mart Supercenter
S: 7-11 Convenience Store*, Dr. Roy Martin Optical, Home Quarters, Jefferson Green Shopping Ctr, N H Northern Hardware Store, Patrick Henry Mall

256AB Oyster Point Road, Poquoson Rd, Victory Blvd

Gas N: Citgo*(CW)

Food N: Burger King, Chesapeake Bagel Bakery, Empress Chinese, Fuddrucker's, Hardees, Kenny Rogers Roasters, Pizzaria Uno, Spaghetti Warehouse, Subway

AServ N: K-Mart, Kramer Tire

ATM N: Cenit Bank, First VA Bank, Old Point Nat'l Bank

Other N: Farm Fresh Grocery, K-Mart (24 Hr. w/ grocery, pharmacy & deli), Mail Boxes Etc,

EXIT VIRGINIA

Village Square Shopping Center

258AB U.S. 17, J Clyde Morris Blvd, Yorktown

FStop S: East Coast*(LP) (Kerosene)

Gas N: Amoco*(D) (Kerosene), Exxon*, Shell*(D)(CW) (Kerosene)
S: Citgo*(D), Exxon*(D)(CW), Texaco*(LP)

Food N: Chanello's Pizza, Domino's Pizza, Glass Pheasant Tea Room, Grandstand Grill, Hauss' Deli, New China Express, Spring Garden (Chinese), Waffle House
S: Belgin Waffle And Family Steakhouse, Bo Dines Hickory Hut Bar-B-Que, Captain's Rail, Don's Bar-B-Que, Egg Roll King (Chinese), El Mariachi Mexican, Hong Kong Restaurant, Joe & Mimma's Pizza, Philly's Sub, Pizza Hut, Rally's Hamburgers, Sammy and Angelo's Steak and Pancake House, Subway, Taco Bell, Wendy's

Lodg N: Budget Lodge, Host Inn, Ramada, Super 8 Motel
S: Motel 6, Omni Hotel

AServ N: Advance Auto Parts, Casey's All State Car Sales, Import Car Service, Shell
S: Car Matic, Casey Cheverolet, Casey Honda, Charlie Faulk Auto Dealer, David's Towing, Dent Doctor, Dunlop Tires, Jiffy Lube, Mike's Q-Lube, Mizer Muffler and Brakes

Med S: ✚ Hospital, ✚ Sentara Urgent Care

ATM N: Central Fidelity Bank, Credit Union
S: First Federal Savings Bank, First Union Bank, Nationsbank (7-11), Peninsula Trust Bank

Other N: 7-11 Convenience Store, Animal Hospital, Bay Berry Village Shopping Center, Coin Laundry, Food Lion Grocery, Kiln Creek Shopping Ctr (UPS Drop Box), Mag's Deli and Ice Cream, Mariners Fine Art Center, Pac-N-Mail, Rips Convience Store
S: 7-11 Convenience Store (24 Hr), Charlie's Car Wash, Coin-Operated Laundry, Newport Square Shoppng Center, Pet Supply Store, Revco Drugs, Rite Aide Pharmacy, Super Fresh Grocery, Tourist Information, Virginia Living Museum

261A Hampton Roads Center Pkwy

EXIT VIRGINIA

Gas N: Citgo*, Exxon*(D)(CW), Shell*
S: Amoco*, Exxon, Texaco*(CW)

Food N: Andrea's Pizza, Applebee's, Blimpie's Subs (Shell), Boston Market, Burger King, Carmela's Pasta Cafe, Chi Chi's Restaurant, Chili's, China Garden, Darryl's Restaurant And Tavern, Daswiener Works, Daybreak Restaurant (Days Inn), Denny's, Dunkin Donuts, East Japanese Restaurant, Fay's Chinese, Golden Corral, Golden Palace Chinese, Holiday Inn, Hooters, IHOP, KFC, Larry's Oyster Bar, McDonalds, Mongolian BBQ, Olive Garden, Olive Garden, Picadilly, Pizza Hut, Rally's Hamburgers, Red Lobster, Schlotzkys Deli, Steak & Ale, Subway, Szechuan Chinese Food, TCBY, Taco Bell, The Grate Steakhouse, Waffle House, Wendy's
S: Arby's, Burger King, Captain George's Seafood, Domino's Pizza, Great Family Restaurant (BBQ, Subs, Pasta), High's Ice Cream, Krispy Kreme Donuts, MiPiseo Mexican Food, Old Country Buffett, Papa John's Pizza, Pizza Hut, Sammy and Nicks Family Steakhouse, Szechuan Pan Mongolian BBQ Beef, The Kettle Restaurant (24 Hr), Tommy's Restaurant, Waffle House

Lodg N: Comfort Inn, Days Inn, Fairfield Inn, Hampton Inn, Holiday Inn, Red Roof Inn
S: Econolodge, La Quinta Inn

AServ N: Advance Auto Parts, Firestone Tire & Auto, Freedom Ford, Hamptons Chevrolet, Joseph Automotive, Napa Auto, Pomoco
S: AAMCO Transmission (Wrecker Service), Amoco, Auto Repair Fast Stop, Big Al's Mufflers and Brakes, Birds Auto Repair, Brakes Parts Auto Speciality, Car Quest Auto Center, Christian Auto Repair, Copeland Auto Plaza, Dr Motor Works Engine Repair, Eurato European Repair (Diesel Service), Jiffy Lube, Midas Mufflers and Brakes, Montgomery Ward, Paul Tysinger Nissan, Pep Boys Auto Center, R & D Carburator, Tread Quarters Tires, Western Auto, Williams Honda Used Cars, Windshields Of America

TServ S: Watsons Petroleum and Repair

Med N: ✚ Hospital
S: ✚ Mercury West (Pharmacy), ✚ Riverside Medical Care Ctr

ATM N: BB&T Banking, Crestar, First Union Bank, First Virginia Bank, First Virginia Bank (Walmart), Langley Credit Union, Nationsbank, Old Point Nat'l Bank, Super Fresh Supermarket
S: Crestar Bank

Other N: Almost A Bank (Check Cashing Services), Car Wash, Casey's Marine Boat Repair, Coliseum Mall (UPS Drop Box), Eye to Eye Optical, Food Lion Supermarket, Langley Air Force Base, Lenscrafter Eyeglasses, National Optical, Pearl Express, Pet World, Phar Mor Drugs, Super Fresh Supermarket, Target Department Store, Wal-Mart Supercenter (Vision Ctr, Pharmacy)
S: 7-11 Convenience Store (24 Hr), Better Vision Ctr, Coliseum Business Center, Farm Fresh Supermarket, Goodman Hardware, Greenwood Shopping Ctr, Mercury Animal Hospital, Sandy Bottom Nature Park, Scrub A Dub Car Wash, Todds Center

262B Hampton Roads Center Pkwy (Westbound)

EXIT VIRGINIA

263AB U.S. 258, VA 134, Mercury Blvd, James River Bridge
- **Gas** N: Exxon*(CW), Shell*(CW)
  S: Amoco*
- **Food** N: Blimpie's Subs, Boston Market, Chi Chi's Mexican, Chili's, China Garden, Days Inn, Denny's, Dunkin Donuts, Golden Corral, Holiday Inn, KFC, McDonalds, Olive Garden, Pizza Hut, Rally's Hamburgers, Red Lobster, Steak & Ale, Taco Bell, Waffle House, Wendy's
  S: La Quinta Inn, Old Country Buffet, Waffle House
- **Lodg** N: Comfort Inn, Courtyard by Marriott, Days Inn, Fairfield Inn, Hampton Inn (see our ad opposite page), Holiday Inn (see our ad opposite page), Quality Inn, Red Roof Inn
  S: Econolodge, La Quinta Inn, Travelodge (see our ad opposite page)
- **AServ** N: Coliseum Mercury, GM Auto Dealership, Thompson Ford
  S: All Tune & Lube, Big Al's Mufflers & Brakes, Pep Boys Auto Center, Western Auto
- **Med** S: **Riverside Medical Care Walk-In/ Pharmacy (Open 7 Days a Week)**
- **ATM** N: First VA Bank, Langley Federal Credit Union
- **Other** N: **Coliseum Mall, Riverdale Shopping Center, Super Fresh Grocery, Wal-Mart**

264 Junction I-664 South, Downtown Newport News, Pembrook Pkwy
- **Gas** S: Citgo*
- **AServ** S: Rod's Transmission
- **ATM** S: Citgo
- **Lodg** S: **Comfort Inn (see our ad this page)**

265C Armistead Ave, Langely AFB

265AB VA 134, VA 167, La Salle Ave
- **Gas** N: Citgo
- **Lodg** N: Super 8 Motel
- **AServ** N: Citgo
- **Med** S: **Hospital**

267 U.S. 60, VA 143, Settlers Lodge Rd
- **Gas** N: Citgo*(D)
- **Food** N: Grill (Citgo), Krispy Kreme (Citgo)
  S: Burger King, Subway Station
- **Med** S: **Hospital**
- **ATM** S: Nations Bank (Collegiate Bookstore)
- **Other** S: **VA Air & Space Museum**

268 VA 169, Ft Monroe, Mallory St
- **FStop** N: **Exxon*, Onmark***
- **Gas** N: Texaco*
- **Food** N: Hardee's, Little Chicago Pizza, McDonalds
  S: Strawberry Banks
- **Lodg** S: Strawberry Banks
- **AServ** N: Texaco
- **Med** S: **Hospital**
- **ATM** N: First Union Bank, Old Point Nat'l Bank, Onmark

(271) Vehicle Inspection - All Vehicles Over 10'6" Wide

272 West Ocean View Ave, Willoughby Spit (No Thru Trucks)
- **Food** S: Fisherman's Wharf
- **Lodg** S: Days Inn

273 4th View St, Ocean View
- **Gas** O: Exxon*(D)(LP)
- **Lodg** O: Chesa-Bay Motel, Econolodge
- **Other** I: **Norfolk Visitor Info Center**
  O: **Sarah Constant Beach**

274 Bay Ave, Naval Air Station

276A Jct I-564, U.S. 460, Granby St, Naval Base
- **Gas** O: Amoco(D), Exxon*(D)(CW)
- **Food** O: Kin's Wok II, McDonalds, Mister Jim's Submarines, Oh! Brian's Restaurant, Papa John's Pizza, Reaino's Restaurant, Saigai Restaurant, Subway, TJ Super Subs, Taco Bell, Wendy's
- **AServ** O: Firestone Tire & Auto, Import Car Wash, Nationwide Safety Brake Service
- **Med** O: **Hospital**
- **ATM** O: BT&T Bank, Central Fidelity (Farm Fresh Supermarket), Centura Bank (Haniford Drugs), Crestar Bank, First Union Bank, Life Savings Bank
- **Lodg** O: Hampton Inn (see our ad opposite page)
- **Parks** I: **Virginia Zoological Park**
- **Other** O: **7-11 Convenience Store, Almost A Bank (Check Cashing), Coin Operated Laundry, Eckerd Drugs, Farm Fresh Supermarket, Haniford Drugs, Locksmith, Optical, Parrott Island (Pets), Pearl Vision Optical, Revco Drugs, Super Fresh Food Store, Wards Corner Mall**

276B Junction I-564, to VA 406

276C VA 165, U.S. 460 West, Little Creek Rd

277AB VA 168, Tidewater Drive
- **Gas** I: Amoco, Citgo*(LP)
  O: Citgo*, Crown(CW), Exxon*, Shell(CW)
- **Food** I: Carolino Seafood, Hunan Express
  O: Arby's, Bamboo Hut, Captain D's Seafood, Fortune Dragon (Chinese Dragon), **Hardee's**, Kings Seafood, **Little Caesars Pizza (K-Mart)**, No Frill Grill, Open House Diner, Philly Cheesesteak House, Rotisserie Lite, TCBY
- **AServ** I: Amoco, Hy-Tech Auto Service, Re-King Tires
  O: Advantage Auto Store, All Tune & Lube, Carburator Clinic, Cinder Tire, Exxon, Maaco, Napa, **Penske (K-Mart)**, Tune-Up Plus
- **Med** I: **Tidewater Walk-In Medical Center**
- **ATM** O: Central Fidelity Bank (Rack-N-Sack Supermarket), NationsBank
- **Other** O: **Coin-Operated Laundry, Eckerd Drugs, K-Mart (Pharmacy), Rack-N-Sack Supermarket,** SSS Car Wash, **Veterinary Clinic**

278 VA 194 South, Chesapeake Blvd

279AB Norview Ave, Norfolk Int'l Airport
- **Gas** O: Shell*(LP) (Wrecker Service)
- **Food** O: Franco's Italian Restaurant & Pizza, New China Restaurant, Pizza Hut
- **AServ** O: J&M Tire Mart
- **Other** O: **Bromlee Shopping Ctr, Eckerd Drugs, Food Lion Supermarket, K-Mart (24 Hr), Norfolk Botanical Gardens**

281 VA 165, Military Highway (Difficult Eastbound Reaccess)
- **Gas** I: Exxon(D)(CW)
  O: Amoco
- **Food** I: Hilton, Yings Chinese & American
  O: Andy's Pizza House, House of Eggs, Pappy's Hacienda, Stone Horse Restaurant, Tidewater Seafood, Wendy's
- **Lodg** I: Hampton Inn, Hilton
  O: Econolodge
- **AServ** I: Calvary Repair, Firestone Tire & Auto
  O: AAMCO Transmission, Atlantic Auto Repair & Service, Cider Tire, ED's Auto Repair, Green Grifford Nissan/ Chrysler/ Plymouth/ Saab, Inco Transmission, Ingrams Used Auto Parts, Planet Cars, Tidewater Transmission, Windshields Of America
- **Other** O: **Airport, Coin Operated Laundry,** Robo Car Wash

282 U.S. 13N, N Hampton Blvd, Chesapeake Bridge Tunnel
- **TStop** O: **Big Charlie's Truck Plaza***(SCALES)
- **Food** O: Quality Inn
- **Lodg** I: Days Inn (see our ad opposite page)
  O: Quality Inn (see our ad opposite page)

284AB Junction I-264, VA 44, Virginia Beach, Newtown Rd
- **Food** O: Quality Inn (see our ad this page)
- **Lodg** O: Clarion Hotel, Comfort Inn, Days Inn, **Holiday Inn (see our ad opposite page)**, LaQuinta Inn, Quality Inn

286AB Indian River Road
- **Gas** I: Amoco*(CW), Citgo*(D), Exxon*(LP), Shell*(CW), Shell*(CW)(LP), Texaco*(D)(CW) (24 Hr)
  O: Citgo, Exxon*(D), Texaco*(LP)
- **Food** I: Catherine's Restaurant, Egg Roll King, Hardee's
  O: Arby's, Captain D's Seafood, China Garden, El Cantine Mexican Restaurant, Famous Uncle Al's Hotdogs & Fries, Golden Corral, IHOP, KFC, Leone's Seafood, McDonalds, Outback Steakhouse, Pizza Hut, Rally's Hamburgers, Riverpoint Deli and Pizzeria, Shoney's, Subway, TCBY, Taco Bell, Waffle House
- **AServ** I: Shell, Texaco, Westview Auto Service and Sal[es]
  O: Exxon, Firestone Tire & Auto
- **Med** O: **Patient First Walk-In**
- **ATM** I: Central Fidelity Bank, Life Savings Bank
- **Other** I: **7-11 Convenience Store, Family Vision Ctr, Indian River Pet Grooming**
  O: **7-11 Convenience Store, Animal Hospital, Farm Fresh Supermarket, K-Mart (Pharmacy) Optical Ctr, Pac-N-Mail Postal Service, Pet Land, Revco, University Shops**

289AB Greenbrier Pkwy (Difficult Reaccess)
- **Gas** I: Amoco*(D)(CW), Citgo*(D)(CW), Exxon(CW)
  O: Amoco, Citgo*, Mobil*(D)
- **Food** I: Bancho Grande Mexican, Beijing Restaurant, Burger King, Chevy's, Cugini's Pizza, Deli, Denny's, Hot Dog Deli, Johnson's BBQ, McDonalds, Oriental Cuisine (Thailand, Philippine, Chinese), Pizza Hut, Subway, Village Grill, Wendy's
  O: Anna's Bar & Grill, KFC, S&D Seafood, Taco Bell
- **Lodg** I: Hampton Inn, Suburban Lodge, **Holiday Inn (see our ad opposite page)**
  O: Econolodge, **Wellesley (see our ad opposite page), Courtyard by Marriott (see our ad opposite page), Comfort Suites (see our ad opposite page), Fairfield Inn (see our ad opposite page)**
- **AServ** I: Amoco, Cavalier Ford, Conoly Phillips Lincoln, Mercury, Greenbriar Chrysler, Jeep, Dodge, Greenbriar Dodge, Kline's Chevrolet, Little Joe's Merchant's Tire & Auto, Monroe Muffler Brake, Napa, **Penke (K-Mart)**, Riddle Acura,Subaru, Stan's Muffler & Brake, Tidewater Auto Electric, Transmission Masters
  O: Advance Auto Parts, Exxon, Firestone Tire & Auto, Highway Service Store, Twin B Auto Parts
- **TServ** I: **Gibbons Honda**
- **Med** O: **Hospital**
- **ATM** I: Heritage Bank & Trust, NationsBank, Signet
- **Other** I: **7-11 Convenience Store, All Pets Pleasures (Pet Supplies), Food Lion Grocery (24 Hr), K-Mart, National Optical, Parkview Shopping Center, Phar Mor Drugs, VA State Police**
  O: **Eckerd Drugs, Food Lion Supermarket (Del[i] Bakery), Midway Shopping Ctr, Rainbow Coi[n] Operated Laundry**

290AB VA 168, Battlefield Blvd, Great Bridg[e]
- **FStop** O: **Amoco***(D)(CW)
- **Gas** O: Texaco*(CW)(LP)
- **Food** O: 2 Mom's Cafe, Anthony's Deli & Grill, Applebee's, Bagel Works, Blimpie's Subs (Texaco), Chanello's Pizza, Chick-fil-A, Chuck E. Cheese's Pizza, Dunkin Donuts (Days Inn), Golden Corral, Hardee's, Honey Glazed Ham & Bread Shop, Joey's Pizza, Maxwell's, Ryan's, Taco Bell, Waffle House, Wendy's
- **Lodg** O: Days Inn, Super 8 Motel
- **AServ** O: **Sam's Club**
- **TServ** I: **Mack's Truck**
- **ATM** O: Bank of Hampton Rd, Crestar Bank, Wal-Ma[rt]
- **Other** O: **7-11 Convenience Store, Animal Clinic, Battlefield Market Place, Sam's Club (Optical) U.S. Post Office, Wal-Mart**

291 Junction I-464, VA 104, to U.S. 17, Norfolk, Elizabeth City
- **Lodg** I: **Econolodge (see our ad opposite page)**

292 VA 190, Dominion Blvd, to VA 104
- **Gas** I: Texaco*
  O: Mobil*
- **Food** I: Hardee's, No. One Chinese Food
- **Other** I: **Food Lion Supermarket (Deli, Bakery), Food Mister Jim's Subs, Pizza & Wings**

296 U.S. 17, Portsmouth, Elizabeth City
- **TServ** I: **Atlantic Coast Equipment, Joseph B Holland Truck Repair, Transmission Engineering Co**
- **RVCamp** O: **Campground**
- **Other** I: **Deep Creek Veterinary Hospital**

297 U.S. 13, U.S. 460, Military Hwy
- **Gas** I: Exxon*
- **AServ** O: Tom and Ruth Auto Serv

299A Junction I-264E, Portsmouth, Norfolk

299B Jct I-664, U.S. 13, U.S.58, U.S. 460, Bowershill, Suffolk

↑VIRGINIA

Begin I-64

Bold red print shows RV & Bus parking available or nearby

EXIT VIRGINIA

Newport News Area

HA2366

Hampton Inn

- 132 Rooms
- Free Deluxe Continental Breakfast
- All Rooms Coffee Maker, Ironing Board and Iron
- Free Local Calls
- Mall and Restaurants, 1 Block

757-838-8484

VIRGINIA ▪ I-64 ▪ EXIT 263B

Newport News, Virginia

R2360

Travelodge

6128 Jefferson Ave.
Newport News, VA 23605

In Room Coffee • Outdoor Pool
Non-Smoking Rooms Available
Pay-Per-View Movies
Free Local Calls • Guest Laundry

757-826-4500

VIRGINIA ▪ I-64 ▪ EXIT 263A

320 ROOMS

HO2366

Holiday Inn

- *Tivoli Gardens* Full Service Restaurant
- Garden Court Lounge
- 16,000 sq.ft. Meeting Space
- Indoor Pool/Exercise Room
- 1 Block from Mall & Restaurants
- Shuttle Service

800-842-9370

VIRGINIA ▪ I-64 ▪ EXIT 263B

Bowers Hill Comfort Inn

On 664S Exit 14 & On 64 East Exit 297 Chesapeake, VA

757-488-7900

- Free Continental Breakfast
- Restaurant Nearby
- Swimming Pool
- Free Cable TV with HBO
- Bus/RV Parking
- Kids Stay Free

CO233C

VIRGINIA ▪ I-64

EXIT VIRGINIA

EC2332

2222 South Military Hwy.

Ad Rate $36.95 and up
Free Morning Coffee & Donuts
Efficiencies
Handicap & Senior Rooms
Special Rates For Tour Groups

757-543-2200 expires 3-31-98

VIRGINIA ▪ I-64 ▪ EXIT 291A

CO233B

757-420-0900

VIRGINIA ▪ I-64 ▪ EXIT 289AB

CO233A

757-420-1600

VIRGINIA ▪ I-64 ▪ EXIT 289AB

DA2345

DAYS INN
The Best Value Under The Sun.

757-460-2205
Virginia Beach AIRPORT

SAVE 15%

5708 Northampton Blvd
Virginia Beach, VA

Free Continental Breakfast

Free Cable TV with HBO
Outdoor Pool • Guest Laundry
Sauna & Exercise Room
Meetings & Banquet Space
Complimentary Airport Shuttle

VIRGINIA ▪ I-64 ▪ EXIT 282

FA2332

757-420-1300

VIRGINIA ▪ I-64 ▪ EXIT 289AB

PR070A.15

Value Never Looked This Good!

Virginia • Exit 289A • 757-366-0100

EXIT VIRGINIA

HA2350

Hampton Inn

SAVE 10%

8501 Hampton Blvd.
Norfolk, VA

757-489-1000

100% Satisfaction Guaranteed

Free Continental Breakfast
Free Local Calls
Free HBO & Cable
Indoor Pool & Jacuzzi

VIRGINIA ▪ I-64 ▪ EXIT 276

Cross south Newton Rd. to Greenwich Road

HO2346

Holiday Inn

- Free Parking • Free Local Calls
- Free Airport Shuttle
- Indoor Heated & Outdoor Pools
- Exercise Room, Sauna + Hot Tub
- Full Service Restaurant + Lounge

757-499-4400 • 800-HOLIDAY

VIRGINIA ▪ I-64E ▪ EXIT 284B

HO2332

Right on Woodlake Dr. Hotel is on right.

Holiday Inn

Chesapeake

- Restaurant • Lounge • Room Service
- Free Parking • Free Phone
- Exercise Room • Sauna
- Indoor Pool with Hot Tub
- Adjacent to Greenbrier Mall
- Special Group Rates Available
- Microwaves & Refrigerators Available

757-523-1500 • 800-HOLIDAY

VIRGINIA ▪ I-64 ▪ EXIT 289A

QU2345

Quality Inn

Virginia Beach

1-800-631-3916

AAA

- Rooms with Micro/Freeze available upon request
- Free Beach Shuttle(Seasonal)
- Free Guest Laundry
- Free Local Calls • Outdoor Pool
- Color Cable Television
- Nearby Golf, Tennis, Fishing
- Restaurant on Premises

757-422-3617 • East on SR 44 to 21st St.

VIRGINIA BEACH ▪ VIRGINIA

Begin I-65

↓INDIANA

262 Junction I-90, Chicago, Ohio (toll)

261 U.S. 6, 15th Ave, Gary
- **Gas** W: 76 Station*

259AB Junction I-80, Junction I-94, Chicago, Ohio

258 U.S. 6 Bus., Ridge Rd.
- **Gas** E: Mobil*, Speedway*(D)
  W: K & G* (24 Hrs), Shell*
- **Food** E: Diner's Choice Family Restaurant
- **AServ** E: Widco Transmission
  W: Jerry's Towing Service, Lou's Auto Repair, Quality Brake & Muffler, USA Muffler Shop

255 61st Ave, Merrillville, Hobart
- **FStop** E: **Speedway***(LP)
- **Gas** E: Amoco*(CW) (24 Hrs), Mobil*(CW), Phillips 66*
  W: Coastal*, Shell*(CW)
- **Food** E: Blimpie's Subs (Mobil), **Cracker Barrel, McDonalds**
  W: Burger King
- **Lodg** E: ◆ Comfort Inn, **Dollar Inn**, Lee's Inn
- **AServ** E: Shaver Chevrolet
  W: Auto Zone Auto Parts, USA Muffler Shops, VIP Brake & Spring
- **Med** E: ✚ **Hospital**
- **ATM** W: Bank One
- **Other** W: **Osco Drugs, Wise Way Supermarket**

253AB U.S. 30, Merrillville, Valparaiso
- **FStop** W: **Amoco* (24 Hrs)**
- **Gas** E: Amoco* (24 Hrs)
  W: Gas City*(LP), Shell*(CW), Speedway*(LP)
- **Food** E: Angelo's Italian American Restaurant, Bakers Square Restaurant, Bob Evans Restaurant, Boston Market, Casa Gallardo Mexican, Dairy Queen, McDonalds (Play Place), Miami Subs, Old Country Buffet, Olive Garden, Popeye's Chicken, Red Lobster, Subway
  W: Arby's, Checkers Burgers, Colorado Steakhouse, Denny's, Hooters, New Moon Chinese, Rio Bravo, White Castle Restaurant
- **Lodg** E: Days Inn, Economy Inns of America, Extended Stay America, Knight's Inn, La Quinta Inn, Motel 6, Super 8 Motel
  W: Dollar Inn, Fairfield Inn (see our ad on this page), Hampton Inn, Holiday Inn Express, Radisson, Red Roof Inn, Residence Inn
- **AServ** W: Amoco
- **Med** W: ✚ **Hospital**, ✚ **Prompt Medical Care**
- **ATM** W: Gas City
- **Other** E: **Southlake Mall**
  W: **Century Consumer Mall, U.S. Post Office**

247 U.S. 231, Crown Point, Hebron
- **Gas** W: Mobil*
- **Med** W: ✚ **Hospital**
- **Other** E: **Vietnam Veterans Memorial**

(241) **Weigh Station - Both Directions**

240 IN 2, Lowell, Hebron
- **TStop** E: **Mobil***(LP)(SCALES)
- **FStop** W: **Marathon***
- **Gas** E: **Citgo*, Shell***
- **Food** E: **Blimpie's Subs (Mobil), Burger King,** Little Caesars Pizza
  W: **USA Interstate Restaurant (Marathon)**
- **Lodg** E: ◆ **Super 8 Motel**

EXIT INDIANA

Lake Michigan
E. Chicago
90 GARY 94
80 94 261 259 80 90
Merrillville 253
Crown Point 247
240
INDIANA
65
Roselawn 230
215 Rensselaer
205
Remington 201
Reynolds
193
188
65
178
175
172
LAFAYETTE
168
65
Frankfort
158
INDIANA
Thorntown 146
141
140
139
138
133
130
129
WI MI IL OH IN KY
Area Detail
N
74

EXIT INDIANA

- **TServ** E: **Mobil**
- **ATM** E: Mobil
- **Other** W: **Indiana State Police Post**

(231) **Rest Area - RR, Phones (Both Directions)**

230 IN 10, Roselawn, Demotte
- **Gas** W: Amoco*, Shell*
- **Food** W: Renfrow's Hamburgers, Subway (Shell)
- **AServ** W: Car Quest Auto Center
- **RVCamp** W: **Oak Lake Campground, Yogi Bear Camp Report**
- **ATM** W: Amoco, Kentland Bank
- **Other** W: **Coin Laundry, Fagen Pharmacy, Roselawn Star Supermarket**

215 IN 114, Morocco, Rensselaer
- **TStop** W: **Tree Trail Truck Stop***
- **FStop** E: **Amoco***
  W: **Phillips 66***(SCALES)
- **Food** E: **Dairy Queen, KFC, McDonalds,** Scotty's Family Restaurant
  W: **Burger King, Grandma's (Phillips 66), Tree Trail Restaurant**
- **Lodg** E: Holiday Inn Express, **Interstate Motel**
  W: **Mid-Continent Inn**
- **TServ** W: **Cooper's Tire Service, Phillips 66, Tree Trail TS**
- **ATM** E: Amoco
  W: Phillips 66
- **Other** W: **Fireworks Supermarket**

205 U.S. 231, Remington, Rensselaer
- **TStop** E: **Travellers Plaza* (Phillips 66)**
- **Gas** E: Marathon*
- **Food** E: **Traveller's Plaza (Restaurant)**
- **Lodg** E: **Knights Inn**
- **AServ** E: Marathon
- **Med** E: ✚ **Hospital**
- **ATM** E: **Travellers Plaza TS**

201 U.S. 24, U.S. 231, Remington, Wolcott
- **TStop** W: **76 Auto/Truck Plaza***(SCALES)
- **FStop** W: **Citgo*, Speedway***(SCALES)
- **Food** W: **76 Auto/Truck Plaza, McDonalds,** Subway
- **Lodg** W: **Holiday Inn**
- **TServ** W: **76**

(195) **Rest Area - RR, Phones, Picnic**

193 U.S. 231, Chalmers, Wolcott
- **Gas** E: Shell*
- **Food** E: **Dairy Queen,** Wayfara Restaurant (Shell)

188 IN. 18, Fowler, Brookston

178 IN 43, West Lafayette, Brookston

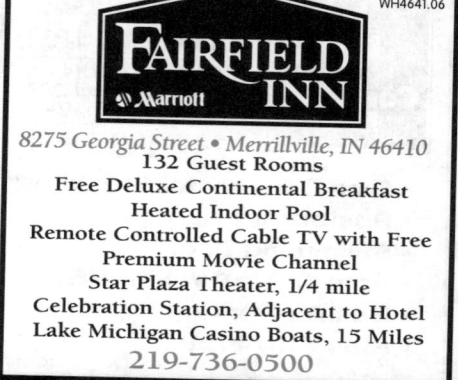

Bold red print shows RV & Bus parking available or nearby

XIT	INDIANA
Gas	**E:** Gas of America*(LP), Phillips 66*(D)(LP), Shell*
Food	**E:** Burger King (Shell), McDonalds (Inside Play Place), Subway (Phillips 66), Taco Bell (Gas of America)
Lodg	**E:** (AAA) Holiday Inn
ATM	**E:** Gas of America, Shell
Other	**E:** Indiana State Police Post
5	IN 25, Lafayette, Delphi
Gas	**E:** Citgo*
RVCamp	**E:** Camping
Med	**W:** ✚ Creekside Animal Hospital
2	IN 26, Lafayette, Rossville
Gas	**E:** Meijer Grocery*(D)(LP)
	W: Amoco*(CW) (24 Hrs), Shell*, Speedway*(LP)
Food	**E:** (Old Country Store) Cracker Barrel, Meijer Food Court, Steak 'n Shake
	W: Arby's, Bob Evans Restaurant, Burger King, Chili's, Damon's Place for Ribs, Denny's, Hour Time (Radisson), McDonalds (Inside Play Place), Mountain Jack's, Olive Garden, Shoney's, Spageddie's Italian Kitchen
Lodg	**E:** Budget Inns of America (see our ad this page), Comfort Suites, ◆ Holiday Inn Express, Lee's Inn
	W: Dollar Inn, ◆ Fairfield Inn, ◆ Knight's Inn, (AAA) Radisson Inn, (AAA) Ramada, ◆ Red Roof Inn, (AAA) Signature Inn
Med	**W:** ✚ Hospital
ATM	**E:** Meijer
	W: Amoco
Other	**E:** Meijer Grocery, Department Store (24 Hrs), Visitor Information
68	IN 38, Lafayette, Dayton
FStop	**E:** Mobil*
58	IN 28, Attica, Frankfort
FStop	**E:** Amoco*
Gas	**E:** Marathon*
RVCamp	**W:** Camping
Med	**E:** ✚ Hospital
50)	Rest Area - RR, Phones, Picnic (Southbound)
48)	Rest Area - RR, Phones, Picnic (Northbound)
46	IN 47, Thorntown, Sheridan
41	U.S. 52, Lafayette Ave. (Left Lane Exit)
40	IN 32, Lebanon, Crawfordsville
FStop	**W:** McClure*
Gas	**E:** Amoco*(CW)
	W: Shell*(CW)
Food	**E:** Ice Cream Paradise, White Castle Restaurant
	W: Arby's, Burger King, KFC, McDonalds, Ponderosa, Steak 'n Shake, Subway, Taco Bell
Lodg	**E:** ◆ Comfort Inn
	W: Dollar Inn, Lee's Inn, ◆ Super 8 Motel
AServ	**E:** Auto Zone Auto Parts, Beason's Mufflers, Goodyear Tire & Auto, Petro's Tires
TServ	**W:** Bob's Towing Recovery
TWash	**W:** Interstate Speed Truck Wash
Med	**E:** ✚ Hospital
ATM	**W:** Shell
Other	**E:** Coin Laundry
39	IN 39, Lizton, Lebanon
Gas	**E:** Gas of America*(D)(LP)
Food	**E:** Choices Restaurant, Wendy's
	W: Old Chicago Family Restaurant
Lodg	**W:** ◆ Holiday Inn

EXIT	INDIANA
ATM	**E:** Gas of America
Other	**E:** Jiffy Wash Coin Car Wash
138	Lebanon (Difficult Reaccess)
Gas	**E:** 76*(D)(LP), Shell*
AServ	**E:** 76
ATM	**E:** 76, State Bank
Other	**E:** Lebanon Bowling Center
133	IN 267, Brownsburg, Whitestown
130	IN 334, Zionsville, Whitestown
TStop	**W:** 76 Auto/Truck Plaza*(SCALES)
FStop	**E:** Crystal Flash*(LP)
Gas	**E:** Phillips 66*
Food	**E:** Stuckey's (Phillips 66), Subway (Crystal Flash)
	W: 76 Auto/Truck Plaza, Shelley's Eatery
TServ	**W:** 76
129	Junction I-465 East, U.S. 52 East
124	71st St
Parks	**W:** Eagle Creek State Park
123	Junction I-465 South, To Indianapolis Inter. Airport
121	Lafayette Road
Gas	**W:** Amoco*(CW), Shell*(CW)
Food	**W:** Sizzling Wok, Subway
Lodg	**E:** Lee's Inn
	W: Dollar Inn
AServ	**W:** Collins Nisson, Discount Tire Co., Indy Lube, McKinny Transmission, NAPA Auto Parts, PEP Boys, Speedway North Auto Parts & Pro. Auto Repair
Med	**W:** ✚ Hospital
ATM	**W:** Amoco
Other	**W:** Coin Car Wash, Georgetown Animal

EXIT	INDIANA
	Hospital
119	38th St. East, State Fairgrounds
117	Martin Luther King Jr. St. (Southbound)
116	29th St., 30th St. (Northbound)
FStop	**E:** Patterson
115	21st St.
Gas	**E:** Shell*
Med	**E:** ✚ Hospital
114	Dr. Martin Luther King Jr. St., To West St.
Food	**W:** Best Taste Chinese Buffet, Donatos Pizza, Hardee's, Papa John's Pizza, Subway, Taco Bell
Med	**W:** ✚ Hospital
113	U.S. 31, IN 37, Meridian St.
AServ	**E:** Midas
Lodg	**E:** Howard Johnson (see our ad this page)
Med	**E:** ✚ Hospital
112A	Junction I-70 East, Columbus, Ohio
111	Michigan St., Ohio St., Fletcher ave.
110A	East St.
Gas	**E:** Citgo*
	W: Speedway*
Food	**W:** Burger King
Other	**W:** Coin Laundry
110B	Junction I-70 West, St. Louis
109	Raymond St
Gas	**E:** Phillips 66*
	W: Speedway*(LP)
Food	**E:** Burger King
	W: Griner's Sub Shop
AServ	**W:** Budggett Tire Service
Med	**E:** ✚ Hospital
Other	**W:** Safe Way Grocery
107	Keystone Ave.
Gas	**E:** Speedway*
	W: Amoco*(CW) (24 Hrs), Speedway*(LP)
Food	**W:** Burger King, McDonalds (Play Place), Subway
Lodg	**E:** Dollar Inn
	W: (AAA) Holiday Inn Express
AServ	**W:** Amoco, Peters Auto Service, Q Lube
Med	**E:** ✚ Hospital
ATM	**W:** Bank One, First of America Bank
Other	**W:** Cub Foods (w/Pharmacy, 24 Hrs), Low Cost Pharmacy
106	Junction I-465, Junction I-74
103	Southport Road
Gas	**E:** Amoco* (24 Hrs), Meijer*(D)(LP), Shell*(CW) (24 Hrs)
	W: Big Foot*, Citgo*, Speedway*(D)(LP)
Food	**E:** Dog n Suds Drive-In, McDonalds (Amoco), Noble Roman's
	W: Bob Evans Restaurant, Burger King, (Old Country Store) Cracker Barrel, KFC, McDonalds (Play Place), Steak 'n Shake, Taco Bell (Speedway), Waffle & Steak, Wendy's
Lodg	**W:** (AAA) Best Western, Dollar Inn, Fairfield Inn, Hampton Inn, (AAA) Signature Inn
AServ	**E:** Q Lube
Med	**E:** ✚ Hospital
ATM	**E:** First of America Bank (Meijer), NBD
	W: Big Foot, Fifth Third Bank
Other	**E:** Meijer Grocery, Department Store (24 Hrs), Mike's Car Wash
99	Greenwood

EXIT INDIANA

TStop E: Phillips 66*(SCALES)
Gas W: 76*(D), Amoco* (24 Hrs), Citgo*, Shell*(CW), Tobacco Road*
Food E: Chester Fried Chicken (Phillips 66)
W: Arby's, Hardee's (Playground), Jonathan Byrd's Cafeteria, McDonalds (Play Place), Shoney's, Subway, TCBY, Taco Bell, Waffle & Steak, White Castle Restaurant
Lodg W: AAA Comfort Inn, Fanta Suite, Greenwood Inn, Lee's Inn
AServ W: Q Lube
TServ E: Phillips 66
RVCamp W: Stout's RV Sales & Service
Med W: + Hospital
ATM W: Bank One, Citgo, Key Bank, NBD (76)
Other W: Low Cost Drug Store, Vale Vista Animal Hospital

95 Whiteland
TStop E: Marathon*(SCALES)
W: Pilot*(SCALES), Speedway*(LP)
Food E: Kathy's Kitchen (Marathon)
W: Arby's (Pilot), Arby's (Speedway)
AServ E: Bleake's Auto
TServ W: Scott Truck Systems Inc. (Trailer Repair, 24 Hr. Wrecker Service)
TWash E: Marathon

90 IN 44, Shelbyville, Franklin
Gas W: Shell*
Food W: Burger King, McDonalds (Play Place), Waffle & Steak
Lodg W: Carlton Lodge, Days Inn AAA Days Inn, Super 8 Motel

80 IN 252, Flat Rock, Edinburgh
Gas W: Shell*

76AB U.S. 31, Taylorsville, Columbus
TStop E: Speedway(SCALES)
FStop E: Shell*
Gas W: Citgo*, Thorntons
Food E: Hoosier Kitchen (Speedway), KFC, Waffle & Steak
W: Arby's, Cracker Barrel Cracker Barrel, Hardee's, McDonalds (Play Place), Snappy Tomato Pizza (Citgo), Subway (Citgo)
Lodg E: Comfort Inn
W: Hampton Inn, Holiday Inn Express, Ramada Inn
AServ W: Taylorsville Tire Co.
TServ W: Kenworth of Columbus, Suburban Tire, White River Truck Repair
RVCamp W: Driftwood RV Park
ATM E: Irwin Union Bank
W: Centra
Other W: Horizon Outlet Center

(74) Rest Area - RR, Phones, Picnic (Southbound)

(72) Rest Area - RR, Phones, Picnic (Northbound)

68 IN 46, Columbus, Bloomington, Nashville
Gas E: Amoco*(D) (24 Hrs), Big Foot*, Shell*, Speedway*(LP)
W: Swifty*
Food E: American Cafe (Ramada), Burger King, McDonalds
W: Bob Evans Restaurant, Bobby G's, Shoney's, Taco Bell

EXIT INDIANA

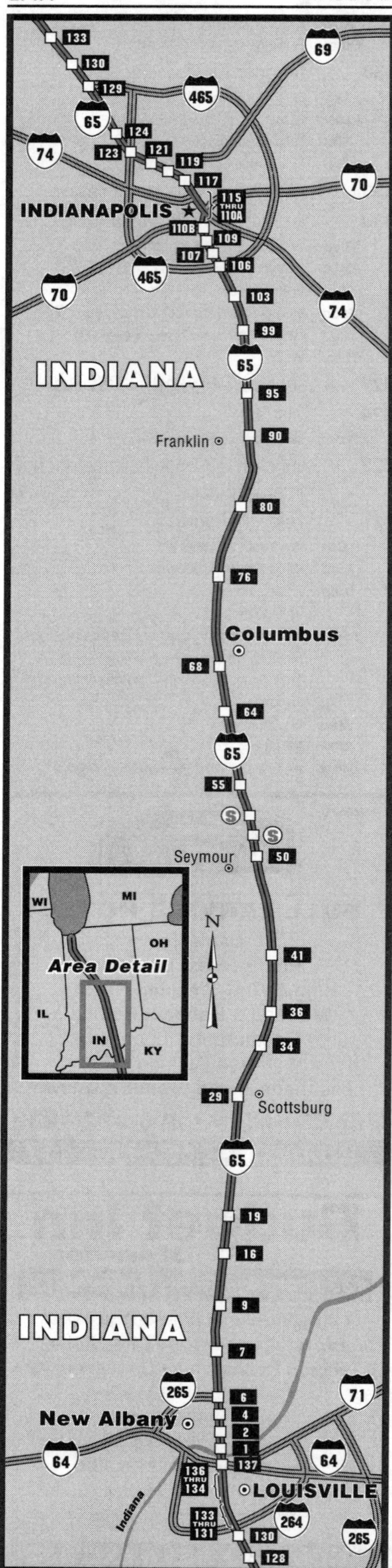

EXIT INDIANA

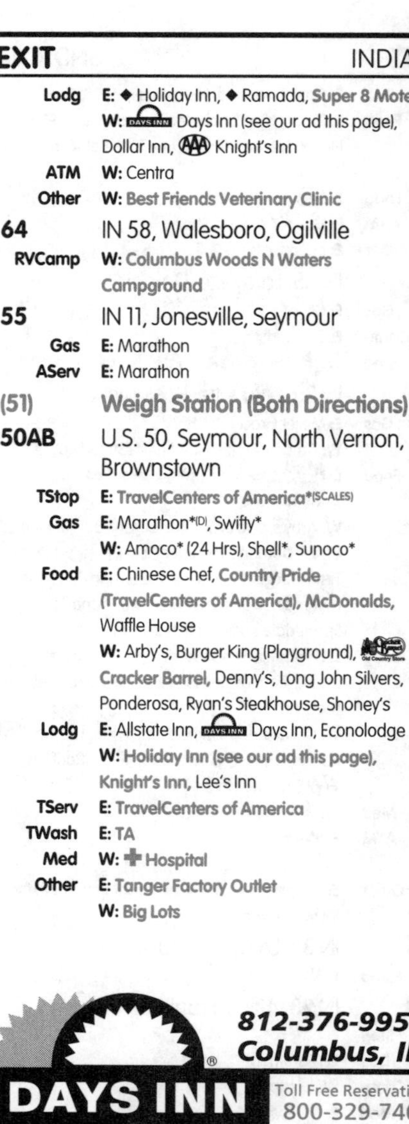

Lodg E: ◆ Holiday Inn, ◆ Ramada, Super 8 Motel
W: Days Inn Days Inn (see our ad this page), Dollar Inn, AAA Knight's Inn
ATM W: Centra
Other W: Best Friends Veterinary Clinic

64 IN 58, Walesboro, Ogilville
RVCamp W: Columbus Woods N Waters Campground

55 IN 11, Jonesville, Seymour
Gas E: Marathon
AServ E: Marathon

(51) Weigh Station (Both Directions)

50AB U.S. 50, Seymour, North Vernon, Brownstown
TStop E: TravelCenters of America*(SCALES)
Gas E: Marathon*(D), Swifty*
W: Amoco* (24 Hrs), Shell*, Sunoco*
Food E: Chinese Chef, Country Pride (TravelCenters of America), McDonalds, Waffle House
W: Arby's, Burger King (Playground), Cracker Barrel Cracker Barrel, Denny's, Long John Silvers, Ponderosa, Ryan's Steakhouse, Shoney's
Lodg E: Allstate Inn, Days Inn Days Inn, Econolodge
W: Holiday Inn (see our ad this page), Knight's Inn, Lee's Inn
TServ E: TravelCenters of America
TWash E: TA
Med W: + Hospital
Other E: Tanger Factory Outlet
W: Big Lots

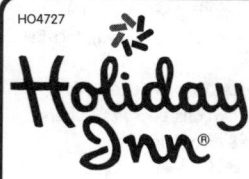

Bold red print shows RV & Bus parking available or nearby

XIT INDIANA

1 IN 250, Uniontown, Crothersville
FStop **W:** Marathon*
Food **W:** Uniontown Restaurant

6 U.S. 31, Crothersville, Austin
Gas **E:** Shell*
W: BP*(CW)(LP)
Food **W:** Beef Boys, Piccadilly Pizza & Subs

4AB IN 256, Austin
FStop **W:** Fuel Mart*(D)
Gas **E:** Big Foot*
Food **E:** Dairy Bar
W: A & W Drive-In, The Home Oven
ATM **E:** Big Foot
Parks **E:** Clifty Falls State Park

9 IN 56, Scottsburg, Salem
Gas **E:** Amoco*(D)(CW) (24 Hrs), Moto Mart* (24 Hrs)
W: Marathon*(D), Shell*, Speedway*(D)(LP)
Food **E:** Burger King (Playground), Mariann Restaurant, Ponderosa, Taco Bell
W: Arby's, Long John Silvers, McDonalds (Inside Play Place), Sweet Tooth, Waffle & Steak, Wendy's
Lodg **E:** Mariann Motel
W: Dollar Inn, Scottsburg Best Western
AServ **E:** Amoco, Grease Pit Oil Change & Lube
W: Wal-Mart
ATM **E:** Bank One
W: Marathon
Other **E:** Touchless Auto Car Wash
W: Wal-Mart (24 Hrs, Pharmacy, Vision Center)

22) Rest Area - RR, Phones, Tourist Info (Both Directions)

9 IN 160, Henryville, Charlestown
Gas **E:** Big Foot*, Citgo*, Shell*
Food **E:** Dairy Queen, Schuler's, Stuckey's (Citgo)
Other **E:** Clark's State Forest

6 Memphis
TStop **E:** Country Style Plaza*(SCALES)
W: Davis Brothers Travel Plaza*(SCALES)
Gas **E:** Marathon*
Food **E:** Country Style Kitchen (Country Style Plaza)
W: Cleo's Buffet (Davis Bro. TS)
TWash **W:** Bay Truck Wash
RVCamp **E:** Customers First RV Inc.
W: Customers First Inc.
ATM **E:** Country Style TS

• IN 311, Sellersburg, New Albany
Gas **E:** Five Star*, Swifty*
W: Dairy Mart*
Food **E:** Arby's, Dairy Queen
W: Burger King (Playground), McDonalds, Taco Bell
AServ **E:** Auto Value Auto Parts, Car America
ATM **W:** Dairy Mart
Other **E:** Adam's IGA Supermarket, Classic Car Wash, Coin Laundry, Disbro's Drugs
W: State Patrol Post

7 IN 60, Salem, Cementville, Hamburg
TStop **E:** Davis Brothers Travel Plaza*(SCALES) (Ashland Gas)

EXIT INDIANA

Gas **E:** BP*(CW) (24 Hrs)
Food **E:** Cleo's Country Kitchen (Davis Bro. TS)
W: KFC
Lodg **W:** DAYS INN ◆ Days Inn (see our ad this page)
ATM **E:** Davis Bro. TS
Other **W:** Care-Pets Animal Hospital

6B I-265 West, to I-64, New Albany

6A IN 265 East, Clark Maritime Ctr.

4 U.S. 31 North, IN 131 S, Clarksville, Cementville
TStop **E:** Big Foot Auto/Truck*(LP)(SCALES) (Shell, Kerosene)
FStop **E:** Shell*(LP)(SCALES)
Gas **E:** Thornton's*(D)
Food **E:** Shell FS, Subway (Thornton's)
W: Arby's, Bob Evans Restaurant, Burger King, Captain D's Seafood, Colonial Inn Restaurant, Denny's, Don Pablo's Mexican, Fazoli's Italian Food, Hooters, Jerry's Restaurant, Logan's Roadhouse, O'Charley's, Penn Station East Coast Subs, Red Lobster, Steak 'n Shake, Wendy's
Lodg **E:** Crest Motel
W: AAA Best Western, Colonial Inn, Dollar Inn
AServ **W:** Carriage Ford, Sears Auto Center
TServ **W:** Goodyear Tire & Auto, Mac
RVCamp **W:** Premier Sales & Service
ATM **E:** Shell FS, Thornton's
W: Community Bank, NBD Bank
Other **E:** Toy Warehouse
W: Bigg's Foods (24 hrs, Pharmacy), CDL Training, Greentree Mall, Office Max, River Falls Car Wash, River Falls Mall, Target Department Store, The Home Depot(LP), Wal-Mart (24 Hrs)

2 Eastern Blvd, Clarksville
Gas **W:** Amoco*(CW), Chevron(D), Dairy Mart*, Sav-A-Step*, Shell*
Food **W:** Restaurant Omelet Shop (24 hrs), Ryan's Family Steakhouse, The Hungry Pelican
Lodg **E:** DAYS INN Days Inn, Motel 6, Super 8 Motel
W: ◆ Econolodge, Rivers Edge Hotel
AServ **W:** AAMCO Transmission, Chevron
Med **W:** ✚ Immediate Care Center (7 Days/Wk, 9am to 9 pm)
ATM **W:** NBD, PNC Bank
Other **E:** U-Haul Center(LP)
W: Cash Advance, Drug Emporium, McClures Drugs, SVS Vision, Value City

1A West New Albany (Northbound)

1 IN 62 E., Stanfifer Ave., Clarksville,

EXIT INDIANA/KENTUCKY

New Albany, U.S. 31 S (Stansifer Ave)
Gas **E:** Thornton's*
Food **E:** China Palace, Dunkin' Donuts, Spaghetti Already
Lodg **W:** AAA Holiday Inn
AServ **E:** Cooper Tires, NAPA Auto Parts, Uniroyal Tire & Auto
RVCamp **W:** KOA Kampground, Tom Stinnett RV
Med **E:** ✚ Clark Memorial Hospital
ATM **E:** Thornton's
Other **E:** All American Car Wash, Daily's 24-Hr Foodmart, Farm Bureau Co-op(LP), WalGreens (Pharmacy, 1-Hr Photo)

0 IN 62 E, Jeffersonville
Gas **E:** Chevron(CW), Thornton's*
Food **E:** China Palace, Hardee's, McDonalds, Parrella's Light Italian, Spaghetti All Ready, Waffle Steak
Lodg **W:** Ramada Inn
AServ **E:** Bales Jeep Eagle Chrysler Plymouth Nissan Hyundai, Chevron, NAPA Auto Parts, Ross Bros. Automatic Transmission Service, Uniroyal Tire & Auto
Med **E:** ✚ Hospital
ATM **E:** NBD, Thornton's
Other **E:** Touchless Auto Car Wash, WalGreens
W: Falls of the Ohio

↑INDIANA

↓KENTUCKY

137 Jct. I-71, I-64, Cincinnati, Lexington, St. Louis

136C Muhammed Ali Blvd.
Gas **W:** Shell*
Food **W:** Backstage Cafe, Miller's Cafeteria, O'Malley's Diner, Subway, White Castle Restaurant
Lodg **W:** DAYS INN Days Inn, The Inn at Jewish Hospital, Travelodge
AServ **W:** Midas
Med **E:** ✚ Jewish Hospital
ATM **W:** Machine (Shell)
Other **W:** Dr. Bizer's Vision World, Visitor Information

136B Brook St. (Northbound)
Gas **E:** Chevron*(D) (24 hrs), Shell*
Food **E:** Chung King Chinese American Restaurant
Lodg **E:** DAYS INN Days Inn, The Inn at Jewish Hospital by Marriot
Med **E:** ✚ Jewish Hospital
ATM **E:** Shell

136A Broadway, Chestnut St. (Northbound)
Gas **W:** Thornton's*(D)
Food **E:** Taco Bell
W: McDonalds, Rally's Hamburgers
Lodg **W:** Holiday Inn
AServ **E:** A-C Brake Company, Firestone Tire & Auto, Monarch Lincoln Mercury, NAPA Auto Parts
W: Cooke's Isuzu of Louisville, Jim Cooke Buick
Med **E:** ✚ Kosair Children's Hospital, ✚ Louisville Medical Center, ✚ University Children's Health Center
Other **W:** Kroger Supermarket, Rapid Auto Wash

135 St. Catherine, Old Louisville (Spalding University)
Gas **E:** Amoco*(CW), Shell*
Food **W:** Dizzy Whiz Hamburgers, Ermin's French

EXIT KENTUCKY

Bakery & Cafe
Other W: Quick Stop Market & Deli

134 KY 61 S, Arthur St (Difficult Reaccess)
Gas W: Amoco*(D)
Lodg W: Days Inn (see our ad this page)
ATM W: Amoco
Other W: Charles Heitzman Bakery

134AB 134A - KY 61N, Jackson St, 134B - Woodbine St (Northbound
Other E: Fire Station, Jerry's Grocery

133AB Alt. US - 60, Eastern Parkway, Univ of Louisville
Gas W: BP*(D)(CW)(LP)
Food E: Dairy Castle, Denny's, Papa John's Pizza, Subway
W: McDonalds
AServ E: Huber Tire
Other E: Sav-A-Step Food Mart
W: U. of L. Thrust Theater, Univ. of Louisville

132 Crittenden Dr., Fair/Expo Ctr. Gates 2, 3, 4 (Southbound, Reaccess Both Directions)
FStop W: BP*(D)(LP)
Food W: Burger King, Clark's Bar-B-Q
Other W: Kentucky Kingdom Fair/Expo Ctr.

131AB 131A - I-264E, W Watterson Expwy Airpt; 131B - Fair Expo Ctr

130 KY 61, Preston Highway, Grade Lane
Gas E: BP*(LP)
Food E: Bob Evans Restaurant, Domino's Pizza, Koreana Restaurant, Papa John's Pizza, Pepper Shaker Chilli, Waffle House
Lodg E: ◆ Red Roof Inn, ◆ Super 8 Motel
AServ E: Big-O Tires, Extend-A-Car, Instant Oil Change, Louisville Brake & Mufflers, Preston Auto Supply
Other E: Save-A-Lot Food Stores, U-Haul Center(LP)

128 KY 1631, Fern Valley Road
Gas E: Amoco* (24 Hrs), BP(D)(LP), Chevron* (24 hrs), Thornton's Food Mart*(D)(LP) (Kerosene)
Food E: Arby's, Bojangles, Frisch's Big Boy, Hardee's, McDonalds, Outback Steakhouse, Shoney's, Subway (Thornton's), Waffle House
Lodg E: ◆ Holiday Inn, AAA Signature Inn, AAA Thrifty Dutchman
AServ E: BP
ATM E: Thornton's Food Mart
Other W: Louisville Airport

127 KY 1065, Okolona, Fairdale
Food E: Texas Roadhouse
W: McDonalds (Playground)

125AB Junction I-265, KY 841 - Gene Snyder Freeway

121 KY 1526, Brooks Road
TStop W: Pilot Travel Center*(SCALES) (24 hrs)
Gas E: Chevron*
W: BP*(D), Shell*
Food E: Arby's, Burger King, Cracker Barrel, Sassy's Subs & Such
W: Subway (Pilot TS), Taco Bell (Pilot TS), Waffle House
Lodg E: ◆ Budgetel Inn, Fairfield Inn by Marriot
W: AAA Comfort Inn, AAA Holiday Inn Express
ATM W: Shell

117 KY 44, Shepherdsville, Mount Washington

EXIT KENTUCKY

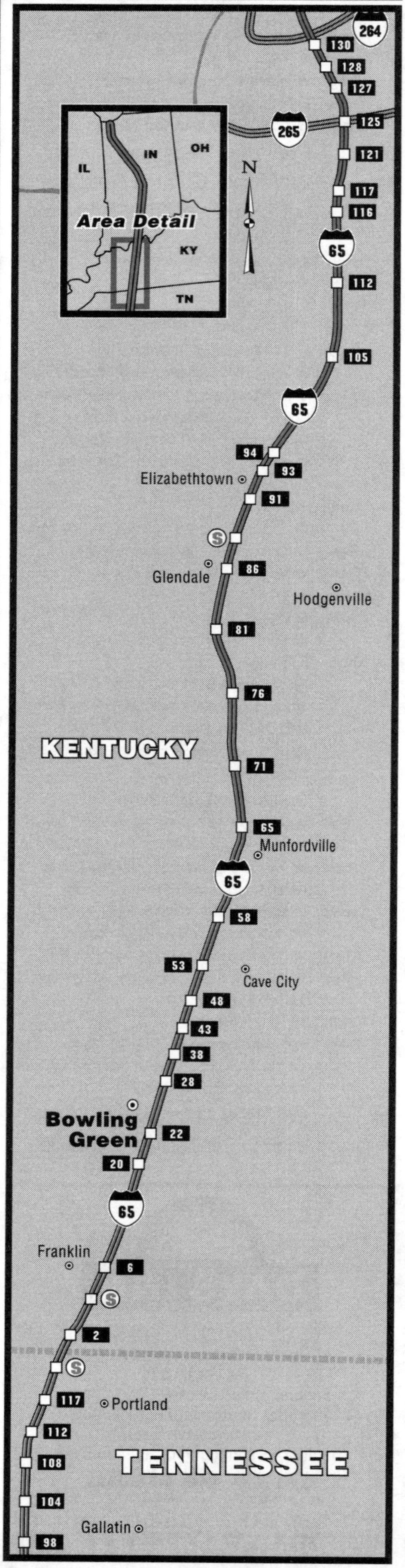

EXIT KENTUCKY

FStop W: Amoco*(D) (24 Hrs), Chevron* (24 hrs)
Gas E: BP*, Shell*
W: Super America*(D)
Food E: Hardee's (Playground), McDonalds (Playground), Pizza Hut (BP), Restaurant (Best Western), The Kitchen Family Rest. (Days Inn)
W: Arby's, Bearno's Little Sicily Pizza, Burger Kir Fazoli's Italian Food, Long John Silvers, McDonalds (Playground), Mr. Gatti's Pizza, Pap John's Pizza, Ponderosa Steakhouse, Shoney's Subway, Taco Bell, Waffle House, Wendy's, White Castle Restaurant
Lodg E: AAA Best Western, Days Inn
W: Motel 6, Super 8 Motel
AServ W: Hardy & Mooney Auto & Tractor Supplies, Price's Auto Parts & Machine Shop
ATM W: Amoco
Parks E: Taylorsville Lake State Park
Other E: Tourist Information
W: ACE Hardware, ACE Rent-All(LP), Kart Kountry (Go Karts, Mini Golf), Rite Aide Pharmacy (1 Hr Photo), Sav-A-Lot Grocery, Speed Queen Laundry (Coin Operated), Tou Information, Winn Dixie Supermarket

116 KY 480 to KY 61
Gas E: Shell*(D)
W: Barnyard*(D) (Kerosene), Citgo* (Kerosene)
RVCamp W: Leisure Life RV
Other E: Buffalo Run Golf Center (Driving Range)

(113) **Rest Area, Picnic, RR, HF Phones (Southbound)**

112 KY 245, Clermont, Bardstown
Gas E: Shell*
ATM E: Shell
Other E: Bernheim Forests

105 KY 61, Lebanon Junction, Boston
TStop W: Davis Brothers Travel Plaza*(SCALES) (Shell Fuel, Travel Store)
Gas W: 105 Plaza*
Food W: Cleo's (Davis Brothers Travel Plaza)
TServ W: Davis Bros. Travel Plaza (24 Hrs)

102 KY 313 to KY 434, Radcliff, Vine Grov

94 U.S. 62, KY 61, Elizabethtown
Gas E: Citgo(D)(CW)(LP), Shell*
W: Chevron*(D) (24 Hrs), SuperAmerica*
Food E: Denny's (Days Inn), KFC, Waffle House, Whit Castle Restaurant (24 Hrs)
W: Burger King (Playground), Cracker Barrel, Ryan's Steakhouse, Shoney's, Texas Roadhouse, Wendy's

Bold red print shows RV & Bus parking available or nearby

XIT KENTUCKY

Lodg E: Days Inn (Playground), ◆ Super 8 Motel
W: Comfort Inn, ◆ Hampton Inn, Holiday Inn, Motel 6, Ramada Limited, Towne Inn
Med W: ✚ Hospital
ATM E: Citgo
W: SuperAmerica
Other W: Kentucky State Police, RW Marine Boating Supplies

3 Bardstown, Bluegrass Parkway, Lexington
RVCamp E: KOA Kampground

1 U.S. 31 West, KY 61, Western KY Pkwy, Elizabethtown, Paducah
TStop E: Big T Truck Stop(SCALES) (Citgo fuel)
Gas E: Chevron* (24 Hrs), Shell*(D)
Food E: Big T TS, High Tide Cafe (Commonwealth Lodge), Long John Silvers, Omelet House
Lodg E: Budget Holiday Motel, Commonwealth Lodge, Heritage Inn, Red Carpet Inn
AServ E: Chevron
TServ E: Big T TS
Other E: Abe Lincoln Birthplace National Marker (13 Miles), Pittstop Car Wash

9) Weigh Station (Both Directions)

6 KY 222, Glendale
TStop W: Country Style Plaza*(SCALES) (Texaco, 24 Hrs)
FStop E: Speedway*(LP)
Gas E: Marathon*(D)
Food W: Country Style Plaza Restaurant
Lodg W: Glendale Economy Inn
AServ E: Glendale Automotive
TServ W: Country Style Plaza TS
RVCamp E: Glendale Campground
ATM W: Country Style Plaza

2) Rest Area - Picnic, RR, HF, Phones (Southbound)

1) Rest Area - Picnic, RR, HF, Phones (Northbound)

1 KY 84, Sonora, Abe Lincoln Birthplace
TStop E: Citgo*(LP), Davis Brothers Travel Plaza(SCALES) (Ashland Gas)
Gas E: BP*
W: Ashland* (Kerosene), Shell* (Kerosene)
Food E: Sammy's (Citgo TS)
AServ E: Horton's Garage 24 Hr Wrecker Service
W: Sonora Garage Inc.
TServ E: Davis Bros. Travel Plaza

EXIT KENTUCKY

TWash E: Blue Beacon (Davis Bros. Travel Plaza)
RVCamp E: Avery's Campground (4 Miles)
ATM E: Davis Bros. Travel Plaza
Other E: Abraham Lincoln Birthplace National Historic Site (13 Miles)
W: US Post Office

76 KY 224, Upton
Gas E: Ashland*, Chevron*(D), Citgo*
Food E: Stuckey's (Citgo)
Lodg E: Sleepy Hollow Motel
AServ E: Chevron

71 KY 728, Bonnieville

65 U.S. 31 W, Munfordville
Gas E: BP*(D), Citgo(D), Texaco*
W: Chevron*(D) (24 Hrs), Shell*
Food E: Dairy Queen, McDonalds (Playground), Pizza Hut, Sheldon's Country Fixens (Texaco), Subway (BP)
W: Cave Country Restaurant
Lodg E: Super 8 Motel
AServ E: Citgo
Other E: Houchens Food Stores, Nolin Lake, Traveler's Food Mart
W: Big Dummy's CB Repair, Farmwald Bulk & Variety

58 KY 218 to KY 335, Horse Cave
FStop W: Chevron Driver's Travel Mart*
Gas W: BP*, Marathon*(D), Shell*
Food W: Budget Host Family Restaurant, Pizza Hut Express (Chevron)
Lodg W: Budget Host Inn
AServ W: Marathon
RVCamp W: KOA Kampground
ATM W: ATM (Chevron)
Other E: American Cave Museum (2.2 Miles), Hidden River Cave (2.2 Miles), Horse Cave Theater (2.2 Miles), Kentucky Caverns, Kentucky Down Under, Tourist Information

(55) Rest Area - Picnic, RR, HF, Phones (Southbound)

53 KY 70, KY 90, Cave City, Glasgow
Gas E: Amoco*(D), BP*, Chevron*(D) (24 Hrs), Jr. Food Stores*, Shell*(D), Super America*
W: Shell*(LP)
Food E: Baker's Dozen Donut & Coffee Shop, Baskin Robbins, Burger King (BP), Country Kitchen (Quality Inn), Hickory Villa Family Restaurant, Hillside Family Restaurant, Jerry's Restaurant, KFC, Long John Silvers, McDonalds (Playground), Oasis Family Restaurant, Pizza Hut, Subway (Jr. Food Stores), Taco Bell, Wendy's
W: Puerto Vallarta Mexican, Watermill Restaurant
Lodg E: Comfort Inn, Days Inn, Executive Inn, Holiday Inn Express, Kentucky Inn (Best Western), Quality Inn, Super 8 Motel
AServ E: Chevron
W: Shell

EXIT KENTUCKY

RVCamp W: Mammoth Cave National Park (10 Miles), Primitive Camping, Singing Hills Camping, Yogi Bear Camp Report(LP)
Med E: ✚ Hospital
Parks E: Barren River Lake State Resort Park
W: Mammoth Cave National Park
Other E: Driving Range Mini Golf
W: Guntown Mountain Visitor Attraction, Kentucky Action Park, Mammoth Cave, Mammoth Cave Wax Museum, Olde Gener'l Store, Onyx Cave, Overnight Film Developing, Smith's Country Store, Ye Olde Fudge Shop

48 KY 255, Park City, Brownsville
FStop E: Shell* (see our ad on this page)
Gas E: Citgo
Lodg E: Parkland Motel
AServ E: Citgo
RVCamp W: Diamond Caverns Camping
Parks W: Mammoth Cave National Park
Other W: Mammoth Cave

43 Cumberland Parkway Toll Road, Glasgow, Somerset
Parks W: Barren River Lake State Resort Park

(40) Rest Area - RR, Phones, Vending, Picnic, 4 - Hr Parking, Dog Walk

38 KY 101, US 68, Scottsville, Smith Grove (Historic Victorian Site On West Side)
TStop W: BP(SCALES)
Gas W: Chevron*(D) (24 Hrs, Kerosene), Kentucky Souvenirs*(D), Shell*
Food W: Buffet Pizza, Donita's Country Diner, McDonalds, Restaurant (BP TS)
Lodg W: Bryce Motel
AServ W: National Tire
TServ W: Lee's Truck Repair & Tire Service (BP)
Other W: Cee Bee Food Store

36 U.S. 68, KY 80, Oakland (Difficult Northbound Reaccess)

(30) Rest Area - RR, Phones, Picnic (Southbound)

28 To U.S. 31 West, Bowling Green
Gas W: BP*, Shell*(CW) (see our ad this page)
Food W: Blimpie's Subs (Shell), Hardee's (Playground), Jerry's Restaurant, Wendy's
Lodg W: Continental Inn (Best Western), Hatfield Inn, Value Lodge
RVCamp W: Camping World (see our ad on this page)
Other W: National Corvette Museum, Tourist Information

EXIT KENTUCKY

22 U.S. 231, Scottsville, Bowling Green
- **FStop** **W:** Texaco*
- **Gas** **E:** Chevron[D], Citgo*, Shell*[CW]
  W: BP*[CW], Exxon*, Marathon*, Racetrac*, Shell[D]
- **Food** **E:** Cracker Barrel, Denny's, Domino's Pizza, Hardee's, Old Kentucky Home Country Hams, Ryan's Steakhouse, Tabatha's Country Inn, Waffle House
  W: Arby's, Bob Evans Restaurant, Burger King (Playground), China Buffet, Fazoli's Italian Food, Hops, McDonalds (Playground), Ponderosa, Shoney's, Waffle House, White Castle Restaurant (24 hrs)
- **Lodg** **E:** Best Western, Comfort Inn, Days Inn, Econolodge, ◆ Fairfield Inn, Quality Inn, ◆ Ramada Inn, Super 8 Motel
  W: ◆ Budgetel Inn, Greenwood Executive Inn, ◆ Hampton Inn, Holiday Inn, Motel 6, News Inn, Scottish Inns
- **AServ** **E:** Chevron
  W: Greenwood BMW, Greenwood Ford Mercury Lincoln, Shell
- **RVCamp** **W:** KOA Kampground
- **ATM** **E:** Citgo
  W: Marathon, Racetrac
- **Other** **W:** Bowling Green Info Center, Clubhouse Golf, Greenwood Car Wash, Greenwood Mall, Houchens Food Store, K-Mart, Three Springs Go-Karts, Tourist Information, Toys "R" Us

20 William H. Natcher Toll Road, Bowling Green, Owensboro

6 KY 100, Scottsville, Franklin
- **TStop** **E:** BP* (Kerosene, 24 hrs)

EXIT KENTUCKY

  W: Mapco Express[SCALES] (see our ad this page)
- **FStop** **E:** Bluegrass Auto/Truck Stop*[SCALES]
  W: Speedway*[SCALES]
- **Food** **E:** Original Ole South Diner (BP TS)
  W: Baskin Robbins (Mapco), Loretta Lynn's Kitchen, Miss Penny's Southern Delicacies, Subway (Speedway), Wendy's (Mapco)
- **Lodg** **W:** Super 8 Motel
- **TServ** **W:** Bluegrass Tire & Truck Repair, Petro Lube
- **RVCamp** **W:** KOA Kampground[LP]
- **Med** **W:** + Hospital
- **ATM** **E:** Bluegrass FStop
  W: Mapco Express
- **Other** **W:** Tourist Information

(4) Weigh Station (Northbound)

2 U.S. 31 West, Franklin
- **TStop** **E:** Flying J Travel Plaza*[LP][SCALES] (CB Shop)
- **FStop** **E:** Keystop* (Texaco)
  W: BP*[D]
- **Gas** **E:** Conoco* (Flying J Travel Plaza)
- **Food** **E:** Burger King (Keystop), Country Market (Flying J TS), Magic Dragon Chinese Eatery (Flying J TS), Pepperoni's the Superslice (Flying J TS)
  W: Cracker Barrel, Huddle House, Lotto Land Market & Grill, McDonalds, Shoney's
- **Lodg** **W:** Comfort Inn (see our ad this page), ◆ Franklin Executive Inn, Holiday Inn Express (see our ad this page), Ramada Limited
- **RVCamp** **W:** Tom Stinnett (see our ad this page)
- **ATM** **E:** Keystop

(1) KY Welcome Center - RR, Phones, Vending, Picnic, Dog Walk (Northbound)

EXIT KENTUCKY/TENNESSEE

↑ KENTUCKY
↓ TENNESSEE

(121) TN Welcome Center - RR, Phones, Picnic, Vending (Southbound)

(119) Weigh Station (Both Directions)

117 TN 52, Orlinda, Portland
- **TStop** **W:** Jiffy Truck-Auto Plaza[SCALES]
- **Gas** **E:** Shell*[D] (Kerosene) (see our ad this page)
  W: Amoco*[D] (Kerosene)
- **Food** **W:** Red River (Jiffy Truck-Auto Plaza)
- **Lodg** **W:** Budget Host Inn (Jiffy Truck-Auto Plaza)
- **AServ** **W:** Portland Collision Center (Jiffy Truck-Auto Plaza)
- **TServ** **W:** Bumper to Bumper Truck Repair (Jiffy Truck-Auto Plaza)
- **Other** **W:** Susie's Fireworks

112 TN 25, Gallatin, Cross Plains, Springfield
- **FStop** **W:** Delta Express*[LP] (Kerosene)
- **Gas** **W:** Exxon* (Kerosene)
- **ATM** **W:** Delta Express
- **Other** **E:** Loco Joe's Fireworks
  W: Uncle Sam's Fireworks

108 TN 76, Springfield, White House
- **Gas** **E:** BP*, Nervous Charlies Market*[D]
  W: Amoco*[D]
- **Food** **E:** Dairy Queen, Dinner Bell Restaurant,

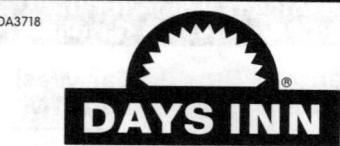

Bold red print shows RV & Bus parking available or nearby

XIT	TENNESSEE
	Hardee's, McDonalds (Playground)
	W: Red Lantern (Days Inn)
Lodg	**E:** Comfort Inn, Holiday Inn Express
	W: AAA Days Inn (see our ad on opposite page)
AServ	**E:** Baskin Body Shop, Xpress Lube
Med	**W:** + Hospital
Other	**E:** United States Post Office
4	TN 257, Ridgetop, Bethel Road, Highland Rim Speedway
TStop	**E:** Phillips 66(LP)(SCALES)
Gas	**W:** Shell*(D), (see our ad on opposite page)
Lodg	**E:** Motel (Phillips 66 TS)
TServ	**E:** Ridgetop 24-Hr Service (Phillips 66 TS)
VCamp	**W:** Owl's Roost Campground(LP)
Other	**E:** Ridgetop Laundry (Motel, Phillips 66 TS)

EXIT TENNESSEE

Springfield · Gallatin · 104 · 98 · 65 · 24 · 97 · 96 · 92 · 90 · 89 · 265 · NASHVILLE · 88B THRU 80 · 40 · 79 · 78A,B · 74A,B · 71 · 69 · 68A,B · 65 · 61 · 24 · 53 · 46 · 65 · 37 · 32 · Lewisburg · 27 · TENNESSEE · 22 · Pulaski · 14 · 65 · 6 · 1 · 365 · 361 · ALABAMA · 354 · 351 · HUNTSVILLE · 65 · 340A,B · Decatur · 334 · 328 · Hartselle · 325 · 322 · 318 · 65

Area Detail: IN · OH · KY · TN · GA · N

EXIT	TENNESSEE
98	U.S. 31 West, Millersville, Springfield
Gas	**E:** Amoco*, Raceway*, Shell*
	W: Amoco*(D) (Kerosene, 24 hrs), Phillips 66*
Food	**E:** Subway (Shell), Waffle House
Lodg	**W:** Economy Inn, Graystone Motel
AServ	**E:** 31W Truck Tire Center, David's Garage (Shell), Goodlettsville Collision & Car Care
TServ	**E:** Fatboy's Truck & Tire Repair Service (24 hrs), Cliff's Speciality Shop (see our ad on this page)
RVCamp	**W:** Graystone Motel & Campground, KOA Campground
ATM	**W:** Amoco
Other	**E:** Music City Trading Post
97	TN 174, Long Hollow Pike, Goodlettsville
FStop	**W:** Citgo*(LP) (Kerosene)
Gas	**E:** BP*(CW), Exxon*(CW), Mapco Express*(D)(LP) (Kerosene)
Food	**E:** Arby's, Captain D's Seafood, China Express, Cracker Barrel, Domino's Pizza, KFC, Little Caesars Pizza, McDonalds (Playground), Shoney's, Subway, Waffle House, Wendy's
	W: Bob Evans Restaurant, Hardee's, Krystal, Ron's Bar-B-Q, Sonic
Lodg	**E:** AAA Comfort Inn (see our ad on this page), AAA Econolodge, ◆ Hampton Inn, Holiday Inn Express, ◆ Red Roof Inn
	W: ◆ Budgetel Inn, Motel 6
AServ	**W:** Xpress Lube
Med	**E:** + Goodlettsville Family Care Center (No Appt. Necessary)
ATM	**E:** Bank of Goodlettsville, Kroger Supermarket (Pharmacy)
	W: Citgo
Other	**E:** Historic Mansker's Station, K-Mart (Pharmacy), Kroger Supermarket, Postal World, US Post Office
	W: City of Goodville Fire Dept., Companion Animal Hospital, Sunshine Car Wash, Volunteer Car Wash
96	Two Mile Parkway, Goodlettsville
Gas	**E:** BP*(D)(CW), Shell*(CW), Texaco*(D)
	W: Amoco* (24 hrs), Phillips 66
Food	**E:** Al Fuente Mexican, Bailey's Sports Grille, Checkers Burgers, Cooker, El Chico Mexican, Hooter's, Las Palmas Mexican, Lee's Famous Recipe Chicken, Mrs. Winner's, McDonalds, McDonalds, Mr. Gatti's Pizza, O'Charley's, Papa John's Pizza, Pargo's, Pen'angelos Italian Kitchen, Pizza Hut, Shoney's, Steak & Ale, Subway, The Honey Baked Ham Co., Uncle Bud's Catfish, Chicken, & Such, Waffle House, Wendy's
Lodg	**E:** Days Inn (see our ad on this page), AAA Rodeway Inn, Super 8 Motel
AServ	**E:** Crest Cadillac, Firestone Tire & Auto, Goodyear Tire & Auto, NTB, Universal Tire
	W: Don Harris Auto Servicenter (Phillips 66)
ATM	**E:** First American, NationsBank, Old Hickory Credit Union, Tennessee Teachers Credit Union, Texaco
Other	**E:** Golf Discount, Pearle Vision, Rivergate Mall, Toys "R" Us
	W: Artrip's Produce Market
95	TN 386, Vietnam Veteran's Blvd., Hendersonville, Gallatin (Limited Access Hwy)
92	TN 45, Madison, Old Hickory Blvd,

EXIT TENNESSEE

Old Hickory Dam

Med **E:** Hospital

90 U.S. 31 W, U.S. 41, TN 155, Briley Pky, Dickerson Pike (Difficult Reaccess)

89 U.S. 31 West, U.S. 41, Dickerson Pike

Gas **E:** Citgo*, Phillips 66*(CW)

Food **E:** Arby's, Burger King, McDonalds (Playground), Mrs Winner's, Lee's Famous Recipe Chicken, Pizza Hut, Shoney's, Taco Bell, Waffle House, Wendy's

Lodg **E:** Days Inn, AAA Days Inn, AAA Econolodge, ◆ Super 8 Motel
W: AAA Econolodge

AServ **E:** Advance Auto Parts

ATM **E:** Citgo, Sun Trust Bank

Other **E:** **H.G. Hill Foodstores, Oak Valley Bowling Lanes, Speed Queen Laundry (Coin Laundry)**

88B Junction I-24 West, Clarksville, Fort Campbell Army Post

87AB U.S. 431, Trinity Lane

TStop **E:** **Pilot Travel Center**(SCALES)

Gas **E:** Circle K*(LP)
W: Amoco*(D)(CW) (Kerosene), BP*(D)(CW), Exxon*(D), Texaco*(D)

Food **E:** **Arby's (Pilot)**, Candle Light Restaurant, Chuch's Chicken, Krystal (24 hrs), Sonic, TJ Restaurant, White Castle Restaurant
W: Burger King, Captain D's Seafood, Chugers Restaurant, Club Paradise, Denny's (24 hrs), Gabe's Lounge, Jack's BAR-B-QUE, Lockers Sports Grill, McDonalds, Ponderosa, Shoney's, Taco Bell, The Broken Spoke Cafe (Ramada Inn), Track 1 Cafe (Rain Tree Inn), Waffle House

Lodg **E:** **Cumberland Inn (Pilot TS)**, Key Motel, Red Carpet Inn, Savoy Motel, Scottish Inns, Trinity Inn
W: Comfort Inn, Days Inn (see our ad on this page), Econolodge, Hallmark Inn, Hampton Inn, **Holiday Inn Express (see our ad on this page)**, Knights Inn, Liberty Inn, Motel 6, Oxford Inn, Rain Tree Inn, Ramada

AServ **E:** Bobby's Tire Service, Gary's, The Tire Store
W: Exxon

RVCamp **E:** **RV and Camper Corral truck Accessories**

ATM **W:** BP

Parks **E:** **Trinity Park (Mobil Home Park)**

Other **E:** Coin Car Wash, National Car Wash, **Sweeney's Food Town Gocery Store, Trinity Gas Co. Inc**(LP), **US Post Office**

86 Junction I-265 Memphis

85B Jefferson St. (East Side Services Are The Same As In Exit 85A)

Gas **W:** Spur

Lodg **W:** Best Western, ◆ Days Inn (see our ad on this page)

85A U.S. 31 E., N. Ellington Pkwy, Spring St.

FStop **W:** **Cone***

Gas **W:** Spur*(D)

Lodg **W:** Best Western (Metro Inn), Days Inn

AServ **W:** Jefferson Street Car Care, NAPA Auto Parts, Nashville Tansmission Parts, Todd's Auto Parts

Other **W:** Burnette's Truck Wash, **U-Haul Center**(LP)

85 James Robertson Parkway, State Capitol (Difficult Reaccess)

TStop **W:** **TravelCenters of America**(SCALES) **(RV Dump)**

Gas **E:** Mapco Express, Shell*(CW)
W: Cone*

EXIT TENNESS

Food **W:** **Country Pride (Truckstop)**, Gersthaus Restaurant, Mrs. C's, Shoney's, **Subway (Truckstop)**

Lodg **W:** Best Host Motel, Econolodge, Ramada Limited

AServ **E:** Amoco, Coin Car Wash
W: Vogely & Todd Collision Repair Experts

TServ **W:** **TravelCenters of America**

ATM **E:** Mapco Express

Other **E:** Main Street Auto Wash
W: **Country Hearth Bread, Bakery Thrift Stor Performing Arts Center, State Capitol, Touris Information**

84 Arena, Shelby Ave

Gas **W:** Exxon*

Food **W:** Gersthaus

Lodg **W:** Econolodge, Ramada Inn (see our ad on th page)

AServ **W:** Napa Auto Parts, Standard Motor Parts

Other **E:** **Lynn Drugs, Martin's Grocery**
W: **Prop Shop & More**

83B I-40 West, South I-65, Memphis, Birmingham

83A I-24 East, I-40 East, Knoxville, Chattanooga

82B I-40 West, Memphis (Left Exit Northbound)

82A North I-65, East 1 - 40, Louisville, Knoxville

81 Wedgewood Ave (Belmont Univ., Fairgrounds, Cumberland Science Museum)

Gas **W:** BP*(CW), Citgo*, Exxon*

Food **E:** Raymond's Barbeque (Raymond's Food Center)
W: Burger King (Playground), Mrs Winner's Chicken & Lee's Famous Recipe Chicken

AServ **E:** Village Tire & Auto Service
W: Hubcap Aanie

ATM **W:** Citgo

Parks **W:** **Reservoir Park**

Other **E:** **Raymond's Food Center, Tennessee State Fair, Wedgewood Station Antique Mall**
W: **Belmont Mansion, Belmont University, Country Music Hall of Fame & Museum, Cumberland Science Museum, U-Haul Center**(LP)

80 Junction I-440, Memphis, Knoxville

79 Armory Drive (Nashville School Of Law & School Of Ballet)

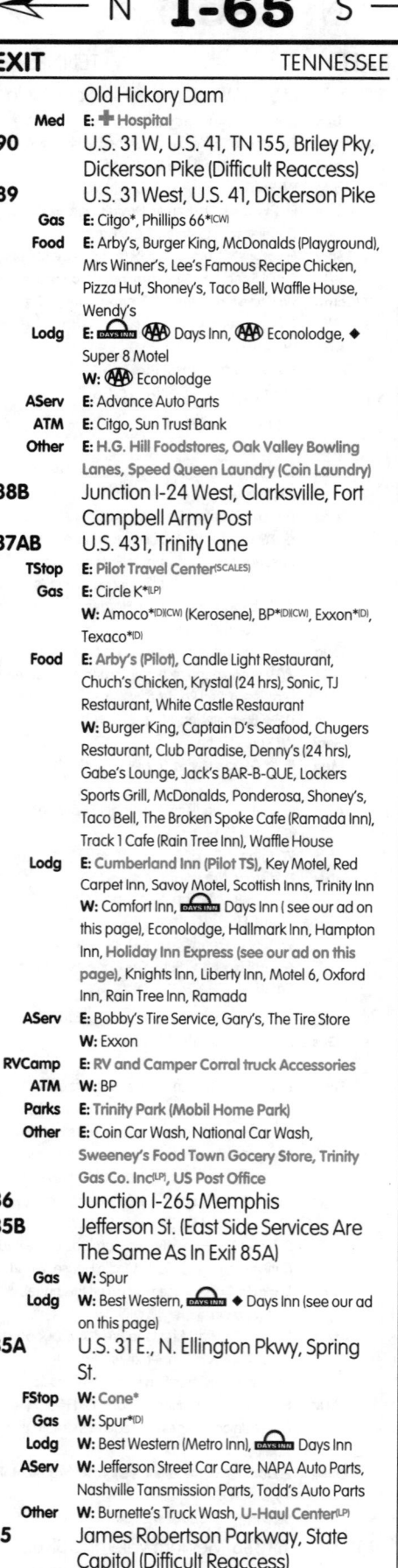

Bold red print shows RV & Bus parking available or nearby

XIT	TENNESSEE
Other	E: 100 Oaks Mall, The Home Depot(LP)
8AB	TN 255, Harding Place
Gas	E: Amoco(D), Exxon, Mapco Express*(LP), Mapco Express*(LP), Texaco*(LP), Texaco
Food	E: Bei Jing Chinese Restaurant, Cracker Barrel, Cuban Cuisine Restaurant, Darryl's, Mama Mia's Italian Restaurant, Santa Fe Cantina & Cattlemen's Club, Waffle House
Lodg	E: La Quinta Inn, Ramada Inn Governor's House, Red Roof Inn
AServ	E: Amoco, Sidco Auto Parts, Texaco
Med	E: ✚ Hospital
Parks	E: Grassmere Wildlife Park
Other	E: Brooks Pharmacy, Ellington Agricultural Center, Fire Dept., Historic Home Travellers Rest W: David Lipscomb University
4AB	TN 254, Old Hickory Blvd, Brentwood
Gas	W: Amoco*(D) (Kerosene), BP*(CW), Phillips 66, Shell*(CW), Texaco*(CW)(LP), Texaco*(D)
Food	E: Captain D's Seafood, Restaurant (Holiday Inn), Shoney's W: August Moon Chinese, Baskin Robbins, Blimpie's Subs, Corky's Bar-B-Q, Heavenly Ham, McDonalds (Playground), Mrs Winner's & Lee's

EXIT TENNESSEE

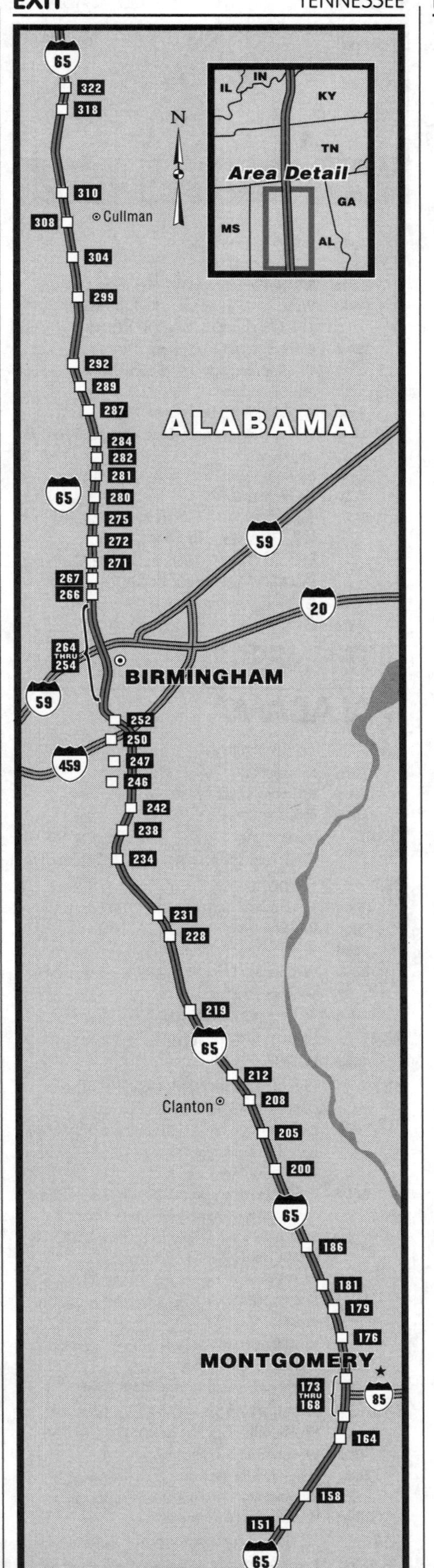

EXIT	TENNESSEE
	Famous Recipe Chicken, Mrs. T's Donut Place, Nick of Thyme, O'Charley's, Papa John's Pizza, Pargo's Restaurant, Phillip's Delicatessen, Quails Up-Scale Dining, Taco Bell, The Flour Shop Bakery, Wendy's
Lodg	E: AmeriSuites (see our ad on this page), Holiday Inn (see our ad on this page), Steeple Chase Inn W: Courtyard (Marriott), Hampton Inn (see our ad on this page), Travelers Rest Inn
AServ	W: Aces Transmission, Amoco, Phillips 66, Valvoline Quick Lube Center
Med	E: ✚ Hospital
ATM	W: AMSouth Bank, NationsBank, SunTrust Bank, Union Planters Bank
Other	W: ACE Hardware(LP), Animal Care Center Brentwood Station, Bradford Animal Hospital, Brentwood Veterinary Clinic, One Hr Photo, Southern Post Mailing Service, US Post Office
71	TN 253, Concord Road, Brentwood
69	TN 441, Moores Lane
Gas	E: Mapco Express*(D)(LP) (Kerosene), Texaco W: Amoco*(D)(CW) (Kerosene), BP*(CW)(LP), Super Clean*(CW)(LP) (Kwik Sak)
Food	E: Applebee's, Baskin Robbins (Mapco Express), Blimpie's Subs* (Mapco Express), Copeland's of New Orleans, Cozymel's, Joe's Crab Shack (Playground), Outback Steakhouse, Tony Roma's Famous for Ribs W: Back Yard Burgers, Boston Market, J. Alexander's Restaurant, McDonalds (Playground), Peking Palace, Red Lobster, Romano's Macaroni Grill, Subway, Taco Bell, The Honey Baked Ham Co.
AServ	E: Texaco
ATM	E: Mapco Express W: NationsBank
Other	E: Bruno's Grocery (Pharmacy), PETsMart, The Home Depot(LP) W: Barnes & Noble, Carmike Cinema, Cool Springs Galleria Mall, Moore's Lane Animal Hospital, Pro Golf Discount, Target Department Store, Toys "R" Us
68AB	Cool Springs Blvd
Food	W: Las Palmas Mexican, Leb's Deli & Grille, Panda Super Buffet
Lodg	W: Hampton Inn & Suites
AServ	W: Saturn of Cool Springs
ATM	W: Kroger Supermarket, SouthTrust Bank
Other	W: Cool Springs Galleria Mall, Golf Tennis Superstore, Kroger Supermarket(LP), Mail Boxes Etc, Recreation World Family Fun Center, Staple's The Office Superstore
65	TN 96, Franklin, Murfreesboro (Historic Site-battle Of Franklin)
Gas	E: Exxon*(CW)(LP), Texaco*(LP) W: BP*(CW)(LP), Mapco Express*(LP), Shell*(CW)
Food	E: Cracker Barrel, Log Cabin Restaurant (Texaco), Sonic W: Back Yard Burgers, Camino Real Mexican, Daylight Donuts, Hardee's, Herbert's Bar-B-Q Restaurant, KFC, McDonalds, Nashville Pizza Co., Nooley's Louisiana Po-Boy's, O'Charley's, Shoney's, Sir Pizza, Taco Bell, Waffle House, Wendy's
Lodg	E: Bugetel Inn, ◆ Comfort Inn, AAA Days Inn W: AAA Best Western (Franklin Inn), AAA Holiday Inn Express
AServ	E: Cain Buick Pont GMC Trucks, Cain GMC, Cain

EXIT — TENNESSEE

Kia, Darrell Waltrip Honda Volvo, Reed's Auto Service, Reed's Quick Lube, Unibody Collision Center, Walker Chev Geo Olds
W: Franklin Nissan Mazda, Hippodrome Dodge Trucks, Valvoline Quick Lube Center

Med E: ✚ Franklin Family Walk-In Clinic, ✚ Williamson Medical Center

ATM W: AMSouth Bank, Bi-Lo Supermarket, First American Bank, First Tennesse Bank, NationsBank

Other E: Pop's Car Wash
W: Allergy Relief Stores, Bi-Lo Supermarket (Pharmacy), Book Den, Coin Laundry, Franklin Aquarium Pet Shop, Franklin Discount Drugs, Franlin Hardware(LP), Harpeth Antique Mall, K-Mart (Pharmacy), One-Hr Photo, Revco Drugs

61 TN 248, Peytonsville Road

TStop E: 76 Auto/Truck Plaza*(SCALES) (BP, FedEx Drop)

FStop E: Speedway*(LP)

Gas W: Cone*(LP) (Kerosene, 24 Hrs), Mapco Express*(LP) (Kerosene)

Food E: Apple Creek Restaurant (76 Auto/Truck Stop)
W: Goose Creek Inn

Lodg W: Goose Creek Inn

TServ E: Truck Lube Center

53 TN 396, Saturn Parkway, Columbia, Spring Hill

(49) Parking Area, Weigh Station - No Facilities (Northbound)

46 U.S. 412, TN 99, Columbia, Chapel Hill (Columbia State Community College)

TStop W: Texaco

Gas E: Chevron*
W: BP*, Exxon*

Food E: Bear Creek Bar-B-Q
W: Burger King (Exxon), Stan's Restaurant (Texaco TS)

Lodg W: (AAA) Econolodge, (AAA) Holiday Inn Express, Relax Inn

Med W: ✚ Hospital

Parks E: Henry Horton State Park (12 Miles)

Other W: James K. Polk Home - Historic Site

37 TN 50, Columbia, Lewisburg

Gas W: Texaco* (Kerosene)

Med E: ✚ Hospital

Other E: TN Walking Horse Association Headquarters
W: James K. Polk Home-Historic Site

32 TN 373, Columbia, Lewisburg, Mooresville Hwy

Gas E: Exxon*

27 TN 129, Lynnville, Cornersville

RVCamp E: Texas Tea Campground

Other E: Hazelburn Golf Cours

(25) Parking Area (Southbound)

(24) Parking Area (Northbound)

22 U.S. 31A, Lewisburg, Cornersville

TStop E: The Tennessean*(SCALES) (Showers, Laundry)

FStop W: BP, Delta Express*(LP) (Kerosene)

Gas W: Exxon*

Food E: The Tennessean TS

Lodg E: (AAA) Econolodge

AServ W: BP Fuel Stop

TServ E: Diesel Shop (The Tennessean TS), Truck Summit Tires
W: Diesel & Truck Repair (BP Fuel Stop)

TWash E: Wayne's Truck Wash (The Tennessean TS)

ATM W: Delta Express Fuel Stop

14 U.S. 64, Pulaski, Fayetteville

FStop E: Texaco*

Gas E: BP*(D) (Kerosene)

Food E: Sarge's Shack, The Sands Restaurant (Super 8

EXIT — TENNESSEE/ALABAMA

Motel)

Lodg E: (AAA) Super 8 Motel

AServ E: Texaco Fuel Stop

RVCamp E: KOA Kampground(LP) (Mini Mart, Laundry)

Parks W: David Crockett State Park (30 Miles)

6 TN 273, Elkton, Bryson Road

TStop E: Shady Lawn Truck Stop*(SCALES) (Phillips 66, CB Shop, FedEx Drop), Mapco Express (see our ad on this page)

Food E: Restaurant (Shady Lawn TS)

Lodg E: Economy Inn, AmeriSuites (see our ad on this page)

TServ E: Shady Lawn TS

ATM E: Shady Lawn TS

(3) TN Welcome Center - Picnic, RR, HF, Phones, RV Dump

1 U.S. 31, TN 7, Pulaski, Lawrenceburg (Martin Methodist College)

Gas E: Chevron*(D) (Kerosene), H.P. Max Fuel*

ATM E: Chevron

↑TENNESSEE

↓ALABAMA

365 AL 53, Ardmore

Gas E: Texaco*

Food E: Granny's Country Restaurant

Lodg E: Budget Inn

(364) Welcome Center - Picnic, RR, HF, Phones, RV Dump (Southbound)

361 Elkmont

TStop W: Charlie's TS*(LP) (Amoco, CB Shop)

FStop W: Exxon*(LP)

Food W: Restaurant (Charlie's TS)

TServ W: Charlie's Truck Stop, Morris Garage (24 Hr Service)

ATM W: Charlie's Truck Stop

354 U.S. 31 South, Athens

Med W: ✚ Hospital

351 U.S. 72, Athens, Huntsville (Athens State College)

Gas E: BP*(D)(CW), Exxon*, Racetrac* (24 Hrs), Shell* (see our ad on this page), Texaco*(D)
W: Chevron*(CW)

Food E: Cracker Barrel, Dunkin' Donuts (Shell), McDonalds (Playground), Shells & Scales Seafood, Pasta, & Salads, Subway (Shell), Waffle House, Wendy's
W: Hardee's (Playground), Krystal, Shoney's

Lodg E: ◆ Comfort Inn, Hampton Inn, ◆ Holiday Inn Express
W: Athens Inn (Best Western), (AAA) Travelodge

ATM E: Exxon, Shell

Other E: Athens Limestone Veterinary Hospital

340AB Junction I-565, AL 20, Decatur, Huntsville (Space & Rocket Center)

Gas W: Racetrac* (24 Hrs), Texaco*(D)(CW)

Lodg E: Comfort Inn (see our ad on this page)
W: Holiday Inn (see our ad on this page)

Other W: Chip Shot Golf Range

334 AL 67, Priceville, Decatur, Somerville

TStop E: Racetrac*, Mapco Express (see our ad on

EXIT — ALABAMA

Bold red print shows RV & Bus parking available or nearby

EXIT	ALABAMA
	this page)
Gas	E: Amoco*(D)
	W: BP*, Chevron*(LP)
Food	W: Breslers Ice Cream (Chevron), Dairy Queen, **Hardee's**, Libby's Catfish & Diner Home Cooking, McDonalds (Playground), Southern Bar-B-Q, Waffle House
Lodg	E: Days Inn Days Inn
TServ	E: **I-65 Truck Shop (RaceTrac TS)**
TWash	E: **Racetrac**
RVCamp	W: **Hood Tractor & RV Center**
ATM	W: BP
Other	W: **Wheeler National Wildlife Refuge, Visitor Center**
328	AL 36, Hartselle
Gas	W: Chevron(D)(LP) (24 Hrs), Shell*
Food	W: Homestyle Bar-B-Q, Huddle House
AServ	W: Chevron
Med	W: ✚ **Hospital**
325	Thompson Road, Hartselle
322	Falkville, Eva
FStop	E: **BP***
318	U.S. 31, Lacon, Vinemont
Gas	E: Texaco*
Food	E: Dairy Queen (Texaco), Stuckey's (Texaco)
Lodg	E: **Lacon Motel**
310	AL 157, Cullman, Moulton, West Point, Florence
Gas	E: Amoco*(CW), Conoco*(D) (Kerosene), Shell*(D)(LP) (24 Hrs)
	W: BP*(D), Exxon*(D)
Food	E: Cracker Barrel **Cracker Barrel**, Delissimo Pizza, Subs, & Breakfast (Shell), Denny's, **McDonalds**, Morrison's Cafeteria, Taco Bell, Waffle House
Lodg	E: AAA Best Western, ◆ Comfort Inn, ◆ Hampton Inn
	W: **Super 8 Motel**
AServ	W: Exxon
RVCamp	W: **Cullman Campground (2.6 Miles)**
Med	E: ✚ **Hospital**
308	U.S. 278, Cullman, Double Springs, Ave Maria Grotto
Gas	W: Chevron*(D) (24 Hrs)
Food	E: **Omelet Shoppe**
	W: **Bryant's Seafood (Howard Johnson)**
Lodg	E: Days Inn ◆ Days Inn
	W: **Howard Johnson**
304	AL 69, Cullman, Good Hope
TStop	E: **Jack's TS*(SCALES) (Fuel-Shell, Showers, Laundromat)**
Gas	E: BP*, Exxon*(D), Texaco*(D)
	W: Citgo*(D) (Kerosene)
Food	E: **Hardee's, Jack's Restaurant (Shell)**, Maxine's (Ramada Inn), Miguel's Mexican, Waffle House
Lodg	E: **Holiday Inn Express (see our ad this page)**, Ramada Inn
AServ	E: Lindsey Motors
TServ	E: **Good Hope Truck & Wrecker Service, Goodyear Tire & Auto, Redding Refrigeration Truck & Trailer Repair**
RVCamp	E: **Good Hope Campground (.5 Miles)**
Med	E: ✚ **Hospital**
Other	W: **Fire Dept**
(302)	**Rest Area - RR, Phones, Picnic, HF, RV Dump (Eastside Available To Both Directions)**
299	AL 69 South, Jasper, Dodge City
FStop	W: **Conoco(D)**

EXIT	ALABAMA
Gas	W: Amoco* (24 Hrs), Texaco*(D), Williams Service*(LP) (Kerosene)
Food	W: Ann's Country Restaurant, **Lee's Cafe (Conoco)**, The Pizza Place
AServ	W: Lee's Tire & Service Shop (Wrecker Service is 24 Hrs)
TServ	W: **Lee's Truck Service (Conoco)**
ATM	W: Amoco, Texaco
Other	W: **CB Shack (Conoco)**
292	AL 91, Hanceville, Arkadelphia
TStop	W: **Shell***
Food	W: **Restaurant (Shell TS)**
TServ	W: **Shell***
TWash	W: **Shell TS**
RVCamp	E: **Country Park RV & Camping (1.4 Miles)**
ATM	W: **Shell TS**

EXIT	ALABAMA
289	Empire, Blount Springs
Gas	W: Texaco*
Food	W: Dairy Queen (Texaco), Stuckey's (Texaco)
Parks	W: **Rickwood Caverns State Park (3 Miles)**
287	U.S. 31 North, Garden City, Blount Springs
Gas	E: Lee's*(D)(LP)
284	U.S. 31 South, AL 160 East, Hayden Corner
Gas	E: Conoco*(D)(LP) (Kerosene), Phillips 66(D), Shell*(LP) (24 Hrs)
AServ	E: Conoco
ATM	E: Community Bank
Other	W: **Rickwood Caverns State Park Scenic Drive**
282	Warrior, Robbins
Gas	E: Chevron*(LP)
	W: Amoco*
Food	E: **Hardee's**, Pizza Hut, **Taco Bell**
ATM	E: Chevron
281	Warrior
280	To U.S. 31, Warrior, Kimberly
275	To U.S. 31, Morris, Kimberly
272	Mount Olive Road, Gardendale
Gas	W: Chevron* (24 Hrs)
RVCamp	E: **Gardendale Campground (.8 Miles)**
271	Fieldstown Road
Gas	E: Chevron(CW) (24 Hrs), Racetrac*
Food	E: Arby's, KFC, Milo's Hamburgers, Shoney's, Subway, Taco Bell, Waffle House
	W: Cracker Barrel **Cracker Barrel**
AServ	E: Serra Chevrolet Geo
Other	E: **Delchamps Supermarket, Wal-Mart (Pharmacy, One-Hr Photo)**
267	Walkers Chapel Road, Fultondale
266	U.S. 31, Fultondale
Gas	E: Chevron* (24 Hrs)
Lodg	E: Days Inn Days Inn, Super 8 Motel
264	41st Ave
263	32nd Ave
Gas	E: Chevron (24 Hrs), Shell*(CW)
	W: Amoco*(LP) (24 Hrs), Exxon*(CW)
Food	E: Hardee's
AServ	E: Chevron
ATM	W: Exxon
262B	Finley Blvd
FStop	E: **South Star Fuel Center*(SCALES) (FedEx Drop, Texaco)**
	W: **Mapco Express*(SCALES)**
Gas	E: Amoco*, Phillips 66*(D) (Racing Fuel)
	W: Chevron* (24 Hrs)
Food	W: Captain D's Seafood, **McDonalds**
AServ	W: Used Tires
TServ	W: **Southern Rubber Truck Tires Truck Service**
ATM	W: Mapco Express
262A	16th Street (Northbound, Reaccess Southbound Only)
Other	E: **Food Fair**, Hand Car Wash
261AB	Junction I-20, Junction I-59, Tuscaloosa, Gadsden, Atlanta
260	6th Ave North, Downtown
259AB	6th St South, 4th Ave South (Southbound, Difficult Reaccess Both Directions)
259	8th Ave South (Northbound, Difficult Reaccess Both Directions)

Bold red print shows RV & Bus parking available or nearby

EXIT ALABAMA

258 Green Springs Ave
256AB Oxmoor Road, Homewood
Lodg W: Comfort Inn (see our ad on page 271)
255 Lakeshore Dr.
254 Alford Ave
252 U.S. 31 Montgomery Highway
Gas E: Amoco, BP*(CW), Chevron*(CW), Shell(CW)
W: Amoco*, Big Green Clean Machine, Chevron*(CW), Exxon, Shell, Texaco*(D)(CW)
Food E: Arby's, Backyard Burgers, Captain D's Seafood, Chuck E. Cheese's Pizza, Comfort Inn, Hardee's, Ichiban Japanese Steak House, Johnny Ray's Restaurant, Milo's Hamburgers, Morrisons Cafeteria, Pizza Hut, Rally's Hamburgers, Ranch House BBQ, Salvador's Italian, Taco Bell, Vestavia Motor Lodge, Waffle House
W: Burger King, Chick-fil-A, El Paletio Mexican, Golden Rule BBQ, Kenny Rogers Roasters, Krispy Kreme Donuts, Krystal, Long Horn Steaks, Manderin Chinese, McDonalds, Outback Steakhouse, Quincy's Family Steakhouse, Sarris Seafood & Steaks, Shoney's, Subway, Waffle House
Lodg E: ◆ Comfort Inn, ◆ Hampton Inn (see our ad on this page), Vestavia Motor Lodge
W: Days Inn, AAA Holiday Inn (see our ad on this page)
AServ E: AAMCO Transmission, Express Lube, Royal Saturn, Vulcan Ford/Lincoln/Mercury
W: Amoco, Crown Nissan, Don Drennen Buick, Express Lube, Exxon, Firestone Tire & Auto, Goodyear Tire & Auto, Hoover Toyota, Ivan Leonard Chevrolet, Mr. Transmission, Shell
Med E: ✚ Hospital
ATM W: AM South Bank, Colonial Bank, First Alabama Bank, SouthTrust Bank
Other E: Anthony's Car Care Center (wash), Big B Drugs, Vestavia Animal Clinic, Vestribge Animal Clinic
W: Bruno's Supermarket (w/pharmacy), Burlington Coat Factory, Coin Car Wash (Big Green Clean Machine), Eckerd Drugs, Green Valley Drug Store
250 to U.S. 280, Junction I-459, Atlanta, Gadsden, Tuscaloosa
247 Valleydale Road
Gas E: BP*
W: Shell*(D)(LP)
Food E: Big Mo's Pizza, Granny's Country Kitchen, Hardee's, Tin Roof BBQ
Lodg W: AAA La Quinta Inn
AServ E: Goodyear Tire & Auto
W: River Chase Complete Auto Service
ATM E: SouthTrust Bank (Food World)
W: First National Bank
Other E: Big B Drugs, Emergency Pet Care, Food World Supermarket
246 AL 119, Cahaba Valley Road
FStop W: Speedway*(LP)
Gas W: BP*(CW)(LP) (24 Hrs), Shell*(CW)(LP)
Food W: Applebee's, Arby's, Captain D's Seafood, Cock of the Walk Restaurant, Cracker Barrel, Dairy Queen, McDonalds (Play Place), O' Charley's, Peaches n Cream, Pizza Hut, Shoney's, Starvin' Marvin (Speedway), Taco Bell, Waffle House, Wendy's
Lodg W: ◆ Best Western, ◆ Comfort Inn, ◆ Holiday Inn Express, ◆ Ramada Limited (see our ad on this page), ◆ Sleep Inn, Travelodge
AServ W: Express Lube
ATM W: Shell
Parks W: Oak Mountain State Park
242 Pelham
FStop E: Exxon*(LP)
Gas E: Chevron*(D)(LP) (24 Hrs)
AServ E: Earl's Auto Repair
TServ E: 242 Tire & Truck Service
ATM E: Chevron
238 U.S. 31 Alabaster
Gas E: BP*(D)
W: Chevron*(D), Shell*, Texaco*(CW)
Food E: Chester Fried Chicken (BP), Icecream Churn (BP)
W: Waffle House
Med W: ✚ Hospital
234 Shelby County Airport
Gas W: Chevron*(LP) (24 Hrs)
ATM W: Chevron
231 U.S. 31, Saginaw
Food W: Kathleen's Restaurant
Lodg E: Holiday Inn Express

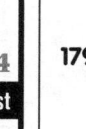

RVCamp E: Burton Camper Sales, Parts & Service
228 AL. 25, Calera
FStop E: Texaco*(CW)(LP)
Gas E: Citgo*(D)
Lodg E: Best Western, Days Inn
219 Jemison, Thorsby
TStop E: Porter's Truckstop*
FStop E: Headco Food Mart*
Gas E: Chevron* (24 Hrs)
W: Shell*(LP)
Food E: Porter's Family Restaurant
W: Icecream Churn (Shell)
TServ E: Porter's Truck & Road Service
RVCamp E: Peach Queen Campground
ATM E: Chevron
(214) Rest Area - Full Facilities, RV Dump (Both Directions)
212 AL. 145, Clanton, Lay Dam
FStop W: Headco Food Mart*
Gas W: BP* (24 Hrs)
Food W: Icecream & Yogurt Shop (Headco Food Mart), Peach Tower Restaurant
Med W: ✚ Hospital
208 Clanton, Lake Mitchell
FStop W: Exxon*
Food W: Heaton Pecan Farm, Shoney's
Lodg W: ◆ Shoney's
ATM W: Exxon
205 U.S. 31, AL 22, Clanton
FStop E: Shell*(LP)
W: BP* (24 Hrs)
Gas E: Amoco*
W: Chevron* (24 Hrs)
Food E: Country's BBQ, McDonalds (Play Place), Tink's Restaurant, Waffle House
W: Burger King, Captain D's Seafood, Hardee's, Heaton Pecan Farm, KFC, Subway, Taco Bell
Lodg E: Best Western, ◆ Holiday Inn, Rodeway Inn
W: ◆ Key West Inn
ATM W: Chevron
200 Verbena
FStop E: BP*(LP)
Gas W: Texaco*
Food W: Dairy Queen (Texaco), Stuckey's (Texaco)
RVCamp W: Holly Hill Plantation Travel Park
186 U.S. 31, Pine Level, Prattville
Gas W: BP*(D), Chevron*, Conoco*(D)
Food W: Days Inn, Icecream Churn (BP)
Lodg W: AAA Days Inn
Parks E: Confederate Memorial Park
181 AL. 14, Prattville, Wetumpka
Gas E: Chevron*(D) (24 Hrs)
W: Phillips 66*(D)(LP)
Food W: Cracker Barrel, Icecream Churn (Phillips 66)
Lodg W: Best Western
Med W: ✚ Hospital
179 Millbrook, Prattville
FStop W: Amoco*
Gas E: National Gas*
W: BP*(D)(CW), Citgo*, Exxon*, Shell*(CW) (24 Hrs) (see our ad on this page)
Food W: Hardee's, McDonalds (Play Place), Shoney's, Waffle House
Lodg W: Econolodge, ◆ Hampton Inn, ◆ Holiday Inn
RVCamp E: K & K Park & Campground (W/ RV Sales & Service)

Bold red print shows RV & Bus parking available or nearby

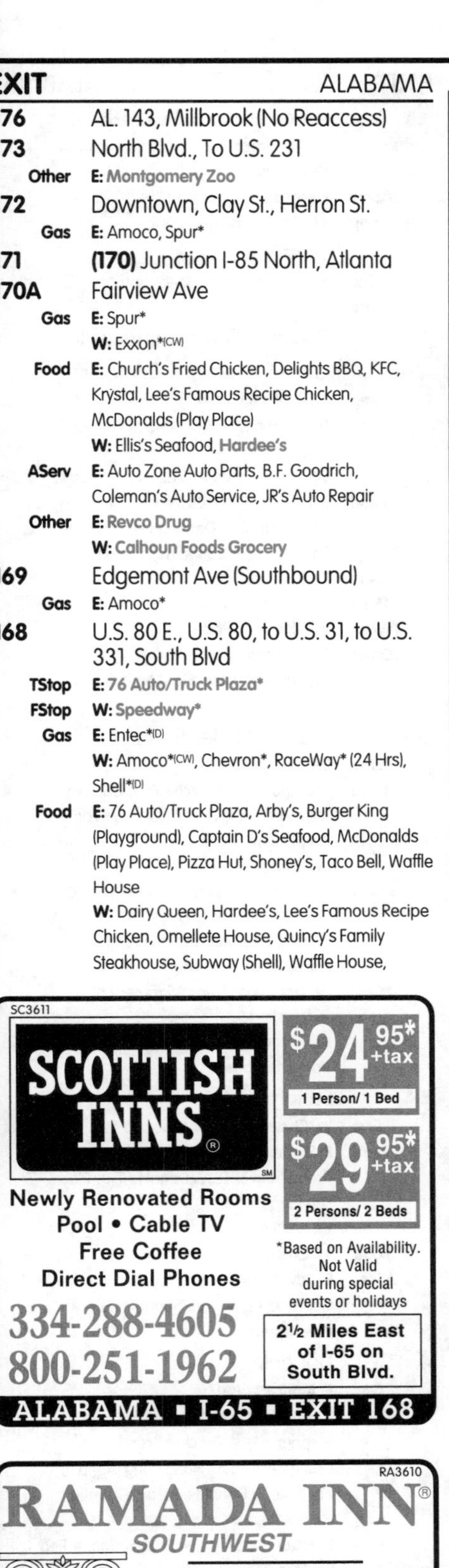

EXIT	ALABAMA
176	AL. 143, Millbrook (No Reaccess)
173	North Blvd., To U.S. 231
Other	**E:** **Montgomery Zoo**
172	Downtown, Clay St., Herron St.
Gas	**E:** Amoco, Spur*
171	**(170)** Junction I-85 North, Atlanta
170A	Fairview Ave
Gas	**E:** Spur*
	W: Exxon*(CW)
Food	**E:** Church's Fried Chicken, Delights BBQ, KFC, Krystal, Lee's Famous Recipe Chicken, McDonalds (Play Place)
	W: Ellis's Seafood, **Hardee's**
AServ	**E:** Auto Zone Auto Parts, B.F. Goodrich, Coleman's Auto Service, JR's Auto Repair
Other	**E:** **Revco Drug**
	W: **Calhoun Foods Grocery**
169	Edgemont Ave (Southbound)
Gas	**E:** Amoco*
168	U.S. 80 E., U.S. 80, to U.S. 31, to U.S. 331, South Blvd
TStop	**E:** **76 Auto/Truck Plaza***
FStop	**W:** **Speedway***
Gas	**E:** Entec*(D)
	W: Amoco*(CW), Chevron*, RaceWay* (24 Hrs), Shell*(D)
Food	**E:** 76 Auto/Truck Plaza, Arby's, Burger King (Playground), Captain D's Seafood, McDonalds (Play Place), Pizza Hut, Shoney's, Taco Bell, Waffle House
	W: Dairy Queen, Hardee's, Lee's Famous Recipe Chicken, Omellete House, Quincy's Family Steakhouse, Subway (Shell), Waffle House,

EXIT	ALABAMA

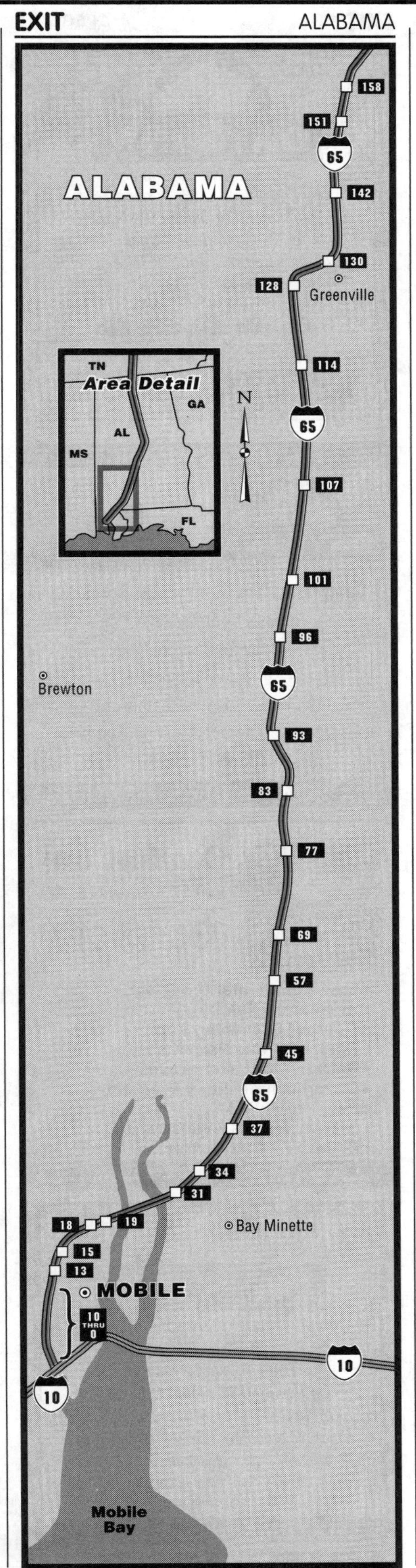

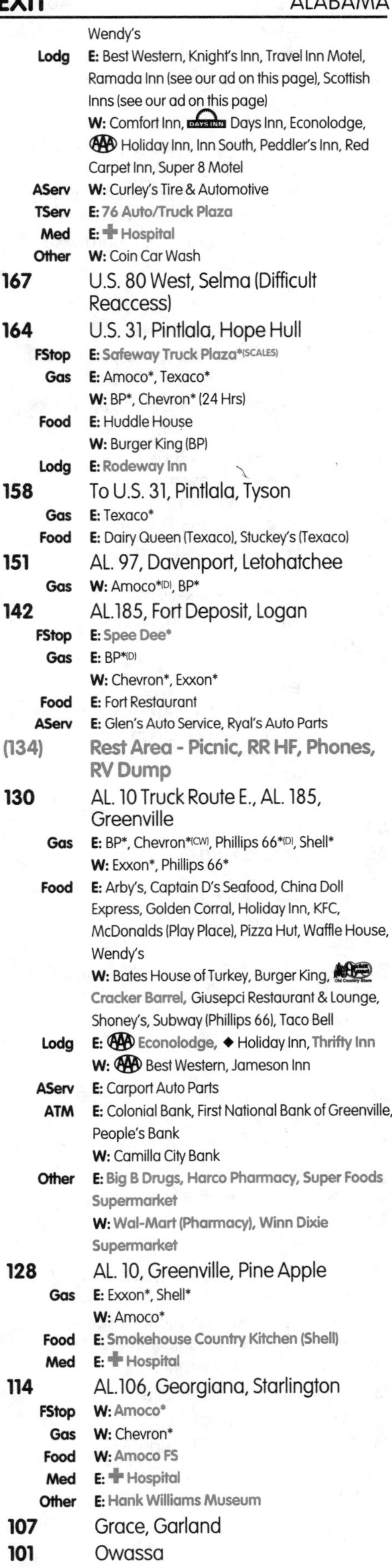

EXIT	ALABAMA
	Wendy's
Lodg	**E:** Best Western, Knight's Inn, Travel Inn Motel, Ramada Inn (see our ad on this page), Scottish Inns (see our ad on this page)
	W: Comfort Inn, Days Inn, Econolodge, (AAA) Holiday Inn, Inn South, Peddler's Inn, Red Carpet Inn, Super 8 Motel
AServ	**W:** Curley's Tire & Automotive
TServ	**E:** **76 Auto/Truck Plaza**
Med	**E:** **✚ Hospital**
Other	**W:** Coin Car Wash
167	U.S. 80 West, Selma (Difficult Reaccess)
164	U.S. 31, Pintlala, Hope Hull
FStop	**E:** **Safeway Truck Plaza***(SCALES)
Gas	**E:** Amoco*, Texaco*
	W: BP*, Chevron* (24 Hrs)
Food	**E:** Huddle House
	W: Burger King (BP)
Lodg	**E:** **Rodeway Inn**
158	To U.S. 31, Pintlala, Tyson
Gas	**E:** Texaco*
Food	**E:** Dairy Queen (Texaco), Stuckey's (Texaco)
151	AL. 97, Davenport, Letohatchee
Gas	**W:** Amoco*(D), BP*
142	AL.185, Fort Deposit, Logan
FStop	**E:** **Spee Dee***
Gas	**E:** BP*(D)
	W: Chevron*, Exxon*
Food	**E:** Fort Restaurant
AServ	**E:** Glen's Auto Service, Ryal's Auto Parts
(134)	**Rest Area - Picnic, RR HF, Phones, RV Dump**
130	AL. 10 Truck Route E., AL. 185, Greenville
Gas	**E:** BP*, Chevron*(CW), Phillips 66*(D), Shell*
	W: Exxon*, Phillips 66*
Food	**E:** Arby's, Captain D's Seafood, China Doll Express, Golden Corral, Holiday Inn, KFC, McDonalds (Play Place), Pizza Hut, Waffle House, Wendy's
	W: Bates House of Turkey, Burger King, **Cracker Barrel**, Giusepci Restaurant & Lounge, Shoney's, Subway (Phillips 66), Taco Bell
Lodg	**E:** (AAA) **Econolodge**, ◆ Holiday Inn, **Thrifty Inn**
	W: (AAA) Best Western, Jameson Inn
AServ	**E:** Carport Auto Parts
ATM	**E:** Colonial Bank, First National Bank of Greenville, People's Bank
	W: Camilla City Bank
Other	**E:** **Big B Drugs, Harco Pharmacy, Super Foods Supermarket**
	W: **Wal-Mart (Pharmacy), Winn Dixie Supermarket**
128	AL. 10, Greenville, Pine Apple
Gas	**E:** Exxon*, Shell*
	W: Amoco*
Food	**E:** **Smokehouse Country Kitchen (Shell)**
Med	**E:** **✚ Hospital**
114	AL.106, Georgiana, Starlington
FStop	**W:** **Amoco***
Gas	**W:** Chevron*
Food	**W:** **Amoco FS**
Med	**E:** **✚ Hospital**
Other	**E:** **Hank Williams Museum**
107	Grace, Garland
101	Owassa

EXIT ALABAMA

FStop E: BP*
Gas W: Exxon*

96 AL 83, Evergreen, Midway
Gas E: Chevron*(CW) (24 Hrs), Shell*
W: BP*(CW)
Food E: Floyd's, Hardee's, McDonalds (Play Place)
W: Waffle House
Lodg W: (AAA) Comfort Inn, Days Inn (see our ad on this page), The Evergreen Inn

93 U.S. 84, Evergreen, Monroeville
Gas E: Exxon*
W: BP*(D)
AServ W: BP*
RVCamp E: Pinecrest Park Camping

(89) Rest Area - RV Dump, Picnic, RR, HF, Phones (Southbound)

(85) Rest Area - RV Dump, Picnic, RR, HF, Phones (Northbound)

83 Castleberry, Lenox
Gas E: Exxon*(D)(LP), Texaco*(D)
Food E: Louise's Restaurant

77 AL 41, Brewton, Repton
FStop W: Amoco*, Shell* (24 Hrs), Texaco*
Gas E: BP*
Food W: Ranch House Restaurant

69 AL 113, Flomaton, Wallace
FStop E: Texaco*(D)
W: Conoco*
Gas E: Phillips 66*
Food W: Hot Stuff Pizza (Conoco), Hudle House

57 AL. 21, Atmore, Uriah
FStop E: Exxon*
Gas W: BP*
Lodg E: Best Western

54 County Hwy 1

45 Rabun, Perdido
Gas W: Chevron*
Food W: Harville's Cafe
AServ W: Rolins Tire Service
TServ W: Rolins Tire Service (24 Hr. Rd. Service)

37 AL. 287, Gulf Shores Parkway, Bay Minette, Rabun
FStop E: Jones Truck Stop* (Spur)
Gas E: Chevron* (24 Hrs)
Food E: Jones Truck Stop

34 AL 59, Bay Minette, Stockton
Med E: + Hospital

31 AL. 225, Stockton, Spanish Fort
FStop W: Conoco*
Parks E: Historic Blakeley State Park

22 Creola
RVCamp W: KOA Campground

19 U.S. 43, Satsuma, Creola
FStop W: Texaco*
Gas E: E-Z Serve*, Northside Bait & Tackle
Lodg E: Days Inn (see our ad on this page)
RVCamp W: I-65 Campground

15 Saraland, Citronelle
Gas W: Circle K*, Citgo*(D)

13 AL. 158, AL. 213, Eight Mile, Saraland
FStop E: Texaco*
Gas E: BP*(CW)
Food E: Shoney's, Waffle House

EXIT ALABAMA

Lodg E: ◆ Comfort Inn (see our ad on this page), Days Inn (see our ad on this page)
AServ E: Henderson's Automotive
RVCamp W: Chickasabogue Park

10 West Lee St
Gas E: Citgo*, Conoco*, Shell*, Spur*
Food E: Subway (Shell)

9 Prichard, Downtown Mobile

8AB U.S. 45, Pritchard, Citronelle
FStop W: Ride With Pride* (24 Hrs)
Gas E: Chevron* (24 Hrs), Conoco*(CW), Gas For Less*
W: Amoco*, Exxon*, RaceWay*, Spur*(D)
Food E: Church's Fried Chicken
W: Burger King, Domino's Pizza, Golden Egg (24 Hrs), McDonalds (Play Place)
AServ E: U-J Chevrolet
TServ E: Peterbilt Dealer
TWash W: Ride With Pride
Other E: Food Tiger Supermarket
W: Coin Car Wash, Eight Mille Animal Clinic

5B U.S. 98, Moffett Road
FStop W: Speedway*(D)(LP)
Food E: Burger King, Church's Fried Chicken, McDonalds (Play Place), Quincy's Family Steakhouse
W: Hardee's
AServ E: Advance Auto Parts, Big Ten Tires, Car Quest Auto Center, Mr. T's
W: Richard's Garage, Southern Auto Repair & Transmission
TServ W: Goodyear Truck Alignment & Retreading
RVCamp W: Seven Oaks RV Park (see our ad on opposite page)

5A Spring Hill Ave.
Gas W: Chevron(D), Exxon*, Texaco*(CW)
Food E: Burger King, Church's Fried Chicken, Dreamland BBQ, McDonalds (Play Place), Sub King
W: Waffle House
Lodg W: Extended Stay America
AServ E: Advance Auto Parts, Big Ten Tires, Ken's Car Tune, L & M Tires, Super-Lube
W: Chevron

4 Dauphin St
Gas E: Amoco*(D), BP*(CW)
Food E: Blimpie's Subs (Amoco), Checkers Burgers, Cracker Barrel, Godfather's Pizza, Hong Kong Island Chinese, McDonalds, Taco Bell, Waffle House, Wendy's
Lodg E: ◆ Comfort Suites, ◆ Econolodge, ◆ Red Roof Inn
AServ E: Mcconnel Pontiac, GMC, Trucks, Mr. Transmission, Super-Lube
Med W: + Spring Hill Medical Complex
ATM E: Ameri South, Amoco, Bank of Mobile, Colonial Bank
Other E: Baptist Bookstore, Del Champs Supermarket
W: Mobile Museum of Art

3AB Airport Blvd
Gas E: Chevron*(CW)

Bold red print shows RV & Bus parking available or nearby

EXIT	ALABAMA
	W: Phillips 66, Texaco*(D)(CW)
Food	**E:** Burger King, Captain D's Seafood, Hooter's, Piccadilly Cafeteria, Wendy's
	W: Chili's, Darryl's Restaurant, Deli Palace Cuisine of India, El Chico Mexican, House of Chin, Lone Star Steakhouse, Mozzarella's, Mulligan's Food Spirits, O' Charley's, Olive Garden, Sub Zone, Waffle House, Weichman's All-Seasons Restaurant
Lodg	**E:** Clarion Hotel
	W: Best Inns of America, Best Western, Bradbury Suites, Courtyard by Marriott, Days Inn, Drury Inn, Fairfield Inn, Family Inn, Holiday Inn, La Quinta Inn, Motel 6, Ramada, Shoney's Inn
AServ	**E:** Auto Air of Alabama, Goodyear Tire & Auto, Master Care Auto Service, Montgomery Ward, Sears Auto Center
	W: Midas Muffler & Brakes, Pep Boys Auto Center, Phillips 66

EXIT	ALABAMA
ATM	**E:** Compass Bank, Regions Bank, SunTrust
	W: Whitney
Other	**E:** Barnes & Noble, Bel Aire Mall, Carmike Cinema's, Springdale Mall
	W: Phar Mor Drugs, Wal-Mart (24 Hrs)
1AB	U.S. 90, Government Blvd
Gas	**E:** Chevron
Food	**E:** All Seasons, Royal Knight
	W: Bamabelles Country Restaurant (Rest Inn), Waffle House
Lodg	**E:** Howard Johnson (see our ad on this page)
	W: Rest Inn
AServ	**E:** AAMCO Transmission, Bay Chevrolet/Geo, Chevron, Grady Buick, Mazda, BMW, Hyundai, Isuzu, Gulf Coast Chrysler, Plymouth, Jeep, Eagle, Lexus of Mobile, Robinson Bros Lincoln/Mercury, Springhill Toyota, Treadwell Honda
	W: All Tune & Lube, Ken's Car Tune
0	Junction I-10 West, Mississippi

↑ALABAMA

Begin I-65

I-66 E →

EXIT	VIRGINIA
	Begin I-66
	↓VIRGINIA
1AB	Jct I-81, Roanoke, Winchester (Westbound)
6	U.S.340, U.S. 522, Winchester, Front Royal
Food	**S:** McDonalds
AServ	**N:** Shenandoah Oldsmobile, GMC
RVCamp	**N:** Fishnet Campgrounds (2.1 Miles)
	S: Gooney Creek Campground (1.4 Miles), Poe's Southfork Campground (1.4 Miles)
Parks	**S:** Shenandoah Nat'l Park
Other	**N:** Cedarville Veterinary Clinic
	S: Skyline Caverns, Skyline Dr
13	VA79, to VA55, Linden, Skyline Dr, Front Royal
Gas	**S:** Mobil*, Texaco*
Food	**S:** The Apple House(D)
Other	**N:** RV Camp KOA, Skyline Caverns
	S: Apple Mountain Lake, Apple Mtn Info Ctr (Texaco)
18	VA688, Markham
23	VA73, to U.S.17N, to VA55W, Delaplane, Paris (Westbound Reaccess)
27	VA 55E, VA 647, Marshall
Gas	**N:** Amoco(D), Chevron(D)
Food	**N:** Main Street Deli, Old Salem Family Restaurant
AServ	**N:** Chevron
28	U.S.17, Marshal, Warrenton,
Gas	**N:** Amoco*
Food	**N:** McDonalds (Amoco)
Med	**S:** ✚ Hospital
31	**(245)** VA 245, Middleburg, The Plains, Old Tavern
40	U.S.15, Haymarket, Leesburg
FStop	**S:** Citgo*(D)(LP) (Kerosene), Sheetz*(LP) (Kerosene)
RVCamp	**N:** Yogi Bear Camp Report
	S: Mountain View Campground
43AB	U.S.29, Gainesville, Warrenton
Gas	**S:** Citgo*, Exxon*(D), Mobil, Racetrac*, Texaco*(D)
Food	**S:** Domino's Pizza, Joe's Pizza & Subs, McDonalds, Subway, Wendy's
AServ	**S:** Exxon, Mobil
RVCamp	**S:** Wildwood Camping Park (1.1 Miles)
ATM	**S:** F&M Bank
Other	**S:** 7-11 Convenience Store (Kerosene), Gainesville Mobile Home & RV Camping Park, Nisson Pavillion
47	VA234, Manassas
Gas	**N:** Texaco*(D)(LP) (Kerosene)
	S: Exxon*, Racetrac*, Shell*, Sunoco*
Food	**N:** Cracker Barrel, Hodiay Inn
	S: Bertucci's Pizzaria, Bob Evans Restaurant, Bruegger's Bagel, Burger King, Checkers Burgers, China Palace, Don Pablo's, Hunan Deli, McDonalds (Playground), Pizza Hut, Pizza Movers House of Uro's, Shoney's, Subway
Lodg	**N:** ◆ Courtyard by Marriott, AAA Holiday Inn
	S: Best Western, Hampton Inn, Roadway Inn,

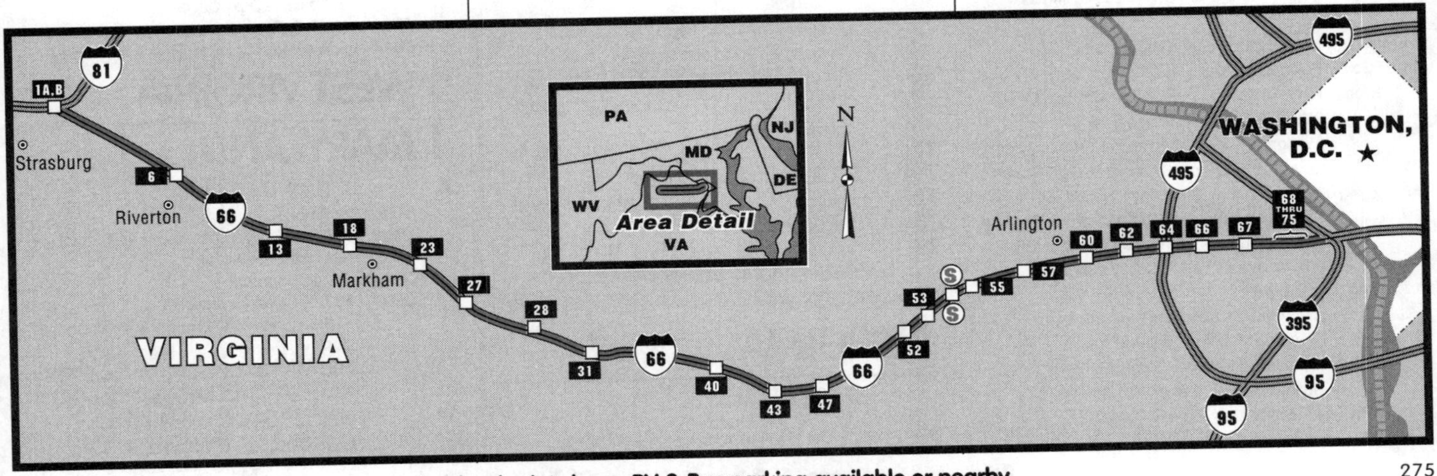

Strasburg
Riverton
Markham
Arlington
WASHINGTON, D.C. ★
VIRGINIA
PA
NJ
MD
DE
WV
VA
Area Detail
N

Bold red print shows RV & Bus parking available or nearby

EXIT	VIRGINIA
	Super 8 Motel
AServ	S: American Truck & Trailer Supply, Koons, Honda, GMC, Pep Boys Auto Center, Shell*, Sunoco
Med	S: ✚ Allmed Medical Center (3 Miles)
Parks	N: Splash Down Water Park
Other	N: Manassas Nat'l Battlefield Park
	S: Fireworks (Wal-Mart), Heba's Grocery & Deli, Wal-Mart (4.5 Miles)
(49)	VA Welcome Center, Rest Area - Picnic, Phones, RR, Pet Rest Area (Westbound)
52	U.S.29, Centreville
Gas	S: Citgo, Mobil*(D)
Food	S: BBQ Country, Bagel Bakery, Baskin Robbins (Centrewood Plaza), Brutopia Coffee Shop, Candy Bouquet, Cio's New York Pizza, El Saborlatino, Hunan Dynasty (Centrewood Plaza), Hunter Mill Deli, Ko Ko Pelli's Pizza, Long Star Steakhouse, McDonalds, Oriental Express, Pizza Hut, Ruby Tuesday, Sho Chiku Japanese Restaurant, Sino's Inn Chinese, Smokehouse Blue Steak, Ribs & Blues, The Shade Tree (Centrewood Plaza), Tippy's Taco (Cenntrewood Plaza), Viet French (Vietnamese Restaurant & Deli), Wendy's
AServ	N: Centreville Automotive (Goodyear)
	S: Centreville Tire & Auto, Citgo, Super Trak
Med	S: ✚ Urgent Medical Care
ATM	S: Chevy Chase Bank, F&M Bank, First Virginia Bank, NationsBank
Other	S: CVS Pharmacy, Centreville Eye Care, Centreville Square Animal Hospital, Centrewood Plaza, Drug Emporium, Fireworks, Giant Food & Pharmacy, Mail Boxes Etc, Optometric Eye Care Assoc, Pet Value, SFW Shoppers Grocery Store, US Post Office
53	VA28, Centreville
Other	S: Walney Visitor Ctr
55	VA7100, Fairfax Cty Pkwy, to US 29, Reston, Herndon, Springfield
Gas	N: Exxon*(CW), Mobil*(CW)
Food	N: Applebee's, Bagel Buddies, Blue Iguana, Burger King, Cantina Italia, Cooker, Crab House, Don Pablo's Mexican Kitchen, Eatery, McDonalds, Olive Garden, Pizza Hut, Red Pepper Chinese, Red Robbin, Sgt Peppers, Star Thai, TCBY Yogurt, Tony's N.Y. Pizza, WattA Rice Bowl, Wendy's
Lodg	N: ◆ Hyatt, Residence Inn
AServ	N: BJ's Tire Ctr, Mobil
Med	N: ✚ Hospital, ✚ Urgent Medical Clinic (Fairlakes Ctr)
ATM	N: George Mason Bank
Other	N: Fairlakes Ctr, Food Lion Grocery (Deli, Bakery), Pets Mart, Unlimited Mailing Service, Wal-Mart
57AB	U.S.50, Fairfax, Winchester
Gas	S: Amoco*, Mobil*, Shell*
Food	N: Nicks (Fair Oaks Shopping Ctr), Red Moon Salon & Steakhouse (Holiday Inn)
	S: Arigato Sushi, Boston Market, Bravo's Italian Cafe, Chilli's, Cookies By Design, Crystal Palace (Chinese), Fanny Mae Candy, Hunter Mill Deli, Manhattan Bagel, McDonalds, Pizza Boli's, Ruby Tuesday, Subway, The Cafe Court in Caldors, The Deli, Thursday's Restaurant, Wendy's
Lodg	N: Holiday Inn
	S: Comfort Inn, Courtyard by Marriott
AServ	N: Sears (Fair Oaks Shopping Ctr)
	S: Amoco, Auto Parts, K-Mart, Mobil, Montgomery Ward, Shell, Ted Britt Ford
ATM	N: Central Fidelity Bank
	S: First Virginia Bank, NationsBank, Southern

EXIT	VIRGINIA
	Financial Bank
Other	N: Fair Oaks Shopping Center, Pender Veterinary Clinic
	S: 7-11 Convenience Store, Bombay Food (Grocery), Caldor, Fairfax Court, Fairfax Optimetric Ctr, Fairfax Plaza, K-Mart, National Rifle Museum
60	VA123, Fairfax, Vienna
Gas	S: Crown*, Exxon*(D), Shell*
Food	N: Bob Evans Restaurant
	S: 29 Tastee Diner, Bambay Bistro Indian, Baskin Robbins, Burger King, Denny's, Dunkin Donuts, Fuddrucker's, Golden Lion Chinese Restaurant, Hooters, Java Express, K-Bob & Steak, Little Panda Chinese Express, Patriot's Cafe, Red Lobster, Roy Rogers, Topkaki, Viet Thi Vietnamese
Lodg	S: Breezeway Motel, ◆ Hampton Inn, AAA Holiday Inn, ◆ Wellesley Inn
AServ	S: Bear Oaks Dodge, Brown Lincoln Mercury, Brown's Buick, Brown's Fairfax & Nissan, Brown's Isuzu, Elm's Mazda, Fairfax Auto Service, Fairfax Honda, Fairfax Pontiac GMC, Fairfax Volvo, Jiffy Lube, Jim McKay Chevrolet, John's Auto Repair, Just Tires, Merchant's Tire & Auto, Midas Muffler, Ourisman Fairfax Toyota, Surburban Body Shop, Windshields Of America
Med	S: ✚ Emergency Medical Care
ATM	S: NationsBank, Patriot Nat'l Bank
Other	S: 7-11 Convenience Store, CVS Pharmacy, Embassy Auto Wash, Fairfax Shopping Ctr, Fed Ex Drop Box (Fairfax Shopping Ctr), Hour Eyes, Pearl Vision, Petco Supplies, Rite Aide Pharmacy
62	VA 243, Nutley St, Vienna
Gas	S: Exxon
Food	S: Domino's Pizza, Lo's Hunan, McDonalds, Pan Am Family Restaurant
AServ	S: A.K. Auto, Exxon
ATM	S: Chevy Chase Bank, First Union Bank (Pan Am Shopping Ctr)
Other	S: CVS Pharmacy, Pan Am Optical, Pan Am Shopping Ctr, Safe Way Grocery
64AB	Jct I-495, to Baltimore, Richmond (No Trucks Beyond This Exit)
66	VA7, Leesburg Pike, Falls Church
Gas	N: Exxon*, Mobil
	S: Xtra*
Food	N: Jerry's Subs & Pizza, Ledo Pizza, New Dynasty Chinese, TCBY Yogurt
	S: Haandi Indian, Long John Silvers, Pizza Hut, Roy Rogers, Starbuck's Coffee
AServ	S: Don Bayer Volvo, Precision Tune & Lube, Speedee Muffler King
ATM	S: First VA Bank
Other	N: Fresh Fields Grocery, Idylwood Shopping Plaza
	S: CVS Pharmacy, Falls Church Animal Hospital, Giant Food Store
68	West Moreland St
69	U.S.29, VA237, Washington Blvd, Lee Hwy
71	VA120, Glebe Rd, VA237, Fairfax Dr
71A	Glebe Rd (Westbound)
72	U.S. 29, Lee Hwy, Spout Run Pkwy
73	Rosslyn, Key Bridge
75	to Jct I-395, to U.S.1, VA110S, Alexandria, Pentagon
76	E Street

↑VIRGINIA

Begin I-66

EXIT	WEST VIRGINIA/MARYLAND

Begin I-68

↓WEST VIRGINIA

0	Jct I-79N, to Washington
1	U.S. 119, University Ave, Downtown
Food	N: Neighbors Food and Spirit, Ramada
Lodg	N: ◆ Comfort Inn, AAA Ramada
Other	N: Fed Ex & UPS Drop Box, Presort Plus Inc
4	WV 7, Sabraton
Gas	N: Amoco*, BP, Exxon*
Food	N: Blimpie's Subs, Classics Restaurant and Lounge, Hardee's, Hero Hut, KFC, Long John Silvers, McDonalds, Pizza Hut, Ponderosa, Rio Grande 3, Subway, Wendy's
AServ	N: Advance Auto Parts, BP, Big A Auto Parts, Campus Ford, Exxon, Full Service Car Wash, Midnight Auto, Napa Auto Parts, Sabraton Chrysler,Plymouth, Dodge
ATM	N: Bruceton Bank (Inside Kroger), One Valley Bank, WesBanco
Other	N: Coin Car Wash, Food Lion Supermarket, Kroger Supermarket (Pharmacy), Revco Drugs, U.S. Post Office, Wash House Coin Laundry
7	WV705, Airport, Pier Pont Rd
Gas	S: 76*
Food	S: Crestpoint 76 Food Mart, Tibeiro's Pasta-Seafood-Steaks
Med	N: ✚ Hospital
	S: ✚ Cheat Lake Urgent Care
ATM	N: Huntington Bank
Other	N: Airport
	S: Coin Car Wash (76)
10	WV 857, Cheat Lake, Fairchance Road (7% Grade On West Ramp)
FStop	N: BP*(LP)
Food	N: Ruby & Ketchys Restaurant
	S: Stone Crab Inn Seafood
ATM	N: Bruceton Bank
Other	S: UPS Drop Box (Near the Stone Crab Restaurant)
15	WV 73/12, Coopers Rock
RVCamp	N: Chestnut Ridge Park (2.1 Miles), Sand Springs Camping Area (2.5 Miles)
Parks	S: Coopers Rock State Forest
23	WV 26, Bruceton Mills
TStop	N: Little Sandy's Truckstop*
FStop	N: BP*
Food	N: Country Fixins, Little Sandy's Restaurant, Tasty Queen, The Mill Place
AServ	N: Mitchel Auto Parts (Coin Car Wash, and Fuel.), Murphy's Garage
RVCamp	N: Glade Farm Campground (6 Miles)
ATM	N: Bruceton Bank, Little Sandy's Truckstop
Other	N: Bruceton Mills Grocery, The Mill Place Hardware
29	WV 5, Hazelton Road
Gas	N: Mobil*(D)
	S: Casteel's*
Food	N: Pine Run Deli (Mobil)
	S: Casteel's Dairy Queen
TServ	N: Mel's Truck Sales
RVCamp	S: Big Bear Lake Campgrounds (3 Miles)

↑WEST VIRGINIA

↓MARYLAND

4	MD 42, Friendsville
Gas	N: Amoco
Food	N: Old Mill Restaurant, Tapers Restaurant and Sunshine Pizza Parlor
RVCamp	S: Campground
Other	N: Trustworthy Hardware
	S: Maryland History Heritage Museum and Library
(6)	MD Welcome Center - RR, Phones, Picnic, Tourist Info (Eastbound)
14	U.S. 219 South, U.S. 40 West, Oakland, Uniontown PA

EXIT	MARYLAND
TStop	N: Keyser Ridge TS
FStop	N: Citgo*(D)
Food	N: Keyser Ridge Truck Stop, McDonalds, Silver Fox Pizza & Ice Cream
Lodg	N: Keyser Ridge Hotel
TServ	N: Menges Trucking and Repair
Parks	N: Deep Creek Lake Recreational Area State Park Ski Area
19	**MD419, Grantsville, Swanton**
Gas	N: Exxon*(D), Mobil*
Food	N: Hey Pizza!
AServ	N: Guy's Tire Service, Sumners Auto Parts Inc.
ATM	N: 1st United Nat'l Bank
Other	N: Beaches Pharmacy, Buckel's Laundromat, HHH Hardware, U.S. Post Office
22	**U.S. 219 North, to U.S. 40 Alt**
TStop	N: BP(SCALES)
FStop	N: Exxon(D)
Gas	S: Amoco*(CW)(LP)
Food	N: BP Restaurant, Burger King, Chester Fried Chicken, Hilltop Delight Restaurant, Subway, Yoders Family Restaurant
Lodg	N: Little Meadows Motel Lake and Campgrounds S: ◆ Holiday Inn
AServ	N: Exxon, Napa Auto Parts
Parks	S: New Germany State Park, Savage River State Forest
Other	N: Food Lend Supermarket, Grantsville Plaza, Hilltop Fruit Market, Pets 4 U, Rite Aide Pharmacy
24	**Lower New Germany Rd, US 40 Alternate**
Food	S: The Avilton Restaurant and Terrace
Parks	S: Big Run State Park, New Germany State Park/Forest, Savage River State Forest
29	**MD 546, to U.S. 40 Alt, Finzel**
RVCamp	N: Mason Dixon Campground (4 Miles)
(31)	**Weight Station (Eastbound)**
33	**Midlothian Road, Frostburg (Dan's Mountain State Park)**
AServ	S: Knieriem Tires
34	**MD 36, Westernport, Frostburg**
Gas	N: Amoco, Sheetz*(LP), Texaco*
Food	N: Comfort Inn, Little Caesars Pizza, McDonalds, Peking House Chinese Restaurant
Lodg	N: ◆ Comfort Inn
AServ	N: Amoco, Big A Auto Parts, Langley's Auto Service
ATM	N: F&M Bank, First Federal Bank, First United Bank, Mercantile Fidelity Bank
Parks	S: Dans Mountain State Park
Other	N: Cinemas123, Food Lion, Mor For Less Food Stores, Nailor's True Value Hardware, Rite Aide Pharmacy, The Medicine Shoppe Pharmacy, Tourist Information
39	**U.S. 40 Alt., La Vale Rd**
40	**Truck Rt 220 South, La Vale (E), Vocke Road (W) (Difficult Reaccess Westbound)**
Gas	N: Amoco(CW), BP*(LP), Citgo*(D), Exxon, Mobil*, Quick and Easy
Food	N: Arby's, Bob Evans Restaurant, Burger King, D'Atri Restaurant, Dairy Queen, Denny's, Gardeners Candies, Gehauf's Restaurant (Best Western), KFC, Kenny Rogers Roasters, Little Caesars Pizza, Long John Silvers, McDonalds, Pizza Hut, Roy Rogers, Sub Express, The Melting Pot Restaurant, Wendy's S: Chi Chi's Mexican, Dragon Chinese, Ponderosa, Roy Rogers, Western Sizzlin'
Lodg	N: Best Western, Scottish Inns, Super 8 Motel
AServ	N: Amoco, Citgo, Exxon, G&G Tire Service, Giant Auto Parts, Midas Muffler & Brakes, Mobil S: K-mart, Kelly Tire, Sears, Trak Auto
Med	N: ✚ Hospital, ✚ Sports Medicine Center
ATM	N: American Trust Bank, First Peoples Credit Union, Home Federal Bank (County Market), Mobil, State Employ's Credit Union of Maryland S: American Trust Bank, First Federal Savings Bank, Martin's Grocery
Parks	N: First Toll Gate House
Other	N: Box Office Video(CW), Braddock Square, Coin Laundry, County Market Grocery, Diamond Shine Automatic Car Wash, La Vale Dry Cleaners, La Vale Pharmacy, La Vale Plaza Shopping Center, Lowes, Naylor's True Value Hardware, One Minute Car Wash, Revco Drugs, State Patrol Post, The Bowler Bowling Lanes, The Mail Room S: AMC 6 Theaters, Country Club Mall, K-Mart, Martin's Grocery (Pharmacy), Thrift Drug (Pharmacy), Vision Center, Wal-Mart (Pharmacy)
41	**Seton Dr, to MD49 (Westbound, Difficult Eastbound Reaccess)**
Med	N: ✚ Sacred Heart Hospital
42	**U.S. 220 South, Greene St, Ridgedale Ave (No Trucks Or Buses)**
Gas	N: Exxon
AServ	N: Exxon
43A	**Johnsoon St , to WV 28Alt, Ridgeley**
Gas	N: Citgo, Sheetz*(LP)
Food	N: Fox's Pizza
AServ	N: Citgo
Other	N: Shopping Mall, Tourist Information
43B	**MD 51, Industrial Blvd, Airport**
Gas	S: Amoco, Citgo*(CW)
Food	S: Roy Rogers, Taco Bell, Wendy's
Lodg	N: Holiday Inn
AServ	S: Amoco, Citgo
ATM	S: America Trust Bank, First United Bank
43C	**Downtown**
Food	S: Dunkin Donuts, Fresh Stuffed Pitas
43D	**Maryland Ave.**
Gas	S: Texaco(D)
Food	S: Subway
Med	S: ✚ Hospital
Other	S: US Post Office
44	**U.S. 40 Alt, Baltimore Ave**
45	**Hillcrest Dr**
FStop	S: BP*(LP)
46	**MD 144, U.S. 220 North, Bedford (Reaccess Via Exit 47)**
Food	N: DaVinci's Pizzaria S: J.B.'s Steak Cellar, L'Osteria, Losterja, Masons Barn
Lodg	N: Cumberland Motel
AServ	N: Premiere Car Wash
TServ	N: CHP Truck Parts
Other	N: Crossroads Animal Hospital
47	**MD 144, Old National Pike**
50	**Pleasant Valley Road**
RVCamp	N: Campground
Parks	N: Rocky Gap State Park
Other	N: Tourist Information, Veterans Cemetery
52	**MD 144 East, National Pike**
56	**MD 144, National Pike, Flintstone**
Gas	S: Citgo*, VanMeters*
Food	S: Jeff's One Stop Pizza To Go
AServ	S: Jeff's One Stop Auto Repair (Citgo)* (ATM), Wrecker Service
Other	S: Helmick's Grocery C-Store, U.S. Post Office
62	**U.S. 40, Scenic Rt, Fifteen Mile Creek Rd**
Other	N: National Reserve Police
64	**M. V. Smith Road**
Parks	S: Green Ridge State Forest Hdqtrs
Other	S: Scenic Overlook (Phone)
68	**Orleans Road**
Gas	N: Citgo*
AServ	N: Bellegrove Auto Service
RVCamp	S: Campground
72	**U.S. 40 Scenic, High Germany Road, Swain Road**
FStop	S: Exxon*(D)
74	**U.S. 40 East Scenic, Mtn Road (Eastbound Reaccess Via US 40)**
(74)	**Sideling Hills Exhibit Center - RR, Phones, Picnic, Vending**
(75)	**Runaway Truck Ramp**
77	**U.S. 40 Scenic, MD 144, Woodmont Rd**
RVCamp	S: Campground
82A	**I-70 West Exit 1B, US 522, South Hancock Winchester**
Gas	S: Gary's, Lowest Gas, Sheetz*
Food	S: Barnyards Ice Cream, Fox's Pizza, Pizza Hut, The Crab Den Seafood Restaurant
Lodg	S: Econolodge
AServ	S: Hancock Auto Supply
RVCamp	S: Happy Hills Campground
Other	S: Home Center Pharmacy, Save-A-Lot Food Stores
82B	**Us 40 West**
82C	**I-70 East Exit**
82ABC	**Jct I-70, U.S. 522, U.S. 40 East, Breezewood, Winchester**

↑MARYLAND

Begin I-68

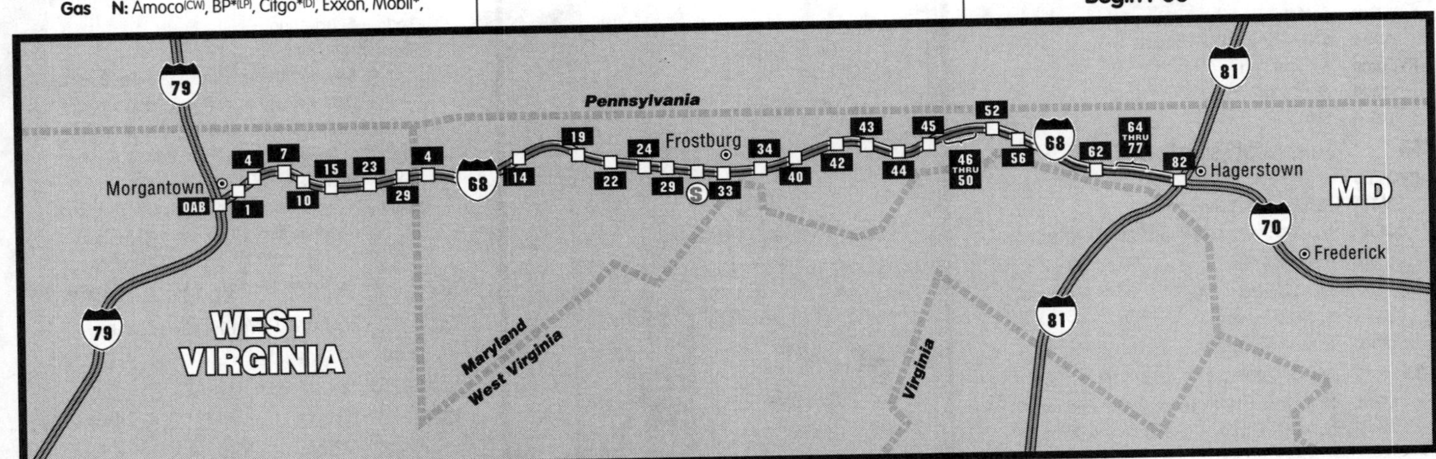

Bold red print shows RV & Bus parking available or nearby

EXIT MICHIGAN

Begin I-69

↓MICHIGAN

199 Business Loop 69, Port Huron

198 Junction I-94W to Detroit, I-94 E to Canada (I-94 W Is Right Lane, I-94 E Is Left Lane)

196 Wadhams Road
- **FStop** N: By-Lo
- **Gas** N: Marathon*, Shell*
- **Food** N: French's Bakery & Cafe, Hungry Howie's Pizza & Subs, McDonalds, Wadhams Country Kitchen, Wadhams House of Pizza
- **RVCamp** N: KOA Campground
- **ATM** N: IGA Supermarket, Old Kent, Speedy Q
- **Other** N: Coin Laundry, IGA Store, Speedy Q Market, Wadhams Pharmacy

194 Barth Road
- **RVCamp** N: Fort Trodd Campground (Good Sam Park)

189 Wales Center Road

184 MI 19, Sandusky, Richmond
- **TStop** N: 76 Auto/Truck Plaza*(LP)(SCALES) (RV Dump)
- **FStop** S: Marathon*(LP) (Kerosene)
- **Food** N: Louie's Restaurant (76)
- **TServ** N: 76
- **ATM** N: 76
  S: Marathon

180 Riley Center Road
- **RVCamp** N: Campground

176 Capac

(174) Rest Area - RR, Phones, Picnic (Southbound)

168 MI 53, Imlay City, Almont
- **FStop** N: Amoco*(CW), Total
- **Food** N: Big Boy, Burger King, Dairy Queen, Little Caesars Pizza, McDonalds, Taco Bell (Total), Wah Wong Chinese, Wendy's
- **Lodg** N: Super 8 Motel
- **AServ** N: Chrysler Auto Dealer, Ford Dealership, GM Dealer, Pennzoil Lube, Total
- **Other** N: Coin Laundry, IGA Supermarket (Pharmacy), Newark Car Wash

163 Lake Pleasant Road

(161) Rest Area - RR, Phones, Picnic (Northbound)

159 Wilder Road

155 MI 24, Pontiac, Lapeer
- **Gas** S: Mobil*(D)
- **AServ** S: Pontiac Buick GM Dealer
- **RVCamp** N: Camping
- **Med** N: ✚ Hospital
- **ATM** S: Mobil

153 Lake Nepessing Road
- **RVCamp** S: Camping
- **ATM** N: Farm Credit Services
- **Other** N: Lake Nepessing Road Convenience Store, Sheriff's Dept

149 Elba Road
- **Food** S: Woody's Pizza
- **Other** S: Country Market

145 MI 15, Davison, Clarkston
- **Gas** N: Marathon* (Kerosene), Shell*, Speedway*
- **Food** N: Arby's, Big Boy, Big John Steak & Onion, Burger King, Chee Kong Chinese, Country Boy, Country Jim's, Cruiser's Drive-Thru, Hungry Howie's Pizza & Subs, Italia Gardens, KFC, Little Caesars Pizza, McDonalds, Sero's, Taco Bell
- **Lodg** N: AAA Comfort Inn
- **AServ** N: Davidson Automotive, Jim Waldron Pontiac Buick GMC Trucks, Minute Lube, Ross Automotive
- **Med** S: ✚ Hospital
- **ATM** N: Citizen's Bank, D&N Bank, NBD Bank
- **Other** N: 7-11 Convenience Store, Car Wash, Coin Laundry, Kessel Grocery, Rite Aide Pharmacy, Robo Car Wash

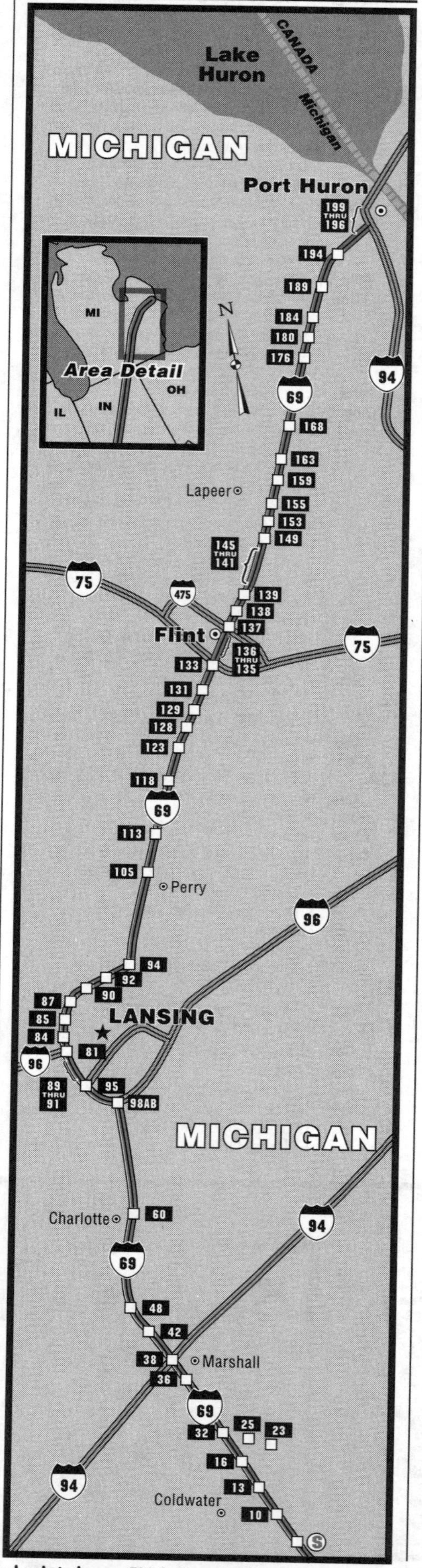

143 Irish Road
- **Gas** S: Marathon*
- **Food** S: Dunkin Donuts, McDonalds (Marathon)
- **ATM** S: 7-11 Convenience Store
- **Other** S: 7-11 Convenience Store, Coin Laundry (Marathon)

141 Belsay Road
- **Gas** N: Shell*
- **Food** N: Little Caesars Pizza (K-Mart), McDonalds, McDonalds (Wal-Mart), Taco Bell
  S: Bootlegger's Bar & Grill
- **AServ** N: K-Mart, Shell
- **Med** S: ✚ Genesy's Medical Bldg
- **Other** N: K-Mart (Pharmacy), Kessel Grocery, Wal-Mart (Pharmacy)
  S: Sparkle Buggy Car Wash

139 Center Road, Flint
- **Gas** N: Total*(D)(LP)
- **Food** N: Applebee's, Big Boy, Boston Market, Halo Burger, Moy Kong Chinese, Old Country Buffet, Olympic Grill, Ponderosa, Subway, Travelodge, Wendy's
  S: Bob Evans Restaurant, China One, Walli's Restaurant
- **Lodg** N: Travelodge
  S: Super 8 Motel, Walli's Motor Lodge
- **AServ** N: Auto Works, Goodyear Tire & Auto
- **Other** N: Car Wash, Courtland Shopping Center, Home Depot, Office Max, VG's Grocery
  S: Farmer Jack Supermarket (Pharmacy), SVS Vision Optical, Staples Office Superstore, Target Department Store

138 MI 54, Dort Highway
- **Gas** N: Amoco, Sunoco
  S: Speedway*, Sunoco
- **Food** N: Sizzler Grill & Bar
  S: Big Boy, Bill Knapp's Restaurant, Empress of China, Hot 'N Now Hamburgers, McDonalds, Subway, Taco Bell
- **Lodg** S: Travel Inn
- **AServ** N: Amoco
  S: Midas Muffler & Brakes, Pontiac Buick GM Dealer, Ross Oil Change & Car Wash, Star Auto Sales & Service, Sunoco, Tuffy Auto Center
- **TServ** S: American Body Truck Parts & Service
- **Med** N: ✚ Hospital
- **Other** N: Am Track Train Station, Eascor Animal Hospital, Rite Aide Pharmacy
  S: E-Z Food Stop*, U-Haul Center(LP)

137 Junction I-475, Saginaw, Detroit

136 Saginaw Street, Downtown
- **Med** N: ✚ Hospital

135 Hammerberg Road

133 Junction I-75, U.S. 23, Saginaw, Ann Arbor

Bold red print shows RV & Bus parking available or nearby

EXIT		MICHIGAN
131		MI 121, Bristol Road
	Gas	**N:** Total(D)
	Food	**N:** Burger King, Halo Burger, Long John Silvers, Subway, Valley Family Dining
	AServ	**N:** Goodyear Tire & Auto, Valley Tire & Service
	ATM	**N:** Chemical Bank, Michigan National Bank
	Other	**S: Airport**
129		Miller Road
	Gas	**S:** Marathon*
	Food	**S:** Arby's, McDonalds
	AServ	**S:** Marathon
128		Morrish Road
	Gas	**S:** Amoco*(CW)
(126)		**Rest Area - RR, Phones, Picnic, Vending (Northbound)**
123		MI 13, Saginaw
	Other	**N: Arnie's Convenient Store**
118		MI 71, Corunna, Durand
	Gas	**N:** Mobil*
		S: Shell*(D), Total*(LP)
	Food	**N:** London's Bakery Shoppe
		S: Crossroads, Great Wall Chinese, Hardee's, McDonalds, Subway
	Lodg	**S:** Crossroads Inn
	AServ	**N:** Mobil
		S: Duran Muffler/Brakes
	RVCamp	**S: Rainbow RV Sales Parts & Service**
	Other	**N: Andy's Food Shop***
		S: Ace Hardware, Carter's Food Center, Coin Laundry, Rite Aide Pharmacy, Train Station
113		Bancroft
	RVCamp	**S: Camping**
105		MI 52, Perry, Owosso
	TStop	**S: Total*(LP) (Kerosene)**
	Gas	**S:** Citgo*(CW), Shell*(CW)
	Food	**S: Burger King,** Cafe Sports, Dunkin Donuts (Citgo), **Joe's (Total),** Taco Bell (Shell), West Side Deli
	Lodg	**S:** Heb's Inn
	RVCamp	**S: Camping**
	ATM	**S:** Joe's
	Other	**N: Perry Animal Clinic**
98		Woodbury Road, Laingsburg
	RVCamp	**S: Moon Lake Campground**
94		East Lansing
	Gas	**S:** Speedway*(LP) (Kerosene)
	Parks	**S: East Lansing Park**
92		Webster Road, Bath
90		U.S. 127 South, East Lansing, Jackson (Westbound)
89		U.S. 127 South, East Lansing, Jackson (Eastbound)
87		U.S. 27, Clare, Lansing
	FStop	**N: Total*(LP)**
	Gas	**N:** Standard*
		S: Speedway*(LP)
	Food	**N:** Arby's, Bob Evans Restaurant, Burger King, Dunkin Donuts (Standard Gas), Little Caesars Pizza, McDonalds, Subway
	Lodg	**N:** ◆ Sleep Inn
	AServ	**N:** GM Auto Dealership
		S: Pennzoil Lube
	RVCamp	**N: RV Sales & Service**
	Med	**N: ✚ Delta Medical Center**
	ATM	**N:** Community First Bank, Michigan National Bank
	Other	**N: Coin Laundry, L & L Food Center**
		S: One Stop Wash & Lube, **Outlet Mall**
85		DeWitt
84		Airport Road
(82)		69 North becomes 69 East
81		Junction I-96 West, Grand Rapids (Southbound)
91		Junction North I-69, U.S. 27, Flint, Clare (Westbound)
93AB		Junction I-69 Business Spur, MI 43, Saginaw Highway
	Gas	**N:** Shell*(CW), Total*(D)
		S: Amoco*(CW)
	Food	**N:** Burger King, Denny's, Hoffman House (BestWestern), McDonalds, T.G.I. Friday's (Holiday Inn)
		S: Cracker Barrel
	Lodg	**N:** Best Western, Fairfield Inn, Hampton Inn, Holiday Inn, Motel 6, Quality Inn, Quality Suites, Red Roof Inn
	AServ	**S:** Regency Olds/ GMC Trucks/ Mazda
	Med	**N: ✚ Hospital, ✚ Westside Medical Center**
	ATM	**N:** Michigan National Bank
	Other	**S: TFC-Farm, Home, Auto**
95		JCT I-496, Downtown Lansing
98AB		Junction I-69 South
72		I-96 East, Detroit
70		Lansing Road
	Food	**E: Mary's Truckstop,** Tom
	Other	**E: State Patrol Post**
66		MI 100, Potterville, Grand Ledge
	Gas	**W:** Amoco*(CW), BP*, Citgo*(D)
	Food	**W:** Charlie's Bar & Grill, McDonalds, Moms Restaurant
	AServ	**W:** Amoco, Auto Value Parts Store
	ATM	**W:** Independent Bank
	Other	**W:** Mr. Clean Car Wash, **Wildern's Pharmacy**
61		Bus Loop 69, Lansing Road, Charlotte
	FStop	**E: Mobil***
	Gas	**E:** Total(D) (Kerosene)
		W: Bay*(D) (Kerosene), Speedway*(LP) (Kerosene)
	Food	**E: Pizza & Deli (Mobil)**
		W: Arby's, Arctic Creamery, Big Boy, Burger King, El Sombrerro, Hot N' Now Hamburgers, KFC, Little Caesars Pizza, Mancino's Pizza, McDonalds, Pizza Hut, Quality*(CW), Subway (Bay), Taco Bell, Tasty Twist, Wendy's
	Lodg	**E:** Sundown Motel
	AServ	**E:** Beacon Chrysler Dodge, Team One Chevrolet Olds, Tiere City
		W: Candy Ford, GM Dealer, Geldof Tire & Auto, Moores Readiator Repair
	Med	**W: ✚ Hospital**
	ATM	**E:** Lafcu
		W: First of America, Independent Bank, Quality
	Other	**E: Carter's Grocery, K-Mart (Pharmacy), RX Optical, Sheriff's Dept, Wal-Mart (Pharmacy)**
		W: Charlotte Car Wash, **Charlotte Vet Hospital, Coast to Coast Hardware, Coin Laundry (Quality),** Out Shiner Car Wash, **Pet Corner, Sav-A-Lot Grocery**
60		MI 50, Charlotte, Eaton Rapids
	Lodg	**W:** (AAA) **Super 8 Motel**
	Med	**W: ✚ Hospital**
	Other	**W: Eaton County Fair Grounds**
57		I-69 Business Loop, Cochran Road
	RVCamp	**E: Campground**
51		Ainger Road, Olivet
	RVCamp	**E: Camping**
48		MI 78, Bellevue
	RVCamp	**E: Camping**
42		N Drive North
	Gas	**W:** Citgo*(D)
	AServ	**W:** J & B Automotive Repair
(41)		**Rest Area - RR, Phones, Picnic (Southbound)**
38		Junction I-94, Chicago, Detroit
36		Bus Loop 94, Michigan Ave, Marshall
	Gas	**E:** Clark*, Mobil*(D) (Kerosene), Shell*(CW)
	Food	**E:** Arby's, Big Boy, Burger King, Cinnamon Street Bakery, Coffee Beanery (Shell), Hot Stuff Pizza, Ice Cream Dream, **Little Caesars Pizza (K-Mart),** Little Caesars Pizza, McDonalds, Pizza Hut, Subway (Shell), Taco Bell, Wendy's, Yin Hi Chinese
	Lodg	**E:** (AAA) AmeriHost Inn
		W: (AAA) Arbor Inn, Bear Camp Inn, Imperial Motel
	AServ	**E:** Caron Chevrolet Olds, Pennzoil Lube
		W: Boshears Ford, Kool Classic Chrysler Plymouth
	Med	**W: ✚ Hospital**
	ATM	**E:** Marshall Savings Bank, Shell
	Other	**E: Big Value Grocery, K-Mart (Pharmacy),** Mission Car Wash, **Rite Aide Pharmacy,** Westside Car Wash
32		F Drive South
	RVCamp	**E: Camping**
(28)		**Rest Area - RR, Phones, Picnic, Travel Directory (Northbound)**
25		MI 60, Three Rivers, Jackson
	TStop	**E: Citgo Truck/Auto Plaza*(SCALES)**
	FStop	**E: 76*(LP), BP* (Kerosene)**
	Food	**E:** Dairy King, **McDonalds, Subway (Citgo)**
	RVCamp	**E: Camping**
	ATM	**E: BP, Citgo Truck Plaza**
	Other	**E: Acme Propane**
23		Tekonsha
	RVCamp	**W: Camping**
16		Jonesville Road
	RVCamp	**W: Camping**
13		U.S. 12, Bus Loop 69, Quincy, Coldwater
	Gas	**E:** Speedway*(D)
		W: 76*, Amoco*(CW), Speedway*, Sunoco*(D)(CW) (Kerosene)
	Food	**E:** Bob Evans Restaurant, China One, Mancino's
		W: Arby's, Burger King, Charlie's, Cold Water Garden Restaurant, Elias Brothers Restaurant, Hot 'N Now Hamburgers, KFC, Little Caesars Pizza, McDonalds, Ponderosa, TCBY Yogurt, Taco Bell, Wendy's
	Lodg	**E:** Little King Motel
		W: Cadet Motor Inn, (AAA) **Quality Inn,** (AAA) **Super 8 Motel**
	AServ	**E:** Fast Lube, Quaker State Lube
		W: Auto Works Auto Parts, Max Larsen Ford Lincoln Mercury, Midas Muffler & Brakes

Bold red print shows RV & Bus parking available or nearby

EXIT MICHIGAN/INDIANA

TServ W: Max Larsen Truck Service
Med W: ✚ Hospital
ATM E: Century Bank
W: Coldwater Banking Center, First of America, Southern Michigan
Other E: Car Wash, Farmer Jack Grocery (Pharmacy), Sav-A-Lot Grocery, Wal-Mart (Pharmacy)
W: K-Mart, Sheriff's Dept, State Patrol Post

10 Business Loop 69, Coldwater

(8) Weigh Station (Northbound)

(6) Welcome Center - RR, Phones, Picnic, Vending (Northbound)

3 Copeland Road, Kinderhook
RVCamp W: Camping

↑MICHIGAN
↓INDIANA

157 Lake George Road, Jamestown, Orland
TStop E: 76 Auto/Truck Plaza*(LP)(SCALES)
W: Shell*(SCALES)
FStop W: Speedway*(LP)(SCALES)
Food E: Baker Street Family Restaurant (76), Subway (76)
W: Hardee's (Speedway), Red Arrow
Lodg E: Lake George Inn (Behind 76)
TServ E: 76
W: Gulick Volvo Truck Dealership

156 Junction I-80/90 (toll), Chicago, Toledo

154 IN 127 to IN 120, Orland, Fremont
Food E: Holiday Inn
Lodg E: Budgeteer Motor Inn, AAA Holiday Inn
W: Pokagon Motel
RVCamp E: Oak Hill Camping, Yogi Bear Camp Report
Parks W: Pokagon State Park

150 County Road. 200 West, Lake James, Crooked Lake
Gas W: Pennzoil Lube*(D)(LP), Shell* (Kerosene)
Food W: Ice Cream Cove
RVCamp W: Campground & Beach

148 U.S. 20, Angola, Lagrange
FStop E: Speedway*
Gas E: Amoco*(LP)
Food E: Subway (Amoco)
W: Stardust (Best Western)
Lodg W: AAA Best Western
RVCamp E: Campground
Med E: ✚ Hospital
ATM E: Amoco

(144) Rest Area - RR, Phones, Picnic (Southbound)

140 IN 4, Hamilton, Ashley, Hudson
Gas W: Marathon*

134 U.S. 6, Kendallville, Waterloo
FStop W: Marathon*(LP)
Lodg W: DAYS INN Days Inn (see our ad on this page)
ATM W: Marathon

129 IN 8, Auburn, Garrett
Gas E: Amoco, BP(D)(LP), Gas America*(LP), Marathon(D), Shell(CW)
Food E: Ambrosia, Arby's, Bob Evans Restaurant, Burger King, Dairy Queen, Dunkin Donuts, Fazoli's Italian Food, KFC, McDonalds, Ponderosa Steakhouse, Richard's Restaurant, Subway, TCBY Yogurt, Taco Bell, Wendy's
Lodg E: Auburn Bed & Breakfast, ◆ Country Hearth Inn, ◆ Holiday Inn Express, Starlite Motel
AServ E: Auburn Motor Sales Ford, Auto Zone Auto Parts, Helmkamp Chrysler Plymouth Dodge, Marathon
RVCamp E: Ben Davis RV Sales

EXIT INDIANA

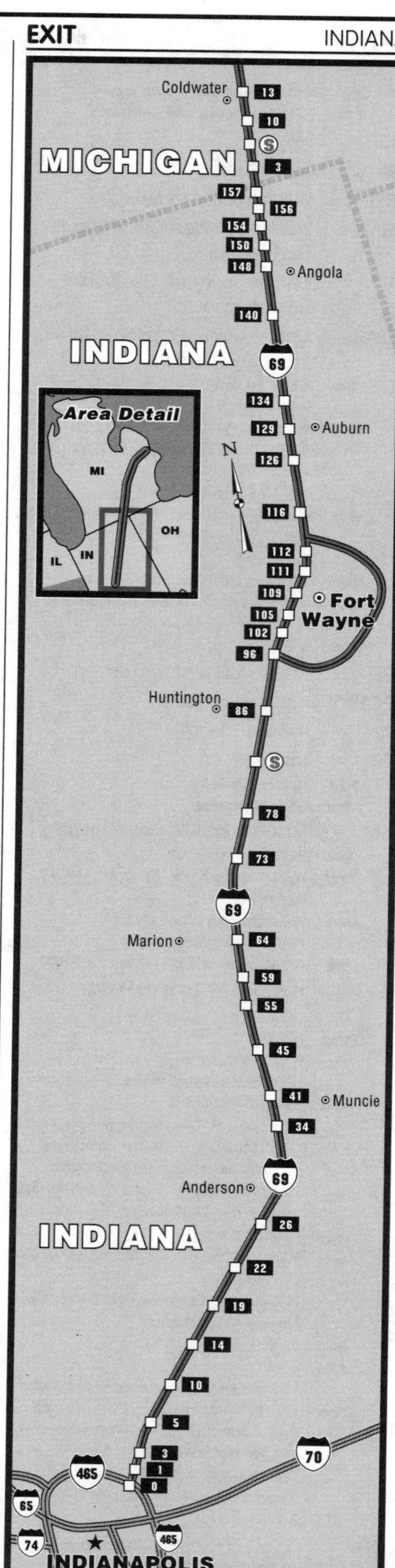

EXIT INDIANA

ATM E: Gas America
Other E: Ace Hardware, Classic Car Wash, Kroger Supermarket, Save-A-Lot Food Stores, Wal-Mart (Pharmacy)

126 County Road. 11A, Garrett, Auburn
RVCamp W: Campground
Other E: Cruse Auction Park

(124) Rest Area - RR, Picnic, Phones, Vending (Both Directions)

116 IN 1 North, Dupont Road
Gas W: BP*(D)(LP), Speedway*(LP) (Kerosene)
Food W: Bob Evans Restaurant

115 Jct I-469, US 30 E

112AB Coldwater Road
Gas E: Amoco*(CW), Marathon
Food E: Arby's, Bill Knapp's, Chi-Chi's Mexican, Cork & Cleaver, Don Hall's Factory Steakhouse, Hunan Chinese, Papa John's, Stuckey Brothers Restaurant, Taco Cabana, Wendy's, Zesto Ice Cream
Lodg E: Marriott, Sumner Suites
AServ E: Marathon
Other E: Red River Steak & BBQ, Wal-Mart

111AB U.S. 27 South, IN 3 North, Fort Wayne
Gas E: Shell*(CW)
W: Amoco, BP*(D)(LP), Meijer*(LP) (Kerosene)
Food E: Dairy Queen, Denny's
W: Burger King, Cracker Barrel, Don's Guest House, Golden China, KFC, McDonalds
Lodg E: DAYS INN Days Inn, Residence Inn
W: Budgetel Inn, Courtyard by Marriott, Dollar Inn, Don Hall's Guest House, Economy Inn, Fairfield Inn, Hampton Inn & Suites, Lee's Inn, Signature Inn, Studio Plus Stay Motel
AServ E: Don Ayres Honda Pontiac, Infiniti of Fort Wayne
Other W: Builders Square, Meijer Supermarket, Mike's Express Car Wash

109AB U.S. 24, U.S. 30, U.S. 33, North, Fort Wayne
TStop E: Fort Wayne Truck Center*(SCALES)
Food E: Azar's Big Boy, McDonalds, Subway (Fort Wayne Truck Plaza), The Point Restaurant
Lodg E: Best Inns of America, Comfort Inn, Dollar Inn, Knight's Inn, Motel 6, Northwest Inn, Plaza Inn, Red Roof Inn
TServ E: Barrys Truck Repair & Body Shop, Cummins Diesel, Fort Wayne Truck Plaza, Goodyear Tire & Auto, Kenworth, Wise International
TWash E: Ray's Truck Wash
Med E: ✚ Hospital
ATM E: Fort Wayne Truck Plaza

105AB IN 14 West, Fort Wayne, Whitley
Gas E: Meijer*(LP), Shell*(CW), Speedway*(LP)
Food E: Ramada Inn, Steak & Shake
Lodg E: Ramada

Bold red print shows RV & Bus parking available or nearby

EXIT INDIANA

AServ E: Hires Auto Parts, Lexus of Fort Wayne, Nissan West, O' Daniel Olds, Poinsatte Chrysler Plymouth Dodge, Tom Kelly Buick
ATM E: First of America Bank, Three Rivers Federal Credit Union
Other E: **Meijer Grocery**, TST Car/ Van Wash

102 U.S. 24 West, Huntington, Fort Wayne
Gas W: Amoco*, Exxon*
Food W: Applebee's, Arby's, Blimpie's Subs (Exxon), Bob Evans Restaurant, Captain D's Seafood, McDonalds, Pizza Hut, Tavern at Coventry, Wendy's, Zesto
Lodg E: Extended Stay America, Hampton Inn
W: AAA Best Western, ◆ Comfort Suites
AServ W: Southwest Automotive Service Center
Med E: ✚ **Hospital**
ATM W: Machine (Amoco), NBD, Northwest Banks
Other W: Perfection Auto Wash, **Scott's Grocery, WalGreens**

99 Lower Huntington Road
96A IN 469, U.S. 24, U.S. 33
96B West LaFayette Center Road, Roanoke
(93) **Rest Area - RR, Phones, Picnic, Vending (Southbound)**
(89) **Rest Area - RR, Phones, Picnic, Vending (Northbound)**
86 U.S. 224, Markle, Huntington
Gas E: Sunco*
Med W: ✚ **Hospital**
(80) **Weigh Station (Both Directions)**
78 IN 5, Warren, Huntington
TStop W: **76 Auto/Truck Plaza**(SCALES) **(RV Dump)**
Gas E: Huggy Bear
W: Clark*(LP) (Kerosene)
Food W: **76 Auto/Truck Plaza**, Hoosier-land Restaurant, **McDonalds**, Subway (Clark)
Lodg E: **Huggy Bear Motel**
TServ W: **76**
ATM W: Clark, First National Bank of Huntington (76 Truck Stop)

73 IN 218, Van Buren, Warren, Berne
64 IN 18, Marion, Montpelier
FStop W: **Marathon***
RVCamp E: **Camping**
Med W: ✚ **Hospital**
59 U.S. 35, IN 22, Gas City
Food E: **Burger King**
Lodg E: Best Western
RVCamp E: **Marbrook Camping**
W: **Post 95 Lake Campground**
55 IN 26, Hartford City, Fairmount
Gas E: Marathon*
(51) **Rest Area - RR, Phones, Picnic, Vending (Southbound)**
(50) **Rest Area - RR, Phones, Picnic, Vending, Tourist Info (Northbound)**
45 U.S. 35 South, IN 28, Alexandria, Albany
TStop E: **Standard Auto Truck Plaza***(SCALES)
Food E: **Restaurant (Standard Truck Plaza)**, Taco Bell
TServ E: **Standard Auto Truck Plaza**
ATM E: Standard Truck Plaza
41 IN 332, Muncie, Frankton
Gas E: Amoco*
AServ W: Jack Smith & Sons RV Service
Med E: ✚ **Hospital**
34 IN 32, Indiana 67 North, Muncie, Anderson
TStop W: **76 Auto/Truck Plaza**(SCALES)
FStop E: **Speedway***(LP)(SCALES)
W: **Gas America***(LP)

EXIT INDIANA

Food E: **Arby's, Denny's, Food Court (Indiana Factory Shops), Hardee's (Speedway)**, Taco Bell
W: **Cleo's Restaurant (76 Truckstop)**, McDonalds, Subway, **Wendy's**
Lodg E: **Budget Inn**
W: ◆ Super 8 Motel
TServ W: **76 Auto/Truck Plaza**
TWash W: **76 Auto/Truck Plaza**
Other E: **Indiana Factory Shops**

26 IN 9 North, Indiana 109, Anderson
Gas E: Meijer*(LP) (Kerosene)
W: Amoco*, Marathon*(CW), Red Barn*, Shell*(CW)
Food E: Ryan's Steakhouse
W: Applebee's, Baskin Robbins, Bob Evans Restaurant, Burger King, Chen Buffet, **Cracker Barrel**, Great Wall Chinese Super Buffet, Grindstone Charley's, Holiday Inn, Lone Star, McDonalds, Noble Romans Pizza, Old Country Buffet, Perkins Family Restaurant, Pizza Hut, Ramada Inn, Rax, Red Lobster, Ruby Tuesday (Holiday Inn), Shants Sports Pub & Eatery, Steak & Shake, Taco Bell, Waffle & Steak, Wendy's, White Castle Restaurant
Lodg E: Dollar Inn
W: Best Inns of America, ◆ Comfort Inn, ◆ Holiday Inn, Lee's Inn, Motel 6, ◆ Ramada Inn
AServ W: Ed Martin Olds, Cadillac, GMC Trucks, Monroe Muffler & Brakes, Quaker State Lube, Shell, Tire Barn
ATM E: First of America (Meijer)
W: Ameriana Bank, Independent Federal Bank, Key Bank, National City Bank, Union Federal Savings Bank
Other E: **Meijer Supermarket, Visitor Information**
W: **Devonshire Veterinary Clinic, GFS Gordon Food Service, Office Max, Optical One, Payless**

EXIT INDIANA

Supermarket (Pharmacy), Target Department Store

22 IN 67 South, Indiana 9 South, Anderson, Pendleton
Med W: ✚ **Hospital**
Other E: **State Patrol Post**
19 IN 38, Pendleton, Noblesville
Gas E: Marathon*(LP)
Food E: **Dairy Queen, McDonalds, Subway**
RVCamp W: **Pine Lake Camping**
14 IN 13, Lapel , Fortville
TStop W: **Pilot Travel Center***(SCALES)
Food W: **Subway (Pilot)**
RVCamp W: **Campground**
ATM W: Pilot
10 IN 238, Noblesville, Fortville
Other E: **Hamilton Commons Outlet Mall**
5 IN 37 North, 116th St, Fishers, Noblesville
Gas W: 76*, Shell*(CW), Speedway*(D)
Food W: **McDonalds, Pantry Pizza (76)**
3 96th Street
Gas E: Amoco*(D)(CW), Meijer Gas & Convenience Store*, Shell*(CW)
W: 76*(D)
Food E: Applebee's, Blimpie's Subs, **Cracker Barrel**, Freshens Yogurt, Golden Wok Chinese, Grindstone Charley's, Ho Lee Chow Chinese, McDonalds, Mega Bites Subs, Muldoon's Grill, Noble Romans Pizza, Ramada Inn, Ruby Tuesday, Steak & Shake, Wendy's
W: Arby's, Burger King, Schlotzkys Deli, Taco Bell
Lodg E: ◆ Holiday Inn Express, Ramada Inn
W: ◆ Residence Inn
AServ E: Wal-Mart
W: Indy Lube, Monroe Muffler & Brake, Q-Lube
ATM E: Amoco, National City Bank, Wal-Mart
Other E: **Marsh Supermarket (Pharmacy), Meijer Supermarket, Sam's Club, VCA Crosspointe Animal Hospital, Wal-Mart (Pharmacy)**
1 82nd Street, Castleton
Gas W: Amoco*, Marathon*, Shell*
Food E: Pizza Hut
W: Arby's, Burger King, Charleston's (Hampton Inn), Denny's, Einstein Brothers Bagels, Fazoli's Italian Food, First China Chinese, Flaky Jake's, Great China Buffet, Hooters, IKE's Deli, Joes Grille, KFC, Krispy Kreem, Laughner's Cafeteria, Le Peep Restaurant, McDonalds, Old Country Buffet, Olive Garden, Pizza Hut, Red Lobster, Sirloin Grill, Sizzler Steak House, Skyline Chili, Steak & Shake, Subway, Taco Bell, The Original Pancake House, Wendy's
Lodg E: American Inn Motel, Dollar Inn, Omni Hotel
W: AAA Best Western, ◆ Days Inn (see our ad on this page), ◆ Fairfield Inn, ◆ Hampton Inn (see our ad on this page)
AServ W: AAMCO Transmission, Car X Muffler & Brakes, Discount Tire Company, Goodyear Tire & Auto, Indy Lube, Indy Tire Center, Meineke Discount Mufflers, Quaker State Lube, Shell, Tire Barn
Med E: ✚ **Hospital**
ATM E: Indiana Fed. Credit Union
W: NBD Bank, National Bank of Indianapolis
Other E: **Dr. Aziz Pharmacy, Friendly Food Supermarket**
W: **Castleton Square Mall**, Coin Car Wash, **Wax Werks**
0 Junction I-465, Indianapolis

↑INDIANA

Begin I-69

EXIT	UTAH
	Begin I-70

↓UTAH

1 Historic Cove Fort
8 Ranch Exit
17 Fremont Indian Museum
- RVCamp S: Castle Rock Campground
- Other N: Visitor Information

23 U.S 89, Panguitch, Kanab (Service More Then 1 Mile From Exit)
- Gas S: Shell

26 Joseph, Monroe, Jct UT 118
- RVCamp S: Flying U Campground

32 Elsinore, Monroe
- Gas S: Chevron*
- Parks S: Mystic Hot Springs 5 Miles

37 Richfield, Bus I-70, Sevier Valley
- RVCamp S: JR Munchie's, KOA Campground
- Med S: ✚ Hospital
- Other S: Highway Patrol Post

40 Business 70, Richfield
- TStop S: Flying J Travel Plaza*(D)(LP) (RV Dump)
- Gas S: Chevron*
- Food S: Arby's, Subway
- Lodg S: DAYS INN AAA Days Inn, ◆ Super 8 Motel
- AServ S: Chevron
- Med S: ✚ Sevier Family Hospital
- Parks S: Capital Reef, Fish Lake

48 CO24, Sigurd, Aurora (Service More Then 1 Mile)
- Gas S: Cedar Ridge Station*
- Parks S: Capital Reef National Park, Fish Lake

54 U.S 89, Business 70, Spur, Salina
- Gas N: Amoco*(D), Chevron, Texaco*
- Food N: Burger King (Chevron), Denny's, Safari Cafe, Subway
- Lodg N: Budget Host/Scenic Hills Motel, AAA Safari Motel
- AServ N: Triple A Towing Service
- TServ N: Triple A Truck & towing Service, Wheeler Tire, Radiator
- RVCamp N: Butch Cassidy RV Park, Salina Creek (Texaco)

61 Gooseberry Road
72 Ranch Exit
(84) Rest Area P (Both Directions)
89 UT72, UT10, Price, Loa

EXIT	UTAH

- Parks N: Capital Reef National Park

97 Ranch Exit
(102) View Area - RR (Both Directions)
105 Ranch Exit
(113) View Area (Both Directions)
114 Moore
(120) View Area (Both Directions)
129 Ranch Exit
(136) Brake Test Area (Eastbound)
(140) View Area - Truck Parking P (Both Directions)
(144) Rest Area - Truck Parking P (Westbound)
147 UT24, Hanksville
- Parks N: Capital Reef National Park, Lake Powell

156 U.S 6, U.S 191, Price, Salt Lake
158 Business 70, Green River, UT 19 (More Services 1 Mile)
- TStop N: Conoco* (RV Dump)
- FStop N: Chevron*(D)
- Gas N: Conoco*
- Food N: Arby's (Conoco TS)
- RVCamp N: KOA Campground

162 UT19, Green River (More Services 1 Mile)
173 Ranch Exit
(179) Rest Area P (Eastbound)
180 U.S 191, Moab
- Gas N: Amoco*(D)
- Food N: Crescent Junction Cafe
- Parks S: Canyon Land Arches National Park

185 Thompson
- FStop N: Texaco- RV Dump*(D)
- Food N: Cook Tent Cafe

(187) Rest Area P (Westbound)
190 Ranch Exit
202 UT128, Cisco
212 Cisco
220 Ranch Exit
225 Westwater

↑UTAH

EXIT	COLORADO

↓COLORADO

2 Rabbit Valley
- Parks N: Dinosaur Quarry Trails

11 Mack
- Gas N: Sinclair*(LP)
- Food N: Colorado Club Restaurant

15 CO139, Loma, Rangely
- Parks S: Highline Lake State Park

(17) Weigh Station (Both Directions)
19 U.S 6, CO340, Fruita
- TStop S: Loco(LP)(SCALES)
- Gas N: Conoco*(D); S: Texaco*(D)(CW)
- Food N: Burger King, Munchie's Pizza & Deli; S: Colorado Line Company Restaurant, McDonalds, Starvin Arvin, Subway, Wendy's (Texaco)
- Lodg N: Balanced Rock Hotel; S: ◆ Super 8 Motel
- AServ N: NAPA Auto Parts
- RVCamp S: Campground
- Med N: ✚ Hospital
- ATM S: Alpine -Texaco
- Parks N: Colorado National Monument
- Other N: Car Wash Napa, City Market Supermarket; S: Colorado Welcome Center - RR, Phones, Picnic, Tourist Info

26 Business 70, U.S 6, U.S 50, Grand Junction
- Gas S: Acorn*(D)
- Food S: Auto's
- Lodg S: Westgate Inn
- RVCamp N: Camping
- Parks S: Walker State Wildlife Area

28 24 Road, Redlands Pkwy (More Services In Town)
- TServ N: Colorado Kenworth
- RVCamp N: Camping; S: Camping

31 Horizon Dr, Grand Junction
- Gas N: Acorn*, Amoco*(CW); S: Conoco*(D)(CW), Phillips 66(CW)
- Food N: Lenny's Family Restaurant, Perks Coffee House, Starvin Arvin's, WW Peppers, Wendy's; S: Applebee's, Burger King, Days Inn, Denny's, Pizza Hut, Taco Bell, The Pour House
- Lodg N: AAA Best Western, ◆ Comfort Inn, Grand Vista Hotel, AAA Holiday Inn, AAA Howard

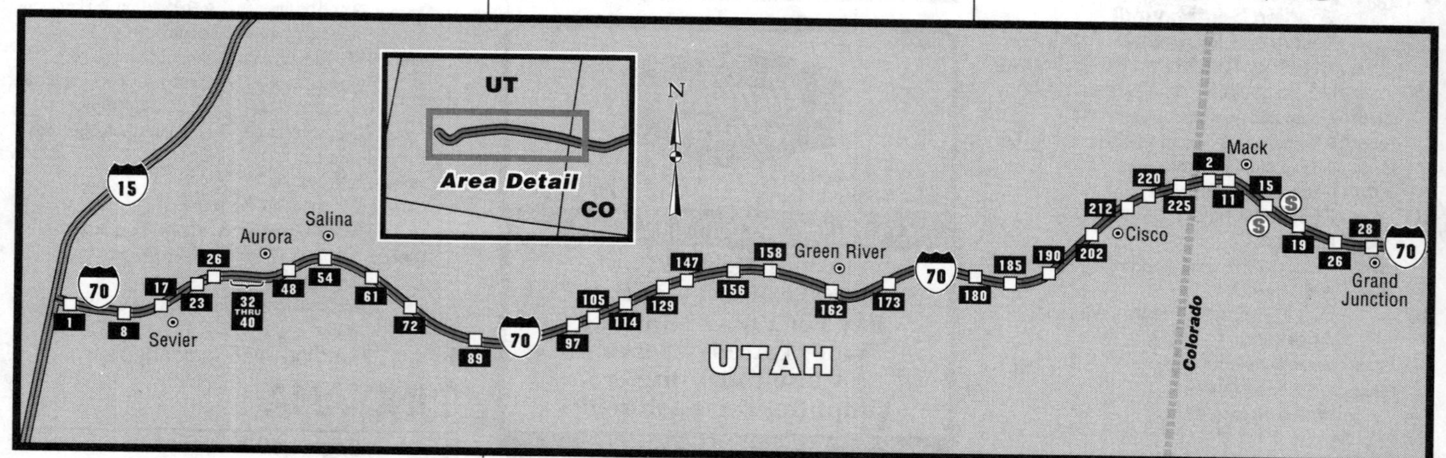

Bold red print shows RV & Bus parking available or nearby

EXIT COLORADO

Johnson, Motel 6
S: AAA Best Western, AAA Budget Host Inn, Country Inn, DAYS INN AAA Days Inn (see our ad on this page), ◆ Hilton, ◆ Super 8 Motel, Travelers Inn
ASers N: Standard Automotive Center
Med S: + Hospital
Parks N: BLM National Forest Information
Other N: US Post Office
S: Visitor Information

37 Business 70, Clifton, Delta (Service More Then 1 Mile)
RVCamp S: Camping

42 U.S 6, Palisade (Eastbound, More Service South 1 Mile)

44 I-70 Bus Loop, Palisades Westbound

46 Cameo

47 Island acres State Rec Area
Gas S: Total*(CW)
Food S: Rosie's
Parks N: Island Acres State Rec Area

49 CO65, CO330, Grand Mesa, Collbran
Other N: Powder Horn Ski Area

(50) Parking Area, View Area (Southbound)

62 DeBeque
AServ N: Wild Horse Service Center(LP)

75 Parachute, Battlement Mesa
FStop N: Total*
Gas N: Sinclair*, Texaco*(LP)
Food N: Hot Stuff Pizza (Texaco), Hungry Mike's Cafe, Jim & Bonnie's Outlaws, Sage Cafe, Smash Hit Subs (Texaco)
S: Taco Bell (Total)
Lodg N: Super 8 Motel
AServ N: Sinclair
S: Valley Car Wash
Other N: Rest Area, Visitor Information

81 Rulison

87 Meeker West Rifle

90 CO13, Rifle, Meeker (North Services More Than Half Of A Mile Away)
Gas N: Amoco, Phillips 66, Texaco*
S: Conoco*, Phillips 66*
Food S: Burger King (Outdoor Playground), McDonalds (Indoor Playground), Red River Restaurant
Lodg S: Red River Inn, AAA Rusty Cannon Motel

EXIT COLORADO

AServ N: Amoco, Phillips 66
Med N: + Hospital
Other N: Rest Area - RR, Phones, Vend, Picnic, Tourist Info, RVDump, Visitor Information (Rest Area)

94 Garfield County Airport Road

97 Silt
Gas N: Conoco, Kim's Tools (Hardware, & Gasoline)
Food N: Pizza Pro, Trail Inn Pizza
Lodg N: Red River Motel (1 Mile)
AServ N: Reed's Auto Service
RVCamp S: Viking RVPark & Camping
Parks N: Harvey Gap State Park

105 New Castle
FStop S: Phillips 66*(LP)
Gas N: Conoco*
RVCamp S: RV Park
ATM N: Alpine Bank
Other N: Car Wash, City Market

107 River Bend Parking Area

109 Canyon Creek

111 South Canyon

114 W Glenwood
Gas N: 7-11 Convenience Store, Conoco(D), Texaco*
Food N: Burger King, Charcoal Burger Drive In, Dairy Queen, Fireside Family Steakhouse, I-70 Cafe, Los Desperados, Marshall Dillons Steakhouse Restaurant
Lodg N: Budget Host Motel, First Choice Inn, Red Mountain Inn
AServ N: Big O Tires, Dodge Plymouth Dealership, Ford Dealership, Glenwood Springs, Henry Taylor's Auto & RV Center, Patterick Tire Company
RVCamp N: Henry Taylor RV Service
Other N: JC Penny, Johnson Park Mini Golf, K-Mart, Radio Shack, Sear's, U-Haul Center

(115) Tourist Info Center, Rest Area (Eastbound)

116 CO82, Glenwood Springs, Aspen
Gas N: Amoco*, Conoco*, Texaco*
Food N: A & W Drive-In, Dairy Kreme, KFC, Mancinelli's Pizza, Ramada, Rosi's Restaurant, Smoking Willie's BBQ
Lodg N: AAA Best Western, Glenwood Motor Inn, AAA Hot Springs, Hotel Colorado, AAA Ramada, AAA Silver Spruce Motel, Starlight Motel
AServ N: Conoco, Toyota Dealership, Volkswagen Dealer
Med N: + Hospital
Other N: Hot Springs Pool, Sunlight Mountain Snow

EXIT COLORADO

Mass Ski Area

119 No name
Other S: Rest Area - RR, Phones, Picnic, Campground

121 Grizzly Creek

(121) Rest Area - RR, Phones, Picnic

123 Shoshone (No Eastbound Reaccess)

125 Hanging Lake (No Reaccess)

129 Bair Ranch (No Vehicles Over 35 Feet)
Other S: Rest Area- RR, HF, Phones, Picnic

(129) Rest Area - RR, Phones, Picnic

133 Dotsero

140 Gypsum
Gas S: Texaco
Food S: Arturo's Meixcan & American
AServ S: SS Auto & Truck Repair
TServ S: SS Auto & Truck Repair

147 Eagle
FStop S: Amoco*(CW)
Gas N: Texaco(CW)
S: Conoco* (Pizza & Subs)
Food N: Bagelopolis, Burger King, Taco Bell (Texaco)
S: Jackie's Olde West Restaurant, Subway
Lodg N: ◆ Holiday Inn Express
S: AAA Best Western
ATM S: Alpine Bank
Other N: City Market
S: Food Town Supermarket

157 CO131, Wolcott, Steamboat Springs

(162) Scenic Overlook

163 Edwards
FStop S: Texaco(CW)
Food S: Jerry's Deli (Texaco), Sacred Grounds Coffee House
AServ S: Texaco

167 Avon (Many Services More Than Half A Mile From The Exit)
Gas N: Coastal
Food N: Pizza Hut
S: Coho Grill, Denny's, Subway
Lodg S: AAA Comfort Inn (see our ad on this page), Lodge at Avon Center
AServ N: Goodyear Tire & Auto
ATM S: First Bank
Other S: Beaver Creek Ski Area

171 U.S 24, Minturn, Leadville
Other N: Cooper Ski Area, National Forest

173 Vail
Gas N: Phillips 66*, Texaco*(D)
S: Conoco*(D)(LP)
Food N: Dairy Queen, Jackalope Cantina, McDonalds, Poppey Seeds Bakery Deli Espresso, Subway, Taco Bell, The Dancing Bear, Wendy's
Lodg N: ◆ Marriott, West Vail Lodge (Restaurant)
AServ N: Phillips 66
Other N: Safe Way Grocery

176 Vail
Gas S: Amoco*
Food S: Craig's Market
Lodg S: ◆ Evergreen Lodge, AAA Holiday Inn, AAA Vail Village Inn
AServ S: Amoco
Med S: + Hospital

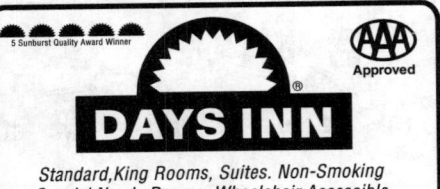

Bold red print shows RV & Bus parking available or nearby

EXIT	COLORADO
ATM	S: First Bank, West Star Bank
Other	S: **Colorado Ski Museum**
179	US 24 E, Minturn, Leadville
180	Vail
(190)	**Rest Area**
195	CO91, Copper Mtn, Leadville
Gas	S: Conoco
Lodg	S: Copper Mountain Ski Resort
Other	S: **Copper Mountain, Ski Cooper, Ski Lodges**
198	Officer's Gulch
201	Frisco, Breckenridge
Food	S: Blue Spruce Restaurant, El Rio
Lodg	S: AAA Blue Spruce Inn, ◆ Creekside Inn, Frisco Bay Inn
Other	S: **Breckenridge Ski Area, Tourist Information, U.S. Post Office**
203	CO9, Frisco
Gas	S: 7-11 Convenience Store, Diamond Shamrock, Texaco*
Food	S: A & W Drive-In, China Szechuan, Claim Jumper Restaurant, Country Kitchen, KFC, Pizza Hut, Smoking Willies, TCBY Yogurt, Taco Bell, Texas Star Cafe
Lodg	S: AAA Alpine Inn, AAA Best Western, AAA Holiday Inn
AServ	S: Big O Tires, NAPA Auto Parts, Wal-Mart
RVCamp	S: **Tiger Run RV Resort**
Med	S: ✚ **Emergency Medical Care**
ATM	S: Community First National Bank
Other	S: Car Wash, **Radio Shack, Safe Way Grocery, Tourist Information, Wal-Mart (Pharmacy)**
(203)	Scenic Overlook - Lake Dillon
205	U.S 6, CO9, Silverthorne, Dillon (Many Services More Than A Half Mile From The Interstate)
Gas	N: 7-11 Convenience Store, Acorn*, Amoco*(CW), Citgo*, Food Mart
	S: Coastal*, Total*
Food	N: Denny's, Good Times Burgers, Mint, Old Dillon Inn Mexican, Quizno's, Silver Thorn Cafe, Village Inn, Wendy's
	S: Arby's, Burger King, Dairy Queen, Dragon Chinese Restaurant, McDonalds, Nick-N-Willy's Pizza, Roberto's Authenic Mexican, Subway, Sweet Peas, The Coffee Cottage
Lodg	N: First Inn, Luxury Inn
	S: Super 8 Motel
Other	N: **Factory Outlet Center**
	S: **City Market Supermarket, Keystone & Rappaho Bassin Ski Area, Silverthorn Factory Stores**
216	U.S 6, Loveland Pass
Other	S: **Loveland Valley Ski Area**
218	No Name
221	Bakerville
226	Silver Plume
Gas	N: Buckley Brothers
AServ	N: Buckley Brothers*
Other	N: **U.S. Post Office**
(226)	Scenic Overlook
228	Georgetown
Gas	S: Conoco*, Georgetown Market, Phillips 66*, Total*
Food	S: Crazy Horse Restaurant, Dairy King, Silver Queen Restaurant, Swiss Inn Restaurant
Lodg	S: **Super 8 Motel**, Swiss Inn (Restaurant)
Other	S: **Tourist Information**
232	U.S 40, Empire, Granby
233	Lawson
Gas	N: Texaco
Food	N: Burger King
(234)	**Weigh Station, Check Station**
234	Downieville, Dumont
FStop	N: **Conoco* (Restaurant)**
Food	N: Burger King, **Subway (Conoco)**
AServ	N: Wesco Supply Auto & Truck Service(LP)
TServ	N: **Wesco Supply Auto & Truck Supply(LP)**
Other	S: **Weigh Station**
238	Fall River Rd.
239	Business 70, Idaho Springs (No Eastbound Reaccess)
Food	N: The Sandwich Mine
Lodg	N: Blair Motel
AServ	N: Silver City Automotive & Towing
RVCamp	S: **Camping**
240	CO103, Mount Evans
Gas	N: Phillips 66, Texaco*
Food	N: Espresso Bar, Picci's Bakery & Pizza
AServ	N: A Okay Auto Clinic
Med	S: ✚ **Chicago Creek Family Clinic**
Other	N: **Coin Laundry, Veterinary Clinic**
	S: **Clear Creek Ranger Station Visitor Center**
241A	Idaho Springs
Gas	N: Conoco*, Phillips 66*(CW), Texaco*(LP), Total*
Food	N: A & W Drive-In, Flip Side Diner, Home on the Range Cafe, King's Derby Restaurant, Marion's Restaurant, Sunrise Donuts, Wild Fire Restaurant
Lodg	N: 6 & 40 Motel, AAA Heritage Inn, JC Suites, AAA National 9 Inn, AAA Peoriana Motel, Tops Motel
243	Hidden Valley
244	U.S 6, Golden, CO119, Blackhawk, Central City (Left Exit)
Other	N: **El Dora Ski Area**
247	Beaver Brook, Floyd Hill (Difficult Reaccess Eastbound)
Food	S: Coffee Factory Outlet, Floyd Hill Grill
251	El Rancho, Evergreen (No Return Eastbound)
Food	S: McDonalds
252	Evergreen, El Rancho
253	Chief Hosa
Lodg	S: ◆ Chief Hosa Lodge
RVCamp	S: **Camping**
254	Genesee Park, Lookout Mtn
Other	N: **Buffalo Bill's Grave**
256	Lookout Mtn, Mother Cabrini Shrine Rd
(257)	Runaway Truck Ramp (Eastbound)
259	Business 70, U.S 40, Golden, Morrison
RVCamp	N: **Camping**
Parks	S: **Red Rocks**
260	CO470, Colorado Springs
261	U.S 6, W 6th Ave
262	Business 70, U.S 40, W Colfax Ave
Gas	N: Sinclair*
	S: Texaco
Lodg	S: Days Inn, AAA Days Inn, AAA Holiday Inn, Mountain View Motel
AServ	N: AC Transmission, Classic Honda
RVCamp	S: **Stevinson RV Sales**
263	Denver West Blvd
Food	N: American Grill, Boston Market, Goldfields, Pizza Hut
Lodg	N: ◆ Marriott
	S: ◆ Marriott
Other	S: **Barnes & Noble, Office Max, Shopping Mall**
264	Youngfield St, W 32nd Ave
Gas	S: Amoco*(CW)(LP), Conoco*, Diamond Shamrock*
Food	N: Country Cafe
	S: Chili's, Chipotle Mexican Grill, Dairy Queen,

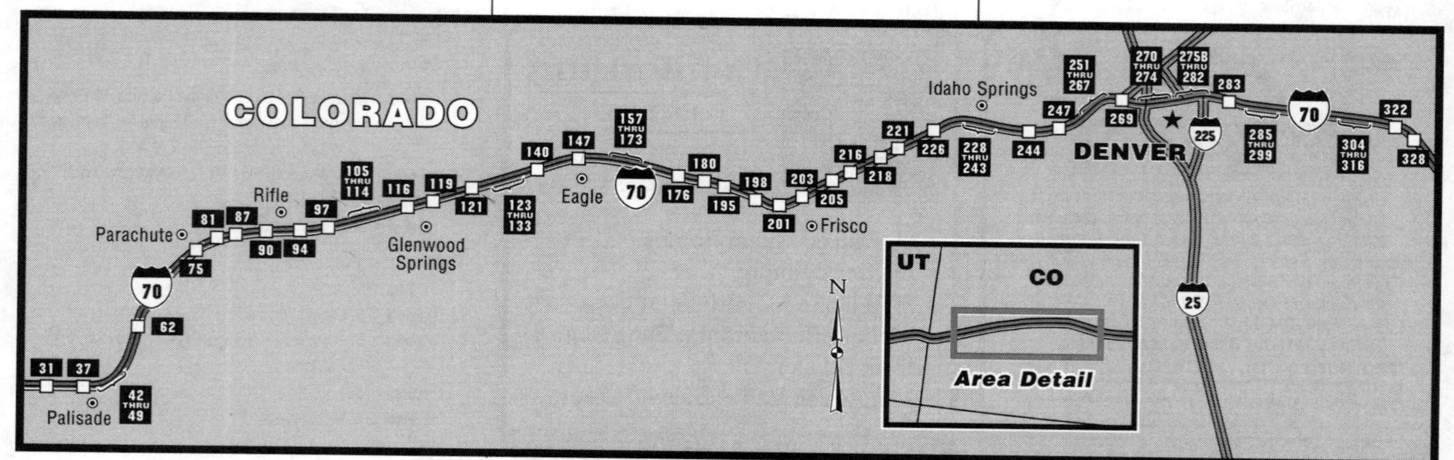

Bold red print shows RV & Bus parking available or nearby

EXIT COLORADO

Juice Shop, Las Carretas Mexican Restaurant, McDonalds, Oriental Kitchen, Starbucks Coffee, Subway, Taco Bell
Lodg N: ◆ La Quinta Inn
AServ S: Amoco
RVCamp N: **Camping World (see our ad on this page)**
ATM S: Northwest Banks
Other S: **King Soopers Grocery, Stacy's Hallmark, Wal-Mart, WalGreens**

66 CO72, Ward Road, W 44th Ave
TStop S: **76 Auto/Truck Plaza(SCALES) (RV Dump)**
Gas S: Total*
Food S: Palancar Reef, Pink Cadillac Restaurant, **Restaurant/76**
Lodg S: Quality Inn
AServ S: JW Brewer Tires
TServ S: **76**
RVCamp S: **The Prospect**

67 CO391, Kipling St
Gas N: Standard, Texaco*(CW)
S: Conoco*(CW)
Food N: Burger King, Cactus Cafe, Denny's, Furr's Family Dining, Pig Skins, Subway
S: Taco Bell, Village Inn
Lodg N: American Motel, Motel 6
S: (AAA) **Holiday Inn Express**, Interstate Inn 17, Motel 6, ◆ Super 8 Motel
AServ N: Standard
Med S: ✚ **Hospital**
ATM N: Foothills Bank
Other N: **Brunswick Bowling Lanes**, Kipling Car Wash, **Lasting Impressions Book Store**

69AB CO121, Wadsworth Blvd, (Jct I-76 to Ft Morgan)
Food N: Applebee's, McDonalds, Schlotzkys Deli, Taco Bell, Winchell's
S: Alamos Varde Mexican Restaurant, IHOP, Red Lobster
AServ N: AT&T Muffler
S: A&G Muffler, Discount Tire Company, Grease Monkey, Pep Boys Auto Center
Med N: ✚ **Hospital**
ATM N: Northwest Bank
S: First Bank
Other N: **Home Depot, Office Depot, Sam's, Walden Books**
S: **Eagle Hdwe**

70 Harlan St

71A **(270)** CO95, Sheridan Blvd

71B Lowell Blvd, Tennyson (Eastbound Reaccess Only)
Gas N: Texaco*(D)(CW)(LP)
Food S: El Paraiso Mexican
AServ N: Texaco
S: Michelin, Montgomery Ward, Seyfer Automotive
ATM S: Bank One, Colorado Nat'l Bank
Other S: **A-Look Optical**, Car Wash, **Lakeside Mall**,

EXIT COLORADO

Target Department Store, WalGreens (Pharmacy)

272 U.S 287, Federal Blvd
Gas N: Amoco*(CW), Sinclair*
S: Amoco, Conoco*(CW)
Food N: Baskin Robbins, Burger King, Good Times Drive Thru Burgers, Luethy Kitchen, McCoy's Family Restaurant, McDonalds, Panda Express Chinese, Pizza Hut, Siagon Vietnamese, Subway, Taco Bell, Village Inn, Wendy's, Winchell's Donuts, Winerschnitzel
S: Los Palmos II Mexican Restaurant (HoJo)
Lodg S: Howard Johnson
AServ N: Amoco, DeLuna Tire Service, Goodyear Tire & Auto, K-Mart, Tire Shop
RVCamp N: **Camping**
ATM N: Mega Bank
Other N: **Coin Laundry**, High Performance Car Wash, **K&M C-Store, K-Mart**

273 Pecos St (Nonaccessible)
Gas S: Circle K*
AServ S: Denver Engine & Transmission Exchange, Pecos Auto Service
ATM S: Circle K
Other N: **Butts Rental(LP), Family Dollar Store, Safe Way Grocery (Pharmacy), Tru-Value Hdwe**
S: **Post Office**

274 Junction I-25, Fort Collins, Colorado Springs

275A Washington St

275B CO265, Brighton Blvd, Coliseum (Reaccess Eastbound Only)
Gas N: Citgo* (7-11 C-Store)
Food N: Cindy Lynn Cafe
AServ N: Auto & Truck Parts Warehouse
ATM N: Citgo
Other N: **Public Library**
S: **Corres Field**

275C York St, Josephine St

276A Steele St, Vasquez Blvd
TStop N: **Pilot Travel Center(SCALES)**
Gas S: 7-11 Convenience Store*
Food N: **Wendy's (Pilot), Western Motor Inn**
S: Burger King
Lodg N: Triple A Motel, **Western Motor Inn**
TServ N: **Ford Dealership, Fruehauf, Goodyear Tire & Auto, Mack Trucks, Peterbilt Dealer, Volvo Dealer**
S: **Automotive Alignment**

EXIT COLORADO

TWash N: **Blue Beacon Truck Wash**
ATM S: 7-11 Convenience Store

276 CO2, to US - 6E, US 85N Colorado Blvd
Gas S: 7-11 Convenience Store*, Conoco*, Diamond Shamrock*, Total*(D)
Food N: El Torro Restaurant
S: Church's Fried Chicken, McDonalds
Lodg S: Budget Inn Motel
AServ S: Rent-A-Heap Jeep
TServ N: **J.W. Brewer Tire Company**
ATM N: Colorado National Bank, The Bank of Cherry Creek
S: 7-11 Convenience Store, Total
Other S: Hi-Performance Car Wash

277 Dahlia St, Holly St, Monaco St (Reaccess Eastbound Only)

278 CO35, Quebec St
TStop N: **Sapp Brothers*(CW)(SCALES)**
FStop N: **Goodyear Tire & Auto**
Gas S: Amoco
Food N: **Burger King (Sapp Brothers)**, Denny's, **Great American Restaurant (Sapp Brothers)**, Red Apple Restaurant (Quality Inn)
S: Capers Bistro (Stapleton Hotel), Morgan's Restaurant (Four Points Motel), Summerfield's Restaurant
Lodg N: **Economy Inn (Sapp Bros TS)**, (AAA) Hampton Inn, (AAA) Quality Inn
S: ◆ Courtyard by Marriott, (AAA) Doubletree, Four Points Motel, ◆ Holiday Inn, Metro Inn, Comfort Inn (see our ad on this page), Renaissance Hotel, (AAA) Stapleton Plaza Hotel
AServ S: Alamo Rent-A-Car, Amoco, Budget Rent A Truck, National Car Rental, Rent-A-Vette, Wyatt Towing
TServ N: **Goodyear Tire & Auto**
TWash N: **Sapp Brothers**
ATM N: Sapp Brothers

279 Jct I-270N, Ft Collins

280 Havana St
Lodg N: (AAA) Embassy Suites
Other N: **JC Penny, Office Depot**

281 Peoria St
Gas N: Conoco*(D)(CW)
S: Amoco*(CW), Total*(D)
Food N: Burger King, Cock Pit Grill (Best Western), McDonalds
S: Bennett's Bar-B-Que Pit, Church's Fried Chicken, Denny's, **IHOP**, KFC, Old Santa Fe Mexican Grille, Pizza Hut, Subway, Taco Bell, The Airport Broker Restaurant, Waffle House
Lodg N: ◆ Best Western, ◆ Drury Inn, Village Inn
S: Airport Value Inn, ◆ La Quinta Inn, (AAA) **Motel 6**, Traveler's Inn
AServ S: Goodyear Tire & Auto, Vazquez Auto Service
ATM N: Citywide Bank, Colorado National Bank
S: Amoco
Other S: **North-East Animal Hospital**

282 Junction I-225S, Colorado Springs (Left Sided Exit)

283 Chambers Road
Gas S: Phillips 66(D)(LP), Total*
Food N: Happy Cafe (Budget Motel)
S: Burger King, Taco Bell (Total)
Lodg N: Budget Motel, Holiday Inn

EXIT		COLORADO
	AServ	S: Phillips 66
284		Pena St.
	Lodg	N: AmeriSuites (see our ad on this page)
285		Airport Blvd.
	Lodg	N: Crystal Inn
286		CO32, Tower Road
288		I-70 Bus, U.S 40, Colfax Ave (Left Sided Exit Westbound)
289		Gun Club Rd
292		CO36, Air Park Rd
	FStop	S: Conoco*
	Food	S: Bev's Kitchen
295		Bus I-70N, Watkins
	TStop	N: Tomahawk*(LP) (Texaco)
	FStop	N: Sinclair*
	Gas	N: Texaco*(SCALES) (Tomahawk)
	Food	N: Lu Lu's (Tomahawk), Restaurant (Tomahawk)
	Lodg	N: Country Manor Motel
	TServ	N: Tomahawk
	Other	N: Coin Operated Laundry (Tomahawk), Post Office (Tomahawk)
299		Manila Rd
	FStop	S: Total*
	Food	S: Taco Bell (Total)
	ATM	S: Total
304		CO79, Bennett
	FStop	N: Hank's Truck Stop*
	Food	N: Restaurant (Hank's Truck Stop)
305		Kiowa, Bennett (Westbound)
	Other	N: Rest Area - RR, Phones, Picnic (Newspaper, Grill)
310		Strasburg
	Gas	N: Amoco*, Ferrell Gas(LP), Texaco(D), Tri Valley Gas(LP)
	Food	N: Cafe, Hank's Steakhouse, The Pizza Shop
	AServ	N: Harv's Diesel Auto Repair, Napa, Texaco
	TServ	N: Harv's Diesel
	RVCamp	N: KAO Campground (RV Dump)
	Other	N: Corner Market, Museum, U.S. Post Office, Western Hdwe
316		U.S 36, Byers
	Gas	N: Sinclair*
		S: Standard*(D)(CW)
	Food	N: Longhorn Restaurant (Longhorn Motel)
		S: Country Burger
	Lodg	N: (AAA) Longhorn Motel

EXIT		COLORADO
		S: Lazy 8 Motel (RV Hook-Up), The Golden Spike
	AServ	S: Standard
	TServ	N: Cummins Diesel
	ATM	N: First National Bank
	Other	N: Byers Super Value Store (Pharmacy)
		S: Car Wash, Post Office, The Wash Tub Coin Laundry
322		Peoria
	Other	S: Racetrac (Greyhound)
328		Deer Trail
	Gas	N: Texaco*
		S: Gas Station*(D)
	Food	S: Deer Trail Cafe
	Lodg	S: Motel
	AServ	N: Texaco
	Other	N: Post Office
(332)		**Rest Area - RR, Phones, Picnic, Vending (Westbound)**
336		Lowland
340		Agate
348		Cedar Point
352		CO86, Kiowa
	Other	S: Scenic Rt to Denver Colorado Rockies
354		No Name
359		Bus 70, to U.S 24, CO71, Limon, Colorado Springs
	TStop	S: Texaco*(CW)(SCALES)
	FStop	S: Total*(D)
	Gas	S: Texaco*(D)
	Food	S: Arby's, Country Fair (Texaco), McDonalds, Subway
	Lodg	S: (AAA) Best Western, ◆ Comfort Inn, ◆ Econolodge, (AAA) Super 8 Motel
	TWash	S: Texaco*
	ATM	S: Texaco TStop*
	Other	S: Tourist Information* (Texaco), Travel Store* (Texaco)
(360)		**Weigh Station - both directions**

EXIT		COLORADO
361		CO71, Limon
	TStop	S: Flying J Travel Plaza*
	FStop	S: Conoco*(LP), Texaco*
	Food	S: Crust Stuff Pitas (Texaco), Dairy Queen, Pizza Hut, Wendy's (Texaco)
	Lodg	S: (AAA) Preferred Motor Inn (Flying J), Travel Inn
	AServ	S: Flying J Travel Plaza
	TServ	S: Flying J Travel Plaza, Texaco
	RVCamp	S: KOA Campground
	ATM	S: Texaco
	Other	S: Colorado Gift Store (Texaco), RV Service (Texaco), State Patrol
363		US40, US 287, Hugo, Kit Carson, US24 to CO71, Limon, Bus 70
	Med	S: ✚ Hospital (13 Miles)
371		Genoa, Hugo
	Food	N: Cafe
	Med	S: ✚ Hospital (9 Miles)
	Other	N: A Point of Interest
376		Bovina
383		Arriba
	FStop	N: Phillips 66*
	Food	N: Hot Deli & Pizza (Phillips 66)
	Lodg	N: Motel
	AServ	N: Phillips 66
	RVCamp	N: Campground
	Parks	N: Old West Campground
	Other	N: Post Office, Tourist Information
		S: Rest Area - RR, Picnic, Phone (Both Directions, Tourist Info, Newspapers), Tourist Information
395		Flagler
	FStop	S: Country Store Gas*
	Gas	N: Total*(D)
	Food	N: Carribbean Ice (Shakes, Ice Cream)
	Lodg	N: Little England Motel
	AServ	N: NAPA Auto Parts
		S: Country Store
	TServ	N: A&J Radiator Repair, NAPA Auto Parts
		S: Country Store
	RVCamp	N: Camping
405		CO59, Seibert
	FStop	S: High Plains Fuel Co, Texaco*
	Food	S: Restaurant (Texaco)
	AServ	S: High Plains Fuel Co
	TServ	S: High Plains Fuel Co
	RVCamp	N: Gortons Campground
	ATM	S: Texaco*

COLORADO
KANSAS
Area Detail

Bold red print shows RV & Bus parking available or nearby

XIT COLORADO/KANSAS

12 Vona

19 Stratton

- **FStop** N: Ampride*, Conoco* (Showers, Laundry)
- **Food** N: Dairy Treat, Restaurant & Lounge
- **Lodg** N: AAA Best Western, AAA Claremont Inn
- **AServ** N: Knox & Sons
- **RVCamp** N: Campground
- **Other** N: Classic on Wheels Museum, Coin Operated Laundry (Conoco), Foods, Foods Grocery Store (Meats & Deli), U.S. Post Office

29 Bethune

37 Bus 70, U.S 385, Rays, Lincoln St

- **FStop** N: Conoco*
- **Gas** N: 7-11 Convenience Store*, Total*(D)
- **Food** N: Arby's, Burger King, Dairy Queen, Interstate House Restaurant (Conoco), McDonalds (Burlington Inn), Pizza Hut, Restaurant (Western Motor Inn), Sonic Drive In
- **Lodg** N: AAA Burlington Inn, Chaparral Motor Inn (Budget Host), Comfort Inn, AAA Sloan's Motel, Super 8 Motel, Westen Motor Inn
- **AServ** N: Goodyear Tire & Auto
- **Med** N: ✚ Hospital
- **ATM** N: Total
- **Other** N: Alco Discount Store, Bonny State Recreation Area, Kar Wash, Western Union

38) Welcome Center, Rest Area - RR, Phones, Picnic, Vending, Mailbox, Tourist Info, Museum, Reservation Hot Line (Both Directions)

38 Bus Loop I-70, U.S 24, Rose Ave.

- **TStop** S: Amoco*
- **FStop** N: Sinclair*, Texaco*(D)
- **Gas** N: Conoco*, Diamon Shamrock, The Red Front Gas
- **Food** N: Dairy Queen
  S: Restaurant (Amoco)
- **Lodg** N: Hi-Lo Motel, Kit Carson Motel, Sleanes Motel
- **AServ** N: Anderson Motors, Big A Auto Parts, Carlin Auto Body & Service Ctr, Ford Dealership, Goodyear Tire & Auto, Kline Tires, NAPA Auto Parts, Pontiac Buick GM Dealer, Ted's Paint & Body, Texaco*
- **TServ** N: Truck Sales
  S: Amoco (24 Hr)
- **RVCamp** N: Camping
- **Med** N: ✚ Hospital
- **Parks** N: Bonny State Rec. Area
- **Other** N: Kit Carson County Carousal, Old Town, Otter Wash Car Wash, Safe Way Grocery, The Red Front Save Market

↑COLORADO

KANSAS

Weigh Station (Eastbound)

KS267, Kanorado

KS Welcome Center, Rest Area - RR, Phones, Picnic, RV Dump, Grills, RV Water P

Ruleton

Caruso

U.S 24, KS27, St Francis, Sharon Springs

- **TStop** S: Texaco*(SCALES)
- **FStop** N: Phillips 66*
  S: Total*(SCALES)
- **Gas** N: Amoco*, Conoco*, Total*
- **Food** N: KFC, McDonalds, Pizza Hut (Phillips 66), Subway, Taco Bell, Wendy's
  S: A & W Drive-In (Total), Apple Trail (Texaco)
- **Lodg** N: Best Western, Comfort Inn, ◆ Howard Johnson, K-Inn, Motel 6, Super 8 Motel
  S: Motel (Texaco)
- **AServ** N: Amoco, Conoco
- **Med** N: ✚ Hospital
- **ATM** S: Texaco, Total
- **Other** N: Gibson's Discount Ctr (Pharmacy), Tourist Information

19 Goodland

- **FStop** N: Conoco* (Picnic Area)
- **TServ** N: Conoco
- **RVCamp** N: KOA Campground
- **Other** N: High Plains Museum

27 (36) KS253, Edson

36 KS184, Brewster

- **Gas** N: Citgo
- **Food** N: Stuckey's (Citgo)

45 U.S 24, Levant

(48) Rest Area - RR, Phones, Picnic, RV Dump, Grills P (Both Directions)

53 KS25, Atwood, Leoti

- **FStop** N: Amoco
- **Gas** N: Phillips 66*
- **Food** N: Arby's, Burger King, KFC, Long John Silvers, McDonalds, Old Depot Restaurant (Bourquins Campground), Ramada Inn, Sirloin Stockade, Subway, Taco Johns
  S: Village Inn
- **Lodg** N: DAYS INN AAA Days Inn, ◆ Econolodge, Ramada, Super 8 Motel
  S: AAA Best Western, AAA Comfort Inn
- **TServ** N: Amoco, Central Detroit Diesel Allison, Cummins Diesel
- **RVCamp** N: Bourquins Campground
- **ATM** N: Phillips 66
- **Other** N: Dillions Grocery (Pharmacy), Prairie Museum of Arts & History, The Health Cottage, Tourist Information, Wal-Mart (Pharmacy)
  S: Whites Factory Outlet Center

54 Country Club Dr

- **RVCamp** N: Camping
- **Med** N: ✚ Hospital

62 Mingo

70 U.S 83, Rexford

- **TStop** S: Phillips 66*
- **FStop** S: Conoco*
- **Food** S: Colonial Steakhouse, Jones Corner Cafe (Phillips 66), Twister Treat Ice Cream
- **Lodg** N: ◆ Inn
- **TServ** S: Phillips 66
- **TWash** S: Blue Beacon Truck Wash (Phillips 66)
- **RVCamp** S: RV Park(LP) (Camp Inn)
- **Med** N: ✚ Hospital
- **Other** S: Prairie Dog Town, Tourist Information

(72) Rest Area - RR, Phones, Picnic, RV Dump P (Both Directions)

76 U.S 40, Oakley, Sharon Springs

- **TStop** S: Texaco(SCALES)
- **FStop** S: Total*
- **Food** S: Texaco Restaurant
- **Lodg** S: 1st Interstate Inn, ◆ Best Western (2 Miles)
- **TServ** S: Dunlop Tires (Texaco)
- **Med** S: ✚ Hospital
- **Other** S: The Fick Museum (4 Miles), Tourist Information

79 Campus Rd

85 KS216, Grinnell

- **Gas** S: Texaco
- **Food** S: Dairy Queen (Texaco), Stuckey's (Texaco)

93 KS23, Grainfield, Gove

- **FStop** N: Conoco*
- **AServ** N: B's Ultimate Finish Car & Truck Service, Conoco
- **TServ** N: B's Ultimate Finish Car & Truck Service, Conoco

95 KS23N, Hoxie, Grainfield

(97) Rest Area - RR, Picnic, RV Dump P (Both Directions)

99 KS211, Park

107 KS212, Castle Rock Road, Quinter

- **FStop** N: Phillips 66*
  S: Coastal* (24 Hr)
- **Food** N: Tepa Mexican American Food (Budget Host Inn)
  S: Dairy Queen
- **Lodg** N: Budget Host Inn
- **AServ** N: Phillips 66
- **RVCamp** N: Camping(LP) (Showers, Water, RV Dump)
- **Med** N: ✚ Hospital
- **ATM** S: Coastal

115 KS198, Banner Road, Collyer

- **FStop** N: Phillips 66*
- **AServ** N: Phillips 66*
- **TServ** N: Phillips 66*

120 Voda Rd

127 U.S 283, WaKeeney, Ness City

- **TStop** S: Amoco*(SCALES)
- **FStop** N: Phillips 66*
- **Gas** S: Conoco*
- **Food** N: Jade Garden Restaurant (Kansas Country Inn), M-W Restaurant, McDonalds, Pizza Hut*
  S: Restaurant (Amoco FS), Subway (Conoco), Wheel (Best Western)
- **Lodg** N: Kansas Country Inn
  S: Best Western
- **AServ** N: Kent's Radiator, Vernie's
- **TServ** S: Amoco
- **RVCamp** S: KOA Campground (RV Dump)
- **Med** N: ✚ Hospital
- **ATM** N: Phillips 66
  S: Amoco
- **Other** N: Picnic Area

128 U.S 283, U.S. 40, WaKeeney, Hill City

- **FStop** N: Conoco* (24 Hr)
- **Food** N: Helen Quality Cafe* (Conoco)
- **Lodg** N: Budget Host Inn
- **Med** N: ✚ Hospital

(131) Rest Area - RR, Picnic, Vending, RV Dump P (Both Directions)

135 KS147, Ogallah

- **FStop** N: Schreiner
- **Food** N: Restaurant (Schreiner)
- **AServ** N: Schreiner

EXIT	KANSAS
TServ	N: Goodyear Tire & Auto (Schreiner)
Parks	S: Cedar Bluff State Park
140	Riga Road
145	KS247, Ellis
Gas	S: Kasey's Gen Store*, Texaco, Total*(D) (RV Dump)
Food	S: Alloway's Restaurant, Ellis Lanes Restaurant, Homemade Pizza & Donuts (Kasey's Gen Store)
Lodg	S: Fischer Motel
AServ	S: Alloway's Ford (Wrecker Service), D & B Body Shop, Texaco
TWash	S: Truck & Car Wash
Other	S: Ellis Railroad Museum, Fire Dept, Truck & Car Wash, Walter P. Chrysler Boyhood Home & Museum
153	Yocemento Ave
157	U.S 183, Hays, LaCrosse (Closed)
Lodg	S: General Hays Inn
Other	S: Fort Hays Sternberg Museum, Tourist Information
159	U.S 183, Hays, Plainville (Ft Hays State University)
FStop	N: Total*
	S: Conoco*, Phillips 66*
Gas	S: Amoco*, Texaco
Food	N: Applebee's
	S: Arby's, Country Kitchen, Golden Corral Family Steakhouse, Golden Ox (24 Hr), Hardee's, KFC, Long John Silvers, McDonalds, Papacito Pizza & Subs, Pheasant Run Pancake Inn, Village Inn Pancake House
Lodg	N: ◆ Comfort Inn
	S: ◆ Comfort Inn, DAYS INN AAA Days Inn, ◆ Hampton Inn, ◆ Holiday Inn, Motel 6, Ramada Ltd, ◆ Super 8 Motel
AServ	N: Ford Dealership, Lewis Paint & Collison, Toyota, Wheels'N Spokes
	S: Firestone Tire & Auto, George's Car Truck Repair, MacDonald Chevrolet & GM Dealer, Wal-Mart, Western Auto
TServ	S: George's
TWash	S: Truck Wash (Conoco)
Med	S: ✚ Hospital
ATM	S: Amoco*
Other	S: Mid West Drug Center (The Mall), Old Historic Ft Hays (4 Miles), The Mall, Wal-Mart (Pharmacy)

Note: I-70 runs concurrent below with I-135. Numbering follows I-135.

EXIT	KANSAS
161	Commerce Pkwy
Med	N: ✚ Hospital

Note: I-70 runs concurrent above with I-135. Numbering follows I-135.

EXIT	KANSAS
(162)	Rest Area - RR, Phones, Picnic P (Both Directions)
163	Toulon Ave
TServ	N: Int'l Lang Diesel Truck & Tractor Repair
168	KS255, Victoria
FStop	S: Ampride*
AServ	S: AAA Mobil Glass & Body, Ampride
Other	S: Historical Cathedral Of The Plains
172	Walker Ave
175	KS257, Gorham
180	Balta Road
184	U.S 281, Russell, Hoisington
FStop	N: Amoco*, Conoco*, Phillips 66*(D) (RV Dump)
Food	N: McDonalds, Meridy's, Pizza Hut (Amoco), Russells Inn, Subway (Conoco)
Lodg	N: Budget Host Inn, Russells Inn, ◆ Super 8 Motel
AServ	N: Heartland Auto, Pennzoil Lube, Phillips 66
RVCamp	N: Campground, Docks Boat & RV Service, Dumler Estates RV Park, Triple J RV Park (Laundry)
Med	N: ✚ Hospital
ATM	N: Amoco
Other	N: Coin Operated Laundry (Triple J), Oil Patch Museum, Tourist Information (Amoco)
(187)	Rest Area - RR, Phones, Picnic, RV Dump, Newspaper P (Both Directions)
189	Russell Pioneer Rd, Bus US40
193	Bunker Hill Road
TStop	N: Total* (Showers)
Food	N: Bearhouse Cafe (Total)

EXIT	KANSA
ATM	N: Total
Other	N: Wilson Lake Wildlife Area
199	KS231, Dorrance
Gas	S: Kerr McGee*(D)(LP)
Parks	N: Wilson Lake (6 miles)
206	KS232, Wilson, Lucas
FStop	S: Texaco*(LP)
Gas	N: Texaco
Food	N: Interstate House Restaurant (Texaco)
Parks	N: Garden of Eden
Other	N: Tourist Information, Wilson Lake (8 Miles)
209	Sylvan Grove
216	Vesper
Gas	S: Texaco
Food	S: Dairy Queen, Stuckey's (Texaco)
219	KS14S, Ellsworth
FStop	S: Conoco*
AServ	S: Conoco*
TServ	S: Conoco*
221	KS14N, Lincoln
(224)	Rest Area - RV Dump, Pet Area, R P (Eastbound)
225	KS156, Ellsworth, Great Bend
FStop	S: Texaco*(LP)
Food	S: Elkhorn Diner
AServ	S: Texaco
Other	S: Ft Larned Nat'l Historic Site (82 Miles)
233	Beverly, Carneiro
238	Brookville, Glendale
244	Hedville, Culver
Lodg	N: Best Western (see our ad on this page)
FStop	S: King Fuel Stop*
Gas	S: Phillips 66*
249	Halstead Rd
250A	Junction I-135, U.S 81S, to Wichita
250B	U.S 81N, Concordia
252	KS143, Nineth St, Salina
TStop	N: Bosselman Travel Center*, Petro(SCALES)
	S: Bosselman's Truck Plaza*
FStop	N: Amoco*
Food	N: Baskin Robbins, Bayard's Cafe, Dairy Queer Denny's, McDonalds, Pizza Hut (Petro), Wendy's (Petro)
	S: Blimpie's Subs, Grandma Mac's (Ramada In Little Caesars Pizza, Mid-America Inn
Lodg	N: DAYS INN ◆ Days Inn, Holiday Inn Express, Mo 6, Salina Inn, AAA Super 8 Motel
	S: AAA Best Western, AAA Mid-America Inn, Ramada Inn
AServ	N: Amoco
TServ	N: Freightliner Dealer, Inland Truck Parts
	S: Bosselman's, Detroit Diesel, International

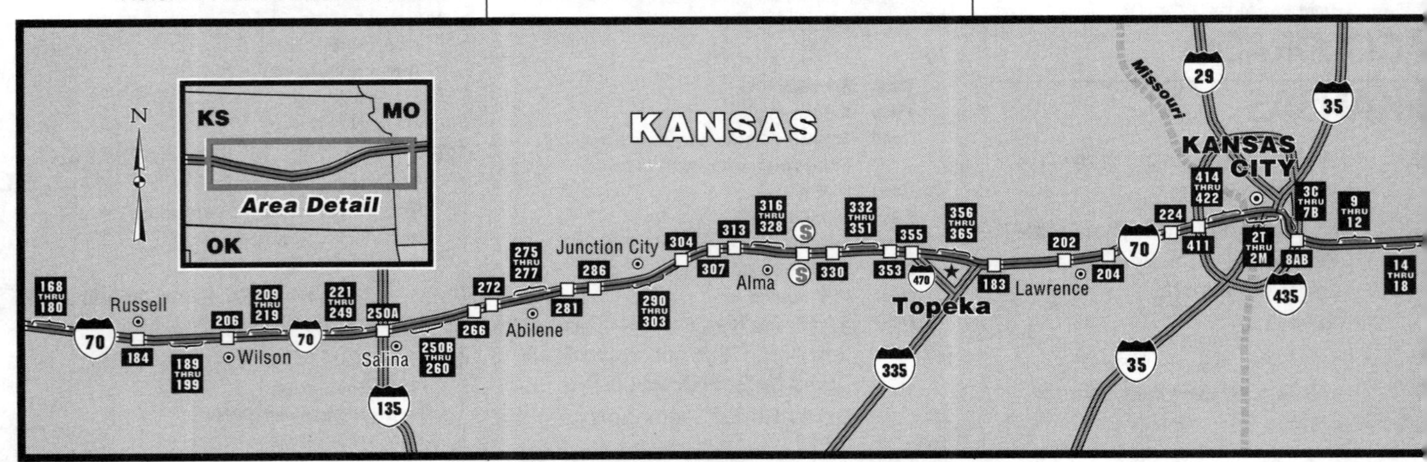

Bold red print shows RV & Bus parking available or nearby

EXIT	KANSAS
	Truck Sales and Service
TWash	N: Petro
53	Ohio St., Bicentennial Center
TStop	S: Flying J Travel Plaza*(SCALES)
Gas	S: Conoco(D)(LP)
Food	S: Country Market Restaurant & Buffet (Flying J), Pepperoni's (Flying J)
TServ	S: Kenworth
60	Niles Road, New Cambria
(65)	Rest Area - RR, Phones, Picnic, RV Dump P (Both Directions)
66	KS221, Solomon
TStop	S: 76 Auto/Truck Plaza(SCALES)
Gas	S: Conoco
Food	S: Restaurant (76 Auto)
TServ	S: 76
ATM	S: 76
72	Fair Rd.
RVCamp	S: RV Park
75	KS215, Abilene Clay Ctr
Gas	S: Amoco*, Phillips 66*(CW), Texaco*(D)
Food	N: Dairy Queen
	S: Baskin Robbins, Blimpie's Subs, Evergreen Chinese Food (Super 8 Motel), Goldmine Restaurant (Best Western), Green Acres Restaurant (Super 8 Motel), McDonalds, Pizza Hut, Subway
Lodg	S: AAA Best Western, Super 8 Motel
AServ	S: Auto Zone Auto Parts, Texaco
ATM	S: UMB Bank
Other	S: Coin Car Wash (2 Miles), Ricco Pharmacy (2 Miles)
77	Jeep Rd.
81	KS43, Enterprise
Gas	N: Apco*
RVCamp	N: Campground, Four Seasons RV Acres Campground RV Sales & Service
86	KS206, Chapman
Gas	S: Citgo*
Food	S: Subway (Citgo)
90	Millford Lake Road
(94)	Rest Area - RR, Picnic, RV Dump, Truck Park P (Both Directions)
95	U.S 77, KS18, Marysville, Herrington
Med	N: Hospital
Other	N: Milford Lake
96	U.S 40 Bus., Washington St
Gas	N: Coastal*, Conoco*(D), Texaco*(D)
Food	N: Country Kitchen, Denny's, McDonalds, Peking Chinese, Sirloin Stockade, Sonic Drive In, Subway
Lodg	N: AAA Budget Host (RV Hook Up), ◆ Comfort Inn, DAYS INN AAA Days Inn, Liberty Inn
ATM	N: Central National Bank South
98	East St, Chestnut St
FStop	N: Texaco*(LP) (Open 24 Hours)
Food	N: Burger King (Texaco), Shoney's, Taco Bell
Lodg	N: Holiday Inn Express, Super 8 Motel
ATM	N: Texaco, Wal-Mart
Other	N: Wal-Mart (Express Lube, Tire Service)
99	J Hill Road, Flinthills Blvd
Gas	N: Texaco, Total*(D)
Food	N: Stacy's
Lodg	N: AAA Best Western, AAA Econolodge, AAA Super 8 Motel, Travelodge
AServ	N: Big A Auto Parts, Texaco
00	U.S 40, KS57, Council Grove
Lodg	N: Dream Land Motel, Sunset Motel
01	Ft Riley, Marshallfield
Other	N: Custer House, First Teritorial Capital, US Calvary Museum

EXIT	KANSAS
303	KS18, Ogden, Manhattan
304	Humboldt Creek Road
307	McDowell Creek Road
(309)	Rest Area - RR, Picnic, RV Dump P (Both Directions)
311	Moritz Rd.
313	KS177, Manhattan, Council Grove
Gas	N: Phillips 66*
316	Deep Creek Road
318	Frontage Road
323	Frontage Road
324	Wabaunsee Road
328	KS99, Wamego, Alma
FStop	N: Gas-N-Shop*(D)
(329)	Weigh Station - both directions
330	KS185, McFarland
332	Spring Creek Road
333	KS138, Paxico
Gas	N: Phillips 66*
RVCamp	N: Camping
335	Snokomo Road
(336)	Rest Area - RR, Picnic, Phones P (Both Directions, Left Lane Exit)
338	Vera Road
Gas	S: Texaco
Food	S: Dairy Queen (Texaco), Stuckey's (Texaco)
341	KS30, Maple Hill, St Marys
FStop	S: Standard*(D)
TServ	S: Standard
ATM	S: Standard
342	Eskridge Road, Keene Road
343	Frontage Road
346	Willard, Rossville, Dover
347	West Union Road
350	Valencia Road
351	Frontage Road
353	KS4, Eskridge
355	Junction I-470E, U.S 75S
Food	S: Cracker Barrel (south on I-470 to Exit 1)

EXIT	KANSAS
356	Wanamaker Road
TStop	S: Topeka Travel Plaza
Gas	S: Citgo, Phillips 66*
Food	S: Burger King, Cracker Barrel, Sirloin Stockade, The Roost Family Restaurant (Texaco FS), Vista Hamburgers
Lodg	N: AAA Amerisuites (see our ad on this page)
	S: Club House Inn, Econolodge, Hampton Inn, Motel 6, Super 8 Motel
AServ	S: Phillips 66
ATM	S: Mercantile Bank
357A	Fairlawn Road
Gas	S: Miller Mart*(D), Phillips 66*(CW)
Lodg	S: Holiday Inn, Motel 6
Other	S: Zoo-Rainforest
357B	Danbury Ln.
358A	JCT US 75 N
358	U.S 75, KS4, Gage Blvd
Gas	S: Conoco*
Food	S: McDonalds, Subway
359	MacVicar Ave
361A	1st Ave (alt. US75) (No Return Access To Eastbound)
361B	Third St, Monroe St
362A	4th St.
Lodg	N: Ramada Inn
362B	Eighth Ave, Downtown Topeka
362C	10th Ave, Madison St., State Capitol
Lodg	N: Holiday Inn
TServ	N: KCR International Trucks
363	Adams St, Branner Trfwy
364A	California Ave
Gas	S: Conoco*(D)
Food	S: Rosa's Mexican, Run Pig Run BBQ
364B	U.S 40, Carnahan Ave, Deer Creek Trfwy
365	21st St, Rice Road
	Note: I-70 runs concurrent below with KSTNPK. Numbering follows KSTNPK.
182	JCT I-70 W, Topeka, Denver, I-470 S Wichita
(183)	Service Plaza - both directions (Left Lane Exit)
FStop	B: Conoco*(D)
Food	B: Hardee's (Playground)
197	Lecompton, Lawrence
202	U.S 59, West Lawrence
Gas	S: Phillips 66*(LP)
Lodg	S: Holiday Inn, Ramada Inn, Super 8 Motel, Best Western (see our ad on this page)
204	U.S 24, U.S 59, East Lawrence
Gas	S: ASAP Texaco*, CCO(D), Citgo*, Total*(D)
Food	S: Burger King (Indoor Playground), Sonic Drive-In, Subway (ASAP Texaco), Taco Bell (ASAP Texaco)
Lodg	S: Bismark Inn
AServ	S: O' Reilly Auto Parts
Other	S: Downtown Outlet Mall, Publishers Warehouse
(209)	Service Plaza - both directions (Left Lane Exit)
224	KS7, to U.S 73, Bonner Springs, Leavenworth (Services One Mile On The Northside)
	Note: I-70 runs concurrent above

Bold red print shows RV & Bus parking available or nearby

EXIT KANSAS/MISSOURI

with KSTNPK. Numbering follows KSTNPK.

410 110th St.

411AB Junction I-435

414 78th St
- Gas **N:** QuikTrip*
  S: Phillips 66*
- Food **N:** Arby's, Burger King, Cracker Barrel, Dairy Queen, Furr's Dinning
- Lodg **S:** American Motel
- Med **N:** ✚ **Hospital**

(414) **Parking Area (Both Directions)**

(415) Kansas State Motor Vehicle Inspection Station (All Trucks Required)

415 U.S. 40, State Ave

417 57th St

418AB Junction I-635 (Exit Closed)

419 Park Dr, 38th St

420A Junction U.S 69, 18th St Expwy

420B US 69, 18th St Exprswy North
- Gas **N:** Conoco*
- Food **N:** Restaurant (Phillips 66)
- Lodg **N:** Eagle Inn Motel
- Med **N:** ✚ **Hospital**

422A U.S 69, U.S 169, 7th St Trfwy

422B US 169, 7th Street Thrwy

422C Service Rd.

422D Central Ave.

423C US 24, Minnesota Ave., JCT US169

423D Fairfax District

↑KANSAS

↓MISSOURI

Note: I-70 runs concurrent below with I-670. Numbering follows I-670.

1B Genesee St, Wyoming St

2N Junction I-29, I-35, U.S 71

2M I-70, U.S 40, U.S 71, end I-670

Note: I-70 runs concurrent above with I-670. Numbering follows I-670.

EXIT MISSOURI

2C US 169, Broadway

2D Maine St, Delaware St, Wyendot St

2E Grand Blvd
- Gas **N:** Conoco
- ATM **N:** NationsBank

2F Oak St (Difficult Reaccess Westbound)

2G JCT I-70 W, I-35 S, US 24, US 40 (Left Hand Exit)

2H US 24, MO 9, Admireable Blvd (Difficult Reaccess Westbound)

2K Harrison St., Troost Ave (Difficult Reaccess)

3A Paseo
- Gas **S:** Amoco*, Total*
- AServ **S:** General Tire

3C U.S 71S, Prospect Ave
- Gas **N:** Service Oil Co.
- Food **N:** Church's Fried Chicken
  S: McDonalds, Taco Bell

4B 18th St (Difficult Reaccess)

4C 23rd St

5C Jackson Ave. (No Reaccess)

6 Van Brunt Blvd
- Gas **S:** Amoco*(CW) (Open 24 Hours)
- Food **S:** KFC, McDonalds, Pizza Hut, Taco Bell
- Other **S:** **Osco Drugs**

7A 31st St (Difficult Reaccess)
- Food **S:** Restaurant (Traveler's Inn)
- Lodg **S:** Traveler's Inn

7B Manchester Frwy (No Reaccess)
- TServ **N:** **Arrow Truck Sales**
  S: **PS Truck Sales, Ryder Truck Sales**

8AB Junction I-435, Des Moines, Wichata

9 Blue Ridge Cutoff
- Gas **S:** Amoco*(CW)
- Food **N:** Denny's
  S: Allison House Restaurant, Taco Bell
- Lodg **N:** Adam's Mark Hotel, ◆ Drury Inn
  S: ◆ Holiday Inn
- Med **S:** ✚ **Humana Medical Health Care**
- ATM **S:** UMB Bank
- Other **S:** **Royal Stadium**

11 U.S 40, Blue Ridge Blvd

EXIT MISSOURI
- Gas **N:** Circle K*
  S: Amoco, Sinclair*, Total*
- Food **N:** Burger King, Long John Silvers, Subway
  S: Chi Chi's Mexican, Winchell's Donuts
- AServ **N:** Quaker State Lube
  S: Amoco
- ATM **S:** First Federal Bank
- Other **S:** **Blue Ridge Mall**

12 Noland Rd
- Gas **N:** Amoco(CW), Texaco*, Total*(D)
  S: Phillips 66*(CW)
- Food **N:** Denny's, Gold China, Golden Corral, Hardee's, Little Caesars Pizza, Magic Wok, Mr. Good Sense Subs & Pasta, Shoney's
  S: Arby's, Burger King, Country Kitchen, Fuddruckers Burgers, KFC, Krispy Kreme Donuts, McDonalds, Red Lobster, Taco Bell, Wendy's
- Lodg **N:** AAA Shoney's, Super 8 Motel
  S: American Inn, AAA Howard Johnson, ◆ Red Roof Inn
- AServ **N:** Amoco, Firestone Tire & Auto, Parts America
- ATM **N:** Bank of Jacomo, Merchantile Bank
- Other **N:** Coin Car Wash, **Osco Drugs, Venture Discount Stores**
  S: **K-Mart**

14 Lee's Summit Road
- Food **S:** **Cracker Barrel**

15AB I-470, MO 291, Liberty, Lee's Summit
- Med **N:** ✚ **Hospital**
- Other **N:** **Mall, Wal-Mart**

18 Blue Springs, Lake Tapawingo, Fleming Park
- Gas **N:** Amoco(CW)
  S: Conoco*(CW), Phillips 66*, QuikTrip*
- Food **S:** Clancy's Cafe, McDonalds, Perkins Family Restaurant, Pizza Hut, Taco Bell, The Diner, Waffle House
- Lodg **N:** **American Inn, Interstate Inn**
- AServ **N:** Amoco
  S: Phillips 66, Pro Lube America
- ATM **S:** UMB Bank

20 MO 7, Lake Lotawana, Blue Springs
- Gas **N:** Kicks Phillips 66*(CW), Sinclair, Total*(CW)
  S: Amoco*(CW), Conoco*, Texaco*
- Food **N:** Bob Evans Restaurant, Country Kitchen,

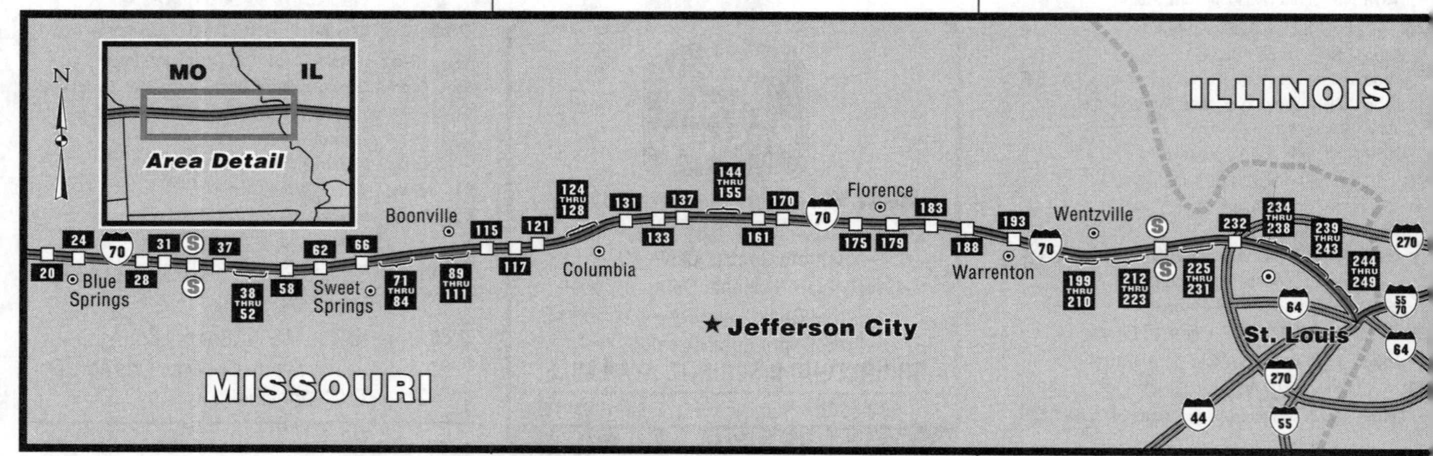

Bold red print shows RV & Bus parking available or nearby

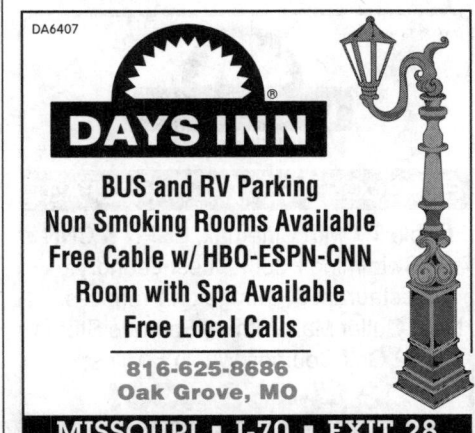

EXIT MISSOURI

Hardee's, Mr. Goodsense Subs & Pasta, T. Rider's
S: Applebee's, China City, Denny's, Godfather's Pizza, Long John Silvers, McDonalds, Subway, Texas Tom's, Wax Italian, Winstead's
Lodg **N:** ◆ Ramada, (AAA) Sleep Inn, Super 8 Motel
S: Holiday Inn Express
AServ **N:** O'reilly Auto Parts, Valvoline Quick Lube Center
ATM **N:** Merchantile Bank
Other **N:** **WalGreens**

Adams Berry Rd.

4 U.S 40, MO AA & BB, Grain Valley, Buckner
TStop **N: Phillips 66*(LP) (RV Dump)**
S: Pilot Travel Center(SCALES)
FStop **N: Mac Stop***
Food **N: Appletrail Restaurant (Phillips 66)**
S: Sonic Drive In, Subway
Lodg **N:** (AAA) Travel Lodge
S: Cozy Inn
RVCamp **N: RV Service**
S: RV Service

Note: I-70 runs concurrent below with I-670. Numbering follows I-670.

T Junction I-35S Wichita

R Central St, Convention Center

Note: I-70 runs concurrent above with I-670. Numbering follows I-670.

8 MO F & H, Levasy, Oak Grove
TStop **N: 76 Auto/Truck Plaza(SCALES)**
S: Gold 70 Truck Stop-Texaco*(SCALES) (RV Dump)
FStop **S: QuikTrip*(SCALES)**
Food **N: Restaurant (76)**
S: Blimpie's Subs (Gold Truck Stop), Hardee's (Texaco), McDonalds, PT's Family Restaurant (Texaco), Restaurant (Texaco), Subway (Texaco), Wendy's (Gold Truck Stop)
Lodg **N:** (DAYS INN) **Days Inn** (see our ad on this page)
S: Econolodge
TServ **N: Buck Truck Sales**
TWash **N: Blue Beacon Truck Wash (76)**
S: Truck o Mat Wash
RVCamp **N: KOA Campground**

1 MO D & Z, Bates City, Napoleon
FStop **S: Citgo**
Gas **S:** Phillips 66 (ATM, Western Union), Total
Food **S:** Taco Bell (Total)
Lodg **N:** Trucker's Motel
TServ **S: Mid America Truck Center**
RVCamp **N: Camping**

36) **Weigh Station - both directions**

7 MO 131, Odessa, Wellington (No Reaccess Eastbound, Must Reacess Exit 38)
Food **N:** Countryside Family Dining

8 MO 131S, to Odessa (Difficult Reaccess, Reaccess To Frontage Rd. Exit 37)
Gas **S:** Amoco(D), Odessa One Stop, Phillips 66*, Sinclair, Total* (ATM)
Food **S:** McDonalds, Morgan's Restaurant, Pizza Hut, Sonic Drive in, Subway, Taco John's, Wendy's
Lodg **S:** Odessa Motorlodge
AServ **S:** Amoco, Sinclair
Other **S: Book Store Outlet, Keene's Super Store**

41 MO M & O, Lexington

45 MO H, Mayview

49 MO 13, Higginsville, Warrensburg
FStop **N: Sinclair**
Gas **N:** Amoco
Food **N: Camelot Restaurant (Sinclair)**
Lodg **N:** ◆ Best Western
S: ◆ **Super 8 Motel**
AServ **N:** Amoco

52 MO T, Aullville

(57) **Rest Area - RR, Phones, Picnic P (Both Directions)**

58 MO 23, Concordia, Waverly
TStop **N: TravelCenters of America(SCALES)**
S: Conoco
Gas **N:** Phillips 66*
S: Casey General Store, MFA Oil*, Texaco* (ATM)
Food **N: Country Pride Restaurant (TravelCenters of America), KFC,** Subway, Taco Bell
S: Creamry Ice Cream, Gambino's Pizza, Hardee's, **Restaurant (Conoco)**
Lodg **S:** Golden Award Motel
AServ **S:** NAPA Auto Parts
TServ **N: TravelCenters of America**
Other **S:** Coin Car Wash

62 MO Y & W, Emma

Food **N: Restaurant (Malfunction Junction Inn)**
Lodg **N: Malfunction Junction Inn**

66 MO 127, Sweet Springs, Mt Leonard
Gas **S:** Break Time MFA Oil*(D), Phillips 66*(D)
Food **S:** Brownsville Station, People's Choice Restaurant
Lodg **S:** (AAA) People's Choice Motel
AServ **S:** Advanced Transmission, Total*(D)

71 MO K, MO EE, Marshall, Houstonia

74 MO YY
TStop **N: Betty's Truck Stop**
S: Standard
Food **N: Restaurant (Betty's Truck Stop)**
S: Restaurant (Standard)
Lodg **N: Motel (Betty's Truck Stop)**
TServ **S: Standard**

78AB U.S 65, Sedalia, Marshall
FStop **S: Breaktime***
Food **S: Cafe,** Countryside Palace
RVCamp **N: Camping**

84 MO J
Gas **N:** Citgo
Food **N:** Dairy Queen, Stuckey's
TServ **N: Fortune Truck & Trailer Repair**

89 MO K North, Blackwater
Gas **S:** Sinclair
Food **S:** Cafe (Sinclair)
Lodg **S:** Rustic Acres Motel
AServ **S:** MFA Oil

98 MO 135, MO 41, Arrow Rock, Pilot Grove
FStop **S: Conoco*(D)**
Gas **N:** Phillips 66
S: Texaco*
TServ **N: Mid Missouri Thermo King**

101 MO 5, Ashley, Boonville, Tipton
Gas **S:** Texaco*
Lodg **N:** ◆ **Comfort Inn**

103 MO B, Boonville, Main St, Bunceton
FStop **S: Amoco*(D), Shell (RV Dump) (see our ad on this page)**
Gas **N:** Phillips 66*, Texaco*
Food **N: McDonalds,** Subway, Taco Bell
S: Bobber Cafe
Lodg **N:** (DAYS INN) Days Inn, ◆ Super 8 Motel
AServ **N:** Phillips 66

(104) **Rest Area - RR, Phones, Picnic, Vending, Truck Parking P (Both Directions)**

106 Bus 70, MO 87, Boonville, Bingham Rd, Prairie Home
Gas **N:** Conoco*(D), Texaco*(D)
Food **N:** Subs & Salads
Lodg **N:** Atlasta Motel

111 MO 179, MO 98, Overton, Wooldbridge
Gas **S:** Phillips 66*(D)
AServ **S:** Phillips 66

115 MO BB, Rocheport
Food **S:** Gregory House Cafe
Lodg **N:** Missouri River Inn

117 MO J, MO O, Harrisburg Huntsdale

121 U.S 40, MO 240, MO UU, Fayette
TStop **N: Midway Travel Plaza(SCALES) (RV Dump)**

Bold red print shows RV & Bus parking available or nearby

EXIT MISSOURI

- **Food** N: Restaurant (Midway Travel Plaza)
- **Lodg** N: Budget Inn
- **TServ** N: Midway Travel Plaza

124 MO 740, MO E Columbia, Stadium Blvd
- **Gas** N: Break Time MFA Oil
  S: Shell*, Texaco*
- **Food** S: Alexandra's Steakhouse, Applebee's, Burger King, Dominio's Pizza, Furr's Family Dining, G-D Family Steaks, Golden Corral, Hardees, KFC, McDonalds, Old Chicago, Pizza Hut, Red Lobster, Ruby Tuesday, TCBY, Taco Bell, The Sub Shop, Wendy's
- **Lodg** S: ◆ Budgetel Inn, DAYS INN AAA Days Inn, ◆ Drury Inn, AAA Holiday Inn (see our ad on this page), Motel 6
- **AServ** S: Jiffy Lube
- **ATM** S: Commerce Bank, First National Bank
- **Other** S: Columbian Mall, K-Mart (Pharmacy), Target Department Store, Wal-Mart

125 West Blvd
- **Gas** S: Citgo*, Conoco*[D], Phillips 66*
- **Food** S: Fazoli's Italian Food, Hardees, Heritage House, Mr. Goodsense Subs & Pasta, Outback Steakhouse, Perkins, Ryan's Steakhouse, Yen Ching Chinese
- **Lodg** S: AAA Econolodge (see our ad on this page)
- **AServ** S: Citgo, Conoco, Firestone Tire & Auto
- **ATM** S: Merchantile Bank
- **Other** S: Car Wash

126 Providence Rd, MS163
- **Gas** N: Amaco*[CW], Conoco*, Texaco*
  S: Brake Time MFA Oil
- **Food** N: Country Kitchen, Shoney's
  S: Burger King, Hong Kong Chinese, Kim's Chinese, Long John Silvers, McDonalds, Pizza Hut, Taco Bell
- **Lodg** N: AAA Holiday Inn, Motel 6, ◆ Red Roof Inn
- **AServ** N: Conoco, McKnight Firstone
  S: Riley's Auto Parts
- **Med** S: ✚ Hospital

127 MO 763, to Range Line St
- **Gas** N: Brake Time MFA Oil
  S: Citgo*, Phillips 66*
- **Food** S: Burger King, Dairy Queen, Everett's, McDonalds, Pizza Hut, Rally's Hamburgers, Sonic, Subway
- **Lodg** N: Budget Host Inn, Motel 6, ◆ Ramada
  S: Super 7 Motel
- **AServ** S: Midas Muffler
- **Other** S: Auto Zone Auto Parts

128A U.S 63, Jefferson City, Moberly
- **Gas** N: QuikTrip*
  S: Break Time MFA Oil*
- **Food** N: Bob Evans, Burger King (Lgr indoor playground), Cracker Barrel, KFC, McDonalds, Steak & Shake, Taco Bell, Wendy's
  S: Haymarket Lounge (Best Western)
- **Lodg** N: Hampton Inn, ◆ Super 8 Motel
  S: AAA Best Western, AAA Comfort Inn, ◆ Holiday Inn Express, AAA Howard Johnson
- **AServ** N: AAMCO Transmission
- **Med** S: ✚ Columbia Regional Hospital, ✚ Keene Medical
- **Other** S: Nowell's Food (Food Mart)

131 Lake of the Woods Road
- **FStop** N: Shell
- **Gas** N: Phillips 66*
  S: Sinclair*
- **Food** N: J J's Cafe (Shell)

EXIT MISSOURI

133 MO Z, Centralia
- **Other** N: Love All's RV's Center

137 MO J & DD, Millersburg
- **Gas** S: Phillips 66*
- **Food** S: Dairy Queen
- **TServ** S: Freightliner Dealer

144 MO M, MO HH, to Hatton
- **RVCamp** S: Crooked Creek RV Park

148 U.S 54, Mexico, Fulton
- **TStop** S: Gasper's Truck Stop[SCALES], Petro[SCALES] (RV Dump)
- **FStop** N: Phillips 66*
  S: Mack Stop of Mid Missouri, Phillips 66
- **Gas** S: Shell*[D]
- **Food** N: Taco Bell
  S: Iron Skillet Restaurant (Gasper's Truck Stop),

EXIT MISSOU

- McDonalds, Restaurant (Gasper's Truck Stop), Subway
- **Lodg** S: ◆ Comfort Inn, DAYS INN Days Inn, ◆ Super 8 Mot
- **TWash** S: Truck Wash (Petro)
- **ATM** N: United Security Bank
- **Other** S: XVIII Wheelers Truck Wash

155 MO A, MO Z, Calwood, Bachelor

161 MO YY & D, Williamsburg
- **TStop** S: Shell
- **TServ** S: Shell

(166) Rest Area - RR, Phones P (Eastbound)

170 MO 161, MO J, Montgomery City, Danville
- **FStop** N: Phillips 66 Fuel Stop
- **Other** N: Grahan Cave State Park

175 MO 19, Hermann, New Florence
- **FStop** N: Amoco*, Shell*[D]
- **Gas** N: Phillips 66*
- **Food** N: Hardee's, Maggie's Cafe, McDonalds
- **Lodg** N: Royal Inn

179 MO F, High Hill
- **FStop** N: MFA Oil*
- **Food** S: Restaurant (Colonial Inn)
- **Lodg** S: Budget Motel, Colonial Inn

183 MO Y, MO NN, MO E, Jonesburg
- **Gas** S: Amoco*, Texaco*[D]
- **AServ** S: Schwartz Automotive
- **RVCamp** N: KOA Campground

188 MO A, MO B, Truxton
- **TStop** S: Flying J Travel Plaza[SCALES] (RV Dump)
- **Gas** S: Conoco
- **Food** S: The Cookery (Flying J Travel Plaza)

193 MO 47, Hawk Point, Warrenton
- **FStop** S: Texaco*[D]
- **Gas** N: Citgo*, Phillips 66*[D], Sinclair[CW]
  S: Amoco*, Coastal*, Conoco*
- **Food** N: Burger King, Dairy Queen, Jack-In-The-Box, McDonalds, Pizza Hut, Ron's Family Restauran, Subway
  S: Hardee's, KFC, Taco Bell
- **Lodg** N: DAYS INN Days Inn (see our ad on this page), Super 8 Motel
  S: Motel 6
- **AServ** N: Sinclair
  S: Auto Zone Auto Parts
- **Med** N: ✚ Doctor's Family Medicine
- **ATM** S: First Bank
- **Other** S: Wal-Mart

(199) Rest Area - RR, Phones, Vending, Picnic, Truck Parking P (Both Directions)

199 MO H, Wright City
- **Food** S: Fast & Company

200 MO J, MO F, Wright City (Difficult Reaccess)
- **Gas** N: Shell*[D][LP] (Kerosene)
  S: Phillips 66
- **Food** S: Big Boy, Restaurant (Phillips 66)
- **Lodg** S: Super 7 Inn

203 MO T & W, Foristell
- **TStop** N: 76 Auto/Truck Plaza[SCALES] (RV Dump)
- **FStop** N: Mr Fuel[SCALES]
  S: Texaco Truck & Auto*
- **Lodg** N: Best Western

208 Pierce Blvd
- **Gas** N: Citgo*, Texaco[D]
  S: Amoco*[CW], Texaco
- **Food** N: Domino's Pizza, KFC, McDonalds, Pizza Hut

EXIT MISSOURI

Stefanina Pizza, Subway, Wendy's
S: Burger King, Hardees, Isabell's Restaurant

Lodg S: AAA Super 8 Motel

AServ N: Goodyear Tire & Auto, NAPA Auto Parts, Phillips 66, Texaco
S: Texaco

Med N: ✚ **Doctor's Hospital**

ATM N: Commerce Bank
S: First Bank

Other S: **Wal-Mart**

9 MO Z, Church St, New Melle (Difficult Reaccess To 70 Westbound)

0AB U.S 40, U.S 61, Wentzville, Hannibal (Difficult Westbound Reaccess)

Gas N: Phillips 66*

Lodg N: Collier Hospitality, **Ramada Limited (see our ad on this page)**

AServ N: Phillips 66

2 CR A

Gas S: Citgo*

Food S: Imo's Pizza, Ponderosa Steakhouse

Lodg S: Holiday Inn

Other S: **A70 Veterinary Hospital**

4 Lake St Louis

Gas S: Phillips 66*, Shell*(CW)

Food S: Cutter's at the Wharf, Denny's, Hardee's, Subway

Lodg S: DAYS INN AAA Days Inn

Med S: ✚ **St. Joseph's Hospital**

ATM S: Mercantile Bank

6 Bryan Road

Food S: Blimpie's Subs

Lodg N: Super 8 Motel

TServ N: **Peter Bilt**

Other N: **Amerigas(LP)**

7 MO K & M, O'Fallon

Gas N: Clark, Huck's*, Mobil*, Vicker's*
S: Citgo*(D), Phillips 66*(D), Shell*(CW)

Food N: Baskin Robbins, Blimpie's Subs, Burger King, Hardee's, Jack-In-The-Box, Ponderosa Steakhouse, Rally's Hamburgers, Taco Bell, Waffle House
S: Fazoli's Italian Food, McDonalds, Papa John's Pizza, Shoney's

AServ N: Firestone Tire & Auto, Jiffy Lube, Valvoline Quick Lube Center
S: Auto Zone Auto Parts, Meineke Discount Mufflers, Midas Muffler & Brakes

ATM N: Nations Bank, Roosevelt Bank
S: Commerce Bank

Other S: **WalGreens**

20 MO 79, Elsberry, Louisiana

Gas S: Amoco*(CW)

Food S: Hardee's

RVCamp N: **Cherokee Lake**

22 MO C

3 MO C, Mid Rivers, Mall Drive, St Peters

FStop N: **QT*(D)**

Gas S: Mobil*(CW)

Food N: Burger King
S: Arby's, Bob Evans, China Wok, Dominio's Pizza, McDonalds, Olive Garden, Ruby Tuesday, Steak & Shake, Taco Bell, Wendy's

Lodg S: Dury Inn

AServ S: B&T Auto Service, Instant Oil Change

Other S: **Mid River Mall**

23) **Weigh Station - both directions**

25 to Ehlmann Road, Cave Springs, Truman Rd.

Gas N: Amoco*(CW), Citgo(D)
S: Citgo*(CW), Conoco, Mobil*, QuikTrip*

Food N: Wendy's
S: Burger King, Dairy Queen, Denny's, Jack in the Box, KFC, McDonalds, Pizza Hut, Ponderosa Steakhouse, Red Lobster, Shoney's, Steak 'N Shake, The Ground Round, The Pasta House, White Castle Restaurant

EXIT MISSOURI

Lodg N: ◆ Hampton Inn, Knight's Inn
S: ◆ Holiday Inn

AServ S: Conoco, Valvoline Quick Lube Center

ATM S: Nations Bank

Other N: **Target Department Store**
S: **Venture Discount Stores**

227 Zumbehl Road

Gas N: Texaco*
S: Amoco*(CW), Citgo(CW), Huck's Food Store*(LP)

EXIT MISSOURI

Food N: Burger King, Rural Route Country Cooking
S: Applebee's, Bob Evans Restaurant, Captain D's Seafood, Chevy's Mexican Restaurant, Fratelli's, Hardee's, Jack-In-The-Box, Judd's BBQ, Kriger's Grill, Lone Star, Luigi Buffet, McDonalds, Mr. Steak, Old Country Buffet, Oriental Palace, Popeye Chicken, Quizno's Subs, St. Louis Bread Company, Subway, Taco Bell, Twister's Frozen Custard

Lodg N: ◆ **Econolodge**
S: Comfort Inn, ◆ Red Roof Inn

AServ S: Jiffy Lube, Wal-Mart

Other S: **Kmart, Sam's Club, Wal-Mart**

228 MO 94, First Capital Dr, Weldon Springs

Gas N: Citgo*
S: Mobil(D)(CW), Shell(CW)

Food N: Arby's, Baskin Robbins, China Town Express, Dairy Queen, Imo's Pizza, Long John Silvers, Papa John's Pizza, Steak & Shake, Victoria's Kitchen, Wendy's
S: Chinese Express, Country Club Place, Fazoli's Italian Food, Gingham Home Style, Grappa Grill, Hot Shot's, Outback Steakhouse, Pizza Hut, Wilker's

AServ N: Auto Zone Auto Parts, Midas Muffler & Brakes, Valvoline Quick Lube Center
S: Dobb's Goodyear

Other N: **Parts America**

229B Bus 70N, 5th St

Gas N: Amoco*(CW), Mobil(D), Moto Mart* (ATM)
S: Phillips 66, QuikTrip*

Food N: Burger King, Denny's, Hardee's, Lee's Chicken, McDonalds, Waffle House
S: Old Country Store Cracker Barrel

Lodg N: ◆ Budgetel Inn, The Charles Inn
S: Fairfield Inn, Noah's Ark Best Western

AServ N: Meineke Discount Mufflers

Other N: **WalGreens**

231B Earth City Expwy

Lodg N: Harley Hotel

232 Junction I-270, Chicago, Memphis

234 MO 180W, St Charles Rock Rd

Gas N: Citgo*

Food N: Applebees, Hatfield & McCoy's Southern Cooking, Imo's Pizza, Long Star

AServ N: Instant Oil Change, Jiffy Lube, Meineke Discount Mufflers, Niehaus

Other N: **K-Mart**, Missouri Pride Car Wash

235A U.S 67S, Lindberg Blvd

235B U.S 67N, Lindbergh Blvd

Gas N: Shell*

Food N: China King, Duffy's (Scottish Inn), Holiday Inn (Scottish Inn), Restaurant (Scottish Inn)
S: Cancun, Hardee's, Lee's Kitchen, Piccadilly Restaurant, Radisson Hotel

Lodg N: Econolodge, Henry the 8th Inn & Lodge, Holiday Inn (see our ad on this page), Howard Johnson, Linair, Scottish Inns, Stanley Motel
S: Bridgeport Motel, Embassey Suites, Radisson Hotel

AServ S: Tire Station Warehouse

Other S: **Mall, Wal-Mart**

235C MO B, to Cypress Road

236 Lambert, St Louis Airport

Lodg S: Holiday Inn (see our ad this page)

237 MO 115

238B Junction I-170

239 Carson Road, Hanley Road

Gas S: Amoco(CW)

Food N: Jack-In-The-Box

EXIT MISSOURI/ILLINOIS

240AB Florissant Rd
- **Gas** **N:** Clark
  S: Q.T.*
- **Food** **N:** Dairy Queen
  S: Wendy's
- **Other** **N:** Coin Laundry, WalGreens

241 Bermuda Ave
- **Gas** **S:** Sinclair*

241B Cty Road U, Lucas and Hunt Road
- **Gas** **N:** Shell

242B Jennings Station Road
- **Gas** **N:** Citgo, Texaco*
  S: Shell*

243 Goodfellow Blvd
- **Gas** **N:** Phillips 66* (Kerosene), Shell*

243B Bircher Blvd, Riverview Blvd

244B Kingshighway

245A Shreve Ave

245B W Florissant Ave

246A Broadway (No Reaccess)
- **FStop** **N:** Phillips 66
- **Gas** **N:** Amoco
- **TServ** **N:** Freightliner Dealer, Phillips 66

246B Adelaide Ave
- **Other** **N:** Safe Way Grocery

247 Grand Ave
- **Gas** **N:** Citgo

248A Salsbury St, McKinley Bridge
- **Gas** **S:** Amoco, Mobil*

248B Branch St (No Reaccess To Interstate)
- **FStop** **N:** North Broadway Truck Stop

249A Madison St

249 Junction I-55, I-70 West

↑MISSOURI

↓ILLINOIS

Note: I-70 runs concurrent below with I-55. Numbering follows I-55.

1 IL 3, Sauget

2B 3rd St (Southbound, Reaccess Northbound Only)

2A ML King Bridge, Downtown St Louis (Southbound Exit, Reaccess Northbound Only)

2 Junction I-64, IL 3, St Clair Avenue, Louisville

3 Exchange Ave (Southbound, Reaccess Northbound Only)

4AB IL 203, Granite City, Colinsville Road
- **TStop** **W:** America's Best Truck Stop*(SCALES), Texaco*(SCALES)
- **Food** **W:** Big Duga's Restaurant(SCALES)
- **TServ** **W:** Texaco
- **Other** **W:** Gateway International Raceway, Volvo Dealer

6 IL 111, Great River Road, Wood River, Washington, Park
- **Gas** **E:** Clark*(D)
- **Food** **E:** Rainbo Motel

EXIT ILLINOIS

- **Lodg** **E:** Rainbo Court Motel
- **AServ** **E:** B&E Auto Repair
- **TServ** **E:** Custom Auto & Truck Accessories
- **RVCamp** **E:** Safari RV Park
- **Other** **E:** Cahokia Mounds Historic Sight, FC Grocery & Pharmacy, Foodland Supermarket (Pharmacy), Venture

9 Black Lane (Northbound)

10 Junction I-255, I-270, Memphis

11 IL 157, Collinsville, Edwardsville
- **Gas** **E:** Amoco*, Mobil*, Shell*(CW)
  W: Moto Mart*(D)
- **Food** **E:** China Palace, Denny's (Pear Tree Inn), Hardee's, Long John Silvers, McDonalds, Pizza Hut, The Pub Lounge & Grill, Waffle House, Wendy's
  W: Arby's, Bob Evans Restaurant, Boston Market, Burger King, Cancun Mexican, Dairy Queen, Ponderosa, Porter's (Holiday Inn), Shoney's, Steak & Shake, White Castle Restaurant
- **Lodg** **E:** ◆ Best Western, Days Inn, Howard Johnson, Motel 6, Pear Tree Inn
  W: Comfort Inn, ◆ Drury Inn, Fairfield Inn, Hampton Inn, Holiday Inn, Ramada Limited, Super 8 Motel
- **AServ** **E:** Amoco, Midas Muffler & Brakes
- **ATM** **W:** Magna Carta
- **Other** **E:** Car Wash
  W: 4 Seasons Auto Wash, Convention Center, State Patrol Post

(14) Weigh Station (Southbound)

15AB IL 159, Collinsville, Maryville
- **Gas** **W:** Conoco*
- **Lodg** **W:** Best Western
- **AServ** **W:** Conoco

17 U.S 40 East, Saint Jacob (Northbound)

18 IL 162, Troy
- **TStop** **E:** Shell(SCALES)
- **Gas** **W:** Mobil*(LP)
- **Food** **E:** Burger King, Dairy Queen, Hardee's, Jack-In-The-Box, McDonalds, Pizza Hut
  W: China Garden, Cracker Barrel, Randy's Restaurant, Taco Bell
- **Lodg** **E:** Relax Inn
  W: Scottish Inns, ◆ Super 8 Motel
- **TServ** **E:** Amoco Auto/Truck Stop*(SCALES), Freightliner Dealer, Speedco Truck Lube
- **TWash** **E:** 18 Wheeler's Truck Wash
- **Med** **E:** ✚ Hospital
- **ATM** **E:** Magna Bank
- **Other** **E:** Super Valu Grocery

Note: I-70 runs concurrent above with I-55. Numbering follows I-55.

15 Junction I-55 North, I-270 West

20AB I-70 E Indiapolis, I-270 W Kansas City

21 IL 4, Lebanon, Staunton
- **FStop** **N:** A.D.R Service
- **Gas** **N:** Mobil
- **TServ** **N:** A.D.R Service
  S: Dorsey Trailor Service

24 IL 143, Marine, Highland
- **Med** **S:** ✚ Hospital

(26) Rest Area - RR, Phones, Picnic P (Eastbound)

EXIT ILLINC

(27) Rest Area - RR, Phones, Picnic, Vending P (Westbound)

30 IL 143, U.S 40, Highland, Pierron
- **Gas** **S:** Shell*
- **Food** **S:** Blue Springs Restaurant
- **RVCamp** **S:** Tomahawk Campground (Seasonal)
- **Med** **S:** ✚ Hospital

36 Pocahontas
- **Gas** **S:** Amoco*(D), Phillips 66(D)
- **Food** **S:** Restaurant (Powhatan)
- **Lodg** **S:** Powhatan Hotel, Tahoe Motel, Wikiup Mote
- **AServ** **S:** Phillips 66, Shuster Repair
- **TServ** **S:** Auto Truck Tires & Service, Shuster Repai Auto Truck & Tire

41 Greenville, East U.S 40 (Eastbound Westbound Reaccess)

45 IL 127, Greenville, Carlyle Lake, Carlyle
- **FStop** **N:** Shell*
- **Gas** **N:** Amoco*
- **Food** **N:** Chang's Chinese Restaurant, KFC, Lu-Bob Restaurant, McDonalds, Southern Edge Restaurant & Lounge
  S: Circle B Steak House
- **Lodg** **N:** 2 Acres Motel, Best Western, Budget Host Inn, Super 8 Motel
- **AServ** **N:** Amoco, Ford Dealership
- **RVCamp** **N:** Campground
- **ATM** **N:** Shell
- **Other** **N:** Classic Touch Car Wash, Grandpa's Groce Store
  S: Airport

52 Mulberry Grove, Keyesport
- **Gas** **N:** Citgo*(D)
- **TServ** **N:** Elliot
- **RVCamp** **N:** Timber Trails Camping
  S: Camping

61 U.S 40, Vandalia
- **Food** **S:** Pondorosa
- **Lodg** **S:** Ramada Limited
- **Other** **S:** Wal-Mart

63 U.S 51, Vandalia, Pana
- **Gas** **S:** Amoco*(CW), Clark*(D), Marathon*(LP)
- **Food** **N:** Chuckwagon Restaurant, Days Inn, Long John Silvers
  S: Dairy Queen, Hardee's, Jay's Family Dinin KFC, McDonalds, Pizza Hut, Subway, Wendy'
- **Lodg** **N:** Days Inn
  S: Jay's Inn, Travelodge
- **AServ** **S:** Amoco
- **Med** **S:** ✚ Hospital
- **ATM** **S:** Citizen's Bank, National Bank
- **Other** **S:** Aldi Supermarket, Coin Laundry, Illinois State Capitol, The Medicine Shop Pharmacy Tourist Information

68 U.S 40, Brownstown, Bluff City
- **RVCamp** **N:** KOA Campground (April-October)

(72) Weigh Station

76 St Elmo
- **Gas** **N:** Phillips 66*
- **Food** **S:** Country Harvest Restaurant
- **RVCamp** **N:** Bales Timber Line Lake

82 IL 128, Altamont
- **FStop** **N:** Speedway*
- **Gas** **N:** Citgo*(CW)

Bold red print shows RV & Bus parking available or nearby

XIT ILLINOIS

S: Phillips 66*(D)
Food N: Dairy Bar, **McDonalds, Restaurant (Best Western)**, Stuckey's (Citgo), Subway (Citgo)
S: **Gilbert's Family Restaurant**
Lodg N: Altamont Hotel, AAA Best Western
S: ◆ Super 8 Motel
ATM N: Citgo

87) **Rest Area - RR, Phones, Picnic, Vending** P

2 Junction I-57 Chicago, Memphis

Note: I-70 runs concurrent below with I-57. Numbering follows I-57.

59 Effingham, Fayette Ave.
TStop E: **76 Auto/Truck Plaza**(LP)(SCALES)
W: **Petro**(SCALES) **(Open 24 Hours), Truck-O-Mat**(SCALES)
FStop E: **Speedway***(SCALES) **(RV Dump)**
Gas E: Amoco*(CW), Clark, Phillips 66*(D)(CW), Shell* (Open 24 Hours), **Speedway***(D)(SCALES)
Food E: Domino's Pizza, G Wilkes Bar and Grill, Golden China Restaurant, **Hardee's**, Little Caesars Pizza, Niemerg's Family Dining, Spaghetti Shop, Subway, The China Buffet
W: Petro
Lodg E: Abe Lincoln Hotel, Comfort Suites, DAYS INN ◆ **Days Inn**, Holiday Inn, **Howard Johnson**, Paradise Inn (see our ad on this page), ◆ **Quality Inn**
W: AAA **Best Western**
AServ E: Amoco, Effingham Tire Center, Firestone Tire & Auto, Pennzoil Lube, Shell
TServ E: **76 Auto/Truck Plaza, Effingham/ International Truck Sales, Firestone Tire & Auto**
W: **Petro**
TWash E: **76 Auto/Truck Plaza**
W: **Truck-O-Mat**
Other E: Car Wash, **Sav-A-Lot Grocery**

160 IL 32, IL 33
TStop W: **Bobbers Truck Plaza***(SCALES) **(RV Dump), Trucks America Truck Plaza***(SCALES)
Gas E: Amoco*, Shell
W: Phillips 66, Shell*(D)
Food E: Dixie Cream Donut Shop, Little Caesars Pizza, Papa John's Pizza, Pizza Hut
W: Arby's, Blimpie's Subs, Bobber Restaurant, Bonanza Steak House, Burger King (Outdoor Playground), **Cracker Barrel**, Denny's, El Rancherito, K Square Food Court, KFC, Long John Silvers, **McDonalds (Inside Wal-Mart)**,

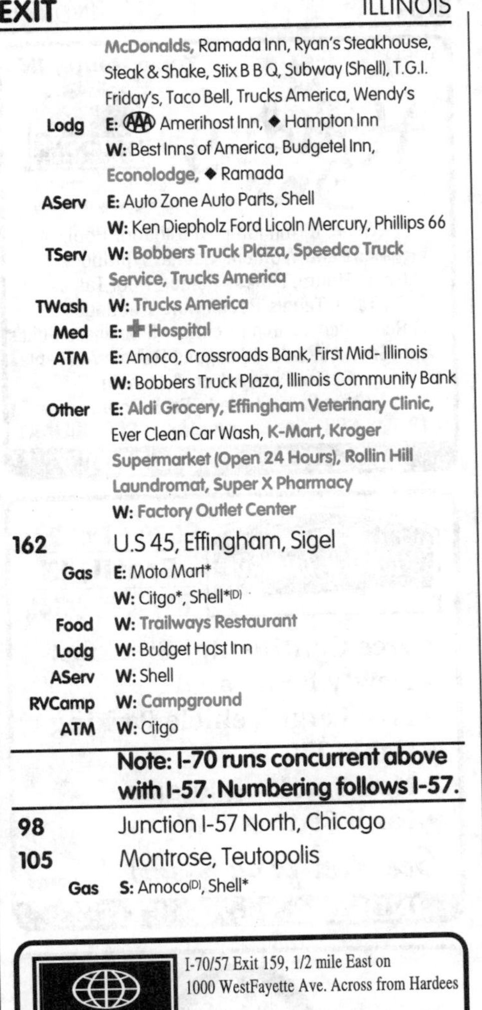

EXIT ILLINOIS

McDonalds, Ramada Inn, Ryan's Steakhouse, Steak & Shake, Stix B B Q, Subway (Shell), T.G.I. Friday's, Taco Bell, Trucks America, Wendy's
Lodg E: AAA Amerihost Inn, ◆ Hampton Inn
W: Best Inns of America, Budgetel Inn, **Econolodge**, ◆ Ramada
AServ E: Auto Zone Auto Parts, Shell
W: Ken Diepholz Ford Licoln Mercury, Phillips 66
TServ W: **Bobbers Truck Plaza, Speedco Truck Service, Trucks America**
TWash W: **Trucks America**
Med E: ✚ **Hospital**
ATM E: Amoco, Crossroads Bank, First Mid- Illinois
W: Bobbers Truck Plaza, Illinois Community Bank
Other E: **Aldi Grocery, Effingham Veterinary Clinic**, Ever Clean Car Wash, **K-Mart, Kroger Supermarket (Open 24 Hours), Rollin Hill Laundromat, Super X Pharmacy**
W: **Factory Outlet Center**

162 U.S 45, Effingham, Sigel
Gas E: Moto Mart*
W: Citgo*, Shell*(D)
Food W: **Trailways Restaurant**
Lodg W: Budget Host Inn
AServ W: Shell
RVCamp W: **Campground**
ATM W: Citgo

Note: I-70 runs concurrent above with I-57. Numbering follows I-57.

98 Junction I-57 North, Chicago

105 Montrose, Teutopolis
Gas S: Amoco(D), Shell*

EXIT ILLINOIS/INDIANA

Lodg S: Montarosa Motel
AServ S: Amoco

119 IL 130, Greenup, Charleston
Gas S: 500 Platolene, 76*(D), Amoco*, Phillips 66*, **Shell***
Food S: Dairy Queen, **Dutch Pantry Restaurant**, Subway (Phillips 66)
Lodg S: BudgetHost Inn, Gateway Inn
AServ S: GM Dealer
RVCamp S: **Camping**
ATM S: Phillips 66
Other S: **Lincoln Log Cabin Home, Tourist Information**

129 IL 49, Casey, Kansas
FStop S: **Citgo**
Gas S: Amoco*(CW)
Food S: **Dairy Queen (Citgo), Hardee's**, Joe's Pizza, **KFC**, McDonalds, Pizza Hut
Lodg S: Casey Motel, ◆ **Comfort Inn**
AServ S: Staley's Complete Tire Service
RVCamp S: **KOA Campground**
Other S: Car Wash

136 Martinsville
RVCamp S: **Camping**

147 IL 1, Marshall, Paris
FStop S: **Phillips 66***(LP)
Gas S: Jiffy* (Kerosene), Shell*
Food N: Ike's Great American Restaurant
S: Burger King, **Hardee's, McDonalds**, Subway, Wendy's
Lodg S: Peak's Motor Inn, Super 8 Motel
AServ N: Randy's Auto
S: Phillips 66
RVCamp S: **Camping**
ATM S: Shell
Parks S: **Lincoln Trail State Park**

(149) **Rest Area - RR, Phones, Picnic, Vending** P **(Westbound)**

(151) **Weigh Station (Westbound)**

154 U.S 40, West

↑ILLINOIS

↓INDIANA

1 U.S 40 East, Terre Haute (Left Exit, Eastbound, No I-70 East)
Lodg N: Stone Lodge Motel

(2) **Rest Area - RR, Phones, Picnic, Tourist Info, Vending (Eastbound)**

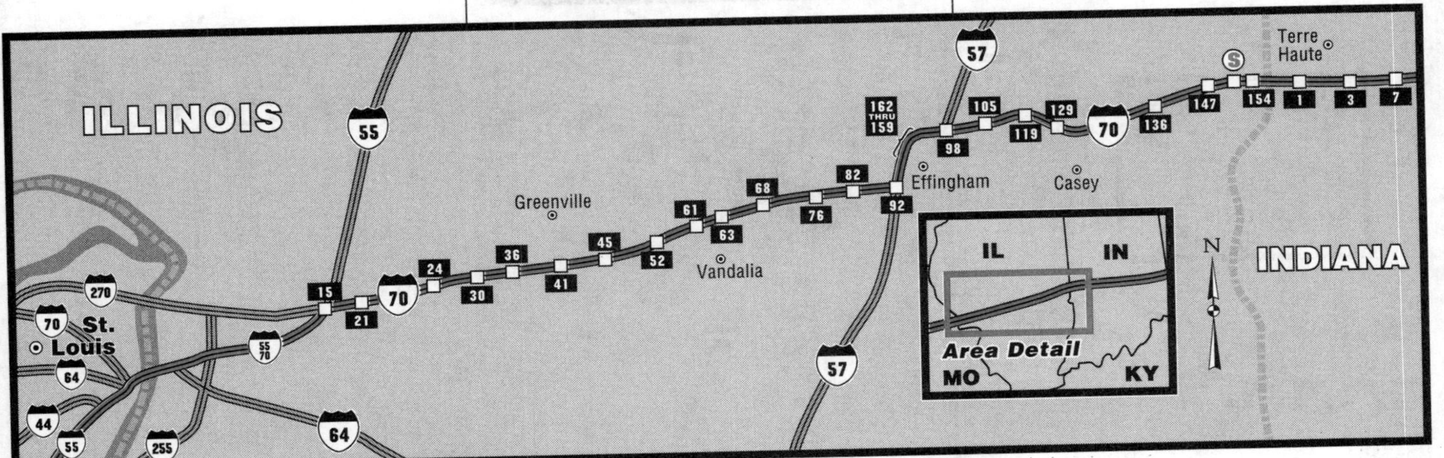

Bold red print shows RV & Bus parking available or nearby

EXIT		INDIANA
3		Darwin Road, West Terre Haute
7		U.S 41, U.S 150, Terre Haute, Evansville
	Gas	N: Amoco*(CW), Big Foot*, Marathon*(D)(CW), Thornton's*(D)(LP) (Kerosene)
		S: Jiffy*(CW), Shell*, Speedway*(LP)
	Food	N: Apple House, Applebee's, Arby's, Bob Evans Restaurant, Cracker Barrel, Dunkin Donuts, Fazoli's Italian Food, Little Caesars Pizza, Peking Chinese, Pizza City, Pizza Hut, Shoney's, Steak & Shake, Texas Road House, Tumbleweed Southwestern Bar & Grill
		S: Be Bops, Can Cun Mexican Restaurant, Chi Chi's Mexican, Dairy Queen, Damon's, Denny's, Garfield's, Hardee's, Holiday Inn, Jade Garden Chinese, Laughner Cafeteria, McDonalds, Olive Garden, Outback Steakhouse, Panda Garden, Pizza Inn, Ponderosa, RAX, Rally's Hamburgers, Red Lobster, Subway, Taco Bell, Wendy's
	Lodg	N: ◆ Comfort Suites, Dollar Inn, ◆ Drury Inn, ◆ Fairfield Inn, ◆ Knight's Inn, Peartree Inn, ◆ Signature Inn, ◆ Super 8 Motel
		S: Best Western, Holiday Inn (see our ad on this page), Motel 6
	AServ	N: Auto Zone Auto Parts, Burger Chrysler Plymouth Dodge, Marathon, Napa Auto Parts, Q-Lube, Southern Indiana Tire
		S: Bowen Olds GMC, Goodyear Tire & Auto, Sears Auto Center
	Med	S: ✚ Ambucare Clinic, ✚ Hospital
	ATM	N: Amoco, Thornton's
		S: 1st Bank & Trust, Citizen's Bank, Merchant's National Bank
	Other	N: Mike's Market
		S: Car Wash, Honey Creek Animal Hospital, Honey Creek Mall, Kroger Supermarket (Pharmacy), Office Max, Osco Drugs, Pet Land, Phar Mor Drugs, Sears Dept Store, Service Merchandise, Toys "R" Us
11		IN 46, Bloomington, Riley, Terre Haute
	TStop	N: Pilot Travel Center(SCALES)
	Gas	N: Thornton's Food Mart*(LP) (Kerosene)
	Food	N: Arby's (Pilot), McDonalds, Subway (Thornton's), TJ Cinnamon's (Pilot)
	AServ	N: Macallistar
	TServ	N: Russ Fisher Truck Parts
	RVCamp	S: KOA Campground
	ATM	N: Thornton's Food Mart
	Parks	N: Hawthorne Park
23		IN 59, Brazil, Linton
	TStop	S: AmBest*(SCALES)
	FStop	S: Speedway*(SCALES)
	Gas	S: Kocolene*, Shell
	Food	S: Brazil 70 Restaurant, Burger King, Ike's Grec American Restaurant, Subway (Speedway)
	Lodg	S: Howard Johnson (see our ad on this page)
	AServ	N: Red Bird Garage
	TServ	N: Red Bird Garage
		S: AmBest, Goodyear Tire & Auto
	Med	N: ✚ Hospital
37		IN 243, Putnamville
	Parks	S: Lieber State Recreation Area
	Other	N: State Patrol Post
41		U.S 231, Greencastle, Cloverdale
	TStop	S: Phillips 66 Travel Plaza*(SCALES)
	Gas	S: Amoco*(D)(CW), Shell*
	Food	N: Steakhouse
		S: Burger King, Chicago Pizza, Hardee's, KFC, McDonalds, Quality Inn, Subway (Phillips 66), Taco Bell, Wendy's
	Lodg	N: Midway Motel
		S: Briana Inn, Days Inn (see our ad on this page), Dollar Inn, Holiday Inn Express, Quality Inn
	AServ	N: Darren's I-70 Service, Thompson's Collision Center
		S: Andy Mohr Chevrolet Trucks, Clover Tire, She
	TServ	S: Phillips 66
	TWash	S: Phillips 66
	RVCamp	N: Cloverdale RV Park
	Med	N: ✚ Hospital
	ATM	S: Amoco, First National Bank
52		CR 1100 West, Little Point
	TStop	S: Phillips 66*
	Gas	S: Coger's Country Mart*
	Food	S: Restaurant (Phillips 66)
	AServ	S: Curtis Garage & Wrecker Service
	TServ	S: Cogers Garage & Wrecker Service, Curtis Garage & Wrecker Service
59		IN 39, Belleville, Monrovia
	TStop	S: 76 Auto/Truck Plaza*(SCALES), Blue & White(SCALES)
	Gas	N: Marathon*
	Food	S: 76 Restaurant, Happ's Place (Blue & White)
	Lodg	N: Canary Motel
	TServ	S: 76, Blue & White
	TWash	S: Blue & White
	ATM	S: 76
(64)		Rest Area - RR, Phones, Picnic, Vending

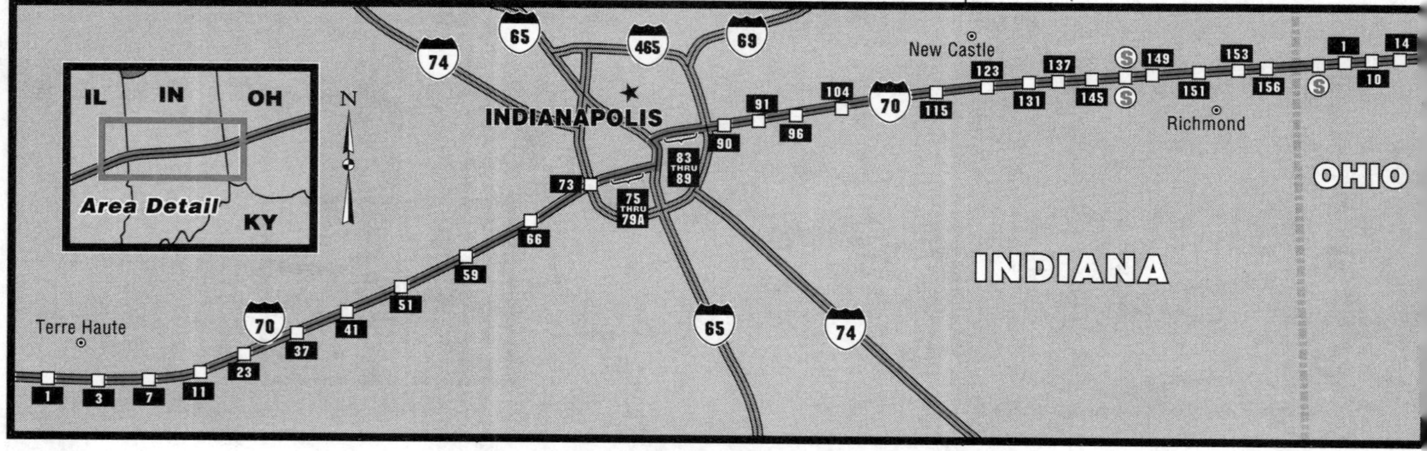

Bold red print shows RV & Bus parking available or nearby

XIT	INDIANA
6	IN 267, Plainfield, Mooresville
Gas	N: Amoco*, Shell*(CW), Speedway*(D)(LP), Thornton's*
Food	N: Arby's, Burger King, Cracker Barrel, Krispy Kreme (Thornton's), McDonalds, Royal Line Pizza, Steak 'N Shake, Subway, The Coachman, Wendy's, White Castle Restaurant
Lodg	N: Ameri Host Inn (see our ad on this page), Days Inn (see our ad on this page), ◆ Holiday Inn Express, Lee's Inn & Suites, Super 8 Motel
ATM	N: Shell
8	I-70 Jct I-465
3A	Junction I-465 South/I-74 East
3B	Junction I-465 North/I-74 West
5	Airport Expwy to Raymond St
7	Holt Road
Food	S: McDonalds
AServ	S: All Pro Auto & Truck Parts
TServ	S: All Pro Auto & Truck Parts, Fruehauf Truck Service
8	Harding St
9A	West St
9B	Illinois St, McCarty St
0	I-65 South, Louisville
3A	Michigan St, Indianapolis (Westbound)
3B	Junction I-65 North
5AB	Rural St, Keystone Ave
Food	S: McDonalds
7	Emerson Ave
Med	S: Hospital
9	Junction I-465 eastbound, Shadeland Ave
Lodg	S: Budget Inn (see our ad on this page)
0	Jct I-465 (Westbound To Ltd Access)
Food	S: JoJo's
Lodg	S: La Quinta Inn
1	Post Road, Fort Harrison
Gas	N: Marathon*, Swifty
	S: Amoco*(CW), Clark*, Shell*(CW)
Food	N: Cracker Barrel, Outback Steakhouse, Steak & Shake, Wendy's
	S: Hardee's, Quality Inn, Taco Bell
Lodg	N: ◆ Budgetel Inn
	S: Dollar Inn, Quality Inn, Super 8 Motel
ATM	N: Marathon
	S: Bank One, Union Federal Savings Bank
Other	N: Logo 7, Lowe's
	S: Car Wash, Kim's Oriental Grocery, Marsh Grocery, Osco Drugs, State Patrol Post
6	Mount Comfort Road
TStop	S: Pilot Travel Center*(SCALES)
FStop	N: Speedway*(SCALES) (Kerosene)
Gas	S: Shell*
Food	S: Grandma's Kitchen (Pilot), McDonalds, Subway (Pilot)
TServ	S: Pilot Travel Center
RVCamp	S: KOA Campground
04	IN 9, Maxwell, Greenfield
Gas	N: Gas America*(LP)
	S: Gas America*, Sunoco*
Food	S: Bob Evans Restaurant, Shoney's, White Castle

EXIT	INDIANA
	Restaurant
Lodg	S: Dollar Inn, Holiday Inn Express, Lees Inn, ◆ Super 8 Motel
RVCamp	N: Camping
	S: Camping
Med	S: Hospital
ATM	N: Gas America
(107)	Rest Area - RR, Phones, Vending, Picnic (Eastbound)
(114)	Rest Area - RR, Phones, Picnic, Vending (Westbound)
115	IN 109, Wilkinson, Knights Town
TStop	N: Gas America*
RVCamp	N: Campground
123	IN 3, New Castle, Spiceland
TStop	S: BP*(SCALES)
FStop	S: Marathon* (Kerosene)
Gas	N: Shell*, Speedway*
Food	N: Denny's
	S: BP Restaurant
Lodg	N: ◆ Days Inn
TServ	N: Hartley Truck Parts
	S: BP
RVCamp	N: Camping
Med	N: Hospital
Other	N: Indiana Basketball Hall of Fame
131	Wilbur Wright Road, New Lisbon
TStop	S: 76*(SCALES)
Food	S: Country Harvest (76 TS), KFC (76 TS)
TServ	S: 76
RVCamp	S: Camping
Other	N: Wilber Wright Birthplace & Museum
137	IN 1, Connersville, Hagerstown
TStop	S: Marathon*
Gas	S: Gas America*(LP), Shell*
Food	N: Dutch Mill Cheese Shoppe (Amish)
	S: Burger King (Shell), Marathon
Lodg	S: Cambridge City Inn*(SCALES)
AServ	N: Road One Towing
TServ	S: Marathon
RVCamp	N: Camp Dakota
ATM	S: Gas America, Marathon, Shell
(144)	Rest Area - RR, Phones, Vending, Picnic, Motorist Service Info
145	Centerville
Gas	N: Amoco
Food	N: Dairy Queen (Amoco), Stuckey's (Amoco)
Lodg	N: Super 8 Motel
AServ	N: Goodyear Truck Service
TServ	N: Goodyear Truck Service
RVCamp	S: Camping
(148)	Weigh Station - both directions
149AB	U.S 35, IN 38, Williamsburg Pike
AServ	N: Pardo's Auto & Service
	S: Westside Auto Service
TServ	N: Pardo's Auto & Truck Service
151AB	U.S 27, Chester Blvd
Gas	S: Sunoco*
Food	N: T-Bird's Cafe
	S: Burger King, Frisch's Big Boy, Pizza Hut, Richard's, Subway, Wendy's
Lodg	S: Comfort Inn, Howard Johnson (see our ad on this page)
RVCamp	N: KOA Campground
Med	S: Hospital

Bold red print shows RV & Bus parking available or nearby

EXIT IDIANA/OHIO

ATM S: Star Bank
Other S: Automatic Car Wash

153 IN 227, Union City
RVCamp N: Campground
S: Campground

156AB U.S 40, Richmond
TStop N: Petro*(CW)(LP)(SCALES)
FStop N: Fuel Mart*
Gas N: Swifty*
S: Shell*
Food N: Iron Skillet (Petro), Subway (Fuel Mart)
S: Bob Evans Restaurant, Cracker Barrel, Fazoli's Italian Food, Holiday Inn, Ice Cream Station, Jerry's Restaurant, KFC, McDonalds, Ponderosa Steakhouse, Red Lobster, Ryan's Steakhouse, TCBY Treats
Lodg N: Fairfield Inn
S: Dollar Inn, Holiday Inn, Knight's Inn, Lee's Inn, Ramada Inn
AServ S: Crown Chrysler Jeep Eagle, Ford Dealership, GM Auto Dealership, Indy Lube, Midas Muffler & Brakes, Shell
TServ N: Petro
TWash N: Blue Beacon Truck Wash (Petro), Interstate 70 Highway 40 Truck Wash
ATM N: Petro
S: Star Bank
Other S: Big Lots, Lowe's, Old National Road Welcome Center, Target Department Store, Thornburg's Quality Foods, U-Haul Center(LP)

↑INDIANA

↓OHIO

(1) Weigh Station (Eastbound)

1 U.S 35 East, Eaton (Eastbound, Reaccess Westbound Only)

(3) Rest Area - RR, Phones, Picnic, Visitor Info, Vending

10 U.S 127, Greenville
TStop N: TravelCenters of America*(SCALES)
FStop S: Pilot Travel Center*(SCALES)
Gas N: BP (TravelCenters of America)
Food N: Country Pride Restaurant (TravelCenters of America), Subway
S: Country Cooker, Dairy Queen (Pilot)
Lodg S: AAA Econolodge
TServ N: TravelCenters of America
ATM S: Pilot Travel Center
Other N: Highway Patrol

14 Ohio 503, Lewisburg, North Alexandria
Gas N: Sunoco*(CW)(LP) (Kerosene)
S: Marathon*(D)(LP)
Food N: Covered Bridge Restaurant, Dari-Twist, Subway (Sunoco)
Lodg N: Super Inn
ATM N: Lewisburg Bank
S: Marathon
Other N: Brennan's Grocery

21 County Road 533, Brookville
FStop S: Speedway*(LP)
Gas S: BP
Food S: Arby's, KFC, McDonalds, Robs Family Dining, Subway (Speedway), Waffle House (Speedway), Wendy's

EXIT OHIO

Lodg S: Days Inn AAA Days Inn
AServ S: BP

24 Ohio 49 North, Clayton, Greenville
RVCamp S: KOA Campground
Other N: Clayton Animal Hospital

26 Ohio 49 South, Trotwood, Hoke Rd

29 Ohio 48, Englewood
FStop N: Cummins Diesel
Gas N: BP*, Sunoco*(CW)(LP) (Kerosene)
S: Meijer Grocery*(D)
Food N: Bob Evans Restaurant, Frisch's Big Boy, Holiday Inn, Perkins Family Restaurant
S: McDonalds, Waffle House
Lodg N: Dollar Inn, ◆ Hampton Inn, ◆ Holiday Inn, Motel 6
S: AAA Cross Country Inn
TServ N: Cummins Diesel
Other S: Meijer Grocery

32 Dayton International Airport

33AB Junction I-75, Toledo, Dayton

36 OH 202, Huber Heights
FStop N: Speedway*(SCALES)
Gas N: Shell, Speedway*
S: BP*(D), Sunoco*(D)
Food N: Applebee's (Mall), Big Boy, Fazoli's Italian Food, Kenny Rogers Roasters, Ruby Tuesday, Skyline Chili, Steak & Shake, Taco Bell, Uno Chicago Bar & Grill, Waffle House, Wendy's
S: Arby's, Bob Evans Restaurant, Boston Market, Burger King, Cadillac Jack Sports Bar & Grill, Cold Beer & Cheeseburgers, Friendly's, Golden Bay, Long John Silvers, McDonalds, Old Country Buffet, White Castle Restaurant
Lodg N: ◆ Super 8 Motel
S: Days Inn ◆ Days Inn
AServ N: Shell, Wal-Mart
S: BP, K-Mart
ATM N: Star Bank (Cub Foods)
S: Citizens Federal, Huntington Bank, National City Bank
Other N: Cub Foods (Pharmacy), Mail Boxes Etc, North Park Mall, PetsMart, Wal-Mart (Pharmacy)
S: K-Mart (Pharmacy), Kroger Supermarket (Pharmacy), Revco Drugs

38 OH 201, Brandt Pike
Gas N: Amoco*
S: Shell*(CW)
Food S: Denny's, Waffle House
Lodg S: Comfort Inn, AAA Travelodge
AServ S: Wilson's Auto Service
Other N: Barney Rental(LP)

41AB OH 4, OH 235, Fairborn, New Carlisle
Gas N: BP*(D)
RVCamp S: Camping
ATM N: BP

44 I-675 South, Fairburn, Cincinnati

47 OH 4 , Springfield (Eastbound, Westbound Reaccess)
RVCamp N: Campground

48 Emon, Donnelsville (Westbound, Eastbound Reaccess)

52AB U.S 68, Urbana, Xenia

54 OH72, Cedarville, Springfield

EXIT OHI

Gas N: BP*, Shell*(CW)(LP), Speedway* (Kerosene), Sunoco*(D)
S: Swifty (Kerosene)
Food N: Bob Evans Restaurant, Cassano's Pizza & Subs, Cracker Barrel, Denny's, KFC, Lee Famous Recipe Chicken, Little Caesars Pizza, Long John Silvers, Mark Pi's Chinese, McDonal Panda Chinese, Perkins Family Restaurant, Ponderosa, Rally's Hamburgers, Subway, Taco Bell, Wendy's
Lodg N: Hampshire Motel, Holiday Inn, Imperial House Motel, Ramada Limited
Med N: ✚ Hospital
Other N: Aldi Grocery, Coin Car Wash, Convenience Mart, Fulmer Grocery, Rite Aide Pharmacy, Super Duper Supermarket

59 OH 41, South Charleston
FStop S: Prime Fuel*
AServ S: Dan's Towing
Med N: ✚ Hospital
ATM S: Prime Fuel
Other S: Zip In Drive Through Convenience Store*

62 U.S 40, Springfield
Gas N: Speedway*(D)(LP) (Kerosene)
Food N: Curly Top Ice Cream
S: Pizza Pause
Lodg N: Harmony Motel
AServ N: Butler's Service, Suburban Automotive
RVCamp S: Beaver Valley Resort (.2 Miles), Crawford's Campground (1.5 Miles)
Parks N: Buck Creek State Park
Other N: State Patrol Post

66 OH 54, Catawba, South Vienna
FStop N: Fuel Mart* (Kerosene)
Gas S: Speedway*
RVCamp S: Crawfords Campground

(71) Rest Area - RR, Phones, Picnic, Vending

72 OH56, Summerford, London
Gas N: Shell*(LP) (Kerosene)
Med S: ✚ Hospital

79 U.S 42, London
TStop S: 76 Auto/Truck Plaza*(SCALES)
FStop N: Circle-K*
S: Speedway*(SCALES)
Gas N: BP*(LP)
S: Sunoco*
Food N: Olde Iron Kettle, Waffle House
S: 76 Auto/Truck Plaza (TS), Blimpie's Subs (Speedway), McDonalds, Pizza Hut (76 Auto TS), Taco Bell (TS), Wendy's
Lodg S: Holiday Inn Express, Trail's Inn
TServ S: 76 Auto/Truck Plaza
TWash S: Red Barren (76 Auto TS)
RVCamp N: RV Headquarters*(LP)
Med S: ✚ Hospital
ATM S: 76 Auto/Truck Plaza

80 OH29, Mechanicsburg, Urbana
Other S: OH State Patrol Post

85 OH 142, West Jefferson, Plain City

91AB Hilliard, New Rome
Gas N: Meijer Grocery*(D), Shell*(LP), Super America*(D)
S: BP*(CW), Marathon*(CW)
Food N: Arby's, Burger King, Cracker Barrel, Danoto's Pizza, Fazoli's Italian Food, Frisch's Big Boy, KFC, McDonalds, Perkins Family Restaurant

Bold red print shows RV & Bus parking available or nearby

XIT		OHIO
		Subway, Wendy's, White Castle Restaurant
		S: Bob Evans Restaurant, Minelli's Pizza, Subway (Marathon)
	Lodg	**N:** Cross Country Inn, Fairfield Inn, Hampton Inn, Motel 6, Red Roof Inn
		S: Best Western
	AServ	**N:** Monroe Muffler & Brake, Wal-Mart
		S: Marathon
	TServ	**N:** Fyda Freightliner
	ATM	**N:** National City Bank
	Other	**N:** Meijer Grocery, PetsMart, Sam's Club, Wal-Mart (Pharmacy)
3AB		JctI-270, Cincinnati, Cleveland
4		Wilson Road, Columbus
	FStop	**S:** Pacific Pride Commercial Fueling*
	Gas	**N:** United Dairy Farmers*
		S: BP*(CW), Shell*(CW), Speedway*
	Food	**N:** Minelli's Pizza*
		S: Waffle House
	Lodg	**S:** AAA Econolodge
	Other	**S:** Anchor Car Wash
5		Hague Ave (Westbound, Eastbound Reaccess)
6		Grandview Ave (Left Exit Eastbound)
7		U.S 40, West Broad St
	Gas	**N:** Amoco*
	Food	**N:** Burger King, Great China Express, McDonalds, Subway, Taco Bell, White Castle Restaurant
	Lodg	**N:** Days Inn AAA Days Inn
	AServ	**N:** AAMCO Transmission
	ATM	**N:** National City Bank
	Other	**N:** Coin Laundry, Cruz Thru Convenience Store, Revco Drugs, U-Haul Center(LP)
8A		U.S 62, Ohio 3, Grove City (Westbound, Eastbound Reaccess)
8B		Mounds St (Westbound, Eastbound Reaccess)
9A		Jct I-71, Cincinnati
9B		OH 315 North
9C		Rich St, Towne St (Westbound)
0A		U.S 23 South, Front St , High St
0B		U.S 23 North, 4th St, Livingston Ave
1A		JctI-71N, to Cleveland (Left Exit Eastbound)
1B		18th St (Eastbound)
	Med	**S:** + Hospital

EXIT		OHIO
102		Miller Ave, Kelton Ave
	Other	**S:** Rite Aide Pharmacy
103A		Main St , Bexley, Alum Creek Dr (Services Same As Exit 103B)
	Gas	**N:** 76, BP*(CW), Thornton's*
		S: Rich, Shell(CW)
	Food	**N:** Domino's Pizza, JC's Bakery, Long John Silvers, Mr Hero, Peking Dynasty, Subway, Taco Bell
		S: McDonalds, Rally's Hamburgers, White Castle Restaurant
	AServ	**N:** Ernie's, Muffler King, Speedee Muffler
	ATM	**N:** National City Bank
	Other	**N:** Car Wash, Conison's Grocery, Deli Delicious C-Store
103B		Alum Creek Dr, Livingston Ave
	Gas	**N:** Speedway*, Thornton's*(LP)
		S: Rich, Shell*(CW)
	Food	**N:** Chinese Cuisine, Danny's Fox, Domino's Pizza, Long John Silvers, Mr Hero, Peking Dynasty, Subway, Taco Bell, Wendy's
		S: McDonalds, Rally's Hamburgers, White Castle Restaurant
	AServ	**N:** Ernie's Brake & Automotive, Muffler King, Speedy Muffler
	ATM	**N:** Bank One, National City Bank
	Other	**N:** Car Wash, Conison's Food Market
105AB		U.S 33, James Road, Lancaster
107		OH 317, Hamilton Road, Whitehall
	Gas	**S:** BP*(CW)(LP), Dairy Mart*, Shell*(D) (Kerosene), Sunoco*(D)(LP)
	Food	**S:** Bob Evans Restaurant, Cookers, Denny's, Holiday Inn, McDonalds, Ponderosa, Red Lobster, Subway, Taco Bell, The Olive Garden
	Lodg	**S:** Holiday Inn (see our ad on this page), AAA Knight's Inn, Ramada Inn, Residence Inn
	AServ	**S:** BP Pro Care, Goodyear Tire & Auto, Tire America
	ATM	**S:** Bank One, Huntington Bank, National City Bank
	Other	**S:** Phar Mor Drugs
108AB		Jct I-270, Cincinnati, Cleveland
110AB		Brice Road, Reynoldsburg
	Gas	**N:** Shell, Speedway*, Sunoco*
		S: BP*(LP), Meijer Grocery*(LP), Super America*(LP)
	Food	**N:** Abner's Country Restaurant, Arby's, Arthur Treacher's Fish & Chips, Best Western, Bob Evans Restaurant, Burger King, Cantina Del Rio, Chi Chi's Mexican, Di Carlo's Pizza, Domino's Pizza, Donato's Pizza, Genji Japanese Steak House, Golden China, Long John Silvers, Max & Erma's, McDonalds, Mi Mexico, Mister Hero, Nick the Greek Restaurant, Pizza Hut, Ryan's Steakhouse, Shoney's, Subway, Tee Jaye's, Waffle House, White Castle Restaurant
		S: Applebee's, Big Boy, Boston Market, Burger King, China Paradise, Home Town Buffet, KFC, McDonalds, Perkins Family Restaurant, Ponderosa, Ruby Tuesday, Shell's Seafood, Subway, Taco Bell, Waffle House, Wendy's, White Castle Restaurant
	Lodg	**N:** Best Western, Cross Country Inn, La Quinta Inn, Red Roof Inn, Super 8 Motel
		S: Days Inn Days Inn, Econolodge, Motel 6
	AServ	**N:** Fast Lube, Goodyear Tire & Auto, Midas Muffler & Brakes, Pro Care, Shell
		S: Batterys Plus, Firestone Tire & Auto, Lindsay Acura, Meineke Discount Mufflers, NTW Tire Center, Precision Tune & Lube, Valvoline Quick Lube Center
	RVCamp	**S:** Farber Recreational Vehicles
	ATM	**N:** Bank 1, First National Bank (Big Bear), Huntington Banks, National City Bank, Star Bank, Sunoco
	Other	**N:** Big Bear Grocery, Dairy Mart Convenience Store
		S: Bryce Outlet Store, Builders Square 2, Cub Foods (Pharmacy), Do-It-Yourself Car Wash, Durg Emporium, Kinko's, Kroger Supermarket (Pharmacy), Meijer Grocery, Office Max, PetsMart, Sam's Club
112AB		OH 256, Pickerington, Reynoldsburg
	Gas	**N:** Shell, Sunoco*(D)

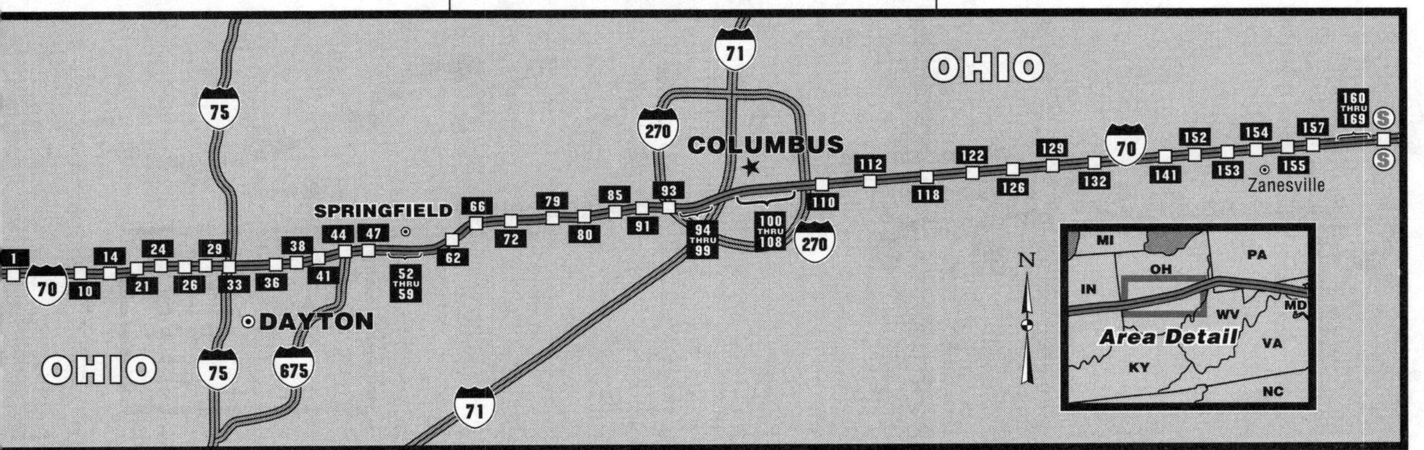

EXIT OHIO

S: Super America*
Food **N:** McDonalds
S: Arby's, Blimpie's Subs, Cracker Barrel, Damon's Ribs, First Wok, Massey's Pizza, TCBY Yogurt, Wendy's
Lodg **N:** Lenox Inn
S: Hampton Inn
AServ **N:** Sunoco
S: Faslube
ATM **S:** Bank One
Other **N:** R&R Car Wash
S: Revco Drugs

118 OH 310, Pataskala
Gas **N:** Sunoco*
S: Duke*
Food **N:** Etna Pizza
S: Red Barn Pizza (Frontage Rd)
AServ **N:** Sunoco
TServ **S:** International Truck Service (Frontage Rd)
ATM **N:** Sunoco
Other **N:** Etna Supermarket

122 OH 158, Kirkersville, Baltimore
TStop **S:** Flying J Travel Plaza
Food **S:** Country Market Restaurant*(LP)(SCALES), Magic Dragon Chinese, Pepperoni's Pizza

126 OH 37, Granville, Lancaster
TStop **N:** Pilot Travel Center*(SCALES)
S: 76 Auto/Truck Plaza*(SCALES)
FStop **N:** Certified*(LP)
S: Truck-O-Mat*(CW)(SCALES)
Gas **N:** Shell*
S: Sunoco*(D) (Kerosene)
Food **N:** Dairy Queen (Pilot)
S: 76 Auto/Truck Plaza
Lodg **S:** Buzz Inn, Motel 76
TServ **S:** 76 Auto/Truck Plaza
TWash **S:** Truck-O-Mat
RVCamp **N:** Campground (1.7 Miles)
S: KOA Campground (3.6 Miles)
ATM **N:** Pilot, Shell

129AB OH 79, Buckeye Lake, Newark
TStop **S:** Duke's Truckstop*(SCALES)
Gas **S:** BP*(LP), Shell
Food **S:** Beechridge Pizza & Subs, Burger King, Dairy Creem, Duke TS, McDonalds, TCBY Treats (Shell), Taco Bell, Wendy's
Lodg **S:** Duke TS
AServ **S:** A1 Auto Parts
TServ **S:** Duke Truckstop
TWash **S:** Beechridge Truck Wash

EXIT OHIO

RVCamp **S:** KOA Campground (1.7 Miles)
Other **S:** Cardinal Grocery(LP), Shoe's CB Shop, Tourist Information

(131) Rest Area - RR, Phones, Picnic

132 OH 13, Newark, Thornville
Gas **S:** BP*, Shell* (Kerosene)
Food **S:** Nyoka's Family Restaurant
AServ **S:** BP
RVCamp **N:** Camping

141 OH 668, Brownsville, Gratiot (Eastbound, Westbound Reaccess)
RVCamp **N:** Campground

142 U.S 40 Gratiot (Westbound, Eastbound Reaccess)
RVCamp **N:** Campground

152 U.S 40, National Road, Zanesville
Gas **N:** Shell*(D)
Food **N:** Big Boy, McDonalds
Lodg **N:** Super 8 Motel
TServ **N:** Hartman's Truck Center
RVCamp **N:** National Trail Campground (10 Miles, U.S.40)
ATM **N:** Shell

153A OH 60 N., Ohio 146 W, State St
Gas **N:** Super America*(D)
S: BP*(D)
AServ **S:** BP
Med **N:** ✚ Hospital
Parks **N:** Dillon State Park
Other **N:** Coin Laundry

153B Ohio 60 N., Ohio 146 W, Maple Ave. (Westbound)
Gas **N:** BP*
Food **N:** Big Boy, Dairy Queen, Picnic Pizza
Med **N:** ✚ Hospital

154 Fifth St
Other **S:** Tourist Information

155 OH 60, OH 146, Underwood St
Gas **S:** Exxon*(D)
Food **N:** Bob Evans Restaurant, Olive Garden, Red Lobster, Shoney's
S: Best Western, Maria Adornetto Italian, Subway, Wendy's
Lodg **N:** AAA Comfort Inn, ◆ Fairfield Inn
S: AAA Amerihost Inn, AAA Best Western, Thrift Lodge
RVCamp **N:** Camping
Med **N:** ✚ Hospital

EXIT OH

Other **N:** Pick 'N Save Supermarket
S: Clay City Outlet Center, Tourist Informatio

157 OH 93, Adamsville, Zanesville
Gas **N:** Duke*(LP)
S: BP, Marathon*, Shell*(D)(LP)
Food **S:** Blimpie's Subs (Marathon), Sub Express (She
AServ **S:** BP, Marathon
TServ **S:** Cat
RVCamp **S:** KOA Campground (.9 Miles)
ATM **S:** Shell
Other **S:** Hittle Roofing(LP), OH State Patrol Post

160 OH 797, Airport, Sonora
Gas **S:** BP*, Shell*(D)
Food **S:** Jake's (Days Inn), McDonalds, Sub Express (Shell), Wendy's
Lodg **S:** Air-Wood Motel, Clara Belle Motel, AAA Days Inn, ◆ Holiday Inn
Other **S:** Air Wood Convenience, Highway Patrol

(163) Rest Area - RR, Phones, Picnic (Westbound)

164 U.S 22, U.S 40, Norwich
Gas **N:** Shell*(D)
Food **N:** Cafe 44
Lodg **N:** Baker's Motel
Other **N:** National Road & Zane Gray Museum

169 OH 83, Cumberland, New Concord
Gas **N:** BP
AServ **N:** BP
RVCamp **N:** Camping

(173) Weigh Station

176 U.S 22, U.S 40, Cambridge
Food **N:** Chal's Ribs & Steak
Lodg **N:** Cambridge Delux Inn, AAA Fairdale Inn
RVCamp **N:** Camping
Other **N:** Ohio State Police

178 OH 209, Cambridge, Byesville
FStop **S:** Speedway* (Kerosene)
Gas **N:** BP*(D), Shell*(D)
Food **N:** Big Boy, Bob Evans Restaurant, Bonanza Steak House, Cracker Barrel, Deer Creek Steak House, KFC, McDonalds, Rax, Spirit Lounge Restaurant, Taco Bell, The Forum
S: Blimpie's Subs (Speedway)
Lodg **N:** ◆ Best Western, ◆ Days Inn, Deer Creek Motel, ◆ Holiday Inn, AAA Travelodge
AServ **N:** Shell
RVCamp **S:** Springvalley Campground (1 Mile)
Other **N:** Kroger Supermarket (Open 24 Hours), More For Less Supermarket, Revco Drugs, Tourist

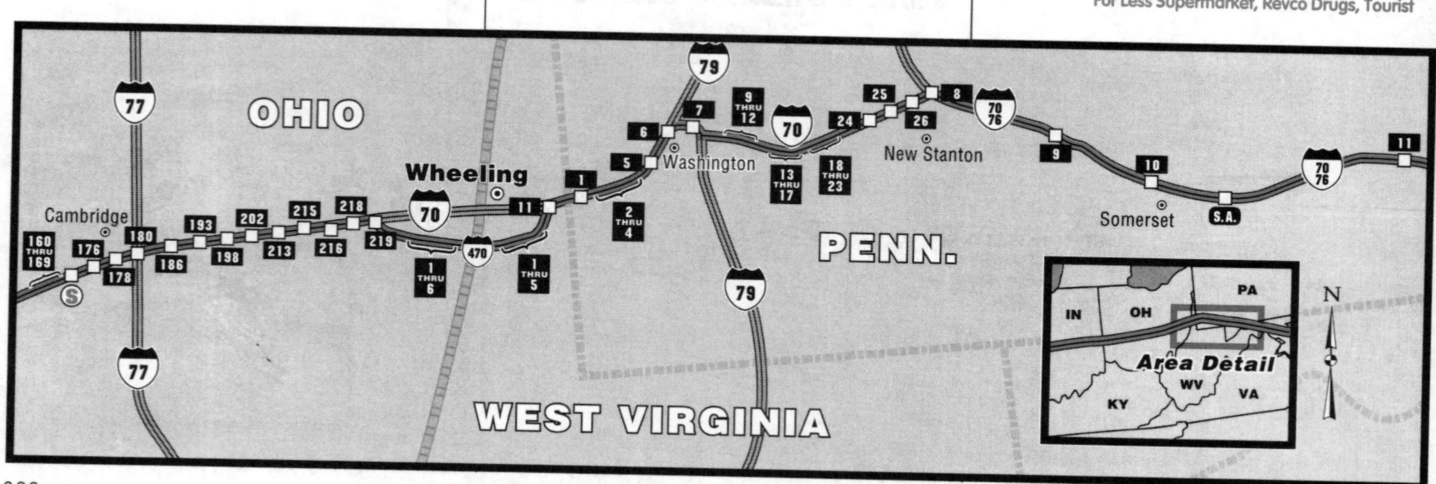

Bold red print shows RV & Bus parking available or nearby

EXIT OHIO

Information
S: K-Mart (Pharmacy)

0AB Jct I-77, Marietta, Cleveland

6 U.S 40, OH 285, Senecaville, Old Washington
- **TStop** S: Shenandoah Truckstop (Citgo)
- **FStop** S: Go-Mart*
- **Gas** N: Shell* (Kerosene)
- **Food** S: Shenandoah TS
- **Lodg** S: Shenandoah Inn
- **TServ** S: Shenandoah Truckstop
- **RVCamp** N: Campground
- **ATM** S: Shenandoah TS

90) Rest Area - RR, Phones, Picnic (Eastbound)

3 OH 513, Quaker City
- **FStop** N: Fuel Mart*
- **Gas** N: BP*, Shell*

8 CR114, Fairview

02 OH 800, Dennison, Barnesville
- **Gas** S: Citgo*(D)
- **AServ** S: Citgo
- **Med** S: + Hospital
- **Other** N: Egypt Valley Wildlife Area

04 U.S 40 East, National Road, CR 100 (Eastbound, Westbound Reaccess)

08 OH149, Morristown, Belmont
- **FStop** N: BP*
- **Gas** S: Marathon*(D)
- **Food** N: Schepp's
  S: Wees' Drive-In
- **AServ** N: Ford Dealership, Morristown Motors
- **RVCamp** N: Jamboree Valley Campground
  S: Camping
- **Parks** S: Barkcamp State Park
- **Other** N: Morristown Pharmacy, State Patrol Post

11) Rest Area - RR, Phones, Picnic

3 U.S 40, OH 331, Flushing
- **Gas** S: Citgo*, Exxon*, Marathon*(LP)
- **Lodg** S: Twin Pines Motel
- **AServ** S: Marathon
- **RVCamp** S: Valley View Campgrounds
- **Other** N: Sheriffs Office
  S: Airport, Coin Car Wash

15 U.S 40, National Rd.
- **Food** N: Burger King, Domino's Pizza, Subway, Wen Wu Chinese
- **ATM** N: Belmont National Bank, Citizens Bank (Riesbeck Supermarket)
- **Other** N: Riesbeck's Supermarket

16 OH 9, St Clairsville
- **Gas** N: BP
  S: Ashland
- **AServ** N: BP
  S: Ashland
- **Other** N: Car Wash

18 Mall Road, Banfield Road
- **Gas** N: Branded Gasoline*(CW)
  S: U.S.A.(D)
- **Food** N: Applebee's, Boston Market, Burger King, De Felice Brothers Restaurant, Denny's, **McDonalds (Wal-Mart)**, Red Lobster, West Texas Roadhouse
  S: Bob Evans Restaurant, Bonanza Steak House, Little Caesars Pizza (K-Mart), Long John Silvers, McDonalds, Rax

EXIT OHIO/WEST VIRGINIA

- **Lodg** N: ◆ Hampton Inn, ◆ Red Roof Inn, Super 8 Motel
- **AServ** N: Auto Zone Auto Parts, GM Auto Dealer, Midas Muffler & Brakes, Wal-Mart
  S: Tire America
- **RVCamp** N: Stewart's RV Center(LP)
- **ATM** N: Steel Valley Bank, WestBanko
  S: Belmont National
- **Other** N: Kroger Supermarket (Pharmacy), Sam's Club, Wal-Mart (Pharmacy, Open 24 Hours)
  S: K-Mart (Pharmacy), Office Max, Ohio Valley Mall, Revco Drugs, Tourist Information

219 Jct I-470E, Bellaire, Washington, PA

220 U.S.40, National Rd, CR 214
- **Gas** N: Citgo (Kerosene), Exxon*(D) (Kerosene)
  S: Marathon*
- **Food** S: Habaneros Grill & Bar
- **Lodg** N: Plaza Motel
  S: Days Inn, AAA Days Inn
- **AServ** N: Citgo
  S: Marathon
- **ATM** N: Exxon
- **Other** N: Hilltop Car Wash

225 U.S.250 W. OH 7, Bridgeport
- **Gas** N: Citgo, Star Fire Express*, Sunoco*
  S: Exxon, Great Gasoline*, Marathon*
- **Food** N: Papa John's Pizza
  S: Domino's Pizza
- **AServ** N: Citgo, Meineke Discount Mufflers
  S: Exxon
- **Med** N: + East Ohio Regional Hospital
- **ATM** N: Belmont National Bank
  S: Citizen's Bank
- **Other** S: Car Wash

Note: I-70 runs concurrent below with I-470. Numbering follows I-470.

1 Banfield Road, Maoo Road (Difficult Reaccess)

3 County Road 214

6 Ohio 7, Bridgeport, Bellaire (Difficult Reaccess)

↑OHIO

↓WEST VIRGINIA

1 U.S 250, WV, Wheeling, Moundsville2

EXIT WEST VIRGINIA

- **Food** S: Best Western
- **Lodg** S: Best Western
- **Med** S: + Hospital

2 WV 88 N, Oglebay Park (Northbound)
- **Gas** N: Exxon*
- **Food** N: Big Boy, Bob Evans Restaurant, Boots Texas Road House, Hardees, Long John Silvers, Rax's, Tj's
- **Lodg** N: Hampton Inn
- **AServ** N: Tire America
- **ATM** N: Progressive Bank, WestBanco
- **Parks** N: Oglebay Park
- **Other** N: Coin Laundry, Minit Car Wash, Revco Drugs
  S: Kroger Supermarket

Note: I-70 runs concurrent above with I-470. Numbering follows I-470.

0 Zane Street, Wheeling Island (Westbound, Eastbound Reaccess)
- **Gas** N: Exxon*(D)
- **Food** N: Abbey's Restaurant, Dairy Queen, Eddie Pepper's, Fabulous Burgers, Hot Stuff Pizza, KFC, Smash Hit Subs

1B U.S.250S, WV 2, South Wheeling, Moundsville (Left Exit Eastbound)

2B Washington Ave
- **Food** S: Sigaretti's
- **Med** S: + Hospital

(4) Weigh Station

4 WV88 S, Elm Grove (Eastbound, Westbound Reaccess)
- **Gas** S: Amoco(D), BP(D), Exxon*(D)
- **Food** S: Dairy Queen, Di Carlo's Pizza, Jaybo's
- **Lodg** S: Grove Terrace Motel
- **AServ** S: Advance Auto Parts, Amoco, Exxon
- **ATM** S: Belmont National Bank, WestBanco
- **Other** S: Stone Church Grocery

5 U.S 40, Elm Grove, Triadelphia
- **Gas** N: Citgo*(D)
- **Food** N: Christopher's, Pizza Hut, Subway, Wendy's
  S: McDonalds
- **Lodg** N: Fort McHenry Bed & Breakfast
- **Other** N: Riesbeck's Supermarket, Rite Aide Pharmacy
  S: State Patrol Post, The Medicine Shoppe

5A Junction I-470, Columbus (Westbound, Left Exit)

11 WV41, Dallas Pike
- **TStop** N: TravelCenters of America*(SCALES)
  S: AmBest* (Dallas Pike TS, Rv Dump)
- **FStop** N: Union 76*(SCALES)
- **Gas** S: BP*(D)(CW), Citgo*
- **Food** N: Country Pride (TravelCenters of America), Windmill Lounge (TravelCenters of America)
  S: Cherokee Trading Post, Dairy Queen (BP), Dallas Pike TS, South Fork (Comfort Inn), Taco Bell
- **Lodg** N: Days Inn (see our ad on this page)
  S: Comfort Inn, Dallas Pike Truckstop, Econolodge, Super 8 Motel
- **TServ** N: TravelCenters of America
- **TWash** N: Americas Truck Wash (Union 76)
- **RVCamp** S: Camping(LP)
- **Other** N: Little Dragon CB Shop (TravelCenters of America)

EXIT WEST VIRGINIA/PENNSYLVANIA

(13) WV Welcome Center - RR, Phones, Tourist Info, Picnic, RV Dump (Westbound)

↑WEST VIRGINIA
↓PENNSYLVANIA

1 West Alexander
- **Gas** S: Texaco*

(5) PA Welcome Center - RR, Phones, Tourist Info, Picnic, Vending (Eastbound)

2 **(6)** PA 231, Claysville
- **TStop** S: **76 Auto/Truck Plaza***(SCALES) **(RV Dump)**
- **Gas** N: Exxon*(LP)
  S: **Citgo (76TS)**
- **Food** S: **I Can't Believe It's Yogurt (76 TS), Kings Family Restaurant (76 TS), Sbarro Italian (76 TS)**
- **TServ** S: **76 Auto/Truck Plaza**

3 **(11)** PA 221, Taylorstown
- **FStop** N: **Amoco***(D)
- **Food** N: Cc's Restaurant
- **AServ** N: **Amoco**

4 **(14)** U.S 40, Chestnut St
- **Gas** S: Amoco, BP*, Exxon*, Sunoco*
- **Food** N: Bonanza Steak House, Hogie Haven, Stone Crab Inn
  S: Big Boy, Blimpie's Subs (BP), Denny's, Donut Connection, Grande Conteninal Cuisine, Hardee's, Long John Silvers, **McDonalds**, Pizza Hut, Taco Bell, Wendy's
- **Lodg** N: Noce Motel, Pugh's Motel
  S: DAYS INN AAA Days Inn (see our ad this page), **Interstate Motel**, AAA Ramada, ◆ Red Roof Inn, AAA Washington Motor Inn
- **AServ** S: Amoco, Sear's
- **ATM** S: Washington Federal Savings
- **Other** N: BJ's Brushless Car Wash, **Chestnut Laundry Depot, Foodland Grocery**
  S: **Crossroads*, Franklin Mall, Thrift Drug, Visitor Information**

5 **(16)** Jessop Place

6 **(17)** PA 18, Jefferson Ave
- **Gas** N: Gulf, Texaco*(D)
  S: Sunoco*, Texaco*, Zappi Oil
- **Food** N: Championship Pizza & Sub, Chinese Kitchen, Dairy Queen, Krency's, McDonalds, Ten Different Hot Dogs
  S: 4-Star Pizza, Burger King, Chicken Charlie's (Texaco), Sub Express
- **AServ** N: All Pro Auto Parts, Ismanger's, Pennzoil Lube
  S: Napa Auto Parts, Sunoco*
- **TServ** N: **Ismingers Auto & Truck Repair**
- **Med** S: ✚ **Hospital**
- **ATM** N: Washington Federal Savings
  S: **Foodland Grocery**, Viehann's Pharmacy
- **Other** N: **Co Go's Convenience Store, Coin Laundry, Rite Aide Pharmacy**
  S: **Foodland Grocery, Viehmann Pharmacy**

(18) Jct I-79 North - Pittsburgh

7AB **(19)** U.S 19, Murtland Ave
- **Gas** S: Amoco*, BP*(D)(CW), Exxon, Sunoco*, Texaco*
- **Food** S: Big Boy, Bob Evans Restaurant, Burger King, China Feast, Donut Connection, **Eat N Park, Jade Inn**, KFC, Long John Silvers, **McDonalds**, Papa John's Pizza, Pizza Hut, Shoney's, Taco Bell, **The Italian Oven**
- **Lodg** S: Hampton Inn, Motel 6
- **AServ** S: Amoco, Baer's Buick, Exxon, Firestone Tire & Auto, Midas Muffler, Monroe Muffler & Brake, Tire America
- **Med** S: ✚ **Hospital**
- **ATM** S: Dollar Bank, Mellon Bank, **National City Bank**, PNC Bank
- **Other** S: **K-Mart, State Patrol Post, Thrift Drug, Wal-Mart (Pharmacy, Vision Center), Washington Mall**

8 **(19)** PA 136, Beau St
- **Lodg** S: **Ramada Inn (see our ad on this page)**
- **RVCamp** N: **KOA Campground**

(21) Jct I-79 South, Waynesburg

9 **(25)** PA 519, Eighty Four, Glyde
- **TStop** S: **BP* (Kerosene)**
- **Gas** S: Amoco* (Kerosene)
- **Food** S: Toot n Scoot Restaurant (BP)
- **TServ** S: **Lou's Auto /Truck Repair (Amoco)**

10 **(28)** Dunningsville
- **Food** S: **Avalon Motor Inn**
- **Lodg** S: **Avalon Motor Inn**

11 **(30)** Kammerer
- **Food** N: Restaurant (Carlton)
- **Lodg** N: Carlton Motel

12A **(33)** PA 917, Ginger Hill

12B **(33)** PA 917, Bentleyville, Ginger Hill
- **TStop** S: **Pilot Travel Center***(SCALES)
- **Gas** S: Amoco*
- **Food** N: King of the Hill Restaurant
  S: Dairy Queen, Hardee's, **McDonalds (Pilot), Subway (Pilot)**
- **AServ** S: Advance Auto Parts, Tregembo Ford
- **TWash** S: **Pilot**
- **RVCamp** N: **Camping (1 Mile)**
- **Other** S: **Giant Eagle Food Store, Rite Aide Pharmacy**

13 **(35)** PA 481, Monongahela , Centerville
- **Med** N: ✚ **Hospital**

14 **(36)** Twin Bridge Rd.

15AB **(38)** PA 43, Toll Road South, CA

16 **(40)** Speers
- **Gas** S: Exxon
- **Food** N: Lorraine's Family
- **AServ** S: Exxon

17 **(41)** PA 88, Charleroi, Allenport
- **FStop** N: **Crossroads***(CW)
- **Gas** N: Amoco*, Pacific Pride Commercial Fueling*(CW)
- **Food** N: KFC (Texaco), Pizza Hut Express (Crossroads), Sub Express (Amoco), Taco Bell (Crossroads)
  S: Campy's Pizza & BBQ, Snooters's

18 **(41)** PA 906, Belle Vernon , Monesson
- **AServ** S: **Wince Auto & Truck Repair**
- **TServ** S: **Wince**
- **Med** N: ✚ **Monn Valley Hospital**

19 **(42)** North Belle Vernon
- **Gas** S: BP*, Sunoco*(D)
- **Food** S: Dairy Queen, Hardees (BP)
- **AServ** S: Pontiac Cadillac Dealer, Sunoco
- **Other** S: **Duritza's Grocery**

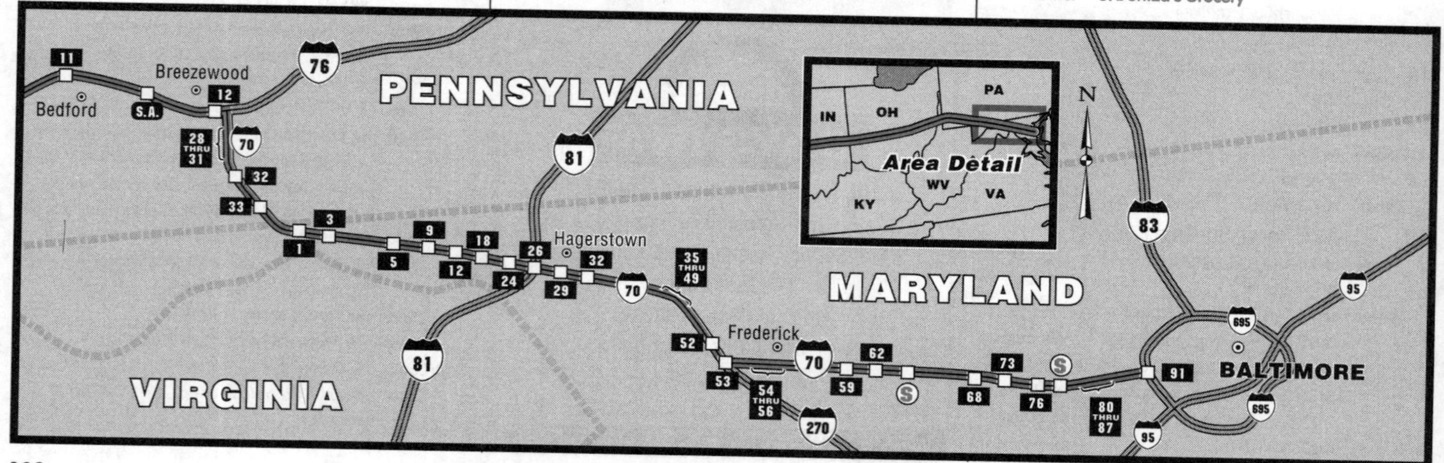

Bold red print shows RV & Bus parking available or nearby

XIT PENNSYLVANIA

0AB **(43)** PA 201, to PA 837, Denora, Fayette City
- **Gas** N: Val's*(CW)
  S: Chevron, Sunoco*, Texaco*
- **Food** S: **Best Val-U Motel, Burger King,** Denny's, Donuts & More, Hoss's Steak and Seafood, K-Mart, KFC, Little Bamboo Chinese, Little Caesars Pizza, Long John Silvers, McDonalds (Wal-Mart), McDonalds, Pizza Hut, Subway, Wendy's
- **Lodg** S: **Best Val-U Motel**
- **AServ** S: Advance Auto Parts, Giant Auto Parts, Kelly Auto Parts and Service, Trak Auto
- **Med** N: **+ Hospital**
- **ATM** S: First Federal Savings, Foodland Grocery, Naional City Bank, PNC Bank
- **Other** N: Coin Car Wash
  S: Aqua Jet Car Wash, **Aqua Jet Express, Foodland Grocery & Drugstore, Giant Eagle Supermarket, K-Mart, Thrift Drug, Union Laundromat, Wal-Mart Supercenter (Supermarket, Vision Center, Pharmacy, Auto Service Center)**

1 **(44)** Arnold City

2AB **(46)** PA 51, Pittsburgh, Uniontown
- **FStop** S: **Texaco***
- **Gas** N: Exxon
- **Food** N: Howard Johnson
  S: **Holiday Inn,** The Main Course (Ron Del Motel)
- **Lodg** N: **Budget Host,** Howard Johnson
  S: 5 M's Hotel, **Holiday Inn,** Knotty Pine Motel, Ron Del Motel
- **AServ** N: Exxon

3 **(49)** Smithton
- **TStop** N: **76 Auto/Truck Plaza*(SCALES) (RV Dump), Smithton Truckstop*(SCALES)**
- **Food** N: **76 Auto/Truck Plaza, Smithton TS**
- **Lodg** N: Smithton Truck Stop
- **TServ** S: **Truck of Western PA Sales,Parts & Service**
- **TWash** N: **Smithton**
- **ATM** N: Smithton Truck Stop

4AB **(51)** PA 31, West Newton, Mt Pleasant

5A **(53)** Yukon

5B **(54)** Madison
- **TStop** S: **Atlantic**
- **FStop** S: **Atlantic*, Citgo(D)**
- **Food** S: **Atlantic**
- **AServ** S: Citgo
- **TServ** S: **Gary's Truck Repair**
- **RVCamp** N: **Camping (.8 Miles)**
- **Other** S: **Larry's CB Shop (Atlantic)**

6 **(57)** New Stanton
- **Gas** N: Exxon*, Sheetz* (Kerosene), Sunoco*
  S: BP*(D), Citgo*(LP), Sunoco*
- **Food** N: Bob Evans Restaurant, **Days Inn,** Donut Chef, Eat N Park, Howard Johnson, KFC, McDonalds, Pagano's Family Restaurant, Pizza Hut, Ramada Inn, Subway, Szechuan Wok, Wendy's
  S: **Cracker Barrel,** TJ's Sport Lounge Restaurant
- **Lodg** N: **Budget Inn,** Comfort Inn, Conley Inn, **Days Inn,** Howard Johnson, Ramada Inn (see our ad this page), **Super 8 Motel**
  S: New Stanton Motel
- **AServ** S: BP
- **TWash** N: **4 Star Truck Wash**
- **RVCamp** N: **Campground**

EXIT PENNSYLVANIA

- **ATM** N: Mellon Bank
- **Other** N: Coin Car Wash
  S: **US Post Office**

Note: I-70 runs concurrent below with I-76. Numbering follows I-76.

8 **(75)** Jct I-70 E, Jct I-76, PATnpk, Pitsburgh, Harrisburg (Toll Rd)

(78) New Stanton Service Plaza (Westbound)
- **FStop** W: **Sunoco***
- **Food** W: **King's Family, McDonalds**
- **ATM** W: **Machine**

9 **(91)** PA711, PA31, Ligonier, Uniontown (Pay Toll To Exit)
- **FStop** S: **Citgo***
- **Gas** S: Exxon*(D), Sunoco, Texaco*
- **Food** N: Laurel Highlands Lodge, Tall Cedars Inn
  S: Candlelight Family, Dairy Queen
- **Lodg** N: Laurel Highlands Lodge
  S: Days Inn, Donegal Motel
- **AServ** S: Sunoco
- **RVCamp** S: **Donegal Campground, Laurel Highland Campgrounds (3 Miles), Mountain Pines Campground (3 Miles), Pioneer Park Campground (11 Miles)**
- **ATM** S: PNC Bank
- **Parks** S: **Kooser State Park, Laurel Hills State Park, Laurel Ridge State Park, Ohiopyle State Park**

(94) Picnic (Westbound)

10 **(110)** U.S 219, Somerset, Johnstown (Pay To Exit)
- **TStop** S: **Jim's Auto/Truckstop***
- **Gas** N: Quick Fill, Sheetz*, Your Car Wash
  S: Exxon*
- **Food** N: Hoss's Steakhouse, Kings Family Restaurant, Pizza Hut, Taco Bell
  S: Arby's, China Garden, Dairy Queen, Dunkin Donuts, Eat N Park, Holiday Inn, **Jim's TS,** KFC, Little Caesars Pizza, Long John Silvers, Maggie Mae's Cafe, McDonalds, Myron's Restaurant, Pine Grill, **Ramada Inn, Subway,** Summit Diner, Wendy's, Yogurt Ice Cream Classics
- **Lodg** N: Dollar Inn, Economy Inn
  S: Best Western, Budget Host Inn, Budget Inn, ◆ **Days Inn,** Economy Inn, **Hampton Inn,** AAA Holiday Inn, ◆ Knight's Inn, AAA **Ramada**
- **AServ** N: Dumbaulds Tire, Ford Dealership, Mardis Chrysler Plymouth, Midas Muffler & Brakes, Speedy Lube

EXIT PENNSYLVANIA

- **TServ** N: **Mulhollen**
  S: **Jim's Auto/Truckstop**
- **TWash** N: **Your Car Wash**
- **Med** S: **+ Hospital**
- **ATM** S: Jim's TS
- **Other** N: Coin Car Wash, **Horizon Outlet Center**
  S: Clean Water Car Wash, **Coin Laundry, Visitor Information**

(112) Somerset Service Plaza (Eastbound)
- **FStop** B: **Sunoco***
- **Food** B: **Bob's Big Boy,** Hershey's Ice Cream, **Roy Rogers, TCBY Yogurt**
  W: **Burger King,** Hershey's Ice Cream, TCBY Yogurt
- **AServ** B: Sunoco
- **ATM** B: ATM

11 **(145)** Jct I-99, U.S. 222, Bedford, Altuna
- **FStop** N: **Amoco*, BP***
- **Gas** N: Sunoco*(D), Texaco*(D)(LP)
- **Food** N: Baskin Robbins, Budget Inn (Sunoco), Burger King, **Chester Fried Chicken (BP),** China Inn, Clara's Place (Best Western), Denny's, Dunkin Donuts, **Ed's Steakhouse,** Hardee's, Hoss's Steak & Sea House, **Long John Silvers, McDonalds (Amoco),** Pizza Hut, Sub Express, The Arena (Quality Inn), **The Country Apple Restaurant (Econolodge)**
- **Lodg** N: **Best Western,** Budget Inn, **Econolodge,** Host Inn, Midway Motel, Quality Inn, **Super 8 Motel**
  S: Motor Town House
- **TServ** N: **Mack Truck**
- **RVCamp** S: **Friendship Village Campground**
- **ATM** N: BP
- **Other** S: **Visitor Information**

(147) North Midway Service Plaza
- **FStop** B: **Sunoco***
- **Food** B: **Hershey's Ice Cream, KFC, Mrs Fields Cookies, Sbarro Italian, TCBY Yogurt**
- **AServ** B: Sunoco
- **ATM** B: PNC Bank

12 **(161)** U.S 30, Baltimore, to I-70 East, Washington, DC (Trucks Avoid U.S. 30E, (3 Mile Hill))
- **TStop** S: **All American 76 Truck Plaza*(SCALES), TA*(SCALES)**
- **Gas** S: **Amoco (TA),** BP*, Citgo*, Exxon*(D), Sheetz*, Sunoco*, Texaco*
- **Food** N: Breeze Manor Restaurant (Quality Inn), **McDonalds, Prime Rib Restaurant (Ramada Inn)**
  S: Arby's, Arthur Treacher's Fish and Chips, Blimpie's Subs (Am Best), Bob Evans Restaurant, Bonanza Steakhouse, Burger King, **Dairy Queen*(SCALES) (TA),** Deli Cafe (Sunoco), **Family House Restaurant (Bus Parking),** Gateway Restaurant (TA), Hardee's, Hershey's Ice Cream (Am Best), KFC, **Little Caesars Pizza (TA),** McDonalds, **Oven Fresh Bakery (TA),** Perkins Family Restaurant (Am Best), Pizza Hut, Post House Cafeteria (Am Best), Stuckey's Express (Exxon), Subway, Taco Bell, Taco Maker (Am Best), Uncle Bud's Pizza (Am Best), Wendy's
- **Lodg** N: **Pan Am Motel,** AAA Quality Inn, **Ramada Inn**
  S: Best Western, Breezewood Motel, Comfort Inn, **Econolodge (TA),** Penn Aire Motel, Wiltshire Motel
- **TServ** S: **Am Best, TA**
- **ATM** S: **Sunoco, TA**

EXIT PENNSYLVANIA

(172) Service Area
- **FStop** B: Sunoco*
- **Food** B: Bob's Big Boy, Burger King, Pretzel Time, TCBY Yogurt
- **AServ** B: Sunoco
- **ATM** B: PNC Bank
- **Other** B: RV Dump Station, Tourist Information

13 Ft Littleton, US 522, McConnellsburg, My Union
- **Gas** N: Amoco*, Exxon*
- **Food** N: Fort Family Restaurant
- **Lodg** N: Downe's Motel #2; S: Downe's Motel
- **AServ** N: Amoco
- **Med** N: Hospital
- **Other** N: State Patrol Post

14 PA 75, Willow Hill, Ft Loudon
- **Food** S: TJ's Restaurant
- **Lodg** S: Willow Hill Motel

15 PA 997, Shipsenburg, Chambersburg
- **Gas** S: Johnnies
- **Food** S: Johnnies
- **Lodg** S: Blue Mountain Brick Motel, AAA Kenmar Motel

Note: I-70 runs concurrent above with I-76. Numbering follows I-76.

28 (149) U.S 30 to I-76, PA Turnpike
- **TStop** S: AmBest Truckstop*, TravelCenters of America*(SCALES)
- **FStop** S: BP*
- **Gas** S: Exxon*, Sunoco*(D), Texaco*(LP)
- **Food** N: Prime Rib (Ramada Inn); S: 76 Auto/Truck Plaza, Arby's, Arthur Treacher's Fish & Chips, Bob Evans Restaurant, Bonanza Steak House, Burger King, China Bamboo, Comfort Inn, Dairy Queen, Deli Cafe (Sunoco), Dunkin Donuts, Family House Restaurant, Gateway Restaurant (TS of America), Hardee's, KFC, Little Caesars Pizza, McDonalds, Pizza Hut, Post House Cafe, Taco Bell, Wendy's
- **Lodg** N: Quality Inn, Ramada; S: Best Western, Breezewood Motel, Comfort Inn, Econolodge (TravelCenters of America), Penn Aire Motel, Wiltshire Motel
- **AServ** N: Sac Inc; S: BP
- **TServ** S: 76 Auto/Truck Plaza*, TravelCenters of America
- **RVCamp** S: Breezewood Campground (4 Miles), Brush Creek Campground (4 Miles), Camping, Crestview Campground (4 Miles)
- **ATM** S: TS of America

29 (150) U.S 30 West, Everett (Low Clearance-14'3")
- **Food** S: The Wildwood Inn
- **Lodg** S: Hi-Way Motel, Panorama Motel, Ritchey's Redwood Motel, Stonewall Jackson Motel, Wildwood Inn
- **RVCamp** S: Hide Away Campground

30 (154) PA 915, Crystal Spring
- **Gas** S: Atlantic
- **Food** N: Dutch Hausn and Restaurant
- **Lodg** N: Motel 70
- **TServ** N: Fischer's Major Repairs

EXIT PENNSYLVANIA/MARYLAND

- **RVCamp** S: Camping
- **Other** S: Country C-Store, Post Office

(156) Rest Area - RR, Phones, Picnic, Vending, Tourist Info (Eastbound)

31 (158) PA 643, Town Hill
- **TStop** N: Town Hill Truckstop- Texaco*(LP)(SCALES) (RV Dump)
- **Food** N: Days Inn
- **Lodg** N: Days Inn

32 (166) PA 731, Amaranth

33 (171) U.S 522 North, Warfordsburg
- **Gas** N: Exxon*
- **Food** N: Nan's Kettle
- **AServ** N: Beatty's

(172) PA Welcome Center - RR, Phones, Picnic, Tourist Info, Vending

↑PENNSYLVANIA

↓MARYLAND

1A Jct I-68, U.S 40 West, Cumberland (Left Exit Westbound)

1B U.S 522 South, Hancock, Winchester (Left Exit Westbound)
- **Gas** S: Gary's, Sheetz*
- **Food** S: Fox's Pizza, Pizza Hut
- **Lodg** S: Comfort Inn, AAA Hancock Motel
- **RVCamp** S: Happy Hills Campground
- **Other** S: Sav-A-Lot Grocery

3 MD 144, Hancock (Left Exit West-

EXIT MARYLAND

bound, No Trucks Over 13 Tons)
- **TStop** S: Hancock Truckstop* (Amoco)
- **Gas** S: BP*, Citgo*(D)(LP)
- **Food** S: Fat Boys Pizza & Subs (BP), Little Sandy's Restaurant, Park-N-Dine
- **Lodg** S: Hancock Budget Inn
- **AServ** S: Kurk Ford
- **TServ** S: Amoco Truck Plaza
- **ATM** S: Home Federal Savings
- **Other** S: C&O Canal Information, Coin Car Wash

5 MD 615 (Left Lane Exit)

9 U.S 40, Indian Springs (Left Lane Exit Eastbound Reaccess Via Exit 5)

12 MD 56, Big Pool, Indian Springs
- **Parks** S: Fort Frederick State Park (1 Mile)

18 MD 68, Clear Spring Ski Area
- **Gas** N: Amoco*, BP*; S: Exxon*(D)
- **Food** N: Al's Pizza (BP), Chester Fried Chicken (BP), McDonalds
- **AServ** N: Thompson Towing; S: Exxon*
- **ATM** N: American Trust Bank, Mac (BP)
- **Other** N: Clearspring Hardware, U.S. Post Office

24 MD 63, Williamsport, Huyett (C&O Canal, Cushwa Basin)
- **AServ** N: Keefer's Towing and Repair Service
- **RVCamp** S: KOA Campground (2 Miles)

26 Jct I-81

29 MD65, Sharpsburg, Hagerstown
- **FStop** S: Shell*
- **Gas** N: Sunoco*(CW)
- **Food** S: Blimpie's Subs (Shell), Burger King, Hot Stuff Pizza, McDonalds (Playland), Wendy's
- **RVCamp** S: Yogi Bear Camp Report
- **Other** S: Antdem Battlefield, Sharpsburg Battlefield

32AB U.S 40, Hagerstown
- **Gas** N: ACT(D)(CW), Citgo*, Exxon
- **Food** N: Bob Evans Restaurant, McDonalds (Playland), Restaurant (Sheraton)
- **Lodg** N: Best Western, Four Points Sheraton, Super 8 Motel (see our ad on this page)
- **AServ** N: Dual Hwy Motors, Exxon (Wrecker Service), Hoffman Dodge, Hoffman's Chevrolet, Younger Auto Service; S: Hagerstown Honda, Sharrett Volkswagon Oldsmobile, Mazda
- **Med** N: Hospital (3.5 Miles)
- **ATM** N: F&M Bank, First United Bank & Trust
- **Other** N: Car Wash, Hagerstown Common Shopping, Martins Food & Pharmacy, Revco Drug, Tourist Information

35 MD 66, Boonsboro, Smithsburg
- **RVCamp** N: Camping; S: Camping
- **Parks** S: George Washington State Park

(39) Welcome Center, Rest Area - RR, Phones, Vending, Picnic, Pet Rest, RV Dump

Bold red print shows RV & Bus parking available or nearby

EXIT	MARYLAND
42	Maryland 17, Myersville, Middletown Road
FStop	S: Amoco*
Gas	N: Exxon*(LP), Sunoco(D) (24 Hr)
AServ	N: Sunoco (Towing)
ATM	N: First United Nat'l Bank (Sunoco)
Parks	N: Gambrill State Park (6 Miles), Greenbrier State Park (6 Miles)
48	US 40 E, Frederick, to US15 & US 340 to Leesburg &Charleston (Difficult Reaccess)
Gas	N: Amoco*(CW), Crown*(D)(CW), Exxon*(D)(LP), Mobil(D), Shell, Sunoco*, Texaco(D)(CW)
Food	N: Barbara Fritchie Fine Dining, Baskin Robbins, Big Chef Chinese Restaurant, Bob Evans Restaurant, Bruegger's Bagels & Bakery, Burger King, Cafe Delight, Casa-Rico Mexican, Chi Chi's Mexican, China King, Dunkin Donuts, Gary Delight (Ice Cream & Subs), Gourmet Express Carry Out, Ground Round, Heavenly Ham, Hillcrest Deli, Holiday Inn, Hot Wok (Chinese), House of Kobe Japanese, Hunan Cheers, Inforno Pizzeria, Jerry's Subs & Pizza, KFC, Kenny Rogers Roasters, Kinklings Donuts, Little Caesars Pizza, Masser's, McDonalds (Playland), Omar Khayam Int'l Foods, Outback Steakhouse, Pizza Hut, Red River, Roy Rogers, Ruby Tuesday, Shanghai Warehouse, Shoney's, TCBY Yogurt, Taco Bell, Village Green Grill, Wendy's
Lodg	N: Holiday Inn, Masser's Motel, Red Horse Motor Inn
AServ	N: Exxon, Forty West Auto Parts, Frederick Ford, Frederick Towne Toyota, Golden Mile, Jiffy Lube, K-Mart, Lube Auto & Truck Repair, Lube Center, Meineke Discount Mufflers, Merchant's Tire & Auto, Midas Muffler & Brakes, Mobil, Montgomery Ward, Pep Boys Auto Center, Rice Auto & Tire Service, Shell, Sunoco, The Lube Ctr, Trak Auto Parts Warehouse
Med	N: Hospital
ATM	N: 7-11 Convenience Store, Chevy Chase Bank, F&M Bank, FCNB Bank, First Bank of Frederick, First Nat'l Bank Of Virginia (Weis Market), Frederick Towne Bank, Frederick Towne Mall, NationsBank
Other	N: 7-11 Convenience Store, Aurorua Car & Van Wash, CVS Pharmacy (Mall), Cellular One (Check Cashing), Coin Operated Laundry, For Eyes Optical, Frederick County Square, Frederick Towne Mall, Giant Food Market (Mall), Grooming Room (Pets), Health Express Food Market, Hillcrest Plaza, K-Mart (Mall, Pharmacy), Martins Food, Revco Drugs, Sam's Convience Store, Shoppers Food Warehouse, Superpetz Pet Owner Warehouse, Visitor Information, Weis Grocery & Pharmacy, West Ridge Shopping Ctr, Westpoint Plaza
49	U.S 40 Alternate, Braddock Heights, Middletown (Difficult Reaccess, Westbound)
Parks	S: Washington Monument State Park
Other	N: MD State Police
52	U.S 340 West, U.S 15 South

EXIT	MARYLAND
53	Jct I-270 to Washington
54	MD 355 to MD 85, Market St, Frederick
TStop	N: Frederick I-70 Truck City*(SCALES)
FStop	S: Mobil*, Southern States*
Gas	N: Amoco*
	S: Exxon*(LP)
Food	S: Burger King, Checkers Burgers, Deli Plus, Dunkin Donuts, El Paso Restaurant, Jerry's Subs & Pizza, Kinklings The Donut Shoppe, McDonalds, Papa John's Pizza, Pizza & Gourmet, Popeye's, Subway, Sunflower's Bakery & Cafe, Wendy's
Lodg	N: I-70 Motor Inn
	S: Days Inn, Knight's Inn (see our ad on this page), Days of Frederick (see our ad on this page)
AServ	S: AAMCO Transmission, Audi, Isuzu, Exxon, Ford Dealership, Ideal Buick, GMC Trucks, Precision Tune & Lube, Saturn Dealership, Tune & Lube, Wal-Mart
TServ	N: Frederick I-70 Truck City*(SCALES)
ATM	N: Truckstop
	S: NationsBank
Other	S: 7-11 Convenience Store, Frederick Veterinary Clinic, Kee Coin Operated Laundry, Police Dept, Wal-Mart (Pharmacy)
55	South St, Frederick
Food	N: Chat 'N Chew Restaurant (Good Grub, north in Frederick)
56	MD144, Patrick St, Frederick

EXIT	MARYLAND
Gas	N: Citgo*, Coastal(LP)
Food	N: Bejing Chinese, Belles' Pit Beef & Ale Hous, Brown's Pizza & Subs, Cheesburgers'N Paradise, Lour's Family Restaurant, McDonalds, Roy Rogers, The Donut Shoppe
TServ	N: Grimes Truck Center, Isuzu Truck
ATM	N: FCNB Bank, S&M Bank
Other	N: Frederick Airport, Plus Mart C- Store
	S: Triangle Outdoor World Sales & Services
59	MD 144 (Westbound, Eastbound Reaccess Only)
62	MD 75, Libertytown, Hyattstown (Services To The Right On Frontage Rd)
Gas	N: Texaco*(D)(CW)
Food	N: Domino's Pizza, Little George's Deli, McDonalds, New Market Carry Out
ATM	N: Farmers & Merchants Bank
Other	N: Animal Care Center, New Market Historic District (Battlefield Site), Pharmacy
(64)	Weigh Station (Eastbound)
(67)	Truck Rest Area - No Facilities (No Cars, Eastbound)
68	MD 27, Damascus, Mount Airy
Gas	N: Amoco*, Citgo*, Texaco*(D)
	S: Exxon(D)(LP)
Food	N: Baskin Robbins* (Amoco), Blimpie's Subs* (Amoco), Bob's Big Boy, Chong Yet Yin Chinese, Christi's Restaurant & Carry Out, Hot Stuff Pizza* (Amoco), KFC, McDonalds, Megan's Bakery, Memories Charcoal House, New York J&P Pizza, Pizza Hut, Roy Rogers, TCBY Treats
AServ	S: Exxon
ATM	N: Citgo, First Nat'l Bank of Maryland, SCNB, Safe Way Grocery
Other	N: Dapper Pet Den (Pet Supplies), Rite Aide Pharmacy, Safe Way Grocery, US Post Office, Visitor Information
73	MD 94, Woodbine, Lisbon
Food	N: Bill's Poultry & Seafood, McDonalds, Mr Teddy's Restaurant & Deli, Pizza Hut
RVCamp	N: Ramblin' Pines (4 Miles)
ATM	N: Sandy Spring Nat'l Bank, Westminster Bank
Other	N: Harvest Fare Supermarket, Woodbine Animal Hospital
76	MD 97, Westminster, Olney, Cookesville
(79)	Weigh Station (Westbound)
80	MD 32, Sykesville, Clarksville
82	U.S 40, Ellicott City (Eastbound, Reaccess Westbound)
83	Marriottsville Road (Westbound, Reaccess Eastbound Only)
87AB	U.S 29, to MD 99, Columbia, Washington
91AB	Jct I-695, to I-95
94	Security Blvd N, Park & Ride

↑MARYLAND

Begin I-70

EXIT OHIO

Begin I-71

↓OHIO

247B Junction I-90 West, Junction I-490

247A West 14th St, Clark Ave (Limited Access)

246 Dennison Ave, Jennings Rd (Left Exit)

245 U.S. 42, Pearl Rd, West 25th St

245A Fulton Rd (Northbound)

244 Denison Ave, West 65th St (Northbound, Left Exit)

242A West 130th St, Bellaire Rd

240 West 150th St
- **Gas** W: BP*(CW)
- **Food** E: Denny's, Marriot
  W: Country Kitchen, Oriental Palace, Winners (Holiday Inn)
- **Lodg** E: Marriot
  W: Budgetel Inn, Holiday Inn
- **AServ** W: All Foreign & Domestic Auto Service
- **Other** W: Full Service Car Wash

239 Ohio 237 South, Airport, Berea

238 Junction I-480, Airport, Toledo, Youngstown

237 Snow Rd, Engle Rd
- **Lodg** E: Best Western, ◆ Fairfield Inn

235 Bagley Rd, Berea, Middleburg Heights
- **Gas** W: BP*(D)(CW)(LP), Shell, Speedway*
- **Food** E: Bob Evans Restaurant
  W: Bruegger's Bagels, Burger King, Capri Pizza, Chinas Best Buffet, Denny's, Dunkin Donuts, Friendly's, Golden Corral, Holiday Inn, Hong Kong Palace, McDonalds, Olive Garden, Penn Station East Coast Subs, Perkins Family Restaurant, Pizza Hut, Ponderosa, Quizno's Classic Subs, Roadhouse, Taco Bell
- **Lodg** E: ◆ Harley Hotel
  W: Comfort Inn, Cross Country Inn, ◆ Holiday Inn, Marriott, Motel 6, Red Roof Inn
- **AServ** E: BP Pro-Care
  W: Gene Norris Olds, K-Mart, Lube Stop, Shell
- **Med** W: ✚ Hospital
- **Other** W: K-Mart (Pharmacy), Sear's Hardware

234 U.S. 42, Parma Heights, Strongsville
- **Gas** E: Shell(CW), Sunoco(D)
- **Food** E: House of Hunan Chinese, Islander Grill, Moondog's Diner, Mr Hero Cheese Steaks, Santo's Pizza, Walt's
- **Lodg** W: La Siesta Motel, Days Inn
- **AServ** E: Hyundai, Honda, Porsche Dealership, Saturn Dealership, Shell, Sunoco

EXIT OHIO

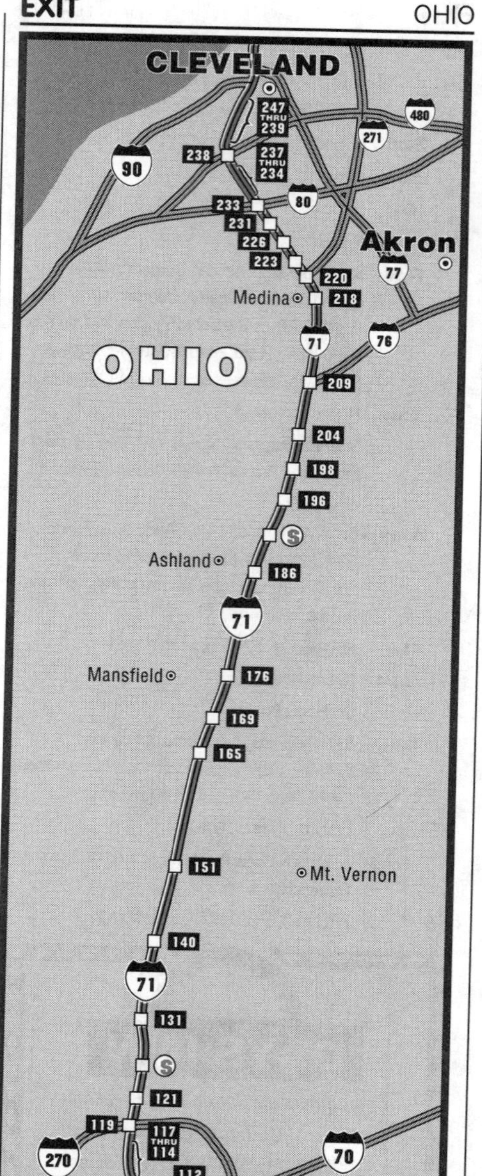

Bold red print shows RV & Bus parking available or nearby

EXIT OHIO

- **Other** E: Coin Laundry, Convenient Food Mart, Pet Supply Store
  W: Builders Square, Coin Car Wash, Wal-Mart (Pharmacy)

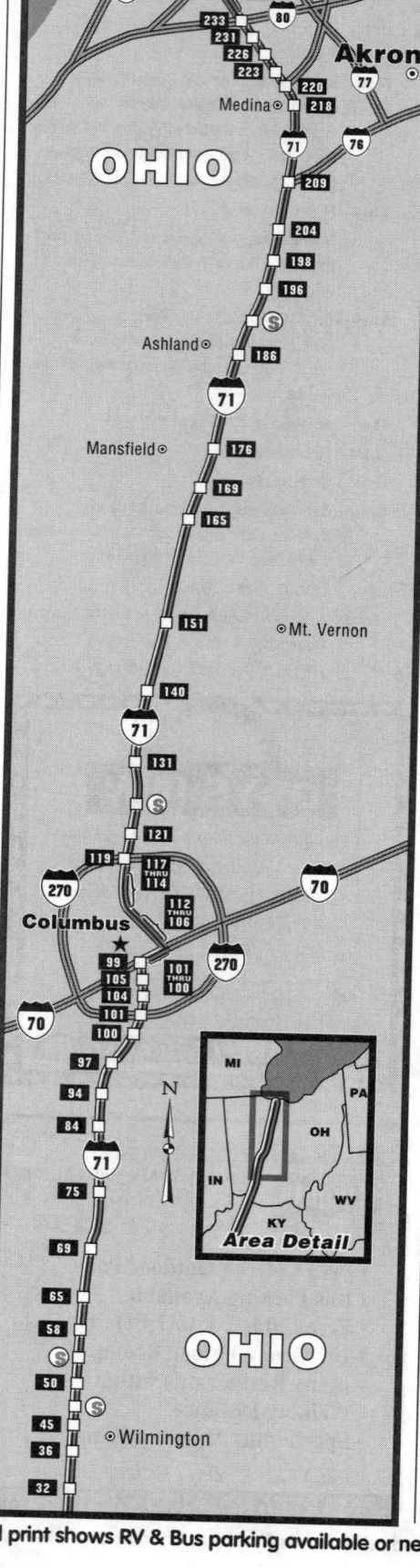

233 Jct I-80, OH Tnpk, Youngstown, Toledo (Toll Rd)

231AB OH 82, Strongsville, Royalton
- **Gas** E: Shell*(CW)
  W: BP*(D)(CW), Marathon*(CW), Sunoco(D)
- **Food** E: Restaurant (Holiday Inn)
  W: Applebee's, Country Kitchen, Ice Cream Ect., Long Horn Steakhouse, Macaroni Romano's Macaroni Grill, Mr Steak, Rio Bravo Cantina
- **Lodg** E: Holiday Inn, Red Roof Inn
- **AServ** E: Shell
  W: BP, Lube Stop, Meuller Tire, Midas Muffler & Brakes, Sear's, Sunoco
- **ATM** W: Key Bank, National City Bank, Star Bank (Rini Rego Grocery)
- **Other** W: Medic Drugs, Rini Rego Grocery, South Park Center Mall

226 OH303, Brunswick
- **FStop** E: Amoco*
- **Gas** W: BP*, Marathon*(D)(CW), Sunoco*
- **Food** E: Denny's, Pizza Hut
  W: Big Boy, Bob Evans Restaurant, Burger King, Friendly's, House of Pearl Chinese, Little Caesars Pizza, McDonalds, Subway, Taco Bell
- **Lodg** W: Howard Johnson
- **AServ** E: Brunswick Toyota/ Chrysler/ Plymouth, Marathon
  W: Midwest Ford
- **RVCamp** E: Camping World (see our ad this page)
- **TServ** E: Goodyear Truck Tire Center
- **ATM** W: Bank One, Marathon
- **Other** W: Automatic Car Wash, Country Counter Food& Drug Store (Pharmacy), K-Mart (Pharmacy, Auto Service)

(225) Rest Area - RR, Picnic, Phones (Northbound)

(224) Rest Area - RR, Phones, Picnic (Southbound)

222 OH 3 , Medina
- **Other** W: Highway Patrol

220 Jct I-271 North, Erie, Pennsylvania

218 OH 18, Akron, Medina
- **Gas** E: 76*(D), BP*(D), Shell*(CW), Speedway*, Sunoco*(D)(LP)
  W: Super America*(D)
- **Food** E: Best Western, Big Boy, Burger King, Cracker Barrel (Sunoco), Dairy Queen, Hunan Dynasty, McDonalds,

EXIT OHIO

Medina Family, On Tap Bar & Grill, Perkins Family Restaurant, Subway (Sunoco)
W: Arby's, BW-3 Grill & Pub, Bob Evans Restaurant, Pizza Hut

Lodg E: AAA Best Western, Days Inn (see our ad this page), Holiday Inn Express, Suburbanite Motel
W: AAA Cross Country Inn

AServ E: 76, Toth Olds/GMC Trucks
W: Chesrown Medina Honda, Norris Pontiac/ Buick/Dodge

Med W: ✚ Hospital

Other E: Finast Grocery Store, Medina Ace Hardware, Windfal Car Wash

209 Jct I-76, U.S. 224, Lodi, Akron (Services On U.S.224)

TStop W: 76 Auto/Truck Plaza*(SCALES), Speedway*(LP)(SCALES), TravelCenters of America*(SCALES)

Food W: 76 Auto/Truck Plaza, Blimpie's Subs (Speedway), Country Pride (TravelCenters of America), McDonalds, Speedway Restauarant (Speedway), Taco Bell (Speedway)

Lodg W: HoJo Inn (TravelCenters of America)

TServ W: 76, TravelCenters of America

TWash W: Blue Beacon Truck Wash, Truck Wash

RVCamp W: Chippewa Campground (7 Miles)

ATM W: 76 Auto/Truck Plaza

204 OH 83, Lodi, Wooster

FStop E: BP*(D), Khalsa's Gas/Diesel*

Gas E: Duke*, Shell*
W: Citgo*

Food E: The Donut Stop

Lodg E: The Plaza Motel

AServ E: Bear's Towing, Burbank Auto Repair

Med W: ✚ Hospital

Other W: Buckeye Factory Shops & Food Court

198 OH 539, West Salem, Congress

RVCamp W: Town & Country Resort (1.9 Miles)

(197) Rest Area - RR, Phones, Vending, Picnic

196 OH301, West Salem (Northbound Only, Southbound Reaccess)

(190) Weigh Station (Southbound)

186 U.S. 250, Ashland, Wooster

TStop W: TravelCenters of America*(SCALES)

Gas E: 76*(D)

Food E: Grandpa's Village Deli Sandwiches, Perkins Family Restaurant

EXIT OHIO

W: Country Pride (TS of America), Ike's Great American Restaurant, McDonalds, Wendy's

Lodg W: Amerihost Inn, Days Inn, AAA Days Inn, AAA Travelodge

TServ W: TravelCenters of America

RVCamp E: Hickory Lake Campground (7.2 Miles)

Med W: ✚ Hospital

ATM W: TravelCenters of America, Wendy's

Other E: Sweeties Jumbo Chocolates
W: Highway Patrol

(180) Rest Area - RR, Phones, Picnic

176 U.S. 30, Mansfield, Wooster

Gas E: Marathon*

Food E: Ike's Great American Restaurant, La City Bar & Grill, The Family Buffet Restaurant
W: Christopher's on the Hill

Lodg E: AAA Econolodge

AServ E: Marathon

Other E: Antique Mall, Country Folks Class General Store

173 OH 39, Mansfield, Lucas

169 OH 13, Mansfield , Bellville

Gas E: BP*(D)
W: 76*

Food E: Cracker Barrel, Daugherty's Family
W: Arby's, Bob Evans Restaurant, Burger King, Denny's, McDonalds, Shoney's, Taco Bell

Lodg E: AAA AmeriHost, ◆ Budgetel Inn
W: ◆ Super 8 Motel, AAA Travelodge

AServ E: BP

Med W: ✚ Hospital

ATM E: BP

Parks E: Mohican State Park (19 Miles)

Other W: OH State Patrol, Tourist Information

165 OH 97, Lexington, Bellville

FStop E: Speedway*

Gas E: BP*(CW), Shell*(D)

Food E: Der Dutchman, Doc's Family Restaurant, McDonalds
W: Dinner Bell, Wendy's

Lodg E: ◆ Comfort Inn, Dollar Motel, ◆ Ramada Limited, Star Inn
W: Mid-Ohio Motel

RVCamp E: Blue Lagoon Campground (8 Miles), Honey Creek Campground (8 Miles), Yogi Bear Camp Report (8 Miles)

Parks E: Mohican State Park (14 Miles)

151 OH 95, Mount Gilead, Fredericktown

TStop E: Duke*(SCALES)

FStop W: BP*(D)

Gas E: Marathon
W: Sunoco*

Food E: Duke, Family Restaurant, McDonalds, Wendy's
W: Leaf Family

Lodg E: AAA Best Western

AServ E: Marathon

TServ E: Duke

RVCamp W: KOA Campground (.4 Miles)

Med W: ✚ Hospital

ATM W: BP

Parks W: Mt Gilead State Park

EXIT OHIO

Other E: 24 Hour Laundromat (Duke), Highway Patrol

140 OH 61, Mount Gilead, Cardington

Gas W: Amoco*, Citgo*

Food E: Sam's Deli
W: Ole Farmstead Inn

AServ E: Brown's

131 Mount Vernon, Wesleyan University

TStop E: Flying J Travel Plaza*(LP)(SCALES) (RV Dump)

FStop E: Speedway*(LP)

Gas W: BP*(D)(CW)(LP), Shell*, Sunoco(D)

Food E: Blimpie's Subs (Speedway), Burger King, Flying J Travel Plaza
W: Arby's, Bob Evans Restaurant, McDonalds, Subway, Taco Bell, Waffle House, Wendy's

Lodg W: Hampton Inn

AServ W: Sunoco

TServ W: North Star Towing & Truck Repair

RVCamp E: Camping

Med W: ✚ Hospital

ATM W: McDonalds

Parks W: The Alum Creek State Park (6 Miles)

(131) Weigh Station (Northbound)

(128) Rest Area - RR, Phones, Picnic, Vending

121 Polaris Parkway

Gas E: BP*(CW), Shell*(CW)(LP)

Food E: Polaris Grill
W: Max & Erma's Restaurant, Wendy's

Lodg E: Wingate Inn

119 Junction I-270, Columbus bypass

117 OH 161, Worthington, New Albany

116 Morse Rd, Sinclair Rd

115 Cooke Rd

114 North Broadway

Gas W: BP*, Sunoco*

Med W: ✚ Hospital

113 Weber Rd

112 Hudson St

Gas E: BP*(CW), Shell*(CW)

Food E: Wendy's

AServ E: John Tires

Med W: ✚ Hospital

Other E: Car Wash

111 17th Ave

110B 11th Ave

110A 5th Ave

Gas E: Delta*(D)
W: BP*, Shell*(CW)

Food E: KFC, White Castle Restaurant
W: Dairy Treat, Wendy's

Med W: ✚ Hospital

109C Spring St (Southbound Left Exit)

109B Leonard Ave, OH 3, Cleveland Ave

109A Junction I-670 West

108B U.S. 40, Broad St

108A U.S. 33, Main St

107 U.S. 70 E (Southbound Left Exit)

106B Ohio 315 N

106A Junction I-70 West

105 Greenlawn Ave

Gas W: Fast and Fair Mini Mart

EXIT OHIO

RVCamp W: Greenlawn RV Sales & Service
Med E: ✚ Hospital

104 OH 104, Frank Rd

101 Jct I-270, Wheeling, Indianapolis

100 Stringtown Rd, Grove City
FStop W: Certified*
Gas E: BP*
W: Speedway*(D)(CW), Sunoco*(D)
Food E: Best Western, Bob Evans Restaurant, Ramada, White Castle Restaurant
W: Captain D's Seafood, China Bell, Cracker Barrel, Daddy-o Express Drive Thru, Fazoli's Italian Food, Gottlieb's, KFC, McDonalds, Millstreet Bagels & Deli, Perkins Family Restaurant, Rally's Hamburgers, Rax, Shoney's, Tee Jaye's, Waffle House, Wendy's
Lodg E: AAA Best Western, AAA Cross Country Inn, ◆ Hampton Inn, Ramada
W: ◆ Comfort Inn, AAA Heritage Inn, ◆ Red Roof Inn, Saver Motel, Value Inn
AServ E: Herb's Auto Service
W: Tuffy Muffler
ATM W: Bank One
Other W: Aldi Grocery, Big Bear Grocery, US Post Office

97 OH 665, London - Groveport Rd
Gas E: Sunoco*(LP)
TServ W: RW Diesel Service &Repair
Other W: Evans Market Convenience Store

94 U.S. 62, OH 3, Grove City, Harrisburg
Gas E: BP*(LP)
W: Shell, Sunoco*
AServ W: Shell
ATM W: Sunoco
Other W: Express Lane

84 OH 56, London, Mount Sterling
Gas E: BP*(LP), Sunoco*(LP)
Food E: Subway (BP)
Lodg E: Royal Inn
ATM E: Sunoco
Parks E: Deer Creek State Park

75 OH 38, Midway, Bloomingburg
Gas W: Sunoco*

69 OH 41, OH 734, Jeffersonville , Washington CH
Gas W: Shell*, Sunoco*
Food W: Arby's, Subway (Shell), TCBY Yogurt, Wendy's, White Castle Restaurant (Sunoco)
Lodg W: AAA Amerihost Inn
RVCamp E: Walnut Lake Campground
Med E: ✚ Hospital
ATM W: Sunoco
Other W: Factory Outlet Center

(68) Rest Area - RR, Vending, Phones, Picnic (Southbound)

65 U.S. 35, Washington Court House, Xenia
TStop E: 76 Auto/Truck Plaza*(SCALES)
W: TravelCenters of America*(SCALES)
FStop E: Amoco*, Circle K*
Gas E: Shell*
W: Royal Mart C-Store
Food E: 76 Restaurant (76 Auto/Truck Stop), Bob Evans Restaurant, Burger King, Dairy Queen, KFC, McDonalds, Shoney's, Taco Bell, Waffle House, Wendy's
W: Country Pride (TC of America)
Lodg E: AAA AmeriHost Inn
W: Dollar Inn
TServ E: 76 Auto/Truck Plaza
W: TravelCenters of America
Med E: ✚ Hospital
Other E: Factory Outlet Center

58 OH72, Jamestown, Sabina

(54) Weigh Station (Southbound)

EXIT OHIO

50 U.S. 68, Wilmington
FStop W: Shell*, Speedway*
Gas W: BP*
Food W: Dairy Queen, L&K Motel, McDonalds, Subway (Speedway)
Lodg W: L&K Motel
AServ W: Sandy's Towing & Tire Service
Med E: ✚ Hospital

(49) Weigh Station (Northbound)

45 OH73, Waynesville, Wilmington
Gas E: BP*, Marathon*
W: Citgo*
RVCamp E: Campground
Med E: ✚ Hospital
Parks W: Caesar Creek State Park (5 Miles)

36 Wilmington Rd
Parks E: Fort Ancient & Campground

(34) Rest Area - RR, Phones, Picnic, Vending, Tourist Info

32 OH123, Morrow, Lebanon
Gas E: Gas*, Marathon*
Food E: Country Kitchen
RVCamp W: Camping
Other W: Fort Ancient State Memorial

28 OH48, South Lebanon
Other W: State Patrol Post

25AB OH 741 North, Kings Mills Rd, Mason
Gas E: Shell*, Speedway*(LP)
W: AmeriStop*
Food E: Bill Knapp's Restaurant, McDonalds, Taco Bell
W: Backstage Bar & Grill, Bob Evans Restaurant, Burger King, Corky's Pizzeria & Deli, Frisch's Big Boy, Perkins Family Restaurant, Skyline Chili, Subway, Waffle House, Wendy's
Lodg E: Comfort Suites
W: Hampton Inn, Holiday Inn Express
AServ W: BP
RVCamp E: Kings Island Jellystone Park Camp Resort (.6 Miles)
ATM W: National Bank & Trust
Other E: Kings Island Amusement Park

24 Western Row Rd, King's Island Drive (Northbound)
Other E: Kings Island Amusement Park

19 Mason Montgomery Rd, Fields Ertel Rd
Gas E: Sunoco*
W: BP*(D)(CW)(LP), Marathon*(CW), Shell*(CW)
Food E: Arby's, Bennigans, Bob Evans Restaurant, Cooker Bar & Grill, Cracker Barrel, Fazoli's Italian Food, Frisch's Big Boy, Grand Oriental, KFC, La Salsa, McDonalds, Olive Garden, Pizza Tower, Ponderosa, T.G.I. Friday's, Taco Bell, White Castle Restaurant
W: Applebee's, Fuddrucker's, Gold Star Chili (Biggs Hyper Mart), Holiday Inn, Lone Star Steakhouse, Marriot, O' Charley's, Skyline Grill, Steak 'N Shake, The Swiss Colony, Tumble Weed, Waffle House, Wendy's
Lodg E: AAA Comfort Inn, AAA Signature Inn
W: ◆ Best Western, ◆ Budgetel Inn, Country Hearth Inn, DAYS INN Days Inn, AAA Holiday Inn, AAA Marriot
AServ E: Firestone Tire & Auto, Williams Auto Superstore
W: BP Pro Care
ATM E: First National Bank, United Dairy Farmers

EXIT OHIO

W: Biggs Hyper Mart, Fifth Third Bank, Marathon
Other E: Kings Veterinary Hospital, Kroger Supermarket (Pharmacy), PetsMart, United Dairy Farmer C-Store
W: Biggs Hyper Mart (Grocery, Pharmacy)

17 Jct I-275, to I-75, OH 32

15 Pfeiffer Rd., Blue Ash
Gas W: BP*(CW)
Food W: Best Western, Bob Evans Restaurant, Subway
Lodg W: Blue Ash Best Western, ◆ Red Roof Inn
Med E: ✚ Hospital
ATM W: BP

14 Ronald Reagan Cross County Hwy.

12 U.S. 22, OH. 3, Montgomery Rd.
Gas E: BP*(D), Shell
W: BP
Food E: Arby's, Bob Evans Restaurant, Lone Star Steakhouse, Outback Steakhouse, Red Lobster, Rio Bravo Cantina, Sigee's (Harley Hotel), T.G.I. Friday's
W: Dick Clark's American Grill, Hotel Discovery, Johnny Rockets
Lodg E: ◆ Harley Hotel
AServ E: Kenwood Dodge, Shell
W: BP, Firestone Tire & Auto
Med W: ✚ Hospital
ATM E: Fifth Third Bank, Provident Bank
W: Key Bank
Other W: Barnes & Noble, Kenwood Towne Centre Mall, Revco Drugs

11 Kenwood Rd (Northbound, Southbound Reaccess)
Gas W: BP*, Chevron*
Food W: Burger King, Graeter's Ice Cream, KFC, McDonalds, Taco Bell, Wendy's, Wok & Roll Chinese
AServ W: BP, Firestone Tire & Auto, Jiffy Lube, Meineke Discount Mufflers, Pep Boys Auto Center, Q Lube
Med W: ✚ Hospital
ATM W: Bank One, PNC, Star Bank
Other W: Kenwood Towne Center Mall

10 Stewart Rd. (Northbound, Southbound Reaccess)
Gas W: Marathon*(D)
ATM W: Marathon
Lodg W: Holiday Inn (see our ad this page)

9 Redbank, Fairfax

8B Ridge Ave. North

8 Ridge Ave. South
Gas W: Speedway*(D)
Food W: Denny's, Long John Silvers, McDonalds, Rally's Hamburgers, Taco Bell, Wendy's
Lodg W: ◆ Howard Johnson
AServ W: Goodyear Tire & Auto, Nolan Ford, Tire Discounters
ATM W: PNC
Other W: Bigg's Grocery, Sam's Club, Wal-Mart

7 Ohio 562, Norwood

6 OH 561, Smith Rd., Edwards Rd.
Gas W: Shell*(CW)
AServ W: Midas
ATM W: Shell

5 Montgomery Ave., Dana Ave.
Gas W: Shell*, United Dairy Farmers*
AServ W: Shell

3 Taft Rd (Southbound, Northbound Reaccess)

2 U.S. 42 North, Reading Rd, Florence Ave

1K Junction I-471 South, U.S. 50, U.S. 52 East, Columbia Parkway

1J Jct I-471 South, Newport, KY

Note: I-71 runs concurrent below with I-75. Numbering follows I-75.

1H Ezzard Charles Dr

Bold red print shows RV & Bus parking available or nearby

EXIT		OHIO/KENTUCKY
1G		To U.S. 50 West, Freeman Ave
	Food	**W:** Frisch's Big Boy, Taco Bell, Wendy's
	Lodg	**W:** Holiday Inn
	ATM	**W:** Fifth Third Bank, Star Bank
1F		Seventh St
	Gas	**W:** Ashland, Sunoco*
	Food	**W:** Chili Company
	Lodg	**W:** Holiday Inn
	ATM	**W:** The Provident Bank
1E		Fifth Street

Note: I-71 runs concurrent above with I-75. Numbering follows I-75.

1D		Main St
1BC		Vine St, Pete Rose Way, Elm St, 3rd St
	Other	**E:** **Cinergy Stadium**
1A		Jct I-75 N, Dayton, 1G - U.S.50 W

↑OHIO

↓KENTUCKY

Note: I-71 runs concurrent below with I-75. Numbering follows I-75.

EXIT		KENTUCKY
192		Fifth St, Covington, Newport
	Gas	**E:** Chevron(D), Shell*(CW), SuperAmerica*(D)(LP) **W:** Deli Direct, JD's Food Mart*
	Food	**E:** Burger King (24 hrs), Dominiques, Frisch's Big Boy, Gold Star Chili, Hardee's, Holiday Inn, KFC, McDonalds, Perkins Family Restaurant, River Front Pizza, Subway, Taco Bell, White Castle Restaurant **W:** Cork & Bottle Cafe, Deli Direct, Willy's Sports Cafe
	Lodg	**E:** Extended Stay America, Holiday Inn, Quality Hotel **W:** Hampton Inn
	AServ	**E:** Chevron, Ford Dealership, Smith Muffler Shop **W:** Mark's Garage
	ATM	**E:** Mac **W:** Deli Direct, JD's Food Mart
	Other	**E:** **Mainstrasse Village & N. KY Visitor Ctr.**
191		Twelfth St, Pike St, Covington
	Gas	**E:** Shell*
	Food	**E:** Anchor Grill, Bakery Outlet, Donut Cafe
	AServ	**E:** Marshall Dodge
	Med	**E:** **Hospital** **W:** **Hospital**
	Other	**E:** **Coin Laundry, Zimmer Hardware**
190		Jefferson Ave
189AB		KY 1072 West, Fort Wright, Kyles Ln
188AB		Dixie Highway, Fort Mitchell
	Gas	**E:** Sunoco* (Kerosene)
	Food	**E:** **Skyline Chili** **W:** Holiday Inn, Indigo Bar & Grill, Pizza Hut, Ramada Inn
	Lodg	**W:** Days Inn, Holiday Inn, Ramada Inn
	ATM	**E:** Fifth Third Bank, PNC Bank **W:** Provident Bank (Walgreen's), Star Bank
	Other	**E:** **Kroger Supermarket** **W:** **Thriftway Food & Drug (24 hrs), WalGreens**
186		KY 371, Buttermilk Pk
	Gas	**E:** Ashland, BP*(CW) (Kerosene) **W:** BP*(CW), Citgo, Marathon, Shell*, Sunoco*(D)(CW)
	Food	**E:** Burbank's Real BBQ & Ribs, Graeter's Ice Cream, Jacqueline's Steak & Ribs, Josh's (Drawbridge Estate), Montoyas Mexican, Oriental Wok Chinese **W:** Arby's, Baskin Robbins, Bob Evans Restaurant, Boston Market, Burger King, Butternut Bakery Outlet, Domino's Pizza, Donatos Pizza, Dunkin Donuts, Fazoli's Italian Food, Goldstar Chili, JB's Canyon Grill, La Rosa's Italian, Marx Hot Bagels, McDonalds, Outback Steakhouse, Papa John's Pizza, Papadinos, Penn Station, Sandwich Block Deli, Subway
	Lodg	**E:** Cross Country Inn, Drawbridge Estate
	AServ	**E:** Ashland, Citgo **W:** Citgo, Jiffy Lube, Marathon

EXIT		KENTUCKY
	ATM	**E:** First National Bank **W:** Fifth Third Bank, PNC Bank, People's Bank
	Other	**E:** **Oldenburg Brewery Tours & Museum, United Dairy Farmers*** **W:** **1 Hr Photo (Walgreens), AmeriStop Food Store, Animal Hospital, Butternut Bakery Outlet, Crestville Drugs, Drug Emporium, Revco Drugs, Service Plus IGA (24 hrs), WalGreens**
185		Jct I-275, Cincinnati Airport
184AB		KY 236, Erlanger
	Gas	**E:** AmeriStop*, BP*, Citgo* **W:** Ashland*, JD's Food Mart*, Speedway*
	Food	**E:** Double Dragon Oriental Cuisine, Rally's Hamburgers (BP) **W:** Subway (Ashland), Waffle House
	Lodg	**W:** Comfort Inn, **Days Inn, Econolodge, HoJo Inn**
	Med	**W:** **Hospital**
	ATM	**E:** Heritage Bank
	Other	**E:** **Cinderella Laundry, US Post Office**
182		KY 1017, Turfway Rd
	Gas	**E:** BP*
	Food	**E:** Blimpie's Subs, Frisch's Big Boy, Krispy Kreme, Lee's Famous Recipe Chicken, Ryan's Steakhouse **W:** Applebee's, **Cracker Barrel,** Fuddrucker's, Italianni's, Longhorn Steakhouse, McDonalds, Ming Garden, O'Charley's, Rafferty's, Shells Seafood, Steak 'N Shake, Tumbleweed, Wendy's
	Lodg	**E:** AAA Courtyard by Marriott, Fairfield Inn, Signature Inn **W:** Commonwealth Hilton, Hampton Inn, Studio Plus
	Med	**W:** **Hospital**
	ATM	**E:** Crawford Insurance **W:** Meijer, Sam's Club, Star Bank
	Other	**E:** **1 Hr Photo, Eye World, Family Fun Center, Pharmacy (Thriftway), Pro Golf Discount, Thriftway Marketplace** **W:** **Advance-U-Cash, Meijer Food Store (24 hrs), Sam's Club, Turfway Park (Horse Races), Wal-Mart (24 hrs)**
181		KY 18, Florence, Burlington
	TStop	**E:** **76 Auto/Truck Plaza***(LP)(SCALES)
	Gas	**E:** Shell(CW), Speedway*, Swifty* **W:** BP*(D)(CW)(LP), Citgo*
	Food	**E:** Fillmore's Dairy Hut, **Restaurant (76 Auto/ Truck Stop),** Waffle House **W:** Espress-way Drive Thru Coffee Bar, Hardee's, La Rosa's, Lonestar
	Lodg	**E:** Cross Country Inn, Florence Best Western
	AServ	**E:** LTD Mobile Glass Auto Glass Specialist **W:** Airport Ford, Dodge Land, Mazda Subaru, Nissan Towne, Precision Tune & Lube, Pro Care (BP), Suber's Auto Repair, Toyota Towne
	TServ	**E:** **76 Auto/Truck Plaza**
	TWash	**E:** **76 Auto/Truck Plaza**
	ATM	**W:** Bank One, Fifth Third Bank, Heritage Bank
	Other	**E:** **AmeriStop*, Angel Animal Hospital,** Classic Car Wash, **Fundome Skating Rink, Speed Queen Coin Laundry** **W:** **K-Mart (Pharmacy), World of Sports Mini Golf**
180		U.S. 42, U.S. 127, Union, Florence
	Gas	**E:** BP*(D)(CW)(LP), Speedway*(D)(LP) **W:** Chevron*, Shell*(CW)
	Food	**E:** Bar-B-Q Pit, Bob Evans Restaurant, Burger King, Captain D's Seafood, Frisch's Big Boy, Jalepeno's, Long John Silvers, Madd Anthony's, McDonalds, Peking Palace Chinese, Penn Station, Pizza Hut, Rally's Hamburgers, Red Lobster, Skyline Chili, Sub Station II, Subway, Warm Ups Sports Cafe, Wendy's **W:** Arby's, Flo's Hot Dogs, KFC, Little Caesars Pizza, Perkins Family Restaurant (24 hrs), Ponderosa, Waffle House, White Castle Restaurant
	Lodg	**E:** Knight's Inn, Motel 6, **Ramada Inn,** AAA Super 8 Motel, The Wildwood Inn **W:** **Budget Host Inn,** Holiday Inn
	AServ	**E:** Jim's Auto Service, Mr. Transmission, Rockcastle Olds, Cad, GM **W:** Car X Muffler & Brakes, Midas Muffler &

Bold red print shows RV & Bus parking available or nearby

← N I-71 S →

EXIT KENTUCKY

Brakes, NTB, Quick Stop Oil Change
ATM E: Fifth Third Bank
Other E: **Check Advance,** Classic Car Wash, **Fantasy Frontier Mini Golf, Go Karts, Lazer Tag,** The Auto Bath Self Serve Car Wash
W: **Ameristops Food Store, Florence Animal Clinic**

178 KY 536, Mt Zion Road
Gas E: Shell*(D)(CW)

(177) Welcome Center - RR, Phones

175 Kentucky 338, Richmond
TStop E: **TravelCenters of America**(SCALES)
FStop E: **Pilot Travel Center***(SCALES)
W: **Pilot Travel Center***(SCALES)
Gas W: BP*, Shell*(D)(CW)
Food E: Arby's, Burger King, **Country Pride Restaurant (TC of America)**
W: **McDonalds,** Snappy Tomato Pizza (BP), **Subway (Pilot),** Waffle House, Wendy's
Lodg W: Days Inn, Econolodge
TServ E: **TravelCenters of America**
ATM W: Huntington Banks
Parks W: **Big Bone Lick State Park**

Note: I-71 runs concurrent above with I-75. Numbering follows I-75.

77 Junction I-75 South, Lexington

(75) **Weigh Station (Southbound)**

72 KY 14, Verona
Gas E: Chevron*(D)
AServ E: Chevron
RVCamp E: **Oak Creek Campground (5.3 Miles)**

62 U.S. 127, Glencoe, Owenton
TStop E: **Exit 62 Truck Plaza***
Gas W: Ashland*(D)
Food E: **Exit 62 TS**

EXIT KENTUCKY

W: Ashland Restaurant
ATM E: **Exit 62 TS**

57 KY 35, Warsaw, Sparta
Gas W: Marathon*
RVCamp E: **Camping**

44 KY 227, Carrollton, Worthville
Gas E: BP*(LP)
W: Ashland*(D), Chevron*(D), Shell*
Food W: Arby's, Burger King, KFC, McDonalds, Taco Bell, Waffle House
Lodg W: Days Inn, ◆ **Holiday Inn Express,** ◆ Super 8 Motel
Med W: **Hospital**
Parks W: **General Butler State Resort Park**

43 KY 389 to KY 55, English, Prestonville

(35) **Weigh Station (Northbound)**

34 U.S. 421, Bedford, Cambellsburg
Gas W: Ashland*(D), Marathon*
Food W: Cody's Burgers & Shakes
AServ W: JR's Auto & Truck Repair
TServ W: **JR's Auto & Truck Repair**

28 KY 153, KY 146, Sligo, New Castle
TStop E: **Davis Brothers Truck Stop***(SCALES)
W: **Pilot Travel Center***(LP)(SCALES)
FStop E: **Marathon***(D)
Food E: **Cleo's (Davis Bro. TS),** Taylor's Dairy Bar
W: Pendelton Food Mart Hickory Smoked BBQ, **Subway (Pilot)**
TWash E: **Davis Brothers TS**
ATM E: **Davis Bro. TS**

22 KY 53, Ballardsville, La Grange
Gas E: Ashland*, BP*, Super America*(D)(LP)
W: Amoco* (24 Hrs), Chevron*(D), Swifty*
Food E: Burger King (Inside Play Place), Lucky Dragon, Papa John's Pizza, Ponderosa, Rally's

EXIT KENTUCKY

Hamburgers (Super America), Shoney's, Waffle House, Wendy's
W: Arby's, **Cracker Barrel,** Dairy Queen, KFC, Long John Silvers, McDonalds (Inside Play Place), Subway, Taco Bell, Taco John's
Lodg E: Best Western, Days Inn
AServ E: Big O Tires, Wal-Mart
W: NAPA Auto Parts, Smiser-Carter Chevrolet, Olds, Buick, Geo
Med E: **Tri County Baptist Hospital**
ATM E: Kroger Supermarket
W: Oldem County, PNC Bank
Other E: **Kroger Supermarket, Wal-Mart (Pharmacy, 24 Hrs)**
W: **La Grange Auto Bath, LaGrange Animal Hospital, Rite Aide Pharmacy**

18 KY. 393, Buckner

17 KY 146, Buckner
Gas W: Shell, Thornton's Fuel Mart*(LP)
AServ E: Tri County Ford Lincoln Mercury
W: Shell
ATM W: PNC Bank
Other W: **U.S. Post Office**

14 KY 329, Crestwood, Pewee Valley
Gas E: Chevron* (24 Hrs), Shell*

(13) **Rest Area - RR, Phones, Picnic**

9AB Jct I-265, KY 841 to I-65

5 Jct I-264, Watterson Exprwy.

2 Zorn Ave.
Gas W: BP*(CW), Chevron*
Lodg W: ◆ Holiday Inn
AServ W: Chevron
Med E: **Hospital**
Other W: **Water Tower Museum**

1A I-64 West, St Louis

1B I-65 Indianapolis, Nashville

I-72 E →

EXIT ILLINOIS

Begin I-72

↓ILLINOIS

(4) JCT US 36, I-172 North to Quincy

10 IL 96, Hull, Payson

20 IL 106, Barry
FStop S: **Phillips 66***(D)
Gas S: Amoco*(D), Shell*
Food S: **Wendy's**
AServ S: Amoco
ATM S: Barry Community Bank, Phillips 66

31 New Salem, Pittsfield

35 US 54, IL 107, Griggsville, Pittsfield
RVCamp S: **Campground**

EXIT ILLINOIS

Other S: **State Patrol Post**

46 IL 100, Bluffs, Detroit

52 IL 106, Winchester

60 US 67, Jacksonville, Berardstown

64 US 67, Alton, Jacksonville

68 IL 104, Jacksonville

76 IL 123, Ashland, Alexander

82 New Berlin

91 Wabash Ave., Loami

93 IL 4, Springfield, Chatham (Left Exit)
Gas N: Huck's*
Food N: Applebee's, Bakers Square Restaurant and Pies, Damon's Clubhouse, Den Chili, Hardee's, Kenny Rogers Roasters, McDonalds (Indoor Playground), Ned Kelly's Steakhouse, Perkin's

EXIT ILLINOIS

Family Restaurant, Subway, Taco Bell
Lodg N: Comfort Inn, Courtyard Inn, Fairfield Inn, Sleep Inn
ATM N: Huck's, Magna Bank
Other N: **Bethel Bookstore,** Full Service Car Wash, **Target Department Store (Pharmacy), Wal-Mart (Tire and Lube Service)**

92AB Sixth Street, Jacksonville
Food N: Burger King, Cancun Mexican Restaurant, Cozy Drive-In, Heritage House Smorgasbord, KFC, Legends Restaurant, McDonalds, New China Restaurant, Subway
Lodg N: Illini Inn, Ramada Inn, Super 8 Motel
AServ N: Fast Lube, Shell
Other N: Car Wash, **WalGreens**

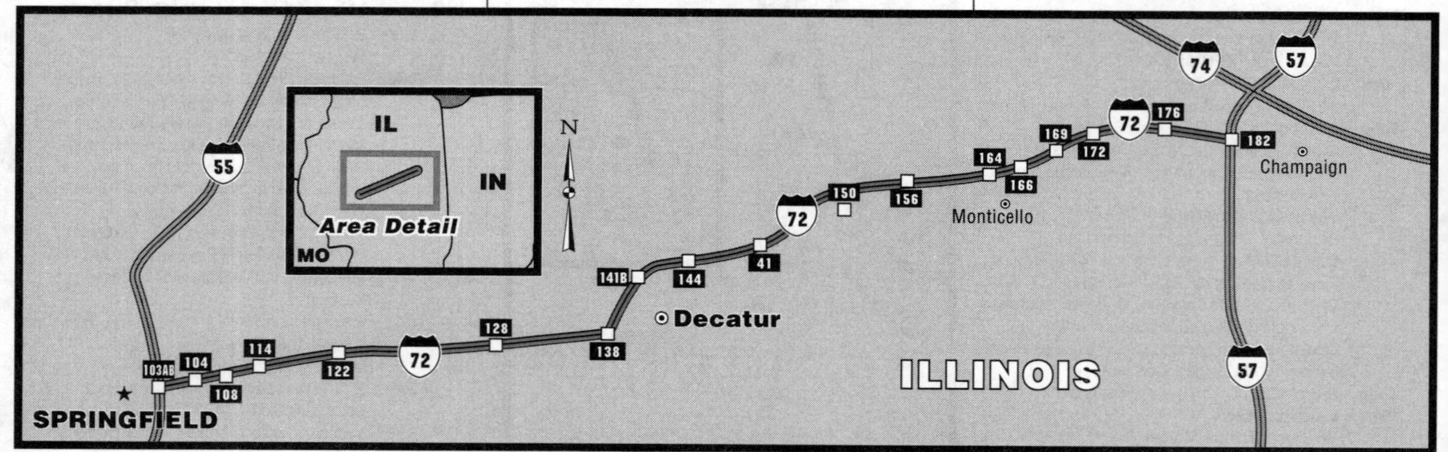

Bold red print shows RV & Bus parking available or nearby

EXIT	ILLINOIS
4	Stevenson Dr., East Lake Dr.
Gas	N: Amoco*(CW), Shell*(D)(CW)
Food	N: 20's Hideout Steakhouse, Arby's, Bob Evans Restaurant, Bombay Bicycle Club, Denny's, Domino's Pizza, **Hardee's**, Long John Silvers, Maverick Family Steakhouse, McDonalds, Red Lobster, Steak & Shake, Subway, Taco Bell, Tai Pan Chinese, Wendy's
Lodg	N: Comfort Suites, Days Inn Days Inn, Drury Inn & Suites, Hampton Inn, Holiday Inn Crown Plaza Hotel, Holiday Inn Motel, Peartree Inn, Signature Inn
ATM	N: First National Bank, Shell
Other	N: **Revco**
6AB	IL 29, Taylorville
Gas	N: Shell*(D)(LP), Union 76*(D)
Food	N: Godfather's Pizza, Nicole's Restaurant
Lodg	N: Knights Inn, Red Roof Inn, Super 8 Motel
AServ	N: Auto Zone Auto Parts
Other	N: **Shop and Save**
03AB	Junction I-55, North - Chicago, South - St. Louis, Jackson
04	Camp Butler
08	Riverton, Dawson
14	Buffalo, Mechanicsburg
22	Mt. Auburn, Illiopolis
28	Niantic
33AB	US 36 E - Decatur, Harristown, US 51 S - Pana (Difficult Reaccess)
Lodg	S: Holiday Inn Select
Med	N: **+ Hospital**
38	I-121 Decatur, Lincoln, Warrensburg
41AB	US 51 S - Decatur, Bloomington
Gas	N: Shell*(LP)
Food	N: Applebee's, Cheddar's, Country Kitchen Restaurant, **Cracker Barrel**, Hardee's, Home Town Buffet, McDonalds (Indoor Playground) S: Arby's, Burger King, China Buffet, El Rodeo, Subway
Lodg	N: Budgetel Inn, Comfort Inn, Country Inn & Suites, Fairfield Inn, Hampton Inn, Ramada Limited
AServ	N: Ten Minute Lube and Oil Change, Western Auto
ATM	N: Mutual Bank (24 Hr Teller) S: First National Bank
Other	N: Coin Car Wash, **Hickory Point Mall** S: **Sam's Club, US Post Office, Wal-Mart Supercenter (Tire Service, Lube)**
44	IL 48 Dacatur, Oreana
FStop	N: **Oasis**
Gas	N: Oasis
Food	N: **Restaurant (Oasis)**
50	Argenta
156)	**Rest Area - RR Phones, Picnic, Vending, Picnic, Playground, Large Truck Parking P**
56	I-48, Sisco, Weldon
RVCamp	S: **Campground**
64	Bridge Street
66	I-105 W, Market Street
Med	W: **+ Hospital**
69	White Heath Road
72	I-10 Clinton
76	I-47 Mahomet
82AB	Junction I-57, South - Memphis, North - Chicago

↑ILLINOIS

Begin I-72

EXIT	IOWA/ILLINOIS
	Begin I-74

↓IOWA

EXIT	
1	53rd Street, Hamilton Tech College
Food	S: Steak & Shake
AServ	N: Big 10 Citgo and Food Mart*(CW)
ATM	S: First Bank
Other	S: **Target Department Store (Super Food Center, Pharmacy)**
2	U.S. 6, Spruce Hill Drive, Kimberly Road
Gas	N: Phillips 66*
Food	N: Old Chicago S: Applebee's, Bob Evans Restaurant, Burger King, China Boy Restaurant, St Louis Bread
Lodg	N: ◆ Courtyard By Marriott, Jumers Castle Inn, AAA Signature Inn S: Days Inn ◆ Days Inn, ◆ Fairfield Inn, ◆ Hampton Inn
AServ	N: Phillips 66 S: The Grease Spot
ATM	S: Merchantile Bank, Norwest Bank
Other	N: **U-Haul Center**(LP) S: **Sam's Club, Wal-Mart Supercenter**
3	Middle Road, Locust Street
4	U.S. 67, Grant Street, State Street, Riverfront (Difficult Reaccess)
Gas	N: Citgo*, Clark, Conoco*(CW), Phillips 66* S: Coastal
Food	N: Conoco Deli, Expressly Ross, Paddle Wheel Restaurant, Ross's 24 Hour Restaurant S: Dairy Queen, Village Inn Restaurant, Waterfront Deli
Lodg	N: Traveler Motel, AAA Twin Bridges Motor Inn S: City Center Motel, Village Inn
TServ	S: **UD Trucks**
ATM	N: Conoco
Other	S: **Coin Laundry**, Full Service Car Wash

↑IOWA

↓ILLINOIS

EXIT	
1	(5) River Dr.
Med	S: **+ Hospital**
2	7th Avenue
3	23rd Avenue
Gas	S: Amoco
Food	N: Hardee's S: Hardee's, Hungry Hobo, Whitey's Ice Cream
4AB	IL 5, John Deere Rd
(5)	**Weigh Station (Eastbound)**
5AB	IL 6, US 6, Moline, Quad City Airport (Exits To Exit Four)
Gas	S: Citgo*
Food	S: McDonalds, Skyline Inn Ribs and Catfish, The Omlet Shop
Lodg	S: Hampton Inn, Holiday Inn, Holiday Inn Express, La Quinta Inn, Ramada Inn
(8)	**Weigh Station (Westbound)**
(14)	Jct I-80
24	IL 81, Kewannee, Cambridge
(28)	**Rest Area - RR, Vending, Phones, Picnic, Playground P (Eastbound)**

EXIT	ILLINOIS
(30)	**Rest Area - RR, Phones, Picnic, Vending, Playground, RV Dump P (Westbound)**
32	IL 17, Woodhull, Alpha
TStop	S: **Mobil**(SCALES)
Gas	N: Citgo(D), Texaco
Food	N: Homestead Restaurant S: **Restaurant (Mobil)**
AServ	S: Mobil
TServ	S: **Mobil, Woodhull Plaza Garage**
Med	N: **+ Woodhull Clinic**
ATM	S: Woodhull Truck Plaza
46AB	US 34, Kewanee, Monmouth
TServ	N: **Nichols Diesel Service**
Med	S: **+ Hospital**
48AB	East Galesburg, Galesburg
Gas	S: Citgo*(CW), Phillips 66*
Food	S: A & W Drive-In, Dairy Queen, KFC, Pizza Hut, Spring Garden Family Restaurant
Lodg	N: Jumers Continental Inn S: Holiday Inn Express
ATM	S: First Mid-West Bank
Other	N: **Eagle Country Market (Grocery Store)** S: Southward's Car Wash
51	Knoxville
FStop	S: **Amoco**
Food	S: **Hardee's, McDonalds**
Lodg	S: Super 8 Motel
ATM	S: Hardee's
54	IL 97, Lewistown
(62)	**Rest Area - Vending, Picnic, Phones, Playground P (Both Directions)**
71	IL 71, Canton, Kewannee
75	Brimfield, Oak Hill
Food	N: Jimalls Restaurant
82	Kickapoo, Edwards Road
Gas	N: Mobil*, Shell
Food	N: **Jubilee Cafe**
AServ	N: Mobil
87AB	Jct I-474, IL 6, Indianapolis, Chillicote
89	U.S. 150, War Memorial Drive
Gas	N: Amoco*(CW), Clark*, Phillips 66(D)
Food	N: Arby's, Burger King, Comfort Suites, Dunkin Doughnuts, Kenny Rogers Roasters, Khoury's Cousine, McDonalds (Outdoor Playground), Mid Kelly Steakhouse, Red Roof Inn, Shakies Pizza, Steak & Shake, Subway, Super 8 Motel, Wendy's
Lodg	N: ◆ Comfort Suites, ◆ Fairfield Inn, ◆ Red Roof Inn, AAA Signature Inn, ◆ Super 8 Motel
AServ	N: Amoco, Auto Zone Auto Parts, Phillips 66
Other	N: **Wal-Mart Supercenter, WalGreens**
90	Gale Ave
91AB	University Street
92A	IL 88, Knoxville Ave. (Knoxville Ave. Left Hand Exit)
92	Glendale Avenue (Difficult Reaccess)
93A	Jefferson Street
93B	U.S. 24, IL 29, Washington Street, Adams Street (No Reaccess)
Lodg	S: Hojo Inn
94	IL 40, Industrial Spur (Difficult

EXIT	ILLINOIS
	Reaccess)
95A	North Main Street
Gas	S: Amoco*(CW)
Food	S: Applebee's, Blimpie's Subs, Bob Evans Restaurant, Hardee's, Long John Silvers, Pizza Hut, Youssef's Deli
Lodg	N: Hampton Inn
	S: Best Western, Mark Twain House, Motel 6
AServ	N: Firestone Tire & Auto
	S: Amoco, Goodyear Tire & Auto
ATM	S: Amoco, Bank One (Kroger)
Other	S: Coin Car Wash, **Kroger Supermarket (Pharmacy), Revco, Wal-Mart Supercenter**
95B	IL 116, Metamora
95C	U.S. 150 East Camp St.
Gas	N: Site
Food	N: Subway
Lodg	N: Super 8 Motel
96	IL 8, East Washington Street
98	Pinecrest Drive
99	Jct I-474
101	Jct I-155 S, Lincoln
102	Morton
Gas	N: Amoco, Phillips 66*
	S: Shell*
Food	N: Blimpie's Subs (Phillips 66), Burger King, Cracker Barrel, Days Inn, Dunkin Doughnuts (Phillips 66), Philllips 66, Taco Bell, Whistles Casual Cafe
	S: McDonalds (Indoor Playground), Subway
Lodg	N: ◆ Comfort Inn, Days Inn, Holiday Inn Express, Knight's Inn
AServ	N: Phillips 66*
ATM	S: Shell
112	IL 117, Goodfield
FStop	N: **Shell***(D)
Gas	N: Amoco
AServ	N: Amoco
RVCamp	N: **Campground**
ATM	N: Goodfield State Bank (Amoco)
(115)	**Rest Area - Vending, RR, Phones, Picnic, Playground, Large Truck Parking, RV Dump P (Both Directions)**
120	Carlock
Gas	N: Amoco
Food	N: Country Side Family Restaurant
AServ	N: Amoco

EXIT	ILLINOIS
(123)	**Weigh Station - both directions**
125	U.S. 150, Mitsubishi Motor Way Diamond - Star Parkway
127	Jct I-55
	Note: I-74 runs concurrent below with I-55. Numbering follows I-55.
160AB	U.S. 150, IL 9, Pekin, Market St
TStop	E: **76 Auto/Truck Plaza*, Pilot Travel Center**(SCALES)
FStop	E: **Speedway***(D)
Gas	E: Amoco*, Freedom*(D)(LP), Phillips 66, Shell*
Food	E: Arby's, Burger King, **Cracker Barrel**, KFC, **McDonalds (Indoor Playground), Restaurant (76)**, Subway, Taco Bell, The Summit, **Wendy's (Pilot Travel Center)**
	W: Country Kitchen, Steak 'N Shake
Lodg	E: Best Inns of America, Comfort Inn, Days Inn, Howard Johnson
AServ	E: Shell
TWash	E: **Pilot Travel Center**
ATM	E: Bank One, Phillips 66
157B	Bus I-55, U.S. 51, Veteran's Pkwy
Gas	E: Shell*
Food	E: McDonalds, Restaurant (Parkway Inn)
Lodg	E: Parkway Inn
	Note: I-74 runs concurrent above with I-55. Numbering follows I-55.
134A	Jct I-55, Chicago, Memphis
134B	I-55 Business, Veterans Parkway
135	U.S. 51, U.S. 51 Bus, Decatur, Bloomington
FStop	S: **Mobil***(D)
Gas	N: Phillips 66*
142	Downs
149	Le Roy
TStop	S: **Shell***(D) **(ATM)**
Gas	N: Amoco*
Food	N: Hardee's
	S: **Woody's Restaurant (Shell)**
Lodg	S: **Super 8 Motel**
152	U.S. 136, Rantoul, Heyworth
(156)	**Rest Area - RR, Phones P (Both Directions)**
159	IL 54, Farmer City, Gibson City
Gas	S: 76*
Food	S: Bon-Aire Rest
Lodg	S: Budgetel Inn, **Days Inn (see our ad this page)**

EXIT	ILLINOIS
AServ	S: NAPA Auto Parts
166	Mansfield
Gas	S: Amoco*
172	IL 47, Gibson City, Mahomet
Gas	S: Apollo*, Clark*
Food	S: Bull Dog Pizza, Hardee's, Hen House, Monical's Pizza, Ricin Garden, Subway, The Treetop Family Dining
Lodg	S: Heritage Inn
AServ	S: NAPA Auto Parts
ATM	S: Busey Bank
Other	S: **Courtesy Coin Laundry, IGA Store**, JR's Car Wash, **Revco Drugs**
174	Prairie View Rd. Lake of the Woods Road
FStop	N: **Mobil***
Gas	N: Amoco*
Food	N: Subway
179AB	Junction I-57 (Difficult Reaccess)
181	Prospect Avenue, Champaign, Urbana
Gas	N: Meijer Grocery*(D)(LP) (Kerosene)
	S: Amoco*(CW), Clark*, Freedom, Mobil*(CW)
Food	N: Burger King, Cheddar's, Chili's, Fazoli's Italian Food, Hardee's, Lone Star Steakhouse, Ryan's Steakhouse, Steak 'N Shake, Subway, Wendy's
	S: Arby's, Chinese Kitchen, Daddy-O's Burger's, Dos Reales, Home Stretch Restaurant, Hong Kong, KFC, Little Caesars Pizza, Long John Silvers, Red Lobster
Lodg	N: Drury Inn & Suites
	S: ◆ **Days Inn (see our ad this page)**, Econolodge

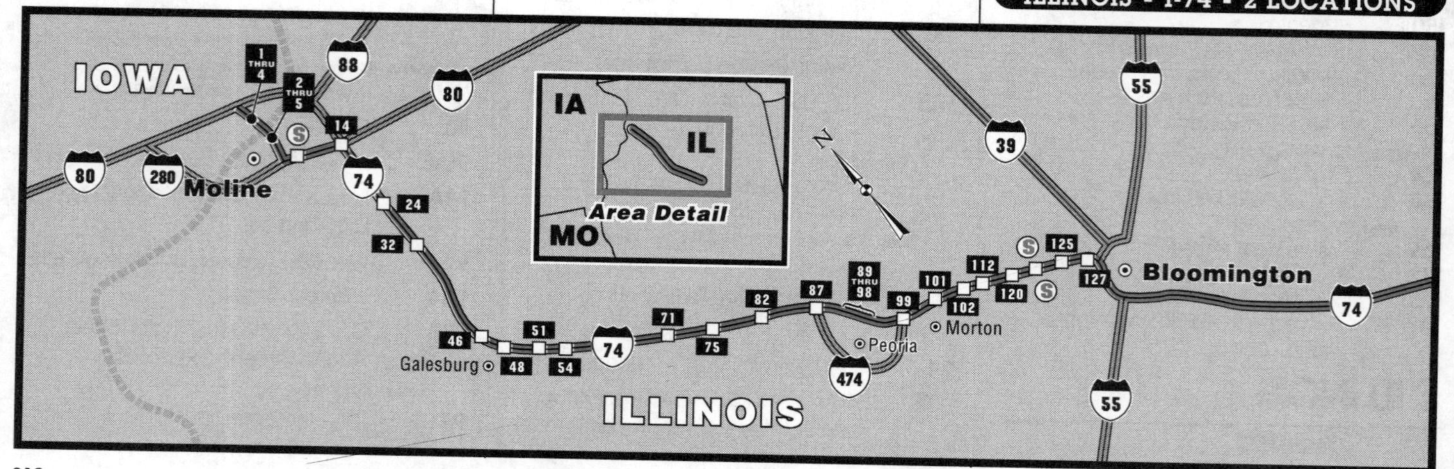

Bold red print shows RV & Bus parking available or nearby

EXIT — ILLINOIS

ASserv N: AAMCO Transmission, Market Town Lube, Prospect Hyundia, Prospect Mitsubishi, Western Auto
S: Car X Muffler & Brakes, Continental Limited, Jiffy Lube, K-Mart, Midas Muffler & Brakes, Tire Barn Warehouse
RVCamp N: Campground
ATM N: Busy Bank, First of America
Other N: Lowe's, Mail Boxes Etc, Meijer Grocery, PetsMart, Sam's Club, Target Department Store (Pharmacy), Wal-Mart
S: Big R Car Wash, K-Mart (Pharmacy, Supermarket, Food Court)

82 Neil Street
Gas S: Mobil*
Food N: Alexander's Steakhouse, Bob Evans Restaurant, Denny's, Grandy's, McDonalds, Olive Garden, Super Delux Chinese Buffet, Taco Bell
S: Howard Johnson, Mountain Jack's Steakhouse, Perkins Family Restaurant
Lodg N: ◆ Budgetel Inn, Comfort Inn, Courtyard by Marriott, Fairfield Inn, La Quinta Inn, Red Roof Inn, Super 8 Motel
S: AAA Howard Johnson
ASserv N: Sear's, Sullivan Chevrolet/Volvo
S: Bickers Auto Repair, JMK Tire & Wheel, Tire America
ATM N: Bank Champaign
Other N: Car Wash, IGA Supermarket, Market Place Mall, Office Depot

83 University of Il. Lincoln Ave.
Gas S: Phillips 66*(CW), Speedway*(LP)
Food S: Holiday Inn, Urbana's Garden
Lodg S: Holiday Inn, Ramada Limited, ◆ Sleep Inn
ASserv N: Atlus Muffler & Brake
TServ N: Interstate Trailor Inc. Service & Parts
Med S: ✚ Hospital
Other N: UPS Drop Box

84AB U.S. 45, Rantoul, Cunningham Ave
Gas N: Amerigas(LP)
S: 76*(D)(LP), Shell*, Speedway*(LP)
Food N: Park Inn
S: Best Western, Bombay Bicycle Club Restaurant & Bar, Old Country Store Cracker Barrel, Domino's Pizza, Ned Kelly's Steakhouse, Steak 'N Shake
Lodg N: Park Inn
S: Best Western, Motel 6
ASserv N: Bernie's Tires
S: Pro Tech Service, Ron's Truck & Auto Repair, Shell, TK Service Center, Tatman's Towing
TServ S: Ron's Truck & Auto Repair
Other S: Sav-A-Lot Supermarket, TJ's Coin Laundry

85 IL 130, University Avenue

92 St Joseph
FStop N: Marathon
Food S: Dairy Queen
ASserv N: Marathon
TServ N: Marathon
Other S: Coin Car Wash

97 IL 49 S, Royal, Ogden
Gas S: Citgo*(D)(LP)
Food S: Ogden Restaurant

00 IL 49 North, Rankin, Fithian

06 Oakwood, Potomac
TStop N: Shell*(SCALES)
S: Knoll's Oakwood Truck Plaza*(SCALES)

EXIT — ILLINOIS/INDIANA

Gas S: Phillips 66*(LP)
Food N: Shell Restaurant
S: Knoll's Oakwood Truck Plaza Restaurant
ATM N: Shell

(208) Rest Area - RR, Phones, Picnic, Vending P (Westbound)

210 U.S. 150, ML King Dr
Gas N: Marathon*
Food N: The Little Nugget
ASserv N: Lynn's Tires
S: Professional Car Care Shop, Tommy House Tire
TServ S: Professional Car Care Shop, Tommy House Tire
RVCamp N: Campground
Med N: ✚ Hospital

214 G Street, Tilton
Gas S: Shell*

215 U.S. 150, Gilbert Street, IL 1, Westville, Georgetown Rd.
Gas N: 76*, Speedway(LP)
S: Speedway*
Food N: Arby's, Central Park Hamburgers, Golden Wok Express, Hardee's, Long John Silvers, McDonalds, Pizza Hut, Steak & Shake, Taco Bell
S: Monical's Pizza, Subway
Lodg N: AAA Best Western, DAYS INN ◆ Days Inn
ASserv N: Bass Tire Co, Care Muffler Shop, Quicklube, Rucker's Auto
S: Royal Buick Pontiac
Med N: ✚ Hospital
ATM S: Palmer Bank
Other N: South Town Pharmacy
S: Aldi Supermarket, Eagle Country Market, Harley Davidson Dealership

216 Bowman Avenue, Perrysville Rd
Gas N: Citgo*(LP), Mobil*(D)(LP)
Food N: Godfathers Pizza (Citgo), Subrageous Subs
RVCamp S: Campground

220 Lynch Road
Gas N: Amoco*
Food N: Big Boy, Ramada Inn, Subway (Amoco)
Lodg N: AAA Best Western, AAA Budget Suites & Inn, ◆ Comfort Inn, ◆ Fairfield Inn, ◆ Ramada Inn, ◆ Super 8 Motel

(220) Welcome Center

↑ILLINOIS
↓INDIANA

(1) Welcome Center - RR, Picnic, Phones, Vending

4 IN 63, West Lebanon, Newport
TStop N: Pilot*(SCALES)
Food N: Beef House, Dairy Queen (Pilot)
TServ N: Pilot

8 Covington
Gas N: Shell*
Food N: Overpass Pizza
ASserv N: Warrick Ford/Mercury
ATM N: Bank of Western Indiana

15 U.S. 41, Attica, Veedersburg
Parks S: Turkey Run State Park

(19) Weigh Station - both directions

EXIT — INDIANA

(23) Rest Area - RR, Phones, Picnic, Vending

25 IN 25, Wingate, Waynetown
RVCamp S: Camping

34 U.S. 231, Linden, Crawfordsville
FStop S: Union 76*
Gas S: Amoco*, Gas America*, Shell*
Food S: Burger King, Holiday Inn, McDonalds
Lodg S: DAYS INN Days Inn (see our ad this page), Dollar Inn, ◆ Holiday Inn, AAA Super 8 Motel
ASserv S: Northridge Auto Service
TServ S: Hoosier Truck Tech
RVCamp S: KOA Campground
Med S: ✚ Hospital
ATM S: Gas America
Other S: Tourist Information

39 IN 32, Crawfordsville

52 IN 75, Advance, Jamestown
RVCamp S: Campground

(57) Rest Area - RR, Phones, Picnic, Vending (Both Directions)

58 IN 39, Lebanon, Lizton
Gas N: Phillips 66*
Food S: Drive-Through Restaurant
ASserv N: Phillips 66
S: Don's Small Engine Service, Scott's Auto Service
Med S: ✚ Hospital
ATM S: State Bank Of Lizton

61 Pittsboro
TStop S: Blue & White*(SCALES)
Food S: Hap's Place (Blue & White)
TServ S: Tires (Blue & White)

66 IN 267, Brownsburg
Gas N: Phillips 66*(D) (Kerosene), Shell*(CW)
S: Amoco*(D), Speedway*(D), Wake* (Kerosene)
Food N: Hardee's
S: Arby's, Blimpie's Subs, Burger King, China's Best Buffet, Elegance Restaurant, Noble Roman Pizza, Papa John's Pizza, TCBY Treats, Taco Bell, Wendy's
Lodg N: Holiday Inn Express
S: Dollar Inn
ASserv N: Texaco Xpress Lube
S: Amoco, Mears Automotive Parts & Service
ATM N: State Bank
S: National City Bank
Other N: Car Wash
S: Kroger Supermarket (Pharmacy), Mail Boxes Etc, Revco Drugs

EXIT INDIANA

Note: I-74 runs concurrent below with I-465. Numbering follows I-465.

14AB Tenth St
- **Gas** **O:** Gas America, Shell*
- **Food** **I:** Pizza Hut Delivery, Wendy's
  O: Subway
- **Med** **I:** ✚ Hospital
- **ATM** **I:** NBD Bank
  O: First Indiana Bank, Gas America
- **Other** **I:** Cub Foods (Pharmacy), Lowe's

13 U.S. 36, Danville, Rockville Rd
- **Gas** **I:** Shell*
  O: Speedway*
- **Food** **I:** Time To Eat Grill
  O: Bob Evans Restaurant, Fabulous 50's Restaurant
- **Lodg** **I:** Comfort Inn
- **AServ** **O:** Bill's Auto Service, Indy Lube
- **ATM** **O:** Union Federal
- **Other** **I:** Convenient Food Mart, Sam's Club

12A U.S. 40 East, Washington St., Plainfield
- **Gas** **I:** Amoco*
  O: Big Foot*, Phillips 66, Shell*(CW)
- **Food** **I:** Fazoli's Italian Food, Taco Bell, White Castle Restaurant
  O: Airport Deli, Arby's, Burger King, Cambridge Inn Cafeterria, Dunkin Donuts, Hardee's, KFC, Little Caesars Pizza (K-Mart), Long John Silvers, McDonalds, Noble Romans Pizza, Number One Wok, Omelet Shoppe, Pizza Hut, Shoney's, Smiley's Pancake & Steak, Steak & Shake, Subway, Wendy's, Western Star, Yogurt
- **Lodg** **O:** Dollar Inn
- **AServ** **I:** Speedway Auto Parts
  O: Bandy's, Burt Nees Tires, Car X Muffler & Brakes, Goodyear Tire & Auto, Midas Muffler & Brakes, Q-Lube, Tire Barn
- **ATM** **I:** Amoco, Bank One
  O: Big Foot, National City Bank
- **Other** **I:** Central Ace Hardware, Coin Laundry, Osco Drugs, U-Haul Center(LP), Village Pantry, WalGreens
  O: Classy Car Wash, Davis Laundromat, K-Mart (Pharmacy), Smiley's Car Wash, Venture Supermarket

11 Airport Expressway
- **Gas** **I:** Marathon*
  O: Speedway*
- **Food** **I:** Denny's
  O: JoJo's Restaurant, The Library Steakhouse & Pub, Waffle Steak

EXIT INDIANA

- **Lodg** **I:** Adam's Mark Hotel, ◆ Courtyard by Marriott, Hampton Inn, Motel 6
  O: Fairfield Inn, La Quinta Inn, Residence Inn
- **AServ** **I:** Marathon

9AB I-70

8 IN 67 South, Kentucky Ave
- **Gas** **O:** Amoco*, Shell*(CW), Speedway*(D), Swifty
- **Food** **O:** Burger King (Shell), Dairy Queen, Hardee's, KFC, McDonalds (Amoco), Subway (Shell)
- **Med** **I:** ✚ Hospital
- **ATM** **O:** NBD Bank
- **Other** **O:** The Cleaning Shop Coin Laundry

7 Mann Rd (Westbound, Reaccess Estbound)

4 IN 37, Harding St
- **TStop** **I:** Pilot Travel Center*(SCALES) (Kerosene)
  O: Flying J Travel Plaza
- **Gas** **O:** Shell*(CW)
- **Food** **I:** Bender's Restaurant (Econolodge), Omelet Shoppe, Restaurant (Pilot), Wendy's (Pilot)
  O: Hardee's, McDonalds, Thads (Flying J Travel Plaza), Waffle & Steak, White Castle Restaurant
- **Lodg** **I:** Dollar Inn, Econolodge, (AAA) Super 8 Motel
  O: Knight's Inn
- **AServ** **O:** Michelin Tire Service
- **TServ** **I:** Peterbilt Dealer
  O: Flying J Travel Plaza, Freightliner Dealer, J & E Tire, Speedco Thirty Minute Truck Lube
- **TWash** **O:** Flying J Travel Plaza
- **Med** **I:** ✚ Hospital
- **Other** **O:** Ferrell Gas(LP)

2AB U.S. 31, IN 37, Indianapolis, East St
- **Gas** **I:** Amoco*(CW), Speedway*
  O: Big Foot*, Shell*(CW), Speedway
- **Food** **I:** Baskin Robbins, Dairy Queen, Dutch Oven, KFC, Laughner's Cafeteria, Noble Roman's Pizza, Old Country Buffet, Pizza Hut, Steak & Ale
  O: Bob Evans Restaurant, Denny's, Heritage Smorgasbord, McDonalds, Red Lobster, Shoney's
- **Lodg** **O:** Comfort Inn, (DAYS INN) Days Inn, Economy Inn, Quality Inn
- **AServ** **I:** Auto Tire Car Care Centers, Ford Dealership, Goodyear Tire & Auto, Jeep Eagle Dealer, Quaker State Lube
  O: Indy Lube
- **ATM** **I:** Amoco, Fifth Third Bank, National City Bank
- **Other** **I:** Coin Laundry, Kroger Supermarket, Office Depot, Revco Drugs, U-Haul Center(LP)

53 Junction I-65 (Limited Access)

52 Emerson Ave., Beech Grove
- **Gas** **I:** Amoco, Marathon*, Shell*(CW)
  O: Citgo*, Shell*(CW), Speedway*(CW)

EXIT INDIANA

- **Food** **I:** Burger King, Domino's Pizza, KFC, Taco Bell
  O: Arby's, Bamboo House, Dairy Queen, Egg Roll, Fazoli's Italian Food, Hardee's, Holiday Inn, Hunan House, McDonalds, Mi Amigos Mexican Restaurant, Pizza Hut, Ponderosa, Ramada, Steak & Shake, Subway, Sunshine Cafe, White Castle Restaurant
- **Lodg** **I:** (AAA) Quality Inn & Suites
  O: ◆ Holiday Inn, ◆ Ramada, ◆ Super 8 Motel
- **AServ** **I:** Marathon
  O: Auto Zone Auto Parts, Brooks Auto Care, Goodyear Tire & Auto, Indy Lube, K-Mart, Quaker State
- **Med** **I:** ✚ Hospital
- **ATM** **I:** Union Federal Savings Bank
  O: Fifth Third Bank, Marsh Supermarket
- **Other** **I:** Automatic & Coin Car Wash, Grime Stopper Car Wash
  O: Automatic Car Wash, Beech Grove Animal Hospital, K-Mart, Kroger Supermarket (Pharmacy), Marsh Grocery Store (Pharmacy), Osco Drugs (Marsh Supermarket), WalGreens

Note: I-74 runs concurrent above with I-465. Numbering follows I-465.

94 Jct I-465, U.S.421 N, I-74 W

96 Post Rd
- **FStop** **N:** Marathon*(LP)
- **Gas** **S:** Amoco, Shell*
- **Food** **N:** Dollar Inn, McDonalds, Subway (Marathon)
  S: Olisgo's Mexican Restaurant
- **Lodg** **N:** Dollar Inn
- **AServ** **S:** Amoco, Mahoney Chevrolet

99 Acton Road

101 Pleasant View Road
- **Gas** **N:** Marathon
- **AServ** **N:** Marathon
- **ATM** **N:** Key Bank
- **Other** **N:** Don's Marine Service, Pleasant View One Stop

103 London Road

109 Fairland Road

113 IN 9, Shelbyville, Greenfield
- **Gas** **S:** Mr T's*, Shell*(CW)
- **Food** **S:** McDonalds, Ramada, Waffle & Steak
- **Lodg** **S:** ◆ Comfort Inn, ◆ Ramada, Super 8 Motel
- **ATM** **S:** Shelby County Savings

116 IN 44, Shelbyville, Rushville
- **FStop** **N:** Big Foot*
- **Gas** **S:** Shell*(CW)
- **Food** **S:** Baskin Robbins, Bavarian Haus, Bob Evans Restaurant, Burger King, Dairy Queen, Dunkin Donuts, McDonalds, New China Buffet,

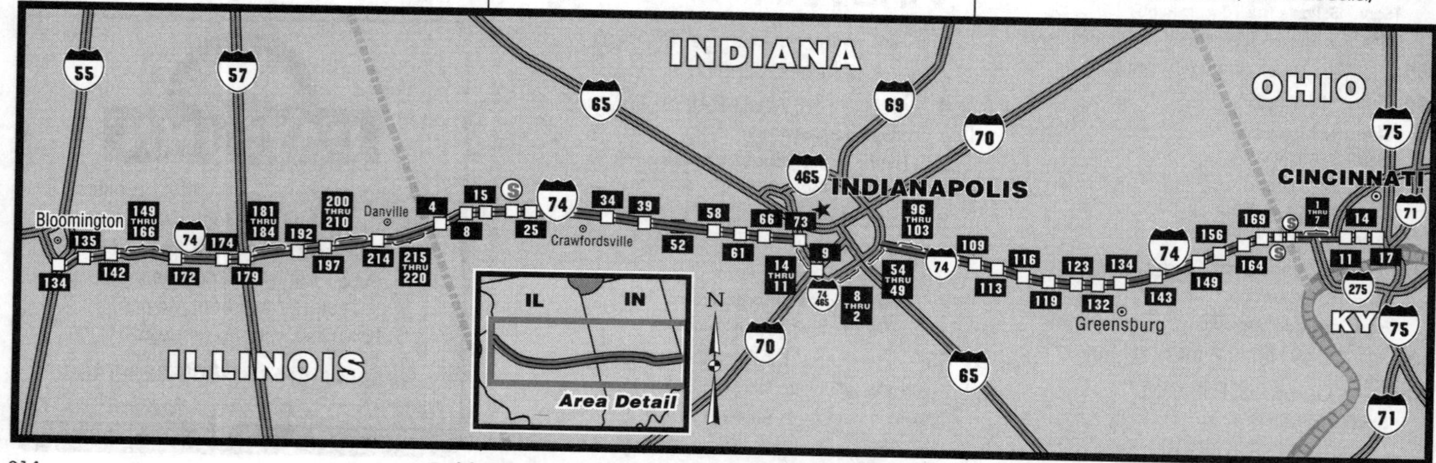

Bold red print shows RV & Bus parking available or nearby

EXIT INDIANA/OHIO

Pasquale's Pizza & Pasta, Pizza Hut, Rax, Shoney's, Subway, Taco Bell
Lodg S: AAA Lees Inn, Rasners Motel
AServ S: Classic Chevrolet Olds Nissan Geo, Hubler Ford/Lincoln/Mercury, Midas Muffler & Brakes
Med S: + Hospital
ATM S: Key Bank, National City Bank
Other S: Ace Hardware, Aldi Food Store, Kroger Supermarket (Pharmacy), Osco Drugs, Revco Drugs, Wal-Mart (Pharmacy)

119 IN 244, Milroy, Andersonville
AServ N: I-74 Auto Center & Towing

123 Saint Paul, Middletown (Westbound Exit, Westbound Reaccess)
RVCamp S: Camping

132 U.S. 421, Greensburg (Eastbound, Westbound Reaccess)

134AB IN 3, Rushville, Greensburg
Gas S: Big Foot*, Shell*(CW)(LP), Speedway*(LP)
Food S: Burger King, Frisch's Big Boy, Great Wall Chinese, Hardee's, KFC, **Little Caesars Pizza (K-Mart)**, Mang's Family, McCamment's Steak House, McDonalds, Papa John's Pizza, Subway, Taco Bell, Waffle House, Wendy's
Lodg S: Belter Motel, Best Western, Lees Inn
AServ S: Meyer Ford
ATM S: Fifth-Third Bank, Peoples Trust Bank
Other S: Aldi Supermarket, Coin Car Wash, Jay C Grocery, K-Mart, Revco Drugs, Wal-Mart (Pharmacy)

143 New Point, St Maurice
TStop N: 76 Auto/Truck Plaza*(SCALES)
Food N: 76 Auto/Truck Plaza
TServ N: 76 Auto/Truck Plaza
RVCamp N: Camping (7 Miles)

149 IN 229, Batesville, Oldenburg
Gas N: Day-Nite*, Shell*
S: Amoco*(CW)
Food N: China Wok, McDonalds, Subway, Wendy's
S: Arby's, Dairy Queen, Hardee's, KFC
Lodg N: ◆ Hampton Inn
S: DAYS INN AAA Days Inn
AServ N: Napa Auto Parts, Pennzoil Lube
S: Amoco
Med S: + Hospital
ATM N: Peoples Bank
S: Home Federal Savings
Other N: Kroger Supermarket (Pharmacy)
S: Revco Drugs

(152) Rest Area - RR, Phones, Picnic, Vending

156 IN 101, Sunman, Milan
AServ S: Todd Wrecker Service & Tire Repair

164 IN 1, Lawrenceburg, St Leon
Gas N: Citgo*, Shell*
S: BP*
Food N: Christina's Family, Tiny Town Pizza
S: Blimpie's Subs (BP)
AServ N: Shell
ATM N: Citgo, Shell
S: BP

169 U.S.52 West, Brookville

(171) Weigh Station (Westbound)

↑INDIANA

↓OHIO

New Haven Road, Harrison
Gas S: Shell*(CW), Speedway*(D), Sunoco*, Super America*

EXIT OHIO

Food N: McDonalds (Bigg' Hypermarket)
S: Arby's, Back's Deli, Burger King, Hardee's, KFC, P.F.C. Tavern & Restaurant, Perkins Family Restaurant, Pizza Hut, Quality Inn, Waffle House, Wendy's
Lodg S: ◆ Quality Inn
AServ N: Kesserling Ford
S: Firestone Tire & Auto, Goodyear Tire & Auto, Harrison Tire & Automotive
ATM N: Harrison Building & Loan, Star Bank (Bigg's Hypermarket)
Other N: Bigg's Hypermarket (Pharmacy)
S: Coin Laundry, Harrison Police Station, Revco Drugs

(3) Weigh Station (Eastbound)

3 Dry Fork Road
Gas N: Citgo*
S: BP*, Chevron*(D), Shell*(D)(CW)
Food S: Burger King (Shell)
Lodg S: Motel Deluxe
AServ S: Hirlinger
ATM S: Chevron, Shell
Other N: Miami Whitewater Forest (1 Mile)
S: Cincinnati S.W. Veterinary Clinic, Suburban Propane(LP)

5 Jct I-275 South , to KY

7 OH 128, Hamilton, Cleves
Gas N: BP*(D), Shell*(D) (Kerosene)
Food N: Wendy's
S: Angelo's Pizza
AServ N: Don's Miami Town

9 Jct I-275 North , to I-75, Dayton

11 Rybolt Road, Harrison Pike
Gas S: Sunoco
Food S: Angilo's Pizza, Dante's Restaurant (Imperial)
Lodg S: Imperial House
AServ S: Sunoco
ATM S: Oak Hills Savings & Loan
Other S: Ameristop Convenience Store

14 North Bend Road, Cheviot
Gas N: Ameristop*, Shell(CW), Speedway*, SuperAmerica*(D)
S: BP*(LP)
Food N: Dairy Queen, Dunkin Donuts, McDonalds, Papa John's Pizza, Rally's Hamburgers, Skyline Chili, Subway, Wendy's
S: Bob Evans Restaurant
Lodg S: Tri Star Motel
AServ N: Jim's Auto Clinic, Midas Muffler & Brakes, Quaker State Q-Lube, Shell, Tuffy Auto Service, Valvoline Quick Lube Center
Med N: + Hospital
ATM N: Fifth Third Bank, Key Bank, The Provident Bank (Thriftway Food & Drug)
Other N: Complete Pet Mart, Sam's Club, Thriftway Food & Drug, WalGreens
S: Monford Heights Animal Clinic

17 Montana Avenue (Westbound, Eastbound Reaccess)
Gas N: BP*(CW)
Other N: Dairy Mart Convenience Store

18 U.S. 27 North, Colerain Ave, Beekman St

19 Elmore St, Spring Grove Ave
Gas N: Sunoco*
ATM N: Sunoco

20 U.S. 27, U.S. 127 South, Central Pkwy (No trucks)

↑OHIO

Begin I-74

EXIT MICHIGAN

Begin I-75

↓MICHIGAN

394 Easterday Ave, Sault Locks, Sault Ste. Marie to Toledo
Gas E: Amoco*
W: Holiday* (Currency Exchange), USA*(D) (Currency Exchange, RV Dump)
Food E: McDonalds
Lodg E: AAA Holiday Inn Express
Med W: + Hospital
ATM W: USA

392 Bus 75, Three Mile Road, Sault Ste. Marie
FStop E: 76*(LP), Mobil*(CW)
Gas E: Amoco*, Holiday*(D), Marathon, Shell*, USA Mini-Mart*(D) (Kerosene)
Food E: Albie's, Ang-gio's Italian Restaurant, Arby's, Indo-China Garden, Jeff's Barbecue, Knife & Fork, La Senorita Mexican, Mancino's Pizza, Robin's Nest, Studebaker's, Wendy's
Lodg E: AAA Best Western, ◆ Comfort Inn, DAYS INN ◆ Days Inn (see our ad on this page), ◆ Hampton Inn, AAA Kewadin Inn, King's Inn, Plaza Motor Motel, Ramada, Super 8 Motel
AServ E: Delta Tire Center, NAPA Auto Parts, Quaker State Lube(CW), Sadler GM Dealership
RVCamp E: Camping, Chippewa Campground
Med E: + Hospital
ATM E: FMB Bank (Glens Grocery), First of America, North Country Bank & Trust
Other E: Car Wash, Coin Laundry, GFS Grocery, Glen's County Market, MI State Police & Sheriff Dept, Mail Boxes Etc (USA Gas), Rite Aide Pharmacy, Sault Tribe Indian Reservation, Tourist Information, Wal-Mart (Pharmacy)

(389) Rest Area - Picnic, RR, Phones, Vending (Northbound)

386 MI 28, Newberry, Munising
Food E: Sharolyn Restaurant
Lodg E: Sharolyn Motel, AAA Sunset Motel
RVCamp W: KOA Campground
Parks W: Brimley State Park

379 Gaines Hwy, Barbeau Area
RVCamp W: Camping

378 Kinross
Gas E: Amoco*
Food E: Frank & Jim's Italian/American
RVCamp E: Campground

EXIT	MICHIGAN
Other	E: Airport
373	MI 48, Rudyard, Pickford
Food	W: Clydes Drive Through
359	MI 134, DeTour Village, Drummond Isand
Lodg	W: Christmas Motel
352	MI 123, Newberry, Tahquamenon Falls
348	Mackinaw Trail, St. Ignace, Cr H 63
Gas	E: Shell*
Lodg	E: Birchwood Motel, Carey's Motel, Cedar's Motel, Great Lakes Motel, Harbor Light Motel, Melody Motel, Northern Air Motel, Pines Motel, Rock View Motel, Sand's Motel, Shores Restaurant
RVCamp	E: **KOA Campground (6 Miles), National Forest Campground (6 Miles)**
	W: **Campground**
Other	E: **Sault Indian Reservation**
	W: **Castle Rock Tourist Attraction**
344AB	Bus I-75, St. Ignace
TStop	W: **St Ignace Truck Stop**
Gas	W: Amoco(CW), Holiday*, Shell*(D)
Food	E: North Country Restaurant, Northern Lights
	W: Big Boy, **Burger King**, Clyde's, McDonalds, St Ignace Truck Stop, Subway, Wally's Restaurant
Lodg	E: Aurora Borealis Motel, Moran Bay Motel, Normandy Motel, Quality Inn, Roadway Inn, St Ignace Inn
	W: **Howard Johnson (see our ad this page)**, Super 8 Motel
AServ	W: Amoco
Med	E: + **Hospital**
ATM	W: First of America
Parks	E: **Straits State Park**
Other	E: **Glen's Market, Mackinaw Island Ferry**
	W: **US Post Office**
(346)	**Rest Area, Scenic Turnout (Southbound)**
(344)	**Rest Area**
343	Bridgeview (Southbound, No East Access)
339	Jamet St (Note: Beware High Winds)
Gas	E: Amoco*, Total*
	W: Shell* (RV Repair), **Sunoco(CW) (RV Repair)**
Food	E: Audie's Family Restaurant, Burger King, Cunningham's Family Restaurant, Dairy Queen, Hendrick's Fudge, Joanne's Fudge, KFC, Mama Mia's Pizzeria, McDonalds, Pancake Chef, Pizza Palace, Subway, TCBY Yogurt
	W: Darrow's Family Dining, Donuts Chef, Hazel's Pasties, I-75 Plaza
Lodg	E: Downing's Downtown Motel, AAA Econolodge, AAA Motel 6, Ramada Inn, AAA Super 8 Motel
	W: Bridgeview Motel, Chalet Motel, Fort Mackinaw, AAA Holiday Inn Express, Trails End Inn, Vin-Del Motel
AServ	W: **Sunoco (RV Repair)**
ATM	E: First of America
Parks	W: **Wilderness State Park**
Other	E: **Colonial Michilimackimac, Gateway to Mackinaw Island, IGA Grocery, U.S. Post Office**
(338)	**Welcome Center, Rest Area - RR, Phones, Picnic, Vending**

EXIT	MICHIGAN
337	MI 108, Nicolet St, Macinaw City
FStop	E: **Citgo*(LP)**
Gas	E: Marathon*(D)
Food	E: Big Boy, Burger King, Chee Peng's Chinese, Days Inn, Embers Motel, Marcino's Pizza, Marios Ristorante, Ponderosa, The Lighthouse Restaurant
	W: Neath The Birches
Lodg	E: Anchor Budget Inn, AAA Best Western, Budgetel Inn, Capri Motel, Days Inn (see our ad on this page), Embers Motel, Friendship Inn, Hampton Inn, Howard Johnson, Huron Motel, Kewadin Inn, Nicolet Inn, Ottawa Motel, Quality Inn, Ramada Limited, Roadway Inn, Starlite Budget Inn, Sundown Motel, Surf Motel, Travelodge
	W: Americana Motel, AAA Val-Ru Motel, White Birches Motel
AServ	E: Citgo
RVCamp	E: **Camping, Macinaw City Campground**
	W: **KOA Campground**
ATM	E: Citizen's National Bank
Parks	W: **Wilderness State Park**
Other	E: **Coin Laundry, IGA Store, Macinaw Island Fairy Tickets**
(328)	**Rest Area - Picnic, RR, HF, Phones (Southbound)**
326	CR 66, Cheboygan, Cross Village
Gas	E: Marathon*
AServ	E: Marathon
Med	E: + **Hospital**
Other	E: **Seashell City Gift Shop**
322	CR 64, Pellston, Cheboygan

EXIT	MICHIGAN
Med	E: ✚ Hospital
Other	E: Blarney Castle Oil Company[LP], State Patrol Post
	W: Airport
(317)	**Rest Area - Picnic, RR, Phones, Vending, Scenic View, Nature Trail (Northbound)**
313	MI 27 North, Topinabee, Cheboygan, Indian River
Lodg	E: Johnson Motel
RVCamp	E: KOA Campground
310	MI 68, Roger City, Indian River
Gas	W: 76*[D], Shell*
Food	W: BC Pizza, Breadeaux Pizza, Burger King, Christopher's Restaurant, Dairy Queen, Don's Burgers & Chicken, Indian River Trading Post, Jan's Donut Shop, Paula's Cafe, Wilson's River Edge Restaurant
Lodg	E: Holiday Inn Express (see our ad on this page)
	W: Coach House Motel, Devoe's Motel, Indian River Motel, Reid's Motor Court
AServ	W: 76, Auto Value Auto Parts, Inland Transmission, Jack's Auto Repair, Pollard's Quick Lube
RVCamp	E: Camping
	W: Camping
ATM	W: Citizen's National Bank, First of America
Parks	W: Burt Lake State Park
Other	W: Big Bear General Store, Coin Laundry, Ken's Market, Tomahawk Trails Canoe, Kayak, & Tube Trips, Tourist Information, Village Pharmacy
301	CR 58, Wolverine

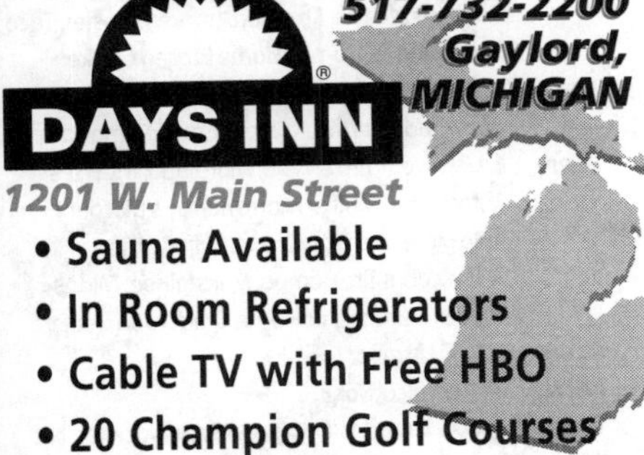

EXIT	MICHIGAN
FStop	E: Marathon*
Food	E: County Line Restaurant
RVCamp	E: Camping
	W: Camping
290	Vanderbilt
Gas	E: 76*
	W: Amoco*
Food	E: Darcy's Bar, Olde Town Pizza & Restaurant
AServ	W: Amoco
(287)	**Rest Area - RR, Phones, Picnic, Vending (Southbound)**
282	Gaylord, Alpena, MI 32
FStop	W: 76, Citgo*, Marathon*
Gas	E: Amoco*[D][CW], Clark*, Holiday*, Phillips 66*, Total*[D][LP]
	W: BP*, Shell*
Food	E: Arby's, Breadeaux Pizza, Burger King, Comfort Inn, Dairy Queen, Dan's Pizza Plus, Dunkin Donuts, Holiday Inn, KFC, La Senorita, McDonalds, Subway, TCBY Yogurt, Wendy's
	W: Albie's, Big Boy, Bob Evans Restaurant, Flap Jack Shack, Little Caesars Pizza, Pizza Hut, Ponderosa, Taco Bell
Lodg	E: AAA Comfort Inn, AAA Holiday Inn
	W: DAYS INN AAA Days Inn (see our ad this page), ◆ Hampton Inn, AAA Super 8 Motel
AServ	E: Gaylord Auto Parts, Muffler Man
	W: Fred's Garage, LJ Ormsbee Dodge Chrysler, Maxi Muffler Service, Midas Muffler & Brakes, Q-Lube, Upper Lakes Tire
RVCamp	W: Gaylord Alpine RV Park & Campground
Med	E: ✚ Hospital
	W: ✚ Medical Care Center
ATM	E: First National Bank of Gaylord, Old Kent Bank, Standard Federal
Parks	E: Bavarian Falls Park
Other	E: Alpine Hardware, Call of the Wild Museum, Coin Laundry, Glen's Market (Pharmacy), Sheriff's Dept, State Patrol Post
	W: Auto Wash, Buy-Low Foods, GFS Grocery, True Value Hardware, Wal-Mart (Pharmacy)
279	Bus Loop 75, Downtown Gaylord, Old 27
FStop	E: Marathon*
Gas	E: 76*, Shell*
Food	E: Dowker's Meat & Deli, Mama Leone's Italian, Willabee's
Lodg	E: Alpine Motel, ◆ Econolodge, The Hamlet Motel
AServ	E: 76, GM Auto Dealership, Pontiac Buick Dealer
RVCamp	E: Gaylord RV Service[LP], Moore's RV Supply
	W: Camping
ATM	E: Old Kent, Shell
Other	E: Coin Car Wash, Sheriff's Dept, State Patrol Post
(277)	**Rest Area - RR, Phones, Picnic, Vending (Northbound)**
270	Waters
TStop	E: 76 Auto/Truck Plaza*
Food	E: 76 Auto/Truck Plaza
	W: Countryside Cafe, McDonalds
TServ	W: Northern Tank Truck Service
RVCamp	W: Camping
Parks	W: Otsego Lake State Park
Other	W: IGA Store
264	Lewiston, Frederic
RVCamp	W: Camping

EXIT	MICHIGAN
Parks	E: Hartwick Pine State Park
(262)	**Rest Area - RR, Phones, Picnic, Vending (Southbound)**
259	Hartwick Pines Road
RVCamp	E: Campground
	W: Campground
Parks	E: Hartwick Pine State Park
256	North Down River (Southbound, North Reaccess)
Med	W: ✚ Hospital
Other	W: Animal Hospital, Sheriff Department
254	Business Loop 75, MI 72, Traverse City, Downtown Grayling (Left Exit Northbound)
Gas	W: Mobil*[D], Phillips 66*, Total*[D]
Food	W: A & W Drive-In, Big Boy, Burger King, China West, Crawford Station Restaurant, Dairy Queen, KFC, Little Caesars Pizza, McDonalds, Patti's Towne House, Pizza Hut, Subway, Taco Bell, Wendy's
Lodg	W: AAA Aquarama Motor Lodge, Holiday Inn
AServ	W: Car Quest Auto Center, Ford Dealership (Auto Parts), NAPA Auto Parts (Auto Parts)
TWash	W: Truck Wash
ATM	W: Chemical Bank, Citizens Bank, Empire National Bank
Other	W: 7-11 Convenience Store, Ace Hardware, Coin Car Wash, Coin Laundry, Glen's Grocery, K-Mart
(252)	**Rest Area - RR, Picnic, Phones, Vending (Northbound)**
251	4 Mile Road
FStop	W: Total*[LP]
Lodg	W: ◆ Super 8 Motel
RVCamp	E: Camping
	W: Campground
ATM	W: Total FS
249	U.S. 27 South, Claire, Lansing (Southbound, No Reaccess)
244	North Higgins Lake State Park
FStop	W: Sunoco*
RVCamp	W: Camping
Other	W: CCC Museum
239	Bus Loop 75, MI 18, South Higgins Lake, Roscommon
RVCamp	E: Camping
	W: Camping
Parks	W: State Park
(235)	**Rest Area - RR, Phones, Picnic, Vending (Southbound)**
227	MI 55 W, Houghton Lake, Cadillac
Food	E: Maple Valley Restaurant
222	Old 76, St Helen
RVCamp	E: Camping
215	Business Loop 75, MI 55 East, West Branch
FStop	E: Total*
Food	E: Restaurant (Total)
AServ	E: Pontiac Buick Dealer
Med	E: ✚ Hospital
212	Bus Loop 75, Cook Road, West Branch
FStop	E: Marathon*

EXIT MICHIGAN

W: BP*
Gas E: Total*
Food E: Arby's, Big Boy, Burger King, McDonalds, Ponderosa Steakhouse, Quality Inn, Subway (Shell), Taco Bell
W: Hot Stuff Pizza (BP Fuel Stop), North Country Junction (BP Fuel Stop)
Lodg E: AAA Quality Inn, AAA Super 8 Motel
RVCamp E: Camping
W: Camping
Med W: Hospital
ATM E: Shell
Other E: Sheriffs Dept, State Patrol Post, Tanger Factory Outlet

(210) Rest Area - RR, Phones, Picnic, Vending (Northbound)

202 MI 33, Alger, Rose City
FStop E: Mobil*
Gas E: Shell*
Food E: Hot Stuff Pizza (Shell), Subway (Shell), Taco Bell (Shell)
RVCamp E: Camping
Other E: Airport

(201) Rest Area - RR, Picnic, Phones, Vending (Southbound)

195 Sterling Road, Sterling
RVCamp E: Camping

190 MI 61, Gladwin, Standish
Gas W: Amoco*, Mobil*
AServ W: John's Auto Repair
Med E: Hospital

188 U.S. 23, Standish, Alpena
RVCamp E: Cedar Springs Campground

181 Pinconning Road, Pinconning
FStop W: Sunoco*(LP)
Gas E: 76*(LP), Mobil*
Food E: Peppermill
AServ E: 76
RVCamp E: Camping

(175) Rest Area - RR, Phones, Picnic, Vending (Northbound)

173 Linwood Road, Linwood
Gas E: Mobil*

168 Beaver Road
Gas W: Mobil*
Parks E: Bay City State Park (5 Miles)

164 to MI 13, Wilder Road, Kawkawlin
Gas E: Meijer Grocery*(D)
Food E: Bill Knapp's Restaurant, Spad's Pizza
AServ E: Hart Tires
Med E: Hospital
Other E: Coin Car Wash, Meijer Grocery, Traxler's Convenience Store

162AB U.S.10 West, Midland, Bay City, MI 25
Med E: Hospital
Other E: Sheriff's Dept, State Patrol Post

160 MI 84, Saginaw Road
Gas E: Mobil*, Shell*
W: Amoco*, Citgo*, Marathon*(D)(CW)
Food E: Dunkin Donuts (Shell), Subway (Mobil)
W: Bergers Family Restaurant, Burger King, Heatherfields Restaurant, McDonalds
Lodg W: AAA Bay Valley Hotel & Resort, AAA Best Western
RVCamp W: International RV World

EXIT MICHIGAN

ATM W: Citgo

(158) Rest Area - RR, Phones, Picnic, Vending (Southbound)

155 Junction I-675, Downtown Saginaw

154 MI 13, Zilwaukee
Gas W: Total*
AServ W: Great Lakes Transmission & Auto Repair

153 MI 13, East Bay City Rd

151 MI 81, Caro, Reese
FStop E: Sunoco*(LP) (Kerosene)
AServ E: Upper Lakes Tire
W: Robinson's Auto Service
TServ E: Michigan CAT, Schaffer Truck Center, Tri-City Suspension & Brake, Upper Lakes Tire
W: Freightliner Dealer
Other W: Express-Mart (Convenience store)

150 Junction I-675 North, Downtown Saginaw

149AB MI 46, Holland Ave, Sandusky
Gas W: BP, Fisca, Sunoco*, Total* (Kerosene)
Food E: Back Street Grill
W: Arby's, Big John Steak & Onion, Burger King, Ern's Seafood, Holiday Inn, Little Caesars Pizza (K-Mart), McDonalds, Quality Inn, Taco Bell, Texan Restaurant, Wendy's
Lodg W: Holiday Inn, Knight's Inn, Quality Inn, Red Roof Inn
Med E: Hospital
ATM W: First of America
Other E: Celebration Square
W: Coin Car Wash (BP), K-Mart (Pharmacy), Kessel Supermarket

144AB Bridgeport, Frankenmuth
TStop W: 76 Auto/Truck Plaza*(SCALES) (RV Dump)
FStop E: Speedway*(LP)(SCALES), Total*
Gas E: Shell*
W: Amoco*
Food E: A & W Drive-In (Total), Blimpie's Subs (Shell), Freeway Fritz, TCBY Yogurt (Shell)
W: Apple Creek Family Restaurant (76 TS), Arby's, Big Boy, Burger King, Cracker Barrel, Dunkin Donuts (Amoco), Little Caesars Pizza, McDonalds, Peking City Chinese, Render's, Sbarro, Subway, Taco Bell, Wendy's
Lodg W: Budgetel Inn, Days Inn, Motel 6
TServ W: 76
ATM W: NBD Bank
Other E: Parker's Propane(LP)
W: IGA Store, Michigan State Police, Rite Aide Pharmacy

(143) Rest Area - Picnic, RR, HF, Phones (Northbound)

(138) Weigh Station (Both Directions)

136 MI 54, MI 83, Birch Run
FStop E: Mobil* (Kerosene)
W: Sunoco*, Total*
Gas E: Citgo*, Marathon*(D), Shell*
W: Amoco*, Citgo*
Food E: Dixie Dave's Restaurant, Dunkin Donuts (Shell), Exit Restaurant, Halo Burger, KFC, Market Street Inn, Subway
W: A & W Drive-In, Arby's, Bob Evans Restaurant, Burger King (Amoco), Christy's, Little Caesars Pizza, McDonalds, Schlotzkys Deli, Shoney's, Spads Pizza, Tony's Restaurant (Truck parking

EXIT MICHIGA

nearby), Wendy's
Lodg E: ◆ Comfort Inn, ◆ Hampton Inn, AAA Holiday Inn Express, Market Street Inn, AAA Super 8 Motel
W: AAA Country Inn & Suites
AServ E: Car Quest Auto Center, HS Auto Service, Uniroyal Tire & Auto
W: GM Auto Dealership
RVCamp W: Campground
ATM W: Citgo
Other E: Coin Laundry
W: Birch Run Drugs, Car Wash, Outlet Mall

131 MI 57, Clio, Montrose
Gas E: Sunoco*
W: Amoco*, Mobil*
Food E: Arby's, Burger King, KFC, McDonalds, Oriente Express (Chinese), Shell*, Subway, Taco Bell
W: Dunkin Donuts (Amoco), Elias Bros Big Boy, McDonalds, Wendy's
Lodg E: AAA Econolodge
AServ E: Clio Chrysler Dealership, Expressway Ford Dealership, GM Auto Dealership
W: Mike's Auto Service
Med E: McLaren Family Care (Urgent Care)
ATM E: NBD Bank
Other E: Pamida Supermarket

(129) Rest Area - RR, Picnic, Phones, Vending, Travel Directory (Both Directions)

126 Mount Morris
TStop E: BP*(LP)(SCALES)
FStop W: Amoco*
Food E: Burger King (BP), Subway (BP)
W: McDonalds (Amoco FS)
AServ W: NAPA Auto Parts
TServ W: Napa Auto Care Center
RVCamp E: Timberwolf Campground (11 Miles), Wolverine Campground (14 Miles)
W: Bell Fork Lift (LP Gas available)
ATM W: Amoco
Other E: Huckleberry Railroad Crossroad Village
W: Cyclone Coin Car Wash

125 Junction I-475, Downtown Flint

122 Pierson Road, Flushing
Gas E: Amoco*(CW), Clark, Marathon*, Total*(D)
W: Meijer Grocery*
Food E: Full House Chinese Restaurant, KFC, McDonalds, Moy Kong Express, Papa's Gyro's, Subway, Walli's Restaurant
W: Arby's, Bill Knapp's Restaurant, Blimpie's Subs, Bob Evans Restaurant, Burger King, Cracker Barrel, Denny's, Halo Burger, Long Joh Silvers, Pizza Hut, Ramada Inn, Red Lobster, Tac Bell, Wendy's, Ya Ya's Flame Broiled Chicken
Lodg E: Super 8 Motel, Walli's Motor Lodge
W: ◆ Budgetel Inn, Knight's Court, Ramada Inn
AServ E: Goodyear Tire & Auto, Marathon, Master's Automotive, Muffler Man, Murray's Discount Auto Store, Rainbow Brake, Tuffy Muffler
W: Discount Tire Company, Instalube, Midas Muffler & Brakes, Valvoline Quick Lube Center
TServ E: Tuffy Muffler
ATM W: Chemical Bank
Other E: Coin Laundry, Double D Super Market, K-Mart (Pharmacy)
W: Meijer Grocery

118 MI 21, Corunna Road, Owosso

Bold red print shows RV & Bus parking available or nearby

Gas E: Sunoco*(CW)
W: MSI*, Mobil*

Food E: Atlas Coney Island, Bada Lebanese West, Big John Steak & Onion, Coffee Haus Donuts, Dairy Delight, Halo Burger, Hardee's, Hungry Howie's Pizza & Subs, King Chinese Buffet, Little Caesars Pizza, Paradise Cafe, Taco Bell, The Whisper Restaurant, Wing Fong Chinese, Ya Ya's Flame Broiled Chicken
W: David's Pizza, Domino's Pizza, Fortune Dragon, Happy Valley Restaurant, Valley Pub

Lodg W: Economy Motel

AServ E: Auto Value (Auto Parts), Auto Works (Auto Parts)
W: Auto World Service, Car Quest Auto Center, Muffler Man, Q Lube & Wash Center

Med E: ✚ Hospital

ATM E: Old Kent

Other E: **Ace Hardware, Asian Supermarket, Harrison's Hardware, IGA Store, Kessel Grocery, Rite Aide Pharmacy, Sheriff's Dept**
W: **Corunna Convenience Store, Fowler's Party Store, Michigan State Police, Sam's Club, Wal-Mart (Pharmacy)**

117B Miller Road

Gas E: Speedway*(D), Sunoco*
W: Amoco*, Marathon*

Food E: Applebee's, Arby's, Bennigan's, Bill Knapp's Restaurant, Chi Chi's Mexican, Cottage Inn Pizza, Don Pablo's Mexican, Fuddrucker's, International Bakery & Pastries, KFC, Laredo Texas Steakhouse, McDonalds, Subway
W: Big Boy Restaurant, Bob Evans Restaurant, Burger King, Chuck E. Cheese's Pizza, Dunkin Donuts (Marathon), Hooters, Mancino's Pizza, McDonalds (Amoco), Old Country Buffet, Olive Garden, Pizza Hut, Salvatore Scallopini, Taco Bell, Wendy's

Lodg E: Comfort Inn, Motel 6, Sleep Inn
W: Howard Johnson, Super 8 Motel

AServ E: K-Mart, Meineke Discount Mufflers, National Tire & Battery, Sunoco, Tuffy Muffler, Valvoline Quick Lube Center
W: All Tune & Lube, Fast Eddie's Oil Change & Car Wash, Midas Muffler & Brakes

Med E: ✚ **Medical Care Center**

ATM E: Citizens Bank
W: Amoco

Other E: **K-Mart (Pharmacy), Pet Supplies Plus, Rite Aide Pharmacy, Sav-A-Lot Grocery**
W: **Animal Hospital, Pet Care Superstore, U-Haul Center(LP)**

117A Junction I-69, Lansing, Port Huron

116AB MI 121, Bristol Road (Southbound)

Gas E: Amoco*, Total*

Food E: Beechtree, Capitol Coney Island Family Restaurant, Days Inn, **Imperial Coney Island, McDonalds**, Taco Bell (Total)

Lodg E: Days Inn

AServ E: Amoco, Auto Zone Auto Parts

Other E: Coin Car Wash, **Super K Convenience Store**
W: **Airport**

114 Junction U.S. 23 southbound to Ann Arbor and Toledo I-475

111 Junction I-475 (Northbound)

109 MI 54, Dort Hwy, Burton

108 Holly Road, Grand Blanc

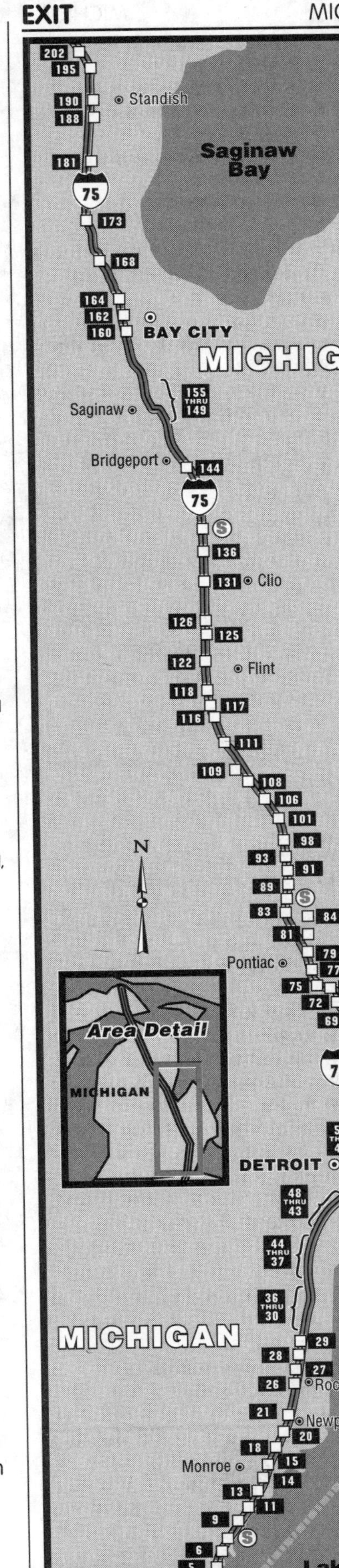

Gas E: Sunoco*

Lodg E: AAA Amerihost Inn

AServ E: Sunoco, Toyota Dealer

Med W: ✚ Hospital

106 Dixie Hwy, Grand Blanc (Southbound)

101 Grange Hall Road, Fenton

RVCamp E: **Yogi Bear Camp Report(LP)**

Parks E: **Groveland Oaks County Park, Holly Recreation Area**
W: **Seven Lakes State Park**

Other E: **State Patrol Post**

98 East Holly Road

Gas E: Mobil*(D)(LP)

(97) **Rest Area - RR, Vending, Picnic, Phones (Northbound)**

(95) **Rest Area - RR, Vending, Phones, Picnic (Southbound)**

93 U.S. 24, Dixie Hwy, Waterford

Food E: Coney Cafe

AServ E: Saturn Dealership
W: Clarkston's Chrysler Plymouth Dealership

Parks W: **Indian Springs Metro Park, Pontiac Lake Recreation Area**

91 MI 15, Clarkston, Davison

Gas E: Citgo*
W: Shell*

Food W: LB's Muffins & Yogurt, Mesquite Creek Restaurant

Lodg W: Millpond Bed & Breakfast

AServ W: Clarkston Muffler & Brake

Med W: ✚ Hospital

89 Sashabaw Road

Gas E: Shell*(CW)
W: Amoco

Food W: Dunkin Donuts, Little Caesars Pizza, Little Dana's Deli & Pizza, Mike's Oasis, Subway, Wai Hong Chinese

AServ E: Shell
W: Amoco, Clarkston Auto Body

ATM W: The State Bank

Parks E: **County Park**

Other W: **Arbor Drugs, Foodtown Grocery, Independence Animal Hospital, State Patrol Post**

(86) **Weigh Station - both directions**

84AB Baldwin Ave

Gas W: Mobil*, Shell*

Food E: Big Boy, Joe's Pub & Grub
W: McDonalds

AServ W: Mobil

83AB Joslyn Ave

Food W: Little Caesars Pizza (K-Mart)

AServ W: K-Mart

ATM W: First of America

Other W: **Food Town Grocery, K-Mart (Pharmacy)**

81 I-75 Business Loop, MI 24, Pontiac, Lapeer (Exit To Divided Hwy, Difficult Reaccess)

Parks E: **Bald Mt. Recreation Area**

Other E: **Sheriff's Dept**

79 University Dr, Rochester

Gas E: Amoco*(CW)
W: Speedway*(D)

Bold red print shows RV & Bus parking available or nearby

EXIT MICHIGAN

Food E: Bristoni's Italian, Domino's Pizza, Dunkin Donuts, Hershel's Hot Bakery Restaurant, Subway, Taste of Thailand
W: **Bistro 75 (Holiday Inn)**, McDonalds, Mountain Jack's, Taco Bell
Lodg W: ◆ AmeriSuites (see our ad this page), Courtyard Inn, ◆ Fairfield, Hampton Inn, ◆ **Holiday Inn**, AAA Motel 6
Med E: ✚ **Hospital**
W: ✚ **Hospital**
Other W: Car Wash

77AB MI 59, Utica, Pontiac (Exit To Divided Hwy, Southbound Reaccess Only)

75 Square Lake Road, Pontiac (Exits To Left)
Med W: ✚ **Hospital**

74 Adams Road
AServ E: Pontiac Silver Dome

72 Crooks Road
Food W: Charley's Crab, Hilton
Lodg W: Doubletree Guest Suites, Hilton
ATM W: NBD Bank

69 **(4)** Big Beaver Road
Gas W: Shell*
Food E: Marriott
W: Big Boy Restaurant, Denny's, Ruth's Chris Steakhouse
Lodg E: Drury Inn, ◆ Marriott

67 Rochester Road
Gas E: Clark, Mobil
Food E: Bangkok Hung Chinese, Big Boy, Burger King, Domino's Pizza, Dunkin Donuts, Leo's Souvlaki Coney Island, Mr Pita, Orchid Thai Cuisine, Pure & Simple Vegetarian Restaurant, Ram's Horn, Taco Bell, Zack's Coney Island
W: Holiday Inn, Mountain Jack's
Lodg W: ◆ Holiday Inn, Red Roof Inn
AServ E: Cottman Transmissions, Discount Tire Company, Fix 'N Go, Mobil, Noelle's
W: Belle Tire, Metro 25 Tire
ATM E: First Federal of Michigan, Michigan National
Other E: **Rite Aide Pharmacy**

65AB 14 Mile Road
Gas E: Shell*
W: Mobil*
Food E: Bob Evans Restaurant, Burger King, Chi Chi's Mexican, Chili's, Denny's, Sign of the Beef Carver, Steak & Ale, Taco Bell
W: Bennigan's, Big Fish Two, McDonalds, New York Coney Island, Outback Steakhouse
Lodg E: Motel 6, Red Roof Inn
W: Extended Stay America, Fairfield Inn, Hampton Inn, Knights Inn, Residence Inn, Troy-Madisson Inn
AServ E: Ford Dealership, Four Seasons Radiator, Goodyear Tire & Auto, Oakland Dodge Dealership, Sears, Shell
ATM E: Michigan National, NBD Bank
W: First Federal of Michigan
Other E: **ABC Eye Care, Doc Eye World, Fannie May Candies, Oakland Mall, Office Depot**

63 12 Mile Road
Gas E: 76*(D), Marathon*, Total*(CW)
W: Marathon*, Mobil*, Total*
Food E: Blimpie's Subs, Golden Wheel Chinese Restaurant, Green Lantern Pizza, Hacienda Azteca Mexican Restaurant, Marienelli's Pizza,

EXIT MICHIGAN

McDonalds, Red Lobster, Seros Family Dining, TCBY Yogurt
W: Denny's, Dunkin Donuts (Marathon)
AServ E: Auto Lab, Ramcharger Automotive, Uncle Ed's Oil Shop
W: GM Auto Dealership, Sparks Tune-up
ATM E: Standard Federal
W: First of America Bank
Other E: **K-Mart (Pharmacy)**
W: **Arbor Drugs, Farmer Jack Grocery**

62 11 Mile Road, 10 Mile Road West
Gas E: Mobil*(CW)
W: Citgo*, Mobil*
Food E: Boodles, Central Station Bar & Grill, Domino's Pizza, Jet's Pizza
W: Chicken Shack, KFC, Pizza Hut, Subway, Taco Bell, Tubby's Sub Shops
AServ E: Foreign Auto Supply, Matt's Tire Center, S & J Auto Service, Tuffy's Auto Service
W: Belle Tire
TServ E: **S & J Diesel Repair**
ATM W: First Bank of America
Other E: **7-11 Convenience Store**
W: Coin Car Wash, **Dairy Mart Convenience Store**

61 Junction I-696, Port Huron, Lansing

60 9 Mile Road, John R Street
Gas W: Marathon*(D), Price, Shell*
Food E: McDonalds, Quality Inn
W: Dairy Park
Lodg E: Quality Inn
AServ W: Rays Tire Center, TNT Complete Auto Repair
ATM W: Marathon

59 8 Mile Road, MI 102
Gas E: Amoco*
W: Clark*(D), Marathon, Shell*
Food E: Arby's, Belmont Coney Island, Burger King, Imperial Superstore, KFC, Little Caesars Pizza, McDonalds, Number 1 Chinese, Subway, Tubby's Submarine, Wendy's
W: Coney Island Restaurant
Lodg E: Bali Motel
AServ E: Brownie's Muffler, Good Wheels Auto Sales & Service, Richards Auto Parts
W: A-1 Kelley Tires, Goodwill Auto Parts, Shell
RVCamp W: **Campers Paradise**
Med W: ✚ **Bi-Country Walk-in Clinic**
ATM E: Michigan National Bank, NBD Bank
Other E: **Coin Laundry, McCrory Grocery, Rite Aide Pharmacy**
W: Car Wash, **Suburban Veterinary Hospital**

58 7 Mile Road
Gas E: 76
W: Amoco*(D)

57 McNichols Road
Gas E: BP*
W: 76*(D)
Food E: KFC, LA Koney Family, Taco Bell
W: Motor City Coney Island

EXIT MICHIGAN

AServ E: D-6 Tire Shop, McCormick Auto & Truck Repa
W: 76

56AB Davison Frwy.

55 Holbrook Ave, Caniff Ave
Gas E: Mobil*
Food W: KFC, Taco Bell
Med E: ✚ **Hospital**

54 East Grand Blvd, Clay Ave
Gas W: Shell*
Food W: Super Coney Island Restaurant
AServ W: Shell

53B Ford Frwy, I-94, Port Huron, Chicag

53A Warren Ave

52 Mack Ave
Gas E: Shell*
Food E: McDonalds
Med W: ✚ **Medical Care Center**

51C Junction I-375, Chrysler Frwy, Flint

51B MI 3, Gratiot Ave (Exits Left)

51A MI1, Woodward Ave

50 Grand River Ave

49B MI 10, Lodge Frwy.

49A Rosa Parks Blvd, Civic Center
Gas E: Shell*
W: Mobil*
Food E: White Castle Restaurant
AServ E: Modern Auto Parts
Other E: **Tiger Stadium**

48 Junction I-96, Jeffries Frwy, Lansing

47B Bridge to Canada, Lafayette Blvd

47A MI 3, Clark Ave
Gas E: Mobil*(CW)

46 Livernois Ave, Downtown Detroit
FStop E: **Pacific Pride Commercial Fueling**
Gas E: Marathon
Food E: KFC, Oscar's Coney Island Restaurant, Taco Bell
ATM E: Comerica

45 Fort St, Springwells Ave
Gas E: Marathon*(D)
W: Mobil*
Food W: Hungarian Village, McDonalds
AServ E: Marathon

44 Dearborn Ave
FStop E: **76* (Kerosene)**
AServ E: Sam's Auto Craft
Other E: **All-American Convenience Store**

43AB MI 85, Fort St, Schaefer Hwy
Gas E: Amoco*, Sunoco*

42 Outer Dr, Melvindale
FStop W: **Quick Fuel***
Gas E: Mobil*
W: Amoco*
Med E: ✚ **Oakwood Downriver Medical Center**
ATM E: Coamerica Bank

41 MI 39, Southfield Hwy
Gas W: Mobil*(D)
Food E: A & W Drive-In, Bangkok Star, Bill's Place
W: Big Boy, China Buffet, Dunkin Donuts, Leon's Dining, Seafood Bay
Lodg E: Economy Motor Inn
AServ E: Murray's Discount Auto Store

Bold red print shows RV & Bus parking available or nearby

KIT | MICHIGAN

W: Mobil
ATM W: Michigan National
Other E: Coin Laundry, Pat's Car Wash
W: Rite Aide Pharmacy

Dix Hwy
Gas E: Citgo*(CW)
W: Shell*(CW), Total*
Food E: Little Caesars Pizza, Pizza Factory, Ponderosa, Toma's Coney Island Restaurant
W: Arby's, Burger King, Dairy Queen, Dunkin Donuts, Long John Silvers, McDonalds, Pizza Hut, Rally's Hamburgers, Spad's Pizza, Taco Bell, Wendy's
Lodg W: Allen Park Motor Lodg, Holiday Motel
AServ E: Dix Auto Parts, Downriver Auto Service, Express Lube, Quiet Zone Muffler, Tuffy Auto Service
W: Downriver Alignment, Firestone Tire & Auto, Jiffy Lube, Lincoln Park Automotive Center, Midas Muffler & Brakes, Perry's Auto Clinic, Pete & Johnnys Automotive, Speedy Muffler, Top Value Muffler Shop
TServ E: Downriver Springs Truck Service
Med E: ✚ Health One Medical Center
ATM W: MI National Bank, NBD Bank
Other E: 7-11 Convenience Store, Arbor Drugs, Car Wash, K-Mart (Pharmacy), Lincoln Park Foriegn Cars
W: Farmer Jack Supermarket, Foodland Grocery, Kroger Supermarket (Pharmacy), Rite Aide Pharmacy

Allen Road, North Line Road
Gas E: Shell*, Speedway*
Food E: The Donut Shop, Yum Yum Donuts
W: Arby's, Burger King, Jonathon B, McDonalds, Nifty 50's, Roaring 90's
Lodg W: ◆ Budgetel Inn, (AAA) Cross Country Inn
AServ E: Lube N More, Pennzoil Lube, Shell
W: Oil & Lube
Med E: ✚ Hospital
Other E: Sam's Club
W: Car Wash, Metro Airport

Eureka Road
Gas E: Shell*, Total*
Food E: Bob Evans Restaurant, Denny's, Ramada Inn, Ryan's Steakhouse, Trovano's Pizzeria
W: Bakers Square Restaurant, Hooters, Mountain Jack's, New York Coney Island, Schlotzkys Deli, Shoney's, Sign of the Beef Carver
Lodg E: (AAA) Ramada Inn, Super 8 Motel
W: ◆ Red Roof Inn
AServ E: Pennzoil Lube, Quality Image Service Center, Shell*
W: Discount Tire Company
Med W: ✚ Medical Care Center
ATM E: Old Kent, Standard Federal, Total
Other E: Southland Animal Hospital
W: Home Depot, Southland Mall, Staples Office Superstore

5 U.S. 24, Telegraph Road (Left Exit, Dificult Reaccess)

AB Sibley Road, Dix Hwy
Gas W: Shell*

2 West Road, Trenton
TStop E: Mobil*(SCALES)
Gas E: Meijer*(LP), Mobil, Total*(LP)
W: Amoco*(CW), Shell*(CW)

EXIT | MICHIGAN

Food E: Birch Tree Cafe (Mobil), Burger King, Church's Fried Chicken, Dunkin Donuts, Good Fellas Pizza, Long John Silvers, PT's Christoff's Restaurant, Pizza Hut, Subway, Sushi Iwa Japanese Restaurant, Taco Bell, Uncle Harry's Family Dining, Wendy's, White Castle Restaurant
W: Amigo's Mexican Restaurant, Best Western, Country Skillet, Domino's Pizza, Kwan's Chop Suey, McDonalds
Lodg W: (AAA) Best Western, Knight's Inn
AServ E: Firestone Tire & Auto, GM Auto Dealership, Gorno Ford, Guardian Car Care, K-Mart, Midas Muffler, Mobil*, Precision Tune
TServ E: Mobil
TWash E: Mobil
ATM E: Old Kent
W: Old Kent Bank, Standard Federal
Other E: Car Wash, Coin Laundry, Customs Info. Center, K-Mart (Pharmacy), Kroger Supermarket, Miejer Foods, Pet Supplies Plus, Sears Hardware, Target Department Store
W: Coin Laundry, Woodhaven Pharmacy

29AB Flat Rock, Gibraltar
Gas W: Marathon*
Lodg W: Sleep Inn
AServ W: Ford Dealership, Marathon
Med E: ✚ Hospital
Parks E: Lake Erie Metro Park

28 MI 85, Fort St

27 N Huron River Dr, Rockwood
Gas E: Total*
W: Speedway*(D)
Food E: Benito's Pizza, Marco's Pizza
W: Riverfront Family Restaurant
ATM E: Old Kent
Other E: Food Town Grocery, Rite Aide Pharmacy, State Patrol Post, US Post Office

26 South Huron River Dr, South Rockwood
Gas E: Sunoco*(D) (Kerosene)
Food E: Dixie Cafe, Drift In
ATM E: Monroe Bank & Trust
Other E: U.S. Post Office

21 Newport Road, Newport
FStop W: Total*

20 Junction I-275 North, Flint

18 Nadeau Road
TStop W: Pilot Travel Center*(SCALES)
Food W: Arby's (Truckstop)

15 MI 50, Dixie Hwy, Monroe
TStop W: BP*
FStop W: Speedway*
Gas E: Shell*
Food E: Bob Evans Restaurant, Burger King, Country Skillet, Red Lobster
W: BP, Big Boy, (Cracker Barrel) Cracker Barrel, Denny's, Holiday Inn, McDonalds, Monroe Lounge, Subway, Wendy's
Lodg E: (AAA) Cross Country Inn, (AAA) Hometown Inn, Motel
W: ◆ Holiday Inn, Knight's Inn
TServ W: Great Lakes Western Star Truck Service
RVCamp E: Camping
Med W: ✚ Hospital
Parks E: State Park

14 Elm Ave

EXIT | MICHIGAN/OHIO

13 Front St, Monroe

11 La Plaisance Road, Downtown Monroe
Gas W: Amoco*(CW), Speedway*
Food W: Burger King, McDonalds
Med W: ✚ Hospital
ATM W: Horizon Outlet
Other W: Horizon Outlet Center, State Patrol Post

(10) Welcome Center - RR, Picnic, Vending, Phones

9 South Otter Creek Road, La Salle
Other W: American Heritage Antique Mall

(8) Weigh Station - both directions

6 Luna Pier
FStop E: Sunoco*(LP)
Gas W: Julie's*
Food E: Baskin Robbins (Sunoco), Gander's, McDonalds (Sunoco), Piasanos Pizza (Sunoco)
ATM E: Monroe Bank & Trust, Sunoco
Other E: Luna Pier Market

5 Erie Road, Temperance

2 Summit St, Erie, Temperance

↑MICHIGAN

↓OHIO

210 OH 184, Alexis Road

209 Ottawa River Road

208 Junction I-280 to I-80/90, Cleveland

207 LaGrange St, Stickney Ave

206 Phillips Ave, to U.S. 24

205B Berdan Ave (No Northbound Reaccess)
Med E: ✚ Hospital

205A Willys Pkwy, Jeep Pkwy

204 Junction I-475 West to U.S. 23, Maumee, Ann Arbor (Left Exit)

203B U.S. 24, Detroit Ave
Gas W: BP*(CW), Shell*
Food W: KFC, McDonalds, Rally's Hamburgers, Wendy's
AServ W: Shell
Other W: Rite Aide Pharmacy, U-Haul Center

203A Bancroft St, Toledo
Med E: ✚ Hospital

202B Collingwood Blvd
Food W: McDonalds
AServ E: Firestone Tire & Auto
Med E: ✚ Hospital
Other E: Art Museum
W: Revco Pharmacy

202A South Washington Street, Downtown

201B OH25N, Downtown (No Northbound Reaccess)

201A to OH 25 South, Collingwood Ave
RVCamp W: Creekside Mobile Village

200 South Ave, Kuhlman Dr

199A OH 65 North

199B OH 65 South, Rossford, Miami St
Food E: Palm's Cafe (EconoLodge)

EXIT OHIO

Lodg	E: Econolodge
198	Wales Road, Oregon Road, Northwood
FStop	E: Shell*
Gas	E: Speedway*
Food	E: Comfort Inn, Coney Island, Kayvon's Grill, Subway (Shell)
Lodg	E: AAA Comfort Inn
197	Buck Road
Gas	E: Buckroad Plaza*
	W: BP*, Sunoco
Food	E: Wendy's
	W: Denny's, Ike's Great American Restaurant, McDonalds
Lodg	W: Knights Inn, Super 8 Motel
AServ	W: Sunoco
195	I-80/90, Ohio Tnpk
193	U.S. 20, U.S. 23, Perrysburg, Fremont
Gas	E: BP*(D)(CW), Sunoco*(D)
	W: Marathon*(CW)
Food	E: Bob Evans Restaurant, Burger King, Cracker Barrel, Croy's Supper Club, Fricker's, Frisch's Big Boy, Holiday Inn, Holiday Inn Express, Jed's BBQ & Brew, McDonalds, Perry Cream Ice Cream Palace, Ralphies Burgers, Ranch Steak House, Stinger's Cafe, Subway, Taco Bell, Wendy's
Lodg	E: AAA Best Western, DAYS INN AAA Days Inn (see our ad on this page), ◆ Holiday Inn, ◆ Holiday Inn Express
	W: ◆ Budgetel Inn
AServ	W: Auto Zone Auto Parts, Marathon*
RVCamp	E: KOA Campground (7 Miles)
ATM	E: Fifth Third Bank, Key Bank
	W: Huntington Bank, Mid-Am Three Meadows Banking Center
Other	E: K-Mart (Pharmacy), Kroger Supermarket (w/pharmacy)
192	Junction I-475, U.S. 23, Maume, Ann Arbor (Left Exit)
187	OH 582, Haskins, Luckey
181	OH 64, OH 105, Bowling Green, Pemberville
Gas	E: Citgo*
	W: 76*(D), BP*(CW), Speedway* (Kerosene), Sunoco*(LP), Super America*
Food	W: Baskin Robbins, Big Boy, Blimpie's Subs (Super America), Bob Evans Restaurant, Burger King, Campus Quarters Sports Bar & Grill, Chi Chi's Mexican, Di Benedetto's Italian Restaurant, Domino's Pizza, Fricker's, Hunan Palace Chinese, Kaufman's Restaurant, Little Caesars Pizza, McDonalds, Pizza Hut, Ranch Steak & Seafood, Subway, Wendy's
Lodg	W: AAA Best Western, Buckeye Budget Motor Inn, DAYS INN AAA Days Inn (see our ad on this page), AAA Quality Inn & Suites
Med	W: ✚ Hospital
ATM	W: MidAm Bank
Other	W: Bowling Green State University, Burlington Optical, Coin Laundry
179	U.S. 6, Napoleon, Fremont
(179)	Welcome Center - RR, Phones, Tourist Info, Picnic, Vending
(176)	Weigh Station (Northbound)

EXIT OHIO

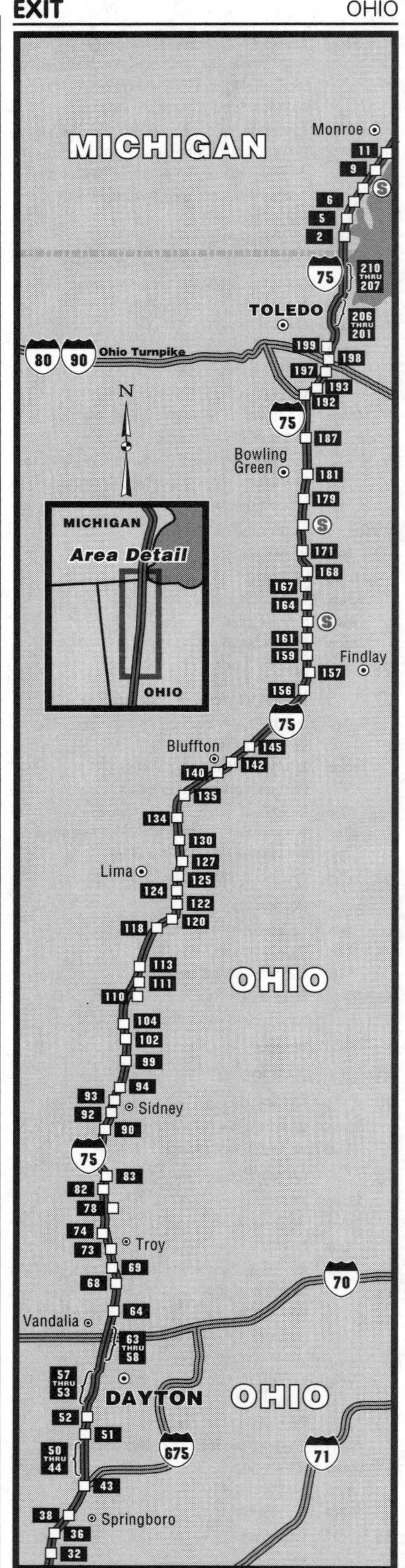

EXIT OHIO

171	OH 25, Cygnet
168	Eagleville, Quarry Road
FStop	E: Fuel Mart*
167	OH 18, North Baltimore, Fostoria
TStop	E: 76 Auto/Truck Plaza*(SCALES) (Travelers Ma
FStop	W: Citgo*
Food	E: McDonalds, Truckstop
	W: Denny's
Lodg	W: Crown Inn
AServ	W: Bernie's
TServ	E: Buckeye Truck Repair
TWash	E: Blue Beacon Truck Wash (76 Truck Stop)
164	OH 613, Fostoria, McComb
TStop	W: Pilot Travel Center*(LP)(SCALES)
Gas	E: Mobil*
Food	W: Dairy Queen (Pilot), Subway (Pilot), Taco Bell (Pilot)
AServ	E: Mobil
RVCamp	E: Camping
ATM	W: Pilot
Parks	E: Van Buren State Park
(162)	Weigh Station (Southbound)
161	Township Road 99
RVCamp	E: Campground
Other	E: Ohio State Patrol Post
	W: Jeffrey's Antique Mall
159	**(160)** OH 224, OH 15, Ottawa, Tiffir
Gas	E: BP*(D)(CW), Marathon*(LP), Swifty* (Kerosene)
	W: Shell*(D)
Food	E: Archie's Drive-in Too, Big Boy Restaurant, Evans Restaurant, Domino's Pizza, Enck's, McDonalds, Ming's Great Wall, Pizza Hut,

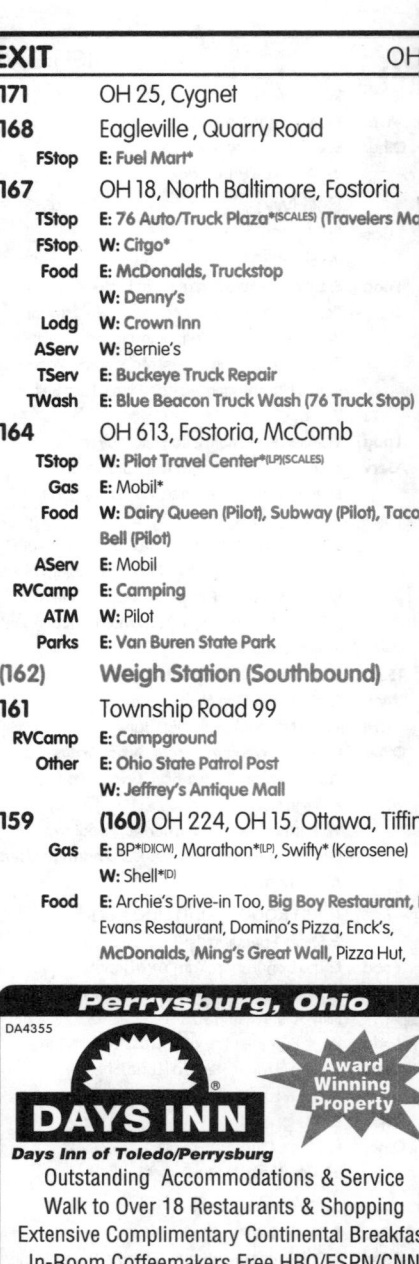

Bold red print shows RV & Bus parking available or nearby

EXIT	OHIO
	Ponderosa, Rally's Hamburgers, Spaghetti Shop, Subway, Taco Bell, The Yogurt Parlor, Wendy's, Willie's Pizza Factory
Lodg	E: AAA Cross Country Inn, ◆ Ramada Inn, Rodeway Inn, Super 8 Motel
	W: ◆ Country Hearth Inn, ◆ Hampton Inn
AServ	E: Rainbow Muffler
	W: Two to Four Auto
TServ	W: Easter Tire, Peterbilt Dealer
ATM	E: Bank One
Other	E: Findlay Car Wash, Great Scott's Food Store, Rite Aide Pharmacy
7	OH 12, Columbus Grove, Findlay
TStop	W: Ohio West Truck Plaza*(SCALES)
Gas	E: Citgo*(D), Marathon*(CW)
	W: BP*
Food	E: Bill Knapp's Restaurant, Blimpie's Subs, Imperial Dining Room (Days Inn), Nathan's (Citgo), TCBY (Citgo)
	W: Fricker's, Pilgrim Family Restaurant, Truckstop
Lodg	E: Days Inn AAA Days Inn (see our ad on this page)
	W: AAA Econolodge
TServ	W: MAC Trucks, McCoi National Lease, Miami Industrial Trucks
ATM	W: Ohio West Truck Plaza
Other	E: Coin Car Wash, Wolfies Convience Store
	W: Findlay Animal Care Center
6	U.S. 68, OH 15 To US 23, Kenton, Columbus
Med	E: ✚ Hospital
Other	E: Airport
3)	Rest Area - RR, Phones, Picnic,

EXIT	OHIO
	Vending
145	OH 235, Mount Cory, Ada
RVCamp	E: Twin Lakes Park Campground
142	(143) OH 103, Arlington, Bluffton
FStop	W: Shell*
Gas	E: Citgo*, Sunoco*(LP)
Food	E: Denny's
	W: Arby's, KFC, McDonalds, Subway
Lodg	E: Howard Johnson
Med	W: ✚ Hospital
Other	W: Car Wash
140	Bentley Road, Bluffton
Med	W: ✚ Hospital
135	(136) U.S. 30, Upper Sandusky, Delphos, Beaver Dam
TStop	E: Beaver Dam Truck Plaza
	W: Flying J Travel Plaza*(LP)(SCALES) (RV Dump)
FStop	E: Speedway*(LP)
Food	E: Blimpie's Subs (Speedway), Nathan's (Beaver Dam Truck Stop), Subway (Beaver Dam Truck Plaza), TCBY Yogurt (Beaver Dam Truck Plaza)
	W: Thads Restaurant (Flying J)
TServ	W: Bridgestone Tire & Auto, Flying J Travel Plaza
TWash	E: Beaver Dam Truck Plaza
ATM	E: Beaver Dam Truck Plaza
134	OH 696, Napoleon Road, Beaverdam (No Northbound Reaccess)
130	Blue Lick Road
Gas	E: Citgo*
Food	E: Mexican Fast Food Express, Ramada Inn, Subway
Lodg	E: Ramada Inn
127AB	(128) OH 81, Lima, Ada
Gas	W: BP*
Food	W: Darrihutt, Econolodge, Waffle House
Lodg	W: Comfort Inn, Days Inn Days Inn (see our ad on this page), Econolodge
AServ	W: BP, General Tire Company, Phoenix Auto Salvage
TServ	E: Truck & Trailer Parts
	W: Tomlinson Truck Service, Western Ohio Truck & Equipment
Other	E: Wonder Bakery Thrift Shop
	W: Big A Auto Parts
125AB	OH 309, OH 117, Kenton, Lima
TStop	W: 76 Auto/Truck Plaza*
Gas	E: BP*(D)(CW), Speedway*(D)(LP)
	W: Citgo* (Kerosene)
Food	E: Arby's, Big Boy Family, Big Owl's Pizza, Bob Evans Restaurant, Burger King, Captain D's Seafood, Cracker Barrel, Fazoli's Italian Food, Hunan Garden, Little Caesars Pizza, McDonalds, Olive Garden, Pat's Donuts & Kreme, Pizza Hut, Ponderosa Steakhouse, Rally's Hamburgers, Red Lobster, Ryan's Steakhouse, Subway, Taco Bell, Wendy's
	W: Kewpee Hamburgers, Shoney's, The Spaghetti Shop
Lodg	E: Hampton Inn, Holiday Inn, Motel 6
	W: East Gate Motel, Economy Inn, Super 8 Motel
AServ	W: Quaker State Q-Lube, Shell
Med	W: ✚ Hospital
ATM	E: AmeriCom, Bank One, Fifth Third Bank
	W: Lima Superior Federal Credit Union

EXIT	OHIO
Other	E: Coin Car Wash, K-Mart, Phar Mor Drugs, Plaza Car Wash, Ray's Grocery Store, Sams Club, Tourist Information, WG Grinders Deli
	W: Coin Laundry, Eastside Coin Laundry, Sav-a-lot Food Store
124	4th St, Lima
Other	E: Ohio State Patrol Post
122	(123) OH 65, Lima, Ottawa
Gas	E: Speedway
	W: Shell*(D)
TServ	E: Freightliner Dealer
	W: Buckeye Truck Center, Volvo Dealer
120	(121) Breese Road, Fort Shawnee
FStop	W: Amoco*
Gas	W: Speedway*
Food	W: Old Barn Deli & Bakery
Lodg	W: Tourest Motel
AServ	W: J&S Auto Service & Towing, Shawnee Motors Service Center
TWash	W: Dixie Truck Wash
Other	W: Amerigas(LP), Bakers Propane, Car Wash
118	(119) Cridersville
FStop	W: Fuel Mart*
Gas	W: Speedway*(LP)
Food	W: Dixie Diner, Home Cooked Food (Fuel Mart), Padrone's Pizza & Subs, Pat's Donuts & Creme, Subway (Fuel Mart), Village Cafe
ATM	W: Bank One
Other	W: Cridersville Laundry, Dave's Market Grocery, Veterinary Clinic
(114)	Rest Area - RR, Phones, Picnic, Vending
113	OH 67, Wapakoneta, Uniopolis
Lodg	E: Catalina Motel
111	(112) Bellefontaine St, Wapakoneta
TStop	E: L & G Auto/Truck Stop*(SCALES)
Gas	W: BP*(CW)(LP), Shell*(CW)
Food	E: Dodge City Restaurant, Mickey's Restaurant (L&G Truck Stop)
	W: Arby's, Burger King, Captain D's Seafood, Dairy Queen, KFC, Lucky Steer, McDonalds, Pizza Hut, Ponderosa, Taco Bell, The Chalet Restaurant (Holiday Inn), Waffle House, Wendy's
Lodg	E: Days Inn Days Inn, Twin Lakes Resort
	W: Dollar Inn, AAA Holiday Inn, ◆ Super 8 Motel
AServ	W: Auto Serviceof Wapak
TServ	E: Cummins Diesel, L & G
TWash	E: L & G
RVCamp	E: KOA Campground
ATM	W: Fifth Third Bank
Other	W: Big Bear Plus Grocery, Car Wash, Neil Armstrong Air and Space Museum
110	U.S. 33, St Marys, Bellefontaine
RVCamp	E: KOA Campground (.9 Miles)
Other	E: Ohio State Patrol Post
104	(105) OH 219, Botkins
Gas	W: Sunoco*
Food	W: Little John Lounge & Restaurant (Budget Host Inn)
Lodg	W: Budget Host Inn
AServ	W: Firestone Tire & Auto
102	(103) OH 274, New Bremen, Jackson Center
99	(100) OH 119, Anna, Minster

EXIT OHIO

TStop E: 99 Truckstop[SCALES], Fuel City*[LP]
Gas W: Sunoco*, Super America*[LP]
Food E: Apple Valley, Cafe 99 (Truckstop)
W: Wendy's
TServ E: L & O Truck Service (99 Truckstop)
ATM W: Super America

94 CR 25A, Sidney

93 OH 29, St Marys, Sidney
RVCamp W: Campground

92 OH 47, Bus Loop 75, Versailles, Sidney
Gas E: Shell*[CW], Speedway*
W: Amoco*[D], BP*, Sunoco*[LP] (Kerosene)
Food E: Arby's, Bakery Outlet, China Garden, East of Chicago Pizza Co., Pub Restaurant, Subway, Taco Bell, Winger's
W: Big Bear Deli, Bob Evans Restaurant, Burger King, Burks Deli, Hardee's, Holiday Inn, KFC, Marco's Pizza, McDonalds, Perkins Family Restaurant, Pizza Hut, Ponderosa, Rally's Hamburgers, Super Subby's, Waffle House
Lodg W: AAA Comfort Inn, DAYS INN AAA Days Inn, AAA Econolodge, AAA Holiday Inn
AServ E: NAPA Auto Parts
W: Pennzoil Lube, Sidney Ford
Med E: + Hospital
ATM W: Bank One, Mutual Federal Savings Bank, Star Bank
Other E: Coin Laundry, Sav-a-lot Food Store, The Pharm Pharmacy
W: Aldi Supermarket, Car Wash, Kroger Supermarket (Pharmacy), Mail Boxes Etc, Revco Drugs, Wal-Mart

90 Fair Road, Sidney
Gas E: Sunoco*[D]
AServ E: G&H Auto Center, Sunoco
W: Edco Automotive
RVCamp W: Campground

83 (84) CR 25A, Piqua
Food W: J J's (Knights Inn)
Lodg W: Knight's Inn
AServ W: Dan Hemm Olds, GM Auto Dealership, Lawsons Auto Service, Paul Sherry Olds/ Cadillac/Chevrolet, Piqua Auto Lube
RVCamp W: Paul Sherry RV Center
Med W: + Hospital
Other W: Mabbitts C-Store

82 U.S. 36, Urbana, Piqua
Gas E: BP*[CW]
W: Speedway*[LP]
Food E: Arby's, China East, Duff's, KFC, Long John Silvers, Ponderosa, Rax, Swag's, Taco Bell, Wendy's
W: Bob Evans Restaurant, Bullies, Burger King (Mall), El Sombrero Mexican Restaurant, McDonalds, Red Lobster
Lodg W: AAA Comfort Inn, Howard Johnson
AServ E: Goodyear Tire & Auto, Pennzoil Lube, Sears Auto Center
Med W: + Hospital
Other E: Piqua Mall
W: Aldi Supermarket, Miami Valley Centre (mall)

(81) Rest Area - RR, HF, Picnic, Phones, RV Dump, Vending (Both Directions)

EXIT OHIO

78 CR 25A

74AB OH 41, Covington, Troy
Gas E: BP*[LP]
W: Meijer Grocery*, Shell*[CW], Speedway*[LP]
Food E: China Garden Buffet, Donatos Pizza, El Sombrero, Little Caesars Pizza, Long John Silvers, McDonalds, Perkins Family Restaurant, Pizza Hut Delivery, Sally's Cafe, Subway
W: Applebee's, Big Boy, Bob Evans Restaurant, Boston Market, Burger King, Dairy Queen, Fazoli's Italian Food, Friendly's, Golden Corral, McDonalds (Wal-Mart), Steak & Shake, Tokyo Peking Restaurant
Lodg W: Hampton Inn, Knight's Inn
AServ E: Goodyear Tire & Auto
W: Auto Zone Auto Parts
Med E: + Hospital
ATM E: National City Bank
W: Fifth Third Bank
Other E: Pearl Vision Center, Super Petz Food Store, The Pharm Pharmacy, Upper Valley Family Care
W: County Market Grocery, Meijer Grocery (Pharmacy), Revco Drugs, Sear's Hardware, Wal-Mart (Pharmacy & Vision)

73 OH 55, Troy, Ludlow Falls
Gas E: Amoco*, BP*[LP]
Food E: Amazing Wok, Mel-O-Dee Restaurant, Papa John's, Waffle House
Lodg E: Microtel, Quality Inn Suites, Super 8 Motel
TServ E: Holt Cat
Med E: + Hospital
ATM E: Fifth Third Bank (Kroger), Star Bank
Other E: Kroger Supermarket (Pharmacy)

69 CR 25A
Gas E: Dairy Mart*, Marathon*, Starfire*[D]
Food E: Taco Bell
AServ E: Buick Dealer, Ford Dealership
ATM E: Marathon

68 OH 571, West Milton, Tipp City
Gas E: BP*[CW], Shell*[CW], Speedway*
W: Citgo*[D] (Kerosene), Super America*
Food E: Burger King, Cassano's Pizza & Subs, Domino's Pizza, Hinders, Hong Kong Kitchen, McDonalds, Osaka Japanese Restaurant, Subway, Taco Bell
W: Arby's, Big Boy, Blimpie's Subs, Tipp O'the Town Family Restaurant, Wendy's
AServ E: Goodyear Tire & Auto, Honda Dealer
W: Citgo
ATM E: Star Bank
W: Bank One
Other E: ACE Hardware[LP], Chmiel's Grocery, Coin Car Wash, Revco Drugs

64 Northwoods Blvd

63 U.S. 40, Vandalia, Donnelsville
Gas E: SuperAmerica*
W: Shell*[CW], Speedway*[D]
Food E: Bunkers Bar & Grill, Dragon China, Frickers, Gliders Soup & Sandwiches
W: Arby's, Hot Air Balloon Lounge, Jim's Donut Shop, KFC, McDonalds, Original Rib House, Pizza Hut (Carry-out only), Subway, TW's Smokehouse, Taco Bell, Wendy's
Lodg E: Crossroads Motel
W: AAA Cross Country Inn
AServ E: Vandalia Auto Clinic

EXIT OH

W: Muffler Brothers, Vandalia Auto Parts
ATM E: Citizens Federal
W: Key Bank, Monroe Federal Savings
Other E: Kroger Supermarket (Pharmacy)
W: Hengers Supermarket, Ken's Pharmacy, Vandalia Animal Clinic, Vandalia Carry Ou C-Store

61AB Junction I-70, Indianapolis, Columbus, Dayton International Airport

60 Little York Road, Stop Eight Road
Gas W: Sunoco* (Kerosene)
Food E: Cooker, Damon's, Olive Garden, Red Lobs Ryan's Steakhouse, Subway
W: Arby's, Bennigan's, Bob Evans Restaura Cracker Barrel, Lone Star Steakhouse, Max & Erma's, Northern Palace Chinese, Outback Steakhouse, Wendy's
Lodg E: ◆ Residence Inn
W: ◆ Comfort Inn, DAYS INN ◆ Days Inn, ◆ Fairfi Inn, Knight's Inn, ◆ Motel 6, ◆ Ramada, ◆ R Roof Inn
AServ E: Frank Z Imports
TServ E: Miami Intenational Trucks
Other E: Kinko's Copies
W: Sams Club

58 Needmore Road/Wright Brothers Pkwy
Gas E: BP*, Shell*[CW]
W: Speedway*[LP] (Kerosene), SuperAmerica[D] (Kerosene)
Food E: Entenmann's Thrift Cake Bakery, Fisch's Big Boy, Hardee's, Quality Inn
W: Friendly's, Long John Silvers, Subway (Speedway), Taco Bell (Super America), Waffle House
Lodg E: Quality Inn
AServ E: Goodyear Tire & Auto
W: Need More Service Center
RVCamp W: Harris Travel Trailer Service

57B Neff Road, Wagner Ford Road, Siebenthaler Road
Gas E: Sunoco*[LP]
W: Speedway*, United Dairy Farmers*
Food E: Holiday Inn
W: Denny's, Little Caesars Pizza, Prescott's Restaurant, Subway
Lodg E: ◆ Holiday Inn
W: AAA Best Western, Econolodge
Other W: Rite Aide Pharmacy

57A Neeva Dr, Dayton (Northbound, Reaccess Southbound)

56AB Stanley Ave (Southbound)
FStop E: Amoco*
Gas W: 76*[D], BP*
Food W: Gold Star Chili, Golden Nugget Pancake, McDonalds, Rally's Hamburgers, Taco Bell, Wendy's
Lodg W: Dayton Motor Hotel, Plaza Motel, Royal Mo
ATM W: 76, Bank One
Other W: Coin Car Wash

55B Keowee St North, Leo St (Northbound)
Gas W: BP*
Food W: Rally's Hamburgers, Taco Bell
AServ E: Big Muffler Shop #4

IT OHIO

ATM W: Bank One

Other W: **Parkside Convenience Store**

C OH 4, Springfield, Webster St

B OH 48, Main St, Downtown Dayton

A Grand Ave

AB OH 49, Salem Ave, First St, Third St

B U.S. 35, Eaton, Xenia

A Albany St, Stewart St

Nicholas Road, Edwin C. Moses Blvd

Gas W: BP*, Shell Express

Food W: **McDonalds**

Lodg W: AAA Econolodge

Med E: ✚ **Hospital**

B OH 741, Springboro Road

Med E: ✚ **Hospital**

A Dryden Road

Gas E: SuperAmerica

W: Sunoco

Food W: Holiday Inn, TJ's Restaurant

Lodg W: ◆ Holiday Inn, ◆ Super 8 Motel

AServ W: Broadway Car Care, Sunoco

Other W: **U-Haul Center(LP)**, **Veterinary Clinic**

Central Ave., West Carrollton

Gas E: Sunoco* (Kerosene)

W: Speedway*

Food E: Domino's Pizza, Frickers, Frisch's Big Boy, Holly's Home Cooking, Waffle House

W: El Meson Restaurant, Krackers, McDonalds

Lodg E: Parkview Inn, Red Horse Motel

AServ E: Precision Tune Auto Care, Slone's Automotive

W: Bill's Transmission, South Point Service Center

VCamp W: **Campground**

Other E: **Don's Convenience Store**

W: **U.S. Post Office, Woody's Supermarket (Pharmacy)**

OH 725, Miamisburg, Centerville, W. Carrollton (Very Congested)

Gas E: BP*(CW), Shell*, Speedway*

W: BP*, Marathon*(CW), Shell

Food E: Amar Indian Restaurant, **Applebee's**, Big Boy, Blimpie's Subs, Bruegger's Bagels, Burger King, **Captain D's Seafood**, Casual Dining Gallery, Chi Chi's Mexican, Chuck E. Cheese's Pizza, Cold Beer & Cheeseburgers, Delphine's Steak, Seafood, & Ribs, **Denny's**, Dunkin Donuts, Friendly's, Grindstone Charlie's Restaurant, Ground Round, Hardee's, India Palace, Interface Cafe, Jokers Comedy Cafe, KFC, Lone Star Steakhouse, Marion's Italian, Max & Erma's, McDonalds, Olive Garden, Peking Express, Penn Station Steak & Subs, Pizza Hut, **Ponderosa**, Rally's Hamburgers, **Red Lobster**, Ruby Tuesday, Shooters, Skyline Chili, Steve Kao's Chinese, TCBY Yogurt, Taco Bell, Wendy's

W: Bob Evans Restaurant, Byers Inn, China Hut, Luciano's Pizza, Perkins Family Restaurant

Lodg E: AAA Holiday Inn, **Motel 6**, Suburban Lodge

W: Byers Inn, DAYS INN AAA **Days Inn**, Knight's Inn, ◆ Red Roof Inn, AAA **Signature Inn**

AServ E: BP, Custom Lube and Oil, Dayton Auto Mall (Nissan, Honda, Mazda), Firestone Tire & Auto, Goodyear Tire & Auto, Jiffy Lube, National Tire & Battery, Sear's, Speedy Muffler King

W: Q-Lube, Shell

Med E: ✚ **The Neighborhood Doctor (Urgent Care Facility)**

EXIT OHIO

W: ✚ **Sycamore Hospital**

ATM E: Bank One, Fifth Third Bank, Key Bank, The Provident Bank

Other E: Bubble Brush Car Wash, **Cub Foods, Dayton Mall, Office Depot, Paws Inn Pet Supplies & Grooming, Super Petz**

W: **M&M Convenience Store**

43 Junction I-675 North, Columbus

38 OH 73, Springboro, Franklin

Gas E: Amoco*(CW), BP, Speedway, **Sunoco(D)**

Food E: **Arby's**, Burger King, La Comedia Dinner Theater, Long John Silvers, McDonalds, Perkin's Family Restaurant, Pizza Hut Carry Out, Skyline Chili, Taco Bell, Wendy's

W: **Frisch's Big Boy**, Village Station Steak & Seafood

Lodg E: ◆ Ramada Limited

W: Econolodge, Knight's Inn

AServ E: BP, Jiffy Lube

ATM E: **First National Bank**, National City Bank

W: Community National Bank

Other E: **Animal Medical Center, Coast to Coast Hardware**, Coin Car Wash, **K-Mart, Kroger Supermarket, US Post Office**

36 OH 123, Franklin, Lebanon

TStop E: **Dayton South Travel Center*(SCALES)**

FStop E: **Speedway*(LP)**

Gas W: BP*(CW)

Food E: Dayton South Travel Center, Hot Dog Construction Co., **McDonalds, Waffle House**

Lodg E: **Royal Inn**, ◆ **Super 8 Motel**

TServ E: **Dayton South Travel Center**

TWash E: **Dayton South Travel Center, Truck Wash**

ATM W: BP

32 OH 122, Middletown

Gas E: BP*(LP), Duke*

W: Marathon*, Meijer*(D)(LP) (Kerosene)

Food E: **McDonalds, Waffle House**

W: Applebee's, Bamboo Garden, Bill Knapps Restaurant, Bob Evans Restaurant, Boston Market, Old Country Store **Cracker Barrel**, Hardee's, KFC, Lone Star Steakhouse, Old Country Buffet, Olive Garden, Ponderosa Steakhouse, Sonic, Steak & Shake, Wendy's

Lodg E: Best Western, ◆ Comfort Inn, ◆ **Super 8 Motel**

W: ◆ Fairfield Inn, ◆ Holiday Inn Express, **Howard Johnson**

AServ E: BP, Williams Jeep/Eagle/Mercury/Lincoln

W: Goodyear Tire & Auto, Guyler Buick GMC Truck, Tire Centers

Med W: ✚ **Doctor's Urgent Care Clinic**

ATM W: Boston Market, Fifth Third Bank (Kroger)

Other W: **Kroger Supermarket (Pharmacy), Meijer Supermarket, Pearl Vision Center, Super Pets, Towne Mall**

29 OH 63, Monroe, Lebanon

TStop E: **AmBest*(SCALES)**

Gas E: Chevron*(D), Shell

W: BP*(D), Sunoco*(D), Super America*

Food E: AmBest

W: Gold Star Chili, **McDonalds**, Perkin's Family Restaurant, Sarah Jane's, Subway (Super America)

Lodg E: DAYS INN Days Inn, **Stony Ridge Inn (Am Best)**

W: Econolodge, ◆ Hampton Inn

AServ W: Goodyear Tire & Auto

TServ E: **AmBest, Bishops Truck Care**

EXIT OHIO

ATM W: Sunoco

(28) Rest Area - RR, Picnic, Phones, Vending, Tourist Info (Both Directions)

22 Tylersville Road, Hamilton, Mason

FStop W: **Speedway*(LP) (Kerosene)**

Gas E: Amoco*(LP), BP*, Sunoco*(LP)

W: Meijer*, Shell*(CW)

Food E: Akira Japanese Restaurant, Arby's, Bob Evans Restaurant, Boston Market, Burger King, Fazoli's Italian Food, Gold Star Chili, KFC, Long John Silvers, McDonalds, Perkins Family Restaurant, Pizza Hut, Pizza Hut Carry Out, Subway, Taco Bell, Waffle House, Wendy's

W: Bruegger's Bagels, Charley's, La Tazza Verde, Steak & Shake

Lodg E: Rodeway Inn

AServ E: Bob Sumerel Tire Company, Firestone Tire & Auto, Goodyear Tire & Auto, Quaker State

W: Oil Express, Tire Discounter, Wal-Mart

ATM E: Fifth Third Bank, Fifth Third Bank (Kroger), First National Bank, PNC Bank

W: Enterprise Federal, Shell, **Star Bank**

Other E: **Kroger Supermarket (Pharmacy)**, Thunderbird Auto Wash, **United Dairy Farmers***

W: **Complete PetsMart, Mail Boxes Etc, Meijer Supermarket, Sear's Hardware, Wal-Mart (Pharmacy)**

21 Cin - Day Road

Gas W: BP, Shell(CW), United Dairy Farmers*(D) (Kerosene)

Food E: Frisch's Big Boy

W: China Fun, La Rosa's, Papa John's Pizza

Lodg E: Holiday Inn Express

W: Knights Inn

AServ W: BP, Shell

ATM W: United Dairy Farmers

Other W: **Pfister Animal Hospital**

16 Junction I-275, to I-74, I-71, Indianapolis, Columbus

15 Sharon Road, Sharonville, Glendale

Gas E: Ashland*, BP*(LP), Chevron*(D), Marathon*(D), Shell

W: Sunoco* (24 hrs)

Food E: Bob Evans Restaurant, Burbank's Real BBQ, Frisch's Big Boy, Pizza Hut, Skyline Chili, Spanky's (Holiday Inn), **Waffle House**

W: Arby's, Bombay Bicycle Club, Captain D's Seafood, Chuck E Cheese's, Long John Silvers, Lotus Buffet Oriental Restaurant, **McDonalds**, Osaka Japanese, Penn Station Steak & Sub, Red Dog Saloon, Shogun, Subway, Taco Bell, Tahan Mongolian BBQ, Texas Roadhouse, Wendy's, **Windjammer Restaurant**

Lodg E: ◆ Fairfield Inn, AAA Hampton Inn, AAA Holiday Inn, ◆ Red Roof Inn, ◆ Woodfield Suites

W: Comfort Inn, Econolodge, Extended Stay America, ◆ Marriott, ◆ Red Roof Inn, Residence Inn, AAA Signature Inn, AAA Super 8 Motel

AServ E: Firestone Tire & Auto, Marathon, Shell

W: Instant Oil Change, Safelite Auto Glass, Top Value Muffler Shops

TServ E: **Dent Spring Truck Service Inc**

ATM W: Fifth Third Bank

Other E: **Malibu Grand Prix**

W: **Maytag Laundry, Princeton Marine Sales & Service**

Bold red print shows RV & Bus parking available or nearby

EXIT		OHIO
14		OH 126, Neumann Way, Woodlawn, Evendale
	Gas	**W:** Swifty (Kerosene)
	Food	**W:** Raffel's (Quality Inn)
	Lodg	**W:** Quality Inn
	TServ	**W:** **Taton Transport Service**
	ATM	**W:** Fifth Third Bank
13		Shepherd Lane, Lincoln Heights
	Gas	**W:** 76
	Food	**W:** Taco Bell, Wendy's
	AServ	**W:** Cooke's Garage
	Other	**W:** **I-75 Check Cash**
12		Wyoming Ave, Cooper Ave, Lockland, Reading (Difficult Reaccess Southbound)
	Gas	**W:** Marathon*
	ATM	**W:** Marathon
10		Galbraith Rd, Arlington Hts, Ronald Reagan Cross County Hwy
	Gas	**E:** BP*(D)(CW), Shell*
		W: Sunoco*(LP), SuperAmerica*
	Food	**E:** Burger King, Howdy's Home Cooking
	AServ	**E:** Car X Mufler & Brake, Midas Muffler & Brakes, Tire Discounters
	ATM	**W:** Huntington Bank (SuperAmerica), Sunoco
9		OH 4, OH 561, Paddock Road, Seymour Ave.
	Food	**W:** White Castle Restaurant
8		Elmwood Place, Towne St (Northbound, No Southbound Reaccess)
7		OH 4, Paddock Rd, to I-71 (Exit To Limited Access Hwy)
6		Mitchell Ave, St Bernard
	Gas	**E:** Marathon*, Shell, Speedway*
		W: BP*
	Food	**E:** KFC, Taco Bell
		W: McDonalds
	AServ	**E:** Jim's Auto Electric, Marathon, Shell
		W: Micheal Tires, Sander Ford
	ATM	**W:** Kroger Supermarket
	Other	**W:** Coin Car Wash, **Kroger Supermarket, Pharmacy (Kroger)**
4		Junction I-74 W, U.S. 52 W, U.S. 27 N, Indianapolis
3		U.S. 52E, U.S. 27S, U.S. 127S, Hopple Street
	Gas	**E:** Marathon
		W: BP*(D), Shell*(CW)
	Food	**E:** Frisch's Big Boy, The Cave, White Castle Restaurant (Interstate Motel)
		W: C.W. Donuts, Camp Washington Chili, Isadore's Pizzeria, Italian Buffet, U.S. Chili
	Lodg	**E:** DAYS INN ◆ Days Inn, Interstate Motel, Travelodge, White Castle Restaurant
	AServ	**E:** Jiffy Lube, Marathon
	Med	**E:** ✚ **Good Samaritan Hospital**
	ATM	**W:** Shell (24 Hrs)
	Other	**E:** **Dairy Mart,** Parkway Auto Wash
		W: **Holmes Pharmacy**
2B		Harrison Ave (Northbound Exit Left)
	Gas	**W:** BP*
	Food	**W:** McDonalds
2A		Western Ave, Liberty St (Southbound)
	Gas	**W:** BP*(D)

EXIT		OHIO/KENTUCKY
	Food	**W:** Gold Star Chili
	AServ	**W:** BP
1H		Ezzard Charles Dr (Northbound)
1G		To U.S. 50 West, Freeman Ave
	Food	**W:** Frisch's Big Boy, Taco Bell, Wendy's
	Lodg	**W:** Holiday Inn (see our ad on this page)
	ATM	**W:** Fifth Third Bank, Star Bank
1F		Seventh St
	Gas	**W:** Ashland, Sunoco*
	Food	**W:** Chili Company
	Lodg	**W:** Holiday Inn
	ATM	**W:** The Provident Bank
1E		Fifth Street
1D		I-71N, I-47, US 50E, Downtown
1C		I-71N, I-47, US 50E, Downtown
1B		I-71N, I-47, US 50E, Downtown
1A		I-71 North, I-471, U.S. 50 E, US 52 E

↑OHIO

↓KENTUCKY

EXIT		KENTUCKY
192		Fifth St, Covington, Newport
	Gas	**E:** Chevron(D), Shell*(CW), SuperAmerica*(D)(LP)
		W: Deli Direct, JD's Food Mart*
	Food	**E:** Burger King (24 hrs), Dominiques, Frisch's Big Boy, Gold Star Chili, Hardee's, Holiday Inn, KFC, McDonalds, Perkins Family Restaurant, River Front Pizza, Subway, Taco Bell, White Castle Restaurant
		W: Cork & Bottle Cafe, Deli Direct, Willy's Sports Cafe
	Lodg	**E:** Extended Stay America, Holiday Inn, Quality Hotel
		W: Hampton Inn
	AServ	**E:** Chevron, Ford Dealership, Smith Muffler Shop
		W: Mark's Garage
	ATM	**E:** Mac
		W: Deli Direct, JD's Food Mart
	Other	**E:** **Mainstrasse Village & N. KY Visitor Ctr.**
191		Twelfth St, Pike St, Covington
	Gas	**E:** Shell*
	Food	**E:** Anchor Grill, Bakery Outlet, Donut Cafe
	AServ	**E:** Marshall Dodge
	Med	**E:** ✚ **Hospital**
		W: ✚ **Hospital**
	Other	**E:** **Coin Laundry, Zimmer Hardware**
190		Jefferson Ave
189		KY 1072, Kyles Lane, Fort Wright, Pa Hills
	Gas	**W:** BP, Chevron, Marathon*(CW), Shell*(D)(CW), SuperAmerica*(LP)
	Food	**W:** D'Andrea's (Ramada Inn), Fort Wright Restaurant, Frisch's Big Boy, Gabriel's Pizza, Grandpa's Gourmet Ice Cream & Coffee Hou Hardee's, Ramada Inn, Sub Station II
	Lodg	**W:** DAYS INN Days Inn, Lookout Motel, Ramada I
	AServ	**W:** BP, Chevron, Marathon, Roger Kuchle Garage & Body Shop
	ATM	**W:** Fifth Third Bank, Guardian Savings Bank, Huntington Banks, People's Bank of Northern
	Other	**W:** **AmeriStop (Food Store), Fort Wright Museum**
188AB		188A - Dixie Hwy S, Ft Mitchell 188E Dixie Hwy N, Ft Mitchell
	Gas	**E:** Sunoco* (Kerosene)
	Food	**E:** **Skyline Chili**
		W: Holiday Inn, Indigo Bar & Grill, Pizza Hut, Ramada Inn
	Lodg	**W:** DAYS INN Days Inn, Holiday Inn (see our ad o this page), Ramada Inn
	ATM	**E:** Fifth Third Bank, PNC Bank
		W: Provident Bank (Walgreen's), Star Bank
	Other	**E:** **Kroger Supermarket**
		W: **Thriftway Food & Drug (24 hrs), WalGree**
186		KY 371, Buttermilk Pk
	Gas	**E:** Ashland, BP*(CW) (Kerosene)
		W: BP*(CW), Citgo, Marathon, Shell*, Sunoco*(D)
	Food	**E:** Burbank's Real BBQ & Ribs, Graeter's Ice Cream, Jacqueline's Steak & Ribs, Josh's (Drawbridge Estate), Montoyas Mexican, Orie Wok Chinese
		W: Arby's, Baskin Robbins, Bob Evans Restaurant, Boston Market, Burger King, Butternut Bakery Outlet, Domino's Pizza, Donc Pizza, Dunkin Donuts, Fazoli's Italian Food, Goldstar Chili, JB's Canyon Grill, La Rosa's Itali Marx Hot Bagels, McDonalds, Outback Steakhouse, Papa John's Pizza, Papadinos, P Station, Sandwich Block Deli, Subway
	Lodg	**E:** Cross Country Inn, Drawbridge Estate
	AServ	**E:** Ashland, Citgo
		W: Citgo, Jiffy Lube, Marathon
	ATM	**E:** First National Bank
		W: Fifth Third Bank, PNC Bank, People's Bank, Star Bank, Star Bank
	Other	**E:** **Oldenburg Brewery Tours & Museum, United Dairy Farmers***
		W: **1 Hr Photo (Walgreens), AmeriStop Food Store, Animal Hospital, Butternut Bakery Outlet, Complete Petmart, Crescent Springs Hardware, Crestville Drugs, Drug Emporium**

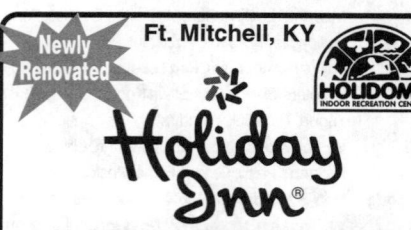

Bold red print shows RV & Bus parking available or nearby

EXIT	KENTUCKY
	Mail Boxes Etc, Revco Drugs, Service Plus IGA (24 hrs), WalGreens
185	Junction I-275, Airport
184AB	184A - KY 236E, Erlanger, 184B - KY 236W, Erlanger
Gas	**E:** AmeriStop*, BP*, Citgo*
	W: Ashland*, JD's Food Mart*, Speedway*
Food	**E:** Double Dragon Oriental Cuisine, Rally's Hamburgers (BP)
	W: Subway (Ashland), Waffle House
Lodg	**W:** Comfort Inn, Days Inn, Econolodge (see our ad on this page), HoJo Inn
Med	**W:** ✚ Hospital
ATM	**E:** Heritage Bank
Other	**E:** Cinderella Laundry, Super Bowl Bowling Lanes, US Post Office
182	KY 1017, Turfway Road

EXIT	KENTUCKY
Gas	**E:** BP*
Food	**E:** Blimpie's Subs, Frisch's Big Boy, Krispy Kreme, Lee's Famous Recipe Chicken, Ryan's Steakhouse
	W: Applebee's, Cracker Barrel, Fuddrucker's, Italianni's, Longhorn Steakhouse, McDonalds, Ming Garden, O'Charley's, Rafferty's, Shells Seafood, Steak 'N Shake, Tumbleweed, Wendy's
Lodg	**E:** AAA Courtyard by Marriott (see our ad on this page), Fairfield Inn, Signature Inn
	W: Commonwealth Hilton, Hampton Inn, Studio Plus
Med	**W:** ✚ Hospital
ATM	**E:** Crawford Insurance
	W: Meijer, Sam's Club, Star Bank
Other	**E:** 1 Hr Photo, Eye World, Family Fun Center, Pharmacy (Thriftway), Pro Golf Discount, Thriftway Marketplace
	W: Advance-U-Cash, Meijer Food Store (24 hrs), Sam's Club, Turfway Park (Horse Races), Wal-Mart (24 hrs)
181	KY 18, Florence, Burlington
TStop	**E:** 76 Auto/Truck Plaza*(LP)(SCALES)
Gas	**E:** Shell(CW), Speedway*, Swifty*
	W: BP*(D)(CW)(LP), Citgo*
Food	**E:** Fillmore's Dairy Hut, Restaurant (76 Auto/Truck Stop), Waffle House
	W: Espress-way Drive Thru Coffee Bar, Hardee's, La Rosa's, Lonestar
Lodg	**E:** Cross Country Inn, Florence Best Western
AServ	**E:** LTD Mobile Glass Auto Glass Specialist
	W: Airport Ford, Dodge Land, Mazda Subaru, Nissan Towne, Precision Tune & Lube, Pro Care (BP), Suber's Auto Repair, Toyota Towne
TServ	**E:** 76 Auto/Truck Plaza
TWash	**E:** 76 Auto/Truck Plaza
ATM	**W:** Bank One, Fifth Third Bank, Heritage Bank
Other	**E:** AmeriStop*, Angel Animal Hospital, Classic Car Wash, Fundome Skating Rink, Speed Queen Coin Laundry
	W: K-Mart (Pharmacy), World of Sports Mini Golf
180	U.S. 42, U.S. 127, Union, Florence 180A - Mall Rd
Gas	**E:** BP*(D)(CW)(LP), Speedway*(D)(LP)
	W: Chevron*, Shell*(CW)
Food	**E:** Bar-B-Q Pit, Bob Evans Restaurant, Burger King, Captain D's Seafood, Frisch's Big Boy, Jalepeno's, Long John Silvers, Madd Anthony's, McDonalds, Peking Palace Chinese, Penn Station, Pizza Hut, Rally's Hamburgers, Red Lobster, Skyline Chili, Sub Station II, Subway, Warm Ups Sports Cafe, Wendy's
	W: Arby's, Flo's Hot Dogs, KFC, Little Caesars Pizza, Perkins Family Restaurant (24 hrs), Ponderosa, Waffle House, White Castle Restaurant
Lodg	**E:** Knight's Inn, Motel 6, Ramada Inn, AAA Super 8 Motel, The Wildwood Inn
	W: Budget Host Inn, Holiday Inn (see our ad on this page)
AServ	**E:** Jim's Auto Service, Mr. Transmission, Rockcastle Olds, Cad, GM
	W: Car X Muffler & Brakes, Midas Muffler & Brakes, NTB, Quick Stop Oil Change
ATM	**E:** Fifth Third Bank
Other	**E:** Check Advance, Classic Car Wash, Fantasy Frontier Mini Golf, Go Karts, Lazer Tag, The Auto Bath Self Serve Car Wash
	W: Ameristops Food Store, Florence Animal Clinic
178	KY 536, Mt Zion Road

Bold red print shows RV & Bus parking available or nearby

EXIT KENTUCKY

Gas E: Shell*(D)(CW)

(177) KY Welcome Center - RR, Phones, Picnic, Vending

(176) **Rest Area - Picnic, RR, Phones, Vending (Northbound)**

175 KY 338, Richwood
- **TStop** E: **TravelCenters of America**(SCALES)
- **FStop** E: **Pilot Travel Center***(SCALES)
  W: **Pilot Travel Center***(SCALES)
- **Gas** W: BP*, Shell*(D)(CW)
- **Food** E: Arby's, Burger King, **Country Pride Restaurant (TC of America)**
  W: **McDonalds**, Snappy Tomato Pizza (BP), **Subway (Pilot)**, Waffle House, Wendy's
- **Lodg** W: Days Inn, Econolodge
- **TServ** E: **TravelCenters of America**
- **ATM** W: Huntington Banks
- **Parks** W: **Big Bone Lick State Park**

173 Jct I-71 South, Louisville

171 KY 14, KY 16, Walton, Verona
- **TStop** W: **Flying J Travel Plaza***(LP)(SCALES) **(RV Dump, Kerosene)**
- **Gas** E: BP*
- **Food** W: **Thads Restaurant (Flying J TS)**
- **AServ** E: Citgo
- **TServ** W: **STI Towing and Truck Repair, Truck Lube Center (24 hrs)**
- **TWash** W: **Magnum Truck Wash**
- **RVCamp** W: **Camping, Delightful Days RV Center**
- **ATM** W: Flying J Travel Plaza, The Bank of Kentucky
- **Other** W: **CB Shop**

(168) **Weigh Station (Southbound)**

166 KY 491, Crittenden
- **Gas** E: 76, Ashland*, BP*
  W: Chevron* (24 hrs), Shell*
- **Food** E: A & W Drive-In (Ashland), Chester Fried Chicken (BP), Taco Bell (Ashland)
  W: **B & E's Log Cabin Restaurant**, Subway (Shell)
- **AServ** E: 76
- **RVCamp** E: **KOA Campground**(LP)
- **ATM** E: Eagle Bank, Grant County Deposit Bank
- **Other** E: **Kentucky State Patrol Post**

159 KY 22, Dry Ridge, Owenton
- **Gas** E: Ashland*, BP*, Shell*(D)(CW)
  W: I-75 Gas, Marathon*
- **Food** E: Arby's, B-52 BBQ, Burger King (BP), Cozy Cottage Restaurant, KFC, McDonalds, Taco Bell, Waffle House, Wendy's
  W: Shoney's, The Country Grill
- **Lodg** E: Dry Ridge Motor Inn, **Super 8 Motel**
  W: Big $avings Motel
- **AServ** W: Dry Ridge Toyota, I-75 Gas
- **ATM** E: Eagle Bank
- **Other** E: **Kentucky State Patrol Post**
  W: **Dry Ridge Outlet Center, Sav-A-Lot Grocery**

154 KY 36, Owenton, Williamstown
- **Gas** E: Ashland*(D)(CW) (Kerosene), Citgo(D)
  W: 76, BP*
- **Food** W: **The Copper Kettle (HoJo Inn)**, Violet's Restaurant (Days Inn)
- **Lodg** W: (AAA) Days Inn, ◆ **HoJo Inn**
- **AServ** E: Citgo
  W: 76
- **Med** E: ✚ **Hospital**

EXIT KENTUCKY

- **ATM** E: Ashland
- **Parks** E: **Kincaid Lake State Park**
- **Other** E: **Corinth Lake (10.5 Miles), Tourist Information**

144 KY 330, Corinth, Owenton
- **TStop** E: **Marathon***
- **Gas** E: Ashland*(D)(LP) (Kerosene), Taylor's Grocery*(LP)
  W: BP*
- **Food** E: **Noble's (Marathon)**
  W: **Free Way Restaurant**, Gameroom Restaurant
- **Lodg** W: K-T Motel
- **TServ** E: **Sechrest Garage (Marathon)**
- **ATM** E: Eagle Bank, Marathon

136 KY 32, Sadieville
- **Gas** E: BP
  W: Chevron
- **AServ** W: Chevron

(131) **Weigh Station (Northbound)**

129 KY 620, Delaplain Road
- **TStop** E: **Pilot Travel Center***(SCALES) **(Showers)**
  W: **Speedway***(LP)(SCALES) **(Kerosene)**
- **Gas** W: Shell*
- **Food** E: Grandma's Kitchen, **Subway (Pilot TS, 24 hrs)**
  W: **Hardee's (Speedway TS)**
- **Lodg** E: (AAA) **Days Inn**, (AAA) Motel 6
- **TServ** E: **American Eagle**(SCALES) **(Pilot TS, 24 hrs)**
- **TWash** E: **Pilot Travel Center**
- **ATM** W: **Speedway**

(127) **Rest Area - RR, Vending, Phones, Dog Walk (Both Directions)**

EXIT KENTUCKY

126 U.S. 62, to U.S. 460, Georgetown, Cynthiana
- **Gas** E: Standard*(LP) (24 hrs)
  W: BP*(CW), Marathon*, Shell*(CW)
- **Food** E: Frisch's Big Boy, **McDonalds**
  W: **Cracker Barrel**, Fazoli's Italian Foo[d], **KFC**, Shoney's, Waffle House
- **Lodg** E: EconoLodge (see our ad on this page)
  W: ◆ Holiday Inn Express, ◆ Shoney's Inn, **Super 8 Motel**
- **Other** W: **Factory Stores of America - Georgetow[n] (see our ad on this page)**

125 U.S. 460, Georgetown, Paris
- **Gas** E: BP*(D)(CW), Shell*(LP)
  W: Dairy Mart*, Swifty*
- **Food** E: Flag Inn Restaurant
  W: Dairy Queen, Heavenly Ham, Little Caesa[rs] Pizza, Long John Silvers, Reno's Roadhouse, Taco Bell, Wendy's
- **Lodg** E: **Econolodge**, Flag Inn
  W: Motel
- **AServ** E: Georgetown Auto Repair
  W: Exhaust Pro, Speedy Lube
- **ATM** E: Shell
  W: Farmer's Bank Square
- **Other** W: **Dr. Clark Cleveland - Veterinarian, K-Mart, Pharmacy (K-Mart), Pharmacy (Winn Dixie), Winn Dixie Supermarket**

120 KY 1973, Ironworks Pike
- **RVCamp** E: **Kentucky Horse Park Campground (.6 Miles)**
- **Med** W: ✚ **Hospital**

118 Junction I-64 West, Frankfort, Louisville

115 KY 922, Bluegrass Pkwy, Lexingto[n]
- **FStop** E: **Exxon***(D)(CW)
- **Gas** E: Shell*(CW)(LP)
  W: Chevron*
- **Food** E: **Cracker Barrel**, McDonalds, Subwa[y] (Shell), Waffle House, Wyndham Garden Hote[l]
  W: Denny's, The Mansion Restaurant, The Pos[t] Restaurant (Holiday Inn)
- **Lodg** E: Knights Inn, La Quinta Inn, Wyndham Gard[en] Hotel
  W: Holiday Inn, Marriott
- **AServ** W: Chevron
- **Other** E: **Kentucky Horse Park**
  W: **Griffin Gate Golf Club (Marriot)**

113 U.S. 27, U.S. 68, Paris, Lexington (Univ. Of KY)
- **Gas** E: BP*(D)(LP), Marathon*(LP), SuperAmerica*(LP) (Kerosene)
  W: Chevron*(D)(LP), Shell*
- **Food** E: Sigee's Restaurant (Harley Hotel), Waffle House
  W: Fazoli's Italian Food, Howard Johnson, Subway (Chevron)
- **Lodg** E: Harley Hotel
  W: **Days Inn**, Howard Johnson
- **AServ** E: BP
- **Med** W: ✚ **Hospital**
- **ATM** E: BP, SuperAmerica, Whitaker Bank
  W: ATM (Shell)
- **Other** E: **Joyland Bowl, Tourist Information**
  W: **Rupp Arena, Tourist Information**

111 Junction I-64 East, Winchester, Ashland

XIT KENTUCKY

0 U.S. 60, Lexington
- **FStop** W: SuperAmerica*(D)
- **Gas** W: Shell*(CW), Thornton's*(LP) (Kerosene)
- **Food** W: Arby's, Bob Evans Restaurant, Cracker Barrel, McDonalds, Shoney's, Waffle House, Wendy's
- **Lodg** W: AAA Best Western, AAA Comfort Inn, ◆ Hampton Inn, HoJo Inn, ◆ Holiday Inn Express, Microtel, Motel 6, ◆ Super 8 Motel, ◆ Wilson Inn
- **ATM** W: Shell, Thornton's

08 Man O' War Blvd
- **Med** W: ✚ Hospital

04 KY 418, Athens, Lexington
- **Gas** E: Exxon*(D)
  W: BP*(D)(CW), Shell*, SuperAmerica*(LP) (Kerosene)
- **Food** E: Baskin Robbins (Exxon), Dunkin' Donuts (Exxon), Waffle House, Wendy's (Exxon)
  W: Jerry's Restaurant, Subway (SuperAmerica), Taco Bell (Super America)
- **Lodg** E: ◆ Comfort Suites (see our ad on this page), DAYS INN ◆ Days Inn, Econolodge (see our ad on this page), ◆ Holiday Inn
- **Med** W: ✚ Hospital
- **ATM** W: Shell, SuperAmerica

9 N. U.S. 25, N. U.S. 421, Clays Ferry

7 South U.S. 25, South U.S. 421
- **TStop** E: Clays Ferry Travel Plaza*(SCALES) (Exxon, Showers)
- **Food** E: Restaurant (Clays Ferry TS), Subway (Clays Ferry TS)
- **TServ** E: Clays Ferry Travel Plaza
- **RVCamp** W: Clays Ferry Landing Campground & RV Park (1.5 Miles)

5 KY 627, Winchester, Boonesborough
- **FStop** W: Shell*(LP)
- **Gas** E: BP*(D) (Kerosene)
- **Food** E: Blimpie's Subs (BP), McDonalds
- **ATM** E: BP
  W: Shell
- **Parks** E: Boonesborough State Park (5 Miles)
- **Other** W: Whitehall State Historic Site (2 Miles)

0BA U.S. 25, U.S. 421, Richmond
- **Gas** E: Shell* (24 Hrs)
  W: BP*(D) (Kerosene), Citgo*, Exxon*, Marathon*(LP), Pennzoil Lube*, Shell*(D)
- **Food** E: Cracker Barrel, Western Sizzlin'
  W: Arby's (Exxon), Dairy Queen, Dunkin' Donuts (Citgo), Frisch's Big Boy, Hardee's, Pizza Hut, Waffle House
- **Lodg** E: Best Western, HoJo Inn, Motel 6, Roadstar Inn
  W: DAYS INN Days Inn, Super 8 Motel
- **AServ** W: NTB, Pennzoil Lube
- **RVCamp** W: Interstate RV Outlet
- **ATM** E: Shell
  W: Bank One, Citgo
- **Other** W: $uper-$ave Food Stores, Auto Wash, Flerlage Marine

37 KY 876, Lancaster, Richmond (Eastern Kentucky University)
- **Gas** E: BP*(D)(CW), Chevron*(LP), Citgo(D) (Kerosene), Dairy Mart*, Shell*(D), Speedway*(D)(LP), SuperAmerica*(D)(LP)
  W: BP*(LP)

EXIT KENTUCKY

- **Food** E: Arby's, Bojangles, Burger King, Casa Cafe Mexican, Denny's, Dunkin' Donuts, Fazoli's Italian Food, Hardee's (Playground), Krystal, Lydia's The Landing Rest., McDonalds, Papa John's Pizza, Pizza Hut, Rally's Hamburgers, Shoney's, Subway, Waffle House, Wendy's, Wok 'n Go Chinese
  W: Fat Jack's Catfish, Chicken, & Ribs
- **Lodg** E: AAA Econolodge, AAA Holiday Inn, Quality Quarters Inn
- **AServ** E: BP, Citgo, Goodyear Tire & Auto, University Tire Center
  W: BP
- **Med** E: ✚ Hospital
- **ATM** E: Bank One, People's Bank of Madison County, Shell, SuperAmerica
- **Other** E: By-Pass Animal Clinic, Madison Optical Co., Paradise Pets, Super One Foods, Tourism Center

(83) Rest Area - Picnic, RR, HF, Phones (Both Directions)

77 KY 595, Berea (Berea College)
- **Gas** W: Shell*
- **Food** W: Denny's
- **Lodg** W: DAYS INN ◆ Days Inn
- **Med** E: ✚ Hospital
- **ATM** W: Shell
- **Other** E: Tourist Information
  W: Mini Golf (Days Inn, Open to Public)

76 KY 21, Berea
- **TStop** W: Spur*
- **Gas** E: BP(CW), Citgo*, Dairy Mart*, Shell*, Speedway*
  W: BP*(LP) (Kerosene), Chevron*(LP) (24 Hrs), Marathon*
- **Food** E: Arby's, Burger King (Shell, Playground), Columbia Steakhouse, Dairy Queen (Citgo), Dinner Bell Restaurant, Hometown Cafeteria, KFC, Long John Silvers, Mario's Pizza, McDonalds (Playground), Papa John's Pizza, Pizza Hut, Sweet Betty's Restaurant, Wendy's
  W: China Sea Restaurant, Lee's Famous Recipe Chicken, Pantry Family Restaurant, Spur Restaurant (Spur TS)
- **Lodg** E: AAA Holiday Motel, AAA Howard Johnson, AAA Super 8 Motel
  W: AAA Econolodge, Mountain View Motel
- **AServ** E: Arvin Muffler & Pipes, BP, Jenning's Auto Parts
  W: Madison County Dodge Chrysler Plymouth
- **TServ** W: Spur Truck Stop
- **RVCamp** W: Covered Wagon RV Park-Full Hookups,

EXIT KENTUCKY

Oh Kentucky RV Park (.5 Miles), Walnut Meadow Campground (.5 Miles)
- **Med** E: ✚ Hospital
- **ATM** E: Berea National Bank, People's Bank of Madison County
- **Other** E: BigValu Discount Foods
  W: Berea True Value Hardware, Southern States Farm Home Garden(LP)

62 U.S. 25 to KY 461, Renfro Valley, Mt. Vernon
- **TStop** E: Derby City Truck Stop (24 Hrs)
- **Gas** W: Ashland*, BP*, Shell*(LP)
- **Food** E: Hardee's, Restaurant (Derby City Truck Stop)
  W: Apollo Pizza, Blimpie's Subs (BP), Dairy Queen, Denny's, Rock Castle Steakhouse, Subway (Shell), Wendy's
- **Lodg** E: Renfro Valley Motel (Renfro Valley RV Park)
  W: DAYS INN ◆ Days Inn, ◆ Econolodge, Super 8
- **TServ** E: Derby City Truck Stop
- **RVCamp** E: KOA Kampground (1.5 Miles), Renfro Valley RV Campground
- **Med** W: ✚ Hospital
- **ATM** E: Derby City Truck Stop
  W: BP
- **Other** W: Big South Fork National River Rec. Area, Lake Cumberland

59 U.S. 25, Livingston, Mt. Vernon
- **FStop** E: Burr Hill*(D)
- **Gas** E: BP*
  W: Peg's Food Mart*(LP)
- **Food** E: Gean's Restaurant, Pizza Hut, Restaurant (Best Western)
- **Lodg** E: Best Western
  W: Holiday Motel
- **AServ** W: Hwy 25 Cleanup Shop
- **RVCamp** E: Nicely Camping (.6 Miles, Full Hookups)

49 KY 909 to US 25, Livingston
- **TStop** W: 49er Fuel Center* (Shell, 24 hrs)
- **Food** W: Kentucky Grub Diner (49er Fuel Center, 24 hrs)

(43) Weigh Station (Both Directions)

41 KY 80, Somerset, London, Daniel Boone Parkway, Toll Rd.
- **TStop** W: London Auto/Truck(SCALES) (BP)
- **FStop** W: Citgo*
- **Gas** E: Chevron* (24 hrs), Marathon
  W: Chevron*, Citgo*(D)(LP), Shell*(CW)
- **Food** E: Arby's, KFC, Rax Gold Star Chili, Zachary's Family Restaurant
  W: Homestyle Cookin, Jerry's Restaurant, Kountry Kookin, Long John Silvers, McDonalds(D) (Chevron), Shiloh Roadhouse, Subway (Shell), Wendy's
- **Lodg** E: AAA Best Western, DAYS INN ◆ Days Inn, Holiday Inn Express, ◆ Sleep Inn
  W: Budget Host Inn
- **AServ** E: Marathon
- **RVCamp** W: Westgate RV Camping (.3 Miles)
- **Med** E: ✚ Hospital
- **ATM** W: Dog Patch Trading Post
- **Parks** E: Levi Jackson State Park
- **Other** E: Kentucky State Police
  W: Dog Patch Trading Post(LP), Tourist Information, Wilderness Rd Info Center

38 KY 192, London, to Daniel Boone Pkwy
- **FStop** E: Shell*(LP)

EXIT KENTUCKY

Gas E: BP*, SuperAmerica*
W: Shell*, Texaco*
Food E: Arby's (Shell), Charcoal House (Ramada Inn), Fazoli's Italian Food, Frisch's Big Boy, Hardee's, Rally's Hamburgers, Rock-A-Billy Cafe
Lodg E: AAA Comfort Suites, ◆ Hampton Inn, AAA Ramada Inn (see our ad on this page)
AServ E: Cook Bros. Auto Parts, Cook Tire Inc., Floyd's Garage, Hesco Muffler Centers, Wal-Mart
TServ E: Martin's Truck Parts & Sales
RVCamp E: Levi Jackson State Park Camping (4 Miles)
Med E: ✚ Hospital
ATM E: Shell
W: Shell
Parks E: Levi Jackson State Park
Other E: Regency Seven (Cinema), US Post Office, Wal-Mart Supercenter (w/pharmacy, 24 hrs), Westside Market
W: Lake Cumberland London Dock (21 Miles), Laurel Lake Holly Bay (18 Miles), Laurel River Lake

29 U.S. 25, U.S. 25 East, Corbin, Barbourville (Union College)
TStop E: 76 Auto/Truck Plaza*(SCALES) (24 hrs), Pilot Travel Center(SCALES) (Kerosene)
FStop E: Speedway*(LP)(SCALES)
Gas E: BP, Exxon*
W: Amoco*(D) (24 hrs), Chevron*(D), Shell*(CW)
Food E: Burger King (Playground), Restaurant (76 Auto/Truck Stop), Shoney's, Subway (Pilot), Western Sizzlin', Western Steer Family Steakhouse
W: Cracker Barrel, Krystal, Sonny's Real Pit Bar-B-Q
Lodg E: Quality Inn, ◆ Super 8 Motel
W: ◆ Budgetel Inn (see our ad on this page), Fairfield Inn, ◆ Knights Inn
AServ E: BP
TServ E: 76 Auto/Truck Plaza, Lens Truck Garage (Pilot Truck Stop)
TWash E: Blue Beacon Truck Wash (76 Auto/Truck Stop), Jiffy Truck Wash
ATM W: Shell
Other E: Antique Mall

25 U.S. 25 West, Corbin
Gas E: Shell*, Speedway*
W: BP*(D), Shell*
Food E: Corbin Burger House, Jerry's Restaurant
W: Arby's, Rino's Roadhouse
Lodg E: ◆ Country Inn & Suites by Carlson, DAYS INN ◆ Days Inn, Red Carpet Inn
W: AAA Best Western, ◆ Holiday Inn
AServ E: Complete Auto & Truck Repair
Parks E: Levi Jackson State Park (15 Miles)
W: Cumberland Falls State Resort Park (15 Miles)
Other W: Grove Recreation Area & Marina (10 Miles)

15 U.S. 25 West, Williamsburg
Gas W: Chevron, Shell*
AServ W: Chevron, Jayne's Tire Company
Other W: Grove Recreation Area & Marina (15 Miles)

11 KY 92, Williamsburg (Cumberland College)
FStop W: Mapco Express*(LP)(SCALES) (RV Dump) (see our ad this page)
Gas E: BP*(D), Dairy Mart*, Direct, Exxon*(LP), Shell*
W: Amoco*, Shell*(D) (Kerosene)
Food E: Arby's, Athenaeum (Cumberland Lodge), Dairy Queen, Hardee's (Playground), KFC, McDonalds (Playground), Pizza Hut, Subway (BP)
W: B J's Restaurant (24 hrs), Baskin Robbins (Mapco Express), Long John Silvers, Mi Pueblo, Restaurant (Williamsburg Motel), Wendy's (Mapco)

EXIT KENTUCKY

EXIT KENTUCKY/TENNESSEE

Lodg E: Cumberland Lodge Conference Center & Museum (Marriot) (see our ad on this page), ◆ Holiday Inn Express
W: Best Western, Williamsburg Motel
AServ E: Kain Chevrolet, Kain Chrysler Plymouth Dodge Geo, NAPA Auto Parts, Paul Steely Ford, Tri County Transmission
Parks W: Big South Fork NRRA
Other E: Car Wash, Riverside Produce, Save-A-Lot Food Stores

(2) KY Welcome Center - Picnic, RR, HF, Phones (Northbound)

↑KENTUCKY

↓TENNESSEE

(161) TN Welcome Center - Full Facilities (Southbound)

160 U.S. 25 West, Jellico
Gas E: Amoco*, Citgo, Exxon*(D), Fina*, Texaco
W: Shell
Food E: Billy's Restaurant (Billy's Motel), Johnny B's Reataurant (Jellico Motel), Subway (Exxon)
W: Arby's (Shell), Gregory's (Days Inn), Hardee's, Heritage Pizza, Taco Village
Lodg E: Billy's Motel, Jellico Motel
W: ◆ Best Western, DAYS INN ◆ Days Inn (see our ad on this page)
AServ W: B&B Auto Supply, Danny's Auto Repair
RVCamp W: Indian Mountain State Park (3 Miles)
Med W: ✚ Hospital
ATM W: Shell
Parks W: Indian Mountain State Park (3 Miles)
Other E: Stuckeys Express
W: Flowers Bakery Thrift Store, Jack's Car

EXIT	TENNESSEE
	Wash, **Sheriff's Dept., Sunbeam Bakery Thrift Store, Tourist Information**
4	Stinking Creek Road
1	TN 63 W, Oneida, Huntsville
FStop	**W:** **Texaco Truck/Auto Center***
Gas	**E:** BP*
	W: Exxon*
Food	**E:** Perkins Family Restaurant, Stuckey's (BP)
	W: **Dairy Queen (Texaco)**
Lodg	**W:** ◆ **Comfort Inn**
Other	**W:** **Big South Fork Recreation Area, Fireworks Superstore, Homemade Mountaintop Fudge**
4	N. U.S. 25 West, E. TN 63, Caryville, La Follette, Jacksboro
Gas	**E:** Exxon*, Shell*(LP)
	W: Amoco*
Food	**E:** Front Door Bar & Grill, The Woods Steakhouse, Tracker Christmas Inn Restaurant, Waffle House
	W: Pizza Palace, Scotty's 55 Cent Hamburgers, Shoney's
Lodg	**E:** Cove Lake Center, Family Inns of America, (AAA) Hampton Inn, (AAA) Holiday Inn, Lakeview Inn
	W: (AAA) Budget Host Inn
AServ	**E:** Talley Tires
Med	**E:** ✚ **Hospital**
ATM	**E:** Exxon, First State Bank, Shell
Parks	**E:** **Cove Lake State Park (.5 Miles)**
Other	**E:** **Fireworks Supercenter, Hill Billy Bob's Cross Fireworks, Laundromat (Cove Lake Center), Russel's Market & Deli, Thunder Mountain Fireworks**
	W: **Caryville Surplus Sales, Fire Station, US Post Office**
30)	**Weigh Station (Both Directions)**
9	U.S. 25 W South, Lake City
Gas	**W:** BP, Citgo* (Kerosene), Exxon*(D), Phillips 66*, Shell*, Texaco
Food	**W:** Cottage Restaurant, (Cracker Barrel logo) **Cracker Barrel,** Domino's Pizza, KFC, Mary J. & Ray's Unicorn Cafe, **McDonalds (Playground), The Lamb's Inn Motel**
Lodg	**W:** Blue Haven Motel, (Days Inn logo) (AAA) Days Inn (see our ad on this page) (Playground), Lake City Motel, **The Lamb's Inn Motel**
AServ	**W:** Auto Salvage Sales, BP
RVCamp	**W:** **Mountain Lake Marina**

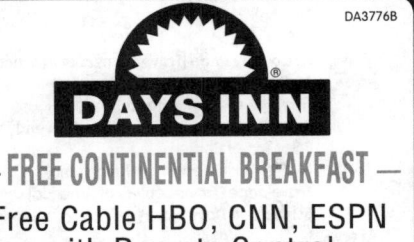

EXIT	TENNESSEE
ATM	**W:** McDonalds
Other	**E:** **Pernell Dist., Inc.(LP)**
	W: **Lamb's Antique Car Museum, Uncle Sam's Fireworks**
128	U.S. 441, Lake City
Gas	**E:** BP*, Phillips 66*
RVCamp	**E:** **Mountain Lake Marina & Campground (1.3 Miles)**
Parks	**E:** **Norris Dam State Park (2.3 Miles)**
Other	**W:** **Hillbilly Market & Deli**
122	TN 61, Norris, Clinton
FStop	**W:** **Citgo*(SCALES)**
Gas	**E:** Phillips 66*, Shell*(LP)
	W: Exxon* (Kerosene), Fina*(LP), **Git 'N Go*(D) (Playground)**
Food	**W:** Git 'N Go Deli & Ice Cream, Golden Girls Restaurant, Hardee's (Playground), Krystal, McDonalds (Playground), Waffle House, Wendy's
Lodg	**W:** (AAA) Best Western, Days Inn, Jameson Inn, ◆ **Super 8 Motel**
RVCamp	**E:** **Fox Inn Campground(LP), TVA Public Campground**
ATM	**E:** Shell
	W: People's National Bank, Playground
Other	**E:** **Twin Gables Antique Mall**
117	Route 170, Raccoon Valley Road
TStop	**E:** **Raccoon Valley Truckstop (Phillips 66)**
FStop	**E:** **Delta Express***
Gas	**E:** BP*(CW), Shell*
Food	**E:** **Travel House Restaurant (24 Hrs, Raccoon Valley TS)**
Lodg	**W:** Valley Inn
TServ	**E:** **Raccoon Valley TS**
TWash	**E:** **Raccoon Valley TS**
RVCamp	**W:** **KOA Kampground (70 ft Pull Throughs), Yogi Bear Camp Report(LP)**
112	TN 131, Emory Road, Powell
Gas	**E:** BP*(CW), Chevron*(LP), Pilot*(LP) (Kerosene)
	W: Phillips 66* (Kerosene), Shell*(LP)
Food	**E:** Aubrey's Restaurant, Buddy's Bar-B-Q (BP), **Dairy Queen (Pilot),** McDonalds (Playground), Pizza Inn Express (Chevron), Steak 'N Shake, TCBY (BP), **Taco Bell (Pilot),** Wendy's
	W: **Hardee's,** Waffle House
Lodg	**E:** Holiday Inn Express
	W: ◆ **Comfort Inn**
ATM	**W:** Phillips 66
Other	**E:** **Ingles**
	W: **Mayes Aviation Airplane Rides & Charter Flights**
110	Callahan Dr
Gas	**E:** Phillips 66*
	W: Amoco*
Food	**E:** The Courtyard Buffet (Quality Inn), The Pizza Eatery
Lodg	**E:** ◆ **Quality Inn, Roadway Inn**
	W: Scottish Inns
ATM	**W:** Amoco
Other	**W:** **United Parcel Service**
(108)	South I-275, East I-640, Knoxville, Asheville (Unnumbered Exit)
108	Merchant Dr
Gas	**E:** BP*(D)(CW)(LP), Citgo*(D), Pilot*(LP), Texaco*(D)

EXIT TENNESSEE

W: Amoco*, Conoco* (Kerosene), Exxon*(D), Pilot*, Shell*

Food E: Applebee's, **Cracker Barrel**, Denny's (Best Western), El Chico Mexican, Monterrey Mexican, O'Charley's, Ryan's Steakhouse, Sagebrush Steakhouse, Sonic, Sugarbakers, Waffle House
W: Bamboo Stix, Baskin Robbins (Pilot), Bob Evans Restaurant, Burger King, Captain D's Seafood, Darryl's Restaurant, Great American Steak & Buffet Co., Mandarin House Chinese Buffet, McDonalds (Playground), Nixon's Deli, Outback Steakhouse, Red Lobster, Subway, T. Ho Vietnamese Oriental Restaurant, TCBY (Exxon), Waffle House

Lodg E: AAA Best Western, AAA Comfort Inn (see our ad this page), ◆ Days Inn, ◆ Hampton Inn, ◆ Ramada Suites, AAA Sleep Inn
W: AAA Econolodge, Family Inns of America, AAA La Quinta Inn, ◆ Red Roof Inn, ◆ **Super 8 Motel (see our ad on this page)**

AServ E: Dunlop, Instant Oil Change
W: **K-Mart**, Pennzoil Lube, Western Auto

RVCamp E: **Campers Corner Too Sales & Services**

Med E: **✚ Prompt Care Medical Center (Walk-In Clinic)**

ATM E: Citgo, Pilot
W: First American, First Tennessee Bank, Home Federal Bank, Pilot, SunTrust

Other E: Car Wash, **Fire Dept, Ingles Supermarket, North Knox Veterinary Clinic, Revco, The Book Case**
W: **Baptist Bookstore, Book Warehouse, K-Mart, Norwood Pharmacy, Revco Drugs,** The Clean Machine Car Wash, **Walgreens, Winn Dixie Supermarket (Pharmacy, One Hr Photo)**

Note: I-75 runs concurrent below with I-640. Numbering follows I-640.

3 TN 25W, Gap Road, Clinton (No Reaccess)

1 TN 62, Western Ave

Gas N: Racetrac*, Texaco*
S: Shell(D)

Food N: Baskin Robbins, Central Park Hamburgers, KFC, Little Caesars Pizza, Long John Silvers, McDonalds (Playground), Panda Chinese Restaurant, Shoney's, Subway, Taco Bell, Wendy's
S: Dad's Donuts & Delights, Domino's Pizza, Hardee's, Krystal, Tracey's Restaurant Home Cookin

AServ S: Advance Auto Parts, Mighty Muffler, Shell(D)

ATM N: First American Bank, **Kroger Supermarket (24 Hrs)**, SunTrust
S: First Tennessee Bank

Other N: **Kroger Supermarket (24 Hrs), Revco, WalGreens**
S: Car Wash, **Cokesbury Books & Church Supplies, Pack n Ship Mail Center,** Super Wash House, **US Post Office**

Note: I-75 runs concurrent above with I-640. Numbering follows I-640.

Note: I-75 runs concurrent below with I-40. Numbering follows I-40.

383 U.S. 11, U.S. 70, Papermill Dr

FStop S: **Pilot*(LP) (Kerosene)**

Gas N: Spur*
S: BP*(CW), Citgo*

Food S: Waffle House

Lodg N: ◆ Holiday Inn
S: ◆ Super 8 Motel

ATM S: Pilot

Other N: **Fire Dept**

380 U.S. 11, U.S. 70, West Hills

Gas S: BP(CW), Conoco*(LP), Pilot*(LP)

Food S: Big Orange Sports Bar & Deli, Cancun Mexican

EXIT TENNESS

Restaurant & Cantina, Chili's, Copper Cellar Seafood & Steaks, Cozymel's, KFC, Korean Restaurant, Pizza Palace Theater & Entertainment Co., Romano's Macaroni Grill, Steak-Ou Char-Broiled Delivery, Subway, Taco Bell

Lodg S: AAA Comfort Hotel (see our ad on this pag Howard Johnson

AServ S: BP, **K-Mart**, NTB

ATM S: Conoco, First American Bank, First Tenness Bank, SunTrust

Other S: **Fire Dept, Food Lion, Jeffrey Photo Lab (1 Photo), K-Mart, Kingston Pike Pet Hospital, Mail Store & More, State Hwy Patrol, West Town Mall**

379 Walker Springs Road

FStop N: **Citgo*(D)(LP)**

Gas N: Exxon*
S: BP*(D)(CW)(LP), Pilot*

Food N: Subway (Exxon)
S: Buddy's Bar-B-Q, Burger King, Can Ton Restaurant, Huck Finn's Catfish Chicken & Ste Joe's Crab Shack, Kyoto Japanese, Logan's Roadhouse, Mrs Winner's Chicken, Old Coun Buffet, Roger's Place, Shoney's, Time-Out Deli Wendy's

Lodg S: Nation's Best Hotel

AServ N: **Sam's Club, Wal-Mart**
S: Auto Zone Auto Parts, Firestone Tire & Auto Goodyear Tire & Auto, Rice Mitsubishi, Rice Oldsmobile Inc GMC Trucks, Rodgers Cadilla

Med S: **✚ Park Med Ambulatory Care Walk-In Medical Center**

ATM N: Citgo
S: First Tennessee Bank, Pilot, Union Planters Bank

Other N: **Sam's Club, Wal-Mart (24 Hrs)**
S: **Books-A-Million, Joshua's Christian Book Music, & Video, Revco, Winn Dixie Supermarket (24 Hrs, Pharmacy)**

378 Cedar Bluff Road

Gas E: Phillips 66
W: Amoco, Pilot Travel Center

Food E: Applebee's, Bob Evans Restaurant, Corkey' BBQ & Ribs, Denny's, Grady's, Outback
W: Arby's, Burger King, Cracker Barrel, KFC, Long John Silvers, McDonalds, Papa Joh Pizza, Pizza Hut, Subway, Taco Bell, Waffle House, Wendy's

Lodg E: AAA La Quinta, ◆ Red Roof Inn, AAA Signature Inn
W: Holiday Inn Express (see our ad this page), Ramada Inn, Scottish Inns

ATM W: First American Bank

376A TN 162, Oak Ridge (Southbound)

Med N: **✚ Hospital**

376B TN 162 North, Mabry Hood Road, Oak Ridge

374 Route 131, Lovell Road

TStop N: **TravelCenters of America*(SCALES)**
S: **Pilot Travel Center*(SCALES)**

Gas N: Amoco*(D) (24 Hrs), BP*(CW), Citgo, Marathon Texaco*(CW)
S: Speedway*

Food N: **Country Pride (TravelCenters of America), McDonalds (Playground), Taco Bell (TravelCenters of America),** Waffle House
S: Arby's, Krystal (24 Hrs), Shoney's, **Wendy's (Pilot Travel Center)**

Lodg N: AAA **Best Western**, ◆ Knights Inn, **Travelodge (TravelCenters of America)**
S: ◆ **Days Inn**, AAA Motel 6

AServ N: Citgo

TServ N: **TravelCenters of America**

Other N: **All Kreatures Pet Supplies, Etc**

373 Campbell Station Road, Farragut

Gas N: Amoco*, Texaco*(D)
S: BP*(CW), Pilot Travel Center*(D), Speedway*

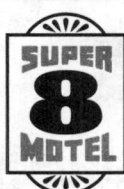

EXIT	TENNESSEE
Food	S: Applecake Tearoom, Applewood, Cracker Barrel, Hardee's
Lodg	N: AAA Comfort Suites, AAA Super 8 Motel
	S: Budgetel Inn, AAA Holiday Inn Express
RVCamp	N: Buddy Gregg Motor Homes
Other	S: Harvest Moon Fresh Fruits & Vegetables
72)	**Weigh Station - both directions**
9	Watt Road
TStop	N: Flying J Travel Plaza*(LP)(SCALES), Knoxville Travel Center, 76*(SCALES) (TFuel-Unocal 76, AFuel-BP)
	S: Petro Travel Plaza*(SCALES)
Food	N: The Cookery (Flying J Travel Plaza)
	S: Burger King (Knoxville Travel Center, 76), Iron Skillet Restaurant (24 Hrs, Petro TS), Perkins (Knoxville Travel Center TS), Pizza Hut (Knoxville Travel Center, 76)
TServ	S: Knoxville Travel Center, 76, Petro TS
TWash	S: Blue Beacon Truck Wash (Petro TS)
Other	N: Fireworks Supermarket
	Note: I-75 runs concurrent above with I-40. Numbering follows I-40.
AB	Junction I-40, Nashville, Knoxville (Northbound)
	U.S. 321, TN 95, Lenoir City, Oak Ridge (Ft Loudon Dam, Great Smoky Mtn. National Park)
FStop	W: Shell* (Kerosene)
Gas	E: Amoco*(LP) (24 Hrs), BP*(D)(CW), Exxon*(D), Phillips 66* (Kerosene), Texaco(D)
	W: Citgo*
Food	E: Baskin Robbins (Exxon), Dinner Bell, KFC, King's Inn, Quincy's Family Steakhouse, Shoney's, Subway (Exxon), Terry's (Crossroads Inn), Waffle House (24 Hrs)
Lodg	E: AAA Crossroads Inn, AAA King's Inn
	W: AAA Econolodge, Ramada Limited
AServ	E: Texaco
	W: Jack Nelson Olds, Pont, Buick, GMC Trucks, Rocky Top Chevrolet
ATM	E: Amoco
Parks	E: Great Smoky Mtns. National Park
Other	E: Fireworks Discounter, Loudon County Visitors Bureau, Tourist Information
	W: Bimbo's Fireworks World (Citgo), Fireworks Factory Outlet, Putt 'N Stuff Mini Golf
	TN 324, Sugar Limb Road
Other	E: River View Golf Course (2.5 Miles)
	TN 72, Loudon
Gas	E: Shell*
	W: Phillips 66*(LP), Shell* (Kerosene)
Food	E: Wendy's (Shell)
Lodg	E: AAA Holiday Inn Express
	W: AAA Knights Inn (see our ad on this page)
RVCamp	W: Express Campground (Water, Electric, & Dump Station)
ATM	W: Shell
Parks	E: Fort Loudon State Park (20 Miles)
Other	E: Sequoia Museum (20 Miles)
	Route 323, Philadelphia
Gas	E: BP* (Kerosene)
Other	E: Crazy Joe's Fireworks (BP)
	W: Cowboy's Dream Ranch
	TN 322, Oakland Road, Sweetwater
Gas	E: Phillips 66* (Dinner Bell)
Food	E: Dinner Bell
RVCamp	W: KOA Kampground (.6 Miles)
Med	E: + Hospital
Other	E: 4 B's Fireworks
	TN 68, Sweetwater, Spring City (Watts Bar Dam)
FStop	W: Exxon* (Kerosene)
Gas	E: Marathon*, Raceway*, Texaco*

EXIT	TENNESSEE
	W: BP*(D), Phillips 66*(CW)
Food	E: Huddle House, McDonalds (Playground)
	W: Blanton's Resturaunt, Blimpie's Subs (Phillips 66), Cracker Barrel, Denny's (Quality Inn)
Lodg	E: AAA Budget Host Inn, AAA Comfort Inn, DAYS INN AAA Days Inn
	W: AAA Quality Inn
TServ	W: 68 Tire & Service Center
RVCamp	W: TN's Largest Flea Market
Med	E: + Hospital
ATM	E: Texaco
Other	E: Joker Joes Fireworks, M-90 Factory Outlet Fireworks
	W: Builderback Animal Clinic, Flea Market Mall
56	TN 309, Niota
TStop	E: Crazy Ed's Fireworks Supercenter*
Food	E: Restaurant (Crazy Ed's TS), Subway (Crazy Ed's TS)
AServ	W: Michael's Wheels & Tires
TServ	E: Jerry's Garage (24 Hr Road Service, Crazy Ed's TS)
RVCamp	E: Country Music Campground
ATM	E: Crazy Ed's Fireworks Supercenter
52	TN 305, Mount Verd Road, Athens
Gas	E: Phillips 66*(D) (Kerosene)
	W: Exxon*(LP)
Lodg	E: Heritage Motel & Campground
	W: ◆ Holiday Inn Express
RVCamp	E: Heritage Motel & Campground, Over-nighter RV Park (.6 Miles)
Other	E: Crazy Ed's Fireworks, McMinn County Living Heritage Museum
49	TN 30, Athens, Decatur (Tennessee Weslyan College)
FStop	E: Texaco*
Gas	E: BP*(D)(CW) (Kerosene), Exxon*(LP) (24 Hrs), Raceway*, Shell*
	W: Phillips 66*(LP) (Kerosene)
Food	E: Applebee's, Burger King (Playground), Clark's Restaurant, Dairy Queen(D)(LP) (Shell), Hardee's, Linda's Country Kitchen, Shoney's, TCBY (Exxon), Waffle House, Wendy's
Lodg	E: DAYS INN AAA Days Inn (see our ad on this page), AAA Homestead Inn, AAA Knights Inn (Playground), AAA Super 8 Motel (see our ad on this page)
	W: AAA Homestead Inn West
AServ	W: Heritage Olds, Pontiac, Buick, Cadillac, GMC Dealer
TServ	E: Roberts Brothers Motors
RVCamp	E: Athens I-75 Campground(LP), KOA Kampground
Med	E: + Hospital
ATM	E: Exxon
Other	E: McMinn County Living Heritage Museum, Tennessee Weslyan College
(45)	**Info Station, Rest Area - RR, HF, Phones, Picnic (Both Directions)**
42	TN 39, Riceville Road
Gas	E: Exxon*(D)
Lodg	E: AAA Relax Inn, AAA Rice Inn
RVCamp	E: Mouse Creek Campground
36	TN 163, Calhoun
33	TN 308, Charleston
TStop	E: Texaco*(SCALES) (Kerosene, FedEx Drop, CB Shop), Mapco Express (see our ad on this page)
Gas	E: Citgo*(LP) (Kerosene)
Food	E: Buffet Style Pizza (Citgo), Texaco Restaurant (Texaco TS)
RVCamp	E: 33 Campground
27	Paul Huff Pkwy, Cleveland
Gas	W: BP*(D), Shell*(LP)
Food	W: Blimpie's Subs (BP), Denny's, Hardee's,

Bold red print shows RV & Bus parking available or nearby

EXIT	TENNESSEE
	Waffle House, Wendy's
Lodg	W: AAA Best Western, AAA Comfort Inn, ◆ Exclusive Quarters, Hampton Inn (see our ad this page), Royal Inn, Super 8 Motel
Med	E: ✚ Cleveland Community Hospital
ATM	W: BP, Shell
Other	E: Tourist Information
	W: Cleveland St. Community Coll.
25	TN 60, Cleveland, Dayton (Lee University, Cleveland St. Comm. College)
Gas	E: Amoco(D), BP*(CW)(LP), Chevron*(D), Citgo*(D)(LP), Racetrac*, Texaco*
	W: Amoco*
Food	E: Big Daddy's Country Buffet, Burger King (Playground), Cracker Barrel, El Meson (Quality Inn), Hardee's, McDonalds (Playground), Roblyn's Steak House, Stadfeld's Family Restaurant, Waffle House
	W: Roy Pepper's
Lodg	E: AAA Colonial Inn, Days Inn AAA Days Inn, AAA Econolodge (see our ad on this page), Heritage Inn, Knight's Inn, Lincoln Inn Motel, AAA Quality Inn, Red Carpet Inn, Travel Inn
	W: ◆ Budgetel Inn, ◆ Holiday Inn
AServ	E: 25th Street Auto Parts, Amoco
Med	E: ✚ Cleveland Community Hospital
ATM	E: Bank of Cleveland, SunTrust
	W: Amoco
Other	E: Car Wash, Cherokee Pharmacy & Medical Supply, Community Animal Hospital, Hiwasee River, Maxwell Munford Home Improvement(LP), Ocoee River, Red Clay State Historic Area, The Laundry Basket "Too", Tourist Information
	W: Roller Coaster, Rolling Hills Golf Course, Sam's Wharehouse Golf
(23)	Truck Inspection Station, Weigh Station (Northbound)
(24)	Weigh Station (Northbound)
20	U.S. 64, Bypass East, Cleveland
Gas	W: Citgo*, Exxon*(LP) (24 Hrs)
Food	W: Stan's Truck Stop(SCALES) (Showers)
Lodg	W: Hospitality Inns Of America
AServ	W: Exxon, Stan's Truck Stop
TServ	W: Stan's Truck Stop
RVCamp	W: KOA Kampground (1 Mile)
Parks	E: Red Clay State Park (16 Miles)
Other	W: Fireworks Supermarket
(16)	Scenic View, Parking Area

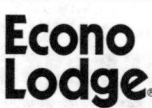

EXIT	TENNESS
	(Southbound)
(13)	Weigh Station, Parking Area - Phones (Southbound)
11	North U.S. 11, East U.S. 64, Ooltew (Southern Adventist University)
FStop	W: Exxon* (Kerosene, FedEx Drop)
Gas	E: Amoco*, Chevron, Citgo* (Kerosene), Racetrac*
Food	E: Arby's, Hardee's, Kreme House Country Cookin', Little Caesars Pizza, McDonalds (Playground), Subway, Taco Bell, Wendy's
	W: TCBY (Exxon), Waffle House
Lodg	W: ◆ Super 8 Motel
AServ	E: Chevron, Ooltewah Auto Center, Sunshine Lube(CW)
ATM	E: Bi-Lo Supermarket, Citgo
Parks	W: Harrison Bay State Park (9 Miles)
Other	E: ACE Hardware, Bi-Lo Supermarket (24 H Pharmacy)
7AB	TN 317, Summit, Collegedale
Gas	W: Texaco*
Food	W: Denny's (Best Western), Waffle House
Lodg	W: Best Western, Comfort Inn, Country Heart Inn, Days Inn Days Inn, Econolodge (see our ad this page), Motel 6
AServ	W: Denton's Garage (24 Hr Wrecker Service)
ATM	W: Texaco
Other	W: Innovative Miniatures Museum, Silverda Confederate Cemetery
5	Shallowford Road
Gas	W: Amoco*, Citgo*(LP), Exxon*, Texaco*
Food	E: Alexander's Restaurant, Arby's, Burger King Central Park Hamburgers, Country Place Restaurant, El Meson, Kanpai of Tokyo, Krysta Outback Steakhouse, Red Lobster, Schlotzkys Deli, Taco Bell, The Acropolis, The Olive Garde
	W: Applebee's, Buckhead Roadhouse, Cancu Mexican, Cracker Barrel, Glen Gene D O'Charley's, Ocean Avenue, Papa John's Pizz Shoney's, Sonic, Subway, Waffle House, Wenc
Lodg	E: AAA Comfort Suites, ◆ Courtyard by Marric
	W: Country Suites by Carlson, Days Inn Days Inn Fairfield Inn, Hampton Inn (see our ad this pag AAA Holiday Inn, AAA Holiday Inn Express, Homewood Suites Hotel, AAA La Quinta Inn, AAA Microtel Inn, Ramada Inn, ◆ Red Roof In ◆ Sleep Inn
AServ	E: Master Care Auto Service
ATM	E: First Tennesee Bank
	W: Citgo, Texaco
Other	E: Barnes & Noble, Bood Browser Christian Bookstore, Hamilton Place Mall, Joshua's Christian Stores, National Knife Museum, Pr Golf Discount, Toys "R" Us, WalGreens (1 Hr Photo), Tenneessee Aquarium (see our ad o this page)
	W: Eyear Optical
4A	Hamilton Place Blvd
Food	E: Grady's Restaurant, Tia's Tex Mex
Med	E: ✚ Physicians Care Walk-In Clinic
ATM	E: Tennessee Valley Federal Credit Union
Other	E: Hamilton Place Mall, Regal Cinemas
4	Junction Hwy 153, Airport, Chickamauga Dam
3AB	Hwy 320 East, E. Brainerd Road
Gas	E: Amoco*(LP) (Kerosene), Exxon*(LP)

Bold red print shows RV & Bus parking available or nearby

KIT — TENNESSEE/GEORGIA

Food E: Baskin Robbins, Subway, TCBY (Exxon)
ATM E: Exxon
Other W: **Animal Hospital Dragon Museum Unique Giftshop**

I-24 West Chattanooga

TN Welcome Center - Picnic, RR, HF, Phones, RV Dump (Northbound)

AB U.S. 41, East Ridge (Northbound)

Gas E: BP*(LP) (ATM), Exxon*
W: Amoco, Conoco*(D)(CW), Texaco*
Food E: **Trip's Seafood**
W: Arby's, Burger King, **Cracker Barrel**, Hardee's, Krystal, Long John Silvers, McDonalds, Porto Fino Italian, Shoney's, Subway, Taco Bell, Uncle Bud's Catfish & Chicken, Waffle House, Wally's Family Restaurant
Lodg E: **Best Western**, Econolodge, **Holiday Inn**, **Ramada**
W: Hospitality Inn of America, Scottish Inns, **Super 8 Motel**, Waverly Motel, World Inn
AServ E: East Ridge Body Shop
W: Amoco
RVCamp E: **Shipp's Jellystone Park* (RV Dump), Shipp's RV Service Center**
W: **Holiday Travel Park**
Other E: **Bi-Lo Supermarket (ATM), Revco Drugs**
W: **Coin Laundry**

↑TENNESSEE

↓GEORGIA

142 **(353)** GA 146, Rossville, Ft. Oglethorpe

Gas E: BP*(LP), Conoco*
W: Exxon*(D) (ATM), Texaco*(LP)
Food E: **Restaurant/Buffet (Howard Johnson)**
Lodg E: **Howard Johnson**
Other W: **Gateway Antique Mall**

(352) **GA Welcome Center - 7-11, Picnic, RR, HF, Phones, RV Dump (Southbound)**

141 **(349)** GA 2, Battlefield Pkwy, Ft. Oglethorpe

FStop E: **Save-a-ton**
Gas E: Exxon*
W: Exxon(D), Racetrac*, Texaco*(LP) (ATM)
Food W: Bar-B-Q Corral, Big O's BBQ
AServ W: Exxon
RVCamp W: **KOA Campground**
Med W: **+ Hospital**
Parks W: **Chickamauga Nat'l Park**

140 **(347)** GA 151, Ringgold, LaFayette

FStop E: **Golden Gallon*, Texaco**
Gas E: **Texaco*(D)**
W: Amoco*(LP), Chevron*, Exxon*(LP)
Food E: **Country Bumpkin**, Hardee's, KFC, Krystal, **McDonalds**, Pizza Hut, Taco Bell, **Waffle House**
W: Wendy's
Lodg E: **Days Inn**, ◆ Hampton Inn, ◆ Holiday Inn Express, (AAA) Super 8 Motel
W: ◆ Comfort Inn
AServ E: Walter Jackson Chevrolet
W: Benny Jackson Ford
TServ W: **Lookout Mtn. Peterbilt**
RVCamp E: **Country Bumpkin**
Other E: Car Wash

EXIT — GEORGIA

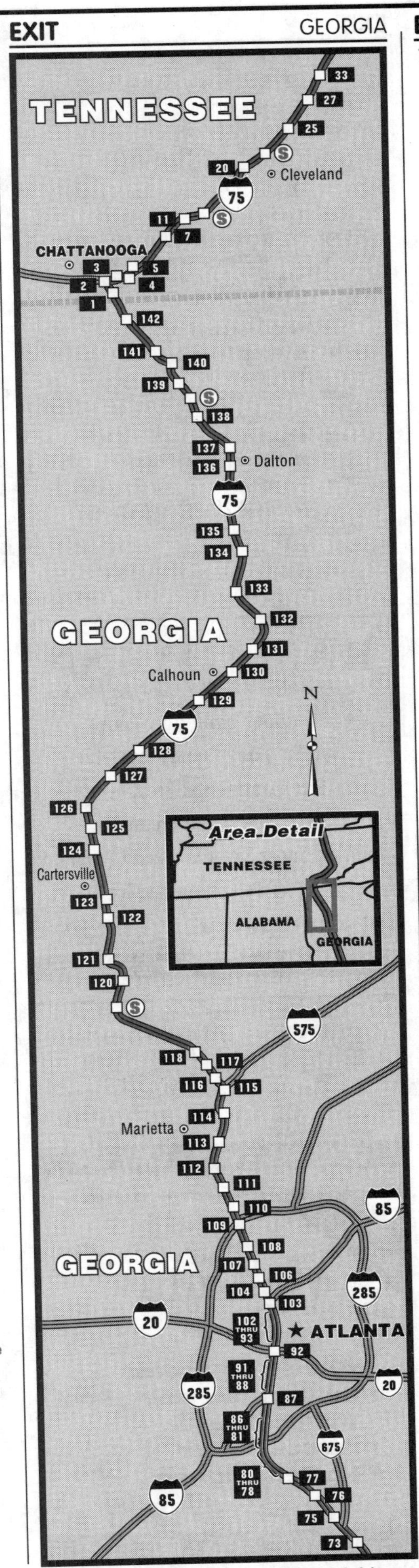

EXIT — GEORGIA

139 **(344)** U.S. 41, U.S. 76, Ringgold, Tunnel Hill

TStop W: **Choo Choo Truck Wash(SCALES), Citgo*(SCALES) (ATM)**
FStop W: **Fuel Mart*(SCALES)**
Gas E: BP*, Cowboy's*(LP)
W: Phillips 66*
Food W: **Waffle House**
Lodg W: **Friendship Inn**
TServ W: **Bell's Truck Parts/Service/Towing**

(343) **Weigh Station - both directions**

138 **(340)** GA 201, Tunnel Hill, Varnell

Gas W: Chevron*(CW), Texaco* (ATM)
Other E: **Knob North Golf Course**

137 **(335)** U.S. 41, U.S. 76, Dalton, Rocky Face

Gas E: BP, Chevron*(D), Racetrac*, Texaco* (ATM)
W: Amoco*(D), Phillips 66*(LP)
Food E: Blimpie's Subs (Chevron), Mr. Biscuit, Tampico Mexican Bar & Grill, **Waffle House**
W: Maggic's Pizza & Deli, Miller Bros Rib Shack, Ta Gin Chinese
Lodg E: ◆ Country Hearth Inn
W: (AAA) **Best Western**, Econolodge, Howard Johnson, Motel 6
AServ E: AAA Tire Store, BP
Med E: **+ Hospital**
Other E: **The Home Depot(LP)**

136 **(333)** GA 52, Walnut Ave, Dalton

Gas E: BP*(LP), Chevron*(CW), Exxon*(CW), Racetrac*
W: Texaco*(LP)
Food E: Applebee's, Burger King, Captain D's Seafood, Chick-fil-A, CiCi's Pizza, **Cracker Barrel**, Dairy Queen, IHOP, KFC, Lizzy's Deli, Long John Silvers, **McDonalds**, O'Charley's, Pizza Hut, Prime Sirloin, Shoney's, Taco Bell, The Cellar, **Waffle House**, Wendy's
W: Red Lobster, Sensations Restaurant & Lounge (Holiday Inn)
Lodg E: (AAA) Best Inns of America, (AAA) Days Inn, ◆ Hampton Inn, Travelodge
W: Comfort Suites, Country Inn & Suites, (AAA) Holiday Inn
AServ E: Brooker Ford, Edd Kirby Chevrolet & Geo, Exxon
ATM E: 1st Union Bank, Harwick's Bank, Wachovia Bank
Other E: **America's Best Contacts & Eyeglasses, Book Warehouse, Dalton Factory Outlet Mall, Kmart (Pharmacy), Kroger Supermarket (Pharmacy), The Chocolate Pot**
W: **Chamber of Commerce, NW Trade & Convention Center**

135 **(328)** GA 3, to U.S. 41

TStop E: **Pilot Travel Center*(SCALES) (ATM & Showers)**
FStop W: **Fuel City***
Food E: **Arby's (Pilot), Southside Cafe (Super 8)**, Waffle House
W: Bill's BBQ
Lodg E: **Super 8 Motel**

134 **(325)** Carbondale Road

FStop W: **Phillips 66**
Gas E: Chevron*(D)(LP)
W: Citgo*(LP) (ATM)
Food E: Annette's Catus Cafe
W: Campsite Restaurant
Lodg E: Country Boy Inn
TServ W: **TCI Truck Tire Center**

Bold red print shows RV & Bus parking available or nearby

EXIT GEORGIA

RVCamp E: Camping (Country Boy Inn), Pa-Paw's Park Campground

133 **(320)** GA 136, Resaca, LaFayette

TStop E: Flying J Travel Plaza(SCALES) (RV Dump)

Food E: Big Daddy Al's Bar-B-Q, Road Runner BBQ

TServ E: A & J Tire & Repair, Dependable Truck Wash, TT Calhoun Truck & Trailer

Other E: Radio Shop

(319) **Rest Area - Picnic, RR, HF, Phones, RV Dump (Southbound)**

132 **(318)** U.S. 41, Resaca

Gas W: Shell*(CW), The Right Stuff*(LP)

Food E: Great Wall Chinese (Best Western), Hardee's, Huddle House

W: Chuckwagon Restaurant

Lodg E: AAA Best Western

W: Burget Inn, AAA Duffy's Motel, AAA Econolodge, AAA Smith Motel, Super 8 Motel

131 **(317)** GA 225, Calhoun, Chatsworth

Lodg W: Express Inn

AServ W: Blackstock's Radiator/Alternator/Starter Repair

Parks E: New Echota Historic Site

Other E: Vann House

130 **(315)** GA 156, Redbud Road, Calhoun

Gas E: Citgo*(D) (ATM), Exxon* (ATM)

W: BP*(CW)(LP), Chevron*(CW), Shell*, Texaco(D)(LP)

Food E: Waffle House

W: Arby's, Shoney's

Lodg E: Scottish Inns

W: Howard Johnson, Ramada Limited

AServ W: Texaco

ATM W: Teller Yourself

Other E: Calhoun Antique Mall

W: Coin Car Wash

129 **(312)** GA 53, Calhoun, Fairmont

Gas E: Amoco*, Texaco*(D)

W: BP*, Exxon* (ATM), Fina*, Shell*

Food W: Arby's, Brangus Cattle Co., Golden Corral, IHOP (Days Inn), KFC, Krystal, Long John Silvers, McDonalds, Pizza Hut, Subway (Fina), Waffle House, Wendy's

Lodg E: Budget Host Inn, Quality Inn

W: Days Inn Days Inn, Guest Inn, ◆ Hampton Inn, The Jameson Inn

AServ W: Mr. Service Chevrolet, Geo

Other E: Calhoun Outlet Center, Ga State Patrol

W: Winn Dixie Supermarket (Pharmacy)

(308) **Rest Area - Picnic, RR, HF, Phones, RV Dump (Northbound)**

128 **(305)** GA 140, Adairsville, Summerville, Rome

TStop E: Patty's Truckstop(SCALES)

W: Citgo Pit Stop*(SCALES)

Gas E: Amoco*(D), Shell*(LP)

W: BP*, Exxon*

Food W: Burger King, Hardee's, Owens Bar-B-Q, Subway (BP), Taco Bell, Waffle House

Lodg W: AAA Comfort Inn, AAA Ramada Limited

AServ E: Shell

TServ E: Cobra Truck Repair

W: A&J Truck Service (Pit Stop)

TWash E: Patty's Truckstop

W: Pit Stop

RVCamp W: Family Leisure Resort

EXIT GEORGIA

Other W: Barnesly Gardens

127 **(296)** Cassville - White Road

FStop E: Speedway*(LP)

Gas E: Amoco*, Conoco*

W: BP*(LP) (ATM), Chevron*, Shell*(D)

Food E: The Flaming Grill

W: Antonino's Grotto (Econo Lodge), Waffle House

Lodg W: AAA Econolodge, AAA Red Carpet Inn

RVCamp W: KOA Kampground (No Trucks)

126 **(293)** U.S. 411, Cartersville, White

FStop E: Exxon*(LP)

W: Coastal* (ATM)

Gas E: Conoco*(LP)

W: BP*(D), Chevron*(LP)

Food W: Denny's (Holiday Inn), Truck Line Cafe (Coastal), Waffle House

Lodg E: Scottish Inns

W: ◆ Holiday Inn, Masters Inn

Other E: Aubery Lake

125 **(290)** GA 20, Rome, Canton

FStop E: Speedway*

Gas E: Chevron*(D) (Subway)

W: BP*(LP), Texaco*(D)

EXIT GEORG

Food E: Arby's, McDonalds, Morrell's BBQ (Econolodge), Wendy's

W: Cracker Barrel, Shoney's, Waffle House

Lodg E: AAA Comfort Inn, Econolodge, ◆ Motel 6, AAA Ramada Limited (see our ad this page), ◆ Super 8 Motel

W: Days Inn AAA Days Inn

124 **(287)** GA 113, Main St, Cartersville

Lodg W: Knights Inn

123 **(285)** Red Top Mountain Road

Gas E: Conoco*(LP)

Parks E: Red Top Mtn State Park

122 **(283)** Emerson - Allatoona Road

121 **(278)** Glade Road, Acworth

Gas E: BP*(CW), Exxon*(LP)

Food E: Subway (BP)

W: Burger King, Country Cafe, His & Her Southern Cookin, KFC, Krystal, Pizza Hut, Subway, Taco Bell, Waffle House

Lodg W: AAA Travelodge

AServ W: Auto Zone Auto Parts

Other W: Ingles Supermarket, K-Mart, Revco Drugs

120 **(276)** GA 92, Woodstock, Acworth

FStop W: Fina*

Gas E: Shell*, Texaco*(D)

W: BP*(CW)(LP), Stuckey's

Food E: Hardee's, Shoney's

W: Bamboo Garden, Dairy Queen, McDonalds, Ricardo's Mexican, Wendy's

Lodg E: Holiday Inn (see our ad on this page), ◆ Ramada

W: AAA Best Western, Days Inn AAA Days Inn, Hometown Lodge, AAA Quality Inn, AAA Super 8 Motel

Other W: Publix Supermarket, Revco

118 **(272)** Wade Green Road, Kennesaw

Gas E: BP*(CW), Chevron*(CW), Racetrac*

W: Texaco*(D) (Blimpie)

Food E: Arby's, McDonalds, Mrs. Winters, Taco Bell, Waffle House

Lodg E: Roadway Inn

AServ E: Georgia Tire

W: Q-Lube

ATM E: NationsBank, Premier Bank

Other E: Eckerd Drugs, Mail Boxes Etc, Winn Dixie Supermarket

117 **(270)** Chastain Road, to North I-575

Gas W: Amoco*, Exxon*, Texaco*(D)(LP)

Food E: Cracker Barrel

W: Arby's, Blimpie's Subs (Exxon), Del Taco, Mrs Winner's Chicken, Waffle House, Wendy's

Lodg E: Best Western, ◆ Fairfield Inn, Residence Inn

W: ◆ Country Inn & Suites, Sun Suites

Other E: Outlets Ltd. Mall

116 **(269)** Barrett Pkwy, to North I-575

Gas W: BP*(CW)

Food E: Baskin Robbins, Fuddrucker's, Fuji Hana, Grady's American Grill, Happy China II, Ippolito's Italian, Long Horn Steakhouse, Mandrian Cafe, Manhattan Bagel, McDonalds, Mellow Mushroom Pizza, Olive Garden, Philly Connection, Pizza Hut, Red Lobster, Rio Bravo Cantina, Rio Mexico, Schlotzkys Deli, Shoney's, Subway, Taco Mac, The Honey Baked Ham Co., Thia Peppers, Three Dollar Cafe, Waffle House

W: Chick-fil-A, Chili's, Cooker, Golden Corral,

Bold red print shows RV & Bus parking available or nearby

EXIT		GEORGIA
		Macaroni Grill, Mayflower Chinese, Outback Steakhouse, Sidelines Sports Grill, Sorrentino N.Y. Pizza, Steak & Shake, T.G.I. Friday's
	Lodg	E: ◆ Econolodge, ◆ Holiday Inn, ◆ Red Roof Inn
		W: ◆ Comfort Inn, Days Inn, ◆ Hampton Inn
	AServ	E: Big 10 Tires
		W: NTB Tire Warehouse
	Med	E: **Physicans Immediate Med**
	ATM	E: Colonial National Bank, Nations Bank, SouthTrust Bank
	Parks	W: **Kennesaw Mountain Battfield National Park**
	Other	E: **Eyesmart Vision Center, Home Depot**(LP), **Hunter's Eye Care, Town Center at Cobb**
		W: **Cobb Place Shopping Mall, Media Play, Target Department Store**
115		**(267)** Junction I-575, GA 5
114AB		**(268)** GA 5, U.S. 41, Marietta (Limited Access)
	Med	W: **Hospital (take 114B)**
113		**(264)** GA 120, Marietta, Roswell
	Med	W: **Hospital**
	Other	E: **White Water/American Adventures Theme Parks**
112		**(262)** GA 120 Loop, South Marietta Pkwy
	Gas	E: Chevron*, Shell*
		W: Fina*(LP), Kwik Trip*
	Food	W: All Star Sports Bar & Grill, Chili's, China Kitchen, Hardee's, International Grocery & Deli, Longhorn Steaks, Pizza Chef, Subway, TJ Applebee's
	Lodg	W: ◆ Hampton Inn, AAA Ramada, AAA Super 8 Motel, ◆ Wyndham Garden Hotel
	AServ	E: Massey Automotive
	RVCamp	W: **La Siseta RV Camp**
	ATM	W: Summit National Bank
111		**(261)** GA 280, Delk Road, Lockheed, Dobbins AFB
	Gas	E: Exxon*(CW), Texaco*(D)(CW)(LP) (ATM)
		W: Amoco(CW)(LP), BP(CW), Chevron*(CW)
	Food	E: China Wok, Denny's (Howard Johnson), Hardee's, KFC, McDonalds (Texaco), Papa John's Pizza, Spaghetti Warehouse, Taco Bell, Texas Bar-B-Q, Waffle House
		W: Cracker Barrel, Gourmet Cafe, Scrub Club, Coin Laundry, Waffle House
	Lodg	E: ◆ Courtyard by Marriott, Dury Inn, Howard Johnson (see our ad this page), AAA Motel 6, Scottish Inns, AAA Sleep Inn, Super 8 Motel (see our ad on this page)
		W: AAA Best Inns of America, AAA Comfort Inn, ◆ Fairfield Inn, AAA **Holiday Inn (see our ad on this page)**, AAA La Quinta Inn, Wingate Hotel
	AServ	E: Q-Lube
	Other	E: **Coin Laundry**
		W: Big A Car Wash, **Fast Trip Convenience Store**
110		**(260)** Windy Hill Road, Smyrna
	Gas	E: Amoco*(CW), BP*
		W: Chevron*(CW), Texaco(D)
	Food	E: Pappadeaux, Pappasito's Cantinia
		W: Arby's, Chick-fil-A, Georgian Cafe, McDonalds, Popeye's Chicken, Three Dollar Cafe, Waffle House, Wendy's
	Lodg	E: Days Inn (see our ad this page), AAA **Travelodge**
		W: ◆ Best Western, ◆ Courtyard by Marriott, AAA Hilton, Master's Inn, ◆ Red Roof Inn
	AServ	W: DeKalb Tire
	Med	W: **Kennestone Hospital**
	ATM	W: NationsBank, Wachovia Bank
	Other	W: **Target Department Store**
109AB		**(261)** Junction I-85, Atlanta Bypass
108		**(259)** Mt. Paran Road, Northside Pkwy, to U.S. 41
107		**(258)** West Paces Ferry Road, Northside Pkwy, to U.S. 41 (Very Congested)
	Gas	E: BP*(D)(CW), Chevron(D)
		W: Amoco (24 Hrs)
	Food	E: Cafe Three, Chick-fil-A, China Moon Restaurant, Gorins Ice Cream, McDonalds, Mrs. Fields Bakery, OK Cafe, Pero's Italian & Pizza Buffet, Starbuck's Coffee, Steak & Shake, Taco Bell
	AServ	E: Chevron
		W: Amoco (24 Hrs)
	Med	E: **Colombia West Paces Medical Center**
	ATM	E: NationsBank, SouthTrust, Suntrust Bank
	Other	E: **A & P Supermarket, Atlanta Historical Center, Big B Drugs, Kinko's Copies**
106		**(257)** Moores Mill Road
	Med	E: **Columbia West Paces Medical Center**
104		**(255)** U.S. 41, Howell Mill Road, Northside Dr (104B Northbound)
	Gas	E: Texaco*(CW)
		W: Shell*(CW)
	Food	E: Chick-fil-A, Chinese Restaurant, Fellinni's Pizza, Hardee's, McDonalds
		W: Arby's, Chinese Buffet, El Amigo, Green Derby Restaurant, KFC, Long John Silvers, Mellow Mushroom Pizza, Picadilly Cafeteria, Popeye's Chicken, Sensational Subs, Subway, Taco Bell, Wendy's
	Lodg	W: Castlegate Hotel & Conference Center, Holiday Inn, Howard Johnson
	AServ	E: D.W. Campbell, Goodyear Tire & Auto, Xpress Lube
		W: J & R Auto Service, Jiffy Lube, Master Care Auto Service, Precision Tune & Lube, Tune Up Clinic
	ATM	E: SouthTrust, Wachovia
		W: Kroger Supermarket, NationsBank
	Other	E: **Coin Laundry, Eckerd Drugs, Howell Mill Pharmacy, Mail Boxes Etc, U.S. Post Office, Winn Dixie Supermarket**
		W: **Kroger Supermarket (Pharmacy)**
104A		**(255)** Northside Dr (Northbound)
	Gas	W: BP(CW)
	Food	W: Dancers II, Oga's Hickory BBQ, Waffle House
	Lodg	W: Days Inn
	AServ	W: BP
	Other	W: **Buckwood Pet Hotel, Northside Drive Pet Hospital**
103		**(252)** Junction I-85 North, Greenville
102		**(251)** Fourteenth St, Tenth St, Techwood Dr (Southbound, Very Congested, Difficult Reaccess)
	Gas	E: Amoco*, BP*(D)(CW)
		W: Citgo*(D)
	Food	E: Dunkin Donuts, Philly Connection, Vini Vidi Vici Italian
		W: Blimpie's Subs, Chinese Buddha Restaurant, City Deli & Bagels, Silver Skillet Restaurant
	Lodg	E: Hampton Inn, Marriott, Occidental Grand Hotel
		W: Courtyard Marriott
	AServ	E: BP
	Other	E: **U.S. Post Office**

EXIT		GEORGIA
		W: Revco Drugs, Wolf Camera & Video
101		**(251)** Fourteenth St, Tenth St, Georgia Institute of Tec (Northbound)
	Gas	E: Amoco*, BP(D)(CW), Chevron* (24 Hrs)
	Food	E: Checkers Burgers, Domino's Pizza, Dunkin Donuts
		W: McDonalds, Papa John's Pizza
	Lodg	E: Regency Suites Hotel, Residence Inn
	AServ	E: BP
	Other	E: Public Transportation (MARTA Midtown Station)
		W: Alexander Memorial Coliseum
100		**(250)** U.S. 78, U.S. 278, North Ave, Georgia Institute of Technology (Difficult Reaccess)
	Gas	E: BP*(D)(CW)
	Food	E: The Varsity (an Atlanta landmark)
	Lodg	E: Days Inn
	Med	E: Crawford Long Hospital
	Other	E: Public Transportation (MARTA)
		W: The Coca Cola Company
99		**(250)** Williams St, World Congress Center, Georgia Dome (Southbound)
	Lodg	W: Days Inn (see our ad this page)
98		**(250)** Pine St, Peachtree St, Civic Ctr
	Lodg	W: Days Inn (see our ad on this page)
97		**(249)** Courtland St, (Southbound, Difficult Reaccess)
	Lodg	W: Hilton, Marriott, AAA Travelodge
	AServ	W: Beaudy Ford
96		**(249)** Ga. 10, International Blvd., Freedom Pwky., Carter Center
	Lodg	W: Courtyard, Fairfield Inn, Radisson
	Med	W: Georgia Baptist Hospital
	Other	W: World Congress Center
95		**(249)** Houston St (Southbound)
94		Edgewood Ave., Auburn Ave
	Med	W: Grady Memorial Hospital

ATLANTA
404-523-1144
DAYS INN
300 Spring Street NW
• Minutes from Underground Atlanta
• Minutes From Fulton County Stadium
• Swimming Pool
• Cocktail Lounge
• Cable TV with Free HBO
• 6 Blocks to GWCC
$58* +tax
From I-75/85 North Exit # 99 Williams St. 3 Blocks on Left.
From I-75/85 South Exit # 96 International Blvd. five blocks right on Spring 2 Blocks, Left on Baker
*Subject to tax. Based on space/rate availability. Not valid during conventions/special events, with any other discounts or group stays.
GEORGIA • I-75/85 • EXIT 99

EXIT		GEORGIA
	Other	E: Martin Luther King Jr. Historic Sight
93		**(248)** Martin Luther King Jr. Blvd, State Capitol, Underground Atlanta
	Other	W: Coca Cola Museum (Atlanta landmark, shopping), Underground Atlanta (Atlanta landmark, shopping)
92		**(248)** Junction I-20, Augusta, Birmingham
	Other	E: Cyclorama (see our ad this page)
91		**(247)** Fulton St. , Central Ave., Georgia Dome
	Parks	W: Zoo Atlanta (see our ad on this page)
90		**(247)** Abernathy Blvd., Capital Ave.
89		**(246)** University Ave, Pryor St
	Gas	E: Exxon*(D)
	Food	W: Brook's Cafeteria, Pool's Take-out
	AServ	E: Anderson Alignment, Charlie's Tire Company, NAPA Auto Parts
	Parks	W: Atlanta City Park (Pittman Park)
	Other	E: Lakewood Amphitheater
88		**(243)** GA 166, East Point, Lakewood Frwy (To Fort McPherson)
87		**(243)** Junction I-85 S
86		**(242)** Cleveland Ave
	Gas	E: Amoco*, Chevron*
		W: Citgo*, Shell*

Bold red print shows RV & Bus parking available or nearby

EXIT GEORGIA

Food E: Checkers Burgers, Church's Fried Chicken, El Progreso, McDonalds
W: Happy Chinese Cafe, Krystal, Pizza Hut Carry-Out, Walter's Cafe, Yasin's Fish Supreme
Lodg W: Days Inn, New American Inn

82 **(240)** U.S. 19, U.S. 41, Henry Ford II Ave
81AB **(239)** Junction I-285 East
80 **(239)** GA 85 South, Riverdale
79 **(239)** Frontage Road
Lodg E: Holiday Inn (see our ad opposite page)

78 **(237)** GA 331, Forest Pkwy
FStop W: BP*
Gas E: Chevron*(CW)
Food E: Waffle House
W: Waffle House
Lodg E: Rodeway Inn
W: Days Inn (see our ad on this page), Ramada (see our ad on this page)
AServ E: Chevron, Universal Tire
W: Auto Parts Machine Shop, Lee Tire Company
RVCamp E: Holiday RV Superstore (see our ad this page)
Other E: State Farmer's Market

77 **(236)** U.S. 19, U.S. 41, Old Dixie Hwy.
FStop W: Fuel Mart*(D)
Gas E: Hess*, Phillips 66*(D)(LP)
W: Amoco*(LP), Citgo*(LP), Shell*(CW)
Food E: Hardee's, Taqueria, Waffle House
W: Applebee's, Benefield's, Blimpie's Subs, Burger King, Captain D's Seafood, Chuck E. Cheese's, Dunkin Donuts, El Meson, Folks, Hot-Wings Cafe, Johnny's Pizza, KFC, Krystal, McDonalds, Monterrey Mexican, O'Hara's, Old Country Buffet, Provino's, Red Lobster, Sonny's Barbecue, Taco Bell, Waffle House, Waffle House (#2), Wendy's
Lodg E: AAA Super 8 Motel, Travelodge
W: AAA Comfort Inn, ◆ Days Inn (see our ad on this page), Econolodge, ◆ Holiday Inn, Shoney's
AServ E: Mark's New & Used Tires, Penske Auto Service
W: Amoco, Jiffy Lube, Landmark Dodge, Precision Tune & Lube, Tune Up Clinic, Xpress Lube
RVCamp W: Sagon Motor Homes (see our ad on this page)
Med W: ✚ Hospital
W: Arrowhead Pharmacy, Car Wash, Cub Foods

76 **(233)** GA 54, Morrow, Lake City
Gas E: BP*(CW), Chevron*(CW), Hess*(D)
W: Exxon*(CW), Texaco*(D)
Food E: Cracker Barrel, Krystal, Mrs Winner's Chicken, Taco Bell, Waffle House, Wendy's
W: KFC, Long John Silvers, McDonalds, Pizza Hut, Shoney's, Waffle House
Lodg E: AAA Best Western, ◆ Drury Inn, ◆ Fairfield Inn, ◆ Red Roof Inn
W: ◆ Hampton Inn, AAA Quality Inn (see our ad on this page)
AServ E: Chevron
Med E: ✚ Medical Care (Minor Emergency Walk-In)
ATM E: First Union Bank

EXIT GEORGIA

W: NationsBank (Texaco)

75A **(231)** Mount Zion Blvd
Gas W: Exxon*(CW)
Food W: Arby's, Blimpie's Subs, Chick-fil-A, Chili's, Del Taco, Los Toribis, McDonalds (Playland), Mrs Winner's Chicken, Papa John's Pizza, Philly Connection, Rio Bravo, Steak N Shake, TCBY Yogurt, Truett's Grill, Waffle House, Wendy's
Lodg W: AAA Sleep Inn (see our ad on this page), Sun Suites
AServ W: Xpress Lube
ATM W: SunTrust (In Publix)

75 **(228)** GA 138, Jonesboro, Stockbridge
Gas E: Racetrac*
W: Amoco* (24 Hrs), Chevron* (24 Hrs), Citgo*
Food E: Applebee's, Baskin Robbins, Chick-fil-A, Cici's Pizza, Damon's Ribs, Emperor's Gourmet Chinese, Frontera Mexican, Gregory's Bar & Grill, IHOP, Kenny Rogers Roasters, Krystal, Long John Silvers, Los Toribis Mexican, Morrison's Cafe, Philly Connection, Subway, The Bistro (Best Western), Waffle House
W: Waffle House
Lodg E: AAA Best Western, AAA Days Inn
Other E: Kroger Supermarket

74 **(227)** I-675 (Northbound)
73 **(224)** Hudson Bridge Road
Gas E: BP*, Citgo*(LP), Exxon*(CW)(LP)
W: Phillips 66*
Food E: Dairy Queen, McDonalds, Subway, Waffle House, Wendy's
W: China Cafe Three, Fifteenth Street Pizza & Calzones, Memories Restaurant
Lodg W: AAA Super 8 Motel
Med E: ✚ Henry General Hospital
ATM E: Citgo
Other E: Eagle's Landing Pharmacy, Hudson Bridge Pharmacy, U.S. Post Office

72 **(222)** Jodeco Road, Flippen
Gas E: Amoco*, Citgo*(LP)
W: Chevron*
Food E: Hardee's
RVCamp W: KOA Campground
Med E: ✚ Hospital

71 **(221)** Jonesboro Road, Lovejoy
FStop E: Mapco Express*(D)(LP)
ATM E: Mapco Express
Other W: Peachtree Peddlers Flea Market

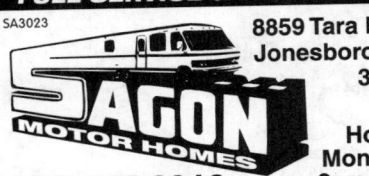

70 **(219)** GA 20, GA 81, McDonough, Hampton
FStop W: Speedway*(LP)
Gas E: Amoco*, BP*
W: Shell*
Food E: Arby's, Burger King (Playground), Dairy Queen, KFC, McDonalds, Mrs Winner's Chicken, Pizza Hut, Sizzling Platter Family Steakhouse, Taco Bell, Waffle House, Wendy's
W: Dusters, Starvin' Marvin (Speedway), Subway, Waffle House

EXIT GEORGIA

Lodg **E:** AAA Brittany Motor Inn, Budget Inn, ◆ Hampton Inn, **Red Carpet Inn**
W: AAA Comfort Inn, AAA Econolodge (see our ad on this page), Masters Inn
AServ **W:** Shell
ATM **W:** Speedway

69 **(216)** GA 155, McDonough
FStop **E: Texaco***(D)(CW)
W: Citgo*(LP)
Gas **E:** Chevron*(D)
W: Amoco, BP*(D)
Food **E:** McGhin's Southern Pit Barbecue, Subway (Chevron), Waffle House
W: Shoney's, **Waffle House**
Lodg **E:** DAYS INN AAA Days Inn, **Sunny Inn**, Welcome Inn
W: AAA Holiday Inn (see our ad on this page)
AServ **E:** Bellamy-Strickland Oldsmobile, Chevrolet, Gmc Trucks
W: Amoco, BP
TServ **E: Perimeter Transport Refrigeration & Truck Repair**

68 **(212)** Locust Grove, Hampton
Gas **E:** Amoco*, BP*(D), Exxon*(D)(CW)(LP), Shell*
W: Chevron*
Food **E:** Hardee's, **Huddle House**, Mom & Pop's, Subway (BP), **Waffle House**
W: B.J.'s
Lodg **E:** Red Carpet Inn
W: Scottish Inns (see our ad on this page), AAA **Super 8 Motel**
AServ **E:** Amoco
Other **E: Tanger Factory Outlet**

67 **(205)** GA 16, Griffin, Jackson
FStop **E: Citgo***
Gas **E:** BP*
W: Amoco*(LP), Chevron*(D), Texaco*(D)
Food **E:** Simmon's BBQ
AServ **W:** Amoco
ATM **W:** Texaco

66 **(201)** GA 36, Jackson, Barnesville
TStop **E: 76 Auto/Truck Plaza*(SCALES) (RV Dump, Citgo Gas)**
W: Flying J Travel Plaza*(D)(LP)(SCALES) (Conoco Gas)
FStop **E: Fuel City*(D) (Smile Gas)**
Gas **W:** The Food Store*
Food **E: Hot Stuff Pizza (76 Travel Center)**, King's Bay Restaurant, **Restaurant (76 Travel Center)**, Subway, Taco Bell
W: Buckner's, **Hardee's (Flying J Travel Plaza)**, **Thads Restaurant (Flying J Travel Plaza)**
TServ **E: Truck Lube Center**
W: Flying J Travel Plaza (24 Hrs)
TWash **E: Blue Beacon Truck Wash (76 Travel Center)**
W: Eagle One Truck Wash

65 **(198)** High Falls Road
RVCamp **W: High Falls Campground**(LP)
Parks **W: High Falls State Park**

64 **(193)** Johnstonville Road
Gas **E:** Amoco*, BP*
AServ **E:** BP

(189) **Weigh Station (Both Directions)**

63 **(187)** GA 42, Forsyth (Northbound)
Gas **E:** Shell*
Food **E: Round-Up (Super 8 Motel)**
Lodg **E:** AAA Best Western, AAA **Super 8 Motel**, Valu Inn
Other **E: Indian Spring State Park**

62 **(187)** GA 83, Forsyth, Monticello
Gas **W:** Amoco*, Citgo*(D), Conoco*(D) (24 Hrs),

EXIT GEORGIA

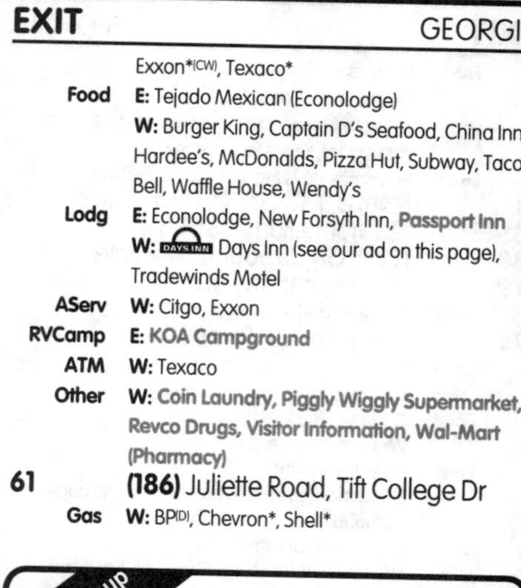

Exxon*(CW), Texaco*
Food **E:** Tejado Mexican (Econolodge)
W: Burger King, Captain D's Seafood, China Inn, Hardee's, McDonalds, Pizza Hut, Subway, Taco Bell, Waffle House, Wendy's
Lodg **E:** Econolodge, New Forsyth Inn, **Passport Inn**
W: DAYS INN Days Inn (see our ad on this page), Tradewinds Motel
AServ **W:** Citgo, Exxon
RVCamp **E: KOA Campground**
ATM **W:** Texaco
Other **W: Coin Laundry, Piggly Wiggly Supermarket, Revco Drugs, Visitor Information, Wal-Mart (Pharmacy)**

61 **(186)** Juliette Road, Tift College Dr
Gas **W:** BP(D), Chevron*, Shell*

Bold red print shows RV & Bus parking available or nearby

EXIT		GEORGIA
	Food	W: Dairy Queen, Hong Kong Palace Chinese, **Lelands Restaurant**, Waffle House
	Lodg	W: Ambassador Inn, ◆ **Hampton Inn**, Holiday Inn
	AServ	W: BP
	RVCamp	E: **KOA Campground**
	Other	E: **Jarrell Plantation Historic Site** W: **Ingle's Supermarket, Nacy Cinema**
60		**(184)** GA 18, Gray
	Gas	W: Amoco* (24 Hrs), Texaco*(D)
	Food	W: Ice Cream Churn (Amoco), Shoney's
	Lodg	W: (AAA) Comfort Inn
	ATM	W: Amoco
	Other	E: **Jarrell Plantation Historical Site** W: **Georgia State Patrol**
59		**(180)** Rumble Road, Smarr
	Gas	E: BP*(D), Shell*
	Food	E: Subway (BP)
	AServ	E: Roy's Automotive & Truck Tire Repair
	TServ	E: **Roy's Truck Tire Service**
	ATM	E: BP
(179)		**Rest Area - RR, Phones, Picnic, Vending, RV Dump, Tourist Info (Southbound)**
58		**(178)** Junction I-475, Valdosta (Southbound)
57		**(175)** Pate Rd, Bowling Broke (Northbound, No Return Access)
56		**(172)** Bass Rd
55		**(171)** U.S.23, GA. 87, Riverside Dr
55A		**(169)** to U.S. 23, Arkwright Rd, Riverside Dr
	Gas	E: Shell* W: Chevron*(LP) (24 Hrs), Conoco*(D)
	Food	E: Carrabba's Italian, Cracker Barrel, Logans Road House, Outback Steakhouse, Waffle House, Wager's W: Applebee's, Baskin Robbins, Burger King, Chick-fil-A, Chili's, Dunkin Donuts, Hooters, KFC, McDonalds, Papa John's, Rio Bravo, Ryan's Steakhouse, Shoney's, Steak 'N Shake, Subway, Taco Bell, Traditions Restaurant (Holiday Inn), Waffle House, What a Pizza
	Lodg	E: ◆ Courtyard by Marriott, (AAA) La Quinta Inn, Residence Inn, (AAA) Sleep Inn, ◆ Wingate Inn W: ◆ Hampton Inn, (AAA) Holiday Inn, ◆ Quality Inn & Suites, Shoney's Inn
	AServ	E: Huckabee Buick Cadillac W: Raffield Tire Master
	ATM	W: First Union Bank, Nations Bank (In Kroger), SunTrust, Wachovia
	Other	W: **Barnes & Noble, K-Mart (Pharmacy), Kroger Supermarket (Pharmacy), Publix Supermarket, Regal Cinemas**
54		**(167)** GA 247, Pierce Ave
	Gas	W: Amoco*, BP*, Chevron* (24 Hrs), Exxon, Fina*(LP), Texaco*
	Food	W: Applebee's, Arby's, Bennigan's, Blimpie's Subs, Captain D's Seafood, Denny's, El Indio, Macon Music City Restaurant, S&S Cafeteria, Steak-Out, Subway (Fina), Texas Cattle Co, Waffle House, Wendy's, Yamato Japanese
	Lodg	E: Days Inn W: Ambassador Inn, (AAA) Best Western, (AAA) Comfort Inn, (AAA) Holiday Inn, Howard Johnson
	AServ	W: Exxon, Goodyear Tire & Auto
	ATM	W: First Liberty Bank, Nations Bank
	Other	W: **Eckerd Drugs**
53		**(165)** Jct I-16 , to Macon, Savannah
52		U.S.41, GA19, Hardman Ave, Forsyth St, Downtown
	Gas	E: Exxon, Fina*, Phillips 66
	Food	E: Sid's Soup, Salad, Sandwich
	Med	E: ✚ **Hospital**
	ATM	E: SunTrust
	Other	E: **Animals & Things Animal Hospital** W: Fountain Car Wash
51		GA. 74 W., Mercer Univ. Dr
	Gas	W: Citgo*, Fina*(D)
	Other	W: **Animal Emergency Care**
50		**(162)** U.S. 80, GA22, Eisenhower Pkwy
	Gas	W: Amoco* (24 Hrs), Chevron*(CW)(LP), Shell*, Speedway*(D)(LP)
	Food	W: Burger King, Captain D's Seafood, Checkers, China King Chinese, Krispy Kreme, Krystal, Lee's Chicken, Little Caesars Pizza, Long John Silvers, McDonalds, Shoney's, Taco Bell, Wendy's
	AServ	W: Goodyear Tire & Auto, Gordon Bush Tire Co., Meineke Discount Mufflers, Penske Auto Service, Pep Boys Auto Center, Super Shops, Sure Brake, Western Auto
	ATM	W: NationsBank
	Other	W: **Butler's Pet Hosptital, Food Max, K-Mart (Pharmacy), Petsmart, Wal-Mart (Pharmacy)**
49AB		**(160)** U.S.41, GA 247, Rocky Creek Rd, Piónono Ave
	Gas	E: Exxon*(LP), Racetrac* W: Citgo, Enmark(D)
	Food	E: Huddle House, Waffle House W: Central Park Hamburgers, China Star, Dairy Queen, Hardee's, Janet's Restaurant, KFC, McDonalds, Waffle House
	Lodg	E: Masters Inn
	AServ	W: BF Goodrich, Jiffy Lube, Raffield Tire Master
	ATM	E: Racetrac W: Citgo, First Macon Bank & Trust, NationsBank, SunTrust
	Other	W: **Eckerd Drugs, Piggly Wiggly Supermarket (24 Hrs), Roses Department**

Note: I-75 runs concurrent below with I-475. Numbering follows I-475.

EXIT		GEORGIA
4		**(15)** Bolingbroke
	Gas	E: Exxon*(D)(LP), Fina*(D)
3		**(11)** Zebulon Road, Wesleyan College, Macon
	Gas	E: BP*(CW) W: Citgo*, Exxon
	Food	E: Buffalo's Cafe, Chen's Wok, Chick-fil-A, Hong Kong Chinese, Subway, Waffle House, Wendy's W: Kuntry Kitchen, Polly's Corner Cafe
	AServ	W: Exxon
	Med	E: ✚ **Hospital**
	ATM	E: NationsBank (In Kroger), Rivoli Bank & Trust, SunTrust
	Other	E: **Kroger Supermarket**
(8)		**Rest Area - RR, Phones, Picnic, Vending (Northbound)**
2		**(5)** GA 74, Macon, Thomaston, Mercer University Dr
	Gas	W: Exxon*, Phillips 66*(D), Shell*
	Food	E: Waffle House W: Speedy Pizza, Subway (Exxon), Taco Bell (Exxon)
	Lodg	W: **Family Inns**
	AServ	E: Goodyear Tire & Auto, Muffler Master
	RVCamp	W: **Lake Tobesofee Recreation Area**
	ATM	W: First Macon Bank & Trust
	Other	W: **Brantley & Jordan Animal Hospital, Food Lion Supermarket**
1		**(3)** U.S. 80, Macon, Roberta, Macon College
	FStop	E: Fina*(LP)
	Gas	E: Citgo*, Racetrac* W: Amoco*, BP*(D)(CW), Shell*
	Food	E: **Cracker Barrel**, JL's BBQ (Citgo), Shoney's, Subway (Citgo), Waffle House W: Blimpie's Subs (Shell), Burger King, El Zarape (Passport Inn), McDonalds
	Lodg	E: (AAA) Best Western, (AAA) Comfort Inn, Hampton Inn, ◆ Holiday Inn, Howard Johnson, Motel 6, (AAA) Ramada, **Red Carpet Inn**, (AAA) Rodeway Inn, (AAA) **Super 8 Motel**, (AAA) Travelodge W: (AAA) Econolodge, (AAA) Knight's Inn, **Passport Inn**
	AServ	W: Amoco, BP
	ATM	E: Citgo, Fina, Racetrac W: Shell
	Other	W: **Sams Club**

Note: I-75 runs concurrent above with I-475. Numbering follows I-475.

EXIT		GEORGIA
48		**(156)** Junction I-475, Macon Bypass (Southbound)
47		**(155)** Hartley Bridge Road
	Gas	E: Fina*(D)(LP), Phillips 66*(D), Shell* W: Auto Masters, Exxon*, Texaco*(LP)
	Food	E: Golden Wok, Popeye's Chicken, Wendy's W: Subway, Waffle House
	Lodg	W: **Ambassador Inn**
	AServ	W: Auto Masters
	ATM	E: NationsBank (In Kroger), Phillips 66
	Other	E: **Kroger Supermarket**
46		**(149)** GA 49, Byron, Fort Valley
	FStop	W: **Citgo*(CW), Fina*, Speedway***
	Gas	E: Shell*, Texaco*(D) W: BP*, Racetrac*
	Food	E: **McDonalds**, Peach Outlet BBQ, Pizza Hut, Shoney's, Waffle House W: **Blimpie's Subs (Citgo)**, Country Cupboard, Dairy Queen, Hardee's, Huddle House, **Icecream Churn (Citgo)**, Papa's Pizza, Subway, Waffle House
	Lodg	E: (AAA) Super 8 Motel W: (AAA) Comfort Inn, (AAA) **Days Inn (see our ad on this page)**, (AAA) Econolodge, **Passport Inn**

EXIT GEORGIA

AServ W: Alan's Alignment & Automotive, BP, Brannen Ford, Butler Chevrolet, Oldsmobile
RVCamp E: **Mid State RV Center**
W: **RV Interstate Camping, Parts & Service**(LP)
ATM W: Fina, Racetrac
Other E: **Peach Festival Outlet Center**

45 **(146)** GA 247, Centerville, Warner Robins
FStop W: **Citgo***
Gas E: Exxon*, Speedway*(D)(LP)
Food E: Subway, Waffle House
Lodg E: Masters Inn
W: **Red Carpet Inn**
AServ E: Exxon
Other E: **Museum of Aviation (East in Warner Robins -- see our ad this page)**

44 **(142)** GA 96, Housers Mill Road
Gas E: Shell*
AServ E: Shell

43A **(138)** Thompson Road, Perry - Fort Valley Airport
FStop E: **Happy Stores***(D)(LP) **(RV Dump, Phillips 66)**
ATM E: Happy Stores

43 **(137)** U.S. 341, Perry, Fort Valley
FStop E: **Speedway***(D)(LP)
Gas E: Amoco*(CW), Chevron*(D)
W: BP*, Conoco*(D)(LP), Racetrac*(LP)
Food E: A.B.'s BBQ, Arby's, Baskin Robbins, Burger King (Playground), Captain D's Seafood, Chick-fil-A, China Moon Restaurant, El Jalisiense, Hardee's, KFC, Kimberly BBQ, Krystal, McDonalds, Pizza Hut, Quincy's Family Steakhouse, Red Lobster, Samantha's Chinese Restaurant, Shoney's, Subway, TCBY Yogurt, Taco Bell, Waffle House, Wendy's
W: Angelina's Italian Cafe, Green Derby Restaurant & Bar
Lodg E: ◆ Fairfield Inn, Hampton Inn, Ramada Inn, Red Gable Inn, Super 8 Motel
W: AAA Comfort Inn, DAYS INN ◆ **Days Inn (see our ad this page)**, Econolodge, ◆ Holiday Inn, Knight's Inn, Passport Inn, AAA Quality Inn
AServ E: Advance Auto Parts, Amoco
W: Wayne Morris Ford
RVCamp E: **Boland's Camping**(LP)
W: **Crossroads of Georgia Travel Park**
Med E: ✚ **Hospital**
ATM E: NationsBank (In Kroger), Speedway
Other E: **K-Mart (Pharmacy), Kroger Supermarket**

42 **(134)** U.S. 41, GA 127, Perry, Marshallville
FStop E: **BP***
Gas E: Exxon*, Shell*, Speedway*(LP)
Food E: **Cracker Barrel**, Dairy Queen, Mandarin House Chinese, Perry Cafe, Waffle House
Lodg E: Crossroads Motel, Red Carpet Inn, Regency Inn, Rodeway Inn, Sandman Motel, Scottish Inns, Thrift Courts, AAA Travelodge
AServ E: Hamby Chrysler Plymouth, Jeep Eagle
Other W: **Georgia State Patrol**

41 **(127)** GA 26, Montezuma, Hawkinsville
FStop W: **Chevron***
Food W: Icecream Churn (Chevron), Judee's Coffee Shop

EXIT GEORGIA

EXIT GEORGIA

RVCamp E: **Twin Oaks RV Camp**

40 **(122)** GA 230, Unadilla, Byromville
FStop E: **Dixie***
Gas E: Fina*
W: Phillips 66*
Lodg W: Red Carpet Inn
AServ E: Brannen Motors Chevrolet, Geo Ford

39 **(121)** U.S. 41, Unadilla, Pinehurst
TStop W: **Citgo***(D)
Gas E: BP*(LP), Shell*, Texaco*
Food E: **Cotton Patch Restaurant (Scottish Inns)**, Dairy Queen, Subway
W: **Citgo**
Lodg E: DAYS INN **Days Inn, Scottish Inns**
W: **Passport Inn**

(118) **Rest Area - RR, Phones, Picnic, RV Dump, Vending (Southbound)**

38 **(117)** Pinehurst
Gas W: BP*(D)
Food W: New Colony Inn
Lodg W: New Colony Inn

37 **(112)** GA 27, Vienna, Hawkinsville
Gas E: Shell
W: Fina*
AServ E: Shell

36 **(109)** GA 215, Vienna, Pitts
Gas E: BP
W: Amoco*(LP), Chevron*, Citgo*(D)
Food W: **Hardee's, Huddle House**, Marise (Knight's Inn)
Lodg W: Knight's Inn
Med W: ✚ **Dooly Medical Center**

(108) **Rest Area - RR, Phones, Picnic, Vending, RV Dump (Northbound)**

35 **(104)** Bus. Loop 75, Farmers Market Road, Cordele
Gas W: Phillips 66(LP)
Lodg E: **Super 8 Motel**
AServ W: Hess Garage, Phillips 66
TServ W: **Hess Garage**
RVCamp W: **Hess Garage**

34 **(102)** GA 257, Hawkinsville, Cordele
Gas E: Country Store*(LP), Shell*
W: Cordele Pecan House
Med W: ✚ **Hospital**

33 **(101)** U.S. 280, GA 90, Cordele, Abbeville
Gas E: Citgo*, Exxon*(D), Texaco*
W: Amoco*(CW), BP(D), Chevron, Racetrac*, Spur
Food E: Denny's, Good Food Restaurant (Ramada), Happy China, Waffle House
W: Burger King, Captain D's Seafood, Dairy Queen, Golden Corral, Hardee's, KFC, Krystal, McDonalds (Play Land), Pizza Hut, Shoney's, T.J.'s Italian, TCBY Yogurt, Taco Bell, Wendy's, Western Steer Family Steakhouse
Lodg E: DAYS INN ◆ Days Inn, AAA **Ramada (see our ad this page)**
W: Athens Motel, AAA Colonial Inn (see our ad this page), ◆ Comfort Inn, AAA Econolodge, Hampton Inn, AAA Holiday Inn, Passport Inn, Rodeway Inn
AServ E: Texaco
W: BP, Chevron
ATM W: Central Bank & Trust, Cordele Banking Company, Racetrac

Bold red print shows RV & Bus parking available or nearby

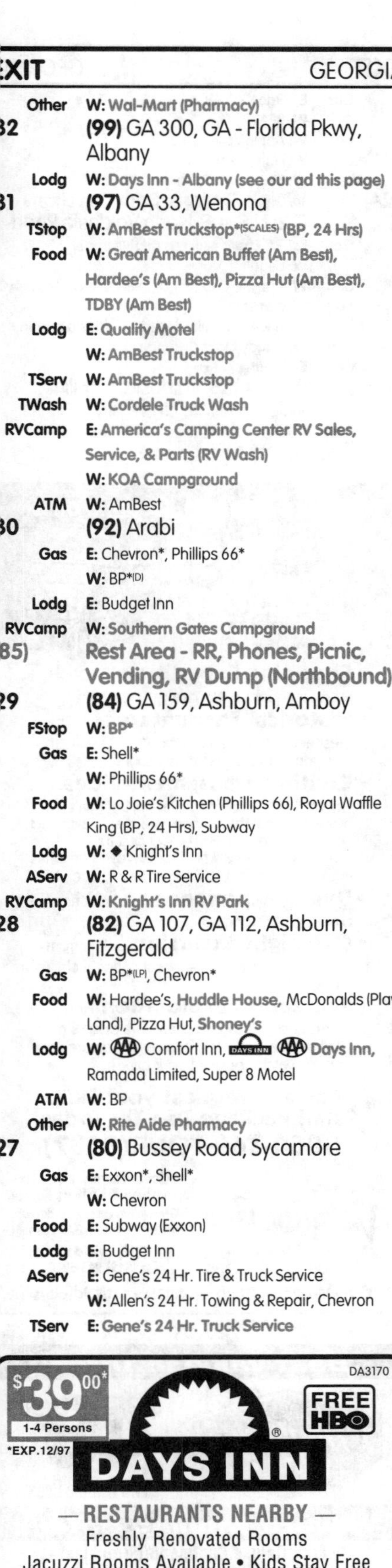

EXIT		GEORGIA
	Other	W: Wal-Mart (Pharmacy)
32		**(99)** GA 300, GA - Florida Pkwy, Albany
	Lodg	W: Days Inn - Albany (see our ad this page)
31		**(97)** GA 33, Wenona
	TStop	W: AmBest Truckstop*(SCALES) (BP, 24 Hrs)
	Food	W: Great American Buffet (Am Best), Hardee's (Am Best), Pizza Hut (Am Best), TDBY (Am Best)
	Lodg	E: Quality Motel
		W: AmBest Truckstop
	TServ	W: AmBest Truckstop
	TWash	W: Cordele Truck Wash
	RVCamp	E: America's Camping Center RV Sales, Service, & Parts (RV Wash)
		W: KOA Campground
	ATM	W: AmBest
30		**(92)** Arabi
	Gas	E: Chevron*, Phillips 66*
		W: BP*(D)
	Lodg	E: Budget Inn
	RVCamp	W: Southern Gates Campground
(85)		Rest Area - RR, Phones, Picnic, Vending, RV Dump (Northbound)
29		**(84)** GA 159, Ashburn, Amboy
	FStop	W: BP*
	Gas	E: Shell*
		W: Phillips 66*
	Food	W: Lo Joie's Kitchen (Phillips 66), Royal Waffle King (BP, 24 Hrs), Subway
	Lodg	W: ◆ Knight's Inn
	AServ	W: R & R Tire Service
	RVCamp	W: Knight's Inn RV Park
28		**(82)** GA 107, GA 112, Ashburn, Fitzgerald
	Gas	W: BP*(LP), Chevron*
	Food	W: Hardee's, Huddle House, McDonalds (Play Land), Pizza Hut, Shoney's
	Lodg	W: AAA Comfort Inn, DAYS INN AAA Days Inn, Ramada Limited, Super 8 Motel
	ATM	W: BP
	Other	W: Rite Aide Pharmacy
27		**(80)** Bussey Road, Sycamore
	Gas	E: Exxon*, Shell*
		W: Chevron
	Food	E: Subway (Exxon)
	Lodg	E: Budget Inn
	AServ	E: Gene's 24 Hr. Tire & Truck Service
		W: Allen's 24 Hr. Towing & Repair, Chevron
	TServ	E: Gene's 24 Hr. Truck Service

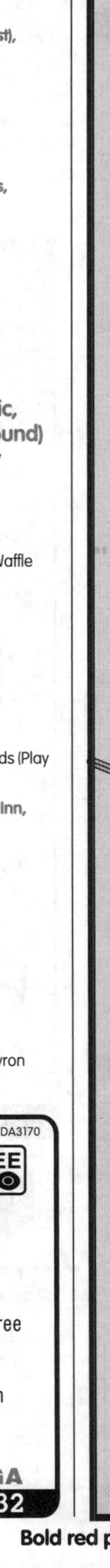

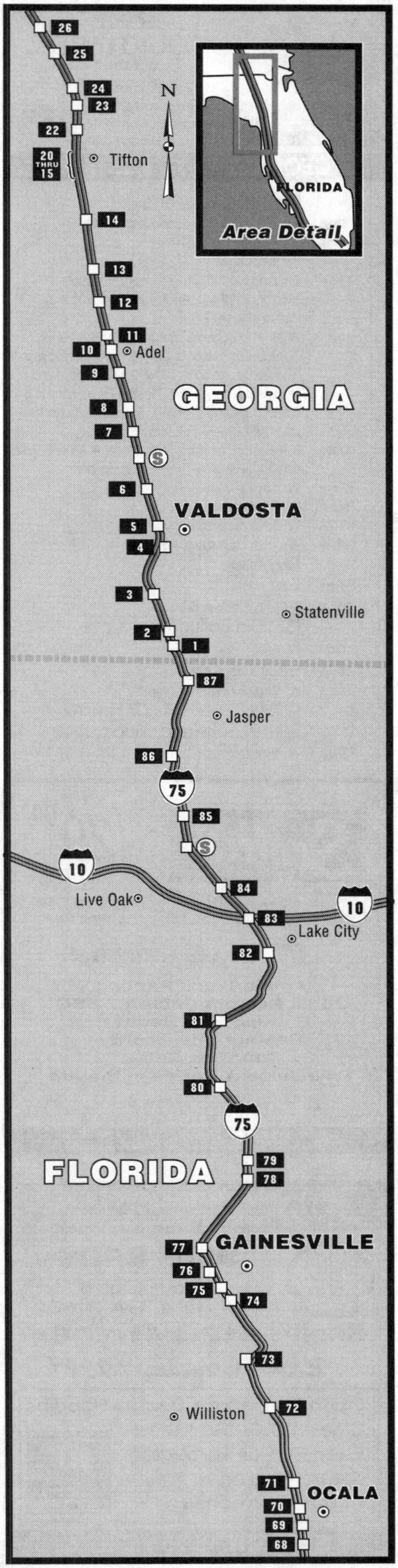

EXIT		GEORGIA
	RVCamp	E: Lakeview Campground
26		**(78)** GA 32, Sycamore, Ocilla
	Parks	E: Jefferson Davis Park & Museum
(76)		Rest Area - RR, Phones, Picnic, Vending, RV Dump (Southbound)
25		**(75)** Inaha Road
	Gas	E: BP*
		W: Citgo*
	Food	W: Dairy Queen (Citgo), Stuckey's (Citgo)
	AServ	E: BP
24		**(72)** Willis Stills Road, Sunsweet
	Gas	W: BP*(D)
	RVCamp	W: Branch Bros. Farm Market RV Camping
23		**(69)** Chula - Brookfield Road
	Gas	E: Phillips 66* (24 Hrs)
	Food	E: Chula Family Restaurant (Red Carpet)
	Lodg	E: Red Carpet Inn
22		**(66)** Brighton Road
21		**(64)** Bus Loop 75, U.S. 41, Tifton ABAC
	Gas	E: Chevron*(D), Fina, M & S(CW)
		W: Citgo*
	AServ	E: Fina
	Med	E: ✚ Hospital
	ATM	E: SunTrust
	Other	E: Food Lion Supermarket
20		**(63)** Eighth St, Tifton
	Gas	E: Exxon*, Texaco*
	Food	E: Hardee's, KFC, Peking House, Subway
		W: Split Rail Grill
	Lodg	E: DAYS INN Days Inn
	Med	E: ✚ Hospital
	ATM	E: NationsBank
	Other	E: Cinema 6, Coin Car Wash, Tifton Mall, Winn Dixie Supermarket
		W: Georgia Agrirama
19		**(62)** Second St, Tifton
	FStop	E: Dixie*
	Gas	E: BP*, Chevron*
	Food	E: Arby's, Baskin Robbins (Chevron), Burger King, Central Park, Checkers Burgers, Chicago Pizza & Pasta, China Garden, Golden Corral, Krystal, Long John Silvers, Los Compadres, McDonalds, O'Neal Country Buffet, Pizza Hut, Red Lobster, Taco Bell
		W: Denny's, Icecream Churn (Citgo), Waffle House
	Lodg	E: Red Carpet Inn, ◆ Super 8 Motel
		W: Best Western, ◆ Comfort Inn, Ramada (see our ad this page)

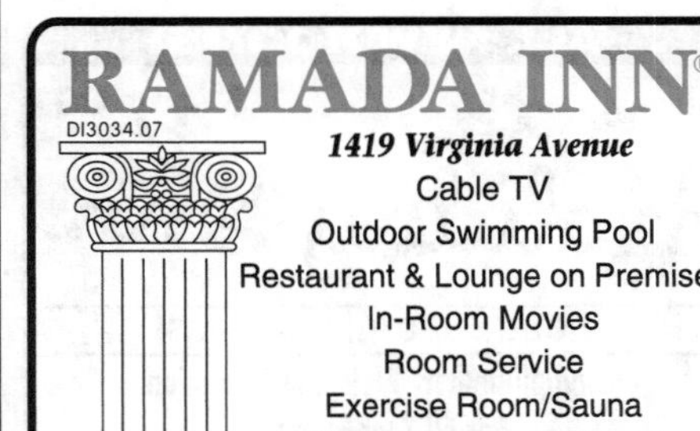

EXIT GEORGIA

AServ E: Advance Auto Parts, Jeff Fender Pontiac, Buick, Cadillac, Super Lube
ATM E: Citizen's Bank, First Community Bank
Other E: Food Max (24 Hrs), K-Mart, Wal-Mart
18 **(62)** to U.S. 82, U.S. 319, Sylvester, Moultrie
FStop W: Texaco*
Gas E: Exxon*(D)(CW), Shell*(D)
W: Amoco*, Chevron* (24 Hrs), Citgo*, Racetrac*, Texaco*
Food E: Charles Seafood Restaurant, Cracker Barrel, Little Caesars Pizza, Sonic, Waffle House, Western Sizzlin'
W: Burger King (Playground , Texaco), Captain D's Seafood, Chick-fil-A, Shoney's, Sonny's BBQ, Subway (Amoco), Waffle House, Wendy's
Lodg E: AAA Courtyard by Marriott (see our ad this page), ◆ Hampton Inn (see our ad this page), Masters Inn, Red Carpet Inn
W: ◆ Holiday Inn (see our ad this page), Ramada Limited, Scottish Inns (see our ad this page)
AServ E: Page Auto & Tire Center
W: Prince Chevrolet,Oldsmobile, Prince Honda, Prince Toyota
Med E: ✚ Hospital
Other E: Food Max, Wal-Mart
17 **(61)** Omega Road
RVCamp E: Covered Wagon RV Park & Campground
16 **(60)** South Central Ave
TStop W: Tifton Travel Center*(D)(SCALES) (RV Dump)
Food E: Granny's Kitchen
W: Steak 'N Shake (Tifton Travel Center)
TServ W: Ten Speed Service Center(SCALES)
TWash W: Express Truck Wash
RVCamp W: Express RV Wash
Med E: ✚ Hospital
15 **(59)** Southwell Blvd
Gas E: Citgo*(D)
14 **(55)** Eldorado, Omega
Gas E: Chevron*
W: Phillips 66*, Shell*
AServ W: Shell
13 **(49)** Kinard Bridge Road, Lenox
FStop E: Dixie*
Gas W: Shell*, Texaco(D)
Food E: Gabi's Cafe & Restaurant, Stella's Diner
W: Ice Cream Churn (Shell), Pit BBQ
Lodg E: Lenox Motel
AServ W: Texaco
(48) Rest Area - RR, Phones, Picnic, Vending (Southbound)
(47) Rest Area - RR, Phones, Picnic, Vending (Northbound)
12 **(45)** Barneyville Road
Lodg E: Red Carpet Inn
11 **(42)** Bus Loop 75, Rountree Bridge Rd
Food W: Plantation Inn Restaurant
Lodg W: ◆ Plantation Inn
Parks W: Reed-Bingham State Park
10 **(40)** GA 37, Adel, Moultrie ,

EXIT GEORGIA

Nashville, Lakeland
TStop W: Citgo Truck & Auto Plaza*(SCALES)
Gas E: BP, Citgo*(D), Shell*(D)
W: BP*, Citgo*
Food E: China House, Dairy Queen, Hardee's, McDonalds (Play Land), Pizza Hut, Subway, Waffle House
W: Baskin Robbins, Burger King (Play Land), Captain D's Seafood, Colombo Yogurt (Days Inn), Huddle House (Citgo), IHOP (Days Inn), King Frog Restaurant, Little Caesars Pizza, Mama's Table, Popeye's Chicken, Stuckey's (Days Inn), Taco Bell, Western Sizzlin'
Lodg E: ◆ Howard Johnson, Scottish Inns, ◆ Super 8 Motel
W: DAYS INN ◆ Days Inn, Hampton Inn
TServ W: Sam's Truck & Auto Repair
Med E: ✚ Hospital
ATM W: Bank of Lenox
Other W: Coin Laundry (Citgo)
9 **(37)** Adel
Gas E: BP
Food E: Rick's Chicken & BBQ
8 **(33)** Old Coffee Road, Cecil
Gas W: Amoco*(LP)
Lodg E: Stagecoach Inn
RVCamp W: Eagles Landing Resort
7 **(29)** U.S. 41N, GA 122, Hahira, Barney, Lakeland, Moody AFB
FStop W: Sav-a-Ton*(SCALES)

EXIT GEORGIA

Gas E: Citgo*
W: BP*(D)(LP)
Food W: Apple Valley Cafe
Lodg W: Hahira Inn
AServ W: BP
(24) Weigh Station - Both Directions
6 **(22)** U.S. 41S, North Valdosta Road
Gas E: BP*, Shell* (see our ad this page)
W: Citgo*(D)
Food W: Burger King (Citgo), Dairy Queen, Stuckey's
Lodg W: DAYS INN ◆ Days Inn
AServ E: Prince Chevrolet, Mazda, Valdosta Lincoln
RVCamp W: Shady Oaks Campground
Med E: ✚ Hospital
5 **(18)** GA 133, Valdosta, Moultrie, Valdosta State University
Gas E: Amoco*(CW)(LP), Citgo*, Shell*, Texaco*(D)
W: BP*(LP), Exxon Gas n Go*

XIT GEORGIA | EXIT GEORGIA | EXIT GEORGIA

Food E: Applebee's, Arby's, Cracker Barrel, Denny's, El Toreo, Fazoli's Italian Food, Hardee's (Playground), KFC, Krystal, Longhorn Steakhouse, McDonalds (Play Land), Ole Time's Country Buffet, Outback Steakhouse, Quincy's Family Steakhouse, Red Lobster, Ruby Tuesday, Taco Bell, Texas Roadhouse, Waffle House, Wendy's

Lodg E: Club House Inn (see our ad on this page), AAA Fairfield Inn (see our ad this page), Hampton Inn, AAA Holiday Inn (see our ad this page), Jameson Inn, Jolly Inn, AAA Quality Inn (see our ad this page), Scottish Inns, Travelodge
W: AAA Best Western

AServ E: Penske Auto Service, Sears
W: BP

RVCamp W: **Riverpark Campground**

Med E: ✚ **Hospital**

ATM W: Gas n Go

Other E: **Visit Valdosta (see our ad this page)**

4 **(16)** U.S. 84, U.S. 221, Ga. 94, Valdosta, Quitman

FStop W: **Texaco***(SCALES)

Gas E: Amoco*(CW)(LP) (24 Hrs), BP*(D), Citgo*, Phillips 66, **Shell*(D) (see our ad this page)**
W: Phillips 66

Food E: Aligatou Chinese & Japanese, Burger King, IHOP, McDonalds (Play Land), Old South BBQ, Pizza Hut, **Shoney's**, Waffle House
W: Austin's Steakhouse, Huddle House (Texaco), Huddle House

Lodg E: Big 7 Motel, Motel 6, ◆ **Quality Inn**, ◆ **Ramada Limited**, AAA Rodeway Inn, **Shoney's**
W: Briarwood Motel, AAA Comfort Inn, Villager Lodge

AServ E: Phillips 66
W: Phillips 66

Med E: ✚ **Hospital**

ATM W: Texaco

3A **(13)** Valdosta, Old Claytville Rd.

3 **(11)** GA 31, Valdosta, Valdosta Airport

FStop E: **Speedway***(LP)(SCALES)

Gas E: Texaco
W: BP*(LP), Texaco

Food E: **Starvin Marvin (Speedway), Subway (Speedway)**

Lodg E: **Villager Lodge**

AServ W: Texaco

2 **(4)** GA 376, Lakes Blvd., Lake Park

FStop E: **Phillips 66***
W: **Citgo Travel Center*(D), Phillips 66***

Gas E: Amoco*(CW), Chevron*(LP), Racetrac*, Shell*, Texaco*

Food E: Chick-fil-A (Playground), Farm House Restaurant, Fine Chinese Restaurant, Hardee's, Shoney's, Subway, Waffle House
W: **Baskin Robbins (Citgo Travel Center), Burger King (Citgo Travel Center), Cracker Barrel, McDonalds**, Pizza Hut, Taco Bell, Wendy's

Lodg E: ◆ Holiday Inn Express, ◆ Shoney's
W: DAYS INN AAA Days Inn (see our ad this page), AAA Travelodge

RVCamp E: **Eagle's Roost Campground, Giant Recreation World (Service & Parts)**

ATM E: Park Avenue Bank
W: Phillips 66

Other E: **Book Warehouse, Lakepark Mill Store Plaza**
W: **Factory Stores of America - Lake Park (see our ad this page)**

(3) **GA Welcome Center - Tourist Info, RR, Phones, Picnic, Vending, RV Dump (Northbound)**

1 **(2)** Bellville (Florida), Lake Park

TStop E: **National Travel Center*(SCALES) (BP Gas)**
W: **Flying J Travel Plaza*(D)(LP)(SCALES) (RV Dump)**

Gas E: Shell*, Texaco*

Food E: **Dairy Queen, Town & Country Restaurant**
W: **Country Market (Flying J)**

Lodg E: Ramada Inn
W: ◆ **Country Hearth Inn**

AServ E: Shell

EXIT GEORGIA/FLORIDA

RVCamp W: **Holiday Campground**
ATM E: National Travel Center
W: Flying J Travel Plaza

↑GEORGIA

↓FLORIDA

(470) FL Welcome Center - RR, Phones, Vending, Picnic, Tourist Info (Southbound)

87 **(468)** FL 143, Jennings
FStop W: **Exxon***(D)
Gas E: Chevron*, Texaco*
W: Amoco*(LP)
Food W: **Burger King (Exxon FS)**
Lodg E: AAA Quality Inn
W: Jennings House Inn, North Florida Inn & Suites
RVCamp W: **Jennings Campground**

86 **(462)** FL 6, Jasper, Madison
TStop W: **Sheffield's TS***
FStop E: **Amoco***(D), **Exxon***
Gas E: Lyman Walker's, RaceWay*
W: Chevron*, Shell*
Food E: **Burger King (Amoco), Huddle House (Exxon FS)**
W: **Catfish House (Sheffield's TStop)**
Lodg E: DAYS INN AAA Days Inn
W: 8 Motel, **Scottish Inns**
AServ W: Tullock & Sons
TServ W: **Sheffield 's**
Parks W: **Suwannee River State Park**
Other W: **Laundry (Sheffield's TS)**

85 **(453)** U.S.129, Live Oak, Jasper
Gas E: Texaco*(CW)
W: BP*, Shell*, Spur*(D)
Food E: Dairy Queen (Texaco)
W: Icecream Churn
RVCamp W: **Spirit of the Suwannee Campground**

(447) **Weigh Station, Agricultural Inspection (Both Directions)**

(445) **Rest Area - RR, Phones, Picnic, Vending (Southbound)**

(442) **Rest Area - RR, Phones, Picnic, Vending (Northbound)**

84 **(441)** FL 136, White Springs, Live Oak
FStop W: **A-1 Fuel Stop* (Citgo Gas)**
Gas E: BP* (24 Hrs), Gate*(D), Texaco*
W: Shell*(LP)
Food E: 3B's Restaurant, **McDonalds**
W: **A-1 Fuel Stop**
Lodg E: U.S. Inn
W: **Colonial House Inn, Scottish Inns**
RVCamp E: **Kelly RV Park, Lee's Country Camping**
Parks E: **Stephen Foster State Culture Center**
Other E: **State Farmer's Market**

83 **(436)** Junction I-10, Jacksonville, Tallahassee

82 **(429)** U.S. 90, Live Oak, Lake City
Gas E: BP*, Chevron*(CW), Exxon, Texaco*(D)
W: Amoco*, BP*, Chevron*, Citgo*(D), Shell*, Spur*, Texaco*(D)
Food E: Applebee's, Arby's, Burger King (Playground), **Cracker Barrel**, Dairy Queen, Fazoli's Italian Food, Hardee's, IHOP, KFC, McDonalds

EXIT FLORIDA

(Play Land), Pizza Hut, Red Lobster, Santilli's Italian Restaurant, Sonny's BBQ, Subway, TCBY Yogurt, Taco Bell, Texas Roadhouse, Waffle House, Wendy's
W: Boarding House, Bob Evans Restaurant, Long John Silvers (Amoco), Shoney's, Waffle House
Lodg E: A-1 Inn, Cypress Inn, DAYS INN AAA **Days Inn**, AAA Driftwood Motel, Executive Inn, Howard Johnson, AAA **Knight's Inn (see our ad this page)**, ◆ Rodeway Inn, Scottish Inns, AAA Villager Lodge
W: Best Western, ◆ Comfort Inn, Econolodge, Hampton Inn, ◆ Holiday Inn, Motel 6, Ramada, AAA Red Carpet Inn, Roadmaster Inn, Travelodge
AServ E: Exxon, Rountree-Moore Ford, Mercury, Lincoln, Toyota, Tire Mart
RVCamp E: **Inn-Out RV Camp**
Other E: Car Wash, **Wal-Mart Supercenter (Pharmacy)**
W: **Addison's Animal Hospital, Florida Sports Hall of Fame, Tourist Information**

81 **(424)** FL 47, Fort White, Lake City
Gas E: Texaco*(D)
W: Amoco*, BP*(D), Chevron*, Citgo*(LP), Express*(D)(LP)
Food W: Anne's Cafe, Cecil's (Super 8 Motel), Icecream Churn (Express), Little Caesars Pizza (Chevron), P.D.Q. Pizza, Subway (Express)
Lodg W: **Motel 8**, ◆ Super 8 Motel
RVCamp W: **Casey Jones Campground**
ATM E: Texaco

80 **(415)** U.S. 41, U.S.441, Lake City, High Springs
TStop W: **L & G Truckstop***(SCALES) **(Spur Gas)**
FStop W: **Amoco***(D), **BP***
Gas E: B & B Gas 'N Go*, Chevron*(D), Mini-Mart*
W: Citgo, Sunshine Food Store
Food W: Huddle House, L & G Restaurant (L & G TS), Subway
Lodg E: **Red Carpet Inn**, Traveler's Inn
W: **Diplomat Motel**, AAA **Econolodge**
AServ W: Citgo
TServ W: **L & G TS 24 Hr. Service & Parts**
RVCamp W: **Wagon Wheel Campground**
Parks W: **O'Leno State Park**

(413) **Rest Area - RR, Phones, Vending, Picnic (Both Directions)**

79 **(406)** CR 236, High Springs, Lake Butler

EXIT FLORID

Gas E: Chevron*, Texaco* (Florida Welcome Station)
Food E: Icecream Churn (Chevron)
RVCamp W: **High Springs Campground**

78 **(400)** U.S. 441, Alachua, High Springs
Gas E: BP*
W: Amoco*(CW), Citgo*, Lil' Champ*, Mobil*
Food E: McDonalds (Play Land), Pizza Hut, Sonny's BBQ, Waffle House
W: **Dairy Queen, Hardee's**, Huddle House (Mobil)
Lodg E: AAA Comfort Inn, Travelodge
W: DAYS INN Days Inn, ◆ Ramada Limited
RVCamp E: **Traveler RV Campground**

77 **(391)** FL 222, Gainesville
Gas E: Chevron*(D), Mobil*(D)
W: Texaco*(D)
Food E: Wendy's
W: Food Court (Texaco)
AServ W: Sal's Auto Service
Other E: **Aalatash Animal Hospital**

76 **(388)** FL 26, Gainesville, Newberry
Gas E: Chevron*, Citgo*, Shell*(CW), Speedway*, Texaco*(D)(CW)
W: BP*(CW), Chevron*(D), Mobil*(CW)
Food E: Bono's Pit BBQ, Boston Market, Burger King, Long John Silvers, McDonalds, Morrison's Cafeteria, Perkins Family Restaurant, Red Lobster, Subway, Wendy's
W: Bagels & Such, Boston Seafood, Domino's Pizza, Hardee's, Jackpot Subs, Little Caesars Pizza, Lucky Lee's, Maui Teriyaki, Napolantano' Pizza Pasta, Pizza Hut, Rocky's Ribs, Shoney's, TCBY Yogurt, Taco Bell, Waffle House
Lodg E: Budget Lodge, AAA La Quinta Inn
W: DAYS INN Days Inn, Econolodge, Fairfield Inn, AAA Holiday Inn
AServ E: Texaco
W: Discount Auto Parts, Jiffy Lube, Pep Boys Au Center
Med E: ✚ **Hospital**
ATM E: First Union
W: Barnett Bank
Other E: **Oaks Mall**
W: **Gainesville Animal Hospital, K-Mart (& Pharmacy), Publix Supermarket, Winn Dixie Supermarket (Pharmacy)**

75 **(385)** FL. 24, Gainesville, Archer
Gas E: Chevron* (24 Hrs)
W: Mobil*
Food E: Bob Evans Restaurant, Burger King, Captain D's Seafood, Gainesville Ale House, Imperial Garden Chinese, Kenny Rogers Roasters, McDonalds, Shoney's, Sonny's BBQ, Texas Roadhouse Steaks, Wendy's
Lodg E: ◆ Cabot Lodge, ◆ Courtyard by Marriott, Hampton Inn, Motel 6, ◆ Ramada Limited, Sup 8 Motel
AServ E: Master Care Auto Service, Midas Muffler & Brakes, Tuffy Auto Service
Other E: **Museum of Natural History, Target Department Store**

74 **(384)** FL 121, Gainesville, Williston
Gas E: Amoco*(CW) (24 Hrs), Citgo*(D)
W: Chevron*(D) (24 Hrs), Lil' Champ*, Shell*(D)
Food W: Icecream Churn (Shell), The Chuckwagon
Lodg W: Briarcliff Inn
AServ E: Amoco

Bold red print shows RV & Bus parking available or nearby

EXIT		FLORIDA

(383) **Rest Area - RR, Phones, Picnic (Both Directions)**

73 **(375)** CR 234, Micanopy
- **FStop** **W:** Citgo*
- **Gas** **E:** Fina*, Shell*(LP)
  - **W:** Amoco, Texaco
- **Food** **E:** Icecream Churn
- **Lodg** **W:** Scottish Inns
- **AServ** **W:** Amoco*
- **ATM** **E:** Shell
- **Parks** **W:** Paynes Ferry State Preserve

72 **(368)** CR 318, Irvine, Orange Lake
- **TStop** **E:** Petro Truck Plaza*(D)(LP)(SCALES) (RV Dump)
- **Gas** **E:** Amoco*, Citgo*
  - **W:** BP(D)
- **Food** **E:** Dairy Twirl (Citgo), Iron Skillet (Petro), Jim's Pit BBQ, Wendy's
- **AServ** **E:** Amoco
  - **W:** BP
- **TServ** **E:** Petro
- **RVCamp** **W:** Callowood Campground

71 **(358)** Fl. 326
- **TStop** **E:** Checkered Flag Truck Plaza*(SCALES) (BP Gas)
- **FStop** **E:** Speedway*(SCALES)
  - **W:** Chevron* (24 Hrs)
- **Gas** **E:** Fina*(D)
- **Food** **E:** Checkered Flag Restaurant, Hardee's (Speedway), Icecream Churn
  - **W:** Dairy Queen, Gator's Restaurant
- **TServ** **E:** Mid-Fla. Truck Repair
- **ATM** **E:** Speedway

70 **(354)** U.S. 27, Ocala, Silver Springs
- **Gas** **E:** Amoco*(CW), Super Test*(D)
  - **W:** BP*, Chevron*(D), Shell*(D), Texaco*(D)
- **Food** **E:** Big Rascal BBQ, Krystal, Raintree Restaurant (Quality Inn)
  - **W:** AJ's (Days Inn), Damon's Restaurant, Waffle House
- **Lodg** **E:** AAA Quality Inn
  - **W:** AAA Budget Host Inn, AAA Days Inn, Howard Johnson, ◆ Ramada, Red Coach Inn
- **AServ** **E:** Amoco
- **TServ** **W:** Raney's Truck Center
- **RVCamp** **W:** Arrowhead RV Camp, Oaktree Village

69 **(353)** FL 40, Ocala, Silver Springs
- **FStop** **E:** BP*(D)
- **Gas** **E:** Chevron*, Citgo*, Racetrac*
  - **W:** Glen Acres*, Texaco*
- **Food** **E:** Comedy House Dinner Theater (Holiday Inn), Gabriel's Restaurant (Holiday Inn), Icecream Churn (Citgo), McDonalds, Mr. Sub (Citgo), Wendy's
  - **W:** Denny's (Horne's), Rocky's Italian, Waffle House
- **Lodg** **E:** ◆ Days Inn, Holiday Inn (see our ad this page), ◆ Motor Inns, Scottish Inns
  - **W:** Comfort Inn, Horne's, Super 8 Motel
- **RVCamp** **W:** Holiday Travel Park
- **ATM** **E:** Citgo, Racetrac
- **Other** **E:** Ocala Breeders Sale Complex (Bloodstock Agents & Auctioneers), Silver Springs Theme Park (Approx. 10 Miles), Wild Waters
  - **W:** Companion Animal Clinic

68 **(350)** FL 200, Hernando, Dunnellon
- **Gas** **E:** Citgo, Racetrac*, Texaco*(D)
  - **W:** Amoco*(CW), Chevron*
- **Food** **E:** Bob Evans Restaurant, Chick-fil-A, Chili's, Olive Garden, Perkins, Po Folks, Quincy's Family Steakhouse, Red Lobster, Ruby Tuesday, Shoney's, T.G.I. Friday's
  - **W:** Cracker Barrel, Dunkin' Donuts, Steak 'N Shake, Waffle House
- **Lodg** **E:** ◆ Hampton Inn, ◆ Hilton
  - **W:** AAA Budgetel Inn, ◆ Courtyard
- **AServ** **E:** Citgo
  - **W:** Don Olson Tire & Auto Center
- **RVCamp** **W:** Camper Village
- **ATM** **E:** Racetrac
  - **W:** SunTrust Bank
- **Other** **E:** Barnes & Noble, Petsmart
  - **W:** Disney Travel Center, Sam's Club

(346) **Rest Area - RR, Phones, Vending, Picnic (Both Directions)**

67 **(341)** CR 484, Belleview
- **TStop** **W:** Pilot Travel Center*(SCALES)
- **Gas** **E:** Chevron*, Citgo*, Exxon*
  - **W:** Amoco
- **Food** **E:** Sonny's BBQ
  - **W:** Arby's (Pilot TS), Dairy Queen (Pilot TS), McDonalds (Play Land), Waffle House
- **AServ** **W:** Amoco
- **TServ** **W:** 484 Tire Service
- **ATM** **W:** Pilot TS
- **Other** **E:** Drag Racing Museum
  - **W:** Ocala Factory Stores

(338) **Weigh Station (Both Directions)**

66 **(329)** FL 44, Wildwood, Inverness
- **TStop** **W:** Wildwood Travel Center*(SCALES)
- **FStop** **E:** Gate*(D)(SCALES)
  - **W:** Speedway*(D)(SCALES), United 500* (24 Hrs)
- **Gas** **E:** Amoco* (24 Hrs), Shell(D), Texaco*
- **Food** **E:** Burger King (Play Land), Dairy Queen, McDonalds, Shoney's, Steak 'N Shake (Gate), Wendy's
  - **W:** Atrium Restaurant (Wildwood Travel Center), KFC, Pizza Hut (Wildwood Travel Center), Starvin' Marvin (Speedway), Subway (Wildwood Travel Center), Waffle House
- **Lodg** **W:** Budget Suites, Days Inn, Knight's Inn, Super 8 Motel
- **AServ** **E:** Amoco
- **TServ** **W:** Tommy's Tire Shop, United 500 Truck & Road Service, Wildwood Travel Center
- **TWash** **W:** Wildwood
- **RVCamp** **E:** KOA Campground

65 **(328)** Florida Tnpk., Orlando, Miami

64 **(321)** CR 470, Sumterville
- **TStop** **E:** Fuel City*
- **Gas** **W:** Chevron*
- **Food** **E:** Smiley's Restaurant (Fuel City)
  - **W:** Pinky's BBQ
- **RVCamp** **W:** Countryside RV Park, Idle Wild Lodge & RV Park, Pan-Vista Lodge & RV Park, Turtle Back RV Resort
- **Other** **W:** Lake Panasoffkee

63 **(314)** FL48, Bushnell
- **Gas** **E:** BP*, Mobil*, Texaco*
  - **W:** Citgo*(D), Shell*(D)
- **Food** **E:** Dairy Queen (Mobil), KFC, Leopard Inn, Stuckey's (Mobil)
  - **W:** McDonalds (Play Land), Waffle House
- **Lodg** **E:** AAA Best Western
- **AServ** **W:** Citgo
- **RVCamp** **E:** Oaks Campground, Red Barn RV Park

62 **(309)** CR 476, Webster
- **RVCamp** **E:** Safire Campground, Sumter Campground

(307) **Rest Area - RR, Phones, Picnic (Both Directions)**

61 **(302)** U.S. 98, FL 50, Dade City, Brookville
- **Gas** **E:** Amoco, Racetrac*, Shell*
  - **W:** Mobil*, Shaw's Service Station(D)
- **Food** **E:** Five Star Pizza Palace, McDonalds (Play Land), River House Pub, Toni's Italian, Waffle House
  - **W:** Subway (Mobil)
- **Lodg** **E:** Days Inn (see our ad this page)
  - **W:** ◆ Hampton Inn, ◆ Holiday Inn
- **AServ** **E:** Amoco, C & F Auto
  - **W:** Shaw's Service

EXIT	FLORIDA
RVCamp	E: **Florida Campland, Tall Pines RV Park**
	W: **Hidden Valley RV Camp**
Med	W: ✚ **Hospital**
ATM	E: SunTrust
Other	E: **Winn Dixie Supermarket (Pharmacy)**
60	**(294)** CR 41, Dade City
59	**(287)** FL 52, Dade City, Newport Richey
TStop	E: **Flying J Travel Plaza***(LP)(SCALES) **(RV Dump)**
FStop	W: **Texaco***(LP)(SCALES)
Gas	W: Shell*
Food	E: **Country Market (Flying J)**
	W: **Blimpie's Subs (Texaco), Waffle House**
Med	E: ✚ **Hospital**
58	**(280)** FL 54, Land O' Lakes, Zephyrhills
Gas	E: Lucky Food Center*, Racetrac*, Shell*, Texaco*(CW)(LP)
	W: Amoco*(D), Circle K*(D), Citgo*(D)
Food	E: ABC Pizza, Brewmaster Steakhouse, **Burger King**, Subway, Waffle House
	W: **Cracker Barrel**, Denny's, McDonalds, Peacock's, Shoney's
Lodg	W: Comfort Inn, **Masters Inn**, Sleep Inn
ATM	E: Racetrac
Other	E: **WalGreens, Winn Dixie Supermarket**
(278)	**Rest Area - RR, Phones, Picnic, Vending (Both Directions)**
57	**(275)** Junction I-275 South, Tampa
Lodg	E: Days Inn (see our ad this page)
56	**(271)** CR. 581, Bruce B. Downs Blvd.
55	**(267)** CR. 582 A., Fletcher Ave
Lodg	W: ◆ Courtyard by Marriott, ◆ Sleep Inn
Med	W: ✚ **Hospital**
54	**(266)** FL. 582, Fowler Ave, Temple Terrace
Food	W: Shoney's
Lodg	W: ◆ Shoney's Inn
RVCamp	E: **Happy Travelers RV Camp**
Other	W: **Busch Gardens, Museum of Science & Industry**
53	**(262)** Jct I-4, Orlando, Tampa, Lakeland
52AB	**(260)** FL. 574 , Mango
Gas	E: Shell (see our ad this page)
Lodg	W: Camberly Plaza
AServ	E: Southern Auto Air, Transworld Transmission
51	**(257)** FL 60, Brandon, Tampa
Gas	W: Shell*
Food	E: Bennigan's, Grady's American Grill, Olive Garden, Red Lobster, Romano's Macaroni Grill, Tia's Tex Mex
	W: Bob Evans Restaurant, Burger King, Hooters, McDonalds, Sonny's BBQ, Subway, Sweet Tomatoes, Villa Rina Pizza, Wendy's
Lodg	W: ◆ Budgetel Inn, **Days Inn (see our ad this page)**, Hampton Inn, ◆ Red Roof Inn, Fairfield Inn (see our ad this page)
Other	E: **Barnes & Noble, Brandon Town Center, Target**
	W: Car Wash
50	**(256)** Cross Town Expy, Tampa
49	**(254)** U.S. 301, Riverview

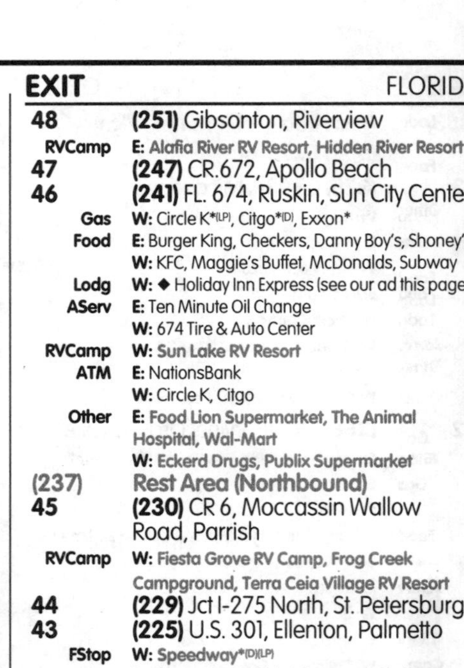

EXIT	FLORIDA
48	**(251)** Gibsonton, Riverview
RVCamp	E: **Alafia River RV Resort, Hidden River Resort**
47	**(247)** CR.672, Apollo Beach
46	**(241)** FL. 674, Ruskin, Sun City Center
Gas	W: Circle K*(LP), Citgo*(D), Exxon*
Food	E: Burger King, Checkers, Danny Boy's, Shoney's
	W: KFC, Maggie's Buffet, McDonalds, Subway
Lodg	W: ◆ Holiday Inn Express (see our ad this page)
AServ	E: Ten Minute Oil Change
	W: 674 Tire & Auto Center
RVCamp	W: **Sun Lake RV Resort**
ATM	E: NationsBank
	W: Circle K, Citgo
Other	E: **Food Lion Supermarket, The Animal Hospital, Wal-Mart**
	W: **Eckerd Drugs, Publix Supermarket**
(237)	**Rest Area (Northbound)**
45	**(230)** CR 6, Moccassin Wallow Road, Parrish
RVCamp	W: **Fiesta Grove RV Camp, Frog Creek Campground, Terra Ceia Village RV Resort**
44	**(229)** Jct I-275 North, St. Petersburg
43	**(225)** U.S. 301, Ellenton, Palmetto
FStop	W: **Speedway***(D)(LP)
Gas	E: Chevron*(CW) (24 Hrs), Racetrac*, Shell*(D)(CW)
Food	E: Checkers Burgers, McDonalds, Wendy's
	W: Shoney's, **Subway (Speedway FS)**, Waffle House
Lodg	W: ◆ **Best Western**, Shoney's Inn
RVCamp	E: **Ellenton Gardens RV Park**
	W: **Bay Palm RV Park**
ATM	E: Barnett Bank, Racetrac, Shell
Other	E: **Gulf Coast Factory Shops, K-Mart, Publix Supermarket, WalGreens**
42AB	**(221)** FL 64, Bradenton, Zolfo Springs, Wauchula
FStop	W: **Circle K***
Gas	W: Chevron*(CW) (24 Hrs), Citgo*(D), Racetrac*, Shell* (24 Hrs)
Food	W: Blimpie's Subs (Citgo), Burger King, **Cracker Barrel, McDonalds (Play Land)**, Subway, Waffle House
Lodg	W: Comfort Inn, **Days Inn (see our ad this page)**, Econolodge, Luxury Inn, Motel 6
RVCamp	W: **Winter Quarters RV Resort**
Med	W: ✚ **Hospital**
ATM	W: Circle K
Other	E: **Little Manetee State Recreational Area**
41	**(218)** FL. 70, Bradenton, Arcadia
Gas	W: Shell*(CW)
Food	W: Bogey's Grill & Pub, China Village, Dakota's Bar & Grill, Denny's, Hungry Howie's, **Publix Cafe**, The Orange Dipper
RVCamp	W: **Horseshoe Cove RV Resort, Pleasant Lake RV, Tropical Gardens RV Park**
ATM	W: Barnett Bank, NationsBank, Shell
Other	W: **Publix Supermarket**
40	**(214)** University Pkwy, Sarasota, International Airport

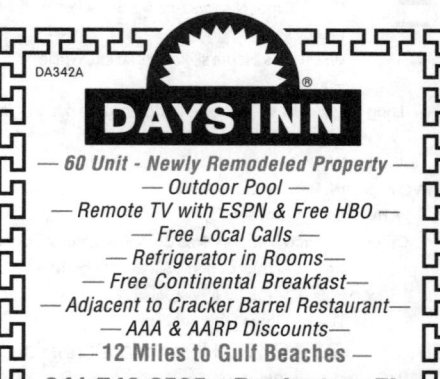

Bold red print shows RV & Bus parking available or nearby

EXIT FLORIDA

Lodg W: Hampton Inn (see our ad this page), Sleep Inn (see our ad this page)
Food W: Applebees, Checkers, KFC, McDonalds, Outback Steak House, Taco Bell
Other W: Publix, Sarasota Outlet Center, Shopping Center, Wal-Mart, Walgreens

39 **(210)** FL. 780, Sarasota, Gulf Beaches
Food E: Applebees, Checkers, Sante Fe Steak house
Lodg W: Wellesley Inns (see our ad this page), Azure Ties Resort
RVCamp E: Sun 'N Fun RV Resort
Other E: Publix, Shopping Center, Target

38 **(208)** Fl. 758, Bee Ridge Road, Sarasota
Gas W: Mobil*(D)
Food W: Arby's, Bagel Cafe, Blimpie's Subs (Texaco), Checkers Burgers, Chili's, Domino's Pizza, McDonalds (Play Land), Sarasota Ale House, Subway (Mobil), Taco Bell

EXIT FLORIDA

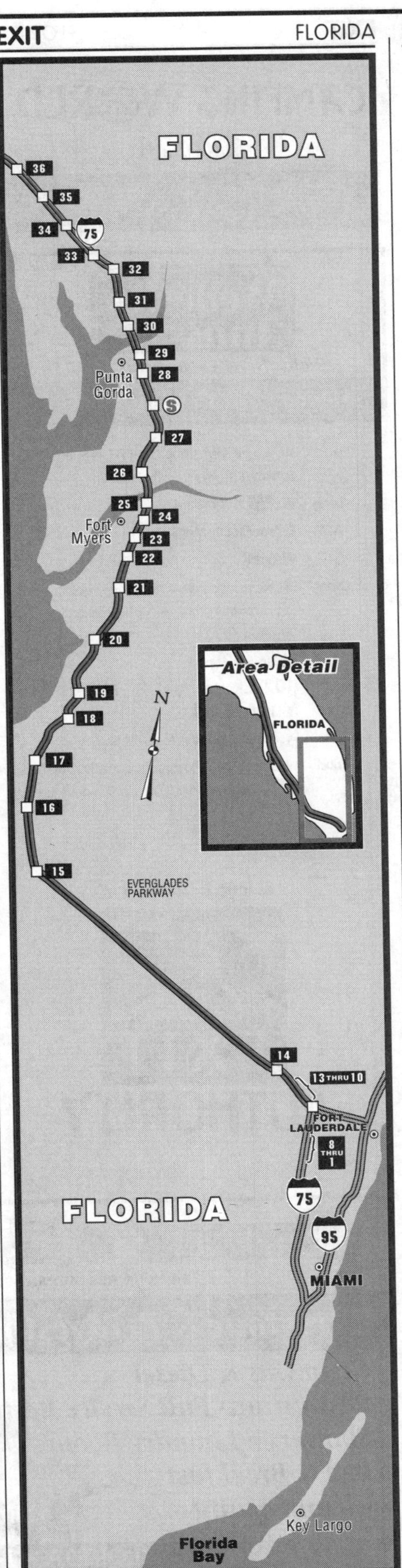

EXIT FLORIDA

Lodg W: Hampton Inn
Med W: Hospital
ATM W: Barnett Bank, SunTrust
Other W: K-Mart, Publix Supermarket (Pharmacy), WalGreens (Pharmacy)

37 **(206)** FL 72, Sarasota, Arcadia
Gas W: Citgo*(D), Mobil*(CW)
Food W: Burger King, Waffle House, Wendy's, Wings-N'-Weenies
Lodg W: Comfort Inn, Ramada Inn
RVCamp W: Gulf Beach Campground, Windward Isle Adult RV Park
ATM W: Citgo, First State Bank of Sarasota, Mobil, Suntrust
Parks E: Myakka River State Park

36 **(200)** FL. 681 South, Venice, Osprey
Med W: Hospital

35A **(195)** Venice, Laurel

35 **(194)** Jacaranda, Venice
Gas W: Hess*
Food W: Cracker Barrel
Med W: Hospital

34 **(191)** River Road, North Port, Englewood
RVCamp E: Ramblers Rest Campground
W: Venice Campground

33 **(182)** North Port

32 **(180)** North Port, Port Charlotte

31 **(171)** CR 769, Arcadia, Port Charlotte
FStop W: Mobil*(D)
Gas W: Hess*(D), Texaco*(D)
Food W: Blimpie's Subs (Mobil), Burger King (Play Land), Cracker Barrel, Dairy Queen (Texaco), Dunkin' Donuts (Mobil), McDonalds (Play Land), Subway, Taco Bell, Zorba's Pizza
ATM W: Mobil
Other W: Eckerd Drugs, Publix Supermarket (Pharmacy)

30 **(167)** CR. 776, Charlotte Harbor

29 **(164)** U.S. 17, Punta Gorda, Arcadia
RVCamp W: KOA Campground

28 **(161)** CR 768, Punta Gorda
TStop W: Speedway*(SCALES)
FStop W: Hess*(D)
Gas W: Amoco*
Food W: Blimpie's Subs (Speedway), Burger King, Church's Fried Chicken (Speedway), Dairy Queen (Amoco), McDonalds, Pizza Hut, Shoney's, Subway (Amoco), Taco Bell, Waffle House, Wendy's
Lodg W: DAYS INN ◆ Days Inn, AAA Motel 6
RVCamp E: KOA Campground
W: Alligator RV Park, Gulf View RV Resort
Other E: Charlotte County Airport
W: Tourist Information

(163) Rest Area - RR, Phones, Picnic
(160) Weigh Station - Both Directions.

27 **(158)** CR 762, Cape Coral, N. Ft Myers
RVCamp E: Sun & Shade
W: Punta Gorda RV Resort, Raintree RV Resort
Other E: Babcock/Webb Wildlife Mgmt

26 **(143)** FL 78, N. Ft Myers, Cape Coral
Gas W: Racetrac* (24 Hrs)
RVCamp E: Seminole RV Campground, Up River Campground
W: Holiday RV Camp (see our ad this page), Pioneer Village RV Mobile Home Resort
ATM W: Racetrac

25 **(141)** FL 80, Fort Myers, La Belle

EXIT FLORIDA

Gas	E: Citgo*(CW) (24 Hrs)
	W: Hess*(D), Racetrac*
Food	E: Cracker Barrel, Waffle House
	W: Chinese Kitchen, Hardee's, Juicy Lucy's Burgers, Perkins Family Restaurant, Pizza Hut, Sonny's BBQ, Subway, Taco Bell
AServ	W: Martin's General Repair
RVCamp	W: North Trail RV Service & Sales
ATM	E: Citgo
	W: First Union Bank
24	**(139)** Fort Myers, Luckett Road
TStop	W: Pilot Travel Center*(SCALES) (RV Dump)
Food	W: Grandma's Kitchen (Pilot), Subway (Pilot)
TServ	W: Pilot Travel Center
RVCamp	W: Gulf Coast RV Service, Lazy J Adventures, Camping World (see our ad this page)
ATM	W: Pilot
23	**(138)** FL. 82, Fort Myers, Immokalee
Gas	W: Racetrac*, Speedway*(LP)
ATM	W: Racetrac, Speedway
Other	W: Edison & Ford Estates, Imaginarium
22	**(136)** FL. 884, Fort Myers, Colonial Blvd, Lehigh Acres
Med	W: ✚ Hospital
Other	W: Nature's Center & Planetarium
21	**(132)** Daniels Pkwy, SW. International Airport, Cape Coral
Gas	W: Citgo*, Hess* (24 Hrs), Racetrac*, Shell*(CW)
Food	W: Arby's, Burger King (Playground), Denny's, McDonalds, Shoney's, Taco Bell, Waffle House, Wendy's
Lodg	W: AAA Comfort Suites, ◆ Hampton Inn, AAA Sleep Inn
Med	W: ✚ Hospital
ATM	W: NationsBank (Shell)
(132)	Rest Area - RR, Phones, Vending (On East Side Of Exit 21)
20	**(128)** Alico Road, San Carlos Park
19	**(123)** Estero , CR 850
RVCamp	W: Covered Wagon RV Park, Shady Acres Travel Park
Other	W: Koreshan State Historic Site
18	**(116)** Bonita Springs, Gulf Beaches
FStop	W: Hess*(LP)
Gas	W: Amoco, Hess*
Food	W: McDonalds, Waffle House
Lodg	W: DAYS INN ◆ Days Inn
RVCamp	W: Bonita Beach Trailer Park, Bonita Lake RV Resort, Citrus Park RV Resort, Imperial Bonita RV Park
ATM	W: Hess, Hess FStop
17	**(112)** CR 846, Naples Park, Immokalee Rd.
RVCamp	W: Lake Sand Marino Resort
Parks	W: Delnor-Wiggins State Park
16	**(107)** CR 896, Naples, Golden Gate
Gas	E: Mobil*
	W: Chevron*(D) (24 Hrs), Shell*(D)(CW) (24 Hrs)
Food	E: McDonalds (Mobil), Patso Cafe

EXIT FLORIDA

	W: Burger King, Cappy's 19th Hole Sports Grill, Subway (Chevron), Waffle House
Lodg	W: AAA Knight's Inn
ATM	E: Amerisouth, Mobil
	W: Shell
Other	E: Crossroads Veterinary Clinic, Publix Supermarket (Pharmacy), WalGreens (Pharmacy)
	W: Naples Airport
15	**(102)** CR. 951, to FL. 84, Naples, Marco Island
Gas	W: Amoco* (24 Hrs), Mobil*(D), Shell*(D)
Food	W: Burger King (Playground), Checkers Burgers, Cracker Barrel, Dunkin' Donuts (Mobil),

EXIT FLORIDA

	McDonalds (Play Land), Subway (Mobil), Waffle House
Lodg	W: ◆ Budgetel Inn, Wellesley Inns (see our ad this page), ◆ Comfort Inn, Super 8 Motel
RVCamp	W: Endless Summer RV Camp, Kountree Kampinn, Naples RV Resort, Silver Lake RV Park
ATM	W: Amoco, Mobil, NationsBank (Shell)
Other	W: Tourist Information
14A	**(80)** FL. 29, Everglades City, Immokalee
Parks	E: Big Cypress National Preserve (17 Miles), Everglades National Park (22 Miles)
Other	E: Ted Smallwood's Store (25 miles)
(63)	Rest Area - RR, Phones, Vending, Picnic (Both Directions)
14	**(50)** Indian Reservation
FStop	E: Shell* (24 Hrs)
(38)	Recreational Area (Southbound)
(35)	Recreational Area
(32)	Recreational Area (Both Directions)
13AB	**(23)** U.S. 27, Miami, South Bay
12	**(22)** Arvida Pkwy.
11	**(21)** FL 84 West, Indian Trace
10	**(17)** Jct I-595, Sawgrass Exprwy, Fort Lauderdale , FL 869
8	**(15)** Arvida Pkwy West
7AB	**(14)** Griffin Road
6AB	**(12)** Sheridan St
Med	E: ✚ Hospital
5AB	**(9)** FL 820, Hollywood Blvd, Pines Blvd
4	**(7)** Miramar Pkwy
3B	**(5)** Florida Tnpk. South, Key West, Homestead
3A	**(6)** New186th St, Miami Gardens Dr
TStop	W: Jimmies Firestone (see our ad this page)
2	Northwest 138th St, Graham Dairy Road
1AB	FL. 826, Palmetto Exprwy

↑FLORIDA

Begin I-75

Bold red print shows RV & Bus parking available or nearby

EXIT	COLORADO
	Begin I-76
	↓COLORADO
1A	NE121, Wadsworth Blvd
1B	NE95, Sheridan Blvd
Food	S: Amici's Pizzeria Italian Restaurant, Great Panda Chinese Restaurant, Mr B's Roadhouse
Other	S: **The Great American Pet Castle**
3	U.S. 287, Federal Blvd
Gas	N: Total
Food	S: McDonalds, Panda Express, Subway, Taco House
Lodg	N: Alpine Rose Motel, North Valley Federal Motel
	S: Joy Motor Motel, Primrose Motel, White Rock Motel
	W: Weekly Motor Inn, Weekly Motor Motel
AServ	S: Munoz Auto Service
RVCamp	S: **Deluxe RV Park**
Other	N: Car Wash
	S: **Old Glory Fireworks**
4	Pecos St
5BA	Jct I-25, U.S. 36, Fort CollinsBoulder
6	Junction I-270E, Aurora
8	CO224, 74th Ave (Westbound Reaccess Only)
Gas	S: Diamond Shamrock*
Food	S: Butcher Block Cafe
ATM	S: Diamond Shamrock
9	Colorado Blvd
10	88th Ave
Gas	N: Conoco*
Food	N: A-Frame Cafe, Brenda Cafe
Lodg	N: ◆ **Super 8 Motel**
	S: ◆ Motel
AServ	N: Conoco
11	96th Ave
12	U.S. 85, Brighton, Greeley
16	CO2W, Commerce City
17	CO51, Brighton
TStop	N: **Texaco*(SCALES)**
Food	N: **Blimpie's Subs (Texaco), Tomahawk Restaurant (Texaco)**
TServ	N: **Texaco**
Med	N: ✚ **Hospital**
ATM	N: Texaco
Other	N: **CB Shop (Texaco), Chiropractic Clinic (Texaco)**
19	Barr Lake
20	136th Ave
RVCamp	N: **Barr Lake Campground(LP), Laundry Facilities**
21	144th Ave
22	Bromley Ln
Parks	S: **Barr Lake State Park**
23	144th St
25	Lochbuie
FStop	N: **Texaco(D)**
Food	N: **Cafe (Texaco)**
31	NE52, Hudson
Gas	S: Amoco*(D)(CW)
Food	S: **Longhorn Family Dining**, McDonalds, **Pepper Pod Restaurant**
AServ	S: L&C Auto Repair
RVCamp	S: **KOA Campground**
ATM	S: Amoco
Other	S: **Andersen Star Grocery, Fire Station, Post Office**
34	Kersey Rd
39	Keenesburg
Gas	S: Phillips 66*(D)
Food	S: **Charlie D's**, Korner Kitchen Cafe
Lodg	S: Keene Motel
ATM	S: Phillips 66
Other	S: **Post Office**
48	Roggen
Gas	N: Texaco*
	S: Amoco*
Lodg	N: **I-76 Motel**
	S: Prairie Lodge
AServ	S: Amoco
57	CR91
60	NE144, Orchard
64	Wiggins
66A	NE39, Goodrich, NE52W, Wiggins (Westbound Reaccess Only)
FStop	S: **Sinclair*(D)(LP) (24 Hr)**
Gas	N: Amoco*(D)
Food	N: **The Trophy Room/Amoco**
Parks	N: **Jackson Lake State Park**
Other	S: **UPS Drop Box**
66B	U.S. 34, Greeley, Estes Park
73	Long Bridge Rd
TServ	N: **Temco Truck, Trailer (Frontage Rd)**
(75)	**Weigh Station - Both Directions**
75A	Bus Loop I-76, US 34E
Food	S: Heinrich Restaurant
Lodg	S: The Madison, Wayward Wind Restaurant
RVCamp	S: **Wayward Wind Camp Ground**
Other	S: **State Patrol**
79	NE144, Weldona
80	NE52, Fort Morgan, New Raymer
Gas	S: Amoco, Conoco*, Phillips 66*(D)
Food	S: A & W Drive-In, **Arby's**, Dairy Queen, Hardee's, KFC, **McDonalds**, Taco John's
Lodg	S: ◆ Best Western, AAA Central Motel, DAYS INN Days Inn, **Super 8 Motel**
Med	S: ✚ **Hospital**
Other	S: **K-Mart, Museum**
82	Barlow Rd
FStop	N: **Texaco***
	S: **Coastal***
Food	N: **Old Fort Restaurant**
	S: **Coastal***
Lodg	N: **Econolodge**
Other	N: **Camping**
86	Dodd Bridge Rd
89	Hospital Rd
Med	S: ✚ **Hospital**
90BA	NE71, Brush, Lyman, to US34 Akron
FStop	N: **Texaco*(LP)**
Gas	S: Conoco*
Food	N: **Pizza Hut, Restaurant (Texaco, 24 Hr), Scotch 'n Steer**, Wendy's
	S: McDonalds
Lodg	N: Best Western
ATM	N: **Texaco**
92BA	U.S. 6E, I-76 Business to U.S. 34
95	Hillrose
102	Merino
(108)	**Rest Area - RR, Picnic, Vending, RV Dump (Both Directions)**
115	NE63, Atwood
FStop	N: **Sinclair*(LP)**
Gas	S: **Standard(D)**
Food	N: **Prairie-Land Cafe (Sinclair)**
	S: **Restaurant (Standard),** The Steak House Fine Dining
AServ	S: **Standard**
Med	N: ✚ **Hospital**
125B	I-76 Bus, US6 Sterling, Holyoke
FStop	S: **Total***
Food	S: Country Kitchen (Ramada), **Jackson's Cafe (Super 8 Motel), Restaurant/Days Inn (24 Hr)**
Lodg	S: DAYS INN Days Inn, Ramada, **Super 8 Motel**
RVCamp	S: **Buffalo Hills Campground(LP)**
Med	N: ✚ **Hospital**
Other	N: **Tourist Information**
134	Iliff
141	Proctor
149	NE55, Crook
FStop	S: **Texaco**
Food	S: **Cafe (Texaco)**
(151)	**Rest Area - Phones, Picnic, Vending, RV Dump, RR**
155	Red Lion Rd
165	Sedgwick
FStop	N: **Conoco*(D)**
Gas	N: Total*
Food	N: **Lucy's Place Cafe (Conoco),** Taco Bell (Total)
ATM	N: Total
172	Ovid
180	U.S. 385, Julesburg
TStop	N: **Flying J Travel Plaza***
FStop	S: **Conoco***
Gas	N: Texaco
Food	N: **Flying J Travel Plaza, Plat Valley Inn,** Sweden Cream (Ice Cream)
	S: **Wagon Wheel* (Conoco)**
Lodg	N: Holiday Motel, **Platt Valley Inn**
AServ	N: Texaco
TServ	N: **Flying J Travel Plaza**
Med	N: ✚ **Hospital**
Other	N: **Colorado Welcom Ctr (RV Dump, Full Facilities)**
	↑COLORADO
	Begin I-76

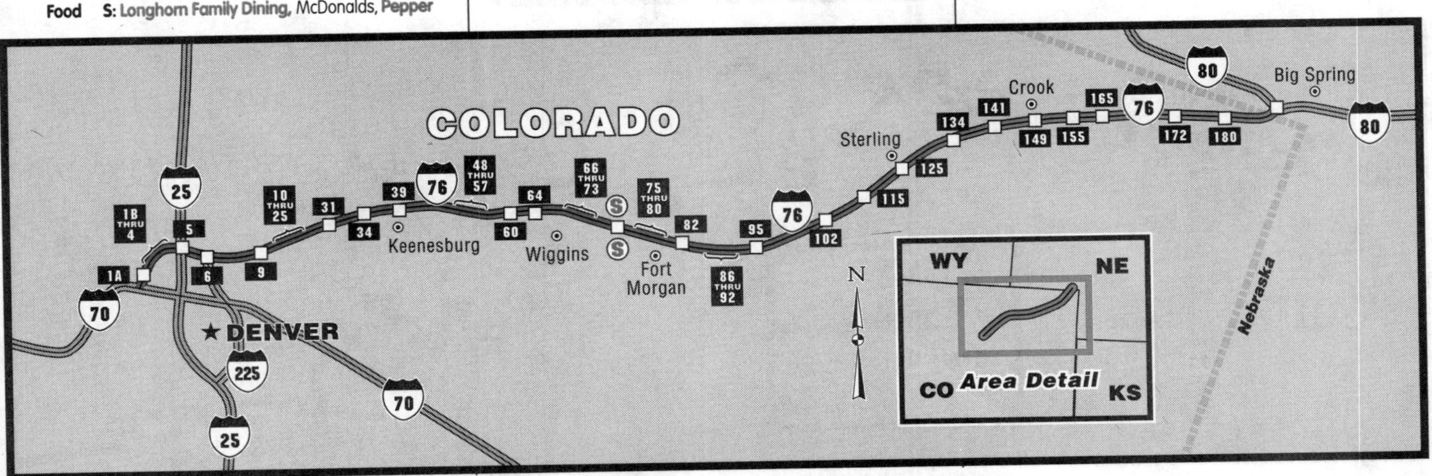

Bold red print shows RV & Bus parking available or nearby

EXIT		OHIO

↓OHIO

2 OH 3, Seville, Medina
- FStop **S: 76*(D)**
- Food N: Dairy Queen, Foxes Pizza, **Hardee's**, Subway
- Lodg N: **Comfort Inn**
- RVCamp N: **Maple Lakes Campground (1.9 Miles)**
  - S: **Camping**

(6) **Weigh Station (Both Directions)**
7 OH 57, Rittman, Medina
9 OH 94, North Royalton, Wadsworth
11 OH 261, Norton
- Gas N: Speedway*
- AServ S: American Tire/Auto

13AB OH 21, Cleveland, Massillon
14 Cleve - Mass Road
16 Barber Road
17A State St (Eastbound, Westbound Reaccess)
17 OH 619, Barberton, Wooster Road
18 Jct I-277, U.S. 224 East, Canton, Barberton (Left Lane Exit)
19 Kenmore Blvd
20 Jct I-77 North, Cleveland (Eastbound Exits Left)
21B Lakeshore, Bowery St
- Food E: AJ's Drive Through
- Med E: ✚ **Hospital**

21C OH 59 East, Downtown Akron (Northside)
22A Main St, Broadway
22B Grant St, Wolf Ledges
- Gas S: BP*
- Food S: McDonalds

23B OH 8 North Cuyahoga Falls (Left Exit Eastbound)
23A Junction I-77, Canton
24 Arlington St, Kelly Ave
25B Martha Ave, General St, Brittan Rd
26 OH 18, East Market St, Mogadore
- FStop N: **Shell***
- Gas S: Speedway*
- Food N: J.D.'s Restaurant, **Tomy's (Shell)**
  - S: East Side Cafe, Lamp Post Restaurant, **McDonalds**, Pizza, Subway
- AServ N: Goodyear Tire & Auto
  - S: Akron Auto Sales
- Other S: **Evergreen Mini-Market, Revco Drugs**

27 OH 91, Gilchrist Road, Canton Road

EXIT		OHIO

29 OH 532, Mogadore, Tallmadge
31 County Road 18, Tallmadge
33 OH 43, Kent, Hartville
- Lodg N: Days Inn (see our ad this page)

38 OH 5, OH 44, Ravenna
43 OH 14, Alliance
(46) **Rest Area - RR, Phones, Picnic**
48 OH 225, Alliance
54 OH 534, Lake Milton, Newton Falls
57 To OH 45, Bailey Road, Warren

Note: I-76 runs concurrent below with OHTNPK. Numbering follows OHTNPK.

15 **(60)** Jct I-80 East, to New York City, I-76 West to TNPK (Toll Road)

EXIT		OHIO/PENNSYLVANIA

16 **(233)** OH 7 Youngstown
- TStop S: **Penn Ohio Plaza*(SCALES) (Amoco)**
- FStop S: **Speedway*(LP)**
- Food N: Dairy Queen, Knights Inn, Smaldino's Italian
  - S: **Giuseppe's (PennOhio TS), Roadhouse Restaurant**
- Lodg N: Budget Inn, **Economy Inn, Knights Inn (see our ad this page)**, Super 8 Motel
  - S: **Davis Motel**, Rodeway Inn
- TServ S: **Penn Ohio Plaza (76)**
- ATM S: Penn Ohio TS
- Other S: **Coin Laundromat (Penn Ohio TS)**

16A **(234)** Jct I-680 North, Youngstown (Westbound)
(237) Glacier Hills Service Plaza
- FStop B: **Sunoco***
- Food B: **McDonalds**
- AServ B: Sunoco
- Other B: **Tourist Information**

(239) Toll Booth

↑OHIO

↓PENNSYLVANIA

Note: I-76 runs concurrent above with OHTNPK. Numbering follows OHTNPK.

(1) Toll Plaza
1A **(9)** Newcastle, PA 60, Pittsburgh (Toll Road , All Traffic Exits To The Northside)
2 **(13)** PA 18, Ellwood City, Beaver Falls (Traffic Exits To The Northside, Toll Booth)
- Food N: Restaurant (Holiday Inn)
  - S: Guiseppe's Italian, **Pino's (Conley Inn)**
- Lodg N: Beaver Valley Motel, Danny's Motel, Hilltop Motel, ◆ Holiday Inn, Lark Motel
  - S: **Conley Inn**
- AServ N: Jim's Service
- RVCamp S: **Camping**

(16) Picnic
(22) Zelienople Service Plaza (Eastbound)
- FStop E: **Sunoco***
- Food E: **Mrs Fields Cookies, Roy Rogers, TCBY Yogurt**
- AServ E: Sunoco

(24) Picnic (Eastbound)
(26) Picnic (Eastbound)
(27) Picnic (Eastbound)
3 **(28)** Jct I-79, U.S. 19, Pittsburgh, Erie

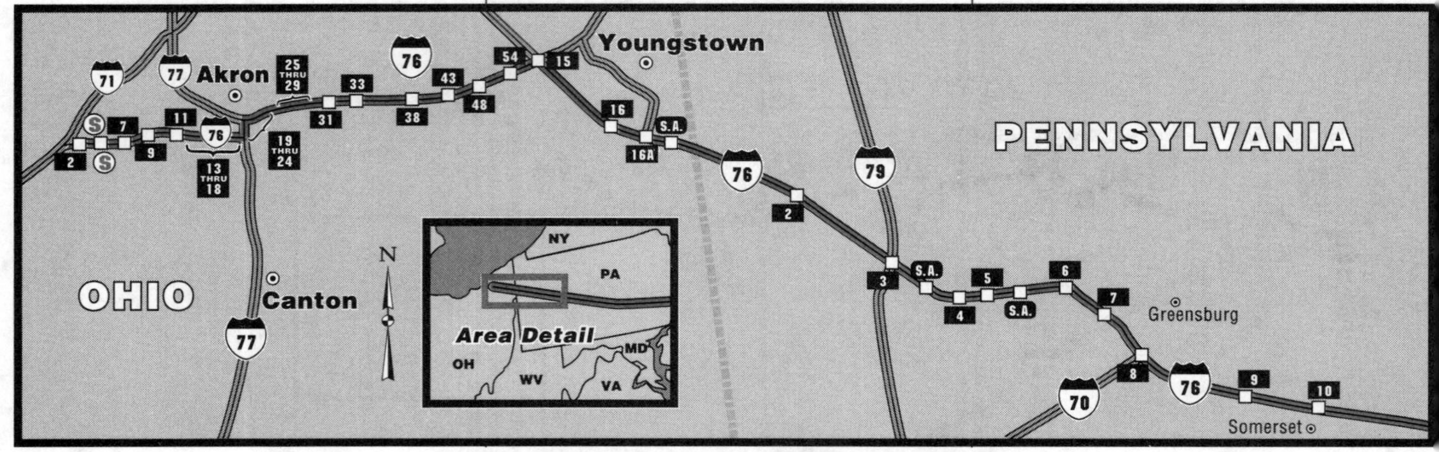

Bold red print shows RV & Bus parking available or nearby

XIT	PENNSYLVANIA
	(Toll Booth)
Gas	**N:** Exxon[D], Sheetz*
Food	**N:** Burger King, Climo's Italian, Denny's, Dunkin Donuts, Kings Family, Long John Silvers, Pizza Outlet, Wendy's
Lodg	**N:** Conley Inn, (AAA) Fairfield Inn, Red Roof Inn, Sheraton
AServ	**N:** Grease Monkey, Meineke Discount Mufflers, Pep Boys Auto Center
RVCamp	**N:** **Campground**
Med	**N:** **+ Hospital**
ATM	**N:** Giant Eagle Supermarket, Mars National Bank, Mellon Bank, Northside Bank, PNC Bank
Other	**N:** **Coin Laundry, Cranberry Mall,** Full Service Car Wash, **Giant Eagle Grocery (Pharmacy), K-Mart (Pharmacy), Office Max, Pet Supplies Plus, Phar Mor Drugs, Shop 'N Save Supermarket, Thrift Drug, U.S. Post Office, Wal-Mart**
31)	Butler Service Plaza (Westbound)
FStop	**W:** **Sunoco***
Food	**W:** **Burger King,** Mrs Fields Cookies, **Popeye's Chicken,** TCBY Yogurt
AServ	**W:** Sunoco
37)	Picnic (Eastbound)
	(39) PA 8, Pittsburgh, Butler, Butler Valley (Toll Booth)
Gas	**N:** BP[LP], Exxon[D], Sheetz*[LP]
Food	**N:** Eat N Park, Empire Gourmet Chinese, Monte Cello's Pizza, Pittsburg North Motor Lodge, Venis Diner
	S: Arby's, Baskin Robbins, Burger King, Donut Connection, Econolodge, McDonalds, Pizza Hut, Pizza Zone
Lodg	**N:** ◆ Comfort Inn, Landmark Manor, Pittsburg North Motor Lodge
	S: Duskey's Motel, Econolodge, Days Inn (see our ad this page)
AServ	**N:** BP, Exxon
	S: BP, Goodyear Tire & Auto, Midas Muffler & Brakes, Parts America, Speedy Muffler, Valvoline Quick Lube Center
ATM	**N:** Sheetz
	S: Northside Bank, PNC Bank
Other	**S:** Full Service Car Wash, **Thrift Drug**
45)	Picnic (Westbound)
	(47) PA 28, New Kensington, Pittsburgh (Toll Booth)
Gas	**N:** Amoco[CW]
	S: Exxon*
Food	**S:** Bob Evans Restaurant, Burger King, Denny's, Eat N Park, Holiday Inn, KFC, **Kings Family Restaurant,** McDonalds, Ponderosa Steakhouse, Taco Bell, Wendy's
Lodg	**S:** Comfort Inn, (DAYS INN) Days Inn, **Holiday Inn,** Super 8 Motel, (AAA) Valley Motel
AServ	**N:** Cooper Tires
	S: Coxon's Ford
Med	**S:** **+ Medi Center**
ATM	**S:** National City Bank
Other	**N:** Car Wash, Harmar Full Service Car Wash
	S: **Giant Eagle Grocery (Pharmacy)**
49)	Oakmont Service Plaza (Eastbound)
FStop	**E:** **Sunoco***
Food	**E:** **Arby's,** Sub Shop
AServ	**E:** Sunoco
	(57) Junction I-376, U.S. 22, Pittsburgh, Monroeville (Toll Booth)
Gas	**N:** Gulf[D]
	S: BP*[CW], Citgo, Sunoco*
Food	**S:** Burger King, Damon's Ribs, Darbar Indian (Sunrise Inn), Dunkin Donuts, Holiday Inn, Kenny Rogers Roasters, Kings Inn, Lone Star Steakhouse, Moio's Italian Pastry Shop, Shanghi Express, Taco Bell
Lodg	**N:** East Exit Motel
	S: (DAYS INN) ◆ Days Inn, ◆ Holiday Inn, King's Motel, Palace Inn, ◆ Red Roof Inn, William Penn Motel
AServ	**N:** Falconi Acura, Mazda, Gulf, Ted McWilliams Toyota
	S: Citgo, Cochran Olds/GMC Trucks/Isuzu/Infinity/Suzuki/Saturn, Tire America, Valley Buick/Jeep-Eagle
Med	**S:** **+ Hospital**
ATM	**S:** Dollar Bank, Mellon Bank
Other	**N:** **Giant Eagle Grocery, Revco Drugs**
	S: **Penn Super Pharmacy, Thrift Drug**
7	**(67)** U.S. 30, Irwin, Greensburg (Toll Booth)
FStop	**S:** **Super America***[LP]
Gas	**N:** BP*, Sheetz*, Texaco
	S: Sunoco*
Food	**S:** Angelo's Italian, Arby's, Baskin Robbins, Bob Evans Restaurant, Burger King, China House, Conley Inn, Dairy Queen, Denny's, Dunkin Doughnuts, KFC, Little Caesars Pizza, Long John Silvers, McDonalds, Pizza Hut, Royal Dragon, Side Show Pizza, Taco Bell, **Teaddy's,** The Italian Oven, The Sandwich Factory, Vincent's Pizza & Pasta, Wendy's
Lodg	**S:** Conley Inn, Holiday Inn Express, Penn Irwin Motel
AServ	**N:** BP, Import Export Tire Company, Texaco, Truzka Foreign Car Parts/Service
	S: Advance Auto Parts, Jiffy Lube, Monroe Muffler & Brake, Speedy Muffler, Sunoco, Toona Auto Parts

EXIT	PENNSYLVANIA
ATM	**S:** Cogo's, Irwin Bank & Trust, National City Bank, PNC Bank
Other	**S:** **Cogo's Convience Store, Giant Eagle Grocery (Pharmacy), Shop 'N Save Supermarket, The Medicine Shoppe, Thrift Drug, Turnpike Laundromat**
(75)	Hempfield Service Plaza (Eastbound)
FStop	**E:** **Sunoco***
Food	**E:** **Breyer's Ice Cream,** Hot Dog Construction Co, **McDonalds**
AServ	**E:** Sunoco
ATM	**E:** PNC Bank
8	**(75)** Jct I-70W, U.S. 119, PA 66 N, New Stanton (Toll Booth)
Lodg	**N:** Hampton Inn (see our ad this page)
(78)	New Stanton Service Plaza (Westbound)
FStop	**W:** **Sunoco***
Food	**W:** **King's Family, McDonalds**
ATM	**W:** **Machine**
9	**(91)** Donegal, PA 31, PA 711, Legonier, Uniontown (Toll Booth)
FStop	**S:** **Citgo***
Gas	**S:** Exxon*[D], Sunoco, Texaco*
Food	**N:** Laurel Highlands Lodge, Tall Cedars Inn
	S: Candlelight Family, Dairy Queen
Lodg	**N:** Laurel Highlands Lodge
	S: (DAYS INN) Days Inn, Donegal Motel
AServ	**S:** Sunoco
RVCamp	**S:** **Donegal Campground, Laurel Highland Campgrounds (3 Miles), Mountain Pines Campground (3 Miles), Pioneer Park Campground (11 Miles)**
ATM	**S:** PNC Bank
Parks	**S:** **Kooser State Park, Laurel Hills State Park, Laurel Ridge State Park, Ohiopyle State Park**
(94)	Picnic (Westbound)
10	**(110)** U.S. 219, Somerset, Johnstown (Toll Booth)
TStop	**S:** **Jim's Auto/Truckstop***
Gas	**N:** Quick Fill, Sheetz*, Your Car Wash
	S: Exxon*
Food	**N:** Hoss's Steakhouse, Kings Family Restaurant, Pizza Hut, Taco Bell
	S: Arby's, China Garden, Dairy Queen, Dunkin Donuts, Eat N Park, Holiday Inn, **Jim's TS,** KFC, Little Caesars Pizza, Long John Silvers, Maggie Mae's Cafe, McDonalds, Myron's Restaurant, Pine Grill, **Ramada Inn, Subway,** Summit Diner, Wendy's, Yogurt Ice Cream Classics
Lodg	**N:** Dollar Inn, Economy Inn
	S: Best Western, Budget Host Inn, Budget Inn, (DAYS INN) ◆ **Days Inn,** Economy Inn, **Hampton Inn,** (AAA) Holiday Inn, ◆ Knight's Inn, (AAA) **Ramada**
AServ	**N:** Dumbaulds Tire, Ford Dealership, Mardis Chrysler Plymouth, Midas Muffler & Brakes, Speedy Lube
TServ	**N:** **Mulhollen**
	S: **Jim's Auto/Truckstop**
TWash	**N:** **Your Car Wash**
Med	**S:** **+ Hospital**
ATM	**S:** Jim's TS
Other	**N:** Coin Car Wash, **Horizon Outlet Center**
	S: Clean Water Car Wash, **Coin Laundry, Visitor Information**
(112)	Somerset Service Plaza - both directions
FStop	**B:** **Sunoco***
Food	**B:** **Bob's Big Boy,** Hershey's Ice Cream, **Roy Rogers, TCBY Yogurt**
	W: **Burger King,** Hershey's Ice Cream, TCBY

EXIT	PNNSYLVANIA
	Yogurt
AServ	B: Sunoco
ATM	B: ATM
11	**(146)** U.S. 220, Bedford, Altoona, Johnstown (Pay Toll To Exit)
FStop	N: Amoco*, BP*
Gas	N: Sunoco*(D), Texaco*(D)(LP)
Food	N: Baskin Robbins, Budget Inn (Sunoco), Burger King, Chester Fried Chicken (BP), China Inn, Clara's Place (Best Western), Denny's, Dunkin Donuts, Ed's Steakhouse, Hardee's, Hoss's Steak & Sea House, Long John Silvers, McDonalds (Amoco), Pizza Hut, Sub Express, The Arena (Quality Inn), The Country Apple Restaurant (Econolodge)
Lodg	N: Best Western, Budget Inn, Econolodge, Host Inn, Midway Motel, Quality Inn, Super 8 Motel S: Motor Town House, Hampton Inn (see our ad this page)
TServ	N: Mack Truck
RVCamp	S: Friendship Village Campground
ATM	N: BP
Other	S: Visitor Information
(147)	North Midway Service Plaza - both directions
FStop	B: Sunoco*
Food	B: Hershey's Ice Cream, KFC, Mrs Fields Cookies, Sbarro Italian, TCBY Yogurt
AServ	B: Sunoco
ATM	B: PNC Bank
12	**(161)** U.S. 30, Everett, Breezewood, to I-70 East, Hagerstown (Pay Toll To Exit)
TStop	S: All American 76 Truck Plaza*(SCALES), TravelCenters of America(SCALES)
Gas	S: Amoco (TA), BP*, Citgo*, Exxon*(D), Sheetz*, Sunoco*, Texaco*
Food	N: Breeze Manor Restaurant (Quality Inn), McDonalds, Prime Rib Restaurant (Ramada Inn) S: Arby's, Arthur Treacher's Fish and Chips, Blimpie's Subs (Am Best), Bob Evans Restaurant, Bonanza Steakhouse, Burger King, Dairy Queen*(SCALES) (TA), Deli Cafe (Sunoco), Family House Restaurant (Bus Parking), Gateway Restaurant (TA), Hardee's, Hershey's Ice Cream (Am Best), KFC, Little Caesars Pizza (TA), McDonalds, Oven Fresh Bakery (TA), Perkins Family Restaurant (Am Best), Pizza Hut, Post House Cafeteria (Am Best), Stuckey's Express (Exxon), Subway, Taco Bell, Taco Maker (Am Best), Uncle Bud's Pizza (Am Best), Wendy's
Lodg	N: Pan Am Motel, AAA Quality Inn, Ramada Inn S: Best Western, Breezewood Motel, Comfort Inn (see our ad this page), Econolodge (TA), Penn Aire Motel, Wiltshire Motel
TServ	S: Am Best, TA
ATM	S: Sunoco, TA
(172)	Sideling Hill Service Area, RV Dump
FStop	B: Sunoco*
Food	B: Bob's Big Boy, Burger King, Pretzel Time, TCBY Yogurt
AServ	B: Sunoco
ATM	B: PNC Bank
Other	B: RV Dump Station, Tourist Information
13	**(179)** U.S. 522, McConnelsburg, Mount Union (Toll Booth)

EXIT	PENNSYLVANIA
Gas	N: Amoco*, Exxon*
Food	N: Fort Family Restaurant
Lodg	N: Downe's Motel #2 S: Downe's Motel
AServ	N: Amoco
Med	N: + Hospital
Other	N: State Patrol Post
14	**(189)** PA 75, Fort Loudon, Willow Hill (Toll Booth)
Food	S: TJ's Restaurant
Lodg	S: Willow Hill Motel
15	**(201)** PA 997, Blue Mountain, Shippensburg, Chambersburg (Toll Booth)
Gas	S: Johnnies

EXIT	PENNSYLVANI
Food	S: Johnnies
Lodg	S: Blue Mountain Brick Motel, AAA Kenmar Motel
(203)	Blue Mountain Service Area (Westbound)
FStop	W: Sunoco*
Food	W: Mrs. Field's, Roy Rogers, TCBY Yogurt
AServ	W: Sunoco
ATM	W: PNC Bank
(214)	Pennsylvania State Police (Westbound)
(215)	Picnic (Eastbound)
(219)	Plainfield Service Area (Eastbound)
FStop	E: Sunoco*
Food	E: Roy Rogers, TCBY Yogurt
AServ	E: Sunoco
(224)	Picnic (Eastbound)
16	**(226)** U.S. 11, to I-81, Carlisle, Harrisburg (Toll Booth)
TStop	N: All American Truck Plaza, AmBest Truckstop*(SCALES), Flying J Travel Plaza*(LP)(SCALES), Gables of Carlisle*(SCALES)
Gas	N: BP*, Citgo, Gables*(LP), Shell, Texaco*(LP)
Food	N: AmBest Truckstop, Arby's, Best Western, Bo Evans Restaurant, Budget Host Inn, Carelli's, Dunkin Donuts, Eat N Park, Embers Inn, Flying J Travel Plaza, Hardee's, Holiday Inn, Iron Kettle Restaurant, McDonalds, Middlesex Diner, Subway, Western Sizzlin' S: Hoss's Steak and Seahouse
Lodg	N: AAA Appalachian Motor Inn, AAA Best Western, Budget Host Inn, Econolodge, Embers Inn, Hampton Inn, Holiday Inn, Quality Inn, AAA Rodeway Inn, Super 8 Motel, Thrift Lodge S: Motel 6
AServ	N: Citgo
TServ	N: AmBest Truckstop (CB sales & service), Flying J Travel Plaza, Soco All American
TWash	N: AmBest Truckstop, Gables TS
Med	S: + Hospital
ATM	N: Farmers Trust
17	**(236)** U.S. 15, Gettysburg Pike, Gettysburg, Harrisburg (Toll Booth)
(237)	Picnic (Eastbound)
18	**(242)** Junction I-83 (Toll Booth)
Lodg	S: Days Inn (see our ad this page)
19	**(247)** Jct I-283, PA 283, Hershey, Harrisburg, Lancaster (Toll Booth)
FStop	N: Citgo*
Food	N: Wendy's
Lodg	N: DAYS INN AAA Days Inn, Doubletree Hotel, Hollywood Motel
AServ	N: Citgo
Med	N: + Hospital
Other	S: Harrisburg Int'l Airport
(250)	Highspire Service Plaza (Eastbound)
FStop	E: Sunoco*
Food	E: Hershey's Ice Cream, Sbarro Italian, TCBY Yogurt
(252)	Parking Area (Both Sides)
(253)	Parking Area (Eastbound)
(254)	Parking Area (Westbound)
(255)	Picnic (Eastbound)

Bold red print shows RV & Bus parking available or nearby

XIT	PENNSYLVANIA
59)	Lawn Service Plaza (Westbound)
Food	N: **Burger King**
63)	Picnic (Westbound)
64)	Picnic (Eastbound)
)	**(266)** PA 72, Lebanon, Lancaster (Northbound, Toll Booth)
FStop	N: **Texaco**(D)
Food	N: Hull's Stagecoach, Little Corner of Germany Cafe, Mt Hope Family Restaurant
Lodg	N: Hull's Stagecoach Motor Inn
	S: Friendship Inn, **Red Carpet Inn, Roadway Inn**
AServ	N: Auto Repair
TServ	N: **Hilltop Truck Service (Texaco)**
RVCamp	S: **Pinch Pond Campground (3.5 to 4.0 Miles)**
Med	N: **VA Medical Center**
Other	S: **Renaissance Fair**
68)	Picnic (Eastbound)
69)	Picnic (Eastbound)
74)	Picnic
	(286) U.S. 222, Reading, Ephrata (Southbound, Toll Booth)
Gas	S: Citgo*(LP)
Food	S: Black Horse Restaurant, Country Pride BBQ, Geoffery's (Holiday Inn), Procopio's Pizza, Zinn's Diner
Lodg	S: Black Horse Lodge, AAA Comfort Inn, AAA **Holiday Inn**, Penn Amish Motel, ◆ Pennsylvania Dutch Motel, Red Carpet Inn, **Days Inn (see our ad this page)**
AServ	S: Scubber Auto (Wrecker Service)
TServ	S: **Dutchman Truck Service**
RVCamp	N: **Dutch Cousins Wooded Campgroung, Shady Grove Campground, Sill's Campground, Sun Valley Campground**
	S: **Hickory Run Campground, KOA Campground, Red Run Campground**
ATM	S: BBNB Nat'l Bank, **Fulton Bank (Weaver Mkt)**
Other	S: **Antique Malls, Doll Express, Tourist Information, Weaver Mkt Supermarket (Pharmacy)**
89)	State Police Outpost
90)	**Bowmansville Service Plaza, Rest Area - RR, Phones (Eastbound)**
FStop	E: **Sunoco***(D)
Food	E: **Bob's Big Boy, Mrs Fields Cookies, TCBY Yogurt**
AServ	E: Sunoco
ATM	E: PNC Bank

EXIT	PENNSYLVANIA
(291)	Picnic (Eastbound)
(294)	Picnic (Westbound)
(295)	Picnic (Eastbound)
(297)	Picnic (Eastbound)
22	**(298)** Jct I-176, PA10 Morgantown, Reading (Northbound, Toll Booth)
Food	N: Heritage
	S: Celebrations (Holiday Inn)
Lodg	N: The Inn at Morgantown
	S: Holiday Inn
Other	N: **Carr's Recreation Park, Hopewell Furnace Nat'l Historic Site**
(300)	Picnic (Westbound)
(305)	Peter J. Camiel Service Plaza

EXIT	PENNSYLVANIA
	(Westbound)
FStop	W: **Sunoco***
Food	W: **Nathan's Famous, Roy Rogers, Sbarro Italian, TCBY Yogurt**
AServ	W: Sunoco
ATM	W: PNC Bank
23	**(312)** PA 100, Pottstown, West Chester (Southbound, Toll Booth)
Gas	S: Gulf(D)(CW), Mobil(LP), Sunoco*(D)
Food	S: Hauss's (Hampton Inn), Holiday Inn
Lodg	S: Comfort Inn, Hampton Inn, Holiday Inn
AServ	S: Car Sense, Chester Cty Hydrolics
TServ	S: **American Truck**
Med	S: **Hospital**
ATM	S: Downing Town Nat'l Bank, Elverton Nat'l Bank
Other	N: **Hopewell Furnace Historical Site**
	S: **Wa Wa C-Store**
(325)	**Valley Forge Service Area - RV Dump, Phones, Rest Area, Fax Services (Eastbound)**
FStop	E: **Sunoco***
Food	E: Burger King, Mrs Fields Cookies, Nathan's, TCBY Yogurt
AServ	E: Sunoco
ATM	E: PNC Bank
24	**(326)** Jct I-76 E, to U.S.202, to I-476, Philadelphia, Valley Forge (Traffic Exits South, Toll Ends)
25	**(326)** Valley Forge, Mall Blvd, Northtown Rd
Gas	N: Exxon*, Mobil*(CW)
Food	N: Bennigan's, Charlie's Place, Chilli's, Denny's, Dick Clark American Bandstand, Doubletree, Dunkin Donuts, Food & Spirits Co, **Food Court at The Plaza King Of Prussia Mall,** Jade Diner, Lone Star Steakhouse, Ruby Diner, Starbuck's Coffee, T.G.I. Friday's, Uno Pizzeria
Lodg	N: Best Western, Doubletree, Fairfield Inn, Holiday Inn, MacIntosh Inn
AServ	N: Jiffy Lube (The Plaza King Of Prussia Mall), Penske, Sears (The Plaza King Of Prussia Mall)
ATM	N: CoreStates Bank
Other	N: **Pearl Express, The Plaza King Of Prussia Mall**
26AB	**(327)** U.S. 202, King of Prussia, West Chester (Northside Services Merge With Exit 25,)
Gas	N: Exxon*, Exxon
	S: Sunoco*

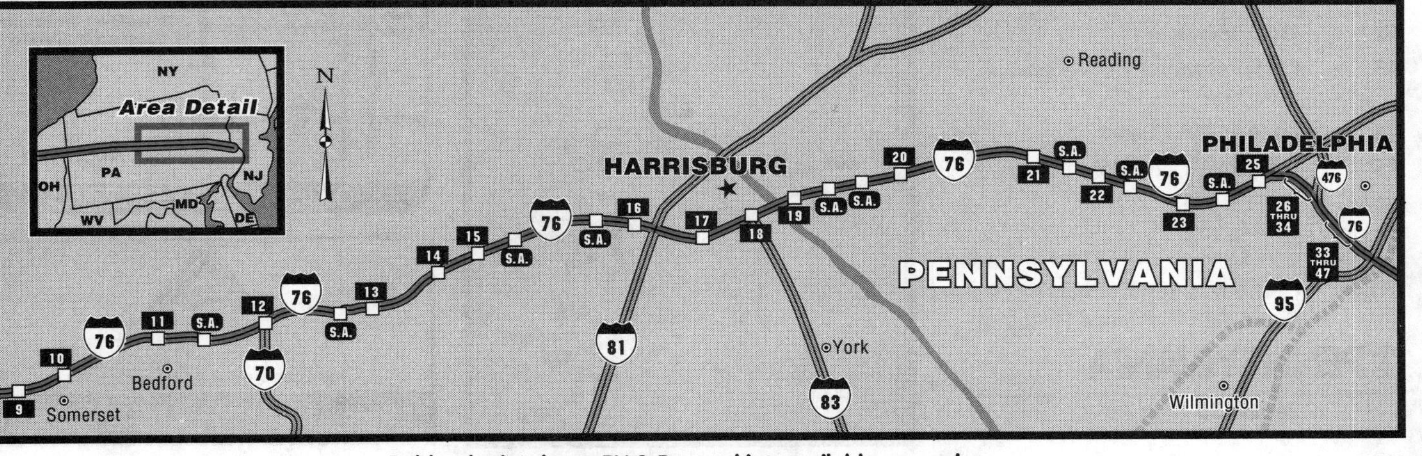

EXIT	PNNSYLVANIA
Food	N: Brandywine Grill, Carlucci's Grill, Chili's, Houlihan's, McDonalds, Pizzaria Uno, Rath Bone
Lodg	N: Holiday Inn, Howard Johnson, **Motel 6**
AServ	N: Exxon, Jiffy Lube
Med	S: **✚ Hospital**
ATM	N: Jefferson, Mellon Bank, Merridian, PCN Bank S: Royal Bank of Pennsylvania
Other	N: **Wa Wa Food Market** S: **Plymouth Mall**
(328)	King of Prussia Plaza - Picnic, RR, HF, Phones
27	**(330)** PA 320, Gulph Mills (Eastbound)
28AB	**(331)** Jct I-476 Chester, Plymouth Meeting
29	**(332)** PA 23, Conshohocken (Westbound, Eastbound Reaccess Only, Toll Booth)
30	**(336)** Gladwyne (Westbound, Eastbound Reaccess)
31	**(337)** Belmont Ave, Green Lane Ave, Manayunk
Gas	S: Sunoco*
AServ	N: Tony's Transmissions
Med	N: **✚ Hospital**
32	**(340)** Lincoln Dr, Kelly Dr (Left Exit Eastbound)
33	**(340)** U.S. 1 South, City Ave
34	**(340)** U.S. 1 North, Roosevelt Blvd
35	**(342)** Montgomery Dr, West River Dr (No Trucks Or Busses)
36	**(343)** U.S. 13, U.S. 30 West, East Fairmount Park, Girard Ave, Philadelphia Zoo
37	**(344)** Spring Garden St, Haverford Ave (Eastbound , Westbound Reaccess)
38	**(344)** Junction I-676, U.S. 30 East, Central Philadelphia (Left Exit Eastbound)
39	**(344)** 30th St, Station Market St
40	**(345)** South St (Left Exit Eastbound)
41	**(346)** Gray Ferry Ave, University Ave
42	**(347)** 28th St
43	**(386)** Passyunk Ave, Oregon Ave, to PA291
44	**(387)** PA 291, Chester
45	**(349)** PA 611, Broad St
46AB	**(349)** to I-95, Paker Ave
Food	S: Holiday Inn
Lodg	S: Holiday Inn
(352)	**Neshaminy Service Plaza - Picnic, RR, HF, Phones, RV Dump**

↑PENNSYLVANIA

Begin I-76

EXIT

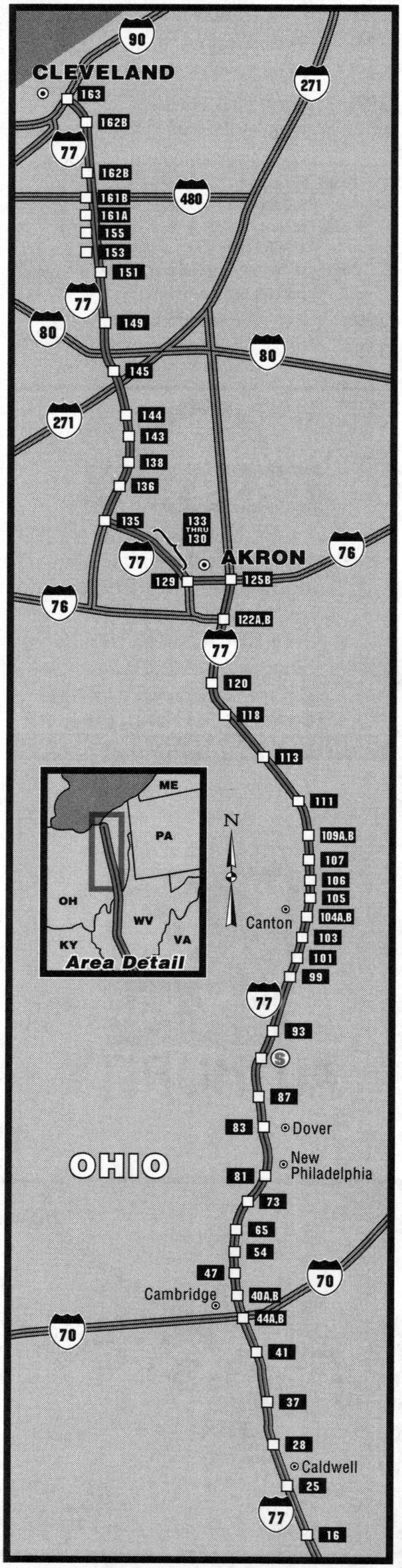

EXIT	OHIO
	Begin I-77

↓OHIO

163	Junction I-90, Toledo, Erie PA
162B	East 22nd, East 14th
162A	Woodland Ave, East 30th St
161B	Junction I-490 West, East 55th St, Toledo
161A	OH 14, Broadway (Northbound, Southbound Reaccess)
160	Pershing Ave
Med	E: **✚ Hospital**
159B	First Ave
159A	Harvard Ave, Newburgh Heights
158	Grant Ave, Cuyahoga Heights
157	OH21, OH17 (Southbound)
156	Junction I-480, Toledo, Youngstown
Lodg	E: Days Inn (see our ad this page)
155	Rockside Rd, Independence
Lodg	W: Amerisuites (see our ad this page)
153	Pleasant Valley Rd, Seven Hills, Independence
151	Wallings Rd
149	OH 82, Brecksville, Broadview Hts
Gas	W: BP*
Food	W: Country Kitchen, Domino's Pizza, Hunan Palace, Mr Hero Subs
Lodg	W: Days Inn (see our ad this page)

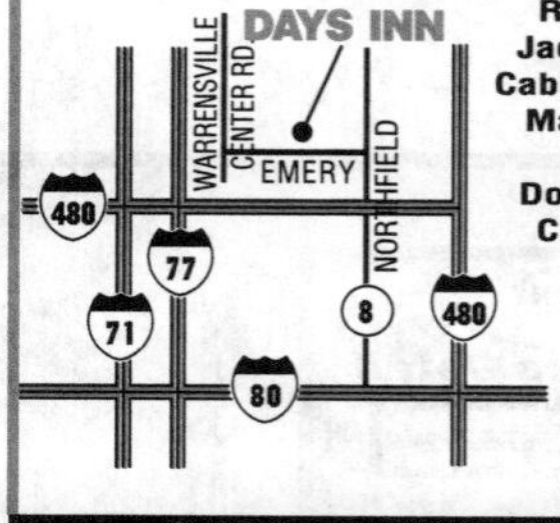

Bold red print shows RV & Bus parking available or nearby

EXIT		OHIO
7		to OH 21, to I-80, Miller Rd (Southbound, Northbound Reaccess)
5		OH 21, to I-80, Brecksville Rd, OH Tnpk (Northbound, Southbound Reaccess)
4		Junction I-271 North, Erie (PA)
3		OH 176, to Jct I-271 South, Richfield
	Gas	W: Sunoco
	Food	W: McDonalds
	AServ	W: Sunoco
	ATM	W: McDonalds
	Lodg	W: Howard Johnson
	Parks	E: Cuyahoga Valley Nat'l Recreation Area
1)		Rest Area - RR, Phones, Picnic
8		Ghent Rd
	Gas	W: BP*
	Food	W: Jimbo's Drive-In, Vaccaro's
7AB		OH 18, Fairlawn, Medina
	Gas	E: 76*(D), Amoco*, BP*(CW)(LP), Marathon*(LP), Speedway*
	Food	E: A Wok, Applebee's, Bob Evans Restaurant, Brubakers Pub, Brueggers Bagel Bakery, Chi Chi's Mexican, Chili's, Cracker Barrel, Friendly's, Holiday Inn, Joe's Bar & Grill, KFC, Lone Star Steakhouse, Macaroni Grill, Max & Erma's, McDonalds, Mustard Seed Market & Cafe, Olive Garden, On Tap Grill & Bar, Red Lobster, Subway, Swenson's, Taco Bell
		W: Burger King, Damon's Ribs, Don Pablo's Mexican, Eastside Mario's, Fuddrucker's, Miss Kitty's Steakhouse, Outback Steakhouse, Shoney's, T.G.I. Friday's
	Lodg	E: Courtyard by Marriott, Fairfield Inn, Hampton Inn, Holiday Inn, Super 8 Motel
		W: Comfort Inn (see our ad this page), Copley Studio Plus, Extended Stay America, Marriott, Radisson
	AServ	E: Marathon, Montrose Ford, National Tire & Battery, Tuffy Auto Center
	ATM	E: Bank One, Fifth Third Bank (Finast), First Merit, Key Bank
		W: First Merit Bank, NationalCity Bank
	Other	E: Acme Supermarket, Builders Square 2, Finast Supermarket, K-Mart (Pharmacy & Supermarket), PetsMart, Sear's Hardware, Wal-Mart (Pharmacy)
6		OH 21 South, Massillon (Left Exit Northbound)
5		Cleve - Mass Rd (Northbound, Southbound Reaccess)
3		Ridgewood Rd, Miller Rd
	Gas	E: 76*(D)(LP)
2		White Pond Dr, Mull Ave
	Food	E: Fontano Restaurant
	Other	E: Mail Boxes Etc
1		OH 162, Copley Rd
	Gas	W: BP*(D)
	Food	E: Dairy Queen, Queens BBQ & Shrimp, The Lantern Restaurant
		W: Charlie's Ribs & Chicken, McDonalds

EXIT		OHIO
	AServ	E: J&H Jaguare Service, Kiehl's Garge
		W: Stan's Towing
	Other	E: Coin Car Wash, Coin Laundry, Full Service Car Wash, IGA Store, U.S. Post Office, WalGreens
		W: Animal Clinic
130		OH 261, Wooster Ave
	Gas	E: 76*(D), BP*
	Food	E: Ann's Place, Burger King, Church's Fried Chicken, New York Style Pizza & Subs, Rally's Hamburgers, The Carriage House Restaurant, White Castle Restaurant
		W: KFC, Primo's Deli
	AServ	E: Auto Parts Center, Auto Zone Auto Parts, Parts America
		W: Bert Greenwald Chevrolet, Geo, Montrose Toyota, Rolling Acres Dodge
	Other	E: Acme Supermarket (Pharmacy), Coin Car Wash, Dairy Mart C-Store, FSO Food Service Outlet Supermarket, Full Service Car Wash
		W: Builder's Square, U-Haul Center(LP)
129		Junction I-76 West, Barberton
		Note: I-77 runs concurrent below with I-76. Numbering follows I-76.
21B		Lake Shore, Bowery St
	Food	E: AJ's Drive Through
	Med	E: Hospital
21A		East Ave
21C		OH 59 E, Downtown Acron (Northbound)
22A		Main Street, Broadway
22B		Grant St, Wolf Ledges
	Gas	S: BP*
	Food	S: McDonalds
23A		JCT I-77, Canton
23B		OH 8 N, Cuyahoga Falls (Eastbound, Left Exit)
		Note: I-77 runs concurrent above with I-76. Numbering follows I-76.
125B		Junction I-76 East, Youngstown
125A		OH 8 North, Cuyahoga Falls (Left Exit Northbound)
124B		Lovers Lane, Coal Ave
124A		Archwood Ave., Firestone Blvd N

EXIT		OHIO
123B		OH 764, Wilbeth Rd, Waterloo Rd
	Food	E: Carl's Jr Hamburgers, Castle Pizza, Dairy Queen
	AServ	E: Texas 10 Minute Oil & Lube
	ATM	E: National City Bank
123A		Waterloo Rd (Northbound Reaccess Only)
	Gas	W: BP*, Speedway*, Waterloo Service Center
	Food	W: Baskin Robbins, Burger King, Guy's Restaurant, House of Hunan, Italo's, John Baha's Waterloo Restaurant, Rally's Hamburgers, Subway
	AServ	W: Eagle Auto & Tire Service, Hank Richard's Automotive Service Center, Waterloo Service Center
	ATM	W: Key Bank
	Other	W: Acme Supermarket (Pharmacy), Car Wash, Finast Supermarket, Giant Eagle Supermarket
122AB		Junction I-277, to I-76, U.S. 224, Mogadore, Barberton
120		Arlington Rd
	Gas	E: Speedway*(LP)
		W: BP*
	Food	E: Denny's, Friendly's, Holiday Inn, Kenny Rogers Roasters, Pizza Hut, Ryan's Steakhouse, Shoney's, White Castle Restaurant
		W: Big Boy, Bob Evans Restaurant, Burger King, McDonalds, Taco Bell, Wendy's
	Lodg	E: AAA Holiday Inn, ◆ Red Roof Inn
		W: Days Inn AAA Days Inn
	AServ	E: K-Mart, Quaker State Q-Lube, Wal-Mart
		W: Doug Chevrolet Geo, Toth Pontiac, Buick
	RVCamp	W: RV Super Center
	ATM	W: Fifth Third Bank
	Other	E: Full Service Car Wash, K-Mart (Pharmacy), Rite Aide Pharmacy, Wal-Mart (Pharmacy)
118		OH 241, to OH619, Masillon Rd
	Gas	E: Speedway*
		W: Citgo*, Duke*
	Food	E: Belgrade Garden South Restaurant, Bobby's Bistro, Gionino's Pizzeria, Subway
		W: Arby's, Lucky Star Chinese, McDonalds, Menches's Brothers, The Lunch Box
	AServ	W: Mack's Transmission
	Med	W: Green Medical Center
	Other	W: Coin Car Wash, Green Animal Medical Center
113		Akron - Canton Airport
	RVCamp	W: Clays Rv Center
	Other	W: Airport
111		Portage St, Canal Fulton, North Canton
	TStop	E: 76 Auto/Truck Plaza*(SCALES)
	FStop	W: BP*
	Gas	E: Shell*(CW), Sunoco
		W: Dairy Mart*, Super America*(CW)
	Food	E: 76 Auto/Truck Plaza, Bourbon Street Grill, Burger King, Chieng's Express, Gaisen Haus, KFC, Palombo's Italian, Rax, Rustee's, Subway, The Brew House
		W: Cracker Barrel, Don Pablo's Mexican Restaurant, Long Horn Steakhouse, McDonalds, Pizza Hut, Wendy's

EXIT OHIO

Lodg W: AAA Best Western, Motel 6
AServ E: Midas Muffler & Brakes, Sunoco
W: Belden Village Towing
TServ **E: 76**
ATM E: 76 Auto/Truck Plaza, National City Bank
W: Star Bank (Giant Eagle Grocery), United Bank
Other E: Royal Car Wash
W: Discount Drug Mart, Giant Eagle Grocery, Wal-Mart (Pharmacy)

109AB Belden Village St, Whipple Ave, Everhard Rd
Gas E: Shell*(CW), Super America*
W: Marathon*, Speedway*
Food E: Blimpie's Subs, Burger King, Dairy Queen, Denny's, Fazoli's Italian Food, McDonalds, Taco Bell
W: Bob Evans Restaurant, Chang Chinese, Chi-Chi's Mexican Restaurant, Damon's Ribs, Friendly's, Holiday Inn, Mountain Jack's, Ponderosa Steakhouse, Red Lobster, Shoney's, The Pub, The Thirsty Dog, Wendy's
Lodg E: Comfort Inn, Fairfield Inn, Hampton Inn, Residence Inn
W: DAYS INN Days Inn, Holiday Inn, Red Roof Inn
AServ E: Mullinax Ford
W: Goodyear Tire & Auto, National Tire & Battery, Sear's, Valvoline Quick Lube Center
ATM E: United Bank
W: First Merit, Key Bank, National City Bank
Other **W: Belden Village Mall, Marc's Supermarket, Pearl Vision Center**

107B U.S. 62, Alliance

107A OH 687, Fulton Rd
Gas W: Citgo
Food E: Italo's Restaurant, Sports Page Bar & Grill, Woody's Root Beer & Sandwiches
W: Kustard Korner
AServ E: All Tune & Lube
ATM W: Kustard Korner
Other **E: Football Hall of Fame, Medicap Pharmacy**
W: Football Hall of Fame, Schnieder's Pet Hospital

106 13th St
Food W: The Stables Restaurant
Med **E: ✚ Hospital**

105 OH 172, Tuscarawas St, Downtown Tuscarawas
Gas E: Gas & Oil One Stop Shop
W: 76, BP*
Food E: Lindsey's Restaurant, McDonalds, Wendy's
W: Harmon's Pub, KFC, Loong Fung
AServ E: Miller's Radiator
W: Auto Zone Auto Parts, Good Deal Auto Mart
Med **W: ✚ Hospital**
Other **W: Dairy Mart***

104AB U.S. 30, U.S. 62, East Liverpool, Massillon

103 OH 800 South, Cleveland Ave
FStop **W: Pennzoil Lube**
Gas E: 76, BP*, Shell, Super America*
Food E: Arby's, Subway, Taco Bell
AServ E: 76, Miracle Automotive
TServ **W: Ziegler Tire & Oil**
Other E: Coin Car Wash, **Coin Laundry, Thurman Munson Memorial Stadium**

EXIT OHIO

101 OH 627, Faircrest St
TStop **E: 77 Gulliver's Travel Plaza*(SCALES)**
Food **E: 77 Gulliver's Travel Plaza, Ice Cream & Yogurt (77 Gulliver's Travel Plaza)**

99 Fohl Rd, Navarre
Gas W: Shell*(LP)
Food W: Subway (Shell)
RVCamp **W: Campground**
ATM W: Shell

93 OH 212, Bolivar, Zoar
Gas E: Dairy Mart*, Shell*
W: BP*, Citgo*
Food E: Der Dutchman (Amish), **McDonalds,** Pizza Hut, Tony D's Pizza, Wendy's, Wilkshire's 19th Hole Grill
W: Dairy Queen (Citgo)
Lodg E: AAA Sleep Inn
AServ E: Napa Auto Parts
W: BP
ATM E: Charter One Bank
Other E: Coin Car Wash, **IGA Supermarket (Pharmacy), Town & Country Vet**

(92) Weigh Station - Both Directions

87 U.S. 250, Strasburg
Gas W: Citgo*
Food W: **Hardee's,** The Manor
Lodg W: Twins Motel
TServ **W: Cummins Diesel**

(85) Rest Area - RR, Phones, Picnic

83 OH 39, OH 211, Sugarcreek, Dover
Gas E: BP(D), Shell*, Speedway
W: Citgo
Food E: Blimpie's Subs (Shell), Bob Evans Restaurant, Denny's, KFC, **McDonalds,** Shoney's, Wendy's
W: Dairy Queen (Citgo)
Lodg E: ◆ Knight's Inn
W: AAA Comfort Inn
AServ E: Flynn Tire, Parkway Nissan Licoln Mercury
Med **E: ✚ Hospital**
ATM W: Citgo
Other **W: Veterinary Clinic**

81 U.S. 250, OH 39, Uhrichsville, New Philadelphia
TStop **W: Eagle Auto Truck Plaza**
Gas E: BP*, Shell*
W: Marathon
Food E: Big Boy, Burger King, **Days Inn,** Denny's, Don Pablo's Tex Mex Cafe, Don Pancho's, Elbys, Goshen, **Holiday Inn, Little Caesars Pizza (K-Mart),** Long John Silvers, McDonalds, **McDonalds (Wal-Mart),** Pizza Hut, Schoenbrunn Inn, Taco Bell, The Spaghetti Shop, Waters Edge Restaurant
W: Family Restaurant (Eagle Auto Truck Plaza)
Lodg E: DAYS INN **◆ Days Inn, AAA Holiday Inn, AAA Motel 6,** AAA Schoenbrunn Inn, ◆ Super 8 Motel, AAA Travelodge
AServ E: Auto Parts Connection, BP, K-Mart, KC's Auto Repair, **Wal-Mart**
W: Marathon
TServ **W: McKnight Trucks**
RVCamp **W: Hobart Propane(LP)**
ATM E: K-Mart, **Wal-Mart**
Other **E: Aldi Food Store, K-Mart (Pharmacy), K-Mart, Wal-Mart (Pharmacy)**

73 OH 751, CR 21, Stone Creek
Gas W: Marathon*

EXIT OH

65 U.S. 36, Newcomerstown, Port Washington
TStop **W: Duke*(LP)(SCALES) (Kerosene)**
Gas W: BP*
Food **W: Duke's Restaurant (Duke's Inn), McDonc**
Lodg **W: Duke's Inn**
TServ **W: Duke**
ATM W: Duke

54 OH 541, CR 831, Kimbolton, Plainfi
Gas W: Shell* (Kerosene)
Food W: Linns, Palace Pizza, **Rocky's Family Restaurant**
RVCamp **E: Campground**

47 U.S. 22, Cambridge, Cadiz
Gas W: BP
AServ W: BP
RVCamp **E: Campground**
Med **W: ✚ Hospital**
Other **W: Degenhart Glass Museum, Tourist Information**

46AB U.S. 40, Old Washington, Cambridge
Gas W: BP*(CW), Speedway*(LP)
Food W: Burger King, Hardee's, House of Hunan, J Restaurant, Long John Silvers, **McDonalds, M Lee's Family Restaurant,** Wally's Pizza Palace
Lodg W: Longs Motel
AServ W: Roger's Radiator & AC
ATM W: Bank One
Other **W: Coin Laundry, Guernsey Veterinary Clini Riesbeck's Grocery, Sav-A-Lot Grocery**

44AB Junction I-70, Wheeling, Columbu

41 OH 209, OH 821, CR 35, Byesville
Gas W: BP
AServ W: BP
ATM W: National Bank
Other **W: Byesville Pharmacy, Byesville Police Dep IGA Store**

(40) Rest Area - RR, Phones, Picnic (Northbound)

37 OH 313, Pleasant City, Senecaville
Gas E: Duke*(LP) (Kerosene), Starfire Express
Food E: Subway
AServ E: Starfire Express
TServ **E: Truck Service**

(37) Rest Area - RR, Phones, Picnic (Southbound)

28 OH 821, Belle Valley
Gas E: Smith's Grocery, Sunoco*(D)
Food E: Marianne's Food Station, Oasis Grill
AServ E: Sunoco
Other **E: Smith's Grocery, U.S. Post Office**

25 OH 78, Caldwell, Woodsfield
Gas E: BP*, Sunoco(D)
Food **E: Lori's Family Restaurant, McDonalds**
Lodg E: ◆ Best Western
AServ E: Cowgill's GM/Geo, Sunoco
Other **E: Tourist Information**

16 OH 821, Macksburg, Dexter City
Food E: Restaurant & Mini Mart Grocery

6 OH 821, Marietta, Lower Salem
FStop **W: BP*(D)(LP)**
Gas E: Exxon*
Food E: Sub Express (Exxon)
RVCamp **W: Campground**

Bold red print shows RV & Bus parking available or nearby

EXIT	OHIO/WEST VIRGINIA
Med	W: Hospital
	Rest Area - RR, Vending, Phones, Picnic (Northbound)
	OH 7, Marietta
FStop	E: **Go Mart***
Gas	E: Ashland
	W: BP*(D), BP(CW), Exxon, Marathon
Food	E: Dairy Queen, Damon's Ribs, Holiday Inn, Ryan's Steakhouse
	W: **Bob Evans Restaurant**, Burger King, China Fun, Hong Kong Chinese American Restaurant, Little Caesars Pizza, Long John Silvers, McDonalds, McHappy's Donuts, Napoli's Pizza, Pizza Hut, Shoney's, Subway, Taco Bell
Lodg	E: AAA Comfort Inn, ◆ Econolodge, AAA Holiday Inn, AAA **Knights Inn**
	W: AAA **Knight's Inn, Super 8 Motel**
AServ	E: Ashland, **Wal-Mart**
	W: Exxon, Marathon, Midas Muffler & Brakes
RVCamp	E: **Camping**
	W: **Campground**
Med	W: **Quick Care Walk In Center**
ATM	W: Bank One, **People's Bank (Kroger)**
Other	E: **Aldi Grocery, Highway Patrol, Wal-Mart (Pharmacy, Vision Center)**
	W: Coin Car Wash, **K-Mart (Pharmacy), Kroger Supermarket (Pharmacy), Marietta Tourist Inconvention Info, Ohio River Museum, Revco Drugs**

↑OHIO
↓WEST VIRGINIA

EXIT	
5	WV 14, WV 31, Williamstown, Vienna
Gas	W: Go Mart*
Food	W: Dutch Pantry Family Restaurant
Lodg	W: DAYS INN AAA **Days Inn**
9	WV 2, WV 68, Emerson Ave, Vienna, North Parkersburg
Gas	E: Exxon* (Kerosene)
	W: BP*(D)
Food	W: Sub Express (BP)
Lodg	W: AAA Expressway Motor Inn
Med	W: **Hospital**
6	U.S. 50, 7th St, Downtown
Gas	W: 76*, Chevron*(D), Exxon*
Food	W: Bob Evans Restaurant, Burger King, Char House, Long John Silvers, **McDonalds**, Mountaineer Family Restaurant, Omelet Shoppe, Shoney's, Subway
Lodg	E: Best Western
	W: Holiday Inn, AAA Ramada Inn, ◆ Red Roof Inn, **The Stables Motor Lodge**
Parks	E: **North Bend State Park**
4	WV 47, Staunton Ave
Other	E: **State Police**
3	WV 95, Camden Ave, Downtown
RVCamp	E: **Campground, RV Service (Travel Trailer - Frontage Rd)**
	W: **Camping**
Med	W: **Hospital**
Parks	W: **Blennerhaffett Historical Park**
0	WV 14, Mineral Wells
TStop	E: **Liberty Truck Stop(SCALES), Parkersburg Truck Stop*(SCALES)**
FStop	E: **BP*(D) (Kerosene)**

EXIT	WEST VIRGINIA

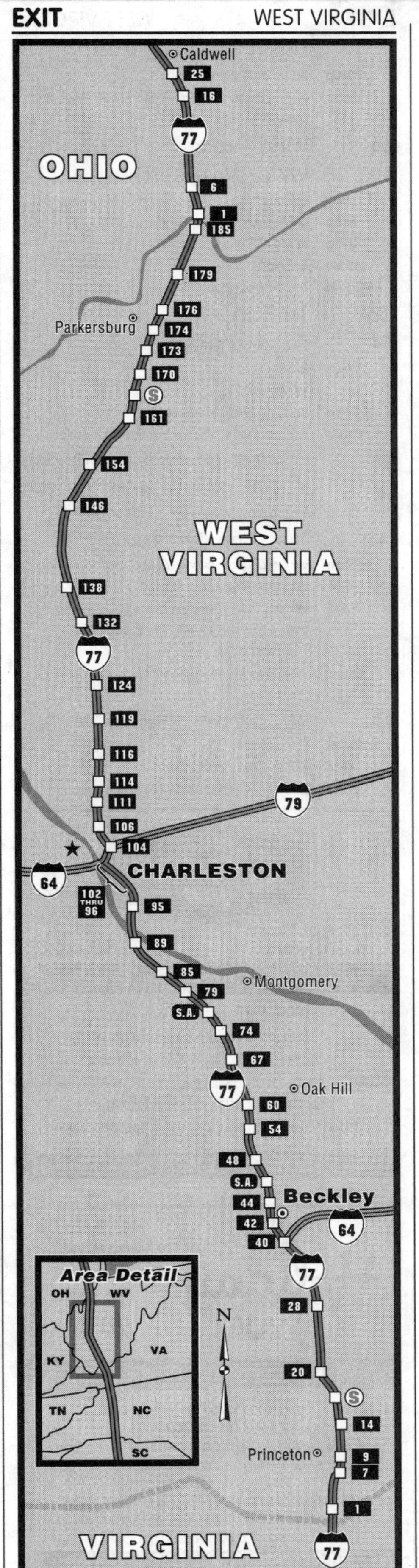

EXIT	WEST VIRGINIA
Gas	E: Chevron* (Kerosene), Exxon*
Food	E: **Liberty Restaurant, McDonalds,** Subway
	W: **Cracker Barrel**
Lodg	E: AAA Comfort Suites, Hampton Inn, **Liberty Motel**
	W: AAA AmeriHost Inn
AServ	E: BP
TServ	E: **Benson Truck Bodies, Parkersburg Truck Stop**
ATM	E: Commercial Bank, Libety Truck Stop
Other	E: **U.S. Post Office**
(169)	**Weigh Station - Both Directions, RV Dump**
(166)	**Rest Area - RR, Vending, Phones, RV Dump**
161	WV 21, Rockport
154	WV 1, Medina Rd
146	WV 2 South, Silverton, Ravenswood
Gas	W: BP*, Exxon, Marathon*
Food	W: Pit' n Git
Lodg	W: Scottish Inns
ATM	W: Exxon
138	U.S. 33, Ripley
FStop	E: **Duke*(D)**
Gas	E: Exxon*(CW), Marathon*(D) (Kerosene)
	W: Exxon*(D)
Food	E: Cozumel, Gabby's Grill & Bar, KFC, Kam Bo Chinese, Long John Silvers, **McDonalds, McDonalds (Wal-Mart)**, Pizza Hut, Wendy's
	W: Ponderosa, Shoney's, Subway
Lodg	E: **McCoy's Conference Center (Same As Best Western)**, ◆ Super 8 Motel
	W: AAA Holiday Inn Express
AServ	E: Napa Auto Parts
RVCamp	E: **Ruby Lake Campground (7 Miles)**
Med	W: **Hospital**
ATM	E: **Jackson (Kroger)**, One Valley Bank, United National Bank
Other	E: **Jackson Animal Clinic, Kroger Supermarket, Rite Aide Pharmacy, Sav-A-Lot Grocery, Wal-Mart (Pharmacy)**
132	WV 21, Ripley, Fairplain
FStop	E: **Go Mart***
Gas	E: BP*(D)
Food	E: Broughton Ice Cream Shop, Cinnamon Street Bakery, Eddie Pepper's Mexican Restaurant, Hot Stuff Pizza (BP)
Lodg	W: **77 Motor Inn**
AServ	E: Denbigh-Garrett Ford Licoln Mercury
124	WV 34, Kenna
Gas	E: Exxon*
Food	E: Patty's Country Cookin
ATM	E: Exxon
119	WV 21, Goldtown
116	WV 2, Haines Bridge Rd, Sissonville
RVCamp	E: **Rippling Waters Campground (4.2 Miles)**
114	WV 622, Sissonville, Pocatalico
ATM	E: **B & B Market**
Other	E: **B & B Market**
111	WV 29, Tuppers Creek Rd
Gas	W: BP*(D)
Food	W: Subway (BP)
106	WV 27, Edens Fork Rd
Gas	W: Exxon*(D) (Kerosene)
Lodg	W: Edens Motel (Exxon)

EXIT WEST VIRGINIA

104 Junction I-79 North (Left Lane Exit)

102 U.S. 119, Westmoreland Rd

101 Junction I-64 West

100 Broad St, Capital St (Difficult Reaccess Northbound)
- Gas **W:** Chevron[D]
- Food **W:** Chesapeake Bagel Bakery, Graziano's Pizza, Lee Street Deli & Lounge, Murfy's Coffee Shop, Pavilion Cafe (Heart O' Town), Ponderosa, Tudor's Biscuit World
- Lodg **W:** Heart O' Town
- AServ **W:** A & S Automotive, Big A Auto Parts, Chevron, Firestone Tire & Auto, Goodyear Tire & Auto, NAPA Auto Parts, Ziebart TidyCar
- Med **W:** ✚ **Charleston Area Medical Center,** ✚ **Columbia St Francis Hosp**
- ATM **W:** Huntington Banks
- Other **W: Capitol Market, Fire Dept, Kanawha County Public Library, Mail Boxes Etc, Park Place Cinema 7, Revco Drugs, Rite Aide Pharmacy, US Post Office**

99 WV 114, Greenbriar St, State Capital
- Gas **W:** Chevron* (711 Store), Exxon*[D]
- Food **W:** Applegarth Cafe, Capital Lounge Rest, Domino's Pizza, New China Restaurant, Rally's Hamburgers, Subway, Wendy's
- ATM **W:** Bank One
- Other **W: US Post Office, West Virginia State Capitol Building**

98 South Park Rd, Charleston (Southbound)

97 WV 60, Kanawha Blvd

96 U.S. 60 East, Belle (Difficult Reaccess)
- Food **E:** Gino's Italian, Tudor's Biscuit House
- Lodg **W:** Budget Host Inn

95 WV 61, McCorkle Ave
- FStop **E: Go Mart***
- Gas **W:** Exxon*[D]
- Food **E:** Bob Evans Restaurant, Graziano's Italian, Lone Star Steakhouse, McDonalds, Wendy's, **West Virginia Fried Chicken (Go Mart)**
  W: Burger King, Cancun Mexican Rest, Captain D's Seafood, Ponderosa Steak House, Taco Bell, Tony's Place Gold Dome BBQ
- Lodg **E:** ◆ **Days Inn,** Knights Inn, Motel 6, ◆ Red Roof Inn
- AServ **E:** Advance Auto Parts, **K-Mart**
  W: W.H. Service Center
- ATM **W:** Bank One, One Valley Bank
- Other **E: K-Mart, Pharmacy (K-Mart)**
  W: Kanawha Mall, Kroger Supermarket (24 hrs), Pharmacy (Kroger)

89 WV 61, WV 94, Marmet, Chesapeake
- FStop **E: Exxon*[D] (Kerosene)**
- Gas **E:** BP[D], Go Mart*
- Food **E:** Gino's Pizza, **Hardees (24 hrs),** Subway, Tudor's Biscuit World
- AServ **E:** BP, Hudson's Auto Repair, NAPA Auto Parts
- Other **E: Kroger Supermarket (24 hrs), Pharmacy (Kroger), Rite Aide Pharmacy**

85 Chelyan (Pay Toll To Exit)
- AServ **E:** Paul White Chevrolet Geo

79 Sharon, Cabin Creek Rd

74 WV 83, Paint Creek Rd

(72) **Travel Plaza - RV Dump (Northbound)**
- FStop **N: Exxon[D] (RV Dump)**
- Food **N: Roy Rogers (Travel Plaza), TCBY Yogurt (Travel Plaza)**

66 WV 15, Mahan

60 WV 612, Mossy, Oak Hill
- Gas **E:** Exxon*[D]
- Food **E:** Miss Ann's Fancy Food
- Lodg **E:** Motel (Exxon)
- AServ **E:** Exxon
- RVCamp **E: Camping (Full Hookup)**

(56) Toll Plaza

54 Pax, Mount Hope
- Gas **E:** BP*
  W: Exxon*[D]
- Food **W:** Long Branch (Kerosene)
- Other **E: Plum Orchard Lake Public Fishing Area**

48 U.S. 19, Summersville, North Beckley (Pay Toll To Exit, Ltd Access Highway)
- Lodg **W:** Holiday Inn (see our ad this page)

(45) Travel Plaza (Southbound)
- FStop **W: Exxon (Tamarack TP, RV Dump)**
- Gas **W:** Exxon (Tamarack TP)
- Food **W: Mrs Fields Cookies (Tamarack TP), Sbarro Italian (Tamarack TP), TCBY, Taco Bell (Tamarack TP)**
- Other **W: Tamarack Huge Craft Center (Tamarack TP)**

44 WV 3, Harper Rd, Beckley
- FStop **E: Go Mart***
- Gas **E:** BP*, Chevron*[D], Exxon*
- Food **E:** Applebees, Baskin Robbins (BP), Beckley

EXIT WEST VIRGIN

Pancake House, Bennetts Smokehouse & Saloon, Dairy Queen, Omelet Shoppe, Pizza H Subway (BP)
W: Bob Evans Restaurant, **Pasquale Mira (Do Inn),** Texas Steakhouse, Wendy's
- Lodg **E:** **Best Western,** Comfort Inn, Fairf Inn, ◆ Holiday Inn, Howard Johnson, **Su 8 Motel**
  W: Country Inn and Suites, **Days Inn,** Hampton Inn (see our ad this page), ◆ **Shone Inn**
- RVCamp **W: Lake Stephens**
- Med **E:** ✚ **Doctors Immedia Care (7 Days/Week, Appt),** ✚ **Hospital**
- Other **E: Beckley Vision Center**

42 WV 16, WV 97, Mabscott (Difficult Reaccess)

40 Junction I-64 East, Lewisburg

(29) Toll Plaza

28 WV 48, Ghent, Flat Top
- Gas **E:** Exxon*[D], Marathon*[D][LP]
- Other **E: US Post Office**

20 U.S. 119, Camp Creek
- FStop **E: Exxon*[LP]**
- Food **E:** Home Cookin'
- RVCamp **W: Camp Creek State Park**

(18) **Weigh Station, Rest Area - RV Dump (Southbound)**

(17) **Travel Plaza, Weigh Station - RV Dump (Northbound)**
- FStop **N: Exxon**
- Food **N: Roy Rogers, TCBY Yogurt (24 hrs)**

14 WV 20, Athens Rd
- Parks **E: Pipestem Resort State Park**

9 U.S. 460, Princeton
- TStop **E: I-77 Truck Stop - 24 hr* (Showers, Laund**
- FStop **W: Marathon[LP] (Kerosene, Propane)**
- Gas **W:** Amoco*[D], Chevron*, Exxon*
- Food **W:** Captain D's Seafood, Cracker Barre Dairy Queen, Hardee's, Ice Cream Churn (Amoco), Johnston's Inn Rest, King's Palace, McDonalds, Omlet Shop, Shoney's, Texas Steakhouse, Wendy's
- Lodg **W:** Comfort Inn, **Days Inn,** ◆ Hampton Inn, Johnston's Inn, Princeton Motel Ramada, **Sleep Inn, Super 8 Motel,** Tov & Country Motel, Turnpike Motel
- AServ **E:** Taylor's Garage
  W: K-Mart
- TServ **E: K.N.T. Truck & Tire Repair (I-77 TS)**
- ATM **W:** Amoco
- Parks **W: Pipestem Resort St Pk**
- Other **E: WV Welcome Center-RR, Vend, Gift, US Postage**
  W: Historical Society Museum, K-Mart, Princeton Mercer County Chamber of Commerce, Stop & Shop Food Mart, WV Sta Police

5 RT 112, Ingleside (No Southbound Reaccess)

7 WV 27, Ingleside Rd. (Southbound Reaccess Only)

1 U.S. 52 North, Bluefield
- Med **W:** ✚ **Hospital**
- Lodg **W:** Holiday Inn (see our ad this page)

Bold red print shows RV & Bus parking available or nearby

(EX)IT WEST VIRGINIA/VIRGINIA

WEST VIRGINIA

VIRGINIA

(6)5 VA 598

(6)4 U.S. 52, VA 61, Rocky Gap
- **Gas** E: BP*
- **Other** E: US Post Office

(6)2 CR 606, South Gap

(6)2) VA Welcome Center - RR, Phones (Southbound)

(5)9) Rest Area - RR, Phones, 2 Hr Parking (Northbound)

(5)8 U.S. 52, Bastian
- **FStop** E: BP* (Kerosene)
  W: Citgo*(LP) (Kerosene)
- **Gas** W: Exxon*(D) (Kerosene)
- **Other** W: Indian Village & Museum

(5)6) Truck Escape Ramp

(5)2 U.S. 52, VA 42, Bland
- **FStop** W: Shell*(D)(LP) (Kerosene)
- **Food** W: Dairy Queen (Shell), The Log Cabin Rest (Big Walker Motel)
- **Lodg** W: ◆ Big Walker Motel
- **AServ** W: Tire Outlet
- **Other** E: Coin Car Wash
  W: A Forest Warden

(5)2) Weigh Station (Northbound)

(4)7 CR 717, CR 601, Deer Trail Park
- **RVCamp** W: Deer Trail Park Family Campground, Stony Fork Campground
- **Other** W: Big Walker Mountain Hwy

(4)1 CR 610, Peppers Ferry Road
- **TStop** W: Wilderness Road Travel Center (24 hrs)
- **FStop** W: Shell*
- **Food** E: Western Steer Family Steakhouse
  W: Country Kitchen (Ramada Inn), Family Rest (Wilderness Road TS), Subway (Wilderness Road TS), Taco Bell (Wilderness Road TS)
- **Lodg** E: Best Western, ◆ Super 8 Motel
  W: ◆ Hampton Inn, (AAA) Ramada
- **TServ** W: 76(SCALES) (Showers)
- **Med** W: + Hospital
- **ATM** W: Wilderness Road TS
- **Other** E: Alliance Tractor & Trailer Training

(4)0 Junction I-81 South, North US 52, Bristol (Southbound)

EXIT VIRGINIA

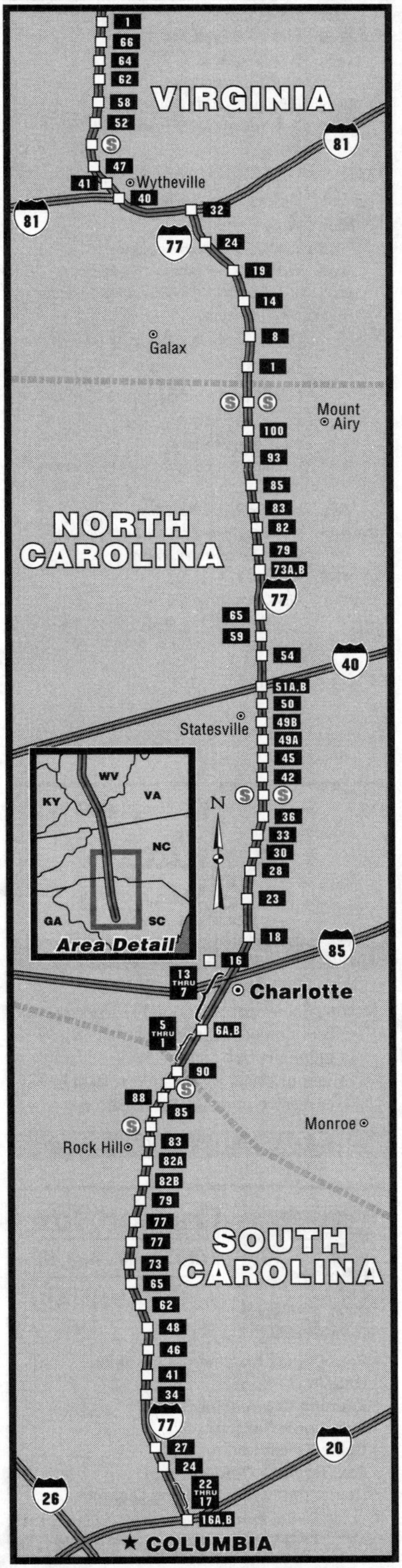

VIRGINIA
81
41 47 52 58 62 64 66 1
Wytheville
40 32 24 19 14 8 1
77
Galax
Mount Airy
100 93 85 83 82 79 73A,B
NORTH CAROLINA
77
65 59 54
40
51A,B 50 49B 49A 45 42 36 33 30 28 23 18 16
Statesville
WV VA KY NC GA SC
Area Detail
N
85
13 THRU 7
Charlotte
5 THRU 1
6A,B 90 88 85 83 82A 82B 79 77 77 73 65 62 48 46 41 34
Rock Hill
Monroe
SOUTH CAROLINA
77
27 24 22 THRU 17 16A,B
26
20
★ COLUMBIA

EXIT VIRGINIA

Note: I-77 runs concurrent below with I-81. Numbering follows I-81.

77 Service Road
- **TStop** E: Flying J Travel Plaza*(LP)(SCALES) (RV Dump)
- **FStop** E: Citgo* (Fuel Stop)
- **Gas** E: Conoco, Texaco*(D)
  W: Exxon* (RV Diesel)
- **Food** E: Burger King, Subway (Citgo), Thads Restaurant(SCALES) (Flying J)
- **RVCamp** E: KOA Campground
- **Other** E: Gale Winds (go-carts, putt-putt, family games)
  W: State Patrol Post

73 U.S. 11, South Wytheville
- **Gas** E: Exxon, Mobil*(LP), Shell*(D)
- **Food** E: Bob Evans Restaurant, El Puerto Mexican Restaurant, KFC, Mom's Country Store, Peking Restaurant, Shoney's, Steak House & Saloon, Waffle House
- **Lodg** E: (DAYS INN) (AAA) Days Inn, (AAA) Holiday Inn, Interstate Motor Lodge, Motel 6, (AAA) Red Carpet Inn, Shenandoah Inn, Wythe Inn
- **AServ** E: Exxon
- **Med** E: + Hospital

80 U.S. 52 South, VA 121 North, Fort Chiswell, Maxmeadows
- **TStop** E: Petro(SCALES)
- **FStop** W: Citgo*(LP)
- **Gas** E: BP*(D), Exxon* (Petro TS)
  W: Amoco
- **Food** E: Buck's Pizza, Iron Skillet Restaurant (Petro TS), Little Caesars Pizza (Petro TS), Wendy's (Petro TS)
  W: McDonalds
- **Lodg** E: Comfort Inn
  W: (AAA) Country Inn
- **AServ** E: Lee's Tire, Brake, & Muffler
- **TServ** E: Petro
  W: Complete Truck Service (24-hour)
- **RVCamp** E: Fort Chiswell RV Campground
- **ATM** E: Petro TS
- **Other** E: Car Wash, Laundromat

Note: I-77 runs concurrent above with I-81. Numbering follows I-81.

32 Junction I-81 North, U.S. 11, Roanoke

24 VA 69, Poplar Camp
- **FStop** W: Citgo* (Kerosene)
- **AServ** E: Poplar Camp Auto Service
- **Parks** E: New River & Shot Tower

19 CR 620, Twin County Airport

14 U.S. 58, U.S. 221, Hillsville, Galax
- **FStop** W: Chevron*(CW), Exxon*
- **Gas** E: Citgo*
  W: BP*
- **Food** E: Peking Palace, Subway (Citgo)
  W: Countryside, Dairy Queen (Chevron), McDonalds
- **Lodg** E: Econolodge
  W: (AAA) Comfort Inn, (AAA) Holiday Inn, Mountain Palace Inn
- **AServ** W: Chevron*(LP) (Kerosene)
- **RVCamp** E: Carroll Wood Campground
- **ATM** W: Chevron
- **Parks** W: Carroll County Recreation Park
- **Other** W: BJ Produce, Southwest VA Farmers Market

8 VA 148, CR 755, Fancy Gap

Bold red print shows RV & Bus parking available or nearby

EXIT — VIRGINIA/NORTH CAROLINA

FStop W: Exxon*
Gas E: Citgo*
W: BP*
Food W: Mayberry Station
Lodg W: Country View Inn, Days Inn, (AAA) Days Inn
AServ E: Allen's Garage
W: David L. Smith Inc.
TServ W: David L. Smith Inc.
RVCamp W: Camping
Other W: VA DOT

1 CR 620
Food E: Wallace Bros BBQ
Other W: Stuart's Creek WMA

(1) VA Welcome Center - RR, Phones, playground (Northbound)

↑VIRGINIA

↓NORTH CAROLINA

(106) NC Welcome Center - RR, Phones (Southbound)

(103) Weigh Station both directions

101 NC 752, Mount Airy
Med E: ✚ Hospital (12 Miles)

100 NC 89, Mount Airy, Galax
TStop E: Brintle's Travel Plaza*(SCALES)
Gas E: Exxon*(D) (Kerosene), Marathon*(D) (Kerosene), Texaco*(LP) (Kerosene)
W: BP* (Kerosene)
Food E: Brintle's Home Style Rest, Wagon Wheel
Lodg E: (AAA) Bryson Inn (Best Western), Comfort Inn (see our ad this page)
AServ W: BP, Blue Ridge Towing
TServ E: Brintle's Truck & Auto
RVCamp W: Camping
Med E: ✚ Hospital (12 Miles)

93 Dobson
FStop E: Exxon*(LP)
Gas E: Citgo*
Food E: Dairy Queen (Exxon), Surry Inn Diner
Lodg E: (AAA) Surry Inn
RVCamp W: Camping

85 SR 1138, C.C. Camp Rd
FStop W: Fuel Stop*(LP)
Med W: ✚ Hospital

83 U.S. 21 Bypass, Sparta (Northbound, Reaccess Southbound)

82 NC 67, Elkin, Jonesville, Boonville
FStop E: Amoco*(LP), Chevron*(D)(LP)
W: Exxon*(LP)
Gas E: BP, Citgo*(LP)
Food E: Arby's, Cracker Barrel, JD's Rest (Holiday Inn), Sally's Joe's Country Kitchen
W: Baskin Robbins (Exxon), Bojangles, Huddle House, McDonalds, Shoney's, Wendy's
Lodg E: ◆ Holiday Inn
W: (AAA) Comfort Inn, Days Inn, (AAA) Days Inn, (AAA) Hampton Inn
AServ E: Chevron
ATM W: Exxon
Other E: 67 Hardware, Holly Ridge Family Campground

79 U.S. 21 South, U.S. 21 Business, Arlington, Jonesville
FStop W: Texaco*(LP) (Kerosene)

EXIT — NORTH CAROLINA

Gas E: Shell*(LP)
Food W: Sally Joe's Kitchen
Lodg E: Super 8 Motel
W: (AAA) Country Inn
AServ E: Bridgestone Tire & Auto, Shell
Med E: ✚ Yadkin County Emerg. Med. Service Station 2

73AB U.S. 421, Winston - Salem, Wilkesboro, Yadkinville
TStop W: Coastal(D)(LP)
FStop E: Amoco*(D) (Kerosene), Exxon*(D)
Food W: Pine View Restaurant (Coastal)
Lodg E: Welborn Motel, Yadkin Inn, Holiday Inn (see our ad this page)

(72) Rest Area - RR, Phones (Northbound)

65 NC 901, Harmony, Union Grove
FStop W: Amoco*(LP), BP* (Kerosene), Texaco*(LP) (Kerosene, 24 Hrs)
Food W: Burger Barn, S&S Cafe (24 hrs), Subway (Texaco)
AServ W: Union Grove Tire & Auto
RVCamp E: Vanhoy Farms RV Park (Full Hookup)
W: Fiddler's Grove Camping
ATM W: Texaco
Other W: ACE Hardware

(63) Rest Area - RR, Phones (Southbound)

59 Tomlin Mill Rd

54 U.S. 21
FStop W: Exxon*(LP)
Gas E: Citgo*

EXIT — NORTH CAROLI

51AB Junction I-40

50 East Broad St, Downtown
Gas E: BP*, Crown(D)(CW), Etna(D)(LP)
Food E: Bojangles, Burger King, Cozumel Mexican, Dragon Golden, El Tio's, Gluttons Rest. & Bar, Golden China, Hardee's, IHOP, Little Pigs, McDonalds, Pizza Hut, Shoney's, Tucker's Bar Grill, Village Inn Pizza Parlor, Wendy's
Lodg E: (AAA) Fairfield Inn, ◆ Red Roof Inn
Med E: ✚ Carolina Primary & Urgent Care (No ap necessary)
W: ✚ Hospital
ATM E: Cash Points, First Citizen's Bank, First Union, Nation's Bank, United Carolina Bank
Other E: Community Police Center, Eckerd Drugs, Gran Prix(CW), Harris Teeter Supermarket, Johnson Cleaners Laundry Mart (Coin Laundry), K-Mart, Mail Boxes Etc, Miniature Golf, New Town Theatre, Signal Hill Mall, W Dixie Supermarket

49B Downtown Statesville
Gas E: PDI*(D) (Kerosene)
W: Amoco*, Citgo*, Exxon
Food E: KFC
Lodg W: Best Stay Inn
AServ E: Carolina Dodge, Everhart Honda, Ford Dealership, Muffler Masters, Toyota West
W: Black Pontiac Buick GMC, Exxon, Lebe's Tir Service
Other E: Animal Shelter

49A U.S. 70, G Bagnal Blvd
FStop E: Etna*, Phillips 66*(LP) (Kerosene)
Gas E: BP*(CW), Circle K*, Texaco*
Food E: Holiday Inn, Waffle House
Lodg E: (AAA) Best Western, Comfort Inn, ◆ Holiday I (AAA) Super 8 Motel
AServ E: BP, Muffler Clinic, Nissan of Statesville, Subu
Other E: La Esperanza Mexican Store, Scana Propane Gas(LP)
W: Statesville Rental(LP)

45 Troutman, Barium Springs
FStop W: Chevron*
RVCamp E: KOA Campground

42 U.S. 21, NC 115, Troutman
TStop E: Wilco Citgo*(LP)(SCALES)
Food E: Restaurant (Wilco), Subway (Wilco), Taco Bell (Wilco)
RVCamp W: Duke Power State Park
ATM E: Wilco Citgo

(38) Rest Area - RR, Phones (Both Directions)

36 NC 150, Mooresville, Lincolnton
Gas E: Exxon*
W: BP*(D)(CW)(LP), Texaco*(D)(CW) (Kerosene)
Food E: Denny's, Fat Boys, La Pizza Cafe, Peking Palace, Taco Bell, Waffle House, Wendy's
W: Arby's, Cracker Barrel, Hardee's, Lew's Fine Food, Pizza, Subway, The Checkere Flag Rest., US Subs
Lodg E: Days Inn, Ramada
W: ◆ Super 8 Motel
Med E: ✚ Hospital
ATM E: Nations Bank
W: Lincoln Bank, SouthTrust Bank, United Carolina Bank
Other E: Auto Bell Car Wash, Harris Teeter Supermarket, K-Mart

Bold red print shows RV & Bus parking available or nearby

EXIT NORTH CAROLINA

W: Food Lion Supermarket, Mail Boxes Etc, NC Auto Racing & Hall of Fame Museum, Revco Drugs, The Country Corner Marine (Boating)

3 U.S. 21 North
- **Gas** **W:** Citgo*(D)

) Davidson, Davidson College
- **Gas** **E:** Exxon*(D)
- **Food** **W:** North Harbour Cafe

3 U.S. 21, NC 73, Lake Norman, Cornelius
- **Gas** **E:** Amoco, Cashion's*(LP), Citgo*(LP)
- **Food** **E:** Bojangles, Dunkin' Donuts (Citgo), Mom's Country Store & Restaurant, Prime Sirloin, Shoney's Prime Bakery/Buffet
  W: Baskin Robbins, Bruegger's Bagel Bakery, Burger King, El Cancun, Fuzion Cafe World Cuisine, Hickory Hams, KFC, Little Caesars Pizza, Lotus 28 Chinese, McDonalds, Papa John's Pizza, Pizza Hut, Queen Bee Palace Chinese, Subway, Taco Bell, The Stock Car Cafe, Uptown Pasta, Wendys
- **Lodg** **E:** ◆ Hampton Inn, ◆ Holiday Inn
  W: ◆ Best Western, ◆ Comfort Inn, (AAA) Microtel Inn
- **AServ** **W:** Fast Lube, Lake Norman Chrysler Plymouth Dodge
- **ATM** **E:** Cashion's, Home Federal Savings
  W: Bi-Lo Supermarket, First Citizen's Bank, Lincoln Bank, United Carolina Bank
- **Other** **E:** Automatic Car Wash, Coin Laundry
  W: 1 Hr Photo, Bi-Lo Supermarket (& Pharmacy), Eckerd Drugs, Harris Teeter Supermarket (24 Hrs), Kerr Drugs, Loan Star Steakhouse, Movies At the Lake, Outdoors Etc Limited, US Post Office, Visitor Information, Wild Willie's Golf Center

5 NC 73, Concord, Lake Norman
- **Gas** **E:** Exxon*, Texaco*(CW)(LP)
  W: BP*(CW)
- **Food** **E:** Bagel Bin & Deli, Burger King, Chili's, Fuddruckers, McDonalds, Northcross Lanes (Bowling), Philly Connection, Pizza Hut (Delivery Only), TCBY, Wendys
  W: Arby's, Bojangles, Dairy Queen, Subway
- **ATM** **E:** Texaco, United Carolina Bank
  W: BB&T, Cash Points
- **Other** **E:** Java House Coffee Store, Mail Boxes Etc, Pet Mania, Target Department Store (Pharmacy), Winn Dixie Supermarket (Pharmacy)
  W: Energy Explorium, Food Lion Supermarket, Golf

3 Huntersville
- **Gas** **E:** Amoco, Texaco*(LP)
- **Food** **E:** Bringuier's Family Rest., Captain's Gallery, Hardee's, Hero Express, Romanello's Pasta, Pizza, Subs, The Palace of China, The Steak & Hogie Shop, Waffle House, Wendys
- **Lodg** **E:** (AAA) Holiday Inn Express
- **AServ** **E:** Amoco, Goodyear Tire & Auto, Lake Lube
- **ATM** **E:** First Charter, First Union, Lincoln Bank, Texaco
- **Other** **E:** Eckerd Drugs, Food Lion Supermarket, Huntersville Animal Care Hosp, Post Office, Revco Drugs

3 Harris Blvd, Reames Rd
- **Gas** **E:** Texaco*(D)
- **Med** **E:** ✚ Hospital

EXIT NORTH CAROLINA

16AB U.S. 21 North, Sunset Rd
- **TStop** **E:** Jake's Red Ball*(SCALES) (Sunset 76), Sunset 76*(SCALES) (Showers)
  W: Jake's Sunset*(LP)(SCALES) (Western Union)
- **Gas** **E:** Circle K*(LP)
  W: BP*(CW), Citgo*(CW), Texaco (Jake's TS)
- **Food** **E:** Captain D's Seafood, Hardee's, KFC, McDonalds, Pagoda Chinese, Pizza Hut Carryout, Subway, Taco Bell, Waffle House, Wendys
  W: Bojangles, Bubba's B.B.Q, Burger King, Denny's, Domino's Pizza, Waffle House
- **Lodg** **E:** Best Stay Inn, (DAYS INN) Days Inn
- **AServ** **E:** Right Way Tires & Serv. Ctr
- **TServ** **E:** Sunset 76 A.G. Boone Sales & Service
- **TWash** **E:** Sunset 76
- **RVCamp** **W:** Independence RV Sales & Services(LP)
- **ATM** **E:** Circle K, Lincoln Bank, Sunset 76 (FedEx Drop)
- **Parks** **W:** Latta Plantation Park
- **Other** **E:** Kerr Drug, Soaphie's, Winn Dixie Supermarket

13 Junction I-85, Greensboro

12 LaSalle St
- **Gas** **W:** Citgo*(D), Texaco*

EXIT NORTH CAROLINA

11B Brookshire Frwy. West

11A Junction I-277 S., NC 16 S., Brookshire Freeway

10C Trade St West, Fifth St, Downtown Charlotte
- **Gas** **E:** Amoco*
- **Food** **E:** Church's Fried Chicken

10B Trade St, Fifth St, Downtown
- **Gas** **W:** Phillips 66*, Servco*(CW)
- **Food** **W:** Bojangles
- **Lodg** **E:** Double Tree Hotel
- **AServ** **W:** Griffin Brothers Tire

10A U.S. 29, NC 27, Morehead St (Difficult Reaccess Southbound)
- **Food** **W:** Alex's Brown Derby, Open Kitchen
- **ATM** **W:** Wachovia Bank

9A NC 160, West Blvd.

9C U.S. 74 West, U.S. 29, NC 27, Wilkinson Blvd (Northbound)

9B Junction I-277, John Belk Frwy., Charlotte convention center
- **Med** **E:** ✚ Hospital Trauma Unit
- **Lodg** **E:** Four Points Hotel (see our ad this page)

9 I-277, U.S. 74, to U.S.29, to N.C.27, John Belk Frwy.
- **Med** **E:** ✚ Hospital Trauma Unit
- **Lodg** **W:** Knights Inn (see our ad this page)

8 NC 160, Remount Rd (Northbound, Reaccess Southbound)

7 Clanton Rd, NC 49
- **FStop** **E:** Citgo*
- **Gas** **W:** Amoco*, Citgo*
- **Lodg** **E:** (AAA) Holiday Inn Express, Motel 6, (AAA) Ramada Inn (see our ad this page), Super 8 Motel
- **Other** **E:** Coin Car Wash

6B NC 49 South, U.S. 521, Billy Graham Pwky. (East Side Services Same As 6A)
- **Gas** **W:** Phillips 66*
- **Food** **W:** McDonalds
- **Lodg** **W:** ◆ Embassy Suites, ◆ Summerfield Suites
- **ATM** **W:** NationsBank

6A U.S. 521 S., Woodlawn Rd (Difficult Reaccess)
- **Gas** **E:** BP*(D)(CW), Citgo*(LP) (24 Hrs), Exxon*(CW), Speedway*(D)(LP), Texaco*(D)(CW)
- **Food** **E:** Bojangles, Captain D's Seafood, Carver's Creek Prime Rib/Steaks, Checkers Burgers, Dragon Inn, Fanz Restaurant, Grill & Pub, Harper's Pizza, Steak,Pasta, House of Hunan (Mongolian BBQ), Krispy Kreme Donuts, Queen City, Shoney's, Steak & Ale
- **Lodg** **E:** (DAYS INN) Days Inn, Howard Johnson, Sterling Inn, Woodland Suites
- **ATM** **E:** BB&T
- **Other** **E:** Revco Drugs

5 Tyvola Rd
- **Gas** **E:** Citgo*(D)
- **Food** **E:** Black-Eyed Pea, Chili's, Kabuto Japanese Steak House and Sushi Bar, McDonalds, Sonny's BBQ
- **Lodg** **E:** (AAA) Comfort Inn, Hampton Inn, (AAA) Hilton, ◆ Marriott, ◆ Residence Inn
- **AServ** **E:** Keffer Buick, Scott Jaguar

EXIT N CAROLINA/S CAROLINA

ATM E: Bi-Lo Supermarket
Other E: **Bi-Lo Supermarket**

4 Nations Ford Rd
Gas E: Amoco*, Citgo*
W: Shell*
Food E: Caravel Seafood, Hardee's, Waffle House
W: Burger King
Lodg E: Cricket Inn, Innkeeper, La Quinta Inn, ◆ Red Roof Inn, Villager Lodge
ATM E: Amoco, Citgo
Other E: **Sam's Club**

3 Arrowood Rd (Southbound, Services Same As Exit 2)
Lodg E: (AAA) Marriot, Amerisuites (see our ad this page)

2 To Arrowood Rd.
Food E: Wendy's
Lodg E: Ameri Suites, Courtyard by Marriott

(1) NC Welcome Center - RR, Phones, Tourist Info (Northbound)

1 Westinghouse Blvd, I-485
FStop E: **Amoco***
Gas W: Texaco*(CW)
Food E: Subway, **Waffle House**
W: Burger King, **McDonalds**
Lodg E: **Super 8 Motel**
AServ W: Star Lube
ATM W: First Union, Texaco

↑NORTH CAROLINA

↓SOUTH CAROLINA

90 U.S. 21, Carowinds Blvd
Gas E: Citgo*(D)(CW)
W: BP*
Food E: Denny's
W: El Cancun Mexican, KFC, Mom's Restaurant, Shoney's, Wendy's
Lodg W: Comfort Inn, Holiday Inn Express (see our ad this page), Ramada, Sleep Inn
AServ E: Fort Mill Chrysler Plymouth, Dodge
RVCamp E: **RV World (Parts & Service)**
ATM E: Citgo
W: Nations Bank
Parks W: **Paramount's Carowinds Theme Park**
Other W: **Outlet Marketplace**

(89) SC Welcome Center - RR, Vending, Phones, Picnic (Southbound)

(88) **Weigh Station (Northbound)**

88 Gold Hill Rd, Pineville

EXIT SOUTH CAROLINA

Gas E: Texaco*
W: Exxon*
RVCamp W: **Lazy Daze Campground, Tracy's-RV Inc. (RV Parts & Service)**
Other W: **Coin Laundry (Lazy Daze)**

85 SC 160, Fort Mill, Taga Cay
Gas W: BP*
Food W: Bojangle's
AServ W: Bob's Automotive
Med W: ✚ **Fort Hill Medical Park**
ATM E: Winn Dixie Supermarket
W: BP, NationsBank
Other E: **Winn Dixie Supermarket (Pharmacy, 24 Hrs)**

EXIT SOUTH CAROLINA

83 Rd 49, Sutton Rd

82AB US 21, SC 161, York, Rock Hill
FStop E: **Phillips 66**
Gas E: Exxon*(D)
W: Citgo*(D), Phillips 66*(LP), Racetrac, Texaco*
Food W: Bojangle's, Burger King, Checkers, Denny's, Godfather's Pizza, H. Dundee Steakhouse, Long John Silvers, McDonalds, Outback Steakhouse, Pizza Hut, Sagebrush Steakhouse, Shoney's, Sonic Drive In, Taco Bell, Tres Amigo's, Waffle House, Wendy's
Lodg E: Holiday Inn
W: Best Way Inn, Best Western, Comfort Inn, (DAYS INN) **Days Inn**, Econolodge, Howard Johnson (see our ad this page), Oak Inn, Ramada Limited, Regency Inn
AServ E: Phillips 66, X Press Lube(CW)
W: Burns Chevrolet, Harrelson Toyota, Midas Muffler, Muffler Masters, Pep Boys Auto Center, Precision Tune
Med W: ✚ **Riverview Med Center (24 Hrs)**
ATM W: NationsBank (Bi-Lo)
Other E: **Museum of York County (see our ad this page)**
W: **Bi-Lo Supermarket**

79 SC 122, Dave Lyle Blvd, Downtown
Gas E: BP*(D)(CW), Fast Way Gas*
Food E: Applebee's, **Cracker Barrel**, Hardee's
Lodg E: ◆ Hampton Inn
AServ E: Honda's Cars of Rock Hill, Meineke Discount Mufflers, Mike's Wrecker Service, Sears
Other E: **Galleria Cinemas, Rock Hill Galleria Mall, Wal-Mart (Pharmacy)**

77 U.S. 21, SC 5, Rock Hill
FStop W: **Phillips 66***(LP)
Gas E: BP*(D), Citgo*(D)(CW)
W: Exxon*(D)(CW)
Food W: **Waffle House**
Med W: ✚ **Hospital**
ATM E: BP
Parks E: **Andrew Jackson State Park**

75 Porter Rd
Gas E: Shell*(D)
Other E: **Rocket Stop Fireworks**

73 SC 901, Rock Hill York
FStop E: **Citgo***
Med W: ✚ **Hospital**

(66) **Rest Area - RR, Vending, Phones, Picnic (Both Directions)**

65 Lancaster, Chester, S.C. 9
FStop E: **Texaco***

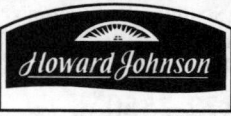

Bold red print shows RV & Bus parking available or nearby

EXIT	SOUTH CAROLINA

	W: Citgo*
Gas	**E:** Amoco*(CW)
Food	**E:** Subway (Amoco), Waffle House
	W: Country Omelet, Front Porch, KFC, McDonalds
Lodg	**E:** Days Inn, AAA Days Inn, AAA Econolodge, AAA Relax Inn
	W: AAA Super 8 Motel
ATM	**E:** Amoco
Parks	**E:** Landsford Canal State Park
	W: Chester State Park
2	Rd 56, Richburg, Fort Lawn
5	SC 97, Great Falls, Chester
FStop	**E:** BP*
Food	**E:** The Home Place
Parks	**W:** Chester State Park
8	SC 200, Great Falls
TStop	**E:** Shell* (see our ad this page)
Food	**E:** Grand Central Station
TWash	**E:** Shell
ATM	**E:** Shell
6	Rd 20, White Oak

EXIT	SOUTH CAROLINA
41	Rd 41, Winnsboro
Parks	**E:** Lake Wateree State Park
34	SC 34, Ridgeway, Winnsboro, Camden
FStop	**E:** BP*(LP)
Gas	**W:** Citgo*(D)
Lodg	**E:** Ridgeway Motel
AServ	**E:** Bryan's Garage, Jerry's Wrecker
RVCamp	**E:** Ridgeway RV Campground
Med	**W:** ✚ Hospital
ATM	**W:** Citgo
27	Blythewood Rd
FStop	**E:** Exxon* (24 Hrs)
Gas	**E:** Citgo*
Food	**E:** Bojangle's (Exxon), McDonalds, Restorante Mexico, W.G.'s Chicken Wings, Waffle House, Wendy's
AServ	**E:** Jim Hall Auto Service
Other	**E:** Blythewood Pharmacy, Foodliner Supermarket
24	U.S. 21, Blythewood
FStop	**E:** Fuel Stop*(D)(LP)
Food	**E:** Subway (Fuel Stop)
AServ	**E:** Abell's Auto Service
Other	**E:** Blythewood Animal Hospital
22	Killian Rd
19	SC 555, Farrow Rd
Gas	**E:** Citgo*(CW)(LP)
Food	**W:** Waffle House
ATM	**E:** Citgo, Wachovia
18	U.S. 277, Columbia, I-20 West, Augusta (Southbound, Northbound Reaccess)
Gas	**E:** BP*(CW)
17	U.S. 1, Two Notch Rd
Gas	**E:** BP*(CW), Exxon*
Food	**E:** Arby's, Burger King, Denny's, Waffle House
	W: Chili's, Fazoli's Italian Food, IHOP, Lizard's Thicket, Mc Kenna's, Outback Steakhouse, Waffle House
Lodg	**E:** Ramada Plaza Hotel
	W: Ameri Suites (see our ad this page), Comfort Inn, Hampton Inn, Holiday Inn, Quality Inn, Red Roof Inn
Med	**E:** ✚ Doctors Care
ATM	**E:** SouthTrust
Parks	**W:** Sesquicentennial Park
Other	**E:** Gregg Animal Hospital, Royal Lanes Bowling, Sparkle Car Wash, U.S. Post Office
16AB	Junction I-20, Atlanta, Florence
15	SC 12, Fort Jackson (Southbound, Reaccess Northbound)
13	Decker Blvd, Dentsville
Gas	**W:** Corner Pantry (see our ad this page)
12	Forest Drive, Forest Acres, Fort Jackson
10	Fort Jackson Blvd, SC760
9	U.S. 76, U.S. 378, Barners Ferry Rd
6	SC 768
5	SC 48, Bluff Rd

↑SOUTH CAROLINA

Begin I-77

EXIT	PENNSYLVANIA
	Begin I-78

↓PENNSYLVANIA

1	**(8)** PA 343, Fredericksburg, Lebanon (No Westbound Reaccess)
Gas	**S:** Redner's(LP)
Food	**S:** Esther's
RVCamp	**S:** Camping (5 Miles)
ATM	**S:** Redner's
Other	**S:** Redner's Grocery (Pharmacy)
2	**(10)** PA 645, Frystown
TStop	**S:** AmBest Truckstop*(SCALES)
FStop	**S:** Gables* (Shell)
Food	**S:** AmBest Truckstop Restaurant
Lodg	**N:** Fairview Motor Court
	S: AmBest Truckstop
TServ	**S:** AmBest Truckstop
ATM	**S:** AmBest Truckstop
3	**(13)** PA 501, Bethel
FStop	**N:** Texaco*
Gas	**S:** Amoco*(D), Exxon (Kerosene)
Lodg	**N:** Midway Motor Lodge
AServ	**S:** Exxon
TServ	**S:** Midway Truck Service
4	**(15)** Grimes (No Tractor Trailers)
Lodg	**S:** Midway Motor Lodge
5	**(16)** Midway
FStop	**N:** Exxon*, Midway*
Food	**N:** Blue Mountain Country Kitchen, Trainer's Midway Diner
Lodg	**N:** ◆ Comfort Inn
AServ	**S:** J & J's Truck & Auto Service
TServ	**S:** J & J's Truck & Auto Service
ATM	**N:** Exxon
6	**(17)** PA 419, Rehrersburg
Gas	**N:** Atlantic (Kerosene)
Lodg	**N:** Seven Star Hotel
AServ	**N:** Atlantic
Other	**N:** Country Bar and Sewing Center
7	**(19)** PA 183, Strausstown
Gas	**N:** Mobil*
	S: Best, Texaco* (Kerosene)
Food	**N:** Coffee Shop Sodas and Sandwiches, Dutch Kitchen
Lodg	**N:** Dutch Motel
AServ	**S:** Best, Texaco
Other	**N:** Pat Garrett Music Park Amphitheater
8	**(23)** Shartlesville
Gas	**N:** Citgo*, Texaco*
Food	**N:** K&M Burger Ranch, Stuckey's
	S: Blue Mountain Family Restaurant, Haag's Pennsylvania Dutch Cooking
Lodg	**N:** Dutch Motel
	S: Fort Motel, Shartlesville Motel
AServ	**N:** Motor Service Co (Texaco), Ted's Garage
RVCamp	**N:** Camping, Mountain Springs Camping Resort (1 Mile)
	S: Camping
ATM	**S:** Bernville Bank
Other	**S:** Post Office
9AB	**(29)** PA 61, Pottsville, Reading
Food	**N:** Cracker Barrel, Wendy's
Other	**N:** State Police Headquarters
10	**(30)** Hamburg
Gas	**S:** Getty*
Food	**S:** Subs and Pizza

EXIT PENNSYLVANIA

ATM S: Getty

11 **(35)** PA 143, Lenhartsville
- Gas S: Sunoco
- Food S: CJ Hummel, Pennsylvanian Dutch Restaurant
- AServ S: Lenhartsville Garage
- RVCamp N: **Blue Rocks Campground (1 Mile)**
  S: **Camping**

12 **(41)** PA 737, Kutztown, Krumsville
- Food N: **The Krumsville Inn**
  S: **Sky View Family Restaurant**
- Lodg N: **The Krumsville Inn,** Top Motel
- RVCamp N: **Robin Hill Campground**

13 **(45)** PA 863, Lynnport, New Smithville
- TStop N: **Bandit Truckstop***
- FStop N: **Texaco***
- Gas N: Gulf*
  S: Coastal*(LP)
- Food N: Terry's Place Restaurant
  S: The Golden Key
- Lodg S: AAA **Super 8 Motel**
- TServ N: **Werley's Truck Service**
  S: **Coastal**

14AB **(50)** PA 100, Fogelsville, Trexler Town
- Gas S: Exxon*, Sunoco*
- Food N: Arby's, Comfort Inn, Cracker Barrel, Flamingo's Frozen Yogurt, Gyros Sandwich Shop, Long John Silvers, Orient Express, Pizza Hut
  S: **Burger King,** Star-Lite Family, Subway (Exxon), Taco Bell, Yocco's Hot Dogs
- Lodg N: Comfort Inn
  S: Cloverleaf Inn, Hampton Inn, **Holiday Inn**
- AServ N: STS Car Service Center
  S: Exxon, Krause Toyota
- ATM N: Core States Bank, First Commonwealth Bank, First Union
- Other N: **Thrift Drug**
  S: **Visitor Information**

15 **(54)** PA 309N, PA Turnpike, Tamaqua
- Lodg N: DAYS INN Days Inn

16AB **(55)** U.S. 222, Hamilton Boulevard
- Gas S: Sunoco(CW)
- Food N: Ambassador Restaurant, Ice Cream World, Limerick Restaurant, O'Hara's Restaurant

EXIT PENNSYLVANIA

(Comfort Suites), Subway
S: Charcoal Drive-in, Pizza Hut, Tom Sawyer Diner
- Lodg N: Comfort Suites, Holiday Inn Express
- AServ S: Becker Wagonmaster Subaru, Queen City Tire Co
- ATM N: Bank of Pennsylvania, Summit Bank
  S: First Union Bank
- Other N: **Dorney Amusement Park (.5 Miles), Wild Water King Theme Park**

17 PA 29, Cedar Crest Boulevard
- Gas N: Texaco*(D)
- Food S: Spice of Life Restaurant
- AServ N: Texaco
- Med N: ✚ **Cedarcrest EmergiCenter**
  S: ✚ **Hosptital**
- ATM N: PNC Bank
- Other S: **Indian Museum**

18A **(57)** Lehigh St.
- Gas N: Hess*(D)
  S: Gettys, Mobil(LP), Texaco
- Food N: Asian Restaurant, Denny's, Willy Jo's
  S: Burger King, Domino's Pizza, Dunkin Donuts, Friendly's, McDonalds, Poncho Sunny's Mexican, Subway, Taco Bell, The Brass Rail
- Lodg N: DAYS INN Days Inn
- AServ N: Dodge Dealer, Haldeman Mercury/Lincoln/Ford, STF Tire Center
  S: A&A Auto Parts, Allentown Mazda/Volvo, Gettys, Jiffy Lube, KNOPF Pontiac, Le High Valley Ford, Midas Muffler & Brakes, Scott Chrysler, TCI Tire Centers, Texaco
- ATM S: Bank of Pennsylvania, First Union Bank, Summit Bank
- Other N: **Hannah Auto Wash, Laneco Grocery & Pharmacy**
  S: Car Wash, **Food 4 Less Grocery, Mail Boxes Etc, Petco, Phar Mor Drugs, South Mall, Weiss Markets**

18B **(58)** Emaus Avenue (No Reaccess)
- Gas N: Hess(D), Texaco
  S: Shell, Sunoco
- Food N: Appleman's Deli
- AServ N: Texaco
  S: Shell, Sunoco
- Other N: **South Mountain Pharmacy**

EXIT PENNSYLVANIA/NEW JERS

S: **Sharon Coin Laundry**

19 **(61)** to PA 145, Summit Lawn (Eastbound)

20AB **(61)** U.S. 309 South, S 4th St., Downtown Allentown, Quakertow

21 **(67)** Hellertown, Bethlehem, PA412 (Le High University)
- Gas N: Citgo*(LP)
  S: Exxon*
- Food N: Vassie's Drive In, Wendy's
  S: Subway (Exxon)
- AServ S: Landis Chevrolet Dealership
- Med N: ✚ **Hospital**
- Other N: **Visitor Information**

22 **(75)** Exit to PA 611, Easton, Philadelphia
- Gas N: Citgo*(LP)
  S: Shell*

(76) **PA Welcome Center - RR, Phones Tourist Info, Truck Parking, RV Parking (Westbound)**

↑PENNSYLVANIA

↓NEW JERSEY

3 **(4)** U.S. 22, U.S. 22 Alt, to PA 33, NJ 173, Phillipsburg, Bloomsbury
- TStop N: **Penn Jersey Truckstop***
- FStop N: **US Gas***
- Food N: **Bob's Big Boy,** Holiday Inn
- Lodg N: Holiday Inn, **Phillipsburg Inn**
- AServ S: GM Auto Dealership, Isuzu, Oldmobile Deal
- Other N: **CR Pharmacy, Home Depot, Lanco Supermarket & Drugstore, Shop Rite (Pharmacy)**

(4) **Weigh Station (Eastbound)**

4 Warren Glen, Stewartsville (Westbound, Reaccess Eastbound)

6 Warren Glen, Asbury (Eastbound, Westbound Reaccess)

7 NJ 173, West Portal, Bloomsbury
- TStop S: **76 Auto/Truck Plaza*(LP)(SCALES) (RV Dump), Pilot(SCALES)**
- Food S: **76 Restaurant, Subway (Pilot)**

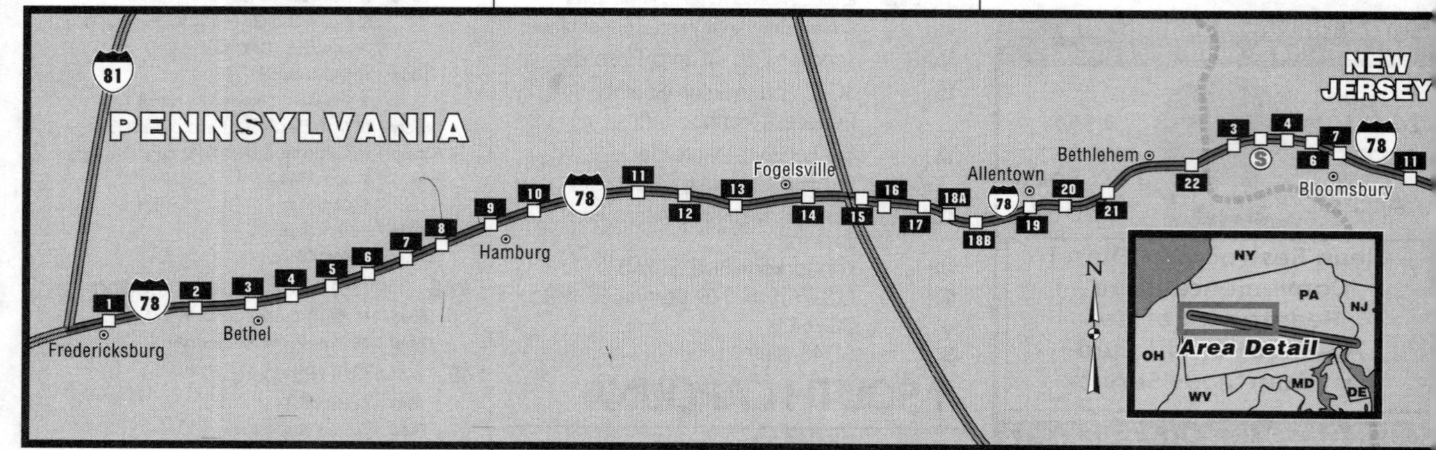

Bold red print shows RV & Bus parking available or nearby

EXIT	NEW JERSEY
TServ	**S:** 76
ATM	**S:** PNC Bank (Truckstop)
Other	**S:** Fed Ex Drop Box (Pilot Truck Stop), UPS Drop Box (Pilot Truck Stop)
8)	Rest Area - Picnic P (Westbound)
1	**(12)** NJ 173, West Portal, Pattenburg
Gas	**N:** Coastal*, Texaco
Food	**N:** Mountain View Shalet, Villa Pizza
AServ	**N:** Texaco
	S: Patrick's Auto Repair
RVCamp	**N:** Camping
Other	**N:** State Patrol Post
2	**(14)** New Jersey 173, Jutland, Norton
TStop	**N:** Citgo
Gas	**N:** Exxon*
	S: Shell
Food	**N:** Truckstop
	S: Bagelsmith's Deli
AServ	**N:** Exxon
	S: Shell
RVCamp	**N:** Spruce Run Recreation Area
Other	**N:** State Patrol Post
3	NJ 173 West, Service Road (Westbound, Eastbound Reaccess Only)
TStop	**N:** Clinton Truckstop*
Food	**N:** Louise's Diner
ATM	**N:** First Union Bank
5	NJ 173 East, Clinton, Pittstown
Gas	**N:** Citgo[CW][LP], Texaco[D]
	S: Clinton Car Care
Food	**N:** Holiday Inn
	S: Frank's Pizza
Lodg	**N:** ◆ Holiday Inn
AServ	**N:** Texaco
	S: Clinton Car Care
Other	**S:** Historical Museum and Art Center, Lanco Supermarket, Wal-Mart
6	NJ 31 North, Washington (Eastbound)
7	**(18)** NJ 31, Clinton, Flemington, Washington DC
Food	**N:** Finnigan's, King Buffet, McDonalds
ATM	**N:** Summit Bank

EXIT	NEW JERSEY
Other	**N:** Amerigas[LP], Coin Car Wash
18	**(19)** U.S. 22, Annandale (Westbound, Difficult Reaccess)
RVCamp	**N:** Campground
20	**(21)** Lebanon, Cokesbury
Gas	**S:** Exxon*[CW], Shell[D]
Food	**S:** Bagelsmith Deli, Dunkin Donuts (Exxon)
AServ	**S:** Shell
RVCamp	**S:** Round Valley Recreation Area
ATM	**S:** Fleetwood
Other	**S:** United States Post Office
24	**(25)** NJ523, to NJ517, Oldwick, Whitehouse
26	**(27)** NJ 523 Spur, North Branch, Lamington (Raritan Valley Community College)
29	**(31)** Jct I-287 to US 202, US 206, Morristown
(33)	Scenic Overlook (No Trailers Or Trucks)
33	**(34)** NJ 525, Bernardsville, Martinsville (Philadelphia College Of The Bible, Golf Museum USGA)
Gas	**S:** Amoco
Food	**N:** Christine's (Somerset Hills Hotel)
Lodg	**N:** ◆ Somerset Hills Hotel
AServ	**S:** Amoco
TServ	**N:** Amoco
Med	**N:** ✚ Hospital

EXIT	NEW JERSEY
	S: ✚ Hospital
ATM	**N:** Summit Bank
36	**(37)** CR527, Spur, Basking Ridge, Warrenville
Gas	**N:** Exxon
AServ	**N:** Exxon
40	**(41)** NJ 531, The Plainsfield, Gillette, Watchung (No Vehicles Over 5 Tons)
Med	**S:** ✚ Hospital
41	Berkeley Heights, Scotch Plains, US22 (Difficult Reaccess)
43	New Providence, Berkeley Heights (Westbound)
44	New Providence, Berkeley Heights
Med	**N:** ✚ Overlook Hospital
45	**(46)** Summit, Glenside Avenue (Eastbound)
48	**(49)** NJ 24, NJ 124W, Springfield, Morristown
49	NJ124, NJ82, Maplewood, Springfield Union
AServ	**S:** Dobson Company (Wrecker Service)
50AB	**(52)** Union, Millburn (Westbound, Reaccess Eastbound)
Gas	**N:** Amoco, Exxon, Mobil
	S: Texaco
AServ	**N:** Exxon, Jiffy Lube
	S: Texaco
Other	**N:** U.S. Post Office
52	**(53)** NJ Garden State Parkway (No Trucks)
54	U.S. 1, U.S. 9, Hillside, Irvington (Eastbound, Eastbound Reaccess)
55	13th Street, Hillside, Irvington (Westbound)
Med	**N:** ✚ Newark Israel Best Medical Center
56	Clinton Avenue, Irvington
Med	**N:** ✚ State Trauma Center
57	U.S. 1, U.S. 9, U.S. 22, Newark Airport
58AB	Junction I-95, U.S. 1, U.S. 9, New Jersey Turnpike, Newark

↑ NEW JERSEY

Begin I-78

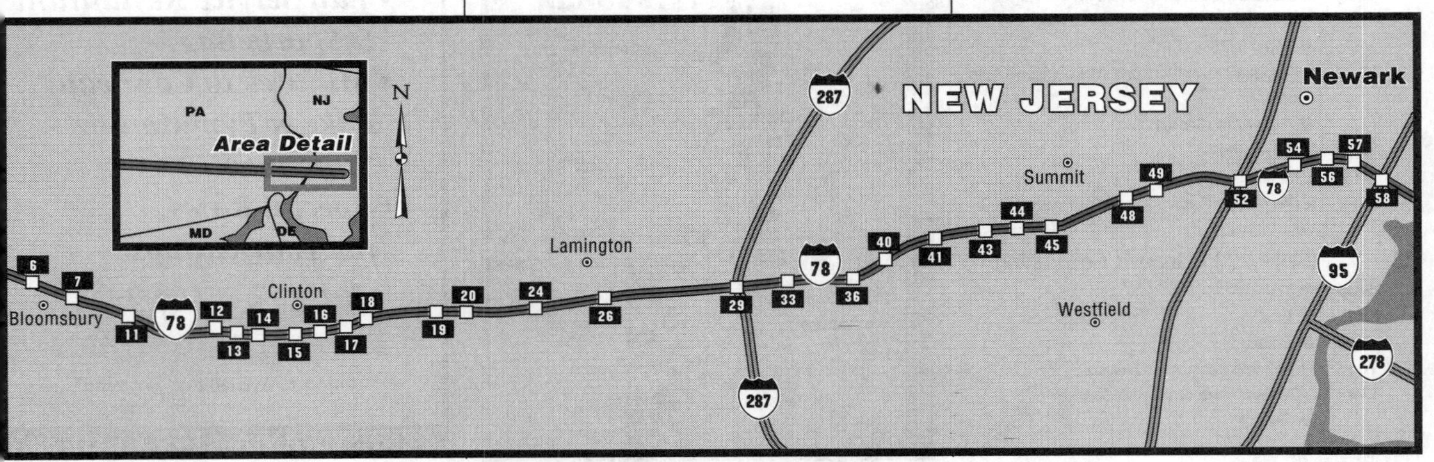

EXIT PENNSYLVANIA

Begin I-79

↓PENNSYLVANIA

44AB **(184)** 44A - PA 5, 12th St, East, 44B - PA 5, 12th St, West
- **Gas** W: BP*(CW)
- **Food** W: Eat N Park, Lin China Buffet
- **AServ** W: Dunn Tires, Monroe Mufflers & Brakes
- **Med** E: ✚ **Hammot Medical Center**
  W: ✚ **Prompt Care**
- **ATM** W: First Western Bank, Mellon Bank, National City Bank
- **Other** W: **Phar Mor Drugs**

43 **(182)** U.S. 20, 26th St
- **Gas** W: Citgo*, Gulf*
- **Food** E: Eric B's Pizza House
  W: McDonalds
- **AServ** W: Auto Land
- **Med** E: ✚ **St. Vincent Medical Center**
- **ATM** W: National City Bank
- **Parks** W: **Presque State Park, Waldemeer Park**
- **Other** W: **K-Mart, Revco**

41 **(180)** U.S. 19, Kearsarge
- **Food** E: Arby's, Chi- Chi's, Ponderosa, Red Lobster, The Brown Derby Roadhouse
- **AServ** E: Firestone Tire & Auto, Sear's
- **RVCamp** W: **Camping**
- **Med** E: ✚ **Hospital**
- **ATM** E: First National Bank, Mellon Bank
- **Other** E: **Mill Creek Mall, Phar Mor Drugs**

40 **(177)** Jct. I-90, Buffalo, Cleveland

39 **(174)** McKean
- **RVCamp** W: **Erie Campground**

38 **(165)** U.S. 6 N, Albion, Edinboro
- **Food** E: Highlander Restaurant
- **RVCamp** W: **Lianas Lake Park Campground**

(163) **Rest Area - RR, Tourist Info, Picnic, Phones, Vending**

37 **(154)** PA 198, Saegertown, Conneautville
- **Gas** W: Citgo
- **AServ** W: Citgo

36AB **(147)** 36A - U.S. 6, U.S. 322, Meadville, 36B - Conneaut Lake
- **Gas** E: Citgo*(D)(LP), Marathon(CW), Sheetz*(D)(LP)
- **Food** E: Big Boy, Dairy Queen, Days Inn, Perkins Family Restaurant, **Sandalini's**
  W: Ponderosa
- **Lodg** E: DAYS INN **Days Inn (see our ad on this page)**, Holiday Inn
  W: Super 8 Motel
- **AServ** E: Kirkpatrick Buick/Pontiac/GM, Robinson's Auto Repair
  W: Classy Auto Service
- **Med** E: ✚ **Hospital**
- **ATM** E: **Mellon Bank**
- **Other** E: **Giant Eagle Supermarket (Pharmacy), Veterinary Clinic**

35 **(141)** PA 285, Geneva, Cochranton
- **TStop** W: **Texaco***
- **FStop** W: **Gulf***
- **Gas** W: Texaco
- **Food** W: Palmer's Restaurant, **Texaco**
- **TServ** E: **Geneva Truck and Equipment**
  W: **Texaco**

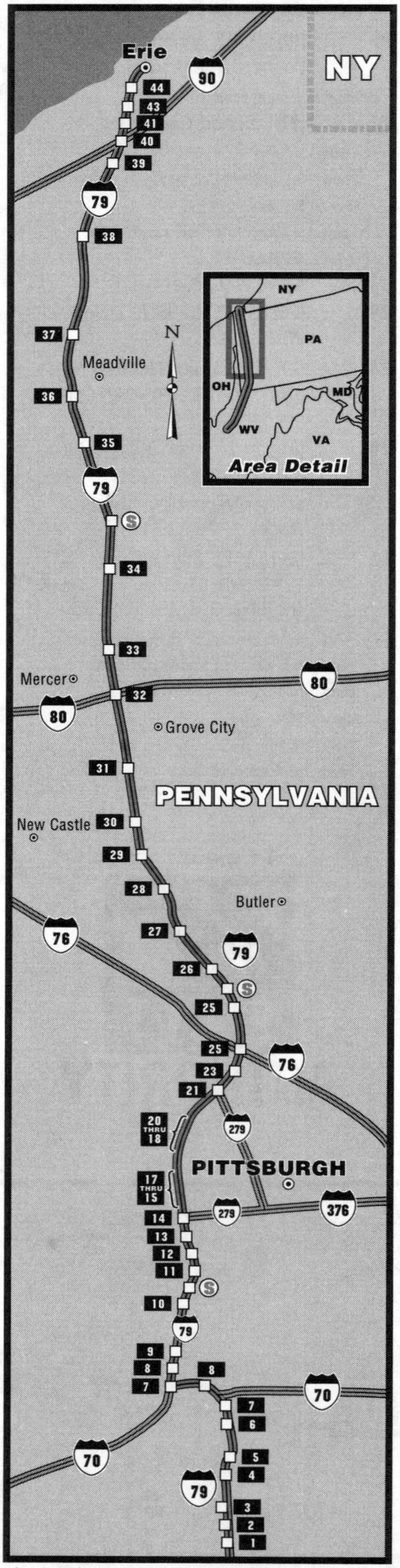

EXIT PENNSYLVANIA

(136) **Weigh Station (Both Directions)**

(135) **Rest Area - RR, Phones, Picnic (Southbound)**

(133) **Rest Area - RR, Phones, Vending, Picnic (Northbound)**

34 **(130)** PA 358, Greenville, Sandy Lake
- **TStop** E: **Lake Way**
- **FStop** W: **Fuelstop***
- **Food** E: Lake Way Truck Plaza
  W: Fuelstop
- **RVCamp** W: **Camp Wilhem, Goddard State Campground**
- **Med** W: ✚ **Hospital (13 Miles)**
- **Parks** W: **MK Goddard State Park, Pymatuing State Park**

33 **(121)** U.S. 62, Mercer, Franklin
- **FStop** W: **Sunoco***
- **Gas** E: Citgo*(CW)
- **ATM** E: Citgo

32 **(116)** Junction I-80, Clarion, Sharon

31 **(113)** PA 208 & 258, Grove City
- **Gas** E: BP*
  W: Pennzoil Lube*, Sunoco*
- **Food** W: Eat N Park, McDonalds, Subway
- **Lodg** W: (AAA) AmeriHost Inn
- **AServ** W: Pennzoil Lube, Sunoco
- **RVCamp** W: **KOA Campground**
- **Med** E: ✚ **Hospital**
- **Other** E: **Mon's Convenience Store***(LP)
  W: **Grove City Factory Shops, Tourist Information**

Bold red print shows RV & Bus parking available or nearby

EXIT	PENNSYLVANIA
09)	Rest Area - RR, Phones, Picnic, Vending (Southbound)
07)	Rest Area - RR, Phones, Picnic, Vending (Northbound)
0	**(105)** PA 108, Slippery Rock
RVCamp	E: Slippery Rock Campground
9	**(99)** U.S. 422, New Castle, Butler
RVCamp	W: Coopers Lake Campground (2 Miles), Rose Point Park Campground (3 Miles)
Parks	E: Moraine State Park
	W: McConnells State Park (2 Miles)
Other	W: Living Treasures Animal Park
8	**(96)** PA 488, Portersville, Prospect
AServ	W: Bill's Auto Towing & Repair
RVCamp	E: Bear Run Canpground
Parks	E: Moraine State Park
	W: McConnells State Park (3 Miles)
7B	**(88)** to U.S. 19, PA 68, Zelienople (Southbound, Northbound Reaccess)
RVCamp	W: Indian Brave Camping Resort
7A	**(87)** PA 68, Zelienople (Northbound, Southbound Reaccess)
Food	W: Thompson's Family Restaurant, Zelienople Restaurant
Lodg	W: Zelienople Motel
AServ	W: Goodyear Tire & Auto
RVCamp	W: Indian Brave Camping Resort
6	**(83)** PA 528, Evans City (No Southbound Reaccess)
80)	Weigh Station, Picnic - Phones (No Facilities)
79)	Weigh Station, Picnic - Phones (No Facilities, Southbound)
5	**(76)** U.S.19 N, to Jct I-76, PA Tnpk, Cranberry (Left Exit Northbound)
Gas	W: Amoco*(CW), BP*(D), Exxon*, Gulf(CW), Sheetz, Sunoco*
Food	W: Arby's, Bob Evans Restaurant, Boston Market, Brighton Hot Dog Shop, Burger King, Climos Pizza and Chicken, De Vos Restaurant, Denny's, Dunkin Donuts, Hardee's, Hartners Restaurant, Jersey Mike's Subs, Kings Family, Lone Star Steakhouse, Long John Silvers, Mac & Erma's, PaPa Don's Italian, Perkins Family Restaurant, Pizza Outlet, Pizza Roma, Sheraton, Shoney's, Subway, TCBY Yogurt, The Italian Oven, Wendy's
Lodg	W: Conley Inn, DAYS INN Days Inn, Fairfield Inn, Hampton Inn, Holiday Inn Express, Junction Inn, Oak Leaf Motel, Red Roof Inn, ◆ Sheraton
AServ	W: Baierl Toyota, Beacon Auto Parts, Goodyear Tire & Auto*, Grease Monkey, Jorden's Service Center, Pep Boys Auto Center
RVCamp	E: Pittsburg North Campground
Med	W: ✚ Hospital
ATM	W: Giant Eagle Grocery, Mars National Bank, National Bank, Northside Bank, PNC Bank
Other	W: Giant Eagle Grocery (Pharmacy), K-Mart, Kinko's Copies, Mail Boxes Etc, Mail Stop Packaging and Shipping, Office Max, Pet Supplies Plus, Phar Mor Drugs, Shop 'N Save Supermarket, Thrift Drug, Wal-Mart
23	**(75)** to U.S. 19 South, Warrendale (Northbound, Southbound Reaccess)
22	**(73)** PA 910, Wexford
Gas	E: BP*
Food	E: King's Family
	W: Carmody's Restaurant
Lodg	E: ◆ Econolodge
ATM	E: Citizens National Bank
Other	E: Shumaker Pharmacy, T-Bones Grocery, VIP Do-It-Yourself Car Wash
21	**(71)** Jct I-279 South, Pittsburgh (Southbound Left Exit, Northbound Reaccess)
20	**(68)** Mount Nebo Rd
Med	W: ✚ Hospital
19	**(66)** PA 65, Emsworth, Sewickley
Med	W: ✚ Hospital
18	**(65)** Neville Island, PA 51
Gas	E: Round Town Gas*(D)
AServ	W: Steel City Tire, Tri State Motors
ATM	E: Machine (Round Town)
Other	W: Neville Island Laundromat
17	**(64)** PA 51, Coraopolis, McKees Rocks
Med	E: ✚ Hospital
16	**(60)** PA 60, Crafton, Moon Run
FStop	E: Exxon*
Gas	W: Sunoco*
Food	E: King's Family
Lodg	E: Econolodge, Motel 6
AServ	W: J&N Foreign Car Service, Sunoco
Med	E: ✚ Hospital
ATM	E: King's Family Restaurant
15	**(58)** U.S. 22, U.S. 30, Airport
14	**(58)** Jct I-279, North to Pittsburgh
13	**(56)** Carnegie
12	**(54)** Heidelberg, Kirwan Heights (All Traffic Exits Eastbound)
Gas	E: BP
Food	E: Arby's, Chuck E. Cheese's, Eat N Park, KFC, Kribel's Bakery, Moon Chinese Buffet, Napoli Pizza, Old Country Buffet, Peter's Place, Pizza Hut, Shanghai Chinese, Subway, Wendy's
AServ	E: Firestone Tire & Auto, Goodyear Tire & Auto, Meineke Discount Mufflers, Procare, Speedee Muffler King, Valvoline Quick Lube Center
Med	E: ✚ Hospital
ATM	E: National City Bank
Other	E: Car Wash, Phar Mor Drugs, Thrift Drug
11	**(52)** PA 50, Bridgeville
Gas	E: BP*(D)(CW), Exxon*(D)
	W: Texaco*(LP)
Food	E: Bo Sue's Ice Cream Shop, Burger King, King's Family (Open 24 Hours), McDonalds, Pop Edward's Famous Roast Beef
Lodg	W: (AAA) Knight's Inn
AServ	E: Burgunder Dodge, Exxon, GM Auto Dealership, Havoline 15 Minute Quick Lube, Midas Muffler & Brakes, Monroe Muffler & Brakes, Napa Auto Parts and Service
	W: Texaco
ATM	E: PNC Bank
Other	W: Car Wash
(50)	Weigh Station
(49)	Rest Area - RR, Phones, Vending, Picnic
10A	**(48)** Southpointe, Hendersonville
10	**(45)** to PA 980, Canonsburg
Gas	W: CoGo's*
Food	W: Hoss's Steak & Sea, KFC, Long John Silvers, McDonalds, Pizza Hut, Wendy's
Lodg	W: Super 8 Motel
AServ	W: Advance Auto Parts, John's Automotive, Tatano Brothers Auto
Med	W: ✚ Hospital
9	**(43)** PA 519, Houston, Eighty Four
Gas	E: Amoco
	W: Sunoco
AServ	E: Amoco
	W: Sunoco
Med	E: ✚ Hospital
	W: ✚ Hospital
8	**(21)** PA 136, Beau St
8B	**(41)** Meadow Lands (No Reaccess Northbound)
Gas	W: BP*(D)(LP)
Food	E: Holiday Inn, McDonalds
	W: Mr. Hungry (Burgers)
Lodg	E: (AAA) Holiday Inn
Other	W: Trolley Museum (3 Miles)
(37)	Jct I-70 East, to New Stanton, West to Washington
7	**(33)** U.S. 40, Laboratory Rd.
RVCamp	W: KOA Campground
(31)	Parking Area, Weigh Station (Southbound)
6	**(30)** U.S. 19, Amity, Loan Pine
Gas	W: Exxon*(D)
RVCamp	W: Lone Pine RV Center(LP)
(28)	Picnic (Northbound)
5	**(23)** Marianna, Prosperity
Gas	W: Texaco*
4	**(18)** to U.S. 19, PA 221, Ruff Creek, Jefferson
Gas	W: Citgo*
Other	W: Hardware Store (Citgo)
3	**(14)** PA 21, Masontown, Waynesburg
Gas	E: Amoco*(D)
	W: BP*(LP), Citgo*(D)(LP), Exxon*
Food	W: Dairy Queen, Golden Corral, Golden Wok Chinese, KFC, Taco Bell (Exxon), Wendy's
Lodg	E: Comfort Inn
	W: (AAA) Econolodge
AServ	E: Amoco
	W: Amerilube, Bortz, Chevrolet, Cadillac, Geo
Med	W: ✚ Hospital
ATM	W: Community Bank
Other	W: PA State Police, Revco Drugs, Shop 'N Save Supermarket
2	**(7)** Kirby, Garards Fort
(5)	PA Welcome Center - RR, Phones, Tourist Info (Northbound)
1	Mount Morris
TStop	E: Texaco*(SCALES)
Gas	W: Ashland, Ashland (Kerosene), BP*(D)
Food	E: Texaco Fuel Stop
AServ	E: Honda Mazda Dealer
RVCamp	W: Campground

EXIT	PENNSYLVANIA/WEST VIRGINIA
ATM	W: BP
Other	W: Coin Car Wash

↑PENNSYLVANIA

↓WEST VIRGINIA

(159) **WV Welcome Center - RR, Phones, Picnic, Vending, RV Dump (Southbound)**

155 U.S. 19, WV 7, West Virginia University

- **Gas** W: Exxon*(D)
- **Lodg** E: Holiday Inn (see our ad on this page), Hampton Inn (see our ad on this page)
- **Med** E: ✚ **Hospital**

152 U.S. 19, Westover, Morgantown

- **Gas** E: BP*(CW), Exxon*
- **Food** E: Donut Connection, Double Play, McDonalds, **Pizza Hut, Western Sizzlin'**
  W: Bob Evans Restaurant, Burger King, Captain D's Seafood
- **Lodg** E: AAA **Econolodge**
- **AServ** E: Midas Muffler & Brakes, The Auto Ranch
- **Med** E: ✚ **Wedgewood Family Practice (Walk-in Clinic)**
- **ATM** E: Shop n Save, West Banco
- **Other** E: Coin Car Wash, **Shop n Save Supermarket, WV State Police**
  W: **Lowes, Morgantown Mall, Office Max, Phar Mor Drugs (Pharmacy, Grocery), Super K-Mart (Pharmacy, Grocery, Open 24 Hours)**

148 Junction I-68 East, Cumberland, MD

146 WV 77, Goshen Rd

(141) **Weigh Station**

139 East Fairmont, WV 33. Prickett's Corner Rd

- **TStop** W: **K and T Truck Stop**
- **Gas** E: Chevron*
  W: Exxon*
- **Food** W: Country Store and Cafe
- **AServ** W: F and B Auto Repair
- **TServ** W: **A1 Truck Service**
- **RVCamp** E: **Campground (.9 Miles)**
- **Other** E: **Mom n Pops Food Mart, Race Zone Nascar Collectables**

137 WV 310, Downtown Fairmont

- **FStop** E: **Exxon*(D)**
  W: **Exxon*(D)**
- **Gas** W: Chevron*(CW)
- **Food** W: Dairy Creme Corner, KFC, McDonalds, Subway, Wendy's
  E: The Simmering Pot (Holiday Inn)
- **Lodg** E: Holiday Inn (see our ad on this page)
- **AServ** E: Exxon
  W: Eastside Garage (Twenty-four hour towing.)

EXIT

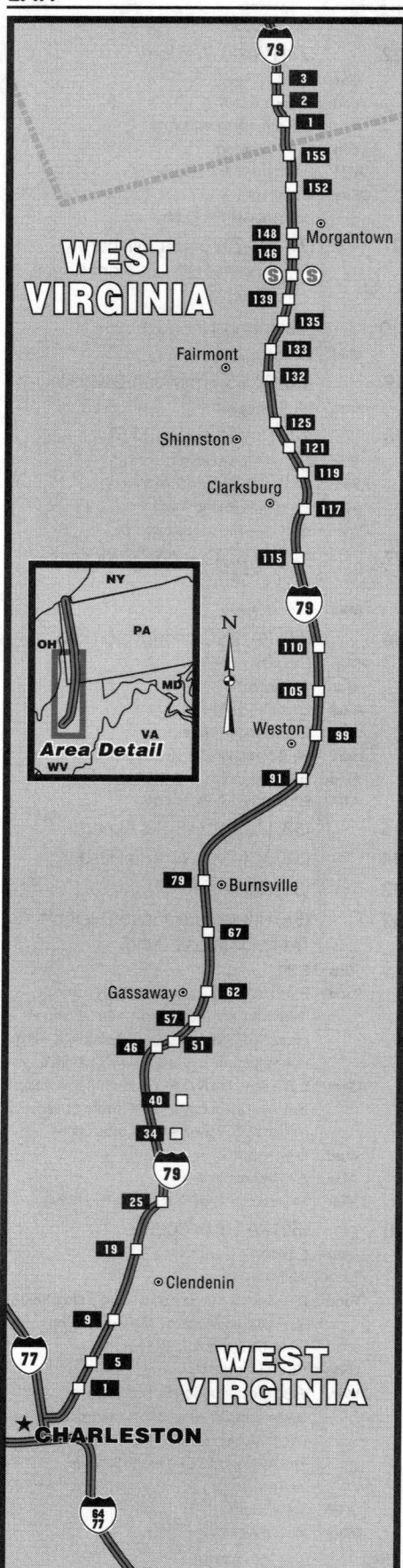

WEST VIRGINIA
Morgantown
Fairmont
Shinnston
Clarksburg
Weston
Burnsville
Gassaway
Clendenin
CHARLESTON
Area Detail
NY PA OH MD VA WV
N

EXIT WEST VIRGINIA

- **ATM** W: Chevron, One Valley Bank
- **Other** E: **Village Pantry Grocery**
  W: **Warehouse Grocery**

135 WV 64, Pleasant Valley Rd

- **Other** W: **Middleton Animal Clinic**

133 WV 64/1, Kingmont Rd

- **FStop** E: **BP*(D)**
- **Food** E: **BFS Foods (BP)**
- **Lodg** E: Super 8 Motel
  W: DAYS INN Days Inn

132 U.S. 250, South Fairmont

- **Gas** W: Exxon*(D), Go Mart*, Sunoco*
- **Food** E: Dutchman's Daughter, Hardee's, Marino's Italian Restaurant, McDonalds, Shoney's, Subway
- **Lodg** E: ◆ Econolodge, ◆ Red Roof Inn
- **AServ** E: Lee's Auto Repair, NAPA Auto Parts, Urse Plymouth/ Chrysler
  W: Astro, Buick, Geo, GMC, Exxon, Mark's Towing, Penn Toyota, Wilson Ford, Lincoln, Mercury
- **ATM** E: OneValley Bank, Wes Banco
- **Parks** E: **Tygart State Park (21 Miles)**
- **Other** E: **Enterprise Rent-a-Car, Eye Care One, Hills Department Store, Lowes, Middletown Mall, Priority Mail and Ship, Sam's Club, Shop 'N Save Supermarket, Tygart Valley Cinemas, Wal-Mart (Pharmacy)**
  W: **Brand Trailer Sales RV Service and Sales**

Bold red print shows RV & Bus parking available or nearby

EXIT	WEST VIRGINIA
125	WV131, Saltwell Rd , Shinnston
Gas	W: Exxon*
Food	W: Subway (Exxon)
ATM	W: Exxon
Other	E: Airport (3 Miles)
124	WV 707, FBI Center Rd
(123)	Rest Area - RR, Phones, RV Dump
121	WV 24, Meadowbrook Rd
Gas	E: Go Mart*
	W: Exxon*(D)
Food	E: Blimpie's Subs (Go Mart), Bob Evans Restaurant, Carl's Cafe (Go Mart)
	W: Burger King, Ponderosa
Lodg	E: ◆ Hampton Inn
	W: Econolodge
AServ	W: Quick Slick Oil Change
ATM	E: Bank One
Other	W: Meadowbrook Mall
119	U.S. 50, Clarksburg, Bridgeport
Gas	E: Chevron*, Exxon*(D)
Food	E: Bonanza Steak House, Daisy Mae's Cafe, Damon's Ribs, Days Inn, Eat N Park, El Rincon Mexican, Elby's Big Boyoy, KFC, Little Caesars Pizza, Long John Silvers, McDonalds, Ming Garden, Pizza Hut, Pizza Man, Shoney's, Taco Bell, Texas Road House, The Italian Oven, The Simmering Pot (Holiday Inn), Western Steer Family Steakhouse
Lodg	E: DAYS INN ◆ Days Inn, AAA Holiday Inn (see our ad on this page), ◆ Knight's Inn, Ramada Inn, ◆ Sleep Inn
AServ	E: Advance Auto Parts, Exxon, Midas Muffler & Brakes, Plaza Service Center, Quick Lube
ATM	E: Huntington Bank, One Valley Bank (Kroger), Wes Banco Bank
Other	E: Bridgeport Eye Care, GNC, Giant Eagle Supermarket (Pharmacy), Hills Grocery Store, Kroger Supermarket (Pharmacy), Lowes, Radio Shack, Supermarket, U.S. Post Office, Wal-Mart (Pharmacy)
117	WV 58, Anmoore
FStop	W: BP*(D) (1.3 Miles)
AServ	W: KC Auto
115	WV 20, Stonewood, Nutter Fort
Gas	E: Chevron*, Texaco*(D)
Food	E: Dee's Diner, Mountaineer Family Restaurant
ATM	E: Texaco
Other	E: Marty's Italian Bakery
110	WV270, Lost Creek
Gas	E: BP*, Citgo*(CW)
Food	E: Subexpress
ATM	E: Harrison County Bank
Parks	W: Watters Smith Memorial State Park
Other	E: Lost Creek Laundromat, U.S. Post Office, Village Pharmacy

EXIT	WEST VIRGINIA
105	WV 7, Jane Lew
TStop	E: I-79 Truck Stop*(SCALES) (RV Dump)
Gas	E: Chevron*(CW)
Food	E: I-79 TS (Scales and RV Dump), The Wilderness Plantation
Lodg	E: The Wilderness Plantation
TServ	E: I-79 Truck Stop
TWash	E: I-79 Truck Stop
ATM	E: Chevron
	W: Citizens Bank-Jane Lew Branch
99	U.S. 33, U.S. 119, Weston, Buckhannon
Gas	E: BP, Exxon(D)(LP), Sheetz*(LP)
Food	E: Comfort Inn, Gino's Pizza & Spaghetti, McDonalds, SheetzaPizza, Subway, Taco Bell (Sheetz Convenience Store), Western Sizzlin'
Lodg	E: AAA Comfort Inn, Super 8 Motel
AServ	E: Advance Auto Parts, Quality Farm & Fleet (Tire and Battery Service.)
	W: Exxon*(CW)
Med	W: ✚ Hospital
ATM	E: Progressive Bank
Parks	E: Black Water Falls (72 Miles), Canaan Valley Park (72 Miles)
Other	E: Kroger Supermarket, Revco Drugs, Wal-Mart (Pharmacy), Weston 4 Theaters
96	WV 30, S Weston
Lodg	W: Western Motor Inn (Nine-tenths of a mile from the Interstate.)
RVCamp	E: Broken Arrow Camp Ground (1.4 Miles)
91	U.S. 19, Roanoke
RVCamp	E: Stonewall Jackson Lake State Park (3.1Miles)
	W: Whisper Mountain Campground (5 Miles)
Other	E: Bull Town Recreation Area
(85)	Rest Area - RR, Phones, Picnic, RV Dump
79	WV 5, Burnsville, Glenville
Gas	E: Exxon*
	W: Go Mart*
Food	E: Burnsville Diner, The Seventy-Niner
Lodg	E: Motel 79 (Coin Laundry)
RVCamp	E: Burnsville Dam Riffle Run (3.5 Miles)
	W: Cedar Creek State Park (24 Miles)
ATM	W: Bank of Gassaway
67	U.S. 19, WV 15, Flatwoods
TStop	E: Go Mart* (ATM)
	W: John Skidmore Truck Stop*
FStop	E: Chevron*(D)
Gas	E: Ashland Convience Store Compress Natural Gas*, Exxon* (ATM)
	W: Ashland Convience Store*, Citgo*
Food	E: Dairy Queen, KFC, McDonalds, Stancati's Pizza (Mid-Mountain Lanes), WV Fried Chicken (ATM), Waffle Hut, Western Steer Family Steakhouse
	W: John Skidmore Restaurant, Shoney's
Lodg	E: DAYS INN Days Inn, Motel and Restaurant
	W: John Skidmore Motel (Pennzoil)
RVCamp	E: Gerald R Freeman, Sutton Lake Campground
Other	E: Coin Car Wash, Coin Laundromat, Dan's Backwood Sporting Supply, Lloyd's Ace Home Center Hardware, Mid -Mountain Bowling Lanes, The Home National Bank (ATM)
	W: Factory Outlet Center
62	WV 4, Sutton, Gassaway
Food	W: Pizza Hut
Med	W: ✚ Hospital
57	U.S. 19 South, Beckley
51	WV 4, Frametown
(49)	Rest Area - RR, Phones, Tourist Info, Picnic, RV Dump
46	WV 11, Servia Rd
40	WV 16, Big Otter
FStop	E: Go Mart*
Gas	W: Exxon*(D) (ATM)
Lodg	W: Country Inn
Other	W: G-Mart Grocery
34	WV 36, Wallback , Clay
25	WV 29, Amma
Gas	E: Exxon*(D)(CW)
19	U.S. 119, WV 53, Clendenin
FStop	E: BP*(D) (.5 Miles)
Food	E: Schaffers Super Stop (.5 Miles), Tutors Biscuit World (2.6 miles from the Interstate.)
9	WV 43, Elkview, Frame Rd
Gas	W: Exxon*(D)(LP), Su perAmerica*
Food	W: Arby's, Baskin Robbins, McDonalds, Pizza Hut Carry Out, Ponderosa, Rainbow Restaurant, Subway
AServ	W: Advance Auto Parts
ATM	W: Machine (SuperAmerica)
Other	W: K-Mart (w/pharmacy), Revco Drugs
5	WV 114, Big Chimney
Gas	E: Exxon (1 Miles)
Food	E: Hardies (Seven-tenths of a mile off of the interstate.)
1	U.S. 119, Mink Shoals
Food	E: Harding's Family Restaurant
Lodg	E: ◆ Sleep Inn
Other	E: Anglers Roost Sport Shop

↑WEST VIRGINIA

Begin I-79

EXIT	CALIFORNIA

Begin I-80

↓CALIFORNIA

4 Treasure Island (Left Hand Exit)
- Lodg N: Days Inn (see our ad on this page)

(7) Toll Plaza (Westbound)

8 West Grand Ave.

9A Powell St., Emeryville
- Gas N: Shell*(D)(CW)
  - S: BP
- Food N: Charlie Brown's Steak & Lobster
  - S: Bettore's Pizza, Burger King, Lyon's (Days Inn), Togo's
- Lodg N: Holiday Inn
  - S: Days Inn
- Other S: **Lens Crafters, Powells St. Plaza (No Trucks)**

9B Ashby Ave., CA. 13

11 University Ave, Berkeley

12A Gilman St

12B Albany

13 Central Ave
- Gas S: Exxon*, Shell*
- Food S: White Knight

14 Carlson Blvd (No Trucks Over 3.5 Tons)
- Gas N: 76 (24 Hrs)
- Lodg N: 40 Flags Motel
  - S: Super 8 Motel
- RVCamp N: **Ralph's RV Service Center**

15A McDonald

15 Cutting Blvd., to I-580, San Rafael Bridge
- Gas N: Alaska(D)
  - S: 76(D) (24 Hrs), Just Cheap
- Food S: Church's Fried Chicken, Jack-in-the-Box
- Lodg S: Freeway Motel
- Other S: **Pet Food Express, Target Department Store**

16 San Pablo Ave., Barrett Ave.
- Food S: Baker's Square, Labamba, Super Wok, Wendy's
- AServ S: Smith Chrysler, Plymouth

17A McBryde Ave

17C San Pablo, Dam Road

17 Solano Ave
- Gas N: BP*
- Food N: **Broiler (Casino San Pablo)**, Denny's (24 Hrs), Mae's Coffee House, Nations, Taco Bell
- RVCamp S: **Dave's Camperland, Windy Flat RV Repair**
- ATM N: Cal Fed
- Other N: **Casino San Pablo, Coin Laundry, Lucky (Pharmacy), Pay Less Drugs**
  - S: Grand Prix Car Wash

18 El Portal Drive, San Pablo

19 Hilltop Drive
- Food N: Chevy's Mexican
- Lodg N: Courtyard by Marriott (Marriott)
- ATM N: Bank Of America, Coast Federal Bank, Home Savings Of America
- Other N: **Hilltop Mall**

20B Richmond Pkwy., Fitzgerald Dr.
- Food N: McDonalds (Play Place), Red Lobster
  - S: Applebee's, Cheese & Steak Shop, In-N-Out Burgers, Taqueria Uno, Togo's
- AServ S: Kragen Auto Parts
- ATM S: Wells Fargo
- Other N: **Barnes & Noble, Office Max, Pearl Vision Center, Petsmart**
  - S: **Mervyn's, Pet Food Express, Pinole Vista Crossing, Target Department Store**

20 Appian Way, El Sobrante
- Gas N: BP
  - S: Exxon*(D)
- Food N: Bangkok Tai Cuisine, McDonalds (Play Place), Peking Garden, Pizza Hut
  - S: Burger King, Carl's Jr Hamburgers, Hometown Buffet, Hot Dog Station, KFC, Long John Silvers, Pinnoli Gourmet Pizza, Sizzler Steak House, Starbuck's Coffee, Taco Bell, Wendy's
- Lodg S: Days Inn (see our ad on this page), Motel 6
- AServ N: BP
  - S: Goodyear Tire & Auto
- Med N: **+ Dr. Hospital**
- ATM N: Bank Of America, Cal Fed
- Other N: Full Service Car Wash, **Long's Drugs, Safe Way Grocery**
  - S: **K-Mart, Lucky's (Pharmacy)**

21 Pinole Valley Road
- Gas S: Beacon(CW), Chevron*, Shell
- Food S: Jack-In-The-Box, Pizza Plenty, Subway, Zip's Restaurant
- AServ S: Beacon, Shell
- ATM S: 7-11 Convenience Store
- Other S: **7-11 Convenience Store, Coin Laundry, Lucky Food Center**

23 Hercules

24 Willow Ave, Rodeo (No Trucks East)
- Gas S: BP*(CW)
- Food S: Adesta Pizza, Donuts, Willow Garden

25 Cummings, Sky Way (Eastbound)

26A Martinez, Concord

26B Crockett Rodeo
- Food N: Yet Wah

27A Toll Bridge (Eastbound)

27B Sonoma Blvd. (Eastbound)

27C Sequoia

27D Maritime Academy Drive (Westbound)
- Gas N: Arco*(D), Chevron*
- Food N: Subway (Arco)

Bold red print shows RV & Bus parking available or nearby

EXIT CALIFORNIA

Lodg N: Motel 6, Rodeway Inn

28 Magazine St

Gas N: Shell

Food N: Rod's Hickory Pit (24 Hrs)
S: McDonalds

Lodg N: Budget Motel, Economy Inn, El Curtola, El Rancho Motel, Golden Penny Inn
S: Thrift Lodge

RVCamp N: **Trade Winds RV Park**

ATM S: 7-11 Convenience Store

Other S: **7-11 Convenience Store**

29 Jct. I-780 & I-680, Benicia, Martinez

30 Georgia St, Centro Vallejo

31A Solano Ave, Spring Road

Gas N: Chevron (24 Hrs)
S: Beacon*

Food N: Burger King, Church's Fried Chicken, Nitti Gritti Restaurant, Taco Bell
S: Burt's Burgers

Lodg N: E-Z 8 Motel, Gateway Motor Hotel

AServ N: Chevron
S: Burt Motor Works

Other N: **Coin Laundry, Lucky's Grocery (Pharmacy), Pay Less**

31B Tennessee St., Mare Island

Gas S: 76, Exxon*(CW)

Food S: Gold Rush Pizza & Potato Co

Lodg S: Quality Inn

AServ S: 76, Minute Man Car Wash

31C Red Wood St (No Trucks)

Gas N: 76*
S: Cheaper*, Shell*(CW)

Food N: Annie's Panda Garden, Denny's
S: Little Caesars Pizza, Papa Murphy's, Royal Jelly Donuts, South Villa Chinese

Lodg N: Days Inn, Motel 6

AServ N: 76
S: Avery Greene Oldsmobile, GMC, Honda, Chief Auto Parts, Oil Changers

Other S: **Redwood Coin Laundry, Redwood Veterinary Clinic, Safe Way Grocery (Pharmacy)**

32 Columbus Pwky.

33 San Rafael, CA. 37

(33) **Rest Area - Tourist Info, RR, Phones, Picnic (Westbound)**

37 American Canyon Rd.

44 Red Top Road

Gas N: 76*(LP) (24 Hrs)

45 County Road 12, Napa, Sonoma

46A Junction I-680, Benicia, San Jose

46B Suisun Valley Road, Green Valley Road

TStop N: **Jimmy's Truck Stop***(SCALES)

Gas S: 76*(D), Arco*, Chevron*, Shell*(D)(CW)

Food N: **Jason's**
S: Arby's, Bravo's Pizza, Burger King (Playground), Carl's Jr Hamburgers, Denny's, Green Bamboo, McDonalds (Playground), Old San Francisco Express Pasta, Subway, Taco Bell, Wendy's

Lodg S: Best Western, Economy Inns of America, Hampton Inn, Overniter Lodge

TServ N: **Jimmy's Truck Stop**

RVCamp S: **Camping World RV Service** **(see our ad on this page)**

Other S: **Scandia Amusement Park/ Shopping Mall**

(47) **Weigh Station (Both Directions)**

49 Rio Vista, CA. 12E., Suisun City

Lodg S: Motel 6

AServ S: Fairfield Toyota, Steve Hopkins Honda

50 West Texas Road, Rockville Road

Gas N: Shell*(D)(CW)

Food N: Chuck E. Cheese's Pizza, Gordito's Restaurant
S: Cenario's Pizza, Fairfield Donut, Jack-In-The-Box, McDonalds, Nations Burgers, Pelayo's

AServ S: Pep Boys Auto Center

Other S: **Chinese Restaurant, Food 4 Less, Target Department Store, Walgreens**

51 Travis Blvd

Gas N: Chevron*(CW)

Food N: Burger King, Denny's, Mary's Pizza Shack, McDonalds, My Cafe, New York Pizza Kitchen, Subway, Taco Bell, Wokman
S: Chevy's Mexican Restaurant, Fresh Choice, Marie Callender's Restaurant, Red Lobster

Lodg N: Holiday Inn Select, Motel 6

AServ N: Fairfield Nisson & Hyundai

Other N: **Kinko's (24 Hrs), Pier One Imports, Ralph Raley's (Drug Store)**
S: **Sears, Solano Mall**

52 Waterman Blvd

Food N: Coffee World, Dynasty Restaurant, Round Table Pizza, Strincs Italian Cafe, TCBY Yogurt, The Hungry Hunter

AServ N: Woodard Chevrolet Dealership

Other N: **Safe Way Grocery (Pharmacy)**

54 Fairfield, North Texas St

Gas S: Arco* (24 Hrs), BP*, Shell*(CW)(LP)

Food S: **Lou's Junction**

Lodg S: E-Z 8 Hotel

56A Cherry Glen Road, Lagoona Valley Road

56B Pena, Adobe

57 Cherry Glen Road

58 Merchant St, Alamo Drive

Gas N: Chevron*(CW), Shell*(D)(CW), Ultramar

Food N: Bakers Square Restaurant, Baskin Robbins, Cenario's Pizza, Digger's Deli, Donut Queen, KFC, Lyon's, Round Table Pizza, Wrenn's Cafe
S: Ho's Donuts & Coffee, Jack-In-The-Box, McDonalds (Playground), Pizza Hut, Port Of Subs, Stir Fry

Lodg N: Alamo Inn, Monta Vista Hotel

ATM N: Eureka Bank

Other N: **Wordsworth Used Books**
S: **Food 4 Less (Grocery Store, 24 Hrs)**

59A Davis St

Gas S: Quik Start Market*

Food S: Grandma's Platter House

AServ N: Bernie's City Garage
S: Advanced Muffler, Chevron, Christian's Auto Service, Gemini Auto Repair, Village Radiator

Other S: **Midvalley Veterinarian**

59B Mason St, Travis AFB

Gas N: Texaco*(CW)
S: Shell

Food N: Chows, Subway
S: Cable Car Coffee Co., Carl's Jr Hamburgers, Domino's Pizza, Donut Wheel, El Azteca, Formosa, Hi-Way, Lisa's Lounge, Solano Baking Co., Sushi Sen, Wah Shine, Wienerschnitzel, Wok-N-Roll

AServ N: Chief Auto Parts, Econo Lube & Tune, Kragen, NAPA Auto Parts, Quaker State Oil Change
S: Agean Tire Service, Ford Dealership, Goodyear Tire & Auto, Shell

Med S: **+ Hospital**

ATM S: West America Bank

Other N: **Albertson's Grocery, WalGreens**
S: Car Wash, **Launder Land, U.S. Post Office**

60 Monte Vista Ave

Gas N: BP*, Chevron*(D)(CW), Shell*(CW)

Food N: Arby's, **Denny's**, IHOP, Java Int'l, McDonalds (Playground), Murillo's, Nations, Round Table Pizza, Taco Bell, Wendy's, Yen King Restaurant
S: Applebee's, Big Apple Bagels, Boston Market, Chili's, Chubby's, Coffee Tree Restaurant (Courtyard), Fresh Choice, In-N-Out Hamburger's, Italian Cafe, Java City, KFC, Pizza Hut, Port Of Subs

Lodg N: Best Western, **Brigadoon Lodge** **(see our ad on this page)**, Super 8 Motel
S: Courtyard by Marriott

AServ N: Firestone Tire & Auto, Grand Auto Supply Tires & Service, Midas Muffler & Brakes

ATM S: Bank Of America

Other N: Vacaville Car Wash
S: **Crown Books Superstore, Nut Tree, Petco, Safe Way Grocery, Vaccaville Commons (Mall)** **(see our ad on this page)**

EXIT CALIFORNIA

61 I-505 North Redding
63 Pleasure Town Road
- Gas S: BP
- Food S: Hick'ry Pit, Jack-In-The-Box, Vaca Joe's
- Lodg S: Quality Inn
- AServ S: BP

64 Meridian Road, Weber Road
66 Midway Road, Lewis Road, Elmira
69 Dixon Ave
- FStop N: **Cheaper*(D)**
- Gas S: Arco*, Shell(CW)(LP) (24 Hrs)
- AServ S: Shell
- ATM S: Shell

70 Pitt School Road
- Gas S: Chevron* (24 Hrs)
- Food S: Arby's, Asian Garden Chinese, Baskin Robbins, Burger King, Chevy's Mexican, Denny's, IHOP (24 Hrs), Jalisco Mexican, Java California, LaBella's Pizza, Marcey's Ice Cream & Yogurt, Mary's Pizza Shack, McDonalds (Playground), Pizza Hut, Solano Baking Co., Taco Bell, Valley Grill
- Lodg S: Best Western
- ATM S: Chevron
- Other S: **Safe Way Grocery (Pharmacy)**

71 CA.113 South, Rio Vista
- Food S: Cattleman's Restaurant, Jack-in-the-Box (Under Construction)

72 Milk Farm Road, Dixon
73 Pedrick Road, County E7
- Gas N: BP
- AServ N: BP

74 Kidwell Road
75 UC Davis
76 Winters Use, CA 113 North
77A EChiles Road (Eastbound)
77B Richards Blvd, Davis
- Gas N: Shell*
  S: Chevron (Under Construction)
- Food N: Cafe Italia (Davis Inn), Murder Burger
  S: Wendy's
- Lodg N: Davis Inn

79 Olive Drive (Westbound)
80 Mace Blvd, El Macero, Liberty Island
- Gas S: BP, Chevron*(CW) (24 Hrs), Shell*(CW)
- Food S: Burger King, Cindy's Restaurant, Denny's, McDonalds, Subway, Taco Bell
- Lodg S: Motel 6
- AServ S: Chrysler Auto Dealer, Courtesy Pontiac, Oldsmobile, Buick, GMC Truck, Ford Dealership, Hanlees Chevrolet, Toyota, Geo Dealership, Nisson Dealership, University Honda Dealership
- RVCamp S: **La Mesa Rv's**

86 Frontage Road (Westbound)
88A West Capital Ave
- FStop N: **Exxon***
- Gas N: Chevron* (24 Hrs), Shell*
  S: Arco* (24 Hrs)
- Food S: Burger King, Denny's
- Lodg N: Granada Inn
- RVCamp S: **KOA RV Campground**

88B Jct. U.S. 50, Business 80 Loop
89 West Sacramento, Reed Ave

EXIT CALIFORNIA

- Gas S: Arco*
- TServ N: **Sacramento Valley Ford (RV Service)**

91 West El Camino Ave
- TStop N: **76 Auto/Truck Plaza(LP)(SCALES) (Shell Gas)**
- Gas N: Chevron*
- Food N: Burger King (Play Area), **I Can't Believe It's Yogurt (76 TStop), Silver Skillet (76 TStop), Subway (Chevron)**
- ATM N: 76 Auto/Truck Plaza

92 Junction I-5
93 Truxel Road
94 Northgate Blvd
- Gas S: Shell*(CW)
- Food S: Carl's Jr Hamburgers, Finnegan's, McDonalds, Taco Bell
- Lodg S: Extended Stay America, **Travelers Inn**

95 Norwood Ave
- Gas N: Arco*
- Food N: Jack-In-The-Box, McDonalds, New Hong Kong Chinese, Subway
- AServ N: Chief Auto Parts
- Other N: **Sav-Max Foods**

97 Raley Blvd, Del Paso Heights, Marysville Blvd
- FStop N: **Bell Gas 'N Diesel*(D)(LP)**
- Gas N: Arco*, Chevron
- Food S: Connie's Drive-in
- AServ N: Chevron
  S: Hooten Tire Co, North Side Tires, Parrish Tire & Wheel

98 Winters St
99 Longview Drive, Light Rail Station
- FStop N: **Pacific Pride**

100 Capital City Freeway, Sacramento, Bus I-80
101 Madison Ave
- Gas N: Beacon*, Shell* (ATM)
  S: 76(D), Arco*
- Food N: Brookfield's Family Restaurant, Cyber Java, Denny's, Foster's Freeze Hamburgers
  S: A & W Drive-In, Boston Market, Eppie's, IHOP, Jack-In-The-Box, Subway
- Lodg N: Motel 6, Super 8 Motel
  S: Holiday Inn
- AServ S: Acura Mitsubishi Porsche Dealer, Larry's Tire and Brake, Suburban Ford
- ATM N: Shell
  S: Home Savings of America
- Other N: **Hillsdale Animal Hospital**

103 Greenback Lane, Elkhorn Blvd
- Gas N: 76
- Food N: Carl's Jr Hamburgers, China Delux, Leyba's, McDonalds, Pizza Hut, Subway, Taco Bell
- AServ N: 76
- ATM N: Bank Of America
- Other N: **Long's Drugs, Safe Way Grocery**

EXIT CALIFORNIA

(104) **Weigh Station - both directions**
106 Antelope Road, Citrus Heights
- Gas N: 76(LP)
- Food N: Bangkok Thai, Carl's Jr Hamburgers, Chubby's, Giant Pizza, Golden China, KFC, Long John Silvers, McDonalds, Round Table Pizza, Subway, Taco Bell
- AServ N: 76
- Other N: **Albertson's Grocery, Payless Drugs, Post Office, Raley's Supermarket and Drug Center, Total Care Vet. Hospital**

107 Roseville, Riverside Ave, Citrus Heights
- Gas N: Arco*
  S: Exxon*(D)(CW), Shell*(CW)
- Food N: Carousel Deli & Creamery
  S: Back 40 Texas BBQ, Blue Moon Pub, California Hamburger, Golden Donuts, Jack-in-the-Box, Jim Boy's Tacos
- AServ N: AAMCO Transmission, Michaels Auto Service, Mitsos Auto Repair, Preformance Muffler
  S: Do-it-Right Transmissions, Economy Garage, K-Mart, Winston Tire
- Other N: Coin Car Wash, **Coin Laundry, Roseville Animal Hospital**
  S: **K-Mart**

109 Douglas Blvd, Sunrise Ave.
- Gas N: 76*(CW)(LP), Arco*, Exxon*
  S: Arco*, Chevron, Shell
- Food N: Baker Ben's Donuts, Baskin Robbins, Burger King, Chubby's, Dairy Queen, Delicias, Jack-In-The-Box, KFC, McDonalds, Mongolian BBQ, Mountain Mike's Pizza, Pizza Guys, Primos, Roseville Gourmet, Sam's Sub Shop, Taco Bell, Yogurt Del
  S: Carl's Jr Hamburgers, Carrows Restaurant, Del Taco, Denny's, Fortune Garden, Lorenzo's Mexican, Round Table Pizza, San Francisco Hot Dog Co., Szechuan, Yoshi Japanese
- Lodg N: Heritage Inn
  S: Oxford Suites
- AServ N: Auto Service of Roseville, Big O Tires, Cragen Auto Plus, Firestone Tire & Auto, Grand Auto Supply, Midas Muffler & Brakes
  S: Chevron, Hyundai, Quality Tune Up, Shell
- ATM N: Bank Of America, Placer Savings Bank, US Bank
  S: American Savings, Western Valley Credit Union
- Other N: **Price Less Drug and Grocery Store**
  S: **Coin Laundry, Payless Drugs**

110 Atlantic St, Eureka Road
- Gas S: Shell*(CW)
- Food S: Black Angus Steakhouse, Brookfield Restaurant, Carver's Steaks and Chops, Taco Bell, Wendy's
- AServ N: Rossville Auto Parts
  S: America's Tire Co.
- Other N: **Atlantic Street Animal Hospital**

111A Rocklin, Taylor Road (Eastbound, Westbound Reaccess Only)
- Food N: Cattlemen's Restaurant
- RVCamp N: **Holiday RV Super Stores (see our ad this page)**

111B CA 65, Lincoln, Marysville

Bold red print shows RV & Bus parking available or nearby

EXIT		CALIFORNIA
113		Rocklin Road
	Gas	N: Exxon*
		S: Arco*
	Food	N: Arby's, Baskin Robbins, Blimpie's Subs, Burger King, Carl's Jr Hamburgers, China Gourmet, Chinese Cuisune, Denny's, Hacienda Del Roble, Ivory Coast Coffee Co., Jack-In-The-Box, Jaspers Giant Hamburgers, KFC, Outrigger, Papa Murphy, Subway, Taco Bell
	Lodg	N: Days Inn (see our ad on this page), First Choice Inn
	Other	N: Camping World (see our ad on this page), Payless Drugs, Safe Way Grocery
114		Sierra College Blvd
	Gas	N: 76(LP), Chevron*(CW) (McDonalds)
	Food	N: McDonalds (Chevron)
	AServ	N: 76(LP)
116		Loomis Horseshoe Bar Rd
	Food	N: Burger King (Playground), Round Table Pizza, Taco Bell
	Other	N: Raley's
117		Penryn
	Food	N: Cattlebarons
	Other	N: Bob's(LP)
121A		Newcastle
	Gas	S: Arco*, Exxon*(D)
	Food	S: Denny's
	AServ	N: Monroe's Transmission
	Other	N: U.S. Post Office
		S: Highway Patrol
121B		CA 193, Lincoln, Taylor Rd
	TServ	S: Wagner's Truck Repair
122		Auburn, Maple St (Eastbound, Reaccess Eastbound Only)
	Gas	S: Shell*
	Food	S: Cafe Delicias, Mary Belle's, Shanghai Restaurant (ATM), Tio Pepe
	ATM	S: Shanghai
123		Ophir Rd. (Westbound, Reaccess Eastbound Only)
123A		Navada St. (Westbound, Reaccess Westbound Only)
124		CA 49, Grass Valley, Placerville
	Gas	N: BP, Shell*
	Food	N: Foster's Freeze Hamburgers, Marie Callender's Restaurant, In-N-Out Burger
	Lodg	N: Holiday Inn
	AServ	N: BP, Chuck's Auto Glass, Placer Smog and Auto Repair
	Other	N: Thrifty Drug and Grocery Outlet
125A		Elm Ave
	Gas	N: Shell*
		S: Rowdy Randy's*, Sierra Superstop
	Food	N: Taco Bell
	Lodg	S: Elmwood Motel
	AServ	S: Sierra Superstop
	ATM	N: Wells Fargo Bank
		S: U.S. Bank
	Other	N: Kinko's, Long's Drugs, Lucky
125B		Lincoln Way, Russell Rd.
	Gas	S: BP, Texaco(D)
	AServ	S: BP
	Other	S: Foothill Market, Post Office
126		Foresthill, Auburn Ravine Road
	Gas	N: Beacon*, First Stop*

EXIT		CALIFORNIA
		S: 76*(CW), Arco*, Chevron*, Shell*
	Food	N: Akatsuki, Arby's, Denny's, Sam's Hof Brau, Sweetpea's Akatsuki, Taco Bell, Wendy's, Wienerschnitzel
		S: Bagel Junction, Bakers Square Restaurant, Baskin Robbins, Burger King, Burrito Shop, China Express, Country Waffles, Dairy Queen, David's Thai Cuisine, Ikeda's, Izzy's Burger Spa, Jack-In-The-Box, KFC, Lou La Bonte's, McDonalds, Pizza Chalet, Subway, Szechuan Food
	Lodg	N: Auburn Inn, Foothills Motel, Sleep Inn, Super 8 Motel (see our ad on this page)
		S: Best Western, Country Squire Inn National 9
	AServ	S: 76, Shell
	Other	N: Foothill Car Wash
		S: Raley's Supermarket, Thrifty Wash Coin Laundry
127		Bowman
	RVCamp	N: Bowman Mobile Home & RV Park
128		Bell Road
	RVCamp	N: KOA Campground (3 Miles)
	Med	N: + Hospital (3 Miles)
129		Dry Creek Road
130		Clipper Gap, Meadow Vista

EXIT		CALIFORNIA
133		Applegate
	Gas	N: Beacon*(D)
	Food	S: Lil' Applegate Saloon and Deli
	Lodg	N: The Original Firehouse Motel
		S: Applegate Motel
	AServ	S: Applegate Garage
	Other	S: Post Office
134		Heather Glen
135		West Paoli Lane
	Gas	S: Weimar Country Store*
	Food	S: Jimmy Inn Restaurant
	Other	S: Campora Propane
136		Weimar, Cross Road
138		Canyon Way, Placer Hills Road
	Food	N: Dingus McGee's
		S: Velarde's California Cantina
	AServ	S: Colfax Garage, Sierra Chevorlet Dealership, Tom's Sierra Tires
	RVCamp	S: Sierra Chevrolet, GM
140		Colfax, Grass Valley, CA 174
	Gas	N: BP*, Chevron
		S: Chevron*(D), Sierra Super Stop(LP)
	Food	N: A & W Drive-In, Chubby's Diner, Little Red Hen, Mr. C's Doughnut Depot, Pizza Factory, Rosy's Cafe, Taco Bell
		S: Colfax Max, Shang Garden Chinese Restaurant, Subway
	Lodg	N: Colfax Motor Inn
	AServ	N: Chevron, Riebes Auto Parts
		S: Sierra Super Stop
	Other	N: Coin Laundry, Sierra Market
		S: AmeriGas(LP)
145		Magra Road, Rollins Lake Road
	RVCamp	N: Camping
146		Secret Town Road, Magra Road
148		Gold Run (Westbound)
	AServ	S: Hi-Sierra Motors
(149)		Rest Area - RR, Phones, Picnic, RV Water, RV Dump (Both Directions)
150		Dutch Flat
	Gas	S: Araco*
	Food	N: Monte Vista Inn
	RVCamp	S: Dutch Flat Campground and RV Resort
	Other	S: CA Highway Patrol
152		Alta
153		Crystal Springs
154		Baxter
	RVCamp	S: Baxter RV Campground
156		Drum Forebay Road
160		Blue Canyon
(161)		Break Check Area (Westbound)
161		Nyack Road
	FStop	S: Shell*(LP)
	Food	S: Nyack Coffee Shop and Restaurant
(162)		Vista Point Observation Point (Westbound, Reaccess Westbound Only)
162		Emigrant Gap
163		Laing Road (Eastbound)
	Food	S: Rancho Sierra Resort
	Lodg	S: Rancho Sierra Resort
166A		Yuba Gap

Bold red print shows RV & Bus parking available or nearby

EXIT CALIFORNIA

Parks S: Snow-Park
166B CA 20 West, Nevada City
169 Eagle Lakes Road
RVCamp N: Campground
170 Cisco Grove
Gas S: Chevron*
Food S: Cisco's
AServ S: Chevron
Parks N: Sno-Park
172 Big Bend (Eastbound)
173 Rainbow Road, Big Bend
Food S: Rainbow Lodge
Lodg S: Rainbow Lodge
Other S: Tahoe National Forest
176 Kingvale
Gas S: Shell*
183 Soda Springs, Norden
Gas S: 76*
Food S: Donner Summit Lodge
Lodg S: Donner Summit Lodge
186 Castle Peak Area, Boreal Ridge Rd.
Lodg S: Boreal Inn
Other S: Western America Ski Sport Museum
(188) Rest Area - RR, Phones, Picnic (Both Directions)
190 Donner Lake
(191) Vista View Point (Westbound)
193 Truckee, Donner State Park
Gas N: Shell*(D)(CW)
S: 76*, Chevron*
Food N: Pat's Diner
S: Donner House, The Beginning Restaurant
Lodg S: Alpine Village Motel
AServ S: 76
Other N: Tahoe Truckee Factory Stores
S: RV Dump (Chevron)
193B Agricultural Inspection Station (Westbound)
194 CA 89 South, Lake Tahoe
Gas S: Shell*(CW)
Food N: Dairy Queen, Little Caesars Pizza, Pizza Junction, Sizzler Steak House, Wild Cherry's Coffee House
S: Burger King, China Garden, KFC, McDonalds, Papa Murphy's Pizza Take and Bake, Pine O's Pizzeria, Subway, Truckee Bagel Company, Wong's Garden
Lodg S: Super 8 Motel
AServ S: Napa Auto Center, Stone's Tire, Truckee Automotive
ATM N: Bank of America
S: Placers Savings Bank
Other N: Allied Auto Parts, Amerigas Propane, Books and Expresso, Payless Drugs, Safe Way Grocery
S: Coin Car Wash, Lucky Food Center, Post Office
195 Central Truckee (Reaccess Westbound Only)
Food S: El Toro Bravo Mexican, Jordan's Restaurant
Lodg S: Cottage Hotel
Med N: + Tahoe Forrest Hospital
ATM N: U.S. Bank
Other N: Tahoe Forest Pharmacy

EXIT CALIFORNIA/NEVADA

196 CA 89 North, CA 267 South, Truckee, Sierraville, Kings Beach
RVCamp N: Coachland RV Park
198 Prosser Village Road
(202) Weigh Station (Westbound)
207 Hirschdale Road
Gas S: United Trails*(LP)
RVCamp S: United Trails
208 Floriston
210 Farad

↑CALIFORNIA

↓NEVADA

1 Verdi, Gold Ranch Road (Westbound, Reaccess Westbound Only)
2 I-80 Business, Verdi
Gas N: Arco*
Food N: The Branding Iron Cafe
S: Rancho Nevada
3 Verdi (Westbound, Reaccess Eastbound Only)
(4) Weigh Station (Eastbound)
4 Garson Road, Boomtown
TStop N: Boomtown Truckstop*(SCALES) (RV Dump)
Gas N: Chevron*(D)(LP)
Food N: Restaurant (Boomtown TStop)
Lodg N: Boomtown Hotel & Casino
TServ N: Boomtown TStop
Other N: RV Park
5 Verdi, Business Loop 80 (Westbound, Reaccess Eastbound Only)
(6) Truck Parking (Both Directions)
7 Mogul
8 West 4th St (Eastbound)
9 Robb Dr. (Westbound)
10 West, McCarran Blvd
Food S: Little Caesars Pizza (K-Mart)
Other S: 7-11 Convenience Store, Super K-Mart (Pharmacy)
12 Keystone Ave
FStop S: Pacific Pride Commercial Fueling
Gas N: 76, Arco*, Shell
S: 76*(D), Chevron*(CW)
Food N: Pizza Hut, The Purple Bean

EXIT NEVADA

S: Baskin Robbins, Burger King, Coffee Grinder Inn, Flavers Espresso, Gold N' Silver Restaurant, Jack-In-The-Box, KFC, McDonalds, Pizza Baron, Port of Subs, Shakeys Pizza, Szechuan Express, Taco Bell, Wendy's
Lodg N: Gateway Inn, Motel 6
S: Gold Dust West Casino and Motel
AServ N: 76, Q-Lube, Shell
S: Allied Auto Parts, Grand Auto Supply Tires and Service, Meineke Discount Mufflers, Midas Muffler & Brakes, Precision Automotive
ATM N: 7-11 Convenience Store, Norwest Bank
S: Wells Fargo
Other N: 7-11 Convenience Store*, Raley's Supermarket, Sav On Drugs
S: Albertson's Grocery, Sundance Bookstore
13 Business 395, Sierra St, Virginia St
Gas N: 7-11 Convenience Store*, Texaco*(D)
S: Texaco*
Food N: Giant Burger, The Break Away
S: Dairy Queen
Lodg N: Capri Motel, CoEd Lodge, Silver Dollar Motor Lodge, Sundance Motel, University Inn
S: Aspen Motel, Circus Circus Hotel, Flamingo Motel, Golden West Motor Lodge, Horseshoe Motel, Ponderosa Motel, Rino Motel, Savoy Motor Lodge, Shamrock Inn, Showboat Inn, Swan Motel, Uptown Motel, White Court, Wonder Lodge
AServ S: Reno Goodyear, Roy Foster's Downtown Service
Med N: + Hospital
S: + Hospital
ATM N: Texaco
S: Bank of America
14 Wells Ave
Gas S: Chevron*(D), Texaco*(D)
Food S: Bavarian World, Carrows Restaurant, Chicago Express, Denny's
Lodg N: Motel 6
S: DAYS INN Days Inn, Holiday Inn (see our ad on this page), Motel 6
AServ S: Fred's Auto Repair, Rankins Auto Service, Texaco Express Lube
TServ S: Purcell's Truck Tire Center Goodyear
Med N: + Hospital
15 U.S. 395, Carson City, Susanville
16 B St, E. 4th St
FStop S: Arco(D)
Gas N: Arco*, Western Mountain*
S: Chevron*
Food N: Jack's Coffee Shop, Taqueria
Lodg N: Motel 6, Mt. Rose Crest Motel, Pony Express Lodge
S: Alejo's Inn, Gold Coin Motel
AServ N: Instant Smog
S: ATW Tire and Wheel, Chevron, D&D Foreign Car
Other N: In&Out Car Wash, Sanders Mobil Home
17 Rock Blvd
Gas N: Arco*, Chevron*, Exxon*
Lodg N: Emerald Motel, Safari Motel, Tarry Motel, Victorian Inn, Wagon Train Motel
AServ N: Jack's Auto Repair, Lock & Glass, Sparks Tire
Other N: A+ Veterinarian Hospital, Coin Laundry, Paperback Depot, Rock & B Coin Laundry
18 NV 445, Pyramid Way (Eastbound)

Bold red print shows RV & Bus parking available or nearby

EXIT NEVADA

Food **S:** Restaurante Orozko (Nugget)

Lodg **N:** Nugget Courtyard, Silver Club Hotel
S: Nugget Hotel

19 E. McCarran Blvd

TStop **N:** 76 Auto/Truck Plaza(LP)(SCALES) (RV Dump)

Gas **N:** Beacon*, Chevron*(D), Western Mountain Oil(D)

Food **N:** Applebee's, Arby's, BJ's Bar-B-Que, Baskin Robbins, Burger King, Craig's, Expresso Plus, Fast Chinese Food, IHOP, Jack-in-the-Box, Jelly Donut, Jerry's, Joe Bob's Chicken Joint, KFC, Little Caesars Pizza, Port of Subs, Sierra Sid's Restaurant, Sinbad's Hotdogs, Wendy's, Wonder Wok
S: Denny's, Juicy's, Pizza Kitchen, The Black Forrest House

Lodg **N:** Budget Motel, Inn Cal, Western Village, Windsor Inn
S: Best Western

AServ **N:** Econo Lube 'n Tune, Sparkling Car Wash
S: Alied Auto Parts

TServ **N:** 76
S: Cummins Diesel, Smith Allison Transmissions

RVCamp **N:** Victorian RV

ATM **N:** Bank Of America, Interwest Bank, Norwest Bank, Wells Fargo

Other **N:** Albertson's Grocery, Long's Drugs, Mervyn's, Safe Way Grocery, Save On Drugs, Target Department Store
S: Crystal Clean Coin Car Wash

20 Sparks Blvd

FStop **N:** Pacific Pride Commercial Fueling

Gas **N:** Texaco*(D)(CW)

TServ **S:** Tyres International

Other **N:** Factory Outlet Center, Wild Island

21 Vista Blvd, Greg St

TStop **S:** Alamo Truck Plaza*(SCALES) (RV Dump)

FStop **S:** Gas Card Club (Card Lock System)

Gas **N:** Chevron (McDonalds)

Food **N:** McDonalds (Chevron)
S: Alamo Truck Plaza

Lodg **S:** Super 8 Motel

TServ **S:** Allison Transmission Center, Silver State International, Sparks Trailer Repair, Worthen Kenworth

Med **N:** ✚ Hospital

22 Lockwood

23 Mustang

Food **N:** Mustang Station

AServ **N:** The Auto Recker

EXIT NEVADA

Other **N:** Sage Trailer Park

(25) **Weigh Station (Westbound)**

28 NV 655, Patrick

32 Tracy, Clark Station

36 Derby Dam

38 Orchard

40 Painted Rock

(41) **Rest Area - RR, Phones, Picnic, HF P (Westbound)**

(42) **Weigh Station (Eastbound)**

43 Wadsworth, Pyramid Lake

Gas **N:** 76*(LP)

RVCamp **N:** I-80 Campground(LP)

46 U.S. 95 Alt, West Fernley

TStop **S:** Pilot Travel Center*(SCALES)

Food **S:** Dairy Queen (Pilot), Wendy's (Pilot)

TWash **S:** Blue Beacon Truck Wash

Parks **N:** Pyramid Lake

48 U.S. 50 Alt, East Fernley

TStop **N:** Truck Inn*(SCALES) (RV Dump)

FStop **S:** Silverado*

Gas **N:** Exxon*(D)
S: Texaco*

Food **N:** Truck Inn
S: McDonalds, Pizza Factory, Silverado Cafe & Cassino

Lodg **N:** Truck Inn
S: AAA Best Western, ◆ Super 8 Motel

AServ **S:** D&D Tire Inc.

TServ **N:** Truck Inn

TWash **N:** Truck Inn

ATM **S:** Norwest Bank

Other **S:** Coin Laundry, Warehouse Market Grocery Store

65 Hot Springs, Nightingale

78 Jessup

83 U.S. 95 South, Fallon, Las Vegas, Rest Area - Phones, Picnic, Grills, RR (Both Directions)

TServ **N:** Trinity Truckstop

93 Toulon

105 Lovelock, 80 Business Loop (Eastbound)

Gas **N:** Beacon, Exxon*(D)

Food **N:** LaCasita Mexican Restaurant

Lodg **N:** Brookwood Motel, Lafon's Motel, Lovelock Inn

EXIT NEVADA

AServ **N:** Beacon, Jim's Tire Shop, Nutter's Tow, Stovall's Auto Clinic, Vonsild's Complete Auto Repair

Med **N:** ✚ Hospital

Other **S:** Suburban Propane

106 Downtown Lovelock

Gas **N:** 76(LP), Two Stiffs*

Food **N:** Mama Jean's Grill, McDonalds, Pizza Factory

Lodg **N:** Covered Wagon Motel, Desert Haven Motel, Sage Motel, Sunset Motel

AServ **N:** 76, California Garage

Other **N:** Coin Car Wash, Coin Laundry, Safe Way Grocery

107 East Lovelock (Westbound)

FStop **N:** Chevron*(D)

Food **N:** Sturgeon's, Wee-B's Subs & Pizza

Lodg **N:** Best Western, Cadillac Inn, Desert Plaza Inn, Sierra Motel, Super 10 Motel

RVCamp **N:** Lazy K Campground Park(LP)

112 Coal Canyon

119 Oreana, Rochester

129 Rye Patch Dam

FStop **S:** Burns Bros Travel Stop*(LP)

RVCamp **N:** Rye Patch Dam Recreational Area

138 Humboldt

145 Imlay

149 NV 400 South, Unionville, Mill City

RVCamp **S:** Starpoint Mobile Home and RV Park(LP)

151 Mill City, Dun Glen

TStop **N:** Burns Bros Travel Stop*(LP)

Food **N:** Burns Bros Travel Stop

Lodg **N:** Super 8 Motel

AServ **N:** Burns Bros Travel Stop(LP)

TServ **N:** Burns Bros Travel Stop

158 **Cosgrave, Rest Area Southbound - RR, Phones, Picnic, Grills, RV Dump**

168 Rose Creek

173 West Winnemucca

TServ **S:** Maga Truck Repair

Other **S:** Humboldt Veterinary Clinic

176 U.S. 95 North, Winnemucca Blvd, Downtown

TStop **S:** Flying J Travel Plaza(LP)(SCALES) (RV Dump)

FStop **N:** Pacific Pride Commercial Fueling

Gas **S:** Chevron*, Shell*

Food **S:** Arby's, Baskin Robbins (Model T), Burger King, China Garden, Denny's, Flying Pig, Jerry's, KFC (Model T), Los Sanchez, McDonalds, Pizza

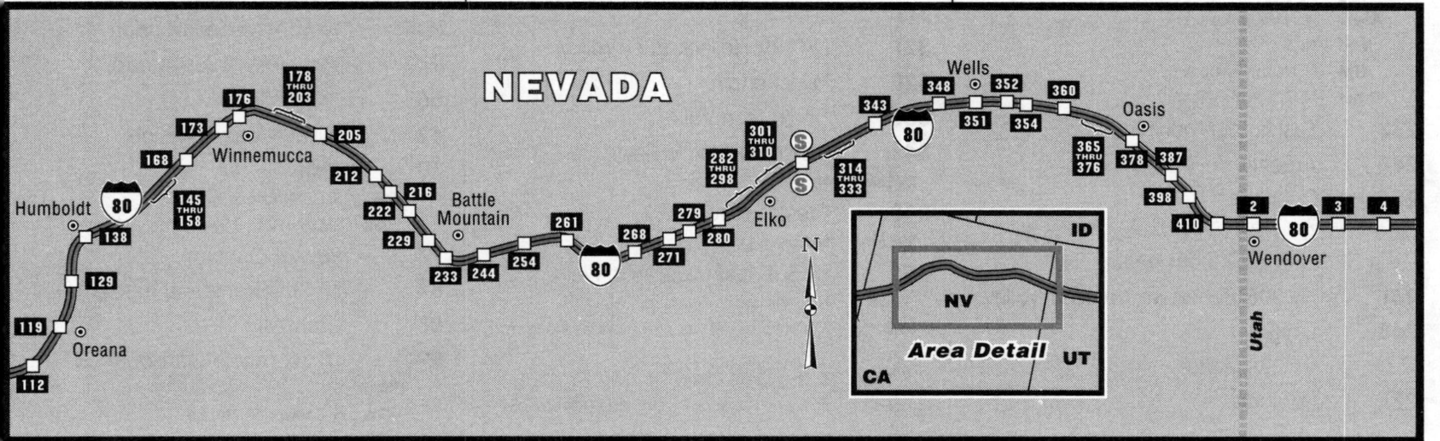

EXIT	NEVADA
	Hut, Round Table Pizza, Taco Time
Lodg	S: AAA Best Western, Holiday Inn Express, **Model T Motel, Motel 6**, Santa Fe Inn, Super 8 Motel
AServ	S: Ford Dealership
TServ	N: **Good Enterprises Welding & Truck Repairs**
	S: **Humboldt Diesel**
RVCamp	S: **Model T RV Park**
ATM	S: Bank of America
Other	N: **Tippin Gas Co. (LP Company)**
	S: **Raley's Supermarket**
178	Winnemucca Blvd., Downtown
Gas	S: Pump 'n Save*
Food	S: Big A Auto Parts, Dave's Dugout, Maverik
Lodg	S: **Bull Head Motel**, Cozy Motel, Downtown Motel, Frontier Motel, Val-U Inn Motel
RVCamp	S: **Westerner Trailer Lodge, Winnemucca RV Park**
Med	S: ✚ **Hospital**
Other	S: Coin Car Wash, **Visitor Information**
180	Winnemucca Blvd East
(187)	**Rest Area - RR, Picnic, Grills, HF** P
194	Golconda, Midas
Food	N: Z Bar & Grill
Lodg	N: **Water Hole #1 Motel**
Other	N: **U.S. Post Office**
(200)	**Golconda Summit (Truck Parking)**
203	Iron Point
205	Pumpernickel Valley
212	Stonehouse
216	**Valmy, Rest Area - Picnic, RR, Grills, RV Dump (Both Directions)**
FStop	N: **Valmey 76 Station***(D)(LP)
Other	N: **U.S. Post Office**
222	Mote
229	NV 305, Downtown Battle Mountain
TStop	N: **Colt Service Center***(LP) **(RV Dump)**
Gas	N: 76*, Chevron
Food	N: **Blimpie's Subs (Colt Service), Colt Inn**
Lodg	N: ◆ Best Western, **Colt Motel**
AServ	N: Chevron, Ed's Tire
TServ	N: **Smith Detroit Diesel-Allison**
RVCamp	N: **Colt RV Park**
ATM	N: Colt Motel Restaurant
231	UT 305 Downtown Battle Mountain
Gas	N: Big R's*(D)
Food	N: Hide-A-Way Steakhouse, McDonalds, **Subway**
Lodg	N: **Super 8 Motel**
AServ	N: NAPA Auto Parts
Med	N: ✚ **Hospital**
ATM	N: Northwest Bank
Other	N: **U.S. Post Office**
233	East Battle Mountain
244	Argenta
254	Dunphy
(259)	**Rest Area - RR, Grills, Picnic, RV Dump** P **(Both Directions)**
261	NV 306, Beowawe, Crescent Valley
268	Emigrant
(271)	**Parking Area**
271	Palisade

EXIT	NEVADA
279	NV 278, West Carlin, Eureka (Eastbound, Reaccess Westbound Only)
280	Central Carlin , Eureka
TStop	S: **Pilot Travel Center***(SCALES) **(RV Dump)**
Food	S: Chins Cafe, **Gear Jammers Cafe**, State Inn, **Subway (Pilot)**
Lodg	S: **Cavalier Motel**
AServ	S: Intermountain Tire
TServ	N: **Anderson Diesel Repair**
Other	S: **Coin Laundry**
282	NV 221, East Carlin (Westbound, Reaccess Eastbound Only)
292	Hunter
298	80 Business loop, Elko West
301	NV 225, Elko Downtown
Gas	N: Maverick*
	S: **Phillips 66***(CW) **(RV Dump)**, Texaco*(D)(LP)
Food	N: 9 Beans & A Burrito, Arby's, Denny's, Hogi Yogi, McDonalds, Papa Murphy's, Round Table Pizza
	S: Cimarron West Family Restaurant (Texaco), **KFC**
Lodg	N: ◆ Shilo Inn
AServ	N: Dirt Busters Car Wash
	S: Checker Auto Parts, Elko Oil-N-Go
Other	N: **Coin Laundry, K-Mart, Raley's Supermarket and Drugs, Wal-Mart**
	S: **Smith's Food and Drug Center**
303	Elko East
Gas	S: Amaco*(CW), Chevron*, Texaco*(D)
Food	S: Baskin Robbins, Burger King, Golden Corral, JR's Bar and Grill, McDonalds, Perfect Subs, Pizza Barn, Pizza Hut, Showboat, Subway, Taco Time, Wendy's
Lodg	S: ◆ Ameritel, AAA Best Western, DAYS INN AAA **Days Inn**, AAA Holiday Inn, Motel 6, Parkview Inn, AAA Red Lion Inn, Super 8 Motel
AServ	S: Chrysler Auto Dealer, Gallagher Ford, Pennzoil Lube(CW)
RVCamp	S: **Gold Country**
Med	S: ✚ **Hospital**
ATM	S: Wells Fargo
Other	S: **JC Penny, Payless Drugs, Tourist Information**
310	Osino
(312)	Check Station (Both Directions, All Trucks With Livestock Must Stop)
314	Ryndon, Devils Gate
RVCamp	S: **KOA Campground**(LP)
317	El Burz
321	NV 229, Halleck, Ruby Valley
328	River Ranch
333	Deeth, Starr Valley
343	Welcome, Starr Valley
RVCamp	N: **Welcome RV Camp**
348	Crested Acres
RVCamp	N: **Camping**
351	80 Business Loop, West Wells
Lodg	N: **Chinatown Motel**
RVCamp	N: **Chinatown Motel, Mountain Shadows RV Park**
Parks	S: **Angel Lake Recreation Area**
Other	N: **Stuart's Foodtown, U.S. Post Office, Wells**

EXIT	NEVADA/UTAH
	Propane INC.
352	Great Basin Hwy, U.S. 93, East Wells
TStop	N: **Texaco***(SCALES)
	S: **Flying J Travel Plaza***(LP)(SCALES)
Gas	N: Chevron(LP)
Food	N: **4 Way Cafe, Burger King**(LP)
	S: **Thads Restaurant (Flying J TStop)**
Lodg	N: **Motel 6, Rest Inn**, Super 8 Motel
AServ	N: Chevron, Intermountain Car & Truck Wash
TServ	N: **Roadway Diesel Goodyear, Texaco**
RVCamp	N: **Moutain Shadows (1 Mile)**
Other	N: Coin Car Wash
(354)	**Parking Area (Eastbound, Reaccess Eastbound Only)**
360	Moor
365	Independence Valley
373	Parking Area (Both Directions)
376	Pequop
378	NV 233, Oasis, Montello
FStop	N: **Chevron***(LP)
Food	N: **Chevron**
AServ	N: Chevron
Other	N: **Post Office**
387	Shafter
398	Pilot Peak
410	I-80 Business Loop, U.S. 93 Alt, West Wendover
Gas	S: Chevron*
Food	S: **Burger Time, Pepper Mill Inn & Casino**
Lodg	S: **Nevada Crossing, Pepper Mill Inn & Casino, Red Garter Hotel**, Super 8 Motel
RVCamp	S: **KOA Campground**
Other	S: **Trucker's Lounge*, Visitor Information**

↑NEVADA

↓UTAH

EXIT	UTAH
2	UT 58, Wendover
3	Port of Entry
4	Bonneville Speedway
TStop	N: **Amaco**(SCALES) **(Restaurant)**
Food	N: **Amaco**
TServ	N: **Amaco**
(10)	**Rest Area - RR, Phones, Picnic** P **(Both Directions)**
41	Knolls
49	Clive
(54)	**View Area (Eastbound)**
(55)	**View Area (Westbound)**
56	Aragonite
62	Military Area, Lakeside
70	Delle
FStop	S: **Delle Auto & Truck***
Food	S: **Delle Auto & Truck**
77	Rowley, Dugway
84	UT 138, Grantsville, Tooele
88	Grantsville
99	UT 36, Tooele, Stansbury
TStop	S: **76 Auto/Truck Plaza***(SCALES)
Gas	S: Chevron*(D), Texaco

Bold red print shows RV & Bus parking available or nearby

EXIT — UTAH

Food — S: McDonalds, Subway (Chevron)
Lodg — S: Oquirrh Motor Inn
AServ — S: Texaco
TServ — S: 76
TWash — S: 76
Other — S: Visitor Information (76 Truck Stop)

102 UT 201, 2100 South Freeway (Eastbound, Reaccess Westbound Only)
Med — S: ✚ Hospital

105 Magna, Saltair Dr., UT 202
Parks — N: Great Salt Lake State Park

111 7200 West

113 56th West
Food — N: Perkins
Lodg — N: Comfort Inn, Fairfield Inn, AAA Quality Inn, ◆ Super 8 Motel
ATM — N: First Security Bank

114 Wright Brothers Dr. (Westbound, Reaccess Eastbound Only)
Food — N: Sardini's Sandwich Villa
Lodg — N: Fairfields Inn, Hilton Inn, La Quinta Inn

115B 40th West

115 Airport

117 I-215, Ogden, Provo

118 UT 68, Redwood Road (Eastbound)
Parks — N: Utah State Fairpark

120 I-15, North Ogden

Note: I-80 runs concurrent below with I-15. Numbering follows I-15.

311 Junction I-80W

310 6th St, City Center
Gas — E: Circle K, Phillips 66*, Sinclair*
Food — E: Cafe Olympus, Denny's, Hilton, McDonalds, Sophi Garcia's Mexican
Lodg — E: Best Western, Crystal Inn, Little America Hotel, Motel 6, Olympus Hotel, Quality Inn
AServ — E: Miller Toyota
ATM — E: Cash Machine (Sinclair)

309 13th St, 21st St, 9th St.
TStop — W: Flying J Travel Plaza(LP)(SCALES)
FStop — E: Cash Saver*
Food — E: Atlantis Burgers
W: Wendy's
AServ — E: Truckland Truck Service
W: Dick's Check 'n Lube, Seiner Chevrolet, Seiner Mitsubishi, Val's

EXIT — UTAH

TServ — E: Cummins Diesel
W: Flying J Travel Plaza, Great Dane Trailer Service, Semi Service, Utility Trailer Service, West Valley Tire
TWash — W: Blue Beacon Truck Wash
ATM — E: Key Bank
Other — E: U-Haul Center(LP)

307 Junction I-80E

Note: I-80 runs concurrent above with I-15. Numbering follows I-15.

124 U.S. 89, State St
Gas — N: Citgo*, Texaco*(D)(LP)
Food — N: Burger King, Busy Bee, Jo Jo's Restaurant, New South Seas Cafe, Sconecutters, Subway, Taco Bell, Wendy's, Zorba's Drive In
S: KFC, Mi Ranchito, Palace Burgers, Salt Lake Doughnuts, Shuchi, Taco Time, Wenchel's Donut House
Lodg — S: Travelodge
AServ — N: Hayes Brothers Buick, GM, Jeep, Hinkley's Dodge Trucks, State Auto Repair, Texaco
S: Transmission Exchange
Med — S: ✚ Hospital
ATM — S: Bank One
Other — N: Denkers Veterinary Hospital

125 UT71, 7th East

126 UT181, 13th East, Sugar House
Gas — N: Chevron*(CW)
Food — N: Hogi Yogi, KFC, Olive Garden, Red Lobster, Sizzler Steak House, Taco Bell, Training Table, Wendy's
AServ — N: Jiffy Lube
Other — N: Barnes & Noble, Coin Car Wash, Desert Book, Shopko Grocery

129 Foothill Drive, Parleys Way
Med — N: ✚ Hospital

130 Junction I-215
Parks — N: Cotton Wood Ski Area

132 Ranch Exit

134 UT65, Emigrations, East Canyons
Parks — N: Mountain Dale Recreation Area

137 Lambs Canyon

(140) Brake Test Area (Westbound)

140 Parleys Summit
Gas — S: Sinclair*
Food — S: Mountain Villgae Coffee Shop
AServ — S: Sinclair

EXIT — UTAH

143 Jeremy Ranch
Gas — N: Amaco* (Blimpie, 24hrs.)
Food — N: Blimpie's Subs (Amaco)

145 UT224, Kimball Junction, Park City
Gas — S: Chevron*, Texaco*(D)(CW)
Food — S: Arby's, Denny's, Kenny Rogers Roasters, McDonalds, Taco Bell
Lodg — S: AAA Best Western
RVCamp — N: Hidden Haven RV Park
ATM — S: First Security Bank
Other — S: Factory Outlet Center, K-Mart, Post Office, Smith's Food and Drug, Wal-Mart

(147) Rest Area - RR, Phones, Picnic, Vending P (Westbound)

148AB U.S. 40, Heber, Provo
FStop — N: Sinclair*
Food — N: Blimpie's Subs (With Sinclair)

152 Ranch Exit

156 UT 32, Wanship, Kamas (Difficult Reaccess)
Gas — S: Sinclair*
Food — N: Spring Chicken Inn
Lodg — N: Spring Chicken Inn
AServ — N: "R" Auto Shop
Parks — N: Rockport State Park, Rail Trail

164 Coalville
FStop — N: Amaco*(D)(LP)
Gas — S: Chevron, Sinclair*(D)
Lodg — S: Blonquist Motel, Moore Motel
AServ — S: Chevron, Napa Auto Parts
RVCamp — N: Holiday Hils RV Park*
Other — S: Key Drugs, U.S. Post Office

166 View Area (Both Directions)

168 Junction I-84W, Ogden

169 Echo (No Eastbound Reaccess)
Lodg — N: Cozy

(170) Rest Area - RR, Phones, Picnic, Vending, Tourist Info P (Both Directions)

180 Emory (Westbound, No Reaccess)

185 Castle Rock

189 Ranch Exit

193 Wahsatch

↑UTAH

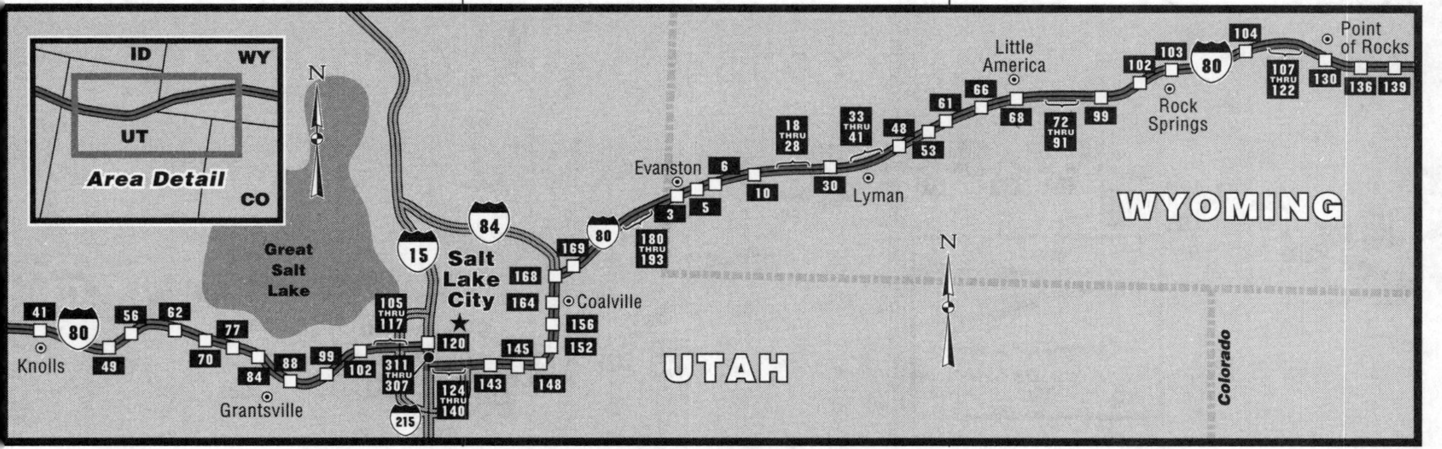

EXIT	WYOMING

↓WYOMING

1 Port of Entry, Weigh Station (Both Directions)

3 Harrison Dr, Evanston
- TStop N: Flying J Travel Plaza*(SCALES)
- Gas N: Amoco, Chevron, Phillips 66*, Shell*(D), Sinclair*, Texaco
  S: Phillips 66*
- Food N: Burger King, Coffee Shop & Fine Family Dining, JB's Restaurant, Lotty's, Taco Time, Wally's Burgers
  S: KFC
- Lodg N: Best Western, DAYS INN AAA Days Inn, Hill Crest Motel, National 9 Inn, Weston Plaza, Weston Super Budget Inn
- AServ N: Amoco, Automotive Plus, Goodyear Tire & Auto, Pontiac Buick GM Dealer
- Med N: ✚ Hospital
  S: ✚ Hospital
- Other N: Hill Crest Laundrymat

(3) Rest Area - Picnic, RR, HF, Phones, RV Dump

5 WY89, WY150, Front St
- Gas N: Amoco*
- Food N: Arby's, Dutch Cowboy Cafe, McDonalds, Shakeys Pizza, Subway, Taco John's, Wendy's
- AServ N: GM Auto Dealership
- Med S: ✚ Hospital
- Other N: Contact Lens Center, IGA Store, Pine Cone Store, Wal-Mart

6 Business 80, U.S. 189, WY89, Bear River Dr, Evanston
- FStop N: Texaco
- Gas N: Amoco*(CW)(LP)
- Food N: The Last Outpost, The Riverside Restaurant
- Lodg N: Alexander Motel, Evanston Inn, Motel 6, AAA Prairie Inn, Super 8 Motel (1.2 Miles), Vagabon Motel
- AServ S: Freeway Tire
- RVCamp N: Phillips RV Camping & Trailer Park, Sunset RV
- Other N: Bear River State Park
  S: Wyoming Welcome Center - RR, Phones, Tourist Info, RVDump (No Trucks)

10 Painter Rd

18 U.S. 189, Kemmerer

21 Coal Rd

EXIT	WYOMING

23 Bar Hat Rd

24 Leroy Rd

(28) Parking Area (Both Directions)

28 French Rd

30 Bigelow Rd
- TStop N: Burns Bros Travel Stop(SCALES)
- Food N: Mrs. B's Restaurant (Burns Stop)

33 Union Rd

34 Fort Bridger
- Lodg S: Wagon Wheel
- RVCamp S: Wagon Wheel

39 WY412, WY414, Carter Mtn View

41 WY413, Lyman
- FStop S: Gas-n-Go*
- Food N: Cowboy Inn Cafe
- TServ N: Terry's Diesel Service
- RVCamp S: KOA Campground (1.2 Miles)
- Other S: Rest Area - RR, Phones, Picnic

(43) Rest Area - RR, HF, Phones, Picnic, RV Dump

48 Loop 80, Lyman, Fort Bridger

53 Church Butte Rd

(60) Parking Area (Both Directions)

61 Cedar Mountain Rd

66 U.S. 30, Kemmerer, Pocatello

68 Little America
- FStop N: Little America(SCALES)
- Food N: Little America
- Lodg N: AAA Little America
- Other N: Campground/Little America

72 Westvaco Rd

83 WY372, LaBarge Rd

85 Covered Wagon Rd

89 Green River
- Gas S: Exxon*, Texaco*
- Other S: Tex's Travel Camp

91 Business 80, U.S. 30, WY530, Green River
- Gas S: Gas-Mat, Mini-Mart*
- Food S: Artic Circle Hamburgers, McDonalds, Pizza Hut, Subway, Taco John's, Wild Horse Cafe
- Lodg S: Coachmen Inn, Mustang Motel
- AServ S: NAPA Auto Parts
- RVCamp S: Adam's RV
- Other S: Tourist Information (2 Miles)

EXIT	WYOMING

99 U.S. 191, East Flaming Gorge Rd
- FStop S: Conoco
- Food S: Restaurant (Conoco)
- RVCamp N: KOA Campground, RV Dealer

102 Dewar Dr
- Gas N: Exxon*(CW), MiniMart*, Texaco*
  S: Citgo, Texaco
- Food N: Denny's (24 Hrs), Pam's Bar-B-Q, Taco Time
  S: Arby's, Baskin Robbins, Burger King, Golden Corral, Hong Kong House, KFC, McDonalds, Pizza Hut, Shakeys Pizza, Subway, Village Inn, Wendy's
- Lodg N: AAA Motel 6, Ramada Limited, AAA The Inn at Rock Springs
  S: AAA Comfort Inn, AAA Holiday Inn, Motel 8, Super 8 Motel
- AServ N: GM Auto Dealership, NAPA Auto Parts
  S: Amoco(D), Texaco
- Other N: K-Mart, Smith's Grocery (Pharmacy), Wal-Mart, White Mountain Mall
  S: Coin Laundry, Visitor Information

103 College Dr, Rock Springs
- Gas S: MiniMart*
- Med N: ✚ Hospital

104 U.S. 191, Elk St
- TStop N: Flying J Travel Plaza(SCALES), Texaco
- FStop N: Exxon*(D)
- Gas N: 7-11 Convenience Store*, Phillipps 66*, Sinclair*, Texaco
  S: Exxon(D)
- Food N: Burger King, Frosty Freeze, Santa Fe Trail Restaurant, Taco Time, The Renegade Cafe (Texaco FS)
  S: Daylight Donuts, De Amigos Restaurant
- Lodg N: AAA Best Western, ◆ Econolodge
  S: DAYS INN AAA Days Inn
- AServ N: Minute Man, Pontiac Buick GM Dealer
- TServ N: Texaco
- Med N: ✚ Hospital
- Other N: Bakery Thrift Shop

107 Business 80, U.S. 30, WY430, Rock Springs, Pilot Butte Ave
- FStop N: Conoco
- Gas N: 7-11 Convenience Store*, Gas For Less(LP), Phillips 66, Texaco*(D)
- Food N: Chinese American, The Sands Motel
- Lodg N: El Rancho, AAA Springs Motel, Thunderbird Motel
- AServ N: 5 Star Towing & Repair, Metrick Motors, Rizzi's 66 Service

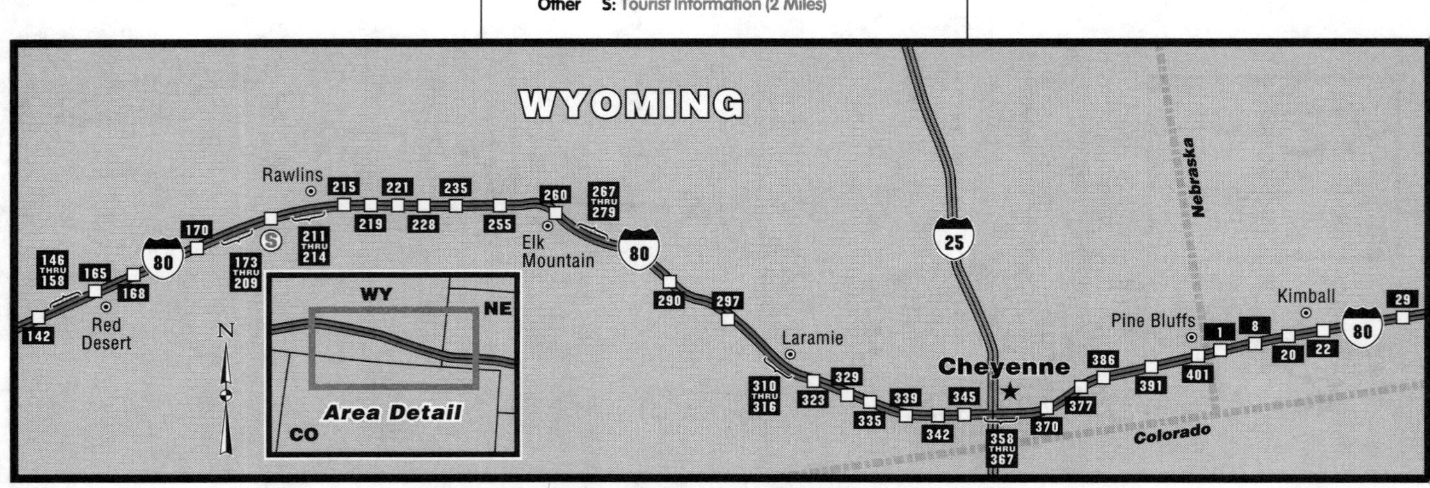

Bold red print shows RV & Bus parking available or nearby

EXIT	WYOMING
11	Airport Rd
22	WY371, Superior
30	Point of Rocks
FStop	N: Conoco
Food	N: Restaurant (Conoco)
Other	N: Coin Laundry
137)	Parking Area (Both Directions)
36	Black Butte Rd
39	Red Hill Rd
42	Bitter Creek Rd
144)	Rest Area - RR, Phones, Picnic, RV Dump (Both Directions)
46	Patrick Draw Rd
50	Table Rock Rd
Gas	S: Sinclair*
Food	S: Sinclair
Other	S: Camping
52	Bar X Rd
54	B L M Rd
56	G L Road
58	Tipton Rd
65	Red Desert
Gas	S: Saveway*
Lodg	S: Motel (Saveway)
66	Booster Rd
68	Frewen Rd
70	Rasmussen Rd
73	Wamsutter
84	Continental Divide Rd
87	WY789, Baggs Rd
89)	Parking Area (Both Directions)
96	Riner Rd
01	Daley Rd
04	Knobs Rd
06	Hadsell Rd
09	Johnson Rd
TServ	N: Flying J Travel Plaza(SCALES)
211)	Weigh Station (Westbound)
11	Spruce St
FStop	N: Sinclair
Gas	N: Conoco, Minit Mart, Texaco
Food	N: Cappy's, Christy's Cafe, JB's Restaurant, The Golden Spike (National 9 Inn)
Lodg	N: (AAA) Best Western, Ideal Motel, LaBella Motel, Motel 7, Super 8 Motel, The Bucking Horse Lodge, The National 9 Inn
AServ	N: Conoco, El Rancho, Terry's Towing
TServ	N: Terry's Towing
RVCamp	N: President's RV Camp, Western Hills Campground
Med	N: ✚ Hospital
Other	N: Gas for Less(LP)
14	Higley Blvd
TStop	S: Texaco(SCALES)
Gas	N: 7-11 Convenience Store
Food	S: Restaurant (Texaco FS)
Lodg	S: The Sleep Inn (Texaco FS)
TServ	S: Texaco
15	Business 80, U.S. 287, WY789,

EXIT	WYOMING
	Rawlins
FStop	N: Conoco*(D)
Gas	N: Coastal*, Phillips 66, Texaco*
Food	N: Country Cookin, Josh's Place, McDonalds, Pizza Hut, Square Shooter's Eating House, Subway, Wendy's
Lodg	N: Bridger Inn, ◆ Days Inn, Key Motel, Weston Inn
AServ	N: American Auto Care, Fast Lube, GM Auto Dealership, Terry's Towing
TServ	N: Terry's Towing, Truck & Radiator Repair
Other	N: Airport, Alco Discount Store, City Market (Pharmacy), Eye Clinic, Pamida Discount Center
219	West Sinclair
221	East Sinclair
TStop	N: Burns Bros Travel Stop
Food	N: Restaurant (Burns Brothers)
228	Unnamed
Gas	S: Fort Steele*
Other	N: Fort Steele Historical Site, Rest Area - RR, Phones, Picnic, RVDump
235	U.S. 30, WY130, U.S. 287, Walcott, Saratoga
FStop	N: Modern Gas*(D)
Other	N: Camping
255	WY72, Hanna, Elk Mountain
FStop	N: Conoco* (Towing)
AServ	N: Conoco (Towing)
260	CR402
Gas	S: Texaco
(262)	Parking Area (Both Directions)
267	Wagonhound Rd
Other	S: Rest Area - RR, Phones, Picnic
272	Arlington
Gas	N: Exxon*
RVCamp	N: Camping
279	Cooper Cove Rd
290	Quealy Dome Rd
FStop	S: Total (ATM)
Food	S: Taco Bell (Total)
297	WY12, Herrick Ln
310	(8) Curtis St
TStop	N: Pilot Travel Center(SCALES)
	S: Petro(SCALES)
FStop	N: Foster's Country Store, Total
Food	N: The Outrider Cafe (Total FS), Wendy's (Pilot FS)
	S: Iron Skillet (Petro)
Lodg	N: ◆ Econolodge, Super 8 Motel
TServ	N: L&W Truck Service
	S: Petro
TWash	S: Blue Beacon Truck Wash (Petro)
Med	N: ✚ Hospital
Other	N: KOA Campground
311	WY130, WY230, Snowy Range Rd, Laramie
FStop	S: Foster's Country Corner(LP)
Gas	S: Conoco*(LP), Phillips 66*(D)
Food	S: Beanry Restaurant, Foster's Country Corner, McDonalds
Lodg	S: (AAA) Best Western, (AAA) Camelot Motel
AServ	S: BF Goodrich
RVCamp	S: Yeao Marine & RV

EXIT	WYOMING
Other	S: Laramie Car Wash
313	U.S. 287, 3rd St, Ft. Collins, CO, Laramie
FStop	N: Pacific Pride
Gas	N: Conoco, Exxon, Granny's Bargin Lots, Robo Wash, Sinclair, Texaco
Food	N: Cafe Ole, Chuck Wagon Restaurant, Denny's, Great Wall Chinese, Motel 8
Lodg	N: Laramie Inn, Motel 8, (AAA) Sunset Inn, Travel Lodge
	S: ◆ Holiday Inn, Motel 6
AServ	N: 24 hour Towing, Walt's Automotive Service
Other	S: Animal Hospital, Snowy Range Ski Area
316	Business 80, U.S. 30, Grand Ave, Laramie
Med	N: ✚ Hospital
323	WY210, Happy Jack Rd
Other	N: Rest Area - RR, Phones, Picnic, RVDump
329	Vedauwoo Rd
Other	S: National Forest Recreation Area
335	Buford
FStop	S: Sinclair
339	Remount Rd
342	Harriman Rd
(344)	Parking Area (Both Directions)
345	Warren Rd
348	Otto Rd
358	W Lincolnway
TStop	N: Sinclair*(LP)
Food	N: Denny's, Little America
Lodg	N: LaQuinta Inn, Little America
359A	Junction I-25, U.S. 87S, Denver
359C	Junction I-25, U.S. 87, Casper
362	Junction I-180, U.S. 85, Central Ave, Greeley Colorado
FStop	S: Total
Gas	N: Conoco*(D)(LP)
	S: Conoco
Food	N: Diamond Horsestop Cafe, Hardees, In-N-Out Hamburgers, Los Amigos Mexican
	S: Burger King, Subway, Taco John's
Lodg	N: Lariat Motel
	S: Holiday Inn, Round Up Motel
AServ	N: Alan's Auto Service
TServ	N: International Dealer
RVCamp	S: RV Service
Med	N: ✚ Hospital
Other	N: Big A Auto Parts, Capitol, Coin Laundry, Frontier Days Park, Tourist Information
	S: Safe Way Grocery, Town & Country Pharmacy
364	Business 80, WY212, Cheyenne, E. Lincolnway
RVCamp	N: RV Camping
Other	N: Outlet Mall
367	Campstool Rd
RVCamp	N: KOA Campground
370	U.S. 30, Archer
TStop	N: Sapp Brothers(SCALES)
Food	N: T-Joe's Saloon (Steak, Pizza & Hamburgers)
Lodg	N: Big G Motel (Sapp Brothers)
TServ	N: Goodyear Tire & Auto, Sapp Brothers
RVCamp	N: T-Joe's RV Park

EXIT	WYOMING/NEBRASKA
(372)	**Weigh & Check Station (Westbound)**
377	WY217, Hillsdale
TStop	N: Burns Bros Travel Stop(SCALES)
Lodg	N: Burns Bros Travel Stop(SCALES)
TServ	N: Burns Bros Travel Stop
Other	N: Campground
386	WY213, 214, Burns, Carpenter
TStop	N: Cenex
TServ	N: Cenex
Other	N: Cenex (Showers, Restaurant, Laundry Mat, ATM)
391	Egbert
401	Pine Bluffs
TStop	N: AmPride
Gas	N: Sinclair, Total
Food	N: Fred's Place, Restaurant (AmPride FS)
Lodg	N: Gator's Travelyn's
AServ	N: Pine's Body Shop, Sinclair
TServ	N: AmPride
Other	N: Camping
	S: Rest Area - RR, Phones, Picnic, Vending, RVDump

↑WYOMING

↓NEBRASKA

EXIT	NEBRASKA
1	Pine Bluffs
8	Link 53C, Bushnell
(19)	**Weigh Station (Eastbound)**
20	NE71, Kimball, Scottsbluff
Gas	N: Phillips 66*(D)
Food	N: Pizza Hut, Subway
Lodg	N: First Interstate Inn (ATM), Super 8 Motel
AServ	N: Phillips 66(LP)
RVCamp	N: Campgound(LP)
Med	N: ✚ Hospital
ATM	N: First State Bank
Other	N: Truck Permit Station
22	Kimball, East Entrance
Lodg	N: Best Western
Other	N: KOA Campground
(25)	**Rest Area - RR, Phones, Picnic (Westbound)**
29	Dix
38	Potter
FStop	N: Cenex(LP), Texaco
Gas	N: Cenex(D)
Food	N: Cafe
Lodg	N: Motel
AServ	N: Texaco
RVCamp	N: Camping
48	Community College
(51)	**Rest Area - RR, Phones, Picnic (Eastbound)**
55	NE19
59	Business 80, U.S. 385, Sydney, Bridgeport
TStop	N: The Sapp Brothers*
FStop	N: Texaco (ATM)
Gas	N: Amoco
Food	N: Arby's, Cabelo's, McDonalds, Runza Restaurant, Taco John's, The High Plains Cache
	S: Country Kitchen
Lodg	N: ◆ Comfort Inn, DAYS INN AAA Days Inn
	S: ◆ Holiday Inn
TServ	N: The Tire Service
TWash	N: Hoff's Truck Wash
RVCamp	N: RV Camping
Med	N: ✚ Hopsital
Other	N: Tourist Information
(61)	**Rest Area - RR, Phones, Picnic (Westbound)**
69	Link 17E, Sunol
76	Link 17F, Lodgepole
(82)	**Rest Area - RR, Phones, Picnic (Eastbound)**
85	Link 25A, Chappell
Gas	S: Texaco
AServ	S: Texaco
RVCamp	N: Creekside RV Park (.6 Miles)
(88)	**Rest Area - RR, Phones, Picnic (Westbound)**
95	NE27, Julesburg, Oshkosh
101	U.S. 138
102	Junction I-76
107	Link 25B, Big Springs
TStop	N: Bosselman Travel Center*(D)(CW)(SCALES)
FStop	N: Total(D)
Food	N: Blimpie's Subs (Bosselman), Food Court (Bosselman Travel Ctr), Grandma Max's Restaurant, Taco Bell (Total), The Char Restaurant
Lodg	N: Budget 8 Motel
TServ	N: Bosselman Travel Ctr
TWash	N: Eagle 1 Truck Wash (Bosselman Travel Ctr)
RVCamp	S: McGreer's Campground
ATM	N: Bosselman
Other	S: Laundry (McGreer's), Walker's CB Repair Shop
117	Link 51A, Brule
FStop	N: Happy Jack's*
RVCamp	N: Happy Trails Campground (Laundry Facilities)
Other	N: Laundry Facilities
(124)	**Rest Area - Phones, RR, HF, Grills, Picnic (Eastbound)**
126	NE61, Ogallala, Grant
TStop	S: 76 Auto/Truck Plaza(SCALES) (RV Dump)
FStop	N: Amoco, Texaco*(D)
Gas	S: Conoco, Phillips 66*
Food	N: Arby's, Cafe Steaks & Seafood (Breakfast), Country Kitchen, McDonalds, Pizza Hut, Valentino's Pizza & Subs
	S: 76 Auto/Truck Plaza, Cassel's Family Restaurant & Pancake House, Conoco (Subway), Dairy Queen, KFC, Wendy's
Lodg	N: DAYS INN AAA Days Inn, Holiday Inn Express, AAA Ramada Limited
	S: ◆ Comfort Inn*, AAA First Interstate Inn, I-80 Motel, Super 8 Motel
TServ	N: Big Mac Trucks
	S: MCM Truck Repair (76 Auto/Truck Stop)
RVCamp	S: Open Corral Camp
Med	N: ✚ Hospital
ATM	N: Texaco
	S: 76 Auto/Truck Plaza
Parks	N: Ogalalla Nature & Outdoor Classroom
Other	N: Ace Hardware, Front Street Museum, Hunting, Fishing, Parking Permits, 24 Hr, RV Dump*(LP) (Amoco), Lake McConaughy, Visitor Information
	S: Meyer Camping, Pamida Grocery Store (Pharmacy), Truck Permit Station (76 Auto/Truck Stop)
(132)	**Rest Area - RR, Phones, Picnic (Westbound)**
133	Link51B, Roscoe
145	Link 51C, Paxton
FStop	N: Texaco*(CW)
Food	N: Ole Big Game Steak
Lodg	N: DAYS INN AAA Days Inn
ATM	N: Texaco*

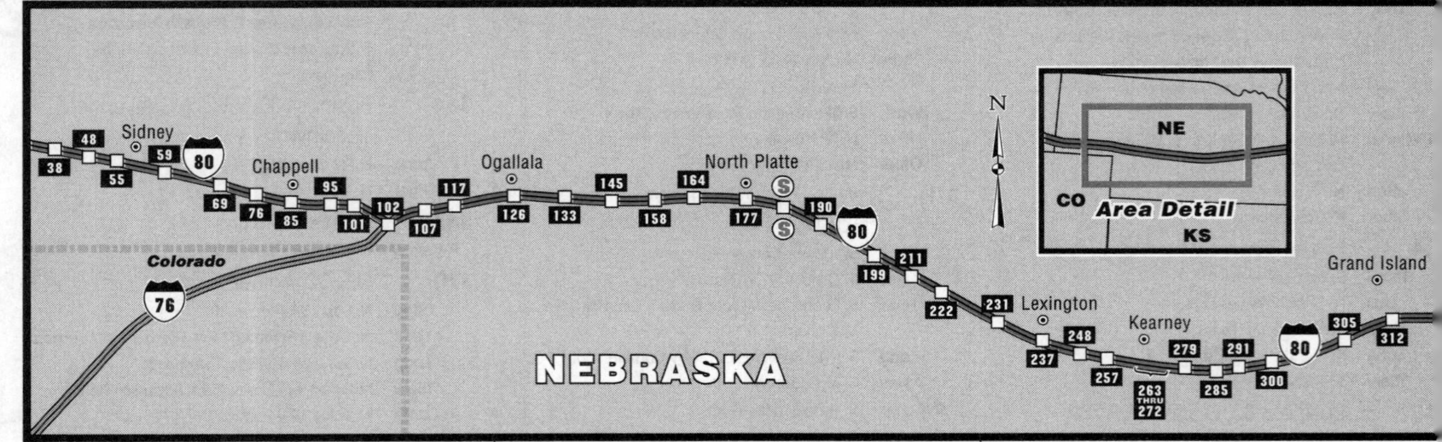

Bold red print shows RV & Bus parking available or nearby

EXIT NEBRASKA

158 NE25, Sutherland
- **FStop** S: Conoco(LP) (24 Hr)
- **RVCamp** S: Camping
- **Other** S: Truck Parking

(160) **Rest Area - Phones, RR, HF, Picnic, Grills, Pet Walk, Tourist Info (Both Directions)**

164 Link56D, Hershey
- **TStop** N: **Texaco*(SCALES) (RV Dump, 24 Hr Road Service)**
- **Gas** N: Sinclair*
- **Food** N: **Restaurant (Texaco)**
- **TServ** N: **Texaco (24 Hr Road Service)**
- **ATM** N: Texaco (24 Hr)
- **Other** N: **Camping, Truck Permit Station* (Texaco)**

177 U.S. 83, McCook, North Platte
- **TStop** S: **Texaco*(D)(LP)(SCALES) (24 Hr)**
- **Gas** N: Amoco, Coastal*, Conoco*, Phillips 66*, Texaco(D), Total(D)
  S: Phillips 66, Total(D)
- **Food** N: A & W Drive-In, Amigo's (Mexican), Applebee's, Arby's, Baskin Robbins, Blimpie's Subs, Branding Iron BBQ, Burger King, Chinese Restaurant, Dunkin Donuts (Phillips 66), Golden Corral, Long John Silvers, McDonalds, Mickey Finn's Sports Cafe, Peking Garden Chinese (Best Western), Perkin's Family Restaurant, Phillips 66, Pizza Hut, Roger's, Stockmann Inn, Subway (Amoco), Subway, Valentino's Pizza Pasta Buffet, Wendy's
  S: **Country Kitchen, Hunan Chinese,** Taco Bell (Total)
- **Lodg** N: AAA Best Western, Blue Spruce Motel, AAA **Camino Inn, First Interstate Inn,** ◆ Hampton Inn, Knight's Inn Motel, Motel 6, Pioneer Motel, ◆ **Sands Motor Inn,** Stockmann Inn
  S: AAA Comfort Inn, DAYS INN AAA **Days Inn,** Holiday Inn Express, **Motel, Super 8 Motel**
- **AServ** N: Dolanburg Motor, Goodyear Tire & Auto, Modern Muffler, Quaker State Lube, Texaco (24 Hr Wrecker Service)
  S: Dodge Dealer, Ford Truck Dealer, GM Auto Dealership, Honda Dealer, Oldsmobile Dealer, Phillips 66
- **TServ** S: **Cat, International Dealer, Nebraska Truck & Equip. Co., Texaco**
- **RVCamp** N: **Holiday Park RV Camp**
- **Med** N: ✚ **Hospital**
- **ATM** N: First Nat'l Bank, Phillips 66, Western Nebraska National Bank
  S: **Texaco (24 Hr)**
- **Other** N: **Cody Go-Karts & Bumper Boats (Water Slide), Factory Outlet Center,** North Majic Spray Car Wash, **Picnic Area, State Patrol Post, The Mall, Wal-Mart (Pharmacy)**
  S: **Ft Cody Museum (No Charge), Laundry Facilities (Texaco 24 Hr), North Platte Veterinary Clinic, Tourist Information, Truck Permit Station**

179 Link56G, to U.S. 30, to North Platte

(182) **Weigh Station - both directions**

190 Spur, Maxwell
- **FStop** N: **Texaco***
- **AServ** N: Texaco
- **RVCamp** S: **Camping**

(194) **Rest Area - Phones, Picnics, RR, HF, Tourist Info (Both Directions)**

EXIT NEBRASKA

199 Brady
- **FStop** N: **Citgo**
- **Food** N: **Dairy Queen (Citgo), Stuckey's (Citgo)**

211 NE47, Gothenburg
- **TStop** N: **Texaco**
- **Gas** N: Total*
  S: Total*(LP)
- **Food** N: **Homestead Cafe (Texaco),** McDonalds, Mirantito Mexican, Pizza Hut, Runza Restaurant, The Snack Shack
- **Lodg** N: **Travel Inn Motel,** Western Motor Inn
- **AServ** N: Big A Auto Parts, Dan's Auto Service, GM Auto Dealership, Walker Auto Repair
- **RVCamp** S: **KOA Campground**
- **Med** N: ✚ **Hospital**
- **ATM** N: **Texaco**
- **Other** N: Car Wash, **Original Pony Express Station, Sod House Museum, Truck Permit Station**

222 NE21, Cozad
- **FStop** N: **Total***
- **Gas** N: Amoco*(D), Conoco*(LP)
- **Food** N: **Burger King, Circle S Restaurant,** Dairy Queen, El Pariso Mexican, McDonalds, PJ's, Subway
- **Lodg** N: **Budget Host Inn,** ◆ **Comfort Inn**
- **AServ** N: Ford Dealership
- **Med** N: ✚ **Hospital**
- **ATM** N: Amoco
- **Other** N: **Alco Discount Store, Robert Henri Museum (1 Mile)**

(227) **Rest Area - Picnics, Phones, RR, HF (Eastbound)**

(228) **Rest Area - Phones, Picnics, RR, HF (Westbound)**

231 Darr

237 U.S. 283, Lexington, Arapahoe
- **TStop** S: **Sinclair***
- **FStop** N: **Ampride*(D)**
- **Gas** N: Phillips 66*(D)
  S: Conoco*(D)
- **Food** N: Arby's, **KFC, McDonalds, Taco Bell, Wendy's**
  S: **Coffee Shop (Sinclair)**
- **Lodg** N: Budget Host Motel, ◆ Comfort Inn, DAYS INN Days Inn, Econolodge, Toddle Inn
  S: ◆ **Super 8 Motel**
- **AServ** N: Key Motors
- **TServ** S: **Goodyear Tire & Auto**
- **Med** N: ✚ **Hospital**
- **ATM** N: Phillips 66
  S: **Sinclair (24 Hr)**
- **Other** N: **Military Museum**
  S: **Johnson Lake Recreation Area, Travel Ctr (Sinclair)**

248 Overton
- **FStop** N: **Burns Bros Travel Stop**
- **Food** N: **Restaurant (Burns Brothers FS)**

257 U.S. 183, Elm Creek, Holdrege
- **TStop** N: **Bosselman Travel Center*(D)(SCALES)**
- **Food** N: **Buffalo Creek Cafe, Subway (Bosselman), Taco Bell (Bosselman)**
- **Lodg** N: **1st Interstate Inn**
- **ATM** N: **Bosselman**
- **Other** N: **Coin Operated Laundry (Bosselman), Travel Ctr (Bosselman)**
  S: **Harlan Cty Reservoir**

263 Odessa

EXIT NEBRASKA

- **TStop** N: **Sapp Bros Truck Stop*(D)**
- **Food** N: **Aunt Lu's Cafe (Sapp Bros)**
- **Lodg** N: **Budget Motel**
- **RVCamp** N: **Campground**
- **ATM** N: **Sapp Bros**
- **Parks** N: **Union Pacific State Wayside Area**

(269) **Rest Area - Phones, Picnics, RR, HF, Tourist Info, Pet Walk (Eastbound)**

(271) **Rest Area - Picnics, Phones, RR, HF (Westbound)**

272 NE44, Kearney
- **Gas** N: Coastal*(CW) (24 Hr), Gas 'N Stop, Phillips 66, Texaco, Total*
- **Food** N: Amigo's Mexican Restaurant, Arby's, Big Apple, Bonanza Steak House, Budget Host Inn, Burger King, Country Kitchen, Golden Dragon Chinese, Grandpa's Steak House, Long John Silvers, McDonalds, Mickey Finn's Sports Cafe, Perkins Family Restaurant, Pizza Hut, Red Lobster, Runza, Taco Bell, Taco John, The Captain's Table (Seafood, Ramada Inn), Valentino's, Wendy's, Whiskey Creek Steakhouse
  S: **Ft Kearny Inn, Grandpa's Steakhouse**
- **Lodg** N: AAA Best Western, Budget Host Inn, AAA **Budget Motel,** ◆ Comfort Inn, Country Inn & Suites, DAYS INN AAA **Days Inn,** ◆ Fairfield Inn, Hampton Inn, ◆ Holiday Inn, ◆ Ramada, Super 8 Motel, AAA Western Inn South
  S: **Fort Kearny Inn**
- **AServ** N: Buick Dealer, Chrysler Auto Dealer, GM Auto Dealership, Goodyear Tire & Auto, Jiffy Lube, Texaco (24 Hr Towing)
- **RVCamp** N: **Clyde & Vi's Campground**
- **Med** N: ✚ **Hospital**
- **ATM** N: Phillips 66, The Platte Valley States Bank
- **Other** N: Buggy Bath Car Wash, Super Wash (Car Wash), **Trails & Rails Museum**

279 NE10, Minden
- **Gas** N: Texaco*(D) (Hunting, Fishing License)
- **Food** N: Picadilly Pizza (Texaco)
- **RVCamp** S: **Pioneer Village (13 Miles)**
- **Other** S: **Pioneer Village (13 Miles)**

285 Link10C, Gibbon
- **Lodg** S: **The Country Inn**
- **AServ** S: Don's Repair
- **Other** N: **R&I RV Sales & Service, The Windmill Recreation Area**

291 Link10D, Kenesaw
- **Gas** S: Kerr McGee*(D)

300 NE11, Wood River
- **TStop** S: **Bosselman* (Texaco)**
- **Food** S: **Mel's Country Cafe (Bosselman), Subway (Bosselman)**
- **Lodg** S: **Motel**
- **ATM** S: **Bosselman**

305 Link40C, Alda
- **TStop** N: **76 Auto/Truck Plaza(D)(SCALES)**
- **FStop** N: **Total* (24 Hr)**
- **Gas** N: Conoco (76 Auto/Truck)
- **Food** N: **Restaurant (76 Auto/Truck)**
- **TServ** N: **76 Auto/Truck Plaza**
- **ATM** N: **76 Auto/Truck Plaza**
- **Other** N: **Travel Ctr (76 Auto/Truck)**

(305) **Rest Area - RR, Phones, Picnic (Westbound)**

312 U.S. 34, U.S. 281, Grand Island, Hastings

Bold red print shows RV & Bus parking available or nearby

EXIT NEBRASKA

TStop N: Bosselman Travel Center*(D)(LP)(SCALES)
Gas N: Conoco*(CW)
S: Conoco* (RV Dump)
Food N: Grandma Max's Restaurant (Bosselman TC), Subway (Bosselman), Sweet Shop, Taco Bell (Bosselman), Tommy's Restaurant (U.S.A. Inns)
S: Restaurant (Holiday Inn)
Lodg N: Days Inn (7 Miles), U.S.A. Inns
S: ◆ Holiday Inn, ◆ Holiday Inn Express
TServ N: CAT Truck Service, Hoffer Truck, Trailer, RV, Bus,Repair
S: Ford Dealership, Ford Truck, Peterbilt Dealer
TWash N: Diamond TW
RVCamp S: KOA Campground (15 Miles)
Med N: ✚ Hospital
ATM N: Bosselman
Other N: Morman Island Rec Area, Stuhr Museum, Tourist Information, Truck Permit Station
S: Hastings Museum

(315) Rest Area - Phones, RR, HF, Picnic, Grills, Pet Walk, Newspaper (Eastbound)

(317) Rest Area - Phones, Picnics, RR, HF (Westbound)

318 NE2, Phillips
RVCamp S: Campground

324 41B, Spur, Giltner

332 NE14, Aurora
FStop N: Ampride*(D)
S: Marlo*
Gas N: Texaco(D)
Lodg N: Budget Host Motel (3 Miles), Hamilton Motor Inn
AServ N: Ford Dealership, Texaco
Med N: ✚ Hospital
Other N: Edgerton Educational Ctr, The Plainsman Museum

338 Link41D, Hampton
FStop S: Kelly's Korner
TServ S: Kelly's Korner

342 Links93A, Henderson, Bradshaw
FStop S: Fuel Mart*
Food S: Dell's Restaurant (Wayfare Motel)
Lodg S: Wayfare Motel
TServ S: Hy-way Trailer
RVCamp S: Western Campground
Med S: ✚ Hospital
Other N: KOA Campground

348 Link93E

(350) Rest Area - Picnics, Phones, RR, HF, Tourist Info, Grills, Pet Walk, Newspaper (Eastbound)

353 U.S. 81, York, Geneva
TStop S: Petro* (Phillips 66)
FStop N: Amoco*(D), Conoco*(SCALES)
S: Conoco, Texaco
Gas N: Texaco*
Food N: A & W Drive-In, Amigo's Mexican, Arby's, Blimpie's Subs (Texaco), Burger King, Country Kitchen (Yorkshire Motel), Deli (Texaco), Golden Gate Express (Chinese), KFC, McDonalds, Pizza Hut, Wendy's
S: Baskin Robbins (Petro), Iron Skillet (Petro), Pizza Hut (Petro), Restaurant (Texaco FS), Tommy's Restaurant (U.S.A. Inns)
Lodg N: Best Western, ◆ Comfort Inn, Days Inn, Super 8 Motel (RV Hook-Ups), The Yorkshire Motel
S: The York Inn, U.S.A. Inns
AServ N: Ford Dealership, GM Auto Dealership, Plymouth Dodge Dealer
S: Conoco
TServ N: Kenworth, Trucks R Us
S: Conoco, Petro
TWash S: Blue Beacon TW
ATM N: Conoco, Texaco
S: Petro, Texaco
Other S: Travel Store (Petro)

(355) Rest Area - Phones, Picnics, RR, HF (Westbound)

360 Link93B, Waco
FStop N: Burns Bros Travel Stop*
Food N: Restaurant* (Burns Bros FS)
RVCamp S: The Double Nickel Campground*
ATM N: Burns Bros

366 Link80F, Utica

369 Link80E, Beaver Crossing

373 Link80G, Goehner
Gas N: Texaco*

(376) Rest Area - Picnics, Phones, RR, HF (Westbound)

379 NE15, Seward, Fairbury
TStop S: Phillips 66*(D)(LP) (24 Hr)
TServ S: The Great Plains Tire Ctr
Med N: ✚ Hospital

(381) Rest Area - Phones, PIcnics, RR, HF (Eastbound)

382 Milford
Gas N: Dahle's I-80 Service*
Food S: Cafe
Lodg N: Milford Inn
AServ N: Dahle's I-80
RVCamp N: Westward Ho Campground

388 NE103, Crete
Gas N: Twin Lakes Gas Station*(LP)
RVCamp N: Campground

395 NW 48th St
TStop S: Texaco*(SCALES)
Food S: Popeye's Chicken, Shoemaker's Restaurant (Texaco TS)
Lodg S: ◆ Cobbler Inn (Texaco), ◆ Super 8 Motel (2 Miles)
TServ S: Cobbler Inn, Texaco
Other N: State Patrol Post, Truck Permit Station

396 US 6, West O St (Eastbound, Westbound Reacess)
Gas S: Total*
Lodg S: Senate Inn Motel, Super 8 Motel
TServ S: Freightliner Dealer
Med S: ✚ Hospital
ATM S: Total

397 US 77S, Beatrice (Divided Hwy, Difficult Reaccess)

399 Lincoln Municipal Airport
Gas N: Amoco*(D)(CW), Phillips 66*
S: Sinclair*(D)
Food N: Best Western, Denny's, Happy Chef, McDonalds, Perkins Family Restaurant
Lodg N: Airport Inn (Best Western), ◆ Comfort I… Days Inn, Hampton Inn, Mot… 6, Quality Inn, ◆ Sleep Inn, Travelodge
S: Econolodge, Inn 4 Less, Ramada
AServ N: Amoco, Honda Dealer

401 Junction I-180 & U.S. 34, 9th St, Downtown Lincoln
RVCamp S: Camping

403 27th St
Other N: Museum, University of Nebraska, Zoo

(405) Rest Area - Phones, Picnics, RR, H… (Westbound)

405 U.S. 77, 56th St
Gas S: Phillips 66*(D)(LP)
TServ S: Freightliner Dealer, Walker Tire
TWash S: Walker Tire
Med S: ✚ Hospital

409 U.S. 6, Waverly
Med S: ✚ Hospital

420 NE63, Ashland, Weigh Station (Both Directions)
TStop N: Amoco*(SCALES) (24 Hr)
Gas S: Phillips 66*(D)
Food N: Restaurant (Amoco TS)
AServ S: Tim's Auto Repair
TServ N: Amoco, Sundowner Trailers
RVCamp N: Campground
ATM N: Amoco
S: Phillips 66
Other S: Platte River State Park (12 Miles)

(425) Rest Area - Max 5 Hr Parking, Pe… Walk, Phones, Picnic, RR, HF (Eastbound)

426 Mahoney State Park
Parks N: Mahoney State Park
Other N: Camping

(432) Rest Area - Phones, Picnics, RR, H… (Westbound)

432 U.S. 6, NE31, Gretna, Louisville
TStop S: Flying J Travel Plaza*(LP) (RV Dump)
Gas N: Texaco*(D)(LP)
S: Conoco (Flying J)
Food N: McDonalds
S: Thads Restaurant (Flying J)
Lodg N: ◆ Super 8 Motel
RVCamp N: KOA Campground
ATM N: 24 Hr Bank (Factory Outlet Mall)
Other N: Nebraska Crossing Factory Stores

439 NE370, Papaillion, Bellevue
Gas N: Phillips 66*(D)
Lodg N: Days Inn, Suburban Inn
Med S: ✚ Hospital
Other S: SAC Museum (13 Miles)

440 NE50, Millard, Springfield.
TStop N: Sapp Bros*(LP)(SCALES)
FStop N: Citgo*(SCALES)
Gas N: Phillips 66*
S: Amoco*
Food N: 24 Hr Cafe (Sapp Bros), Catfish Charlie's Restaurant, Hardee's, McDonalds, Sandwic… Shop (Citgo), Subway (Sapp)
S: El Bee's Mexican Food
Lodg N: American 9 Inn, Ben Franklin Motel, Comfort Inn, Park Inn, ◆ Ramada Limite…
AServ N: Ford Dealership, Madza

Bold red print shows RV & Bus parking available or nearby

XIT	NEBRASKA
TServ	**N:** Ford Dealership, Midland Radiator, Sapp Bros, Volvo Dealer **S:** Fleet Truck Sales, Wick's Trailer Serv, Young Truck Trailer Inc
TWash	**N:** T/Wash(LP)
Med	**S:** ✚ Hospital (7 Miles)
ATM	**N:** Sapp Bros (24 Hr)
Other	**N:** Travel Store*(SCALES) (Sapp Bros), Truck Permit Station
42	126th, Harrison St
44	Q St, L St
45	U.S. 275, NE92, L St
Gas	**S:** 24 Hr*, Amoco*(CW), Conoco(D)(CW), Phillips 66*(CW), QuikTrip*(D)
Food	**N:** Abigail's (Sheraton), Austin's Steaks **S:** Arby's, Burger King, Godfather's Pizza, Golden Corral, Hardee's, Hong Kong Cafe, Hunan Garden (Chinese), Little King, Long John Silvers, McDonalds, Shoney's, TCBY Yogurt, Taco Bell, The Gold Coast Lounge, Three Dollar Cafe, Valentino's (Italian), Village Inn, Wendy's
Lodg	**N:** ◆ Sheraton **S:** (AAA) Best Western, ◆ Budgetel Inn, Clarion Motel, Comfort Inn, ◆ Hampton Inn, Hawthorne Suites Hotel, Holiday Inn Express, Motel 6, ◆ Super 8 Motel
AServ	**S:** Checker Auto Parts
TServ	**N:** Int'l Dealer
RVCamp	**N:** A. C. Nelson Camper World, RV Service
ATM	**S:** Albertson's Grocery, Conoco, First Bank
Other	**N:** Sam's Club **S:** Albertson's Grocery (Pharmacy), Bag-n-Save Grocery Store, Cat Clinic, Fed Ex Drop Box, State Patrol, Suds City Car Wash, Vision Trends Optical, Wash World Coin Laundry
46	Jct I-680N, Downtown
48	84th St
Gas	**N:** Amoco*, Citgo*, Ginn (Full Service) **S:** Phillips 66*(D), Sinclair*(CW), Super Gas*
Food	**N:** Club Paradise, Denny's, Gators, Long John Silvers, McDonalds, Taco Bell **S:** Aki Oriental, Arby's, Bucky's Dexter, Schlotzkys Deli, Subway, Wendy's
Lodg	**N:** Econolodge
AServ	**N:** Master Tune-Up, NAPA Auto Parts, Shopko Grocery **S:** A Auto Parts, Acura Dealer, Bancroft Body Shop, GM Auto Dealership, Goodyear Tire & Auto, House of Mufflers & Brakes, Michelin, Sinclair, Spark's

EXIT	NEBRASKA
ATM	**N:** Amoco, First American Savings Bank, First Nat'l Bank **S:** Norwest Banks, Phillips 66
Other	**N:** Bakers, Kohlls Drug, Omaha Fire Dept, Pearl Vision **S:** Foot Care Clinic, Rent-All(LP)
449	La Vista, Ralston, 72nd St
Gas	**N:** Amoco*(CW) **S:** Amoco*(CW)
Food	**N:** Burger King, Grandmother's (Holiday Inn), Great American Diner, La Strada 72 Cafe (Italian), Perkins Family Restaurant **S:** Anthony's, Paddock
Lodg	**N:** (AAA) Central (Best Western), ◆ Hampton Inn, Holiday Inn, Homewood Suites Hotel, (AAA) Ramada, Roadway Inn, Super 8 Motel
AServ	**S:** Amoco
Med	**N:** ✚ Hospital
ATM	**N:** Amoco
450	60th St
FStop	**N:** Phillips 66*
Gas	**N:** Total*(CW)
Lodg	**S:** (AAA) Satellite Motel
AServ	**N:** Alfred Tire, NAPA Auto Parts, Phillips 66 **S:** Allpro Muffler & Brakes, Burns Body Shop
TServ	**N:** Phillips 66
ATM	**N:** Total
Other	**N:** Herman Nut House, Omaha Fire Dept
451	42nd St
Gas	**N:** 76*(CW)(LP), Phillips 66, Texaco*(CW) **S:** Phillips 66
Food	**S:** McDonalds, Taco Bell
AServ	**N:** 76, Phillips 66 **S:** Phillips 66
ATM	**N:** 76, Texaco
Other	**N:** Don's Service, Family Dentistry, Prescription Center
452	Jct I-480N, Eppley Airfield
453	24th St (Eastbound, Westbound Reaccess Only)
Food	**N:** Asiana Cafe, Changi Hi Garden Chinese Mexican Buffet, KFC **S:** Taqueiro (Mexican)
AServ	**N:** Car Quest Auto Center, Motorwest, Muckey's Service
ATM	**N:** Fast Bank
Other	**N:** Bakers Grocery, Self Serve Car Wash, WalGreens (Pharmacy) **S:** South O Laundry

EXIT	NEBRASKA/IOWA
454	13th St
Gas	**N:** Total* **S:** Auto Gas Club(D)(LP), Phillips 66*(CW)
Food	**N:** Famous Goodrich Ice Cream (Subs & Pizza) **S:** Zesto's (Ice Cream & Chicken)
ATM	**N:** Total **S:** Phillips 66
Other	**N:** Welcome Center **S:** Henry Doorly Zoo, Rosenblatt Stadium

↑NEBRASKA

↓IOWA

EXIT	IOWA
1A	Junction I-29N, Sioux City (Left Sided Exit)
	Note: I-80 runs concurrent below with I-29. Numbering follows I-29.
1B	South 24th St, Council Bluffs
3	IA 192, Council Bluffs, Lake Manawa
	Note: I-80 runs concurrent above with I-29. Numbering follows I-29.
2	Junction I-29S, to Kansas City
5	Madison Ave, Council Bluffs
Gas	**N:** Amoco* **S:** Phillips 66* (ATM), Texaco*(CW)
Food	**N:** Burger King, Great Wall, Pizza Hut, Subway, The Garden Cafe **S:** Dairy Queen, Shoney's
Lodg	**N:** ◆ Heartland Inn **S:** (AAA) Western Inn
ATM	**N:** First Bank
Other	**N:** Coin Laundry, Council Bluffs Medical Mall, Hy-Vee Food Store (Pharmacy), Mall of the Bluffs, Target Department Store **S:** Scrub & Dub Auto Wash
8	U.S. 6, Council Bluffs, Oakland
Gas	**N:** Total Convenience Store
AServ	**N:** K-Mart
Med	**N:** ✚ Hospital
Other	**N:** K-Mart **S:** Iowa State Patrol
17	CR G30, Underwood
TStop	**N:** Phillips 66
Food	**N:** Restaurant (Phillips 66)
Lodg	**N:** I-80 Inn, Sunshine Motel
(20)	**Rest Area - RR, Picnic, Phones, RV Dump (Both Directions)**

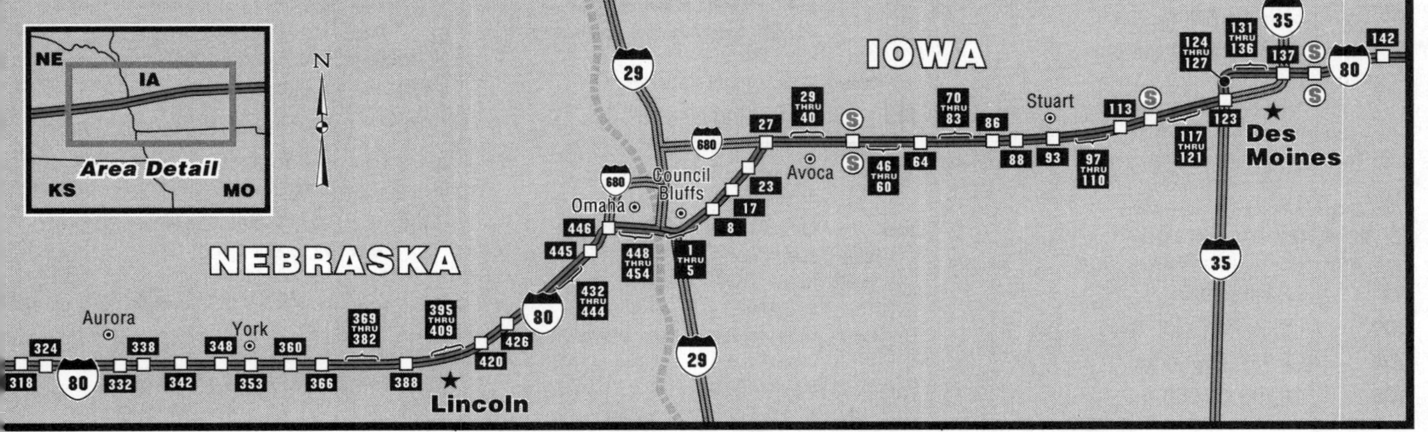

EXIT		IOWA
23		IA244, CR L55, Neola
	FStop	S: Phillips 66* (ATM)
	Parks	S: Arrowhead State Park
27		Junction I-680W
29		CR L66, Minden
	FStop	S: Conoco
	Gas	S: Total*
	Food	S: A & W Drive-In (Total), Kopper Kettle (Conoco)
(32)		Rest Area - No Services
34		CR M16, Shelby
	FStop	N: Texaco
	Gas	N: Phillips 66
	Food	N: Branded Steakhouse (Shelby ML), Dairy Queen, The Cornstalk (Texaco)
	Lodg	N: Sheby Motel Lodge
	AServ	N: Phillips 66
40		U.S. 59, Avoca, Harlan
	FStop	N: Phillips 66*(D)
	Food	S: The Embers Restaurant
	Lodg	S: Avoca Motel
(44)		Weigh Station - Both Directions
46		CR M47, Walnut Antique City Dr
	Gas	N: Amoco*
		S: Kum & Go*, Phillips 66*
	Food	N: McDonalds, The Villager
	Lodg	N: AAA Super 8 Motel
		S: Walnut Creek Inn
	RVCamp	S: Walnut Creek Inn
51		CR M56, Marne
54		IA173, Elk Horn, Kimballton
57		CR N16, Atlantic
	Med	S: Hospital
60		U.S. 6, U.S. 71, Atlantic, Audubon
64		CR N28, Wiota
70		IA148, Anita, Exira
75		CR G30
76		IA925, CR N54, Adair
	FStop	N: Phillips 66
	Gas	N: Amoco*(D) (AServ)
	Food	N: Happy Chef
	Lodg	N: Adair Guest Inn, AAA Best Western
(80)		Rest Area - RV Dump (Both Directions)
83		CR N77, Casey
	RVCamp	N: Camping
86		IA25, Guthrie Center, Greenfield
	Gas	S: Conoco*
88		CR P20, Menlo
93		CR P28, Stuart, Panora
	FStop	N: Conoco
		S: Phillips 66*(D)
	Gas	N: Amoco
	Food	N: Burger King, Harris House Restaurant, McDonalds, Subway
	Lodg	S: New Edgetowner Motel
	AServ	N: Wallace Auto Supplies
	ATM	N: Security State Bank
97		CR P48, Dexter
100		U.S. 6, Redfield, Dexter
104		CR P57, Pearlham
106		CR F90, CR P58

EXIT		IOWA
	RVCamp	N: KOA Campground
110		U.S. 169, Desoto, Adel
	Gas	S: Amoco, Casey's
	Food	S: Rattler's
	Lodg	S: Desoto Motor Inn, ◆ Edgetowner Motel
	AServ	S: Highway Auto & Tire
113		CR R16, Van Meter
(114)		Weigh Station (Westbound)
117		CR R - 22, Waukee, Booneville
	Gas	S: Kum and Go*
	Food	S: Taco Johns, Weeping Radish
(119)		Rest Area - RV Dump (Both Directions)
121		74th St, West Des Moines
	Gas	S: Kum & Go*
	Food	S: Arby's, Blimpie's Subs, Burger King, McDonalds, Taco Johns
	Lodg	N: ◆ Hampton Inn
		S: Marriott
123B		I-35 & I-80 split
124		University Ave, Clive
	Gas	N: Texaco*(CW)
	Food	S: Chili's
	Lodg	S: ◆ Courtyard by Marriott, ◆ Fairfield Inn, ◆ Holiday Inn, ◆ Residence Inn
	AServ	N: Texaco*(CW)
125		U.S. 6, Hickman Road, Adel
	Food	S: Abigails (Best Western), Iowa Machine Shed (Comfort Suites)
	Lodg	S: Best Western, AAA Comfort Suites
	AServ	S: Goodyear Tire & Auto
126		Douglas Ave, Urbandale
	TStop	N: Pilot Travel Center(SCALES)
	Gas	S: Phillips 66*(CW)
	Food	S: Dragon House
	Lodg	S: DAYS INN Days Inn, Econolodge
	TServ	N: Pilot Travel Center(D)(SCALES) (ATM)
	TWash	N: Pilot Travel Center
	ATM	S: Magna Bank
127		IA 141, Grimes, Perry
	FStop	N: Phillips 66*(D)
	Food	N: Subway
131		IA 28, Merle Hay Road, Urbandale
	Gas	N: Coastal*(D), QuikTrip*
		S: Amoco*(CW), QuikTrip*, Sinclair*
	Food	N: North Inn Diner (The Inn)
		S: Burger King, Denny's, Embers, Hostetler's BBQ, McDonalds, Perkins Family Restaurant, Wendy's
	Lodg	N: AAA Best Inns of America, AAA The Inn
		S: ◆ Comfort Inn, DAYS INN AAA Days Inn, Holiday Inn, AAA Howard Johnson, Roadway Inn, ◆ Super 8 Motel
	Other	S: Touchless of Merle Hay(CW)
135		IA 415, 2nd Ave, Polk City
	Gas	S: Coastal*, QuikTrip*
	TServ	N: Interstate Detroit Diesel
		S: Freightliner Dealer
136		U.S. 69, East 14th St, Ankeny
	FStop	S: QuikTrip*(SCALES)
	Gas	N: Amoco*(D)(CW), Phillips 66*(CW), Sinclair*(D)
		S: QuikTrip
	Food	N: Bonanza Steak House, Country Kitchen, Okoboji
		S: Burger King
	Lodg	N: AAA Best Western, Motel 6

EXIT		IOWA
		S: 14th Street Inn
	AServ	N: Sinclair
	TServ	S: Housby Mack Trucks
	Other	N: K-Mart
137B		Jct. I-35, Minneapolis
137A		Jct. I-235
(141)		Weigh Station - Both Directions
141		US 6, US 65, Pleasant Hill, Des Moines
142		U.S. 65, Hubbel Ave, Des Moines, Bondurant, Altoona
	TStop	S: Bosselman Travel Center*
	Gas	S: Cenex*(D)
	Food	S: Adventureland Inn, Burger King, Godfathers Pizza, McDonalds (Indoor Playground), Pizza Hut, Restaurants (Bosselman Travel Center), Taco John's, Viking Dining and Lounge
	Lodg	S: Adventure Land, Country Inn Motel, ◆ Heartland Inn, Holiday Inn Express
	TServ	S: Bosselman Travel Center, Peterbilt Dealer
	TWash	S: Bosselman Travel Center
143		Bondurant, Altoona
	TStop	N: 76 Auto/Truck Plaza(SCALES)
	Food	N: 76
	TServ	N: 76
	TWash	N: 76
(148)		Rest Area - RR, Picnic, Phones, Vending, RV Dump (Both Directions)
149		Mitchellville
(151)		Weigh Station (Westbound)
155		IA 117, Colfax, Mingo
	FStop	S: Texaco*(D) (Open 24 Hours)
	Gas	S: Phillips 66*(D)(CW)(LP)
	AServ	S: Phillips 66
	ATM	S: Texaco
159		Cnty Road F48, Baxter
164		IA 14, US 6, Newton, Monroe
	Gas	N: Amoco*, Conoco*, Phillips 66
	Food	N: Country Kitchen, Golden Corral, KFC, Perkins Family Restaurant, Sizzlin Sams, Subway
	Lodg	N: DAYS INN ◆ Days Inn, Holiday Inn Express, Ramada Inn, Super 8 Motel
		S: AAA Best Western
	AServ	N: Amoco, Conoco
168		Southeast Beltline Drive
	RVCamp	N: Camping
173		IA 224, Kellogg, Sully
	FStop	N: 76*(D)
	Food	N: Iowa's Best Burger Cafe
	RVCamp	N: Camping
179		Lynnville, Oakland Acres
(181)		Rest Area - RR, Picnic, Vending, Phones, Playground (Both Directions)
182		IA 146, Grinnell, New Sharon
	Gas	N: Coastal
	Food	N: City Limits Restaurant
	Lodg	N: AAA Best Western, DAYS INN Days Inn, Super 8 Motel
	AServ	N: Rick's 66
	Med	N: Hospital
191		U.S. 63, Tama, Montezuma

Bold red print shows RV & Bus parking available or nearby

EXIT	IOWA
FStop	S: Fuelmart
Gas	S: Texaco*(D)
Food	S: Dinner Bell, Star Resturaunt
AServ	S: Texaco
7	Brooklyn
FStop	N: Amoco
Food	N: Brooklyn - 80
TServ	N: Amoco
01	IA 21, Balle Plaine, What Cheer
TStop	S: Kwik Star*(D)(SCALES) (Conoco)
FStop	N: Texaco* (Overnight RV Parking, RV Hook-ups)
Gas	N: 76
Food	N: Nick's
TServ	S: Kwik Star
05	Victor
09)	Rest Area - RR, Picnic, Phones, Vending (Both Directions)
1	Millersburg, LaDora
6	Marengo, North English
FStop	N: Texaco*
Food	N: Texaco
20	IA 149, Iowa County Road V 77, Williamsburg, Parnell
FStop	N: Phillips 66*(D)
Gas	N: Conoco*
Food	N: Arby's, Landmark Restaurant, McDonalds, Pizza Hut, Rocky Mountain Chocolate, Subway
Lodg	N: (AAA) Crest Motel
Other	N: Factory Outlet Center, Publishers Warehouse
25	U.S. 151, Amana Colonies
Gas	S: Phillips 66*, Standard*
Food	S: Colony Haus Restaurant, Colony Village Restaurant, Little Amana Family Style Restaurant, Seven Village Restaurant
Lodg	N: (AAA) Comfort Inn
	S: Colony Haus Motor Inn, (Days Inn) (AAA) Days Inn, (AAA) Holiday Inn, Super 8 Motel
30	Johnson Cnty. W38, Oxford, Kalona Village Museum
Gas	N: Sinclair*(CW)
Food	N: Dairy Queen, Sleepy Hollow Restaurant*
RVCamp	N: Sleepy Hollow Campground
37)	Rest Area - RR, Vending, Phones, Picnic, Playground, RV Dump (Both Directions)
37	Tiffin

EXIT	IOWA
239A	Jct. U.S. 218 S, Mt. Pleasant, Keokuk
239B	Jct. I-380, Cedar Rapids, Waterloo
240	IA 965, Coralville, North Liberty
Gas	N: Texaco*
Lodg	N: (AAA) Best Western, Express Way Motel
242	Coralville
FStop	S: QuikTrip*(D), Texaco*(D)
Gas	S: Amoco*, Coastal
Food	N: Weatherby's
	S: Arby's, Cancun Mexican Restaurant, Country Kitchen, Hawk I (Texaco), KFC, McDonalds, Perkins Family Restaurant
Lodg	N: ◆ Clarion Hotel, ◆ Hampton Inn
	S: (AAA) Best Western, Big Ten Inn, ◆ Comfort Inn, Econolodge, ◆ Fairfield Inn, Motel 6, ◆ Super 8 Motel
AServ	S: Amoco
TWash	S: Texaco
Med	S: ✚ Coral Center Medical
ATM	S: Texaco
244	Dubuque St, Coralville Lake
246	IA 1, Dodge St, Mount Vernon
Gas	N: Phillips 66*(CW)
	S: Sinclair*
Food	N: Highlander Inn
Lodg	N: (AAA) Highlander Inn
249	Herbert Hoover Highway
254	Cedar Cnty. X30, West Branch
Gas	N: Amoco*(CW)
	S: Phillips 66*
Food	S: McDonalds
Lodg	S: Presidential Motor Inn
ATM	S: Phillips 66
259	West Liberty
FStop	S: Amoco* (RV Dump)
Food	S: Rockitz Diner
Lodg	S: Econolodge
RVCamp	S: KOA Campground(LP)
265	Atalissa
FStop	N: Phillips 66*(D) (No Auto Fuel. Diesel And Home Heating Fuel Only.)
Food	N: Birdy's Restaurant
267	IA 38, Tipton, Moscow
Gas	N: Sinclair*(LP)
Food	S: The Cove Restaurant
RVCamp	N: Mini Farm Acres
(268)	Weigh Station - Both Directions
(270)	Rest Area - RV Dump, Phones,

EXIT	IOWA/ILLINOIS
	Picnic, Vending (Both Directions)
271	U.S. 6, IA 38, Wilton, Muscatine
277	Durant, Bennett
280	Scott Cnty. Y30, New Liberty, Stockton
FStop	S: Burns Bros Travel Stop
284	Scott Cnty. Y40, Walcott, Plainview
TStop	N: Iowa 80 Truck Stop*(CW)(SCALES) (Amoco, ATM)
FStop	S: Pilot Travel Center*
Gas	N: Phillips 66*(D)
Food	N: Granma's Kitchen, Restaurants* (80 Truck Stop)
	S: McDonalds, Subway
TServ	N: Iowa 80 Truck Stop
290	Jct. I-280, US 6 Left Exit, Rock Island, Moline
292	IA 130, Maysville
FStop	N: Flying J Travel Plaza*(D)(SCALES) (RV Dump)
Gas	S: Citgo*
Food	N: Thads Restaurant (Flying J)
	S: Farmers Market Restaurant
Lodg	S: ◆ Comfort Inn
AServ	S: Citgo
295	U.S. 61, Brady St., Elridge, Dewitt
Gas	S: Amoco*(CW)
Food	S: Bonanza Steakhouse, Burger King, Country Kitchen, (Cracker Barrel) Cracker Barrel, Hardee's, Thunder Bay Grill, Village Inn
Lodg	S: Best Western, Budgetel Inn, Country Inn and Suites, Economy Inn, Exel Inn, Heartland Inn, Holiday Inn, Motel 6, Ramada, ◆ Residence Inn, Super 8 Motel
ATM	S: Norwest Bank
Other	S: N. Brady Animal Hospital
298	I-74E, Peoria
(300)	Rest Area - RV Dump (Both Directions)
301	Middle Road
306	U.S. 67, Le Claire, Bettendorf
Food	N: Slagles Bakery & Deli, Steventon's
Other	N: Camping, Iowa Welcome Center

↑IOWA

↓ILLINOIS

1	IL 84, East Moline, Savanna
(1)	Rest Area (P) (Eastbound)

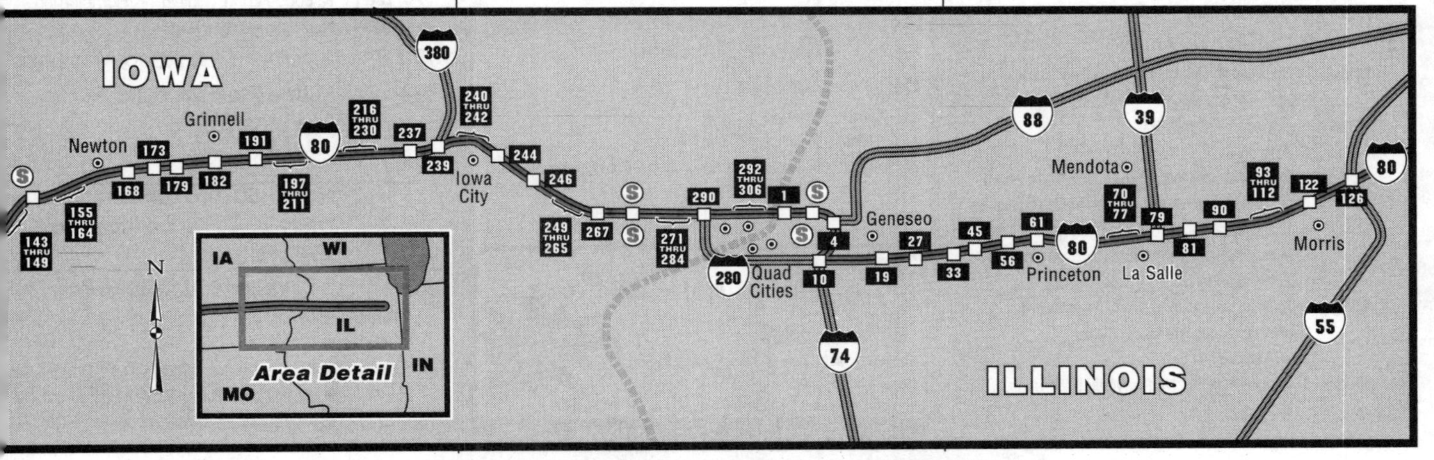

EXIT ILLINOIS

(2) **Weigh Station - both directions**

4AB Jct. I-88, IL 5, Silvis

7 Colona, Green Rock

9 U.S. 6

10 Jct. I-74, I-280

19 IL 82, Geneseo, Cambridge
- **FStop** N: **Phillips 66**
- **Food** N: Hardee's, Subway
  S: **KFC**
- **Lodg** N: AAA Deck Plaza
- **TServ** N: **Phillips 66**
- **Other** N: **Wal-Mart**

27 Galva, Atkinson
- **TStop** N: **Shell**
- **Gas** N: 76*(D)(CW)(LP)
- **TServ** N: **Wash Out**

33 IL 78, Prophets Town
- **Gas** S: Amoco(CW), Phillips 66*(D)
- **Food** S: The Loft
- **AServ** S: Amoco

45 IL 88, Peoria, Sterling
- **Gas** N: Amoco
- **Food** N: Farm House, Denny's (see our ad on this page)
- **Lodg** N: Days Inn

(51) **Rest Area - RV Dump P (Both Directions)**

56 IL 26, Dixon, Princeton
- **TStop** N: **Princeton Auto/Truck Plaza***(SCALES)
- **Gas** S: Amoco*(CW), Phillips 66, Shell*
- **Food** S: Country Kitchen, KFC, **McDonalds**, Subway, Taco Bell, Wendy's
- **Lodg** N: Super 8 Motel
  S: Comfort Inn, Days Inn, ◆ Princeton Motor Lodge
- **AServ** S: Auto Zone Auto Parts, Penzoil Quick Lube
- **TServ** N: **Phillips 66**
- **Other** S: **Animal Clinic, Wal-Mart**

61 Jct. I-180, Hennepin

70 IL 89, Ladd, Spring Valley
- **Gas** N: Casey's Store* (ATM)
- **Lodg** S: Motel Riviera

73 Plank Road
- **TStop** N: **Sapp Brothers***(LP) **(ATM)**
- **Food** N: Burger King, **Great American Buffet (Sapp Brothers)**, Interstate Pancakes, Subway
- **TServ** N: **Goodyear Tire & Auto, Kenworth, Unitz Truck Service**
- **TWash** N: **Sapp Brothers**

75 IL 251, Peru, Mendota
- **TStop** N: **76 Auto/Truck Plaza**(SCALES)**, Tiki Truck Stop**(SCALES) **(RV Dump)**
- **Gas** N: Amoco*(LP), Shell*
  S: Amoco(CW)
- **Food** N: Arby's, **McDonalds, Pine Cone Restaurant (Tiki Truck Stop), Shakeys (Howard Johnson's)**
  S: Bob Evans Restaurant, Dairy Queen, Dunkin Donuts, Oogie's, Red Lobster, **Subway**, Wendy's, Willaby's Sports
- **Lodg** N: Comfort Inn, Days Inn, **Howard Johnson**, Motel 6, **Super 8 Motel**
  S: ◆ Fairfield Inn
- **AServ** S: Amoco, K-Mart, Midas Muffler & Brakes,

EXIT ILLINOIS

  NAPA Auto Parts, Pennzoil Lube
- **TServ** N: **Tiki**
- **RVCamp** N: **Tiki RV Park**
- **ATM** S: First State Bank
- **Other** S: **K-Mart, Peru Mall**, Super Car Wash, **Target Department Store, Wal-Mart, WalGreens**

77 IL 351, La Salle

79AB Jct. I-39, U.S. 51

81 Il 178, Utica
- **Gas** S: Amoco* (ATM)
- **Food** S: Restaurant/Bait Shop
- **Lodg** S: Starved Rock Gateway Motel
- **AServ** S: Greg's Automotive
- **RVCamp** N: **Camping**
- **Other** S: **Camping**

90 IL 23, Ottawa, Dekalb
- **Gas** N: Amoco*
  S: Shell*
- **Food** N: Subway
  S: Dunkin Donuts, Hardee's, KFC, Little Caesars Pizza, Ponderosa
- **Lodg** N: AAA Holiday Inn Express
  S: Comfort Inn, Ottawa Inn, Super 8 Motel, The Surrey Motel
- **AServ** S: Penzoil
- **TServ** S: **International Truck Service**
- **ATM** S: First National Bank of Ottawa
- **Other** S: Car Wash, **K-Mart (Pharmacy), Kroger Supermarket (Pharmacy), Wal-Mart (Pharmacy)**

93 IL 71, Ottawa, Oswego
- **FStop** N: **J&L Gas, Shell**(SCALES)
- **Food** N: **PJ's Family Restaurant (Shell)**
  S: New Chiam Chinese
- **TServ** N: **Shell**

97 Marseilles
- **RVCamp** S: **Glen Wood RV Resort**

105 Seneca

112AB IL 47, Morris, Yorkville
- **TStop** N: **Standard**(SCALES)
- **Gas** S: Amoco*, Clark, Shell*
- **Food** N: **R-Place**
  S: Bozo Hot Dogs, Burger King, KFC, Marias Restaurant, McDonalds, Pizza Hut, Taco Bell, Wendy's
- **Lodg** N: Comfort Inn, Holiday Inn
  S: Morris Motel, Park Motel, Super 8 Motel
- **AServ** S: Pennzoil Lube
- **TServ** N: **Standard**
- **Other** S: **Wal-Mart**

(116) **Rest Area - RR, Phones, Vending P (Both Directions)**

122 Minooka
- **FStop** N: **Citgo**
- **Gas** S: Amoco* (ATM)
- **Food** S: Big Henry's Diner, McDonalds, Subway, Wendy's

EXIT ILLINOIS

- **ATM** S: Tri-County Bank
- **Other** S: **SuperValu Foods (Pharmacy)**

126AB Jct. I-55

127 Houbolt Rd
- **Gas** N: Amoco
- **Food** N: **Cracker Barrel**, Wendy's
- **Lodg** N: ◆ Fairfield Inn, ◆ Hampton Inn, Ramada Inn

130AB IL 7, Larkin Ave, Joliet
- **Gas** N: 76*, Clark*, Marathon*, Shell*(CW), Speedway*(D), Thornton's Food Mart*
- **Food** N: Bob Evans Restaurant, Boston Market, Bulle[...] BBQ, Burger King, Dunkin Donuts, Family Table Restaurant, McDonalds, Pizza Hut, Steak & Shake, Subway, Taco Bell, Wendy's, White Cast[...] Restaurant, White Hen Pantry
- **Lodg** N: Comfort Inn, Holiday Inn, Motel 6, Red Roof I[...]
- **AServ** N: Car X Muffler, Goodyear Tire & Auto, J&J Aut[...] Parts Warehouse, Meineke Discount Mufflers, Pep Boys Auto Center, Tuffie's Auto Center
- **TServ** N: **International Trucks Sales & Service Repai[...]**
- **ATM** N: First American
- **Other** N: **Sam's Club, Wal-Mart**

131 Meadow Ave, Center St

132AB Chicago St, U.S. 52 East, IL 53
- **Gas** N: King Gas*
- **Food** N: Burger King

133 Richards St

134 Briggs St
- **Gas** N: Speedway*
  S: Amoco*, General Store*(D)
- **ATM** S: Bank of Joliet, Machine (Amoco)

137 U.S. 30, Maple St, New Lenox
- **Gas** S: Speedway(D)
- **Food** N: Les Brothers Restaurant
  S: AJ Gyros, Hardee's, KFC, Kirby's Bakery, McDonalds, New Lenox Restaurant, Pizza Hut, Pizza Restaurant, Taco Bell
- **ATM** S: NDB Bank, NISB Bank
- **Other** N: **K-Mart**
  S: **Animal Hospital, Eagle Country Market**, Super Wash, **WalGreens**

(143) **Weigh Station (Eastbound)**

145AB U.S. 45, 96th Ave

(148) **Weigh Station (Westbound)**

148AB IL 43, Harlem Ave
- **FStop** N: **Speedway***
- **Food** N: Burger King, **Cracker Barrel, Subway (Speedway)**, Wendy's
- **Lodg** N: Budgetel Inn, Fairfield Inn, Hampton Inn
- **ATM** N: Machine (Speedway)

151AB Junction I-57, Chicago, Memphis

154 Kedzie Ave (Eastbound, No Reaccess)

155 I-294 (Toll)

Note: I-80 runs concurrent below with I-294. Numbering follows I-294.

3 IL 1, Halsted St (Pay Toll To Exit)
- **FStop** N: **Quick Fuel**
- **Gas** N: 76*(D), Clark
- **Food** N: Alf's Pub, Burger King, Hot Spot Drive-In, Mr Philly, Wendy's, Yellow Ribbon Restaurant
  S: Arby's, Dunkin Donuts, Hardee's, IHOP,

Bold red print shows RV & Bus parking available or nearby

EXIT	ILLINOIS

	McDonalds, Shooter's Buffet, Taco Bell, Washington Square Family Restaurant
Lodg	**N:** Budgetel Inn, Hampton Inn (see our ad this page), Junction Inn, Ramada, Red Roof Inn
	S: Days Inn, Homewood Hotel, Motel 6
AServ	**S:** Auto Dynamic's
TServ	**N:** **Road Ready**
ATM	**N:** Mutual Bank
Other	**S:** **K-Mart**
	Dixie Highway (Eastbound, No Reaccess)

Note: I-80 runs concurrent above with I-294. Numbering follows I-294.

0A	IL 394, Danville
60)	Lincoln Oasis
0B	Junction I-94 West, Chicago
1	U.S. 6, IL 83, Torrence Ave
Gas	**N:** Amoco*(CW) (ATM)
	S: Clark*, Gas City*, Mobil*(D), Shell
Food	**N:** Arby's, Bob Evans Restaurant, Checkers Burgers, Chili's, Lansing House Buffet, Little Caesars Pizza, Outrigger's Fish, That Oriental Place (Chinese), Wendy's
	S: Al's Hamburgers, Brown's Chicken, Burger King, Cafe Borgia (Roman Food), China Chef II, Golden Crown Restaurant, McDonalds, Pappy's Gyros, Vienna Beef Hot Dogs
Lodg	**N:** Best Western, ◆ Fairfield, Holiday Inn, Red Roof Inn, Super 8 Motel
	S: Pioneer Motel
AServ	**N:** Car X Muffler & Brakes, Firestone Tire & Auto,

EXIT	ILINOIS/INDIANA
	Goodyear Tire & Auto
	S: Auto Clinic Muffler, Jiffy Lube, Meineke Discount Mufflers, Shell
ATM	**N:** US Bank
Other	**N:** **Fannie Mae Candies, K-Mart**
	S: **Komo's Grocery Store, Pets Mart, Sam's Club, Speedwash Coin Laundry**

↑ILLINOIS

↓INDIANA

1	Calumet Ave, U.S. 41 North
Gas	**N:** Amoco(CW), Gas Center*, Gas City*, On The Go
	S: Marathon*, Shell
Food	**N:** Barton's Pizzeria, Dunkin Donuts, Kalipsos Seafood, Porky's Pizza, Rick's Grill, Subway (Gas Ctr)
	S: Arby's, Boston Market, Burger King, Edwardo's Natural Pizza, Fat Guy's BBQ, Taco Bell
AServ	**N:** Levin, Precision Tune & Lube
	S: Shell, Ward's Auto Express
Med	**N:** **Hospital**
ATM	**N:** Mercantile Bank
Other	**N:** **Calumet Laundromat, WalGreens,** Wash Station - Car Wash
	S: **McCoroy Grocery Store, Sterks Grocery**
2AB	U.S. 41 South, IN 152 North, Indianapolis Boulevard
Gas	**N:** Gas Center*(CW), Shell*(CW), Witham's*
	S: Speedway
Food	**N:** Arby's, Chop Suey, Dunkin Donuts, McTavern's Deli, Papa John's Pizza, The Wheel, Woodmar Restaurant
	S: **Burger King**
AServ	**N:** Apex Muffler & Brake, Car X Muffler & Brakes, Midas Muffler & Brakes, Shell
	S: K-Mart
Med	**N:** **Hospital**
ATM	**N:** Mercantile Nat'l Bank of Indiana, People's Bank
Other	**N:** **Woodmar Animal Clinic**
	S: **K-Mart**
3	Kennedy Ave
5AB	IN 912, Cline Ave
Gas	**S:** Mobil(CW)
Food	**S:** Bob Evans Restaurant, Burger King, Dunkin Donuts, Subway
Lodg	**S:** Holiday Inn, Motel 6, Super 8 Motel
AServ	**S:** Michelin Tire Service
Med	**N:** **Hospital**

EXIT	INDIANA
6	Burr St, Gary
9AB	Grant St
10AB	Broadway
Gas	**N:** Amoco*
	S: Amoco*
Food	**N:** Broadway BBQ, Coney Island Hot Dogs
Med	**N:** **Hospital**
Other	**N:** Full Service Car Wash
11	I-65 South, Indianapolis (Eastbound)
12AB	Junction I-65 North, Gary
13	Central Avenue (No Eastbound Reaccess)
AServ	**S:** Frank's Auto Repair, S & F Tire
Other	**S:** **7 Elephants Deli & Foodmart**
15AB	U.S. 6 East, U.S. 20, IN 51, Ripley St
TStop	**N:** **Petro*(SCALES), Travel Port*(SCALES)**
FStop	**N:** **Speedway*(SCALES)**
	S: **Speedway*(SCALES)**
Gas	**S:** 76*, Marathon*(CW)
Food	**N:** **Dunkin Donuts (Speedway), McDonalds (Petro)**
	S: Arman's Polish Sausage et. al, Long John Silvers, Miami Subs, Papa G's Gyros, Ruth & Bud's Grill, Snak Time Family Restaurant
AServ	**N:** Petro
	S: Discount Transmission
TServ	**N:** **Travel Port, Weber's**
TWash	**N:** **Blue Beacon Truck Wash, Murray (Travel Port), Red Baron (Speedway)**
ATM	**N:** **Machine (Petro)**
16	Junction I-90/I-80 (Toll)
17	Junction I-65, U.S. 12, U.S. 20, Dunes Hwy., Indianapolis
Med	**S:** **Hospital**
21	I-80, I-94, U.S. 6, IN. 51 W., Des Moines (Difficult Reaccess)
TStop	**N:** **Dunes Plaza* (24 Hrs), Petro*(SCALES) (24 Hrs), TravelPort*(SCALES)**
FStop	**N:** **Speedway*(LP)(SCALES)**
Food	**N:** **Buckhorn Family Restaurant (TravelPort),** Burger King, **Iron Skillet Restaurant (Petro),** McDonalds (Play Land), Ponderosa, **Subway (Travelport),** Wing Wah
AServ	**N:** Jiffy Lube
TServ	**N:** **Petro, Weber's Truck Repair**
TWash	**N:** **Blue Beacon Truck Wash (Petro), Murray Truck Wash (Travelport), Red Baron**
Other	**N:** Coin Car Wash

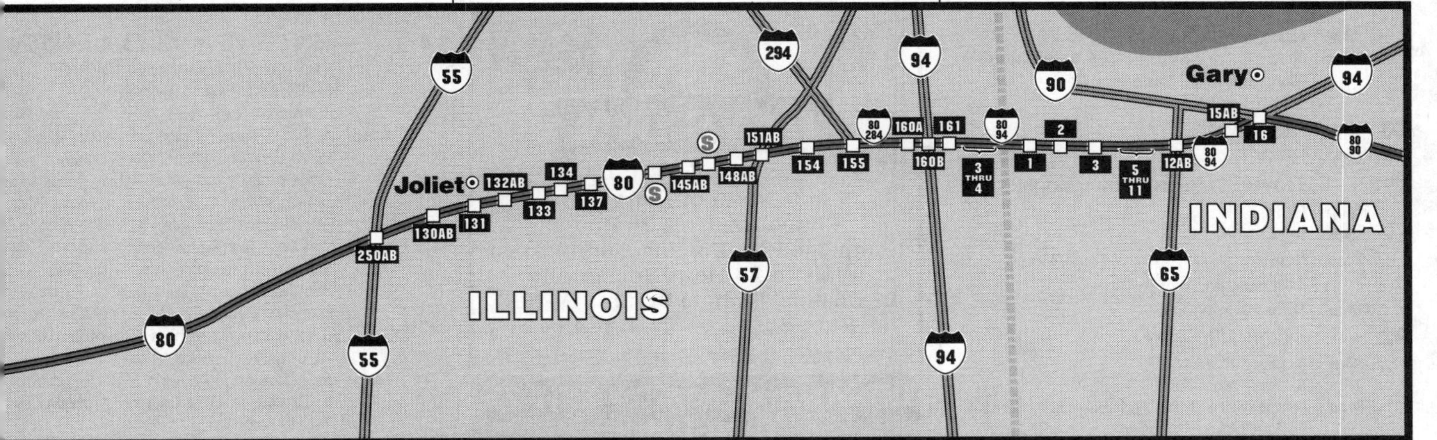

Bold red print shows RV & Bus parking available or nearby

EXIT INDIANA

(22) Travel Plaza (Both Directions)
- FStop N: SC(LP)(SCALES)
- Gas N: BP*(D)
- Food N: Baskin Robbins, Fazoli's Italian Food, Hardee's
- Other N: Tourist Information

23 Portage
- Gas S: Amoco, Marathon*
- Food S: Burger King, Dunkin' Donuts, KFC, McDonalds (Play Place), Subway, Wendy's
- Lodg N: Lee's Inn
- AServ S: Amoco, Marathon, Muffler Shops, Portage Quick Change Oil
- ATM N: Pennacle Bank
  S: First National, Indiana Federal Bank
- Other S: Town & Country Grocery (24 Hrs), WalGreens (Pharmacy, 24 Hrs)

(24) Indiana Toll Plaza

31 Indiana 49, Chesterton, Valparaiso
- Med S: Hospital
- Parks N: Indiana Dunes State Park
- Other N: Indiana Dunes National Lakeshore, Tourist Information (2 Miles)

39 U.S. 421, Michigan City, Westville

49 Indiana 39, La Porte
- Lodg S: Cassidy Motel
- TServ N: Tomenko Tire & Truck Service
- Other S: Kingsbury State Fish & Wildlife Area

(56) Travel Plaza & Service Area
- FStop N: SC
- Gas N: BP*(D)
- Food N: Baked Goods, Dairy Queen, McDonalds
- Other N: Travel Emporium*

72 U.S. 31 By - Pass, South Bend, Plymouth, Niles
- FStop N: Speedway*(LP)
- Food N: Hardee's (Speedway)
- ATM N: Speedway
- Parks S: Potato Creek State Park

77 U.S. 33, Bus. U.S. 31, South Bend, Notre Dame University
- Gas S: Amoco(CW) (24 Hrs), Phillips 66*(D)
- Food N: Burger King, Denny's, Fazoli's Italian Food, J & N Restaurant (24 Hrs), Marco's Pizza, McDonalds, Pizza Hut, Ponderosa, Steak & Ale, Subway
  S: Bennitt's Restaurant, Bill Knapp's Restaurant, Bob Evans, Colonial Pancake House, Donut Delight, King Gyro's, Perkins Family Restaurant, Schlotzkys Deli, Shoney's, Wendy's
- Lodg N: Days Inn (see our ad on this page), ◆ Hampton Inn & Suites, Motel 6, AAA Ramada, Super 8 Motel
  S: AAA Best Inns of America, Holiday Inn, Howard Johnson, Knights Inn, AAA Signature Inn, St. Mary's Inn
- AServ N: Giant Auto Supply
  S: Amoco, Q Lube
- ATM N: Standard Federal
- Other N: All Star Car Wash, Key Bank, North Village Mall, WalGreens (Pharmacy)
  S: Rose Land Animal Hospital

83 Mishawaka
- RVCamp N: KOA Campground

(90) Travel Plaza (Both Directions)
- FStop N: SC
- Gas N: BP*(D)
- Food N: Arby's, Dunkin' Donuts, South Bend Chocolate Co.
- Other N: Tourist Information

92 Indiana 19, Elkhart
- Gas N: Citgo*, Phillips 66*(D)
  S: Clark*, Marathon*(D)
- Food N: Andini Fine Dining, Applebee's, Lee's Famous Recipe Chicken (Phillips 66), Steak 'n Shake
  S: Blimpie's Subs, Bob Evans Restaurant, Burger King, Callahan's (24 Hrs), DaVinci's Pizza, King Wha Chinese, McDonalds (Inside Play Place), Olive Garden, Perkins Family Restaurant, Red Lobster, Weston Restaurant (Weston Plaza Hotel)
- Lodg N: Comfort Inn, AAA Diplomat Motel, AAA Econolodge, Hampton Inn, Holiday Inn Express, AAA Knight's Inn, ◆ Shoney's, Turnpike Motel
  S: Budget Inn, Ramada, ◆ Red Roof Inn, AAA Signature Inn, Super 8 Motel, Weston Plaza Hotel
- RVCamp N: American Trailer Supply, Dan's Service Center (Hitches & Trailers), Elkhart Campground, Traveland RV Service, Worldwide RV Sales & Service
- ATM N: NBD
  S: Key Bank, Marathon
- Other N: Aldi Supermarket, K-Mart (Pharmacy), Martin's Supermarket, Revco Drugs, Visitor Information
  S: Car Wash World

101 IN 15, Bristol, Goshen
- RVCamp N: Eby's Pines Camping

107 U.S. 131, Indiana 13, Constantine, Middlebury
- FStop N: Mobil*
- Lodg N: Plaza Motel
- AServ N: Dick's Auto Parts

121 Indiana 9, Howe, LaGrange
- Gas N: J & M Service Center(D)
- Food N: Golden Buddha
- Lodg N: Green Briar Inn, Travel Inn Motel
  S: AAA Super 8 Motel
- AServ N: J&M Service Center
- Med N: Hospital

(126) Service Area (Both Directions)
- FStop N: SC
- Gas N: BP*(D)
- Food N: Baskin Robbins, Fazoli's Italian Food, Hardee's
- Other N: Tourist Information

144 Junction I-69, U.S. 27, Angola, Ft. Wayne, Lansing (Pay Toll To Exit)
- TStop N: 76 Auto/Truck Plaza*(SCALES)
- FStop S: Pioneer, Speedway*(SCALES)
- Gas S: Marathon*, Shell*
- Food N: Baker St Family Restaurant (76 TS), Subway (76 TS)
  S: Deli Mart (Marathon), Hardee's (Speedway), Red Arrow Restaurant (24 Hrs)
- Lodg N: Lake George Inn
  S: E&L Motel, Hampton Inn, ◆ Holiday Inn Express, Redwood Lodge
- TServ N: 76 Auto/Truck Plaza, Gulick Trucks & Parts Service
  S: Volvo Dealer
- RVCamp S: Yogi Bear Camp Report
- Parks S: Pokagon State Park

EXIT INDIANA/OH

- Other S: Country Meadows Golf Resort, Horizon Outlet Center

(146) Service Area (Both Directions)
- FStop N: BP*(D)
- Food N: Baked Goods, Dairy Queen, McDonalds

(153) Indiana Toll Plaza

↑INDIANA

↓OHIO

Note: I-80 runs concurrent below with OHTNPK. Numbering follows OHTNPK.

1 Ohio49
- Gas N: Mobil*(LP)
- Food N: Burger King, Sub Shop (Mobil)
- ATM N: Mobil
- Other N: Ohio Tourist Center

(2) Ohio Toll Plaza

2 (10) Ohio 15, Bryan, Montpelier (Pay Toll To Exit)
- FStop S: Pennzoil Lube*
- Gas S: Marathon*(D)
- Food S: Country Fare, Subway (Marathon)
- Lodg S: Econolodge, Holiday Inn, Rainbow Motel
- TServ S: Hutch's Tractor & Trailer Repair
- Med S: Hospital
- ATM S: Pennzoil FS

(21) Service Plaza - RV Dump (Both Directions)
- FStop N: Sunoco*
- Food N: Hardee's
- AServ N: Sunoco
- Other N: Tourist Information

3 (31) Ohio 108, Wauseon (Pay Toll To Exit)
- FStop S: Hy-Miler* (Shell)
- Food S: Smith's
- Lodg S: Arrowhead Motor Lodge, Del-Mar Best Western, Super 8 Motel
- AServ S: Wood Truck & Auto Service
- TServ S: Wood Truck & Auto Service
- RVCamp S: Executive Travelers Sales & RV Service

3B OH. 109, Delta, Lyons (Pay Toll To Ex

(49) Service Area (Both Directions)
- FStop N: Sunoco*
- Food B: Charlie Brown's Family Restaurant & General Store
- Other B: Tourist Information

3A (59) Ohio 2, Toledo Airport, Swanton (Pay Toll To Exit)
- Lodg S: Toledo Airport Motel
- TServ S: Express Auto & Truck Service

4 (63) U.S. 20, to U.S. 23, to I-475 (Pay Toll To Exit), Maumee, Toledo
- Gas N: BP*, Fast Check*, Speedway*(LP)
  S: Amoco*, Speedway*
- Food N: Arby's, Beast of Chicago Pizza Co., Bob Evans Restaurant, China Buffet, Connie Mac's Bar and Grill, Dominic's Family Italian, Little Caesars Pizza, Mark Pi's China Gate Restaurant, Max's Diner, McDonalds, Nick's Cafe, Pizza Hut, Ramada, Tandoor Indian Restaurant
  S: Baverian Brewing Company, Big Boy, Brandie's Diner, Chi Chi's Mexican, Fricker's, Friendly's, Gourmet of China, Popoff's Pizza and Lebanese Food, Ralphie's Burgers, Red Lobster, Rib Cage, Yes Solid Rock Cafe
- Lodg N: Budget Inn, Holiday Inn, Motel 6, Ramada
  S: Comfort Inn, Cross Country Inn, Days Inn, Hampton Inn

EXIT OHIO

AServ N: Auto Express, Goodyear Tire & Auto, K-Mart, Napa Auto Parts, The Car Doctor, Tom's Tire
S: Bob Schmidt GM, Hatfield GM
ATM S: Huntington Bank
Other N: K-Mart (Pharmacy), Southwyck Mall
S: DJ's Car Wash, Meijer Grocery

5A **(72)** Exit 4A, Jct I-75, Perrysburg, Moline - Martin Rd

5 **(73)** Junction I-280, Ohio 420, to I-75, to Stony Ridge, Toledo (Pay Toll To Exit)

TStop N: Flying J Travel Plaza*(LP)(SCALES), Petro*(SCALES)
S: 76 Auto/Truck Plaza, TravelCenters of America*(SCALES)
FStop S: Speedway*(SCALES)
Gas S: Mobil (76 Auto Truckstop)
Food N: Dad's (Flying J), Howard Johnson, Iron Skillet (Petro), Metro Inn, Pizza Hut (Petro)
S: 76 Auto/Truck Plaza, McDonalds, Sbarro Pizza, Toledo 5 TS, TravelCenters of America, Wendy's
Lodg N: Budget Inn, Howard Johnson, Knights Inn, Metro Inn
TServ N: Petro
S: 76 Auto/Truck Plaza, Fleet Tire Center, Perkins Detroit Diesel, TravelCenters of America*(SCALES)
TWash N: Blue Beacon Truck Wash (Petro)
S: Stony Ridge Truck Wash (TS)
ATM N: Flying J Travel Plaza
S: 76 Auto/Truck Plaza

(77) Service Plaza - RV Dump (Both Directions)

FStop B: Sunoco
Food B: Fresh Fried Chicken, Hardee's
AServ B: Sunoco
Other B: Tourist Information

6 **(91)** Ohio 53, Fremont, Ft. Clinton (Pay Toll To Exit)

FStop S: Shell*
Food N: Days Inn, Sneaky Fox Steak House, Z's Diner
S: Holiday Inn
Lodg N: Best Budget Inn, Days Inn
S: Fremont Turnpike Motel, Holiday Inn
Med S: Hospital

(100) Commodore Perry Service Plaza (Both Directions)

FStop B: Sunoco
Food B: Rax
AServ B: Sunoco
Other B: Travel Information

7 **(119)** U.S. 250, Sandusky, Norwalk (Pay Toll To Exit))

Gas N: Marathon*(D) (Kerosene), Speedway*
Food N: Dick's Place (take-out), Subway, Super 8 Motel
S: Homestead Diner
Lodg N: Comfort Inn, Days Inn, Hampton Inn,

EXIT OHIO

Homestead Inn, Ramada Limited, Super 8 Motel
S: Crown Motel, Homestead Farm
AServ S: Dorr Chevrolet/Geo
RVCamp N: Milan Travel Park
Other N: Lake Erie Factory Outlet Center, State Patrol Post

(139) Service Plaza - RV Dump (Both Directions)

FStop B: Sunoco
Food B: Bob's Big Boy, Burger King, TCBY Yogurt

8A **(143)** Jct. I-90, Ohio 2 (Pay Toll To Exit)

8 **(147)** Ohio 57 (Pay Toll To Exit)

Gas N: BP*(D), Speedway*
S: Shell*
Food N: Arby's, Bob Evans Restaurant, Delphine Restaurant, Holiday Inn, McDonalds, Mountain Jack's Restaurant, Pizza Hut, Red Lobster, Rubin's Restaurant, Wendy's, White Castle Restaurant
S: Mario's Restaurant (Ramada Inn)
Lodg N: Comfort Inn, Days Inn, Econolodge, Holiday Inn (see our ad on this page)
S: Howard Johnson, Journey Inn, Ramada Inn
AServ N: Firestone Tire & Auto, Goodyear Tire & Auto, Sears (Midway Mall)
Med S: Hospital
ATM N: National City Bank, Star Bank
Other N: Midway Mall, Rini-Rigo Food Mart, Wal-Mart
S: Ohio State Patrol Post

9A **(151)** Junction I-80, I-480, North Ridgeville, Cleveland (Pay Toll To Exit)

9 **(158)** Ohio State Penitentiary, To North Olmstead, Cleveland (Difficult Reaccess)

10 **(162)** Junction I-71, U.S. 42, Strongsville, Cleveland

(170) Towpath Service Plaza

FStop B: Sunoco
Food B: McDonalds, Sunoco
Other B: Tourist Information

11 **(176)** Ohio 21, to I-77

FStop S: Speedway*
Gas N: Clark*
S: BP
Food N: Demetrio's Kitchen (Scottish Inn), Lake Motel, Valley Forge Restaurant (Howard Johnson)
S: Dairy Queen, Holiday Inn, Richfield Family Restaurant
Lodg N: Howard Johnson, Lake Motel, Scottish Inns
S: Brushwood Motel, Holiday Inn (see our ad on this page), Super 8 Motel, Days Inn (see our ad on this page)
AServ S: BP, Richfield Radiator Repair

12 **(180)** Ohio 8, to I-90

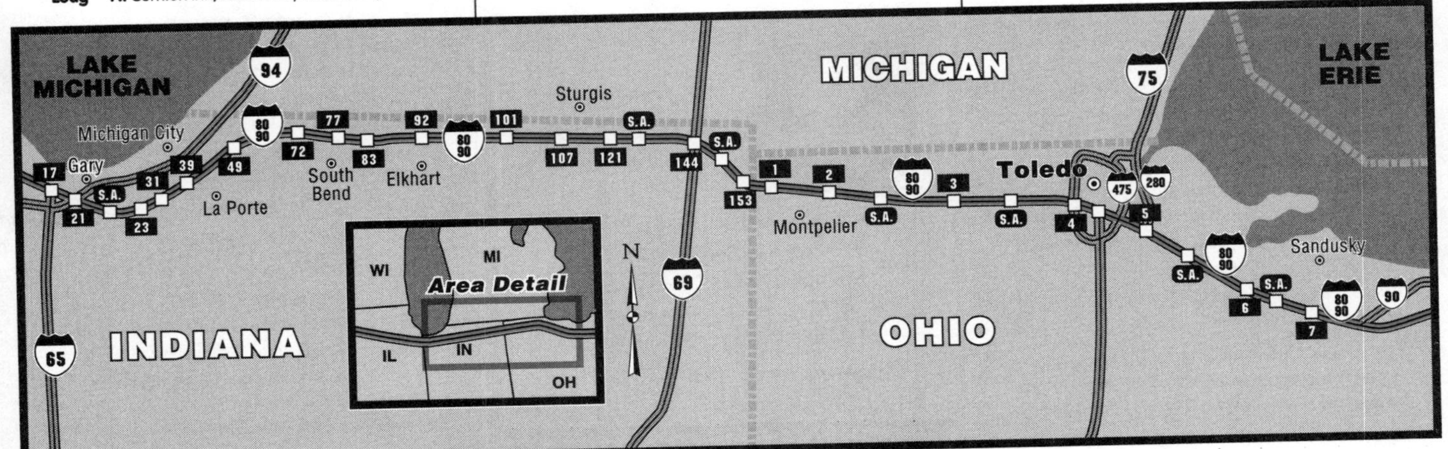

Bold red print shows RV & Bus parking available or nearby

EXIT OHIO

13 **(187)** Junction I-480, Ohio 14, Streetsboro (Pay Toll To Exit)
- **Lodg** **N:** Comfort Inn (see our ad on this page)

13A Ravenna Rd., OH 44

(197) **Brady's Service Plaza, RV Dump**
- **FStop** **B:** Sunoco
- **Food** **B:** Dunkin Donuts, Hot Dog City, Popeye's Chicken, TCBY Yogurt, Taco Bell
- **AServ** **B:** Sunoco

14 **(208)** Ohio 5, Warren
- **Gas** **S:** Marathon
- **Food** **S:** Wrangler's (Marithon)
- **Lodg** **N:** Budget Lodge, Scottish Inns; **S:** Roadway Inn
- **TServ** **S:** Wayne's Truck & Auto Service (Marithon TStop)
- **RVCamp** **S:** Camping

14A **(209)** Lordstown (Eastbound, Westbound Reaccess)

14B **(210)** Lordstown (Westbound, Eastbound Reaccess)

15 **(219)** Junction I-76, I-80 East

Note: I-80 runs concurrent above with OHTNPK. Numbering follows OHTNPK.

223AB Ohio 46, Niles
- **TStop** **N:** Universal Truck Plaza(SCALES); **S:** 76 Auto/Truck Plaza(SCALES) (RV Dump)
- **FStop** **S:** Fuel Mart*, Speedway*(LP)
- **Gas** **N:** Citgo*(LP); **S:** BP*(D), Sunoco*
- **Food** **N:** Bob Evans Restaurant, Burger King, Country Kitchen (Universal Truck Plaza), NT Mugs (Budget Inn); **S:** Antone's, Arby's, Cracker Barrel (Bus Parking), McDonalds, Perkins Family Restaurant, Subway (Sunoco), Taco Bell, The Ranch Family Restaurant, Wendy's, Winston's Tavern (Best Western)
- **Lodg** **N:** Budget Luxury Inn (Universal Truck Plaza), TraveLodge; **S:** Best Western, Hampton Inn, Knight's Inn, Super 8 Motel
- **TServ** **N:** CB Repair, Universal Truck Plaza; **S:** 76 Auto/Truck Plaza, Freightliner Dealer
- **TWash** **N:** Universal Truck Plaza; **S:** 76 Auto/Truck Plaza
- **ATM** **N:** Citgo
- **Other** **N:** Coin Car Wash

224 Ohio 11 South, Canfield

224B I-680, Youngstown

226 Salt Springs Rd, to McDonald
- **TStop** **S:** Petro(SCALES), Pilot Travel Center(SCALES)
- **FStop** **S:** Mr. Fuel
- **Gas** **N:** Shell*(D)(CW)
- **Food** **N:** McDonalds, Summit Carry-Out & Deli; **S:** Arby's (Pilot Travel Center), Baskin Robbins, Iron Skillet (Pilot Travel Center)
- **TServ** **S:** Petro, Pilot Travel Center
- **TWash** **S:** Eagle One (Pilot Travel Center), Frank's Truck Wash (Petro TStop)

227 U.S. 422, Girard
- **Gas** **N:** Amoco*, McQuaids*, Shell*(D)(CW)
- **Food** **N:** Burger King, Dairy Queen, Jab Hotdog Shop, Rocco's Pizza
- **ATM** **N:** Charter One Bank

228 Ohio 11, Warren, Ashtabula (Eastbound Exit Left)

229 Ohio 193, Belmont Ave
- **Gas** **N:** Shell, Speedway*; **S:** Amoco*, Rich, Speedway*
- **Food** **N:** Days Inn, Granny's Home Cooking, Handel's Ice Cream, Ramada Inn, Station Square, Tally Ho Tel; **S:** Antone's Restaurant, Arby's, Armando's Italian, Arthur Treacher's Fish & Chips, Bob Evans Restaurant, C.R. Berry's Burgers, Cancun Restaurant, Inner Circle Pizza, Long John Silvers, McDonalds, Perkins Family Restaurant, Pizza Hut, Taco Bell, Western Sizzlin'
- **Lodg** **N:** ◆ Days Inn, Holiday Inn, Motel 6, ◆ Ramada Inn, Super 8 Motel; **S:** ◆ Comfort Inn, (AAA) Econolodge
- **AServ** **N:** Pennzoil Lube; **S:** Goodyear Tire & Auto, Monroe Mufflers & Brakes, Super Shop Automotive Performance Centers
- **Med** **S:** ✚ Hospital
- **ATM** **N:** Charter One Bank; **S:** Bank One, Home Savings and Loan Company, Mahoning National Bank, Metropolitan Savings Bank
- **Other** **N:** Hi Land Foods*; **S:** Big John's Car Wash, Phar Mor Drugs

(232) **Weigh Station (Westbound)**

234B U.S. 62, Ohio 7, to Hubbard, Sharon
- **TStop** **N:** Truck World(D)(SCALES)
- **Gas** **N:** Shell*(D); **S:** Clark*
- **Food** **N:** Arby's, Jab Hotdog Shop (Truck World), McDonalds, Restaurant (Truck World); **S:** Castaway's BBQ Pitt, Katie's Korner
- **Lodg** **N:** Truck World; **S:** The Superette Motel*
- **AServ** **S:** Hi-Tech Automotive, Morrow's
- **TServ** **N:** Truck World
- **TWash** **N:** Blue Beacon Truck Wash (Truck World)
- **ATM** **N:** Truck World

(236) **OH Welcome Center - RR, Phones, Picnic (Westbound)**

EXIT OHIO/PENNSYLVANIA

↑OHIO
↓PENNSYLVANIA

(1) PA Welcome Center - RR, Phones, Picnic (Eastbound)

1 **(5)** PA 18, PA 60, Sharon, Hermitage

2 **(15)** U.S. 19, Mercer
- **Gas** **N:** Amoco, Citgo
- **Food** **N:** Howard Johnson, McDonalds
- **Lodg** **N:** (AAA) Howard Johnson
- **AServ** **N:** Amoco, Exit 2 Tire & Service
- **RVCamp** **S:** KOA Campground, The Junction 19-80 Campground
- **Other** **N:** PA. State Police Patrol Post, Tourist Information

2A **(19)** Junction I-79, Erie, Pittsburgh

3A **(24)** PA 173, Sandy Lake, Grove City (Grove City College)
- **Med** **S:** ✚ Hospital
- **Other** **S:** Wendall August Forge

3 **(29)** PA 8, Barkeyville, Franklin
- **TStop** **N:** Phoenix Auto Truck Plaza*; **S:** Kwick Fill*(SCALES), TravelCenters of America*(SCALES)
- **FStop** **S:** Citgo* (Kerosene)
- **Food** **N:** Burger King, Kings Family Restaurant, Phoenix Auto Truck Plaza; **S:** Country Pride (TC of America), Kimberly's Restaurant, Roadhouse Restaurant (Kwik Fill), Subway (TC of America)
- **Lodg** **N:** ◆ Days Inn
- **TServ** **N:** Diesel Injection, Phoenix Auto Truck Plaza; **S:** Brown's Truck Service, Kwik Fill, TravelCenters of America
- **TWash** **S:** Kwik Fill
- **ATM** **N:** Kings Restaurant, Phoenix Auto Truck Plaza
- **Other** **N:** Airport

(30) **Rest Area - RR, Phones, Picnic, Vending (Eastbound)**

(31) **Rest Area - RR, Phones, Vending, Picnic (Westbound)**

4 **(35)** PA 308, Clintonville
- **FStop** **N:** Gulf*(LP)

5 **(42)** PA 38, Emlenton
- **TStop** **N:** Emlenton Truck Plaza*(SCALES)
- **FStop** **N:** Texaco*
- **Gas** **N:** Exxon*(D)(LP)
- **Food** **N:** Emlenton TS, Subway
- **Lodg** **N:** Emlenton Motor Inn
- **TServ** **N:** Emlenton Truck Plaza, Snyder Bros Towing & Service
- **TWash** **N:** Interstate Truck Wash
- **RVCamp** **N:** Gaslight Campground
- **ATM** **N:** Emlenton TS

6 **(46)** PA 478, Emlenton, St Petersburg

7 **(54)** to PA 338, Knox
- **FStop** **N:** Gulf*
- **Food** **N:** B.J.'s Eatery, Wolf's Den
- **Lodg** **N:** Wolf's Den Bed and Breakfast
- **AServ** **N:** Major Brand Tire Company; **S:** GW Dollar, Good Tire Service
- **TServ** **S:** Good Tire Service
- **RVCamp** **N:** Wolf's Camping Resort

EXIT PENNSYLVANIA

7) **Weigh Station (Westbound)**

(60) PA 66 North, Shippenville
- **Gas** N: Citgo*
- **Food** N: Bonanza (see our ad on this page), Citgo
- **ATM** N: Citgo
- **Other** N: Airport, Pennsylvania State Patrol Post

(62) PA 68, Clarion
- **Gas** N: BP*, Exxon*, Kwick Fill*(D)
- **Food** N: Arby's, Burger King, Days Inn, Dominick's Pizza & Subs, Holiday Inn, Long John Silvers, Perkins Family Restaurant, Pizza Hut, Subway
- **Lodg** N: Comfort Inn, Days Inn, ◆ Holiday Inn, AAA Super 8 Motel
- **AServ** N: Western Auto
- **Med** N: ✚ Hospital
- **ATM** N: National City Bank
- **Other** N: Clarion Mall, County Market Grocery (Pharmacy), Exit 9 Car Wash, K-Mart (Pharmacy), Rite Aide Pharmacy, Tourist Information, Wal-Mart

0 **(65)** PA 66 South, Clarion, New Bethlehem (Clarion University)

(71) U.S. 322, Strattanville
- **TStop** N: Keystone Short Way*(SCALES)
- **FStop** N: Exxon*
- **Gas** N: Shell (Keystone)
- **Food** N: Keystone TS
- **TServ** N: Keystone TS
- **ATM** N: Keystone Truckstop
- **Other** N: Fed Ex Drop Box (Keystone)

2 **(73)** PA 949, Corsica
- **Parks** N: Clear Creek State Park (12 Miles)
- **Other** S: U.S. Post Office

3 **(78)** PA 36, Sigel, Brookville
- **TStop** N: TravelCenters of America*(SCALES) (RV Dump)
- **FStop** N: Agway*
- **Gas** N: BP* (TC of America)
  S: BP*, Country Fair*, Exxon*, Sunoco*(LP)
- **Food** N: Country Pride (TC of America), Dairy Queen, KFC, McDonalds, Pizza Hut, Taco Bell (TC of America)
  S: American Hotel, Arby's, Burger King, Days Inn, Plyer's Pizza, Subway, Tony's Lil- Roma Italian Restaurant
- **Lodg** N: Howard Johnson (TC of America), ◆ Super 8 Motel
  S: Days Inn, AAA Gold Eagle Inn, Holiday Inn Express
- **AServ** N: Napa Auto Parts, Stultz Pontiac, Buick, Cadillac
- **TServ** N: TravelCenters of America
- **Med** S: ✚ The Brookville Hospital, ✚ The Charles Medical Center
- **ATM** S: Country Fair, S & T Bank
- **Other** N: Barber Shop (TC of America), Coin Laundry (TC of America)
  S: Coin Car Wash, The Medicine Shop

14 **(81)** PA 28, Hazen

15 **(87)** PA 830, Reynoldsville
- **TStop** S: Diamond J's Truck & Auto Stop(SCALES)
- **Gas** S: Diamond J's*(CW)
- **Food** S: Diamond J's
- **Lodg** S: Diamond J's
- **AServ** S: Diamond J's
- **TServ** S: Diamond J's
- **TWash** S: Diamond J's
- **ATM** S: Diamond J's
- **Other** N: Airport
  S: Coin Laundry (Diamond J's)

(88) **Rest Area - RR, Phones, Picnic, Vending (Westbound)**

16 **(97)** U.S. 219, Du Bois, Brockway
- **TStop** S: Pilot*, Sheetz Truckstop*(SCALES)
- **Gas** S: BP*, Citgo*
- **Food** S: Arby's (Pilot), Dutch Pantry, Holiday Inn, Sheetz
- **Lodg** S: ◆ Holiday Inn, Miller's Motel
- **AServ** S: Citgo, Pennzoil Lube
- **ATM** S: Sheetz
- **Other** N: Allagheny National Forest
  S: Airport, State Patrol Post

17 **(101)** PA 255, Du Bois, Penfield
- **Gas** N: Amoco*
- **Food** N: Jessie's
  S: Ramada Inn
- **Lodg** S: AAA Ramada
- **RVCamp** N: Campground
- **Med** S: ✚ Hospital
- **Other** N: Supermart C-Store
  S: Pennsyvlania State Patrol Post

18 **(111)** PA 153, Pennfield
- **Med** S: ✚ Hospital
- **Parks** N: Parker Dam State Park, S.B. Elliott State Park

19 **(120)** PA 879, Clearfield, Shawville
- **TStop** N: Sapp Brothers*(SCALES)
- **FStop** S: Keystone*
- **Gas** S: Amoco*, BP*
- **Food** N: Aunt Lou's (Sapp Brothers), Best Western, Chester Fried Chicken
  S: Arby's, Burger King, Days Inn, Dutch Pantry, McDonalds, Pickens Ice Cream, Subway (Inside BP)
- **Lodg** N: AAA Best Western, Sapp Brothers
  S: Comfort Inn, AAA Days Inn, ◆ Super 8 Motel
- **AServ** S: Fullington Buick
- **TServ** N: Sapp Brothers
- **Other** S: Wal-Mart (Pharmacy, Vision, Grocery, 24 Hrs)

20 **(123)** PA 970, Woodland , Shawville
- **FStop** S: Pacific Pride Commercial Fueling* (self serve)
- **Other** N: Woodland Campground (1 Mile)

21 **(133)** PA 53, Kylertown, Philipsburg
- **Gas** N: Citgo* (Kerosene), Sunoco*(LP) (Kerosene)
- **Food** N: Kwik Fill TS, Napoli's Pizza (Sunoco)
- **Lodg** N: Kwik Fill TS
- **AServ** N: Citgo
- **TServ** N: Truckstop 21 Garage
- **Med** S: ✚ Hospital
- **Parks** N: Black Moshannon State Park
- **Other** N: Coin Laundry, Hardware Service Star, Mountain View Market, U.S. Post Office

(146) **Rest Area - RR, Phones, Picnic, Vending**

22 **(147)** PA 144, Snow Shoe
- **TStop** N: Snow Shoe 22
- **FStop** N: Exxon*
- **Gas** N: Citgo (Snow Shoe Truck Plaza)
- **Food** N: Snow Shoe 22 Restaurant, Snow Shoe Sandwich Shop
- **AServ** N: Exxon
- **TServ** N: Snow Shoe 22 Truck Service

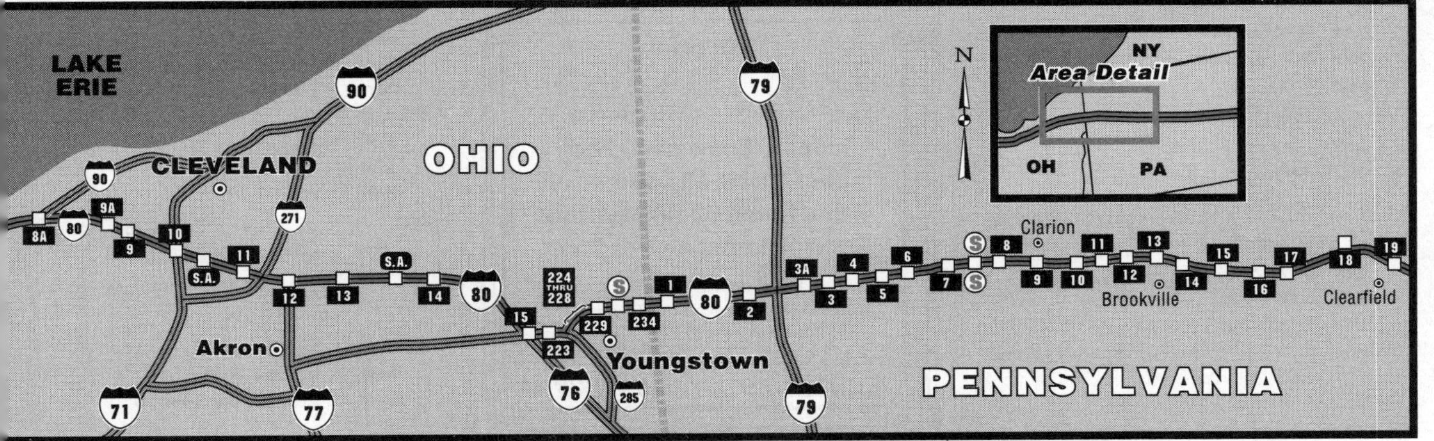

Bold red print shows RV & Bus parking available or nearby

EXIT	PENNSYLVANIA
ATM	N: IGA, Machine (Exxon)
Other	N: IGA Store
23	**(156)** U.S. 220 South, PA 150, Altoona, Milesburg
TStop	N: Best Way*(SCALES), Travel Port*(SCALES)
Gas	N: Amoco*(LP), Mobil*(D)
Food	N: Bald Eagle Restaurant (Travel Port), Best Way Truckstop, Holiday Inn, McDonalds, Subway
Lodg	N: Best Way Motel, ◆ Holiday Inn
AServ	S: Brownson's Auto Center
TServ	N: Best Way
RVCamp	N: Bald Eagle State Park
ATM	N: Travel Port
Other	S: Pennsylvania State Patrol Post
24	**(161)** PA 26, Bellefonte (Penn State University)
Gas	S: Exxon, Texaco*
Food	S: Catherman's, Centre Villa Family Rest.
Lodg	N: Hampton Inn (see our ad on this page)
AServ	S: Exxon, Texaco
RVCamp	N: KOA Campground (2 Miles)
(171)	Picnic (Eastbound)
25	**(173)** PA 64, Lamar
TStop	S: TravelCenters of America*(SCALES)
FStop	N: Texaco*
Gas	N: Gulf S: Citgo*
Food	N: Comfort Inn, McDonalds, The Cottage Family Restaurant S: Country Pride (Truckstop), Sub Express (Truckstop), Subway (Truckstop)
Lodg	N: Comfort Inn, Traveler's Delite Motel
AServ	N: Gulf S: LMR Tires (Citgo)
TServ	S: TravelCenters of America
ATM	N: Texaco
Other	N: Pennsylvania State Patrol Post
26	**(178)** U.S. 220, Lock Haven (Lock Haven University)
Gas	S: Exxon(D), Sunoco(D)
AServ	S: Exxon, Sunoco
27	**(186)** PA 477, Loganton
Gas	N: Gulf S: Citgo
Food	N: Homemade Ice Cream S: Hershey's Ice Cream
AServ	N: Gulf
RVCamp	N: Holiday Pines Campground (2 Miles)
28	**(192)** PA 880, Jersey Shore
TStop	N: Citgo Truck Plaza*
Gas	S: BP*
Food	N: Pit Stop Restaurant (Inside Citgo)
TServ	S: Bressler's Garage Towing and Recovery
Med	S: ✚ Hospital
(194)	Weigh Station, Rest Area - RR, Phones, Picnic, Vending
29	**(199)** Mile Run
30AB	**(210)** U.S. 15, 30A - Lewisburg, 30B - Williamsport
Gas	S: Citgo* (Kerosene), Shell(D)
Food	S: Bonanza
Lodg	S: Comfort Inn
RVCamp	S: Nintony Mountain Campground (5 Miles), Willow Lake Campground
Med	S: ✚ Hospital (U.S.15 South)

EXIT	PENNSYLVANIA
31AB	**(212)** Junction I-180, PA 147, Williamsport, Milton
32	**(215)** PA 254, Limestoneville
TStop	N: Milton 32 Truck Plaza*(SCALES) (RV Dump) S: AmBest Truckstop*(SCALES)
Gas	N: Shell S: Texaco(D)
Food	N: Milton 32 S: AmBest Truckstop
TServ	N: Milton 32 Truck Plaza S: AmBest Truckstop
TWash	S: AmBest Truckstop
ATM	N: Milton 32 Truck Plaza S: AmBest Truckstop
(219)	Rest Area - RR, Phones, Picnic, Vending (Eastbound)
(220)	Rest Area - RR, Phones, Picnic, Vending (Westbound)
33	**(224)** PA 54, Danville
FStop	N: Amoco*(D), Mobil* S: Texaco*
Gas	S: Citgo*
Food	N: Ming's (Howard Johnson), Subway (Howard Johnson), Subway (Amoco) S: Days Inn, Dutch Pantry, Friendly's, McDonalds
Lodg	N: Howard Johnson S: DAYS INN ◆ Days Inn, Red Roof Inn, Travelodge
Med	S: ✚ Hospital (3 Miles)
34	**(232)** PA 42, Buckhorn
TStop	N: Travel Port*(SCALES)
Gas	N: Amoco, Mobil*
Food	N: Buck Horn Restaurant, Burger King, KFC,

EXIT	PENNSYLVANIA
	Perkins Family Restaurant, Quality Inn, Subway (Amoco), Subway (Travel Port), Wendy's, Western Sizzlin'
Lodg	N: AAA Econolodge, AAA Quality Inn
AServ	N: Sear's
TServ	N: Travel Port
Other	N: Columbia Mall
35AB	**(236)** PA 487, Lightstreet, Bloomsburg (Bloomsburg University)
Gas	S: Coastal*(LP)
Food	S: Denny's
Lodg	S: The Inn at Turkey Hill
Med	S: ✚ Hospital
ATM	S: Columbia Country Farmers Nat'l Bank
Other	S: Tourist Information
36	**(241)** U.S. 11, Lime Ridge, Berwick
37	**(242)** PA 339, Mifflinville, Mainville
TStop	N: Brennan's Auto/Truck Plaza*
FStop	N: Gulf* S: Citgo*
Food	N: Brennan's, McDonalds (Brennans), Spur's Steak & Ribs
Lodg	N: ◆ Super 8 Motel
(246)	Rest Area - RR, Phones, Vending, Picnic
(247)	Weigh Station
38	**(256)** PA 93, Conyngham, Nescopeck
TStop	N: Pilot Travel Center*(SCALES)
Gas	N: Sunoco S: Texaco*
Food	N: Subway (Pilot)
Lodg	S: DAYS INN AAA Days Inn
AServ	N: Sunoco
TServ	S: Drumm's Truck Service
RVCamp	N: Council Cup Campground, Moyers Grove Campground
Med	S: ✚ Hospital
38A	**(260)** Junction I-81, Harrisburg, Wilkes - Barre
39	**(263)** PA 309, Mountain Top, Hazleton
TStop	N: Truckstop 39*
FStop	S: Amoco*
Gas	N: Texaco*
Food	N: Kisenwether's Store & Deli, Mountain View Family Restaurant (Truckstop 39)
Lodg	N: AAA Econolodge
AServ	N: Kisenwether's Auto & Truck Repair
TServ	N: Kisenwether's Truck Service
RVCamp	N: KOA Campground (.5 Miles)
Med	S: ✚ Hospital
ATM	N: Truckstop 39
Other	S: State Patrol Post
(270)	Rest Area - RR, Phones, Tourist Info, Picnic (Eastbound)
40	**(273)** PA 940, PA 437, White Haven, Freeland
Gas	N: Mobil*, Shell*
Food	N: Subs Now (Shell)
AServ	S: Schlier's Towing & Auto Service

XIT PENNSYLVANIA

TServ S: Schlier's
VCamp S: Sandy Valley Campground

(274) PA 534
TStop N: Bandit Truckstop*, Hickory Run Plaza*
Gas N: Sunoco*
Food N: Hickory Run Plaza
Parks S: Hickory Run State Park

2 **(277)** PA Tnpk, PA 940
Gas N: Exxon, Shell*, Texaco*
Food N: Arby's, Burger King, David's Pizza Cafe, Howard Johnson, McDonalds
Lodg N: DAYS INN ◆ Days Inn, Howard Johnson, Mountain Laurel Resort, Ramada (see our ad on opposite page)
Other N: Tourist Information

3 **(284)** PA 115, Blakeslee
Gas S: Exxon*, Sunoco(D)
Food S: Pizz-Nut
AServ S: Monaghan's New Tech Auto Towing and Tires, Sunoco
Other N: State Patrol Post

5A **(293)** Junction I-380 North, Scranton

295) Rest Area - RR, Phones, Vending, Tourist Info, Picnic (Eastbound)

44 **(299)** PA611, Scotrun (Westbound Reaccess Exit 45)
Gas N: Sunoco*, Texaco
Food N: Lou's Deli, Scotrun Diner, The Plaza Deli
Lodg N: Scotrun Motel
AServ N: Sunoco, Texaco

45 **(298)** PA 715, Tannersville (Camelback Ski Area)
Gas N: BP*, Mobil*
S: Amoco, Exxon*(CW)
Food N: Billy's Pocono Diner
S: Best Western, Tannersville Diner, Train-Coach Restaurant
Lodg N: Pocono Lodge
S: Best Western
AServ S: Amoco, Schlier
TServ S: Schlier
ATM N: Mellon Bank
Parks N: Big Pocono State Park
Other N: Coin Laundry (BP), The Crossings Factory Outlet
S: Soft Touch Car Wash (Amoco)

EXIT PENNSYLVANIA

46B **(303)** PA 611, Bartonsville
TStop N: 76 Auto/Truck Plaza*(SCALES), Pocono Mountain Travel Center
FStop N: Texaco*(SCALES)
Gas N: Mobil
Food N: Great American Restaurant, Lee's Japanese Restaurant, Pizza Hut, Subway, TCBY Yogurt, Taco Bell
Lodg N: Comfort Inn, ◆ Holiday Inn
AServ N: Gulf
TServ N: 76
TWash N: Texaco

46A **(303)** to South PA 33, to South U.S. 209, Snydersville

47 **(304)** Ninth St, Bushkill (Eastbound, Westbound Reaccess)
Food N: Arby's, Beef and Ale Restaurant, Boston Market, Burger King, Dunkin Donuts, Fat Cat Pizza 2, Fulay Chinese Buffet, McDonalds, Pic'n Lic'n Rotisserie Chicken, Pizza Hut, Taco Rio
AServ N: Abeloff Pontiac, Grey Chevrolet
ATM N: Mellon Bank
Other N: Ames, Stroud Mall

48 **(305)** U.S. 209 Business, Main St
Gas N: Mobil*
S: Texaco* (Kerosene)
Food N: Perkins Family Restaurant
S: Big Star Drive-In (Texaco), India Gate Restaurant (Texaco), Jonny B's Pizza (Texaco), Mommies' Coffee Shop (Texaco)
Lodg N: Pocono Plaza Hotel
S: Colony Motor Lodge
AServ N: Gulf
S: Bud Warners Auto Garage, Cost Auto Service & Muffler, Cottman Transmission
TServ S: Claude Cyphers Truck Parts
Other S: Coin Laundry (Texaco), Steinhauer's Office Supplies and Hardwares

49 **(306)** Dreher Ave (Westbound, Reaccess Eastbound)

50 **(307)** PA 191, Broad St, Park Ave
Gas N: Gulf
S: Sunoco*
Food N: CJ's Drive In, Fu Lay Chinese, KFC, Main Street Bagels and Deli, McDonalds, Panda Kitchen Chinese, Roma Italian Restaurant
S: Angelo's Oyster House, Compton's Pancake House, Days Inn
Lodg N: Best Western, Pocono Inn
S: DAYS INN AAA Days Inn

EXIT PENNSYLVANIA/NEW JERSEY

AServ N: Gulf, Napa Auto Parts
S: Glen Edinger Auto Repair
Med N: ✚ Hospital

51 **(308)** East Stroudsburg
Gas N: Texaco*
Food N: Casa's Pizzeria
Lodg S: AAA Budget Motel, ◆ Super 8 Motel
AServ N: Schlier's Service Center
Med N: ✚ Hospital
ATM N: Texaco
Other N: East Stroudsburg Animal Hospital, Pocono Prescription Center, Wawa Food Market C-Store

52 **(309)** U.S. 209, PA 447, Marshalls Creek
Gas N: Exxon, Gulf
Lodg N: Shannon Inn
AServ N: Exxon, Gulf
RVCamp N: Happy Campers RV Service
Med N: ✚ Hospital
Other N: Mosier's Grocery

53 **(310)** PA 611S, Delaware Water Gap
Gas S: Gulf, Shell*
Food S: Ramada, Water Gap Diner
Lodg S: Ramada
AServ S: Gulf
Other S: Delaware Gap Recreation Area, Tourist Information

↑PENNSYLVANIA

↓NEW JERSEY

(1) Rest Area - Picnic P (Eastbound)

(2) Weigh Station (Eastbound)

4AB NJ 94, Columbia, Portland
TStop N: TravelCenters of America*(SCALES)
Gas N: BP (Truckstop)
Food N: Country Pride (Truckstop), McDonalds, Taco Bell
Lodg N: Daystop Inn
AServ S: Hummel's Garage
ATM N: Instant Teller (TravelCenters of America)

4C NJ 94 North, Blairstown

(7) Welcome Center, Rest Area - Picnic P

(8) Scenic Overlook (Westbound, Cars Only)

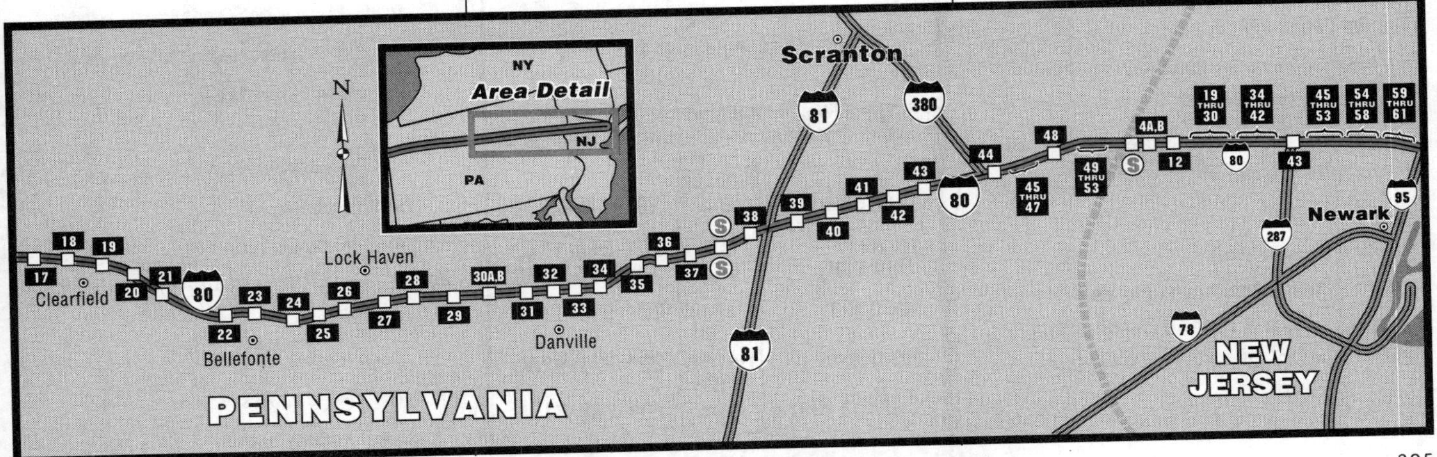

Bold red print shows RV & Bus parking available or nearby

EXIT	NEW JERSEY
12	(10) NJ 521, Hope, Blairstown
RVCamp	S: Campground (see our ad on this page)
Other	N: State Patrol Post
	S: Land of Make Believe
19	NJ 517, Hackettstown, Andover
Med	S: Hospital
(19)	Scenic Overlook (No Trucks, Westbound)
(20)	Rest Area - Picnic P
25	U.S. 206, Newton, Stanhope (International Trade Center, Waterloo Village)
26	U.S. 46, Budd Lake, Hackettstown (Westbound, Reaccess Eastbound)
27	U.S. 206, NJ 183, Netcong, Sommerville
Gas	N: Mobil
	S: Texaco*(D)
Food	N: El Coyoto Mexican Restaurant, Joseph's Family Reataurant
	S: Days Inn
Lodg	S: Days Inn ◆ Days Inn
AServ	N: Mobil
	S: Family Ford
RVCamp	S: Campground
Parks	N: State Park
28	to NJ 10, Ledgewood, Lake Hopatcong
Gas	S: G&N(D), Vantage(D)
Food	S: Cliff's Ice Cream, Days Inn
Lodg	S: Days Inn Days Inn
AServ	S: G&N, Ledgewood Transmission, Meineke Discount Mufflers, Towne Toyota, Vantage
ATM	S: The Bank of New York
Other	S: Car Wash
30	Howard Blvd, Mount Arlington
Gas	N: Exxon(D)
Food	N: Davy's Hot Dog Deli, Four Points Motel (Formerly Sheraton), Sports Authority Restaurant
Lodg	N: Four Points Motel (Formerly Sheraton) (see our ad this page)
AServ	N: Exxon
ATM	N: Summit Bank
(32)	Truck Rest Area - No Vehicles Under 5 Tons P (Westbound)
34	NJ 15, Wharton, Dover, Jefferson, Sparta

EXIT	NEW JERSEY
Gas	N: Exxon(D)
Food	S: Dunkin Donuts, Lorenzo's Pizza, Townsquare Diner, Vincent's Pizza
Lodg	N: Old Lafayette (see our ad on this page)

EXIT	NEW JERS
AServ	N: Exxon
Other	N: Coin Laundry, Rite Aide Pharmacy, U.S. P Office
	S: Car Wash, Coin Laundry
35AB	Mount Hope, Dover
Food	S: Peking Garden Chinese, Sizzler, Villa Napol Restaurant
AServ	S: Sears
Med	N: Hospital
Parks	N: Mount Hope Historical Park
Other	S: Acme, Discovery Zone, Pearle Express Vision Center, Rockaway Mall
37	NJ 513, Hibernia, Rockaway
Gas	N: Exxon(D), Shell
	S: Mobil
Food	N: Hibernia Diner, Howard Johnson (see our a this page)
	S: Fresh Start Deli
Lodg	N: Howard Johnson
AServ	S: Mobil
Med	S: Hospital
ATM	N: The Bank of New York
38	U.S. 46 East, to NJ 53, Denville (Eastbound)
Gas	N: Exxon, Sunoco
	S: Texaco
Food	N: Burger King, Charlie Brown's Restaurant & Lounge
AServ	N: Firestone Tire & Auto, Sunoco
	S: Texaco
Med	N: Hospital
Other	S: Coin Laundry, Denville Animal Hospital
39	U.S. 46, NJ 53, Denville (Westboun Eastbound Difficult Reaccess)
Gas	N: Exxon
Food	N: Burger King, Charlie Brown's, Wendy's
	S: Casa Bella Fine Dining, Hunan Taste
AServ	N: American Car Care, Denville Alignment & Service, Exxon
	S: Firestone Tire & Auto
Med	N: Hospital
Other	N: Denville Police Station
42	U.S. 202, U.S. 46, Parsippany, Mor Plains
Gas	N: Exxon
Food	N: Fuddrucker's, McDonalds, Wendy's
Lodg	N: Days Inn

Bold red print shows RV & Bus parking available or nearby

XIT — NEW JERSEY

AServ	**N:** Exxon
Other	**N: Morris Hill Shopping Center**
3	Jct I-287, Morristown, Boonton, U.S. 46
5	U.S. 46, Whippany, Lake Hiawatha (Eastbound, Reaccess Difficult)
Gas	**N:** Amoco*
Food	**N:** Friendly's, Gourmet Kitchen Chinese, Holiday Inn, International House of Pancakes, McDonalds, Ramada Limited, Wendy's
Lodg	**N:** Holiday Inn, Howard Johnson, Ramada Limited, Red Roof Inn
RVCamp	**N: Campground**
ATM	**N:** Valley National Bank
Other	**N: K-Mart (Pharmacy), Shop Rite**
7A	Jct I-280, Newark, The Oranges
7	U.S. 46 West, Parsippany
Lodg	**N:** Radisson (see our ad opposite page)
8	The Caldwells, Lincoln Park, Montville, Pinebrook (Westbound, Difficult Reaccess)
Food	**N:** Wendy's
	S: Ramada Inn
Lodg	**N:** Holiday Inn, Howard Johnson
	S: Ramada Inn
Med	**N:** ✚ **Chilton Memorial Hospital**
3	U.S. 46, NJ 23, Wayne Clifton, Lincoln Tunnel
Gas	**S:** Sunoco
Food	**S:** Dunkin Donuts, Red Lobster, The Anthony Wayne, Wendy's
Lodg	**S:** ◆ Holiday Inn, Howard Johnson (see our ad opposite page)
AServ	**S:** Speedee Muffler King, Wayne Ford
ATM	**S:** First Union Bank

EXIT — NEW JERSEY

Other	**S: Home Depot**
54	Minnisink Rd, Totowa, Little Falls
Food	**S:** Applebee's, Belmora Pizzaria, Classic Deli, Wo Lee Chinese
ATM	**S:** Great Falls Bank
55AB	Union Blvd, Totowa, Little Falls (Difficult Reaccess)
Gas	**N:** Texaco(D)
Food	**N:** Sponzilli Deli
	S: Holiday Inn, The Bethwood
Lodg	**S:** Holiday Inn
AServ	**N:** Texaco, Totowa Tires
Other	**N:** Valley Car Wash
56AB	Squirrelwood Rd, West Paterson
Gas	**S:** Mobil
AServ	**S:** Mobil
ATM	**S:** PNC Bank
57AB	NJ 19, Main St, Downtown Paterson, Clifton
57C	Main St, Paterson (Westbound, Reaccess Exit 57AB)
58AB	Madison Ave, Patterson, Clifton
Food	**S:** The Madison IV Diner
AServ	**S:** B & C Radiator Service
Med	**S:** ✚ **Hospital**
59	Market St, Patterson (All Traffic Exits Northbound)
AServ	**N:** Mercedes Dealership, Reiman Pontiac

EXIT — NEW JERSEY

60	NJ 20, to U.S. 46, Hawthorne, Passaic
Gas	**N:** Exxon*, Texaco
61	NJ 507, Garfield, Elmwood Park
Gas	**N:** QAC Quality Care
AServ	**N:** QAC Quality Care
62A	To Garden State Pkwy - No Trucks, Saddlebrook
Gas	**N:** Texaco
Food	**N:** Allies, Oasis (The Howard Johnson)
Lodg	**N:** Howard Johnson, Marriott, Ramada (see our ad this page)
AServ	**N:** Texaco
62B	**(63)** to NJ17, to NJ4, Lodi, Fair Lawn
FStop	**N: Hess***
Gas	**N:** Amoco, Citgo, Gulf
Food	**N:** Carmella's Italian, The Rusty Nail
Lodg	**N:** Holiday Inn (see our ad on this page)
Med	**N:** ✚ **Hackensack Medical Center**
63	North NJ17, Rochelle Park
FStop	**N: Hess***
Gas	**N:** Amoco*, Citgo, Gulf*
Food	**N:** Carmella's Italian, The Rusty Nail
AServ	**N:** P&A Auto Parts
Med	**N:** ✚ **Hospital**
64	NJ 17 , to NJ4, Hasbroock Heights, Newark
Food	**N:** The Crows Nest Restaurant and Pub
Lodg	**N:** Crowne Plaza (see our ad opposite page)
65	Green St, Peterboro, South Hackensack
Food	**S:** Club 80
66	Hudson St, Hackensack, Little Ferry
Other	**N: New Jersey Naval Museum**
67	Bogota, Ridgefield Park (Eastbound, Westbound Reaccess)
68A	I-95 South, U.S. 46, New Jersey Tnpk South (toll)
68B	I-95 North, George Washington Bridge, New York
70	U.S. 93, Leonia
70B	Teaneck
Lodg	**N:** Marriott
71	Broad Ave, Leonia, Inglewood

↑NEW JERSEY

Begin I-80

EXIT	NEW YORK

Begin I-81

↓NEW YORK

52 **(183)** De Wolf Point, Island Road

51 **(180)** Island Road, Fineview
- **Gas** E: Citgo*
  W: Fineview*
- **Food** E: Island Cafe
- **Lodg** W: Torch Light Motel
- **RVCamp** **E: Campground**
  W: Campground
- **Parks** **E: DeWolf State Park**
  W: Golf Course State Park, Wellsly State Park
- **Other** **E: U.S. Post Office**

(178) NY Welcome Center - RR, Phones (Southbound)

50 **(178)** NY 12, Clayton, Ogdensburg
- **FStop** **E: Mobil***
  W: Citgo*
- **Food** E: Bonnie Castle Downs Restaurant & Resort, Kountry Kottage, Subway
  W: Yalvell's Restaurant
- **Lodg** E: Green Acres River Motel & Cottages, PineHurst
  W: Bridgeview Motel, RiverEdge Resort
- **AServ** W: NAPA Auto Parts
- **Med** **E: ✚ Hospital**
- **Other** **E: Tourist Information**

(174) Rest Area - RR, Phones (Northbound)

49 **(169)** NY 411, Theresa , LaFargeville
- **FStop** **E: Sunoco***
  W: Exit 49 Truckstop
- **Food** **W: Exit 49 Diner**
- **AServ** E: Sunoco

(167) Rest Area - Picnic (Southbound)

(161) Rest Area - Picnic (Northbound)

48 **(158)** U.S. 11, NY 37, NY 342, Black River , Fort Drum
- **TStop** **E: Long Ways*(D)**
- **FStop** **E: Sunoco*(D)**
- **Gas** E: Citgo*(D)(LP)
- **Food** E: Long Ways Diner
- **Lodg** **E: Allen's Budget Motel, Long Ways Motel**
- **AServ** E: Citgo
- **TServ** **E: Bandag Tire**
- **Other** **E: NY State Police**

(156) Parking Area

47 **(155)** NY 12, Bradley St
- **Gas** E: Citgo(LP)
- **Lodg** W: Motel 47, The Maples Motel
- **TServ** **E: Walsh Equipment**
- **Med** **E: ✚ Hospital**

46 **(154)** NY 12F, Coffeen St
- **FStop** **E: Mobil***
- **Gas** W: Citgo*(LP)
- **Food** E: Chappy's 50 Classic Diner, Cracker Barrel
- **TServ** **W: Beam Mac Trucks**
- **ATM** W: Watertown Savings Bank

45 **(153)** NY 3, Arsenal St
- **Gas** E: BP*, Citgo*, Mobil*, Sunoco*(CW)
- **Food** E: Apollo Restaurant, Arby's, Benny's Steak House, Bickfords Family Restaurant, Burger King, China Cafe, Denny's, Dunkin Donuts, Friendly's,

EXIT	NEW YORK

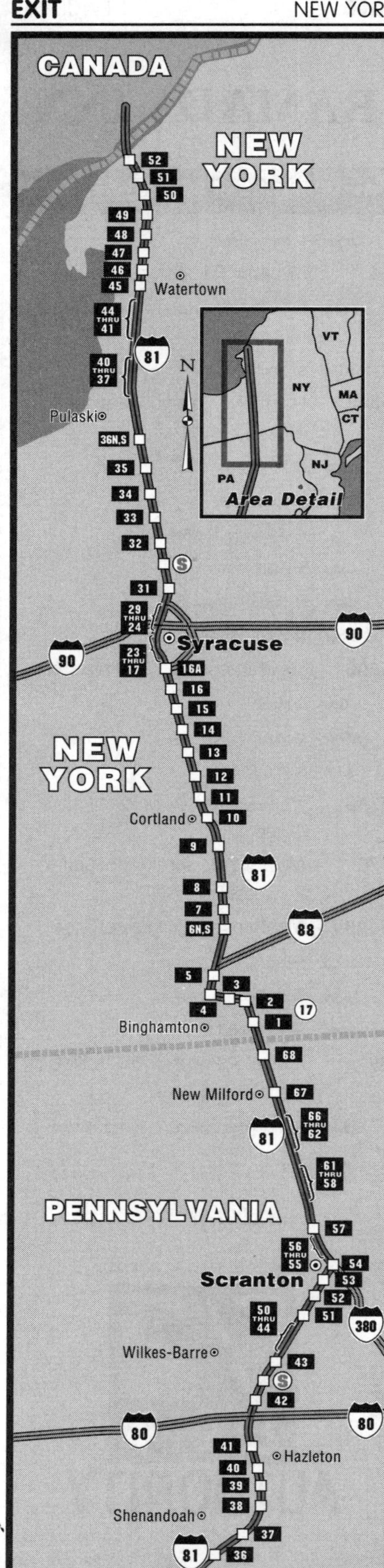

EXIT	NEW YOR

KFC, Long John Silvers, McDonalds, Ponderos Steakhouse, Smiley's Bagel Bites, Taco Bell, Thialand Restaurant, Wendy's
W: Ann's Restaurant, Bob Evans Restaurant, R Lobster
- **Lodg** E: Arsenal Street Motel, ◆ Days Inn, Econolodge, Quality Inn
  W: Ramada (see our ad on this page)
- **AServ** E: Auto Palace, Coal Muffler, Kost Tires, Midas Muffler, Monroe Muffler & Brake, Parts Americ Pennzoil Lube
  W: Auto Express
- **Med** **E: ✚ Hospital**
- **ATM** E: Community Bank, Marine Midland Bank
  W: Key Bank
- **Other** **E: Alde's Supermarket,** Coin Car Wash, **Ecker Drugs, Empire Vision Center, Pharmhouse Discount Pharmacy, Post Office, Price Chop Supermarket, State Way Plaza**
  W: Hannaford Supermarket, K-Mart, Salmo Run Mall, Sam's Club (Tire Center), Wal-Mar

(149) Parking Area - Phones

44 **(148)** NY 232, Watertown Center
- **Med** **E: ✚ Hospital**

(147) Rest Area - RR, Phones, Picnic (Southbound Closed)

43 **(146)** U.S. 11, Kellogg Hill

42 **(144)** NY 177, Smithville, Adams Center
- **FStop** **E: Mobil(D)(LP)**

41 **(141)** NY 178, Adams, Henderson
- **FStop** **E: Citgo*(LP)**

40 **(135)** NY 193, Ellisburg, Pierrepont Manor
- **Parks** **W: Southwick Beach State Park**

(134) Parking Area

39 **(133)** Mannsville

38 **(131)** U.S. 11
- **Lodg** E: 81-11 Motel

37 **(128)** Sandy Creek, Lacona
- **FStop** **W: Gas Mart***
- **Gas** W: Citgo
- **Food** W: Cathy's Deli, Nelson's Cafe, The Snackery
- **Lodg** **W: Angler's Roost Bed/Breakfast**
- **AServ** W: Citgo, Northern Repair & Rebuilding, Ontari Tire Service
- **RVCamp** **W: Angler's Roost Campground, Colonial Campground**

Bold red print shows RV & Bus parking available or nearby

XIT NEW YORK

Other W: Coin Laundry, Sandy Creek Market[LP], U.S. Post Office

6 **(122)** NY 13, Pulaski (Difficult Reaccess)

FStop E: Agway*

Gas E: Citgo*
W: Citgo*[LP], Mobil*[LP]

Food E: C & M Diner, Dunkin Donuts, Ponderosa
W: Arby's, Burger King, McDonalds

Lodg E: Redwood Motel, Simon Acres Lodge, Whitaker's Sport Store & Motel[LP]
W: Pulaski Country Pizza & Motel, Salmon Country Motel, Super 8 Motel

AServ W: NAPA Auto Parts

Other E: Rite Aide Pharmacy
W: Kinney Drugs, Selkirk Shores State Park, Tourist Information

5 **(118)** to U.S. 11, Tinker Cavern Road

4 **(115)** NY 104, Mexico

TStop E: Ezze Auto and Truckstop[LP]

Food E: Ezze Restaurant (Ezze)

RVCamp W: KOA Campground

ATM E: OnBank (Ezze)

3 **(111)** NY 69, Parish

FStop E: Sunoco*

Gas W: Citgo*[LP], Mobil*

Food E: Grist Mill

RVCamp E: East Coast Campground, UpCountry Campground

Other E: Wayne's Drugs

2 **(103)** NY 49, Central Square

TStop E: Penn-Can Truckstop*[LP]

FStop E: Citgo*
W: BP*

Food E: Wilborn's
W: Arby's, Burger King, Sundae Best Ice Cream

AServ W: Ford Dealership, NAPA Auto Parts

Other W: IGA Store[LP], Rite Aide Pharmacy

01) Rest Area - RR, Phones, Picnic, Vending (Southbound)

00) Weigh Station (Southbound)

1 **(99)** Bartell Road, Brewerton

Gas W: Citgo*, Mobil*[D] (ATM)

Food W: Burger King, Jay R. Restaurant, Little Caesars Pizza, McDonalds, Sam's Lakeside Restaurant, Subway

ATM W: Fleet Bank

Other W: Brewerton Plaza, Kinney Drugs, Purrfect Pet, Smith's Supermarket (ATM), U.S. Post Office

0 **(95)** NY 31, Cicero

FStop E: Sunoco*

Gas E: Hess*[D], Mobil[D]
W: Kwik-Fill*

Food E: Arby's, Cracker Barrel, Dickenson's Pizza, Dunkin Donuts, Fish Haven, Genio's & Jay Pizza, Great Wall Chinese, McDonalds, Subway

AServ E: Dan's Auto Repair, Mobil, Sunoco

RVCamp W: Campers by Ganlen

ATM E: Skaneatoes Bank

Other E: Casey's Drugs, Hometown Supermarket
W: Byrne C-Store

9 **(93)** Jct I-481, NY 481, Dewitt, Oswego

8 **(91)** Taft Road, North Syracuse

Gas E: Sunoco*[LP]
W: Hess*[D], Mobil*[CW]

Food W: Fresno

AServ E: Ballard's Auto Repair, Lonze's Radiator, Lots Fine Auto Repair
W: Buick Mitsubishi Isuzu Dealer, Cole Muffler, Miles Auto Supply

EXIT NEW YORK

Other E: Eckerd Drugs
W: U.S. Post Office

27 **(91)** Syracuse Airport

26 **(89)** U.S. 11, Mattydale

Gas E: Mobil*, Sunoco*
W: Gulf*[D], Kwik Fill, Sunoco

Food E: Doug's Fish Fry, Friendly's, Hunan Chinese, Peking Duck House, Pizza Hut, Zebb's Grill
W: Applebees, Arby's, Burger King, Carnikee Cafe, Carvelle Ice Cream Bakery, Crittela's, DeWitt Bakery, Denny's, Fresno's Southwest, McDonalds, Ponderosa, Subway, Taco Bell, Wendy's

Lodg E: Alma Motel
W: Airflite Motel, Royal Motel

AServ E: Auto Parts America, Courtesy Ford, Goodyear Tire & Auto, K-Mart, Mobil, Valvoline Quick Lube Center
W: AAMCO Transmission, Kost Tire, Midas, Roger Burdick Mazda, Sunoco, Toyota

ATM W: Fleet Bank, Skaneateoes Bank

Other E: Coin Car Wash, Eckerd Drugs, K-Mart, Mattydale Animal Hospital, Northern Lights Shopping
W: Eckerd Drugs, Empire Vision Center, P&C Grocery Store, Sweetheart Market

25A **(88)** JctI-90, Rochester, Albany, NYTHWY

25 **(88)** 7th St, North St, Liverpool

TStop E: Pilot Travel Center*[SCALES]

Gas W: Mobil* (ATM), Sunoco

Food E: KFC (Pilot), Subway (Pilot)
W: Bob Evans Restaurant, Burger King, Colorado Mine Steak House, Denny's, Friendly's, Italian Carry-Out, Jreck Subs, The Ground Round

Lodg W: DAYS INN Days Inn, Econolodge, Hampton Inn, Holiday Inn, Knights Inn (see our ad this page), ◆ Quality Inn, Ramada (see our ad on this page), Super 8 Motel

EXIT NEW YORK

AServ W: Sunoco

TServ E: Cummins Diesel, Hino Diesel Trucks

TWash E: Express Truck Wash

ATM W: Chase Bank

24 **(87)** Liverpool, NY 370W

23 **(86)** Hiawatha Blvd

Gas W: Hess*[D]

Food E: Ball Park Deli

ATM E: Key Bank

Other W: Carousel Center Mall

22 **(86)** NY298, Court St

21 **(86)** Spencer St, Cataba St (Southbound)

20 **(85)** Franklin St, West St (Southbound)

19 **(85)** Clinton St, Salina St (Southbound)

(85) Jct I-690, E Syracuse, Fair Grounds, Baldwinsville

18 **(84)** Harrison St, Adams St

Med E: ✚ Hospital

17 **(82)** South Salina St, Brighton Ave

Gas W: Kwik Fill*, Mobil

Med W: ✚ Hospital

Other W: Expressway Market C-Store

16A **(81)** Junction I-481N, DeWitt

16 **(78)** U.S. 11, Onondaga Nation Territory

Other W: Freem's Place C-Store

15 **(73)** U.S. 20, LaFayette

Gas E: Citgo[D], Mobil*

Food E: LaFayette Diner, Old Tymes Cafe, Quinto's NY Pizza
W: McDonalds

AServ E: LaFayette Garage, Mike Amidon's Auto Service, NAPA Auto Parts

ATM E: OnBank

Other E: LaFayette IGA Supermarket, State Patrol Post, U.S. Post Office

(70) Parking Area - Phones

14 **(66)** NY 80, Tully

FStop E: Mobil* (ATM)

Food E: Best Western
W: Burger King

Lodg E: AAA Best Western

ATM E: First National Bank of Courtland

13 **(63)** NY 281, Preble

FStop E: Mobil*

AServ E: Mobil, Walburgers Service (Towing)

TServ E: Walburgers Service (Towing)

(60) Rest Area - RR, Phones, Picnic, Vending

12 **(54)** U.S. 11, NY 281, Homer (Closed)

Med W: ✚ Hospital

11 **(52)** NY 13, Cortland

Gas W: Mobil*

Food E: Denny's
W: Arby's, Bob Evans Restaurant, Friendly's, Little Caesars Pizza, McDonalds, Panda Wok, River Junction, Subway, Taco Bell, W.B. Goff Ice Cream, Wendy's

Lodg E: ◆ Comfort Inn, ◆ Super 8 Motel
W: Holiday Inn

AServ W: AA 1 Transmission & Brakes, Jiffy Lube, Kost Tire, Mobil, Parts America

Other E: NY State Police, Tourist Information
W: Eckerd Drugs, P & C Grocery, Riverside Shopping Plaza

10 **(50)** U.S. 11, NY 41, Cortland, McGraw

EXIT NEW YORK

FStop W: Agway[LP], Citgo*, Mobil*, Sunoco*
Food W: Burger King, Little Treat Shop, Lori's Diner, Suburban Skyliner Diner, Subway (Mobil)
Lodg W: Econolodge
AServ W: Cooper Tires, Courtland Saab

(48) Rest Area (Northbound, Closed)

(44) Parking Area - Picnic (Northbound)

9 **(40)** U.S. 11, NY 221, Marathon
Gas W: Citgo (Gregg's), Sunoco*
Food W: Kathy's Diner, New York Pizzeria, Taco Bell (Sunoco)
Lodg W: Three Bear Inn
AServ W: Napa Auto Parts
RVCamp W: Country Hill Campground
ATM W: First National Bank of Courtland
Other W: Coin Car Wash, Coin Laundry, Gregg's Supermarket, Pet Connection, State Police, US Post Office

(33) Rest Area - RR, Phones, Picnic, Vending (Southbound)

8 **(31)** to U.S. 11, NY 26, to NY 79, Whitney Point, Lisle
Gas E: Hess*, Kwik Fill*[LP], Mobil*[D]
Food E: Aiello's Restaurante, Arby's, Country Kitchen Diner, Subway, The Sundae Shoppe
Lodg E: (AAA) Point Motel
AServ E: Napa Auto Parts, Parts Plus
RVCamp E: Campground
W: Campground
ATM E: Chase Bank (Hess), First National Bank of Courtland
Other E: Car Wash, Eckerd Drugs, Greg's Supermarket[LP]

7 **(21)** U.S. 11, Castle Creek
FStop W: Citgo*
Gas W: Mobil*
AServ E: Sickel's Garage

6B **(17)** U.S. 11, NY 12, Chenango Bridge, to Jct I-88 (Southbound)

6A **(16)** U.S. 11, Norwich, NY12, Chenango Bridge (Northbound)
FStop E: Hess*
Gas E: BP*[D], Citgo*, Hess*[D], Kwik Fill
W: Mobil*
Food E: Burger King, Denny's, Dunkin Donuts, Empire Wok, Grande Pizza, Kelly's Ice Cream, McDonalds, Morrie's, PapaG's, Pizza Hut, Ponderosa, Spiedie & Rib Pit, Subway, The Bull's Head, Wendy's
W: Casey's, Chicken Chalet, Friendly's, International Deli, McDonalds, Nirchi's Pizza, Spot's Diner, Subway, The Indian Restaurant
Lodg E: (DAYS INN) Days Inn, Motel 6
W: Banner Motel, Comfort Inn (see our ad on this page)
AServ E: Cory North Gate Ford, Mazda, Jiffy Lube, Kost Tire, Midas Muffler & Brakes, Monroe Muffler & Brake, Valvoline Quick Lube Center
W: O'Shay Auto Parts
ATM E: BSB Bank & Trust, Marine Midland Bank, NBT Bank of Binghamton
W: Vision Federal Credit Union
Other E: CVS Pharmacy, Eckerd Drugs, Empire Vision Center, Giant Supermarket, Northgateway Speedway, OptiKal Eye, Pet World
W: Auto Magic Car Wash, Cars R Us, Coin Car

EXIT NEW YORK

EXIT NEW YORK/PENNSYLVAN

Wash, Nimmonsburg Square, Taylor Rental Center[LP]

(15) Jct I-88E (Northbound)

5 **(14)** U.S. 11, Front St
Gas W: Mobil*[D]
Food W: Howard Johnson
Lodg W: Howard Johnson (see our ad on this pag
Super 8 Motel, Ramada Inn (see our ad on this page)

4 **(12)** NY 7, Binghamton, Hill Crest
TServ E: Cook Brothers Truck Parts, Driveline Repai

3 **(12)** Broad Ave (Northbound, Southbound Reaccess Only)
TStop W: Travel Port*[D][SCALES]
Gas W: BP*, Citgo*[D]
Food W: Country Bob's, Subway
Lodg W: (AAA) Super 8 Motel (see our ad on this page)
AServ W: Citgo[LP]
TServ W: Universal Joint Sales
TWash W: Travel Port
Other W: Car Wash

2 **(8)** U.S. 11, NY 17E , Industrial Park, New York City
ATM W: Chase Bank
Other W: Schneider's Supermarket

1 **(4)** U.S.11, NY 7, Kirkwood, Conklin
Gas W: Mobil*[D]
Other W: State Patrol Post

(2) NY Welcome Center - RR, Phones, Picnic (Northbound)
Other E: State Patrol Post

(0) Weigh Station (Northbound)

↑NEW YORK

↓PENNSYLVANIA

68 **(231)** PA 171, Great Bend, Susquehanna
FStop W: Exxon*
Gas E: Mobil*
W: Sunoco[D]
Food W: Arby's, Beaver's Restaurant, Dobb's Countr Kitchen, McDonalds, Subway, Tedeschi's Pizza Pasta
Lodg W: (AAA) Colonial Brick Motel
AServ W: B&D Auto Center, Sunoco
RVCamp E: Lakeside Campground
ATM W: CoreState Bank, People's Bank
Other W: Coin Car Wash, Coin Laundry, Insalaco's Grocery Store, Reddon's Rexall Drugs, Rob's Big M C-Store

67 **(224)** PA 492, New Milford, Lakesid
Gas W: Shell*
Lodg W: All Season Motel
RVCamp W: All Season Campground[LP]

66 **(219)** PA 848, Gibson
TStop E: Amoco*[SCALES]
Food E: Amoco Truckstop
TServ E: Gibson Truck & Tire Service
RVCamp E: April Valley Campground
Other E: State Patrol Post

65 **(217)** PA 547, Harford
TStop E: Liberty Auto/Truck Stop* (ATM), Penn-Can Truckstop*
Food E: Liberty Auto, Penn-Can Truckstop (ATM)

64 **(211)** PA 92, Lenox
FStop W: Exxon*, Texaco*

Bold red print shows RV & Bus parking available or nearby

XIT PENNSYLVANIA

Food W: Bingham's Restaurant, Lenox Dairy Bar, Lenox Restaurant, Mam's Bakery
ATM W: CoreState Bank
Other W: Lenox Pharmacy
208) PA Welcome Center - RR, Vending, Phones (Southbound)
3 **(206)** PA 374, Glenwood, Lenoxville
Med E: Hospital
203) **Rest Area - RR, Phones, Visitor Info, Picnic (Northbound)**
2 **(202)** PA 107, Fleetville, Tompkinsville
1 **(200)** PA 438, East Benton
Food W: East Benton Dairy Bar, Restaurant, The North 40
AServ W: Exxon (Towing, ATM)
0 **(199)** PA 524, Scott
TStop E: Amaco* (ATM), Scott 60 Plaza*
Food E: Scott 60 Plaza
W: Elaine's Cafe & Deli
Lodg W: ◆ Motel 81
9 **(194)** PA 632, Waverly
Gas W: Sunoco*
Other E: Mr. Z's Food Mart, Rite Aide Pharmacy
8 **(194)** U.S. 6, U.S. 11, PA Tnpk, Clarks Summit
Lodg E: Ramada Plaza Hotel (see our ad on this page)
7 **(192)** U.S. 6 East, U.S. 11 South, Carbondale
Food E: American Buffet, Arby's, Burger King, China Palace, Denny's, Donut Connection, Fresno, Grace's Hogie, Heavenly Ham, Long John Silvers, Long Star Steakhouse, McDonalds, Perkins Family Restaurant, Pizza Chef, Pizza Hut, Ponderosa, Rita's Italian Ices, Ruby Tuesday, TCBY, Wendy's
AServ E: Firestone Tire & Auto, Kmart, Sears
Med W: Hospital
ATM E: First National Community Bank, PNC Bank
Other E: Kmart, Mall, Optical, Petco
6 **(190)** Dickson City, Main Ave
Gas E: Gulf*
5 **(188)** PA 347, Blakely St, Throop
Gas E: Sunoco*(D)
W: Mobil*
Food E: Big Boy, Boston Market, China World Buffet, Donut Connection, McDonalds, Wendy's
W: Burger King, Friendly's
Lodg E: Days Inn, Dunmore Inn, Econolodge
AServ E: Midas Muffler & Brakes, Monroe Muffler, Parts America
W: Bour's Motor Mazada
TServ E: Kenworth Motor Co.
ATM E: CoreState Bank, Fidelity Bank
Other E: CVS Pharmacy, Price Chopper Supermarket, State Patrol Post
W: Mose Taylor Hospital
54 **(187)** Jct I-84, I-380, PA 435, Drinker St (No Northbound Reaccess)
Gas W: Citgo(D)
Food W: Kuzzins Cafe, Tallo's Ristorante
AServ W: Citgo
53 **(184)** Central Scranton Expwy
52 **(184)** PA 307, River St, Moosic
Gas W: Citgo*(LP), Exxon*, Shell*, Texaco(D)
Food W: Chick's Diner, House of China, Profera's Pizza, Ramada Inn, Stampy's Pizza and Hoagies
Lodg W: Ramada
AServ W: Jiffy Lube, Texaco
Med W: CMC Trauma Service
ATM W: CoreStates Bank, Penn Security Bank
Other W: CVS Pharmacy, Gerrity Supermarket
51 **(182)** Montage Mtn Road, Davis St

EXIT PENNSYLVANIA

(Lackawanna Stadium)
FStop W: Sunoco*(D)
Gas W: Exxon*
Food E: Marvelous Muggs Restaurant and Pub
W: Econolodge, Valentino's Restaurant and Lounge
Lodg E: Comfort Inn, Courtyard Inn, ◆ Hampton Inn
W: AAA Econolodge
Other E: Lackawanna Coal Mine Tour, PA Anthrosite Heritage Museum
W: U.S. Post Office
50 **(179)** U.S. 11, PA 502, Moosic (Northbound Left Exit, Difficult Reaccess)
Gas E: Mobil(D)
W: Shell*(D)
Food E: Grande Pizza, Joey's Pizza
Lodg E: Days Inn, Trotters Motel

HA1850
Hampton Inn
717-342-7002
Montage Mountain Rd. & Davis St.
Scranton, PA
Complimentary Continental Breakfast
Indoor Pool & Fitness Center
Free HBO & Free Local Calls
Complimentary Shuttle Service
Gameroom & Complimentary Beverage Bar
Work Center in Lobby
100% Satisfaction Guarantee
PENNSYLVANIA • I-81 • EXIT 51

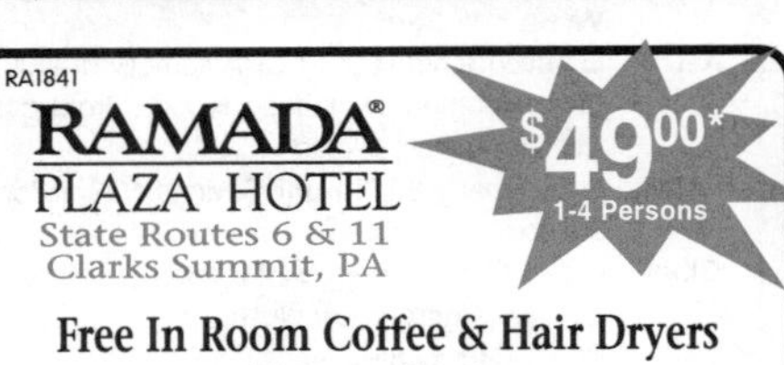

RA1841
RAMADA®
PLAZA HOTEL
State Routes 6 & 11
Clarks Summit, PA
$49.00*
1-4 Persons
Free In Room Coffee & Hair Dryers
Fitness Center
Free Local Phone Calls
Restaurant & Cocktail Bar
Close to Skiing Elk & Montage Mountain
Historic Steamtown National Park
*Show this ad at check-in to receive the special rate.
Based on availability. Not valid with any other discounts.
717-586-2730 • Toll Free Reservations: 1-800-2-RAMADA
PENNSYLVANIA • I-81 • EXIT 58

Howard Johnson
I-81 • Exit 48
Pittston, PA
717-654-3301
• Meeting Room Up to 300
• Non-Smoking Rooms
• Outdoor Pool • Restaurant
• Cable TV (Free HBO)
• Free Continental Breakfast
• Guest Coin Laundry
• Free Local Calls
HO1864
PENNSYLVANIA • I-81 • EXIT 48

EXIT PENNSYLVANIA

AServ E: Domiano Auto Service, Jack Williams Tire and Auto Service
49AB **(177)** Avoca, Wilkes Barre, 49A - Wilkes Barre Airport, 49B - Avoca
FStop W: Petro*(D)(SCALES)
Food E: Damon's Ribs (Holiday Inn Express)
W: Iron Skillet*(SCALES) (Petro)
Lodg E: Holiday Inn Express
ATM W: Petro
48 **(176)** PA Tnpk, PA 9, PA 315, Dupont, Clark Summit, Allentown
TStop W: Skyliner Truck Plaza*(SCALES)
FStop W: Shell*
Gas E: Exxon, Mobil*, Sunoco
Food E: Bonanza Steak House, Howard Johnson
W: Slyliner TS
Lodg E: Howard Johnson (see our ad on this page), Knight's Inn
W: Skyliner Truck Plaza, Victoria Inn
AServ E: Abraham Geo, Exxon, Sunoco
W: Goodyear Tire & Auto, Vullo Auto Repair
TServ W: Goodyear Tire & Auto, Skyliner, Vullo Truck Repairs
TWash W: Skyliner
ATM E: Core States Bank
47AB **(170)** PA 115, Bear Creek, Wilkes Barre, 47A - 115 S, 47B - 309 N
Gas E: Amoco*
Food E: Carriage Stop Inn (Family Restaurant)
Lodg E: Best Western, Melody Motel
W: Holiday Inn
Med E: Hospital (1 Mile)
W: VA Medical Center/General Hospital
45 **(165)** PA 309, Mountain Top, Wilks Barre (Exits Left Northbound)
Gas W: Shell*(CW), Shell(D), Texaco
Food W: Dunkin Donuts, Mark 2 Restaurant, McDonalds, Milazzo's Pizza Subs and Wings, Perkins Family Restaurant, Taco Bell, Tin Tin Exotic Chinese Food
AServ W: Auto Parts America, K-Mart, M.J. Auto Service, Monroe Muffler and Brake, Texaco
TServ W: M.J. Truck Service
ATM W: PNC Bank
Other W: K-Mart (Pharmacy), Thrift Drug Pharmacy
44 **(165)** PA 29, Nanticoke
43 **(159)** Nuangola
Gas W: Shell*
Food W: Godfather's Pizza (Shell Gas Station)
RVCamp W: Counsel Cup Camp Grounds
ATM W: Shell
(158) **Rest Area - RR, Phones, Vending, Picnic (Southbound)**
(157) **Weigh Station (Southbound)**
(156) **Rest Area - RR, Phones, Vending, Picnic (Northbound)**
(155) **Weigh Station (Northbound)**
42 **(155)** Dorrance
FStop W: Blue Ridge Plaza*(D)
Gas E: Sunoco(D)
AServ E: Sunoco(D)
RVCamp E: KOA Campground (2 Miles)
W: Moyer's Grove Campground & Country RV (7.3 Miles)
(151) Junction I-80, Bloomsburg, Stroudsburg
41 **(145)** PA 93, West Hazleton, Conyngham (Penn State University - Hazelton Campus)

Bold red print shows RV & Bus parking available or nearby

EXIT PENNSYLVANIA

Gas E: Mobil, Sunoco*(D)
Food E: Perkins Family Restaurant, Rossi's
W: Hampton Inn
Lodg E: ◆ Comfort Inn, AAA Forest Hill Inn
W: ◆ Hampton Inn
AServ E: Mobil
Other E: Airport, State Patrol Post

40 **(143)** PA 924, Hazleton (Northbound)
Gas W: Texaco
Food E: Carmen's Family Restaurant
Lodg E: Hazelton Motor Inn

39 **(138)** PA 309, McAdoo , Tamaqua

38 **(134)** Delano

(132) Parking Area

37 **(131)** PA 54, Hometown, Mahoney City
Gas W: Citgo*, Texaco*(D)
AServ W: Texaco
Parks E: State Park

36 **(124)** PA 61, Frackville, St Clair (Must Take Mall Rd To East Services)
FStop W: Gulf*(LP)
Gas W: BP(D), Exxon*(LP), Getty*(LP), Shell*(CW)
Food E: McDonalds
W: Blimpie's Subs (Getty Gas Station.), Cesari's, Dutch Kitchen Restaurant, Hardee's, Subway, The Pizza Place
Lodg W: Central Hotel (AAA), Econolodge, Granny's Motel
AServ E: A&A Auto Stores, K-Mart, Sears
W: Goodyear Tire & Auto, Rinaldi's Dodge
Med W: ✚ Hospital
ATM E: First Federal Savings
W: Core States Bank, Getty, Pennsylvania National Bank
Other E: K-Mart, Phar Mor Drugs, Schuylkill Mall, Weiss Markets
W: Ace Hardware, Rite Aide Pharmacy, State Patrol Post, Thomas Animal Hospital

35 **(116)** PA 901, Minersville
Food E: 901 Pub and Restaurant
Other E: Airport (1.5 Miles)

34 **(112)** PA 25, Hegins
RVCamp W: Camp-A-While

33 **(107)** U.S. 209, Tremont, Tower City, Pottsville

32 **(104)** PA 125, Ravine
FStop E: Mobil*
Food E: Raceway Family Restaurant
ATM E: Mobil

31 **(100)** PA 443, Pine Grove
TStop W: All American Ambest Truckstop*(SCALES)
FStop W: Texaco*
Gas E: Citgo*
Food E: Arby's, McDonalds, Pizza-Beer Restaurant, Ulsh's Family Restaurant
W: All American Family Rest. (All American TS), Subway (All American TS)
Lodg E: Colony Lodge, ◆ Comfort Inn, AAA Econolodge
AServ E: Citgo
W: Motters Wrecker Service
TServ W: All American Truck Plaza*

30 **(91)** PA 72, Lebanon
Gas E: Exxon*(D), Shell*(D)(LP)
AServ E: Exxon
RVCamp E: KOA Campground (5 Miles), Lickdale Campground

(89) Junction I-78E, Allentown

29AB **(85)** PA 934, Annville, Fort Indiantown Gap (29A- PA 934 South Annville 29B- Fort Indiantown Gap)
Gas W: Texaco*(D)
Food E: Harper's Tavern Restaurant, Swatara Creek Inn Bed and Breakfast
W: Funck's Family Restaurant

EXIT PENNSYLVANIA

Lodg E: Swatara Creek Inn Bed and Breakfast
RVCamp E: Campground
Parks W: Indiantown National Cemetery, Memorial Lake State Park

28 **(80)** PA 743, Grantville, Hershey
Gas E: Mobil*(D)
W: Exxon*, Texaco*(LP)
Food W: Holiday Inn
Lodg E: AAA Econolodge, Hampton Inn
W: ◆ Holiday Inn

(79) Weigh Station both directions

(78) Rest Area - RR, Phones, Vending, Picnic, Weather Info

27 **(77)** PA 39, Hershey, Manada Hill
TStop E: Travel America
W: 76 Auto/Truck Plaza(SCALES), Travel America, TravelCenters of America*(SCALES)
FStop E: Pilot Travel Center*(SCALES)
W: Shell*(LP)(SCALES)
Gas E: BP*, Texaco
Food E: Travel America Truck Stop
W: 76 Auto/Truck Plaza, Country Pride (Truckstop)
Lodg E: AAA Sleep Inn
W: Daystop Inn (TravelCenters of America)
AServ E: Texaco
TServ W: 76 Auto/Truck Plaza, Goodyear Truck Service, Travel America
TWash W: Travel America

26AB **(72)** Paxtonia, Linglestown, A - US 22, B - Linglestown
Gas E: Citgo(D), Hess*, Sunoco, Texaco*(D)
Food E: Burger King, Dutch Pantry, Geo's Family Restaurant, Great Wall Chinese, Malley's, McDonalds, The Harrisburg Inn, The Sandwich Man
W: Country Oven (Best Western)
Lodg E: Budgetel Inn, Reese's Motel
W: Best Western
AServ E: Advance Auto Parts, C&P Automotive Repair, Citgo, Hartman Toyota Jeep, Monroe Brake and Muffler, Sunoco, Texaco
ATM E: Commerce Bank, Harris Savings, Mellon Bank, Sunoco
Other E: CVS Pharmacy (Open 24 Hours), Festival Grocery, Hartman Car Wash
W: Ace Hardware

(70) Junction I-83 South, U.S. 322, PA Tnpk, Hershey, York

24 **(69)** Progress Ave
Gas E: Gulf(D)
Food W: Best Western, Red Roof Inn, Your Place
Lodg W: AAA Best Western, Inn of the Dove Luxury Motel, ◆ Red Roof Inn
AServ E: Gulf (Auto Towing)
ATM W: Fulton Bank
Other E: State Patrol Post

23 **(67)** U.S. 322w, PA 230, Cameron St, Lewistown, U.S. 22
TStop N: Clark's Ferry All American*(SCALES) (in Duncannon on U.S. 322)

22 **(66)** Front St
Gas W: Amoco*(CW)(LP), Exxon*, Sunoco
Food W: Chopsticks House, JR's (Super 8), McDonalds, Pizza Hut, Ponderosa, Taco Bell, The Arches (Super 8), Wendy's, White Mountain Ice Cream
Lodg W: DAYS INN AAA Days Inn, AAA Super 8 Motel
AServ W: Sunoco
Med E: ✚ Hospital
ATM W: Amoco

21 **(65)** U.S. 11, U.S. 15, Enola, Marysville
Gas E: Mobil
Food E: Eden Park, Hardee's, McDonald's, Wendy's
Lodg E: Quality Inn

20 **(61)** PA 944, Wertzville Road
Food E: Pizza Ect.

EXIT PENNSYLVANIA

19 **(59)** PA581 E, to U.S. 11, Camp Hill, Gettysburg

18 **(57)** PA 114, Mechanicsburg
Food E: McDonald's, Pizza Hut, Shoney's

17 **(52)** U.S. 11, to I-76, PA Tnpk, Middlesex, Kingstown
TStop E: Flying J Travel Plaza*(LP) (RV Dump)
W: All American Truck Plaza, Gables*(SCALES), Soco's All American Truckstop*(SCALES)
Gas E: Citgo(D)
W: BP*, Texaco*(LP)
Food E: Bob Evans Restaurant, Duffy's Restaurant (Holiday Inn), Flying J Travel Plaza, Hardee's, Middlesex Diner
W: All American Truck Plaza, Arby's, Best Western, Budget Host Motel, Carelli's Deli, Dunkin Donuts, Eat N Park, Iron Kettle Restaurant, McDonalds, Subway, Western Sizzlin'
Lodg E: Appalachian Trail Inn, Comfort Inn, ◆ Econolodge, AAA Embers Inn, ◆ Holiday Inn, AAA Super 8 Motel
W: Best Western, Budget Host Motel, Hampton Inn, Quality Inn, Rodeway Inn, Thriftlodge
AServ E: Citgo
TServ W: Soco Garage
TWash W: All American Truck Plaza, Gables
ATM E: Flying J Travel Plaza
W: All American Truck Plaza

16 **(48)** PA 641, High St (Southbound, Reaccess Northbound Exit 15)
Gas W: Getty*, Hess*
Food W: Burger King, Golden China Restaurant, Little John's Family Restaurant, Long John Silvers, McDonalds, Pizza Hut, Scalles Warehouse Restaurant, Taco Bell
AServ W: Jiffy Lube, Midas Muffler & Brakes, Patriot Auto Parts, Ralph Peipers Auto Parts, Trak Auto
ATM W: Core States Bank, Farmers Trust Company
Other W: Carlisle Plaza Mall, Pharmacy CVS, Weiss Market

15 **(48)** PA 74, York Road (Northbound Vai Exit 16)
Gas W: Gulf(D), Kwik Fill
Food W: Farmers Market Restaurant
AServ W: Gulf, Royer's Towing Service (Gulf)

14 **(47)** PA 34, Hanover St
FStop E: Gulf(D)
Gas E: Texaco(D)
W: Exxon(D)
Food W: Genova Restaurant, Papa John's Pizza, Wendy's
AServ E: Carlisle Car and Truck, Gulf, Mullen's Towing
W: Carlisle's Expert Tire, Exxon, Hanover Auto Works
Other W: M.J. Carlisle Mall, Soft Cloth Car Wash, Tritt's General Store

13 **(46)** College St (Dickinson School Of Law)
Gas E: Mobil, Texaco*(LP)
Food E: Alfredo Pizza, Bonanza Steak House, Carnelli's Pizza, Dunkin Donuts, McDonalds, Shoney's
Lodg E: DAYS INN ◆ Days Inn, Super 8 Motel
AServ E: Best Auto Supply, Mobil, Monroe Muffler & Brakes
RVCamp E: Western Village Campground (2 Miles)
Med W: ✚ Hospital
ATM E: Farmers Trust, Orrstown Bank
Other E: Family Hardware, Mail Boxes Etc, Stonehedge Square, Thrift Drug

12 **(44)** PA 465, Plainfield
Lodg E: DAYS INN Days Inn, Super 8 Motel
RVCamp W: Carlisle Campground (5.5 Miles)
Other E: State Patrol Post

(39) Rest Area - RR, Phones, Picnic, Vending (Southbound)

(38) Rest Area - RR, Phones, Picnic (Northbound)

Bold red print shows RV & Bus parking available or nearby

XIT PENNSYLVANIA

1 **(37)** PA 233, Newville
- **Parks** E: Pine Grove Furnace State Park (Eight Miles)
- W: Colonel Denning State Park (Thirteen Miles)
- **Other** E: Kings Gap EE&T Center (Five Miles)

0 **(28)** PA 174, King St
- **TStop** E: Pharo's Truck Stop(SCALES)
- **Food** E: Pharo's Truckstop
- **Lodg** E: Budget Host Inn
- W: Ameri Host Inn, Budget Host University Lodge
- **AServ** W: Interstate Ford
- **TServ** E: DJ's Truck Repair, Pharo's

(24) PA 696, Fayette St (Shippensburg University)
- **Gas** W: Mobil*(LP)

(20) PA 997, Scotland
- **FStop** E: Shell*(D)
- W: Amoco*(D)
- **Food** E: Bonanza, McDonalds
- **Lodg** E: AAA Comfort Inn, Super 8 Motel
- **AServ** E: Sears
- **ATM** W: Chambersburg Trust Bank
- **Other** E: Chambersburg Mall, Cinema 4, Thrift Drugs
- W: Coin Car Wash

(16) U.S. 30, Chambersburg, Gettysburg
- **FStop** W: Hess*
- **Gas** E: Exxon*, Sheetz*
- **Food** E: Boston Market, Chris's Country Kitchen, KFC, Perkins, Shoney's, Two Brothers Pizzaria and Restaurant
- W: Burger King, Copper Kettle Prime Rib, Golden China, Hardee's, Howard Johnson, Long John Silvers, McDonalds, Pizza Hut, Ponderosa, Roadster Diner, Taco Bell
- **Lodg** E: Days Inn AAA Days Inn (see our ad on this page)
- W: AAA Howard Johnson
- **AServ** E: Battery Warehouse, Fitzgerald Toyota, Goodyear Tire & Auto, Meineke Discount

EXIT PENNSYLVANIA

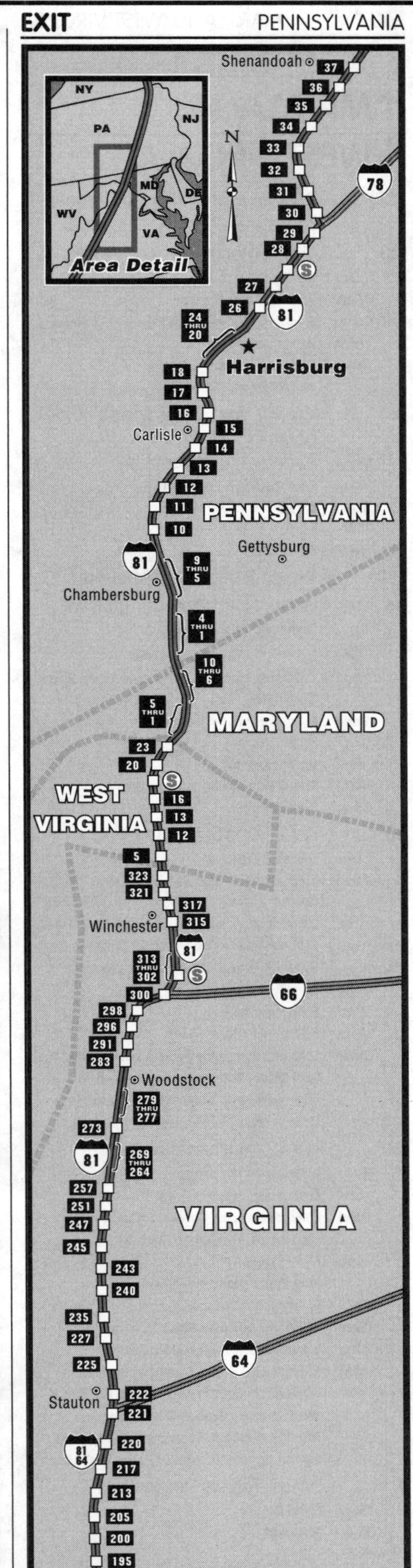

EXIT PENNSYLVANIA/MARYLAND

Mufflers, Midas Muffler & Brakes
- W: Forresters Mercury Lincon
- **Med** W: Hospital
- **ATM** E: Sheetz*(LP)
- **Other** E: Big Lots, Food Lion, Franklin Center, Lowes, Tourist Information
- W: Coin Laundry, State Patrol Post

5 **(14)** PA 316, Wayne Ave
- **Gas** W: Exxon*, Quick Fuel*, Sheetz*, Sunoco, Texaco*(D)
- **Food** E: Bob Evans Restaurant, Cracker Barrel
- W: Applebee's, China Wok, Denny's, Little Caesars Pizza, Mario's Italian, Pizza Hut, Razzle's (Econolodge), Red Lobster, September's (Holiday Inn), Subway, Wendy's
- **Lodg** E: ◆ Fairfield Inn (see our ad on this page), AAA Hampton Inn
- W: AAA Econolodge, AAA Holiday Inn
- **AServ** W: Jiffy Lube, Monroe Muffler & Brakes
- **TServ** E: A&B Garage
- **ATM** W: Chambersburg Trust, F&M Trust Bank, Patriot Federal Credit Union, Sheetz
- **Other** W: CVS Pharmacy, Coin Laundry, Giant Supermarket (Pharmacy), K-Mart, Mail Boxes Etc, Pugh's Petcetera Pet Shop, Rain Tunne Car Wash, Staples Office Store, Wayne Plaza

(11) **Parking Area - Picnic, Phones (Southbound)**

4 **(10)** PA 914, Marion

(7) **Weigh Station (Northbound)**

3 **(5)** PA 16, Greencastle, Waynesboro
- **TStop** E: Travel Port(SCALES) (RV Dump)
- **FStop** E: Texaco*(D)
- **Gas** E: Mobil*
- W: Exxon(D)
- **Food** E: Arby's, McDonalds, Subway (Travel Port), Travel Port Restaurant
- W: Green Castle Motel
- **Lodg** E: Econolodge, Roadway Inn (Travel Port), Travel Port Inn
- W: Green Castle Motel
- **AServ** W: Exxon
- **TWash** E: John Wayne Truck Wash and Detailing
- **ATM** E: Machine (Travel Port)
- W: Chambersburg Trust
- **Other** E: The Wash House Laundromat

2 **(3)** U.S. 11, Molly Pitcher Hwy
- **Gas** E: Exxon*
- W: Sunoco
- **Food** E: Brother's Pizza 2, Chef's Kitchen Family Restaurant
- **Lodg** E: AAA Comfort Inn
- **AServ** E: Bill Bower's Auto Service
- W: Carbaugh's Garage, Sunoco
- **Other** E: Blue Chip Mini-Mart, Coin Laundry

(1) **PA Welcome Center - RR, Phones, Vending, Tourist Info, RV Parking**

1 **(0)** PA 163, Mason - Dixon Road
- **Food** W: Black Steer Steak House
- **Lodg** W: Econolodge, State Line Motel
- **TServ** W: Expedited Services Inc., Mason-Dixon Repair and Service Inc.
- **RVCamp** W: Keystone RV Center

↑PENNSYLVANIA

↓MARYLAND

10AB Showalter Road, A - Eastbound, B - Westbound
- **Other** E: Airport

9 **(10)** Maugens Ave
- **FStop** E: Texaco*(D)
- **Gas** E: Amoco

Bold red print shows RV & Bus parking available or nearby

EXIT MARYLAND

Food	E: McDonalds
	W: Family Time
AServ	E: Amoco, Battery Warehouse, Mobil
	W: Warren Martin Auto Service
TServ	W: **Truck Enterprises of Hagerstown**
Other	E: Coin Car Wash

8 Maugansville Road (Southbound, Reaccess Northbound)

7AB **(8)** Hagerstown , MD58, Cearfoss, A - 58 East, B - 58 West

TServ	W: **Grimes Truck Center, Interstate Truck Equiptment Inc., MAC Trucks**

6AB **(7)** U.S. 40, Hagerstown , Huyett, A - US 40 East, B - US 40 West

Gas	E: Texaco
Lodg	E: Wellsley Inns
AServ	E: Texaco
Med	E: **Hospital (2 Miles)**
Other	E: **Visitor Information**

5 Halfway Blvd

TStop	W: **AC&T*(SCALES)**
FStop	E: **AC&T*(D)**
Food	E: Applebee's, Boston Market, China Town, Crazy Horse Steakhouse, Howard Johnson, Little Caesars Pizza, McDonalds, Pizza Hut, Red Lobster, Rick's China Garden, Roy Rogers, Ruby Tuesday, Shoney's, Taco Bell, The Ground Round, Wendy's
Lodg	E: AAA Howard Johnson, AAA Motel 6
AServ	E: Firestone Tire & Auto, Hagerstown Ford, Massey Ford/ Mercury/ Lincoln/ Hyundai, Montgomery Ward
TServ	W: **Freightliner Dealer, Higman Truck and Trailor Repair (Truck Wash)**
ATM	E: First National Bank of Maryland, First United Bank (Martin's Supermarket.), Hagerstown Trust
Other	E: **K-Mart, Lowes, Martin's Supermarket, Optical Service (JC Penny), Pet Land, Revco Drugs, Sam's Club, Staples Office Superstore, Valley Mall, Valley Mall Theater 123, Valley Park Commons, Value Vision Center, Wal-Mart (Pharmacy), Weiss Market**

4 Jct I-70, Hancock, Frederick

3 MD 144, Hancock

TStop	E: **Amoco Truck Plaza**
FStop	E: **Citgo*(D)(LP)**
Gas	E: BP*
Food	E: Fatboy's Pizza and Subs, Little Sandy's Restaurant, **Park N Dine**
AServ	E: Kurk Ford
TServ	E: **Amoco Truck Plaza**
ATM	E: Home Federal Savings Bank
Other	E: **C&O Canal Information,** Coin Car Wash

2 U.S. 11, Williamsport

FStop	W: **Sunoco***
Gas	E: AT&C*(D)
	W: Exxon(D), Shell*(D)
Lodg	W: DAYS INN Days Inn
AServ	W: Exxon, Sunoco
RVCamp	W: **KOA Campground (4 Miles)**
ATM	W: American Trust Bank
Other	W: **Tourist Information**

1 MD 63, MD 68, Williamsport, Boonsboro

AServ	W: Napa Auto
RVCamp	E: **Jelly Stone Camp, Safari Campground**

EXIT MARYLAND/WEST VIRGINIA

Other	E: **C&O Canal Historical Site**
	W: **AC Ban Locksmith**

↑MARYLAND

↓WEST VIRGINIA

(25) WV Welcome Center - RR, Phones, Tourist Info, Picnic (Southbound)

23 U.S. 11, Marlowe, Falling Waters

Gas	W: Citgo*, Texaco*
TServ	W: **Ruds Truck Repair**
RVCamp	E: **Falling Waters Campground(LP) (.8 Miles)**
ATM	W: Huntington Banks
Other	E: **Handy Shopper C-Store**
	W: **RV Express (Sales & Service)**

20 WV 901, Spring Mills Road (Southbound)

TStop	E: **76 Auto/Truck Plaza*(D)(SCALES)**
Food	E: **Econolodge, Truckstop**
Lodg	E: AAA Econolodge
TServ	E: **76**
ATM	E: Truckstop

(20) **Weigh Station (Southbound)**

16 WV 9E, North Queen St, Berkeley Springs, Hodgville

Gas	E: Exxon*(D), Sheetz* (Kerosene)
Food	E: Denny's, Hardee's, Hoss's, Long John Silvers, Taste of Italy, Yams Chinese/American
Lodg	E: Comfort Inn, Knight's Inn, Stone Anchor Inn, Super 8 Motel
Med	W: **Hospital**
ATM	E: Blue Ridge Bank, OneValley Bank
Other	E: **State Police**

13 WV 15, King St, Downtown

Gas	E: Sheetz*, Texaco*
Food	E: Applebee's, Asian Mkt, Burger King, **Cracker Barrel**, El Ranchero Mexican, Fazoli's Italian Food, Holiday Inn, Hungry Howies Pizza, Outback Steakhouse, Pizza Hut, Shoney's, Subway (Texaco), TCBY (Texaco), Wendy's
Lodg	E: AAA Holiday Inn
Med	E: **Hospital**
ATM	E: OneValley Bank, Sheetz
Other	E: **Martinsburg Mall (Frontage Rd), Outlet Mall, Post Office (Roger's County Market), Roger's County Market Grocery (Frontage Rd, Post Office), Tourist Information**

12 WV 45, Winchester Ave

FStop	E: **Qualls 24 Hr Fueling**
Gas	E: Southern States*(LP), Texaco*
Food	E: **Hardee's, Little Caesars Pizza (K-Mart),** McDonalds, **Ponderosa,** Taco Bell
Lodg	E: ◆ Hampton Inn, Krista Lite Motel, AAA Scottish Inns, AAA Windewald Motel
AServ	E: American Car Care Ctr, Battery Mart, **Sears**
TServ	E: **Ridgeway Trailer Sales**
RVCamp	E: **Nahkeeta Campground (3 Miles)**
ATM	E: Huntington Banks, OneValley Bank
Other	E: **Food Loin Supermarket (Deli & Bakery), K-Mart, Martin's Grocery (Pharmacy), Martinsburg Mall, Sport Center Camping Equipment, VA Veteran's Center, Wal-Mart**

8 WV 32, Tablers Station Road

TServ	E: **MS Carrier**
Other	E: **Airport**

EXIT WEST VIRGINIA/VIRGINI

5 WV 51, Inwood, Charles Town

Gas	E: Citgo*, Exxon*, Mobil(D), Texaco(D)(CW)(LP) (Kerosene)
Food	E: Luwegee's Pizza, McDonalds, Pack's Frozen Yogurt, Pizza Hut
AServ	E: Mobil
RVCamp	W: **Lazy A Campground (9 Miles)**
Med	E: **First Med Walk-In Clinic**
ATM	E: Huntington Banks of WV, OneValley Bank
Parks	E: **Harper Fairy**
Other	E: **Barnhart's Supermarket, Berkley Pharmac Coin Laundry, Fish Hatchery & Fishing Lakes, Food Lion Supermarket, Inwood Animal Ctr, Patterson Pharmacy**

(2) WV Welcome Center - RR, Phones, Picnic, Vending (Northbound)

↑WEST VIRGINIA

↓VIRGINIA

323 U.S. 11, VA 669, Whitehall

Gas	E: Exxon*(LP)
Other	E: **Fireworks**

321 VA 672, Clearbrook, Brucetown

FStop	E: **Olde Stone***
Food	E: **Restaurant* (Olde Stone FS)**
Other	E: **Kingdom Animal Hospital**

(320) VA Welcome Center - RR, Phones, Picnic, Vending (Southbound)

317 U.S. 11, to VA 37, Stephenson, to U.S.522N, to U.S.50 W

FStop	W: **Mobil***
Gas	W: Amoco*(D)(CW), Exxon*(D), Quick Stop*, Shell*(D)
Food	W: Blimpie's Subs, Burger King, Denny's (Rodeway Inn), Godfathers Pizza, McDonalds
Lodg	W: AAA Econolodge, Rodeway Inn
AServ	W: Exxon
Med	W: **Hospital**
ATM	W: First VA Bank

315 VA 7, Winchester, Berryville

Gas	E: Mobil*, Sheetz*(D) (Kerosene)
	W: Chevron(D), Citgo*, Exxon, Mobil, Shell*(D), Texaco(D)
Food	E: 220 Seafood, Blind Pig BBQ (Mobil), Dairy Corner Frozen Custard
	W: Arby's, Camino Real Mexican(SCALES), Captain D's Seafood, China Town, Deli (Citgo), Domino's Pizza, George's Pizza, Hardee's, KFC, Long John Silvers, McDonalds, Pizza Hut, Shoney's, Wendy
Lodg	W: Shoney's
AServ	E: Clark Volvo/Dodge
	W: Chevron, Exxon, Texaco
Med	W: **Hospital**
ATM	W: F&M Bank, First Union Bank, Jefferson National Bank, Marathon Bank
Other	E: **Revco, Winchester Shopping Ctr**
	W: **Coin Laundry, Food Lion Supermarket, Martin's Grocery Store, PharmHouse, Revco Drugs, Veterinary Clinic**

313 U.S. 17, U.S. 50, U.S. 522, Winchester

Gas	E: Chevron*(D)(LP), Shell*, Texaco*(D)
	W: Chevron*, Sheetz*(LP)
Food	E: Asian Garden, Bagels & Pizza, Baskin Robbins, Belle Starr, Chason's Country Buffet, China Town Buffet, **Cracker Barrel,** Dunkin Donuts, Hoss' Steakhouse, Jenny's

Bold red print shows RV & Bus parking available or nearby

EXIT	VIRGINIA
	Restaurant (Holiday Inn), Texas Steak House
	W: Best Western, Blue's Pit BBQ, Bob Evans Restaurant, KFC, Pargo's, Ponderosa, Taco Bell
Lodg	E: AAA **Comfort Inn,** AAA Holiday Inn, Super 8 Motel, AAA Travelodge
	W: AAA Best Western, ◆ Budgetel Inn, AAA Hampton Inn, AAA Quality Inn
AServ	E: Texaco
	W: Chevron, Jiffy Lube, Sears
Med	W: **✚ Hospital**
ATM	E: Southern Financial Bank
	W: Bank Of Clark County, F&M Bank
Other	E: **Fireworks (Shell), Food Lion Supermarket, Harley Davison Motorcycle**
	W: **Apple Blossom Mall, Bell Tone Hearing Ctr, Best Western, K-Mart, Pearl Vision Center, Visitor Information, Wal-Mart**
310	VA 37, U.S. 50 West, U.S. 522, Romney, Berkley Springs (Exit US11)
Lodg	W: Eco Village Budget Motel, AAA Howard Johnson
AServ	W: Miller Suzuki, Napa Auto, Snappy Lube Plus
TServ	W: **Donald B Rice Goodyear Tires**
RVCamp	W: **Candy Hill Campground**
Med	W: **✚ Hospital**
ATM	W: Marathon Bank
Other	W: **State Police Office**
307	VA 277, Stephens City , to RT 340
FStop	E: **Texaco***(D)
Gas	E: Chevron*(LP) (Kerosene), Shell*
	W: Exxon*, Sheetz*(LP) (Kerosene)
Food	E: Bella Vista Italion Restaurant, Domino's Pizza, KFC, Lilly Garden (Chinese), **McDonalds,** Pack's Frozen Yogurt, Roma Restaurant & Pizzeria, Subway (Shell), Wendy's, **Western Steer Family Steakhouse**
	W: Dunkin Donuts* (Exxon), **Tastee Freeze (Exxon)**
Lodg	E: AAA Comfort Inn, AAA Holiday Inn Express
AServ	E: Texaco (Wrecker Service)
Med	E: **✚ Urgent Care**
ATM	E: Bank of Clark County, Jefferson National Bank
Other	E: Coin Car Wash, **Food Loin Supermarket, James Way Shopping Center, Rite Aide Pharmacy, Stephens City Animal Hospital**
(304)	**Weigh Station (Both Directions)**
302	VA 627, Middletown, Cedarville
FStop	E: **Exxon***(LP)
	W: **Amoco***
Gas	W: Mobil*
Food	W: **Blimpie's Subs (Amoco), Godfathers Pizza**

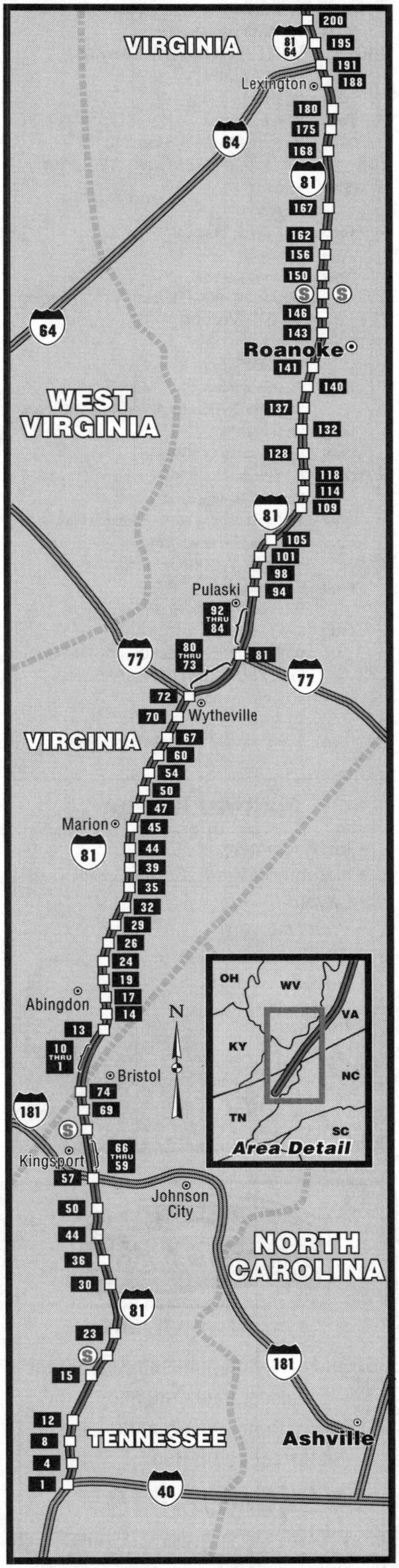

EXIT	VIRGINIA
	(Amoco), Stuckey's (Mobil)
AServ	E: Exxon
300	Junction I-66, Washington D.C., Front Royal
298	U.S. 11, Strasburg
Gas	E: Mobil
Food	E: Burger King
AServ	E: Mobil
RVCamp	W: **Battle of Cedar Creek Campground (1.4 Miles)**
Other	E: **Strasburg Museum**
	W: **Belle Grove Plantation**
296	VA 55, Strasburg
Other	E: **Hupps Battlefield Museum**
291	VA 651, Toms Brook
TStop	W: **The Virginian Truck Center*, Wilco Travel Plaza***
Food	W: **Dairy Queen (Wilco), Milestone* (Wilco), The Virginian, Wilco Travel Plaza**
Lodg	W: **The Virginian**
TServ	W: **The Virginian**
TWash	W: **Truck Wash**
283	VA 42, Woodstock
Gas	E: Chevron*(CW), Texaco*
	W: Coastal*, Exxon(D)
Food	E: **Hardee's, KFC,** McDonalds (Playland), Pizza Hut, Ramada, TCBY Yogurt (Texaco), Taco Bell, **Wendy's**
Lodg	E: ◆ Ramada (see our ad on this page)
AServ	W: Exxon (Wrecker Service)
Med	E: **✚ Hospital**
ATM	E: First Virginia Bank
Other	E: **7-11 Convenience Store**
279	VA 185, VA 675, Edinburg
Gas	E: Amoco*(D), Shell*
Food	E: Subway (Amoco)
AServ	E: Cook's Auto Sales
ATM	E: Crestar Bank, Shell
Other	E: **Coin Operated Laundry**
277	VA 614, Bowmans Crossing (Northbound, Reaccess Southbound Only)
273	VA 292, VA 703, Mt Jackson, Basye
TStop	E: **Sheetz*(SCALES), Shenandoah Truck Center*(SCALES)**
Gas	E: Citgo*
Food	E: **Baskin Robbins (Sheetz),** Burger King, **Denny's (Best Western), Pizza (Sheetz), Subs & More (Sheetz)**
Lodg	E: AAA **Best Western**
AServ	E: Jeff's Auto Body
ATM	E: Shenandoah
269	VA 730, Shenandoah Caverns
Gas	E: Chevron*(D) (Kerosene)
Other	W: **Shenandoah Caverns, Tuttle & Spice Museum***
264	U.S. 211, New Market, Timberville Luray
Gas	E: Chevron*(D), Exxon, Shell*(D)(LP)
	W: Citgo*
Food	E: Blimpie's Subs, **Johnny Appleseed Restaurant (Quality Inn),** McDonalds
	W: Jane's Cafe
Lodg	E: ◆ Blue Ridge Inn, AAA Budget Inn, AAA **Quality Inn**
	W: DAYS INN Days Inn

EXIT — VIRGINIA

AServ — E: Exxon

RVCamp — E: Rancho Campground

ATM — E: Crestar Bank, F&M Bank, Shell

Parks — E: Shenandoah Nat'l Park

Other — E: Applecore Village, Fireworks, Laundromat, Luray Caverns, Skyline Drive

W: Calvary Museum, New Market Civil War Battlesite, Tourist Information

(262) Rest Area - RR, Phones, Vending, Picnic

257 U.S. 11, VA 259, Mauzy, Broadway.

RVCamp — E: Endless Cavern Campground (5.2 Miles), KOA Campground (3.4 Miles)

251 U.S. 11, Harrisonburg

Gas — W: Shell*(D)

Food — W: Bar-B-Q Ranch

Lodg — W: Scotish Inn

Other — W: Fed Ex Drop Box (Shell), Fireworks

247AB U.S. 33, Elkton, Harrisonburg

Gas — E: Royal*

W: Amoco*, Chevron*(D), Etna*(D)(LP), Exxon*(D) (FedEx Box), Royal*, Texaco*

Food — E: Barney's Cafe, Boston Beanery, Captain D's Seafood, Chili's, China Jade, El Charro Mexican, Guiseppe's Pizza, Little Caesars Pizza, Long John Silvers, Mr J's Bagels, Pargo's Southern Food, Ponderosa Steakhouse, Red Lobster, Ruby Tuesday, Taco Bell, Texas Steakhouse, Waffle House, Wendy's

W: Arby's, Blimpie's Subs (Amoco), Carasel Frozen Treats, Chang House, China Inn, Golden China, Hardee's, KFC, L'Itilia Restaurant, McDonalds, Mr Gatty's Pizza, Papa John's, Pizza Hut, Sigon Vietnamese, Subway

Lodg — E: Comfort Inn, Econolodge, Hampton Inn, Motel 6, Sheraton Four Points

W: Motel Marvilla, Shoney's

AServ — E: Jiffy Lube, Nisson, Speedee Oil Change

W: Advance Auto Parts, Blue Ridge Tire Co

RVCamp — E: RV Camping

Med — W: ✚ Hospital

ATM — E: Central Fidelity, Crestar Bank, F&M Bank, First Union, First Virginia Bank, Jefferson National Bank, NationsBank

W: Amoco, NationsBank

Other — E: CVS Pharamacy, Food Lion Supermarket, K-Mart (Pharmacy), Kroger Supermarket (Pharmacy), Rac-N-Sac Groceries, Valley Mall, Wal-Mart Supercenter

W: Car Wash, Coin Laundry, Rite Aide Pharmacy, Save-a-Center Grocery

245 VA 659, Port Republic Road

Gas — E: Chevron*(D), Exxon*(D), Texaco*

Food — E: Blimpie's Subs, Citgo*(D) (Kerosene), Dairy Queen, Howard Johnson, J Willowby's Road house (Days Inn), Subway (Exxon)

Lodg — E: DAYS INN ◆ Days Inn, Howard Johnson

Med — W: ✚ Hospital

ATM — E: Exxon

243 U.S. 11, Harrisonburg (West Side Services Accessible Pleasant Valley Rd)

TStop — W: Travel Center*(D)(LP)

Gas — W: Amoco(D)(CW), Citgo*, Exxon*, Mobil*(D) (Kerosene)

Food — W: Crab Apple Restaurant (Ramada Inn), Cracker Barrel, Deli (Exxon), Double Happiness (Chinese), Pano's Seafood, Southside Diner (Travel Center), Waffle House

Lodg — W: AAA Ramada, Red Carpet Inn, AAA Super 8 Motel

AServ — E: Big L Tires

W: Amoco, Bob Wade Isuza, Carr's Auto, Exxon, Mobil, Toyota

TServ — E: Big L Tire, Freeman Trucking

W: ThermoKing, Truck Enterprises (Kenworth)

EXIT — VIRGINIA

TWash — W: Travel Truckstop*(D)(LP) (Kerosene)

ATM — W: F&M Bank, Travel Ctr

240 VA 257, VA 682, Mt Crawford, Bridgewater

Gas — W: Exxon*(D) (FedEx)

Food — W: Arby's (Exxon)

Other — W: Fed Ex Drop Box (Exxon)

235 VA 256, Weyers Cave, Mt Sidney

FStop — E: Mobil*(D)

W: Amoco*

Food — W: Subway (Amoco)

ATM — W: Amoco

Other — E: Grand Caverns

(232) Rest Area - RR, Phones, Picnic, Vending

227 VA 612, Verona

Gas — E: Amoco*(D)

W: Exxon*

Food — E: Subway (Amoco), Waffle Inn

W: China City (Take-Out), Hardee's, McDonalds

Lodg — W: Scottish Inns

AServ — E: Verona Car Care Center

RVCamp — E: Waynesboro North Forty Campground (11.6 Miles)

W: KOA Campground (2.9 Miles)

Other — W: Coin Laundry, Factory Antique Mall, Food Lion Grocery, Revco Drugs

225 VA 275, Woodrow Wilson Pkwy

Food — E: Innkeeper

W: Holiday Inn, The Host Inn

Lodg — E: ◆ Innkeeper

W: AAA Holiday Inn (Golf Course), The Host Inn

222 U.S. 250, Staunton, Fishersville

Gas — E: Exxon*, Texaco*(D)(LP)

W: Augusta*(LP)

Food — E: McDonalds (Exxon), Mrs Rowes Family

EXIT — VIRGINIA

Restaurant, Shoney's, Texas Steak House

W: Burger King, Shorties Diner, Waffle House

Lodg — E: AAA Best Western

W: AAA Comfort Inn, AAA Econolodge, AAA Super 8 Motel

AServ — W: Augusta*(LP), McDonnah Truck Service, Tire Mart, Toyota

ATM — W: Planters Bank

Other — W: Animal Care, Wal-Mart Supercenter

221 Junction I-64 East (Southbound)

220 VA 262, U.S. 11, Staunton

217 VA 654, Mint Spring, Stuarts Draft

213 U.S. 11, Greenville

205 VA 606, Raphines, Steeles Tavern

200 VA 710, Fairfield

(199) Rest Area - RR, Phones (Southbound)

195 U.S. 11, Lee Hwy

TStop — W: Shell*(D)

Gas — W: Citgo

Food — E: Maple Hall

W: Aunt Sara's

Lodg — E: ◆ Maple Hall

W: AAA Howard Johnson (see our ad on this page), AAA Ramada, Red Oaks Inn

191 Junction I-64 West

188 U.S. 60, Lexington, Buena Vista

Gas — W: Exxon*(D)

AServ — W: Exxon

Med — W: ✚ Hospital (2.5 Miles)

Other — W: Tourist Information (3 Miles)

180 U.S. 11, Natural Bridge, Glasgow

Gas — E: Fergusons Grocery* (Fireworks), Shell (Kerosene)

W: Chevron (Kerosene), Texaco*(D)

Food — E: Fancy Hill

W: Westmoreland Restaurant

Lodg — E: ◆ Fancy Hill Motel, Natural Bridge Inn & Conference Center (see our ad this page)

W: AAA Budget Inn

AServ — E: Shell

W: Chevron

RVCamp — W: KOA Campground

Other — E: Enchanted Castle Studio Tours

W: The Wax Museum & Factory Tour of Natural Bridge

175 U.S. 11, Natural Bridge

Gas — E: Exxon* (Kerosene)

Lodg — E: Natural Bridge Inn & Conference Center (see our ad on this page)

AServ — E: Exxon

RVCamp — E: Campground by Natural Bridge

168 CR 614, Arcadia

Gas — E: Shell*

Food — E: Spice House Restaurant

Lodg — E: Wattstull Motel

AServ — E: Shell

RVCamp — E: Yogi Bear Camp Report (6.1 Miles)

RVCamp — E: Caverns of Natural Bridge, Wax Museum of Natural Bridge

167 U.S. 11, Buchanan (Southbound Reaccess Via Frontage Rd On West Side)

162 U.S. 11, Buchanan

Gas — W: BP(D), Texaco*(LP)

AServ — W: BP

(158) Rest Area - RR, Phones (Southbound)

156 CR 640, Troutville

150AB 150A - US 220 to US 460, US 11

Bold red print shows RV & Bus parking available or nearby

EXIT VIRGINIA

Troutville, Cloverdale, Lynch, 150B - Clifton Forge, Daleville, Fincastle, US 220N
FStop E: Citgo*(D)(LP), Pilot Travel Center*(SCALES)
Gas E: TravelCenters of America, Dodge's Store*(D) (Kerosene), Exxon(LP)
W: Amoco*(D) (24 hrs), Express*, Exxon(D)(LP) (Kerosene)
Food E: Burger King (TC of A), Country Pride Restaurant ((TC of A), Country-Cookin, Cracker Barrel, Dodge's Store, Hardee's, Italian Bella, McDonalds, Shoney's, Subway (Pilot), Taco Bell, Waffle House
W: Pizza Hut, Western Sizzlin'
Lodg E: Comfort Inn, Daystop ((TC of A), Holiday Inn, Travelodge
W: Best Western, Howard Johnson
AServ E: Exxon
TServ E: Carter CAT
TWash E: Truck Wash U.S.A
RVCamp E: Berglund Scott Cooper
ATM E: Central Fidelity Bank, TravelCenters of America
W: Bank of Botetourt, Bank of Fincastle
Other E: Revco Drugs, Winn Dixie Supermarket
W: Coin Car Wash, Pet Health Clinic
49) Weigh Station (Both Directions)
46 VA 115, Hollins, Roanoke
Gas E: Exxon*, Shell*(D)
Food E: Country Kitchen (Country Inn & Suites), McDonalds, The Peaks
Lodg E: AAA Country Inn & Suites, ◆ Days Inn
ATM E: First Union
43 Junction I-581, U.S. 220, Downtown Roanoke
41 VA 519, to VA North 311, Salem, New Castle
Gas E: Texaco*
Food E: Burger King (Texaco)
Lodg E: Budgetel Inn, AAA Quality Inn
W: Holiday Inn (see our ad on this page)
Med E: ✚ Veterinary Clinic
40 VA 311, Salem, Newcastle
37 VA 112, CR 619, Salem
Gas E: Amoco*, Shell*(D)
W: Chevron
Food E: Burger Boy, Denny's, El Rodeo Mexican, Mamma Maria, Omelet Shoppe, Shoney's
Lodg E: AAA Comfort Inn, AAA Knights Inn, Super 8 Motel
W: AAA Holiday Inn
AServ E: Anderson's Auto Repair
W: Chevron
RVCamp E: Snyder's Sales Part & Services
ATM E: Shell
Other E: Food Lion Supermarket
32 CR 647, Dixie Caverns, Elliston, Shawsville
Food E: Blue Jay Restaurant, Wilson Restaurant (see our ad on this page)
Lodg E: Blue Jay Motel, AAA Budget Host
RVCamp E: Dixie Caverns & Campground
29) Rest Area - RR, Phones (Northbound)
28 U.S. 11, CR 603, Ironto
FStop W: Citgo*
Gas E: Mobil*
ATM E: Mobil
W: Citgo
18 U.S. 11, U.S. 460, Christiansburg, Blacksburg
FStop E: Shell*(LP)

EXIT VIRGINIA

W: Crown*(LP)
Gas W: Amoco, Chevron(LP), Exxon*, Racetrac*, Shell*(LP), Texaco*(D)
Food E: Cracker Barrel, Huckleberry Restaurant
W: Hardee's, McDonalds, Pizza Hut, Shoney's, The Outpost Restaurant, Waffle House, Wendy's, Western Sizzlin'
Lodg E: ◆ Days Inn, Hampton Inn
W: AAA Econolodge, HoJo Inn
AServ W: Advance Auto Parts, Amoco, Chevron, Exxon, Homer Cox Ford, Shelor Chevrolet
RVCamp E: Camping
Med W: ✚ Hospital
114 VA 8, Christiansburg, Floyd
Gas W: Citgo*, Deli Mart*(LP)

EXIT VIRGINIA

AServ W: A & M Auto Repair, Jack's Garage
109 VA 177, CR 600, Radford
TStop W: BP*(LP)
Gas W: Marathon* (Kerosene)
Food W: A Diesel Cafe
Med W: ✚ Hospital
(108) Rest Area - RR, Phones (Both Directions)
105 VA 232, CR 605, Radford
101 VA 660, Claytor Lake State Park
Gas W: Chevron (Kerosene), Citgo*(D) (Kerosene)
Food W: Mr Burger (No Sleeping), VA Rest & Deli
Lodg W: Claytor Lake Inn
AServ W: Chevron
Parks W: Claytor Lake State Park
Other W: K & K Marine
98 VA 100 North, Dublin, Pearisburg
FStop W: Texaco*
Gas E: Exxon(D), Mobil*(LP)
W: Marathon* (Kerosene)
Food E: Bon Fire Restaurant (Comfort Inn)
W: Blimpie's Subs (Texaco), Burger King, McDonalds, Waffle House, Wendy's
Lodg E: AAA Comfort Inn
AServ E: Exxon Auto Care Center
Med W: ✚ Hospital
ATM W: Premier Bank
Other E: Wilderness Road Regional Museum
W: Army Reserve Center, Community Animal Hospital
94 VA 99, Pulaski, Service Road
Gas W: Exxon*(LP) (Kerosene)
Food E: Rib & Sirloin Rest Sandwich Shop (Red Carpet Inn)
Lodg E: Red Carpet Inn
AServ E: Jimmy's Roll Back Service & Auto Repair
Parks W: New River Trail State Park
92 CR 658, Draper
Gas E: BP*
RVCamp E: Horseshoe Campground
Parks E: New River Trail State Park
89AB 89A - VA 100S, Hillsville, 89B - US 11N, Pulaski (US 11N Not Recommended For Trucks)
Lodg E: Skyline Motel (Frontage Road)
86 CR 618, Service Road
TStop W: I-81 Auto/Truck Plaza (Chevron)
Food W: The Appletree Restaurant (I-81 Auto/Truck Plaza)
Lodg W: AAA Gateway Motel #2 (Frontage Road), I-81 Auto/Truck Plaza
TServ W: I-81 Auto/Truck Plaza
84 CR 619, Grahams Forge
FStop W: Shell*
Food W: Fox Mountain Inn
Lodg W: Fox Mountain Inn
Other W: Grahams Forge Foodette*
81 Junction I-77 South, Charlotte
80 U.S. 52 South, VA 121 North, Fort Chiswell, Maxmeadows
TStop E: Petro(SCALES)
FStop W: Citgo*(LP)
Gas E: BP*(D), Exxon* (Petro TS)
W: Amoco
Food E: Buck's Pizza, Iron Skillet Restaurant (Petro TS), Little Caesars Pizza (Petro TS), Wendy's (Petro TS)
W: McDonalds
Lodg E: Comfort Inn
W: AAA Country Inn
AServ E: Lee's Tire, Brake, & Muffler
TServ E: Petro
W: Complete Truck Service (24-hour)
RVCamp E: Fort Chiswell RV Campground
ATM E: Petro TS
Other E: Car Wash, Factory Merchants - Fort Chiswell (see our ad this page), Laundromat

EXIT VIRGINIA

77 Service Road
- **TStop** E: Flying J Travel Plaza*(LP)(SCALES) (RV Dump)
- **FStop** E: Citgo* (Fuel Stop)
- **Gas** E: Conoco, Texaco*(D)
  W: Exxon* (RV Diesel)
- **Food** E: Burger King, Subway (Citgo), Thads Restaurant(SCALES) (Flying J)
- **RVCamp** E: KOA Campground
- **Other** E: Gale Winds (go-carts, putt-putt, family games)
  W: State Patrol Post

73 U.S. 11, South Wytheville
- **Gas** E: Exxon, Mobil*(LP), Shell*(D)
- **Food** E: Bob Evans Restaurant, El Puerto Mexican Restaurant, KFC, Mom's Country Store, Peking Restaurant, Shoney's, Steak House & Saloon, Waffle House
- **Lodg** E: AAA Days Inn (see our ad this page), AAA Holiday Inn, Interstate Motor Lodge, Motel 6, AAA Red Carpet Inn, Shenandoah Inn, Wythe Inn
- **AServ** E: Exxon
- **Med** E: Hospital

72 Junction I-77 North, Bluefield

70 U.S. 21, U.S. 52, Wytheville
- **Gas** E: BP*(D)
  W: Exxon*, Shell*(D)(CW)(LP) (Kerosene)
- **Food** E: Arby's, McDonalds (Playground)
  W: Scrooge's Restaurant
- **Lodg** W: AAA Comfort Inn
- **AServ** E: Fas-T-Lube
  W: Shell
- **Med** E: Hospital
- **ATM** E: First Virginia Bank
- **Other** E: Food Lion Supermarket, Revco, Tourist Information

67 U.S. 11, Wytheville (Northbound)

(61) Rest Area - RR, Phones (Northbound)

60 VA 90, Rural Retreat
- **FStop** E: Citgo*(LP)
- **Food** E: McDonalds, The Sub Hub
- **Other** E: Putt Putt Mini Golf, Trout Pond, Rural Retreat Lake

(54) Rest Area - RR, HF, Phones, Picnic (Southbound)

54 CR 683, Groseclose (Mountain Empire Airport, Settlers Museum Of SW Virginia)
- **FStop** E: Texaco*
- **Gas** E: Exxon*
- **Food** E: Cumbow's Restaurant, Dairy Queen (Exxon), Village Motel
- **Lodg** E: Village Motel
- **Other** E: Settler's Museum of SW Virginia

50 U.S. 11, Atkins
- **FStop** W: Citgo*
- **Food** W: Country Cookin Restaurant
- **Lodg** W: ◆ Days Inn
- **Other** W: Cullop's Old Stone Tavern

47 U.S. 11, Marion
- **RVCamp** W: Hungry Mother Family Campground (4.1 Miles)
- **Med** W: Hospital
- **Parks** W: Hungry Mother State Park

45 VA 16, Marion
- **Gas** E: Shell*
- **Food** E: Apple Tree Restaurant
- **Med** W: Hospital
- **Parks** E: Grayson Highlands State Park (33.5 Miles)
- **Other** E: Camping (7 Miles), Mount Rogers Wreck Area Visitors Center

44 U.S. 11, Marion
- **AServ** W: Marion Tire Dealer Inc.
- **Other** W: Smyth County Animal Hospital

39 U.S. 11, CR 645, Seven Mile Ford Road
- **Lodg** E: Ford Motel
- **RVCamp** W: Interstate Campground

35 VA 107, Chilhowie
- **Gas** E: Gas-haus, Texaco*
  W: Exxon*, Rouse Fuel Center(D)(LP)
- **Food** E: Mountainside Restaurant (Mount Rogers Inn)
  W: McDonalds (Playground), Shuler's Pizza Factory
- **Lodg** E: Mount Rogers Inn
- **Other** W: Greever's Drugstore Inc.

32 U.S. 11, Chilhowie

29 VA 91, Damascus, Glade Spring, Saltville
- **Gas** E: Shell*(D)(LP)
  W: Chevron*(D) (24 hrs, Kerosene), Texaco*
- **Food** E: Subway (Shell)
- **Lodg** E: Economy Inn & Restaurant, AAA Swiss Inn
- **AServ** E: Apache Towing, Interstate Auto Parts & Service
  W: Hick's Tire & Alignment
- **TServ** E: Apache Towing
- **Other** W: Jiff-e-Jack

26 CR 737, Emory (Emory And Henry College)

24 VA 80, Meadowview

22 CR 704, Enterprise Rd
- **Gas** E: Chevron(D) (Kerosene)
- **AServ** E: Chevron

19 U.S. 11, U.S. 58, Abingdon, Damascus
- **FStop** E: Texaco* (Kerosene)
  W: Citgo*
- **Gas** W: Chevron*(CW), Exxon*(D)
- **Food** E: Cherokee Restaurant (Cherokee Motel), Pizza Plus, Subway (Texaco FS), The Ice Cream Stop
  W: Bella's Pizza & Subs, Burger King, Old Country Store Cracker Barrel, Harbor House Seafood, Omelette Shop, The Bonfire Restaurant
- **Lodg** E: Cherokee Motel, Holiday Lodge Motel
  W: AAA Alpine Motel, Empire Motor Lodge, ◆ Holiday Inn Express
- **AServ** W: Chevron
- **RVCamp** E: Calleb's Cove Campground
- **ATM** E: Highlands Union Bank
  W: First Bank & Trust Co.
- **Other** E: Highlands Animal Hospital, Pet Shop
  W: The Arts Depot, Tourist Information

EXIT VIRGINI

17 U.S. 58 alt, VA 75, Abingdon, South Holston Dam
- **Gas** E: Amoco*(CW)
  W: Exxon*, Shell*(D)(LP)
- **Food** E: Domino's Pizza, Long John Silvers
  W: Arby's, Bonanza Steak House, Hardee's, Hunan Chinese, KFC, Little Caesars Pizza, Los Arcos Authentic Mexican, McDonalds (Playground), Pizza Hut, Shoney's, Subway, Sun Chinese, TCBY, Taco Bell, Wendy's
- **Lodg** W: ◆ Super 8 Motel
- **AServ** E: Xpress Lube
  W: Advance Auto Parts
- **ATM** W: Central Fidelity, First American, First Virginia Bank, Food City
- **Other** E: Anchor Bookshop, Mr. Klean Kar Wash, Put In Mini Golf
  W: Auto Wash, Blue Ridge Photo One Hr Lab, Food City, Food Lion, K-Mart (Pharmacy), Kroger Supermarket (24 Hrs, Pharmacy), Magic Mart, Revco, Super K Discount Drugs, The Art Depot, Tourist Information, Wm. King Reg. Art Center

14 U.S. 19 North, VA 140, Abingdon
- **Gas** W: Chevron*(CW) (24 Hrs), Texaco*
- **Food** E: VA Creeper Bakery
- **Lodg** W: AAA Comfort Inn
- **AServ** W: Empire Ford
- **RVCamp** W: Riverside Campground
- **ATM** W: Highlands Union Bank
- **Other** W: Tourist Information

(13) Rest Area - RR, HF, Phones, Picnic

13 CR 611, Lee Highway (VA Highland Airport)
- **Gas** W: Texaco* (Kerosene)
- **TServ** W: Blueridge Kenworth
- **Other** W: Lee Highway Animal Hospital, Tourist Information

10 U.S. 11, U.S. 19, Lee Highway (VA Intermont College, N. Campus)
- **FStop** W: Conoco* (24 Hrs, Kerosene)
- **Gas** W: Chevron*(CW) (24 Hrs), Coastal*
- **Lodg** W: Beacon Motel, Evergreen Motor Court, Red Carpet Inn, Robert E. Lee Motel, Scottish Inns, Skyland Inn, Thrifty Inn Motel
- **ATM** W: Coastal
- **Other** W: Fleenor Mobile Home Park (Overnight Parking)

7 Old Airport Road, Bonham Road
- **Gas** E: Amoco, Texaco*(D)
  W: BP*, Citgo*, Conoco*
- **Food** E: Crossings Pizza
  W: Damon's, Fazoli's Italian Food, McDonalds (Wal-Mart), Perkins Family Restaurant, Prime Sirloin, Subway, Taco Bell, Texas Steakhouse & Saloon, Wendy's (Citgo)
- **Lodg** E: AAA La Quinta Inn
- **AServ** E: Amoco

Bold red print shows RV & Bus parking available or nearby

XIT VIRGINIA

W: Advance Auto Parts, Exit 7 Muffler Man, Wal-Mart
TServ W: Good Pasture
RVCamp W: Sugar Hollow
Med W: ✚ Mountain Spring Family Care (Walk-Ins Welcome)
ATM E: Highlands Union Bank, Texaco
W: Citgo, First Union
Other W: Food Country USA, Wal-Mart Supercenter (24 Hrs, One-Hr Photo)

U.S. 11, U.S. 19, Lee Highway (King College, VA Intermont College)
Gas E: Amoco*, Shell*(CW)
W: Exxon*(LP)
Food E: Arby's, Burger King (Playground), Hardee's, KFC, Long John Silvers, McDonalds (Playground), Shoney's
Lodg E: Budget Inn, Crest Motel, Siesta Motel, ◆ Super 8 Motel
W: AAA Comfort Inn (see our ad on this page)
AServ E: Parts Plus
W: Blevins Tire, Crabtree Buick Pontiac
ATM E: Commonwealth Community Bank, First Bank & Trust
Other E: Food Lion (24 Hrs), Kroger Supermarket, Revco Drugs, VA Intermont College

Junction I-381 South, Bristol (Bristol Motor Speedway)
Gas E: Chevron, Conoco
Food E: Omelette Shop, Pizza Hut
Lodg E: ◆ Econolodge

West U.S. 58, U.S. 421, Bristol, Gate City

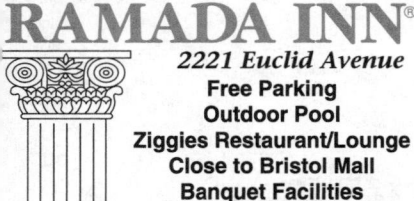

EXIT VIRGINIA/TENNESSEE

Lodg E: Ramada Inn (see our ad on this page)
(0) VA Welcome Center - RR, HF, Phones, Picnic, No Trucks

↑VIRGINIA

↓TENNESSEE

(75) TN Welcome Center - RR, Phones (Southbound)
74AB U.S. 11W, Kingsport, Bristol (King College)
FStop W: One Stop Fuel Stop*(D)
Food E: Scooter's Restaurant (Quality Inn)
W: Chicago Bar & Grill, Pizzeria Uno
Lodg E: Hampton Inn, Quality Inn
W: Holiday Inn
TServ W: W&E Truck Service
Med E: ✚ Bristol Regional Medical Center
69 TN 37, Blountville
Gas E: BP*(D)
Food E: Arby's, Subway (BP, Kerosene)
ATM E: BP
Other E: Appalachian Caverns
66 TN 126, Kingsport, Blountville
Gas E: Amoco* (24 Hrs, Kerosene), Chevron*(CW) (24 Hrs), Exxon*
Food W: McDonalds (Playground)
Other W: Factory Stores of America (see our ad on this page)

EXIT TENNESSEE

63 TN 357, Tri - City Airport
FStop W: Phillips 66*(LP)
Gas E: Exxon*(D)(LP)
W: Amoco*
Food E: Baskin Robbins (Exxon), Cracker Barrel, Taco Bell (Exxon), Wendy's
Lodg E: AAA La Quinta Inn
W: Red Carpet Inn
AServ W: Sam's Club
RVCamp W: KOA Kampground (1.5 Miles), Rocky Top Campground (1.2 Miles)
ATM E: Exxon
W: Phillips 66
Other W: Sam's Club, The Antique Mall
59 TN 36, Kingsport, Johnson City
Gas E: Amoco*
W: Texaco*(D)(CW)(LP) (Kerosene)
Food W: Arby's, Burger King, Domino's Pizza, Hardee's, Huddle House, La Carreta Authentic Mexican, Little Caesars Pizza, McDonalds (Playground), Moto Japanese Restaurant, Motz's Italian Restaurant, Pal's Sudden Service, Perkins Family Restaurant, Raffael's Pizza, Pasta, etc., Subway, The Plum Tree Oriental, Wendy's
Lodg E: AAA Super 8 Motel
W: AAA Comfort Inn, Holiday Inn Express
AServ W: Advance Auto Parts, Express Lube
ATM W: First Bank of Tennessee, First Tennessee Bank, Ingles, Texaco
Parks W: Warrior's Path State Park
Other E: Green Spring Marine
W: Colonial Heights Hardware, Colonial Heights Pharmacy, Fire Dept., Food Lion, Golf Unlimited, Ingles Supermarket, One-Hr Photo, Revco Drugs, Soft Cloth Auto Wash, US Post Office, Wrap-it Mail
57AB Junction I-181, South U.S. 23, Johnson City, Kingsport (Rocky Mount And ETSU)
Lodg W: Days Inn (see our ad this page)
50 TN 93, Jonesborough, Fall Branch (Highway Patrol Hq)
44 Jearoldstown Road
Gas E: Amoco*
(41) Rest Area (Southbound)
(38) Rest Area - RR, HF, Phones, Picnic (Northbound)
36 TN172, Baileyton Road
TStop W: Davy Crockett Auto/Truck Stop*(LP)(SCALES) (Marathon, 24 Road Service)
FStop W: Texaco*(CW)
Gas W: BP*(LP) (24 hrs)
Food W: Subway (Texaco)
Lodg W: 36 Motel
AServ E: Rader's Garage (24 Hr Road Service)
TServ W: Davy Crockett Auto/Truck Stop
TWash E: B&J Truck Wash
RVCamp W: Baileyton Camp Inn (1.7 Miles)
ATM W: Davy Crockett Auto/Truck Stop, Texaco

← N I-81

EXIT TENNESSEE

Other **E:** Andrew Johnson Visitor Center (13 Miles)

30 TN 70, Greeneville, Rogersville

Gas **E:** Exxon*(D)(LP)

Food **E:** Dairy Queen (Exxon)

23 U.S. 11 East, Mosheim, Greeneville, Bulls Gap (Jonesborough-TN's Oldest Town, Tusculum College)

FStop **W:** Phillips 66(SCALES)

Gas **E:** Amoco*, BP*

W: Citgo*

Food **E:** Pearl's Diner, TCBY (Amoco), Wendy's (Amoco)

W: McDonalds (Playground), New York Pizza & Italian Restaurant, Taco Bell, Tony's Bar-B-Que

Lodg **W:** AAA Comfort Inn (see our ad on this page), ◆ Super 8 Motel

(21) Weigh Station (S.B.)

(20) Parking Area - No Facilities, (N.B.)

15 TN 340, Fish Hatchery Road

FStop **E:** Amoco*

12 TN 160, Morristown, Lowland

Gas **E:** Citgo*, Streamline*

EXIT TENNESSEE

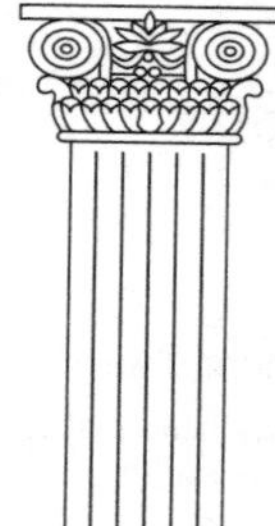

EXIT TENNESSEE

W: Shell* (24 Hrs, Kerosene)

8 U.S. 25 East, Morristown, White Pine

Gas **E:** Shell(D) (Kerosene)

W: BP*(D)(CW), Phillips 66*

Food **E:** Catawba Restaurant

W: Hardee's, Ramada Inn Restaurant

Lodg **W:** Parkway Inn, AAA Ramada Inn (see our ad on this page), ◆ Super 8 Motel

AServ **E:** Shell

4 TN 341, White Pine Road

TStop **E:** Pioneer Travel Center*(LP)(SCALES) (BP, CB Shop)

W: Pines Truck Plaza

Gas **E:** Exxon(D)

W: Amoco*, Citgo, Texaco*(D) (Kerosene)

Food **W:** Huddle House (24 hrs)

Lodg **E:** Crown Inn

W: Days Inn AAA Days Inn, Hillcrest Inn

(3) Rest Area (Southbound)

1AB 1A - I-40 E, Asheville, 1B - I-40 W, Knoxville

↑TENNESSEE

Begin I-81

I-82 E →

EXIT WASHINGTON

Begin I-82

↓WASHINGTON

3 WA 821S, Thrall Road

(7) Scenic View - Truck Parking, RV Parking

11 Military Area

(23) Rest Area - RR, Phones, RV Dump P (Westbound)

(25) Rest Area - RR, Phones, Free Coffee, Picnic, RV Dump P

EXIT WASHINGTON

(Eastbound)

26 WA 821N., To WA 823 Canyon North, Selah

29 East Selah Road

30AB Rest Haven Road, Selah

31 First Street, Yakima, N.16th Ave, N. 40th Ave.

FStop **S:** ARCO*

Gas **S:** Arco*, Exxon*, Texaco

Food **S:** Artic Circle Hamburgers, Country Harvest Buffet, Dairy Queen, Denny's, Espinoza's Mexican & American, Golden Scenic Buffet, La

EXIT WASHINGTON

Tiendita Del Mexicano, Red Lobster, Settlers Inn, Sizzler, The Lariat BBQ, Wendy's

Lodg **S:** Allstar Motel, Big Valley Motel, Double Tree Hotel, Econolodge, May Wood Lodge, Motel 6, Nendels Inn, Pepper Tree Motel, AAA Red Lion Inn, AAA Sun Country Inn, Tourist Motor Inn, Twin Bridges Inn

AServ **S:** Dick's Auto, Texaco

RVCamp **S:** The Trailer Inns

Med **S:** ✚ Hospital

Other **S:** Coin Laundry

33AB Yakima Ave., Terrace Heights,

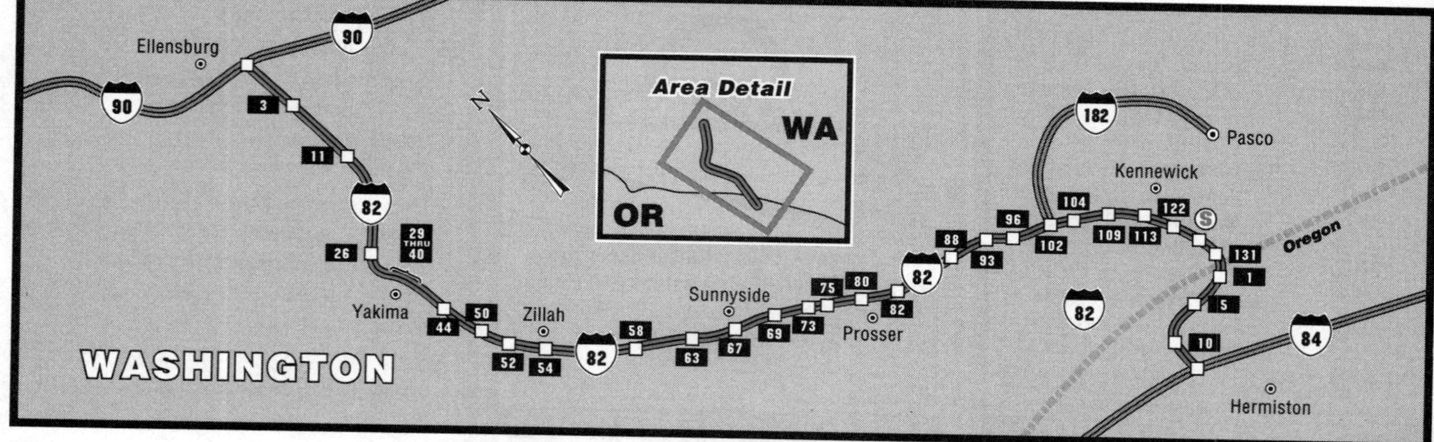

Bold red print shows RV & Bus parking available or nearby

EXIT | WASHINGTON

Lincoln Ave., Fair Ave.
Food S: Best Western
Lodg S: Best Western, Holiday Inn
AServ S: Buick Olds, Honda Dealer, Wal-Mart
Other N: Office Max, Pavillion Foods, Target Department Store
S: Wal-Mart (Pharmacy)

4 WA 24E, Moxee, Nob Hill Blvd
Gas S: 7-11 Convenience Store*, Arco*, Chevron*[D][CW], Exxon[D], Shell*[D]
Food S: Subway (Chevron)
AServ S: Al's Auto Supply, Carlton's Transmission Center, Goodyear Tire & Auto (Frontage Rd.), Paul's Autobody Shop
TServ N: Woodpecker Truck Service
S: Coffey Truck & Equipment, Cummins Diesel, Freightliner Dealer (Frontage Rd.), Kenworth Northwest, Peterbilt Dealer (Frontage Rd.), Volvo Dealer (Frontage Rd.)
RVCamp S: Circle Ranch RV Park
ATM S: 7-11 Convenience Store, Chevron
Parks N: State Park (2 Miles)
Other S: Family Foods

6 Union Gap, Valley Mall Blvd
TStop S: GearJammer Truck Plaza (RV Dump)
Gas S: Arco*, Texaco*
Food S: Denny's, IHOP, JB's Restaurant, KFC, Old Town Station Family Restaurant, Sea Galley, Shangrila Restaurant
Lodg S: Days Inn, (AAA) Quality Inn, Super 8 Motel
AServ S: Frank's Tire's, Sears, Valley Mall Auto Plaza
TServ S: GearJammer Truck Plaza
RVCamp S: Canopy RV Service
Med S: + Valley Medi-Center
ATM S: Seafirst Bank
Other S: Shopko Grocery (Pharmacy), Valley Mall

7 U.S. 97 S., Goldendale, Bend Or (Divided Hwy)

8 Union Gap (Westbound, Reaccess Eastbound)

0 Thorp Road, Parker Road

4 Wapato
Gas N: Texaco*
Other N: Donald Fruit & Mercantile

0 WA 22E, Toppenish, Buena
RVCamp S: RV Park
Med S: + Hospital
Other S: Mural of Museums, Yakima Cultural Station (4 Miles)

2 Zillah
Gas N: ARCO, BP*[D], Chevron* (24 Hrs)
Food N: Hyatt's Restaurant, McDonalds, Subway
Lodg N: ◆ Comfort Inn
Other N: Car Wash, Fruit Store

4 Zillah, Division Road
Gas S: Gas[D][LP] (Frontage Rd.)
Food N: El Ranchito
Other N: Tourist Information

8 WA 223S, Granger
AServ S: Eddie Post OK Tires, Eddie's Muffler's
RVCamp S: Camping
Other S: Mezas Market, Village Market

3 Sunnyside, Outlook
RVCamp N: RV Park (3 Miles)

7 Sunnyside, Mabton
Med N: + Hospital

9 WA 241N, Vernita Blvd., Mapton
FStop N: Texaco*
Gas N: Arco*
Food N: Arby's, Burger King (Arco), Dairy Queen, McDonalds, Pizza Hut, Taco Bell (24 Hrs)
Lodg N: Rodeway Inn
Med N: + Hospital

3 Grandview, Stover Road
AServ S: Chrysler Auto Dealer (RV Service)

EXIT | WASHINGTON/OREGON

RVCamp N: Camping (RV Park)
S: Chrysler Auto Dealer

75 County Line Road
FStop S: Pacific Pride Commercial Fueling
Gas S: Conoco*[D][LP], Exxon*
Food S: L & L
Lodg S: Grandview Motel
AServ S: Auto Mania Body Shop, Conoco, Dayton Tires
RVCamp S: Long's Home Center (RV Service)
Other S: Star Food City, Visitor Information

80 Gap Road, Prosser
TStop S: Texaco*[LP][SCALES]
Food S: Grand Slam Pizza, McDonalds, Northwoods Family Restaurant
Lodg S: ◆ Best Western, Prosser Motel
AServ S: Ford Dealership
RVCamp S: RV Park
Med S: + Hospital
Other S: Visitor Information

(80) Rest Area - RR, Phones, Picnic (P) (South Side Exit 80)

82 WA 22, WA 221, Mabton, Patterson
Other S: Museum, Visitor Information

88 Gibbon Road
FStop S: Four-P Truck Stop*

93 Yakitat Road

96 WA 224E, Benton City, West Richland, WA 225N
Gas N: BP*[D][LP]
Food N: Cactus Jack's Cafe (BP)
RVCamp N: RV Park
ATM N: BP

102 Junction I-182, U.S. 12E, Richland, Pasco
Med N: + Hospital

104 Goose Gap Road, Dallas Rd
Gas N: Conoco*
Food N: Hot Stuff Pizza (Conoco), Smash Hit Subs (Conoco)

109 Badger Road
RVCamp N: Campground

113 U.S. 395N to I-82, Kennewick, Spokane
Med N: + Hospital
Other N: State Patrol

114 Locust Grove Road

122 Coffin Road

(130) Weigh Station (Westbound)

131 WA 14W, Plymouth, Vancouver
RVCamp S: Campground

↑WASHINGTON

↓OREGON

1 U.S. 395, U.S. 730, Umatilla, Hermiston
TStop S: Crossroads*[SCALES]
FStop S: Arco*
Food S: Restaurant (Crossroads Motel), The Sub Shop
Lodg S: Crossroads Motel, Rest-A-Bit Motel
RVCamp N: Hat Rock Campground
Med N: + Hospital
S: + Umatilla Medical Clinic
ATM S: Crosssroads Truck Stop
Other S: Oregon Welcome Center, Red Apple Market, U.S. Post Office, Western Union (Crossroads TS)

5 Power Line Road

10 Westland Road

↑OREGON

Begin I-82

EXIT | PENNSYLVANIA

Begin I-83

↓PENNSYLVANIA

(50) Jct I-81, Hazeltown, Allentown, U.S. 322W, Carlisle (Left Exit)

30 **(49)** U.S. 22, Colonial Park, Progress
Gas E: Atlantic*, Mobil*, Sunoco, Texaco*[CW][LP]
W: Citgo*, Exxon*, TLC*
Food E: Applebee's, Colonial Park Diner, El Rodeo, Long John Silvers, McDonalds, Red Lobster, Taco Bell
W: Dairy Queen, Dunkin Donuts, Friendly's, KFC, Pasta & Pizza
AServ E: Avellino's Auto Center, BF Goodrich, Goodyear Tire & Auto, Meineke Discount Mufflers, Penbrook Auto Repair, Sears, Sunoco, Tire America, Wimmer Tire
W: Bob's Auto & Tire Service, Jiffy Lube
ATM E: Commerce Bank, MAC
Other E: Colonial Park Mall
W: Coin Laundry

29A **(48)** Union Deposit Rd

29 **(47)** Deery St

28 **(46)** U.S. 322E, to Hershey

(46) Jct I-283S, to I-76, PA Tnpk Airport, Lancaster (Left Exit)

26 Paxtang (Southbound Reaccess Exit 25)

25 **(44)** 19th St

24 **(43)** PA230, 13th St

23 **(43)** Capital, 2nd St

22 **(42)** Lemoyne (Eastside Same As Exit 21)

21 **(42)** Highland Park
Gas E: Exxon*[CW], Mobil*, Turkey Hill*
W: Mobil*, Texaco*
Food E: KFC
W: Rascal's, Royal Sub Shop
AServ E: Mobil
ATM W: Dauphin Deposit Bank
Other W: Thrift Drug, Weis Market

20 **(41)** PA581W, Camp Hill

19 **(41)** New Cumberland

18A **(40)** Limekiln Rd
Gas E: BP*
W: Exxon
Food E: Best Western, Bob Evans Restaurant, Eat N Park, Holiday Inn, Pizza Hut
Lodg E: Best Western, Fairfield Inn, Holiday Inn, MacIntosh Inn
W: Knight's Inn, Motel 6
AServ E: BP
W: Exxon
Other W: West Shore Veterinary Hospital

(39) Jct I-76, PA Tnpk, to Phil, Harrisburg, Hershey, Lancaster

18 **(39)** PA114, Lewisberry Rd
Food E: Captain Wolf Restaurant (Days Inn)
Lodg E: Highland Motel, Keystone Days Inn
AServ E: Richards Auto Repair

17 **(38)** Reesers Summit (Southbound)
AServ W: Ken's Service Ctr

EXIT		PENNSYLVANIA
16		**(37)** PA262, Fishing Creek
	Gas	**E:** Shell*(LP)
		W: Texaco(D)(LP)
	Food	**W:** Culhane's Steakhouse
	AServ	**W:** Texaco
	ATM	**E:** Dauphin Bank
	Other	**E: Fairview Pharmacy**
15		**(36)** PA177, Lewisberry
	Gas	**W:** Mobil* (Kerosene)
	Food	**E:** Hillside Steak House, Jalapeno's Mexican
		W: Francesco's ItaItian, **Hardees**
	AServ	**E:** Gross General Repair
	TServ	**W: Interstate Truck Park**
	Other	**W: Skiing**
14A		**(35)** PA392, Yocumtown
	TStop	**E: Henry's***
	FStop	**E: Exxon*(D)**
	Gas	**E:** Rutters*(LP)
	Food	**E:** 2 Brothers Italian, **Alice's Restaurant (Truckstop)**, Maple Donuts, McDonalds, New China, USA & I Pizza
	Lodg	**E: Super 8 Motel**
	TServ	**E: Truck Service**
	ATM	**E:** Harris Savings Bank, Super Fresh Supermarket, The York Bank
	Other	**E:** Coin Car Wash, **Live Bait (Rutters), Newberry Common, Park Away Park(LP), Super Fresh Supermarket, Thrift Drug**
14		**(35)** PA392, Yocumtown (Under Construction Northbound)
	RVCamp	**E: Park Away Park(LP)**
(35)		**Weigh Station (Closed Due To Construction Southbound)**
(33)		**Weigh Station (Northbound)**
13		**(33)** PA382, Newberrytown
	FStop	**W: Mobil***
	Gas	**E:** Exxon*, Sunoco(D)
	AServ	**E:** RW Auto Repair, Sunoco (Wrecker Service)
	TServ	**E: Detroit Diesel**
	ATM	**E:** Exxon
12		**(28)** PA295, Strinestown
	Gas	**W:** Exxon*(LP)
	Food	**W: I-83 FamilyRestaurant**, Wendy's
	Lodg	**E:** Charles Tone Motel
	AServ	**W:** York Auto Repair
	ATM	**W:** Exxon
11		**(24)** PA238, Emigsville
	Food	**W: Sybil's Log Cabin Steak House**
	TServ	**E: C Earl Brown Inc (Frontage Rd)**
	Other	**W: Animal Emergency Clinic**
10		**(22)** Bus Rt 83, PA181, to U.S.30W, N George St
	Gas	**W:** Sunoco
	AServ	**W:** Sunoco
9		**(21)** U.S. 30, Arsenal Rd, Lancaster, Gettysburg (Difficult Reaccess)
	Gas	**W:** Exxon, Mobil(CW)
	Food	**E:** Big Boppers Drive In, Bob Evans Restaurant, Holiday Inn, Round the Clock Diner, San Carlo's, The Hop
		W: Arby's, Boston Market, Bubbles Bagels, Burger King, Friendly's, Hardees, Italian Oven, Long John Silvers, McDonalds, Ponderosa Steakhouse, Subway, Taco Bell, Wendy's
	Lodg	**E:** DAYS INN Days Inn, Holiday Inn, Ramada Inn, Red Roof Inn
		W: Motel 6, Super 8 Motel
	AServ	**E:** Giambaldo
		W: Econolube & Tune, Exxon, Fisher Auto Parts, Mobil, NTB, Pep Boys Auto Center

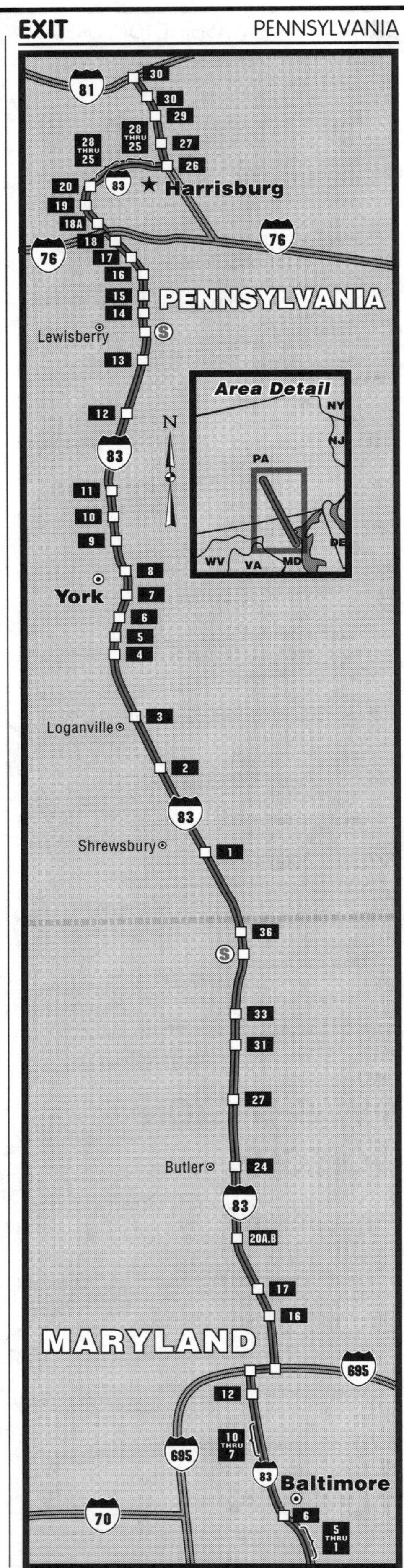

EXIT		PENNSYLVANI
	Med	**W: ✚ Hospital**
	ATM	**W:** CoreState Bank, Mellon Bank, York Bank
	Other	**E: Harley Davison, Tourist Information**
		W: Bell Tone Hearing Aid Ctr, CVS Pharmacy, County Market, Crossroads Shopping Center, Thrift Drug, Weiss Market
8		**(19)** PA462, Market St
	Gas	**W:** Sunoco*
	Med	**W: ✚ Memorial Hospital**
	Other	**W:** Automated Car Wash
7		**(18)** PA124, Mt Rose Ave
	FStop	**E: Pacific Pride Commercial Fueling***
	Gas	**E:** Exxon*, Sunoco*
	Food	**E:** Big Apple Bagels, Burger King, Denny's, Lee Gardens, Penrod's, Pizza Hut (Sunoco), Pizza H Subway, Viet Garden
	Lodg	**E:** AAA **Budget Host Inn**
	AServ	**E:** Exxon (Wrecker Service)
	ATM	**E:** Farmers Bank, Sunoco, York Bank
	Other	**E: Brothers C-Store, K-Mart, Revco Drugs, W Food Mart**
6		**(16)** E&W PA74, Queen St
	Gas	**E:** Mobil
		W: Mobil*
	Food	**E:** Chinese Restaurant, **Cracker Barrel**, Hardees, Orzo's Pizza, Prime Time Restaurant, Your Pet & Fish
		W: Donut Delite, McDonalds, Ponderosa Steakhouse, Subs to Go, Taco Bell, Wonderful Garden
	AServ	**E:** Mobil, Square Deal Garage, York Lincoln, Mercury
		W: Griffith Cadillac, BMW, Griffith Honda of Yor Monroe Muffler & Brake, Shaull Olds
	TServ	**E: Snyder's Volvo**
	ATM	**E:** Connor's Bank, Drovers Bank
		W: CoreState Bank
	Other	**E: Auto Hearing Prof Ctr, Giant Food & Pharmacy**
		W: Phar Mor Drugs, Queens Gate Shopping Center, Veterinary Clinic
5		**(15)** South George St
4		**(14)** PA182, Leader Heights
	Gas	**E:** Exxon(D)
		W: Exxon*, Sunoco*
	Food	**E:** Domino's Pizza, First Wok, Mr Bill's Quarterback Lounge, Subway
		W: McDonalds, Pizza Hut (Exxon), Rutter's Deli (Exxon)
	AServ	**E:** Exxon(D)
	RVCamp	**W: Indian Rock Campground**
	ATM	**W:** Exxon
	Parks	**W: Kain Park**
	Other	**E: Eye Care, Help Thy Self (Health Food Store**
3		**(11)** PA214, Loganville
	Food	**W:** Lee's Drive In, Mama's Pizza
	Lodg	**W:** Midway Motel
	ATM	**W:** The Glen Rock State Bank
	Parks	**W: Kain, Nixon**
	Other	**W: Ace Hardware,** Coin Car Wash, **PA Police Station, Tru-Value Hdwe**
2		**(8)** PA216, Glen Rock, Winterstown
	Lodg	**E: Rocky Ridge Motel**
	RVCamp	**E: Rocky Ridge Motel, Spring Valley Camp**
1		**(4)** PA851, Shrewsbury
	FStop	**E: Citgo*(D), Mobil***
	Food	**E: TCBY Treats (Mobil), Tom's Pizza & Subs (Mobil)**
		W: Coach Lite Family Restaurant, Hardees, McDonalds

Bold red print shows RV & Bus parking available or nearby

EXIT	PENNSYLVANIA/MARYLAND
ATM	E: Drovers Bank
	W: York Bank
Other	E: True Value Hdwe
	W: Giant Food Store, K-Mart, Revco Drugs, Saubels Grocery, Shewsbury Veterinary Clinic(LP)
)	PA Welcome Center, Rest Area - Picnic, RR, Phones, Vending, Pets (Northbound)

↑PENNSYLVANIA

MARYLAND

EXIT	
6	MD439, Maryland Line, Bell Air
RVCamp	W: Morris Meadows Tra-V-L Park
5)	Weigh Station (Southbound)
3	Parkton, MD 45
Gas	E: Exxon*(D)
AServ	E: Exxon*(D) (Wrecker Service)
	Middletown Rd, Parkton
7	MD137, Mt Carmel Rd, Hereford
Gas	E: Exxon*(D)(LP)
AServ	E: Exxon (Wrecker Service)
ATM	E: First National Bank of MD, Signet
Other	E: Graul's Market Food Store, Hereford Pharmacy
4	Bellfast Rd, Butler, Sparks (No Trucks West)
0AB	Shawan Rd, Cockeysville
Food	E: Marriot, Turtle Bay Seafood
Lodg	E: Courtyard Of Marriott, Embassy Suites, Hunt Valley Inn Marriot, Residence Inn
AServ	E: Sears
ATM	E: First National Bank
Parks	W: Oregon Ridge State Park
Other	E: Hunt Valley Mall
	Warren Rd, Cockeysville (Northbound, Reaccess Southbound Only)
	Padonia Rd, Cockeysville
Food	E: Days Inn, Denny's (Days Inn), Dinner Theatre, Domino's Pizza, Peerce's Diner, Ralphie's Diner
Lodg	E: ◆ Days Hotel
AServ	E: AAMCO Transmission, Precision Alternator & Starter, Radiator Warehouse
ATM	E: First National Bank of MD
AB	Timonium Rd
Gas	E: Amoco*(CW), Exxon*(D), Sunoco*(D)
Food	E: Chi Chi Mexican, McDonalds, Michael's Cafe, Nashville's (Holiday Inn), Steak & Ale, TCBY Treats, Uncle Teddy's Soft Pretzels
Lodg	E: Holiday Inn Select, Red Roof Inn
AServ	E: Auto World, Jiffy Lube, Nationwide Infinity, Nationwide Service Center, Nissan
ATM	E: NationsBank, Signet Bank
Other	E: Coin Operated Laundry, Lykos Pharmacy, Royal Farms C-Store
	Jct I-695, to Towson, Pikesville, to I-95S
	Ruxton Rd (Northbound, Southbound Reaccess Only)
AB	Northern Pkwy
AB	Cold Springs Ln
	MD25N, Falls Rd (Downtown)
	28th St, Druid Park, Lake Dr (Downtown)
	U.S. 1, Truck Rt U.S. 40, North Ave (Downtown)

↑MARYLAND

Begin I-83

EXIT	OREGON

Begin I-84

↓OREGON

EXIT	
1A	Lloyd Boulevard (Westbound, Reaccess Eastbound Only)
Food	N: Stanford's
	S: 1500 Deli East
Lodg	N: Double Tree Hotel, Marriott
AServ	S: Oldmobile Dealer
Other	N: Lloyd's Center
1B	33rd Ave (Eastbound)
Gas	N: Albina Fuel Co., BP, Texaco*(D)(CW)
Food	N: Burger King, Domino's Pizza, The Blind Onion
	S: Poncho's, Shaughnessy's Bar and Grill, Wendy's
Lodg	S: Banfield Motel, Hollywood Motel
AServ	N: BP, Sparks Auto Repair
	S: Breslin Pontiac, GMC Trucks, Jiffy Lube, Les Schwab Tires, Wallis Buick
2A	39th Ave (Eastbound, Reaccess Eastbound Only)
Food	N: Baskin Robbins, Burger King, McDonalds, My Canh, Poor Richard's Restaurant, Sweet Indulgence, The Pagoda, Tio Loco's, Winchell's Donuts, Wing Wah Chinese Restaurant
Lodg	N: Howard Johnson
AServ	N: Perry Thieman's Auto Service
ATM	N: Bank of America
Other	N: Hollywood Vision Center, Paulsen's Pharmacy, Wrigley-Cross Books
2B	43rd Ave (Westbound, Reaccess Eastbound Only)
Gas	N: Chevron*
Food	N: Baskin Robbins, Chin's Kitchen, The Pagoda, Winchell's, XO
Lodg	N: Roadway Inn
ATM	N: Bank of America
3	58th Ave (Eastbound, Reaccess Westbound Only)
Gas	S: BP*, Texaco
Food	S: Sandwich Depot Deli
AServ	S: Texaco
Med	S: ✚ Hospital
4	68th Ave (Eastbound, No Reaccess)
Food	N: Blaze Jose's Mexican
AServ	N: Danny's Auto Repair
Other	N: Coin Laundry
5	OR 213, 82nd Ave (Eastbound, Difficult Reaccess)
Gas	S: Chevron*, Shell, Texaco*(D)
Food	N: Dragon Garden Chinese Restaurant, Gems
	S: Burgerville USA, Elmer's, Pappy's Family Restaurant, Pho Van Vietnamese, Pizza Hut, Subway, Taco Bell, Utopia's, Wendys
Lodg	N: Capri Motel, Motel Cabana
AServ	N: Les Schwab Tires
	S: AAMCO Transmission, Car Quest Auto Center, Jiffy Lube
Other	N: Animal Hospital
	S: Full Service Car Wash, Safe Way Grocery, Walgreens
6	I-205 South, Salem
7	Halsey Street, Gateway District (Eastbound, Difficult Reaccess)
Gas	S: Arco*
Food	S: Applebee's, Boston Market, Carl's Jr Hamburgers, Izzy's Pizza, J.J North's Grand Buffet, McDonalds, Skewer's, Subway
AServ	S: Q-Lube
Med	S: ✚ Hospital
ATM	S: Key Bank, Wells Fargo
Other	S: Cub Foods, Fred Meyer Grocery, Mervyn's, Tower Books
8	I-205 North, Seattle, Portland Airport (Eastbound)
9A	102 ND Ave, Parkrose (Eastbound, Reaccess Eastbound Only)
9B	I-205, Salem, Seattle (Westbound)
10	122 North Ave (Eastbound, Reaccess Eastbound Only)
13	Gresham, Fairview
Gas	S: 76, Arco*, BP*(CW)(LP)
Food	S: Burger King (Playground), Dotty's, Jung's Dynasty Restaurant, Plumtree Restaurant, Sari's Restaurant, Wendy's
Lodg	N: (AAA) Hampton Inn, Pioneer Motel
	S: (AAA) Holiday Inn Express, Sleep Inn
AServ	S: 76
ATM	S: 7-11 Convenience Store, Wells Fargo
Other	S: 7-11 Convenience Store* (ATM), Safe Way Grocery, Wildwood Animal Hospital
14	207th Ave., Fairview
Food	N: Wind Cove
RVCamp	N: American Dream RV Sales and Service, Portland Fairview RV Park, Rolling Hills
16A	Wood Village, Gresham
TStop	N: Krueger's Auto/Truckstop(SCALES)
Gas	N: Arco*, Chevron(CW)
	S: Chevron*
Food	N: Good Buddies Restaurant, Krueger's Restaurant, Royal Chinese
Lodg	N: (AAA) TraveLodge
AServ	N: Chevron
16B	Crown Point Hwy (Difficult Eastbound Reaccess)
17	Troutdale, Airport
TStop	S: Burns Bros Travel Stop*(LP)(SCALES), Flying J Travel Plaza*(SCALES) (RV Dump)
Gas	S: Chevron(LP)
Food	S: Burger King, McDonalds, Mrs. B's (Burns Bros TStop), Sharis, Subway (Burns Bros), Taco Bell
Lodg	S: Burns West Motel, Motel 6, (AAA) Phoenix Inn
	N: Inn America (see our ad this page)
AServ	S: Chevron

EXIT OREGON

TServ S: Burns Bros Travel Stop
Other S: Columbia Gorge Factory Stores, Visitor Information
18 Lewis & Clark State Park, Oxbow County Park
22 Corbett
Food S: Royal Chinook Inn
RVCamp S: Crown Point RV Park (1 3/4 Miles)
25 Rooster Rock State Park (No Trucks)
28 Bridal Veil (Eastbound, Reaccess Eastbound Only)
29 Dalton Point (Westbound, Reaccess Westbound Only)
30 Benson State Park (Eastbound)
31 Multnomah Falls (Exit To The Left)
35 Historic Hwy, Ainsworth Park
37 Warrendale (Westbound, Reaccess Eastbound Only)
40 Fish Hatchery, Bonneville Dam
41 Fish Hatchery, Eagle Creek Recreation Area (Eastbound, Reaccess Eastbound Only)
44A U.S. 30 West, Cascade Locks, Stevenson
Gas N: Big D's[D], Chevron[LP], Shell*, Texaco
Food N: Cascade Inn, Charburger, Eastwind Drive-In, Gum Oak, Salmon Row Pub
Lodg N: Best Western, Bridge of the Gods, Econo Inn, Scenic Winds Motel
AServ N: Chevron
RVCamp N: KOA Campground
Other N: Columbia Market, U.S. Post Office
S: Weigh Station
47 Forest Lane, Herman Creek (Westbound, Reaccess Eastbound Only)
RVCamp S: Campground
51 Wyeth
RVCamp S: Camping
(54) Weigh Station (Westbound)
(55) Rest Area - RR, Phones, Starvation Creek State Park P (Eastbound)
56 Viento Park
58 Mitchell Point Overlook (Eastbound, Reaccess Eastbound Only)
60 Service Road (Westbound, Reaccess Eastbound Only)
62 US 30, West Hood River, Westcliff Drive
Gas S: Chevron*, Texaco*[D][LP]
Food N: Charburger Country
S: Dairy Queen, Red Carpet Inn, Stonehedge Inn, Taco Bell
Lodg N: Columbia Gorge Hotel, Meredith Motel, AAA Vagabond Lodge
S: AAA Comfort Suites
AServ S: Cliff Smith Chevrolet, GM, Les Schwab Tires
Med S: + Hospital
ATM S: Columbia River Bank
Other N: The Fruit Tree
S: Wal-Mart
63 Hood River City Center
Food S: Big Horse Brew Pub, Chilis Cantina, Golden Rose Chinese, Hood River Restaurant, Pietro's Pizza, River View Deli, Sage's Cafe
Lodg S: Hood River Hotel
ATM S: Bank Of America, Wells Fargo

EXIT OREGON

Other S: Artifacts Used Books
64 U.S. 30, OR 35, Mt Hood Hwy, White Salmon Gov't Camp
Gas N: Texaco*
Food N: McDonalds, River House Gourmet Pizza, Riverside Grill (Best Western), Taco Time, The Gorge Cafe
S: Cafe Winddance, China Gorge Restaurant
Lodg N: AAA Best Western
Other N: Hood River Toll Bridge
(66) Rest Area P (Westbound)
69 U.S. 30, Mosier
(73) Info Center, Rest Area - RR, Phones, RV Water, RV Dump P (Both Directions)
Other N: Memalose State Park
76 Rowena, Mayer State Park
Parks N: Mayer State park
82 Chenoweth Area (Eastbound, Reaccess Westbound Only)
FStop S: Pacific Pride Commercial Fueling
Gas S: Texaco*
Lodg S: Oregon Trail Motel
AServ S: Texaco
83 West The Dalles
FStop N: Shell*
Gas S: Chevron[LP], Exxon*[LP]
Food N: Orient Cafe
S: Arby's, Burger King, Circle C, Cousins Restaurant, Denny's, KFC, Lindo's Mexico, McDonalds, Papa Murphy's Pizza, Skipper's, Slugger's Pizzaeria, Subway, Taco Bell, Wendy's
Lodg S: DAYS INN AAA Days Inn, AAA Quality Inn
AServ N: Burling's Auto Repair, Nelson Tire
S: Chevron, GM Auto Dealership, Ray Schultens, Schuck's Auto Supply
Other S: Albertson's Grocery, Fred Meyer Grocery, K-Mart, Payless Drugs, Shop'n Kart, Wash N Shop Laundry
84 West The Dalles (Difficult Eastbound Reaccess)
FStop S: Pacific Pride
Gas S: Chevron*, Texaco[D]
Food S: Burgerville USA, Casa El Marado
Lodg S: AAA Best Western, Oregan Motor Motel, Shamrock Motel
AServ S: Auto Electric Supply Co., Dale's Auto Parts, Napa Auto Parts, Texaco
Med S: + Hospital
Other S: Old Fort Dalles Museum, Safeway Drugs
85 City Center, The Dalles
Gas S: BP
Food S: Domino's Pizza, Guadalajara, Holesteins Coffee Co.
AServ S: Precision Auto Repair
Med S: + Hospital
87 U.S. 30, U.S. 197, Dufur Bend
Gas N: BP*, Texaco*[D][LP]
Food N: Lone Pine Restaurant, McDonalds, O'Calahans (Shilo Inn), Taco Bell
S: Big Jim's Drive-in
Lodg N: AAA Lone Pine Motel, AAA Shilo Inn
S: The Inn at the Dalles
RVCamp N: Lone Pine Park
88 The Dalles Dam
97 OR 206, Celilo Park
104 U.S. 97, Yakima Bend
TStop S: Biggs Auto/Truckstop[D][LP]

EXIT OREGON

Gas S: Chevron, Texaco
Food S: Biggs Cafe, Jack's Fine Food, Linda's Restaurant
Lodg S: AAA Best Western, Biggs Motel, Dinty Motor Inn
AServ S: Chevron, Texaco
TServ S: Biggs Auto/Truckstop
RVCamp S: Biggs RV Park
ATM S: Mini Market
Other S: Mini Market (ATM)
109 Rufus, John Day Dam
FStop S: Pacific Pride Commercial Fueling
Gas S: BP* (ATM)
Food S: Bob's Texas T-Bone, Restaurant
Lodg S: Tyee Motel
ATM S: BP
(112) Parking Area (Both Directions, Dam Overlook)
114 John Day River Recreation Area
123 Philippi Canyon
129 Blalock Canyon
131 Woelpern Road (Eastbound, Reaccess Westbound Only)
136 Viewpoint (Westbound)
137 OR 19, Arlington, Condon
Gas S: BP*[D][LP], Chevron*
Food S: Happy Canyon, Pheasant Grill Drive-in, Village Inn Restaurant
Lodg S: Village Inn
RVCamp S: Port of Arlington
Other S: Post Office
147 OR 74, Ione, Heppner
151 Threemile Canyon
159 Tower Road
(161) Rest Area - Phones, RR, Picnic P (Both Directions)
164 Boardman
FStop N: Texaco*[D]
S: BP*[D] (Taco Bell)
Gas N: Chevron[LP]
Food N: C&D Drive-in, Longbranch Room (Dodge City Motel)
S: Nomad Restaurant, Pizza & Subs, Taco Bell
Lodg N: Dodge City Motel, Riverview Motel
S: The Nugget Inn
AServ N: Chevron
S: Boardman Auto Repair, NAPA Auto Parts
RVCamp S: Driftwood (2 Miles)
Other N: Coin Car Wash, Post Office
S: Century Foods
165 Port of Morrow
FStop S: Pacific Pride Commercial Fueling
168 U.S. 730, Irrigon
171 Paterson Ferry Road
177 Umatilla Army Depot
179 I-82 West, Umatilla, Kennewick, Hermiston
180 Westland Road, Hermiston
TServ S: Barton Industries-Truck Body & Trailer Repair
182 OR 207, Hermiston, Lexington
FStop N: Buffalo Junction*
Food N: Buffalo Junction, McDonalds, Shari's

Bold red print shows RV & Bus parking available or nearby

XIT OREGON

Restaurant
Lodg N: Sands Motel
RVCamp N: Buttercreek RV Park
Med N: ✚ Hospital

86) Rest Area - RR, Phones, Picnic, Pet Excercise, Tourist Info P (Both Directions)

8 U.S. 395 North, Stanfield, Echo, Hermiston
TStop N: Pilot*(LP)(SCALES) (24 Hrs)
Food N: Country Cooker (Pilot TS), Subway (Pilot TS)
RVCamp N: Camping
S: Fort Henrietta RV Park (1 1/4 Miles)

3 Echo, Lexington

8 Lorenzen Road, McClintock Rd
TServ N: PJ's Truck Repair

9 Yoakum Road, Stage Gulch

02 Stage Gulch, Barnhart Road
TStop S: Floyd's Truck Ranch*
Food S: Ranch Cafe
Lodg S: 7 Inn, Floyd's Truck Ranch Motel
TServ N: International Dealer, Woodpecker Truck
S: Floyd's Truck Ranch, Steve's Truck Body Repair

07 Airport, West Pendleton, US 30
Gas N: Chevron, Texaco*(D)(LP)
AServ N: Chevron, Pacific Power, Texaco
RVCamp N: Brooke RV West
Other N: Cash & Carry United Grocery

09 U.S. 395, OR 37, John Day Pendleton
Gas N: Town Pump*
S: Exxon*(D)(LP), Sunshine Gas And Wash*(CW), Texaco*
Food N: KFC
S: Burger King, Denny's, Klondike Pizza, McDonalds, Rooster's, Subway, Wendy's
Lodg N: Traveller's Inn
S: AAA Chaparral Motel
AServ N: Battery X-Change
S: Honda Dealer, Kube Lube, Les Schwab Tires, My Own Auto Sales, Pontiac, Oldsmobile, Buick, Cadillac, GMC
VCamp S: Thompson RV Service
ATM N: Pacific One Bank
Other N: Liberty Laundrymat
S: Golden Dragon Fireworks, K-Mart, Southgate Mini Mart

0 Pendleton, OR 11, Milton - Freewater
FStop N: BP*
Gas S: Texaco*(D)(LP)
Food S: Kopper Kitchen
Lodg S: AAA Best Western, Double Tree Hotel, Motel 6, ◆ Super 8 Motel
VCamp S: Camping (1.25 Miles)
Med N: ✚ Hospital
Other N: State Police
S: Bi-Mart (Pharmacy)

3 U.S. 30, OR 11, Pendleton City Center, Milton - Freewater (Difficult Reaccess Westbound)

EXIT OREGON

216 Milton - Freewater, Walla Walla
TStop N: Arrowhead Truck Plaza*(SCALES)
Food N: Arrowhead Truck Plaza
TServ N: Arrowhead Truck Plaza

(222) View Point (Eastbound)

224 Poverty Flat Road, Old Emigrant Hill Road

(227) Weigh Station (Westbound)

228 Deadman's Pass

(228) Rest Area - RR, Phones, Picnic P

234 Emigrant Park, Meacham

238 Meacham, Kamela

243 Mt Emily Road, Summit Rd.

248 Kamela, Spring Creek Road
Parks N: Oregon Trail Interpretive Park (3 Miles)

252 OR 244, Starkey
Parks S: Hilgard State Park

256 Perry

(259) Weigh Station (Eastbound)

259 U.S. 30, La Grande (Eastbound, Westbound Reaccess Only)

261 OR 82, La Grande, Elgin
FStop N: BP*
S: Exxon*(CW)
Gas N: Texaco*(D)
S: Conoco*
Food N: Denny's, Pizza Hut, Subway
S: Dairy Queen, Klondike Pizza, McDonalds (Playground), Skippers Seafood, Taco Time, Wendy's
Lodg N: Best Western
S: Broken Arrow Lodge, AAA Comfort Inn, Mr. Sandman Motel, Royal Motor Inn, AAA Super 8 Motel
AServ N: Ford Dealership, Lincoln
S: Blue Mountain Radiator, Transmissions Unlimited
TServ N: Eagle Truck & Machine Co, Freightliner Dealer
TWash N: Truck Wash
RVCamp N: Camping, RV Resort
Med S: ✚ Hospital
ATM N: Pioneer Bank
S: Western Bank
Other N: Coin Car Wash
S: Albertson's Grocery, Coin Laundry, Drug Emporium, Payless Drugs, Visitor Information, Vista Optical

265 OR 203, La Grande Union
TStop S: Flying J Travel Plaza*(SCALES)
Food S: Flying J Travel Plaza, TCBY Yogurt (Flying J Travel Plaza)
RVCamp N: Hot Lake RV Park (5 Miles)

268 Foothill Road

(269) Rest Area - RR, Phones, Picnic, RV Dump P (Both Directions)

270 Ladd Creek Road (Eastbound, Reaccess Westbound Only)

273 Ladd Canyon

EXIT OREGON

278 Clover Creek

283 Wolf Creek Ln.

285 U.S. 30, OR 237, North Powder, Haines
FStop N: Cenex*(LP)
Food N: Cafe
Lodg N: Powder River Motel
Other S: Anthony Lakes (19 Miles), Wildlife Viewing Area

(295) Rest Area - RR, Picnic, Phones P (Both Directions)

298 OR 203, Medical Springs, Haines, Bakers City Airport

302 OR 86 East, Richland, North Baker City
RVCamp S: Oregon Trails West RV Park(LP)
Med S: ✚ Hospital
Other N: Hells Canyon (64 Miles), Oregon Trail Interpretive Center (5 Miles)

304 OR 7 South, Baker City Center, Geiser Grand Hotel
TStop S: Baker's Truck Corral*(LP)(SCALES)
FStop N: Conoco*
S: Texaco*(CW)(LP)
Gas S: Sinclair*
Food N: Burger King, TCBY Yogurt (Conoco FS)
S: Baker Truck Corral, Best Western, Diner, Fong's American Chinese, McDonalds, Pizza Hut, Subway, Sumpter Junction, Taco Time
Lodg N: AAA Super 8 Motel
S: AAA Best Western, Eldorado Inn, AAA Quality Inn
AServ S: Discount Auto Parts & Auto Service, Paul's Automatic Transmission and Repair, Quaker State Quality Lube
TServ S: Baker Truck Corral
TWash S: Baker Truck Corral
RVCamp S: Mountain View RV Park (2 Miles)
ATM S: Baker Truck Corral
Other N: Oregon Trail Interpretive Center
S: Baker City Center Historic District, Baker City Coin Laundry, Humble's Car Wash, Oregon Trail Antique Mall, Payless Drugs (Pharmacy), Safe Way Grocery (Pharmacy), Tourist Information

306 U.S. 30 West, Baker City
Other N: State Patrol

315 Pleasant Valley

327 Durkee
FStop N: Oregon Trail Travel Plaza*(LP)
Food N: Wagon Wheel Restaurant
AServ N: Interstate Battery (Oregon Trail Travel Plaza)

330 Plano Road, Cement Plant Rd.

335 Weatherby
Other N: Rest Area - RR, Picnic, Phone

(335) Rest Area - RR, Phones, Picnic P

338 Lookout Mountain

340 Rye Valley

342 Lime (Eastbound, Reaccess

Bold red print shows RV & Bus parking available or nearby

EXIT	OREGON
	Westbound Only)
345	Lime, U.S. 30 Bus., Huntington
Other	N: Snake River Area
353	Farewell Bend State Park
TStop	N: Farewell Bend
Food	N: Farewell Bend Motel (Farewell Bend TS)
Lodg	N: Farewell Bend Motel (Farewell Bend TS)
AServ	N: Farewell Bend TS
TServ	N: Farewell Bend TS
RVCamp	N: Farewell Bend State Park
Parks	N: Farewell Bend State Park
Other	N: Tourist Information
(353)	Weigh Station (Westbound)
(354)	Weigh Station (Eastbound)
356	Or 201, Weiser
362	Moores Hollow Road
371	Stanton Boulevard
374	OR 201, Ontario, U.S. 30 Bus
FStop	S: Conoco*(LP)
Food	S: Denny's
Lodg	S: Budget Inn, Easy Access Motel
AServ	S: Doersch Engine & Machine Repair, Freeway Texaco
TServ	N: Royals Truck Diesel Repair
	S: Ontario Truck Park
ATM	S: Conoco
Parks	N: Ontario State Park
Other	N: Lake Owyhee
376AB	U.S. 30, Payette, Ontario
TStop	S: Texaco*(LP)(SCALES)
FStop	S: Conoco*
Gas	N: Chevron*(CW)
	S: Phillips 66*
Food	N: Burger King, Country Kitchen, Dairy Queen, Denny's, McDonalds, Taco Time
	S: DJ's Family Restaurant, Domino's

EXIT	OREGON/IDAHO
	Pizza, Far East Chinese Restaurant, Klondike Pizza, Rusty's Pancake & Steakhouse, Sizzler, Taco Bell, Wendy's
Lodg	N: Best Western, Colonial Inn, Holiday Inn, Motel 6, Sleep Inn, Super 8 Motel
	S: Holiday Motel, Oregon Trail Motel, Regency Crest Motel, Stockman's Motel, V-Hotel
AServ	N: Chrysler Auto Dealer
	S: Commercial Tire, Dewey Lube, Fred's Auto Repair, Les Schwab Tires, Miller's Repair, NAPA Auto Parts
TServ	S: Ontario Diesel
RVCamp	S: Rocking R Campers RV Service (Frontage Rd.)
ATM	N: Akins Supermarket
	S: Phillips 66, Texaco TS
Other	N: Akins Supermarket (24 Hrs), Golden Dragon Fireworks, K-Mart, State Police, Wal-Mart (Pharmacy)
	S: Jim's Laundry Mat
(377)	OR Welcome Center - Tourist Info., RR, Phones, Picnic (Westbound)

↑OREGON

↓IDAHO

(1)	ID Welcome Center - Tourist Info, RR, Phones, Picnic (Eastbound)
3	U.S. 95, Payette , Parma

EXIT	IDAHO
Food	N: Palisades Bar and Grill
RVCamp	N: Curtis' Neat Retreat
Other	N: Hells Canyon Recreation Area (79 Miles)
9	U.S. 30, New Plymouth, Emmett
13	Black Canyon Junction
TStop	S: Stinker Fuelstop(SCALES)
Food	S: Black Canyon Restaurant (Stinker TS)
Lodg	S: Stinker Motel
ATM	S: Stinker TS
17	Sand Hollow
Gas	N: Sinclair*(D)
Food	N: Sand Hollow Country Cafe
25	ID 44, Middleton
26	U.S. 20, U.S. 26, Notus, Parma
RVCamp	N: Camp Caldwell Campground
27	ID 19, Homedale, Caldwell, I-84 Bus Loop
Other	S: Tourist Information
28	10th Ave., City Center
Gas	N: Maverik Country Store*
	S: Amoco*, Chevron* (24 Hrs), Citgo*, Conoco*(CW)
Food	S: Dairy Queen, Holiday Motel Cafe, Jack-In-The-Box, Mr V's Restaurant, Pizza Hut, Wendy's
Lodg	N: Budget Motel, I-84 Motor Inn
	S: Holiday Motel
AServ	S: Bruneel Tire & Auto Service
Med	S: Hospital
ATM	S: Citgo
Parks	N: City Park
Other	N: Hands Off Coin Car Wash
	S: Paul's Market Grocery Store, Tourist Information
29	Franklin Road, US 20, US 26
TStop	S: Sinclair*(LP)
Food	S: Perkins Family Restaurant, Sage Cafe (Sinclair TS)
Lodg	S: AAA Comfort Inn, Desert Inn
ATM	S: Sinclair
35	ID 55, Marsing, Nampa Blvd

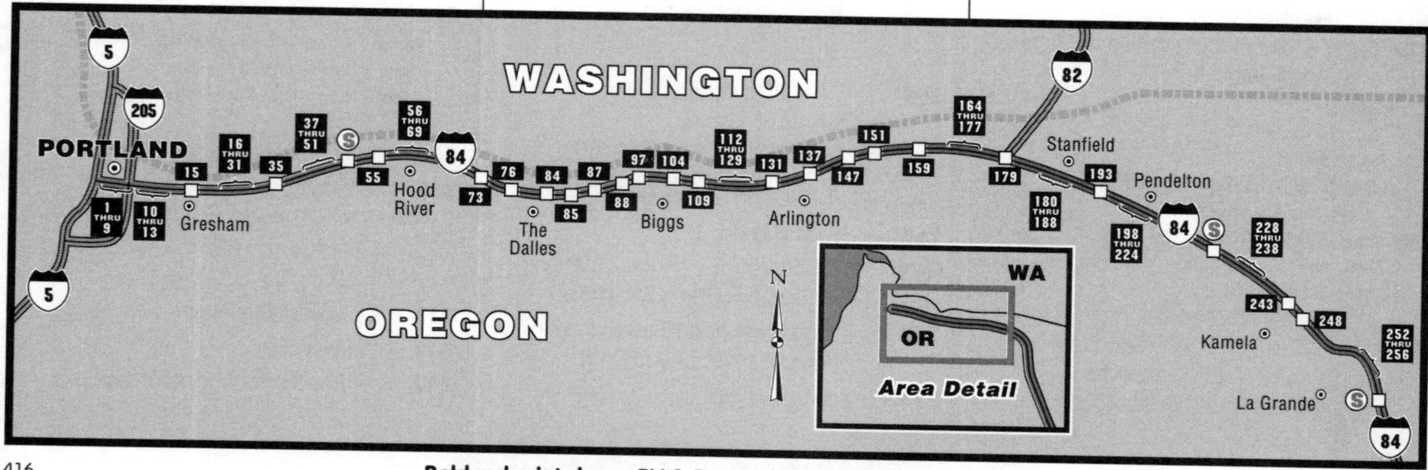

Bold red print shows RV & Bus parking available or nearby

XIT IDAHO

TStop S: Conoco*(CW)(LP)(SCALES)
Food S: Denny's
Lodg S: ◆ Inn America (see our ad this page), Shilo Inn Motel, Super 8 Motel
TServ S: Conoco TS, Triangle Truck Center
TWash S: Conoco TS
Med S: ✚ Hospital
ATM S: Conoco TS
Other S: Trip Permit

6 Franklin Boulevard
FStop S: Chevron*(LP), Texaco*(SCALES)
Gas S: Western Marine(LP)
Food N: Callahan's (Shilo Inn), Noodles Restaurant (Pizza, Pasta)
S: Chester Fried Chicken (Chevron FS), Hot N Now (Texaco), Taco Bell (Texaco)
Lodg N: Shilo Inn Suites Motel
S: ◆ Sleep Inn
RVCamp S: Mason Creek RV Park, Minor's RV Sales & Service
ATM S: Texaco FS

3 Business 84, Garrity Blvd.
RVCamp S: RV Furniture Center

4 ID 69, Meridian, Kuna
Gas N: Chevron*(D)(CW), Sinclair*(D)
S: Husky*(D)(LP), Texaco*(CW)
Food N: Blimpie's Subs, Bolo's Pub and Eatery, Dairy Queen, Godfather's Pizza, KFC, McDonalds, Pizza Hut, Shari's Restaurant, Taco Bell, Taco Time
S: JB's Restaurant
Lodg N: AAA Best Western
S: Mr. Sandman Motel

EXIT IDAHO

AServ N: Les Schwab Tires
S: Meridian Ford
RVCamp N: Regional RV Center Service
S: Playground Sports and RV Park (1 Mile)
Med N: ✚ Mercy Medical Center (Family Dr.)
ATM N: Wells Fargo
Parks N: City Park (Playground)
Other N: Tourist Information
S: Snake River Birds of Prey (23 Miles)

46 ID 55, Eagle, McCall
Gas N: Texaco*(D)
Food N: Taco Bell (Texaco)
RVCamp N: Fiesta RV Park (3.5 Miles)
Med N: ✚ St. Lukes Meridian Medical Center
ATM N: Texaco

49 Jct. I-184, City Center (Left Lane Exit)

50 Overland Road, Cole Road
TStop S: Flying J Travel Plaza*(SCALES) (RV Dump)
Gas N: Chevron* (24 Hrs)
Food N: Buster's Grill, Cancun Mexican Restaurant, Eddie's Restaurant, McDonalds (Playground), Outback Steakhouse, Pizza Hut, Pizzaz Pizza, Subway, TCBY Yogurt, Taco Bell
S: McDonalds (Wal-Mart), Restaurant (Flying J TS)
Lodg S: Best Rest Inn
AServ N: Economy Transmission
S: Wal-Mart
TServ S: Cat Truck Service (Flying J TS), Commercial Tire (Frontage Rd.), Goodyear Tire & Auto, New Holland, Northwest Equipment, Volvo Dealer (Frontage Rd.)
TWash S: Flying J Travel Plaza
ATM N: Chevron, Wells Fargo
S: Flying J Travel Plaza
Other N: Coin Car Wash, Coin Laundry, Deseret Book Store, Dollar Store
S: Wal-Mart (Optical, Pharmacy)

52 Orchard Street
Gas N: Texaco*(D)
Food N: Taco Bell (Texaco)
AServ N: Oldmobile Dealer
ATM N: Texaco

53 Vista Ave., Air Terminal
Gas N: Chevron*, Citgo*, Conoco*(CW), Sinclair, Texaco*(D)(CW) (24 Hrs)
S: Chevron*(CW)
Food N: Taco Bell (Texaco)
S: Denny's (24 Hrs), Kopper Kitchen, McDonalds (Chevron)
Lodg N: Extended Stay America, ◆ Fairfield inn, ◆ Hampton Inn, AAA Holiday Inn (see our ad this page), ◆ Holiday Inn Express, AAA Quality Inn, ◆ Super 8 Motel
S: AAA Best Western, AAA Comfort Inn, AAA Inn America (see our ad this page), Motel 6, AAA Sleep Inn

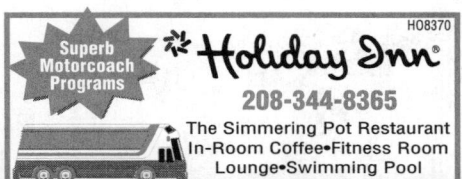

EXIT IDAHO

AServ N: Chevron, Parts n' Stuff, Sinclair
S: Chevron
ATM N: Citgo
S: U.S. Bank
Other N: State Capital Museum, State Police, Zoo
S: Post Office, Tourist Information

54 Broadway Ave.
TStop N: Flying J Travel Plaza*(LP)
S: Burns Bros Travel Stop*(SCALES)
Food N: Hong Chieu, Pepperoni's (Flying J TS)
S: Burns Bros TS
Lodg N: Skyline Motel
S: ◆ Shilo Inn
AServ N: Cummins Diesel, Dowdy's Automotive, NAPA Auto Parts
TServ N: Flying J Travel Plaza
S: Burns Bros Travel Stop*(SCALES), Freuhaus, Kenworth Dealer, Lake City International, Utility Trailer
TWash S: Burns Bros TS
RVCamp S: Camping
Med N: ✚ Hospital
Other N: Office Max, Shopko Grocery (Pharmacy, Optical)

57 ID 21, Idaho City, Gowen Rd.
Gas S: Chevron*(CW)
TServ N: Heno Diesel Trucks Sales and Services, Idaho Peterbilt, Volvo Dealer
S: Jack's Tire & Oil
Other S: Boise Factory Outlets

(62) Rest Area - RR, Picnic P (Both Directions)

EXIT		IDAHO
64		Blacks Creek, Kuna Road
(66)		**Weigh Station - Both Directions**
71		Mayfield, Orchard
	TStop	**S:** Boise Stage Stop*(SCALES)
	Food	**S:** Stage Stop Restaurant (Boise Stage Stop)
	TServ	**S:** Boise Stage Stop
	Other	**S:** Trip Permits
74		Simco Road
90		Mountain Home, Bus. I-84, ID 51, ID 67, Bruneau
	FStop	**S:** Three Ole Gals*(LP)
	Lodg	**S:** Towne Center Hotel
	AServ	**S:** Three Ole Gals
	RVCamp	**S:** KOA Campground (3 Miles)
	Med	**S:** ✚ Hospital
95		U.S. 20, Mountain Home, Fairfield
	TStop	**N:** Pilot Travel Center*(SCALES)
	Gas	**N:** Chevron* (24 Hrs)
	Food	**N:** Arctic Circle, Dairy Queen (Pilot TS), Great American Food (Pilot TS), JB's Restaurant, Subway (With Pilot)
		S: McDonalds
	Lodg	**N:** (AAA) Best Western, Sleep Inn
	RVCamp	**S:** KOA Campground (2 1/2 Miles)
	Med	**S:** ✚ Hospital
	Other	**S:** K-Mart (Pharmacy), Tourist Information
99		Bennett Road
	RVCamp	**S:** Camping
112		Business I-84, ID 78, Hammett
114		Business I-84, ID 78, Cold Springs Road (Westbound, Eastbound Reaccess Only)
120		Business 84, Glenns Ferry (Eastbound, No Reaccess)
	Lodg	**S:** Radford Motel
	Parks	**S:** 3 Islands State Park
121		Business 84, Glenns Ferry, King Hill
	Gas	**S:** Amoco*, IGA Store*(D)
	Food	**S:** Expresso Station, Hanson's Cafe
	Lodg	**S:** Hanson's Hotel
	AServ	**S:** Main St. Garage, NAPA Auto Parts, Uniroyal Tire & Auto(LP)
	Med	**S:** ✚ Medical Care Center
	Parks	**S:** Three Island State Park
	Other	**S:** Coin Laundry, IGA Grocery
125		Paradise Valley
129		King Hill
(133)		**Rest Area - Picnic, RR (P) (Both Directions)**
(134)		**Weigh Station - Both Directions**
137		Bliss, Pioneer Road
	RVCamp	**S:** Camping
141		U.S. 26, U.S. 30, Hagerman, Gooding

EXIT		IDAHO
	TStop	**S:** Texaco(LP)
	FStop	**S:** Phillips 66*, Sinclair*
	Food	**S:** Ox-Bow Cafe, Pizza Bar, Roadrunner Cafe (Texaco), Royal Cafe (Phillips 66 FS)
	Lodg	**S:** Amber Inn Motel, V-Motel
	TServ	**S:** Texaco TS
	RVCamp	**S:** Hagerman RV Village (8 Miles)
	Med	**N:** ✚ Hospital
	ATM	**S:** Texaco
	Other	**S:** Coin Laundry, Post Office
147		Tuttle
	Parks	**S:** Malad Gorge State Park
155		Wendell, ID 46
	RVCamp	**N:** Intermountain RV Camp
157		ID 46, Wendell
	FStop	**S:** Texaco*
	Gas	**N:** Maverick Country*, The Pit Stop*
	Food	**N:** Hogie's Restaurant, Wendell's Snack Bar
		S: Farmhouse Restaurant
	AServ	**N:** BF Goodrich, Hub City Auto Parts Big A, Miller Brother's, Wendell OK Tire
	TServ	**S:** K & D General Truck Repair
	RVCamp	**N:** Intermountain Motor Homes Inc. RV Sales and Service (Frontage Rd.), Intermountain RV Camp
	ATM	**N:** U.S. Bank
	Parks	**N:** City Park
	Other	**N:** Blue Bird Cold Storage(LP), Simerly's Grocery, Winslow's Wendell Department Store
165		ID 25, Jerome
	Gas	**N:** Sinclair*
	AServ	**N:** Number One Auto Part Auto Service
	TServ	**N:** Centennial Truck Repair
	RVCamp	**N:** Campground
	Med	**N:** ✚ Hospital
	Other	**N:** Sawtooth Veterinary Hospital, Suburban Propane(LP)
168		ID 79, Jerome
	TStop	**N:** Sinclair*(LP)
	Gas	**N:** Chevron*(CW)
	Food	**N:** Cindy's Restaurant, McDonalds
		S: Subway
	Lodg	**N:** Best Western, Crest Motel, Holiday Motel
	TServ	**N:** Honker's Diesel, Kenworth Dealer (Frontage Rd.)
	RVCamp	**N:** Brockman's RV Sales and Service
	ATM	**N:** Sinclair
(171)		**Rest Area - RR, Picnic, Phones (P) (Eastbound)**
(172)		**Weigh Station - (Eastbound)**
173		U.S. 93, Twin Falls, Wells, Nevada
	TStop	**N:** Petro*(LP)(SCALES)
	Food	**N:** Quick Skillet* (Petro)
	Lodg	**N:** Sleep Inn
	TWash	**N:** Blue Beacon Truck Wash
	RVCamp	**N:** KOA Campground (1 Mile)

EXIT		IDAH
	Med	**S:** ✚ Hospital
	Other	**S:** Sho Shone Falls (8 Miles), Tourist Information (3 Miles)
(175)		**Rest Area (P)**
(176)		**Weigh Station**
182		ID 50, Kimberly, Eden
	TStop	**S:** Texaco*(SCALES)
	Food	**S:** Blimpie's Subs (Texaco TS), Restaurant (Texaco TS)
	Lodg	**S:** Amber Inn Motel
	RVCamp	**N:** Anderson's Trav-L-Park, Gary's RV Sales
188		Valley Road, Hazelton
194		Ridgeway Road
	FStop	**S:** Amoco*
	RVCamp	**S:** Camping
201		ID 25, Kasota Road, Paul
208		ID 27, Burley, Paul, Bus. I-84
	TStop	**N:** Phillips 66*
	Gas	**S:** Amoco*, Chevron*(D), Sinclair*, Texaco*
	Food	**N:** Connor's Cafe (Phillips 66), Hot Stuff Pizza, Smash Hit Subs (Phillips 66)
		S: Burger King, CJ's Billiards & Burgers, George K's, JB's Restaurant, Jalisco Mexican, KFC, Little Caesars Pizza (K-Mart), McDonalds (Playground), Melina's Mexican, Perkins Family Restaurant, Sod Buster, Subway (Chevron), Taco Bell, Taco Time, Wendy's
	Lodg	**S:** (AAA) Best Western, ◆ Budget Motel
	AServ	**S:** Clete's Auto Repair, Commercial Tires, Pontiac Buick, GMC Dealer, Ray's Muffler's
	TServ	**N:** International Dealer
	Med	**S:** ✚ Hospital
	ATM	**N:** Phillips 66
		S: D.L. Evans Bank, U.S. Bank
	Other	**S:** JC Penny, K-Mart (Pharmacy), Stokes Grocery, Wal-Mart (Pharmacy)
211		U.S. 30, ID 24, Heyburn, Rupert
	FStop	**N:** Stinker*
	Food	**N:** Wayside Cafe (24 Hrs)
	Lodg	**N:** Tops Motel
	Med	**N:** ✚ Hospital
		S: ✚ Hospital
	Parks	**N:** Lake Walcott State Park (16 Miles)
216		ID 77, ID 25, Declo, Albion
	FStop	**N:** Phillips 66*(LP)
	Food	**N:** The River View Cafe (Phillips 66)
	RVCamp	**N:** Snake River Campground
222		Jct. I-86, U.S. 30 East, Pocatello
228		ID 81, Yale Road, Malta
(229)		**Weigh Station, Rest Area - Phones, RR, Picnic, Vending (P) (Both Directions)**
(230)		**Weigh Station - Both Directions**
237		Idahome Road
245		Sublett Road, Malta

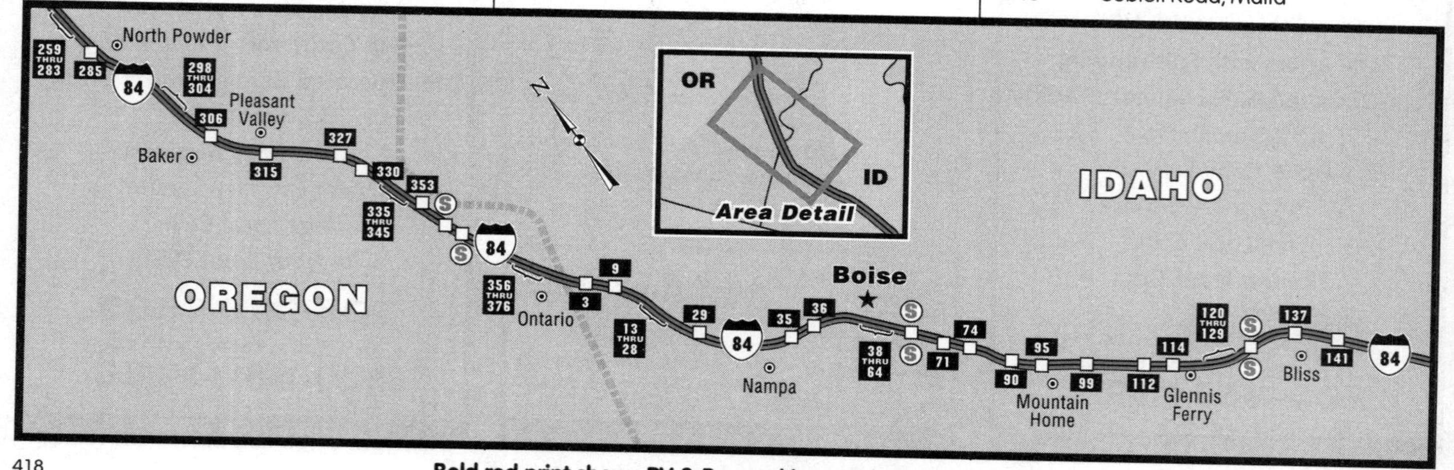

Bold red print shows RV & Bus parking available or nearby

EXIT	IDAHO/UTAH
FStop	N: C-Store* (24 Hrs)
Food	N: Restaurant (C -Store, 24 Hrs)
RVCamp	S: Camping
54	Sweetzer Road
63	Juniper Road
(69)	**Rest Area - RR, Phones, Picnic P (Both Directions)**

IDAHO

UTAH

	UT 30, Park Valley
	Snowville
FStop	N: Flying J Travel Plaza*(D)(LP)
Gas	N: Chevron*
Food	N: Restaurant (Flying J TS), Round-Up Cafe
Lodg	N: Outsiders Inn
RVCamp	N: Lottie-Dell Campground (.5 Miles)
2	Ranch Exit
	Hansel Valley
	Ranch Exit
0	Blue Creek
4	Valley
	UT 83, Howell, Thiokol
Other	S: Golden Spike National Historic Sight
2	Ranch Exit
9	Garland, Bothwell
Med	N: Hospital
	UT 102, Tremont, Bus. I-84
FStop	N: Chevron* (Burger King), Texaco*(LP)
Food	N: Burger King (Chevron FS), Denny's, McDonalds (Playground)
Lodg	N: AAA Western Inn
AServ	N: Chevron
TServ	N: Chevron
TWash	N: Semi Wash
Med	N: Hospital
Other	N: Car Wash S: Rocket Display, Golden Spike National Historic Sight
	I-15, North Pocatello

Note: I-84 runs concurrent below with I-15. Numbering follows I-15.

92	I-84 West, Tremonton, Boise
37	UT 30 East, Riverside, Logan
RVCamp	E: Camping

EXIT	UTAH
383	Business 15, Tremonton, Garland
Med	E: Hospital
379	UT 13, Business 15, Business 84, Tremonton
375	UT 24 to UT 13, Honeyville, Bear River
RVCamp	E: Crystal Hot Springs Camping
(370)	**Rest Area P (Southbound)**
368	UT 13, Brigham City, Corinne
Parks	W: Bear River Bird Refuge
364	U.S. 91, U.S. 89, Brigham City, Logan
Med	E: Hospital
(363)	**Rest Area P (Northbound)**
361	Port of Entry
360	UT 315, Willard, Perry
FStop	E: Flying J Travel Plaza*(D)(LP) (Country Market)
Food	E: Country Market (Flying J)
RVCamp	W: Camping
354	UT 126, U.S. 89, Willard
352	UT 134, Farr West, Pleasant View
Gas	E: Maverick* (Playground)
Food	E: McDonalds, Subway
ATM	E: Maverick
349	Defense Depot, Harrisville
Gas	W: Texaco*
TServ	W: JM Bodily & Sons TService
347	UT 39, 12th Street
TStop	W: Pilot Travel Center*(SCALES)
FStop	E: Flying J Travel Plaza*
Gas	E: Phillips 66*(CW)
Food	E: Jeremiah's Restaurant W: Country Kitchen (Pilot), Dairy Queen (Pilot), Subway (Pilot), Taco Bell (Pilot)
Lodg	E: AAA Best Western W: ◆ Sleep Inn
AServ	E: Steve's Car Care
TServ	W: General Diesel
Parks	E: Ogden Canyon Recreation Area
346	UT 104, 21st Street, Wilson Lane
TStop	E: Flying J Travel Plaza(SCALES) (RV Dump)
FStop	E: Chevron*(CW) W: Texaco*(D) (RV Dump)
Food	E: Arby's (Chevron), Big Z Restaurant, Cactus Reds (Comfort Suites), The Cookery Buffet W: Freeway Cafe
Lodg	E: Best Rest Inn, AAA Big Z Motel, ◆ Comfort Suites W: ◆ Super 8 Motel

EXIT	UTAH
TServ	E: Flying J Travel Plaza, Lake City International, Ogden Diesel
RVCamp	W: Century Home & RV Park
ATM	E: Flying J Travel Plaza
Other	E: Flying J Coin Laundry
345	UT 53, 24th Street (Northbound, Reaccess Southbound Only)
FStop	E: Sinclair*
Gas	E: Texaco*
344	UT 79, Ogden, 31st Street
343	I-84 East, Cheyenne (Southbound)

Note: I-84 runs concurrent above with I-15. Numbering follows I-15.

81	UT26, I-15, Riverdale
Gas	N: Conoco*(D)(CW)(LP)
Food	N: Applebee's, Boston Market, Chili's, Einstein Bros. Bagels, La Salsa S: Palermo's Italian
Lodg	S: Motel 6
AServ	N: Petersen Nissan, Pontiac, Buick, GMC
Other	N: Pearle Vision, Super Target, Wal-Mart
85	Uintah, South Weber
87AB	US89, Ogden, Salt Lake
(91)	**Rest Area - RR, Picnic P (Eastbound)**
92	UT 167, Mount Green, Huntsville (Eastbound, Westbound Reaccess)
(94)	**Rest Area - RR, Picnic P (Westbound)**
96	UT167, Peterson, Mount Green, Stoddard
Gas	S: Phillips 66*
AServ	S: Phillips 66
Other	S: Coin Car Wash
103	UT66, Morgan
Food	S: Buzzys Grill, Country Cafe
AServ	S: Heiner Ford Dealership, Little Lube and Tire, Nelson Auto Repair
Other	S: Morgan Drugs, Visitor Information
106	Ranch Exit
108	Taggert
111	Croydon
112	Henefer
115	Henefer, Echo

↑ UTAH

Begin I-84

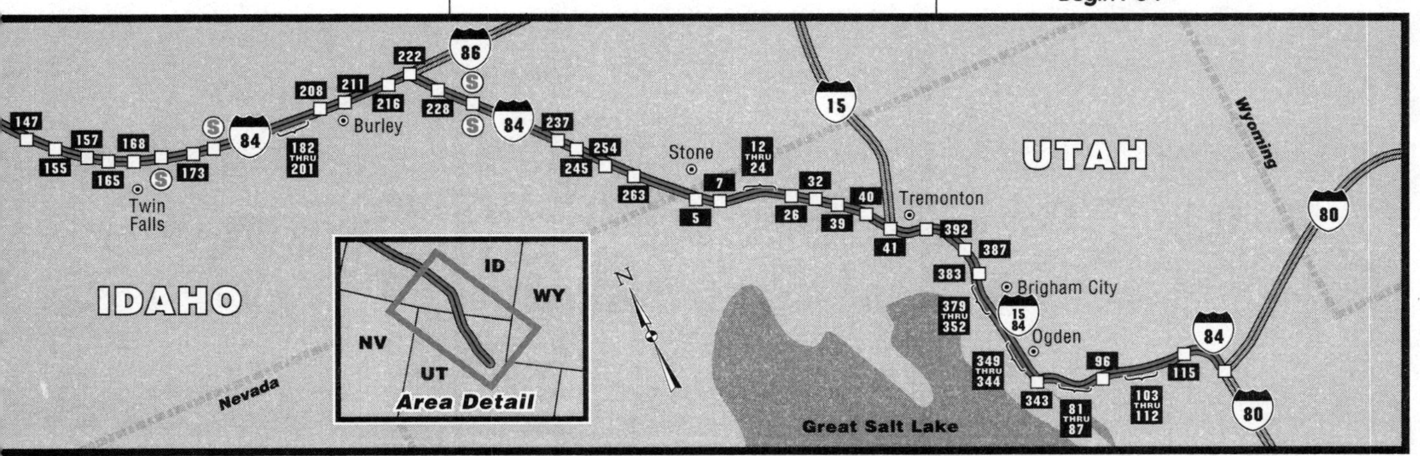

EXIT	PENNSYLVANIA

↓PENNSYLVANIA

(0) Jct I-81, to Wilks - Bear, PA, Binghamton, NY

1 Tigue St, Dunmore
- FStop S: **Texaco(D) (Kerosene)**
- Food N: Holiday Inn
- Lodg N: Holiday Inn
- AServ S: Texaco

2 PA 435, Elmhurst

(4) Junction I-380, Scranton, Mt Pocono

4 **(8)** PA 247, PA 348, Mount Cobb (PA247 - No Trucks Over 10.5 Tons)
- FStop N: **Mobil*(D)**
- Gas N: Citgo*
- ATM N: LA Bank
- Other N: **Four Star Super Mkt**

5 **(14)** PA191, Newfoundland, Hamlin
- TStop N: **Howe's Auto/Truck Plaza(D)(SCALES) (Exxon Gas)**
- FStop N: **Exxon(D)**
- Food N: **Comfort Inn (Howe's Truckstop), Twin Rocks (Howe's Truckstop)**
- Lodg N: AAA **Comfort Inn (Howe's Truckstop)**
- TServ N: **Howe's Truckstop**

6 **(20)** PA 507, Green Town Lake, Lake Wallenpaupack
- FStop N: **Exxon(D)**
- Gas N: Mobil*
- Food N: **John's Italian Restaurant**
- Other N: **Claws N'Paws Animal Park**

(24) **Weigh Station, Rest Area - Pet Area, RR, Phones, Picnic, Vending, Tourist Info**

7 **(27)** PA 390, Tafton, Promised Land State Park
- Parks N: **Promised Land State Park**

8 **(30)** PA 402, Porter's Lake, Blooming Grove

9 **(35)** PA 739, Dingman's Ferry, Lord's Valley
- Gas S: Sunoco*
- Food S: **McDonalds**
- AServ N: Lord's Valley Towing

10 **(46)** U.S. 6, Milford
- Gas N: Mobil*(D)

EXIT	PENNSYLVANIA/NEW YORK

- S: Citgo*
- Food S: Red Carpet Inn Restaurant (Red Carpet Inn)
- Lodg S: AAA Red Carpet Inn
- AServ N: My Place Auto Service (Towing Service)

11 **(54)** U.S. 6, U.S. 209, Matamoras (No Trucks 7PM To 7AM)
- Gas N: 24 Hr Gas & C-Store*, Citgo*, Shell*
  S: Mobil*
- Food N: Landmark, Polar Bear Ice Cream Cafe
  S: Best Western, **Coffee Bar (Lazy River Books)**, McDonalds, Peking Garden, Perkins, Pizza House Family Restaurant, Riverview Landing, Village Inn Diner, Wendy's
- Lodg N: West Falls Motel
  S: AAA Best Western (see our ad this page), ◆ Blue Spruce Motel, Village Inn
- AServ N: Shell
- TServ S: **A&M Truck Ctr(LP)**
- RVCamp S: **Hickory Grove Campground, Tri-State Canoe Campground**
- ATM N: CoreState Bank
  S: Wal-Mart
- Other N: **Tri-State Canoe**
  S: **Eckerd Drugs, Grand Union Grocery, K-Mart (Pharmacy), Lazy River Books (Coffee Bar), Wal-Mart (Pharmacy), Westfall Ctr**

↑PENNSYLVANIA

↓NEW YORK

1 U.S. 6, NJ 23, Port Jervis, Sussex
- FStop N: **Mobil***
  S: **Citgo*, Gulf*(D)(LP), Xtra***

EXIT	NEW YOR

- Food N: Arlene & Tom's Diner, Dunkin Donuts
  S: Dairy Queen, McDonalds, Village Pizza
- Lodg N: **Deer Dale Motel, Painted Aprons Motor Lodge,** Shady Brook Motel
  S: Comfort Inn
- AServ S: Gulf
- RVCamp N: **Campground (2 Miles)**
- Med N: **Hospital**
- Other N: **Erie Depot, Ft Decker Historic Site**
  S: **Rite Aide Pharmacy, Shop Lite Grocery Store, Tri- State Mall**

(3) **Scenic Overlook, Parking Area (N Services)**

2 **(5)** Mountain Road

3 **(15)** U.S. 6, NY 17M, Middletown, Goshen
- FStop S: **Sunoco*(D)(LP)**
- Gas N: Citgo*(D), Hess(D), Mobil*(D), Texaco*, Wally Mart(D)
  S: **84 Quick Stop***
- Food N: 6-17 Diner, Blimpie's Subs, Boston Market, Bradley Corner's Restaurant, Burger King, Cancun Inn (Mexican & Spanish), Carvel Ice Cream, China Town Restaurant, Dunkin Donut Ground Round, McDonalds, Momma Pina Pizz Peking Restaurant, Perkin's Family Restaurant, Ponderosa Steakhouse, Taco Bell, Wendy's
  S: 84 Quick Stop Deli, Salt & Pepper Deli, Pizza, Tudy's Coffee Shop
- Lodg S: DAYS INN Days Inn, Global Budget Inn Of Amer
- AServ N: Express Lube, Feder's Subaru/Acura, Fulton Cadillac/Chevrolet/Geo, Middletown Madza, Monroe's Muffler, Pit Stop Oil Change, Pontiac, GMC
  S: Al Scala Suzuki, Johnson Toyota, Ken's Auto Serv, Nissan Of Middletown, Plymouth, Jeep, Dodge, Sunoco, Village Lincoln Mercury
- Med N: **Hospital**
- ATM N: M&T Bank, MSB Bank
- Other N: **Campbell Shopping Ctr,** Crystal Clean Car Wash, **Middletown Commons Shopping Center, Redner's Warehouse Groc, Rite Aide Pharmacy, Shop-Rite Grocery**
  S: **U.S. Post Office**

(17) **Rest Area - Picnic, RR, Phones, Pe Walk, Vending (Eastbound)**

4 **(19)** NY 17, Binghamton, New York
- Med S: **Hospital**
- Other S: **State Police Headquarters**

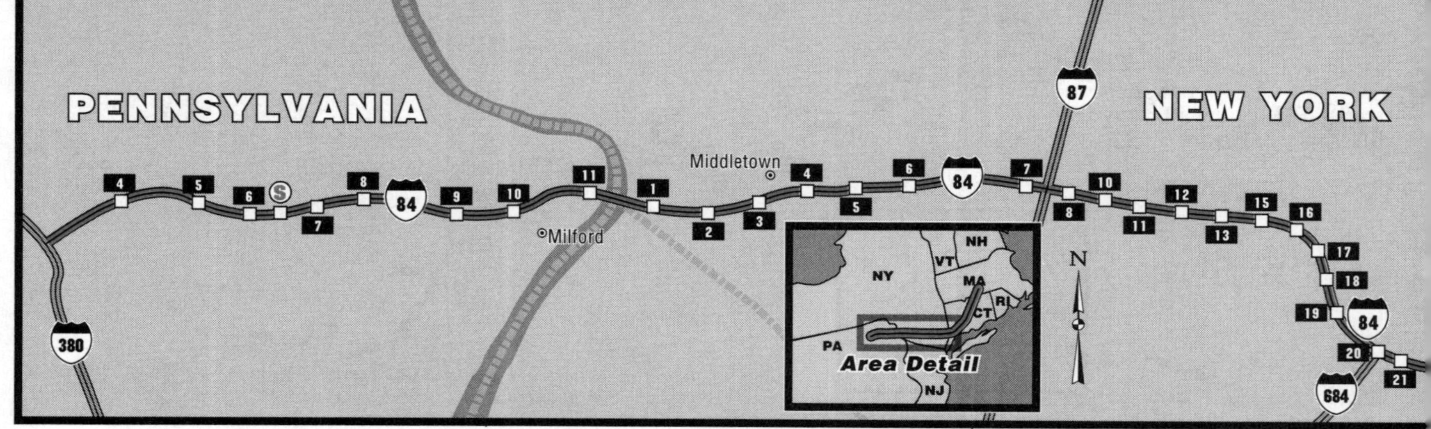

Bold red print shows RV & Bus parking available or nearby

EXIT NEW YORK

3) **Rest Area - RR, Phones, Picnic, Vending (Westbound)**

(29) NY 208, Maybrook, Walden, New Paltz
- **TStop** **S:** Travel Port*(SCALES)
- **Gas** **N:** Exxon, Mobil
  S: Sunoco (Travel Port)
- **Food** **N:** Blazing Bagels, Burger King, Gold Chain Chinese Restaurant, Leaning Tower III (Pizza), McDonalds
  S: Buckhorn Retaurant (Travel Port), Pizza Hut (Travel Port), The Roadside Inn
- **Lodg** **N:** Super 8 Motel
  S: Super 8 Motel
- **AServ** **N:** Exxon, Mobil (Wrecker Service)
- **TServ** **S:** Travel Port(SCALES)
- **TWash** **S:** Blue Beacon Truck Wash (Travel Port)
- **RVCamp** **N:** Winding Hill Campground
- **ATM** **N:** Orange County Trust Company, Shop Rite Grocery
- **Other** **N:** Eckerd Drugs, Shop Rite Grocery
  S: Variety Farms C-Store

(34) NY 17K, Montgomery, Newburgh
- **Gas** **N:** Mobil*(CW)(LP)
  S: Exxon*
- **Food** **N:** Stewart Airport Diner
  S: Courtyard By Marriott, Deli (Exxon)
- **Lodg** **N:** AAA Comfort Inn
  S: ◆ Courtyard By Marriott
- **Other** **S:** Stewart Int'l Airport

(37) Junction I-87 Thruway, NY 300, Union Ave

(37) NY 300, Union Ave
- **Gas** **N:** Exxon*, Mobil*(CW)
  S: Getty*, Sunoco*
- **Food** **N:** Ground Round, King Buffet, McDonalds, Roy Rogers, Subway, Taco Bell, Wendy's
  S: Burger King, Cafe International (Ramada Inn), Denny's, Holiday Inn, Neptune Diner, Yobo Oriental
- **Lodg** **S:** Hampton Inn, Holiday Inn, Howard Johnson, Ramada Inn (see our ad this page), Super 8 Motel
- **AServ** **N:** Auto Palace, Exxon (Wrecker Service), Mavis Discount Tire, Midas Muffler & Brakes, Pitstop Oil Change, Sears
  S: Getty, Newburgh Nissan
- **RVCamp** **S:** KOA Camp (8 Miles)
- **ATM** **N:** All Bank, Fleet Bank
  S: Fleet Bank
- **Other** **N:** Dairy Mart C-Store, Newburgh Mall, Weis Supermarket
  S: Tourist Information

(37) NY 52, Walden
- **Gas** **N:** Citgo*(D)
  S: Gulf*
- **AServ** **S:** Century Auto Service, Gulf
- **Med** **S:** ✚ Hospital

(39) U.S. 9 W, NY 32, Newburgh, Highland
- **Gas** **N:** Mobil(CW), Texaco*
  S: Exxon*(D), Sunoco*
- **Food** **N:** Burger King, KFC, Lexus Diner, New York Bagel & Bean's Coffee Shop, Perkins Family Restaurant, Roma Imperial Restaurant &

EXIT NEW YORK

Pizza, Safari Joe's
S: North Plank Rd Tavern & Restaurant
- **AServ** **N:** Firestone Tire & Auto, Texaco
- **Med** **S:** ✚ Hospital
- **ATM** **N:** First Union Bank, Price Chopper Supermarket
- **Other** **N:** Knox's Headquarters Historic Site, Medical Arts Pharmacy, Price Chopper Supermarket (Pharmacy), RX Place (Pharmacy)
  S: Dairy Mart Convenience Store, New Windsor Cantonment Historic Site, Washingtons Headquarters Historic Site

11 **(41)** NY 9D, Wappinger Falls, Beacon
- **Gas** **N:** Sunoco*
- **Med** **N:** ✚ Hospital
- **Other** **N:** Mt Gulian Historic Site, Tourist Information

12 **(45)** NY 52, Fishkill
- **Gas** **N:** Coastal(LP) (Kerosene)
  S: Shell*, Sunoco
- **Food** **S:** Fishkill Frostee Hot Dog Stand, Home Town Deli, I-84 Diner
- **AServ** **S:** Brownell Mercury/Lincoln/Saab, Shell, Sunoco

13 **(46)** U.S. 9, Poughkeepsie, Peekskill
- **Gas** **N:** Mobil*(CW)(LP)

EXIT NEW YORK/CONNECTICUT

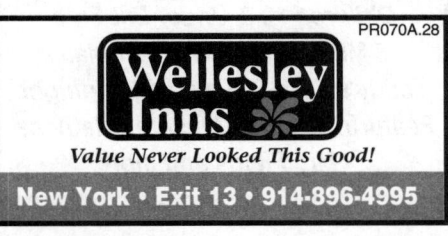

- **Food** **N:** Boston Market, Denny's, New Deli News, Pizza Hut, Stanley's Eatery (Holiday Inn)
  S: Cattleman's, McDonalds, Moog's Farm
- **Lodg** **N:** ◆ Courtyard by Marriott, ◆ Holiday Inn (see our ad this page), ◆ Residence Inn, Wellesley Inn (see our ad this page)
- **AServ** **N:** Sam's (Tire & Battery)
- **ATM** **N:** Poughkeepsie Savings Bank
  S: Al Bank, Hudson Valley Credit Union
- **Other** **N:** Factory Outlet Center, Sam's Club, State Patrol Post, Wal-Mart (Pharmacy)
  S: Drug World, Dutchess Mall Shopping Center, Lens Crafters Optical, Vanwick Historic Site

15 **(51)** Lime Kiln Road

16 **(53)** Taconic State Pkwy, Albany (No Trucks)

(55) **Rest Area - RR, Phones, Picnic, Pet Walk, Vending**

17 **(59)** Ludingtonville Road
- **FStop** **S:** Hess*(D)
- **Gas** **S:** Sunoco
- **Food** **S:** Blimpie's Subs (Hess), Lou's Restaurant & Deli
- **AServ** **N:** Metric Motors
  S: Sunoco

18 **(62)** NY 311, Lake Carmel, Patterson

19 **(65)** NY 312, Carmel
- **Med** **S:** ✚ Hospital

20 **(68)** Jct I-684, US 6, US 202, US 22, White Plains, Brewster, Pawling

21 **(69)** U.S. 6, U.S. 202, NY 121, North Salem, Brewster (Westbound)

↑NEW YORK

↓CONNECTICUT

1 Saw Mill Road

(2) **Weigh Station, Rest Area - RV Dump (Eastbound)**

2AB **(1)** U.S. 6, U.S. 202, Mill Plain Road, Old Ridgebury Rd
- **Food** **S:** Hilton
- **Lodg** **S:** Hilton

3 **(4)** U.S. 7, Norwalk (Left Exit)
- **AServ** **S:** Jiffy Lube, Sears
- **ATM** **S:** First Union Bank
- **Other** **N:** Danbury Airport
  S: Danbury Fair Mall, Military Museum

4 U.S. 6, U.S. 202, Lake Ave
- **Gas** **N:** Amoco*, Gulf(D), Texaco*(D)(CW)
- **Food** **N:** Dunkin Donuts, Ethan Allen Inn, McDonalds
- **Lodg** **N:** ◆ Ethan Allen Inn, Super 8 Motel
- **AServ** **N:** Gulf

5 CT 37, CT 39, CT 53, Downtown Danbury
- **Gas** **N:** Texaco*
  S: Mobil*
- **Food** **S:** Deli & Snack Shop, Pizza Hut, Taco Bell
- **Lodg** **N:** Exit 5 Motel

Bold red print shows RV & Bus parking available or nearby

EXIT		CONNECTICUT
	AServ	**S:** Avis Lube
6		CT 37, New Fairfield
	Gas	**N:** Shell[CW], Texaco
		S: Amoco*
	Food	**N:** Bella Italia Restaurant, Burger King, Carvelle Ice Cream Bakery, McDonalds, Mykono's
		S: KFC
	AServ	**N:** Shell
		S: A&S Auto Sales (Wrecker Service 24 hour), Axel's Foreign Auto Parts, Bob's Auto Supply
	ATM	**N:** First Union Bank
	Other	**N:** **A & P Supermarket, Brooks Pharmacy**
		S: **Deep's Market, Railway Museum**
7		**(8)** U.S. 7, U.S. 202, New Milford, Brookfield
8		**(9)** U.S. 6, Newtown Road, Bethel
	Gas	**N:** Mobil*[D], Texaco[D][CW]
		S: Shell*[CW], Texaco*[D]
	Food	**N:** Mykonos Pizza, The Ground Round (AAA)
		S: Bangkok, Blimpie's Subs, Burger King, Chili's, Danbury Vience II, Denny's, Dunkin Donuts, Finley's, Italian & Pizza, McDonalds, Mike's Pizza, Ponderosa Steak House, Taco Bell
	Lodg	**N:** AAA Ramada (see our ad this page)
		S: AAA Best Western, AAA Holiday Inn (see our ad this page)
	AServ	**S:** Action GMC Trucks, Mohawk Tire & Auto Goodyear, Roberts Pontiac, Buick, Texaco
	Med	**S:** + **Hospital**
	ATM	**S:** Chase Manhattan Bank, First Union, Fleet Bank, Union Saving's Bank
	Other	**N:** **A&P Supermarket**
		S: **Animal House, CVS Pharmacy, Coin Laundry, Eye Care Plus, Nutmeg Square Mall, Super Stop Grocery (ATM)**
9		**(12)** CT 25, Brookfield, New Milford
10		**(15)** U.S. 6 West, Newtown, Sandy Hook (Low Bridge On U.S. 6W, 12'7")
	Gas	**S:** Amoco*, Shell*
	Food	**S:** Blue Colony Diner, Pizza Palace
	ATM	**S:** Chase Manhattan Bank
	Other	**S:** **Newton Hardware & Drug Center**
11		**(17)** CT 34, Derby, New Haven
13		**(19)** River Road (Eastbound)
14		**(20)** CT 172, South Britain
	Gas	**N:** Texaco*
	Food	**N:** Thatcher's Restaurant
		S: Tartuso Restaurant
	Other	**S:** **Colonial Vision, State Patrol Post**

EXIT		CONNECTICUT
15		**(22)** U.S. 6, CT 67, Southbury
	Gas	**N:** Mobile
	Food	**N:** Carnival Ice Cream, Dunkin Donuts, Europeon Shop, Jordan's Restaurant, Market Cafe, McDonalds, Nut's & Such, The Bakery
	ATM	**N:** People's Bank, Webster Bank
	Parks	**S:** **Kettletown State Park**
	Other	**N:** **CVS Pharmacy, Grand Union Grocery, K-Mart, OptiCare**
16		**(25)** CT 188, Middlebury
	Gas	**N:** Mobil*
	Food	**N:** Patty's Pantry & Deli
	Lodg	**N:** ◆ Hilton
17		**(29)** CT 63, CT 64, Naugatuck, Watertown
	Gas	**S:** Mobile*
	Food	**S:** Dunkin Donuts (At Mobile), I Can't Believe Its Not Yogurt, Java Coast Fine Coffee, Leo's Deli Buffet, Maple's Restaurant
	Other	**N:** **Amusement Park**
		S: **Primrose Square**
18		**(32)** West Main St, Highland Ave (Difficult Reaccess)

EXIT		CONNECTIC
	Gas	**N:** Texaco
	Food	**N:** New Moon Chinese, Robinwood Luncheon The Cheese Cakery
	Med	**N:** + **Waterbury Hospital**
	Other	**N:** **CVS Pharmacy**
19		**(32)** CT 8 South, Naugatuck
20		**(32)** CT 8 North, Torrington (Left Exi
21		**(32)** Meadow St, Banks St
	Gas	**S:** Exxon*[D], Mobile
	Food	**S:** Sun Kee Kitchen
22		**(33)** Union St, Downtown Waterbu
	Gas	**N:** Gulf
	Lodg	**N:** ◆ Courtyard by Marriott, AAA Quality Inn
	Med	**N:** + **St.Mary's Hospital**
	ATM	**N:** First Union, Webster Bank
	Other	**N:** **Post Office**
23		**(34)** CT 69, Wolcott, Prospect, Hamilton Ave (No Reaccess Westbound)
	Med	**N:** + **Hospital**
24		**(37)** Harpers Ferry Road (No Reaccess)
	Gas	**S:** Texaco*
25		**(41)** East Main St, Scott
	Gas	**N:** Exxon*
	Food	**N:** Bob's Grocery, China Buffet, Dunkin Donuts Friendly's, Nardelli's, Sandwich Shop Plus
		S: Bagel Mania, Burger King, Carvel Ice Crean Frankie's, Friendly's, Golden Wok, McDonalds, Neno's Restaurant
	Lodg	**S:** ◆ Super 8 Motel
	AServ	**N:** Gulf[D]
		S: Loehman Geo, Chevrolet
	Other	**S:** **CVS Pharmacy, Coin Laundry**
26		**(38)** CT 70, Cheshire, Prospect
	Food	**N:** Sheraton
	Lodg	**N:** Sheraton
27		**(41)** Jct I-691 East, Meridan
28		**(41)** CT 322, Marion, Wolcott
	TStop	**S:** **76 Auto/Truck Plaza***[SCALES]
	Gas	**N:** Gulf*
		S: Mobil*
	Food	**N:** Extra Mart
		S: **76 Auto/Truck Plaza, American Eagle (At 76)**, Burger King, China, Dunkin Donuts
	Lodg	**S:** DAYS INN ◆ Days Inn
	AServ	**N:** Mantz Auto Sales & Repair
		S: Quick Call Check

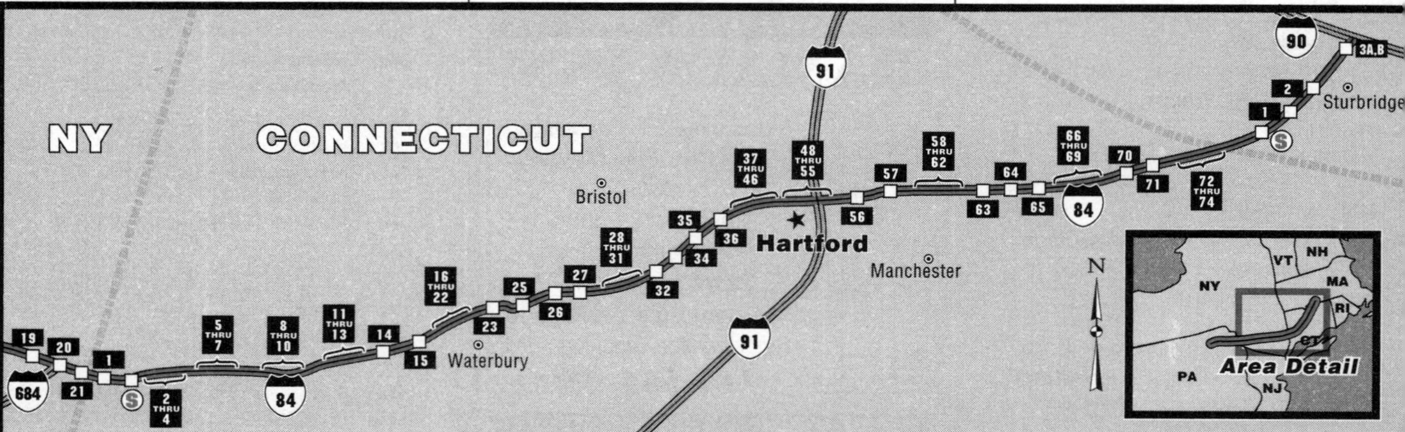

Bold red print shows RV & Bus parking available or nearby

EXIT CONNECTICUT

RVCamp S: Bell Camper Land(LP)
(42) Rest Area - RR, Phones, RV Dump (Eastbound)
9 (42) Milldale, CT 10
0 (43) West Main St, Marion Ave, Southington
Gas S: Mobil*
AServ S: Mobil
TServ S: Aszklar
Med S: ✚ Bradley Memorial Hospital
Other N: Ski Area
1 (44) CT 229, West St, Bristol
FStop N: Sunoco*(D)
Gas N: Mobil*
S: Citgo*, Gulf*
Food N: Dunkin Donuts (Mobil)
S: Scooter's Ice Cream, Vallendino Pizza, Westview Seafood
AServ S: Gulf*
Other S: Country Time Store
2 (46) CT 10, Queen St
Gas N: Shell*(CW)
S: Citgo*(D), Mobil*(CW), Sunoco*
Food N: Beijang, Bruegger's Bagel, Burger King, Carvell Ice Cream, Chili's, Chip's Ice Cream, D'Angelo, Denny's, Dunkin Donuts, Friendly's, JD's Great Food & Drinks, Kenny Rogers Roasters, Luen Hop Chinese Kitchen, Manhattan Bakery, McDonalds, Outback, Pizza Hut, Taco Bell
S: Bickfords Family Restaurant, Blimpie's Subs, Dunkin Donuts, Friendly's, Little Caesars Pizza, Ponderosa, Subway, The Whole Donut, Wendy's
Lodg N: Holiday Inn Express, Motel 6, Motel 6
S: Red Carpet Inn, Susse Chalet
AServ N: Auto Zone Auto Parts
S: Firestone Tire & Auto, Southington Geo, Speedy
ATM N: First Union
S: Webster Bank
Other N: CVS Pharmacy, Coin Laundry, Pet Supply Store
S: K-Mart, Southington Car Wash
3 (49) CT 72, Plainville, Bristol (Left Exit Westbound)
4 (49) CT 372, Crooked St, Plainville (Left Exit)
FStop N: Gasoline Alley*
Food N: Chung's Buffet, Howard Johnson, Manhattan Bagel
Lodg N: ◆ Howard Johnson
AServ N: Crowley's Vokeswagon
S: Tire Repair
ATM N: Peoples Savings and Trust
Other N: Big Y Supermarket, Rite Aide Pharmacy
S: Jim's Grocery
5 (50) CT 72, to CT 9, New Britain, Middletown (Left Exit Westbound)
Med S: ✚ Hospital
6 (51) Slater Road
7 (53) US 6, Fienemann Road, Farmington
Lodg N: ◆ Marriott
Med S: ✚ Hospital
8 (54) U.S. 6 West, Bristol

EXIT CONNECTICUT

39 (54) CT 4, Farmington
Med N: ✚ Hospital
39A (55) CT9, Newington, New Britain
9A (55) CT9S, Newington, New Britain
40 (56) CT 71, New Britain Ave, Corbins Corner (Difficult Reaccess Westbound)
Food S: Joe's American Bar & Grill, Wendy's
AServ S: Jiffy Lube, Sears Auto Center
ATM S: Fleet Bank
41 (57) South Main St, Elmwood
Food S: Charley's Place, China Buffet, IHOP
ATM S: Webster Bank
Other N: Noah's Webster's Home Historical Site
42 (58) Trout Brook Dr, Elmwood (Left Exit)
43 (58) Park Road, West Parkford Center (Left Exit Eastbound)
Other N: Police Dept, Science Center of Connecticut
44 (59) Prospect Ave
Gas N: Exxon*(CW), Texaco*(D)
Food N: Bess Eaton Donuts, Burger King, Carvel Ice Cream Bakery, D'Angelos/Chips, Hometown Buffet, McDonalds, Pizza Hut, Prospect Pizza, Wendy's
ATM N: Fleet Bank
Other N: Shaw's Food & Drug, Vision Corner
45 (60) Park Road, Flatbush Ave (Left Exit Westbound)
46 (59) Sisson Ave
47 (59) Sigourney St
48 (60) Asylum St, Capital Ave
Med N: ✚ Hospital
49 (60) Civic Center, Ann St, High St
50 (61) Main St
51 (62) Junction I-91 North, Springfield
52 (62) Junction I-91 South, New Haven, U.S. 44, Main St
53 (63) U.S. 44, East Hartford
54 (64) CT 2 W, Downtown Hartford (Left Exit Westbound)
55 (64) CT 2, Norwich, New London
56 (64) Governor St, East Hartford (Left Exit Eastbound)
57 (65) CT 15, to I-91, Charter Oak Bridge, NY City (Left Exit Westbound)
58 (65) Roberts St, Silver Lane, Burnside Ave
Lodg N: ◆ Holiday Inn, AAA Wellesley Inn (see our ad this page)
TServ N: Freightliner Dealer
ATM N: BSW

EXIT CONNECTICUT

59 (66) Junction I-384, Providence
60 (67) U.S. 44W, Middle Tnpk W, U.S. 6, Manchester
Food S: Chez Ben Diner (American/Canadian)
AServ S: Final Inspection
Med S: ✚ Hospital
61 (69) Jct 291W, Windsor
62 (70) U.S. 44, Buckland St
FStop N: Exxon
Gas S: Mobil*
Food N: Between Round Bagel, Boston Market, Bugaboo Creek Steak House, Chili's, Friendly's, John Harvard's Stew House, Olive Garden, Pizza Hut Express, Starbuck's Coffee, Taco Bell
S: Chowder Town Seafood, Chuck E. Cheese's Pizza, Dunkin Donuts, Friendly's, Golden Dragon, Ground Round Restaurant, Manhattan Bagel, McDonalds, Subway, Wooster Pizza Shop
AServ N: Exxon
S: Firestone Tire & Auto, Mobile
ATM S: Eagle Federal Savings Bank, SBM
Other N: Buckland Hills Mall, Sam's Club
S: Mail Boxes Etc, Waldbaum's Food Mart
63 (72) CT 30, CT 83, South Windsor
FStop N: Texaco
Gas S: Exxon, Sunoco*
Food N: Chicago Bar & Grill, McDonalds, Old Country Buffet, Outback Steak House, Ramono's Macaroni Grill, T.G.I. Friday's
S: King Buffet, Palace Restaurant & Lounge, Roy Rogers
Lodg S: Connecticut Inn Motor Lodge
AServ S: Marnde Ford, Lincoln, Mazada
Med S: ✚ Hospital
ATM S: Savings Bank of Rockville
Other N: Christmas Tree Shops, Wal-Mart
S: Big Y Supermarket
64 (73) CT 30, CT 83, Vernon Center
Gas N: Sunoco
Food N: Anthony's Pizza, Between The Rounds, Congress Rotisserie, Damion's, Denny's, Dunkin Donuts, Friendly's, J Copperfield Restaurant, KFC, Kevin Coffee, Kim's Oriental, McDonalds, Papa Geno's, Taco Bell, The Polish Bakery
Lodg N: Colonial Vernon Budget Motor Inn, Holiday Inn Express
AServ N: Auto Parts America, Goodyear Tire & Auto, Jiffy Lube, Sunoco
ATM N: First Federal Savings, Fleet Bank, People's Bank, Savings Bank of Rockville
Other N: Adam's Superfood Store, Animal City, CVS Pharmacy, Mail Boxes Etc, Vernon Walk-in Medical Center
65 (73) CT 30, Vernon Center
Gas N: Mobile*, Shell, Texaco*(CW)
Food N: Bickfords Family Restaurant, Burger King, Joy Wok Chinese, KFC, Lotus Restaurant
AServ N: Firestone Tire & Auto, Mobile, Post Road Plaza, Shell*
ATM N: Holland Bank, SBM Bank
Other N: K-Mart, Vernon Drug
66 (75) Tunnel Road, Vernon, Bolton
67 (77) CT 31, Rockville, Coventry
Gas N: Texaco*(D)
Food N: Bess Eatin Donuts, Blimpie's Subs, McDonalds, Wongs Too Chinese
RVCamp N: Campground

EXIT	CONNECTICUT
Med	N: Hospital
ATM	N: The Savings Bank of Rockville
Other	N: Coin Laundry, Medicine Shop, Red Apple Supermarket
68	(81) CT 195, Tolland, Mansfield
Gas	N: Gulf
	S: Citgo*, Getty*
Food	N: Poppa T's Family Restaurant, Subway
	S: Lee's Garden Chinese Restaurant, Rhodo's Pizza, Toland Pizza
AServ	N: Gulf, Martha's Auto Parts
	S: Texaco*, Toland's Auto Parts
ATM	S: Toland Bank
Other	S: 7-11 Convenience Store, IGA Store, Toland Pharmacy
69	(84) CT 74 to U.S. 44, Willingham
RVCamp	S: Campground
Other	N: State Patrol Post
(84)	Rest Area - Tourist Info, RR, Picnic, Phones, RV Dump.
(85)	Weigh Station
70	(86) CT 32, Willington, Stafford Springs
Gas	S: Mobil*, Sunoco*(D)
RVCamp	N: Rainbow Acres Campground (1.5 Miles)
	S: Campground
Med	N: Hospital
71	(88) CT 320, Ruby Road
TStop	S: PA Truck Stop (Travel Store)
Gas	N: Citgo(D)
Food	S: Burger King, Country Pride Restaurant, Dunkin Donuts
Lodg	S: Sleep Inn

EXIT	CONNECTICUT/MASSACHUSETTS
AServ	N: Citgo
TServ	S: PA Truck Stop(SCALES)
RVCamp	S: RV Dump Station
72	(92) CT 89, Westford, Ashford
Lodg	N: Ashford Motel
RVCamp	N: Campground
	S: Campground
73	(93) CT 190, Stafford Springs, Union
RVCamp	S: Campground
74	(97) CT 171, Union, Holland MA
Gas	N: Citgo*
	S: BP
Food	N: Lucy's Place
	S: Traveler Restaurant
AServ	S: BP, Goodhall's Garage, Jerry Yost's Chrysler, Plymoth, Jeep
TServ	S: Goodhall's Garage, Truck Service International
RVCamp	S: Campground

↑CONNECTICUT

↓MASSACHUSETTS

EXIT	MASSACHUSETTS
(1)	Rest Area - Phones, Picnic (Eastbound)
(3)	Weigh Station (Both Directions)
1	(4) Mashapaug Road, Southbridge
FStop	N: Sturbridge Isle
	S: Mobil
Gas	N: Texaco*(D)
	S: Mobil*
Food	N: Boston Pizza & Deli, Coffee Isle, Country Kitchen Restaurant, Ice Cream
	S: Roy Rogers, Sbarro Pizza
Med	S: Hospital
Other	N: Tourist Information
(4)	Picnic (Northbound)
2	(5) to MA 131, Sturbridge, Southbridge
Food	S: Herbert Candy & Ice Cream
RVCamp	N: Yogi Bear Camp Report
	S: Campground
Other	N: Olde Sturbridge Village
3AB	(7) U.S. 20, Worcester, Palmer
TStop	S: New England Truck Stop(SCALES)
Gas	N: Citgo*, Mobil*, Texaco (Towing)
Food	N: Admiral PJ's O'Brien Restaurant & Pub, Bagel Express, Bob's Homemade Ice Cream, Burger King, Friendly's, La Petiete French Bakery, McDonalds, Oxhead Tavern (Sturbridge Hotel), Piccadilly Pub & Restaurant, Steamer's Grill, Sturbridge Host Hotel, Sturbridge's Pizza House
	S: Heritage Family Restaurant
Lodg	N: American Lodging, Sturbridge Host Hotel, Super 8 Motel, Thistle Inn, Village Motel
	S: Quality Inn, Sturbridge Inn
AServ	N: Texaco
	S: Sturbridge Motors
TServ	N: New England Truck Stop(SCALES)
ATM	N: Fleet Bank
Other	N: State Patrol Post, Tourist Information, Uncle Wallen's Country Store

↑MASSACHUSETTS

Begin I-84

I-85 N →

EXIT	VIRGINIA

Begin I-85

↓VIRGINIA

EXIT	VIRGINIA
68	Junction I-95S, U.S.460E, Rocky Mt, NC
65	Squirrel Level Road
63AB	U.S. 1
FStop	E: Chevron*(CW), Exxon*(LP), Texaco*(D)
Gas	E: Citgo*(LP)
	W: Amaco
Food	E: Burger King (Citgo), Exxon, Hardee's, Waffle House
	W: Dunkin Donuts, McDonalds
AServ	E: Citgo
	W: Greyline
Med	W: State Hospital
Parks	E: Pamplin Civil War Park
61	U.S. 460, Blackstone
FStop	E: East Coast*(D) (ATM)
Gas	W: Texaco(D)(LP)
Food	W: Bullets, Dunkin Donuts
RVCamp	E: Picture Lake Camping (2.1 Miles)
(55)	Rest Area - RR, Phones, Picnic, Vending
53	VA 703, Dinwiddie
FStop	W: Exxon*(D)
Food	W: Thats A Burger
AServ	W: Exxon (Towing)
Parks	W: Petersburg National Battlefield
48	VA 650, DeWitt
42	VA 40, McKenney
Gas	W: Citgo*(D), Exxon*(D)(LP)
Food	W: Dairy-Freeze
AServ	W: Lafonna's Auto Service
Other	W: Saunder's Gas, Oil, Propane, Wallace's Market
39	VA 712, Rawlings
FStop	W: Chevron*
Gas	W: Citgo(D) (Nottaway Motel)
Food	W: Nottaway Restaurant
Lodg	W: Nottaway Motel
Other	E: VA Battlerama
34	VA 630, Warfield
Gas	W: Exxon*(D)
AServ	W: Warfield Service Center
(32)	Rest Area - RR, Phones, Picnic, Vending
28	U.S. 1, VA 46, Alberta, Lawrenceville
FStop	E: Amoco*
Gas	W: Citgo*
Other	W: Synergy Gas(LP)
27	VA 46, Blackstone, Lawrenceville (Northbound, Southbound Reaccess)
24	VA 644, Meredithville
(22)	Weigh Station - Both Directions
15	U.S. 1, South Hill, Kenbridge
Gas	W: Shell*(D)
Food	W: Kahill's Restaurant, Lake Country BBQ (Shell)
Med	W: Hospital
12	U.S. 58, VA 47, Norfolk, South Hill
FStop	W: Exxon*
Gas	E: Racetrac*, Shell*(CW)
	W: Amoco(D)(CW), Chevron*, Petrol*(LP), Texaco*(D)(CW)
Food	E: Bagel's Plus, Bojangles
	W: Brian's Steak House & Lounge, Burger King, Denny's, Golden Corral, Hardee's, KFC, McDonalds, New China Restaurant, Taco Bell, The Medicine Shop, Wendy's
Lodg	E: (AAA) Hampton Inn
	W: ◆ Best Western, Comfort Inn, (AAA) Econolodge, Super 8 Motel
AServ	E: Amerilube
	W: Chevron (Towing), Colony Tire, Frank Jackson Ford, Lincoln, Mercury
Med	W: Hospital
ATM	E: Signet Bank
	W: First Citizens Bank, First Virginia Bank

EXIT	VIRGINIA/NORTH CAROLINA
Other	E: Wal-Mart (w/pharmacy)
	W: Car Wash, Coin Laundry, Farmer's Foods, Fresh & Friendly Food Store, Revco Drugs, Rite Aide Pharmacy, Winn Dixie Supermarket
	VA 903, Bracey, Lake Gaston
TStop	E: Simmons Auto/Truck Terminal*(SCALES)
Gas	E: Amoco*(D)(LP)
Food	E: Bracey's Junction, Dairy Queen, Simmons Truckstop
	W: Countryside Family Restaurant
Lodg	W: Days Inn, AAA Days Inn
TServ	E: Simmons
ATM	E: Signet Bank (Simmons)
2)	Welcome Center - RR, Phones, Picnic, Vending, Tourist Info (Northbound)

↑ VIRGINIA

↓ NORTH CAROLINA

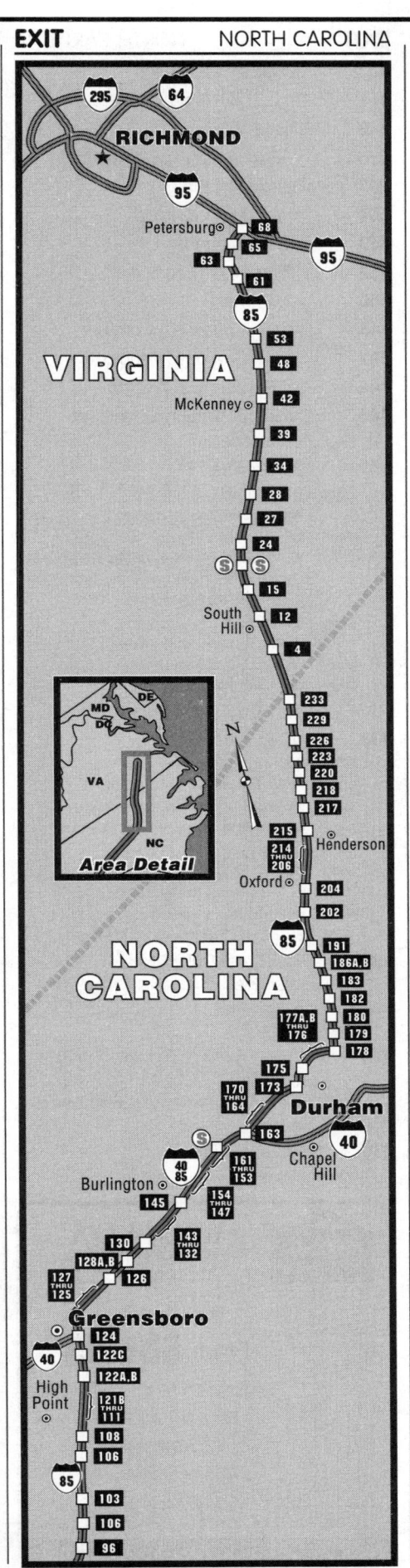

EXIT	NORTH CAROLINA
33	U.S. 1, Wise
TStop	E: Wise Truck Stop*
Gas	E: Shell*(D)
Food	E: Wise Truck Stop
Lodg	E: Budget Inn
AServ	E: Shell*
231)	Welcome Center - RR, Phones, Vending, Tourist Info (Southbound)
29	Oine Road
26	Ridgeway Road
Other	W: State Recreation Area
23	Manson Road
RVCamp	E: BP
20	U.S. 1, U.S. 158, Norlina, Fleming Road
TStop	W: Chex*(SCALES)
Gas	E: Exxon*(D)
Food	W: Floyd's Grill
Lodg	W: Chex
AServ	W: Floyd's Auto Repair
TWash	W: Chex
ATM	W: Chex Truckstop
18	U.S. 1 South, Raleigh (Left Exit, Southbound)
17	Satterwhite Point, Nutbush Bridge
Gas	W: Exxon*(D)
15	U.S. 158 East Bypass

EXIT	NORTH CAROLINA
FStop	E: Winoco*(D)
Gas	E: Shell*(D)
Food	E: 220 Seafood, Burger King, Dockside Seafood, Freeze Made Dairy Bar, Golden China, Nunnery-Freeman Barbacue, PD Quix Fast Food, Subway, Taste Freeze, Waffle & Pancake House
Lodg	E: AAA Budget Host Inn, Comfort Inn, AAA Howard Johnson, Quality Inn, Scottish Inns
AServ	E: A&E Imported Auto Parts, East Carolina Tire
ATM	E: Nations Bank
Other	E: Coin Laundry, Food Lion Grocery, Revco Drugs
214	NC 39, Henderson
Gas	E: BP*
	W: Amoco*(LP), Shell*(LP)
Food	E: Dabney Road Restaurant, The Sandwich Shop
	W: Hot Stuff Pizza (Amoco)
AServ	E: King Tire
	W: Doug's Towing & Recovery
Other	E: Little River LP(LP), U.S. Post Office
	W: Kerr Lake Recreation Area
213	Dabney Rd
Gas	E: Citgo*(D), Shell*(LP)
	W: Exxon*, Shell (ATM), Speed-EZ*(LP)
Food	E: Bamboo Garden Chinese, Birds Deli & Bakery, Dairy Queen, Denny's, Lans Chinese, Little Caesars Pizza, McDonalds, Pino's Italian, Pizza Inn, Subway, Wendy's
	W: Gators, Golden Corral, Krispy Kreme Donuts
Lodg	W: Holiday Inn Express
AServ	E: Goodyear Tire & Auto
	W: Boyd's Chevrolet, Buick, Chrysler Auto Dealer, Henderson's Motors, Simmons Parts & Service, Wal-Mart
Med	W: ✚ Hospital
ATM	E: Centura Bank, Fidelity Bank, First Citizens Bank
Other	E: Eckerd Drugs, Food Lion Grocery, Henderson Mall, Opti Eye Care Center, Postal Plus, Revco Drugs, Winn Dixie Supermarket
	W: Henderson Square (Pharmacy), Wal-Mart (Pharmacy)
212	Ruin Creek Road
Gas	E: Shell*
	W: Amoco* (ATM)
Food	E: Cracker Barrel, Mazatlan Mexican (Days Inn), The Silo Steaks & Seafood
	W: Baskin Robbins, Burger King, Western Sizzlin'
Lodg	E: Days Inn (see our ad this page)
Med	W: ✚ Maria Parham Hospital
209	Poplar Creek Road
206	U.S. 158, Oxford
Gas	E: Citgo*(D)
	W: Texaco*(D)
AServ	E: Citgo
204	NC 96, Oxford
FStop	E: Amoco*(D)
Gas	W: Shell*
Food	W: Burger King, China Wok Chinese, KFC, Little Caesars Pizza, McDonalds, Peter's Bakery, Pizza Hut, Ramada, Rena's Family Restaurant, Subway, Taco Bell, Yogurt Depot
Lodg	E: King's Inn Motel
	W: Ramada
AServ	E: Boyd Oldsmobile/GMC Trucks
Med	W: ✚ Hospital
ATM	W: Central Carolina Bank
Other	W: Byrd's Supermarket, Coin Laundry, Wal-

EXIT	NORTH CAROLINA
	Mart
202	U.S. 115, Oxford, Clarksville
(199)	**Rest Area - RR, Phones, Picnic, Vending**
191	NC 56, Butner, Creedmoor
FStop	**E:** **Amoco*(CW), Trade Mart***
	W: **Shell***
Gas	**E:** BP*
	W: Exxon*(LP)
Food	**E:** Bob's Barbecue, **Bojangles**, Burger King, China Taste, KFC, **McDonalds**, Pizza Hut, Subway, Taco Bell, Wendy's
	W: **Hardee's, The Depot (Exxon)**
Lodg	**E:** ◆ Comfort Inn
	W: **Econolodge, Holiday Inn Express,** AAA **Sunset Inn**
AServ	**E:** M&H Tire
	W: JR Tire & Auto
ATM	**E:** Carolina Central Bank, Fidelity Bank
Other	**E:** Car Wash, **Coin Laundry, Dutch Eye Center, Eckerd Drugs, Food Lion Supermarket**
189	Butner
186	U.S.15, Creedmoor, Butner
183	Redwood Road
Gas	**E:** Days Inn
Food	**E:** **Toddle House Diner (Days Inn)**
Lodg	**E:** DAYS INN AAA **Days Inn**
AServ	**E:** Yarbo Automotive
182	Red Mill Road
FStop	**E:** **Exxon*(D)**
TServ	**E:** **Kenworth**
180	Glenn School Road
FStop	**W:** **Mobil*(D)**
AServ	**E:** Mark's Auto Service
179	East Club Blvd, to U.S. 70 E, Raleigh
Gas	**W:** BP(D), Exxon*, Mobil*
AServ	**W:** BP, Kennedy Auto Service
178	Roxboro Road, U.S. 70 (Northbound)
177B	NC 55 E, Avondale Dr. (Westside Merges With Exit 177A, Difficult Reaccess)
177A	Downtown Durham
176	U.S. 501, Gregson St
175	Guess Road
174	Hillandale Road
174B	Bypass U.S. 15, U.S. 501 (Southbound)
173	U.S. 15, U.S. 501 South
172	U.S. 70
170	To NC 751, U.S. 70, Duke University
TStop	**E:** **Dixie Truckstop***
Food	**E:** **Dixie Truckstop**, Harbor Bay Seafood, Skyland (Best Western)
Lodg	**E:** AAA Best Western (Skyland Inn), Scottish Inns
AServ	**W:** Bull City Radiator
Parks	**W:** **Eno River State Park**
165	NC 86, Chapel Hill, Hillsborough
164	Hillsborough
Gas	**E:** Exxon*(LP)
Food	**E:** **McDonalds**
Lodg	**E:** Holiday Inn Express
ATM	**E:** Exxon

EXIT	NORTH CAROLINA
163	Junction I-40, Raleigh
161	To U.S. 70, NC 86
160	Efland
(159)	**Weigh Station - both directions**
157	Buckhorn Road
154	Mebane - Oaks Road
153	NC 119, Mebane
152	Trollingwood Road
150	Haw River
148	NC 54, Chapel Hill, Carrboro
147	NC 87, Graham, Pittsboro
145	NC 49, Liberty, Burlington
143	NC 62, Burlington, Downtown Alamance
141	Elon College
Gas	**E:** Amoco*, BP*
	W: Phillips 66*
Food	**E:** IHOP
	W: Applebee's, Arby's, Bojangles, Burger King, Chick-fil-A, **Cracker Barrel**, Golden Corral, The Summit, The Village Grill
Lodg	**E:** Hampton Inn
	W: **Best Western**, Super 8 Motel
AServ	**W:** Jiffy Lube, Phillips 66
Other	**W:** **K-Mart, Wal-Mart**
(140)	**Rest Area - RR, Phones**
138	NC 61, Gibsonville
TStop	**W:** **TravelCenters of America(SCALES)**
Gas	**W:** **BP (TravelCenters of America)**
Food	**W:** **Burger King (TravelCenters of America), Country Pride Restaurant (TravelCenters of America)**
Lodg	**W:** **Day Stop (TravelCenters of America)**
TServ	**W:** **Truck Service (TravelCenters of America)**
TWash	**W:** **Piedmont Truck Wash**
ATM	**W:** **Machine (TravelCenters of America)**
135	Rock Creek Dairy Road
Gas	**W:** Citgo*, Exxon*
Food	**W:** McDonalds
AServ	**W:** Exxon
132	McLeansville, Mtn Hope Church Road
Gas	**W:** Handi Pik*, Shell*(CW)(LP), Texaco*(LP) (Towing)
AServ	**W:** Texaco*
130	McConnell Road
Gas	**E:** Texaco*

EXIT	NORTH CAROLINA
AServ	**E:** Texaco* (Towing)
128	NC 6, East Lee St
Gas	**W:** BP*(CW)
Food	**W:** Blimpie's Subs
Lodg	**W:** Holiday Inn Express
127	U.S. 29 North, U.S. 70 East, U.S. 220, U.S. 421, Reidsville, Danville (No Northbound Reaccess)
126	U.S. 421 South, Sanford
Gas	**E:** Exxon*
Food	**E:** Burger King, Domino's, McDonalds, Pizza Pronto, Szechuan Kitchen, Wendy's
AServ	**E:** Advantage Auto Stores, Goodyear Tire & Auto Hall Tire Company
ATM	**E:** BB & T
Other	**E:** **A-1 Convenience Store, Bi-Lo Supermarket, Buchman's Discount Drugs, Food Lion Supermarket, Post Office, Revco Drugs**
125	East Elm - Eugene St
Gas	**E:** Amoco*(D)(CW), Texaco*(D)
	W: Crown*
Food	**W:** BT Sandwich Express, VIP Express Chinese
Lodg	**E:** **Cricket Inn,** DAYS INN AAA Days Inn, Howard Johnson, Super 8 Motel
	W: Homestead Lodge, **Ramada Limited (see our ad this page)**
AServ	**W:** Auto Zone Auto Parts, Superior Auto Parts Store
ATM	**W:** Cash Point
Other	**W:** **Coint Laundry, Food Lion Supermarket, Kerr Pharmacy, Visitor Information**
124	To Jct. I-40 West, Randleman Road
Gas	**E:** BP*(CW), Exxon, Texaco*(LP)
	W: Amoco* (ATM), Shell*(LP)
Food	**E:** Cafe 212, Cookout Hamburgers, Mayflower Seafood, Quincy's Family Steakhouse, **Waffle House**, Wendy's
	W: Arby's, Ben's Diner, Burger King, Captain D's Seafood, China Town Express, Dairy Queen, Jed's BBQ, KFC, McDonalds, Pizza Hut, Sub Station 2, Taco Bell
Lodg	**W:** Budget Motel
AServ	**E:** Exxon
	W: Grease Monkey, Meineke Discount Mufflers, Precision Tune & Lube
ATM	**E:** Machine (Texaco)
	W: Wachovia Bank
Other	**E:** **The Optical Place**
	W: **Dry Clean America, Eckerd Drugs**, New Glow Car Wash
122C	Rehobeth Church Road, Vandalia Road
Gas	**W:** Citgo*
Food	**W:** Shannon Hills Cafe
Lodg	**E:** Motel 6
122AB	U.S. 220 South, Asheboro (Difficult Reaccess)
121B	Holden Road
Gas	**E:** Texaco*
Food	**E:** Arby's, Burger King, Denny's, Hooligan's, **K & W Cafeteria**
	W: Niblick Pub
Lodg	**W:** Americana Hotel, Howard Johnson, Traveler's Express
ATM	**E:** Cash Points
Parks	**W:** **Gilford Courthouse National Park**

Bold red print shows RV & Bus parking available or nearby

XIT	NORTH CAROLINA
Other	E: **Camping(LP), K-Mart, Winn Dixie Supermarket**
	W: **Camping**
1A	Holden Road, Coliseum Area
Food	W: American Motel, Howard Johnson
Lodg	W: American Motel, Howard Johnson, Scottish Inns
RVCamp	E: **Field's RV Campground**
Other	W: **Emerald Pointe Water Park**
0	Groometown Road
FStop	W: **Phillips 66***
8	U.S. 29 South, U.S. 70 West
3	NC 62
1	U.S. 311, Archdale, High Point
8	Hopewell Church Road
6	Finch Farm Road
3	NC 109, Thomasville
Lodg	W: **Ramada Limited (see our ad this page)**
2	Lake Road
Gas	W: Texaco*
Lodg	W: **Days Inn**
Med	W: **Hospital**
00)	**Rest Area - RR, Phones (Southbound)**
9)	**Rest Area - RR, Phones (Northbound)**
6	U.S. 64, Asheboro
Gas	E: Mobil*
	W: Chevron*(D)
AServ	E: Modern Tire & Service
	W: Chevron
ATM	E: Machine (Mobile)
4	Old U.S. 64
Gas	E: Texaco*
1	NC 8, Lexington, Southmont
Gas	E: Amoco*(D), Citgo*(LP), Phillips 66*, Texaco*(D)
	W: Exxon*(D), Quality Mart*(LP)
Food	E: Bisquit King, Jimmy's BBQ, McDonalds, Sonic Drive In, Stephen's Lakeside Seafood, The Pizza Oven, Wendy's
	W: Arby's, Burger King, **Cracker Barrel,** Hardee's, Hunan Chinese, LaFunete Mexican, Long John Silvers, Taco Bell
Lodg	E: ◆ Comfort Suites, AAA Super 8 Motel
RVCamp	E: **High Rock Lake Campground**
Med	W: **Hospital**
ATM	E: BBT Bank
	W: First Union
Other	E: Car Wash, **Food Lion**
	W: **Ingles Supermarket, Wal-Mart (w/ pharmacy)**
8	Linwood
Gas	W: BP*
Food	W: Country Cousin Cafe (BP)
Med	W: **Hospital**
7	U.S. 29, U.S. 52, U.S. 70, Lexington, High Point, Winston - Salem
6	Belmont Road, NC 97
TStop	E: **Phillips 66*(D)(SCALES)**
Food	E: **Bill's Restaurant (Phillips 66)**
TServ	E: **Phillips 66**
5	Clark Road
Food	E: Tracks End Restaurant

EXIT	NORTH CAROLINA
Lodg	E: Tracks End Motel
83	NC 150 W, Spencer
Gas	W: Traveller's
Other	W: Coin Car Wash (Traveller's)
82	Southbound U.S. 29 (Southbound)
81	Spencer
Gas	E: Amoco*
	W: Texaco*
79	Spencer
Lodg	E: Chanticleer Motel
76AB	U.S. 52, Albemarle, Salisbury
Gas	E: Amoco, Racetrac*, Speedway*
	W: BP*(CW)(LP), Exxon(D), Shell*(LP)
Food	E: City View BBQ, Denny's, Golden Palace Chinese, Lighthouse Family Seafood, Little Caesars Pizza, Pizza Hut, Shoney's, Winks BBQ
	W: Bojangles, Burger King, Captain D's Seafood, China Garden, Dunkin Donuts, Hardee's, KFC, Krispy Kreme Donuts, McDonalds, Waffle House, Wendy's
Lodg	E: Happy Traveler Inn
	W: Econolodge, Harold's Motel, Roadway Inn
AServ	E: Big M Muffler Service, Meineke Discount Mufflers, Yost & Crowe Auto Service
	W: Exxon, Firestone Tire & Auto
Med	W: **Hospital**
ATM	E: CCB
Other	E: **Eckerd Optical, Food Lion Supermarket, Revco**
	W: **K-Mart (w/pharmacy), Post Office,** Sam's Full Service Car Wash
75	U.S. 601, Jake Alexander Blvd,

EXIT	NORTH CAROLINA
	Rowan
Gas	W: Amoco*, Citgo*, Exxon*
Food	E: **Arby's,** Farmhouse Restaurant, **Ramada**
	W: **Holiday Inn,** Ichiban Japanese Steakhouse, **Ryan's Steakhouse,** Sage Brush Steakhouse, Stadium Club, Subway, **Waffle House,** Wendy's
Lodg	E: Ramada Limited (see our ad this page)
	W: AAA Days Inn, ◆ Hampton Inn, **Holiday Inn**
AServ	W: Brad Farrah Pontiac, GMC, Ford Dealership, Gerry Wood Honda, Chrysler, Team Chevrolet
RVCamp	W: **Bull Hill Campground**
Other	W: **Wal-Mart**
74	Julian Road
72	Peach Orchard Road
71	Peeler Road
TStop	E: **Derrick 76 Auto/Truck Stop*(D)(SCALES)**
	W: **Wilco Citgo Truck Stop*(D)**
Food	W: **Express Cafe (Wilco), Taco Bell (Wilco)**
70	Webb Road
Other	W: **State Patrol Post**
68	NC 152, China Grove, Rockwell
63	Kannapolis
TStop	E: **Speedway Truckstop*(D)(LP)(SCALES)**
60	Earnhardt Road
Gas	E: Exxon*
Food	E: Burger King, **Cracker Barrel**
Lodg	E: ◆ Hampton Inn
Med	E: **Hospital**
(59)	**Rest Area - RR, Phones**
58	U.S. 29, U.S. 601, Kannapolis, Concord
Gas	E: BP*, Crown*(CW), Texaco*(D)(LP)
Food	E: Burger King, Captain D's Seafood, Chick-fil-A, China Orchad, El Vallarta, Golden Corral, KFC, Little Caesars Pizza, Mayflower Seafood, McDonalds, Shoney's, Sonic, Subway, Taco Bell, Waffle House, Wendy's
	W: Ryan's Steakhouse
Lodg	E: Colonial Inn, Holiday Inn Express, Mayfair Motel, Roadway Inn
	W: Comfort Inn, AAA Fairfield Inn
AServ	E: Brave Xpert Muffler Shop, Chrysler Auto Dealer, Sears
Med	E: **Hospital**
Other	E: **Contact Lens & Glasses Shop, Eckerd Drugs**
	W: **Drug Emporium**
55	NC 73, Davidson
FStop	W: **Phillips 66***
Gas	E: Exxon*
	W: BP*
Food	W: Days Inn, Huddle House
Lodg	W: Days Inn
52	Poplar Tent Road
Gas	E: Texaco*(D)(LP)
48	To Speedway Blvd.
46	Mallard Creek Church Road
45AB	Harris Blvd
Food	E: Applebee's, Bojangles, Chick-fil-A, Chili's, Max & Erma's, Shoney's, T.G.I. Friday's, Taco Bell
Lodg	E: Courtyard by Marriott, Dury Inn, Hampton Inn, Hilton, Holiday Inn, Quality Inn, Residence Inn
AServ	E: Tire and Lube Express
Med	E: **Hospital**

Bold red print shows RV & Bus parking available or nearby

EXIT	NORTH CAROLINA

ATM E: First Union, Home Federal Bank, NationsBank, SouthTrust Bank
Other E: **Hanaford Supermarket, Sam's Club, Vision Works, Wal-Mart**

43 To U.S. 29, to NC 49 (Northbound)

41 Sugar Creek Road
Gas E: Racetrac*, Texaco*(D)(CW)
W: BP*(CW), Exxon*
Food E: Bojangles, Dunkin Donuts (RaceTrac), McDonalds, Taco Bell, Wendy's
W: Shoney's, Texas Ranch Steakhouse, Waffle House
Lodg E: Best Western, AAA Continental Inn, Crickett Inn, AAA Econolodge, ◆ Red Roof Inn
W: AAA Comfort Inn, DAYS INN **Days Inn (see our ad this page)**, ◆ Fairfield Inn, ◆ Holiday Inn Express, Roadway Inn, Super 8 Motel
TServ W: **Ameri-Truck Goodyear, Freightliner Dealer**

40 Graham St
Food E: Hardees, Hereford Barn Steak House
Lodg E: Econolodge, Quality Inn, Travel Lodge
TServ E: **Tar Hill Ford Truck**
W: **Adkinson Truck Service**

39 Statesville Ave
FStop E: **Pilot Travel Center**(SCALES)
Gas W: Shell*(CW) (seeour ad on this page)
Food E: **Subway (Pilot FS)**
W: BoJangles
Lodg W: Knight's Inn
AServ E: Car Quest Auto Center, Statesville Avenue Garage
W: J&K Auto Repair, Super Transport
TServ E: **Adams International, Cummins Diesel, Custom Hydraulics, Peterbilt Dealer, Stone Heavy Vehicle Service, Volvo Dealer**
W: **Bailey's Truck Service, Truck Pro's Parts Warehouse**

38 Junction I-77, U.S. 21, Statesville, Columbia

37 Beatties Ford Road
Gas E: Petro Express* (Kerosene), Shell*(D) (see our ad this page), Texaco*(CW)
W: Coastal*
Food E: Burger King, McDonalds
W: **McDonalds Cafeteria (McDonalds Inn)**
Lodg W: **McDonalds Inn**
AServ W: Coastal
Med W: ✚ **Mecklenburg County Health Dept**
ATM E: NationsBank
Parks W: **Hornets Nest Park**
Other E: **Charlotte Coin Laundry, Fire Station, Library**
W: **Beatties Ford Hardware & Beauty Salon(LP) (Kerosene)**

36 NC16 to U.S. 76E, Brookshire Blvd. Downtown
Gas E: Amoco*(D), Auto Spa(CW)
W: Speedway*(LP)
Food E: China City
W: Bojangles, Burger King, Quincy's Family Steakhouse, Rogers Lunch, Wendy's
Lodg E: Hornet's Rest Inn
AServ W: Brake Experts
Med E: ✚ **Hospital (Trauma Center)**
Other W: **Golf Etc**

35 Glenwood Dr
Gas E: Citgo* (Kerosene)

EXIT	NORTH CAROLIN

Lodg E: **Innkeeper**
AServ E: Citgo

34 NC 27, Freedom Drive, Tuckaseege Road
Gas E: Amoco*(D), Exxon*
Food E: Burger King, Chad's Family Buffet, IHOP, Pap John's Pizza, Pizza Hut, Ruby Palace Chinese, **Subway**, Taco Bell, Tung Hoi Chinese, Wendy's
W: Dave's Grill
Lodg W: Apartment Inn, **Howard Johnson (see our ad on this page)**, Ramada Limited (see our ad on this page)
AServ E: Amoco, Ashley Road Auto Center, Exxon, Midas Muffler & Brakes, Western Auto
W: Discount Brake & Auto Center, Jiffy Lube
Med E: ✚ **Pro-Med Minor Emergency Center**
ATM E: State Employees Credit Union
Other E: **Action Clean Coin Laundry, Bi-Lo Supermarket, Checks Cashed,** EZ Car Wash, **Eckerd Drugs, Freedom Animal Hosp, Freedom Mall, Super Coin Laundry, U-Haul Center**(LP)
W: **Food Lion**

33 U.S. 521, Billy Graham Pkwy
Gas E: BP*(CW)
W: Exxon*(D)(LP) (Kerosene)
Food W: Cracker Barrel, Prime Sirloin Steakhouse, Waffle House
Lodg E: DAYS INN AAA **Days Inn,** ◆ Fairfield Inn, Microt Inn, ◆ Sheraton
W: Hampton Inn, AAA La Quinta Inn, ◆ Red Ro Inn, Save Inn
Other E: **Post Office**

32 Little Rock Road
Gas W: Circle K*(LP), Exxon*(D)
Food E: Waffle House
W: Arby's, Hardee's, Hickory House Restauran Lantana Rest. (Bradley Motel), Little Caesars Pizza, Little Rock Deli, **Quincy's Family Steakhouse, Shoney's**, Subway
Lodg E: AAA **Best Western,** AAA Courtyard by Marriot, Econolodge
W: Bradley Motel, Comfort Inn, **Country Inn & Suites**
Other W: **Eckerd Drugs, Food Lion Supermarket**

29 Sam Wilson Road
Gas W: Texaco*(D)

(28) **Weigh Station - Both Directions**

27 NC 273, Mount Holly, Belmont
FStop W: **Texaco* (Western Union)**
Gas E: Exxon*
W: BP*(D)
Food E: Arby's, Burger King, KFC, Pizza Hut, Subway, Taco Bell (24 Hrs), The Captain's Cap Restauran Waffle House
AServ E: Exxon
ATM E: First Union, NationsBank, Wachovia Bank
Other E: **College Park Pharmacy, Eckerd Drugs, Foo**

Bold red print shows RV & Bus parking available or nearby

EXIT NORTH CAROLINA

EXIT NORTH CAROLINA

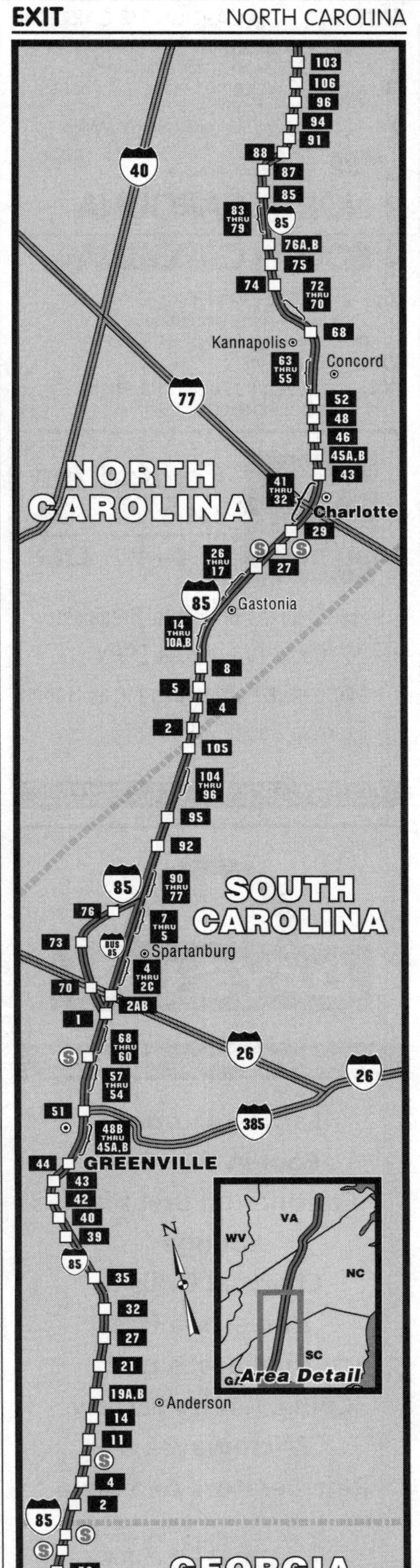

EXIT NORTH CAROLINA

Lion Supermarket, Harris Teeter Supermarket

26 Belmont, Mount Holly
- Food — E: Bagels, Bojangles, Happy China, Hardee's, McDonalds, Papa John's Pizza, Quincy's Family Steakhouse
- Other — E: Bi-Lo Supermarket, College Coin-Up Laundry, Winn Dixie Supermarket

23 NC 7, McAdenville
- Gas — W: Shell* (see our ad on this page)
- Food — W: Hardees, Hillbilly's BBQ

22 Cramerton, Lowell
- AServ — E: Auto Tech & Parts, Haygood Mercury Lincoln, McKenney Salinas Mitsubishi, McKenney Salinas Suzuki, Sonny Hancock Hyundai, Sonny Hancock Mazda
- Other — E: Pop-a-Top Grocery, Wilkinson Animal Hospital

21 Cox Road, Ranlo
- Gas — E: Citgo* (Kerosene)
  W: Exxon*, Phillips 66*(CW)
- Food — E: China House Restaurant, Jackson's Cafeteria, Longhorn Steaks, Max Mexican Eatery, Pizza Inn
  W: IHOP
- Lodg — E: Caravan Motel
  W: Regal Inn, Villager Lodge
- AServ — E: Sonny Hancock Cadillac, Oldsmobile, Speedy Oil Change & Tune-Up, Tire Kingdom
  W: Cox Road Auto Repair, Exxon
- Med — E: + Hospital
- ATM — E: Bi-Lo Supermarket, First Union Bank
  W: First Citizen's Bank, State Employees Credit Union
- Other — E: Batting Cages, Bi-Lo Supermarket, Gaston Mall, Harris Teeter Supermarket (24 Hrs)
  W: Medical Center Pharmacy

20 NC 279, New Hope Road
- Gas — E: Amoco*, Texaco(CW)
- Food — E: Captain D's Seafood, Red Lobster, The Italian Oven
  W: Bojangles, Cracker Barrel, Hickory Hams Cafe, KFC, Outback Steakhouse, Waffle House
- Lodg — W: AAA Fairfield Inn, ◆ Hampton Inn (see our ad on this page), Innkeeper
- AServ — E: Amoco, Firestone Tire & Auto, Nappa Auto Parts
- Med — W: + Hospital
- Other — E: East Ridge Mall, Shiele Museum & Planetarium

19 NC 7, East Gastonia
- Gas — E: Shell* (see our ad on this page)
  W: Servco (Kerosene)
- AServ — E: Shell
- Other — E: Amtrac
  W: Shirley's Fresh Food & Groceries

17 U.S. 321, Gastonia, Lincolnton
- Gas — E: Exxon*(D)(LP)
  W: Shell*(D) (see our ad on this page)
- Food — E: Hardee's, Pancake House
  W: Waffle House, Wendy's, Western Sizzlin'
- Lodg — E: Days Inn
  W: AAA Motel 6, Howard Johnson (see our ad on this page)

EXIT		NORTH CAROLINA

	AServ	E: Auto Service Brake Products, Kelly Tires
	RVCamp	E: **Allen's Mobile Home Parts & Supplies**
	Other	E: **Fire Dept**
14		NC 274, Bessemer City, West Gastonia
	FStop	W: **Citgo* (Kerosene)**
	Gas	E: Phillips 66*(D)
		W: BP*
	Food	E: Burger King
		W: **Bojangles**, Subway, Waffle House
	Lodg	W: **Econolodge**
	Other	E: Super Clean Coin Car Wash
13		Bessemer City, Edgewood Rd.
	Gas	W: Amoco*, Shell*(D) (see our ad on this page)
	Food	W: **Master's Inn**
	Lodg	W: **Master's Inn**
	AServ	W: Shell*
	Parks	E: **Crowders Mountain State Park**
10B		U.S. 74 West, Kings Mountain, Shelby
	Lodg	W: Hampton Inn (see our ad on this page)
10A		U.S. 29, U.S. 74 East
8		NC 161, King's Mountain
	Gas	E: Exxon(D)(LP)
		W: Amoco* (24 Hrs)
	Food	W: Burger King, Waffle House
	Lodg	W: AAA **Comfort Inn**, Ramada
	AServ	W: McKenney Chevrolet
	RVCamp	W: **Camping**
	Med	W: ✚ **Hospital**
	ATM	W: Amoco
(6)		**Rest Area - RR, Phones (Southbound)**
5		Dixon School Road
	TStop	E: **Kings Mountain Travel Center*(SCALES) (RV Dump)**
	Gas	E: BP*
	Food	E: **Family Restaurant, Subway (Kings Mountain TC)**
	AServ	E: BP

EXIT		N CAROLINA/S CAROLINA
	TServ	E: **Kings Mountain TC**
4		U.S. 29 South (Southbound)
(3)		NC Welcome Center - RR, Phones (Northbound)
2		NC 216, Kings Mountain, Military Park
	Gas	W: Chevron* (Kerosene)

↑NORTH CAROLINA

↓SOUTH CAROLINA

106		U.S. 29, Grover
	Gas	W: Crown*, Exxon* (Kerosene)
	Other	W: **Grover Car Wash & Coin Laundry**
104		Road 99
(103)		Welcome Center - RR, Phones (Southbound)

EXIT		SOUTH CAROLIN
102		SC 198, Blacksburg, Earl
	TStop	W: **Flying J Travel Plaza(SCALES) (RV Dump)**
	FStop	E: **Exxon*(LP)**
	Gas	E: Phillips 66* (Fireworks)
		W: Texaco*
	Food	E: **Hardee's**
		W: **McDonalds, Thads Restaurant (Flying J), Waffle House**
100		SC 5, Blacksburg, Rock Hill
	TStop	W: **Speedway*(SCALES)**
	Gas	W: Texaco*
	Food	W: Subway (Speedway), Wilson's Country Store Cafe
	ATM	W: Speedway
98		Frontage Road (Northbound, Reaccess Via Frontage Rd)
	TStop	E: **Broad River TruckStop**
	Food	E: **Broad River TruckStop**
	TServ	E: **Broad River TruckStop (Tire & CB repair)**
96		SC 18, Shelby
	FStop	W: **Exxon*(CW)**
	TServ	W: **INS Diesel Service Inc**
95		SC 18, Gaffney, Shelby (Reaccess Vi Frontage Road)
	TStop	W: **Norma's Truck Stop**
	FStop	E: **Citgo*(D) (Overnight Parking), Mr. Petro*(D) (2 hrs)**
	Food	E: **Aunt M's Good Cookin'**, Blackbeard's Arsen & Grill, Breakfast Point U.S.A., Fatz Cafe, Mr. Waffle (24 Hrs)
		W: **Norma's Truck Stop Restaurant**
	Lodg	E: Shamrock Inn
	AServ	E: Wallace White Pontiac, Buick, GMC Truck
	Med	E: ✚ **Hospital**, ✚ **Hospital**
	Other	W: **CB Repair**
92		SC 11 to SC 150, Gaffney, Boiling Springs
	Gas	W: Exxon*, Phillips 66*, Texaco*
	Food	E: Bojangles, Burger King, **McDonalds**, New China II, Pizza Hut, Quincy's Family Steakhouse, Shoney's, Subway, Wendy's
		W: **Top of the Hub (Days Inn), Waffle House**
	Lodg	E: The Jameson Inn
		W: AAA **Comfort Inn (see our ad on this page)**
		DAYS INN **Days Inn (see our ad on this page)**
	AServ	W: Exxon
	ATM	E: NationsBank, Wal-Mart
	Other	E: **Ingles Supermarket, Mail Boxes Etc, Wal-Mart**
		W: **Cowpens Battlefield**
90		SC 105, Road 42, Gaffney
	Gas	W: Citgo*
	Food	W: Baskin Robbins (Citgo), Burger King (Citgo)
	ATM	W: Citgo
	Other	W: **Carolina Factory Shops, Sunny Hill Farms**
(89)		**Rest Area - RR, Phones**
87		Road 39
	RVCamp	E: **Camping**
	Other	W: **Antique Mall**
83		SC 110, Cowpens
	TStop	W: **Cowpens Truck Stop(SCALES), Mr. Waffle*(SCALES)**
	Gas	E: Poor Paul's Fireworks
	Food	W: **Mr. Waffle Restaurant**
	Lodg	W: **Mr. Waffle**
	AServ	W: Mr. Waffle
	TServ	W: **Mr. Waffle**
	TWash	W: **Mr. Waffle TW**
	Other	E: **Abbott Farms Peaches, Red Star Fireworks**
81		Frontage Road (Northbound, Reaccess Frontage Rd)
80		Road 57, Gossett Rd
78		U.S. 221, Chesnee, Spartanburg
	Gas	E: Shell*(D)

Bold red print shows RV & Bus parking available or nearby

EXIT SOUTH CAROLINA

W: BP*[CW], Exxon*[D]
Food E: Hardee's, Waffle House
W: Burger King (Exxon), McDonalds, Southern BBQ
Lodg E: Motel 6
AServ E: Bojan's Tire & Lube Service, Cooper Tires
W: Advance Auto Parts
Other W: Ingles, Red Star Fireworks

77 I-85 Business Loop South

75A S.C. 9, Spartanburg, Boiling Springs
Gas W: BP*[D][LP], Phillips 66*
Food W: Burger King (BP), Long John Silvers (BP), McDonalds
AServ W: Advance Auto Parts
Other W: Ingles, Revco Drugs

72 U.S. 176 to I-585, Spartanburg, Inman
Med E: ✚ Hospital

70 Jct. I-26, Ashville, Columbia

Note: I-85 runs concurrent below with I-85B. Numbering follows I-85B.

7 Bryant Road
Gas E: Amoco*[LP]

6 Converse & Wofford Colleges, SC 9 Boiling Springs Rd
Food E: Waffle House
Lodg E: Comfort Inn, Days Inn
W: Best Western, Travelers Inn
RVCamp E: Winfield Co Inc Mobile Home and RV Supply

5AB Junction I-585 South, U.S. 176, Pine Street, Downtown (Difficult Reaccess)
Food W: Fatz Cafe, Steak & Ale
Lodg W: Ramada, Super 8 Motel

4B SC 56, Asheville Hwy
FStop E: Texaco*
Gas E: Exxon*[LP], Speedway*[LP]
W: Shell*
Food E: Arby's, Bojangles, Burger King, Dominoe's Pizza, Hardee's, KFC, Miami Subs Grill, Pizza Hut, Quincy's Family Steakhouse, Taco Bell, Waffle House
W: Carolina BBQ, Darie Dream, McDonalds, Steak & Ale, Waffle House, Wendy's
Lodg E: Courtyard by Marriott, Extended Stay America, Fairfield Inn
W: Quality Hotel, Spartanburg Motor Lodge
AServ E: Exxon
ATM E: Bi-Lo Supermarket, NationsBank
W: Wachovia Bank
Other E: Bi-Lo Supermarket, Eckerd Drugs

4A Viaduct Road

3 SC 295, New Cut Road
Food E: So Then Diner
AServ E: Superior Parts

2C (3) Fair Forrest Rd (Reaccess Via Frontage Rd)
Gas E: Phillips 66* (Kerosene)
Food E: Cracker Barrel, Wilson World
W: J.D. Peaches Rest. (Holiday Inn)
Lodg E: Budget Inn, Residence Inn, Wilson World
W: Holiday Inn

2B Junction I-26 West, Asheville

2A Junction I-26 East, Columbia

EXIT SOUTH CAROLINA

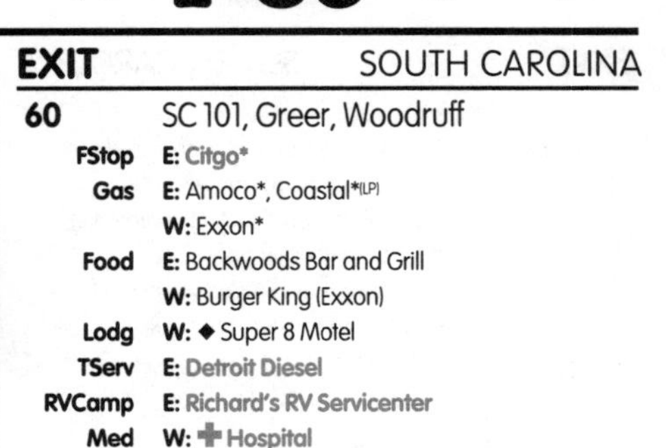

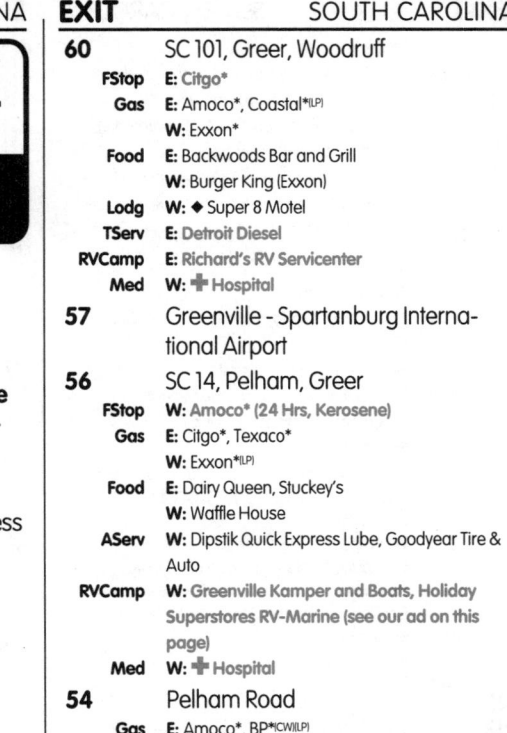

1 Road 41, North Blackstock Rd (Southbound, Reaccess Northbound)
Lodg W: Hampton Inn, Tower Motel

Note: I-85 runs concurrent above with I-85B. Numbering follows I-85B.

69 I-85N, Spartanburg

68 SC 29, Greer (Southbound, Reaccess Northbound)

66 U.S. 29, Wellford, Spartanburg - West, Lyman

63 SC 290, Duncan, Moore
TStop W: TravelCenters of America*[SCALES]
FStop W: Pilot Travel Center*[SCALES] (Kerosene)
Gas E: Citgo*[D], Exxon*
W: Amoco*, BP (TravelCenters of America)
Food E: Arby's, Aunt M's Good Cookin', Denny's, Pizza Inn
W: Bojangles, Country Pride (TravelCenters of America), Dairy Queen (TravelCenters of America), Demetre's Gourmet Grill & Bar, Hardee's, McDonalds, Waffle House, Wendy's (Pilot FS)
Lodg W: ◆ Comfort Inn, ◆ Days Inn, Travelodge (TravelCenters of America)
TServ W: USA Truck Wash
RVCamp W: Sonny's Camp & Travel
Med E: ✚ Hospital
ATM E: Exxon, Nations Bank

(62) Rest Area - RR, Phones

EXIT SOUTH CAROLINA

60 SC 101, Greer, Woodruff
FStop E: Citgo*
Gas E: Amoco*, Coastal*[LP]
W: Exxon*
Food E: Backwoods Bar and Grill
W: Burger King (Exxon)
Lodg W: ◆ Super 8 Motel
TServ E: Detroit Diesel
RVCamp E: Richard's RV Servicenter
Med W: ✚ Hospital

57 Greenville - Spartanburg International Airport

56 SC 14, Pelham, Greer
FStop W: Amoco* (24 Hrs, Kerosene)
Gas E: Citgo*, Texaco*
W: Exxon*[LP]
Food E: Dairy Queen, Stuckey's
W: Waffle House
AServ W: Dipstik Quick Express Lube, Goodyear Tire & Auto
RVCamp W: Greenville Kamper and Boats, Holiday Superstores RV-Marine (see our ad on this page)
Med W: ✚ Hospital

54 Pelham Road
Gas E: Amoco*, BP*[CW][LP]
W: Exxon[CW]
Food E: Brady's Biscuits & Burgers, Burger King, Joe's Crab Shack, Luigi's Pizzeria
W: Applebee's, Bertolo's Pizza, Boston Market, California Dreaming, Happy China, Hardee's, J & S Cafeteria, Macaroni Grill, McDonalds, On the Border Mexican Cafe, Ruby Tuesday, Tony Roma's
Lodg E: ◆ Holiday Inn Express
W: Extended Stay America, ◆ Hampton Inn, AAA Microtel Inn, ◆ Wingate Inn, Comfort Suites (see our ad on this page
ATM W: BB&T, First Union, Wachovia
Other E: Fire Station
W: Bi-Lo Supermarket, Mail Express USA, Revco Drugs

51B Junction I-385

51 Junction I-385, SC 146, Columbia, Greenville, Woodruff Road
Gas E: Hess*, Red Robin*
W: Racetrac*
Food E: Marcelino's
W: Capris Italian, Cracker Barrel, McDonalds, Waffle House
Lodg W: ◆ La Quinta Inn
AServ E: Goodyear Tire & Auto
ATM E: Red Robin, Wachovia
Other E: All Creatures Animal Hospital
W: Greenville Mall, The Home Depot

48AB U.S. 276, Mauldin, Greenville
FStop E: Speedway*[LP]
Gas W: BP*[LP] (24 Hrs), Exxon
Food E: Charlie T's Buffalo Wings & Subs, Waffle House
W: Burger King, Olive Garden
Lodg E: Intown Suites Apartment Hotel, Red Roof Inn
W: Howard Johnson, Value-Lodge
AServ W: Century BMW, GM Auto Dealer

46 Mauldin Road
Gas W: Amoco*[LP], Texaco*
Food W: Arby's, Steakout, Subway
Lodg W: Comfort Inn, Ramada Limited, Suburban

Bold red print shows RV & Bus parking available or nearby

EXIT SOUTH CAROLINA

Lodge
ASserv W: AAMCO Transmission, Consolidated Auto Tire & Serv, Japan Auto Parts, Jiffy Lube
ATM W: Carolina First, Harris Teeter Supermarket, United Carolina Bank (Bi-Lo)
Other W: Bi-Lo Supermarket (w/pharmacy), Harris Teeter Supermarket, Revco Drugs

45AB U.S. 25, SC 291, Greenwood, Greenville
Gas E: Hess, Li'l Cricket*
Food E: Bojangles, Holiday Inn Restaurant, Waffle House
W: Dixie Family Restaurant
Lodg E: Camelot Inn, Holiday Inn, Motel 6
W: Cricket Inn, Days Inn
AServ E: Used Tires
Other E: Augusta Road Grocery, Nick's Car Washes

44 U.S. 25, White Horse Road
FStop E: Exxon*(D)
Gas W: Amoco*(D), Texaco*(D) (Kerosene)
Food E: El Cactus, Masters Inn Restaurant, Subway (Exxon)
W: Captain D's Seafood, McDonalds, Quincy's Family Steakhouse, Shoney's, Waffle House
Lodg E: Masters Inn
AServ W: L & L Discount Muffler
ATM E: Exxon

43 SC 20, Piedmont

42 Junction I-185, U.S. 29, Greenville

40 SC 153
Gas E: Exxon*
W: Texaco*(D)
Food E: Waffle House
W: Arby's, Burger King, Hardee's, KFC
Lodg W: Executive Inn, Super 8 Motel
ATM E: Exxon

39 Road 143
Gas E: BP*
W: Texaco*(LP)
AServ E: BP

35 SC 86, Piedmont, Easley
TStop E: Speedway Truck/Auto Plaza*(SCALES)
Food E: Speedway Truck/Auto Plaza

34 U.S. 29, Williamston, South Anderson

32 SC 8, Pelzer, Easley
Gas E: Citgo*
AServ E: Citgo

27 Anderson College, SC 81
TStop W: Anderson Truck Plaza*(SCALES)
FStop E: Phillips 66*
Gas E: Citgo, Exxon*
Food E: Arby's, McDonalds (Playground), Waffle House
W: Anderson Truck Plaza
Lodg E: AAA Holiday Inn
ATM E: Exxon
Other E: Big Zack's 10,000 Fireworks

(23) Rest Area - RR, Phones (Southbound)

21 U.S. 178, Anderson, Liberty
FStop E: Texaco*(D)

19AB U.S. 76, SC 28, Anderson, Clemson
Gas E: Exxon*(D)
W: Racetrac*
Food E: Charlie T's Original Buffalo Wings, Hardee's, Katherine's Kitchen
W: Buffalo's Cafe, Carrabba's Italian Grill, Cracker Barrel, Outback Steakhouse, Waffle House, Wendy's
Lodg E: Days Inn, Park Inn, Royal American Motor In, AmeriSuites (see our ad on this page)
W: Hampton Inn, Jameson Inn
ATM E: Exxon
Other W: S.C. Botanical Gardens

(18) Rest Area - RR, Phones, Vending (Northbound)

14 SC 187, Pendleton, Clemson
FStop E: Fuel Mart*

EXIT SOUTH CAROLINA/GEORGIA

Gas W: Amoco*(D)(LP) (ATM)
Food E: Huddle House (24 Hrs)
Lodg W: Economy Lodge
RVCamp E: KOA Campground

11 SC 24, SC 243, Anderson, Townville
FStop E: Speedway*(LP)
Gas W: Texaco*
Food W: Townville Station Rest.
RVCamp W: Hartwell Four Seasons Campground
Other E: Fireworks Superstore, The Racin' Station Conv. Store*

(9) Weigh Station (Northbound)

4 SC 243, Road 23, Fair Play
TStop E: Cherokee Run*(LP)(SCALES)
Gas E: Exxon*(D)(LP)
Food E: Cherokee Run, Glenn's Diner
RVCamp E: Camping

2 SC 59, Fair Play, Seneca
Gas E: Econolodge
Food E: Econolodge
Lodg E: Econolodge
RVCamp E: Lakeshore Campground
Other E: Crazy Steve's Fireworks* (ATM)

1 SC 11, Walhalla, Westminster
FStop W: Amoco*(D)
Food W: Gazebo Restaurant & Deli
RVCamp W: Camping
Parks W: Lake Hartwell State Park
Other W: I-85 Fireworks Outlet

(1) SC Welcome Center - RR, Phones, Vending (Northbound)

↑SOUTH CAROLINA

↓GEORGIA

(177) GA Welcome Center - RR, Phones, RV Dump (Southbound)

59 **(177)** GA 77, Elberton, Hartwell
Gas E: BP*
Food E: Dad's (BP)
Other E: Dad's Boat Storage, Hilltop Marine

58 **(173)** GA 17, Lavonia, Toccoa, Hartwell, Helen
FStop E: Texaco*(D) (Kerosene)
Gas E: BP*(LP)
W: Exxon*(LP) (ATM), Shell
Food E: Fernside Home Cooking Buffet, KFC, McDonalds, Subway, Waffle House
W: Arby's, Burger King (Playground), Hardee's, Pizza Hut, Shoney's, Waffle House, Wendy's
Lodg E: Days Inn, ◆ Sleep Inn
W: Shoney's Inn
Med E: ✚ Hospital
ATM E: People's Bank
Other E: Lavonia Foods, Rite Aide Pharmacy

57 **(166)** GA 106, GA 145, Toccoa
TStop W: Echo*(LP)(SCALES) (Restaurant 24 Hrs)
Gas E: Shell*
Med E: ✚ Hospital

56 **(164)** GA 320, Carnesville
TStop E: Sunshine Travel Center*(SCALES) (Restaurant)
Food E: Hardees (Sunshine Travel Center)
AServ E: Harper's Auto Truck Plaza, Truck Tires
Med E: ✚ Hospital

(161) Rest Area - RR, Phones, RV Dump

EXIT GEORGIA

(Northbound)

55 **(160)** GA 51, Royston, Elberton
TStop E: Shell*(LP)(SCALES) (RV Dump, ATM, Restaurant)
W: Petro*(LP)(SCALES) (RV Dump)
FStop W: Amoco (Petro)
Gas W: Amoco*(LP) (RV Dump)
Food E: Subway (Shell)
W: Iron Skillet (Petro)
TServ W: Petro(LP)(SCALES) (Travel Plaza Restaurant, shopping center, 24 Hrs)
TWash W: Blue Beacon (Petro)
Parks E: Victoria Bryant State Park (Camping)

54 **(154)** GA 63, Martin Bridge Road

53 **(149)** U.S. 441, GA 15, Commerce
TStop E: 76 Auto/Truck Plaza*(SCALES)
FStop E: Citgo(LP) (ATM)
W: BP*(D)
Gas E: Amoco
W: Racetrac*
Food E: BJ's Grill (The Pottery), Captain D's Seafood, China Doll Chinese, McDonalds, Shoney's, Sonny's Real Pit Barbecue, South Fork Steak House, Stringers Seafood, Taco Bell, Waffle House
W: Cracker Barrel, Arby's, Burger King, Checkers Burgers, Dairy Queen, KFC, La Hacienda Mexican, McDonalds (Playground), Pizza Hut, RJ T-Bones Steakhouse, Ryan's Steakhouse, Subway, Waffle House, Wendy's
Lodg E: ◆ Days Inn, Guest House Inn, Hampton Inn, AAA Holiday Inn Express (see our ad on this page)
W: AAA Comfort Inn, Dollar Wise, Howard Johnson (see our ad on this page), Ramada Limited, The Jameson Inn
AServ E: Amoco* (24 Hrs)
W: Jimmy McBride (Pontiac, Buick, GMC Trucks)
RVCamp E: The Pottery Campground
W: Campground

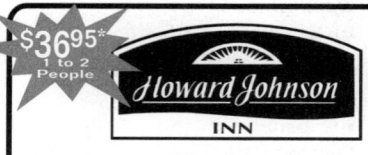

Bold red print shows RV & Bus parking available or nearby

EXIT GEORGIA

ATM E: Regions Bank
Other E: Commerce Factory Stores, Swimwear Outlet, Tanger Factory Outlet, The Pottery
W: Tanger Factory Outlet

52 **(147)** GA 98, Commerce, Maysville
FStop E: Fuel Mart*, Speedway*
Gas W: Shell*
AServ E: Holman's Tire & Auto
Med E: + Hospital

51 **(140)** GA 82, Dry Pond Road

50 **(136)** U.S. 129, GA 11, Gainesville
FStop E: Phillips 66*(D)(LP)
Gas E: BP*, Shell* (ATM) (see our ad on this page)
Food E: Arby's, Hardee's, McDonalds (Playground), Waffle House
W: Katherine's Kitchen, Waffle House
Other E: Crawford W. Long Museum
W: Georgia's Largest Flea Market (Has Food Court)

49 **(129)** GA 53, Braselton, Hoschton, Lanier Raceway, Winder
FStop E: Shell*(D)(LP) (24 Hrs), Texaco*(D)(LP) (Kerosene)
W: BP*(D), Fina*(D)(LP) (Full Service), Speedway*(LP)(SCALES)
Food E: Waffle House
W: Subway (BP)
Lodg E: Best Western
AServ W: Fina
ATM E: Northeast Georgia Bank
Other W: Braselton Hardware(LP)

48 **(125)** GA 211, Winder
Gas W: BP*(LP) (ATM, Kerosene)
Lodg W: ◆ Days Inn (see our ad on this page)
Parks E: Fort Yargo State Park (Camping, Picknicking)
Other W: Chateau Elan (Winery, Spa, Inn, Conference Center, Golf Club and Villas)

47 **(120)** Hamilton Mill Parkway, Hamilton Mill Road
FStop E: Circle K*(D)(LP) (ATM)
Gas E: BP*
W: Shell (see our ad on this page)
Food E: Buffalo's Cafe, Dos Copas Mexican Grill, McDonalds (Playground), Mr. Edd's Pizza and Grill, Subway
AServ W: Hamilton Mill Auto Repair Center
ATM E: First Commerce Bank, SunTrust Bank
Other E: Publix Supermarket

46 **(115)** GA 20, Lawrenceville, Buford Dam, Lake Lanier Islands
Med E: + Hospital

45 **(113)** Junction I-985N, Gainesville, Lake Lanier Parkway (Left Lane Exit)

(112) Rest Area - RR, Phones, Picnic, HF, RV Dump (Northbound)

(114) Rest Area (Southbound)

44 **(111)** GA 317N, Suwanee
Gas E: Amoco*(CW), BP*(CW)(LP), Phillips 66*(LP)
W: Chevron* (24 Hrs), Exxon
Food E: Arby's, Baskin Robbins, Blimpie's Subs, Burger King, Checkers Burgers, Chick-fil-A, Cracker Barrel, Del Pizzo's Rest. and Pizzaria, Del Taco, Dunkin' Donuts, Mrs Winner's Chicken, Outback Steakhouse, Philly Connection, Pizza Hut Delivery, Subway, Taco Bell, The Orient Garden Chinese, Waffle House, Wendy's
W: Denny's, McDonalds (Playground), Waffle House
Lodg E: AAA Comfort Inn, AAA Holiday Inn (see our ad on this page), Howard Johnson, Sun Suites, Days Inn (see our ad on this page)
W: ◆ Days Inn (see our ad on this page), AAA Falcon Inn (Best Western), Ramada Limited (see our ad on this page)
AServ E: Kauffman Tires
W: Exxon
ATM E: Citizen's Bank, NationsBank, SunTrust Bank
Other E: Ace Hardware, Ingle's Supermarket, Mail

Bold red print shows RV & Bus parking available or nearby

EXIT GEORGIA

Boxes Etc, Publix Supermarket, Revco Drugs, Ski Nautiques (Marine and Boating Supply)
W: Falcons' Training Camp

43 **(109)** Old Peachtree Road
- **Gas** **E:** QuikTrip* (24 Hrs)

42 **(107)** GA 120, Duluth, Lawrenceville
- **Gas** **E:** Speedway*(LP) (ATM)
  W: Chevron*(CW) (24 hr service)
- **Lodg** **E:** Days Inn (see our ad on this page)
- **AServ** **W:** BP(CW)
- **Med** **E:** + Hospital

41 **(106)** GA 316E, Lawrenceville, Athens
40 **(104)** Pleasant Hill Road
- **FStop** **E:** Shell*(D)(CW)
- **Gas** **E:** Chevron*(CW), Circle K*(LP) (ATM), Phillips 66*(LP)
  W: Amoco*, Phillips 66*(CW)(LP), QuikTrip*, Texaco*(CW)
- **Food** **E:** Baskin Robbins, Blimpie's Subs, Buffalo's Cafe, Burger King, Carrabbas Italian Restaurant, Chesapeake Bagel Bakery, Chick-fil-A, Combo Express Chinese, Cooker, Corkey's BBQ & Ribs, Dunkin' Donuts, Georgia Diner (24 Hrs), Geronimo's Cafe, Grady's, Guacamole's, Hardee's, Hickory Hams, Kinyobee, Krispy Kreme Donuts, Los Loros, Mandarin House III, McDonalds, Monterrey Mexican, New York Pizza, O'Charley's, Philly Connection, Poona Indian Gourmet, Popeye's Chicken, Red Garlic, Romano's Macaroni Grill, Ruby Tuesday, S & S Cafeteria, Schlotzkys Deli, Shiki, Subway, T.G.I. Friday's, The Original Pancake House, Waffle House, Wendy's
  W: A Fondue Restaurant, Applebee's, Barnacle's Seafood, Baskin Robbins, Black-Eyed Pea, Bruegger's Bagel, Burger King, Chili's, Circus Pizza World, County Seat Cafe, Cripple Creek, Devito's Pizza, El Torero, Golden China Restaurant, Hooters, IHOP, Johnnie's Pizza & Subs, KFC, Long Horn's Steak House, Manhatten Bagel, Marchello's Italian Rest., Matsuri Restaurant, McDonalds, Mick's, Mother India Restaurant, Mrs Winner's Chicken, My Friend's Place, Olive Garden, On The Border, Philly Connection, Pizza Hut, Provino's Iltalian Restaurant, Red Lobster, Rio Bravo Cantina, Roaster Rotisserie, Rugby Sports Bar and Grill, Ryan's Steakhouse, Sho Gun, Shoney's, Skeeter's, Starbucks Coffee, Steak & Shake, Subway, TCBY, Taco Bell, Tung Sing Chinese, Waffle House, Wangs Chinese, Wendy's
- **Lodg** **E:** ◆ Comfort Suites, Hampton Inn, ◆ Holiday Inn Express, ◆ Marriott
  W: Ameri Suites (see our ad on this page), ◆ Courtyard by Marriott, ◆ Fairfield Inn, Ramada Limited, Suburban Lodge, Summer Suites, Wyndham Garden Hotel
- **AServ** **E:** Eurasain Motor Sports, Goodyear Tire & Auto, K-Mart, Tune Up
  W: Atlanta Toyota Trucks, Batteries Plus, Brake-O Brake Shop, Econo Lube & Tune, Firestone Tire & Auto, Goodyear Tire & Auto, Midas Muffler & Brakes, Precision Tune & Lube, Q-Lube
- **TServ** **E:** Nalley Motortrucks
- **Med** **E:** + Med Plus (Walk-Ins welcome)
  W: + Any Lab Test, + Physicians Immediate Med (Minor Emerg, 9am-9pm, no appt necessary)
- **ATM** **E:** First Union Bank, SunTrust Bank
  W: NationsBank, SouthTrust Bank
- **Other** **E:** Eckerd Drugs, K-Mart (Pharmacy), Kinko's Copies (24 Hrs), Mail Boxes Etc, Office Depot, Pro Golf Discount, Publix Supermarket, The Home Depot, United Mail Services, Wal-Mart (24 Hrs), Winn Dixie Supermarket
  W: Ace Checks Cashed, Carmike Cinemas, Drug Emporium, General Cinema, Gwinnett Place Mall, Island Mania, Kroger Supermarket (ATM, 24 Hrs), Mail Boxes Etc, OfficeMax, PETsMART, Pearle Express Vision Center, Pets Etc., Pirates Cove Adventure Golf, Q-Zar (Lazer Game), Target Department Store, The

EXIT GEORGIA

Aviarium (A Bird Pet Store), US Post Office, Wolf Camera

39A **(103)** Steve Reynolds Blvd (Northbound, Reaccess Southbound)
- **Gas** **W:** QuikTrip*, Texaco*(CW)
- **Food** **E:** Roadhouse Grill
  W: Han Gang, Waffle House
- **Lodg** **E:** Comfort Suites, Sun Suites Extended Stay Hotel
- **ATM** **E:** SunTrust Bank, Wachovia
  W: NationsBank
- **Other** **E:** Sportsman Warehouse, The Home Depot
  W: Boater's World, Cinema 12 Theaters, Costco Wholesale, Outlet's Limited Mall, PETsMART, Sam's Club, Waccamaw

39 **(102)** GA 378, Beaver Ruin Road, Lilburn
- **FStop** **E:** Shell*(D)(LP)
  W: Texaco*(D)
- **Gas** **E:** Racetrac*
  W: Amoco*(CW)
- **AServ** **E:** Timmers Chev-GM

38 **(101)** Indian Trail - Lilburn Road
- **FStop** **W:** BP*(D)(CW)
- **Gas** **E:** QuikTrip* (ATM), Texaco*(CW)
  W: Chevron*(CW) (ATM, 24 Hrs), Conoco*(LP)
- **Food** **E:** Blimpie's Subs, Burger King, Dunkin Donuts, McDonalds, Shoney's, Taco Bell, Waffle House
  W: Arby's, Dairy Queen, La Pantera Rosa, La Tapatia, Lee's Golden Budda, Lupita's Mexican, Mega Tacos, Mrs Winner's Chicken, Papa John's Pizza, Pinoy, Taylor's, The Eating Exchange, The Grill Cafe, Waffle House, Wendy's, Yoko
- **Lodg** **E:** (AAA) Shoney's Inn, Suburban Lodge, Super 8 Motel
  W: ◆ Red Roof Inn, Villager Lodge
- **AServ** **W:** Carmax Chrysler-Plymouth Jeep-Eagle, Jiffylube
- **RVCamp** **E:** Jones Mobile Home Estates and RV Park
- **ATM** **E:** Atlantic States Bank, NationsBank, Wachovia Bank
- **Other** **W:** Big B Drugs, Check Express Western Union Money Centers, Patient's Pharmacy, The Laundry Express (Coin Laundry), Winn Dixie Supermarket

37 **(99)** GA 140, Jimmy Carter Blvd.
- **Gas** **E:** Amoco*(CW), Chevron, Texaco* (ATM)
  W: Chevron*(LP), Phillips 66*(D)(CW), QT*
- **Food** **E:** Atlanta Bread Co, Bennigan's, Burger King, Cafe Atlanta, Checkered Parrot, Checkers Burgers, Chili's, China Wok, Costa del Sol, Cracker Barrel, Denny's, Dunkin Donuts, El Grill, Fujita Sushi Bar, KFC, Krystal's, Long John Silvers, Mandarin Garden Chinese, McDonalds, Melvin & Elmo's Bagels, North China Super Buffet, Papa John's Pizza, Pastorcito, Pizza Hut, Steak & Ale, Taco Bell, The Heritage Steak & Seafood, The Mughals, The Varsity, Waffle House, Wendy's, Wing's House

EXIT GEORG

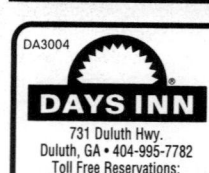

W: Arby's, Barnacle's, Blimpie's Subs, Chop Ch Chinese Food, Hooter's, Papa Beaux, Shoney's, Taste of Thai, Waffle House, Waffle House, Wendy's
- **Lodg** **E:** Amberly Suite Hotels, ◆ Best Western, Club House Inn, ◆ Courtyard by Marriott, (AAA) La Quinta Inn, Motel 6, ◆ Quality Inn
  W: (AAA) Comfort Inn, ◆ Drury Inn, Guest House Inn
- **AServ** **E:** A&B Carburetor & Electric(CW), Advance Auto Parts, Brake-O, Chevron
  W: Auto Zone Auto Parts, NTW Tire Center, Pep Boys Auto Center
- **ATM** **E:** First Union Bank, McDonalds, SouthTrust Ban Wachovia Bank
  W: NationsBank, SunTrust
- **Other** **E:** Al-Madina Pakistanian Grocery, Cub Food Oriental Supermarket, Revco Drugs, Shalima Supermarket, Village Laundromat
  W: Gwinnett Co. Fire Station

36A **(97)** Pleasantdale Road
- **FStop** **W:** QuikTrip*(D) (24 Hrs)
- **Food** **E:** Burger King, Pleasantdale Chinese Fast Food
  W: B J Country Cafe
- **Lodg** **W:** Holiday Inn Express
- **RVCamp** **E:** Morgan's RV
- **ATM** **W:** Wachovia Bank

36B **(96)** Northcrest Road (Accessible Vi Pleasantdale Exit)
- **Gas** **W:** BP(D)(CW)
- **Food** **W:** Waffle House
- **Lodg** **E:** Econolodge
- **AServ** **W:** BP
- **TServ** **W:** GMC Trucks of Atlanta
- **RVCamp** **E:** Northcrest RV Storage
- **Other** **E:** E&B Boat Gear, Old Sport Golf

35AB **(96)** I-285, Chattanooga, Augusta

34 **(94)** Chamblee - Tucker Road
- **Gas** **E:** Amoco*, Fina*
  W: QuikTrip*, Texaco*
- **Food** **E:** Nan's Chinese, Oriental Restaurant (Chinese)
  W: Dairy Queen, Selena's Mexican, Waffle Hous
- **Lodg** **E:** Masters Inn
- **AServ** **E:** Chamblee-Tucker Automotive

33 **(93)** Shallowford Road, Doraville
- **Gas** **E:** Shell*(LP)
  W: Circle K*(LP), Texaco*(LP)
- **Food** **E:** La Playa, Waffle House
  W: Anthony's Pizza Express, Chicken Plaza
- **Lodg** **W:** Quality Inn
- **AServ** **W:** Apollon Auto Repair
- **ATM** **W:** Circle K, Texaco
- **Other** **E:** Big Food Supermarket, Coin Laundry

32 **(91)** U.S. 23, GA 155, Clairmont Road, Decatur
- **Gas** **E:** Chevron(CW), Speedway*(D)(LP)
  W: Amoco*(D)(CW)
- **Food** **E:** IHOP, Madras Indian Cuisine, Mo's Pizza & Sandwich Shop, Popeye's Chicken, Waffle House
  W: Hattie's Restaurant & Grill, Waffle House
- **Lodg** **W:** (AAA) Days Inn (see our ad on opposite page)
- **AServ** **E:** Chevron, Firestone Tire & Auto, Robert Bosch Parts & Service, Tune-Up Clinic
  W: K-Mart, Penske Auto Service
- **ATM** **E:** SouthTrust Bank, Speedway, Wachovia Bank
- **Other** **E:** Barwick's Pharmacy, Coin Laundry, IGA Grocery Store, Revco, U.S. Post Office
  W: K-Mart

Bold red print shows RV & Bus parking available or nearby

XIT	GEORGIA

1 **(90)** GA 42, North Druid Hills Road

Gas E: Amoco, Crown*[CW], QuikTrip*
W: BP[D][CW], Chevron, Crown*[D][CW][LP], Exxon*, Hess[D]

Food E: Boston Market, Burger King, Chick-fil-A, El Torero, Grady's, Le Peep Restaurant, McDonalds, Morrisons Cafeteria, St. Charles Deli, TCBY Yogurt, Taco Bell, Wallstreet Pizza
W: Atlanta Diner, Cafe Lawrence's, Captain D's Seafood, Denny's, Dunkin Donuts, Folks, Fuzzy's Place, Havana Restaurant, International Cafe, Krystal, Lotus Garden Chinese, Red China, Rusty Nail, Waffle House

Lodg W: AAA Hampton Inn, ◆ Radisson, ◆ Red Roof Inn

AServ E: Amoco
W: BP, Chevron, Gordy Tire Service

Med E: ✚ Egleston Chilren's Hospital

ATM E: NationsBank, SunTrust
W: Crown, Wachovia Bank

Other W: Coin Laundry, High Speed Car Wash, Lawrence Animal Hospital, Revco

30 **(89)** to GA 400 North, Cheshire Bridge Road, Lenox Road

Gas E: BP*[D][CW], Citgo*, Fina*[D][LP], Shell*[CW], Spur*

Food E: Bamboo Luau Chinese, Dunk'n Dine, Happy Herman's Deli, Original Pancake House, Seasiam, South of France, Varsity Jr Drive-In, Waffle House

Lodg W: ◆ Buckhead Bed & Breakfast, Courtyard by Marriott (see our ad on this page)

AServ E: Highland Automotive, Precision Tune

ATM E: SouthTrust Bank, Wachovia Bank

Other E: Patient's Pharmacy, Tara Movie Theater

29 **(87)** GA 400 North, Buckhead, Cumming (toll) (Northbound)

28 **(86)** US 13 (Difficult Reaccess)

Gas E: BP

Food E: Chef's Cafe (Comfort Inn), Denny's, Don Juan's Spanish Restaurant, Javaco Restaurant, Mrs Winner's Chicken, Touch of India, Wendy's

Lodg E: Comfort Inn

AServ E: BP, Brake-O Brake Shop
W: Meineke Discount Mufflers

Note: I-85 runs concurrent below with I-75. Numbering follows I-75.

103 **(252)** Junction I-85 North, Greenville

102 **(251)** Fourteenth St, Tenth St, Techwood Dr (Southbound)

Gas E: Amoco*, BP*[D][CW]

EXIT	GEORGIA

W: Citgo*[D]

Food E: Dunkin Donuts, Philly Connection, Vini Vidi Vici Italian
W: Blimpie's Subs, Chinese Buddha Restaurant, City Deli & Bagels, Silver Skillet Restaurant

Lodg E: Hampton Inn, Marriott, Occidental Grand Hotel
W: Courtyard Marriott

AServ E: BP

Other E: U.S. Post Office
W: Revco Drugs, Wolf Camera & Video

101 **(251)** Fourteenth St, Tenth St, Georgia Insitute of Tech

Gas E: Amoco*, BP[D][CW], Chevron* (24 Hrs)

Food E: Checkers Burgers, Domino's Pizza, Dunkin Donuts
W: McDonalds, Papa John's Pizza

Lodg E: Regency Suites Hotel, Residence Inn

AServ E: BP

Other E: Public Transportation (MARTA Midtown Station)
W: Alexander Memorial Coliseum

100 **(250)** U.S. 78, U.S. 278, North Ave, Georgia Institute of Technology

Gas E: BP*[D][CW]

Food E: The Varsity (an Atlanta landmark)

Lodg E: Days Inn

Med E: ✚ Crawford Long Hospital

Other E: Public Transportation (MARTA)
W: The Coca Cola Company

99 **(250)** Williams St, Downtown Atlanta, Georgia Dome

Lodg W: Days Inn

98 **(250)** Pine St, Peachtree St (Northbound)

97 **(249)** Courtland St, Georgia State University (Southbound, Difficult Reaccess)

Lodg W: Hilton, Marriott, AAA Travelodge

AServ W: Beaudy Ford

96 **(249)** Boulevard, Carter Center

Lodg W: Courtyard, Fairfield Inn, Radisson

Med W: ✚ Georgia Baptist Hospital

Other W: World Congress Center

95 **(249)** Butler St, Houston St

95A **(249)** Piedmont St, Baker St

94 **(249)** Edgewood Ave, Auburn Ave

Med W: ✚ Grady Memorial Hospital

EXIT	GEORGIA

Other E: Martin Luther King Jr. Historic Sight, World Congress Center

93 **(248)** Martin Luther King Jr. Blvd, State Capitol, Underground Atlanta

Other W: Coca Cola Museum (Atlanta landmark, shopping), Underground Atlanta (Atlanta landmark, shopping)

92 **(248)** Junction I-20, Augusta, Birmingham

91 **(247)** Fulton St

90 **(247)** Georgia Ave, Capitol Ave

89 **(246)** University Ave

Gas E: Exxon*[D]

Food W: Brook's Cafeteria, Pool's Take-out

AServ E: Anderson Alignment, Charlie's Tire Company, NAPA Auto Parts

Parks W: Atlanta City Park (Pittman Park)

Other E: Lakewood Amphitheater

88 **(243)** GA 166, Lakewood Frwy, East Point

87 **(243)** Junction I-85 S. Atlanta Airport, Montgomery (Southbound)

Note: I-85 runs concurrent above with I-75. Numbering follows I-75.

24 **(77)** GA 166, Lakewood Freeway (Northbound)

23 **(76)** U.S.19, U.S. 41, Stewart Ave (Southbound, Difficult Reaccess)

Gas E: Conoco*[LP], Hess[D]

Food E: Arby's, Blimpie's Subs, Burger King, Del Taco, Jade Buddha Chinese, Mrs Winner's Chicken, Popeye's Chicken, West Texas Steakhouse

Lodg E: Colonial Motor Lodge
W: Town & Country Hotel Courts

AServ E: Auto Zone Auto Parts
W: Nalley Chevrolet, Nalley Honda & Kia

ATM E: NationsBank

Other E: Big B Drugs, Kroger Supermarket (Pharmacy)

22 **(75)** Cleveland Ave, East Point (Northbound)

Gas E: Conoco*[LP], Hess[D]
W: Amoco*[CW]

Food E: Blimpie's Subs (Conoco), Burger King, Mrs Winner's Chicken
W: Chick-fil-A, Church's Fried Chicken, KFC, Taco Bell

AServ W: Sylvan & Cleveland Auto Service

Med W: ✚ Hospital

ATM E: Kroger Supermarket

Other E: Big B Drugs, Kroger Supermarket (Pharmacy)
W: Dekalb Car Wash

21 **(74)** Sylvan Road, Central Ave, Hapeville

Gas E: Fina*[LP], Texaco*[D], Virginia St. Auto Service

Food E: Chester Fried Chicken

Lodg E: In Town Suites
W: Mark Inn

Other E: Papa Foodmart

20 **(73)** Loop Rd. (Southbound)

19AB **(72)** Virginia Ave.

Gas E: Citgo*[D][CW]
W: Chevron*[CW], Texaco[D]

Food E: Malone's, McDonalds, Morrison's Cafeteria, Waffle House

EXIT	**GEORGIA**
	W: BBQ Kitchen, Blimpie's Subs, Happy Buddha Chinese, Hardee's, Hartsfield's Bar & Grill, KFC, La Fiesta, Steak & Ale, Waffle House
Lodg	**E:** Courtyard by Marriott, Hilton Hotel, Renaissance Hotel
	W: Crowne Plaza Hotel, Econolodge, Holiday Inn, Howard Johnson (see our ad page 435), Ramada
AServ	**E:** Citgo
	W: Texaco
Med	**E:** ✚ **Hospital**
18A	**(72)** Camp Creek Pkwy, Atlanta Airport
18B	**(72)** Riverdale Road, GA 139
Food	**E:** Ruby Tuesday
Lodg	**E:** Comfort Suites, Hampton Inn, Sheraton Gateway, Sleep Inn, Sumner Suites, Super 8 Motel
	W: DAYS INN Days Inn, Embassy Suites, Marriott, Quality Inn, Travelodge, Westin Hotel
Other	**E: Georgia International Convention Center**
17	**(71)** Junction I-285, Macon, Birmingham (Southbound)
16AB	**(70)** GA 279, Old National Hwy, Atlanta Int'l Airport
Gas	**E:** Exxon*(D), Hess(D), Racetrac*, Texaco*
	W: Chevron*
Food	**E:** Arby's, Blimpie's Subs, Burger King, Captain D's Seafood, Checkers Burgers, China Cafeteria, China Dragon, Church's Fried Chicken, El Ranchero Mexican, Howard Johnson, KFC, Krystal, McDonalds, Pizza Hut, Sizzler Steak House, Steak & Ale, Subway, Waffle House, Wendy's
	W: Denny's, Dipper Dan Ice Cream, Waffle House
Lodg	**E:** Budgetel Inn, Holiday Inn, Howard Johnson, Red Roof Inn, Red Roof Inn
	W: La Quinta Inn
AServ	**E:** Car Wash & Lube, Midas Muffler & Brakes, Tune-Up Clinic
	W: Goodyear Tire & Auto
Med	**N:** ✚ **Med-First Physician**
ATM	**E:** NationsBank, Trust Company Bank
Other	**E: Big B Drugs, Kroger Supermarket**
16A	**(69)** GA 14, GA 279
15	**(68)** Junction I-285, Atlanta Bypass
14	**(67)** Flat Shoals Road
FStop	**W: Speedway***
Gas	**W:** Amoco* (24 Hrs)
Food	**W:** Hardee's, Waffle House
Lodg	**W:** Knight's Inn
13	**(66)** GA 138, Union City, Jonesboro
Gas	**E:** BP*
	W: Chevron*(CW) (24 Hrs), Racetrac*, Texaco*, USA Gasoline(D) (24 Hrs)
Food	**E:** Waffle House
	W: Burger King, Captain D's Seafood, China King Restaurant, Church's Fried Chicken (Texaco), Corner Cafe Hot Wings, Cracker Barrel, Donna Marie's Pizza & Pasta, KFC, Krystal, Pizza Hut, Shoney's, Subway, TCBY Treats, Taco Bell, Wendy's
Lodg	**E:** AAA Econolodge, Ramada Limited
	W: DAYS INN ◆ Days Inn (see our ad on this page)
AServ	**E:** BP, Gene Evans Ford, Steve Rayman Pontiac,

EXIT GEORGIA

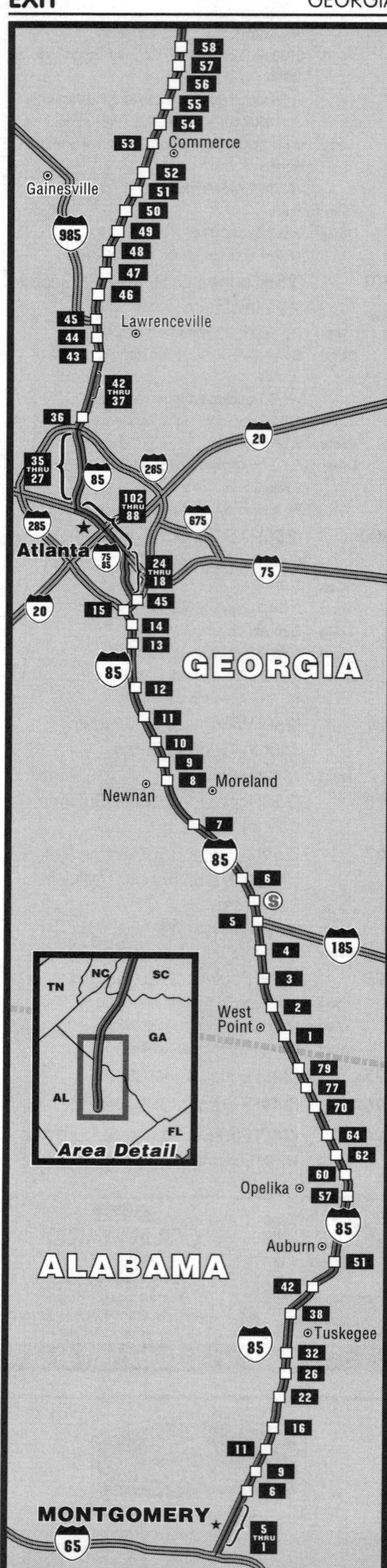

EXIT	**GEORGIA**
	Buick, GMC Truck
	W: Firestone Tire & Auto, Goodyear Tire & Auto, Jiffy Lube, Precision Tune & Lube
ATM	**W:** NationsBank, NationsBank, Racetrac, SunTrust, Tucker Federal Bank
Other	**W: Coin Laundry, Drug Emporium, Eckerd Drugs, Ingles Supermarket, Kroger Supermarket (24 Hrs, Pharmacy), Shannon Theatres, South Park Mall, United Artist Theaters, Wal-Mart (Pharmacy)**
12	**(61)** GA 74, Fairburn
Gas	**E:** Amoco*, Chevron*(D)(CW)(LP), Racetrac*
	W: BP*(D)(CW)
Food	**E:** Subway (Chevron), **Waffle House**
Lodg	**W: Efficiency Motel**
ATM	**E:** Chevron, Racetrac
11	**(56)** Collinsworth Road, Palmetto, Tyrone
Gas	**W:** Fina*(LP)
Food	**W:** Frank's Family Restaurant
10	**(51)** GA. 154, McCollum, Sharpsburg Road
FStop	**E: Money Back*(LP) (24 Hrs), Phillips 66*(D) (24 Hrs)**
	W: Chevron*(CW)(LP) (24 Hrs)
Gas	**W:** Shell*
Food	**E:** Hardee's
ATM	**E:** Money Back FS
9	**(46)** GA 34, Newnan, Peachtree City
FStop	**E: Texaco***
Gas	**E:** Chevron*(D)(LP), Citgo*(LP), Petro*(D)(CW), Shell*
	W: Amoco(CW)(LP), Phillips 66*(LP), Racetrac*

Bold red print shows RV & Bus parking available or nearby

EXIT		GEORGIA

	Food	**E:** Applebee's, Atlanta Bread Co., Dunkin' Donuts (Citgo), Hooters, Hot Stuff Pizza (Petro), Miano's Pasta Express, Philly Connection, Ryan's Steakhouse, Sprayberry's BBQ, Subway (Citgo), Waffle House, Wendy's
		W: Hardee's, IHOP, Waffle House
	Lodg	**E:** Hampton Inn
		W: Admiral Benbow Inn, AAA Holiday Inn Express, Shenandoah Inn Best Western
	AServ	**E:** Wal-Mart
		W: Patton Chevrolet, Oldsmobile, Chrysler, Jeep, Southtown Pontiac, Buick, Cadillac, GMC Truck
	Med	**W:** ✚ Hospital
	ATM	**E:** Petro
		W: Amoco, First Union Bank, Racetrac
	Other	**E:** Peachtree Factory Outlet Stores, Wal-Mart Supercenter (Pharmacy)
		W: Rainbow Car Wash
8		**(41)** U.S. 27, U.S. 29, Newnan
	FStop	**W:** Amoco*(D) (24 Hrs)
	Gas	**W:** BP, Phillips 66* (ATM), Speedway*(D)
	Food	**W:** Denny's (Days Inn), Huddle House, McDonalds (Play Place), Waffle House
	Lodg	**W:** ◆ Comfort Inn, DAYS INN AAA Days Inn, Ramada Limited
	AServ	**W:** BP
	ATM	**W:** Phillips 66
7		**(35)** U.S. 29 , Grantville, Moreland
	FStop	**W:** Amoco*, Phillips 66*
	ATM	**W:** Amoco FS
6		**(28)** GA 54, GA 100, Hogansville
	TStop	**W:** Nobel Auto Truck Plaza*(SCALES) (BP Gas, Travel Store)
	FStop	**E:** Shell*(D)(LP)
		W: Amoco*(LP)
	Gas	**W:** Citgo*(CW)
	Food	**W:** Buford's Hickory Grill, Hardee's, Janet's Country Cooking (Nobel TS), KFC (Citgo), McDonalds (Play Place), Subway, Waffle House, Wayback Steakhouse & Seafood
	Lodg	**W:** Hummingbird Inn, AAA Key West Inn
	TServ	**W:** Crockett Bros. (Nobel TS)
	RVCamp	**W:** Flat Creek Campground
	ATM	**W:** Citiizens Bank
	Other	**W:** Ingles Supermarket
(22)		Weigh Station (Both Directions)
5		**(20)** Junction I-185 South, Columbus
4		**(18)** Georgia 109, Greenville, Warm Springs
	Gas	**W:** Amoco*(D)(LP), Racetrac*, Shell*(LP), Spur*(D)(LP)
	Food	**W:** Blimpie's Subs (Shell), Bonzai Japanese Steakhouse, Knickers, Ryan's Steakhouse, Subway, The Inland Restaurant, Waffle House
	Lodg	**E:** AAA La Grange Inn (Best Western)
		W: AAA AmeriHost Inn, Jameson Inn, Ramada Inn, Super 8 Motel

EXIT		GEORGIA/ALABAMA
	Parks	**E:** F.D.Roosevelt State Park
	Other	**W:** Carmike Cinemas
3		**(14)** U.S. 27, LaGrange
	Gas	**W:** Amoco*(LP)
	Other	**E:** Georgia State Patrol Post
2		**(13)** GA. 219, LaGrange
	TStop	**E:** La Grange Auto/Truck Plaza*(LP)(SCALES) (Travel Store, Texaco Gas)
	FStop	**W:** Amoco*(D)(LP), Speedway*(LP)(SCALES)
	Food	**E:** **Days Inn (see our ad on this page)**, Huddle House, LaGrange Truck Plaza Restaurant, Waffle House
		W: Hardee's, McDonalds (Play Place)
	Lodg	**E:** Admiral Benbow Inn, DAYS INN ◆ Days Inn
1		**(2)** GA 18, West Point, Callaway Gardens
	FStop	**E:** Shell*
	Gas	**E:** Amoco* (24 Hrs)
	Lodg	**E:** Travelodge
	ATM	**E:** Amoco
(1)		GA Welcome Center - RV Dump, Picnic, RR, HF, Phones (Northbound)

↑GEORGIA

↓ALABAMA

79		U.S. 29, Lanett, Valley
	Gas	**E:** Amoco*(D)(CW)(LP), Racetrac*
		W: Exxon* (24 Hrs), Phillips 66*(D)
	Food	**E:** Burger King (Playground), Captain D's Seafood, KFC, Krispy Kreme (Amoco), Krystal, McDonalds (Play Place), Subway, Taco Bell (Amoco), Waffle House, Wendy's
		W: Shoney's
	Lodg	**W:** DAYS INN Days Inn (see our ad on this page), Econolodge
	AServ	**E:** Advance Auto Parts, Bryan's Transmission Service
		W: Auto Zone Auto Parts
	Med	**E:** ✚ Hospital
	ATM	**E:** Colonial Bank
	Other	**E:** Food Max Supermarket, Revco Drugs
		W: Riverside Veterinary Clinic
(78)		Rest Area - Picnic, RR, HF, Phones, RV Dump (Southbound)
77		CR 208, Valley, Huguley
	FStop	**E:** Amoco*(D)(LP)
	Gas	**E:** Texaco*
	Food	**E:** Krispy Kreme (Amoco FS)
	AServ	**E:** King Chevrolet, King Ford, Mercury, Lincoln, Chrysler, Plymouth, Dodge
	Med	**E:** ✚ Hospital

EXIT		ALABAMA
	Other	**E:** Jolly Joe's Fireworks
70		CR. 388, Cusseta
	TStop	**E:** Perlis*(SCALES) (Texaco Gas)
	Gas	**E:** BP*
	Food	**E:** Country Pride (Perlis TS), Subway (Perlis TS)
	AServ	**E:** Perlis TS
	TServ	**E:** Perlis
	RVCamp	**E:** B&B RV Park
64		U.S. 29, Opelika
	Gas	**E:** Amoco*(CW)
		W: Exxon*
	AServ	**E:** Amoco
62		U.S. 280, U.S. 431, Opelika, Phenix City
	FStop	**E:** Texaco*(D)
	Gas	**E:** BP*(D)(CW), Chevron*
		W: Exxon*, Shell*(LP)
	Food	**E:** Church's Fried Chicken (Texaco FS), Denny's, Krispy Kreme (Texaco FS), La Herradura Mexican, McDonalds (Play Place), Subway
		W: Cracker Barrel, Waffle House, Western Sizzlin'
	Lodg	**E:** DAYS INN Days Inn, ◆ Holiday Inn (see our ad on this page), Knight's Inn, Motel 6, Ramada Limited, Shoney's Inn
		W: AAA Best Western (Mariner Inn), ◆ Comfort Inn
	Other	**W:** Food World Supermarket, USA Factory Stores (see our ad on this page)
60		AL. 51, AL. 169, Opelika, Hurtsboro
	Gas	**E:** Amoco*(CW), Leco*
		W: Chevron* (24 Hrs)

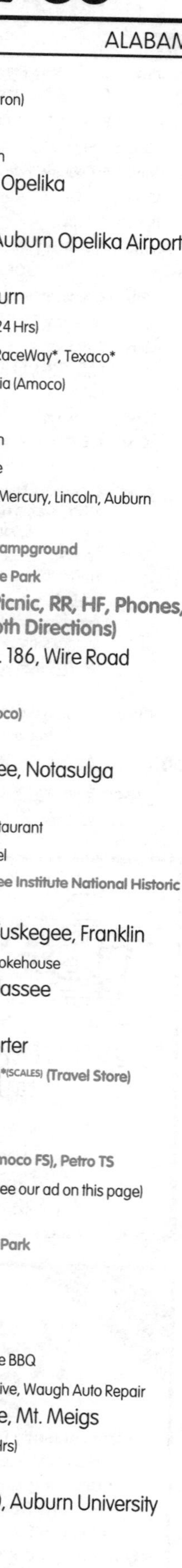

EXIT ALABAMA

Food E: Hardee's
W: Krystal (Chevron)
Med W: Hospital
Other E: Coin Car Wash

58 U.S. 280 W., Opelika
Gas W: Chevron*(D)

57 Glenn Ave, Auburn Opelika Airport
Gas W: Phillips 66*

51 U.S. 29, Auburn
Gas E: Amoco*(D)(CW) (24 Hrs)
W: BP*, Exxon*, RaceWay*, Texaco*
Food E: Hot Stuff Pizzaria (Amoco)
W: Waffle House
Lodg E: ◆ Hampton Inn
W: ◆ Econolodge
AServ W: Auburn Ford, Mercury, Lincoln, Auburn Mitsubishi
RVCamp E: **Leisure Time Campground**
Parks E: **Chewacla State Park**

(44) **Rest Area - Picnic, RR, HF, Phones, RV Dump (Both Directions)**

42 To U.S. 80, AL. 186, Wire Road
FStop W: **Amoco*(D)**
Food W: **Torch 85 (Amoco)**
ATM W: Amoco FS

38 AL.81, Tuskegee, Notasulga
Gas E: Texaco*(D)
Food E: Western Inn Restaurant
Lodg E: Western Inn Hotel
Other E: **Airport, Tuskegee Institute National Historic Site**

32 AL. 49 North, Tuskegee, Franklin
Food W: J&S Country Smokehouse

26 AL. 229 N., Tallassee
Med W: Hospital

22 To U.S. 80, Shorter
TStop E: **Petro 2-Chevron*(SCALES) (Travel Store)**
FStop E: **Amoco***
Gas E: Exxon*
Food E: **Krispy Kreme (Amoco FS), Petro TS**
Lodg E: Days Inn (see our ad on this page)
TServ E: **Petro 2-Chevron**
RVCamp E: **Wind Drift Travel Park**
ATM E: Exxon

16 Cecil, Waugh
Gas E: BP*
Food E: Fuller's Char House BBQ
AServ E: Interstate Automotive, Waugh Auto Repair

11 U.S. 80, Mitylene, Mt. Meigs
Gas W: Chevron*(CW) (24 Hrs)
ATM W: Chevron

9 AL. 271 to AL. 110, Auburn University at Montgomery
Med W: Hospital

6 U.S. 80, U.S. 231, AL 21, East Blvd
Gas E: Chevron*(CW) (24 Hrs), Citgo*, Exxon*(D),

EXIT ALABAMA

EXIT ALABAMA

Racetrac*
W: Conoco*, Shell*
Food E: American Pie (Ramada), Burger King (Playground), Cracker Barrel, Cuco's Mexican, Fifth Quarter Steak & Seafood, Luigi's Pizzaria, Schlotzkys Deli, Shogun Japanese, Taco Bell, Waffle House
W: Church's Fried Chicken, Denny's, KFC, Lone Star Steakhouse, McDonalds (Play Place), Outback Steakhouse, Shoney's, Waffle House
Lodg E: AAA Best Suites, ◆ Budgetel Inn, ◆ Courtyard by Marriott, ◆ Fairfield Inn, ◆ Hampton Inn, AAA La Quinta Inn, Quality Inn, ◆ Ramada Inn, ◆ Residence Inn
W: Best Western, ◆ Comfort Suites (see our ad on this page), ◆ Holiday Inn, Motel 6, Wynfield Inn
AServ E: Exxon, Xpert Tune
W: Capitol Chevrolet, Conoco, Express Lube, Reinhardt Toyota, Royal Chrysler/Plymouth, Jeep/Eagle
ATM E: AmSouth Bank
Other E: **Carmike Cinemas,** Coin Car Wash, **Sam's Club**

4 Perry Hill Road
Gas W: Chevron*(CW) (24 Hrs), Citgo*(D), Texaco*(LP)
Food E: Chappy's Deli
W: Hardee's, New China
ATM E: Whitney Bank
Other E: **Del Champs Supermarket, Harco Super Drug, Hour Glass Optical**
W: **K & B Drugs**

3 Ann St
Gas E: BP*(D)(CW) (24 Hrs), Citgo*(LP)
W: Amoco*(CW) (24 Hrs), Exxon*
Food E: Arby's, Captain D's Seafood, Country's BBQ, Domino's Pizza, Down The Street Cafe, Great Wall Chinese, Hardee's, Lee's Famous Recipe Chicken, McDonalds (Play Place), Taco Bell, Wendy's
W: Burger King (Exxon), Quincy's Family Steakhouse
Lodg E: Days Inn
W: Villager Lodge
AServ E: Big 10 Tires
ATM E: SouthTrust Bank

2A Mulberry St.
Med W: **Jackson Hospital**

2 Forest Ave. (Northbound)
Med W: **Hospital**
Other E: **Revco Drug**

1 Court St.
Gas E: Exxon*, Pacecar*(CW)
AServ E: Ragan Auto Repair
W: Mr. Transmission
Med W: **Hospital**
ATM W: Colonial Bank

↑ALABAMA

Begin I-85

Bold red print shows RV & Bus parking available or nearby

XIT IDAHO

Begin I-86

↓IDAHO

	I-84 East, Ogden (Westbound)
)	**Weigh Station - Both Directions**
5	Raft River Area
FStop	**S:** Sinclair*
9)	**Rest Area - RR, Phones, Picnic** (P) **(Eastbound)**
1	Coldwater Area
8	Massacre Rocks State Park
RVCamp	**N:** Camping
Parks	**N:** Massacre Rocks State Park
1)	**Rest Area - RR, Phones, Picnic** (P) **(Westbound)**
3	Neeley Area
6	Bus. I-86, ID 37, American Falls.
RVCamp	**S:** Indian Springs RV Resort (2 Miles)
Med	**N:** ✚ Hospital
0	ID 39, American Falls

EXIT IDAHO

TStop	**S:** Amoco*(LP)
FStop	**S:** Lakeview
Food	**S:** Hilltop Cafe (Amoco), Lakeview Cafe (Lakevies FS)
Lodg	**S:** Hillview Motel
TServ	**S:** Amoco
RVCamp	**N:** Willow Bay Recreation Area
ATM	**S:** Amoco TS
Other	**N:** American Falls Recreation Area
	S: Coin Laundry (Amoco)
44	Seagull Bay
49	Rainbow Road
52	Arbon Valley
FStop	**S:** Sinclair*
56	Pocatello, Air Terminal
58	U.S. 30, West Pocatello
TServ	**S:** Kenworth Dealer (3 Miles)
61	U.S. 91, Chubbuck, Yellowstone Ave
Gas	**N:** Exxon*, Sinclair*, Vicki's Hitchin Kitchen*(D)(LP)
	S: Chevron*, Flying J Travel Plaza*(D)(CW), Phillips

EXIT IDAHO

	66*(D)
Food	**N:** Arctic Circle Hamburgers, Baskin Robbins (Vicki's Hitchin Kitchen), Burger King, Cafe (Vicki's Hitchin Kitchen), Casa Ole Mexican, Days Inn, Johnny B. Goode, Pizza Hut, Subway
	S: Denny's, JJ North, McDonalds, Taco Bandido, Taco Bell (Flying J)
Lodg	**N:** DAYS INN Days Inn, Motel 6
	S: Nendels Inn, Pine Ridge Inn
AServ	**N:** Denny's Wrecker Service, Quaker State Cowboy Lube Barn(CW)
	S: Chevron, Les Schwab Tires, Wal-Mart
RVCamp	**N:** Budget RV Park, Days Inn
	S: Herb's RV Service
ATM	**N:** Circle K
Other	**N:** Eddy's Bakery Outlet, Parrish Grocery
	S: Grocery Outlet, K-Mart (Pharmacy), Mall, Shopko Grocery (Pharmacy, Optical), Wal-Mart (Pharmacy)
63	Jct. I-15, Idaho Falls, Salt Lake City

↑IDAHO

Begin I-86

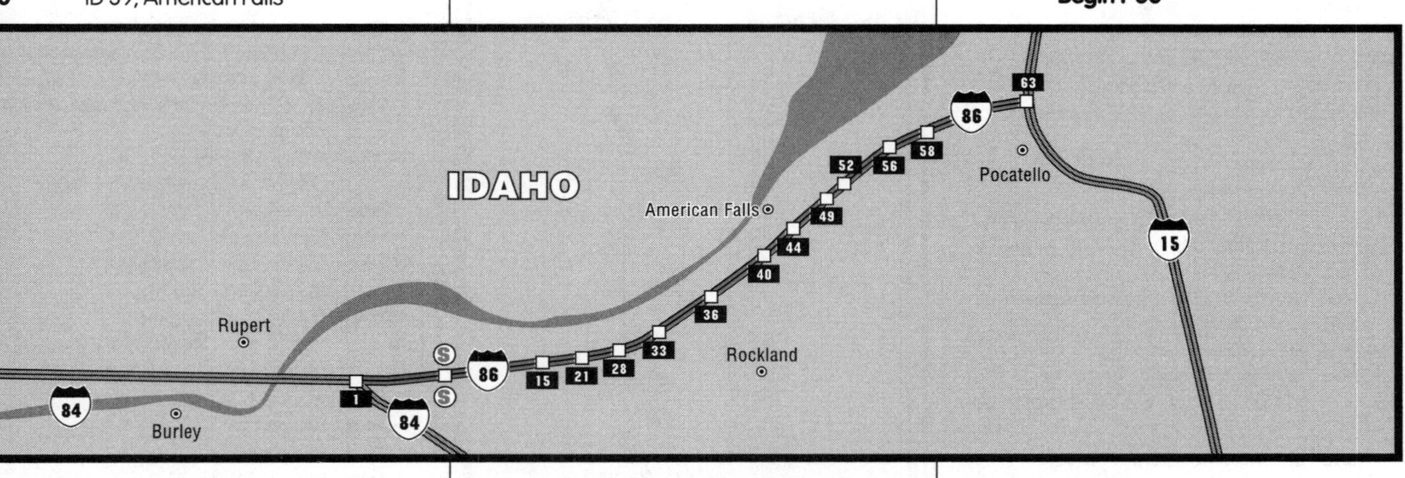

I-87 S ⟶

XIT NEW YORK

Begin I-87

↓NEW YORK

3	**(175)** U.S. 9, Champlain
TServ	**W:** Champlain Peterbilt Service
2	**(174)** U.S. 11, Moors, Rouses Point
TStop	**W:** 11-87 Truck Plaza*(LP)(SCALES)
FStop	**W:** Sunoco*
Gas	**E:** Citgo*, Mobil*
	W: Neverett Brothers*(LP)(SCALES), Petro Canada
Food	**E:** Art's Place, Choices Chinese, Peppercorn Family Restaurant, Pizza Plus (Mobil)
	W: Dunkin Donuts (Petro), Main Street Eatery, McDonalds
TServ	**W:** 11-87 Truck Plaza
ATM	**E:** AlBank, Key Bank
Other	**E:** Car Wash, Corner Market, Grand Union Supermarket, Rite Aide Pharmacy
	W: Miromar Factory Outlet Center, Tourist Information

EXIT NEW YORK

41	**(168)** NY 191, Chazy, Sciota
Other	**E:** State Patrol Post
(162)	NY Welcome Center - RR, Phones, Picnic
40	**(160)** NY 456, Beekmantown, Point au Roche
Gas	**E:** Mobil*
Food	**E:** Deli (Mobil), Stonehelm Motel
Lodg	**E:** AAA Stonehelm Motel
39	**(156)** NY 314, Cumberland Head, Moffitt Rd
Gas	**E:** Mobil*, Stewart's Shops*
Food	**E:** Broadview Deli, Gus' Red Hots
Lodg	**E:** Chateau Motel, Landmark Hotel, Rip Van Winkle Hotel, Sundance Motel, AAA Super 8 Motel
AServ	**E:** Mobil
RVCamp	**E:** Campground, Plattsburg RV Park
Parks	**E:** State Park Beach
Other	**E:** Stewart Shop

EXIT NEW YORK

38	**(155)** NY 22N, Saranac Lake Dennemorre, NY 374W, Plattsburg
Med	**E:** ✚ Hospital
Other	**W:** Sherriff's Department
37	**(153)** NY 3, Plattsburgh
Gas	**E:** Exxon*, Petro Canada*, Stewart's*, Sunoco*
	W: Sunoco*
Food	**E:** Barkin' Dog Restaurant, Bok Ben, Bootleggers, Dairy Queen, Dunkin Donuts, Friendly's, Ground Round, Jay's Buffet, Jreck Subs, KFC, Little Caesars Pizza, Mangia, Maxwell's, McDonalds, Pizza Hut, Red Lobster, Ron's Family Restaurant, Subway, Wendy's, What's Cookin, Wong's
	W: Anthony's Steaks & Seafood, Butcher Block, Ponderosa
Lodg	**E:** ◆ Budgetel Inn, ◆ Comfort Inn, ◆ Howard Johnson, ◆ Ramada Inn
	W: DAYS INN ◆ Days Inn, Econolodge, Travelers Motor Inn
AServ	**E:** Auto Palace, Auto Parts America, Bill Santa Pontiac, Drew Isuzu, Buick, Express Lube,

EXIT NEW YORK

Firestone Tire & Auto, Knight GMC, Midas Muffler & Brakes, Monroe Muffler & Brake, Montogomery Ward, Sears, Sunoco
W: Donna's Auto Sales & Service, Jiffy Lube, McGee's Used Car & Servcie
Med **E:** ✚ Hospital
W: ✚ Urgent Care Walk-in
ATM **E:** Adarondike Bank, AllBank, Key Bank, Marine Midland Bank
W: Albank
Other **E:** Ames Shopping Center, Consumer's Square, Empire Visions Center, Grand Union Grocery Store, K-Mart, Northern Pride Car Wash, P&C Grocery, Pearl Vision Center, Pet Shop, Price Chopper Supermarket (Pharmacy), Sam's Club, Super Store, Wal-Mart (Pharmacy)
W: AAA Office, Nuway Car Wash

36 **(150)** NY 22, Air Force Base, Lake Shore
TStop **E:** Mobil
Food **E:** Mobil
AServ **W:** Garage
TServ **E:** Charbroille, Ma Jerry & Co.
Other **W:** New York State Police

35 **(144)** NY 442, Peru, Port Kent
Food **E:** Leaning Tower, McDonalds
RVCamp **E:** Ausabale Chasm KOA, Ausabale Pines, Iroquois Campground
W: Birchwood Campground
Other **E:** Post Office

34 **(138)** NY 9 North, Ausable Chasm, Ausable Forts
Gas **E:** Sunoco
Food **E:** Pleasant Corners Restaurant, Tastee Freeze
AServ **E:** Sunoco
RVCamp **W:** Ausable River Camp Sites
Other **E:** Keeseville Veterinary Clinic

33 **(135)** U.S. 9, NY 22, Willsboro, Essex Ferry
FStop **E:** Power Champs (Closed Sundays)
Lodg **E:** Chesterfield Motel
AServ **E:** Power Champs
RVCamp **E:** Campground

32 **(123)** Lewis, Willsboro
TStop **W:** Andre's Truck Stop
FStop **W:** Sunoco
Food **W:** Andre's TruckStop
AServ **W:** Sunoco
RVCamp **E:** Spruce Mill Campsite

(123) **Rest Area - Phones, Tourist Info, Picnic, NO Trucks, NO RR (Northbound)**

31 **(117)** NY 9 North, Elizabethtown, Westport
Gas **E:** Mobil*
Food **E:** Goff's (Mobil)
Lodg **E:** Hilltop Motel
Med **W:** ✚ Hospital
Other **E:** State Patrol Post

(112) **Rest Area - RR, Phones, Grills (Closed)**

30 **(104)** U.S. 9, NY 73, Keene Valley, N Hudson

(99) **Rest Area - RR, Phones, Picnic (Southbound)**

29 **(94)** Newcomb, North Hudson

EXIT NEW YORK

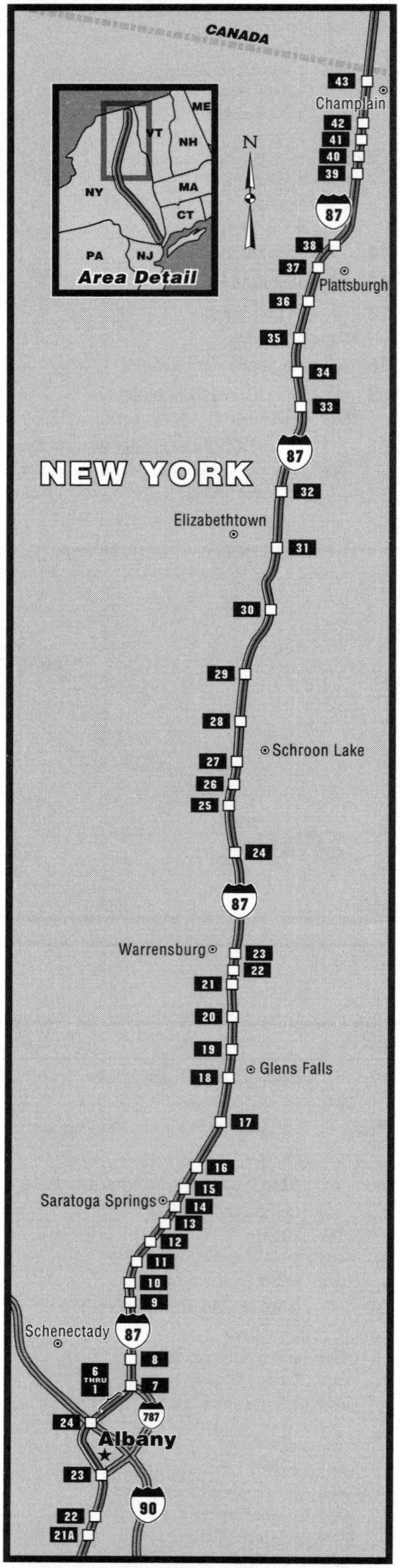

CANADA
Champlain
Plattsburgh
NEW YORK
Elizabethtown
Schroon Lake
Warrensburg
Glens Falls
Saratoga Springs
Schenectady
Albany
Area Detail

EXIT NEW YOR

FStop **E:** Mobil*
Food **E:** Frontier Town Restaurant, McDonalds, Smo Pit
Lodg **E:** Frontier Town Motel
RVCamp **E:** Paradise Pines Camping
W: Blue Ridge Falls Campground
Other **E:** Frontier Town, North Hudson Grocery, Pos Office, The General Store
W: Visitor Information

28 **(89)** NY 74, Ticonderoga - Ferry, Schroon Lake
Gas **E:** Sunoco*[D]
Food **E:** Chesapeake Fine Dining
Lodg **E:** Schroon Lake Bed & Breakfast
AServ **E:** Sunoco
Other **E:** State Patrol Post

(83) **Scenic Overview, Rest Area - RR, Phones, Picnic**

27 **(82)** U.S. 9, Schroon Lake (Northbound, Southbound Reaccess Onl

RVCamp **E:** Campground

26 **(79)** U.S. 9, Pottersville, Minerva (Difficult Reaccess)
FStop **W:** Mobil*
Food **E:** Black Bill, The Family Deli, The Taste of Heave
Lodg **E:** Spoon Lake Cottages, Wakondia Campgrou
AServ **E:** Great Northern Car & Truck Supply, Pottersvi Garage
RVCamp **E:** Ideal Campsite

25 **(73)** NY 8, Chestertown, Hague
FStop **W:** Sunoco*
Gas **E:** Gulf*, Sunoco
RVCamp **E:** Riverside Pines Campground

24 **(67)** Bolton Landing
RVCamp **E:** Good Sam Campground

(64) **Rest Area - No RR (Northbound)**

23 **(60)** Warrensburg, Diamond Point
FStop **W:** Citgo
Gas **W:** Mobil*
Food **W:** McDonalds
Lodg **W:** Super 8 Motel
AServ **W:** Ford Dealership
TServ **W:** Lake Shore Garage
RVCamp **W:** Rainbow View

22 **(55)** U.S. 9, NY 9 N, Lake George Village, Diamond Point
Gas **E:** Mobil*
Food **E:** Adarondik Bar & Grill, Ben & Jerry's, Betty's Place, Boardwalk, Bob's Ice Cream, Capri Pizzeria, Carmen Corner Deli, Chowder's Ice Cream, Claire's, Duffy's Tavern, Golden Wok, LaRoma, Luigi's Italian, Mario's, Micheal's Townhouse, New York Pizza, Oasis Motel, Pizz Hut, RJ Annie Candy, Rosie's, Scillion Spathetti House, Subway, Wagner's
Lodg **E:** Admiral Motel, Bamoral Motel, Blue Moon Motel, Brookside Motel, Cedar Hurst, (AAA) Cole Motel, (AAA) Econolodge, Georgian Luxury Resc Green Acres, Hilside, Lake Crest Motel, Lake Haven Motel, Lake Haven Motel, Lake Motel, Marine Village Motel, ◆ Mohawk Motel, (AAA) Motel Montreal, Nortic Motel, O'Sullivan's, Oasi Motel, S&J Motel, Scotty's Motel, Seven Dwarfs Motor Court, Sundowner Motel, Sunrise, Surfsid on the Lake, The Warehouse Cottages
Other **E:** Rexal Drugs
W: Scenic Overview

Bold red print shows RV & Bus parking available or nearby

EXIT	NEW YORK
1	**(53)** NY 9 N, Lake Luzerne
FStop	**W:** Mobil*
Gas	**E:** BP, Citgo*, Stewarts*, Sunoco, Sunoco
Food	**E:** A & W Drive-In, Avanti Italian, Barnsider's Motel, Comfort, Dairy Queen, Lake George Pancake House, McDonalds, Micheal Anthony's Seafood, Mountaineer, Mr. B's Subs, Pizza Hut, Zachery's
Lodg	**E:** AAA Best Western, Blue Bell, Colonial Inn, Colonial Williams Motor Inn, Comfort Inn, Congo Motor Court, Do-Rest Court, Domonio's Motor Inn, Howard Johnson, Lee's Motel, Lin Airet Motel, Naussa Motel, Rosey's Country Cabine, Travelodge
	W: Kathy's Motel
AServ	**E:** Sunoco(D) (Towing), Sunoco
RVCamp	**E:** King Phillips, Lake George RV Park, Wipperowill Camp Sites
Other	**E:** Lake George Action Park, Magic Forrest Park, Tourist Haunted Attraction, Wild West Family Park
0	**(49)** NY 149, Fort Ann, Whitehall
Gas	**E:** Shell*
Food	**E:** Small World Shoppes Restaurant & Gifts, The Meetings Place, Trading Post
Lodg	**E:** Gray Court Motel, Kay's Motel, Martha's Motel, Samoset Cabins
RVCamp	**E:** Lake George, Lake George Sales(LP)
ATM	**E:** OnBank
Other	**E:** Factory Outlet Center, Outlet Center, Small World Shoppe, State Patrol Post
	(47) U.S. 9, NY 254, Glens Falls, Hudson Falls
Gas	**E:** Citgo*, Hess*, Mobile*, Sunoco*
	W: Mobil*
Food	**E:** Blacksmith Stop, Burger King, Dunkin Donuts, Friendly's, Howard Johnson, KFC, McDonalds, North Country Eatery, Old China Buffet, Red Lobster, Seven Steers Western Grill, Subway, Taco Bell, The Lighthouse Fish Fry, The Olive Garden, The Silo Restaurant & Gifts
Lodg	**E:** Alpenus House Motel, ◆ Econolodge, ◆ Howard Johnson
	W: Ramada Inn (see our ad this page)
AServ	**E:** Monroe Muffler, Parts America, Quick Lube, Sears, Warren Tire Service
RVCamp	**E:** Lake George Campsite (2 Miles)
Med	**E:** ✚ Convenient Care Walk-in
ATM	**E:** All Bank, Troy Savings Bank
Other	**E:** CVS Pharmacy, Pearl Vision Center, Post Office, Queensberry Plaza
	W: State Patrol Post
	(45) Glens Falls, Corinth
FStop	**E:** Hess*
Gas	**E:** Citgo, Gulf*, Mobil*
Food	**E:** Carl R's Cafe, Lox of Bagels & Moor, Pizza Hut, Subway
	W: McDonalds
Lodg	**E:** ◆ Susse Chalet
	W: ◆ Super 8 Motel
AServ	**E:** Citgo (Towing)
Med	**E:** ✚ Hospital
3)	Rest Area - RR, Phones, Picnic
	(40) U.S. 9, South Glens Falls
TStop	**E:** Buddy Beaver's Truck Stop
FStop	**E:** Mobil*
Gas	**E:** Gulf*
Lodg	**E:** Landmark Motor Inn, Swiss American Motel
	W: Terry's Motel

EXIT	NEW YORK
TServ	**E:** Hillman Brothers Transp. Ser., Truck Dealership
RVCamp	**E:** American Campgrounds
Parks	**W:** Moreau Lake State Park
16	**(36)** Ballard Rd, Gansevoorit, Willton
TStop	**W:** Sunoco*(SCALES)
Gas	**W:** Mobil*, Stewart's Shops*
Food	**E:** Cattone's Pizza
	W: Scotty's (Sunoco)
RVCamp	**E:** Cole Brook Campsite (1 Mile)
	W: Freedom RV Outlet (Sales & Service)
Other	**W:** State Patrol Post
15	**(30)** NY 50, Saratoga Springs, Gansevoort, NY29, Schuylerville
Gas	**E:** Hess*(D), Mobil*
Food	**E:** Burger King, McDonalds, Ponderosa Steakhouse, Ruby Tuesday
AServ	**E:** Sears
ATM	**E:** 1st National Bank, Key Bank, Midland Bank, Trustco
Lodg	**W:** Sheraton (see our ad this page), Ramada Inn (see our ad this page), Wellesley Inn (see our ad this page), Howard Johnson (south in Ramsey-see our ad this page)
Other	**E:** Eckerd Drugs, K-Mart, Price Chopper Supermarket, Saratoga Mall, Shop 'N Save Supermarket, Wal-Mart (Pharmacy, Tire Center), Wilton Mall
14	**(28)** NY 9P, NY 29, Saratoga Springs, Schuylerville
13	**(24)** U.S. 9, Saratoga Lake, Saratoga Springs
Gas	**W:** Mobil*, Sunoco*(D)
Food	**E:** Andy's Pizza, Chez Sophie, The Leprechaun
	W: Packhorse Restaurant
Lodg	**E:** Locust Grove Motel, Maggiore's Inn, The Post Road Lodge
	W: Coronet Motel, Roosevelt Suites, Thorobred Motel
12	**(21)** Malta, Ballston Spa
FStop	**E:** Mobil*(CW), Sunoco
Food	**E:** Bishop Gate Pub, Briarwood, Malta Diner, McDonalds, McGoo's Bagel Shop, Tasty's Chinese
Lodg	**E:** Coco's Motel, Murphy's Motel, Riveria Motel
AServ	**E:** Croteu Auto Repair, Sunoco
RVCamp	**E:** Northway Travel Trailers, RV & Camping Accesories
Med	**E:** ✚ Malta Walk-in Clinic
ATM	**E:** Adirondock Truck Co., Key Bank, Tructco Bank
Parks	**E:** Saratoga National Historic Park
Other	**E:** Bonfare Convenience Store, Coin Laundry, Grand Union Supermarket, Malta Commons, Revco Pharmacy, Shops of Malta, State Patrol Post, Stewarts Shops
11	**(19)** Burnt Hill, Round Lake
FStop	**W:** Mobil*(LP)
Gas	**W:** Sunoco*(LP)
Food	**W:** Gran-Prix
Lodg	**W:** Gran-Prix
AServ	**W:** Sunoco
Other	**W:** Stewart's Shops
10	**(16)** Ushers Rd , Jonesville
FStop	**E:** Sunoco*
Food	**E:** Ferretti's Italian
(14)	Rest Area - RR, Phones, Picnic (Northbound)

Bold red print shows RV & Bus parking available or nearby

EXIT		NEW YORK
9		**(13)** NY 146, Rexford, Halfmoon
	Gas	**E:** Hess*[D], Raizada Enterprises*
		W: Exxon, Mobil*[CW], Sunoco
	Food	**E:** Burger King, Carvelle Ice Cream Bakery, Clifton Pot Pizza, Fortune Wok Chinese, Mr. Sub, Panda Garden, Pizza Barron, Pizza Hut, The Grill
		W: Applebee's, Boston Market, Denny's, Dominio's Pizza, Dragon Buffet, Dunkin Donuts, East Wok, Friendly's, KFC, Little Caesars Pizza, Manhattan Bagel, McDonalds, Mr. Sub, Neal's Cafe & Deli, Persnickety Coffee Bar, Taco Bell, Venevis Pizza & Subs, Wendy's
	Lodg	**W:** Best Western
		E: Comfort Inn (see our ad page 441)
	AServ	**E:** B&J Auto Supply, Jiffy Lube, Midas Muffler & Brakes, Quick Lube, Valvoline Quick Lube Center, Warren Tire Service
		W: Exxon, Sunoco (Towing)
	ATM	**E:** Key Bank, Troy Savings Bank
		W: Amsterdam Savings Bank, Fleet Bank, Marine Midland Bank, Trustco Bank
	Other	**E:** **Empire Vision Center, Grand Union Grocery (ATM)**, Hoffman Car Wash, **Laundry Mat, Revco Drugs, St. John Plaza, The Crossing Shopping Plaza**
		W: **CVS Pharmacy**, Car Wash, **Clifton Country Mall**, Express Car Wash, **K-Mart, Price Chopper Supermarket**
8A		**(12)** Grooms Rd, Waterford
8		**(10)** Vischer Ferry, Crescent
	Gas	**W:** Mobil*, Sunoco
	Food	**E:** McDonalds
		W: Mr Sub (Mobil)
	ATM	**W:** Key Bank, Trustco Bank
	Other	**W:** **CVS Pharmacy**, Raindancer Car Wash, **Stewart's Shops**
7		**(6)** U.S. 9, NY 7E, Troy, Cohoes
	Gas	**E:** Mobil
	Food	**E:** Dunkin Donuts, Mr. Sub, Subway
	Lodg	**E:** Century House Inn, ◆ Hampton Inn, ◆ Holiday Inn Express, Monte Mario Motel
	AServ	**E:** Cotton Transmission, Discount Muffler, Ford Dealership, NAPA Auto Parts, Service Parts Auto Store, Volvo Dealer
	Med	**E:** ✚ **Hospital**
	ATM	**E:** Trustco Bank
	Other	**E:** **Crossroads Shopping Plaza, Eckerd Drugs, Factory Outlet Center, Price Shopper**
6		NY 7 West, NY 2, Schenectady, Watervliet
	Gas	**E:** Mobil*
		W: Mobil*
	Food	**E:** Boston Market, Car Wash, Coco's Bar & Grill, Dakota Steak & Seafood, Ginza Japanese Cuisine, Golden Wok, Ground Round, Papa Genio's, Pizzaria Uno
		W: Friendly's, King's Buffet, Sebastian's
	Lodg	**E:** Thunderbird Motel
		W: Clarion Inn, (AAA) Microtel, Super 8 Motel
	AServ	**E:** Goodyear Tire & Auto, Nemer VW/Jeep/Eagle, Quick Lube
		W: Goodyear Tire & Auto, Jiffy Lube
	ATM	**E:** Marine Midland Bank, OnBank, Trustco Bank
	Other	**E:** **Latham Circle Shopping Mall (Lathams Farms), Latham Farms Shopping Center, Sam's Club, Wal-Mart (Lathams Farms)**
5		NY 155, Latham
	Food	**E:** Vintage Sub's & Wings

EXIT		NEW YORK
	Other	**W:** **Colony Police Dept.**
4		NY 155, Wolf Rd, Albany Airport
	Gas	**E:** Hess*, Sunoco*
	Food	**E:** Arby's, Ben & Jerry's, Burger King, Denny's, Java's Coffee House, Lexington's, Maxie's Bar Grill, McDonalds, Olive Garden, Pizza Hut, Rea Seafood Company, Wednesday's Deli, Wolf Ro Diner
		W: The Desmond
	Lodg	**E:** (AAA) Best Western, ◆ Courtyard by Marriott, Hampton Inn (see our ad this page), ◆ Holiday Inn (see our ad this page), ◆ Marriott, ◆ Red R Inn
		W: ◆ The Desmond
	AServ	**E:** Ford Dealership, GM Auto Dealership
	ATM	**E:** Fleet Bank, OnBank
	Other	**E:** **Pearl Vision Center, Pets & More**
2		**(1)** NY 5 East, Wolf Rd, Albany
	Gas	**E:** Citgo, Mobil*
		W: BP*, Mobile* (ATM)
	Food	**E:** Applebees, Bangkok Thai Restaurant, IHOP, Kansas City Grill, Kuhveh Korean, Maggie's Co Perkins, Wendy's
		W: Bruno's Pizza, Butcher Block Grill, Domino's Pizza, Houlihan's, L-Ken's Sandwich, Mr Sub, Northway Inn, Red Lobster
	Lodg	**E:** Cocca's Motel, (DAYS INN) Days Inn, Susse Chale
		W: ◆ Ambassador Motor Inn, Comfort Inn, Econolodge, Howard Johnson, Northway Inn, Ramada Limited, (AAA) Super 8 Motel
	AServ	**E:** Goodyear Tire & Auto, Jiffy Lube, Kost Tire & Muffler, Sears
		W: Auto Palace, Goodyear Tire & Auto, Monro Muffler & Brakes, Valvoline Quick Lube Center

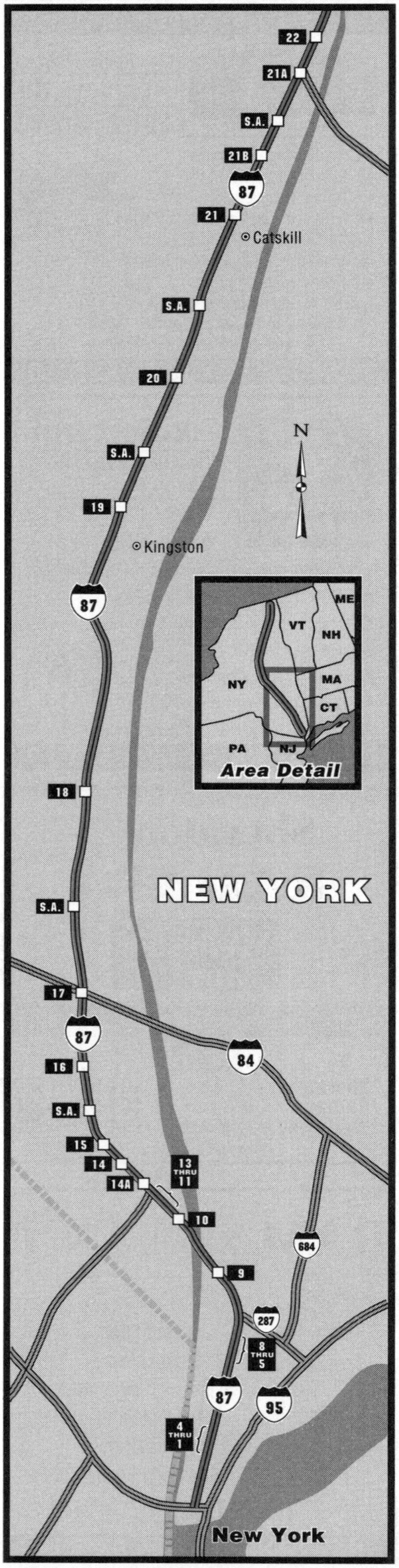

Bold red print shows RV & Bus parking available or nearby

EXIT NEW YORK

ATM **E:** All Bank, Trustco Bank
Other **E:** CVS Pharmacy, Car Wash, Coin Laundry, Colonie Center Mall, Northway Mall

(0) Junction I-90 , New York , Buffalo , Boston

B **(0)** to U.S. 20, Western Ave
Gas **W:** Exxon*, Sunoco
Food **E:** Bountiful Bread, Bruegger's Bagel, Burger King, Cafe Londonbury, Coco's, Cowan Lobel Gourmet Food, Denny's, Mangia Mexican, Peach'n Cream Coffee Shop, Star Buck's Coffee, T.G.I. Friday's, TCBY
W: China Buffet, Jade Fountion, KFC, McDonalds, Pizza Hut, Ponderosa, Sidedoor Cafe, Wendy's
Lodg **E:** Holiday Inn Express
AServ **W:** Tire Warehouse
TServ **W:** H.L.Gage Sales International
ATM **E:** All Bank, Fleet Bank, Trustco Bank
W: Key Bank
Other **E:** Albany County Veterinary Hospital, Dinapoli Opticians, Post Office, Revco Drugs, Shopping Plaza
W: Crossgate Mall, Sam's Club

Note: I-87 runs concurrent below with NYTHWY. Numbering follows NYTHWY.

24 **(148)** Junction I-90, Albany
Other **N:** Crossgates Mall

23 **(142)** Junction I-787, U.S 9W Albany, Rensselaer
TStop **E:** Big M Truckstop*
Gas **E:** Mobil*
Food **E:** Big M Truckstop, Dominio's Pizza, Reuters Sports Bar & Grill, TCBY
Lodg **E:** Howard Johnson
AServ **E:** Buff O Matic II, Glenmont Family Tire & Auto, Mobil
TServ **E:** Roberts Towing & Recovery Center
Med **W:** + Hospital

(139) Truck Inspection, Parking Area - Picnic, Phones (Southbound)

22 **(135)** NY 144, NY 396, Selkirk
AServ **E:** Valley Auto & Tire Service
Other **E:** State Patrol Post

21A **(134)** to I-90 East, MATNPK, Boston (No Vehicles Over 10'6" Wide)

(127) New Baltimore Service Area
FStop **B:** Mobil
Food **B:** Bob's Big Boy, Mrs Fields Cookies, Roy Rogers, TCBY Yogurt
AServ **B:** Mobil
ATM **B:** On Bank

21B **(125)** U.S. 9 W, NY 81, Coxsackie
TStop **W:** Fox Run Truckstop(LP)(SCALES)
FStop **W:** Sunoco*
Lodg **W:** Fox Run (Fox Run Truckstop)
AServ **W:** Dr How's Automotive
Other **W:** New Baltimore Animal Hospital

21 **(114)** NY 23, Catskill, Cairo
Gas **W:** Mobil*
Food **W:** 21 Restaurant, Days Inn, Log Cider Cafe, Rip Van Winkle Motor Lodge
Lodg **W:** DAYS INN Days Inn, Green Lake Resort, Rip Van Winkle Motor Lodge

EXIT NEW YORK

AServ **E:** 10 Min Quick Lube
RVCamp **W:** Indian Ridge Camp Sites
Other **E:** State Patrol Post
W: Tourist Information

(104) Parking Area - Phones (Southbound)

(103) Mauldin Service Area
FStop **E:** Mobil
Food **E:** Carvelle Ice Cream Bakery, McDonalds
AServ **E:** Mobil

20 **(101)** NY 32, Saugerties, Woodstock
FStop **E:** Getty
W: Hess
Gas **E:** Mobile, Stewarts Shop*
W: Hess*(D)
Food **E:** Dairy Queen, McDonalds, Star Pizza, The Stairway Cafe
W: Casey's, Howard Johnson, LaCucina Family Restaurant
Lodg **W:** Cloverleaf Hotel, Comfort Inn, Howard Johnson
Other **E:** CVS Pharmacy, Grand Union Grocery Store

(99) Rest Area - Phones (Northbound)

(96) Ulster Service Area (Southbound)
FStop **E:** Mobil
Food **E:** Nathan's Restaurant, Roy Rogers, Strathmore's Bagel & Deli, TCBY Yogurt
AServ **E:** Mobil
ATM **E:** Onbank

19 **(92)** NY 28, Kingston, Rhinecliff Bridge
Gas **E:** BP
Food **E:** B'NBagels, Blimpie's Subs, Chic's Restaurant, Elaine's Diner, Plaza Pizza
W: Kingston's Family Restaurant (Travelodge), Ramada, Skycop Steak House
Lodg **E:** Holiday Inn, Super 8 Motel
W: Ramada, Skycop Motel, Super Lodge, Travelodge
AServ **E:** Auto Plaza & Tire, BF Goodrich, BP, Parts America
W: Johnson Ford/Nissan, Throughway Nissan
RVCamp **W:** Johnson's RV Camper Barn
Med **E:** + Hospital
ATM **E:** Bank of New York, Fleet Bank, Ulster Savings Bank
Other **E:** Coin Laundry, Grand Union Grocery, Kindgston's Plaza Shopping Center, Tourist Information, WalGreens
W: State Patrol Post, Tourist Information

18 **(76)** NY 299, Poughkeepsie, New Paltz
Gas **E:** Citgo, Mobil
W: Sunoco*
Food **E:** Austrian Village Restaurant, China Buffet, College Diner, La Bella Pizza & Pasta
W: Bertone's Deli, Blue Jeans Steak House, Burger King, Dunkin Donuts, Gadaletl's Seafood, Great Wall Kitchen Chinese, McDonalds, Napali Pizza, New Paul's Hot Bagels, Pasquale's Pizza & Italian, Pizza Hut, Plaza Diner, TCBY
Lodg **E:** 87 Motel, DAYS INN Days Inn, Econolodge
W: Super 8 Motel
AServ **E:** Citgo, Mobil
W: Auto Parts, Midas Muffler & Brakes
RVCamp **W:** KOA Campground (9 Miles), Yogi Bear Camp Report (9 Miles)

EXIT NEW YORK

ATM **W:** Fleet Bank
Other **E:** Cumberland Farms Convenient Store, Diamond Car Wash
W: Adair Winery (6 Miles), Baxter's Pharmacy, Coin Laundry, Eckerd Drugs, New Paltz Plaza, Rite Aide Pharmacy, Shoprite Grocery

(66) Modena Service Area (Southbound)
FStop **W:** Mobil
Food **W:** Arby's, Carvelle Ice Cream Bakery, Mama Ilardo's, McDonalds
AServ **W:** Mobil
ATM **W:** Onbank

(65) Platekill Service Area (Northbound)
FStop **W:** Mobil
Food **E:** Bob's Big Boy, Nathan's Restaurant, Roy Rogers
AServ **E:** Mobil
ATM **E:** Onbank

17 **(60)** Junction I-84, NY 17K, Newburgh, Stewart Airport
Gas **E:** Getti's*, Sunoco*
Food **E:** Banta's Steak & Stein, Burger King, Denny's, Monroe's Restaurant, Neptune Diner, Ramada Inn, Windmill Cafe, Yobo's Oriental
W: Cake Bins, Diana's Pizza, Flower of the Orient, Pizza Hut, Plaza Deli, Shop Rite, Siam Restaurant
Lodg **E:** DAYS INN Days Inn, Hampton's Inn, Holiday Inn, Ramada Inn, Super 8 Motel
AServ **E:** Meineke Discount Mufflers, Rizzo's
W: Colandrea Pontiac, Buick, NAPA Auto Parts, Newburg Auto Parts, STS Tire & Auto, Sunshine Ford
RVCamp **E:** KOA Camping
ATM **E:** Hudson Valley Credit Union, Key Bank, Marine Midland Bank
W: Bank of New York, First Hudson Valley Bank
Other **E:** Flannery's Animal Hospital, Lloyd's Convenience Store, Tourist Information, Wal-Mart
W: Ames Shopping Center, Rite Aide Pharmacy, Steve's Coin Laundry, Vision City

16 **(45)** U.S. 6, NY 17, Harriman
Gas **W:** Exxon*(D)
Food **W:** Brookside Restaurant, Triangle Deli
Lodg **W:** American Budget Inn
AServ **W:** Rallye Chev, Buick
Other **W:** Coin Laundry, Tourist Information, Woodbury Commons Factory Outlet

(33) Ramapo Service Area
FStop **B:** Sunoco
Food **E:** Burger King, Dunkin Donuts, Sbarro Italian, TCBY Yogurt
W: Breyer's Ice Cream, McDonalds
AServ **B:** Sunoco
Other **B:** Tourist Information

15A **(31)** NY 17N, NY59, Sloatsburg, Suffern
15 **(30)** NJ17S, Jct I-287S, New Jersey
14B **(28)** Airmont Rd, Montbello
14A **(23)** Garden State Pkwy
14 **(22)** NY 59, Springvalley, Nanuet
13 **(21)** Palisades Pkwy, New Jersey, Bear Mountain (Passenger Cars Only)
12 **(19)** NY 303, W Nyack

← N **I-87** S →

EXIT	NEW YORK
11	**(18)** U.S. 9W, Nyack, S Nyack (Southbound)
Gas	**W:** Exxon, Texaco*
Food	**W:** KFC, McDonalds
Lodg	**W:** Super 8 Motel
AServ	**W:** Action Nissan, Exxon, J&L Auto & Tire Center, Jean's Auto Repair & Sales, Jiffy Lube, Midas Muffler & Brakes
Med	**W:** ✚ **Hospital**
(19)	Tappan Zee Bridge Toll (Pay Toll Southbound Only)
9	**(13)** U.S. 9, Tarrytown
8	**(12)** Jct I-287E, Cross Westchester Expressway, White Plains
7A	**(11)** Saw Mill River Pkwy (Passenger Cars Only)
7	**(8)** NY9A, Ardsley (Northbound, Southbound Reaccess Only)
(8)	Service Area (Northbound)
FStop	**B: Sunoco**
Food	**B:** Burger King, Roy Rogers, TCBY Yogurt
(5)	Toll Plaza
6A	**(5)** Cooporate Dr
6	**(4)** Tuckahoe Rd, Yonkers, Bronxville
4	**(3)** Cross County PKWY, Mile Square Rd
Gas	**W:** Sunoco, Texaco*(D)
Food	**W:** Roy Rogers, Wild Cactus Cafe
AServ	**W:** Firestone Tire & Auto, Sunoco
ATM	**E:** Chase Manhattan Bank
Other	**E: Mall at Cross Country**
3	Mile Square (Northbound)
2	Yonkers Ave, Raceway (Northbound)
1	**(0)** Hall Place, McLean Ave

↑NEW YORK

Begin I-87

I-88 Illinois →

EXIT	ILLINOIS
	Begin I-88 / ILLINOIS

↓ILLINOIS

EXIT	ILLINOIS
1AB	I-80, Des Moines, Peoria
2	Former IL 2
6	IL92E, Joslin
Other	**S: Camping**
10	Hillsdale, Port Byron
FStop	**S: Citgo*(D)**
Other	**S: Hillsdale Dental**
18	Erie, Albany
26	IL 78, Morrison, Prophetstown
36	US 30W, Rock Falls, Sterling
Med	**N:** ✚ **Hospital**
Other	**N: Camping**
41	IL 40 Rock Falls, Sterling
Gas	**N:** Amoco, Clark*, Mobil*
Food	**N:** Arthur's Deli, Bennigan's, Long John Silvers, **McDonalds**, Subway, The Red Apple Family
Lodg	**N:** All Seasons Motel, Holiday Inn, ◆ **Super 8 Motel (see our ad this page)**
AServ	**N:** Amoco, Auto Zone Auto Parts, Goodyear Tire & Auto, Quaker State Express Lube
TServ	**N: Brad Ragin Goodyear**
Med	**N:** ✚ **Hospital**
ATM	**N:** Community State Bank
Other	**N:** Coin Car Wash, **Wal-Mart, WalGreens**
44	Joilette
60	US 30, Rock Falls
69	IL 88
70	IL 26, Dixon
Lodg	**N:** Motel 8
92	IL 251, Rochelle, Mendota
FStop	**N: Citgo* (ATM)**
Gas	**N:** 76*(D), Shell*
Food	**N:** Blimpie's Subs (Citgo), Taco John (Citgo)
ATM	**N:** Amcore Bank
Other	**N:** Coin Car Wash
94	Jct I-39, US 51, Rockford, Bloomington/Normal (Difficult Reaccess)
(106)	Dekalb Toll Plaza
107	IL38, IL23, Annie Glidden Rd., Dekalb
Lodg	**N:** Super 8 Motel
(108)	**Rest Area** 🅿
110	IL 38, Piece Rd. (Difficult Westbound Reaccess)
129	IL 56W, US 30, IL 47, Sugar Grove (Difficult Reaccess)
130	Orchard Rd. (Difficult Westbound Reaccess)
AServ	**N:** Chrysler Auto Dealer, Fox Valley Ford Dealer, Saturn Dealership
132	IL31 Aurora, Batavia
Gas	**S:** Mobil*, Thornton Food Mart
Food	**S:** Best Bet Gyro's, Bilias Restaurant, **Denny's**
Lodg	**N:** Asbury Court
	S: Howard Johnson
AServ	**S:** Firestone Tire & Auto
Med	**S:** ✚ **Mercy Health Care Center**

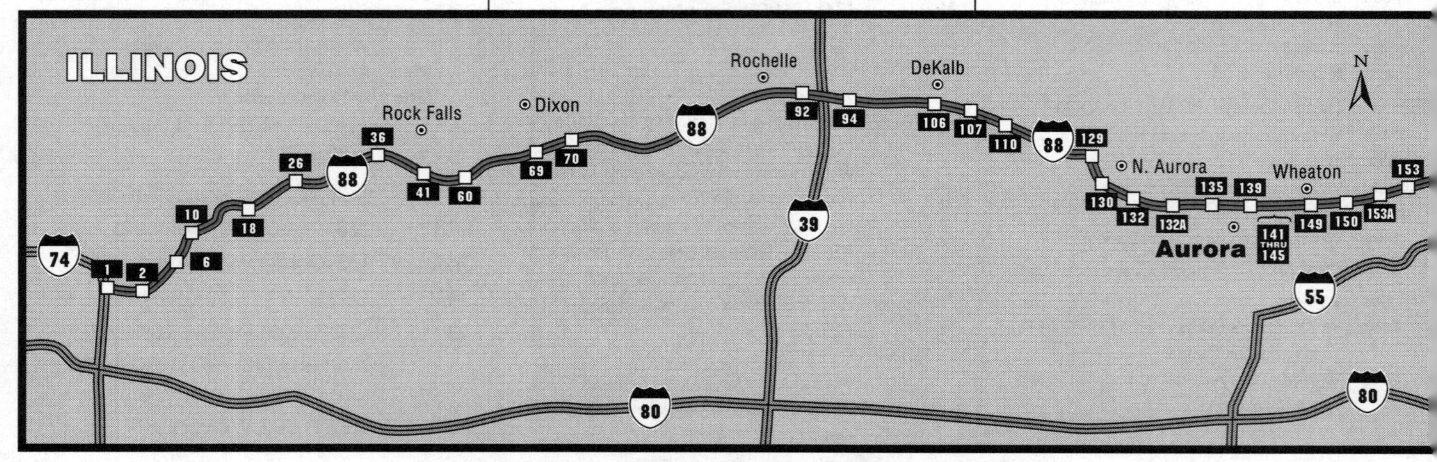

Bold red print shows RV & Bus parking available or nearby

EXIT	ILLINOIS
Other	**S: Lincoln Way Dental Center**
132A	Aurora Plaza Station
135	Farnsworth Ave.
Gas	**N:** Amoco*
	S: Marathon*, Shell
Food	**N:** Pappa Bear Restaurant
Lodg	**N:** Motel 6
AServ	**S:** Shell
Other	**S: White Hen Pantry**
139	IL59
Gas	**N:** 76*, Amoco*(CW)
	S: Mobil*(D)
Food	**N:** Cafe 59
	S: Adamo's Pizza, **Cracker Barrel,** Danny's Grille, Golden Wok, Steak & Shake, Subway, Wendy's
Lodg	**S:** Red Roof Inn, Sleep Inn
AServ	**N:** Mr. Jim's Auto Care (76)
Other	**S: White Hen Pantry***
141	Winfield Rd.
Gas	**N:** Amoco*(CW), Mobil*(CW)
	S: Mobil
Food	**N:** McDonalds (Amoco)
Lodg	**S:** AmericSuites
143	Naperville Rd.
Gas	**S:** Mobil*(CW)
Food	**N:** Rio Bravo
	S: Arby's, Bertucci's Pizza, Bob Evans Restaurant, Buona Beef, Casa Lupita, McDonalds, Old Peking, T.G.I. Friday's, Wendy's
Lodg	**N:** Radisson Hotel
	S: Courtyard by Marriott, Days Inn, Exel Inn, Fairfield Inn, Hampton Inn, Holiday Inn Select, Travelodge, Wyndham Garden Hotel
145	IL 53, Bryant Avenue (Difficult Westbound Reaccess)
Gas	**S:** Amoco*(CW), Shell
Lodg	**S:** Hyatt
AServ	**S:** Amoco, Honda Volvo Dealer, Infiniti Dealer
149	Jct. I-355, Downer's Grove, Joilet
150	IL 56, Highland Ave, Downer's, Lombard, Butterfield Rd. (Difficult Reaccess)
Food	**N:** Bakers Square Restaurant, Diamond Back Char House, Olive Garden
	S: Highland Grill
Lodg	**N:** Red Roof Inn
Med	**S: + Hospital**
Other	**N: Mall**
153A	IL83S, Oakbrook, Hindale (Difficult Reaccess)
154	Jct I-294
153	Oakbrook, Cermak Rd., IL83N

↑ILLINOIS

Begin I-88 / ILLINOIS

EXIT	ILLINOIS
	Begin I-88 / NEW YORK

↓NEW YORK

EXIT	
1	**(0)** NY7W, Binghamton
2	NY12A West, Chenango Bridge
Gas	**N:** Red Barrell*
AServ	**N:** Bridge Auto Care, By the Book Car Care
ATM	**N:** Key Bank
3	**(4)** NY369, Port Craine
Gas	**S:** Discount Petroleum, Kwik Fill*(LP)
Parks	**N: Chenango State Park**
4	**(8)** NY17E, Sanitaria Springs
Parks	**S: Nathanial Cole Park**
5	**(12)** Martin Hill Rd, Belden
RVCamp	**N: Belden Manor**
6	**(16)** NY79, Harpursville, Nineveh
RVCamp	**S: Hawkins RV Jayco Dealer**
7	NY41, Aston
8	**(29)** NY206, Bainbridge, Masonville
FStop	**N: Sunoco*(CW)**
Food	**N:** Dairy Cream, Taco Bell
Lodg	**N:** Susquehana
AServ	**N:** Parts Plus, Scoville Chevrolet, Geo
RVCamp	**N: Riverside Camping (Good Sam Park)**
ATM	**N:** NBT Bank
Parks	**S: Oquaga Creek State Park**
9	**(33)** NY8, Sidney
FStop	**N: Citgo***
Food	**N:** Burger King, China Buffet Center, **Little Caesars Pizza (Kmart),** McDonalds, Pizza Hut
Lodg	**N:** Super 8 Motel
RVCamp	**N: Tall Pines Campground**
Med	**N: + Hospital**
Other	**N: AmeriGas(LP), Coin Laundry, Grand Union Supermarket (Pharmacy), Kmart, Post Office**
10	**(37)** NY7, Unadilla
AServ	**S:** RC Sales & Service (Towing, AAA)
Other	**N: State Patrol**
(39)	**Rest Area - Picnic, Phones, RR, Tourist Info (Eastbound)**
11	**(40)** NY357, Unadilla, Franklin
RVCamp	**N: KOA Campground (3mi)**
(42)	**Rest Area - Picnic, RR, Phones (Westbound)**
12	**(47)** NY7, Otego, Wells Bridge
FStop	**S: Sunoco***
Food	**S: Country Kitchen (Sunoco)**
13	**(53)** NY205, to NY23W, Oneonta, Morris
Gas	**N:** Citgo*
AServ	**N:** Country Club Chevrolet, Jeep, Eagle, Honda, Oldsmobile, GMC, Pontiac, Parts Plus
TServ	**N: Hanna Fleet Service**
RVCamp	**N: Campground, Gilbert Lake State Park**
Other	**N: AAA Office, Coin Laundry, Oneonta Airport**
14	**(56)** NY28S, Main St (Eastbound, Southbound Reaccess)
Gas	**N:** Citgo, Stewart's Shop*, Sunoco*
Food	**N:** Alfresco's Italian, Golden Guernsey Ice Cream, India House of Tandoori, Little Panda Take-Out
AServ	**N:** Citgo (Towing)
Med	**N: + Hospital**

← W I-88

EXIT — NEW YORK

Other N: CVS Pharmacy, Coin Laundry, National Soccer Hall of Fame

15 **(56)** NY23, NY28S, Davenport
FStop S: Getty*(CW)
Gas N: Getty*
Food N: Dunkin Donuts, Friendly's, KFC, Puggie's Pizza
S: Holiday Inn, McDonalds, Neptune Diner, Perkins, Sabatini's Little Italy
Lodg N: Mama Nino's, South Pizzeria, Town House Motor Inn
S: Christopher's Country Lodge, ◆ Holiday Inn, Riverview Motel, ◆ Super 8 Motel
AServ S: Goodyear Tire & Auto, Kahn's Ford, Lincoln, Mercury, Midas Muffler & Brakes
Med N: ✚ Hospital
ATM N: Wilber National Bank
S: Amsterdam Federal (Shop N'Save)
Other N: National Soccer Hall of Fame, U.S. Post Office
S: Aldi Grocery Store, Empire Vison Center, K-Mart, Shop 'N Save Supermarket, Southside Mall, State Patrol Post, Wal-Mart

16 **(58)** Emmons, West Davenport
Food N: Arby's, Brook's Barbecue, Farm House Restaurant, Morey's Family, Perrucci Pizza
Lodg N: Master Host Inns

EXIT — NEW YORK

AServ N: Parts America
Other N: Eckerd Drugs, Price Chopper Supermarket

17 **(61)** to NY7, NY28N, Colliersville, Copperstown
Other N: Baseball Hall of Fame (18 Miles)

18 **(70)** Schenevus

(74) **Rest Area - RR, Picnic, Phones (Eastbound)**

19 **(76)** to NY7, Worcester, East Worcester
FStop N: Sunoco*(CW)
Gas N: Stewart's Shop*
Food N: Chris' Pizzeria
AServ N: NAPA Auto Parts
Other N: Motorhead Car Wash

(79) **Rest Area - Phones, RR, Picnic (Westbound)**

20 **(88)** NY7, NY10S, Richmondville
Gas S: Mobil*(LP)
Lodg S: 88 Motel
RVCamp S: Hi-View Campsites

21 **(90)** NY7, NY10N, Warnerville, Cobbleskill
Food N: Pee Wee's

EXIT — NEW YORK

AServ N: Murray's, Warnerville Garage
Med N: ✚ Hospital
Other N: Post Office

22 **(95)** NY7, NY145, Cobbleskill, Middleburgh
AServ N: Roosevelt Towing
TServ N: Roosevelt Towing
RVCamp S: Twin Oaks Campground
Med N: ✚ Hospital
Other N: Howes Caverns, Iroquois Museum
S: Police Station

23 **(101)** NY7, NY30, NY30A, Schoharie Central Bridge
Food S: Dunkin Donuts
RVCamp N: Hideway Campground

24 **(112)** U.S. 20, NY7, Duanesburg
Gas N: Mobil*
RVCamp S: Frosty Acres Campground
Other N: State Patrol Post

25 **(117)** NY7, Rotterdam, Schenectady

(25) Toll Booth

↑ NEW YORK

Begin I-88 / NEW YORK

I-89 S →

EXIT — VERMONT

Begin I-89

↓ VERMONT

22 **(130)** U.S. 7 South, Highgate Springs
Other E: Duty Free Shop

21 **(123)** VT 78, to U.S. 7, Swanton
FStop E: Exxon
W: Mobil*(D)
Gas W: Sunoco(D), Texaco*
Food E: Champlains Farms (Exxon)
W: Dunkin Donuts, McDonalds, Pam's Place, Reedie's Diner, The Old My-T-Time Creamery
AServ W: Sunoco
Other W: Grand Union Grocery, Travel Information

20 **(118)** U.S. 7, VT 207, St. Albans
Gas W: Mobil*, Shell*
Food W: Burger King, Diamond Jim's Grill, Dunkin Donuts, McDonalds, Pizza Hut, Royal Dynasty
AServ W: Cob Web, Ford Dealership, Handy Chevrolet/ Olds
ATM W: Franklin Lamoille Bank, Key Bank
Other W: Kinney Drug Store, Mail Boxes Etc, Travel Information

19 **(114)** U.S. 7, VT 36, VT 104, St. Albans
TStop W: Gulf*
Lodg W: Cadallic Motel, ◆ Comfort Inn, Econolodge
AServ W: Ford Dealership
TServ W: Gulf
Med W: ✚ Hospital
Other W: Vermont State Police

(111) **Rest Area - No Services (Southbound)**

(110) **Rest Area - No Services (Northbound)**

18 **(106)** U.S. 7, VT 104A, Milton, Fairfax
FStop E: Mobil, Texaco*(LP)
Gas E: Citgo*
Food E: Georgia Farm House, Michealangelo's Snack Bar

EXIT — VERMONT

AServ E: Blake's, Georgia Auto Service, Mobil
RVCamp E: Campground
ATM E: People's Trust Bank

17 **(98)** U.S. 2, U.S. 7, Lake Champlain Islands , Milton
FStop E: Texaco*
Gas E: Mobil
Food E: Simon's Deli (Texaco)
AServ E: Mobil, S&B Auto Center
RVCamp E: Campground
W: Campground

(96) **Weigh Station (Both Directions)**

16 **(92)** U.S. 2, U.S. 7, to VT 15, Coalchester, Winooski
Gas W: Go Go Gas, Texaco*(D)(CW)
Food E: Coal Chester Beef Lighthouse, Shoney's
W: Junior's Pizza, Kenny's, Libby's Blue Line Diner, McDonalds, Rathskeller Pizza
Lodg E: ◆ Hampton Inn
W: ◆ Fairfield Inn, Hi-Way Motel, Solid Comfort Motel
AServ W: Pectro (Towing)
TServ E: J&B Truck & Sales, RR Chaleibos Freightliner Trucks
ATM W: Merchants Bank
Other E: Shaw's Supermarket
W: Main Street Laundry, McGregor's Pharmacy

15 **(90)** VT 15, Winooski, Essex Jct (Northbound, Southbound Reaccess Only)
Gas W: Exxon*, Mobil*
Food E: Bagel Factory
W: Westside Deli
Lodg E: DAYS INN ◆ Days Inn

14 **(89)** U.S. 2, Burlington, S Burlington
Gas E: Citgo, Coastal*, Mobil*, Shell*(D), Sunoco(CW), Texaco(D)
W: Mobil
Food E: #1 Chinese, Al's Burgers, Al's Ice Cream, Burger King, Cheese Trader's, Dunkin Donuts,

EXIT — VERMONT

Econolodge, Friendly's, Howard Johnson, KFC, Lee Zachery Pizzaria, Margo's Pizza, McDonal[ds], Silver Palace Chinese
W: Sheraton
Lodg E: Anchor's Bank, Comfort Inn (see our ad opposite page), Econolodge, Holiday Inn, Howard Johnson (see our ad opposite page), Ramada Inn, Swiss Host
W: Sheraton
AServ E: Citgo, Midas Muffler & Brakes, Texaco
W: Mobil
ATM E: Critton Bank, Howard Bank, VT Federal Bank
W: Chitteden Bank
Other E: Bed & Bath Plaza, Coin Laundry, Grand Union Grocery Store, Lens Crafters, Universit[y] Mall
W: Price Chopper Supermarket

13 **(88)** Junction I-189, U.S. 7, Shelburn[e] Burlington
Lodg E: Howard Johnson (see our ad opposite page)

12 **(84)** VT 2A, to U.S. 2, Williston, Essex Jct
FStop E: Sunoco
Gas E: Mobil*
Food E: Better Bagel, Evergreen Eddie's, Friendly's, I Can't Believe Its Yogurt, Kalval Coffee Shop, M[e] at Wok Chinese, Mexicali Mexican, Pepperoni Pizza & More
Lodg E: ◆ Susse Chalet
W: ◆ Marriott
AServ E: Imported Car Center, Sunoco
TServ E: Vermont Mac
ATM E: Crittendon Bank, Howard Bank, Vermont National Bank
Other E: Hanaforde Grocery Store, Mail Boxes Etc, Simons Deli & Grocery, TLC Laundrymat, VT State Police, Wal-Mart

(82) **Rest Area - RR, Phones, Picnic, D[og] Walk, Vending**

11 **(78)** U.S. 2 to VT 117, Richmond, Williston

EXIT VERMONT

NY
CANADA
22
21
Swanton
89
20
St. Albans
19
18
89
Milton
17
Burlington
16
15
14
13
12
11
Richmond
N
NY
Vermont
NH
Area Detail
ME
89
VERMONT
Waterbury
10
9
Montpelier
8
7
6
Barre
5
Brookfield
89
4
Randolph
3
Royalton

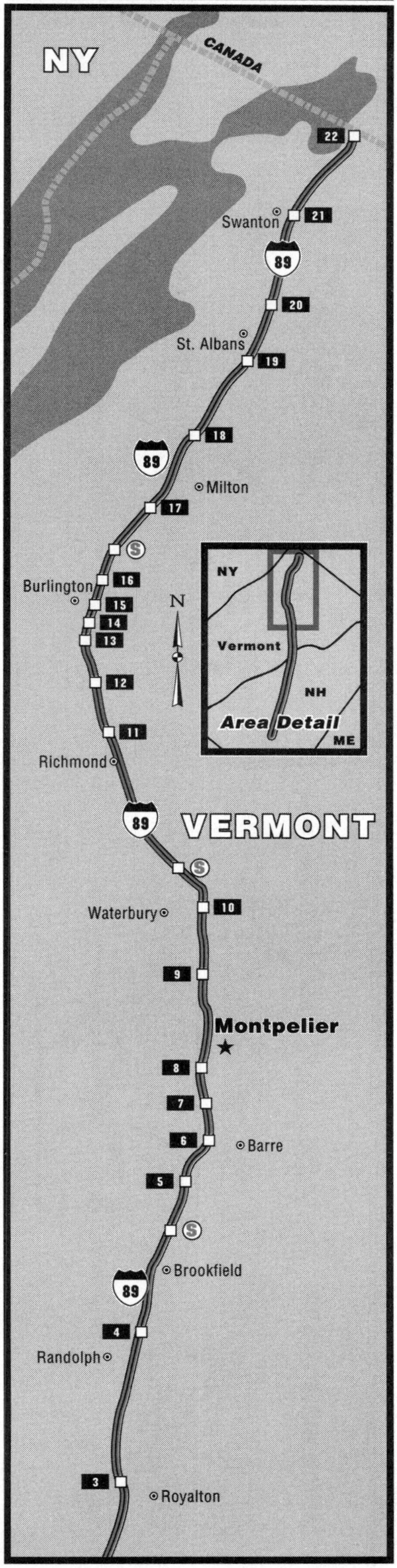

EXIT VERMONT

Gas	E: Citgo*(D)
	W: Mobil*
Food	E: Chequered Restaurant
Lodg	E: Chequered House Motel
Other	E: **Blue Flame Corp.**(LP)
(68)	**Weigh Station (Both Directions)**
(67)	**Rest Area (Southbound)**
(66)	**Rest Area - Picnic (Northbound)**
10	**(64)** to U.S. 2, VT 100, Waterbury, Stowe
Gas	E: Exxon*(D)
	W: Citgo*(D), Gulf*
Food	W: Crossroads Beverage & Deli (Citgo), Lee Zachary Pizza House
Lodg	E: Batchor Brooks, AAA Holiday Inn
	W: Stage Coach Inn
Other	E: **Area Information, Ben & Jerry Ice Cream Factory, Casoria Market,** Lots O'Suds Car Wash
	W: Car Wash, **Ideal Market, Post Office**
9	**(59)** U.S. 2 to VT 100B, Middlesex, Moretown
Gas	W: Citgo (Towing)
Food	W: Camp Meade Motor Inn
Lodg	W: Camp Meade Motor Inn
AServ	W: Citgo (Towing)
Other	W: **Vermont State Patrol Post**
8	**(53)** U.S. 2, VT 12, Montpelier
RVCamp	E: **Campground**
7	**(50)** to U.S. 302, VT 62, Barre
FStop	E: **Mobil***
Food	E: Shoney's
Lodg	E: ◆ Comfort Inn
AServ	E: Town & Country Honda
Med	E: ✚ **Hospital**
Other	E: **Shaw's Supermarket (ATM)**
6	**(47)** VT 63, VT 14, South Barre
RVCamp	E: **Campground**
Food	E: Days Inn)
Lodg	E: Days Inn (see our ad this page)
5	**(43)** VT 64, to VT 12, to VT 14, Northfield
RVCamp	E: **Campground**
(34)	**Rest Area - RR, Vending, Phones, Picnic**
(33)	**Weigh Station (Both Directions)**
4	**(31)** VT 66, Randolph, to VT 12, VT 14
Food	W: McDonalds
RVCamp	E: **Campground**
	W: **Campground**
Other	W: **Tourist Information (ATM)**
3	**(22)** VT 107, Bethel, Royalton
FStop	E: **Texaco***(D)(CW)

EXIT VERMONT/NEW HAMPSHIRE

Food	**E:** Eaton's, Village Pizza
Lodg	**E:** ◆ Fox Stand Inn
Other	**W:** VT Police

2 **(13)** VT 14, VT 132, Sharon

Gas	**W:** Citgo (ATM)
Food	**W:** Brooksie's Steak & Seafood
Lodg	**E:** Half-Acre Motel
	W: The Columns Motor Lodge
Other	**W:** Post Office

(11) **Rest Area - RR, Phones (Southbound)**

(10) VT Welcome Center - RR, Phones, Vending, Picnic

(9) **Weigh Station (Both Directions)**

1 **(4)** U.S. 4, Woodstock, Rutland

Gas	**W:** Citgo*, Mobil*
	E: Shell (see our ad this page)
Lodg	**E:** Comfort Inn (see our ad this page)
RVCamp	**W:** Pine Valley(LP)

(1) Junction I-91, Brattleboro, Whote River Jct

↑VERMONT

↓NEW HAMPSHIRE

20 **(60)** NH 12A, West Lebanon, Claremont

Gas	**E:** Mobil*(LP), Sunoco*(CW)
	W: Citgo(D)
Food	**E:** Bangkok Garden, Boston Market, Del Taco, House of Pizza, KFC, Lui Lui Pizza & Pasta, McDonalds, Men at Wok, Shorty's Mexican Roadhouse, Subway
	W: Burger King, China Lite Restaurant, D'Angelo's, Denny's, Dragon Island Chinese, Friendly's, McDonalds, Pizza Hut, TCBY Yogurt, The Seven Barrel Brewery, Weathervane Seafood, Wendy's
Lodg	**W:** AAA Economy Inn, ◆ Radisson
AServ	**E:** Jiffy Lube
	W: Auto Palace, Citgo, Midas Muffler & Brakes
ATM	**E:** First Savings of NH, Vermont National Bank
	W: Citizen's Bank
Other	**E:** Brooks Pharmacy, Car Wash, K-Mart, Lebanon Pet & Aquarium Supply, Mail Boxes Etc, Pearle Vision, Powerhouse Mall, Shaw's Grocery
	W: Airport, CVS Pharmacy, Car Wash, Eye Glass Outlet, Shaw's Grocery

EXIT NEW HAMPSHIRE

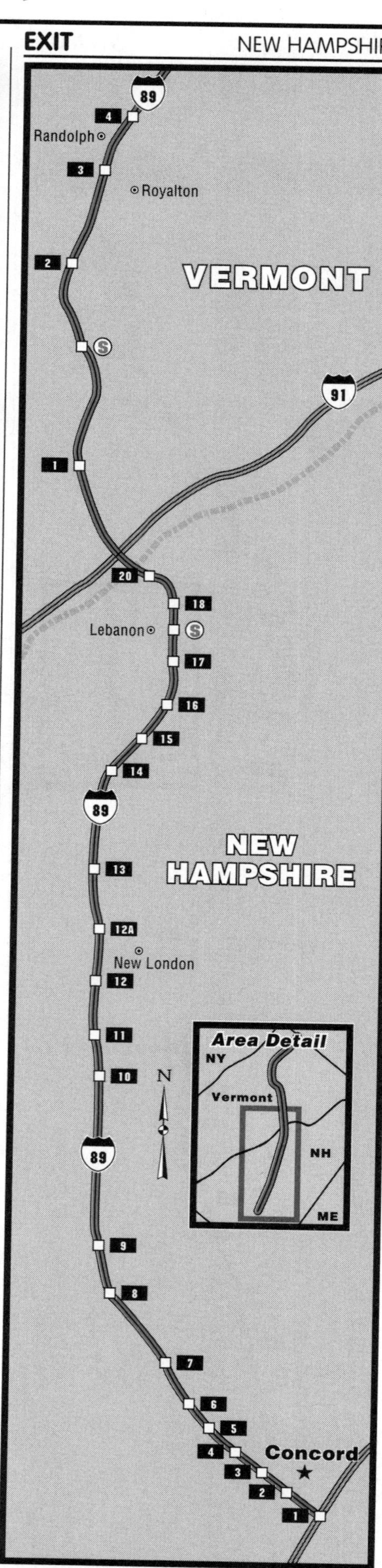

EXIT NEW HAMPSHIRE

19 **(58)** U.S. 4, NH 10, Lebanon, Hanover

Gas	**E:** Exxon*, Mobil*
	W: Sunoco
Food	**E:** Blimpie's Subs, China Station
AServ	**E:** Flander's & Pump Ford, Honda
	W: Dave's Starter & Alternator, Robert's Auto Service (Towing), Sunoco
Other	**E:** America Mile Plaza, Coin Car Wash, Coin Laundry, Grand Union Grocery

(57) NH Welcome Center - RR, Phones, Picnic, HF (Northbound)

(56) **Weigh Station (Southbound)**

18 **(56)** NH 120, Lebanon, Hanover

TStop	**E:** Coastal*(LP) (RV Dump)
Gas	**W:** Citgo*(D), Exxon*
Food	**W:** Bagel Basement, Lebonnan Bakery
Lodg	**E:** ◆ Holiday Inn Express
AServ	**E:** Dodge Mazda Dealer, Smith Buick/Olds
TServ	**E:** Freightliner Dealer
Med	**E:** ✚ Hospital
	W: ✚ Hospital
ATM	**E:** Lake Sunapee Bank
	W: Lake Sunapee Bank

17 **(54)** U.S. 4, NH 4A, Enfield, Canaan (Trucks Advised To Use Low Gear)

Food	**E:** Riverside Grill
AServ	**E:** Northern States Tire
RVCamp	**E:** Campground, RV and Truck Services
Other	**E:** Shaker Museum

16 **(52)** Eastman Hill Rd, Purmort

FStop	**E:** Evans Exit 16 Truck Stop* (Exxon)
	W: Mobil
Food	**E:** Dunkin Donuts (Exxon), Pizza (Exxon)
	W: Burger King (Mobil)
TServ	**E:** Evans Truck Stop

15 **(50)** Montcalm

14 **(48)** North Grantham (Southbound, Reaccess Northbound)

13 **(43)** NH 10, Grantham, Croydon

FStop	**E:** Gulf
Gas	**W:** Mobil* (Towing)
Food	**E:** Doodels Diner
AServ	**W:** Mobil
ATM	**W:** Lake Sunapee Bank, Sugar River Savings Bank

(40) **Tourist Information Center -**

EXIT	NEW HAMPSHIRE
	Phones, RR, Picnic, HF
2A	**(37)** Georges Mills, Springfield
Parks	**E:** Sunapee State Park
2	**(35)** NH 11 West, New London, Sunapee
Lodg	**E:** Maple Hill Farm Bed/Breakfast
Med	**E:** ✚ Hospital
Other	**E:** The Fells Historic Site
1	**(31)** NH 11 East, King Hill Rd, New London
0	**(27)** to NH 114, Sutton
Parks	**E:** Winslow State Park
	W: Wadleigh State Park
26)	Rest Area - RR, HF, Phones, Picnic (Southbound)
	(20) NH 103, Warner, Bradford
FStop	**E:** Exxon*, Mobil
Food	**E:** McDonalds, Subway (Exxon), Warner Market Place (Exxon)
8	**(17)** NH 103, Warner (Northbound)
FStop	**E:** Exxon*, Mobil
Food	**E:** McDonalds, Subway, Warner Market Place
Other	**E:** Market Basket Supermarket
	W: Indian Museum
7	**(15)** NH 103, Davisville, Contoocook
RVCamp	**E:** Campground
6	**(11)** NH 127, Contoocook, W Hopkinton
5	**(9)** U.S. 202, NH 9, Hopkinton, Henniker (Southbound, Difficult Southbound Reaccess)
4	**(7)** to NH 103, Hopkinton (Northbound, Difficult Northbound Reaccess)
3	**(4)** Stickney Hill Rd (Northbound, Southbound Reaccess Only)
2	NH 13, Clinton St
Med	**E:** ✚ Hospital
Other	**E:** Cilly Veterinary Clinic
1	Junction I-93, Concord, Bow, Logging Hill Rd
Gas	**E:** Mobil*
Food	**E:** The Grist Mill
Lodg	**E:** Hampton Inn

↑ NEW HAMPSHIRE

Begin I--89

-90 E →

EXIT	WASHINGTON
	Begin I-90

↓ WASHINGTON

EXIT	WASHINGTON
BC	Junction I-5 Portland, Vancouver
AB	WA 900, Rainier Ave. S. (No Reaccess)
Gas	**N:** Texaco*(CW)
	S: 7-11 Convenience Store*
Food	**N:** Mi La Cay Oriental Restaurant, Stan's Fish & Chips
	S: Baskin Robbins, McDonalds, Teriyaki, Vietnamese
AServ	**N:** Budd & Co. Complete Auto Repair, Electronic Tune Up, GMC Dealer
	S: Midas Muffler & Brakes
Med	**N:** ✚ Family Dr. Medical Clinic, ✚ Pacific Medical Center
Other	**N:** Rainier Veterinary Hospital, Vision Center
	S: Dental Clinic
	West Mercer Way
	Island Crest Way
Gas	**S:** 76, Chevron*, Texaco*
Food	**S:** Bruger's Bagel, Denny's, Roberto's Pizza, Subway, Thai On Mercer
Lodg	**S:** (AAA) Travelodge (see our ad this page)
AServ	**S:** 76, Simba's Auto Service(LP), Texaco
ATM	**S:** U.S. Bank
Other	**S:** WalGreens (Pharmacy)
	E. Mercer Way
	Bellevue Way
0AB	I-405, Bellevue, Renton
1AB	161st Ave. SE., 156th Ave, 150th Ave
Gas	**N:** 7-11 Convenience Store*, Texaco*
	S: 76, Chevron*
Food	**N:** Cascade's Grill, Chinese Food, Dairy Queen, Gulliver's, India Gate, Lil John Restaurant, McDonalds
	S: Baskin Robbins, Buon Jorno, Denny's, Domino's Pizza, Outback Steakhouse, Pizza Hut
Lodg	**N:** (DAYS INN) Days Inn, Embassy Suites
AServ	**N:** Subaru Dealer, Texaco
	S: 76
RVCamp	**N:** RV Park
Other	**N:** Safe Way Grocery, State Patrol
	S: Albertson's Grocery (24 Hrs), Bosley's Pet Food, Payless Drugs (Pharmacy)
13	Southeast Newport Way, West Lake Sammamish Pkwy. SE
15	WA. 900 South, Renton
Gas	**N:** Arco* (24 Hrs)
	S: Texaco*(CW)
Food	**N:** IHOP, Tully's Coffee
	S: Acapulco Fresh Mexican, Baskin Robbins, Burger King, Cascade Garden Chinese, Georgio's Subs, Godfather's Pizza, Issaquah Cafe, McDonalds, Sushiman Japanese, Tully's Coffee
Lodg	**N:** (AAA) Holiday Inn, (AAA) Motel 6
AServ	**N:** Honda Dealer (Frontage Rd.)
	S: Al's Auto Supply, Firestone Tire & Auto, Ford Dealership
RVCamp	**S:** Camping (4 Miles), RV Service & Sales
ATM	**S:** U.S. Bank, Wells Fargo
Parks	**N:** Lake Sammamish State Park
Other	**N:** Crown Book Superstore, Office Depot
	S: Ark Pet Supplies, Issaquah Dental Care, Looks Pharmacy, Meadows Cat Hospital, Quality Food Grocery Store, Western Union, Zoo
17	East Lake Sammamish Pkwy.
Gas	**N:** 76*(LP)
	S: Arco*, Chevron*, Shell*, Texaco*(D)(CW)
Food	**S:** Skipper's
AServ	**N:** 76
	S: Arco
RVCamp	**N:** Issaquah RV Park
ATM	**S:** Seafirst Bank
Other	**S:** Front Street Grocery & Deli, Tourist Information
18	E Sunset Way (Westbound, Reaccess Eastbound Only)
20	High Point Way
22	Preston, Fall City
Gas	**N:** Chevron*(D)
Food	**N:** Coffee Too, Savannah's Burger's
RVCamp	**N:** Snoqualmie River (4.5 Miles)
	S: Blue Sky Campground
25	WA 18 West, Auburn, Tacoma, Weigh Station
TServ	**N:** Weigh Station
27	Snoqualmie, North Bend (Eastbound, Reaccess Westbound Only)
31	WA 202 West, North Bend, Snoqualmie
Gas	**N:** Chevron*(D)(CW), Texaco*(D)
Food	**N:** Arby's, Blimpie's Subs, Dairy Queen, McDonalds, Mitzel's American Kitchen, Romio's Pizza & Pasta, Starbuck's Coffee, Taco Time
Lodg	**N:** Northbend Motel, Sunset Motel
ATM	**N:** Chevron
Other	**N:** Factory Outlet Center, GNC, Museum, Safe Way Grocery (24 Hrs, Pharmacy), State Patrol Post, Visitor Information
32	436th Ave. Southeast
Other	**N:** North Bend Ranger Station
34	Edgewick Road
TStop	**N:** Seattle East Auto Truck Plaza*(SCALES)
FStop	**N:** Texaco*(LP) (24 Hrs)
Gas	**N:** BP*(D)
Food	**N:** Ken's (Seattle East TS), Pizza Hut (BP), Taco Bell (BP)
Lodg	**N:** (AAA) Edgewick Inn, Norwest Motel
TServ	**N:** Seattle East TS
TWash	**N:** Seattle East Auto Truck Plaza(SCALES)

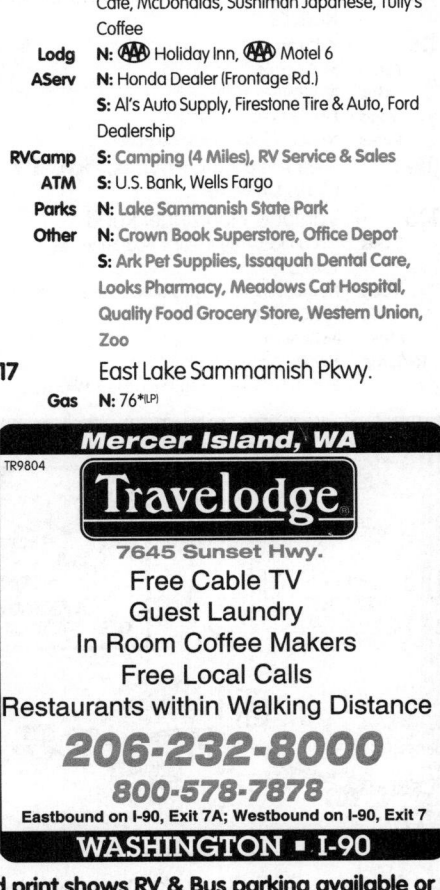

EXIT	WASHINGTON
RVCamp	N: Norwest RV Campground
ATM	N: Seattle East TS, Texaco
38	Fire Training Center
42	Tinkham Road
45	Lookout Point Road, USFS Rd 9030
47	Denny Creek, Asahel Curtis
52	West Summit (Eastbound)
53	Snoqualmie Pass Rec Area
Gas	S: Chevron*
Food	S: Family Pancake House
Lodg	S: Snoqualmie Path Inn, Summit Inn
Other	S: Time Wise Grocery & Deli, Visitor Information
54	Hyak Gold Creek
62	Stampede Pass, Lake Kachess
63	Cabin Creek Road
RVCamp	S: Cabin Creek RV Park
70	Easton, Sparks Road
Gas	N: Texaco*(LP)
Food	N: Parkside Cafe
AServ	N: Cascade Towing & Auto Repair
RVCamp	N: Camping
	S: Lake Easton Campground
Parks	S: Lake Easton State Park
71	Easton
Gas	S: CB's Cafe Grocery*
Food	S: CB's Cafe
Lodg	S: Hotel, Rex's Country Inn
Parks	S: Iron Horse State Park
Other	S: John Wayne Trail Access, U.S. Post Office
74	West Nelson, Sliding Road
78	Golf Course Rd.
RVCamp	S: Sun Country RV Park
(80)	Weigh Station - Both Directions
80	Roslyn, Salmon, La Sac
84	Cle Elum
Gas	N: Chevron*(D)(CW), Conoco, Texaco*
Food	N: Buttercup Restaurant, Cavallini's Restaurant, Dairy Queen, El Caporal, Mama Vallone's Steakhouse, Mug Rootbeer, Sunset Cafe
Lodg	N: Ramblin Rose Bed & Breakfast, (AAA) Stewart Lodge
AServ	N: Able Machine & Repair, Brake & Muffler, Conoco, NAPA Auto Parts, Texaco
TServ	N: Pacific Pride
RVCamp	N: Whispering Pines RV Park(LP) (RV Service)
ATM	N: Seafirst Bank
Other	N: Cascade East Animal Clinic, Cavallini's Pharmacy, Cle Elum Drugs, Fairway Grocery, Museum, Safe Way Grocery, Tourist Information, U.S. Post Office, Valley Laundromat
85	WA 970 N., WA 903, Cle Elum, Wenatchee

EXIT	WASHINGTON
Gas	N: BP*
Food	N: Homestead Restaurant BBQ
Lodg	N: Wind Blew Inn
RVCamp	N: Trailer Corral RV Park, Twin Pines RV Park
Other	N: Safe Way Grocery (2 Miles)
(89)	Rest Area - RR, Phones, Picnic, RV Dump (P) (Both Directions)
93	Elk Heights Road
101	Thorp Hwy, Thorp
Other	N: Thorp Fruit Store
106	U.S. 97 N, Wenatchee
TStop	N: Pilot Travel Center(SCALES)
Gas	N: BP*, Chevron*(CW)(LP), Conoco*
Food	N: Blue Grouse Restaurant, Dairy Queen, Perkins Family Restaurant, Roswell Cafe, Subway (Pilot TS)
Lodg	N: (AAA) I-90 Inn Motel
RVCamp	S: KOA Campground
Other	S: Washington State Patrol
109	Canyon Road, Ellensburg
TStop	S: Flying J Travel Plaza(LP)(SCALES)
FStop	N: Exxon*
Gas	N: BP*, Chevron* (24 Hrs), Shell*, Texaco*, Texaco*(D) (RV Dump)
Food	N: Appleseed Inn Restaurant, Arby's, Baskin Robbins, Bruchi's Cheesesteak's & Subs, Burger King, Casa De Blanca Restaurant, KFC, McDonalds, Ranch House Restaurant, Skippers Seafood, Subway, Taco Bell (24 Hrs)
	S: Leaton's Family Restaurant, Sacks Restaurant (Flying J)
Lodg	N: (AAA) Best Western, Comfort Inn, Nites Inn Motel, Super 8 Motel, The Inn At Goose Creek
AServ	N: GM Auto Dealership, Hills Auto Repair & Quick Lube, Les Schwab Tires
RVCamp	S: Leaton's RV Park
Med	N: + Hospital
ATM	N: Shell
Other	N: Museum
110	I-82 East & U.S. 97 S, Yakima (Difficult Reaccess)
115	Kittitas
FStop	N: Texaco*
Gas	N: BP*(LP)
Food	N: RJ's Cafe
Parks	N: Olmstead Place State Park
(126)	Rest Area - RR, Phones, Picnic (P) (Both Directions)
136	Vantage, Huntzinger Road
Gas	N: Conoco*(LP) (Auto Parts, Boat Supplies), Texaco*(D)
Food	N: Golden Harvest Restaurant, The Wanapum
Lodg	N: KOA Motel
AServ	N: Conoco
RVCamp	N: KOA Campground
	S: Getty's Cove Campground (4.1 Miles)

EXIT	WASHINGTON
ATM	N: Conoco
Parks	S: Wanapum State Park (3 Miles)
Other	N: The Store (Grocery Store)
	S: Ginkgo Petrified Forest
137	WA 26 East, Othello, Richland
139	Scenic View (Westbound)
143	Silica Rd
Other	N: Gorge Amphitheatre
149	WA 281 North, George
FStop	S: BP*(LP), Exxon*
Food	S: Eat, Martha's Inn Cafe, Subway (BP)
TServ	S: Bassetts Truck Repair
Med	N: + Hospital (12 Miles)
ATM	S: BP, Exxon
Other	N: Gorg Amphitheater
	S: Visitor Information
151	WA 281 N., Quincy, Wenatchee
FStop	N: Texaco* (24 Hrs)
Food	N: Hot Stuff Pizza (Texaco FS), Subs (Texaco FS
RVCamp	N: Shady Tree RV Park
Med	N: + Hospital
ATM	N: Texaco
154	Adams Road
(162)	Rest Area - RR, Phones, Picnic, RV Dump (P) (Both Directions)
164	Dodson Road
RVCamp	N: RV Park
169	Hiawatha Road
174	Mae Valley
Gas	S: Conoco*(D)
Food	S: Barbara Ann's Inn
RVCamp	N: Suncrest Resort RV Park
	S: RV Park & Camping
Other	S: Mae Valley State Patrol
175	Moses Lake State Park, Mae Valley (Westbound)
Parks	N: Moses Lake State Park
176	Moses Lake
Gas	N: BP*(D), Texaco*
Food	N: Cade's (Best Western), Kayoji's Japanese, Perkins Family Restaurant
Lodg	N: (AAA) Best Western, Interstate Inn, Motel 6, Oasis Budget Inn, Super 8 Motel
	S: Lakeshore Motel
RVCamp	N: Big Sun Resort
Med	N: + Hospital
179	WA 17, Othello, Moses Lake, Ephrata
TStop	N: Ernie's Truck Stop(SCALES)
Gas	N: Arco*, Chevron (Ernie's TS), Conoco*(D), Exxon*(LP), Shell*
Food	N: Bob's Cafe, Denny's, McDonalds, Restaurant (Shilo Inn), Shari's Restaurant
Lodg	N: ◆ Holiday Inn Express, Shilo Inn Motel
AServ	N: B&S Engine Repair
RVCamp	S: Burnham RV Service, Mar Don Campground

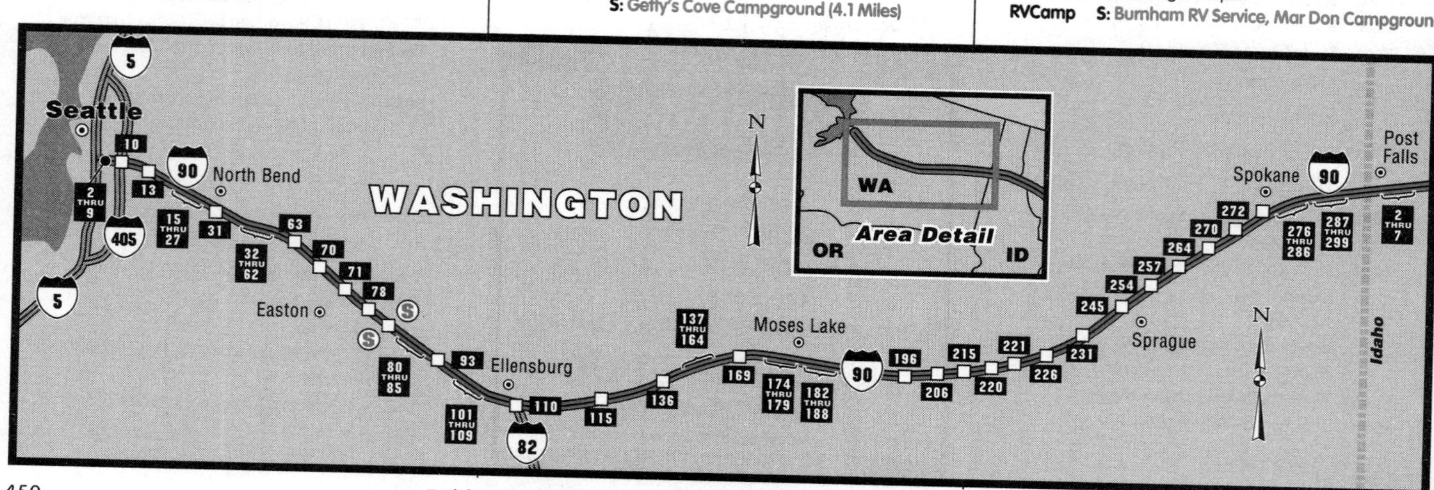

Bold red print shows RV & Bus parking available or nearby

XIT WASHINGTON

(15 Miles), Sun Country RV Service, Willows Campground (2 Miles)
Med N: Hospital
ATM N: Conoco
Other N: Coin Car Wash

30) Rest Area - Picnic, RR, Phones P

2 Wheeler, "O" Road, Northeast, Southeast

4 "Q" Road, Northeast

8 Warden, "U" Road

6 Batum, Schrag, Deal Rd

98) Rest Area - RR, Phones, Picnic P (Both Directions)

06 WA 21, To U.S. 395, Lind, Odessa

15 Paha, Packard

20 U.S. 395 South, Ritzville, Pasco
FStop N: Cenex, Exxon*(D), Vista
Food N: Texas John's Restaurant
Lodg N: AAA Colwell Motor Inn, Westside Motel
AServ N: Cooper Tires, Les Schwab Tires, NAPA Auto Parts
ATM N: Vista
Other N: State Patrol

21 WA 261 South, Ritzville, Washtucna
Gas N: Chevron*, Texaco*(D)(LP)
Food N: McDonalds, Perkins Family Restaurant, Zip's
Lodg N: AAA Best Western, Empire Motel
RVCamp N: Best Western (Best Western)
Med N: Hospital
ATM N: Texaco
Other N: Visitor Information

26 Coker Road
RVCamp S: Camping

31 Tokio, Weigh Station - Southside
Gas S: Exxon*
Food S: Templin's Cafe

42) Rest Area - RR, HF, Phones, Picnic, RV Dump, Weather Info P (Both Directions)

45 WA 23, Sprague, Harrington
Gas S: Chevron*(D), Viking Drive In
Lodg S: Last Roundup Motel, Purple Sage Motel
AServ S: Highway Garage, Tire Service (Sprague Grain Supply)
RVCamp S: Four Season Campground, Sprague Lake Resort
Other S: Sprague Grain Supply(LP)

54 Fishtrap
RVCamp S: Fishtrap RV Camp

57 WA 904, Tyler, Cheney

64 Cheney, Medical Lake, WA 902
RVCamp N: Barber's Resort Campground, Clear Lake Campground, Connie's Cove Campground
Parks N: Clear Lake Recreation Area

70 WA 904, Four Lakes, Cheney
Gas S: Exxon*
Lodg S: Saddle Inn
RVCamp S: Peaceful Pines (7 Miles)

72 WA. 902, Medical Lake
Gas N: Texaco*
Lodg S: ◆ Super 8 Motel
TServ S: Cummins Diesel
RVCamp N: Overland Station Campground
S: Smokey Trails Campground

76 Geiger
TStop N: Flying J Travel Plaza*(LP)(SCALES)
Food N: Denny's, Restaurant (Flying J Travel Plaza), Subway (Flying J Travel Plaza)
Lodg N: ◆ Best Western, The Starlite Motel
RVCamp S: Hideaway Campground, Sunset Campground
Other N: State Patrol

77A Garden Springs

EXIT WASHINGTON

277B Airport, U.S. 2 W., Davenport (Difficult Reaccess)
Food N: Spokane House (Friendship Inn)
Lodg N: Ciera Hotel, Friendship Inn, Hampton Inn, Motel 6, ◆ Ramada Inn, Rodeway Inn

279 U.S. 195 South, Colfax, Pullman (Divided Hwy)
Lodg W: Quality Inn (see our ad this page)

280A Maple Blvd.

280B Lincoln St (Westbound)
Gas N: Chevron*(CW), Conoco*, Exxon*
Food N: Burger King, Coyote Cafe, Godfather's Pizza, Hot Stuff Pizza (Exxon), IHOP, Jack-In-The-Box, McDonalds, TCBY Yogurt (Exxon), Taco Bell
Lodg N: Hotel Carlyle, Rodeway Inn
AServ N: Barton Oldsmobile, Big O Tires, Chevron, Mechanic Pride Auto Service, Saturn, Toyota & Honda Dealer
Med N: Hospital

281 U.S. 2 Newport, U.S. 395 Colville
Gas N: Conoco*, Shell, Texaco(D)
Food N: Azteca Restaurant, Best Western, Dick's Hamburger's, Frankie Doodle's Restaurant, Jack-in-the-Box, McDonalds, Perkins Family Restaurant, Pizza Hut, Rancho Restaurant, Red Lion BBQ, Subway, Taco Time, Thadeus T. Thudpucker's, Waffles n More
Lodg N: AAA Best Western, Double Tree Hotel, ◆ Fairfield Inn, ◆ Suntree Inn, The Courtyard Marriott, Tiki Lodge, ◆ Travelodge
AServ N: Carburetor & Electric, Firestone Tire & Auto, Goodyear Tire & Auto, Les Schwab Tires, Meineke Discount Mufflers, Nationwide, Shell
Med S: Sacred Heart Medical Center
ATM N: Sterling Saving Bank
Parks N: Riverside State Park (10 Miles)
Other N: Arena, Carriage Car Wash, Visitor

EXIT WASHINGTON

Information

282AB Second Ave, Trent Ave, Hamilton St (No Reaccess)
Gas N: Cenex
Lodg N: Shilo Inn
Other S: Office Depot

283A Altamont St
Gas S: Circle K*

283B Freya St., Thor St. (No Westbound Reaccess)
TStop N: Safe-way Pacific Pride Truck Stop
Gas N: Chevron*, Conoco*(D), Texaco*
S: Texaco*
Food N: Natural Cafe & Grocery, Otter Restaurant, Subway
S: Taco Bell (Texaco)
AServ N: NAPA Auto Parts

284 Havana St (Reaccess Westbound Only)
Lodg N: Park Lake Motel & RV Park
TServ N: Bocan Freightliner
RVCamp N: Park Lake RV Park

285 Sprague Ave
Gas S: Exxon*
Food N: Denny's, McDonalds
S: Mandarin House, Puerto Vallarta Mexican, Subway, Taco Time, Zip's Hamburger's
AServ S: Al's Auto Supply, Alton's Tire, Chrysler Auto Dealer, Hyundai Dealer, Nissan
RVCamp N: By-Rite RV, Quality RV Service, Ray's RV Service
S: L & L RV Service, Milestone RV Service Center
Other N: Eastwind Pet Wellness Clinic

286 Broadway Ave, Interstate Fairgrounds
TStop N: Flying J Travel Plaza*(LP)(SCALES)
Gas N: Ferrell(LP)
Food N: Flying J Travel Plaza, My Place Grill, Zip's Burgers
Lodg N: Broadway Motel, AAA Comfort Inn
AServ N: Les Schwab Tires
S: Kimpson's Collision & Alignment Center
TServ N: Detroit Diesel, Goodyear Tire & Auto, International Trucks, J&R Rigging & Supply, Kenworth, Peterbilt Dealer, Volvo Dealer, Western State CAT, Williams Equipment Co., Younker Bros.
ATM N: Flying J Travel Plaza

287 Argonne Road, Millwood
Gas N: Holiday*(D)
S: Chevron, Conoco*(D), Exxon*
Food N: 4B's Restaurant, Boston Market, Burger King, Jack-In-The-Box, Longhorn BBQ, Marie Callender's Restaurant, McDonalds, Subway, Taco Bell (Holiday), Tidyman's, Wolffy's Rockin' 50's
S: Casa Da Oro, Godfather's Pizza, Isaac's Frozen Yogurt, Little Caesars Pizza, Perkins Family Restaurant
Lodg N: DAYS INN Days Inn, ◆ Super 8 Motel
S: ◆ Holiday Inn Express, AAA Quality Inn
AServ N: Martin's Auto Service, Shuck's Auto Supply
S: Q-Lube
ATM N: Seafirst Bank
Parks N: State Park
Other N: Albertson's Grocery (Pharmacy), Super Saver Drugs
S: Book Exchange, Payless Drugs, Safe Way Grocery (24 Hrs)

289 WA. 27 South, Pines Road.
Gas N: 7-11 Convenience Store*
S: BP*(CW), Shell*(D)
Food N: Derringer's Grill, Trendd
S: Applebee's, Brown Bag, Jack-in-the-Box, Nature's Kitchen

EXIT WASHINGTON/IDAHO

Lodg S: Best Western
AServ S: Shell, Standard Battery
TServ N: Detroit Diesel
Med S: Hospital
ATM N: 7-11 Convenience Store
Other N: Pilgrim Nutrition Center
S: Hander Eye Clinic

291 Sullivan Road, Veradale
Gas S: Chevron*(CW), Conoco*, Texaco*(LP)
Food S: Bruchi's Cheesesteaks & Subs, Dairy Queen, Jack-in the-Box, McDonalds, Noodle Express, Pizza Hut, Red Lion Motor, Schlotzkys Deli, Shari's Restaurant (24 Hrs), Subway, Taco Bell
Lodg S: Comfort Inn, Double Tree Hotel
AServ S: Wal-Mart
Med S: Valley Med Center
ATM S: Texaco, Washington Trust Bank
Other N: Valley Mall
S: Fred Meyer Grocery, Wal-Mart (Pharmacy)

293 Barker Road, Greenacres
TStop N: GTX Truckstop
Food N: GTX TS
Lodg N: AAA Alpine Motel
RVCamp N: Alpine RV Park, KOA Campground (1 Mile)
S: Northwest RV Service

294 Sprague Ave, Business Rte (Westbound, Reaccess Eastbound Only)

296 Liberty Lake, Otis Orchards
Gas N: Shell*(D)
S: BP*
Food S: Burger King, McDonalds, Papa Murphy's
ATM S: Albertson's Grocery, BP
Other S: Albertson's Grocery (Pharmacy), GNC

299 State Line

(299) Weigh Station, Rest Area - Tourist Info, Picnic, RR, Phones, Picnic P (Both Directions)

↑WASHINGTON

↓IDAHO

2 Pleasant View Road
TStop N: Flying J Travel Plaza*(LP)(SCALES)
Gas N: Texaco*
S: Exxon*
Food N: Burger King, Flying J Travel Plaza, McDonalds, Subway, Tora Viejo
S: Jack-In-The-Box, Wichel's Restaurant
Lodg N: AAA Suntree Inn
S: ◆ Riverbend Inn, ◆ Sleep Inn
TServ N: Truck & Tire Repair
TWash N: Splash N Dash Truck/RV Wash
RVCamp N: Splash n' Dash RV Wash, Suntree RV Park
Other S: Post Falls Factory Outlet, Tourist Information

5 Spokane St
FStop S: Pacific Pride
Gas N: BP*(D)(CW), Exxon(LP), Texaco*
S: Handy Mart*
Food N: Andy's Pantry Family Dining, Bagel Works, Bobby's Cafe, Little Rascal's Pizza, Mallard's, Rob's Seafood & Burger's, Sub Shop
Lodg S: AAA Best Western
AServ N: Cooper Tires, Les Schwab Tires, Parts Plus
S: Align Brake & Tire Master
RVCamp S: RV Service
ATM N: BP, Bank of America, Wells Fargo
Parks S: Falls Park, Treaty Rock Park
Other N: Coin Laundry, Excel Foods, Gentle Dental Care, Interstate Office Supply, St. Vincent Books, Treaty Rock Historical Site
S: Centennial Trail, Post Falls Eye Care, Post Falls Police Department, Tourist Information

6 Bus Loop 90, Seltice Way, City Center (Reaccess Eastbound Only)
Gas N: 7-11 Convenience Store* (24 Hrs), Exxon* (RV Dump), Texaco*(D)

EXIT IDAHO

EXIT IDAHO

Food N: La Cabana Mexican, Pizza Hut
S: Arby's, Breeze's Famiy Restaurant, Godfather's Pizza, Little Caesars Pizza, McDonalds, Papa Murphy's Pizza, Subway
AServ N: NAPA Auto Parts
RVCamp N: Camping
S: Traveland RV Service
ATM N: Bank of America, Mountain West Bank, U.S. Bank
S: Washington Trust Bank
Other N: Animal Center West, Excel Foods Supermarket, N. Idaho Immediate Care Center, Super Foods (24 Hrs)
S: Coin Car Wash, Little Mo's Optical, Tidyman's

7 ID.41, Rathdrum, Spirit Lake
Gas N: Exxon*(D)(LP)
S: Chevron*(D)
Food N: Finney's Pub
S: Casey's Grill, Dairy Queen, KFC
Lodg E: Holiday Inn (see our ad this page)
AServ N: Ros Point Auto Repair, Valley Cars
TServ N: Astro Truck Electric
S: Truck Parts
RVCamp N: Camping, Valley RV Repair
S: Seltice RV Sales (Frontage Rd.), Truck Parts

(8) Weigh Station E, Rest Area - RR, Phones, Picnic, Newspapers P (Both Directions)

(9) Weigh Station (Eastbound)

11 Northwest Blvd
Gas N: Country Qwik Shop*
S: Exxon*, Qwik Stop*
Food N: Players
S: Doc & Crockett's Grill
Lodg S: Boulevard Hotel, DAYS INN ◆ Days Inn, Garden Motel
AServ N: Ford Dealership
S: Honda Dealer
TServ N: Industrial Service
RVCamp S: RV Service
ATM S: First Bank
Other N: Car Wash
S: Camping, Tourist Information

12 U.S. 95, Sandpoint, Moscow
Gas N: Chevron*, Exxon*(D), Holiday*(D), Shell*(D)(LP)
S: Texaco*(D)
Food N: Arby's, Back Door Grill, Burger King, Coeur D' Alene, Domino's Pizza, Dragon House Restaurant, JB's Restaurant, Log Cabin Restaurant, McDonalds, Monarch's Fishery, Paddy's, Perkins Family Restaurant, Pizza Hut, Pizza Shoppe, Red Lobster, Taco Bell, Thai Palace, Tomato Street Italian Restaurant, Village Inn
S: Mr Steak, Sea of Subs, Shari's Restaurant, Starbuck's Coffee, TCBY
Lodg N: AAA Coeur D'Alene Inn, AAA Comfort Inn, Motel 6, Shilo Inn, ◆ Super 8 Motel, Travelodge
S: AmeriTeln, Inn America (see our ad this page)
AServ N: Al's Service Center, Alton's Tire & Automotive, Big O Tires
Med S: Hospital
ATM N: Holiday, Rosauer's Grocery, Shell
S: Albertson's Grocery, First Security Bank
RVCamp N: Foretravel (see our ad this page)
Other N: Golden Dragon Fireworks, K-Mart, Rosauer's Grocery (Pharmacy, 24 Hrs), Safe Way Grocery
S: Albertson's Grocery (24 Hrs), GNC, Payless Drugs, Shop Co (Pharmacy, Optical), Staples Office Supply Store, Vision Center, Wild Water Waterslide Park

13 4th St
Gas N: 7-11 Convenience Store*, Conoco*(D), Exxon*(D)
S: Exxon*
Food N: Baskin Robbins, Burger King, Dairy Queen, Denny's, Godfather's Pizza, IHOP, KFC, Little

Bold red print shows RV & Bus parking available or nearby

EXIT	IDAHO
	Caesars Pizza, McDonalds, Muffins & Donuts, Pizza Hut, Taco Bell, Taco John's, Wendy's
	S: Hunter's Grill, Subway
Lodg	**N:** ◆ Fairfield Inn
AServ	**N:** Alpine Tire, Alton's Tires, Atlas Automotive, Kelly Tires, Lake City Auto, Motor Tech, NAPA Auto Parts, Schucks Auto Parts
	S: Beaugry Motors, Car Quest Auto Center, Chrysler Auto Dealer, Coeur D' Alene Lube Center, GM Auto Dealership
VCamp	**N:** Erickson's RV Service
ATM	**N:** Well's Fargo
Other	**N:** Bakery Thrift Shop, Hastings Bookstore & Music Store, Kinko's Copies (24 Hrs), Montgomery Ward, Western Union
	15th St
	Bus Loop I-90, City Center, Sherman Ave.
FStop	**S:** Exxon*(LP)
Gas	**S:** Cenex*, Conoco*, Piggie's Gas Mart*
Food	**S:** Casa Maria Mexican, Chuck Wagon Cafe, Coffee House, Cove Bowl, Down the Street Restaurant, Eduardo's Mexican Restaurant, Mad Mary's Oriental Express, Roger's Home Made Icecream
Lodg	**S:** Bate's Motel, Cedar Motel, Coeur D' Allene Budget Saver Motel, AAA El Rancho Motel, AAA Holiday Inn Express, Holiday Motel, Lake Drive Motel, Monte Vista Motel, Sandman Motel, Star Motel, State Motel, Sundowner Motel
AServ	**S:** GM Dealer
TServ	**S:** Robideaux Truck Service
VCamp	**S:** Cedar RV Camp, Monte Vista RV Park
Other	**N:** Forrest Information, Ranger Station
	S: Animal Medical Center, Coin Laundry, Sherman IGA Grocery (24 Hrs), Tourist Information
2	ID. 97, Harrison, St. Maries
Lodg	**N:** Wolf Lodge Inn
VCamp	**N:** Wolf Lodge Campground
	S: KOA Camping (.5 Miles), Squaw Bay Resort (7.5 Miles)
Other	**N:** Coeur D' Alene Scenic Byway
8	4th of July Pass Recreation Area
Other	**S:** The Mullin Tree Historical Point
2)	Parking Area (Westbound)
4	ID.3, Rose Lake, St. Maries
FStop	**S:** Conoco*(LP)
Gas	**S:** General Store*
Food	**S:** Country Chef Cafe
Parks	**S:** White Pines Scenic Route
Other	**S:** Scenic By-way Info.
9	Old Mission State Park
Parks	**S:** Cataldo Mission National Historic Site, Old Mission State Park
0	Cataldo
Lodg	**N:** Cataldo Inn
RVCamp	**S:** Coeur D'Alene River RV Park

EXIT	IDAHO
43	Kingston
Gas	**N:** Texaco*(D)
Other	**N:** National Forest
45	Pinehurst
Gas	**S:** Chevron*(D), Kanoko*(D)
AServ	**S:** Chevron
RVCamp	**S:** KOA Campground, Pinehurst RV Service(LP)
Other	**S:** Laundry Express
48	Smelterville
FStop	**N:** Silver Valley Car Truck Stop
Food	**N:** Buffalo Nickel Bar & Grill
RVCamp	**N:** Buffalo Nickle RV Park, White's Buffalo RV Park
49	Bus Loop 90, Bunker Ave, Silver Mtn
Gas	**N:** Chevron*
Food	**N:** McDonalds, Restaurant (Silver Horn), Sam's, Sea of Subs, Subway
Lodg	**N:** Silver Horn Motor Inn, Silver Ridge Mountain Lodge
AServ	**N:** Cooper Tires
Med	**N:** ✚ Kellogg Medical Clinic, ✚ Shoshone Medical Center
ATM	**N:** Chevron
Other	**N:** Greyhound Bus Station, Historic Site
	S: Museum, Tourist Information
51	Bus Loop 90, Division St, Wardner
Gas	**N:** Conoco*(D), Conoco*, Jack's*
Food	**N:** Broken Wheel Restaurant, Humdigger Drive In, Sunshine Restaurant
	S: Kopper Keg Pizza, McKinnley Inn(SCALES), Wah Hing Restaurant (Chinese), Zany's Pizza
Lodg	**N:** Motel 51, Sunshine Inn, Trail Motel
	S: McKinnley Inn
AServ	**N:** Auto Dealers, Cooper Tires, Les Schwab Tires, NAPA Auto Parts, Parts Plus, Shoshone Glass
	S: Reco's Muffler & Repair
TServ	**N:** Truck Tire Service
ATM	**N:** IGA
	S: First Security Bank, US Bank
Other	**N:** Jack's Bookshop, Kellogg Animal Hospital, Pet Grooming & Supply, Stien's IGA Store, Sunnyside Drug's, Western Union (IGA)
	S: Coin Laundry, Museum, Pack'n Save Foods Grocery, Penney's Dept Store, Todd's Office Supply, Tourist Information, Western Union (Todd's Office Supply)
54	Big Creek
Other	**N:** Elk Creek Store C-Store (Frontage Rd), Historic Site
57	Bus Loop 90, Osburn
Gas	**S:** Texaco*
AServ	**S:** Gary's Lube & Oil, Silver Auto
RVCamp	**S:** Blue Anchor RV Park
60	Bus Loop 90, Silverton, Osburn
Lodg	**S:** Silver Leaf Motel
RVCamp	**S:** Camping
Med	**N:** ✚ Hospital
Other	**N:** National Forrest Ranger Station

EXIT	IDAHO/MONTANA
61	Bus Loop 90, Wallace
Gas	**S:** Conoco*(D)
Food	**S:** JC's Restaurant
Lodg	**S:** Motel, Super Valley Inn, AAA Wallace Inn (Best Western)
RVCamp	**S:** Camping
Other	**S:** Tourist Information
62	Bus Loop, ID4, Wallace Burke
Food	**S:** BJ's Pizza, Historical Saloon & Grill, The Pizza Factory
Lodg	**S:** Brooks Hotel
RVCamp	**S:** Down-by-the-Depot RV Park
Med	**S:** ✚ Family Practice Doctor
Other	**S:** Excel Foods Grocery, Police Station, Tourist Information, Wallace Historical District, Western Union
64	Golconda District
65	Compressor District
66	Gold Creek (Eastbound)
(67)	Parking Area (Both Directions)
67	Morning District
RVCamp	**N:** Camping
68	Bus Loop, Mullan (Eastbound)
69	Bus Loop 90, Mullan
Gas	**N:** Exxon*(D)(LP)
AServ	**N:** Cooper Tires
Parks	**N:** Shoshone Park (3 Miles)
Other	**N:** Museum
(73)	Historical Site (Westbound)

↑IDAHO

↓MONTANA

EXIT	MONTANA
0	Local Access
(5)	Rest Area - Full Facilities P (Both Directions)
5	Taft Area
10	Saltese
Lodg	**N:** 4-D's Motel, Motel
Other	**N:** Post Office
(15)	Weigh Station (Both Directions, Left Lane Exit)
16	Haugan
FStop	**N:** 10,000 Silver Dollars
Gas	**N:** 10,000 Silver Dollars
Food	**N:** 10,000 Silver Dollars
Lodg	**N:** 10,000 Silver Dollars
RVCamp	**N:** 10,000 Silver Dollars
ATM	**N:** 10,000 Silver Dollars
18	DeBorgia
Gas	**N:** Cenex*(D)(LP)
Food	**N:** Restaurant-Chicken & Hamburgers

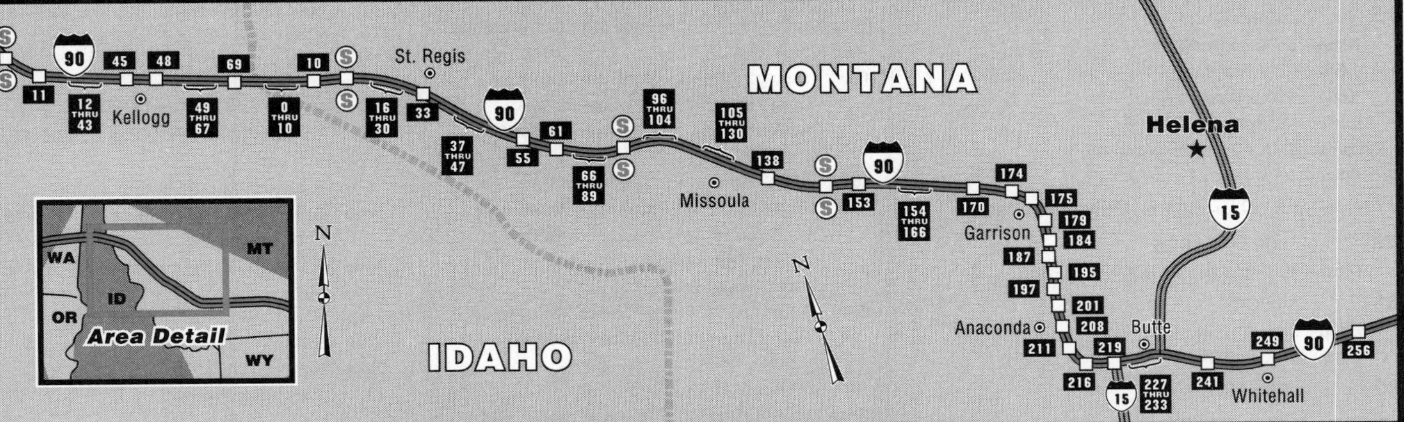

Bold red print shows RV & Bus parking available or nearby

EXIT MONTANA

(Pinecrest Motel)

Lodg N: ◆ Hotel Albert, Pinecrest Lodge Motel

Other N: Whole Fireworks

22 Henderson

RVCamp N: Camping

25 Drexel

29 Fishing Access (Westbound, Reaccess Westbound Only)

Other S: Fishing

30 Two Mile Road

33 MT 135, St Regis

Gas N: Cenex(LP), Conoco*(D), Exxon*

Food N: Chester Fried Chicken (Conoco), Frosty Drive-in, Hot Stuff Pizza (Exxon), Huckleberry's Family Restaurant (Conoco), Jasper's Restaurant

Lodg N: Little River Motel, Super 8 Motel

AServ N: Schober's Truck/RV & Auto Service

TServ N: Schober's Truck/RV & Auto Service

RVCamp N: KOA Campground (1 Mile), Schober's Truck/RV & Auto Service, St Regis Camp

Other N: Visitor Information

37 Sloway Area

RVCamp S: Camping

43 Dry Creek Road

RVCamp N: National Forest Campground

47 CR 257, Superior

FStop S: Exxon*

Gas N: BP*, Cenex*(CW), Conoco*

Food N: Breadboard Restaurant, Durango's Restaurant

Lodg N: Belleview Motel, Budget Host Motel

AServ N: Cenex, Schenider Auto Service

S: Carl's Auto Repair

RVCamp N: National Forest Campground (7 Miles)

Med N: ✚ Hospital

ATM S: Exxon

Other N: IGA Store, Mineral County Law Enforcement Center, Mineral Pharmacy, Museum, National Forest Ranger Station, US Post Office

S: Lucky Lil's Casino, Tourist Information

55 Lozeau Quartz

(58) Rest Area - Phones, Picnics, RR, HF P (Both Directions)

61 Tarkio

66 Fish Creek Road

70 Cyr

(72) Parking Area P (Eastbound)

(73) Parking Area P (Westbound)

75 Alberton

FStop S: River Edge Motel*(LP)

Gas N: Panther Express*

Lodg S: River Edge Motel

AServ N: Alberton Mechanical

RVCamp S: River Edge Campground

Other S: Fishing

77 CR 507, Alberton, Petty Creek Road

82 Nine Mile Road

Other N: Historic Ranger Station

85 Huson

FStop S: Huson Mercantile*(D)

Other S: US Post Office

89 Frenchtown

EXIT MONTANA

Gas S: Conoco*, Sinclair*

Food S: French Connection Casino & Grill, Frenchtown Club, Smash Hit Subs (Conoco), The Alcar Cafe, The Coffee Cup Restaurant

AServ S: Pete's Garage

ATM S: Conoco

Parks N: Frenchtown Pond State Park

Other S: Bronc's Grocery, Coin Laundromat, US Post Office

(93) Weigh Station (Both Directions)

96 U.S. 93 North, MT 200 West, Kalispell

TStop N: Conoco(SCALES) (RV Dump)

S: Cross Road's Travel Center(LP) (Sinclair, RV Dump)

Food N: Muralt's Restaurant (Conoco)

Lodg N: DAYS INN (AAA) Days Inn

S: (AAA) Redwood Lodge

TServ N: Conoco, Ford Dealership, Freightliner Dealer, Peterbilt Dealer

S: Cross Road's, Kenworth

RVCamp N: Jellystone RV Park

Parks N: Glacier National Park

101 CR 430, Reserve Street, Hamilton

FStop S: Cenex*(LP) (24 Hrs, RV Dump), Exxon* (RV Dump)

Gas N: Conoco*(D)

S: Conoco*(CW)

Food S: 4B's Restaurant, Hot Stuff Pizza (Exxon), Joker's Wild Casino & Restaurant, McDonalds, McKenzie River Pizza, Restaurant (Exxon), Smash Subs (Exxon), Taco Time

Lodg N: (AAA) Best Western

S: (AAA) 4 B's Inn, ◆ Comfort Inn, ◆ Hampton Inn, (AAA) Ruby's Reserve Street Inn, ◆ Super 8 Motel, Traveler's Inn

TServ S: CAT Truck Service, Cummins Diesel, Detroit Diesel, Okie's Electric, Onon Truck & RV Service

RVCamp N: Camping

S: Onon Truck & RV Service

ATM N: Conoco

S: Cenex

Other N: Skiing

104 U.S. 93 South, Orange Street

Gas S: Conoco*, Sinclair(D)(LP)

Food S: Pagoda Chinese, Subway

Lodg S: (AAA) Budget Motor Inn

AServ S: Conoco, Sinclair

Med S: ✚ Hospital

ATM S: Conoco

Other S: Coin Laundry

105 Business 90, U.S. 12 West, Van Buren Street

FStop S: Conoco*(D), Pacific Pride Commercial Fueling

Gas S: Sinclair*

Food S: Burger King, Finnegans Family Restaurant, Goldsmith Ice Cream, Little Caesars Pizza, McDonalds, Pizza Hut, Ponderosa, Press Box, Taco Bell

Lodg S: (AAA) Campus Inn, Canyon Motel, (AAA) Creekside Inn, (AAA) Doubletree Hotel, ◆ Family Inn, Goldsmith Bed & Breakfast, ◆ Holiday Inn Express, (AAA) Ponderosa Lodge, Thunder Bird Motel, Village Motor Inn

AServ S: Champion Auto Parts, Conoco, Pennzoil Lube, Quaker State, Unocal 76, Wholesale Battery Supply

ATM S: Sinclair

Other S: Broadway Market, Buttrey Grocery Store,

EXIT MONTANA

Coin Car Wash, Coin Laundromat, Tourist Information, University Of Montana, Vietnam Vets Memorial

107 East Missoula

FStop N: Conoco*

Gas N: BP* (Coin Laundry), Sinclair

Food N: Dixie's Cafe (Conoco), Kolbs Cafe

Lodg N: OK Motel, The Motel

AServ N: Bill's Tranmission, Carl's Auto

TServ N: Bill's Transmission, Nick's Diesel

ATM N: Conoco

Other N: Coin Laundry (BP), Ski Area

109 MT 200 East, Bonner, Great Falls

TStop N: Exxon Travel Plaza*(LP)

Gas N: Conoco*, Sinclair*(LP)

Food N: Arby's (Exxon), Finky's Foods (Exxon), Hot Stuff Pizza* (Sinclair), Lucky Lil's Casino (Exxon), Restaurant (Exxon), Subway

AServ N: Interstate Truck & Auto, Milltown Garage

TServ N: Brian Motors, Interstate Truck & Auto

ATM N: Exxon

Other N: Historic Point, US Post Office

113 Turah

Food N: Other Place

RVCamp S: Turah Campground

120 Clinton

Gas S: Conoco*(LP)

Other S: Post Office

126 Rock Creek Road

Gas S: Rock Creek Lodge

Food S: Rock Creek Lodge

Lodg S: Rock Creek Lodge

Other S: Recreation Area

(128) Parking Area P (Both Directions)

130 Beavertail Road

RVCamp S: Campground

Parks S: Beavertail State Park (.25 Miles)

Other S: Fishing

138 Bearmouth Area

Gas N: Chalet Bearmouth

Food N: Chalet Bearmouth

Lodg N: Chalet Bearmouth (RV Dump)

RVCamp N: Bearmouth Campground

Other N: Chalet Bearmouth C-Store

(143) Rest Area - Phones, Picnics, RR, HF, Historic Site, Pet Walk P (Both Directions)

(151) Weigh Station (Both Directions)

153 MT 1, Drummond, Philipsburg (Eastbound, Reaccess Westbouund Only, See Exit 154 Southbound Services)

154 MT 1, Drummond, Philipsburg (Westbound, Reaccess Eastbound)

FStop S: Exxon*

Gas S: Cenex(LP), Sinclair*(CW)

Food S: D-M Restaurant, Frosty Freeze, The Corner Cafe, Wagon Wheel

Lodg S: Drummond Hotel, Sky Motel, Wagon Wheel Motel

RVCamp N: The Good Time Camping & RV

ATM S: Exxon

Other N: Pentler Scenic Loop

S: Front Street Market

Bold red print shows RV & Bus parking available or nearby

XIT MONTANA

62 (646) Jens

66 Gold Creek

67) Rest Area - Phones, Picnics, RR, HF, Pet Walk P (Westbound)

69) Rest Area - Phones, Picnics, RR, HF P (Eastbound)

70 Phosphate

74 U.S. 12 East, Garrison, Helena (Eastbound, Reaccess Westbound)

75 U.S. 12 East, Garrison, Helena (Westbound, Reaccess Eastbound)

RVCamp N: RV Park

79 Beck Hill Road

84 Business 90, Deer Lodge

FStop S: Conoco(CW) (RV Dump)
Gas S: Sinclair*
Food S: 4B's Restaurant (Steak & Prime Rib), McDonalds, Restaurant (Sinclair)
Lodg S: AAA Super 8 Motel
RVCamp S: Indian Creek Campground(LP) (Laundry Facilities)
ATM S: Conoco
Other S: Historic Site, I-90 Powerwash (Car Wash), Laundry Facilities (Indain Creek Campground)

87 Business 90, Deer Lodge (Reaccess Eastbound Only)

RVCamp N: KOA Campground (Open May 1st To October 15th)
Med N: + Hospital
Other N: Grant-Kohrs Ranch Nat'l Historic Site

95 Racetrack

97 CR 273, Galen

01 Warm Springs

Gas S: Sinclair*
Med S: + Montana State Hospital
Other S: Coin Laundry (Sinclair)

08 MT 1, Anaconda (Eastbound Reaccess Only)

Med S: + Hospital
Other S: Scenic Loop

10 Scenic Loop Information Turnout (Westbound)

11 CR 441, Gregson, Fairmont, Hot Springs

16 Ramsay

19 I-15 South, Dillon, Idaho Falls (Port Of Montana Trasportation)

Note: I-90 runs concurrent below with I-15. Numbering follows I-15.

21 Jct I-15S, Idaho Falls

22 Rocker, Weigh Station (All Trucks Must Exit When Weigh Station Is Open)

TStop E: Conoco*(LP)
W: Flying J Travel Plaza(LP)
Food E: 4 B's Restaurant (Conoco), Arby's (Conoco TS)
W: Flying J Travel Plaza
Lodg W: AAA Rocker Inn
AServ E: Rocker Repair (24 Hr)
TServ E: Rocker Repair (24 Hr)

EXIT MONTANA

ATM E: Conoco
W: Cash Machine (Flying J TS)

124 I-115, City Center (Northbound, Southbound Reaccess Only)

126 Montana Street

Gas E: Conoco*, Exxon*(D)
W: Cenex*, Sinclair
Food E: Muzz & Stan's Food
W: Jokers Wild Casino & Restaurant (Chicken/ BBQ Ribs), Winter Garden Lanes
Lodg W: Eddy's Motel
AServ W: Sinclair
RVCamp W: KOA Campground
Med W: + Hospital
ATM E: Conoco
Other W: Dental Clinic, Safe Way Grocery, Tourist Information

127AB Bus15, Bus I-90, Harrison Ave

Gas E: Conoco* (#1 of 1), Conoco* (#2 of 2), Exxon*(D)(CW)
W: Cenex*(D)(CW), Conoco*, Sinclair* (RV Dump)
Food E: 4 B's Restaurant (24Hr), Arby's, Burger King, Godfather's Pizza, Joey's Seafood, KFC, McDonalds, Perkins Family Restaurant, Pizza Hut, Plaza Royale Restaurant, Ray's Place, Red Rooster Supper Club, Restaurant (Motel), Silver Bow Pizza, Subway, TCBY Yogurt, Taco Bell, The Ponderosa Cafe, Uno's, Wendy's
W: Arctic Circle Hamburgers, Dairy Queen, Denny's, Derby Steakhouse, Domino's Pizza, El Taco Mexican Food, Hardee's, Hot Stuff Pizza (Cenex), John's, Little Caesars Pizza, Restaurant (War Bonnet Inn), Smash Hit Subs (Cenex), Taco John's, Top Deck
Lodg E: Best Western, Comfort Inn, DAYS INN Days Inn, Mile High, Motel, Super 8 Motel
W: Holiday Inn Express, War Bonnet Inn
AServ E: American Car Care Centers, Checker Auto Parts, Ford Dealership, GM Auto Dealership, Glenn's, Honda Dealer, Pennzoil Lube, Wal-Mart
W: Bob's Fast Lube, Champion Auto Parts Store, Nissan Dealer, Uniroyal Tire & Auto
ATM E: American Federal Savings Bank, Conoco (#1 of 1), Conoco (#2 of 2), Exxon, First Citizens Bank, First National Bank, Norwest Bank
W: Cenex, Conoco
Other E: Butte Plaza Mall, Buttrey Food Grocery (Pharmacy), K-Mart, Optical, Wal-Mart (Pharmacy)
W: Downey Drug, Natural Healing, Safe Way Grocery (24 Hrs, W/Pharmacy)

Note: I-90 runs concurrent above with I-15. Numbering follows I-15.

227 I-15 North, Helena, Great Falls

228 CR 375, Continental Drive

233 Homestake

(235) Rest Area - Picnic, Phones, RR, HF P (Both Directions, All Trucks Must Stop For Grade Information)

241 Pipestone

RVCamp S: Piipestone Campground (Open April 15th to October 1st)
Other N: Delmoe Lake

249 MT 55, to MT 69, Whitehall

EXIT MONTANA

TStop S: Exxon*(D)
Food S: Lucky Lil's, Subway (Exxon)
Lodg S: Chief Motel (1 Mile), Super 8 Motel
AServ S: Quaker State, Tobaccoroot
ATM S: Exxon

256 CR 359, Cardwell, Boulder

FStop S: Conoco*(D)(LP)
RVCamp S: Camping
Parks S: Lewis & Clark Caverns State Park

267 Milligan Canyon Road

274 U.S. 287, Helena, Ennis

FStop S: Exxon*(LP)
Gas N: Conoco*
Food N: Cafe, Steer In Casino & Restaurant, Wheat Montana Bakery
S: Lucky Lil's Casino & Restaurant (Exxon), Subway (Exxon)
Lodg N: AAA Three Forks Motel
TServ N: Ron's Diesel Repair
RVCamp N: Silos RV Park (39 Miles)
S: KOA Campground (Open April 15th To October 15th)
ATM S: Exxon
Parks S: Lewis & Clark Caverns State Park, Yellowstone National Park
Other N: Canyon Ferry Lake

278 MT 2, Three Forks, Trident, CR205

Parks N: Missouri Headwaters State Park

283 Logan, Trident

Parks S: Madison Buffalo Jump State Monument (7 Miles)

288 CR 346, CR 288, Manhattan, Amsterdam

Gas N: Conoco*(D)
Food N: Cafe On Broadway, Garden Cafe, Manhattan Cafe
AServ N: Gallatin Repair
RVCamp N: Camping(LP) (RV Dump, Laundromat)
ATM N: Manhattan State Bank
Other N: Food Farm Grocery, Laundry Facilities (RV Camp), Museum, Picnic Area

298 CR 291, MT 85, Amsterdam Belgrade, West Yellowstone

TStop S: Conoco(SCALES)
FStop N: Cenex*(LP)
Gas N: Conoco*, Exxon*(D)(CW)
S: Chalet Market Gas*
Food N: Charlie's Deli & Coffee Shop, Lucky Lil's Casino & Restaurant, McDonalds, Rosa's Pizza, Subway (Exxon), Taco Time
S: Cafe (Motel), Country Kitchen, Restaurant (Conoco)
Lodg S: Homestead Motel, Kelly Inn, Motel, Super 8 Motel
AServ N: Don's Auto Repair, Lube-it
S: Small Car Clinic
TServ S: Goodyear Tire & Auto (Frontage Rd)
TWash S: Rabbit Car-Truck Wash
RVCamp S: Camping
ATM N: Cenex, Exxon, Lee & Dad IGA
Other N: Lee & Dad's IGA Store, NAPA Auto Parts, Trust Worthy Hardware (RV Dump)

305 CR412, N 19th Ave, Springhill

Other S: Montana State University

306 Bus Loop I-90, U.S. 191, Bozeman, MT 205 N 7th Ave

Bold red print shows RV & Bus parking available or nearby

EXIT	MONTANA
Gas	**N:** Conoco*
	S: Conoco*(D), Exxon*, Holiday*, Sinclair
Food	**N:** Apple Tree Restaurant, McDonalds
	S: Applebee's, Best Western, Ferraros Fine Italian Food, Hardee's (Conoco), Restaurant (Holiday Inn), Santa Fe Red Cantina, Village Inn Pizza
Lodg	**N:** ◆ Fairfield Inn, AAA Prime Rate Motel, ◆ Ramada Limited, Sleep Inn, ◆ Super 8 Motel
	S: AAA Best Western, AAA Bozeman Inn, Comfort Inn, DAYS INN ◆ Days Inn, ◆ Holiday Inn, AAA Rainbow Motel, Sunset Motel
AServ	**S:** Car Quest Auto Center, K-Mart, Sinclair, Wal-Mart
TServ	**N:** Bridgestone Tire & Auto, Michelin
RVCamp	**N:** C&T Trailor Supply Store, Camping
ATM	**S:** American Federal Savings Bank, Conoco
Other	**N:** Ski Area
	S: Optical House, Price Rite Drugs, Scrubby's Car Wash, Tourist Information, Van's IGA Store, Wal-Mart (Pharmacy)
309	U.S. 191, Main Street, Bozeman
Gas	**S:** Conoco*, Exxon*, Sinclair*, Town Pump*(D)
Food	**S:** Family Restaurant
Lodg	**S:** Alpine Lodge, Blue Sky Motel, Continental Motor Inn, Ranch House Motel, Western Heritage Inn
AServ	**N:** Dick Motors
	S: Rocky Mountain RV & Auto, Straight Away Auto Repair
RVCamp	**N:** Sunrise Campground
	S: Rocky Mountain RV & Auto Service
Med	**S:** ✚ Hospital
Parks	**S:** Lindley Park
Other	**S:** Buggy Bath Car Wash, Skiing, Visitor Information
313	Bear Canyon Road
RVCamp	**S:** Camping
316	Trail Creek Road
319	Jackson Creek Road
324	Ranch Access
330	Bus Loop I-90, Livingston, Local Access
TStop	**N:** Yellowstone Truck Stop*
Food	**N:** Restaurant (Yellowstone Truck Stop)
TServ	**N:** Yellowstone Truck Stop
333	U.S. 89S, City Center, Yellowstone Park
FStop	**S:** Conoco*(CW)
Gas	**N:** Conoco, Gas Station, Holiday*, Sinclair(D)(LP) (RV Dump)
	S: Cenex*, Exxon*
Food	**N:** Best Western, Crazy Coyote Mexican, Dairy Queen, Domino's Pizza, Papa Murphy's Pizza, Paradise Inn Motel, Pizza Hut
	S: Hardee's, Lucky Lil's Casino (Conoco), McDonalds, Subway
Lodg	**N:** Best Western, Del-Mar Motel, Livingston Inn, Paradise Inn Motel
	S: ◆ Comfort Inn, Super 8 Motel
AServ	**N:** Conoco, Gas Station, Kimo RV & Auto Service, Livingston Ford Mercury, Sinclair
RVCamp	**N:** Camping, Kimo RV & Auto Service, Paradise Oasis, Windmill RV Park
	S: KOA Campground (Open April 15th thru October 31st)
ATM	**S:** Buttrey Food (Buttrey Food), Cenex, Conoco
Parks	**S:** Yellowstone National Park
Other	**N:** County Market Grocery, Pamita Discount Ctr (Pharmacy), Visitor Information, Western Drugs, Yellowstone Drug
	S: Buttrey Food & Drug, Colmey Veterinary Hospital, Ranger Station, We Buy Antlers
337	I-90 Bus Loop, Livingston, Local Access
RVCamp	**N:** Camping
340	U.S. 89 North, White Sulphur Springs
343	Mission Creek Road
350	East End Access
352	Ranch Access
354	CR 563, Springdale
362	DeHart
367	Bus Loop I-90, U.S. 191, Big Timber, Harlowton
FStop	**N:** Exxon*
Gas	**N:** Conoco*
Food	**N:** Crazy Jane's Family Eatery (24Hr), Fried Cafe
Lodg	**N:** Russell Motel, ◆ Super 8 Motel
AServ	**N:** Car Quest Auto Center
RVCamp	**N:** Spring Creek Camp & Trout Ranch
ATM	**N:** Conoco, Exxon
Other	**N:** Big Timber Visitor Information, Historic Point, Victorian Village Museum
370	U.S. 191, Business 90, Big Timber, Harlowton
RVCamp	**N:** Spring Creek Camp & Trout Ranch
377	Greycliff
TStop	**S:** Truck Stop
Food	**S:** Ranch House Cooking (Truck Stop)
Lodg	**S:** Four Winds Inn
TServ	**S:** H&H Service & Repair (Truck Stop)
RVCamp	**S:** KOA Campground
Parks	**S:** Prairie Dog Town State Park & Monument
Other	**S:** Big Timber Water Slide Park
(381)	Rest Area - Phones, Picnic, RR, HF, P (Both Directions)
384	Bridger Creek Road
392	Reedpoint
Gas	**N:** Sinclair*(D)(LP)
Food	**N:** Waterhole Saloon Family Dining
Lodg	**N:** Hotel Montana
RVCamp	**N:** Cedar Hills Campground (Laundromat)
Other	**N:** Laundry Facilities (Cedar Hills Campground), US Post Office
396	Ranch Access
400	Springtime Road
408	MT 78, Absarokee, Columbus
FStop	**S:** Exxon (RV Dump)
Food	**S:** Apple Village Cafe, KFC, Lucky Lil's Casino, McDonalds, Mountain Expresso, Taco Bell
Lodg	**S:** Big Sky Motel, AAA Super 8 Motel
TServ	**S:** JC Tire
RVCamp	**N:** Mountain Range RV Park
Med	**S:** ✚ Hospital
Parks	**S:** Granite Peak Park
Other	**S:** Visitor Information
(419)	Rest Area - Picnic, RR, HF, Phones, P (Both Directions)
426	Park City
FStop	**S:** Cenex*
Gas	**S:** Kwik Stop*
Food	**S:** Restaurant (Cenex)
Lodg	**S:** Lazy Motel
433	Bus Loop I-90, West Laurel (Eastbound, Westbound Reaccess)
434	U.S. 212, U.S. 310, Laurel, Red Lodge
FStop	**N:** Exxon*
Gas	**N:** Cenex*(D)(CW)(LP), Conoco*, Exxon*
Food	**N:** Burger King, Hardee's, Little Big Men Pizza, Locomotive Inn Restaurant, Pizza Hut, Subway, Taco John's
Lodg	**N:** Best Western
AServ	**N:** Expert Lube & Wash, GM Auto Dealership, Laurel Ford, Rapid Tire
TServ	**N:** By Pass Truck Repair & Welding (2 Miles)
ATM	**N:** Conoco, Exxon Fuel Stop
Parks	**S:** Yellowstone National Park

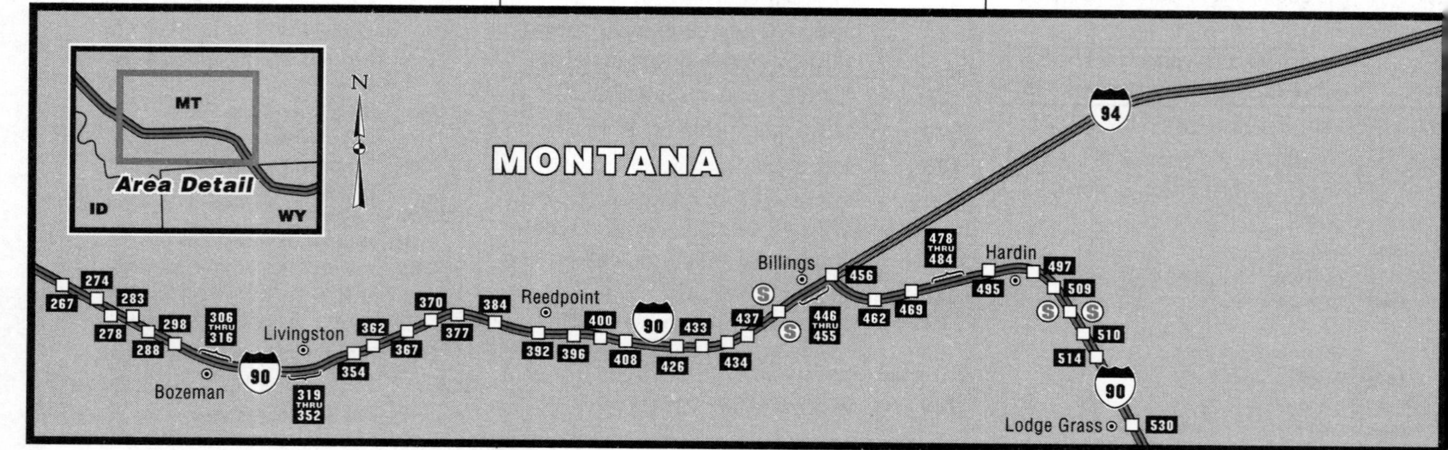

Bold red print shows RV & Bus parking available or nearby

XIT	MONTANA
Other	N: Car Wash (Expert Car & Wash), IGA Store, Western Drug
37	South Frontage Road, Bus Loop 90, E. Laurel
TStop	S: Sinclair TS[SCALES] (RV Dump)
Food	S: Chester Fried Chicken (Sinclair TS), Restaurant (Sinclair TS)
Lodg	S: Motel (Sinclair TS)
RVCamp	S: Camping
Other	S: Animal Lodge Kennels, Laurel East Veterinary Clinic
(39)	Weigh Station (Both Directions)
46	Bus Loop I-90, City Center, King Ave W., Mulloney W
Gas	N: Conoco*[CW], Holiday*[LP] S: Conoco*[D] (24Hr)
Food	N: Applebee's, Burger King, Country Harvest, Denny's, Fuddrucker's, Golden Corral, Lucky Lady Gambling Hall & Food, Olive Garden, Perkins Family Restaurant, Pizza Hut, Red Lobster S: Apple Tree Restaurant, Miyajima Garden Japanese, Silver Dollar Restaurant & Casino, Torres Cafe
Lodg	N: ◆ C'mon Inn, ◆ Comfort Inn, DAYS INN Days Inn, ◆ Fairfield, ◆ Quality Inn, Super 8 Motel S: ◆ Best Western, Clairon, Holiday Inn, Kelly Inn West, Motel 6, Ramada Limited
AServ	N: Auto Parts Store, Conoco, Costco Tire Sales, Dodge Dealer, Ford Dealership, Hyundai, Subaru Dealer, Incredible Auto Sales, NAPA Auto Parts, Wal-Mart
TServ	N: CAT Service S: D&D Truck Service (Frontage Rd)
ATM	N: Holiday, Norwest Bank S: Conoco
Other	N: Barnes & Noble, K-Mart (Pharmacy), Office Max, Optical (ShopKo), Shopko Grocery (Pharmacy & Optical), Toys "R" Us, Wal-Mart (Pharmacy) S: Big Splash Water Park
47	S Billings Blvd
Food	N: McDonalds
Lodg	N: DAYS INN ◆ Days Inn (1 Mile), ◆ Sleep Inn, ◆ Super 8 Motel (1 Mile)
TServ	S: Kenworth Dealer
Parks	N: Geyser Park Family Fun Center S: Chief Plenty Coups State Park
50	MT 3, 27th Street, City Center
Gas	N: Conoco*, Exxon, Sinclair*[CW][LP]
Food	N: Blondy's, Lucky Cuss Casino & Restaurant, Pizza Hut
Lodg	N: ◆ Howard Johnson, Sheraton, War Bonnet Inn
AServ	N: Car Quest Auto Center, Sinclair
RVCamp	S: Camping, KOA Campground
Med	N: ✚ Hospital
ATM	N: Cash Machine (Conoco)
52	U.S. 87 North, City Center, Roundup
FStop	N: Exxon* S: Cenex*
Gas	N: Conoco*
Food	N: Big Dipper Drive In, Little Caesars Pizza (Conoco), Second Shift Casino & Restaurant S: V.R. Grill
Lodg	N: Best Western (3 Miles)
TServ	N: Fruehauf
Parks	S: Pictograph Cave State Park

EXIT	MONTANA
Other	S: Chief Plenty Coups State Park
455	Johnson Lane
TStop	N: The Truck Stop (Frontage Rd) S: Flying J Travel Plaza*[LP][SCALES] (RV Dump)
Gas	S: Exxon*
Food	S: Burger King, Thads Restaurant
TServ	S: Fly-N-Lube
TWash	S: Fly-N-Lube
RVCamp	S: Tour America RV Service
ATM	S: Cash Machine (Exxon), Security Bank
456A	Jct I-94, Hardin, Sheridan
456B	Jct I-94E, Miles City, Bismarck (Westbound)
462	Pryor Creek Road
469	Arrow Creek Road
(476)	Rest Area - Picnic, Phones, RR, HF P (Both Directions)
478	Fly Creek Road
484	Toluca, Frontage Road
495	Bus Loop I-90, MT 47, City Center
FStop	N: Texaco* S: Conoco*[D], Exxon*, Sinclair*[LP] (RV Dump, Casino)
Gas	S: Cenex*
Food	N: Purple Cow Family Restaurant (Texaco) S: Dairy Queen, Far West Restaurant, McDonalds, Pizza Hut, Subway (Conoco), Taco John's
Lodg	S: Americann Inn, Lariat Motel, Super 8 Motel, ◆ Western Motel
RVCamp	S: Grandview Campground, Sunset Village RV Park
Med	S: ✚ Hospital
ATM	S: Conoco, Exxon, Sinclair
Parks	S: Bighorn Canyon Recreational Area
Other	S: Sinclair Casino
497	Bus Loop I-90, MT 313, Third Street, Hardin
Other	N: Fireworks Store S: Visitor Information, Yellowtail Dam
503	Dunmore
509	Crow Agency
Gas	N: Conoco*
Parks	S: Big Horn Canyon National Recreation Area
Other	N: Crow Laundromat, Crow Mercantile Grocery Store, U.S. Post Office
(510)	Weigh Station (Both Directions, Left Lane Exit)
510	U.S. 212 East, Little Big Horn Battlefield, Broadus
Gas	N: Sinclair*[D] S: Little Big Horn
Food	N: Little Big Horn Casino, Little Bighorn Casino, Pizza's -N- Cream (Sinclair), The Custer Battlefield Trading Post & Cafe
Lodg	S: Little Big Horn Motel
RVCamp	S: Camping
Med	N: ✚ Hospital
Other	N: Little Big Horn Nat'l Historic Site S: Coin Laundry
514	Garryowen
Gas	N: Conoco (Historic Site & Museum)
Other	N: Post Office, World's Largest Custer Museum

EXIT	MONTANA/WYOMING
530	CR 463, Lodge Grass
FStop	S: Cenex*[LP] (1 Mile)
AServ	S: Cenex (1 Mile)
Other	S: Stevinsons IGA
544	Wyola
549	Aberdeen

↑MONTANA

↓WYOMING

EXIT	MONTANA/WYOMING
1	Parkman
9	U.S. 14 West, Ranchester, Grey Bull - Lovell
Gas	S: Big Country Oil* (1 Mile), Conoco*, Texaco* (1 Mile)
Food	S: Good Earth Eatery, Kelly's Kitchen, Ranch House
Lodg	S: Western Motel (1 Mile)
AServ	S: Big Country Oil (1 Mile), Tongue River Auto
RVCamp	S: Camping (1 Mile)
ATM	S: Ranchester State Bank (1 Mile)
Parks	S: Big Horn Canyon National Recreation Area, Teton National Park, Tongue River Creative Playground, Yellowstone National Park
Other	S: Coin Laundry, Connor Battlefield, Ranchester Fire Dept, Tongue River Health Center
14	Acme Road
16	Decker Mount
(16)	Parking Area (Westbound)
20	Main Street (Wyoming Port Of Entry)
TStop	S: Exxon*
Gas	S: Cenex*, Conoco*, Gas For Less*
Food	S: Country Kitchen, Little Caesars Pizza, McDonalds, Restaurant (Exxon TS), Subway (Conoco), Trails End Motel
Lodg	S: Bramble Motel, Stage Stop Motel, Trails End Motel
AServ	S: A Towing & Repair, Bridgestone Tire & Auto
TServ	S: 10-4 Diesel Repair, Steve's Truck & RV Service
RVCamp	S: KOA Campground
Other	S: Car Wash, K-Mart (Pharmacy), Ron's[LP], Wyoming Port Of Entry
23	WY 336, Fifth St
Gas	S: Texaco*
Lodg	S: Evergreen Inn, Super Saver Inn
Med	S: ✚ Hospital
Other	N: Wyoming Welcome Center
(23)	WY Welcome Center, Rest Area - RV Dump, Playground, Pet Walk (Both Directions)
25	Bus Loop I-90, U.S. 14 East, Sheridan, Big Horn, Ucross
FStop	S: Sinclair[LP]
Gas	S: Bison Oil[LP], Holiday*[D], Texaco[D]
Food	S: Arby's, Burger King, Carl's Corner, Golden China Restaurant, Taco Bell, Taco John's
Lodg	N: ◆ Comfort Inn S: DAYS INN Days Inn, AAA Holiday Inn, Lariat Motel, AAA Mill Inn, Parkway Motel
AServ	S: Cooper Tires, Firestone Tire & Auto, GM Dealer, NAPA Auto Parts, Sheridan Motor, Toyota Dealer
RVCamp	S: Sheridan RV Park (1.1 Miles)
ATM	S: Buttery's Foods, Holiday, Sheridan State Bank

Bold red print shows RV & Bus parking available or nearby

EXIT WYOMING

Other S: Buggy Bath Car Wash, **Buttery's Food Grocery (Pharmacy), Carl's Super Store, IGA Store, National Forest Info, Wal-Mart (Pharmacy)**

(31) **Parking Area (Eastbound)**

33 Meade Creek Road, Big Horn
Other S: **Bradford Brinton Memorial**

37 Story, Prairie Dog Creek Road

(39) **Parking Area, Scenic Turnout (Westbound)**

44 U.S. 87N, Piney Creek Road, Story, Banner (Difficult Reaccess)
Other N: **Site of Fort Phil Kearny (Visitors Center & Museum)**

47 Shell Creek Road

51 Lake De Smet
RVCamp N: **Camping**

53 Rock Creek Road

56A Bus Loop I-90, Business 25, Buffalo (Eastbound, Westbound Reaccess Only)

56B I-25, U.S. 87 South, U.S. 16, Buffalo, Casper (Left Lane Exit)

58 Bus Loop I-90, U.S. 16, Buffalo, Ucross
FStop S: **Sinclair***(LP)
Gas S: Big Horn Petroleum(LP)
Food S: **Cowbow Restaurant**
Lodg S: **Bunkhouse Motel**
RVCamp S: **Big Horn Industries**(LP)
Med S: **Hospital**
ATM S: Sinclair
Other S: **Information Center, National Historic District**

(61) **Parking Area (Eastbound)**

65 Red Hills Road, Tiperary Road

(68) **Parking Area (Westbound)**

69 Dry Creek Road

73 Crazy Woman Creek Road

77 Schoonover Road

82 Indian Creek Road

88 Powder River Road
Other N: **Rest Area (Full Facilities, Pet Walk)**

(88) **Rest Area (Both Directions)**

91 Dead Horse Creek Road

102 Unnamed

106 Unnamed

110 Parking Area (Both Directions)

113 Wild Horse Creek Road

116 Force Road

124 Bus Loop I-90, U.S. 14, U.S. 16 West, Gillette
Gas N: 7-11 Convenience Store* (24Hr), Conoco*, Texaco
S: 7-11 Convenience Store*(D)
Food N: Best Western, Fireside Cafe & Lounge, Granny's Kitchen, **Hong Kong Restaurant**, Long John Silvers, Pizza Hut
Lodg N: Best Western, Motel 6, **Super 8 Motel**
AServ N: Texaco

EXIT WYOMING

S: Big Horn Tire & Brake
TServ N: **Texaco**
RVCamp N: **Camping**
Med N: **Hospital**
ATM N: Cash Machine (Conoco)
Other N: **Airport**, Car Wash, **Decker's Grocery (Pharmacy), Northside Laundromat, Western Union (Decker's Grocery)**

126 WY 59, Gillette, Douglas
TStop S: **Flying J Travel Plaza***
FStop N: **Cenex**
Gas S: Exxon*, Texaco
Food N: Daylight Donuts, McDonalds, Polar Bear Frozen Yogurt, Subway, The Prime Rib Restaurant
S: Arby's, Burger King, China Buffet, Dairy Queen, Golden Corral, Good Times, Holiday Inn, KFC, Las Margaritas Mexican Family Restaurant, Ole's Pizza, Pappa Murphy's (Pizza), Perkins Family Restaurant, Pizza Hut, **Restaurant* (Flying J Travel Plaza)**, Taco Bell, Wendy's
Lodg S: Days Inn, ◆ Holiday Inn
AServ N: C&F Repair, Plains Alignment, Plains Tire Co
S: Big O Tires, Midas, Plymouth Dodge Dealer
ATM N: Buttrey's Food, First Bank, Key Bank
S: First Interstate Bank
Parks N: **Lasting Legacy Park (Playground/Picnic)**
Other N: **Buttrey's Food & Drugs, Optical Store, Water Slide**
S: **Albertson's Grocery, Dan's Supermarket**, E-Z100 Auto Wash, EZ Auto Wash, **K-Mart (Pharmacy), Tourist Information, Wal-Mart (Pharmacy)**

128 Bus Loop I-90, U.S. 14, U.S. 16, Gillette
FStop N: **Texaco**(D)
Gas N: 7-11 Convenience Store*
Food N: **Mona's Cafe (Mexican)**
Lodg N: Arrowhead Motel, Econolodge
AServ N: Texaco
RVCamp S: **High Plains Campground**
Other S: **Port Of Entry**

129 Garner Lake Road
RVCamp S: **High Plains Campground, RV Service & Repair**

132 Wyodak Road

138 Parking Area (Both Directions)

141 Rozet

153 Bus Loop I-90, U.S. 14 East, U.S. 16, Moorcroft, Newcastle
Other N: **Rest Area (Full Facilities)**

(153) **Rest Area - RR, HF, Phones, Picnic (Both Directions)**

154 Bus Loop I-90, U.S. 14, U.S. 16 East, Moorcroft
FStop S: **Conoco***(LP)
Gas S: Texaco*
Food S: Hub Cafe
Lodg S: Cozy Motel, Moorcourt Motel
ATM S: **Cash Machine (Conoco FS)**

160 Wind Creek Road

163 Parking Area (Both Directions)

165 Pine Ridge Road

171 Parking Area (Both Directions)

172 Inyan Kara Road

177 Parking Area (Both Directions)

EXIT WYOMING/SOUTH DAKOT

178 Coal Divide Road

185 To WY 116, Sundance
Gas S: Texaco*(D)
AServ S: Texaco
Other N: **Devil's Tower National Monument**

187 WY 585, Sundance, New Castle
Gas N: Amoco, Conoco*
Food N: Flo's Place
Lodg N: Best Western
AServ N: Amoco
S: Pennzoil Lube
ATM N: Conoco
Other N: **Devil's Tower National Monument**, Hoppy Car Wash

189 U.S. 14W, Sundance (Wyoming Port Of Entry)
RVCamp N: **Camping**
Med N: **Hospital**
Other N: **Devil's Tower National Monument**
S: **Rest Area - Visitor's Information Center (Fu Facilities, RV Dump), Weigh Station**

(189) **Rest Area - RR, HF, Phones, Picnic, RV Dump (Both Directions)**

(190) **Weigh Station - Both Directions**

191 Moskee Road

199 WY 111, Aladdin

205 Beulah
FStop N: **Conoco***(D)(LP)
Other N: **Tourist Information, U.S. Post Office**

↑WYOMING

↓SOUTH DAKOTA

(1) **Rest Area - RR, HF, Phones, Picnic, RV Dump (Eastbound)**

2 McNenny State Fish Hatchery
Other N: **McNenny State Fish Hatchery**

10 U.S. 85 North, Spearfish, Belle Fourche
Lodg S: Days Inn
RVCamp S: **KOA Campground**

12 Spearfish
Gas S: Amoco*(CW), Kwik Mart*, Sinclair*, Texaco*(CW)
Food S: Arby's, Burger King, Dairy Queen, Domino's Pizza, Golden Dragon Restaurant, Lown House Restaurant, McDonalds, Pizza Hut, Stadium Sports Grill, The Millstone Family Restaurant, Wendy's
Lodg S: Best Western, Days Inn, ◆ Kelly Inn, Rancho Motel, Sherwood Lodge, Spearfish Mote
AServ S: Big A Auto Parts
Med S: **Hospital**
ATM S: Pioneer Bank
Other S: Car Wash, **Tourist Information**

14 Bus Loop I-90, U.S. 14 A, Spearfish
Gas S: Amoco*(D)
Food N: Happy Chef
S: Guadalajara Mexican, KFC, Perkins Family Restaurant
Lodg N: (AAA) Comfort Inn, ◆ Fairfield Inn, (AAA) Holida Inn
S: All Americann Inn, Best Western, Super 8 Motel
AServ S: Johnson Ford Mercury

Bold red print shows RV & Bus parking available or nearby

EXIT SOUTH DAKOTA

RVCamp S: Camping, Spearfish Mobil Homes
Other N: Black Hills National Forest
S: High Plains Heritage Center, K-Mart (Pharmacy), Spearfish Canyon Scenic Byway, Ziegler Building Ctr(LP)

U.S. 85 South, Deadwood, Lead
RVCamp S: KOA Campground
Other S: Deadwood National Historic Landmark, Winter Recreation Area

3 SD 34 West, Whitewood, Belle Fourche
FStop S: Sinclair*
Gas S: Conoco*(D)
Food S: Maggi's Diner, Pizza (Conoco), Subs (Conoco)
Lodg S: Tony's Motel
AServ S: Performance Plus
Other S: Coin Laundry (Casino)

0 Business 90, U.S. 14 A, SD 34, Sturgis, Deadwood - Lead
FStop S: Phillips 66*
Gas N: Cenex*, Conoco*, Sinclair*(D)
S: Texaco*(D) (Casino), Trailside General Store*
Food N: McDonalds, Pizza (Cenex), Subs (Cenex), Subway
S: Boulder Canyon Restaurant, Burger King
Lodg S: Canyon Inn, DAYS INN ◆ Days Inn, Super 8 Motel
AServ N: Big A Auto Parts
S: Hersrud's Chevrolet
RVCamp N: Camping
Other N: Coin Car Wash, Coin Laundry, Pamida Grocery (Pharmacy)

2 Bus Loop I-90, SD 79 North, Sturgis
Gas N: Conoco*, Exxon*, McPherson(LP)
Food N: Subway (Conoco), Taco John's
Lodg N: AAA Best Western, Star Lite Motel
AServ N: Ford Dealership, Uniroyal Tire & Auto
RVCamp N: Camping
Med N: ✚ Fort Mead VA Hospital
Other N: National Motorcycle Hall Of Fame, Northern Hills Eyecare, Southside Coin Car Wash, The Drug Store of Sturgis, Tourist Information, Wonderland Cave

4 Black Hills National Cemetery
Other S: Black Hills Nat'l Cemetery

7 Pleasant Valley Road
RVCamp S: Bull Dog Campground (.3 Miles)

38) Weigh Station (Westbound)

0 Tilford Road

EXIT SOUTH DAKOTA

(42) Rest Area - RR, HF, Phones, Picnic, RV Dump, RV Water

44 Bethlehem Road

46 Piedmont, Elk Creek Road
Gas S: Conoco*(LP)
Food N: Elk Creek Steakhouse & Lounge
S: Subway (Conoco)
Lodg N: AAA Elk Creek Resort
RVCamp N: Camping
ATM S: Conoco
Other N: Lantis Fireworks, Petrified Forest

48 Stagebarn Canyon Road
Gas S: Sinclair*, Stage Stop*
Food N: Cattleman's Club (Steaks)
S: Classics Restaurant, Mike's Pizza
AServ N: Interstate Auto (Frontage Rd)
TServ N: Northwest Peterbilt (Frontage Rd)
RVCamp S: Covered Wagon Resort (1.5 Miles)
Med S: ✚ Piedmont Medical Center
Other N: Black Hawk Marine (Boat Service)
S: Coin Laundry, Stagebarn Coin Car Wash, Stagebarn Crystal Cave, Stagebarn Dental Care

51 Bus Loop I-90, SD 79, Black Hawk Road (Difficult Reaccess)
RVCamp S: Camping

55 Deadwood Ave
TStop S: Sinclair*(D)(LP)(SCALES) (RV Dump, Casino)
Food S: Restaurant (Sinclair TS), Subway (Sinclair TS)
TServ N: Butler CAT, Diesel Machinery Inc. (Frontage Rd)
S: L&L Truck & Auto Electric, Sinclair, Terex Parts, West River Int'l Truck Service
TWash S: Superior Truck Wash
ATM S: Sinclair TS
Other S: Black Hills National Forest Information, Casino (Sinclair TS)

57 I-190, U.S. 16 West, Rapid City, Mt Rushmore (Exit Left Side)
Other S: Black Hills Nat'l Forest, Crazy Horse Monument, Mt. Rushmore, Tourist Information

58 Haines Ave
Gas N: Conoco*
S: Gas n' Snax*
Food N: Applebee's, Hardee's, La Costa Mexican Food, Royal Fork Buffet
S: Taco John's
Lodg N: AAA Sunburst Inn
AServ N: Big O Tires, Tires Plus

EXIT SOUTH DAKOTA

Med S: ✚ Hospital, ✚ Rapid Care Medical Center
ATM N: Norwest Bank
Other S: Area Dental Associates (& Pharmacy/ Optical), Office Max, Shopko Grocery (Pharmacy & Optical)

59 LaCrosse Street
Gas N: Amoco*(CW), Phillips 66*, Texaco
S: Cenex*(CW), Exxon*
Food N: A & W Drive-In (Phillips 66), Fuddrucker's (Frontage Rd), Happy Chef, Howard Johnson, Red Lobster (Frontage Rd)
S: Golden Corral, McDonalds (Wal-Mart), Millstone Family Dining, Perkins Family Restaurant
Lodg N: ◆ Econolodge, AAA Howard Johnson, ◆ Super 8 Motel
S: AmericInn Inn, ◆ Comfort Inn, DAYS INN ◆ Days Inn, AAA Fair Value Inn, Foothills Inn, AAA Motel 6, Quality Inn, ◆ Ramada, AAA Rushmore Motel, AAA Thrifty Motor Inn, ◆ Travelodge
AServ N: Sears
S: Wal-Mart
Other N: Rushmore Mall, Target Department Store, Toys "R" Us
S: Optical (Wal-Mart), Pirate's Cove Mini Golf, Wal-Mart (Pharmacy & Optical)

60 Bus Loop I-90, SD 79, North Street, Mount Rushmore (Westbound Exits Left Side Of Highway, Eastbound Reaccess)
AServ S: Ed's Towing & Repair, Mid-America
TServ S: Ed's Towing & Repair
RVCamp S: Berry Patch Campground

61 St Patrick Street
TStop N: Conoco*(LP)(SCALES) (RV Dump)
Food N: Restaurant (Conoco)
TServ N: Conoco
S: Freuhauf Truck Services
TWash N: Conoco
RVCamp N: The Longland Trailer Service
S: KOA Campground

63 Ellsworth A.F.B Commercial Entrance, Box Elder (Eastbound, Reaccess Westbound Only)

66 Ellsworth A.F.B Main Entrance, Box Elder
FStop N: Texaco(CW)
Gas N: Texaco*
Food N: KFC, McDonalds, Pizza Hut, Taco Bell, Taco John's

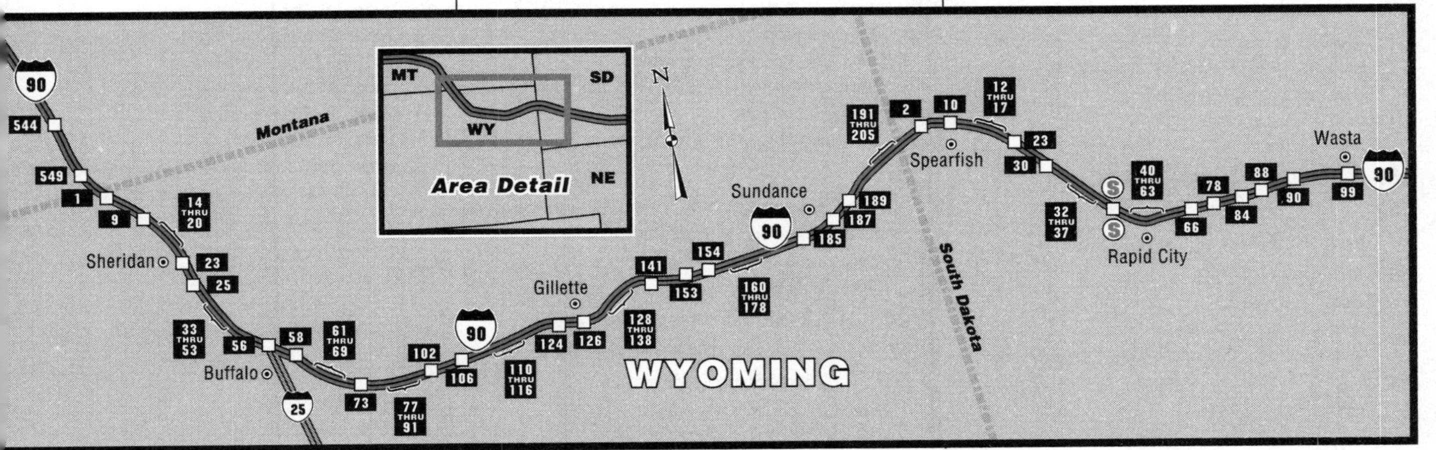

EXIT SOUTH DAKOTA

AServ **S:** Interstate Engine Service
TServ **S:** Interstate Engine Service
Other **N:** Anderson Dental, Fireworks, South Dakota Air & Space Museum

(69) Parking Area (Both Directions)

78 New Underwood
Gas **S:** Frontier*
Food **S:** Diamond Cafe Family Dining, Restaurant (Frontier)
Lodg **S:** Motel

84 CR 497
Other **N:** Old Glory Fireworks

88 CR 473

90 Unnamed

99 Wasta
Lodg **N:** Redwood Motel
AServ **N:** Watsu Tire
RVCamp **N:** Camping (1 Mile, Laundromat)
Other **N:** U.S. Post Office

(99) Rest Area - RR, HF, Phones, Picnic, RV Dump, Tourist Info (Both Directions, Highway Patrol Station)

101 CR T - 504, Jensen Road

107 Cedar Butte Road

109 Bus Loop I-90, Wall

110 Bus Loop I-90, SD 240, Badlands Loop, Wall
Gas **N:** Amoco*(D)(CW), Conoco*, Exxon*, Texaco
Food **N:** Dairy Queen, Elkton House Restaurant, Subway, Wall Drug Restaurant
Lodg **N:** Ann's Motel, Best Western, ◆ Days Inn, Elk Motel, Hillcrest Motel, Kitchen Coast Motel, Super 8 Motel, The Sands Motor Inn, Wall Motel, Welsh Motel
AServ **N:** Badlands Automotive, Big A
RVCamp **N:** Arrow Campground
ATM **N:** Conoco
Parks **S:** Badlands National Park (7 Miles)
Other **N:** National Grasslands Visitor Center, Tourist Information, Wall Drug, Wall Factory Outlet, Wild West Wax Museum
S: Fireworks

112 Unnamed

116 Unnamed

121 Unnamed

127 Unnamed

131 Badlands Loop, SD 240, Badlands Interior
Gas **S:** Amoco*
Food **S:** Circle 10 Cafe (Circle 10 Motel)
Lodg **S:** Circle 10 Motel
RVCamp **S:** Circle 10 Campground
Other **S:** Badlands Trading Post, Prairie Homestead

138 Scenic Overlook (Westbound)

143 SD 73 North, Philip
Med **N:** ✚ Hospital (15 Miles)

150 SD 73 South, Kadoka
FStop **S:** Conoco*
Gas **S:** Amoco*, Texaco*
Food **S:** Happy Chef, Pizza (Conoco), Restaurant (Texaco), Subs (Conoco)
Lodg **N:** Kadoka Inn

EXIT SOUTH DAKOTA

S: Best Western, Sundowner Motor Inn (Budget Host), Super 8 Motel
RVCamp **S:** Camping
ATM **S:** Conoco

152 Bus Loop I-90, Kadoka, South Creek Road
TStop **N:** Burns Bros Travel Stop*
Food **N:** Restaurant* (Burns Bros Travel Stop)
Lodg **S:** Best Western (1.6 Miles), Budget Host Motel (2 Miles)
TServ **N:** Burns Bros Travel Stop*

163 SD 63 South, Belvidere
FStop **S:** Amoco*, Phillips 66*
Food **S:** Cafe

(165) Rest Area - RR, HF, Phones, Picnic, RV Dump (Eastbound, Closed Due To Construction)

(166) Rest Area - RR, HF, Phones, Picnic, RV Dump (Westbound, Closed Due To Construction)

170 SD 63 North, Midland
FStop **N:** Texaco*
Food **N:** 50's Train Diner
RVCamp **N:** KOA Campground (.5 Miles)
Other **N:** South Dakota's Original 1880 Town (Props From "Dances With Wolves")

172 Unnamed

177 Unnamed

183 Okaton
Gas **S:** West Lakes Ghost Town* (Secnic Overlook, Rock Store)
Other **S:** Petting Farm, Senic Overlook (West Lakes Ghost Town)

188 Parking Area (Both Directions)

191 Bus Loop 90, Murdo

192 Bus Loop I-90, U.S. 83 South, Murdo, White River
TStop **N:** Texaco*(LP)
FStop **N:** Sinclair*(D)
Gas **N:** Amoco*, Kwik Mart*, Phillips 66*
Food **N:** Homemade Ice Cream, KFC, Murdo Drive-In, Restaurant (Texaco), Star Restaurant, Sub Station, The Teepee Restaurant, Triple H Restaurant (24Hr)
Lodg **N:** Anchor Inn, Best Western, Chucks Motel, Country Inn, Country Inn, Hospitality Inn, Lee Motel, Sioux Motel, Super 8 Motel, The Teepee Motel (Campground)
AServ **N:** Phillips 66, Sinclair
TServ **N:** Sinclair, Texaco
RVCamp **N:** Camping, Teepee Motel
Other **N:** Pioneer Auto Museum, Super Value Grocery

(194) Parking Area (Both Directions)

201 Draper
FStop **N:** Total*
Food **N:** Restaurant (Total)
AServ **N:** Total
TServ **N:** Total

208 Unnamed

212 U.S. 83 North, SD 53, Fort Pierre, Pierre State Capitol
FStop **N:** Phillips 66*(LP) (Casino)

EXIT SOUTH DAKO

Food **N:** Vivian Jct Restaurant
Med **N:** ✚ Hospital (34 Miles)
ATM **N:** Phillips 66
Other **N:** Casino (Phillips 66), Cultural Heritage Ctr (33 Miles), OAHE Dam (40 Miles)

214 Vivian

(218) Rest Area - RR, HF, Phones, Picni RV Dump, Pet Walk (Eastbound)

220 Exit to unpaved highway

(221) Rest Area - RR, HF, Phones, Picni RV Dump, Tourist Info (Westboun

225 Bus Loop I-90, Presho
FStop **N:** Conoco*, Standard*(LP)
Gas **N:** Texaco*
Food **N:** Hutch's Motel
Lodg **N:** Coachlight Inn, Hutch's Motel
AServ **N:** National Tire
RVCamp **N:** Camping
Other **N:** Pioneer Museum

226 Bus Loop I-90, U.S. 183, Winner, Presho

235 SD 273, Kennebec
FStop **N:** Conoco*(LP)
Gas **N:** King's
Lodg **N:** Budget Host Motel, Gerry's Motel, King's Motel
AServ **N:** Conoco Fuel Stop, King's
RVCamp **N:** KOA Campground (.4 Miles)

241 Unnamed

248 SD 47 North, Reliance
FStop **N:** Standard
AServ **N:** Standard
Parks **N:** Big Bend Dam Recreation Area

251 SD 47 South, Gregory, Winner

260 Oacoma
FStop **N:** Amoco*(CW)(LP), Conoco*
Food **N:** Al's Oasis Restaurant, Burger King (Amoc Taco John's (Conoco)
Lodg **N:** Comfort Inn, ◆ Days Inn, Econolodg Oasis Inn
TServ **N:** Conoco
RVCamp **N:** The Familyland Campground(LP) (Showers
ATM **N:** Cash Machine (Al's Oasis)
Other **N:** Cedar Boat Ramp & Recreation Area (3 Miles), Old West Museum, Wildlife Adventur Museum
S: Coin Laundry

263 Chamberlain
FStop **N:** Sinclair*(LP)
Food **N:** A & W Drive-In, Casey's Cafe, Pizza Hut, Tac John's
Lodg **N:** Bel-Air Motel, Super 8 Motel
AServ **N:** Sinclair FS
TServ **N:** Sinclair FS
Other **N:** Casey's Drugs, Crow/Creek & Sioux Tribal Headquarters (20 Miles), Super Value (Mini Grocery Store), The American Creek Recreation Center (2 Miles)

(264) Scenic Overlook, Rest Area - RR, HF, Phones, Picnic, RV Dump , Grills, Tourist Info (Both Directions)

265 Bus Loop I-90, SD 50, Chamberlain
FStop **S:** Conoco*
Gas **N:** Amoco*

Bold red print shows RV & Bus parking available or nearby

XIT SOUTH DAKOTA

Food N: Dairy Queen
AServ N: Amoco, Gary's Body Shop
TServ N: A&R Truck Service (24Hr)
RVCamp N: Happy Campers Campground
S: KOA Campground
Med N: Hospital
Other N: Mid River Veterinary Clinic, St Joseph Akta LaKota (4 Miles)

72 SD 50, Pukwana
Parks S: Platte Creek Recreation Area (32 Miles), Snake Creek Recreation Area (38 Miles)

84 SD 45 North, Kimball
FStop N: Phillips 66*
Gas N: Amoco*(CW)
Food N: Doo-Wah-Ditty's Diner, Frosty King Drive In, Restaurant (Phillips 66)
Lodg N: ◆ Super 8 Motel, Travelers Motel
S: Kimball Motel
AServ N: Amoco
RVCamp N: Camping

89 SD 45 South, Platte

293) Parking Area (Both Directions)

96 White Lake
FStop N: Cenex*(D)(LP) (24 Hr Automated Fueling)
Gas N: Texaco*
Food N: Sportsman's Pub
Lodg N: White Lake Motel
S: Motel
AServ N: Cenex
TServ N: Cenex
RVCamp S: Camping
Other N: Frontier Foods Grocery

301) Rest Area - RR, HF, Phones, Picnic, RV Dump (Both Directions)

08 Bus Loop I-90, Plankinton
FStop N: Cenex*(LP), Phillips 66*
Gas N: AJ's Mini Mart*
Food N: Cafe (Phillips 66)
Lodg N: I-90 Motel, ◆ Super 8 Motel
AServ N: AJ's, Cenex, Exhaust Master, Skip's Auto Service
RVCamp N: Gordy's Camping (Laundry)
Other N: Laundry Facilities (Gordy's Camping)

10 U.S. 281, Stickney, Aberdeen
FStop S: Conoco*
Other S: Fort Randall Dam (59 Miles)

19 Mt Vernon

25 Betts Road

EXIT SOUTH DAKOTA

RVCamp S: Camping

330 Bus Loop I-90, Mitchell, Corn Palace
Gas N: Texaco*
Food N: Brandy's (Holiday Inn), Dairy Queen, Happy Chef Restaurant, Shakes-N-Stuff
Lodg N: AAA Anthony Motel, Coach Light Motel, AAA Econolodge, ◆ Holiday Inn (see our ad this page), Motel 6, Siesta Motel, Wheel Inn Motel
RVCamp N: Goldies Shady Acres Campground, Jack's Campers (RV Service)
S: Dakota Campground
Med N: Hospital (3 Miles)
Other N: Balloon & Airship Museum, County Fair Food Store, Enchanted World Doll Museum, Friend of the Middle Border Museum, Prehistoric Indian Village Museum, Wayne & Mary Nutrition Ctr, Weigh Station

332 Bus Loop I-90, SD 37, Mitchell, Parkston
TStop N: Texaco*(SCALES) (RV Dump)
FStop N: Cenex*
Gas N: Amoco* (24Hr), Phillips 66*, Texaco*
Food N: Arby's, Bonanza Steak House, Burger King (Amoco), Country Kitchen, Domino's Pizza, Happy Chef, Hardee's, Hot Stuff Pizza (Cenex), McDonalds, Smash Hit Subs, Subway (TS), Truck Haven Restaurant (Texaco TS), Twin Dragon Chinese
Lodg N: AmericInn Motel, Best Western, Chief Motel, ◆ Comfort Inn, Corn Palace Motel, DAYS INN ◆ Days Inn, Super 8 Motel, Thunderbird Motel
AServ N: Dale's A-1 Transmission, Pennzoil Lube, Texaco

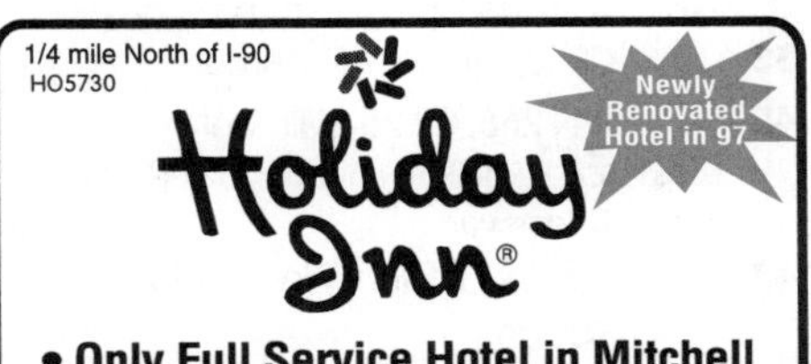

EXIT SOUTH DAKOTA

TServ N: Texaco
S: Trail King Truck Service
TWash N: Texaco, Truck/Car Wash
RVCamp N: R&R Campground
Med N: Hospital (2 Miles)
ATM N: Cenex
Other N: Car Wash, Friends Of Middle Border Museum, K-Mart (& Pharmacy)

335 Riverside Road
RVCamp N: KOA Campground
Other N: Fireworks

(336) Parking Area (Westbound)

344 SD 262 East, Fulton, Alexandria
FStop S: Cenex*, Texaco*(CW)
Food S: Restaurant (Texaco FS)
AServ S: Cenex

350 SD 25, Emery, Farmer
Other N: De Smet - Home of Laura Ingalls Wilder (55 Miles)

353 Spencer, Emery
TStop S: Burns Bros Travel Stop*(LP)
Food S: Deli Mart Restaurant* (Burns Bros TS), Subway* (Burns Bros TS)

357 Bridgewater, Canova

(362) Rest Area - RR, HF, Phones, Picnic, Highway Patrol, RV Dump (Both Directions)

364 U.S. 81, Salem, Yankton
Food N: Edith's Cafe
RVCamp N: Camp America

368 Canistota

374 Montrose
RVCamp S: Camping (5.5 Miles)
Parks S: Lake Vermillion Recreation Area (5 Miles)
Other N: Montrose Veterinary Clinic

379 SD 19, Humboldt, Madison

387 Hartford
FStop N: Phillips 66*
AServ N: Phillips 66 FS
TServ N: Phillips 66 Fuel Stop

390 SD 38, Buffalo Ridge, Hartford
Gas N: Phillips 66 (2.9 Miles)
S: Buffalo Ridge*
Food S: Coffee Shop (Buffalo Ridge FS)
RVCamp N: The Camp Dakota Campground
Other S: Buffalo Ridge Ghost Town

396 Jct I-29

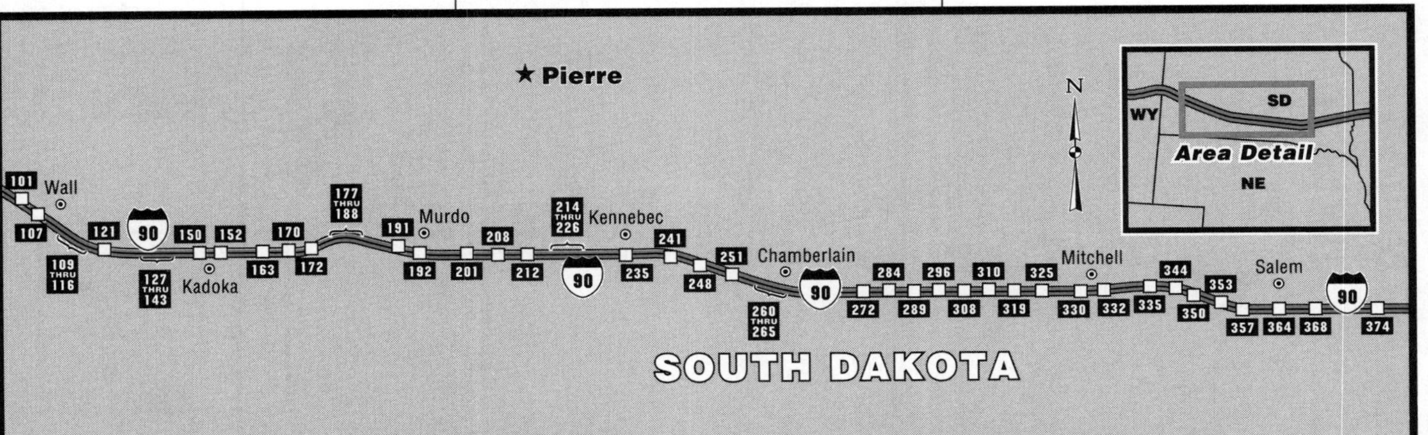

Bold red print shows RV & Bus parking available or nearby

EXIT	SOUTHDAKOTA/MINNESOTA
399	Bus Loop I-90, SD 115, Cliff Ave
TStop	**N:** **Sinclair*(SCALES) (CB Shop)**
	S: **Pilot Travel Center*(SCALES)**
Gas	**N:** Total*(D)
	S: Holiday*(D), Phillips 66*(D)(CW)
Food	**N:** **Cody's Restaurant (Sinclair TS)**
	S: Arby's, Burger King, Perkins Family Restaurant, **Restaurant (Pilot Travel Center), Subway (24 Hr, Pilot Travel Center), The Grain Bin & Lucky Horseshoe**
Lodg	**S:** **Cloud 9 Motel**, AAA Comfort Inn, DAYS INN ◆ **Days Inn**, ◆ **Super 8 Motel**
AServ	**S:** Goodyear Tire & Auto
TServ	**S:** **American Rim & Brake, Dakota Volvo, Dorsey Trailers, Goodyear Tire & Auto, Kenworth Dealer, Peterbilt Dealer, Pilot Travel Center, Precision Industries**
RVCamp	**N:** **KOA Campground (.4 Miles), Spader Camper Center (Frontage Rd)**
Med	**N:** ✚ **Hospital**
ATM	**S:** Dakota State Bank, Holiday, Phillips 66, Pilot Travel Center
400	Jct I-229
402	EROS Data Center
RVCamp	**N:** **Camping**
Other	**N:** **Fireworks**
406	SD 11, Corsan, Brandon
Gas	**S:** Amoco* (24Hr, Casino), Ampride*, Texaco
Food	**S:** Dairy Queen, Hardee's, Subway
Lodg	**S:** **Holiday Inn Express**
AServ	**S:** Brandon Motors, Exhaust & Lube
Parks	**S:** **Big Sioux Recreation Center (4 Miles)**
Other	**S:** **Casino (Amoco), Jubilee Foods Grocery (Pharmacy)**, Super Wash Car Wash
410	Valley Springs, Garretson
Parks	**S:** **Beaver Creek Nature Area (6 miles south), Palisades State Park (7 Miles)**
(412)	**SD Welcome Center, Weigh Station - Full Facilities, Tourist Info, RV Dump (Westbound)**

↑SOUTH DAKOTA

↓MINNESOTA

EXIT	MINNESOTA
(1)	MN Welcome Center - Phones, Picnics, HF, RR (Northbound)
1	MN 23, CR 17, Jasper, Pipestone
Other	**N:** **Pipestone National Monument (30 Miles)**
3	CR 4, Beaver Creek
5	CR 6, Beaver Creek, Hills
Gas	**N:** Texaco(D)
12	U.S. 75, Luverne, Rock Rapids
Gas	**N:** Amoco*, Casey's General Store*, Cenex*, Co-Op(D)(LP), Ferrell Gas(LP), Phillips 66
Food	**N:** Country Kitchen, Hardee's, Homemade Donuts (Casey's General Store), JJ's Tasty Drive In, McDonalds, Pizza Hut, Scotty's Bar & Grill, Subway, Taco John's
	S: **Magnolia Steak House**
Lodg	**N:** Comfort Inn, Hillcrest Motel
	S: **Super 8 Motel**
AServ	**N:** Chrysler Auto Dealer, Exhaust Pros
RVCamp	**N:** **Camping**
	S: **Camping**
Med	**N:** ✚ **Hospital**
Parks	**N:** **Blue Mounds State Park**
Other	**N:** Coin Car Wash, **Jubilee Foods Grocery, Pipestone Nat'l Monument, Tru-Value Hdwe**
18	CR 3, Magnolia, Kanaranzi
(24)	**Rest Area - Phones, Picnics, HF, RR, Pet Walk, Playground (Both Directions)**
26	MN 91, Lake Wilson, Adrian, Ellsworth
FStop	**S:** **Cenex*(LP) (24Hr)**
Gas	**S:** Amoco*
Food	**S:** Restaurant (Amoco)
AServ	**S:** Cenex
TServ	**S:** **Cenex**
RVCamp	**S:** **Adrian Campground (Laundry Facilities)**
Other	**S:** **Laundry Facilities (Adrian Campground)**
33	CR 13, Wilmont, Rushmore
42	MN 266, CR 25, Wilmont
Lodg	**S:** ◆ Budget Host Inn, **Super 8 Motel**
Other	**S:** **Camping**
43	U.S. 59, Slayton, Worthington, Flayton
FStop	**S:** **Phillips 66*(D)(CW)**
Gas	**S:** Cenex*(D), Conoco, Texaco*(D), Total*
Food	**S:** Dairy Queen, Hardee's, KFC, McDonalds, Perkins Family Restaurant, Pizza Hut, Ruttles 50's Grill, Subway, Taco Bell (Cenex), Taco John's
Lodg	**S:** Best Western, Budget Inn, ◆ Ramada Inn
AServ	**S:** Car Quest Auto Center, Conoco, GM Auto Dealership, Joe's Exhaust Pros, NAPA Auto Parts, RBS Transmission
RVCamp	**S:** **Camping**

EXIT	MINNESOT
Med	**S:** ✚ **Hospital**
ATM	**S:** Cenex, First State Bank
Other	**S:** Car Wash, **Colonial Coin Laundry, EconoFood, Optical (Shopko), Shopko Groce (Pharmacy & Optical)**
45	MN 60, Windom
FStop	**S:** **Texaco*(SCALES)**
Food	**S:** **Hot Stuff Pizza (Texaco)**, Smash Hit Subs
Lodg	**S:** AAA Best Western
(46)	**Weigh Station (Eastbound)**
47	CR 53
50	MN 284, CR 1, Brewster, Round Lak
RVCamp	**S:** **Camping**
57	CR 9, Heron Lake
64	MN 86, Lakefield
RVCamp	**N:** **Camping**
	S: **Camping**
Parks	**N:** **Kilen Woods State Park (12 Miles)**
(69)	**Rest Area - Phones, Picnics, RR, H (Eastbound)**
(73)	**Rest Area - Phones, Picnics, RR, HF, Pet Walk, Playground (Westbound)**
73	U.S. 71, Jackson, Windom
FStop	**N:** **Conoco***
Gas	**S:** Amoco*
Food	**N:** **Best Western, Hardee's**
	S: Burger King (Amoco), Santa Fe Crossing Mot
Lodg	**N:** **Best Western**
	S: Santa Fe Crossing Motel
AServ	**N:** Goodyear Tire & Auto
	S: GM Auto Dealership
TServ	**N:** **Goodyear Tire & Auto**
RVCamp	**N:** **Camping, KOA Campground (May thru September)**
Med	**S:** ✚ **Hospital**
ATM	**N:** **Conoco Fuel Stop (Conoco FS)**
Parks	**N:** **Kilen Woods State Park**
80	CR 29, Alpha
87	MN 4, St James, Sherburn
Gas	**S:** Texaco*
Food	**S:** Homemade Pizza (Texaco), Ma Faber's Hom Cookin
RVCamp	**N:** **Caverns Landing Campground, Everett Campground (5 Miles)**
93	MN 263, CR 27, Welcome, Caylon
Gas	**S:** Welcome Campground
RVCamp	**S:** **Welcome Campground (Laundry Services,**

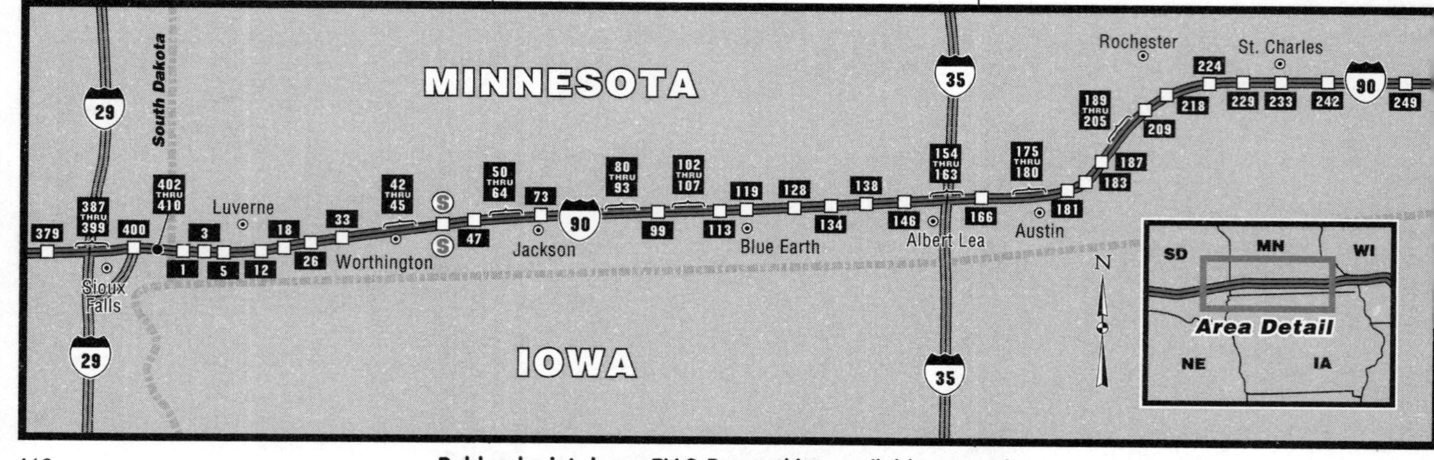

Bold red print shows RV & Bus parking available or nearby

XIT	MINNESOTA
	Gas)
Other	S: Laundry (Welcome Campground)
9	CR 39, Fairmont
02	Bus Loop I-90, MN 15, Madelia, Fairmont
FStop	S: Cenex*, Super America*
Gas	S: Phillips 66*(CW)
Food	S: China Restaurant, Happy Chef, McDonalds, Perkins Family Restaurant (24 Hr), Pizza Hut, Taco Bell (Cenex)
Lodg	S: ◆ Comfort Inn, ◆ Holiday Inn, ◆ Super 8 Motel
AServ	S: Car Quest Auto Center, GM Auto Dealership
TServ	N: Truck Service S: Chesley Freightliner
TWash	N: Truck Wash
RVCamp	S: Camping
Med	S: ✚ Hospital
Parks	S: Lincoln Park
Other	S: K-Mart
07	MN 262, CR 53, Granada, East Chain
RVCamp	S: Flying Goose Camp Ground (1 Mile, Open May to Oct)
13	CR 1, Huntley, Guckeen
19)	Rest Area - Phones, Picnics, RR, HF, Pet Walk (Both Directions)
19	U.S. 169, Blue Earth
FStop	S: Sinclair*(LP)
Gas	S: Texaco*
Food	S: Country Kitchen, Dairy Queen, Hardee's, Hot Stuff Pizza (Texaco), McDonalds, Pizza Hut, Subway, Taco Bell (Texaco)
Lodg	S: ◆ AmericInn Motel, ◆ Super 8 Motel
AServ	S: Sinclair FS
TServ	S: Sinclair FS
RVCamp	S: Camping
Med	S: ✚ Hospital
ATM	S: Texaco
Other	S: Super Wash Automatic & Coin Car Wash, Wal-Mart (Pharmacy)
28	MN 254, CR 17, Easton, Frost
34	MN 253, CR 21, Minnesota Lake, Bricelyn
38	MN 22, Wells, Kiester
RVCamp	S: Camping
46	MN 109, CR 6, Alden, Mankato
FStop	S: Amoco*
Gas	S: Cenex*(LP)
Food	N: Main St Grill (Steaks) S: Club 90 Restaurant
AServ	S: Alden Truck & Auto Repair
TServ	S: Alden Truck & Auto Repair, Amoco Fuel Stop
Parks	N: Moran Park (Swimming)
Other	N: Coin Car Wash, U.S. Post Office
54	MN 13, U.S. 69, Waseca, Albert Lea
FStop	N: Super America*
57	CR 22, Albert Lea
FStop	S: Citgo*(CW)(LP)
Gas	S: Conoco*(CW)
Food	S: Deli & Cafe (Hy-Vee Grocery), McDonalds, Subway* (Citgo FS)
Lodg	S: Americ Inn
AServ	S: Indy Lube
Med	S: ✚ Hospital

EXIT	MINNESOTA
ATM	S: Citgo Fuel Stop (Conoco), Conoco
Other	S: Coin Operated Car Wash, Hy-Vee Grocery (24 Hrs), Northbridge Mall, Optical (Shopko), Shopko Grocery (Pharmacy & Optical)
159AB	Jct I-35, Twin Cities, Albert Lea
(162)	Rest Area - RR, Picnic, Phones, Vending (Eastbound)
163	CR 26, Hayward
FStop	S: Amoco
Food	S: Netts Diner (Amoco FS)
AServ	S: Auto Service (Amoco FS)
RVCamp	S: Camping
Parks	S: Myre-Big Island State Park
166	CR 46, Oakland Road
RVCamp	N: KOA Campground
(171)	Rest Area - RR, Phones, Picnic, Vending (Westbound)
175	MN 105, CR 46, Oakland Road
FStop	N: Phillips 66*(LP) S: Conoco*
Gas	S: Amoco*(CW)
Food	N: Restaurant (Phillips 66 FS), Sports Family Dining
Lodg	N: AAA Rodeway Inn
AServ	S: Amoco, Austin Ford Mercury
177	U.S. 218, 14th St NW, Owatonna
FStop	S: Total*
Food	N: KFC S: Hardee's
Lodg	S: Super 8 Motel
AServ	N: Indy Lube
Other	N: K-Mart (Pharmacy), Oak Park Mall
178A	4th St NW
Gas	S: Amoco*(CW)(LP), Conoco*(D)(CW)(LP), Sinclair(D)
Food	N: Holiday Inn, Perkins Family Restaurant S: Burger King
Lodg	N: DAYS INN ◆ Days Inn, ◆ Holiday Inn
Other	N: I-Vee Food Store
178B	6th St NE Downtown, Austin
179	11th St NE
TStop	N: Citgo*(LP)
Food	N: Citgo Truck Stop
TServ	N: Citgo Truck Stop
180B	U.S. 218S, 21st St NE, Lyle
Gas	S: Texaco* (ATM)
Lodg	S: Austin Motel
181	28th St NE
183	MN 56, Brownsdale, Rose Creek
FStop	S: Cenex*(LP)
AServ	S: Cenex FS
187	CR 20
RVCamp	S: Beaver Trails Campground
189	CR13, Elkton
193	MN 16, Dexter, Preston
TStop	S: Amoco*
Food	S: Wind Mill Restaurant (Amoco TS)
Lodg	S: Mill Inn
ATM	S: Cash Machine (Amoco TS)
(202)	Rest Area - RR, Phones, Picnic, Vending (Eastbound)
205	CR 6
209AB	U.S. 63, MN 30, Rochester,

EXIT	MINNESOTA/WISCONSIN
	Stewartville
TStop	S: Texaco
Food	S: Restaurant (Texaco TS)
TServ	S: Texaco Truck Stop
TWash	S: Texaco Truck Stop
RVCamp	S: Camping
ATM	S: Cash Machine (Texaco TS)
218	U.S. 52, Chatfield
RVCamp	N: KOA Campground
(222)	Rest Area - RR, Phones, Picnic, Vending (Westbound)
224	CR 7, Eyota
229	CR 10, Dover
233	MN 74, Chatfield, St Charles
TStop	S: Texaco*(LP) (RV Dump)
Food	S: The Amish Oven
242	CR 29, Lewiston
(244)	Rest Area - RR, Phones, Picnic, Vending (Eastbound)
249	MN 43, Rushford
TServ	N: Peterbilt Dealer
252	MN 43, Winona
Med	N: ✚ Hospital
257	MN 76, Houston
RVCamp	S: Camping
266	CR12, Nodine
TStop	S: Amoco*(SCALES)
Food	S: Subway (Amoco TS)
TServ	S: Amoco Truck Stop
TWash	S: Amoco
Parks	N: O. L. Kipp State Park
270	U.S. 61N, Dakota
272A	Dresbach
Gas	S: Mobil*
AServ	S: Mobil
272B	Dresbach
Gas	S: Mobil(D)
AServ	S: Mobil
275	U.S. 14, U.S. 61, La Crescent, La Crosse
Med	N: ✚ Hospital
Other	N: Minnesota Welcome Center (Info, RR, vending, picnic, phones)

↑MINNESOTA

↓WISCONSIN

EXIT	WISCONSIN
(1)	WI Welcome Center - RR, Phones, Vending, Picnic, RV Dump
2	CR B, French Island
Gas	S: Citgo*(D)
Lodg	S: DAYS INN AAA Days Inn
3AB	U.S. 53, WI 35, La Crosse, Onalaska
Gas	S: Amoco*(CW), QuikTrip*, Super America*(D)
Food	S: Burger King, Chee Peng Palace Chinese, Embers Restaurant & Bakery, KFC, McDonalds, Pizza Hut, Ponderosa
Lodg	S: Excel Inn, Hampton Inn, Road Star Inn
AServ	S: Amoco
ATM	S: First Federal
Other	S: Bridgeview Car Wash
4	U.S. 53N, WI 157, to WI 16, La Crosse,

EXIT WISCONSIN

Onalaska
- **Gas** S: QuikTrip*(D)(LP)
- **Food** S: Bakers Square Restaurant, Burger King, Rocky Rococo Pizza, Shakeys Pizza, Taco Bell
- **Lodg** S: ◆ Comfort Inn
- **AServ** S: Tires Plus, Zzip Lube
- **ATM** S: First Bank
- **Other** S: Full Serve & Coin Car Wash, Sam's Club, Wal-Mart (Pharmacy & Optical)

5 WI 16, Onalaska
- **Gas** N: Woodman's
- **Food** S: Applebees, ChiChi's Restaurant, Chuck E Cheese Pizza, Fazoli's Italian Food, McDonalds, Old Country Buffet, Olive Garden, Subway
- **Lodg** N: ◆ Budgetel Inn
  S: Holiday Inn Express
- **AServ** N: Woodman's
- **Other** N: Woodman's Supermarket
  S: Barnes & Noble, Best Buy Discount, Shopko Grocery, Target Department Store, Valley View Shopping Mall

(10) Weigh Station (Eastbound)

12 CR C, West Salem
- **RVCamp** N: Camping, Coulee Region RV Service

15 WI 162, Bangor, Coon Valley
- **TServ** S: Wehrs Used Trucks

(20) Rest Area - RR, Phones, Picnic, Vending (Eastbound)

(22) Rest Area - RR, Phones, Picnic, Vending (Westbound)

25 WI 27, Sparta, Melvina
- **Gas** N: Citgo*
- **Food** N: Happy Chef
- **Lodg** N: (AAA) Country Inn, (AAA) Super 8 Motel

28 WI 16, Sparta

41 WI 131, Tomah, Wilton
- **Lodg** N: (AAA) HoJo Inn
- **Med** N: ✚ VA Medical Center
- **Other** S: State Patrol Post

43 U.S. 12, WI 16, Tomah
- **Lodg** N: Rest Well Motel

45A Junction I-94

45B Jct I-94, St Paul

(48) Weigh Station (Westbound)

49 CR PP, Oakdale
- **FStop** N: Speedway*(LP)
- **Gas** S: Mobil*(LP)

EXIT WISCONSIN

- **RVCamp** N: Kamp Dakota

(55) Rest Area - RR, Phones, Picnic, Grills, No Running Water (North At Exit 55)

55 CR C, Camp Douglas, Volkfield
- **FStop** S: Amoco*
- **Gas** S: Mobil*(LP)
- **Food** S: Mobil, Subway (Mobil), Target Bluff Restaurant
- **Lodg** S: Walsh's K&K Motel
- **ATM** S: Mobil

61 WI 80, Necedah, New Lisbon
- **TStop** N: Citgo, New Lisbon 76 Auto/Truck Stop*(SCALES)
- **Food** N: 76, Grandma's Kitchen, Subway
- **Lodg** N: Edge of the Wood Motel, Rafter's Motor Inn
- **TServ** N: 76 Auto/Truck Plaza

69 WI 82, Mauston, Necedah
- **FStop** N: Cennex Travel Mart*(D)
  S: Kwik Trip - Amoco*(D)
- **Gas** N: Phillips 66*, Shell(D)
  S: Citgo*
- **Food** N: Country Kitchen, Phillips 66, The Family Restaurant
  S: Culver's Frozen Custard & Burgers, Garden Valley Restaurant, Hardee's, McDonalds, Roman Castle Italian, The Alaskan Restaurant
- **Lodg** N: (AAA) Park Oasis Inn, Super 8 Motel
  S: The Alaskan Motel
- **TServ** S: Interstate Truck Repair
- **ATM** S: Bank of Mauston
- **Other** S: Car Wash, Cheese Mart

(76) Rest Area - RR, Picnic, Vending,

EXIT WISCONSIN

Phones, RV Dump, Truck Parking (Both Directions)

79 CR HH, Lyndon Station
- **FStop** S: Shell*
- **Food** S: Shell
- **RVCamp** N: Camping, Dreamfield RV Sales & Service

85 U.S. 12, WI 16, Wisconsin Dells
- **Lodg** S: (AAA) Day's End Motel
- **TServ** S: G & S Inc
- **RVCamp** N: Camping
  S: Camping
- **Parks** S: Rocky Arbor State Park

87 WI 13 North, Wisconsin Dells
- **FStop** N: Amoco(D) (RV Dump)
- **Gas** N: Citgo, Mobil*, Shell*
- **Food** N: Burger King, Country Kitchen, Denny's, Dino' Dinosaur Buffet Pizza, Gia's Cuisine, Paul Bunion's Cook Shanty, Perkins Family Restaurar Taco Bell, Wendy's
- **Lodg** N: (AAA) Best Western, (AAA) Comfort Inn, Dells Eagle Motel, (AAA) Holiday Inn, ◆ Super 8 Mote The Polynesian Motel
- **AServ** N: Amoco

89 WI 23, Lake Delton, Reedsburg
- **FStop** N: Phillips 66
- **Gas** N: Terry's 76
- **Food** N: Pioneer Family Restaurant (Phillips FS)
- **Lodg** N: Malibu Inn, Olympia Motel, Roadway Inn, Sahara Motel
- **RVCamp** N: Yogi Bear Camp Report

92 U.S. 12, Baraboo, Lake Delton
- **FStop** N: Mobil*
- **Gas** N: Citgo*(CW)
- **Food** N: Burger King, Danny's Diner, Market Square Cheese Store, McDonalds, Ponderosa, Subway Wintergreen Restaurant
- **Lodg** N: Alakai Hotel & Suites, Black Wolf Lodge, Country Squire Motel, Del Rancho, Grand Marquis Inn, (AAA) Ramada Limited, ◆ Wintergreen Resort
  S: Motel 6, Vagabond Motel
- **AServ** N: Parkway Auto Parts
- **RVCamp** S: Red Oaks Camping, Scenic Traveler RV Sales & Service, Yogi Bear Camp Report
- **Other** S: Pioneer Camping

106 WI 33, Portage, Baraboo
- **FStop** N: Mobil*
- **Gas** S: 76*
- **Food** N: Quackers Restaurant & Coffee Shop (Mobi FS)

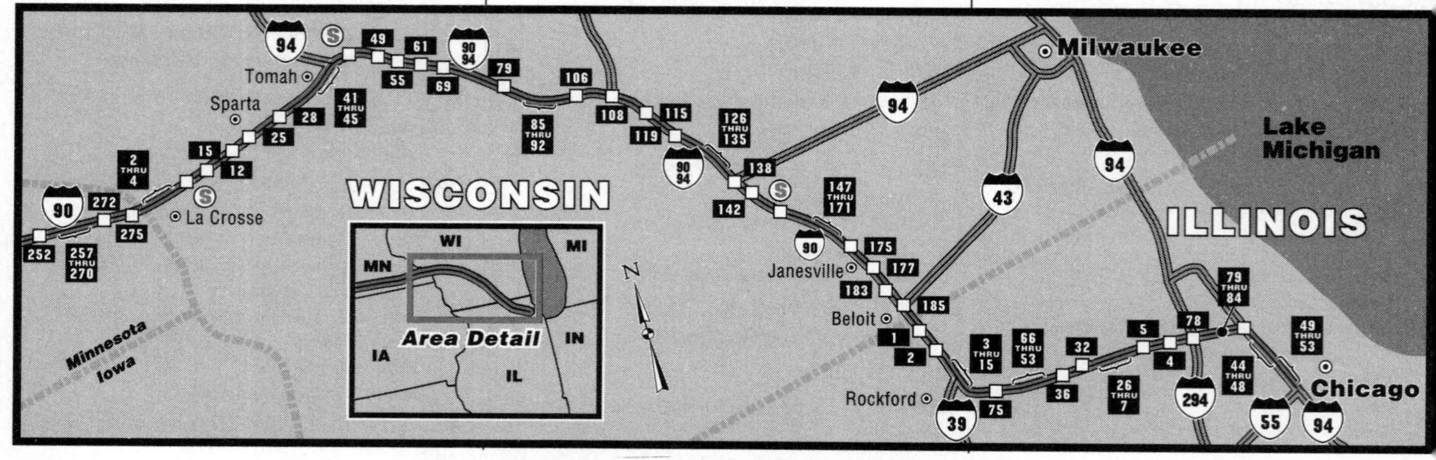

Bold red print shows RV & Bus parking available or nearby

EXIT		WISCONSIN
	Med	N: Hospital
108		WI 78, U.S. 51, Wausau, Merrimac
	TStop	S: Petro*(LP)(SCALES)
	Gas	S: Phillips 66*
	Food	S: Dairy Queen (Petro), Little Caesars Pizza (Petro), The Iron Skillet (Petro)
	Lodg	S: Days Inn (see our ad this page)
	AServ	S: Phillips 66
	TServ	S: Petro
(114)		Rest Area - RR, Vending, Phones, Picnic, RV Dump (Both Directions)
115		CR CS, Poynette, Lake Wisconsin
	FStop	N: Citgo*(D)
	Food	N: McDonalds, Subway
	TWash	N: Citgo
119		WI 60, Lodi, Arlington
126		CR V, Dane, DeForest
	Gas	N: Amoco*(CW)
		S: Phillips 66*
	Food	N: Cheese Chalet, Culver's Frozen Custard & Hamburgers, McDonalds, Subway
	Lodg	N: Holiday Inn Express
	AServ	S: Phillips 66
	RVCamp	N: KOA Campground
131		WI 19, Waunakee, Sun Praire
	Gas	N: Mobil, Quick Trip*(D), Super America*
	Food	N: A & W Drive-In, McDonalds, Mouse House
	Lodg	N: Days Inn, AAA Super 8 Motel
	AServ	N: Mobil
132		U.S. 51, Madison, DeForest
	TStop	N: Marathon*(LP)(SCALES), Shell*(SCALES)
		S: 76 Auto/Truck Plaza*(SCALES) (RV Dump)
	FStop	S: Pumper Truck Stop
	Food	N: The Copper Kitchen (Marathon)
		S: 76 Auto/Truck Plaza, Subway
	TServ	N: Diesel Specialists of Madison
		S: Brad Ragin Tire Service, Peterbilt Dealer
	RVCamp	S: WI RV World RV Service
135AB		U.S. 151, Madison, Sun Prairie
	Gas	S: Amoco* (ATM), Shell* (Towing), Sinclair*
	Food	S: Applebee's, Bread Smith, Carlos O'Kely, Chili's, Country Kitchen, Dunkin Donuts, Hardee's, IHOP, KFC, McDonalds, Mountain Jack's Prime Rib, Olive Garden, Perkins Family Restaurant, Pizza Hut, Ponderosa, Red Lobster
	Lodg	S: Comfort Inn, Crown Plaza, Hampton Inn, Select Inn
	ATM	S: Bank One, First Federal, Firststar Bank
	Other	S: Essex Square, Half Price Books
138A		I-94 East Milwaukee (Difficult Reaccess)
138B		WI 30, Madison (Difficult Reaccess)
142AB		U.S. 12, U.S. 18, Madison, Cambridge
	Gas	N: Mobil*
	Food	N: McDonalds, Subway (Mobil)
	Lodg	N: Motel 6, Ramada Inn
		S: Quality Inn
(147)		Weigh Station (Eastbound)
147		CR N, Stoughton, Cottage Grove
	FStop	S: Mobil Travel Center*
	Gas	S: Amoco
	Food	S: Amoco, Burger King, Country View Restaurant, Cousin's Subs

EXIT		WISCONSIN
156		U.S. 51 North, Stoughton
	Food	S: Coachman's Inn
	Lodg	S: ◆ Coachman's Inn
160		U.S. 51 South, WI 73, WI 106, Edgertown, Deerfield
	TStop	S: Shell(LP)
		W: Edgertown Oasis(SCALES)
	Food	S: Restaurant (Shell)
	TServ	S: Edgerton Oasis
	RVCamp	S: Camping
	Med	S: Hospital
	ATM	S: Cash Machine (Shell TS)
163		WI 59, Milton, Edgerton
	Gas	N: Mobil*(CW), Shell*
	Food	N: Burger King, Cheryls Restaurant, Cousin's Subs, McDonalds, Newville Drive In, Red Apple Restaurant
	Other	N: Red Apple Car Wash
(169)		Rest Area - RR, Picnic, Phones, Vending (Both Directions)
171AB		WI 26, Janesville, Milton, U.S. 14W
	FStop	N: Coastal*(D)
	Gas	S: Amoco*(CW), Kwik Trip*, Shell*
	Food	N: Alfresco Cafe, Campi's Prime Rib's, Cracker Barrel
		S: Amazon Station, Applebee's, Arby's, Burger King, Chi Chi's Mexican, Cornerstone Grill, Country Kitchen, Fazoli's Italian Food, Hardee's, Hoffman House, Jessica's Restaurant, KFC, McDonalds, McRaven's, Peking China Restaurant, Perkins Family Restaurant, Shakeys Pizza, Taco Bell, The Ground Round, Uno Pizzeria
	Lodg	N: Best Western (see our ad this page), Hampton Inn, Motel 6
		S: Country Inn, Oasis Motel, Ramada Inn, Select Inn, Super 8 Motel
	AServ	S: BF Goodrich, Jiffy Lube, K-Mart
	TServ	S: Transport America
	ATM	S: First Financial, FirstStar
	Other	S: Janesville Shopping Mall, K-Mart, Target Department Store, Wal-Mart, Wonder Car Wash
171C		U.S. 14, Janesville, Milton (Southside Services Accessible From Exit 171AB)
	TStop	N: Mobil*(SCALES)
	Food	N: Damon's, Shoney's, Wendy's
	Lodg	N: ◆ Holiday Inn Express, Microtel
	AServ	N: Batteries Plus
	RVCamp	N: Kamp Dakota

EXIT		WISCONSIN/ILLINOIS
175		WI 11, Delavan, Janesville
	FStop	N: Shell*
	Food	N: Denny's, Subway
	Lodg	N: AAA Budgetel Inn
177		WI 351, Avalon, Janesville
183		CR S, Shopiere Road
	FStop	S: Citgo*
	AServ	S: L & C Automotive
	RVCamp	N: Turtle Creek Campsite
185A		WI 81, Beloit, Milwaukee
	TStop	W: Pilot Travel Center(SCALES)
	FStop	W: Citgo*, Super America*
	Gas	W: Amoco*, Phillips 66*, Shell*, Super America*(D)(LP)
	Food	W: Arby's, Country Kitchen, Dairy Queen, Fazoli's Italian Food, McDonalds, Shirley's Home Cooking (Super America FS), Subway (Super America FS), Taco Bell (Super America FS), Wendy's
	Lodg	W: Comfort Inn, Econo Lounge, Fairfield Inn, Holiday Inn Express, Super 8 Motel
	ATM	W: 76
	Other	W: Wal-Mart
185B		Jct I-43N, Milwaukee
(187)		Rest Area - RR, Phones, Vending, Tourist Info, Picnic

↑WISCONSIN

↓ILLINOIS

EXIT		ILLINOIS
(1)		IL Welcome Center - RV Dump (Eastbound)
1		U.S. 51N, IL 71, S. Beloit
	Lodg	S: Tollway Motel, Holiday Inn (see our ad this page)
	AServ	S: Beloit Ford Lincoln Mercury
3		Rockton Road, CR 9
12		IL76, Poplar Grove
15		U.S. 20, Downtown Rockford
16		I-39, Rockford

Note: I-90 runs concurrent below with NWTLWY. Numbering follows NWTLWY.

EXIT		ILLINOIS
(75)		Toll Plaza
66		East Riverside Blvd
	Gas	S: Mobil*

EXIT ILLINOIS

63 U.S. 20 Bus, State Street
- **Gas** N: Phillips 66*(D)(CW)
  - S: Citgo*(LP), Mobil*
- **Food** N: Subway
  - S: Blimpie's Subs, Bombay Bicycle Club, Burger King, Country Kitchen, Steak & Shake, The Machine Shed
- **Lodg** N: Best Western, Excel Inn
  - S: Courtyard by Marriott, Fairfield Inn, **Hampton Inn**, Holiday Inn, Red Roof Inn
- **Other** S: **Sam's Club, Wal-Mart**

61 Jct I-39, U.S. 20, Rockford

(47) Toll Plaza

54 Belvidere Oasis - RR, HF, Phones
- **FStop** N: **Mobil**
- **Food** N: McDonalds

53 Genoa Road, Belvedere, Sycamore

(41) Toll Plaza

36 U.S. 20, Marengo, Hampshire
- **TStop** N: **76 Auto/Truck Plaza(SCALES) (RV Dump), Elgin Truck Stop*(SCALES)**
- **FStop** N: **Mobil(SCALES)**
- **Food** N: **Dairy Queen**, Dunkin Donuts, Subway, **Wendy's**
- **TServ** N: **76 Auto/Truck Plaza, Elgin Tire Service, Shell**

32 IL 47, Woodstock (No Westbound Reaccess)

26 Randall Road (8 Ton Per Axle Wt Limit)

(25) Toll Plaza

23 IL 31, Elgin
- **Gas** N: Amoco*(CW)(LP) (ATM)
- **Food** N: Alexanders Restaurant, **Cracker Barrel**, Wendy's
- **Lodg** N: Budgetel Inn, Hampton Inn, Super 8 Motel
- **Other** N: **Factory Card Outlet**

22 IL 25, Elgin
- **Gas** S: Amoco(CW), Checkers
- **Food** S: Arby's, George's Family Restaurant, Subway
- **Lodg** N: DAYS INN Days Inn
- **AServ** S: Amoco, Midas Muffler & Brakes
- **Med** S: ✚ **Hospital**

11 IL 59, Sutton Road

10 Barrington Road (No Westbound Reaccess)

9 Roselle Road (No West Reaccess)

8 Jct I-290

7 Arlington Heights Road
- **Lodg** N: Amerisuites (see our ad this page)

5 Elmhurst Road (No Westbound Reaccess)

4 Jct IL 72, Lee St
- **Gas** N: Mobil
- **Food** N: **McDonalds**

Note: I-90 runs concurrent above with NWTLWY. Numbering follows NWTLWY.

77 Jct I-294N

78 Jct I-294, I-190, River Road, Mannheim Road

EXIT ILLINOIS

79A Cumberland Ave South

79B Cumberland Ave North

80 Canfield Road

81A IL 43, Harlem Ave
- **Gas** S: Amoco*, Shell*(CW)
- **Food** S: La Scala, Mr. K's, Sally's Restaurant, Skylark Restaurant, Wendy's
- **AServ** S: Midas Muffler

81B Sayre Ave

82A Nagle Ave.
- **Gas** N: Mobil(CW)

82B Bryn Mawr Ave

82C Austin Ave. (Difficult Reaccess)

83AB Central Ave, Foster Ave

84 Lawrence Ave
- **Gas** N: Amoco
- **Food** S: Baskin Robbins, Dunkin Donuts, Gale Street Inn, Hunan Chinese, Jefferson Restaurant, Krakus Polish Deli, Little Spain, Magnolia Restaurant, McDonalds, Rey's Restaurant, Theresa & Magnolia Polish Restaurant, Zona Rosa Mexican Cuisine
- **AServ** N: Amoco*
  - S: Firestone Tire & Auto
- **Other** S: **Dental Clinic**

85 I-94, Skokie, Chicago

Note: I-90 runs concurrent below with I-94. Numbering follows I-94.

43D Kostner Ave (No Reaccess)

44A Keeler Ave, Irving Park Road
- **Gas** S: Amoco*, Mobil*, Shell*
- **AServ** S: Amoco
- **ATM** S: Midwest Bank

44B IL19, Pulaski Rd, Irving Park Rd (Difficult Reaccess)
- **Gas** N: Amoco*, Mobil*, Shell*
- **Food** S: La Villa Pizzaria, Las Brisas
- **AServ** N: Amoco, Mobil
- **ATM** N: Midwest Bank
- **Other** S: **7-11 Convenience Store**

45 Addison Street
- **Gas** S: Citgo*
- **Food** N: Little Caesars Pizza
  - S: Subway (Citgo)
- **Other** N: **K-Mart**

45B Kimball Ave

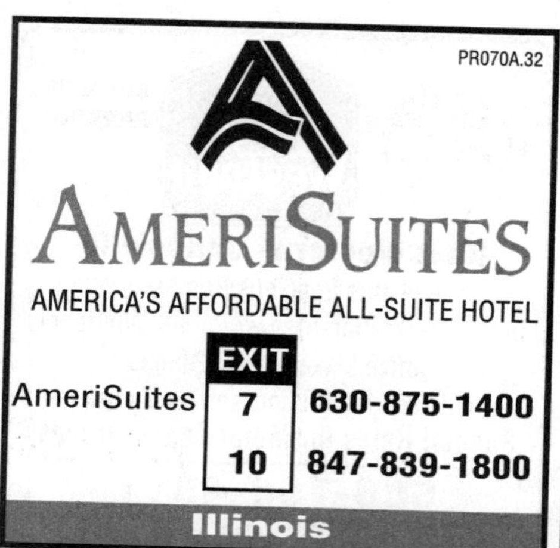

EXIT ILLINOIS

- **Gas** N: Marathon*(D)
  - S: Amoco
- **Food** S: Dunkin Donuts, Subway, Subway
- **Other** S: **Dominick's Grocery, WalGreens**

45C Belmont Ave. (Reaccess At Exit 45B)
- **Gas** S: Amoco
- **Food** S: Dunkin Donuts, Subway, Wendy's
- **Other** S: **Domick's Grocery (Pharmacy), WalGreens**

46A California Ave
- **Gas** N: Cabwerks
  - S: Mobil*
- **Food** N: KFC
  - S: IHOP, Popeye's Chicken
- **AServ** N: James Auto Repair
  - S: Mobil

46B Diversey Ave. (Difficult Reaccess)
- **Gas** N: Mobil*
- **Food** N: IHOP, Popeye's Chicken
- **AServ** N: Mobil

47A Fullerton Ave
- **Gas** S: 76, Amoco
- **Food** N: Chuck E Cheese Pizza, Dixie Que Restaurant, Dunkin Donuts, Popeye's Chicken, Scoops Ice Cream, Subway
  - S: Domino's Pizza
- **AServ** S: 76, Amoco (24hr towing)
- **ATM** N: Midtown Bank
- **Other** N: Express Car Wash
  - S: **Fullerton Western Pharmacy**

47B Damen Ave., 2000W (Difficult Reaccess To 94, I-90W)
- **Food** N: Popeye's Chicken
  - S: Baba Luci Italian Eatery, Bravo's Pizza, Flo's Bakery Shop
- **Med** S: ✚ **Medical & Dental Center**
- **Other** N: Squeaky Clean Hand Car Wash, **U-Haul Center(LP)**

48A Armitage Ave
- **Gas** S: Amoco*

48B IL 64, North Ave
- **Gas** N: Amoco*(CW)
  - S: Amoco*, Citgo, Shell
- **Food** N: Art's Drive In, Hollywood Grill, Tripp's Chicken
- **AServ** N: Goodyear Tire & Auto, Nortown Automotive
  - S: American Transmission, Shell
- **ATM** N: Cash Machine (Amoco)

49A Division Street
- **Gas** S: Amoco*, Shell*
- **AServ** S: Amoco*, Shell

50A Ogden Ave (No Eastbound Reaccess)
- **AServ** S: Firestone Tire & Auto
- **TServ** S: **International Transmission**

50B Ohio Street (No Return Access From Ohio Street)
- **Gas** S: Marathon*

51 Jct I-290

51A Lake St. 200 N

51B West Randolph Street (Southbound)
- **Food** N: Jim Ching's Restaurant, Perez Restaurant, S & S Restaurant
  - S: New Star Restaurant (Chinese)

51C Washington Blvd (Difficult Reaccess)

51D Madison Street (Exit Closed)

Bold red print shows RV & Bus parking available or nearby

XIT	ILLINOIS
51E	Monroe St.
51F	Adams Street
Food	S: Greek Island, Pegasus Restaurant, San Torini Restaurant
51G	Jackson Blvd
Food	N: Greek Town Gyros, Sorbes Restaurant
	S: Mitchell's Restaurant
Lodg	N: New Jackson Hotel
AServ	S: Toyota Dealer
51H	Jct I-290, Chicago Loop
52A	Taylor Street, Roosevelt Road
Gas	N: Amoco
Food	N: Eppel's Restaurant
AServ	N: Midas Muffler & Brakes
52C	18th Street
53	Canal Port Ave.
Gas	N: Shell*
Food	N: Ken Tones (Italian Beef & Sausage)
AServ	N: Joe's Auto Repair
53B	I-55 South, St. Louis
53C	I-55, St Louis, Lake Shore Drive
54	31st Street
Gas	N: Shell*
Food	N: Fat Albert's Italian Food, Maxwell Street Depot
Med	N: ✚ **Hospital**
55A	35th Street
55B	Pershing Road
56A	43rd Street
56B	47th Street
57A	51st Street
Food	N: McDonalds
57B	Garfield Boulevard
Gas	N: Amoco(CW)
	S: Shell*, Speedway*
Food	N: Checkers Burgers, KFC, Mr. Pizza King
	S: Wendy's
Other	N: **WalGreens**
58A	59th Street
Food	N: Church's Fried Chicken
58B	63rd Street (No Reaccess To I-94)
60A	75th Street
Gas	N: Shell*
Food	S: KFC
Other	N: **Wal Green's**
60B	76th Street
Gas	N: Amoco, Mobil*(CW)

EXIT	ILLINOIS/INDIANA
Other	N: **WalGreens**
60C	79th Street
Gas	N: Shell
	S: Amoco*, Shell(CW)
Food	S: Church's Fried Chicken, Dock's Fish, Dunkin Donuts, Subway
Other	S: **WalGreens**
61B	87th St.
Gas	N: Amoco*, Shell
Food	N: Burger King, McDonalds
	S: 87th Street BBQ, Grand Chinese Kitchen, Reggio's Pizza, Subway
AServ	S: Lube Pro's
Other	S: **Jewel Grocery, Osco Drugs**

↑ILLINOIS

↓INDIANA

Note: I-90 runs concurrent above with I-94. Numbering follows I-94.

EXIT	ILLINOIS/INDIANA
0	U.S. 12, U.S. 20, U.S. 41, Indianapolis Blvd (Southbound Difficult Reaccess)
Gas	S: 76*
Food	S: Burger King, KFC, McDonalds, Vienna Hot Dogs
3	Cline Ave, Indiana 912 E (Eastbound)
5	U.S. 41, Hammond
10	Indiana 912, Cline Ave
13	Grant St
15	Broadway, Illinois 53
17	Junction I-65, U.S. 12, U.S. 20, Dunes Hwy, Indianapolis (Pay Toll To Exit)

Note: I-90 runs concurrent below with I-80. Numbering follows I-80.

EXIT	ILLINOIS/INDIANA
21	Junction I-80 (Illinois tollway), Junction I-94, U.S. 6 West, Indiana 51, Des Moines (Pay Toll To Exit, Services On IN 51)
TStop	N: **Dunes Plaza* (24 Hrs), Petro*(SCALES) (24 Hrs), TravelPort*(SCALES)**
FStop	N: **Speedway*(LP)(SCALES)**
Food	N: **Buckhorn Family Restaurant (TravelPort)**, Burger King, **Iron Skillet Restaurant (Petro)**, McDonalds (Play Land), Ponderosa, **Subway (Travelport)**, Wing Wah
AServ	N: Jiffy Lube

EXIT	INDIANA
TServ	N: **Petro, Weber's Truck Repair**
TWash	N: **Blue Beacon Truck Wash (Petro), Murray Truck Wash (Travelport), Red Baron**
Other	N: Coin Car Wash
(22)	Service Area (Both Directions)
FStop	N: **SC(LP)(SCALES)**
Gas	N: BP*(D)
Food	N: Baskin Robbins, Fazoli's Italian Food, Hardee's
Other	N: **Tourist Information**
23	Portage (Pay Toll To Exit)
Gas	S: Amoco, Marathon*
Food	S: Burger King, Dunkin' Donuts, KFC, McDonalds (Play Place), Subway, Wendy's
Lodg	N: **Lee's Inn**
AServ	S: Amoco, Marathon, Muffler Shops, Portage Quick Change Oil
ATM	N: Pennacle Bank
	S: First National, Indiana Federal Bank
Other	S: **Town & Country Grocery (24 Hrs), WalGreens (Pharmacy, 24 Hrs)**
(24)	Indiana Toll Plaza
31	Indiana 49, Chesterton, Valparaiso (Pay Toll To Exit)
Med	S: ✚ **Hospital**
Parks	N: **Indiana Dunes State Park**
Other	N: **Indiana Dunes National Lakeshore, Tourist Information (2 Miles)**
39	U.S. 421, Michigan City, Westville (Pay Toll To Exit)
49	Indiana 39, La Porte (Pay Toll To Exit)
Lodg	S: **Cassidy Motel**
TServ	N: **Tomenko Tire & Truck Service**
Other	S: **Kingsbury State Fish & Wildlife Area**
(56)	Service Area (Both Directions)
FStop	N: **SC**
Gas	N: BP*(D)
Food	N: Baked Goods, Dairy Queen, McDonalds
Other	N: **Travel Emporium***
72	U.S. 31 By - Pass, South Bend, Plymouth, Niles (Pay Toll To Exit)
FStop	N: **Speedway*(LP)**
Food	N: **Hardee's (Speedway)**
ATM	N: Speedway
Parks	S: **Potato Creek State Park**
77	U.S. 33, U.S. 31, South Bend, Notre Dame University (Pay Toll To Exit)
Gas	S: Amoco(CW) (24 Hrs), Phillips 66*(D)
Food	N: Burger King, Denny's, Fazoli's Italian Food, J & N Restaurant (24 Hrs), Marco's Pizza,

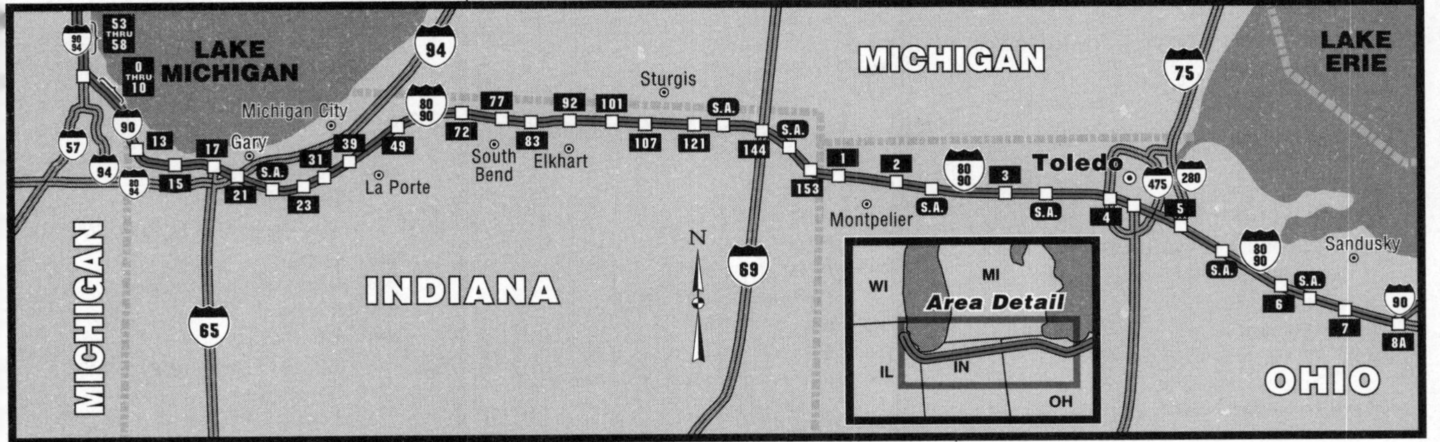

EXIT INDIANA

McDonalds, Pizza Hut, Ponderosa, Steak & Ale, Subway
S: Bennitt's Restaurant, Bill Knapp's Restaurant, Bob Evans, Colonial Pancake House, Donut Delight, King Gyro's, Perkins Family Restaurant, Schlotzkys Deli, Shoney's, Wendy's
Lodg **N:** Days Inn, ◆ Hampton Inn & Suites, Motel 6, AAA Ramada, Super 8 Motel
S: AAA Best Inns of America, Holiday Inn, Howard Johnson, Knights Inn, AAA Signature Inn, St. Mary's Inn
AServ **N:** Giant Auto Supply
S: Amoco, Q Lube
ATM **N:** Standard Federal
Other **N:** All Star Car Wash, Key Bank, North Village Mall, WalGreens (Pharmacy)
S: Rose Land Animal Hospital

83 Mishawaka (Pay Toll To Exit)
RVCamp **N:** KOA Campground

(90) Service Area - RV Dump (Both Directions)
FStop **N:** SC
Gas **N:** BP*(D)
Food **N:** Arby's, Dunkin' Donuts, South Bend Chocolate Co.
Other **N:** Tourist Information

92 Indiana 19, Elkhart (Pay Toll To Exit)
Gas **N:** Citgo*, Phillips 66*(D)
S: Clark*, Marathon*(D)
Food **N:** Andini Fine Dining, Applebee's, Lee's Famous Recipe Chicken (Phillips 66), Steak 'n Shake
S: Blimpie's Subs, Bob Evans Restaurant, Burger King, Callahan's (24 Hrs), DaVincci's Pizza, King Wha Chinese, McDonalds (Inside Play Place), Olive Garden, Perkins Family Restaurant, Red Lobster, Weston Restaurant (Weston Plaza Hotel)
Lodg **N:** Comfort Inn, AAA Diplomat Motel, AAA Econolodge, Hampton Inn, Holiday Inn Express, AAA Knight's Inn, ◆ Shoney's, Turnpike Motel
S: Budget Inn, Ramada, ◆ Red Roof Inn, AAA Signature Inn, Super 8 Motel, Weston Plaza Hotel
RVCamp **N:** American Trailer Supply, Dan's Service Center (Hitches & Trailers), Elkhart Campground, Traveland RV Service, Worldwide RV Sales & Service
ATM **N:** NBD
S: Key Bank, Marathon
Other **N:** Aldi Supermarket, K-Mart (Pharmacy), Martin's Supermarket, Revco Drugs, Visitor Information
S: Car Wash World

101 IN 15, Bristol, Goshen (Pay Toll To Exit)
RVCamp **N:** Eby's Pines Camping

107 U.S. 131, Indiana 13, Constantine, Middlebury (Pay Toll To Exit)
FStop **N:** Mobil*
Lodg **N:** Plaza Motel
AServ **N:** Dick's Auto Parts

121 Indiana 9, Howe, LaGrange (Pay Toll To Exit)
Gas **N:** J & M Service Center(D)
Food **N:** Golden Buddha
Lodg **N:** Green Briar Inn, Travel Inn Motel
S: AAA Super 8 Motel
AServ **N:** J&M Service Center

EXIT INDIANA/OHIO

Med **N:** ✚ Hospital

(126) Service Area (Both Directions)
FStop **N:** SC
Gas **N:** BP*(D)
Food **N:** Baskin Robbins, Fazoli's Italian Food, Hardee's
Other **N:** Tourist Information

144 Jct I-69, U.S. 27, Angola, Ft Wayne, Lansing
TStop **N:** 76 Auto/Truck Plaza*(SCALES)
FStop **S:** Pioneer, Speedway*(SCALES)
Gas **S:** Marathon*, Shell*
Food **N:** Baker St Family Restaurant (76 TS), Subway (76 TS)
S: Deli Mart (Marathon), Hardee's (Speedway), Red Arrow Restaurant (24 Hrs)
Lodg **N:** Lake George Inn
S: E&L Motel, Hampton Inn, ◆ Holiday Inn Express, Redwood Lodge
TServ **N:** 76 Auto/Truck Plaza, Gulick Trucks & Parts Service
S: Volvo Dealer
RVCamp **S:** Yogi Bear Camp Report
Parks **S:** Pokagon State Park
Other **S:** Country Meadows Golf Resort, Horizon Outlet Center

(146) Booth Tarkington/ James Witcomb Riley Service Plaza
FStop **N:** BP*(D)
Food **N:** Baked Goods, Dairy Queen, McDonalds

153 Indiana Toll Plaza

↑INDIANA

↓OHIO

Note: I-90 runs concurrent above with I-80. Numbering follows I-80.

Note: I-90 runs concurrent below with OHTNPK. Numbering follows OHTNPK.

1 **(2)** OH 49
Gas **N:** Mobil*(LP)
Food **N:** Burger King, Sub Shop (Mobil)
ATM **N:** Mobil
Other **N:** Ohio Tourist Center

(3) Toll Plaza

2 **(13)** OH 15, Montpelier, Bryan
FStop **S:** Pennzoil Lube*
Gas **S:** Marathon*(D)
Food **S:** Country Fare, Subway (Marathon)
Lodg **S:** Econolodge, Holiday Inn, Rainbow Motel
TServ **S:** Hutch's Tractor & Trailer Repair
Med **S:** ✚ Hospital
ATM **S:** Pennzoil FS

(21) Indian Meadow/Tiffin River Service Plaza
FStop **N:** Sunoco*
Food **N:** Hardee's
AServ **N:** Sunoco
Other **N:** Tourist Information

3 **(35)** OH 108, Wauseon
FStop **S:** Hy-Miler* (Shell)
Food **S:** Smith's
Lodg **S:** Arrowhead Motor Lodge, Del-Mar Best Western, Super 8 Motel

EXIT OHIO

AServ **S:** Wood Truck & Auto Service
TServ **S:** Wood Truck & Auto Service
RVCamp **S:** Executive Travelers Sales & RV Service

(49) Oak Openings/Fallen Timbers Service Plaza
FStop **N:** Sunoco*
Food **B:** Charlie Brown's Family Restaurant & General Store
Other **B:** Tourist Information

3A **(53)** Ohio 2, Toledo Airport, Swanton
Lodg **S:** Toledo Airport Motel
TServ **S:** Express Auto & Truck Service

4 **(59)** U.S. 20, to U.S. 23, to I-475 Maumee, Toledo
Gas **N:** BP*, Fast Check*, Speedway*(LP)
S: Amoco*, Speedway*
Food **N:** Arby's, Beast of Chicago Pizza Co., Bob Evans Restaurant, China Buffet, Connie Mac's Bar and Grill, Dominic's Family Italian, Little Caesars Pizza, Mark Pi's China Gate Restaurant, Max's Diner, McDonalds, Nick's Cafe, Pizza Hut, Ramada, Tandoor Indian Restaurant
S: Baverian Brewing Company, Big Boy, Brandie's Diner, Chi Chi's Mexican, Fricker's, Friendly's, Gourmet of China, Popoff's Pizza and Lebanese Food, Ralphie's Burgers, Red Lobster, Rib Cage, Yes Solid Rock Cafe
Lodg **N:** Budget Inn, Holiday Inn, Motel 6, Ramada
S: Comfort Inn, Cross Country Inn, Days Inn, Hampton Inn
AServ **N:** Auto Express, Goodyear Tire & Auto, K-Mart, Napa Auto Parts, The Car Doctor, Tom's Tire
S: Bob Schmidt GM, Hatfield GM
ATM **S:** Huntington Bank
Other **N:** K-Mart (Pharmacy), Southwyck Mall
S: DJ's Car Wash, Meijer Grocery

4A JCT I-75 to Perrysburg, Toledo

5 **(72)** Jct I-280, OH 420, to I-75, to Stony Ridge, Toledo
TStop **N:** Flying J Travel Plaza*(LP)(SCALES), Petro*(SCALES)
S: 76 Auto/Truck Plaza, TravelCenters of America*(SCALES)
FStop **S:** Speedway*(SCALES)
Gas **S:** Mobil (76 Auto Truckstop)
Food **N:** Dad's (Flying J), Howard Johnson, Iron Skillet (Petro), Metro Inn, Pizza Hut (Petro)
S: 76 Auto/Truck Plaza, McDonalds, Sbarro Pizza, Toledo 5 TS, TravelCenters of America, Wendy's
Lodg **N:** Budget Inn, Howard Johnson, Knights Inn, Metro Inn
TServ **N:** Petro
S: 76 Auto/Truck Plaza, Fleet Tire Center, Perkins Detroit Diesel, TravelCenters of America*(SCALES)
TWash **N:** Blue Beacon Truck Wash (Petro)
S: Stony Ridge Truck Wash (TS)
ATM **N:** Flying J Travel Plaza
S: 76 Auto/Truck Plaza

(77) Blue Heron Service Plaza/Wyandot
FStop **B:** Sunoco
Food **B:** Fresh Fried Chicken, Hardee's
AServ **B:** Sunoco
Other **B:** Tourist Information

5A **(82)** OH 51, Elmore, Woodville, Gibsonburg

Bold red print shows RV & Bus parking available or nearby

EXIT OHIO

(91) Ohio 53, Fremont, Port Clinton
FStop S: Shell*
Food N: Days Inn, Sneaky Fox Steak House, Z's Diner
S: Holiday Inn
Lodg N: Best Budget Inn, DAYS INN Days Inn
S: Fremont Turnpike Motel, Holiday Inn
Med S: + Hospital

00) Erie Island Service Plaza/Comodore Perry
FStop B: Sunoco
Food B: Rax
AServ B: Sunoco
Other B: Travel Information

6A (110) OH 4, Sandusky, Bellevue

(119) U.S.250, Sandusky, Norwalk
Gas N: Marathon*(D) (Kerosene), Speedway*
Food N: Dick's Place (take-out), Subway, Super 8 Motel
S: Homestead Diner
Lodg N: Comfort Inn, DAYS INN Days Inn, Hampton Inn, Homestead Inn, Ramada Limited, Super 8 Motel
S: Crown Motel, Homestead Farm
AServ S: Dorr Chevrolet/Geo
RVCamp N: Milan Travel Park
Other N: Lake Erie Factory Outlet Center, State Patrol Post

7A (136) Baumhart Rd, Vermillion

39) Service Plaza
FStop B: Sunoco
Food B: Bob's Big Boy, Burger King, TCBY Yogurt

3A (142) Jct I-90, OH 2 (Eastbound)

Note: I-90 runs concurrent above with OHTNPK. Numbering follows OHTNPK.

44 OH 2, Sandusky

45A OH 57, To I-80, Elyria, Lorain
FStop S: Speedway*
Gas S: BP*(D)
Food S: Bob Evans Restaurant, Country Kitchen, Fun Times, Holiday Inn, McDonalds, Mountain Jacks, Pizza Hut, Red Lobster, Rubins Restaurant, Wendy's
Lodg S: Best Western, Camelot Inn, Comfort Inn, DAYS INN Days Inn, Econolodge, Holiday Inn
AServ S: Sears, Tuffy Service Center
TServ N: Goodyear Tire & Auto
Med S: + Hospital
ATM S: Lorain Nat'l Bank, National City Bank, Star Bank
Other N: U-Haul Center(LP)
S: Automatic Car Wash, Car Wash, K-Mart, Midway Mall, Sams Club, State Patrol Post

48 Sheffield, Avon, OH 254
Gas S: Speedway*(LP)
Food S: Burger King, KFC, Marco's Pizza, McDonalds, Pizza Hut, Sips & Nibbles, Subway, Sugar Creek Restaurant, Taco Bell
AServ N: Goerlich Mufflers, Mike Bass Ford
ATM S: First Merit Bank, Premier Bank and Trust (Rini Rego)
Other S: Aldi Grocery, Car Wash, Revco Drugs, Rini Rego Grocery, Sear's Hardware

51 OH 611, Sheffield, Avon
FStop N: BP*(CW)(LP), Speedway*(SCALES)
Food N: Church's Fried Chicken, McDonalds, Subway (BP)
AServ S: Ray's
TServ S: Ray's
RVCamp N: Avon RV Superstore

53 OH 83, Avon

EXIT OHIO

156 Bassett Rd , Crocker Road
Gas N: BP, Shell*(CW)
Food N: Holiday Inn
S: Bob Evans Restaurant, Max & Erma's, McDonalds, T.G.I. Friday's, Wendy's
Lodg N: ◆ Holiday Inn, ◆ Red Roof Inn, ◆ Residence Inn
S: AAA Hampton Inn (see our ad this page)
AServ N: BP
S: K-Mart, Procare
ATM S: Key Bank, Star Bank
Other S: K-Mart, Marc's Grocery, Prominade of Westlake, Rini Rego Grocery

159 OH 252, Columbia Road
Gas S: Amoco, BP*(CW), Shell
Food N: Cooker Bar and Grill
S: Houlihan's, KFC, McDonalds, Taco Bell
Lodg N: Cross Country Inn
AServ S: Lube Stop, Mueller Tire, Shell
Med N: + Lakewood Medical Center
S: + Hospital
ATM S: Charter One Bank, Key Bank, National City Bank, Strongsville Savings Bank
Other S: Finast Supermarket, Revco Drugs

160 Clague Rd (Eastbound, Westbound Reaccess)

161 OH 2, OH 254, Detroit Rd

162 Hilliard Blvd, Rocky River

163 OH 237, Rocky River Drive

164 McKinley Ave

165 Warren Road, Lakewood

165B West 140th St, Bunts Rd
Gas N: Sunoco*
Other N: Dairy Mart C-Store

166 West 117th St

167A West 98th St, West Blvd

167 Lorrain Ave, OH 10

169 West 41st St, West 44th St

170A OH 14, Broadway, US 42

170B Jct I-71 South to Columbus, Jct I-90 W to Toledo (Left Exit Westbound)

171 West 14th St, Abbey Ave

172A Jct I-77S, to Akron

172C Downtown Cleveland

173A Prospect Ave
Med N: + Hospital

173B Chester Ave

173C Superior Ave, St Clair Ave
Gas N: BP*

174A Lakeside Ave

174B OH 2 West, Downtown Cleveland
Other N: Science Museum

EXIT OHIO

175 East 55th St, Marginal Road

176 East 72nd St

177 University Circle, MLK Jr Dr (No Trucks)
Med S: + Hospital

178 Eddy Road, Bratenahl

179 OH 283, Lake Shore Blvd, Bratenahal (Eastbound)

180A East 140th St

180B East 152nd St, East 140th St

181 East 156th St (No Trucks)
Gas N: Clark*
AServ N: Century Tire Service

182A East 185th St

182B East 200th St
Gas S: BP(CW), Sunoco*
AServ N: Case Honda
Med N: + Hospital
Other S: Car Wash

183 East 222nd St
Gas N: Sunoco*(D)
AServ N: Flickinger Goodyear Tire & Auto
S: Euclid Auto Service
Other N: Coin Laundry

184B OH 175, East 260th St, Babbitt Rd
Gas N: Shell
Food N: Draw Bridge Family Restaurant
AServ N: Clay Matthews Pontiac, Shell
S: Euclid Transmission, Meuller's Tire
Other S: Veterinary Clinic

184 OH 175, East 260th St

185 OH 2 East, Painesville, OH 2 W Cleveland

186 U.S. 20, Euclid Ave
Gas N: Sunoco*
Food N: Denny's, Four Points Sheraton, McDonalds
Lodg N: Four Points Sheraton, ◆ Hampton Inn
S: Invoy Motel
AServ N: Glavic Olds, Subaru, Kaufman's Radiator, Mullinex East Ford, Saturn Dealership, Universal Mechanics
S: The Lube Stop
ATM S: Huntington Bank
Other N: Same Drug Store, Wickliffe Food Mart
S: Coin Laundry, Convenient Food Mart

187 OH 84, Bishop Road
Gas S: BP*(CW), Fast Check*, Shell*(CW)
Food S: Bakers Square Restaurant, Burger King, Friendly's, Golden Mountain Restaurant, Mama Lauro's Pizza, Manhattan Deli, McDonalds, Subway
Lodg S: Quality Inn
AServ S: Mazda/VW Dealer, Rainbow Muffler, Tony La Riche Chevrolet/ Geo
Med S: + Hospital
ATM S: Bank One, First Merit, Great Lakes Bank, National City Bank
Other S: Coin Laundry, Dairy Mart C-Store, Marc Grocery, Revco Drugs, Rini Rego Grocery, Sam's Club, Sear's Hardware, Willoughby Hills Veterinarian

188 Jct I-271S, Local Lanes (Left Exit Westbound)

189 OH 91, Willoughby Hills
Gas N: BP*(D)(CW), Shell*(CW)
Food N: Bob Evans Restaurant, Bruegger's Bagels, Cafe Europa, Harley Hotel, Peking Chef, Pizza and Restaurant
S: Fazzio's

EXIT		OHIO
	Lodg	N: ◆ Fairfield Inn, ◆ Harley Hotel, ◆ TraveLodge
	Med	N: Hospital
	Other	N: Convenient Mart, Revco Drugs
(190)		Weigh Station (Eastbound)
190		Jct I-271S to Columbus, Express Lanes (Left Exit Westbound)
193		OH 306, Kirtland, Mentor
	Gas	N: BP*
		S: Marathon*(CW)
	Food	N: McDonalds, Subway (BP)
		S: Bluey's Cafe, Brown Derby, Burger King
	Lodg	N: ◆ Knight's Inn
		S: ◆ Days Inn (see our ad this page), ◆ Red Roof Inn
(198)		Rest Area - RR, Phones, Picnic, Vending
200		OH 44, Chardon, Painesville
	Gas	S: BP*(D)
	Food	S: McDonalds, Quail Hollow Resort, Subway (BP)
	Lodg	S: AAA Quail Hollow Resort
	Med	N: Hospital
205		Vrooman Road
	Gas	S: BP*(D)
	Food	S: Frary's Restaurant
212		OH 528, Madison, Thompson
	Gas	S: Marathon*(D)
	Food	N: McDonalds, Potbelly's Family Restaurant
	AServ	S: Jonny Ray's Radiator Service, Mace Auto Service Repair
	TServ	S: Hercules Tires
218		OH 534, Geneva
	TStop	S: Kwik Fill Auto Truck Plaza*(SCALES)
	Gas	N: BP*, Sunoco (Kerosene)
	Food	N: Geneva Country Kitchen, Geneva Inn Restaurant, McDonalds
		S: Applewood Family Restaurant (Kwik Fill Truck Stop)
	Lodg	N: ◆ Howard Johnson
	AServ	N: Sunoco
	TServ	N: Goodyear Tire & Auto
	Med	N: Hospital
	Parks	N: Geneva Lake Park (9 Miles)
	Other	S: Country Mall C-Store
223		OH 45, Warren
	FStop	S: Speedway*(LP)
	Gas	S: BP*
	Food	N: Holiday Inn, Mr C's Restaurant
		S: Burger King, Clay Street Inn, JD's on the Freeway, McDonalds, Subway (Speedway)

EXIT		OHIO
	Lodg	N: ◆ Holiday Inn, AAA Travelodge
		S: ◆ Hampton Inn
	RVCamp	S: Indian Creek (10 Miles)
	Other	S: Niagara Falls Info Center (BP)
229		OH 11, Ashtabula, Youngstown
	Med	N: Hospital
235		OH 84, OH 193, Ñ Kingsville
	TStop	S: 76 Auto/Truck Plaza*(SCALES)
	FStop	S: Speedway*(SCALES)
	Gas	N: Citgo*
		S: Amoco
	Food	N: Roesch's Natural Foods
		S: Jonathan's Family Dining, Subway (Speedway)
	Lodg	N: Dav-Ed Motel
		S: Kingsville Motel
	AServ	S: Amoco, Mr Boltz
	TServ	S: Mr Boltz
	RVCamp	N: Campground
	Other	N: Glad-E-O C-Store
241		OH 7, Conneaut, Andover
	Food	N: Burger King
		S: Beef and Beer Family Restaurant
	Lodg	N: AAA Days Inn
	AServ	N: K-Mart
	RVCamp	N: Evergreen Lake Park
	Med	N: Hospital
	Other	N: Giant Eagle Supermarket, K-Mart, Revco Drugs
(242)		OH Welcome Center - RR, Phones, Picnic, Vending (Westbound)
(243)		Weigh Station (Westbound)

EXIT		OHIO/PENNSYLVANIA

↑OHIO

↓PENNSYLVANIA

(1)		PA Welcome Center - RR, Phones, Picnic, Vending (Eastbound)
(2)		Weigh Station
1		(3) U.S. 6 N, Cherry Hill, West Springfield
	TStop	S: BP
	Food	S: Hitchen Post Restaurant
	TServ	S: Bartone's Service (24 Hr Towing)
	ATM	S: BP
2		(6) PA 215, Albion, East Springfield
	Gas	S: Sunoco*
	Lodg	S: Miracle Motel
	AServ	S: Morley's
3		(10) PA 18, Platea, Girard
	TServ	N: Keystone Diesel Engine Co
	Other	N: State Patrol Post
4		(16) PA 98, Franklin Center, Fairview
	RVCamp	S: Folly's End Campground
5		(18) PA 832, Sterrettania, Presque Isle
	TStop	S: AmBest Truckstop*(SCALES)
	FStop	N: Citgo*, Shell*
		S: Shell*(SCALES)
	Food	N: Burger King
		S: Best Western, Green Shingle (Am Best)
	Lodg	S: Best Western
	TServ	S: CB Repair (Am Best), Poplar Thruway Service Garage (Am Best)
	TWash	S: AmBest
	RVCamp	N: Hills Family Campground
		S: Erie KOA Campground
	ATM	S: AmBest Truckstop
	Parks	N: Walameer Park (8 Miles)
	Other	N: Coin Laundry (Citgo)
(22)		Junction I-79, Pittsburgh, Erie
6		(24) U.S. 19, Waterford, Peach St
	Gas	N: Citgo*(D), Kwik Fill*
		S: Exxon*(D)
	Food	N: Applebee's, Burger King, Chuck E. Cheese's Pizza, Damon's Rib House, Eat N Park, Howard Johnson, Kenny Rogers Roasters, McDonalds, Old Country Buffet, Sub Fare(LP), Taco Bell
		S: Bob Evans Restaurant
	Lodg	N: Howard Johnson

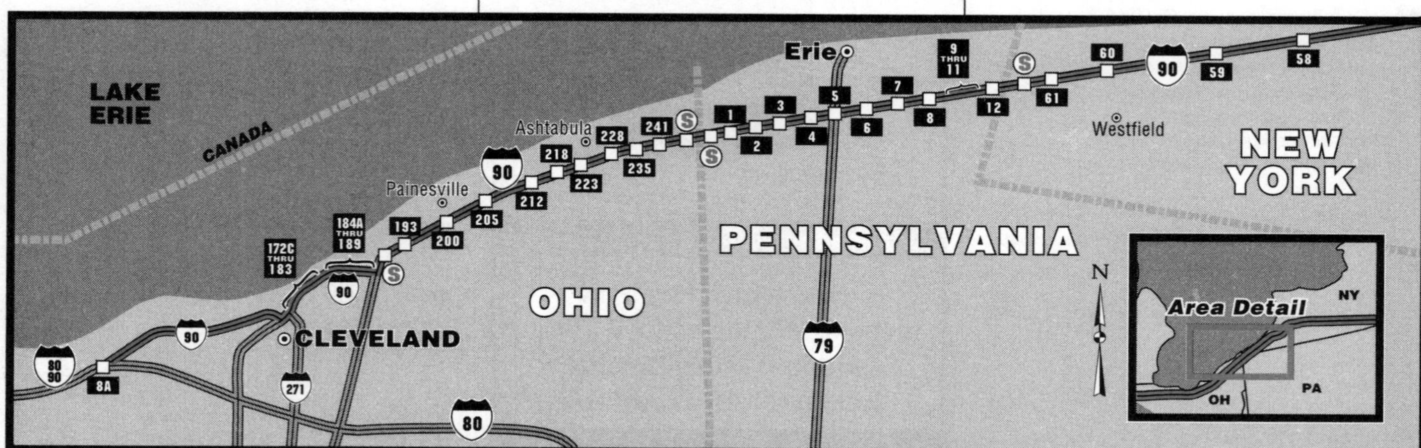

Bold red print shows RV & Bus parking available or nearby

EXIT PENNSYLVANIA

S: ◆ Comfort Inn, ◆ Econolodge, AAA Microtel, ◆ Residence Inn
AServ N: Wal-Mart
ATM N: Country Fair, National City Bank, PNC Bank
Other N: Builders Square, Country Fair*, Giant Eagle Grocery (Pharmacy), K-Mart (Pharmacy), Lowe's, Petsmart, Sam's Club, Thrift Drug, Towne Center Summit, Wal-Mart (Pharmacy)

7 **(27)** PA 97, Waterford, State St
FStop S: Pilot*(D), Shell*(D)
Gas N: BP*(D), Citgo*, Kwik-Fill
Food N: Arby's, Big Boy, McDonalds
S: Dairy Queen (Pilot), Holiday Inn, Subway (Pilot)
Lodg N: DAYS INN ◆ Days Inn, Knight's Inn, ◆ Red Roof Inn
S: AAA Holiday Inn
ATM N: Machine (Citgo)
S: Shell
Other S: Big Al's Auto Wash

8 **(29)** PA 8, Hammett, Parade St
Gas N: Citgo*, Shell*
Food N: Sub Fare, Wendy's
S: Ramada Inn
Lodg S: AAA Ramada
AServ N: Bob's Radiator, Jack Cooney's
TServ S: International Truck Service, Lake Erie Ford Trucks
ATM N: Country Fair (Citgo)

9 **(32)** PA 430, Wesleyville
FStop S: Sunoco*
Gas N: Citgo* (Kerosene)
Food N: Sub Fare
ATM N: Country Fair

10 **(35)** PA 531, Harborcreek, Phillipsville
TStop N: Travel Port*(SCALES) (RV Dump)
Gas N: Sunoco (Travel Port)
Food N: Buckhorn Family Restaurant (TS), Pizza Hut (TS)
Lodg N: Rodeway Inn (TS)
TServ N: Robert's Truck Service, Travel Port
ATM N: Travel Port

10A **(36)** PA 17, Jamestown

11 **(41)** PA 89, North East
Gas N: Shell
Food N: Orton's Ice Cream, The Harvest Inn (Super 8)
Lodg N: ◆ Super 8 Motel
AServ N: Shell
RVCamp S: Family Affair Campground

12 **(45)** U.S. 20, State Line
TStop N: Kwik Fill Auto Truck Plaza*(SCALES), North East Truck Plaza*(LP)(SCALES) (RV Dump)
FStop S: BP*
Food N: McDonalds, North East TS, Pizza Hut (TS), Taco Bell
S: Subway (BP)
Lodg S: Red Carpet Inn
TWash N: North East Truckstop
ATM S: BP
Other S: Tourist Information

(46) Weigh Station (Westbound)

(47) PA Welcome Center - RR, Phones, Vending (Westbound)

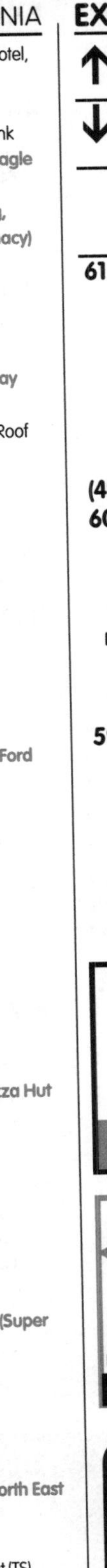

EXIT PENNSYLVANIA/NEW YORK

↑PENNSYLVANIA

↓NEW YORK

Note: I-90 runs concurrent below with NYTHWY. Numbering follows NYTHWY.

61 **(495)** Shortman Road, Ripley
TStop N: 76 Auto/Truck Plaza*(SCALES)
Gas N: Shell (76 TS)
Food N: 76 Auto/Truck Plaza
S: Colonial Squire Restaurant (Budget Host Inn)
Lodg S: Colonial Squire Motel
TServ N: 76
Other S: Tourist Information

(494) Toll Booths

60 **(485)** NY 394, Westfield, Mayville
Gas N: Keystone*
S: Mobil* (Kerosene)
Lodg S: Thruway Holiday Motel
AServ S: Mobil
RVCamp N: KOA Campground
Med S: ✚ Hospital
Parks N: Lake Erie State Park
Other N: Tourist Information

59 **(468)** NY 60, Dunkirk, Fredonia
FStop S: Keystone*
Gas S: BP*, Citgo*, Sunoco
Food S: BP, Bob Evans Restaurant, Burger King, McDonalds, Perkins Family Restaurant, Pizza Hut, Ponderosa, The Vinyard, Wendy's
Lodg S: Comfort Inn, DAYS INN Days Inn, Quality Vineyard Inn, Holiday Inn (see our ad this page)
AServ N: Napa Auto Parts

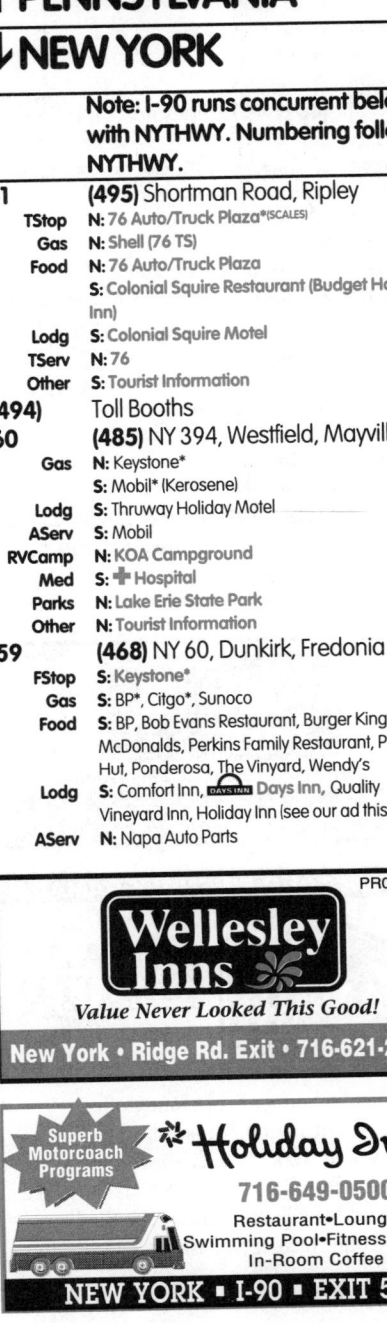

H01407
Superb Motorcoach Programs
Holiday Inn®
716-649-0500
Restaurant•Lounge
Swimming Pool•Fitness Center
In-Room Coffee
NEW YORK ▪ I-90 ▪ EXIT 57

Jamestown, New York
Route 17 Exit 12
Holiday Inn®
of Jamestown
H01404
• Full Service Restaurant & Lounge
• Indoor Pool
• AAA/AARP Rates
• Chautauqua Institution 15 miles
716-664-3400 For Reservations: 800-528-8791
NEW YORK ▪ I-90 ▪ EXIT 59

EXIT NEW YORK

S: Lakeshore Ford Mercury Lincoln, Midas Muffler & Brakes, Monroe Mufflers and Brakes, Quality Farm & Fleet Auto Service, Sunoco
Med S: ✚ Hospital
ATM S: Fleet Bank, M&T Bank
Other S: Car Wash, Fay's Drugs, K-Mart (Pharmacy), Tops Grocery & Pharmacy, Wal-Mart (Pharmacy)

58 **(456)** U.S. 20, NY 5, Silver Creek, Irving
Gas N: Citgo*, Kwik Fill*(LP)
Food N: Burger King

(447) Angola Service Plaza (Eastbound)
FStop B: Mobil(D)
Food B: Denny's, McDonalds
AServ B: Mobil
ATM B: OnBank
Other B: Tourist Information

57A **(445)** Eden, Angola

(443) Rest Area - Phones, Picnic

57 **(436)** NY 75, Hamburg, East Aurora
TStop N: 57 Truck Plaza*(SCALES)
Gas N: Mobil*(D)
S: Kwik Fill, Mobil
Food N: Bozanna's Pizzeria, Exit 57 TS, Perkins Family Restaurant, Snack Shop Truck Plaza, Wendy's
S: Arby's, Burger King, Camp Road Diner, Holiday Inn, Pizza Express, Pizza Hut, Subway
Lodg N: DAYS INN Days Inn (Exit 57 TS), Howard Johnson
S: Holiday Inn (see our ad this page)
AServ N: Austin Pontiac, Continental Transmission, Meineke Discount Mufflers, Mobil, S & S Service, Towne Chrysler Jeep, West-Herr Ford, West-Herr Mitsubishi
S: All Systems Go, Automotive Plus, Basil Chevrolet, Instalube, Mobil, Monroe Mufflers and Brakes
TServ N: Exit 57 Truck Plaza
ATM N: Mobil
S: Rochester Community Bank
Other S: Camp Road Pharmacy, Convenient Food Mart, Small Animal Clinic

56 **(432)** NY 179, Mile Strip Road, Blazedall, Orchard Park
Gas N: Getty, Sunoco*
Food N: Blasdell Pizza, Odyssey Family Restaurant
S: Applebee's, Boston Market, McDonalds, Pizza Hut, Ruby Tuesday
Lodg N: Econolodge
AServ N: Gregorio's Auto Service, Mike Scora's Collision, Orchard Auto Parts
ATM S: NYCE ATM
Other N: CVS Pharmacy
S: Builders Square, McKinley Mall, Petsmart

(430) Toll Booths

55 **(430)** U.S. 219, Ridge Road, Lackawanna, West Seneca
Gas N: Mobil*, Sunoco
S: Coastal*(CW)
Food S: Perkins Family Restaurant
Lodg N: Wesley Inn (see our ad this page)
AServ N: Cole Muffler, Midas Muffler & Brakes, Napa Auto Parts, Sunoco, Valvoline Quick Lube Center
S: Monroe Muffler & Brakes
Med N: ✚ Hospital
ATM S: M&T Bank, Rochester Community Bank
Other N: Car Wash, Community Care Animal Hospital, WalGreens
S: Home Depot, Mail Boxes Etc, Topps Grocery (Pharmacy)

54 **(428)** NY 16, NY 400, West Seneca, East Aurora

Bold red print shows RV & Bus parking available or nearby

EXIT	NEW YORK
53	**(426)** Junction I-190, Niagra Falls, Downtown Buffalo
52A	**(425)** Williams St
52	**(424)** Walden Ave, Buffalo, Cheektowaga
TStop	**S:** Jim's Truck Plaza*(SCALES)
Gas	**S:** Sunoco*(D)
Food	**N:** Arby's, Bob Evans Restaurant, T.G.I. Friday's, Wendy's **S:** Jim's Truck Plaza, Olive Garden, Ruby Tuesday (Walden Gallaria Mall), Sheraton
Lodg	**N:** Hampton Inn **S:** Sheraton
AServ	**N:** Dunlop Tires, Four City Auto Parts, Paul Batt Buick **S:** K-Mart, Sear's
TServ	**N:** Buffalo Spring **S:** Jim's Truck Plaza
Med	**N:** + St. Joseph's Hospital
ATM	**N:** Marine Midland Bank, Rochester Community Bank **S:** Jim's Truck Plaza Restaurant
Other	**N:** Target Department Store (Pharmacy) **S:** K-Mart (Pharmacy), Walden Gallaria Mall, Wegmans Food & Pharmacy
51	**(422)** NY 33 , Buffalo
Lodg	**S:** Quality Inn
Other	**N:** Airport
50A	**(421)** Cleveland Dr (Eastbound, Westbound Reacess)
Gas	**S:** Coastal
Food	**N:** Lunetta's
AServ	**S:** Coastal
ATM	**S:** Tile Pharmacy
Other	**S:** Tile Pharmacy
50	**(420)** Junction I-290, Toll Booth
49	**(417)** NY 78, Depew, Lockport (Buffalo Airport)
Food	**N:** Cracker Barrel, Mighty Taco, Picasso Pizza, Protocol Restaurant, Wendy's **S:** Bob Evans Restaurant, Garden Place Hotel, Howard Johnson
Lodg	**N:** Fairfield Inn, Holiday Inn Express, Lancaster Motor Inn, Microtel, Ramada (see our ad this page), Wellesley Inn (see our ad this page) **S:** Garden Place Hotel, Hospitality Inn, Howard Johnson, Red Roof Inn
AServ	**N:** Al Maroone Ford, Dan Georger's, The Battery Warehouse

EXIT	NEW YORK
(412)	Clarence Service Plaza (Westbound)
FStop	**W:** Sunoco
Food	**W:** Mrs. Fields Cookies, Nathan's
AServ	**W:** Sunoco
ATM	**W:** OnBank
48A	**(402)** NY 77, Pembroke, Medina
TStop	**S:** 76 Auto/Truck Plaza*(SCALES) (RV Dump)
Food	**S:** 76 Auto/Truck Plaza, Apple Creek Family Restaurant, Exit 48A Diner
Lodg	**S:** Econolodge
TServ	**S:** 76
RVCamp	**S:** Darien Lakes Campground
Other	**S:** Darien Lake Theme Park & Camping Resort
(397)	Pembroke Service Plaza (Eastbound)
FStop	**E:** Sunoco
Food	**E:** Burger King, Mrs Fields Cookies, Popeye's Chicken, TCBY Yogurt
AServ	**E:** Sunoco
ATM	**E:** OnBank
Other	**E:** Tourist Information
48	**(390)** NY 98, Batavia, Albion, Attica
Food	**S:** Best Western, Bob Evans Restaurant, Days Inn, Holiday Inn
Lodg	**S:** Best Western, Days Inn, Holiday Inn,

EXIT	NEW YORK
	Park-Oak Hotel, Roadway Inn, Super 8 Motel
Other	**N:** State Patrol Post
47	**(379)** Junction I-490, NY 19, Leroy, Rochester
(376)	Ontario Service Plaza (Westbound)
Food	**W:** Ben & Jerry's Ice Cream, McDonalds
AServ	**W:** Mobil
ATM	**W:** OnBank
(366)	Scottsville Service Plaza (Eastbound)
FStop	**E:** Mobil
Food	**E:** Burger King, Dunkin Donuts, TCBY Yogurt
AServ	**E:** Mobil
Other	**E:** Tourist Information
46	**(362)** Jct I-390, Rochester, Corning
Lodg	**N:** Ramada Inn
(353)	Parking Area - Picnic, Phones (Eastbound)
45	**(351)** Junction I-490, Rochester
Lodg	**S:** Microtel
(350)	Seneca Service Plaza (Westbound)
FStop	**W:** Mobil*
Food	**W:** Burger King, Mrs Fields Cookies, TCBY
AServ	**W:** Mobil
Other	**W:** Tourist Information
44	**(347)** NY 332, Victor, Canandaigua
Food	**S:** Best Western
Lodg	**S:** Best Western, Economy Inn
RVCamp	**S:** KOA Campground
Other	**S:** State Patrol Post
43	**(340)** NY 21 Manchester, Palmyra
FStop	**S:** Mobil*
Food	**S:** McDonalds, Steak-Out Restaurant
Lodg	**S:** Abbott's Motel, Roadside Inn
(337)	Clifton Springs Service Plaza (Eastbound)
FStop	**E:** Sunoco
Food	**E:** Roy Rogers, Sbarro Italian, TCBY Yogurt
AServ	**E:** Sunoco
ATM	**E:** OnBank
42	**(327)** NY 14, Geneva, Lyons
FStop	**S:** Mobil*
Food	**S:** Daystop Inn
Lodg	**S:** Daystop Inn
(324)	Junius Ponds Service Plaza (Westbound)
FStop	**W:** Sunoco*
Food	**W:** Roy Rogers, TCBY Yogurt

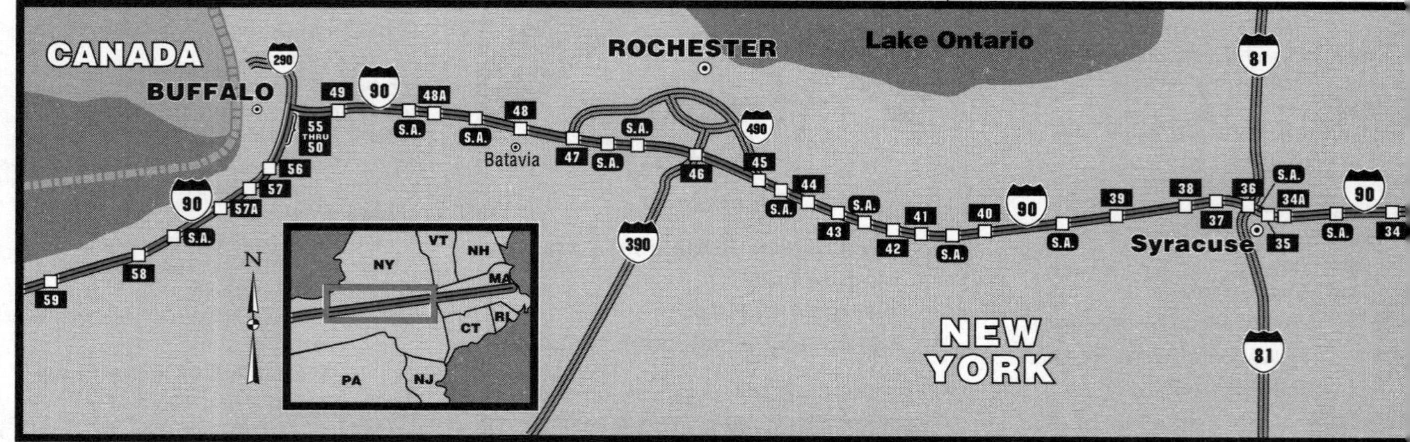

Bold red print shows RV & Bus parking available or nearby

XIT	NEW YORK
ASserv	W: Sunoco
ATM	W: OnBank
Other	W: State Patrol Post
1	**(320)** NY 414, Waterloo, Clyde
FStop	S: Mobil*(LP)
Food	S: Magee Country Diner
Other	S: Wildlife Refuge, Women's Hall of Fame
318)	Parking Area - Picnic, Phones (Westbound)
310)	Port Byron Service Plaza (Eastbound)
FStop	E: Mobil
Food	E: Ben & Jerry's, Mama Ilardo's Pizza, McDonalds
AServ	E: Mobil
ATM	E: On Bank
40	**(304)** NY 34 Weedsport, Auburn
FStop	S: Mobil*, Sunoco*(LP)
Gas	S: Quik Fill*
Food	S: Arby's, Arnold's Family Restaurant, DB's Drive In, Jrek Subs, New York Pizzeria
Lodg	S: Best Western, Days Inn
AServ	S: NAPA Auto Parts
TServ	S: Pullen's Truck Repair
RVCamp	N: River Forest Park
Med	S: Hospital
ATM	S: Cayuga Bank
Other	S: Ace Hardware(LP), Big M Convenience Store, Coin Laundry, Post Office
(292)	Warners Service Plaza (Westbound)
FStop	W: Mobil
Food	W: Ben & Jerry's, Mama Ilardo's Pizza, McDonalds
AServ	W: Mobil
ATM	W: On Bank
39	**(289)** Junction I-690, NY 690, Syracuse, Fulton
Food	N: Denny's, The Simmering Pot (Holiday Inn)
Lodg	N: Comfort Inn, Holiday Inn (see our ad this page)
TServ	N: Pen Truck Detroit Diesel Allison
38	**(286)** CR 57, Liverpool, Syracuse
Food	S: Houglin's
37	**(284)** Electronics Pkwy, Liverpool, Syracuse
Gas	S: Hess*
Food	S: Four Points Hotel
Lodg	S: Four Points Hotel, Knight's Inn
36	**(283)** Junction I-81, Watertown, Binghamton
(280)	DeWitt Service Plaza (Eastbound)
FStop	N: Sunoco
Food	N: Ben & Jerry's Ice Cream, McDonalds
AServ	N: Sunoco
ATM	N: OnBank
35	**(279)** NY 298, Syracuse, East Syracuse
Gas	S: Mobil*, Sunoco* (Natural Gas Fuel)
Food	S: Anthony's, Denny's, Fred Grimaldi's Chop House, Holiday Inn, Howard Johnson, Hub Diner, Joey's Fine Italian, Marriot, McDonalds, Pronto's Pizza
Lodg	S: Comfort Inn, Courtyard by Marriott, Days Inn, Embassy Suites, Extended Stay America, Fairfield Inn, Hampton Inn, Holiday Inn (see our ad this page), Howard Johnson, John Milton Inn, Marriott, Microtel, Motel 6, Ramada

EXIT	NEW YORK
	Limited Inn (see our ad this page), Red Roof Inn, Residence Inn
AServ	S: Mobil
ATM	S: OnBank
34A	**(277)** Junction I-481, Chittenango
(266)	Chittenango Service Plaza (Westbound)
FStop	N: Sunoco
Food	N: Dunkin Donuts, Sbarro Italian, TCBY Yogurt
AServ	N: Sunoco
ATM	N: OnBank
34	**(262)** NY 13, Canastoca, Oneida
Gas	S: Mobil*(D)(CW)
Food	S: Arby's, Graziano's Motel, McDonalds
Lodg	S: Days Inn, Graziano's Motel
AServ	S: S & J Associates Auto Service
RVCamp	S: Ta-Ga-Soke Campground (7 Miles), The Landing Campground (7 Miles), Verona Beach State Park/Camping (7 Miles)
Other	S: Canal Museum, International Boxing Hall of Fame
(256)	Parking Area - Phones, Picnic (Westbound)

EXIT	NEW YORK
33	**(253)** NY 365, Verona, Rome
FStop	N: Save On*(D)
Gas	N: Citgo*, Mobil*(LP)
Food	N: Joel's Front Yard Steak House, Mini Mason Jar (Sunoco)
Lodg	N: Super 8 Motel, Verona Motor Inn
RVCamp	N: Verona Beach State Park S: The Villages (2 Miles)
Med	S: Hospital
(250)	Parking Area - Picnic, Phones (Eastbound)
(244)	Oneida Service Plaza (Eastbound)
FStop	S: Sunoco
Food	S: Burger King, Sbarro Italian, TCBY Yogurt
AServ	S: Sunoco
ATM	S: OnBank
32	**(243)** NY 233, Westmoreland
Food	S: Carriage Motor Inn
Lodg	S: Carriage Motor Inn
Other	N: Delta Lake (10 Miles), Erie Canal Village (10 Miles), Ft Stanwix National Monument (10 Miles), Orisknay Battlefield (10 Miles) S: Airport, US Post Office
31	**(233)** Junction I-790, NY 8, NY 12, Utica
FStop	S: Sunoco*
Gas	N: Citgo*(D)(CW), Mobil* S: Hess(D), Mobil*
Food	N: Franco's Pizza & Pasta, Good Friend Chinese Restaurant, Lota'burger, Rosario's Pizzeria S: Best Western, Friendly's, Howard Johnson, Jack Appleseeds Grill, McDonalds, Pizza Hut, Reck Subs, Taco Bell, Wendy's
Lodg	S: A-1 Motel, Best Western, Howard Johnson, Motel 6, Red Roof Inn, Super 8 Motel
AServ	N: Papandrea's Automotive S: Genesee Automotive, Monroe Mufflers & Brakes
ATM	N: Marine Midland Bank
Other	N: Price Chopper Supermarket, Rite Aide Pharmacy
(227)	Schuyler Service Plaza (Westbound)
FStop	N: Sunoco
Food	N: Breyer's Ice Cream, McDonalds
AServ	N: Sunoco
ATM	N: OnBank
Other	N: State Patrol Post
30	**(220)** NY 28, Herkimer, Mohawk
Gas	N: Citgo*
Food	N: Chef's Home Cooking, Denny's, Friendly's, Mr Shake Ice Cream, Subway S: Mohawk Station
Lodg	N: Herkimer Motel, Mohawk Valley Motor Inn
AServ	N: Skinner Ford
Med	S: Hospital
Other	S: Big M Supermarket, Tourist Information
29A	**(211)** NY 169, Little Falls, Dolgeville
Med	N: Hospital
Other	N: General Herkimer Home
(209)	Indian Castle Service Plaza
FStop	B: Sunoco
Food	E: Bob's Big Boy, Mrs Fields Cookies, Roy Rogers W: Burger King, Dunkin Donuts, TCBY Yogurt
AServ	B: Sunoco
ATM	B: OnBank

Bold red print shows RV & Bus parking available or nearby

EXIT NEW YORK

29 **(194)** NY 10, Canajoharie, Sharon Springs
- **Gas** N: Gulf*(LP)
  S: Sunoco*
- **Food** N: Little Buffet Chinese, McDonalds, Pizza Hut
  S: Rosita's Place Bakery, The Pub Restaurant, Tony's Pizzaria, Village Restaurant
- **ASer**v N: Roosevelt's Auto Service Towing
- **ATM** S: Central National Bank
- **Other** N: Rite Aide Pharmacy
  S: Dievendorf Pharmacy, Tourist Information (No Trucks), U.S. Post Office

(184) Parking Area - Phones, Picnic (Both Directions)

28 **(182)** NY 30A, Fultonville, Fonda
- **TStop** N: Citgo*, Glen Travel Plaza*(LP), Travel Port Truckstop*(LP)(SCALES)
- **Gas** N: Gulf, Mobil (Glen Travel Plaza), Sunoco*(LP) (Glen Travel Plaza)
- **Food** N: Citgo Truck Stop, Glen Travel Plaza, McDonalds, Sugar Shack Bar and Grill, The Poplars Inn, Travel Port
- **Lodg** N: Glen Travel Plaza, Riverside Motel, The Poplars Inn, Travel Port
- **AServ** N: Gulf
- **TServ** N: Citgo, Travel Port
- **TWash** N: Citgo Travel Center, Glen Travel Plaza
- **Med** N: Hospital
- **ATM** N: Glen Travel Plaza
- **Other** N: State Patrol Post

27 **(174)** NY 30, Amsterdam
- **FStop** N: Getty*
- **Gas** N: Mobil*
- **Food** N: Sundowner
- **Lodg** N: Super 8 Motel, Valley View Motor Inn
- **Med** N: Hospital
- **Other** N: Coin Laundry (Valley View Inn)

(172) Mohawk Service Plaza (Eastbound)
- **FStop** E: Sunoco
- **Food** E: Breyer's Ice Cream, McDonalds
- **AServ** E: Sunoco
- **ATM** E: ONBank

(168) Pattersonville Service Plaza (Westbound)
- **FStop** W: Sunoco
- **Food** W: Bob's Big Boy, Mrs Fields Cookies, Roy Rogers
- **AServ** W: Sunoco
- **ATM** W: OnBank

EXIT NEW YORK

- **Other** W: Tourist Information

26 **(162)** Junction I-890, NY 5 S, Schenectady

25A **(159)** Junction I-88, NY 7, Schenectady, Binghamton

25 **(154)** Junction I-890, NY 7, NY 146, Schenectady

(153) Guilderland Service Plaza (Eastbound)
- **FStop** E: Sunoco
- **Food** E: Ben & Jerry's Ice Cream, McDonalds, Mr Subb
- **AServ** E: Sunoco
- **ATM** E: OnBank

24 **(148)** Jct I-87N, Jct I-90 E, Albany, Montreal (Left Exit Eastbound, All Traffic Exits To Northside)
- **Other** N: Crossgates Mall

Note: I-90 runs concurrent above with NYTHWY. Numbering follows NYTHWY.

1 **(147)** Westbound Junction I-87

2 **(146)** Washington Ave, Fuller Road

3 **(145)** Albany, State Offices

4 **(144)** NY 85, Albany, Slingerlands

5 **(143)** Everett Road

5A Cooporate Woods Blvd

6 NY9, Loudonville, Arbor Hill, Jct I-787

7 **(139)** Rensselaer, Defreestville, Washington Ave (Eastbound, Westbound Reaccess)

9 U.S. 4, Greenbush Road, Rensselaer, Troy

10 **(13)** Miller Road, E Greenbush

11 **(133)** U.S. 9, U.S. 20, Nassau
- **Gas** N: Citgo*(D), Hess*(D)
- **Food** S: Burger King
- **Lodg** S: AAA Four Seasons Motel
- **ATM** N: Key Bank
- **Other** N: State Patrol Post
  S: Coin Laundry, Grand Union Grocery, Rite Aide Pharmacy

(18) Rest Area

12 **(19)** U.S. 9, Hudson

EXIT NEW YORK/MASSACHUSETTS

B1 **(1)** Jct I-87 to New York City, Jct I-90 E Boston

B2 **(123)** NY 295, Taconic State Pkwy

B3 **(18)** NY 22, Austerlitz, New Lebanon
- **FStop** N: Citgo*(SCALES) (Grocery, Deli), Mobil*
- **Gas** S: Sunoco*(D)
- **Food** N: Racing Cafe
- **Lodg** S: Berkshire Spur Motel
- **TServ** N: Mobil

↑NEW YORK

↓MASSACHUSETTS

1 **(3)** MA 41, West Stockbridge

(8) Service Plaza Pinic Area
- **FStop** E: Mobil*
  W: Mobil*
- **Food** E: Burger King
- **Other** E: Welcome Center

2 **(11)** U.S. 20, Lee, Pittsfield , Adams
- **TStop** N: Diesel Dan's Truck Stop*
- **Gas** N: Texaco
  S: Shell*(D)
- **Food** N: Cathy's Kitchen (Dan's TS)
  S: Burger King, Friendly's, McDonalds
- **Lodg** N: Hunter's Motel (Diesel Dan's)
  S: AAA Pilgrim Motel, Sunset Motel, Super 8 Motel
- **AServ** N: Bob's Auto Service, NAPA Auto Parts, Texaco (Wrecker Service)
- **Other** N: Outlet Village

(29) Service Plaza
- **FStop** B: Mobil*
- **Food** W: Burger King
- **ATM** B: ATM

3 **(40)** U.S. 202, MA 10, Westfield, Northampton
- **Gas** N: Mobil*
  S: Texaco
- **Food** N: Friendly's
  S: Bickfords Family Restaurant, Dunkin Donuts
- **Lodg** S: Westfield Motor Inn
- **AServ** S: Texaco(D)
- **Med** N: Urgent Care Center
  S: Hospital

4 **(46)** Jct I-91, U.S. 5, West Springfield, Holyoke
- **Gas** N: Shell*(CW)

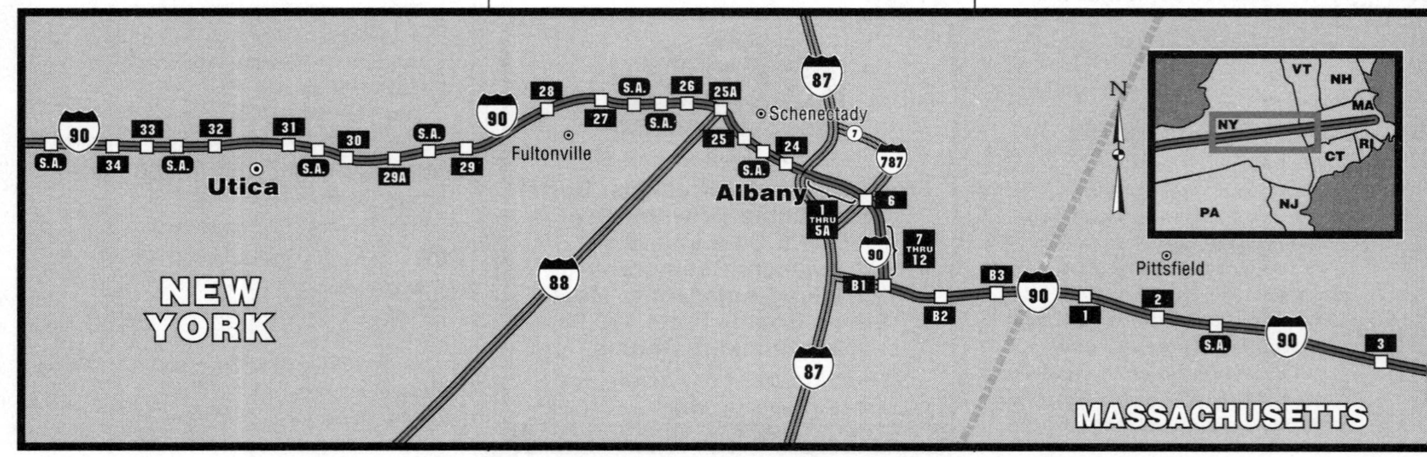

Bold red print shows RV & Bus parking available or nearby

EXIT MASSACHUSETTS

S: Citgo
Food N: Dunkin Donuts
S: B'shara's Restaurant, Bickfords Family Restaurant, Chili's, Donut Dip, Friendly's*, Kenny Rogers Roasters, Pizza Hut, Subway
Lodg S: Corral Motel, Hampton Inn, Knight's Inn, Motel 6, Quality Inn, Ramada Hotel, Red Roof Inn, Super 8 Motel
AServ S: BMW, Blade Collision Repair, Century Auto Services, Express Lube
Med N: Providence Hospital
Other S: Mall

(49) MA 33, Chicopee, Westoverfield
Food S: Admiral DW's, Burger King, Denny's, IHOP, Pizza Hut, Trumpet's (Comfort Inn), Wendy's
Lodg S: AAA Best Western, ◆ Comfort Inn
AServ S: Bob Pion Pontiac, Buick, GMC, Chicopee Cadillac, Olds, Midas Muffler, Monroe Muffler & Brake, Strauss Discount Auto
ATM S: Bank of Boston, Chicopee Savings, Fleet Bank
Other S: Big Y Supermarket, CVS Pharmacy, East Coast Market Place, Fairfield Mall, Super Stop Shop (ATM)

(51) JctI-291, Springfield, Hartford (CT)
FStop S: Pride*(D)(SCALES)
Gas S: Pride*
Food S: Fifty's Diner, McDonalds, Ramada Inn, Subway
Lodg S: Motel 6, ◆ Plantation Inn, Ramada Inn
TServ S: Dave's Truck Repair

(55) MA 21, Ludlow, Belchertown
Gas N: Cumberland Farms*, Gulf*, Mobil, Pride*(D), Sunoco*(CW)
Food N: Burger King, McDonalds
S: Dunkin Donuts, Friendly's, Joy's Country Kitchen, Subway
AServ N: Express Lube, Ludlow Tire Center, Mobil
ATM N: Bank Boston, Bay Bank, Fleet Bank
Other N: Big Y Supermarket, CVS Pharmacy, Rocky Bay Hardward, Serv-U Auto
S: Bat Ring Beach

56) Service Plaza
FStop B: Mobil*
Food B: Roy Rogers

(63) MA 32, Palmer, Amaherst
Gas S: Mobil*
Food N: McDonalds
S: Dunkin Donuts (Mobil)
AServ N: Auto Gallery, Baldyag's Auto Repair, GM Auto Dealership, Goodyear Tire & Auto, Palmer Auto Mall, The Car Store
S: A-Plus Transmission Center, Doorman's Garage
RVCamp S: Camping
Other N: Big Y Supermarket, Brooks Pharmacy, Car Wash

9 **(78)** Jct I-84, U.S. 20, Sturbridge

(80) Service Plaza (Eastbound)
FStop E: Mobil
Food E: Mrs. Fields Cookies, Roy Rogers, TCBY Yogurt
Other E: Tourist Information

(84) Service Plaza (Westbound)
FStop W: Mobil*
Food W: Burger King
Other W: Tourist Information

10 **(90)** Jct I-395, I-290, MA 12, Auburn, Worcester
Gas N: Texaco*
S: Shell*
Food N: Piccadilly Pub, Ramada Inn
S: D'Angelo's, Dunkin Donuts, Friendly's, Quality Pizza, Wendy's
Lodg N: AAA Ramada Inn
AServ S: Shell
ATM N: First MA Bank
Other N: CAT Hospital
S: AAA office, CVS Pharmacy, Honey Farms Mini Mart, Park N Shop Grocery

11 **(96)** MA 122, Millbury, Worcester, Providence, RI
Gas N: Getty
AServ N: Getty, Red's

(104) Service Plaza (Westbound)

11A **(106)** Jct I-495, NH, to Maine, Cape Cod

12 **(111)** MA 9, Framingham

(115) Service Plaza (Westbound)

13 **(117)** MA 30, Natick, Framingham

(118) Service Plaza (Eastbound)

(120) State Patrol Headquarters

14 **(123)** Junction I-95, MA 128, MA 30

15 **(124)** Junction I-95, MA 30, Weston

16 **(125)** West Newton

17 **(128)** Newton, Watertown
Lodg N: DAYS INN Days Inn (see our ad this page)

18 **(131)** Allston, Cambridge (Eastbound)

19 **(131)** Allston, Cambridge (Eastbound)

20 **(131)** Allston, Cambridge

21 **(132)** Massachusetts Ave, Boston (Eastbound)

22 **(133)** Prudential Center, Copley Square (Eastbound)

23 **(134)** Downtown Boston, South Station

24 **(134)** Junction I-93, Central Artery

↑MASSACHUSETTS

Begin I-90

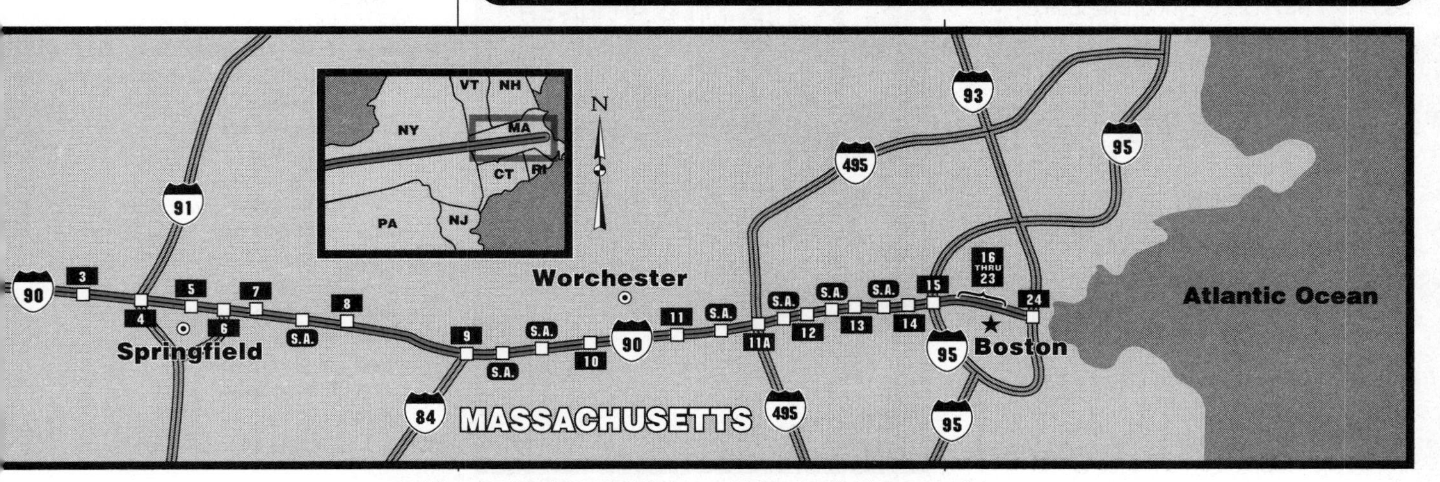

EXIT VERMONT

Begin I-91

↓VERMONT

29 **(177)** To U.S. 5, Derby Line

(176) **Welcome Center, Rest Area (Southbound)**

28 **(172)** U.S. 5, VT 105, Newport

- **FStop** **W:** **BP***
- **Gas** **E:** Petrol King*
  - **W:** Gulf*
- **Food** **W:** Dunkin Donuts, McDonalds, Village Pizza
- **Lodg** **W:** Pepin's Motel, ◆ **Super 8 Motel**
- **AServ** **E:** Turners Tires
  - **W:** Mulkin's Chevrolet, Sumner Tire
- **RVCamp** **E:** **RV Camping**
- **Med** **E:** ✚ **Hospital**
- **Other** **E:** **State Patrol Post, Tourist Information**
  - **W:** **Derby Veterinary Hospital, Rite Aide Pharmacy, Shop 'N Save Supermarket**

27 **(170)** VT 191, to U.S. 5, VT 105, Newport

- **Other** **W:** **Green Mountain Gas(LP), State Patrol Post**

(169) **Rest Area - RR, Phones (Northbound)**

(168) **Rest Area - Weigh Station (Southbound)**

(167) **Rest Area - Picnic (Both Directions)**

26 **(162)** U.S. 5, VT 58, Orleans, Irasburg

- **Gas** **E:** Irving, Sunoco*(LP)
- **Food** **E:** Applejack's, Henry's Pizza, Village Pizza
- **Lodg** **E:** Valley House Inn
- **AServ** **E:** Sunoco
- **RVCamp** **E:** **Campground**
- **ATM** **E:** Access
- **Other** **E:** **Austin Rexall Drugs, Cole's Market, Orleans General Store**

25 **(156)** VT 16, Barton, Hardwick

- **Gas** **E:** Gulf*(LP)
- **RVCamp** **E:** **RV Camping**
- **Other** **W:** **Travel Information**

(154) **Rest Area - No Services (Northbound, Steep Climb)**

(143) Scenic Overlook (Northbound)

(141) **Rest Area - RR, Phones (Southbound)**

24 **(140)** VT 122, Wheelock, Sheffield

23 **(137)** U.S. 5, Lyndonville, Burke

- **Gas** **E:** Gulf*(D)(LP), Mobil*
- **Food** **E:** Dunkin Donuts (Mobil), McDonalds
- **Lodg** **E:** DAYS INN ◆ Days Inn
  - **W:** ◆ Lyndon Motor Lodge
- **ATM** **E:** Passumpic Bank
- **Other** **E:** **Amerigas(LP), Rite Aide Pharmacy, Shop 'N Save Supermarket**

22 **(133)** to U.S. 5, St Johnsbury (Exit On Steep Grade)

- **Med** **E:** ✚ **Hospital**

21 **(131)** U.S. 2 to VT 15, St Johnsbury, Montpelier

20 **(129)** U.S. 5, St Johnsbury, to U.S. 2E

- **FStop** **E:** **BP***
- **Gas** **E:** Mobil*

EXIT VERMONT

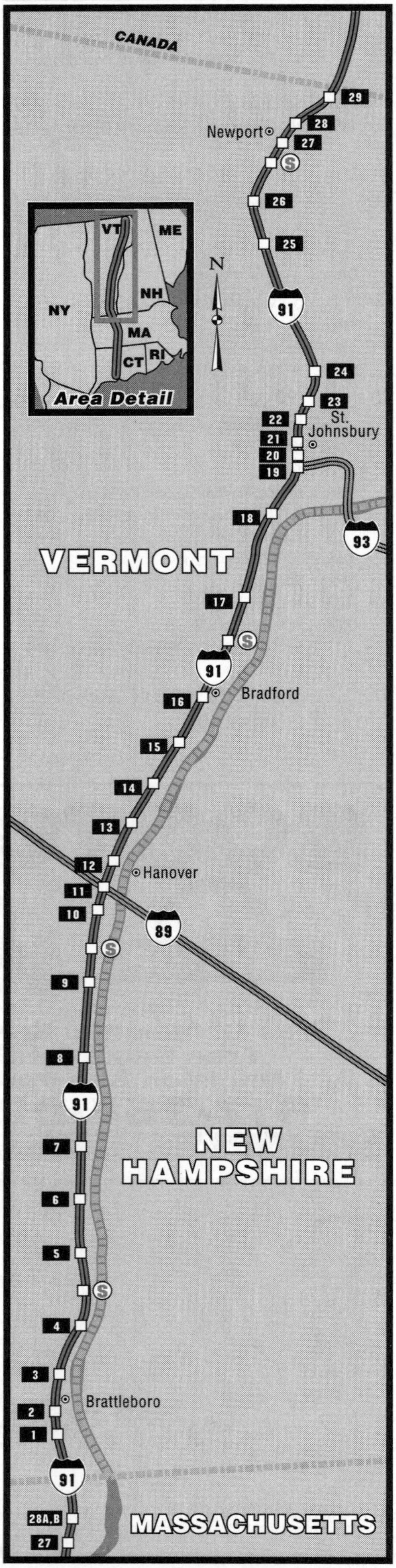

EXIT VERMON

- **Food** **E:** McDonalds, Subway
- **AServ** **E:** Mobil
- **Other** **E:** Car Wash

19 **(128)** Junction I-93 South

18 **(120)** U.S. 5, Barnet, Peacham

- **RVCamp** **E:** **Campground**
  - **W:** **Campground**

(115) **Rest Area - No Services (Southbound)**

(113) **Rest Area - No Services (Northbound)**

17 **(110)** U.S. 302, Wells River, Woodsville NH

- **TStop** **W:** **P & H Truck Stop(SCALES)**
- **FStop** **E:** **Mobil***
- **Food** **E:** Warners Gallery Restaurant
- **Med** **E:** ✚ **Hospital**
- **ATM** **W:** P&H Truckstop

(101) **Rest Area - Southbound - No Services, Northbound - RR, Phones**

(100) **Weigh Station (Both Directions, Closed)**

16 **(97)** U.S. 5, to VT 25, Bradford

- **Other** **W:** **State Patrol Post**

15 **(97)** U.S. 5, Fairlee, Orford NH

- **Gas** **E:** Mobil*, Texaco*(D)
- **Food** **W:** Lake Morey Inn
- **Lodg** **W:** Lake Morey Inn
- **AServ** **E:** Texaco
- **Other** **E:** **Fairlee General Store, Post Office**

14 **(84)** VT 113 to U.S. 5, Thetford

- **RVCamp** **E:** **Campground**
  - **W:** **Campground**

13 **(75)** U.S. 5, VT 10A, Norwich, Hanover NH

- **RVCamp** **E:** **Campground**
- **Med** **E:** ✚ **Hospital**

12 **(72)** U.S. 5, Wilder, White River Junction

- **Gas** **E:** Cumberland Farms* (Gulf)
- **Food** **E:** Blood's Seafood
- **Lodg** **E:** Wilder Motel (2 Miles)
- **Other** **E:** **Taft Optical Shop, White River Animal Hospital**

11 **(70)** U.S. 5, White River Junction

- **FStop** **W:** **Texaco***
- **Gas** **E:** Gulf*, Sunoco*
  - **W:** Citgo(LP)
- **Food** **E:** AJ's, Canton House Chinese, Cross Roads Cafe, **McDonalds**, Mustard Seed Deli, **William Tally House**
  - **W:** Howard Johnson
- **Lodg** **E:** Best Western, Coach An' Four Motel, AAA Comfort Inn, **Holiday Inn**
  - **W:** Best Western, Hampton Inn, Super 8 Motel
- **AServ** **E:** Big A Auto Parts, Gateway Motors, Mobil, White River Toyota, Wilder Auto & Tip Top Tire
  - **W:** Citgo
- **TServ** **E:** **Gateway Ford Trucks**
- **RVCamp** **W:** **Campground**
- **Med** **W:** ✚ **VA Hospital**
- **ATM** **E:** Mascoma Savings Bank
- **Other** **E:** Car Wash, **U.S. Post Office**

10 **(70)** Junction I-89, to New Hamp-

Bold red print shows RV & Bus parking available or nearby

XIT VERMONT

shire

8) **Rest Area - RR, Phones, Picnic (Both Directions)**

7) **Weigh Station (Both Directions)**

(61) U.S. 5, VT12, Hartland, Windsor
- **Med** **E:** ✚ **Hospital**

(51) U.S. 5, VT 12, VT 131, Ascutney, Windsor
- **FStop** **E:** **Citgo, Sunoco*, Texaco***
- **Gas** **E:** Mobil
- **Food** **E:** Mr. G's Restaurant
- **AServ** **E:** Citgo, Sunoco
- **Med** **E:** ✚ **Hospital**
- **Other** **E:** **Max's Country Village Store, Tourist Information**

(42) U.S. 5, VT 11, VT 106, Springfield VT, Charlestown NH
- **FStop** **W:** **Mobil(LP), Texaco***
- **Food** **W:** **HoJo**
- **Lodg** **W:** (AAA) **Holiday Inn Express,** Howard Johnson
- **AServ** **W:** Mobil
- **RVCamp** **W:** **Campground**
- **Med** **W:** ✚ **Hospital**

9) **Rest Area (Both Directions)**

(35) U.S. 5, VT 103, Rockingham, Bellows Falls
- **Gas** **E:** Citgo*(LP)
  W: Sunoco*(LP)
- **Other** **W:** **State Patrol Post**

(29) U.S. 5, Westminster, Walpole NH

4) **Rest Area - RR, Phones (Southbound, Closed)**

3) **Rest Area - RR, Phones (Northbound, Closed)**

2) **Weigh Station (Both Directions)**

0) **Weigh Station (Northbound)**

(18) U.S. 5, Putney
- **FStop** **W:** **Sunoco* (ATM)**
- **Gas** **W:** Mobil
- **Food** **E:** Putney Inn
  W: Curtis BBQ
- **Lodg** **E:** (AAA) Putney Inn
- **AServ** **W:** Mobil
- **ATM** **W:** Chittenden Bank, Putney Cafe, River Valley Credit Union, Vermont National Bank
- **Other** **E:** **Tourist Information**
  W: **Community Grocery & Deli, Post Office**

(12) U.S. 5, VT 9 East, Brattleboro, Keene (Bridge Clearance 13'2", Trucks Use Exit #2)
- **FStop** **E:** **Citgo**
- **Gas** **E:** Agway, Mobil*, Texaco(LP)
- **Food** **E:** Bickfords Family Restaurant, Chung Yan Chinese, Dunkin Donuts, Friendly's, Haun Chinese, KFC, Lamp Lighter Inn, McDonalds, N. End Market, Pizza Hut, Steak Out Restaurant, Village Pizza
- **Lodg** **E:** (AAA) Colonial Motel, (Days Inn) ◆ Days Inn (AAA), Lamp Lighter Inn, Motel 6, (AAA) Quality Inn, Riverside Motel, Super 8 Motel
- **AServ** **E:** Central Auto Parts, Dodge, Chrysler, Ford Dealership, Pontiac, GMC

EXIT VERMONT/MASSACHUSETTS

- **ATM** **E:** First Vermont Bank
- **Other** **E:** **Hanaford Supermarket, Mail Boxes Etc, Rite Aide Pharmacy, U.S. Post Office**

2 **(9)** VT 9, to VT 30, Brattleboro, Bennington
- **Gas** **E:** Sunoco
  W: Mobil*, Texaco*
- **Food** **W:** Gourmet Kitchen
- **Lodg** **E:** The Tudor Bed/Breakfast
- **AServ** **E:** Sunoco
- **Other** **W:** **Tourist Information, Vermont State Patrol Post**

1 **(8)** U.S. 5, Brattleboro
- **FStop** **E:** **Texaco*(D)(CW)**
- **Gas** **E:** Coastal*(CW), Mobil*, Sunoco*
- **Food** **E:** Burger King, Dunkin Donuts, Exit 1 Village Pizza, Vermont Inn Pizza
- **Lodg** **E:** Econolodge
- **AServ** **E:** Roberts Jeep/Chevrolet/Olds/Cadillac
- **RVCamp** **E:** **Campground**
- **Med** **E:** ✚ **Hospital**
- **ATM** **E:** Vermont National Bank
- **Other** **E:** **Brooks Pharmacy, Factory Outlet Center, Price Chopper Supermarket, VT State Police**

(0) VT Welcome Center - RR, Vending, Phones (Northbound)

↑VERMONT

↓MASSACHUSETTS

(54) **Parking Area (Both Directions)**

28AB **(50)** MA 10, Bernardston, Northfield
- **Gas** **W:** Citgo*, Sunoco
- **Food** **E:** Andiamo Italian
  W: Falls River Inn, Four Leaf Clover, Thunder Lodge Vegetarian
- **Lodg** **E:** Fox Inn
  W: Falls River Inn
- **RVCamp** **E:** **Campground**
  W: **Campground**
- **Other** **W:** **Marshall's Country Corner**

27 **(46)** MA 2 East, Greenfield, Boston
- **Gas** **E:** Citgo, Cumberland Farms*, Mobile, Sunoco*(CW)
- **Food** **E:** Burger King, Denny's, Friendly's, Little Caesars Pizza
- **AServ** **E:** Citgo, Dodge, Chrysler, Jeep, Meineke Discount Mufflers, Mobile, Pontiac, Oldsmobile, GMC,
- **TServ** **E:** **Goly Mac Truck Service**
- **Med** **E:** ✚ **Hospital**
- **Other** **E:** **CVS Pharmacy, Dairy Mart**

26 **(43)** MA 2 W, MA 2A E, Greenfield Center, N Adams
- **Gas** **E:** Mobil*
  W: Texaco*
- **Food** **E:** Howard Johnson, McDonalds
  W: Bickfords Family Restaurant, Bricker's, Friendly's, McDonalds, New Fortune Chinese, Red Spot Pizza, Turnbull's Family Restaurant
- **Lodg** **E:** Howard Johnson
  W: Candlelight Resort Inn, Motor Inn, ◆ Super 8 Motel
- **AServ** **W:** Nissan Dealer
- **Med** **E:** ✚ **Hospital**
- **ATM** **W:** United Bank

EXIT MASSACHUSETTS

- **Other** **W:** **Big Y Supermarket, Brooks Pharmacy, Waldbaum's Grocery Store**

(37) **Weigh Station (Both Directions)**

25 **(36)** MA 116, Deerfield, Conway (Southbound, Northbound Reaccess Only)
- **Food** **E:** New China Restaurant
- **RVCamp** **W:** **White Birch Campground**

24 **(34)** U.S. 5, MA 10, MA 116, Whately (Difficult Northbound Reaccess)
- **TStop** **W:** **BP Truckstop & Restaurant**
- **Gas** **E:** Citgo*
- **Food** **E:** New England Country Sampler, Sugarloaf Delif
- **Lodg** **E:** **Motel 6**
- **ATM** **E:** BankBoston
- **Other** **E:** **Tourist Information**

(33) **Parking Area - No Services**

23 **(32)** U.S. 5, Whately, MA10, N Hatfield (Southbound, Reaccess Northbound Only)
- **Food** **E:** Jenny's Long Steamed Dogs, Tom's Long HotDog & Burgers
- **Lodg** **E:** Colony Motel, Rainbow Motel
- **RVCamp** **E:** **Rainbow Campground**

22 **(30)** U.S. 5, MA 10, N Hatfield, Whately (Northbound, Reaccess Southbound Only)

21 **(27)** U.S. 5, MA 10, Hatfield
- **Gas** **W:** Sunoco
- **Lodg** **W:** Country View Motel, The North King Motel
- **AServ** **W:** Sunoco
- **RVCamp** **W:** **Long View RV Service(LP)**
- **Other** **W:** **State Patrol Post**

20 **(26)** U.S. 5, MA 9, MA 10, Northampton, Hadley (Southbound, Difficult Southbound)
- **Gas** **W:** BP, Mobil*, Pride*(D)
- **Food** **W:** Bluebonnet Diner, Burger King, Chip's Ice Cream, D'Angelo Sandwich Shop, Hunan Chinese Garden, Papa Gino's, Ponderosa
- **AServ** **W:** Burke Whitaker Cadillac/ Olds, Chrysler Auto Dealer, Dana Chevrolet/VW, Firestone Tire & Auto, Ford Dealership, Monroe Muffler & Brake, NAPA Auto Parts, Town Fair Tire
- **Med** **W:** ✚ **DVA Medical Center**
- **ATM** **W:** BankBoston, SIS Bank

19 **(25)** MA 9, N Hampton, Amhurst (Northbound, Difficult Northbound Reaccess)
- **Gas** **E:** Citgo, Getty*
- **Food** **E:** Banana-Rama Frozen Yogurt, Webster's Fish Hook
- **AServ** **E:** Citgo

18 **(23)** U.S. 5, N Hampton, E Hampton
- **Gas** **E:** Mobil
  W: Shell*
- **Food** **E:** The Inn at North Hampton
  W: 5-91 Diner
- **Lodg** **E:** The Inn at North Hampton
  W: (Days Inn) Days Inn
- **AServ** **E:** Mobil
  W: Pro-Lube
- **Other** **W:** Pleasant Street Car Wash

(19) Scenic View (Southbound)

EXIT MASSACHUSETTS

17AB **(16)** MA 141, S Hadley
- Gas E: Citgo[D], Mobil*[D]
- Food E: Bess-n-Eaten Donuts, Real China, Subway, Taco Bell
- Lodg E: Motel 8, Ramada (see our ad this page)
- AServ E: Joe Leisur's Motor Clinic, Magna Buick, Reardon's Auto Service
- ATM E: Bank of Western MA, Fleet Bank
- Other E: **Brooks Pharmacy, Dairy Mart Convenience Store**

16 **(14)** U.S. 202, Holyoke, Westfield
- Food E: The Yankee Peddler
- Lodg E: Peddler Court
- Med E: **✚ Hospital**
- ATM E: United Corporate Bank
- Parks E: **Hertiage State Park**

15 **(12)** Ingleside
- Gas E: Shell*
- Lodg E: ◆ Holiday Inn
- AServ E: Sears
- Med E: **✚ Hospital**
- ATM E: Fleet Bank
- Other E: **Holyoke Mall, Petco (Pet Supply)**

14 **(11)** Junction I-90, MA Tnpk (toll) to Boston, Albany NY

13AB **(9)** U.S. 5, West Springfield, Riverdale St
- Gas E: Citgo*, Shell*
  W: Mobil*[CW], Sunoco*[CW]
- Food E: B'Shara's, Bigford's Pancakes & Family Fair, Donut Dip, Dunkin Donuts, Kenny Rogers Roasters, Piccadilly Pub & Restaurant, Subway, Wendy's
  W: Abbow's Family Restaurant, Burger King, Chi Chi's Mexican, Chili's, Chip's Ice Cream, D'angelo's, Debbie Wong, Friendly's, Hoots Bar & Grill, Intal Buffet, KFC, Old Country Buffet, Pizza Hut
- Lodg E: Arrowhead Motel, **Capri Motel**, Cyril's Motel, Knight's Inn, Motel 6, Super 8 Motel
  W: Elsie's Motel, Hampton Inn, Quality Inn, Ramada
- AServ E: Balise Lexus, Honda, Toyota, Red's Quick Lube Center, Wagner BMW
  W: Town Center Tire
- ATM W: BankBoston, SIS Bank, West Bank
- Other E: **AAA Office**
  W: **Riverdale Shopping Center, Super Stop Shop (Pharmacy)**

12 **(8)** Junction I-391 North, Chicopee

11 **(7)** U.S. 20 West, Birnie Ave, W Springfield
- Med E: **✚ Bay State Trauma Center**

10 **(7)** Main St, Springfield (Northbound)
- Gas E: Mobil*[CW]
- Med E: **✚ Bay Street Trauma Center**
- ATM E: BankBoston
- Other E: **Post Office**

9 **(7)** U.S. 20 West, West Springfield, MA 20 E (Northbound, Northbound Reaccess Only)

8 **(7)** Junction I-291, U.S. 20 East, Boston

7 **(6)** Columbus Ave, Springfield Center (Southbound)
- Other W: **Basketball Hall of Fame**

6 Springfield Center (Northbound)
- AServ E: Brake King
- ATM E: Bank of W. MA

EXIT MASSACHUSETTS/CONNECTICUT

5 Broad St, Springfield
- FStop W: **Sunoco**
- Gas W: Sunoco*[CW], Todd
- AServ E: Houser Hyaundi, Buick, Houser Mitsubishi

4 **(5)** Main St, MA83
- Gas E: Mobil, Texaco*[D]
- AServ E: Houser Hyaundai/ Mitsubishi
  W: Balise Chevrolet, Olds

3 **(5)** U.S. 5 N, Columbus Ave
- Gas E: Citgo

2 **(4)** MA 83, Forest Park, East Longmeadow (Northbound)

1 **(4)** U.S. 5 South, Forest Park, Longmeadow

↑MASSACHUSETTS

↓CONNECTICUT

49 **(58)** U.S. 5, Longmeadow MA, Enfield St
- Gas E: Citgo*, Mobil*
- Food E: Friendly's, Twister's Ice Cream & Yogurt
  W: Cloverleaf Cafe, Dairy Queen, McDonalds
- Lodg E: ◆ Harley Hotel
- AServ E: Meineke Discount Mufflers, State Line Service
- Other W: **Taylor Rentals[LP]**

48 **(56)** CT 220, Elm St, Thompsonville
- Gas E: Mobil*
- Food E: Boston Seafood, Burger King, Chi Chi's Mexican, China Buffet, Denny's, Dunkin Donuts, Friendly's, Kenny Rogers Roasters, McDonalds, Pumpernickel Pub & Restaurant, Ruby Tuesday, Wendy's
- AServ E: Auto Palace, Sears
- ATM E: Webster Bank
- Other E: **AAA Office, CVS Pharmacy, Enfield Square Mall, Super Food Mart**

47 **(55)** CT 190, Hazardville, Summers
- Gas E: Gulf[CW], Mobile*[D]
- Food E: Bickfords Family Restaurant, Carvel Ice Cream, Chang's Garden, D'Angelo's, Dunkin Donuts, Ground Round, HomeTown Buffet, KFC, Little Caesars Pizza, McDonalds, Olive Garden, Pizza Hut Express, Taco Bell
- Lodg E: Motel 6, Red Roof Inn
- AServ E: Enfield Ford, Goodyear Tire & Auto
- Med E: **✚ Hospital (10 Miles)**
- ATM E: Bank Boston, New England Bank, People's Bank, Savings Bank of Manchester
- Other E: **Mail Boxes Etc, RX Pharmacy, Shaw's Supermarket (ATM), Super Stop & Shop, Vision Corner, WalGreens**

46 **(53)** U.S. 5, King St
- Gas E: Mobil*[CW]
- Food E: Astro's Plaza
  W: Crystal
- Lodg W: (AAA) Super 8 Motel
- Other E: King St Car Wash
  W: **Post Office**

45 **(51)** CT 140, Ellington, Warehouse Point
- FStop W: **Sunoco*[D]**
- Gas E: Shell*[CW]
- Food E: Best Bagels, Burger King, Dunkin Donuts, Giorgio's Restaurant & Deli, Kowloon Chinese, Puffins Ice Cream Palor, Sofia's Seafood/Pasta/ Pizza, Sweet Sensations, Wings Bar & Grill
  W: **Best Western**
- Lodg W: (AAA) **Best Western**
- AServ E: **Cammon Care (RV Service)**, Radiators & Air Conditioning
- Med E: **✚ East Windsor Medical Center**
- ATM E: SBM Bank
- Other E: **Coin Laundry**

44 **(50)** U.S. 5 South, East Windsor
- FStop E: **Citgo***
- Gas E: Citgo
- Food E: Dunkin Donuts, **East Windsor Restaurant**, LaNotte Fine Dining, Primavera Restaurant
- Lodg E: ◆ Holiday Inn Express
- AServ E: Citgo

EXIT CONNECTICU

- Other E: **Wal-Mart**

42 **(49)** CT 159, Windsor Locks
- Gas W: Gulf
- Food W: Charles-Ten
- AServ W: Gulf (towing)

41 **(49)** Center St
- Food E: Ad Pizzaria
- RVCamp E: **Longview RV Super Store**
- Other W: **K-Mart**

40 **(48)** Bradley Int'l Airport, CT 20

39 **(48)** Kennedy Rd (Northbound)
- Gas W: Shell*[CW]
- Food W: Charkoon
- ATM W: Fleet Bank
- Other W: **K-Mart (Pharmacy), Super Stop & Shop**

38 **(46)** CT 75, Poquonock, Windsor
- Gas E: Mobil*
- Food E: China Sea, Domino's Pizza, McDonalds, Subway, The Beanery Bistro
- Lodg W: ◆ Courtyard by Marriott

37 **(45)** CT 305, Bloomfield Ave, Windsor Center
- Gas E: Mobil*
  W: Citgo*
- Food E: McDonalds
- Lodg W: ◆ Residence Inn

36 **(44)** CT 178, Park Ave

35B **(45)** Junction I-291, Windsor, Bloomfield

35 **(45)** Junction I-291, Manchester

34 **(41)** CT 159, North Main St, Hartford

33 **(40)** Jennings Rd

32B **(39)** Jct I-84 West, Trumball St, Waterbu

- AServ W: Goodyear Tire & Auto

30 **(38)** Junction I-84 East (Northbound)

29A **(37)** Capital Area
- Med W: **✚ Hospital**

28 **(36)** U.S. 5, CT 15, Berlin TNPK, Wethersfield

27 **(35)** Airport Rd, Brainard
- FStop E: **Texaco*[CW]**
- Gas W: Citgo*, Save A Step*
- Food E: USS Chowder Pot IV Seafood & Prime Rib
  W: Baker's Dozen Donuts, Burger King, Carmicheal's, Dunkin Donuts, Tasalama, Wendy's
- Lodg E: (DAYS INN) Days Inn, Susse Chalet Inn
- AServ W: Meineke Discount Mufflers, Super Lube
- TServ E: **International, Interstate Ford**
- ATM W: Webster Bank

26 **(34)** Marsh St, Old Wethersfield

25 **(34)** CT 3, Wethersfield, Glastonbury

24 **(32)** CT 99, Wethersfield, Rocky Hill
- Gas E: Mobil*
  W: Mobil*, Texaco
- Food E: Bickfords Family Restaurant, Boston Market, China Pavilion, J. Copperfield, McDonalds, Subway
  W: Bagel n Boo, Bombay Bicycle Club, D'Angelo's, **Denny's (see our ad on opposite page)**, KFC, Ming Dynasty, Old Country Buffet, **Panda King Chinese (Goff Brook Shops)**, Red Lobster, The Ground Round, Town Line Diner
- Lodg E: Great Meadow Inn, Susse Chalet, Travelers Budget Lodge
  W: Motel 6, Ramada
- AServ E: Midas Muffler, Monroe Muffler & Brake
  W: Auto Palace, Texaco
- ATM W: Bank of Boton, People's Bank
- Other W: **Goof Brook Shops, Mail Boxes Etc, Medicine Shoppe, Walbums Food Mart**

23 **(29)** To CT 3, West St, Rocky Hill
- Gas W: Citgo, Mobile
- Food E: Marriott
  W: Bagelz, D'Angelo's, Manhattan Bagel, McDonalds, Michelangelo's Pizzeria, Pizza Hut, Westside Pizza
- Lodg E: ◆ Marriott
- AServ W: Citgo, Mobile
- Med E: **✚ VA Hospital**

Bold red print shows RV & Bus parking available or nearby

EXIT		CONNECTICUT
	ATM	**W:** Farmers & Merchants Bank, Fleet Bank, Money Center
	Parks	**E:** Dinosaur State Park
	Other	**W:** Coin Car Wash
2		**(28)** CT 9, New Britain, Middletown
		(27) CT 372, Cromwell, Berlin
	Gas	**E:** Sunoco(D)
		W: Citgo(D)(LP), Mobil*(D), Shell
	Food	**E:** Radisson, Wooster Street Pizza
		W: Blimpie's Subs, Burger King, Cromwell's Diner, Holiday Inn, McDonalds, Sherlock's Cafe
	Lodg	**E:** (AAA) Comfort Inn, ◆ Radisson
		W: (AAA) Holiday Inn, Super 8 Motel
	AServ	**E:** Sunoco
		W: Bishop's Garage, Citgo, Firestone Tire & Auto, Shell
	Med	**W:** ✚ Walk In Care
	ATM	**E:** Webster Bank
	Other	**W:** Companion Animal Hospital, Wal-Mart (Pharmacy)
		(23) Country Club Rd, Middle St
3)		Rest Area - RR, Phones, RV Dump (Northbound)
2)		Weigh Station - RV Dump (Northbound)
		(21) Baldwin Ave, Preston Ave (Southbound Only, Reaccess Northbound Only)
		(20) Junction I-691W, Meridian, Waterbury
		(19) CT 15 South, West Cross Pkwy
	Lodg	**E:** Ramada Plaza Hotel (see our ad on this page)
		(19) East Main St, Shelton (Northbound)
	Gas	**E:** Gulf, Mobil*, Texaco*
		W: Amoco, Exxon*, Getty, Gulf, Sunoco
	Food	**E:** American Steak House, Dunkin Donuts, King Garden's Chinese, Little Caesars Pizza, Royal Guard Fish & Chips
		W: Bess Eaton Donuts, Boston Market, Burger King, Friendly's, KFC, Manhattan Bagel, McDonalds, Olympus Diner, Taco Bell
	Lodg	**E:** Hampton Inn, Meriden Inn
		W: Regents Motel
	AServ	**E:** Ford Dealership
		W: Car Quest Auto Center, Gulf
	Med	**W:** ✚ Hospital
	ATM	**W:** Dines Savings Bank, First Union Bank
	Other	**E:** CVS Pharmacy

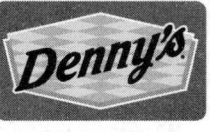

EXIT CONNECTICUT

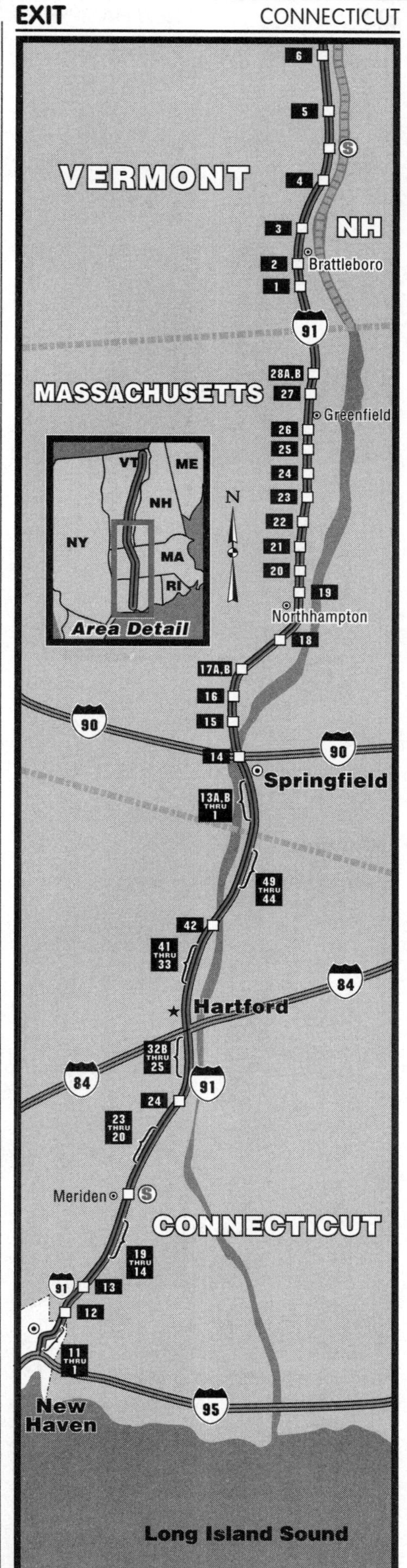

EXIT		CONNECTICUT
		W: Brooks Pharmacy, Hancock's Pharmacy
15		**(16)** CT 68, Yalesville, Durham
	Lodg	**W:** ◆ Courtyard by Marriott, ◆ Susse Chalet
(14)		Rest Area RR, Phones, Picnic, Tourist Info, Canteen
14		**(13)** CT 150, East Center St (Southbound, Difficult Southbound Reaccess)
	Food	**E:** Hong Kong
	Other	**E:** Coin Laundry, Dairy Mart, East Center Market
13		**(11)** U.S. 5 Wallingford, North Haven
	Parks	**W:** Whorton Brook State Park
12		**(8)** U.S. 5, Washington Ave
	Gas	**E:** Exxon*(D), Shell, Sunoco*, Texaco
	Food	**E:** Angelo's Pizza, Boston Market, Burger King, China Buffet, D'angelo's, Dunkin Donuts, Friendly's, McDonalds, Pizza House, Roy Rogers, The Rustic Oak
		W: Antonio's Sub Shop, Athena Diner (Greek), Athena's Diner, Holiday Inn
	Lodg	**W:** (AAA) Holiday Inn
	AServ	**E:** Insty-Lube (Texaco), Shell
	Other	**E:** WalGreens, Washington Center Shopping Plaza
		W: Washington Avenue Car Wash
11		**(8)** CT 22, New Haven (Northbound)
	Gas	**E:** Shell, Sunoco*, Texaco
	Food	**E:** Carbones Deli, Carvel Ice Cream, China Buffet, Dunkin Donuts, Hunan, Java Jive, Pizza House, Subway, The Bagel
	AServ	**E:** Auto Palace, Shell, Texaco
	ATM	**E:** First Union Bank, People's Bank, Webster Bank
	Other	**E:** Boston Market, CVS Pharmacy, Optical, Penny's Outlet, Super Stop & Shop, Town Fair, WalGreens, Walbaum Supermarket
10		**(7)** CT 40, Hamden, Cheshire
9		**(5)** Montowese Ave
	FStop	**W:** Berkshire*
	Gas	**E:** Citgo
		W: Sunoco
	AServ	**W:** Sunoco
	Other	**E:** Dairy Mart
8		**(3)** CT 17, CT 80, Middletown Ave, N Branford
	Gas	**E:** Amoco*(CW), Exxon*, Shell*(CW), Sunoco*
		W: Mobil
	Food	**E:** Burger King, Carvel Ice Cream, D'Angelo's Sandwich Shops, Dunkin Donuts, Exit 8 Diner, McDonalds, Pizza Hut
	Lodg	**E:** Motel 6
	AServ	**E:** Castro Oil Express, Valvoline Quick Lube Center
		W: Mobil
	Other	**E:** Car Wash, K-Mart
7		**(3)** Ferry St, Fair Haven (Southbound, Difficult Reaccess)
6		**(2)** U.S. 5, Willow St, Blatchly Ave
5		**(1)** U.S. 5, State St (Northbound, Very Difficult Reaccess)
4		**(1)** State St
3		**(1)** Trumbull St
2		**(0)** Hamilton St
1		**(0)** CT 34, Downtown New Haven
	Lodg	**W:** Holiday Inn (see our ad on this page)

↑CONNECTICUT

Begin I-91

EXIT	VERMONT/NEW HAMPSHIRE
	Begin I-93

↓VERMONT

(11) I-91S to Whitewater Jct, I-91N to St Johnsbury

1 **(7)** to U.S. 2, Vermont 18, St. Johnsbury
- **Food** **E:** Aime's Motel
- **Lodg** **E:** (AAA) Aime's Motel (AAA)
  - **W:** The Moon Star Inn
- **RVCamp** **E:** Moose River Campground
- **Other** **E:** Pettio Country Store*

(1) VT Welcome Center - RR, Phones, HF, Picnic

↑VERMONT

↓NEW HAMPSHIRE

(130) NH Welcome Center - RR, Phones

44 **(130)** NH 18, NH 135, Monroe, Waterford CT
- **Other** **W:** Rest Area & Scenic View (RR, Picnic Phones, Tourist Information)

43 **(126)** NH 135, to NH 18, Littleton, Dalton
- **RVCamp** **W:** Crazy Horse Campground

42 **(124)** U.S. 302, NH 10
- **FStop** **W:** Irving
- **Gas** **E:** Sunoco*
  - **W:** Gulf*, Mobil*
- **Food** **E:** Burger King, Cafe Munchies, Clam Hut, Jing Fong Italian, McDonalds, Pizza Hut, Subway, The Clam Shell
  - **W:** Continental Motor Inn
- **Lodg** **W:** Continental Motor Inn
- **AServ** **E:** Convenient Lube
  - **W:** GM Auto Dealership
- **RVCamp** **W:** KOA Campground
- **ATM** **W:** Berlin National Bank
- **Other** **E:** Coin Laundry, Rite Aide Pharmacy, The Medicine Shoppe Drugs
  - **W:** Stop N Save Grocery Store

41 **(123)** U.S. 302, NH 18, NH 116, Littleton
- **Gas** **E:** Irving*[D][LP] (ATM)
- **Food** **E:** Bishop's Ice Cream
- **Lodg** **E:** (AAA) Eastgate Motor Inn
- **TServ** **W:** Commercial Tire
- **Med** **E:** ✚ Hospital

40 **(121)** U.S. 302, NH 10 East, Bethlehem, Twin Mtn
- **Gas** **E:** Exxon
- **Food** **E:** Adair Country Inn
- **Lodg** **E:** Adair Country Inn
- **AServ** **E:** Exxon
- **RVCamp** **E:** Snowy Mountain Campground
- **Med** **W:** ✚ Hospital

39 **(119)** NH 18, NH 116 , N Franconia, Sugar Hill (Southbound)

38 **(117)** NH 18, NH 116, NH 117, NH 142, Franconia, Sugar Hill, Lisbon (Difficult Reaccess)
- **Gas** **W:** Mobil
- **Food** **E:** Red Coach Inn

EXIT	NEW HAMPSHIRE

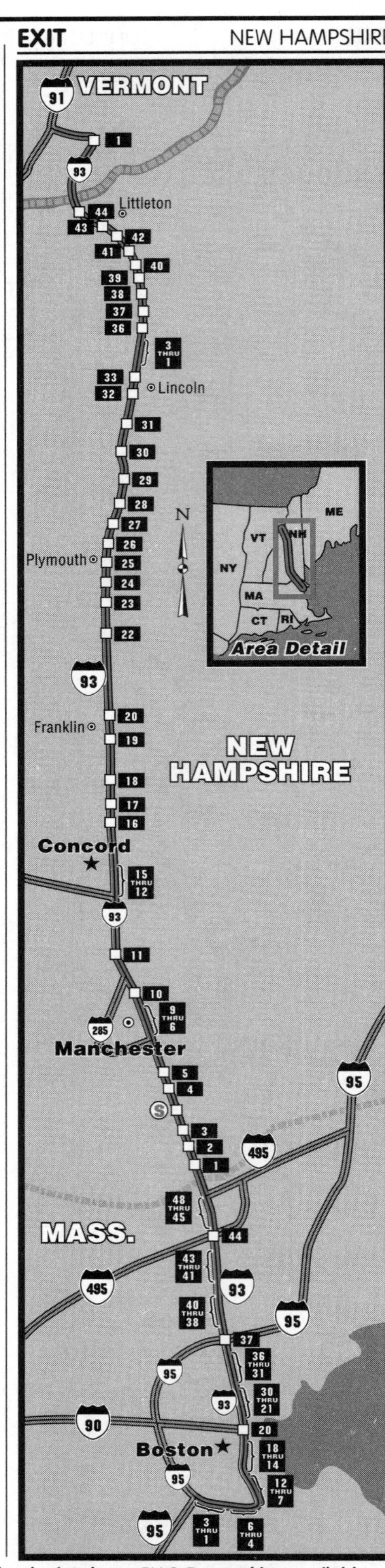

EXIT	NEW HAMPSHIRE

- **W:** Cannon Ball Pizza, Dutch Treat Restaurant, Hillwinds Motel, Quality Bakery
- **Lodg** **E:** (AAA) Red Coach Inn
  - **W:** Hillwinds Motel
- **ATM** **W:** People's Bank
- **Other** **E:** Robert Front Museum
  - **W:** Franconia Food Store, Kelly's Food Town, Tourist Information

37 **(114)** NH 142, NH 18, Franconia, Bethlehem (Northbound)
- **Lodg** **E:** ◆ The Inn
  - **W:** Hill Winds, Raynor's Motor Lodge
- **RVCamp** **W:** Franstead Family Campground

36 **(113)** NH 141, South Franconia

35 **(112)** U.S. 3, Twin Mountain, Lancaster
- **Other** **E:** White Mountain National Forest

Note: I-93 runs concurrent below with FNPKWY. Numbering follow FNPKWY.

3 **(111)** NH 18, Echo Lake Beach, Peabody Lodge
- **Other** **E:** Govenor Gallen Memorial, Viewing Area
  - **W:** Echo Lake Swimming & Parking Area

2 **(110)** Cannon Mountain Tramway, Old Man Viewing
- **Other** **W:** Park Information Center

(110) Trail Head Parking Area (Southbound)

(109) Boise Rock Parking Area (No Trucks, Northbound)

1A **(108)** Lafayette Place Campground (No Busses)

(108) Trail Head Parking Area - RR, Phones

(107) Rest Area/The Basin - RR

1 **(105)** U.S. 3, The Flume Visitor Cente
- **Other** **E:** Applachian Trail, Flumes Gorge & Visitor Center

Note: I-93 runs concurrent above with FNPKWY. Numbering follow FNPKWY.

33 **(103)** U.S. 3, North Woodstock, Nort Lincoln
- **Gas** **E:** Irving*, Mobil[D]
- **Food** **E:** Dad's Lounge, Fresolone's Italian, Longhorn Palace, Mountaineer Motel, Mr. Freeze, Notch View Country Kitchen, Woodward Motor Inn
  - **W:** Georgiana Falls
- **Lodg** **E:** Beacon North, (AAA) Drummer Boy Motor Inn, (AAA) Indian Head Resort, (AAA) Mount Coolidge Motel, Mountaineer Motel, Pemmi Motor Ct., (AAA) Profile Motel, (AAA) Red Doors Motel, SanTonio's Notch Motel, The Beacon Resort, Village Green by the River, Woodward Motor Ir
  - **W:** Cozy Cabins, Mount Liberty Motel, White Mountaineer Motel
- **AServ** **E:** Dale's Radiator, Mobil
  - **W:** Ted's White Mountain Garage
- **RVCamp** **W:** Cold Springs Camp, Country Bumpkins Campground & Cottages[LP], Lost River Gorg
- **Other** **E:** Whale's Tale Water Park
  - **W:** Clark's Trading Post (RV Dump)

Bold red print shows RV & Bus parking available or nearby

XIT NEW HAMPSHIRE

2 **(101)** NH 112, North Woodstock, Lincoln

Gas **E:** Citgo*(D), Irving*, Mobil*(CW)
Food **E:** Bill & Bob's Famous Roast Beef, Bishops Homemade IceCream Outlet, Burger King, Chieng Garden, Dragon Light, Dunkin Donuts, Earl of Sandwich, Elvio Pizzeria, GH Pizza & Greek Restaurant, Kings Korner, McDonalds, Micheal's, Mr W 's House of Pancakes, New England Yogurt Co., Old Timber Mill, Pizza Hut, Skoop's, The Country Mile, The Italian Garden, The Mill House Inn, White Mountain Chowder House, White Mtn Bagel Company
Lodg **E:** River Green Restort Hotel, The Lincoln Motel, The Mill House Inn
AServ **E:** Nosworthy Automotive Supply
RVCamp **W:** **KOA Campground**
Med **E:** ✚ **Winwood Medical Center**
ATM **E:** Citizen's Bank, Fleet Bank
Other **E:** **Coin Laundry, Grand Union Grocery Store, Lincoln Police, Post Office, Rite Aide Pharmacy, White Mountain Information Center**

1 **(98)** To NH 175, Tripoli Rd

RVCamp **E:** **Russel Pond Campground**
W: **Campground**

0 **(95)** U.S. 3, Woodstock, Thornton

Food **E:** Frannie's Place, Jack-O-Lantern Resort
Lodg **E:** Jack-O-Lantern Resort, Pioneer Motel

9 **(89)** U.S. 3, Thornton

Lodg **W:** Gilcrest Motel
RVCamp **E:** **Pemi River Campground***

8 **(88)** NH 49, Campton, Waterville Valley, to NH175

Gas **E:** Citgo*, Mobil*
Food **E:** The Lost Sailor Grill
W: Scandanavi Inn
Lodg **W:** Scandanavi Inn
AServ **E:** Citgo, Mobil
RVCamp **W:** **RV Campground**
Other **E:** **National Forest Information, Post Office, Quik Pik Grocery, Tourist Information, True Value Hardware(LP)**

7 **(84)** Blair Bridge, W Campton , to U.S. 3

Food **E:** The Covered Bridge Restaurant
Lodg **E:** AAA Best Western (AAA), Red Sleigh Chalets & Motel

6 **(82)** U.S. 3, NH 25, NH 3A, Plymouth, Rumney

Food **E:** **McDonalds**
Lodg **E:** Pilgrim Inn, ◆ Susse Chalet
RVCamp **E:** **Campground**
Med **E:** ✚ **Hospital**
Other **E:** **Pilgrim Police**

5 **(81)** NH 175A, Holderness Rd, Plymouth (Northbound, Southbound Reaccess Only)

FStop **W:** **Irving***
Gas **W:** Mobil
Food **W:** Bridgeside Diner
AServ **W:** Mobil
Med **W:** ✚ **Hospital**

4 **(76)** U.S. 3, NH 25, Ashland, Holderness, Squam Lake Region

FStop **E:** **Irving***(D)
Gas **E:** Mobil(CW)

EXIT NEW HAMPSHIRE

Food **E:** Burger King, Dunkin Donuts, Subway
Lodg **E:** ◆ **Comfort Inn**
AServ **E:** Ashland Auto Parts, Mobil
RVCamp **E:** **Squam Lakes Resort**
ATM **E:** Pemigewass National Bank

23 **(70)** NH 132, NH 104, Meredith, New Hampton

FStop **E:** **Citgo*, Irving**
Food **E:** Dunkin Donuts, Norm's Ice Cream, Rossi's Italian & Pizza
RVCamp **E:** **Ames Brook Campground, Clearwater Campground, Yogi Bear Camp Report**
ATM **E:** Franklin Savings Bank

22 **(62)** NH 127, Sanbornton, W Franklin

Med **W:** ✚ **Hospital**

(61) **Rest Area - RR, HF, Phones, Picnic (Southbound)**

20 **(57)** U.S. 3, NH 11, NH 140, Tilton, Laconia

FStop **E:** **Irving***
Gas **E:** Exxon*, Mobile
Food **E:** Burger King, Dunkin Donuts, **McDonalds,** Oliver's Restaurant, Ree Garden, Taco Bell, Tilt'N Diner, Upper Crust Pizzeria
W: **McDonalds (Wal-Mart)**
Lodg **E:** ◆ Super 8 Motel
W: Wayland Hotel
AServ **E:** VIP Discount Auto Service
W: Tilton,Ford, Chrysler,Plymouth, Dodge, Wal-Mart
ATM **E:** Concord Savings Bank
W: Concord Savings
Other **E:** **Factory Outlet Center**
W: **Post Office, Wal-Mart**

19 **(55)** NH 132, Northfield, Franklin (Reaccess Southbound Only)

Gas **W:** Mobil*
Food **W:** Jim's Drive-In
Med **W:** ✚ **Hospital**
ATM **W:** Franklins Savings Bank
Other **W:** **Northfield Building(LP), Police Dept, Visitor Information**

(52) **Rest Area - RR, Phones, Picnic, Tourist Info (Northbound)**

18 **(49)** NH 132, Canterbury

Gas **E:** Sunoco*

17 **(46)** U.S. 4 to U.S. 3, NH 132, Penacook, Boscawen

16 **(41)** NH 132, East Concord

Gas **E:** Sunoco(D)
Food **E:** Sunflower's Bakery
AServ **E:** Sunoco
Other **E:** **Mill Stream Market**

15 **(39)** Jct I-393, W U.S.2, to U.S. 3, Main St

FStop **W:** **Mobile**
Gas **W:** Citgo, Exxon*, Getty*, Gulf*, Merritt*
Food **W:** Friendly's
AServ **W:** Citgo
Med **W:** ✚ **Hospital**
Other **W:** Car Wash, **Cumberland Farms Convenience Store**

14 **(39)** NH 9, Loudon

FStop **E:** **Texaco***
Gas **W:** Citgo*, Exxon*, Getty*, Gulf*, Merritt*

EXIT NEW HAMPSHIRE

Food **E:** Boston Market, Einstein Bagels, Pizzeria Uno, Week's Family Restaurant
W: Friendly's, Holiday Inn, Tea Garden Restaurant, The Gas Lighter
Lodg **W:** ◆ Holiday Inn
AServ **E:** Auto Palace, Midas Muffler & Brakes
W: Concord Car Care
Med **W:** ✚ **Hospital**
ATM **W:** Bank of NH
Other **E:** Car Wash, **Osco Drugs, Shaw's Grocery Store, Shop 'N Save Supermarket, U.S. Post Office, Vision Center**
W: **Mango's Market, NH Museum of Natural History**

13 **(38)** U.S. 3, Manchester St

FStop **E:** **Sunoco***
Gas **E:** Sunoco*
W: Mobil(D)
Food **E:** Dunkin Donuts, Egg Shell, Skuffy's
W: Al's Capitol City Diner, Burger King, DeAngelo's Sandwich Shop, Dunkin Donuts, Hawaiian Isle II, KFC, McDonalds, Miami Subs, Pizza Hut
Lodg **W:** ◆ Comfort Inn, AAA Econolodge
AServ **E:** Auto House, Carlson's Chrysler/Plymouth, Ford Dealership
W: Goodyear Tire & Auto, Meineke Discount Mufflers, Patsy's GMC Trucks
TServ **W:** **Patsy's GMC /Kenworth**
RVCamp **E:** **RV Sports Center**
Med **W:** ✚ **Hospital**

12 **(37)** NH 3A, South Main St

Gas **E:** Exxon* (ATM), Irving*(D)
Food **E:** Dunkin Donuts, Subway
Lodg **E:** Brick Tower Motor Inn, DAYS INN Days Inn, Hampton Inn
AServ **E:** Ford Dealership
TServ **E:** **John Grapone (RV Repair)**
Med **W:** ✚ **Hospital**
ATM **E:** Fleet Bank

(36) Jct I-89, Dartmouth, Lake Sunapee, White River Junction

(31) **Rest Area - Both Directions, RR, Phones, Tourist Info**

11 **(29)** NH 3A, Hooksett (Toll Plaza)

(27) Junction I-293, Everett Tnpk, Manchester

10 **(26)** NH 3A, Hooksett

FStop **E:** **Fuel Stop***
Gas **E:** Mobil* (ATM)
W: Exxon*
Food **W:** Big Cheese Pizza, Dunkin Donuts
Other **W:** **Riverside Park**

9 **(24)** U.S. 3, NH 28, Hooksett, Manchester

Gas **E:** Exxon*
W: Exxon*(D), Mobil*, Sunoco(D)
Food **E:** Chantilly's, Hooksett Bagel & Deli
W: Boston Market, Burger King, Cheung Kee Chinese, DeAngelo's Sandwich Shop, Happy Garden, Japanese Steak House, KFC, Lui's, Max's Deli, Papa Genio's, Shorty's Mexican Road House
Lodg **E:** Firebird Motel
AServ **W:** Dodge,Jeep, Eagle, Ford Dealership, Sunoco, VIP Auto Center
ATM **E:** Fleet Bank

← N I-93 S →

EXIT NEW HAMPSHIRE

W: BankBoston, St.Mary's Bank
Other E: Car Wash, Colonial Ace Hardware(LP)
W: Maple Tree Mall, Optical, Partside Convenience Store, Shop 'N Save Supermarket

8 **(22)** Wellington Bridge, Bridge St
Med W: Hospital
Other W: NH Performing Arts Center

7 **(21)** NH 101 East, Portsmith, Seacoast

6 **(21)** Candia Rd, Hanover St (Difficult Northbound Reaccess)
AServ E: Transmission Man
Med W: Hospital
Other E: Candia Rd Convenient Store & Sub Shop

(19) Junction I-293, NH 101, West Manchester, Bedford

5 **(15)** NH 28, North Londonderry
FStop E: Sunoco
Gas W: Exxon*(D)
Food E: Dunkin Donuts, Poor Boy's Family Dining, Subway

4 **(12)** NH 102, Derry, Londonderry
FStop E: Citgo*
Gas E: Citgo*(D), Gulf(D), Mutual, Shell*(CW) (ATM), Sunoco
W: Citgo*, Texaco*(CW)
Food E: Burger King, Derry
W: China Garden, Church's Fried Chicken, Domino's Pizza, Honey Dew Donuts, Maple Garden Chinese, McDonalds, Papa Gino's, Steven's Cafe, USA Subs, Wendy's, Whipper Snappers
AServ E: Gulf (Towing), Sunoco
W: Gladstone's Ford/Dodge/Chrysler/ Plymouth, Instant Lube
ATM W: BankBoston
Other E: Coin Laundry
W: Coin Car Wash, K-Mart, Market Basket Supermarket, WalGreens

(7) Weigh Station - both directions

3 **(6)** NH 111, Windham, Canobie Lake, N Salem
Gas E: Citgo*(D)
W: Exxon*(D), Sunoco*
Food E: Capri Pizza, Windham House of Pizza
W: Dunkin Donuts, Subway
ATM E: Citizen's Bank, Southern NH Bank
W: First NH Bank

2 **(3)** To NH 38, NH 97, Salem, Pelham
Gas W: Citgo*
Food W: Adam's, Holiday Inn, Millstone Manor
Lodg E: ◆ Red Roof Inn
W: ◆ Holiday Inn, ◆ Susse Chalet
AServ W: Uniroyal Tire & Auto
ATM W: Southern NH Bank
Other W: Lucy's Country Store

1 **(2)** NH 24, Rockingham Park Blvd, Salem
Gas E: Citgo(D), Exxon*, Getty
Food E: Bickfords Family Restaurant, Burger King, Denny's, Findeisen's Ice Cream, Friendly's, Grand China, Papa Genio

EXIT NEWHAMPSHIRE/MASS.

Lodg E: AAA Park View Inn
AServ E: Citgo, K-Mart, Sears
Other E: Coin Car Wash, K-Mart, Lens Crafters, Market Basket Supermarket, Osco Drugs, Rockingham Mall, Salem Animal Hospital, Salem Police, Shaw's Grocery, WalGreens

(1) Welcome Center - RR, Phones, Picnic, Vending, Pet Walk (Northbound)

↑NEW HAMPSHIRE

↓MASSACHUSETTS

48 **(47)** MA 213, Methuen, Haverhill

47 MA 213, Pelham St, Pelham, Methuen
Gas E: Shell, Sunoco*
W: Galloway's Grocerette(D)(LP), Getty
Food W: Days Inn, Fireside Restaurant
Lodg W: DAYS INN AAA Days Inn
AServ E: Shell
W: Clark Chrysler,Plymouth, Jeep, Eagle, Getty
Med E: Hospital

46 **(44)** MA 110, MA 113, Lawrence, Dracut
Gas E: Getty, Mobil*, Shell*(CW)
W: Coastal, Gulf*
Food E: Burger King, China Cafe, DeAngelo's Sandwich Shop, Dunkin Donuts, McDonalds, Papa Gino's, Pizza Hut
W: Bea's Sandwiches, Dunkin Donuts, Jackson's, Royal House Of Pizza
AServ E: Getty, Mobil, Shell
W: Gulf
Med E: Hospital
ATM E: Andover Bank, BankBoston, First Ethics Bank, Fleet Bank
Other E: Coin Car Wash, Coin Laundry, Market Basket Supermarket, Merimak Plaza, Osco Drugs
W: Variety Convenience Store

45 **(43)** River Rd, South Lawrence
Gas W: Mobil*
Food E: Andover Marriot
W: Dunkin Donuts, Grill 93
Lodg E: Andover Marriot, ◆ Courtyard by Marriott
W: AAA Tage Inn
AServ W: Mobil
Parks E: Lawrence River Front State Park

44AB **(41)** Junction I-495, Lawrence, Lowell

43 **(39)** MA 133, North Tewksbury, Andover
Gas E: Mobil*
Food E: Dunkin Donuts (Mobil), Ramada
Lodg E: ◆ Ramada

42 **(38)** Dascomb Rd, Tewksbury

41 **(35)** MA 125, Andover, North Andover

EXIT MASSACHUSETTS

40 **(34)** MA 62, North Reading, Wilmington

39 **(33)** Concord St

38 **(32)** MA 129, Reading, Wilmington
Gas W: Mobil
Food W: Burger King
AServ W: Mobil

37AB **(29)** Junction I-95, Waltham, Peabody

36 **(27)** Montvale Ave, Stoneham, Woburn
Gas E: Mobil
W: Citgo*, Exxon*, Getty, Shell
Food E: Al La Kitchen Pizza & Subs, Dunkin Donuts, Monty's Restaurant
W: Bigford's Family Restaurant, Friendly's, McDonalds, Primo's Italian, Spud's
Lodg W: Howard Johnson
AServ E: Mobil, Rite Way
W: Getty, Shell
ATM W: BankBoston, US Trust
Other E: Car Wash, Coin Car Wash

35 Winchester Highlands (Southbound, Reaccess Northbound)

34 **(26)** MA 28 North, Stoneham, Melrose (Northbound)
Gas E: Mobil
Food E: Friendly's
Lodg E: Spot Pond Motel
AServ E: Mobil
Med E: Boston Regional Medical Center
Other E: The Stone Zoo

33 MA 28, Fellsway, Winchester
Med E: Hospital
W: Hospital

32 **(23)** MA 60, Medford Square, Malden
Food W: Carvel Ice Cream, Donuts, Jin House Chinese, McDonalds
AServ E: Gem Auto Parts, Ulta Tech
Med W: Hospital
ATM E: Medford Cooperative Bank
Other E: Macy's Convenience
W: CVS Pharmacy

31 MA 16, Mystic Valley Pkwy

29 MA 28, MA 38, Somerville, Everett
Food E: Dunkin Donuts, McDonalds
W: MtVernon
AServ E: K-Mart
W: Broadway Brake Supply
ATM E: BankBoston
Other E: K-Mart

27 U.S. 1 North, Tobin Bridge

26 Storrow Drive, Cambridge

25 Causeway St, North Station

24 Callahan Tunnel, Logan Airport

23 High St, Congress St

22 Atlantic Ave, Northern Ave

21 Kneeland St, Chinatown

Bold red print shows RV & Bus parking available or nearby

EXIT	MASSACHUSETTS
0	Junction I-90, MA Turnpike, Chinatown
3	Massachusetts Ave, Roxbury (Northbound Exits Left)
Med	S: University Hospital
17	East Berkeley St, Broadway
16	South Hampton St, Andrew Square
Gas	W: Shell*
Food	E: Dunkin Donuts
	W: Bickford Family
Lodg	W: South Bay Hotel
TServ	W: Boston Tire
Med	W: Boston Medical Center
Other	W: K-Mart (Pharmacy)
15	Columbia Rd
14	Morrissey Blvd (Northbound)
Other	W: JFK Library
13	Freeport St, Dorchester
Food	W: Carvel Ice Cream, Dunkin Donuts, Pizza
Lodg	W: Swiss Chalet
AServ	W: Jiffy Lube, Toyota & GM Dealer
ATM	W: Citizen's Bank
12	MA 3A, Neponset, Quincy
11AB	To MA203, Granite Ave, Ashmont
10	Squatum St, Milton (Southbound, No Reaccess)
9	Bryant Ave, W Quincy (Difficult Reaccess)
8	Furnace Brook Pkwy, Quincy (No Reaccess)
7	MA 3 South, Braintree, Cape Cod (Exit To Ltd Access)
6	MA 37, Holbrook, Braintree
Gas	W: Mobil*
Food	W: Molly Malone's (Sheraton), Uno Chicago Bar & Grill
Lodg	W: Sheraton
AServ	E: Dave Dinger Ford, Quirk GMC ,Olds
	W: Sears & Jiffy Lube
Other	W: South Shore Shopping Plaza
5AB	MA 28, Randolph, Milton
Gas	E: Citgo(CW), Shell*(D), Texaco(CW)
Food	E: D'angelo's, Dunkin Donuts, Holiday Inn, Lombard's, Pizza Hut, Sal's House Calzone, Wong's
Lodg	E: Holiday Inn
AServ	E: Shell
ATM	E: BankBoston
Other	E: Store 1
4	MA 24 South, Brockton, New Bedford (Exit To Ltd Access)
3	Houghton's Pond, Ponkapoag Trail
Other	W: Picnic Area
2AB	MA 138, Milton, Stoughton
Gas	W: Gulf(D), Mobil*, Sunoco, Texaco(D)
AServ	W: Sunoco
1	Junction I-95

↑ MASSACHUSETTS

Begin I-93

EXIT	MONTANA

Begin I-94

↓ MONTANA

EXIT	MONTANA
0	Jct I-90 East, Hardin, Sheridan
6	CR 522, Huntley
FStop	N: Pryor Creek
Food	N: Pryor Creek
RVCamp	N: Camping
14	Ballantine, Worden
FStop	N: Conoco
	S: Sinclair
Food	S: The Long Branch Cafe
Lodg	N: Motel
Other	N: Coin Laundry, Post Office
23	Pompeys Pillar
Other	N: Pompeys Pillar Monument
36	Waco
(38)	Rest Area - RR, Picnic P (Eastbound)
(41)	Rest Area - RR, Picnic P (Westbound)
47	Custer
Gas	S: Conoco*
Food	S: D & L Cafe, Junction City Saloon
Lodg	S: D & L Motel
AServ	S: Conoco
Other	S: Custer Food Market (C-Store), Post Office
49	MT 47, Hardin
Food	S: Fort Custer Restaurant
Parks	S: The Little Big Horn Battlefield
Other	S: Camping
53	Bighorn
63	Ranch Access
(65)	Rest Area - RR, Picnic, Phones P (Both Directions)
67	Hysham
72	CR 384, Sarpy Creek Road
82	Reservation Creek Road
87	MT 39, Colstrip
93	U.S. 12 West, Forsyth, Roundup
FStop	N: Conoco(LP)
Gas	N: Sinclair*
Lodg	N: Howdy Hotel, Pat's Motel, Shade Tree Inn, AAA Westwind Motor Inn
AServ	N: Sinclair
Med	N: Hospital
Other	N: Amerigas(LP)
95	Forsyth
FStop	N: Exxon (ATM)
Gas	N: Conoco
Food	N: Dairy Queen, M & M Cafe
Lodg	N: AAA Best Western, Hillside Inn, Howdy Hotel, Montana Inn, Rails Inn, Rest Well Motel
AServ	N: Conoco, Haberle Ford, King's Muffler & Brake
RVCamp	N: Campground Wagon Wheel
Med	N: Hospital
Parks	S: Rosebud Recreation Area
Other	N: Car Wash
(99)	Weigh Station - Both Directions
103	CR 447, CR 446, Rosebud
106	Rosebud, Butte Creek Road
(113)	Rest Area - RR, Picnic, Phones, RV Dump P (Westbound)
(114)	Rest Area - RR, Picnic, Phones, RV Dump P (Eastbound)
117	Hathaway
Gas	N: Hathaway
Other	N: U.S. Post Office (Hathaway)
126	Moon Creek Road
128	Local Access
135	Bus Loop I-94, Miles City, Jordan
RVCamp	N: KOA Campground
Other	N: Tourist Information
138	MT 59, Miles City, Broadus
FStop	N: Cenex* (RV Dump), Exxon*(CW)
Gas	N: Conoco*, Sinex*
Food	N: 4B's Family Dining, Boardwalk Restaurant, Dairy Queen, Gallagher's Family Restaurant, Hardee's, McDonalds, Pizza Hut, Subway, Taco Bell, Taco John's, Varsity Bar & Grill, Wendy's
	S: The New Hunan Restaurant
Lodg	N: AAA Best Western, Budget Host Inn, Days Inn AAA Days Inn, Motel 6
	S: ◆ Comfort Inn, Holiday Inn Express, Super 8 Motel
AServ	N: Exxon, Frank's Body Shop, Glader Electric Service
RVCamp	N: KOA Campground (7 Miles)
Med	N: Hospital
ATM	N: Cash Machine (Cenex FS)
Other	N: Buttrey Food & Drug, County Market Grocery, K-Mart, Osco Drugs
141	U.S. 12, Miles City, Baker
TStop	N: Flying J Travel Plaza
Food	N: Flying J Travel Plaza
Lodg	N: Star Motel
AServ	N: Flying J Travel Plaza
TServ	N: Flying J Travel Plaza
RVCamp	N: Big Sky Campground
ATM	N: Flying J Travel Plaza
148	Valley Access
159	Diamond Ring
169	Powder River Road
176	CR 253, Terry
Gas	N: Cenex*
Food	N: Gaopland Cafe, Overland Restaurant
Lodg	N: Campton Motel, Diamond Motel
RVCamp	N: Terry RV Oasis
Med	N: Hospital
185	CR 340, Fallon
(192)	Rest Area - RR, Picnic, Phones P (Bad Route Road)
198	Cracker Box Road
204	Whoopup Creek Road
206	Pleasant View Road
210	Bus Loop I-94, MT 200S, Glendive, Circle

EXIT — MONTANA

Parks S: Makoshika State Park
Other S: Camping

211 MT 200 S, Circle (Westbound, No Westbound Reaccess)

213 MT 16, Glendive, Sidney
TStop S: Sinclair
FStop N: Exxon(LP)
Gas S: Cenex*, Conoco*, Sinclair (RV Dump)
Food S: McDonalds, Pizza Hut, Sinclair, Subway
Lodg S: Budget Host Inn, Parkwood Motel
AServ N: Exxon
S: D & R Repair, Goodyear Tire & Auto, Mercury, Ford, Lincoln, NAPA Auto Parts, Sinclair, Smith Auto Sales
TServ S: D&R Repair
RVCamp N: Green Valley Campground
ATM S: Cenex
Other N: MT Highway Patrol
S: Anthony's Dept Store, Buttrey Food & Drug, Econowash Coin Laundry, K-Mart

215 Glendive, City Center
Gas N: Conoco*
S: Exxon*(D), Sinclair(D)(LP)
Food N: CC's Family Cafe, Cafe (Kings Inn)
S: Hardee's
Lodg N: DAYS INN Days Inn, Kings Inn (Casino), Super 8 Motel
S: AAA Best Western, AAA El Centro Motel, Glendive Budget Motel
AServ S: Bob's Body Shop, Sinclair
RVCamp N: Glendive Campground
Med S: + Glendive Medical Center
Parks S: Makoshika State Park
Other N: Frontier Museum, True Value Hardware(LP)

224 Griffith Creek, Frontage Road

231 Hodges Road

236 Ranch Access

(240) Weigh Station - Both Directions, RV Dump

241 MT 7, CR 261, Wibaux, Baker (Eastbound, No Reaccess Eastbound)
FStop S: CoOp*
Gas S: Conoco*(D)
Lodg S: Super 8 Motel, W-V Motel, Wibaux Motel
AServ S: CoOp
TServ S: CoOp
Other S: Visitor Information

EXIT — MONTANA/NORTH DAKOTA

242 MT 7, CR 261, Wibaux, Baker (Westbound, No Reaccess Westbound)

248 Carlyle Road

↑MONTANA

↓NORTH DAKOTA

(1) Weigh Station, Rest Area (Both Directions)

2 ND 16, Beach
TStop S: Flying J Travel Plaza
FStop S: Amoco*
Food S: Flying J Travel Plaza
Lodg N: Outpost Motel
S: Buckboard Inn
Other N: Camping

7 Home on the Range

10 Sentinel Butte, Camel Hump Lake

(13) Rest Area - RR, Phones, Picnic (Eastbound)

(15) Rest Area - RR, Phones, Picnic (Westbound)

18 Buffalo Gap
RVCamp N: Camping

(21) Scenic View (Eastbound)

23 W River Road (Westbound, Reaccess Eastbound Only)

24 Historic Medora
RVCamp S: Camping
Parks S: Theodore Roosevelt National Park
Other S: Medora Visitor Center

27 Historic Medora (Eastbound, No Eastbound Reaccess)

32 Painted Canyon, Rest Area - RR, Phones, Vending, Info
Other N: Painted Canyon Visitor Center

36 Fryburg

42 U.S. 85, Belfield, Grassy Butte
FStop S: Amoco(LP)
Gas S: Conoco* (ATM)
Food S: Dairy Queen, Rendevous, Trapper's Kettle
Lodg S: Trapper's Inn Motel
RVCamp S: Campground
Parks N: Theodore Roosevelt National Park - North

EXIT — NORTH DAKO

Unit
Other S: Tourist Information

51 South Heart

59 Bus Loop I-94, Dickinson, City Cente
RVCamp S: Camping
Parks S: Patterson Lake Recreation Area

61 ND 22, Dickinson, Kill Deer
Gas N: Cenex*(D), North Hill*, Sinclair*(D)
S: Amoco*(CW)(LP), Cenex, Conoco(D)(LP), Holiday*
Food N: Applebee's, Arby's, Bonanza Steak House, Burger King, Dairy Queen, Happy Joe's, Hospitality Inn, Sergio's Mexican Restaurant, TCBY, Taco Bell, Taco John's, The Donut Hole, Wendy's
S: Army's West, China Doll Chinese, Country Kitchen, Domino's Pizza, KFC, McDonalds, Perkins Family Restaurant, Pizza Hut, Subway
Lodg N: American Motel, AAA Comfort Inn, AAA Hospitality Inn
S: AAA Best Western, Budget Inn, Select Inn, ◆ Super 8 Motel
AServ N: Goodyear Tire & Auto, Midas Muffler & Brakes, NAPA Auto Parts, Sears, Season's Auto Care
S: Conoco
RVCamp S: RV Camping
Med S: + Hospital
ATM N: Norwest Bank
Other N: Buttrey Food & Drug, Coin Laundry, Dan's Supermarket (ATM), K-Mart, Prairie Hills Ma Wal-Mart
S: Ace Hardware(LP), Dinosaur Museum, Law Enforcement Center, Tourist Information

64 Dickinson
FStop S: Amoco*
AServ S: Budget Battery, Ford Dealership
TServ S: Diamond Truck Equipment, Schmidt Repa (Amoco)
RVCamp S: Camping
ATM S: Amoco

(69) Rest Area - RR, Phones, Picnic, RV Dump (Both Directions)

72 Gladstone, Lefor

78 Taylor

84 ND 8, Richardton, Mott
FStop N: Cenex*
Med N: + Hospital
Other N: Schnell Ranch Recreation Area

90 Antelope

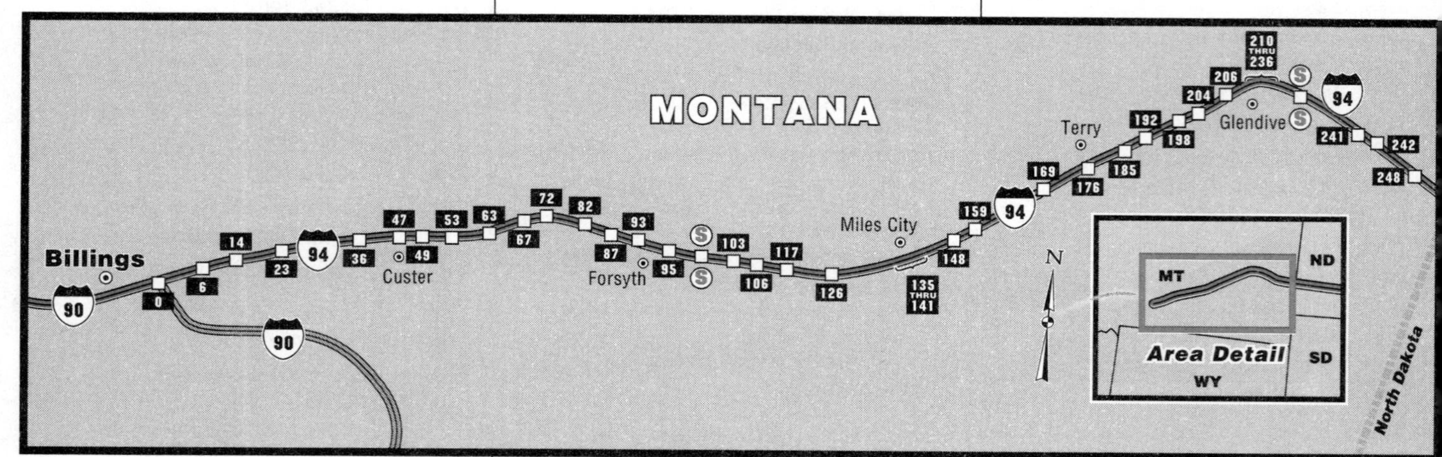

Bold red print shows RV & Bus parking available or nearby

XIT NORTH DAKOTA

4) Rest Area - RR, Phones, Picnic, RV Dump (Both Directions)

7 Hebron

)2 Glen Ullin, Lake Tschida, Hebron

)8 Glen Ullin

RVCamp N: Camping

0 ND 49, Glen Ullin, Beulah

3 Geck Township

7 Dengate

19) Rest Area - RR, Phones, Picnic, RV Dump (Both Directions)

20 Blue Grass

23 Almont

27 ND 31, New Salem, Hannover

FStop S: Amoco(LP) (ATM), Conoco*

Food S: Hot Stuff Pizza (Amoco), RJ's Restaurant (Amoco), Sunset Inn Cafe (Conoco), The Golden West Restaurant

Lodg S: Sunset Inn

AServ S: Amoco

Med S: ✚ New Salem Clinic

ATM S: Conoco

Other N: Knife River Indian Villages (35 Miles)

S: Gaebe Pharmacy, RJ's Laundry Mat, RV Dump (Amoco), Randy's Super Value Grocery, The Golden West Shopping Center, World's Largest Cow

34 Judson, Sweet Briar Lake

40 Crown Butte, Crown Butte Dam

47 Bus Loop I-94, ND 25, to ND 6, Center, Mandan

FStop S: Amoco(LP)

Food S: Amoco Fuel Stop

TServ S: Amoco Fuel Stop

TWash S: Amoco Fuel Stop

151) Scenic View (Eastbound)

52 (4) Sunset Drive, Mandan

Gas N: Conoco*

S: Amoco*(CW)

Food S: Los Amigos Mexican

Lodg N: ◆ Best Western, ◆ River Ridge Inn

RVCamp S: Camping

Med S: ✚ Med Center One

Parks S: Fort Lincoln State Park

Other S: Tourist Information

53 Mandan Ave

EXIT NORTH DAKOTA

FStop N: Conoco* (Auto & Truck Service)

Parks S: Dakota Park, Fort Lincoln State Park

155 ND 6, Bus Loop 94 (Westbound)

156 Bus 46, Bismarck Expwy, West Bismarck, East Mandan

Other N: Bismarck Zoo, Camping

157 Divide Ave

FStop N: Conoco*(LP)

Gas S: Cenex*(D)(CW)(LP) (RV Dump), Conoco*, Gas Food Station

Food N: McDonalds

S: Cafe (Conoco)

TServ N: Conoco

TWash N: Conoco

ATM N: Conoco

Other N: Post Office

S: Camping, Econo Foods (& Pharmacy)

159 U.S. 83 N Bismarck, Minot

Gas N: Mini Mart*, Sinclair*(D)

S: Amoco*(CW), Conoco*(D)(LP)

Food N: A&B Pizza, Arby's, Burger King, Country House, HongKong Chinese American, KFC, Kroll's Kitchen, McDonalds, O'Brians, Paradiso Mexican Restaurant, Perkins Family Restaurant, Red Lobster, Rockin 50's Cafe, Royal Fork Buffet, State's Alien Grill

S: Bonanza Steak House, Dairy Queen, Donut Hole, Hardee's, International Restaurant, Kerry's Family Restaurant, Mineriva Restaurant, Pizza Hut, Subway, Taco Bell, Taco John's, The Woodhouse Restaurant, Wendy's

Lodg N: American Motel, ◆ Fairfield Inn, Motel 6

S: ◆ Best Western, (Days Inn logo) Days Inn, Hillside Motel, ◆ Kelly Inn, (AAA logo) Select Inn, ◆ Super 8 Motel

AServ N: GM Auto Dealership, K-Mart, Mr. Muffler, Northwest Tire

S: Conoco

Med S: ✚ Hospital

ATM N: Gate City Federal, Kirkwood Bank & Trust, Norwest Bank

S: First Bank of Bismark

Other N: Coin Car Wash, Coin Laundry, Dan's Supermarket, Gateway Mall, Gateway Supervalue, Interstate Eye Care, Osco Drugs, Tourist Information

161 Lincoln, Bismarck Expwy

TStop S: Amoco(SCALES)

FStop N: Cenex* (RV Dump)

Food N: Hot Stuff Pizza (Cenex FS)

EXIT NORTH DAKOTA

S: Oasis (Amoco)

Lodg S: Ramada Limited

TServ S: Butler CAT Engine Service, Freightliner Dealer, Goodyear Tire & Auto, International, Johnson Trailer Sales, Kenworth

TWash S: Truck Wash

RVCamp N: RV Dump (Cenex FS)

S: Capital RV Service

ATM N: Cash Machine (Cenex)

(169) Rest Area - RR, Phones, Picnic, RV Dump (Both Directions)

170 Menoken

176 McKenzie

182 U.S. 83S, ND 14, Sterling, Wing

FStop S: Cenex*(LP) (RV Dump)

Food S: Tops Cafe (Cenex FS)

Lodg S: Tops Motel

190 Driscoll

195 Long Lake

200 ND 3, Dawson, Steele, Tuttle

FStop S: Conoco*

Gas S: Amoco*(LP)

Food S: Lone Steer Restaurant

Lodg S: Lone Steer Motel, OK Motel

AServ S: Amoco

TWash S: Conoco Fuel Stop

RVCamp S: Campground

205 Robinson

208 ND 3S, Dawson

Gas S: Midway Service

RVCamp N: LaQua Dakota RV Campground

S: Camping

Other S: Post Office, Yankee Doodle Diner

214 Tappen

Gas S: Amoco*

Food S: I-94 Cafe

AServ S: Amoco

217 Pettibone

(221) Rest Area - RR, Phones, Picnic, RV Dump (Eastbound)

221 CR 70, Crystal Springs

(224) Rest Area - RR, Phones, Picnic, RV Dump (Westbound)

228 ND 30 South, Streeter

230 Medina

RVCamp N: Camping

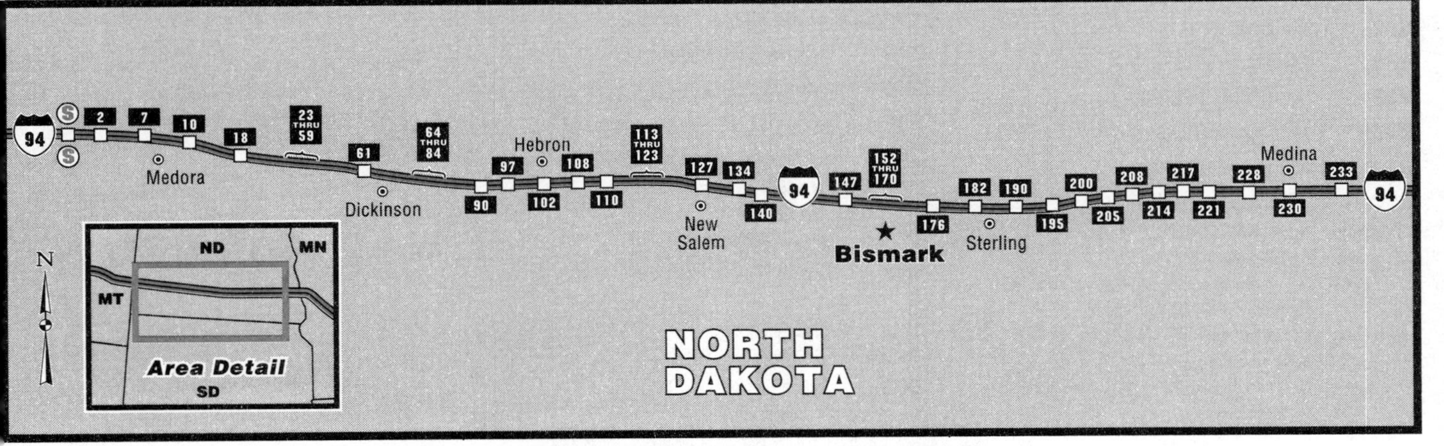

EXIT NORTH DAKOTA

233 Halfway Lake

238 Cleveland, Gackle
- Gas N: Super K Service
- AServ N: Super K Service

242 Windsor

245 Oswego

248 Lippert Township

251 Eldridge

(254) **Rest Area - RR, Phones, Picnic, RV Dump (Both Directions)**

256 Woodbury
- TServ S: **Wiest Truck & Trailer Repair**
- RVCamp S: **KOA Campground**

257 Bus Loop I-94, Jamestown (Exits Left Side Eastbound, No Eastbound Reaccess)

258 U.S. 281, Jamestown
- FStop N: **Sinclair***
  S: **Conoco***
- Gas N: Amoco*(D)
- Food N: Arby's, McDonalds, Wagon Masters Restaurant
  S: **Big Jim Cafe (Conoco FS)**, Burger King, Little Caesars Pizza, Perkins Family Restaurant, Sergio's Mexican Restaurant, **Subway (Conoco)**, The Kings House Buffet
- Lodg N: Buffalo Motel, ◆ Comfort Inn, DAYS INN Days Inn, ◆ Holiday Inn Express, Select Inn
  S: Best Western, **Interstate Motel**
- AServ N: Goodyear Tire & Auto, Jamestown Body Shop, Klein Collision Center, Master Care Auto Service
  S: Valvoline Quick Lube Center
- TServ N: **Jamestown Truck & Diesel Service**
- RVCamp S: **Snows Appliance & RV Sales**
- Med N: ✚ **Hospital**
- Other N: **Big A Auto Part Store**, Car Wash, **National Buffalo Museum, Tourist Information**
  S: **Bakery Outlet, Buffalo Mall, K-Mart, Wal-Mart**

260 U.S. 52 West, Jamestown
- TStop N: **Amoco***(LP)(SCALES)
- Gas N: Stop 'N Go*
- Lodg N: Starlite Motel
- AServ N: Dodge Dealer, GM Auto Dealership
- TServ N: **Uniroyal Tire & Auto**
- Med N: ✚ **Hospital**
- Other N: **Camping, Prairie Veterinary Hospital**

262 Bloom

269 Spiritwood

272 Urbana

276 Eckelson

281 Sanborn, Litchville

283 ND 1 North, Rogers

288 ND 1 South, Oakes

290 I-94 Bus Loop, Valley City
- FStop N: **Sinclair***
- Gas N: Conoco(D)(CW)
- Food N: Hardees, Kenny's Family Restaurant, Pizza Hut, Roby Family Dining, Subway
- Lodg N: Bel-Air Motel
  S: Flickertail Inn
- AServ N: Miller Chrysler, NAPA Auto Parts, Penzoil
- Med N: ✚ **Hospital**

EXIT NORTH DAKOTA

- Other N: **Barnes Historical Museum, Pimida Discount Center**

292 Valley City
- FStop N: **Amoco***
- Lodg N: AAA Wagon Wheel Inn
- AServ S: Dietrich's
- TServ S: **Dietrich's**
- Med N: ✚ **Hospital**
- ATM N: Amoco
- Parks S: **Fort Ransom State Park**
- Other N: **Camping**

294 I-94 Business Loop, Valley City
- RVCamp N: **Camping**

296 Peak

298 Cuba

302 ND 32, Oriska, Fingal

(304) **Rest Area - RR, Phones, Picnic, Tourist Info (Both Directions)**

307 Tower City
- FStop N: **Mobil**
- Food N: **Cafe & Bakery (Mobil FS)**
- Lodg N: Tower Motel
- TServ N: **Mobil**
- RVCamp N: **Camping**

310 Hill Township

314 ND 38, Buffalo, Alice

317 AYR

320 Embden

322 Absaraka

324 Wheatland, Chaffee

(327) **Rest Area - RR, Phones, Picnic, RV Dump (Eastbound)**

328 Lynchburg

331 ND 18, Casselton, Leonard
- FStop N: **Phillips 66**(CW)
- Food N: **Cafe (Phillips 66)**, Club 94
- Lodg N: **Shamrock Motel**
- AServ N: Phillips 66
- TServ N: **Gordy Service Center**(CW), **Phillips 66**
- TWash N: **Phillips 66**
- ATM N: Phillips 66
- Other N: Car Wash (Phillips 66)

(337) **Rest Area - RR, Phones, Picnic, RV Dump (Westbound)**

338 CR 11, Durbin, Mapleton
- FStop N: **AmPride***
- AServ N: Ampride
- TServ N: **Ampride**

340 Kindred, Davenport

(342) **Weigh Station - both directions**

342 Raymond

343 Bus Loop I-94, U.S. 10, Fargo
- Food N: **Highway Host Restaurant**
- Lodg N: Hi-Lo Inn
- RVCamp N: **Campground**

346 West Fargo, Horace (Local Trucks Only)
- Gas S: Conoco*(D)
- AServ S: Conoco

348 45th Street
- TStop N: **Petro**(SCALES)

EXIT NORTH DAKOTA/MINNESO

- Food N: **Iron Skillet (Petro)**, McDonalds
- Lodg N: Come On In, ◆ Sleep Inn
- TServ N: **Petro**
- TWash N: **Blue Beacon**
- Other N: **Tourist Information**

349 I-29, U.S. 81, Sioux Falls, Grand Fork Fargo

350 25th Street
- Gas N: Stop 'N Go*
  S: Mini Mart*
- ATM N: Cash Machine (Stop-N-Go)
  S: Mini Mart

351 Bus U.S. 81, Downtown Fargo
- Gas N: Mobil(CW), Sinclair*, Stop 'N 'Go*(LP)
  S: Amoco*(CW), Phillips 66*(D)(CW), Stop 'N Go*(D)
- Food N: Baskin Robbins, Duane's House Of Pizza, Godfather's Pizza, Great Wall Chinese Food, Hornbacher Express (24 Hr), Little Caesars Pizz Pizza Hut, Taco Shop
  S: A & W Drive-In, Burger King, Denny's, Domino's Pizza, Expressway Inn, Happy Joe's Pizza, KFC, McDonalds, Pepper's Cafe, Quality Bakery, Randy's Family Restaurant, Roger's, Subway, Taco Bell
- Lodg S: ◆ Expressway Inn, Roadway Inn
- AServ N: Mobil
  S: Amoco, Gateway Chevrolet, K-Mart, Phillips 66
- RVCamp N: **Lindwood Park**
- Med N: ✚ **Heartland Hospital**
- ATM N: First Bank, Gate City Federal, Hornbacher's Express
  S: K-Mart, State Bank of Fargo, Stop 'N Go
- Other N: **Ace Hardware, Animal Health Clinic, Ben Franklin, Natural Pet Ctr, Sinkler Optical, Thri Drugs, Vincent Locksmith**
  S: **7-11 Convenience Store**, Fargo Coin Car Wash, **K-Mart**

↑NORTH DAKOTA

↓MINNESOTA

1A U.S. 75, Moorehead, Breckenridge
- Gas N: Phillips 66, Sinclair(CW)
  S: 76*, Amoco*(CW)
- Food N: Burger King, Debbie's Homestyle Kitchen, Domino's Pizza, Fryn' Pan, Sia Gon Cafe
  S: Courtney's Restaurant, Golden Phoenix Asia Cuisine, Hardee's, Red River Cafe (Best Westerr Speak Easy Restaurant, Village Inn
- Lodg N: Regency Inn
  S: ◆ Best Western, AAA Motel 75, ◆ Super 8 Motel
- AServ N: Champion Auto Service, Phillips 66, Sinclair
  S: Amoco, Pontiac Dealer
- ATM N: Norwest Bank
- Other N: **Family Dentistry, Holiday Mall, Tourist Information**
  S: **Moorehead Marine (Boat Service), Osco Drugs, Sunmart Grocery**, Whale Of A Wash

1B 20th Street (Eastbound, Reaccess Westbound)

(2) **Rest Area - Phones, Picnics, RR, HF, Tourist Info, Playground (Eastbound)**

2 Moorhead

Bold red print shows RV & Bus parking available or nearby

XIT MINNESOTA

RVCamp N: KOA Campground, Larry's RV Service

) **Weigh Station (Eastbound)**

MN 336, CR 11, Sabin
- TStop N: Phillips 66*(LP)(SCALES)
- TServ N: Phillips 66
- TWash N: Phillips 66
- RVCamp N: Camping

6 CR 10, Downer

2 MN 9, Barnesville

4 MN 34, Barnesville, Detroit Lakes
- Food N: PJ's Drive Inn
- Other N: Butch's Boat's

2 MN 108, CR 30, Lawndale, Pelican Rapids

8 Rothsay
- FStop S: Amoco*(LP)
- Food S: Cafe (Amoco), Tower House
- TServ S: J's Tire & Service
- ATM S: Amoco
- Other S: Picnic Area (Prarie Chicken Statue)

0 CR 88, CR 52, Fergus Falls
- FStop N: Texaco*(LP) (RV Dump)
- Food N: Texaco
- RVCamp N: Camping
- Med N: ✚ Hospital (6 Miles)
- ATM N: Texaco

4 MN 210, Breckenridge, Fergus Falls
- FStop N: Cenex*
- Gas N: Amoco(CW), Spur*(CW)
- Food N: Burger King, Debbie's Homestyle Kitchen, Hardee's, KFC, King Buffet, McDonalds, Mr C's Family Dining (BBQ Ribs), Perkins Family Restaurant, Ponderosa, Subway
  S: Mabel Murphy's Food & Drinks
- Lodg N: Americinn Inn, Best Western, AAA Comfort Inn, DAYS INN ◆ Days Inn, Motel 7, Super 8 Motel
- AServ N: Ford Dealership, Pennzoil Lube, Spur
- RVCamp N: Pine Plaza
- Med N: ✚ Hospital
- ATM N: Community First Nat'l
- Other N: 4 Seasons Coin Car Wash, Tourist Information, West Ridge Mall
  S: Wal-Mart (Pharmacy)

55 CR 1, Fergus Falls, Wendell
- Other N: County Museum, Interstate Ink(LP)

57 MN 210E, CR25, Furgus Falls

(60) **Rest Area - Phones, Picnics, RR, HF (Eastbound)**

EXIT MINNESOTA

61 U.S. 59S, CR 82, Elbow Lake
- FStop N: Citgo*
- Food N: Big Chief Cafe (Citgo)
- RVCamp N: Camping
  S: Camping
- Med N: ✚ Hospital (6 Miles)

67 CR 35, Dalton
- RVCamp N: Camping
  S: Camping

(69) **Rest Area - Phones, Picnics, RR, HF (Westbound)**

77 MN 78, CR 10, Barrett, Ashby
- RVCamp N: Camping
  S: Camping
- Med S: ✚ Hospital

82 MN 79, CR 41, Evansville

90 CR 7, Brandon
- RVCamp N: Camping
  S: Camping
- Other S: Ski Area

97 MN 114, CR 40, Garfield, Lowry
- RVCamp N: Camping
  S: Camping

(100) **Rest Area - Full Facilities, Playground, Pet Walk (Eastbound)**

100 MN27, Alexandria

103 MN 29, Alexandria, Glenwood
- FStop N: Texaco*
  S: Conoco*(CW)(LP)
- Gas N: Amoco*(CW), Citgo*(D)
  S: BNH*(D), Mobil
- Food N: Flagship Restaurant, Hardee's, McDonalds, Perkins Family Restaurant, Subway (Amoco), Taco Bell
- Lodg N: ◆ Americinn Motel, DAYS INN ◆ Days Inn, ◆ Super 8 Motel
  S: AAA Country Inn & Suites, ◆ Holiday Inn
- AServ N: GM Auto Dealership
  S: Lee Buick
- TServ S: Alexandria Diesel
- RVCamp N: Campground
  S: Alexandria RV Service, Camping
- Med N: ✚ Hospital
- ATM N: Amoco
  S: BNH
- Other N: Target Department Store (Pharmacy), Tourist Information, Wal-Mart (Pharmacy)

(105) **Rest Area - Phones, Picnics, RR, HF**

EXIT MINNESOTA

(Westbound)

114 MN 127, CR 3, Westport, Osakis
- RVCamp N: Camping

119 CR 46, West Union

124 Sinclair Lewis Ave. (Eastbound, Westbound Reaccess)

127 U.S. 71, MN 28, Glenwood, Sauk Centre, Willmar
- TStop S: Texaco*(SCALES)
- FStop N: Amoco*, Holiday*
- Gas N: Super America*
- Food N: Dairy Queen, Hardee's, McDonalds, Subway, Taco John's (Amoco)
  S: Restaurant (Texaco)
- Lodg N: Americinn Motel, AAA Econolodge, Gopher Prairie Motel, AAA Hillcrest Motel
  S: Super 8 Motel
- AServ N: Boyer Motors, Ford Dealership, Kawy's Auto & Truck Serv, S/W Exhaust
  S: GM Auto Dealership
- TServ N: Kawy's Auto & Truck Serv
  S: Felling Trailers, Sauk Centre Tire, Texaco Truck Stop
- RVCamp N: Campground (2 Miles)
  S: MPG RV Service
- Med N: ✚ Hospital
- ATM N: Amoco FS, Super America
  S: Texaco TS
- Parks N: Sauk Center Park (Picnic Area)
- Other N: Rest Area - RR, Phone, Playground, Picnic, Sinclair Lewis Interpretive Ctr
  S: Sauk Centre Veterinary Clinic

131 MN 4, Meire Grove, Paynesville

135 CR 13, Melrose
- FStop N: Mobil(LP)
  S: Conoco*(CW)
- Gas N: Phillips 66* (24 Hr)
- Food N: Burger King, Hardee's
  S: Countryside Family Restaurant, Dairy Queen, Hot Stuff Pizza (Conoco FS)
- Lodg S: Super 8 Motel
- AServ N: Mobil(LP), Tire One
  S: Loren Collison Ctr
- TServ N: Ford Truck Service, Mobil
  S: Truck Service
- RVCamp N: Camping
- Med N: ✚ Hospital (1 Mile)
- Other N: Melrose Fire & Ambulance
  S: Free-Way Foods, Melrose Veterinary Clinic, Veterinary Clinic

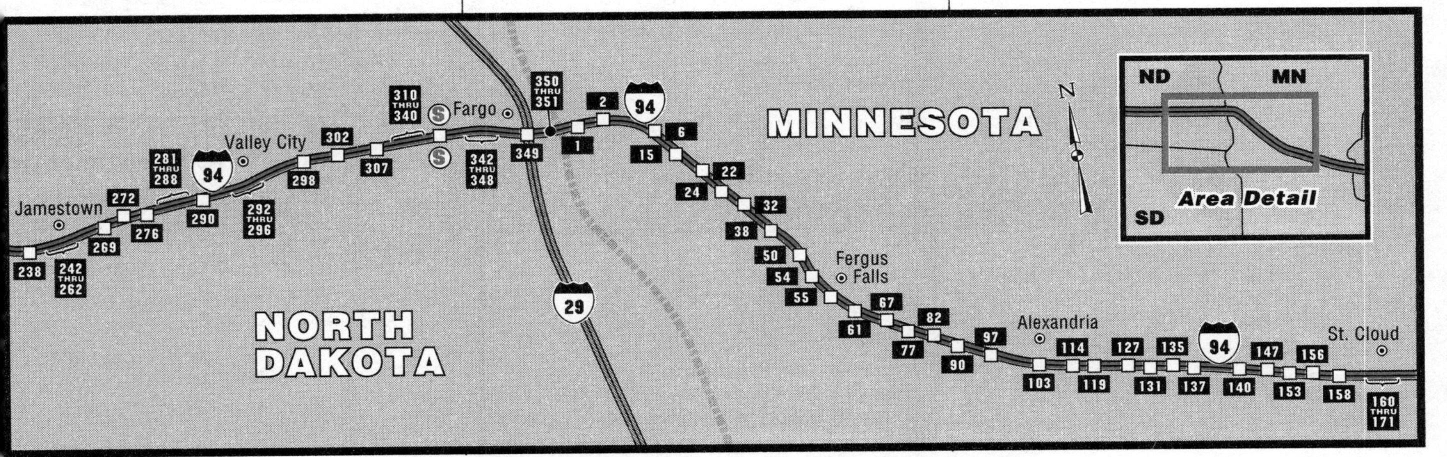

Bold red print shows RV & Bus parking available or nearby

EXIT | MINNESOTA

137 MN 237, CR 65, New Munich

140 CR 11, Freeport
Gas N: Conoco*(D), Mobil*(D)(LP)
Food N: Charlie's Cafe
AServ N: Amoco Motor Club (24 Hr Towing), Cooper Tire
Other N: Corner Store, Jay's Car Wash, U.S. Post Office

147 MN 238, CR 10, Albany
FStop N: Holiday* (RV Dump)
Gas N: Ashland*, Texaco*(D)(LP), Thellen
Food N: Dairy Queen, Sands Restaurant (Sands Motel)
S: KFC
Lodg N: Sands Motel
AServ N: Auto Value Parts Store, Joel Snyder, Plymouth, Thelen
S: Syl's Ford
TWash N: Albany's
Med N: ✚ Hospital
ATM N: Cash Machine (Holiday)
Other N: Albany Coin Car Wash, Amby's IGA Food Store

(152) Rest Area - Phones, Picnics. RR, HF (Both Directions)

153 CR 9, Avon
Gas N: Phillips 66*(LP)
Food N: Burger Treat, Rascals Restaurant
Lodg N: Americinn Inn
RVCamp S: Camping
Other N: U.S. Post Office

156 CR 159, St John's University

158 CR 75, St Cloud (Exit Left Side Of Hwy, Eastbound, Westbound Reaccess)

160 CR 2, St Joseph, Cold Springs

164 MN 23, St Cloud, Rockville
Food N: Antique Mall Coffee Shop
Lodg N: Comfort Inn (5 Miles), Country Inn & Suites (5 Miles), Motel 6 (4 Miles)

167 MN 15, Kimball, St Cloud (Difficult Reaccess, Divided Hwy)
Med N: ✚ Hospital

171 CR 7, CR 75, St Augusta, St Cloud
FStop N: Holiday*(CW)(LP) (RV Dump)
Food N: Hot Stuff Pizza (Holiday), McDonalds (Holiday), Smash Hit Subs (Holiday)
Lodg N: ◆ Americinn Motel, Raddison
TServ S: Ziegler CAT, Zips Diesel Service
Med N: ✚ Hospital
ATM N: Holiday
Other S: County Park

(177) Rest Area - Phones, Picnics, RR, HF (Westbound)

178 MN 24, Annandale, Clearwater
TStop N: Clearwater Travel Plaza(CW)(LP) (24Hr, Coin Laundry)
Gas N: Holiday*(LP)
S: Phillips 66*
Food N: Clearwater Travel Plaza (24Hr), Dairy Queen, Hot Stuff Pizza (Holiday), Subway, The Ole Kettle Restaurant
S: Brigitte's Cafe (Phillips 66)
Lodg N: ◆ Budget Inn
AServ N: Valvoline Quick Lube Center
S: Mathison Motors Auto Service
TServ N: Clearwater Travel Plaza(CW)(LP) (24 Hr), Mr

EXIT | MINNESOTA

Tire, Valvoline Quick Lube Center
TWash N: Clearwater Truck Plaza
RVCamp N: KOA Campground
S: A-J Acres Campground
ATM N: Cash Machine (Holiday), Clearwater Travel Plaza (24 Hr)
Other N: Centre Drug, Cobom's Grocery, Coin Laundry (Clearwater Truck Plaza), Pet Clinic, The Clearwater Clinic, True Value Hdwe

183 CR 8, Silver Creek, Hasty
TStop S: Conoco*(LP)
Food S: Pump House Restaurant(SCALES) (Conoco)
TServ S: Conoco
RVCamp N: Whole Service RV
S: Camping
Parks S: Lake Maria State Park

(187) Rest Area - Phones, Picnics, RR, HF, Playground, Pet Walk (Eastbound)

193 MN 25, Buffalo, Monticello
FStop S: Amoco*(LP)
Gas N: Ferrell Gas(LP), Holiday*
S: Super America*(LP), Total*(D)(CW)(LP)
Food N: Burger King, Country Grill, Dairy Queen, KFC, Perkins Family Restaurant, Skillet Restaurant, Taco Bell
S: McDonalds, Subway, Tree House (Comfort Inn), Wendy's
Lodg N: Americinn Inn
S: (AAA) Comfort Inn
AServ N: Auto Value
S: Champion Auto Stores, Conoco Ultra Lube, Ford Dealership, GM Auto Dealership, Goodyear Tire & Auto, Total
RVCamp N: Campground
S: Monticello RV Center
Med N: ✚ Hospital
ATM N: Cash Machine (Holiday), First Bank, First National Bank, Marquette Bank
S: Amoco, Super America, Total
Other N: K-Mart (Pharmacy), Lake Maria (8 Miles), Maus Foods Grocery (Pharmacy), Monticello Mall (& Pharmacy), Pet Center

195 CR 75, Monticello (Westbound, Difficult Reaccess)
Med N: ✚ Hospital

201 CR 19, Albertville, St Michael (Eastbound)

202 CR 37, Albertville
FStop S: Phillips 66*

EXIT | MINNESO

Gas N: Conoco*(D)(LP)
S: Amoco*(LP)
Food S: Hot Stuff Pizza (Phillips 66), Stephen's Br
AServ S: Amoco, Car Quest Auto Center
ATM N: Cash Machine (Conoco)
S: Highland Bank
Other S: Car Wash

205 MN 241, CR 36, St Michael
Lodg N: Crow River Motel

207 MN 101, Rogers, Elk River
TStop N: 76 Auto/Truck Plaza*(SCALES) (RV Dump)
Gas S: Amoco*(CW), Holiday(LP), Sinclair*(D)(LP), Super America*
Food N: McDonalds
S: Dairy Queen, Domino's Pizza, Happy Chef, Kernel Restaurant, Subway
Lodg N: Super 8 Motel
S: Americ Inn
AServ S: Miller Chevrolet, Sinclair
TServ N: 76 Auto/Truck Plaza
S: Glen's Truck Center
TWash N: 76 Auto/Truck Wash
RVCamp N: Camping World (see our ad on this page)
Med S: ✚ Northwest Family Clinic
ATM N: 76 Auto/Truck Plaza
S: Holiday, State Bank of Rogers, Super Americ
Other S: Rogers Pet Clinic, Super Value Grocery, T Rogers Fire Dept, True Value Hdwe, Value R Drug

213 CR 30, 95th Ave N
Gas N: Conoco*(D), Super America*(D)
RVCamp S: KOA Camping
ATM N: Conoco, Super America

(215) Rest Area - Phones, Picnics, RR, HF, Indian Mound (Eastbound)

215 MN 109, Weaver Lake Road
Gas N: Super America*(D), Texaco*(CW)
Food N: Bakers Square Restaurant, Boston Market, Broadway Pizza, Burger King, Caribou Coffee, Chin Yon Restaurant, Hot Wok, May Inn (Chinese), McDonalds, Old Country Buffet, Pizz Hut, Subway, Taco Bell, Wendy's
S: Applebee's, Fuddrucker's
Lodg N: Raddison
AServ N: Goodyear Tire & Auto, Midas, Texaco, Tires Plus
ATM N: Cub Foods, Norwest Bank, Texaco
Other N: Cub Foods (Pharmacy), Dental Clinic, K-Mart (Pharmacy), Kohl's Dept Store, Maplegrove Eye Clinic, Optical Center, Pearl Vision, Pet Hospital, Petco, WalGreens

216 Jct. I-494
Lodg I: AmeriSuites (see our ad on this page)

225 MN 252, I-694 East

Note: I-94 runs concurrent below with I-694. Numbering follows I-694.

28 CR 61, Hemlock Lane

29AB U.S.169, Osseo

30 Boone Ave
Food O: Northlind Inn
Lodg O: Northlind Inn
Other I: Home Depot, Petsmart

31 CR 81, Downtown Minneapolis
Gas I: Amoco*(CW)

Bold red print shows RV & Bus parking available or nearby

EXIT MINNESOTA

O: 76*
Food I: Wings & Ribs
O: Best Western, Dairy Queen, Wendy's
Lodg I: Budget Host Inn
O: Best Western
AServ I: Kara Collision & Glass, Kennedy Transmission
O: 76
TServ I: **ABM Truck Service**
ATM I: Amoco

3 CR 152, Brooklyn Blvd
Gas O: Total*
AServ O: Car X, Dodge Truck Dealer, GM Auto Dealership
RVCamp O: **Truck/RV Service**

4 MN 100, Shingle Creek Pkwy
Gas O: Mobil*(CW)
Food I: Ground Round, Lee Ann Chin Chinese (Carry Out), Subs Ect
O: Barnacle Bill's, Chi Chi's Mexican, **Cracker Barrel**, Denny's, Hardee's, Olive Garden, T.G.I. Friday's
Lodg I: Country Inn & Suites
O: Budgetel Inn, Comfort Inn, The Hilton
AServ I: Tires Plus
O: Mobil
Other I: **Police Station, Target Department Store**

5A MN 100, Robbinsdale

Note: I-94 runs concurrent above with I-694. Numbering follows I-694.

26 49th Ave, 53rd Ave
Gas S: Texaco
AServ S: Texaco

28 Dowling Ave

29 CR 81, Washington Ave, W Broadway Ave

30 MN55

21A **(231)** Jct. I-394, U.S. 12W

31A Jct. I-394, U.S. 12W

31B Lyndale Ave, Hennepin Ave (Hazmat Vehicles Prohibited Thru Tunnel, Must Exit Here)

33 Jct. I-35

34A MN 55, Hiawatha Ave.

34B 5th Street (Difficult Reaccess)

34C Cedar Ave (Westbound, No

EXIT MINNESOTA

Westbound Reaccess)

235A 25th Ave, Riverside Ave (Difficult Reaccess)

235B Huron Blvd

236 MN 280 (Left Lane Exit, Very Congested, Difficult Reaccess)

237 Cretin Ave, Vandalia St

238 MN 51, Snelling Ave
Gas S: 76(CW), Phillips 66*, Total(D)
Food N: Applebee's, Burger King, McDonalds
S: Pizza Central, The Malt Shop
AServ N: Firestone Tire & Auto
S: Freeway Auto Body, Phillips 66, Tires Plus
ATM S: Liberty Motor Bank
Other N: **Metro Dental, Rainbow Foods Supermarket, WalGreens**
S: Car Wash

239A Hamline Ave (Westbound, No Westbound Reaccess)

239B Lexington Pkwy

240 Dale Street
Gas N: Clark*
Food N: Hickory Chips, Popeye's Chicken, Wendy's, Wings & Ribs
AServ N: American Auto Radiator & Air Conditioning Service, Meineke Discount Mufflers, Tires Plus
ATM N: Western Bank
Other N: **Dental Clinic, Unidale Mall**

241A Marion St, Kellogg Blvd

241B 5th St, 10th St (Difficult Reaccess)

241C Jct I-35E South

242B Jct I-35E North, U.S. 10W (Left Lane Exit)

242A 12th St State Capitol

242C 7th Ave (Difficult Reaccess)

242D U.S. 52

243 Mounds Blvd (Left Lane Exit, Difficult

EXIT MINNESOTA/WISCONSIN

Reaccess)

244 U.S. 10, U.S. 61

245 White Bear Ave

246A Ruth St. (No Eastbound Reaccess)

246B McKnight Road

247 MN 120, Century Ave

249 I-494 South & I-694 North
Lodg S: AmeriSuites (see our ad on this page)

250 CR 13, Radio Drive, Inwood Ave
Food S: Angeto's, Blimpie's Subs, Champs, Sunsets, T.G.I. Friday's, The Boston Market, Wendy's
AServ S: Rapid Oil Change, Tires Plus

251 CR 19, Woodbury Drive, Keats Ave
Gas S: SuperAmerica*(D) (ATM)
Lodg N: Countryside Motel (Frontage Rd)
S: Holiday Inn Express
Other S: **Horizon Outlet Center, The Book Store**

253 CR 15, Manning Ave
RVCamp S: **KOA Campground**
Parks S: **Afton State Park**

(256) **Rest Area - RR, Info, Picnic, Vending, Phones (Westbound)**

(257) **Weigh Station (Westbound)**

258 MN 95, Hastings, Stillwater
Food N: Bungalow Inn
Lodg N: Bungalow Inn

↑ MINNESOTA

↓ WISCONSIN

1 WI 35, Hudson
Gas N: Auto Stop*, Freedom Value Center*(CW)(LP)
Food N: Dairy Queen, Hickory Chicken & Ribs
ATM N: First Federal

2 CR F, Carmichael Road
Gas S: Amoco*(CW), Phillips 66*
Food S: Arby's, Best Western (Phillips 66), Burger King, Country Kitchen (Phillips 66), McDonalds, Shoney's, Subway, Taco Bell, Wendy's
Lodg S: Best Western, ◆ Fairfield Inn, Holiday Inn Express
ATM S: Amoco, Bank St Croix, First National Bank of Hudson
Other S: **K-Mart (& Pharmacy), NAPA Auto Parts, Wal-Mart (& Pharmacy)**

3 WI 35S, River Falls

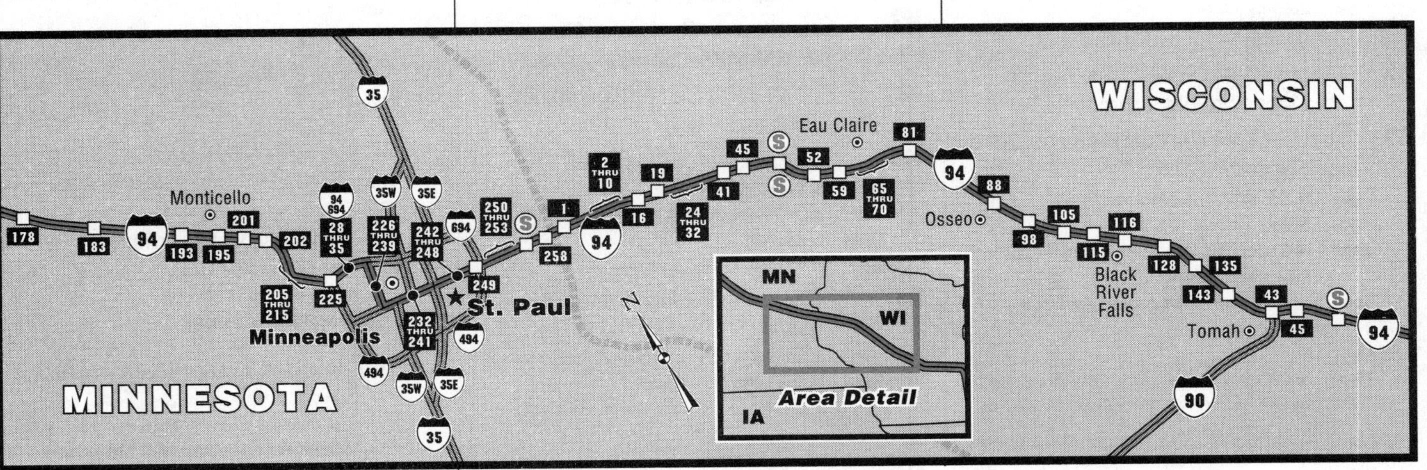

Bold red print shows RV & Bus parking available or nearby

EXIT WISCONSIN

4 U.S. 12, CR U, Somerset
- **TStop** N: 76(SCALES)
- **Gas** N: Citgo*(D)(LP)
- **Food** N: 76 TS, Ranch Motel
- **Lodg** N: Ranch Motel
- **ATM** N: Citgo (76 TS)

(8) Weigh Station (Eastbound)

10 WI 65, New Richmond, Roberts, River Falls

16 CR T, Hammond
- **Other** N: Sheepskin Outlet

19 U.S. 63, Baldwin, Amery, Ellsworth
- **FStop** N: Phillips 66*(D)
  S: Conoco*
- **Gas** N: Freedom*(D)(LP)
- **Food** N: A & W Drive-In, Hardee's, Phillips 66
  S: The Coachman Supper Club
- **Lodg** N: Colonial Motel

24 CR B, Woodville, Spring Valley
- **FStop** N: Mobil*(LP)
- **Lodg** N: Woodville Motel

28 WI 128, Wilson, Glenwood City, Spring Valley, Elmwood
- **FStop** N: Kwik Trip (ATM)

32 CR Q, Knapp

41 WI 25, Menomonie, Barron
- **Gas** N: Cenex*(CW)
  S: Amoco(CW), Citgo*, Super America*(D) (ATM)
- **Food** N: Burger King
  S: Cheesy Pizza & Pasta, Country Kitchen, Hardee's, Kernel Restaurant, McDonalds, Perkins Family Restaurant, Pizza Hut, Taco Bell, Taco John's
- **Lodg** S: (AAA) Americinn, (AAA) Best Western, Bolo Country Inn, Super 8 Motel
- **AServ** S: Pennzoil Lube
- **RVCamp** N: KOA Campground
- **ATM** N: Mutual Savings Bank
  S: First American Bank of Wisconsin
- **Other** N: Wal-Mart (Pharmacy)
  S: K-Mart

(43) Rest Area - RR, Phones, Info, Picnic, Vending (Both Directions)

45 **(44)** CR B, Menomonie
- **TStop** N: 76 Auto/Truck Plaza(LP)
  S: Amoco*
- **Food** N: 76
  S: Heckel's Family Restaurant (Amoco), Subway (Amoco)

(48) Weigh Station (Westbound)

52 U.S. 12, WI 29, WI 40, Elk Mound, Chippewa Falls, Colfax

59 U.S. 12, WI 124, County EE, Eau Claire, Chippewa Falls
- **FStop** N: Cenex*(LP) (ATM), Holiday*(LP)(SCALES)
- **Gas** N: 76(D)
- **Food** N: Burger King, Charcoal Grill Family Restaurant, McDonalds, Subway
- **Lodg** N: ◆ Americinn, (DAYS INN) (AAA) Days Inn, Super 8 Motel
- **AServ** N: 76
- **TServ** N: O'Claire Mac Sales & Service, Riverstates Truck & Trailer Freightliner Cumings, Thermal King Refrigeration

EXIT WISCONSIN

- **ATM** N: Holiday

65 WI 37, WI 85, Eau Claire, Mondovi (Services Approx 1 Mile)
- **Gas** N: Amoco*, Cenex*(D), Conoco*, Phillips 66*(D)(CW)
- **Food** N: Arby's, Best Western, Chi Chi Mexican, China Buffet, Fireside Restaurant, Godfather's Pizza, Hardee's, Heckle's Steakhouse, Italian Slice, McDonalds, Randy's Family Restaurant, Red Lobster, Subway, Sweetwater's, TCBY, Taco Bell, Wendy's, Yen King Chinese Restaurant
- **Lodg** N: Best Western, Comfort Inn, Hampton Inn, Highlander Inn, Holiday Inn, Quality Inn, Ramada, Road Star Inn
- **AServ** N: Amoco
- **TServ** N: International & Cumings Diesel
- **Med** N: ✚ Hospital
- **ATM** N: Firstar National Bank

68 WI 93, Eleva, Eau Claire
- **Gas** N: 76*(CW)(LP), Kwik Trip*(LP), Super America*(CW)(LP)
- **Food** N: Burger King, Dairy Queen, Marks Brothers Bagels
- **ATM** N: People National Bank, Super America

70 U.S. 53, Eau Claire, Chippewa Falls (Southside Accessible From Exit 68)
- **Gas** N: Conoco*(D)
- **Food** N: Fazoli's Italian Food, McDonalds, Shoney's
  S: A & W Drive-In, Applebees, Baker's Square, Garfield's, Mancino's, Olive Garden
- **Lodg** N: ◆ Country Inn
  S: ◆ Heartland Inn
- **Other** N: Sam's Club, Wal-Mart
  S: Best Buys, Border's Books, Oak Wood Shopping Mall, Target Department Store

81 CR HH, CR KK, Foster, Fall Creek
- **TStop** S: Amoco
- **FStop** S: Cenex*
- **Food** S: Amoco

88 U.S. 10, Osseo, Fairchild
- **TStop** N: Holiday(SCALES) (RV Dump)
- **Gas** N: Mobil*, Phillips 66*
  S: Amoco*
- **Food** N: Dairy Queen, Hardee's, Heckel's Big Steer Restaurant
  S: McDonalds, Subway
- **Lodg** N: Budget Host Inn
  S: Roadway Inn
- **AServ** S: Amoco
- **ATM** N: Mobil

(91) Rest Area - RR, Phones, Picnic, Vending (Eastbound)

(94) Rest Area - RR, Phones, Picnic, Vending (Westbound)

98 WI 121, Alma Center, Pigeon Falls, Northfield

105 WI 95, Hixton, Alma Center
- **FStop** S: Cenex*(D), Phillips 66*
- **Food** S: Jeffrey's Cafe (Phillips 66)
- **Lodg** N: Motel 95
- **RVCamp** N: Camping

115 U.S. 12, WI 27, Black River Falls, Merrillan
- **Gas** S: Amoco*(D), Conoco*, Holiday*, Phillips 66
- **Food** N: The Pines Motor Lodge
  S: Country Kitchen, Dairy Queen (Phillips 66), Hardee's, KFC, Subway, Tasty Treat

EXIT WISCONS

- **Lodg** N: The Pines Motor Lodge
- **Other** S: Coin & Automatic Car Wash

116 WI 54, Black River Falls, Wisconsin Rapids
- **TStop** S: Citgo*(SCALES)
- **FStop** N: Cenex*(LP)
  S: Kwik Trip*
- **Food** N: Best Western, Perkins Family Restaurant
  S: Burger King, McDonalds, Pizza Hut, Taco Time
- **Lodg** S: (AAA) American Heritage Inn
- **TServ** S: Power Brake Wisconsin
- **TWash** S: Citgo
- **RVCamp** N: Parkland Village Campground
- **ATM** S: Jackson County Bank
- **Other** S: Wal-Mart (Pharmacy)

128 CR O, Millston
- **Gas** S: Phillips 66*(D)
- **Lodg** S: Millston Motel
- **RVCamp** N: State Forest Campground

(132) Rest Area - RR, Phones, Picnic, Truck Parking, RV Parking (Both Directions)

135 CR E, Warrens

143 U.S. 12, WI 21, Tomah, Necedah
- **TStop** S: Amoco*(LP)(SCALES) (RV Dump)
- **Gas** N: Mobil*(D)
  S: Citgo*, Shell*(D)
- **Food** N: Country Kitchen
  S: Culver's Frozen Custard & Burgers, Hardee's, KFC, Luigi's Subs, McDonalds, Pizza Inn, Subwa
- **Lodg** N: Americ Inn, (AAA) Holiday Inn, ◆ Super 8 Mo
  S: ◆ Comfort Inn, Cranberry Suites Motel, (AAA) Econolodge
- **ATM** S: Shell

144 Jct I-90, La Crosse

Note: I-94 runs concurrent below with I-90. Numbering follows I-90

49 CR PP, Oakdale
- **FStop** N: Speedway*(LP)
- **Gas** S: Mobil*(LP)
- **RVCamp** N: Kamp Dakota

55 CR C, Camp Douglas, Volkfield
- **FStop** S: Amoco*
- **Gas** S: Mobil*(LP)
- **Food** S: Mobil, Subway (Mobil), Target Bluff Restaurant
- **Lodg** S: Walsh's K&K Motel
- **ATM** S: Mobil

(51) Weigh Station (Eastbound)

61 WI 80, Necedah, New Lisbon
- **TStop** N: Citgo, New Lisbon 76 Auto/Truck Stop*(SCALES)
- **Food** N: 76, Grandma's Kitchen, Subway
- **Lodg** N: Edge of the Wood Motel, Rafter's Motor Inn
- **TServ** N: 76 Auto/Truck Plaza

69 WI 82, Mauston, Oxford
- **FStop** N: Cennex Travel Mart*(D)
  S: Kwik Trip - Amoco*(D)
- **Gas** N: Phillips 66*, Shell(D)
  S: Citgo*
- **Food** N: Country Kitchen, Phillips 66, The Family Restaurant
  S: Culver's Frozen Custard & Burgers, Garden Valley Restaurant, Hardee's, McDonalds,

Bold red print shows RV & Bus parking available or nearby

XIT	WISCONSIN
	Roman Castle Italian, **The Alaskan Restaurant**
Lodg	N: AAA **Park Oasis Inn, Super 8 Motel**
	S: **The Alaskan Motel**
TServ	S: **Interstate Truck Repair**
ATM	S: Bank of Mauston
Other	S: Car Wash, **Cheese Mart**
6)	**Rest Area - RR, Picnic, Vending, Phones**
9	CR Huddle House, Lyndon Station
FStop	S: **Shell***
Food	S: **Shell**
RVCamp	N: **Camping, Dreamfield RV Sales & Service**
5	U.S. 12, WI 16, Wisconsin Dells
Lodg	S: AAA Day's End Motel
TServ	S: **G & S Inc**
RVCamp	N: **Camping**
	S: **Camping**
Parks	S: **Rocky Arbor State Park**
7	WI 13 North, Wisconsin Dells
FStop	N: **Amoco(D) (RV Dump)**
Gas	N: Citgo, Mobil*, Shell*
Food	N: Burger King, Country Kitchen, Denny's, Dino's Dinosaur Buffet Pizza, Gia's Cuisine, Paul Bunion's Cook Shanty, Perkins Family Restaurant, Taco Bell, Wendy's
Lodg	N: AAA Best Western, AAA Comfort Inn, Dells Eagle Motel, AAA Holiday Inn, ◆ **Super 8 Motel,** The Polynesian Motel
AServ	N: Amoco
9	WI 23, Lake Delton, Reedsburg
FStop	N: **Phillips 66**
Gas	N: Terry's 76
Food	N: **Pioneer Family Restaurant (Phillips FS)**
Lodg	N: Malibu Inn, Olympia Motel, Roadway Inn, Sahara Motel
RVCamp	N: **Yogi Bear Camp Report**
2	U.S. 12, Baraboo, Devils Lake
FStop	N: **Mobil***
Gas	N: Citgo*(CW)
Food	N: Burger King, Danny's Diner, Market Square Cheese Store, McDonalds, Ponderosa, Subway, Wintergreen Restaurant
Lodg	N: Alakai Hotel & Suites, Black Wolf Lodge, Country Squire Motel, Del Rancho, Grand Marquis Inn, AAA Ramada Limited, ◆ Wintergreen Resort
	S: Motel 6, Vagabond Motel
AServ	N: Parkway Auto Parts
RVCamp	S: **Red Oaks Camping, Scenic Traveler RV Sales & Service, Yogi Bear Camp Report**
Other	S: **Pioneer Camping**
106	WI 33, Portage, Baraboo
FStop	N: **Mobil***
Gas	S: 76*
Food	N: **Quackers Restaurant & Coffee Shop (Mobil FS)**
Med	N: **+ Hospital**
108	WI 78, Stevens Pt., Wausau
TStop	S: **Petro*(LP)(SCALES)**
Gas	S: Phillips 66*
Food	S: **Dairy Queen (Petro), Little Caesars Pizza (Petro), The Iron Skillet (Petro)**
Lodg	S: **DAYS INN Days Inn**
AServ	S: Phillips 66
TServ	S: **Petro**
(114)	**Rest Area - RR, Info, Picnic,**

EXIT	WISCONSIN
	Vending, Phones, RV Dump
115	CR CS, Poynette, Lake Wisconsin
FStop	N: **Citgo*(D)**
Food	N: **McDonalds, Subway**
TWash	N: **Citgo**
119	WI 60, Lodi, Arlington
126	CR V, Dane, DeForest
Gas	N: Amoco*(CW)
	S: Phillips 66*
Food	N: Cheese Chalet, Culver's Frozen Custard & Hamburgers, **McDonalds,** Subway
Lodg	N: Holiday Inn Express
AServ	S: Phillips 66
RVCamp	N: **KOA Campground**
131	WI 19, Waunakee, Sun Praire
Gas	N: Mobil, Quick Trip*(D), Super America*
Food	N: A & W Drive-In, **McDonalds,** Mouse House
Lodg	N: DAYS INN Days Inn, AAA Super 8 Motel
AServ	N: Mobil
132	U.S. 51, Madison, DeForest
TStop	N: **Marathon*(LP)(SCALES), Shell*(SCALES)**
	S: **76 Auto/Truck Plaza*(SCALES) (RV Dump)**
FStop	S: **Pumper Truck Stop**
Food	N: **The Copper Kitchen (Marathon)**
	S: **76 Auto/Truck Plaza,** Subway
TServ	N: **Diesel Specialists of Madison**
	S: **Brad Ragin Tire Service, Peterbilt Dealer**
RVCamp	S: **WI RV World RV Service**
135AB	U.S. 151, Madison, Sun Prairie
Gas	S: Amoco* (ATM), Shell* (Towing), Sinclair*
Food	S: Applebee's, Bread Smith, Carlos O'Kely, Chili's, Country Kitchen, Dunkin Donuts, Hardee's, IHOP, KFC, McDonalds, Mountain Jack's Prime Rib, Olive Garden, Perkins Family Restaurant, Pizza Hut, Ponderosa, Red Lobster
Lodg	S: **Comfort Inn,** Crown Plaza, **Hampton Inn,** Select Inn
ATM	S: Bank One, First Federal, Firststar Bank
Other	S: **Essex Square, Half Price Books**
138	Junction I-94 (I-94 Left Lane)
	Note: I-94 runs concurrent above with I-90. Numbering follows I-90.
244	CR N, Sun Prairie, Cottage Grove
Gas	N: Citgo Super Store*, Phillips 66*
AServ	N: Citgo Super Store*(D), Phillips 66 (Towing)
(245)	**Weigh Station (Eastbound)**
250	WI 73, Marshall Deerfield
259	WI 89, Lake Mills, Waterloo, Marshall
FStop	N: **Phillips 66*(D)**
	S: **Kwik Trip**
Gas	S: 76*, Amoco*
Food	S: Harrington's Country Chef Restaurant, **McDonalds,** Subway, The Pizza Pit
Lodg	N: ◆ **Lake Country Inn**
	S: Pyramid Motel
Other	S: Car Wash
(261)	**Rest Area - RR, Phones, Picnic, Vending (Eastbound)**
(264)	**Rest Area - RR, Phones, Picnic, Vending, Truck Parking (Westbound)**
267	WI 26, Watertown, Johnson Creek
TStop	N: **Shell*(SCALES) (ATM & Showers)**
Gas	S: Citgo*(D)

EXIT	WISCONSIN
Food	N: **Pine Cone Restaurant (Shell)**
	S: **Hardees (Citgo)**
Lodg	N: DAYS INN AAA Days Inn
TServ	N: **Goodyear Tire & Auto**
275	CR F, Sullivan, Ixonia
RVCamp	S: **Camping**
277	Willow Glen Road (Eastbound, No Eastbound Reaccess)
282	WI 67, Dousman, Oconomowoc
Food	N: Mr Slow's Sandwiches, Subway
ATM	N: M&I Bank
Other	N: **Century Foods, K-Mart, Pick & Save Grocery**
283	CR P Sawyer (Difficult Reaccess)
285	CR C, Delafield
Gas	N: Amoco, Mobil*
AServ	N: Amoco
287	WI 83, Hartland, Wales
Gas	S: Amoco*(CW), PDQ*(D)
Food	N: Cousins Subs, Hardee's, **McDonalds,** Shoney's
	S: Burger King, Dairy Queen, First Star, Heidi's Cafe, Marty's Pizza, Rococo Pizza, Subway
Lodg	N: Holiday Inn Express
	S: Budgetal Inn
ATM	N: M&I Bank, Wakesha State Bank
Other	S: **Little Professor's Book Center, Shop Rite Grocery Store, Target Department Store, Wal-Mart**
290	CR SS
ATM	N: M & I Bank
291	CR G, Pewaukee
Gas	S: Amoco*
Food	N: Country Inn Hotel
Lodg	N: Country Inn Hotel
293AB	CR T, Waukesha, Pewaukee
294	CR J, Waukesha
Gas	N: Mobil*(CW)
ATM	N: M & I Bank
295	WI 164, Waukesha, Sussex
Med	S: **+ Hospital**
297	U.S. 18, CR Y, Waukesha, Blue Mound Road
Food	N: Annie's American Cafe, Applebee's, KFC, Zorba's Greek Cuisine
Lodg	N: ◆ Budgetel Inn, AAA Fairfield Inn, ◆ **Hampton Inn,** Motel 6
301AB	Moorland Road
FStop	N: **Amoco*(CW)**
Gas	N: Mobil
Food	N: McDonalds, Whitney's Dining
	S: Maxwell's Restaurant, Midway Hotel
Lodg	N: Marriott
	S: Best Western, Embassy Suites, Residence Inn
AServ	N: Michelin Tire Service, Sears Auto Center
Other	N: **Brookfield Square Mall, Mall**
304AB	WI 100
Gas	N: Shell*(D)(CW)
	S: Phillips 66*, Super America*(D)
Food	N: Bagel Tracks, Edwardo's Natural Pizza, Giuseppis Pizza, Manderian Chinese, Pizza Hut, Taco Bell, The Ground Round, Tosa Deli
Lodg	N: 40 Winks Inn, Best Western, Camelot Inn, Exel Inn, Ramada Inn
AServ	S: Quaker State Lube

EXIT		WISCONSIN
	ATM	N: Tri City National Bank
	Other	S: Rider Rent A Truck
305A		Jct I-895, U.S. 45, Chicago
306		WI 181, 84th Street
	Med	N: Hospital
	Other	N: Scrub-A-Dub Car Wash
		S: Milwaukee Speedway, Olympic Training Center
307A		68th Street, 70th Street
	AServ	S: Pennzoil Lube
	Other	S: Automatic Car Wash
305B		U.S. 45 N Split
307B		Hawley Road
308A		VA Center, Milwaukee County Stadium (Exits Left)
308B		U.S. 41 South
308C		U.S. 41 North
309A		35th Street
309B		26th Street, St Paul Ave
310A		Jct I-43N, Green Bay
310B		Jct I-43 North, Green Bay
310C		Jct I-794, Downtown Milwaukee
311		WI 59, National Ave., 6th St.
313		Lapham Ave.
314A		Holt Ave.
	Gas	N: Citgo*(D)
	Other	N: Pik & Save
314B		Howard Ave.
	Gas	N: Clark*, Mobil*, Union 76(CW)
	Food	N: Copper Kitchen
	AServ	N: Mobil
	ATM	N: Mutual Savings Bank
	Other	N: WalGreens
316		Jct I-43, I-894W, Beloit
317		Layton Ave
	Gas	N: Clark*
	Food	N: Annie's Cafe, Arby's, Burger King, Dunkin Donuts, KFC, McDonalds, Napoli Pizza, Pizza Hut, Ponderosa Steakhouse, Porterhouse, San Dong Express, Sizzler's, Taco Bell
		S: Big City Pizza
	Lodg	S: Howard Johnson
	AServ	N: Meineke Discount Mufflers, Speedy Lube
	ATM	N: Norwest Bank
	Other	N: Car Wash

EXIT		WISCONSIN
318		General Mitchell Int'l Airport
319		CR ZZ, College Ave
	Gas	N: Shell, Super America*(LP)
		S: Amoco*, Citgo*
	Food	N: Georgy's Restaurant (Ramada Inn), McDonalds, Shoney's
	Lodg	N: AAA Exel Inn, ◆ Hampton Inn, AAA Holiday Inn, Ramada Inn, ◆ Red Roof Inn
320		CR BB, Rawson Ave
	Gas	N: Amoco, Mobil*(CW)
	Food	N: Burger King
	Lodg	N: ◆ Budgetel Inn
322		WI 100, Ryan Road
	TStop	S: Milwaukee 76 Auto/Truck Stop*(SCALES), Speedway*(D)(SCALES)
	Gas	S: Citgo*
	Food	N: Hardees, McDonalds, Wendy's
		S: Country Kitchen (Citgo), Jalisco Mexican, Perkins Family Restaurant, Subway (Citgo)
	Lodg	S: ◆ Knight's Inn, Traveler's Motel
	AServ	N: Goodyear Tire & Auto
	TServ	N: Cummins Diesel
		S: Freightliner Dealer, Kenworth
	TWash	S: Blue Beacon Truck Wash
	ATM	S: Tri City National Bank
326		Seven Mile Road
	Gas	N: Amoco*
		S: Mobil*
	RVCamp	N: Jellystone Park Campground
327		CR G
(328)		Weigh Station (Eastbound)
329		CR K, Racine
	FStop	N: Pilot Travel Center*(SCALES)
	Gas	S: Union 76

EXIT		WISCONSIN
	Food	N: Arby's (Pilot FS)
		S: A & W Drive-In
	TServ	S: Eastern Star Trucks
333		WI 20, Racine, Waterford
	FStop	S: Citgo*, Mobil*
	Gas	N: Kwik Trip
	Food	N: Burger King, McDonalds
		S: Wendy's (Citgo FS)
	Lodg	N: 94 Motel, Ramada
		S: Rodeway Inn
	TServ	S: International Service & Sales
	ATM	S: BankOne
335		WI 11, Racine, Burlington
	Lodg	S: Motel
	RVCamp	S: Camping
337		CR KR, County Line Road
	Food	S: The Apple Holler Restaurant
339		CR E
	FStop	N: Speedway*
340		WI 142, CR S, Kenosha, Burlington
	Gas	N: 76*, Mobil
	Food	S: Mars Cheese Castle, Star Restaurant
	Lodg	S: Easter Day Motel
	AServ	N: Mobil (Towing)
342		WI 158, Kenosha
344		WI 50, Kenosha, Lake Geneva
	FStop	S: Speedway
	Gas	N: Shell*(D)
		S: Amoco*
	Food	N: Annie's American Cafe, Dunkin Donuts, Shoney's
		S: Bratstop Cheese & Restaurant, Burger King, Chefs Table, Cracker Barrel, Denny's, KFC, Long John Silvers, McDonalds, Perkins Family Restaurant, Taco Bell, Taste of Wisconsin Family Dining, Wendy's
	Lodg	N: Budgetel Inn
		S: Best Western, Days Inn (see our ad on this page), ◆ Knight's Inn
	Other	S: Action Teritory Family Fun Pk.
345		CR C
	FStop	N: Phillips 66* (RV Dump)
	AServ	N: Phillips 66
	RVCamp	S: RV Sales
(347)		Rest Area - Picnic, RR, HF, Phones
347		WI 165, CR Q, The Lakeview Pkwy
	Gas	N: Amoco*
	Food	N: McDonalds
	Lodg	N: Radisson
	Other	N: Wisconsin Welcome Center (Rest Area & Tourist Information)

Bold red print shows RV & Bus parking available or nearby

EXIT	WISCONSIN/ILLINOIS

↑WISCONSIN
↓ILLINOIS

A Russell Road
- **FStop** W: Marathon*
- **Gas** W: Mobil
- **RVCamp** E: Sky Harbor RV

3 U.S. 41, Waukegan

Note: I-94 runs concurrent below with TSTLWY. Numbering follows TSTLWY.

'4) Toll Plaza

4 IL 132, Grand Ave, Waukegan, Gurnee
- **Gas** N: Speedway*(D)
  S: Mobil*, Shell*(CW)
- **Food** N: Burger King, Cracker Barrel, Hot Diggity Dawgs, IHOP, Ichiban Steakhouse, Little Caesars Pizza, McDonalds, Ming's of China Restaurant, Outback Steakhouse, TCBY
  S: Applebees, Baker's Square, Boston Market, Chili's, Denny's, Holiday Inn, Lone Star Steakhouse, Max's and Erma's Restaurant, McDonalds, Pizza Hut, T.G.I. Friday's, Taco Bell, Wendy's, White Castle Restaurant
- **Lodg** N: Budgatel Inn, Hampton Inn (see our ad on this page)
  S: Comfort Inn, Fairfield Inn, Holiday Inn (see our ad on this page)
- **AServ** S: Fast Lane Lube, Midas Muffler & Brakes
- **Med** S: ✚ Acute Care Center
- **ATM** N: First Federal Bank
  S: First Chicago, Harris Bank
- **Other** N: Piggly Wiggly Supermarket, Six Flags Theme Park
  S: Gurnee Mills Mall, Jewell Osco Drugs, Sam's Club, Spot Not Car Wash, Target Department Store, Wal-Mart

9 IL 21, Milwaukee Ave (No Eastbound Reaccess)

8 Calumet Park
- **Lodg** W: Super 8 (see our ad this page)

5 IL 120, Belvidere Road
- **Med** S: ✚ Hospital

60) Tollway Oasis
- **Gas** E: Mobil
- **Food** E: Baskin Robbins, Wendy's

9 IL 60, Town Line Road
- **Lodg** W: AmeriSuites (see our ad this page)

7 IL 22, Half Day Road

54) Toll Plaza

53 Jct I-294 South, O'Hare

51 IL 43, Waukegan Road
- **Gas** E: Amoco*(CW), Shell
- **Food** E: Applebees, Chili's, Chung Chopsuey, Cooker's Hot Dogs, Full Slab Ribs, Japanese Steakhouse
- **Lodg** E: Red Roof Inn

EXIT	ILLINOIS

- **AServ** E: Just Tires
- **Other** E: Border's Bookstore, Jewell Osco Phamacy, Shopping Mall

Note: I-94 runs concurrent above with TSTLWY. Numbering follows TSTLWY.

30A Dundee Rd. (No Westbound Reaccess)
- **Gas** S: 76, Amoco(CW)

EXIT	ILLINOIS

- **AServ** S: 76

31 Tower Road

33A Willow Rd (No Westbound Reaccess)
- **Gas** N: Northfield Auto Clinic

34 U.S. 41S, Lake Avenue (No Reaccess)
- **Gas** N: Amaco (ATM)
  S: Amoco, Shell
- **Food** N: Akai Hana Japanese, LaMadeleine Cafe, LouMalnati's, Sea Ranch Fresh Fish
  S: Dairy Queen
- **AServ** N: Amoco
  S: Amoco, Shell
- **ATM** N: Eden's Bank
- **Other** N: Borders Bookstore

35 Old Orchard Road
- **Gas** N: Amoco, Shell
- **Food** N: Boston Market, California Pizza Kitchen
- **AServ** N: Amoco, Shell (Towing)
- **Med** N: ✚ Hospital
- **ATM** N: CitiBank, LaSal Bank

37AB IL 58, Dempster Street
- **Gas** N: Amoco, Mobil*
  S: Amoco, Hillerich's, Shell*
- **Food** N: Hong Kong Chinese, Kaufman's Bakery, Maggie's Hot Dogs & Subs, Mazzini's Pizza, McDonalds, Mr. B's Subs
  S: Baskin Robbins, Golden Chopstick
- **AServ** N: A Tire & Auto Center, Amoco, Car X Muffler & Brakes, Firstone, Skokie Auto Parts, Value Transmission Center
  S: Amoco, Midas Muffler & Brakes
- **Med** N: ✚ Eden's Medical Center
- **ATM** N: La Salle Bank

39AB Touhy Ave
- **Gas** N: Amoco*(D), Shell
  S: Amoco, Citgo*, Mobil, Shell
- **Food** N: Canton Express, Psistaria Greek Tavern
  S: Applebees, Chicago Style Pizza, Chili's, Chuck E Cheese Pizza, Dunkin Donuts, Jak's Restaurant, McDonalds, Pete's Fast Food Italian, Sander's Restaurant & Pancake House
- **Lodg** N: Radisson, Holiday Inn (see our ad on this page)
- **AServ** N: Uniroyal Tire & Auto
  S: Amoco, Lee Auto Parts, Mobil
- **Other** N: Dominick's Grocery
  S: Book Store, Jewell Osco Drugs

41B U.S. 14, Peterson Ave
- **Food** N: Campeche Mexican, Sauganash Restaurant
- **Lodg** N: Edens Motel
- **ATM** N: LaSalle Bank
- **Other** N: Tarpey Drugs

41C IL 50, Cicero Ave (No Eastbound Reaccess)

42 Foster Ave
- **Gas** S: Amoco
- **AServ** S: Auto Repair Shop
- **Other** S: Eden's Foods

43A Wilson Ave

44A Keeler Ave, Irving Park Road, IL 19
- **Gas** S: Amoco*, Mobil*, Shell*
- **AServ** S: Amoco
- **ATM** S: Midwest Bank

45 Addison Street
- **Gas** S: Citgo*
- **Food** N: Little Caesars Pizza
  S: Subway (Citgo)
- **Other** N: K-Mart

45B Kimball Ave
- **Gas** N: Marathon*(D)
  S: Amoco
- **Food** S: Dunkin Donuts, Subway, Subway
- **Other** S: Dominick's Grocery, WalGreens

46A California Ave
- **Gas** N: Cabwerks
  S: Mobil*
- **Food** N: KFC
  S: IHOP, Popeye's Chicken

EXIT ILLINOIS

AServ N: James Auto Repair
S: Mobil

47A Fullerton Ave
Gas S: 76, Amoco
Food N: Chuck E Cheese Pizza, Dixie Que Restaurant, Dunkin Donuts, Popeye's Chicken, Scoops Ice Cream, Subway
S: Domino's Pizza
AServ S: 76, Amoco (24hr Towing)
ATM N: Midtown Bank
Other N: Express Car Wash
S: **Fullerton Western Pharmacy**

48A Armitage Ave Ashland Ave
Gas S: Amoco*

48B IL 64, North Ave
Gas N: Amoco*(CW)
S: Amoco*, Citgo, Shell
Food N: Art's Drive In, Hollywood Grill, Tripp's Chicken
AServ N: Goodyear Tire & Auto, Nortown Automotive
S: American Transmission, Shell
ATM N: Cash Machine (Amoco)

49A Division Street
Gas S: Amoco*, Shell*
AServ S: Amoco*, Shell

50A Ogden Ave (No Eastbound Reaccess)
Lodg N: Days Inn (see our ad on this page)
AServ S: Firestone Tire & Auto
TServ S: **International Transmission**

50B Ohio Street (No Return Access)
Gas S: Marathon*

51 Jct I-290

51B West Randolph St (Difficult Reaccess)
Food N: Jim Ching's Restaurant, Perez Restaurant, S & S Restaurant
S: New Star Restaurant (Chinese)

51C E Washington Blvd (Difficult Reaccess)

51D Madison Street

51E Monroe St.

51F Adams Street
Food S: Greek Island, Pegasus Restaurant, San Torini Restaurant

51G Jackson Blvd
Food N: Greek Town Gyros, Sorbes Restaurant
S: Mitchell's Restaurant
Lodg N: New Jackson Hotel
AServ S: Toyota Dealer

51H Jct I-290W, West Suburbs

52A Taylor Street, Roosevelt Road
Gas N: Amoco
Food N: Eppel's Restaurant
AServ N: Midas Muffler & Brakes

52C 18th Street

53 Jct I-55, Chicago
Gas N: Shell*
Food N: Ken Tones (Italian Beef & Sausage)
AServ N: Joe's Auto Repair

53B Junction I-55 South, I-90/94 West

54 31st Street
Gas N: Shell*
Food N: Fat Albert's Italian Food, Maxwell Street Depot
Med N: **+ Hospital**

55B Pershing Road

56A 43rd Street

56B 47th Street

57A 51st Street
Food N: McDonalds

EXIT ILLINOIS

57B Garfield Blvd
Gas N: Amoco(CW)
S: Shell*, Speedway*
Food N: Checkers Burgers, KFC, Mr. Pizza King
S: Wendy's
Other N: **WalGreens**

58A 59th Street
Food N: Church's Fried Chicken

58B 63rd Street (No Reaccess To I-94)

59A Junction I-90, Indiana (Toll)

59C 71st Street

60A 75th Street
Gas N: Shell*
Food S: KFC
Other N: **Wal Green's**

60C 79th Street
Gas N: Shell
S: Amoco*, Shell(CW)
Food S: Church's Fried Chicken, Dock's Fish, Dunkin Donuts, Subway
Other S: **WalGreens**

61A 83rd Street
Gas N: Shell
Other N: **State Patrol Post**

61B 87th Street
Gas N: Amoco*, Shell
Food N: Burger King, McDonalds
S: 87th Street BBQ, Grand Chinese Kitchen, Reggio's Pizza, Subway
AServ S: Lube Pro's
Other S: **Jewel Grocery, Osco Drugs**

62 U.S. 12, U.S. 20, 95th Street (Difficult Reaccess)
Gas N: Shell*
S: Amoco
Food N: Top Dog
S: Expressway 95th Hot Dogs
AServ S: Precision Tune & Lube

EXIT ILLINOIS

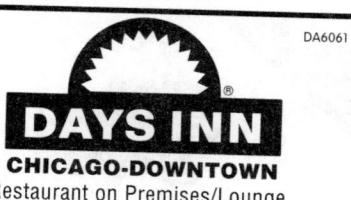

63 Junction I-57 South, Memphis

65 95th Street, 103rd Street, Stony Island Ave

66A 111th Street

66B 115th Street

68AB 130th Street
Lodg N: Super 8 (see our ad on this page)

69 Beaubien Woods (Eastbound, Westbound Reaccess)

70AB Dolton Ave

71AB Sibley Blvd, IL 83
Gas N: 76*, Minuteman*, Mobil*(D)(CW)
S: Amoco(CW), Shell*(CW)
Food N: Dock's Sea Food, McDonalds, Nicky's Gyros, Popeye's Chicken, Subway
S: Arby's, Dunkin Donuts, Long John Silvers, Wendy's, White Castle Restaurant
AServ S: Ultimate Penn
Other N: **Dominick's Grocery & Drugstore**
S: **Fair Play Grocery,** Ultimate Car Wash

73AB U.S. 6, 159th St

74A IL 394 South, Danville

74B Junction I-80 West, I-294 (Toll)

Note: I-94 runs concurrent below with I-80. Numbering follows I-80.

161 U.S. 6, IL 83, Torrence Ave
Gas N: Amoco*(CW) (ATM)
S: Clark*, Gas City*, Mobil*(D), Shell
Food N: Arby's, Bob Evans Restaurant, Checkers Burgers, Chili's, Lansing House Buffet, Little Caesars Pizza, Outrigger's Fish, Thai Oriental Place (Chinese), Wendy's
S: Al's Hamburgers, Brown's Chicken, Burger King, Cafe Borgia (Roman Food), China Chef II, Golden Crown Restaurant, McDonalds, Pappy's Gyros, Vienna Beef Hot Dogs
Lodg N: Best Western, ◆ Fairfield, Holiday Inn, Red Roof Inn, Super 8 Motel
S: Pioneer Motel
AServ N: Car X Muffler & Brakes, Firestone Tire & Auto, Goodyear Tire & Auto
S: Auto Clinic Muffler, Jiffy Lube, Meineke Discount Mufflers, Shell
ATM N: US Bank
Other N: **Fannie Mae Candies, K-Mart**
S: **Komo's Grocery Store, Pets Mart, Sam's Club, Speedwash Coin Laundry**

Note: I-94 runs concurrent above with I-80. Numbering follows I-80.

Note: I-94 runs concurrent below with I-294. Numbering follows I-294.

3 IL 1, Halsted St, Harvey, Chicago Heights
FStop N: **Quick Fuel**
Gas N: 76*(D), Clark
Food N: Alf's Pub, Burger King, Hot Spot Drive-In, Mr Philly, Wendy's, Yellow Ribbon Restaurant
S: Arby's, Dunkin Donuts, Hardee's, IHOP, McDonalds, Shooter's Buffet, Taco Bell, Washington Square Family Restaurant
Lodg N: Budgetel Inn, Junction Inn, Ramada, Red Roof Inn
S: Days Inn, Homewood Hotel, Motel 6
AServ S: Auto Dynamic's
TServ N: **Road Ready**
ATM N: Mutual Bank
Other S: **K-Mart**

Bold red print shows RV & Bus parking available or nearby

XIT	ILLINOIS/INDIANA
	Dixie Highway (Eastbound, No Reaccess)

↑ILLINOIS

↓INDIANA

Note: I-94 runs concurrent above with I-294. Numbering follows I-294.

	Calumet Ave, U.S. 41 North
AB	U.S. 41 South, IN 152 North, Indianapolis Blvd
Gas	N: Gas Center*(CW), Shell, Witham's*
	S: Speedway*
Food	N: Arby's, Chop Suey, Dunkin Donuts, Papa John's Pizza, The Wheel, Woodmar Restaurant
	S: **Burger King**, Little Caesars Pizza
AServ	N: Apex Muffler & Brake, Car X Muffler & Brakes, Midas Muffler & Brakes, Shell
	S: K-Mart
Med	N: ✚ **Hospital**
ATM	N: Mercantile Nat'l Bank of Indiana
Other	N: **Woodmar Animal Clinic**
	S: **K-Mart**
3	Kennedy Ave
Lodg	N: Fairfield Inn (see our ad on this page)
5AB	IN 912, Cline Ave
Food	S: Abigail's (Holiday), Bob Evans Restaurant, Burger King
Lodg	S: Holiday Inn, Motel 6, Super 8 Motel
6	Burr St, Gary

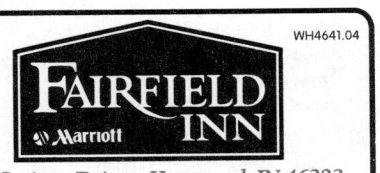

EXIT	INDIANA
9AB	Grant St
10AB	Broadway, IN53 S
11	I-65 South, Indianapolis
12AB	Junction I-65 North, Gary
13	Central Ave
15AB	U.S. 6 East, U.S. 20, IN 51, Lake Station
16	IN. 51 N., Ripley St., Jct. I-90, I-80 (Toll Rd)
(22)	Service Plaza
19	Indiana 249, Portage, Port of Indiana
Gas	S: Amoco* (24 Hrs), Shell*(CW) (24 Hrs)
Food	S: **Angel's Garden (Ramada)**, D.D. Crandalls Grill, **Drifter's (Super 8)**, McDuffy's (24 Hrs), Shoney's
Lodg	S: **Days Inn (see our ad on this page), Dollar Inn**, Knights Inn, ◆ **Ramada**, ◆ **Super 8 Motel**
TServ	N: **Great Lakes Peterbilt**
22AB	U.S. 20, Porter, Burn's Harbor
TStop	N: **Travel Port***(D)(SCALES)
FStop	N: **Steel City Express***(LP)(SCALES)
Gas	S: Shell*
Food	N: **Restaurant (Travel Port), Subway (Travel Port)**
AServ	N: Cleveland Tire Co
	S: Arnell Chevrolet, GMC, Lakeshore Ford/ Mercury/Toyota, Roger's Repair
TServ	N: **Cleveland Tire, Travel Port**
	S: **Roger's Repair**

EXIT	INDIANA
TWash	N: **Blue Beacon Truck Wash (Mobil)**
Other	N: **Sheldon Discount Fireworks**
	S: **Club Fireworks**
26AB	Indiana 49, Chesterton, Indiana Dunes
Gas	S: Amoco*, Mobil*(CW), Shell* (24 Hrs)
Food	S: Arby's, Bert's Bagels, Burger King, Dunkin' Donuts, Gelsosomo's Pizzeria, Jade East Chinese, KFC, Little Caesars Pizza, Long John Silvers, McDonalds (Play Place), Pizza Hut, Subway, Taco Bell, Wendy's, Wingfield's Restaurant (Indian Oak)
Lodg	S: Econolodge, Indian Oak Inn, **Super 8 Motel**
AServ	S: Oil Works (10 Min. Oil Change)
Med	S: ✚ **Duneland Health Center of St. Anthony's Hospital**
ATM	S: First National Bank, First Source Bank
Parks	N: **Indiana Dunes State Park**
Other	N: **Indiana Dunes National Lakeshore**
	S: Coin Car Wash, **Coin Laundry, Jwele Osco Supermarket (24 Hrs, Pharmacy), K-Mart (Pharmacy), Paper Back Book Exchange**
(29)	**Weigh Station - Both Directions**
34AB	U.S. 421, Michigan City, Westville
Gas	N: Clark*, Meijer*(D)(LP), Mobil*(D)(CW), Speedway*(LP)
Food	N: Baskin Robbins, Bob Evans Restaurant, Boston Market, Chili's, Dune Land Brew House, Dunkin' Donuts, **Food Court (Meijer)**, Jade's Buffet, KFC, Mino's, Peking Restaurant Chinese, Pizza Hut, Red Lobster, Steak n Shake
Lodg	N: City Manor, **Comfort Inn, Dollar Inn**, Hampton Inn, Holiday Inn, Knight's Inn, **Red Roof Inn**, Super 8 Motel
AServ	N: AA Quality, Auto Zone Auto Parts, **Wal-Mart**
RVCamp	S: **Michigan City Campground**
Med	N: ✚ **Hospital**
ATM	N: **First America (Meijer)**, NBD
Parks	N: **Creek Ridge County Park**
Other	N: **Aldi, Meijer Supermarket, Department Store (24 Hrs)**, Spot-Not Car Wash, **Wal-Mart (Pharmacy, 24 Hrs)**
40AB	U.S. 20, U.S. 35, Michigan City, La Porte
FStop	N: **Amoco***
Parks	N: **Dunes National Park**
Other	N: **Indiana Welcome Center (Tourist Info.), Lighthouse Place Visitor Center, Museum & Zoo**
(43)	**Rest Area - RR, Phones, Picnic (Westbound)**

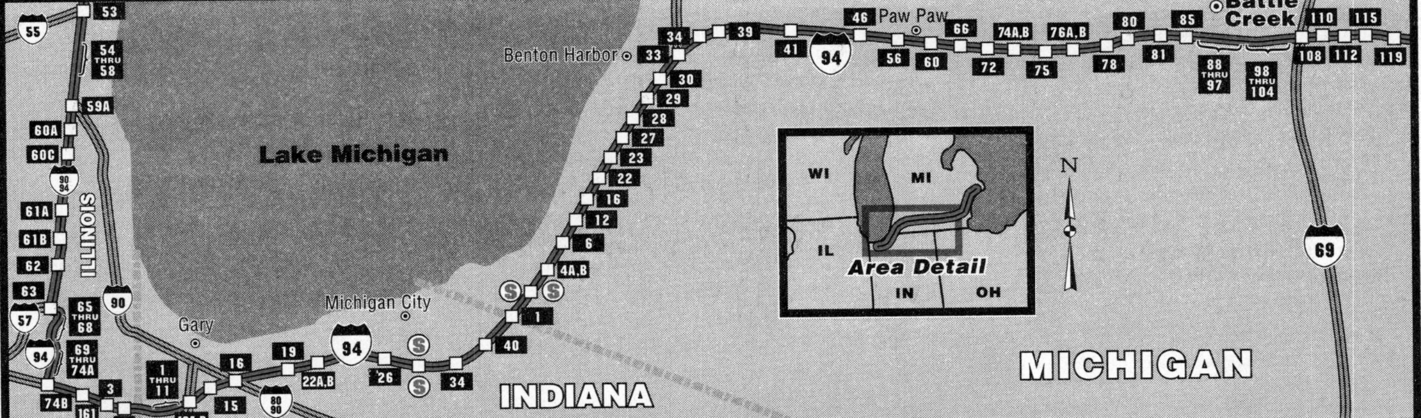

Bold red print shows RV & Bus parking available or nearby

EXIT	INDIANA/MICHIGAN
77	to South Bend, Notre Dame, to U.S.33, U.S. 31 Bus
83	Mishawaka
(90)	Service Plaza
92	IN19, to Elkhart
Gas	N: Citgo*
	S: Clark*, Marathon
Food	N: Applebee's, Cracker Barrel, Food Lin Chinese Buffet, Shoney's, Steak & Shake
	S: Bob Evans Restaurant, Burger King, Callahan's, Davinti's, King Wha Chinese, McDonalds, Olive Garden, Perkins, Red Lobster
Lodg	N: Comfort Inn, Diplomat Motel, Econolodge, Knight's Inn, Shoney's Inn, Turnpike Motel
	S: Budget Inn, Days Inn, Ramada Inn, Red Roof Inn, Signature Inn, Super 8 Motel, Weston Plaza
RVCamp	N: **Elkhart Campground, Travel World, Worldwide Recreation**
	S: **RV Sales, Service & Parts**
Med	S: **+ Hospital**
ATM	N: NBD Bank
	S: Key Bank
Other	N: Coin Car Wash, **K-Mart, Martin's Supermarket, Revco Drugs, Visitor Information**
101	to IN 15, Bristol, Goshen
107	to U.S. 131, IN13, Middlebury, Constantine
121	IN 9, La Grange, Howe
(126)	Service Plaza
144	To I-69, U.S. 27, Angola, Fort Wayne, Lansing

↑INDIANA

↓MICHIGAN

EXIT	MICHIGAN
(1)	MI Welcome Center - RR, Phones, Vending (Eastbound)
1	MI 239, LaPorte Road
TStop	S: **New Buffalo Plaza***(SCALES) **(24 Hrs, Phillips 66 Gas)**
FStop	S: **Gas & Snacks***
Gas	N: Marathon*(D)
Food	S: Arby's (24 Hrs), **Plaza One (New Buffalo Plaza)**, Wendy's
Lodg	N: **Edgewood Motel**
	S: AAA **Comfort Inn**
TServ	S: **J & A Tire**
(2)	**Weigh Station - Both Directions**
4AB	U.S. 12, New Buffalo, Niles
FStop	N: **BP***
Gas	S: Amoco*(D)
Food	S: Expressway Stop (Amoco)
AServ	S: Dale's Repair Service, Roger's Wrecker Service
TServ	S: **Dale's Repair Service, Roger's Wrecker Service**
6	Union Pier
RVCamp	S: **Bob-A-Ron Camping**
Parks	S: **Warren Woods State Park**
12	Sawyer
TStop	N: **Citgo***(SCALES)
	S: **76 Auto/Truck Plaza***(SCALES) **(RV Dump, Mobil Gas)**
Food	N: **Citgo TS**
	S: **Restaurant (76 TStop)**
TServ	N: **Dunes Truck Service**
	S: **76 Auto/Truck Plaza**
TWash	N: **Dunes Truck Wash**
RVCamp	N: **Camping**
ATM	N: Citgo
16	Bridgman
FStop	S: **Speedway***(LP)(SCALES)
Food	S: **McDonalds**
Lodg	S: Bridgman Inn
Parks	N: **Warren Dunes State Park**
Other	S: Car Wash
22	John Beers Road
Parks	S: **Grand Mere State Park**
23	Red Arrow Hwy., Stevensville
Gas	N: Amoco*, Shell*
Food	N: Big Boy, Burger King, **Cracker Barrel**, Dunkin' Donuts, Long John Silvers, McDonalds, Popeye's, Taco Bell (Amoco)
	S: Schuler's
Lodg	N: ◆ Budgetel Inn
	S: ◆ Hampton Inn
ATM	N: Shell
27	MI. 63, Niles Ave.
FStop	S: **Total***(D)
Gas	N: Amoco*
Med	N: **+ Hospital**
28	U.S. 31, MI 139 N., Niles
FStop	N: **Total***(D)
Gas	N: Citgo*(D)
	S: Marathon*
Food	N: Bill Knapp's Restaurant, Burger King, Country Kitchen, KFC, McLaughlins, Red Rose Restaurant (Ramada)
	S: Quality Inn
Lodg	N: AAA **Days Inn**, AAA **Ramada**
	S: AAA Quality Inn
AServ	N: Firestone Tire & Auto, Mike's Radiator Service, Pro Transmission
	S: Marathon
ATM	N: Shoreline Bank
Other	N: **Blossom Lanes Bowling, Freier Animal Hospital**
29	Pipestone Road
FStop	S: **Citgo***
Gas	N: Meijer Grocery*(D) (24 Hrs)
Food	N: Applebee's, Burger King, Busch Garden, Hardee's, Mancino's Pizza & Grinders, McDonalds (Inside Play Place), Pizza Hut, Shoney's, Steak n Shake, Subway
	S: **Blimpie's Subs (Citgo)**, Bob Evans Restaurant, **Taco Bell (Citgo FS)**
Lodg	N: ◆ Comfort Inn, ◆ Courtyard by Marriott, **Motel 6**, ◆ Red Roof Inn
AServ	N: Best Ford/Lincoln/Mercury, Goodyear Tire & Auto, Quick Lube, **Wal-Mart**
Med	N: **+ Orchards Urgent Care Walk-In Clinic**
ATM	N: Shoreline Bank
Other	N: **K-Mart (Pharmacy), Meijer Grocery, Department Store (24 Hrs), Orchards Mall, Wal-Mart (Pharmacy, Optical)**
	S: **Tourist Information**
30	Napier Ave
TStop	N: **Petro2***(D)(CW)(LP)(SCALES) **(Mobil Gas, 24 Hrs)**
Gas	N: Shell*(D) (24 Hrs)
Food	N: **Wendy's (Petro)**
Lodg	N: **Super 8 Motel**
TServ	N: **Petro 2**
TWash	N: **Blue Beacon (Petro 2)**
Med	N: **+ Hospital**
ATM	N: Petro
33	Bus. Loop 94, Downtown, Benton Harbor, St. Joseph (Westbound)
34	Junction I-196, U.S. 31, Holland, Grand Rapids
(36)	**Rest Area - RR, Phones, Picnic (Eastbound)**
39	Millburg, Coloma
Gas	N: Mobil*(CW)
Food	N: **McDonalds (Inside Play Place)**
Other	N: **Hardings Grocery**
41	MI 140, Watervliet, Niles
Gas	N: Amoco*(CW) (24 Hrs), Citgo*
Food	N: **Burger King, Chicken Coop Restaurant**, Waffle House of America
AServ	N: Amoco
RVCamp	N: **Camping**
Med	N: **+ Community Hospital (W/ Walk-In Clinic)**
(43)	**Rest Area - RR, Phones, Picnic (Westbound)**
46	Hartford
Gas	N: Shell*(D)
Food	N: Panel Room Restaurant
Other	N: **Vietnam War Memorial**
52	Lawrence
Food	N: Waffle House
56	MI 51, Decatur, Dowagiac
TStop	S: **Road Hawk Travel Center***(SCALES) **(Total Gas)**
FStop	S: **Citgo***
Food	S: **Nibbles Home Style Food (Road Hawk TS)**
ATM	S: **Road Hawk**
Other	N: **State Patrol Post**
60	MI 40, Lawton, Paw Paw
Gas	N: Amoco*(CW) (24 Hrs), Crystal Flash*(D), Speedway*(D)
Food	N: Big Boy, Burger King, Chicken Coop, Little River Cafe, McDonalds (Inside Play Place), Pizza Hut, Subway, Taco Bell, Wendy's
Lodg	N: **Mroczek Inn**
AServ	N: Paw Paw Chrysler/ Plymouth/Dodge/Jeep Eagle
Med	N: **+ Hospital**
Other	N: **Sheriff's Dept, Village Market (24 Hrs, Pharmacy)**
66	Mattawan
FStop	N: **Mobil***
Gas	S: Shell*
Food	N: Main St. Icecream, Samuel Mancino's Italian Eatery
AServ	N: Rossman Auto
RVCamp	N: **R&S RV Service**
	S: **Camping**
ATM	N: **Mobil FS**
Other	N: **Mattawan Pharmacy**
	S: **Formula K Fun Park**
72	Oshtemo, 9th St.
TStop	N: **Road Hawk* (Phillips 66)**
FStop	N: **Citgo***
Gas	N: Total*(D)(CW)
Food	N: Burger King, **Hot Stuff Pizza (Citgo FS)**, McDonalds (Play Place), **The Rock (Road Hawk**

Bold red print shows RV & Bus parking available or nearby

XIT MICHIGAN

TS)
S: Cracker Barrel

ATM N: Citgo FS, Road Hawk TS

73) Rest Area - RR, Phones, Picnic (Eastbound)

'4AB U.S. 131 North, Bus 94, Three Rivers, Kalamazoo, Grand Rapid

'5 Oakland Dr.

'6AB Westnedge Ave.

Gas N: Admiral*, Clark*, Meijer Grocery*(D)(LP), Speedway*(D)(LP)
S: Shell*

Food N: Arby's, Lee's Famous Recipe Chicken, McDonalds (Meijer), Outback Steakhouse, Papa John's Pizza, Pappy's Place Mexican, Pizza Hut Carry Out, Samuel Mancino's Italian Eatery, Steak 'n Shake, Subway
S: Fazoli's Italian Food, KFC, Little Caesars Pizza, Peking Palace, Schlotzkys Deli, Southside Warehouse, Wendy's

AServ N: Discount Tire Co., Goodyear Tire & Auto, Midas Muffler & Brakes
S: Pep Boys Auto Center, Quick Lube

Med S: ✚ Westnedge Clinic

ATM N: Michigan National Bank

Other N: Meijer Grocery, Department Store (24 Hrs), Rite Aide Pharmacy
S: Kinko's Copies

'8 Portage Road, Kilgore Road

Gas N: Citgo*
S: Shell*, Total

Food N: Dane's Buffet, Hungry Howie's Pizza & Subs, Subway, Uncle Ernie's Pancake House
S: Bill Knapp's Restaurant, Gum Ho Chinese, Olympia Family, Pizza King, Propeller Club (Days Inn), Subs Plus, Taco Bell, Theo & Stacy's Family Restaurant

Lodg N: ◆ Hampton Inn
S: AAA Days Inn, Lee's Inn

AServ N: Allen's Service Center, Pete's B-Line Service, Uncle Ed's Oil Shoppe
S: Airport Auto Service, Parts Plus Auto Store, Total

ATM N: Old Kent

Other N: Milwood Car Wash
S: Tourist Information

30 Sprinkle Road, Cork St.

Gas N: Clark*, Mobil*(CW), Wesco*(D)
S: Amoco*(CW) (24 Hrs), Speedway*(D)

EXIT MICHIGAN

Food N: Burger King, Cork 'N Cleaver Steak & Seafood, Denny's, Godfather's Pizza, Hot 'n Now, Perkins Family Restaurant, Taco Bell
S: Country Kitchen, Holly's Landing (Holiday Inn), McDonalds, Subway

Lodg N: Clarion, ◆ Fairfield Inn, Kelly Inn Best Western, ◆ Red Roof Inn
S: ◆ Holiday Inn, ◆ La Quinta Inn, Motel 6

AServ N: D & S Auto Inc., Goodyear Tire & Auto, Lentz Mufflers, Pennzoil Lube

TServ N: Great Lakes Truck Center

ATM N: Old Kent

Other N: Sprinkle Road Veterinary Clinic
S: Continental Lanes Bowling, Wing's Stadium

(84) Rest Area - RR, Phones (Westbound)

85 Galesburg

Gas N: Shell* (24 Hrs)

Food N: Burger King (Shell), McDonalds

88 Climax

92 Business Loop 94, Battle Creek, Springfield

TStop N: Arlene's Truck Stop*(D) (Citgo Gas)

Gas N: Shell*

Food N: Arlene's TS

TServ N: Glen's Truck & Farm Center (1 Mile)

RVCamp N: B & B's Camper Sales & RV Service
S: Timberlake Camping

Parks N: Fort Custer Recreation Area

95 Helmer Road, Springfield

FStop N: Marathon*(D)

Food N: Miller's Time-Out

Other N: State Patrol Post

(96) Rest Area - RR, Phones, Picnic (Eastbound)

97 Capital Ave

Gas N: Amoco*(CW), Clark*
S: Shell* (24 Hrs)

Food N: Arby's, Lone Star Steakhouse, McDonalds, Red Lobster, Welcome Inn Family Restaurant
S: Bill Knapp's Restaurant, Bob Evans Restaurant, Canton Buffet, Cracker Barrel, Denny's, Donut Mill, Fazoli's Italian Food, Little Caesars Pizza, Mr. Cribbins (Battle Creek Inn), Pizza Hut, Subway, Taco Bell, Wendy's

Lodg N: AAA Knights Inn
S: AAA Battle Creek Inn, ◆ Budgetel Inn, ◆ Hampton Inn, AAA Motel 6, ◆ Super 8 Motel

AServ S: Goodyear Tire & Auto, Uncle Ed's Oil Shop

EXIT MICHIGAN

ATM N: Amoco, First of America
S: Southern Michigan Bank & Trust

Other N: State Police Post (3 Miles)
S: Coin Car Wash, K-Mart, Mission Car Wash, Target Department Store

98AB MI 66, Sturgis, Downtown Battle Creek

Gas S: Citgo*(CW)

Food S: Blimpie's Subs (Citgo), McDonalds (Wal-Mart), Old Country Buffet, Schlotzkys Deli, Steak 'n Shake, TCBY Yogurt (Citgo)

AServ S: Firestone Tire & Auto, Wal-Mart

ATM S: Michigan National Bank

Other S: Binder Park Zoo, Lakeview Square Mall, Sam's Club, Target Department Store, Wal-Mart (Pharmacy & Optical)

100 Beadle Lake Road

FStop S: Phillips 66*

Food N: Moonraker

AServ S: Auto Value Parts Store, Phillips 66

RVCamp N: Camping

Parks S: Binder Park

Other S: Binder Park Zoo

104 11 Mile Road, Michigan Ave.

TStop N: Te-Khi Truck Stop*(SCALES) (Phillips 66)

FStop N: Total*(D)(LP)
S: Mobil*

Food N: Four Seasons Restaurant (Te-Khi TS), Taco Bell (Total FS)
S: Roadrunner Restaurant (Quality Inn), Subway (Mobil)

Lodg S: Quality Inn & Suites

TServ N: Te-Khi TS

ATM N: Te-Khi TS
S: Mobil FS

Other N: C.O. Brown Stadium

108 Junction I-69, U.S. 27, Lansing, Ft. Wayne

110 Old 27, Marshall

TStop N: Shell* (24 Hrs)

Food N: Hillcrest Coffee Shop (24 Hrs), Subway (Shell)

TServ N: Shell

Med S: ✚ Hospital

ATM N: Shell TS

112 Partello Road

Gas S: Shell* (24 Hrs)

Food S: Sports Rock Bar & Grill

(113) Rest Area - RR, Phones (Westbound)

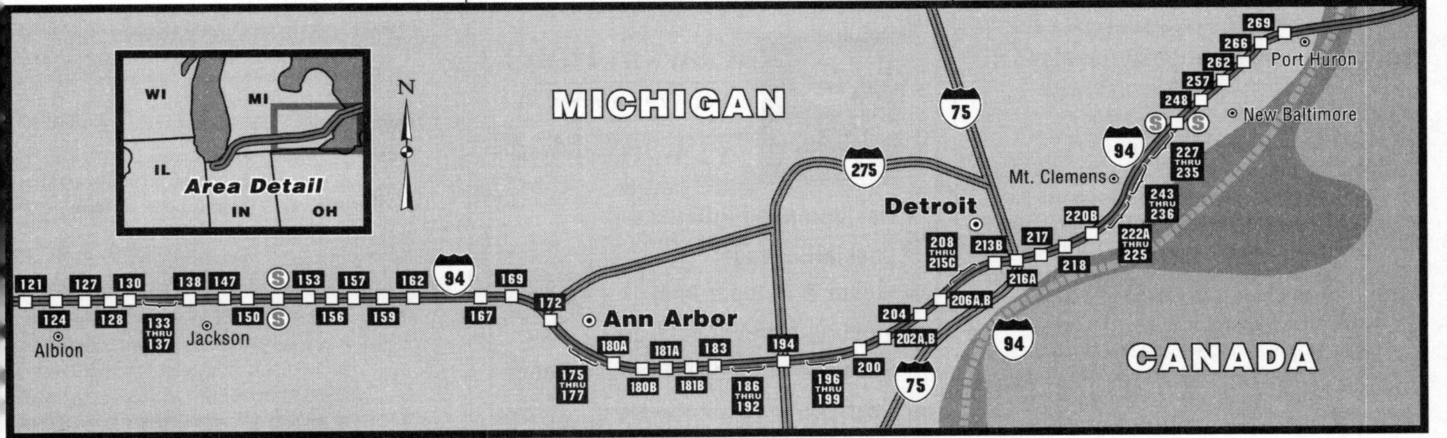

Bold red print shows RV & Bus parking available or nearby

EXIT MICHIGAN

115 22 1/2 Mile Road
- TStop N: 115 Truck Stop* (Total Gas)
- Food N: 115 Truck Stop Restaurant
- TServ N: 115 TS

119 26 Mile Road

121 Business Loop 94, Albion
- Gas N: Mobil*
  S: Amoco*(D)(CW), Shell*(CW), Speedway*(D)(LP)
- Food N: Arby's
  S: A & W Drive-In, Big Boy, Frosty Dan's, McDonalds, Paradise Inn Chinese, Pizza Hut, Ponderosa
- Lodg N: Days Inn (AAA) Days Inn
  S: Adams Arms Best Western
- AServ S: Albion Ford/ Mercury, Auto Zone Auto Parts, Bob Frahm GMC, Chevrolet, Buick, Pontiac, Cooper Tires
- RVCamp N: Camping
- ATM S: Chemical Bank South, Great Lakes Bank Court, Shell
- Other S: Car Wash, Felpausch Pharmacy, Felpausch Supermarket, K-Mart, Rite Aide Pharmacy

124 Michigan 99, Eaton Rapids

127 Concord

128 Michigan Ave
- FStop N: Sunoco*(SCALES)
- Gas N: Amoco*(D)
- Food N: Burger King (Sunoco)
- RVCamp N: Camping

130 Parma
- TStop S: 76 Travel Express*
- Food S: Cafe Cracker Hill (76 TS)
- AServ N: Oil Zone
- TServ N: J&T Repair

133 Dearing Road, Spring Arbor

(135) Rest Area - RR, Phones (Eastbound)

136 MI. 60, Bus Loop 94, Jackson

137 Airport Road
- Gas N: Meijer Grocery*(D)(LP), Shell*
  S: Amoco*(CW)
- Food N: Burger King, Denny's, Dunkin' Donuts (Shell), Food Court (Meijer), Greek Villa, Hot 'n Now Hamburgers, McDonalds, Subway, Wendy's
  S: Cracker Barrel, Lone Star Steakhouse, Olive Garden
- AServ S: K-Mart
- ATM N: Comerica, Shell, Standard Federal
- Other N: Meijer Grocery, Department Store (Pharmacy, 24 Hrs), Paul's Auto Wash
  S: K-Mart (Pharmacy), Sam's Club

138 U.S. 127 North, MI 50, Lansing, Jackson (Difficult Reaccess For Northside)
- Gas S: Marathon*, Shell* (24 Hrs), Total*
- Food N: Bill Knapp's, Gilbert's, Red Lobster, Yen King
  S: Arby's, Big Boy, Bob Evans Restaurant, Davis' Restaurant & Deli, Fazoli's Italian Food, Ground Round Restaurant, Hot N' Now Hamburgers, KFC, Long John Silvers, McDonalds, Old Country Buffet, Outback Steakhouse, Pizza Hut, Ponderosa
- Lodg N: Budgetel Inn, ◆ Fairfield Inn, Hampton Inn, Holiday Inn, Holiday Motel, ◆ Super 8 Motel
  S: Best Motel, Country Hearth, (AAA) Motel 6

EXIT MICHIGAN

- AServ S: Jackson Ford, Meineke Discount Mufflers, Midas Muffler & Brakes, Muffler Man
- ATM S: Comerica Bank, Republic Bank, Shell
- Parks S: Cascade Falls Park, Ella Sharp Park
- Other S: DOC Optical, Ella Sharp Museum, Jackson Crossing Mall, Jay's Book Mart, Target Department Store

139 MI 106, Downtown Jackson, Cooper Street
- FStop N: Commercial Fuel Inc. (Members Only)
- Gas S: Citgo*, Marathon* (24 Hrs)
- Food S: Cherry's Cafe, Dunkin' Donuts (24 Hrs, Marathon)
- TServ S: Vischer Tire
- Med S: ✚ Hospital
- Other S: Bowlorama

141 Elm Road
- Lodg N: (AAA) Rodeway Inn
- Med S: ✚ Hospital

142 U.S. 127 South, Hudson

145 Sargent Street
- TStop S: Citgo*(SCALES) (24 Hrs)
- Gas S: Sunoco*
- Food S: 145 Restaurant (Citgo TS), Ambassadore Diner, Jackson Brewing Co., McDonalds, White Castle Restaurant (Sunoco)
- Lodg S: Colonial Inn
- TServ S: Citgo

147 Race Road
- RVCamp N: Oaks Camping, Sherwood Forest Camp
  S: Greenwood Acres
- Other N: Waterloo Recreation Area

(149) Rest Area - RR, Phones (Westbound)

150 Grass Lake
- RVCamp S: Camping
- Other N: Waterloo Recreation Area (4 Miles)

(151) Weigh Station - Both Directions

153 Clear Lake Road
- Gas N: Total*(D)(LP)

156 Kalmbach Road

157 Old U.S.12, Pierce Road

159 MI 52, Chelsea, Manchester
- Gas N: Amoco*(D)(CW), Mobil*(D)(CW)
- Food N: A & W Drive-In, Big Boy, Chinese Tonite, Little Caesars Pizza, McDonalds, Schumm's Restaurant, Subway, Taco Bell
- AServ N: Amoco, Faist Moorow Buick, Chevrolet,

EXIT MICHIGAN

Oldsmobile, Palmer Ford, Village Motors Plymouth, Dodge, Jeep Eagle
- RVCamp N: Lloyd Bridges Sales, Rental, Service & Parts
- Med N: ✚ Hospital
- ATM N: Amoco, Chelsea State
- Other N: Chelsea Lanes, Chelsea Pharmacy

162 Old US 12, Jackson Road
- FStop S: Clark*
- Food S: Stivers, Subway (Clark)

167 Baker Road, Dexter
- TStop S: 76 Auto/Truck Plaza*(SCALES), Pilot Travel Center*(SCALES)
- FStop N: Speedway*
- Food S: Arby's (Pilot), Wolverine's (76 TS)
- TServ S: 76
- TWash S: Blue Beacon Truck Wash (Pilot)

(168) Rest Area - RR, Phones (Eastbound)

169 Zeeb Road
- Gas N: Amoco* (24 Hrs)
  S: Mobil*
- Food N: Baxter's Party Store & Deli, McDonalds (Inside Play Place)
  S: Arby's, Pizza Hut, Taco Bell, Wendy's, Westside Grill
- AServ N: Amoco
  S: Mobil
- ATM S: First of America

171 MI 14, Plymouth (Left Lane Exit)

172 Bus Loop I-94, Ann Arbor
- Gas N: Amoco* (24 Hrs), Marathon*(LP), Shell*(CW)
  S: Sunoco*(D)
- Food N: Bill Knapp's Restaurant, Dunkin' Donuts, KFC, Papa Romano's Pizza, Schlotzkys Deli, Subway
  S: Chee Peng Restaurant, City Limits (Clarion), Weber's
- Lodg S: (AAA) Clarion (see our ad on this page), Michigan Inn
- AServ N: Amoco, K-Mart, Midas Muffler & Brakes, Tuffy Auto Service, Worldwide Transmission
  S: Muffler Man
- Med S: ✚ Hospital
- ATM N: Kroger Comerica Bank, NBD, Standard Federal
- Other N: Fox Village Theatres, K-Mart (Pharmacy), Kroger Supermarket, Little Professor Book Co., Rite Aide Pharmacy
  S: Super Suds Laundry

175 Ann Arbor, Saline Road
- Gas N: Shell*(CW)
  S: Meijer Grocery*(D)
- Food N: Applebee's, Lone Star Steakhouse, Old Country Buffet, Pizza Republic
  S: Food Court (Meijer), McDonalds, T.G.I. Friday's
- ATM N: Old Kent
- Other N: Parkway Animal Clinic
  S: Meijer Grocery, Department Store (24 Hrs, Pharmacy), Target Department Store

177 State Rd
- Gas N: Mobil*
  S: Total*(D)
- Food N: Bill Knapp's Restaurant, Bombay Backyard, Burger King, Romano's Macaroni's Grill
  S: Chi-Chi's Mexican, Mark's Midtown Coney Island, McDonalds
- Lodg N: ◆ Crowne Plaza, ◆ Sheraton Inn, Wolverine

EXIT	MICHIGAN
	Inn Best Western
	S: AAA Motel 6
Med	N: Hospital
Other	N: Briarwood Mall
180A	U.S. 23 South, Toledo, Flint
181AB	Michigan Ave, U.S. 12 West, Ypsilanti
Gas	N: Speedway*, Total*(D)
Food	N: Burger King, Little Saigon Restaurant, Round Tree, Taco Bell
AServ	N: Firestone Tire & Auto, The Brake Shop, Victory Lane Oil Changes
Med	N: Hospital
ATM	N: Standard Federal
Other	N: Big Lots, Busch's Valu-Land Supermarket, Coin Laundry, Rite Aide Pharmacy, Wal-Mart (Pharmacy)
183	Business U.S. 12, Huron St, Downtown Ypsilanti
Gas	N: BP*(D)
Food	N: Hawkins BBQ
	S: McDonalds
Lodg	S: AAA Marriott
AServ	N: Maynard's Muffler & Brake
	S: GM Auto Dealer
Med	N: Hospital
Parks	S: Rolling Hills County Park (3.5 Miles)
Other	S: State Patrol Post
185	U.S. 12, Michigan Ave, Willow Run Airport (Left Lane Exit)
187	Rawsonville Road
Gas	S: Mobil*(D), Speedway*(D)(LP)
Food	S: Burger King, Denny's, KFC, Little Caesars Pizza, Maria's Pizzeria, McDonalds (Play Place), Pizza Hut, Tom Horton's, Wendy's
RVCamp	S: KOA Camping
ATM	S: First of America, NBD
Other	S: K-Mart (Pharmacy), Mr. Bubble Car Wash, Rite Aide Pharmacy
(188)	Rest Area - RR, Phones, Picnic (Westbound)
190	Belleville Road, Belleville
Gas	N: Amoco*(CW) (24 Hrs), Marathon*
	S: Shell*
Food	N: Bamboo Garden Chinese, Big Boy, Cracker Barrel, Hungry Howie's Pizza, McDonalds, Tin Lizzie Casual Dining, Wendy's
	S: Burger King, China King, Demetri's Kitchen, Domino's Pizza, Dos Tesos, Subway, Uncle Joe's Coney Island
Lodg	N: ◆ Red Roof Inn
	S: ◆ Super 8 Motel
AServ	N: Auto Works, Marathon
	S: Firestone Tire & Auto, Pennzoil Lube
ATM	N: Amoco, Marathon
	S: Comerica Bank, Shell
RVCamp	N: Camping World (see our ad this page)

EXIT	MICHIGAN
Other	N: Arbor Drugs
	S: Belle Coin Laundry, U.S. Post Office
192	Haggerty Road
Gas	N: Mobil*
Parks	S: Lower Huron Metro Park
194AB	Junction I-275S., Toledo
196	Romulus, Wayne
Gas	N: Shell*(CW)
	S: Mobil*(D), Total*

EXIT	MICHIGAN
Food	N: McDonalds
	S: Burger King
AServ	N: Shell
	S: Potter Tire Center
ATM	N: First of America
198	Metropolitan Airport, Middle Belt Road (Services Are Straight Off Exit)
Lodg	N: Clarion Inn (see our ad this page)
200	Ecorse Road
202AB	U.S. 24, Telegraph Road
204	Southfield Freeway, MI 39, Pelham Road
206AB	Oakwood Blvd
208	Schaefer Road
209	Rotunda Dr
210A	U.S. 12, Michigan Ave
210B	Addison Ave, MI 153, Ford Rd
211A	Lonyo Ave
211B	Cecil Ave, Central Ave
212A	Livernois Ave
212B	Warren Ave
213A	West Grand Blvd, Warren Ave
213B	Junction I-96
214A	Grand River Ave, Linwood Ave
214B	Trumbull Ave
215AB	MI 10, Lodge Freeway
215C	MI 1, Woodward Ave, Brush St
216A	Junction I-75, Chrysler Fwy, Flint
Lodg	N: Holiday Inn Select of Auburn Hills (take I-75 North -- see our ad this page)
216B	Russell St
217B	Mount Elliott Dr
217A	Mount Elliot Ave, East Brand Ave
Gas	S: Mobil*, Shell*
Food	S: Legends Coney Island Restaurant, Taco Bell, Young's BBQ
Other	S: White Cross Pharmacy
218	Van Dyke Ave, MI 53
Gas	N: 76*, Amoco*
	S: Sunoco*
AServ	S: Sunoco
219	MI 3, Gratiot Ave
Gas	S: BP*
Food	N: Elmo's Fine Food, KFC
	S: Burger King
Other	N: Farmer Jack Grocery
220A	French Road
Gas	S: Shell*
220B	Conner Ave, City Airport
Gas	N: BP*, Marathon*(CW)
Lodg	N: Regency Inn
Med	S: Hospital
Other	N: Airport, Sam's Market

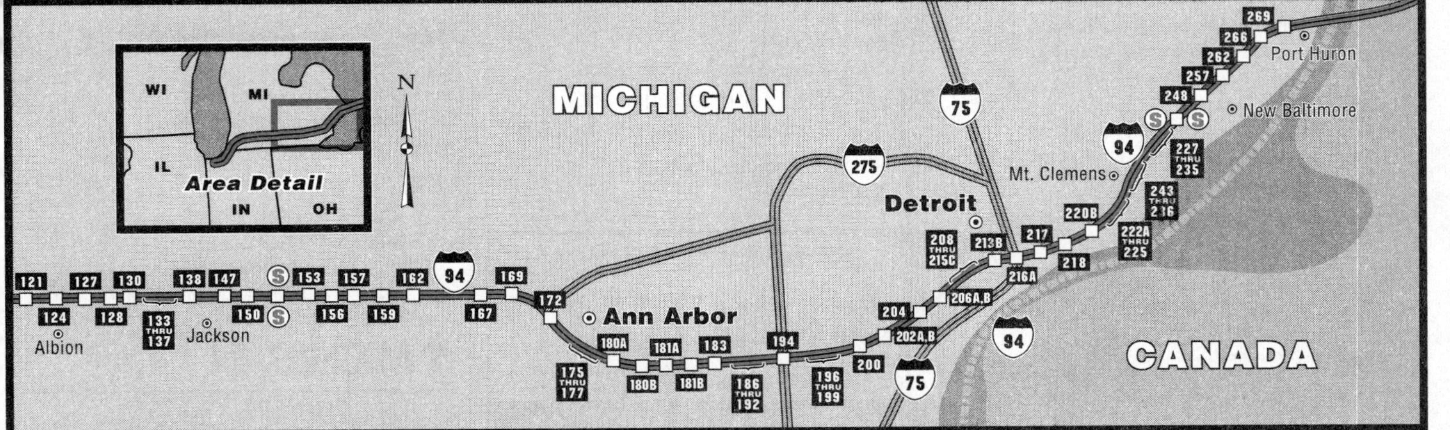

Bold red print shows RV & Bus parking available or nearby

EXIT MICHIGAN

222A Outer Dr, Chalmers Ave
- Gas N: 76*, Amoco*, BP*, Shell
- Food N: Alabama Style Chicken, KFC, Little Caesars Pizza, Olympia Coney Island, Universal Restaurant, White Castle Restaurant
- AServ N: Santoro Auto Service, Shell
- TServ N: **Santoro Truck Service**
- Other N: **Coin-O-Matic Laundromat, Sam's Discount Auto Parts**

222B Harper Ave (Eastbound, Difficult Reaccess)

223 Cadieux Ave
- Gas S: Amoco*, Citgo*, Mobil*, Shell*, Sunoco*
- Food S: Boston Market, McDonalds, Rally's Hamburgers, Taco Bell, Tubby's Submarine, Wendy's
- AServ S: Mobil, Valvoline Quick Lube Center
- Other N: **Vogue Convenience**
  - S: Auto Wash, **Rite Aide Pharmacy**

224A Moross Road
- Gas S: Shell*
- Other S: **Farmer Jack Grocery**

224B Eastwood Drive, Allard Ave
- Gas S: Shell*
- Other N: **Police Station**
  - S: **Farmer Jack Grocery**

225 MI 102, 8 Mile Road, Vernier Road (Difficult Reaccess)
- Gas S: Mobil*
- Food S: KFC, Round Table, Wendy's
- AServ S: Cadillac Dealer, Mobil
- Other S: **Kroger Supermarket**

227 9 Mile Road
- Gas N: Marathon, Speedway*(D), Sunoco*
  - S: 76 Express, Mobil*
- Food N: China Wok, Dolly's Pizza, Elaine's Bagels, Karan's Restaurant, Marc's Pub & Grill, McDonalds, Olympic Family Dining, Porky's BBQ, Taco Bell, Wendy's
  - S: Stooges
- AServ N: Marathon, Minit-lube
- ATM N: First State Bank, Standard Federal
  - S: First of America
- Other N: **Arbor Drugs, Coin Laundry, F & M Drugstore, Farmer Jack Grocery, Office Depot, U.S. Post Office (Farmer Jack Grocery)**

228 10 Mile Road
- Gas N: Amoco*(D), Mobil, Shell*(CW)
- Food N: 3D's Pizza, Baskin Robbins, Coney Island Restaurant, East Wind Chinese, Fairway Sports Bar & Grill, Jet's Pizza, RJ's Vault Grill & Bar
- AServ N: Mobil
- Other N: **3D's Convenience Store, Arbor Drugs, Doctor's Pharmacy, Fairway Drugs, Pet Supply Store, Rite Aide Pharmacy**

229 Junction I-696 West, 11 Mile Road

230 12 Mile Road
- Gas N: Marathon*(D), Mobil*(D)
  - S: Marathon*
- Food N: Burger King, Dunkin Donuts, Outback Steakhouse
- AServ N: Arnold Ford/Lincoln/Mercury
  - S: Marathon
- ATM N: Marathon, Mobil, NBD Bank
- Other N: **Arbor Drugs**, Car Wash, **Embassy Market, Farmer Jack Grocery, Service Merchandise, Super Petz**

231 Mi 3, Gratiot Ave (Left Exit, Eastbound, Westbound Reaccess Only)

232 Little Mack Ave
- Gas N: Speedway*
  - S: Amoco*, Meijer Grocery*(D)(LP), Total*(D)
- Food N: Applebee's, Arby's, Boston Market, Chuck E. Cheese's Pizza, Denny's, Lone Star Steakhouse, Pizza Hut
  - S: Cantonese Restaurant, Dunkin Donuts (Amoco), IHOP, Pearl City
- Lodg N: (AAA) Eastin Hotel, Econolodge, ◆ Red Roof Inn, Super 8 Motel
  - S: ◆ Budgetel Inn
- AServ S: K-Mart
- Other N: **Builders Square 2, Macomb Mall, Murray's Discount Auto Store, Sams Club, Staples Office Supply**
  - S: **Home Depot, K-Mart (Pharmacy), Meijer Grocery, PetsMart**

234AB Harper Ave
- Gas N: Speedway*(LP), Speedy Q Mart*, Sunoco*
- Food N: Carmen's Deli & Pizza, Cosmo's Family Restaurant, LA's Careless Cafe, Sorrento Pizza, Teddy's Tavern
  - S: China Moon, Golden Donuts, Little Caesars Pizza, Sub Company, Subway, Travis Restaurant, **Winner's Sports Cafe**
- AServ N: Jiffy Lube, Stan & Mike Graphic Auto Care
- Med N: ✚ **Hospital**
  - S: ✚ **Convenient Walk-in Clinic**
- ATM S: Old Kent Bank
- Other N: Car Wash, **Patterson Vet Hospital, Police Department, Sandy's Coin Laundry**
  - S: CK Car Wash

235 Shook Rd. (Westbound, Eastbound Reaccess)

236 Metropolitan Parkway
- Food S: Fong's Chinese, Little Caesars Pizza, McDonalds, Subway
- Med S: ✚ **Hospital**
- Other S: **Arbor Drugs, Kroger Supermarket, Sear's Hardware**

237 North River Road, Mount Clemens
- Gas N: Amoco*(D), Mobil*
- Food N: Subway (Mobil)
- AServ N: Mobil(D)
- Other S: **The Captains Convenience Store**

240 MI 59, Utica, Self Ridge ANG (Eastbound, Westbound Reaccess)

241 21 Mile Road, Selfridge
- Gas N: Marathon*
- Other N: **Sheriff Dept.**

243 MI 3, MI 29, MI 59, Utica, New Baltimore
- FStop S: **Speedway***
- Gas N: Citgo* (Kerosene), Meijer Convenience Store*, Sunoco*
  - S: Mobil*(D)
- Food N: Applebee's, Arby's, Baskin Robbins, Burger King, Coney Island, **Father & Son Pizzeria**, Horn of Plenty, Le Grand Chinese Buffet, Little Caesars Pizza, McDonalds, New Riviera Restaurant, Papa Ramon's Pizza, Ponderosa, Score Card Bar & Grill, Wendy's, White Castle Restaurant
  - S: Big Boy, Buscemis, Hot 'N Now, TCBY Treats (Mobil), Taco Bell
- Lodg N: Chesterfield Motor Inn
  - S: **Lodge Keeper**
- AServ N: Chesterfield Auto Parts, Discount Tires, Goodyear Tire & Auto, Jiffy Lube, Midas Muffler & Brakes
  - S: Mobil
- ATM N: NBD Bank, Old Kent
  - S: Mobil
- Other N: Clearview Car Wash, **Farmer Jack Grocery, K-Mart (Pharmacy), Meijer Grocery, Pet Supplies Plus, Rite Aide Pharmacy, Sear's Hardware, Staples Office Superstore, Target Department Store**
  - S: Auto SPA Car Wash

(246) **Weigh Station - both directions**

247 MI 19, Richmond, New Haven (Eastbound, Reaccess Westbound)
- Other N: **State Patrol Post**

248 26 Mile Road, New Haven, Marine City
- Food S: My Place Coffee Donut Shop

(250) **Rest Area - RR, Picnic, Phones, Vending**

(255) **Rest Area - RR, Phones, Picnic, Vending (Eastbound)**

257 St Clair, Richmond
- Other N: **State Patrol Post**

262 Wadhams Road
- FStop S: **Total***(SCALES)
- RVCamp N: **Camping**
- ATM S: **Total**

266 I-94 Loop, Gratiot Road, Marysville
- Lodg N: Days Inn (see our ad on this page)
- Med S: ✚ **Hospital**
- Other N: **Airport**

269 Range Road, Dove St
- FStop N: **Total***(D)(CW)
- Food N: Burger King, **Subway (Total)**
- Other N: **Factory Outlet Center**

271 Junction I-69

274 Water St
- Gas S: By-Lo Convenience Store, Total*
- Food N: **Cracker Barrel (on Water St)**
  - S: Bob Evans Restaurant
- Lodg S: ◆ Comfort Inn, ◆ Fairfield Inn, Knights Inn
- RVCamp N: **Campground**

↑MICHIGAN

Begin I-94

Bold red print shows RV & Bus parking available or nearby

EXIT MAINE

Begin I-95

↓MAINE

3 **(298)** U.S. 2, Houlton Airport (Northbound Only To Can, Southbound To I-95)
TServ E: Shannon's Repair
2 **(295)** U.S. 1, Houlton, Presque Isle
TStop W: Irving Truckstop*(LP)
FStop W: Exxon(LP)
Gas E: Exxon*, Irving*(D)
W: Citgo*
Food E: Asian Palace, Burger King, Dunkin Donuts, McDonalds, Pizza Hut
W: Shiretown Motor Inn, York's Dairy Bar
Lodg W: AAA Ivey's Motor Lodge (AAA), AAA Shiretown Motor Inn
AServ E: VIP Discount Auto Parts
W: York's Chev/Olds/Pontiac/Buick
TServ W: Irving Truckstop(SCALES)
RVCamp W: Campground
Med E: ✚ Hospital
ATM E: Citizen's Bank
W: City Bank, Peoples Bank
Other E: Rite Aide Pharmacy
W: Rest Area - RR, Picnic, Phone, Shop & Save, Wal-Mart (Pharmacy)
1 **(284)** U.S. 2, Smyrna
Food E: Brookside Restaurant
Lodg E: Brookside Motel
0 **(279)** to ME 11, Oakfield Rd, Smyrna Mills
FStop W: Irving*(D), Mobil*
Food W: Chester Fried Chicken, Country Corner Cafe, Crossroads Cafe, Hilton's Farm Dairy Bar
Other W: Coin Laundry, U.S. Post Office
9 **(269)** ME 159, Island Falls
FStop E: Mobil*(D)(LP)
Gas E: Citgo*
Food E: Pipe's Family Restaurant
RVCamp E: Campground
Parks W: Baxter State Park
Other E: Bishop's Market, Porter's Oil Co.(LP)
8 **(257)** to ME 11, to ME 158, Shermon, Patten
FStop E: Mobil(LP)
W: Irving*
Food W: Shermon Dairy Bar
Lodg W: Katahdin Valley Motel, Shermon Bed/ Breakfast, Summit Farms
AServ W: Tom's Garage
Parks W: Baxter State Park
7 **(252)** Benedicta (No Northbound Reaccess)
245) Scenic View - Picnic (Northbound)
6 **(237)** ME 157, Medway, Millinocket
TStop W: Irving Truckstop*
Food W: Irving
Lodg W: ◆ Gateway Inn
RVCamp W: Campground
Med W: ✚ Hospital
Other E: Alabaster State Park
237) Rest Area - RR, Phones
5 **(221)** to U.S. 2, ME 6, ME 116, Lincoln, Mattawamkeag
RVCamp E: Campground
Med E: ✚ Hospital
4 **(210)** ME 6, ME 155, Howland, Lagrange
Gas E: Irving*
Food E: D & J's Ice Cream
AServ E: 95'er Towing Service
RVCamp E: King's Camping Park
Other E: Car Wash
3 **(193)** ME 16, LaGrange, Milo (Northbound, Reaccess Southbound)

EXIT MAINE

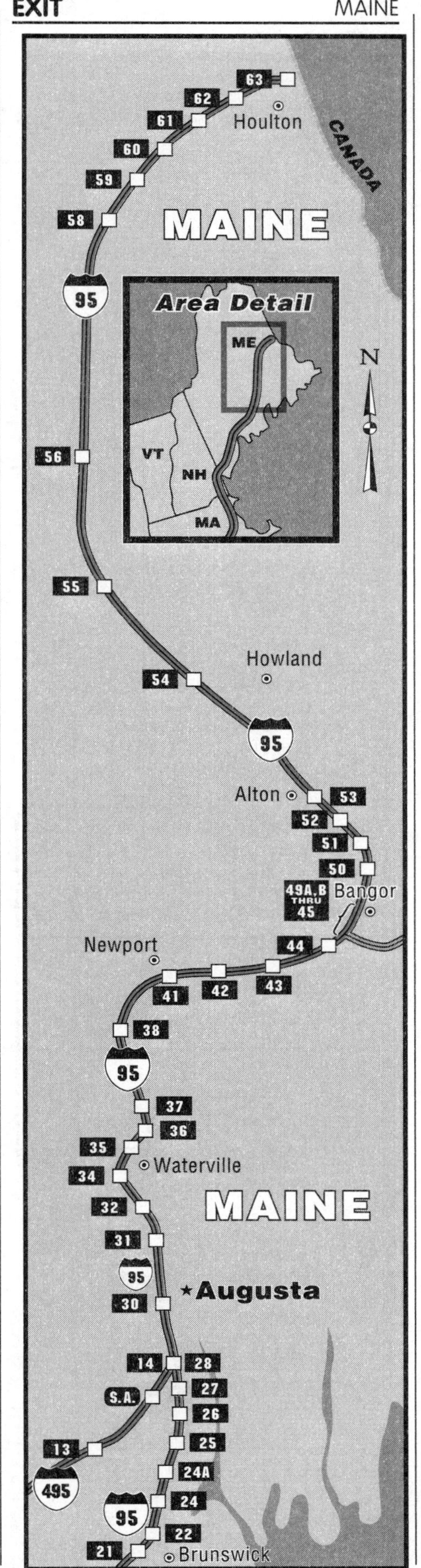

EXIT MAINE

(192) Weigh Station - both directions (Closed)
52 **(190)** ME 43, Old Town, Hudson
51 **(186)** Stillwater Ave, Old Town
FStop E: Citgo*(D)
Gas E: Citgo*(D), Irving*(D), Texaco*(D)
Food E: Burger King
Lodg E: ◆ Best Western
50 **(184)** Kelly Road, Orono
49 **(181)** Hogan Road, Bangor, Veazie
Gas E: Citgo*
W: Citgo*(D)(CW), Exxon(D)(CW)(LP)
Food W: Arby's, Asian Palace, Burger King, China Wall, Colonial House of Pancakes, KFC, McDonalds, Olive Garden, Oriental Jade, Paul's Steak & Seafood, Pizza Hut, Red Lobster, Wendy's
Lodg W: AAA Bangor Motor Inn, ◆ Comfort Inn, Country Hospitality Inn, ◆ Hampton Inn
AServ E: Bangor Ford/ VW/ Plymouth/ Dodge/ Saturn/ Mitsubishi/ Hyundai, Bangor Mercedes, Citgo*, Firestone Tire & Auto, Honda Nissan Dealer, Saturn Dealership, Varney Izusu,GMC, Village Car & Truck Center, Village Subaru
W: Goodyear Tire & Auto, K-Mart, McQuick's Oil Change, Sears, VIP Discount Auto Service, Van Syckle Ford,Mercury,Lincoln
TServ E: Peterbilt Dealer
ATM E: Bangor Federal Credit Union
W: E. Maine Federal Credit Union, Fleet Bank, Heritage Bank, Merrill Merchant Bank
Other E: Sam's Club, Vision Center
W: Bangor Shopping Center, Shaw's Supermarket, Wal-Mart
48 **(179)** ME 15, Broadway
Gas E: Irving
W: Exxon, Mobil*
Food E: Tri-City Pizza
W: Baldacci, BoBo's, China Lite, Dairy Queen, Friendly's, Governor's Restaurant, Jane's Java & Deli, KFC, King's Kitchen, Little Caesars Pizza, McDonalds, Pizza Hut
AServ W: Kelly Pontaic, Prompto Auto Care, Vernie's Auto Supply
Med E: ✚ Hospital
ATM W: Bangor Savings Bank, Fleet Bank
Other E: Larry Barron's(LP)
W: Christy's Convenience Store, Rite Aide Pharmacy, Super Shop & Save
47 **(178)** ME 222, Ohio St, Airport , Union St
Gas E: Citgo, Exxon
AServ E: Citgo (Towing)
RVCamp W: Campground
ATM E: Merrill Merchant Bank
Other W: Airport
46 **(177)** U.S. 2, ME 100, Hammond St, Airport
Gas E: Exxon
Food E: Napoli Pizza, Subway
AServ E: Exxon, NAPA Auto Parts
ATM W: Bangor Savings Bank
Other E: Corner Store, Price's Fairmont Market
W: Airport
45AB **(176)** Junction I-395, to U.S. 1A, to U.S. 2, to ME 9, Bangor, Breweraccess)
FStop W: Citgo*, Mobil(CW)
Gas W: Irving(D)
Food W: Barnaby's, Dunkin Donuts, Ground Round, Holiday Inn, Howard Johnson
Lodg W: DAYS INN Days Inn, Econolodge, Fairfield Inn, Holiday Inn, Howard Johnson, Ramada Inn, Roadway Inn, Super 8 Motel
TServ E: Brewer's Truck & Parts
RVCamp W: Campground
ATM W: Fleet Bank
Other W: Blackbeards Family Fun Park
44 **(173)** Coldbrook Road, Hermon,

Bold red print shows RV & Bus parking available or nearby

EXIT MAINE

Hampden
TStop W: Dysart's Truck Stop*(D)(LP)(SCALES)
FStop E: Citgo*
Lodg W: AAA Best Western
TServ W: Dysart's, Kenworth International

(172) **Rest Area - RR, Phones, Picnic, Tourist Info (Southbound)**

(169) **Rest Area - RR, Phones, Picnic, Tourist Info (Northbound)**

43 **(167)** ME 69, Carmel, Winterport
Gas E: Citgo*(CW)
AServ E: Max Paint & Body Shop
RVCamp W: Campground
Other E: Butterfield's Car Wash, Butterfield's Convenient Store

42 **(161)** ME 69, ME 143, Etna, Dixmont
RVCamp W: Campground

41 **(154)** U.S. 2, ME 7, East Newport, Plymouth
RVCamp W: Campground(LP)

39 **(151)** to ME 7, ME 11, ME 100, Newport, Dexter
TStop W: Irving*
FStop W: Mobil*(D)
Gas W: Citgo*(D)
Food W: China Way, Irving Restaurant, McDonalds, Popeye's Chicken, Subway
Lodg W: AAA Lovley's Motel
AServ W: Big A Auto Parts, Muffler King, Plymouth Dodge Dealer
RVCamp W: Campground
Med E: + Hospital
ATM W: Merrill Merchant Bank
Other W: Bud's Shop & Save, Coin Laundry, Newport Plaza, Rite Aide Pharmacy, Shop 'N Save Supermarket, Triangle Plaza, Visitor Information, Wal-Mart

38 **(144)** ME 11, ME 100, ME 152, Hartland, Somerset Ave, Pittsfield
Food E: Subway
Lodg E: Ponderosa Motel
AServ E: Michelin Tire Service, Uniroyal Tire & Auto, Wright Chev/GEO
RVCamp E: Campground
Med E: + Hospital
Other E: Bag-n Shop Grocery, Java Shop, Rite Aide Pharmacy, Scrub A Dub Car Wash, Shop 'N Save Supermarket

(141) **Rest Area - RR, Phones, Picnic (Both Directions)**

37 **(132)** Hinckley Road, Clinton, Burnham
Gas W: Citgo*(D)(LP) (Towing)
AServ W: Citgo

36 **(127)** U.S. 201, Fairfield, Skowhegan
Other E: Coin Car Wash, Coin Laundry

35 **(126)** ME 139, Fairfield , Benton
TStop W: Citgo*(CW)(SCALES)
Gas E: Texaco*(LP)
TServ W: Citgo(LP)

34 **(124)** ME 104, Main St, Waterville
Gas E: Citgo*, Exxon
Food E: Arby's, Family Dining, Friendly's, Killarney's, McDonalds, TCBY Yogurt, The Governor's, Wendy's
Lodg E: AAA Best Western, ◆ Holiday Inn, The Atrium Motel
AServ E: Citgo, Thompson VW/Audi/Mazda, VIP Discount Auto Center
ATM E: Fleet Bank
Other E: Coin Laundry, K-Mart, Shop & Save, Vision Express

33 **(121)** ME 11, ME 137, Waterville, Oakland

EXIT MAINE

FStop E: Irving*
W: Coastal*, Exxon(D)
Gas E: Citgo*(CW), Coastal*
Food E: Angelo's, Burger King, Classics Restaurant, Dunkin Donuts, Pizza Hut, Subway, Weathervane Seafood
W: China Express, Coastal, Pine Acres
Lodg E: Budget Host Hotel, AAA Econolodge, Motor Lodge
AServ E: GM Auto Dealership, Pontiac Buick GM Dealer, Texaco Auto Lube
W: A&A Auto Parts, Pullen's Ford/Lincoln/Mercury/Mercedes
Med E: + Hospital
ATM E: Citi Bank
Other E: Coin Laundry, Lucyen Car Wash, Rite Aide Pharmacy, Shaw's Supermarket, Wal-Mart

32 **(114)** Lyons Road, Sidney

(110) **Rest Area - RR, Phones (Southbound)**

(107) **Rest Area - RR, Phones (Northbound)**

31AB **(106)** ME 8, ME 11, ME 27, Augusta, Belgrade
Gas E: Citgo*, Getty*(LP)
W: Irving*(D)
Food E: Captain Cote's BBQ, Civic Center Pizza, Sandwiches, TCBY, The Ground Round
W: Alfred's
Lodg E: Holiday Inn (see our ad on this page)
W: Comfort Inn (see our ad on this page)
AServ W: Davis Motors
ATM E: Augusta Federal Savings, Fleet Bank
Other E: Augusta Civic Center, Sam's Club, Vision Express (Eye Wear), Wal-Mart

30 **(106)** U.S. 202, ME 100, ME 17, Augusta, ME11

EXIT MAIN

Gas E: Citgo*(D), Gulf*(LP), Irving*(D), Texaco*(LP)
W: Exxon*, Getty*
Food E: Arby's, Bagels Mania, Best Western, Charlie's Pizza & More, Dairy Queen, Daman's Italian Sandwiches & Pizza, Hong Kong Isle Chinese, KFC, McDonalds, Oyster Bar & Grill
W: Bonanza Steak House, Margarita's
Lodg E: Best Western, Senator Motel
W: Augusta Hotel, Motel 6, ◆ Super 8 Motel, Th Susse Chalet
AServ E: Citgo, Prompto Lube
W: Charlie's Motor Mall, Charlie's Toyota, Exxon Hyundai,Chryser, Jeep, Eagle, Sears Auto Cente
ATM E: Fleet Bank
Other E: Coin Car Wash, Rite Aide Pharmacy, Shaw's Supermarket, Smart Eye Care
W: AAA Travel Center, Shop & Save

28 **(96)** Jct I-95 S, Me Tnpk, Litchfield, Gardiner

27 **(94)** U.S. 201, Gardiner
AServ E: Daniel's Auto Service (24 hr Towing)

26 **(88)** ME 197, Richmond, Litchfield

25 **(81)** ME 125, ME 138, Bowdoinham, Bowdoin

24AB **(75)** ME 196, Topsham, Lisbon
Food E: Arby's, McDonalds, Romeo's Pizza
AServ E: Meineke Discount Mufflers, NAPA Auto Parts, Woody's Performance Center
ATM E: Coastal Bank, Garner Saving
Other E: Coin Laundry, Rite Aide Pharmacy, Super Shop & Save, Topsham Fare Mall

22 **(72)** U.S. 1, Coastal Route, Brunswick (Eastbound)
Gas E: Citgo*, Irving*, Mobil*(D)
Food E: Dunkin Donuts, Hong Kong Island, MacLean's
Lodg E: Comfort Inn , ◆ Econolodge,
AServ E: Goodwin's Chevrolet, Texaco Express Lube
RVCamp E: KOA Campground
Med E: + Hospital

21 **(68)** U.S. 1, Freeport (Northbound, Difficult Reaccess Southbound))

20 **(66)** ME 125, ME 136, Durham
Gas E: Exxon*
Food E: Arby's, Broad Arrow Tavern, China Rose, Friendly's, McDonalds
Lodg E: Brewster House Bed/Breakfast, Harraseektn, Mains St Bed/Breakfast, Nicholson Inn, Village Inn Bed & Breakfast, White Cedar Inn Bed/Breakfast
RVCamp W: RV Campground
ATM E: Coastal Bank
Parks W: Bradbury State Park
Other E: Coin Laundry, Outlet Stores, Pet Pantry, U.S Post Office

19 **(65)** U.S. 1, Desert Road, Freeport
Gas E: Citgo*
Food E: Blue Onion, Cricket's, Gritty's, Subway, Thia Garden
Lodg E: Cindy's Motel, ◆ Coastline Inn, Dutch Village Motel, Freeport, ◆ Super 8 Motel
ATM E: Bass Savings
Other E: Desert Rose Camp Ground(LP) (AAA), Factory Outlet Center, Freeport Car Wash

17 **(62)** U.S. 1, Yarmouth
FStop W: Texaco*(CW)
Food E: Day's, Muddy Rudder Fine Dining
W: Bill's Pizza & Pasta, McDonalds, Pat's Pizza
Lodg E: AAA Freeport Inn (AAA)
W: Down-East Village, Red Wagon Motel
AServ E: Casco Bay Ford
W: Texaco
ATM W: People's Heritage Bank, Texaco
Other E: Rest Area - RR, Picnic, Phone
W: Coin Laundry, Royal River's Natural Food

Bold red print shows RV & Bus parking available or nearby

EXIT MAINE

6 **(59)** U.S. 1, Cumberland, Yarmouth
Gas **W:** Exxon, Mobil*(D)
Food **W:** Birchwood Restaurant, Lox's & Bagels, Main Roaster's Coffee Shop, Romeo's Pizza
Lodg **W:** ◆ Brookside Motel
AServ **W:** Exxon
ATM **W:** Bass Saving
Other **W:** **AAA Office, Rite Aide Pharmacy, Yarmouth Market Place**

5 **(55)** U.S. 1, Falmouth

Note: I-95 runs concurrent below with METNPK. Numbering follows METNPK.

4AB **(99)** Junction I-95 South, ME 9, ME 126

98) Toll Plaza

97) Service Area (Northbound)
FStop **E: Citgo**
Food **E: Burger King**, TCBY

81) Service Area (Southbound)
FStop **W: Citgo**
Food **W: Burger King**

3 **(78)** Lewiston
Gas **W:** Mobil(D), Sunoco*, Texaco*(D)
Food **E:** Fast Breaks
W: Cathay Hut Polenesian, Dunkin Donuts, Gendrum's Seafood, KFC, McDonalds, Taco Bell, The Governor's, Wendy's
Lodg **E:** Super 8 Motel
W: Ramada Inn, The Chalet
AServ **E:** Double Discount Auto Service, Tire Warehouse, VIP Auto Service
W: Mobil, Prompto 10 Min Oil Change
Med **W:** ✚ **Hospital**

2 **(73)** U.S. 202, Auburn, ME 4, ME100
FStop **W: Bundy's*(D)**
Food **W: Auburn Inn**, Taco Bell
Lodg **E:** Sleepy Time Motel
W: Auburn Inn
RVCamp **W: Campground**
Med **E:** ✚ **Hospital**
Parks **W: Bradbury Mtn State Park**
Other **W: Airport**

65) Toll Plaza

1 **(61)** U.S. 202, ME 115, ME 16, Gray
Gas **E:** Exxon, Gulf*(LP), Mobil, Texaco*
Food **E:** Dunkin Donuts, McDonalds, New England Pizzeria, Subway, T-Jay's Ice Cream, The Village Kitchen
AServ **E:** Exxon, Mobile (Towing)
RVCamp **E: Campground**
W: Campground
Other **E: IGA Supermarket, Rite Aide Pharmacy**

57) Service Area both directions
FStop **B: Citgo**
Food **B: Burger King**, TCBY Yogurt

0 **(50)** Portland, West Falmouth
Med **E:** ✚ **Hospital**

• **(49)** Junction I-95, U.S. 1 Coastal, Falmouth, Brunswick

3 **(46)** ME 25, Westbrook, Portland
FStop **W: Global***
Gas **E:** Mobil*(D), Texaco*(CW)
W: Exxon*(D)
Food **E:** Denny's, Governor's Restaurant, KFC, McDonalds, Pizza Hut, Subway, Zachery's
W: Exit 8 Diner (Howard Johnson), Verrillo's Restaurant
Lodg **E:** Holiday Inn, Susse Chalet, Westbrook Inn
W: Howard Johnson (see ou r ad on this page), Super 8 Motel
AServ **E:** NAPA Auto Parts, VIP Auto Center
W: Auto Tire Warehouse, Double Discount Auto Parts

EXIT MAINE

TServ **E: Roe Ford Truck**
Med **E:** ✚ **Hospital**
Other **E: Super Shop & Save**

7 **(43)** to U.S. 1, Maine Mall Road, Airport
Gas **W:** Citgo*(LP), Mobile*, Texaco
Food **E:** Bean Sprouts Chinese, Ground Round, Hot Shots Grill, Pizza Plus, Subway
W: Burger King, Chicago Bar & Grill, Chili's, Friendly's, IHOP, McDonalds, Old Country Buffet, Pizza Hut, Pizzeria Uno, The Silver Shell, Weather Vane, Wendy's
Lodg **E:** Comfort Inn, Fairfield Inn, Hampton Inn
W: DAYS INN Days Inn, Sheraton
AServ **E:** American Brake Service, Express Lube, Honda
W: Sears
ATM **W:** Atlantic Bank
Other **E: Maine Mall, Pet Quarters, Sam's Club, Wal-Mart (Pharmacy), Wassamki Springs Campground**
W: Airport, Main Mall

6A **(42)** Junction I-295 North, South Portland, Downtown Portland

6 **(40)** U.S. 1, Scarborough

5 **(33)** Junction I-195, U.S. 1, Saco, Old Orchard

4 **(30)** ME 111, Biddeford
FStop **E: Irving***
Food **W:** Wendy's
AServ **W:** Tom's Auto Repair, **Wal-Mart**
Med **E:** ✚ **Hospital**
ATM **W:** People's Heritage Bank
Other **W: Shaw's Food & Drugs, Wal-Mart (Pharmacy)**

(24) Service Area - both directions
FStop **B: Citgo**
Food **B: Burger King, Popeye's Chicken, TCBY Yogurt**

EXIT MAINE

AServ **B:** Citgo

3 **(24)** Kennebunk
FStop **E: Citgo**
Food **E:** Burger King, Popeyes, TCBY Yogurt
Lodg **W: Turnpike Motel (AAA)**
Other **E: Kennebunk Beach, Kennebunk Veterinary Hospital**

2 **(18)** Wells, Sanford, ME 9, ME109
RVCamp **E: RVCamp**
W: Gregoires Campground
Med **W:** ✚ **Hospital**

Note: I-95 runs concurrent above with METNPK. Numbering follows METNPK.

(5) **Weigh Station (Northbound)**

(4) **Weigh Station (Southbound)**

4 **(6)** to ME 91, to U.S. 1, Yorks, Ogunquit
Gas **E:** Citgo*, Irvin's*(D), Mobil*
Food **E:** Breakfast Plus, Fooders Pizza, Green Leaves Chinese, Ice Cream & More, Norma's, Sadie's Cafe, Stonewall Kitchen, Vincent Subs & Pizza
AServ **E:** Frank Goodwin's Chevrolet
Med **E:** ✚ **Hospital**
ATM **E:** Mobil
Other **E:** Kleen City Car Wash, **Leather Outlet, Mail Boxes Etc, Rite Aide Pharmacy, Visitor Information**

(3) Maine Wlcm. Ctr. - RR, Phones, fax, Vending, Playground, Picnic

3 **(2)** U.S. 1 North, ME 236 (Northbound)
Gas **E:** Getty*
Food **E:** Sunrise Grill
Other **E: Outlet Mall, Police Station**

2 **(1)** U.S.1 S, ME 236, Kittery, S Burwick
TStop **E: Howell's Travel Stop*(D)(SCALES)**
FStop **E: Citgo(SCALES)**
Gas **E:** Citgo*
W: Mobil
Food **E:** Bagel Caboose, Christy's Subs, Dairy Queen, **Lazy 8 Wishbone Steak House**, Payrin Thai, Simple Subs, Sunrise Grill, Taco Bell
Lodg **E:** Blue Roof Motel, DAYS INN AAA Days Inn, Rex Motel, **Super 8 Motel**
AServ **W:** Mobil
TServ **E: Northeast Hydraulics**
ATM **E:** Citgo
Other **E: Groceries on the Go, Kittery Animal Hospital, Kittery Historical & Naval Museum**

1 ME 103, Dennett Road (No Northbound Reaccess)
Gas **E:** Citgo

↑MAINE

↓NEW HAMPSHIRE

(20) **Rest Area - Picnic, Phones, RR, HF**

7 **(19)** Bus. District Portsmouth, Waterfront Historic Sites, Newington
Lodg **E:** Sheraton
W: Courtyard Hotel
Other **E: Tourist Information, USS Albacor Submarine**

6 **(16)** Woodbury Ave, Portsmouth
Food **E:** Anchorage Inn, Bickfords Family Restaurant, Meadowbrook Inn
Lodg **E:** Anchorage Inn, Holiday Inn, Howard Johnson, Meadowbrook Inn, Portsmouth Inn, The Port Motor Inn

5 **(14)** U.S. 1, Portsmouth
FStop **W: Exxon*, Texaco***
Gas **W:** Gulf
Food **E:** Bickfords Family Restaurant

EXIT NEW HAMPSHIRE/MASSACHUSETTS

Lodg E: ◆ Holiday Inn, ◆ Howard Johnson, Medow Brook Inn, Swiss Chalet, (AAA) The Port Motor Inn (AAA)
W: Portsmouth Inn
AServ E: Pontiac, Oldsmobile, GMC, Cadillac
Med E: ✚ Hospital

4 **(14)** U.S. 4, NH 16, Newington, Dover (northbound) NH Lakes

3 **(12)** NH 101, Greenland, Portsmouth
FStop W: **Sunoco***
Food W: **Dunkin Donuts (Sunoco)**
Med E: ✚ Hospital

2 **(6)** NH 51, to NH 101, Hampton, Manchester (Exit To Toll)

1 NH 107, Seabrook, Kingston
FStop E: **A1 Farms, Getty**
Gas E: Getty(D), Jack's*, Richdale*, Sunoco*
W: Citgo*
Food E: 99 Restaurant & Pub, Burger King, DeAngelo's Sandwich, Dunkin Donuts, Hananah's, Honey Bee Donuts, Johnny Dale's Italian, Moe's Italian Sandwiches, Papa Gino's, Road Kill Cafe
W: Captain K's Steak & Seafood, Master MacGrath's
Lodg E: Hampshire Motor Inn
W: (AAA) Best Western, Cinnamon Suites
AServ E: Getty, Jiffy Lube, Midas Muffler & Brakes, Pronto 10 Min Oil Change
W: Stravton
TServ E: **Seabrook Truck Center**(SCALES)
ATM E: Fleet Bank
Other E: **Market Basket, Science & Nature Center Seabrook Station, Seabrook South Gate Plaza**

(0) NH Welcome Center - RR, Phones, Picnic, HF (Northbound)

↑ NEW HAMPSHIRE

↓ MASSACHUSETTS

(61) MA Welcome Center - RR, Phones (Southbound)

60 MA 286, Beaches, Salisbury
FStop E: **Mobil***
Food E: Buster's Breakfast, Chubby's Diner, Hodgie's Too Ice Cream, Jabe's Subs & Pizza, Lena's Seafood, Martha's Seafood
TServ E: **AER Sales & Truck Service**
RVCamp E: **Blackbear Campground**
Other E: **Marte-L Convenience Store**

58AB **(87)** MA 110, to I-495, Salisbury, Amesbury
Gas W: Mobil*, Sunoco*
Food E: Simon's Roast Beef & Subs
W: Burger King, Dunkin Donuts, Friendly's, McDonalds
Lodg W: Susse Chalet
AServ E: **Auto Mall**, Dodge, Meineke Discount Mufflers, Pro Lube Auto Repair, Sea Coast Motors Plymouth, Jeep
W: Yeo Volkswagen
ATM E: BankBoston
Parks E: **Maudslay State Park**
Other E: Car Wash (Auto Mall)

57 MA 113, West Newbury, Newburyport
Gas E: Mobil*, Shell, Sunoco
Food E: Cafe Bagel, Dunkin Donuts, Giufepps Pasta, McDonalds, Ming Jade Polynesian, Papa

EXIT MASSACHUSETTS

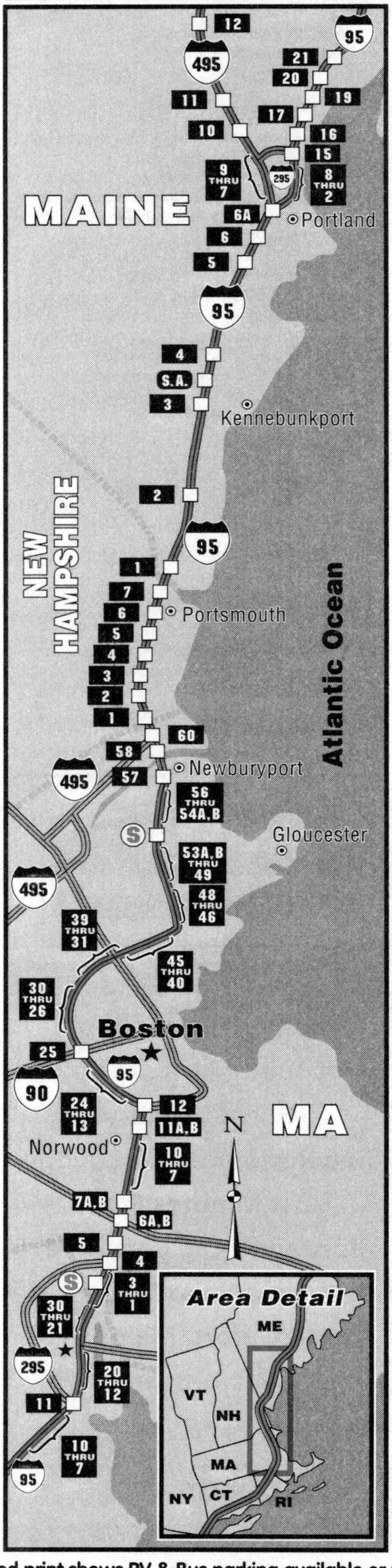

EXIT MASSACHUSETTS

Genio's, White Hen Pantry Deli
AServ E: Midas Muffler & Brakes, Shell, Sunoco
Med E: ✚ **Hospital**
ATM E: BankBoston, Fleet Bank
Parks E: **Maudslay State Park**
Other E: **Brook's Pharmacy, K-mart, Port Plaza, Shaw's Grocery, Village Pet Shop**
W: **Newburyport Veterinary Clinic**

56 Scotland Road, Newbury
Other E: **State Patrol Post**

55 Central St, Byfield , Newbury
Gas W: Prime(D)
Food E: Buddy's Fine Food, Buddy's Ice Cream
Other E: **Byfield General Store, Pearson Hardware**
W: **Expo Market Place, Tourist Information, U.S. Post Office**

54AB MA 133, Rowley , Georgetown
(54) **Weigh Station - Both Directions (Facility Closed)**

53AB **(75)** MA 97, Topsfield , Georgetown

52 **(73)** Topsfield Road, Topsfield , Boxford

51 Endicott Road, Topsfield

50 U.S. 1, Topsfield (Difficult Reaccess)
Gas E: Exxon*, Mobil*
Food E: Dunkin Donuts, Four 66 Pub & Grill, Newbury Street Deli & Bakery, Pechi's Italian
Lodg W: Fern Croft Hotel

49 MA 62, Danvers, Middleton (Difficult Reaccess)
Food W: Putnam Pantry (Candy, Ice Cream), Subway, Supino's Italian
Med E: ✚ **Hospital**
Other E: **Danvers Animal Hospital**
W: **CVS Pharmacy, MA State Police, Mail Boxes Etc, Stop N Shop Grocery**

48 Centre St, Danvers (Southbound, Difficult Reaccess)
Food W: Cabris Italian, Vito Brothers Italian
Lodg W: Comfort Inn

47A MA 114, Peabody, Middleton (Difficult Reaccess)
Gas E: Exxon*(D), Shell(CW), Sunoco
Food E: Dunkin Donuts, Jenny's, **McDonalds**, Papa Gino's Pizza
W: Jake's Grill
Lodg W: Motel 6, Residence Inn
AServ E: Herb Chambers Dodge, NTW Tire Center, Pac Hyundai/Pontiac, Sunoco

46 U.S. 1 South (Southbound, Difficult Reaccess)
Gas E: BP*(D), Citgo*(D)
W: Citgo
Food E: Burger King, Dunkin Donuts, Newbury Street Cafe, Royal Garden Chinese, Sri Thai Restaurant, Sunrise Pizza & Subs, Union Jack British Food
W: Honey Dew Donuts, Roadside Restaurant, T.G.I. Friday's
Lodg E: Holiday Inn, Sir John Motel
W: Marriott, Motel Mario's, Sir John Motel
AServ E: Citgo, Courtsey Jeep Eagle, Midas Muffler & Brakes
ATM E: BankBoston

45 MA 128 North, Gloucester

44 U.S. 1, MA 129, Boston, Everett, Danvers (Difficult Reaccess)
Gas W: Mobil, Shell, Texaco
Food W: Bennigan's, Bertucci's Brick Oven Pizza, Bickfords Family Restaurant
Lodg W: Econo Motor Inn, Holiday Inn
AServ W: 4-Star Service Center
Med E: ✚ **Hospital**
Other E: **State Patrol Post**

Bold red print shows RV & Bus parking available or nearby

EXIT MASSACHUSETTS

43 Walnut St, Lynnfield Ctr, Saugus (Difficult Reaccess)
- **Food** W: Colonial Hilton Resort
- **Lodg** W: Colonial Hilton Resort

42 Salem St, Montrose Ave, Wakefield
- **Gas** E: A Prime
  W: Sunoco
- **Food** E: Sub Stop
  W: Brother's Pizza
- **Lodg** W: Hilton
- **AServ** E: A Prime
  W: Sunoco

41 Main St , Lynnfield Ctr, Wakefield

40 (58) MA 129, Wakefield Center, North Reading
- **Gas** E: Exxon*
  W: Cumberland Farms, Gulf
- **Food** E: Honey Dew Donuts, Lanai Island (Chinese & Polynesian)
  W: Dunkin Donuts
- **AServ** E: Exxon
- **TServ** E: **Detroit Diesel**
- **Med** E: ✚ **Wakefield Medical Center**
- **Other** E: **Animal Hospital of Wakefield**

39 North Ave, Reading, Wakefield
- **Gas** W: Exxon, Texaco*(D)
- **Food** E: Oscar's Food & Spirits
  W: Honey Dew Donuts
- **Lodg** E: (AAA) Best Western
- **AServ** E: 128 Ford/Volvo/Saab, 128 Olds/Mazda/Isuzu
  W: 128 Ford, Exxon
- **Med** E: ✚ **Hospital**
- **ATM** E: Savings Bank

38AB MA 28, Stoneham, Reading
- **Gas** E: Gulf, Texaco
  W: Exxon*, Mobil*, Shell
- **Food** E: Boston Market, Denny's, Dunkin Donuts, The Ground Round
  W: Friendly's, Gregory's Subs & Deli, Harrow's, Mr Softie Ice Cream
- **AServ** E: Texaco
  W: Meineke Discount Mufflers, Shell
- **Med** E: ✚ **Hospital**
- **ATM** E: Eastern Bank, Stone & Savings
- **Other** E: **CVS Pharmacy, Lens Crafters, Petsmart, Red Stone Shopping Center, Shaw's Supermarket**
  W: **JK Market**

37AB Junction I-93, Boston

36 Washington St, Winchester, Woburn, Reading
- **Gas** E: Getty
- **Food** W: 99 Restaurant & Pub, Bertucci's Pizzaria, Chicago Bar & Grille, DeAngelo's Sandwich Shop, J.C. Hillary's Restaurant, McDonalds, On The Border Mexican, Papa Genio's, Pizza Hut, T.G.I. Friday's, Weylu's Thai, Wonder Wok
- **Lodg** W: ◆ Comfort Inn, Courtyard Inn, Hampton Inn, ◆ Red Roof Inn
- **AServ** E: Crest GMC/ Pontiac/ Buick/ Nissan, Getty
  W: NTB Tire Warehouse
- **ATM** W: Woburn National Bank
- **Other** W: **CVS Pharmacy, Market Basket Supermarket, Woburn Mall**

35 MA 38, Woburn, Wilmington
- **Gas** W: Mobil*, Shell
- **Food** E: Juliet's Restaurant
  W: Baldwin's Restaurant, Dunkin Donuts, Roast Beef Round Up
- **Lodg** E: ◆ **Ramada**
- **AServ** W: Mobil, Shell
- **Med** W: ✚ **Hospital**
- **ATM** W: Fleet Bank, Northern Bank & Trust
- **Other** W: **Super Stop Pharmacy**

34 Winn St, Woburn, Burlington

EXIT MASSACHUSETTS

- **Food** E: Delicious Desserts, Domino's Pizza, Maria's Pizza, Midtown Deli
- **Med** E: ✚ **Hospital**
- **Other** E: **Coin Laundry**

33 U.S. 3 South, MA 3A North, Winchester, Burlington
- **Food** E: Bickfords Family Restaurant, Chuck E. Cheese's Pizza, Outback Steakhouse, Papa Razzi's
  W: Marriott
- **Lodg** W: Marriott
- **AServ** E: Herb Chamber's Honda
- **Other** E: **CVS Pharmacy, Lens Crafters, Petsmart**

32B Middlesex Tnpk, Arlington, Burlington
- **Gas** E: Mobil*, Shell
- **Food** E: Burger King, Chip's Ice Cream, D'angelo's Subs & Sandwiches, McDonalds
  W: Chili's, Jenny's Steak & Seafood, Johnny Rockets, Pizzeria
- **Lodg** W: Howard Johnson
- **AServ** E: Lexington Auto Parts, Midas Muffler & Brakes, Ned's Auto Service, Shell
  W: Burlington Dodge, Sears & Jiffy Lube
- **Med** W: ✚ **Hospital**
- **Other** E: **Market Basket, Middlesex Mall**
  W: **Burlington Mall**

Two Locations
Holiday Inn
Golf Course & Indoor Tennis
Full Service Restaurant
Lounge
Swimming Pool
Health Club Facilities
Spectra vision
Newly Renovated Spacious Guest Rooms
HO0204
Holiday Inn - Mansfield
31 Hampshire Street
Mansfield, MA 02048
508-339-2200
I-95 • Exit 7A
HO0216
Holiday Inn - Newton
399 Grove Street
Newton, MA 02162
617-969-5300
I-95 • Exit 22
MASS • I-95 • 2 LOCATIONS

EXIT MASSACHUSETTS

32A U.S. 3 North, Lowell, Nashua NH

31 MA 4, MA 225, Hanscom Field, Carlisle, Lexington, Bedford
- **Gas** E: Mobil
- **Food** E: Lexington Lobster & Seafood Co.
  W: Denny's
- **Lodg** W: Holiday Inn Express, Travelodge (see our ad on this page)
- **AServ** E: Mobil (Wrecker)
- **Med** W: ✚ **Beth Israel Childrens Hospital**
- **Other** W: **Veterinary Clinic**

30 MA 2A, Concord, East Lexington
- **Gas** E: Shell
- **Lodg** W: Sheraton Tara Lexington Inn
- **AServ** E: Shell
- **Parks** W: **Minuteman National Historic Park**

(30) Service Area (Northbound)
- **FStop** E: **Sunoco**
- **Food** E: **Roy Rogers**
- **ATM** E: Fleet Bank

29B MA 2W, Acton, Fitchburg

29 MA 2 East, Arlington, Cambridge

28AB Trapelo Road, Belmont, Lincoln
- **Gas** E: Exxon

27AB Totten Pond Road, Waltham, Wyman, Winter
- **Gas** E: Shell
- **Food** E: Ben & Gusko Japanese, De'Angelo's Sandwich Shop, Lanai Restaurant, Pizza Hut Carry Out, Thackery's
  W: Bertucci's Restaurant, Gourmet Bagel, The Grill (Doubletree)
- **Lodg** E: Best Western, Home Suites Hotel, Susse Chalet, Westin Hotel, Wyndham Garden Hotel
  W: Doubletree Guest Suites, The Green Papaya
- **AServ** E: Shell
- **Other** E: **Coffee & More**

26 U.S. 20, MA 117, Waltham, Weston (Difficult Reaccess)
- **Gas** E: Texaco
  W: Mobil
- **AServ** E: NTB Warehouse, Texaco
- **Med** E: ✚ **Hospital**

25 Junction I-90, MA Tnpk (Toll)

24 MA 30, MA 28, Newton, Waltham
- **Gas** E: Mobil
- **Food** E: Marriott, S & S Livestock Company
- **Lodg** E: ◆ Marriott
- **AServ** E: Mobil
- **Med** E: ✚ **Hospital**

23 Recreation Road

22 Grove St , MBTA Station
- **Lodg** E: Holiday Inn (see our ad on this page)

(22) Rest Area (Soutbound)
- **FStop** W: **Sunoco**
- **Food** W: **Roy Rogers**
- **ATM** W: Fleet Bank

21 MA 16, Newton, Wellesley
- **Gas** W: Mobil
- **Food** W: Pillar House
- **AServ** W: Mobil
- **Med** E: ✚ **Hospital**

EXIT	MASSACHUSETTS
20	**(36)** MA 9, Boston, Brookline, Wooster
Gas	**W:** Mobile
Food	**W:** Pillar House, Wok Chinese
AServ	**W:** Dodge, Mobile* (Wrecker Service)
Med	**W:** ✚ **Hospital**
19AB	**(35)** Highland Ave, Newton Highlands
Gas	**E:** Charter, Gulf
Food	**E:** DeAngelo's Sandwich Shops, The Ground Round
Lodg	**E:** Sheraton
AServ	**E:** Berejik Olds, Gulf, Midas Muffler & Brakes
	W: Muzi Ford/ Chevrolet/ Geo
ATM	**E:** BankBoston, MetroBank
Other	**E: AAA Office**
18	**(33)** Great Plain Ave, West Roxbury
Med	**W:** ✚ **Hospital**
17	**(32)** MA 135, Needham, Natick
Med	**W:** ✚ **Hospital**
Other	**W: Norfolk County Sheriff Office**
16	**(31)** MA 109, Dedham, Westwood
(18)	**Rest Area - RR, Phones**
15	U.S. 1, to MA 1A, Dedham (Difficult Reaccess)
Gas	**W:** Shell
Food	**E:** Bickfords Family Restaurant, Boston Market, Chili's, Friendly's Ice Cream, J.C. Hillarie's Food & Drink, Jenny's, T.G.I. Friday's
	W: Dunkin Donuts, Patricia's Subs
Lodg	**E:** Comfort Inn, Holiday Inn
	W: Westwood Motor Lodge
AServ	**E:** Dedham Nissan, Goodyear Tire & Auto, Midas Muffler & Brakes, Owen Mercury/Lincoln
	W: Shell
TServ	**E: Cummins Diesel**
Med	**E:** ✚ **Faulkner Hospital**
ATM	**E:** Dedham Savings
Other	**E: Brooks Pharmacy, Lens Crafters, Pet Supply Store**
	W: Lambert's Market
14	**(29)** East St, Canton St
Lodg	**W:** ◆ Hilton
TServ	**E: Cumming's Truck Service**
(8)	**Rest Area (Southbound)**
13	**(28)** Railway Station, University Ave
12	**(26)** Junction I-93 North, Braintree, Boston
11AB	**(23)** Neponset St, Canton, Norwood
Gas	**E:** Citgo, Sunoco
Lodg	**E:** Ramada (see our ad on this page)
AServ	**E:** Citgo, Sunoco
Med	**W:** ✚ **Hospital**
10	**(23)** Coney St, Sharon, Walpole (Southbound, Northbound Reaccess Only)
Food	**W:** 99 Restaurant, Becon's, Dunkin Donuts, McDonalds, Taco Bell
AServ	**W:** Acura/Lexus
ATM	**W:** BankBoston
Other	**W: CVS Pharmacy, Walpole Mall**
9	**(19)** U.S. 1, to MA 27, Walpole
Gas	**E:** Exxon*, Mobil*, US Petroleum
Food	**E:** Clyde's Smokehouse Saloon, Pizza Hut, The Ground Round
	W: Bigford's Family Fair
Lodg	**E:** Boston View Motel
	W: The Sharon Inn
AServ	**E:** US Petroleum
RVCamp	**E: Campers Head Quarters**(LP)
Other	**E: Wal-Mart**
8	**(16)** South Main St, Sharon, Foxboro
Food	**E:** Bliss Ice Cream Restaurant & Deli, Tony Lema's Sandwich & Pizza
Other	**E: Osco Drugs, Shaw's Grocery**
7AB	**(13)** MA 140, Mansfield, Foxboro
Gas	**W:** Shell
Food	**E:** 99 Restaurant, Fran's Apple Pie Bakery, Piccadilly Pub & Restaurant, PJ's Place Chicken & Ribs
	W: Papa Genio's
Lodg	**E:** Courtyard by Marriott, Days Inn, Holiday Inn (see our ad page 505), **Motel 6**
AServ	**W:** Shell
Other	**E: Foxfield Plaza**
	W: Mail Boxes Etc
6AB	**(11)** Junction I-495, Worcester, Taunton
(10)	MA Welcome Center - RR, Phones, Picnic
(9)	**Rest Area - Phones (Southbound)**
5	**(7)** to MA 152, Attleboro
Gas	**W:** Gulf(D)
Food	**W:** Bill's Pizza & Subs, Bliss Ice Cream Restaurant & Deli
AServ	**W:** Gulf
Med	**E:** ✚ **Hospital**
	W: ✚ **Walk-in Clinic**
ATM	**W:** Bank of Boston, Fleet Bank
Other	**E: Police Station**
	W: Brooks Pharmacy, Coin Laundry, Shaw's Grocery (Citizen's Bank ATM)
4	**(5)** Junction I-295, Woonsocket
3	**(4)** MA 123, Norton, Attleboro
Gas	**E:** Texaco*(D)
Med	**E:** ✚ **Hospital**
Other	**E: Capron Zoo**
(2)	**Weigh Station - both directions**
(1)	**Rest Area (Northbound)**
2AB	**(1)** MA1A, Pawtuket, S Attleboro
Gas	**E:** Mobil*(CW)

EXIT	MASSACHUSETTS/RHODE ISLAND
Food	**E:** Honey Dew Donuts, McDonalds, Olive Garden, Spumoni's Italian, Tony V's Deli
AServ	**E:** Country Pontiac, **K-Mart**
ATM	**E:** Bank of Boston
Other	**E: K-Mart, Pearl Vision Center, Shaw's Supermarket (ATM)**
1	**(0)** U.S. 1 S, Broadway, Pawtucket (Southbound, Difficult Reaccess)
Food	**E:** Super Dragon Chinese, Vista Donuts
Lodg	**E:** Days Inn (AAA) Days Inn (AAA)
AServ	**E:** Royal Tire Servcie
Med	**E:** ✚ **Hospital**
Other	**E: Brook's Pharmacy**

↑ MASSACHUSETTS

↓ RHODE ISLAND

EXIT	RHODE ISLAND
30	**(43)** East St, Central Falls
Food	**E:** Dunkin Donuts
29	**(43)** U.S. 1, Broadway, Cottage St (Difficult Reaccess Northbound)
Food	**W:** DeAngelo's Sandwich Shop, Pizza Hut
AServ	**W:** Crown Collision, Division Brakes, Firestone Tire & Auto
Other	**W: Children's Museum of RI**
28	**(42)** RI 114, School St (Northbound, Difficult Reaccess)
Gas	**E:** Sunoco*
Food	**W:** Newport Creamery
AServ	**E:** Tire Pro's
	W: Apex Auto Service
Med	**E:** ✚ **Memorial Hospital**
Other	**W: Apex Shopping Center**
27	**(42)** Downtown Pawtucket (Difficult Reaccess)
Gas	**W:** Shell, Sunoco(D)
Food	**W:** Burger King, Dunkin Donuts, The Ground Round
Lodg	**W:** ◆ Comfort Inn
AServ	**W:** Shell, Sunoco
26	**(41)** RI 122, Lonsdale Ave, Main St (Northbound, Southbound Reaccess Only)
Med	**E:** ✚ **Memorial Hospital of RI**
Other	**E: Shaw's Plaza**
25	**(40)** U.S. 1, RI 126, North Main St, Smithfield Ave (Difficult Reaccess Eastbound)
Food	**W:** Chello's, Chili's, Hunan Garden
AServ	**W:** AAMCO Transmission, Auto Parts Plus, Luigi Auto Repair
24	**(39)** Branch Ave
Gas	**W:** Mobil*
Food	**E:** KFC
AServ	**E:** Action Auto Parts
	W: Mobil
Med	**E:** ✚ **Miriam Hospital**
Other	**W: Super Stop N Shop**
23	**(38)** RI 146, RI 7 Woonsocket, Downtown Providence (Difficult Northbound Reaccess)
Gas	**W:** Shell(D)(CW)
Food	**W:** Dunkin Donuts (Shell Station), Marriot
Lodg	**W:** ◆ Marriot

Bold red print shows RV & Bus parking available or nearby

EXIT RHODE ISLAND

AServ **W:** Shell
Med **W:** ✚ Miriam Hospital
Other **W:** Post Office

2 **(38)** U.S. 6, RI 10

1 **(37)** Broadway (Difficult Reaccess)

0 **(37)** U.S. 6, East Providence, Cape Cod

9 **(36)** Eddy St, Allens Ave (Southbound, No Return Reaccess)
Med **W:** ✚ Hospital

8 **(35)** U.S. 1A, Thurbers Ave (Difficult Eastside Reaccess)
Gas **E:** Diesel Snacks*[D]
Food **E:** Stacy's Restaurant
W: Burger King
Med **W:** ✚ Hospital

7 **(34)** U.S. 1, Elmwood Ave (Difficult Reaccess)
AServ **E:** Dick & Sons Auto Repair
W: Tire Warehouse
Other **W:** Coin Laundry

6 **(34)** RI 10, to U.S. 1, to RI 2
Med **W:** ✚ St. Joseph's Hospital
Other **E:** Roger Williams Park & Zoo

5 **(32)** Jefferson Blvd
Gas **E:** Getty*[D]
Food **E:** Bigford's Family Restaurant, Dunkin Donuts
Lodg **E:** Motel 6, ◆ Susse Chalet Inn
AServ **E:** Getty*
TServ **W:** Coastal International Trucks, Colony Ford Trucks

4 **(31)** RI 37, to U.S. 1, to RI 2, Pranston, Warwick

3 **(30)** T.F. Green Airport
Lodg **W:** ◆ Residence Inn

2A **(28)** RI 113 East, Warwick
Gas **E:** Sunoco*, Texaco*
Lodg **E:** ◆ Holiday Inn (2 Restaurants)
AServ **E:** Sunoco, Texaco

1 **(28)** Junction I-295, Woonsocket (Northbound)

0AB **(27)** RI 117, Warwick
Med **W:** ✚ Hospital
Other **W:** Triple A Offices

(25) RI 4, N Kingstown

AB **(24)** RI 2, to RI 4, West Warwick, East Greenwich (Difficult Reaccess)
FStop **E:** Texaco*[D]
Food **E:** McDonalds, Outback Steakhouse, Ro-Jack Market Place, Ruby Tuesday
W: Little Chopsticks Chinese, Papa Gino's Pizza, Pinelli Deli, Wendy's
Lodg **W:** Open Gate Motel
AServ **W:** Jennings Car Care
RVCamp **W:** Arlington RV Super Center
Other **E:** Discovery Zone, E. Greenwich Square, Harvard Walk-in Medical Center, WalGreens
W: Greenwich Animal Clinic

(21) Coventry, West Warwick

EXIT RHODE ISLAND/CONNECTICUT

FStop **E:** Mobil[D]
AServ **E:** Mobil

6A **(20)** Hopkins Hill Road

6 **(18)** RI 3, West Greenwich, Coventry
Gas **W:** Sunoco[D], Texaco*
Food **W:** Bess Eaton Coffee Shop & Bakery, Dunkin Donuts, Mark's Grille, Salvator's
Lodg **W:** (AAA) Best Western, Congress Inn
AServ **W:** Sunoco*

5AB **(14)** RI 102, W Greenwich, Exeter (B Exit To Truck Stop)
TStop **W:** RI's Only 24 Hr T/A Plaza*[SCALES]
Food **W:** Exit 5 Coffee Shop
Lodg **W:** Classic Motor Lodge
TServ **W:** T/A Plaza
ATM **W:** Centreville Bank

(11) Rest Area - Phones Ⓟ

4 **(10)** RI 3, to RI 165, Arcadia, Exeter (Northbound, Reaccess Southbound)

3AB **(8)** RI 138, Kingston, Newport, Wyoming, Hope Valley
Gas **W:** Exxon, Gulf*, Mobil*, Sunoco*
Food **E:** Bess Eaton Coffee Shop & Bakery, Dunkin Donuts, McDonalds
W: Bali Village Chinese, Bess Eaton Coffee Shop & Bakery, Nick's Pizza, Pizza King, Village Pizza
Lodg **W:** Sun Valley Motel
AServ **E:** NAPA Auto Parts
W: Exxon
ATM **W:** Washington Trust
Other **W:** Ocean Pharmacy, Shop & Stop (Pharmacy), Tourist Information

(9) RI Welcome Center - RR, Phones, Picnic, Vending, Pet Area (Northbound)

2 **(5)** Hope Valley, Alton
RVCamp **W:** RV Camping

1 RI 3, Hopkinton, Westerly
RVCamp **E:** Campground
Med **E:** ✚ Hospital
Other **E:** Tourist Information

↑RHODE ISLAND

↓CONNECTICUT

93 **(111)** CT 216, Clark Falls, Ashaway RI, CT184
TStop **W:** Texaco[SCALES]
FStop **W:** Citgo*[D][LP]
Gas **E:** Shell*
Food **W:** Bess Eaton, McDonalds, Republic Family Restaurant
Lodg **W:** Stardust Motel
AServ **E:** Shell*
TServ **W:** Republic
Other **E:** Burlingange Park

(108) Connecticut Wlcm. Ctr. - RR, Phones

92 **(108)** CT 2, CT 49, North Stonington, Powcatuck (Difficult Reaccess Southbound)
RVCamp **E:** Highland Orchards RV Resort Park & Service
W: Highland Orchards Park Inc.
Other **E:** Tourist Information

EXIT CONNECTICUT

91 **(104)** CT 234, North Main St, Stonington, Borough

90 **(101)** CT 27, Mystic Aquarium, Mystic Port
Gas **W:** Texaco*[D] (ATM)
Food **E:** Friendly's, Go Fish, Howard Johnson, Jamm's, McDonalds, McQuade's Market Place, Moorings (Hilton), Mystic Hilton, Newport Creamery, Steak Loft
W: Ashby's Restaurant, Best Western, Copperfield's, Days Inn, Dunkin Donuts, Mystic Bagel, Subway
Lodg **E:** Howard Johnson, Mystic Hilton, Old Mystic Motor Lodge
W: ◆ Comfort Inn, (DAYS INN) ◆ Days Inn, (AAA) Marriott
RVCamp **W:** Campground
Med **W:** ✚ Seaport Walk In Medical Center
ATM **E:** Liberty Bank
W: Society Bank
Other **E:** Mystic Factory Outlet, Mystic Marinelife Aquarium, Old Mystic Village Shopping Center, Tourist Information
W: Carosal Museum, Coin Laundry, Mystic Aquarium, Nature Center

(101) Scenic Overlook (Northbound, No Trucks)

89 **(100)** Allyn St, Noank, Groton Long Point

88 **(98)** CT 117, Noank, Groton Point
Med **E:** ✚ Emergency Medical Treatment

87 **(96)** CT 349, to U.S. 1, Clarence B Sharp Hwy

86 **(96)** CT 184, CT 12 (Northbound Exits Left)
Gas **W:** Cory's*, Merit*, Mobile
Food **E:** Gee Willikers, Rosie's 24-Hr Diner, Skybox Cafe
W: Cafe DelMar, Chinese Kitchen, Domino's Pizza, Famous Famigla Pizza, Flanagan's, IHOP, JB's Fun 'N Food Clam Bar, KFC, Peking Chinese, Russell's Ribs, Taco Bell
Lodg **E:** (AAA) Clarion, Morgan Inn
W: Best Western, Super 8 Motel (RV Hook-Ups)
AServ **E:** Volkswagen Dealer
W: Cardinal Honda, Gerard Mitsubishi, Midas Muffler & Brakes
ATM **W:** Liberty Bank, Society Savings
Other **E:** Super Stop & Shop, USS Nautilus Memorial & Mûseum, Wal-Mart (Pharmacy & Eyecare)

85 **(95)** U.S. 1 North, Thames St, Downtown Groton (Northbound, Difficult Reaccess)

Bold red print shows RV & Bus parking available or nearby

EXIT CONNECTICUT

Gas E: Citgo

Food E: Norm's Country Lounge Cafe, Norm's Diner

AServ E: Citgo, NAPA Auto Parts

84 **(94)** CT 32, Downtown New London, Norwich (Difficult Reaccess)

Other E: **Eugene O'Neil Homestead, Nathan Hale School House, Shaw Mansion**

83 **(93)** CT 32, New London, U.S. 1, Frontage Rd (Northbound)

Gas W: Mobile

Food W: Chuck E. Cheese's Pizza, PaPa Pizza, Red Lobster

Lodg W: Holiday Inn

Med W: **+ Hospital**

Other W: **New London Mall**

82A **(92)** New London, Frontage Road, Colchester, Hartford (Northbound)

Food E: Pizza Hut
W: American Steak House, Golden Wok Chinese

Lodg W: ◆ Red Roof Inn

AServ E: Auto Palace, Town Fair Tire
W: Goodyear Tire & Auto

ATM E: Citizen's Bank, Fleet Bank

Other E: **IGA Supermarket, New London Shoppping Center**

82 **(92)** CT 85, Broad. St, Waterford

Gas W: Mobil*

Food W: Bee Bee's Dairy Restaurant

AServ W: **Sears**

Other W: **Crystal Mall**

(91) **Weigh Station - Both Directions**

81 **(90)** Cross Road

Lodg W: AAA Lamplighter Motel

Other W: **Wal-Mart**

80 **(89)** Oil Mill Road (Difficult Reaccess, Southbound Only)

77 **(2)** CT85, Colchester, Waterford

76 **(88)** Junction I-395 North, Norwich, Plainfield (Northbound)

75 **(88)** U.S. 1, Waterford

Food E: Flander's Seafood, Lou Lou's

RVCamp W: **Camping**

ATM W: Citizen's Bank

74 **(87)** CT 161, Flanders, Niantic

Gas E: Exxon*, Mobil
W: Shell*(CW)

Food E: Bickfords Family Restaurant, Burger King, Fat Friday's (Connecticut Yankee Inn), Starlight Motor Inn (Connecticut Yankee Inn), The Bootlegger Country Steakhouse, The Shack
W: BeeBee Dairy, Flander's House Cafe, Flanders Donuts & Bagels, Flanders Pizza, Kathy's Corner, McDonalds, Scott's Quality Meats, The Shack Restaurant

Lodg E: Connecticut Yankee Inn, DAYS INN Days Inn, East Lymn Inn, Motel 6, Starlight Motor Inn

AServ E: Mobil, National Tire
W: Plaza Ford

RVCamp W: **Campground**

Med W: **+ Walk In Medical Center**

ATM W: Citizen's Bank

Other E: **Children's Museum, Tourist Information**
W: **A2Z Medical Supply, Brooks Pharmacy, Cumberland Farms*, IGA Market, Proud Parrot Pet Shop, Tru Value Hardware**

EXIT CONNECTICUT

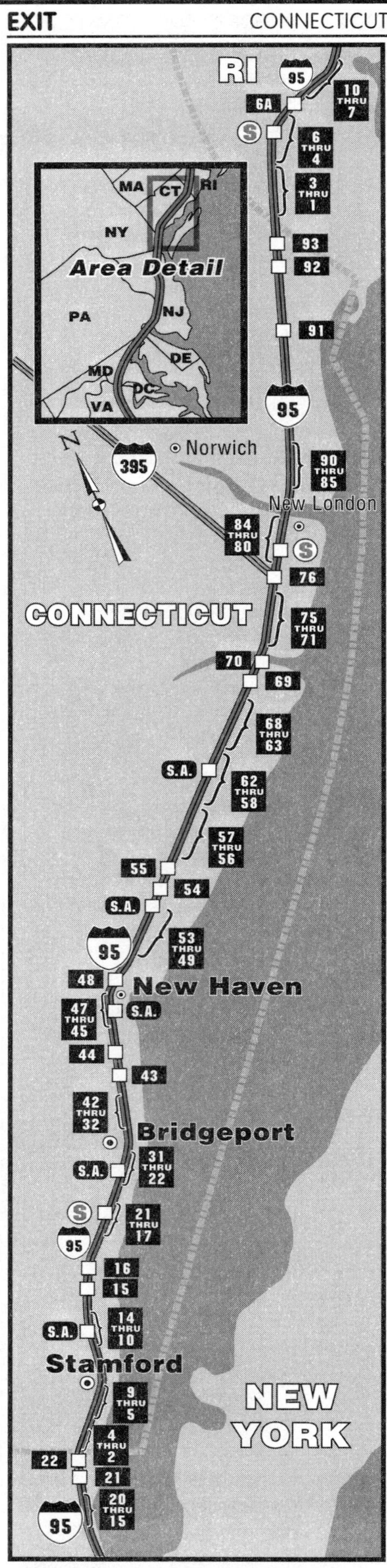

EXIT CONNECTICU

73 **(86)** Society Road (Northbound, Northbound Reaccess Only)

(85) **Rest Area (Southbound)**

72 **(84)** Rocky Neck State Park

RVCamp W: **Camp Niantic**(LP)

71 **(84)** Four Mile River Road

RVCamp E: **Campground**

Other E: **Rocky Neck State Park**

70 **(79)** U.S. 1, CT 156, Old Lyme (Difficul Reaccess Northbound)

Gas W: Texaco(D)

Food W: Anne's Kitchen, Bess Eaton Donuts, Mama Mia Pizza, Old Lymn Chinese Food, The Grist M

AServ W: Texaco

ATM W: Fleet Bank

Other W: **A&P Super Food Mart, Old Lyme Shoppin Center, Old Lyme Pharmacy, Post Office, Shoreline Eye Care**

69 **(78)** CT 9 North, Essex, Hartford

68 **(77)** U.S. 1, Old Saybrook (Southbound)

FStop E: **Citgo**(CW)(LP)

AServ E: Isuzu, Saybrook Motors, Stanley Motors

Other E: Classic Carriage Car Wash

67 **(76)** CT 154, Old Saybrook, Elm St (Northbound, Difficult Northbound Reaccess)

Gas E: BP*, Citgo*(CW), Mobil, Sunoco*

Food E: Andriana's Seafood & Steaks, Bagel Bakery, Pizza Hut, Pizza Works, Subway, Sully's Seafoo The Leaning Tower Takeout, Two Brothers Fami Pizza, Whistle Stop, Zhang's

AServ E: Chrysler Auto Dealer, Computer Tune & Lube Old Saybrook, Saybrook Ford, Zip Lube

ATM E: New Haven Savings Bank

Other E: **Country Farms Dairy Store, Pharmacy, Saybrook Junction Shopping Center**

66 **(74)** CT 166, Spencer Plain Road

Gas E: Citgo*

Food E: Aleia's, Alforno Brick Oven Pizza, Cookoo's Nest Mexican, Mike's Deli, Sol-E-Mar Cafe, TNT Family Restaurant

Lodg E: DAYS INN ◆ Days Inn, Saybrook Motor Inn, Supe 8 Motel

AServ E: Benny's, Mal's Auto & Truck Repair (24 hr Towing), Smith Bros Transmission

Other E: **DMK Animal Hospital**

(74) **Rest Area - RR, Phones, Picnic (Northbound)**

65 **(73)** CT 153, Westbrook

Gas E: Exxon*

Food E: Andy's Steak & Seafood, Highliner Cafe

AServ E: Highlander Cafe, Westbrook Honda

Med W: **+ 24 Hour Emergency Medical Center (3 Miles)**

Other E: **Westbrook Factory Stores (Food Court)**

64 **(71)** CT 145, Horse Hill Road

63 **(69)** CT 81, Clinton, Killingworth

Gas E: Shell*

Other W: **Clinton Crossings Premium Outlet (Food Court)**

62 **(66)** Hammonasset State Park

RVCamp W: **Campground**

Parks E: **Hammonasset State Park (1.5 Miles)**

Bold red print shows RV & Bus parking available or nearby

EXIT — CONNECTICUT

(66) Service Area - Picnic, ATM, RR (Both Directions)
- FStop — **B:** Mobil
- Food — **B:** McDonalds

61 **(65)** CT 79, Madison
- Med — **E:** ✚ East Shore Medical Center
- ATM — **E:** Webster Bank
- Other — **E:** Madison Police

60 **(64)** CT 79, Madison, North Madison, Mungertown Rd (Southbound)

59 **(61)** Guilford, Goose Lane
- Gas — **E:** Citgo*, Mobil*, Texaco(D)
- Food — **E:** Boss Pizza & Subs, McDonalds, Seabreeze Family Restaurant, Shepard's, Wendy's
- Lodg — **E:** Tower Motel
- AServ — **E:** Gary's Transmission, Goodyear Tire & Auto, Instant Change Auto, Texaco, Town Line Auto
- ATM — **E:** The Gilford Savings Bank
- Other — **E:** Pepper's General Store

58 **(60)** CT 77, Guilford, North Guilford
- Other — **E:** Tourist Information
  W: Guilford Police Station

57 **(59)** U.S. 1, Boston Post Road
- AServ — **W:** Saab Dealer

56 CT 146, Leetes Island Road, Stony Creek
- TStop — **W:** 76 Auto/Truck Plaza(SCALES) (RV Dump)
- FStop — **W:** Berkshire*(D)(CW)(LP)
- Gas — **W:** Mobil*(CW)
- Food — **W:** Friendly's
- Lodg — **E:** (AAA) Advanced Motel
  W: (AAA) Ramada Limited
- AServ — **E:** New Age Motor
- TServ — **W:** 76
- Other — **W:** Super Stop & Shop

55 U.S. 1, North Branford, East Main St
- Gas — **E:** Mobil
  W: Texaco*
- Food — **E:** 280 Pub, McDonalds, My Dad's Place, Paula's Candy Kitchen
  W: Lyn's Deli, Margarita's Mexican, Steak Loft, Su Casa Mexican, The Parthenon Diner
- Lodg — **E:** Knight's Inn, (AAA) Motel 6
  W: (DAYS INN) (AAA) Days Inn (see our ad on this page)
- AServ — **E:** Mobil, Wilson Ford
- Other — **E:** WalGreens

54 Branford, Cedar St
- Gas — **E:** Citgo, Mobil*(CW), Petro Plus
- Food — **E:** Brandford Townhouse, Dunkin Donuts
  W: Born in America, The Lion City Chinese
- AServ — **E:** Citgo, Petro Plus
  W: Eastshore Auto Body
- Other — **E:** AAA Office
  W: Brandford's Farmers Market, Brushy Hill Shopping Center, Crousner's Conveniece Store

(53) Service Area - Both Directions

53 **(52)** U.S. 1, CT 142, CT 146, Short Beach (Northbound, Difficult Reaccess Northbound)
- Gas — **E:** Getty*
- Food — **E:** Bagel Connection, Moon Star Chinese

52 **(50)** CT 100, Easthaven , N High St (Southbound)
- FStop — **B:** Mobil
- Food — **B:** McDonalds
- ATM — **B:** ATM

51 **(50)** U.S. 1, Frontage Road , E Haven
- Gas — **E:** Merit(D), Sunoco*
  W: Shell*(CW), Sunoco*
- Food — **E:** Boston Market, Friendly's
  W: Dunkin Donuts, The Bon China Buffet, Wendy's
- Lodg — **E:** (AAA) Holiday Inn Express
- AServ — **E:** GM Auto Dealership
  W: Auto Palace
- ATM — **E:** New Haven Bank
- Other — **E:** Coin Laundry, Tourist Information, X-pect Pharmacy

50 **(49)** Woodward Ave, Lighthouse Point (Northbound, Difficult Northbound Reaccess)
- Gas — **E:** Texaco
  W: Forbes, Gulf*
- Food — **W:** BJ's Lunch, Dunkin Donuts, Fireside, Planet Mars Cafe
- AServ — **W:** DeNatto & Son Garage, Gulf
- Other — **E:** Airport
  W: Quality Mart

49 Stiles St (Difficult Reaccess Northbound)
- Food — **W:** Lou's Lunch
- AServ — **W:** A&A Towing & Auto Repair, DeMufis Radiator Service, Fountions Garage, Sas-t Service
- TServ — **E:** PC Truck Repairs
  W: Mikes Truck & Trailer Repair
- Other — **W:** Forbes Car Wash, Pay Rite Food Market

48 **(47)** Junction I-91 North, Hartford (Left Exit)

47 CT 34, Downtown New Haven
- Lodg — **E:** Holiday Inn (see our ad on this page)

46 **(49)** Long Wharf Dr, Sargent Dr
- FStop — **W:** Mobil*(CW)(LP)
- Food — **W:** Brazi's Pizza & Pasta, Dunkin Donuts

(Mobil), Howard Johnson
- Lodg — **W:** Howard Johnson
- AServ — **W:** Mobil
- Med — **W:** ✚ Hospital
- ATM — **W:** Fleet Bank
- Other — **E:** Tourist Information

45 **(48)** CT 10, Blvd. (Southbound)

44 **(46)** CT 10, Kimberley Ave
- Gas — **W:** Getty*
- Food — **W:** Dairy Queen, Dunkin Donuts, McDonalds, O-Boy Lunchenette, Tropical Krust Chicken (Jamaican)
- AServ — **W:** Catapano Brothers, Meineke Discount Mufflers, Tony Long Warf

43 **(45)** CT 122, First Ave, Downtown (No Trucks)
- Gas — **W:** Amaco*, Mobile
- Food — **E:** Duchess Diner, Gaetano's Italian, Matthews Tavern, Pizza Works
  W: Crousers Food Market, Richard's Chik N' Ribs
- AServ — **E:** Bob's Auto Center, Greco's Auto Parts, Insti-Lube
  W: Andre Towing, Kelly Tire, Marshal's Garage, Mobile*
- TServ — **E:** Black Diamond Truck Repair
- Med — **W:** ✚ Veterns Hospital
- Other — **E:** Cheer Deli Mart, Coin Laundry
  W: Animal Clinic

42 **(44)** CT 162, Saw Mill Road
- Gas — **E:** Mobil*
  W: Shell*
- Food — **E:** Pizza Hut, West Haven Pizza Place
  W: American Steak House, D'Angelo's Subs & Salads, Days Hotel, Denny's, Dunkin Donuts, Friendly's
- Lodg — **E:** Econolodge
  W: ◆ Days Hotel
- AServ — **E:** Mobil
- Other — **E:** Coin Laundry, Saw Mill Shopping Plaza
  W: 7-11 Convenience Store

41 **(42)** Marsh Hill Road, Orange
- Food — **W:** Outback Steakhouse
- Lodg — **W:** Courtyard by Marriott
- Other — **W:** Mall, Movie Cinema

(41) Service Area - RR
- FStop — **B:** Mobil
- Food — **B:** McDonalds
- AServ — **B:** Mobil
- ATM — **B:** ATM
- Other — **B:** Gift Shop

40 Old Gate Lane, Woodmont Road
- TStop — **E:** Gulf*(LP)
- Gas — **E:** Getty(D), Sunoco*, Texaco*
- Food — **E:** Bennigan's, (Old Country Store) Cracker Barrel, D'angelo's Sandwich Shop, Duchess Burgers, Gipper's, Gulf, Mayflower Diner
- Lodg — **E:** Comfort Inn, Mayflower Motel, Milford Inn
- AServ — **E:** Brake Center of America, Getty
- TServ — **E:** Mayflower Kenworth, Secondi Bros (Gulf)
- TWash — **E:** Secondi Bros
- ATM — **E:** Bank of New Haven

39AB U.S. 1
- FStop — **E:** Gulf
- Gas — **W:** Mobile*(CW)
- Food — **E:** Athnia Diner, Howard Johnson, Mama Teresa Pizza, Pizzeria Uno, The Gathering Steaks
  W: Alexander's Buffet, Baskin Robbins, Burger

EXIT CONNECTICUT

King, Dunkin Donuts, Honey Baked Ham Co., McDonalds, Miami Grill, Richie's Deli
Lodg **E:** Howard Johnson, Milford Motel, The Connecticut Turnpike Motel
AServ **E:** Firestone Tire & Auto, Interprize Volvo, Mazada, Oldsmobile
ATM **W:** Fleet Bank, New Haven Savings Bank
Other **E:** Car Wash
W: **Pearl Vision Center, Post Mall**

38 **(37)** CT 15, Merritt Pkwy, W Cross PKWY (No Commercial Vehicles)

37 High St (Northbound, Southbound Reaccess Only)
Gas **E:** Gulf, Texaco
Food **E:** Kimberly Diner
AServ **E:** Colonial Toyota, Gulf, Meineke Discount Mufflers, Texaco
Other **E:** **7-11 Convenience Store, Expect Discount Pharmacy, Mobile Vet Clinic**

36 Plains Road, Stratford
Gas **E:** Exxon*[D]
Food **E:** Gusto Fine Italian Cuisene
Med **E:** ✚ **Hospital**
ATM **E:** Fleet Bank

35 **(36)** School House Road, Bic Dr
Food **E:** Wendy's
Lodg **E:** The Susse Chalet
W: ◆ Red Roof Inn
AServ **E:** Ford Dealership, NAPA Auto Parts
Med **E:** ✚ **Milford Medical Center**

34 **(35)** U.S. 1, Milford
Gas **E:** Getty, Gulf[D], Shell
Food **E:** BelAir Seafood, Boston Market, China City, Denny's, Domonio's, Dunkin Donuts, Gourmet Buffet International, Manhattan Bagel, McDonalds, New Haven Pizza, TCBY Yogurt, Taco Bell, The Bakery
Lodg **E:** Devon Motel
AServ **E:** Action Auto Supply, Brake Masters USA, Brangaccio Bros. Transmission, Getty, Gulf, Lodus Auto Service
Med **E:** ✚ **Milford Medical Hospital**
Other **E:** **K-Mart, Milk Plus Food Stores, Rite Aide Pharmacy, Walbaum's Supermarket**

33 **(34)** U.S. 1, CT 110, Ferry Blvd, Devon (Difficult Reaccess)
Gas **E:** Sunoco
W: BP
Food **E:** Danny's Drive-In, Fagan's Restaurant
W: House of Tong, Peter Pan Pizza, Ponderosa
AServ **E:** Sunoco
W: 5 Star Muffler, BP
Other **E:** **Pet Supply Store**
W: **Post Office, Strattford Square, Super Stop Pharmacy (ATM & Pharmacy)**

32 **(33)** West Broad St, Stratford
Gas **W:** Gulf
Food **W:** Dunkin Donuts, Leonardo's A Pizza, Palarimo Pizza, Pepin's Restaurant
Med **W:** ✚ **Hospital**
Other **W:** **Pet Superbowl, Quick Convenience Store**

31 **(32)** Honeyspot Road
Gas **E:** Gulf[D]
W: Citgo
Lodg **E:** Honey Spot Motor Lodge
AServ **E:** Gulf

EXIT CONNECTICUT

Other **W:** Car Wash

30 **(31)** CT 113, Lordship Blvd (Northbound)
FStop **W:** **Citgo***[D]
Gas **E:** Shell*
Food **E:** Ramada
Lodg **E:** ◆ Ramada
AServ **E:** Shell
Other **E:** **Airport**

29 **(30)** CT 130, Stratford Ave (Difficult Reaccess)
Med **W:** ✚ **Hospital**

28 **(30)** Main St
Med **E:** ✚ **Eastport Medical Center**

27A **(29)** CT25, CT8, Trumball, Waterbury

27 **(29)** Lafayette Blvd, Downtown
Lodg **E:** **Holiday Inn (see our ad on this page)**
Other **W:** **State Police, The Barnum Museum**

26 **(28)** Wordin Ave

25 **(27)** Fairfield, State St, Commerce Dr (Difficult Northbound Reaccess)
Gas **W:** Gulf[D]
Food **W:** McDonalds
AServ **E:** Continental Motors
W: Gulf, Kim Mufflers
TServ **W:** **Dialing Fleet Service, Don Stevens Tires**

24 **(26)** Black Rock Tnpk
Gas **W:** Gulf[D]
Food **E:** D'angelo Chips Deli, Fairfield Diner
Lodg **E:** Bridgeport Motor Inn
AServ **E:** Lexus
W: Gulf

23 **(26)** U.S. 1, Kings Highway

22 **(25)** Round Hill Road

(25) Service Area - Phones, RR
FStop **B:** **Mobil**
Food **B:** **McDonalds**
AServ **B:** Mobil
ATM **B:** ATM

21 **(24)** Mill Plain Road

20 **(23)** Bronson Road (Southbound, Difficult Reaccess)
Food **E:** Callaghan's Cafe

19 **(23)** U.S. 1, Center St, Southport (Difficult Northbound Reaccess)
Gas **W:** Sunoco, Texaco*

EXIT CONNECTICU

Food **W:** Athena Diner, Friendly's, S&S Dugout
Lodg **W:** Piquot Motor Inn
AServ **W:** Sunoco, Town Fair Tire
Other **W:** Southport Car Wash, **Vision Center Associates**

18 **(20)** Sherwood Island State Park (East)

(20) **Weigh Station (Southbound)**

17 **(18)** CT 33, CT 136, Westport, Saugatuck
Food **E:** Jazaman Sushi Bar

16 East Norwalk
Gas **E:** BP[D], Texaco[D][CW]
Food **E:** Baskin Robbins, Donut Den, Dunkin Donuts, East Avenue Pizza Company, Golden Million Chinese Food, Mike's Deli, Penney's III Diner, Vertelli Pizza
AServ **E:** BP, Ohara's Brake & Alignment, Texaco
Med **W:** ✚ **Norwalk Medical Center**
Other **E:** **Coin Laundry, Mike's Deli & Grocery, Rite Aide Pharmacy**, Star Wash

15 U.S. 7, Norwalk, Danbury
Gas **W:** Mobile, Sunaco*[D], Texaco*
Food **W:** Dunkin Donuts
AServ **W:** Maritime Oldsmobile, Chevrolet, Mobile (Towing)
Other **W:** **Police Station**

14 U.S. 1, Connecticut Ave, South Norwalk
Gas **W:** Amoco*, Gulf*, Shell, Texaco
Food **W:** Angelo's Pizzaria, Bagel's Bakery, Cosmos Grill Family Restaurant, Pizza Hut, Silver Star Restaurant, Speedy Donuts, Swanky Frank's Diner
AServ **W:** Firestone Tire & Auto, Gulf, John's Auto Rep Shell, Texaco
Med **W:** ✚ **Hospital**
Other **W:** **New England Eye Care, Police Station**

13 U.S. 1, Post Road (Difficult Reaccess Northbound)
Gas **W:** Exxon, Mobil, Shell*
Food **W:** American Steak House, Bertucci's House of Pizza, Burger King, Carvelle Ice Cream Bakery, Dutchess Hamburgers & Hot Dogs, IHOP, KFC, McDonalds, Red Lobster, Wendy's
Lodg **W:** Ramada Hotel
AServ **W:** Auto Service Town Fair, Darien Car Clinic[CW] Mobil, Sand Oval of Darien
ATM **W:** Citi Bank, Fairfield County Savings Bank, Firs Union Bank
Other **W:** Exit 13 Car Wash, **Norwalk Animal Hospita Stop and Shop Food Mart**

(12) Service Area (Northbound)
FStop **W:** **Mobil**
Food **W:** Lavava Cafe, **McDonalds**
AServ **W:** Mobil
ATM **W:** ATM
Other **W:** **Postal Service, RV Dump, Tourist Information**

12 CT 136, Tokeneke Road, Rowayton (Northbound, Southbound Reacces Only)

11 U.S. 1, Darien, Rowayton
Gas **E:** Amoco, Sunoco
W: Exxon

Bold red print shows RV & Bus parking available or nearby

XIT CONNECTICUT

Food **W:** Black Foot Goose Grill, Howard Johnson, Sugar Bowl Luncheonette, TCBY, Tom E Toes, Uncle's Deli
Lodg **W:** ◆ Howard Johnson
AServ **E:** H&L Chevrolet & Geo, Jaguar Ford Lincoln Dealer
W: Citgo, Exxon
ATM **W:** First Union, Fleet Bank
Other **E:** Darien Car Wash, **Darien Veterinary Hospital**
W: **CVS Pharmacy, Pharmacy**

) Noroton
FStop **W:** **Texaco***(CW)
Gas **W:** Getty, Mobil
AServ **W:** Getty, Mobil

I) Service Area (Southbound)

U.S. 1, CT 106, Glenbrook
Gas **W:** Gulf(CW)(LP)
Food **W:** Blimpie's Subs, Buster's Barbacue, Exit 9 Bagels, Giovani's Steak House, McDonalds, Mi Terruno, Pizza Hut, Stamford Motor Inn, Stamford Pizza
Lodg **W:** Stamford Motor Inn
AServ **W:** Midas Muffler
Other **W:** **Fairlawn Pharmacy, Just Cats Veterinary Hospital**

Elm St, Atlantic St
Lodg **W:** Marriot Hotel, Holiday Inn (see our ad on this page)
Med **W:** ✚ **Hospital**

CT 137, Greenwich Ave (Nothbound, Reaccess Northbound Only)
Food **E:** Sheraton Stanford Hotel
Lodg **E:** Sheraton Stanford Hotel

Harvard Ave
Gas **E:** BP(CW)
W: Shell*
Lodg **W:** ◆ Super 8 Motel
AServ **W:** Marc Service Center
TServ **E:** **CT Truck and Trailer**
Med **W:** ✚ **Standford Hospital**
Other **E:** **Maxi Discount Pharmacy**

U.S. 1, Riverside, Old Greenwich
Gas **W:** Mobil*, Shell(D)
Food **W:** Bang Cafe, DiMare Pastry Shop, Fuji Mart, Hunan Cafe Chinese, Hustler Diner, McDonalds, Riverside Deli, Starbucks Coffee
Lodg **W:** ◆ Howard Johnson
AServ **W:** Precision Tune & Lube, Riverside Auto Parts
ATM **W:** Chase Bank, Peoples Bank, Putman Trust
Other **W:** **A & P Supermarket, CVS Pharmacy,**

EXIT CONNECTICUT/NEW YORK

Greenwich Care Center, Optic Care, Post Office

4 Indian Field Road, Cos Cob

3 Arch St, Greenwich
Gas **W:** Mobil, Texaco(D)
AServ **W:** Lexus, Mobil, Texaco
Med **W:** ✚ **Greenwich Hospital**

(2) **Weigh Station (Northbound)**

2 **(1)** Delavan Ave, Bryam

↑CONNECTICUT

↓NEW YORK

22 **(14)** Midland Ave, Port Chester, Rye

21 **(13)** Junction I-287, U.S. 1 North, Tappenzee Bridge (Toll), Port Chester, White Plains

20 **(14)** U.S. 1 South, Rye
Gas **E:** Shell, Texaco
Food **W:** Bagel Shop, Carvel Ice Cream, Clino's Pizza, Maselli's Deli, McDonalds, Szechuan Empire II
AServ **E:** Ford Rye, Shell, Texaco
ATM **W:** Chase Bank, Fleet Bank
Other **E:** **A&P Supermarket, Police Station**
W: **Rockbottom Pharmacy, Vision World, Walbum's Supermarket**

19 **(12)** Playland Pkwy, Rye, Harrison (Difficult Reaccess)

18B **(11)** Mamaroneck Ave, White Plains
Gas **E:** Hess(D), Shell
Food **E:** Hillside Deli, Jenny's Pizza
AServ **E:** Shell
Other **E:** **A&P SuperMarket, Auto Service Expressway,** Dippin Car Wash, **Mavis Discount Tire**

18A **(9)** Fenimore Road, Mamaroneck (Difficult Reaccess)
Gas **E:** Sunoco
Food **E:** Cinamore Deli
AServ **E:** A&N Garage, Centre Auto Service (Wrecker Service, Transmission Repair), Sunoco, Vincent's Garage
Other **E:** **Deli, Convenient Store**

17 **(7)** Chatsworth Ave, Larchmont (NY State Thruway Begins Northbound- Toll - Difficult Reaccess Northbound)

16 **(6)** North Ave, Cedar St, New Rochelle
Gas **E:** Gatti's
Food **E:** Del Pote's, Luke's Hong Kong Garden, Pizza Hut, R&B Deli, Taco Bell
Lodg **E:** (AAA) Ramada
AServ **E:** Toyota Dealer
Med **E:** ✚ **Hospital**
ATM **E:** Chase Bank, Fleet Bank
Other **E:** **ANC Veterinary Hospital, Coin Laundry**

15 **(5)** U.S. 1, New Rochelle, The Pelhams
Gas **E:** Shorco*
Food **E:** Atostino's, Remi Deli, The Thru Way Diner
AServ **E:** Auto Care Center, Empire Discount Transmission, Thanbo Auto Repair

14 **(3)** Hutchison PKWY S, Whitestone

EXIT NEW YORK

Bridge (Southbound, No Commercial Traffic)

13 **(2)** Conner St, Baychester Ave
FStop **E:** **Gasateria***(CW)
Gas **W:** Amoco*, City gas(D)
Food **W:** Dunkin Donuts, McDonalds, Wonder Bakery
Lodg **W:** Holiday Motel
AServ **E:** New England Auto Repair, Shell(D)
W: Auto Parts Super Store, Eagle Tire, Gatti, Lee Myles Auto Service, Michelin, V&L
TServ **E:** **Frank's Truck & Automotive, New England Auto Repair**
Med **E:** ✚ **Lady of Mercy Medical Hospital**

12 Baychester Ave (Left Exit, Northbound)

11 Bartow Ave, Coop City Blvd
Gas **E:** Mobil
W: Shell
Food **E:** Bartow Pizza, Baskin Robbins, Boston Market, Burger King, Champion Chicken, Chinese Cuisine, Dunkin Donuts, Garden Bakery, Little Caesars Pizza, Magic Wok, Ponderosa, Red Lobster, Seven Seas
AServ **W:** Shell
ATM **E:** Algamated Bank of NY, Marine Midland Bank, Republic National Bank
Other **E:** **Bay Plaza Mall, General Vision Eyewear, Lens Lab Optical, Pathmark Pharmacy, Petland**

10 Gun Hill Road (Left Exit)

9 Hutchinson Pkwy (No Commercial Traffic)

8C Pelham Pkwy (No Trucks)
Food **W:** Pelham Bay Diner, Walbum's Supermarket
Lodg **W:** Pelham Garden Motor Lodge
AServ **W:** Speedy Muffler King
ATM **W:** City Bank

8B Orchard Beach, City Island

7D East Tremont Ave (Difficult Reaccess)

7C Country Club Road, Pelham Bay Park
Food **E:** Best Bagels in Town, Cross Town Diner
AServ **E:** Mobil

6B Junction I-295 South, Throgsneck Bridge

6A Junction I-678 South, Whitestone Bridge (Difficult Reaccess)

5B Castle Hill Ave
Food **E:** Big Market West Indian Grocery, Castle Hill Bake Shop, House of Hero's and Deli
W: Brisas-Delcaribe (Mexican), Castle Hill Grocery
AServ **W:** R&S Strauss, Senico

5A White Plains Road, Westchester

4B Rosedale Ave , Bronx River Pkwy
Other **E:** **Bronx Zoo**

4A Jct I-895 South, Sheridan Expwy, Triboro Bridge
Other **E:** **Yankee Stadium**

2B Webster Ave
Food **E:** McDonalds

2A Jerome Ave

1C Junction I-87, Deegan Expwy,

EXIT	NEW YORK/NEW JERSEY
	Upstate Triboro Road (Exit Only)
1B	Harlem River Dr, Amsterdam Ave
1A	**(122)** 181 St, NY9A, H Hudson Pkwy

↑NEW YORK

↓NEW JERSEY

73	**(122)** Palisades Pkwy (Cars Only)
72	**(122)** NJ67, U.S. 9W, Fort Lee
Lodg	**W:** Days Inn, Hilton
71	**(122)** U.S. 1, U.S. 9, U.S.46
(122)	NJ4, NJ17
70	to NJ93, Leonia, Tenack
Gas	**W:** Mobile, Shell
Lodg	**W:** Country Inn, Days Inn, Hilton, Marriott
69	**(119)** Jct I-80, to NJ17
68A	**(118)** U.S. 46, New Jersey Tnpk
68B	**(118)** Junction I-95 North, George Washington Bridge, New York
68	**(118)** Jct I-80 West, to NJ17
(69)	Clara Barton/John Fenwick Service Area

Note: I-95 runs concurrent below with NJTNPK. Numbering follows NJTNPK.

18	**(114)** U.S. 46, Hackensack
(113)	Vince Lombardi Service Area
17	**(117)** NY 3, Lincoln Tunnel
Food	**E:** Ladelkaribe (At Holiday Inn)
Lodg	**E:** Holiday Inn
	W: Radisson Suites (see our ad this page)
Other	**E:** **Mall**
16	**(113)** NY 3, Secaucus, Rutherford
Gas	**E:** Amoco
Food	**E:** Bagels Plus, Chi Chi's Mexican, Grettle's Pretzel, Haagen-Dazs, Herbert Billard, Hulahan's, McConkey Big City Grill (Holiday Inn), Players Steakhouse (Embassy Suites), Red Robbin Restaurant (Courtyard By Marriott), Sizzler, Subway
	W: Burger King, Cafe, Pizzeria Uno, TCBY Treats, Tuttaeene Italian
Lodg	**E:** Ameri Suites, Courtyard By Marriott, Embassy Suites, Holiday Inn (see our ad on this page), Howard Johnson (see our ad this page), Ramada Inn (see our ad this page)
	W: Radisson Suites
ATM	**E:** Summit Bank
Other	**E:** **Meadowland Convention Ctr**
	W: **Six Flags Factory Outlet (see our ad on this page), Hear X (Hearing Aid)**
15	**(109)** Jct I-280, Newark, The Oranges
15E	**(107)** Newark, Jersey City
14C	**(105)** Holland Tunnel
14B	**(105)** Jersey City
14A	**(105)** Bayonne
14	**(105)** Junction I-78, U.S. 1, U.S. 22
(102)	Admiral W.M. Halsey Service Area (Southbound)

EXIT NEW JERSEY

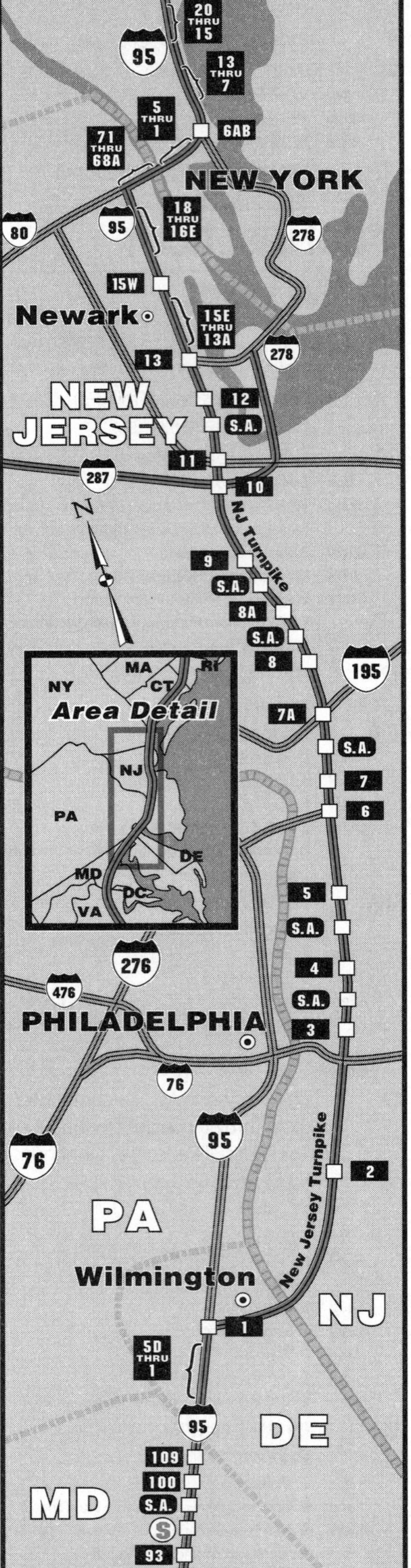

EXIT	NEW JERSEY
13A	**(101)** Newark Airport, Elizabeth Seaport
13	**(100)** Junction I-278, Elizabeth, Goethals Bridge, Verrazano
12	**(96)** Carteret, Rahway
Food	**E:** Blimpie's Subs, Burger Express, Burger King, Carlo's Pizza & Subs, Chopsticks Kitchen Chinese, Hang Hing Chinese, **Holiday Inn,** McDonalds, Riuniti Pizza, The Bus Stop
Lodg	**E:** **Holiday Inn**
AServ	**E:** George Lucas Chevrolet/Geo, Goodyear Tire Auto
Med	**E:** **Dr's Medi Center (Open 7 Days a Week**
ATM	**E:** First Union Bank

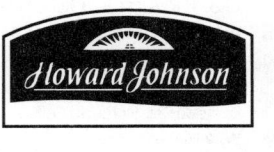

Bold red print shows RV & Bus parking available or nearby

XIT	NEW JERSEY
Other	**E:** Coin Laundry, Revco Drugs, Shoprite Grocery, Short Stop C-Store, WalGreens
3)	Thomas A. Edison Service Area
	(91) U.S. 9, Woodbridge, Garden State Pkwy
	(88) Junction I-287, NJ 440, Metuchen
	(84) U.S.11, NJ 18, New Brunswick
9)	Joyce Kilmer Service Area (Northbound)
A	**(74)** Jamesburg, Cranbury
2)	Molly Pitcher Service Area (Southbound)
	(68) NJ 33, Hightstown, Freehold
A	**(61)** Jct I-195, Trenton, Hamilton
9)	Richard Stockton Service Area (Southbound)
	(54) U.S. 206, Bordentown, Trenton
	(51) Jct PA Tnpk
	(44) Burlington, Mount Holly, Willingboro
Lodg	**E:** Howard Johnson (see our ad on this page)
9)	J. Fenimore Cooper Service Area (Northbound)
	(34) NJ 73, Philadelphia, Camden
9)	Walt Whitman Service Area (Southbound)
FStop	**E:** Sunoco
Food	**E:** Nathan's Famous, Roy Rogers
AServ	**E:** Sunoco
	(26) NJ 168, Camden, Woodbury
FStop	**E:** Gulf*
Gas	**W:** Coastal(D), Texaco(D), Xtra*
Food	**W:** Bagal Wiches, Club Diner, Dunkin Donuts, Italia Pizza, Pulcinell's, Wendy's
Lodg	**E:** Comfort Inn, Holiday Inn
	W: Bellmawr Motor Inn, Econolodge, Howard Johnson
AServ	**E:** Gulf
	W: Coastal
ATM	**W:** Midlantic Bank
Other	**W:** NJ State Aquarium
	(13) U.S. 322, Swedesboro , Chester PA
5)	Clara Barton Service Area
A	**(0)** NJ49E, Pennsville, Salem
	Note: I-95 runs concurrent above with NJTNPK. Numbering follows NJTNPK.
B	CR583N, Princeton Pike
A	CR583S, Princeton Pike
AB	U.S.206, Lawrenceville, Princeton, Trenton
Gas	**E:** Mobil
Food	**W:** Lawrence Deli
AServ	**E:** Mobil
ATM	**W:** TNC Bank
Parks	**W:** Mercer County Park, Rosedale Park
AB	**(6)** Federal City Rd (Southbound, No

EXIT	NEW JERSEY
	Trucks Over 10 Tons)
Lodg	**W:** Howard Johnson (see our ad this page)
4	NJ31, Ewing, Pennington
Parks	**W:** Washington Crossing Nat'l Park
3	Scotch Rd
Lodg	**E:** Comfort Inn (see our ad on this page)
	W: Howard Johnson (see our ad on this page)

EXIT	NEW JERSEY/PENNSYLVANIA
ATM	**E:** CoreState Bank
2	CR 579, Harbourton, W Trenton, Airport
Parks	**W:** South Park At Ewing
Other	**E:** Mercer Airport
1	NJ29, Trenton, Lambertville
Lodg	**W:** Wellesley (see our ad this page)
Parks	**W:** DNR Canal State Park & Scutters Fall
Other	**W:** NJ State Police Headquarter & Museum, Washington Trail Of History

↑NEW JERSEY

↓PENNSYLVANIA

EXIT	PENNSYLVANIA
31	**(50)** New Hope
Other	**W:** Washington Crossing Park (3 Miles)
(50)	Weigh Station, Rest Area - Phones, RR, Picnic, Pet Walk (Southbound)
30	**(48)** PA332, Newtown, Yardley
29	**(45)** U.S. 1, Morrisville, Langhorne
Lodg	**E:** Radisson Hotel (see our ad on this page)
Med	**W:** + Delaware Hospital
28	**(43)** PA413, U.S. 1 Bus, Penndel, Levittown
Gas	**E:** Texaco*(CW)(LP)
	W: Mobil(D)
Food	**E:** Ballpark Pizza, Chuck E. Cheese's Pizza, Friendly's, Hong Kong, Jack's Bagels, Mario's Pasta
	W: Denny's, McDonalds
AServ	**E:** Dearden Buick, Goodyear Tire & Auto, McCafferty Hyundai
	W: Collison Repair, Combustion Engineering, Simon's Garage & Towing, Team Toyota
TServ	**W:** Bucks Cty Int
Med	**E:** + Delaware Valley Medical Center
ATM	**E:** CoreState, First Nat'l Bank & Trust
	W: Mobil
Other	**E:** Drug Emporium, Langhorn Square
26	**(39)** PA 413, Bristol, Jct I-276, PA Tnpk
Med	**E:** + Lower Bucks Hospital
25	**(36)** PA 132, Street Rd, to US13, Bristol Pike
Gas	**W:** Amoco, Coastal*, Sunoco*
Food	**W:** Burger King, China Sun, Dunkin Donuts, Fisher's, KFC, Lee's Hoagie House, Manhattan Bagel, McDonalds, Tudor House Restaurant
AServ	**E:** DP Brown
	W: Coastal (Wrecker Service), Cottman

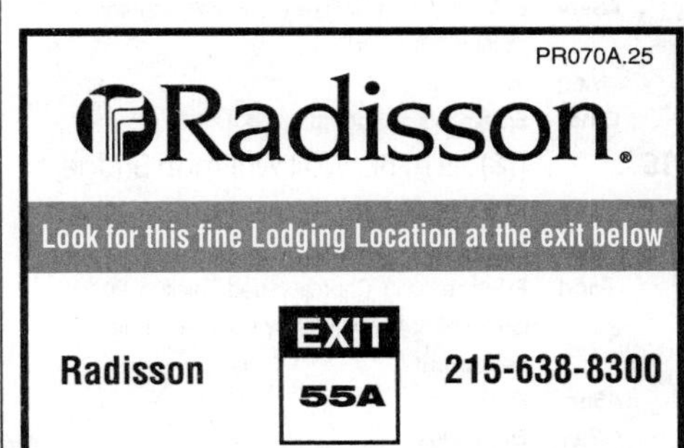

EXIT PENNSYLVANIA

Transmission, K-Mart, Keystone Discount Tire Co, Pep Boys Auto Center
ATM **W:** 7-11 Convenience Store, Keystone Bank, Mellon
Parks **E:** State Park
Other **W:** 7-11 Convenience Store, K-Mart, Philadelphia Eye Glass Lab, Shelly's Pharmacy

24 **(34)** PA 63W, to U.S.13, Woodhaven Rd, Bristol Park
Lodg **W:** Hampton Inn (Frontage Rd)
AServ **W:** NTB Service
Med **W:** ✚ Hospital

23 **(32)** Academy Road, Porresdale Ave
Gas **W:** Mobil*
Food **W:** Dunkin Donuts, KFC
Lodg **W:** Holiday Inn (see our ad on this page)
AServ **W:** Parts USA
Med **W:** ✚ Hospital
Other **W:** Coin Operated Laundry, Philadelphia NE Airport

22 **(30)** PA73, Cottman Ave (Northbound Reaccess, Difficult Southbound Reaccess)
FStop **E:** Sunoco*(D)
Food **E:** Inna G's
Lodg **W:** Doubletree Club Hotel (see our ad on this page)
AServ **E:** Rossetti's Collision, Sunoco
TServ **E:** Foster Michelin Big Trucks (Frontage Rd)

21 **(26)** Bridge St
Gas **W:** Getty's*
AServ **W:** Pratt Auto Repair
Med **W:** ✚ Hospital

20 **(25)** Pennsauken NJ, Betsy Ross Bridge

19 **(25)** Allegheny Ave
Gas **W:** Sunoco*
Food **W:** Syrenka (Polish Food)

18 **(23)** Girard Ave, Lehigh Ave (Northbound, No Reaccess)
Gas **E:** Sunoco*(D)
Food **E:** Dunkin Donuts
AServ **E:** Ft Richmond Village, Pep Boys Auto Center
ATM **E:** Sunoco
Other **E:** Thrift Drug

17 **(22)** Jct I-676, U.S. 30, Central Philadelphia, Callohill St (Left Exit)

16 **(20)** Columbus Blvd, Washington Ave (Difficult Reaccess)
Gas **E:** Mobil*(D)(CW)
Food **E:** Engine 46 Steakhouse, Margarita Cafe, Philadelphia Bagel Co, Sicilly's Pizza, Silly Bar & Grill, Wings To Go, Wok & Roll Chinese
AServ **E:** All Day Long Oil Filter & Lube(CW), Schafer Muffler & Brakes
Med **W:** ✚ Hospital
Other **E:** River View Shopping Center

15 **(18)** Jct I-76E, Walt Whitman Bridge, Packer Ave
Gas **E:** Sunoco*(D), Texaco*
Food **E:** Burger King, Church's Fried Chicken, Pizza (Sunoco), Subway, Wendy's, White Castle Restaurant
AServ **E:** R&S
Other **E:** Car Wash

EXIT PENNSYLVANIA

14 **(16)** PA 611, Broad St, Pattison Ave
Lodg **W:** Airport Tower (see our ad on this page)
Other **W:** Stadium

13 Enterprise Ave, Island Ave (Southbound)
Lodg **E:** Days Inn (see our ad on this page)

EXIT PENNSYLVANIA

12 Bartram Rd, Essington Ave (Southbound)

11 to Jct I-76, PA291W, Central Philadelphia
FStop **E:** Exxon*(D)
Food **E:** Atrium Cafe (Doubletree Cafe), Season's Cafe (Days Inn)
Lodg **E:** Courtyard by Marriott, DAYS INN AAA Days Inn, AAA Doubletree Guest Suites, AAA Hilton, ◆ Residence Inn

10 PA291E, Luster, Cargo (Northbound)
Food **E:** Embassy Suites, Tropix
W: Marriott
Lodg **E:** ◆ Courtyard by Marriott, Embassy Suites
W: ◆ Marriott
Other **E:** John Hines Nat'l Refuge
W: Philadelphia International Airport

9 PA420, Essington
FStop **E:** Coastal*
Gas **E:** Exxon(D)
Food **E:** Denny's, Holiday Inn, Jen's Place Philly Cheese Steak, Jim's Place, Miller's Luncheonette, Preston Diner, Radisson, Ramada Inn, S&S Pizza, Shoney's
Lodg **E:** Comfort Inn, Econolodge, Holiday Inn, Motel 6, Radisson, Ramada Inn, Red Roof Inn
AServ **E:** Exxon
ATM **E:** County Savings Assoc., First Union Bank
Parks **E:** Governor Print State Park
W: Morton Homestead Historic Site, Prospect Park
Other **E:** US Post Office

8 Ridley Park
Med **W:** ✚ Taylor Hospital
Other **W:** Chester, Penn Terminal, Water Front

7 Junction I-476, Plymouth Meeting

6 PA352, PA320 Edgmont Ave, Providence (Difficult Northbound Reaccess)
Food **W:** Howard Johnson
Lodg **W:** Howard Johnson
Med **W:** ✚ Clozer-Chester Hospital

5 Kerlin St, to DE 291 (Difficult Northbound Reaccess)

4 U.S. 322 East, to 9th St, Bridgeport

3 Highland Ave, 322, Wilmington, W Chester, NJ Bridge (Difficult Northbound Reaccess)
Gas **E:** Sunoco*
AServ **E:** Goodyear Tire & Auto, Lions 24 Hr Towing, Murphy Lincoln, Mercury
TServ **E:** EPC Truck Service
Other **W:** Garden Food Market C-Store

2 DE 452, Market St
Gas **E:** Getty
W: Citgo*, Exxon*
Food **E:** Abe's Place (Hoagies, Steaks & Chicken), Dairy King, DiCostanza's Sandwiches
W: Linwood Restaurant, Ye Old Restaurant
AServ **E:** Getty
W: Interstate Auto Electric & Repair
TServ **W:** Nationwide Equipment Carrier

1 Chichester Ave
Gas **E:** Sunoco
W: Amoco*(CW)(LP)

Bold red print shows RV & Bus parking available or nearby

XIT	PENNSYLVANIA/DELAWARE
AServ	**E:** Rick's Auto Repair, Sunoco
Other	**W:** WaWa C-Store
)	**Weigh Stations, PA Welcome Center - RR, Phones, Picnic (Northbound)**

↑ PENNSYLVANIA
↓ DELAWARE

	(23) Jct I-495, DE 92, Naamans Road, Hambys Corner, Naamans Corner
Food	**E:** Callhan's, Crownery, Eatery Family Retaurant (K-Mart), Food Court (Tri-State Mall), Shop & Bag Bakery, Deli **W:** Evergreen's Restaurant (Hilton), Howard Johnson, Taco Bell
Lodg	**W:** AAA Hilton
AServ	**E:** Goodyear Tire & Auto **W:** Gulf, Jiffy Lube
ATM	**E:** Wilmington Trust Bank **W:** PNC Bank
Other	**E:** Discount Eye Glasses, K-Mart, Tri-State Mall, US Optical, Wa-Ma C-Store **W:** Acme Grocery Store, Eckerd Drugs, James Way Shopping
0	**(22)** Harvey Road, to Philadelphia Pike (Northbound, Southbound Reaccess Only)
	(19) DE 3, Marsh Road
Parks	**E:** Bellevue State Park
Other	**E:** DE State Police
	(17) U.S. 202, Concord Pike, Wilmington, Westchester
Med	**E:** + Hospital **W:** + E I Dupont Children's Hospital
	(16) DE 52, Delaware Ave
Gas	**W:** Shell*
Food	**E:** Sheraton, Willoughby Restaurant
Lodg	**E:** AAA Sheraton Suites
Med	**E:** + Hospital **W:** + Hospital
ATM	**W:** Shell
	(15) DE 4, to DE 100, Maryland Ave, Downtown Wilmington, MLK Blvd
D	U.S. 13, Dover, Baltimore, Norfolk
C	Junction I-295 (Left Exit)
AB	**(10)** DE 141 N, Newport, New Castle
AB	**(8)** DE 7, Christiana , DE1, DE58, Metroform
Food	**E:** Food Court (Christiana Mall), Ruby Tuesday **W:** Chili's, Hilton, Michael's Family Restaurant, Shoney's
Lodg	**W:** Courtyard by Marriott, Fairfield Inn, Hilton, Red Roof Inn, Shoney's Inn
Other	**E:** Christina Mall
	DE 273, Newark, Dover
Gas	**E:** Amoco, Exxon, Getty*, Shell*
Food	**E:** Authenic Chinese, Bob Evan's Restaurant, Boston Market, Country Made Deli, Denny's, Donut Connection, Oliver's Restaurant (Holiday Inn), Pizza City, Pizza Hut, Ristoranetrevi Restaurant (Best Western), Santa Fe, Wendy's
Lodg	**E:** Best Western, ◆ Hampton Inn, ◆ Holiday Inn, MacIntosh Inn
AServ	**E:** Exxon
ATM	**E:** PNC Bank, WSFS Bank
Other	**E:** 7-11 Convenience Store **W:** Acme Grocery, Coin Operated Laundry, Happy Holly Discount Drugs, University Plaza
2)	Service Area - Gift Shop, Vending, PP Phonecards, Phones, RR (Both

EXIT	DELAWARE/MARYLAND
	Directions)
FStop	**E:** Exxon, Mobil
Food	**E:** Bob's Big Boy, Gourmet Bean Coffee, Mrs Field's Cookies, Roy Rogers, Sbarro Italian, TCBY Yogurt, Taco Bell
AServ	**E:** Exxon, Mobil
ATM	**E:** Exxon
Other	**E:** US Mailbox
1AB	**(2)** DE 896, Newark, Middletown
FStop	**W:** Texaco*
Gas	**W:** Exxon*, Gulf*(D), Mobil*, Shell*
Food	**W:** Boston Market, Friendly's, Ground Round, McDonalds, Mother's Kitchen, Pepper's Pizza,Subs & Steaks, Pit O' Scotland Bakery
Lodg	**W:** Comfort Inn, Howard Johnson
AServ	**W:** BF Goodrich, Exxon, Gulf, Shell*
RVCamp	**E:** Camping
Other	**W:** White Glove Car Wash

↑ DELAWARE
↓ MARYLAND

109	MD 279, Elkton, Newark DE
TStop	**E:** Petro*(LP)(SCALES) **W:** 76 Auto/Truck Plaza(D)(SCALES) (RV Dump)
FStop	**E:** Texaco*
Gas	**E:** Texaco*(D)
Food	**E:** Aunt Nanny's Restaurant, Friendly's, Iron Skillet (Petro), McDonalds **W:** 76 Auto/Truck Plaza
Lodg	**E:** Elkton Lodge, Howard Johnson, Knight's Inn, Motel 6
TServ	**W:** 76
TWash	**E:** Blue Beacon Truck Wash
Med	**E:** + Hospital (3 Miles)
100	MD 272, North East, Rising Sun
TStop	**E:** Flying J Travel Plaza*(LP) (RV Dump, FedEx, Fax)
FStop	**W:** Citgo*
Food	**E:** Crystal Inn, Schroeder's Deli (Flying J's), Thads Restaurant (Flying J's)
Lodg	**E:** ◆ Crystal Inn
Parks	**E:** Elk Neck State Park
Other	**E:** State Patrol Post, Upper Bay Museum **W:** Plumpton Park Zoo
(97)	Chesapeake House Service Area - Phones, PP Phonecards. RR, Picnic, FedEx Box (Both Directions, Exit Left)
FStop	**B:** Exxon, Texaco
Gas	**B:** Sunoco(D)
Food	**B:** Burger King, Gourmet Bean, Hot Dog City, Mrs Fields Cookies, Popeye's Chicken, TCBY

EXIT	MARYLAND
	Yogurt, Taco Bell
AServ	**B:** Exxon, Sunoco
ATM	**B:** Machine
(95)	**Weigh Station**
93	**(94)** MD222, Perryville, Port Deposit
TStop	**E:** Perryville Travel Plaza*(D)(SCALES)
FStop	**E:** Exxon*
Gas	**E:** Chevron*
Food	**E:** Dairy Queen (Exxon), Denny's, Subway (Exxon)
Lodg	**E:** AAA Comfort Inn
TServ	**E:** River City Truck & Tire
Other	**E:** Chesapeake Mall Village, Perryville Outlet Center, State Patrol Post
(93)	Toll Booths (Northbound)
89	MD155, Churchville, Havre de Grace
Med	**E:** + Hospital
Parks	**W:** State Park
Other	**W:** Stepping Stone Museum
85	MD 22, Aberdeen, Churchville
FStop	**E:** Crown*(D)(CW)
Gas	**E:** Amoco, Citgo, Sunoco*(D), Texaco*(D)(CW)
Food	**E:** Bob Evans Restaurant, Carvel Ice Cream, Charcoal Pit BBQ, Durango's (Sheraton), Eagle's Nest, Golden Corral, Grumpy's (Econolodge), Holiday Inn, Hunan, Japan House, KFC, Kings Chinese, Little Caesars Pizza, McDonalds, Olive Tree Italian, Oriental Food Express, Papa John's, Subway, Taco Bell, Villa Cafe, Wendy's
Lodg	**E:** DAYS INN AAA Days Inn, Econolodge, AAA Holiday Inn, Howard Johnson, ◆ Red Roof Inn, AAA Sheraton, Super 8 Motel
AServ	**E:** Aberdeen Muffler Works, Amoco, Harco Olds/ Pontiac/ Cadillac, Sunoco*(D), Texaco
ATM	**E:** Aberdeen Fed Credit Union, County Bank, Mars Grocery, Merchantile Cty Bank, Nations Bank
Other	**E:** Aberdeen Market Place, Beards Hill Plaza, Cal Ripken Museum, Coin Operated Laundry, House of Pets, K-Mart, Kleine's Supermarket, Mars Grocery, Optical, Pet Value, Rite Aide Pharmacy
(82)	**Service Area, Rest Area - Vending, Phones (Both Directions, Left Exit)**
FStop	**B:** Exxon, Sunoco*
Food	**B:** Bob's Big Boy, Gourmet Bean Coffee, Hot Dog City, Mrs Fields Cookies, Roy Rogers, Sbarro Italian, TCBY Yogurt
AServ	**B:** Exxon, Sunoco
ATM	**B:** Machine
Other	**B:** Travel Mart C-Store, Travel Mart C-Store
80	MD 543, Riverside, Churchville
Gas	**E:** Amoco*, Mobil*(D)(CW)
Food	**E:** All In One (Amoco), China Moon, Cracker Barrel, McDonalds, Riverside Pizza
RVCamp	**E:** Bar Harbor RV Park
ATM	**E:** Forest Hill Bank, Mobil
Other	**E:** Kline's Supermarket, Mail Call Shipper, Thrift Drug
77AB	MD 24, Edgewood, Bel Air
FStop	**E:** Texaco*(D)
Gas	**E:** Exxon*(D), Shell
Food	**E:** Burger King, Comfort Inn, Denny's (see our ad on this page), McDonalds **W:** King's Chinese, Little Caesars Pizza, Subway
Lodg	**E:** Best Western, Comfort Inn, DAYS INN Days Inn, Sleep Inn

EXIT | MARYLAND

AServ	**E:** Shell
ATM	**W:** Maryland Fed Credit Union, Nations Bank
Other	**E:** Full Service Car Wash, **Weis Market**
	W: **Rite Aide Pharmacy**

74 MD 152, Joppatown, Fallston

Gas	**E:** Citgo*(D)
Med	**W:** + **Hospital**
Other	**E:** **Hall's C-Store**

67AB MD 43, to U.S. 1, Whitemarsh Blvd, to U.S. 40

Gas	**W:** Exxon
Food	**E:** Cassamea Carry Out
	W: McDonalds, Olive Garden, Ruby Tuesday, T.G.I. Friday's
Lodg	**W:** Hampton Inn
AServ	**W:** Exxon
ATM	**E:** Family Alliance Fed Credit Union
	W: Merchantile Bank
Other	**E:** **7-11 Convenience Store, Byron's Station, Gunpowder State Park, Warner Bros. Outlet Store**
	W: **US Post Office, Whitemarsh Mall**

64AB Junction I-695 Beltway, Towson, Essex

62 Jct I-895 South, Fort McHenry Tunnel, Annapolis, Harbor Tunnel Thruway (Southbound Left Exit)

Lodg	**E:** Holiday Inn (see our ad on this page)

61 U.S. 40, Pulaski Highway (Northbound)

60 Moravia Road (Northbound)

59 MD 150, Eastern Ave

Gas	**W:** Amoco(D), Exxon*(CW)
Food	**E:** Eastwood Inn
AServ	**W:** Eastern Auto Service
TServ	**W:** **Jag's Auto/Truck Service**
Other	**W:** **C-Store, Eastern Animal Hospital**

58 Dundalk Ave (Northbound)

57 O'Donnell St

TStop	**E:** **Baltimore Travel Plaza(SCALES), Port Travel Plaza*(SCALES)**
FStop	**E:** **Mobil*(D)**
Food	**E:** Buckham Family Restaurant, KFC, **McDonalds**, Peter Pan, Sbarro Italian, Subway, Taco Bell, **Tradewinds (Best Western)**, Travel Port Plaza
Lodg	**E:** (AAA) **Best Western**, Port Travel Plaza
AServ	**E:** **Mobil Fuel Stop(D)**
TServ	**E:** **Port Travel Plaza*(SCALES)**

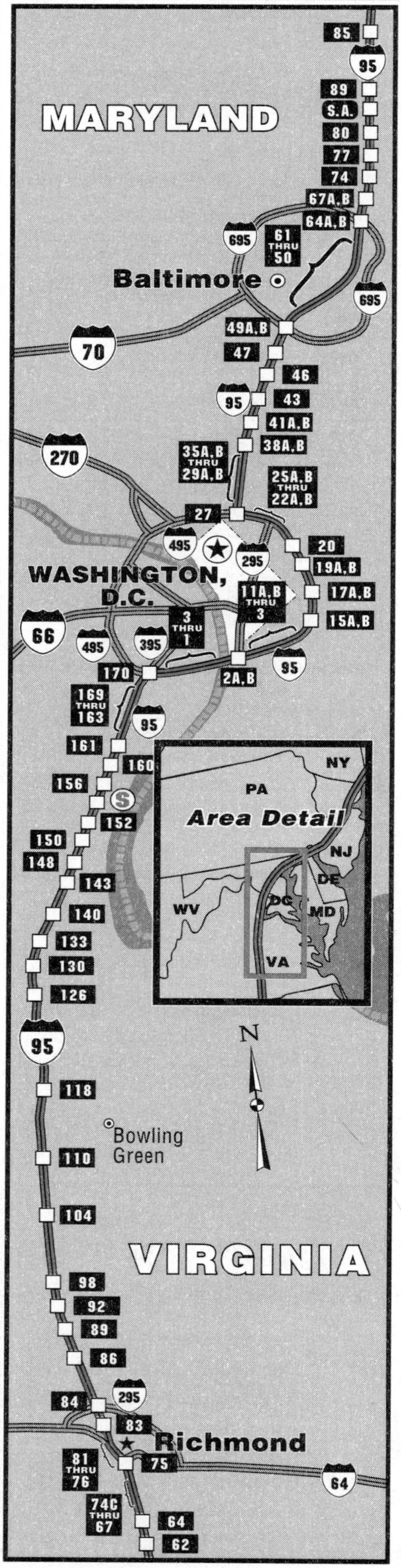

56 Keith Ave

55 Key Highway

Med	**W:** + **Hospital**
Other	**W:** **Ft McHenry Historic Site**

54 MD 2, Hanover St (Closed Due To Construction)

53 MD395, Baltimore Downtown

Lodg	**W:** Baltimore Marriott (see our ad on this page)

52 MD 295 South, Baltimore, Washington Pkwy, Int'l Airport (Closed Due To Construction)

51 Mount Clare, Washington Blvd (Closed Due To Construction)

50 Caton Ave, West to Wilkens Ave, East Washington Blvd

Gas	**E:** Shell*
Food	**E:** Caton House Restaurant, Pargos
Lodg	**E:** Holiday Inn Express
AServ	**E:** Toyota Express Lube Collison Ctr
ATM	**E:** NationsBank (Joes Ave), Shell

49AB Jct I-695, Pawson, Annapolis, Glen Burnie

47AB Jct I-195, MD 166, Catonsville, Airpo

46 Jct I-895 North, Harbor Tunnel (Northbound)

43 MD 100, Glen Burnie, Ellicott City

41AB MD 175, MD 108, Columbia, Jessup Ellicott City

TStop	**E:** **TravelCenters of America*(SCALES)**
Gas	**E:** Citgo*(D), Exxon*(D)(CW)
Food	**E:** **Burger King, Country Pride Restaurant(SCALE) (TravelCenters of America)**, Crab House, Holiday Inn, JR Chicken, Jimmy G's Pizzeria, McDonalds, Steak Subs Burgers Deli, Subway
Lodg	**E:** Holiday Inn (see our ad on this page), Suesse Chalet

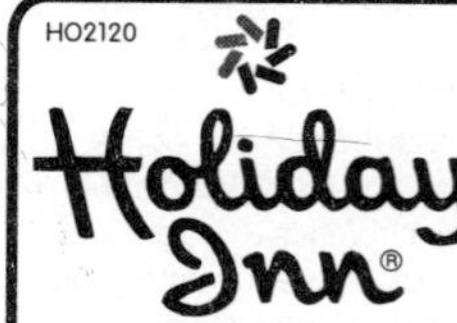

Bold red print shows RV & Bus parking available or nearby

XIT — MARYLAND

AServ E: Citgo*(D), Shell (Wrecker Service)
TServ E: **TravelCenters of America**
TWash E: **TravelCenters of America (Tank Wash Center)**
Med W: ✚ **Hospital**
ATM E: Citizens Nat'l Bank, NationsBank
Other E: **Convenience Store (Deli), Jessups Plaza, State Patrol Post**

8AB MD 32, Fort Meade, Columbia, Savage, Annapolis
Other E: **Laurel Park Racecourse**

36) **MD Welcome Center - Pet Area, RR, Phones, Picnic, Vending, RV Dump, 24 Hr Attendant**

5AB MD 216, Laurel, Scaggsville

3AB MD 198, Burtonsville, Laurel, Sweitzer Ln
Gas E: Exxon*(CW)
W: Exxon*(D)
Food E: Domino's
W: Blimpie's Subs*(D) (Exxon), Glass Duck (Best Western)
Lodg W: Best Western
AServ W: Exxon*(D)
Med E: ✚ **Hospital**
ATM E: Tower Federal Credit Union
W: Best Western
Other E: **Hughes Dairy Store**

9AB MD 212, Beltsville, Calverton
Gas W: Amoco*(D)
Food W: Baskin Robbins, Holiday Inn, McDonalds, Peking Delight, Plata Grande Grill (Ramada), Sierra's Grill (Mexican), TJ's of Calverton Restaurant, Wendy's
Lodg W: **Holiday Inn (see our ad on this page)**, Ramada
AServ W: Amoco(D)
RVCamp W: **Cherry Hill Park (see our ad on this page)**
ATM W: Crestar Bank
Other W: **CVS Pharmacy, Calverton Shopping Ctr, County Police, Doggie's Depot (Pet Grooming), Giant Grocery**

27 Jct I-495, I-95N, Baltimore, College Park, Silver Spring
Other O: **Baltimore, Washington International Airport**

Note: I-95 runs concurrent below with I-495. Numbering follows I-495.

27 Junction I-95 North, Baltimore

EXIT — MARYLAND

Other O: **Baltimore Washington Int'l Airport**

25AB U.S. 1, Baltimore Blvd, Laurel, College Park
Gas I: Amoco, Shell
O: Shell*
Food I: Days Inn (Korean), Lasieks Seafood Buffet, Pizza Hut, University Donut Shop
O: Jonesie's Steak, Mooses Creek Steak House
Lodg I: Comfort Inn, DAYS INN Days Inn, Econolodge, Royal Pine Inn, Super 8 Motel
O: Holiday Inn
AServ I: College Park Auto Repair, College Park Volkswagen, Mazda, College Part Hyndai, Honda, Jiffy Lube, RT 1 Auto Repair, Shell
O: Shell
RVCamp I: **Campground, Queenstown Motors**
ATM I: Commerce Bank, State Employees Credit Union
Other I: **College Park Animal Hospital, Colllege Park Boat & RV Center, Hardware Store**

24 Greenbelt Station (Commuters Only, No Return Southbound)
Other I: **Metro Train Station**

23 MD 201, Kenilworth Ave, Bladensburg, Greenbelt
Gas I: Amoco*(CW), Mobil, Shell, Texaco*(CW), Xtra Fuel
Food I: **Baskin Robbins (Giant Supermarket)**, Bennigan's, Boston Market, Checkers Burgers, Dunkin Donuts, Fireside Beef House, Ledo Pizza, McDonalds, Muashi Japanese Steakhouse, Pizza Hut, Popeye's, Roy Rogers, Sir Walter Raliegh, TJ's Roadhouse Grill, Wendy's
O: T.G.I. Friday's
Lodg I: Sir Walter Raliegh Inn
O: Courtyard
AServ I: Greenbelt Auto, Shell, Texaco
RVCamp O: **Greenbelt Park Campground**
Med I: ✚ **Med Serve**
ATM I: Chevy Chase Bank, Liberty Nat'l Bank, Maryland Federal Bank, NationsBank
Parks O: **Greenbelt Park**
Other I: **Beltway Keys & Locks, Beltway Plaza*, CVS Pharmacy**, Fennigan's Car Wash, **Giant Supermarket (Pharmacy), NASA, Optical, Pearl Vision Center, Postal Stand (Beltway Plaza)**

22AB Baltimore Washington Parkway (No Trucks)

20 MD 450, Annapolis Road, Lanham, Bladensburg
Gas I: Citgo*(D), Sunoco*
Food I: Bob's Big Boy, Cheasapeak Bay Seafood, Cuisine Of China, Dunkin Donuts, Good Earth Chinese, KFC, Pappa John's, Popeye's Chicken, Ramada
O: Days Inn, Horn & Horn Buffet, Jerry's Sub Shop, Red Lobster
Lodg I: Ramada
O: Best Western, DAYS INN Days Inn, Red Roof Inn
AServ I: Amoco(LP), Bob Banning Pontiac, Citgo, Grease-N-Oil Xpress, Just Tires, Merchant's Tire & Auto, Nissan, Shell, Sunoco
O: Midas Muffler & Brakes, Speedy Muffler King
Med O: ✚ **Hospital**
ATM I: Crestar Bank, Maryland Federal Credit Union
Other I: **Ames Shopping Ctr, CVS Pharmacy, Carrollton Mall, Checks Cashed, Hinson Optical, Safe Way Grocery, Shoppers Food Warehouse**
O: **Harley Davison Motorcycle Shop, Lanham Shopping Center**

19AB U.S. 50, Annapolis, Washington
Lodg O: Cambridge Inn (see our ad this page)

17AB MD 202, Landover Road, Bladensburg
Food I: IHOP
O: Holiday Inn
Lodg O: Holiday Inn
AServ I: Just Tires, Sears, Tollison Buick
Other I: **County Police, Landover Mall, Sam's Club**
O: **US Air Center (Capitol Arena)**

15AB MD 214, Central Ave
Gas I: Crown*(D)(CW), Exxon*(D), Shell*, Sunoco
Food I: Daybreak Restaurant (Days Inn), Jerry's Pizza & Subs, KFC, McDonalds, Pizza Hut, Taco Bell
Lodg I: DAYS INN Days Inn, Motel 6
O: Hampton Inn
AServ I: Hampton Auto Body (24 Hr Towing), Jiffy Lube, Midas, Shell, Traks
Other I: **7-11 Convenience Store**, Full Service Car Wash
O: **Adventure World, US Air Arena**

11AB MD 4, Pennsylvania Ave, Washington, Upper Marlboro
Gas I: Amoco, Exxon*(CW), Sunoco
O: Texaco*(D)
Food I: Hunan House Restaurant, Villa Sara Pizza Parlor
O: Murray's Hot Food To Go
AServ I: DJ's One Stop Auto Service, Michelin, Napa

EXIT	MARYLAND/VIRGINIA
	Auto Parts, Shell*
TServ	I: Goodyear Tire & Auto
ATM	I: Industrial Bank
Other	I: Ames Shopping Ctr, Coin Laundry, Forestville Business Ctr (Commercial Truck Services), Friendly Food Mart Coffee & Deli, Furniture Outlet Shopping Ctr, Goodman Eye Care Ctr, State Patrol Post
9	MD 337, Allen Town Road, Andrews AFB (Difficult Reaccess)
Gas	I: Mobil*(D)(CW)
Other	O: U-Haul Center(LP)
7AB	MD 5, Branch Ave, Waldorf, Silver Hill
Lodg	I: Days Inn (see our ad on this page)
4AB	MD 414, Saint Barnabas Road, Oxen Hill, Marlow Heights
Gas	I: Exxon*(D), Shell O: Exxon(D), Xtra
Food	I: China Best, McDonalds, Pizza Hut Carry Out O: Black Eyed Pea, Burger King, CJ's Restaurant, Checkers, Hunan's Restaurant, KFC, Little Caesars Pizza, McDonalds, Roy Rogers, Sharky's Seafood & Crabhouse, Subway, Wendy's
Lodg	O: Red Roof Inn
AServ	I: Exxon, Jiffy Lube, Shell O: Quaker State Oil & Lube
Med	I: + Hospital
ATM	I: First VA Bank O: Chevy Chase Bank (Safeway Grocery), First Union, Industrial Bank, NationsBank
Other	I: Coin Laundry, Sunrise Plaza At St Barnabas O: 7-11 Convenience Store, CVS Pharmacy, Car Wash, K-Mart (Pharmacy), Rivertowne Commons Marketplace, Rose Craft Raceway, Safe Way Grocery
3AB	MD 210, Indian Head Hwy, Forest Heights
Gas	O: Crown*, Mobil
Food	O: KFC, Ramada
Lodg	O: Ramada
AServ	O: Goodyear Tire & Auto, Mobil
Other	O: Coin Operated Car Wash
2AB	Junction I-295, Washington

↑MARYLAND

↓VIRGINIA

EXIT	VIRGINIA
1AB	U.S. 1, Fort Belvoir, Alexandria
Gas	I: Exxon, Merrit(D) (Kerosene) O: Amoco, Mobil*(D), Sunoco, Texaco
Food	O: Domino's Pizza, Howard Johnson, Kyoto Japanese Restaurant, Oh's Palace (Chinese), Western Sizzlin'
Lodg	O: Brookside Motel, Howard Johnson, Red Roof Inn, Statesman Motel, Travelers Motel, Virginia Lodge
AServ	I: Exxon, Heritage Chevrolet & Jeep O: Ourisman Dodge/ Suzuki/ Chrysler/ Plymouth, Texaco
Med	I: + Herrin Medical O: + Urgent Medical Care Walk-In
ATM	O: Central Fidelity, Crestar Bank
Other	O: Fireworks, Huntington Parkway Shops, Optical (Huntingon Parkway Ctr)
2AB	VA 241, Telegraph Road, Alexandria
Gas	O: Citgo(D)(CW), Exxon*, Hess(D)
Food	I: Mansenies Deli & Bakery O: Honolulu Restaurant
Lodg	I: Courtyard by Marriott, Holiday Inn
AServ	I: Nissan O: Citgo
Med	I: + Hospital
ATM	I: NationsBank O: 7-11 Convenience Store, Burke & Herbert Bank & Trust
Other	I: Alexandria Animal Hospital, George Washington Mason Hall O: 7-11 Convenience Store
3	VA 613, Van Dorn Street, Franconia
Lodg	O: Comfort Inn
4B	Junction I-395, Arlington, Washington
4A	Junction I-495 North, Rockville

Note: I-95 runs concurrent above with I-495. Numbering follows I-495.

EXIT	VIRGINIA
170	Junction I-495, I-395, Rockville
Lodg	W: Days Inn (see our ad on this page)
169	VA 644, Old Keen Rd, Springfield, Franconia
Gas	E: Mobil W: Citgo*(D), Mobil, Shell*
Food	E: Bennigan's, Bertucci's Pizza, Blackie's Steak, Blue Parrot (Hilton), Chesapeake Bagel Bakery, Daybreak Restaurant (Days Inn), Makati Deli & Donuts, Mozzarella's Cafe, Osaki Japanese Restaurant, Ruby Tuesday, Silver Diner, Starbucks Coffee W: Baskin Robbins, Chi Chi's Mexican, Ding Ho Carry-Out (Chinese), Donut Masters, Dunkin Donuts, Family Restaurant & Pizzeria, Generous George's Pasta & Pizza, Houlihan's, Hunan's Springfield Restaurant, MaKong Groc & Deli, Malek's Pizza Palace, McDonalds, Mike's American Grill, Pasta Peddler Restaurant, Payton Place Restaurant, PeKing Garden, Pizza Hut Delivery, Popeye's Chicken, Rocco' Italian Restaurant, Roy Rogers, Royal Restaurant, Shoney's, Subway, T.K. Restaurant (Taiwan), The Bug-A-Boo Creek, Village Chicken
Lodg	E: Comfort Inn, Days Inn, Hilton W: Holiday Inn Express, Motel 6
AServ	E: Firestone Tire & Auto, Mobil, Montgomery Ward, Sheehy Ford W: Advantage Auto, Citgo, Goodyear Tire & Auto, Jerry's Dodge, K-Mart, Midas Mufflers & Brakes, Mobil, Shell, Springfield Toyota, Toyota Express, Volkswagen Dealer
Med	W: + Inove Urgent Care Ctr
ATM	E: Central Fidelity Bank, NationsBank W: Crestar Bank, F&M Bank, First Union
Other	E: 7-11 Convenience Store, Animal Magic Pet Grooming, CVS Pharmacy, For Eyes Opti (Optical), Petsmart, Shopping Mall W: 7-11 Convenience Store, CVS Pharmacy, Car Wash, Concord Centre, Fireworks, Fisher's Hdwe(LP), Giant Drugs, K-Mart, Petco Pet Supplies, Pro Food Pet Supply, South Paws Veterinary Referral, Springfield Animal Hospital, Springfield Plaza, Super Mart C-Store, Tower Center, Vision Works Opitcal
167	VA 617, Backlick Road, Fullerton Road (Difficult Reaccess)
AServ	W: Action Auto Works
166AB	VA 790, Ft Belvoir, Newington, Alban Rd, Backlick Rd
FStop	E: Quill's
Gas	E: Exxon*(D)(LP)
Food	E: Delly's Deli & Grill, Fun Cafe, Hunter Motel, Ollie's Deli, Terminal Rd Deli (Commerce Ctr) W: Benjamin's, Tex-Mex
Lodg	E: Hunter Motel
AServ	E: All Tune & Lube, Corvette Service, Exxon, Fairfax Auto Parts, Fullerton Auto Body, International Automotive, Japanese & Domestic Auto Care, Kline, Meineke Discount Mufflers, NTW Tire Center, Pallone Chevrolet, Springfield Auto, US Auto Service W: Alban Tire, Central, Perfection Auto Body & Paint, T.K. Small Engine Repair
TServ	E: Driveline Repair Of Virginia, Red Man, Wilson Trucking Co W: Kenworth
ATM	E: F&M Bank
Other	W: Mini Mart
163	VA 643, Lorton
Gas	E: Citgo*(LP), Shell W: Texaco*(D)(CW)
AServ	E: Shell (Wrecker Service)
Other	E: Pohick Bay Park

Bold red print shows RV & Bus parking available or nearby

EXIT	VIRGINIA

161 U.S. 1, Ft Belvoir, Mt Vernon (Reaccess Exit 160)
Other E: Gunston Hall, Woodlawn Plantation

160 VA 123 North, Woodbridge to Occoquan
Gas E: Amoco*(D), Exxon, Shell, Texaco(D)
W: Exxon*(D), Mobil*, Shell* (24 Hr)
Food E: Astorra Pizza, Dixie Bones BBQ, Dunkin Donuts, Hunan Restaurant, Joe's Pizza & Deli, Kowloon Restaurant, Little Caesars Pizza (K-Mart), Lobster Farm Mkt, Lum's, Manni's Italian Restaurant, Market Street Buffet & Bakery, McDonalds, Roy Rogers, Shoney's (Woodbridge Ctr), Stow Away Restaurant, Sushui Bar & BBQ, Taco Bell, Tex-Mex Chilli Parlor
W: KFC
Lodg E: Comfort Inn, Econolodge, Inns Of Virginia
AServ E: All Tune & Lube, C&E Auto Service, Cowles Ford, Dunivin's Corner, Jiffy Lube, K&W Auto Body, K-Mart, Midas Mufflers & Brakes, Morton's Service Ctr, Rainbow Auto Service, Shell, Trak Auto Service
W: Exxon
TServ E: Holly Acres Truck Ctr(LP)
ATM E: Crestar Bank, Jefferson Nat'l Bank
Other E: 7-11 Convenience Store, Car Wash, Fireworks, Food Lion Supermarket, Food Mart (Woodbridge Ctr), Gordon Plaza, Handy Dandy Mkt, K-Mart, Laundry Salon (Coin Operated Woodbridge Ctr), Patomic Plaza, Pets (Gordon Plaza), Tourist Information, Woodbridge Animal Hospital, Woodbridge Ctr
W: Prince William Marina

158 VA3000, Prince William Pkwy, Manassas, Woodbridge
Gas W: Mobil*(CW), Shell*(CW)
Food W: Red River Crossing BBQ, Wendy's
Lodg W: Fairfield Inn
Other W: Target Department Store (Pharmacy)

156 VA 784, Dale City, Rippon Landing
Gas W: Exxon, Mobil*, Shell*, Texaco*(D)
Food W: Black Eyed Pea, Bob Evans Restaurant, Carlos O'Kelly's Mexican, Cheaspeake Bay Seafood, Chili's, Denny's, Domino's Pizza, El Charro Mexican, Fu Kien Gourmet Chinese, Jerry's Subs & Pizza, Lone Star Steakhouse, McDonalds, Olive Garden, Palm Tree Chinese, Popeye's, Red Lobster, Seoul BBQ (Korean), Wendy's
Lodg W: Days Inn
AServ W: Advantage Auto, Exxon, Firestone Tire & Auto, K-Mart, NTW Tire Center, Napa Auto, Precision Tune & Lube, Shell, Texaco, Windshields Of America
Med W: ✚ Hospital, ✚ Urgent Care
ATM W: F&M Bank, Navy Federated Credit Union, Rigs Bank, Signet Bank
Other W: Ashdale Plaza, Dale City Animal Hospital, K-Mart, Potomic Mills Shopping Center, Prince Wiliam Square

(157) Rest Area - RR, Phones, Picnic, Grills

(154) Weigh Station - Truckers Rest Area, RR, Phones, Picnic

EXIT	VIRGINIA

152 VA 234, Dumfries, Manassas
Gas E: Amoco*(D), Sunoco, Texaco*(D)
W: Citgo*, Exxon*(LP), Shell
Food E: Golden Corral, Happy Eatery, KFC, McDonalds, Taco Bell
W: Cracker Barrel, Don Pepe (Mediterranian), Mont Clair Family Restaurant, Waffle House
Lodg E: ◆ Super 8 Motel
AServ E: American Car Care & Ctr, Beatty's Service Ctr, Goodyear Tire & Auto, Grease Monkey, Gumphry's Auto Body, Meineke Discount Mufflers
W: Exxon
ATM W: Citgo
Other E: Coach House Plaza Shopping Ctr, Weens-Botts Museum

150AB VA 619, Triangle, Quantico
Gas E: Amoco(D), Exxon*, Shell*
Food E: Burger King, Dent's Seafood, McDonalds, Ralph's Ice Cream & BBQ, Tru Grit Restaurant, U.S. Inn (Chinese), Wendy's
Lodg E: Ramada, U.S. Inn
AServ E: ASE Auto Service, Amoco, Howard's Triangle Camping
Parks E: Forest Greens Park, Locust Shade Park
W: Prince William Forest Park
Other E: Post Office
W: Quantico Nat'l Cemetary

148 Quantico, Marine Base
Other E: US Marine Airground Museum

143AB VA 610, Aquia, Garrisonville
FStop W: BP*(CW)
Gas E: Amoco*, Exxon*(CW), Shell*(D)
W: Amoco*(D)(CW), Citgo*, Crown*(D)(CW)(LP) (Kerosene), Texaco*(D)(CW)
Food E: Carlos O'Kelly's Mexican, Dairy Queen, Davazo's Cafe, Gargoyles Coffee Bar, Hunan's Chef Restaurant, Imperial Garden, KFC, Little Caesars Pizza, Mediterranian Cafe & Market, Pizza Hut, Roy Rogers, Ruby Tuesday, Shoney's (Days Inn)
W: Baskin Robbins, Botta Bagel & Deli, Buffett King, Burger King, Dad's Deli, Dunkin Donuts, Formosa Chinese Restaurant, Golden Corral, Hardee's (BP), Kobe Japanese Restaurant, McDonalds, Popeye's Chicken, Taco Bell, Tony's Pizza & Deli, Wendy's
Lodg E: Days Inn
W: Comfort Inn, Country Inn, Super 8 Motel
AServ W: American Automotive & Tire, Jiffy Lube, Wal-Mart
RVCamp W: Aquia Pines Campground
ATM E: Central Fidelity Bank, Crestar Bank, Nationsbank (Amoco)
W: BP, Crestar Bank, First Union Bank, Jefferson National Bank, Texaco, Virginia First Savings Bank
Other E: Aqua Towne Center, Coin Laundry, Optical, Petmania, Rite Aide Pharmacy
W: Bafferton Center, CVS Pharmacy, Car Wash, Giant Food & Drugs, North Stafford Plaza, Wal-Mart Supercenter (Pharmacy)

140 VA 630, Stafford
FStop W: Texaco*
Gas E: Mobil*, Texaco
W: Shell*(D)
Food E: Chix Subs & Pizza, McDonalds
AServ E: Texaco
Other W: Fireworks

133AB U.S. 17 North, Warrenton, Falmouth
TStop W: Servicetown Travel Plaza*(SCALES) (Truck Wash)
FStop E: Racetrac*
W: East Coast*(D)(LP) (24 Hr)
Gas E: Mobil*(D)(CW)
W: Amoco*(D), Citgo*, Shell*, Texaco*(D)
Food E: Aliby's Restaurant, Arby's, Family Tie's Pizzeria, Majestic's Restaurant
W: Anthony's Pizza Restaurant, Burger King, Hardee's, Holiday Inn, Johnny Appleseed Restaurant (Best Western), McDonalds, Pizza Hut, Ponderosa, Servicetown Travel Plaza (24 Hr), Taco Bell, Waffle House, Wendy's
Lodg E: Howard Johnson (see our ad on this page), Motel 6
W: Best Western, Comfort Inn, Days Inn, Holiday Inn, Ramada Limited, Sleep Inn, Travelodge
AServ E: Darvin's Transmission, Woodard's Automotive
W: Falls Run Auto Repair, Progressive Auto, Texaco (Wrecker Service), Xpress Lube
Med E: ✚ Hospital
ATM W: F&M Bank, Nat'l Bank of Fredricksburg
Other E: 7-11 Convenience Store, Coin Laundry
W: Fireworks (Citgo), Fireworks, Food Lion Supermarket, Inglewood Shopping Ctr, Revco Drugs

(131) VA Welcome Center - RR, Phones, Picnic (Southbound)

130AB VA 3, Fredericksburg, Culpepper
Gas E: Amoco(D) (24 Hr), Mobil(CW), Pep Boys Auto Center, Racetrac*, Shell*(D)(LP) (Kerosene)
W: Amoco*, Crown*(CW)(LP), Exxon*, Shell*
Food E: Arby's, Bagel Station, Bob Evans Restaurant, Carlos O'Kelly's Mexican, Chesapeake Bay Seafood, DeVancZo Italian Cafe, Dunkin Donuts, Formosa Restaurant, Friendly's, Golden Rail Restaurant, Heavenly Ham, KFC, Lone Star Steakhouse, Long John Silvers, Market Street Buffet & Bakery, McDonalds, Popeye's Chicken,

Bold red print shows RV & Bus parking available or nearby

EXIT VIRGINIA

Shoney's, Top's China Buffet
W: Applebee's, Aunt Sara's Pancakes House, Baskin Robbins, Boston Market, Burger King, Checkers, Chi-Chi's Restaurant, Denny's, Dragon Inn Chinese & Seafood, Dunkin Donuts, Einstein Bagels, Fuddrucker's, Hardee's, IHOP, Italian Oven, Kenny Rogers Roasters, Little Caesars Pizza, McDonalds, Morrison's Cafeteria, Old Country Buffet, Outback Steakhouse, Pancho Mexican Restaurant, Pizza Hut, Red Lobster, Sheraton, Starbucks Cafe, Subway, Taco Bell, Tia's Tex Mex

Lodg **E:** Best Western, Hampton Inn
W: Best Western, Econolodge, Ramada, Sheraton, Super 8 Motel

AServ **E:** Battery Plus, Pep Boys Auto Center
W: ATA Tire, BJ's Tire Ctr, Jiffy Lube, Mineke Mufflers & Brakes, Montgomery Ward, Precision Tune, Sear's

Med **E:** ✚ **Hospital**

ATM **E:** First VA Bank, Signet Bank of Virginia
W: Crestar Bank, First Union Bank, First Union Bank, Jefferson Nat'l Bank, Virginia First Savings Bank

Other **E: 7-11 Convenience Store, Chancelorsville Battlefield, Fed Ex Drop Box, Fredericksburg Museum, Fredricksburg Area Museum & Cultural Center, Gateway Village Shopping Center, Greenbriar Shopping Ctr, Kenmore Plantation (James Monroe Museum), Maico Hearing Aid Service, Optical, Westwood Shopping Center**
W: CVS Pharmacy, Car Wash, **Fireworks, Giant Food & Drug Store, K-Mart, Mail House Plus, Petsmart, Spotsylvania Crossing Ctr, Spotsylvania Mall, Target Department Store (Pharmacy), Village Square, Wal-Mart Supercenter (Optical, Pharmacy)**

126 U.S. 1, U.S. 17, Massaponax

FStop **W: Racetrac***

Gas **E:** BP, Chevron*(D), **Citgo*(D) (Fireworks), Exxon*,** Fas Mart*, Mobil, Shell*(CW), Texaco*, U-Save
W: Exxon*(CW)

Food **E:** Arby's, China Max, Dairy Queen, **Days Inn,** Denny's, El Charro Mexican, Hardee's, Holiday Inn, Jimmy's Diner, **McDonalds,** Pat's Ice Cream, Pizza Connection, Pizza Hut, Rally's Hamburgers, Subway (Exxon), Subway (Lees Hill Center), Taco Bell, **Waffle House,** Wendy's, Western Sizzlin'
W: Aunt Sara's Pancakes, Blimpie's Subs (Exxon), Burger King, Cousin John's Chicken & Steaks, **Cracker Barrel,** Damon's Ribs, Edy's Grand Ice Cream, Fantastic Italian Restaurant, Golden China, Hot Stuff Pizza (Exxon), **KFC, McDonalds**

Lodg **E:** AAA Econolodge, Fairfield Inn, Heritage Inn, AAA Holiday Inn (see our ad on this page), **Howard Johnson (see our ad this page),** AAA Royal Inn, Super 8 Motel
W: AAA Comfort Inn, AAA WhyteStone's Inn

AServ **E:** BP, Bill Britt Mazda Volkswagon, Bonnett Pontiac, Buick, Goodyear Tire & Auto, Grease Monkey Oil & Lube, Hunda, Olds, Cadillac, Jim's Auto Repair, Merchant's Tire & Auto, Mobil*, Rowe's Automotive Clinic, Safford Dodge, Texaco, U-Save

ATM **E:** Crestar, First Virginia Bank
W: The National Bank of Fredericksburg

Other **E: CVS Pharmacy,** Car Wash, **Country Corner C-Store (Fireworks), Fireworks, Food Lion Supermarket (Deli, Bakery), Lees Hill Center, Maru Pet Shop, Pharmahouse Drugs, Revco Drugs, Safe Way Grocery, Visitor Information**
W: Dept of Motor Vehicles, Fed Ex Drop Box, Massaponax Outlet Center, South Point Shopping Ctr

118 VA 606, Thornburg

FStop **E: Shell*(D)**
W: Citgo*(D)

Gas **W:** Exxon*, Shell*(LP) (24 Hr)

Food **W: Burger King, McDonalds**

Lodg **W: ◆ Holiday Inn Express**

AServ **E:** Shell (Towing)
W: Campbell's Wrecker (Citgo), Citgo

TServ **E: Atkins Truck & Equipment Sales**

RVCamp **W: Campground**

ATM **W:** Citgo

Parks **E: Lake Anna State Park**

Other **E: Stonewall Jackson's Shrine**
W: Fireworks (Shell)

110 VA 639, Ladysmith

Gas **E:** Shell(D)
W: Exxon* (Kerosene)

AServ **E:** Fletcher's Auto Service (Towing)
W: Exxon, Speedy's Auto Repair & Lube

(108) **Rest Area - Tourist Info, RR, Phones, Picnic, Vending**

104 VA 207 to 301, Carmel Church, Bowling Green, Ft. AP Hill

TStop **E: Petro*(CW)(LP)(SCALES), Pilot Travel Center***
W: Flying J Travel Plaza*(LP)(SCALES)

FStop **E: Mr Fuel* (Showers Available), Petro, Texaco***

Gas **E:** Amoco*(D)(LP) (Kerosene), Chevron*(D), Exxon*(D), Texaco*(D)
W: Exxon*(D)(CW) (Kerosene)

Food **E: Dairy Queen (Pilot), Iron Skillet Restaurant (Petro), McDonalds,** Perkee's Pizza, **Subway (Pilot), Wendy's (Petro)**
W: Aunt Sara's Pancakes (Comfort Inn), Country Mkt Restaurant & Buffett, **Days Inn,** Magic Dragon Chinese, Maria's Cafe Pizza & Subs, Pepperoni's Pizza, **Waffle House**

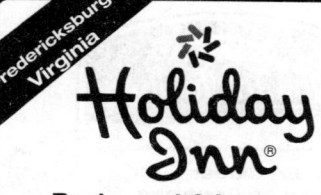

Lodg **E:** ◆ Holiday Inn Express, ◆ Howard Johnson
W: Comfort Inn, DAYS INN ◆ **Days Inn**

AServ **W:** Retallic Auto Parts

TServ **E: Petro**

TWash **E: Petro/ Blue Beacon**

RVCamp **E: KOA Campground (13.5 Miles)**

ATM **E:** Exxon, Petro

Other **E: Fireworks**
W: Car Wash, **Carmel Church Hdwe, Coin Operated Laundry, Fed Ex Drop Box (Flying J Travel), Fireworks, U.S. Post Office, UPS Drop Box (Flying J Travel)**

98 VA 30, Doswell , Westpoint

TStop **E: Doswell All American Travel Plaza*(SCALES) (RV Dump, 24 Hr)**

Gas **E:** Citgo*, Texaco*

Food **E: All American Truck Plaza (Best Western), Burger King**

Lodg **E:** Best Western (Kings Dominion), **Econolodge (All American Truckstop)**

TServ **E: All American**

TWash **E: All American**

RVCamp **E: All American Truck Plaza***

ATM **E: All American Truckstop,** Machine (Citgo)

Other **E: Factory Fireworks Outlet, Kings Dominion Theme Park**

92AB VA 54, Hanover, Ashland

TStop **W: TravelCenters of America*(SCALES)**

FStop **W: East Coast*(D)(LP) (Kerosene), Pilot Travel Center*(D)**

Gas **E:** Mobil*
W: Amoco*(CW)(LP), BP*, Chevron*, Exxon*(D), Shell*, Texaco*(CW)(LP)

Food **W:** Anthony's Italian Pizza, Arby's, Aunt Sara' Pancakes (Ramada), Blimpie's Subs (East Coast), Burger King, Captain D's Seafood, **Country Pride Restaurant (TStop),** **Cracker Barrel,** Dunkin Donuts (East Coast), Hardee's, **KFC,** Little Caesars Pizza, McDonalds, McDonalds, Omelet Shoppe, Pizza Hut, Ponderosa Steakhouse, **Popeye's Chicken,** Season's Restaurant (Holiday Inn), Shoney's, Subway (Exxon), Taco Bell, The Smokey Pig Restaurant, **TravelCenters of America,** Wendy's

Lodg **W:** Comfort Inn, Econolodge, HoJo, Holiday Inn, Palm Leaf Motel, Ramada, Super 8 Motel, **Travelodge,** Twin Oaks Motel

AServ **E:** Mobil
W: Advance Auto Parts, Amoco, Chevron, Feild's Auto (Towing), Patrick Pontaic, Buick, GMC, Trak Auto, Tuffy's Mufflers & Brakes

ATM **W:** Central Fidelity Bank, First Virginia Savings Bank, Texaco

Other **W: Bumper To Bumper Auto Service, CVS Pharmacy, Coin Operated Laundry, Food Lion Supermarket, Pearl Vision, Rite Aide Pharmacy, Tower Optical, UKrop's Grocery (Pharmacy)**

89 VA 802, Lewistown Road

TStop **E: Speed & Briscoe 76(SCALES) (RV Dump)**

Gas **E: Mobil* (Truckstop),** Shell*

Food **E:** I Can't Believe It's Not Yogurt, Pizza Hut, **Taco Bell (Truckstop), Truckstop (Truckstop)**

Bold red print shows RV & Bus parking available or nearby

XIT	VIRGINIA
AServ	E: Shell
TServ	E: Speed & Briscoe 76
RVCamp	E: Americamps Campground[LP]
	W: Kosmo Campground
ATM	E: Speed & Briscoe
Other	E: County Airport
6	VA 656, Atlee, Elmont
Gas	W: Amoco*[D], Mobil, Texaco*[CW]
Food	W: Best Western, Dairy Queen (Amoco), Sbarro (Amoco, Picnic), Subway (Texaco)
Lodg	W: Best Western
AServ	W: Mobil
TServ	E: Western Branch Diesel (24 Hr)
Other	E: Hanover Airport
4AB	Junction I-295, Williamsburg, Charlottesville (Left Exit Southbound)
3AB	VA 73, Parham Road
FStop	W: Watch Car Fleet Fuel Management
Gas	W: Amoco, East Coast*[D][LP] (Kerosene), Exxon, Texaco
Food	W: Aunt Sara's Pancake House (Quality Inn), Denny's, El Paso Mexican Restaurant, Hardee's, Little Caesars Pizza, McDonalds, Mings Dynasty (Chinese), Stuffys Subs and Salads, Subway, Theresa'a Italian Village, Wendy's
Lodg	W: Broadway Motel, Cavilier, Econolodge, HoJo, Quality Inn
AServ	W: Amoco, Exxon, Francis Motor Company, Martins Car Care (Firestone), Texaco, Wal-Mart Supercenter
ATM	W: Crestar Bank, Haniford Supermarket
Other	W: B. Lewis Genter Botanical Garden, CVS Pharmacy, Food Lion Supermarket (24 Hr), Haniford Supermarket, Revco Drugs, Wal-Mart Supercenter
82	U.S. 301, Chamberlayne Ave
FStop	E: Exxon*[D]
Gas	E: BP[LP], Mobil*[D]
Food	E: Bojangle's, Buger King, Kentucky Fried Chicken, McDonalds, Mello Buttercup and Ice Cream, Red House American and Chinese, Seafood House, Subway, Taco Bell, The Virginia Inn
Lodg	E: Chamberlaine Motel, Super 8 Motel, The Virginia Inn
AServ	E: Azalea Auto and Tire, Lube Check Oil Change, Mobil, Tuffy Mufflers
ATM	E: Consolidated Bank and Trust, Creststar Bank
Other	E: Coin Operated Laundry, Farmer Jacks Supermarket, Southern Express Convience Stores, Strawberry Hill
81	U.S. 1 (Northbound)
80	VA 161, Lakeside Ave, Hermitage Road (Northbound, Southbound Reaccess Only)
79	Junction I-64, I-195 South, Charlottesville
78	Boulevard
Gas	E: Citgo
	W: Amoco*[D], Lucky*[LP]
Food	E: Holiday Inn, Zippy's BBQ
	W: Bill's Virginia BBQ, Taylor's Family Restaurant (Days Inn)

EXIT	VIRGINIA
Lodg	E: Diamond Lodge and Suites, Gadnes Restaurant, Holiday Inn
	W: Days Inn (see our ad on this page)
TServ	W: Dolan International
ATM	W: Jefferson National Bank (Lucky)
Other	W: Tourist Information, US Marine Museum
76B	U.S. 1, U.S. 301, Belvidere
Med	W: + Belvidere Medical Center
Other	W: Maggie Walker Historical Site, VA War Memorial
76A	Chamberlayne Ave (Northbound)
Gas	E: Citgo
Food	E: Burger King, Captain D's Seafood, Dunkin Donuts, Hawks Bar-B-Que And Seafood, McDonalds
Lodg	E: Belmont Motel
AServ	E: Emrick Chevrolet, Napa Auto Parts, Texaco*[D] (Kerosene)
Other	E: 7-11 Convenience Store, Easters Convience Store
75	Junction I-64 East, Williamsburg Norfolk
74C	U.S. 33, U.S. 250, Broad St
Gas	E: Exxon*[CW]
Food	E: McDonalds
Med	W: + VA Commonwealth Medical Center
Parks	E: Richmond Battlefield Park
Other	E: Museum of the Confedercy, Valentine Museum
74B	Franklin St (Re-Entry 74A)

EXIT	VIRGINIA
74A	VA 195, Downtown Expwy, to Jct I-195N (Toll Rd)
73	Maury St, Commerce Road
Food	E: Sonny's Grill
AServ	E: Napa Auto Parts
TServ	E: James River Automated Fuel Service, Standard Parts Corporation
Med	E: + VA Hospital
Other	E: Denis Truck and Trailer Repair
69	Bells Road, VA161
FStop	W: East Coast, Texaco* (24 Hr)
Gas	W: Exxon*[D]
Food	W: Hardee's, Holiday Inn
Lodg	W: ◆ Holiday Inn, ◆ Red Roof Inn
AServ	W: Exxon, Texaco
TServ	E: Cumming Truck Service, Great Dane, Peterbilt Dealer
Other	W: Port of Richmond
67	VA 150, Chippenhan Pkwy, to U.S. 60 & U.S. 360 (Divides To VA 613 At Willis Rd)
64	VA 613, Willis Road
FStop	W: Texaco*[D]
Gas	E: Exxon*, Shell*[CW]
	W: Citgo*, Crown*[D][LP], Mobil*
Food	E: Aunt Sara's Pancake House (Ramada), Subway (Exxon), Waffle House
	W: Almonds Pit Cooked Bar-B-Que, Dunkin Donuts (Texaco), McDonalds, Pancake House Restaurant (VIP Inn)
Lodg	E: ◆ Econolodge, ◆ Ramada
	W: Economy House Motel, Sleep Inn, VIP Inn
AServ	E: Shell
TServ	E: Kenworth Truck Enterprises, Tamahawk Truck And Trailer
62	VA 288, Chesterfield to Powhite Pkwy
Food	W: Pietro's Italian Restaurant, Raven Restaurant
AServ	E: Advance Auto Parts
	W: Bennett Ford, White's Automotive
Other	W: 7-11 Convenience Store
61AB	VA 10, Chester, Hopewell
FStop	E: Racetrac* (24 Hr, UPS Drop)
	W: Exxon*[LP]
Gas	W: Amoco, Citgo*[D][LP], Crown*[D][CW], East Coast*[D]
Food	E: Comfort Inn, Hardee's, Imperial Food Buffet
	W: Burger King, Captain D's Seafood, Cracker Barrel, Days Inn, Denny's, Friendly's, Hardee's, KFC, McDonalds, Piasano's Italian Restaurant, Pizza Hut, Rosa's Italian Restaurant, Shoney's, Subway, Taco Bell, Waffle House, Wendy's, Western Sizzlin'
Lodg	E: Comfort Inn, Hampton Inn, Holiday Inn (see our ad on this page)
	W: Days Inn, Fairfield Inn, Howard Johnson, Super 8 Motel
AServ	W: Amoco, Auto World, Heritage Chevrolet, Trak Auto
TServ	W: Best Truck Repairing
Med	W: + Med-Care Family Practice
ATM	E: Fibers Federal Credit
	W: Citgo, Crestar Bank, Exxon, First VA Bank,

EXIT VIRGINIA

Virginia First Savings Bank
Other **E:** Surburban Propane(LP)
W: Bermuda Square Shopping Center, Breckenridge Center, CVS Pharmacy, Car Wash, K-Laundry, K-Mart, Paradise Pets, Rite Aide Pharmacy, Target Department Store, Ukrops Supermarket, Winn Dixie Supermarket

58 VA 746, VA 620, Walthall
FStop **E:** Texaco*(LP)
W: Chevron*
Gas **W:** Amoco*(D)(LP) (Picnic Area)
Food **E:** Travelodge
W: Bullets (Amoco), Dunkin Donuts (Amoco), Interstate Inn
Lodg **E:** (AAA) Travelodge
W: (DAYS INN) Days Inn (see our ad on this page), Interstate Inn

54 VA 144, Temple Ave
FStop **W:** Texaco*(D)(CW)(LP)
Gas **E:** Chevron*(LP)
Food **E:** Applebee's, Arby's, Fuddrucker's, Golden Corral, Lone Star Steakhouse, McDonalds, Morrison's Cafeteria, Old Country Buffett, Red Lobster, Subway
W: Hardee's
AServ **E:** Sears
ATM **E:** First VA Bank
Other **E:** Colonial Heights C-Store, Sam's Club, South Park Crossing, South Park Mall
W: Colonial Heights Police Dept

53 Southpark Blvd
Gas **E:** BP*(CW), Citgo*(CW)
Food **E:** China Man's Buffett, Koreana Korean Food, La Carreta Authentic Mexican, Padow's Hams and Deli, Sagebrush Steakhouse, TCBY (BP), Wendy's
AServ **E:** Wal-Mart
TServ **W:** Goodyear Tire & Auto (24Hour)
ATM **E:** BP, Citgo
Other **E:** Demmock Center Shopping Mall, K-Mart (Pharmacy), One-Hour Optical, Petsmart, Southgate Square, Southlake Shopping Center, Target Department Store (Pharmacy), Wal-Mart Supercenter

52B Downtown, Washington St, Wythe St
FStop **E:** Texaco*(LP)
Gas **E:** Amoco
W: Crown*(D)
Food **E:** Aunt Sara's Pancakes, Pelican Restaurant (Best Western), Ramada Host, Steak & Ale, Subway
W: Church's Fried Chicken, Dairy Queen, Hawks Bar-B-Que, Nick's Pancake and Steakhouse (Howard Johnson)
Lodg **E:** Best Western, Travel Lodge
W: Countryside Inn, Econolodge, Howard Johnson, King Motel, Knight's Inn, Star Motel, Super 8 Motel
AServ **W:** B & B Auto, Blanford 66
Med **E:** + Hospital
ATM **E:** Cignet Bank, VA First Savings Bank
Other **E:** Police Station, Roses Drugstore, S & J Handwash Car Wash, Tourist Information

52A Wythe St, Washington St - (Southbound)

51 Junction I-85 South, U.S.460W, to Durham, to Atlanta

50 U.S. 301, U.S. 460, Crater Rd, County

EXIT VIRGIN

Dr
Gas **E:** Amoco, Citgo*, Racetrac
Food **E:** Hardee's
Lodg **E:** Countryside Inn, Flagship Inns
AServ **E:** Budget Muffler, Express Lube, Jim Whelans Service Center, Texaco
Med **W:** + Hospital
Other **E:** Car Wash

50BC U.S. 301, Crater Rd (Northbound)

48AB Wagner Road
Gas **W:** Chevron(LP), Shell*, Texaco*(CW)
Food **W:** Anabelles' Restaurant and Pub, Arby's, Burger King, Golden Elephant Garden Restaurant (Chinese Buffet), McDonalds, Pizza Hut, Sal's Italian Restaurant, Subway, Taco Bell, The Big Scoop Ice Cream, The Mad Italian Pasta and Steakhouse, Yesterday's Restaurant
Lodg **W:** Crater Inn
AServ **W:** 301 Auto Parts, Advance Auto Parts, Auto Zone Auto Parts, Expert Brakes, Exxon, Maggie Lube and Oil Change, Mazda, Merchants Tire and Auto, Pep Boys Auto Center, Quality Ford, Triangle Dodge
ATM **W:** Bank of Southside Virginia, Central Fidelity Bank, Community Bank, Signet Bank, Texaco
Other **W:** CVS Pharmacy, E-Z Check Cashing, Food Lion Supermarket, Petersburg Animal Hospital, Revco Drugs, Rite Aide Pharmacy, Ukrops Supermarket (Pharmacy), Walk-In Medical Ctr

47 VA 629, Rives Road
Gas **W:** Texaco*
Lodg **W:** Heritage Motor Lodge
AServ **W:** Texaco*
Other **W:** US SSA Softball Hall of Fame Museum

46 Jct I-295N, to Washington (Left Exit Southbound)

45 U.S. 301
Gas **E:** Texaco(D)
W: Exxon*(LP)
Food **W:** Days Inn, Nanny's Family BBQ, Steven Ker Restaurant (Quality Inn)
Lodg **W:** (AAA) Comfort Inn, (DAYS INN) (AAA) Days Inn (see our ad this page), (AAA) Holiday Express (see our ad on this page), (AAA) Quality Inn (see our ad on this page)
AServ **E:** Texaco

41 U.S. 301, VA 35, VA 156, Courtland
FStop **E:** Chevron* (24 Hr)
Food **E:** Econolodge, Nino's North Italian Restaurant (KOA Campground)
W: Rose Garden Inn
Lodg **E:** Econolodge
W: Rose Garden Inn, Super 8 Motel
RVCamp **E:** KOA Campground
ATM **E:** Chevron

(40) Weigh Station - both directions

37 U.S. 623, Carson
Gas **W:** Amoco*(D), Texaco*

(36) Info Center, Rest Area - Picnic, RR, Phones, Vending

33 VA 602
TStop **W:** Davis Truck Plaza* (RV Dump)
FStop **W:** Chevron
Gas **W:** Exxon*, Shell(D)
Food **W:** Burger King, Denny's

Bold red print shows RV & Bus parking available or nearby

XIT	VIRGINIA
Lodg	**W:** Stony Creek Motel
AServ	**W:** Brooks Brothers Collison Center, Johnny's 24 Hour Trailer/Truck Repair (Wrecker Service)
TServ	**W:** Jimmy Matthew's Towing & Repair
ATM	**W:** Davis Truck Plaza
1	VA 40, Stony Creek, Waverly
FStop	**W:** Texaco*
Food	**W:** Stoney Creek Tastee BBQ, Texaco
AServ	**W:** Carters Car & Truck Repair, JR's Towing, Texaco
4	VA 645
0	VA 631, Jarratt
FStop	**W:** Citgo*
Gas	**W:** Exxon*(LP) (24 Hr)
Food	**W:** Blimpie's Subs (Exxon)

EXIT VIRGINIA

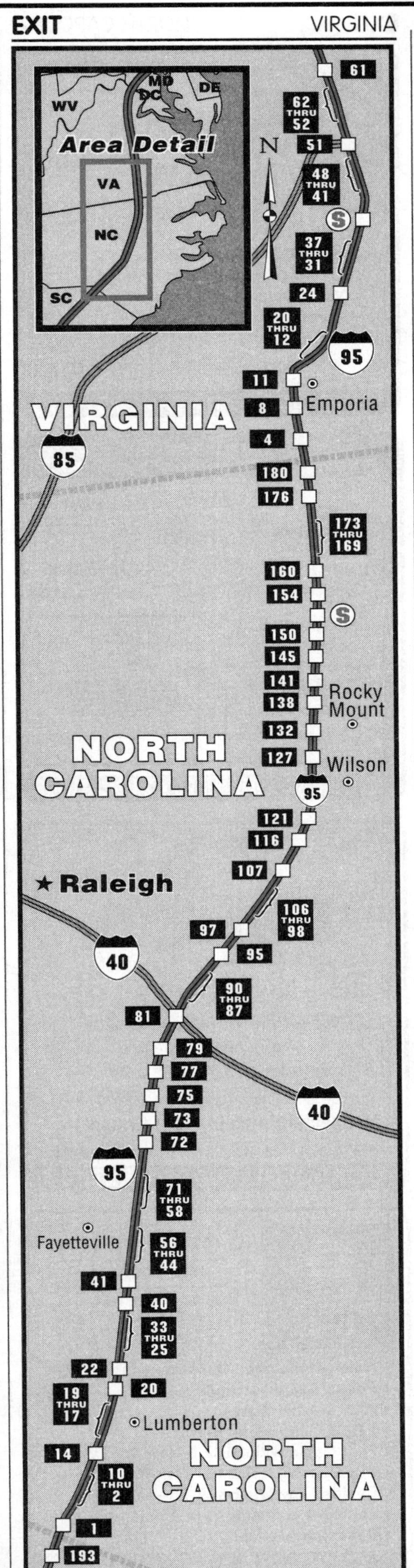

EXIT	VIRGINIA/NORTH CAROLINA
AServ	**W:** Owen Ford
ATM	**W:** Exxon
17	U.S. 301
Food	**E:** China Star
Lodg	**E:** Econolodge, ◆ Reste' Motel
RVCamp	**E:** Jellystone Park Campground(LP) (1.7 Miles, RV Dump Station)
13	VA 614, Emporia
FStop	**E:** Shell* (Kerosene)
Gas	**E:** Texaco*
Food	**E:** Pit Cooked BBQ
Lodg	**E:** Dixie Motel
11AB	U.S. 58, Emporia, South Hill, Norfolk
TStop	**W:** Sadler Travel Plaza*
Gas	**E:** Amoco*(LP), Citgo* (RV Dump), Phillips 66*, Shell*, Texaco*(D)(CW) **W:** Citgo* (Kerosene), Shell* (Sadler Travel Plaza), Texaco*(D)
Food	**E:** Arby's, Burger King (Citgo), China Dragon Chinese, Dairy Queen, Hardee's, Holiday Inn, KFC, McDonalds, Pizza Hut, Subway, Taco Bell (Amoco), Wendy's, Western Sizzlin' **W:** Days Inn, Johnson Peanut Company, Shoney's, T.J.'s Family Restaurant
Lodg	**E:** Holiday Inn (see our ad on this page), Comfort Inn (see our ad on this page) **W:** Best Western, Days Inn, Hampton Inn
AServ	**E:** Advance Auto Parts, Auto Mart, Freeman Auto Parts **W:** Exxon*
TServ	**W:** Sadler
Med	**E:** ✚ Hospital
ATM	**E:** First Citizens Bank
Other	**E:** CVS Pharamacy, Food Lion Grocery, Revco Drugs, Winn Dixie Supermarket
8	U.S. 301, Emporia
TStop	**E:** Amoco*(D)(SCALES)
Gas	**E:** Citgo, Exxon* (Kerosene)
Food	**E:** Baskin Robbins, Denny's (Comfort Inn)(see our ad on this page), Homestyle Cooking (Cooking), Marie's Restaurant, Red Carpet Inn
Lodg	**E:** AAA Comfort Inn (see our ad on this page), Red Carpet Inn
AServ	**E:** Citgo
TServ	**E:** Redwine Garage
ATM	**E:** F&M Bank (Amoco)
4	VA 629, Skippers
Gas	**W:** Citgo*(D)
Food	**W:** Econolodge
Lodg	**W:** Econolodge
RVCamp	**W:** Cattail Creek Campground (2.9 Miles)
(1)	VA Welcome Center - RR, Phones, Picnic, Vending

↑VIRGINIA

↓NORTH CAROLINA

(181)	NC Welcome Center - RR, Phones
180	NC 48, Gaston
TStop	**W:** McElroy Truck Lines, Speedway*(LP)(SCALES)
176	NC 46, Gaston, Garysburg
Gas	**E:** Texaco* **W:** Shell* (24 Hr)
Food	**E:** Stuckey's (Texaco) **W:** Aunt Sarah's Pancakes (Comfort Inn), Burger King

Bold red print shows RV & Bus parking available or nearby

EXIT NORTH CAROLINA

Lodg W: AAA Comfort Inn
TServ W: Redwines Parts and Sales Truck Service

173 U.S. 158, Roanoke Rapids, Weldon
FStop E: Speedway*(LP)
W: Phillips 66*(CW)(LP) (RV/ Truck Wash)
Gas E: Shell* (see our ad on this page), Texaco*(D)(CW) (Picnic Area)
W: Amoco*(D)(CW) (Picnic Area), Amoco*(LP), BP*(CW), Exxon, Racetrac*, Shell* (Kerosene)
Food E: Blimpie's Subs, Frazier's, Ralph's BBQ, Trigger's Steak House, Waffle House
W: Blimpie's Subs (Shell), Burger King, China King Chinese Restaurant, Country Porch Restaurant (Roses Dept. Store), Cracker Barrel, Dino's Pizza, Fisherman's Paradise, Hardee's, Holiday Inn, KFC, McDonalds, New China Chinese Restaurant, Piccolowe's, Pizza Hut, Pizza Inn, Ribeye Steakhouse, Ryan's Steakhouse, Shoney's, Subway, Taco Bell, Waffle House, Wendy's
Lodg E: Days Inn AAA Days Inn (see our ad on this page), Interstate Inns, Orchard Inn
W: ◆ Hampton Inn, Holiday Inn (see our ad on this page), AAA Motel 6, Sleep Inn/Comfort Inn
AServ E: Nenc Service Center
W: Advance Auto Parts, Alan Vester Ford Lincoln, Mercury, Auto Mart, Auto Zone Auto Parts, Draper Wrecker Serv. (24 Hr), Exxon, Firestone Tire & Auto, Honda
RVCamp E: Camping (Interstate Inn)
Med W: ✚ Hospital
ATM E: Speedway
W: Cashpoint (BP), Centura Bank
Other W: Chockoyotte Park, Coin Car Wash (Phillips 66), Discount Drug Barn, Eckerd Drugs, Food Lion Supermarket, Piggly Wiggly Supermarket, Roses Dept. Store, Wal-Mart(LP) (W/ Pharmacy, Open 24 Hours)

171 NC 125, Roanoke Rapids
Other W: State Patrol

169 **(168)** NC 903, Halifax, Littleton
TStop W: Citgo(LP) (Auto Repair, Truck Wash)
FStop E: Exxon*(D), Speedway*(LP) (Kerosene)
Food W: Citgo
ATM E: Speedway

160 NC 161, Louisburg, Roanoke Rapids
Gas E: Exxon
W: Citgo*
AServ E: 561 Auto Repair*

154 NC 481, Enfield
Gas E: Mobil*
RVCamp W: KOA Campground(LP) (8/10 Mile, RV Dump Station)

(151) Weigh Station - Both Directions

150 NC 33, Whitakers
Gas W: Texaco*
Food W: Dairy Queen, Stuckey's (Texaco)
Other W: Exit 150 Travel Center (No Trucks, No Turn around)

145 to NC4, to NC48, Goldrock
FStop E: Amoco*, Shell*(D)(LP)
Gas E: BP*(CW), Texaco*
Food E: Denny's, Hardee's, Howard Johnson, Quality Inn, Shoney's, Waffle House, Wendy's (Days Inn)
Lodg E: AAA Best Western, ◆ Comfort Inn, Days Inn ◆ Days Inn, Deluxe Inn, AAA Howard Johnson, Masters Inn, Motel 6, ◆ Quality Inn, Red Carpet Inn, Scottish Inns, AAA Shoney's Inn
ATM E: Cash Point

(142) Rest Area - RR, Vending, Phones, Picnic (No Overnight Parking)

141 NC 43, Red Oak, Dorches
FStop E: I-95 Food Stop*(D)(LP), Texaco*(D)(CW)
Gas W: BP*
Food W: River Restaurant (Holiday Inn)
Lodg W: AAA Holiday Inn
ATM E: Texaco

138 U.S. 64, Tarboro, Nashville, Rocky Mount, Winstead Ave
Food E: Double Back's Old World Eatery (Holiday Inn), Outback Steakhouse
Lodg E: Comfort Inn (see our ad on this page), ◆ Hampton Inn, AAA Holiday Inn (see our ad on this page)
AServ E: Davenport Honda, Isuzu, Pontiac, Cadillac
ATM E: Centura Bank, Triangle Bank

132 to NC 58
FStop E: Citgo*
AServ E: 24-Hr Towing & Repair

127 NC 97, Airport, Zebulon, Rocky Mount
Gas E: Exxon*(D)

121 U.S. 264, Wilson, Greenville, Zebulon
FStop E: Bullets East Coast*(LP) (Kerosene, UPS Boxes)
Gas E: Amoco*(LP), Exxon*(D), Shell*(LP)
W: BP*
Food E: Aunt Sara's Pancakes, Bullets, Burgers, Chicken And More (East Coast), Hardee's, Subway (Exxon)
W: Blimpie's Subs (BP), Cracker Barrel, McDonalds
Lodg E: Comfort Inn
Med E: ✚ Hospital (5 Miles)
ATM E: Amoco, Nationsbank
W: Cashpoint (BP)

116 NC 42, Wilson, Rock Ridge, Clayton
FStop W: Texaco*(D)(LP)
Gas E: Shell*(D)(LP) (Kerosene)
W: Mobil*(D)
AServ W: Bunn's Auto Repair And Sales
RVCamp W: Rock Ridge Campground(LP) (1.9 Miles)
Med E: ✚ Hospital (6 Miles)
ATM W: Texacp

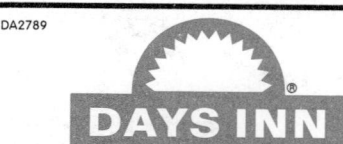

Bold red print shows RV & Bus parking available or nearby

EXIT NORTH CAROLINA

07 U.S. 301, Kenly, Wilson
FStop E: Citgo*
Gas E: Amoco*, BP*(LP), Coastal, Exxon*(D), Texaco*
Food E: Burger King, Golden China, McDonalds, Moore's Bar-B-Que, Nik's Pizza, Patrick's Cafeteria, Subway (Coastal), Waffle House
Lodg E: Budget Inn, Deluxe Inn, AAA Econolodge
AServ E: Big A Kenly Service Center(D) (Exxon), Durham's Garage, Kenly Ford Dealer, Texaco, Top Dog Auto Center
ATM E: Citgo
Other E: 1870 Farmstead, Big A Auto Parts, Food Lion Supermarket, Gov. Charles B. Accott, Tobacco Farm Life Museum

06 Truck Stop Road
TStop W: TravelCenters of America*(SCALES) (Truck Wash), Wilco Travel Plaza*(SCALES)
Gas W: Texaco*(D)
Food W: TravelCenters of America (Country Prod), Waffle House, Wilco Travel Plaza
Lodg W: AAA Best Western, DAYS INN AAA Days Inn
TServ W: TravelCenters of America*(SCALES)
ATM W: TravelCenters of America

05 Bagley Road
TStop E: Citgo Truck and Auto Plaza*(LP)(SCALES) (Kerosene)
Food E: Stormin Norman
TServ E: Bunns Mobil Truck and Trailer Repair
Other E: Big Boys Travel Store (Citgo)

02 Micro
Gas E: BP
AServ E: BP

01 Pittman Road
TStop E: Truck Stop(D)

99) Rest Area - RR, Phones, Picnic, Vending

98 Selma
RVCamp E: KOA Campground(LP) (Dumpsite)

97 U.S. 70, NC39, Selma, Pine Level
FStop E: Citgo*(LP)
W: Amoco*(D), Texaco*(D)(LP)
Gas W: Exxon*(D)
Food E: Denny's, Kathy's*
W: Bojangles, KFC, McDonalds, Mucho's Mexico (Mexican Restaurant), Oliver's Steakhouse, Pizza Hut, Royal Inn, Shoney's, Waffle House
Lodg E: AAA Holiday Inn Express
W: AAA Comfort Inn, DAYS INN Days Inn, Luxury Inn, Masters Inn, Regency Inn, Royal Inn
AServ W: Exxon
TServ W: I-95 Truck Center
Med W: Hospital
ATM E: Citgo
W: Amoco
Parks E: Cape Lookout National Seashore
Other E: J&R Outlets
W: RV & Truck Parking

95 U.S. 70, Smithfield, Goldsboro
FStop W: Speedway*(LP)
Food E: Howard Johnson, Log Cabin Motel, Village Motor Lodge
W: Burger King, Cracker Barrel, Waverly Inn
Lodg W: Waverly Inn

EXIT NORTH CAROLINA

AServ W: Valvoline Quick Lube Center
Other O: Visitor Information
W: Ava Gardner Museum, Factory Stores of America - Smithfield (see our ad on this page)

93 Smithfield, Brogden Road
Gas W: Shell*

90 U.S. 301, U.S. 701, NC 96, Newton Grove
FStop E: Citgo*(D)
W: Phillips 66(D)
Food E: Roz's Country Buffet
Lodg E: Travelers Inn
AServ E: Rick's Towing
W: Ormond's Auto Sales
RVCamp E: Holiday Travel Park
Other E: Bentonville Battle Ground (15 Miles)
W: Highway Patrol

87 Four Oaks (Northbound Reaccess Via Frontage Rd)
Gas W: BP
AServ W: BP

81 Junction I-40, Raleigh, Wilmington

79 NC 50, NC 242, Benson, Newton Grove
FStop E: BP*
Gas E: Blackmon's Gas*, Citgo*
W: Amoco*(D)(CW)(LP), Coastal*(D), Exxon*, Phillips 66*(CW)
Food E: Brothers Famous Subs & Pizza, Golden Corral, Olde South BBQ, Waffle House
W: Burger King (Exxon), China Eight, Hardee's, KFC, McDonalds, Subway (Coastal)
Lodg E: Dutch Inn
W: DAYS INN Days Inn
AServ W: Lube Xpress, Phillips 66
ATM W: BB&T Bank, Coastal, First Citizens Bank
Other E: Food Lion Supermarket
W: Byrd's Supermarket, Coin Car Wash, Kerr Drugs

77 Dennings Road

EXIT NORTH CAROLINA

FStop E: Speedway*(SCALES)
TWash E: Speedway
RVCamp E: Speedway Rite Wash RV Wash
ATM E: Speedway

75 Jonesboro Road
TStop W: Sadler Travel Plaza*(SCALES)
Gas W: Citgo*
Food W: Sadler Restaurant*, Tart's Diner
TServ W: Sadler Travel Plaza

73 US 421, NC 55, Newton Grove
Gas W: Chevron Servico*, Exxon*(D), Texaco*
Food E: Wendy's
W: Bojangle's, Burger King, Dairy Freeze, Taco Bell, Triangle Waffle, Western Steer Family Steakhouse
Lodg W: Day Inn, ◆ Ramada
Med W: Hospital
ATM W: United Carolina Bank

72 Pope Road
Gas E: BP*
W: Amoco*, Chevron*(D)
Food E: Best Western
W: Brass Lantern Steakhouse, Gym's Restaurant
Lodg E: Best Western
W: Econolodge, Howard Johnson

71 Long Branch Road
Gas E: Shell*(D)
Food E: Hardee's (Shell)
TServ E: G&L Truck & Tire Service

70 SR 1811
Lodg E: Relax Inn
AServ E: Interstate Auto
TServ E: Interstate Truck Service

65 NC 82, Godwin, Falcon
Gas W: Citgo*(D), Sam's*
AServ W: Smith's Garage

61 Wade
FStop E: Citgo*
Gas W: BP*
Food E: Dairy Queen (Citgo)
Lodg W: Queen Anne's Motel
AServ E: Citgo
RVCamp E: KOA Campground

(60) Parking Area

58 US 13, Newton Grove
FStop E: Texaco*
Gas E: Citgo*(D) (24 Hrs)
Food E: Tasty World (Days Inn)
Lodg E: DAYS INN ◆ Days Inn
ATM E: Texaco

56 Bus I-95 S, US 301, Fayetteville, Ft. Bragg, Pope AFB (Southbound)
Lodg W: Economy Inn (see our ad on this page)

55 SR 1832
Gas W: Exxon*(D), Texaco*(LP)
Lodg W: Budget Inn

52 NC 24, Fayetteville, Clinton (Difficult Reaccess)
Other W: Botanical Gardens, Military Museum, Museum

49 NC 53, NC 210, Fayetteville
Gas E: Amoco*(D), Exxon*(D), Texaco*(D)(LP)

Bold red print shows RV & Bus parking available or nearby

EXIT NORTH CAROLINA

W: Amoco*(D), Exxon*(D), Shell*(D)

Food E: Burger King, **Lilly's Restaurant & Lounge**, McDonalds, Pizza Hut, Waffle House
W: Beaver Dam Seafood, **Cracker Barrel**, Shoney's, Subway (Amoco)

Lodg E: Days Inn, AAA **Days Inn**, Deluxe Inn, Motel 6, **Quality Inn**
W: AAA Comfort Inn, AAA Econolodge (see our ad on this page, ◆ Hampton Inn (see our ad on this page), AAA **Holiday Inn**, AAA Howard Johnson, Innkeeper, AAA Sleep Inn (see our ad on this page), Super 8 Motel

AServ E: Amoco

ATM E: Amoco
W: Exxon

(48) **Rest Area - RR, Phones, Picnic, Vending (Both Directions)**

46 NC 87, Elizabethtown, Fayetteville

Med W: ✚ **Hospital**

44 Airport, Snow Hill Road

RVCamp W: **Lazy Acres Campground**

41 NC 59, Hope Mills, Parkton

Gas E: Texaco*
W: BP*

RVCamp W: **Spring Valley Park Campground**

40 Bus. I-95N, US 301, Fayettville, Fort Bragg, Pope Air Force (Northbound)

33 U.S. 301, N.C. 71, Parkton

Gas E: Amoco(D)

AServ E: Amoco

31 NC 20, St Paul, Raeford

FStop E: **Shell***

Gas E: Amoco*(LP) (24 Hrs), BP*
W: Citgo*(LP), Exxon*(D)

Food E: Burger King, **Joe Don Danny's (Shell)**

Lodg E: Days Inn Days Inn

TServ E: **Shell**

25 US 301, Local Traffic

(24) **Weigh Station - Both Directions**

22 U.S. 301, Local Traffic

TStop W: **Shell*(D) (see our ad on this page)**

FStop W: **BP*(LP)**

Gas E: Exxon*

Food E: Denny's (see our ad opposite page), **Hardee's**, Huddle House, Ryan's Steakhouse, Waffle House
W: Uncle George's Restaurant

Lodg E: AAA Best Western, ◆ Comfort Suites (see our ad on this page), AAA Hampton Inn, AAA

EXIT NORTH CAROLINA

EXIT NORTH CAROLINA

Holiday Inn, AAA Super 8 Motel

AServ E: Carolina Tire Co.

20 NC 211, to NC 41, Lumberton, Red Springs

Gas E: Amoco*(D), Citgo* (24 Hrs), Exxon*(D)
W: Texaco*(D)(LP)

Food E: Bojangle's, Golden City Chinese, Hardee's, Lane's Icecream Cafe, **Little Caesars Pizza**, McDonalds, New China Restaurant, Shoney's, Subway, Village Station Restaurant, Waffle House, Western Sizzlin'
W: **Cracker Barrel**, Kelly Green's Cafe, Lung Wah Chinese

Lodg E: Howard Johnson, Quality Inn & Suites, ◆ Ramada Limited
W: Days Inn, Econolodge, ◆ Fairfield Inn

Med E: ✚ **Hospital**

Other E: **Food Lion Supermarket, K-Mart (Pharmacy)**, **Winn Dixie Supermarket**

19 Carthage Road

Gas W: Citgo*, Exxon*(D)(CW)

Food E: San Jose

Lodg E: Travelers Inn
W: **Knight's Inn**, Motel 6

AServ E: McGrit's Auto

17 NC 72, NC 711, Lumberdon, Pembroke

FStop E: **Go Gas**

Gas E: BP*, Mobil*(D), Texaco*

Food E: Burger King, Central Park, **Hardee's**, Hot Stuff Pizza, Little China Chinese, Little Sparkie's Burgers, **McDonalds**, Subway, Taco Bell, Waffle House

Lodg E: Southern Inn

AServ E: Advance Auto Parts, BP, Mobil (AServ 24 Hrs), O.K. Auto Service

ATM E: BP, United Carolina Bank

Other E: **Food Lion Supermarket, Revco Drugs**

14 US 74, Maxton, Laurinburg, Wilmington, Whiteville, NC beaches

FStop W: **BP***

Lodg W: Exit Inn

RVCamp W: **Sleepy Bear Campground**

10 US 301, Fairmont

Gas W: Hunts 301*

7 McDonald, Raynham

(5) NC Welcome Center - RR, Phones, Picnic, Vending (Northbound)

2 NC 130, NC 904, Rowland, Fairmont

Bold red print shows RV & Bus parking available or nearby

EXIT NORTH/SOUTH CAROLINA

AServ W: Rowland Motor Co. Chevrolet, Pontiac, Buick

AB US 301, US 501, Rowland, Laurinburg
- **TStop** E: **Porky's Truck Stop (Citgo)*(SCALES)**
- **FStop** W: **Amoco* (24 Hrs), Texaco*(D)**
- **Gas** E: BP, Exxon, Shell*(D); W: Citgo*
- **Food** E: Cafe of the Pink Turtle, Hot Tamale, Peddler Steakhouse, Pedro's Coffee Shop, Pedro's Diner, Pedro's Icecream, Sombrero Restaurant; W: Denny's, **Family Inns, Hardee's,** Waffle House
- **Lodg** E: Family Inns, **South of the Border Motel**; W: (Days Inn logo) Days Inn, Holiday Inn
- **RVCamp** E: **Pedro's Campground**
- **ATM** E: First Citizens Bank, Porky's TStop
- **Other** E: **El Drug Store, Fort Pedro Fireworks, Pedro Land Amusement Park, Rocket City Fireworks, South of the Border, U.S. Post Office**

↑NORTH CAROLINA
↓SOUTH CAROLINA

(195) SC Welcome Center - RR, Phones, Picnic, Vending

193 SC 9, SC 57, Dillon, Little Rock
- **FStop** E: **Speedway*(D)(LP)**
- **Gas** E: Amoco*(CW); W: Amoco*, BP*(D)
- **Food** E: Burger King, Golden Corral, **Huddle House, Shoney's,** Wendy's; W: Hubbard House, **Waffle House**
- **Lodg** E: (AAA) Comfort Inn, (Days Inn logo) (AAA) Days Inn, ◆ Hampton Inn, ◆ Holiday Inn Express; W: Econolodge, (AAA) **Super 8 Motel**
- **TServ** W: **Cottingham Trailor Service**
- **RVCamp** W: **Bass Lake Campground**
- **Other** E: **JABS Fireworks**

190 SC 34, Dillon
- **Gas** E: Citgo*
- **Food** E: Dairy Queen (Citgo), Stuckey's (Citgo)
- **AServ** E: Peewee's Auto Service

EXIT SOUTH CAROLINA

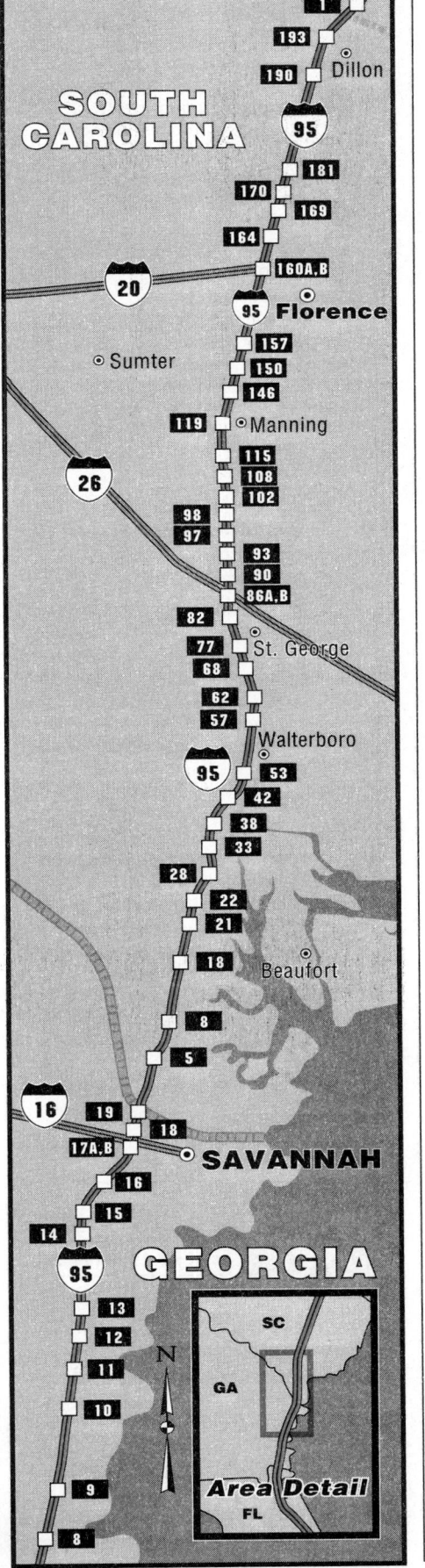

SOUTH CAROLINA — GEORGIA — Dillon — Florence — Sumter — Manning — St. George — Walterboro — Beaufort — SAVANNAH — Area Detail: SC, GA, FL

EXIT SOUTH CAROLINA

181 S.C.38, Marion, Bennetsville, Lattia.
- **TStop** E: **Flying J Travel Plaza*(LP)(SCALES) (RV Dump)**
- **Gas** E: BP*, Exxon*, Texaco*; W: Exxon*
- **Food** E: Subway (Texaco), **Thads Restaurant (Flying J Plaza)**
- **AServ** E: Auto Repair Shop (BP); W: Exxon
- **TServ** E: **Little Chief Complete Truck Repair**
- **TWash** E: **Little Chief**
- **ATM** E: Texaco
- **RVCamp** W: **Camping World (see our ad on this page)**
- **Other** E: **Fireworks Exxon**

(172) **Rest Area - RR, Phones, Picnic, Vending (Northbound)**

170 SC 327, Marion, Myrtle Beach
- **FStop** E: **Speedway***; W: **Shell Auto Truck Plaza***
- **Gas** E: Amoco*
- **Food** E: **Waffle and Egg Family Restaurant**; W: Restaurant (Shell)

169 TV Road, Quimby, Florence
- **TStop** W: **Petro*(CW)(SCALES) (RV Dump)**
- **FStop** W: **Advantage***
- **Gas** W: Amoco*, BP, Texaco*(D)(CW)
- **Food** W: **Iron Skillet Restaurant, Petro**
- **Lodg** W: (AAA) **Quality Inn**
- **AServ** W: BP
- **TServ** E: **Peterbilt Dealer**; W: **Petro**
- **TWash** W: **Petro**
- **RVCamp** E: **KOA Campground**
- **ATM** W: Petro Travel Plaza

164 U.S. 52, Florence, Darlington
- **FStop** W: **Pilot Travel Center*(SCALES)**
- **Gas** E: Exxon*(D)(CW), Racetrac* (24 Hrs), Texaco*(D); W: Amoco*, BP*
- **Food** E: (Cracker Barrel logo) Cracker Barrel, Denny's, Hardee's, McDonalds, Perkins Family Restaurant, Pizza Hut, Quincy's Family Steakhouse, Waffle House, Wendy's; W: Arby's, Bojangles, Burger King, **Dairy Queen (Pilot),** Jose Jose, Shoney's, **Subway (Pilot)**
- **Lodg** E: (AAA) Best Western, (AAA) Comfort Inn, (AAA) **Econolodge,** ◆ Hampton Inn, (AAA) Holiday Inn, (AAA) Motel 6, ◆ Super 8 Motel, (AAA) Travelers Inn; W: (Days Inn logo) Days Inn, ◆ Radisson Inn, (AAA) Ramada, Shoney's Inn, (AAA) **Sleep Inn,** Thunderbird Inn
- **AServ** E: Efird Buick, Pontiac, Jeep Eagle
- **ATM** E: Wachovia

160AB I-20 Business, Florence
- **FStop** E: **Shell*(D)**
- **Food** E: Baskin Robbins, Burger King, Chick-fil-A, Huddle House, Outback Steakhouse, Red Lobster, Ruby Tuesday, Waffle House, Western Sizzlin'
- **Lodg** E: Fairfield Inn, Hampton Inn, Holiday Inn Express, Red Roof Inn
- **AServ** E: Sears, Western Auto
- **ATM** E: NationsBank, Shell
- **Other** E: **Carmike Cinema, Crossroads Center Mall, Magnolia Mall, Wal-Mart (Pharmacy)**

157 U.S. 76, Florence, Timmonsvile
- **Gas** E: BP(D), Exxon*(D), Miller's Amoco(D), Texaco*(D); W: Shell*(D) (24 Hrs)
- **Food** E: Carolina BBQ, McDonalds, Miz Jenny's,

EXIT SOUTH CAROLINA

Swamp Fox Diner, Waffle House
W: Magnolia Dining Room (Young's Plantation Inn), The Tree Room
Lodg E: AAA Days Inn, AAA Howard Johnson, Swamp Fox Inn, Villager Lodge (Extended Stay)
W: Econolodge, AAA Young's Plantation Inn
AServ E: Amoco, BP
RVCamp W: Swamp Fox Camping
Other E: Swamp Fox Fireworks

150 SC 403, Sardis
FStop E: Sardis Auto/Truck Plaza*(D) (BP)
Gas W: Exxon*(D)
Food E: Steak & Waffle (Sardis FS)
TServ E: Sardis Auto Truck Plaza (24 Hr. road service)
RVCamp W: Lake Honeydew
ATM E: Sardis Auto Truck Plaza

146 SC 341, Lake City, Lynchburg, Olanta
Gas E: Exxon*(D)
Lodg E: Relax Inn

141 SC 53, SC 58, Shiloh
FStop E: Exxon*
Gas W: Texaco*
RVCamp E: Don Mar RV Sales & Service
Parks E: Woods Bay State Park

(139) Rest Area - RR, Phones, Picnic, Vending (Both Directions)

135 U.S. 378, Turbeville, Sumter
Gas E: BP*(D), Citgo* (24 Hrs)
Food E: Compass Restaurant, Days Inn
Lodg E: Days Inn, Exit Inn
Other W: Pineview Golf

132 SC 527, Sardinia, Kingstree

122 SC 521, Alcolu, Manning
Gas W: Exxon*(D)(CW)
Food W: Shrimper Sea Food

119 SC 261, Manning, Paxville
TStop E: 76 Auto/Truck Plaza*(LP)(SCALES)
Gas E: Shell*, Texaco*
W: BP*, Exxon (24 Hrs)
Food E: Blimpie's Subs (76 Truck Stop), Huddle House, Long John Silvers, Shoney's, Taco Bell, Waffle House, Wendy's, Western Steer Family Steakhouse
W: Lyle's Subs & More (BP)
Lodg E: AAA Comfort Inn, AAA Economy Inn
W: Alpha Inn
AServ E: Stokes-Craven Ford
W: Clarendon Auto Repair & Sales, Exxon, Spry & Sons Auto Repair
TWash E: 76 Auto/Truck Plaza
RVCamp E: Camper's Paradise (Propane Filling Station)
ATM E: 76 Auto/Truck Plaza
Other E: Crazy Fireworks

115 U.S. 301, Summerton, Manning
FStop W: Texaco*
Gas E: Exxon(D)
Food E: Shanghai Restaurant (Travelers Inn)
W: Days Inn, Georgio's Pizza
Lodg E: Carolina Inn, Travelers Inn
W: Colonial Inn, ◆ Days Inn, Sun Set Inn

EXIT SOUTH CAROLINA

AServ E: Exxon
W: Ram-Bay Auto Repair

108 Road 102, Summerton
Gas E: Citgo, Shell*
W: BP
Food E: Dairy Queen (Citgo), Stuckey's (Citgo)
W: Family Folks Restaurant, Summerton Restaurant, Super 6 Subs
Lodg W: Econolodge, Economy Inn, Knight's Inn, Summerton Inn, Travelers Inn
AServ W: BP
Other E: Loco Joe's Fireworks

102 North Santee
TStop E: Lake Marion Truck Stop*(D) (BP)
Gas E: Texaco*(D)
W: Exxon*
Food E: Restaurant (Santee Resort)
Lodg E: Santee Resort
AServ E: Frankie's Auto Service, Frankie's Truck Repair
RVCamp E: Cooper's Campground, Santee Lakes Campground
Other W: Santee National Wildlife Refuge

(99) Rest Area - RR, Phones, Tourist Info, Vending, Picnic (Both Directions)

98 SC 6, Santee, Eutawville
Gas E: Amoco*, Chevron*(LP), Citgo*, Texaco*(D)(CW)
W: BP*(D), By-Lo*(D), Citgo*, Exxon*(LP), Shell*(D)
Food E: Georgio's House of Pizza, Huddle House, Jake's Steak & Ribs, KFC, Randazzo's Italian, Shoney's, Subway, Verandah, Western Steer Family Steakhouse
W: Burger King, Clark's Family Rest (Clark's Inn), Cracker Barrel, Denny's, Hardee's, McDonalds, Peking Chinese, TCBY Yogurt (Shell), Tastee Food Shop, The Taco Maker, Waffle House, Ziggy's Restaurant (Holiday Inn)
Lodg E: Ashley Inn, AAA Best Western, Carolina Lodge, ◆ Days Inn, AAA Ramada, ◆ Super 8 Motel, Tara Inn
W: Budget Motel, AAA Clarks Inn, AAA Comfort Inn (see our ad on this page), AAA Economy Inn, ◆ Holiday Inn, ◆ Lake Marion Inn, Mansion Park Motor Lodge
AServ W: BP
ATM W: Citgo, First National Bank, Shell, The Bank Of Clarendon
Parks W: Santee State Park

EXIT SOUTH CAROLINA

Other E: Lake Marion Golf Course, Santee Factory Stores
W: Food Lion Supermarket, Revco, The Rivers Country Store

97 U.S. 301, Orangeburg (Southbound, Northbound Reaccess)
AServ W: Avingers Auto

93 U.S. 15, Holly Hill

90 U.S. 176, Holly Hill, Cameron
Gas W: Exxon*

86AB Junction I-26

82 U.S. 178, Harleyville, Bowman
TStop E: Wilco Travel Center*(SCALES) (Exxon, RV Dump)
W: BP*(LP)
Gas E: Amoco*
W: Texaco*(D)
Food E: Dairy Queen (Exxon TStop), Krispy Kreme (Exxon), Stuckey's (Exxon TStop), Wendy's (Exxon TStop)
W: Jerry's Truck Stop Restaurant (BP TStop)

77 U.S. 78, St George, Branchville
Gas E: Exxon*, Shell*(CW)
W: BP*, Texaco*(D)(CW)
Food E: Georgio's Pizza & Restaurant, Griffin's Family Restaurant, Hardee's, KFC (Exxon), McDonalds, Subway (Shell), TCBY Yogurt (Shell), Waffle House, Western Steer Family Steakhouse
W: Huddle House
Lodg E: ◆ Comfort Inn, AAA Economy Inns of America, ◆ Holiday Inn, AAA St George Economy Inn
W: AAA Best Western, Econolodge, Southern Inn, Super 8 Motel
ATM E: Shell
Other E: Crazy Bob's Fireworks, Food Lion Supermarket, Revco Drugs, U.S. Post Office
W: Crazy Bob Fireworks, Food Lion Supermarket, Revco Drugs

(74) Parking Area (Both Directions)

68 SC 61, Canadys, Bamberg
TStop E: Texaco*(SCALES) (RV Dump)
Gas E: Amoco*, Citgo*
Food E: Necee's Restaurant(D) (Texaco), Subway (Texaco)
TServ E: Texaco
ATM E: Texaco
Parks E: Colleton State Park

62 Road 34
Gas W: Exxon*
RVCamp E: Lakeside Campground, Lakeside Used RV Parts

57 SC 64, Walterboro, Lodge
FStop E: Texaco*(D)(CW)
Gas E: BP*, Exxon*, Shell*
W: Amoco*(D)
Food E: Blimpie's Subs (Texaco), Burger King, Dairy Queen (Shell), Huddle House, Olde House Restaurant, Stuckey's (Shell), Subway (BP), TCBY Yogurt (BP), Taco Maker (BP), Waffle House
Lodg E: Carolina Lodge, AAA Howard Johnson, AAA Southern Inn
W: AAA Super 8 Motel
ATM E: Exxon
Other E: Sad Sam's Fireworks

Bold red print shows RV & Bus parking available or nearby

EXIT SOUTH CAROLINA

53 S.C. 63, Walterboro, Varnville, Hampton
- Gas: E: Amoco* (24 Hrs), Citgo*, Econo Mart*(D)(CW), Exxon*, Shell, Texaco* W: BP*
- Food: E: Burger King, Carolina Fried Chicken, Dairy Queen, Huddle House, Joint Venture Seafood, Keith's BBQ, **Longhorn Steaks**, McDonalds, Shoney's, Waffle House
- Lodg: E: AAA **Best Western**, AAA Comfort Inn (see our ad on this page), Econolodge, ◆ Holiday Inn, AAA Rice Planter's Inn, AAA Thunderbird Inn, Town & Country Inn W: DAYS INN ◆ Days Inn, **Deluxe Inn**, ◆ Hampton Inn
- RVCamp: W: **Green Acres Campground**
- ATM: E: Citgo, Exxon W: BP
- Other: E: **Fireworks Supermarket, Hobo Joe's Fireworks, Jolly Joe's Fireworks**

(47) **Rest Area - RR, Phones, Vending, Picnic (Both Directions)**

42 U.S. 21, Yemassee, Beaufort

38 SC 68, Yemassee, Hampton, Beaufort
- TStop: W: **Simon's Truck Stop* (Fina)**
- Gas: E: Chevron* W: BP*, Shell*, Texaco*(D)
- Food: W: **Diner (Simon's Truck Stop)**, Home Cooking, Round House Restaurant, Subway (BP), TCBY Yogurt (BP)
- Lodg: W: **Palmetto Lodge**, ◆ **Super 8 Motel**
- TServ: W: **Simon's Truck Stop**
- TWash: W: **Simon's TStop**
- ATM: E: Chevron

33 U.S. 17 N., Charleston, Beaufort
- Gas: E: BP*(D), Exxon*, Shell*, Texaco*
- Food: E: Daybreak Restaurant, Denny's, McDonalds (Exxon), Point Cafe, Waffle House
- Lodg: E: **Best Western**, DAYS INN Days Inn, ◆ Holiday Inn Express, **Knight's Inn**
- RVCamp: E: **KOA Campground**
- ATM: E: BP, Exxon

28 SC 462, Coosawhatchie, Hilton Head Island
- TStop: W: **Texaco Travel Center*(LP)(SCALES)**
- Gas: W: Amoco*, BP*, Chevron*(D), Citgo*
- Food: W: Dairy Queen, **Restaurant (Texaco Travel Center)**, Stuckey's (Citgo)

EXIT SOUTH CAROLINA

- ATM: W: Texaco Truck Stop

22 U.S. 17 South, Ridgeland
- Gas: W: Shell*
- Food: W: Plantation Restaurant
- Lodg: W: Plantation Motel
- AServ: W: Little T's Wrecker Service
- Med: W: ✚ **Hospital**

21 S.C. 336, Hilton Head Island, Ridgeland
- Gas: W: Chevron*(D), Exxon*, Texaco*(D)
- Food: W: Blimpie's Subs, Depot Restaurant, **Hardee's**, Huddle House, Pizza Station 2, Scoops & Slices, Subway, Waffle House
- Lodg: W: Carolina Lodge, AAA Comfort Inn, ◆ **Econolodge**, The Station Inn
- Other: W: **Eckerd Drugs, Food Lion Supermarket**

EXIT SOUTH CAROLINA/GEORGIA

18 Road 13, Switzerland
- Med: W: ✚ **Hospital**

(17) **Parking Area (Both Directions)**

8 U.S. 278, Hardeeville, Hilton Head
- TStop: E: **Joker Joe's Truck Stop*(D)(SCALES) (BP)**
- Gas: E: Exxon* W: Shell*(LP)
- Food: E: Joker Joe's Truck Stop
- Lodg: W: AAA **Ramada Inn**
- TServ: E: **Joker Joe's**
- Parks: E: **Sgt. Jasper State Park, Hilton Head Island (see our ad on this page)**

5 U.S. 17, U.S. 321, Hardeeville, Savannah
- FStop: W: **Speedway*(D)(LP)**
- Gas: E: Exxon*(LP) W: Amoco*, BP*(D)(CW), Exxon
- Food: E: Blimpie's Subs, Denny's, TCBY Yogurt (Exxon), **Waffle House** W: Budget Inn, **Burger King**, Harry's Sports Bar & Grill, Jasper's Restaurant (Howard Johnson), McDonalds, Shoney's, Subway, The Pizza & Sub Station, **Wendy's**
- Lodg: E: DAYS INN AAA **Days Inn (see our ad on this page)** W: Budget Inn, Carolina Inn Express, AAA Comfort Inn, ◆ Holiday Inn Express, AAA Howard Johnson, Magnolia Motel, Super 8 Motel (see our ad on this page), Thunderbird Lodge (see our ad on this page)
- AServ: W: Exxon, Parts City Auto Parts, Western Auto
- ATM: W: Amoco, Speedway
- Other: E: **Alamo Wholesale Fireworks, Crazy Ed's Fireworks, Fireworks Papa Joe's, Tourist Information** W: **Crazy Joe's Fireworks**

(5) SC Welcome Center - RR, Phones, Vending, Picnic (Northbound)

(3) **Weigh Station - both directions**

↑SOUTH CAROLINA

↓GEORGIA

(111) **GA Welcome Center, Weigh Station - RR, Phones, Vending, Picnic, RV Dump, Tourist Info (Southbound)**

19 **(108)** Georgia 21, Port Wentworth, Savannah

Bold red print shows RV & Bus parking available or nearby

EXIT GEORGIA

TStop E: Speedway*(SCALES) (24 Hrs)
FStop E: Enmark*(D)
W: Smile Gas*(D)(LP)
Gas W: BP*
Food E: Waffle House
W: Big Boy
Lodg E: ◆ Hampton Inn
W: Park Inn (see our ad on this page), Ramada, ◆ Sleep Inn
TServ E: Enmark, Freightliner Dealer, Peterbilt Dealer, Speedway Mechanic & Tire Service
TWash E: Savannag Truck Wash
RVCamp W: Camping

18A **(104)** Savannah International Airport

18 **(102)** U.S. 80, Pooler, Garden City
Gas E: Amoco*(LP), Enmark*(D)(CW)
W: Amoco*(LP), Chevron, Gate*(D)
Food E: Baldino's Subs (Amoco), Blimpie's Subs, Hardee's, Huddle House, KFC, McDonalds, Waffle House
W: Burger King, Don's Famous BBQ, Hardee's, Hong Kong Chinese, Lovezzola's Pizza, Spanky's, Subway (Gate gas), Wendy's, Western Sizzlin'
Lodg E: AAA Comfort Inn, Microtel Inn
W: AAA Econolodge, Ramada
AServ W: Quick Change Quick Lube
ATM W: NationsBank, Wachovia Bank
Other E: Air Force Heritage Museum, Carter's Pharmacy, Food Lion Supermarket, Fort Pulaskie National Monument
W: Coin Laundry

17AB **(99)** Junction I-16, Savannah, Macon

16 **(94)** Georgia 204, Savannah, Pembroke
Gas E: Amoco*(LP), Exxon*, Shell*
W: Chevron* (24 Hrs)
Food E: Cracker Barrel, Denny's (Best Western), Hardee's, McDonalds, Shoney's
W: Giovanni's Italian, Huddle House, Subway, The Shell House Seafood, Waffle House
Lodg E: AAA Best Western, DAYS INN ◆ Days Inn, ◆ Hampton Inn, AAA Holiday Inn (see our ad on this page), AAA La Quinta Inn, AAA Quality Inn & Suites, AAA Shoney's Inn, ◆ Sleep Inn, AAA Travelodge
W: Econolodge, Red Carpet Inn, Super 8 Motel
RVCamp W: Bellaire Woods Camping
Med E: ✚ Hospital
Parks E: Skidaway State Park
Other E: Savannah Festival Outlet Center, Wormsloe Historic Site

15 **(90)** Georgia 144, Fort Stewart, Richmond Hill
FStop W: Texaco*(D)(LP)
Gas E: Amoco*, Exxon*
Food E: Millie's BBQ
TServ W: Roberts Truck Center
RVCamp E: Waterway RV Camp
Parks E: Fort McAllister State Park & Historic Site

14 **(87)** U.S. 17, Coastal Highway
TStop W: Savannah 76 Auto/Truck Stop*(SCALES)
FStop W: Chevron*(LP), Speedway*(LP) (RV Dump), Texaco*(D)
Gas E: Amoco*, Racetrac*, Shell*(D)(CW)
W: BP*(CW)
Food E: Dairy Queen, Denny's, Huddle House, Mike's Pizza, Sandra Dee's Steaks & Seafood, Subway

EXIT GEORGIA

EXIT GEORGIA

(Amoco), Waffle House
W: Arby's, Burger King, Courtesy House Restaurant, KFC, Long John Silvers (76 TStop), McDonalds, Pizza Hut (76 TStop), Taco Bell, Waffle House, Wendy's
Lodg E: DAYS INN Days Inn (see our ad on this page), Motel 6, Royal Inn, Scottish Inns, Travelodge
W: AAA Econolodge (see our ad on this page), ◆ Holiday Inn, Ramada Inn
TServ W: 76
TWash W: Louis' Truck Wash & RV
RVCamp W: KOA Campground
ATM E: Amoco, Racetrac
W: Speedway
Other E: Richmond Hill Georgia Welcome Center

13 **(76)** U.S. 84, Georgia 38, Midway, Sunbury
FStop W: El Cheapo*(SCALES)
Gas W: Amoco*(LP), Shell*(D)
Food W: Holton's Seafood, Huddle House
RVCamp W: Glebe Plantation Camping
Med W: ✚ Hospital
ATM W: El Cheapo
Other E: Fort Morris Historical Site
W: Fort Stewart Museum, Midway Museum

12 **(67)** U.S. 17, South Newport
Gas E: BP*, Chevron, Shell*, Texaco*(D)
Food E: McDonalds (Shell)
RVCamp E: Riverfront RV Camp, South Newport Campground
ATM E: Heritage Bank

11 **(58)** Georgia 57, Georgia 99, Eulonia, Townsend Road
Gas E: Amoco*
W: Amoco*(D), Chevron* (24 Hrs), Shell*(D), Texaco (24 Hrs)
Food E: The Georgia Cracker
W: Eulonia Cafe (Days Inn), Huddle House, JR's Kitchen
Lodg W: DAYS INN Days Inn, Ramada Limited
RVCamp E: Lake Harmony
W: McIntosh Lake Camp

(55) Weigh Station (Both Directions)

10 **(49)** Business 95 Loop, Georgia 251, Darian
FStop W: Texaco*(SCALES)
Gas E: Chevron*, Citgo*(D)
W: Amoco*, Mobil*(D), Shell*(D) (24 Hrs)
Food E: Dairy Queen, McDonalds, Waffle House
W: Burger King, Huddle House, KFC (Texaco)
Lodg W: ◆ Holiday Inn Express, AAA Super 8 Motel
AServ W: Interstate Auto Repair
TServ W: Interstate Truck Repair
RVCamp E: Inland Harbor, Tall Pine Camping
Other W: Magnolia Bluff Factory Shops

9 **(42)** Georgia 99

(40) Tourist Info Center, Rest Area - RR, Phones, Vending, Picnic, RV Dump (Southbound)

8 **(38)** U.S. 17, Georgia 25, Brunswick, North Golden Isles Pkwy
Gas W: Amoco* (24 Hrs), Shell*(D)(LP)
Lodg W: ◆ Econolodge, Guest Cottage, AAA Quality Inn
RVCamp E: Golden Isles Campground

Bold red print shows RV & Bus parking available or nearby

EXIT GEORGIA

Med E: Hospital

7AB **(36)** U.S. 25, U.S. 341, Jesup, Brunswick

FStop E: Shell*(D)

Gas E: Amoco*, Chevron*(CW) (24 Hrs), Exxon*(D), Racetrac*, Texaco*(D)
W: BP*(D), Mobil*(D), Phillips 66*(LP)

Food E: Burger King, Cracker Barrel, International House Of Pancake, KFC, Krystal, McDonalds, Pizza Hut, Shoney's, Taco Bell, Waffle House, Wendy's
W: Captain Joe's Seafood, Denny's, Huddle House, Matteo's Italian, Minh-Sun (Chinese), Quincy's Family Steakhouse, Sonny's BBQ, Waffle House

Lodg E: Budgetel Inn, Days Inn, Hampton Inn, Howard Johnson, Knight's Inn, Ramada
W: Best Western, Comfort Inn, Holiday Inn (see our ad on this page), Motel 6, Sleep Inn, Super 8 Motel

Med E: Hospital

ATM E: Racetrac, Shell
W: Mobil

Other W: Coin Laundry, Revco Drugs, Winn Dixie Supermarket

6 **(29)** U.S. 17, U.S. 82, Georgia 520, Brunswick, S. Ga. Pkwy.

TStop E: Pilot Travel Center*(D)(SCALES)
W: Flying J Travel Plaza*(LP)(SCALES) (RV Dump), TravelCenters of America(SCALES) (BP)

FStop E: Texaco*(D)
W: El Cheapo*(D)(SCALES) (Shell Gas)

Gas E: Amoco*(D), Exxon*
W: Citgo*, Phillips 66*(LP)

Food E: Huddle House, Oyster Box, Steak 'N Shake (Pilot), Subway (Pilot)
W: Country Market (Flying J), Country Pride (TC of America), Dairy Queen (TC of America), Magic Dragon, Super Slice, Waffle House

Lodg E: Holiday Inn (see our ad on this page)
W: Daystop Inn (TravelCenters of America), Super 8 Motel

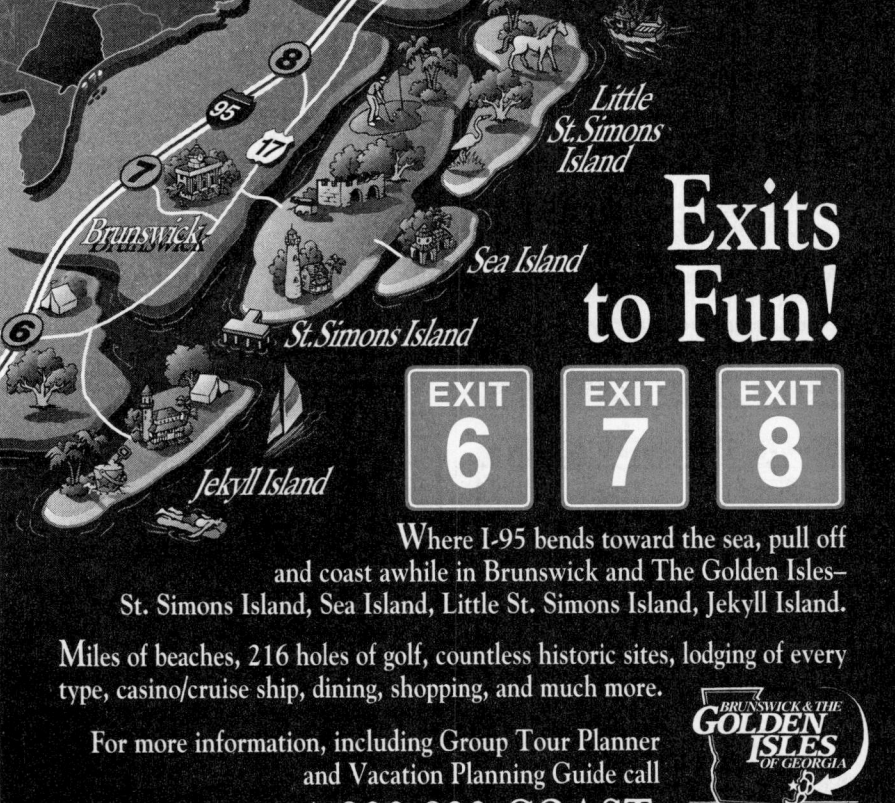

TServ W: Five Star Truck Quick Lube, Truckstop Of America

TWash W: Five Star Truck Wash, TravelCenters of America

Other E: Chamber of Brunswick (see our ad on this page)

5 **(26)** Dover Bluff Road

Gas E: Shell*(D)(LP)

Food E: Choo Choo BBQ

RVCamp E: Ocean Breeze Camp (2 Miles)

Other E: Rawl's Gift Shop (Shell)

4 **(14)** Georgia 25, Woodbine

FStop W: Sunshine Plaza*(D)(SCALES)

Gas W: Exxon*

Food W: Jack's BBQ, Sunshine Grill

3 **(7)** Harriett's Bluff Road

FStop E: Texaco*(D)

Gas E: Shell*(LP)

Food E: Angelo's Italian, Huddle House, Jack's Famous BBQ, Taco Bell (Texaco)

TServ E: Texaco

Other E: King George RV

2A **(6)** Colerain, St Mary's Road

FStop E: Cone*

TServ E: Cisco

ATM E: Cisco

Other E: Arby's (Cisco), Cisco Travel Plaza*(D)(SCALES) (BP Gas)

2 **(3)** Georgia 40, Kingsland, St Mary's

FStop E: Enmark(D)

Gas E: BP*, Chevron*, Mobil*, Shell*
W: Exxon*(D)

Food E: Applebee's, Burger King, Dairy Queen, Dynasty Chinese, Hardee's, KFC, McDonalds, Osaka Japanese, Ponderosa, Rose Garden Chinese, Shoney's, Shorty's BBQ, Sonnys BBQ, Subway (Shell), Waffle House, Wendy's
W: Bennigan's

Lodg E: AAA Comfort Inn, AAA Days Inn, AAA Hampton Inn, Peachtree Inn, Super 8 Motel
W: ◆ Econolodge, AAA Holiday Inn

AServ E: Xpress Lube

Med E: Hospital

ATM E: BP, Chevron

Parks E: Crooked River State Park

Other E: Revco Drugs, Winn Dixie Supermarket

1 St.Mary's Road

Gas W: BP*

RVCamp W: KOA Campground

EXIT		GEORGIA/FLORIDA
(1)		GA Welcome Center - RR, Phones, Picnic, Vending, RV Dump, Tourist Info (Northbound)

↑GEORGIA

↓FLORIDA

(380)		Weigh Station, Agricultural Inspection (Both Directions)
130		**(379)** U.S. 17
	Gas	W: Amoco*, Shell* (24 Hrs)
	Food	W: Amandeo's Italian (Days Inn)
	Lodg	W: Days Inn, ◆ Holiday Inn Express
	AServ	W: Shell
	RVCamp	E: Hance's RV Park(LP)
(378)		FL Welcome Center - RR, Phones, Vending, Picnic (Southbound)
129		**(373)** FL A1A, Callahan, Fl. 200, Fernandina Beach
	TStop	W: Sunshine Truck Plaza*(SCALES)
	Gas	E: Citgo*, Shell*
		W: BP*, Citgo*, Exxon*(D)
	Food	E: Dairy Queen, McDonalds, Stuckey's (Citgo)
		W: Sunshine Restaurant (Sunshine Truck Plaza), Waffle House, Wayfarer Restaurant (Citgo)
	AServ	E: Shell
	TServ	W: Sunshine Truck Plaza
	RVCamp	E: Bow & Arrow Campground, Fort Clinch State Park
	ATM	W: Citgo
128		**(365)** Pecan Park Road
	FStop	E: Citgo*
	Gas	E: BP*(D)
	Food	E: Ice Cream Churn
127AB		**(363)** Duval Road, Jacksonville International Airport
	Gas	E: Mobil*(D)
		W: BP*, Chevron*(D), Texaco*(D)
	Food	W: Denny's, Dunkin Donuts, Subway, Taco Bell, Waffle House
	Lodg	W: Admiral Benbow Inn, Airport Motor Inn, Courtyard by Marriott, Days Inn, Hampton Inn, Holiday Inn, Jax Airport Hotel, Red Roof Inn, Valu-Lodge
	AServ	E: Mobil
126A		**(362)** FL 9A, Blount Island
126B		**(362)** Junction I-295, to I-10
125		**(359)** FL 104, Dunn Ave, Busch Drive
	FStop	E: Gate*(D)
	Gas	W: Hess*, Shell*(CW) (24 Hrs), Texaco* (24 Hrs)
	Food	E: Applebee's, Hardee's, Riverview Seafood & Steaks, Shane's Sandwich Shop, Waffle House
		W: Arby's, Bono's BBQ, Burger King, Captain D's Seafood, Dunkin Donuts, KFC, Krystal, Long John Silvers, McDonalds, Pizza Hut, Popeye's Chicken, Quincy's Family Steakhouse, Rally's Hamburgers, Shoney's, Sonny's BBQ, Taco Bell, Wendy's
	Lodg	E: Admiral Benbow Inn
		W: (AAA) Best Western, Jaguar Inn, (AAA) La Quinta Inn, Motel 6, Super 8 Motel
	AServ	W: Discount Auto Parts, Pep Boys Auto Center
	ATM	E: First Union, Gate
		W: Barnett Bank, First Union
	Other	E: Sam's Club
		W: Eckerd Drugs, Publix Supermarket, Winn Dixie Supermarket
124B		**(358)** Broward Road, FL. 105, Heckscherd Dr.
	Food	W: Red Horse Restaurant
	Lodg	W: Days Inn

EXIT		FLORIDA
124A		**(358)** FL 105, Heckscher Dr (Exit To Ltd. Access)
123		**(356)** FL 111, Edgewood Ave
	Gas	W: Amoco(CW) (24 Hrs), Citgo*
	AServ	W: Amoco
	ATM	W: Amoco
122AB		**(356)** FL 115, Lem Turner Road
	Gas	W: Amoco*(CW) (24 Hrs), Hess
	Food	E: Hardee's
		W: Famous Sandwiches, Golden Eggroll, Krystal, Popeye's
	AServ	W: Discount Auto Parts, Goodyear Tire & Auto, Jiffy Lube, Meineke Discount Mufflers, Midas Muffler & Brakes, One Stop Auto Parts, Wilbun's Garage
	Other	E: Gateway Animal Hospital
		W: Star Coin Car Wash, WalGreens
121		**(354)** Golfair Blvd
	Gas	E: Shell* (24 Hrs), Texaco*
		W: Amoco* (24 Hrs), Exxon*
	Lodg	W: Valu-Lodge
120AB		**(354)** U.S. 1, 20th St (Difficult Reaccess)
119		**(353)** FL 114, 8th St
	Gas	E: Amoco*
	Food	E: McDonalds
	Med	E: ✚ University Medical Center
118		**(353)** U.S. 23 North, Kings Road
117		**(352)** Union St (Northbound)
116		**(352)** Church St., Myrtle Ave., Forsyth St.
115		**(352)** Monroe St
114		**(352)** Myrtle Ave
113		**(352)** Stockton St
112		**(351)** Margaret St

EXIT		FLORIDA
111		**(351)** Junction I-10 West
110		**(351)** College St (Closed For Construction)
109		**(351)** U.S. 17, Riverside Dr.
108		**(350)** San Marco Blvd
107		**(350)** Prudential Dr, St. Mary's St., Main St.
	Med	E: ✚ Wolfson Children Hospital
	Other	E: Museum of Science & History, Southbank River Walks & Museum
106		**(349)** U.S. 90 East, Beach Blvd (Eastbound)
105		**(349)** U.S. 1 South, Phillips Highway
	Gas	W: Amoco*(CW)
	Lodg	W: City Center Motel, Scottish Inns
	AServ	W: Jerry Hamn Chevrolet, O' Steen Volvo
104		**(347)** FL.126, Emerson St
	Gas	E: Chevron, Shell* (24 Hrs)
		W: Amoco*(CW) (24 Hrs), BP*(D), Exxon*(CW)
	Food	W: McDonalds, Taco Bell
	Lodg	W: Comfort Inn, Emerson Inn, Howard Johnson (see our ad on this page)
	AServ	E: Chevron, Discount Auto Parts, Shell
	ATM	W: First Union
	Other	E: Food Lion Supermarket, U.S. Post Office, WalGreens (Pharmacy)
		W: Skate World
103AB		**(347)** FL 109, University Blvd. (Southbound)
	Gas	E: Amoco* (24 Hrs), Texaco*
		W: BP*, Racetrac*
	Food	E: Captain D's Seafood, Happy Garden Chinese, Krystal, Ying's Chinese
		W: Buckingham Grill, Burger King, Dunkin Donuts, IHOP, Mel's Ocean Bay Seafood, Ryan's Steakhouse, Shoney's, Sonny's BBQ, Taco Bell, Waffle House
	Lodg	W: Comfort Lodge, Econolodge, Ramada, Red Carpet Inn (see our ad on this page)
	AServ	E: Midas Muffler & Brakes, One Stop Auto Parts, Tire Kingdom
	ATM	E: SunTrust, Texaco
	Other	W: University Blvd. Animal Hospital
102		**(345)** FL109, Bowden Road, University Blvd (Northbound)
	Gas	E: Chevron*, Gate*(D)
	Food	E: Bono's BBQ, Larry's Giant Subs, Wedge Cafe
	AServ	E: Edward's Automotive
101		**(343)** FL.202, Butler Blvd, Jacksonville
	Food	E: Vito's (Quality Inn)
	Lodg	E: ◆ Marriott, (AAA) Quality Inn, Comfort Inn (see our ad on this page)
	Med	E: ✚ St Luke's Hospital

Bold red print shows RV & Bus parking available or nearby

EXIT FLORIDA

100 **(341)** FL 152, Baymeadows Road

Gas E: Chevron*(CW) (24 Hrs), Gate*(D)(CW), Shell(D), Texaco*(D)(CW)
W: Shell*(CW) (24 Hrs), Texaco*(D)

Food E: Arby's, Hardee's, Quincy's Family Steakhouse, T.G.I. Friday's, Waffle House
W: Bennigan's, Bombay Bicycle Club, Burger King, Denny's, KFC, McDonalds (Play Land), Pizza Hut, Red Lobster, Rio Bravo, Steak & Ale

Lodg E: AAA AmeriSuites (see our ad on this page), AAA Embassy Suites, Fairfield Inn, ◆ Holiday Inn
W: AAA Best Inns, Comfort Inn, ◆ Homewood Suites, AAA La Quinta Inn, Motel 6, ◆ Residence Inn, ◆ Travelodge

AServ E: Shell, Xpress Lube

ATM E: NationsBank, SunTrust
W: Compass Bank, First Union Bank

99 **(339)** Southside Blvd, Fl. 115

98 **(338)** U.S. 1

Food E: Arby's, Burger King (Play Land), McDonalds (Play Land), Olive Garden, Taco Bell, Waffle House

Other E: The Avenues Mall

97 **(338)** Jct I-295 N, Orange Park

(331) **Rest Area - Picnic, RR, HF, Phones (Both Directions)**

96 **(329)** County Road 210

TStop E: G & M 76 Auto Truck Plaza*(SCALES) (BP)

Gas E: Citgo*, Exxon*
W: Amoco*, Chevron*(D), Texaco*

Food E: G & M TS, Waffle House

AServ W: Amoco, Chevron

TServ E: G & M TS

RVCamp E: KOA Campground

Other W: Phantom Fireworks

95A International Golf Pwky.

95 **(317)** FL 16, St. Augustine

FStop E: Speedway*(D)(SCALES)

Gas E: Amoco* (24 Hrs), BP*, Citgo*, Shell*
W: Exxon*, Texaco*(D)

Food E: Burger King (Playground), McDonalds (Play

EXIT FLORIDA

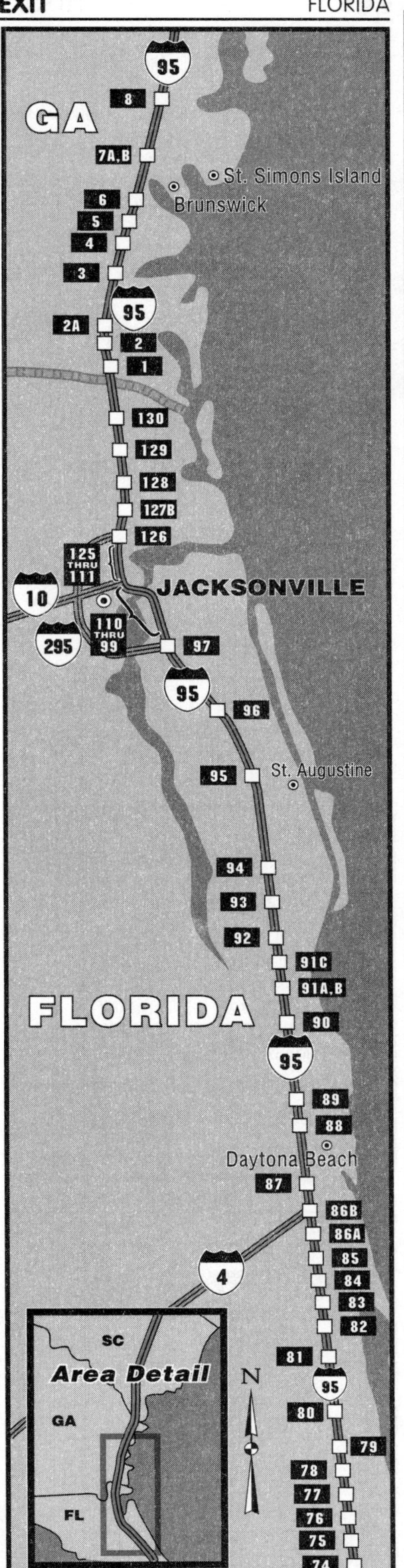

EXIT FLORIDA

Land), Ocean Palace Chinese Buffet, Subway, Waffle House
W: Cracker Barrel, Denny's, KFC, Shoney's, Wendy's

Lodg E: Guesthouse Inn, Holiday Inn Express, Thriftlodge (see our ad on this page)
W: AAA Best Western, Days Inn ◆ Days Inn, AAA Econolodge, Scottish Inns, AAA Super 8 Motel

AServ E: Amoco, Shell
W: Exxon

RVCamp W: Stagecoach RV Park

Other W: St. Augustine Outlet Center (see our ad on this page)

94 **(311)** FL.207, St. Augustine

Gas E: Chevron*(LP) (24 Hrs), Gas*
W: Gas For Less*(D), Texaco*

Lodg W: AAA Comfort Inn

AServ E: Gas
W: Gas For Less, Texaco

RVCamp E: St. John's RV Park

Med E: ✚ Hospital

Parks E: Anastasia State Park

93 **(305)** FL.06, Hastings, Crescent Beach

Gas E: Texaco*(D)

AServ W: Continental Auto Service (24 Hr. Towing)

TServ W: Continental Truck Service (24 Hr. Towing)

RVCamp W: Continental RV Service (24 Hr. Towing)

Other E: Marineland of Florida (13 Miles)

(303) **Rest Area - Picnic, RR, HF, Phones (Southbound)**

(302) **Rest Area - Picnic, RR, HF, Phones (Northbound)**

92 **(298)** U.S. 1, St. Augustine

TStop W: Charlie T's Truck Stop*(SCALES) (76)

Gas E: BP*(D), Citgo*, D & L Groves*, Texaco*(D)
W: Shell*(D)

Food W: Charlie T's TS, Dairy Queen (Shell), Stuckey's (Shell), Waffle House

Lodg W: Charlie T's

AServ E: Texaco

Parks E: Favor Dykes State Park

91C **(289)** Palm Coast

Gas E: BP*, Exxon*(D), Shell*(CW)
W: BP, Chevron*(D)(LP)

Food E: Caruso's Family Restaurant & Pizzaria, Cracker Barrel, Denny's, KFC, McDonalds (Play Land), Shoney's, The Sub Base
W: China One, Perkins Family Restaurant, Sonny's Pit BBQ, Steak 'N Shake, Subway, Taco

EXIT FLORIDA

Bell, Taco Bell (Chevron)

Lodg **E:** ◆ Hampton Inn, ◆ Sleep Inn

AServ **E:** Exxon
W: BP, Discount Auto Parts, Jiffy Lube, Tire Kingdom

ATM **E:** Barnett, First Union Bank, SouthTrust
W: NationsBank (Winn Dixie)

Other **E:** Book Rack, Eckerd Drugs, Publix Supermarket
W: K-Mart, Wal-Mart Supercenter, WalGreens, Winn Dixie Supermarket (Pharmacy)

(286) Weigh Station - both directions

91AB **(284)** FL. 100, Bunnell, Flagler Beach

Gas **E:** BP*, Chevron*(D)(CW)

Food **E:** Burger King (Playground), McDonalds (Playground), Oriental Garden, Subway, Taco Bell (BP)

AServ **W:** Flagler Chrysler, Plymouth, Dodge, Jeep, Eagle

Med **W:** ✚ Hospital

ATM **E:** Barnett Bank

Other **E:** Animal Hospital, Winn Dixie Supermarket

90 **(279)** Old Dixie Highway

Gas **W:** Texaco*

Lodg **W:** AAA Best Western

RVCamp **E:** Bulow Resort & Campground
W: Holiday Trav-L-Park

Parks **E:** Tomaka State Park

Other **E:** Bulow Plantation State Historic Site

89 **(273)** U.S. 1

FStop **E:** Fina*(SCALES), Mobil*

Gas **E:** Amoco*, BP*, Citgo*
W: Exxon*(D)

Food **E:** Baskin Robbins, Denny's, McDonalds (Play Land), Waffle House, Wendy's (Mobil FS)
W: Dairy Queen, J & J Barn (Ramada)

Lodg **E:** AAA Comfort Inn
W: Budget Host Motel, DAYS INN AAA Days Inn, ◆ Ramada Inn, AAA Super 8 Motel

AServ **E:** Citgo

TServ **E:** Fina Truck, Tire, CB Service

RVCamp **E:** Sunshine RV Park

Med **E:** ✚ Hospital

ATM **E:** Mobil

88 **(268)** FL 40, Ormond Beach

Gas **E:** BP*, Citgo*(D)
W: Amoco*, Chevron*(CW), Mobil*(D)

Food **E:** Applebee's, Boston Market, Chili's, Denny's, Little Caesars Pizza, Regent's Court Cafe, Steak 'N Shake, Subway, Taco Bell (Citgo), Waffle House
W: Burger King (Amoco)

Lodg **E:** Sleep Inn

AServ **E:** Xpress Lube

Med **E:** ✚ Hospital

ATM **E:** Barnett Bank, SunTrust Bank

Parks **E:** Tomoka State Park

Other **E:** K-Mart (& Pharmacy), Publix Supermarket (Pharmacy), Wal-Mart Supercenter

87C LPGA Blvd., Holly Hill

EXIT FLORIDA

Other **W:** LPGA International

87AB **(262)** U.S. 92, Daytona Beach, DeLand

Gas **E:** Chevron*(D) (24 Hrs), Hess*(D), Mobil*, Racetrac*
W: BP*(CW), Citgo, Super 8 Motel

Food **E:** Blimpie's Subs (Hess), Burger King, Fazoli's Italian Food, Krystal, Pizza Hut Express (Hess), Red Lobster, S & S Cafeteria, Shoney's, Waffle House
W: Denny's, McDonalds (Play Land)

Lodg **E:** Golden Eagle Motel, Days Inn (see our ad on this page)Holiday Inn, La Quinta Inn
W: DAYS INN Days Inn (see our ad on this page)

RVCamp **W:** Town & Country RV Campground

Other **W:** Mark Martin's Klassix Auto Museum

86AB **(258)** FL. 400 East, South Daytona, Jct. I-4, West Orlando

Lodg **E:** Ramada Limited

85 **(256)** FL. 421, Port Orange

Gas **E:** Amoco*(CW)
W: Citgo*, Shell*(LP)

Food **E:** Spruce Creek Pizza
W: McDonalds (Play Land), Subway

AServ **W:** Xpress Lube

ATM **W:** Barnett Bank, Commercial National Bank

Other **W:** Publix Supermarket (W/ Pharmacy)

(255) Rest Area - RR, Phones, Vending, Picnic (Southbound)

(254) Rest Area - RR, Phones, Vending, Picnic (Northbound)

EXIT FLORIDA

84AB **(249)** FL. 44, DeLand, New Smyrna Beach

Gas **E:** Shell*
W: Chevron*

RVCamp **E:** KOA Campground

Med **E:** ✚ Hospital

83 **(244)** FL. 442, Edgewater, Oak Hill

Gas **E:** Chevron* (24 Hrs)

TServ **E:** Florida Shores Truck Center

82 **(231)** CR. 5A, Oak Hill, Scottsmoor

FStop **E:** BP*(D)

Gas **W:** Citgo*(D)

Food **E:** Stuckey's (BP)
W: Dairy Queen (Citgo)

RVCamp **E:** Crystal Lake Campground

(227) Rest Area - RR, Phones, Vending, Picnic (Southbound)

(225) Rest Area - RR, Phones, Vending, Picnic (Northbound)

81 **(223)** FL. 46, Sanford, Mims

FStop **W:** Mobil*

Gas **W:** Shell*

Food **E:** McDonalds

AServ **E:** Precision Auto Care
W: Shell

Med **E:** ✚ Hospital

ATM **W:** Mobil

80 **(220)** FL. 406, Titusville

FStop **W:** Texaco*

Gas **E:** BP*(D), Shell*

Food **E:** Taste of China (Travelodge)

Lodg **E:** Travelodge

Med **E:** ✚ Hospital

Other **E:** Marritt Island National Wildlife Refuge, National Seashore Wildlife Refuge

79 **(215)** FL. 50, Titusville, Orlando, Kennedy Space Center

FStop **E:** Space Shuttle Fuel*(D)

Gas **E:** Amoco*(D)(LP), Chevron*, Circle K*, Shell*, Texaco*

Food **E:** Denny's, McDonalds (Playground), Shoney's, Waffle House, Wendy's
W: Cracker Barrel, Fama's Italian Bistro (Days Inn)

Lodg **E:** AAA Best Western, AAA Ramada
W: DAYS INN AAA Days Inn, Luck's Way Inn

AServ **E:** Shell

ATM **E:** Texaco

Other **E:** Wal-Mart (Pharmacy, & Optical Department)

78 **(211)** FL. 407, Kennedy Space Center

(209) Parking Area - Picnic (Both Directions)

77 **(205)** FL 528, Fl. 528 Toll, Canaveral, Cape - Port A.F.S., Orlando

Lodg **E:** Howard Johnson

76 **(202)** FL.524, Canaveral, Cape - Port - A.F.S.

Gas **E:** Citgo*(D)
W: Amoco*(D) (24 Hrs)

Lodg **W:** DAYS INN AAA Days Inn

75 **(201)** FL. 520, Cocoa, Orlando

FStop **E:** BP*(D), Speedway*(LP)

Gas **E:** Fina*

EXIT	FLORIDA
	W: Chevron*(D), Shell*
Food	**E:** IHOP, **Subway (Speedway FS)**, Waffle House
	W: McDonalds
Lodg	**E:** (AAA) Best Western, Budget Inn, (AAA) Cocoa Inn
	W: **Holiday Inn (see our ad on this page)**
Med	**E:** + **Hospital**
ATM	**E:** BP
Other	**W:** **Cocoa Stadium**
74	**(196)** Fiske Blvd, Rockledge
Gas	**E:** Citgo*
RVCamp	**E:** **Space Coast RV Resort**
73	**(191)** Satellite Beach, Patrick AFB, CR. 509
Gas	**E:** Citgo*
	W: Texaco*(D)
Food	**E:** Denny's, McDonalds (Playground), Miami Sub's, Wendy's
	W: Burger King, **Cracker Barrel**, Long John Silvers (Texaco)
Lodg	**E:** (AAA) Comfort Inn (see our ad on this page)
	W: ◆ Budgetel Inn
Other	**E:** **Brevard Zoo**
(189)	**Rest Area - Picnic (Both Directions)**
72	**(183)** FL. 518, Melbourne, Indian Harbor Beach
Gas	**E:** Amoco*(D)(LP), Citgo*, Racetrac*
ATM	**E:** Citgo, Racetrac
71	**(180)** U.S.192, West Melbourne
Gas	**E:** Circle K*, Citgo*, Mobil*(CW), Speedway*(LP)
	W: Texaco*(D)
Food	**E:** Denny's, IHOP (York Inn), Shoney's, Waffle House
Lodg	**E:** Holiday Inn, Shoney's, Melbourne Beach Hilton (see our ad this page), Travelodge, York Inn, Qualtiy Suites (see our ad on this page)
AServ	**E:** Mobil, Space Coast Saturn
Med	**E:** + **Hospital**
ATM	**E:** Circle K*
Other	**E:** **International Airport, Sam's Club**
70A	**(176)** CR. 516, Palm Bay
Gas	**E:** Fina*(LP)
Food	**E:** Denny's, Taco Bell (Fina)
ATM	**E:** Albertson's Grocery, Fina
Other	**E:** **Albertson's Grocery (& Pharmacy)**
70	**(173)** FL. 514, Palm Bay, Malabar
FStop	**W:** **Speedway*(LP)**
Gas	**E:** Cumberland Farms*, Speedway*(D)
	W: Shell* (see our ad on this page)
Food	**E:** McDonalds (Playground), Starvin' Marvin
	W: Arby's, IHOP, Starvin' Marvin (Speedway),

EXIT	FLORIDA
	Subway, Taco Bell, **Waffle House**, Wendy's
AServ	**E:** Meineke Discount Mufflers, Mobil 10 Min. Oil Change
	W: Discount Auto Parts, Tire Kingdom
Med	**E:** + **Hospital**
ATM	**E:** First Federal Osceola
	W: Barnett Bank, Riverside National Bank
Other	**W:** **Eckerd Drugs, Publix Supermarket (Pharmacy)**
(169)	**Rest Area - RR, Phones, Vending, Picnic (Southbound)**
(168)	**Rest Area - RR, Phones, Vending, Picnic (Northbound)**
69	**(156)** Fellsmere Road, CR. 512
Gas	**E:** Amoco*, Citgo*(D)
Food	**E:** Dairy Queen (Citgo), McDonalds, Stuckey's (Citgo)
RVCamp	**E:** **KOA Campground, Sunshine RV Park, Whispering Palms RV Park**
Med	**E:** + **Hospital**
68	**(146)** FL 60, Vero Beach, Lake Wales
TStop	**E:** **Amoco Vero Beach Travel Center*(D)(SCALES)**
FStop	**E:** **Citgo***
Gas	**E:** Chevron, Citgo*, Mobil*, Shell*
Food	**E:** **Courtesy Restaurant**, Waffle House, Wendy's
	W: **Cracker Barrel**, McDonalds
Lodg	**E:** (AAA) **Days Inn, Holiday Inn, Howard Johnson, Super 8 Motel**
AServ	**E:** Chevron, Fulmer Brothers Auto, Mobil, NAPA Auto Parts
TServ	**E:** **Amoco Truck Stop, Fulmer Brothers Truck Service**
ATM	**E:** Citgo, Shell
Other	**W:** **Horizon Outlet Center**
67	**(137)** FL. 614, Indrio Road
(133)	**Rest Area - RR, Phones, Vending, Picnic (Both Directions)**
66AB	**(131)** FL. 68, Orange Ave.
Gas	**E:** The Grove*(D)
AServ	**E:** NAPA Auto Parts, St. Lucy Battery & Tire
TServ	**E:** **Peterbilt Dealer**
RVCamp	**W:** **Roadrunner Travel Resort RV Park**
Parks	**E:** **Fort Pierce Inlet State Park**
65	**(129)** FL. 70, Okeechobee Road
FStop	**W:** **Amoco*(SCALES)**
Gas	**W:** Chevron*(CW), Exxon*, Mobil* (24 Hrs), Texaco*(CW)
Food	**E:** Applebee's, Golden Corral
	W: Burger King, **Cracker Barrel**, Denny's, Dunkin' Donuts, KFC, McDonalds, Miami Subs, Perkins Family Restaurant, Red Lobster, **Shoney's**, Subway (Mobil), Taco Bell, **Waffle House**
Lodg	**W:** **Days Inn, Econolodge**, Holiday Inn Express, Motel 6, Treasure Coast Inn
AServ	**E:** Goodyear Tire & Auto
TServ	**E:** **Goodyear Tire & Auto**
Med	**E:** + **Hospital**
Other	**E:** **Wal-Mart**
	W: **R Manufacturers Outlet Center**
64	**(125)** County Road 712, Midway Road
Gas	**E:** Fina*(D)
Food	**E:** Subway (Fina)
Other	**E:** **Fairwinds Public Golf Course**
63C	**(121)** St. Lucie West Blvd.

EXIT		FLORIDA
	Gas	E: Chevron*(CW) (24 Hrs), Shell*(D)
	Food	E: McDonalds, Subway (Shell), Wendy's
	Other	E: **St. Lucie Sports Complex (NY Mets Training Camp)**
63AB		**(117)** Gatlin Blvd
	Med	E: ✚ **Hospital**
62		**(110)** CR. 714, Fl. 714, Martin Hwy.
(106)		**Rest Area - RR, Phones, Vending (Both Directions)**
61C		**(102)** CR. 713, Stuart, Palm City
61		**(100)** FL. 76, Stuart, Indiantown
	Gas	E: Chevron* (24 Hrs), Mobil*(D)
	Food	E: **Cracker Barrel**, McDonalds
	Lodg	E: Holiday Inn (see our ad on this page)
	Med	E: ✚ **Hospital**
60		**(96)** CR. 708, Hobe Sound
59AB		**(87)** FL. 706, Jupiter, Okeechobee
	Gas	E: Chevron*(CW) (24 Hrs), Mobil*, Shell*(CW), Texaco*(CW)
	Food	E: Applebee's, Bresler's Icecream & Yogurt, China Dragon Chinese, IHOP, KFC, Little Caesars Pizza, **McDonalds (Play Land)**, Shoney's
	Med	E: ✚ **Hospital**
	ATM	E: Bank Atlantic, Barnett Bank, First Union
	Other	E: **Eckerd Drugs, Winn Dixie Supermarket**
58		**(84)** Donald Ross Road
57C		**(79)** CR. 809, Military Trail (Southbound, Difficult Reaccess)
57		**(80)** FL. 786, PGA Blvd.
	Gas	E: Mobil*
	Food	E: China Wok Garden
	Lodg	E: ◆ Marriott
	Med	E: ✚ **Hospital**
56		**(77)** North Lake Blvd
	Gas	E: Shell, Texaco*
		W: Amoco*, Chevron*(CW), Hess*, Mobil*
	Food	E: Arby's, Checkers, McDonalds, Taco Bell, Wendy's
		W: Pizza Hut, RJ Gator's Restaurant, Subway
	Lodg	W: (AAA) Inns of America
	AServ	E: Ed Morse Chrysler, Plymouth, Jeep, Eagle, Oldsmobile, Isuzu, Shell, Wallace Mercury, Lincoln
		W: Don Olson Tire & Auto Center
	ATM	E: SunTrust
		W: Great Western Bank
	Other	W: **Gardens Animal Hospital, Publix Supermarket, Winn Dixie Supermarket**
55		**(75)** Blue Heron Blvd.
	Gas	E: Amoco*(D)(CW) (24 Hrs), Shell*(D) (24 Hrs)
		W: Cumberland Farm, Mobil*(D)(CW), Racetrac*, Texaco*
	Food	W: Blimpie's Subs (Texaco), Burger King (Playground), Denny's, McDonalds
	Lodg	E: **Motel 6**, Days Inn (see our ad on this page)
		W: ◆ Super 8 Motel
	AServ	E: Bennett Auto Supply, J&M Tire & Auto Service, Tidewater Transmission
	TServ	E: **Mack Trucks**
	ATM	E: First Union
54		**(74)** County Road 702, 45th St
	Food	E: Burger King, Hong Kong Cafe, IHOP
		W: **Cracker Barrel**, Wendy's
	Lodg	E: ◆ Days Inn, Knight's Inn

Cape Canaveral
Cocoa Beach
Area Detail
FL
N
95
74 73 72 71 70A 69 68 67 66 65 64 63C 63A,B 62 61 60 59A,B 58 57A,B 56
Vero Beach
Fort Pierce
FLORIDA
55 THRU 48
West Palm Beach
47 THRU 42
41 THRU 38
37A,B
36 THRU 29
27 26A 26
Fort Lauderdale
75
595
25 THRU 17
16
14 THRU 1
MIAMI
Miami Beach

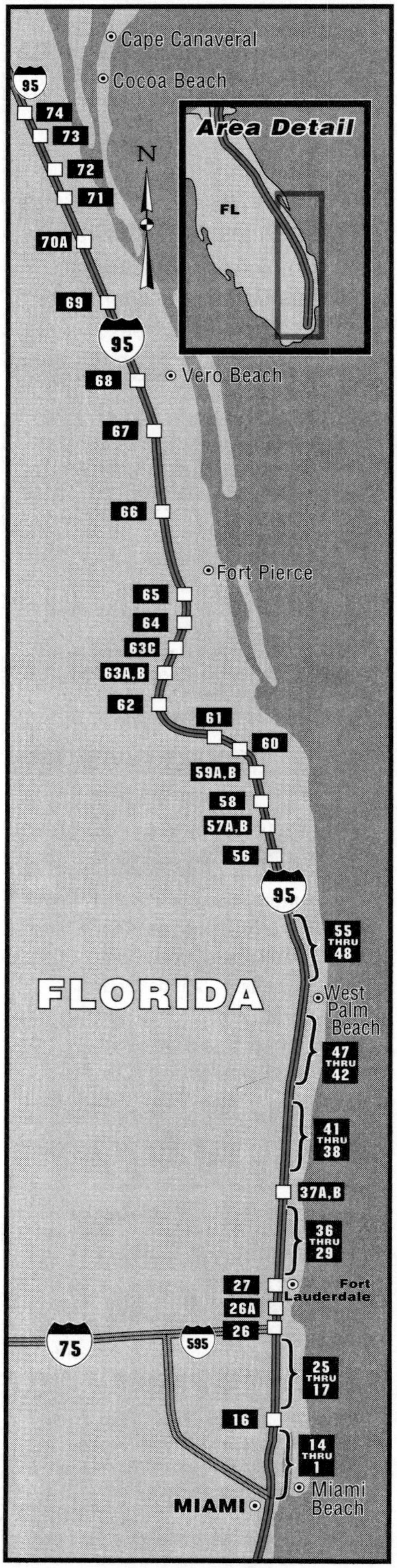

EXIT		FLORIDA
		W: ◆ Courtyard by Marriott, ◆ Red Roof Inn
	AServ	E: Schooley Cadillac
	Med	E: ✚ **Hospital**
53		**(71)** Palm Beach Lakes Blvd.
	Food	E: King Chinese Buffet
		W: Bancock House Thai Restaurant, Durango Steakhouse, Jerry's Family Restaurant, Morrison Cafeteria, No Anchovies, Olive Garden
	Lodg	E: (AAA) Best Western (Palm Beach Lakes)
		W: (AAA) Comfort Inn, (AAA) Wellesley Inn (see our ad on this page)
	AServ	E: Firestone Tire & Auto, Jiffy Lube
	ATM	E: First Union, Great Western Bank
		W: Barnett Bank, Capital Bank
	Other	E: **Palm Beach Mall, Target Department Store, The Mall Cinema**
		W: **Vision Works**

Bold red print shows RV & Bus parking available or nearby

EXIT FLORIDA

52AB **(70)** Fl. 704
- **Gas** W: Chevron*[D][CW] (24 Hrs), Good Way Oil, Shell*, Texaco*
- **Food** W: Aleyda's, McDonalds, Shells Seafood, Subs
- **Lodg** E: Sheraton (see our ad on this page)
- **AServ** W: Arrigo Dodge Dealership, Oil Connection Fast Change Oil, Palm Beach Lincoln, Mercury, Pat's Tire, Roger Dean Chevrolet, Texaco, West Palm Mitsubishi
- **Other** E: **Museum & Fine Arts Center**
  W: Car Wash, **United Artist Theaters**

51 **(68)** Belvedere Road
- **Gas** E: Amoco* (24 Hrs), Fina*[D], Texaco*
  W: Shell*
- **Food** E: McDonalds
  W: Denny's, Phillips Seafood, Shoney's
- **Lodg** W: ◆ Hampton Inn, (AAA) Holiday Inn
- **AServ** E: Congress Auto Parts, Fina

50 **(68)** U.S. 98, Southern Blvd, International Airport
- **Gas** E: Chevron*[D], Texaco*
- **Food** E: Capri
  W: Veranda (Hilton)
- **Lodg** W: ◆ Hilton
- **AServ** E: Chevron, Oil Well
- **Med** E: ✚ **Hospital**
- **ATM** E: First Union
- **Other** E: **DeSoto Pharmacy, Eckerd Drugs, Publix Supermarket, Science Museum, U.S. Post Office, Zoo**

49 **(66)** Forest Hill Blvd

EXIT FLORIDA

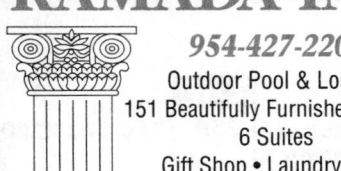

48 **(65)** 10th Ave North
- **Gas** W: BP*, Texaco*[D][LP]
- **Food** W: 10th Ave. Cafe, Boomer's Cafe, Dunkin' Donuts, Julio's Cafe, Lindburgers, Roxy's Cafe, Wendy's
- **AServ** W: Joe's Auto Parts, Wayne Akers Ford
- **Med** W: ✚ **Hospital**
- **ATM** W: Great Western Bank
- **Other** W: **Eckerd Drugs, Winn Dixie Supermarket**

47 **(63)** 6th Ave. S.
- **Med** W: ✚ **Hospital**

46 **(64)** Cty Road 812, Lantana Road
- **Gas** E: Shell*[D]
- **Food** E: Family Rotisserie Chicken, McDonalds, Subway
- **Lodg** E: Motel 6
- **ATM** E: First Union, Great Western Bank
- **Other** E: **Coin Laundry, Eckerd Drugs, Florida Hwy. Patrol Station, Publix Supermarket**
  W: **Costco**

45 **(60)** Hypoluxo Road
- **FStop** W: **Shell***[D]
- **Gas** E: Amoco* (24 Hrs), Mobil*, Texaco*[D]
- **Food** E: Pizza Hut, Shoney's, Taco Bell
- **Lodg** E: (AAA) Inns of America, Super 8 Motel
- **AServ** E: Don Olson Tire & Auto Center, Palm Beach Tire Inc., Tire Kingdom

44C **(59)** Gateway Blvd

44 **(57)** Boynton Beach Blvd
- **Gas** E: Texaco[LP]
  W: Texaco[LP]
- **Food** E: Captain Franks Seafood (take-out)
  W: Checkers Burgers, Waffle House, Wendy's
- **Lodg** E: (AAA) Holiday Inn Express
- **AServ** E: Texaco
  W: Texaco*
- **Other** E: **Coin Laundry**

43 **(56)** Woolbright Road
- **Gas** W: Racetrac*
- **Food** W: Burger King, **Cracker Barrel**
- **Med** E: ✚ **Hospital**
- **ATM** W: Racetrac

42AB **(53)** FL. 806, Atlantic Ave.
- **Gas** W: Chevron*[CW]
- **Food** W: Subway
- **AServ** W: Bennett Auto Supply, Eastern Auto Care
- **Med** W: ✚ **Hospital**
- **Other** W: **Atlantic Animal Hospital**

41 **(51)** Linton Blvd
- **Gas** W: Shell*
- **Food** E: Dairy Queen, McDonalds, Outback Steakhouse
  W: Blimpie's Subs (Shell), Bono's BBQ
- **AServ** E: Wallace Chrysler Plymouth, Nisson
  W: Goodman's Auto Service, NAPA Auto Parts
- **ATM** W: First Union Bank
- **Other** E: **Target Department Store**
  W: **Winn Dixie Supermarket**

40C **(50)** Congress Ave
- **Lodg** W: ◆ Residence Inn

40 **(48)** FL. 794, Yamato Road
- **Gas** E: Mobil*[CW]
- **Lodg** W: ◆ Embassy Suites

EXIT FLORIDA

- **Other** E: **Pharmacy**

39 **(45)** FL. 808, Glades Road
- **Gas** W: Amoco* (24 Hrs)
- **Food** W: California Pizza Kitchen, Hooters, Houston's, Macaroni Grill, Mario's Italian, Nick's, Ruby Tuesday
- **Lodg** W: ◆ Courtyard by Marriott, (AAA) Holiday Inn, Sheraton Inn
- **Med** E: ✚ **Hospital**
- **ATM** W: Amoco, Barnett Bank, NationsBank

38 **(44)** Palmetto Park Road
- **Gas** E: Chevron*[D]
- **Food** E: Denny's, Pizza Time, Subway
- **ATM** E: Sun Trust Bank
- **Other** E: **K-Mart, Publix Supermarket, U.S. Post Office**

37 **(43)** FL 810, Hillsboro Blvd
- **Gas** E: Amoco*[CW] (24 Hrs), Texaco*[D]
  W: Chevron*, Mobil*[CW]
- **Food** E: Clock Family Restaurant
  W: Denny's
- **Lodg** E: (AAA) Hilton, (AAA) La Quinta Inn
  W: Ramada Inn (see our ad on this page), Village Lodge, Wellesley Inns
- **Parks** E: **Tivoli Nature Park**

36C **(45)** FL. 869 (Toll), Southwest 10th St., to I-75 Naples
- **Lodg** W: (AAA) Comfort Suites, Quality Inn

36 **(43)** FL 834, Sample Road
- **Gas** E: Shell*[D], Texaco*[D]
  W: Chevron, Mobil*, Texaco*[D][CW]
- **Food** E: Bejing Express Oriental Cooking, Four Corners of Europe Restaurant, Hong Kong House
  W: McDonalds (Play Land), Metro Pizzeria, Sub's Miami
- **AServ** E: Shell
  W: Chevron, Mobil
- **Med** E: ✚ **North Broward Medical Ctr**
- **ATM** E: Commercial Bank of Florida, First Union
- **Other** E: **Swifty Coin Laundry**
  W: **Eckerd Drugs**

35AB **(39)** Copans Road
- **Gas** E: Amoco* (24 Hrs)
  W: Amoco*[D][CW]
- **AServ** E: Champion Audi
- **Other** E: **Pep Boys Auto Center, Wal-Mart**

34AB **(36)** FL. 814, Atlantic Blvd
- **Food** E: KFC, Miami Sub's, Taco Bell
- **Med** E: ✚ **Hospital**
  W: ✚ **Hospital**

33B **(34)** CR. 840W., Cypress Creek Road
- **Gas** E: Amoco*[CW]
  W: Hess*[D], Shell*
- **Food** W: Bennigan's
- **Lodg** E: Hampton Inn
  W: Doubletree Suites, ◆ Marriott
- **AServ** E: Jiffy Lube
  W: Shell
- **Med** E: ✚ **Hospital**
- **ATM** W: First Union
- **Other** E: **Coin Laundry**
  W: **Pearl Vision, Regal Cinema's**

33A CR. 840E., Cypress Creek Rd.

EXIT		FLORIDA
32		(36) FL. 870, Commercial Blvd
	Gas	W: Amoco*(CW), Coastal*(D), Mobil*(D), Shell*
	Food	E: Subway
		W: Dunkin Donuts, KFC, McDonalds, Sub's Miami, Waffle House
	Lodg	E: Holiday Inn (see our ad on this page)
		W: Holiday Inn (see our ad on this page), ◆ Red Roof Inn
	AServ	W: Brake World, Discount Auto Parts, Kelly Tires, Precision Tune & Lube
	Med	E: Hospital
31B		FL. 816W., Oakland Park Blvd.
	Gas	E: Chevron*(CW) (24 Hrs), Citgo(D), Mobil*
		W: BP*, Hess*(D)
	Food	E: Denny's, Sub's Miami, Wendy's
		W: Dunkin Donuts
	Lodg	W: Days Inn
	AServ	E: Citgo, Goodyear Tire & Auto, Meineke Discount Mufflers, Oakland Center Auto
	ATM	E: Chevron
		W: Capital Bank
	Other	E: Swifty Coin Laundry
31A		Fl. 816E., Oakland Park Blvd.
30B		(29) FL. 838W., Sunrise Blvd
	Gas	E: Amoco*(CW) (24 Hrs), Hess*(D)(CW), Texaco*
		W: Shell*(CW)
	Food	E: Burger King, Popeye's Chicken
		W: Captain Crab's, Church's Fried Chicken, McDonalds
	AServ	E: Discount Auto, O.K.'s Complete Auto Center, Smalley's Tire & Auto
	ATM	E: Amoco
		W: Shell
	Other	E: Coin Laundry, Hugh Taylor Birch State Recreation Area

EXIT		FLORIDA
30A		Fl. 838E., Sunrise Blvd.
29		(28) FL. 842, Broward Blvd, Downtown
	Gas	E: Amoco*
		W: Texaco*
	Food	E: Daybreak (Days Inn), Jack's Grill
		W: Checkers Burgers
	Lodg	E: Days Inn
	AServ	E: Precision Tune & Lube
	Med	W: Hospital
	Other	W: Coin Laundry, Eckerd Drugs
28		FL. 736, Davie Blvd
	Gas	E: Speedy Food Store*
		W: Amoco*, Hess(D), Mobil*(CW)
	Food	W: Shanghai Gardens, Subs Miami, Subway
27C		(29) Junction I-595, Point Everglades (Ft Lauderdale Int'l Airport)
27		(29) FL 84
	Gas	E: Chevron*, Citgo*(D), Coastal*(D), Mobil*, Texaco
	Food	E: Li'l Red's Cookin', McDonalds
	Lodg	E: Best Western, Budget Inn, Motel 6
		W: Cross Roads Motor Park, Ramada Inn, Red Carpet Inn
	AServ	E: Mobil, Texaco
	ATM	E: Coastal
26A		(29) Junction I-595 to Fl. Turnpike I-75
	Lodg	E: AmeriSuites (see our ad on this page)
26		FL. 818, Griffin Road
	Gas	W: Amoco*(D), Citgo(D)(CW)
	Food	E: Moon Dance (Wyndam Hotel)
	Lodg	E: ◆ Hilton, Wyndham Garden Hotel
	ATM	E: Amoco

EXIT		FLORIDA
25		(26) Stirling Road (Cooper's City & Grand Prix)
	Food	E: Burger King (Play Land), Little Caesars Pizza, Taco Bell
		W: Juicy Lucy's Restaurant, Subs Miami, Subway
	Lodg	E: AAA Comfort Inn, AAA Hampton Inn
	AServ	E: Discount Auto Parts
	Parks	E: John U. Lloyd State Park
	Other	E: Barnes & Noble, K-Mart (Pharmacy), PetsMart
24		(25) FL. 822, Sheridan St
	Gas	E: Mobil*(D)
		W: Shell*(CW) (24 Hrs)
	Food	E: Blimpie's Subs (Mobil)
	Lodg	W: ◆ Days Inn, ◆ Holiday Inn (see our ad on this page)
	ATM	W: Shell
	Parks	W: Topeekeegee Yugnee State Park
23		FL. 820, Hollywood Blvd
	Gas	E: Shell*
	Food	E: IHOP, McDonalds, Subs Miami, The Old Steakhouse
	Lodg	E: HO JO Inn
	AServ	E: Goodyear Tire & Auto
	Med	W: Hospital
	Other	E: Eckerd Drugs
22		FL. 824, Pembroke Road
	Gas	E: Shell*
	AServ	E: Vinnie's Tires
21		FL. 858, Hallandale Beach Blvd
	Gas	E: Amoco*, Citgo*(D), Hess*, Shell*
		W: Racetrac*
	Food	E: Blimpie's Subs (Shell), Burger King, Denny's, Dunkin Donuts, IHOP, KFC, Little Caesars Pizza,

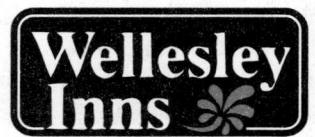

Bold red print shows RV & Bus parking available or nearby

:XIT FLORIDA

Long John Silvers, McDonalds, Smokehouse BBQ
Lodg E: ◆ Holiday Inn Express
AServ E: Authority Muffler & Brake, Precision Tune & Lube
W: Discount Auto Parts
ATM E: Citgo, Racetrac
Other E: **Coin Laundry, Winn Dixie Supermarket (Pharmacy)**

20 **(19)** Ives Dairy Road, Northeast 203rd St.
Gas W: Amoco* (24 Hrs)
AServ W: Joel's Auto Service
Other W: **Ives Plaza Animal Clinic**

9 **(16)** Miami Gardens Dr., FL. 860

8 **(15)** U.S. 441
Med E: **✚ Hospital**

7 **(15)** North 167th Street (Reaccess Northbound Only)
Gas E: Amoco*, Chevron*, Citgo*, Shell* (24 Hrs)
Food E: Dunkin Donuts, KFC
Lodg E: Holiday Inn, Howard Johnson
AServ E: Chevron
Med E: **✚ Parkway Medical Plaza**

6 **(15)** FL. 826 W., Tnpk. N. (Reaccess Northbound Only)

EXIT FLORIDA

15 **(13)** Northwest 151st St. (Reaccess Northbound Only, Difficult Reaccess)
Food W: Jah-Nets Jamaican, May Fu Chinese Restaurant, McDonalds
AServ W: Discount Auto Parts
Other W: **Super Saver Discount Drugs, Winn Dixie Supermarket**

14 **(12)** Northwest 135th St., Opa Locka Blvd., FL 916
Gas W: Amoco(CW), BP*, Chevron*, Mobil*
Food W: Dunkin' Donuts, Pizza Hut, Subway
Lodg W: Motel 7
AServ W: Amoco, BP, Damian Tires, Goodyear Tire & Auto, Howie's Tires, Rose Auto Stores

13 **(11)** Northwest 125th St, North Miami, Bal Harbor
Gas W: Shell*
Food W: Burger King, Royal Castle, Wendy's
Lodg E: Howard Johnson (see our ad on this page)
AServ W: Shell
ATM W: Great Western Bank

12 Northwest 119th St., Fl. 924
Gas W: Amoco*(CW)
Food W: BBQ Barn, Jimmy's Place, Popeye's
Other W: Car Wash, **Eckerd Drugs**

11 **(9)** Northwest 103rd St , Miami Shores
Gas E: Shell*(CW)
W: Amoco, Chevron*, Mobil*
Food W: Cesar Discount Cafe, Dunkin Donuts
Lodg E: Ramada (see our ad on this page)
AServ W: Amoco, Mad Hatter Mufflers, Perfect Auto Parts

10 **(8)** Northwest 95th St
Gas W: Mobil*, Shell* (24 Hrs)
Food W: Burger King
AServ W: Discount Auto Parts
Med W: **✚ Hospital**
Other W: **WalGreens (Pharmacy)**

9 **(7)** NW. 79th St., NW. 81st St., FL. 934
Gas E: BP*, Chevron*(D) (24 Hrs)
W: Shell*
Food E: Waffle House
W: Cafe China, Checkers Burgers
Lodg W: Days Inn (see our ad on this page)
AServ E: Midas Muffler & Brakes, NAPA Auto Parts
W: Colonial Chrysler, Jeep, Eagle, Miami Lincoln Mercury

8 **(6)** Northwest 62nd St., Northwest 54th St.
Food W: McDonalds

EXIT FLORIDA

Other W: **Winn Dixie Supermarket**

7 **(5)** Junction I-195, FL.112, Airport
Lodg W: Days Inn (see our ad on this page)

6 **(4)** FL. 836, Airport
Med W: **✚ Hospital**

5A **(3)** Northwest 8th St, Orange Bowl, Port of Miami

5 **(3)** I-395, Miami Beach
Lodg W: Howard Johnson (see our ad on this page)

4 **(2)** Northwest 2nd St

3A Downtown, Miami Ave. (Southbound)

3 **(2)** U.S. 1, Biscayne Blvd, Downtown

2 Southwest 8th St, U.S. 41
Gas E: Chevron
W: Shell*
Food E: McDonalds, Wendy's

1 Rickenbacker Causeway, Biscayne Blvd
Lodg W: Holiday Inn (see our ad on this page), Days Inn (see our ad on this page)

↑FLORIDA

Begin I-95

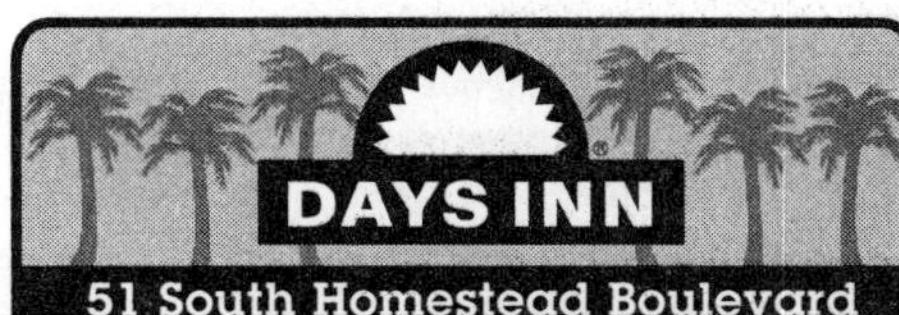

Bold red print shows RV & Bus parking available or nearby

EXIT	MICHIGAN
	Begin I-96

↓MICHIGAN

1AB U.S. 31, Ludington, Grand Haven
- Gas — **N:** Mobil*

4 Airline Road
- Gas — **S:** Wesco*(D)
- Food — **S:** Burger Crest, Pizza Reaction, Subway
- AServ — **S:** Fruitwood Automotive
- ATM — **S:** First of America
- Parks — **S:** **PJ Hoffmaster State Park**
- Other — **S:** **Farmer Dave's General Store, Fruitport Foods Grocery, Pleasure Island Amusement**

5 Fruitport (Westbound, Eastbound Reaccess)

(8) **Rest Area - RR, Phones, Picnic (Westbound)**

9 MI 104, Spring Lake, Grand Haven (Eastbound, Westbound Reaccess)
- Parks — **S:** **State Park**

10 CR B31, Nunica (Difficult Reaccess)

16 CR B35, Eastmanville
- FStop — **N:** **Speedway***; **S:** **Pacific Pride Commercial Fueling**
- Gas — **N:** Amoco*(CW), Shell*(D)
- Food — **N:** **Arby's**, Burger King, **McDonalds**, Subway, Taco Bell
- Lodg — **N:** (AAA) AmeriHost Inn
- AServ — **N:** Amoco
- RVCamp — **N:** **Fun N Sun RV Center (Parts & Service)**
- ATM — **N:** Comerica Bank
- Other — **N:** **Rite Aide Pharmacy, Shop Rite Super Market**

19 Lamont, Coopersville
- Gas — **N:** 76*(D)
- Food — **S:** Sam's Joint
- RVCamp — **N:** **Prime Time RV**; **S:** **Vanandell**(LP)

23 Marne
- Gas — **N:** Shell*(D) (Kerosene)
- Food — **S:** Rinaldi Pizza
- AServ — **N:** Schneider Tire Service, Shell
- ATM — **S:** Comerica Bank

25 Eighth Ave, 4 Mile Road (Westbound, Eastbound Reaccess)
- Gas — **S:** Marathon*

(25) **Rest Area - RR, Phones, Picnic (Eastbound)**

26 Fruit Ridge Avenue
- FStop — **S:** **Amoco***(LP)
- Gas — **N:** Sunoco*; **S:** Pacific Pride Commercial Fueling(D)
- AServ — **S:** Mr. Bill's Tire & Auto
- ATM — **S:** Amoco

28 Walker Avenue
- Gas — **S:** Meijer*(D)
- Food — **S:** Blimpie's Subs (Meijer), McDonalds
- Lodg — **S:** (AAA) AmeriHost

30AB MI 37 North, Alpine Avenue, Newaygo
- Gas — **N:** Amoco*(CW), Marathon*(D), Shell*; **S:** Admiral*, **Meijer Grocery***(D)(LP), Total*(D)
- Food — **N:** Blimpie's Subs, Chuck E. Cheese's Pizza, Clock FIne Food, Cooker, **Cracker Barrel**, First Wok Chinese, Little Caesars Pizza, McDonalds, Old Country Buffet, Olive Garden, Outback Steakhouse, Paunchy Pete's Mexican, Perkins Family Restaurant, Russ, Ryan's Steakhouse, Steak & Shake, Subway, Taco Bell, Three Happiness Chinese, Village Inn Pizza Parlor; **S:** Arby's, Burger King, KFC, Labate's Ristorante, Long John Silvers, McDonalds (Meijer), McDonalds, Ole Tacos, Pizza Hut (Carry-Out), Ponderosa, Wendy's
- Lodg — **S:** Motel 6
- AServ — **N:** Alpine Automotive Supply Clinic, Belle Tire, Car Quest Auto Center, Keller Ford, Pep Boys Auto Center; **S:** Goodyear Tire & Auto, Midas Muffler & Brakes, Valvoline Quick Lube Center
- ATM — **N:** Ameribank, Comerica Bank, First of America; **S:** Comerica Bank, FMB Bank, Michigan National Bank, Old Kent Bank
- Other — **N:** **Builders Square 2, K-Mart (Pharmacy), Office Depot, Office Max, Wal-Mart (Pharmacy)**; **S:** **Alpine Coin Laundry, Meijer Grocery*, U-Haul Center**(LP)

31AB U.S. 131, Kalamazoo, Cadillac

33 MI 44 Connector, Plainfield Avenue
- Gas — **N:** Four Star, Total*; **S:** Amoco*(CW)(LP)
- Food — **N:** Arby's, Burger King, Charlie's Grill, Domino's Pizza, McDonalds, Pizza Hut, Pizza Hut Express, Taco Boy, Wendy's; **S:** Bill Knapp's Restaurant, Denny's
- Lodg — **N:** Lazy T Motel, Plainfield Inn, President Inn
- AServ — **N:** Ace's Transmission, Auto Lab (Total), Goodyear Tire & Auto, Meineke Discount Mufflers, Midas Muffler & Brakes, NAPA Auto Parts, NTB, Pennzoil Lube, Speedy Quik Oil Change, Top Value Muffler Shops, Valvoline Quick Lube Center, West Michigan Transmission
- RVCamp — **N:** **Camps & Cruise**
- Med — **N:** **+ Family Physicians Urgent Care, + Plainfield Medical Center**
- Other — **N:** Great Lakes Car Wash, **Rite Aide Pharmacy**, Touch of Class Car Wash, **U-Haul Center**(LP)

36 Leonard Street
- Other — **N:** **Sheriff's Dept**

37 I-196 Junction (left lane exit) Gerald R Ford Pkwy

38 MI 37 South, MI 44, MI 21, East Belt Line Ave
- Gas — **N:** Amoco*(CW)
- Food — **S:** Dulba's Fine Dining
- RVCamp — **N:** **Camping**
- ATM — **N:** Old Kent Bank
- Other — **N:** **Ski Area**

40AB Cascade Road
- Gas — **N:** BP*, Marathon*; **S:** Shell, Speedway*(D)
- Food — **N:** Forest Hills Inn, Hud's Italian; **S:** Sigee's Restaurant (Harley)
- Lodg — **S:** Harley Hotel
- AServ — **N:** BP; **S:** Shell
- Med — **S:** **+ Hospital**
- ATM — **N:** Old Kent Bank, United Bank
- Other — **N:** **7-11 Convenience Store**

43AB MI 11, 28th Street, Kent County Airport, Cascade
- Gas — **N:** Marathon*; **S:** MSI*, Mobil*(CW)
- Food — **N:** Big Boy, Brann's Steakhouse, Burger King, Gippers, Shanghai Garden, Sundance Grill; **S:** Applebee's, Arby's, Bob Evans Restaurant, Carloso' Kelly's, McDonalds, Perkins Family Restaurant, Rio Bravo, Spinnaker (Hilton)
- Lodg — **N:** Budgetel Inn, Country Inn Suites, Crowne Plaza Hotel, (Days Inn) Days Inn, Lexington Suites, New England Suites Hotel; **S:** Econolodge, Excel Inn, Hampton Inn, Hilton, Red Roof Inn
- AServ — **S:** Car Quest Auto Center, Pennzoil Lube
- ATM — **N:** First of America
- Other — **N:** **Meijer Grocery, Wal-Mart**

(45) **Rest Area - RR, Phones, Picnic, Vending (Westbound)**

52 MI 50, Lowell
- RVCamp — **S:** **Camping**

59 Clarksville

(63) **Rest Area - RR, Phones, Picnic (Eastbound)**

64 Lake Odessa, Saranac
- Parks — **N:** **Ionia State Park**

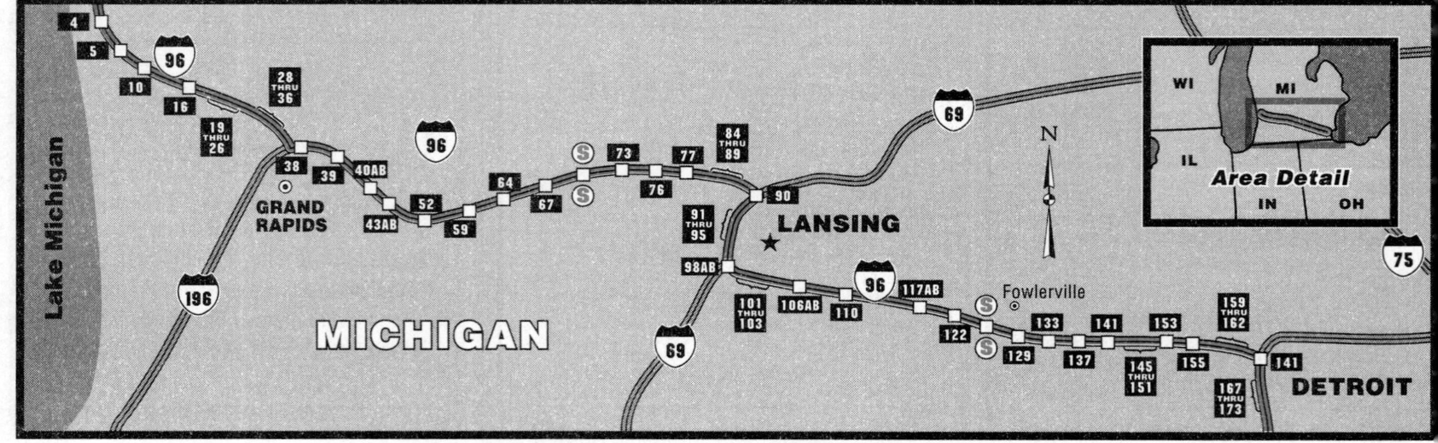

Bold red print shows RV & Bus parking available or nearby

EXIT MICHIGAN

67 MI 66, Ionia, Battle Creek
- **FStop** N: Total*(SCALES)
- **Gas** N: Amoco*
- **Food** N: Scalehouse Restaurant, Subway (Total)
- **Lodg** N: Midway Motel, (AAA) Super 8 Motel
- **TServ** S: I-96 Towing & Recovery
- **TWash** S: I-96 Truck Wash
- **RVCamp** N: Camping
- **Med** N: ✚ Hospital
- **ATM** N: Scalehouse
- **Other** N: Sheriff's Dept, State Patrol Post

(69) Rest Area (Both Directions)

73 Lyons - Muir, Grand River Avenue

76 Kent Street (No Trucks)

77 Grand River Avenue, Junction Business 96, Portland
- **FStop** N: Speedway*(LP)
- **Gas** N: Amoco*(CW), Shell*(CW), United*(LP) (Kerosene)
- **Food** N: Arby's, Burger King, Diana's Italian, McDonalds, Subway (Shell), Tommie's
- **Lodg** N: (AAA) Best Western
- **AServ** N: Quicklube, Sawyer's Ford/Mercury
- **ATM** N: FMB Bank, Independent Bank
- **Other** N: Coin Car Wash, Rite Aid Pharmacy, Tom's Grocery, Tom's Hardware

(79) Rest Area - RR, Phones, Picnic (Westbound)

84 Westphalia, Eagle

86 MI 100, Right Road
- **FStop** S: Total*
- **Food** S: Taco Bell (Total)
- **ATM** S: Total

(87) Rest Area - RR, Phones, Picnic (Eastbound)

90 Grand River Avenue
- **Other** N: Airport

91 Junction I-69 North, U.S. 27, Flint, Clare

93AB Junction I-69 Business , MI 43, Saginaw Highway
- **Gas** N: Shell*(CW), Total*(D)
  S: Amoco*(CW)
- **Food** N: Burger King, Denny's, Frank's Press Box, Hoffman House (Best Western), McDonalds, T.G.I. Friday's (Holiday)
  S: Cracker Barrel, Dunkin Donuts (Amoco)
- **Lodg** N: Best Western, Fairfield Inn, Hampton Inn, Holiday Inn, Motel 6, Quality Inn, Red Roof Inn, Residence Inn
- **AServ** S: Regency Olds/ GMC Trucks/Mazda
- **Med** N: ✚ Hospital, ✚ Westside Medical Center
- **ATM** N: Michigan National Bank
  S: Amoco

95 Junction I-496, Downtown Lansing

97 Junction I-69 South, Charlotte, Ft. Wayne

98AB Lansing Road
- **TStop** S: Citgo*
- **Food** S: Windmill Restaurant (Citgo)
- **Lodg** S: Citgo Hotel, Windmill Motel
- **AServ** S: Cooper Tires
- **TServ** S: Citgo
- **TWash** N: General
  S: Citgo
- **Other** S: State Patrol Post

101 MI 99, Logan Street, Eaton Rapids
- **Gas** N: Bay*(D), QD*
- **Food** N: Los Gringo's, Poppa Leo's Pizza
  S: McDonalds, Pizza To Go
- **Parks** S: Grand River Park

EXIT MICHIGAN

- **Other** N: Mario's Market, Pleasant Grove Car Wash
  S: Rich's Country Store

104 Business Loop 96, Cedar Street, Holt
- **Gas** N: Meijer*, Shell*, Speedway*
  S: Speedway*(D), Total*
- **Food** N: Bill Knapp's Restaurant, Blimpie's Subs, Bob Evans Restaurant, Brewsters (Best Western), Country Skillet, Days Inn, Denny's, Finley's Family Dining, Flapjack, KFC, Kewpee Restaurant, Long John Silvers, McDonalds (Meijer), Mr. Taco, Pizza Hut, Rally's Hamburgers, Wendy's
  S: A & W Drive-In, Burger King, Cedar Park Restaurant, Frank's Press Box, McDonalds, Ponderosa, Stop's Family Restaurant
- **Lodg** N: ◆ Best Western, Days Inn(see our ad on this page), Knight's Inn, Regent Inn, Super 8

EXIT MICHIGAN

  Motel
  S: (AAA) Holiday Inn (see our ad on this page), ◆ Howard Johnson
- **AServ** N: Capitol Cadillac, Miller Jeep/Eagle, NTB, Snethkamp Dodge/Saab, University Olds/GMC
  S: Chapman Car Care, Eddie's Quick Lube, Muffler Hanger, Muffler Man
- **RVCamp** N: Campground
- **Med** N: ✚ Hospital
  S: ✚ Redi Care Walk-in Urgent Care
- **ATM** N: Community First Bank, First of America, Michigan National Bank, NBD, Old Kent Bank
  S: Dart National Bank, Total*, Total
- **Other** N: Aldi Supermarket, Meijer Grocery, Office Max, Sam's Club, Target Department Store
  S: Car Wash, Kroger Supermarket (Pharmacy), Rite Aide Pharmacy

106AB Junction I-496, U.S. 127, Jackson, Downtown Lansing

110 Okemos, Mason
- **Gas** N: Amoco*, Marathon*(CW), Mobil*(CW)
- **Food** N: Big Boy, Burger King, Cafe Oriental, Dunkin Donuts (Amoco), Little Caesars Pizza, McDonalds, Subway, Tubby's, Yo Cone Deli & Cafe
- **Lodg** N: Best Western, (AAA) Comfort Inn, ◆ Fairfield Inn, (AAA) Holiday Inn Express
- **ATM** N: First of America, Michigan National Bank
- **Other** N: 7-11 Convenience Store, BGPA Pharmacy, Michigan State Spartan Stadium

(111) Rest Area - RR, Phones, Picnic, Vending (Westbound)

117AB Dansville, Williamston
- **AServ** N: J&C Auto
  S: Bill's Wrecker & Radiator Service

122 MI 43, MI 52, Webberville, Stockbridge
- **FStop** N: Mobil*
- **Food** N: West Side Deli
- **ATM** N: Mobil

(126) Weigh Station - both directions

129 Fowlerville
- **Gas** N: Amoco, Shell*(CW)
  S: Mobil*
- **Food** N: Big Boy, Fowlerville Farms, McDonalds, Shooters Grill, Taco Bell, Wendy's
  S: Dunkin Donuts (South Mobil)
- **Lodg** N: ◆ Best Western
- **AServ** N: GM Auto Dealership, Shell
  S: Plymouth Dodge Jeep Dealer
- **ATM** N: First National Bank

133 Highland Road, MI 59
- **RVCamp** N: Camping
- **Other** N: Kensington Valley Factory Shops, Sheriff's Dept

(135) Rest Area - RR, Phones, Picnic, Vending (Eastbound)

137 Howell, Pinckney, CR D19
- **Gas** N: Mobil*, Shell*(CW)(LP), Speedway*(LP), Total*
- **Food** N: Pizza by the Slice, Time Out Grill
  S: Country Kitchen (Best Western)
- **Lodg** N: Knight's Inn, Park Inn
  S: (AAA) Best Western
- **AServ** S: Howell Auto Repair Center
- **Med** N: ✚ Hospital
- **Parks** S: Recreation Area
- **Other** N: True Value Hardware

(140) Rest Area - RR, Phones, Picnic, Vending (Westbound)

141 Junction I-96 Business Loop, Howell

EXIT MICHIGAN

(Westbound, No Reaccess)
Gas N: Sunoco*(D)(LP) (Kerosene)
AServ N: Champion Chevrolet
145 Grand River Avenue
Gas N: Amoco*, Shell*
S: Clark*, Sunoco*
Food N: Arby's, Cracker Barrel, KC's Cookery, O'Connor's Deli, Pizza Hut
S: Big Boy, Border Cantina, Burger King, Chili's, Dunkin Donuts, KFC, Lil' Chef, Little Caesars Pizza, McDonalds, Ponderosa, Subway (Clark), Wendy's
Lodg N: Woodland Lake Motel
AServ N: Brighton Ford/Mercury/Lincoln, Superior Olds Cadillac
S: Auto Works Auto Parts, Midas Muffler & Brakes
Med S: Hospital
ATM N: Comerica Bank, First of America
S: Michigan National Bank, Old Kent Bank (Farmer Jack), Standard Federal
Other N: **GFS Market**
S: **Farmer Jack Grocery (Pharmacy), Home Depot, K-Mart (Pharmacy), Meijer Grocery, Rite Aide Pharmacy, Ski Area, Target Department Store, US Post Office**
147 Brighton
Parks N: **Brighton Recreation Area**
Other N: **State Patrol Post**
148AB U.S. 23, Ann Arbor, Flint
150 Pleasant Valley Road (Westbound)
151 Kensington Road
Med S: Hospital
Parks N: **Island Lake Recreation Area, Kensington Metro Park**
153 Kent Lake Road, Kensington
Gas S: Mobil*
Parks N: **Kensington State Park**
155 New Hudson, Milford
159 Wixom
Gas S: Meijer*(LP), Mobil*(CW), Shell*(CW)
Food S: Arby's, McDonalds, Subway (Shell), Taco Bell
AServ S: Goodyear Tire & Auto, Varsity Ford/Mercury/Lincoln
Med S: Hospital
Parks N: **Proud Lake Recreation Area**
Other S: **Meijer Grocery**
160 Beck Road
Med N: Hospital
Parks N: **Mayberry State Park**
162 Novi, Walled Lake
Gas N: Amoco*(CW)
S: Mobil*
Food N: Denny's, Kerby's Coney Island, McDonalds, New Bangkok Thie Restaurant, Oaks Grill (Double Tree), Pizza Hut, Red Lobster, Subway
S: Bates Burgers, Big Boy, Bob Evans, Boston Market Restaurant, Coney Island Inn, Fuddruckers, Grady's, Kims Gardens Chinese, Kosch's Eatery, Maisono's Italian, Olive Garden, Oxford Steaks & Ribs, Red Robin, T.G.I. Friday's
Lodg N: Double Tree Hotel, Hotel Baronette
AServ N: Midas Muffler & Brakes, Sear's
S: Dan's Auto Repair, Discount Tire Company, NAPA Auto Parts, Tommy's Tire & Auto
Med S: Hospital
ATM N: Comerical Bank, Standard Federal
S: NBD Bank, Old Kent Bank
Other N: **12 Oaks Mall, K-Mart (Pharmacy), Mail Boxes Etc**

S: **Novi Veterinary Clinic,** Soft Shine Auto Wash
165 Junction I-696, I-275 ends
167 8 Mile Road
Gas S: Meijer Grocery*(D), Speedway*
Food S: Big Boy, Chili's, Hilton, Kirby Coney Island, Kyoto Japanese, McDonalds, Taco Bell
Lodg S: Hampton Inn, Hilton, Travelodge
Med S: Hospital
ATM S: Comerica Bank, Michigan Heritage Bank, NBD
Parks S: **State Park**
Other N: **Veterinary Clinic**
S: **Meijer Grocery**
169AB 7 Mile Road
Food N: Cascade Restaurant (Embassy), Lone Star Steakhouse, Rio Bravo
S: Cooker
Lodg N: Embassy Suites
Other S: **State Patrol Post**
170 6 Mile Road
Gas S: Amoco*, Mobil*(CW)
Food N: Akasa Japanese Restaurant, Big Boy, Bill Knapp's Restaurant, Boston Market, Bruegger's Bagel, Holiday Inn, Kerby's Koney Island, Marriott, Max & Erma's, Papa Romano's Pizza, The Ground Round, Wing Yee's
S: Charley's Grille, McDonalds, Papa Vino's, Standard Federal, Wendy's
Lodg N: Best Western, Courtyard by Marriott, Holiday Inn, Marriott
Med N: Hospital
ATM N: Comerica Bank, First Federal of Michigan, Michigan National Bank, NBD Bank
S: Old Kent
Other N: **Laural Park Place Mall, Rite Aide Pharmacy**
S: **Arbor Drugs, Farmer Jack Supermarket, Office Depot, PetsMart**
172 I-275 South, MI 14 West
173A Newburgh Road
173B Levan Road
Med N: Hospital
174 Farmington Road
Gas N: Mobil*(D), Total*
S: Amoco*
Food N: Looney Baker Cafe
S: KFC, Mason's Lounge
ATM S: First of America
Other S: **Livonia Veterinary**
175 Merriman Road
Gas N: Mobil*, Total*
S: Sunoco*
Food S: Blimpie's Subs, Mountain Jack's Steakhouse, Royal Coney Island Restaurant
176 Middlebelt Road
FStop S: **Express Fueling**
Food N: Bob Evans Restaurant, Chi Chi's Mexican, French Epi Bakery/Cafe, Olive Garden
S: Derby, McDonalds
Lodg N: Comfort Inn
AServ S: Auto Village, Oil Dispatch
Other S: **Discount Pharmacy, HQ**
177 Inkster Road
Food N: Baskin Robbins, Murphy's Restaurant, Subway
ATM S: Comerical Bank
Other N: **7-11 Convenience Store**
S: **Home Depot, Jerusalem Food Market, U-Haul Center**
178 Beech Daly Road
Gas N: Sunoco*

179 U.S. 24, Telegraph Road
Gas N: Marathon*(D), Sunoco*
Food N: Arby's, Church's Fried Chicken, Larry's Dining, Lucky Lau Chinese, McDonalds, Rally's Hamburgers, Subway, Taco Bell, White Castle Restaurant
Lodg N: Tel-96 Inn
AServ N: Dodge Dealer, Matic Chevrolet
Other N: **Express Pharmacy, Pet Supplies Plus, Rite Aide Pharmacy**
180 Outer Dr
Gas N: BP*(D)
AServ N: BP
Other N: **Coin Laundry**
182 Evergreen Road
Gas N: Sunoco*
Food N: Sonny's Hamburgers
AServ S: Firestone Tire & Auto, Metro 25 Tire
Other S: **Rite Aide Pharmacy**
183 MI 39, Southfield Fwy
184 Greenfield Road
Gas N: Amoco*, BP*, Sunoco*
S: BP*, Shell*
Food N: Greenfield Seli & Soul Food, Rikshaw Chinese, Subway (Amoco)
S: KFC, McDonalds, Rally's Hamburgers, Super Coney Island
AServ N: Greenfield, Midas Muffler & Brake, Sunoco, Tuffy Muffler
S: Gardner's Auto Service, Greenfield Auto
Other N: **Giant Supermarket,** Mac Full-Service Car Wash
S: **Rite Aide Pharmacy**
185 Grand River Ave., Schaefer Hwy
Gas N: Amoco*, Mobil*, Shell*(CW)
Food N: Capital Coney Island, China Dragon, McDonalds
AServ N: Shell
S: H & R Tires, Joy's Tires, Parkes Garage, Three Stars Auto Repair & Service
TServ S: **American Motor Coach**
Other N: **Grand Auto Parts,** Hand Car Wash
S: **Grand Price Food Center**
186A Wyoming Ave
Gas N: Marathon*, Shell*
Food N: Asian Corned Beef Chinese & American Food, Tubby's Grilled Subs
AServ N: Shell
186B Davison Ave
Gas N: Davison Ewald
AServ N: Davison Ewald
188A Livernois Avenue
Gas N: Mobil*, Shell*
S: BP*
Food N: Burger King, KFC, McDonalds, Spad's Pizza, Wendy's, Young's BBQ
AServ N: Midas Muffler & Brake, Sonny's Truck Repair
S: Hughes Auto Service, Westfield Auto Service
Other S: Full Service Car Wash
188B Joy Road
Food N: Capital One Coney Island, Church's Fried Chicken, Famous Pizza
189 West Grand Blvd, Tireman Ave
Gas N: Amoco, Mobil*
Other N: **Rite Way True Value Hardware**
190B Warren Ave

↑ **MICHIGAN**

Begin I-96

Bold red print shows RV & Bus parking available or nearby

EXIT	CALIFORNIA
	Begin CA99

↓CALIFORNIA

316B Bus.80, Reno

316 CA99, I-5 Jct., Redding, US50, South Lake Tahoe

315 12th Ave., Sutterville Rd.
- Gas E: Oak Park Market*
  W: Shell* (ATM)
- Food W: Adalberto's, Casa de los Ninos, China Wok Chinese, Yum Yum Donuts
- ATM W: Shell
- Other W: **Coin Laundry, Lee's Food King Grocery Store**

313B Fruitridge Rd. W.
- Gas E: Hites*, Shell*(CW)
  W: Exxon*
- Food E: Taco Bell
  W: Caballo Ballanca Mexican, Hacienda Mexican, Jim Boy's Mexican Fiesta, Wienerschnitzel
- AServ E: Lube Express, Quality Tune Up Shop
  W: A&E Auto Parts, L&M Tires, Parks USA, Sierra Transmission
- Other E: Trifty Wash Coin Car Wash
  W: **South Sacramento Pharmacy**

312 47th Ave.
- Gas W: 76*, Orbit*, Star*(D)
- Food E: Steakhouse (Southpointe)
  W: Pitts Stop Restaurant
- Lodg E: Southpointe Inn and Suites
- AServ E: Quality Muffler and Brake
  W: AAMCO Transmission, Clutch Specialist Parts and Service, Napa Auto Parts, Number 1 Auto Repair, Sacremento Radiator Service, Unitied Tire Co.
- RVCamp W: **Stillman RV Park**

310 Florin Rd.
- Gas W: Chevron
- Food E: Dennys
  W: IHOP
- AServ E: Montgomery Ward
- RVCamp E: **RV Travel World & Repair**
- Other E: **Circuit City, Florin Mall, Montgomery Ward, Toys "R" Us**
  W: **Bank of America, Home Depot, Southgate Shopping Center**

308 Mack Rd., Stockton Blvd., Bruceville Rd.
- Gas W: 76, Arco*, Shell*(CW)
- Food E: Chinoy Cuisine, Jack-in-the-Box, Taco Loco
  W: Burger King, Carl's Jr Hamburgers, Del Taco, Denny's, Jim Boy's, KFC, Little Caesars Pizza (K-Mart), Lucky Donuts, McDonalds, Mr. Perry's, Super Taco, Teriyaki Chicken Express
- Lodg E: Best Western, Gold Rush Inn, Motel 6
- AServ E: Jo Hill's Transmission, Q-Lube
  W: 76, Kragen Auto Parts
- RVCamp E: **B&L Trailor Supply**
- Med W: **✚ Hospital**
- ATM W: Wells Fargo
- Other E: **Hwy Patrol, Sam's Club (Members Only)**
  W: **Food 4 Less, K-Mart, Long's Drugs, Target Department Store**

306 Calvine Rd.

305 Sheldon Rd.

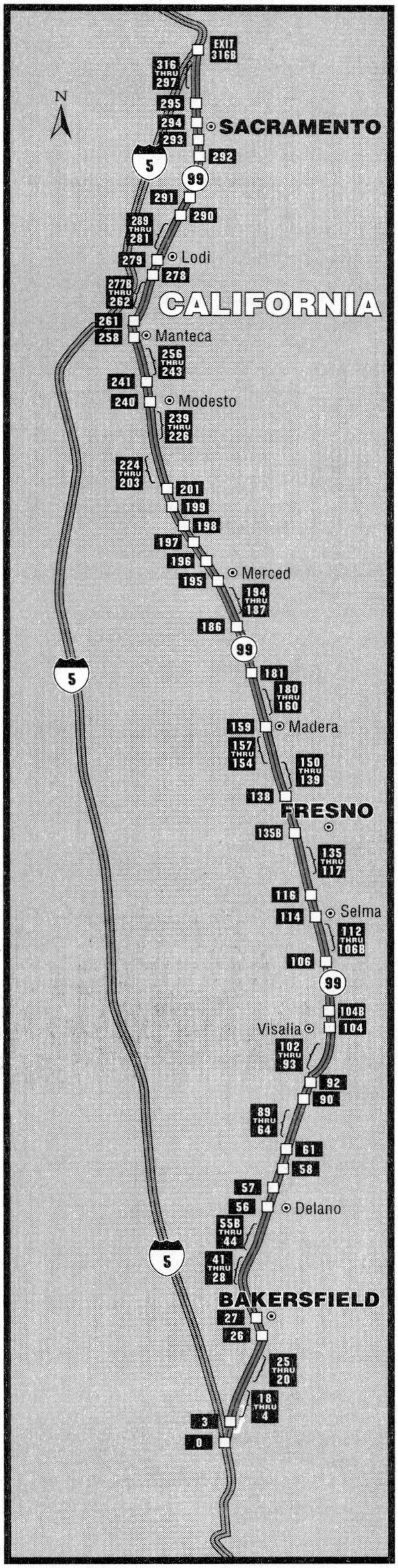

- RVCamp W: **Camping**

304 Laguna Blvd., Bond Rd.
- Gas E: Shell*(CW)
- Food E: Applebee's, Burger King, IHOP, In-N-Out Hamburgers, Marie Callender's Restaurant, Taco Bell
- Other E: **Marketplace 99 Shopping Mall, Raley's**

302 Elk Grove Blvd.
- Gas E: Exxon*, Shell*(CW)
  W: 76*(LP), Arco*
- Food E: Burger King, Cafe La Bou, Carl's Jr Hamburgers, Cheezer's Pizza, Denny's, Hunan Garden, Jo's Restaurant, KFC, Long John Silvers, Lucky, McDonalds, Mountain Mike's Pizza, Papa Murphy's, Pizza Barn, Pizza Bell, Puerto Vallerta, Stagecoach Restaurant, Subway, Taco Bell, ToGo's
  W: Casa Gomez, Continental Cafe, Lyon's, **McDonalds (Wal-Mart)**, Pizza Hut, Sunflower, UC Sub
- AServ E: Goodyear Tire & Auto, Kragen Auto Supply, Maita Chevrolet, Spee Dee
- ATM E: Home Savings
  W: The Golden 1
- Other E: **Elk Grove Veterinary Hospital**, Golden State Express Car Wash, **Payless Drugs**
  W: **Almost Perfect Books, Pak'N Save Grocery Store, Wal-Mart**

299 Grant Line Rd.
- Gas E: Arco*
- Other E: **Kamp's Propane**

298 Eschinger

297 Dillard Rd.

295 Arno Rd.

294B Frontage Rd. (Southbound)

294 Mingo Rd. (Northbound)

293 Jackson, CA104
- Gas E: Exxon*(LP)
- TServ E: **Royer's Trailor, Truck Repair**

292 Walnut Ave

291 Ayers Ln., Pringle Ave.
- Gas W: Cheaper*(D), Galt (RV Dump)
- Lodg W: Holiday Inn Express, **Royal Delta Inn**
- AServ W: Galt
- Other W: **Country Oaks Vet. Hospital, U.S. Post Office**

290 Simmerhorn Rd. (Northbound)
- Gas E: Exxon
- AServ E: Cain Brothers Auto Service, Exxon

289 Central Gath
- Gas W: Exxon(D) (Baskin Robbins)
- Food W: Baskin Robbins (Exxon), Carl's Jr Hamburgers, Chubby's, Denny's, Fancy Doughnuts, **McDonalds**, Polar Bear Yogurt, Round Table Pizza, Subway, Taco Bell, Valley Pharmacy, Wholey Ravoli
- AServ W: Chief Auto Parts
- ATM W: Farmers and Merchants Bank
- Other W: Coin Car Wash, **Coin Laundry, Payless Drugs, Save Mart Grocery Store**

287C Fairway Dr. (Southbound)
- AServ W: Mastercraft Tires

287B Crystal Way, Boessow Rd. (Northbound)
- Food E: Golden Acorn

Bold red print shows RV & Bus parking available or nearby

EXIT	CALIFORNIA
287	Liberty Rd.
286	Collier Rd.
Gas	**W:** Chevron*
Food	**W:** Lay's
Other	**W:** **Collierville Country Store**
285	Jahant Rd.
TStop	**W:** **Texaco*(LP)**
Food	**W:** **Airport Cafe (With Texaco)**, Johnsons Drive-In
TServ	**W:** **Texaco**
Other	**W:** **Lodi Airport**
284	Peltier Rd
283	Acampo Rd.
282	Woodbridge Rd.
281B	Frontage Rd.
281	Turner Rd.
279	CA12E, Central Lodi, San Andreas
FStop	**E:** **CFN**
Gas	**E:** Shell*(LP)
	W: BP*, Beacon*
Food	**E:** El Papagallo Mexican
	W: Burger King, Cherokee Auto Center, Express Doughnuts, Felten's Topaz, The Back Bay, UJ's Family Restaurant
Lodg	**W:** Comfort Inn, Del Rancho Inn, Lodi El Rancho Inn
AServ	**E:** Shell
	W: Big O Tires, Lodi Tire
RVCamp	**E:** **Richards RV Repairs**
Other	**W:** **Coin Laundry**
278	Fairfield, CA12W, Kettleman Ln.
FStop	**W:** **Paul's Mini Mart(D)(LP)**
Gas	**W:** Arco*
Food	**E:** **McDonalds**
	W: Carrows Restaurant, Denny's, Subway, Wendy's
Lodg	**W:** Holiday Inn Express
AServ	**E:** Buick Olds Dealer, Geweke Toyota
	W: Knowles Auto Parts, Midas Muffler and Brakes, Paul's Mini Mart, Q-Lube, Sanborn Chevrolet
RVCamp	**W:** **Geweke RV Service**
Other	**W:** **The Book Lady**
277B	Lodi (Northbound)
Gas	**W:** Quick Stop*
Food	**W:** Omega Restaurant
AServ	**W:** Lodi Honda
277	Harney Ln.
276	Armstrong Rd., Micke Grove Park
Parks	**W:** **Micke Grove Park**
274	Eight Mile Rd
Gas	**E:** BP*(D)
Food	**E:** Chicken Kitchen, Ernie's Pasta Barn
273	Morada Ln
271	Hammer Lane
Gas	**E:** Arco*
	W: Shell*(LP)
Food	**E:** Denny Boy's Restaurant, Wienerschnitzel
	W: Taco Bell
Lodg	**E:** Arbor Motel, **El Rancho Motel**, St. Francis Motel
	W: Sunshine Inn
Other	**W:** **Petmart, Wal-Mart**
270	Frontage Rd.

EXIT	CALIFORNIA
269C	Wilson Way, Central Stockton (Sothbound)
269B	Cherokee Rd.
TStop	**W:** **Cherokee Truck Stop*(SCALES)**
Food	**W:** **Carl's**, Donut Time
TServ	**W:** **Cherokee Truck Parts, Cherokee Truck Stop**
TWash	**W:** **Cherokee Truck Stop**
269	Waterloo Rd., Jackson, CA88
FStop	**W:** **Cardlock Fuels**
Gas	**E:** Chevron, Ernie's*, Shell*
	W: Waterloo*(LP)
Food	**E:** Burger King, Denny's, McDonalds, Perko's Cafe, Subway, Taco Bell
Lodg	**E:** Best Western, Comfort Inn, Guest Inn, Sixpence Inn
AServ	**E:** Chevron
	W: Advanced Tune Up, Knowles Auto Parts
Other	**W:** **Centro Mart**
268B	Freemont St., CA26 , Linden
Gas	**E:** Grewal's Gas*(D)
Food	**E:** The Blues Cafe
AServ	**E:** Drive Line Service of Stockton, Kamita Automotive
TServ	**W:** **Mack Diesel Truck Service**
268	CA4, W.Jct.I-5, Downtown Stockton
267B	Charterway West (Southbound)
267	Main St. (Northbound, Difficult Reaccess)
Gas	**E:** Exxon, QuickStop*
	W: Beacon
Food	**E:** Original 39cent Hamburger Stand, T's Drive-In
AServ	**E:** Exxon, Mike's Service Center
	W: Quality Tires
Other	**E:** Buggy Bath Coin Car Wash, **Coin Laundry**
	W: **Post Office, Save-Mart**
266	Farmington Rd., CA4
265	Escalon, Mariposa Rd.
Gas	**W:** USA Gas
Food	**W:** Sam's Donuts
Lodg	**W:** Mission Motel
AServ	**W:** California Smog
TServ	**W:** **Complete Diesel Repair**
Other	**W:** Coin Car Wash, **K-Mart, Wash Time Coin Laundry**
263B	Clark Dr. (Northbound)
263	Arch Rd.
FStop	**W:** **Gas Card**
Gas	**E:** Citgo*
Food	**E:** **Burger King (24 Hrs, Playground)**, Denny's, Jack-In-The-Box, Taco Bell
Other	**W:** **U.S. Post Office**
262	Frontage Rd.
261	French Camp
258B	Lathrop Rd.
Gas	**W:** Exxon*
AServ	**W:** Rick's Automotive
258	Manteca
Gas	**W:** Citgo*
Food	**W:** Casa Herrera, Taco Bell (24 Hrs)
AServ	**W:** Country Nissan
Other	**W:** **K-Mart**
256	CA120, Sonora, Yosimite Ave., Oakdale
Gas	**E:** 76*, Arco*

EXIT	CALIFORNI
	W: Exxon*
Food	**E:** Bock's Ranch, Burger King, Lyon's, Wendy's
	W: Jimmy's Restaurant (24 hrs), Lu Lu's Mexican McDonalds (Indoor Playground), Rick's Donuts, Taqueria Mexican, Three Flames Pizza, Tsing T
Lodg	**E:** Comfort Inn
	W: Best Western, **Manteca Inn**
AServ	**E:** Kurt Hughes Dodge, Piskel's Auto Air and Radiator, Vern's Auto Repair
	W: Tradeway Chevrolet, GM
Other	**W:** **Big Boy**, Indy's Car Wash, **Rancho Pharmacy, Wash Time Coin Laundry**
255	CA120W, San Francisco, Manteca (Difficult Reaccess)
254	Austin Rd.
Parks	**W:** **Caswell State Park**
252	Jack Tone Rd.
TStop	**E:** **Flying J Travel Plaza*(SCALES) (Country Marke**
Food	**E:** Country Market (Flying J)
251B	Milgeo Ave (Northbound)
TStop	**E:** **Jimco*(LP)(SCALES) (Restaurant)**
FStop	**E:** **Pacific Pride(SCALES)**
Food	**E:** **Giovanni's Pizza and Restaurant**, Jimco
TServ	**E:** **Howard's Truck Tire, Jimco*(CW)(SCALES)**
251	Ripon
Gas	**E:** Chevron*, Shell*(CW)
	W: Kwick Serve*(D)
Food	**E:** The Barnwood Restaurant and Deli
Lodg	**E:** Blue Light Motel
AServ	**E:** Chevron
	W: Ripon Auto Parts
249	Hammett Rd.
247B	Riverbank Rd., Salida, CA219
FStop	**W:** **Boyett Petrolium, Pacific Pride Commercia Fueling*(D)**
Gas	**E:** BP*(D)(CW)
Food	**E:** Burger King (Playground)
	W: Big Rig, La Hacienda Mexican, Salida Donut
Lodg	**W:** Don Pedro Motel
AServ	**W:** Salida Auto Parts
RVCamp	**W:** **Valley RV Center**
ATM	**W:** Union Safe Deposit Bank
Other	**W:** **Salida Vet. Hospital, U.S. Post Office**
247	Pelandale Ave.
Gas	**E:** Chevron* (24 hrs), Exxon*(LP)
	W: Arco*
Food	**E:** Carl's Jr Hamburgers, Fosters Freeze Jr., In-N Out Hamburgers, La Fiesta Mexicana, Me-N-Ed Pizza Parlor, Pizza Connection, Subway, Taco B
	W: Del Taco (Indoor Playground), **McDonalds (Indoor Playground)**
Lodg	**W:** Holiday Inn Express
RVCamp	**W:** **Dan Gamel RV Center**
Other	**E:** **Heavenly Place Bookstore, One Stop Pet Care, Payless Drugs, Save Mart Grocery Store**
245	Beckwith Rd., Standiford Ave.
Gas	**E:** Unocal 76*
Food	**E:** Chuck E. Cheese's Pizza, Coco's, Family Garcia's Restaurant, Garcia's Jo Jo's, Jade Garden, KFC, Red Lobster, Sizzler Steak House, The Hungry Hunter, Wendy's
AServ	**E:** Firestone Tire & Auto, Pro 10 Minute Oil Change, Sears Auto Center
ATM	**E:** Home Savings Of America
Other	**E:** **Julie's Books, Longs Drugs, Pier 1 Imports, Sears, Vintage Fair Shopping Mall**

Bold red print shows RV & Bus parking available or nearby

XIT CALIFORNIA

43 Carpenter Rd., Briggsmore Ave.
- **Gas** E: Arco*, Chevron(D), Shell(CW)
  W: BP*(LP) (ATM)
- **Food** E: Albertos Molcasalsa, Bakers Square Restaurant, Black Angus, Burger King, Denny's, Domino's Pizza, El Pollo Loco, Fresh Choise, Hometown Buffett, IHOP, Imperial Garden, Jack-In-The-Box, Kirin, McDonalds (Wal-Mart), McDonalds, Olive Garden, Outback Steakhouse, Taco Bell, Taco Shop, Teriyaki King, Togo's
  W: Wendy's
- **Lodg** E: Best Western, Holiday Inn, **Motel 6**, Ramada Inn, Super 8 Motel
- **AServ** E: Chevron, Precision Tune
- **Med** E: **+ Hospital**
- **ATM** W: BP
- **Other** E: **Food Maxx, Kinko's,** Prime Shine Express Car Wash, **Wal-Mart**

241 Kansas Ave.
- **Gas** W: Chevron*
- **Food** E: Cafe Orleans, The Sandwich Shop
  W: Jack-in-the-Box, Pho Viet Chinese, TheEarly Dawn Cattle
- **Lodg** E: **Econolodge**
  W: Econo Inn
- **AServ** E: Northern Tire
  W: Kansas Ave. Auto Center

240 CA108, CA132, Mays Blvd. (One Way Streets)
- **Gas** W: Arco*
- **Food** E: St. Stans
  W: Sun Sun
- **Lodg** E: Double Tree Hotel
- **AServ** E: Madesto Engine Renew
  W: Express Lube, May's Animal Hospital, Trojan Batteries
- **TServ** E: **Madesto Truck and Brake**

239 Tuolomne Blvd., B Street

238 Crows Landing Rd.
- **Food** W: El Marinero, Josefina's Bakery
- **AServ** W: A&P Wheels and Tires, Bul's Auto Service, Economy Tire Co.

237 South 9th St. (Northbound, Difficult Northbound Reaccess)
- **Gas** E: Gas'N Shop*(D)(LP)
- **Lodg** E: Budget Inn, Sea Breeze Motel
  W: Driftwood Motor Hotel, Holiday Motel, Sahara Motel
- **AServ** E: California Auto Parts, Quiet Masters Mufflers, Willie's Auto Electric
  W: Modesto Auto Service
- **TServ** W: **A&B Truck Repair, AFI Diesel Fuel Injection, California Equipment, Independent Fleet Services Truck and Trailer Repair(CW), S & S Diesel**

236 Hatch Rd.
- **Gas** E: Chevron*
  W: Rocket*, USA*
- **Food** E: Baskin Robbins, Burger King (Playground), La Morenita Mexican, Long John Silvers, McDonalds, Scotty's Donuts, Taco Bell, Teriyaki King, Weinerschnitchel, Wendy's
  W: El Tapatio
- **Lodg** E: Howard Johnson
- **AServ** E: Big O Tires, Kragen Auto Parts, Triple A Transmission Service, USA Parts
  W: Lary's Tire Mart

EXIT CALIFORNIA

- **ATM** E: 7-11 Convenience Store
- **Other** E: **7-11 Convenience Store (ATM), Food 4 Less, K-Mart, Long's Drugs**

235 Whitmore Ave, Hughson
- **Gas** E: Chevron*
- **Food** E: KFC

233B Ceres
- **Gas** E: Moon Gas*, Shell*, Texaco
- **Food** E: Alfonso's
- **AServ** E: Texaco

233 Mitchel Rd.

230 Keyes Rd.

229 Taylor Rd.
- **FStop** E: **Texaco*(D)(LP)**
- **Gas** E: Arco*
- **Food** E: Eppie's
- **Lodg** E: Best Western
- **AServ** E: Patchetts Ford, Lincoln, Mercury
- **TServ** E: **Bonander Truck Parts and Service**

228 Denair, Monte Vista Ave.
- **Med** E: **+ Hospital**

227 Fulkerth Rd., Pedretti Park
- **Gas** E: Shell*(CW)
- **Food** E: Baskin Robbins, Burrito Vitta, Chubby's, Hollywood Chicken, IHOP, Rico's Pizza, Subway, Top Dog Hotdogs
- **AServ** E: Chief Auto Parts
- **Other** E: **Food Maxx (24 Hrs), Wal-Mart**

226 Patterson, W. Main St.
- **FStop** E: **CFN**
- **Gas** E: BP*
  W: Arco*
- **Food** E: Burger King (Playground), Lyon's Family Dining
  W: **Carl's Jr Hamburgers (Playground),** Cindy's Restaurant, Dairy Queen, Golden Dragon, McDonalds (Indoor Playground), Taco Bell
- **Lodg** E: Western Budget Motel
  W: **Motel 6**
- **AServ** W: Toby's Auto Repair
- **TServ** W: **Turlock Tire Co.**
- **Med** E: **+ Hospital**

224 Los Banos, Lander Ave., CA 165
- **Gas** E: Chevron*(CW), Quick Stop*
  W: Arco*(D)
- **Food** E: Denny's, Jack-In-The-Box, Roundtable Pizza, Subway, Tequila Cafe
  W: Almond Tree (24 Hrs)
- **Lodg** W: **Comfort Inn**
- **AServ** E: Kragen Auto Parts
- **TServ** W: **Morado Tire**
- **Other** E: **Long's Drugs, Save Mart,** Touchless Coin Car Wash

(222) **Rest Area - Picnic, RR, Vending, Phones (Both Directions)**

222 Golden State Blvd., Central Turlock (Northbound, Reaccess Southbound Only)

220 Merced Ave.

219 Delhi, Shanks Rd.

(218) **Weigh Station (Southbound)**

216C Hammatt Ave.
- **Food** W: **Almond Tree Restaurant**
- **TWash** W: **Rocket Truck Wash**
- **Other** W: **Livingston Animal Clinic**

EXIT CALIFORNIA

216B Winston Pkwy.

216 Collier Rd.

215 Bloss Ave., Hilmir (Westbound)

214 Robin Ave.

(214) **Weigh Station (Northbound)**

212 Hunter Rd.

211B Arena Way

211 Peach Ave., N. Saultana Ave.

210 Cressey Way

209 Stein Rd.

208 Westside Blvd., Central Ave.

207 Gross Ave., Olive Ave.

206 Atwater (Southbound)

205 Applegate Rd.
- **Gas** E: 76, Chevron*, Exxon*(LP), Gas N Save
- **Food** E: Almond Tree, KFC, Los Panchos
- **Lodg** E: **Super 8 Motel, Valley Motel**
- **AServ** E: 76, Chevron, Jensen's
- **Med** E: **+ Hospital**
- **Other** E: **Bertelli's Drugs, Coin Laundry, U.S. Post Office**

203 Buhach Rd. (Left Hand Exit For Southbound Traffic)
- **RVCamp** E: **H&H Campers RV Parts and Service(LP)**
- **Other** E: **Atwater Veterinary Clinic**

201 Franklin Rd. (Northbound)

200 16th Street
- **Food** W: Nagame Japanese
- **AServ** W: Leo's Garage

199 V Street, West CA140
- **Gas** E: 76*(D), Shell
  W: Arco*
- **Food** E: Burger King, Carl's Jr Hamburgers
  W: Guss and Nick's Deli, Jack-In-The-Box, Pine Cone
- **Lodg** E: Gateway Motel, Motel 6, San Joaquin Motel, Siesta Motel, Slumber Motel
  W: Best Western
- **AServ** E: Advanced Transmission, Buick Pontiac Jeep Dealer, CNG Tire Service, Del's Auto Parts, Midas Muffler & Brakes, Pennzoil Lube, Razzari Chrysler, Plymouth, Razzari Ford
  W: Condell Radiator

198 R Street
- **FStop** E: **Pacific Pride Commercial Fueling**
- **Gas** E: Exxon*(LP), Gas-N-Save*
  W: Beacon*, Shell*
- **Food** E: Apple Annie's Donuts, Su Casa
  W: Denny's, McDonalds
- **Lodg** W: Motel 6
- **AServ** E: Chief Auto Parts, Winston Tires
- **Other** E: **Wal Green's**

197 Martin Luther King Blvd
- **Gas** E: Shell(LP)
  W: Beacon*, World Gas*(LP)
- **Food** E: In-N-Out Hamburgers, KFC, Taco Bell, Wendy's
  W: Star Garden
- **AServ** E: Shell
  W: A-1 Auto Repair, Jerry's Tire Shop, Napa Auto Parts
- **Med** W: **+ Hospital**
- **Other** W: **Wash-N-Dry Laundry Mat**

195 Mariposa, Yosemite, CA 140

Bold red print shows RV & Bus parking available or nearby

EXIT	CALIFORNIA
Gas	E: BP
Food	E: Carrows Restaurant, Domino's Pizza, Sir James, Victoria's Mexican
Lodg	E: Best Western, Days Inn, Happy Inn, Holiday Inn Express, Sandpiper Motel, Sierra Lodge
AServ	E: BP, Bob's Auto Works
Med	W: Hospital
194	**Child's Ave.**
Gas	E: Beacon*, Chevron*
Food	E: Eagle's Nest (Ramada), McDonalds, Mi Casa Mexican
Lodg	E: Ramada, Super 8 Motel
193	**Gerard Ave.**
192	**Helly Rd., Vassar Ave.**
191	**Harvard Ave., Yale Ave.**
190	**McHenry Rd.**
189	**Mariposa Way**
188	**Lingard Rd.**
187	**Pioneer Rd.**
186	**Worden Ave.**
185	**Le Grand**
183	**Athlone Rd.**
182	**Bachanan Hollow**
181	**Sandy Mush Rd.**
180	**Yosemite Plainsburg Rd.**
179	**Harvey Pettit Rd.**
TStop	W: **Diesel Country Truckstop* (Chuckwagon Coffee Shop)**
Food	W: Chuckwagon Coffee Shop
178	**Vista Ave.**
(178)	**Weigh Station (Northbound)**
177	**Le Grand Ave**
176	**Road 15**
175	**Avenue 26, CA233, Robertson Blvd., Chowchilla**
FStop	W: **The Way Station***
Gas	W: Beacon*(D), Chevron*(CW), Texaco*(D)(CW)
Food	E: Taco Bell
	W: Burger King, **Los Tejanos, McDonalds**
Lodg	W: **Days Inn**
AServ	E: Tom DuBose
Other	W: Coin Car Wash
173	**Avenue 24.5**
172	**Avenue 24**
171	**CA152, Los Banos Gilroy**
170	**Avenue 22.5 Fairmead**
169	**Road 19.5**
168	**Road 21.5**
166	**Avenue 20, Avenue 20.5**
164	**Avenue 18.5, Road 23**
TStop	W: **Pilot Travel Center*(D)(LP)(SCALES)**
Food	W: Dairy Queen, Great American Buffet, Subway, **Wendy's**
Lodg	W: **Liberty Inn**
TServ	W: **Kenworth, Schoettler**
TWash	W: **Pilot Travel Center**
162	**Avenue 17**
Other	W: **Airport, Highway Patrol**
161	**Avenue 16**
TStop	E: **49er Truck Stop*(D)(LP)**
Gas	W: Shell(LP)
Food	E: **Cafe (49er)**
	W: Farnesi's
Lodg	W: Gateway Inn
AServ	E: Whitaker's 4X4 Heaven
	W: Shell
Other	E: **Suburban Propane**
160	**Cleveland Ave**
Gas	E: Mobile*(D)(LP), Texaco*
	W: Chevron*
Food	E: Baskin Robbins, Burger King, Carl's Jr Hamburgers, Eppie's 24 Hour, Hong Kong Chinese, Jack-In-The-Box, KFC, Long John Silvers, Mei Wah, Subway, Wendy's
	W: Chubby's, International House of Pancakes, Little Caesars Pizza, Perko Family Restaurant, Red Dragon, TCBY, Taco Bell
Lodg	W: Economy Lodges of America
AServ	E: GM Auto Dealership, Kragen Auto Parts
	W: Chief Auto Parts, Winston Tires
Other	E: **Longs Drug Store, Lucky Food Center**
	W: **Food 4 Less, Pak-N-Sav, Wal-Mart**
159	**Central Madera, Fourth St.**
Gas	E: 76 Buggy Shower(CW), BJ's Gas*
Food	E: Lucca's Restaurant, Piccolo's Pizza, Taco Shop, Tex Mex Taco Shop, The Village Chinese American, Yum Yum Donuts
Lodg	E: Best Western, M & R Motel
Other	E: **Police Department Headquarters**
157	**CA 145, Firebaugh, Kerman**
Gas	E: Gas-N-Save*, Madera Fast Stop(LP), Texas Gold*
	W: Texaco
Food	E: El Ranchero Tacos, Mejia Taco Shop, Sun Sun Chinese
	W: Burrito King, Carl's Jr Hamburgers, DiCicco's Italian, The Vineyard
Lodg	E: B&Z Motel
AServ	E: 10 Minute Oil Change, Madera Cars Unlimited ,Lube Center, Madera Ford, Mercury, Monterrey Auto
	W: Donovan, Texaco
Med	E: **Hospital**
ATM	W: 7-11 Convenience Store
Other	E: Gateway Coin Car Wash
	W: **7-11 Convenience Store, MK Medical, Madera Vet. Clinic, Thrifty Drug**
156	**Madera, Gateway Dr. (Northbound)**
FStop	E: **Tesei Propane(LP)**
Gas	E: Alliance*, Beacon*(D)
Food	E: Rancho Madera
Lodg	E: Dixie Motel
AServ	E: Madera Automatic Transmission, Madera Front End and Brake Service(D), Paul's Auto Repair
Med	W: **Madera Community Hospital**
154	**Ave 12, Rd.29**
Lodg	W: Casa Grande
150	**Ave 9, Rd 311/2**
148	**Ave 7, RD 33**
146	**Herndon, Grantland Ave**
TStop	E: **Klein's*(SCALES)**
FStop	W: **Shell*(D)(LP)**
Gas	E: Chevron*, Texaco(LP) (Burger King)
Food	E: Burger King (Texaco), **Klein's**
TServ	E: **Klein's, Trucker's Air**
Other	E: **Hardin Scale(SCALES), MP Truck Stop**
144	**Shaw Ave, Biola**
Gas	E: Chevron*, Shell, Texaco*(D)(CW)
	W: BP*, Parkway Mini Mart
Food	E: **Carl's Jr Hamburgers,** In-N-Out Hamburgers, McDonalds
Lodg	E: Economy Inns of America, Formosa Inn
AServ	E: Car Quest Auto Center, Shell
142B	**Ashlan Ave**
Gas	W: Chevron, Citgo* (ATM), Mobile*
Food	E: Bon Appetit, Dairy Queen, Jack-in-the-Box
	W: **Brook's Ranch Restaurant**
Lodg	W: **Brook's Ranch Inn**
AServ	E: Johnson's Transmission Service
	W: Chevron
ATM	W: Citgo
Other	W: **Coin Laundry**
142	**North Golden State Blvd. (Northbound)**
141C	**Dakota Ave. (Southbound)**
Lodg	W: **Astro Motel,** Lite Inn
RVCamp	W: **Sunset West**
141B	**Schields Ave. (Southbound)**
FStop	W: **Beacon(LP)(SCALES)**
Gas	W: Cheaper*
TServ	W: **Beacon, Shoettler**
141A	**Princeton Ave. (Southbound)**
141	**Clinton Ave**
Gas	E: Beacon*(LP), Exxon*, U Save*
	W: Arco*, Chevron*
Food	E: DiCicco's Italian, Me N Ed's Pizza
	W: Gaslight Room Steakhouse, Grandma's Kitchen (24 Hrs, Travelodge)
Lodg	W: Best Western, Travelodge
AServ	E: Goodyear Tire & Auto
Other	E: Coin Car Wash
140	**McKinley Ave (Northbound, Difficult N.bound Reaccess)**
AServ	E: AR Tansmissions
139	**Olive Ave**
Gas	E: Chevron* (24 Hrs)
	W: 76, Exxon
Food	E: Taco Bell (24 Hrs), Tiny's
	W: Denny's, KFC, **McDonalds,** Wendy's
Lodg	E: Best Western
	W: Days Inn, London Motel, **Motel 6,** Parkway Inn, **Plaza Inn,** Roadway Inn, Super 8 Motel, Villa Motel, Welcome Inn
AServ	W: Bruce's, Exxon
RVCamp	W: **Parkview**
Other	E: **Frezno's Zoo, K-Mart**
138	**Belmont Ave., Pine Flat Dam**
Gas	W: Arco*, Texaco(D)
Food	E: Judi-Ken's Drive-in
	W: Charbroiled Burgers
Lodg	W: **Best Budget Inn,** Econolodge, London Inn, Palm Court Inn, Villa Inn, Welcome Inn
AServ	W: Texaco
Parks	E: **Roeding Park**
135B	**CA180W, Fresno St., Mendota (Watch One Way's)**
Gas	W: BP*, Oasis, Shell
Food	E: Chihuahua, Rally's Hamburgers
	W: KFC, Wendy's
AServ	E: Gennuso's
	W: Chief Auto Parts
ATM	E: Bank Of America
Other	W: **Coin Laundry**
135	**CA180E, Ventura St., Kings Canyon**
Gas	E: Beacon
134	**CA 41N, Yosemite, Millerton Lake National Park (Difficult Northbound Reaccess)**
133	**Sanger, CA41S, Jenson Ave**
FStop	W: **Texaco*(D) (Subway)**
Gas	E: BP*, Shell*
Food	E: Carl's Jr Hamburgers, Denny's, **In-N-Out Hamburgers,** KFC, McDonalds, Taco Bell, Wendy's
	W: Subway (Texaco)

Bold red print shows RV & Bus parking available or nearby

EXIT	CALIFORNIA
Lodg	E: Travelers Inn
AServ	W: Valley Tire Company
TServ	E: Central Valley Trailor Repair
31	Cedar Ave, North Ave
TServ	E: Sahara Nevada Truck Repair W: Ry-Den Diesel Inc., Valley Diesel
TWash	E: Western Truck Wash
29	Chestnut Ave, Malaga, Central Ave
FStop	E: Beacon(SCALES) W: Chevron*(SCALES)
Gas	E: Arco*, Shell*, Texaco*
Food	E: Pop's, The Brook's Ranch
TServ	E: Beacon Service and Wash, Fresno Truck Center, G & H Diesel Service, Golden State Peterbilt*, Goodyear Tire & Auto W: Rogers Sales and Service Complete Truck Repair
TWash	E: Truck and RV Wash
RVCamp	W: Dan Gamel's RV Service, Paul Evert's Sales and Service
Other	E: Central Car Wash
28	American Ave.
Gas	W: Arco*, Texaco*
Food	E: Aldo's, El Unico, Judi-Ken's Drive In, Triangle Burger, Will's Texas BBQ
Lodg	W: Motel 6
AServ	W: Custom Tech, Texaco
26	Clovis Ave
Gas	W: Texaco*(D)(LP)
25B	Adams Ave, Fowler (Southbound)
25	Merced St., Fowler
FStop	W: Wright Oil Company(LP)
Gas	E: Exxon*, Zip-N-Go* W: 76*
Other	E: Coin Car Wash
23	Manning Ave, Reedley, San Joaquin
FStop	E: Texaco*(D)(SCALES)
20	Floral Ave, Selma Ave., CA 43S, Hanford, Corcoran
TStop	W: Selma Truck Stop*(SCALES)
Gas	E: BP*, Shell*(CW), Texaco*(CW) W: Mobil*(D)
Food	E: Ann's Donuts, Arthur's, Brooks Ranch Coffee Shop, Carl's Jr Hamburgers, El Conquiestidor, McDonalds, Paradise Pizza and Subs, Piccolo's Pizza, Subway, Wendy's W: Baskin Robbins, Big Road (Selma Truck Stop), Burger King, McDonalds (Wal-Mart), Pea Soup Andersen's (Holiday Inn), Pizza Hut
Lodg	E: Best Western, Super 8 Motel W: Holiday Inn
AServ	E: Chief Auto Parts, Kragen, Selma Toyota, Swanson-Fahrney Ford, Buick W: Wal-Mart
TServ	W: Selma
Other	E: Pay Less W: Wal-Mart
119	Second St.
Gas	E: Beacon* W: Exxon*, PDQ*
Lodg	E: Villager Inn
AServ	E: Tire Shop Tune Up Service
Med	E: + Hospital
117	Mt. View Ave, Caruthers, Dinuba
FStop	E: Darling Oil And Tire
Gas	W: Arco(D)(LP)
AServ	E: Darling
TServ	E: Darling
116	Bethel Ave, Kamm Ave
AServ	W: Gonzales Tires

EXIT	CALIFORNIA
RVCamp	E: Viking Trailor Park
114	Conejo, CA 201, Kingsburg
Gas	E: 76(LP), Chevron*(LP) (TCBY) W: Arco*, Texaco*
Food	E: Denny's, TCBY (Chevron) W: Bobby Salazar's Mexican, Brenda Jo's, Burger King, Jack-in-the-Box, McDonalds, Me N Ed's Pizza, Subway, Taco Bell
Lodg	E: The Valley Inn W: Kingsburg Swedish Inn
AServ	E: 76, Village Tire Sales
Other	E: Coin Car Wash W: K-Mart, Kings Market and Deli Grocery Store
112	Kingsburg, Sanger, Mendecina Ave.
Gas	W: Shell*
Food	W: Los Pepe's
111	Ave 384
FStop	E: Exxon*(D)
Food	E: A & W Drive-In
Lodg	E: Kings Inn Motel
RVCamp	W: Riverland RV Park
Other	E: Rest Area (Picnic Area, Phones, Vending, RR, RV Dump)
108	Traver, Merritt Dr.
Gas	E: Shell*
Other	E: Post Office
106B	Goshen Elder Street
FStop	W: Arco*(D)
Gas	E: Exxon* W: Texaco*(D)(CW)
Lodg	W: Goshen Motel
TWash	W: Truck and Car Wash
RVCamp	W: The Wooden Shoe
106	Goshen, Ave 304
Food	W: Depot Cafe
TServ	E: Goodyear Tire & Auto
Other	W: Greyhound Bus Terminal
104	CA 198E, Visaila, Sequoia Park, Dinuba, Reedley
102	Ave 280, Exeter, Farmersville
98	Tagus
Lodg	E: Friendship Inn
97B	Tulare, Jay Street
97	Ots St., Cartmill Rd.
94	Hillman St., Prosperity Ave
Gas	E: Chevron*(CW) W: Exxon*(LP)
Food	E: Carl's Jr Hamburgers, Long John Silvers, McDonalds (Wal-Mart) W: Apple Annie's, Baskin Robbins, Burger King, Denny's, KFC, Mandarin House, McDonalds, Senor Taco, Subway
Lodg	E: Green Gable Inn W: Best Western, Inns of America, Motel 6
AServ	E: Wal-Mart, Winston Tires W: BP, Kragen Auto Parts
Med	W: + Tulare Medical Center
ATM	E: Great Western Bank W: Union Bank of California
Other	E: K-Mart, Longs Drug Store, Wal-Mart (McDonalds) W: Lucky, Thrifty Drugs
93	Lindsay, Central Tulare, CA 137, Visalia
Gas	W: Shell(LP), Texaco*(D)
Food	E: Cafe Launda, D'Oliveiras International Cuisine, Donut Factory, Foster's Freeze, Hong Kong Chinese, Rosa's Italian Restaurante, Wimpy's W: Ryan's Place, Wendy's

EXIT	CALIFORNIA
AServ	E: Auto Oil Changers W: Howell's Service Center, Shell
Other	E: Coin Laundry
92	Bardsley Ave
FStop	W: Exxon
Gas	E: Circle K*, Texaco*
AServ	W: Exxon, Parts Plus Auto Store
TWash	E: BJ's Car and Truck Wash
90	Paige Ave.
FStop	E: Exxon* W: Roche Oil
Lodg	E: Tulare Inn
AServ	W: Bryant Ross Dealership
RVCamp	E: Tulare RV Park
89	Tulare, K - Street (Left Hand, Northbound)
TWash	W: Truck Wash Truck Tub
85	Ave 200
Gas	W: Chevron* (24 Hrs)
Food	W: Country Cafe, Lynn's Cafe
Lodg	W: Agri-Center Lodge
RVCamp	W: Sun N Fun
83	Ave 184
(82)	Rest Area - RR, Picnic, Vending
79	Ave 152
Gas	E: Chevron*
78	Jct.CA190, Tipton, Porterville
75	Ave. 120
74B	Pixley (Souhtbound)
Gas	W: Exxon*
Food	W: Coffee Shop, Ritchie Z's
Other	W: Coin Car Wash
74	Ave.124 (Norhthbound)
Gas	E: Texaco*
Food	E: Three Brother's Burgers
AServ	E: Miranda'a Tire Shop
73	Unnamed (Northbound)
FStop	E: USA*
72	Court St.
FStop	E: Bob's*(LP)
Gas	W: Exxon*
Food	E: El Sarape W: Ritchie Z's
AServ	W: Pixley Auto Parts
71	Pixley, Terra Bella (Northbound)
Gas	E: Texaco* W: Shell*(LP)
Food	E: Mr. Suds Burger Bunch
Lodg	W: Butler Motel
AServ	E: J&W Tire Service W: Shell
Other	W: Pixley Laundry Mat
68	Ave 80, Ave 76 (Southbound)
67B	72
67	Ducor, Aplough, Erlimart
Gas	W: Terrible Herbst
Parks	W: Colonel Allansworth State Historic Park (9 Miles)
64	Earlimart, Ave 48
Gas	E: Bazea, Chevron
Food	E: Mendozas Bakery, Taqueria Jalisco
AServ	E: Bazea, Chevron
61	Ave 24
60	Ave. 16 (Southbound)
FStop	W: Pacific Pride(SCALES) (Subway)
Food	W: Subway (Pacific Pride)
Lodg	W: Motel
58	County Line Rd.

Bold red print shows RV & Bus parking available or nearby

EXIT — CALIFORNIA

FStop W: Exxon*(D)
Gas E: Arco*
Food E: Burger King, Fruit Tree Cafe, Jack-In-The-Box, Kong's Dynasty
Lodg E: Comfort Inn, Shilo Inn
AServ W: JC's Auto Truck Repair
Other E: K-Mart

57 Cecil Ave.
Gas E: USA*
W: Arco*, Chevron*
Food E: Donuts to Go, KFC, McDonalds, Pizza Hut, Rocky's Pizza, Taco Bell, Taco Factory, Wendy's
W: Taco Mex
Lodg W: Sundance Inn
AServ E: Daffern's Oil, Delano's Repair Service, Delany Motors Dodge, Chrysler, Plymouth, Kragen, R&S Tire Shop
W: Texaco Express Lube
Other E: Thrifty Drug Store
W: Coin Car Wash'

56 CA 155, Central Delano, Glennville
Gas W: Texaco*
Food E: Carmen's Mexican
W: Rosa's Bakery
Other W: Coin Car Wash, Komoto Pharmacy

55B First Ave. (Northbound)
AServ E: Singh Pontiac, Buick, GMC

55 Delano Ave, Woollomes Ave.
FStop W: Texaco*
Food E: Aldo's
W: Pioneer Restaurant
Lodg W: Pioneer Motel
Med E: Hospital

53 Pond Rd., Lake Woollomes

50 Perkins Ave., Elmo Hwy. (Southbound)
Gas W: Texaco*(D)
Food W: La Paloma
Lodg W: National 9 Inn
AServ W: LA Tires, Rameriz Muffler Shop
TServ W: Dave's Truck and Tire Repair(SCALES)

49 McFarland, Sherwood Ave. (Difficult Northbound Reaccess)
Gas W: Chevron
Food W: Golden Oven, McDonalds, Snow White Drive-In
AServ W: Napa Auto Parts

47 Whisler Rd.

44 CA 46, Wasco, Paso Robos, Famoso
Gas E: Chevron*, Mobil*
Food E: Chinese Cafe
W: Idle Spur Cafe
Lodg E: Famoso Inn

41 Kimberlina Rd.

39 Merced Ave.
TStop E: Flying J Travel Plaza*(D)(LP)(SCALES)
Food E: Flying J Travel Plaza

37 Shafter, Lerdo Hwy.
RVCamp W: Camping

31 7th Standard Rd.
Food W: Bob's Cafe
TServ W: 7th Standard Tire Company, Bakersfield Truck Center, Freightliner Dealer, Glenn's Truck Refrigeration

30 CA 65, Porterville, Sequoia Park (Northbound)
Gas E: Arco*, Texaco*
Other E: Bakersfield Airport

EXIT — CALIFORNIA

29 Roberts Rd., Oildale (Southbound)

28 Olive Dr.
Gas W: Chevron*(CW), Shell
Food E: Little Caesars, Los Tacos, Sonic Drive-in, Subway
W: Burger King (Playground), Hodel's, Jack-In-The-Box, Milt's Coffee Shop, Mom's Donut Shop, Rusty's Pizza Parlor, Taco Bell, The Old Hacienda
Lodg W: E-Z 8 Motel, Economy Motels of America, Motel 6
AServ W: Shell
ATM W: Bank of America
Other E: Animal Hospital, Lucky, Payless Drugs
W: Coin Car Wash, Vons Grocery

27 Oildale, Airport Dr., Pierce Rd.
Lodg E: Quality Inn
TServ E: Electric Motor Diesel Repair

26 Jct. 58, CA178, Downtown, Rosedale Hwy.
FStop E: Beacon
W: Texaco*(D)(LP)
Gas E: Arco*, Shell, Texaco(D)
W: Shell*
Food E: Arby's, Burger King, Denny's, IHOP, Tanuki Japanese, The Junction Dinner Lounge, Zingo's Cafe
W: Benji's French-Basque, Black Angus, Carl's Jr Hamburgers, Hungry Hunter, Sushi Kato, Taco Bell
Lodg E: Best Western, E-Z 8 Motel, La Quinta Inn, Road Runner Motel
W: DAYS INN Days Inn, Double Tree Hotel, Marriot Courtyard, Ramada Inn
AServ E: Chuck's Automotive, Shell
TServ E: Kenworth
Other E: Coin Car Wash

25 California Ave., Civic Center
Gas E: Chevron*(D), Circle K*
W: Arco*, Shell*, Texaco (Baskin Robbins)
Food E: Carrows Restaurant, Jack-In-The-Box, John's Burgers, Siagon, Taco Bell
W: Baskin Robbins (Texico), Burger King, Carl's Jr Hamburgers, Carrows Restaurant, Fresh Choice, Marie Callender's Restaurant, McDonalds, Pizza Hut, Regency Lanes (Travelodge), Roxanne's, Sizzler Steak House, Taco Fresco, Wendy's
Lodg E: Best Western, Extended Stay America, Hampton Inn, Quality Inn
W: California Inn, Motel 6, Radisson Suites Inn, Super 8 Motel, Travelodge
AServ E: Jiffy Lube, Sir Lube
W: Shell, Three Way Chevrolet, GM, Xpress Lube
Med E: Hospital
ATM W: Home Savings of America
Other E: Camelot Amusement Park
W: Barnes & Noble, Long's Drugs, Vons Grocery

24B Stockdale Hwy., Brundage Ln.
Gas E: BP, Handy Major
W: Citgo*
Food E: KFC
W: Foster's Freeze

EXIT — CALIFORNI

AServ E: Econo Lube and Tune, Midas Muffler & Brake, Tire Man Bridgestone, Windshields Of America
W: Stockdale Auto Electric
ATM W: Citgo
Other W: Crown Car Wash

24 Jct. 58E, Tehachapi, Mojave (Southbound)

23 Ming Ave.
Gas E: Arco*
W: 76*, Shell*, Texaco(D)
Food E: IHOP, Jack-In-The-Box, Taco Rey
W: Baker's Square, Foster's Donuts, King Chwa, Long John Silvers, Magoo's Pizza, Teriyaki Bowl, The Bagel, Wendy's
AServ E: Sears
W: Kragen, Sangera Auto Haus, Mercedes, Texaco
ATM W: Bank of America, Great Western Bank
Other E: Payless Drugs, Target Department Store

21 White Ln., Wible Rd.
Gas E: Chevron*, Mobil*, Shell(CW), Texaco*(D)
W: Arco*
Food E: Burger King, California Pizzaria and Chicken, Carl's Jr Hamburgers, Denny's, Donut Queen, E Chili Verde, J's Pancake House, McDonalds, Motel 777, Rally's Hamburgers
Lodg E: Comfort Inn, Holiday Inn Express, Motel 6
W: Travelodge
AServ W: Discount Tire Center, Southern Auto Supply, Victory Lane Oil Change
Other E: Wal-Mart
W: Rainbow Hand Car wash (With Mobil)

20 Panama Ln.
Gas E: Chevron*(LP), Texaco*(D)
W: Arco*
Food E: Denny's, In-N-Out Hamburgers, Jack-In-The-Box
Lodg E: Economy Motels of America
ATM E: Texaco
Other W: Holiday RV Super Store (see our ad on this page)

18 Jct. 119W, Taft, Lamont
FStop W: Mikuls Diesel*(D)(SCALES)
Gas W: Mobil*(D)(LP), Texaco*(LP)
RVCamp E: Leisure Times RV Repair, Southland RV Park

16 Houghton Rd., Weed Patch

13 Bear Mountain Blvd., Arvin, CA223
TStop W: Bear Mountain Truck Stop*(SCALES)
Food E: Beryl's Cafe
W: Harvest Steak House

10 Union Ave., Greenfield, Rt.99 Bus. (Northbound, Southbound Reacces Only, Difficult Reaccess)

9 Herring Rd.

7 Sandrini Rd.

5 David Rd., Copus Rd.

4 Mettler (Southbound)
FStop W: Pacific Pride
Gas W: Wadkins*
Food W: Day and Night Market Restaurant

3 Jct. CA166, Maricopa, Taft
FStop W: Chevron*(D)(SCALES), Texaco*(SCALES) (Subway)
Food W: Cantina Laminita Mexican, Subway (Texaco)
AServ W: Chevron

0 Jct. I-5

↑CALIFORNIA

Begin CA99

Bold red print shows RV & Bus parking available or nearby

EXIT	PENNSYLVANIA
	Begin I-99

↓PENNSYLVANIA

	(51) PA 350, Bald Eagle
Gas	**W:** Amoco*
Food	**W:** Subway (Amoco)
	(48) PA 453, Tyrone
Food	**W:** Burger King, Main Moon Chinese
ATM	**W:** Burger King
	(45) Tipton, Grazierville
AServ	**W:** Chrysler Auto Dealer
Med	**W:** **+ Hospital**
Parks	**W:** **Bland's Park (1 Mile)**
Other	**W:** **Visitor Information**
	(41) PA 865, Bellwood
Gas	**E:** Citgo*(D), Sheetz*
Parks	**E:** **Bland's Park**
Other	**E:** **Fort Roberdeau Historic Site**
	(39) PA 764 S, Pinecroft
	(33) 17th St.
Food	**W:** Chicken Express Restaurant
AServ	**W:** D&M Chrysler Plymouth
	(32) PA 36, Frankstown Rd.
Food	**W:** Chuck E. Cheeses Pizza, Em's Subs, Pizza Hut, Red Lobster, Shoney's, The Gingerbread Man Restaurant, The Olive Garden, Wendy's
Lodg	**W:** DAYS INN Days Inn, Super 8 Motel
AServ	**W:** Barts Auto Center, Big A Auto Parts
Med	**W:** **+ Hospital**
ATM	**W:** Mellon Bank
Other	**W:** **Thrift Drugs**
	(31) Plank Rd.
Food	**E:** Boston Market, Jethro's Italian, Kings Family Restaurant, Ramada Inn, Robbie's Pizza, T.G.I. Friday's
	W: Arby's, Best Way Pizza, **Denny's**, Eat N Park, KFC, Long John Silvers, Mrs. G's Country Dining, Ponderosa Steakhouse, Taco Bell
Lodg	**E:** Ramada Inn
	W: HoJo Inn
AServ	**E:** Firestone Tire & Auto
	W: Jiffy Lube, Sear's
ATM	**E:** Laurel Bank
Other	**E:** **Food Lion, Sam's Club, State Police, Wal-Mart (Pharmacy)**
	W: **Bi-Lo Supermarket, Giant Eagle Supermarket, K-Mart (Auto Service, Pharmacy), Logan Mall, Pet Owners Warehouse, Phar Mor Drugs, Tourist Information, Weis Supermarket (Pharmacy)**
	(28) Ebensburg, Hollidaysburg, US 22, PA 764
RVCamp	**W:** **764 North Campground**
	(23) US 22, PA 36, PA 164, Hollidaysburg, Roaring Spring Portage
TStop	**E:** **Mobil Travel Plaza***
Food	**E:** Blimpie's Subs (Mobil), Creek Side Inn, Rax (Mobil)
AServ	**E:** Shuster Chrysler Plymouth Dodge Jeep Eagle
Med	**E:** **+ Hospital**
Other	**E:** **Airport**
	(15) Claysburg, King
	(10) Blue Knob State Park
Food	**W:** 220 Sporting Goods Dairy Delight
Parks	**W:** **Blue Knob State Park**
	(07) Osterburg, St Clairsville
	(03) PA 56, Fishertown, Cessna
FStop	**E:** **Amoco* (Kerosene)**
AServ	**E:** Jenkins Plymouth Dodge

↑PENNSYLVANIA

Begin I-99

EXIT	KANSAS
	Begin I-135

↓KANSAS

95AB		Junction I-70, US 40E to Topeka, Hays
93		KS140, State Road
92		Crawford Street
	FStop	**W:** **Phillips 66*(SCALES)**
	Gas	**E:** Amoco*, Conoco*(D), KwikShop*, Texaco*
	Food	**E:** Anchor Room (Mexican), BBQ, Beijing Chinese, Braum's, Chaser's (Holiday Inn), Hardee's, Panda's Chinese Restaurant, Russell's Restaurant, Spangles, Subway, Taco Bell, Wood Grill Buffet
		W: Red Coach Restaurant
	Lodg	**E:** Best Western, ◆ Comfort Inn, ◆ Fairfield Inn, ◆ Holiday Inn, ◆ Super 8 Motel
		W: AAA Red Coach Inn
	AServ	**E:** Royal Tire
	TServ	**E:** **Royal Tire Truck Town**
	Med	**E:** **+ Hospital**
	ATM	**E:** First Bank Kansas
		W: Phillips 66
	Other	**E:** **Pet Life**
88		9th Street, Schilling Road
89		Schilling Road
	Food	**E:** Applebee's, **McDonalds (Wal-Mart)**, Pizza Hut, Red Lobster
	AServ	**E:** GM Auto Dealership, **Wal-Mart**
	Med	**E:** **+ Hospital**
	ATM	**E:** First Bank Kansas
	Other	**E:** **Office Max (Frontage Rd), Sam's Club, Target Department Store, Wal-Mart (Pharmacy)**
86		Mentor, Smolan
82		KS4, Assaria, Herington
78		Lindsborg, Bridgeport
	Gas	**E:** Texaco
	Food	**E:** Dairy Queen (Texaco), Stuckey's (Texaco)
	RVCamp	**W:** **Camping**
	Parks	**W:** **Coronado Heights Park**
72		Bus US 81 Lindsborg, Roxbury
	Other	**W:** **Old Mill Museum, Sandzen Art Gallery**
(68)		**Rest Area - RR, Picnic, RV Dump P (Both Directions)**
65		**(61)** Pawnee Road
	Other	**E:** **Maxwell Wildlife Refuge, McPherson Fishing Lake**
60		U.S. 56, McPherson, Marion
	FStop	**W:** **Conoco***
	Food	**W:** Arby's, **Best Western**, Braum's, Golden Dragon (Chinese), Hunan Chinese, KFC, Long John Silvers, McDonalds, Pizza Cafe, **Pizza Hut**, Red Coach Inn, Sirloin Stockade, Subway, Taco Tico
	Lodg	**W:** ◆ **Best Western**, AAA Red Coach Inn, Super 8 Motel
	AServ	**W:** Auto Zone Auto Parts, Ford Dealership, John's Motor, **Wal-Mart**

EXIT		KANSAS
	TWash	**W:** **Car & Truck Wash**
	Med	**W:** **+ Hospital**
	Other	**E:** **Kansas Kampers Service(LP)**
		W: Car & Truck Wash, **Wal-Mart (Pharmacy)**
58		U.S. 81, KS61, McPherson, Hutchinson
54		Elyria (Difficult Access)
48		KS260, Moundridge
46		KS260, Moundridge
	Other	**W:** U-Do-It Coin Car Wash
40		Hesston
	FStop	**W:** **Phillips 66*(SCALES) (RV Dump)**
	Food	**W:** Hesston Heritage Inn, **Pizza Hut (Phillips 66)**, Subway
	Lodg	**W:** Hesston Heritage Inn
	TWash	**W:** **Truck Wash**
	RVCamp	**E:** **Cottonwood Grove Campground**
	ATM	**W:** Phillips 66
	Other	**W:** Car Wash, Coin Car Wash
34		KS15, N Newton, Abilene
	FStop	**E:** **Phillips 66*(LP)**
	Food	**E:** Mid-Kansas RV Park Restaurant
	AServ	**E:** Phillips 66
	RVCamp	**E:** **Mid-Kansas RV Park(LP)**
	Other	**E:** **Coin Operated Laundry (Mid-Kansas RV Park)**
		W: **Kauffman Museum**
33		U.S. 50 E, Peabody, Emporia (Closed Due To Construction, Difficult Reaccess)
32		Broadway Ave
	AServ	**E:** Chrysler Dealer
31		First Street
	TStop	**E:** **Texaco*(SCALES) (RV Dump)**
	FStop	**E:** **Conoco*(D)**
	Gas	**E:** Phillips 66*
		W: Phillips 66*(D)(LP)
	Food	**E:** CJ's Pancake House, Charlie's (Texaco), KFC
		W: Braum's Ice Cream Shop, Red House Restaurant (Best Western), Sirloin Stockade
	Lodg	**E:** 1st Inn, DAYS INN ◆ Days Inn, **Super 8 Motel**
		W: AAA Best Western, AAA Red Coach Inn
	AServ	**E:** Hill's Auto (Phillips 66)
		W: Phillips 66*
	TServ	**E:** **NCH Truck Service, Texaco**
	TWash	**E:** **Texaco**
	ATM	**E:** Texaco
30		US 50, KS15, Newton, Hutchinson (Divided Hwy, Difficult Reaccess)
28		McLains Road
	Gas	**W:** Total*(CW)
	Food	**W:** Burger King, Chisholm BBQ, Food Court (Newton Factory Outlet), Gambino's Pizza, Hong Kong Garden (Chinese), Ice Cream Factory, Rocky Mtn Choc Factory, Subway
	ATM	**W:** Nationsbank, Total
	Other	**W:** **Newton Factory Outlet (Food Court), Vitamin World**

Bold red print shows RV & Bus parking available or nearby

EXIT		KANSAS
25		KS196, Whitewater, El Dorado
(23)		**Rest Area - RR, Phones, Picnic, Vending, RV Dump P (Both Directions)**
22		Sedgwick, 125th Street
19		101st Street
	Other	W: **Camping**
17		85th Street
	Other	E: **Kansas City Colliseum**
16		77th Street
	Parks	E: **Wichita Greyhound Park**
14		Park City, Kechi, 61st Street
	TStop	W: **Coastal**(SCALES)
	Gas	E: QuikTrip*(D)
		W: Total*
	Food	E: Casa Grande (Mexican), Wendy's (QT)
		W: **Hen House Restaurant (Coastal), McDonalds**
	Lodg	W: **Super 8 Motel**
	AServ	W: GM Auto Dealership, Garnett Auto Supply, Goodyear Tire & Auto
	TServ	W: **Coastal**
	TWash	W: **Coastal**
	ATM	E: QT
		W: Chisholm Trail Bank
	Other	W: **Ark Valley Chiropractor, Family Vision Ctr, Leeker's Family Food Grocery Store**
13		53rd Street
	Gas	W: Phillips 66*(D)(CW)
	Food	W: The Red Coach Restaurant (Best Western)
	Lodg	W: AAA Best Western, DAYS INN AAA Days Inn
	Other	W: Car Wash, **Fire Dept**
11B		Junction I-235W
11A		KS254, El Dorado (Divided Hwy, Difficult Reaccess)
10B		Hydraulic Ave, 29th Street
	Gas	W: Conoco*(D)
10A		Kansas Rt 96E (Difficult Reaccess - Eastside Only)
9		Wichita State University, 21st Street
	Gas	E: Amoco*(D)
	Food	E: Burger King
	Other	E: **Turner Family Dentistry**
8		13th Street
	Gas	E: Total*
		W: Conoco*(D)
	Parks	W: **Emerson McAdams Park**
	Other	E: **13th St Coin Operated Laundry**
7B		8th Street, 9th Street
	Other	E: **University of Kansas Medical School**
6AB		1st Street, 2nd Street (Access To 6A Southbound Only, Exit For Kellogg Ave)
	AServ	E: Import Auto Service Center
5B		U.S. 54, U.S. 400, KS96, Kellogg Ave (Divided Hwy, Difficult Reaccess)
5A		Lincoln Street

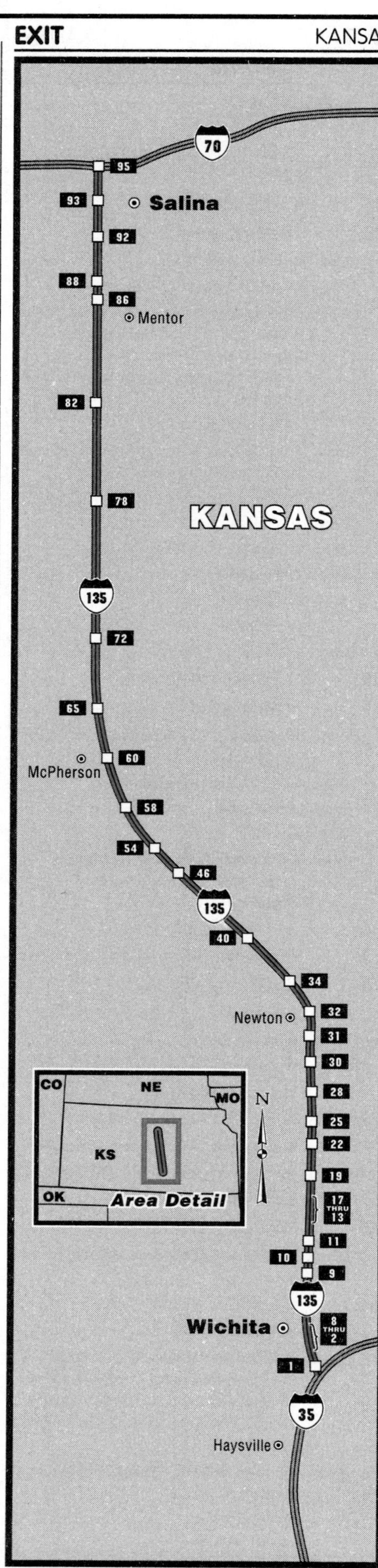

EXIT		KANSAS
	Gas	W: QuikTrip*
	Food	E: Dairy Queen
	AServ	W: Tire Center
	ATM	E: Commerce Bank
	Other	W: **Dillions Grocery (24 Hr)**
4		Harry Street
	Gas	E: QuikTrip*
		W: Phillips 66*(CW)
	Food	E: Angelo's Italian, Barb's Carry Out BBQ, Dog N Shake, Domino's, Dunkin Donuts, India Emporium, McDonalds, Taco Tico's, Toc's Teehouse
		W: Sport Burger
	AServ	E: Budget, Magic Touch
		W: Auto Truck Ind, Johnson Bros Auto Supply, Mike's Custom, Phillips 66*(CW)
	Med	E: **+ Hospital**
	ATM	E: QuikTrip
	Other	E: **Dopp's Chiropractor, Fabric Care Center Coin Operated Laundry**
		W: **Quick'n Easy Wash-O-Mat, SOS Optical**
3B		Pawnee
	Gas	W: Conoco*(D)
	Food	E: Grandma's Diner
		W: Burger King, Dog n Shake, KFC, Li'l Mexico, Pizza Hut, Rice & Roll Express (Chinese), Spangl
	AServ	W: Jiffy Lube, Little Giant Muffler Shop, Mola Aut Repair
	ATM	W: Emprise Bank
	Other	E: **American Family Dental**
		W: **Checker's Grocery Store,** Richard's Car Wash, **TG&Y Hdwe**
3E		Kansas Route 15S SE Blvd (Southbound)
	AServ	E: Walker Truck Service
2		Hydraulic Ave
	Gas	E: QuikTrip
	Food	W: McDonalds
1C		Junction I-235
1AB		47th Street E
	Gas	E: Coastal*, QT*, Total*(D)
		W: Phillips 66*(CW)
	Food	E: Pot Bellys, Super Wok (Chinese Food), Taco Tico's
		W: Applebee's, Braum's, Domino's, Godfather's Pizza, Gourmet Food Court, KFC, Little Caesars Pizza, Long John Silvers, McDonalds, Mr Goodcents Subs & Pasta, Pizza Hut, Spaghetti Jack's, Spangles, Sub & Stuff Sandwich Shop, Subway, Taco Bell
	Lodg	E: Holiday Inn
		W: Red Carpet Inn
	AServ	E: **K-Mart,** Tucker's Tire Service
		W: Auto Parts, Whitlock Auto Supply
	RVCamp	W: **R&D Camperland**
	ATM	W: The Garden Plain State Bank
	Other	W: **Big Lots, Checkers Grocery Store, Dillions Grocery, Goodwill Store, K-Mart (Pharmacy)**

↑KANSAS

Begin I-135

Bold red print shows RV & Bus parking available or nearby

I-155

EXIT ILLINOIS

Begin I-155

↓ ILLINOIS

JCT I-55, Chicago, St Louis
Hartsburg
0 US 136, Havana, McLean
5 IL 122 West, Delavan
9 IL 122 East, Hopedale
2 Townline Rd.
5 IL 9, Peakin, Bloomington
8 Broadway Rd.
0 Queenwood Rd.
- **Gas** E: Casey's General Store, Citgo*
- **Food** E: Baskin Robbins, Godfather's Pizza, Hoppers, Pizza Hut, Villager Lodge
- **Lodg** E: **Villager Lodge**
- **ATM** E: Citgo
- **Other** E: Car Wash

1 IL 98, Birchwood St.
- **Gas** E: Speedway*(LP)

↑ ILLINOIS

Begin I-155

I-172

EXIT ILLINOIS

Begin I-172

↓ ILLINOIS

9 US 24, IL 96, Mt Sterling, Keokuk
5 Columbus Rd., Wismann Ln
4 IL 104, Broadway
- **Gas** W: Shell*(D), Texaco Food Mart*(D)
- **Food** W: Applebee's, Bagels, Baskin Robbins, Bejing Garden Chinese, Burger King, Domino's, Fazoli's Italian Food, Golden Corral, Hardee's, KFC, Lakeview Restaurant, Little Caesars Pizza (K-Mart), Made Right Sandwich Shop, McDonalds (Indoor Playground), Pizza Hut, Steak & Shake, Subway, Taco Bell, Wendy's
- **Lodg** W: Comfort Inn, ◆ Fairfield Inn
- **AServ** W: Auto Zone Auto Parts, Big A Auto Parts, Goodyear Tire & Auto, Sear's, Shell, Western Auto
- **ATM** W: First Bankers Trust, State Street Bank
- **Other** W: **K-Mart (Pharmacy, Auto Service), Sear's, Vision Center, Wal-Mart**

10 IL 96, Thirty - Sixth Street, Payson
4C I-36 West To Hannibal
2 IL 57, Marblehead

↑ ILLINOIS

Begin I-172

I-185

EXIT GEORGIA

Begin I-185

↓ GEORGIA

15 **(46)** Big Springs Rd.
14 **(41)** U.S. 27, Pine Mt.
- **Gas** W: Amoco*(LP) (24 Hrs), Shell*(D)
- **Food** W: Waffle House
- **Parks** W: **Callaway Gardens, Roosevelt State Park**
- **Other** W: **Ga. State Patrol Post, Little White House Historic Stie**

13 **(34)** Ga. 18, West Point
12 **(30)** Hopewell Church Rd., Whitesville
- **FStop** W: **Amoco*(LP)**

11 **(26)** Ga. 116, Hamilton
- **Food** W: Hunter's Pub & Steakhouse

10 **(19)** GA. 315, Mulberry Grove
- **FStop** W: **Chevron* (24 Hrs)**

9 **(15)** Smith Rd.
2 **(3)** St. Mary's Rd.
- **Gas** E: Quick Change*
  W: Amoco*(LP) (24 Hrs), BP*(D)(LP)
- **Food** E: Catfish Country, Domino's Pizza
  W: Golden Chopsticks Korean Restaurant, Hardee's, Hong Kong Chinese Restaurant, Lula's Diner
- **RVCamp** E: **Ford's RV Center Sales & Service**
- **ATM** W: Barnett Bank, First Union
- **Other** E: Goo-Goo Coin Car Wash
  W: **Piggly Wiggly Supermarket**

1 Ga. 520, U.S. 27, U.S. 280, U.S. 431, Victory Dr.
- **Gas** W: Conoco*(D)(LP), Racetrac*
- **Food** W: Country Crossing Buffet, Hardee's, Smokin Branch BBQ
- **Lodg** W: **Econolodge**
- **Parks** E: **Florence Marina State Park, Providence Cayon State Park**
- **Other** E: **Fort Benning Military Reservation**

8 Williams Rd.
(12) GA Welcome Center - Full Facilities, Tourist Info (West Side Of Exit 8)
7 **(10)** Ga. 22, U.S. 80, Phenix City Alabama, Macon
6 **(9)** Airport Thruway, Columbus Airport
- **Gas** W: Amoco*(D)(LP), Crown*(CW)(LP)
- **Food** E: Blimpie's Subs, China Moon, Shoney's
  W: A Bagel Cafe, Al Who's Place, Applebee's, Burger King (Playground), Cafe Di Italia, Captain D's Seafood, Folks, Hardee's, KFC, Los Amigo's, Mandarin House, McDonalds (Play Place), Outback Steakhouse, Pargo's, Taco Bell, Texas Steakhouse & Saloon
- **Lodg** W: Comfort Suites, ◆ Hampton Inn(See our ad this page), ◆ Sheraton
- **AServ** E: Western Auto
- **Med** W: ✚ **Hospital**
- **ATM** E: Barnett Bank
  W: SouthTrust Bank, SunTrust Bank
- **Other** E: **Columbus Metro Airport, Sam's Club, Wal-Mart (Pharmacy)**
  W: **Carmike Cinemas, Food Max Supermarket (Pharmacy), K-Mart (Pharmacy)**

5 **(8)** U.S.27, Ga. 85, Columbus
- **Gas** W: BP*(D)(LP), Chevron*(CW), Crown*(LP) (24 Hrs)
- **Food** E: Applebee's, J. McDaniel's, Krystal, Rio Bravo, Ruby Tuesday
  W: Arby's, Dunkin' Donuts, Krystal River Seafood, Logan's Roadhouse, Subway, Waffle House, Wendy's
- **Lodg** E: ◆ Super 8 Motel
  W: Holiday Inn
- **AServ** W: Victory Auto
- **Med** W: ✚ **Hospital**
- **ATM** E: CB&T Bank, First Union, SunTrust, SunTrust
- **Other** E: **Peachtree Cinemas, Peachtree Mall**
  W: **Animal Emergency Center, Civil War Navel Museum**

4 **(6)** Ga. 22, Macon Rd.,
- **Gas** W: Amoco*, Chevron* (24 Hrs)
- **Food** E: Burger King (Playground), China Buffet, Choctaw Grill, Dairy Queen, Domino's Pizza, KFC, Pizza Hut, Taco Bell, Waffle House, Western Sizzlin'
  W: Captain D's Seafood, Central Park Hamburgers, China Star Restaurant, Chuck E. Cheese's, Cisco's Mexican Grill, Country's BBQ, Denny's, El Vaquero, Golden Donut, Longhorn Steakhouse, McDonalds (Play Place), Shoney's, Subway
- **Lodg** E: AAA Comfort Inn, DAYS INN ◆ Days Inn
  W: AAA La Quinta Inn
- **AServ** E: Carl Gregory Hundai, Chrysler, Jeep, Eagle, Freeway Ford, Jay Mazda, Jay Toyota, Suzuki, Mike Collins Nissan
  W: Carl Gregory Honda, Firestone Tire & Auto, Goodyear Tire & Auto, Mr. Transmission
- **Med** W: ✚ **Hospital**
- **ATM** W: Amoco, Columbus Banking & Trust Co., SouthTrust, SunTrust Bank
- **Other** E: **Lewis Jones Foodmarket**
  W: **Big B Drugs, Books-A-Million, Columbus Museum, Columbus Square Mall, K-Mart, Phar Mor Drugs, Publix Supermarket**

3 **(4)** Buena Vista Rd.
- **Gas** E: BP*(LP), Shell*
  W: Chevron, Speedway*(D)
- **Food** E: Burger King (Playground), Captain D's Seafood, Chef Lee's Peking Chinese, Church's Fried Chicken, McDonalds (Playground), Pizza Hut, TCBY Treats, Waffle House
- **AServ** E: A&A Transmission, Firestone Tire & Auto, King Lube, Tires N Terms, Tune-Up Center, Victory Auto
  W: Chevron
- **Other** E: **Lewis Jones Food**
  W: **Steve Parker Lock Smith**

↑ GEORGIA

Begin I-185

EXIT RHODE ISLAND/MASSACHUSETTS

Begin I-195

↓RHODE ISLAND

1 Downtown Providence

2 U.S. 44W, South Main St (One Way Street, Difficult Reaccess)

Med S: ✚ **Hospital**

3 Gano St, India Point

Lodg S: Days Inn

6 To RI 103, U.S. 44, E Providence

Food N: Amanda's Coffee Shop, Churrasqueria, Santoria Pizza

AServ N: Replacement Auto Parts, Speedee Muffler King, Vargas

S: Driver's Seat

ATM S: Bankeritz

Other S: **Coin Laundry**

(0) Transfer to RI

↑RHODE ISLAND

↓MASSACHUSETTS

1 MA114A, Seekonk, Barrington RI

Gas N: Shell*

S: Sunoco(D)

Food N: 99 Restaurant, Newport Creamery

S: Bagel Station, Bigford's Family Restaurant, Bug A Boo Steak House, Burger King, China Wok, Cisco's Pizza, Coffee Spoons, Friendly's, McDonalds, New Bagels, Old Country Buffett, Ramada Inn, Subway, T.G.I. Friday's, Taco Bell

Lodg N: Motel 6

S: Gateway Motor Inn, Mary's Motor Lodge, ◆ Ramada Inn, ◆ Susse Chalet

AServ N: Shell*

S: Grease Monkey Oil Change, Jiffy Lube

ATM S: Bank of Boston, Fleet Bank

Other S: **Lo-Jacks Supermarket, Sam's Club, Wal-Mart**

(3) **Weigh Station (Westbound)**

2 **(5)** MA136, Warren RI, Newport RI

Gas S: Mobil*, Shell*(CW)

Food S: Cathay Pearl Polonesian, Dunkin Donuts, Eskimo King Ice Cream, Quikava, Subway, The Nut House

ATM S: Slade's Ferry Bank

EXIT MASSACHUSETTS

Other S: **Pegas Building(LP)**

(6) **Rest Area - Parking Area Only**

3 **(7)** U.S. 6, MA 118, Swansea, Rehoboth

Gas N: Charter*, Texaco(D)

S: Cumberland Farms*

Food N: Bess Eaton, Burger King, DeAngelo's Sandwich Shop, Dunkin Donuts, Friendly's, Hoy Tin, Island Candy, King Pizza, McDonalds, Plaza Pizza, Ponderosa, Thai Taste

Lodg S: Swansea Motor Inn

AServ N: Broadway Auto Service, Firestone Tire & Auto, Midas Muffler & Brakes, Swansea Auto Center, Texaco

S: NAPA Auto Parts

Med S: ✚ **Walk In Clinic**

ATM N: Compass Bank, Fleet Bank

Other N: Kwik'N Kleen Car Wash, **Swansea Mall**

4AB **(10)** MA103, Somerset, Ocean Grove

Gas N: Getty

S: Shell*

Food N: Famous Pizza, Roger's

Lodg S: Comfort Inn, Quality Inn

AServ N: Getty, Lee Auto Repair

Other N: **Somerset Animal Hospital**

5 **(12)** MA79, MA138, Taunton, N Tiverton RI

Gas S: Gulf(D), Mutual

Food S: Honey Dew Donuts

AServ S: Gulf

Med S: ✚ **St. Anne's Hospital**

6 **(13)** Pleasant St (Westbound, Difficult Westbound Reaccess)

Food S: 99 Restaurant & Pub

7 **(13)** MA81, Plymouth Ave

Gas N: Getty, Merrit(D)

S: Shell*

Food N: 99 Restaurant & Pub, Burger King, Carvelle Ice Cream Bakery, China Room, DeAngelo's Sandwich Shop, Dunkin Donuts, KFC

S: Applebee's, McDonalds

AServ N: Firestone Tire & Auto, Getty, Poirier Buick, Pontiac, GMC, Ford, Lincoln, Mercury

S: Goodyear Tire & Auto

Med N: ✚ **Charleton Memorial Hospital**

S: ✚ **St. Anne's Hospital,** ✚ **Walk-in Clinic**

Other N: **Factory Outlet Center**

EXIT MASSACHUSETT

S: **WalGreens**

8A **(14)** MA24S, Tiverton RI, Newport RI

Other S: **Outlet Center**

8B **(15)** MA24N, Taunton, Boston

Lodg S: Hampton Inn

10 **(16)** MA88, Westport, Horseneck Beach, to U.S.6

11AB **(19)** Reed Rd, Hixville, Dartmouth

12AB **(22)** Faunce Corner, N Dartmouth

Food S: McDonalds, Newport Creamery

AServ S: Firestone Tire & Auto, Hart Toyota

Med S: ✚ **Dartmouth Walk-in Clinic**

ATM S: Fleet Bank

Other S: **National Discount Pet Center**

13AB **(24)** MA140, Taunton, Dartmouth

15 **(25)** MA18S, Downtown, New Bedford

17 **(26)** Coggeshall

Gas N: Mutual*, Shell*(CW), Sunoco

Food N: Dunkin Donuts

AServ N: Roland's, Shell*(CW), Sunoco*

18 **(27)** MA240S, Fairhaven

Gas S: Exxon*, Fast Gas

Food S: Burger King, Dunkin Donuts, Great Wall China, McDonalds, Taco Bell, The Pasta House, Wendy

Lodg S: Hampton Inn

AServ S: Alden Pontiac, Buick, Auto Palace, Fairhaven Mazda, Jiffy Lube, Midas Muffler

Med S: ✚ **Family Medical Walk-in Clinic**

ATM S: Bank of Fall River, Citizen's Bank, Edgewater Credit Union

Other S: **Brooks Pharmacy, K-Mart (Pharmacy), Optical Eyewear, Shaw's Supermarket, Super Stop, Wal-Mart**

19AB **(31)** North Rochester, Mattapoisett

Gas S: Citgo

Food S: Uncle John's Coffee

20 **(35)** MA105, Marion, Rochester

Food S: The Wave Family Restaurant

21 **(39)** MA28

Food N: Zeadey's

Med S: ✚ **Hospital**

Other S: **Wearham Police**

↑MASSACHUSETTS

Begin I-195

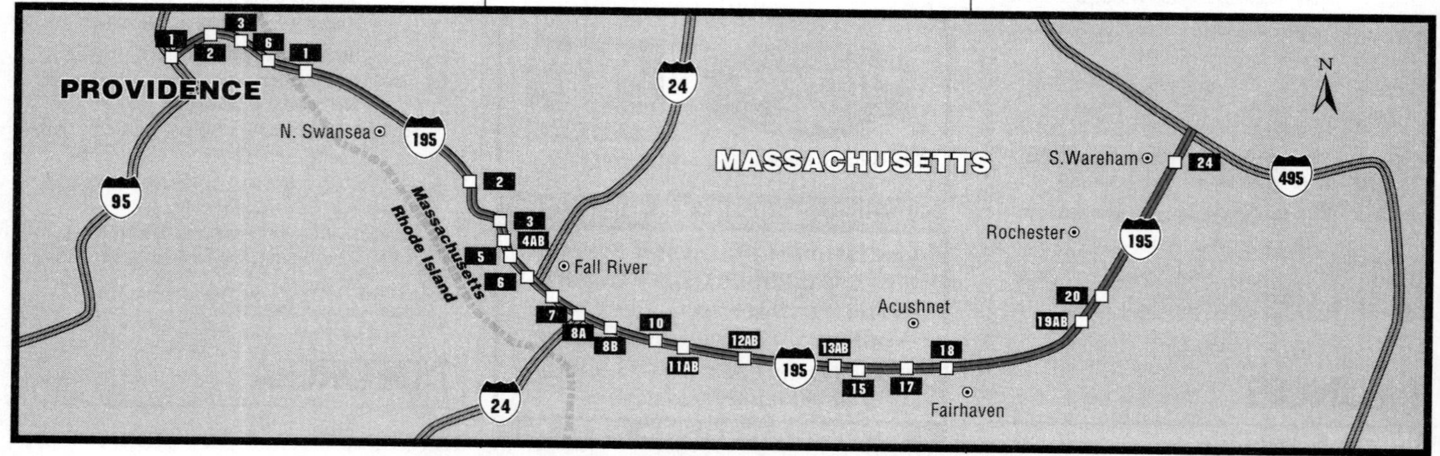

Bold red print shows RV & Bus parking available or nearby

XIT MICHIGAN

Begin I-196

↓MICHIGAN

9 Fuller Ave
- **Gas** N: Marathon*, Super Clean*
  S: Shell*, Speedway*(LP)
- **Food** N: Elbow Room Bar & Grill, Palermo Pizza, Red Lion, Spad's Pizza
  S: Bill's Family Restaurant, Burger King, Checkers Burgers, Dairy Cone, Russo's Pizza & Sub, Subway, Wendy's
- **AServ** N: Tuffy Auto Service
  S: Al's Tamminga Garage, Shell
- **Med** N: ✚ **Kent Community Hospital Complex**
- **ATM** N: FNB Bank
  S: Grand Rapids Teachers Credit Union, NBD, Old Kent Bank
- **Other** N: **Family Fair Supermarket (Pharmacy), Sheriff's Dept**
  S: Auto Spa Car Wash, Coin Car Wash, **Duthler's Grocery, Michigan St. Laundromat, Northeast Cat & Dog Hospital**

8 College Ave
- **Gas** S: Dairy Mart* (24 Hrs)
- **Food** S: Bagel Beanery, McDonalds, Pizza Hut
- **AServ** S: Chuck's Auto Repair
- **Med** S: ✚ **Hospital**
- **Other** S: **Coin Laundry, Rite Aide Pharmacy**

77C Ottawa Ave., Downtown

77AB Junction U.S. 131, Cadillac, Kalamazoo
- **Other** S: **Van Andel Arena**

76 MI. 45 East, Lane Ave.
- **AServ** S: Cowdin's Auto Repair
- **ATM** S: BankWest
- **Other** S: **Coin Laundrymat**

75 MI 45 West

73 Market Ave
- **Other** N: **Van Andel Arena**

70 MI 11, Grandville, Walker (Left Lane Exit)
- **Gas** S: Shell*
- **Food** S: New Beginnings
- **Lodg** S: Land's End Suites
- **AServ** S: Brake Shop, Hub Cap Center, Kenowa Auto Supply, Shell
- **ATM** S: First American Bank
- **Other** S: **Pet Vet Clinic**

69AB Chicago Dr.
- **Gas** N: Meijer Grocery*(D)(LP) (24 Hrs), Phillips 66*, Shell, Total*(LP)
  S: Clark
- **Food** N: Hot n Now, Hungry Howie's Pizza, KFC, McDonalds (Inside Play Place), McDonalds, Mr. Fable's, Subway, TCBY
  S: Arby's, Brann's Steakhouse, Burger King, Get-Em-N-Go, Jerry's Country Inn, Little Caesars Pizza, Ole' Tacos, Pizza Hut, Russ' Family Restaurant, Spad's Pizza, Wendy's
- **Lodg** S: Best Western, Fountain Motel
- **AServ** N: 10 Min. Oil Change, Lentz Muffler's & Brakes, Muffler Man, Phillips 66, Quick Lube, Shell, Total, Valvoline Quick Lube Center
  S: NAPA Auto Parts
- **ATM** N: **First America Bank (Meijer)**, Michigan National Bank
- **Other** N: Baldwin Car Wash, **Coin Laundry**, Jenison Car Wash, **Meijer Supermarket, Department Store (24 Hrs)**
  S: A+ Auto Car Wash, **Fairlanes Bowling**, Grand Village Car Wash, **Tarry Hall Skating, Village Green Miniature Golf**

67 44th St
- **Gas** N: Mobil*(D)(LP)
- **Food** N: Burger King, **Cracker Barrel**
- **Med** S: ✚ **Butterworth Health Campus Med + Center**
- **ATM** S: Byron Center State Bank

62 32nd Ave, Hudsonville
- **Gas** N: Amoco*(D)(CW), BP*(D)(CW)(LP)
  S: Mobil*(D)(CW)
- **Food** N: **Burger King, McDonalds**
  S: Hot Stuff Pizza (Mobil)
- **Lodg** N: Amerihost Inn
  S: Rest All Inn
- **AServ** N: Fast Trak Quick Lube, Timmer Chevrolet
- **TServ** S: **Hudsonville Trailer**
- **RVCamp** N: **Camping**
- **ATM** N: Byron Center State Bank

(58) **Rest Area - RR, Phones, Picnic**

55 Bus. Loop 196, Zeeland, Holland
- **Gas** N: Citgo*
- **Food** N: **McDonalds (Inside Play Place)**
- **RVCamp** N: **Camping**
- **Med** N: ✚ **Hospital**
- **Parks** N: **Holland State Park**

52 16th St., Adams St.
- **Gas** S: Mobil*(D)
- **Food** S: **Burger King**, Subway (Mobil)
- **Med** N: ✚ **Hospital**
  S: ✚ **Hospital**

49 MI 40, Allegan
- **TStop** S: **Tulip City Truck Stop*(SCALES) (Total Gas)**
- **Food** S: Rock Island Family Restaurant
- **TServ** S: **CB Shop Tulip City TS, K & R Trailer Repair, Tulip City TS (Wrecker Service)**
- **TWash** S: **Tulip City TS**
- **ATM** S: **Tulip City TS**

44 U.S. 31, Holland, Bus. Loop 196, Muskegon (Eastbound)

(43) **Rest Area - RR, Phones, Picnic (Westbound)**

41 County Rd A2, Saugatuck, Douglas
- **Gas** N: Shell*(D), Total*(D)(LP)
- **Food** N: Burger King, Subway (Shell)
- **RVCamp** N: **West WinnCamping**
- **ATM** N: Shell
- **Parks** N: **Saugatuck Dunes Park**

36 County Rd A2, Ganges

34 MI 89, Fennville
- **Gas** S: Shell*
- **AServ** S: Fleming's Collision
- **Parks** N: **Westside County Park**

30 County Rd. A2, Glenn

(29) **Rest Area - RR, Picnic, HF, Phones (Eastbound)**

26 109th Ave., Pullman

22 North Shore Dr.
- **Food** N: Cousins' Restaurant
- **RVCamp** N: **Cousins' Camping, Jensen's RV Camp**
- **Parks** N: **Kal-Haven Trail State Park**

20 Bus Loop I-196, MI. 43, South Haven, Bangor
- **Gas** N: Amoco*(D)(CW), Marathon*(D)
- **Food** N: Hot n Now Hamburgers
  S: Big Boy, **McDonalds (Wal-Mart)**, Sherman Dairy Bar
- **Lodg** N: South Haven Motel
- **AServ** S: Wal-Mart
- **RVCamp** N: **Camping**
  S: **Camping**
- **Med** N: ✚ **Hospital**
- **ATM** N: Amoco
- **Other** N: Car Wash, **State Police, Village Coin Laundry**
  S: **Sears, Wal-Mart (Pharmacy & Optical)**

18 MI 140, Watervliet
- **FStop** N: **Shell***
- **Gas** N: Checker*
- **Food** N: **Ma's Coffee Pot (24 Hrs)**
- **AServ** N: McFadden Chrysler Plymouth Dealer
  S: Randy's Transmission Service

13 Covert
- **RVCamp** N: **Camping**
- **Parks** N: **Van Buren State Park**

7 MI 63, Benton Harbor, St. Joseph
- **Food** N: DiMaggio's Pizza, Flo & Jerry's Beachside Restaurant, Icecream Vault, Vitale's Market/Deli
- **AServ** N: Skip's Ace Garage, Stefani's Truck & Auto Parts
- **TServ** N: **Stephani's Truck Parts**

4 Coloma, Riverside
- **Gas** S: Marathon*(D) (24 Hrs)
- **RVCamp** S: **KOA Camping**
- **ATM** S: Marathon

1 Red Arrow Hwy
- **TServ** N: **ABC Truck Repair**

(0) Junction I-94, Chicago, Detroit

↑MICHIGAN

Begin I-196

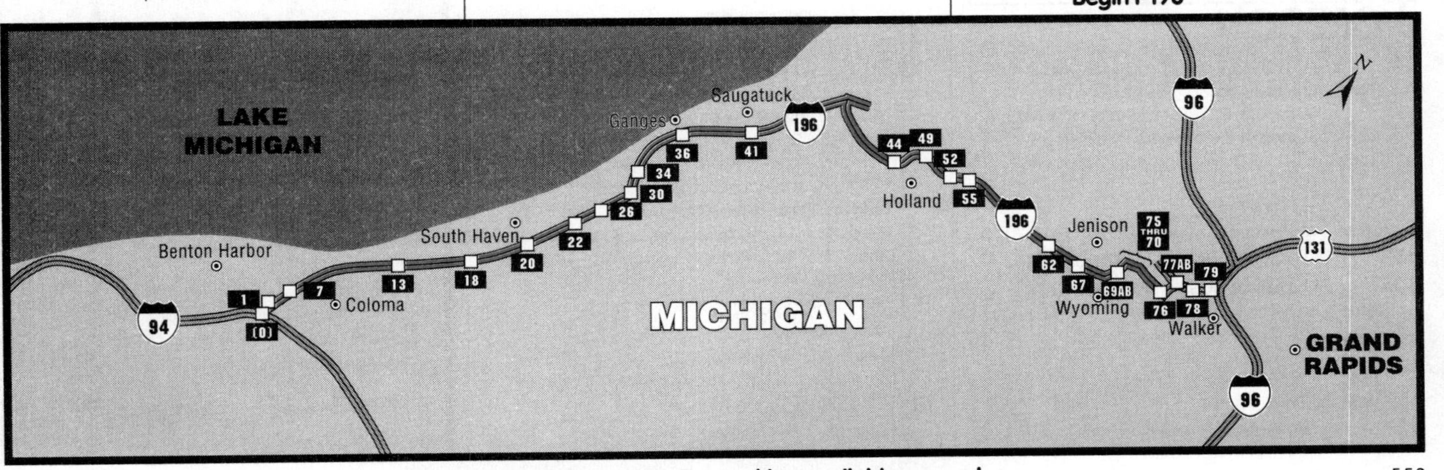

Bold red print shows RV & Bus parking available or nearby

EXIT WASHINGTON/OREGON

Begin I-205

↓WASHINGTON

36 NE 134th St.
- **Gas** E: Vista Mart*[D]
  - W: 76, Citgo*, Trail Mart*[D]
- **Food** W: Burger King (Playground), Burgerville USA, J.B's Roadhouse, McDonalds (Playground), Round Table Pizza, Taco Bell
- **Lodg** W: Comfort Inn, Shilo Inn
- **AServ** W: 76
- **RVCamp** W: **Spring's RV Service**
- **ATM** E: Vista Mart
  - W: Citgo, Trail Mart
- **Other** E: **Albertson's Grocery, Salmon Creek Vision Center**
  - W: **Hi-School Pharmacy**

32 NE 83rd St. Andresen Rd.

30 WA 500 Vancouver Orchards (No Southbound Reaccess)
- **Gas** E: PDQ Market*
  - W: Arco*, Chevron*
- **Food** E: Dairy Queen, Denny's, Imperial Palace, Orchard's Subs, Orchards Teriyaki, Smokey's Hot Oven Pizza, Suprema Pizzaeria
- **Lodg** W: (AAA) Comfort Suites, ◆ Holiday Inn Express, Sleep Inn
- **AServ** E: Napa Auto Parts, Orchards Automotive, Orchards Tire Service, Ron's Automotive & RV Service, Vancouver Toyota
  - W: Master Tech Automotive (Chevron), Sears (Chevron)
- **Other** E: **E-Z Coin Laundry, Hi-School Pharmacy, Spencer's Books, Thriftway**
  - W: **JC Penny, Vancouver Mall**

28 Mill Plain Rd.
- **Gas** E: BP*, Chevron[D], Texaco*[CW][LP]
  - W: Arco*, Texaco*
- **Food** E: Baskin Robbins, Brandy's, Burger King, Burgerville USA, Cascade Lantern, Dairy Queen, El Charrito, Elmer's, Gingiss Kahn, Mings, Pizza Hut Carryout, Saris Restaurant, Smokey's Pizza, Starbuck's Coffee, Taco Bell, Taco King
  - W: Happy Family Restaurant
- **Lodg** E: Extended Stay America, Roadway Inn, The Guest House Motel, The Travel Lodge
- **AServ** E: Les Schwab Tires, Schuck's
  - W: Bob Kendall's Chevrolet, Geo, Hyundai, Jiffy Lube, Ricki's Tire Factory, The Brake Shop, Windshield Express
- **Med** W: **✚ Hospital**
- **ATM** E: First Independent Bank, Northwest National Bank, Seafirst Bank
- **Other** E: **Fred Meyer Grocery, Petco**

27 Washington 14 W. Vancouver

↑WASHINGTON

↓OREGON

24 Airport Way, Portland Airport
- **Food** E: Burger King (Playground), China Wok, McDonalds (Playground), Restaurant (Shilo Inn), Sharis, Sidelines Restaurant, Subway
- **Lodg** E: Comfort Suites, Fairfield Inn, Holiday Inn Express, Marriott, Shilo Inn, Silver Cloud Inn, Super 8 Motel
- **Other** W: **Portland International Airport**

23B U.S 30 Bypass W. Columbia Blvd.

30A U.S. 30 Bypass E. Sandy Blvd.
- **FStop** E: **Shell***[D]
- **Gas** E: Leather's Fuel*[LP], Texaco*[D][CW]
- **Food** E: Bill's Steakhouse, Elmer's
- **Lodg** E: Best Western, Travelodge
- **ATM** E: Bank Of America

22 The Dalles. E. 184 & U.S. 30

EXIT OREGON

21B West I-84, US 30, West Portland

21A Stark St., Glisan St.

20 Washington Street, Stark St. (Northbound, Watch One Ways)
- **Food** E: Arby's, Bagel Sphere, Baskin Robbins, Burger King, Coffee's On, Elmers, McMenamins, Newport Bay, Subway, Tony Roma's, Wallstreet Pizza
  - W: Stark Street Pizza Co., Taco Bell
- **Lodg** E: Chestnut Tree Motel, Holiday Inn Express, Montavilla
  - W: Roadway Inn
- **AServ** W: Small 205 Transmissions, Tanner's Automotive
- **ATM** W: 7-11 Convenience Store
- **Other** E: **Kinko's, Mall 205**
  - W: **7-11 Convenience Store (ATM)**

19 U.S. 26 Division St.
- **Gas** E: Exxon
  - W: Chevron, Texaco*
- **Food** W: Burgerville, Chuck E. Cheese's Pizza, McDonalds
- **AServ** E: Exxon
  - W: Chevron, Jiffy Lube
- **ATM** W: 7-11 Convenience Store
- **Other** W: **7-11 Convenience Store (ATM)**

17 Foster Rd.
- **Gas** W: Chevron*
- **Food** W: New Copper Penney Restaurant

16 Johnson Creek Blvd.
- **Gas** W: Texaco
- **Food** W: McDonalds (Playground), Ron's Restaurant, Taco Bell, Tortilla Flats
- **AServ** W: Texaco
- **RVCamp** W: **Fred's RV World, Olinger Travel Homes, Spencer RV Sales & Service, Town and Country RV Park**
- **ATM** W: 7-11 Convenience Store, Wells Fargo
- **Other** W: **7-11 Convenience Store (ATM),** Coin Car Wash, **Fred Meyer Grocery, Petsmart**

14 Sunnyside Rd.
- **Gas** E: 76[LP]
- **Food** E: Coffee Biestro, Gustav's Pub & Grill, Izzy's Pizza, KFC, Menamins, Subway, Thai House
  - W: Chevy's, Macheezmo Mouse, Olive Garden, Red Robins, Rockin Robins, Spaghetti Factory, Starbuck's Coffee, ToGo's Eatery
- **Lodg** E: (AAA) Best Western, (DAYS INN) (AAA) Days Inn
  - W: Monarch Motel
- **AServ** E: 76
  - W: Montgomery Ward
- **Med** E: **✚ Hospital**
- **ATM** E: U. S. Bank
  - W: Bank of America
- **Other** E: **Kinko's**
  - W: **Barnes & Noble, Clackamas Town Mall, JC Penny, Montgomery Ward, Target Department Store**

13 OR. 224W Milwaukee

12AB E. OR. 212 & OR. 224, Clackamas Estacada, Webster Rd., Johnson City
- **Gas** E: Chevron*, Texaco*[CW]
- **Food** E: 2 Scoops, Clackadeli, Denny's, KFC, McDonalds (Playground), New Cathay Deli, Sunshine Pizza Exchange, Taco Bell, Wendy's
- **Lodg** E: Clackamas Inn, Hampton Inn
- **ATM** E: 7-11 Convenience Store, US Bank, Wells Fargo
- **Other** E: **7-11 Convenience Store (ATM), Fred Meyer, Post Office**

11 Gladstone
- **Gas** W: Arco*
- **AServ** W: The Golden Wrench
- **Other** W: **Safe Way Grocery (Grocery Store)**

10 Park Place Molalla
- **FStop** E: **Pacific Pride Commercial Fueling**

EXIT OREGO

- **Med** E: **✚ Hospital**

9 OR. 99E. Oregan City Gladstone
- **Gas** E: 76*, Arco
- **Food** E: Friendship Chinese, KFC
  - W: Edgewater Inn, La Hacienda, McDonalds (Playground), Restaurant (Val-U Inn), Shari's, Subway
- **Lodg** W: (AAA) Val-U Inn
- **AServ** E: Arco, Clackamas Auto Parts, John Link Pontia GMC, Mazda, Ming's Auto Repair, Napa Auto Parts
  - W: Firestone Tire & Auto
- **ATM** E: US Bank
  - W: Wells Fargo
- **Other** E: **McLoughlin House**
  - W: **Book Store, Payless Drugs**

(8) **Rest Area - RR, HF, Picnic, Phones, RV Dump (P)**

8 West Linn Lake Oswego
- **Gas** E: BP
  - W: Astro[D][LP], Texaco[D][LP]
- **Food** W: Bagel Basket, Pasta's, Starbuck's Coffee
- **AServ** E: BP
  - W: Astro, Texaco
- **ATM** W: Key Bank
- **Other** W: **Post Office, Thriftway**

6 West Linn 10th St
- **Gas** E: Chevron*[LP]
- **Food** E: McDonalds, McMenamins, Sharis

3 Stafford Rd. Lake Oswego
- **Med** W: **✚ Hospital**

2 Tigard Salem. Jct. I-5

↑OREGON

Begin I-205

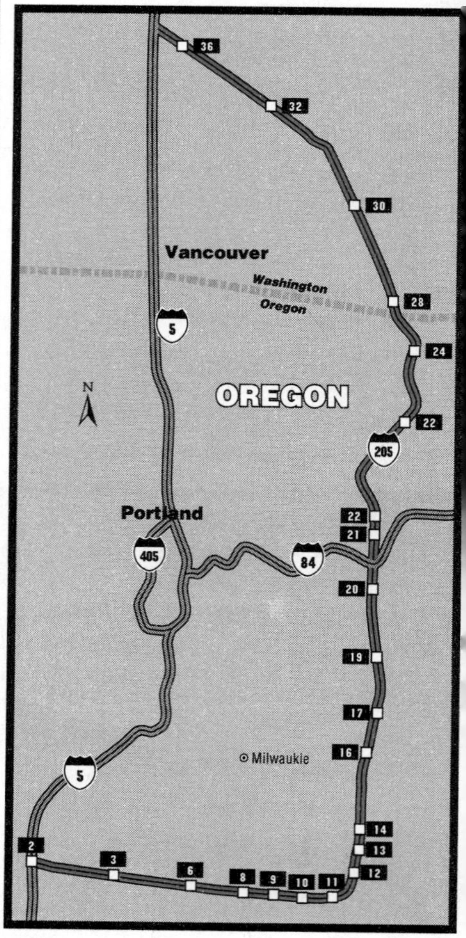

Bold red print shows RV & Bus parking available or nearby

XIT	LOUISIANA

Begin I-220

↓LOUISIANA

7B I-20 West Shreveport, I-20 East Monroe

7A US 79, US 80, Racetrack
- **Gas** O: Circle K*[LP]
- **Food** O: Waffle House
- **ATM** O: Circle K
- **Other** O: Louisiana Downs

5 Shed Rd

3 Swan Lake Rd
- **Other** I: Food Lion Grocery

2 LA 3105, Airline Drive, Bossier City
- **Food** I: Burger King (Playground), McDonalds (Playground), Taco Bell
- **Med** O: ✚ Bossier Health Center
- **Other** I: Total* (No Gas)

1 LA 3, Benton, Bossier City
- **Gas** I: Chevron*[CW] (24 Hrs)
- **AServ** O: John Harvey Toyota, Lexus, Mike Morgan Pontiac, Buick
- **RVCamp** O: Maplewood RV Park (2.8 Miles)
- **Med** O: ✚ Bossier Health Center

7AB US 71, LA 1, Shreveport, Texarkana
- **Gas** I: Chevron*[CW], Racetrac*, Texaco*[D][CW]
  O: Conoco*, Exxon*[D], Fina*[D], Mobil*
- **Food** I: McDonalds (Playground), O'Malley's Deli, Podnuh's BBQ, Taco Bell, Trejo's (Howard Johnson)
  O: Domino's Pizza, Frosty Express, Sonny's Catfish, Seafood, Subway, What A Burger (24 Hrs)
- **Lodg** I: Howard Johnson
- **AServ** I: AAA Auto Clinic
  O: NAPA Auto Parts, Nelson's Kar Kare, Tri-State Battery Supply
- **RVCamp** O: Coleman RV Mobile Home Park
- **ATM** I: Hibernia
  O: Regions Bank
- **Other** I: Eckerd Drugs, North Shreveport Animal Hospital, Pet Lodge & Supply, Speed Queen Coin Laundry
  O: Brookshire's Grocery (Pharmacy), Northern Hills Pharmacy, The Soap Opera (Coin Laundry)

5 LA 173, Blanchard Rd.

2 Lakeshore Drive

1D LA 511, West 70th St. (Westbound)

1C I-20 East, Monroe (Westbound)

1B West I-20, Dallas (Westbound)

1A Jefferson Paige Rd.

↑LOUISIANA

Begin I-220

EXIT	IOWA

Begin I-235

↓IOWA

(0) JCT I-35, I-80

(1) Valley West Dr.
- **Gas** I: Amoco*[D][CW]
- **Food** I: T.G.I. Friday's
- **ATM** I: Iowa Savings Bank, Magna Bank
- **Other** I: West Valley Mall

(2) 22nd Street
- **Food** I: Chuck E. Cheese's Pizza, Gino's West, McDonalds, Taco Bell
- **AServ** I: Goodyear Tire & Auto
- **ATM** I: First Bank, Mercantile Bank
- **Other** I: Half Price Grocery Store, Osco Drugs

(3) 8th Street, West Des Moines, Windsor Heights
- **Gas** I: Amoco*, Coastal*, Kum & Go (Open 24 Hours), Quick Trip*
- **Food** I: B-Bops Hamburgers, Blimpie's Subs, Burger King, Cabo San Lucas (Best Western), Jimmy's American Cafe
- **Lodg** I: Best Western
- **AServ** I: Amoco
- **ATM** I: Amerius Bank, Kum & Go
- **Other** I: Sam's Club, Wal-Mart

(4) IA 28, 63rd Street

13 56th Street

12 42nd Street

11 31st Street

10 Cottage Grove

9 Martin Luther Jr. Pkwy, Airport
- **Lodg** S: Hampton Inn Airport

8 Keo Way
- **Lodg** I: Holiday Inn

7 3rd Street, 5th Ave
- **Lodg** O: Best Western

6 East 6th Street

5 Penn Ave. (Difficult Reaccess)

4 US 65 South, US 69 South, East 15th Street, East 14th (Difficult Reaccess)
- **Gas** I: Sinclare

3 US 65 North, IA 163 West, East University Ave.

2 Guthrie Ave.
- **FStop** I: Coastal*[D]

1 US 6, Euclid
- **Gas** O: Amoco, Coastal*, Phillips 66*
- **Food** I: Denny's
  O: Burger King (Indoor Playground), Perkins Family Restaurant
- **ATM** O: Firststar Bank
- **Other** O: Vision Center (Leetown Center), WalGreens

↑IOWA

Begin I-235

EXIT	TENNESSEE

Begin I-240

↓TENNESSEE

25AB 25A - I-55 S, Jackson MS, 25B - I-55 N, St Louis MO

24 Millbranch Rd
- **Lodg** O: ◆ Budgetel Inn, ◆ Hampton Inn
- **TServ** O: Kenworth

23AB Airways Blvd International Airport
- **Gas** I: Texaco*, Total*
- **Food** I: Big Dip Dairy Bar, Church's Fried Chicken, Dixie Queen Ice Cream & Hamburgers, Subway
- **Lodg** I: Airways Inn, Economy Inn
- **AServ** I: Danny Dunn Transmission
- **Other** I: Cash It Plus, Comfort Coin-Op Laundry, Easy Quick #2*

21 US 78, Lamar Ave., Birmingham
- **Gas** I: Circle K*, Citgo*, Citgo, Phillips 66*
- **Food** I: Big D Dairy Bar
- **Lodg** I: Liberty Inn
  O: ◆ Econolodge, Holiday Inn Express
- **AServ** I: Citgo, Park Lane Service
  O: Bill King's Brake-o
- **Other** I: Coin-Op Laundry & Car Wash, Rook Sales, Supervalu Foods, U-Haul Center[LP]

20 Get Well Rd
- **Gas** I: BP*[CW]
  O: Amoco*[CW], Coastal*[D], Texaco*[D]
- **Food** I: McDonalds
  O: Emilio's, Rice Bowl, Subway, Wendy's
- **Lodg** O: Red Roof Inn
- **AServ** I: Bluff City Nissan Isuzu Jaguar Buick Pontiac, Coleman Taylor Transmissions, Infiniti of Memphis
  O: Jiffy Lube
- **ATM** O: Kroger Supermarket
- **Other** I: Sam's Club
  O: Coin-Op Laundry, Kroger Supermarket (w/ drug store), ProGolf, Save Shop Grocery Store, Z National Market*

18 Perkins Rd
- **Gas** O: Ace, Circle K*, Exxon, Gas, Texaco*[D][CW], Total*[LP]
- **Food** O: Arby's, Captain D's Seafood, Danver's Restaurant, Golden Corral, IHOP, Kenny Rogers Roasters, Pancho's Taco, Pizza Hut, S & S Cafeterias, Shoney's, Taco Bell, The Olive Garden
- **Lodg** O: ◆ Hampton Inn, Marriot Hotel
- **AServ** O: Ace, Express Lube, Exxon, Master Care Auto Service
- **ATM** O: First Tennessee Bank, Union Planter's Bank
- **Other** O: Car Wash, Mall of Memphis, Office Max, Pet Inns, Toys "R" Us

17 Mt Moriah Rd
- **Gas** I: BP*[LP]
  O: Pump and Save, Total*[LP]

EXIT	TENNESSEE
Food	**I:** Crumpy's Hot Wings, Three Little Pigs Bar-B-Q **O:** Denny's, El Torero Mexican, Lone Star Steakhouse, Picadilly Cafeteria, The Flaming Grille, Wendy's
Lodg	**O:** ◆ Holiday Inn
AServ	**O:** Automax, Chuck Hutton Olds, Chev, Chuck Hutton Trucks, Dobb's Ford, Liberty Chrysler Plymouth Subaru, Midas, NTW Tire Center, Performance Toyota, **Wal-Mart**
ATM	**O:** Megamarket, Union Planter's Bank
Other	**I:** **Billy Hardwick's All Star Lanes (Bowling), WalGreens, Yorkshire Launderette Coin Laundry** **O:** **Buck and Bass Sports Center, Delta Square Shopping Mall, Megamarket,** Mr. Pride Car Wash, **Police Station, Wal-Mart**
16	TN 385 East, Nonconnah Pkwy (Limited Access Hwy)
15	US 72, Poplar Ave, Germantown, St Francis Hosp. Ex. 15E
Gas	**I:** Texaco*(LP)
Food	**I:** Cafe Expresso (The Ridgeway Inn), Garibaldi's Pizza, Wendy's, Yates Avenue Grocery & Deli **O:** Cockeyed Camel (Comfort Inn)
Lodg	**I:** ◆ The Ridgeway Inn **O:** Adam's Mark Hotel, ◆ Comfort Inn, ◆ Holiday Inn, Homewood Suites Motel, Hampton Inn (see our ad page 555)
Med	**O:** **✚ Hospital**
Other	**I:** **The Dive Shop**
13	Walnut Grove Rd
Med	**O:** **✚ Baptist Memorial Hospital East**
12C	I-40 East, Nashville
31B	I-40 West, Little Rock
26	Norris Rd
28AB	S Parkway
Gas	**O:** Amoco, BP*, Exxon*
Food	**I:** Coletta's Italian, McDonalds **O:** Fish Market
Lodg	**I:** Parkway Inn
AServ	**O:** Amoco, Auto Factory, BP, Miles of Service
Other	**I:** **Reed's Sunry*** **O:** **Kerr Avenue Supermarket**
29	Lamar Avenue
Gas	**I:** Total* **O:** Citgo*
Food	**I:** China Doll, Rally's Hamburgers **O:** Nate's Fish & Wings
Lodg	**I:** Lamplighter Motor Inn
Med	**I:** **✚ Hospital**
Other	**I:** **Alpha Express Food Mart** **O:** **Kim's Grocery Store**
30	Union Ave, Madison Ave
Gas	**I:** Exxon*, Exxon*(CW)
Food	**I:** Admiral Benbow Inn, Arby's, Backyard Burgers, Burger King, Burger King, C K's Coffee Shop, Denny's, Krystal, McDonalds, Minh-Chau, Shoney's, Tops Bar-B-Q, Western Steak House & Lounge
Lodg	**I:** Admiral Benbow Inn, ◆ Hampton Inn, ◆ Red Roof Inn **O:** AAA La Quinta Inn
AServ	**I:** Goodyear Tire & Auto, Schilling Lincoln Mercury **O:** Master Care Auto Service
Med	**I:** **✚ Baptist Memorial Hospital, ✚ Methodist Hospitals of Memphis** **O:** **✚ Hospital**
ATM	**I:** First Tennessee Bank
Other	**I:** **Ed's Camera Shop, Southern College of Optometry Eye Institute, Stuart Bros Hardware Co**

↑TENNESSEE

Begin I-240

EXIT ILLINOIS

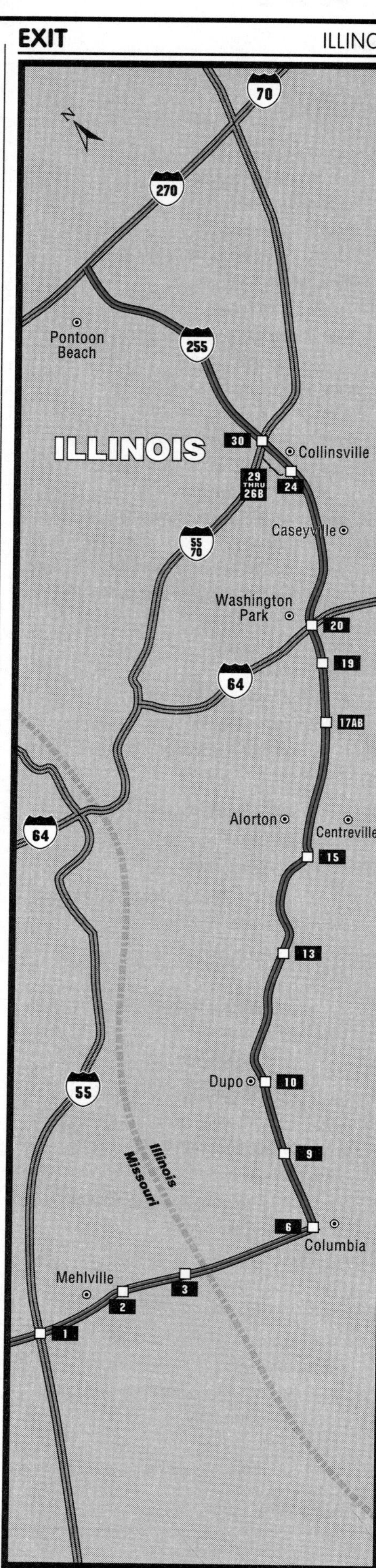

EXIT	MISSOURI/ILLINOIS

Begin I-255

↓MISSOURI

1	I-55 Jct
(5)	**Rest Area - Picnic, RR, HF, Phones** P

↑MISSOURI

↓ILLINOIS

30	JCT I-70, Kansas City
29	**(30)** IL. 162 Glen Carbon, Granite City
26	**(28)** Horseshoe Rd
26B	Jct I-55 S I-70 W
24	Collinsville Rd.
20	**(23)** I-64 Jct.
19	**(20)** State St. E. St. Louis
17AB	E. Saint Louis, Belleville
Lodg	**I:** Budget Motel, Deluxe Hotel, Lakeside Motel
Med	**I:** **✚ Hospital, ✚ Hospital**
15	Mousette Ln.
13	**(15)** IL. 157 Cahokia
Gas	**I:** Amoco*(CW) **O:** Clark*
Food	**I:** Burger King, Captain D's Seafood, China Express, Hardee's, KFC, Pizza Hut Delivery, Rally Hamburgers **O:** Norman's Family, Smokie's BBQ
AServ	**I:** Goodyear Tire & Auto **O:** General Auto Service
RVCamp	**I:** **Camping**
Other	**I:** **Auto Zone Auto Parts, Car Quest Auto Center, Pharmacy, Quaker State Lube, Schnucks Grocery Store, Scotchwash Coin Laundry, Wal-Mart** **O:** Car Wash, **Coin Laundry, Fisher Pharmacy**
10	**(13)** IL. 3 N Cahokia E. Saint Louis
9	Dupo
6	**(9)** IL. 3 South Columbia

↑ILLINOIS

↓MISSOURI

3	Koch Rd
2	**(3)** MO 231, Telegraph Rd
1C	U.S. 61, U.S. 67, MO 267 (No Reaccess Westbound)
Gas	**I:** Amoco, Amoco*, Phillips 66*(CW), Shell
Food	**I:** Arby's, Arby's, Chinese Buffet, Chuck E. Cheese's Pizza, Dairy Queen, Gingham's Homestyle Restaurant, Hooter's, KFC, LJS, McDonalds, Po Folks, R.T. Furr's, Steak 'N Shake, Subway, Taco Bell, Uncle Bill's Pancake & Dinner House, Wendy's
AServ	**I:** Amoco, Chrysler Auto Dealer, Dobbs Tire & Auto, Dodge Dealer, Easton Tire Co., Ford Dealership, GM Auto Dealership, Meineke Discount Mufflers, Midas Muffler & Brakes, Tire America, Valvoline Quick Lube Center
ATM	**I:** Allegiant Bank, Mark Twain Bank
Other	**I:** **A to Z Auto Part Store, K-Mart, Mall, Office Max,** South County Car Wash, **Western Auto**

↑MISSOURI

Begin I-255

XIT KENTUCKY

Begin I-264

↓KENTUCKY

AB I-64, US 150, Lexington, St Louis
Bank Street, NW Parkway (Southbound, Counterclockwise)
Gas **I:** Loto*

River Park Dr., Muhammad Ali Blvd.
AServ **I:** Tune-Up Center
O: Alignment & Brake Service
Other **O:** **Fire Station**

Dumesnil St., Virginia Ave.

Bells Lane, Algonquin Pkwy

Ralph Ave., Cane Run Rd.
FStop **I:** **Marathon***(D)
Gas **I:** Speedway
O: Amoco*, Dairy Mart* (Kerosene), Thornton's*(D)(LP) (Kerosene)
Food **O:** Dunkin Donuts (Thornton's)
AServ **I:** Alvin's Auto Service, Ferguson's Truck Stop (Car & Light Truck Repair), J.J. Exhaust Systems, Mike Hall Auto Parts
O: Hi-Way Muffler Sales & Service, Ken Towery's Firestone
ATM **O:** Dairy Mart, Thornton's
Other **O:** **Cane Run Hardware, Coin Laundry**

AB US 31W, US 60, Ft Knox, Shively
Gas **I:** Amoco*
O: Citgo*, Speedway* (Kerosene), Sunoco* (Kerosene), Thornton's* (24 hrs)
Food **O:** Bojangles, Chi-Chi's Mexican, Dunkin Donuts (Thornton's), KFC, McDonalds, O'Charley's, Restaurant (Royal Inn Motel), Taco Bell, The Banquet Table Buffet, The Italian Oven Restaurant
Lodg **I:** Holiday Inn
O: Louisville Manor, Royal Inn Motel, Toad Stool Inn
AServ **I:** Auto Radiator Co., Byerly Ford, Byerly Nissan
O: Auto Glass of Louisville, Jim Vincent Body Shop, Midas Muffler & Brakes
RVCamp **O:** **Otter Creek Park (25 Miles)**
Med **O:** ✚ **Hospital**
ATM **I:** Citizen's Bank, Great Financial Bank
O: Fifth Third Bank, Thornton's
Other **I:** **Rite Aide Pharmacy**
O: **Check Advance,** Rainbow Car Wash, Regal Auto Wash, Van & Truck Wash (light trucks)

9 Taylor Blvd, Church Hill Downs
Gas **O:** Ashland* (Kerosene), Dairymart*
Food **I:** Cactus Jack's Bar & Grill
O: Dairy Queen, Restaurant, Sassy's Subs & Stuff
AServ **O:** NAPA Auto Parts
Med **O:** ✚ **Hospital**
Other **I:** **Church Hill Downs, Kentucky Derby Museum**
O: **Hazelwood Coin Laundry, Reasor's Supermarket**

10 KY1020, Southern Pkwy, 3rd St.
Gas **I:** Dairy Mart* (Kerosene)
O: Thornton's*(D)(LP) (Kerosene)
Food **O:** Dunkin Donuts (Thornton's), Long John Silvers
ATM **I:** Dairy Mart
O: Thornton's

(10) Crittenden Dr., Fair/Expo Ctr. Gates 2, 3, 4

11 Crittenden Dr., Airport, Fair/Expo Ctr

12 I-65, KY 61, Nashville, Indianapolis

EXIT KENTUCKY

14 KY 864, Poplar Level Rd.
Gas **I:** SuperAmerica*(D)(LP)
Food **I:** Bamboo House Chinese American Restaurant, Charles Hertzman Bakery, Donut Factory, Frisch's Big Boy, Lolita's Mexican
O: Arby's
AServ **I:** Art's Service Center, Auto Truck & Fiberglass of Louisville, Consolidated Tire, Franck Bros. Auto Shop, K-Mart, The Precision Body Shop of Louisville, WIP Imported Auto Parts
O: Veh-I-Cycle Auto Repair
TServ **O:** **Kenworth, Peterson GMC Trucks**
Med **I:** ✚ **Hospital**
ATM **I:** Bank One
Other **I:** **Abram's Animal Hospital, K-Mart, Kroger Supermarket (w/ Pharmacy), Louisville Zoo**

15AB Newburg Rd., KY1703
Gas **O:** BP*(CW)(LP), SuperAmerica*(LP)
Food **O:** Horsefeathers (Ramada), Long John Silvers, Oriental Star, Subway, Waffle House
Lodg **O:** Ramada, Red Roof Inn, Wilson Inn
AServ **O:** Cross Pontiac GMC Trucks (Light Trucks), Motor Works Inc. Auto Repair, NAPA Auto Parts
ATM **O:** National City Bank, PNC Bank, SuperAmerica
Other **O:** **Animal Emergency Center & Hospital**

16 Bardstown Rd, US31E, US150
Gas **I:** BP*(CW)
O: Shell, Thornton's Food Mart*(LP)
Food **I:** Buckhead, Darryl's, Krispy Kreme, Thai-Saim Restaurant, The Bakery, The Daily Fare Deli & Market
O: India Palace (Quality Inn), KFC, McDonalds, Mr. Gatti's, New World Chinese Restaurant, Rally's Hamburgers, Steak 'N Shake, Wendy's
Lodg **O:** (AAA) Holiday Inn, Junction Inn, Quality Inn
AServ **I:** Goodyear Tire & Auto
O: Midas, Shell
ATM **I:** National City Bank
Other **I:** **Mail Boxes Etc, Markwell's Supermarket, Rite Aide Pharmacy (24 hrs)**
O: **Coin Laundry, Rose Bowl (Bowling)**

17AB KY155, Taylorsville Rd.
Gas **O:** Thornton's*(D)

18 18A - KY1932 S, Breckenridge Ln 18B - KY1932 N, Breckenridge Ln
Food **I:** China Town, Dooley's Hot Bagels, Mamma Grisanti's Italian, O'Charley's, Rafferty's, Red Lobster, Skyline Chili, Texas Roadhouse, The Italian Oven Restaurant, Wendy's
O: Rally's Hamburgers
Lodg **O:** Breckinridge Inn
AServ **I:** Neil Huffman Nissan, Pep Boys Auto Center, St Matthews Automotive Center, Subaru, Tri City Olds
O: **K-Mart**
Med **I:** ✚ **Columbia Suburban Hospital**
Other **I:** Car Wash, **Dr. Bizer's Vision World, Korrect Optical, Springer & Lee Optical, Village 8 Theatres, Winn Dixie Supermarket (w/ Pharmacy)**
O: **K-Mart**

19AB I-64, Lexington, Louisville

20AB Jct. US 60, Shelbyville Rd., Middle Town, St. Matthews
Gas **O:** BP*(LP), Chevron(LP)
Food **I:** Frisch's Big Boy, Honey Baked Ham Co., Logan's Roadhouse, McDonalds, Moby Dick's, Outback Steakhouse, Remington's, Rollo Pollo

EXIT KENTUCKY

Chicken, Taco Bell
O: Shoney's, Taco Bell
Lodg **I:** (DAYS INN) Days Inn
AServ **I:** Goodyear Tire & Auto, Precision Tune Auto Care
O: Acura at Oxmoor, Budget, Chevron, Ken Towery Firestone, Star Ford
Other **I:** **Drug Emporium, Mall St. Matthews,** Sparkle Brite Car Wash, **Ten Pin Lanes**
O: **Oxmoor Center**

22 Brownsboro Rd., US 42
Gas **I:** Marathon
O: BP(CW), Chevron*(CW), Swifty
Food **I:** Brewed Awakening Coffee Shop, Gast Haus, Mike Best's Meat Market
O: Arby's, Baskin Robbins, Bruegger's Bagel, Emperor of China, KFC, Kingsley Meat & Seafood, Pizza Hut & Subs, Taco Bell, Thai Cafe
Lodg **I:** ◆ Ramada Inn
AServ **I:** Marathon
O: BP, Chevron
ATM **I:** PNC Bank
O: Bank One, Bank of Louisville, Kroger Supermarket, Republic Bank
Other **I:** **Brownsboro Hardware**
O: **AAA Auto Club KY, Golf USA, Kroger Supermarket (24 hrs), Perfect Photo, Revco Drugs, Springer & Lee Optical**

23AB 23A - I-71N, Cincinatti, 23B - I-71S, Louisville

↑KENTUCKY

Begin I-264

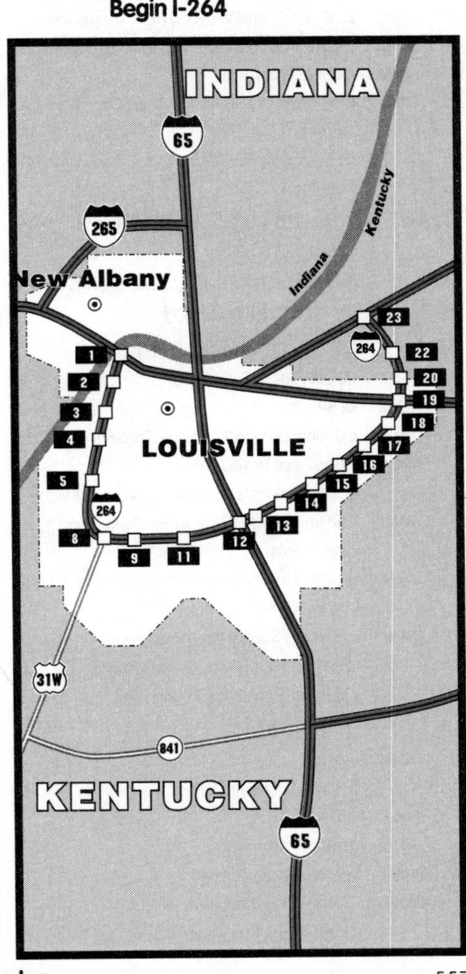

EXIT KENTUCKY

Begin I-265

↓KENTUCKY

35AB 35A - N. I-71, Cincinatti, 35B - S. I-71, Louisville (Gene Snyder Hwy)

34 KY 22, Brownsboro Rd, Crestwood
- **Gas** O: BP*(LP)
- **Food** I: 888 Great Wall, Philly Connection, The Italian Oven Restaurant
- **ATM** I: Kroger Supermarket, Stockyards Bank
- **Other** I: **Feeders Supply Pet Foods, Kroger Supermarket (w/pharmacy, 24 hrs), One-Hr. Moto Photo**

32 KY 1447, Westport Rd, Chamberlain Lane
- **Gas** I: BP*(CW), Thornton's*(LP)
  O: SuperAmerica(D)(CW)
- **Food** I: Dunkin Donuts (Thornton's)
  O: Burger King, Frisch's Big Boy, HomeTown Buffet, Moby Dick, Subway (SuperAmerica), Waffle House
- **AServ** I: Bob Smith Chevrolet
- **Parks** I: **E.P. "Tom" Sawyer State Park**
- **Other** O: **Wal-Mart Supercenter**(LP) **(w/ pharmacy, one-hr photo)**

30 KY 146, La Grange Rd, Anchorage, Pewee Valley

27 US 60, Shelbyville Rd, Middletown, Eastwood
- **Gas** I: BP(D)(CW), Thornton's Food Mart* (Kerosene)
- **Food** I: Captain D's Seafood, Waffle House, Wendy's
- **ATM** I: Bank of Louisville
- **Other** I: **The Park at Middleton's Golf, Cars, Boats, & Games**
  O: **Bluegrass Mushroom Farm, Shelbyville Rd Veterinary Clinic**

25AB 25A - I-64 East, Lexington, 25B - I-64 West, Louisville

23 KY 155, Taylorsville, Jeffersontown

19 KY 18, 19; Billtown Rd

17 US 31East, US 150, Bardstown, Louisville, Fern Creek
- **Gas** O: BP*(CW)(LP)
- **Food** I: Double Dragon II, Subway, Taco Bell
- **AServ** I: Goodyear Tire & Auto
  O: Don Newton's Corvette Classic Trucks (light)
- **Med** I: **✚ Immediate Care Center (9am-9pm, 7 days/week)**
- **ATM** I: Bank of Louisville, Kroger Supermarket, Nation City Bank, PNC Bank
- **Other** I: **Feeder's Supply Pet Food, Kroger Supermarket (24 hrs, w/ Pharmacy)**

15 KY 864, Beulah Church Rd

14 Smyrna Rd
- **Gas** I: Citgo*
  O: Key Supermarket*, Marathon*(D)
- **Food** O: Barrel of Fun
- **ATM** O: Marathon
- **Parks** O: **McNeely State Park**
- **Other** O: **Smyrna Hardware**(LP)

12 KY 61, Preston Hwy

EXIT KENTUCKY/INDIANA

- **Parks** O: **McNeely Lake Park**

10AB 10A - N.I-65, Louisville, 10B - S.I-65, Nashville

↑KENTUCKY

↓INDIANA

0 I-64W, St Louis, I-64E Louisville

1 State Street
- **Gas** I: Thornton's Food Market*
- **Food** I: Burger King, Chinese Restaurant, Dooley's Bagels & Deli, Dunkin Donuts (Thornton's), Hardee's, Hoosier Pizza, Long John Silvers, Subway
- **AServ** I: Big O Tires, Greg Andre's New Albany Auto Body
- **Med** I: **✚ Hospital**
- **ATM** I: Community Bank, Kroger Supermarket, National City Bank
- **Other** I: **Check Into Cash, Dahlem Animal Hospital, Hoosier Bowling Lanes, Kroger Supermarket (24 hrs, w/Pharmacy), Mail Boxes Etc, Rite Aide Pharmacy, Target Department Store**
  O: **Just for Pets, Laundry Basket Coin Laundry**

3 IN 111, Grant Line Rd
- **Gas** I: Dairy Mart*, Thornton's Food Mart*(D) (Kerosene)
  O: Sav-A-Step*(D) (Kerosene)
- **Food** I: Burger King, Lee's Famous Recipe Chicken, Little Caesars Pizza, Mancino's Pizza & Grinders, Nifty's, Papa John's Pizza, Pizza Hut & Subs, Rich O's BBQ, Sportstime Pizza, Subway, You-A-Carry-Out-A
  O: Frisch's Big Boy, McDonalds
- **AServ** I: Mortons Auto Repair Center
- **ATM** I: NBD, PNC Bank, Regional Bank
  O: Sav-A-Step
- **Other** I: **Aldi Grocery, Feeder's Supply Pet Food,** G.T. Clean Car Wash, **K-Mart (w/Pharmacy), Kroger Supermarket, Plaza Laundry Coin Laundry**
  O: **Bill's 5 Lakes, Indiana University S.E.**

4 IN 311, Charlestown Rd
- **Gas** I: Convenient Gasoline*
  O: Marathon*(D)
- **Food** I: A Nice Restaurant, Sam's
  O: Big Bubba's Bub-Ba-Q
- **AServ** I: Pennzoil Lube, Specialty Automotive
  O: All Tune & Lube Automotive Specialists, Floyd Cornett's Auto Body
- **Med** I: **✚ S. IN Rehabilitation Hospital**
- **Other** O: **Especially Pets Small Animal Hospital, Prosser School of Technology**

7 I-65, Indianapolis, Louisville

10AB 10A - West Jeffersonville, 10B - IN 62E, Charlestown

↑INDIANA

Begin I-265

EXIT OHIO

Begin I-270

↓OHIO

2 U.S. 62, OH3, Grove City
- **Gas** O: Shell*(CW), Sunoco*, Super America*(CW)
- **Food** O: Frisch's Big Boy, Johnny D's Long Branch (Relax Inn)
- **Lodg** O: Relax Inn
- **ATM** O: Shell

5 Georgesville Rd
- **Gas** I: Amoco*, Sunoco*(D)(LP)
- **Food** O: Blocks Bagel & Cafe (Kroger), Lone Star Steakhouse, Wendy's
- **AServ** O: Hatfield Hundai/ Isuzu/ Subaru/ Toyota, Toyota West, Volkswagen/ Jeep-Eagle
- **TServ** O: **Fruehauf**
- **ATM** I: National City Bank, Sunoco
  O: Huntington Bank (Kroger)
- **Other** I: **Georgesville Road Animal Hospital**
  O: **Kroger Supermarket (Pharmacy)**

7 U.S. 40, Broad St
- **Gas** I: BP*(CW), Shell*(CW)
  O: Speedway*(D), Sunoco*, Super America*(D), Thorntons*
- **Food** I: Bob Evans Restaurant, Boston Market, Chi Chi's Mexican, Golden Chopsticks, Peacock West (Chinese/American), Red Lobster, Subway, Super China Buffet, The Ground Round, Wendy's, York Steak House
  O: Arby's, China Inn, Friendly's, Frisch's Big Boy, KFC, Little Caesars Pizza, Long John Silvers, McDonalds, O'Tooles, Pizza Hut, Pizza Hut Carry Out, Rally's Hamburgers, Ramada, Roosters, Subway (Thorntons), Waffle House
- **Lodg** O: Holiday Inn Express, **Ramada**
- **AServ** I: Firestone Tire & Auto, Haydocy GMC Trucks Pontiac, Sears Auto Center, Tuffy Muffler
  O: BP Pro Care, Goodyear Tire & Auto, Rickart-West, Speede Muffler King
- **Med** I: **✚ Urgent Care Medical Care**
- **ATM** I: Star Bank
  O: **Bank One (Big Bear Supermarket)**, Bank One
- **Other** I: **Phar Mor Drugs, Westland Mall**
  O: **20/20 Vision Center, Ace Hardware, Big Bear Grocery, My Cleaners & Coin Laundry, Pet Land, Revco Drugs**

8 Junction I-70, Columbus, Indianapolis

10 Roberts Rd
- **Gas** I: BP*(D)(CW)(LP)
- **Food** I: Subway
  O: Holiday Inn, **Waffle House**
- **Lodg** O: ◆ Holiday Inn, ◆ Royal Inn
- **TServ** O: **Center City International**

13AB Cemetary Rd, Fishinger Rd
- **Gas** I: Shell*(CW)
  O: BP*(CW), Speedway*(LP)
- **Food** I: Chili's, Damon's Ribs, Home Town Buffet, Lone Star Steakhouse, Spageddie's, T.G.I. Friday's
  O: Bob Evans Restaurant, Evergreen Chinese, **McDonalds,** Roadhouse Grill, Tim Horton's, Wendy's
- **Lodg** I: Comfort Suites, Homewood Suites Hotel
  O: Motel 6
- **AServ** I: National Tire & Battery, Saturn Dealership
  O: Buckeye Nissan
- **TServ** O: **Kenworth**
- **Other** I: **K-Mart (Pharmacy)**

15 Tuttle Crossing Blvd
- **Gas** I: BP*(CW), United Dairy Farmers* (Kerosene)
  O: Shell*(CW)
- **Food** I: Bob Evans Restaurant, Boston Market, Cozymel's Mexican, Grady's American Grill, Macaroni Grill, Taco Bell, Wendy's

EXIT		OHIO
		O: McDonalds (Wal-Mart)
	Lodg	I: Budgetel Inn, ◆ Sumner Suites
	AServ	O: Wal-Mart
	ATM	I: Star Bank
	Other	I: **The Mall at Tuttle Crossing**
		O: **Wal-Mart (Pharmacy, Vision Center)**
17AB		U.S. 33, OH 161, Dublin, Marysville
	Gas	I: Amoco, Shell*
	Food	I: Baskin Robbins, Bob Evans Restaurant, Chi-Chi's Mexican, China Garden, Hyde Chophouse, Little Caesars Pizza, McDonalds, Subway
	Lodg	I: Courtyard by Marriott, Cross Country Inn, Red Roof Inn
		O: AmeriHost Inn (West on U.S. 33 to U.S. 36 -- see our ad on this page)
	AServ	I: BP Pro Care, Midwestern Group BMW
	ATM	I: Heartland, Huntington Bank, National City, State Savings Bank
	Other	I: Coin Car Wash, **Kroger Supermarket (Pharmacy), Revco Drugs, Roush Hardware**
20		Sawmill Rd
	Gas	I: BP*(CW), Dairy Mart*(LP), Shell*(CW), Sunoco*
		O: BP*(CW)(LP), Marathon*
	Food	I: Applebee's, Bob Evans Restaurant, Boulevard Grill, Damon's, Doubles, Einsein Brothers Bagels, Four Bakers, KFC, Long Horn Steaks
		O: McDonalds, Olive Garden, Taco Bell, Wendy's
	AServ	I: AAMCO Transmission, BP Pro Care, Crestview Cadillac, Precision Tune
		O: Krieger Ford, Lincoln, Mercury, Marathon, Patrick Subaru , Mazda, Ruhl Ford Dealership, Tire America
	ATM	I: Bank One, The Loan Zone
	Other	I: **Phar Mor Drugs**
		O: **New Market Mall**
22AB		OH 315, Olentangy River Rd.
23		U.S. 23, Delaware, Worthington (Outer Services Take First Right)
	Gas	I: Marathon, Sunoco*(D)
	Food	I: Bruegger's Bakery, Chili's, Dalts Classic American Grill, Darci's Cafe & Bakery, Franco's Pizza & Pasta, Frank & Patty's, Godfry's, McDonalds, Michael Dominics Steak & Seafood, Omaha Steaks
		O: 55 at Crossroads, Alexander's, Bob Evans Restaurant, Bravo Italian Kitchen, Casa Fiesta, Champps, Dick Clark's American Band Stand Grill, Fuddrucker's, Holiday Inn, Rio Bravo
	Lodg	I: (AAA) Econolodge
		O: (AAA) AmeriSuites (see our ad on this page), ◆ Courtyard by Marriott, (AAA) Holiday Inn (see our ad on this page), (AAA) Homewood Suites, (AAA) Microtel, ◆ Red Roof Inn, Travelodge
	AServ	I: Marathon
	ATM	I: Bank One, National City Bank, State Savings Bank
	Other	I: **Kroger Supermarket, Rite Aide Pharmacy, Worthington Mall**
26AB		Jct I-71, Columbus, Cleveland
27		OH 710, Cleveland Ave
	Gas	O: 76*, Shell*(CW)
	Food	I: Antolino Pizza, Cheddar's Cafe, Cooker Grill, Fuddrucker's, Golden House Chinese Restaurant, McDonalds, Monte Carlo, T.G.I. Friday's, Yanni's Greek Grill
		O: Grape Vine, KFC, Schmidt's Restaurant, Wendy's
	AServ	O: Faslube, Goodrich Auto, Lee Miles Transmissions, Tuffy Muffler & Brake, Valvoline Quick Lube Center
	Med	O: + **Hospital**
	ATM	O: National City Bank (Big Bear Grocery)
	Other	I: Coin Car Wash
		O: **Big Bear Grocery (Pharmacy), Revco Drugs**
29		OH 3, Westerville
	Gas	I: Shell*(CW), Speedway*(LP), Sunoco* (Kerosene)
		O: BP*, Shell*
	Food	I: Carsonie's Italian, China House, Domino's,

COLUMBUS

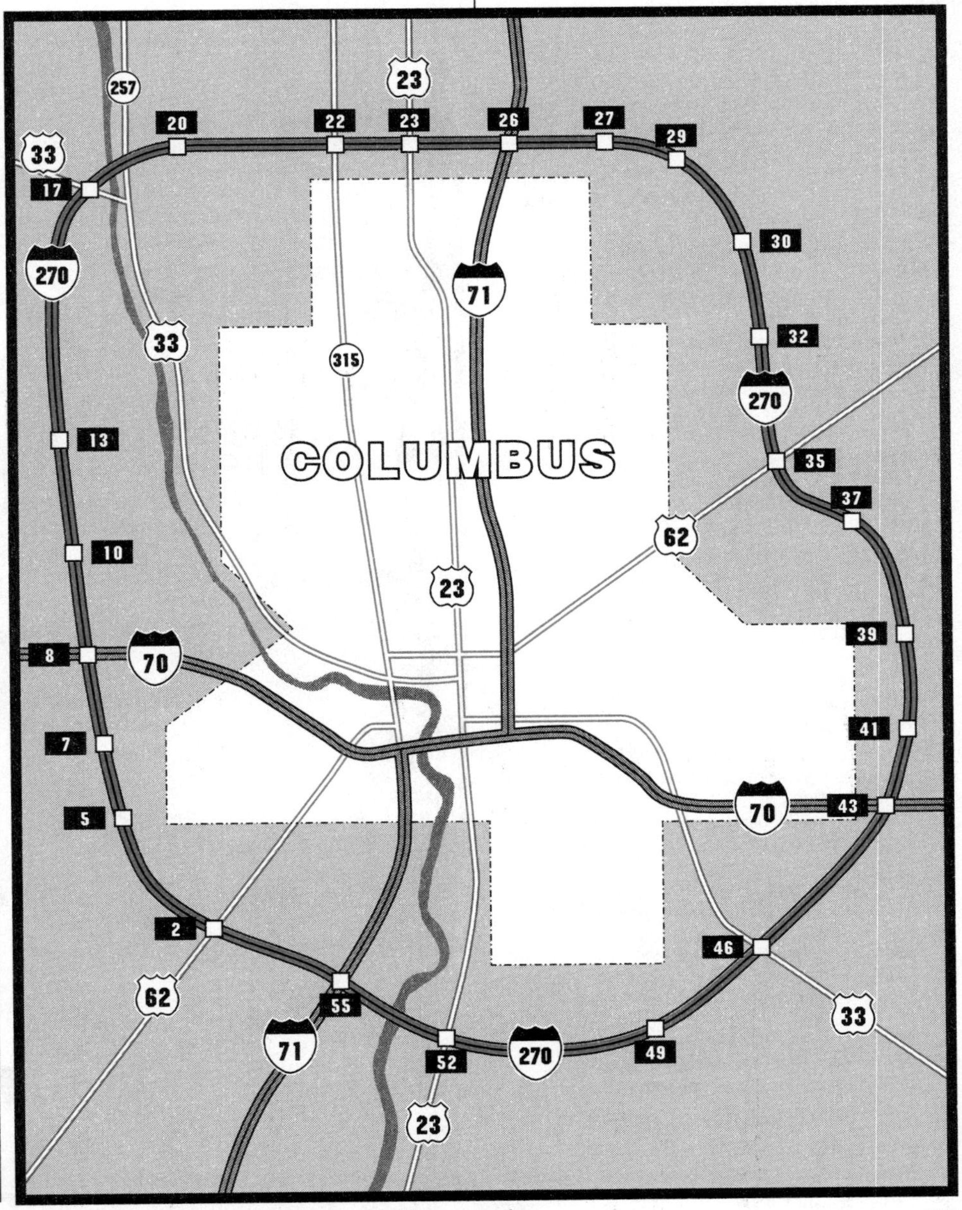

EXIT	OHIO
	Subway
	O: Arby's, Bob Evans Restaurant, Burger King, Denny's, McDonalds, Pizza Hut, Ponderosa, Rally's Hamburgers, Tee Jay's Country, Wendy's
Lodg	**O:** AAA Cross Country Inn, ◆ Knight's Inn
AServ	**I:** Grismer Tires, Jan's Auto Repair, Midas Muffler & Brakes
	O: BP Pro Care, Car Quest Auto Center, Firestone Tire & Auto, Quick Oil Change
ATM	**O:** Bank One, Fifth Third Bank (Kroger), Huntington National Bank, Star Bank, State Savings Bank
Other	**I:** **Aldi Grocery,** Miracle Car Wash, **Revco Drugs, U.S. Post Office**
	O: **Kroger Supermarket, Office Max**
30AB	OH 161, Minerva Park, New Albany
32	Morse Rd
Gas	**I:** BP*, Shell*(CW)
	O: Citgo*, Speedway*(D)
Food	**I:** McDonalds (Wal-Mart), Subway
	O: Donato's Pizza
AServ	**I:** Kelly Motors, Wal-Mart
	O: Toyota Direct
RVCamp	**I:** **Jerry Greer Airstream**
Other	**I:** **Sam's Club, Wal-Mart (Pharmacy)**
35AB	I-670, U.S. 62, Airport, Gahanna
37	OH 317, Hamilton Rd
Gas	**O:** Shell*(CW), Speedway*(LP) (Kerosene)
Food	**O:** Bob Evans Restaurant, Burger King, China Garden, **Damon's Ribs,** Frisch's Big Boy, KFC, Luck's Cafe, Subs, Yogurts, and Muffins, Taco Bell
AServ	**O:** Firestone Tire & Auto, K-Mart, Shell
ATM	**O:** Huntington Bank
Other	**O:** **Abbott Food Service Outlet Supermarket,** C & B Car Wash, **K-Mart (Pharmacy)**
39	Ohio 16, Broad St
Med	**O:** ✚ **Mt Carmel East Hospital**
41AB	U.S. 40, Main Street, Whitehall, Reynoldsburg
Gas	**O:** Shell*(CW)
Food	**I:** Don Pablo's Mexican Kitchen, Spageddie's
	O: Blocks Bagel, Bob Evans Restaurant, Denny's, **Hooter's,** Manhattan Bagel, Outback Steakhouse, Pizza Hut, Rally's Hamburgers, Skyline Chili, The Italian Oven, The Spaghetti Shop
AServ	**O:** Shell
Med	**O:** ✚ **Hospital**
Other	**O:** **Revco Drugs**
43AB	Jct I-70, Wheeling, Columbus
46AB	U.S. 33, Bexley, Lancaster
49	Alum Creek Dr, Obetz
Gas	**I:** Sunoco*(D)(LP)
	O: BP*(D)(LP)
Food	**I:** Donato's Pizza, **Subway**
	O: J.R. Valentine's, McDonalds, Taco Bell, Wendy's
Other	**I:** Coin Car Wash
52	U.S. 23, High St, Circleville
Gas	**I:** Speedway*(LP) (Kerosene), Sunoco*(CW)
	O: BP*(D)
Food	**I:** Abner's Country, Arby's, Blimpie's Subs (Sunoco), Bob Evans Restaurant, Golden China Town, Little Caesars Pizza, Long John Silvers, McDonalds, Ohio Deli & Restaurant, Pizza Hut, Ponderosa, RAX, Taco Bell, Wendy's, Wings & Rings
AServ	**I:** Firestone Tire & Auto, K-Mart
ATM	**I:** Bank One, Bank One (Big Bear), Fifth Third Bank, Huntington Bank, National City Bank, Star Bank (Kroger)
Other	**I:** **Big Bear Grocery (Pharamcy), K-Mart (Pharmacy), Kroger Supermarket, Pearl Vision Center, Pet Center, Rite Aide Pharmacy**
55	Jct I-71, Columbus, Cincinatti

↑OHIO

Begin I-270

EXIT	ILLINOIS

Begin I-270

↓ILLINOIS

12	IL - 159, Collinsville, Edwardsville
Other	**S:** **Dental Clinic**
9	IL 157, Collinsville, Edwardsville
7	I-255 S Jct I-270 E
6AB	IL. 110 Wood River, Pontoon Beach
Gas	**O:** 76*, Amoco*
Food	**I:** McDonalds
	O: **Hen House Restaurant**
Lodg	**I:** **Ramada Limited**
	O: Apple Valley Motel, Best Western
AServ	**I:** Mick's Garage
TServ	**O:** **Cities Service**
Med	**I:** ✚ **Hospital**
4	**(3)** IL. 203 Granite City

EXIT	ILLINOIS/MISSOURI
3AB	Alton, Granite City

↑ILLINOIS

↓MISSOURI

(34)	**Rest Area - Picnic, RR, HF, Phones** P
33	Lilac Ave
32	Bellefontaine Rd.
Lodg	**I:** Econo Lodge (see our ad on this page)

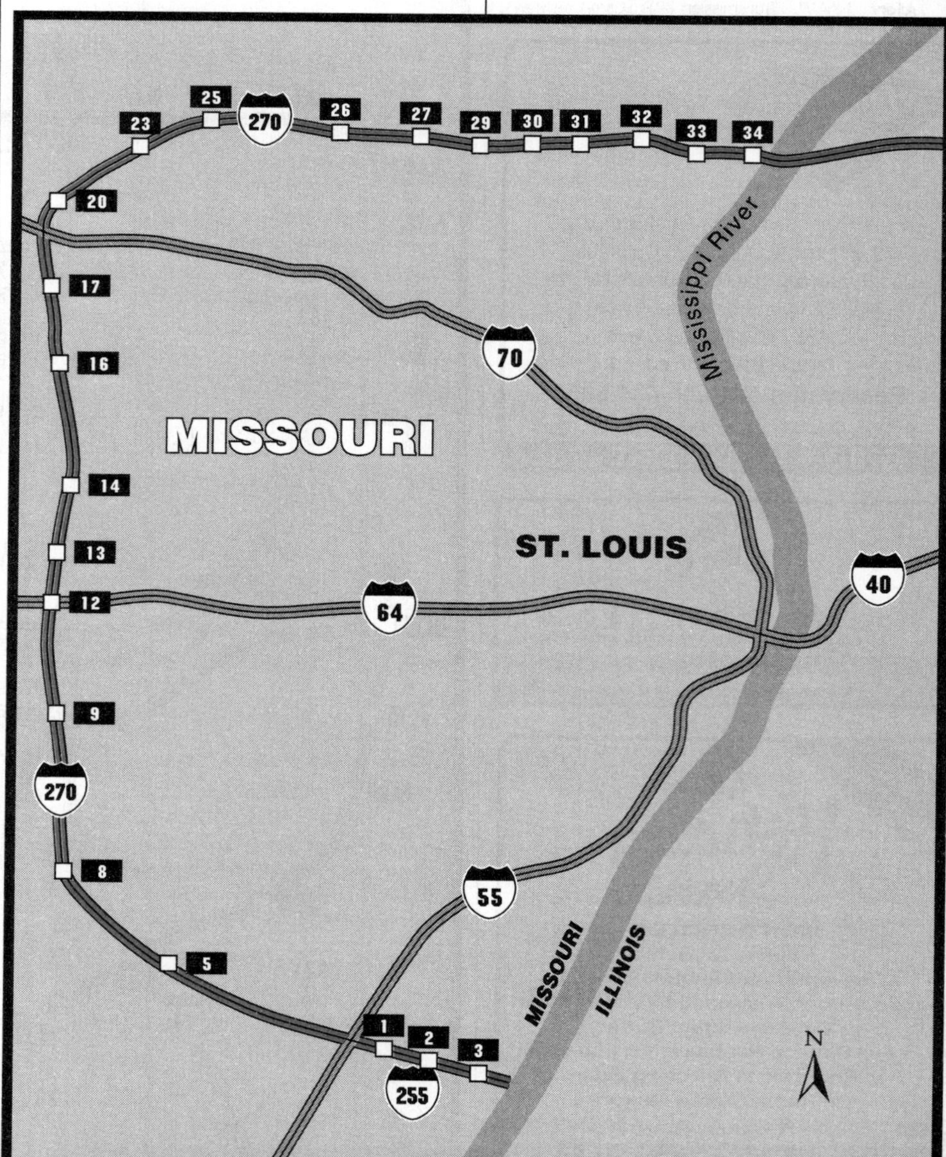

Bold red print shows RV & Bus parking available or nearby

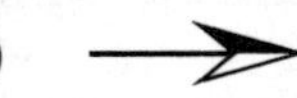

EXIT MISSOURI

31AB MO 367 St. Louis

30 CR AC, Halls Ferry Rd
- **Gas** I: Amoco*(CW), Hucks*, Shell*(CW)
  - O: Mobil(D), Phillips 66, Q.T*, Shell*
- **Food** I: Arby's, Buffet Place, Church's Fried Chicken, Emperor's Wok, KFC, McDonalds
  - O: Applebee's, Captain D's Seafood, Casa Gallard, Casa Gallardo, Red Lobster, The Olive Garden, Wendy's, White Castle Restaurant
- **Lodg** O: ◆ Super 8 Motel
- **AServ** I: Auto Zone Auto Parts, Buick Dealer, Car Care Center, Firestone Tire & Auto, Rapid Lube
  - O: Actra Transmisson, Ford Dealership, Jiffy Lube, Shell, Tire America
- **ATM** I: UMB Bank
  - O: Boatmen's Bank, Mercantile Bank
- **Other** I: Owner's Pride Car Wash
  - O: **Hydro Jets, Target Department Store**

29 West Florissant Ave

28 Washington St., Elizabeth Ave.

EXIT MISSOURI

27 CR N, New Florissant Rd

26B N. Hanley Rd, Graham Rd.

26A I-170 S

25AB **(24)** U.S. 67 Lindbergh Blvd

23 McDonnell Blvd.
- **Gas** O: Q.T.*
- **Food** I: Denny's, McDonalds
  - O: Hardee's, Jack-In-The-Box, Lion's Choice, Roast Beef, Steak 'N Shake
- **Lodg** I: (AAA) La Quinta Inn
- **AServ** O: Pontiac Buick GM Dealer, Tony's Auto Repair
- **ATM** I: Mercantile Bank

22 **(21)** MO. 370 W. Missouri Bottom Rd

20B **(18)** St. Charles Rock Rd, MO 180, Natural Bridge Rd.
- **Lodg** I: Holiday Inn (see our ad on this page)

20A **(18)** I-70 Jct.

17 Dorsett Rd
- **Lodg** I: Hampton Inn (see our ad on this page)

16A **(15)** Page Ave, CR D East

14 MO. 340 Olive Blvd.
- **Gas** I: Amoco(CW), Phillips 66(CW)
- **Food** I: Dairy Queen, Denny's, Dunkin Donuts, KFC, Meyer's Deli, Padrino's Pizza, Peking Inn Chinese Restaurant
- **AServ** I: Cadillac Dealer, GM Auto Dealership, Jeep Eagle Dealer, Phillips 66
- **Med** O: ✚ **Hospital**
- **ATM** I: First Bank
  - O: Commerce Bank, Mark Twain
- **Other** I: **Asian Market**

13 Ladue Rd

12 **(11)** I-64. U.S. 40 U.S. 61 Jct. (Difficult Reaccess)
- **Other** S: **Dental Clinic**

9 MO. 100 Manchester Rd.

5A Dougherty Ferry Rd.

5B **(4)** I-44 U.S. 50 Jct.

3 MO. 30 Gravois Rd

2 MO21

↑ MISSOURI

Begin I-270

EXIT MARYLAND

Begin I-270

↓ MARYLAND

31AB Buckeystown, Frederick
- **FStop** E: **Mobil***
- **Gas** E: Lowest Price*, Sheetz*(LP), Shell*(CW), Southern States(D)
- **Food** E: Applebee's, Bob Evans Restaurant, Burger King, Checkers, Deli Plus, Dunkin Donuts, El Paso Restaurant, Golden Corral, Holiday Inn, Italian Oven, Jerry's Subs & Pizza, KFC, King & Queens Donut Shop, McDonalds, Olive Garden, Papa John's, Pargo's, Pizza & Gourmet, Pizza Hut, Popeye's, Roy Rogers, Subway, Taco Bell, Wendy's
  - W: **Cracker Barrel**, Hampton Inn, Marriott
- **Lodg** E: Days Inn, Holiday Inn, Holiday Inn Express, Knights Inn
  - W: Fairfield Inn, Hampton Inn, Marriott
- **AServ** E: AAMCO Transmission, Antedem Auto Parts, British Auto Service, FSK Ford, Lincoln, Mercury, Audi, Goodyear Tire & Auto, Ideal GMC Buick, Ideal Hyundai, Precision Tune, S&S Tire, Sam's Club (Tire), Saturn, Sears (Mall), Tate Chrysler Plymouth, Tube & Lube, Wal-Mart
  - W: Shackley Honda
- **RVCamp** E: **Scotty Camper Trailers**
- **ATM** E: FCM Bank, First Nat'l Bank of Maryland, NationsBank, Sheetz
- **Other** E: **7-11 Convenience Store, Evergreen Square Shopping Ctr, Francis Scott Key Mall, Frederick Veterinary Ctr (Grove Rd), Kee Laundro-Mat (Self Service), Sam's Club (Tires), Scotty Camper Trailer Repair, Sheriffs Dept, Wal-Mart (Pharmacy, Auto)**

26 MD80, Urbana, Buckeystown
- **Gas** E: Exxon*
- **Other** E: **Countryside Animal Clinic**

22 MD109, Bondsville, Hyattstown
- **Other** W: **Sugarloaf Mtn**

23 Weigh Station

18 MD121, Boyds, Clarksburg
- **RVCamp** E: **Little Bennett Regional Park (From April to October)**
- **Parks** E: **Little Bennett Regional Park**
  - W: **Black Hill Regional Park, Boyds Nat'l State Park**

16 MD27, Father Hurley Blvd, Damascus

15AB MD118, MD355, to Germantown
- **Gas** W: Amoco, Exxon*(CW)
- **Food** E: Season's Cafe (Hampton Inn)
  - W: Pizza Hut
- **Lodg** E: Hampton Inn
- **AServ** W: Amoco, D Brake Shop, Discount Tire Auto Service, Exxon*(CW), Germantown Service Ctr
- **Med** E: ✚ **Shady Grove Medical Ctr**
- **Other** W: **County Police, Tourist Information**

11AB MD124, Montgomery Fair Grds, Quince Orchard Rd, Gaithersburg
- **Gas** E: Exxon*(D)(CW)
  - W: Shell*, Texaco*(D)
- **Food** E: Boston Market, Bruegger's Bagel, Chuck E. Cheese's Pizza, Cookies By Design, David's Ice Cream, Golden Bull Grand Cafe, Heavenly Ham, Hilton, Holiday Inn, KFC, McDonalds, Omaha Steaks, Paisaons Bar & Grill, Ricky's Rice Bowl,

EXIT		MARYLAND
		Roy Rogers, Starbuck's Coffee, Subway, Wendy's **W:** Baskin Robbins, Chilli's, Deli, Denny's, Einstein Bros Bagel, Ernie's Restaurant, Friendly's, Haloha Inn (Polonesian, Chinese, Continental Cuisine), Jerry's Subs & Pizza, Lone Star Steakhouse, Pastry Shop, Pat & Mike, PeKing Cheers Restaurant, Roy Rogers, Shakey Pizza & Buffet, Taste Of Europe, Toyko Lighthouse
	Lodg	**E:** Hilton, Holiday Inn **W:** Red Roof Inn, Red Roof Inn
	AServ	**E:** Exxon*(D)(CW), Montgomery Ward, **Sam's Club** **W:** Chris Well Chevrolet, Texaco*(D), Trak Auto
	Med	**E:** + **Urgent Care Med Clinic**
	ATM	**E:** Chevy Chase Bank, City Bank, NationsBank, Patomic Valley Bank, Sandy Spring Nat'l Bank **W:** Chevy Chase Bank, Crestar Bank
	Parks	**W: Seneca State Park**
	Other	**E: CVS Pharmacy, Gathersburg Square Shopping Ctr, Pearl Vision Ctr, Sam's Club** **W: Giant Grocery Store (Pharmacy), Godman Square Shopping Ctr, Pet Center, Quince Orchard Plaza**
9		Leads to 370
8		Shadygrove Rd, Omega Dr. (Difficult Reaccess)
	Gas	**E:** 270 Center, Crown*, Texaco
	Food	**E:** Bakery Cafe, Bug-A-Boo Creek Steakhouse, Burger King, Chesapeake Bagel Factory, Chicken Out, Claude's Bistro, Dai-loc Oriental Mkt, Days Inn, Deli (7-11 C-Store), Hunan Palace, India Kitchen, Philidelphia Mike, Red Lobster, Tokyo Rice Bowl, Zan-Zan Mkt House of Halal, Zios **W:** Marriott
	Lodg	**E:** DAYS INN Days Inn **W:** ◆ Marriott
	AServ	**E:** Texaco
	ATM	**E:** Crestar Bank, NationsBank **W:** NationsBank
	Other	**E: 7-11 Convenience Store, Animal Dermatology & Eye Care Ctr, Fed Ex Drop Box, Gather Center, Grove Ctr Veterinary Hospital, Mail Boxes Etc, Montgomery County Airport, Shadygrove Ctr** **W: John's Hopkins University, Shadygrove Pharmacy**
6AB		MD28, W Montgomery Ave.
	Gas	**W:** Shell*
	Food	**W:** Callaway's (Best Western)
	Lodg	**W:** Best Western
	ATM	**W:** Crestar Bank
	Other	**W: Dames Town, Rockville Town, Ctr**
5		Falls Rd, Potomac
	Other	**W: Police Station**
4AB		Montrose Rd
1		JCT495, Democracy Blvd
	Gas	**W:** Exxon(D), Texaco*(D)
	Food	**E:** Marriott **W:** Boston Market, Cuisine China
	Lodg	**E:** Marriott
	AServ	**W:** Exxon(D), Jim Coleman, Ourisman Ford, Sears, Texaco(D)
	ATM	**W:** First Union Bank, Maryland Federal
	Other	**W: County Police Facility (Montgomery Mall), Montgomery Mall (On Bethesda), Westlake Crossing**

↑MARYLAND

Begin I-270

EXIT		OHIO
		Begin I-271

↓OHIO

EXIT		OHIO
3		OH 94, Wadsworth, North Royalton, I-71 N, Cleveland
	AServ	**E:** M&M Towing and Recovery
	TServ	**E: M&M Towing and Recovery**
1		JCT I-71 S, Columbus
36		Wilson Mills Rd
	Gas	**E:** Shell* **W:** BP, Marathon, Sunoco
	Food	**E:** Austin's Steakhouse **W:** Denny's, Wellington's Restaurant
	Lodg	**E:** Holiday Inn
	AServ	**W:** BP, Marathon, Sunoco
	ATM	**W:** First Merit Bank, Sunoco
	Other	**W: Builders Square 2**
34		Mayfield Hieghts
	Gas	**E:** BP*, Sunoco **W:** Shell
	Food	**E:** Domino's Pizza, Eastside Mario's, Hunan Garden Gourmet Chinese, Master Pizza, Rio Bravo, Subway, Tony Roma's **W:** Arrabibta's Italian Restaurant, Bob Evans Restaurant, Burger King, China Express, Jeppetto's Pizza, Long Horn Steakhouse, Longo's Pizza, McDonalds, Moose O' Malley's, T.G.I. Friday's
	AServ	**E:** BP Pro Care, Sunoco, Tire & Wheel Service Center **W:** Goodyear Tire & Auto, Marshall Ford, Midas Muffler and Brakes, Shell, Speedy Muffler King
	Med	**E:** + **Hospital**
	ATM	**E:** National City Bank (Rini Rego Supermarket), Sunoco **W:** Huntington Bank, Star Bank
	Other	**E: Coin Laundry, Gary Mart C-Store, Revco, Rini Rego Supermarket** **W:** Automatic Car Wash, **Convenience Store, PetsMart**
32		Brainard Rd., Cedar Rd.
29		US 422 W, OH 87, Chagrin Blvd
	Gas	**E:** Gulf, Marathon, Shell*(CW) **W:** 76*, BP*
	Food	**E:** Coffee & Creations, Domino's, Hong Kong Restaurant, McDonalds, Red Lobster, The Olive Garden, Uno Pizzeria Restaurant and Bar, Village Square Pizza, Wendy's **W:** Charley's Crab, Marriott, Radisson Inn
	Lodg	**E:** Holiday Inn, The Courtyard Marriott, Travelodge **W:** Embassy Suites, Marriott, Radisson Inn
	AServ	**E:** Gulf **W:** Crestmont Cadillac, Don Jordon Chrysler Plymouth
	ATM	**E:** National City Bank
	Other	**E: Gale's Supermarket, Revco, Rite Aide Pharmacy**, Village Auto Wash (Marathon)
28		OH 175, Richmond Rd., Emory Rd
	Gas	**E:** BP*
	Food	**E:** Astoria Restaurant, Country Kitchen
	Med	**E:** + **Hospital**
	ATM	**E:** Star Bank
	Other	**E: Emory Food Mart***
27AB		JCT I-480, US 422 E

EXIT		OHIO
26		Rockslide Rd.
	Gas	**E:** Speedway*, Sunoco*
	Food	**E:** Burger King, Cugino's Pizza, Donut Works, Ramada Inn, Subway
	Lodg	**E:** Ramada Inn
	Med	**W:** + **Hospital**
	Other	**E: Dairy Mart Convience Store, Stop N Go Supermarket**

Note: I-271 runs concurrent below with I-480. Numbering follows I-480.

EXIT		OHIO
23		OH 14 W, Forbes Rd., Broadway Ave.
	Gas	**E:** Marathon* **W:** BP*
	Food	**E:** McDonalds, Pizza Hut, Subway, Taco Bell, Wendy's, Wing Kee Chinese
	Lodg	**E:** Holiday Inn Express
	AServ	**W:** Lube Stop
	Med	**W:** + **Hospital**
	Other	**E: Builders Square, Oakwood Village Hardware, PetsMart, Sam's Club** **W:** Car Wash
21		OH 14 E, Youngstown
	Food	**N:** Zech Inn
	Med	**N:** + **Hospital**
	Other	**N: Convenient Food Mart**

Note: I-271 runs concurrent above with I-480. Numbering follows I-480.

EXIT		OHIO
19		OH 82, Macedonia, Twinsburg (Difficult Reaccess Southbound)
	Gas	**E:** Speedway* **W:** Sunoco
	Food	**E:** Blue Willow, Zachary's Restaurant **W:** Applebee's, Boston Market, Burger King, Dairy Queen, Fuji Japanese Restaurant, KFC, Little Caesars Pizza, McDonalds, Old Country Buffet, Outback Steakhouse, Pizza Hut, Taco Bell, The Ground Round, Wendy's, Winking Lizard
	AServ	**E:** Henchick's Auto Service **W:** Instant Oil Change, K-Mart, Sunoco
	ATM	**E:** First National Bank **W:** Ohio Savings Bank
	Other	**W: Country Counter Supermarket, Finast Pharmacy, K-Mart (Pharmacy), Revco, Wal-Mart (Pharmacy, Vision Center, Auto Service)**
18		OH 8, Boston Heights, Acran
	Gas	**E:** Amoco, BP*(CW), Speedway*
	Food	**E:** Bob Evans Restaurant, Dolphin Family Restaurant, Donut Tree **W:** Applebee's, KFC, McDonalds, The Ground Round Steakhouse
	Lodg	**E:** Budgetel Inn, Knights Inn, Motel 6, Travelodge
	AServ	**E:** Amoco
	ATM	**E:** BP
	Other	**W: Wal-Mart (Pharmacy, Vision Center, Auto Service)**
12		OH 303, Richfield, Peninsula
10		JCT I-77, Acron, Cleveland
(8)		**Rest Area - RR, Picnic, Phones**

↑OHIO

Begin I-271

Bold red print shows RV & Bus parking available or nearby

EXIT	KENTUCKY/OHIO
	Begin I-275, Cincinnati

↓KENTUCKY

84 JCT I-71, I-75

83 U.S. 25, U.S. 42, U.S. 127, Dixie Hwy
- **ATM** I: People's Bank
- **Other** O: **Crestview Hills Mall**

82 KY 1303, Turkeyfoot Rd
- **Food** O: Applebee's, T.G.I. Friday's
- **Med** O: ✚ **Hospital**
- **ATM** O: Star Bank
- **Other** O: **Crestview Hills Mall**

80 KY 17, Covington, Independence
- **Gas** I: Super America*(LP), United Dairy Farmers*
- **Food** I: Arby's, Bob Evans Restaurant, Burger King, Frisch's Big Boy, Taco Bell (Super America), Wendy's
- **AServ** I: Eagle Tire & Car Care, Quick Stop 10 Minute Oil Change
- **TServ** I: **Freightliner Dealer**
- **ATM** I: United Dairy Farmers
- **Other** I: Classic Car Wash

79 Taylor, Mill Rd, Covington
- **Gas** O: BP*(D)(CW)
- **Food** O: KFC, McDonalds, Oriental Wok, Snappy Tomatoe Pizza, Subway, Taco Bell
- **ATM** O: Star Bank
- **Other** O: **Emergency Veterinary & Specialty Clinic, Remke's Supermarket, Rite Aide Pharmacy**

77 KY 9, Wilder, Maysville

76 Three Mile Rd, Northern Kentucky University (No Reaccess Eastbound)

74AB Jct I-471 N, to U.S. 27, Newport, Cincinnati, Alexandria
- **Med** I: ✚ **Hospital**

↑KENTUCKY

↓OHIO

72 U.S. 52, Kellogg Ave , New Richmond
- **Food** I: Lebo's, River Sea Restaurant
- **Other** I: **Bill's Carry Out Grocery, Steve's California Carry Out Convenience Store**
  O: **Coney Island Water Park**

71 U.S.52 E, New Richmond

69 Five Mile Rd
- **Med** I: ✚ **Hospital**

65 OH 125, Amelia, Beechmont Ave
- **Gas** I: BP*, Clark, Super America*(LP)
  O: Shell*(CW), Super America*, United Dairy Farmers*
- **Food** I: Bob Evans Restaurant, Burger King, Chi Chi's Mexican, Ditto's Grill, Frisch's Big Boy, McDonalds, Montgomery Inn East, Olive Garden
  O: Arby's, BW-3 Grill, Denny's, Red Lobster, Ryan's Steakhouse, Szechuan House, Wendy's
- **Lodg** I: (AAA) **Cross Country Inn**, (DAYS INN) Days Inn, Montgomery Inn East, Red Roof Inn
  O: (AAA) Motel 6
- **AServ** I: AAMCO Transmission, Firestone Tire & Auto, Honda
  O: Beechmont Ford
- **ATM** I: Super America

EXIT	OHIO

- O: Huntington Bank, PNC Bank, Provident Bank
- **Other** O: Coin Car Wash, **WalGreens**

63 OH 32, Batavia, Newtown
- **Gas** I: Ashland*, Speedway*(D)
  O: Marathon*
- **Food** I: Gramma's Pizza (Ashland)
  O: Bob Evans Restaurant, Burger King, Frisch's Big Boy, Holiday Inn, Krispy Kreme Donuts, McDonalds, O'Charley's, Penn Station, Perkins Family Restaurant, Pizza Hut, Rally's Hamburgers, Sizzling Wok, Subway, Taco Bell, Wendy's
- **Lodg** O: Holiday Inn
- **AServ** I: Eastern Hills Tire, Midas Muffler & Brakes
  O: Firestone Tire & Auto (Eastgate Mall), National Tire & Battery, Sears (Eastgate Mall), Tire America, Wal-Mart
- **ATM** O: Clairmont Savings, Fifth Third Bank
- **Other** I: Coin Car Wash
  O: **Builders Square Hardware, Eastgate Mall, Kroger Supermarket (Pharmacy), PetsMart, Wal-Mart (Pharmacy)**

EXIT	OHIO

(60) **Rest Area - RR, Phones (Closed)**

59 U.S. 50, Hillsboro

57 OH 28, Milford, Blanchester
- **Gas** I: BP*, Shell*
  O: Ameristop*, USA Gasoline
- **Food** I: Ponderosa, R.W. Roosters
  O: Arby's, Berryman's Tastee Treat, Burger King, Dunkin Donuts, **KFC**, Long John Silvers, Old Town Ice Cream, Papa John's Pizza, Perkins Family Restaurant, Taco Bell, White Castle Restaurant
- **AServ** I: Auto Parts Warehouse, BP Pro Care Center, Chrysler Auto Dealer
  O: Bob Sumerel Tire, Car X Muffler & Brakes, Firestone Tire & Auto, Jiffy Lube, K-Mart, Meineke Discount Mufflers, Midas Muffler & Brakes, Milford Dodge, NAPA Auto Parts, Tuffy Auto Care Center
- **Med** O: ✚ **Walk-In Medical Clinic**
- **ATM** I: Bank One, Star Bank
  O: Fifth Third Bank, Key Bank, **Provident Bank**
- **Other** O: **Coin Laundry**, Eastern Hills Car Wash, **K-Mart (Pharmacy), Medicine Shop Pharmacy, Pearl Vision Center, Thriftway Food & Drug**

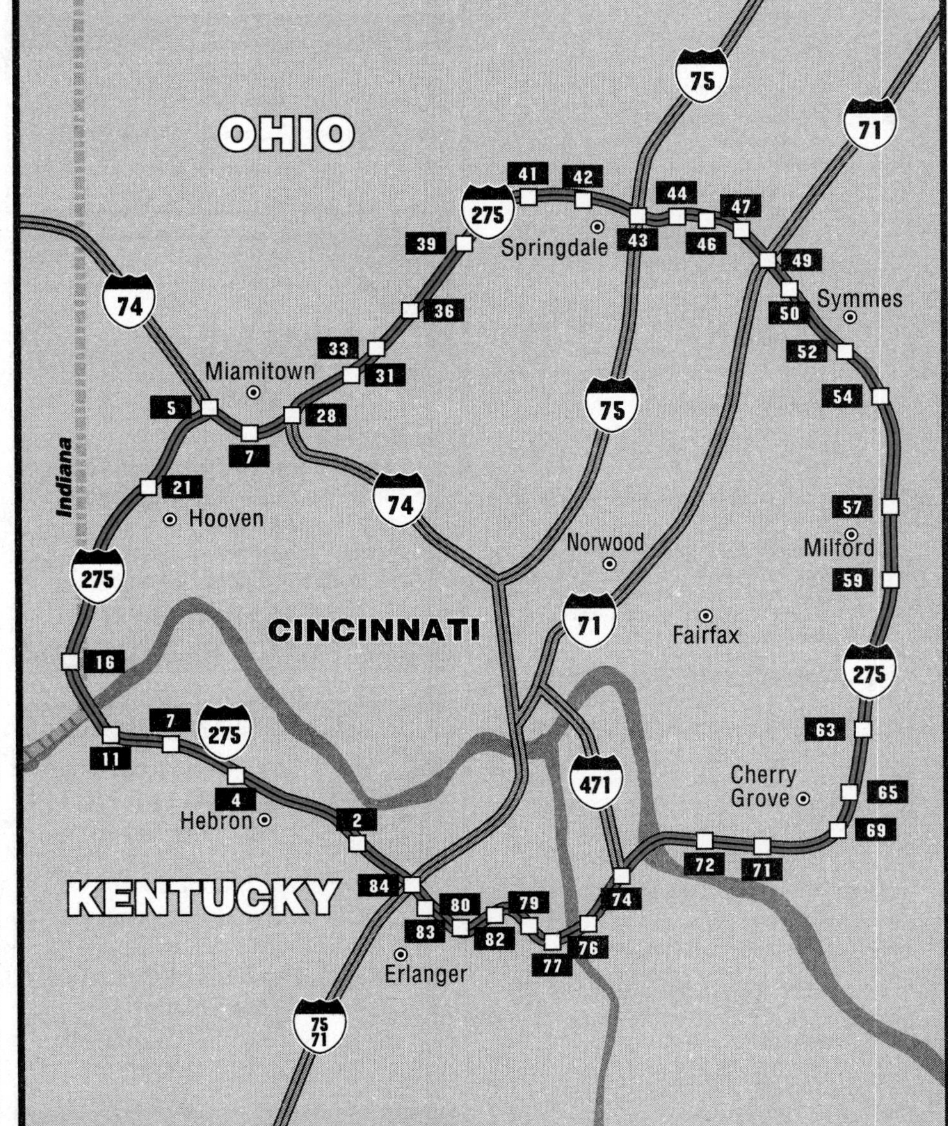

EXIT OHIO

54 Wards Corner Rd

Gas **I:** United Dairy Farmers*

O: BP*

Food **I:** Frisch's Big Boy, Gold Star Chili, Subway

Other **I:** **East Hills Veterinary Clinic**

52 Indian Hill, Loveland

Gas **O:** Marathon[D], Shell*[CW], Super America*[D], Thornton's*

Food **O:** Arby's, Bruegger's Bagels, Skyline Chili, Subway, Taco Bell, Wendy's

AServ **O:** Century Motors VW

ATM **O:** Fidelity Federal, Machine (Super America), Star Bank

Parks **I:** **Park at the Lake Isabela Park**

Other **O:** **WalGreens**

50 U.S. 22, OH 3, Montgomery, Morrow

Gas **I:** BP*[D][LP], Shell*[CW]

O: Speedway*[D][LP]

Food **I:** Crown's Bar & Grill, Gold Star Chili, La Rosa's Pizzeria, McDonalds, P.J.'s Restaurant & Lounge, Skyline Chili, Village Wok, Wendy's

O: The Melting Pot (Fondue)

AServ **O:** Columbia Acura

Med **I:** ✚ **Hospital**

ATM **I:** Bank One, Star Bank, Suburban Federal

Other **I:** **Ameristops C-Store, Hanson Animal Hospital, Montgomery Health Mart & Pharmacy**

O: **Kyle Veterinary Hospital**

49AB Junction I-71, Columbus, Cincinnati

47 Reed - Hartman Hwy, Blue Ash

Lodg **I:** AmeriSuites (see our ad this page, AAA Doubletree Guest Suites

46 U.S. 42, Sharonville, Mason

Gas **I:** Ashland*[D][LP], Marathon*[CW][LP], United Dairy Farmers* (Kerosene)

O: BP*[D][CW]

Food **I:** Arby's, Kitchen Express Restaurant, Szechuan House, Waffle House

O: Burger King, Damon's Ribs, Dunkin Donuts, Hardee's, **Holiday Inn**, House of Sun Chinese, KFC, McDonalds, Mountain Jacks, Perkins Family Restaurant, Skyline Chili, Wendy's, **White Castle Restaurant**

Lodg **I:** DAYS INN ◆ **Days Inn**

O: ◆ **Holiday Inn (see our ad on this page)**, Motel 6

AServ **I:** Car X Muffler & Brakes, Jiffy Lube, Marathon, Midas Muffler & Brakes, Sharon Woods Auto Care, Tire Discounters

O: BP Pro Care, Goodyear Tire & Auto, **K-Mart**

ATM **I:** Fifth Third Bank

O: PNC Bank

Other **I:** Coin Car Wash, **Indian Grocers Food Market**

O: **K-Mart (w/pharmacy), Kroger Supermarket, Revco Drugs**

44 Mosteller Rd

Lodg **I:** Homewood Suites Hotel

TServ **O:** **Tri State Ford**

EXIT OHIO

43AB Jct I-75, Dayton, Cincinnati

42AB OH 747, Springdale, Glendale

FStop **O:** **Amoco***

Gas **I:** BP*[CW], Shell*[CW]

O: Thornton's*

Food **I:** Frisch's Big Boy, Hardee's, Jade Garden, KFC, La Rosa, Long Horn Steaks, Norka Futon, Panda Chinese, Ponderosa, Red Squirrel, Schlotzkys Deli, Steak 'N Shake, T.G.I. Friday's, TCBY Yogurt, Wendy's

AServ **I:** BP Pro Care, Jake Sweeney Chevrolet, Princeton Tire Co, Saturn of Tri-County

O: Gateway Tire Co.

ATM **I:** Bank One, PNC Bank, Star Bank

41 OH 4, Springfield Pike

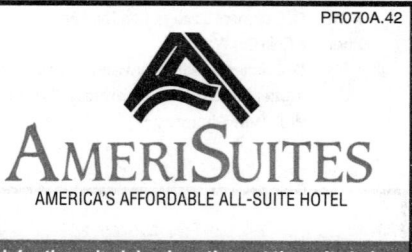

EXIT OHIO

Gas **I:** BP*

O: Cocolene*, Shell*[CW]

Food **I:** Applebee's, Chi Chi's Mexican, Dairy Queen, New York Deli, Outback, Penn Station, Perkins, Wok & Roll

O: Bennigan's, Bob Evans Restaurant, Burger King, Camel Won, East-West International Buffet, Hooters, Remington's, Sheraton, Taste of China, Wendy's

Lodg **I:** AAA Cross Country Inn, Extended Stay America, ◆ Howard Johnson, Best Western (see our ad on this page), AAA Ramada Inn

O: ◆ **Budgetel Inn**, Rodeway Inn, Sheraton

AServ **O:** Midas Muffler & Brakes, Tire Discounters

Med **I:** ✚ **Doctor's Urgent Care (8:30am-11pm)**

ATM **O:** Shell

Other **O:** **Veterinary Clinic**

39 Fairfield, Forest Park, Winton Rd

Gas **I:** Marathon*, Shell*[CW], United Dairy Farmers*

Food **I:** Bill Knapp's Restaurant, **Cracker Barrel, Frisch's Big Boy**, Fuddrucker's, **Old Country Buffet, Peking Wok, Ponderosa**, Skyline Chili, Wendy's

O: Burbanks, Cheyenne Cattle Company, Little Caesars Pizza (K-Mart), McDonalds, Ruby Tuesday

Lodg **I:** AAA **AmeriSuites (see our ad on this page)**

AServ **I:** Goodyear Tire & Auto, McQuick's

ATM **I:** United Dairy Farmers*

O: Meijer Grocery

Other **I:** **Builders Square, Revco Drugs**

O: **Forest Fair Mall, K-Mart (Pharmacy), Meijer Grocery (Grocery, Pharmacy)**

36 U.S. 127, Mount Healthy

Gas **I:** BP[D], Shell*, Sunoco*

O: Dairy Mart*, Super America*[LP] (Kerosene)

Food **I:** **Arby's, Burger King**, China Inn, **Frisch's Big Boy**, Golden Chain, **Little Caesars Pizza, McDonalds, Rally's Hamburgers, Subway, Taco Bell**

AServ **I:** Auto Supply, Pennzoil Lube

ATM **I:** Provident Bank

Other **I:** Car Wash, **County Sheriff's Office, Thriftway Supermarket (Pharmacy)**

33 U.S. 27, Colerain Ave

Gas **I:** BP*[D], Dairy Mart, Shell*[CW]

O: BP*, Super America*

Food **I:** Arby's, Bob Evans Restaurant, Denny's, Domino's Pizza, Frisch's Big Boy, Frontier, Hardee's, Italian Oven, KFC, Kenny Rogers Roasters, Long John Silvers, McDonalds, Olive Garden, Oriental Gardens, Outback Steakhouse, P&S Restaurant, Pizza Hut, Prescotts, Red Lobster, T.G.I. Friday's, TCBY Yogurt, The Ground Round, White Castle Restaurant

O: Burger King, Wendy's

AServ **I:** BP Pro Care, Bob Sumerel Tire, Car Quest Auto Center, Castrucci's Jeep, Eagle, Isuzu, Firestone Tire & Auto, Goodyear Tire & Auto, North Gate Transmission, Sear's, Warehouse Outlet Auto Parts

O: Wal-Mart

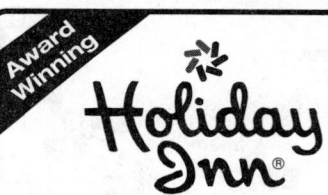

Bold red print shows RV & Bus parking available or nearby

EXIT OH/IN/KY

ATM I: Fifth Third Bank, Huntington Bank, Key Bank, PNC Bank, Provident Bank
O: First National Bank, Provident BAnk (Thriftway)

Other I: **Kroger Supermarket, Northgate Shopping Center, Pearl Vision Center, Revco Drugs**
O: **Office Max, Petsmart, Thriftway Drug, Wal-Mart (Pharmacy)**

31 Blue Rock Rd, Ronald Reagan Cross Cnty Hwy
Med I: ✚ **Hospital**

Note: I-275 runs concurrent below with I-74. Numbering follows I-74.

9 N I-275, Dayton (Left Exit)
7 OH 128, Cleves, Hamilton
Gas N: BP*(D), Shell*(D) (Kerosene)
Food N: Wendy's
S: Angelo's Pizza
AServ N: Don's Miami Town
5 I-275S to KY

Note: I-275 runs concurrent above with I-74. Numbering follows I-74.

25 JCT I-74, US 52, Cincinnati, Indianapolis (Left Exit)
21 Kilby Rd
Other I: **Mitchell Memorial Forest, Shawnee Look Out**

↑OHIO

↓INDIANA

16 U.S. 50, Lawrenceburg, Greendale, Aurora
TStop O: **Kinnett Truckstop***
Gas O: Ameri-Stop*, Marathon, Shell*
Food O: Dunkin Donuts, Empress Chili (Kinnett TS)
Lodg O: Holiday Inn Express, The Wishing Well Motel
AServ O: Andee Chevrolet, Olds, Kidd Dodge, Tri State Tire Co
TServ O: **Kinnett Truck & Trailer Repair (Kinnett TS)**
ATM O: American State Bank

↑INDIANA

↓KENTUCKY

11 to KY 20, Petersburg
7 KY 237, Hebron
Gas O: Shell*(D)(CW)
Food O: Subway (Shell)
ATM O: Huntington Bank (Shell)
4 KY 212, to KY 20, Cincinnati - No. Kentucky Int'l Airport
Gas I: ValAir
Lodg O: Radisson Inn
Med O: ✚ **Hospital**
Other O: **US Post Office**
2 Mineola Pike
Food I: **Holiday Inn**
Lodg I: Budgetel Inns & Suites, **Holiday Inn**
O: Marriott

↑KENTUCKY

Begin I-275, Cincinnati

EXIT FLORIDA

Begin I-275, Tampa

↓FLORIDA

36 **(52)** Bearss Ave
Gas E: Citgo*
W: Amoco*, Chevron*(D), Shell*
Food W: Baby Cakes BBQ, Burger King (Playground), McDonalds (Play Land), Subway, Wendy's
Lodg W: Holiday Inn
ATM W: Barnett Bank, NationsBank, NationsBank Shell
Other W: **Albertson's Grocery (Pharmacy), Eckerd Drugs, Publix Supermarket**
35 **(51)** Fletcher Ave
Gas E: Exxon*, Fina*(D)
W: BP*, Chevron*, Citgo*(CW)
Lodg E: DAYS INN ◆ Days Inn
W: Americana Inn
AServ E: Tire Kingdom
34 **(50)** Flower Ave, FL. 582
Gas E: Shell*
Food E: Burger King, Hops Grill & Bar, Luvy's, McDonalds, Pizza Hut, Ponderosa Steak House, Shoney's, Subway, Waffle House
W: Boston Grill, Cypress Wood

EXIT FLORIDA

Lodg E: Howard Johnson
W: Discovery Inn, Interchange Motor Inn, Motel 6, Sleep Rite
AServ E: Auto Air, Discount Auto Parts, Shell
W: AAMCO Transmission, Bob Wilson Dodge, Chrysler, Ferman Oldsmobile, Napa Auto Parts, Reeves Porsche, Audi, Reeves Subru, Roger Whitley Chevrolet
RVCamp W: **Happy Traveler RV Park**
Other E: **Eckerd Drugs, Varsity Theaters**
W: **U.S. Post Office**
33 **(48)** Busch Blvd., FL. 580
Gas E: Chevron*(CW)
W: Amoco (24 Hrs), Exxon*
Food W: Canton Chinese Restaurant, KFC
Lodg E: AAA Best Western
W: University Motel
AServ W: Amoco, Goodyear Tire & Auto, Master Care Auto Service
Other E: **Adventure Island, Busch Gardens,** Tropical Car Wash
32 **(48)** Bird St
31 **(466)** Sligh Ave
Gas E: Amoco*(CW), Coastal*, Spur(D)
W: BP*(D)

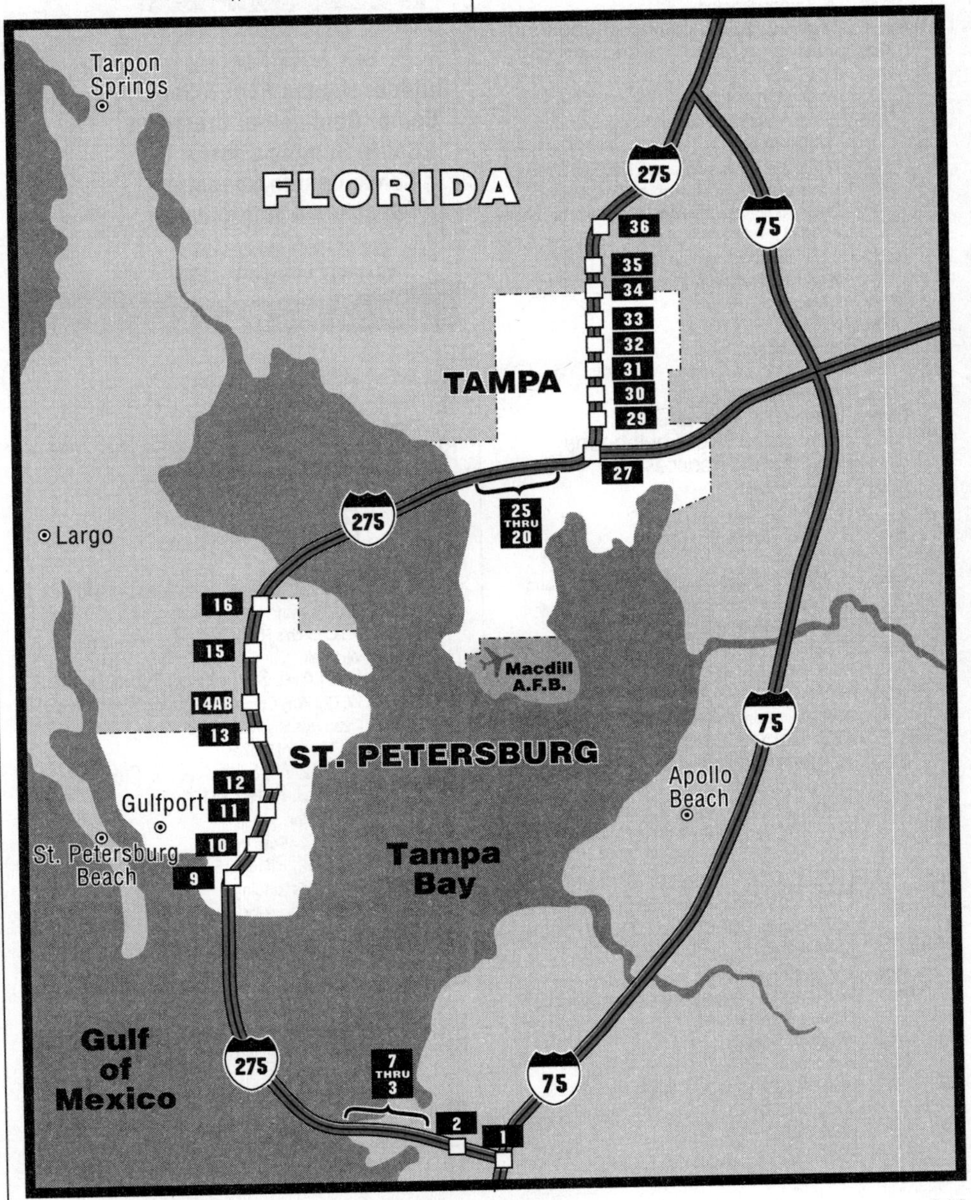

Bold red print shows RV & Bus parking available or nearby

EXIT MICHIGAN

AServ E: Amoco, Spur
Other W: **Lowry Park Zoo**
30AB U.S. 92, Hillsborough Ave
Gas E: Citgo(CW), Mobil(CW)
W: Amoco*
Food E: High Tide Fish & Chips
AServ E: Citgo, Discount Auto Parts, Mobil
W: Brake World
29 FL. 574, MLK Jr. Blvd
Gas E: Amoco*, Chevron*
W: Cumberland Farms*
Food W: McDonalds
28 Florbraska St.
27 I-4, Orlando
26 Downtown, E. Jefferson St.
25 Downtown, Ashley Dr., Tampa St.
Lodg E: AAA Holiday Inn Select
Med E: + **Hospital**
Other E: **Performing Arts Center, Tampa Convention Center**
W: **Police Station**
24 **(41)** Armenia Ave, Howard Ave
Gas W: Citgo*, Texaco*
Food W: Popeye's
23C Himes Ave.
Gas W: Supertest* (24 Hrs)
Other W: **Tampa Stadium**
23AB U.S. 92, Dale Marby (B Northbound)
Gas E: Exxon*(D)(CW) (24 Hrs), Mobil(CW), Shell*(CW) (24 Hrs)
W: Citgo(CW), Exxon*(D)
Food E: Alexander's Restaurant, Carrabba's, Pizza Hut, Rio Bravo
W: Blimpie's Subs, Checkers, Denny's, Dunkin Donuts, Longhorn Steakhouse, McDonalds, Sweet Tomatoes, Taco Bell, Tia's Mexican, Waffle House, Wendy's
Lodg E: Courtyard by Marriott
W: Howard Johnson
AServ E: Mobil
ATM E: Shell
Other W: **Wal-Mart**
22 Lois Ave.
Gas W: Radiant*(D)
Other W: **Woolf Animal Hospital**
21 West Shore Blvd. (Southbound, Northbound Reaccess)
Gas E: Citgo*, Shell*
W: Shell*
Food E: Steak & Ale, Waffle House
W: Subway
Lodg E: AAA Embassy Suites, AAA Sheraton Grand
W: Crowne Plaza Hotel, Doubletree Hotel, ◆ Marriott, Quality Hotel
AServ E: Shell
W: Shell
ATM E: Citgo
W: Barnett Bank, NationsBank
20A **(38)** Tampa Airport, Clearwater
Lodg N: AmeriSuites (see our ad on this page, Days Inn (see our ad on this page)
19 S. to FL. 687, to U.S. 92, 4th St. N.
18 FL. 688 W., Largo, Ulmerton Rd.
17 M. L. King St., Nineth St. N.
16 **(30)** FL. 686, Roosevelt Blvd.
15 **(27)** FL.694 W., Pinellas Park, Seminole, Gandy Blvd, Indian Shore
14AB Kenneth City, 54th Ave N
Gas W: Citgo*, Racetrac*(LP)
Food W: Nishu's Family Restaurant, Son's Cafe Chinese & Vietnamese
Lodg W: DAYS INN Days Inn
ATM W: Racetrac
13 38th Ave N, Reddington Beaches, Maderia Beach
12 **(23)** 22nd Ave. N.

EXIT MICHIGAN

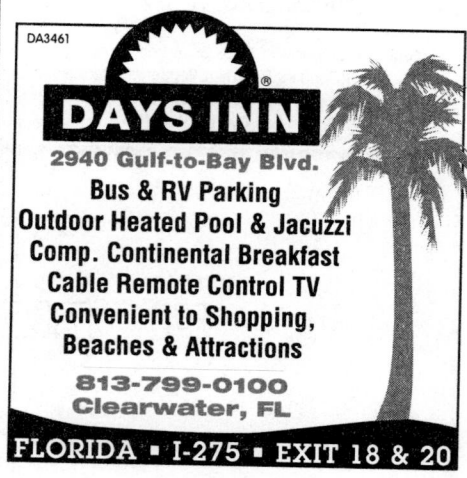

11 **(12)** Fl. 595, Fifth Ave. N.
Med W: + **Hospital**
Other W: **Coin Laundry**
10 Jct I-375 E., The Pier (Southbound, Northbound Reaccess)
9 Jct I-175, Dr. M.L.K. St.
8 **(21)** 28th St. S. (Southbound, Northbound Reaccess)
7 31st St S
6 **(20)** 22nd Ave. S. (Southbound)
Gas W: Chevron*, Citgo*, Shell*
Food W: Church's Fried Chicken
AServ W: Shell
5 26th Ave. S., Gulfport (Northbound)
Food W: Cat Way Chinese, KFC, King Wah Chinese, Pizza Hut, Sky Way Jack's
AServ W: Discount Brakes
4 FL. 682 W., 54th Ave. S., Pinellas Bay Way, St. Pete Beach
Lodg W: Park Inn
Parks W: **Fort DeSoto Park**
3 34th St., Pinellas Point Dr.
Food E: Parker's Landing
Lodg E: DAYS INN Days Inn
Parks W: **Maximo Park**
2B Sky Way Fishing Pier, Rest Area
2A Sky Way Fishing Pier S., Rest Area - RR, Phones, Picnic, Vending
2 U.S. 19 S., Palmetto, Bradenton
1 **(2)** Port Manatee, Palmetto, Bradenton

↑FLORIDA

Begin I-275, Tampa

EXIT MICHIGAN

Begin I-275, Detroit

↓MICHIGAN

Note: I-275 runs concurrent below with I-96. Numbering follows I-96.

167 8 Mile Rd, Northville
Gas S: Meijer Grocery*(D), Speedway*
Food S: Big Boy, Chili's, Hilton, Kirby Coney Island, Kyoto Japanese, McDonalds, Taco Bell
Lodg S: Hampton Inn, Hilton, Travelodge
Med S: + **Hospital**
ATM S: Comerica Bank, Michigan Heritage Bank, NBD
Parks S: **State Park**
Other N: **Veterinary Clinic**
S: **Meijer Grocery**
169AB 7 Mile Road
Food N: Cascade Restaurant (Embassy), Lone Star Steakhouse, Rio Bravo
S: Cooker
Lodg N: Embassy Suites
Other S: **State Patrol Post**
170 6 Mile Rd
Gas S: Amoco*, Mobil*(CW)
Food N: Akasa Japanese Restaurant, Big Boy, Bill Knapp's Restaurant, Boston Market, Bruegger's Bagel, Holiday Inn, Kerby's Koney Island, Marriott, Max & Erma's, Papa Romano's Pizza, The Ground Round, Wing Yee's
S: Charley's Grille, McDonalds, Papa Vino's, Standard Federal, Wendy's
Lodg N: Best Western, Courtyard by Marriott, Holiday Inn, Marriott
Med N: + **Hospital**
ATM N: Comerica Bank, First Federal of Michigan, Michigan National Bank, NBD Bank
S: Old Kent
Other N: **Laural Park Place Mall, Rite Aide Pharmacy**
S: **Arbor Drugs, Farmer Jack Supermarket, Office Depot, PetsMart**

Note: I-275 runs concurrent above with I-96. Numbering follows I-96.

29 I-96 E Junction - Detroit, Ann Arbor
28 Ann Arbor Rd, Plymouth, Livonia
Gas E: Amoco*, Shell*
Food E: Denny's, Dunkin Donuts (Amoco), Water Club Seafood
W: Bill Knapp's, Burger King, Steak & Ale
Lodg W: AAA Quality Inn
AServ E: Shell
W: Cadillac, K-Mart, Lincoln Mercury Dealer
ATM E: First of America
W: Michigan Nat'l Bank, Standard Federal
Other W: **K-Mart (Pharmacy)**
25 MI 153, Ford Rd, Westland, Garden City
Gas W: Amoco*(CW), Shell*, Speedway*, Sunoco*(D) (Kerosene)
Food E: Don Pablo's
W: BJ Bowery's, Bob Evans Restaurant, Bruegger's Bakery, Buck Jones Deli, Chuck E. Cheese's Pizza, Cooker, Dunkin Donuts, Hunan Empire, Johnson's Restaurant, K-Mart, KFC, Little Caesars Pizza, Outback Steakhouse, Platos, Red Oak Steakhouse, Shark Club, T.C. Gators,

Bold red print shows RV & Bus parking available or nearby

EXIT		MICHIGAN
		The Donut Scene, The Roman Forum, Tim Hortons, Toarmina's Pizza, Wendy's, White Castle Restaurant
	Lodg	W: ◆ Budgetel Inn, ◆ Fairfield Inn, AAA Motel 6
	AServ	E: Auto Nation USA
		W: Discount Tire, Midas Muffler & Brakes, Pennzoil Lube, Speedy
	Med	W: ✚ Hospital
	ATM	W: First Federal of MI, First of America
	Other	E: Andy Plc BMW Motorcycles
		W: Builder Square 2, Good Food Company, K-Mart, Lighthouse Car Wash, Pet Supply Store, Sears Hardware
(24)		Rest Area - RR, Phones, Picnic (Northbound)
22		US 12, Michigan Ave, Ypslanti, Dearborn
	Gas	E: Amoco*, Shell*, Speedway*(D)
	Food	E: Dawn Donuts (Amoco), McDonalds, Subway
	Lodg	E: ◆ Country Hearth Inn, Fellows Creek Motel, Maple Lawn Motel, ◆ Super 8 Motel, Willo Acres Motel
20		Ecorse Rd, Romulus
	Other	W: Airport
17		I-94 Junction, Detriot, Chicago
15		Eureka Rd
	Gas	E: Shell*(D)
13		New Boston, Sibley Rd
	Parks	W: Lower Heron Metro Park
11		South Huron Rd
	Gas	W: Sunoco*
	Food	W: Burger King
	TServ	W: Jim's Heavy Equip Repair
	Parks	E: Willow Metro Park
8		Will Carleton, Flat Rock
	Other	E: State Patrol Post
		W: Cross Winds Marsh
5		Carleton , South Rockwood
(4)		Rest Area - RR, Phones, Picnic (Southbound)
2		U.S. 24, Telegraph Rd
	Lodg	E: Glee Motel
1		Junction I-75, Detroit, Toledo

↑MICHIGAN

Begin I-275, Detroit

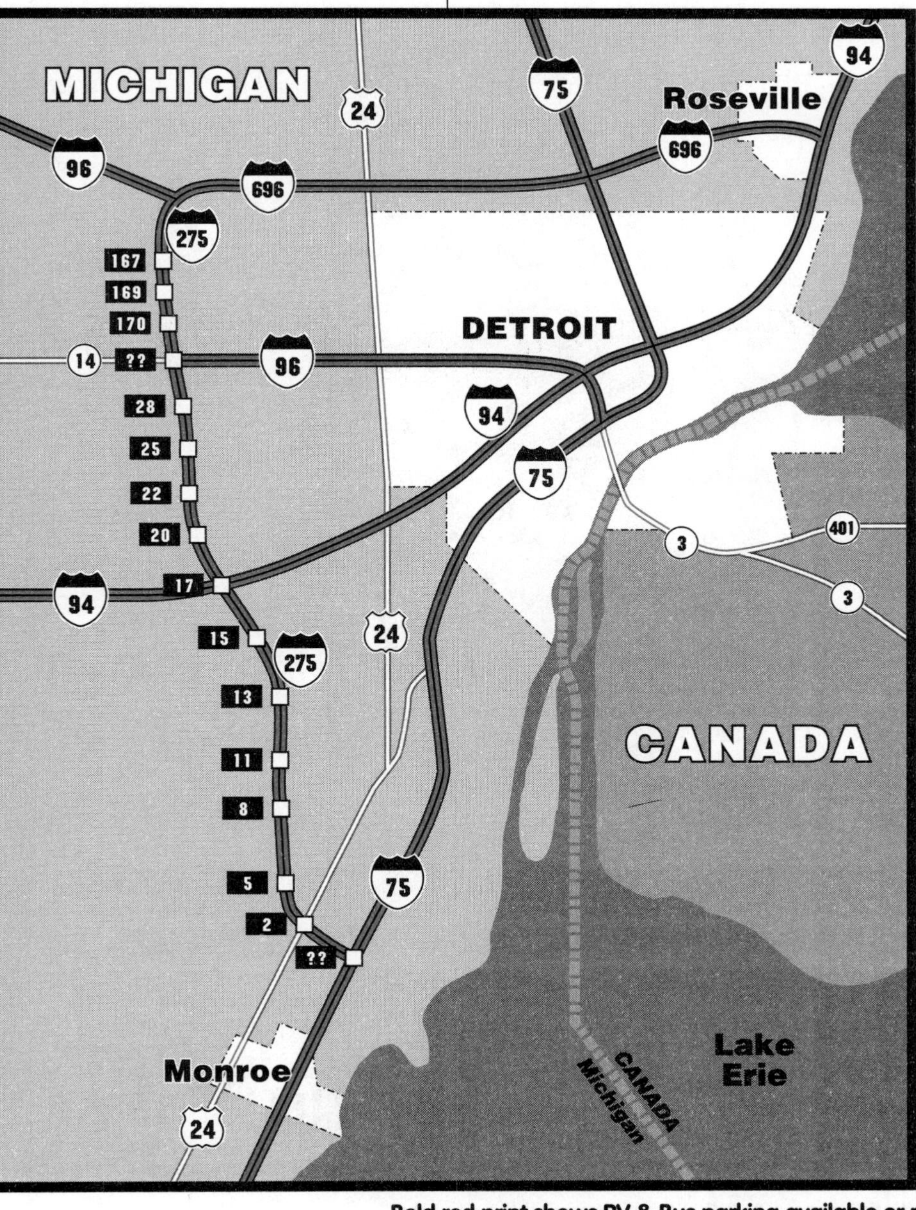

EXIT		PENNSYLVANIA
		Begin I-276

↓PENNSYLVANIA

EXIT		PENNSYLVANIA
26		**(339)** PA 309, Ft Washington, Philadepphia, Ambler, Toll Rd (Highland Ave, To Piketown Rd
	Gas	N: Exxon(CW), Mobil*
	Food	N: Friendly's, Holiday Inn, Koners Ye Old Beef & Ale, Palace of Asia (Indian Cuisine, Ft Washington Inn), Park Place Deli, Subway
	Lodg	N: Clarion, Ft Washington Inn, Holiday Inn
	AServ	N: Exxon, Mazda, Volvo, Mercedes, Old Ft Pontiac GMC
	ATM	N: CoreStates Bank, PSFS Bank
	Other	N: Car Wash, Post Office
27		PA611, Willowgrove, to Doylestown
	Gas	N: Amoco*(D) (Kerosene), Hess*(D), Mobil*, Mobil*(D)(CW), Texaco*
	Food	N: Bagels Plus, Bennigan's, China Garden, Domino's Pizza, Donut Haven, Nino Pizza-Rama, The Bakers Inn, Williamson Restaurant
		S: Dunkin Donuts, Giulo's Italian
	Lodg	S: Hampton Inn
	AServ	N: Adams Pittstop, Texaco
		S: J&J Automotive, Pep Boys Auto Center
	Other	N: 7-11 Convenience Store, United Check Cashing
		S: Regency Square, Thrift Drugs
28		**(350)** US1, Philadelphia, Trenton
	FStop	S: Sunoco*(D)
	Gas	S: Amoco(CW), Exxon(D), Getty*(LP), Mobil, Sunoco*
	Food	S: Baskin Robbins, Bumpers (Radisson Inn), Dumino's, Dunkin Donuts, Miami Grill, Steak & Ale, Subs
	Lodg	S: Howard Johnson, Knights Inn, Neshaminy Inn, Penn Motel, Raddison Inn, Red Roof Inn, The Lincoln Motel
	AServ	S: Faulkner GM, Faulkner Toyota, Sunoco
	ATM	S: Summit Bank
	Other	N: Neshaminy Mall
		S: Neshaminy State Park
(352)		Neshaminy Service Center - RR, Phones, Vending, Arcade, Phone Cards
	Gas	N: Sunoco
	Food	N: Bakery, Bryer's Ice Cream, Hot Dogs Under Construction, McDonalds (Playland), Salads
		S: Burger King, Nathan's
	Other	N: C-Store
29		US13, Delaware Valley, Bristol, to I-95
	FStop	S: Bristol Fuel(D), Coastal(LP)
	Gas	N: Citgo*(D)
		S: Mobil, Sunoco*(D)
	Food	N: Dallas Diner, Divas (Econolodge), Edgley
		S: Big Apple Grill, Big Daddy's Italian Style Sandwiches, Boston Market, Burger King, Dari-Deli, Days Inn, Italian Family Pizza, McDonalds, Pizza Hut, Subway, The Grand Family Diner, The Original Eagle
	Lodg	N: Comfort Inn, Econolodge
		S: DAYS INN Days Inn, The-Vow Motel
	AServ	N: AAMCO Transmission, Al Rue's Auto Serv, Bruce's Auto Service, H&H Auto Repair, Pays Auto Repair
		S: AAA Cooling Specialities, Dieckhaus GM, Low Cost Exhaust & Brakes, Meineke Discount Mufflers, Mobil, Percision Performance, Quick Lube, STS Tire & Auto, Sunoco*(D), Zebart Tidy Car
	Med	S: ✚ Hospital
	Other	N: 7-11 Convenience Store, Pets Best Friend Veterinary Hospital, U-Haul Center(LP)
		S: Action Safe & Locksmith, Coin Operated Laundry, Self Service Car Wash, Super Fresh Food Market (Pharmacy), Thrift Drug

↑PENNSYLVANIA

Begin I-276

Bold red print shows RV & Bus parking available or nearby

EXIT	PENNSYLVANIA

Begin I-279

↓PENNSYLVANIA

21A	Jct I-79
21	**(15)** Camp Horne Rd.
20	**(14)** Bellevue, West View
Med	**W:** **Hospital**
19	**(12)** Perrysville Ave. (Northbound)
18	Truck US 19 North, McKnight Rd, Evergreen Rd (Northbound)
17	Venture St (Southbound)
16	**(10)** Hazlett St
15	PA 28 North, East St
14	Intersection 579 South Veterans Bridge
13	PA 28 North, Chestnut St, East Ohio St, Etna
12	Three Rivers Stadium
11	PA 65, Ohio River Blvd
10	Fort Duquesne Blvd Convention Center
9	Boulevard of Allies, Liberty Ave, Civic Arena
Food	**W:** McDonalds
ATM	**W:** Mellon Bank, PNC Bank
8	Jct I-376 East, Monroeville
6	US 19 South, Banksville Rd. South Only
5	Parkway Mall Center
Food	**W:** Chi Chi's Mexican Restaurant, Parkway Center Inn Best Western, Sam and Tony's Cafe
Lodg	**W:** Parkway Center Inn
Other	**W:** **Giant Eagle Supermarket, K-Mart, Parkway Center Mall, Phar Mor Drugs, US Post Office**
7AB	**(5)** South Truck US 19, PA 51, Uniontown. (No Northbound Reaccess)
Gas	**E:** Amoco*
Food	**E:** Georgie's Diner
AServ	**E:** Amoco
4	PA 121, Green Tree, Mount Lebanon
Gas	**E:** BP*, Citgo*, Exxon*
Food	**E:** Aracri's Greentree Restaurant, Boston Market, Claudio's Pizza, Einstein Brothers Bagels, The Olive Garden
	W: Holiday Inn, Pizza Pub
Lodg	**E:** The Greentree Inn
	W: ◆ Hampton Inn, AAA Holiday Inn, Pittsburg Marriott Greentree
Med	**E:** **Primary Care Center**
ATM	**W:** PNC Bank
Other	**E:** **Kinko's Copies**
3	PA 50 West, Carnegie
Gas	**E:** BP*, Sunoco
Food	**E:** Donut Connection, Pizza Outlet, The Main Dinning Room
AServ	**E:** Batey Chevrolet, Napa Auto Parts, Sunoco, Wright Pontiac
Other	**E:** **Co Go's Supermarket, The Medicine Shoppe**
2	Rosslyn Farms (Southbound, Northbound Reaccess)
Med	**W:** **Hospital**

↑PENNSYLVANIA

Begin I-279

EXIT	CALIFORNIA

Begin I-280

↓CALIFORNIA

(53)	**Rest Area - Picnic, RR, HF, Phones**
40	McLaughlan Ave.
39	7th Street, CA 82
37	Guadalupe Pky., CA 78
38	Bird Ave.
36	Meridian Ave.
Gas	**W:** Chevron
Food	**W:** Mr. Chau's, Taco Bell, Wienerschnitzel
AServ	**W:** Chevron
35	Bascom Ave. & Leigh Ave.
34	Junction I-880 & CA 17
33	Winchester Blvd. Campbell
Gas	**E:** Exxon
	W: Arco, Gas & Shop*(D)(LP)
Food	**E:** Burger King, Chili's, Flames, Lyon's, Mandarian House, Mr. Chau's, Ocean Harbor, Rock'n Tacos, Sweet Treats Cafe, ToGo's
	W: Florentine, Marie Callender's Restaurant, Tho Saigon
AServ	**E:** Courtesy Chevrolet Geo Dealership, Exxon
	W: Arco
ATM	**E:** Great Western Bank
32	Saratoga Ave.
Gas	**E:** Arco*, Chevron*
	W: BP, Exxon, Shell
Food	**E:** Azabu, Bankok Garden, Bijan Backery & Cafe, Black Angus, Burger Factory, Burger King, Coco's, Family BBQ, Happi House, Harry's Hofbrau, High Thai, Le Papillon, McDonalds, Tasty Doughnuts, U-Bake Pizza
	W: Baskin Robbins, Chinese, Denny's, Round Table Pizza, Tony Roma's
Lodg	**N:** Ho Jo (see our ad on opposite page)
AServ	**E:** Chevron, Dayton Tire, Jiffy Lube
	W: BP, Exxon, Shell
ATM	**E:** 7-11 Convenience Store, Bank Of The West
	W: 7-11 Convenience Store
Other	**E:** **7-11 Convenience Store (ATM), Lucky Food**

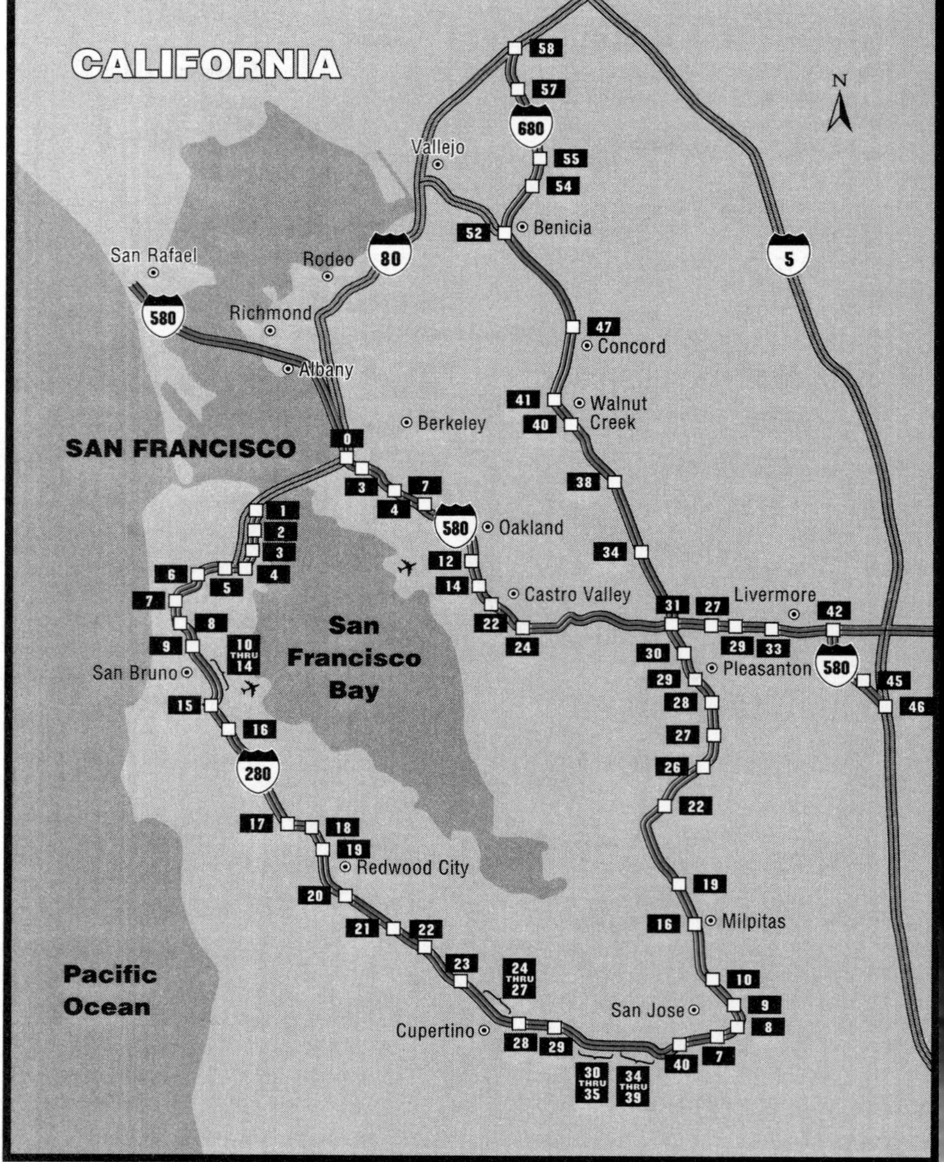

Bold red print shows RV & Bus parking available or nearby

EXIT CALIFORNIA

Center, Magic Touch Car Wash
W: **7-11 Convenience Store (ATM)**

31 Stevens Creek Blvd.
- Gas W: 76, Rotten Robbie(D)
- Food E: Boston Market, McDonalds
  W: IHOP, Subway
- Lodg W: Howard Johnson
- AServ W: 76
- Other E: **Marchall's, Payless Drugs, Safeway Drugs**

30 Wolfe Rd.
- Gas E: Arco*
  W: BP (AServ)
- Food E: Cafe Gourmet, Duke Of Edinburch, Samkee, Sushi Depot, Wolfe Cafe
  W: Azuma, El Torito, Erik's Deli Cafe, Fresh Choice, Harry's Hofbrau, McDonalds, Pizza Hut, T.G.I. Friday's, Taco Grill
- Lodg E: Marriott
- AServ W: BP, Sear's Auto Center
- ATM E: Bank Of The West
  W: Great Western Bank
- Other W: **Sear's, Stacy's Book Store, Valco Mall**

29 De Anza Blvd. Saratoga Rd. Cupertino
- Gas E: Chevron*(CW)
  W: Arco (AServ)
- Food E: Carl's Jr Hamburgers, Manley's Donuts
  W: Outback Steakhouse, Peppermill, Santa Barbara Grill
- Lodg E: Cupertinon
- AServ E: Carlos', Goodyear Tire & Auto
  W: Arco
- Other E: **Oasis Coin Laundry**
  W: **World Headquarter of Apple Computer**

28 CA 85

27 Grant Rd., Foothill Expwy.
- Gas E: Chevron*
- Food E: China Shuttle, Hickory Pitt, Pacific Steamer, Red Pepper
- AServ E: Chevron
- Parks W: **Rancho San Antonio County Park**
- Other E: **Payless Drugs**

26B Jct. Ca. 85 S., Gilroy

26 Grant Rd., Foothill Expwy.

25 Magdalena Rd.

24 El Monte Rd. Moody Rd.
- Other W: **Foothill College**

23A Page Mill Rd. Arastradero Rd.

23 Alpine Rd.

22 Sand Hill Rd. Menlo Park
- Med E: **Hospital**

21 Woodside Rd. CA 84

20 Farm Hill Blvd. (No Trucks)

19 Edgewood Rd., Canada Rd., San Carlos

EXIT CALIFORNIA

- Med E: **Hospital**

18B La Vista Point

18 San Mateo, Hayward, Belmont, Ralston, E. Ca. 92

17 CA 30, CA 35, CA 92W, Bunker Hill

(16) **Rest Area - RR, Call Box, Phones (Northbound)**

15 Black Moutain Rd, Hayne Rd.
- Parks W: **Sawyer Camp Trail**

14 Trousdale Dr.
- Med E: **Hospital**

13 Larkspur Dr. Hillcrest Blvd.
- Gas E: Chevron
- AServ E: Chevron

12 Crystal Springs
- Parks E: **Junipero Serra County Park**

11 San Bruno Ave., Sneath Ave. (Norhtbound)
- Gas W: Chevron (24 Hrs)
- Food E: Baskin Robbins, Burger Saloon, Carl's Jr Hamburgers, Dicino Deli, Manilla Eatery, Taco Bell, Vietnam Village
- AServ W: Chevron
- ATM W: 7-11 Convenience Store
- Other E: **Long's Drugs**
  W: **7-11 Convenience Store**

10B I-380 to U.S. 101, San Franscisco International Airport

10A Sneath Ln. (Difficult Southbound Reaccess)
- Gas W: BP*(D), RWA
- Food W: Bakers Square Restaurant
- AServ W: RWA

9B Abalone Dr.

9A Westborough Blvd.
- Gas W: Arco, Exxon(D)
- Food W: McDonalds, Paradise Restaurant, West Borough Deli
- AServ W: Arco, Exxon
- ATM W: Sanwa Bank
- Other W: **WalGreens**

8 Hickey Blvd.
- Gas E: Chevron*(D)(CW) (24 Hrs), Shell(CW)
  W: Shell*(CW) (24 Hrs)
- Food W: El Torito, Hungry Hunter, Sizzler
- ATM W: Bank of America, Cal Fed

7 Serramonte Blvd. (No Southbound Reaccess)
- Gas E: Chevron
  W: 76, Olympic*
- Food E: Fresh Choice, Round Table Pizza, Sizzler
  W: McDonalds, Toones
- AServ E: Chevron, Serramonte Nisson
  W: 76, Montgomery Ward
- ATM W: Wells Fargo
- Other E: **Complete Eyeglasses, Target Department Store**
  W: **Mervyn's, Sierra Monte Center**

6 Pacifica, Ca. 1
- Gas W: Arco, Exxon, Shell
- AServ W: Arco, Broadmoor Auto Serv., Exxon, M & N Car Wash
- Med W: **Hospital**

EXIT CALIFORNIA

5 Eastmore Ave., Mission Ridge
- Gas E: BP*
  W: Arco*, Exxon(D), Shell
- AServ W: Arco, Exxon
- Med W: **Hospital**
- ATM E: Bank of America
- Other E: **Lucky Food Center**

4 Daily City, West Lake Ave.

3 Mission St., Daily St., CA. 1, CA. 82

2 Ocean Ave. & Geneva Ave. (Difficult Reaccess)

1 Jct. I-280 & U.S. 101

↑ CALIFORNIA

Begin I-280 at San Jose

Begin I-580 at Jct I-80, kBerkeley

↓ CALIFORNIA

46 Jct. I-580 & I-5

45 Jct. CA. 132

44 Corral Hollow Rd.

43 Patterson Pass Rd.
- Gas S: **Arco***

42 Jct I-205 Tracy & Stockton

41 Grant Line Rd. Byron

39 North Flynn Rd. (Brake Check)

38 North Greenville Rd.

37 Weigh Station (Both Directions)

36 Vasco Rd.
- FStop N: **Shell(CW) (Gino's)**
  S: **BP***
- Food N: **Gino's (Shell FStop)**
  S: Jack-In-The-Box

34 1st St. CA 84 Springtown Blvd.
- Gas N: Springtown*(D)
  S: 76, BP*, Shell*(CW)
- Food N: Chinese Buffet
  S: Applebee's, Arby's, Bagels, Baskin Robbins, Blimpie's Subs, Burger King, Happy Juice, Home Style Buffet, Kenny Rogers Roasters, McDonalds, Me N Ed's Pizza, Peking Restaurant, Starbuck's Coffee, Strings, TCBY Yogurt, Taco Bell, Togo's
- Lodg N: Holiday Inn, Motel 6, **Springtown Motel**
- AServ S: 76, American Tire Co.
- ATM N: 7-11 Convenience Store
- Other N: **7-11 Convenience Store (ATM)**
  S: **Animal Health, Long's Drugs, Safe Way Grocery, Target Department Store**

33 N. Livermoore Ave.

32 Portola Ave. (No Eastbound Reaccess)
- FStop S: **Coast*, Grafco**
- Gas S: Chevron, Shell*
- Lodg S: Palace Motel
- AServ S: Chevron, Oil Changers
- Other S: Coin Car Wash

31 Colier Canyon Rd. Airway Blvd.
- Gas N: Exxon*(D)(CW)
- Food N: Baskin Robbins, Wendy's
  S: Beeb's Sports Bar and Grill, Cattleman's
- Lodg N: Comfort Inn, Hamption Inn, Marriott

Bold red print shows RV & Bus parking available or nearby

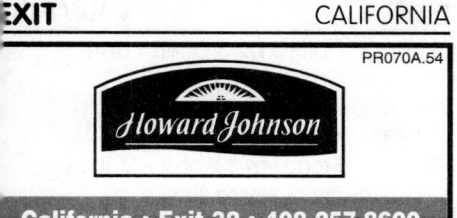

EXIT	CALIFORNIA
30	El Charro Rd. Fallon Rd.
29	Sante Rita Rd. , Tassajara
Gas	**S:** Shell*(D)(CW)
Food	**S:** Bagels, Bakers Square Restaurant, Baskin Robbins, Boston Market, California Burger, Doughnuts and Yogurt, Erik's Deli Cafe, McDonalds, Nations, Subway, T.G.I. Friday's, Taco Bell
AServ	**S:** Infinity, Lexus, Acura, Volvo, Oil Changer
Med	**S:** + Hospital
Other	**S:** Long's Drugs
28	Hacienda Dr.
Med	**S:** + Hospital
27	Hopyard Rd., Pleasanton (No Trucks Over 3 Tons)
Gas	**N:** 76*(CW), BP*
	S: Chevron*(CW), Shell*(D)
Food	**S:** Arby's, Burger King, Buttercup Pantry Restaurant, Denny's, El Molino, Hungery Hunter, N-N-Out Burger, Nations Hamburgers, Pleasant Asian Cuisine, Taco Bell
Lodg	**S:** Motel 6, Super 8 Motel
AServ	**N:** Dublin Pontiac, Toyota
RVCamp	**N:** El Monte RV Center
26B	Junction I-680
26	San Ramon Rd. Foothill Rd. (No Trucks Over 3 Ton)
Gas	**N:** Chevron*, Shell*(D)
Food	**N:** Burger King, Coco's, Frankie Johnnie & Luigi Too, Starbuck's Coffee, Willow Tree
	S: Black Angus
Lodg	**N:** Best Western
	S: Holiday Inn Select, Wyndham Garden Hotel
AServ	**N:** Grand Automobiles
ATM	**N:** Cal Fed
25	Eden Canyon Rd., Balomares Rd.
24B	Center Street, Crow Canyon Road
Gas	**S:** 76(D)(CW), Arco*, Chevron*
Food	**N:** Casper's, Domino's Pizza
24	Redwood Rd. Castrol Valley (Difficult Reaccess)
Gas	**N:** Chevron
Food	**N:** Chinese Cooking, Dino's Italian, Genghis Khan, Hof Brau, Java Bob's, KFC, Rudy's Doughnut House, Taco Bell
AServ	**N:** Chevron
Other	**N:** Long's Drugs, Safe Way Grocery
23	Strobridge
Gas	**N:** 76, BP
Food	**N:** Carrows Restaurant, Donut Express, McDonalds, Wendy's
Lodg	**N:** Holiday Inn Express
AServ	**N:** 76, Big Tires, Jiffy Lube
Med	**N:** + Hospital
Other	**N:** Veterinary Clinic
22B	Junction I-523 & I-880 (Westbound)
22	CA 238 Hayward
21	164th Ave. Miramar
Gas	**N:** Beacon(D), Chevron
Food	**N:** Grand View, Uncle Ben's Pizza
Lodg	**N:** 580 Motel, Budget Inn
AServ	**N:** Beacon, Chevron
20	105th Ave. Fairmont Dr.
Gas	**S:** Arco*, Shell
Food	**S:** Denny's

EXIT	CALIFORNIA
AServ	**S:** Shell
Other	**S:** Coin Laundry
19	Dutton Ave. Estudillo Ave. (Difficult Reaccess)
Gas	**S:** Payless Drugs
Food	**S:** Cafe Encore, Sabino's, The Cheesecake Creamery, The Lyon Restaurant
Other	**S:** Payless Drugs, Quick Stop
18	106th Ave. Foothilll Blvd (No Eastbound Reaccess)
17	Golf Links Rd. 98th Ave.
Gas	**N:** Shell*
	S: BP
AServ	**S:** BP
Other	**N:** Knowland Park, Oakland Zoo
16	Kelller Ave., Mt. Blvd.
15	Edwards Ave.
14	Seminary Ave. Warren Frwy. CA.13
13	McCarther Blvd. (No Eastbound Reaccess)
Gas	**S:** 76 (AServ)
Lodg	**N:** Holiday Motel, Mills Motel, Sage Motel
AServ	**S:** 76
ATM	**N:** 7-11 Convenience Store
Other	**N:** 7-11 Convenience Store* (ATM)
12	High St. (No Eastbound Reaccess)
Gas	**N:** 76, Shell
	S: BP
Food	**N:** Daniel's Palace, Giant Burgers, Subway, Yummy Yogurt
	S: Dick's Doughnuts
AServ	**N:** 76, Kragen Auto Parts
	S: BP
Other	**N:** Coin Car Wash, High St. Pharmacy
	S: Coin Laundry, WalGreens
11	35th Ave. (Eastbound, No Reaccess)
Gas	**N:** 76 (24 Hrs, AServ), Chevron
	S: BP*(LP), Quik Stop*
Food	**N:** Glen's Hot Dog, Golden City Restaurant, Taco Bell
AServ	**N:** 76, Chevron
9	Fruitvale Ave., Coolidge Ave. (Difficult Eastbound Reaccess)
Gas	**N:** 76
	S: 76*
Food	**N:** Mel's Pancakes
ATM	**N:** Great Western Bank
Other	**N:** Lucky
8	14th St. (Westbound)
Med	**S:** + Hospital
7	Park Blvd.
Gas	**N:** Shell*
	S: Arco*
AServ	**S:** Oil Changers
6	Grand Ave., LakeShore Ave.
Gas	**N:** 76* (AServ), Chevron(D) (AServ)
	S: Chevron (24 Hrs)
Food	**N:** 4 Star Pizza, Boston Market, Chinese Cuisine, Colonial Donuts, Domino's Pizza, Good Nature, KFC, Milano, Mimosa Cafe, Peet's Coffee & Tea, Subway, Szechuan, The Burrito Shop, The Coffee Mill, Yogurt Deluxe
AServ	**N:** 76, 76, Cheveron
ATM	**N:** America Savings Bank, Bank Of America
Other	**S:** Launderville Coin Laundry

EXIT	CALIFORNIA
5	Harrison St. & Oakland Ave. (Eastbound)
Gas	**N:** 76, BP
AServ	**N:** 76, BP
4	Broadway & Webster St. (Hwy. Patrol, Hospital)
3	Jct. I-980 - I-880 S.
2	Market St. (Westbound)
1	McCarther Blvd. San Pableo Ave. (Difficult Reaccess)
00	Jct. I-580E & I-580W (No Trucks Over 4 Tons)

↑CALIFORNIA

Begin I-580 at Jct I-5 near Tracy

Begin I-680 at San Jose

↓CALIFORNIA

EXIT	CALIFORNIA
58	Cordelia Rd., GreenValley Rd.
57	Gold Hill Rd.
Gas	**W:** Cheaper*(D)
56	Marshview Rd.
55	Parish Rd.
54	Lake Herman Rd., Vista Point
FStop	**W:** Shell*(D)(CW) (TCBY Yogurt)
Food	**W:** Carl's Jr Hamburgers, TCBY Yogurt
Other	**W:** Vista Point
53	Bayshore Rd.
(52)	I-780 Toll Plaza
51	Marina, Vista, Martinez
50B	Contra Costa Blvd. (Southbound)
Gas	**W:** 76, Exxon, Shell*(D)(CW)
Food	**W:** Burger King, Carrows, Denny's, KFC, McDonalds, The Velvet Turtle
AServ	**W:** Mark Morris
Other	**W:** Barnes & Noble, Full Service Car Wash, K-Mart, Target Department Store
50	Pacheco Blvd. (No Northbound Reaccess)
FStop	**W:** Shell*(D) (24 Hrs)
Gas	**W:** BP* (24 Hrs)
49	W. CA 4, Martinez, Richmond
48	CA 4, Martinez, Pittsburg, Antioch
47	Concord Ave., Burnett Rd.
Lodg	**E:** Holiday Inn
AServ	**E:** Diablo Chrysler Plymouth, Hyundai & Volkswagon, Infiniti Acura Chevrolet Dealer
46	Willow Pass Rd.
Food	**E:** Denny's, El Torito, Red Lobster, Sizzler Steak House
Lodg	**E:** Hilton
45	Concord, Pittsburg, CA 242
44	Contra Costa Blvd.
Gas	**E:** Chevron*(D)(CW)(LP)
Food	**E:** Ming Wah
	W: Boston Market, Confetti Restaurant, Lyons
Lodg	**W:** Marriott, Sunvalley Inn
AServ	**E:** Montgomery Ward
	W: Grand Auto Supply
ATM	**W:** Great Western Bank
Other	**E:** Montgomery Ward
	W: Payless Drugs

Bold red print shows RV & Bus parking available or nearby

EXIT CALIFORNIA

43 Treat Blvd., Geary Rd, Weigh Station Northbound
- Gas W: Chevron, Shell(D)
- Food W: Burger King, Golden City Restaurant, Magic Garlic Restaurant, Wendy's
- Lodg E: Embassy Suites
  W: Holiday Inn
- AServ W: Chevron, Clutchmart, Joe's Foreign Auto Service, Walnut Creek Car Wash
- Other W: **Holiday Coin Laundry**

42 North Main St.
- Gas E: Chevron
  W: 76(D)
- Food E: Baja Grill, Jack-In-The-Box, La Virage, Minerva's Cottage, Subway (Chevron), Taco Bell
  W: Domino's Pizza
- Lodg E: Marriott, Motel 6
- AServ E: GMC Truck, Oldsmobile Dealership, Something Special, Wayne Stead Cadillac
  W: Broadway Mufflers, Michael Stead Nissan
- ATM W: 7-11 Convenience Store
- Other W: **7-11 Convenience Store**

Note: I-680 runs concurrent below with I-670. Numbering follows I-670.

41 Ygnacio Valley Rd. (No Northbound Reaccess)
- Gas E: Shell(D) (24 Hrs)
- Med E: **Hospital**

Note: I-680 runs concurrent above with I-670. Numbering follows I-670.

40 Walnut Creek, South Main Street, Junction CA 24 (Difficult Northbound Reaccess)
- Food E: Johnny Love's, The Original Hick'ry Pit
- Med E: **Hospital**

39 Livorno Rd.

38 Stone Valley Rd.
- Gas W: Chevron, Shell(D)(CW)
- Food W: Alamo Cafe, Bagel St. Cafe, Baskin Robbins, Donuts, Hi Tech Burrito, Jitr Thai, Kona Cafe, Round Table Pizza, Taco Bell, The Brass Bear Deli, The Whole Grain Co.
- AServ W: Chevron
- ATM W: Bank Of America, Wells Fargo Bank
- Other W: **Post Office, Safe Way Grocery, Thrifty Drug & Discounts**

37 El Cerro Blvd, El Pintado Rd.

36 Diablo Rd.
- Gas E: 76*(LP)
- Food E: Sun's Chinese Cuisine, Taco Bell
  W: CoCo's
- AServ E: 76
- Other E: **Lucky (Grocery)**

35 Sycamore Valley Rd., Danville
- Gas E: Shell(D)
  W: BP, Exxon
- Food E: Denny's
  W: Cierra Yogurt Club, Giorgio's Coffee and Backery, Hi Tech Burrito, Patrick David's Cafe, Piatti, Tony Romans, Woody's Burger & Pizza
- Lodg E: Danville Inn
- ATM W: Eureka Bank

34 Crow Canyon Rd., San Ramon (No Trucks Over 15 Ton)
- Gas E: Shell*
  W: BP*, Exxon*, Shell(CW)
- Food E: Baskin Robbins, Blue Tatoo, Burger King, Carl's Jr Hamburgers, Chili's, Hopsing's, Max's Diner, Pasta Primavera, Round Table Pizza, Uncle Joe's Pizzaria
  W: Bagel Company, Big Horn Grill, Boston Market, Diablo Deli, In and Out Burger, Maestro's, McDonalds, Subway, Taco Bell
- AServ W: BP, Exxon
- Med E: **Hospital**
- ATM E: Bank Of America, Wells Fargo
  W: Bay Bank
- Other E: **Lucky's, Payless Drugs**
  W: **Long Drugs, Safe Way Grocery**

33 Bollinger Canyon Rd.
- Lodg E: Marriott
- Other E: **Target Department Store**

32 Alcosta Blvd.
- Gas E: BP
  W: Chevron, Shell(D)(CW)
- Food W: Chatillon Restaurant, Chubby's, Great Wall Of China, Hometown Donuts, Little House Thai Cuisine, McDonalds, Mountain Mike's Pizza, New Joe's, PaPa Murphy's Pizza, Royal Gourmet Coffee, SunSun Garden, Taco Bell, Taylor Made Pizza
- AServ E: BP
  W: Chevron
- ATM E: 7-11 Convenience Store
- Other E: **7-11 Convenience Store**
  W: **Walgreens, Wash and Dry Coin Laundry**

31 I-580, Oakland, Stockton

30 Stoneridge Dr.

29 Bernal Ave.
- Food E: Chinese Garden, Mexico Linda, Rings Super Burgers, Vic's

28 Sunol Blvd, Pleasanton

27 E. CA 84 , Livermore

26 Calveras, Sunol, W. CA 84

25 Andrade Rd.
- Gas E: Sunol-Tree*

(24B) **Weigh Station (Northbound Only)**

24 Sheridan Rd.

23 Vargas Rd.

22 Mission Blvd, CA 238
- Gas E: Shell*
- Food E: McDonalds

21 Washington Blvd. (No Trucks)
- Gas E: Quick Stop Markets*
- Food E: Coffee Roasting, Mission Jarrito, Mission Pizza, Yen Ching
- Med W: **Hospital**
- Other E: **Pet Hospital**

20 Durham Rd.
- Gas W: 76*(CW) (Subway)
- Food W: Subway (76)

19 Mission Blvd.
- Gas W: Exxon
- Food W: Burger King, Carl's Jr Hamburgers, Denny's, Domino's Pizza, Doughnut House, Jack-In-The-Box, Pizza Hut, Round Table Pizza, Subway, Taco Bell, Taqueria, The Better Bagel, Wock City Diner, Zorba's Deli Cafe
- AServ W: Exxon
- Other W: **Long's Drugs, Lucky**

18 Scott Creek Rd.

17 Jacklin Rd.
- Gas W: Shell(CW)
- Food E: Cafe Romeo (Chinese), Hung Wu, The Pizza Box

16 Calaveras Blvd, Milpitas, Jct. 237
- Gas E: 76*, Shell
  W: Shell(CW)
- Food E: Chin's Take-Out, Mr. Tung's, Pizza Hut, Round Table Pizza, Subway
  W: Hong Kong Gardens, Lyon's, Red Lobster
- Lodg E: Inn Call
  W: Embassy Suites
- AServ E: Shell
- ATM E: 7-11 Convenience Store
- Other E: **7-11 Convenience Store (ATM), Book Garden, Victoria Laundry**

15 Landess Ave. Montague Expressway
- Gas E: 76(LP), Arco, Chevron
- Food E: Burger King, Dong Ba, Jack-in-the-Box, McDonalds, My Chinese Restaurant, Pho Mai, Royal Taco, Straw Hat Pizza, Taco Bell, Wienerschnitzel, Winchell's
- AServ E: 76, Arco, Chevron, Firestone Tire & Auto
- ATM E: Wells Fargo
- Other E: **Lucky, Park Town Vet. Clinic, Target Department Store, Thrifty, Walgreens**

14 Capital Ave.

13 Hostetter Rd. (No Northbound Reaccess)
- Gas E: Shell*(CW)
- Food E: Italo's Pizza

12 Berryessa Rd.
- Gas E: Arco*, Chevron
- Food E: Baskin Robbins, Denny's, New Capital Egg Roll, Phu Yen, Taco Bell
- AServ E: Chevron, Exxon
- Other E: **Long's Drugs, Safe Way Grocery**

11 McKee Rd.
- Gas E: BP(CW), Shell*(CW)
  W: World Gas* (24 Hrs)
- Food E: Burger King, Country Harvest Buffet, Royal Taco, Starbuck's Coffee, ToGo's, Wienerschnitzel
  W: Baskin Robbins, China Express, Foster's Freeze, McDonalds, Winchell's Doughnuts
- AServ E: Chevron, Montgomery Ward
  W: Penske
- ATM E: Wells Fargo
- Other E: **Capital Square Mall, Lucky's Grocery Store, Montgomery Ward, WalGreens**
  W: **Andy Pharmacy, K-Mart**

10 Alum Rock Ave.

9 Capital Expressway

8B Jackson Ave.

8A King Rd.

7 U.S. 101, San Francisco

6 McLaughlin Ave.

5B 10th & 11th St.
- Other E: **7-11 Convenience Store**
  W: **Spartan Stadium**

5 7th St. CA. 82 Virginia St.
- Gas W: Shell
- AServ W: Shell

Note: I-680 runs concurrent below with I-280. Numbering follows I-280.

4 Vine St. Almaden Ave.

Note: I-680 runs concurrent above with I-280. Numbering follows I-280.

3 Guadalupe Pkwy . CA 87
- Gas E: 76*(D), Chevron
- AServ E: Chevron

2 Median Ave

1 Bascom Ave. Leigh Ave.

0 Jct. 880 CA 17 Oakland

↑CALIFORNIA

Begin I-680 at Jct I-80

Bold red print shows RV & Bus parking available or nearby

EXIT	GEORGIA

Begin I-285

↓GEORGIA

1 GA 279, GA 14, Old National Hwy, South Fulton Pkwy
- **Gas** I: Chevron*
  O: Citgo*[CW], Exxon*[D], Hess[D], Racetrac* (Open 24 Hours)
- **Food** I: Denny's, Tropical Restaurant, Waffle House
  O: Arby's, Blimpie's Subs, Burger King, Checkers, China Cafeteria, China Dragon, Church's Fried Chicken, El Nopal, El Ranchero, Formosa Chinese Buffet, KFC, Krystal, McDonalds, Owens Fine Food, Piccadilly Cafeteria, Pizza Hut, Red Lobster, Steak & Ale, Subway, Taco Bell, Waffle House, Wendy's
- **Lodg** I: La Quinta Inn
  O: Best American Inn, Budgetel Inn, Days Inn, Fairfield Inn, Holiday Inn, Red Roof Inn
- **AServ** I: Goodyear Tire & Auto, Mustang Auto Parts
  O: Hess, Midas Muffler & Brakes, Tune-Up Clinic
- **Med** O: ✚ **The Med First Physician**
- **ATM** I: Chevron
  O: NationsBank, Racetrac, SunTrust Bank, Wachovia
- **Other** I: **Big B Drugs, Lock Surgeon, Service Merchandise**
  O: **Animal Clinic, Coin Laundry, Kroger Supermarket (Pharmacy), Target Department Store**

2 (1) Washington Road
- **Gas** O: Chevron*
- **Lodg** O: **Good Nite Inn**
- **Other** O: **Best Supermarket,** Coin Car Wash, **Coin Laundry**

3 Camp Creek Pkwy, US Army Reserve, Atl. Airport
- **Gas** I: BP*[CW], Citgo*
- **Food** I: Checkers, McDonalds, Mrs Winner's Chicken
- **Lodg** I: ◆ Sheraton (see our ad on this page)
- **AServ** I: Park & Go
- **Other** I: Coin Car Wash

4A GA 166, Lakewood Fwy, GA.154 (Norhtbound Reaccess Only, Difficult Reaccess)

4B Campbelton Rd.
- **Gas** I: BP*[CW], Texaco*[CW]
  O: Amoco*[CW], Racetrac*, Shell*[D][LP], Texaco*
- **Food** I: Burger King, Captain D's Seafood, Checkers Burgers, Dairy Queen, Taco Bell, Wendy's
  O: Church's Fried Chicken, Mrs Winner's Chicken
- **AServ** I: B&B Auto, Midas Muffler, Rich's Auto Center
- **ATM** O: Racetrac
- **Other** I: **Greenbriar Mall, Kroger Supermarket (Pharmacy)**
  O: **Big B Drugs,** Coin Car Wash, **Greenbriar Animal Hospital**

5 Cascade Rd
- **Gas** I: Chevron*[CW], Coastal
  O: Amoco*, BP*[CW]
- **Food** O: KFC, Mrs Winner's Chicken
- **AServ** I: Coastal
- **Med** O: ✚ **Southwest Hospital**
- **ATM** I: Chevron, NationsBank
- **Other** I: **Bruno's Supermarket**

6 GA 139, Martin Luther King Jr. Dr, Adamsville
- **Gas** I: Amoco*[CW], Fina*
  O: Chevron (Open 24 Hours)
- **Food** I: Mrs Winner's Chicken
- **AServ** I: One Stop Auto Shop
  O: Chevron

7A Junction I-20 East, Atlanta, Abernathy Hwy.

EXIT	GEORGIA

7B I-20 West, Birmingham

8 (11) U.S. 78, U.S. 278, Bankhead Hwy
- **TStop** I: **Petro Travel Plaza*[LP][SCALES] (Truck Wash)**
- **FStop** I: **Citgo[LP]**
- **Gas** I: Amoco*[D]
  O: Amoco*[CW], Texaco*
- **Food** I: Iron Skillet (Petro Travel Plaza), Mrs Winner's Chicken

9 Bolton Rd. (Southbound, No Reaccess)

10 GA 280, South Cobb Dr, Smyrna
- **Gas** O: Amoco*, BP[D], Racetrac*, Shell*[D][CW]
- **Food** O: Arby's, Blimpie's Subs, Checkers Burgers, Church's Fried Chicken, Krystal, McDonalds, Monterrey Mexican, New China Chinese, Taco Bell, Waffle House, Wendy's
- **Lodg** O: Knight's Inn
- **Med** O: ✚ **Hospital**
- **ATM** O: Racetrac
- **Other** I: **U-Haul Center**

11 South Atlanta Rd
- **TStop** I: **Pilot Travel Center*[D][SCALES]**
- **Gas** I: Shell*[D][CW]
- **Food** I: **KFC (Pilot), Subway (Pilot)**
  O: Waffle House
- **Lodg** O: (AAA) Holiday Inn Express

12 (17) Paces Ferry Rd
- **Gas** I: BP*[LP], Texaco*[CW][LP]
- **Food** I: Blimpie's Subs (Texaco), Mrs. Winner's Chicken, Subway
- **Lodg** O: ◆ Fairfield Inn
- **AServ** I: Goodyear Tire & Auto
- **ATM** I: Texaco
- **Other** I: **Home Depot, Mail Boxes Etc, Publix Supermarket (Pharmacy)**

13 (19) U.S. 41, Cobb Pkwy, Dobbins Airforce Base

EXIT	GEORGIA

- **Gas** O: Amoco*[CW], Chevron*[CW], Citgo*, QT*
- **Food** I: Buffalo's Cafe, Chick-fil-A, El Toro, Haagen-Dazs Icecream, Hooters, Johnny Rocket's Hamburger, Krispy Kreme Donuts, Longhorn Steakhouse, Malone's, May's Chinese, Philly's Connection, Pizza Hut, Schlotzkys Deli
  O: Applebee's, Arby's, Black-Eyed Pea, Blimpie's Subs (Citgo), Carrabba's Italian Grill, Chuck E Cheese's, Denny's, Dunkin Donuts, Haveli Indian Cuisine, Jade Palace, Jock's & Jill's Sports Bar, KFC, Old Hickory House, Olive Garden, Papa John's Pizza, Pizza Hut, Red Lobster, Steak & Shake, Subway, Taco Bell, Waffle House, Wendy's
- **Lodg** I: ◆ Courtyard by Marriott, ◆ Embassy Suites, ◆ Homewood Suites, Renaissance Waverly Hotel, Sheraton Suites
  O: Double Tree Guest Suites, Holiday Inn, Residence Inn, Sumner Suites
- **AServ** I: Sears
  O: Chevron
- **ATM** I: First Union, NationsBank, SunTrust Bank
  O: SouthTrust Bank
- **Other** I: **Cinema, Cumberland Mall, Franklin's Printing and Copies, Galleria Mall, Revco, U.S. Post Office**
  O: **Pearl Vision Center, PetsMart**

14 Junction I-75, Atlanta, Chattanooga

15 Northside Dr, New Northside Dr, Powers Ferry Rd.
- **Gas** I: Amoco*[CW], Chevron*[CW] (Open 24 Hours)
  O: Texaco*[CW]
- **Food** I: Blimpie's Subs, McDonalds, Rio Bravo Cantina, Wendy's
- **Lodg** I: Harvey Hotel
- **AServ** O: Texaco Express Lube
- **ATM** I: NationsBank, Sun Trust Bank, Wachovia Bank
  O: NationsBank
- **Other** I: **Big B Drugs, Powers Ferry Animal Clinic, Publix Supermarket**

16 (24) Riverside Dr.

17 (25) U.S. 19 South, Roswell Rd, Sandy Springs
- **Gas** I: Chevron, Majik Market*[LP], Texaco*[CW]
  O: Amoco*[CW], BP*[CW], Chevron[CW], Shell*[D]
- **Food** I: China Cooks, Country Cork Food, El Taco, Frankie's, Il Frno Pizza & Pasta, Jilly's Ribs, Old Hickory House, Southern Style Cooking, Tagueria Deldado Mexican
  O: American Pie, Billy McHale's, Boston Market, Burger King, Cabo Wobo Grill, Capri, Chick-fil-A, Dunkin Donuts, El Toro, Good Ol' Day's, Lettuce Suprise You, Palsanos Italian, Ruth's Chris Steakhouse, Subway, Taco Mac, Wendy's
- **Lodg** I: Days Inn
  O: (AAA) Country Hearth Inn
- **AServ** I: Chevron
  O: BF Goodrich, BP, Chevron, K-Mart, Tune-Up Clinic
- **ATM** O: SunTrust Bank
- **Other** I: **Coin Laundry**
  O: **Fed Ex Drop Box, K-Mart (w/pharmacy)**

18 Glenridge Dr., Glenridge Connector (No Reaccess)

19 (27) U.S. 19, GA 400, Dahlonega, Alpharetta (Southbound, Toll Road)
- **Lodg** O: Hampton Inn (see our ad on this page)

20 Peachtree - Dunwoody Rd (Westbound, Reaccess Eastbound Only)
- **Lodg** I: ◆ Courtyard by Marriott
  O: (AAA) Comfort Suites, Doubletree Hotel
- **Med** I: ✚ **Northside Hospital,** ✚ **Scottish Rite,** ✚ **St. Joseph**
- **ATM** I: NationsBank

21 Ashford - Dunwoody Rd
- **Gas** O: Exxon*

Bold red print shows RV & Bus parking available or nearby

EXIT	GEORGIA
Food	O: California Pizza Kitchen, Houlihan's, McKendrick's Steakhouse, Mick's
Lodg	O: Crowne Plaza Hotel, The Marque Hotel & Suites
ATM	O: Fidelity National Bank, First Union Bank, Nations Bank, SouthTrust Bank, SunTrust Bank, Wachovia Bank
Other	O: **Cinema 1,2,3,4, Perimeter Expo and Perimeter Malls**
22	**(28)** Chamblee - Dunwoody Rd, N. Shallowford Rd, N. Peachtree Rd
Gas	I: BP[CW], Citgo*, Conoco*[LP], Texaco*[CW]
	O: Amoco*[D][CW], BP*[CW], Chevron*[CW]
Food	I: Arby's, Blimpie's Subs, Mad Italian, Malone's Bar & Grill, Pizza Inn, Savoy Restaurant, Steak & Ale, Taco Bell, Wendy's
	O: Bagel Co. & Deli, Dairy Queen, Denny's, Mrs Winner's Chicken, Popeyes, Subway, Waffle House
Lodg	I: ◆ Holiday Inn, ◆ Residence Inn
	O: ◆ Four Points Hotel
AServ	I: BP
Med	I: ✚ **Peachford Hospital**
	O: ✚ **Chamblee-Dunwoody Medical Center**
ATM	I: Texaco
	O: Kroger Supermarket, Nations Bank, SunTrust Bank
Other	O: **Big B Drugs, Kroger Supermarket, Mail Boxes Etc**
23A	GA 141 South, Peachtree Industrial Blvd, Chamblee
Food	I: Waffle House
AServ	O: Jim Ellis Audi, Mitsubishi Motors
23B	GA 141 North, Peachtree Industrial Blvd
Food	I: Oga Restaurant, Piccadilly Cafeteria
AServ	I: Kelley Buick, Tune Up Clinic

EXIT	GEORGIA
24	Tilly Mill Rd, Flowers Rd (Northbound, No Reaccess)
25	U.S. 23, Doraville, Buford Hwy.
Gas	I: Crown*[D][CW][LP], Fina*[LP]
	O: Amoco*[LP]
Food	I: Captain D's Seafood, Dunkin Donuts, El Charrua Mexican, First China Restaurant, House of Peking, McDonalds, Mi Casa Mexican, Subway, Szechuan Restaurant, Taco Bell, Waffle House
	O: Arby's, Baldinos Subs, Burger King, Checkers Burgers, Chick-fil-A, El Azetec Mexican, KFC, Krystal, Mrs Winner's Chicken, Steak & Shake, Sushi & Hibachii Japanese, Taqueria Mexican
Lodg	I: Atlanta Inn, AAA Comfort Inn(see oura d on this page)
AServ	I: Auto Value, Brake-O Brake Shop, Meineke Discount Mufflers, Midas Muffler & Brakes
	O: Big 10 Tires, Goodyear Tire & Auto, Penske
ATM	O: First Union, SouthTrust Bank, Wachovia Bank
Other	I: **Big B Drugs,** Coin Car Wash, **Doraville Animal Hospital, Doraville Police Station,** Hand Car Wash, **U.S. Post Office**
	O: **Coin Laundry, Fed Ex Drop Box, K-Mart (& Pharmacy), U.S. Post Office**

EXIT	GEORGIA
26A	**(33)** Junction I-85 South, Chamblee - Tucker Rd
26B	Junction I-85 North, Greenville
27	**(34)** Chamblee - Tucker Rd (Northbound)
Gas	I: Phillips 66*[LP]
	O: BP*[CW][LP], Chevron, Chevron, Citgo*[LP], Phillips 66*
Food	I: Lone Star Steakhouse, McDonalds, Mexicali Restaurant, Philly Connection, Pizzeria, Round Dragon Chinese Restaurant, TCBY Treats, Waffle House
	O: Arby's, Blimpie's Subs, Deli Express, Hot Wings, Hunan Inn Chinese, KFC, S & S Cafeteria, Taco Bell
Lodg	O: Motel 6
AServ	O: Bon Auto Service, Chevron, Warbington's Service Center
ATM	I: NationsBank
	O: Embry National Bank (24 hour banking), First Union Bank, SunTrust Bank (Drive-thru location)
Other	I: **Kroger Supermarket (& Pharmacy), Pet**

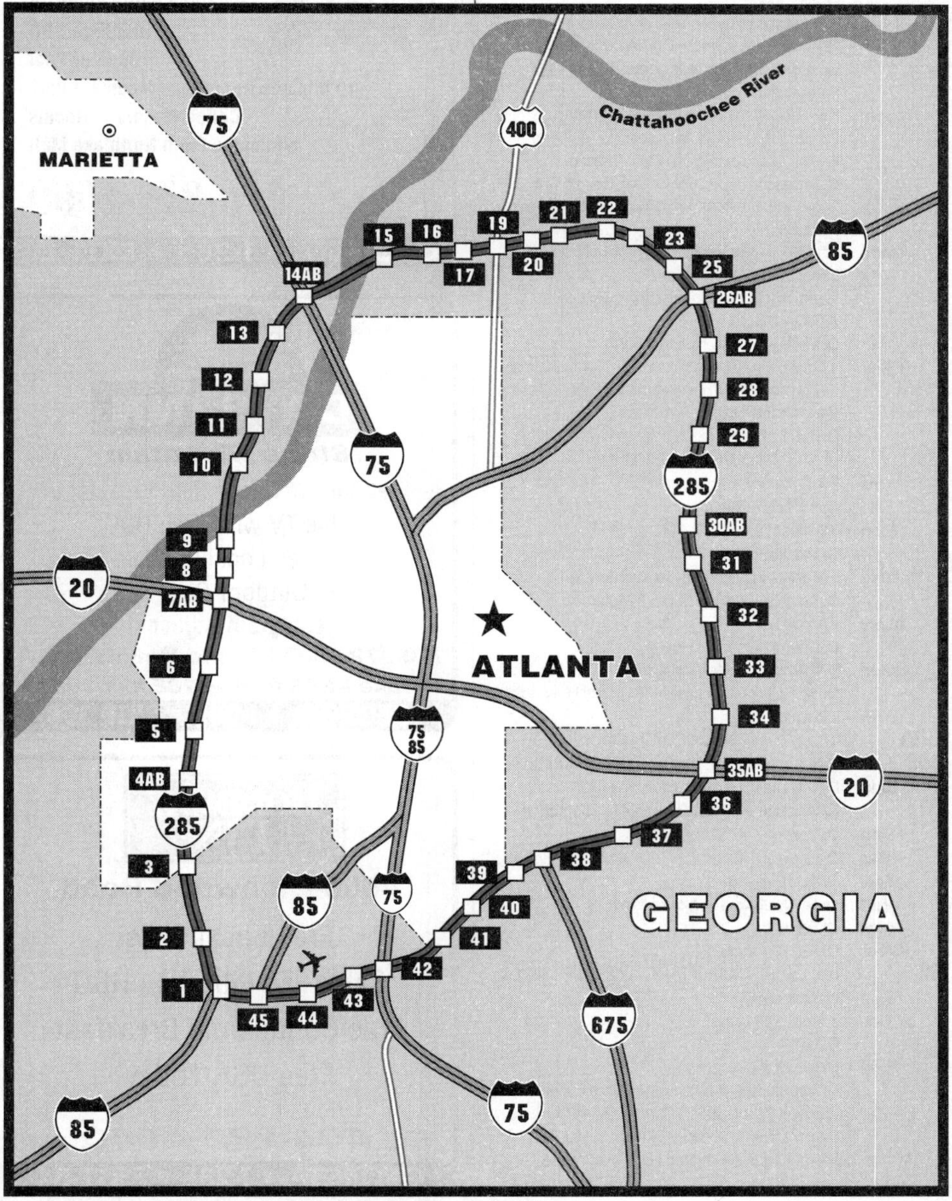

EXIT GEORGIA

Clinic, Animal Hospital, Post Net (UPS, Fax, Copy)
O: Coin Laundry, Eckerd Drugs, Post Office, Tracy's Medicine Center, Winn Dixie Supermarket

27A Northlake Parkway
Lodg I: Hampton Inn (see our ad onpage 573)

28 GA 236, LaVista Rd, Tucker
FStop I: Hess(D)
Gas I: Amoco*(CW)(LP), BP*(CW), Texaco*(CW)
O: Chevron*(CW), Circle K*(LP)
Food I: Arby's, Bagle Boulevard, Black-Eyed Pea, Blimpie's Subs, Blue Ribbon Grill, Captain D's Seafood, Carvel Ice Cream Bakery, Dairy Queen, Domino's Pizza, Dunkin Donuts, Freshens Yogurt, Fuddrucker's, Golden Dragon Chinese Restaurant, LA Bonne Cuisine, Lucky Key Chinese Restaurant, Maharaja Indian Cuisine, Manderin Palace, McDonalds, Mellow Mushroom, Montrrey Mexican Restaurant, Pizza Hut, Premier Indian Restaurant, Red Lobster, St. Louis Bread Company, Subway, Taco Bell, The Italian Oven, The Plaza Diner, Wendy's
O: Checkers Burgers, Chili's, Folks, IHOP, O'Charley's, Old Hickory House, Olive Garden, Piccadilly Cafeteria, Schlotzkys Deli, Steak & Ale, Waffle House
Lodg I: ◆ Courtyard by Marriott, ◆ Fairfield Inn, ◆ Holiday Inn, Wyndham Garden Hotel
O: ◆ Best Western, ◆ Bradbury Suites, AAA Ramada (see our ad on this page)
AServ I: BP, Goodyear Tire & Auto, McNamara Pontiac,GMC Trucks, Isuzu, Midas Muffler & Brakes, Sears (Northlake Mall), Texaco
O: American Sports Car Center, Dekalb Tire Company, Firestone Tire & Auto, RWB Mercedes Benz, BMW
Med I: ✚ Northlake Regional Hospital (2-3 Miles)
ATM I: First Union, NationsBank (Kroger Supermarket), SouthTrust Bank, SunTrust Bank, Texaco, Wachovia Bank
O: Circle K, Fidelity National Bank
Other I: Eckerd Drugs, Eye First Vision Center, Kroger Supermarket (w/pharmacy), Northlake Mall, Pearl Vision Center, PetsMart
O: Fed Ex Drop Box

29 U.S. 29, Lawrenceville Hwy
Gas I: Amoco*(CW)(LP), Racetrac*(LP)
O: BP*(CW)(LP), Chevron
Food I: Waffle House
O: Waffle House
Lodg I: Masters Inn Economy, ◆ Red Roof Inn
O: Knight's Inn, AAA Super 8 Motel
AServ I: Amoco
O: Cheron (Open 24 Hours)
Med I: ✚ Montreal Medical Center
O: ✚ Columbia Lakeside Regional Hospital
ATM I: Racetrac

30A U.S. 78 West, Decatur, Atlanta

30B **(38)** U.S. 78 East, Athens, Stone Mountain
Lodg O: Days Inn (see our ad on this page)

31 **(39)** East Ponce de Leon, Clarkston
FStop O: Texaco*(D)(LP)
Gas O: Chevron*, Citgo*
AServ O: God Blessed
Med O: ✚ Northlake Regional Medical
ATM O: Texaco
Other O: Coin Laundry Mat

32 **(41)** GA 10, Memorial Dr, Avondale Estates
FStop O: Texaco*(CW)
Gas I: Amoco*(CW), Fina*
O: QT*
Food I: KFC, Waffle King
O: Applebee's, Arby's, Burger King, Canyon BBQ, Caravan's Crab Shack, Church's Fried Chicken, Denny's, Dunkin Donuts, Hardee's, Mama's Restaurant & Buffet, Steak & Shake, Waffle House, Wendy's
AServ O: Continental General Tire Care, European Specialists of Atlanta, Tune-Up Clinic
ATM O: First Union Bank, NationsBank, QT, SouthTrust Bank, SunTrust, Texaco, Wachovia
Other I: Dekalb County Department of Public Safety Police Fire
O: Coin Laundry, Dunaire Pharmacy

33 U.S. 278, Covington Hwy
FStop O: Crown*(D)(CW)(LP)
Gas I: Amoco*(CW), BP*, Texaco*(CW)(LP)
O: Chevron*(CW), Citgo*(LP)
Food I: Checkers Burgers, KFC, Wendy's
O: Waffle House
AServ I: BP, D&R Auto Service, Firestone Tire & Auto, Tri-US
ATM O: Crown
Other I: Big B Drugs, Cub Foods, Target Department Store

32A **(42)** Transit Parking Marta Only (Northbound, Southbound Reaccess Only)

34 **(46)** GA 260, Glenwood Rd
Gas I: Racetrac*, Texaco*
Food I: Glenwood Diner, Miss Dossey's, Mrs. Winner's Chicken
Lodg I: AAA Glenwood Inn
O: Rodeway Inn (see our ad on this page)
ATM I: Racetrac
Other I: Coin Laundry, Dekalb Police Station

35 **(46)** Junction I-20, Atlanta, Augusta

36 **(50)** GA 155, Flat Shoals Rd, Candler Rd
FStop I: Shell*(D)
Gas I: Circle K*, Conoco*(CW)
O: Citgo*, Fina*
Food I: Checkers Burgers, Dairy Queen, Homebox Restaurant, KFC, McDonalds, Pizza House, Taco Bell, WK Wings, Waffle King
Lodg I: Efficiency Lodge, Gulf American Inns
AServ O: A's Tires, Fina
ATM I: Circle K, Conoco, First Union, NationsBank
Other I: Coin Laundry, Holland & Knight Pharmacy

37 **(51)** Bouldercrest Rd
TStop I: Pilot Travel Center*(D)(SCALES)
Gas I: Amoco* (Open 24 Hours)
O: Chevron*(LP)
Food I: Dairy Queen, Domino's Pizza, Hardee's, W.K. Wings, Wendy's
Lodg I: Dekalb Inn
Other I: Big B Drugs, Fed Ex Drop Box, Wayfield Foods

38 **(53)** Junction I-675 South, Macon

39 **(53)** U.S. 23, Mooreland Ave., Ft. Gillam
TStop O: TA(D)(SCALES) (Truck Wash)
FStop I: Conoco* (Open 24 Hours), Speedway*(LP)
O: Amoco(D)(LP)
Food O: Charlie's, Country Pride Restaurant, Rio Vista (Catfish and Hush-puppies), Waffle House
Lodg O: AAA Econolodge, AAA TA
AServ O: Goodyear Tire & Auto
TServ I: William's Detroit Diesel Allison
O: Sunbelt Cat Parts & Services, TA, Woods Truck Tires
Med O: ✚ Mooreland Medical Clinic
ATM I: Speedway
Other I: Southern Towing Services

40 **(55)** GA 54, Jonesboro Rd, Forest Park
Gas O: Amoco*
Food O: Arby's, McDonalds, Shoney's
Lodg I: AAA Comfort Inn
AServ O: Metro Tire and Auto Service
TServ O: Trucks of Atlanta

41 **(58)** Junction I-75, Atlanta, Macon

42 Clark Howell Hwy, Air Cargo (Difficult Reaccess)

43 GA 139, Riverdale Rd, Ga. International Covention Center
Gas I: Hess*(D), Racetrac*
O: BP*(CW), Speedway*(D)(LP), Texaco*(D)(CW)
Food I: McDonalds, Waffle House
O: Waffle House
Lodg O: AAA Comfort Inn, DAYS INN ◆ Days Inn, Ramada Inn
AServ I: ACC Auto Parts & JT Auto Specialists, National Tire Warehouse
ATM I: Racetrac
Other I: Coin Laundry

44 **(63)** Junction I-85 South, Atlanta, Montgomery

↑GEORGIA

Begin I-285

Bold red print shows RV & Bus parking available or nearby

EXIT — NEW JERSEY

Begin I-287

↓NEW JERSEY

66 **(67)** NJ17S, to Mahwah
TStop **E:** Travel Port Express*(D)(SCALES)
FStop **E:** International FuelStop*, Mobil*
Gas **E:** Dean(D), Getti*(D), Gulf, U-Save
Food **E:** Burger King, Mason Jaw Family Restaurant, McDonalds, Ramada Inn, State Line Diner, Subway, Vido's Pizza
Lodg **E:** Comfort Inn, ◆ Courtyard by Marriott, ◆ Ramada Inn, ◆ Sheraton (see our ad on this page)
AServ **E:** Gulf
ATM **E:** First Union Bank, Summit Bank

59 **(60)** NJ208S, Franklin Lakes

58 **(59)** U.S.202, Oakland
Gas **W:** Exxon
Food **E:** Baskin Robbins, Blimpie's Subs, Bread Basket Deli, Jr's Pizza & Subs, KFC, Linda's Chicken, Oakland Bagels & Pastry, Oakland Bakery, Sun Yuan, The Pepper Mill, Topps China
W: Cafe L'Amore, Hot Bagels
AServ **E:** CBR Tires
W: Exxon, Oakland Auto Parts
ATM **E:** Fleet Bank, Hudson City Savings Bank, The Bank of New York, Valley National Bank
Other **E:** CVS Pharmacy, Coin Laundry, Drug Fair, Grand Union Grocery, Krauzer's C-Store, Oakland Drugs, Oakland Vision Center

57 **(58)** Skyline Dr, Ringwood
Parks **W:** Ringwood State Park

53 **(54)** CR694, Alt CR511, Bloomingdale, Pompton Lakes
Gas **E:** Coastal*(D), Sunoco
Food **E:** Bella Italia, Blimpie's Subs, Charcoal Grill Cheese Steak, Dough House, Riverdale Lunchenette, Rosemary Lunch Dinner & Sage
AServ **E:** Coastal
Other **E:** Mike's Feed Farm, Post Office, Riverdale C-Store, Riverdale Police, Urgent Care Center

52AB **(53)** NJ23, Riverdale, Wayne, Butler
Gas **E:** Amoco(CW)
Food **E:** Venny's Pizza
AServ **E:** Circle Auto Parts, Maroon Jeep, Pontiac
Med **E:** ✚ Chilton Hospital

47 U.S. 202, Montville, Lincoln Park
Gas **E:** Exxon*
AServ **E:** Exxon
Med **W:** ✚ Hospital
ATM **E:** South Burgeon Savings Bank

45 **(46)** Myrtle Ave, Boonton (Southbound, Northbound Reaccess)
Gas **W:** Citgo*, Exxon*(D)
Food **W:** Boonton Pharmacy, McDonalds
AServ **W:** Boonton Tire Supply & Auto Repair, National Auto Parts, Scerbo Buick,Chev., Geo
Other **W:** Jack's IGA Supermarket

43 Intervale Rd, Mountain Lakes (Difficult Reaccess Northbound, No Southbound Access)
Gas **E:** Texaco*(D)
Food **E:** Bevac Qua's
AServ **W:** Coriglino's Auto Service

42 U.S.46, U.S.202, Dover, Clifton
Gas **W:** Exxon
Food **W:** Embassy Suites, Fuddrucker's, Lucky's Star Kitchen, McDonalds, The Diner, The Great Wazu, Wendy's
Lodg **W:** DAYS INN ◆ Days Inn, Embassy Suites
AServ **W:** Exxon
RVCamp **E:** Brookwood Swim & Tennis Club, Camping
Med **E:** ✚ Hospital
ATM **W:** Summit Bank

EXIT — NEW JERSEY

Other **W:** Coin Laundry, Sav On Pharmacy, US Post Office

41AB Jct I-80, Delaware Water Gap, New York City

40AB CR511, Parssipany Rd, Whippany (B - Lake Parssipany, Lake Shore Dr)
Gas **W:** Mobil
Food **W:** Frankinstein Pizza, Lake Parssipany Luncheonette, Woks Chinese Kitchen
AServ **W:** Mobil (Wrecker Service)
ATM **W:** First Union Bank
Other **W:** Krauser's C-Store

39AB NJ10, Dover, Whippany
Gas **W:** Mobil
Food **W:** Demiamo, Hilton, Howard Johnson (Chinese)
Lodg **W:** Hilton, Howard Johnson
AServ **W:** Mobil, Norman Gayle Oldsmobile

37 NJ24E, Springfield, Columbia (Difficult Reaccess)

36 CR510, Morris Ave, Lafayette, Ridgedale
Gas **W:** Amoco
Food **E:** Governor Morris Hotel
W: Ridgedale Lunch
Lodg **E:** Governor Morris Hotel
AServ **W:** Amoco
Other **E:** Washington's Headquarters (1,000 Ft)

35 NJ124, Old NJ124, Madison Ave
Food **W:** Ceede's Restaurant, Hamilton Coffee Shop
Med **E:** ✚ Morristown Memorial Hospital (Trauma Ctr)
ATM **E:** Nations Bank
W: Chase Bank, TNC Bank
Other **W:** Kings Grocery

33 **(34)** Harter Rd (No Trucks Over 5 Tons, Difficult Reaccess)

EXIT — NEW JERSEY

Other **E:** Morriston Police

(15) Rest Area - RR, HF, Phones, Picnic, P

30AB U.S.202, Bernardsville, Basking Ridge
Food **W:** The Grain House (Old Mill Inn)
Lodg **W:** Old Mill Inn
Other **W:** Jockey Hollow Nat'l Autobon Site

26 Mount Airy Rd, Liberty Corner, VA Hospital
Med **E:** ✚ VA Hospital
Other **E:** USGA Golf Museum

22 U.S.202, U.S.206, Bedminster, Pluckemin
Gas **E:** Amoco
Food **E:** Blimpie's Subs, Golden Palace (Chinese), McDonalds, Nina's Pizza, TCBY Treats
ATM **E:** Peapack Gladstone Bank, TNC Bank
Other **E:** Kings Supermarket, Mail Boxes Etc (Village At Bedminster), Revco Drugs, Village At Bedminster

21AB Jct I-78, to Clinton, Newark

13 **(18)** U.S.202, U.S.206S, U.S.22W, Summerville, Flemington

11A **(15)** U.S.22E, to New York City

(14) NJ28, Bound Brook, Summerville
Gas **E:** Amoco*, Mobil
Food **E:** Big Daddy's, Carvelle Ice Cream Bakery, Costa del Sol, Pompeii Italian, Subway, The Fill In Station Deli
AServ **E:** Amoco, Royal Chevrolet
Med **W:** ✚ Hospital
ATM **E:** Summit Bank
Other **E:** Coin Laundry

7 **(12)** Weston Canal Rd, Manville, South Bound Brook

6 **(11)** CR527, New Brunswick
Gas **W:** Amoco
Food **E:** Marriot
W: Cafe Alfredo, Hong Kong Chinese, Quality Inn, Vinny & Sons Pizzaria, Zee Best Bagel
Lodg **E:** ◆ Marriot
W: AAA Quality Inn
ATM **W:** Summit Bank
Other **W:** Cedar Lane Animal Clinic, Drug Fare, Kwik Pik C-Store

5 **(11)** Bound Brook, Highland Park

(11) Centenial Ave, Highland Park, Middlesex

(10) Weigh Station (Northbound)

4 Centennial Ave (Southbound Reaccess)

3 **(7)** South Randolphville Rd, Piscataway
Gas **E:** Mobil
AServ **E:** Mobil

2 **(6)** CR529, Eddison, Dunellen

1 CR529, Dunellen, Eddison

1A **(4)** Durham Ave, South Plainfield (Northbound, Southbound Reaccess)
Med **E:** ✚ Hospital

(3) CR501, Metuchen, New Durham (Southbound, Reaccess Northbound)

(2) NJ27S, to New Brunswick (Northbound, Southbound Reaccess)

(1) U.S.1, CR531

(0) CR514W, Jct I-95, NJ Tnpk, (Toll Rd)Woodbridge, Bonhampton

↑NEW JERSEY

Begin I-287

EXIT	MASSACHUSETTS
	Begin I-290

↓MASSACHUSETTS

9 Swanson Rd & Auburn
- **Gas** S: Getty*, Shell*
- **Food** S: Addow's, Arby's, Auburn Town Pizza, Burger King, Federal Deli, McDonalds, Papa Genio's, Yong Shing
- **Lodg** S: Auburn Motel, Buget Motel, Days Inn
- **AServ** S: Acura, Auto Palace, Firestone Tire & Auto, Midas, Sears, Speedy Muffler King
- **ATM** S: Fleet Bank, MCU Federal Credit Union, Webster's Savings
- **Other** S: **Auburn Mall, Auburn Police,** Car Wash, **Post Office, Shaw's Supermarket**

11 Worchester, College Sqaure
- **Gas** N: Texaco*(CW)
- **Food** N: Culpeppers Bakery & Cafe, Pizzeria Delight, Wendy's
- **ATM** N: Commerence Bank
- **Other** N: **G&M Grocery Mart, Pet Center**

12 Millbury
- **Food** S: Dunkin Donuts, Market Deli
- **ATM** S: Safety Fine Bank

13A Vernon Street, Kelly Square
- **Gas** N: Exxon*(CW), Merritt
- **Food** N: Broadway, Dunkin Donuts(CW), Kelly's Square Pizza, Lederman's Coffee Bar & Bakery, Tony's Pizza & Subs, Weintrauf's Deli
- **AServ** N: General Automotive Supply, Hardin Tire Firestone
- **Med** S: **✚ Hospital**

14 Grafton, Posner Square
- **Gas** S: Citgo*, Mobil*
- **Food** N: Bill's Sandwich Shop
  S: Auntie Dot's Pizza, Campy's Deli, Dunkin Donuts, Roberts Fish & Chips
- **AServ** N: Firstone, Harry's Radiator Service, International Discount Muffler, K&K Garage
  S: Thomas Auto Service
- **ATM** S: BankBoston, Fleet Bank
- **Other** S: **Grafton Street Market, Honey Farms Convenience**

15 Shrewsbury St., E. Central
- **Gas** S: Mobile
- **Food** N: Treet-me-Donuts
  S: Black Orchid, Boulevard Diner, Brew City Grill, Food Caffe, Leo's, Mac's Diner, Poscano Market & Deli
- **Lodg** N: Crowne Plaza, Hampton Inn
- **AServ** S: Chandler, Edward's Buick, GMC,, Gentile's Auto Service, Mobile
- **Med** N: **✚ Fallon Medical Center**
- **Other** N: **Post Office, Worchester Fashion Outlet Mall**
  S: **Boulevard Pharmacy Drugs,** Minute Car Wash

16 Central St.

17 MA 9, Belmont St.

EXIT	MASSACHUSETTS/ILLINOIS

- **Food** S: Dimare Deli, Jim's Pizza, Roberts Fish & Chips, Roberts Seafood
- **Lodg** N: Crown Plaza Motel, Honey Farms
- **Med** S: **✚ Memorial Hospital**
- **ATM** N: Fleet Bank
- **Other** N: **Worchester Airport**
  S: **24 Convenience Store**

18 Lincoln St., Worchester Blvd

19 I-190, Lincoln, MA1

20 MA70, Lincoln, Burncoat St.
- **Gas** N: Exxon*
- **Food** N: Baskin Robbins, Bickfords Family Restaurant, Denny's, Dunkin Donuts, Subway, Wendy's
- **Lodg** N: Day's Lodge, ◆ Holiday Inn
- **AServ** N: Auto Plaza, Lincoln Auto Plaza
- **Med** N: **✚ Hahnemann Hospital**
- **ATM** N: BankBoston, Fleet Bank
- **Other** N: **Cambridge Eye Care, Lincoln Plaza, Post Office, Shaw's Supermarket, WalGreens**

21 Plantation St., Worchester (Difficult Reaccess)
- **Food** N: Christie's Pizza, Mick's
- **Lodg** N: Days Inn
- **AServ** N: Clearview Auto, Firestone Tire & Auto
- **Med** S: **✚ Hospital**
- **Other** N: **Premier Optical**

22 Main St., Shrewsbury, Worchester

23AB MA140, Shrewsbury, Boyalston
- **Gas** N: Citgo*, Mobile*
- **Food** N: Dragon 88, Dunkin Donuts
- **AServ** N: Mobile

↑MASSACHUSETTS

Begin I-290, Massachusetts
Begin I-290, Illinois

↓ILLINOIS

1A Jct I-90

1B IL72, Higgins Rd. , IL58 Gulf Rd.
- **Gas** W: Shell(CW)
- **Food** W: Benhana, Bennigan's, Bertucci's, Denny's, Wendy's
- **Lodg** W: Drury Inn, Hampton Inn, LaQuinta Inn, Marriott
- **Other** W: **Shopping Mall, Target Department Store**

4 IL53, Biesterfield Rd.
- **Food** W: Marino's Pizza

5 Thorndale
- **Gas** W: Marathon(D)
- **Food** W: Dunkin Donuts, Ma's Chinese, Rosati's Pizza, Tasty Dawg, Wendy's
- **Lodg** E: ◆ Wyndham Garden Hotel
  W: Excel Inn
- **AServ** W: Marathon
- **ATM** W: First Star Bank
- **Other** W: Monarch Car Wash

7 Jct I355

EXIT	ILLINOIS

10AB I183

12 York Rd., US20 W
- **Gas** W: Amoco, Shell*
- **Food** E: Bagels & Things, Christopher's, Hoxie's Bar-B-Que, Mimmo's Pizza
  W: Angelo's, Arby's, Burger King, Little Caesars Pizza, Rainbow Pancakes, Steven's Steakhouse
- **Lodg** E: ◆ Holiday Inn
- **AServ** W: Amoco, Firstone, Jiffy Lube
- **Med** W: **✚ Elmhurst Hospital**
- **Other** W: **Dental Clinic, Eye Care**

13A IL64 North Avenue.

14AB St. Charles Rd.
- **Gas** E: Citgo, Union 76(CW)
- **Food** E: Dunkin Donuts, LosMarichi's, McDonalds
- **AServ** E: Citgo, Kraft Auto Center, Union 76
- **ATM** E: National Bank of Commerence

15AB I-290W Rockford, I-294 Tollway, I-88

16 Wolf Rd.
- **Lodg** N: Holiday Inn, Renanissance Inn

17B US12, US20, US45, Mannheim

18 25th Ave

19A 17th Ave
- **Gas** S: Citgo*

20 Herst Ave.
- **Gas** N: Amoco*, Shell*(CW)
  S: Union 76
- **Food** N: Dunkin Donuts, KFC
  S: Burger King, Checker's, Dunkin Donuts, Poor Boy Sandwiches
- **Med** S: **✚ Hospital**
- **Other** S: **Brookfield Zoo**

21B IL43, Harlem Avenue
- **Med** N: **✚ Hospital**

23A Austin Blvd

23B Central Avenue
- **Med** N: **✚ Loretto Hospital**

24A Laramie Ave.
- **Food** N: Chop Suey
- **Med** N: **✚ Hospital**

25 Kostner
- **Gas** N: Amoco

29B Morgan Street

26A Independance Blvd
- **Gas** S: Amoco

26B Homan Avenue
- **Gas** S: Minute Man Citgo

27B California Ave (Difficult Westbound Reaccess)

28B Ashland Blvd
- **Gas** N: Amoco(CW)
- **Lodg** S: Hyatt
- **Med** S: **✚ St. Lukes Hospital**

27C Oakley Blvd, Western Ave

28A Damen Ave.
- **Med** S: **✚ Hospital**

↑ILLINOIS

Begin I-290

Bold red print shows RV & Bus parking available or nearby

EXIT ILLINOIS

Begin I-294

↓ILLINOIS

52 Willow Rd
48 Golf Rd
47 Jct I-190, Kennedy Expwy, (No Reaccess Southbound)
Gas E: Clark*, Marathon(D)
Food W: Mirage Restaurant
Lodg W: Howard Johnson
AServ E: Marathon
38 O'hare Oasis - RR, HF, Phones
37 Jct I-290 W
36 Jct I-290 E, Eisenhower Exprwy
35 I-88 W Jct.
34 Cermak St.
29 U.S. 34 Ogden Ave.
Gas W: Amoco*, Shell*
Food W: Dunkin Donuts
AServ W: Firestone Tire & Auto
Med W: ✚ Hospital
ATM W: First Chicago Bank
26 Jct. I-55 to Saint Louis
24 Jct. I-55, Joliet Rd, Willowbrook
Gas E: Mobil
Food E: Baskin Robbins (Hinsdale Oasis), Wendy's (Hinsdale Oasis)
23 75th St. Willow Springs Rd
22 **(18)** US 12, US 20, 95th St., 76 Ave.
Gas E: Keene, Shell
W: 76*, Shell*, Super America*(D) (ATM)
Food E: Bennigan's, Billy Boy's Restaurant, Dunkin Donuts, Oak Ridge Restaurant
W: Delphian, Denny's, Less Brothers, Sghoops Hamburgers, Wendy's
Lodg E: Holiday Inn (see our ad on this page)
AServ E: Car X Muffler & Brakes, On The Go, Shell
W: Penzoil Oil Change
ATM E: Lasall Bank
Other E: Oak Lawn Medical Center
W: Osco Drugs
16 **(13)** IL 50, Cicero Ave.
FStop W: Gas City
Gas E: 76
Food E: Bob Evans Restaurant, Dunkin Donuts
W: Boston Market, Einstein Brothers Bagels, Hardee's, IHOP, Portillo's Hot Dog, Rosewood West Restaurant, Subway
Lodg E: Budgetel Inn, Delux Budget Motel
W: Hampton Inn, Radisson
12 **(6)** US 6, 159th St.
Gas E: Shell*(CW)
Food W: Burger King, McDonalds, Taco Bell, White Castle
AServ W: Firestone Tire & Auto
ATM W: Southwest Financial Bank
Other W: Tower Car Wash, Walgreens
11 Jct I-80W

↑ILLINOIS

Begin I-294

EXIT ILLINOIS

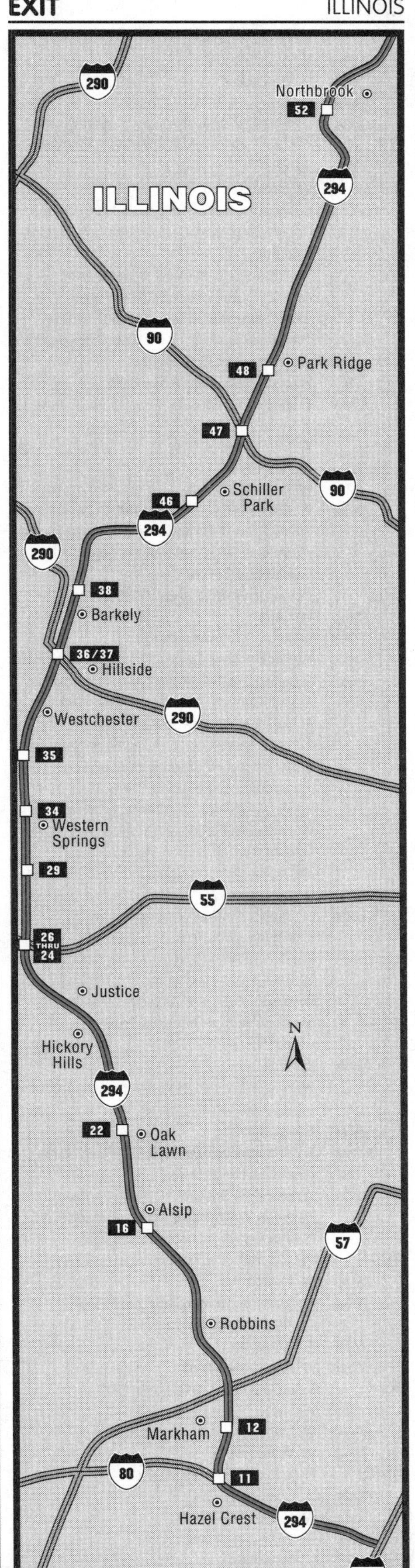

EXIT NEW JERSEY

Begin I-295, Trenton

↓NEW JERSEY

(5) Clara Barton Service Area
(1) Penns Grove
1A NJ Tnpk (Northbound)
1B U.S. 40, Atlantic City
1C CR551, Salem
Lodg E: White Oak Inn
2B U.S. 40E, NJ Tnpk
2C U.S. 130, Deepwater
TStop E: Pilot Travel Center*(SCALES)
W: TA Truckstop*(SCALES)
Food E: Landmark Inn, Popeye's Chicken, Subway (Truckstop)
W: Blimpie's Subs
Lodg E: Friendship Inn, Landmark Inn
Other E: Turnpike Inn C-Store
(3) Rest Area - Phones, RR, Picnic, RV Dump, Vending, Tourist Info P (Northbound)
(4) Weigh Station (Northbound)
4 NJ 48, Pennsgrove, Woodstown
Food W: Roman Pantry Pizza
7 Auburn, Pedricktown
TStop E: 295 Auto/Truck Plaza*(SCALES)
Gas E: 295 Fuel
Food E: 295 Truck Plaza
Lodg E: 295 Truck Plaza
10 Center Square Rd, Swebesboro
Gas E: Texaco*(D)(LP) (Kerosene)
Food E: China Wok, Holiday Inn Select, McDonalds, Momma Ciconte's Italian Eatery, Spyros Pizza, Village Ctr Restaurant
Lodg E: ◆ Holiday Inn Select
AServ E: Texaco
RVCamp W: Camping World RV Center (Frontage Rd) (see our ad on this page)
Med W: ✚ Urgent Care (Pure Land Ind Complex)
ATM E: Woodstown National Bank & Trust
Other E: Beckett Shopping Ctr, Eckerd Drugs, Village Ctr, Wa Wa C-Store
11 **(12)** U.S. 322E, Mullica Hill, Bridgeport
13 **(14)** U.S.130 S, U.S. 322W, Bridgeport
14 CR 684, NJ44, Repaupo Town
15 CR 607, Harrisonville
16A CR 653, Swebesboro, Paulsboro
16B CR 551 Spur, Gibbstown, Nickleton
17 CR 680, Gibbstown, Nickleton
Food W: Boar's Head Meats & Hogies, Burger King, Disantillo
Lodg W: Dutch Inn
TServ E: Truck Service
Other W: Coin Laundry, Funari Thriftway Supermarket, Rite Aide Pharmacy
18 CR 678, CR 667, Paulsboro, Clarksboro
TStop E: Travel Port*(SCALES)
FStop E: Amoco(D)
Food E: Brothers Pizza (Travel Port), Buckhorn Family Restaurant (Travel Port), Dragon Nest Chinese, KFC, McDonalds (Playland), Taco Bell, The Starting Gate Restaurant, Wendy's (Travel Port)
AServ E: Amoco(D) (Wrecker Service)
TServ E: Travel Port
TWash E: Travel Port
RVCamp E: Campground
Other E: CB Specialist (Travel Port), Coin Operated

EXIT NEW JERSEY

Laundry, Fed Ex Drop Box (Travel Port), Royal C-Store (Travel Port), UPS Drop Box (Travel Port)

18B CR 667, Clarksboro, CR678, Mt Royal, Paulsboro
- FStop E: **Texaco*(D) (Kerosene)**

19 CR 656, Mantua

20 CR 643, to CR660
- FStop E: **Amoco***
- Food E: Angelina's (Best Western)
- Lodg E: AAA Best Western
- Parks W: **Thorofare Nat'l Park**

21 NJ44S, to CR534E, Woodbury
- TStop W: **Crown Point Truckstop* (Frontage Rd)**
- Food E: Country House Restaurant
- W: Amazing Wok I, Vinny's Pizza (Crown Pt Plaza)
- Lodg W: **Westwood Motor Lodge**
- AServ W: Little Joe's Auto Repair
- Other W: **Police Dept, Wa Wa C-Store**

22 Jct CR 644, Woodbury, Red Bank
- Gas E: Mobil
- TServ E: **Freedom Int'l**
- Med E: **Hospital**
- Other E: **One Stop Shoppe C-Store**

23 U.S. 130 N, to Westville, Gloucester
- Lodg W: Cundey Motel
- Parks W: **National Park**

24AB CR551, NJ45, Westville, Woodbury

25AB NJ47, Deptford, Glassboro.
- Gas E: Amoco*
- Food W: Baby Loona Pizza
- TServ E: **TS Auto & Truck Repair (Towing)**
- Other W: Coin Car Wash

26 Jct I-76, to Jct I-676, Philadelphia

28 NJ 168, to NJ Turnpike, Bellmawr, Runnemede, Mt Ephraim
- FStop E: **Xtra C-Store*(D) (Kerosene)**
- Gas E: Shell
- W: Amoco(CW), Exxon
- Food E: Bakery Thrift Shop, Burger King, Club Diner
- W: Carvel Ice Cream, Chik-N-Lickin, Del Buono's Bakery, Good Friend Chinese, McDonalds, Pop's Homemade Italian Ice, Ravioli Company, Seafood Gallery, Taco Bell, Tu Sei Della Family Restorante
- AServ E: Ben Tilla Foreign Truck, Brakes & Mufflers, DeAngelis Auto, Shell
- W: American Battery, Exxon, Grease N Go
- ATM E: Commerce Bank, PNC Bank
- Other E: Car Wash

29 **(30)** U.S. 30, Berlin, Collingswood, Hadden Heights, Lawnside
- FStop E: **US Gas(D)**
- Gas E: Amoco, Texaco
- W: Shell*
- Food E: Burger King, Church's Fried Chicken, Dunkin Donuts, KFC, Oriental Inn, Wendy's
- Lodg E: Barrington Motel
- AServ E: Barrington Transmission, Jiffy Lube
- Med E: **Hospital**
- ATM E: **Pathmark Food**
- W: Jefferson Bank
- Other E: Car Wash, **Clean Machine Coin Operated Laundry, Drug Emporium, Pathmark Food & Drug**

30 Warwick Rd, Lawnside, Haddenfield (Southbound, Northbound Reaccess Only)
- Med E: **Hospital**

31 Woodcrest Station (Park & Ride)
- Other E: **Echelon Mall**

32 CR 561, Haddonfield, Gibbsboro
- Gas E: Mobil*(D)
- Food E: Bagel Place, Herman's Deli, Lotus Oriental

EXIT NEW JERSEY

Chinese & Japanese, Nopoli Pizza, Vito's Pizza
- AServ E: Mobil, Trio Tire
- Med E: **Hospital**
- ATM E: Commerce Bank, PNC Bank
- Other E: **Talk of the Town Grocery, Thrift Drug**

34 NJ 70, Cherry Hill, Marlton, Camden
- Gas E: Exxon*
- W: Mobil, Texaco*(D)
- Food E: Big John's Deli & Pizza, Burger King, Dunkin Donuts, Empress of China, Korea Garden, Uno Pizzeria
- W: Alkhimah Middle Eastern Restaurant, Androtti's Vienese Cafe, Boston Market, Manhattan Bagel, Morgan's (Four Points by Sheraton), Norma's Middle Eastern Restaurant, Pizza Time, Pracacci Gourmet
- Lodg W: Four Points Hotel by Sheraton
- AServ E: Bargain Brakes & Mufflers, FTS Tire, Grease Monkey
- W: Mobil, Texaco*(D)
- Med W: **Hospital**
- ATM W: Hudson City, Jefferson Bank, Summit Bank
- Other W: **Cherry Hill Animal Hospital,** Cherry Hill Car Wash, **Cherry Hill Food Store, Covered Bridge Pharmacy, Dogs Dynamic Pet Salon, Lynchwood Optical**

36 NJ 73, Berlin, to NJ Tnpk
- FStop W: **Citgo(D)**
- Gas E: Exxon, Gulf*(D), Mobil*
- W: Exxon(D), Shell, Texaco*(D)
- Food E: Bennigan's, Bob Evans Restaurant, Boston Market, Denny's, G G's (Doubletree), Giavani's Pasta, Jake's Boardwalk (Ramada), Jin Ching Chinese, Kelly's Court (Quality Inn), **McDonalds,** Pastry Palace, Raddison, Regency Palace (Ramada), Sage Restaurant Diner, The Garden Room (Quality Inn), The Greenery Cafe, Wendy's
- W: Bertucci's, Burger King, Dunkin Donuts, German Kitchen (Sharon Motor Inn), Golden Eagle Diner (Prime Rib Seafood), KFC, Nathan's (Texaco), Wendy's
- Lodg E: Courtyard by Marriott, Doubletree, Econolodge, MacIntosh Inn, Quality Inn, Raddison, Ramada, Red Roof Inn, Travelodge
- W: Bel-Air Motor Lodge, Motel 6, Motel Clover, Rodeway Inn, Sharon Motor Inn, Track & Turf Motor Hotel, Trav-Lers Motel (Truckers Welcome)
- AServ E: Mobil
- W: Ford Dealership, Haddon Transmission, Jiffy Lube, Lexus, Monroe Muffler, NTB, Sears, Shell
- ATM E: First Union
- Other E: **7-11 Convenience Store, Garden Of Eden Groceries, Home Depot**
- W: **East Gate Square Shopping Center, Petsmart (Eastgate), Shop Rite Groceries & Pharmacy,** White Car Wash & Detail Ctr

40AB NJ 38, Mount Holly, Morrestown
- FStop W: **Texaco***
- Food W: Food Court (Morrestown Mall), Ground Round, **Subway (Texaco)**
- Med E: **Hospital**
- Other W: **Morrestown Mall**

43 Rancocas Woods, Delran
- Gas W: Texaco*
- AServ W: Texaco
- Med W: **Hospital**

45AB Mount Holly, Willingboro
- FStop W: **Exxon(D)**
- Gas W: Mobil
- AServ W: Brake & Go, Exxon
- Med E: **Hospital**

EXIT NEW JERSE

- W: **Hospital**

47AB CR 541, Mount Holly, Burlington
- Gas E: Exxon*(D)
- W: Gulf, Hess(D), Shell*
- Food E: Silver Star Chinese, Tarshishi's Deli
- W: Checkers Burgers, French Country Bread Bakery, Heavenly Ham, Holiday Ice Cream & Restaurant, Kumfong Chinese, Little Caesars Pizza, Pizza Village, Shoney's, Subway, Villa Pizza, Wedgewood Farm Family Restaurant
- AServ E: Sphere's Auto Ctr (Burlington Ctr Shopping)
- W: ABC Discount Auto Parts, Goodyear Tire & Auto, Gulf, Jiffy Lube, **K-Mart, Wal-Mart**
- ATM W: Andrews Fed Credit Union, Burlington Count Bank, Farmers & Merchants Bank, First Union Bank
- Other E: **Burlington Center Shopping Mall, Wonder Hostess Bakery & Thrift Shop**
- W: **Hear RX, K-Mart (Pharmacy), Thrift Drug, Wal-Mart Supercenter (Pharmacy, Optical)**

(50) **Rest Area - RR, HF, Phones, Picnic, RV Dump P (Overnite Truck Parking)**

52AB Columbus, Florence

56 U.S. 206, to NJ Tnpk, Fort Dix, McGuire AFB (Northbound)
- TStop E: **Petro Truck Serv***
- Food E: Iron Skillet Restaurant (Petro)
- TServ E: **Truck Repair**

57 U.S. 130, US206 Burlington, Bordentown
- Gas E: Amoco, Exxon, Mobil, Raceway, Shell*
- Food E: Burger King, Cannie's Restaurant (European Cuisine), Chinese Restaurant, Denny's, Ground Round, McDonalds, Rosario's Pizza, Taco Bell
- Lodg E: Econolodge
- AServ E: Bob McGuire Chevrolet, Bordentown Auto Parts, Challenger Transmissions, Exxon, Mobil (Wrecker Service), Raceway, STS Tire, Saturn, Shell*, Town Ford
- ATM E: CoreState Bank, People's Saving Bank, Summ Bank
- Other E: **Acme Grocery, Quik Chek C-Store, Zenith RV's**
- W: **State Patrol Post**

(58) Scenic Overlook (Vehicles Over 5 Ton-Prohibited)

60 Jct I-195N, Jct I-95N, NJ29, NJ129, Trenton, Shore Points

61A Arena Dr E, Whitehorse Ave
- AServ W: Bob Hope's Auto Repair
- Other W: **7-11 Convenience Store**

61B **(62)** Olden Ave N (Southbound, Northbound Reaccess Only)
- Gas W: Delta

63 **(64)** NJ33, Trenton, Hamilton Ave, Nottingham
- Gas E: Mobil
- W: Exxon*
- Food E: Lobster Dock, Popeye's, Steakhouse, Vincent's Pizza & Hoagie's
- W: AJ's Cafe, Dunkin Donuts, Joey's Place, Peter Pan Diner
- AServ E: Ford Dealership, Mufflex, Walt's Place Auto Repair
- W: Hamilton Transmission, Parts America, Tune-Up Center Plus
- Med W: **Hospital**
- Other E: **A-1 Safe & Lock, Acu-Vision, Episcopo Pharmacy**
- W: **Bill's Pet Supply, Wa Wa Food Market & C-Store**

64 **(65)** CR 535N, to NJ 33E, Mercerville

Bold red print shows RV & Bus parking available or nearby

XIT | NEW JERSEY/DELAWARE

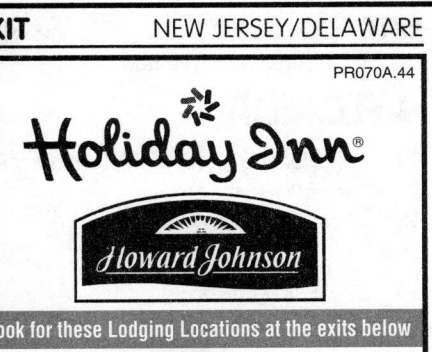

(Southbound, Northbound Reaccess Only)

55AB Sloan Ave
- **Gas** E: Exxon
- **Food** E: Brookwood Restaurant & Catering, Burger King, China Buffet, Dunkin Donuts, Hoagie Shack, Lenny & Haps Pizza, Taco Bell
- **AServ** E: Car Performance Auto Repair, Quaker Ridge Auto, RW Tire
- **ATM** E: Yardsill Nat'l Bank
- **Other** E: **Animal Veterinary Care Center Hospital, Roslidi Thriftway Groc, Speed Wash Coin Laundry**
  W: **Fed Ex Drop Box**

(35) **Rest Area - RR, HF, Phones, Picnic, RV Dump** P

67 **(68)** U.S.1, New Brunswick, Trenton
- **FStop** E: **Amoco***(D)
- **Gas** W: Exxon*(LP), Shell
- **Food** E: Applebee's, Bagel Brothers, Ground Round, Red Lobster, Season's Restaurant & Bar
  W: Domino's Pizza, Golden Empire Restaurant, Michael's, Swiss Bakery
- **Lodg** E: McIntosh Inn, Red Roof Inn
  W: Howard Johnson (see our ad on this page), Motel Mount's, Sleep-ee Hollow Motel
- **AServ** E: Midas Muffler, Sears (Mall)
  W: Greenfield Dodge, Jiffy Lube, Just Mercedes Benz, Lawrence Lincoln Mercury, Lawrence Toyota, STS Tire, Volvo Dealer
- **Other** E: **Quaker Bridge Mall**

↑NEW JERSEY

↓DELAWARE

9N DE 9, New Castle Ave, New Castle, Wilmington
- **FStop** W: **Gulf***
- **Gas** E: Amoco*, Conoco
  W: Texaco*
- **Food** E: Dragon Palace Chinese, Giavani's Pizza
  W: Gordan's Seafood, McDonalds
- **Lodg** E: **Country Inn**
  W: Days Inn, **Motel 6**, Travelodge
- **AServ** E: Big A, Conoco (Wrecker Service), Firestone Tire & Auto, Penn/Jersey Tire & Serv Ctr
- **Other** E: Car Wash, **Crossroads Hdwe, Food & Fresh Grocery & Pharmacy, Laundry, Rite Aide Pharmacy**
  W: C's Car Wash

↑DELAWARE

Begin I-295, Wilmington

EXIT | VIRGINIA

Begin I-295, Richmond

↓VIRGINIA

51AB Nuckols Rd
- **Other** S: **US Post Office**

49AB US 33, to Richmond, Montplier
- **Other** S: **Meadow Farm Museum**

45AB Woodman Rd

43 Junction I-295, I-95 & US 1 to Richmond

41AB US301, VA2, to Hanover and Richmond (Exit To 64 & 95 To Richmond)
- **FStop** W: **Exxon**(LP)
- **Gas** E: Amoco*(D), Texaco*(D)(LP)
- **Food** E: Brunetti's Restaurant, Burger King, China Kitchen, Dunkin Donuts (Texaco), McDonalds
- **AServ** E: Atlee Auto Service
- **ATM** E: First Virginia Bank, Signet (Amoco)
- **Other** E: **Patrick Henry's Birthplace, Revco Drug**

38 VA 627, Polgreen Rd, Meadowbridge Rd
- **AServ** W: Stubbs Auto
- **Med** W: ✚ **Hanover Medical Mall**

37AB US360, Tappahannock, Mechanicsville
- **Gas** E: Texaco*(CW)(LP)
  W: Amoco*(LP), Citgo*, Mobil(D)
- **Food** E: Baskin Robbins, **Cracker Barrel**, Gus' Italian Cafe, Mexico Restaurant, Taco Bell, Ukrop's Cafe'
  W: Mechanicville Seafood, Prairie Schooner, Rendezvous Cafe, Sno-cone Shack
- **AServ** W: Bruce Auto Parts, Carquest Auto Parts, Don's Auto Care, Napa Auto Parts, RJ Automobile and Truck Repair
- **ATM** W: Central Fidelity Bank, Creststar Banking, First Virginia Bank, Nationsbank
- **Other** E: **Ukrop's Supermarket and Pharmacy, Wal-Mart (Pharmacy)**
  W: **Colonial Pharmacy, Mechanicsville Drug Store, Old Dominion Locksmith Co., Revco Drugs, Richmond Battlefield, Winn Dixie Supermarket**

34AB VA615, Kreighton Rd

31AB VA156, Highland Springs

28 Jct I-64, U.S. 60, Norfolk, Richmond, Williamsburg

22AB VA5, Charles City, Varina

15AB VA10, Hopewell, Chester
- **Gas** E: Amoco*(D)(LP), Citgo*(LP)
  W: Chevron*(CW)(LP)
- **Food** W: **McDonalds**
- **Med** E: ✚ **Hospital**
- **ATM** W: Chevron

9AB VA36, Hopewell, Ft Lee, Colonial Heights
- **Gas** E: Petrol*(LP) (Kerosene)
  W: Amoco*(D)(LP), Exxon*(D), Pilot Travel Center*(D)(LP) (Kerosene), Texaco*
- **Food** E: Honey Bee's, Hong Kong Chinese Restaurant, Mexican Restaurant, Rosa's Italian Food
  W: Bullets (Exxon), Days Inn, Denny's, Gary's Eatery and Tavern (Comfort Inn), Kanpai Japanese Steakhouse, Leone's Italian Restaurant, McDonalds, Papa John's, Pizza Hut, Shoney's, Splash Seafood, Subway, Taco Bell, Top's Chinese, Waffle House, Wendy's, Western Sizzlin', Zero's Subs
- **Lodg** E: Innkeeper
  W: Comfort Inn, Days Inn, Hampton Inn
- **AServ** W: Strosnder Chevrolet, Geo
- **ATM** W: First Colonial Savings Vank, First VA Bank
- **Other** E: **Army Museum, City Point National Historical, Super Fresh Supermarket, Tourist Information**
  W: **Amoco (Coin Operated Laundry), Food Lion Grocery, Petersburg Historical, Revco Drugs, Rite Aide Pharmacy, Winn Dixie Supermarket (24 Hr)**

3AB U.S. 460, Norfolk, Petersburg
- **FStop** E: **East Coast***(LP)
- **Gas** E: Texaco* (Kerosene)
  W: **Exxon***
- **Food** E: **Dunkin Donuts* (East Coast), Subway (East Coast)**
  W: **McDonalds**
- **AServ** E: Firestone Tire & Auto
- **ATM** W: Exxon
- **Parks** W: **Virginia Motor Sports Park**

2 Jct I-95S, to Rocky Mount, NC (Left Exit Southbound)

1 **(46)** Jct - 95 to I-85 Petersburg (Southbound)

↑VIRGINIA

Begin I-295, Richmond

Begin I-295, D.C. & Maryland

↓MARYLAND

41A Ellicott City, Glen Burnie

46 JCT I-895N, Harbor Tunnel (North-bound)

47AB JCT I-95, MD166 Catonsville, VWI Airport.

19 JCT695, Glen Burnie, Pawson, Annapolis

18 West Nursery Rd
- **Lodg** E: AmeriSuites (see our ad on this page)

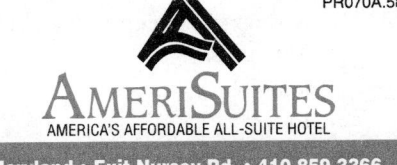

17 Jct I-195, Catonsville, Odenton, Lithicum, Aviaton Blvd (Services On Belkridge Landing Rd Frontage Rd)
- **Gas** E: Convenience Store
- **Food** E: Charlie's Cafe (Airport), Holiday Inn (Ridge Rd), Marriott, McDonalds
- **Lodg** E: Empire Airport North, Hampton Inn, Holiday Inn (Ridge Rd), Marriott, Susse Chalet
- **AServ** E: Dasis Auto Service, Goodyear Tire & Auto

EXIT		MARYLAND
	ATM	E: First Nat'l Bank
	Other	E: VWI Airport
16		MD100, Glen Burnie, I-95
	Food	W: 3 Aces Restaurant
	Lodg	W: Ramada Inn
15		MD175, East Odenton, West Jessup
	Gas	E: Shell* (Kerosene), Shell
	Food	E: Parkway Inn
	Lodg	E: Parkway Inn
	AServ	E: Hare's Service Ctr, Shell
14		MD32, East Ft Meade, West Columbia
	Gas	W: Shell
	AServ	W: Shell (Wrecker, Auto)
	Other	W: National Cryptologic Museum
13		MD198, Ft Meade
	AServ	E: Laurel Auto Body, Tischer Volkswagon
12		MD197, Laurel, Bowie
	Gas	E: Exxon
		W: Getty's(D), Texaco
	Food	E: China Diner, Little Caesars Pizza
		W: Chintao Chinese Restaurant, Roy Rogers, Subway, TCBY Treats
	Lodg	W: Red Rood Inn
	AServ	W: Getty's (Towing), Whip's Towing
	Med	W: + Hospital
	ATM	W: Citizens Nat'l Bank
	Other	E: 7-11 Convenience Store, E-Z Suds Coin Operated Laundry
		W: Babies Are Us Plaza, Montplier Mansion & Art Ctr, US Post Office
11		Powder Mill Rd, Beltsville
	Other	E: Nassa Goddard Visitor Ctr, Nat'l Wildlife Visitor Ctr
10		MD193, Greenbelt, Nassa, Goddard Ctr (No Trucks)
	Gas	E: Exxon
		W: Citgo*
	Food	E: Chesapeake Bagel Bakery, Chevys Mexican, Chi Chi's Mexican, Denny's, Hunan Treasure, Jasper's, Subway, Wendy's
	AServ	E: Exxon, Super Trak
		W: Citgo
	ATM	E: Chevy Chase Bank, First Nat'l Bank of Maryland, Safe Way Grocery
		W: Citgo
	Other	E: CVS Pharmacy, For Eyes Optical, Greenway Ctr, Maryland History Greenbelt Mueseum, Parcel Plus Mail, Safe Way Grocery (Deli Bar, ATM)
		W: Park Police Station
9		Baltimore Washington Pkwy, Jct 495. (No Trucks)
8		Riverdale Dr
	Gas	E: Exxon*(D)
	Food	E: A Wok Restaurant, Pizza Oven

EXIT		MARYLAND
	AServ	E: Exxon
	Other	E: Check Cashing, Coin Operated Laundry, East Pine Shopping Ctr, Super C-Store
7		MD450, Bladensburg, Annapolis
	Gas	E: Amoco, Capitol Plaza Auto Center*, Econoway, Lowest Price, Mobil, Super Trak
	Food	E: Bob's Big Boy, Burger King, Dunkin Donuts, Eastern Restaurant (Chinese), Howard Johnson (Italian), Italian Inn, McDonalds, Pizza Hut, Shakey's Pizza & Buffet, Wendy's
	Lodg	E: Bladensburg Motel, Comfort Inn, Howard Johnson
	AServ	E: Amoco, Econoway, Mobil, Montgomery Ward, Pep Boys Auto Center, Rosenthall Toyota, Trak Auto Service
	ATM	E: Chevy Chase Bank (Safeway Supermarket), Chevy Chase Bank (The Capital Plaza Mall), Crestar Bank
	Other	E: Ace Cash Express (Check Cashing), Bubbluz Coin Laundry, Capitol Corner Shopping Ctr, Car Wash, Eko Food Store, Safe Way Grocery (Pharmacy), The Capital Plaza Mall
		W: Bladensburg Animal Hospital
6		(202) US 1, Bladensburg, Cheverly
	Gas	W: Exxon
	Food	W: Fertelli Italian Restaurant
	Lodg	W: Howard Johnson
	AServ	W: Goodyear Tire & Auto
	Other	E: Prince Valley Medical Ctr
		W: Coin Operated Laundry, Drug Emporium
5		Pennsylvania Ave, & Andrews Airforce Base, Frederick Douglas
	Gas	E: Amoco, Citgo*, Mobil*, Shell*
	Food	W: McDonalds
	AServ	E: Citgo
	Med	W: + Hospital
	Other	E: Southeast Animal Hospital
4		Navy Yard, Downtown Washington, I-395
3AB		South Capitol St, Downtown, Sutland Pkwy, US Naval Station (Difficult Reaccess)
	Other	E: Frederick Douglas Home
2		Portland St, Boling Airforce Base
5AB		JCT I-95.
6AB		US 1, Washington Blvd, to Laurel, Elk Ridge
	FStop	W: Quarrels Card Fueling (24 Hr)
	Gas	W: Extra Fuel*(D)(LP)
	Lodg	W: Blvd Motel
	AServ	W: All Pros, All Tune & Brake, World Class Auto & Truck (Napa Dealer)
1		US Naval Research Lab
	Med	W: + Hospital

↑MARYLAND

Begin I-295, Maryland

EXIT		FLORIDA
		Begin I-295, Florida

↓FLORIDA

EXIT		
13		Flamingo Lake, FL 115, Lemturner Rd
	Other	E: Wal-Mart
12		FL104, Dunn Ave
11		U.S. Hwy 1, U.S. Hwy 23
10		Pritchard Rd
9		Commonwealth Ave
8AB		**(20)** Jct I-10, Lake City, Tallahassee, Jacksonville
7		**(19)** FL228, Normandy Blvd
6		**(18)** FL208, Wilson Blvd.
5		**(16)** FL134, 103rd St
4		**(12)** FL21, Clanding Blvd, Orange Pa
	Lodg	W: Hampton Inn (see our ad on this page)
3		**(10)** U.S.17, FL15, Nas Jack, Orange Park
2		**(5)** San Jose Blvd, FL13
1		**(3)** Old St. Augustine Rd
	Gas	E: Amoco, BP*(CW), Texaco
		W: Chevron
	Food	E: Denny's, Taco Bell, Wendy's
	ATM	E: AM Bank

↑FLORIDA

Begin I-295, Florida

Bold red print shows RV & Bus parking available or nearby

EXIT PENNSYLVANIA

Begin I-376

↓PENNSYLVANIA

1 I-279 North, Three Rivers Stadium
2 Stan Wix Left Lane Exit.
3 Grant St. Left Exit
4 Second Ave.
5 Forbes Ave, Oakland
6 Boulevard of the Allies
7AB PA 885, Oakland, Glenwood, 7A - PA 885 N, 7B - PA 885 S
- Lodg **N:** Best Western

8 **(4)** Squirrel Homestead
9 **(6)** Edgewood, Swiss Vale
- Gas **N:** Citgo, Sunoco
  S: BP*
- Food **N:** Food For Thought Deli
  S: Applebee's, Coral Garden Chinese, Eat N Park, Subway, Taco Bell, **The Italian Oven**
- AServ **N:** Citgo, Sunoco
- ATM **S:** Mellon Bank, National City
- Other **N:** **Pittsburgh Zoo**
  S: **Coin Laundry, Giant Eagle Supermarket, K-Mart, Mail Boxes Etc, Phar Mor Drugs**

10 US 30 East, Forest Hills
- Gas **N:** Amoco*, Texaco
- Food **N:** Holiday Inn
- Lodg **N:** Holiday Inn
- AServ **N:** Lincoln Garage, Parkway East Ford, Texaco

11 PA 8, Wilkinsburg
12 Greensburg Pike
13 PA 130, Churchill (Westbound)
14 US 22 Business, Monroeville
- Gas **N:** BP*[LP]
- Food **N:** Anthony's Restaurant, Harley Hotel
- Lodg **N:** Harley Hotel
- AServ **N:** Sovak and Miller Auto Center

15 PA 791 North, Penn Hills
16 PA 48, Monroeville, Plum, 16A - 48 South, 16B - 48 North
- Gas **N:** Gulf, Sunoco*
  S: BP*[CW], Citgo, Exxon*, Sunoco
- Food **S:** Arby's, Baskin Robbins, Big Boy, Burger King, Dar Bar Indian Restaurant, Denny's, Dunkin Doughnuts, Holiday Inn, **Kenny Rogers Roasters,** Kings Motel, Kings Restaurant, Lone Star Steakhouse, McDonalds, Nick Marie's Esta Esta Italian Restaurant, Palace Inn, Shanghai Express, **Taco Bell,** Wendy's
- Lodg **N:** East Exit Motel
  S: Days Inn, Kings Motel, Palace Inn, Penn Motel, Red Roof Inn
- AServ **N:** Gulf, Ted McWilliams Toyota
  S: Citgo, Import Export Tire Company, Procare, Sunoco, Tire America, Valley Buick
- Med **S:** ✚ **Hospital**
- ATM **S:** Mellon Bank, **PNC Bank**
- Other **N:** **Giant Eagle Supermarket, Revco**
  S: **Mosside Animal Clinic, Penn Super Pharmacy, Thrift Drug**

↑PENNSYLVANIA

Begin 376

EXIT IOWA

Begin I-380

↓IOWA

72 San Marnan
- Gas **W:** Conoco*, Holiday*
- Food **W:** Bonanza Steakhouse, Boston Market, Burger King, Carlos O Kelly's Mexican Restaurant, Country Kitchen, Genghis Khan Chinese, Godfathers Pizza, Goldcoast Dogs, Great Dragon Buffet, Long John Silvers, McDonalds, Olive Garden, Pizza Hut, Subway, Taco Bell, Taco John's, The Ground Round Restaurant, Wendy's
- Lodg **W:** ◆ Comfort Inn, ◆ Fairfield Inn, ◆ Heartland Inn, ◆ Super 8 Motel
- AServ **W:** Jiffy Lube, Midas Mufflers and Brakes, Tire Plus, Western Auto
- ATM **W:** Magna Bank
- Other **W:** Car Wash, **K-Mart, Shopping Mall, Target Department Store, Venture, Wal-Mart Supercenter**

71 US 218, Waterloo North, Airport
70 River Forrest Rd
68 Elkrun Hieghts, Evansdale Dr.
- TStop **E:** **Elkrun Hieghts Truck Plaza***
- Food **E:** Blimpie's Subs, Jumie's (Elkrun Hieghts), **McDonalds (Indoor Playground)**
- Lodg **E:** AAA **Ramada Limited**
- TServ **E:** **Elkrun Hieghts Truck Plaza**

66 IA 297, Raymond, Gilbertville
65 US 20 East, Dubuque
62 Gilbertsville, Blackhawk County Road D 38
55 LaPorte City, Jesup, Buchanan County Road V 55
(54) **Weigh Station (Southbound)**
(51) **Weigh Station (Northbound)**
49 IA 283, Brandon
43 IA 150, Independence, Venton
41 IA 363, Urbana
- FStop **W:** **Conoco***
- Food **W:** Dashboard Subs (Conoco)

35 Cedar Point, Linn County Road W36
- Gas **E:** Standard

28 Toddville, Robbins, Linn County Road 34
25 Boyson Rd.
- Gas **E:** Phillips 66*[CW]
- Food **E:** Blimpie's Subs (Phillips 66), Oscars Restaurant

24B Blairs Ferry Rd
- Gas **E:** Quick Shop*
  W: QT*
- Food **E:** Hardee's, Papa Juans Mexican Restaurant, Shoney's, Zio Johno's Spaghetti Of Course
  W: Burger King, Egg Roll House Chinese, Godfathers Pizza, Pizza Hut, Royal Fork Buffet Restaurant, Subway, Taj Mahal Restaurant
- Lodg **E:** AAA Shoney's Inn

EXIT IOWA

- AServ **E:** Jiffy Lube
  W: Auto Zone Auto Parts
- Other **W:** **Sam's Club, Wal-Mart**

23 42nd Street
- Gas **E:** Amoco[CW], Texaco*
- Food **E:** Baskin Robbins, Domino's Pizza, Doughnut Land, Hardee's, KFC, Little Caesars Pizza, Pizza Hut, Tasty Freeze, The Red Lion, The Springhouse Restaurant, Wendy's
- AServ **E:** Amoco, Napa Auto Parts
- Other **E:** **Clarks Pharmacy, Coin Laundry**

22 Coldstream, 29th Street, Glass Rd, 32nd Street
- Gas **E:** Texaco*
- AServ **E:** Kevin's Auto Repair

21 H Ave., J Ave.
20B 7th Street East, Five Season Center Hospital
19A IA 94, US 151, Fifth Ave Southwest, Diagonal Dr. Downtown (Difficult Reaccess Northbound)
18 Wilson Ave. Southwest
17 Thirty - Third Ave. South West, Hawkeye Downs
- Gas **E:** Quick Stop*
  W: 66 Handi Mart*, Amoco*[CW]
- Food **W:** Cancun Restaurant, Denny's, Happy Chef Restaurant, Hot Stuff Pizza (66 Handi Mart), McDonalds, Pei's Manderine Restaurant, Perkins Family Restaurant, Subway, Wendy's, Zazza's Supper Club
- Lodg **E:** ◆ **Hampton Inn**
  W: ◆ Comfort Inn, ◆ Days Inn, AAA Econolodge, AAA Exel Inn, ◆ Fairfield Inn, Four Points Hotel Sheritan, ◆ Heartland Inn, HoJo Inn, ◆ Red Roof Inn, ◆ Super 8 Motel
- AServ **E:** Big A Auto Parts
  W: Goodyear Tire & Auto
- ATM **W:** Norwest Bank

16AB US 30, US 151, US 218, Mt Vernon, Tama
- Gas **E:** Phillips 66*
- Food **E:** Blimpie's Subs, Hardee's
- Other **E:** Coin Car Wash

13 Cedar Rapids Airport, Ely
- FStop **W:** **Coastal*[D] (Open 24 Hours)**
- TServ **W:** **Cedar Rapids Truck Center**

(12) **Rest Area - RR, Picnic, Phones, Vending, RV Dump (Both Directions)**
10 County Road F12, Swisher, Shueyville
- Gas **E:** Citgo*, Swisher Sinclare[LP]

4 County Road F28. North Liberty
- Gas **E:** Amoco*
- Food **E:** Subway
- RVCamp **E:** **Campground**

↑IOWA

Begin I-380

Bold red print shows RV & Bus parking available or nearby

EXIT SOUTH CAROLINA

Begin I-385

↓SOUTH CAROLINA

42 U.S. 276, Stone Ave., Traveler's Rest
- Gas W: Amoco(D)
- Food E: Rugby's Sports Grill, The Big Clock Drive-in
  W: Zorba Restaurant & Lounge
- AServ E: Acne Radiator, Car Quest Auto Center
- Other W: **Greenville zoo**

40AB SC 291, Pleasantburg Dr.
- Gas E: Exxon, Hess*, Speedway*(D)(LP), Texaco*(CW)
- Food E: Bojangle's, Hardee's, Pizza Hut, S&S Cafeteria, Steak & Ale, Subway, Taco Casa, Wendy's
- AServ E: Exxon
- ATM E: Summit National Bank
- Other W: **Bob Jones University, Furman University, Greenville Tech., Palmetto Expo Center**

39 Haywood Rd.
- Gas E: BP*(CW)(LP), Exxon*(D)
  W: Crown*(CW)
- Food E: Domino's Pizza, Gourmet Kitchen, Long Horn Steakhouse, Outback Steakhouse, Philly Connection
  W: Arby's, Black-Eyed Pea, Burger King, Carrabba's, Chuck E Cheese's, CiCi's Pizza, Don Pablo's Mexican, El Pollo Loco, Italian Market & Grill, Kanpai Japanese Steakhouse, O' Charley's, Quincy's, Waffle House, Wendy's
- Lodg E: Ameri Suites, ◆ Courtyard by Marriott, ◆ Hilton, Quality Inn, ◆ Residence Inn
  W: ◆ Hampton Inn
- AServ E: Jiffy Lube
  W: Sears, Tire America
- Med E: **✚ Doctors Care**
- ATM E: American Federal, Carolina First, Exxon, NationsBank
  W: BB&T Bank, First Union, Wachovia
- Other W: **Haywood Mall, Kinko's Copies, Movies At The Mall, Phar Mor Drugs, The Pet Motel**

37 Roper Mountain Rd.
- Gas W: BP*(CW), Exxon*(LP)
- Food E: Pizza Express (Harris Teeter)
  W: Olympian Greek & Italian Food, Rio Bravo, Waffle House
- Lodg W: DAYS INN ◆ Days Inn, ◆ Fairfield Inn, AAA Holiday Inn Select
- ATM E: First Citizens Bank (Harris Teeter)
- Other E: **Harris Teeter Supermarket (Pharmacy), Roper Mountain Science Center (Planitarium & Observatory)**

36AB Junction I-85, Spartanburg, Atlanta

35 Woodruff Rd., SC 146
- Gas W: Red Head*
- Food E: Applebee's, Arizona Steakhouse, Bojangle's, Boston Pizzaria, Hardees
  W: Acropolis, Best BBQ, Cheers Food & Drink, Fuddrucker's, Hong Kong Chinese, IHOP, Marcelino's Italian, Midori's Steak & Seafood, Stock Car Cafe
- AServ W: Smith-Davidson Tire
- ATM E: Carolina First Bank
  W: Red Head, Wachovia
- Other W: **All Creatures Animal Hospital**

34 Butler Rd.

33 Bridges Rd.

31 Laurens Rd., Simpsonville

30 U.S. 276, Standing Springs Rd. Maudlin
- AServ E: Bradshaw Ford, Love Chevrolet

29 Georgia Rd., Simpsonville
- FStop E: **Amoco* (24 Hrs)**

EXIT SOUTH CAROLINA

- Gas E: BP*(LP)

27 Fairview Rd., Simpsonville
- Gas E: Exxon*(D)(LP), Sav-Way*(D)(LP), Short Stop*, Texaco*(D)
  W: Amoco*(LP), BP*(CW)(LP)
- Food E: Christy's Cafe, Coach House Restaurant, Hardees, J.B.'s BBQ, Little Caesars Pizza, Manzanillo Mexican, McDonalds, Subway
  W: Applebee's, Arby's (Amoco), Pizza Hut, Ryan's Steakhouse, Taco Bell, Wendy's
- Lodg E: Palmetto Inn
  W: ◆ **Comfort Inn (see our ad this page),** Jameson Inn
- AServ E: Advance Auto Parts, Auto Zone Auto Parts
  W: Tire & Lube Express (Wal-Mart)
- Med E: **✚ Hospital**
- ATM E: BB&T Banking, Short Stop
  W: Amoco, NationsBank
- Other E: **Bi-Lo Supermarket,** Coin Car Wash, **Coin Laundry, Eckerd Drugs, Fairview Pointe Animal Hospital, Revco Drugs, Winn Dixie Supermarket**
  W: **Fairview Station, Ingles Supermarket, K-Mart (Pharmacy), Wal-Mart (Pharmacy)**

24 Fairview
- Gas E: Exxon, Phillip 66*(D)(LP)
- Food E: **Hardees,** Waffle House
- AServ E: Berry Woods Motors

23 SC 418, Fountain Inn, Pelzer
- Gas E: Exxon*(D)
- Food E: Subway
- ATM E: Exxon

22 S.C. 14 West, Fountain Inn.

19 SC14 East, Owings

16 SC 101, Woodruff, Gray Court

10 Rd. 23, Barksdale.

9 Enoree, Laurens, U.S. 221
- FStop E: **Exxon***
- Food E: Subway (Exxon)
- Lodg E: Southern Economy Inn
- ATM E: Exxon
- Other E: **Fed Ex Drop Box (Exxon)**

(4) **Rest Area - RR, Phones, Vending, Picnic (Both Directions)**

5 SC 49, Laurens

2 SC 308, Clinton, Ora
- Med W: **✚ Hospital**

1 Junction I-26

↑SOUTH CAROLINA

Begin I-385

EXIT NEW YORK

Begin I-390

↓NEW YORK

20AB **(76)** Jct I-490, to Rochester, Buffalo

19 **(75)** NY33A, Chili Ave

18 **(74)** Brooks Ave, Airport, NY204

17 **(73)** NY383, Scottsville Rd

16AB **(72)** NY15, W Henrietta Rd, River Rd

16B E Henrietta Rd
- Lodg W: Wellesley (see our ad on this page)

15 **(71)** Jct I-590N

14AB **(69)** NY252, NY15A, E Henrietta Rd

13 **(68)** Hylan Dr
- Gas W: Mobil*
- Food W: Ciau Italian, McDonalds, Pizzaria Uno, Wendy's
- Other W: **Market Place Mall, Wegmans Grocery Store**

12AB **(66)** NY253, Lehigh Station Rd, Jct I-90
- TStop W: **Sugar Creek Travel Plaza***
- Gas W: Hess*(D), Sunoco*(CW)
- Food W: Days Inn, McDonalds, Peppermints Family Restaurant, Perkins, Wendy's
- Lodg W: DAYS INN **Days Inn,** Fairfield Inn, Highland Motor Inn, Microtel, Red Roof Inn, **Super 8 Motel**

11 **(62)** NY15, NY251, Rush
- AServ E: Jim Bucci's Towing

10 **(55)** U.S. 20, NY5, Avon, Lima
- FStop W: **Citgo**
- Gas W: Mobil (ATM)
- Food W: Avon

9 **(52)** NY15, Lakeville, Conesus Lake
- FStop E: **Hess***
- Food E: **Manhattan Bagel (Hess), Subway (Hess)**

8 **(49)** NY20A, Geneseo
- RVCamp E: **Conesus Lake Campground**

7 **(39)** NY63, NY408, Morris
- Gas W: Mobil*(LP)
- Food W: Yesteryears Ice Cream

(38) **Rest Area - Picnic, RR, Phones (Both Directions)**

6 **(34)** NY36, Sonyea

5 **(24)** Dansville, NY36
- TStop W: **Travelport*(LP)(SCALES) (RV Dump)**
- Gas E: Kwik Fill*, Mobil*
- Food E: Arby's, Burger King, Ice Cream Island, McDonalds, Pizza Hut, Subway
  W: **Buckhorn Restaurant (Travelport)**
- Lodg W: AAA **Daystop (Travelport)**
- Other E: **Airport, Revco Drugs, Rite Aide Pharmacy, Topps Grocery Store**

4 **(23)** NY36, Dansville, Hornell
- RVCamp E: **Goodwin's RV Center**
  W: **Stony Brook Park Campground, Sugar Creek Glen Campground**
- Med E: **✚ Hospital**

3 **(17)** NY15, NY21, Wayland
- RVCamp E: **Holiday Hills Campground**
- Other E: **State Patrol Post**

2 **(11)** NY415, Cohocton

(9) Scenic Area (Northbound)

1 **(2)** NY415, Avoca

(0) Jct NY 17

↑NEW YORK

Begin I-390

Bold red print shows RV & Bus parking available or nearby

EXIT MASSACHUSETTS/CONNECTICUT

Begin I-395

↓ MASSACHUSETTS

7 **(0)** Jct I-90 MA TNPK, MA12S, Auburn
- **Food** **E:** Golden Lion
  - **W:** Abdow's, Arby's, Burger King, Papa Gino's
- **Lodg** **W:** Budgetel Inn, Days Inn
- **AServ** **W:** Firestone Tire & Auto, Midas Muffler & Brakes, Ron Buchard Acura, Sears, Speedee Muffler King
- **ATM** **E:** Bank of Boston
  - **W:** Fleet Bank, Webster Five Cents Savings Bank
- **Other** **W:** Auburn Mall, Auburn Police

6 **(12)** Jct U.S. 20, Auburn, Charlton
- **Gas** **W:** Shell
- **Food** **E:** KFC
  - **W:** Chuck's Steak House, D'Angelo's, Friendly's, Holiday Pizza, Quality Pizza, Wendy's
- **Lodg** **E:** Queen Elizabeth Motel
- **AServ** **W:** Interstate Transmission, Millbury Motors, Shell
- **Other** **W:** CVS Pharmacy, Honey Farms, Park & Shop

5 **(9)** Depot Rd, N Oxford

4 **(6)** Sutton Ave, Oxford, Sutton
- **Gas** **W:** Citgo*
- **Food** **W:** Dairy Express Pizza, Dunkin Donuts, McDonalds, New England Pizza, Subway
- **AServ** **W:** Cahill's Tire Center
- **ATM** **W:** Fleet Bank
- **Other** **W:** Cumberland Farms, Post Office

3 **(4)** Cudworth Rd, N Webster, S Oxford
- **Food** **W:** Uncle Tannou's

2 MA16, Webster, Douglas
- **Gas** **W:** Exxon*(CW), Getty*, Hess, Mobile, Shell
- **Food** **W:** Burger King, Dunkin Donuts, E. Main Street Cafe, Friendly's, Gil's Seafood Deli, Honey Dew Donuts, Ma Ling's Chinese, Men at Wok, Papa Gino's, Stuffy's Express, Webster Donuts, Yum-Yum Chicken
- **AServ** **W:** NAPA Auto Parts, Place Motors Ford, Shell* (Towing)
- **RVCamp** **E:** Campground
- **ATM** **W:** Commerence Bank, Fleet Bank
- **Other** **W:** CVS Pharmacy, Coin Laundry, Price Chopper Supermarket

1 MA193, Webster
- **Gas** **W:** Citgo
- **Food** **W:** Wind Tiki Chinese
- **AServ** **W:** Citgo, Goodyear Tire & Auto, Rich & Sons Auto Repair
- **Med** **E:** ✚ Hospital
- **Other** **E:** Webster Lake Veterinary Hospital

↑ MASSACHUSETTS

↓ CONNECTICUT

100 **(54)** Wilsonville, E Thompson

99 **(50)** CT200, N Grosvenor Dale, Thompson
- **Lodg** **E:** Lord Thompson Bed & Breakfast
- **RVCamp** **E:** Campground

98 **(49)** Grosvenor Dale, CT12 (Northbound, Southbound Reaccess Only)
- **Gas** **W:** Sunoco
- **AServ** **W:** Sunoco
- **Med** **W:** ✚ Hospital
- **Other** **W:** Franklin Pharmacy

EXIT CONNECTICUT

97 **(47)** U.S.44, Woodstock, E Putnam
- **Gas** **W:** Shell, Sunoco*
- **Food** **E:** Dunkin Donuts, KFC, McDonalds, Subway, Wendy
  - **W:** McDonalds (Wal-Mart)
- **AServ** **E:** K-Mart, Meineke Discount Mufflers
  - **W:** Pontiac, Buick, Shell*
- **ATM** **E:** New London Trust
- **Other** **E:** K-Mart
  - **W:** Wal-Mart(LP) (ATM)

96 **(46)** to CT12, Putnam
- **Food** **W:** Kings Inn
- **Lodg** **W:** Kings Inn
- **Med** **W:** ✚ Hospital

95 **(45)** Kennedy Dr
- **Med** **W:** ✚ Hospital
- **Other** **W:** Animal Hospital of Putnam, Eye Care

94 **(43)** Attawaugan, Ballouville
- **TServ** **W:** International Truck Parts & Maintanence

93 **(41)** CT101, Dayville, E Killingly
- **FStop** **W:** Mobil*
- **Gas** **E:** Royal Guardian, Shell
- **Food** **E:** Burger King, Carvel Ice Cream, China Garden, Dunkin Donuts, Fitzgerald's Pub & Grill, Joker's Cafe, McDonalds, Menagerie, Subway, Zip's Diner
- **Lodg** **E:** Budget Inn
- **AServ** **E:** Firestone Tire & Auto, Sam Wibberly Tires, Shell*
- **RVCamp** **E:** Campground
  - **W:** Campground
- **ATM** **E:** Fleet Bank, New London Trust, Savings Bank of Manchester
  - **W:** The Savings Inst.
- **Other** **E:** A&P Grocery, Cumberland Farms
  - **W:** Airport, Bell Park Pharmacy

92 **(38)** S Killingly
- **Food** **W:** Giant Pizza
- **ATM** **W:** Fleet Bank, Putnam Savings
- **Other** **W:** Bonneville Pharmacy, Coin Laundry, State Patrol Post

91 **(38)** U.S. 6W, Danielson, Hartford

90 **(35)** to U.S. 6E, Providence

(35) Rest Area - Both Directions
- **FStop** **B:** Mobil*

89 **(32)** CT14, Central Village, Sterling
- **Gas** **W:** Citgo*(D), Sunoco(D)
- **Food** **W:** Subway
- **Lodg** **W:** Plainfield Motel
- **AServ** **W:** Sunoco
- **RVCamp** **E:** Campground
  - **W:** Campground (RV)
- **Other** **W:** Vision Care Center

88 **(30)** CT14A, Plainfield, Oneco
- **Gas** **W:** Mobile
- **Lodg** **W:** Renaissance Bed & Breakfast
- **RVCamp** **E:** Camp-N-Canoe at Riverbend, Sterling Park
- **Med** **W:** ✚ Walk-In Clinic
- **Other** **W:** Coin Laundry

87 **(28)** Lathrop Rd
- **Gas** **W:** Citgo*, Sunoco*
- **Food** **E:** Dunkin Donuts, Great Oak Pizza, Hong Kong Star
  - **W:** Baker's Donuts, Golden Eagle, Golden Greek, McDonalds
- **Lodg** **E:** Plainfield Yankee Motor Inn
- **ATM** **E:** Savings Society

EXIT CONNECTICUT

- **Other** **E:** CVS Pharmacy, Super Big Y Grocery Store
  - **W:** Boyl's Drugs, Car Wash

86 **(24)** CT201, Hopeville
- **RVCamp** **E:** Campground
- **Parks** **E:** Hopeville State Park
- **Other** **E:** Park & Ride (Telephone)

85 **(22)** CT164, CT138, Preston City, Pachaug
- **Food** **E:** NAPA Auto Parts
- **RVCamp** **E:** Hidden Acres Campground (1 Mile)
- **ATM** **E:** Savings Society
- **Other** **E:** Slater Shopping Center

84 **(21)** CT12, Griswold, Jewett City
- **FStop** **W:** Mobil*
- **Gas** **W:** Citgo*, Shell*(CW)
- **Food** **W:** McDonalds, Taco Bell
- **RVCamp** **W:** Campground
- **ATM** **W:** Yankee
- **Other** **W:** Better Value Supermarket, Supervalue Pharmacy

83A **(19)** CT169, Lisbon (Northbound, Southbound Reaccess Only)

83 **(18)** CT97, Occum, Taftville
- **Gas** **E:** Getty*
  - **W:** Citgo*(D)
- **RVCamp** **W:** Camping

82 **(14)** Yantic, Norwichtown
- **FStop** **W:** Mobil*, Shell, Texaco
- **Gas** **W:** Texaco*(CW)
- **Food** **E:** Friendly's Ice Cream
  - **W:** Bagel Co., Norwich Motel, Rena's Pizza
- **Lodg** **W:** Marriott, Norwich Motel
- **AServ** **E:** Norwich Tire Service
  - **W:** Texaco
- **ATM** **E:** Dime Bank
- **Other** **E:** True Value Hardware(LP)
  - **W:** Car Wash, Total Vision

81 **(14)** CT2, CT32, Norwich, Hartford
- **Med** **E:** ✚ Hospital

80 **(11)** CT82, Downtown, Norwich, Salem
- **Gas** **E:** Mobil*, Shell
- **Food** **E:** Bess Eaton, Burger King, KFC, McDonalds, Pizza Hut, Subway
- **Lodg** **W:** ◆ Ramada Hotel
- **AServ** **E:** Shell, Town Center Tire
  - **W:** Wal-Mart
- **ATM** **E:** Liberty Bank
  - **W:** Norwich Federal Credit Union
- **Other** **W:** Big Y Market, Tourist Information, Wal-Mart

79A **(9)** CT2A E, Preston, Ledyard
- **Other** **E:** Mohegan Sun Casino

79 **(6)** CT163, Uncasville, Montville

78 **(5)** CT32, New London (Northbound)

(8) Rest Area - RR, Phones, Picnic
- **FStop** **E:** Mobil*
- **Other** **E:** State Police

77 **(2)** CT85, Colchester, Waterford
- **Gas** **W:** Citgo
- **Food** **W:** Dunkin Donuts
- **Lodg** **W:** Oakdale Motel

75 **(1)** U.S.1, Waterford, I-395S (Southbound)

↑ CONNECTICUT

Begin I-395

Bold red print shows RV & Bus parking available or nearby

EXIT VIRGINIA

Begin I-395

↓VIRGINIA

9 Pentagon, Crystal City, Richmond, Washington Blvd, Ridge Rd
- **Food** **E:** California Pizza, Chevys Mexican Restaurant, **Crystal City Mall**, Nell's Carry Out, Starbuck's Coffee
- **Lodg** **E:** Doubletree Hotel, Residence Inn
- **Other** **E:** **Best Buy Ctr, Pentagon City Mall, Tourist Information**

8AB VA 27, Washington Blvd, Ft Myer, & Memorial Bridge
- **Food** **W:** Marvelous Bagels & Deli, Pizza Movers, Sheraton
- **Lodg** **W:** Sheraton
- **AServ** **W:** A-1 Auto Clinic

7AB VA 120, Glebe Rd.
- **Gas** **E:** Exxon
- **Food** **E:** Auggie's Crabbie Pig, Pizza Hut
  W: Best Western, Brandy's Restaurant, China Village Restaurant (Econolodge)
- **Lodg** **W:** Best Western, Econolodge
- **AServ** **E:** Exxon
- **Med** **E:** ✚ **Hospital**
- **ATM** **E:** First Virginia Bank
- **Other** **E:** **Giant Foods, Rite Aide Pharmacy**

6 Shirlington, Quaker Ln
- **Gas** **W:** Exxon, Texaco
- **Food** **W:** Best Buns, Bistro-Bistro, California Pizza Kitchen, Caryle Gran, Charlie Chang, Damon's, Deli Cafe, Honey Baked Ham, St Pepper's, Thai Shirlington
- **AServ** **W:** Exxon, Texaco
- **Med** **E:** ✚ **Hospital**
- **ATM** **W:** Virginia Bank
- **Other** **W:** **Subway, The Village At Shirlington, Village Market Grocery**

5 King St, Falls Church, Alexandria
- **Gas** **W:** Exxon*
- **Food** **E:** Atlanta's Pizza, Bagel Bakery, I Can't Believe Its Not Yogurt, Pastry Shop, Roy Rogers, Starbuck's Coffee, Subway, The Cafe
  W: Copeland's (Parks Ctr)
- **ATM** **E:** Chevy Chase Bank, Crestar Bank
- **Other** **E:** **Giant Supermarket, Post Office, Rite Aide Pharmacy, Tourist Information, United Optical**
  W: **Parks Center**

4 Seminary Rd
- **Gas** **E:** Exxon
- **Food** **E:** Hunan Restaurant, Roma Pizza Deli & Subs, Steak & Ale
  W: Raddison Hotel
- **Lodg** **E:** Ramada Inn
  W: Raddison Hotel
- **Med** **E:** ✚ **Hospital**
- **Other** **E:** **CVS Pharmacy, Safe Way Grocery**

3AB VA 236, Duke Street, Little River Turnpike, Lincolnia

EXIT VIRGINIA

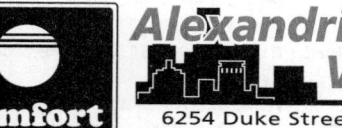

- **Gas** **E:** Extra Fuel*(CW), Exxon*(D), Texaco(D)
  W: Exxon*, Mobil*, Texaco*
- **Food** **E:** Akasaka Japanese Restaurant, Anette's BBQ Haven, Charlie's Chiang Restaurant (Mangolian BBQ), Chesapeake Bagel Bakery, Chi Chi's, Dunkin Donuts, Four Seasons Diner, Friendly's, Good Taste Chinese Restaurant, Ho King, Jerry's Subs & Pizza, Lupita's Mexican Restaurant, McDonalds, Mozarella Cafe, Pappa John's, Red Lobster, Ruby Tuesday, Santa Rosa Seafood, Savio's, Scechuan Express, Subway, Thi Hut Restaurant, Ti Lemon Grass Restaurant, Wendy's
  W: Arby's, Bennigan's, Boston Market, Casa Fiesta Mexican & Salvadorian Food, Einstein Brothers Bagels, **Happy Eatery Buffet & Bakery**, Hee Been Japanese, IHOP, McDonalds, Pamekay's Donuts, Pizza Hut, Roy Rogers, Subway, Szechuan Delight Chinese Carry-Out
- **Lodg** **W:** **Comfort Inn (see our ad on this page)**, DAYS INN Days Inn
- **AServ** **E:** Amoco, Exxon, Goodyear Tire & Auto, Jiffy Lube, Jiffy Lube, K&B Auto Service, Kelly Springfield Tires, NTW Tire Center, Passport Chrysler Plymouth, Sears
  W: Mobil, Super Trak, Texaco
- **ATM** **E:** Crestar Bank
- **Other** **E:** **7-11 Convenience Store, CVS Pharmacy, Coin Operated Laundry, Giant Supermarket & Pharmacy, Landmark Shopping Mall, Pets, Vandorn Safeway Grocery**
  W: **7-11 Convenience Store, 7-11 Convenience Store, For Eyes Optical, Latin African Grocery Store (Check Cashing), Plaza At Landmark Shopping Ctr, Rite Aide Pharmacy**

2A VA 648, Edsall Rd.
- **Gas** **I:** Amoco*, Crown(D)(CW), Exxon
  O: Mobil*
- **Food** **O:** **Denny's**
- **Lodg** **O:** Home-Style Inn
- **AServ** **I:** Amoco, Exxon
- **ATM** **O:** NationsBank

1 Jct I-95, Jct I-495

↑VIRGINIA

Begin I-395

EXIT CALIFORNIA

Begin I-405

↓CALIFORNIA

75 Lancaster

74 JCT I-210

73 Rexford St.

72 Rinaldi Street
- **Gas** **E:** Arco
  W: Shell
- **Lodg** **W:** Grenada Motel

71A San Fernando Mission Blvd, San Fernando
- **Gas** **E:** 76, Arco
- **Food** **E:** Donuts, Subway, Sutters Mill

71B Junction CA118

70 Devonshire Street
- **Gas** **E:** 76, Mobil, Shell, Thrifty
- **Food** **E:** Buon Ristorante, Holiday Burger, Milly's, Safari Room

69 Nordhoff Street
- **Gas** **E:** Arco
  W: Thrifty*
- **Food** **E:** ABC Donuts, China Palace, El Patio
  W: Chips And Salsa
- **Lodg** **E:** Redwood Inn Motel, Sepulvada Motel, Tahiti Motel, Vacation Inn
- **AServ** **E:** Frontier Auto Parts

68 Roscoe Blvd, Panorama City
- **Gas** **E:** 76
  W: Shell, Texaco
- **Food** **E:** Burger King, Carl's Jr Hamburgers, Denny's, Jack-In-The-Box, Shakeys Pizza, Subway, Taco Bell
  W: Coco's, Daily Donuts, Tommy's Hamburgers
- **Lodg** **E:** Holiday Inn Express
  W: Motel 6, Panorama Motel
- **AServ** **E:** Midas Muffler & Brakes

66 Sherman Way, Reseda
- **Gas** **E:** 76, Chevron, Mobil, Shell
  W: 76, Mobil, Texaco
- **Food** **E:** ABC Chinese, Bangkok Express, Foster's Freeze, Peter's Donuts
  W: Beep's Hamburgers, Subway
- **AServ** **E:** X-pert Tune
  W: John's Auto Service
- **Other** **W:** **U.S. Post Office**

65 Victory Blvd, Van Nuys
- **Gas** **E:** 76, Mobil
  W: Arco
- **Food** **E:** Carl's Jr Hamburgers, Fast Taco, Lido Pizza, Subway
- **Med** **E:** ✚ **Hospital**

64 Burbank Blvd
- **Gas** **E:** 76, Chevron, Shell
- **Food** **E:** Carriage Inn Coffee Shop, Happy Family Chinese
- **Lodg** **E:** Carriage Inn Motor Hotel

63 Ventura Blvd, Sherman Oaks, U.S. 101
- **Gas** **W:** 76
- **Food** **E:** Carl's Jr Hamburgers, Sizzler Steak House, Subway
  W: California Chicken Cafe, IHOP, McDonalds, Tony Roma's, Valley Inn
- **Lodg** **W:** Heritage Motel, Radisson, Country Inn (see our ads opposite page), Danish Country Inn (see our ad opposite page), Quality Inn (see our ad opposite page)

61 Mulholland Dr, Skirball Center Dr

Bold red print shows RV & Bus parking available or nearby

EXIT CALIFORNIA

60 Getty Center Dr
58 Moraga Dr
57 Sunset Blvd
56 Montana Ave
55A Wilshire Blvd, Westwood
55B CA2, Santa Monica Blvd
ASERV E: Spee Dee Oil Change
53A Junction I-10
53B National
Gas W: 76
Food E: Bagels, Habit Hamburgers, McDonalds, Pizza Hut
W: McDonalds
Other E: **Thrifty Drugs**

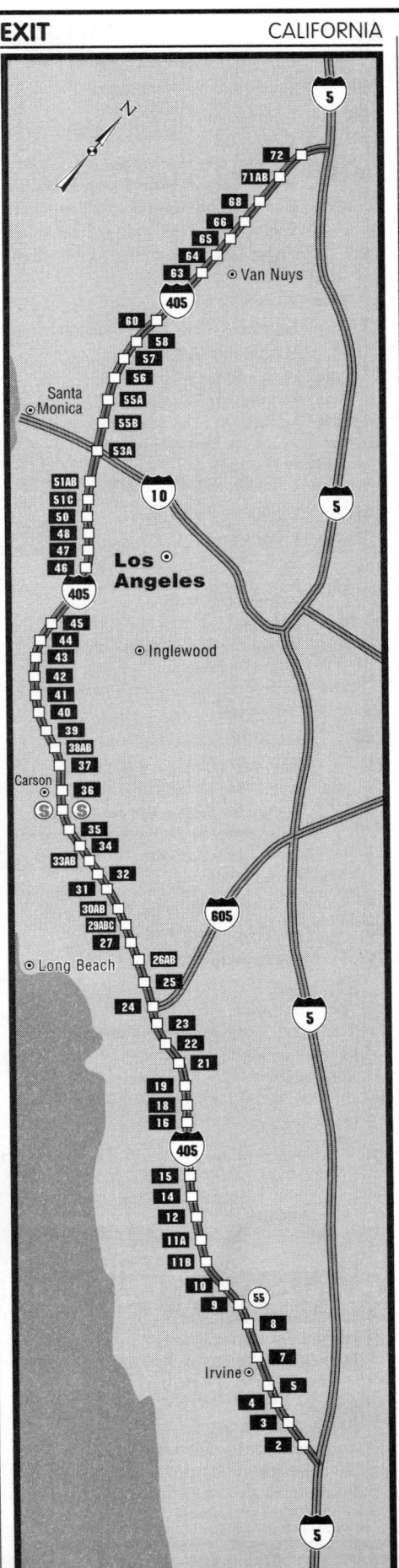

EXIT CALIFORNIA

EXIT CALIFORNIA

51A Washington Blvd, Venice Blvd
Gas E: Chevron
W: Chevron, Mobil
Food E: Metro Diner, Tito's Tacos
W: Baskin Robbins, Campos Donuts, Spudnuts
Lodg E: Travelodge
AServ E: Culver Auto Center, K&S Automotive Repair, Washington Place
51B Culver Blvd, Culver City
Gas W: 76, Thrifty
51C Junction CA90, Jefferson
50 Slauson Ave, Sepulvada
Gas E: Arco, Mobil
W: 76, Mobil
Food E: Del Taco, Royal Chinese, Sizzler Steak House
W: Joseph's Restaurant, Viva Fresh Mexican
Lodg E: Ramada, Windam Garden Hotel
W: Red Lyon Hotel
AServ E: Goodyear Tire & Auto
ATM E: Wells Fargo Bank
Other E: **Fox Hills Plaza**
48 La Tijera Blvd, Loyola Marymount
Gas E: Chevron, Mobil
W: 76, Arco, Chevron
Food E: Baskin Robbins, Burger King, China Town Express, El Pollo Loco, Fat Burger, Hong Kong Express, McDonalds, Numero Uno Pizzaria
W: Buggy Whip Restaurant, Marie Callender's Restaurant
ATM E: Wells Fargo Express
Other E: **Ralph's Grocery**
47 Manchester, La Cienga Blvd, CA42
Gas E: 76
W: Fast Fuel, Shell
Food E: Carl's Jr Hamburgers, Don Amigos, Tam's Jr. Hamburgers, The Green Burrito
W: Arby's, Randy's Donuts, Tinos Pizza
Lodg E: Best Western, Econolodge, Lotus Motel
W: Days Inn, Sand Man Motel
46 Century Blvd, L.A. Airport
Gas E: Mobil
W: Arco, Chevron, Texaco
Food E: China Express, Flower Drum Chinese, Jim's Taco Fiesta, Mellow Burgers, Rally's Hamburgers
W: Denny's, McDonalds, Taco Bell
Lodg E: Best Western, Comfort Inn, Motel 6, Tivoli Hotel
W: Doubletree Hotel, Holiday Inn, Quality Hotel, Travelodge
Med E: **Centela Hospital**
45 Imperial Hwy, Century Blvd
44 El Segundo Blvd, El Segundo
Gas E: Arco, Chevron, Thrifty
Food E: Campos Tacos, China House, Jack-In-The-Box, Los Chorros, Pizza Hut, Taco Bell
W: Chappie's Restaurant, Denny's
Lodg E: El Segundo Inn
W: Ramada
AServ E: Arco
43 Rosecrans Ave, City of Hawthorne
Gas E: 76, Mobil, Shell
W: Thrifty
Food E: Amigo's Donuts, Chicken Madras, Denny's, El Pollo Loco, Falcon Inn, Subway
W: Carl's Jr Hamburgers, Christie's Coffee Shop, Luigi's, Soup Exchange
AServ E: Bay Tires, Shell
42 Inglewood Ave
Gas E: Arco
W: Mobil, Shell, Texaco
Food E: In-N-Out Hamburgers, Mom's Donuts, Taco Bueno

Bold red print shows RV & Bus parking available or nearby

EXIT CALIFORNIA

W: El Gauchos, La Salsa, Leo's Mexican, Manhattan Diner
ASery E: Tune Dynamics
W: Texaco, Texaco

41 CA107, Hawthorne Blvd, Lawndale
Gas W: Arco
Food E: Golden China, Jack-In-The-Box, McDonalds, Spires
W: Subway
Lodg E: Best Western
W: Days Inn (see our ad on this page)
AServ W: Winston Tires

40 CA91, Artesia Blvd, Redondo Beach
Gas W: 76
Food W: Sizzler Steak House
ATM W: Wells Fargo Bank

39 Crenshaw Blvd, Torrance
Gas W: Arco, Mobil, Texaco
Food W: Denny's

38A Western Ave
Gas E: 76, Shell
W: Mobil
Food E: Del Taco, Denny's, Fu Shing, Winchell's Donuts
W: Millie's
AServ W: New Tomato Auto Service
ATM W: Union Bank

38B Normandie Ave, Gardena
Gas W: Texaco
Lodg E: Comfort Inn

37 Junction I-110, San Pedro, Los Angeles

36 Main Street

(35) **Weigh Station (Both Directions)**

35 Avalon Ave N
Gas E: Shell
W: Arco
Food E: Chuck E. Cheese's Pizza, Jack-In-The-Box, Pizza Hut, Sizzler Steak House, Tony Roma's Ribs
Lodg W: Ramada Inn
AServ E: Tire Station
Other E: **Southbay Pavilion**

34 Carson Street, Avalon Blvd S
Gas W: 76
Food E: Rosario's Italian
W: Carl's Jr Hamburgers
Lodg E: Comfort Inn
W: Hilton

33A Wilmington Ave
Gas E: Arco, Chevron
Food E: Carson Burgers
W: Subway, Taco Bell

33B Alameda Street

32 Santa Fe Ave

31 Junction I-710, Pasadena, Long Beach

30A Long Beach Blvd
Gas W: Mobil, Shell

30B Atlantic Ave
Gas E: Arco, Bixby Knolls Carwash, Texaco
Food E: Arby's, Baskin Robbins, Dave's Burgers, Denny's, El Torito, Jack-In-The-Box, Jenny's Donuts, New York Deli, Polly's, Subway, The Green Burrito

EXIT CALIFORNIA

Other E: Car Wash, **Mail Boxes Etc, Western Union**

29A Orange Ave (Southbound)

29B Cherry Ave
Gas E: Mobil, Shell
Food E: Fantastic Charbrioled Burgers
W: Charley Brown's Steaks & Seafood, The Rib
AServ E: A&A Towing & Repair, Advantage Auto Glass, C&F

29C Cherry Ave, Signal Hill (Northbound)
Gas E: Mobil
Food E: John's Hamburgers

27 Lakewood Blvd, CA19, Access To Long Beach Airport
Gas W: Chevron, Shell
Food W: Spires, Taco Bell
Lodg E: Motel Suites
AServ W: Just Tires, Oldmobile Dealer
Med W: **+ Hospital**
Other E: **Sky Link Public Golf Course**

26A Bellflower Blvd, Lakewood
Gas E: Chevron
W: 76, Shell
Food E: Tracy's
W: Burger King, Wendy's
AServ E: Chevron
ATM W: Bank of America, California Federal Bank, Glendale Federal Bank
Other W: **Save-On Drugs, Target Department Store, Thrifty Drugs**

26B Woodruff Ave

25 Palo Verde Ave
Gas W: 76
Food W: China Star, Del Taco, Domino's Pizza, Dr. Yi Donuts, Fish n Chips, Marri's Pizza, Pizzamania, Ryan's, Subway, Taco Bell
AServ W: 76, The Mechanic

24 Junction I-605

23 7th Street, Longbeach, CA22W

22 Los Alamitos, Seal Beach Blvd, Los Alamitos Blvd
Gas E: Chevron
Food E: KFC, Spaghettini, Tortilla Beach
AServ E: Chevron
ATM E: Farmers & Merchants Bank

21 Valleyview Street, Garden Grove, CA22E

19 Westminster Ave

EXIT CALIFORNIA

Gas E: 76, Thrifty
W: Chevron, Shell
Food E: Cafe Westminster, Country Harvest Buffet, Dino's Italian Restaurant, In-N-Out Hamburgers, Jades Chinese, KFC, Taco Bell
W: Casa Gamino Mexican, Chinese Jades, Donut Storr, Hong Kong Express, Pizza Hut, Spaghetti Eddies, Subway, The Fireside
Lodg E: Executive Suites Inn, Motel 6, Travelodge
W: Best Western, Days Inn
AServ E: AJ's Automotive Repair & Service, Chief Auto Parts, Jiffy Lube, Kragan Auto Supply, Tune up & Lube
W: Jiffy Lube
Other E: **Pick & Save,** Westminister Car Wash
W: **Coin Laundry**

18 Goldenwest Street, Bolsa Ave
Gas W: Mobil, Shell
Food E: Super Submarine
W: Coco's, El Torito, IHOP, Jack-In-The-Box, KD's Donuts, Kiku Japanese Restaurant, Togo's Eatery
AServ W: Sears Auto Center
ATM W: Wells Fargo Bank
Other W: **Castle For Brides Wedding Preparation, Westminister Mall**

16 CA39, Beach Blvd, Westminster (Many More Services At This Exit)
Gas E: Mobil
W: 76, Shell
Food E: Hof's Hut, Hong Kong Express Chinese, Little Caesars Pizza, Marie Callender's Restaurant, Mei's Chinese
W: Arby's, El Torito, Jack-In-The-Box, Los Castillos, Macaroni Grill
Lodg E: Super 8 Motel, Westminster Beach West Inn
W: Princess Inn
AServ E: Just Tires, Midas Muffler & Brakes, Toyota Dealership
W: Lynn's Auto Air, Pep Boys Auto Center
ATM E: Wells Fargo
Other E: **K-Mart**

15 Magnolia Street, Warner Ave
Gas E: Texaco
W: Mobil
Food E: Japanese Steakhouse, Jean's Donuts, Pizza Deli
W: Bullwinkles Family Restaurant, Carro's, Jeffery's, Magnolia Cafe, Taste Of India, Tommy's Hamburgers, Winchell's Donuts
Lodg W: Ramada Limited
AServ E: Texaco
Med W: **+ Emergicare Family Medicine, + Medical Care Center**
ATM W: Downey Savings
Other W: **Family Fun Center, Ralph's Grocery, Save-On Pharmacy**

14 Brookhurst Street, Fountain Valley
Gas E: Arco, Thrifty
W: Chevron
Food E: Coco's, Del Taco, Donut City, KFC
W: Big City Bagels, Black Angus Steakhouse, Borrelli Ristarante, Claim Jumper, Funashin Japanese Restaurant, Juice It Up, Pick Up Stix, Pizza & Pasta, Pogo's Eastery, Sports Stop Pizza, Sushi, Thia Cousine, Wendy's

Bold red print shows RV & Bus parking available or nearby

EXIT		CALIFORNIA
	Lodg	E: Courtyard by Marriott, Residence Inn
	AServ	E: Parnelli Jones Express Tire
	Med	W: ✚ **Orange Coast Medical Center**
	ATM	E: Bank of America, Bank of Orange County, Glendale Federal Bank, Wells Fargo W: First Bank & Trust
	Other	W: **Albertson's Grocery, Sams Club, US Post Office**
12		Euclid Street
	Food	E: George's Burgers
	AServ	E: Rancho Auto Wash
11A		Harbor Blvd
	Gas	W: Arco, Chevron, Mobil
	Food	E: Coco's W: Burger King, Denny's, IHOP, Kaplan's Deli/ Bakery, Long John Silvers, McDonalds
	Lodg	E: La Quinta Inn W: The Inn at Costa Mesa, The Vagabond Inn
	AServ	W: Big O Tires, Mobil
11B		Fairview Rd
	Gas	E: Mobil(CW), Shell(CW)
	Food	E: Sam Woo Oriental Seafood
10		Bristol Street, Costa Mesa
	Gas	W: 76, Chevron, Shell
	Food	E: Carl Jr's Hamburgers, Corner Bakery, Maggiano's Little Italy, Scott's Seafood W: Costa Mesa Super Burgers, Del Taco, El Pollo Loco, McDonalds, Mother India
	Lodg	E: The Westin Southcoast Plaza W: Holiday Inn
	AServ	E: Sear's
	Med	E: ✚ **Hospital**
	ATM	E: Wells Fargo
	Other	E: **South Coast Plaza, Southcoast Plaza Mall, Town Shopping Center**
9		CA55, Costa Mesa Fwy, Newport Beach, Riverside
8		MacArthur Blvd, Main Street E, John Wayne Airport
	Gas	W: Chevron, Mobil
	Food	E: Carl's Jr Hamburgers, Chicago Joe's, KooKooMoo's, McDonalds, Schlotzkys Deli, Sensations, Southeast Deli, Spires Burgers W: El Torito, Gullivers, IHOP
	Lodg	W: Atrium Hotel
	ATM	E: Union Bank of CA
7		Jamboree Rd
	Food	E: Chow Mein W: Chris Steakhouse, KooKooRoo
	Lodg	E: Courtyard by Marriott, Hyatt
5		Culver Dr
	Gas	W: Chevron
4		University Dr, Jeffrey Rd
	Gas	W: Mobil
	Food	W: Ben & Jerry's, Donut Star, Hudson's Grill, IHOP, Knots, Pick Up Stix, Pogos Eatery, Taco Bell
3		Sand Canyon Ave
2		CA133S, Laguna Beach
1		Irvine Center
	Other	E: **Wild River Water Park**
0		Jct I-5

↑CALIFORNIA

Begin I-405

EXIT		WASHINGTON

Begin I-405, Washington

↓WASHINGTON

27		I-5 Jct.
26		WA 527, Bothell, Mill Creek
	Gas	W: Texaco*, Texaco
	Food	E: Canyon's, McDonalds, Thairama W: Baskin Robbins, Bruegger's Bagels, Burger King, Godfather's Pizza, KFC, Mitzel's, Mongolian Grill, Papa Murphy's Pizza, Starbuck's Coffee, Taco Bell, Taco Time, Teriyaki Plus
	Lodg	W: Comfort Inn
	AServ	W: Texaco
	RVCamp	W: **Camping**
	Med	E: ✚ **Evergreen Medical Clinic**
	ATM	E: Seafirst Bank W: QFC, U.S. Bank, Wells Fargo
	Other	W: **7-11 Convenience Store, Albertson's Grocery, Bartell Drugs, Canyon Park Dental, Cat Clinic, King Books, Payless Drugs, QFC Grocery (24 Hrs), Vision Clinic**
24		NE. 195th St., Beardslee Blvd.
	Lodg	E: Residence Inn, Wyndham Garden Hotel
23		WA 522E., to Wa. 202, Monroe.
22		NE 160th St.
	Gas	E: Chevron* (24 Hrs), Texaco*(D)
	Food	E: Cafe, Domino's Pizza
20AB		NE 124th St.
	Gas	E: BP*(D)(CW), Chevron*, Texaco* W: Arco*, BP
	Food	E: A & W Drive-In (Texaco), Denny's, Evergreen China Restaurant, Old Country Buffet, Pizza Hut W: Baskin Robbins, Burger King, Hunam Wok Chinese, McDonalds, Mong Thai Restaurant, Taco Del Mar, Taco Time, Wendy's
	Lodg	E: Best Western (Frontage Rd.), Motel 6, Silver Cloud Inn
	AServ	E: Discount Tire Co., Gibson Auto Center, Q Lube W: Al's Auto Supply, BP
	Med	E: ✚ **Hospital**
	ATM	E: Key Bank, Seafirst Bank, Texaco
	Other	E: **Office Max, Payless Drugs** W: **Drug Emporium (Pharmacy)**
18		WA 908, NE 85th St. Kirkland

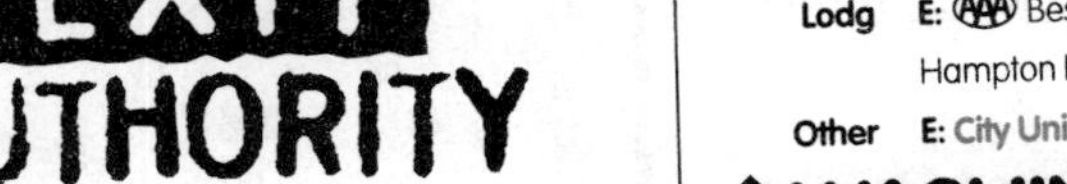

EXIT		WASHINGTON
	Gas	E: Arco*, BP*
	Food	E: Burger King, Domino's Pizza, Outback Steakhouse, Pizza Hut, Teriyaki
	AServ	E: GM Auto Dealership, Honda Dealer, Meineke Discount Mufflers, Pennzoil Lube, Premium Tune & Lube, Speedy
	Med	E: ✚ **Hospital**
	Other	E: **Maytag Laundry, Safe Way Grocery, U-Haul Center(LP)**
17		NE. 70th
14		Redmond, WA 520, Seattle
13AB		NE 8th St, NE 4th St.
	Gas	E: Arco*, Chevron*, Chevron*(CW), Texaco*
	Food	E: Burger King, Chinese American Orders To Go, Denny's, Hunan Garden Chinese Food, Pumphouse, Taste of Tokyo W: Azteca Mexican (Frontage Rd.), Dairy Queen
	Lodg	E: Westcoast Bellevue Hotel W: Bellevue Inn (Frontage Rd.), Doubletree Hotel, Hilton
	AServ	E: A-1 Clutch, Brake, Transmission, Chevron, HC Auto Care, Wagon Shop
	Med	E: ✚ **Hospital** W: ✚ **Hospital**
	ATM	E: Texaco
	Other	E: **Eye Care Center, Kinko's Copies**
12		**(11)** SE 8th St.
11		I-90 Jct., Seattle, Spokane
10		Coal Creek Pkwy., Factoria
9		112th Ave. Southeast, New Castle
7		NE. 44th St.
	Food	E: Denny's, Dino's Beefstro, McDonalds, Subway
	Lodg	E: Traveler's Inn
	Other	E: **I-405 Express MiniMart**
6		NE 30th St
	Gas	E: Arco* W: Chevron*(CW), Texaco*
5		WA 900 East, Park Ave. N., Sunset Blvd. Northeast
4		WA 169 S., Maple Valley, Enumclaw, WA 900 W. (Difficult Reaccess)
	Gas	W: Texaco*(D)
	Food	E: Shari's Restaurant W: Burger King, Yoko's Teriyaki
	Lodg	E: Silver Cloud Inn W: Dona Lisa Motel
	AServ	W: Express Tune
	TServ	W: **International Dealer**
	Other	W: **7-11 Convenience Store, Dollar Mart**
2		WA 167, Auburn, Rainier Ave.
1		WA 181 S Tukwila
	Food	E: Barnaby's, Best Western, Jack-In-The-Box, Taco Bell
	Lodg	E: (AAA) Best Western, (AAA) Embassy Suites, (AAA) Hampton Inn, ◆ Residence Inn
	Other	E: **City University**

↑WASHINGTON

Begin I-405, Washington

Bold red print shows RV & Bus parking available or nearby

EXIT	TEXAS

Begin I-410

↓TEXAS

53 I-35, San Antonio, Laredo
51 Somerset Rd., FM 2790, Somerset
49 TX 16S, Spur 422, Palo Alto Rd., Poteet
- **Food** I: Church's Fried Chicken
- **Med** I: ✚ Hospital
- **Other** I: **Palo Alto College**

48 Zarzamora St.
46 Moursund Blvd.
44 S. US 281, S. Spur 536, Roosevelt Ave., Pleasanton
- **Gas** I: Texaco*(D)
- **AServ** O: U-Save Tire Shop
- **RVCamp** I: **Camping**
- **Other** I: **Stinson Airport**
  O: **Mission San Francisco De La Espada**

42 Spur 122, S. Presa St., Southton Rd
- **FStop** O: **Whiteside's Gas & Diesel**
- **TServ** I: **Truck Palace Paint & Body**
- **RVCamp** O: **Camping**

41 I-37, N US 281, San Antonio, Corpus Christi, Lucian Adams Fwy
39 Spur 117, W. W. White Rd
37 Southcross Blvd, Sulphur Springs Rd.
- **FStop** O: Fina*(D)
- **Med** I: ✚ Hospital

35 Rigsby Ave., Victoria , US 87, Sinclair Ave
- **FStop** I: **Datafleet (Credit Card Sales Only)**
- **Gas** I: Chevron*
  O: Diamond Shamrock*(D), Exxon (24 Hrs)
- **Food** I: Beef & Bourbon Steakhouse, **Bill Miller BBQ**, El Tipico, **Luby's Cafeteria**, Subway, **What A Burger (24 Hrs)**
  O: **Jack-in-the-Box (24 Hrs)**, McDonalds (Playground), Taco Bell
- **Lodg** I: **Days Inn**
- **AServ** I: AAMCO Transmission, Mendez & Son Auto Body, Sielo Action Center, Xpress Lube
  O: Exxon, Ramco Transmission
- **RVCamp** I: **Sielo Action Center**
- **Other** I: Car Wash, **Rigsby Veterinary Clinic**

34 FM 1346, E. Houston St.
33 US 90, I-10, San Antonio, Houston
32 Dietrich Rd.
32A FM 78, Kirby
31 I-35, South US 81, San Antonio, Binz - Englemann Rd
31B Loop 13, W. W. White Rd.
31A FM 78, Kirby
30A Space Center Dr.
30 Binz - Englemann Rd
27 South I-35, South I-410 San Antonio
26B Interchange Pkwy, Perrin Creek Dr
26A South Loop 368, Alamo Heights
- **Food** I: Ada's Seafood, Wang Goong

25B FM 2252, Perrin - Beitel Rd.
- **Gas** O: Chevron(CW) (24 Hrs), Diamond Shamrock*(D)(CW), Mobil
- **Food** I: China Doll Restaurant, Jim's Restaurant
  O: Denny's, Rally's Hamburgers, Schlotzkys Deli, Taco Bell, Taco Cabana (24 Hrs), Wendy's
- **Lodg** O: Comfort Inn
- **AServ** I: All Tune & Lube
  O: Mobil
- **ATM** I: Bank of America, NationsBank, San Antonio Federal Credit Union
  O: Diamond Shamrock
- **Other** I: **Animal Hospital, Galaxy Theatres**
  O: **Academy (Outdoors Store)**

25A Star Crest Dr.
- **Gas** I: Conoco*, Phillips 66

EXIT	TEXAS

- O: Exxon, Texaco*
- **Lodg** O: Family Gardens Suites
- **AServ** I: Autobahn Import & Domestic, Phillips 66
  O: Alamo Toyota, Exxon
- **Med** O: ✚ **Northeast Baptist Hospital**
- **ATM** I: Conoco
  O: Texaco
- **Other** O: **Stopn Go***

24 Harry Wurzbach Rd, Ft. Sam Houston
- **Gas** I: Texaco*
- **Food** I: The BBQ Station
- **AServ** I: Texaco
- **ATM** O: Jefferson State Bank
- **Parks** O: **MacArthur Park (Playgrounds)**
- **Other** I: **Northwood Animal Hospital, Stopn Drive**

23 Nacogdoches Rd.
- **Gas** I: Texaco
  O: Circle K*
- **Food** O: Au Bon Pain (Club Hotel), Baskin Robbins, Bill Miller BBQ, Church's Fried Chicken, Domino's Pizza, Formosa Garden, IHOP, Jack-in-the-Box (24 Hrs), Luby's Cafeteria, Mama's Cafe, Miami International Cafe & Bakery, Pizza Hut, Schlotzkys Deli, Sonic, Subway, Taco Cabana, Veladi Ranch Steakhouse, Wendy's
- **Lodg** O: Club Hotel (DoubleTree)
- **AServ** I: Rod East Volkswagon Audi, Volvo Dealer
  O: Kwik Kar
- **ATM** I: Frost Bank, Guaranty Federal Bank, International Bank of Commerce
  O: Circle K
- **Other** O: **Forest Oaks Animal Clinic, Sun Harvest Grocery**

22 Broadway
- **Gas** I: Diamond Shamrock*(D), Exxon*(CW), Stopn Go*
  O: Texaco*(D) (24 Hrs)
- **Food** I: 410 Diner, Burger King, Chester's Hamburgers, Finkel's Salad Bar, Hsiu Yu Chinese, Jim's, Magic Time Machine Restaurant
  O: Capparelli's Pizza, Empire Grill, McDonalds (Playground), Rooty's All American Eatery, Zito's
- **Lodg** I: Best Western, Residence Inn (Marriott)
  O: Courtyard by Marriott
- **AServ** I: Brake Check, Carquest Auto Parts, Earl Scheib

EXIT	TEXAS

- Paint & Body, Gunn Oldsmobile, Q Lube
  O: Alamo Body & Paint, Prime Wheel, Tire Station
- **ATM** I: Stopn Go, Texas Commerce Bank, Wells Fargo
- **Other** I: **Many Mansions**
  O: **Emergency Pet Clinic, Lutheran Bookstore**

21B US 281 S, Airport Blvd, Wetmore Rd
- **Gas** I: Texaco*(D)(CW) (24 Hrs)
- **Food** I: What A Burger
- **Lodg** I: Days Inn
- **ATM** I: Texaco
- **Other** I: **Pearle Vision, Target Department Store**

21A Jones Maltsberger Rd.
- **Food** I: Pappadeaux Seafood Kitchen, Red Lobster, Texas Land & Cattle Steakhouse
- **Lodg** I: Country Inn (Carlson), Courtyard (Marriott), Fairfield Inn (Marriott)
- **AServ** I: Metro Mitsubishi
- **Other** I: **Wal-Mart (24 Hrs, Pharmacy, Optical, 1-Hr Photo)**

20B McCullough Ave
- **Food** O: Bean Sprout Chinese, Cascabell (DoubleTree Hotel), Jason's Deli, T.G.I. Friday's
- **Lodg** O: DoubleTree Hotel, Holiday Inn Select
- **Other** O: **Office Depot**

20A US 281N, San Pedro Ave.
- **Food** I: Arby's, Bennigans, Church's Fried Chicken, IHOP (24 Hrs), Sea Island Shrimphouse, Taco Bell, Taco Cabana, Teaka Molino
  O: BBQ's Galore, Coffee Beanery, PT's, Souper Salad, Texas Grill & Sports Bar
- **Lodg** O: Hilton
- **AServ** I: K.C. International Centre, Master Care Auto Service (Firestone), North Star Dodge Sales, Sears
  O: Benson Honda, Benson Mazda, North Park Lexus, North Park Lincoln Mercury
- **ATM** O: Smith Barney
- **Other** I: **Central Park Mall, Eyemart Express, Northstar Mall**
  O: **Baptist Bookstore, Barnes & Noble, Pearle Vision, Petco Supplies**

19B FM 1535, Military Highway, Blanco Rd.
- **Gas** I: Exxon*
- **Food** I: Aldino Cucina Italiana, Burger King (Exxon), Casa de Martha's, Charlotte's, Denny's, Hook Line & Sinker, Jim's, Luby's Cafeteria, Wang's

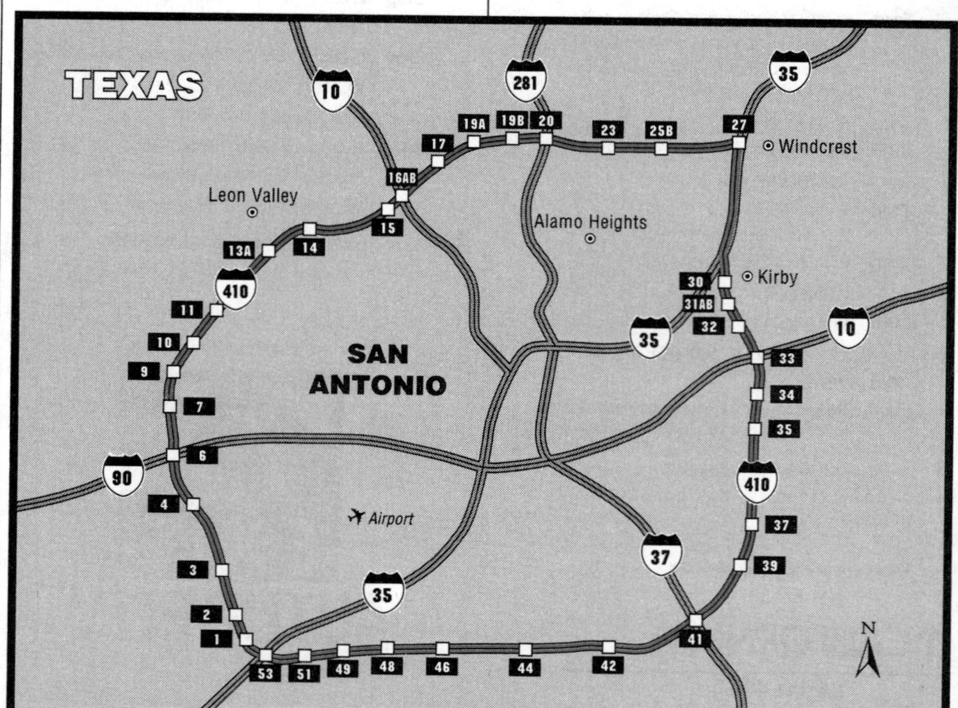

EXIT TEXAS

Garden Chinese
ATM I: Compass Bank, Frost Bank
Other I: **Central Park Mall, West Marine**

19A Honeysuckle Lane, Castle Hills

18 Jackson - Keller Rd, West Ave
Food I: Bill Miller BBQ, Dairy Queen, Little Caesars Pizza, My Luck Chinese, Northloop Molino, Subway, The Cookie Lady
AServ I: Dean Coley Auto Service, Dean Coley Imports, Larry's Super Tune Up, Tire Station
Other I: **H-E-B Grocery (Pharmacy), Mail Center, My $39.95 Optical 1-Hr Lab, Wiseman Animal Hospital**

17 Vance Jackson Rd. (Counterclockwise Exit Is Exit 17 & 17B)
Gas I: Exxon
O: Diamond Shamrock*(D), Mobil, Phillips 66*(CW)
Food O: Burger King, Jack-in-the-Box (24 Hrs), KFC, McDonalds (Playground), Olly's Steaks, Lobster, & Music, Steak & Ale, Taco Bell, Taco Cabana, Teriyaki & More Sushi, Tom's Ribs
Lodg O: Holiday Inn
AServ I: Exxon
O: Carquest Auto Parts, **Discount Tire Company**, Kustom Klean Lube Center(CW), Mobil
ATM I: Fas Stop
O: Diamond Shamrock
Other I: **Fas Stop**
O: **Malibu Fun Center, Target Department Store**

16B West I-10, N. US 87, El Paso, San Angelo (Clockwise Is A Left Exit)

16A East I-10, South US 87, San Antonio, Houston (Counterclockwise Is A Left Exit)

15 Loop 345, Fredericksburg Rd., Balcones Hts
Gas O: Thornton's Food Mart*(D)
Food I: Babe's Old Fashion Food, Bonanza, Jack-in-the-Box, Jim's (24 hrs), Kettle Restaurant (24 Hrs), Luby's Cafeteria, North China Buffet, Ports O' Call, Shoney's, Simi's India Cuisine, Wendy's, What A Burger
O: La Fonda
Lodg I: Motel Travis, Sumner Suites
AServ I: Auto Express (Montgomery Ward), BMW Center, Western Auto
O: **K-Mart**
ATM O: Frost Bank, Guaranty Federal Bank
Other I: **Crossroads Malls, Crossroads Theatres, Fox Photo 1-Hr Lab, Jumbo Sports, Stopn Go**
O: **K-Mart (Pharmacy)**

14 Callaghan Rd, Babcock Rd
Gas I: Chevron*(CW) (24 Hrs), Diamond Shamrock* (24 Hrs), Exxon*(CW), Texaco*
Food I: Catalinas, Zito's Deli
ATM I: Diamond Shamrock, Exxon

14A Summit Parkway
Food O: Burger King (Playground), Chili's, Drake's Restaurant, El Chico, French Quarter Grill, Golden Corral, Landry's Seafood House, Shoney's, Souper Salad, U.R. Cooks Steakhouse, Wendy's
AServ O: **Wal-Mart**
ATM O: Instant Cash (Sam's Club Parking Lot)
Other O: **Half Price Books, Sam's Club, The Home Depot(LP), Wal-Mart (24 Hrs, Pharmacy)**

14B Callaghan Rd
Gas O: Diamond Shamrock*, Phillips 66*
Food O: Ding How Chinese, Pizza Hut, Red Lobster, What A Burger
ATM O: Diamond Shamrock, International Bank of Commerce

14C Babcock Rd.
Food O: China Star, Massimo Ristorante Italiano, Nadler's Bakery, Old World Deli, Subway

EXIT TEXAS

Med O: Hospital
ATM O: Bank of America, International Bank of Commerce
Other O: **Foto First 1-Hr, Oak Hills Veterinary Hospital, Postal Center**

13BB Rolling Ridge Dr.
Food I: Marie Callender's Restaurant
Lodg I: Comfort Inn, Hampton Inn, Travelodge Suites
AServ I: McCombs Hyundai Pontiac GMC Trucks

13B Evers Rd.
Gas O: Conoco*(D) (24 Hrs), Diamond Shamrock* (24 Hrs), Diamond Shamrock*
Food O: Capparelli's Pizza, Gin's Chinese Restaurant, HomeTown Buffet, Jack-in-the-Box, KFC, Las Palapas, Subway, Thai Taste
AServ O: Schumann Auto Repair
ATM O: Diamond Shamrock, Diamond Shamrock
Other O: **Booketeria, Eckerd Drugs, Glen Oaks Laundromat, Kenwick Veterinary Hospital, PETsMART, Quick Wash Coin Laundry**

13A Spur 421, Bandera Rd, Evers Rd, TX 16 N, Leon Vly, San Antonio
Gas I: H-E-B Marketplace
O: Diamond Shamrock*(D), Exxon(D)
Food I: Cha Cha's, EZ's Pizza & Burgers, Outback Steakhouse, The "M" Grill (H-E-B)
O: Bill Miller BBQ, Black-Eyed Pea, IHOP, Jim's Restaurant, McDonalds (Playground), Schlotzkys Deli, Taco Cabana
AServ I: Alamo Body & Paint, Cavender Oldsmobile Dealer, Cavender Toyota, Q-Lube, Saturn of San Antonio
O: Exxon, NTB, Sparks Computerized Car Care
ATM I: Albertson's Grocery
Other I: **Albertson's Grocery (Pharmacy, 24 Hrs), Animal Hospital, Eye Care Eyewear, H-E-B Grocery Store, National Outdoors, Office Depot, OfficeMax, Pear Apple County Fair (The Family Fun Park), Petco Supplies, Target Department Store**
O: **Discovery Zone, Kenwick Animal Hospital**

12 Exchange Parkway
Food O: Applebee's, IHOP, Jason's Deli, Sea Island Shrimp House, Starbucks Coffee (Barnes & Noble), The Black-eyed Pea
Lodg O: Super 8 Motel
AServ O: NTB
ATM O: Security Service Federal Credit Union
Other O: **Barnes & Noble, Discovery Zone, Santikos Bandera 6 (Cinema)**

11 Ingram Rd.
Gas I: Texaco*(D)(CW)
O: Citgo*
Food O: Burger King, Chinatown Restaurant, Long John Silvers, Olive Garden, Tink-A-Tako, Villa Maria Prado's, What A Burger (24 Hrs)

EXIT TEXAS

Lodg I: Econolodge, Red Roof Inn
AServ I: Benson Mazda, Benson Nissan, Ingram Park Chrysler
O: Q-Lube, Sears Auto Center
ATM O: Lackland Federal Credit Union, NationsBank, San Antonio Federal Credit Union
Other O: **Ingram Park Animal Hospital Animal Emergency Clinic, Ingram Park Mall, My $39.95 Optical, One-Hr Photo Lab & Portrait Studio, The Shepherd's Shoppe, Toys "R" Us, Trinity Vision Center**

10 FM 3487, Culebra Rd., St. Mary's Univ.
Gas O: Exxon*(D)(CW), Phillips 66*(CW)
Food I: Bill Miller BBQ, Denny's, McDonalds (Playground), Wendy's
O: Blimpie's Subs, Casa Real, Chuck E. Cheese's, Denny's, Fuddruckers, Ming Garden Patio Cafe
Lodg I: Holiday Inn Express, La Quinta Inn
O: Best Western
AServ O: Firestone Tire & Auto, Gillespie Ford, Slim's Paint & Body, Westloop Mitsubishi
Other I: **Fire Station, Northside Aquatics Center**
O: **Eyemart, Ingram Sq. 8 Santikos (Cinema), Sharp Care Animal Hospital**

9 Military Drive, TX 151, Sea World
Food O: **McDonalds (Wal-Mart)**
AServ I: MAACO Auto Painting
O: **Wal-Mart**
Other O: **Wal-Mart(LP) (Pharmacy, Optical, 1-Hr Photo, 24 Hrs)**

7 Marbach Rd.
Gas I: Exxon*(CW), Texaco*
O: Chevron*(D)(CW), Texaco*(D) (24 Hrs)
Food I: Apetito's, Beijing Express, Church's Fried Chicken, Gilbert's
O: Asia Kitchen & Market, Burger King (Playground), Golden Wok, Jack-in-the-Box, Jim's, KFC, Lily's Bakery, Little Caesars Pizza, Long John Silvers, Luby's Cafeteria, McDonalds (Playground), Pizza Hut, Red Lobster, Sirlion Stockade, Sonic, Subway, Taco Bell, What A Burger (24 Hrs)
AServ I: Pep Boys Auto Center, Sparks Computerized Car Care
O: Discount Tire Co., Econo Lube N Tune, Firestone Tire & Auto, Midas
ATM O: Bank of America
Other I: **Arbach Coin-Op Laundry & Cleaners**
O: **Eckerd Drugs, Fox Photo 1-Hr Lab, H-E-B Grocery Store (1 Hr Photo, 24 Hr Pharmacy), Mail Center USA, Westlakes Mall**

6 US 90, San Antonio, Del Rio
Lodg I: Best Western (see our ad on this page)

4 Valley Hi Drive, Lackland A.F.B.
Gas I: Diamon Shamrock* (Stop-N-Go), Diamond Shamrock*, H-E-B, Mobil*, Phillips 66*(CW)
O: Texaco*(D)(CW)
Food I: Church's Fried Chicken, McDonalds (Playground), Medina County Line Cafe, Pizza Hut
AServ I: Auto Zone Auto Parts
ATM I: Diamond Shamrock, Norwest Banks
Other I: **H-E-B Grocery Store, US Post Office, Washateria Self-Service (Coin-Op Laundry)**

3 Ray Ellison Dr., Medina Base Rd.
Gas O: Diamond Shamrock*
ATM O: Diamond Shamrock
Other O: **Public Library**

2 FM 2536, Old Pearsall Rd.

1A Frontage

1 Frontage Rd.

↑TEXAS

Begin I-410

I-435 →

EXIT	MISSOURI
	Begin I-435
	↓MISSOURI
67	Gregory Blvd
Other	**I:** Swope Park Zoo
66	MO 350, Lee's Summit (Left Exit)
65	Eastwood Frwy
Gas	**I:** Q.T.*
Food	**I:** KFC, McDonalds, Pizza Hut, Subway
Other	**I:** Russell's Car Wash
63C	Ray Town Rd., Stadium Dr., Sports Complet (Northbound)
63AB	I-70 Jct
61	MO 78, 23rd St.
Gas	**E:** Fisca*(D)
60	MO 12, 12th St, Truman Rd.
Gas	**O:** Amoco*(CW), Phillips 66
AServ	**I:** Eastside Auto Parts
	O: Amoco
59	US 24, Winner Rd., Independence Ave
Gas	**O:** Q.T.*
ATM	**O:** QT
57	**(56)** Front St.
TStop	**O:** Conoco* (Open 24 Hours)
FStop	**O:** Q.T.*(D) (Open 24 Hours)
Gas	**I:** Gas Station*(D)(CW), Phillips 66*(D)(CW)
Food	**I:** KFC, McDonalds, Pizza Hut, Shoney's, Taco Bell, Waffle House, Wendy's
	O: Burger King
Lodg	**I:** Hampton Inn, Park Place Hotel
TServ	**O:** Goodyear Tire & Auto, International Truck Service, Midwest Kenworth
TWash	**O:** TNT Truck Wash
Med	**I:** ✚ Compcare Clinics
ATM	**I:** UMB Bank
	O: Quick Trip
55	MO. 210, Richmond N. Kansas City (Difficult Reaccess)
Gas	**O:** King Super Store*(D)(LP)
Lodg	**O:** Red Roof Inn
TServ	**O:** Midway Truck Center
54	48th St. Parvin Rd
Gas	**I:** QuikTrip*
Food	**I:** Ponderosa
Lodg	**I:** ◆ Best Western, ◆ Holiday Inn, Super 8 Motel
AServ	**O:** Freightliner Dealer
52	I-35 S Jct, U.S. 69, Kansas City Claycomo
51	Shoal Creek Dr.
49AB	MO. 152
47	NE 96th St.
46	NE 108th St.
45	**(43)** MO. 291, to I-35 N.
42	N. Woodland Ave
41AB	U.S. 169, Riverside, Smithville

EXIT	MISSOURI
40	NW Cookingham Dr
37	MO C, Skyview Ave.
36	**(35)** NW Cookingham Dr.
Gas	**O:** Citgo, Total*
Lodg	**I:** Best Western, Club House Inn, Hampton Inn, Holiday Inn
14	I-435 E. Split
15	Mexico City Ave.
31	I-29 Jct U.S. 71 ST. Josephs Kansas City
29	**(25)** CR D NW 120th ST
24	CR N MO 152 NW Barry Rd.
22	**(19)** MO. 45 Weston Parkville
18	**(16)** State Rd 5 Wolcott Dr
15AB	KS 5, Leavenworth Rd
14AB	Parallel Pkwy
Med	**I:** ✚ Hospital
13B	US 24, US 40, US 70, State Ave. West
13A	U.S 24, U.S. 40 State Ave
12AB	I-70 Jct, KS Turnpike

EXIT	MISSOURI
11	Kansas Ave
9	Rt.32 Kansas City, Edwardsville, Bonnersprings
8B	Woodend Rd.
8A	Holliday Dr.
6C	Johnson Dr.
Other	**I:** Old Shawnee Town
6A	Shawney Mission Pkwy (Difficult Reaccess)
5	**(4)** Midland Dr.
Gas	**I:** Circle K, Texaco*(CW)
Food	**I:** Blimpie's Subs (Texaco), Donuts, Mexicalli Alley, Par T Golf Grill, Pizza Stop, Wendy's
3	87th St
Food	**I:** McDonalds
Other	**I:** K-Mart
2	**(1)** 95th St.
1B	KS 10 JCT, Lawrence (Difficult Reaccess)
1A	Lackman Rd (Difficult Reaccess)

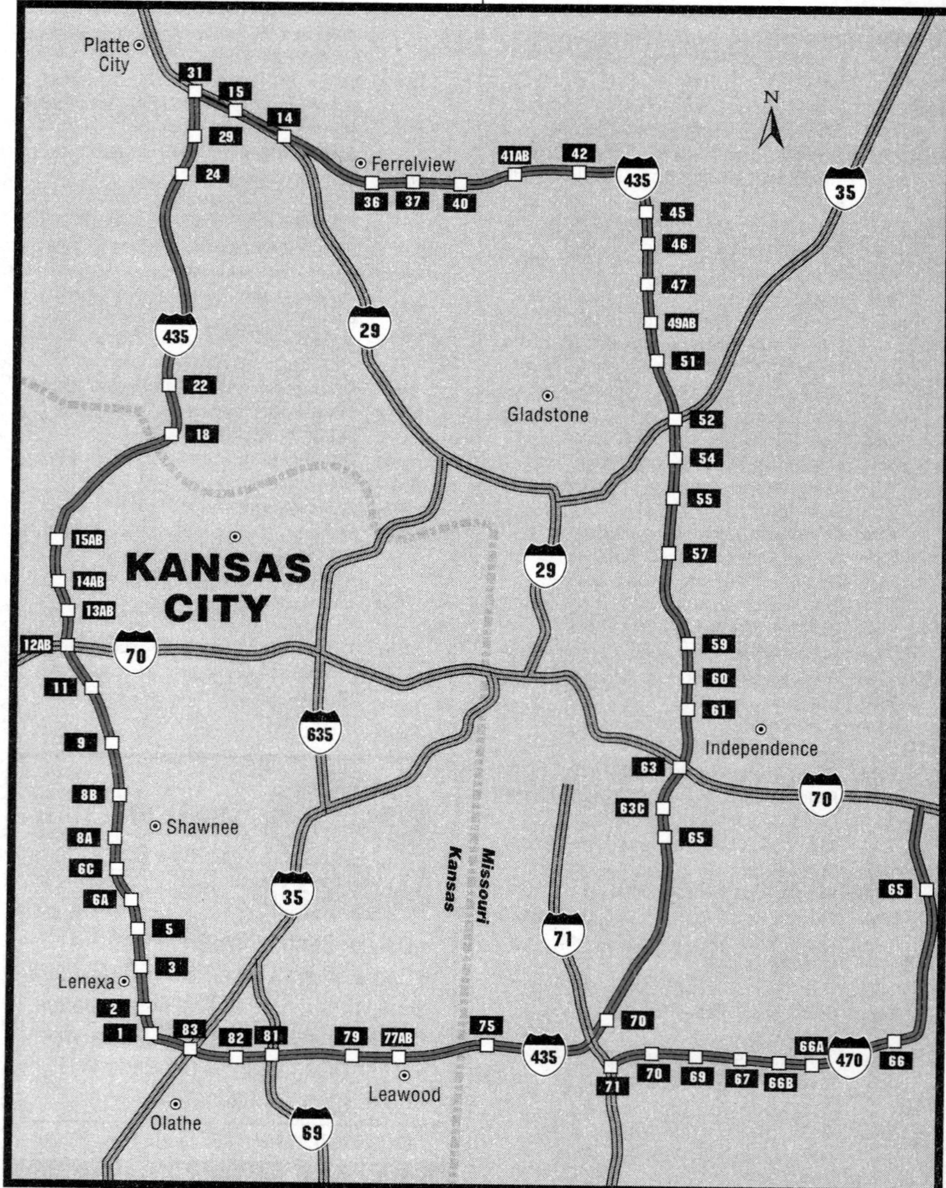

Bold red print shows RV & Bus parking available or nearby

EXIT KANSAS/MISSOURI

↑MISSOURI

↓KANSAS

83 I-35 Jct Witichia, DeMoine
82 Quivira Rd
Food O: McDonalds
AServ O: Instant Oil Change
Med I: ✚ **Overland Park Regional Medical Hospital**
81 **(80)** U.S. 69, Fort Scott
Gas I: Texaco*
Food I: McDonalds, Taco Bell
79 US 169, Metcalf Ave.
Gas I: Amoco(CW), Texaco(LP)
Food I: Chi-Chi's Mexican, Chuck E. Cheese's Pizza, Denny's, Hardee's
O: American Bandstand, Applebee's,

McDonalds, St Louis Bread Company, Tippin's
Lodg I: Atrium by Holiday Inn, Club House Inn, Embassy Suites, Holiday Inn Express, (AAA) Wyndham Garden Hotel, Hampton Inn (see our ad on this page), AmeriSuites (see our ad on this page)
O: ◆ Drury Inn, ◆ Marriott
AServ I: Amoco, Texaco
ATM I: Amoco, Auto World Tire
O: UMB Bank
77AB Roe Ave, Nall Ave
Gas I: Amoco(CW)(LP), Texaco*
Food I: A & W Drive-In (Texaco), KFC (Texaco), Mr. Goodcents Subs & Pasta, Winstead's
O: Wendy's
AServ I: Amoco
Med O: ✚ **Foxhill Medical Building**
ATM O: UMB Bank

↑KANSAS

↓MISSOURI

75B State Line Rd
Gas I: Conoco*
O: Amoco(CW)
Food I: Applebee's, Blimpie's Subs, Chopstix, Fuzzy's, Gates Bar B Que, Guadalajara Cafe, McDonalds

EXIT MISSOURI

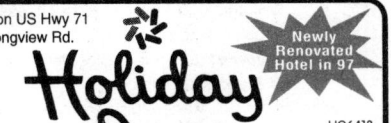

(Outdoor Playground), Waid's Restaurant, Wendy's
O: EBT Restaurant (UMB Bank)
AServ I: Goodyear Tire & Auto, Jiffy Lube, Midas Muffler & Brakes
O: Amoco
Med O: ✚ **Hospital**
ATM I: NationsBank
O: UMB Bank
Other I: **Drug Emporium, Price Shopper Food Store**
75A Wornall Rd
Gas I: Amoco(CW), Texaco*
Food I: Sub Station, Taco Bell
AServ I: Amoco
Med O: ✚ **Hospital**
74 Holmes Rd
Gas I: Gas Station*
Food I: Andy's Wok Chinese, Domino's Pizza, Fin and Pete's Grill, Gomer's Chicken, Patnikios Mexican Restaurant, Subway
ATM I: Mercantile Bank
O: Mark Twaine Bank
Other I: **24 Hour Laundromat**
73 103rd St (No Westbound Reaccess)
71 I-470 Jct, US 50, US 71
Lodg I: **Holiday Inn (see our ad on this page)**
70 Bannister Rd
Gas O: Q.T.* (Open 24 Hours)
Food I: Long John Silvers, Old Country Buffet, Taco Bell
O: Bennigan's, Burger King, McDonalds, Red Lobster
AServ I: Firestone Tire & Auto
ATM O: Commerce Bank, Hillcrest Bank, Midwest United Credit Union
Other I: **Venture Store**
O: **K-Mart (Pharmacy), Mall, Michael's**
69 **(68)** 87th St.
Gas I: Amoco*(CW)
O: Conoco*, Total
Food O: Benjamin Ranch Cafe, Darryl's, Denny's, Mr. Goodsence Subs and Pasta, Tippin's
Lodg I: ◆ Budgetel Inn
O: (AAA) Benjamin Hotel and Suites, (AAA) Motel 6
Other O: **America's Contact and Eye Glasses, Bannister Mall (Dept Store), Best Buy, Family Bookstores, Wal-Mart Supercenter (Open 24 Hours)**
66B Blue Parkway
66A 63rd St. Raytown (Northbound)

↑MISSOURI

Begin I-435

EXIT ALABAMA

Begin I-459

↓ALABAMA

33A I-59S, Birmingham
33B I-59N, Gadsden
32 US11, Trussville
Gas E: Chevron*(D)(CW)(LP), Exxon*, RacTrac*
Food E: Jack's, Waffle House
AServ E: Courtesy Pontiac, GMC Trucks
31 Derby Parkway
29 I-20, Birmingham, Atlanta
27 Grant Mill Road
23 Liberty Parkway
19 US280, Mountain Brook, Childersburg
Food E: Morrison Cafeteria, Ruby Tuesday
Lodg E: Hampton Inn (see our ad this page)
Other E: **Ralph & Kocoo Seafood, The Shops of Colonade**
17 Acton Road
Gas W: Texaco*(D)(CW)(LP)
Food W: Krystal, McDonalds
Other W: **AAA Auto Club**
15 I65, Montgomery, Birmingham
13 US31, Hoover, Pelham
Food E: Fuddrucker's, Grady's Grill, Manny's NY Bagels, McDonalds, The Olive Garden
W: Ali Baba Persian Food, Guadalajara
Lodg E: Winfrey Hotel
AServ E: Kmart
Other E: **Galleria Mall, Information Center, K-Mart**
W: **Hoover Public Library**
10 AL150, Hoover, Bessemer
Gas E: BP*(CW)
Other E: **The Met Vet**
6 Helana, Bessemer
Gas E: BP*(LP), Chevron*(CW)(LP), Texaco*(D)(LP)
Food E: Johnny Ray's BBQ, McDonalds, Pizza Hut, Waffle House, Wendy's
Lodg E: Sleep Inn
RVCamp E: **Cherokee Beach Camper Village**
Other E: **Revco, Winn Dixie Supermarket**
1 Bessemer, McCalla
Gas E: Amoco*(D)
Parks E: **Tannerhill State Park**
Other E: **Animal Hospital**
0 Jct. I-59, I-20

↑ALABAMA

Begin I-459

EXIT INDIANA

Begin I-465

↓INDIANA

53 Junction I-65

52 Emerson Ave., Beech Grove

Gas **I:** Amoco, Marathon*, Shell*(CW)
O: Citgo*, Shell*(CW), Speedway*(CW)

Food **I:** Burger King, Domino's Pizza, KFC, Taco Bell
O: Arby's, Bamboo House, Dairy Queen, Egg Roll, Fazoli's Italian Food, Hardee's, Holiday Inn, Hunan House, McDonalds, Mi Amigos Mexican Restaurant, Pizza Hut, Ponderosa, Ramada, Steak & Shake, Subway, Sunshine Cafe, White Castle Restaurant

Lodg **I:** AAA Quality Inn & Suites
O: ◆ Holiday Inn, ◆ Ramada, ◆ Super 8 Motel

AServ **I:** Marathon
O: Auto Zone Auto Parts, Brooks Auto Care, Goodyear Tire & Auto, Indy Lube, K-Mart, Quaker State

Med **I:** ✚ **Hospital**

ATM **I:** Union Federal Savings Bank
O: Fifth Third Bank, Marsh Supermarket

Other **I:** Automatic & Coin Car Wash, Grime Stopper Car Wash
O: Automatic Car Wash, **Beech Grove Animal Hospital, K-Mart, Kroger Supermarket (Pharmacy), Marsh Grocery Store (Pharmacy), Osco Drugs (Marsh Supermarket), WalGreens**

49AB Junction I-74 East, U.S. 421 South

47 U.S. 52, Brookville Rd., New Palestine

Gas **O:** Shell*

Food **O:** Burger King (Shell)

Med **I:** ✚ **Hospital**

46 U.S. 40, Washington St.

Gas **I:** Amoco, Thornton's*
O: Marathon, Shell(CW)

Food **I:** Al Green's Famous Food, Applebee's, Bob Evans Restaurant, Burger King, Chi-Chi's Mexican Restaurant, Dan Pablo's, Fazoli's Italian Food, Hardee's, Mark Pi's Express, McDonalds, Pizza Hut, Steak 'N Shake
O: Arby's, China Buffet, Chuck E. Cheese's Pizza, Flakey Jake's Burgers, Grindstone Charley's, Laughners Cafeterria, Old Country Buffet, Olive Garden, Shoney's, Skyline Chili, Steak & Shake, Subway

Lodg **I:** Signature Inn

AServ **I:** Amoco, Car Quest Auto Center, K-Mart, Pep Boys Auto Center, Wood Pontiac/GMC/Mazda
O: Bandy's Auto, Q-Lube, USA Muffler & Brakes

Med **I:** ✚ **Hospital**

ATM **I:** First of America Bank (Cub Foods)

Other **I:** **Builders Square, Cub Foods, East Gate Consumer Mall, K-Mart (Pharmacy),** Mike's Express Car Wash, **Service Merchandise**
O: **Animal Medical Center, Osco Drugs**

44 I-70 Junction, West to Indianapolis, East to Columbus

42 U.S. 36 East, IN 67 North, Pendleton Pike

Gas **I:** Clark, Crystal Flash Convenience Store, Speedway*(LP)
O: Amoco*(CW)

Food **I:** Arby's, Baskin Robbins, Denny's, Foon Ying Restaurant, Heidelberg Cafe, LJS, Little Caesars Pizza (K-Mart), Los Tapatios, McDonalds, Rally's Hamburgers, Rayzr's, Subway, Taco Bell, Tony's Char & Grill, White Castle Restaurant

EXIT INDIANA

O: Burger King, Domino's Pizza, Hardee's, Papa's Pancake, Pizza Hut

Lodg **I:** American Inn Motel, Motor 8 Inn, USA Inn
O: Econolodge, Ramada

AServ **I:** Bandy's, Meineke Discount Mufflers, Precision Tune & Lube
O: Indiana Car & Truck, Quaker State Q-Lube, Tom Wood East

Med **I:** ✚ **Hospital**

Other **I:** **IGA Store, K-Mart (Pharmacy), Mail Boxes Etc, Pet Supplies Plus**

40 Shadeland Ave., 56th St.

Gas **O:** Marathon

AServ **O:** Marathon

ATM **O:** Finance Center

37A IN 37 South, Indianapolis

37B IN 37 North, I-69, Fort Wayne

35 Allisonville Rd.

Gas **I:** Shell*(LP), Speedway

Food **I:** 24 Hour Restaurant, Bob Evans Restaurant, Great Wall Chinese, Hughies Subs, Salads, & Pizza, Perkins Family Restaurant, Subway, White Castle Restaurant

EXIT INDIANA

O: BW-3 Grill & Pub, Bravo Italian, Hardee's, MCL Cafeteria, Max & Erma's, Outback Steakhouse

Lodg **I:** AAA Signature Inn
O: ◆ Courtyard by Marriott

AServ **I:** Precision Tune & Lube, Q-Lube
O: Car Quest Auto Center

ATM **I:** Union Federal Savings Bank
O: NBD Bank (Cub Foods)

Other **I:** **Allisonville Animal Hospital, Kroger Supermarket (Pharmacy),** Mike's Express Car Wash
O: **Cub Foods (Pharmacy), PetsMart**

33 IN 431, Keystone Ave.

Gas **O:** Amoco(CW), Marathon*(D), Shell*

Food **I:** California Pizza Kitchen, Cooker, Keystone Grill
O: Arby's, Bob Evans Restaurant, Burger King, Luca Pizza, Ruth's & Chris Steakhouse, Steak & Ale, Subway

Lodg **I:** AAA AmeriSuites (see our ad on this page), AAA Radisson
O: ◆ Knight's Inn

AServ **O:** Butler Hyundia, Butler Toyota, Dan Young Chevrolet/GM, Ed Martin Pontiac GMC Acura, Nissan Dealership, Tom Wood Ford

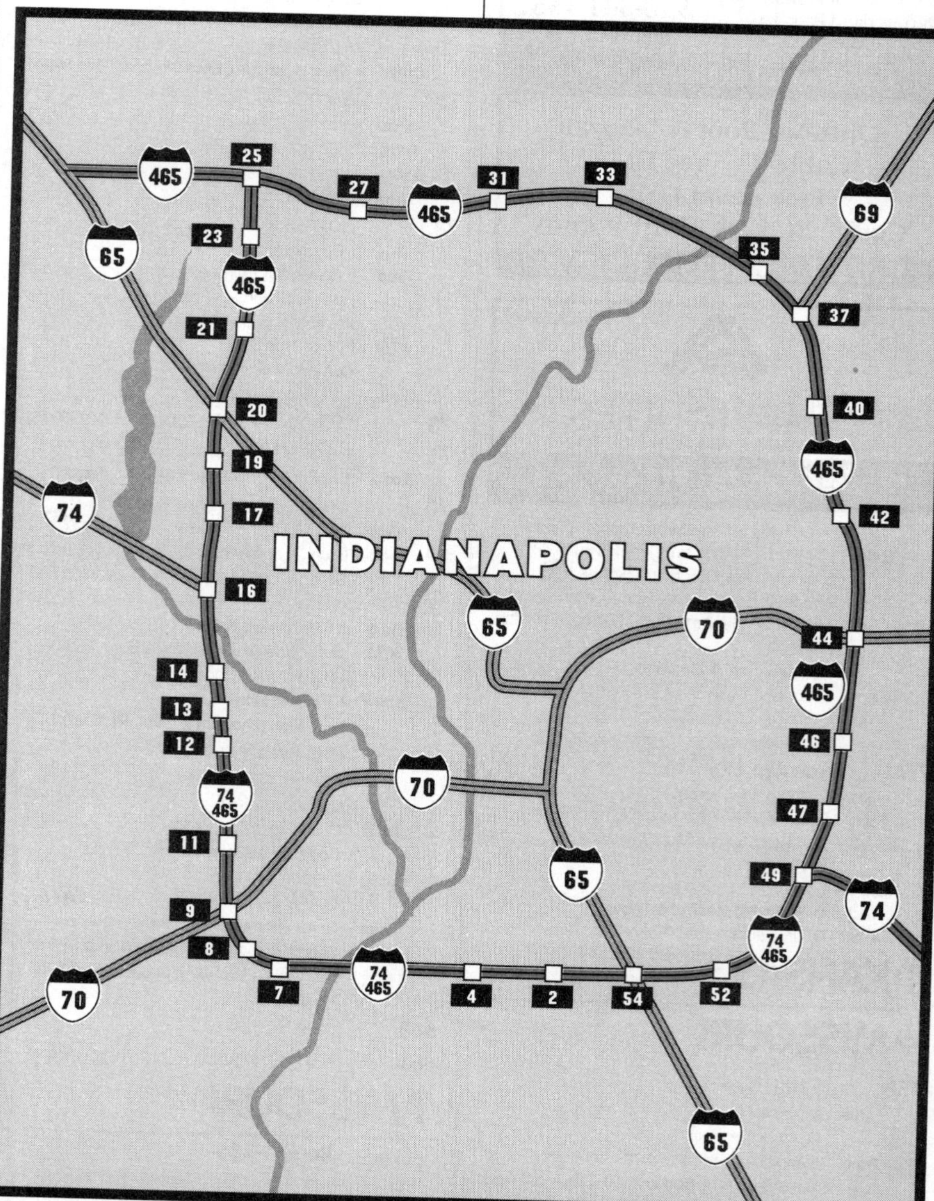

Bold red print shows RV & Bus parking available or nearby

EXIT INDIANA

ATM O: Indy Bank, National City Bank, Shell
Other I: The Fashion Mall
O: Castle Clean Car Wash

31 U.S. 31, Meridian St.
Gas I: Marathon, Shell*(D)
Food I: Deli Jack's, Kona Jack's Fish Market, McDonalds
Lodg O: ◆ Courtyard by Marriott, ◆ Wyndham Garden Hotel
AServ I: Marathon
O: Tutwiler Cadillac
Med I: Hospital
ATM I: Credit Union
Other I: **Kroger Supermarket, Revco Drugs (Kroger)**

27 U.S. 421 North, Michigan Rd.
Gas I: Amoco*, Marathon*, Shell*(CW)
Food I: Arby's, Arnis, Blimpie's Subs, Bob Evans Restaurant, Burger King, Chi Chi's Mexican, Dante's Pizza, Daruma Japanese Restaurant, Denny's, Einstein Brothers Bagels, Embassy Suites, Fortune House Chinese, Holiday Inn Select, IHOP, Little Caesars Pizza (K-Mart), McDonalds, Mi Amigo's Mexican Restaurant, Moscow Nights, O' Charley's, Outback Steakhouse, Pizza Hut, Rally's Hamburgers, Ruby Tuesday, Steak 'N Shake, Subway, Wendy's, Wild Cat Brewing Co.
O: Dairy Queen, McDonalds
Lodg I: ◆ Comfort Inn, Dollar Inn, ◆ Drury Inn, ◆ Embassy Suites, ◆ Fairfield Inn, Holiday Inn Select, New England Suites, AAA Quality Inn, ◆ Residence Inn, AAA Signature Inn, Studio Plus Motel
O: ◆ Red Roof Inn
AServ I: Bridgestone Tire & Auto, Car X, Goodyear Tire & Auto, Indy Lube Express, K-Mart, Marathon, Meineke Discount Mufflers, Midas Muffler & Brakes, Precision Tune & Lube, Wal-Mart
O: Bill Estes Chevrolet Dealership, NAPA Auto Parts
Med I: Hospital
ATM I: Bank One
Other I: Car Wash, **K-Mart (Pharmacy)**, Mike's Express Car Wash, **Office Max, PetsMart, Revco Drugs, Sam's Club, Wal-Mart (Pharmacy)**

25 I-65, Chicago, US 52 W (Left Lane Exit)

24 I-465 West

23 86th St.
Gas I: Shell*(CW), Speedway*
Food I: Arby's, Burger King, Charcoal Steakhouse, Claude & Annie's, Derby Grill, Smiley's Pancake & Steakhouse
Lodg I: Main Stay Suites, Suburban Lodge
Med I: Hospital
ATM I: Bank One, National City Bank

21 71st St.
Gas I: Amoco*(CW)
Food I: DK Brunchies Deli, Galahad's Cafe, **McDonalds**, Steak 'N Shake, Subway
Lodg I: Clarion Inn, ◆ Hampton Inn
ATM I: First of America

20 I-65 North Junction, Chicago

EXIT INDIANA

19 56th St. (No Northbound Reaccess)
Gas I: Speedway*
Med I: Hospital

17 38th St.
Gas I: Amoco*(CW), Marathon*(CW), Shell*(CW)
Food I: Bettie's Seafood, Chi Chi's Mexican, Dairy Queen, Dunkin Donuts (Amoco), Just Pizza, Little Caesars Pizza, New Hoss Pizza & Deli, Pizza Hut, Red Lobster, Shoney's, Steak 'N Shake, Subway, The Olive Garden, The Sports Page Bar & Grill, Thirty- Seventh Bar & Grill, Thoroughbred's Steakhouse
O: Burger King, Chili's, **Cracker Barrel**, Don Pablo's Mexican, Harvest Rotisserie Chicken, McDonalds, Mountain Jack's Restaurant, Ruby Tuesday, T.G.I. Friday's, Taco Bell
Lodg I: Days Inn
O: ◆ Country Hearth Inn, AAA Signature Inn
AServ I: Marathon
Med I: Hospital
ATM I: Huntington Bank
O: Bank One, Union Federal Savings (Marsh Supermarket)
Other I: Car Wash, **Coin Laundry, K-Mart, Osco Drugs, Revco Drugs**
O: **Marsh Supermarket & Pharmacy**

16A U.S. 136, Speedway, Clermont
Gas I: Big Foot*, Shell*(CW)
Food I: Arby's, Burger King, Charlie Browns Pancake & Steakhouse, Denny's, Dog n Suds, Long John Silvers, McDonalds, Sirloin Grill, Taco Bell, Wendy's, White Castle Restaurant
Lodg I: Dollar Inn, Howard Johnson, Motel 6, Red Roof Inn
AServ I: Goodyear Tire & Auto, Jiffy Lube
ATM I: Bank One, Big Foot
Other I: **Kroger Supermarket, Marsh Grocery Store, PetsMart, Village Pantry Grocery**

16B I-74 West, Peoria, IL

14AB Tenth St.
Gas O: Gas America, Shell*
Food I: Pizza Hut Delivery, Wendy's
O: Subway
Med I: Hospital
ATM I: NBD Bank
O: First Indiana Bank, Gas America
Other I: **Cub Foods (Pharmacy), Lowe's**

13 U.S. 36, Danville, Rockville Rd.
Gas I: Shell*
O: Speedway*
Food I: Time To Eat Grill
O: Bob Evans Restaurant, Fabulous 50's Restaurant
Lodg I: Comfort Inn
AServ O: Bill's Auto Service, Indy Lube
ATM O: Union Federal
Other I: **Convenient Food Mart, Sam's Club**

12A U.S. 40 East, Washington St., Plainfield
Gas I: Amoco*
O: Big Foot*, Phillips 66, Shell*(CW)
Food I: Fazoli's Italian Food, Taco Bell, White Castle Restaurant
O: Airport Deli, Arby's, Burger King, Cambridge Inn Cafeterria, Dunkin Donuts, Hardee's, KFC, Little Caesars Pizza (K-Mart), Long John Silvers, McDonalds, Noble Romans Pizza, Number One Wok, Omelet Shoppe, Pizza Hut, Shoney's, Smiley's Pancake & Steak, Steak & Shake, Subway, Wendy's, Western Star, Yogurt

EXIT INDIANA

Lodg O: Dollar Inn
AServ I: Speedway Auto Parts
O: Bandy's, Burt Nees Tires, Car X Muffler & Brakes, Goodyear Tire & Auto, Midas Muffler & Brakes, Q-Lube, Tire Barn
ATM I: Amoco, Bank One
O: Big Foot, National City Bank
Other I: **Central Ace Hardware, Coin Laundry, Osco Drugs, U-Haul Center(LP), Village Pantry, WalGreens**
O: Classy Car Wash, **Davis Laundromat, K-Mart (Pharmacy)**, Smiley's Car Wash, **Venture Supermarket**

11 Airport Expressway
Gas I: Marathon*
O: Speedway*
Food I: Denny's
O: JoJo's Restaurant, The Library Steakhouse & Pub, Waffle Steak
Lodg I: Adam's Mark Hotel, ◆ Courtyard by Marriott, Hampton Inn, Motel 6
O: Fairfield Inn, La Quinta Inn, Residence Inn
AServ I: Marathon

9AB I-70

8 IN 67 South, Kentucky Ave.
Gas O: Amoco*, Shell*(CW), Speedway*(D), Swifty
Food O: Burger King (Shell), Dairy Queen, Hardee's, KFC, McDonalds (Amoco), Subway (Shell)
Med I: Hospital
ATM O: NBD Bank
Other O: **The Cleaning Shop Coin Laundry**

7 Mann Rd. (Westbound, Reaccess Estbound)

4 IN 37, Harding St.
TStop I: **Pilot Travel Center*(SCALES) (Kerosene)**
O: **Flying J Travel Plaza**
Gas O: Shell*(CW)
Food I: **Bender's Restaurant (Econolodge), Omelet Shoppe, Restaurant (Pilot), Wendy's (Pilot)**
O: **Hardee's, McDonalds**, Thads (Flying J Travel Plaza), **Waffle & Steak**, White Castle Restaurant
Lodg I: **Dollar Inn, Econolodge, AAA Super 8 Motel**
O: **Knight's Inn**
AServ O: Michelin Tire Service
TServ I: **Peterbilt Dealer**
O: **Flying J Travel Plaza, Freightliner Dealer, J & E Tire, Speedco Thirty Minute Truck Lube**
TWash O: **Flying J Travel Plaza**
Med I: Hospital
Other O: **Ferrell Gas(LP)**

2AB U.S. 31, IN 37, Indianapolis, East St.
Gas I: Amoco*(CW), Speedway*
O: Big Foot*, Shell*(CW), Speedway
Food I: Baskin Robbins, Dairy Queen, Dutch Oven, KFC, Laughner's Cafeteria, Noble Roman's Pizza, Old Country Buffet, Pizza Hut, Steak & Ale
O: Bob Evans Restaurant, Denny's, Heritage Smorgasbord, McDonalds, Red Lobster, Shoney's
Lodg O: Comfort Inn, Days Inn, Economy Inn, Quality Inn
AServ I: Auto Tire Car Care Centers, Ford Dealership, Goodyear Tire & Auto, Jeep Eagle Dealer, Quaker State Lube
O: Indy Lube
ATM I: Amoco, Fifth Third Bank, National City Bank
Other I: **Coin Laundry, Kroger Supermarket, Office Depot, Revco Drugs, U-Haul Center(LP)**

↑INDIANA

Begin I-465

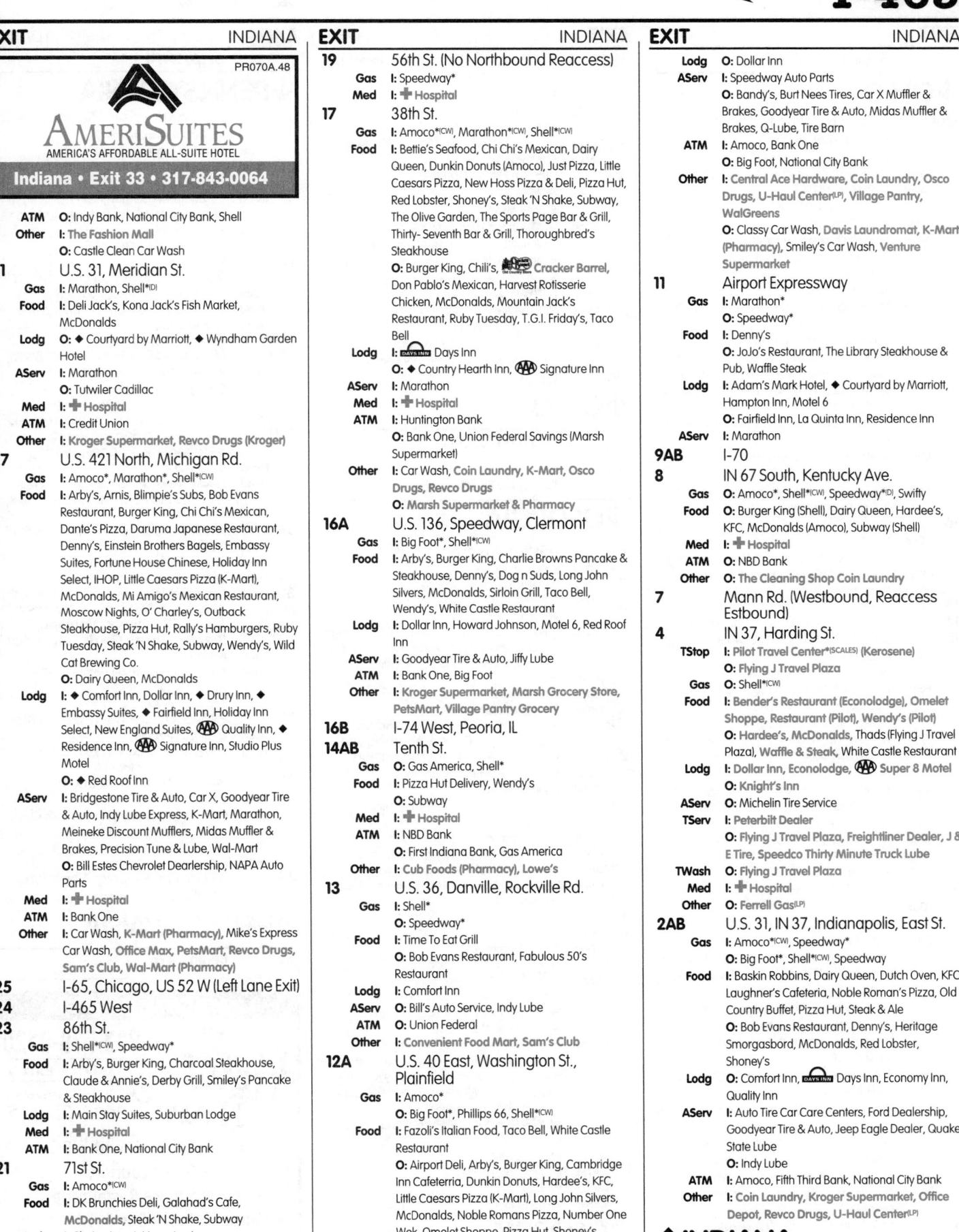

EXIT OHIO

Begin I-475, Ohio

↓OHIO

(20) JCT I-75 N Detroit, I-75 S Daton
19 Central Ave
- **Gas** S: Amoco*
- **AServ** N: Throne Auto Service
- **Med** S: ✚ Hospital

18A OH 51, Monroe St.
- **Gas** S: 76*, Clark*
- **Food** N: Jojo's Original Pizzeria, Wendy's
  S: Chuck's Hungry i, Monroe Street Diner

17 Secor Rd.
- **Gas** N: Amoco*, Clark*
- **Food** N: Betsy Ross Restaurant, Boston Market, KFC, Monroe Street Diner
  S: Al Smith's Place, Big Boy, Breakfast at Marie, China Express, Chinese Restaurant, Claudia's, Denny's, Long John Silvers, McDonalds, Meal Time Express, Mr. Philly, Papa's Pizza, Pizza Hut, Wendy's
- **Lodg** S: (AAA) Clarion Hotel, (AAA) Comfort Inn, ◆ Red Roof Inn
- **AServ** N: Jiffy Lube, Toledo Radiator Co.
  S: Sear's, Valvoline Quick Lube Center
- **Other** N: **Kroger Supermarket (Pharmacy, Open 24 Hours)**
  S: **Town Plus Supermarket (Pharmacy)**

16 Talmitch Rd.
- **Gas** S: BP*(LP), Speedway*
- **Food** S: Charcoal House, Chi Chi's Mexican Restaurant
- **Other** S: **Franklin Park Mall, Orchard Drugs, Pet Supplies Plus**

15 Corey Rd. (Westbound Exit)
14 US 23 N, Sylvania (Left Lane Exit)
13 US 20, OH 120, Central Ave
- **Gas** E: Speedway*
  W: Amoco*, BP*(CW), Speedway*
- **Food** E: Big Boy, Bob Evans, Burger King, Friendly's, Magic Rock, McDonalds, Rally's Hamburgers, Subway, The Greek Restaurant
  W: Alexander's Pizza, Christopher's Grill, Fortune Inn Chinese, Gourmet Ice Cream Shop, Marco's Pizza, TCBY Yogurt
- **AServ** E: Auto Fair, Brown Honda, Saturn, Toyota, Lexus, Family Automotive, **K-Mart**, Michel Tire Co., Midas Muffler and Brake, Tiffy, York Olds
- **Med** W: ✚ **Afterhours Medical Center (Walk-in)**
- **Other** E: Car Wash, **K-Mart**
  W: **Shales Pharmacy**

8 OH 2, Airport Hwy
- **Gas** E: BP*(D)
  W: Amoco*, BP*, Speedway*, Sunoco*
- **Food** E: Don Pablo's Mexican
  W: Arby's, Arthur Treachers Fish and Chips, Bob Evans, Boston Market, Chili's, Coffee Mill Cafe, Cooker, Dairy Queen, Frisch's Big Boy, Frisco's Deli, Magic Rock, Marco's Pizza, McDonalds, New Empire Chinese, Pizza Hut, Rally's Hamburgers, Ranch Steak and Seafood, Subway, TCBY Yogurt, Wendy's
- **Lodg** E: ◆ Extended Stay America, ◆ Red Roof Inn, ◆ Residence Inn
  W: Courtyard Inn, (AAA) Cross Country Inn, ◆ Fairfield Inn
- **AServ** W: Firestone Tire & Auto, National Tire and Battery, The Andersons Tire Man Auto Center
- **Other** W: Express Car Wash, **Kroger Supermarket (Pharmacy), Office Max, Rite Aide Pharmacy, Sam's Club**

6 Salisbury Rd., I-80/90 Trnpk
- **Gas** E: BP*(CW)
- **Food** E: Applebee's, Bill Knapp's, Marie Cafe,

EXIT OHIO

McDonalds, Sam's Diner, Subway, Toledo Big Apple Bagels, Wendy's
  W: Chinese Connection
- **Lodg** E: Country Inn & Suites, Studio PLUS, ◆ Tharaldson Inn & Suites

4 US 24, Napoleon, Maumee
- **Med** E: ✚ **Hospital**

2 OH 25, Perrysburg
- **Gas** N: 76*(D), BP*
  S: Speedway*(LP)
- **Food** N: Alexanders Pizza, Berry Bagels, Charlie's Restaurant, Hungry Howie's Pizza, La Bella Restaurante, McDonalds, Subway, Wendy's
  S: The China Place
- **Lodg** S: (AAA) Red Carpet Inn
- **AServ** N: Tom's Tire
  S: Garrett's Auto Works
- **Other** N: **Churchill's Supermarket, Mail Depot, Shale's Pharmacy, Young's RV Center**

192 US 23, Maumee (Left Exit)

↑OHIO

Begin I-475, Ohio

Begin I-475, Georgia

↓GEORGIA

4 **(15)** Boling Broke
- **Gas** E: Exxon*(D)(LP), Fina*(D)

3 Zebulon Rd.
- **Gas** E: BP*(CW)
  W: Citgo*, Exxon
- **Food** E: Buffalo's Cafe, Chen's Wok, Chick-fil-A, Hong Kong Chinese, Subway, Waffle House, Wendy's
  W: Kuntry Kitchen, Polly's Corner Cafe
- **AServ** W: Exxon
- **Med** E: ✚ **Hospital**
- **Other** E: **Kroger Supermarket**

(8) **Rest Area - RR, Phones, Picnic, Vending, RV Dump (Northbound)**

2 Georgia 74, Macon, Thomaston
- **Gas** W: Exxon*, Phillips 66*(D), Shell*
- **Food** E: Waffle House
  W: Speedy Pizza, Subway (Exxon), Taco Bell (Exxon)
- **Lodg** W: **Family Inns**
- **AServ** E: Goodyear Tire & Auto, Muffler Master
- **RVCamp** W: **Lake Tobesofee Recreation Area**
- **Other** W: **Brantley & Jordan Animal Hospital, Food Lion Supermarket**

1 **(4)** U.S. 80, Macon, Roberta
- **FStop** E: **Fina***(LP)
- **Gas** E: Citgo*, Racetrac*
  W: Amoco*, BP*(D)(CW), Shell*
- **Food** E: **Cracker Barrel**, JL's BBQ (Citgo), Shoney's, Subway (Citgo), Waffle House
  W: Blimpie's Subs (Shell), Burger King, El Zarape (Passport Inn), McDonalds
- **Lodg** E: (AAA) Best Western, (AAA) Comfort Inn, Hampton Inn, ◆ **Holiday Inn(see our ad this page)**, Howard Johnson, Motel 6, (AAA) Ramada, **Red Carpet Inn**, (AAA) Rodeway Inn, (AAA) **Super 8 Motel**, (AAA) Travelodge
  W: (AAA) Econolodge, (AAA) Knight's Inn, **Passport Inn**
- **AServ** W: Amoco, BP
- **ATM** E: Citgo, Fina, Racetrac
  W: Shell
- **Other** W: **Sams Club**

↑GEORGIA

Begin I-475, Georgia

EXIT PENNSYLVANIA

Begin I-476

↓PENNSYLVANIA

37 I-81, PA315, Pittston, Scranton
- **TStop** W: **Truck Plaza Citgo***(D)
- **Gas** E: Mobil*, Sunoco
- **Food** E: Bonanza
  W: **Skyliner Restaurant (Truck Plaza)**
- **Lodg** E: Howard Johnson, Knights Inn
  W: **Skyliner Motel (Truck Plaza)**, Victoria Inn
- **Med** E: ✚ **Hospital**

(91) Jct I-81, Toll Plaza
(90) Picnic
36 **(85)** Wilkes - Barre, Bear Creek (Toll Booth)
- **Gas** E: Mobil*(D), Shell*, Texaco* (Kerosene)
- **Food** E: JB Pizzeria & Subs (Shell), Subway (Mobil)
- **Med** E: ✚ **Hospital**

(79) **Emergency Parking**
35 **(74)** I-80, PA940, Hazelton (Toll Booth)
- **Gas** W: Exxon, Shell*, Texaco*
- **Food** E: **Burger King**
  W: **Arby's**, McDonalds, Mt Laurel Resort
- **Lodg** W: Mountain Laurel Inn Resort, **Pokeno Mtn Lodge**

(66) **Hickory Run Service Plaza - Vending, RR, Phones, Picnic**
- **FStop** B: **Sunoco***(D)
- **Food** B: **Aunt Annie's Pretzel, Breyer's Ice Cream, Hot Dog Construction, McDonalds, Pizza & Salad Stand**

34 Mahonie Valley, Lehighton (Toll Booth)
- **Gas** W: Exxon*
- **Food** W: Platz Restaurant, Subway
- **RVCamp** W: **Campground, Sales, Service**(LP)

33 Leighi Valley, I-78, US22 (Toll Booth)
(32) **Allentown Service Center, Rest Area - RR, Phones, Vending, Prepaid Phone Cards, Snack Bar**
- **FStop** B: **Sunoco***(D)
- **Food** B: **Bob's Big Boy, Gourmet Bean Coffee, Hershey's Ice Cream Stand, Mrs. Field's Cookies, Roy Rogers, TCBY Yogurt**

32 PA663, Quakertown (Toll Booth)
- **Food** E: Farazo's Pizzeria
- **Lodg** E: (AAA) **Econolodge**, (AAA) Rodeway Inn
- **Med** E: ✚ **Quakerpoint Med Ctr**

31 PA63, Lansdale (Toll Booth)
- **Gas** E: Citgo*
- **Food** E: Carlo's Pizza, Holiday Inn, Kulpsville Kitchen (Days Inn), Sandwich Mill, Swiss Chalet (Days Inn), Tigers Restaurant
- **Lodg** E: Days Inn, ◆ Holiday Inn

25 Jct295, Philadelphia, Allentown (Northeast Extention, toll Booth)

↑PENNSYLVANIA

Begin I-476

Bold red print shows RV & Bus parking available or nearby

EXIT OHIO

Begin I-480

↓OHIO

1 JCT I-80

1A Oberlin, OH 10 W, Norwalk (Left Exit)

1B I-80, OH 10 E, OH Tnpk, Lorain Rd.
- Gas S: Citgo, Marathon, Speedway*, Sunoco(LP) (Kerosene)
- Food S: Kartels Restaurant, McDonalds
- Lodg S: Super 8 Motel, **Travelers Inn**
- AServ S: Citgo, Marathon, Sunoco

3 Stearns Rd.

6BA OH 252, N Olmsted, Great Northern Blvd.
- Food N: Applebee's, Chili's, Lone Star Steakhouse, McDonalds (Wal-Mart), Pizza Uno, Red Robin, Romano's Macaroni Grill, T.G.I. Friday's, The Spaghetti Company and Deli Too, Tony Roma's
- Lodg N: Hampton Inn, Radisson Inn
- AServ N: BP Pro Care, Goodyear Tire & Auto, NTB National Tire and Battery, Sear's
- Other N: **Great Northern Mall, PetsMart, Rini Rego Supermarket, Wal-Mart (Pharmacy)**

7 Claque Rd., West Lake, Fairview Park

9 OH 237, Grayton Rd., Airport
- Med N: ✚ **Hospital**
- Other S: **Airport**

10 Oh 237, Airport, Berea

11 JCT I-71, Cleveland, Columbus

12 West 150th St., West 130th St., Brook Park
- FStop S: **Marathon**
- Food S: Budget Inn, Harpo's Sports Club, Hot Rock Concert Club, Ramada Inn, Rusty's Family Restaurant
- Lodg S: Budget Inn (see our ad this page), Parkbrook Inn, Ramada Inn
- AServ S: Metro Toyota
- TServ S: **Internation Truck Service, R&B Truck Repair**
- TWash S: **R&B Truck Service**
- Other S: **ABC Check Cashing**

13 Tiedeman Rd., Brooklin
- Gas S: BP*(CW), Speedway*
- Food S: A Wok Chinese, Blimpie's Subs (BP), **Burger King**, Long John Silvers, McDonalds, Perkins Family Restaurant, Pizza Hut, Subway
- AServ S: Wal-Mart
- Other S: **Aldi Supermarket, Sams Club, Wal-Mart (Pharmacy)**

EXIT OHIO

15 US 42, Ridge Rd
- Gas S: Super America*
- Food N: Applebee's, Baskin Robbins, Boston Market, Dunkin Donuts, Golden Corral, Manhattan Bagel, McDonalds, Mr. Hero, Rockne's, Skyline Chili
- S: Arby's, Colonial Eatery Deli Bakery, Denny's, Wendy's
- AServ S: Axelrod Chrysler Plymouth, K-Mart, Lube Depot, Lube Shop, Mazda Dealership, Parma Transmission, Speedy Muffler King
- Med S: ✚ **Hospital**
- ATM S: Bank One, Home Bank
- Other N: **Finast Supermarket (Pharmacy), Mail Boxes Etc, Marks Supermarket**
- S: **ABC Check Cashing**, Coin Car Wash, **K-Mart, Rini Rego Supermarket**

16 OH 94, State Rd.
- Gas S: BP*(CW), Sunoco*
- Food N: Sports Inn
- S: Ictoria's Delight, The Goal Post
- AServ N: Lenny's Auto Service, Russo Tire and Service
- S: Myron's Automotive, Transmissions by Bruce
- Med N: ✚ **Hospital**
- S: ✚ **Hospital**
- Other N: **Cleveland Zoo, Convenient Food Mart**

20AB JCT I-77, Cleveland, Acron

21 Transportation Blvd., East 98th St
- Food N: Zech Inn
- Med N: ✚ **Hospital**
- Other N: **Convenient Food Mart**

22 OH 17, Granger Rd. (No Eastbound Reaccess)
- Gas S: Speedway*, Sunoco
- Food S: Dairy Whip
- AServ S: Sunoco
- Med N: ✚ **Hospital**

Note: I-480 runs concurrent below with I-271. Numbering follows I-271.

23 OH 14, Broadway Ave.

Note: I-480 runs concurrent above with I-271. Numbering follows I-271.

24 Lee Rd., Maple Heights (Westbound)
- Gas S: Marathon
- Food S: Jade Garden Chinese Restaurant
- AServ N: General Transmissions
- S: Friend-Lee Automotive, Marathon
- Other S: **WalGreens**

26 I-271 N, Erie PA, US 422

25ABC Oh 8, OH 43, Northfield, Warrensville Rd.
- Gas N: BP*, Shell*(CW)
- S: Marathon*
- Food N: Bakers Square, Bob Evans, Gallop Inn, Long John Silvers, Popeye's Chicken
- AServ N: Firestone Tire & Auto, Goodyear Tire & Auto, Meineke Discount Mufflers, Sear's, Speedy Muffler King, Valvoline Quick Lube Center
- S: Central Auto Repair, Mike's Automotive
- ATM N: Key Bank
- Other N: Car Wash, **Randall Park Mall**
- S: **Henry's Bi-Rite Supermarket**

↑OHIO

Begin I-480

EXIT NEW YORK

Begin I-490

↓NEW YORK

4 NY259, N. Chili (Closed)

5 NY386, Chili Center

6 NY204, County Airport

7AB Gates Center
- Gas E: Sunoco*
- Food E: Burger King, Denny's, Dunkin Donuts, Ground Round, Jilly Beans Coffee Shop, Marks Pizzeria, No. 1 Chinese Restaurant, Perkins, Winni's, Wokery Chinese
- W: Abbott's Frozen Custard, Delta House Restaurant
- AServ E: Goodyear Tire & Auto
- ATM E: Citi Bank, Fleet Bank, MNT Bank, OnBank
- Other E: **Eckerd Drugs, Top Super Food Market**

8 NY531, Spencerport, Brockport

9AB I-390, NY31 Wyelle Ave.

10 Mt. Read Blvd.
- Gas N: Gulf
- Food N: Combo's Deli
- Lodg N: 490 Motel

11 Ames St.
- Other N: **Police Station**

12 Brown St., Broad St.

13 St. Paul Ave., Clinton Ave.

14 Broad Street, Plymouth Avenue

15 South, NY15

16 **(23)** Clinton Ave Downtown
- Lodg N: Hyatt, Sheraton
- Other N: **Eastman Theatre, Rochester Public Market**

17 Goodman St.
- Food W: Highland Park Diner, Imperial Restaurant, USA Pizzeria
- AServ W: Mt. Hope Auto Parts

18 NY31, Monroe Avenue
- Gas N: Hess
- Food N: Beacho's, Pizza Peddler, Plumhouse Japanese, River City Subs, The Food Store, The Hub
- W: Pizza World, The Rose & Crown
- AServ N: Hess, Penstad
- Med N: ✚ **Hospital**
- Other W: **Wilson Convenience Store**

19 Clover Road

20 University Avenue, Winton Rd.
- Gas N: Hess*
- Food N: Coffees & Cafe, Ioncone's, Liberty Family Restaurant, McDonalds, Wendy's, Wong's Kitchen
- AServ N: Myttie Muffler, Vogel Collison Center
- ATM N: M&T Bank, Marine Midland Bank, OnBank
- Other N: **Post Office, Tops Supermarket (Pharmacy), Wegmans Food & Pharmacy**

21 Jct. 590,

23 **(27)** NY441, Lindon Avenue, Pennfield

(29) West Commercial St.

25 **(29)** NY31S, Fairport, East Rochester

26 **(32)** Pittsford

27 **(34)** NY96, Bushnell Basin

28 **(35)** NY96
- Other S: **Eastview Mall**

29 **(37)** NY 96, Bickter
- Gas N: Citgo*, Mobil*(LP)
- Food N: Brubaker Bagels, Mark's Pizzeria, Oven Door Bakery, Pontillo's Pizza, Sophia's Cafe, Subway, Subway
- Lodg N: Brookwood Inn
- AServ N: Mobil*(LP), Sears (Mall)
- ATM N: Marine Midland Bank
- Other N: **Animal Hospital, Eastview Mall, Eckerd Drugs**

↑NEW YORK

Begin I-490

EXIT MINNESOTA

Begin I-494

↓MINNESOTA

26 CR10, Bass Lake Rd
- **Gas** I: Conoco*(CW)
  O: Amoco(CW), Sinclair*(LP)
- **Food** I: McDonalds, Subway
- **AServ** O: Amoco
- **ATM** I: Norwest Bank
  O: Amoco
- **Other** I: **Maple Grove Mall**

23 CR 9, Rockford Rd (Closed)
- **Gas** I: Amoco(CW), Holiday*(CW)
  O: PDQ*
- **Food** I: Baker's Square Restaurant, Broadway Station (Pizza), Chili's, Domino's, Dufner's Soup & Sandwiches
  O: Bruegger's Bagel Bakery, Dairy Queen, Pizza Hut Carry-Out, Subway
- **AServ** I: Ford Dealership
- **ATM** I: Amoco, Anchor Bank, Holiday
  O: PDQ
- **Other** I: **Dental Clinic, Domino's Pizza, Petsmart, Rainbow Food Grocery (24 Hr), Rockford Rd Animal Hospital, WalGreens (Pharmacy)**

22 MN 55
- **Gas** O: Holiday*
- **Food** I: Best Western, Denny's
  O: Arby's, Burger King, Mulligan's, New Dynasty Restaurant, Perkins Family Restaurant
- **Lodg** I: ◆ Best Western, Radisson Hotel
  O: DAYS INN AAA Days Inn
- **AServ** O: Roger's Body Shop, Tires Plus
- **ATM** I: Franklin National Bank
  O: Holiday, Norwest Banks
- **Other** O: Car Wash

21 CR6
- **Other** I: **Home Depot**

20 Carlton Pkwy
- **Food** I: Einstein Bros Bagels
  O: Country Kitchen, Italianni's (Country Inn)
- **Lodg** O: AAA Country Inn & Suites

19A U.S.12 W. Wayzata

19B I-394 Jct

17 Minnetonka Blvd

16AB MN. 7
- **Gas** O: Total*(LP)

13 **(16)** MN 62, CR 62

12 CR 39, Valley View Rd (Northbound Reaccess Only)
- **AServ** O: GM Auto Dealership

11C MN 5 W.

11AB U.S. 212
- **Lodg** I: AmeriSuites (see our ad page 488)

10 U.S. 169 N to CR 18

7AB MN 100, CR 34 Normandale Blvd
- **Gas** I: Phillips 66, Wave*
  O: 76*(D)(CW) (1), 76*(CW) (11)
- **Food** I: Burger King, Chili's, Embers, T.G.I. Friday's
  O: Olive Garden, Tony Roma Rib's
- **Lodg** I: Radisson Motel, Select Inn
  O: Best Western, DAYS INN Days Inn, Holiday Inn
- **AServ** O: 76, Ford Dealership
- **ATM** O: Highland Bank

6AB CR 17, CR 32 Penn Ave
- **Gas** I: 76*, Mobil
- **Food** I: Fuddrucker's, McDonalds, Perkins Family Restaurant
  O: Applebee's, Bruegger's Bagels, Denny's, Dragon Jade Restaurant, Edwardo's, Grandma's Deli, Joe Cense's Bar & Grill, KFC, Lincoln Deli, McDonalds, Red Lobster, Starbuck's Coffee, Steak & Ale, Subway, Tortilla Ria (Mexican), Wendy's
- **Lodg** I: Best Western
  O: Doubletree Inn, Hampton Inn
- **AServ** I: 76, ABRA Body & Glass, Hyundai Dealer, Mazda Dealer, Mobil, Wally McCarthy's Olds, Walser Buick
  O: Chrysler Auto Dealer, Dodge Dealer, Goodyear Tire & Auto, Jiffy Lube, South Town Tire Center
- **Med** I: ✚ **Hospital**
- **ATM** I: Mobil
  O: TCF Bank
- **Other** I: **Office Depot**
  O: **Kohl's Dept, Montgomery Ward, Target Department Store (Pharmacy), Toys "R" Us**

5AB I-35 W, Jct Albert Lea, Minneapolis

4B Lyndale Ave
- **Gas** I: Amoco(CW), Super America*
  O: Phillips 66(D)
- **Food** I: Boston Market, Dairy Queen, Don Pablo's, El Jalape Mkt (Latin American, Mexican), Hope's Chow Mein (Chinese), Pappa John's Pizza, Vietnamese Restaurant
- **Lodg** I: ◆ Hampton Inn
- **AServ** I: Amoco, Champion Auto Stores, Honda Dealer, Tire Plus
  O: Acura Dealer, Ford Dealership, Phillips 66, Subaru Dealer
- **ATM** I: Amoco, Super America
- **Other** I: **Grocery Of The Orient, Petsmart, Veterinary Clinic**
  O: **Home Depot**

4A Nicollet Ave
- **Gas** I: Super America*(D)
  O: Total*(D)(LP)
- **Food** I: Burger King, Chi Chi's Mexican, Ember's
  O: Chubby Monroe's (79th St Grill), McDonalds
- **Lodg** O: ◆ Budgetel Inn, Super 8 Motel
- **AServ** I: RPM Automotive
  O: Auto Tech, Bromley Car Care
- **ATM** I: Super America
  O: Total

3 Portland Ave, 12th Ave
- **Gas** I: Sinclair*(D)
  O: Amoco*(D)(CW)
- **Food** I: Arby's, Mongolian BBQ Restaurant
  O: Denny's (Frontage Rd), First Wok (Chinese), Ming Cou Chow Mein (Chinese), Pizza Gallery, Pizza Hut Carry Out, Subway
- **Lodg** O: Americinn Hotel (Frontage Rd), ◆ Comfort Inn (Frontage Rd), AAA Holiday Inn Express (Frontage Rd)
- **AServ** I: Elsen's Service Garage, Sinclair
  O: Amoco
- **Other** I: **Kwik Way Foods (C-Store)**
  O: **Pilgrim Coin Laundry, Snyder Drugs, Super Value Foods, Wal-Mart Supercenter, WalGreens**

2BC MN 77
- **Other** O: **Malls Of America**

2A 24th Ave
- **Gas** O: Amoco*(CW), Super America*(CW)
- **Food** O: Marriott
- **Lodg** O: AAA Best Western, AAA Doubletree Grand, Excel Inn, Fairfield Inn, ◆ Marriott, ◆ Sheraton Inn, AAA Thunderbird Hotel
- **Other** O: **Malls Of America**

1B 34th Ave, HHH Terminal
- **Lodg** O: Ameri Suites, ◆ Embassy Suites, ◆ Holiday Inn Select, The Hilton

1A MN 5, Main Terminal

71 CR 31, Pilot Knob Rd
- **Lodg** I: Courtyard Marriott
  O: AAA Holiday Inn Select

70 I-35 E Jct. Albert Lee, St. Paul

69 MN 149, MN 55 Dodd Rd

67 MN3, Robert St
- **Gas** O: PDQ*
- **Food** I: White Castle Restaurant
- **AServ** I: Ford Dealership, Nissan Dealer, Toyota Dealer
- **Other** I: **Cub Foods, Office Max, Sam's Club, Southview Animal Hospital**

66 U.S. 52, Rochester, St. Paul (Divided Hwy)

65 7th Ave, 5th Ave

64B MN 156, Concord St.
- **FStop** O: **Super America***
- **Food** I: Best Western, Burger King
  O: **Blimpie's Subs (Super America)**, Golden Steer Motor Inn, **Taco John's (Super America)**
- **Lodg** I: ◆ Best Western
  O: Golden Steer Motor Inn
- **AServ** I: Grimley Auto Repair, Linford Transmissions
  O: MPS Motor Part Service
- **TServ** I: **Ford Truck Dealer**
  O: **Peterbilt Dealer**
- **ATM** I: First American Bank
- **Other** O: **ABC Rentals**(LP), Car Wash

64A Hardman Ave
- **TStop** I: **76***(SCALES), **Conoco***(SCALES)
- **Food** I: **Restaurant (Conoco), Restaurant (76)**
- **TServ** I: **ABN Truck/Trailer Parts & Service, Conoco, Goodyear Tire & Auto, Peterbilt Dealer, Pumps Tire**
- **TWash** I: **76**
- **ATM** I: **Conoco**

63C **(64)** Maxwell Ave

63AB U.S. 10, U.S. 61, St. Paul, Hastings (Divided Hwy)

60 Lake Rd

59 Valley Creek Rd
- **Gas** I: PBQ*
  O: Amoco*(CW)(LP), Super America*(CW)
- **Food** I: Burger King, China City, Old Country Store **Cracker Barrel**, Keys Cafe and Bakery, Maggiore Ristorante, McDonalds
  O: Applebee's, Broadway Pizza, Cheasapeake Bagel Bakery, Ciatti's Italian, Marguerita Murphy's, Old Country Buffet, Oriental Restaurant, Perkin's Family Restaurant, Starbuck's Coffee, Yang's Chinese
- **Lodg** I: Hampton Inn
  O: Red Roof Inn
- **AServ** I: Goodyear Tire & Auto
  O: Amoco
- **ATM** I: PBQ
  O: Amoco
- **Other** I: **Ace Hardware, Valley Creek Mall**
  O: **Coin Laundry, Kohl's Dept Store, Pearl Vision, Pet Food Outlet, Rainbow Foods Grocery, Target Department Store (Pharmacy), WalGreens**

55AB I-94 Junction to Madison, St. Paul

↑MINNESOTA

Begin I-494

Bold red print shows RV & Bus parking available or nearby

EXIT	NEW JERSEY

Begin I-495

↓NEW JERSEY

14 to25A, 21st St, Long Island City

15 Van Dam St., South Brooklyn, North Queens Blvd

16 Queensboro Bridge, Green Point Ave
- **Lodg** I: Ramada Plaza Suite (see our ad on this page)

17 Brooklyn, Queens, Expway278, Stanten Island, LaGuardia Airpt

↑NEW JERSEY

↓NEW YORK

19 Woodhaven Blvd, Rockaway, Queens Blvd, NY25
- **Gas** N: Amoco*, Getty, Mobil
- **Food** S: Church's Fried Chicken, White Castle Restaurant
- **AServ** N: Getty, Mobil
- **Med** S: ✚ St John's Queens Hospital
- **ATM** S: Emmigrant Bank
- **Other** S: Queens Ctr Mall

20 Junction Blvd.

21 108th St.
- **Food** S: Main St Bagel Rolls
- **Other** S: Alternate Parking for Shay Staduim, Car Wash, Hall Of Science, New Speedy Laundry Mat, Queens Zoo, Quik Stop C-Store

22 Zanwick Expwy, College Point Blvd
- **Gas** S: Mobil*
- **AServ** S: Mobil
- **Med** S: ✚ Hospital
- **Other** N: Hall Of Science

23 Main St
- **Food** N: Ill Cimine Italian, Kanwah, Palace Diner, Pioneer Pub Sports Cafe
- **AServ** N: Auto Depot
- **Med** N: ✚ NY Hospital of Queens
- **Other** N: Queens Botanical Garden

24 Kissena Blvd
- **Gas** N: Exxon*(D)
  S: Mobil*(D)
- **Food** N: Blimpie's Subs, Dunkin Donuts, Kissena Deli, Lum Wok, Wok's Famous Chinese
- **AServ** S: Kristy's, Mobil
- **Med** S: ✚ Hospital
- **Other** N: 59/24 C-Store, Coin Operated Laundry, Superior Pharmacy

25 Utopia Pkwy, 188th St
- **Gas** N: Mobil*, Sunoco*
  S: Amoco(CW) (Brushless CW), Shell*, Sunoco
- **AServ** N: Mobil
  S: A-1 Tires & Repair, Carriage House Collison, Goodyear Tire & Auto

EXIT	NEW JERSEY

26 Francis Lewis Blvd

27 I-295, Clearview Expwy, Ny25 Hillside Ave, Throgs.
- **Gas** N: Gulf
- **Food** N: Blue Made Diner, Bon Gusto Pizza
- **AServ** N: Gulf
- **ATM** N: Jamaica Savings Bank
- **Other** N: 7-11 Convenience Store, Mall, Rock Bottom Pharmacy, Waldhaun Supermarket

28 Oceania St, Francis Lewis Blvd

29 Springfield Blvd.
- **Gas** S: Exxon*
- **Food** S: Anthony's Fish, Bayside Deli, Burger King, Coffee Shop, Empire Garden, Empire House, Gino's Pizza, Hot Bagels, Imperial Wok (Chinese Take-Out), Joe's Pizza, Mario's Deli, McDonalds, Nathan's, New York Style Bagel, Papoung Restaurant, Slim's Bagels, Slim's Coffee Shop, Subway
- **ATM** S: Bank Of New York, City Bank, FS Bank
- **Other** N: Belmont Raceway
  S: Bayfield Pharmacy, Check Cashing, Circle Pt Optical, Dime's Savings Bank, Douglaston Plaza, Eric's Deli & Grocery, Genovese Pharmacy, Key Food Grocery, Petland Discount

30 Cross Island Parkway, White Stone Bridge

31 Douglaston Pkwy
- **Gas** S: Amoco
- **Food** S: Bakery, Burger King, Fish Market, Golden Imperial Chinese, Joe's Pizza Place Too, Orange Farm & Deli, Persian Tea Room, Ralph Potton's Nut House, TJK Cafe, Tia King Chinese
- **AServ** S: Amoco
- **Other** S: Specticle Shop, US Post Office

32 Little Neck Pkwy
- **Gas** N: Gulf(D)
- **Food** N: Boston Market, Center Pizza, Navurs Fresh Baked Bread, Slim's Hot Bagel
- **AServ** N: Gulf
- **Med** N: ✚ Little Neck Comm Hospital
- **ATM** N: Roosevelt Savings Bank (Open Sunday's)
- **Other** N: Le Drug Store, Vision World

33 Great Neck, Manahasset, Lakeville Rd, Community Dr
- **Med** N: ✚ Hospital
- **Other** N: Saddlerock Grist Mill

34 New Hide Park Rd
- **Med** N: ✚ Hospital

35 Shelter Rock Rd, Manahasset

36 Port Washington, Manhasset
- **Med** N: ✚ Hospital

37 Roslyn Rd, Willis Ave, Mineola
- **Gas** N: Exxon, Shell
  S: Mobil
- **Food** N: #1 Chinese Kitchen, Attilios Italian Eatery, Domino's, Hot Bagels, Ilsatore Italian, Mexican Grill, Northside Bagels, Roslyn Diner
  S: Tofu Chinese & Japanese Restaurant
- **AServ** N: Shell
  S: Mobil
- **ATM** N: Exxon, First Bank Of Long Island
- **Other** N: Hills Shoppes, International Food Market, Veterinary Clinic

38 Northern Pkwy E, to Meadowbrook Pkwy, Jones Beach

39 S Glenco Rd, Hanstead

EXIT	NEW JERSEY

- **Gas** N: Mobil*(D)
- **Food** N: Kitchen Caberea & Deli
- **AServ** N: Mobil
- **ATM** N: AT&T Bank, Chase Bank
- **Other** N: CVS Pharmacy, The Opticians of East Hills, Village Of East Hills

40 Westberry, Synosset, Jericho
- **Gas** S: Amoco, Concord, Exxon*, Hess*(D)
- **Food** S: Angelo's Restaurant, Bagels, Burger King, Cafe Doraldo, Chinatown Chef, Frank's Steaks, Friendly's, IHOP, Long Island Internet Cafe (Westberry Inn), McDonalds, Nagashima, Nathan's, Wendy's
- **Lodg** S: Edgewood Motel, Host Way Motor Inn, Howard Johnson, Westberry
- **AServ** S: Amoco, Exxon
- **ATM** S: EAB Bank
- **Other** S: 7-11 Convenience Store, Jericho Shopping Plaza, Village Animal Hospital

41 Ny106, Ny107, to Hicksville, Oyster Bay
- **Gas** S: Amoco, Sunoco
- **Food** N: Ben's Kosher Restaurant, Starbuck's Coffee
  S: Broadway Diner, Burger King, Caravella, Deli, Dunkin Donuts, Hot Bagels, McDonalds, Mulberry St Pizza & Pasta, Sun-Up Pizza
- **AServ** S: Goodyear Tire & Auto
- **ATM** N: Fleet Bank, State Bank Of Long Island
  S: Chase Bank, Marine Midland Bank
- **Other** N: CVS Pharmacy, Waldhaum's Grocery
  S: Path Mart Super Ctr & Groc, Pearl Vision Ctr, Petco

42 Northwestern Parkway (Westbound, Passenger Cars Only)

43 South Oyster Bay Rd
- **Gas** N: Mobil*
- **ATM** N: Mobil*

44 NY 135, Seafort, Syosset

45 Manneto Hill Rd, Plainsview, Woodbury

46 Sunnyside Blvd, Plainview
- **Food** N: Holiday Inn
- **Lodg** N: Holiday Inn
  S: Plainview Plaza Hotel

48 Round Swamp Rd, Old Bethpage, Farmingdale
- **Gas** S: Mobil*(D)
- **Food** S: Comfort Zone (Comfort Inn), Renaissance Gallaxy Cators
- **Lodg** S: Comfort Inn
- **RVCamp** S: Camping
- **ATM** S: Mobil
- **Other** S: Old Village Restoration

49 Farmingdale, Pinelawn, Amityville, Huntington
- **Food** N: Marriott
- **Lodg** N: Marriott

49B NY 110, Huntington, Amityville
- **ATM** S: Marine Midland Bank, North Fork Bank
- **Other** S: Republic Airport, Walt Whitman Historic Site

50 Wyandanch, Bagatelle Rd, Wheatley Heights, Dix

51 231 to Northport, Babylon

52 NY4, CR Comic, North Babylon, Bay Shore
- **FStop** N: Gas(D)
- **Gas** N: Mobil(D), Shell*

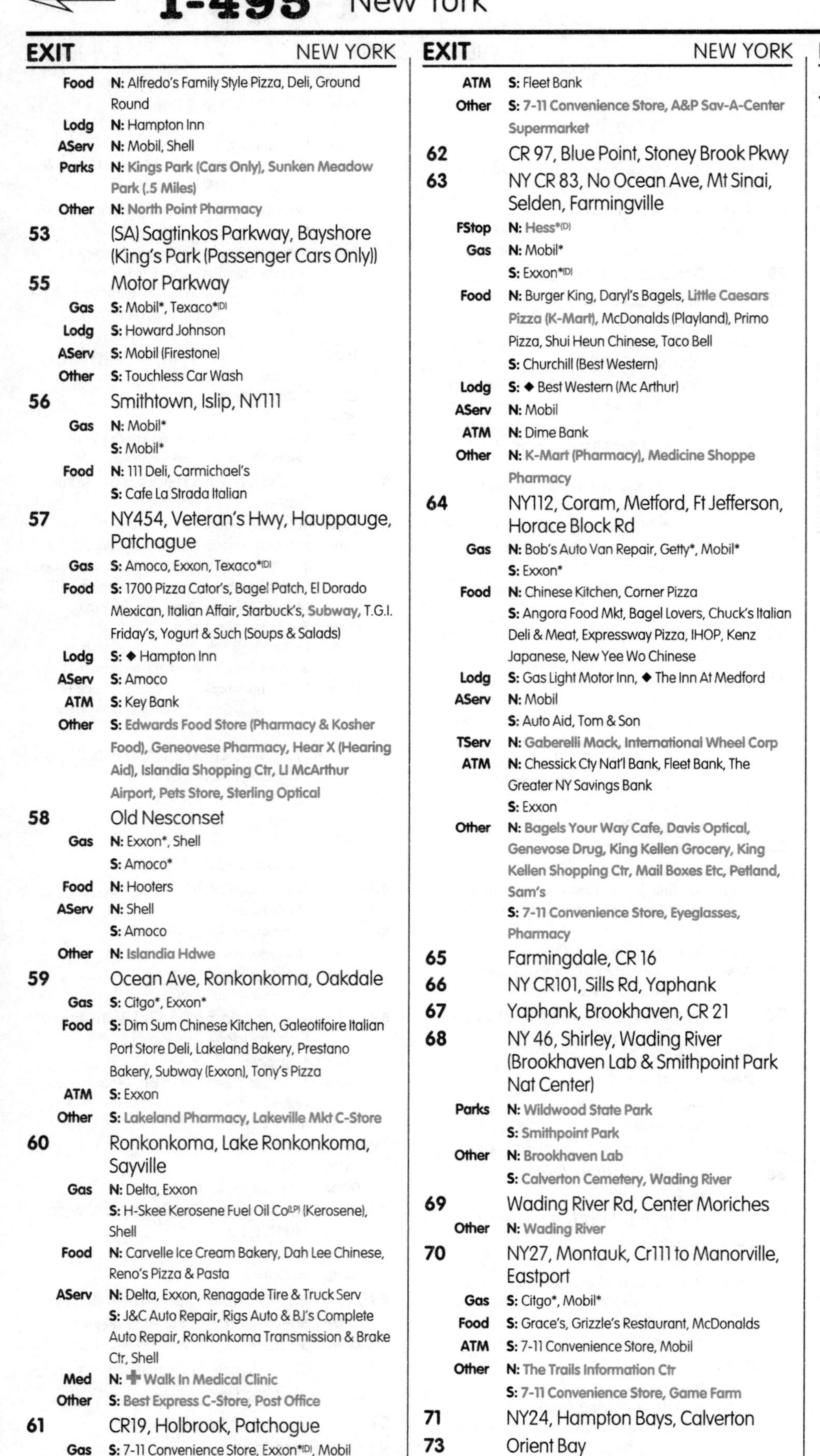

EXIT NEW YORK

Food **N:** Alfredo's Family Style Pizza, Deli, Ground Round
Lodg **N:** Hampton Inn
AServ **N:** Mobil, Shell
Parks **N:** **Kings Park (Cars Only), Sunken Meadow Park (.5 Miles)**
Other **N:** **North Point Pharmacy**

53 (SA) Sagtinkos Parkway, Bayshore (King's Park (Passenger Cars Only))

55 Motor Parkway
Gas **S:** Mobil*, Texaco*(D)
Lodg **S:** Howard Johnson
AServ **S:** Mobil (Firestone)
Other **S:** Touchless Car Wash

56 Smithtown, Islip, NY111
Gas **N:** Mobil*
S: Mobil*
Food **N:** 111 Deli, Carmichael's
S: Cafe La Strada Italian

57 NY454, Veteran's Hwy, Hauppauge, Patchague
Gas **S:** Amoco, Exxon, Texaco*(D)
Food **S:** 1700 Pizza Cator's, Bagel Patch, El Dorado Mexican, Italian Affair, Starbuck's, **Subway**, T.G.I. Friday's, Yogurt & Such (Soups & Salads)
Lodg **S:** ◆ Hampton Inn
AServ **S:** Amoco
ATM **S:** Key Bank
Other **S:** **Edwards Food Store (Pharmacy & Kosher Food), Geneovese Pharmacy, Hear X (Hearing Aid), Islandia Shopping Ctr, LI McArthur Airport, Pets Store, Sterling Optical**

58 Old Nesconset
Gas **N:** Exxon*, Shell
S: Amoco*
Food **N:** Hooters
AServ **N:** Shell
S: Amoco
Other **N:** **Islandia Hdwe**

59 Ocean Ave, Ronkonkoma, Oakdale
Gas **S:** Citgo*, Exxon*
Food **S:** Dim Sum Chinese Kitchen, Galeotifoire Italian Port Store Deli, Lakeland Bakery, Prestano Bakery, Subway (Exxon), Tony's Pizza
ATM **S:** Exxon
Other **S:** **Lakeland Pharmacy, Lakeville Mkt C-Store**

60 Ronkonkoma, Lake Ronkonkoma, Sayville
Gas **N:** Delta, Exxon
S: H-Skee Kerosene Fuel Oil Co(LP) (Kerosene), Shell
Food **N:** Carvelle Ice Cream Bakery, Dah Lee Chinese, Reno's Pizza & Pasta
AServ **N:** Delta, Exxon, Renagade Tire & Truck Serv
S: J&C Auto Repair, Rigs Auto & BJ's Complete Auto Repair, Ronkonkoma Transmission & Brake Ctr, Shell
Med **N:** **✚ Walk In Medical Clinic**
Other **S:** **Best Express C-Store, Post Office**

61 CR19, Holbrook, Patchogue
Gas **S:** 7-11 Convenience Store, Exxon*(D), Mobil
Food **S:** Carvel Ice Cream, Holbrook House Of Bagels, Hung Chinese, Joe's Pizza
AServ **S:** 7-11 Convenience Store, Holbrook Auto Serv

EXIT NEW YORK

ATM **S:** Fleet Bank
Other **S:** **7-11 Convenience Store, A&P Sav-A-Center Supermarket**

62 CR 97, Blue Point, Stoney Brook Pkwy

63 NY CR 83, No Ocean Ave, Mt Sinai, Selden, Farmingville
FStop **N:** **Hess*(D)**
Gas **N:** Mobil*
S: Exxon*(D)
Food **N:** Burger King, Daryl's Bagels, **Little Caesars Pizza (K-Mart)**, McDonalds (Playland), Primo Pizza, Shui Heun Chinese, Taco Bell
S: Churchill (Best Western)
Lodg **S:** ◆ Best Western (Mc Arthur)
AServ **N:** Mobil
ATM **N:** Dime Bank
Other **N:** **K-Mart (Pharmacy), Medicine Shoppe Pharmacy**

64 NY112, Coram, Metford, Ft Jefferson, Horace Block Rd
Gas **N:** Bob's Auto Van Repair, Getty*, Mobil*
S: Exxon*
Food **N:** Chinese Kitchen, Corner Pizza
S: Angora Food Mkt, Bagel Lovers, Chuck's Italian Deli & Meat, Expressway Pizza, IHOP, Kenz Japanese, New Yee Wo Chinese
Lodg **S:** Gas Light Motor Inn, ◆ The Inn At Medford
AServ **N:** Mobil
S: Auto Aid, Tom & Son
TServ **N:** **Gaberelli Mack, International Wheel Corp**
ATM **N:** Chessick Cty Nat'l Bank, Fleet Bank, The Greater NY Savings Bank
S: Exxon
Other **N:** **Bagels Your Way Cafe, Davis Optical, Genevose Drug, King Kellen Grocery, King Kellen Shopping Ctr, Mail Boxes Etc, Petland, Sam's**
S: **7-11 Convenience Store, Eyeglasses, Pharmacy**

65 Farmingdale, CR 16

66 NY CR101, Sills Rd, Yaphank

67 Yaphank, Brookhaven, CR 21

68 NY 46, Shirley, Wading River (Brookhaven Lab & Smithpoint Park Nat Center)
Parks **N:** **Wildwood State Park**
S: **Smithpoint Park**
Other **N:** **Brookhaven Lab**
S: **Calverton Cemetery, Wading River**

69 Wading River Rd, Center Moriches
Other **N:** **Wading River**

70 NY27, Montauk, Cr111 to Manorville, Eastport
Gas **S:** Citgo*, Mobil*
Food **S:** Grace's, Grizzle's Restaurant, McDonalds
ATM **S:** 7-11 Convenience Store, Mobil
Other **N:** **The Trails Information Ctr**
S: **7-11 Convenience Store, Game Farm**

71 NY24, Hampton Bays, Calverton

73 Orient Bay

↑NEW YORK

Begin I-495

EXIT MARYLAND

Begin I-495

↓MARYLAND

27 Jct I-95 N, to Baltimore
Other **O:** **Baltimore Washington Int'l Airport**

25AB U.S. 1, Baltimore Ave, College Park, Laurel
Gas **I:** Amoco, Shell
O: Shell*
Food **I:** Days Inn (Korean), Lasieks Seafood Buffet, Pizza Hut, University Donut Shop
O: Jonesie's Steak, Mooses Creek Steak House
Lodg **I:** Comfort Inn, DAYS INN Days Inn, Econolodge, Royal Pine Inn, Super 8 Motel
O: Holiday Inn
AServ **I:** College Park Auto Repair, College Park Volkswagen, Mazda, College Part Hyndai, Honda, Jiffy Lube, RT 1 Auto Repair, Shell
O: Shell
RVCamp **I:** **Campground, Queenstown Motors**
ATM **I:** Commerce Bank, State Employees Credit Union
Other **I:** **College Park Animal Hospital, College Park Boat & RV Center, Hardware Store**

24 Greenbelt Station, MD 201, Kenilworth Ave. (No Reaccess)
Other **I:** **Metro Train Station**

23 MD 201, Kenilworth Ave, Bladensburg, Greenbelt
Gas **I:** Amoco*(CW), Mobil, Shell, Texaco*(CW), Xtra Fuel
Food **I:** Baskin Robbins (Giant Supermarket), Bennigan's, Boston Market, Checkers Burgers, Dunkin Donuts, Fireside Beef House, Ledo Pizza, McDonalds, Muashi Japanese Steakhouse, Pizza Hut, Popeye's, Roy Rogers, Sir Walter Raliegh, TJ's Roadhouse Grill, Wendy's
O: T.G.I. Friday's
Lodg **I:** Sir Walter Raliegh Inn
O: Courtyard
AServ **I:** Greenbelt Auto, Shell, Texaco
RVCamp **O:** **Greenbelt Park Campground**
Med **I:** **✚ Med Serve**
ATM **I:** Chevy Chase Bank, Liberty Nat'l Bank, Maryland Federal Bank, NationsBank
Parks **O:** **Greenbelt Park**
Other **I:** **Beltway Keys & Locks, Beltway Plaza*, CVS Pharmacy**, Fennigan's Car Wash, **Giant Supermarket (Pharmacy), NASA, Optical, Pearl Vision Center, Postal Stand (Beltway Plaza)**

22AB Baltimore Washington Parkway

20 MD 450, Annapolis Rd, Lanham, Bladensburg
Gas **I:** Citgo*(D), Sunoco*
Food **I:** Bob's Big Boy, Cheasapeak Bay Seafood, Cuisine Of China, Dunkin Donuts, Good Earth Chinese, KFC, Pappa John's, Popeye's Chicken, Ramada
O: Days Inn, Horn & Horn Buffet, Jerry's Sub Shop, Red Lobster
Lodg **I:** Ramada
O: Best Western, DAYS INN Days Inn, Red Roof Inn
AServ **I:** Amoco(LP), Bob Banning Pontiac, Citgo, Grease-N-Oil Xpress, Just Tires, Merchant's Tire & Auto, Nissan, Shell, Sunoco
O: Midas Muffler & Brakes, Speedy Muffler King
Med **O:** **✚ Hospital**
ATM **I:** Crestar Bank, Maryland Federal Credit Union

Bold red print shows RV & Bus parking available or nearby

EXIT		MARYLAND

Other I: **Ames Shopping Ctr, CVS Pharmacy, Carrollton Mall, Checks Cashed, Hinson Optical, Safe Way Grocery, Shoppers Food Warehouse**
O: **Harley Davison Motorcycle Shop, Lanham Shopping Center**

19AB U.S. 50, Annapolis, Washington

17AB MD 202, Landover Rd, Bladensburg, Upper Marlboro
- **Food** I: IHOP
  O: Holiday Inn
- **Lodg** O: Holiday Inn
- **AServ** I: Just Tires, Sears, Tollison Buick
- **Other** I: **County Police, Landover Mall, Sam's Club**
  O: **US Air Center (Capitol Arena)**

15AB MD 214, Central Ave
- **Gas** I: Crown*[D][CW], Exxon*[D], Shell*, Sunoco
- **Food** I: Daybreak Restaurant (Days Inn), Jerry's Pizza & Subs, KFC, McDonalds, Pizza Hut, Taco Bell
- **Lodg** I: DAYS INN Days Inn, Motel 6
  O: Hampton Inn
- **AServ** I: Hampton Auto Body (24 Hr Towing), Jiffy Lube, Midas, Shell, Traks
- **Other** I: **7-11 Convenience Store,** Full Service Car Wash
  O: **Adventure World, US Air Arena**

11AB MD 4, Pennsylvania Ave, Washington, Upper Marlboro
- **Gas** I: Amoco, Exxon*[CW], Sunoco
  O: Texaco*[D]
- **Food** I: Hunan House Restaurant, Villa Sara Pizza Parlor
  O: Murray's Hot Food To Go
- **AServ** I: DJ's One Stop Auto Service, Michelin, Napa Auto Parts, Shell*
- **TServ** I: **Goodyear Tire & Auto**
- **ATM** I: Industrial Bank
- **Other** I: **Ames Shopping Ctr, Coin Laundry, Forestville Business Ctr (Commercial Truck Services), Friendly Food Mart Coffee & Deli, Furniture Outlet Shopping Ctr, Goodman Eye Care Ctr, State Patrol Post**

9 MD 337, Allentown Rd, Andrews AFB (Difficult Reaccess)
- **Gas** I: Mobil*[D][CW]
- **Other** O: **U-Haul Center**[LP]

5 MD 5, Branch Ave, Waldorf, Silver Hill
- **Gas** I: Texaco*[D]
- **Food** I: Days Inn
- **Lodg** I: DAYS INN Days Inn
- **AServ** I: Capitol Cadillac, Sheehy Ford, Texaco, Volvo Dealer, Wilson Powell Lincoln Mercury
- **Med** O: ✚ **Hospital**

4AB MD 414, Saint Barnabas Rd, Oxon Hill, Marlow Heights
- **Gas** I: Exxon*[D], Shell
  O: Exxon[D], Xtra
- **Food** I: China Best, McDonalds, Pizza Hut Carry Out
  O: Black Eyed Pea, Burger King, CJ's Restaurant, Checkers, Hunan's Restaurant, KFC, Little Caesars Pizza, McDonalds, Roy Rogers, Sharky's Seafood & Crabhouse, Subway, Wendy's
- **Lodg** O: Red Roof Inn
- **AServ** I: Exxon, Jiffy Lube, Shell
  O: Quaker State Oil & Lube
- **Med** I: ✚ **Hospital**

EXIT		MARYLAND/VIRGINIA

- **ATM** I: First VA Bank
  O: Chevy Chase Bank (Safeway Grocery), First Union, Industrial Bank, NationsBank
- **Other** I: **Coin Laundry, Sunrise Plaza At St Barnabas**
  O: **7-11 Convenience Store, CVS Pharmacy,** Car Wash, **K-Mart (Pharmacy), Rivertowne Commons Marketplace, Rose Craft Raceway, Safe Way Grocery**

3AB MD 210, Indian Head Hwy, Forest Heights
- **Gas** O: Crown*, Mobil
- **Food** O: KFC, Ramada
- **Lodg** O: Ramada
- **AServ** O: Goodyear Tire & Auto, Mobil
- **Other** O: Coin Operated Car Wash

2AB Junction I-295, Washington

↑MARYLAND

↓VIRGINIA

1AB U.S. 1, Fort Belvoir, Alexandria
- **Gas** I: Exxon, Merrit[D] (Kerosene)
  O: Amoco, Mobil*[D], Sunoco, Texaco
- **Food** O: Domino's Pizza, Howard Johnson, Kyoto Japanese Restaurant, Oh's Palace (Chinese), Western Sizzlin'
- **Lodg** O: Brookside Motel, Howard Johnson, Red Roof Inn, Statesman Motel, Travelers Motel, Virginia Lodge
- **AServ** I: Exxon, Heritage Chevrolet & Jeep
  O: Ourisman Dodge/ Suzuki/ Chrysler/ Plymouth, Texaco
- **Med** I: ✚ **Herrin Medical**
  O: ✚ **Urgent Medical Care Walk-In**
- **ATM** O: Central Fidelity, Crestar Bank
- **Other** O: **Fireworks, Huntington Parkway Shops, Optical (Huntingon Parkway Ctr)**

2AB VA 241, Telegraph Rd, Alexandria
- **Gas** O: Citgo[D][CW], Exxon*, Hess[D]
- **Food** I: Mansenies Deli & Bakery
  O: Honolulu Restaurant
- **Lodg** I: Courtyard by Marriott, Holiday Inn
- **AServ** I: Nissan
  O: Citgo
- **Med** I: ✚ **Hospital**
- **ATM** I: NationsBank
  O: 7-11 Convenience Store, Burke & Herbert Bank & Trust
- **Other** I: **Alexandria Animal Hospital, George Washington Mason Hall**
  O: **7-11 Convenience Store**

3 VA 613, Van Dorn Street, Franconia
- **Lodg** O: Comfort Inn

4 Junction I-95 South, Richmond; Junction I-95 North (East Beltway), Alexandria; Junction I-395, Arlington, Washington

4A Junction I-495 North, Rockville

4B Junction I-395, Arlington, Washington

5 VA 620, Braddock Rd
- **Gas** O: Mobil[D]
- **Food** O: Chesapeake Bagel Bakery, Kilroy's

EXIT		VIRGINIA

- **AServ** O: Mobil
- **ATM** O: Chevy Chase Bank, NationsBank
- **Other** O: **Patriotic Center For The Arts, Ravensworth Shopping Center, Rite Aide Pharmacy, Safe Way Grocery**

6 VA 236, Little River Tnpk, Annandale, Fairfax
- **Gas** I: Citgo[D][CW], Exxon
- **Food** I: McDonalds, PeKing
- **AServ** I: Citgo, Exxon, Trail Auto
- **ATM** I: Crestar Bank, First Union
- **Other** I: **7-11 Convenience Store, Coin Laundry, Rite Aide Pharmacy, Safe Way Grocery**

7 VA 650, Gallows Rd
- **Gas** O: Exxon*
- **AServ** O: Exxon
- **Med** O: ✚ **Inova Fairfax Hospital**
- **Other** O: **7-11 Convenience Store**

8 U.S. 50, Arlington Blvd, Arlington to Lee Hwy
- **Gas** O: Amoco[CW], Mobil*[D][CW], Shell*
- **Food** O: Baskin Robbins, CVS Pharmacy, Cheasapeake Bagel, Dominion Deli, Grevey's Restaurant, Hunan Gate, McDonalds, Sir Walter Raliegh Inn, Starbucks Coffee, Uno Chicago Bar & Grill, Wendy's, York Towne Center
- **Lodg** I: ◆ Marriott
  O: Sir Walter Raliegh Inn
- **AServ** O: Midas, Shell (Wrecker Service)
- **ATM** O: Chevy Chase Bank
- **Other** O: **Giant Food Supermarket**

9 Capitol Beltway to the North, JCT I-66, Vienna, Front Royal (No Trucks Allowed To The East, B Exit Right, C To Ft Royal)

10AB VA 7, Leesburg Pike, Falls Church, Tysons Corner
- **Gas** I: Crown[D], Exxon, Mobil[CW], Shell
- **Food** I: Aubonpain Bakery & Cafe (Doubletree Hotel), Bertucci's Pizza, Busara Ti Restaurant, Chilli's, Elli's Deli, JR's Stockyard Inn, Marketplace Cafe, Marriot, Olive Garden, On The Border Mexican Cafe, Roy Rogers, Silver Diner, The Bagel Market
- **Lodg** I: Doubletree Hotel, JR's Stockyard Inn, Marriott
- **AServ** I: Exxon, Tysons Ford
- **ATM** I: Central Fidelity, Crestar, George Mason Bank, NationsBank, Signet Bank (Crown)
- **Other** I: **7-11 Convenience Store, Petco**

11 VA 123, Chain Bridge Rd, McLean, Tysons Corner
- **Gas** O: Mobil, Shell[D]
- **Food** O: Dadomenico Italian, Hilton, Holiday Inn, La Madeline, Legal Seafood, Maggie's Corner Bakery, Ritz Carlton
- **Lodg** O: Hilton, Holiday Inn, Marriott, Ritz Carlton
- **AServ** O: Coon's Pontiac GMC, Shell
- **ATM** O: Shell
- **Other** O: **Pearl Vision Center, The Galleria at Tysons II, Tysons Corner Center**

12 VA 267, Dulles Airport Exit No toll (Left Exit West A Toll Rd)

13 VA 193, Georgetown Pike, Langley Air Force Base, Great Falls

EXIT VIRGINIA/MARYLAND

14 George Washington Memorial Pkwy (No Trucks)

↑VIRGINIA

↓MARYLAND

41 Clara Barton Pkwy, Carderock, Glen Echo (No Trucks)

39 MD 190, River Rd, Washington, Potomac

38 Junction I-270, Rockville, Frederick (Left Exit)

36 MD 187, Old Georgetown Rd, Bethesda, Rockville
- **Med** I: ✚ **Hospital (2,1/2 Miles)**

34 MD 355, Wisconsin Ave, Bethesda
- **Food** I: Bellomondo & American Grill (Marriott)
- **Lodg** I: Marriott

33AB MD 185, Connecticut Ave, Chevy Chase, Kensington (No Thru Trucks Southside)
- **Gas** I: Amoco, Exxon, Sunoco*
- **Food** I: Einstein Brothers Bagels, Eurogrill, Starbucks Coffee
- **AServ** I: Amoco, Exxon, Sunoco
- **ATM** I: Chevy Chase Bank
- **Other** I: **Chevy Chase Lake Shopping Center, Chevy Chase Supermarket, Hippchen Optical, Lake Care Pharmacy**

31AB MD 97, Georgia Ave, Silver Spring, Wheaton
- **Gas** I: Amoco*, Exxon*, Shell*, Texaco*(D)
  O: Exxon*(D)
- **Food** I: Armand's Chicago Pizzeria, Bagel's DeLox, Domino's Pizza, Dunkin Donuts, Hunan City, Kerston's Cafe, Mayflower Chinese Restaurant, Montgomery Donuts, Tropical Ice Cream, Woodside Deli
- **Lodg** I: Holiday Inn (see our ad on this page)
- **AServ** I: Amoco, Exxon, Texaco
- **Med** O: ✚ **Forest Glenn Hospital**
- **ATM** I: Citibank
- **Other** I: **Ace Hardware, CVS Pharmacy,** Car Wash, **Cemenary Place Shopping Ctr, Silver Spring Animal Hospital**

30 U.S. 29, Colesville Rd, Silver Spring
- **Gas** O: Amoco, Shell, Texaco*
- **Food** O: China House Restaurant, Jerry's Pizza &Subs, Pappa John's Pizza, Roy Rogers, The Corner Pub & Restaurant
- **AServ** O: Shell, Texaco
- **ATM** O: Chevy Chase Bank

EXIT MARYLAND/MASSACHUSETTS

- **Other** O: **7-11 Convenience Store, Four Corner Animal Hospital, Safe Way Grocery**

29B MD 193 East, University Blvd, Langley Park

28AB MD 650, New Hampshire Ave, Takoma Park, White Oak
- **Gas** O: Amoco, Exxon*(D), Shell*
- **Food** O: Big Boy, China Restaurant, Domino's, Eastern Carry -Out, KFC, Mike's Cheese Steak & Pizza
- **AServ** O: Amoco, Shell
- **ATM** O: Chevy Chase Bank
- **Other** O: **CVS Pharmacy, Hillandale Shopping Center, Hillandale Pharmacy, Piney Branch Optical, Safe Way Grocery**

↑MARYLAND

Begin I-495, D.C.

Begin I-495, Boston

↓MASSACHUSETTS

54 **(117)** MA150, Amesbury
- **TStop** E: **Gulf***
- **Gas** W: Getty
- **Food** W: Bagel Express
- **AServ** W: Getty
- **RVCamp** W: **Campground**
- **Other** E: **Coin Laundry (Gulf)**
  W: **Liddia Homestead (1.5mi)**

53 **(115)** Broad St , Merrimac, Merrimacport
- **Gas** W: Texaco
- **AServ** W: Texaco
- **ATM** W: BankBoston

(52) **Rest Area - Phones, Picnic**

52 **(111)** MA110, Haverhill, Merrimac
- **Gas** W: Mobil*(D), Mobile*
- **Food** W: Biggart's Ice Cream, Dunkin Donuts
- **Med** E: ✚ **Hospital**

51AB **(109)** MA125, Plaistow, NH, Haverhill
- **Gas** E: Mobil (Towing), Richdale*
- **Food** E: Brothers Pizza, Dunkin Donuts
- **AServ** E: Mobil
- **Med** E: ✚ **Hospital**
- **ATM** E: First Essex Bank
- **Other** E: **Coin Laundry, Richfield Convenience Store**

50 **(108)** MA97, Haverhill, Salem, NH
- **Food** W: Cafe 97, Dunkin Donuts
- **AServ** W: Regan Ford
- **Other** W: **One Stop Market**

49 **(107)** MA110, MA113, River St
- **Gas** E: Gulf*, Sunoco*
- **Food** E: 99 Restaurant, Athens Pizza, Bigford's Family Restaurant, Dunkin Donuts, McDonalds, Oriental Garden, Papa Gino's Pizza &More
- **Lodg** E: (AAA) **Best Western,** Comfort Inn
- **AServ** E: Tire Warehouse
- **Med** E: ✚ **Hale Hospital**
- **ATM** E: BankBoston, Fleet Bank, Pentucke Bank
- **Other** E: **CVS Pharmacy, Coin Laundry, K-Mart, Market Basket**

48 **(106)** MA125, Ward Hill, Bradford

47 **(104)** MA213, Methuen, Salem, NH (Difficult Reaccess)
- **Med** W: ✚ **Hospital**

46 **(103)** MA110, Merrimac St, Pleasant Valley
- **Gas** E: Gulf, Sunoco*
- **Food** E: Carr's Variety Deli, **Heav'nly Donuts,** Super Swirl Ice Cream
- **AServ** E: Gulf, Valley Towing, Walls Ford, Lincloln, Mercury
- **ATM** E: First Essex

EXIT MASSACHUSETTS

45 **(102)** Marston St, Lawrence
- **Food** E: Anthony's Pizza & Subs
- **AServ** E: Commonwealth Geo, Isuzu,Chevrolet, Honda, Ferro's Auto Service
- **TServ** E: **B&B Truck & Equipment Repair, Coady's Garage, Northeast Truck Service**
- **Other** E: **Tourist Information**

44 **(102)** Merrimac St, Sutton St (Difficult Reaccess Southbound)
- **Gas** E: Citgo
  W: Gulf*(D)
- **Food** E: Cafe Bakery, Honey Dew Donuts, Sutton Square Grill
  W: Italian Deli, J&J Super Sub, Paisano's Pizza, The Frosty Mug
- **AServ** E: Trombly Towing
  W: Merrimac Auto Technician, NAPA Auto Parts
- **Other** E: **Belmont Propane(LP), Richdale Food Store**
  W: **Car Cleaning Service, Dog Gone It, Food**

(87) **Rest Area - Phones (Both Directions)**

43 **(101)** Massachusetts Ave, N Andover
- **Gas** W: Mobil, Shell*
- **AServ** W: Mobil, Shell
- **ATM** W: Lawrence Savings Bank

42B **(101)** MA114W. St Lawrence
- **Gas** E: Exxon*, Gulf*, Mobil*(D)
  W: BP, Texaco*(D)
- **Food** E: Boston Market, Friendly's, Pasta Palazzi, Pizza Hut
  W: Denny's, Dunkin Donuts, Garden House, KFC, McDonalds, Pizzaria Bravo, Royal Dragon, Wendy's
- **Lodg** E: Hampton Inn
- **AServ** E: Gulf, Mobil
  W: Auto Pep Plaza, BP, Speedy Muffler King, Texaco, VIP Auto Center
- **Med** W: ✚ **Hospital**
- **ATM** E: Fleet Bank
  W: BankBoston, First Essex's, US Trust Bank
- **Other** E: **7-11 Convenience Store, Andover Mall, CVS Pharmacy, Market Basket**
  W: **Osco Drugs, Parelli Optical, Stadium Plaza Shopping**

42A **(100)** MA114E, Middleton

41 **(99)** MA28, Lawrence, Andover
- **Gas** W: Mobil
- **Food** E: Courtyard Restaurant, Dunkin Donuts
  W: Dunkin Donuts
- **Lodg** E: Shawsheen Motel
- **AServ** E: GM Auto Dealership, Shawsheen Garage
  W: Mobil
- **ATM** E: Lawrence Savings Bank
- **Other** E: **Letourneau's, U.S. Post Office**

40AB **(97)** Jct I-93 to Concord, NH, Boston

39 **(95)** MA133, Dracut, Andover
- **Gas** E: Charter
- **Food** E: McDonalds
  W: Holiday Inn, Wendy's
- **Lodg** E: Ramada Inn (see our ad opposite page)
  W: ◆ Holiday Inn, ◆ Residence Inn, ◆ Susse Chalet
- **Med** W: ✚ **Hospital**

38 **(92)** MA38, Lowell, Tewksbury
- **Gas** W: Citgo, Mobil*, Sunoco, Texaco, USA Petroleum(D)
- **Food** E: Applebee's, Burger King, Econolodge, Waffle House
  W: DeAngelo's Sandwich, Dunkin Donuts, Jillie's Roast Beef, Little Caesars Pizza, McDonalds, Milan Pizza, Pizza Hut
- **Lodg** E: Econolodge
- **AServ** E: Atamian Honda, VW
  W: Bornivale Chevrolet, Citgo, Jiffy Lube, Jim Witt Pontiac, Buick, GMC, Sunoco
- **Med** W: ✚ **Hospital**
- **ATM** W: Fleet Bank

Bold red print shows RV & Bus parking available or nearby

EXIT	MASSACHUSETTS
Other	E: Wal-Mart (Pharmacy,Optical)
	W: CVS Pharmacy, Covent Farm Convenient Store, K-Mart (Pharmacy), Market Basket
37	**(91)** Woburn St, S Lowell, N Billerica
Gas	W: Exxon*
Food	W: Stefano's Italian
36	**(90)** Lowell, Connector
35AB	**(90)** U.S. 3, Nahua, NH, Burlington
34	**(89)** MA4, MA110, Chelmsford
Gas	E: Ampet, Mobil*, Sunoco*
	W: Shell*
Food	E: Cafe Bake Shoppe, Dunkin Donuts, Hong Kong Chinese, Jimmy's Subs & Pizza, Skips, Skips Ice Cream, Town Meeting, Townsford Creamery
	W: Ground Round
Lodg	E: ◆ Radisson Heritage Hotel
	W: ◆ Best Western
AServ	E: All Tune & Lube
Other	E: Coin Laundry, Palmart Plaza
32	**(84)** Boston Rd, Westford
Gas	E: Exxon*, Gulf, Mobil*
Food	E: Applebees, Bel-Deli, Burger King, Colonial Bakery, DeAngelo's Sandwich Shop, Dunkin Donuts, McDonalds, Papa Genio's Pizza, Pizza Express, Pizza Hut
AServ	E: Exxon (Towing), Gulf
ATM	E: BankBoston, Middle Bank
Other	E: CVS Pharmacy, Eye Care, Market Basket, Osco Drugs, Pets-Pets-Pets, WalGreens, Westford Convenience Store
31	**(80)** MA119, Groton, Acton
Gas	E: Mobile*, Shell*
Food	E: Dunkin Donuts, Ken's American Cafe, Littleton Sub Shop
AServ	E: Mobile*, Thrifty Thompson Auto Parts
Med	W: ✚ Hospital
Other	E: CVS Pharmacy, Cumberland Farms, Post Office, Woodstove & Fireplace Shop(LP)
30	**(79)** MA2A, MA110, Littletown, Ayer
RVCamp	W: Campground
Med	W: ✚ Hospital
29AB	**(78)** MA2, Leominster, Boston
28	**(75)** MA111, Boxboro, Harvard
Gas	E: Exxon
Food	E: Jerry's (Holiday Inn)
Lodg	E: Holiday Inn (see our ad on this page)
AServ	E: Exxon
Other	W: Shepard's Vet Clinic
27	**(70)** MA117, Bolton, Stowe
Food	W: Herbert's Candy, Ice Cream
RVCamp	W: Campground
ATM	W: Quinton Savings Bank
Other	E: Country Cuboard Store
26	**(67)** MA62, Berlin, Hudson
Gas	E: Exxon
	W: Texaco(D)
AServ	E: Certified Repair Shop, Exxon
	W: Texaco
25B	**(65)** Jct I-290
25A	**(65)** MA85, Marlboro, Worcester
Med	E: ✚ Hospital
24AB	**(63)** U.S. 20, Northboro, Marlboro
Gas	E: C&N
	W: Exxon*, Shell*(CW)
Food	E: Holiday Inn
	W: Best Western, Boston Market, Brubakers Bagels, China Case, Hunan Cuisine, McDonalds, Pizzeria, Radisson Inn, Starbucks Cafe, Subway, Wendy's
Lodg	E: Holiday Inn
	W: Best Western, Embassey Suites, Radisson Inn
AServ	E: Meineke Discount Mufflers
	W: Post Road Buick
Med	E: ✚ Hospital
	W: ✚ Walk-In Clinic
Other	W: Mail Box, Marlboro Animal Hospital, Victory Supermarket

EXIT	MASSACHUSETTS
23AB	**(60)** MA9, Shrewsbury, Framingham
Gas	E: Exxon(D), Exxon*(D)
Food	E: Wendy's
	W: Marriot
Lodg	E: Red Roof Inn
	W: Marriot
AServ	E: Watson Auto Body
22	**(58)** Jct I-90, MA Trnpk, Boston, Albany, NY
21AB	West Main St, Upton, Hopkinton
Gas	E: Gulf*, Mobil*
Food	E: Dino's Pizza, Subs & More, Dunkin Donuts, Dynasty Chinese, Golden Spoon Coffee Shop, Jelly Donuts
AServ	E: Lumber Street Auto Repair
Med	E: ✚ Hopkinton Walk In Clinic
ATM	E: BankBoston
20	**(50)** MA85, MA16, Mildord, Holliston
Gas	W: Gulf(D)
Food	W: Marriot, Wendy's
Lodg	W: ◆ Marriot
AServ	W: Gulf(LP)
19	**(49)** MA109, Milford, Medway
Gas	W: Mobil*, Shell*
Food	W: Burger King, Lil Cigna, McDonalds, Papa Genio's, Radisson Hotel
Lodg	W: Pagen, ◆ Radisson Hotel
AServ	W: Meineke Discount Mufflers, Plaza Discount Auto Parts, Shell
ATM	W: BankBoston, Milford Federal Bank
Other	E: 495 Rental Center(LP)
	W: Coin Laundry, K-Mart
(18)	Weigh Station (Southbound)
18	MA126, Medway, Bellingham
Gas	W: Charter*, Mobil*, Sunoco
Food	E: McDonalds
	W: Chicago Bar & Grill, Dunkin Donuts
AServ	W: Sunoco
RVCamp	W: Campground
ATM	E: BankBoston
Other	E: Market Basket Supermarket, Stallbrook Market Place, Wal-Mart
	W: Petsmart
17	**(43)** MA140, Franklin, Bellingham
Gas	E: Mobil*
Food	E: Applebee's, Burger King, Dunkin Donuts, Hunan Gourmet II, PaPa Genio's Pizza, Pizza Hut, TCBY Yogurt, Taco Bell, The Caravan, Village Cafe
AServ	E: Jiffy Lube
Med	W: ✚ Hospital
ATM	E: Bank of Boston, Summit Bank
Other	E: Cambridge Optical Wear, Mail Boxes Etc, Stop & Shop
16	**(41)** King St, Franklin, Woonsocket, RI
Gas	E: Sunoco*
Food	E: Lambert's Gourmet Deli
15	**(38)** MA 1A, Wrentham, Plainville

EXIT	MASSACHUSETTS
RVCamp	W: KOA Campground(LP)
Med	E: ✚ Hospital
14AB	**(36)** U.S. 1, N Attleboro, Wrentham
Food	E: Arbor Inn, Luciano's
Lodg	E: Arbor Inn
AServ	E: Morrocco Brothers
RVCamp	E: Campground
13AB	**(33)** Jct I-95, Providence, RI, Boston
12	**(30)** MA140N, Mansfield
(11)	Rest Area - Parking Area Only
11	**(30)** MA140S, Norton
AServ	E: Goodyear Tire & Auto
(10)	Rest Area - Parking Area Only
10	**(27)** MA123, Norton, Easton
Food	E: Kricket's Corner, Mino's Pizza
RVCamp	E: McLaughlin's Campground
Med	W: ✚ Hospital
Other	E: Quick Stop
9	**(25)** Bay St, Taunton, Easton
Food	W: Dunkin Donuts, Golden Flower Chinese, Holiday Inn, Pizza Hut
ATM	W: Tebeschi
8	**(22)** MA138, Raynham, Taunton
Gas	E: Mobil*(CW)
	W: Exxon*(D)
Food	E: Christopher's, Honey Dew Donuts
	W: China Garden, Dunkin Donuts, Honey Dew Donuts, McDonalds, Pepperoni's Pizza
AServ	W: Exxon(LP)
Med	W: ✚ Hospital
ATM	W: Bristol County Savings Bank
Other	W: Bristol County Animal Clinic, Smart Mart
7AB	**(19)** MA24, Fall River, Boston
6	**(15)** U.S.44, Middleboro, Plymouth
Gas	E: Circle Farm*(D)
Food	E: Burger King, Dunkin Donuts, Fireside Grille, Friendly's, Frosty's Place, Panda Inn, The Lobster Shack, USA Pizza
Lodg	W: Susse Chalet
AServ	E: Kelly's Tire Mart
	W: Cannon Buick,Pontiac
ATM	E: Bridgewater Savings Bank
Other	E: Buggy Bath Car Wash
	W: Kahian's Propane
5	MA18, Lakeville, New Bedford
Gas	E: Circle Farm*(D)
Food	E: Dunkin Donuts, Fireside Grill, Friendly's, Frosty's Place
RVCamp	E: Rousseau's RV Center (2 Miles)
	W: Campground
Other	E: Buggy Bath Car Wash
4	**(12)** MA105, Middleboro Center, Lakeville
Gas	E: Exxon*(D), Shell(CW), Sunoco*, Texaco
Food	E: Dairy Queen, Dunkin Donuts, McDonalds, Papa Timmy's Pizza
Lodg	E: DAYS INN ◆ Days Inn
AServ	E: LH Chace & Sons Chevrolet, Shell, Texaco
ATM	E: Rockland Trust
	W: Bridgewater Savings Bank
Other	E: Brooks Pharmacy, Coin Laundry, Osco Drugs
3	**(8)** MA28, Rock Village, S Middleboro
FStop	E: Citgo*(D)
Food	W: Huckleberry's Fine Dining
AServ	E: Fred's Repair
TServ	E: Exit 3 Truck Center
2	**(3)** MA58, Carver, W Wareham
RVCamp	W: Campground
Other	E: Bell Haven Camp Ground
	W: Roby's Propane
1	**(0)** Jct I-195W, Wareham, New Bedford

↑MASSACHUSETTS

Begin I-495, Boston

I-526 SC

EXIT SOUTH CAROLINA

Begin I-526

↓SOUTH CAROLINA

32A US 17 S., Charleston
- **Other** I: **K-Mart**

32 Isle of Palms, US 17 N., Georgetown (Westbound Reaccess)
- **Gas** O: Amoco*, Hess*[D], Speedway*[LP]
- **Food** O: Blimpie's Subs, Fazoli's Italian Food, Gullah Cuisine, McDonalds, Ryan's Steakhouse, Ye Old Fashioned Icecream & Sandwich Cafe
- **AServ** O: Advance Auto Parts, Gerald's Tire & Brake, Midas, Precision Tune, The Car Shop, Transmission Clinic
- **ATM** O: First Union, NBSC
- **Other** O: **Bookland, Piggly Wiggly, Revco Drugs, Rite Aide Pharmacy, Wal-Mart**

30 Long Point Rd.
- **Food** O: Waffle House
- **Other** O: **Charles Pinckney National Historic Site**

23AB Road 33

20 Virginia Ave. (Eastbound, Reaccess Westbound Only)

19 North Rhett Ave.
- **Gas** I: Texaco*
  - O: Citgo*[D], Hess*[D]
- **AServ** O: Bouchillon Auto Service
- **Other** O: **Eckerd Drugs, Food Lion**

18AB US 52, US 78, Rivers Ave.
- **FStop** O: **Amoco***
- **Gas** I: Exxon*
  - O: Phillips 66
- **Food** I: Captain Don's Seafood, Mayflower Chinese Restaurant
  - O: Alex's Restaurant
- **Lodg** I: Catalina Inn, Economy Inn
- **AServ** I: Goodyear Tire & Auto, Low Country Muffler & Brakes
  - O: Meineke Discount Mufflers, Parks Auto Parts, Parts America, Phillips 66
- **ATM** I: Exxon
- **Other** O: **Piggly Wiggly Supermarket**

17AB Jct. 26, Charleston, Columbia

17 Montague Ave.
- **Gas** O: BP*[CW], Speedway*[LP]
- **ATM** O: BB&T
- **Other** O: **Rite Aide Pharmacy**

15 Paramont Dr., Dorchester Rd.
- **Gas** I: Exxon*
- **Food** I: Burger King, Checkers, Domino's, Huddle House, Pizza Hut
- **Lodg** I: Super 8 Motel
- **AServ** I: Sentury Buick, Pontiac, GMC
  - O: Palmetto Tire Service
- **Other** I: Water Works Car Wash

13 Leeds Ave
- **Med** I: ✚ **Hospital**

11AB Ashley River Rd.

↑SOUTH CAROLINA

Begin I-526

I-575 GA

EXIT GEORGIA

Begin I-575

↓GEORGIA

13 **(26)** Ga. 5, Howell Branch Rd., Ball Ground

12 Hwy 372

11 Ga. 5, Canton
- **FStop** W: **Exxon***[LP]
- **Gas** E: Amoco*
  - W: Tim's Auto Buff
- **Food** E: Cafe Cantina, Chick-fil-A (Playground), Gondolier, **McDonalds (Wal-Mart)**
  - W: **Waffle House**
- **Lodg** E: Comfort Inn
- **AServ** W: S&H Tire Service
- **ATM** E: First Market Bank (Wal-Mart)
- **Other** E: **Wal-Mart Supercenter (Pharmacy)**

10 **(19)** Ga. 20 E., Canton, Cumming

8,9 8 - Ga. 5 Bus. N., Ga. 20W., Ga. 140W. 9 - Ga. 140E. Canton

7 **(14)** Holly Springs, Canton
- **Gas** W: Chevron*[CW] (24 Hrs), Racetrac*[LP]
- **ATM** W: Racetrac
- **Other** W: **Kroger Supermarket (Pharmacy)**

Note: I-575 runs concurrent below with I-595. Numbering follows I-595.

6 **(11)** Sixes Rd.
- **Gas** E: Chevron*[D][CW][LP] (24 Hrs)
  - W: Fina*[D]

Note: I-575 runs concurrent above with I-595. Numbering follows I-595.

5 **(8)** Towne Lake Pkwy., Woodstock
- **Gas** E: BP*[CW][LP], Chevron*[LP]
- **Food** E: Hot Rod Cafe, Waffle House, Waffle King
- **AServ** E: Cherokee Ford
- **ATM** E: SunTrust (BP)

4 **(6)** Ga. 92, Wooodstock
- **Gas** E: Racetrac* (24 Hrs), Shell*[D][CW] (24 Hrs), Texaco*[D][CW][LP]
  - W: Amoco*[CW]
- **Food** E: Buffalo's Cafe, Captain D's Seafood, Checkers, Chick-fil-A Dwarf House, China Bay, Dairy Queen, Folks, KFC, Manhattan Bagel, McDonalds (Play Place), Monterrey Mexican, Ruby Tuesday, Taco Bell, Waffle House
  - W: Blimpie's Subs (Amoco), IHOP
- **AServ** E: American Lube Fast, Goodyear Tire & Auto, Master Care Auto Service, Xpress Lube
  - W: Hennessy Honda
- **ATM** E: Bank of Canton, Bank of North Georgia, First National Bank, Racetrac, SunTrust, Texaco, Wachovia
  - W: Etowah Bank
- **Other** E: **Big B Drug, Ingles Supermarket, Pearl Vision Center, Wal-Mart, Winn Dixie Supermarket (24 Hrs, Pharmacy)**
  - W: **Cub Food, Wolf Camera,** Woodstock Car Wash

3 Bells Fairy Rd.
- **Gas** W: Chevron*[CW], QT*, Texaco*
- **Food** W: Arby's, Burger King (Playground), Dunkin' Donuts, Subs, Subway, Waffle House
- **AServ** W: Xpress Lube
- **ATM** W: NationsBank
- **Other** W: **A&P Supermarket (Pharmacy), Big B Drugs, Eckerd Drugs, Suburban Veterinarian**

2 Chastain Rd.
- **Gas** E: BP*[CW][LP] (24 Hrs)

1 To I-75, Barrett Pkwy.
- **Gas** E: Chevron*[CW][LP] (24 Hrs)
  - W: Texaco*[CW]
- **Food** E: Burger King (Playground), Waffle House
  - W: 3 Dollar Cafe, Applebee's, Fuddruckers, Fuji Hana Japanese, Grady's, Happy China II, Ippolito's Italian, Longhorn Steakhouse, Mandarin Village Chinese, Manhattan Bagel, McDonalds (Play Place), Mellow Mushroom Pizza, Mick's, Olive Garden, Philly Connection, Pizza Hut, Red Lobster, Rio Mexico, Schlotzkys Deli, Shoney's, Starbucks Coffee, Subway, Taco Mac, Thai Peppers, Waffle House
- **Lodg** W: Econolodge, Holiday Inn Express, Red Roof Inn, Shoney's Inn
- **AServ** E: Tune-Up Clinic, Xpress Lube
  - W: Big 10 Tires, Firestone Tire & Auto, Midas Muffler & Brakes
- **Med** W: ✚ **Physician's Immediate Med.**
- **ATM** E: Premier Bank
  - W: Colonial Bank, NationsBank, SouthTrust, Texaco
- **Other** E: **Barnes & Noble, Carmike Cinemas, Drug Emporium, Mountasia Family Fun Center, Publix Supermarket (Pharmacy)**
  - W: **Town Center Mall At Cobb**

↑GEORGIA

Begin I-575

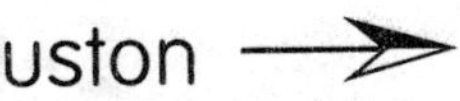

EXIT FLORIDA

Begin I-595

↓FLORIDA

2 Flamingo Rd.
- **Gas** N: Mobil*(D)(CW), Shell*(D)(CW)
  - S: Cumberland Farms*
- **AServ** N: Discount Auto Parts
- **Other** S: **Coin Laundry**

1 SR. 84, SW. 136 Ave. (Westbound)
- **Food** S: Antonio's Pizza-Rant, China Island Restaurant, Subway
- **Lodg** N: ◆ Budgetel Inn, (AAA) Wellesley Inn
- **AServ** S: Tire Kingdom
- **ATM** S: NationsBank
- **Other** S: **Publix Supermarket, WalGreens**

3 Hiatus Rd.
- **Gas** S: Mobil*(CW)
- **Food** S: 84 Diner, Bobby Rubino's Ribs, Deli-Boy Subs
- **Other** S: **Animal Hospital, U.S. Post Office**

4 Nob Hill Rd.
- **Food** S: Burger King (Play Land)

5 Pine Island Rd.
- **Gas** S: Citgo*, Mobil*
- **Food** S: Bellanotte Italian, China Town, Ciro's Deli & Pizzaria, Garcia's Mexican, KFC, Little Caesars Pizza, The Diner
- **ATM** S: First Union
- **Other** S: **Animal Hospital, Publix Supermarket**

6 University Dr.
- **Gas** N: Chevron* (24 Hrs)
  - S: Chevron*(CW) (24 Hrs)
- **Food** N: Bagel Whole, IHOP, King China Buffet, McDonalds, Truly Scrumptious
  - S: Arby's, Burger King (Playground), Davie Ale House, Dunkin' Donuts, Longhorn Steakhouse, Papa Roni Pizza, Roadhouse Grill, Subs Miami, Subway, Taco Bell, The Shell, Vinnie's
- **AServ** N: Chevron, Plantation Tire Center
  - S: Pep Boys Auto Center, Spee Dee Oil Change
- **Med** N: ✚ **First Med (Walk-In Med Center)**
- **ATM** N: Barnett Bank
  - S: Regent Bank
- **Other** S: **K-Mart, WalGreens**

7 Davie Rd.
- **Gas** S: Racetrac*
- **Food** S: Bagel Express & Deli, Juicy Lucy's, Kenny Rogers Roasters, Shoney's, Subs Miami, Waffle House
- **AServ** S: Kovac Automotive

8 Florida Turnpike, Fl. 91

9AB U.S. 441

10 Fl. 84E.

11AB I-95

12A U.S.1

12B International Airport

12C Port Everglades, Fort Lauderdale

↑FLORIDA

Begin I-595

EXIT TEXAS

Begin I-610, Houston

↓TEXAS

1A FM 521 Almeda Rd.
- **FStop** I: **Chevron*(D) (24 Hrs)**
- **Gas** O: Texaco*(CW)
- **AServ** I: Inner City Tire
  - O: AAMCO Transmission, American & Foreign Motors

1B Fannin St.
- **Gas** I: Conoco
- **Food** O: McDonalds
- **AServ** I: 4 Wheel Parts of Texas, Conoco, Cunningham Auto Service
  - O: Sam's Club
- **Other** I: **The Astrodome (Houston Astro's Baseball)**
  - O: **Sam's Club**

1C Kirby Dr.
- **Food** I: Restaurant (Sheraton), Shoney's (Shoney's Inn)
  - O: Joe's Crab Shack, Pappa Sito's Cantina
- **Lodg** I: (DAYS INN) Days Inn, (AAA) Sheraton Astrodome, (AAA) Shoney's Inn
- **AServ** O: Davis Chevrolet Trucks, Mike Calvert Toyota, NTB
- **Other** I: **Astrodome, Six Flags Astroworld**

2 US 90 Alt, Buffalo Speedway, S. Main St.
- **Gas** I: Exxon*, Shell*
  - O: Phillips 66
- **Food** I: Arby's, Bennigan's, Brisket Bar B.Q., Church's Fried Chicken, Denny's, KFC, Main Street Chinese Buffet, Ruchis Taqueria El Rincon De Mexico, Taco Bell, Wendy's
  - O: Joe's Crab Shack, Steak 'N Egg Kitchen (24 Hrs), What A Burger (24 Hrs)
- **Lodg** I: Astro Motor Inn, Howard Johnson
  - O: 7 Tweety's Diamond Inn, (AAA) La Quinta Inn, Motel 6, Red Coach Motel, **Super 8 Motel**
- **AServ** I: Russell & Smith Ford, Russell & Smith Honda, Tire Station
  - O: Baker/Jackson Nissan, NTW Tire Center, Nationwide

EXIT TEXAS

- **RVCamp** I: **Camping (RV Dump)**
  - O: **South Main Trailor Village RV Park**
- **Other** I: **Pathfinder Bookstore, Texas Dept. of Agriculture, Tropical Fish Pet Shop**

3 Stella Link Rd.
- **Gas** I: Chevron
  - O: Conoco*, Exxon*(D)
- **Food** I: Domino's Pizza
  - O: Super Fried Chicken & Taqueria
- **AServ** I: All Tune & Lube, Chevron, Discount Tire Co., Q-Lube, Tuneup...Etc.
  - O: Brake Check
- **ATM** I: Chevron
- **Other** I: **All Cats Veterinary Clinic**
  - O: **Stella Link Animal Clinic, Washateria**

4A Braeswood Blvd
- **Gas** I: Chevron(D)(CW), Exxon
- **Food** I: Alice's Diner, Boston Market, IHOP, McDonalds (Playground)
  - O: Austin Coffeehouse, Bandana's Grill, Cafe Express, Doc's Real Italian Ices, Los Tios Mexican Restaurant, Salt Grass Steakhouse, Souper Salad, Starbucks Coffee, Szechuan Empress
- **AServ** I: Chevron, Exxon
- **ATM** I: Bank United, First Prosperity Bank
  - O: Bank of America, Compass Bank
- **Other** I: **Builders Square II(LP), Police Station**
  - O: **Borders (Bookstore), Dr. Panzer Optometrist, Meyerland General Cinema, OfficeMax, Pets America, Texas State Optical, Venture, Wolf Camera**

4B South Post Oak Rd. (Limited Access Hwy)

5A Beechnut St. (Southbound)
- **Gas** O: Chevron(D)(CW)
- **Food** O: Alice's Diner, Bandana's Grill, Boston Market, Cafe (Borders), Doc's Real Italian Ices, IHOP, Los Tios Mexican Restaurant, McDonalds, Souper Salad, Starbucks Coffee, Szechuan Empress
- **AServ** O: Chevron
- **ATM** O: Bank of America, Compass Bank
- **Other** O: **Borders (Books, Music, & Cafe), Dr. Panzer Optometrist, Texas State Optical, The Optical**

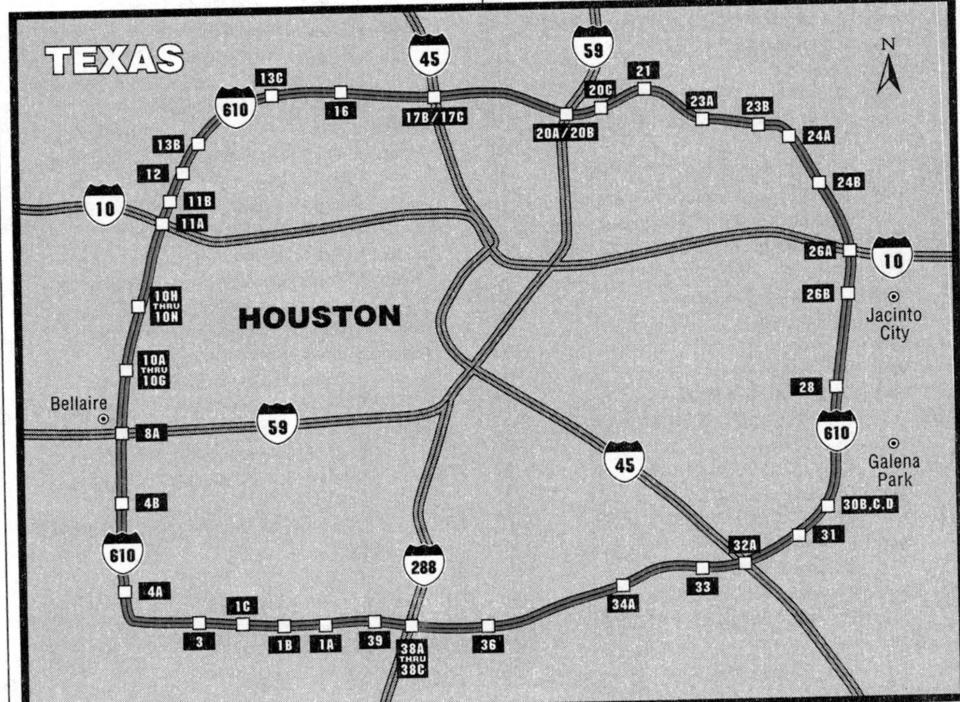

EXIT		TEXAS
		Place
5B		Evergreen St., Bellaire Blvd.
6		Bellaire Blvd. (Southbound)
7		Fournace Place, Bissonnet St.
	Gas	O: Texaco*(D)(CW)
	Med	O: ✚ Hospital
	ATM	O: Texas Commerce Bank
8A		U.S. 59 South, Victoria; U.S. 59 North, Downtown (Limited Access Hwy)
8B		Richmond Ave (Southbound)
	Food	I: Bayou City's Burger Cafe, Bayou City's Seafood & Pasta, Carmines Big Tomato, Fu's Garden Hunan Cuisine, Luling City Market Bar-B-Que & Bar, Pizza Hut Delivery
		O: Chinese Cafe, Dimassi's Middle Eastern Cafe, Marie Callender's Restaurant, Smoothie King, Steak & Ale, Subway, The Mason Jar
	Lodg	I: Courtyard by Marriott, Fairfield Inn by Marriott
	AServ	I: NTB
	ATM	O: American Bank, Bank One, Texas Commerce Bank
	Other	I: **Super K Food Store**
		O: **Mail Boxes Etc, OfficeMax, Sears Optical**
8C		FM 1093 West Heimer Rd (Northbound)
	Gas	I: Texaco
		O: Shell*
	Food	I: Capt'n Benny's Seafood & Oyster Bar, Le Peep Restaurant, Tokyo Gardens Japanese Restaurant, Westcreek Market & Deli
		O: Lunes Uptown Burgers
	Lodg	I: Red Lion Hotel
	AServ	I: Central Ford, Texaco
	Other	I: **Joe's Golfhouse**
		O: **The Galleria**
9		San Felipe Rd. (Northbound)
	Gas	I: Exxon*(D)
	Food	I: Jack-in-the-Box
	Lodg	I: ◆ Hampton Inn, ◆ The Ritz-Carlton
	AServ	I: Exxon
	ATM	I: Heritage Bank
	Other	I: **Super Crown Bookstore, Target Department Store**
9A		FM 1093, San Felipe Rd., Westheimer Rd. (Southbound)
	Food	O: California Pizza Kitchen, Domino's Pizza, Luby's Cafeteria, Rancho Tejas, Subway
	Lodg	O: Marriott
	ATM	O: Comerica Bank
	Other	O: **Duke's Golf, Fox Photo 1-Hr Lab, Toys To Love**
9B		Post Oak Blvd. (Southbound)
	Food	O: McDonalds (Playground), Ninfa's Fine Mexican Food, Willie G's Seafood & Steakhouse
10		Woodway Drive, Memorial Drive
	Gas	O: Exxon*(D), Shell(D)(CW)
	Food	O: Swiss Chalet Restaurant
	Lodg	O: Omni Hotel
	AServ	O: Exxon, Shell
	Parks	I: **Arboretum Memorial Park**
	Other	O: **Houston Mounted Police Station**
11A		I-10 West, San Antonio; I-10 East, Downtown
11B		Katy Rd. (Southbound)
	Other	O: **Malibu Grand Prix**
12		W. 18th St., Hempstead Rd
	Food	I: Tea's Bar-B-Q & Broiler, What A Burger
		O: Demeris Bar-B-Q
	AServ	I: American Car Care Centers
		O: American Wheel & Tire, HTC Tires
	Other	I: **VIP Convenience Stop**
		O: **Blockbuster, Boat US Marine Center,**

EXIT		TEXAS
		Northwest Mall, The Box Store
13B		U.S. 290 W. Austin (Northbound Is Left Exit)
13C		W. TC Jester Blvd. & E. TC Jester Blvd.
	Gas	I: Phillips 66*(D)(CW)
		O: Chevron(CW), Texaco*
	Food	O: Antone's, Atchafalaya River Cafe, Baskin Robbins, Denny's, Golden Gate, Supreme Sandwiches
	Lodg	O: Courtyard by Marriott
	AServ	O: Chevron, Rock Chip Repair
	Other	O: **Sunny's Foodstore**
14		Ella Blvd
	Gas	I: Shell*(CW)
		O: Exxon*, Shell*, Texaco(CW) (Dr. Gleem Car Wash)
	Food	I: Tecate Mexican Restaurant, Thomas Bar-B-Q
		O: Burger King (Playground), KFC, McDonalds (Playground), Popeye's, Taco Bell, Wok Buffet
	AServ	I: Brake Check
		O: Alex's Service, All Tune & Lube, Complete Auto & Truck Parts, Q Lube, Texaco
	Med	I: ✚ **Memorial Hospital Northwest**
	ATM	I: Bank One
	Other	I: **Circle Pone Food Store, Cunningham Pharmacy**
		O: Dr. Gleem Car Wash, Rick's Handwash & Detail (Car Wash)
15		Spur 261, North Durham Dr., North Shepherd Dr.
	Gas	I: Chevron*(D)(CW) (24 Hrs), Mobil*, Shell*, Texaco
		O: Texaco
	Food	I: What A Burger (24 Hrs)
		O: Gabby's Genuine Barbecue, Mi Sombrero, Sonic, Taco Cabana
	AServ	I: Auto Glass Center, Auto Upholstery (Emissions Tests), CV Joint Specialist, Hubcap City(CW), Pep Boys Auto Center, Shell, Texaco, Transmission Exchange
		O: Kim's Complete Service
	ATM	I: Mobil, Shell
	Other	I: **The Home Depot(LP)**
		O: **Brink's Hardware, Sunny's Grocery Store**
16		N. Main St., Yale St.
	Gas	I: Mobil*, Shell*(CW), Shop & Save*
	Food	I: Church's Fried Chicken
		O: Bar-B-Que Plate Lunch
	AServ	I: Bluebell Imports, Opersall Kindall Diesel & Auto Repair, P.Q. Tire Shop
	ATM	I: Mobil, Shell
	Other	I: **Sunset Heights Food Market, Wing Fong Food Market**
		O: Hand Car Wash, **Harley Davidson Dealer**
17A		Airline Drive (Eastbound)
	Gas	I: Mobil*(D), Texaco*(D)
	Food	I: Connie's Market Seafood, Golden Seafood House, Jack-in-the-Box
	Lodg	I: Fiesta Inn Motel, **Western Inn Motel**
	AServ	I: Transinternational Transmissions, Vulcanizadora Rey New & Used Tires
		O: High Speed Transmission
	Other	I: **Farmer's Market, Jurica's Egg & Meat Co., Kalil Fruit & Vegetable Inc.**
17B		I-45 North, Dallas (Eastbound Is A Left Exit)
17C		I-45 Downtown (Westbound Is Left Exit)
18		Irvington Blvd., Fulton Dr.
	Gas	I: Texaco* (24 Hrs)
		O: Chevron(CW) (24 Hrs)
	Food	O: Speedy's Burgers
	ATM	I: Stop N Go, Texaco

EXIT		TEXAS
		O: Chevron
	Other	I: **Stop N Go**
		O: **Lindale Grocery**
19A		Hardy Street, Elysian Street
19B		North Hardy Toll Rd.
20AB		20A - US 59 North, Cleveland 20B - US 59 South, Downtown
20C		Hirsch Rd. (Westbound)
21		Lockwood Dr., Hirsch Rd.
	Gas	I: Conoco*(D)
		O: Chevron*(CW) (24 Hrs), Exxon, Gas*, Hopcus Mini Mart*
	Food	I: Snowflake Donut
		O: Burger King, C N Seafood Market & Restaurant, Church's Fried Chicken, Popeye's Chicken, Tasty Fast Foods, Triple J's Smokehouse Bar-B-Q
	AServ	I: First Klass Tire Shop & Auto Supply
		O: High Performance Auto Center, Mose Stephens
	Parks	I: **Hutcheson Park**
	Other	I: **B. B. Pharmacy, C & J Supermarkets Co. of Houston,** Family Detailing Car Wash, **Public Library**
		O: **Fire Station, Price Buster Foods**
22		Homestead Rd
	FStop	O: **Texaco*(D)**
	Food	O: What A Burger
	AServ	I: **Goodyear Tire & Auto**
	TServ	I: **Goodyear Tire & Auto, Houston Peterbilt Inc., Lone Star GMC Trucks**
		O: **Performance Kenworth, Western Star Trucks**
23A		Kirkpatrick Blvd
	TServ	I: **Gulf Coast Truck & Trailor**
		O: **Al Tucker Trailors, Inc.**
23B		N. Wayside Dr., Liberty Rd.
	Gas	O: Exxon* (24 Hrs)
24A		U.S. 90 E. McCarty Dr., Wallisville Rd. (Eastbound Exit Is Listed As Only Exit 24)
	TStop	O: **A-1 Truck Stop (Phillips 66)**
	FStop	I: **Texaco*(D)**
	Gas	O: Chevron*
	Food	O: **Luby's Cafeteria (A-1 Truck Stop)**, Wendy's
	AServ	I: Hart Radiators
	TServ	I: **Olympic International Trucks Inc.**
	ATM	I: **Texaco**
		O: Texas Commerce Bank
24B		Wallisville (Westbound)
	FStop	O: **Mobil***
	Gas	O: Texaco*(D)(CW)
	Food	O: **McDonalds (Playground)**
	ATM	O: Pinemont Bank
25		Gellhorn Drive (Eastbound)
	TServ	I: **CCC Truck Parts Specialists, Houston Freightliner Inc.**
	Med	I: ✚ **Hospital**
26A		I-10 West, Downtown, I-10 East, Beaumont
26B		Market St. (Northbound)
	TStop	I: **Fu-Kim Restaurant & Truck Stop* (24 Hrs)**
	Food	I: **Tiger Sports Bar (Fu-Kim TS)**
	ATM	I: **Fu-Kim TS**
27		Turning Basin Drive (Southbound)
28		Clinton Dr.
	Gas	O: Mobil*(D)
29		Port of Houston Main Entrance
30A		Manchester St. (Southbound, No Southbound Reaccess)
	AServ	O: A & B Auto Electric, Guajardo's

Bold red print shows RV & Bus parking available or nearby

EXIT TEXAS

Other O: Moonlight Grocery
30B East Tex 225, Pasadena, La Porte, Deer Park (San Jacinto Battlepark)
30C Texas 225 W., Lawndale Ave. (Limited Access Hwy)
30D Lawndale Ave. (Northbound)
31 Broadway Blvd.
Gas I: Conoco*(D)
O: Diamond Shamrock*(D) (24 Hrs), Texaco*(D)
Food I: Mexico Tipico
O: Ostioneria Villa del Sol, Taqueria Reyes
ATM I: Conoco
O: Diamond Shamrock
Other I: **Colony Express Market, Dry Clean Washateria**
32A I-45 South Galveston
32B I-45 North Downtown, TX 35 South, Alvin
33 Phones Rd. & Woodridge Dr.
Gas O: Phillips 66, Shell*(CW)
Food I: Piccadilly Cafeteria, Wendy's
O: Frank's Grill, KFC, Spanky's Pizza, Pasta, Salads, Burgers, What A Burger
AServ I: Auto Air Conditioning Center, Carburetor Shop, Davila's, Leon's Tire Shop
O: Houston Car Clinic, Phillips 66
ATM I: Savings of America, Texas Commerce Bank
Other I: **Gulf Gate Shopping City**
O: Family Wash Center Washateria
34A Long Dr., S. Wayside Dr.
Gas I: Chevron, Exxon*, Texaco* (24 Hrs)
O: Coastal*(D), Conoco*, Diamond Shamrock*
Food I: **Wendy's**
AServ I: Chevron
O: Allstate Transmissions, Dickson Auto Supply, Doreck's Automotive Service, Holstien's Auto, Radiator's Plus, Republic Battery
ATM I: Texaco
O: Conoco, Diamond Shamrock
35 Crestmont St., M L King Blvd, Mykawa Rd.
FStop I: **Mobil***(D)
Gas I: Conoco*(D)
Food I: Burger King (Playground)
O: Best Seafood Market
Other I: **Mrs. Baird's Thrift Store, Trading Fair II**
36A Calais Rd., Holmes Rd., FM 865, Cullen Blvd., M L King Blvd
Gas O: 5-11 Convenient Store*(D)
AServ I: King Center Auto Parts, Onazena Used Car & Auto Repair, Toyota Specialists, Willie's Tire Service
Other I: **Donovan Marine**
O: **Welding Supplies**(LP)
36B FM 865, Cullen Blvd (Westbound)
Gas O: Chevron (24 Hrs), Diamond Shamrock* (24 Hrs), Mobil*
Food O: Tri's Bar-B-Que (Diamond Shamrock)
Lodg I: Crystal Inn
O: Crown Plaza Inn, Cullen Inn
AServ I: Marshall "Genuine" Autos
37 Scott Street
Gas I: Shell*, Texaco*
Food I: Bunky's Cafe, Pappa's Bar-B-Q, Seafood King
Lodg I: Motel
AServ I: Dave's Foreign Auto Parts Inc.
Other I: **Grocery Save & Save, Joseph's Food Store, St. Theresa's Meat Market**
38A TX 288 N, Downtown, Zoo TX 288 S. Lake Jackson, Freeport (Limited Access Highway)
38C S. TX 288 Lake Jackson Freeport
39 Court Of Houston. Main Entrance

↑TEXAS

Begin I-610, Houston

EXIT TEXAS

Begin I-635

↓TEXAS

35A TX121S, Dallas - Ft.Worth Airport, TX121N, Bethel Rd.
35 Royal Lane
34 Freeport Pkwy

EXIT TEXAS

33 Belt Line Rd.
31 MacArthur Blvd.
30 TX161S, Valley View Ln
29 Luna Rd.
27BC I-35E West, Dallas - Ft.Worth Airport. I-35E North, Denton
27A Denton Drive, Harry Hines Blvd.
Food S: Kore a House Restaurant & Club, Mama's & Daughter's Diner
Lodg N: Holiday Inn Select
Other S: **Pensky Truck Rental & Leasing**
25 Webb Chaple Rd., Josey Lane
Med N: **✚ Hospital**
24 Marsh Lane
AServ S: Master Care Auto Service
23 Midway Rd, Welch Rd
Gas N: Exxon
S: Mobil*(D)(CW)
Food N: Luby's Cafeteria, McDonalds, Owens, Taco Bell, Wendy's
Lodg N: Hilton
Other N: **Freed's Home Furnishings, The Home Place, Wal-Mart**
22CD Toll Way
22B Dallas Parkway, Inwood Rd, Welsh Rd.
22A Montfort Dr.
Gas N: Shell(CW)
Food N: Arby's, Burger King, Taco Bueno
AServ N: Goodyear Tire & Auto
Other N: **Target Department Store**
21 TX289, Preston Rd.
Lodg N: Best Western (see our ad on this page)
20 Hillcrest Rd.
19B US75S, Coit Rd.
19A Richardson Plano, Sherman, US175
Lodg N: Hampton Inn (see our ad on thispage),

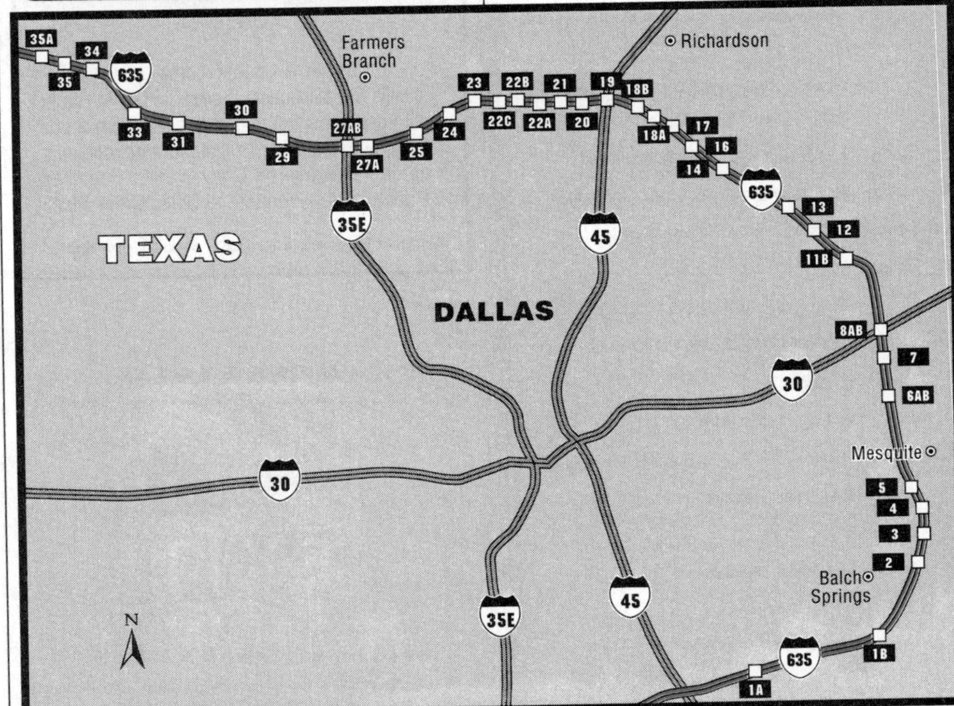

EXIT	TEXAS
	Holiday Inn (see our ad on this page)
18B	Floyd Rd
18A	Greenville Ave.
Gas	**E:** Texaco*(D)
	W: Exxon*(CW)
Food	**W:** Cheddar's, Ci Ci's Pizza, Don Pablo's, Olive Garden, Outback Steakhouse, Red Lobster, Shoney's, Sushi Ichiban
Other	**W:** Animal Clinic
17	Forest Lane, Abrams Rd.
Gas	**W:** Chevron*(CW), Fina, Mobil, Texaco
Food	**W:** Colter's BBQ, KFC, Long John Silvers, McDonalds, Sonic, Steak & Ale, Taco Bell
Lodg	**W:** Royal Inn
ATM	**W:** Bank of America, Mobil
Other	**W:** Tom Thumb Grocery (Pharmacy)
16	Skillman St., Audelia Rd.
Food	**E:** Denny's, Italian Oven
RVCamp	**W:** Foretravel (see our ad on this page)
14	Plano Rd
Gas	**W:** Texaco
Food	**W:** BoBo China
13	Jupiter Rd, Kingsley Rd
Gas	**W:** Texaco*
Lodg	**E:** (AAA) Sleep Inn
Other	**E:** Travel Center
11B	Northwest Hwy, Shiloh Road
Gas	**E:** Citgo* (ATM), Racetrac* (ATM)
	W: Chevron*
Food	**E:** Arby's, Chili's, Chinese Buffet, Colter's BBQ, Denny's, ElChico, Furr's, Grandy's, KFC, Steak & Ale
	W: Bennigan's, Sonic, Taco Cabana, What A Burger
Lodg	**E:** (AAA) La Quinta Inn, Ramada Limited
Lodg	**W:** Hampton Inn (see our ad on this page)
RVCamp	**W:** Morgan's RV's Sales & Service
Other	**E:** Pep Boys Auto Center
	W: Incredible Universe, Sam's Club
11A	Centerville Rd., Ferguson Rd.
Gas	**E:** Shell(CW), Total
	W: Citgo, Mobil* (ATM)
Food	**E:** Jack-In-The-Box, McDonalds, Wendy's
Lodg	**E:** (AAA) Best Western
ATM	**E:** Bank One
Other	**E:** Mervin's, Office Depot
	W: McKinney Animal Clinic
9A	Oates Dr.
Gas	**E:** Diamond Shamrock*(CW), Exxon*
	W: Citgo*, Exxon*(CW), Mobil*, Swifty
Food	**E:** Burger King, Pizza Hut
	W: Cracker Barrel, Hunan Dragon Chinese
AServ	**W:** Kwik Kar Lube & Tune
ATM	**E:** Northeast Bank
Other	**E:** Albertson's Grocery
8AB	I-30W Dallas, I-30E Texarkana
7	Town East Blvd.
Food	**E:** Chili's, Grady's Bar & Grill, Owens, Tia's Tex Mex
	W: Hooters, Olive Garden, Outback Steakhouse, Red Lobster
Other	**E:** Town East Mall
6AB	US80W Dallas, I-80E Terrell (Difficult Reaccess)
5	Gross Rd.
Gas	**E:** Diamond Shamrock*(D)(CW)
	W: Overland*
Food	**W:** Hoff Brau Steaks
Lodg	**E:** Spanish Trails Inn
4	TX352, Military Parkway, Scyene Rd
Gas	**W:** 1 Stop, Chevron*
Food	**W:** Dairy Queen, Jack-In-The-Box, McDonalds, Taco Bell, Waffle House, What A Burger
3	Bruton Road
Gas	**E:** Mobil*
Food	**E:** Mesquite Arena
2	Lake June Rd.
Gas	**E:** Exxon*(CW)
	W: Citgo*, Total*
Food	**W:** Arby's
1B	Elam Road
Gas	**E:** Shell*(CW)
	W: Texaco*(D)
Food	**E:** Captain D's Seafood, KFC, Sonic
	W: Burger King (Outdoor Playground), Checkers Burgers, Grandy's, McDonalds, Taco Bell, Taco Bueno
AServ	**W:** Chief Auto Parts, Kwik Kar
ATM	**E:** Gateway Bank
Other	**E:** One Stop Food Store
	W: K-Mart, Kroger Supermarket
1A	Seagoville Rd

↑ TEXAS

Begin I-635

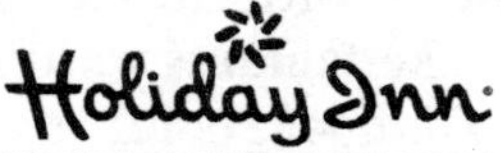

PLANO
700 East Central Parkway
Plano, Texas 75074
972-881-1881
• 161 guest rooms
• NEW! Pin & Pockets Electronic Game Room with golf simulator, video games and billiards
• Bristol's Restaurant • Outdoor pool and spa
• In-room hair dryers • Convenient to SouthFork Ranch, historic Plano and Collin Creek Mall
Independently owned and operated by Bristol Hotels & Resorts™
HO7507

Bold red print shows RV & Bus parking available or nearby

EXIT TENNESSEE

Begin I-640

↓TENNESSEE

0 Junction I-40, US 25, Asheville, Nashville

8 Washington Pike, Mall Rd South, Millertown Pike
- **Gas** S: Conoco*
- **Food** S: Little Caesars Pizza, Subway
- **AServ** S: Pep Boys Auto Center
- **Med** S: ✚ Walk-In Healthcare Center
- **ATM** S: First American, First Tennessee Bank
- **Other** N: Farmer's Market, Three Ridges Golf Course
  S: Food Lion (24 Hrs), Good Shepherd Christian Bookstore, Regal Cinema's, The Home Depot[LP]

6 US 441, Broadway
- **Gas** N: Conoco*
- **Food** N: Ci Ci's Pizza, Fazoli's Italian Food, Hardee's, Mustard Seed Cafe
  S: Buddy's bar-b-q, Hobo's Family Style Restaurant, Little Caesars Pizza, Louis' Inn, The Original Louis Drive-In
- **AServ** N: Adam's Auto Repair, Classic Car Supplies, Prestige Performance
  S: Alert Transmissions, Eddie's Auto Parts, Xpert Tune
- **ATM** N: SunTrust Bank
- **Other** N: Companion Animal Hospital, J B's Car Wash, Mail Boxes Etc, Revco, Strictly Feathers
  S: 30 Min. Alignment, Food City, K-Mart, Revco, Wharehouse Golf

3 N I-75, S I-275, Knoxville, Lexington

3B US 25 W, North Gap Rd, Clinton (Eastbound)

TN 62, Western Ave
- **Gas** N: Racetrac*, Texaco*
  S: Shell[D]
- **Food** N: Baskin Robbins, Central Park Hamburgers, KFC, Little Caesars Pizza, Long John Silvers, McDonalds (Playground), Panda Chinese Restaurant, Shoney's, Subway, Taco Bell, Wendy's
  S: Dad's Donuts & Delights, Domino's Pizza, Hardee's, Krystal, Tracey's Restaurant Home Cookin
- **AServ** S: Advance Auto Parts, Mighty Muffler, Shell[D]
- **ATM** N: First American Bank, Kroger Supermarket (24 Hrs), SunTrust
  S: First Tennessee Bank
- **Other** N: Kroger Supermarket (24 Hrs), Revco, WalGreens
  S: Car Wash, Cokesbury Books & Church Supplies, Pack n Ship Mail Center, Super Wash House, US Post Office

↑TENNESSEE

Begin I-640

EXIT OHIO

Begin I-675

↓OHIO

26AB I-70

24 OH 444, Enon

22 OH 235, Fairborn

20 Dayton, Yellow Springs Rd.
- **Other** W: Fairborn Animal Hospital

17 North Fairfield Rd. Wright Patterson Air Force Base Area A&C
- **Gas** E: BP*[CW]
  W: Speedway*
- **Food** E: Bagel Cafe, Boston Market, Burger King, Chili's, Cooker, Don Pablo's Mexican Restaurant, Max & Erma's Restaurant, McDonalds (Wal-Mart), Olive Garden, Red Lobster, Steak & Shake, T.G.I. Friday's
  W: Arby's, Blimpie's Subs, Bob Evans Restaurant, Chi-Chi's Mexican Restaurant, Cold Beer & Burgers, Greek Isle Deli, Holiday Inn, Kim's Chinese, McDonalds, Taco Bell, The Bagel Shoppe, The Flying Pizza
- **Lodg** W: ◆ Fairfield Inn, ◆ Hampton Inn, ◆ Holiday Inn, AAA Homewood Suites, ◆ Red Roof Inn
- **Other** E: Drug Emporium, PetsMart, Sam's Club, The Mall of Fairfield, Wal-Mart (Pharmacy)

15 Wright Patterson Airforce Base

13AB Dayton, Xenia, US 35

10 Dorothy Ln., Indian Ripple Rd.
- **Gas** W: BP*, Shell[CW]
- **Food** W: Bill Knapp's Restaurant, Burger King, China House, Fulmer Deli & Bakery, Martino's, Pizza Hut Carry Out
- **Other** W: K-Mart (Pharmacy), Revco Drugs

7 Wilmington Pike, Bellbrook
- **Gas** E: BP*[CW], Shell*[CW]
  W: Super America*
- **Food** E: Arby's, Beijing Garden Chinese, Boston Market, Burger King, Fazoli's Italian Food, Frisch's Big Boy, Kenny Rogers Roasters, McDonalds, Ponderosa Steakhouse, Steak & Shake, Subway, Taco Bell, Wendy's
  W: Bob Evans Restaurant
- **ATM** E: National City Bank, Star Bank (Cub Foods)
- **Other** E: Aldi Supermarket, Cub Foods Supermarket, Drug Emporium, Wal-Mart (Pharmacy)

4AB OH 48, Kettering, Centerville
- **Food** E: Double Day's Grill and Tavern, My Favorite Muffin & Bagel, Outback Steakhouse, Subby's, TCBY Yogurt
  W: Bill Knapps Restaurant
- **Other** E: Pearl Vision Center, Sear's Hardware

2 Miamisburg, Centerville
- **Gas** E: BP*
  W: Amoco[CW]
- **Food** E: Boston Market, Carvers Steaks and Chops, Damon's, J Alexanders Restaurant, Long John Silvers, McDonalds, The Chop House
  W: Arby's, Baskin Robbins, Bravo Italian Kitchen, Jade Garden, T.G.I. Friday's
- **Lodg** E: ◆ Hampton Inn
  W: ◆ Quality Inn
- **AServ** E: Lexus of Dayton
  W: K-Mart, Midas Muffler
- **Med** W: ✚ Hospital
- **ATM** E: Bank One
  W: Citizens Federal
- **Other** E: Car Wash
  W: K-Mart (Pharmacy)

↑OHIO

Begin I-675

EXIT OHIO

Begin I-694

↓MINNESOTA

27 I-94 split from I-694, to St Cloud

28 Hemlock Lane

29AB U.S. 169 (Divided Hwy)

30 (31) Boone Ave
- **Food** O: Northlind Inn
- **Lodg** O: Northlind Inn
- **Other** I: Home Depot, Petsmart

31 County Road 81
- **Gas** I: Amoco*[CW]
  O: 76*
- **Food** I: Wings & Ribs
  O: Best Western, Dairy Queen, Wendy's
- **Lodg** I: Budget Host Inn
  O: Best Western
- **AServ** I: Kara Collision & Glass, Kennedy Transmission
  O: 76
- **TServ** I: ABM Truck Service
- **ATM** I: Amoco

33 Brooklyn Blvd
- **Gas** O: Total*
- **AServ** O: Car X, Dodge Truck Dealer, GM Auto Dealership
- **RVCamp** O: Truck/RV Service

34 Shingle Creek Pkwy
- **Gas** O: Mobil*[CW]
- **Food** I: Ground Round, Lee Ann Chin Chinese (Carry Out), Subs Ect
  O: Barnacle Bill's, Chi Chi's Mexican, Cracker Barrel, Denny's, Hardee's, Olive Garden, T.G.I. Friday's
- **Lodg** I: Country Inn & Suites
  O: Budgetel Inn, Comfort Inn, The Hilton
- **AServ** I: Tires Plus
  O: Mobil
- **Other** I: Police Station, Target Department Store

35A Minnesota 100

35B I-94 Jct East

35C 252

36 East River Rd

37 Minnesota 47, University Ave
- **Gas** I: 76, Amoco[CW]
  O: Super America*
- **Food** O: Burger King, Hardee's, McDonalds, The Cattle Company
- **AServ** I: 76, Amoco
  O: Valvoline Quick Lube Center
- **ATM** O: Super America
- **Other** O: Holiday Plus Grocery, Petsmart

38AB (39) Minnesota 65, Central Ave
- **Gas** I: Ashland*, Super America*
  O: Citgo, Sinclair
- **Food** I: Boston Market, Denny's, Dunkin Donuts, Embers, Hardee's, KFC, La Casita Mexican, Taco Bell, The Ground Round Restaurant, Tycoon's Tavern & Grill, White Castle Restaurant
  O: Joe DiMaggio's Sports Bar & Grill, Shorwood, Subway
- **Lodg** I: Best Western
- **AServ** I: Jiffy Lube, Tires Plus
- **ATM** I: First Bank, TCF Bank
- **Parks** O: City Park
- **Other** I: All Pets Animal Hospital, Car Wash, Dental Clinic, K-Mart, Menard's, Petco, Skywood Mall, Target Department Store (Pharmacy)
  O: Brown Berry Bakery Outlet, Dental Clinic

39 Silver Lake Rd
- **AServ** O: Ford Dealership

40 Long Lake Rd, Tenth St NW

41AB I-35 Junction

EXIT	MINNESOTA
42B	U.S. 10 W, Anoka Ave (Divided Hwy)
42A	Minnesota 51, Hamline, Snelling (Eastbound Reaccess)
43A	Lexington Ave
Gas	I: Amoco*(D)(LP), Sinclair*, Total*
Food	I: Blue Fox Bar & Grill, Burger King, Perkin's Family Restaurant
Lodg	I: Emerald Inn O: Hampton Inn
AServ	I: Abra Auto Body & Glass, Amoco, Goodyear Tire & Auto, Kennedy Transmission, Sinclair
ATM	I: Amoco O: Firststar Bank
Other	I: **Target Department Store**
43B	Victoria Street
45	**(46)** Minnesota 49, Rice Street
Gas	O: Phillips 66*(CW), Total*(D)(LP)
Food	I: Burger King, Hardee's, Taco John's O: Taco Bell
ATM	O: Phillips 66
Other	O: **Dental Clinic, Pet Junction**
46	I-35 E Jct S to St. Paul (Left Lane Exit)
47	I-35 East, N to Duluth
48	U.S. 61
AServ	O: Buick/Honda Dealer, Cook's, Dodge Dealer, Ford Dealership, GM Auto Dealership, K&K Transmission, Subaru Acura Dealer
50	**(49)** White Bear Ave (Closed Due To Construction)
Gas	I: Amoco* O: Phillips 66*(CW), Super America*
Food	I: Applebee's, Arby's, Bakers Square Restaurant, Best Western, Chi Chi's Mexican, Ciatti's Italian, Denny's, Godfather's Pizza, Hardee's, North China, Old Country Buffet, Perkin's Family Restaurant, Pizza Hut, Pizza Hut Carry-Out, Taco Bell, Wendy's
Lodg	I: (AAA) Best Western
AServ	I: AAMCO Transmission, Amoco, Firestone Tire & Auto, Goodyear Tire & Auto, Midas Muffler & Brakes O: K-Mart
ATM	I: TCF Bank
Other	I: **Cole's Department Store**, Hurricane Carwash, **Maplewood Mall, Pet Food Warehouse** O: **K-Mart (Pharmacy)**
51	Minnesota 120
Gas	I: Conoco*(LP) O: Super America*(D)
Food	O: Blimpie's Subs, Bridgeman's Soda House, Jethro's Char-house
ATM	O: Super America
Other	I: **Pet Images**
52AB	Minnesota 36, Stillwater, St. Paul (Divided Hwy)
55	Minnesota 5
Gas	I: Amoco*(CW) (24 Hr), Holiday*(D)(CW)
AServ	I: Amoco
ATM	I: Holiday
Other	I: **Menards**
57	County Rd 10 , Tenth St
Gas	I: Conoco* O: Super America*(D)(CW)(LP)
Food	I: Brick Hearth Pizza, Brothers Subs, Burger King, China House Buffet, Hunan Palace, KFC, Little Caesars Pizza
ATM	I: Conoco, Norwest Bank, **Rainbow Food Mkt**, Western Bank O: Super America
Other	I: **K-Mart (Pharmacy), Oakdale Mall, Oakdale Optical, Pet Food Warehouse, Rainbow Foods Market**
58AB	I-94 Junction

↑ MINNESOTA

Begin I-694

EXIT	MARYLAND

Begin I-695

↓ MARYLAND

1	MD173, Hawkins Point Rd, Pennington Ave, Ft Smallwood Rd
2	MD10S, Severna Parkway (Exit 3AB Outer Loop Services Accessible From This Exit)
Food	I: Roy Rogers, Sbarro Pizza
AServ	I: Bay Meadow Isuzu
ATM	I: First National Bank Of Maryland
Other	I: **Petsmart, Price Club Plaza**
3AB	MD2S, Ritchie Hwy, Glen Burnie
Gas	O: Texaco*
Food	I: Days Inn O: China Spring (Chinese), Chuck E. Cheese's, Denny's, Golden Corral, Holiday Inn, Hunan Inn, KFC, Old Country Buffet, Pizza Hut, **Pizza Hut (Wal-Mart)**, Taco Bell, Tail Gators
Lodg	I: (DAYS INN) Days Inn, Holiday Inn O: Hampton Inn (see our ad opposite page), Holiday Inn
AServ	I: AAMCO Transmission, Bill's Transmission O: Just Tires, Montgomery Ward, Parts America, Saturn
Med	I: ✚ **Glen Burnie Medical Clinic**
ATM	O: First Union, Giant Supermarket, **Mars Supermarket, Morrow's Supermarket**
Other	O: **America's Best Contact Lens & Eyeglasses, Barenburg Eye Assoc, Expressway Shopping Ctr, For Eyes Optical, Giant Supermarket & Pharmacy, Glen Burnie's Mall, Governor Plaza, Haus Dairy Store, K-Mart, Mail Boxes Etc, Mars Supermarket, Morrow's Supermarket, Optical Center**, Self Service Car Wash, **Wal-Mart Supercenter (Pharmacy, Optical)**
4	Jct I-97S, to Annapolis, Bay Bridge (Left Exit)
5	MD648, Baltimore - Annapolis Blvd, Ferndale
Gas	O: Amoco*
Food	O: Kwong's Chinese, Little Caesars Pizza, Roy

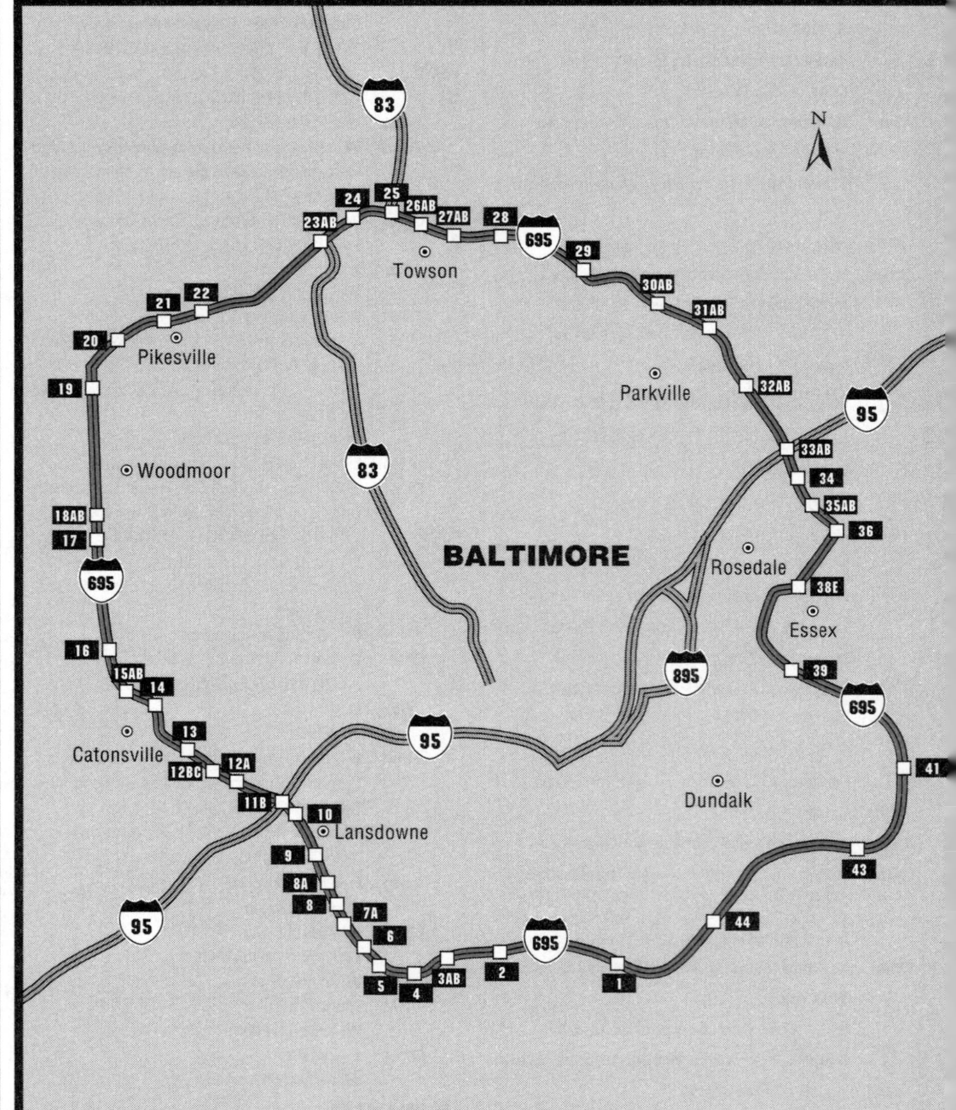

Bold red print shows RV & Bus parking available or nearby

EXIT MARYLAND

Rogers
AServ I: Mike's Auto Sales & Repair
ATM O: Crestar Bank
Other I: County Police
O: 7-11 Convenience Store, Basics Supermarket, Burrwood Plaza, Revco Drugs

6 MD170, Camp Meade Rd, Linthicum (Outerloop, No Eastbound Reaccess)
Gas I: Shell*
Food I: Bogie's Hoagies, Boli's Pizza, Checkers Burgers, The Rose (Comfort Inn)
Lodg I: Comfort Inn
AServ I: All State Motors, JS Lee's Body Shop, John's Auto Body, M&D Auto Repair, North Lin Auto Parts

7A MD295S, Baltimore Washington Pkwy

8 Hammonds Ferry Rd, MD168, Nursery Rd
Gas I: Exxon*, Shell*(CW)
O: Amoco*, Citgo*(D)(CW)
Food I: Burger King, KFC, McDonalds (Playland), Taco Bell, Wendy's
O: Charlie's Donuts & Subs, G&M, Philadelphia Style Pizza & Steaks, Snyder's Willow Grove Restaurant, Szechuan Hot Wok
Lodg I: Motel 6
AServ I: Shell*(CW)
TServ I: Baltimore Mack Truck Service
ATM O: Signet Bank
Other O: Whitei's Grocery Store

8A Jct I-895N Tunnel, Toll Rd

9 Hollins Ferry Rd, Lansdown
Gas I: Exxon*(D), Mobil
Food I: Victor's Place Deli*
AServ I: Exxon, Mobil
TServ I: Hooe's Truck/Auto Service
O: Goodyear Tire & Auto
Other I: Royal Farms C-Store

10 U.S. 1 Alt, Washington Blvd (Innerloop)

11B Jct I-95, Baltimore, Washington, Sulfer Rd

12A U.S. 1, Southwestern Blvd, Arbutus (Outerloop, Innerloop Reaccess Only)

12BC MD372, Wilkens Ave

EXIT MARYLAND

13 MD144, Fred Rd, Catonsville
Gas O: Amoco(LP), Cisco*, Crown(LP)
Food O: Dunkin Donuts, Indian Delight Restaurant, Roy Rogers
AServ O: Amoco, Jiffy Lube
ATM O: Amoco, Crestar Bank, F&M Bank, Signet Bank
Other O: 7-11 Convenience Store, Hi's C-Store

14 Edmonson Ave
Gas I: Exxon
O: Amoco
Food I: Catonsville Bakery
O: Opie's Snoballs, Ice Cream, Soup & Sandwiches, Papa John's Pizza
AServ I: Catonsville Auto Care Center, Exxon
O: Amoco (Wrecker Service)
Other I: Belt Way Drug Store, Royal Farms C-Store
O: Beltway Animal Clinic

15AB U.S. 40, Ellicott City, Baltimore
Gas I: Amoco(CW)
O: Amoco(D), Crown*, Shell*
Food I: Bejing Chinese, Burger King, Chi Chi's Mexican, Chuck E. Cheese's Pizza, IHOP, KFC, Kings Court, McDonalds, Mr G's Fast Lane Hamburgers, Ronda's Diner (Howard Johnson), Tiovoli Restaurant, Tivoli Garden Restaurant (Days Inn)
O: Doubletree Diner, Dunkin Donuts, Mariah's, McDonalds, McDonalds (Wal-Mart), Sorrento's Pizza, Taco Bell
Lodg I: Days Inn, Howard Johnson
AServ I: Firestone Tire & Auto, Just Tires, Musselman's Dodge, Speedee Muffler King, Westview Motors
O: Caton Auto Clinic, Firestone Tire & Auto, Fisher Auto, Goodyear Tire & Auto, Jiffy Lube, Nanik Mufflers, Precision Tune, Shell, Windshields Of America
ATM I: First Nat'l Bank of Maryland, Merchantile Bank & Trust, NationsBank
O: Crown, Signet Bank
Other I: Ingleside Shopping Ctr, Optical Fair, Rite Aide Pharmacy, Safe Way Grocery, Water Works Car Wash, West View Mall
O: Car Wash, Giant Food & Pharmacy, Super First Food Mkt, Wal-Mart (Pharmacy, McDonald's)

16 Jct I-70, Frederick

17 Security Blvd, Woodlawn, Rolling Rd
Gas I: Amoco*, Shell*(CW)
O: Exxon*(D), Mobil*(CW)
Food I: Chih Yuan, Dunkin Donuts, McDonalds,

EXIT MARYLAND

Pargo's (Comfort Inn), Pizza Bolli's, Red Lobster, Wendy's
O: Applo's Carry Out, Bennigans, Burger King, Holiday Inn, IHOP, McDonalds, Szechuan Express
Lodg I: AAA Comfort Inn, Days Inn ◆ Days Inn, Motel 6
O: ◆ Holiday Inn (Belmont Ave, Frontage Rd) (see our ad on this page)
AServ I: Amoco, GM Auto Dealership, Jiffy Lube, Security Nissan
O: Firestone Tire & Auto, Koons Ford, Montgomery Ward
ATM I: NationsBank
O: Merchantile Bank, Mobil, NationsBank, Super Fresh Food Mkt, Susquehanna Bank
Other I: Food King Grocery Store, Food King Shopping Ctr, Rite Aide Pharmacy
O: Chadwick Manor Shopping Ctr, Coin Operated Laundry, Rite Aide Pharmacy, Security Square Mall, Super Fresh Food Mkt

18AB MD26, Randallstown, Lochearn
Gas I: Crown(D)
O: Exxon*(D)
Food I: Domino's Pizza
O: Baskin Robbins, Burger King, Dunkin Donuts, Golden Dragon Chinese, American, Rice Express (Milford Shopping Ctr), Sea Pride 2 Crab, Subway, Taco Bell
AServ O: D&S Auto Repairs, Exxon, Jiffy Lube, Parts America
Med O: ✚ Hospital
ATM O: Signet Bank
Other I: Revco Drugs
O: 7-11 Convenience Store, Giant Food & Drugs, Milford Self Serv Car Wash, Milford Shopping Ctr, Pearl Vision

19 Jct I-795, NW Expressway, Owings Mills, Reisterstown

20 MD140, Reisterstown Rd, Garrison, Pikesville
Gas I: Amoco(D), Mobil, Shell*
O: Exxon(D)
Food I: Brasserie, Fuddrucker's, Holiday Inn, McDonalds, San Marco's Restaurant (Hilton), Shoney's
O: Al Petchino Cafe, Donna's Coffee Bar (Woodholme Shopping Ctr), Jasper's, Kenny Rogers Roasters, Starbuck's Coffee, Sutton's Place Gourmet, TCBY Yogurt
Lodg I: AAA Comfort Inn, Hilton, ◆ Holiday Inn
AServ I: Mobil, Shell
ATM I: Crestar Bank, Hopkins Federal Savings Bank, Provident Bank
O: First Union, Merchantile Bank & Trust (Woodholme Shopping Ctr), Signet Bank
Other I: Old Court Animal Hospital, Pamona Square, Royalty Hand Car Wash, Shopping At The Hilton
O: Neighbor Care Pharmacy, Woodholme Shopping Ctr

21 Stevenson Rd, to Park Hgts Ave (Difficult Reaccess)

22 Greenspring Ave
Other O: Irvine Natural Science Center

23AB JctI-83S, Baltimore, MD25, Falls Rd, Greenspring Ave
Gas O: Exxon(D)
Food O: Food Court At Greenspring Station, Harvey's

Bold red print shows RV & Bus parking available or nearby

EXIT		MARYLAND
		Restaurant, Joey Chiu Restaurant, **Windy Valley Homemade Ice Cream (General Store)**
	Lodg	**O:** Greenspring Inn
	AServ	**O:** Exxon
	ATM	**O:** Merchantile Bank & Trust, NationsBank
	Other	**O: Greensprings Station, Post Office (Greenspring Station), Windy Valley General Store (Ice Cream)**
24		Jct I-83 N, Timonium, York PA
25		MD139S, Charles St (No Trucks)
	Med	**I: Hospital**
26AB		MD45, York Rd, Lutherville, Towson
	Gas	**O:** Exxon*
	Food	**I:** McDonalds
		O: Friendly's, Subway, The Peppermill
	AServ	**I:** Goodyear Tire & Auto, Kimmel Car Care, Midas Muffler & Brakes, Towson Ford
		O: Lutherville Car Care
	ATM	**I:** Provident Bank of MD, Signet Bank
	Other	**I:** Car Wash
		O: Fire Museum of MD
27AB		MD146, Dulaney Valley Rd, Towson
	Food	**I:** Boston Market, Carnegie's Restaurant (Sheraton), Fisherman's Warf, Silver Diner, T.G.I. Friday's (Towson Town Center)
	Lodg	**I:** Sheraton
	ATM	**I:** NationsBank (Towson Town Center), NationsBank
	Other	**I: Dulaney Plaza, Howard & Morris Pharmacy, Mail Boxes Etc, Super Fresh Food Market, Towson Town Center Mall**
		O: Hampton National Historic Site
28		MD 45, to York Rd, Lutherville, Towson
	Gas	**I:** Citgo*(D), Crown*(D)(CW)
		O: Amoco(D), Exxon*(D), Shell
	Food	**I:** Bubba's Breakaway, McDonalds, The Diner Restaurant
		O: Carvella's, Chesapeake Gourmet Seafood, Friendly's, Ocean Pride Restaurant, Peppermill Restaurant, Pizza Hut, Steak House Of Japan, Subway, Szechuan House Restaurant
	AServ	**I:** Brooks Tire & Auto Ctr, Citgo, Griffith, Kimmel Tire, Merchant's Tire & Auto, Midas Muffler, Star Auto, Towson Ford, Tyre's Auto Repair
		O: Exxon*(D), Rehax Lutherville Car Care, Shell
	ATM	**I:** Harbor Federal Savings, Royal Farms C-Store
		O: 7-11 Convenience Store, Columbia Bank, Farmers Bank Of Maryland
	Other	**I:** Car Wash (Full Service), **Royal Farms C-Store, United Optical, Vinson Animal Hospital**
		O: 7-11 Convenience Store (ATM), Fire Museum Of Maryland, Galleria, Rite Aide Pharmacy
29		MD542, Loch Raven Blvd, Cromwell Bridge Rd
	Gas	**I:** Amoco(D), Crown*, Hess(D)
	Food	**I:** Bel Loc Diner, Crystal Restaurant (Days Inn), Don's Crab, Dunkin Donuts, Enroy*, Fisherman Island Restaurant (Ramada Inn), Joppa Seafood Market, McDonalds, Papa John's, Pastori's Delly, Philadelphia Style Pizza, Pizza Hut, Sarando's Restaurant, Vit's Pizzeria, Wendy's
	Lodg	**I:** DAYS INN Days Inn, Ramada Inn, Towson East Motel, Welcome Inn
		O: Holiday Inn
	AServ	**I:** All Tune & Lube, Amoco, Hollenshade's, Koller Auto Service, Mike's Auto Supply, Pep Boys Auto Center, Precision Tune, Savco Tune & Lube, Speedee Muffler King, The Pit Stop Oil Change, Towson Dodge
	ATM	**I:** Carrollton Bank
	Other	**I: 7-11 Convenience Store, Beltway Animal Hospital, Midas Muffler,** Will's Car Wash (Hand Wash & Dry)
30AB		MD41, Perring Pkwy
	Gas	**O:** Texaco*
	Food	**O:** Burger King, Carney Crab House, Denny's, Dunkin Donuts, Fishermans Exchange Seafood Buffet, KFC, Kelly's Deli House, McDonalds, Milanos Restaurant (Italian, American, Greek), Momma Lucia Expresso & Capaccino, Old Country Buffet, Pizza Bollie, Popeye's, Poulet Chicken, Roy Rogers, Subway, The Ground Round
	AServ	**O:** Bob Davison Ford, Hillen Tire, Jeff's Tires, Jerry's, Jiffy Lube, K-Mart, National Tire & Battery, Texaco
	ATM	**O:** Chesapeake Federal Bank, First Nat'l Bank, Metro Food Market, NationsBank
	Other	**O: Giant Food & Drug Store, Hives Dairy Store, K-Mart, Metro Food Market, North Plaza Mall, Pering Plaza, Valu Food Grocery Store**
31AB		MD147, Harford Rd, Carney, Parkville
	Gas	**O:** Amoco*, Citgo*, Mobil, Texaco(D)
	AServ	**I:** MB Auto Supply
		O: BMW Auto Electric, Carney Tire & Car Care, Griffeths Dodge,Plymouth, J&J Auto Supply, Mobil, Texaco
	ATM	**O:** Citgo
	Other	**I: Carney Animal Hospital, Coin Laundry**
		O: CVS Pharmacy, Carney Professional Eye Care
32AB		U.S.1, Bel Air, Overlea
	Gas	**I:** Crown*, Getty
		O: Exxon*(CW)
	Food	**I:** Cookies By Design, McDonalds, Mr Crab Seafood, Schonners, Szechuan Taste, The Canopy BBQ
		O: Burger King, Caravel Ice Cream & Bakery, Denny's, Dunkin Donuts, Fisherman's Warf, IHOP, Kenny Rogers Roasters, PeKing House Chinese, Philedelphia Style Pizza & Subs, Roy Rogers, Taco Bell, Tully's
	AServ	**I:** Battery Warehouse, Melvin's Tire, Papa John's Pizza
		O: Fullerton Auto Service, Jerry's Toyota, **K-Mart,** Kimmel Tire & Auto, Merchant's Tire & Auto, Perry Hall, Salvo Auto Parts, Strohminger
	ATM	**O:** 7-11 Convenience Store, Eastern Savings Bank, NationsBank
	Other	**I: Animal Medical Hospital, Bookstore, United Optical, Valu Food**
		O: 7-11 Convenience Store, Fullerton Animal Hospital, Giant Food & Drug Store, K-Mart (Pharmacy), Pet Nook (Pet Grooming)
33AB		Jct I-95, North to New York, South to Baltimore, Washington (Southbound Exits Left)
34		MD7, Philadelphia Rd, Rosedale (Services Also Accessible From Exit 35)
	FStop	**I: Exxon*(D)**
	Food	**I:** Pizza Factory Carry-Out
		O: Baskin Robbins, George's Open Pit BBQ, Wendy's (Golden Ring Plaza), Your Way Cafe
	Lodg	**O:** ◆ Susse Chalet
	AServ	**O: Goodyear Tire & Auto (Golden Ring Plaza)**
	Med	**O: Hospital, Rosedale Group Walk-In Clinic**
	ATM	**O:** Point Breeze Credit Union, Royal Farms Stores
	Other	**O: Eastside Animal Hospital, Giant Food & Drug Store, Golden Ring Plaza, Haus Dairy Store, Royal Farms Stores**
35AB		U.S. 40, Pulaski Hwy, Aberdeen, Baltimore (Services Also Accessible From Exit 34)
	Food	**O:** Baskin Robbins, Dunkin Donuts, Gourmet Chinese Restaurant, McDonalds, Wendy's (Golden Ring Plaza)
	Lodg	**O:** Christlen Motel, Continental Inn, Susse Chalet
	AServ	**O:** Bepz Garage, Goodyear Tire & Auto, Montgomery Ward, NTB Tires & Batteries, Sunny's Auto & Truck Repair
	Med	**O: Franklin Square Hospital**
	ATM	**O: NationsBank (Golden Ring Mall),** Point Breeze Credit Union, Provident Bank Of Maryland, Signet Bank
	Other	**O: Coin Operated Laundry, Golden Ring Mall, U-Haul Center(LP), US Post Office**
36		MD702 S, to Essex
38E		MD150 E, Eastern Blvd, Essex
	Gas	**O:** Enroy*
	Food	**I:** Golden Corral, Horn & Horn Smorgasbord, KFC, Kings Court Chicken Steak & Seafood (Southport Mall), McDonalds (Southport Mall)
		O: Chuck E. Cheese's Pizza, Roy Rogers
	AServ	**I:** Bob Bell Chevrolet , Nissan, Sears (Southport Mall), Speedee Muffler King
		O: Sam's Club, Treadway's Auto Service
	ATM	**I:** NationsBank (Southport Mall)
		O: Enroy*
	Other	**I: Metro Supermarket, Revco Drugs, Rite Aide Pharmacy (Southport Mall), Southport Mall**
		O: Diamond Point Plaza, Sam's Club
39		Merrit Blvd, Dundalk (Innerloop)
40		MD151, North Point Blvd N, MD150 Eastern Blvd
41		Cove Rd, MD20, MD151, Dundalk
	Gas	**I:** Citgo*(D) (Kerosene)
	Food	**I:** Pizza Roma
	ATM	**I:** 7-11 Convenience Store, First Nat'l Bank Of Maryland, Provident Bank Of Maryland
	Other	**I: 7-11 Convenience Store, Coin Operated Laundry, Mickey Village,** Self Service Car Wash
43		Bethleham Blvd, Sparrows Point (Difficult Reaccess)
44		Broening Hwy, Dundalk (Outerloop, Innerloop Reaccess Only)
(47)		Toll Plaza - Key Bridge

↑MARYLAND

Begin I-695, Baltimore

Bold red print shows RV & Bus parking available or nearby

EXIT	MICHIGAN

Begin I-696

↓MICHIGAN

17 Bermuda, Mohawk
16 MI 1, Woodward Ave, Main St.
- **Gas** I: Mobil*
- **Food** I: Anna's Coffee Shop
  O: Amici's Pizza, Main Donut Stop, Oxford
- **AServ** I: Mobil, Pennzoil Lube, Wetmore's Tires
- **Other** O: **Detroit Zoo, Holiday Supermarket, Westcott Veterinary Care Center**

14 Coolidge Rd., Ten Mile Road
- **Gas** I: Total*(CW)
- **Food** I: Caesar Kabob, Det. Bagel, Hungry Howie's, Jade Palace, Little Caesars Pizza, Subway
  O: A Taste of the Orient, Ramatari Vegetarian
- **Med** I: **✚ Oak Park Medical Clinic**
- **Other** I: **All Pets, Arbor Drugs, Farmer Jack Supermarket, Wash Quarters Coin Laundry**
  O: **Lincoln Drugs, Rite Aide Pharmacy**

13 Greenfield Rd.
- **Gas** I: Marathon(CW), Mobil, Sunoco*
- **Food** I: Classic Coney Island Family Restaurant, Dunkin Donuts, New York Bagel, Oriental City, Pita Cafe, Royal Kubo Manila Cousine, Schlotzkys Deli, Silverman's, Taste of Israel
  O: Bagel's Plus, Baskin Robbins, Dot & Etta's Shrimp Hut, Euro Food, Giorgio's, McDonalds, New York Pizza World, Pasta To Go, Ponderosa, Ribs & Pizza, Subway, Wendy's
- **Other** I: **Efros Drugs, Greenfield Animal Hospital, VCA Animal Hospital**
  O: **7-11 Convenience Store, Farmer Jack Supermarket, K-Mart (Pharmacy)**

5 Orchard Lake Rd.
- **Gas** O: Marathon, Mobil*(D), Shell*, Sunoco(CW)
- **Food** O: Arby's, Bill Knapps Restaurant, Jets Pizza, Kerby's Koney Island, Orchard Grille Restaurant (Best Western), Quizno's Subs, Roosevelts Grill, Ruby Tuesday, Silverman's Deli Restaurant, Steak & Ale, Steamers Seafood & Grill, Subway, Sushi Ko, Wendy's, Zia's Italian
- **Lodg** O: Best Western, Comfort Inn, ◆ Extended Stay America

12 Southfield Rd., 11 Mile Rd.
- **Gas** I: Shell*
- **Food** O: St Louis Bread

11 Evergreen Rd.
- **Food** I: BW-3 Grill & Pub, T.G.I. Friday's
  O: Eastern American Cousine, La Fendi, Papa Romano's Pizza, Yaldoo's Grocery

10 US 24, Telegraph Rd., Mi 10, Northwest Hwy, Lahser Rd.
- **Gas** I: Marathon*
  O: Mobil*, Shell
- **Food** I: Sunrise Donuts
  O: Empress Gardens, Gateway Deli, Pearl City
- **Lodg** O: Hampton Inn, Marriott
- **Med** O: **✚ Henry Ford Medical Center**
- **Other** O: **Arbor Drugs, Farmer Jack Supermarket**

9 US 24, Telegraph Rd
- **Gas** I: Mobil*, Sunoco
  O: Sunoco
- **Food** I: Bob Evans Restaurant, Holiday Inn
  O: China City Restaurant, Denny's, Food Court (Mall), Little Caesars Pizza (K-Mart), New Seoul Garden, Old Country Buffet, Pearl City Chinese & American Restaurant, Ruby Tuesday
- **Lodg** I: Holiday Inn
  O: Hampton Inn
- **Other** O: Jax Car Wash, **K-Mart (Pharmacy), Office Max, PetsMart, Tell Twelve Mall**

↑MICHIGAN

Begin I-696

EXIT	TEXAS

Begin I-820, Fort Worth

↓TEXAS

34A Jct I-20E, US287S, Waxahachie, Jct I-20W, Abaline
33C Sun Valley Dr.
- **Gas** E: Al's Quick Stop & Grill*
- **Lodg** W: Ecomony Inn

32A US287N
32 Wilberger St. (Difficult Southbound Reaccess)
31 E. Berry St.
30C Ramey Ave.
30BB TX180, TX303, Rosedale St. (Same Exit As 30B)
30B TX180, TX303, Lancaster Ave.
30A Craig St.
29 Meadowbrook Dr.
28AB I-30W Dallas Ft. Worth
28C Brentwood, Stair Rd.
27 John T. White Bridge St.
- **Gas** E: Texaco*
  W: Exxon*(CW), Mobil*
- **Food** W: Blackeye Pea, Lubby's Cafeteria, Subway
- **AServ** W: Discount Tire Co., Firestone Tire & Auto, Kwick Kar Lube and Tune
- **Other** W: **Albertson's Grocery, Animal Clinic, Kroger Supermarket, Wal-Mart**

26 Randol Mill Rd.
- **Gas** E: Texaco*(D)(LP)
  W: Conoco*(CW), Racetrac(D)
- **Food** W: Waffle House
- **ATM** E: Texaco

25 Trinity Blvd.
24B TX121S, Downtown Ft.Worth
24A TX10E, TX183W, Richland Hills
- **Gas** E: Racetrac*
  W: Chevron(D)
- **Food** W: Dairy Queen, El Chico Mexican
- **Lodg** W: Best Western
- **Med** W: **✚ Hospital**

EXIT	TEXAS

- **ATM** E: Racetrac
- **Other** W: Car Wash, **Pet Clinic, Sams Club**

23 Pipeline Rd, Glenview Dr.
- **Gas** E: Mobil*(D)
- **Food** E: McDonalds
- **Other** E: **Discount Tire Company, K-Mart (Pharmacy)**

22B TX121N, TX183E, Dallas, Ft.Worth Airport, Grapevine
22A TX26, Colleyville
- **Gas** S: Racetrac*
- **Food** N: Chuck E. Cheese's Pizza, Pizza Hut, Red Lobster
  S: Arby's, China Cafe, Golden Corral, IHOP, Long John Silvers, Owens Restaurant, Ruby's Cafeteria, Waffle House
- **Lodg** N: Motel 6
  S: Budget Inn
- **AServ** S: Muffin Muffler
- **Med** S: **✚ Hospital**
- **Other** N: **Eckerd Drugs**
  S: **Best Buys**

21 Holiday Lane
- **Food** N: Ryans Steakhouse

20 Rufe Snow Drive
- **Gas** I: Exxon, Mobil*(D)
  O: Chevron(CW), Citgo, Diamond Shamrock
- **Food** O: Applebees, Arby's, Chick-fil-A, CiCi's Pizza, Don Pablo's, Grandy's, Hill Top Chinese, Italian Oven, McDonalds, Pippin's, Ryan's Steakhouse, Taco Bueno, Wendy's, What A Burger (24hr)
- **AServ** I: Exxon, Mobil, National Tire & Battery, Pep Boys Auto Center
  O: Pep Boys Auto Center
- **ATM** I: Central Bank & Trust
  O: First National Bank of Texas
- **Other** O: **Wal-Mart**

19 US377, Denton Hwy
- **Gas** I: Diamond Shamrock, Stop 'N Go*, Texaco
  O: Texaco*(CW)
- **Food** I: Haltom Chinese
  O: Braum's, Chicken Express
- **Lodg** O: Warren Inn
- **AServ** O: G&B Auto Repair, Quaker State
- **Other** I: Car Wash, **Diamond Oak Vet**

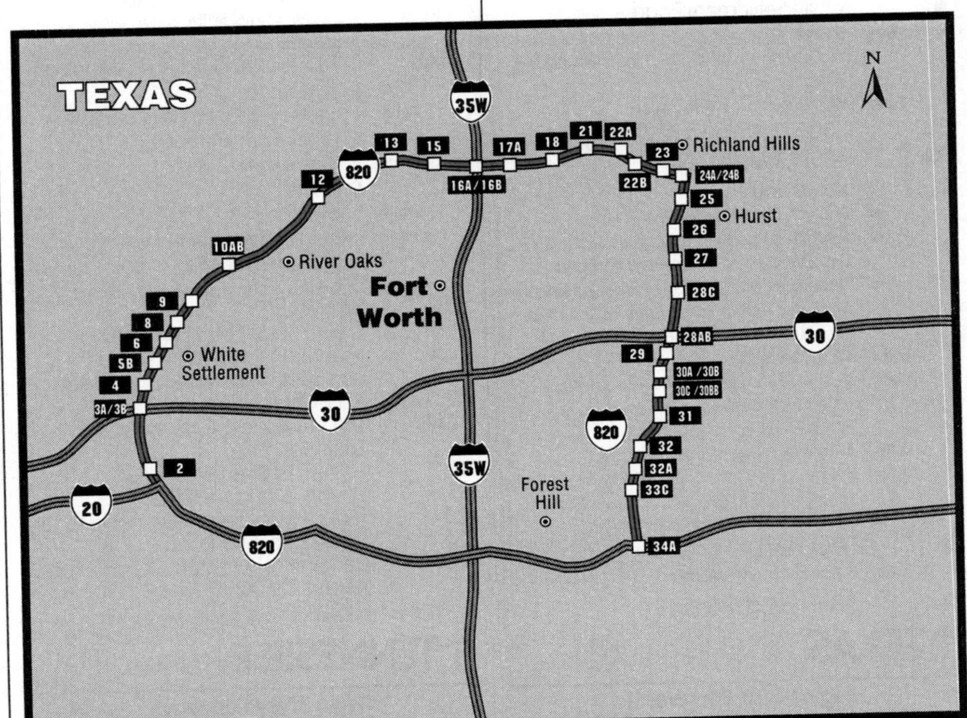

EXIT TEXAS

O: Car Wash

18 Haltom Rd.
- **Lodg** O: Great Western Inn

17A N. Beach St.
- **Gas** O: Mobil*(CW)
- **Food** I: IHOP
- O: Lubby's Cafeteria
- **Lodg** O: Courtyard by Marriott, Fairfield Inn, Lubby's

16A I-35W North, US 287 Denton

16B I-35W South, US287 Downtown Ft.Worth

16C Mark IV Parkway

15 TX156, Blue Mound
- **Gas** S: Citgo*(D), Texaco
- **Food** S: **Buffy's**, Handsome Hank's Fresh Baked Pizza, **McDonalds**, Taco Bell, Wendy's, What A Burger
- **Lodg** S: **Great Western Inn**
- **TServ** S: **Brucner Truck Sales**
- **ATM** S: Norwest Bank

13 Bus US287, Saginaw Main Street

12 Marine Creek Parkway, Old Decatur Road

10AB TX 199, Jacksboro Hwy, Azle Ave
- **Gas** N: Citgo, Diamond Shamrock, Racetrac, Texaco
- S: Chevron*(CW), Texaco*(D)(CW)
- **Food** N: Burger King, Jack-In-The-Box, Sack & Snack (Citgo), Sonic, Taco Bell, Taco Bueno, Waffle House
- S: Long John Silvers, Pizza Hut, Wendy's, What A Burger
- **Lodg** S: Great Western Inn
- **AServ** N: Hi-Li Auto Supplies, Pennzoil Lube, Pep Boys Auto Center, Texaco
- **Med** S: **+ Family Clinic**
- **Other** N: **Alberton's**, Car Wash, **Wal-Mart**, **WalGreens**
- S: **Kroger Supermarket**

9 Quebec Street

8 Navajo Tr., Cahoba Dr.
- **Gas** N: Texaco*(D)
- **Food** N: Tommy's Hamburger's (Texaco)

6 Las Vegas Tr., Heron Dr

5B Silver Creek Road

5A Clifford St

4 White Settlement Road
- **Gas** E: Fina*
- W: Chevron*(CW) (ATM), Diamond Shamrock*(CW) (ATM)
- **Food** W: Mancusso's, Waffle House

3B I-30E, Downtown Ft. Worth

3A I-30W, Weatherford

2 Route 580
- **Gas** E: Citgo*, Citgo, Texaco(CW)
- **Food** E: Arby's, Burger King, Captain D's Seafood, KFC, Lim's Donut, Little Caesars Pizza, McDonalds, McDonalds, Mr. Jim's Pizza, Pancho's Mexican Buffet, Red Lobster, Spring Creek BBQ, Subway, Taco Bueno, Tokyo Steak House, Wendy's
- **Lodg** W: La Mirage Inn
- **AServ** E: All Tune and Lube, GM Auto Dealership, Q-Lube, Toyota Dealer
- W: G&S Auto
- **ATM** E: Bank One, Central Bank and Trust
- **Other** E: Classic Car Wash, Colonial Car Wash, **Eckerd Drugs**, **Kroger Supermarket**, **Speed Queen Laundry Mat**, **Town and Country Drug**

1B Link Crest Dr
- **Gas** E: Chevron, Fina(D), Mobil(D)

1A Team Ranch Road

↑TEXAS

Begin I-820, Fort Worth

EXIT TENNESSEE

Begin Briley Parkway

↓TENNESSEE

6AB I-40, Nashville, Knoxville

7 Elm Hill Pike
- **Gas** I: Circle K*(LP)
- O: Cone*, Mapco Express*(LP)
- **Food** I: **Hardee's**
- O: Waffle House
- **Lodg** O: ClubHouse Inn, DoubleTree Guest Suites, Embassy Suites, Hampton Inn, Holiday Inn Select, La Quinta Inn, Marriot, Ramada Inn, Residence Inn by Marriot, Wilson Inn
- **ATM** O: Mapco Express
- **Other** O: **Bridal Path Wedding Chapel**

8AB US 70, Lebanon Pike, Donelson
- **Food** O: Burger King (Playground), Taco Bell
- **Other** O: **ACE Hardware**, **Fire Station**, **Food Lion**, **Library**, **One Hour Moto Photo**, **The Hermitage**

10 Two Rivers Pky
- **Parks** O: **Two Rivers Park**
- **Other** I: **Gaylord Entertainment**
- O: **Two Rivers Golf Course**, **Wave Country (Kids' Country Playground - Free)**

11 Opryland USA
- **Other** I: **Opryland USA**

12B McGavock Pike Music Valley Drive, Opryland Hotel Conven. Ctr
- **Gas** I: Citgo*(D)
- **Food** I: Bob Evans Restaurant, Captain D's Seafood, Cracker Barrel, John Andretti's Car-B-Que Grill & Ribs, Luby's Cafeteria, McDonalds, Ramada Inn Restaurant (Ramada Inn), Santa Fe Cantina & Cattleman's Club, Schlotzkys Deli, Shoney's, The Country Skillet
- O: Waffle House
- **Lodg** I: AmeriSuites, Best Western, Budget Host Inn, Courtyard by Marriot, Econolodge, Fairfield Inn, Fiddlers Inn, Holiday Inn Express, Opryland Hotel, Ramada Inn, Shoney's Inn
- **ATM** I: First Union (Ramada Inn)
- **Other** I: **1 Hr Photo**, **Cars of the Stars**, **Convention Center**, **Factory Outlet Center**, **Grand Old Golf - 36 Hole Putting Golf Course**, **Music Valley Village (Shops)**, **Music Valley Wax Museum of the Stars**, **Toy Museum**, **Willie Nelson & Friends Museum & Showcase**

14AB US 31E North, Gallatin Rd, Madison
- **Gas** I: Circle K*(LP), Texaco(D)
- **Food** I: Daniel's Diner, Es Fernando's, Expressway Great American Hamburgers
- **AServ** I: Supreme Muffler, Texaco
- O: **K-Mart**
- **Med** I: **+ Hospital**
- **Other** I: **Brush Pharmacy**, **Colonial Bakery Store**, **Mobley Veterinary Clinic**
- O: **K-Mart**, **Nashville National Cemetery**

15A Briarville Rd

15B US 31E South, Ellington Pkwy

16A I-65 North, Louisville

16B I-65 South, Nashville (Clockwise)

17 Brick Church Pike

18AB 18A - I-24 West, Clarksville, 18B - I-24 East, Nashville

19 US 431, Whites Creek Pike

21 US 41A, Clarksville Pike

24 TN 12, Hydes Ferry Pike, Ashland City

25 County Hospital Road

26AB Centennial Blvd, 26B - John C. Tune Airport, TN DOT
- **Gas** O: Texaco*(D)

↑TENNESSEE

Begin Briley Parkway

EXIT TENNESSEE

EXIT AUTHORITY

TRAVEL SAVERS

Save Money on Your Favorite Travel Stops & Services!

Interstate America

Save Money on your favorite travel stops & services!

Dear *Exit Authority* traveler,

Enjoy these Travel Savers coupons, compliments of Interstate America (publisher of *Exit Authority*) and our advertising merchants. We are excited about giving you another opportunity to save time and money with *Exit Authority*. Inside this supplement, you'll find coupons for hotel discounts, free or discounted merchandise, and restaurant discounts — up to $1,000 in savings! Next year, we plan to offer even more Travel Savers coupons for our readers.

Travel Savers coupons are an example of Interstate America's ongoing commitment to *Exit Authority* readers and to expanding our value and service to the traveling public. We welcome your ideas and comments on Travel Savers Coupons - simply write to us at the address below.

Thank you again for your purchase of *Exit Authority*. We hope you enjoy safe and pleasant travels — now for less money with Travel Savers!

Sincerely,

Bonnie Elmore

Bonnie Elmore
Marketing Director

PS Relax - you have plenty of time to take advantage of these money-saving coupons! All coupons are good through Dec. 31, 1998.

5695-G Oakbrook Pkwy ▪ Norcross, GA 30093

FREE VIP Coupon Book!

Just redeem this voucher at any of our stores to receive your FREE VIP coupon book.

US 65 to Hwy 248 & Gretna Rd or Hwy 76 to Roark Valley Rd & Gretna Rd in Branson Meadows • Branson, MO

Call 1-800-SHOP-USA for tenant list and hours • http://www.factorystores.com

Expires: 12/31/98

FREE VIP Coupon Book!

Just redeem this voucher at West Point Pepperell to receive your FREE VIP coupon book.

I-10, Exit 194 (Rt. 287 Florence Blvd.) • Casa Grande, Arizona

Call 1-800-SHOP-USA for tenant list and shopping hours
http://www.factorystores.com

FREE VIP Coupons!

Just redeem this voucher at VF Factory Outlet to receive your FREE VIP coupon sheet.

I-45, Exit 229 to Hwy 287 • Corsicana, Texas

Call 1-800-SHOP-USA for tenant list and shopping hours
http://www.factorystores.com

FREE VIP Coupon Book!

Just redeem this voucher at VF Factory Outlet to receive your FREE VIP coupon book.

I-40, Exit 320 • Crossville, TN

Call 1-800-SHOP-USA for tenant list and shopping hours
http://www.factorystores.com

Expires: 12/31/98

FREE VIP Coupon Book!

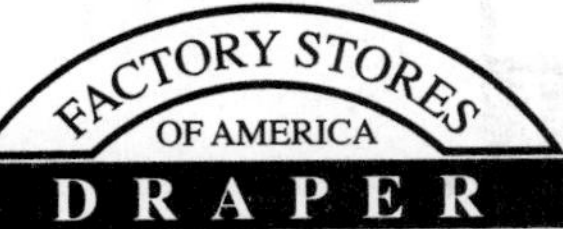

Just redeem this voucher at VF Factory Outlet to receive your FREE VIP coupon book.

I-15, Exit 294 • Draper, UT

Call 1-800-SHOP-USA for tenant list and shopping hours
http://www.factorystores.com

Expires: 12/31/98

FREE VIP Coupon Book!

Just redeem this voucher at Carolina Pottery to receive your FREE VIP coupon book.

I-75, Exit 126 • Georgetown, KY

Call 1-800-SHOP-USA for tenant list and shopping hours
http://www.factorystores.com

Expires: 12/31/98

FREE VIP Coupons!

Just redeem this voucher at VF Factory Outlet to receive your FREE VIP coupon sheet.

US Hwy 290, Exit Factory Outlet Drive • Hempstead, Texas

Call 1-800-SHOP-USA for tenant list and shopping hours
http://www.factorystores.com

Expires: 12/31/98

FREE VIP Coupons!

Just redeem this voucher at Carolina Pottery to receive your FREE VIP coupon sheet.

I-75, Exit 2 • Lake Park, GA

Call 1-800-SHOP-USA for tenant list and shopping hours
http://www.factorystores.com

Expires: 12/31/98

FREE VIP Coupon Book!

Just redeem this voucher at West Point Pepperell to receive your FREE VIP coupon book.

I-45, Exit 13 • La Marque, TX

Call 1-800-SHOP-USA for tenant list and shopping hours
http://www.factorystores.com

Expires: 12/31/98

Attention Merchants!

For more information on how you can participate in the Exit Authority Travel Savers Coupon Program,

Call today, 1-800-683-3948

KNOW THE EXIT BEFORE YOU EXIT

KNOW THE EXIT BEFORE YOU EXIT

KNOW THE EXIT BEFORE YOU EXIT

KNOW THE EXIT BEFORE YOU EXIT

KNOW THE EXIT BEFORE YOU EXIT

KNOW THE EXIT BEFORE YOU EXIT

KNOW THE EXIT BEFORE YOU EXIT

KNOW THE EXIT BEFORE YOU EXIT

KNOW THE EXIT BEFORE YOU EXIT

KNOW THE EXIT BEFORE YOU EXIT

Follow the Sun to Days Inn.

Now you can save even more at any of our 1,800 Days Inns throughout the United States and internationally. Just present this coupon upon check-in and we'll take 10% off our regular rate for your entire length of stay! Advance reservations recommended so call now! For reservations and location information call: ***1-800-DAYS INN***

10% SPECIAL DISCOUNT

Available at participating properties. This coupon cannot be combined with any other special discount offer. Not valid during blackout periods or special events. Void where prohibited. No reproductions accepted. Limit one coupon per stay. Expires 12/31/98

Save 30% at Comfort Inns, Hotels & Suites!

If you're 50 or over, call 1-800-228-5150 and ask for the 30% Prime Rate discount.

You'll save 30% off each night's stay.

Discount is based on availability at participating hotels & can't be used with any other discount. Advance reservations through 1-800-228-5150 required. You must request the 30% Prime Rate Discount. Discount available to those 50 years of age or older. **Expires 12/31/98**

Travelodge®

Call 1-800-578-7878

for reservations. Upon check-in at any Travelodge location, present this coupon and get 10% off the regular room rate. Be sure to ask about Travelodge Miles Program.

Offer good at participating properties. 10% off the reg. rate. Void where prohibited. Offer can not be combined with any other special discount offer. Must submit coupon upon check-in. No reproductions accepted Black-out periods may apply. Expires 12/31/98.

10% STAY WITH US & SAVE! 10%

We invite you to stay at any of our over 900 Ramada Limited, Inn or Plaza Hotel locations throughout the United States. Present this coupon at check-in, and we'll take 10% off your room rate.

For reservations call your travel consultant or 1-800-228-2828 and mention this Ramada 10% discount coupon.

This coupon cannot be combined with other promotions, group rates, or packages. No photocopies of coupon accepted. Availability may be limited and advance reservations are required, so call now.

10% **RAMADA®** Expires 12/31/98 10%

RAMADA LIMITEDS • INNS • PLAZA HOTELS

Save 10% at Quality Inns, Hotels & Suites!

Traveling soon?? Save10% off each night's stay at participating Quality Inns, Hotels & Suites.

Just Call 1-800-228-5151 & ask for "The Travelers" Discount ID # 00004444

Discount is based on availability at participating hotels & can't be used with any other discount. Advance reservations through 1-800-228-5151 required. You must ask for The Travelers Discount & quote ID number at time of reservation. **Expires 12/31/98**

FREE! 4-Roll Pak of SEALAND TISSUE

on any purchase of $25 or more. Reg. $2.48 #6667

Present coupon at start of transaction. One coupon per customer. May not be used in combination with any other offer except regular President's Club 10% discount. Expires 12/31/98

Your RV Traveling Companion Since 1966

Offer Good at all Camping World Locations

Installation/Service/Repair

Discount Code: 8819

10% OFF GUEST CHECK

Iron Skillet Restaurant.

Valid at participating locations only. Not valid with any other discount. One coupon per transaction per customer. Expires 12/31/98.

Kids Stay Free at Clarion Hotel & Suites!

Kids 18 and under stay free when they share the same room as a parent or grandparent.

Call 1-800-252-7466 to reserve a room today & treat the kids... it's on us!

Kids 18 & under stay free at participating hotels in same room as long as no additional bedding is required. Advanced reservations through 1-800-252-7466 required. **Expires 12/31/98**

10% STAY WITH US AND SAVE! 10%

You're invited to stay at one of nearly 600 Howard Johnson Hotels and Inns throughout the United States, Canada and Mexico. And if you present this certificate upon check-in, we'll take 10% off the regular room rate for your stay up to 7 nights. Rates start as low as $29 and kids always stay free*. So make your reservation today!

For reservations, call 1-800-I-GO-HOJO

. . . and ask for the TRAVELDEAL Discount

Expires 12/31/98

Advance reservations required. This offer cannot be combined with other discounts or offers. Restrictions may apply.
*Children 17 and under stay free in their parent's room with existing bedding.

10% 10%

KNOW THE EXIT BEFORE YOU EXIT

KNOW THE EXIT BEFORE YOU EXIT

KNOW THE EXIT BEFORE YOU EXIT

KNOW THE EXIT BEFORE YOU EXIT

KNOW THE EXIT BEFORE YOU EXIT

KNOW THE EXIT BEFORE YOU EXIT

KNOW THE EXIT BEFORE YOU EXIT

KNOW THE EXIT BEFORE YOU EXIT

KNOW THE EXIT BEFORE YOU EXIT

KNOW THE EXIT BEFORE YOU EXIT

FREE VIP Coupon Book!

Just redeem this voucher at West Point Pepperell to receive your FREE VIP coupon book.

I-15 South, Exit 27 • Blue Diamond Rd to Las Vegas Blvd. South • Las Vegas, NV

Call 1-800-SHOP-USA for tenant list and shopping hours
http://www.factorystores.com Expires: 12/31/98

FREE VIP Coupons!

Just redeem this voucher at VF Factory Outlet to receive your FREE VIP coupon sheet.

US 59 Loop South • Factory Outlet Drive • Livingston, TX

Call 1-800-SHOP-USA for tenant list and shopping hours
http://www.factorystores.com Expires: 12/31/98

FREE VIP Coupon Book!

Just redeem this voucher at VF Factory Outlet to receive your FREE VIP coupon book.

Intersection of Power & Baseline Roads • Mesa, Arizona

Call 1-800-SHOP-USA for tenant list and shopping hours
http://www.factorystores.com Expires: 12/31/98

FREE VIP Coupons!

Just redeem this voucher at VF Factory Outlet to receive your FREE VIP coupon sheet.

Hwy 180 to Mineral Wells, TX

Call 1-800-SHOP-USA for tenant list and shopping hours
http://www.factorystores.com Expires: 12/31/98

FREE VIP Coupon Book!

Just redeem this voucher at Reading China & More to receive your FREE VIP coupon book.

I-24, I-40 or I-65 to Briley Pkwy, Exit 12B • Enter at McGavock Pike or Music Valley Drive • Across from Opryland Hotel • Nashville, TN

Call 1-800-SHOP-USA for tenant list and shopping hours
http://www.factorystores.com Expires: 12/31/98

FREE VIP Coupon Book!

Just redeem this voucher at Corning Revere to receive your FREE VIP coupon book.

I-90, Exit 31 • North Bend, Washington

Call 1-800-SHOP-USA for tenant list and shopping hours
http://www.factorystores.com Expires: 12/31/98

FREE VIP Coupon Book!

Just redeem this voucher at Carolina Pottery to receive your FREE VIP coupon book.

I-95, Exit 95 • Smithfield, NC

Call 1-800-SHOP-USA for tenant list and shopping hours
http://www.factorystores.com Expires: 12/31/98

FREE VIP Coupons!

Just redeem this voucher at VF Factory Outlet to receive your FREE VIP coupon sheet.

I-30, Exit 124 • Sulphur Springs, TX

Call 1-800-SHOP-USA for tenant list and shopping hours
http://www.factorystores.com Expires: 12/31/98

FREE VIP Coupons!

Just redeem this voucher at Carolina Pottery to receive your FREE VIP coupons.

I-81, Exit 66 • Tri-Cities, TN

Call 1-800-SHOP-USA for tenant list and shopping hours
http://www.factorystores.com Expires: 12/31/98

KNOW THE EXIT BEFORE YOU EXIT

KNOW THE EXIT BEFORE YOU EXIT

KNOW THE EXIT BEFORE YOU EXIT

KNOW THE EXIT BEFORE YOU EXIT

KNOW THE EXIT BEFORE YOU EXIT

KNOW THE EXIT BEFORE YOU EXIT

KNOW THE EXIT BEFORE YOU EXIT

KNOW THE EXIT BEFORE YOU EXIT

KNOW THE EXIT BEFORE YOU EXIT

KNOW THE EXIT BEFORE YOU EXIT

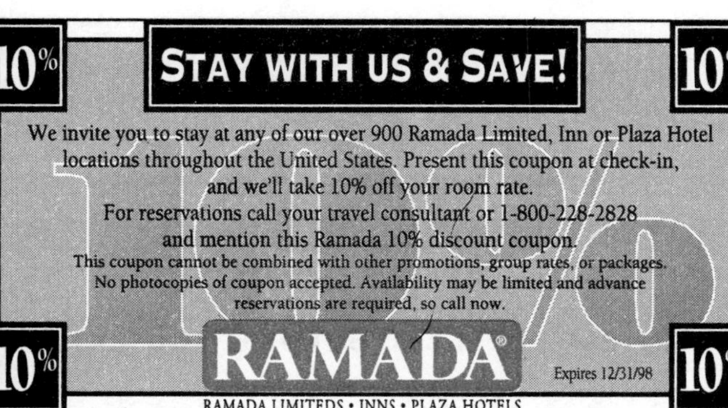

Join us for a Free Deluxe Continental Breakfast!

Coffee & donuts are on us! Get a Free Deluxe Continental Breakfast at Comfort Inns, Hotels & Suites!

Free Deluxe Continental Breakfast available at participating hotels. Advance reservations through 1-800-228-5150 required. **Expires 12/31/98**

Save 10% at Rodeway Inn!

Traveling soon?? Save10% off each night's stay at participating Rodeway Inns

Just Call 1-800-228-2000 & ask for "The Travelers" Discount ID # 00004444

Discount is based on availability at participating hotels & can't be used with any other discount. Advance reservations through 1-800-228-2000 required. You must ask for The Travelers Discount & quote ID number at time of reservation. **Expires 12/31/98**

Save 30% at Clarion Hotel & Suites!

If you're 50 or over, call 1-800-252-7466 and ask for the 30% Prime Rate Discount.

You'll save 30% off each night's stay.

Discount is based on availability at participating hotels & can't be used with any other discount. Advance reservations through 1-800-252-7466 required. You must request the 30% Prime Rate Discount. Discount available to those 50 years of age or older. **Expires 12/31/98**

Enjoy a 100% Satisfaction Guarantee!

We guarantee your satisfaction at Sleep Inns!

If you are dissatisfied with your room or service, inform the front desk staff. If they cannot correct the situation in a timely manner and to your satisfaction, you will not be charged for your room.

Call 1-800-753-3746 Today! **Expires 12/31/98**

2¢ PER GAL. FUEL DISCOUNT

Auto fuel island only.

Valid at participating locations only. Not valid with any other discount. One coupon per transaction per customer. Expires 12/31/98.

Follow the Sun to Days Inn.

Now you can save even more at any of our 1,800 Days Inns throughout the United States and internationally. Just present this coupon upon check-in and we'll take 10% off our regular rate for your entire length of stay! Advance reservations recommended so call now! For reservations and location information call: ***1-800-DAYS INN***

10% SPECIAL DISCOUNT

Available at participating properties. This coupon cannot be combined with any other special discount offer. Not valid during blackout periods or special events. Void where prohibited. No reproductions accepted. Limit one coupon per stay. Expires 12/31/98

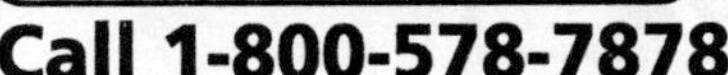

Call 1-800-578-7878

for reservations. Upon check-in at any Travelodge location, present this coupon and get 10% off the regular room rate. Be sure to ask about Travelodge Miles Program.

Offer good at participating properties. 10% off the reg. rate. Void where prohibited. Offer can not be combined with any other special discount offer. Must submit coupon upon check-in. No reproductions accepted Black-out periods may apply. Expires 12/31/98.

Kids Stay Free at Quality Inns, Hotels & Suites!

Kids 18 and under stay free when they share the same room as a parent or grandparent.

Call 1-800-228-5151 to reserve a room today & treat the kids... it's on us!

Kids 18 & under stay free at participating hotels in same room as long as no additional bedding is required. Advanced reservations through 1-800-228-5151 required. **Expires 12/31/98**

KNOW THE EXIT BEFORE YOU EXIT

KNOW THE EXIT BEFORE YOU EXIT

KNOW THE EXIT BEFORE YOU EXIT

KNOW THE EXIT BEFORE YOU EXIT

KNOW THE EXIT BEFORE YOU EXIT

KNOW THE EXIT BEFORE YOU EXIT

KNOW THE EXIT BEFORE YOU EXIT

KNOW THE EXIT BEFORE YOU EXIT

KNOW THE EXIT BEFORE YOU EXIT

KNOW THE EXIT BEFORE YOU EXIT

KNOW THE EXIT BEFORE YOU EXIT

KNOW THE EXIT BEFORE YOU EXIT

KNOW THE EXIT BEFORE YOU EXIT

KNOW THE EXIT BEFORE YOU EXIT

KNOW THE EXIT BEFORE YOU EXIT

KNOW THE EXIT BEFORE YOU EXIT

KNOW THE EXIT BEFORE YOU EXIT

KNOW THE EXIT BEFORE YOU EXIT

KNOW THE EXIT BEFORE YOU EXIT

KNOW THE EXIT BEFORE YOU EXIT

Kids Stay Free at Rodeway Inn!

Kids 18 and under stay free when they share the same room as a parent or grandparent.

Call 1-800-228-2000 to reserve a room today & treat the kids... it's on us!

Kids 18 & under stay free at participating hotels in same room as long as no additional bedding is required. Advanced reservations through 1-800-228-2000 required. **Expires 12/31/98**

Save 30% at Econo Lodge!

If you're 50 or over, call 1-800-553-2666 and ask for the 30% Prime Rate Discount.

You'll save 30% off each night's stay. Try our new Econo Lodge Senior Room!

Discount is based on availability at participating hotels & can't be used with any other discount. Advance reservations through 1-800-553-2666 required. You must request the 30% Prime Rate Discount. Discount available to those 50 years of age or older. **Expires 12/31/98**

Save 10%!

Spend Less Here. Spend More There.

Call 1-800-TheKnights for reservations and ask about the Royalty Club. Upon check-in at any Knights Inn location, present this coupon for 10% off an already great value.

Be sure to visit our web site at www.knightsinn.com.

Available at participating properties. This offer cannot be combined with any other special discount offer. Limit one per room, per stay. Not valid during blackout periods or special events. Void where prohibited. No reproductions accepted. Expires 12/31/98

Save 30% at Quality Inns, Hotels & Suites!

If you're 50 or over, call 1-800-228-5151 and ask for the 30% Prime Rate Discount.

You'll save 30% off each night's stay.

Discount is based on availability at participating hotels & can't be used with any other discount. Advance reservations through 1-800-228-5151 required. You must request the 30% Prime Rate Discount. Discount available to those 50 years of age or older. **Expires 12/31/98**

Save 10% at Clarion Hotel & Suites!

Traveling soon?? Save10% off each night's stay at participating Clarion Hotel & Suites

Just Call 1-800-252-7466 & ask for "The Travelers" Discount ID # 00004444

Discount is based on availability at participating hotels & can't be used with any other discount. Advance reservations through 1-800-252-7466 required. You must ask for The Travelers Discount & quote ID number at time of reservation. **Expires 12/31/98**

10% OFF GUEST CHECK

Iron Skillet Restaurant.

Valid at participating locations only. Not valid with any other discount.
One coupon per transaction per customer. Expires 12/31/98.

10% **STAY WITH US & SAVE!** 10%

We invite you to stay at any of our over 900 Ramada Limited, Inn or Plaza Hotel locations throughout the United States. Present this coupon at check-in, and we'll take 10% off your room rate.

For reservations call your travel consultant or 1-800-228-2828 and mention this Ramada 10% discount coupon.

This coupon cannot be combined with other promotions, group rates, or packages. No photocopies of coupon accepted. Availability may be limited and advance reservations are required, so call now.

10% RAMADA® Expires 12/31/98 10%

RAMADA LIMITEDS • INNS • PLAZA HOTELS

Kids Stay Free at Sleep Inns!

Kids 18 and under stay free when they share the same room as a parent or grandparent.

Call 1-800-753-3746 to reserve a room today & treat the kids... it's on us!

Kids 18 & under stay free at participating hotels in same room as long as no additional bedding is required. Advanced reservations through 1-800-753-3746 required. **Expires 12/31/98**

10% **STAY WITH US AND SAVE!** 10%

You're invited to stay at one of nearly 600 Howard Johnson Hotels and Inns throughout the United States, Canada and Mexico. And if you present this certificate upon check-in, we'll take 10% off the regular room rate for your stay up to 7 nights. Rates start as low as $29 and kids always stay free*. So make your reservation today!

For reservations, call **1-800-I-GO-HOJO**

. . . and ask for the TRAVELDEAL Discount

Expires 12/31/98

Advance reservations required. This offer cannot be combined with other discounts or offers. Restrictions may apply.
*Children 17 and under stay free in their parent's room with existing bedding.

10% 10%

Attention Merchants!

For more information on how you can participate in the Exit Authority Travel Savers Coupon Program,

Call today, 1-800-683-3948

KNOW THE EXIT BEFORE YOU EXIT

KNOW THE EXIT BEFORE YOU EXIT

KNOW THE EXIT BEFORE YOU EXIT

KNOW THE EXIT BEFORE YOU EXIT

KNOW THE EXIT BEFORE YOU EXIT

KNOW THE EXIT BEFORE YOU EXIT

KNOW THE EXIT BEFORE YOU EXIT

KNOW THE EXIT BEFORE YOU EXIT

KNOW THE EXIT BEFORE YOU EXIT

KNOW THE EXIT BEFORE YOU EXIT

Save 10% at Rodeway Inn!

Traveling soon?? Save10% off each night's stay at participating Rodeway Inns

Just Call 1-800-228-2000 & ask for "The Travelers" Discount ID # 00004444

Discount is based on availability at participating hotels & can't be used with any other discount. Advance reservations through 1-800-228-2000 required. You must ask for The Travelers Discount & quote ID number at time of reservation. **Expires 12/31/98**

Save 30% at Quality Inns, Hotels & Suites!

Inn & Suites

If you're 50 or over, call 1-800-228-5151 and ask for the 30% Prime Rate Discount.

You'll save 30% off each night's stay.

Discount is based on availability at participating hotels & can't be used with any other discount. Advance reservations through 1-800-228-5151 required. You must request the 30% Prime Rate Discount. Discount available to those 50 years of age or older. **Expires 12/31/98**

10% 10%

STAY WITH US AND SAVE!

MAKES YOU FEEL AT HOME℠

You're invited to stay at one of nearly 600 Howard Johnson Hotels and Inns throughout the United States, Canada and Mexico. And if you present this certificate upon check-in, we'll take 10% off the regular room rate for your stay up to 7 nights. Rates start as low as $29 and kids always stay free*. So make your reservation today!

For reservations, call 1-800-I-GO-HOJO

. . . and ask for the TRAVELDEAL Discount

Expires 12/31/98

Advance reservations required. This offer cannot be combined with other discounts or offers. Restrictions may apply.

*Children 17 and under stay free in their parent's room with existing bedding.

10% 10%

FREE BAG OF ICE

With 25 gallon fuel purchase. Auto fuel island only.

Valid at participating locations only. Not valid with any other discount. One coupon per transaction per customer. Expires 12/31/98.

Save 30% at Clarion Hotel & Suites!

If you're 50 or over, call 1-800-252-7466 and ask for the 30% Prime Rate Discount.

You'll save 30% off each night's stay.

Discount is based on availability at participating hotels & can't be used with any other discount. Advance reservations through 1-800-252-7466 required. You must request the 30% Prime Rate Discount. Discount available to those 50 years of age or older. **Expires 12/31/98**

Kids Stay Free at Econo Lodge!

Kids 18 and under stay free when they share the same room as a parent or grandparent.

Call 1-800-553-2666 to reserve a room today & treat the kids... it's on us!

Kids 18 & under stay free at participating hotels in same room as long as no additional bedding is required. Advanced reservations through 1-800-553-2666 required. **Expires 12/31/98**

10% 10%

STAY WITH US & SAVE!

We invite you to stay at any of our over 900 Ramada Limited, Inn or Plaza Hotel locations throughout the United States. Present this coupon at check-in, and we'll take 10% off your room rate.

For reservations call your travel consultant or 1-800-228-2828 and mention this Ramada 10% discount coupon.

This coupon cannot be combined with other promotions, group rates, or packages. No photocopies of coupon accepted. Availability may be limited and advance reservations are required, so call now.

10% RAMADA® Expires 12/31/98 10%

RAMADA LIMITEDS • INNS • PLAZA HOTELS

Follow the Sun to Days Inn.

Now you can save even more at any of our 1,800 Days Inns throughout the United States and internationally. Just present this coupon upon check-in and we'll take 10% off our regular rate for your entire length of stay! Advance reservations recommended so call now! For reservations and location information call: ***1-800-DAYS INN***

10% SPECIAL DISCOUNT

Available at participating properties. This coupon cannot be combined with any other special discount offer. Not valid during blackout periods or special events. Void where prohibited. No reproductions accepted. Limit one coupon per stay. Expires 12/31/98

Save 10%!

Spend Less Here. Spend More There.

Call 1-800-TheKnights for reservations and ask about the Royalty Club. Upon check-in at any Knights Inn location, present this coupon for 10% off an already great value.

Be sure to visit our web site at www.knightsinn.com.

Available at participating properties. This offer cannot be combined with any other special discount offer. Limit one per room, per stay. Not valid during blackout periods or special events. Void where prohibited. No reproductions accepted. Expires 12/31/98

Attention Merchants!

For more information on how you can participate in the Exit Authority Travel Savers Coupon Program,

Call today, 1-800-683-3948

KNOW THE EXIT BEFORE YOU EXIT

KNOW THE EXIT BEFORE YOU EXIT

KNOW THE EXIT BEFORE YOU EXIT

KNOW THE EXIT BEFORE YOU EXIT

KNOW THE EXIT BEFORE YOU EXIT

KNOW THE EXIT BEFORE YOU EXIT

KNOW THE EXIT BEFORE YOU EXIT

KNOW THE EXIT BEFORE YOU EXIT

KNOW THE EXIT BEFORE YOU EXIT

KNOW THE EXIT BEFORE YOU EXIT

KNOW THE EXIT BEFORE YOU EXIT

KNOW THE EXIT BEFORE YOU EXIT

KNOW THE EXIT BEFORE YOU EXIT

KNOW THE EXIT BEFORE YOU EXIT

KNOW THE EXIT BEFORE YOU EXIT

KNOW THE EXIT BEFORE YOU EXIT

KNOW THE EXIT BEFORE YOU EXIT

KNOW THE EXIT BEFORE YOU EXIT

KNOW THE EXIT BEFORE YOU EXIT

KNOW THE EXIT BEFORE YOU EXIT

Save 10% at Clarion Hotel & Suites!

Traveling soon?? Save10% off each night's stay at participating Clarion Hotel & Suites

Just Call 1-800-252-7466 & ask for "The Travelers" Discount ID # 00004444

Discount is based on availability at participating hotels & can't be used with any other discount. Advance reservations through 1-800-252-7466 required. You must ask for The Travelers Discount & quote ID number at time of reservation. **Expires 12/31/98**

Special offer to Exit Authority readers from

WOODALL'S®

America's #1 choice for campground information and accommodations-

Woodall's 1998 North American Edition Campground Directory- a $21.95 value can be yours for only **$17.00** Plus $4.50 P&H

Simply call Woodall's at ***1-800-323-9076*** and ask for Department ***3263*** (VISA, Mastercard and Discover accepted.)
Or send this coupon along with your check or money order to:
Woodall's 13975 W. Polo Trail Drive, Lake Forest, IL 60045. Expires 12/31/98

10% **STAY WITH US & SAVE!** 10%

We invite you to stay at any of our over 900 Ramada Limited, Inn or Plaza Hotel locations throughout the United States. Present this coupon at check-in, and we'll take 10% off your room rate.
For reservations call your travel consultant or 1-800-228-2828 and mention this Ramada 10% discount coupon.
This coupon cannot be combined with other promotions, group rates, or packages. No photocopies of coupon accepted. Availability may be limited and advance reservations are required, so call now.

10% **RAMADA®** Expires 12/31/98 10%

RAMADA LIMITEDS • INNS • PLAZA HOTELS

Save 10% at Econo Lodge!

Traveling soon?? Save10% off each night's stay at participating Econo Lodges.

Just Call 1-800-553-2666 & ask for "The Travelers" Discount ID # 00004444

Discount is based on availability at participating hotels & can't be used with any other discount. Advance reservations through 1-800-553-2666 required. You must ask for The Travelers Discount & quote ID number at time of reservation. **Expires 12/31/98**

Save 10%!

Spend Less Here. Spend More There.

Call 1-800-TheKnights for reservations and ask about the Royalty Club. Upon check-in at any Knights Inn location, present this coupon for 10% off an already great value.

Be sure to visit our web site at www.knightsinn.com.

Available at participating properties. This offer cannot be combined with any other special discount offer. Limit one per room, per stay. Not valid during blackout periods or special events. Void where prohibited. No reproductions accepted. Expires 12/31/98

Follow the Sun to Days Inn.

Now you can save even more at any of our 1,800 Days Inns throughout the United States and internationally. Just present this coupon upon check-in and we'll take 10% off our regular rate for your entire length of stay! Advance reservations recommended so call now! For reservations and location information call: ***1-800-DAYS INN***

10% SPECIAL DISCOUNT

Available at participating properties. This coupon cannot be combined with any other special discount offer. Not valid during blackout periods or special events. Void where prohibited. No reproductions accepted. Limit one coupon per stay. Expires 12/31/98

Save 30% at Comfort Inns, Hotels & Suites!

If you're 50 or over, call 1-800-228-5150 and ask for the 30% Prime Rate discount.

You'll save 30% off each night's stay.

Discount is based on availability at participating hotels & can't be used with any other discount. Advance reservations through 1-800-228-5150 required. You must request the 30% Prime Rate Discount. Discount available to those 50 years of age or older. **Expires 12/31/98**

Save 30% at Sleep Inns!

If you're 50 or over, call 1-800-753-3746 and ask for the 30% Prime Rate discount.

You'll save 30% off each night's stay.

Discount is based on availability at participating hotels & can't be used with any other discount. Advance reservations through 1-800-753-3746 required. You must request the 30% Prime Rate Discount. Discount available to those 50 years of age or older. **Expires 12/31/98**

Travelodge® **SAVE 10%**

Call 1-800-578-7878

for reservations. Upon check-in at any Travelodge location, present this coupon and get 10% off the regular room rate. Be sure to ask about Travelodge Miles Program.

Offer good at participating properties. 10% off the reg. rate. Void where prohibited. Offer can not be combined with any other special discount offer. Must submit coupon upon check-in. No reproductions accepted Black-out periods may apply. Expires 12/31/98.

Attention Merchants!

For more information on how you can participate in the Exit Authority Travel Savers Coupon Program,

Call today, 1-800-683-3948

KNOW THE EXIT BEFORE YOU EXIT

KNOW THE EXIT BEFORE YOU EXIT

KNOW THE EXIT BEFORE YOU EXIT

KNOW THE EXIT BEFORE YOU EXIT

KNOW THE EXIT BEFORE YOU EXIT

KNOW THE EXIT BEFORE YOU EXIT

KNOW THE EXIT BEFORE YOU EXIT

KNOW THE EXIT BEFORE YOU EXIT

KNOW THE EXIT BEFORE YOU EXIT

KNOW THE EXIT BEFORE YOU EXIT

Save 30% at Rodeway Inn!

If you're 50 or over, call 1-800-228-2000 and ask for the 30% Prime Rate Discount.

You'll save 30% off each night's stay. Try our new Rodeway Choice Room!

Discount is based on availability at participating hotels & can't be used with any other discount. Advance reservations through 1-800-228-2000 required. You must request the 30% Prime Rate Discount. Discount available to those 50 years of age or older. **Expires 12/31/98**

Save 30% at Econo Lodge!

Econo Lodge®

If you're 50 or over, call 1-800-553-2666 and ask for the 30% Prime Rate Discount.

You'll save 30% off each night's stay. Try our new Econo Lodge Senior Room!

Discount is based on availability at participating hotels & can't be used with any other discount. Advance reservations through 1-800-553-2666 required. You must request the 30% Prime Rate Discount. Discount available to those 50 years of age or older. **Expires 12/31/98**

Enjoy a 100% Satisfaction Guarantee!

We guarantee your satisfaction at Comfort Inns, Hotels & Suites!

If you are dissatisfied with your room or service, inform the front desk staff. If they cannot correct the situation in a timely manner and to your satisfaction, you will not be charged for your room.

Call 1-800-228-5150 Today! Expires 12/31/98

10% STAY WITH US AND SAVE! 10%

Howard Johnson®

MAKES YOU FEEL AT HOME℠

You're invited to stay at one of nearly 600 Howard Johnson Hotels and Inns throughout the United States, Canada and Mexico. And if you present this certificate upon check-in, we'll take 10% off the regular room rate for your stay up to 7 nights. Rates start as low as $29 and kids always stay free*. So make your reservation today!

For reservations, call 1-800-I-GO-HOJO

. . . and ask for the TRAVELDEAL Discount

Expires 12/31/98

Advance reservations required. This offer cannot be combined with other discounts or offers. Restrictions may apply.
*Children 17 and under stay free in their parent's room with existing bedding.

10% 10%

Follow the Sun to Days Inn.

Now you can save even more at any of our 1,800 Days Inns throughout the United States and internationally. Just present this coupon upon check-in and we'll take 10% off our regular rate for your entire length of stay! Advance reservations recommended so call now! For reservations and location information call: ***1-800-DAYS INN***

10% SPECIAL DISCOUNT

Available at participating properties. This coupon cannot be combined with any other special discount offer. Not valid during blackout periods or special events. Void where prohibited. No reproductions accepted. Limit one coupon per stay. Expires 12/31/98

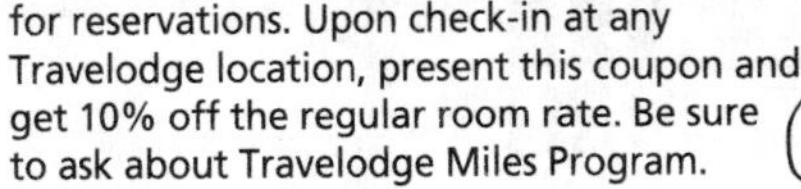

SAVE 10%

Call 1-800-578-7878

for reservations. Upon check-in at any Travelodge location, present this coupon and get 10% off the regular room rate. Be sure to ask about Travelodge Miles Program.

Offer good at participating properties. 10% off the reg. rate. Void where prohibited. Offer can not be combined with any other special discount offer. Must submit coupon upon check-in. No reproductions accepted Black-out periods may apply. Expires 12/31/98.

Enjoy a 100% Satisfaction Guarantee!

We guarantee your satisfaction at Quality Inns, Hotels & Suites!

If you are dissatisfied with your room or service, inform the front desk staff. If they cannot correct the situation in a timely manner and to your satisfaction, you will not be charged for your room.

Call 1-800-228-5151 Today! Expires 12/31/98

Save 30% at Sleep Inns!

If you're 50 or over, call 1-800-753-3746 and ask for the 30% Prime Rate discount.

You'll save 30% off each night's stay.

Discount is based on availability at participating hotels & can't be used with any other discount. Advance reservations through 1-800-753-3746 required. You must request the 30% Prime Rate Discount. Discount available to those 50 years of age or older. **Expires 12/31/98**

Save 10%!

Spend Less Here. Spend More There.

Call 1-800-TheKnights for reservations and ask about the Royalty Club. Upon check-in at any Knights Inn location, present this coupon for 10% off an already great value.

Be sure to visit our web site at www.knightsinn.com.

Available at participating properties. This offer cannot be combined with any other special discount offer. Limit one per room, per stay. Not valid during blackout periods or special events. Void where prohibited. No reproductions accepted. Expires 12/31/98

Attention Merchants!

For more information on how you can participate in the Exit Authority Travel Savers Coupon Program,

Call today, 1-800-683-3948

KNOW THE EXIT BEFORE YOU EXIT

KNOW THE EXIT BEFORE YOU EXIT

KNOW THE EXIT BEFORE YOU EXIT

KNOW THE EXIT BEFORE YOU EXIT

KNOW THE EXIT BEFORE YOU EXIT

KNOW THE EXIT BEFORE YOU EXIT

KNOW THE EXIT BEFORE YOU EXIT

KNOW THE EXIT BEFORE YOU EXIT

KNOW THE EXIT BEFORE YOU EXIT

KNOW THE EXIT BEFORE YOU EXIT

Recreational Vehicle Dealer's Association

RETAIL DEALERS:

Murphy's RV, Inc.	PO Box 202063	Anchorage	AK	99520	907-276-0688
A & M Motors, Inc.	2225 E. 5th Avenue	Anchorage	AK	99501	907-279-5508
Anchorage Nissan / Itasca, Inc.	4748 Old Seward Highway	Anchorage	AK	99503	907-561-1750
Alaskan Adventurers RV Center, Inc.	918 E. 73rd Avenue	Anchorage	AK	99523	907-344-2072
Murphy's RV, Inc.	PO Box 202063	Anchorage	AK	99520	907-276-0688
Reed RV	3901 South Cushman	Fairbanks	AK	99701	907-452-1701
Fairbanks Motorsports	3285 South Cushman	Fairbanks	AK	99701	907-452-2050
Colonial RV Center	6400 1st Ave. South	Birmingham	AL	35212	205-591-3500
Dandy R.V., Inc.	7834 1st Avenue North	Birmingham	AL	35206	205-833-1469
Goodlett RV	27 US 278 East/ Apt. 8	Cullman	AL	35055	205-734-5161
Hood Tractor & RV Center	3504 Deere Road	Decatur	AL	35603	205-353-8712
Waylon Jones Motor Co., Inc.	1501 East Park Avenue	Enterprise	AL	36330	334-347-4302
Madison RV Center	1707 Jordan Lane NW	Huntsville	AL	35816	205-837-3882
Bankston Motor Homes, Inc.	2191 Jordan Lane, NW	Huntsville	AL	35816	205-533-3100
Bankston Motor Homes, Inc.	2191 Jordan Lane, NW	Huntsville	AL	35816	205-533-3100
Webb's R.V. Center, Inc.	5300 Rangeline Road	Mobile	AL	36619	334-661-5400
Marlin Ingram RV	4504 Troy Hwy.	Montgomery	AL	36116	334-288-0331
Alabama Motorcoach	3410 Birmingham Highway	Montgomery	AL	36121	334-270-1100
Millican RV America, Inc.	36115 US HWY 280	Sylacauga	AL	35150	205-249-3773
Trussville RV	175 Railroad Avenue	Trussville	AL	35173	205-655-4467
R.V. City II	18925 I-30	Benton	AR	72015	501-536-2353
Moix RV	1213 Collier Drive	Conway	AR	72032	501-327-2255
Crabtree Motor Sales	2925 Midland Blvd.	Fort Smith	AR	72904	501-783-6129
Mashburn's RV Center, Inc.	2304 E. Nettleton Ave.	Jonesboro	AR	72401	501-932-7989
Camper Capps	9801 Interstate 30	Little Rock	AR	72209	501-568-0338
Pflug's RV Center, Inc.	l0800 Interstate 30	Little Rock	AR	72209	50l-455-2127
Moix RV	12903 I-30	Little Rock	AR	72209	501-455-3333
Dale's Camping Center	3000 West Pullen	Pine Bluff	AR	71601	501-536-8300
R.V. City I	3418 Highway 65 South	Pine Bluff	AR	71601	501-534-3532
R.V. City I	3418 Highway 65 South	Pine Bluff	AR	71601	501-534-3532
Ron Blackwell RV Center	2537 S. Honeysuckle	Rogers	AR	72758	501-621-9300
Outdoor Living Center	Hwy. 7 South Box 1081	Russellville	AR	72801	501-968-7706
National Travelers, Inc.	6027 Warden Road	Sherwood	AR	72116	50l-835-1704
National Travelers, Inc.	6027 Warden Road	Sherwood	AR	72116	50l-835-1704
McGaugh RV Center	2650 Wagon Wheel Road	Springdale	AR	72762	501-751-2174
RV's and More	3819 W. Sunset	Springdale	AR	72762	501-756-3636
Biddulph R.V.	4611 W. Glendale Avenue	Glendale	AZ	85301	602-934-5211
Best West RV Center	4648 N.W. Grand Avenue	Glendale	AZ	85301	602-934-5295
USA Auto & R.V. Sales	951 N. Lake Havasu Avenue	L. Havasu City	AZ	86403	520-453-8721
Cruise America, Inc.	11 West Hampton Avenue	Mesa	AZ	85210	602-464-7300
Arizona State Trailer Sales	2038 N. Country Club Drive	Mesa	AZ	85201	602-834-9581
RV Traders	3619 East Main	Mesa	AZ	85205	602-830-0961
Earnhardt's RV	2222 E. Main St.	Mesa	AZ	85213	602-964-6616
Kempton Travel Town	3335 E. Main Street	Mesa	AZ	85213	602-832-2222
Robert Crist & Company	2025 E. Main	Mesa	AZ	85213	602-834-9410
Dillon's Van City	10310 E. Apache Trail	Mesa	AZ	85201	602-898-9749
American Pride RV Center	2145 East Main St.	Mesa	AZ	85213	602-644-1500
Tveten RV South	3431 East Main St.	Mesa	AZ	85213	602-924-9002
RV Trader	2501 East Main	Mesa	AZ	85213	602-464-9724
Robert Crist & Company	2025 E. Main	Mesa	AZ	85213	602-834-9410
FamilyVan RV, Inc.	9665 N. Cave Creek Road	Phoenix	AZ	85020	602-944-1147
York R.V. Service & Sales	3197 Willow Creek Road	Prescott	AZ	86301	502-445-7910
Many Trails RV	6850 Highway 69	Prescott Valley	AZ	86314	520-775-5770
RV Lifestyle	P.O. Box 12009-167	Scottsdale	AZ	85252	520-927-4001
Beaudry RV	3200 E. Irvington	Tucson	AZ	85706	520-889-6000
Sandy's West RV Center	1451 W. Miracle Mile	Tucson	AZ	85705	520-884-8866
AAA RV	2700 N. ORACLE ROAD	Tucson	AZ	85705	520-791-2700
Magic Carpet RV, Inc.	6718 E. Highway 80	Yuma	AZ	85365	520-726-5477
Second Home Park Model Sales	1562 W. 12 Lane	Yuma	AZ	85364	520-783-0658

R.V. World Recreation Vehicle Center	5875 E. Gila Ridge Road	Yuma	AZ	85365	520-726-6600
Anaheim Coach & Trailer	2222 E. Howell Avenue	Anaheim	CA	92802	714-533-7765
Holiday RV Superstores West, Inc.	2701 Auto Mall Drive	Bakersfield	CA	93313-3208	805-831-5451
Stier's Leisure Vehicles, Inc.	500 S. Union Avenue	Bakersfield	CA	93307	805-323-8000
Altmans Winnebago	1201 Baldwin Park Blvd.	Baldwin Park	CA	91706-5877	818-960-1884
Bob Miller RV	1463 E. Sixth St.	Beaumont	CA	92223	909-845-4611
Bob's Camping World	1365 E. 6th Street	Beaumont	CA	92223	909-769-0185
Hitching Post Trailer Sales	711 W. Imperial Hwy.	Brea	CA	92621	714-529-8766
John Tarter's Holiday World	311 Daily Drive	Camarillo	CA	93010	805-383-6981
Altman's RV Center	1313 R.V. Center Drive	Colton	CA	92324	909-422-0311
Romer's RV Center, Inc.	2295 Arnold Ind. Way.	Concord	CA	94520	510-689-5300
Romer's RV Center, Inc.	2295 Arnold Ind. Way.	Concord	CA	94520	510-689-5300
Dinuba RV Center	391 South Alta Ave.	Dinuba	CA	93618	209-591-0220
Dinuba RV Center	391 South Alta Ave.	Dinuba	CA	93618	209-591-0220
RV Trailerland, Inc.	12803 Old Hwy. 8	El Cajon	CA	92021	619-441-0665
American Sales	5400 N Peck Road	El Monte	CA	91732	818-444-1867
Vehicles Ink, Inc.	P.O. Box 460298	Escondido	CA	92046-0298	619-744-8042
Foretravel of Nevada	1700 Mission Road	Escondido	CA	92029-1111	619-747-4800
Roger's Camping Trailers	4038 Irvington Ave.	Fremont	CA	94538	510-657-5218
Pan-Pacific RV Center	PO Box 1300	French Camp	CA	95231	209-948-8305
Paul Evert's RV Country, Inc.	3633 South Maple	Fresno	CA	93725-2482	209-486-1000
Paul Evert's RV Country, Inc.	3633 South Maple	Fresno	CA	93725-2482	209-486-1000
Bonessa Brothers	8595 Monterey Hwy.	Gilroy	CA	95050	408-842-8418
Pat Meeks RV Center, Inc.	28168 Mission Blvd.	Hayward	CA	94544	510-888-9010
STI, Inc.	847 Industrial Parkway West	Hayward	CA	94544	510-247-1119
STI, Inc.	847 Industrial Parkway West	Hayward	CA	94544	510-247-1119
Hemet Valley R.V.	41491 E. Florida Avenue	Hemet	CA	92544	909-765-5075
Range Vehicle Center, Inc.	11626 Mariposa Road	Hesperia	CA	92345	760-948-1923
Range Vehicle Center, Inc.	11626 Mariposa Road	Hesperia	CA	92345	760-948-1923
Range Vehicle Center, Inc.	11626 Mariposa Road	Hesperia	CA	92345	760-948-1923
Saddleback RV	6441 Burt Rd. Lot #1	Irvine	CA	92720	714-552-0056
Golden Way RV, Inc.	6441 Burt Road #38	Irvine	CA	92720	714-551-3125
Canyon Coach & Trailer, Inc.	6441 Burt Road #7 & 8	Irvine	CA	92720	714-551-9218
Irvine R.V. Center	6441 Burt Road #54	Irvine	CA	92720-3900	714-733-3368
Southwest Coaches	6441 Burt Road #46 & 47	Irvine	CA	92720	714-551-8597
Steve Austin's Motor Home Safari, Inc	6441 Burt Road, Lot 16	Irvine	CA	92720	714-857-8914
Saddleback RV	6441 Burt Rd. Lot #1	Irvine	CA	92720	714-552-0056
Young's RV Center	2337 West Avenue I	Lancaster	CA	93536	805-942-8447
Geweke Ford RV	P.O. Box 1210	Lodi	CA	95241	209-334-0987
Brown's Recreation Center, Inc.	PO Box 1150/ 9702 Hwy 53	Lower Lake	CA	95457	707-994-9418
Manteca Trailers & Campers	P.O. Box 2036/1990 E. Yosemite	Manteca	CA	95336	209-239-1267
Alpine Recreation Sales	19380 Monterey Road	Morgan Hill	CA	95037	408-779-4511
Niel's Motor Homes	8646 Sepulveda Blvd.	North Hills	CA	91343	818-891-0786
Carl's Acres of Trailers	1200 West Mission Blvd.	Ontario	CA	91762	909-983-8411
RV Parts Outlet, Inc.	P.O. Box 50575	Palto Alto	CA	94303	408-730-9991
Hansel RV Center	1215 Auto Center Drive	Petaluma	CA	94952	707-769-2390
The Trailer Hitch	444 S. Dolliver	Pismo Beach	CA	93449	805-773-5448
Coastal RV Brokers	460 South Dolliver	Pismo Beach	CA	93449	805-773-1117
Porterville Trailer Center	1085 Linda Vista Ave.	Porterville	CA	93257	209-784-6036
C.V.L. RV Specialists	3479 Sunrise Blvd.	Rancho Cordova	CA	95742	916-635-4545
Cousin Gary's RV Center	3000 Park Marina Drive	Redding	CA	96001	916-241-3545
Redding RV Center	4850 South Hwy. 99	Redding	CA	96001	916-243-3461
Richardson's RV Center, Inc.	10717 Indiana Avenue	Riverside	CA	92503	909-354-2288
Holiday RV Superstores West, Inc.	2020 Taylor Road	Roseville	CA	95678-1901	916-782-3178
Village RV, Inc.	1039 Orlando Avenue	Roseville	CA	95661	916-757-1155
Happy Daze R.V.	1199 El Camino Avenue	Sacramento	CA	95815	916-920-8255
Total Camper Sales	1925 El Camino Avenue	Sacramento	CA	95815	916-925-5643
Romer's RV Center, Inc.	1064 El Camino Ave.	Sacramento	CA	95815	916-925-3539
Mike Daugherty Chevrolet & RV	2449 Fulton Avenue	Sacramento	CA	95825	916-482-1600
Fox Trailer Sales, Inc.	6630 4th Avenue	Sacramento	CA	95817	916-456-4669
Romer's RV Center, Inc.	1064 El Camino Ave.	Sacramento	CA	95815	916-925-3539
La Mesa RV Center	7430 Copley Park Place	San Diego	CA	92111	619-874-8000
C & D Motorhomes	4530 Convoy Street	San Diego	CA	92111	619-292-8700
RV's of Merritt	1605 Soquel Avenue	Santa Cruz	CA	95062	408-458-1777
El Monte RV Center	12818 Firestone Blvd.	Santa Fe Spring	CA	90670-5404	818-443-6158
El Monte RV Center	12818 Firestone Blvd.	Santa Fe Spring	CA	90670-5404	818-443-6158
Mike Thompson's RV Center	13940 Firestone Blvd.	Santa Fe Springs	CA	90670	714-522-0250
Rainbow's End Trailer Sales	3167 Santa Rosa Avenue	Santa Rosa	CA	95407	707-542-0428
Santa Rosa RV Center	3122 Santa Rosa Avenue	Santa Rosa	CA	95407	707-527-9347
Guaranty RV Center	5015 Scotts Valley Road	Scotts Valley	CA	95066	408-438-0323
Domiano RV Center	12210 Beach Blvd.	Stanton	CA	90680	714-897-5954
South Coast RV	12392 Beach Blvd.	Stanton	CA	90680	714-894-6500
Sky River RV	1039 E. El Camino Real	Sunnyvale	CA	94087	408-749-0117
Denis R.V. Center	2419 W. Monte Vista	Turlock	CA	95382	209-634-9046
Giant Inland Empire RV Center	1090 W. Foothill Blvd.	Upland	CA	91786	909-981-0444

Weslos RV Center	1069 E. Monte Vista Ave.	Vacaville	CA	95688	707-448-1075
Trailer City Sales	2500 Springs Road	Vallejo	CA	94591	707-552-1875
Barber RV	P.O. Box 3318	Ventura	CA	93006	805-642-0276
Crown Dodge	6500 Leland Avenue	Ventura	CA	93003	805-656-6669
Kamper's Korner	14615 Palmdale Road	Victorville	CA	92392	760-241-7351
Amack's Recreation, Inc.	1256 East Main	Woodland	CA	95776	916-662-9658
All Seasons RV	3300 Colusa Hwy.	Yuba City	CA	95993	916-671-9070
All Seasons RV	3300 Colusa Hwy.	Yuba City	CA	95993	916-671-9070
Mountain States RV, Inc.	14300 East Colfax Avenue	Aurora	CO	80011	303-360-0252
Pikes Peak Traveland	4815 E Platte Avenue	Colorado Springs	CO	80915	719-596-2716
K & C RV	136 E. Garden of the Gods Rd.	Colorado Springs	CO	80907	719-528-6337
Dee's RV	2330 Naegele Road	Colorado Springs	CO	80904	719-634-7606
Pikes Peak Traveland	4815 E Platte Avenue	Colorado Springs	CO	80915	719-596-2716
Nolan's RV Center	6935 North Federal Blvd	Denver	CO	80221	303-429-6114
Nolan's RV Center	6935 North Federal Blvd	Denver	CO	80221	303-429-6114
Tarpley RV	25871 Hwy. 160	Durango	CO	81301	970-247-8700
Tarpley RV	25871 Hwy. 160	Durango	CO	81301	970-247-8700
Powder River RV, Inc.	1926 SE Frontage Rd.	Fort Collins	CO	80525	970-493-9193
Stevinson RV Center, Inc.	I5000 W. Colfax Avenue	Golden	CO	80401	303-277-0550
Gavin's RV & Marine	2980 Hwy. 50	Grand Junction	CO	81503	970-245-1800
K & C RV, Inc.	14504 I-25 Frontage Rd.	Longmont	CO	80504	303-776-1309
Loveland RV Service, Inc.	900 E. Colorado Hwy. 402	Loveland	CO	80537	970-669-7465
Casey's Recreational Sales	4120 Youngfield	Wheat Ridge	CO	80033	303-422-2001
Ketelsen Campers of Colorado	9870 S I70 Service Road	Wheatridge	CO	80033	303-43l-2211
Gustine's RV Sales	71 Mott Hill Rd.	East Hampton	CT	06424	860-267-5300
Hemlock Hill R.V. Sales, Inc.	22 Thomaston Road	Litchfield	CT	06759	860-482-0085
Fairchild Sales, Inc.	804 Bridgeport Avenue	Milford	CT	06460	203-874-2561
Highland Orchards	P.O. Box 222	N. Stonington	CT	06359	860-599-5101
Hi-Way Campers, Inc.	992 Norwich Road	Plainfield	CT	06374	860-564-0141
Van's Leisure Living	417 John Fitch Blvd.	South Windsor	CT	06047	860-528-9800
Alexander RV Center, Inc.	1414 N. Dupont Hwy.	Dover	DE	19901	302-674-9240
Brinton's R.V.	Box 220M, R.D. 5	Lewes	DE	19958	302-684-4576
Slicer's Camping Trailers	773 So. Dupont Highway	New Castle	DE	19720	302-836-4110
Kevin's Trailer Sales	RD 1 Box 192, Cedar Neck Road	Ocean View	DE	19970	302-539-3244
Parkview RV Center	5511 Dupont Parkway	Smyrna	DE	19977	302-653-6619
Parkview RV Center	5511 Dupont Parkway	Smyrna	DE	19977	302-653-6619
Golden Sunset RV Resort	2881 US 27 North	Avon Park	FL	33825	810-449-2770
Dusty's Camper World	2835 State Rd. 60 East	Bartow	FL	33830	941-533-2458
Conley Buick Travel Center	800 Cortez Rd. West	Bradenton	FL	34205	941-755-8531
Harberson Swanston, Inc.	17028 US 19 North	Clearwater	FL	33764	813-539-8714
Sundance Motor Homes	5200 South State Road 7	Ft. Lauderdale	FL	33314	954-587-7220
Broward RV Sales	2000 S. State Road 7	Ft. Lauderdale	FL	33317-6720	305-583-3382
Sundance Motor Homes	5200 South State Road 7	Ft. Lauderdale	FL	33314	954-587-7220
North Trail RV Center, Inc.	5270 Orange River Blvd	Ft. Myers	FL	33905	941-693-8200
Gulf Coast Easy Livin' Country	6061 Hamilton Drive	Ft. Myers	FL	33905	941-694-2665
North Trail RV Center, Inc.	5270 Orange River Blvd	Ft. Myers	FL	33905	941-693-8200
Top Notch Marine & RV	2450 North U.S.1	Ft. Pierce	FL	34946	561-466-3119
J.D. Sanders, Inc.	4400 NW 6th Street	Gainesville	FL	32609	352-373-4400
Emerald Coast RV Center, Inc.	6240 Gulf Breeze Pky.	Gulf Breeze	FL	32561	904-939-3484
Nick Nicholas Ford, Inc.	2901 Hwy. 44 West	Inverness	FL	34451	352-726-1231
Revels Nationwide RV Sales	1726 Cassat Avenue	Jacksonville	FL	32210	904-388-9400
Dick Gore's RV World	14590 Duval Place West	Jacksonville	FL	32218	904-741-5100
Suncoast RV Center	9012 Beach Blvd.	Jacksonville	FL	32216	904-642-1600
Dick Gore's RV World	14590 Duval Place West	Jacksonville	FL	32218	904-741-5100
Suncoast RV Center	9012 Beach Blvd.	Jacksonville	FL	32216	904-642-1600
Holiday RV, Inc.	P.O. Box I546	Key Largo	FL	33037	305-45l-4555
Holiday World, Inc.	PO Box 490580	Leesburg	FL	34749	352-787-7744
Preferred RV, Inc.	1532 E. Main Street	Leesburg	FL	34746	352-728-0565
Leisure Tyme RV	1490 Hwy 98 West	Mary Esther	FL	32569	904-581-0880
Holiday RV Superstores, Inc.	16901 N. Cleveland Ave.	N. Fort Myers	FL	33903	941-977-7111
Sun R.V. Center	4023 US 19	New Port Richey	FL	34652	813-842-9735
R.V. World	2110 Tamiami Trail North	Nokomis	FL	34275	941-966-2182
R.V. World	2110 Tamiami Trail North	Nokomis	FL	34275	941-966-2182
Tri-Am R.V. Center, Inc.	544l N.E. Jacksonville Road	Ocala	FL	34479	352-732-6269
Tradewinds R.V.	7677 South US Hwy. 441	Ocala	FL	34480	352-622-7733
County Line RV	3040 N.W. Gainesville Rd.	Ocala	FL	34470	352-351-5255
Holiday RV Superstores, Inc.	7851 Greenbriar Parkway	Orlando	FL	32819-8926	407-363-9211
Leisure Time RVs	3898 W. Colonial Drive	Orlando	FL	32808	407-299-0120
Holiday RV Superstores, Inc.	5001 Sand Lake Road	Orlando	FL	32819	407-351-3096
Holiday On Wheels, Inc.	4100 W. 23RD St.	Panama City	FL	32405	904-785-1566
Carpenter's Camper City, Inc.	8450 Pensacola Blvd.	Pensacola	FL	32534	904-477-6666
Hill Kelly Dodge, Inc.	6171 Pensacola Blvd.	Pensacola	FL	32575	904-476-9078
Leisure Tyme RV	6428 Pensacola Blvd.	Pensacola	FL	32505	904-476-6848
Camper Corral RV Sales	7406 US 27 North	Sebring	FL	33870	941-385-1558
Lazy Days RV Center, Inc.	6130 Lazy Days Boulevard	Seffner	FL	33584-2968	813-246-4333

Lazy Days RV Center, Inc.	6130 Lazy Days Boulevard	Seffner	FL	33584-2968	813-246-4333
Heritage RV Center, Inc.	1150 S. Federal Hwy.	Stuart	FL	34994	561-223-2111
Pepco RV Center	7130 W. Tennessee Street	Tallahassee	FL	32304	904-576-8822
Foretravel of Florida, Inc.	4321 US Hwy. 301 North	Tampa	FL	33610	813-621-9644
Holiday RV SuperStores, Inc.	2910 Overpass Road	Tampa	FL	33619-1318	813-622-8777
Holiday RV Superstores, Inc.	2910 Overpass Road	Tampa	FL	33619-1318	813-622-8777
Giant Recreation World of Tampa	3315 N. US Highway 301	Tampa	FL	33619-2247	813-623-6383
Cathey-RVs, Inc.	2870-A Forest Hill Blvd.	West Palm Beach	FL	33406	407-967-2445
Giant Recreation World, Inc.	13906 West Colonial Dr.	Winter Garden	FL	34787	407-656-6444
Luke Potter Winnebago	12201 West Colonial Drive	Winter Garden	FL	34787	407-877-8558
Devencrest RV Sales & Service	1833 Liberty Expressway SE	Albany	GA	31705-5913	912-888-7880
R.V. Specialties, Inc.	1218 Liberty Expressway S.E.	Albany	GA	31705	912-435-5852
Mid-State RV Center	Hwy. 49, Peachtree Park	Byron	GA	31008	912-956-3654
Cleveland Campers, Inc.	1764 Hwy. 129 South	Cleveland	GA	30528	706-865-6900
Crown R.V. Center	2102 Iris Drive	Conyers	GA	30094	770-483-5057
Coach & Campers	1936 Iris Drive	Conyers	GA	30207	770-483-9080
Super One RV	1774 Iris Drive	Conyers	GA	30207	770-760-0484
John Bleakley RV Center	4411 Bankhead Hwy. East	Douglasville	GA	30134	770-949-4500
John Bleakley RV Center	4411 Bankhead Hwy. East	Douglasville	GA	30134	770-949-4500
Holiday RV Superstores, Inc.	5814 Frontage Road	Forest Park	GA	30050	404-362-9559
Sagon Motorhomes	8859 Tara Blvd.	Jonesboro	GA	30236	770-477-2010
Bankston Motor Homes	8870 Tara Blvd.	Jonesboro	GA	30236	770-477-8095
C.S.R.A. Camperland, Inc.	3844 Washington Road	Martinez	GA	30907	706-863-6294
Camping Time, Inc.	4643 Smithson Blvd.	Oakwood	GA	30566	770-532-9620
Bill Waits RV World	15002 Abercorn St. Ext.	Savannah	GA	31419	912-925-5726
A & H Camper Mart, Inc.	2912 Main Street West	Snellville	GA	30078	770-972-2737
Lasso E Camper Sales	RR 3/Box 131	Anamosa	IA	52205	319-462-3258
Ace Fogdall, Inc.	5424 University Avenue	Cedar Falls	IA	50613	319-277-2641
Leach Camper Sales, Inc.	1629 W. S. Omaha Bridge Rd.	Council Bluffs	IA	51501	712-366-2581
Cylinder Camper Sales	507 Main Street	Cylinder	IA	50528	712-424-3274
Thompson & Sons Trailer Sales	14040 110th Avenue	Davenport	IA	52804-9515	319-381-1023
Ed Garner's Autorama RV Center, Inc.	2227 SE 14th Street	Des Moines	IA	50320	515-282-0443
Kamper Korner	412 1st Street/Box 193	Dewitt	IA	52742	319-659-9343
Paines RV Sales	3573 Lafayette Road	Evansdale	IA	50707	319-234-3039
Brown's Sales & Leasing	Highway 52 South	Guttenburg	IA	52052	319-252-1611
Mobility RV	PO Box 85/Jct. I-35 & Iowa Rt 9	Hanlontown	IA	50444-0085	515-896-2222
Herold Trailer Sales	1806 West 2nd	Indianola	IA	50125	515-961-7405
Sun & Fun, Inc.	4821 Highway 6 SE	Iowa City	IA	52240-8309	319-337-4996
Harrison R.V. Land	P.O. Box 189	Jefferson	IA	50129	515-386-2121
Johnny Ketelsen RV, Inc.	598 57th Street	Marion	IA	52302	319-377-8244
Agency RV Sales & Service	674 East Hwy. 30	Mechanicsville	IA	52306	319-432-7724
RV Central, Inc.	3011 S. 2nd Avenue	Sheldon	IA	51201	712-324-5395
Hutton's RV Center, Inc.	PO Box 155	Urbana	IA	52345	319-443-4386
Kueens Trailer Ranch, Inc.	1120 E. 2nd Box 340	Webster City	IA	50595	515-832-5715
Northwest Recreation Vehicles, Inc.	4033 Chinden Blvd.	Boise	ID	83714	208-345-6644
Nelson's RV's, Inc.	5309 Chinden Blvd.	Boise	ID	83714	208-322-4121
Lesh's Travel Center	3987 Chinden Boulevard.	Boise	ID	83704	208-342-8679
Treasure Valley RV Center	4033 Chinden Blvd.	Boise	ID	83714	208-336-7500
Seventh Heaven Recreation	3880 Chinden Blvd.	Boise	ID	83714	208-343-6203
D.G. Recreational Sales	3800 E. Cleveland/Box 623	Caldwell	ID	83605	208-454-1482
Mike Lesh RV	5500 Cleveland Blvd.	Caldwell	ID	83605	208-459-7778
Cross Roads RV & Sports, Inc.	174 Circle Inn Street	Chubbuck	ID	83202	208-237-8900
Ray's R.V.	North 3340 Atlas Road	Coeur D' Alene	ID	83814	208-666-1478
Foretravel Northwest	7416 North Government Way	Coeur d'Alene	ID	83814	208-772-8561
R & L RV Sales & Service, Inc.	N. 10789 Hwy. 95	Hayden	ID	83835	208-772-7634
Magic Carpet RV Center	N11300 Center Road	Hayden Lake	ID	83835	208-772-2500
Romrell RV Sales	3912 Yellowstone	Idaho Falls	ID	83401	208-529-4682
High Country RV, Inc	3788 North 5th East	Idaho Falls	ID	83401	208-524-5353
Bish's Outdoors	3911 N. 5th East	Idaho Falls	ID	83401	208-529-4386
Lewiston Autobody & RV Sales	1720 21st Street / P. O. Box 606	Lewiston	ID	83501	208-746-8632
Tom Scott Regional RV Center	203 E. 1st Street	Meridian	ID	83642	208-888-4241
Bob's RV Center	3318 Caldwell Blvd.	Nampa	ID	83651	208-467-2533
Minor's RV	1414 Franklin Ave.	Nampa	ID	83651	208-466-7844
Speed's R.V., Inc.	2417 Caldwell Blvd.	Nampa	ID	83651	208-467-2137
OK Trailer Sales	480 North State	Shelley	ID	83274	208-357-7477
Hathaway's, Inc.	2155 S. Yellowstone	St. Anthony	ID	83445	208-624-3141
Gary's Freeway RV	1427 Blue Lakes Blvd. North	Twin Falls	ID	83301	208-733-1823
Mirrielees RV Center	1400 Lock Drive/P.O.Box 427	Bradley	IL	60915	815-933-2251
Whispering Pines Campground Inc.	940 N. Division	Braidwood	IL	60408	815-458-2191
Kamper's Supply	Rt. 2, Box 176	Carterville	IL	62918	618-985-6959
Casey KOA Kampground & Camper Sales	P.O. Box 56	Casey	IL	62420	217-932-5319
Randy's Trailer Town	2012 Mall Road	Collinsville	IL	62234	618-345-7755
Crystal Valley RV	4220 Northwest Hwy.	Crystal Lake	IL	60014	815-459-6611
Ehrhardt's Trailer Sales	776 W. Oakton	Des Plaines	IL	60018	847-437-3421
J & J Camper Sales, Inc.	1501 Il. Rte. 5	East Moline	IL	61244-9606	309-792-2795

Motorhomes Unlimited, Inc.	Route I Box I28	Elgin	IL	60120	847-741-3000
All Seasons Camping, Inc.	260 West Grand Avenue	Elmhurst	IL	60126-1123	630-832-0800
Colman's Country Campers	#2 Fun Street	Hartford	IL	62048	618-254-1180
Quality RV & Minnesota Tool	RR1 Box 107	Heyworth	IL	61745	309-473-2413
Rick's RV Center	4360 W. Jefferson St	Joliet	IL	60431	815-725-4061
Recreation Plantation, Inc.	19660 Stoney Island Ave.	Lynwood	IL	60411	708-895-0660
R & S Sales & Service	218 S. Lake of the Woods Road	Mahomet	IL	61853	217-586-2055
Fourwinds of America	RR 1 Box 135	Maroa	IL	61756	217-794-2292
Whispering Pines R.V.	6245 N. Seneca Road	Morris	IL	60450-0827	815-942-0922
Terry's Leisure Living	9513 West 143rd	Orland Park	IL	60462	708-349-3480
H & M Camper Sales	1500 S. 2nd	Pekin	IL	61554	309-346-8242
Myers Cycle Sales & R.V. Center	PO Box 245/1002 W. Pine	Percy	IL	62272	618-497-2241
Pontiac RV	RR 3, Box 16A	Pontiac	IL	6l764	8l5-844-5000
Mark Wiker Inc. dba Hank Bright RV	2l09 Emmons Road	Rock Falls	IL	61071	815-625-4343
Winnebago Motor Homes	6841 Auburn Road	Rockford	IL	61103	815-964-5591
Collier RV Center	3510 Merchandise Drive	Rockford	IL	61109	815-874-7188
Sky Harbor RV	I-94 & Russell Rd./PO Box 219	Russell	IL	60075	847-395-9500
Trailer Masters	1560 Recreation Drive	Springfield	IL	62707-9469	217-787-7900
Kramer's Kampers	3840l North Sheridan Road	Waukegan	IL	60087	847-623-3989
Larry's Trailer Sales, Inc.	P.O. Box 98, Highway 148	Zeigler	IL	62999	6l8-596-6414
Modern Trailer Sales, Inc.	5123 Pendleton Avenue	Anderson	IN	46011	765-644-4497
Lee's RVs	7247 State Road 46E.	Batesville	IN	47006	812-934-3210
Hal-Dar	U.S. 31 /P.O. Box 277	Bunker Hill	IN	46914	765-689-9193
Tom Stinnett R.V. Center	520 Marriott Drive	Clarksville	IN	47129	812-282-7718
Premier RV, Inc.	1400 Leisure Way	Clarksville	IN	47130	812-284-1400
Tom Stinnett R.V. Center	520 Marriott Drive	Clarksville	IN	47129	812-282-7718
Gillilands Trailer Sales	2710 N. S R 9	Columbus	IN	47203	812-546-5432
Holiday World, Inc.	2345 Cassopolis	Elkhart	IN	46514	2l9-264-0678
Michiana Easy Livin'	2503 Cassopolis	Elkhart	IN	46515	219-262-3658
Hart City RV Sales, Inc.	2300 S. Nappanee Street	Elkhart	IN	46517	219-295-5793
Michiana Easy Livin'	2503 Cassopolis	Elkhart	IN	46515	219-262-3658
Basden RV Center, Inc.	1015 E. Columbia Street	Evansville	IN	47711	812-423-2820
Berning Trailer Sales, Inc.	5220 New Haven Avenue	Ft. Wayne	IN	46803	219-749-9415
Coplen's Coleman Camper Center	9810 Lima Road	Ft. Wayne	IN	46818	219-489-9811
Indy RV Center, Inc.	P. O. Box 447	Greenwood	IN	46142	317-881-0300
Stout's R.V. Sales, Inc.	303 Sheek Road	Greenwood	IN	46143	317-881-7670
Mark's R.V. Sales, Inc.	9702 Pendleton Pike	Indianapolis	IN	46236	317-898-6676
South Side RV Capital	64944 US 31 South	Lakeville	IN	46536	219-291-3000
Whitewater Valley RV's	853 South Road #101	Liberty	IN	47353	765-458-7414
Camp-Land, Inc.	400 W. 81st Avenue	Merrillville	IN	46410	219-769-8496
Monroe City RV	RR1 PO Box 46	Monroe City	IN	47557	812-743-5267
Larry Mayes RV Sales, Inc.	1099 US 31 North	New Whiteland	IN	46184	317-535-5973
Tom Raper, Inc.	2250 Williamsburg Pk.	Richmond	IN	47374	765-966-836l
Rollin On, Inc.	500 US 30	Schererville	IN	46375	219-865-1656
Wetnight RV Sales and Service	P.O. Box 5197	Terre Haute	IN	47805	812-466-3961
Four Seasons RV Acres	2502 Mink Road	Abilene	KS	67410	913-598-2221
B & B Traveland	PO Box 247	Andover	KS	67002	316-733-2200
Augusta RV, Inc.	159th Street East & Hwy. 59	Andover	KS	67002-0378	316-733-8300
Lydia Craig RV & Vanland, Inc.	2617 E. Kellogg Drive	Andover	KS	67002	316-733-6454
Hawley Brothers	1900 E. Wyatt Earp	Dodge City	KS	67801	316-225-5452
Olathe Ford RV Center, Inc.	19310 Gardner Road	Gardner	KS	66030	913-856-8145
Avs Camper Center	1511 W 4th	Hutchinson	KS	67501	316-663-4293
Ned Hiatt's Country Sales, Inc.	Highway 268	Lyndon	KS	66451	913-828-3596
Kansas Kampers Sales & Service	2475 East Kansas	McPherson	KS	67460	316-241-8062
Wilcox Homes & RV Center	835 NE Hwy. 24	Topeka	KS	66608	913-357-5111
Rusty Eck RV Center	9331 West Kellogg	Wichita	KS	67209	316-721-1333
Summit R.V. Sales, Inc.	69l7 U.S. 60	Ashland	KY	41101	606-928-6795
Camp-A-Rama Sales	US Hwy. 68/ Route 9	Benton	KY	42025	502-527-1374
Skagg's Custom Company	7170 South Wilson	Elizabethtown	KY	42701	502-765-7245
Hall's Camper & Motorhomes	975 Beasley	Lexington	KY	40555	606-233-1777
Arrowhead Camper Sales, Inc.	2574 State Route 80 E.	Mayfield	KY	42066	502-247-8187
Youngblood's RV Center, Inc.	2132 State Rt. 45 N.	Mayfield	KY	42066	502-247-8591
Murphy's RV's, Inc.	1619 Dawson Rd./Hwy. 62 East	Princeton	KY	42445	502-365-5082
Interstate RV Outlet	233 North Keeneland Drive	Richmond	KY	40475	606-625-0600
Leisure Life RV	450 Sparrow Drive	Shepardsville	KY	40165	502-543-6005
Cenla Camping Center, Inc.	1916 North Bolton	Alexandria	LA	71303	318-445-6981
Jimmy Walker's RV Sales	3712 Coliseum Blvd.	Alexandria	LA	72303	318-442-6691
Blanchard Trailer Sales	6632 Airline Hwy.	Baton Rouge	LA	70805	504-355-4449
American RV, Inc.	15150 Florida Blvd.	Baton Rouge	LA	70819	504-272-2222
Millers RV	12912 Florida Blvd.	Baton Rouge	LA	70815	504-275-2940
Baton Rouge R.V. Sales & Service, Inc.	13616 Florida Blvd.	Baton Rouge	LA	70819	504-275-4446
Millers RV	12912 Florida Blvd.	Baton Rouge	LA	70815	504-275-2940
Friendly Campers, Inc.	6325 Paris Road	Chalmette	LA	70043	504-277-5186
Dixie Motors RV, Inc.	318 N. Morrison Blvd.	Hammond	LA	70401	504-345-0321
Gauthiers' RV Center, Inc.	P.O. Box 2206	Lafayette	LA	70502	318-235-8547

Steve's Mobile Home & RV Center	3208 E. Judge Perez Drive	Meraux	LA	70075	504-271-2836
Cajun Camping, Inc.	7531 Chef Menteur Hwy.	New Orleans	LA	70126	504-242-9039
Camper's R.V. Center	7700 West 70th St.	Shreveport	LA	71129	318-938-5441
Campers Inn Of Ayer	4 Littleton Rd.	Ayer	MA	01432	508-772-4577
Stearns RV	71 Mechanic Street/RT. 140	Bellingham	MA	02019	508-966-1220
U.S.R.V., Inc.	71 County St.	Berkley	MA	02779	508-823-7571
Bradford Trailer Sales, Inc.	1906 Main Street	Brockton	MA	02401	508-583-1440
Dufour's RV Center	650 John Fitch Hwy.	Fitchburg	MA	01420	508-345-6047
Diamond RV Centre, Inc.	164 West Street/ PO Box 86	Hatfield	MA	01088	4l3-247-3144
Rousseau Recreation, Inc.	150 Bedford Street	Lakeville	MA	02347	508-947-7700
Macdonald's RV Center, Inc.	l7l Washington St.	Plainville	MA	02762	508-695-0066
Long View R.V., Inc.	Rt 5 & 10	W. Hatfield	MA	01088	413-247-9756
Bob's Camper & R.V., Inc.	2810 Hancock Road, Route 43	Williamstown	MA	01267	413-458-3093
Recreation World, Inc.	119 Ferguson Road/ US Rt. 50	Annapolis	MD	21401	410-626-8226
Chesaco Motors, Inc.	1413 Fuselage Avenue	Baltimore	MD	21220	410-687-0858
Chesapeake Recreation, Inc.	9740 Ocean Gateway	Easton	MD	21601	410-820-9178
Triangle Outdoor World	P.O. Box 1127/1501 E. Patrick St.	Frederick	MD	21702	301-662-5722
Endless Summer RV's	7633 Devilbiss Bridge Road	Frederick	MD	21701-1900	301-898-9848
Leo's Vacation Center, Inc.	729 MD Route 3 N. Lane	Gambrills	MD	21054	410-987-4793
Four Seasons RV Service & Sales	790 Potomac Avenue	Hagerstown	MD	21742-3843	301-733-8827
La Grande RV Sales	Rt. 1 Box 118A	Leonardtown	MD	20650	301-475-8550
Charlie's Camping Center, Inc.	10100 Liberty Road	Randallstown	MD	21133	410-655-5200
L.P. Safford, Inc.	3110 Automobile Boulevard	Silver Spring	MD	20904	301-890-3900
Beckley's Camping Center	11109 Angleberger Road	Thurmont	MD	21788	301-898-3300
Brooks Ramsey Motors	11345 Pulaski Highway	White Marsh	MD	21162	410-335-1502
Motor Home Center	747 Minot Avenue	Auburn	ME	04210	207-783-1169
J & M Camper Sales	RR6 New Belgrade Road	Augusta	ME	04330	207-623-4047
Don's Campers, Inc.	340 Main Road	East Holden	ME	04429	207-989-3851
McKay's RV Center	260 Main Road	East Holden	ME	04429	207-989-4300
Good Times Unlimited	Rt. 5/ Box 2121	Farmington	ME	04938	207-778-3482
Boucher's RV Sales, Inc.	1773 Alfred Road	Lyman	ME	04002	207-499-7713
Call of the Wild, Inc.	Rt. 26	Oxford	ME	04270	207-539-4410
Moon's Camper Sales	855 Atlantic Hwy, Rte 1	Waldoboro	ME	04572	207-832-4444
Lee's Family Trailer Sales & Ser	Route 302/408 Roosevelt Trail	Windham	ME	04062	207-892-8308
Circle Star Trailer Sales	5300 Shepherd Rd.	Adrian	MI	49221	517-265-8779
Ranch R.V. Sales, Inc.	6825 M-68	Alanson	MI	49706	616-548-5443
A & S Sales Center, Inc.	2375 Opdyke	Auburn Hills	MI	48326	810-373-5811
George Ewing, Inc.	4251 W. Columbia Ave.	Battle Creek	MI	49017	616-965-0597
Beaverton Outdoor Center	3340 S. M-18	Beaverton	MI	48612	517-435-7761
Les Stanford R.V. Center	44700 I-94 N. Serv Dr.	Belleville	MI	48111	313-565-6000
Parshallburg Campers	15775 Oakley Road	Chesaning	MI	48616	517-845-3189
Roseville RV Center	49685 Gratiot Ave.	Chesterfield	MI	48051-2527	810-949-6400
Leisure Days Travel Trailer	1354 E. Vienna Road	Clio	MI	48420	810-686-2090
Ferris RV Sales & Service	690 Marshall Rd.	Coldwater	MI	49036	517-278-5691
Annie Rae RV Sales	P.O. Box 150	Dewitt	MI	48820	517-669-2755
Annie Rae RV Stores	13200 US 27 North	Dewitt	MI	48820	517-669-2021
Gillette Trailer Center	7210 E. Saginaw Hwy.	East Lansing	MI	48823	517-339-8271
Hilltop Camper, Inc.	2905 North Lincoln Road	Escanaba	MI	49829	906-786-7986
Gaylord RV Sales & Service	271 West Johnson	Gaylord	MI	49735	517-732-6141
Howard Veurink Trailers	4822 S. Division Avenue	Grand Rapids	MI	49548	616-538-2910
Woodland Travel Center, Inc.	5l90 Plainfield Ave.NE	Grand Rapids	MI	49505	616-363-9038
Midway Motorhomes, Inc.	5590 S Division	Grand Rapids	MI	49548	616-534-9641
Modern Trailer Sales, Inc.	3449 S Division	Grand Rapids	MI	49548	616-241-2925
Terrytown Travel Center	7145 S. Division	Grand Rapids	MI	49548	616-455-5590
American RV Sales & Service, Inc.	201 76th Street, SW	Grand Rapids	MI	49548-7228	616-455-3250
America's Choice RV Sales	7834 S. Division Avenue	Grand Rapids	MI	49548	800-732-6236
Betsie Rae RV Stores	11851 E. Michigan	Grass Lake	MI	49240	517-522-8437
Campers Paradise	20804 John R.	Hazel Park	MI	48030	248-545-9026
Hillsdale Travel Center	5100 S. Hillsdale Rd.	Hillsdale	MI	49242	517-437-3859
Holland Motor Homes	670 East 16th Street	Holland	MI	49423	6l6-396-l46l
Holland Motor Homes	670 East 16th Street	Holland	MI	49423	6l6-396-l46l
Burnside Brothers R.V. Rental	2735 West Houghton Lake Drive	Houghton Lake	MI	48629	517-366-8988
Dolney's RV Sales	1171 S. Huron Rd.	Kawkawlin	MI	48631	517-686-6291
After The Fact RV Center	103 Tone Road	Kincheloe	MI	49788	906-495-7242
Dennis Trailer Sales	5226 N. Grand River	Lansing	MI	48906	517-321-1805
Circle K RV's, Inc.	1879 N. Lapeer Rd.	Lapeer	MI	48446	810-664-1942
Raymond's Camper Sales, Inc.	720 N. Mill St/P.O. Box 258	Marion	MI	49665	616-743-2278
Marshall R V Sales & Service	14805 W. Michigan Avenue	Marshall	MI	49068	616-781-2851
Silver Eagle Trailer Sales & Service	3700 Harper Road	Mason	MI	48854	517-676-8835
Raymond's Camper Sales, Inc.	2420 E. Broomfield Road	Mt. Pleasant	MI	48858	517-773-2223
Fun In Motion RV	8639 Mason Drive	Newaygo	MI	49337	616-924-2450
Northern RV Center, Inc.	P.O. Box 539/Hwy. US 2	Quinnesec	MI	49876	906-774-1052
Richard's Motor Sales	749 S. Main Street	Reading	MI	49274	517-283-2114
Northtown Motor Homes	10947 Northland Dr. NE	Rockford	MI	49341	616-866-4300
Hamilton's RV of Saginaw, Inc.	1580 Tittabawassee	Saginaw	MI	48604	517-752-6262

TC-RV, Inc.	705 N. US 31 South	Traverse City	MI	49684	616-943-4050
M & M Camping Center	2960 W. Jefferson	Trenton	MI	48183	313-676-2383
Lowery Trailer Sales, Inc.	21000 Van Dyke	Warren	MI	48089	810-755-9620
West Branch R.V. Center	2268 S. M-76	West Branch	MI	48661	517-345-2033
Haus of Trailers	8285 Highland Rd. (M-59)	White Lake	MI	48386	810-666-2270
General Trailer	48500 12 Mile	Wixom	MI	48393	248-349-0900
Albert Lea RV & Marine	Hwy. 65 South, P.O. Box 58	Albert Lea	MN	56007	507-373-3559
Dalles RV World	7103 Hwy 10 NW	Anoka	MN	55303	612-422-4171
Shorewood R.V.	8390 Highway 10 NW	Anoka	MN	55303	612-421-2505
North Country RV, Inc.	11290 Hwy. 65 N.E.	Blaine	MN	55434	612-757-0550
Blaine Winnebago	10061 Highway 65	Blaine	MN	55434	612-784-2701
Lewie's RV Center, Inc.	1649 Highway 371 North	Brainerd	MN	56401	218-829-3695
Northwest Campers	P.O. Box 158	Byron	MN	55920	507-775-2361
Wold's R.V. Sales	Rt.#1 Box D17	Detroit Lakes	MN	56501	218-847-6333
Bullyan Trailer Sales, Inc.	4945 Miller Trunk Hwy.	Duluth	MN	55811	218-729-9111
Leisureland RV	1819 Central Avenue	East Grand Forks	MN	56721	218-773-7464
Pine Plaza Campers	1701 W. Lincoln Avenue	Fergus Falls	MN	56537-1082	218-736-5135
Hilmerson Motors	S.Hwy 10/P.O.Box 354	Little Falls	MN	56345	612-632-4065
Hart Trailer Sales, Inc.	Hwy. 71 South/ Rt. 2 Box 296	Long Prairie	MN	56347	320-732-6106
Gag's Camper Way, Inc.	Route 9 Box 242	Mankato	MN	56001	507-345-5858
Keepers RV Center	Route 81 Box 91	Mankato	MN	56001	507-625-4647
Niemeyer Trailer Sales	7918 Troy Lane	Maple Grove	MN	55369	612-420-2727
Aarcee Rental Center	2900 Lyndale Avenue, South	Minneapolis	MN	55408	612-827-5746
Hilltop Trailer Sales, Inc.	4560 Central Avenue NE	Minneapolis	MN	55421	612-571-9103
Monticello RV Center	1150 South Hwy. 25	Monticello	MN	55362	612-295-3434
Imperial Camper Sales	1156 Hastings Ave.	Newport	MN	55055	612-459-1804
Modern Trailer Sales & Service, Inc.	219 County Rd. 81	Osseo	MN	55369	612-425-8000
Curtis Camper Sales	7050 - 11th Avenue, SW	Rochester	MN	55902	507-252-1481
Minneapolis Trailer Sales	14490 Northdale Blvd.	Rogers	MN	55374	612-428-2201
Can Am Recreational Vehicles	6220 Highway 101 South	Shakopee	MN	55379	612-445-1252
Hulbert's Camper Corral	Route 2 Box 122	Spring Valley	MN	55975	507-346-2355
PleasureLand RV Center, Inc.	37th Avenue & Division St.	St. Cloud	MN	56302-0669	320-251-7588
Coates Rental & Trailer Sales, Inc.	509 Como Avenue	St. Paul	MN	55103	612-488-0234
Hemmingsen Auto Sales Company	700 Craig Avenue/P.O.Box 1155	Tracy	MN	56175	507-629-4160
Maple Leaf Trailer Sales	14 Meramec Heights Shop Ctr.	Arnold	MO	63010-3201	314-296-6818
Shoemaker Travel Trailers	Box 6 Highway 36	Bevier	MO	63532	816-773-5313
Shoemaker Travel Trailers	Box 6 Highway 36	Bevier	MO	63532	816-773-5313
Bourbon RV Center, Inc.	Old Springfield Road	Bourbon	MO	65441	573-732-5100
Bill Thomas Camper Sales, Inc.	5217 North Lindbergh	Bridgeton	MO	63044	314-731-2217
Bison Campers	Highway 65 South	Buffalo	MO	65622	417-345-2325
Capetown RV	P.O. Box 1985	Cape Girardeau	MO	63702-1985	573-334-7152
Coachlight RV Sales, Inc.	Route 4 Box 515/I44 & 71A	Carthage	MO	64836	417-358-7444
Four State RV	Route 4 Box 515AA	Carthage	MO	64836	417-358-0700
Colaw RV Sales	Route 4 Box 515A	Carthage	MO	64836	417-358-4640
Wilder RV's	1011 E. Clark Street	Clinton	MO	64735	816-885-6177
Loveall RV's of Columbia	8877 E. I-70 Drive N.E.	Columbia	MO	65202	573-474-9584
Trailside R.V. Center	700 R.D. Mize Road	Grain Valley	MO	64029	816-229-2257
G.R. Milner Ford Sales, Inc.	P.O. Box 445/Jct. 291 & 71	Harrisonville	MO	64701	816-380-3251
Apache Village	9001 Dunn Road	Hazelwood	MO	63042	314-895-4567
U S Rents-It RV Center	1513 Industrial Drive	Jefferson City	MO	65102	573-635-6171
Kansas City Trailer Sales	11530 S 71 Hwy.	Kansas City	MO	64137	816-761-3322
Loveall's RV of Kansas City	8753 East Highway 40	Kansas City	MO	64129	816-921-0065
Beilstein Camper Sales	Hwy. 61 South	La Grange	MO	63448	573-655-2254
Vernon Griffon's Trailer Town	#10 East Woodlawn Drive	Leadington	MO	63601	573-431-2728
Scott's Holiday Corral	504 S. Hwy 13 Box 158	Lowry City	MO	64763	417-644-2246
Byerly Trailer Sales	13988 Manchester Road	Manchester	MO	63011	314-227-1552
Elder RV Sales & Accessories	808 Illinois	Montrose	MO	64770	816-693-4743
Loveall RV's of St. Louis	2462 E. Pitman Street	O'Fallon	MO	63366-4112	314-928-1124
Smith Boys, Inc.	Rt. 4 Box 2310	Osage Beach	MO	65065	573-348-3148
Factory Outlet RV Center	309 Flier Drive	Pacific	MO	63069	314-257-7878
Branson West Sales Company	22415 Main Street Road	Sedalia	MO	65301	417-338-5755
Sunshine RV Sales	3536 E. Sunshine	Springfield	MO	65809	417-881-2844
Town & Country RV, Inc.	3250 W. Sunshine	Springfield	MO	65807	417-882-4539
Bass Pro Shops	1935 South Campbell	Springfield	MO	65807	417-887-1915
Morgan Buildings & Spas, Inc.	3550 West Clay	St. Charles	MO	63301	314-949-9495
Howard Motor Homes	4060 Gravois	St. Louis	MO	63116	314-772-4242
Morris B. Thomas Auto Sales, Inc.	275 Lemay Ferry Road	St. Louis	MO	63125	314-631-5600
Van City Sales	3100 Telegraph Rd.	St. Louis	MO	63125	314-894-3905
Harney's RV Center	1195 Dunn Road-Front	St. Louis	MO	63138	314-868-9441
Behlmann GMC Trucks	820 McDonnell Blvd.	St. Louis	MO	63042	314-895-1600
Covered Wagon RV Sales	Box Q	Waynesville	MO	65583	314-336-5111
Country Creek R.V. Center	13 Bullock Lane	Columbia	MS	39429	601-731-2738
Johnny Bishop, Inc.	3352 Highway 45 N	Columbus	MS	39705	601-327-6552
Reliable RV Sales, Inc.	727 Pass Rd.	Gulfport	MS	39501	601-868-1000
Davis Camper Sales	7696 Hwy. 49 North	Hattiesburg	MS	39402	601-268-1800

S & S Apache Camping Center, Inc.	1820 Terry Road	Jackson	MS	39204	601-372-6426
Bus Supply Company	Highway 98 East	McComb	MS	39648	601-684-2900
Ethridge RV Center, Inc.	400 N. Frontage Road	Meridian	MS	39301	601-483-4301
National Travelers	7400 Craft Goodman Road	Olive Branch	MS	72120-6098	601-895-4711
Desoto RV Center	8150 New Craft Road	Olive Branch	MS	38654	601-893-3040
Paw Paw's Camper City	123 Elm Street	Picayune	MS	39466	601-799-0696
George R. Pierce, Inc.	518 St. Johns	Billings	MT	59107	406-252-9313
Magic Carpet RV Sales	5124 Laurel Road	Billings	MT	59101	406-252-6855
Tour America RV Center, Inc.	2220 Old Hardin Road	Billings	MT	59101	406-248-7481
Al's RV/Boat Service	5041 Harrison Avenue	Butte	MT	59701	406-494-2902
McCollum Modern RV's	4200 l0th Ave. South	Great Falls	MT	59405	406-761-3520
Travel Time RV's	4035 10th Avenue South	Great Falls	MT	59405	406-454-0777
Bretz RV & Marine	2045 Mullan Road	Missoula	MT	59802	406-721-4010
Rangitsch Brothers LLC	2001 W. Broadway	Missoula	MT	59807	406-728-4040
Amunrud's RV	P.O. Box 1288	Sidney	MT	59270	406-482-2226
Charles H. Jenkins & Company	720 E. Memorial Drive	Ahoskie	NC	27910	919-332-2191
Americamp RV Sales	P. O. Box 1671	Asheboro	NC	27204-1671	910-824-4600
Bob Ledford's RV Super Market	328 New Leicester Hwy.	Asheville	NC	28806	704-253-1681
Crisp RV Center	2042 Hwy 178	Chocowinity	NC	27817	919-946-0311
Allsport RV Center	PO Box 64337	Fayetteville	NC	28306	910-323-3888
Carolina Coach & Camper	P.O. Box 2166	Hickory	NC	28603	704-322-9009
Hawley's Camping Center	Rt. 3 Box 2HCC	Hope Mills	NC	28348	910-423-5200
Out of Doors Mart, Inc.	Drawer 799/8510 Norcross Rd.	Kernersville	NC	27284	910-993-4518
Young Motor Sales, Inc.	101 Oakgrove Rd.	Kings Mountain	NC	28086	704-734-0595
Tom Johnson's Camping Center	2695 Hwy. 70 West	Marion	NC	28752	704-724-4l05
Wray Frazier Camping Center	P.O. Box 87	Newton	NC	28658	704-464-4521
College Park	4208 New Bern Avenue	Raleigh	NC	27610	919-231-8710
Randy's RV's	5322 Fayettville Road	Raleigh	NC	27603	919-779-5445
Van & RV Sales	1300 S. Wesleyan Blvd.	Rocky Mount	NC	27803	919-446-6126
College Park Div Service/Sales	2550 North Church Street	Rocky Mount	NC	27804	919-446-7166
Bill Plemmons, Inc.	6725 University Parkway	Rural Hall	NC	27045	910-377-2213
Terrell Camping Center, Inc.	P.O. Box 385	Terrell	NC	28682	704-478-2651
Howard RV Center	6811 Market Street	Wilmington	NC	28405	910-791-5371
Oldtown Camper Sales, Inc.	5109 North Causeway Drive	Winston Salem	NC	27106	910-924-9864
Capital R.V. Center, Inc.	1900 N. Bismark Expressway	Bismarck	ND	58501	701-255-7878
Roughrider Mobile Homes	Route 2 Box 207	Dickinson	ND	58601	701-225-9844
Budget Auto & RV, Inc.	5200 Gateway Drive	Grand Forks	ND	58201	701-772-7233
Don Solberg Camper Sales	Highway 2 East/ Box 401	Lakota	ND	58344	701-247-2800
Riverwood RV	3700 Memorial Hwy.	Mandan	ND	58554-4651	701-663-0050
Key RV	2305 2 & 52 Bypass West	Minot	ND	58701	701-838-4343
Apache Camper Center, Inc.	1120 Fort Crook Road 9	Bellevue	NE	68005-2937	402-292-1455
Mobile World, Inc.	P.O. Box l665	Grand Island	NE	68801	308-382-3866
Rich & Sons Camper Sales	5112 S. Antelope Drive	Grand Island	NE	68803	308-384-2040
Hastings Motor Sales, Inc.	901 West J	Hastings	NE	68901	402-463-1338
Transportation Equipment Company	l0l0 West J	Hastings	NE	68901	402-463-4402
Leach Camper Sales of NE, Inc.	2727 Cornhusker Hwy.	Lincoln	NE	68504	402-466-858l
Apache Camper Center, Inc.	P.O. Box 22578	Lincoln	NE	68542-2578	402-423-3218
Larry's RV Sales/Service	1802 E 4th St.	North Platte	NE	69101	308-532-5474
A.C. Nelsen Enterprises, Inc.	11818 L Street	Omaha	NE	68137	402-333-1122
Gary's Sports & RV Center, Inc.	P.O. Box 328	Chichester	NH	03234	603-798-4030
Hill's RV's	Conway Outlet Ctr Rt 16	Conway	NH	03818	603-447-5101
Campers Inn of Kingston	146 Route 125	Kingston	NH	03848	603-642-5555
C.H. Dana R.V. Sales & Service	R.R.1 Box 304	Monroe	NH	03771	603-638-2200
Campers Inn, Inc.	550 Amherst St.	Nashua	NH	03063	603-883-1082
CampAmerica, Inc.	222 Plaistow Road	Plaistow	NH	03865	603-382-9296
Stone's Camping World	121 Route 73	Berlin	NJ	08009	609-767-5422
Cedar Ridge Trailer Sales Inc.	109 Route 206 South	Branchville	NJ	07826	201-948-3800
Bridgeton Travel Trailer Center	588 Shiloh Pike Rt. 49	Bridgeton	NJ	08302	609-451-7585
Driftwood RV Center	1955 Route 9	Clermont	NJ	08210	609-624-1899
Germaine's RV Headquarters, Inc.	407 Route 539	Cream Ridge	NJ	08514	609-758-1919
Sunbird Recreational	1170 Hwy. 36	Hazlet	NJ	07730	908-888-9400
Scott Motor Coach Sales, Inc.	1133 Route 88	Lakewood	NJ	08701	908-370-1022
Scott Motor Coach Sales, Inc.	1133 Route 88	Lakewood	NJ	08701	908-370-1022
Joe Bennett Chevrolet	Rt. 72 P.O. Box 610	Manahawkin	NJ	08050	609-597-2501
Campers of America, Inc.	3405 State Hwy. 33	Neptune	NJ	07753	201-922-2500
Garick RV	3134 Route 23 North	Oak Ridge	NJ	07438	201-208-9200
Garick RV	3134 Route 23 North	Oak Ridge	NJ	07438	201-208-9200
Ocean View Trailer Sales, Inc.	2555 Route 9, P.O. Box 607	Ocean View	NJ	08230	609-624-0370
Rayewood Trailer Sales	120 Clove Road	Sussex	NJ	07461	201-875-4961
Hitcharama Recreational Vehicles	4121 Route 42	Turnersville	NJ	08012	609-629-7400
A & B Travel Trailers	2063 North Blackhorse Pike	Williamstown	NJ	08094	609-629-0444
Woodbridge Dodge & R.V. Center	450 King George Road	Woodbridge	NJ	07095	908-826-1220
American R.V. & Marine	11810 Central S.E.	Albuquerque	NM	87123	505-293-1983
Myers RV Center, Inc.	12024 Central SE	Albuquerque	NM	87123	505-298-7691
Suburban Recreational Vehicles	6022 2nd St. N.W.	Albuquerque	NM	87107	505-344-3583

American Holiday R.V.	9999 Central N.E.	Albuquerque	NM	87123	505-299-6838
Aloha RV, Inc.	11200 Central Avenue, SE	Albuquerque	NM	87123	505-298-8444
Holiday Travel Trailers	11100 Central Ave. SE	Albuquerque	NM	87123	505-294-8280
Rocky Mountain RV World	11109 Central N.E.	Albuquerque	NM	87123	505-255-6441
On The Road Again RV's	4305 Lomas NE	Albuquerque	NM	87110	505-266-3363
Myers RV Center, Inc.	12024 Central SE	Albuquerque	NM	87123	505-298-7691
Bison R. V. Center	1200 East First Street	Clovis	NM	88101	505-762-9506
Lowe's RV	3124 N. Lovington Hwy.	Hobbs	NM	88240	505-392-3412
Holiday RV Superstores of	1285 Avenida De Mesilla	Las Cruces	NM	88005-3904	505-523-0715
American R.V. & Marine	200 N. Telshor Blvd.	Las Cruces	NM	88001	505-522-5512
S & H RV & Mobile Home Center	3100 N. Main St.	Las Cruces	NM	88005	505-524-8057
R.V. Sales	2702 West Central/P.O.Box 1788	Moriarty	NM	87035-1788	505-832-2400
Carson R.V.	4550 North Carson	Carson City	NV	89706	702-882-8333
McKinnish RV	5130 Highway 50 East	Carson City	NV	89701	702-883-7834
Michael Hohl RV Center	3700 S. Carson St.	Carson City	NV	89701	702-883-5777
Johnnie Walker RV's	3700 Boulder Hwy.	Las Vegas	NV	89121	702-458-2092
Sahara R.V.	1518 Scotland Lane	Las Vegas	NV	89102	702-383-7039
Wheeler's Las Vegas RV	13175 Las Vegas Blvd. South	Las Vegas	NV	89124	702-896-9000
Findlay RV Center Inc.	4530 Boulder Highway	Las Vegas	NV	89121	702-435-2500
Tom Poling RV's	3673 Boulder Hwy.	Las Vegas	NV	89121	702-432-9115
Sahara R.V.	1518 Scotland Lane	Las Vegas	NV	89102	702-383-7039
Mountain Family RV	PO Box 18216/16300 S. Virginia	Reno	NV	89511	702-849-1005
Liberty Recreation Vehicles	9125 S. Virginia Street	Reno	NV	89511	702-851-1204
RV Sales & Rentals of Albany	76 Exchange St.	Albany	NY	12205	518-459-4695
Alpin Haus	Route 30	Amsterdam	NY	12010	518-843-4400
Northway Travel Trailers, Inc.	P.O. Box 2110 (Rt. 9 Malta)	Ballston Spa	NY	12020	518-899-2526
Lei-Ti RV Sales & Service	9979 Francis Road	Batavia	NY	14020	716-343-8600
Walker's Four Seasons, Inc.	Route 17 Box 265	Big Flats	NY	14814	607-562-8731
Ozzie's Camping Center, Inc.	1607 Lakeland Avenue	Bohemia	NY	11716	516-589-5295
Walton H. Bull, Inc.	Route 374 Box 405	Cadyville	NY	12918	518-492-7007
Meyer's Campers	3338 State Road Rt 5	Caledonia	NY	14423	716-538-6848
Lent's RV Sales	7 Main St.	Dansville	NY	14437	716-335-3800
Skyline RV & Home Sales, Inc.	10933 Townline Road	Darien Center	NY	14040	716-591-2021
Crestwood RV Center, Inc.	29 Lakeshore Drive	East Chester	NY	10709	914-961-4400
Trailer Land, Inc.	7120 Seneca St.	Elma	NY	14059	716-652-4500
Paradise Bay RV	3039 Route 426	Findley Lake	NY	14736	716-769-7584
The Great Outdoors Recreation Center	1122 County Route 57	Fulton	NY	13069	315-695-5020
City Garage Company of Jamestown NY	Rt. 60 P.O. Box 38	Gerry	NY	14740	716-985-4623
Skyway RV Center	Box 194	Greenfield Park	NY	12435	914-647-3100
Fairway R.V., Inc.	5788 Camp Road	Hamburg	NY	14075	716-649-9400
Wilkins Recreational Vehicles	1099 Almond Rd/ RD2 Rt 21	Hornell	NY	14843	607-324-1313
Twin Trailer Sales, Inc.	1720 Foote Ave/RD # 5	Jamestown	NY	14701	716-487-0011
Campers Barn of Kingston	124 Route 28	Kingston	NY	12401	914-338-8200
Campers Barn of Kingston	124 Route 28	Kingston	NY	12401	914-338-8200
Seven O's Rec. Sales & Service	7917 DeVaul Road	Kirkville	NY	13082	315-687-9342
State Line Camping Center	P.O. Box 93/ 600 State Hwy	Lindley	NY	14858	607-523-7396
Mantelli Trailer Sales, Inc.	6865 S. Transit Road	Lockport	NY	14094	716-625-8877
Ward's Travel Trailers	Rt. 11/ RD 1 Box 2A	Marathon	NY	13803	607-849-3994
Grand Am RV, Inc.	147 Mastic Road	Mastic Beach	NY	11951	516-395-3377
Grand Am RV, Inc.	147 Mastic Road	Mastic Beach	NY	11951	516-395-3377
Price Rite Trailer Sales	1128 Route 17 K	Montgomery	NY	12549	914-561-4580
Gary Johnston's RV Center, Inc.	2311 Marion Road	Palmyra	NY	14522	315-597-5388
Holiday On Wheels, Inc.	Route 22 & Robin Hill Corp. Pk.	Patterson	NY	12563	914-878-9400
Mike's RV Center, Inc.	7865 Scenic Highway	Pulaski	NY	13142	315-298-4232
Herbert S. Hyde, Inc.	64 Blue Barns Road	Rexford	NY	12148	518-399-2880
Interlake Trailer Sales	RD2 Box 354	Rhinebeck	NY	12572	914-266-5387
Bob McKerrow Sons, Inc.	RT 98	Sandusky	NY	14133	716-492-3196
Journeytime Trailers, Inc.	940 Middle Country Road	Selden	NY	11784	516-698-0055
Butler Mobile Homes, Inc.	360 Rt. 97/ P.O. Box 277	Sparrowbush	NY	12780	914-856-6639
Ballantyne/KL Trailers	7447 Route 96	Victor	NY	14564	716-924-3264
W E S Trailer Sales	Route 25	Wading River	NY	11792	516-727-5852
Freedom RV Outlet	One Gordon Lane	Wilton	NY	12831	518-581-7701
Freedom RV Outlet	One Gordon Lane	Wilton	NY	12831	518-581-7701
Boulevard Trailers, Inc.	P.O. Box 62	Yorkville	NY	13495	315-736-5851
Price-Rite Boat & Camper Sales	7450 State/Rt. 52	Youngsville	NY	12791	914-292-5943
Sirpilla RV Center, Inc.	1005 Interstate Parkway	Akron	OH	44312	330-645-1991
Duncan's Trailers	860 S. Arlington St.	Akron	OH	44306	216-773-7474
Sirpilla RV Center, Inc.	1005 Interstate Parkway	Akron	OH	44312	330-645-1991
Avon RV Superstore	38525 Chester Road	Avon	OH	44011	216-934-7500
Holman Motors, Inc.	4387 Elick Lane & Hwy. 32	Batavia	OH	45103	513-752-3123
Ron Gist RV & Auto Sales	169 G St. Claridon	Caledonia	OH	43314	614-389-2516
Stoney's RV Sales & Service	11510 East Pike	Cambridge	OH	43725	614-439-7285
Brocks RV Center	6101 S.R. 219	Celina	OH	45822	419-268-2025
Colerain Travel Trailer Center	3491 Struble Road	Cincinnati	OH	45251-4945	513-923-3600
Colerain Travel Trailer Center	3491 Struble Road	Cincinnati	OH	45251-4945	513-923-3600

Post Industries, Inc.	4330 Westerville Rd.	Columbus	OH	43231	614-471-0551
Farber Motors, Inc.	5858 Scarborough Blvd.	Columbus	OH	43232	614-864-7878
Post Industries, Inc.	4330 Westerville Rd.	Columbus	OH	43231	614-471-0551
Brandt Trailer Sales	1100 Brandt Pike	Dayton	OH	45404	937-236-0200
Alum Creek RV, Inc.	5742 E. State Rt. 37	Delaware	OH	43015	614-548-4068
McNulty Motors	547 South Main	Englewood	OH	45322	937-836-9911
Young's R.V. Center	1450 Dickinson	Fremont	OH	43420	419-334-2648
Kenisee's Grand River RV Sales	4680 Route 307 E	Geneva	OH	44041	216-466-6320
Hartville RV Center	540 South Prospect Avenue	Hartville	OH	44632	330-877-3500
Hartville RV Center	540 South Prospect Avenue	Hartville	OH	44632	330-877-3500
RCD Sales Company	I990 Hebron Road/Box 267	Hebron	OH	43025	6I4-928-6836
Homestead RV Sales & Service	1436 ST. RT.7, SE	Hubbard	OH	44425	330-448-2938
Adena Ridge Recreation	505 West 6th Avenue	Lancaster	OH	43130	614-653-1581
Avalon RV - Marine	1604 Medina Road	Medina	OH	44256	330-239-2131
Avalon RV - Marine	1604 Medina Road	Medina	OH	44256	330-239-2131
Ralph's Motorhome Sales/Service	521 Youngstown-Warren	Niles	OH	44446	216-652-1333
Winn's RV Center, Inc.	6075 Dressler Road NW	North Canton	OH	44720	216-494-3811
Clay's RV Center, Inc.	3063 Greensburg Road	North Canton	OH	44720	330-896-8977
Clay's RV Center, Inc.	3063 Greensburg Road	North Canton	OH	44720	330-896-8977
Moore's Travel Trailers	34155 Lorain Road	North Ridgeville	OH	44039	216-327-6911
Johnny's Nimrod Sales/Service	10593 Old Lincolnway	Orrville	OH	44667	330-682-9261
Paul Sherry RV's	P.O. Box 742	Piqua	OH	45356	937-778-5250
Cecil Caudill Trailer Sales, Inc.	5737 Gallia	Portsmouth	OH	45662	614-776-2151
Stewart's RV, Inc.	50863 National Road East	St. Clairsville	OH	43950	614-695-0918
Jerry Greer RV	6700 E. St. Rt. 37	Sunbury	OH	43074	614-369-3922
Jerry Greer RV	6700 E. St. Rt. 37	Sunbury	OH	43074	614-369-3922
Jerry Greer RV	6700 E. St. Rt. 37	Sunbury	OH	43074	614-369-3922
All American Coach Company	5080 Alexis Road	Sylvania	OH	43560	419-885-4601
All American Coach Company	5080 Alexis Road	Sylvania	OH	43560	419-885-4601
Beggs Motor Homes	11197 Cleveland Ave. NW	Uniontown	OH	44685	330-499-9755
Clare-Mar Lakes RV Sales & Service	PO Box 226	Wellington	OH	44090	216-647-3318
Camperland RV Center	9260 Cincinnati-Dayton Hwy.	West Chester	OH	45069	513-777-8700
Camperland RV Center	9260 Cincinnati-Dayton Hwy.	West Chester	OH	45069	513-777-8700
All Seasons RV Center	37200 Vine	Willoughby	OH	44094	216-951-2011
Ruff's RV Center	28920 Chardon Road	Willoughby Hills	OH	44092	216-944-9224
Barlow's Auto & RV Sales	2817 State Rt. 73 S.	Wilmington	OH	45177	937-382-5514
Bell Camper Sales	PO Box 1142	Bartlesville	OK	74005	918-333-5333
Bell Camper Sales	PO Box 1142	Bartlesville	OK	74005	918-333-5333
Thurman Traveland	I9I23 E. Admiral Place	Catoosa	OK	74015	9I8-266-IIII
Glass R.V. Center	South Hwy. 81	Chickasha	OK	73023	405-224-5773
Glass R.V. Center	South Hwy. 81	Chickasha	OK	73023	405-224-5773
Dave's Claremore RV, Inc.	1400 S. Lynn Riggs Blvd.	Claremore	OK	74017	918-341-0114
Sherrard RV & KOA	Route 1 Box 500	Colbert	OK	74733-9735	405-296-2373
Sherrard RV & KOA	Route 1 Box 500	Colbert	OK	74733-9735	405-296-2373
Kerr Kountry RV's	Highway 70 South	Madill	OK	73446	405-795-3337
David R.V.	Route 3 Box 79A	Marlow	OK	73055	405-658-6899
David R.V.	Route 3 Box 79A	Marlow	OK	73055	405-658-6899
C & L Mobile Home & RV Center	3101 South 32nd Street	Muskogee	OK	74401	918-687-9537
Walker RV Center	1559 N. Main, Hwy 62	Newcastle	OK	73065	405-387-3322
Floyd's R.V.'s	2501 O. J. Talley	Norman	OK	73072	405-288-2355
RV General Store, Inc.	I707 Topeka Drive	Norman	OK	73069	405-366-7934
I-35 RV Sales	5797 N. Interstate Drive	Norman	OK	73069	405-447-2626
Floyd's R.V.'s	2501 O. J. Talley	Norman	OK	73072	405-288-2355
Lewis RV Center	1600 East Reno	Oklahoma City	OK	73117	405-670-4865
Lee's RV City, Inc.	P.O. Box 6250	Oklahoma City	OK	73130	405-733-1753
McClain's RV Superstore	7110 West Reno	Oklahoma City	OK	73128	405-789-4773
McClain's Mr. Motorhome	3333 S. Interstate 35	Oklahoma City	OK	73129-6761	405-943-3366
Mike's RV Sales	9820 Grimsbey Court	Oklahoma City	OK	73159	405-691-2208
Lee's RV City, Inc.	P.O. Box 6250	Oklahoma City	OK	73130	405-733-1753
Mitchell Coach Mfg. Company	P.O. Box 339	Pryor	OK	74362-0339	918-825-7000
Dean's RV Superstore	9955 East 21st Street	Tulsa	OK	74129	918-664-3333
Nichols RV World	8347 East 11th Street	Tulsa	OK	74112	918-836-6606
Nichols RV World	8347 East 11th Street	Tulsa	OK	74112	918-836-6606
Curtis Trailers, Inc.	21525 SW Tualitin Valley Hwy.	Aloha	OR	97006-1330	503-649-8528
All Seasons RV & Marine	P.O. Box 5699	Bend	OR	97708	541-382-5009
Patterson's RV Sales, Inc.	PO Box 6445	Bend	OR	97708	541-382-3I86
Larry's RV, Inc.	2115 NE Hwy. 20	Bend	OR	97701	541-388-7552
Jim Smolich RV Center	63695 N. Hwy 97	Bend	OR	97701	541-330-2495
Gib's RV's, Inc.	1845 Ocean Blvd	Coos Bay	OR	97420	541-888-3424
Porter's R.V.'s	971 S Broadway	Coos Bay	OR	97420	541-269-5121
Romania's RV Centers	P.O. Box 3216	Eugene	OR	97403-0216	541-465-3222
Paul Harris RV	4000 Franklin Blvd.	Eugene	OR	97403	541-747-4099
George M. Sutton RV/Truck Sales	2400 W. 7th	Eugene	OR	97402	541-686-6296
The RV Corral, Inc.	1890 Hwy. 99 North	Eugene	OR	97402	541-689-9204
Daryl's RV Village	160 S.W. Redwood Hwy.	Grants Pass	OR	97527	541-476-0897

Siskiyou World	225 Northeast Terry Lane	Grants Pass	OR	97526	541-535-5551
River City RV's	111 Union Avenue	Grants Pass	OR	97527	541-476-1541
Gresham RV Center Inc.	325 NE Hogan Rd.	Gresham	OR	97030	503-661-5946
Olinger Travel Homes	6503 Alexander	Hillsboro	OR	97123	503-649-2141
Guaranty RV Centers	20 Hwy. 99 S.	Junction City	OR	97448	541-998-2333
Valley RV Center, Inc.	11160 Durham Lane & Hwy 18W	McMinnville	OR	97128	503-472-9567
Triple A RV Center, Inc.	938 Chevy Way	Medford	OR	97504	541-772-1938
Thompson RV, Inc.	1220 Southgate/P.O.Box 1459	Pendleton	OR	97801	541-276-4836
Apache Camping Center, Inc.	P.O. Box 66208	Portland	OR	97290-6208	503-659-5166
Curtis Trailers, Inc.	10177 SE Powell Blvd.	Portland	OR	97266	503-760-1363
Olinger Travel Homes	9401 SE 82nd Avenue	Portland	OR	97266	503-771-2121
Gresham RV	5111 NE 82nd Avenue	Portland	OR	97220	503-254-7015
Wagers Trailer Sales	3282 Silverton Rd. NE	Salem	OR	97303	503-585-7713
Highway Trailer Sales	4560 Portland Rd. NE	Salem	OR	97305	503-393-2400
Roberson R.V. Center	3100 Ryan Drive, S.E.	Salem	OR	97301	503-363-4117
Mitchell RV	16800 SE 362nd Drive	Sandy	OR	97055	503-668-7461
Sweet Home RV Center	4691 Hwy 20 E	Sweet Home	OR	97386	541-367-4293
Kamper Korner, Inc.	5719 NE Stephens	Winchester	OR	97495	541-673-1258
Stoltzfus RV's & Marine	Rt. 272 Box 564	Adamstown	PA	19501	717-484-4344
Harold's RV Center	7514 Bath Pike	Bath	PA	18014	610-837-9880
W.T. Family RV Sales & Service, Inc.	Route 115, Box 1486	Blakeslee	PA	18610	717-646-4040
Heritage Motors	526 Montour Blvd	Bloomsburg	PA	17815	717-784-5388
Starr's Trailer Sales	RD 3 Box 137	Brockway	PA	15824	814-265-0632
Outdoor World Corporation	P.O. Box 447	Bushkill	PA	18324	717-588-6661
Cumberland Valley Camping Center	7050 Carlisle Pike	Carlisle	PA	17013	717-766-2103
Glasgow Recreational Vehicles	1650 Lincoln Way East	Chambersburg	PA	17201	717-264-9551
Rhone's Travel Trailers	4368 Lycoming Creek Road	Cogan Station	PA	17728	717-494-1364
Unger Recreation Vehicles	800 Flaugherty Run Road	Coraopolis	PA	15108	412-264-8100
Unger Recreation Vehicles	800 Flaugherty Run Road	Coraopolis	PA	15108	412-264-8100
Kieffer's RV Sales	244 Ben Franklin Hwy.	Douglassville	PA	19518	610-326-5998
Hopewell R.V. Center, Inc.	290 Corner Ketch Road	Downingtown	PA	19335	610-269-3445
Ansley & Lewis, Inc.	P.O. Box 239/1280 Route 764	Duncansville	PA	16635-0239	814-695-9817
Dewalt's RV, Inc.	270 Country Club Rd.	Easton	PA	18045	610-258-0486
Clem's Trailer Sales Inc.	1580 State Route 65 South	Ellwood City	PA	16117	412-752-1541
Farnsworth Camping Center, Inc.	51 North Market Street	Elsburg	PA	17824-9624	717-672-2332
Boyer RV Center	8495 Peach St.	Erie	PA	16509	814-868-7561
Brien's Trailer Sales	168 Lincoln Hwy. Rt. 1	Fairless Hills	PA	19030	215-946-9530
Huffy's Trailer Sales, Inc.	RD 2 Rt. 88	Finleyville	PA	15332	412-348-5353
Keystone RV Center	15799 Young Road	Greencastle	PA	17225	717-597-0939
Keystone RV Center	15799 Young Road	Greencastle	PA	17225	717-597-0939
Reichart's Camping Center	2100 Baltimore Pike	Hanover	PA	17331	717-637-2882
Reichart's Camping Center	2100 Baltimore Pike	Hanover	PA	17331	717-637-2882
Grumbine's R.V. Center, Inc.	7501 Allentown Blvd.	Harrisburg	PA	17112	717-657-3747
Media Camping Center	1651 Bethlehem Pike	Hatfield	PA	19440-1302	215-822-1345
Ashley Chevrolet & Cadillac	206 Willow Avenue	Honesdale	PA	18431	717-253-3030
Turner Airstream, Inc.	472 Lincoln Hwy West	Jeannette	PA	15644	412-523-6545
Lerch RVs	R.D. #4 Box 399A	Lewistown	PA	17044	717-242-1789
Martin Camper Sales	6785 Carlisle Pike	Mechanicsburg	PA	17055	717-766-5569
Quality Coach, Inc.	Stump Road & Commerce Drive	Montgomeryville	PA	18936	215-643-2211
Spitler, Inc.	Old Rt 220/ PO Box 267	Montoursville	PA	17754	717-368-1771
Gayle Kline R.V. Center, Inc.	444 East Main Street	Mountville	PA	17554	717-285-3159
Ray Wakley's Car Care Center	10261 West Main Road	North East	PA	16428	814-725-9608
Northwood RV Center	P.O. Box 398	Oaks	PA	19456-0398	610-253-1300
Jim's Marine & Travel, Inc.	302 South Keystone Avenue	Sayre	PA	18840	717-888-2353
Susquehanna Valley RV	RD #1 Box 131-S	Selinsgrove	PA	17870	717-374-2267
Tom Schaeffer's Camping & Trailer	1236 Pottsville Pike/US Rt. 61	Shoemakersville	PA	19555	610-562-3071
Fretz Enterprise	3479 Bethlehem Pike	Souderton	PA	18964	215-723-8118
Indian Valley Camping Center	3400 Old Bethlehem Pike	Souderton	PA	18964	215-723-4852
Stoltzfus RV's & Marine	1335 Wilmington Pike	West Chester	PA	19382	610-399-0628
Stoltzfus RV's & Marine	1335 Wilmington Pike	West Chester	PA	19382	610-399-0628
Lehigh Gorge RV Sales & Campground	Box 6A/HRC #1	White Haven	PA	18661	717-443-9191
Wide-World R.V. Center, Inc.	1570 Route 315	Wilkes-Barre	PA	18702	717-825-6673
Mellott Brothers Trailer Sales	2718 Willow Street Pike	Willow Street	PA	17584	717-464-2601
Ben's RV Center	1590 Whiteford Road	York	PA	17402	717-755-9669
Arlington RV Super Center, Inc.	966 Quaker Lane	E. Greenwich	RI	02818	401-884-7550
Arlington RV Super Center, Inc.	966 Quaker Lane	E. Greenwich	RI	02818	401-884-7550
Holiday On Wheels, Inc.	8015 Sumter Hwy.	Columbia	SC	29209	803-776-3752
Holiday Kamper and Boats	3630 Fernandina RD.	Columbia	SC	29210	803-798-0450
Sonny's Camp & Travel	333 Frontage Road	Duncan	SC	29334	864-433-0887
Jerry Lathan RV World	2744 Carowinds Blvd.	Fort Mill	SC	29716	803-548-6550
Astro Sales Center	2400 N. Pleasantburg Drive	Greenville	SC	29609	864-242-6716
Masters RV Centre, Inc.	104 Hwy. 246 South	Greenwood	SC	29649	864-223-2267
Holiday RV Superstores, Inc.	198 Bob Ledford Drive	Greer	SC	29651-9206	864-877-8218
Carolina R.V. Sales	3590 Savanna Hwy.	Johns Island	SC	29455	803-556-3449
Carolina R.V. Sales	3590 Savanna Hwy.	Johns Island	SC	29455	803-556-3449

John's RV Sales/Service	242 Glassmaster Road	Lexington	SC	29072	803-359-2957
Don Mar RV of South Carolina	265 Pudding Swamp Road	Lynchburg	SC	29080	803-453-5011
Brown's RV's	Rt. 2 Box 61 Hwy. 151	McBee	SC	29101	803-335-8829
Brown's RV's	Rt. 2 Box 61 Hwy. 151	McBee	SC	29101	803-335-8829
Camping Sales & Service	6003 Hwy. 17 South	Myrtle Beach	SC	29575	803-238-5532
Camping Sales & Service	6003 Hwy. 17 South	Myrtle Beach	SC	29575	803-238-5532
Goodman Chevrolet Coachmen	100 Main St./ P.O. Drawer G	New Ellenton	SC	29809	803-652-2233
The Trail Center	5728 Dorchester Road	North Charleston	SC	29418	803-552-4700
Sonny's Camp & Travel	8155 Rivers Avenue	North Charleston	SC	29406	803-553-6633
The Trail Center	5728 Dorchester Road	North Charleston	SC	29418	803-552-4700
Sonny's Camp & Travel	8155 Rivers Avenue	North Charleston	SC	29406	803-553-6633
Marine One	8745 Asheville Highway	Spartanburg	SC	29303	864-578-8158
Happy Trails RV Sales	4070 Broad St. Ext.	Sumter	SC	29154	803-494-2055
Camper Country	P.O. Box 14328	Surfside Beach	SC	29587	803-238-5678
Tebben's Campers	47504 271st Street	Harrisburg	SD	57032-8103	605-743-2166
Jack's Campers Sales	1313 W Norway	Mitchell	SD	57301	605-996-3268
Langland Trailer Sales	4300 N. I-90 Service Road	Rapid City	SD	57701	605-343-6675
Green Star Campers	P.O. Box 534/120 Cambell St.	Rapid City	SD	57701	605-343-6877
Mid-State Camper Sales	5900 W. Hwy 44	Rapid City	SD	57702	605-348-0623
Spader Camper Center	5300 Hwy. 77 N.	Sioux Falls	SD	57101	605-339-3230
Schaap's Traveland, Inc.	3100 W Russell	Sioux Falls	SD	57107	605-332-6241
Midway Trailer Sales, Inc.	P.O. BOX 6/E Highway 14	St. Lawrence	SD	57373	605-853-2103
Chilhowee Trailer Sales, Inc.	P.O. Box 236	Alcoa	TN	37701	423-970-4085
Chilhowee Trailer Sales, Inc.	P.O. Box 236	Alcoa	TN	37701	423-970-4085
Shipp's RV	6728 Ringgold Road	Chattanooga	TN	37412	423-892-8275
Clarksville RV	550 Thun RD	Clarksville	TN	37040	615-648-1800
Country Roads Campers	2525 Highway 48 & 13 South	Clarksville	TN	37040	615-552-7820
Crowder RV Center, Inc.	4533 Bristol Highway	Johnson City	TN	37601	423-282-5011
Crowder RV Center, Inc.	4533 Bristol Highway	Johnson City	TN	37601	423-282-5011
Tri-City Travel Trailer	3377 E. Stone Dr.	Kingsport	TN	37660	423-288-3131
Campers Corner, Inc.	4723 Clinton Highway	Knoxville	TN	37950-0728	423-688-4733
Bob Cox Camper Country, Inc.	6014 Clinton Hwy.	Knoxville	TN	37912	423-688-0881
Buddy Gregg Motor Homes	P.O. Box 23470	Knoxville	TN	37933	423-675-1986
America's Choice	4223 Airport Hwy.	Louisville	TN	37777	423-970-7080
Brown RV Center, Inc.	4050 North Thomas	Memphis	TN	38127	901-353-1999
Cullum & Maxey Camping Center	2614 Music Valley Dr.	Nashville	TN	37214	615-889-1600
Nashville Easy Livin' Country	I5I6 Murfreesboro Road	Nashville	TN	37217	423-361-3867
Chad's Camping Center	336 Welch Rd.	Nashville	TN	37211	615-833-0254
Nashville Easy Livin' Country	I5I6 Murfreesboro Road	Nashville	TN	37217	423-361-3867
Jack Sisemore Traveland	4341 Canyon Drive	Amarillo	TX	79110	806-358-4891
Marshall's Traveland	I0704 IH 35 South	Austin	TX	78748	5I2-282-5524
Ancira GMC Trucks & Motor Homes, Inc.	30500 IH 10 W/PO Box 1887	Boerne	TX	78006	210-981-9000
Crestview RV Center	PO Box 1028/ Exit #220 IH 35 S.	Buda	TX	78610	512-295-2308
Marshall's RV Centers, Inc.	Interstate 20 at Exit 521	Canton	TX	75103	903-865-1130
Traylor Motor Homes, Inc.	P.O. Box 1720	Cedar Hill	TX	75104	972-526-1224
Traylor Motor Homes, Inc.	P.O. Box 1720	Cedar Hill	TX	75104	972-526-1224
Fun Time RV	P. O. Box 790	Cleburne	TX	76033-0790	817-517-2200
Morgan Building Systems, Inc.	P.O. Box 660280	Dallas	TX	75266	972-840-1200
Vogt Motor Homes	5624 Airport Freeway	Ft. Worth	TX	76117	817-831-4222
Kerr Kountry RV's	P.O. Box 1079	Gainesville	TX	76240	817-458-7010
R & K Camping Center, Inc.	725 South Jupiter Road	Garland	TX	75042-7702	972-276-7637
Camper Capitol USA	4233 Forest Lane	Garland	TX	75042	972-276-3323
Foretravel of Texas, Inc.	4213 Forest Lane	Garland	TX	75042	972-276-3673
Bennett's Camping Center	2708 Hwy. 377 East	Granbury	TX	76049	817-573-3665
Eastex Camper Sales, Inc.	P.O. Box 60632	Houston	TX	77205	281-441-2138
Holiday World, Inc.	8224 North Freeway	Houston	TX	77037	281-448-0035
Demontrond Automotive Group	14101 N Freeway	Houston	TX	77090	281-872-3898
Topper Sales, Inc.	17930 US 290 West	Houston	TX	77065	713-896-8441
P.P.L. Motorhomes	10777 Southwest Freeway	Houston	TX	77074	713-988-5555
Dar Fun Time RV Sales	P.O. Box 29	Keene	TX	76059	817-641-2534
Marshall's RV Center	9474 E US HWY 175	Kemp	TX	75143	903-498-3711
McClain's RV Superstore	North I-35 East, Exit 460	Lake Dallas	TX	75065	817-497-3300
Avion Sales & Service	702 S. Stemmons Fwy.	Lewisville	TX	75067	972-436-3525
Hayes Trailer Sales, Inc.	5009 Judson Rd.	Longview	TX	75605	903-663-3488
Pharr RV's, Inc.	320 N Loop 289	Lubbock	TX	79403	806-765-6088
Billy Sims Trailer Town	1615 S Loop 289	Lubbock	TX	79423	806-745-8791
Holiday World RV Center	4630 I 30 East	Mesquite	TX	75150	214-327-5858
Blakely RV Complex, Inc.	P.O. Box 61207	Midland	TX	79711-1207	915-561-9551
Gooding RV Center	1601 E Expressway 83	Mission	TX	78572	210-585-4481
Billy Sims Trailer Town	520 E 2nd Street	Odessa	TX	79761	915-580-3000
Superior RV Center, Inc.	1019 Alcock/PO Box 2036	Pampa	TX	79065	806-665-3166
Lloyd's Trailer Sales	350 Twin City Hwy	Port Neches	TX	77651	409-727-1666
Ron Hoover Company	P.O. Box 747/1510 W. Market	Rockport	TX	78382	512-729-9695
Camper Clinic, Inc.	302 West Market	Rockport	TX	78382	512-729-0031
Crestview RV Superstore	P.O. Box 612	Schertz	TX	78154	210-651-6300

Name	Address	City	State	Zip	Phone
Cliff Jones, Inc.	1629 South Circle Dr.	Sealy	TX	77474	409-885-3554
W & W Marketing	1620 North 123 Bypass	Seguin	TX	78155	210-379-4100
Shady Pines Trailer Sales, Inc.	Hwy. 67 South, Box 418R	Texarkana	TX	75501	903-838-5486
Otis Thomas Sales	2606 Jacksboro Hwy	Wichita Falls	TX	76302	817-767-1234
Stewarts RV	854 East 1100 South	American Fork	UT	84003	801-492-1428
Blaine Jensen RV Center	220 North 650 West	Kaysville	UT	84037-2476	801-544-4298
Terry's RV Center	5545 South State	Murray	UT	84107	801-262-2486
Motor Sportsland	4001 S. State	Salt Lake City	UT	84107	801-262-2921
Ardell Brown R.V.	9200 S State	Sandy	UT	84070	801-255-9200
Miller Camper-Trailer	950 E 800 N	Spanish Fork	UT	84660	801-798-7447
Painter's RV	1500 South Hilton Drive	St. George	UT	84770	801-673-1500
Trips Auto Sales	P.O. Box 342	Berryville	VA	22611	540-955-1367
Mt. Joy RV Sales & Service	P.O. Box 249, 2740 Main Street	Buchanan	VA	24066	540-254-2360
Camp-A-Rama, Inc.	1107 N. Geo. Washington Hwy	Chesapeake	VA	23323	757-487-8113
Reines R.V. Center, Inc.	9711 Lee Highway	Fairfax	VA	22031	703-591-2700
Koogler Sales & Service, Inc.	Rt. 2 Box 193	Fishersville	VA	22939	540-942-5556
Fredericksburg RV, Inc.	1132 Jefferson Davis Highway	Fredericksburg	VA	22405	540-659-5800
Fredericksburg RV, Inc.	1132 Jefferson Davis Highway	Fredericksburg	VA	22405	540-659-5800
Rainbow Acres	Route 2 Box 16	King & Queen Cthse.	VA	23085	804-785-9441
Restless Wheels, Inc.	8104 Centreville Road	Manassas	VA	22111-2220	703-257-1067
Dixie R.V. Superstore	Jefferson Ave. & Muller Lane	Newport News	VA	23606	757-249-1257
Cheek & Shockley Auto Trailer	2600 Mechanicsville Pike	Richmond	VA	23223	804-649-7508
Rolling Hills RV Superstore, Inc.	6200 W. Broad Street	Richmond	VA	23230	804-285-9071
Southern Recreation Vehicles	7001 Jeff Davis Hwy	Richmond	VA	23237	804-275-8345
Snyder's R.V.	2011 West Main Street	Salem	VA	24153	540-389-8000
Rule, Inc.	1611 Greenville Avenue	Staunton	VA	24401	540-886-2357
Snyder's R.V.	5632 Virginia Beach Blvd	Virginia Beach	VA	23462	757-499-3300
Virginia RV Sales, Inc.	7023 Rt. 17	Yorktown	VA	23692	757-898-5700
Green Mountain Campers	P.O. Box 7 Historic Route 7A	Arlington	VT	05250	802-375-9661
Mekkelsen Trailer Sales	Route 2	E. Montpelier	VT	05651	802-223-3684
Mekkelsen Trailer Sales	Route 2	E. Montpelier	VT	05651	802-223-3684
Ehler's RV, Inc.	70 Upper Main Street	Essex Junction	VT	05452	802-878-4907
Pete's RV Center	4016 Williston Road	South Burlington	VT	05403	802-864-9350
Auburn Mobile Homes	2536 Auburn Way North	Auburn	WA	98002	206-833-3368
Vacationland R.V. Sales	1400 Iowa Street	Bellingham	WA	98226	360-734-5112
Uhlmann RV	P.O. Box 1366, 173 Hamilton Rd.	Chehalis	WA	98532	360-748-6658
J & D RV Sales	165 Hamilton Road	Chehalis	WA	98532	360-748-3692
Ponderosa Hill Park Model Sales, Inc.	7520 South Thomas Mallen Road	Cheney	WA	99004	509-358-6050
Krueger's 1st Stop R.V., Inc.	1427 Bridge Street	Clarkston	WA	99403	509-758-6454
Holiday World RV Center	12620 Highway 99 South	Everett	WA	98204	425-355-5944
5th Wheel Travel Homes, Inc.	11308 Hwy 99	Everett	WA	98204-4815	206-355-9146
Rainbow RV Center	13210 Hwy 99 S	Everett	WA	98204	206-745-4730
Campers Paradise	12800 Hwy. 99 S.	Everett	WA	98204-6225	425-347-1214
Apache Camping Center, Inc.	13304 Hwy. 99 South	Everett	WA	98204	206-745-8810
Travel Time RV Center, Inc.	12517 Highway 99 South	Everett	WA	98204	206-353-9377
Signal Trailer Sales	1871 Ross Avenue, Suite E	Everett	WA	98205	206-745-5060
Continental RV, Inc.	27454 Pacific Hwy S	Federal Way	WA	98063	206-941-5200
Richard's RV, Inc.	4942 Pacific Highway	Ferndale	WA	98248	360-380-2003
Holiday World, Inc.	4902 Pacific Hwy. East	Fife	WA	98424	206-473-2080
Great American RV	4902 Pacific Hwy. East	Fife	WA	98424	206-926-2626
RV's Northwest, Inc.	E 18919 Broadway	Greenacres	WA	99016	509-891-5854
Valley I-5	23051 Military Road So.	Kent	WA	98032	206-852-0150
Longview RV Center, Inc.	915 Tennent Way	Longview	WA	98632	360-577-1919
Wholesale Travel Sales, Inc.	8165 Guide Meridian Road	Lynden	WA	98264	360-354-4477
Western Motorhome Rentals	19303 Hwy. 99	Lynnwood	WA	98036	206-774-1414
Northwest Trailer Center	15703 Hwy. 99	Lynnwood	WA	98037	206-742-2100
Seaview Chevrolet	P.O. Box 1976	Lynnwood	WA	98046	206-742-1920
Western Motorhome Rentals	19303 Hwy. 99	Lynnwood	WA	98036	206-774-1414
Roy Robinson Chevrolet	PO Box 168	Marysville	WA	98270	360-659-6236
Xplorer Motor Homes NW	7200 Pacific Hwy. E	Milton	WA	98354	206-243-4440
Tveten RV/Airstream	7700 Pacific Hwy. E	Milton	WA	98354	206-922-7770
Blade Chevrolet, Inc.	1100 Freeway Drive	Mt. Vernon	WA	98273	360-424-3231
Valley RV	315 Freeway Drive	Mt. Vernon	WA	98273	360-336-3164
Russ Dean's Family RV Center	3201 W. Octave	Pasco	WA	99301	509-545-9501
Chief's RV Center	1120 North 28th Avenue	Pasco	WA	99301-3970	509-547-1198
Poulsbo RV, Inc.	19705 Viking Avenue NW	Poulsbo	WA	98370	360-697-4445
Courtesy RV	20081 Viking Ave NW	Poulsbo	WA	98370	360-697-2700
Korum Ford/Mitsubishi/RV	P.O. Box 538/812 N. Meridian	Puyallup	WA	98371	206-841-9600
U-Neek R.V. Center	17611 NE Union Road	Ridgefield	WA	98642	360-574-4422
U-Neek R.V. Center	17611 NE Union Road	Ridgefield	WA	98642	360-574-4422
Carl North Company, Inc.	14061 Lake City Way NE	Seattle	WA	98125	206-364-7500
Helgeson's Trailer Exchange, Inc.	21050 Pacific Hwy. S.	Seattle	WA	98198	206-824-0224
Selah Trailer Camper Sales, Inc.	518 South 1st Street	Selah	WA	98942	509-697-7156
A-1 RV Center	145 Silverdale Way	Silverdale	WA	98383	360-692-1098
Clearview RV Repair and Sales	17104 Hwy. 9	Snohomish	WA	98290	206-668-9595

Ray's RV's	4808 East Sprague	Spokane	WA	99212	509-535-6727
R'N'R Holiday R.V., Inc.	North 108 Vista Road	Spokane	WA	99212	509-927-9000
Sumner RV Center	4309 E. Valley Highway	Sumner	WA	98390	206-863-5644
Baydo's Trailer Sales	7230 South Tacoma Way	Tacoma	WA	98409	206-475-1411
Pacific Travel Center, Inc.	10211 Pacific Avenue	Tacoma	WA	98444	206-531-4774
South Side Motors	7202 S Tacoma Way	Tacoma	WA	98409	206-474-9421
Tacoma RV Center	8909 S. Tacoma Way	Tacoma	WA	98499	206-581-7703
Apache Camping Center, Inc.	9402 Pacific Avenue	Tacoma	WA	98444	206-535-6522
Family Fun RV	6722 Pacific Hwy. East	Tacoma	WA	98424-1541	206-472-5040
Family Fun RV	6722 Pacific Hwy. East	Tacoma	WA	98424-1541	206-472-5040
Aubrey's RV Center Inc.	2010 Landon Ave.	Yakima	WA	98903-1479	509-453-4709
Canopy Country RV	2904 South Main St.	Yakima	WA	98903	509-248-7050
Appleton Camping Center	2100 N. McCarthy Road	Appleton	WI	54914-7048	414-757-6112
Finnegan RV	902 Broad Street	Beloit	WI	53511	608-365-2306
Burlington Camping & Travel, Inc.	3145 Wegge Drive	Burlington	WI	53105	414-763-9595
North Point RV	5955 S. Prairie View Road	Chippewa Falls	WI	54729	715-723-5380
Lundmark Camper Sales	2236 Hwy. 63	Cumberland	WI	54829	715-822-8714
Quinnette's RV Center	840 North 9th Street	DePere	WI	54115-1598	414-339-1000
Dick's RV & Sport Shop, Inc.	406 West Main	Durand	WI	54736	715-672-4218
Schiek's Camping Center	406 East Main	Eden	WI	53019	414-477-4561
De Haan RV Sales	9 Deer Road	Elkhorn	WI	53121	414-723-2260
Merz RV Center, Inc.	1140 Highway 151 South	Fond du Lac	WI	54935	414-921-1164
Van Boxtel RV, Inc.	1010 S. Military Ave	Green Bay	WI	54307-1567	414-499-3131
Van Boxtel RV, Inc.	1010 S. Military Ave	Green Bay	WI	54307-1567	414-499-3131
FMB, Ltd.	P.O. Box 97	Holmen	WI	54636-0097	608-526-3336
Camperland Sales	5498 Co. Hwy. CV	Madison	WI	53704	608-241-1636
Wisconsin RV World	P.O. Box 8218	Madison	WI	53708-8218	608-244-6228
Roskopf's RV Center, Ltd.	West 164 N 9306 Water Street	Menomonee Falls	WI	53051	414-255-2240
KOA Trailer Sales	2501 Broadway Street N.	Menomonie	WI	54751	715-235-0641
Advance Camping Sales	6606 W. Layton Avenue	Milwaukee	WI	53220	414-281-6330
Camper Corral	1922 Co Trunk MM	Oregon	WI	53575	608-835-5398
Hubert Trailers	1842 County Trunk Hwy. MM	Oregon	WI	53575	608-835-3002
Horn's RV	Route 3, 143 South	Sheboygan	WI	53082	414-564-2381
Horn's RV	Route 3, 143 South	Sheboygan	WI	53082	414-564-2381
Coulee Region RV Center, Inc.	Route 1, Highway B	West Salem	WI	54669	608-786-2244
Greeneway, Inc.	8220 Hwy. 13 South	Wisconsin Rapids	WI	54494	715-325-5170
Setzer's World of Camping, Inc.	5840 Davis Creek Road	Barboursville	WV	25504	304-736-5287
Setzer's World of Camping, Inc.	5840 Davis Creek Road	Barboursville	WV	25504	304-736-5287
Brand Trailer Sales	2045 Fairmont Ave.	Fairmont	WV	26554	304-366-7104
Roy's Travel Trailers	P.O. Box 146	Harman	WV	27270	304-227-4100
Burdette Camping Center, Inc.	3749 Winfield Rd.	Winfield	WV	25213	304-586-3084
Stalkup's RV Superstore	501 W. Yellowstone	Casper	WY	82601	307-577-9350
Max Auto & Marine Sales, Inc.	550 East 1st Street	Casper	WY	82601	307-577-9333

RENTAL DEALERS:

Affordable New Car Rental	4707 Spenard Road	Anchorage	AK	99517	907-243-3370
Clippership Motorhome Rentals	5401 Old Seward Highway	Anchorage	AK	99518	907-562-7051
Great Alaskan Holidays	3901 W. Internat'l. Airport Rd.	Anchorage	AK	99502	907-248-7777
Sweet Retreat, Inc.	6820 Arctic Blvd.	Anchorage	AK	99518	907-344-9155
POW Island Getaway R.V. Rentals	P.O. Box 421	Craig	AK	99921	907-826-4150
Fireweed R.V. Rentals	3401 Peger Road	Fairbanks	AK	99706	907-452-4949
Last Frontier RV Adventures	P.O. Box 32466	Juneau	AK	99803-2466	907-789-1982
Alaska Recreational Rentals	Box 592/ Mile 102 Sterling Hwy.	Soldotna	AK	99669	907-262-2700
Grand Travel Services	7020 NW Grand Avenue	Glendale	AZ	85301	602-939-6909
Orangewood RV & Marine	7520 N. 67th Avenue	Glendale	AZ	85301	602-939-2521
Bates Motor Home Network of Arizona	15112 North 73rd Avenue	Peoria	AZ	85381	602-878-4930
Freeway Easy Livin' Country	4500 W. Ramsey St.	Banning	CA	92220	909-849-6785
Metro RV	160 W. Olive Avenue	Burbank	CA	91502	818-841-2441
All Valley Rentals	36510 Cathedral Canyon	Cathedral City	CA	92234	619-324-0454
Quality Motorhome Rentals	129 E. Grand Blvd.	Corona	CA	91720-5419	909-278-8868
Moturis, Inc.	400 W. Compton Blvd.	Gardena	CA	90248	310-767-5988
R.V. Rentals	11265 Lime Kiln Road	Grass Valley	CA	95949	916-268-0822
Bea Rentals	1500 Crestfield	Irwindale	CA	91010	818-359-0068
Global Motorhome Travel, Inc.	1147 Manhatten Avenue	Manhattan Beach	CA	90266	310-796-5665
RVStore & RVRent	105 Chalk Creek Ct.	Martinez	CA	94553	510-935-8260
California RV Rentals	5933 N McHenry Avenue	Modesto	CA	95356	209-523-2131
Adventure Rentals, Inc.	1200 West Mission Blvd.	Ontario	CA	91762	909-983-2567
America Traveler RV Rental	770 W. Carson Mesa Road	Palmdale	CA	93550	805-273-7479
Norm's R.V. Rentals	12538 Poway Road	Poway	CA	92064	619-679-2250
Vacation Bound Inc.	1315 Antrim Dr.	Roseville	CA	95747	916-783-8473
Adventure R.V. Rental	P.O. Box 254791	Sacramento	CA	95865	916-485-0282
The Motorhome Club	3803 Convoy Street	San Diego	CA	92111	619-492-9500
California Campers	141 Behr Avenue	San Francisco	CA	94131	415-665-1558
Family R.V.	2828 Monterey Road	San Jose	CA	95111	408-365-1991
Moturis, Inc.	420 San Leandro Blvd.	San Leandro	CA	94577	510-562-8504

RV America	2905 San Pablo Dam Road	San Pablo	CA	94803	510-669-9005
Bates Motor Home Rental Of San Jose	2005 De La Cruz Blvd. Suite 131	Santa Clara	CA	95050	408-988-1711
Crest RV	9483 San Fernando Rd.	Sun Valley	CA	91352	818-252-3300
Road Bear International	972 Calle Brusca	Thousand Oaks	CA	91360	818-865-2925
California R.V. Rentals, Inc.	2455 Sepulveda, Blvd.	Torrance	CA	90501-4325	310-518-4487
Travel Time of Vacaville	941 Merchant Street	Vacaville	CA	95688-5315	707-447-7548
All Tent Trailer Rentals	895 Via Arroyo	Ventura	CA	93003	805-644-2450
T & L RV Rental	2606 North Ventura Avenue	Ventura	CA	93001	805-653-0714
Club Travel Motorhome Rentals	7600 Westminster Blvd.	Westminster	CA	92683-3920	714-775-2730
Colorado RV Vacations	5024 S. Uravan Court	Aurora	CO	80015	303-617-3705
Adventures In R.V.'s	775 North Murray Blvd.	Colorado Springs	CO	80915	719-473-2583
Aspen RV Rentals	5635 Country Heights Drive	Colorado Springs	CO	80917	719-380-7950
Aspen RV Rentals	3137 Silverwood Drive	Fort Collins	CO	80525	970-206-0475
Sportsmen Rentals	3197 Hall Avenue	Grand Junction	CO	81504	970-523-0611
Kathy's Kampers	709 Foulk Road	Wilmington	DE	19803	302-328-9417
1st State Camping Center	6611 Governor Printz Blvd.	Wilmington	DE	19809	302-798-4033
America On The Move, Inc.	1800 West State Road 84	Ft. Lauderdale	FL	33315	305-523-4334
Palm RV Centers	2441 South State Rd 7	Ft. Lauderdale	FL	33317-6999	954-584-3200
Ft. Myers Truck & Auto Land	16065 S. Tamiami Trail	Ft. Myers	FL	33908	941-482-4511
Rover Rentals	13997 Beach Blvd.	Jacksonville	FL	32224	904-992-8222
Blue Lagoon Resorts International	99096 Overseas Hwy.	Key Largo	FL	33037	305-453-0094
Moturis USA, Inc.	3901 N.W. 16th Street	Lauderhill	FL	33313	305-587-6450
Budget Rentals, Inc.	7254 65th Drive	Live Oak	FL	32060	904-364-4909
House on Wheels	9921 Bellville Road	Miami	FL	33157	305-278-8350
Florida RV World, Inc.	4260 U.S. 92 East	Plant City	FL	33566	813-754-6171
Camptown RV Country	23905 SW 132nd Ave.	Princeton	FL	33032	305-258-1783
M & S Rec. Rentals	3490 Green Hill Road	Gainesville	GA	30506	770-535-2562
Family R.V. Center	5463 B. Goshen Springs Road	Norcross	GA	30093	404-279-7117
Andrews RV Rental, Inc.	4145 North Jones Avenue	Boise	ID	83704	208-377-5490
Erickson's R.V. Rentals, Inc.	425 E. Borah Avenue	Coeur D'Alene	ID	83814	208-664-8902
Just Motor Homes, Inc.	5625 W. 107th Street	Chicago Ridge	IL	60415	708-371-9570
Wheel-Go Camping, Inc.	13515 West 159th Street	Lockport	IL	60441	708-301-9110
Traveltime USA	11678 Crockett Road	Roscoe	IL	61073	815-335-7031
The R.V. Oasis	24245 County Road 6	Elkhart	IN	46514	219-264-7748
Brick Road Enterprises, Inc.	1450 Kirklin Brick Road	Frankfort	IN	46041	317-654-9060
In-A-Pinch? Rent-A-Car	491 West Main Street	Greenwood	IN	46142	317-888-2215
Runyon Enterprises	16555 North Gray Road	Noblesville	IN	46060	317-896-9276
Heartland R.V.	P. O. Box 285	Woodburn	IN	46797	219-632-4815
Grant County Implement, Inc.	P.O. Box 40	Ulysses	KS	67880	316-356-3460
Northlake Motorcoach Rental, LLC	2022 Tamvest Court	Mandeville	LA	70470-1753	504-727-0022
Bates Motorhome Rental of New Orleans	40 Tennyson Place	New Orleans	LA	70131	888-392-0300
Fuller Motorhome Rental	92 Diamond Hill Avenue	Boylston	MA	01505-1316	508-869-2905
Trip Makers, Inc.	P.O. Box 9132	Foxboro	MA	02035	508-660-5000
Rent 'N Roam RV Rental, Inc.	796 Hartford Turnpike	Shrewsbury	MA	01545	508-842-1400
AAA Wickers Recreational Rentals	236 Boston Street	Topsfield	MA	01983	508-887-7336
A to B Charters	P.O. Box 606	Grassville	MD	21638	410-819-7800
H.W. Motor Homes, Inc.	107 N. Canton Center Road	Canton	MI	48187	313-981-1535
Hay-Lett's Auto & RV	891 East Chicago	Coldwater	MI	49036	517-278-5196
Krenek Leasing/Motors, Inc.	PO Box 615/ 6542 Ryno Road	Coloma	MI	49038	616-468-7900
Holiday Fun Motorhome Rentals, Inc.	1075 Lakeshore Drive	Columbiaville	MI	48421	810-793-6898
Adventure Motorhomes, Inc.	32430 Northwestern Hwy.	Farmington Hills	MI	48334	810-851-8120
Ridgeway RV, Inc.	A 4220 Blue Star Highway	Holland	MI	49423	616-396-5575
Midland RV Sales	607 S. Saginaw	Midland	MI	48640	517-631-1231
Redmond Rental	7613 Gratiot	Saginaw	MI	48609	517-781-1800
Recreational Rentals & Sales	12445 N. Hwy. U.S. 131	Schoolcraft	MI	49087-9401	616-668-3627
West Motor Home Leasing	52074 Nancy Lane	Three Rivers	MI	49093	616-496-7647
Jules, Inc.	10145 Dogwood Street	Coon Rapids	MN	55448	612-784-6460
Quality RV, Inc.	11044 - 167th Avenue	Elk River	MN	55330	612-441-6657
Winjum's Shady Acre Resort	17759 W. 177th Street	Faribault	MN	55021	507-334-6661
Majestic RV, Inc.	16527 Highway 65 Northeast	Ham Lake	MN	55304	612-434-0350
Vacation Time RV, Inc.	2727 E. Minnehaha Avenue	Maplewood	MN	55119	612-731-4932
Brambilla's, Inc.	PO Box 37/ 550 Valley Park Drive	Shakopee	MN	55379	612-445-2611
Rent-N-Travel	970 Hwy. 10 NE	Spring Lake Park	MN	55432	612-783-8873
Happy Times RV Center	1112 E. 6th Street	Winona	MN	55987	507-452-8916
AAA RV Sales	330 North 291 Hwy.	Liberty	MO	64068	816-781-7081
Gary's R.V. Repair & Rental	8030 N.E. 69 Hwy.	Pleasant Valley	MO	64068	816-452-8788
Camper Rentals	275 Lemay Ferry Road	St. Louis	MO	63125	314-631-5600
C & T Trailer Supply/Service	2000 N 7th Ave.	Bozeman	MT	59715	406-587-8610
J & L Motor Home Rentals, Inc.	P.O. Box 2443	Kalispell	MT	59901	406-752-4515
Bates Mtrhm Rntl Network of W. Montana	2230 N. Reserve Street # 250	Missoula	MT	59802	406-329-9195
Paradise RV Rentals, L.L.P.	1001 Hiberta	Missoula	MT	59806	406-721-6729
Colfax Country RV Sales	8615 Triad Drive	Colfax	NC	27235	910-996-6661
S & S RV Rentals	1105 South 2nd Street	Lillington	NC	27546	910-893-6725
Riverside R.V. Rentals	1108 27th St. NW	Mandan	ND	58554	701-663-2154
Choice Rental	81 Ross Avenue	Manchester	NH	03103	603-647-1007

Kastco RV	100 Walnut Avenue	Clark	NJ	07066	908-232-3161
Wesley Company	7 Aldrich Dr.	Howell	NJ	07731	908-370-3432
Roamin' Holiday, Inc.	P.O. Box 40182	Albuquerque	NM	87196	505-255-6611
Worldwide Motorhome Rentals, Inc.	1700 N. Gateway Rd., Lot E-9	Las Vegas	NV	89115	702-452-8031
Bates Motor Home Rental Network	3690 S.Eastern Ave. Suite 220	Las Vegas	NV	89109	702-737-9050
Hitchin' Post RV Sales	4784 Boulder Hwy.	Las Vegas	NV	89121	702-431-4434
American Motorhome Rentals	P.O. Box 406	Logandale	NV	89021	702-398-7592
Sierra RV Rentals, Inc.	324 Vine St.	Reno	NV	89503	702-324-0522
Proudfoot RV, Ltd.	Abbey Field/Farm to Market Rd	Brewster	NY	10509	914-279-7055
Great Outdoors Recreation Center	1122 Route 57	Fulton	NY	13069	315-695-5020
Bill's Vacation Trailers Inc.	100 W Sunrise Hwy	Lindenhurst	NY	11757	516-957-8810
Four Seasons Motor Home Rental	724 Old Liverpool Road	Liverpool	NY	13088	315-457-4746
Colton Auto Inc.	3176 Niagara Falls Blvd.	N. Tonawanda	NY	14120	716-694-0188
Transatlantic RV Rentals	1245 Park Street	Peekskill	NY	10566	914-739-8314
Bright Star Coach Rentals	35 Chestnut Street	Poughkeepsie	NY	12601	914-473-8463
Stacey's Camper Rental	932 Howard Road	Rochester	NY	14624	716-247-3255
Greece Rentals	950 Edgemere Drive	Rochester	NY	14612	716-865-8868
All Seasons RV Rentals, Inc.	18 North Drive	Saugerties	NY	12477	914-336-6975
Countryside RV	75 Barnstead Drive #13	Springville	NY	14141-1065	716-592-2802
Roaming Roads R.V. Rentals	6941 Greenway-New London Rd	Verona	NY	13478	315-339-4495
Holman Chevrolet Oldsmobile, Inc.	3075 State Route 125	Bethel	OH	45106	513-734-2206
S.A.L. Enterprises	5000 Akron-Cleveland Road	Peninsula	OH	44264	216-656-4014
Arbuckle R.V. Rental	301 West Benton	Davis	OK	73030	405-369-5206
Back Home RV Rental	8232 East 38th Street	Tulsa	OK	74145	918-599-0090
Coast To Coast RV Rentals	9205 SE Clackamas Road #127	Clackamas	OR	97015	503-513-0289
El-Mar Enterprises, Inc.	15555 S.E. McLoughlin Blvd.	Milwaukee	OR	97268	503-654-1002
Adventure On Wheels	17985 SW Pacific Hwy.	Tualatin	OR	97062	800-601-rent
Mike Ziegler's Motorhome Rentals	455 Auburn Street	Allentown	PA	18103	610-435-6920
General Rental & Sales Center	Old Rt. 1 & 41	Avondale	PA	19311	610-268-2825
Bradco Industries, Inc.	3293 Thornwood Drive	Bethel Park	PA	15102-0354	412-835-0449
Miley RV Sales	23 Chestnut Street	Carnegie	PA	15106	412-279-6200
Gordon's Auto & RV Rental, Inc.	317 Hadley Road	Greenville	PA	16125	412-588-2209
M & M RV Rental	2076 County Line Rd. #245	Huntingdon Valley	PA	19006-1739	215-355-7035
K Rentals	508 Route 30	Irwin	PA	15642-4502	412-864-8533
Wright's RV Rentals	P.O. Box 327	Orwigsburg	PA	17961	610-562-3749
Chaffee's Motorhome Rentals	RR1 Box 114A	Rome	PA	18837	717-247-2267
Royal RV Center	240 Greenbush Street	Scranton	PA	18508	717-344-3025
Lockman Motorhome Rentals	4189 Old Spartanburg Hwy.	Moore	SC	29369	803-574-2616
D & N Camper Sales	1160 East Brooks Road	Memphis	TN	38116-1706	901-345-2267
Lost Pines RV Rentals	1293 Lei Court	Bastrop	TX	78602	512-303-1391
Quality RV	990 South Highway 5	McKinney	TX	75069	214-542-7406
Marsh Motorhome Rentals	26910 Maplewood	Spring	TX	77386	713-367-7922
Bates Motorhomes	21520 I-45 North	Spring	TX	77373	713-353-7336
Bridgerland R.V. & Rental	315 North 100 West	Mendon	UT	84325-0308	801-755-6067
Pappy's Motorhome Rentals, Inc.	8201 South State	Midvale	UT	84047	800-888-2230
Access RV Rental Group, Inc.	200 So. Orchard Dr.	North Salt Lake	UT	84054	801-550-9666
Vacation World RV Center	960 South Bluff	St. George	UT	84770	801-673-7283
Bates Motorhome Rental of S. Utah	784 South River Road, Suite 195	St. George	UT	84790	801-688-2525
Freedom R.V. Rentals, Inc.	4908 Embassy Drive	Richmond	VA	23230	804-798-3379
Travel Rite Motorhome Rentals	4016 Williston Blvd.	S. Burlington	VT	05403	800-639-5093
Ken Do Kamping Rentals	RR2 Box 487 Alpine Drive	Underhill	VT	05489	802-899-4115
Executive RV Charters, Inc.	30624 5th Place South	Federal Way	WA	98003	206-529-0905
Eastside Motor Home Rentals	6011 East Lake Sammamish SE	Issaquah	WA	98027	206-392-9226
All Seasons Recreation Rentals	17224 162nd SE	Monroe	WA	98272	360-794-4386
Dallas Leasing	6035 State Route 12	Oakville	WA	98568	360-273-2003
Michael's Luxury Motorcoach Rentals	P.O. Box 9958	Spokane	WA	99209	509-468-8719
Saunders RV Rental, Inc.	3864 North 100 Street	Milwaukee	WI	53222	414-438-0799
CMC Leasing	13370 W. Maple Ridge Road	New Berlin	WI	53151-6997	414-641-0700
Interstate Travel Inc.	612 13th Avenue South	Onalaska	WI	54650	608-781-9094
Outdoor Express, Inc.	P.O. Box 100/Landis Lane	Falling Waters	WV	25419	304-274-9114
A & A Auto & R.V. Rentals, Inc.	1200 E. 2nd/Box 3254	Gillette	WY	82717	307-686-8250

PARTS & SERVICE DEALERS:

E-Z Camper Sales/Supply	417 E Main	Avondale	AZ	85323	602-932-2990
Handy Man Mobile RV Service	2350 Miracle Mile #188-365	Bullhead City	AZ	86442	502-846-1685
Norris RV	P.O. Box 10553	Casa Grande	AZ	85230	520-836-7921
Camelot RV Center	651 N. Main Street	Cottonwood	AZ	86326	520-634-3011
Road Runner R.V. Parts & Service, Inc.	750 N. Lake Havasu City Avenue	Lake Havasu	AZ	86403	520-453-1213
Midway Mobile	1411 S. Anaheim Blvd.	Anaheim	CA	92805	714-774-7012
Travelmaster R.V. Center	960 South G Street	Arcata	CA	95521	707-822-4833
Shull & Ford RV	169 Borland Avenue	Auburn	CA	95603	916-885-4416
Valley Palms Trailer Supply	8401 East Hobsonway	Blythe	CA	92225	619-922-7335
Dr. Scott's RV Clinic	2424 Bates Ave.	Concord	CA	94520	510-827-3855
Century Service Center	1807 Truesdale Street	Eureka	CA	95503-3836	707-445-8411
California Trailer & RV Supply	1914 Twin View Blvd.	Redding	CA	96003	916-241-7746

Custom Equipments Sales, Inc.	330 Keyes Street	San Jose	CA	95112	408-294-1977
Barrett Enterprises	1724 Country Oak Lane	Thousand Oaks	CA	91362-1900	805-494-6847
Bill's RV Service	2811 Beene Road	Ventura	CA	93003-7203	805-339-0882
Quality RV	2021 Live Oak Blvd.	Yuba City	CA	95991	916-755-4036
Lone Oak, Inc.	360 Norfold Road	East Canaan	CT	06024	203-824-7051
Pioneer Insurance Agency, Inc.	406 Farmington Avenue	Farmington	CT	06032	860-676-7716
Park Garage RV Cycle	PO Box 3223	Wilmington	DE	19804	302-652-03l7
Space Coast R.V. Center	4101 U.S. Hwy. 1 N	Cocoa	FL	32926	407-639-4883
Lakeside R.V.	37936 Hwy. 19	Umatilla	FL	32784	352-669-3223
Dalton Camper Sales, Inc.	1617 Murray Avenue	Dalton	GA	30720	706-226-5142
Gause RV Center, Inc.	P.O. Box 1067	Hiawassee	GA	30546	706-896-4846
Rainmaker Camp Store	North Litchfield Township	Litchfield	IL	62056	217-532-6370
Northwest RV & Marine, Inc.	540 S. Rand Road	Wauconda	IL	60084	847-526-5151
Bill's RV Appliance	57628 Hawthorne Street	Elkhart	IN	46517	219-522-1569
Major's RV Service Center	150 McArthur Blvd. Rt. 28	Bourne	MA	02532	508-759-2833
Charles Holden Associates, Inc.	2205 Pheasant Creek Lane	Peabody	MA	01960-4751	508-535-4020
Tim's Service Center	1319 So. Philadelphia Blvd.	Aberdeen	MD	21001	410-272-4772
The Custom Coach Company	8332 Pulaski Hwy.	Baltimore	MD	21237	410-687-7200
Cherry Hill Park	9800 Cherry Hill Road	College Park	MD	20740-1210	301-937-7116
J.M.F. Trailer & Accessories	Rt. 3 Bedford Rd. Box 5	Cumberland	MD	21502	301-724-1530
B & M Recreational Vehicle	P.O. Box 268	Jarrettsville	MD	21084-1268	410-692-9525
W. R. Bashaw & Associates	5381 Sands Road	Lothian	MD	20711	301-574-0368
VIP Systems International	2103 W. Dixon Lake Drive	Gaylord	MI	49735	517-732-3767
Standby Power, Inc.	7580 Expressway Drive/ S.W.	Grand Rapids	MI	49548	616-281-2211
L & A RV Service	8358 Alpine NW	Sparta	MI	49345	616-887-8241
Rec-Rest, Inc.	6900 E. 14 Mile Road,	Warren	MI	48090	810-977-2770
Crystal Welding, Inc.	17601-113th Avenue N.	Maple Grove	MN	55369	612-428-8281
Power Solutions	6507 Olde Savannah Road	Charlotte	NC	28227	704-536-4540
Blair Creek RV & Marine Products	964 NC 69, Suite 4	Hayesville	NC	28904	704-389-4145
Osburn's RV Service, Inc.	6001 Holly Ave. NE	Albuquerque	NM	87113	505-821-0543
Bogart Enterprises	2210 S. Valley Dr.	Las Cruces	NM	88005	505-524-0881
NE Rec Vehicle Camping & Equip	529 North Street	Middletown	NY	10940	914-343-2772
Flying W Caps, Inc.	135 W. Campbell Road	Schenectady	NY	12306	518-393-1301
Valley Toppers	86 North Division Street	St. Johnsville	NY	13452	518-568-3000
Diederich Mobile Home Park Sales	Route 9 West Box 101	West Coxackie	NY	12192	518-735-6492
Valley View R. V. Supply	1815 Limbach Road	Clinton	OH	44216	216-882-3226
Paradise Lake Park	6940 Rochester Road	East Rochester	OH	44625	330-525-7726
Country Boy Enterprises	6487 North St. N.W.	Granville	OH	43023	614-587-4675
The Awning Outlet	6487 North Street, N.W.	Granville	OH	43023	614-587-4735
Mobile RV & Marine Accessories	5881 N. State Rt. 590	Oak Harbor	OH	43449	419-898-0483
Adventure RV Service	220 Tahlequah Trail	Springboro	OH	45066	513-746-1871
J & J RV Repair	1708 Lakeview Drive	Newcastle	OK	73065	405-387-4751
Motley RV Repair	8300 W. Reno	Oklahoma City	OK	73127	405-789-4848
McKay Truck & RV Center	6225 Old Salem Road	Albany	OR	97321	541-928-3331
Central Point RV Service Center	900 S. Front	Central Point	OR	97502	541-664-5207
Andy Anderson Enterprises	4182 E. Evans Creek Road	Rogue River	OR	97537	541-582-4521
The RV Water Filter Man	P.O. Box 2040	Roseburg	OR	97470	
Oak Park R.V. Supplies & Service	4180 Silverton Rd. NE	Salem	OR	97305	503-581-5407
S & S Enterprises	1135 Baltimore Pike	Gettysburg	PA	17325	717-359-7856
Richard's R.V. Service Center	470 Mayfield Road	Duncan	SC	29334	803-879-2067
R & R RV	P.O. Box 687	Goodrich	TX	77335	409-365-2940
RV Service of VA, Inc.	P.O. Box 6443	Ashland	VA	23005-6443	804-798-1433
Richard's RV Service	421 Butternut Dr.	Fredericksburg	VA	22401	540-898-1172
Capital Camper Supply	2830 Gallows Road	Vienna	VA	22180	703-560-6424

TRAVEL NOTES

TRAVEL NOTES

TELEPHONE DIRECTORY

Name	Phone	Name	Phone

TELEPHONE DIRECTORY

Name	Phone	Name	Phone

INDEX

Highway	State	Pages
I-4	Florida	5-9
I-5	Washington	9-15
	Oregon	15-20
	California	20-33
I-8	California	33-35
	Arizona	35-36
I-10	California	36-41
	Arizona	41-45
	New Mexico	45-46
	Texas	46-57
	Louisiana	57-61
	Mississippi	61-62
	Alabama	62-63
	Florida	63-68
I-12	Louisiana	69-70
I-15	Montana	70-72
	Idaho	72-73
	Utah	73-78
	Arizona	78
	Nevada	78-79
	California	79-83
I-16	Georgia	83-84
I-17	Arizona	84-86
I-19	Arizona	86-87
I-20	Texas	87-95
	Louisiana	95-98
	Mississippi	98-101
	Alabama	101-103
	Georgia	103-108
	South Carolina	108-109
I-24	Illinois	110
	Kentucky	110-111
	Tennessee	111-116
I-25	Wyoming	116-118
	Colorado	118-123
	New Mexico	123-125
I-26	North Carolina	126
	South Carolina	126-130
I-27	Texas	130-131
I-29	North Dakota	132-133
	South Dakota	133-135
	Iowa	135-136
	Missouri	136-137
I-30	Texas	137-141
	Arkansas	141-143
I-35	Minnesota	143-147
	Iowa	147-149
	Missouri	149-150
	Kansas	150-152
	Oklahoma	152-155
	Texas	155-171
I-37	Texas	172-173
I-39	Illinois/Wisconsin	174-175
I-40	California	175-176
	Arizona	176-178
	New Mexico	178-181
	Texas	181-183
	Oklahoma	183-187
	Arkansas	187-190
	Tennessee	190-201
	North Carolina	201-206
I-43	Wisconsin	207-208
I-44	Texas	208-209
	Oklahoma	209-212
	Missouri	212-215
I-45	Texas	215-222
I-49	Louisiana	222-223
I-55	Illinois	224-227
	Missouri	227-229
	Arkansas	229-230
	Tennessee	230-231
	Mississippi	231-235
	Louisiana	235-236
I-57	Illinois	236-240
	Missouri	240
I-59	Georgia	240
	Alabama	240-243
	Mississippi	243-246
	Louisiana	246
I-64	Illinois	246-247
	Indiana	247-248
	Kentucky	248-250
	West Virginia	250-253
	Virginia	253-258
I-65	Indiana	260-263
	Kentucky	263-266
	Tennessee	266-270
	Alabama	270-275
I-66	Virginia	275-276
I-68	West Virginia	276
	Maryland	276-277
I-69	Michigan	278-280
	Indiana	280-281
I-70	Utah	282
	Colorado	282-287
	Kansas	287-290
	Missouri	290-294
	Illinois	294-295
	Indiana	295-298
	Ohio	298-301
	West Virginia	301-302
	Pennsylvania	302-304
	Maryland	304-305
I-71	Ohio	306-309
	Kentucky	309-310
I-72	Illinois	310-311
I-74	Iowa	311
	Illinois	311-313
	Indiana	313-315
	Ohio	315
I-75	Michigan	315-321
	Ohio	321-326
	Kentucky	326-330
	Tennessee	330-335
	Georgia	335-346
	Florida	346-350
I-76	Colorado	351
	Ohio	352
	Pennsylvania	352-356
I-77	Ohio	356-359
	West Virginia	359-361
	Virginia	361-362
	North Carolina	362-364
	South Carolina	364-365
I-78	Pennsylvania	365-366
	West Virginia	366-367
I-79	Pennsylvania	368-370
	West Virginia	370-371
I-80	California	372-376
	Nevada	376-378
	Utah	378-379
	Wyoming	380-382
I-80	Nebraska	382-385
	Iowa	385-387
	Illinois	387-389
	Indiana	389-390
	Ohio	390-392
	Pennsylvania	392-395
	New Jersey	395-397
I-81	New York	398-400
	Pennsylvania	400-403
	Maryland	403-404
	West Virginia	404
	Virginia	404-409
	Tennessee	409-410
I-82	Washington	410-411
	Oregon	411
I-83	Pennsylvania	411-413
	Maryland	413
I-84	Oregon	413-416
	Idaho	416-419
	Utah	419
	Pennsylvania	420
	New York	420-421
	Connecticut	421-424
	Massachusetts	424
I-85	Virginia	424-425
	North Carolina	425-430
	South Carolina	430-432
	Georgia	432-437
	Alabama	437-438
I-86	Idaho	439
I-87	New York	439-444
I-88	Illinois	444-445
	New York	445-446
I-89	Vermont	446-448
	New Hampshire	448-449
I-90	Washington	449-452
	Idaho	452-453
	Montana	453-457
	Wyoming	457-458
	South Dakota	458-462
	Minnesota	462-463
	Wisconsin	463-465
	Illinois	465-467
	Indiana	467-468
	Ohio	468-470
	Pennsylvania	470-471
	New York	471-474
	Massachusetts	474-475
I-91	Vermont	476-477
	Massachusetts	477-478
	Connecticut	478-479
I-93	Vermont	480
	New Hampshire	480-482
	Massachusetts	482-483
I-94	Montana	483-484
	North Dakota	484-486
	Minnesota	486-489
	Wisconsin	489-493
	Illinois	493-495
	Indiana	495-496
	Michigan	496-500
		422-424
I-95	Maine	501-503
	New Hampshire	503-504
	Massachusetts	504-506
	Rhode Island	506-507
	Connecticut	507-511
	New York	511-512
	New Jersey	512-513
	Pennsylvania	513-515
	Delaware	515
	Maryland	515-518
	Virginia	518-523
I-95	North Carolina	523-527
	South Carolina	527-529
	Georgia	529-532
	Florida	532-539
I-96	Michigan	540-542
CA99	California	543-548
I-99	Pennsylvania	549
I-135	Kansas	549-550
I-155	Illinois	551
I-172	Illinois	551
I-185	Georgia	551
I-195	Rhode Island	552
I-195	Massachusetts	552
I-196	Michigan	553
I-205	Washington	554
I-205	Oregon	554
I-220	Louisiana	555
I-235	Iowa	555
I-240	Tennessee	555-556
I-255	Missouri	556
I-255	Illinois	556
I-264	Kentucky	557
I-265	Kentucky	558
I-270	Ohio	558-560
I-270	Illinois	560
I-270	Missouri	560-561
I-270	Maryland	561-562
I-271	Ohio	562
I-275	Kentucky	563
I-275	Ohio	563-565
I-275	Florida	565-566
I-275	Michigan	566-567
I-276	Pennsylvania	567
I-279	Pennsylvania	568
I-280	California* (*S.F. Bay Area includes 580 & 680)	568-571
I-285	Georgia	572-574
I-287	New Jersey	575
I-290	Massachusetts	576
I-290	Illinois	576
I-294	Illinois	577
I-295	New Jersey	577-579
I-295	Delaware	579
I-295	Virginia	579
I-295	Maryland	579-580
I-295	Florida	580
I-376	Pennsylvania	581
I-380	Iowa	581
I-385	South Carolina	582
I-390	New York	582
I-395	Massachusetts	583
I-395	Connecticut	583
I-395	Virginia	584
I-405	California	584-587
I-405	Washington	587
I-410	Texas	588-589
I-435	Missouri	590-591
I-459	Alabama	591
I-465	Indiana	592-593
I-475	Ohio	594
I-475	Georgia	594
I-476	Pennsylvania	594
I-480	Ohio	595
I-490	New York	595
I-494	Minnesota	596
I-495	New Jersey	597
I-495	New York	597-598
I-495	Maryland	598-599
I-495	Virginia	599-600
I-495	Massachusetts	600-601
I-526	South Carolina	602
I-575	Georgia	602
I-595	Florida	603
I-610	Texas	603-605
I-635	Texas	605-606
I-640	Tennessee	607
I-675	Ohio	607
I-694	Minnesota	607-608
I-695	Maryland	608-610
I-696	Michigan	611
I-820	Texas	611-612
Briley Pkw	Tennessee	612

*Note: "Travel Savers" Coupon Book begins on page 613; Recreational Vehicle Dealer's Association Membership Directory begins on page 635